			IIIA	IVA	VA	VIA	VIIA	0
								2 **He** 4.003
			5 **B** 10.81	6 **C** 12.011	7 **N** 14.007	8 **O** 15.9994	9 **F** 18.998	10 **Ne** 20.18
	IB	IIB	13 **Al** 26.98	14 **Si** 28.09	15 **P** 30.97	16 **S** 32.06	17 **Cl** 35.453	18 **Ar** 39.95
28 **Ni** 58.71	29 **Cu** 63.55	30 **Zn** 65.38	31 **Ga** 69.72	32 **Ge** 72.59	33 **As** 74.92	34 **Se** 78.96	35 **Br** 79.90	36 **Kr** 83.80
46 **Pd** 106.4	47 **Ag** 107.87	48 **Cd** 112.40	49 **In** 114.82	50 **Sn** 118.69	51 **Sb** 121.75	52 **Te** 127.60	53 **I** 126.90	54 **Xe** 131.30
78 **Pt** 195.09	79 **Au** 196.97	80 **Hg** 200.59	81 **Tl** 204.37	82 **Pb** 207.2	83 **Bi** 208.98	84 **Po** (210)	85 **At** (210)	86 **Rn** (222)

| 63 **Eu** 151.96 | 64 **Gd** 157.25 | 65 **Tb** 158.93 | 66 **Dy** 162.50 | 67 **Ho** 164.93 | 68 **Er** 167.26 | 69 **Tm** 168.93 | 70 **Yb** 173.04 | 71 **Lu** 174.97 |

| 95 **Am** (243) | 96 **Cm** (247) | 97 **Bk** (247) | 98 **Cf** (251) | 99 **Es** (254) | 100 **Fm** (253) | 101 **Md** (258) | 102 **No** (255) | 103 **Lr** (256) |

General Chemistry
Principles and Modern Applications

General Chemistry
Principles and Modern Applications
Fourth Edition

Ralph H. Petrucci
California State University, San Bernardino

Macmillan Publishing Company
New York

Collier Macmillan Publishers
London

Copyright © 1985, Ralph H. Petrucci

Printed in the United States of America

Earlier editions copyright © 1972, 1977, and 1982 by Ralph H. Petrucci.

Macmillan Publishing Company
866 Third Avenue, New York, New York 10022

Collier Macmillan Canada, Inc.

Library of Congress Cataloging in Publication Data

Petrucci, Ralph H.
  General chemistry.

  Includes index.
  1. Chemistry. I. Title.
QD31.2.P48 1985   540   84-14958
ISBN 0-02-394790-X (Hardcover Edition)
ISBN 0-02-946550-8 (International Edition)

Printing: 1 2 3 4 5 6 7 8   Year: 5 6 7 8 9 0 1 2 3

ISBN 0-02-394790-X

Preface

Each edition of this text has differed in some important respects from those preceding it, but all have been written with this central idea in mind: The majority of students who study general chemistry have career interests in fields other than chemistry—biology, medicine, engineering, and the agricultural sciences, to name but a few. Not only should a general chemistry text provide such students with the modern chemical background they need for their specialized studies, but also it should be as readily understandable as possible, for these students have limited amounts of time available to study general chemistry. As with previous editions, I have attempted to strike the balances between principles and applications, between qualitative and quantitative discussions, and between rigor and simplification that seem most appropriate for these typical general chemistry students. I have been guided in my revision by comments from users of previous editions and reviewers of the present one; and I have given special attention to all suggestions, even minor ones, that might in some way improve the clarity of the presentation.

Perhaps the major changes between the previous edition and this one are the increased coverage of applications and descriptive inorganic chemistry and the more extensive integration of these topics throughout the text. Numerous applications are presented at the points where their underlying principles are considered. In addition, a new feature of this edition is the inclusion of special "Focus" sections at the ends of certain chapters (thirteen in all). For example, a discussion of the key chemicals of industry is included in Chapter 1, aspects of industrial chemistry in Chapter 4, semiconductors in Chapter 10, liquid crystals in Chapter 11, and polymers in Chapters 9 and 26. Descriptive inorganic chemistry is introduced in the middle of the text (Chapter 13) with a consideration of the first 20 elements. The topics chosen for this discussion are those that can be presented through principles developed in the first twelve chapters. The chemistry of these elements is revisited in Chapters 21 and 22 in a discussion of the representative elements that employs the full range of principles developed in the first twenty chapters of the text. Chapter 23 deals with the transition elements, Chapter 24 with coordination chemistry, and Chapter 25 with nuclear chemistry. As in earlier editions, one chapter (Chapter 26) is devoted to organic chemistry and one (Chapter 27) to biochemistry. Topics related to qualitative analysis of cations appear in several chapters (18, 19, 21, 23, and 24), each at a point appropriate to the principle involved.

Less apparent than those just described, but also important, are some changes made from a pedagogical standpoint. The discussion of oxidation states has been moved forward (from Chapter 9 to Chapter 3) to permit a more thorough early introduction of nomenclature. Basic concepts of oxidation–reduction and the oxidation state change method of balancing oxidation–reduction equations are now introduced in Chapter 4. This permits the use of oxidation–reduction chemistry as one of the principal themes in the introductory chapter on descriptive chemistry. As in the third edition, solution stoichiometry is introduced early (Chapter 4) to coordinate with typical laboratory programs. Also as in the third edition, the subject of thermodynamics is divided between an early chapter on thermochemistry (Chapter 6) and a later one on additional aspects of thermodynamics (Chapter 16). The first law of thermodynamics has been moved from the first to the second of these chapters, so that most of the formalism of thermodynamics is now found in the later chapter. Solubility and complex ion equilibria have been combined into a single chapter (Chapter 19) that now follows the two chapters on acid–base equilibria. Several topics have been added to the chapter on coordination chemistry, the principal one being optical isomerism (previously introduced in the biochemistry chapter). In a few cases (e.g., crystal structures) sections have been rewritten to offer a new approach to the subject that may be more understandable to students. Toward this same end, some new in-text examples have been added and some existing ones have been worked out in greater detail.

I continue to believe that there exists no single ideal organizational scheme for a general chemistry text. This edition, like its predecessors, is so structured that a number of alternative orders of presentation are possible. For example, the "Focus" features can be studied as encountered, or some features may be deferred to later contexts (such as considering the features on key chemicals and industrial chemistry together at the close of Chapter 4). Also, some of these features can be made optional, for although they relate closely to fundamental principles from the portions of the text that precede them, these features introduce no material that is crucial to the development of later subject matter. Instructors who favor dealing with systematic descriptive inorganic chemistry as a single unit will find that Chapter 13 can easily be deferred and taken up together with Chapters 21–23. None of the material introduced in Chapter 13 is fundamental to the subject matter of Chapters 14–20. On the other hand, those who prefer to disperse descriptive topics more widely will find that Chapter 13 can be approached section by section in conjunction with other chapters. Instructors who wish to combine the material on thermodynamics into a single presentation may defer Chapter 6 until Chapter 16 is reached. Those who wish to limit organic chemistry to a description of bonding, structure, and nomenclature can take up these matters at any time after Chapters 9 and 10 on chemical bonding. In the separate *Instructor's Manual* that accompanies this text, several alternative organizational schemes are described.

As in the third edition, each chapter concludes with a number of study aids. First is a brief chapter summary, which is followed by a set of learning objectives. These objectives are stated in fairly general terms, and individual instructors may choose to emphasize some more than others or to add specific objectives of their own. The third end-of-chapter feature is a set of definitions of important new terms introduced in the chapter. Reference to each end-of-chapter definition is made through a **boldface** page number in the Index. The collection of these Index listings and end-of-chapter definitions constitutes a glossary of the entire text. New to this edition is a listing of a small number of references for each chapter. These "Suggestions for Further Study" are typically of several sorts: Some offer an alternative, often more elementary, approach to important concepts considered in the chapters. Some carry the subject matter beyond the level of the text. Many of the references deal with interesting applications of topics discussed in the chapter. A few provide historical background. Practically all of these references are from journals and magazines, rather than monographs or texts, in the belief that students are more likely to refer to supplementary materials if they are rather brief.

Each chapter has four categories of exercises. The first category, "Review Problems," is new to this edition. These problems require straightforward applications of principles introduced in the chapter, each problem usually involving just a single concept. In the category "Exercises," exercises are grouped by subject matter and are of a broader nature than the "Review Problems"; those that either are more difficult or require an extension beyond the concepts presented in the chapter are designated by a star ★. The

"Additional Exercises" are not grouped by type. The final category, "Self-Test Questions," presents a group of multiple-choice items, together with brief essay questions and/or problems, that are typical of examination questions. Answers to most of the "Review Problems" and the "Self-Test Questions" are provided at the end of the book. About one half of the "Exercises" are similarly answered, but answers for the "Additional Exercises" are not given. Complete solutions to all but the "Additional Exercises" are provided in a separate *Solutions Manual*. Solutions to the "Additional Exercises" are available in the *Instructor's Manual*.

Also available to accompany this text are a *Student Study Guide* and a laboratory manual, *Experiments in General Chemistry*. The study guide was written by Professor Robert K. Wismer and the laboratory manual by Professors Gerald S. Weiss, Robert K. Wismer, and Thomas G. Greco, all of Millersville University of Pennsylvania. The study guide is organized around the "Learning Objectives" in the textbook and features brief discussions of these objectives, drill problems, self-quizzes, and sample tests. The laboratory manual contains thirty-seven experiments that parallel the text, including a final group of six experiments on qualitative analysis.

If I have succeeded in improving this edition over previous ones, it is largely a result of the diligence with which reviewers have studied and commented on what I have written and of their willingness to share some of their own pedagogical ideas. The following have provided critiques of the third edition: R. Kent Murmann, University of Missouri, Columbia; Saul I. Shupack, Villanova University; Joseph Topich, Virginia Commonwealth University; and Mary S. Vennos, Essex Community College of Baltimore County (Maryland). Charles W. J. Scaife of Union College supplied helpful commentary on the Answers to Exercises in the third edition. Those who read and commented on portions of the manuscript of this edition are O. T. Beachley, SUNY, Buffalo; Billy L. Stump, Virginia Commonwealth University; and Carl Trindle, University of Virginia. Commenting on the entire manuscript were Michael F. Golde, University of Pittsburgh; Philip S. Lamprey, University of Lowell; and William H. McMahan, Mississippi State University. The following colleagues from my campus helped by reviewing manuscript, passing along student comments, and being available for consultation on various matters that arose during the preparation of this edition: Dennis Pederson, Arlo Harris, Kenneth Mantei, Lee Kalbus, and James Crum. As with the third edition, I owe a special debt of gratitude to Robert K. Wismer who, in addition to his own efforts in producing the accompanying study guide and contributing to the laboratory manual, commented extensively on the manuscript of the fourth edition, read proof, shared the rather onerous task of preparing the Solutions Manual, and was the willing recipient of countless long-distance (and sometimes long-winded) telephone calls.

Michael S. Smith and John Schultz helped to procure a number of the photographs in the text. In addition to many of the color plates, the photographs on the following pages were taken for this edition, some in association with Arlo Harris, by Carey B. Van Loon: 114, 234, 246, 312, 313, 319, 352, 383, 682, 817.

I have continued to receive assistance from many individuals associated with Macmillan Publishing Company, including a number, I am sure, who are not even known to me. Several individuals deserve special mention, however. Elisabeth Belfer supervised the conversion of a piecemeal manuscript into a bound volume in her usual calm professional manner. My editor, Peter Gordon, undertook several tasks on my behalf that really helped out in the pinch. Kate Aker, who contributed so much to the development of the third edition, was also on call when needed. And offering encouragement at appropriate times were Gregory Payne, Gary Ostedt, and John Snyder.

Top prize for forbearance goes to my wife, Ruth, and to members of my immediate and extended family. They have fashioned lives for themselves that include a special niche for a preoccupied textbook author. I am especially grateful to them for this.

San Bernardino, CA   R. H. P.

Contents

10   Chemical Bonding II: Additional Aspects   257

11   Liquids, Solids, and Intermolecular Forces   285

12   Solutions   328

13   An Introduction to Descriptive Chemistry: The First 20 Elements   364

14   Chemical Kinetics   406

15 Principles of Chemical Equilibrium   443

16 Thermodynamics and Chemistry   474

17 Acids and Bases   504

25   Nuclear Chemistry   768

26   Organic Chemistry   797

1

Matter—Its Properties and Measurement

Fire, known from earliest times, is a powerful agent for producing change. Fire was first used for cooking foods. Later it was used for baking pottery, making glass, and smelting ores to produce metals—first copper and later lead, tin, and iron. Other processes known since ancient times include making butter and cheese from milk, wine from grapes, beer from grains, leather from hides, and soap from fats.

All of the processes just cited are familiar examples of chemical change. A useful, albeit limited, definition of chemistry is that it is the science whose central concern is to discover how the "stuff" or matter of the universe can be changed from existing forms with certain sets of characteristics or properties to other forms with different properties. For example, chemistry has provided the knowledge with which it has become possible to transform the natural resource petroleum into a variety of fuels and countless numbers of plastics, drugs, and pesticides.

Early applications of chemistry were discovered by chance or by trial and error, but most modern uses require the careful application of the fundamental principles of chemistry. Also of recent origin is the discovery that certain chemical processes (e.g., the production of smog) have detrimental effects on the environment. An important challenge facing chemists today is to develop the processes and materials needed by modern society and simultaneously to minimize their environmental impacts. Both of these objectives require a firm understanding of chemical principles, and this is one reason why principles are emphasized together with applications in this text.

1-1 Properties of Matter

It is easier to describe matter intuitively than to define it precisely. Let us say that **matter** is any object or material that occupies space, and that the quantity of matter is measured by a property called **mass** (described further in Section 1-5). Mass is only one of many properties or characteristics by which samples of matter can be identified and distinguished from one another. Properties of matter can be grouped into two categories: physical and chemical.

Physical Properties and Physical Change. Color, luster, and hardness are among the many **physical properties** that may be used to describe the appearance of an object. A process in which an object changes its physical appearance but not its basic identity is called a **physical change.** A cube of copper metal can be flattened into a very thin foil;

1

FIGURE 1-1
A classification scheme for matter.

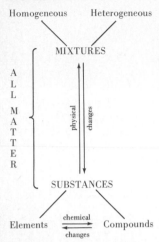

Every sample of matter is either a substance or a mixture. If a substance, it is either an element or a compound; if a mixture, either homogeneous or heterogeneous. Transformations between elements and compounds involve chemical changes; conversions between substances and mixtures, physical changes.

At times scientific definitions may be at variance with practical definitions. For instance, if a sample of homogenized milk is viewed through a microscope, globules of fat can be seen dispersed in a watery medium. Homogenized milk is a heterogeneous mixture.

copper is malleable. Copper can also be drawn out into a fine wire; it is ductile. The melting of ice and the boiling of water are further examples of physical change.

Chemical Properties and Chemical Change.   Paper burns, iron rusts, and wood rots. In each case the object changes not only in physical appearance but also in its basic identity. In a **chemical change** a sample of matter is transformed into completely different materials. The types of chemical changes that a material can undergo are determined by its **chemical properties.**

1-2 Classification of Matter

Matter exists in countless forms, and it is necessary to develop broad categories for its description. The scheme in Figure 1-1 classifies matter into the categories *substance* and *mixture*. Particularly important are those substances that cannot be made into simpler materials either by physical or chemical changes, that is, not by heating, crushing, exposing to acids, and so on. These substances are the **chemical elements.** At present, 106 different elements are known. They range from such common materials as iron, copper, silver, and gold to uncommon ones such as lutetium, promethium, and thulium. About 90 of the elements can be extracted from natural sources; the rest have been created through nuclear processes. A complete listing of the elements is presented on the inside back cover. A special tabular arrangement known as the periodic table (discussed in Chapter 8) is shown on the inside front cover.

Chemical compounds comprise a second class of substances. These are chemical combinations of two or more elements. Clearly, the potential number of different combinations of the 106 elements is enormous. The number of chemical compounds now known is in the millions, ranging in complexity, for example, from ordinary water to the protein hemoglobin. Chemical compounds retain their identities during physical changes but can be separated into their component elements by appropriate *chemical* changes.

The composition and properties of an element or a compound are uniform throughout a given sample *and from one sample to another*. Elements and compounds are said to be *pure* and are called **substances.** Some mixtures of substances also have compositions and properties that are uniform throughout a given sample *but variable from one sample to another*. These are **homogeneous mixtures** or **solutions.** A homogeneous mixture can be separated into its components by appropriate *physical* changes. Ordinary air is a solution of several gases, principally the *elements* nitrogen and oxygen. Seawater is a solution of the *compounds* water, sodium chloride (salt), and a host of others.

In some mixtures—sand and water, for example—the components separate into physically distinct regions. As a result the composition and physical properties vary from one part of the mixture to another. Such mixtures are said to be **heterogeneous.** Samples of matter ranging from a glass of iced tea to a slab of concrete to the leaf of a plant are heterogeneous.

1-3 The Scientific Method

What most distinguishes science from other intellectual activity is the manner in which scientific knowledge is acquired together with the way in which this knowledge can be used to *predict* future events. The time required for a rocket to reach the moon can be predicted even more accurately than can the time required to drive an automobile from New York City to Washington. This is because the scientific basis of rocket propulsion is so well understood.

The Greek approach to acquiring knowledge was based on **deduction.** Starting from certain basic premises or assumptions, the Greeks developed means for establishing conclusions that must follow. This is the logical method that they employed with such success in the study of geometry. However, in the deductive method the validity of the basic assumptions always remains in doubt.

In the seventeenth century the importance of experimentation in the discovery of scientific facts was first appreciated. Rather than start with certain assumptions, seventeenth-century scientists, such as Galileo, Francis Bacon, Descartes, Boyle, Hooke, and Newton, made careful observations of phenomena and then formulated natural laws to summarize them. This process of formulating a general statement, a **natural law,** from a series of observations is called **induction.**

To test a natural law a scientist designs a controlled situation, an **experiment,** to see if the conclusions deduced from the natural law agree with actual experience. If a natural law stands the test of repeated experimentation, confidence in it grows. If agreement between predicted and observed behavior is imperfect, the natural law must be modified or limited in scope. The success of a natural law then is judged by how effective it is in summarizing observations and in predicting new phenomena. However, no natural law can be accepted as an *absolute* truth because there is always the possibility that some experiment may be devised that would refute it.

A tentative explanation of a natural law is called a **hypothesis.** A hypothesis can be tested by experimentation, and if it survives this testing it is often referred to as a **theory.** The term *theory* can also be used in a broader sense: It is a conceptual framework or model (a way of looking at things) that can be used to explain and to make further predictions about natural phenomena. Sometimes, differing or conflicting theories are advanced to explain the same phenomenon. The theory chosen is usually the one that is most successful in its predictions. Also, the theory that involves the smallest number of assumptions, the simplest theory, is usually preferred. Over a period of time, as new evidence accumulates, most scientific theories undergo modification; some are discarded.

By way of illustration, consider the development of Dalton's atomic theory. Dalton's conception or model was that all matter is composed of minute, indivisible particles called atoms. He based his theory on two natural laws of chemical combination that had been discovered previously. One was the law of conservation of mass: matter is neither created nor destroyed in a chemical reaction. The other was the law of constant composition: the proportions in which elements are combined in a compound are independent of the source of the compound. And Dalton's theory was useful in explaining laws of chemical combination discovered subsequently. These laws and Dalton's theory are discussed in Chapter 2.

The sum of all the activities described above—observations, experimentation, and the formulation of laws, hypotheses, and theories—is called the **scientific method.** Single-minded devotion to the scientific method cannot alone guarantee success in scientific investigations, however. Occasionally, it is necessary for someone to break away from established patterns of thinking to discover the key to a scientific puzzle. These are the developments called scientific breakthroughs (see, for example, the discussion of the quantum theory in Chapter 7). And always, it is necessary to be alert to unexpected observations. A number of great scientific discoveries (e.g., x rays, radioactivity, penicillin) were made by accident in the course of other investigations. But accidental discoveries do not usually happen by accident. As noted by Louis Pasteur (1822–1895). "Chance favors the prepared mind."

There is an inherent problem in the method of induction. When have enough observations been made to justify a generalization (a natural law)? A city dweller has observed only a few dozen sheep in her lifetime and all have had a white coat. Is she justified in making the statement "All sheep are white"?

1-4 The Need for Measurement

In scientific work observations that can be assigned numerical values, **quantitative** observations, are generally preferred over verbal qualitative statements. The use of mathematics in describing the laws of nature enhances the scope of these laws and the precision with which they can be applied.

Even in nonscientific work, and from earliest times, there has existed a need to measure quantities and to express these measurements in convenient units. People needed to determine when to plant crops, how to exchange goods, and how to formulate recipes for primitive manufacturing. And, some of the quantities they most needed to measure—mass, length, and time—are also the basic properties of concern to scientists. Of course, the precision of these measurements and the units for expressing them have changed greatly over the years.

Some early units of measurement were based on human features and some on common objects. For example, the yard was introduced in the fifteenth century as the distance from a person's nose to the tip of the middle finger of the extended arm. An inch was defined as the length of "three barleycorns, round and dry, when laid together." Obviously, such quantities cannot be reproduced exactly, since the physical dimensions of human beings (and barleycorns) vary widely. Unless a unit of measurement can be precisely defined, it cannot be used for scientific work.

The yard, in England, was based on the arm length of the reigning monarch. Thus, it varied by as much as several inches from time to time. It was finally standardized in 1592 during the reign of Queen Elizabeth I.

1-5 The English and Metric Systems of Measurement

One system of measurement once widely used in English-speaking countries but now limited mainly to the United States has as its basic unit of mass, the standard **pound (lb),** and as its basic unit of length, the standard **yard (yd).** The **English system** is sufficiently precise to be used in modern manufacturing and commerce, but it is not particularly useful in scientific work. Primarily, this is because there is no regularity in the different units that may be used to express a measured quantity. The number of inches in 1 foot (12) is not the

FIGURE 1-2
A comparison of the metric and English systems.

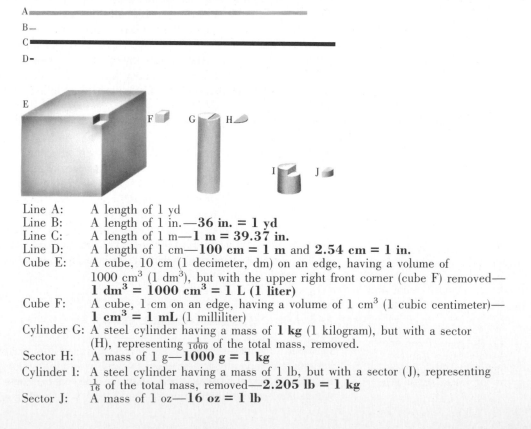

Line A: A length of 1 yd
Line B: A length of 1 in.—**36 in. = 1 yd**
Line C: A length of 1 m—**1 m = 39.37 in.**
Line D: A length of 1 cm—**100 cm = 1 m** and **2.54 cm = 1 in.**
Cube E: A cube, 10 cm (1 decimeter, dm) on an edge, having a volume of 1000 cm³ (1 dm³), but with the upper right front corner (cube F) removed—**1 dm³ = 1000 cm³ = 1 L (1 liter)**
Cube F: A cube, 1 cm on an edge, having a volume of 1 cm³ (1 cubic centimeter)—**1 cm³ = 1 mL** (1 milliliter)
Cylinder G: A steel cylinder having a mass of **1 kg** (1 kilogram), but with a sector (H), representing $\frac{1}{1000}$ of the total mass, removed.
Sector H: A mass of 1 g—**1000 g = 1 kg**
Cylinder I: A steel cylinder having a mass of 1 lb, but with a sector (J), representing $\frac{1}{16}$ of the total mass, removed—**2.205 lb = 1 kg**
Sector J: A mass of 1 oz—**16 oz = 1 lb**

TABLE 1-1
English and metric
equivalents

Metric		English
mass		
1 kg	=	**2.205 lb**
453.6 g	=	**1 lb**
length		
1.609 km	=	**1 mi**
1 m	=	39.37 in.
2.54 cm	=	**1 in.**
volume		
3.785 L	=	**1 gal**
1 L	=	1.057 qt

same as the number of feet in 1 yard (3) or the number of yards in 1 mile (1760). As a result, calculations requiring a conversion between units (e.g., inches to miles) are sometimes difficult to perform.

Prior to the French Revolution the system of measurement in France varied from province to province, making commercial transactions difficult to conduct. The need for a uniform system was clear, and in 1790 a commission of scientists was established to propose one. The commission chose as the unit of length, called a **meter (m)**, 1/10,000,000 of the distance at sea level from the North Pole to the equator along a meridian passing through Paris. The fact that the original measurement was in error is interesting but not significant. The actual standard was the distance between two marks on a certain platinum-iridium metal bar kept in the International Bureau of Weights and Measures in Sèvres, near Paris. The system of measurement based on the meter is called the **metric system.** It was adopted officially in France in 1840. Interest in the metric system in the United States originated in 1821 with John Quincy Adams, then Secretary of State, and continues to this day. Although the metric standards have been adopted as the *official* standards in the United States, the English system is still used widely for measurements in commerce and industry. Today, the United States is an "island in a metric world." Eventual adoption of the metric system in the United States does seem likely, however.

The metric system is a decimal system. The several different multiple and submultiple units by which a measured property may be expressed differ from one another by factors of ten. For example,

kilo means *one thousand times* the base unit

centi means *one hundredth of* the base unit

milli means *one thousandth of* the base unit

Thus, 1 *kilo*meter (km) is 1000 m (about 0.6 mile); 1 *centi*meter (cm) is $\frac{1}{100}$ of 1 m (about 0.4 inch); and 1 *milli*meter is $\frac{1}{1000}$ of 1 m.

Some of the important relationships in the metric system and a comparison to the English system are presented in Figure 1-2. A few equivalent quantities in the two systems are listed in Table 1-1 and used in calculations in Section 1-10.

Mass. Mass describes the quantity of matter in an object. The **kilogram (kg)** was originally defined as the mass of 1000 cubic centimeters (cm^3) of water at 4°C and normal atmospheric pressure. It is now taken to be the mass of a bar of platinum-iridium metal kept at the International Bureau of Weights and Measures. The kilogram is a fairly large unit for most applications in chemistry, so the unit **gram (g)** is more commonly employed (see Figure 1-2).

Weight, which describes the force of gravity on an object, is directly proportional to mass, as represented through a simple mathematical equation.

$$W \propto m \qquad \text{and} \qquad W = g \cdot m \qquad\qquad (1.1)$$

The symbol $\propto$ means "proportional to." It can always be replaced by an equality sign and a proportionality constant. The proportionality constant in equation (1.1), g, is called the acceleration due to gravity. Its significance is explored further in Appendix B.

Although a given quantity of matter has a fixed mass (m), regardless of where or how the measurement is made, its weight (W) may vary because g varies slightly from one point on earth to another. Thus, an object weighs about 0.4% more in Leningrad than in Panama, even though its mass remains constant. The terms *weight* and *mass* are used interchangeably, but only mass is a measure of the quantity of matter. Figure 1-3 illustrates how mass is measured by the principle of weighing.

Volume. Volume is an important property, but it is not as fundamental as mass because volume varies with temperature and pressure. Mass does not. Volume has the unit (length)^3. The basic metric unit of volume is the **cubic meter (m^3).** Another commonly used unit is the **cubic centimeter (cm^3),** and still another is the **liter (L).** One liter is

FIGURE 1-3
Determination of mass
by the principle of
weighing.

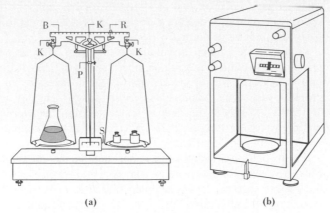

(a)   (b)

(a) The principle of weighing. A balance condition is achieved when the beam (B) is in a horizontal position and the pointer (P) rests at the center of the scale (S). When this occurs, the masses of the unknown and the standards are equal. The final adjustment of the balance point is made by moving a light metal wire called a rider (R) along the beam. The beam and pans pivot about fulcrums or pivot points called knife edges (K).

(b) A modern single-pan balance differs in several ways from the two-pan balance in (a). (1) Imagine that the balance in (a) is viewed "end on" so that only one pan is visible. (2) Now imagine that the pan out of view is replaced by a single large constant weight. (3) An appropriate set of weights is added *above* the single pan to just balance the constant weight. (4) When an unknown is placed on the pan, the balance condition is upset. This condition can be restored, however, by *removing* from the set above the pan a number of weights with mass equal to the unknown. The total mass removed is registered on the dials on the face of the balance.

defined as a volume of 1000 cm^3, which means that one **milliliter (1 mL)** is equal to one cubic centimeter (1 cm^3). The liter is also equal to one **cubic decimeter (1 dm^3).** These volume units are illustrated in Figure 1-2.

1-6 SI Units

The metric standards described in Section 1-5 have been modified to overcome certain difficulties. First is the difficulty in comparing objects with a standard when the standard is one of a kind (such as the standard meter in Sèvres). Furthermore, such standards are subject to change. For example, the metal bar that has served as the standard meter changes in length as the temperature changes.

Such difficulties can be overcome by basing standards of measurement on *natural* universal constants. Then, any scientist, at any time, may set up standards and be assured that they will be the same as those obtained by others. This has been provided for in an international agreement adopted in 1960 and called the International System of Units or **SI units** (from the French, Le Système International d'Unités). In the SI system the unit of length corresponding to 1 meter is defined as a length equal to the distance traveled by light in a vacuum during a time interval of 1/299,792,458 of a second. The unit of time, the second, is defined as the duration of 9,192,631,770 periods of a particular radiation emitted by cesium-133 atoms. The unit of mass, the kilogram, has not been defined in terms of a basic physical constant, so it remains the mass of a cylindrical bar of metal maintained at Sèvres.

Another aspect of the SI system is that to facilitate communication among scientists in

In the past physicists have used a version of the metric system called the mks (meter–kilogram–second) system, whereas chemists have preferred the cgs (centimeter–gram–second) system. This variability is eliminated in the SI system.

different disciplines and in different nations, certain base units and derived units are preferred over others. In time it is expected that the SI convention will be adopted and used consistently in all scientific work. However, as with any new development affecting the activities of large numbers of individuals, a transition period is necessary. In this text both familiar metric units and SI units are employed, and where they differ this fact is mentioned. Of the familiar units introduced to this point, the liter and milliliter are not SI units. Their SI counterparts are the cubic decimeter (dm^3) and the cubic centimeter (cm^3), respectively (recall Figure 1-2). A more complete description of the SI system is presented in Appendix C, to which you will need to refer from time to time. Some common SI prefixes are listed in Table 1-2.

1-7 Significant Figures

A number such as 299,792,458, which appears in the SI definition of the meter, is extremely precise: It is stated to one part in 300 million. On the other hand, a measured quantity such as 51 g is not so precisely known, only to about one part in 50. When a measured or calculated quantity is written down, some indication of the **precision** of the measurement should be given as well. For example, suppose that the same object is weighed on two different balances—one a relatively crude platform balance and the other a sophisticated analytical balance. Typical results might be

	Platform balance	Analytical balance
measured quantity	10.3 g	10.3107 g
uncertainty	±0.1 g	±0.0001 g
mass	10.3 ± 0.1 g	10.3107 ± 0.0001 g
precision	low or poor	high
	(one part in 103)	(one part in 103,107)

Precision refers to the degree of reproducibility of a measured quantity, that is, the closeness of agreement among the values obtained when the same quantity is measured several times. The precision of a series of repeated measurements is high if each of the individual measurements deviates from the average for the series by only a small amount. Conversely, if there is wide deviation among the measurements the precision is poor. Measurements that yield results close to the "correct" or most probable value are said to be **accurate.**

On a platform balance the determination of mass is reproducible only to the nearest one-tenth gram (±0.1 g), whereas on an analytical balance the measurement is reproducible to the nearest one-tenth *milli*gram (±0.0001 g). The notations "10.3 ± 0.1g" and "10.3107 ± 0.0001 g" indicate the precision of these measurements very clearly. This is a type of notation frequently used in laboratory notebooks and reports in scientific journals. But this notation is a bit cumbersome to write and to use in numerical calculations.

An alternative approach is to assume that when a number is written down, all the digits preceding the last are known with certainty and that there is an *uncertainty of about one unit in the last digit shown.* Thus, the number 10.3 is "between 10.2 and 10.4," whereas the number 10.3107 is "between 10.3106 and 10.3108." The number 10.3 is said to consist of *three* significant figures, whereas 10.3107 consists of *six* significant figures. To designate how many significant figures in a number is to indicate the confidence with which the number is known. The greater the number of significant figures, the smaller the uncertainty (and the greater the precision) of a measurement.

Determining the numbers of significant figures in 4.006, 12.012, and 10.070 is not difficult: They are *4*, *5*, and *5*, respectively. The applicable rules are

TABLE 1-2
Some common SI prefixes

Multiple	Prefix
10^9	giga (G)
10^6	mega (M)
10^3	kilo (k)
10^{-1}	deci (d)
10^{-2}	centi (c)
10^{-3}	milli (m)
10^{-6}	micro (μ)[a]
10^{-9}	nano (n)

[a] The Greek letter μ (pronounced "mew").

1. All nonzero digits are *significant* (i.e., 4.006, 12.012, and 10.070).
2. Zeros placed between nonzero digits are *significant* (i.e., 4.006, 12.012, and 10.070).
3. Zeros at the end of a number to the *right* of the decimal point are *significant* (i.e., 10.070).

How many significant figures are there in 0.00002 and 0.000020? The number 0.00002 has only *one* significant figure, since

4. Zeros to the left of the first nonzero digit are *not significant*. (They simply locate the decimal point.)

The number 0.000020 has *two* significant figures, based on rules 3 and 4.
Finally, how many significant figures are there in 750 and 20,000? We cannot be certain whether the number 750 is meant to indicate 750 ± 10 (in which case there are *two* significant figures) or 750 ± 1 (in which case there are *three* significant figures). This ambiguity can be stated as follows:

5. Zeros appearing at the *end* of a number and to the *left* of the decimal point *may or may not be significant*.

A way of resolving this difficulty is to use exponential notation in expressing a number (see Appendix A). Thus, the number 20,000 is expressed differently depending on the precision with which it is known.

One significant figure	Two significant figures	Three significant figures
2×10^4	2.0×10^4	2.00×10^4

Precision can be neither gained nor lost during arithmetic operations. This requirement is generally met by this simple rule for multiplication and division of numbers: *The result may carry no more significant figures than the least precisely known quantity involved in the calculation.* In the chain multiplication that follows, the result should be rounded off to *three* significant figures.*

$$14.80 \quad \times \quad 12.10 \quad \times \quad 5.05 \quad = 904.\cancel{354000} = 904 = 9.04 \times 10^2$$
(4 sig. fig.)   (4 sig. fig.)   (3 sig. fig.)   (3 sig. fig.)

To prove this, multiply 14.80 by 12.10 and then multiply their product first by 5.04, then by 5.05, and finally by 5.06 (because the number 5.05 is really 5.05 ± 0.01). The results are 902.5632, 904.3540, and 906.1448. Since the variation in these three numbers begins with the third digit, only the first three digits are significant—904. In adding or subtracting numbers, the uncertainty in the sum or difference is the same as that of the least precisely known quantity. Consider the sum

```
115.016  g
 12.0     g
  3.5182 g
```
130.5$\cancel{342}$ g = 130.5 g

Although the least precisely known quantity, "12.0," carries only three significant fig-

---

*The rule to follow in "rounding off" is to increase the final digit by one unit if the digit dropped is greater than 5 and to leave the final digit unchanged if the digit dropped is less than 5. For example, to three significant figures, 15.56 rounds off to 15.6, and 15.54 rounds off to 15.5. If the digit dropped is 5, the final remaining digit is increased by one unit if necessary to make it *even;* otherwise, it is left unchanged. Thus, to three significant figures, 15.55 is rounded off to 15.6 and 15.45 is rounded off to 15.4.

ures, the sum has four. The limitation here is not on significant figures but on the fact that the sum cannot be expressed any more precisely than ±0.1, the absolute precision with which "12.0" is stated.

There are two situations when a number appearing in a calculation may be *exact*. This may occur by definition (3 ft = 1 yd; 2.54 cm = 1 in.) or as a result of counting rather than measurement (*two* hydrogen atoms in a water molecule). Exact numbers can be considered to have an unlimited number of significant figures.

1-8 Density

Density is obtained by dividing the mass of an object by its volume.

$$\text{density } (d) = \frac{\text{mass } (m)}{\text{volume } (V)} \tag{1.2}$$

A property whose magnitude depends on the quantity of material is an **extensive** property. Both mass and volume are extensive properties. Any property that is independent of the quantity of material is an **intensive** property. Density, which is the ratio of mass to volume, is an intensive property. Intensive properties are generally preferred for scientific work because of their independence of the quantity of matter being studied.

The mass of 1000 cm^3 of water at 4°C and normal atmospheric pressure is almost exactly (but slightly less than) 1 kg. The density of water under these conditions is 1000 g/1000 cm^3 = 1.000 g/cm^3. Because volume varies with temperature while mass remains constant, density is a function of temperature. At 20°C the density of water is 0.998 g/cm^3. Densities at 20°C for two other liquids are ethyl alcohol, 0.789 g/cm^3, and carbon tetrachloride, 1.59 g/cm^3; and for some representative solids, aluminum, 2.70 g/cm^3; iron, 7.86 g/cm^3; lead, 11.34 g/cm^3; gold, 19.3 g/cm^3. Numerical calculations involving the density concept are presented in Section 1-10.

There is an old riddle that goes: "What weighs more, a ton of bricks or a ton of feathers?" The correct answer is that they weigh the same. One who gives this answer has demonstrated insight into the meaning of mass—a measure of the quantity of matter. One who answers that the bricks weigh more than the feathers has confused mass and density. Matter in a brick is more concentrated than in feathers, that is, confined to a smaller volume; bricks are more dense than feathers.

The SI units of density are kg/m^3 or g/cm^3, but you may occasionally encounter density expressed in g/mL or, for gases, in g/L.

1-9 Temperature

Temperature is a property that is difficult to define, even though we have an intuitive idea of what it is. To say that **temperature** is the degree of "hotness" of an object is not very precise, yet it does convey a certain meaning. If two objects of different temperature are brought into contact, the warmer object becomes colder and the colder one becomes warmer. Eventually, both objects come to the same degree of "hotness"—the same temperature.

Temperature can be measured by its effect on some other measurable property, for example, length. One common temperature-measuring device, a thermometer, is based on the length of a liquid column in a thin capillary bore in a glass tube. As the temperature of the thermometer changes, so does the length of the liquid column, increasing as the temperature increases.

To set up a scale of temperatures requires establishing certain fixed points of temperature and a degree of temperature change. Two commonly used fixed points are the temperature at which ice melts (the ice point) and the temperature at which water boils (the steam point), both under normal atmospheric pressure.

On the Fahrenheit temperature scale the ice point is 32°F, the steam point is 212°F, and

FIGURE 1-4
A comparison of
temperature scales.

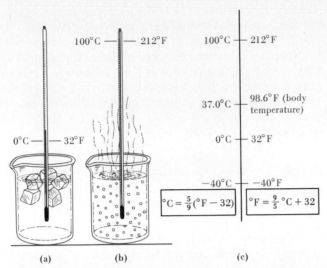

(a) Ice point. (b) Steam point. (c) Comparison of Fahrenheit and Celsius (formerly centigrade) temperature scales. Also shown are equations for converting between °C and °F. For example, to determine the Celsius equivalent of 68°F,

$$°C = \tfrac{5}{9}(°F - 32) = \tfrac{5}{9}(68 - 32) = \tfrac{5}{9}(36) = 20$$

The Fahrenheit equivalent of 30°C is

$$°F = \tfrac{9}{5}°C + 32 = \tfrac{9}{5}(30) + 32 = 86$$

the interval between is divided into 180 equal parts, called degrees Fahrenheit. On the Celsius (centigrade) scale the ice point is 0°C, the steam point is 100°C, and the interval is divided into 100 equal parts, called degrees Celsius. Figure 1-4 compares the Fahrenheit and Celsius temperature scales. Also presented in Figure 1-4 are equations used to convert temperatures between these two scales. The SI unit of temperature, the kelvin (K), is introduced in Section 5-3.

1-10 Problem Solving

The Conversion Factor (Factor-Label) Method. Where possible we will emphasize properties that can be expressed through numbers—quantitative measurements. But a number by itself is usually meaningless. A measured quantity must be accompanied by a **unit.** The unit indicates the standard against which the measured quantity is to be compared. A metal rod 9 m in length is nine times as long as the standard meter.

A number of scientific calculations require a conversion of measured quantities from one set of units to another, through the use of conversion factors. Consider this well-known fact, expressed as a simple mathematical equation.

1 yd = 36 in.

Divide each side of the equation by 1 yd.

$$\frac{1 \; \cancel{yd}}{1 \; \cancel{yd}} = \frac{36 \; in.}{1 \; yd}$$

The numerator and denominator on the left side are identical; they cancel.

$$1 = \frac{36 \text{ in.}}{1 \text{ yd}} \tag{1.3}$$

The numerator and denominator on the right side of equation (1.3) represent the *same length*. It is for this reason that the ratio of the numerator (36 in.) and the denominator (1 yd) is equal to 1. *A conversion factor must always have the numerator and denominator representing equivalent quantities.*

Consider the question: How many inches are there in 6 yd? The measured quantity is 6 yd and multiplying this quantity by 1 does not change its value.

6 yd × 1 = 6 yd

Now replace the 1 by its equivalent—the conversion factor (1.3). Cancel the unit, yd, and carry out the required multiplication.

$$6 \cancel{\text{yd}} \times \underbrace{\frac{36 \text{ in.}}{1 \cancel{\text{yd}}}}_{\substack{\text{this factor} \\ \text{converts} \\ \text{yd to in.}}} = 216 \text{ in.}$$

Next consider the question: How many yards are there in 540 in.? We cannot use exactly the same factor (1.3) as previously; the result would be nonsensical.

$$540 \text{ in.} \times \frac{36 \text{ in.}}{1 \text{ yd}} = 19{,}400 \text{ in.}^2/\text{yd}$$

Factor (1.3) must be rearranged to 1 yd/36 in.

$$540 \cancel{\text{in.}} \times \underbrace{\frac{1 \text{ yd}}{36 \cancel{\text{in.}}}}_{\substack{\text{this factor} \\ \text{converts} \\ \text{in. to yd}}} = 15 \text{ yd} \tag{1.4}$$

This second illustration emphasizes two important points.

1. There are two ways of writing a conversion factor—in one form or its reciprocal (inverse). Since a conversion factor is equivalent to 1, its value is not changed by inversion, but
2. A conversion factor must be used in such a way as to produce the desired cancellation of units.

A convenient way to think about calculations involving conversion factors is that

information sought = information given × conversion factor(s) (1.5)

Consider how expression (1.5) is used to answer this question: What is the length of 22 in., expressed in cm? The answer must consist of *two* parts—a number and a unit. The required unit is suggested in the statement of the problem—cm. The numerical part of the answer must be determined by calculation, and we refer to this as ''number'' or ''no.'' The *information sought* is *no. cm*. The *information given*, that is, the quantity that is to be multiplied by a conversion factor, is also determined by a close reading of the problem; it is *22 in*. Thus, expression (1.5) takes the form

no. cm = 22 in. × conversion factor

Sometimes the statement of the problem includes the necessary conversion factor(s), and sometimes you are expected to know or to be able to derive the factor(s) you need. *The key to problem solving by the conversion factor method lies in knowing where to find and how to use conversion factors.*

In the present case the necessary relationship is 1 in. = 2.54 cm; the unit *in.* must cancel and the unit *cm* must remain.

$$\text{no. cm} = 22 \ \cancel{\text{in.}} \times \underbrace{\frac{2.54 \text{ cm}}{1 \ \cancel{\text{in.}}}}_{\substack{\text{this factor} \\ \text{converts} \\ \text{in. to cm}}} = 56 \text{ cm}$$

Example 1-1   What is the distance 15 miles (mi), expressed in kilometers?

Solution.   The starting point is a form of expression (1.5). Information is sought in the unit kilometers (km) and the given information is the distance, 15 mi.

no. km = 15 mi × conversion factors

From this point the problem can be solved in a number of ways, depending on the conversion factors that are known, for example, miles → yards → inches → meters → kilometers. An alternative is miles → feet → inches → meters → kilometers.

$$\text{no. km} = 15 \ \cancel{\text{mi}} \times \frac{5280 \ \cancel{\text{ft}}}{1 \ \cancel{\text{mi}}} \times \frac{12 \ \cancel{\text{in.}}}{1 \ \cancel{\text{ft}}} \times \frac{1 \ \cancel{\text{m}}}{39.37 \ \cancel{\text{in.}}} \times \frac{1 \text{ km}}{1000 \ \cancel{\text{m}}} = 24 \text{ km}$$

$$(\text{mi} \longrightarrow \text{ft} \longrightarrow \text{in.} \longrightarrow \text{m} \longrightarrow \text{km})$$

SIMILAR EXAMPLES: Review Problems 1, 2, 3; Exercises 15 through 18.

Three new ideas are illustrated through Example 1-2.

(a) At times, if units appear to higher than the first power (e.g., squared or cubed), it may be necessary to raise conversion factors to higher powers.
(b) Often the solution to a problem can be seen more clearly by drawing a sketch of the physical situation, as in Figure 1-5.
(c) There are generally alternative ways of solving a problem, and any logical method is acceptable.

Example 1-2   How many square feet (ft^2) correspond to an area of 1.00 square meter (m^2)?

Solution.   An area of 1.00 m^2 is represented in Figure 1-5; it can be thought of as a square with sides 1 m long. Also depicted are the length, 1 ft, and the area, 1.00 ft^2. There are somewhat more than 9 ft^2 in 1 m^2.
    Expression (1.5) is written as follows:

$$\text{no. ft}^2 = 1.00 \text{ m}^2 \times \underbrace{\left(\frac{39.37 \text{ in.}}{1 \text{ m}}\right)\left(\frac{39.37 \text{ in.}}{1 \text{ m}}\right)}_{\substack{\text{to convert} \\ \text{m}^2 \text{ to in.}^2}} \times \underbrace{\left(\frac{1 \text{ ft}}{12 \text{ in.}}\right)\left(\frac{1 \text{ ft}}{12 \text{ in.}}\right)}_{\substack{\text{to convert} \\ \text{in.}^2 \text{ to ft}^2}}$$

This is the same as writing

$$\text{no. ft}^2 = 1.00 \ \cancel{\text{m}^2} \times \frac{(39.37)^2 \ \cancel{\text{in.}^2}}{1 \ \cancel{\text{m}^2}} \times \frac{1 \text{ ft}^2}{(12)^2 \ \cancel{\text{in.}^2}} = 10.8 \text{ ft}^2$$

Another way to look at the problem is to convert the length 1.00 m to feet

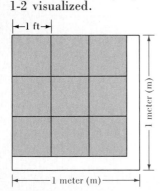

FIGURE 1-5
A comparison of one square foot and one square meter—Example 1-2 visualized.

|←1 ft→|

1 meter (m)

1 meter (m)

From this illustration can you see that 1 m is slightly longer than 3 ft, and that 1 m^2 is somewhat larger than 9 ft^2? The answer obtained in Example 1-2 (10.8 ft^2) is consistent with this observation.

$$\text{no. ft} = 1.00 \text{ m} \times \frac{39.37 \text{ in.}}{1.00 \text{ m}} \times \frac{1 \text{ ft}}{12 \text{ in.}} = 3.28 \text{ ft}$$

and square the result

$$\text{no. ft}^2 = 3.28 \text{ ft} \times 3.28 \text{ ft} - 10.8 \text{ ft}^2$$

SIMILAR EXAMPLES: Review Problem 4; Exercises 20, 21.

Two more new ideas are introduced through Example 1-3.

(d) In some calculations the answer may need to be expressed as a *ratio* of units. This, in turn, may require that one series of conversions be applied to the numerator and another series to the denominator.

(e) The number of conversion factors appearing in a setup may at times be rather large. Even in these cases the conversion-factor method works perfectly well, but it may be easier to follow the logic of the solution by breaking it down into several steps.

Example 1-3 What is the speed of 55.2 mi/h expressed in meters per second (m/s)?

Solution. What is required here is a conversion from miles to meters in the *numerator* and hours to seconds in the *denominator*. The necessary conversion factors must be set up with care to ensure that each factor has a numerator and denominator that are equivalent, and that each factor produces the desired cancellation of units.

$$\text{no. } \frac{m}{s} = \frac{55.2 \text{ mi}}{1 \text{ h}} \times \frac{5280 \text{ ft}}{1 \text{ mi}} \times \frac{12 \text{ in.}}{1 \text{ ft}} \times \frac{1 \text{ m}}{39.37 \text{ in.}} \times \frac{1 \text{ h}}{60 \text{ min}} \times \frac{1 \text{ min}}{60 \text{ s}}$$

$$= 24.7 \frac{m}{s}$$

Here is an alternative way of looking at the problem: Speed is a ratio of distance traveled to the time required. Basically what we must do here is to (1) convert 55.2 miles to a distance in meters, (2) convert one hour to a time in seconds, and (3) express the speed as the ratio of the two new quantities.

Step 1

$$\text{no. m} = 55.2 \text{ mi} \times \frac{5280 \text{ ft}}{1 \text{ mi}} \times \frac{12 \text{ in.}}{1 \text{ ft}} \times \frac{1 \text{ m}}{39.37 \text{ in.}} = 88,800 \text{ m}$$

Step 2

$$\text{no. s} = 1 \text{ h} \times \frac{60 \text{ min}}{1 \text{ hr}} \times \frac{60 \text{ s}}{1 \text{ min}} = 3600 \text{ s}$$

Step 3

$$\text{no. } \frac{m}{s} = \frac{88,800 \text{ m}}{3600 \text{ s}} = 24.7 \frac{m}{s}$$

SIMILAR EXAMPLES: Exercises 19, 22.

The next four examples illustrate the density concept. In Example 1-4 a geometric formula is used to calculate volume; in Example 1-5 volume is measured by the displacement of a liquid; in Example 1-6 the volume of a liquid is calculated from its mass, with density as a conversion factor; and in Example 1-7 a precise density determination with a pycnometer is illustrated.

FIGURE 1-6
Measurement of the
volume of irregularly
shaped objects.

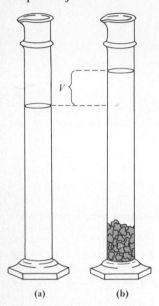

(a)   (b)

(a) A graduated cylinder
is filled to a certain level
with a liquid.
(b) Irregularly shaped
objects are added to the
liquid. They sink to the
bottom of the cylinder and
cause the liquid level to
rise to an extent *V* that
corresponds to their total
volume. (If you are familiar with cooking tech-
niques, you may recognize
this as a common method
of measuring the volumes
of solid fats, such as butter.)

Example 1-4   A block of wood having the dimensions 105 cm × 5.1 cm × 6.2 cm
weighs 2.72 kg. What is the density of the wood, expressed in grams per cubic centimeter?

Solution.   The volume of a rectangular block (a parallelepiped) is simply the product
of the length (*l*), the width (*w*), and the height (*h*).

$$V = 105 \text{ cm} \times 5.1 \text{ cm} \times 6.2 \text{ cm} = 3300 \text{ cm}^3$$

The mass of the block must be expressed in grams.

$$m = 2.72 \text{ kg} \times \frac{1000 \text{ g}}{1 \text{ kg}} = 2720 \text{ g}$$

The density of the wood is

$$d = \frac{m}{V} = \frac{2720 \text{ g}}{3300 \text{ cm}^3} = 0.82 \text{ g/cm}^3$$

SIMILAR EXAMPLES: Review Problem 6; Exercises 27, 28.

Example 1-5   Several irregularly shaped pieces of zinc, weighing 30.0 g, are
dropped into a graduated cylinder containing 20.0 cm³ of water. The water level rises
to 24.2 cm³. What is the density of the zinc?

Solution.   The volume of metal is the difference of the two water levels in Figure 1-6.

volume of zinc = 24.2 cm³ − 20.0 cm³ = 4.2 cm³

$$\text{density} = \frac{\text{mass}}{\text{volume}} = \frac{30.0 \text{ g}}{4.2 \text{ cm}^3} = 7.1 \text{ g/cm}^3$$

SIMILAR EXAMPLE: Exercise 32.

Example 1-6   What is the volume, in liters, occupied by 50.0 kg of ethanol at 20°C?
The density of ethanol at 20°C is 0.789 g/cm³.

Solution.   The information given is *50.0 kg of ethanol* and what is sought is *number
of liters of ethanol*. The series of conversions required is kg → g → cm³ → L. Density
can be thought of as a conversion factor between mass and volume: 1.00 cm³ of
ethanol = 0.789 g of ethanol.

$$\text{no. L ethanol} = 50.0 \text{ kg ethanol} \times \frac{1000 \text{ g ethanol}}{1 \text{ kg ethanol}} \times \frac{1.00 \text{ cm}^3 \text{ ethanol}}{0.789 \text{ g ethanol}}$$
$$\times \frac{1 \text{ L ethanol}}{1000 \text{ cm}^3 \text{ ethanol}}$$
$$= 63.4 \text{ L ethanol}$$

An alternative method involves solving the density equation for volume, *V* = *m/d*,
and substituting the appropriate information.

$$V = \frac{50,000 \text{ g}}{0.789 \text{ g/cm}^3} = 63,400 \text{ cm}^3 = 63.4 \text{ L}$$

SIMILAR EXAMPLES: Review Problem 7; Exercises 29, 31.

Example 1-7   The density of methanol at 20°C is determined by the series of measurements pictured in Figure 1-7. Express this density in the appropriate number of
significant figures.

Solution.   The mass of water required to fill the pycnometer at 20°C is

FIGURE 1-7
Determination of density
with a pycnometer.

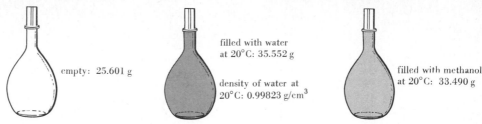

empty: 25.601 g

filled with water
at 20°C: 35.552 g

density of water at
20°C: 0.99823 g/cm³

filled with methanol
at 20°C: 33.490 g

The calculations required in this density determination are the subject of Example 1-7.

$$35.552 \text{ g} - 25.601 \text{ g} = 9.951 \text{ g}$$

The volume of water, and hence that of the pycnometer, is

$$V = \frac{m}{d} = \frac{9.951 \text{ g}}{0.99823 \text{ g/cm}^3} = 9.969 \text{ cm}^3$$

The mass of methanol required to fill the pycnometer at 20°C is

$$33.490 \text{ g} - 25.601 \text{ g} = 7.889 \text{ g}$$

The density of methanol at 20°C is

$$d = \frac{m}{V} = \frac{7.889 \text{ g}}{9.969 \text{ cm}^3} = 0.7914 \text{ g/cm}^3$$

SIMILAR EXAMPLES: Exercise 29; Additional Exercise 8.

FIGURE 1-8
The concept of
equivalence.

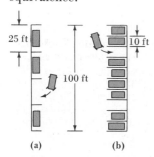

(a)   (b)

(a) For a parallel parking
arrangement we can say
that each automobile *is
equivalent to* 25 ft. That
is, 1 automobile $\backsimeq$ 25 ft.

no. automobiles

$$= 100 \text{ ft} \times \frac{1 \text{ automobile}}{25 \text{ ft}}$$

$$= 4 \text{ automobiles}$$

(b) In this parking arrangement (perpendicular)
each automobile is equiv-
alent to 10 ft of curb
space. How many automo-
biles can be parked along
the 100-ft section of
curb?

Equivalence and Equality. An additional insight into the use of conversion factors is suggested by Figure 1-8, which establishes an equivalence between an automobile and the curb space required to park it. In Figure 1-8a we need a conversion factor based on the relationship 1 automobile $\backsimeq$ 25 ft, where an equivalence sign ($\backsimeq$) has been used rather than an equality sign (=). An automobile is not literally the same thing as 25 ft of curb space. *For the purposes of this calculation* we consider them to be *equivalent:* For every 25 ft of curb space we are able to park one automobile (regardless of its actual length). A different equivalence between the automobile and curb space exists for the perpendicular parking arrangement in Figure 1-8b. We will use the sign = when an equality exists and $\backsimeq$ for an equivalence, but conversion factors are set up and used in the same way in either case.

Percent as a Conversion Factor. The term **percent** means, literally, parts per hundred, that is, parts of one constituent to 100 parts of the whole. For example, the statement that a seawater sample contains 3.5% sodium chloride, by mass, means that in a 100-g sample of the seawater there will be found 3.5 g of sodium chloride.

3.5 g sodium chloride $\backsimeq$ 100 g seawater   (1.6)

A common difficulty encountered by students in using the notion of percent can be put as a question: When do you multiply and when do you divide by percent? Expressing percentages as conversion factors should help to clarify this matter. The factors are always set up to require multiplication, as illustrated in Examples 1-8 and 1-9.

Example 1-8 325 g of seawater containing 3.5% sodium chloride, by mass, is evaporated to dryness. How much sodium chloride is present in the solid residue?

Solution. Expression (1.6) is written as a conversion factor with "g sodium chloride" in the numerator and "g seawater" in the denominator.

$$\text{no. g sodium chloride} = 325 \text{ g seawater} \times \frac{3.5 \text{ g sodium chloride}}{100 \text{ g seawater}}$$

$$= 11 \text{ g sodium chloride}$$

SIMILAR EXAMPLES: Review Problem 8; Exercise 34.

Example 1-9 87 g of sodium chloride is to be produced by evaporating to dryness a quantity of seawater containing 3.5% sodium chloride, by mass. The seawater has a density of 1.03 g/cm³. How many *liters* of seawater must be taken for this purpose?

Solution. In this example we need a conversion factor with "g seawater" in the numerator and "g sodium chloride" in the denominator. Also required are conversion factors for g seawater → cm³ seawater → L seawater.

$$\text{no. L seawater} = 87 \text{ g sodium chloride} \times \frac{100 \text{ g seawater}}{3.5 \text{ g sodium chloride}}$$

$$\times \frac{1 \text{ cm}^3 \text{ seawater}}{1.03 \text{ g seawater}} \times \frac{1 \text{ L seawater}}{1000 \text{ cm}^3 \text{ seawater}}$$

$$= 2.4 \text{ L seawater}$$

SIMILAR EXAMPLES: Review Problem 9; Exercise 35.

Algebraic Solutions to Problems. Several types of problems encountered in this text cannot be solved by the conversion-factor method. They require instead the solution of algebraic equations. The simplest situation is one in which the necessary equation is given and a solution is obtained by substituting a numerical value for one of the variables, as in Example 1-10. Other times it is necessary to *rearrange* the equation to solve for the desired unknown; this is the case in Example 1-11. The most demanding situation is one in which the necessary equation must be *derived* for the particular problem and then solved. Applications of this type will arise later in the text. A brief review of basic algebra is provided in Appendix A.

A detailed discussion of the properties of gases, e.g., pressure, is provided in Chapter 5, together with a more useful method of calculating gas densities. What we are illustrating here is not so much the behavior of gases as the use of algebraic equations.

Example 1-10 A handbook gives the following equation for determining the density (d) of dry air (in g/cm³) at standard atmospheric pressure as a function of Celsius temperature (t): $d = 0.001293/(1 + 0.00367t)$. What is the density of dry air at 25°C?

Solution. Substitute $t = 25$ and solve for d.

$$d = \frac{0.001293}{1 + (25 \times 0.00367)} = \frac{0.001293}{1 + 0.092} = \frac{0.001293}{1.092} = 0.001184 \text{ g/cm}^3$$

SIMILAR EXAMPLES: Review Problem 15; Exercise 36.

Example 1-11 Use the equation of Example 1-10 to determine the temperature at which the density of dry air at standard atmospheric pressure is 0.001200 g/cm³.

Solution. Rearrange the density equation to solve for t.

$$d = \frac{0.001293}{(1 + 0.00367 \, t)} \qquad 1 + 0.00367 \, t = \frac{0.001293}{d}$$

$$0.00367 \, t = (0.001293/d) - 1 \qquad t = \frac{(0.001293/d) - 1}{0.00367}$$

Substitute $d = 0.001200$.

$$t = \frac{(0.001293/0.001200) - 1}{0.00367} = \frac{1.078 - 1}{0.00367} = \frac{0.078}{0.00367} = 21°\text{C}$$

SIMILAR EXAMPLES: Exercises 37, 38.

FOCUS ON
Some Key Chemicals

The known chemical substances number about five million. Of this large number, several tens of thousands are commercially important, but only a relatively small number are key chemicals. These are the ones used to synthesize other chemicals and to produce materials important to modern society. In turn, these key chemicals are derived from a small number of basic raw materials, such as air, water, limestone, salt (sodium chloride), sulfur, phosphate rock, coal, petroleum, and natural gas. We will refer to basic raw materials and key chemicals throughout this text. Specifically, we will consider how the key chemicals are produced from raw materials—nitrogen and oxygen from air (Chapter 13); sulfuric acid from sulfur, air, and water (Chapter 14); ammonia from air and natural gas (Chapter 15); etc. Moreover, we will discover that many of the key chemicals nicely illustrate fundamental chemical principles.

Table 1-3 lists some of the key chemicals, their relative order in terms of quantity produced, and their important uses. The entries in this table are referred to by name

The production of key chemicals usually requires a large, complex, and costly industrial facility. One of the first steps in constructing such a plant is to build a model that can be used to visualize the plant operation. [Courtesy of The M. W. Kellogg Company, a subsidiary of Kellogg Rust, Inc., one of The Signal Companies, Inc.]

TABLE 1-3
Some of the top 50 chemicals produced in the United States (1982)

Rank	Billions of pounds	Chemical	Formula	Principal uses and end products
1	65.36	sulfuric acid	H_2SO_4	fertilizers (70%); metallurgy (5%); petroleum refining (5%); manufacture of chemicals (5%)
2	35.07	nitrogen	N_2	inert atmospheres for manufacture of chemicals (25%); electronics manufacturing (15%); petroleum recovery (15%); metals processing (5%); freezing foods (5%)
3	31.56	ammonia	NH_3	fertilizers (80%); manufacture of plastics and fibers (10%) and explosives (5%)
4	28.81	oxygen	O_2	fabricating metals (35%); steel-making and other metallurgical processes (30%); medical uses (15%)
5	28.15	calcium oxide (lime)	CaO	metallurgy (40%); manufacture of chemicals (10%); water treatment (10%); sewage and pollution control (10%); construction (5%); manufacture of pulp and paper (5%)
6	24.50	ethylene	C_2H_4	manufacture of plastics (65%), antifreeze (10%), fibers (5%), solvents (5%)
7	18.77	sodium hydroxide	$NaOH$	manufacture of chemicals (50%), pulp and paper (20%), soaps and detergents (5%); oil refining (5%)
8	18.35	chlorine	Cl_2	manufacture of organic chemicals and plastics (70%), pulp and paper (10%)
9	16.52	phosphoric acid	H_3PO_4	fertilizers (85%); detergent components (5%); animal feed (5%)

17

TABLE 1-3 (continued)

Rank	Billions of pounds	Chemical	Formula	Principal uses and end products
10	15.64	sodium carbonate	Na_2CO_3	manufacture of glass (55%), chemicals (25%), pulp and paper (5%), soaps and detergents (5%)
13	13.04	urea	$CO(NH_2)_2$	fertilizers (80%); animal feed (10%); manufacture of adhesives and plastics (5%)
17	7.56	methyl alcohol	CH_3OH	manufacture of adhesives, fibers and plastics (50%), solvents (10%)
18	7.36	carbon dioxide	CO_2	refrigeration (30%); carbonated beverages (25%); manufacture of urea (15%)
24	4.90	vinyl chloride	CH_2CHCl	manufacture of plastics (~100%)
35	2.30	carbon black	C	tires (65%); other rubber goods (25%); pigment for plastics and ink (10%)
37	2.02	phenol	C_6H_5OH	manufacture of adhesives (60%), plastics (20%), fibers (10%)
38	1.92	butadiene	C_4H_6	synthetic rubber (70%); fibers (10%)
42	1.69	acetone	$(CH_3)_2CO$	solvents (60%); manufacture of plastics (30%)
46	1.31	titanium dioxide	TiO_2	paints (50%); paper filler and coatings (25%); plastics and rubber filler (15%); ceramics (5%)
50	1.02	ethyl alcohol	C_2H_5OH	solvents (80%); food and beverages (10%)

Sources: *Chemical and Engineering News:* May 7, 1984; Oct. 24, 1983; Sept. 19, 1983; June 20, 1983; May 2, 1983; Mar. 21, 1983; Jan. 24, 1983; Oct. 25, 1982; Aug. 2, 1982; Apr. 26, 1982; Mar. 1, 1982; Nov. 9, 1981.

and by symbolic designations known as chemical formulas. We will begin the task of relating names and formulas—referred to as chemical nomenclature—in the next chapter and gradually expand our knowledge of chemical nomenclature throughout the text. Most of the first 10 chemicals in Table 1-3 (ethylene, C_2H_4, is the exception) are derived from inanimate mineral sources and are called inorganic chemicals. Most of the remaining chemicals in Table 1-3 are derived from a source (coal, petroleum, natural gas) that was living matter at one time. These carbon-and-hydrogen-based compounds are called organic chemicals. We will comment further on these two broad categories of compounds in later chapters. In later chapters we will also discover other important ways of categorizing compounds to help us understand their behavior. Some of these key chemicals will be featured in those discussions.

The chemical industry is that segment of industry that produces chemicals and products obtainable from them. Certain other sectors of industry deal with materials that are not usually thought of as chemicals but which are chemicals nevertheless, such as gasoline and liquefied petroleum gases (LPG) in the petroleum industry and iron and steel, aluminum, magnesium, copper, and other metals in the metallurgical industry. Our study of chemistry will deal with chemicals in a very broad sense.

Summary

Chemistry is a study of matter, and one of the first needs in this study is to classify matter into useful categories. The scheme introduced in this chapter considers matter as being either a substance—element or compound—or a mixture—homogeneous or heterogeneous. Of interest to chemists are the distinctive characteristics or properties of matter and how these may be changed by physical and chemical means. Like other branches of modern science, chemistry makes use of the scientific method. This consists of a series of activities that starts with observations and culminates in theories to explain and predict natural phenomena.

The need for precise measurement in chemistry is essential, and a uniform system for expressing the results of measurements is also important. The SI system provides this uniformity, but because SI units have not yet been universally adopted, other units of measurement are also used in this text. Still another requirement is for methods of performing calculations involving measured quantities. A method introduced in the chapter requires recognizing relationships between quantities and using these relationships to establish conversion factors. The desired result is obtained by multiplying a given quantity by one or more conversion factors. In this method cancellation of units is used as a guide to ensure that conversion factors are properly formulated. Some chemical calculations, however, require that algebraic equations be solved. Finally, measurements cannot be performed with certainty. Measurements should be expressed in such a way as to indicate the uncertainties that exist in them. This is done through significant figures and certain simple rules for handling significant figures in arithmetic operations.

Learning Objectives

As a result of studying Chapter 1, you should be able to

1. Use the terms *element, compound, homogeneous mixture,* and *heterogeneous mixture* to describe common materials.

2. Write the names and chemical symbols of about one half the elements, including the first 20 (see the inside front and back covers).

3. Distinguish between physical and chemical properties and simple physical and chemical changes.

4. Describe the principal features of the scientific method and its limitations.

5. State the basic units of mass, length, and volume and the common prefixes in the metric system.

6. Relate at least one unit each of mass, length, and volume in the English system to a corresponding unit in the metric system.

7. Describe the relationship of the SI to the older metric system, particularly the ways in which the systems resemble and differ from one another.

8. State the number of significant figures in a numerical quantity.

9. Express the result of a calculation with the appropriate number of significant figures.

10. Convert Fahrenheit temperature to Celsius, and Celsius to Fahrenheit.

11. Write a conversion factor from a relationship between two quantities.

12. Express density and percent composition in the form of conversion factors.

13. Use conversion factors in a general problem-solving method.

14. Solve algebraic equations that arise in the course of working chemistry problems.

Some New Terms

A **chemical change** is a transformation of one or more substance(s) into one or more new substance(s).

A **chemical property** describes a type of chemical change a substance can undergo: for example, to combine with oxygen or to dissolve in an acid.

A **compound** is a substance made up of two or more elements. It does not change its identity in physical changes but can be broken down into its constituent elements by chemical changes.

Density is a physical property obtained by dividing the mass of a substance or object by its volume (i.e., mass per unit volume).

An **element** is one of a group of fundamental substances that cannot be broken down into simpler substances.

An **extensive property** is one, like mass or volume, whose value depends on the quantity of matter observed.

In a **heterogeneous mixture** the components separate into physically distinct regions of differing properties.

A **homogeneous mixture (solution)** is a mixture of elements and/or compounds that has a uniform composition within a given sample but a varying composition from one sample to another.

An **intensive property** is *independent* of the quantity of matter involved in the observation. Examples of such properties are density and temperature.

Mass is a measure of the inertia possessed by an object. (Inertia is the tendency to remain at rest or in constant motion unless acted upon by an external force. The more mass in an object, the greater its inertia.)

Matter is anything that occupies space and has mass.

A **mixture** is any sample of matter that is not pure, i.e., not an element or compound. The composition of a mixture, unlike that of a substance, can be varied. Mixtures are either *homogeneous* or *heterogeneous*.

A **natural law** is a general statement that can be used to summarize observations of natural phenomena.

A **physical change** is one in which the physical appearance of a substance changes but its basic identity, that is, its chemical composition, remains unchanged.

A **physical property** is a characteristic that a substance can display without undergoing a change in its identity.

Scientific method refers to the general sequence of activities—observation, experimentation, and the formulation of laws and theories—that lead to the advancement of scientific knowledge.

Significant figures are those digits in an experimentally measured quantity that establish the precision with which the quantity is known.

A **substance** has constant composition and properties throughout a given sample and from one sample to another; all substances are either elements or compounds.

A **theory** is a conceptual framework with which one is able to *explain* one or a group of related natural laws.

Weight refers to the force exerted on an object when it is placed in a gravitational field (the "force of gravity"). The terms *weight* and *mass* are often used synonomously.

Suggestions for Further Study

BATTINO, R., and A. G. WILLIAMSON, "Single-Pan Balances, Buoyancy, and Gravity, or a 'Mass of Confusion'," *J. Chem. Educ., 61,* 51 (1984).

C&EN Staff, "C&EN's Top 50 Chemical Products and Producers," *Chem. & Eng. News,* first issue in May, annually.

National Bureau of Standards, *A Metric America: A Decision Whose Time Has Come,* Special Publication 345, July 1971.

ROBINSON, A. L., "Using Time to Measure Length," *Science,* **220,** 1367 (1983).

WERTIME, T. A., "Pyrotechnology: Man's First Industrial Use of Fire," *American Scientist,* **61,** 670 (1973).

WITTCOFF, H., and W. C. FERNELIUS, "The Chemical Industry: What Is It?," *J. Chem. Educ.,* **56,** 253 (1979).

Review Problems

1. Perform the following conversions within the metric system.
   (a) 1.17 kg = _____ g
   (b) 7115 mm = _____ m
   (c) 621 mg = _____ kg
   (d) 673 mL = _____ L
   (e) 17.3 cm = _____ mm
   (f) 0.482 L = _____ cm^3
   (g) 2.07 g = _____ mg
   (h) 0.481 km = _____ m

2. Perform the following conversions within the English system.
   (a) 21.18 ft = _____ in.
   (b) 416 oz = _____ lb
   (c) 14.0 yd = _____ in.
   (d) 2.5 h = _____ s
   (e) 2721 ft = _____ yd
   (f) 3.70 mi = _____ ft

3. Peform the following conversions between the English and metric system.
   (a) 21 in. = _____ cm
   (b) 22 ft = _____ m
   (c) 11 oz = _____ g
   (d) 63 kg = _____ lb
   (e) 31.3 m = _____ ft
   (f) 5500 mg = _____ oz

4. Determine the number of
   (a) square meters (m^2) in 1 square kilometer (km^2);
   (b) square feet (ft^2) in 1 square mile (mi^2);
   (c) square meters in 1 square mile;
   (d) cubic inches ($in.^3$) in 1 cubic foot (ft^3);
   (e) cubic centimeters (cm^3) in 1 cubic foot.
   (f) cubic nanometers (nm^3) in 1 cubic centimeter (cm^3).

5. Use the equations in Figure 1-4 to perform the following conversions between temperature scales.
   (a) 35°C = _____°F
   (b) 86°F = _____°C
   (c) 832°F = _____°C
   (d) −163°C = _____°F

6. A 2.50-L sample of pure glycerol has a mass of 3153 g. What is the density of glycerol?

7. Ethylene glycol, an antifreeze, has a density of 1.11g/cm^3 at 20°C.
   (a) What is the mass, in grams, of 3.50×10^2 cm^3 of the liquid?

   (b) What is the volume, in liters, occupied by 2.50 kg of the liquid?
   (c) What is the mass, in pounds, of 4.00 gal of the liquid?

8. A particular fertilizer is listed as containing 6.6% phosphorus by mass. What mass, in grams, of phosphorus is contained in a 25.0-lb bag of the fertilizer?

9. A sample of solid sodium chloride is said to be 99.2% pure. How much of this solid sample must be taken to contain 135.0 g of sodium chloride?

10. Express the following numbers in exponential notation (see Appendix A).
    (a) 3800
    (b) 482,000
    (c) 0.100
    (d) 6,212
    (e) 211,100
    (f) 0.0000065
    (g) 0.0087
    (h) 0.0600
    (i) 22

11. Express the following numbers in common decimal form (see Appendix A).
    (a) 6.18×10^3
    (b) 8.12×10^{-1}
    (c) 4.613×10^{-4}
    (d) 43×10^{-3}
    (e) 3.47×10^{-6}
    (f) 670×10^{-2}
    (g) 6.2×10^0
    (h) 2.98×10^{10}
    (i) 0.00168×10^{-4}

12. How many significant figures are shown in each of the following numbers? If indeterminate, give the range of significant figures possible.
    (a) 317
    (b) 420
    (c) 0.051
    (d) 620.03
    (e) 9.0062
    (f) 4620.0
    (g) 88,000
    (h) 0.01070
    (i) 0.0007

13. Rewrite each of the following numbers to consist of *four* significant figures.

(a) 3162.3   (b) 3.2×10^3

(c) 218.51   (d) 0.065045

(e) 60×10^{-5} (f) 327.251

(g) 186,000   (h) 22×10^4

(i) 14.7050

14. Perform the following calculations, expressing each number and the answer in exponential form and with the correct number of significant figures.

(a) $200 \times 4000 =$ (b) $32 \times 30 \times 6100 =$

(c) $0.087 \times 0.0040 =$ (d) $0.0070 \times 612 =$

(e) $\dfrac{5300}{0.0070} =$ (f) $\dfrac{140 \times 600 \times 0.10}{0.030 \times 3.3} =$

15. The following equation can be used to relate the density of liquid water to Celsius temperature (in the range from 0°C to about 20°C).

$$d \ (\text{g/cm}^3) = \frac{0.99984 + 1.6945 \times 10^{-2}t - 7.987 \times 10^{-6}t^2}{1 + (1.6880 \times 10^{-2}t)}$$

To *four* significant figures determine the density of water at (a) 5°C and (b) 10°C.

Exercises

Properties and classification of matter

1. State whether each property is physical or chemical.

(a) An iron nail is attracted to a magnet.

(b) A silver spoon is tarnished in air.

(c) Ice floats on liquid water.

(d) Rubber objects disintegrate in smog-filled air.

2. Indicate whether each sample of matter listed is a substance or a mixture, and, if a mixture, whether homogeneous or heterogeneous.

(a) a cube of sugar   (b) clam chowder

(c) unleaded gasoline   (d) mayonnaise

(e) iodized salt   (f) tap water

(g) ice   (h) white paint

3. What type of change—physical or chemical—is necessary to bring about each of the following separations? (*Hint:* Refer to the inside back cover for a listing of elements.)

(a) hydrogen and oxygen gases from water

(b) pure water from seawater

(c) nitrogen and oxygen gases from air

4. Suggest physical changes by which the following mixtures can be separated.

(a) salt and sand

(b) iron filings and wood chips

(c) mineral oil and water

5. Indicate which of the following are extensive and which are intensive quantities.

(a) the mass of air in a balloon

(b) the temperature of boiling water

(c) the length of time required to melt ice

(d) the color of light given off by a neon lamp

Scientific method

6. Is it possible to predict how many experiments are required to verify a natural law? Explain.

7. What are the principal reasons why one theory might be adopted over a conflicting one?

8. An important premise of science is that there exists an underlying order to nature. Einstein described this belief in the words "God is subtle but He is not malicious." Explain more fully what he probably meant by this remark.

Exponential arithmetic (see Appendix A)

9. A variety of measured or estimated quantities follow. Express each value in exponential form.

(a) speed of light in vacuum: 186 thousand miles per second

(b) mass of air in the atmosphere: 5 to 6 quadrillion tons

(c) solar radiation received by the earth: 173 thousand trillion watts

(d) diameter of a typical aerosol smog particle: one millionth of a meter

(e) average diameter of a human cell: ten millionths of a meter

10. Express the result of each of the following calculations in exponential form.

(a) $0.048 + (62 \times 7.0 \times 10^{-4}) =$

(b) $\dfrac{31 + 283 + (1.60 \times 10^2)}{2.3 \times 10^{-1}} =$

(c) $\dfrac{(1.7 \times 10^{-3})^2}{0.060 + (2.0 \times 10^{-2})} =$

(d) $\dfrac{[(6.0 \times 10^3) + (4.2 \times 10^4)]^2}{(2.2 \times 10^3)^2 + 180,000} =$

Significant figures

11. Indicate whether each of the following is an exact number or a measured quantity subject to uncertainty.

(a) the number of oranges in one dozen

(b) the number of gallons of gasoline to fill an automobile gas tank

(c) the distance between the earth and the sun

(d) the number of days in the month of January

(e) the area of a city lot

12. Perform the following calculations, retaining the appropriate number of significant figures in each result.
   (a) $411 \times 183 =$
   (b) $12.30 \times 10^4 \times 3.5 \times 10^5 =$
   (c) $\dfrac{5.11 \times 10^2}{1.7 \times 10^5} =$
   (d) $35.24 + 36.3 + 1.08 =$
   (e) $(1.561 \times 10^3) - (1.80 \times 10^2) + (2.02 \times 10^4) =$

13. To determine the mass of an object by the method of Figure 1-3a, the following standard masses are required to achieve a balance condition: 20 g, 1 g, 500 mg, 200 mg, 100 mg, 50 mg, 20 mg, 5 mg, and 2 mg. Assuming that each mass is accurate to the nearest 1 mg, what is the mass of the object, expressed with an appropriate number of significant figures?

*14. According to the rules on significant figures, the product 99.9×1.008 should be expressed to three significant figures—101. Yet, in this case, it would be appropriate to express the result to *four* significant figures—100.7. Explain why this is so. (*Hint:* The relative precision of the product can be no greater than that of the least precisely stated number.)

Systems of measurement

15. The English unit, the rod, is equal to 16.5 ft. What is this length expressed in meters?

16. A certain brand of coffee is offered for sale at $7.26 for a 3-lb can or $5.42 for a 1-kg can. Which is the better buy?

17. A sprinter runs the 100-yd dash in 9.3 s. What would be his time for a 100-m run if he ran at the same rate?

18. The unit of length, the furlong, is used in horseracing. The units of length, the chain and the link, are used in surveying. There are 8 furlongs in 1 mi, 10 chains in 1 furlong, and 100 links in 1 chain. To three significant figures, what is the length of 1 link in inches?

19. An English unit of mass used in pharmaceutical work is the grain (gr). 15 gr = 1.0 g. An aspirin tablet contains 5.0 gr of aspirin. A 145-lb person takes two aspirin tablets.
   (a) What is the quantity of aspirin taken, expressed in milligrams?
   (b) What is the dosage rate of the aspirin, expressed in milligrams of aspirin per kilogram of body weight?

20. A block of ice measures 24 in. $\times$ 36 in. $\times$ 18 in.
   (a) What is the volume of this block in cubic meters?
   (b) What is the total surface area of the block in square centimeters?

*21. In the English system of measurement the *acre* is an important unit for measuring land area. The corresponding unit in the metric system is the *hectare*. There are 640 acres in 1 square mile (mi^2), and 1 hectare is defined as 1 square hectometer (hm^2); 1 hm = 100 m.
   (a) How many square feet are there in 1 acre?
   (b) How many hectares are there in 1 acre?

*22. An airplane flying at the speed of sound is said to be at Mach 1. Mach 1.5 is 1.5 times the speed of sound; and so on. If the speed of sound in air is given as 1130 ft/s, what is the speed of an airplane, in km/h, flying at Mach 1.38?

*23. Table 1-1 lists that 1 in. = 2.54 cm and that 1 m = 39.37 in.
   (a) Explain why only one of these statements can be *exact*.
   (b) The accepted basic relationship between the English and metric systems is that 1 in. = 2.54 cm, *exactly*. How many inches are there in one meter, expressed to six significant figures?

Temperature scales

24. A table of climatic data lists the highest and lowest temperatures on record for San Bernardino, California, as 118°F and 17°F, respectively. What are these temperatures on the Celsius scale?

25. A class in home economics is given an assignment in candy making. The candy recipe calls for a sugar mixture to be brought to a ''soft ball'' stage (234 to 240°F). A student borrows a thermometer of range -10 to 110°C from the chemistry laboratory to do this assignment. Will this thermometer serve the purpose?

26. The absolute zero of temperature, introduced in Section 5-3, is -273.15°C. What is the absolute zero of temperature on the Fahrenheit scale?

Density

27. To determine the density of a liquid, a 125-mL volumetric flask is weighed when empty (110.4 g) and again when filled to the mark with liquid (209.4 g). What is the density of the liquid?

28. A barrel contains 42.0 gal of petroleum having a mass of 298 lb. What is the density of this petroleum in g/cm^3?

29. It is desired to determine the volume of liquid that can be contained in an irregularly shaped glass vessel. The vessel is weighed when empty and found to have a mass of 121.3 g. Filled with liquid carbon tetrachloride (density = 1.59 g/cm^3), the vessel weighs 283.2 g. What is the volume capacity of the vessel?

30. A European cheese-making recipe calls for 2.50 kg of whole milk. An American who wants to make the recipe has no scale and only volume-measuring containers marked in quarts, pints, cups, and fractions thereof. Assuming that the density of milk is 1.03 g/cm^3, what volume of milk is needed? (1 L = 1.06 qt; 1 qt = 2 pt; 1 pt = 2 cups)

31. The following densities are given at 20°C: water, 0.998 g/cm^3; iron, 7.86 g/cm^3; aluminum, 2.70 g/cm^3. Arrange the following items in terms of *increasing* mass.
   (a) A rectangular bar of iron, 162 cm $\times$ 1.1 cm $\times$ 0.70 cm.
   (b) A sheet of aluminum foil, 165.0 cm $\times$ 23.0 cm $\times$ 0.980 mm.
   (c) 1.00 L of water.

32. To determine the approximate mass of a small spherical shot of copper metal, the following experiment is performed. 100 pieces of the shot are counted out and added to 8.4 mL of water in a graduated cylinder; the total volume becomes 8.8 mL. The density of copper metal is 8.92 g/cm^3. Determine the approximate mass of a single piece of shot, assuming that all the pieces are of nearly the same dimensions.

Percent composition

33. In a class of 55 students the results of a particular examination were 7 A, 14 B, 24 C, 7 D, and 3 F. What was the percent distribution of grades, that is, % A, % B, and so on?

34. A water solution that is 9.5% ethanol, by mass, has a density of 0.981 g/cm^3 at 25°C. What mass, in grams, of ethanol is contained in 5.75 L of this solution?

35. A solution containing 12.0% sodium hydroxide, by mass, has a density of 1.131 g/cm^3. What volume, in liters, of this solution must be used in an application requiring 1.65 kg of sodium hydroxide?

Algebraic equations

36. A tabulation of data lists the following equation for the densities (d) of solutions of naphthalene in benzene (at 30°C) as a function of the mass percent naphthalene (% N).

$$d \text{ (g/cm}^3) = \frac{1}{1.153 - 0.00182 \, (\% \, N) + 1.08 \times 10^{-6} \, (\% \, N)^2}$$

What is the density of a solution (at 30°C) with 1.15% naphthalene in benzene, by mass, i.e., with % N = 1.15?

37. For a benzene solution containing 6.38% para-dichlorobenzene, by mass, the solution density as a function of temperature (over the range 15 to 65°C) is given by the equation

$$d \text{ (g/cm}^3) = 1.5794 - 1.836 \times 10^{-3}(t - 15°)$$

At what temperature (t) will this solution have a density of 1.543 g/cm^3?

★**38.** It is desired to construct a cube with length l that will have the same volume as a sphere of radius r. What must be the ratio l/r?

★**39.** Use the equations given in Figure 1-4 to find an *algebraic* solution to this question: At what single temperature are the numerical values on the Celsius and Fahrenheit scales equal? (*Hint:* A trial-and-error method is not an algebraic solution.)

★**40.** Refer to the equation given in Review Problem 15 for the density of water as a function of Celsius temperature, and show that the density of water passes through a maximum somewhere in the temperature range in which the equation applies (i.e., from 0°C to about 20°C).

Additional Exercises

1. Human behavior cannot be studied quite as readily as the phenomena of natural science. Nevertheless, there are certain "laws" that are applicable to a variety of human activities. Explain what is meant by the "law of averages" and the "law of diminishing returns."

2. Use the concept of significant figures to criticize the manner in which the following information was stated in an official report: "In 1974, for example, there were approximately 1,573,006 students enrolled in California's four sectors of higher education."

3. Perform the following conversions between the English and metric systems of measurement.
- (a) 55 mi/h = _____ km/h
- (b) 62 ft/s = _____ km/h
- (c) 315 in.^{3} = _____ cm^3
- (d) 17.5 lb/in.^{2} = _____ g/cm^2
- (e) 57.5 lb/ft^3 = _____ g/cm^3

4. It is necessary to determine the density of a solution to *four* significant figures. The volume of solution can be measured to the nearest 0.1 mL.
- (a) What is the minimum volume of sample that can be used for the measurement?
- (b) Assuming the minimum volume sample determined in part (a), how accurately must the sample be weighed (i.e., to the nearest 0.1 g, 0.01 g, . . .) if the density of the solution is greater than 1.00 g/cm^3? If the density is less than 1.00 g/cm^3?

5. A solution used to chlorinate a home swimming pool contains 7% chlorine, by mass. An ideal chlorine level in the pool is about one part per million (1 ppm). (Think of 1 ppm as being 1 g chlorine per one million grams of water.) Assuming densities of 1.10 g/cm^3 for the chlorine solution and 1.00 g/cm^3 for the water, what volume of the chlorine solution is required to produce a chlorine level of 1 ppm in a 20,000-gallon swimming pool?

★**6.** Use the equation of Review Problem 15 to determine the temperature at which the density of water is 0.99916 g/cm^3. (*Hint:* Use the quadratic formula; see Appendix A.)

★**7.** A Fahrenheit and a Celsius thermometer are immersed in the same medium, whose temperature is to be measured. At what Celsius temperature will the reading on the Fahrenheit thermometer be
- (a) twice that on the Celsius thermometer?
- (b) four times that on the Celsius thermometer?
- (c) one-sixth that on the Celsius thermometer?
- (d) 200° more than that on the Celsius thermometer?

*8. A pycnometer weighs 25.60 g empty and 35.55 g when filled with water at 20°C. The density of the water at 20°C is 0.998 g/cm^3. When 10.20 g of lead is placed in the pycnometer and the pycnometer again filled with water at 20°C, the total mass is 44.83 g. What is the density of lead?

*9. The standard kilogram mass pictured in Figure 1-2 was cut from a cylindrical bar of steel with a diameter of 1.50 in. The density of the steel is 7.86 g/cm^3. How long was the section that was cut?

*10. The volume of seawater on earth is estimated to be 330,000,000 mi^3. Assuming that seawater is 3.5% sodium chloride, by mass, and that the density of seawater is 1.03 g/cm^3, what is the approximate mass of sodium chloride dissolved in the seawater on earth, expressed in tons?

*11. The diameter of metal wire is often referred to by its American wire gauge number. A 16-gauge wire has a diameter of 0.05082 in. What length of wire, in meters, is there in a 1.00-lb spool of 16-gauge copper wire? The density of copper is 8.92 g/cm^3.

*12. The principal source of magnesium metal is seawater, from which it is extracted by the Dow process (see Section 22-2). It occurs in seawater to the extent of 1.4 g of magnesium per kilogram of seawater. The annual production of magnesium in the United States is approximately 10^5 tons. What volume, in mi^3, of seawater must be processed to yield this much magnesium? (Density of seawater = 1.03 g/cm^3.)

*13. The Antarctic, Greenland, and other ice caps contain approximately 7.2 million mi^3 of ice.
(a) Given that the density of ice is 0.92 g/cm^3, together with other appropriate conversion factors, determine the mass of this ice, in tons.
(b) When ice is melted, its volume decreases by about 10%. If all the polar ice were to melt completely, estimate the increase in sea level that would result from the additional liquid water entering the oceans. The oceans of the world cover about 1.4×10^8 mi^2.

*14. A typical rate of deposit of dust (''dustfall'') from air that is not significantly polluted might be 10 tons per square mile per month. What is this dustfall, expressed in milligrams per square meter per hour?

*15. When water is used for irrigation purposes, its volume is often expressed in acre-feet. One acre-foot is a volume of water sufficient to cover 1 acre of land to a depth of 1 ft (640 acres = 1 mi^2). The principal lake in the California State Water Project is Lake Oroville, whose water storage capacity is listed as 3.54×10^6 acre-feet. Express this volume in (a) ft^3; (b) m^3; (c) gal.

Self-Test Questions

For questions 1 through 8 select the single item that best completes each statement.

1. Of the following masses, that which is expressed to the nearest milligram is (a) 14.7 g; (b) 14.72 g; (c) 14.721 g; (d) 14.7213 g.

2. The greatest length of the following group is (a) 4.0 m; (b) 140 in.; (c) 12 ft; (d) 0.001 km.

3. The highest temperature of the following group is (a) −250°F; (b) 20°C; (c) 217°F; (d) 105°C.

4. Of the following substances, the greatest density is that of
(a) 1000 g of water at 4°C
(b) 100.0 cm^3 of chloroform, which weighs 148.9 g
(c) a 10.0-cm^3 piece of wood weighing 9.50 g
(d) an alcohol–water mixture of density 0.83 g/cm^3

5. The largest volume of the following group is that of
(a) 380 g of water at 4°C
(b) 600 g of chloroform at 20°C (density = 1.5 g/cm^3)
(c) 0.50 L of milk
(d) 100 cm^3 of steel (density = 7.86 g/cm^3)

6. Of the following numbers, the one with three significant figures is (a) 16.07; (b) 0.0140; (c) 1.070; (d) 0.016; (e) 200.

7. A fertilizer contains 20% nitrogen, by mass. To provide a fruit tree with an equivalent of 1 lb of nitrogen, the quantity of *fertilizer* required is (a) 20 lb; (b) 0.20 lb; (c) 0.05 lb; (d) 5 lb.

8. An example of a homogeneous mixture or solution is (a) acid rain; (b) liquid oxygen; (c) sucrose (sugar); (d) salad dressing.

9. Describe briefly the distinction between the following pairs of terms.
(a) element and compound
(b) homogeneous and heterogeneous mixture
(c) mass and density.

10. Use exponential notation and the appropriate number of significant figures to express the result of the following calculation.

$(19.541 + 1.05 - 3.6) \times 651 = ?$

11. A 55.0-gal drum weighs 75.0 lb when empty. What will be its total mass when filled with ethyl alcohol (density = 0.789 g/cm^3; 1 gal = 3.78 L; 1 lb = 454 g)?

12. A carpet material is listed as costing \$18.50 per square yard (yd^2), installed. How much would it cost to carpet a rectangular room 6.52 m $\times$ 4.18 m with this material?

2 Development of the Atomic Theory

One of the oldest scientific concepts is that all matter can be broken down into particles that cannot be subdivided further. The Greek philosopher Democritus (ca. 460–370 B.C.) considered these particles to be in constant motion but able to fit together into stable combinations. Supposedly, the particular properties of a material resulted from the different sizes, shapes, and arrangements of these particles. Today, we refer to these ultimate particles—these building blocks of all matter—as **atoms** [Gr. *atomos* (*a*, not + *tomos*, to cut)—uncut, undivided, or indivisible]. Many scientific discoveries have resulted from attempts to learn more about the fundamental nature of atoms.

2-1 Dalton's Atomic Theory

One name that we associate particularly with the origin of modern atomic theory is that of the English schoolteacher-chemist John Dalton (1776–1844). Dalton's contribution was unique in two ways: He was the first to make use of *chemical* as well as physical evidence, and he based assumptions on *quantitative* data, not merely qualitative observations or speculation.

Experimental Basis. Two types of experimental evidence—two natural laws—serve as the basis of Dalton's atomic theory.

In the course of a series of investigations of combustion and related processes, Antoine Lavoisier (1743–1794) proved that in these processes oxygen in the air combines with the materials undergoing change. Figure 2-1 depicts one of his experiments, in which liquid mercury was combined with oxygen to form red mercuric oxide (mercury calx). When the calx was recovered and reheated, it decomposed to produce liquid mercury and a quantity of oxygen gas equal in volume to that of the air consumed in the formation of the calx. Lavoisier established the **law of conservation of mass:** *the total mass of materials present after a chemical reaction is the same as before the reaction.*

A second law of chemical combination is the **law of constant composition** (also known as the **law of definite proportions**). A chemical compound, no matter what its origin or its method of preparation, always has the same composition, that is, the same proportions by mass of its constituent elements. For example, in 1799 Joseph Proust (1754–1826) found that copper carbonate, whether from natural sources or synthesized in the laboratory, has a fixed composition.

FIGURE 2-1
Lavoisier's experiment
on heating mercury with
air.

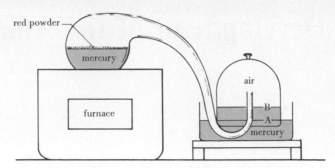

The original mercury level in the air container was at A, but after a number of days it rose to B and remained there. The difference in level between A and B represented the volume of air consumed by the mercury in forming the red powder (mercury calx). As proof of this fact Lavoisier collected the mercury calx and reheated it. The red powder decomposed into liquid mercury and a volume of gas (oxygen) equal to that of the air consumed in the original experiment.

Example 2-1   The following data were obtained by heating strips of magnesium metal in oxygen gas to produce the white powder, magnesium oxide. Show that these data are consistent with the law of constant composition.

	Before heating, g magnesium	After heating, g magnesium oxide	Ratio: g magnesium/g magnesium oxide
strip 1	0.62	1.02	0.62/1.02 = 0.61*
strip 2	0.48	0.79	0.48/0.79 = 0.60*
strip 3	0.36	0.60	0.36/0.60 = 0.60*

Solution.   According to the law of constant composition, the ratio of mass of magnesium to mass of magnesium oxide should have a constant value. The last column of values in the table demonstrates that it does. Within the precision of the measurements used (masses determined to ±0.01 g), the law of constant composition is verified.

SIMILAR EXAMPLES: Review Problem 2; Exercises 3, 4.

Dalton's Assumptions.   Dalton's atomic theory was developed during the period 1803–1808 and was based on three principal assumptions.

1. Each chemical element is composed of minute indivisible, indestructible particles called atoms. Atoms can be neither created nor destroyed during a chemical change.
2. All atoms of an element are alike in mass (weight) and other properties, but the atoms of one element are different from those of all other elements.
3. In chemical compounds, atoms of different elements combine in simple numerical ratios: for example, one atom of A to one of B (AB), one atom of A to two of B (AB$_2$).

If the atoms of an element are indestructible (assumption 1), then the *very same* atoms must be present after a chemical reaction as were present before the reaction. The total

*These ratios, multiplied by 100, yield the percent magnesium in magnesium oxide, that is, 61%, 60%, and 60%, respectively.

FIGURE 2-2
The "atomic weight problem."

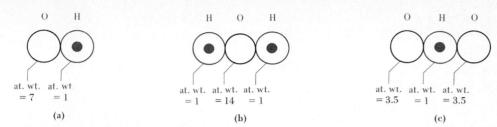

Analyses available to Dalton (87% O and 13% H) suggested that water contained seven times the mass of oxygen as hydrogen. Dalton assumed the simplest formula (a), and as a result assigned an atomic weight of 7 to oxygen. If he had assumed formula (b), his result would have been 14; if formula (c), 3.5.

$$\frac{\text{Mass of oxygen}}{\text{Mass of hydrogen}}: \qquad \overset{\text{Compound (a)}}{\frac{7}{1} = 7} \qquad \overset{\text{Compound (b)}}{\frac{14}{2 \times 1} = 7} \qquad \overset{\text{Compound (c)}}{\frac{2 \times 3.5}{1} = 7}$$

Modern data indicate (b) is correct and that the atomic weight of oxygen is 16, not 14.

mass of reactants and products must be the same. *Dalton's theory explains the law of conservation of mass.*

If all atoms of an element are alike in mass (assumption 2), and if atoms unite in *fixed* numerical ratios (assumption 3), the percentage composition of a compound must have a unique value, regardless of the size of the sample analyzed or its origin. *Dalton's theory also explains the law of constant composition.*

The characteristic masses of the atoms of an element required by assumption 2 above became known as **atomic weights,** and Dalton tried to establish a set of *relative* atomic weights. For example, if an atom of hydrogen is arbitrarily assigned a mass of 1 unit, what must be the mass of an oxygen atom by comparison? The composition of water is 88.81% oxygen and 11.19% hydrogen, by mass, i.e., the mass of oxygen in water is about *eight* times that of hydrogen.* This suggests a relative atomic weight of 8 for oxygen. But there is an assumption here: that hydrogen and oxygen combine in the ratio 1:1—one atom of hydrogen for each oxygen atom.

Figure 2-2 shows how the relative atomic weight of oxygen depends on the assumption made about the ratio in which hydrogen and oxygen atoms combine (i.e., the chemical formula assumed for water). The assignment of atomic weights to the elements remained a confusing matter until an important clarifying suggestion was made by Cannizzaro in 1860 (see Section 5-4).

Law of Multiple Proportions. Despite its early difficulties, Dalton's atomic theory did provide a basis for predicting the **law of multiple proportions** (1805): *If two elements form more than a single compound, the masses of one element combined with a fixed mass of the second are in the ratio of small whole numbers.* Dalton observed that twice as much hydrogen is combined with a given mass of carbon in methane gas (carburetted hydrogen) as in ethylene gas (olefiant gas). He assigned the formula CH_2 to methane and CH to ethylene. (The correct formulas based on present knowledge are CH_4 and C_2H_4.) Dalton's reasoning is outlined in Figure 2-3.

*The analysis of water reported in Dalton's time was 87% oxygen and 13% hydrogen, by mass. This corresponds to a mass ratio of oxygen to hydrogen of 7:1, and suggested that the atomic weight of oxygen should be 7.

FIGURE 2-3
The law of multiple proportions illustrated—with Dalton's symbols and atomic weights.

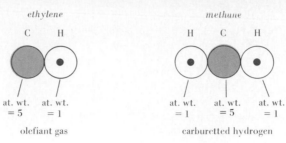

Per gram of hydrogen in olefiant gas there is 5 g of carbon, that is,

$$\frac{5 \text{ g carbon}}{1 \text{ g hydrogen}}$$

Per gram of hydrogen in carburetted hydrogen there is 2.5 g of carbon, that is,

$$\frac{5 \text{ g carbon}}{2 \text{ g hydrogen}} = \frac{2.5 \text{ g carbon}}{1 \text{ g hydrogen}}$$

$$\text{ratio} = \frac{5 \text{ g carbon/1 g hydrogen}}{2.5 \text{ g carbon/1 g hydrogen}} = \frac{2}{1}$$

2-2 Cathode Rays

In 1838, Michael Faraday (1791–1867) reported on his studies of electric discharge through gases. He used an apparatus similar to that pictured in Figure 2-4. Metal plates called electrodes are sealed into the ends of a glass tube having a side-arm opening. One electrode, the **cathode,** is connected to the negative terminal of a source of electric current at high voltage (several thousand volts); the other, the **anode,** is connected to the positive terminal. As long as the tube is filled with air, no electric current flows. Air is a very poor conductor of electricity. The glass tube can be evacuated of air by connecting it to a vacuum pump. The phenomena indicated in Figure 2-4 occur when the pressure of the residual air in the tube is reduced to about one thousandth of normal atmospheric pressure, the limit to which mechanical vacuum pumps could be operated in Faraday's time.

With the advent of better vacuum pumps it became possible to reduce the pressure within discharge tubes to about one millionth of atmospheric pressure. Additional phenomena were observed at these greatly reduced pressures. In 1858, Plucker reported that the dark region enlarged as the air pressure was reduced, that the region of the cathode glow was extended, and that the glass tube itself emitted a phosphorescent glow. In 1869, Hittorf mounted an object in a discharge tube and observed it to cast a shadow. This suggested that the glowing of the tube was caused by rays emanating from the cathode and traveling in straight lines. Thus began a long and exciting series of experiments into the nature of cathode rays.

Properties of Cathode Rays. Here are some of the more significant properties of cathode rays determined by Plucker, Hittorf, Crookes, and others.

1. Cathode rays are emitted from the cathode in an evacuated tube when electric current is passed. (Electric current is essential.)
2. Cathode rays travel in straight lines.
3. The rays, upon striking glass or certain other materials, cause these materials to fluoresce (give off light); the rays themselves are invisible.
4. Cathode rays are deflected by electric and magnetic fields in the manner expected for *negatively charged* particles (see Figure 2-5).

FIGURE 2-4
Electric discharge in an evacuated chamber.

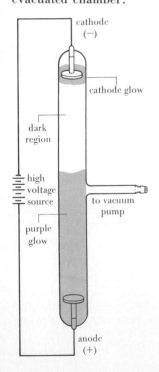

FIGURE 2-5
Deflection of cathode
rays in a magnetic field.

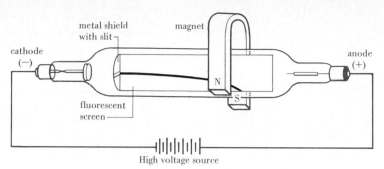

Cathode rays are invisible. Only through their impact on a fluorescent material can they be detected. The beam of cathode rays is deflected as it enters the field of the horse-shoe magnet. The deflection of the beam corresponds to that anticipated for negatively charged particles. (See also, Color Section A.)

5. The properties of the rays are independent of the electrode material (i.e., whether iron, platinum, etc.).

Investigations of J. J. Thomson.   During the period 1894–1897, J. J. Thomson (1856–1940) conducted a series of investigations that established the particle nature of cathode rays. He found the speeds of cathode rays to be only a small fraction of the speed of light, thus ruling out the possibility of their being electromagnetic radiation. Thomson also determined the ratio of electric charge (e) to mass (m), that is, e/m for cathode rays. His experimental design and a brief analysis of it are presented in Figure 2-6.

FIGURE 2-6
Thomson's apparatus
for determining the
charge-to-mass ratio,
e/m, for cathode rays.

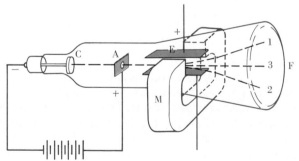

Code: C, cathode; A, anode (perforated to allow the passage of a narrow beam of cathode rays); E, electrically charged condenser plates; M, magnet; F, fluorescent screen.

Path 1: In the presence of an electric field only, the cathode ray beam is deflected upward, striking the end screen at a point such as 1.

Path 2: In the presence of a magnetic field only, the cathode ray beam is deflected downward, striking the screen at point 2.

Path 3: The cathode ray beam can be made to strike the end screen at point 3, undeflected, if the forces on the particles exerted by the electric and magnetic fields are just counterbalanced.

Determination of e/m: From a knowledge of the strengths of the electric and magnetic fields producing path 3 and the radius of curvature of path 2, a value of e/m can be obtained. Precise measurements yield a result of

-1.759×10^8 coulombs per gram

(Because cathode rays carry a negative charge, the sign of the charge-to-mass ratio is also negative.)

FIGURE 2-7
Millikan's oil drop
experiment.

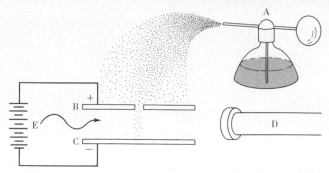

A spray of oil droplets is produced by the atomizer (A). These enter the apparatus through a tiny hole in the top plate of an electrical condenser. The motion of the droplets is observed with a telescope equipped with a micrometer eyepiece (D). Ions are produced by ionizing radiation, such as x rays, from a suitable source (E). Some of the oil droplets acquire an electric charge by adsorbing ions (attaching ions to their surface).

The fall of a droplet between the charged condenser plates (B and C) is either speeded up or slowed down to an extent that depends on the sign and magnitude of the charge on the droplet. By analyzing data from large numbers of droplets, Millikan concluded that the magnitude of the charge, q, on a droplet is always an integral multiple of the electronic charge, e. That is, $q = n \cdot e$ (where $n = 1, 2, 3$, and so on).

The coulomb (C) is the SI unit of electrical charge.

The average value Thomson obtained for e/m for cathode rays was about 2×10^8 coulombs per gram. This value is about 2000 times greater than the e/m ratio calculated for hydrogen liberated in the electrolysis of water. For various reasons Thomson assumed cathode rays to have about the same electrical charge as that associated with hydrogen atoms in the electrolysis of water. This meant that they should possess only about $\frac{1}{2000}$ of the mass of a hydrogen atom. The assumed small size of cathode ray particles and the fact that the value of e/m was independent of the cathode material led Thomson to the following conclusion: Cathode ray particles are negatively charged *fundamental* particles of matter that must be found in *all* atoms. Cathode ray particles are the fundamental units of negative electric charge for which, in 1874, Stoney had proposed the term **electron.**

Charge on the Electron.   Thomson provided a precise *ratio* of charge to mass for an electron. He also speculated on probable values of e and m, but neither the charge nor the mass could be determined by his method alone. An independent evaluation of either e or m was clearly needed. From a precise measurement of one of these and from a knowledge of e/m, the value of the other could be determined. The electronic charge e was chosen as the quantity most amenable to measurement, and the definitive measurements were made by Robert Millikan (1868–1953) at the University of Chicago during the period 1906–1914.

"Here, then, is direct, unimpeachable proof that the electron is not a "statistical mean," but that rather the electrical charges found on ions all have either exactly the same value or else small exact multiples of that value."
ROBERT MILLIKAN

Millikan's famous "oil drop" experiment is suggested by Figure 2-7. Millikan found that the electric charge on all oil droplets that had acquired a charge could be expressed as $n \times e$, where n is a positive or negative integer and e represents the smallest observable unit of electric charge. The currently accepted value of the electronic charge e is -1.60219×10^{-19} C. Combining the results of Millikan and Thomson, one obtains the mass of an electron: 9.110×10^{-28} g.

2-3 Canal Rays (Positive Rays)

In addition to characterizing the electron, cathode ray research also yielded evidence of a fundamental unit of *positive* charge. In 1886, Eugen Goldstein discovered a new type of particles called **canal rays** or positive rays, with the following properties:

FIGURE 2-8
The "plum pudding"
model of atomic
structure.

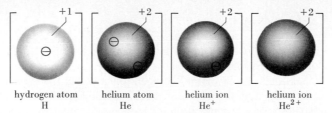

hydrogen atom
H

helium atom
He

helium ion
He$^+$

helium ion
He^{2+}

Thomson's conception was of a "cloud" of positive electric charge with a sufficient number of electrons embedded within to neutralize the positive charge. Thus, a hydrogen atom was thought to consist of a positive cloud of charge $+1$, containing one electron (-1); helium, a positive cloud of charge $+2$, containing two electrons (-2); and so on. Normal atoms are electrically neutral. If an atomic species carries a *net* electric charge it is said to be an **ion.** The loss of one electron by a helium atom, through a collision with cathode rays, for example, results in the production of an ion with a net charge of $+1$, that is, He$^+$. The loss of both electrons results in the ion He^{2+} with a net charge of $+2$.

1. The particles are deflected by electric and magnetic fields in a way that reveals their positive charge.
2. The ratio of charge to mass, e/m, for positive rays is considerably smaller than for electrons.
3. The e/m ratio of the positive rays depends on the nature of the gas in the tube. The highest e/m ratio is obtained with hydrogen gas. For other gases, e/m is an integral fraction (e.g., $\frac{1}{4}$, $\frac{1}{20}$) of the ratio for hydrogen.
4. The e/m ratio of positive rays produced from hydrogen gas is identical to the e/m ratio for hydrogen produced by the electrolysis of water.

These observations can be explained by an atomic model proposed by J. J. Thomson and illustrated in Figure 2-8. An explanation of the production of positive rays (ions) is provided by Figure 2-9. A conclusion consistent with the properties of canal rays is that all atoms carry fundamental units of positive charge, one in the hydrogen atom and larger numbers in other atoms. The fundamental unit of positive charge is now called the **proton.**

FIGURE 2-9
A canal ray tube.

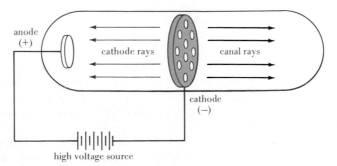

anode
(+)

cathode rays

canal rays

cathode
(−)

high voltage source

The distinctive feature of this tube is the perforated cathode. Cathode rays stream toward the anode. Their collisions with residual gas atoms dislodge electrons from the atoms, producing positively charged ions. These ions are attracted to the cathode (−), but some of the ions pass through the holes in the cathode and appear as a stream of particles (black arrows) on the other side. These beams of positive ions are called positive rays or canal rays.

FIGURE 2-10
The production of
x rays.

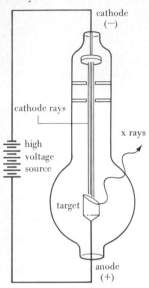

2-4 X Rays

A number of cathode ray experimenters, including apparently J. J. Thomson, occasionally observed objects *outside* cathode ray tubes to fluoresce or glow when the tubes were run. It remained for Wilhelm Roentgen to show that the impact of cathode rays (electrons) on a surface produces a type of radiation which, in its turn, can cause certain substances to fluoresce at a distance from the cathode ray tube. Because of the unknown nature of this radiation, the term *x ray* was coined, a name still in common use.

Within a few weeks of his initial discovery in 1895, Roentgen achieved a rather complete characterization of x rays. He found them to be undeflected by electric and magnetic fields and to have a very high penetrating power through matter. All properties of x rays suggested them to be electromagnetic radiation with wavelengths of approximately 1 angstrom unit (1 Å $= 10^{-10}$ m).*

Practical applications followed closely on the heels of the discovery of x rays. Because x rays have different penetrating powers for different types of matter, they can be used to photograph the interior of objects. Roentgen's original announcement of x rays was made on December 28, 1895. On January 20, 1896, in Dartmouth, New Hampshire, x rays were used to assist in setting a person's broken arm. This was something of a record time for turning a scientific discovery into a practical application.

Roentgen found that an especially effective source of x rays resulted from the impact of cathode rays on a dense metal anode called a *target*. Figure 2-10 is a diagram of an x ray tube based on this principle.

2-5 Radioactivity

The discovery of x rays in turn brought about the discovery of radioactivity. This discovery, by the French physicist Henri Becquerel, came within a few months of Roentgen's. In Roentgen's experiments the impact of cathode rays upon glass produced a fluorescence of the glass as well as the emission of x rays. Becquerel associated the emission of x rays with fluorescence and asked the question: Will naturally fluorescent materials produce x rays? To answer this question he proceeded as follows.

A photographic plate was wrapped in black paper. The paper was covered with a layer of a particular crystalline substance (a double sulfate of uranium and potassium), and the assembly was placed in sunlight. As expected, the photographic plate became exposed. Becquerel thought that sunlight caused the substance to fluoresce or glow and that some of this fluorescent radiation (x rays) penetrated the paper and exposed the photographic plate. On one occasion when he attempted to repeat this experiment, the sky became overcast, and Becquerel put the assembly into a drawer for several days. Before resuming the experiment he replaced the photographic plate, expecting it to be slightly exposed because of the length of time that had elapsed. To his surprise he found that the plate was *strongly* exposed, as much so as in his initial experiments. Becquerel hypothesized that the radiation responsible for exposing the photographic plate was not associated with fluorescence at all, and he designed experiments to test this hypothesis. He found that the radiation was emitted continuously by the substance used in his experiments, in particular by the element uranium. Thus, radioactivity was discovered.

Radioactive Elements and Their Radiations.   Ernest Rutherford identified two types of radiation. One type, which he called **alpha rays (α rays),** has a high ionizing power but a low penetration through matter. Alpha rays can be stopped by a sheet of ordinary paper. The other type is of a lower ionizing power but greater penetrating power. These rays can

---

*A complete discussion of wavelength and other properties of electromagnetic radiation is presented in Chapter 7.

Radiation emanating from radioactive materials can dislodge electrons when it strikes atoms of matter; ions are produced. Ionizing power refers to the number of ions produced by a given quantity of radiation.

pass through aluminum foil up to 3 mm thick. Rutherford called this radiation **beta rays (β rays).**

Alpha rays are *particles* carrying two fundamental units of *positive* charge and having a mass equal to that of a helium atom; thus, an α particle is identical to the ion He^{2+} (see, again, Figure 2-8). Beta rays are *negatively* charged particles with the same charge-to-mass ratio, *e/m,* as electrons. They are indistinguishable from electrons. A third form of radiation has an extremely high penetrating power and is not deflected by electric and magnetic fields. This electromagnetic radiation is known as **gamma rays (γ rays).**

By the early 1900s several additional radioactive elements were discovered (e.g., thorium, radium, and polonium), principally through the work of Marie and Pierre Curie in France. And, in collaboration with Frederick Soddy, Rutherford made another profound discovery concerning radioactivity: The chemical properties of a radioactive element *change* as it undergoes radioactive decay. This observation could only be explained by assuming that radioactivity involves fundamental changes at the *subatomic* level—that in radioactive decay one element replaces another. Thus, with this discovery of *transmutation* one of Dalton's basic assumptions—that atoms of an element are indivisible and unchangeable—toppled. Additional aspects of radioactivity and nuclear chemistry are considered in Chapter 25.

2-6 The Nuclear Atom

Cathode ray production, light emission, and ionization all could be explained, after a fashion, by Thomson's plum pudding model (1898). However, this model could not be used to make quantitative predictions. Neither was the model consistent with other observations being made at about the same time.

Scattering of Alpha Particles. In 1909, at Rutherford's suggestion, Hans Geiger and Ernest Marsden undertook experiments in which very thin foils of gold and other metals (10^{-4} to 10^{-5} cm thick) were bombarded with α particles derived from a radioactive substance. The experimental apparatus used is suggested by Figure 2-11.

The radioactive substance was enclosed in a lead block in such a way that only a narrow beam of α particles could escape. The presence of α particles was detected by the scintillations or flashes of light they produced on a zinc sulfide screen mounted on the end of a telescope. Geiger and Marsden observed that

1. The majority of the α particles penetrated the metal foil undeflected.
2. A few (about one in every 20,000) suffered rather serious deflections as they penetrated the foil.
3. A similar number did not pass through the foil at all but "bounced back" in the direction from which they had come.

Using Thomson's atomic model, Rutherford reasoned that the positive charge of the

FIGURE 2-11
The scattering of alpha particles by metal foil.

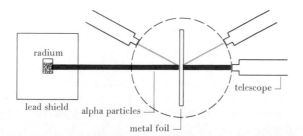

radium

lead shield   alpha particles

metal foil

telescope

FIGURE 2-12
Rutherford's
interpretation of the
scattering of alpha
particles by metal foil.

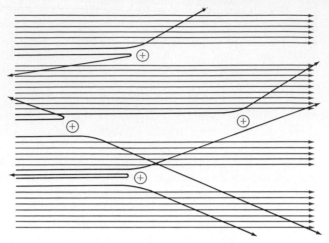

Each arrow represents an α particle. The symbol $\oplus$ represents an atomic nucleus.

"It is about as incredible as if you had fired a 15-in. shell at a piece of tissue paper and it came back and hit you."
ERNEST RUTHERFORD

atom was so diffuse that α particles should pass through this weak electric field largely undeflected. The severe deflections experienced by some of the particles, particularly those that bounced off the foils, astounded Rutherford. This behavior would be expected only if the positive charge and mass of an atom were highly concentrated in a small region, a **nucleus.** The approach of an α particle to a nucleus of high positive charge and mass would lead to repulsive forces strong enough to reverse the direction of the α particle while leaving the nucleus practically unmoved (see Figure 2-12). In this way the idea of a *nuclear* atom originated.

Rutherford's Atomic Model. The main features of the atom postulated by Rutherford were these.

1. Most of the mass and all of the positive charge of an atom are centered in a very small region called the nucleus—the atom is mostly empty space.
2. The magnitude of the charge on the nucleus is different for different atoms and is approximately one half the atomic weight of the element.
3. There must be a number of electrons outside the nucleus of an atom that is equal to the number of units of nuclear charge (to account for the fact that the atom is electrically neutral.)

2-7 Isotopes

In 1912, J. J. Thomson determined the charge-to-mass (e/m) ratios of positive ions produced in a canal ray tube. The results he obtained with neon gas were quite unexpected. He was able to explain them in only one way: In ordinary neon gas about 91% of the atoms have a "normal" mass and about 9% of the atoms are 22/20 heavier than this. Thomson's discovery was that atoms of the same element *may differ slightly in mass.* These differing atoms are called **isotopes.** Some years earlier Soddy proposed this term for atomic species which, although having different radioactive properties, have identical chemical properties; but Thomson discovered the existence of isotopes among nonradioactive elements as well. The discovery of radioactivity required modification of one of the assumptions of Dalton's atomic theory—that atoms of an element are unchangeable. The discovery of isotopes required modification of a second—that all atoms of an element are alike in mass.

FIGURE 2-13
The nuclear atom—
illustrated by the
helium atom.

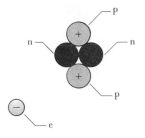

2-8 Protons and Neutrons

In 1913, H. G. J. Moseley, one of a group of brilliant scientists whose careers were launched under Rutherford's direction, reported on experiments in which different elements or their compounds were used as targets in an x ray tube. He found that the x ray wavelength varied with the target material. Moseley was able to correlate these wavelengths through a mathematical equation in which it was necessary to assign to each element a unique integral (whole) number. These integers, which he called atomic numbers, proved to be identical to the nuclear charges that Rutherford had described.

In 1919, Rutherford established the existence of particles carrying a fundamental unit of positive charge, **protons,** through studies on the passage of α particles through air. Rutherford found that scintillations could be detected on a zinc sulfide screen at distances from a radium source considerably greater than α particles were expected to travel before being absorbed. He concluded that upon striking the nuclei of nitrogen atoms in air α particles ejected protons and it was these that reached the zinc sulfide screen.

The concept of the constitution of the nucleus favored by Rutherford and a number of other physicists was this: The nucleus contains a number of protons equal to the atomic number and a sufficient number of *neutral* particles, called neutrons, to account for the observed mass of an atom.

In the early 1930s, in experiments in which beryllium and boron were bombarded with α particles, a very penetrating radiation was obtained. It seemed to have properties similar to γ rays, but was more energetic. In 1932, Chadwick showed that the properties of this radiation were more readily explained by assuming a beam of neutral particles having masses slightly greater than protons. In this way **neutrons,** first postulated 12 years earlier, were finally discovered. An atomic model involving protons, neutrons, and electrons is pictured in Figure 2-13.

2-9 Summary of the Properties of Fundamental Particles

Many fundamental particles are now known in addition to the three described in this chapter, but the chemical behavior of an element is derived just from these three—the proton, the neutron, and the electron. Electric charges and masses of these three particles are presented in Table 2-1 in two sets of units. One is the metric unit, and the other a special atomic unit with which the properties of fundamental particles can be related to one another.

Since the electric charge carried by an electron is the smallest unit of electric charge that can exist, we might call it an atomic unit of electricity. The proton also carries an atomic unit of electric charge, but positive in sign. It would be convenient also to think of protons and neutrons as possessing an atomic unit of mass. The atomic unit of mass, however, is not defined in terms of the mass of a proton or a neutron (for one thing, these two masses are not equal). Instead, the **atomic mass unit** (referred to by the abbreviation,

TABLE 2-1
Properties of three fundamental particles

	Electric charge		Mass	
	Metric (C)	Atomic	Metric (g)	Atomic (u)[a]
proton	$+1.602 \times 10^{-19}$	$+1$	1.673×10^{-24}	1.0073
neutron	0	0	1.675×10^{-24}	1.0087
electron	-1.602×10^{-19}	-1	9.110×10^{-28}	0.00055

[a]u is the SI symbol for atomic mass unit (abbreviated amu).

amu, and the symbol, **u**) is taken to be exactly $\frac{1}{12}$ of the mass of an atom of the isotope of carbon known as carbon-12 (see page 39). As can be seen in Table 2-1, the proton and neutron masses are just slightly greater than 1 u. By comparison the mass of an electron is seen to be extremely small.

Example 2-2 From the data given in Table 2-1 calculate the *e/m* ratio for the electron and compare this with the value listed in Figure 2-6.

Solution

charge: -1.602×10^{-19} C

mass: 9.110×10^{-28} g

charge-to-mass ratio: $e/m = \dfrac{-1.602 \times 10^{-19} \text{ C}}{9.110 \times 10^{-28} \text{ g}} = -1.759 \times 10^{8}$ C/g

This ratio has the same value as that listed Figure 2-6.

SIMILAR EXAMPLES: Exercises 16, 17.

___

The number of protons in the nucleus of an atom is called the **atomic number,** or proton number, **Z**. The number of electrons in an electrically neutral atom is also equal to the atomic number Z. The total mass of an atom is determined very nearly by the total number of protons and neutrons in its nucleus. This total is called the **mass number** A. The number of neutrons in an atom, the neutron number, is given by the quantity $A - Z$.

2-10 Chemical Elements

Atoms of an element are all of a single kind. To the chemist the kind of atom is specified by its atomic number, since this is the property most closely related to chemical behavior. At present all the atoms from $Z = 1$ to $Z = 106$ are known: There are 106 chemical elements. Each chemical element has been given a name and a distinctive symbol. For most elements the symbol is the abbreviated form of the English name, consisting of one or two letters. The first (but never the second) letter of the symbol is capitalized. For example,

oxygen = O   nitrogen = N   neon = Ne   magnesium = Mg

Some elements that have been known for a long time have symbols based on their Latin names, for example,

iron = Fe (ferrum)   copper = Cu (cuprum)   lead = Pb (plumbum)

A few elements have symbols based on the Latin name of a compound, the elements themselves having been discovered only relatively recently, for example,

sodium = Na (natrium = sodium carbonate)

potassium = K (kalium = potassium carbonate)

The symbol for tungsten, W, is based on its German name, wolfram. A complete listing of the elements may be found on the inside back cover.

Information on the composition of atomic nuclei may be given through the symbolism $^{A}_{Z}X$,

$^{12}_{6}C$, $^{14}_{7}N$, $^{16}_{8}O$, $^{24}_{12}Mg$, $^{56}_{26}Fe$, $^{238}_{92}U$,   and so on.

It is possible to arrange four numbers as subscripts and superscripts about a chemical symbol, that is $^{a}_{b}X^{c}_{d}$. In this scheme a is the mass number, b is the atomic number, and c is the net electric charge, as in $^{20}_{10}Ne^{2+}$. The fourth number, the subscript to the right (d), represents the number of atoms in a molecule. Its use is introduced in Chapter 3.

The symbol $^{A}_{Z}X$ represents an atomic species called a **nuclide** of the element X, having an atomic number Z and a mass number A. One of the nuclides shown above, that of the element carbon, is an atom with six protons and six neutrons in its nucleus and six electrons outside the nucleus.

All atoms of an element must have the same atomic number, but they may have different mass numbers. The different nuclides of an element are referred to collectively as isotopes of the element. In Section 2-7, it was stated that one type of neon atom has a mass $\frac{22}{20}$ as great as the predominant atomic species. Actually three different nuclides exist; there are three isotopes of neon. By symbol, these are $^{20}_{10}Ne$, $^{21}_{10}Ne$, and $^{22}_{10}Ne$. Of all Ne atoms in the earth's crust, 90.9% are $^{20}_{10}Ne$; the percentages, by number, of $^{21}_{10}Ne$ and $^{22}_{10}Ne$ are 0.3 and 8.8%, respectively. These percentages—90.9%, 0.3%, 8.8%—are called the percent natural abundances of the neon isotopes. Sometimes the mass numbers of isotopes are incorporated into the names of elements, such as neon-20, carbon-12, and oxygen-16.

In a neutral atom the number of electrons is equal to the number of protons, Z. But if an atom either loses or gains electrons, it acquires a net electric charge; it becomes an **ion**. The species $^{20}_{10}Ne^{+}$ and $^{20}_{10}Ne^{2+}$ are ions. The first one has ten protons, ten neutrons, and nine electrons; the second, ten protons, ten neutrons, and eight electrons.

Example 2-3 Indicate the numbers of protons, neutrons, and electrons in (a) $^{35}_{17}Cl$ and (b) $^{80}_{35}Br^{-}$.

Solution. If the species is a neutral atom, the number of electrons is equal to the number of protons, which in turn is equal to Z, the subscript numeral in the symbol $^{A}_{Z}X$. An atom becomes an ion only by losing or gaining *electrons*. An ion has the same number of protons as the atom from which it is formed, but a different number of electrons.

(a) $^{35}_{17}Cl$: $Z = 17$, $A = 35$, a neutral atom.

number of protons = 17; number of electrons = 17; number of neutrons = $A - Z = 35 - 17 = 18$.

(b) $^{80}_{35}Br^{-}$: $Z = 35$, $A = 80$, an ion with a net charge of -1.

number of protons = 35; number of electrons = 36 (one electron must be *gained* to yield an ion having a charge of -1); number of neutrons = $A - Z = 80 - 35 = 45$.

SIMILAR EXAMPLES: Review Problems 5, 6.

Example 2-4 Write appropriate symbols for (a) an atom of strontium-90; (b) the species containing 29 protons, 34 neutrons, and 27 electrons.

Solution
(a) Look up the element strontium on the inside back cover. It has the symbol Sr and the atomic number 38. The mass number is 90.

$^{90}_{38}Sr$

(b) The element with $Z = 29$ is copper (Cu). The mass number is equal to the number of protons plus the number of neutrons: $29 + 34 = 63$. Because the species has only 27 electrons, it must be an ion with a net charge of $+2$.

$^{63}_{29}Cu^{2+}$

SIMILAR EXAMPLES: Review Problems 5, 6.

FIGURE 2-14
The Bainbridge mass
spectrograph.

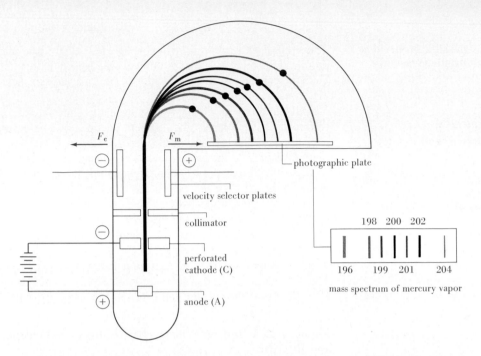

mass spectrum of mercury vapor

In passing from the cathode (C) to the anode (A), cathode rays strike gaseous atoms and produce positive ions (as illustrated in Figure 2-9). The positive ions are attracted to the cathode (C), but some of them pass through the hole in the cathode and enter the space between the velocity selector plates.

One of the velocity selector plates carries a negative charge and the other, a positive charge. Between the plates is an electric field, and positive ions entering this field experience a force that attracts them to the negative plate (to the left in this drawing). The magnitude of this force on a particular ion depends on the strength of the electric field (how strongly the plates are charged) and on the amount of net positive charge carried by that ion. At the same time, a magnetic field is imposed perpendicular to the plane of the page. (The poles of the magnet, not pictured here, are located above and below the page in an arrangement similar to Figure 2-6.) Positive ions in this magnetic field experience a force that drives them to the right. The magnitude of this force on a particular ion is proportional to the magnetic field strength, the amount of net positive charge carried by that ion, *and* the velocity (speed) of the ion.

Thus, every positive ion between the selector plates experiences two forces simultaneously, one that tends to draw it to the left and the other, to the right. For ions of low velocity, the magnetic field force is weak, and the ions are drawn to the left into a velocity selector plate. For ions of high velocity, the magnetic field force is strong, and the ions are drawn into the velocity selector plate on the right. For some particular intermediate velocity, ions experience equal forces in the electric and magnetic fields and are not deflected at all. These are the ions that appear in the mass spectrograph.

Beyond the selector plates only a magnetic field is maintained, and because all the positive ions at this point have the same velocity all ions carrying the same net positive charge experience the same force in the magnetic field. However, the lighter ions are bent into circular paths of smaller radii than are the heavier ones. As a result, the photographic plate is struck at the left by the lightest ions and at the right by the heaviest. Furthermore, the more ions of a particular mass present, the more intense the corresponding spot of the photographic plate. The photographic record consists of a number of exposed regions or "lines." This collection of lines is called the **mass spectrum** of the sample. The mass spectrum of mercury is shown here.

2-11 Atomic Weights

By international agreement a single atom of the nuclide $^{12}_{6}C$ is *arbitrarily* assigned a mass of 12.00000 u (which explains the definition of one atomic mass unit, 1 u, as equal to $\frac{1}{12}$ the mass of a carbon-12 atom).* As discussed in Section 25-6, the mass of an individual atom, known as a **nuclidic mass,** cannot be obtained just by totaling the masses of its fundamental particles—a small quantity of matter is converted to energy in the creation of an atomic nucleus from protons and neutrons. However, the ratio of the mass of any other atom to that of carbon-12 can be established with a **mass spectrometer.** In this device a beam of gaseous ions is separated into components of differing mass by passage through electric and magnetic fields. The separated components are focused on a measuring instrument, where their presence is detected and recorded. (If the record is photographic, the device is called a **mass spectrograph.**) The principle of mass spectrometry is illustrated in Figure 2-14.

Example 2-5 With mass spectral data it can be established that the ratio of the mass of $^{16}_{8}O$ to $^{12}_{6}C$ is 1.3329. What is the mass of the $^{16}_{8}O$ atom?

Solution. The ratio of the masses is $^{16}_{8}O/^{12}_{6}C = 1.3329$. The mass of the $^{16}_{8}O$ atom is 1.3329 times the mass of $^{12}_{6}C$.

mass of $^{16}_{8}O = 1.3329 \times 12.00000$ u $= 15.9948$ u

SIMILAR EXAMPLES: Review Problems 8, 9.

In a table of international atomic weights,† the atomic weight listed for carbon is 12.011; yet our atomic weight standard was taken as 12.00000. It is important to note that the atomic weight standard is based on the pure nuclide $^{12}_{6}C$, whereas naturally occurring carbon contains small amounts of $^{13}_{6}C$ (and traces of $^{14}_{6}C$). The existence of these heavier isotopes causes the atomic weight to be greater than 12. Tabulated atomic weights of the elements are *weighted averages* for *mixtures* of isotopes in their naturally occurring abundances. These atomic weights are pure numbers, relative to 12.00000 for $^{12}_{6}C$, and are calculated from mass spectrometric data as illustrated in Example 2-6.

Example 2-6 The mass spectrum of carbon shows that 98.892% of carbon atoms are $^{12}_{6}C$ with a mass of 12.00000 u and 1.108% are $^{13}_{6}C$ with a mass of 13.00335 u. Calculate the atomic weight of naturally occurring carbon.

Solution. The *simple average* of the masses of C-12 and C-13 is (12.00000 + 13.00335)/2 = 12.50168. However, since in a sample of carbon there are many more C-12 atoms than C-13, we must weigh the contribution of C-12 to the average atomic weight more heavily than that of C-13. We need to calculate a *weighted average*. One way to do so is in terms of a *hypothetical* carbon atom that has 98.892% of the mass of a C-12 atom plus 1.108% of the mass of a C-13 atom. Or, we can say that each nuclide contributes to the average atomic weight in accordance with its nuclidic mass *and* the fraction of all the atoms that are of that nuclide.

---

*Oxygen was formerly taken as the atomic weight standard. However, because physicists assigned an atomic weight of 16.00000 to the nuclide $^{16}_{8}O$, whereas chemists assigned this value to the naturally occurring *mixture* of oxygen isotopes, a number of troublesome discrepancies arose. The definition of atomic weights based on $^{12}_{6}C$ is unambiguous.

†From the discussion of mass and weight in Section 1-5, it is clear that the term atomic mass is more appropriate than atomic weight. However, atomic weight has become so engrained in the vocabulary of chemists that its use is likely to continue for some time. A comparison of several terms that deal with the masses of atoms is made in Table 3-1.

$$\begin{aligned}
\begin{matrix}\text{contribution to} \\ \text{atomic weight} \\ \text{by } {}^{12}_{6}\text{C}\end{matrix} &= \left(\begin{matrix}\text{fraction of} \\ \text{all C atoms} \\ \text{that are } {}^{12}_{6}\text{C}\end{matrix}\right) \times \left(\begin{matrix}\text{mass of} \\ \text{a } {}^{12}_{6}\text{C atom}\end{matrix}\right) \\
&= \quad 0.98892 \quad\quad \times \quad 12.00000 \quad = 11.867
\end{aligned}$$

$$\begin{aligned}
\begin{matrix}\text{contribution to} \\ \text{atomic weight} \\ \text{by } {}^{13}_{6}\text{C}\end{matrix} &= \left(\begin{matrix}\text{fraction of} \\ \text{all C atoms} \\ \text{that are } {}^{13}_{6}\text{C}\end{matrix}\right) \times \left(\begin{matrix}\text{mass of} \\ \text{a } {}^{13}_{6}\text{C atom}\end{matrix}\right) \\
&= \quad 0.01108 \quad\quad \times \quad 13.00335 \quad = 0.1441
\end{aligned}$$

$$\begin{aligned}
\begin{matrix}\text{atomic weight} \\ \text{of naturally} \\ \text{occurring carbon}\end{matrix} &= \left(\begin{matrix}\text{contribution} \\ \text{by } {}^{12}_{6}\text{C}\end{matrix}\right) + \left(\begin{matrix}\text{contribution} \\ \text{by } {}^{13}_{6}\text{C}\end{matrix}\right) \\
&= \quad 11.867 \quad\quad + \quad 0.144 \quad\quad = 12.011
\end{aligned}$$

SIMILAR EXAMPLES: Review Problem 10; Exercise 33.

Example 2-7   Bromine has *two* naturally occurring isotopes. The mass and percent natural abundance of one of them, ${}^{79}_{35}\text{Br}$, are 78.9183 u and 50.54%. What must be the mass and percent natural abundance of the second isotope, ${}^{81}_{35}\text{Br}$?

Solution.   One fact that we can put to immediate use is that *the sum of the percent natural abundances of all the isotopes of an element must total 100.00%.* The percent natural abundance of ${}^{81}_{35}\text{Br}$ is $100.00\% - 50.54\% = 49.46\%$. And, as in Example 2-6, we can break down the average atomic weight into the respective contributions of each isotope.

at. wt. = contribution by ${}^{79}_{35}\text{Br}$ + contribution by ${}^{81}_{35}\text{Br}$

$$\text{at. wt.} = \left(\begin{matrix}\text{fraction of atoms that are} \\ {}^{79}_{35}\text{Br} \times \text{ nuclidic mass } {}^{79}_{35}\text{Br}\end{matrix}\right) + \left(\begin{matrix}\text{fraction of atoms that are} \\ {}^{81}_{35}\text{Br} \times \text{ nuclidic mass } {}^{81}_{35}\text{Br}\end{matrix}\right)$$

The average atomic weight is obtained from the table on the inside back cover, and the unknown in the equation below is mass ${}^{81}_{35}\text{Br}$.

$$79.904 = (0.5054 \times 78.9183) + (0.4946 \times \text{mass } {}^{81}_{35}\text{Br})$$

$$79.904 = 39.89 + (0.4946 \text{ mass } {}^{81}_{35}\text{Br})$$

$$\text{mass } {}^{81}_{35}\text{Br} = \frac{79.904 - 39.89}{0.4946} = \frac{40.01}{0.4946} = 80.89$$

To four significant figures, the percent natural abundance and the nuclidic mass of ${}^{81}_{35}\text{Br}$ are 49.46% and 80.89 u.

SIMILAR EXAMPLES: Exercises 28, 29, 30.

FIGURE 2-15
Visibility of individual atoms.

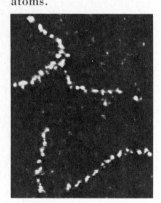

This photograph was obtained with a high resolution scanning electron microscope at the University of Chicago by A. V. Crewe, J. Wall, and J. Langmore [*Science,* 168:1340 (1970)], and is reproduced with Professor Crewe's permission. The bright spots result from individual thorium atoms, arranged in chains, in a complex substance containing thorium. The blurring of the spots is probably caused by the vibrational motion of the thorium atoms.

2-12 Postscript: Do Atoms Exist?

We have traced certain aspects of the development of the atomic theory, from the speculations of the ancient Greeks to some sophisticated experiments of modern times. We will continue to pursue additional aspects of the subject of atomic structure, but we will also come to the realization that the final word on the atomic theory can never be written. So, we may ask, when will there be sufficient evidence to prove the existence of atoms? One type of experimental observation which seems to go a long way toward substantiating the atomic theory is suggested by Figure 2-15. Do atoms exist? At this point the evidence is most convincing. (See also, Color Section A.)

Summary

The most important of the early discoveries concerning chemical change were the laws of conservation of mass and constant composition. These laws led to Dalton's atomic theory. Dalton's theory did not meet with immediate success, however, because of difficulties in assigning chemical formulas and atomic weights.

More fruitful were investigations of the discharge of electricity through gases—cathode ray research. Cathode rays (electrons) were found to be fundamental particles of all matter and also fundamental units of negative electrical charge. Cathode ray research was directly responsible for the discovery of x rays, which in turn led to the discovery of radioactivity. Experiments with positive ions formed in cathode ray tubes yielded evidence of a fundamental unit of positive charge, led to the discovery of isotopes, and served as the basis of mass spectrometry. From studies of the scattering of α particles by thin metal foils, Rutherford and his coworkers proposed the concept of the nuclear atom. This concept, followed by later discoveries of the proton and neutron, permitted atomic structure to be described in terms of protons, neutrons, and electrons.

By assigning an atomic mass of 12.00000 u to carbon-12, the masses of other atoms can now be determined with a mass spectrometer. From the masses of the different isotopes of an element and their percent natural abundances the atomic weight of an element can be determined.

Learning Objectives

As a result of studying Chapter 2, you should be able to

1. Describe and illustrate the laws of conservation of mass, constant composition, and multiple proportions.

2. State the basic assumptions of Dalton's atomic theory.

3. List some of the characteristic properties of cathode rays and of positive rays (canal rays).

4. Describe the production of x rays, the phenomenon of radioactivity, and the characteristics of α, β, and γ radiation.

5. Describe J. J. Thomson's determination of the charge-to-mass ratio of cathode rays, Millikan's determination of the charge on an electron, and Rutherford's studies of α particle scattering by thin metal foils.

6. Outline the general features of Rutherford's model of the nuclear atom.

7. Do simple calculations involving the masses and charges of protons, neutrons, and electrons.

8. Use the symbol $^A_Z X$ to list the numbers of protons, neutrons, and electrons present in atoms and ions.

9. Explain how a mass spectrometer works and calculate atomic masses from experimentally determined mass ratios.

10. Relate the atomic weight of an element to the masses and percent natural abundances of its isotopes, and perform calculations involving these quantities.

Some New Terms

An **atom** is the smallest particle of matter that characterizes an element.

An **atomic mass unit (u)** is a basic unit for expressing the masses of individual atoms. One u is $\frac{1}{12}$ of the mass of a $^{12}_6 C$ atom; the masses of the proton and of the neutron are just slightly greater than 1 u.

The **atomic number, Z,** is the number of protons in the nucleus of an atom. It is also the number of electrons outside the nucleus of an electrically neutral atom.

The **atomic weight** of an element is the mass of the naturally occurring mixture of isotopes of the element, relative to an arbitrarily assigned mass of 12.00000 for carbon-12.

Cathode rays are negatively charged particles (electrons) emitted at the negative electrode (cathode) in the passage of electricity through gases at low pressures.

Chemical symbols are abbreviations consisting of one or two letters assigned to the chemical elements (e.g., N = nitrogen and Ne = neon).

Electrons are particles carrying the fundamental unit of negative electric charge and found outside the nuclei of all atoms.

An **ion** is an electrically charged species consisting of a single atom or a group of atoms. It is formed when a neutral atom or a group of atoms either gains or loses electrons.

Isotopes of an element are atoms with different numbers of neutrons in their nuclei and thus, different masses.

The **law of conservation of mass** states that matter can neither be created nor destroyed in ordinary physical or chemical processes.

The **law of constant composition (definite proportions)** states that a chemical compound has a unique composition in terms of its constituent elements, that is, regardless of the source of the compound.

The **law of multiple proportions** deals with the proportions in which two elements combine when they are able to form more than a single compound: For each compound the ratio of the mass of one element to a fixed mass of the second is formulated. Then, for any pair of compounds, these ratios are themselves in the ratio of small whole numbers, such as $2.66 : 1/1.33 : 1 = 2 : 1$.

The **mass number** A is the total of the number of protons and neutrons in the nucleus of an atom.

A **mass spectrograph (mass spectrometer)** is a device used to separate and to measure the quantities and masses of the different ions in a beam of positively charged gaseous ions.

Neutrons are electrically neutral fundamental particles of matter found in all atomic nuclei except that of the simple hydrogen atom, protium, ^1_1H.

Nuclide is a term used to designate a specific atomic species, as represented by the symbolism $^A_Z X$.

Nuclidic mass is the mass, in atomic weight units u, of an individual atom, relative to an arbitrarily assigned value of 12.00000 u as the nuclidic mass of $^{12}_6C$.

Positive rays are beams of positively charged gaseous ions produced through collisions of cathode rays with residual gas atoms in cathode ray tubes.

Protons are fundamental particles carrying the basic unit of positive electric charge and found in the nuclei of all atoms.

Radioactivity is a phenomenon in which small particles of matter (α or β particles) and/or electromagnetic radiation (γ rays) are emitted by unstable atomic nuclei.

Suggestions for Further Study

CREWE, A. V., J. WALL, and J. LANGMORE, "Visibility of Individual Atoms," *Science,* **168,** 1340 (1970).

KELLER, E., "Man and the Universe, Part II: World of the Atom," *Chemistry,* **45**[7], 9 (1972).

KOLB, D., "What Is an Element?," *J. Chem. Educ.,* **54,** 696 (1977).

MORROW, B. A., "On the Discovery of the Electron," *J. Chem. Educ.,* **46,** 584 (1969).

ORNA, M. V., "On Naming the Elements with Atomic Number Greater Than 100," *J. Chem. Educ.,* **59,** 123 (1982).

PETT, V. B., "Whole Numbers and Atomic Theory," *Chemistry,* **51**[1], 16 (1978).

Review Problems

1. An 8.50-g sample of zinc dust is mixed with 2.20 g of powdered sulfur and the mixture is heated. All the sulfur is used up and 6.69 g of white zinc sulfide is the only product. What mass of zinc remains unreacted?

2. Samples of pure carbon weighing 1.07, 1.96, and 2.03 g, respectively, were burned in an excess of air. The masses of carbon dioxide obtained (the sole product in each case) were 3.92, 7.18, and 7.44 g, respectively.

    (a) Do these data establish that carbon dioxide has a fixed composition?

    (b) What is the composition of carbon dioxide, expressed in % C and % O, by mass?

3. Use modern atomic weights and the method of Figure 2-3 to show that the law of multiple proportions applies to each of the following pairs of compounds: **(a)** SO_2 and SO_3; **(b)** H_2O and H_2O_2; **(c)** PCl_3 and PCl_5.

4. Use data from Table 2-1 to determine the *net* charge, in coulombs, associated with 1.50×10^{15} **(a)** hydrogen atoms (^1_1H); **(b)** bromide ions (Br^-); **(c)** $^{22}_{10}Ne^{2+}$ ions.

5. Complete the table at the right. What minimum amount of information is required to characterize completely an atomic species (nuclide)?

6. Arrange the following species in order of increasing **(a)** number of electrons; **(b)** number of neutrons; **(c)** mass.
$^{112}_{50}Sn$, $^{40}_{18}Ar$, $^{122}_{52}Te$, $^{59}_{29}Cu$, $^{120}_{48}Cd$, $^{58}_{27}Co$, $^{39}_{19}K$

7. For the nuclide $^{133}_{55}Cs$, express

    (a) the percentage, *by number,* of the fundamental particles in the nucleus that are neutrons.

    (b) the approximate percentage of the mass of the atom contributed by protons.

8. The following data on atomic masses are given in a handbook. **(a)** ^9_4Be, 9.01218 u; **(b)** $^{31}_{15}P$, 30.97376 u; **(c)** $^{94}_{40}Zr$, 93.9061 u. What is the ratio of each of these masses to that of $^{12}_6C$?

9. The following ratios of masses were obtained with a mass spectrometer. **(a)** $^{19}_9F : ^{12}_6C = 1.5832$; **(b)** $^{35}_{17}Cl : ^{19}_9F = 1.8406$; **(c)** $^{81}_{35}Br : ^{35}_{17}Cl = 2.3140$. Determine the mass of an $^{81}_{35}Br$ atom in atomic mass units.

10. In naturally occurring uranium, 99.27% of the atoms are $^{238}_{92}U$ with mass 238.05 u; 0.72%, $^{235}_{92}U$ with mass 235.04 u; and 0.006%, $^{234}_{92}U$ with mass 234.04 u. Calculate the atomic weight of naturally occurring uranium.

Name	Symbol	Number protons	Number electrons	Number neutrons	Mass number
sodium	$^{23}_{11}Na$	11	11	12	23
silicon	—	—	—	14	—
—	—	37	—	—	85
—	^{40}K	—	—	—	—
—	—	—	33	42	—
—	$^{20}Ne^{2+}$	—	—	—	—
—	—	—	—	—	80
—	—	—	—	126	—

Exercises

Law of conservation of mass

1. When an iron object rusts, its mass increases. When a match is burned, its mass decreases. Do these observations violate the law of conservation of mass? Explain.

2. By calculation, show whether the law of conservation of mass is obeyed, within the limits of experimental error, in the following experiment. 10.00 g of calcium carbonate is dissolved in 100.0 cm^3 of hydrochloric acid (density = 1.148 g/cm^3). The products are 120.40 g of solution (a mixture of hydrochloric acid and calcium chloride) and 2.22 L of carbon dioxide gas, which has a density of 1.9769 g/L.

Law of constant composition

3. In one experiment 1.76 g of sodium metal was allowed to react with 13.21 g of chlorine gas. All the sodium was used up, and 4.47 g of sodium chloride (salt) was produced. In a second experiment 1.00 g of chlorine was allowed to react with 10.00 g of sodium. All the chlorine was used up, and 1.65 g of sodium chloride was produced. Show that these results are consistent with the law of constant composition.

4. When 2.10 g of hydrogen gas was allowed to react with an excess of oxygen, 18.77 g of water was obtained. In a second experiment a sample of water was decomposed by electrolysis, resulting in 1.45 g of hydrogen and 11.51 g of oxygen. Are these results consistent with the law of constant composition? Demonstrate why or why not.

Dalton's atomic theory

5. A compound of hydrogen and oxygen unknown in Dalton's time is hydrogen peroxide, consisting of 94% oxygen and 6% hydrogen, by mass. Had Dalton known of this compound, what formula do you think he would have assigned to it, having assigned a formula of OH to water?

6. Estimate the atomic weight of magnesium by using 16.0 for the atomic weight of oxygen, information presented in Example 2-1, and Dalton's idea that if there is but a single compound of two elements it should have the formula AB.

Law of multiple proportions

7. The formulas used by Dalton in establishing the law of multiple proportions were CH_2 for methane and CH for ethylene (see Figure 2-3). Show that the law is just as well established with modern formulas for these gases (methane CH_4 and ethylene C_2H_4) and modern atomic weights.

8. Dalton knew of three oxides of nitrogen. To one (nitrous gas) he assigned the formula NO. Following his rule of adopting the *simplest formulas possible*, what formulas do you suppose Dalton assigned to the other two? Show that the formulas for the three compounds are consistent with the law of multiple proportions.

9. Mercury and oxygen form two compounds. One contains 96.2% mercury, by mass, and the other 92.6%. Show that these data conform to the law of multiple proportions. Speculate on the formulas of these two oxides.

Fundamental particles

10. Cite the evidence that establishes most convincingly that electrons are fundamental particles of all matter.

11. List several significant differences between cathode rays and canal rays, stressing the manner of production, electric charge, mass, and so on.

12. Investigation of electrical discharge in gases led to the discovery of both negative and positive particles of matter. Explain why the negative particles proved to be fundamental particles of matter but the positive particles did not.

13. Why couldn't the same methods that had been used to characterize electrons be used to isolate and detect neutrons?

Fundamental charges and charge-to-mass ratios

14. These observations are made for a series of 10 oil drops in an experiment similar to Millikan's (see Figure 2-7). Drop 1 carries a charge of 1.28×10^{-18} C; drops 2 and 3 each carry $\frac{1}{2}$ the charge of drop 1; drop 4 carries $\frac{1}{4}$ the charge of drop 1; drop 5 has a charge four times that of drop 1; drops 6 and 7 have charges three times that of drop 1; drops 8 and 9 have charges twice that of drop 1; and drop 10 has the same charge as drop 1. Are these data consistent with the value of the electronic charge given in the text? Could Millikan have inferred the charge on the electron from this particular series of data? Explain.

15. It is now known that static electric charges are caused by the transfer of electrons.
   (a) How many excess electrons are present on an object with a charge of -3.8×10^{-14} C?
   (b) How many electrons are deficient from an object with a net charge of $+1.7 \times 10^{-11}$ C?

16. Use data from Table 2-1 to verify the following statements made in this chapter.
   (a) The mass of electrons is about $\frac{1}{2000}$ that of hydrogen atoms.
   (b) The charge-to-mass ratio, e/m, for positive rays is considerably smaller than for electrons.

17. Arrange the following species in order of *increasing* absolute magnitude of charge to mass, e/m: proton, electron, neutron, α particle, the atom $^{40}_{18}Ar$, the ion $^{37}_{17}Cl^-$. (The absolute magnitude refers to the value of the e/m ratio without regard to its sign. Note also whether any two of the species listed have identical e/m ratios.)

Atomic models

18. Use the atomic model of J. J. Thomson (see Figure 2-8) to draw pictures of the following gaseous atoms and ions. **(a)** He; **(b)** O; **(c)** N^+; **(d)** F^-.

19. Represent the species given in Exercise 18 by the Rutherford model of the atom (see Figure 2-13).

Atomic number, mass number, nuclides, and isotopes

20. Describe the significance of each term in the symbol $^A_Z X$.

21. For the atom, $^{107}_{47}Ag$, with a mass of 106.90509 u, determine **(a)** the numbers of protons, neutrons, and electrons in the atom; **(b)** the ratio of the mass of this atom to that of an atom of $^{12}_6C$; **(c)** the ratio of the mass of this atom to that of an atom of $^{16}_8O$ (refer to Example 2-5).

22. Explain why the symbols $^{35}_{17}Cl$ and ^{35}Cl actually convey the same information. Do the symbols $^{35}_{17}Cl$ and $_{17}Cl$ have the same meaning?

23. The isotopes of hydrogen are called protium, deuterium, and tritium and have mass numbers 1, 2, and 3, respectively. What are the basic differences among these three types of hydrogen atoms? Which do you think occurs in greatest natural abundance? Explain.

Atomic mass units, atomic masses

24. What is the mass, in grams, corresponding to 1.000 u? (*Hint:* Refer to Table 2-1.)

25. What is the mass, in grams, of each of the following. (*Hint:* Use the result of Exercise 24.)
  (a) 1.00×10^{15} atoms of bromine-79, having individual masses of 78.9183 u.
  (b) 1.00×10^{15} atoms of bromine-81, having individual masses of 80.9163 u.
  (c) 1.00×10^{15} atoms of a mixture of bromine-79 and bromine-81 in their naturally occurring abundances, 50.54% $^{79}_{35}Br$ atoms and 49.46% $^{81}_{35}Br$ atoms.

Atomic weights

26. Which statement is probably true concerning the masses of *individual* copper atoms: that *all*, *some*, or *none* have a mass of 63.546 u? Explain your reasoning.

27. Carbon has an atomic weight of 12.011 and its principal isotope is $^{12}_6C$. Oxygen has an atomic weight of 15.9994 and its principal isotope is $^{16}_8O$. Does it necessarily follow that there will

exist a naturally occurring nuclide with a mass number closest to the tabulated atomic weight and that this will be the most abundant of the isotopes of an element? Explain. (*Hint:* Can you find exceptions to this idea in the text?)

28. There are *three* naturally occurring isotopes of magnesium. The masses and percent natural abundances of two of the isotopes are (24.98584 u, 10.13%) and (25.98259 u, 11.17%). Determine for the third isotope **(a)** its percent natural abundance; **(b)** its mass number; **(c)** its nuclidic mass in atomic mass units.

29. The two principal isotopes of lithium have masses of 6.01513 and 7.01601 u. The atomic weight of lithium is 6.941.
  (a) Which of these two isotopes is more abundant?
  (b) What is the approximate ratio of atoms of the more abundant to the less abundant isotope, i.e., $2:1, 3:1, \ldots$?
  ⋆ **(c)** Calculate the percent natural abundances of the two.

⋆ **30.** The two naturally occurring isotopes of nitrogen have masses of 14.0031 and 15.0001 u, respectively. Use the atomic weight listed for nitrogen to determine the percentage of the atoms in naturally occurring nitrogen that are $^{15}_7N$.

⋆ **31.** Silicon has one major isotope, $^{28}_{14}Si$ (27.97693 u, 92.21% natural abundance) and two minor ones, $^{29}_{14}Si$ (28.97649 u) and $^{30}_{14}Si$ (29.97376 u). What are the percent natural abundances of the two minor isotopes? Comment on the limitation of the precision of this calculation.

Mass spectrometry

32. The three isotopes of hydrogen described in Exercise 23 can all combine with chlorine to form simple diatomic molecules, HCl. The naturally occurring isotopes of chlorine and their percent natural abundances are $^{35}_{17}Cl$, 75.53%, and $^{37}_{17}Cl$, 24.47%.
  (a) How many different HCl molecules are possible?
  (b) What are the mass numbers of these different molecules (i.e., the sum of the mass numbers of the two atoms in the molecules)?
  (c) Which is the most abundant of the possible HCl molecules? Which is the second most abundant?
  (d) Sketch the mass spectrum you would expect to obtain for HCl molecules if all the positive ions obtained are $(HCl)^+$. (Refer to Figure 2-14.)

33. The masses and percent natural abundances of the mercury isotopes referred to in Figure 2-14 are as follows: Hg-196, 195.9658 u, 0.146%; Hg-198, 197.9668 u, 10.02%; Hg-199, 198.9683 u, 16.84%; Hg-200, 199.9683 u, 23.13%; Hg-201, 200.9703 u, 13.22%; Hg-202, 201.9706 u, 29.80%; and Hg-204, 203.9735 u, 6.85%. Use these data to calculate the atomic weight of mercury.

Additional Exercises

1. All of these radioactive nuclides have applications in medicine. Write their symbols in the form $^A_Z X$. **(a)** cobalt-60; **(b)** phosphorus-32; **(c)** iodine-131; **(d)** sulfur-35.

2. Given the following species: $^{24}Mg^{2+}$; ^{47}Cr; $^{59}Co^{2+}$; $^{35}Cl^-$; $^{124}Sn^{2+}$; ^{226}Th; ^{90}Sr. Which of these species
  (a) has equal numbers of neutrons and electrons?
  (b) has protons contributing more than 50% of the mass?
  (c) has a number of neutrons equal to the number of protons plus one-half the number of electrons?

3. Determine the approximate value of the charge-to-mass ratio, e/m, in coulombs per gram, for the ions $^{35}_{17}Cl^+$ and $^{32}_{16}S^{2-}$. Why are these values only approximate?

4. A nuclide of silver has a mass that is 6.68374 times that of O-16. What is the mass (in u) of this nuclide? What is the ratio of its mass to that of C-12? (*Hint:* Refer to Example 2-5.)

5. Use 1×10^{-13} cm as the approximate diameter of the spherical nucleus of the ^1_1H atom, together with data from Table 2-1, to estimate the density of matter in a proton.

6. The highest density encountered is about 22 g/cm³ (osmium metal). What does a comparison of this value and the density of the proton calculated in Additional Exercise 5 suggest about the amount of empty space in matter?

★7. The mass of a $^{12}_6C$ atom is taken to be exactly 12.00000 u. Are there likely to be any other nuclides with an exact integral (whole number) mass, expressed in u? Explain.

★8. The atomic weight of oxygen listed in this book is 15.9994. A textbook printed 30 years ago lists a value of 16.0000. How do you account for this discrepancy? Would you expect other atomic weights listed in the older text to be the same, generally higher, or generally lower than in this text? Explain.

★9. Suppose the atomic weight scale was redefined by choosing as an arbitrary standard having an atomic weight of 35.00000 the naturally occurring *mixture* of chlorine isotopes.
  (a) What would be the atomic weights of helium, sodium, and iodine on this new atomic weight scale?
  (b) Why do these three elements have nearly integral (whole number) atomic weights based on $^{12}_6C$ but not based on naturally occurring chlorine?

★10. From the densities of the lines in the mass spectrum of krypton gas, the following observations were made.

(1) Somewhat more than 50% of the atoms were $^{84}_{36}Kr$.
(2) The numbers of $^{82}_{36}Kr$ and $^{83}_{36}Kr$ atoms were essentially equal.
(3) The number of $^{86}_{36}Kr$ atoms was 1.50 times greater than the number of $^{82}_{36}Kr$ atoms.
(4) The number of $^{80}_{36}Kr$ atoms was 19.6% of the number of $^{82}_{36}Kr$ atoms.
(5) The number of $^{78}_{36}Kr$ atoms was 3.0% of the number of $^{82}_{36}Kr$ atoms.

The nuclidic masses are $^{78}_{36}Kr$, 77.9204 u; $^{80}_{36}Kr$, 79.9164 u; $^{82}_{36}Kr$, 81.9135 u; $^{83}_{36}Kr$, 82.9141 u; $^{84}_{36}Kr$, 83.9115 u; $^{86}_{36}Kr$, 85.9106 u. The atomic weight of Kr is 83.80. Use the data presented here to calculate the percent natural abundances of the six isotopes.

Self-Test Questions

For questions 1 through 8 select the single item that best completes each statement.

1. Of these assumptions or results of Dalton's atomic theory, the only one that remains essentially correct is
(a) All atoms of an element are identical in mass.
(b) Atoms are indivisible and indestructible.
(c) Oxygen has an atomic weight of 7.
(d) Atoms of elements combine in the ratios of small whole numbers to form compounds.

2. One oxide of rubidium has 0.187 g O per gram Rb. A possible O-to-Rb mass ratio for a second oxide of rubidium is (atomic weights, O = 16.0; Rb = 85.5) (a) 16:1; (b) 85.5:16; (c) 0.374:1; (d) all of these.

3. Cathode rays (a) may be positively or negatively charged; (b) have properties identical to β particles; (c) are a form of electromagnetic radiation; (d) have masses that depend on the material that emits them.

4. The scattering of α particles by thin metal foils established that (a) the mass and positive charge of an atom are concentrated in a nucleus; (b) electrons are fundamental particles of all matter; (c) all electrons have the same charge; (d) atoms are electrically neutral.

5. The species that has the same number of electrons as $^{32}_{16}S$ is (a) $^{35}_{17}Cl^-$; (b) $^{34}_{16}S^+$; (c) $^{40}_{18}Ar^{2+}$; (d) $^{35}_{16}S^{2-}$.

6. All of the following masses are possible for an *individual* carbon atom except one. That impossible one is (a) 12.00000 u; (b) 12.01115 u; (c) 13.00335 u; (d) 14.00324 u.

7. There are *two* principal isotopes of indium (atomic weight = 114.82). One of these, $^{113}_{49}In$, has an atomic mass of 112.9043 u. The second isotope is most likely to be (a) $^{111}_{49}In$; (b) $^{112}_{49}In$; (c) $^{114}_{49}In$; (d) $^{115}_{49}In$.

8. The mass of the nuclide $^{84}_{36}Kr$ is 83.9115 u. If the atomic weight scale were *redefined* so that $^{84}_{36}Kr$ = 84.00000 u, exactly, the mass of $^{12}_6C$ would be (a) 11.9115 u; (b) 11.9874 u; (c) 12.1027 u; (d) 12.0885 u.

9. When a strip of magnesium metal is burned in air, it glows brilliantly and produces a white powder that weighs more than the original metal. When a strip of magnesium metal is ignited in a photoflash bulb, a brilliant glow is also produced, but in this case the bulb weighs the same before and after it is flashed. Explain the difference in these observations.

10. Cite some probable reasons why the discovery of neutrons came so much later than the discovery of electrons.

11. What is the charge-to-mass ratio, e/m, in C/g, of a chloride ion, $^{37}_{17}Cl^-$? The mass of $^{37}_{17}Cl$ is 36.966 u. The electronic charge is -1.602×10^{-19} C, and the relationship between the units, u and g, is 1.673×10^{-24} g = 1.0073 u.

12. There are two principal isotopes of silver, $^{107}_{47}Ag$ and $^{109}_{47}Ag$. The atomic weight of silver is 107.87, with 51.82% of the atoms being $^{107}_{47}Ag$. The mass of an atom of $^{107}_{47}Ag$ is 106.9 u. What is the mass, in u, of an atom of $^{109}_{47}Ag$?

3 Stoichiometry I: Elements and Compounds

The word stoichiometry is derived from the Greek *stoicheion,* meaning element. Literally, stoichiometry means to measure the elements. The term is generally used more broadly, however, to include a wide variety of relationships involving substances and mixtures of chemical interest. This chapter is concerned with the stoichiometry of elements and compounds and with certain related subjects. The ideas discussed here are crucial to an understanding of many of the topics that will arise later in the text.

3-1 Avogadro's Number and the Concept of the Mole

The most fundamental relationships among chemical quantities, as we will learn in this and later chapters, involve relative *numbers* of atoms, ions, or molecules. However, although we recognize the importance of numbers of chemical units, we do not count them in the usual sense. We must resort to some other measurement that is related to numbers of atoms, and one of the most convenient measurements is that of mass. Thus, we need to establish a relationship between the *measured* mass of an element and some *known* but *uncountable* number of atoms contained in that mass. Practical examples of substituting mass for a desired number of items are numerous. For example, when we plant a lawn, even though what is required is a certain number of grass seeds, we do not count them out—we buy them by the pound or kilogram.

Atomic weights can be established by comparing the masses of a large number of atoms of one kind with an *equal* number of atoms of the atomic weight standard, $^{12}_{6}C$. The number that is used for this purpose is the number of atoms present in exactly 12.00000 g of $^{12}_{6}C$. This number, called **Avogadro's number,** N_A,* has the value, 6.02205×10^{23} (usually rounded off to 6.02×10^{23}). A term that is nearly synonymous with Avogadro's number is the **mole** (abbreviated **mol**).

A mole of substance is an amount of substance that contains the same number of elementary units as there are $^{12}_{6}C$ atoms in 12.00000 g $^{12}_{6}C$.

If a substance contains atoms all of a single nuclide, we may write

1 mol $^{12}_{6}C$ consists of 6.02205×10^{23} $^{12}_{6}C$ atoms and weighs 12.00000 g; 1 mol $^{16}_{8}O$ consists of 6.02205×10^{23} $^{16}_{8}O$ atoms and weighs 15.9948 g; and so on.

Avogadro's number is an enormous number. Suppose that it were desired to pile up 6.02205×10^{23} garden peas (each having a volume of about 0.1 cm^3). The required pile would cover the entire United States to a depth of about 6 km.

---

*Although Avogadro recognized the significance of this numerical relationship (see Section 5-4), he did not evaluate this number himself. This was first accomplished by Loschmidt. Avogadro's number is also referred to as Loschmidt's number L. One method of evaluating Avogadro's number is presented in Additional Exercise 19. Another is discussed in Section 11-13.

FIGURE 3-1
An attempt to picture a mole of atoms.

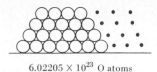
6.02205×10^{23} O atoms
$= 15.9994$ g O

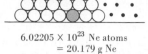
6.02205×10^{23} Ne atoms
$= 20.179$ g Ne

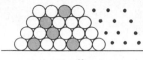
6.02205×10^{23} Cl atoms
$= 35.453$ g Cl

If atoms could be piled up in the manner suggested here, it would take an enormously large pile to contain one mole of atoms. In this representation, because the relative abundances of ^{17}O and ^{18}O are so small, the oxygen atoms are all shown to be alike. In the case of neon, about one of every ten atoms is of the heavier isotope, ^{22}Ne. In chlorine roughly three fourths of the atoms are ^{35}Cl and one fourth are ^{37}Cl.

Most elements are composed of mixtures of two or more isotopes. The atoms to be "counted out" to yield one mole are not all of the same mass. As we learned in Section 2-11, they must be taken in the proportions in which they occur naturally. Thus, in 1 mol of carbon, most of the atoms are carbon-12, but some are carbon-13 (and a very few are carbon-14).

1 mol of carbon consists of 6.02205×10^{23} C atoms and weighs 12.011 g; 1 mol of oxygen consists of 6.02205×10^{23} O atoms and weighs 15.9994 g; and so on.

The mass of one mole of atoms, called the **molar mass,** is easily obtained from a table of atomic weights, e.g., 6.941 g Li/mol Li. Figure 3-1 is a pictorial analogy of the meaning of 1 mol of atoms.

3-2 Calculations Involving the Mole Concept

Example 3-1 How many atoms are present in 2.80 mol of iron metal?

Solution. This problem can be solved by a single-step application of the general problem-solving method (recall equation 1.5). The conversion factor is derived from the fact that one mole of atoms is equivalent to Avogadro's number of atoms: 1 mol Fe $\backsimeq 6.02 \times 10^{23}$ Fe atoms.

From this point on in the text the cancellation of units will not be routinely shown in calculations. In every case, though, you should assure yourself that the proper cancellation will occur.

$$\text{no. Fe atoms} = 2.80 \text{ mol Fe} \times \frac{6.02 \times 10^{23} \text{ Fe atoms}}{1 \text{ mol Fe}}$$
$$= 16.9 \times 10^{23} \text{ Fe atoms} = 1.69 \times 10^{24} \text{ Fe atoms}$$

SIMILAR EXAMPLES: Review Problem 1; Exercise 2(a).

Example 3-2 How many moles of magnesium are represented by a collection of 3.05×10^{20} Mg atoms?

Solution. Again the required conversion factor is based on the definition of a mole. Here, however, the factor is inverted from that of Example 3-1 to provide for the proper cancellation of units. As further indication that we should divide rather than multiply by Avogadro's number, note that 3.05×10^{20} atoms is only a fraction of one mole. Our answer must be less than 1.

$$\text{no. mol Mg} = 3.05 \times 10^{20} \text{ Mg atoms} \times \frac{1 \text{ mol Mg}}{6.02 \times 10^{23} \text{ Mg atoms}}$$
$$= 0.507 \times 10^{-3} \text{ mol Mg} = 5.07 \times 10^{-4} \text{ mol Mg}$$

SIMILAR EXAMPLES: Review Problem 2(a); Exercise 3(a).

In the two examples just considered

1. The solutions did not depend on the elements chosen (that is, the answers would have been 1.69×10^{24} and 5.07×10^{-4} regardless of the elements of choice).
2. The situations described were only hypothetical—atoms cannot be counted directly. Some other property must have been measured.

Examples 3-3 and 3-4 indicate how molar mass enters into calculations involving the mole concept. Example 3-5 illustrates that various other factors may also be required.

Example 3-3   What is the mass of 6.12 mol Ca?

Solution.   The molar mass of Ca is expressed as 40.08 g Ca/mol Ca and can serve as a conversion factor between the amount of substance, in moles, and its mass, in grams.

$$\text{no. g Ca} = 6.12 \text{ mol Ca} \times \frac{40.08 \text{ g Ca}}{1 \text{ mol Ca}} = 245 \text{ g Ca}$$

SIMILAR EXAMPLES: Review Problems 2(b), 2(c); Exercise 5.

Example 3-4   How many Na atoms are present in 15.5 g Na?

Solution.   Here we can think in terms of a two-step approach. First, convert the mass of Na to the number of moles of Na. This requires using the molar mass in the inverse manner to Example 3-3.

$$\text{no. mol Na} = 15.5 \text{ g Na} \times \frac{1 \text{ mol Na}}{23.0 \text{ g Na}} = 0.674 \text{ mol Na}$$

Then convert the number of moles of Na to number of Na atoms (as in Example 3-1).

$$\text{no. Na atoms} = 0.674 \text{ mol Na} \times \frac{6.02 \times 10^{23} \text{ Na atoms}}{1 \text{ mol Na}} = 4.06 \times 10^{23} \text{ Na atoms}$$

Although we think in terms of the two steps above, the steps can be combined into a single setup.

$$\text{no. Na atoms} = 15.5 \text{ g Na} \times \frac{1 \text{ mol Na}}{23.0 \text{ g Na}} \times \frac{6.02 \times 10^{23} \text{ Na atoms}}{1 \text{ mol Na}}$$
$$= 4.06 \times 10^{23} \text{ Na atoms}$$

SIMILAR EXAMPLES: Review Problem 2(d); Exercise 6.

Example 3-5   How many iron atoms are present in a stainless steel ball bearing having a radius of 2.00 mm? The stainless steel contains 85.6% Fe, by mass, and has a density of 7.75 g/cm³.

Solution.   We must reason in a stepwise fashion and use several conversion factors.
Step 1.   Determine the volume of the steel ball in cubic centimeters. (Use the formula for the volume of a sphere.)

$$V = \frac{4}{3}\pi r^3 = \frac{4}{3}\pi \left(2.00 \text{ mm} \times \frac{1 \text{ cm}}{10 \text{ mm}} \right)^3 = \frac{4}{3}(3.14)(0.200)^3 \text{ cm}^3$$
$$= 0.0335 \text{ cm}^3$$

Step 2.   Use the conversion factors indicated in parentheses to convert successively from
(a) cubic centimeters to grams of stainless steel (density);

(b) grams of stainless steel to grams of iron (percent composition);
(c) grams of iron to moles of iron (molar mass of iron);
(d) moles of iron to number of iron atoms (Avogadro's number).

Proceed by solving for a numerical answer to Step 2a; use this value to obtain the answer to Step 2b; and so on. Or, use the setup for each step (*without* solving for a numerical answer) as the starting point for the next step, as shown below.

$$\text{no. Fe atoms} = 0.0335 \text{ cm}^3 \text{ steel} \times \frac{7.75 \text{ g steel}}{1.00 \text{ cm}^3 \text{ steel}} \times \frac{85.6 \text{ g Fe}}{100 \text{ g steel}}$$

$$(\text{cm}^3 \text{ steel} \longrightarrow \quad \text{g steel} \longrightarrow \quad \text{g Fe}$$

$$\times \frac{1 \text{ mol Fe}}{55.8 \text{ g Fe}} \times \frac{6.02 \times 10^{23} \text{ Fe atoms}}{1 \text{ mol Fe}}$$

$$\longrightarrow \quad \text{mol Fe} \longrightarrow \quad \text{Fe atoms})$$

$$= 2.40 \times 10^{21} \text{ Fe atoms}$$

SIMILAR EXAMPLES: Exercises 7, 8, 9.

3-3 Chemical Compounds

Chemical compounds are denoted by combinations of symbols called chemical formulas. A **chemical formula** represents

1. The elements present in a compound.
2. The relative numbers of atoms of each element.

A **formula unit** is the *smallest* collection of atoms from which the formula can be derived. In the following formula the elements present are denoted by their symbols and the relative numbers of atoms by *subscript* numerals (where no subscript is written, the number 1 is understood).

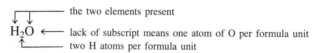

the two elements present
H_2O ← lack of subscript means one atom of O per formula unit
two H atoms per formula unit

Here are three additional formula units.

NaCl	MgCl$_2$	CCl$_4$
sodium chloride	magnesium chloride	carbon tetrachloride

A **molecule** is a group of bonded atoms that exists as a separate entity and has characteristic physical and chemical properties. From Figure 3-2 we see that the formula unit of CCl_4 is a molecule. The formula unit of NaCl, on the other hand, is simply a pair of atoms (ions) selected from a much larger array of atoms (ions). It is inappropriate to speak of a molecule of solid sodium chloride. The situation with $MgCl_2$ is similar to that of NaCl.

The case of the compound hydrogen peroxide is again somewhat different. The smallest collection of atoms that can exist, a *molecule,* contains two atoms of hydrogen and two of oxygen: H_2O_2. But the smallest collection of atoms that discloses the combining ratio (relative numbers of atoms) of hydrogen and oxygen is HO. If this combination of an H atom and an O atom is taken as the formula unit, then the molecule contains *two* formula units.

A chemical formula based on the formula unit is called the **simplest** or **empirical**

FIGURE 3-2
Formula units of ionic and covalent compounds.

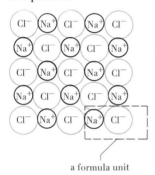

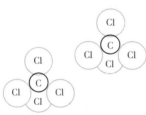

a formula unit

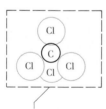

a formula unit (molecule)

NaCl illustrates the situation encountered with solid ionic compounds; CCl_4 is an example of a covalent compound. An ionic compound forms between a metal and a nonmetal; covalent compounds result from combinations of nonmetals. The distinction between metals and nonmetals is introduced in Section 3-5 and is explored more fully in Chapter 8.

formula. The formula based on an actual molecule is called the **molecular formula.** There are three possible relationships to consider.

1. The empirical and molecular formulas may be identical, as with CCl_4.
2. The molecular formula may be a *multiple* of the empirical formula. (The molecular formula H_2O_2 is twice the empirical formula HO.)
3. A compound may have an empirical formula and no molecular formula (such as the solids NaCl, $MgCl_2$, and $NaNO_3$).

Formula Weight and Molecular Weight.   Once the formula unit has been identified, it is a simple matter to establish the **formula weight** of a compound. *It is the mass of the formula unit relative to an assigned mass of* 12.00000 *for* $^{12}_6C$. Since atomic weights are also relative to $^{12}_6C$, formula weights can be determined by adding up atomic weights. Thus, for sodium chloride, NaCl:

one formula unit of NaCl consists of one Na^+ and one Cl^-

formula weight NaCl = at. wt. Na + at. wt. Cl
$$= 22.99 + 35.45 = 58.44$$

And for magnesium chloride, $MgCl_2$:

formula weight $MgCl_2$ = at. wt. Mg + (2 × at. wt. Cl)
$$= 24.30 + (2 × 35.45) = 95.20$$

The atomic weights of an atom and its ion are practically identical. This is because the atom and ion differ only in the number of electrons they contain, and electrons contribute very little to the mass of an atom.

If a compound consists of discrete molecules, it is also appropriate to define a **molecular weight.** *It is the mass of a molecule relative to an assigned mass of* 12.00000 *for* $^{12}_6C$. To determine the molecular weight of carbon tetrachloride, CCl_4, for example, we note that

1 molecule of CCl_4 consists of one C atom and 4 Cl atoms

molecular weight CCl_4 = at. wt. C + (4 × at. wt. Cl)
$$= 12.01 + (4 × 35.45) = 153.8$$

Although it is always acceptable to speak of the formula weight of a compound, the term *molecular weight* is valid only if discrete molecules of the compound exist. When the term is applied to compounds such as NaCl, $MgCl_2$, and $NaNO_3$ in the solid state, what is actually meant is formula weight. If the formula unit and a molecule are identical (as in CCl_4), the formula weight and the molecular weight are identical. Where molecules of a compound consist of two or more formula units, the molecular weight is a corresponding *multiple* of the formula weight.

Mole of a Compound.   The concept of a mole can be applied to any species—atoms, ions, formula units, molecules . . . . As a consequence we can describe a mole of compound as an amount of compound containing Avogadro's number of formula units or molecules. The term molar mass can also be extended to moles of formula units or molecules, leading to relationships of the type

1 mol $MgCl_2$ ≎ 95.20 g $MgCl_2$ ≎ 6.02205 × 10^{23} $MgCl_2$ formula units

and

1 mol CCl_4 ≎ 153.8 g CCl_4 ≎ 6.02205 × 10^{23} CCl_4 molecules

Example 3-6   How many Cl^- ions are present in 50.0 g $MgCl_2$?

Solution.   This calculation is similar to Example 3-4, in that we use molar mass to form a factor to convert from mass to number of moles of $MgCl_2$. This is followed by a factor based on Avogadro's number to convert from moles to number of formula

FIGURE 3-3
What is a "mole of sulfur"?

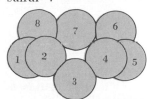

Experimental evidence indicates that in solid sulfur atoms are joined together into puckered rings with eight members. A molecule of sulfur is composed of eight atoms, represented as S_8. In a sample of solid sulfur, there exist eight times as many atoms as there are molecules. If by a "mole of sulfur" we mean 1 mol of sulfur atoms, we can measure out $\frac{1}{8}$ mol of the S_8 molecules and describe this amount as 1 mol S. On the other hand, 1 mol S_8 means that we measure out 1 mol of the S_8 molecules and obtain thereby 8 mol of individual sulfur atoms.

units of $MgCl_2$. Finally, we need an additional factor based on the fact that there are two Cl^- ions per formula unit of $MgCl_2$ (abbreviated f.u. below).

$$\text{no. } Cl^- \text{ ions} = 50.0 \text{ g } MgCl_2 \times \frac{1 \text{ mol } MgCl_2}{95.2 \text{ g } MgCl_2} \times \frac{6.02 \times 10^{23} \text{ f.u. } MgCl_2}{1 \text{ mol } MgCl_2}$$
$$\times \frac{2 \ Cl^- \text{ ions}}{1 \text{ f.u. } MgCl_2}$$
$$= 6.32 \times 10^{23} \ Cl^- \text{ ions}$$

SIMILAR EXAMPLE: Review Problem 2(e).

Example 3-7 How many liters of liquid CCl_4 ($d = 1.59$ g/cm^3) must be measured out to contain 3.58×10^{26} CCl_4 molecules?

Solution. This calculation will require the series of conversions: molecules → mol → g → cm^3 → L. Perhaps the best approach is in this stepwise fashion.

Step 1

$$\text{no. mol } CCl_4 = 3.58 \times 10^{26} \ CCl_4 \text{ molecules} \times \frac{1 \text{ mol } CCl_4}{6.02 \times 10^{23} \ CCl_4 \text{ molecules}}$$
$$= 5.95 \times 10^2 \text{ mol } CCl_4$$

Step 2

$$\text{no. g } CCl_4 = 5.95 \times 10^2 \text{ mol } CCl_4 \times \frac{153.8 \text{ g } CCl_4}{1 \text{ mol } CCl_4} = 9.15 \times 10^4 \text{ g } CCl_4$$

Step 3

$$\text{no. cm}^3 \ CCl_4 = 9.15 \times 10^4 \text{ g } CCl_4 \times \frac{1 \text{ cm}^3 \ CCl_4}{1.59 \text{ g } CCl_4} = 5.75 \times 10^4 \text{ cm}^3 \ CCl_4$$

Step 4

$$\text{no. L } CCl_4 = 5.75 \times 10^4 \text{ cm}^3 \ CCl_4 \times \frac{1 \text{ L } CCl_4}{1000 \text{ cm}^3 \ CCl_4} = 57.5 \text{ L } CCl_4$$

SIMILAR EXAMPLES: Exercises 9, 10.

Mole of an Element—A Second Look. In Section 3-1 we defined a mole of an element as Avogadro's number of *atoms*. This is the only definition possible when describing certain elements, like iron, magnesium, sodium, and copper, in which enormous numbers of individual spherical atoms are clustered together, much like marbles in a can. But with some elements—hydrogen, oxygen, nitrogen, fluorine, chlorine, bromine, iodine, phosphorus, and sulfur, for example—atoms of the same kind are joined together to form molecules, and bulk samples of the elements are composed of collections of molecules. For these elements it is appropriate to speak of 1 mol of molecules and the molecular weight of the element. Molecular forms of the elements just cited are

$$H_2 \quad O_2 \quad N_2 \quad F_2 \quad Cl_2 \quad Br_2 \quad I_2 \quad P_4 \quad S_8$$

The distinction made here between 1 mol H and 1 mol H_2 is very much like the distinction between one dozen socks and one dozen pairs of socks (i.e., the H atom is analogous to a single sock and the H_2 molecule to a pair of socks).

To say a mole of hydrogen is ambiguous; one should say either a mole of hydrogen atoms or a mole of hydrogen molecules. Better still is to write 1 mol H or 1 mol H_2. Furthermore, we may write the molar masses, 1.008 g H/mol H and 2.016 g H_2/mol H_2. The situation for the element sulfur is pictured in Figure 3-3.

Atomic Weights, Molecular Weights, . . . : A Final Word. We have introduced a number of important terms related to the stoichiometry of elements and compounds, beginning in Chapter 2 and continuing through this section. A summary of the meanings of these terms and the relationships among them is provided in Table 3-1 for ready reference.

TABLE 3-1
A summary of terms used in stoichiometry

Term	Definition or usage
atomic weight standard	An atomic weight of 12.00000 is arbitrarily assigned to $^{12}_{6}C$.
nuclidic mass (isotopic mass)	The mass, in atomic mass units, u, of a single atom of a specific nuclide, on a scale in which the mass of an atom of $^{12}_{6}C$ is arbitrarily defined as 12.00000 u (e.g., $^{14}_{7}N$ has a nuclidic mass of 14.00307 u).
atomic weight (relative atomic weight)	A dimensionless (pure) number that expresses the mass of the naturally occurring mixture of isotopes of an element, relative to an arbitrarily assigned atomic weight of 12.00000 for carbon-12. These are the values listed in a table of atomic weights (e.g., atomic weight of Cl = 35.453).
formula weight (relative formula weight)	A dimensionless (pure) number that expresses the mass of a formula unit of a compound, relative to the atomic weight standard, carbon-12 (e.g., formula weight of NaCl = 58.45). Tabulated atomic weights are used in computing formula weights.
molecular weight (relative molecular weight)	A dimensionless (pure) number that compares the mass of a molecule to the atomic weight standard, carbon-12 (e.g., molecular weight of CS_2 = 76.13). Tabulated atomic weights are used in computing molecular weights. Use of this term should be limited to situations where discrete, identifiable molecules actually exist.
mole	An amount of substance containing the same number of elementary units (6.02205×10^{23}) as there are $^{12}_{6}C$ atoms in 12.00000 g $^{12}_{6}C$.
molar mass (molar weight, mole weight)	The mass of one mole of a substance, whether the substance is composed of individual atoms (e.g., 55.85 g Fe/mol Fe), formula units (e.g., 166.0 g KI/mol KI), or molecules (e.g., 32.00 g O_2/mol O_2). Often the terms atomic, formula, or molecular weight are used, even though molar mass is what is intended.

3-4 Composition of Chemical Compounds

We can derive a great deal of information about the composition of chemical compounds from chemical formulas and atomic weights. Consider, for example, the once commonly used insecticide DDT.* It has the formula $C_{14}H_9Cl_5$ and a molecular weight of 354.5. *Per mole* of DDT there are 14 moles of C atoms, 9 moles of H atoms, and 5 moles of Cl atoms. That is,

$$1 \text{ mol } C_{14}H_9Cl_5 \backsimeq 14 \text{ mol C} \backsimeq 9 \text{ mol H} \backsimeq 5 \text{ mol Cl} \tag{3.1}$$

Formulas, as we have been writing them, do not tell us which atoms in a compound are joined or bonded together. That is, there is nothing in the formula $C_{14}H_9Cl_5$ to suggest that the H atoms and the Cl atoms are bonded to C atoms and not to each other. This kind of information about compounds is revealed through *structural formulas,* a topic explored

*Full name: *di*chloro*di*phenyl*tri*chloroethane.

fully in Chapters 9 and 10. For present matters this kind of detailed knowledge about the structure of a compound is not required. We just need to keep in mind that all the atoms represented in a formula are *combined* atoms and that there are no free or uncombined atoms in a compound.

The equivalencies written in expression (3.1) are used to write conversion factors to answer the questions posed in Example 3-8.

Example 3-8   (a) How many H atoms are present in 75.0 g DDT ($C_{14}H_9Cl_5$)? **(b)** What is the mass ratio of Cl to C in DDT? (That is, how many grams of Cl are present *for every gram* of C?)

Solution

(a) To find the number of atoms in a quantity of substance implies that Avogadro's number will be required. But first the quantity of substance must be expressed in moles, and this implies that a molar mass will also be required. The additional factor needed in the series of conversions, g DDT $\to$ mol DDT $\to$ mol H atoms $\to$ no. H atoms, and shown in color below, is a factor derived from expression (3.1) to convert from mol DDT to mol H atoms.

$$\text{no. H atoms} = 75.0 \text{ g C}_{14}\text{H}_9\text{Cl}_5 \times \frac{1 \text{ mol C}_{14}\text{H}_9\text{Cl}_5}{354.5 \text{ g C}_{14}\text{H}_9\text{Cl}_5} \times \frac{9 \text{ mol H}}{1 \text{ mol C}_{14}\text{H}_9\text{Cl}_5}$$

$$\times \frac{6.02 \times 10^{23} \text{ H atoms}}{1 \text{ mol H}}$$

$$= 1.15 \times 10^{24} \text{ H atoms}$$

(b) One approach is to start with 1.00 g C and convert this to mol C. Then convert from mol C to mol Cl, and, finally, from mol Cl to g Cl.

$$\text{no. g Cl} = 1.00 \text{ g C} \times \frac{1 \text{ mol C}}{12.0 \text{ g C}} \times \frac{5 \text{ mol Cl}}{14 \text{ mol C}} \times \frac{35.45 \text{ g Cl}}{1 \text{ mol Cl}} = 1.06 \text{ g Cl}$$

That is, the mass ratio is 1.06 g Cl/g C. An easier approach, perhaps, is to start with the *mole* ratio, 5 mol Cl/14 mol C, and convert both numerator and denominator to mass, in grams.

$$\text{no. g Cl/g C} = \frac{5 \text{ mol Cl} \times \dfrac{35.45 \text{ g Cl}}{1 \text{ mol Cl}}}{14 \text{ mol C} \times \dfrac{12.0 \text{ g C}}{1 \text{ mol C}}} = \frac{177.25 \text{ g Cl}}{168 \text{ g C}} = 1.06 \text{ g Cl/g C}$$

SIMILAR EXAMPLES: Review Problems 3, 4; Exercise 13.

Calculating Percent Composition from a Chemical Formula.   In a manner similar to that of Example 3-8(b), we can start with the mole ratio, 5 mol Cl/1 mol $C_{14}H_9Cl_5$, and convert this to the mass ratio, g Cl/g $C_{14}H_9Cl_5$. This result represents the fractional part of the mass of DDT that is contributed by Cl atoms. Then, if we multiply this fraction (i.e., the mass ratio) by 100, the expression obtained is stated as parts of Cl per 100 parts of DDT, by mass. But this is the definition of *percent;* our result is the percent chlorine (% Cl) in DDT, by mass. If this type of calculation is performed for each element in a compound, the combination of all these percentages represents the percent composition of the compound, as illustrated in Example 3-9.

Example 3-9   What is the percent composition, by mass, of $C_{14}H_9Cl_5$?

Solution.   Following the steps outlined above, for the % Cl in DDT we obtain

$$\% \text{ Cl, by mass} = \frac{5 \text{ mol Cl} \times \dfrac{35.45 \text{ g Cl}}{1 \text{ mol Cl}}}{1 \text{ mol C}_{14}\text{H}_9\text{Cl}_5 \times \dfrac{354.5 \text{ g C}_{14}\text{H}_9\text{Cl}_5}{1 \text{ mol C}_{14}\text{H}_9\text{Cl}_5}} \times 100$$

$$= \frac{177.25 \text{ g Cl}}{354.5 \text{ g C}_{14}\text{H}_9\text{Cl}_5} \times 100 = 50.00\%$$

This procedure can be simplified somewhat by writing, for % C,

$$\% \text{ C, by mass} = \frac{(14 \times 12.01) \text{ g C}}{354.5 \text{ g C}_{14}\text{H}_9\text{Cl}_5} \times 100 = 47.43\%$$

To calculate % H to three significant figures in this step requires carrying four significant figures in the calculation of % Cl and % C.

To determine % H we could proceed as outlined above. Also, however, since H is the remaining element in the compound, its percentage must be consistent with the fact that the percentages of all the elements in a compound total 100.00%.

$$\% \text{ H, by mass} = 100.00\% - \% \text{ C} - \% \text{ Cl} = 100.00\% - 47.43\% - 50.00\% = 2.57\%$$

SIMILAR EXAMPLES: Review Problem 5; Exercises 14, 16.

Establishing an Empirical Formula from the Experimentally Determined Composition of a Compound.   We have learned that the chemical formula of a compound contains a wealth of information, but how is the formula established in the first place? The method is the same that Dalton tried to employ—deduce the formula from the experimentally determined composition of the compound. Our advantage over Dalton is that we now have a well-established table of atomic weights with which to work.

Percent composition establishes the relative proportions of the elements of a compound on a mass basis. A chemical formula requires that these proportions be in terms of *numbers* of atoms, i.e., on a mole basis. The principle of the method used in Example 3-10 is this: The relative numbers of atoms of each type are independent of whether a single formula unit, a mole, or any arbitrary mass of compound is chosen for study. The sample size selected, 100.0 g, allows the easiest conversion of percentages to actual masses of the elements.

The formula obtained by the method of Example 3-10 is the simplest possible formula—the *empirical formula*. The empirical formula can be used to calculate the formula weight of a compound. From a separate experiment the molecular weight of a compound can be measured (with methods to be introduced in Chapters 5 and 12). The molecular weight will prove to be either equal to or some multiple of the formula weight. The *molecular formula* can be obtained by multiplying all the subscripts in the empirical formula by the same multiplier that relates the formula weight to the molecular weight.

Example 3-10   The compound methyl benzoate, used in the manufacture of perfumes, consists of 70.58% C, 5.93% H, and 23.49% O, by mass. Also, by experiment, the molecular weight is found to be 136. What are the empirical formula and the molecular formula of methyl benzoate?

Solution

Step 1.   Determine the mass of each element in a 100.0-g sample. In 100.0 g of compound (100 parts) there are 70.58 g C (70.58 parts), and so on. The masses are

70.58 g C;   5.93 g H;   23.49 g O

Step 2.   Convert the mass of each element in the 100.0-g sample to number of moles.

$$\text{no. mol C} = 70.58 \text{ g C} \times \frac{1 \text{ mol C}}{12.0 \text{ g C}} = 5.88 \text{ mol C}$$

$$\text{no. mol H} = 5.93 \text{ g H} \times \frac{1 \text{ mol H}}{1.01 \text{ g H}} = 5.87 \text{ mol H}$$

$$\text{no. mol O} = 23.49 \text{ g O} \times \frac{1 \text{ mol O}}{16.0 \text{ g O}} = 1.47 \text{ mol O}$$

Step 3. Write a tentative formula based on the numbers of moles just determined.

$C_{5.88}H_{5.87}O_{1.47}$

Step 4. Attempt to convert the subscripts of Step 3 to small whole numbers. This requires dividing each of the subscripts by the smallest one (1.47).

$$C_{\frac{5.88}{1.47}}H_{\frac{5.87}{1.47}}O_{\frac{1.47}{1.47}} = C_{4.00}H_{3.99}O_{1.00}$$

Step 5. If possible to do so at this point, round off the subscripts obtained in Step 4 to small whole numbers.

C_4H_4O (empirical formula)

The formula weight of the compound is $[(4 \times 12.0) + (4 \times 1.01) + 16.0] = 68.0$. Since the experimentally determined molecular weight is twice the formula weight, the molecular formula is $C_8H_8O_2$.

SIMILAR EXAMPLES: Review Problems 6, 7; Exercises 18, 19.

Sometimes it is not possible to achieve small integral subscripts at the point in the procedure labeled Step 5 in Example 3-10. In these cases an additional step is required, as explained in Example 3-11.

Example 3-11 What is the empirical formula of a compound found to consist of 59.53% C, 5.38% H, 10.68% N, and 24.40% O, by mass.

Solution
Step 1. Determine the mass of each element in a 100.0-g sample.

59.53 g C;   5.38 g H;   10.68 g N;   24.40 g O

Step 2. Convert these masses to numbers of moles. Since these conversions take the same form as in Example 3-10, they are not shown here. The results are

4.96 mol C;   5.33 mol H;   0.763 mol N;   1.52 mol O

Step 3. Write a tentative formula based on the numbers of moles above.

$C_{4.96}H_{5.33}N_{0.763}O_{1.52}$

Step 4. Attempt to convert the subscripts to small whole numbers by dividing each subscript by the smallest (0.763).

$$C_{\frac{4.96}{0.763}}H_{\frac{5.33}{0.763}}N_{\frac{0.763}{0.763}}O_{\frac{1.52}{0.763}} = C_{6.50}H_{6.99}N_{1.00}O_{1.99}$$

Step 5. Round off the subscripts to whole numbers, *where possible*. A useful rule is that a subscript can be rounded off if it is within a few hundredths of a whole number, but certainly no more than 0.1.

$C_{6.50}H_7NO_2$

In Step 4 the deviations of the subscripts of H and O from whole numbers can be attributed to experimental error, but not that of C.
Step 6. Multiply all subscripts by a small whole number chosen to make all subscripts integral. Since all subscripts except that of C are already integral, we focus on it. Multiplication by 2 serves our purpose (i.e., $6.50 \times 2 = 13$).

$C_{6.50 \times 2}H_{7 \times 2}N_{1 \times 2}O_{2 \times 2} = C_{13}H_{14}N_2O_4$

SIMILAR EXAMPLE: Exercise 20.

FIGURE 3-4
Apparatus for
combustion analysis.

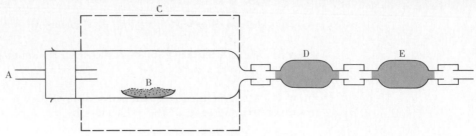

Oxygen gas (A) is passed into the combustion tube containing the sample to be analyzed (B). This portion of the apparatus is enclosed within a high-temperature furnace
(C). Products of the combustion are absorbed as they leave the furnace—water vapor
by magnesium perchlorate (D) and carbon dioxide gas by sodium hydroxide (E) (to produce sodium carbonate).

Combustion Analysis.   How can we determine the percent composition of a compound
by experiment? This can sometimes be done by combustion analysis. As illustrated in
Figure 3-4, a weighed sample of the compound is heated in a stream of oxygen gas in a
furnace. The water vapor and carbon dioxide gas produced in the combustion are absorbed
by appropriate substances. The increases in mass of the absorbers correspond to the
masses of water and carbon dioxide. The principle of combustion analysis is suggested by
Figure 3-5. Example 3-12 illustrates that the empirical formula of a compound can be
established from combustion analysis data without first converting the data to percent
composition by mass.

FIGURE 3-5
Principle of combustion
analysis—Example 3-12
visualized.

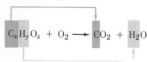

The basis of combustion
analysis is in being able
to trace what happens to
all the atoms in the com-
pound being analyzed
(represented here as
$C_xH_yO_z$). All of the C
atoms in the compound
appear as C atoms in CO_2
(and there is no other
source of C atoms for the
CO_2). All of the H atoms
in the compound appear
as H atoms in H_2O (and
again, there is no other
source of H atoms). Oxy-
gen atoms in the CO_2 and
H_2O come both from the
compound being analyzed
and from oxygen gas con-
sumed in the combustion.
The quantity of oxygen in
the compound must be
determined indirectly.

Example 3-12   Combustion of an 0.2000-g sample of the carbon–hydrogen–oxygen
compound vitamin C (an agent that may be involved in preventing colds) produces
0.2998 g CO_2 and 0.0819 g H_2O. **(a)** What is the empirical formula of vitamin C?
(b) Determine the elemental percent composition of vitamin C.

Solution.   We need to establish both the masses (in grams) and the numbers of moles
of C, H, and O in the 0.2000-g sample of vitamin C in order to solve the two parts of
this example. This will be our first step. Recall that all the C in 0.2000 g of vitamin C
appears in 0.2998 g CO_2, and that all the H appears in 0.0819 g H_2O.

$$\text{no. mol C} = 0.2998 \text{ g } CO_2 \times \frac{1 \text{ mol } CO_2}{44.01 \text{ g } CO_2} \times \frac{1 \text{ mol C}}{1 \text{ mol } CO_2} = 0.006812 \text{ mol C}$$

$$\text{no. g C} = 0.006812 \text{ mol C} \times \frac{12.01 \text{ g C}}{1 \text{ mol C}} = 0.08181 \text{ g C}$$

$$\text{no. mol H} = 0.0819 \text{ g } H_2O \times \frac{1 \text{ mol } H_2O}{18.02 \text{ g } H_2O} \times \frac{2 \text{ mol H}}{1 \text{ mol } H_2O} = 0.00909 \text{ mol H}$$

$$\text{no. g H} = 0.00909 \text{ mol H} \times \frac{1.008 \text{ g H}}{1 \text{ mol H}} = 0.00916 \text{ g H}$$

For oxygen we obtain these quantities in the reverse order. That is,

$$\begin{aligned}
\text{no. g O} &= \text{no. g cpd.} - \text{no. g C} - \text{no. g H} \\
&= 0.2000 \text{ g} - 0.08181 \text{ g} - 0.00916 \text{ g} \\
&= 0.1090 \text{ g O}
\end{aligned}$$

$$\text{no. mol O} = 0.1090 \text{ g O} \times \frac{1 \text{ mol O}}{16.00 \text{ g O}} = 0.006812 \text{ mol O}$$

(a) The numbers of moles of C, H, and O in 0.2000 g of vitamin C provide us with some trial subscripts for the empirical formula.

$$C_{0.006812}H_{0.00909}O_{0.006812}$$

Next we divide each subscript by the smallest—0.006812—to obtain

$$CH_{1.33}O_{1.00}$$

Finally, we multiply each subscript by 3 (since $3 \times 1.33 = 3.99 \simeq 4.00$). The empirical formula of vitamin C is $C_3H_4O_3$.

(b) From the experimentally determined masses of C and H in 0.2000 g vitamin C we can calculate the percentages of these two elements. That of oxygen can be obtained by difference.

$$\% \text{ C} = \frac{0.08181 \text{ g C}}{0.2000 \text{ g cpd.}} \times 100 = 40.90\%$$

$$\% \text{ H} = \frac{0.00916 \text{ g H}}{0.2000 \text{ g cpd.}} \times 100 = 4.58\%$$

$$\% \text{ O} = 100.00\% - 40.90\% - 4.58\% = 54.52\%$$

[An alternative method for this example would have been to first determine the percent composition of the compound, as done here in part (b), followed by the method of Example 3-11 to establish the empirical formula from the percent composition.]

SIMILAR EXAMPLES: Exercises 31, 32.

Precipitation Analysis. Combustion analysis has been applied mostly to establishing the compositions of compounds of carbon and hydrogen with oxygen, nitrogen, and a few other elements (organic compounds, discussed in Chapter 26). For other compounds a useful method has been precipitation analysis. In this method a component of the sample being analyzed deposits from solution as an insoluble material, a precipitate. This precipitate is then treated in such a way as to yield a *pure* solid of *known composition*. From the measured masses of this solid and of the original sample, the percentage of the component in the sample may be determined. The determination of the percent tin in a sample of brass is illustrated through Figure 3-6 and Example 3-13. (See also, Color Section C.)

FIGURE 3-6
Determination of tin in brass—Example 3-13 visualized.

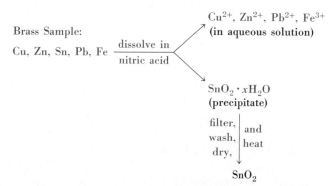

Brass is an alloy of copper and zinc with small amounts of tin, lead, and iron. When a weighed sample of brass is treated with nitric acid (a water solution of HNO_3), the copper, zinc, lead, and iron dissolve and appear in aqueous solution in their ionic forms. Tin is converted to an insoluble oxide with an unknown amount of water associated with it ($SnO_2 \cdot x \, H_2O$). This precipitate is filtered off from the solution, washed, dried, and then heated to drive off all the water. The result is *pure* SnO_2, which is weighed.

Example 3-13   A 2.568-g sample of brass, when treated in the manner outlined in Figure 3-6, yields 0.1330 g SnO_2. What is the % Sn in the brass?

Solution.   To determine the mass of tin in 0.1330 g SnO_2, proceed through these three steps.

Step 1.   Convert 0.1330 g SnO_2 to the number of moles of SnO_2.

$$\text{no. mol } SnO_2 = 0.1330 \text{ g } SnO_2 \times \frac{1 \text{ mol } SnO_2}{150.7 \text{ g } SnO_2} = 8.825 \times 10^{-4} \text{ mol } SnO_2$$

Step 2.   Convert 8.825×10^{-4} mol SnO_2 to mol Sn.

$$\text{no. mol Sn} = 8.825 \times 10^{-4} \text{ mol } SnO_2 \times \frac{1 \text{ mol Sn}}{1 \text{ mol } SnO_2} = 8.825 \times 10^{-4} \text{ mol Sn}$$

Step 3.   Convert 8.825×10^{-4} mol Sn to g Sn.

$$\text{no. g Sn} = 8.825 \times 10^{-4} \text{ mol Sn} \times \frac{118.7 \text{ g Sn}}{1 \text{ mol Sn}} = 0.1048 \text{ g Sn}$$

An alternate route from g SnO_2 to g Sn is based on the conversion factor 118.7 g $Sn/150.7$ g SnO_2. There is one mole of Sn (with mass, 118.7 g) for every mole of SnO_2 (with mass, 150.7 g).

$$\text{no. g Sn} = 0.1330 \text{ g } SnO_2 \times \frac{118.7 \text{ g Sn}}{150.7 \text{ g } SnO_2} = 0.1048 \text{ g Sn}$$

The percent tin in the brass sample is

$$\% \text{ Sn} = \frac{0.1048 \text{ g Sn}}{2.568 \text{ g brass}} \times 100 = 4.08\%$$

SIMILAR EXAMPLES: Exercises 35, 36.

Atomic Weight Determinations.   Historically, precipitation analysis has featured prominently in the determination of atomic weights, although as we learned in Chapter 2 mass spectrometric measurements are now the most accurate for this purpose. Examples 3-14 and 3-15 show how precipitation analysis might be used to establish the atomic weight of an element.

Example 3-14   By precipitation analysis it is found that there is 0.6454 g Cl present in a 0.7718-g sample of the compound XCl. What must be the atomic weight of X?

Solution.   The atomic weight is equal numerically to the molar mass of X. To determine the molar mass requires us to know, for the given sample, the mass of X and the number of moles of X. The mass of X in the sample is

$$\text{no. g X} = 0.7718 \text{ g XCl} - 0.6454 \text{ g Cl} = 0.1264 \text{ g X}$$

Because the elements X and Cl combine in the mole ratio 1:1, the number of moles of X is equal to the number of moles of Cl.

$$\text{no. mol X} = \text{no. mol Cl} = 0.6454 \text{ g Cl} \times \frac{1 \text{ mol Cl}}{35.453 \text{ g Cl}} = 0.01820 \text{ mol Cl}$$

The molar mass is the ratio no. g X/no. mol X or

$$\text{molar mass} = \frac{0.1264 \text{ g X}}{0.01820 \text{ mol X}} = 6.945 \text{ g X/mol X}$$

The atomic weight of X = 6.945.

SIMILAR EXAMPLES: Review Problem 9; Exercise 39.

Example 3-15   The chlorine present in an 0.5250-g sample of the compound XCl_2 is precipitated as 0.5070 g AgCl. What is the atomic weight of X?

Solution.   The situation here differs from Example 3-14 in two ways.
- The mass of Cl in the sample is not given; it must be obtained from the mass of precipitate.
- The mole ratio of X to Cl in the compound is 1:2, not 1:1.

Step 1.   Determine the number of grams of Cl in the AgCl.

$$\text{no. g Cl} = 0.5070 \text{ g AgCl} \times \frac{35.453 \text{ g Cl}}{143.3 \text{ g AgCl}} = 0.1254 \text{ g Cl}$$

Step 2.   The number of grams of Cl in the XCl_2 is also 0.1254 g.

Step 3.   Determine the number of grams of X in the XCl_2.

$$\text{no. g X} = 0.5250 \text{ g } XCl_2 - 0.1254 \text{ g Cl} = 0.3996 \text{ g X}$$

Step 4.   Determine the number of moles of Cl in the XCl_2.

$$\text{no. mol Cl} = 0.1254 \text{ g Cl} \times \frac{1 \text{ mol Cl}}{35.453 \text{ g Cl}} = 3.537 \times 10^{-3} \text{ mol Cl}$$

Step 5.   Determine the number of moles of X in the XCl_2.

$$\text{no. mol X} = 3.537 \times 10^{-3} \text{ mol Cl} \times \frac{1 \text{ mol X}}{2 \text{ mol Cl}} = 1.768 \times 10^{-3} \text{ mol X}$$

Step 6.   Determine the molar mass of X.

$$\text{molar mass} = \frac{0.3996 \text{ g X}}{1.768 \times 10^{-3} \text{ mol X}} = 226.0 \text{ g X/mol X}$$

Step 7.   The atomic weight of X = 226.0.

SIMILAR EXAMPLES: Exercises 40, 41.

3-5 The Need to Name Chemical Compounds—Nomenclature

Throughout this chapter we have referred to compounds mostly by their formulas, not their names. After all, it is the formula that conveys quantitative information about the composition of a compound. Yet, we also need to identify compounds by name. This allows us to look up properties in a handbook, locate a chemical on a storeroom shelf, or just discuss the results of a laboratory experiment with a colleague. A name is the simplest means of calling to mind the general properties of a substance. Later in the text, we will encounter situations in which different compounds have the same formula, and we will find it essential to distinguish among them by name.

To assign symbols and names to the elements is basically simple. But to name chemical compounds is an entirely different matter. It is vital that no two substances have the same name, yet at the same time there should be some similarities in the names of similar substances. Otherwise, the task of naming millions of different compounds would be next to impossible, as would be the case if, for example, all compounds were referred to by a common or trivial name such as water (H_2O) or ammonia (NH_3).

What is needed is a *systematic* method of assigning names—a system of nomenclature. Actually several such systems exist, and each is introduced at an appropriate point in the text. For the present we will restrict ourselves to some of the simpler aspects of naming compounds, and begin by introducing two new concepts that are useful in nomenclature.

3-6 Metals and Nonmetals—An Introduction to the Periodic Table

Although each chemical element has its own set of properties, each element also bears similarities to certain other elements. The elements can be grouped into categories according to these similarities. This subject is explored in some detail in Chapter 8, through the tabular arrangement known as the periodic table of the elements. A periodic table is found on the inside front cover and is reproduced in Figure 3-7.

For now we note that the elements can be grouped into two broad categories—metals and nonmetals. All metals (except mercury, a liquid) are solids at room temperature. Metals generally share the physical properties of being able to conduct heat and electricity, of being malleable (capable of being flattened into thin sheets), of being ductile (capable of being drawn into fine wires), and of having a lustrous or shiny appearance. Nonmetals generally have the "opposite" properties of metals, e.g., they are nonconductors of heat and electricity. Several of the nonmetals are gases at room temperature (e.g., N_2 and O_2); some are brittle solids (e.g., S and Si); one, bromine, is a liquid.

From a chemical standpoint, metal atoms display a tendency to lose one or more electrons when entering into compound formation with nonmetal atoms. The nonmetal atoms in these combinations show a tendency to gain one or more electrons. Since the loss of electron(s) produces a *positive* ion (**cation**) and the gain of electron(s), a *negative* ion (**anion**), the combination of a metal and a nonmetal produces an **ionic compound.** Compounds formed by combinations of nonmetal atoms with one another involve the *sharing* of electrons, not a loss and gain. These compounds are called **covalent compounds.** Elements in the upper right hand portion of the periodic table are nonmetals. Most of the

FIGURE 3-7
Metals, nonmetals, and the periodic table.

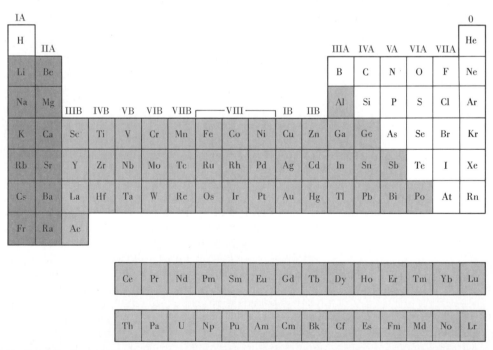

In this tabular arrangement the nonmetals are shown in black. They are found mostly in the right-hand portion of the table. The metallic elements are shown in color; the two groups of elements at the extreme left of the table are the most metallic of all. The special case of hydrogen (a nonmetal grouped with the metals) and further refinements of the classification scheme are discussed in Chapter 8. Refer to the periodic table on the inside front cover for the atomic numbers and atomic weights of the elements.

A few elements, located between the metals and nonmetals, have some of the properties of each and are called metalloids.

rest of the elements are metals, with those in the first two columns on the far left being most metallic of all.

Another feature of the periodic table with which we should become acquainted at this point is the arrangement of similar elements into vertical groups or families. For example, the elements in group IA (except H) are called the alkali metals; the elements in group IIA are the alkaline earth metals; the nonmetals of group VIIA are the halogens; group VIA, the chalcogens (oxygen family); and group VA, the nitrogen family.

3-7 Oxidation States

Oxidation states (oxidation numbers)* are numerical values assigned to atoms that reflect, in a general way, how their electrons are involved in compound formation. Consider the case of NaCl. In this compound a Na atom, a metal, loses one electron to a Cl atom, a nonmetal. The compound consists of the ions, Na^+ and Cl^-, as pictured in Figure 3-2. The Na is said to be in the oxidation state $+1$ and the Cl, -1. *The oxidation states of atoms in their ionic forms are equal to the ionic charges.*

In $MgCl_2$, an Mg atom loses two electrons to become Mg^{2+}, and each Cl atom gains one electron to become Cl^-. The oxidation state of the Mg is $+2$ and of the Cl, -1. If we take the *total* of the oxidation states of all the atoms (ions) in a formula unit of $MgCl_2$, we obtain $+2 - 1 - 1 = 0$. *The total of the oxidation states of all the atoms or ions in a molecule or formula unit is zero.*

In the covalent compound H_2O, if we arbitrarily assign H the oxidation state $+1$, then according to the rule just stated the oxidation state of oxygen must be -2. In another covalent molecule, Cl_2, in order for each Cl atom to be in the same oxidation state and to have the total of these oxidation states be zero, each oxidation state must itself be zero. *The oxidation state of an atom in the free element is zero.*

From the examples just given we can see that some conventions or rules must be followed in assigning oxidation states. The six rules listed below are sufficient to deal with all cases in this text, with this important qualification: *Whenever two rules appear to contradict one another (as they often will), follow the rule that appears **highest** in the list.*

1. The oxidation state of an atom in a free (uncombined) element is 0.
2. The total of the oxidation states of all the atoms in a molecule or formula unit is zero. For an *ion* this total is equal to the charge on the ion, both in magnitude and sign, and regardless of whether the ion consists of a single atom (monatomic) or two or more atoms (polyatomic).
3. In their compounds the alkali metals (group IA of the periodic table, i.e., Li, Na, K, Rb, Cs, Fr) have an oxidation state of $+1$ and the alkaline earth metals (IIA), $+2$.
4. In its compounds the oxidation state of hydrogen is $+1$; that of fluorine is -1.
5. In its compounds oxygen has an oxidation state of -2.
6. In their binary compounds with metals, the elements of group VIIA have an oxidation state of -1; those of group VIA, -2; and those of group VA, -3. (Binary compounds are comprised of *two* elements, e.g., KI, $MgBr_2$, and BaS.)

Example 3-16 What is the oxidation state of the underlined atom in each of the following? (a) $\underline{P}_4$; (b) $\underline{Al}_2O_3$; (c) $\underline{Mn}O_4^-$; (d) $Na\underline{H}$; (e) $H_2\underline{O}_2$; (f) $K\underline{O}_2$; (g) $\underline{Fe}_3O_4$.

Solution
(a) P_4: This formula represents a molecule of the element phosphorus. For an atom of a free element the oxidation state is 0 (rule 1). The oxidation state of P in P_4 is 0.

---

*Because oxidation state refers to a number, the term *oxidation number* is often used synonymously. We will use the two terms interchangeably. Another term whose meaning is similar to oxidation state is valence.

(b) Al_2O_3: The total of the oxidation numbers of all the atoms in a formula unit is 0 (rule 2). The oxidation state of O is -2 (rule 5). The total for three O atoms is -6. The total for two Al atoms is $+6$. The oxidation state of Al is $+3$.

(c) MnO_4^-: The total of the oxidation numbers of all the atoms in this ion must be -1 (rule 2). The total of the oxidation numbers of the four O atoms is -8. The oxidation state of Mn is $+7$.

(d) NaH: Rule 3 states that Na should have an oxidation state of $+1$. Rule 4 indicates that the oxidation state of H should also be $+1$. If both atoms had an oxidation number of $+1$, the total for the formula unit would be $+2$. This would violate rule 2. *Rules 2 and 3 take precedence over rule 4.* The oxidation state of Na is $+1$; the total for the formula unit is 0; and the oxidation state of H in NaH is -1.

(e) H_2O_2: Rule 4, stating that H has an oxidation state of $+1$, takes precedence over rule 5. The sum of the oxidation numbers of the two H atoms is $+2$ and that of the two O atoms is -2. The oxidation state of O in H_2O_2 is -1.

(f) KO_2: Rule 3 (requiring that the oxidation state of K $= +1$) takes precedence over rule 5. The sum of the oxidation numbers of the two O atoms is -1. The oxidation state of each O atom in KO_2 is $-\frac{1}{2}$.

(g) Fe_3O_4: The total of the oxidation numbers of the four O atoms is -8. For three Fe atoms the total is $+8$. This requires an assignment of $+2\frac{2}{3}$ for each Fe atom.

SIMILAR EXAMPLES: Review Problem 12; Exercises 22, 23.

Examples (f) and (g) of Example 3-16 bring out a certain artificiality in the concept of oxidation state. What is the significance of a nonintegral oxidation state? In (f), a molecule of O_2 can be considered to gain an electron from a K atom, forming K^+ and O_2^-. In (g), two of the Fe atoms can be considered to have the oxidation state $+3$ and the third, $+2$. This results in an average of $+2\frac{2}{3}$. We will gain more insight into the significance of the oxidation state concept in later chapters, but for present purposes we should be well served by what we have learned so far.

3-8 Systematic Nomenclature of Inorganic Compounds

Compounds formed by carbon and hydrogen in combination with oxygen, nitrogen, and a few other elements are sufficiently different from other compounds that they are generally considered a special branch of chemistry—organic chemistry (see Chapter 26). At this time we will consider only the naming of the "other" compounds—inorganic compounds. In later chapters we will have much more to say about each of the several categories of inorganic compounds that follow.

Binary Ionic Compounds. **Binary compounds** are those formed between *two* elements. A binary *ionic* compound is formed between a metal and a nonmetal. To name a binary ionic compound, first write the *unmodified* name of the metal, and follow this by the name of the nonmetal, modified to end in "ide."

NaCl = sodium chloride

name unchanged ; *"ide" ending*

$MgBr_2$ = magnesium bromide

Ionic compounds, though comprised of positive and negative ions, must be electrically neutral. That is, the formula unit must contain positive and negative ions in such

TABLE 3-2
Some simple ions[a]

Name	Symbol	Name	Symbol
Positive ions (cations)		copper(II) or cupric	Cu^{2+}
lithium	Li^+	zinc	Zn^{2+}
sodium	Na^+	silver	Ag^+
potassium	K^+	mercury(I) or mercurous[b]	Hg_2^{2+}
rubidium	Rb^+	mercury(II) or mercuric	Hg^{2+}
cesium	Cs^+	tin(II) or stannous	Sn^{2+}
magnesium	Mg^{2+}	lead(II) or plumbous	Pb^{2+}
calcium	Ca^{2+}	**Negative ions (anions)**	
strontium	Sr^{2+}	hydride[c]	H^-
barium	Ba^{2+}	nitride	N^{3-}
aluminum	Al^{3+}	oxide	O^{2-}
chromium(II) or chromous	Cr^{2+}	sulfide	S^{2-}
chromium(III) or chromic	Cr^{3+}	fluoride	F^-
iron(II) or ferrous	Fe^{2+}	chloride	Cl^-
iron(III) or ferric	Fe^{3+}	bromide	Br^-
copper(I) or cuprous	Cu^+	iodide	I^-

[a] The arrangement of ions in this table is similar to the arrangement of the elements in the periodic table (see Figure 3-7).
[b] Unlike other common ions, which are monatomic, the mercurous ion is *diatomic*. The ion is referred to as mercury(I) because two ions with *single* positive charges (Hg^+) are joined together to form the species Hg_2^{2+}. Formulas and names thus take the form: Hg_2Cl_2 = mercury(I) chloride and $HgCl_2$ = mercury(II) chloride.
[c] More common is a *positive* ion of hydrogen, but this can only be found in water solutions and other mixtures, not in pure compounds. Moreover, the positive ion is not simple H^+, although this symbol is often used. The nature of hydrogen ion in aqueous solutions is explored in Section 17-2. (See also, footnote on page 65.)

numbers that their net (total) charge is *zero*. This means *one* Na^+ to *one* Cl^-; *one* Mg^{2+} to *two* Br^-; *one* Mg^{2+} to *one* O^{2-}; and so on. With the ionic charges presented in Table 3-2, it is possible to write formulas for and then to name a host of binary ionic compounds.

Table 3-2 indicates that some metals can exist in more than one ionic form. To write distinctive formulas and names of compounds in these cases, we refer to the oxidation state of the metal. In $FeCl_2$, with one Fe^{2+} ion and *two* Cl^- ions in a formula unit, iron is in the oxidation state +2. In $FeCl_3$, the oxidation state of iron is +3. In the Stock system, oxidation states are written as Roman numerals, as in the names iron(II) chloride and iron(III) chloride. In an older system of nomenclature that is still in use, Fe^{2+} is called ferr*ous* ion and Fe^{3+}, ferr*ic* ion. This system uses the "ous" ending for a cation of lower oxidation state and the "ic" ending for a cation of higher oxidation state. Also, this system uses Latin rather than English names for some metals (e.g., iron = ferrum). Several additional examples are

Formula	Systematic (Stock) name	Old name
$CrCl_2$	chromium(II) chloride	chromous chloride
$CrCl_3$	chromium(III) chloride	chromic chloride
Cu_2O	copper(I) oxide	cuprous oxide
CuO	copper(II) oxide	cupric oxide
$FeBr_2$	iron(II) bromide	ferrous bromide
$FeBr_3$	iron(III) bromide	ferric bromide

By the Stock system we might call NaCl sodium(I) chloride and $MgCl_2$, magnesium(II) chloride. We do not generally do so, however, because the metals of groups IA and IIA of the periodic table have *only one ionic form*. Later in the text you will acquire a better understanding of the factors that determine whether a metal can exist in several oxidation states. For now, use the information in Table 3-2 as a guide.

Binary Covalent Compounds.   The basic method for naming a binary covalent compound is similar to that for an ionic compound. That is,

HCl = hydrogen chloride

H_2O = hydrogen oxide (water)

In both the formula and the name, we write first the element with the positive oxidation state (hence, HCl and *not* ClH).

Some pairs of nonmetals form more than a single binary compound, and we need to be able to distinguish among them. Although the Stock system can be used for this purpose, it is not as generally useful as a system based on prefixes to designate the relative numbers of atoms of each type in a molecule.

mono = 1;   di = 2;   tri = 3;   tetra = 4;   penta = 5;   hexa = 6

Thus, for the two principal oxides of sulfur we can write

SO_2 = sulfur dioxide = sulfur(IV) oxide

SO_3 = sulfur trioxide = sulfur(VI) oxide

Prefixes define the relationship between name and formula precisely. The Stock system does not always do so, as illustrated by this boron–bromine compound.

B_2Br_4 = diboron tetrabromide = boron(II) bromide $[BBr_2(?)]$

Several other binary covalent compounds are named in Table 3-3. Note that although

TABLE 3-3
Naming binary covalent compounds

Formula	Name[a]	
	Using prefixes	Stock system
BCl_3	boron trichloride	boron(III) chloride
CCl_4	carbon tetrachloride	carbon(IV) chloride
CO	carbon monoxide	carbon(II) oxide
CO_2	carbon dioxide	carbon(IV) oxide
NO	nitrogen oxide	nitrogen(II) oxide
NO_2	nitrogen dioxide	nitrogen(IV) oxide
N_2O	dinitrogen oxide	nitrogen(I) oxide
N_2O_3	dinitrogen trioxide	nitrogen(III) oxide
N_2O_4	dinitrogen tetroxide	nitrogen(IV) oxide[b]
N_2O_5	dinitrogen pentoxide	nitrogen(V) oxide
SF_6	sulfur hexafluoride	sulfur(VI) fluoride

[a]Where the prefix ends in an "a" or "o" and the element name begins with an "a" or "o," the final vowel of the prefix is often *dropped for ease of pronunciation.* For example, carbon monoxide (*not* monooxide) and dinitrogen tetroxide (*not* tetraoxide). However, PI_3 is phosphorus triiodide (*not* triodide) and SI_4 is sulfur tetraiodide (*not* tetriodide).

[b]To distinguish it from NO_2, N_2O_4 can be called the dimer of nitrogen(IV) oxide. A dimer (N_2O_4) is formed from two simpler monomers (NO_2).

the prefix mon(o)- is used to name CO, general practice is to use *no* prefix when there is just one atom of a given type per molecule. Thus, NO is nitrogen oxide, not *mono*nitrogen *mon*oxide. Finally, it should also be noted that several substances have common names that are so well established that these substances are not often called by their systematic names. For example, H_2O = water; NH_3 = ammonia; N_2O = nitrous oxide; NO = nitric oxide.

Binary Acids. A special kind of binary covalent compound is one that contains hydrogen, and from which, under appropriate conditions, hydrogen ions (H^+)* can be obtained. Such compounds are called acids, and the principal condition under which H^+ ions are obtained is when the compound is dissolved in water. (In fact, the term *acid* generally is taken to mean a water solution of such a compound.) HCl is an acid. When dissolved in water it produces hydrogen ions (H^+)* and chloride ions (Cl^-). NH_3 in water solutions is *not* an acid; it shows very little tendency to produce H^+ under any conditions. The important binary acids are limited enough in number that they can be listed. They are named by a combination of the prefix "hydro" and the nonmetal name modified to have an "ic" ending.

HF = *hydro*fluor*ic* acid

HCl = *hydro*chlor*ic* acid

HBr = *hydro*brom*ic* acid

HI = *hydro*iod*ic* acid

H_2S = *hydro*sulfur*ic* acid

Polyatomic Ions. The ions listed in Table 3-2 (with one exception) are monatomic ions; they consist of a single atom. Ions in which two or more atoms are bonded together, polyatomic ions, are also commonly encountered, especially among the nonmetals. A number of polyatomic ions and representative compounds containing these ions are listed in Table 3-4, from which we can infer that

1. Polyatomic anions occur more frequently than polyatomic cations. A common polyatomic cation is NH_4^+.
2. Very few polyatomic anions carry the "ide" ending in their names. Of those listed only OH^- (hydroxide ion) and CN^- (cyanide ion) do. Instead, the common endings are "ite" and "ate," and some names carry prefixes (e.g., "hypo" and "per").
3. An element common to many polyatomic anions is *oxygen*. The oxygen is combined with another nonmetal; such ions are called **oxoanions.**
4. Certain nonmetals (e.g., Cl, N, P, and S) form a series of oxoanions containing different numbers of oxygen atoms. Their names are related to the oxidation state of the nonmetal atom to which the O atoms are bonded, according to the scheme

—Increasing oxidation state →

hypo——ite ——ite ——ate per——ate   (3.2)

—Increasing number of oxygen atoms →

5. All the common oxoanions of Cl carry a charge of -1; of S, -2.
6. Some series of oxoanions contain varying numbers of H atoms also and are named accordingly. For example, HPO_4^{2-} is the *hydrogen phosphate* ion and $H_2PO_4^-$ is the *dihydrogen phosphate* ion.

*The species obtained in water solution is actually more complex than the simple ion H^+. In Chapter 17 we will refer to this species as hydronium ion H_3O^+. For present purposes we will not pursue this distinction and will use the simpler notation H^+.

NH_3 belongs to a complementary category of substances called **bases.** These substances yield hydroxide ion (OH^-) in water solutions, either because they contain hydroxide ion—as do NaOH, KOH, and $Ca(OH)_2$, for example—or because they produce OH^- through a reaction with water, as does NH_3.

TABLE 3-4
Some common polyatomic ions

Name	Formula	Typical compound
Cation		
ammonium	NH_4^+	NH_4Cl
Anions		
acetate	$C_2H_3O_2^-$	$NaC_2H_3O_2$
carbonate	CO_3^{2-}	Na_2CO_3
hydrogen carbonate[a] (or bicarbonate)	HCO_3^-	$NaHCO_3$
hypochlorite	ClO^-	$NaClO$
chlorite	ClO_2^-	$NaClO_2$
chlorate	ClO_3^-	$NaClO_3$
perchlorate	ClO_4^-	$NaClO_4$
chromate	CrO_4^{2-}	Na_2CrO_4
cyanide	CN^-	$NaCN$
hydroxide	OH^-	$NaOH$
nitrite	NO_2^-	$NaNO_2$
nitrate	NO_3^-	$NaNO_3$
permanganate	MnO_4^-	$NaMnO_4$
phosphate	PO_4^{3-}	Na_3PO_4
hydrogen phosphate[a]	HPO_4^{2-}	Na_2HPO_4
dihydrogen phosphate[a]	$H_2PO_4^-$	NaH_2PO_4
sulfite	SO_3^{2-}	Na_2SO_3
sulfate	SO_4^{2-}	Na_2SO_4
hydrogen sulfate[a] (or bisulfate)	HSO_4^-	$NaHSO_4$
thiosulfate	$S_2O_3^{2-}$	$Na_2S_2O_3$

[a]These anion names are sometimes written as a single word, i.e., hydrogencarbonate, hydrogenphosphate, etc.

7. The prefix "thio" signifies that a sulfur atom has been substituted for an oxygen atom. (The sulfate ion has *one* S and *four* O atoms; thiosulfate ion has *two* S and *three* O atoms.)

Oxoacids. Binary acids consist of hydrogen and a single nonmetal. The majority of acids are compounds consisting of *three* different elements (**ternary** compounds)—hydrogen, *oxygen*, and another nonmetal. These are called **oxoacids.** Another view of oxoacids is that they result from the combination of hydrogen ions (H^+) and oxoanions. For every series of oxoanions there is a series of oxoacids. The scheme for naming oxoacids is similar to that outlined in expression (3.2), except that the ending "ous" is used instead of "ite" and "ic" instead of "ate." The names, formulas, and oxidation states of the nonmetal atom in several oxoacids are listed in Table 3-5. Also listed in the table are the names and formulas of the compounds that result when the hydrogen of an oxoacid is replaced by sodium. Although we will encounter more precise definitions later, we can think of any compound in which an H atom of an acid is replaced by a metal as a **salt.** The acid is a covalent compound and the salt is ionic.

Example 3-17   Name the compounds (a) $CuCl_2$; (b) NI_3; (c) HIO_4; and (d) $Ca(H_2PO_4)_2$.

Solution
(a) The oxidation state of Cu is $+2$. Since Cu can also exist in the oxidation state $+1$,

TABLE 3-5
Nomenclature of some oxoacids and their salts[a]

Oxidation state	Formula of acid[b]	Name of acid	Formula of salt	Name of salt
+1	$HClO$	*hypo*chlor*ous* acid	$NaClO$	sodium *hypo*chlor*ite*
+3	$HClO_2$	chlor*ous* acid	$NaClO_2$	sodium chlor*ite*
+5	$HClO_3$	chlor*ic* acid	$NaClO_3$	sodium chlor*ate*
+7	$HClO_4$	*per*chlor*ic* acid	$NaClO_4$	sodium *per*chlor*ate*
+3	HNO_2	nitr*ous* acid	$NaNO_2$	sodium nitr*ite*
+5	HNO_3	nitr*ic* acid	$NaNO_3$	sodium nitr*ate*
+4	H_2SO_3	sulfur*ous* acid	Na_2SO_3	sodium sulf*ite*
+6	H_2SO_4	sulfur*ic* acid	Na_2SO_4	sodium sulf*ate*

[a]In general the "ic" and "ate" names are assigned to compounds in which the central non-metal atom has an oxidation state equal to the periodic group number. Halogen compounds are exceptional in that the "ic" and "ate" names are assigned to compounds in which the halogen has an oxidation state of +5 (even though the group number is 7, that is, VIIA).

[b]In all these acids H atoms are bonded to O atoms, not the central nonmetal atom. Often formulas are written to reflect this fact, such as HOCl instead of HClO and HOClO instead of $HClO_2$.

we should use the Stock system to distinguish clearly between the two possible chlorides. $CuCl_2$ is copper(II) chloride. Note that it is unnecessary to call this compound copper(II) dichloride. The composition of the formula unit is established by the charges on the ions. Prefixes are not used to name binary ionic compounds.

(b) Both N and I are nonmetals. NI_3 is a binary covalent compound called nitrogen triiodide.

(c) The oxidation state of I is +7. By analogy to the chlorine-containing oxoacids in Table 3-5, we should name this compound periodic acid.

(d) The polyatomic anion $H_2PO_4^-$ is dihydrogen phosphate ion. Two of these ions are present for every Ca^{2+} ion in the compound calcium dihydrogen phosphate.

SIMILAR EXAMPLES: Review Problem 14; Exercises 25, 26.

Example 3-18 Write acceptable formulas for the following compounds: **(a)** tetranitrogen tetrasulfide; **(b)** iron(III) oxide; **(c)** ammonium chromate; **(d)** bromic acid; **(e)** calcium hypochlorite.

Solution

(a) This is a covalent molecule containing *four* N atoms and *four* S atoms. The correct formula is N_4S_4.

(b) In an electrically neutral formula unit there are *two* Fe^{3+} ions $[2 \times (+3) = +6]$ and *three* O^{2-} ions $[3 \times (-2) = -6]$. The formula is Fe_2O_3.

(c) Two ammonium ions (NH_4^+) must be present for every chromate ion (CrO_4^{2-}). The parentheses around the NH_4, followed by the subscript 2, signify *two* NH_4^+ ions in a formula unit. The correct formula is $(NH_4)_2CrO_4$. (This formula is read as "N—H—4, taken twice, C—R—O—4.")

(d) The "ic" acid for the oxoacid series of the halogens (group VIIA) has the halogen in the oxidation state +5. Bromic acid is $HBrO_3$.

(e) Here there are *one* monatomic cation, Ca^{2+}, and *two* oxoanions, ClO^-, in a formula unit, leading to the formula $Ca(ClO)_2$. Note the importance of the parentheses here. If they were omitted, one would have $CaClO_2$, an *incorrect* formula both for calcium hypochlorite and for calcium chlorite.

SIMILAR EXAMPLES: Review Problem 13; Exercise 27.

Compounds of Greater Complexity. Let us note a few additional ideas represented through the following formulas, which are somewhat more complex than those we have been considering.

$CuSO_4 \cdot 5H_2O$. This compound is of a type known as a hydrate. A **hydrate** is a substance in which a formula unit has associated with it a certain number of water molecules. The formula shown here signifies *five* H_2O molecules per formula unit of $CuSO_4$. The compound is called copper(II) sulfate *penta*hydrate. Its formula weight is that of $CuSO_4$ *plus* that associated with five H_2O: 160 + 90 = 250. We can speak of the percent water in a hydrate. For $CuSO_4 \cdot 5H_2O$ this would be

$$\% \ H_2O = \frac{(5 \times 18.0) \ \text{g} \ H_2O}{250 \ \text{g} \ CuSO_4 \cdot 5H_2O} \times 100 = 36.0\%$$

In Section 24-13 we will learn of the different ways in which water molecules may be incorporated in a solid compound.

$K_4[Fe(CN)_6]$. This compound, called potassium ferrocyanide or potassium hexacyanoferrate(II), belongs to a class known as coordination compounds. It consists of simple ions—K^+—and complex ions—$[Fe(CN)_6]^{4-}$. The atoms within the square brackets are those present in the complex ion. In $[Fe(CN)_6]^{4-}$ there are six CN^- ions bonded to one Fe^{2+} ion. The nature of complex ions and coordination compounds is explored fully in Chapter 24. For the present all that we need to recognize is that

1. The total number of atoms in a formula unit of $K_4[Fe(CN)_6]$ is 17, i.e., 4 K, 1 Fe, 6 C, and 6 N atoms.
2. The formula weight of the compound is $(4 \times 39.1) + 55.85 + (6 \times 12.0) + (6 \times 14.0) = 368.2$.

Summary

The mole concept is central to the study of chemistry. The mole describes an amount of substance in terms of the *number* of individual units present. The mole concept requires that the units being counted be clearly established—atoms, ions, formula units, molecules. Moreover, since amounts of substances must still be measured in terms of mass or volume, it is necessary to know how a mole of substance is related to these other quantities.

The percentage of each element present in a compound can be determined readily from the formula of the compound. Also, the formula can be derived from an experimental measurement of the percent composition of a compound. Formulas obtained in this way are empirical formulas—the simplest formulas that can be written. In some cases the empirical formula also represents the composition of a molecule of a substance; in others, the molecular formula is an integral multiple of the empirical formula.

Chemical compounds are designated by names as well as by formulas. To establish a systematic basis for relating names and formulas—nomenclature—three important ideas are introduced in this chapter, although each is discussed more fully later. These are the concepts of metals and nonmetals, the periodic classification of the elements, and oxidation states.

Learning Objectives

As a result of studying Chapter 3, you should be able to

1. Write Avogadro's number and explain the meaning of a mole of an element and a mole of a compound.

2. Write formulas for the molecular forms of some common nonmetallic elements and distinguish between a mole of atoms and a mole of molecules.

3. Distinguish between the empirical and molecular formulas of a compound.

4. Calculate the numbers of atoms, ions, formula units, or molecules in a substance from a given mass, or vice versa.

5. Use chemical formulas as a source of conversion factors for stoichiometric calculations.

6. Determine the percent composition of a compound from its formula.

7. Establish the empirical formula of a compound from its experimentally determined percent composition.

8. Establish the percent composition of a compound from the results of a combustion analysis.

9. Determine the quantity of a constituent in a compound or mixture from the results of a precipitation analysis.

10. Use appropriate experimental data to establish the atomic weight of an element.

11. State the oxidation state conventions or rules and apply them to assigning oxidation states.

12. Write the names and symbols of common cations and anions and the names and formulas of binary ionic and binary covalent compounds.

13. Name some of the more common polyatomic ions, and write names and formulas of compounds containing these ions.

14. Write names and formulas of binary acids and oxoacids.

Some New Terms

An **acid** is a hydrogen-containing compound that, under appropriate conditions, can produce hydrogen ions H^+. (More general definitions are given in Chapter 17.)

Avogadro's number, N_A, has a value of 6.02205×10^{23}, usually rounded off to 6.02×10^{23}. It is the number of individual units in one mole.

Binary acids are certain combinations of hydrogen with another nonmetallic element.

A **binary compound** is comprised of *two* elements.

A **chemical formula** represents the relative numbers of atoms of each of the elements present in a compound.

Combustion analysis is a procedure that relates the percent composition of a compound to the amounts of CO_2 and H_2O produced when the compound is burned in oxygen.

An **empirical formula** is the simplest formula that can be written for a compound, that is, has the smallest integral subscripts possible.

A **formula unit** is the smallest collection of atoms from which the formula of a compound can be established.

Formula weight is the mass of a formula unit of a compound relative to that of the atomic weight standard $^{12}_{6}C$.

A **hydrate** is a compound in which a certain number of water molecules are associated with each formula unit, for example, $CuSO_4 \cdot 5H_2O$.

Molar mass is the mass (usually in grams) of one mole of atoms, formula units, or molecules.

A **mole** is an amount of substance that contains Avogadro's number of individual units, that is, 6.02×10^{23} atoms, formula units, or molecules.

A **molecular formula** denotes the numbers of the different atoms present in an actual molecule. In some cases the molecular formula is the same as the empirical formula; in others it is an integral multiple of that formula.

Molecular weight is the mass of a molecule of a compound relative to that of the atomic weight standard $^{12}_{6}C$.

A **molecule** is a combination of atoms that can exist as an individual identifiable unit possessing a unique set of properties.

Nomenclature refers to the writing of chemical names and formulas by some systematic method.

An **oxoacid** is an acid containing hydrogen, oxygen, and another nonmetal.

An **oxoanion** is a polyatomic anion containing a nonmetal such as Cl, N, P, or S in combination with some number of oxygen atoms.

A **polyatomic ion** contains two or more atoms.

Salts are ionic compounds in which hydrogen atoms of acids are replaced by metal ions.

Stoichiometry refers to measurements and relationships involving substances and mixtures of chemical interest.

A **ternary compound** is comprised of *three* elements.

Suggestions for Further Study

KOLB, D., "The Mole," *J. Chem. Educ.,* **55,** 758 (1978).

KOPPERL, S. J., "Theodore Williams Richards: America's First Nobel Laureate Chemist," *Chemistry,* **46**[6], (1973).

TODD, J. F. J., "Modern Aspects of Mass Spectrometry," *Educ. in Chem.,* **10,** 89 (1973).

Review Problems

1. What is the number of atoms in each of the following samples of matter? **(a)** 3.85 mol Cu; **(b)** 0.0163 mol Ne; **(c)** 3.4×10^{-9} mol Pu.

2. Calculate the quantities indicated.
   (a) the number of moles represented by 8.21×10^{24} Al atoms

(b) the mass, in grams, of 4.18 mol Cl_2

(c) the mass, in kg, of 6.15×10^{27} Zn atoms

(d) the number of atoms in 35.3 cm^3 Fe (density of Fe = 7.86 g/cm^3)

(e) the number of Li^+ ions in 1.51 kg Li_2S

3. The amino acid methionine has the molecular formula $C_5H_{11}NO_2S$. Determine (a) the molecular weight of methionine; (b) the number of moles of H in 3.17 mol methionine; (c) the number of C atoms in 1.53 mol methionine; (d) the number of grams of O per gram of N in the compound.

4. The compound trinitrotoluene (TNT) has the formula $C_7H_5N_3O_6$. Determine (a) the total number of atoms in one formula unit; (b) the ratio of H atoms to N atoms; (c) the ratio, by mass, of O to C in the compound; (d) the percent N, by mass.

5. Determine the percent, by mass, of each of the elements in the antimalarial drug quinine, $C_{20}H_{24}N_2O_2$.

6. An oxide of cobalt contains 71.06% Co and 28.94% O. What is the empirical formula of the oxide?

7. Rubbing alcohol, isopropyl alcohol, is a carbon–hydrogen–oxygen compound with 59.96% C and 13.42% H. What is its empirical formula?

8. A compound of C, H, and O, known as terephthalic acid, is used in the manufacture of Dacron. Its molecular weight is 166.1, and by combustion analysis it is found to have 57.83% C and 3.64% H. What is the *molecular* formula of terephthalic acid?

9. The iodide ion in a 1.552-g sample of the ionic compound XI

is removed through precipitation. The precipitate is found to contain 1.186 g I. What is the element X? (*Hint:* What is its atomic weight?)

10. Name the following compounds: (a) KI; (b) $CaCl_2$; (c) KCN; (d) $Mg(NO_3)_2$; (e) ICl_3; (f) ClO_2; (g) PCl_5.

11. From the information given about one compound, supply the missing information about the other in each of the following pairs:

(a) SnF_2, _____ $SnCl_4$, stannic chloride

(b) $Pb(NO_3)_2$, lead(II) nitrate   _____, lead(IV) oxide

(c) $CoSO_4$, cobalt(II) sulfate   _____, cobalt(III) sulfide

(d) KIO_3, potassium iodate KIO_4, _____

(e) AuCl, _____ $AuCl_3$, auric chloride

12. Indicate the oxidation state of the underlined element in each of the following: (a) <u>Al</u>; (b) $K_2\underline{S}$; (c) <u>N</u>O_2; (d) Br$\underline{F}_5$; (e) H<u>N</u>O_3; (f) $K_2\underline{Mn}O_4$; (g) <u>Co</u>$^{3+}$; (h) $\underline{Cr}_2O_7^{2-}$.

13. Write correct formulas for the following compounds: (a) calcium oxide; (b) strontium fluoride; (c) magnesium hydroxide; (d) cesium carbonate; (e) mercury(II) nitrate; (f) iron(III) sulfide; (g) magnesium perchlorate; (h) potassium hydrogen sulfate.

14. Supply the name or formula of each of the following acids:

(a) _____ = hydrobromic acid   (b) $HClO_2$ = _____

(c) _____ = iodic acid   (d) HNO_2 = _____

(e) _____ = sulfuric acid   (f) H_2S = _____

15. What is the percent, by mass, of water in the hydrate $ZnSO_4 \cdot 7H_2O$?

Exercises

Terminology

1. Explain the distinction between the terms in each pair: (a) formula unit and molecule; (b) empirical formula and molecular formula; (c) cation and anion; (d) binary and ternary compound; (e) oxoacid and oxoanion; (f) hypo_____ous and per_____ate.

Avogadro's number and the mole

2. How many S atoms are present in each of the following samples? (a) 3.85 mol S; (b) 0.0163 mol S_8; (c) 3.4×10^{-9} mol H_2S; (d) 0.162 mol CS_2.

3. In a collection of 4.15×10^{24} molecules of C_3H_7OH, what is the number of moles of (a) C_3H_7OH molecules; (b) C atoms; (c) H atoms; (d) O atoms?

4. In 0.355 mol of the compound Li_2S
   (a) What fraction of the total *number* of ions is S^{2-}?
   (b) What fraction of the total *mass* is contributed by S^{2-}?
   (c) What is the total *number* of Li^+ ions?

5. An alloy contains Sn, Pb, and Bi atoms in the ratio 2:4:3,

respectively. What is the mass of a sample of this alloy containing a total of 5.75×10^{25} atoms?

6. How many Ag atoms are present in a piece of sterling silver jewelry weighing 65.2 g? Sterling silver contains 92.5% Ag, by mass.

*7. How many Cu atoms are present in a 1.00-m length of 20-gauge copper wire? (A 20-gauge wire has a diameter of 0.03196 in.; density of Cu = 8.92 g/cm^3.)

8. During a severe air pollution episode the concentration of lead in air was observed to be 3.01 μg Pb/m^3. How many Pb atoms would be present in a 500-mL sample of this air (the approximate lung capacity of a human adult)?

9. In rhombic sulfur, S atoms are joined into the molecules S_8 (see Figure 3-3). If the density of rhombic sulfur is 2.07 g/cm^3, determine for a crystal of volume 6.15 mm^3: (a) the number of moles of S_8 present; (b) the total number of S atoms.

10. A public water supply was found to contain 1 part per billion (ppb) by mass of chloroform, $CHCl_3$. (Consider this to be essentially 1.00 g $CHCl_3$ per 10^9 g water.)

(a) How many $CHCl_3$ molecules would be present in a glassful of this water (250 mL)?

(b) If the $CHCl_3$ found in (a) could be isolated, would this quantity be detectable on an ordinary analytical balance that measures mass to about ± 0.0001 g?

Chemical formulas

11. Explain which of the following statements (is) are correct concerning the compound $C_6H_{12}O_6$.
   (a) The percentages, by mass, of C and O are the same as in CO.
   (b) The ratio of number of H atoms to number of O atoms is the same as in water.
   (c) The element present in highest percent by mass is O.
   (d) The proportions of C and O, by mass, are equal.

12. Each of the following formulas represents an actual substance. Which are molecular formulas? Can you tell whether the others are empirical or molecular formulas? Explain. (a) C_2H_6; (b) Cl_2O; (c) CH_4O; (d) N_2O_4

13. For the compound $Ge[S(CH_2)_4CH_3]_4$, determine
   (a) the total number of atoms in one formula unit
   (b) the ratio, by number, of C atoms to H atoms
   (c) the ratio, by mass, of Ge to S
   (d) the number of g S in 1 mole of the compound
   (e) the mass of compound required to contain 1.00 g Ge
   (f) the number of C atoms in 25.00 g of the compound

Percent composition of compounds

14. Determine the % O, by mass, in the mineral malachite, $Cu_2(OH)_2CO_3$.

15. Without performing detailed calculations explain which of the following compounds has the greatest % S, by mass: SO_2, SO_3, $MgSO_4$, Li_2S.

16. All of the materials listed below are of value in fertilizers because they supply the element nitrogen. Which of these is potentially the richest source of nitrogen on a mass basis? urea, $CO(NH_2)_2$, ammonium nitrate, NH_4NO_3; or guanidine, $HNC(NH_2)_2$.

17. Ammonium sulfate, $(NH_4)_2SO_4$, is a commonly used fertilizer. Instructions for fertilizing a mature avocado tree call for 1 lb of actual nitrogen per year. How many pounds of ammonium sulfate would be required per year?

Chemical formulas from percent composition

18. A compound of carbon and hydrogen consists of 93.71% C and 6.29% H, by mass. The molecular weight of the compound is found to be 128. What is its molecular formula?

19. Selenium forms two oxides. One has 28.8% O, by mass, and the other, 37.8% O. What are the formulas of these oxides? Propose acceptable names for them.

20. What is the empirical formula of

(a) the rodenticide Warfarin, which has the composition: 74.01% C, 5.23 % H, and 20.76% O;

(b) citric acid, which has the composition: 37.51% C, 4.20% H, and 58.29% O.

★21. Freons are compounds containing C, Cl, F, and possibly H. They are used as refrigerants, aerosol propellants, and solvents. They are derived from simple hydrocarbons by replacing some or all of the H atoms with F and Cl (e.g., $CHCl_2F$ from CH_4). What is the formula of a freon derived from CH_4 that has 31.43% F, by mass? (*Hint:* The formula unit contains *five* atoms, *one* carbon and *four* others.)

Oxidation states

22. Indicate the oxidation state of the underlined element in each of the following: (a) $\underline{C}H_4$; (b) $\underline{S}F_4$; (c) $Na_2\underline{O}_2$; (d) $\underline{C}_2H_3O_2{}^-$; (e) $\underline{Fe}O_4{}^{2-}$; (f) $\underline{S}_4O_6{}^{2-}$.

23. Arrange the following sulfur-containing anions in order of *increasing* oxidation state of the S atom: $SO_3{}^{2-}$; $S_2O_3{}^{2-}$; $S_2O_8{}^{2-}$; $HSO_4{}^-$; HS^-; $S_4O_6{}^{2-}$.

24. Nitrogen forms five different compounds with oxygen. Write proper formulas for these compounds if the oxidation state of N in them is $+1$, $+2$, $+3$, $+4$, and $+5$, respectively.

Nomenclature

25. Name the following compounds: (a) MgS; (b) ZnO; (c) K_2CrO_4; (d) Cs_2SO_4; (e) Cr_2O_3; (f) $FeSO_4$; (g) $Ca(HCO_3)_2$; (h) K_2HPO_4; (i) NH_4I; (j) $Cu(OH)_2$; (k) HNO_2; (l) $HBrO_3$; (m) $KClO_3$; (n) KIO.

26. Assign plausible names to the following compounds: (a) ICl; (b) ClF_3; (c) SF_4; (d) BrF_5.

27. Write correct formulas for the following: (a) aluminum sulfate; (b) ammonium chromate; (c) silicon tetrafluoride; (d) zinc acetate; (e) iron(II) oxide; (f) tricarbon disulfide; (g) chromium(II) chloride; (h) lithium sulfide; (i) chlorine dioxide; (j) calcium dihydrogen phosphate; (k) tin(IV) oxide; (l) chlorous acid; (m) hydrobromic acid.

Hydrates

28. Without performing detailed calculations, indicate which of the following hydrates has the greatest % H_2O, by mass.
   (a) $CuSO_4 \cdot 5H_2O$ (b) $Cr_2(SO_4)_3 \cdot 18H_2O$
   (c) $MgCl_2 \cdot 6H_2O$ (d) $LiC_2H_3O_2 \cdot 2H_2O$

29. A sample of $MgSO_4 \cdot xH_2O$ weighing 5.018 g is heated until all the water of hydration is driven off. The resulting anhydrous compound, $MgSO_4$, weighs 2.449 g. What is the formula of the hydrate?

30. Anhydrous sodium sulfate, Na_2SO_4, absorbs water vapor and is converted to the *deca*hydrate, $Na_2SO_4 \cdot 10H_2O$. How much would the mass of 1.00 g of dry Na_2SO_4 increase if exposed to sufficient water vapor to be converted completely to the decahydrate?

Combustion analysis

31. An 0.1510-g sample of a hydrocarbon produces 0.5008 g CO_2 and 0.1282 g H_2O in combustion analysis. Its molecular weight is found to be 106. For this hydrocarbon, determine **(a)** its percent composition; **(b)** its empirical formula; **(c)** its molecular formula.

32. An 0.4590-g sample of the carbon–hydrogen–oxygen compound, *p*-cresol, yields 1.3077 g CO_2 and 0.3061 g H_2O in combustion analysis. What is the empirical formula of this compound?

★33. The substance dimethylhydrazine is a carbon–hydrogen–nitrogen compound used in rocket fuels. When burned completely, a 0.208-g sample yields 0.305 g CO_2 and 0.249 g H_2O. From a separate 0.350-g sample, the nitrogen content is converted to 0.163 g N_2. What is the empirical formula of dimethylhydrazine?

★34. A hydrocarbon, C_xH_y, is burned and produces 1.955 g CO_2 for every 1.000 g H_2O. What is the empirical formula of this hydrocarbon?

Precipitation analysis

35. A 0.3518-g sample of KI is dissolved in water, and all the iodide present is precipitated as AgI. How many grams of pure, dry AgI are obtained?

36. A particular type of brass contains the elements Cu, Sn, Pb, and Zn. A sample weighing 1.713 g is treated in such a way as to convert the Sn to 0.245 g SnO_2, the Pb to 0.115 g $PbSO_4$, and the Zn to 0.246 g $Zn_2P_2O_7$. What is the percent, by mass, of each element in the brass?

★37. A 1.013-g sample of $ZnSO_4 \cdot xH_2O$ is dissolved in water and the sulfate ion is precipitated by adding an excess of $BaCl_2$ solution. The mass of pure, dry $BaSO_4$ obtained is 0.8223 g. What is the formula of the zinc sulfate hydrate?

Atomic weight determinations

38. Two compounds of Cl and X are found to have molecular weights and % Cl, by mass, as follows: mol. wt. = 137, 77.5% Cl; mol. wt. = 208, 85.1% Cl. What is the element X? What is the formula of each compound?

39. A sample of the compound MSO_4 weighing 0.1131 g reacts with barium chloride and yields 0.2193 g $BaSO_4$. What must be the atomic weight of the element M?

40. The metal M forms the sulfate $M_2(SO_4)_3$. A sample of this sulfate weighing 0.738 g is converted to 1.511 g $BaSO_4$. What is the atomic weight of M?

★41. An 0.622-g sample of a metal oxide with the formula M_2O_3 is converted to the sulfide, MS, yielding 0.685 g. What is the atomic weight of the metal M?

Additional Exercises

1. If a sample of $MgBr_2$ is to contain 3.64×10^{24} Br^- ions:
   (a) How many Mg^{2+} ions will the sample contain?
   (b) How many formula units of $MgBr_2$ will be present?
   (c) What will be the mass of the sample?

2. Spodumene has the formula $Li_2O \cdot Al_2O_3 \cdot 4SiO_2$. Given that the percentage of ^6_3Li atoms in naturally occurring lithium compounds is 7.40%, how many ^6_3Li atoms are present in a 185.0-g sample of spodumene?

3. What is the percent, by mass, of boron in the mineral axinite, $HCa_3Al_2BSi_4O_{16}$?

4. The food flavor-enhancer monosodium glutamate (MSG) has the composition 13.6% Na, 35.5% C, 4.8% H, 8.3% N, 37.8% O, by mass. What is the empirical formula for MSG?

5. Three different brands of "liquid chlorine" for use in purifying water in home swimming pools all cost $1 per gallon and are water solutions of NaOCl. Brand A contains 10% OCl by mass; brand B, 7% available chlorine (Cl) by mass; and brand C, 14% NaOCl by mass. Which of the three brands would you buy?

6. The substance chlorophyll contains magnesium to the extent of 2.72% by mass. Assuming one Mg atom per chlorophyll molecule, what is the molecular weight of chlorophyll?

7. A certain hydrate is found to have the composition: 20.3%

Cu, 8.95% Si, 36.3% F, and 34.5% H_2O, by mass. What is the empirical formula of the hydrate?

8. The element X forms the chloride XCl_4 containing 75.0% Cl, by mass. What is the atomic weight of X? What is the element X?

9. A 1.562-g sample of the hydrocarbon C_7H_{16} is burned in an excess of oxygen. What masses of CO_2 and H_2O should be obtained?

10. What is the percent composition of a carbon–hydrogen–oxygen compound if 1.5181 g of the compound yields 3.830 g CO_2 and 0.6721 g H_2O? What is the empirical formula of this compound?

11. A certain brand of lunch meat contains 0.10% sodium benzoate, $NaC_7H_5O_2$, by mass, as a preservative. If a person eats 2.52 oz of this meat, how many mg Na will that person consume?

12. How many Cl atoms are present in a 3.50-L sample of chloroform, $CHCl_3$? (Density of $CHCl_3$ = 1.48 g/cm^3.)

13. Write a formula for
   (a) a sulfate of iron, with Fe in the oxidation state +3.
   (b) an oxoacid of nitrogen, with N in the oxidation state +3.
   (c) carbonic acid, which has carbon in the oxidation state +4.

(d) an oxide of chlorine in which Cl is in the oxidation state $+7$.

*14. When 2.750 g of the oxide of lead Pb_3O_4 is heated strongly, decomposition occurs, producing 0.0640 g of oxygen gas and 2.686 g of a second oxide of lead. What is the empirical formula of the second oxide?

*15. A hydrocarbon mixture consists of 60.0% by mass of C_3H_8 and 40.0% of C_xH_y. When 10.0 g of this mixture is burned, it yields 29.0 g CO_2 and 18.8 g H_2O as the only products. What is the formula of the unknown hydrocarbon?

*16. A 0.732-g mixture of methane, CH_4, and ethane, C_2H_6, is burned, yielding 2.064 g CO_2. What is the percent composition of this mixture **(a)** by mass; **(b)** on a mole basis?

*17. A thoroughly dried 1.271-g sample of Na_2SO_4 is exposed to the atmosphere and found to gain 0.387 g in mass. What is the percent, by mass, of $Na_2SO_4 \cdot 10\ H_2O$ in the resulting mixture of hydrate and Na_2SO_4?

*18. The atomic weight of Bi is to be determined by converting the compound $Bi(C_6H_5)_3$ to Bi_2O_3. If 5.610 g $Bi(C_6H_5)_3$ yields 2.969 g Bi_2O_3, what is the atomic weight of Bi?

*19. To deposit exactly one mole of Ag from an aqueous solution containing Ag^+ requires that 96,487 coulombs of electric charge be passed through the solution. The electrodeposition requires that each Ag^+ ion converted to an Ag atom gain one electron. Use this information and other relevant data from the text to obtain a value of Avogadro's number, N_A.

Self-Test Questions

For questions 1 through 7 select the single item that best completes each statement.

1. One *mole* of fluorine gas, F_2 (a) weighs 19.0 g; (b) contains 6.02×10^{23} F atoms; (c) contains 1.20×10^{24} F atoms; (d) weighs 6.02×10^{23} g.

2. Three of the following formulas might be either empirical or molecular formulas, but one of the four must be a molecular formula. That one is (a) N_2O; (b) N_2O_4; (c) NH_3; (d) Mg_3N_2.

3. The compound $C_7H_7NO_2$ (a) contains 17 atoms per mole; (b) contains equal percentages of C and H, by mass; (c) contains twice the percent by mass of O as of N; (d) contains twice the percent by mass of N as of H.

4. The greatest number of N atoms is found in (a) 50.0 g N_2O; (b) 17 g NH_3; (c) 150 cm^3 of liquid pyridine, C_6H_5N ($d = 0.983$ g/cm^3); (d) 1 mol N_2.

5. XF_3 is found to consist of 65% F, by mass. The *atomic weight* of X must be (a) 8; (b) 11; (c) 31; (d) 35.

6. The oxidation state of I in the ion $H_4IO_6^-$ is (a) -1; (b) $+1$; (c) $+7$; (d) $+8$.

7. The correct formula for calcium chlorite is (a) $Ca(ClO_3)_2$; (b) $CaClO_2$; (c) $Ca(ClO_2)_2$; (d) $Ca(ClO_4)_2$.

8. The liquid $CHBr_3$ has a density of 2.89 g/cm^3. What volume of this liquid should be measured out to contain a total of 3.40×10^{24} molecules of $CHBr_3$?

9. Supply the missing name or formula.
   (a) CaI_2 = _____
   (b) _____ = iron(III) sulfate
   (c) _____ = sulfur trioxide
   (d) _____ = bromine pentafluoride
   (e) NH_4CN = _____
   (f) $Ca(ClO_2)_2$ = _____
   (g) _____ = lithium hydrogen carbonate

10. An important copper-containing mineral is malachite, $CuCO_3 \cdot Cu(OH)_2$.
   (a) What is the percent Cu, by mass, in this mineral?
   (b) When malachite is heated strongly, carbon dioxide and water are driven off, yielding copper(II) oxide. What mass of copper(II) oxide is produced *per kilogram* of malachite?

11. Hexachlorophene, used in making germicidal soaps, has the percent composition, by mass: 38.37% C, 1.49% H, 52.28% Cl, and 7.86% O. What is the empirical formula of this compound?

12. A hydrate of sodium sulfite contains almost exactly 50% H_2O, by mass. What is the formula of this hydrate?

4 Stoichiometry II: Chemical Reactions

Chemical reactions are the central concern of chemistry, and our focus in this chapter is on some practical questions about chemical reactions: What is a chemical reaction? What are some important types of reactions? How do we describe a reaction in the symbolic form known as a chemical equation, and what kinds of calculations are possible with these equations? How are chemical reactions used to analyze and synthesize materials? Later in the text we will study the underlying principles of chemical reactions.

As a working definition we can say that a chemical reaction is a process in which new chemical substances, **products,** are produced from a set of original substances, **reactants.** Usually, a chemical reaction is accompanied by physical evidence, such as a color change, formation of a precipitate, or evolution of a gas. At times, however, chemical analysis, sometimes employing sophisticated instruments, may be required to prove that a reaction has occurred.

4-1 The Chemical Equation

In the symbolic representation of a chemical reaction known as a **chemical equation,** formulas of the reactants are written on the left side and those of products on the right. The two sides are joined by an equal sign ($=$) or an arrow ($\rightarrow$). Writing a chemical equation is usually a three-step procedure, although often the first step is thought about but not explicitly written down.

1. The names of reactants and products are written, resulting in a word expression. For example,

$$\text{nitrogen oxide} + \text{oxygen} \longrightarrow \text{nitrogen dioxide} \tag{4.1}$$

2. Chemical formulas are substituted for names, resulting in a formula expression. For example,

$$NO + O_2 \longrightarrow NO_2 \tag{4.2}$$

3. The formula expression is *balanced,* resulting in a *chemical equation.* * For example,

$$2\,NO + O_2 \longrightarrow 2\,NO_2 \tag{4.3}$$

---

*An equation—whether mathematical or chemical—must have the left and right sides equal. Strictly speaking, a formula expression cannot be called an equation until it is balanced; and the term "chemical equation" signifies that this balance has been achieved. Nevertheless, to stress the importance of the balanced condition, the somewhat contradictory statement "unbalanced chemical equation" and the somewhat redundant statement "balanced chemical equation" are both frequently used.

In expression (4.2) there is a total of *three* O atoms on the left and only *two* on the right. This situation is corrected by representing *two* molecules of NO on the left and *two* of NO_2 on the right. The result is *two* N atoms and *four* O atoms on *each* side. In writing a chemical equation, appropriate stoichiometric coefficients (numbers) are placed in front of formulas so that

The total number of atoms of each type remains unchanged in the chemical reaction; atoms are neither created nor destroyed in a chemical reaction.

To achieve a balanced condition, only coefficients may be adjusted, *never* the subscripts in formulas. That is, it would be *incorrect* to write $NO + O_2 \rightarrow NO_3$ in attempting to balance expression (4.2). Nitrogen dioxide can only have the formula NO_2.

Example 4-1   Propane gas, C_3H_8, is easily liquefied, stored and transported for use as a fuel. Write a balanced chemical equation to represent its complete combustion.

Solution.   We learned in Section 3-4 that when a hydrocarbon (carbon-hydrogen compound) is burned in an excess of oxygen the sole products are CO_2 and H_2O.

Physical evidence of a combustion reaction is the evolution of heat. (A reaction that gives up heat to the surroundings is an exothermic reaction; see Chapter 6.)

word expression: propane + oxygen $\longrightarrow$ carbon dioxide + water

formula expression: $C_3H_8 + O_2 \longrightarrow CO_2 + H_2O$

The elements C, H, and O can be balanced in any order that we choose. When one of the reactants or products of a reaction is an element in its free state, however, it is generally best to balance that one *last*, meaning oxygen in this case. As each coefficient is established, it is kept fixed while the next one is being set, and so on, until a final balance results.

Balance C: $C_3H_8 + O_2 \longrightarrow 3\,CO_2 + H_2O$

Balance H: $C_3H_8 + O_2 \longrightarrow 3\,CO_2 + 4\,H_2O$

Balance O: $C_3H_8 + 5\,O_2 \longrightarrow 3\,CO_2 + 4\,H_2O$ (balanced)   (4.4)

SIMILAR EXAMPLES: Review Problems 2, 4(a); Exercises 2, 4(a).

Example 4-2   Triethylene glycol, $C_6H_{14}O_4$, is used as a solvent and plasticizer. Write a balanced chemical equation for its complete combustion.

Solution.   Carbon-hydrogen-oxygen compounds, like hydrocarbons, yield CO_2 and H_2O upon burning in oxygen gas.

formula expression: $C_6H_{14}O_4 + O_2 \longrightarrow CO_2 + H_2O$

Balance C: $C_6H_{14}O_4 + O_2 \longrightarrow 6\,CO_2 + H_2O$

Balance H: $C_6H_{14}O_4 + O_2 \longrightarrow 6\,CO_2 + 7\,H_2O$ (4.5)

The right side of expression (4.5) has 19 O atoms. To obtain 19 on the left side, we start with 4 in $C_6H_{14}O_4$ and add 15 more. This requires a fractional coefficient of $\frac{15}{2}$ for O_2.

Balance O: $C_6H_{14}O_4 + \frac{15}{2}O_2 \longrightarrow 6\,CO_2 + 7\,H_2O$ (balanced)   (4.6)

Final Adjustment of Coefficients.   Although fractional coefficients are acceptable in many circumstances, general practice is to remove them by multiplying all coefficients in a chemical equation by the same whole number—in this case "2."

$2\,C_6H_{14}O_4 + 15\,O_2 \longrightarrow 12\,CO_2 + 14\,H_2O$ (balanced)   (4.7)

SIMILAR EXAMPLES: Review Problems 2, 4(b); Exercises 2, 4.

The symbol (c) is sometimes used to represent the crystalline form of a substance, but the symbol (s) serves essentially the same purpose.

The state of matter or physical form in which reactants and products occur can also be represented in a chemical equation, through the following symbols.

(g) = gas   **(l)** = liquid   **(s)** = solid   **(aq)** = aqueous (water) solution

Thus, for the reaction of hydrogen and oxygen gases to form liquid water

$$2 H_2(g) + O_2(g) \longrightarrow 2 H_2O(l) \tag{4.8}$$

Net Ionic Equations.   The reaction of water solutions of silver nitrate and sodium chloride can be represented by the equation

$$AgNO_3(aq) + NaCl(aq) \longrightarrow AgCl(s) + NaNO_3(aq) \tag{4.9}$$

Physical evidence of reaction (4.9) is the appearance of a voluminous white precipitate (AgCl) from a colorless solution.

What are the actual species present in the aqueous solutions of $AgNO_3$, NaCl, and $NaNO_3$ described in equation (4.9)? Figure 3-2 shows that pure NaCl is comprised of ions—Na^+ and Cl^-. When this compound is dissolved in water, the ions become dissociated from one another. It is appropriate to think of Na^+ and Cl^- in water solution as if they were separate constituents. The compounds $AgNO_3$ and $NaNO_3$ are also ionic, and their ions are dissociated from one another in water solution. Thus, we may write an *ionic* equation.

$$Ag^+(aq) + \cancel{NO_3^-(aq)} + \cancel{Na^+(aq)} + Cl^-(aq) \longrightarrow AgCl(s) + \cancel{Na^+(aq)} + \cancel{NO_3^-(aq)} \tag{4.10}$$

A still further refinement is to note that any species that appears on both sides of an equation is not directly involved in the reaction. The "spectator" ions in equation (4.10) can be eliminated (noted by cancellation signs), yielding a *net ionic equation*.

$$Ag^+(aq) + Cl^-(aq) \longrightarrow AgCl(s) \tag{4.11}$$

The following expression shows that copper metal displaces silver metal from a solution containing silver ions.

$$Cu(s) + Ag^+(aq) \longrightarrow Cu^{2+}(aq) + Ag(s) \quad \text{(not balanced)} \tag{4.12}$$

Physical evidence of reaction (4.13) is the deposition of shiny crystals of silver and the development of a blue color (Cu^{2+}) in an originally colorless solution (see also Color Section G).

Although this ionic expression has the same number of atoms of each type on the two sides, *it is not balanced*. There must be a balance of electric charge as well. *Electric charge can neither be created nor destroyed in a chemical reaction.* In expression (4.12) one unit of positive charge is shown on the left and two on the right. This situation is corrected in equation (4.13).

$$Cu(s) + 2 Ag^+(aq) \longrightarrow Cu^{2+}(aq) + 2 Ag(s) \quad \text{(balanced)} \tag{4.13}$$

Writing and balancing net ionic equations is an important skill. This activity will take on more meaning as we learn how to distinguish ionic from covalent compounds, dissociated from undissociated molecules in aqueous solutions, and soluble from insoluble compounds. The concept of net ionic equations is explored further in Chapter 19.

Example 4-3   When hydrogen sulfide gas is passed into a water solution containing the ion Bi^{3+}, a dark brown precipitate of bismuth sulfide, Bi_2S_3, forms, accompanied by an increase in the number of H^+ in solution. Write a balanced chemical equation for this reaction.

Solution.   Electric charges are shown in the following expression (it is written in ionic form). The final equation must show a balance both in numbers of atoms *and in electric charges*. The stepwise balancing is suggested below.

$$Bi^{3+}(aq) + H_2S(aq) \longrightarrow Bi_2S_3(s) + H^+(aq)$$

Balance Bi: $2 Bi^{3+}(aq) + H_2S(aq) \longrightarrow Bi_2S_3(s) + H^+(aq)$

An aqueous solution is always electrically neutral. The reason that the solution here appears to carry a net charge of $+6$ is that spectator anions (e.g., 6 Cl^-) are not included in the net ionic equation.

Balance S: $2 \ Bi^{3+}(aq) + 3 \ H_2S(aq) \longrightarrow Bi_2S_3(s) + H^+(aq)$

Balance H: $2 \ Bi^{3+}(aq) + 3 \ H_2S(aq) \longrightarrow Bi_2S_3(s) + 6 \ H^+(aq)$ (balanced)

Proof of balance of electric charge

$$\underbrace{2 \times (+3)}_{\text{charge on } Bi^{3+}} = \underbrace{6 \times (+1)}_{\text{charge on } H^+}$$

SIMILAR EXAMPLES: Review Problem 3; Exercise 3.

Reaction Conditions.   To write a chemical equation, it is not necessary to know the conditions under which a reaction occurs. However, this information is necessary in the laboratory or chemical plant. Reaction conditions are often written above or below the arrow. For example, the capital Greek letter delta, Δ, means that an elevated temperature is required; that is, the reaction mixture must be heated.

$$2 \ Ag_2O(s) \xrightarrow{\Delta} 4 \ Ag(s) + O_2(g) \tag{4.14}$$

Gas pressure is discussed in Chapter 5; the function of a catalyst in a chemical reaction, in Chapter 14.

Reaction conditions may be stated even more explicitly, as in the BASF (Badische Anilin- & Soda Fabrik) process for the synthesis of methanol from carbon monoxide and hydrogen. This reaction occurs at 350°C, under a total pressure of 340 atm, and on the surface of a mixture of ZnO and Cr_2O_3 (acting as a catalyst).

$$CO(g) + 2 \ H_2(g) \xrightarrow[\substack{340 \ atm \\ ZnO, \ Cr_2O_3}]{350°C} CH_3OH(g) \tag{4.15}$$

4-2 Types of Chemical Reactions

We can describe chemical reactions more readily, and perhaps better understand them, by establishing some characteristic types of reactions. One classification scheme that covers all the reactions considered to this point in the chapter uses the following terms.

1. Combustion—a reaction in which an element or compound combines with oxygen to produce simple oxygen-containing compounds such as CO_2, H_2O, and SO_2. The reaction of propane with oxygen (4.4) and triethylene glycol with oxygen (4.7) are combustion reactions.

$$C_3H_8(g) + 5 \ O_2(g) \longrightarrow 3 \ CO_2(g) + 4 \ H_2O(l) \tag{4.4}$$

$$2 \ C_6H_{14}O_4(l) + 15 \ O_2(g) \longrightarrow 12 \ CO_2(g) + 14 \ H_2O(l) \tag{4.7}$$

2. Combination (or synthesis)—a reaction in which a more complex substance is formed from two or more simpler substances (either elements or compounds). Reaction (4.8) is the synthesis of water from its elements; reaction (4.15), methanol from CO and H_2.

$$2 \ H_2(g) + O_2(g) \longrightarrow 2 \ H_2O(l) \tag{4.8}$$

$$CO(g) + 2 \ H_2(g) \longrightarrow CH_3OH(g) \tag{4.15}$$

3. Decomposition—a reaction in which a substance is broken down into simpler ones. Reaction (4.14) is the decomposition of silver oxide.

$$2 \ Ag_2O(s) \longrightarrow 4 \ Ag(s) + O_2(g) \tag{4.14}$$

4. Displacement (or single replacement)—a reaction in which one element replaces another in a compound. In reaction (4.13) Cu displaces Ag^+ from an aqueous solution (formed, for example, by dissolving $AgNO_3$ in water).

$$Cu(s) + 2\,Ag^+(aq) \longrightarrow Cu^{2+}(aq) + 2\,Ag(s) \tag{4.13}$$

5. Metathesis (or double replacement)—a reaction in which an exchange occurs between two reactants. In reaction (4.9) NO_3^- and Cl^- are exchanged between Ag^+ and Na^+. When combined, Ag^+ and Cl^- form insoluble AgCl.

$$AgNO_3(aq) + NaCl(aq) \longrightarrow AgCl(s) + NaNO_3(aq) \tag{4.9}$$

The above terms have a long-standing tradition in chemistry, and some of them (e.g., combustion, decomposition) are used regularly. These categories are not as generally useful, however, as are other types discussed later in the text. For example, some reactions occur because certain combinations of ions (such as Ag^+ and Cl^- in reaction 4.11) cannot be maintained in aqueous solutions and deposit from solution as precipitates. *Precipitation reactions* are discussed in Chapter 19. Another type of reaction occurs as a result of the transfer of protons (H^+) from one species (an acid) to another species (a base). *Acid–base reactions* are discussed in Chapters 17 and 18. Still another type involves the transfer of electrons among reacting species. This type of reaction, *oxidation–reduction,* is presented briefly in Section 4-5 and considered more fully in Chapter 20.

4-3 Quantitative Significance of the Chemical Equation

The coefficients in the chemical equation

$$2\,H_2(g) + O_2(g) \longrightarrow 2\,H_2O(l)$$

mean that

$$2 \text{ molecules } H_2 + 1 \text{ molecule } O_2 \longrightarrow 2 \text{ molecules } H_2O$$

or that

$$2x \text{ molecules } H_2 + x \text{ molecules } O_2 \longrightarrow 2x \text{ molecules } H_2O$$

Suppose that we let $x = 6.02205 \times 10^{23}$—Avogadro's number. Then x molecules represents *1 mol*. Thus, the chemical equation also means that

$$2 \text{ mol } H_2 + 1 \text{ mol } O_2 \longrightarrow 2 \text{ mol } H_2O$$

The chemical equation allows us to write the following expressions.

(1) $2 \text{ mol } H_2O \Leftrightarrow 2 \text{ mol } H_2$

(2) $2 \text{ mol } H_2O \Leftrightarrow 1 \text{ mol } O_2$

(3) $2 \text{ mol } H_2 \Leftrightarrow 1 \text{ mol } O_2$

The literal meaning of these expressions is that

1. *two* moles of H_2O are *produced* for every *two* moles of H_2 *consumed*.
2. *two* moles of H_2O are *produced* for every *one* mole of O_2 *consumed*.
3. *two* moles of H_2 are *consumed* for every *one* mole of O_2 *consumed*.

These expressions (and hence the chemical equation from which they are derived) are the source of conversion factors in the two examples that follow.

Example 4-4 How much H_2O, in moles, results from burning an excess of H_2 in 3.3 mol O_2?

Solution.   The statement "an excess of H_2" signifies that there is more than enough H_2 available to permit the complete conversion of 3.3 mol O_2 to H_2O. The necessary conversion factor is derived from the expression, 2 mol $H_2O \backsim$ 1 mol O_2.

$$\text{no. mol } H_2O = 3.3 \text{ mol } O_2 \times \frac{2 \text{ mol } H_2O}{1 \text{ mol } O_2} = 6.6 \text{ mol } H_2O$$

SIMILAR EXAMPLE: Review Problem 5.

Example 4-5   What mass of H_2 must react with excess O_2 to produce 5.40 g H_2O?

Solution.   In this problem, (a) O_2 is in excess instead of H_2; (b) the unknown is a quantity of one of the *reactants* (H_2) instead of the product (H_2O); and (c) information is given and sought in the unit *gram* rather than mole. Even though the actual calculation can be performed through a single setup, we should think in terms of three steps.

Step 1.   Convert the quantity of H_2O from grams to moles. This requires a conversion factor based on the molar mass of H_2O.

Step 2.   From the amount of H_2O in Step 1, calculate the amount of H_2 consumed. This requires a factor from the chemical equation.

Step 3.   Convert the amount of H_2 in Step 2 to mass, in grams, using the molar mass of H_2 as a conversion factor.

$$\text{no. g } H_2 = 5.40 \text{ g } H_2O \times \frac{1 \text{ mol } H_2O}{18.0 \text{ g } H_2O} \times \frac{2 \text{ mol } H_2}{2 \text{ mol } H_2O} \times \frac{2.02 \text{ g } H_2}{1 \text{ mol } H_2} = 0.606 \text{ g } H_2$$

$$(\text{g } H_2O \longrightarrow \text{mol } H_2O \longrightarrow \text{mol } H_2 \longrightarrow \text{g } H_2)$$

SIMILAR EXAMPLES: Review Problems 6, 7(b).

Now we shift our attention to the reaction pictured in Figure 4-1.

$$2 \text{ Al(s)} + 6 \text{ HCl(aq)} \longrightarrow 2 \text{ AlCl}_3\text{(aq)} + 3 \text{ H}_2\text{(g)} \tag{4.16}$$

FIGURE 4-1
The reaction

$2 \text{ Al(s)} + 6 \text{ HCl(aq)} \rightarrow$
$\quad 2 \text{ AlCl}_3\text{(aq)} + 3 \text{ H}_2\text{(g)}$

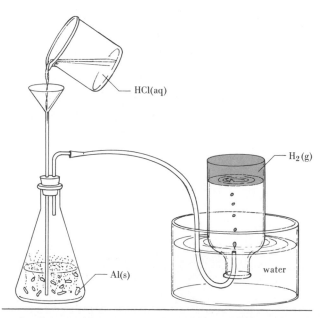

HCl(aq) is introduced to the flask on the left through a long funnel. The reaction of HCl(aq) and Al(s) occurs within the flask. The liberated H_2(g) is conducted to a gas collection apparatus where it displaces water. Hydrogen is only very slightly soluble in water.

Example 4-6 A piece of pure Al(s) having a volume of 0.842 cm^3 reacts with an excess of HCl(aq). What is the mass of H_2 obtained? (The density of Al is 2.70 g/cm^3.)

Solution
Step 1. Use density to convert from volume to mass.

$$\text{no. g Al} = 0.842 \text{ cm}^3 \text{ Al} \times \frac{2.70 \text{ g Al}}{1 \text{ cm}^3 \text{ Al}} = 2.27 \text{ g Al}$$

Step 2. Express the quantity of Al from Step 1 in moles.

$$\text{no. mol Al} = 2.27 \text{ g Al} \times \frac{1 \text{ mol Al}}{27.0 \text{ g Al}} = 0.0841 \text{ mol Al}$$

Step 3. Use a factor from the chemical equation to determine the amount of H_2 that will be produced.

$$\text{no. mol } H_2 = 0.0841 \text{ mol Al} \times \frac{3 \text{ mol } H_2}{2 \text{ mol Al}} = 0.126 \text{ mol } H_2$$

Step 4. Convert the amount of H_2 from Step 3 to mass in grams.

$$\text{no. g } H_2 = 0.126 \text{ mol } H_2 \times \frac{2.02 \text{ g } H_2}{1 \text{ mol } H_2} = 0.255 \text{ g } H_2$$

SIMILAR EXAMPLE: Exercise 9.

Figure 4-2 may prove helpful in analyzing the stepwise solutions of the preceding example and of the two examples that follow.

Example 4-7 An alloy consisting of 95.0% Al and 5.0% Cu, by mass, is used in reaction (4.16). Assuming that all the Al and none of the Cu reacts, what mass of the alloy is required to produce $1.75 \text{ g } H_2$?

Solution. Here the reactant whose quantity we are seeking is not pure. A conversion factor based on the percent composition of the alloy is required in the last step of this four-step solution.

FIGURE 4-2
Visualizing a calculation based on the chemical equation—Examples 4-6, 4-7, and 4-8 illustrated.

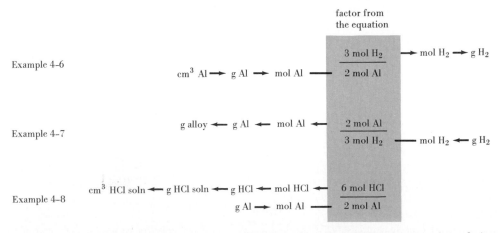

Basically, each of these examples starts with information about one substance (in color) and seeks information about another (in black). The chemical equation provides the factor to convert from one substance to the other, *on a mole basis*. Other conversions use molar mass, density, or percent composition, as necessary. The starting point of a calculation may be on either the left or the right side of the equation, as may the ending point, depending on the specific situation.

Step 1. Describe the H_2 produced as an amount in moles.

$$\text{no. mol } H_2 = 1.75 \text{ g } H_2 \times \frac{1 \text{ mol } H_2}{2.02 \text{ g } H_2} = 0.866 \text{ mol } H_2$$

Step 2. Convert from mol H_2 to mol Al with a factor from the equation.

$$\text{no. mol Al} = 0.866 \text{ mol } H_2 \times \frac{2 \text{ mol Al}}{3 \text{ mol } H_2} = 0.577 \text{ mol Al}$$

Step 3. Express the required quantity of Al as a mass, in grams.

$$\text{no. g Al} = 0.577 \text{ mol Al} \times \frac{27.0 \text{ g Al}}{1 \text{ mol Al}} = 15.6 \text{ g Al}$$

Step 4. If the Al were pure, the quantity required would be 15.6 g; but the sample is only 95.0% pure—the mass of alloy is *greater* than 15.6 g.

$$\text{no. g alloy} = 15.6 \text{ g Al} \times \frac{100.0 \text{ g alloy}}{95.0 \text{ g Al}} = 16.4 \text{ g alloy}$$

SIMILAR EXAMPLES: Exercises 7, 8.

Example 4-8 A hydrochloric acid solution consists of 28.0% HCl, by mass, and has a density of 1.14 g/cm^3. What is the volume of this solution required to dissolve 2.35 g Al in reaction (4.16)?

Solution.   The several steps outlined in Figure 4-2 are performed below.
Step 1.   Convert 2.35 g Al to no. mol Al. *Result:* 0.0870 mol Al
Step 2.   Determine the number of moles of HCl required to dissolve the Al.

$$\text{no. mol HCl} = 0.0870 \text{ mol Al} \times \frac{6 \text{ mol HCl}}{2 \text{ mol Al}} = 0.261 \text{ mol HCl}$$

Step 3.   Determine the mass of 0.261 mol HCl. *Result:* 9.53 g HCl.
Step 4.   Calculate the mass of the acid solution to contain 9.53 g HCl.

$$\text{no. g HCl soln} = 9.53 \text{ g HCl} \times \frac{100.0 \text{ g HCl soln}}{28.0 \text{ g HCl}} = 34.0 \text{ g HCl soln}$$

Step 5.   Use density as a factor to convert from mass to volume of solution.

$$\text{no. } cm^3 \text{ HCl soln} = 34.0 \text{ g HCl soln} \times \frac{1 \text{ } cm^3 \text{ HCl soln}}{1.14 \text{ g HCl soln}} = 29.8 \text{ } cm^3 \text{ HCl soln}$$

This example, as all others that we have considered in stepwise fashion, can also be solved through a single setup, if you are able to visualize each step within the setup.

$$\text{no. } cm^3 \text{ HCl soln} = 2.35 \text{ g Al} \times \frac{1 \text{ mol Al}}{27.0 \text{ g Al}} \times \frac{6 \text{ mol HCl}}{2 \text{ mol Al}} \times \frac{36.5 \text{ g HCl}}{1 \text{ mol HCl}}$$

$$(\text{g Al} \longrightarrow \text{mol Al} \longrightarrow \text{mol HCl} \longrightarrow \text{g HCl}$$

$$\times \frac{100 \text{ g HCl soln}}{28.0 \text{ g HCl}} \times \frac{1 \text{ } cm^3 \text{ HCl soln}}{1.14 \text{ g HCl soln}}$$

$$\longrightarrow \text{g HCl soln} \longrightarrow cm^3 \text{ HCl soln})$$

$$= 29.9 \text{ } cm^3 \text{ HCl soln}*$$

SIMILAR EXAMPLE: Exercise 10.

---

*Note that the answer obtained in the single-setup calculation (29.9 cm^3) differs slightly from that of the stepwise calculation (29.8 cm^3). These differences are caused by the multiple rounding off of numbers required in the stepwise approach.

4-4 Chemical Reactions in Solutions

We need to explore more fully this aspect of chemical reactions: *Some of the reactants and/or products may exist in solution.* One component, the one that determines whether the solution exists as a solid, liquid, or gas, is called the **solvent.** The other component(s) are called **solute(s).** NaCl(aq), for example, describes a solution in which water is the solvent and NaCl, the solute. In seawater, water is again the solvent, but there are many solutes, of which NaCl is simply the most abundant.

The quantity of solute that can be dissolved in a solvent varies widely. As as result, it is necessary to specify the exact composition of a solution if calculations are to be made on chemical reactions in solution. In Example 4-8 the composition of a hydrochloric acid solution was described in terms of the solution density and its percent composition by mass. More useful is a description based on the concept of the mole.

In the SI system the term liter (L) is discouraged, and its equivalent, the cubic decimeter (dm^3), has been adopted. Several other concentration units and their use are introduced in Chapter 12.

Molar Concentration (Molarity).   The composition or concentration of a solution expressed as the *number of moles of solute per liter of solution* is called the **molar concentration** or **molarity (M).**

$$\text{molar concentration } (M) = \frac{\text{number of moles solute}}{\text{number liters solution}} \tag{4.17}$$

If both terms on the right side of equation (4.17) are divided by 1000, the value of the molarity does not change. The unit, mol/1000, is a **millimole (mmol),** and the unit, L/1000, is a milliliter (mL). Thus, an alternate definition of molarity is

$$\text{molar concentration } (M) = \frac{\text{no. mol/1000}}{\text{no. L/1000}} = \frac{\text{no. mmol solute}}{\text{no. mL solution}} \tag{4.18}$$

For example, 0.500 mol of urea, $CO(NH_2)_2$, dissolved in 1.000 L of water solution corresponds to a molar concentration of

$$\frac{0.500 \text{ mol } CO(NH_2)_2}{1.000 \text{ L soln}} = 0.500 \text{ } M \text{ } CO(NH_2)_2$$

and 3.52 mmol (0.00352 mol) of ethanol, C_2H_5OH, dissolved in 100.0 mL (0.1000 L) of aqueous solution corresponds to a molar concentration of

$$\frac{3.52 \text{ mmol } C_2H_5OH}{100.0 \text{ mL soln}} = 0.0352 \text{ } M \text{ } C_2H_5OH$$

Of course, molar quantities cannot be measured out directly; they must be related to other measurements, usually mass or volume.

Example 4-9   A solution is prepared by dissolving 25.0 cm^3 ethanol, C_2H_2OH ($d = 0.789$ g/cm^3), in a sufficient quantity of water to produce 250.0 mL of solution. What is the molarity of C_2H_5OH in this solution?

Solution.   As a first step we must calculate the number of moles of ethanol in a 25.0-cm^3 sample. Density and molar mass provide the necessary conversion factors.

$$\text{no. mol } C_2H_5OH = 25.0 \text{ cm}^3 \text{ } C_2H_5OH \times \frac{0.789 \text{ g } C_2H_5OH}{1 \text{ cm}^3 \text{ } C_2H_5OH} \times \frac{1 \text{ mol } C_2H_5OH}{46.1 \text{ g } C_2H_5OH}$$

$$= 0.428 \text{ mol } C_2H_5OH$$

Now, we use the definition of molar concentration, expressed either through equation (4.17) or (4.18). Note that 250 mL = 0.250 L.

$$\text{molarity} = \frac{0.428 \text{ mol } C_2H_5OH}{0.250 \text{ L soln}} = \frac{428 \text{ mmol } C_2H_5OH}{250 \text{ mL soln}} = 1.71 \text{ } M \text{ } C_2H_5OH$$

SIMILAR EXAMPLES: Review Problem 8; Exercise 11.

FIGURE 4-3
Preparation of 0.250 *M*
Na₂SO₄—Example 4-10
illustrated.

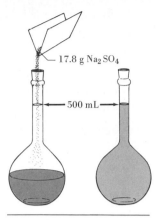

The desired solution requires that 0.125 mol Na₂SO₄ (17.8 g) be dissolved in exactly 500 mL of water solution. One type of container used in the preparation of solutions is the volumetric flask. The flask pictured here contains 500.0 ± 0.2 mL when filled to the mark. The procedure is to dissolve the solute in less than 500 mL of water and, when dissolving is complete, to fill the container exactly to the mark with water.

Example 4-10 It is desired to prepare exactly 0.5000 L (500.0 mL) of an 0.250 *M* Na₂SO₄ solution in water. What is the mass of Na₂SO₄ required for this purpose (see Figure 4-3)?

Solution. One method has as its first step rearranging equation (4.17) to solve for the amount of solute.

no. mol Na₂SO₄ = molarity (mol/L) × volume (L)

$$= \frac{0.250 \text{ mol Na}_2\text{SO}_4}{1 \text{ L}} \times 0.5000 \text{ L}$$

$$= 0.125 \text{ mol Na}_2\text{SO}_4$$

Then the mass of solute is computed.

$$\text{no. g Na}_2\text{SO}_4 = 0.125 \text{ mol Na}_2\text{SO}_4 \times \frac{142 \text{ g Na}_2\text{SO}_4}{1 \text{ mol Na}_2\text{SO}_4}$$

$$= 17.8 \text{ g Na}_2\text{SO}_4$$

Another approach uses solution molarity as a conversion factor between volume of solution and number of moles of solute. That is, for the solution here, 1.00 L solution ⇌ 0.250 mol Na₂SO₄.

$$\text{no. g Na}_2\text{SO}_4 = 0.5000 \text{ L soln} \times \frac{0.250 \text{ mol Na}_2\text{SO}_4}{1 \text{ L soln}} \times \frac{142 \text{ g Na}_2\text{SO}_4}{1 \text{ mol Na}_2\text{SO}_4}$$

(L soln ⟶ mol Na₂SO₄ ⟶ g Na₂SO₄)

$$= 17.8 \text{ g Na}_2\text{SO}_4$$

SIMILAR EXAMPLES: Review Problem 9; Exercise 12.

Many practical applications require that equation (4.17) or (4.18) be used two (or more) times. This is the case, for example, if two solutions are mixed and the final concentration is to be calculated. More common is the situation described in Example 4-11, in which a desired solution is prepared by adding water to a more concentrated solution. This procedure is often used in the laboratory, where stock solutions of fairly high concentrations are stored and other solutions prepared by an appropriate dilution.

FIGURE 4-4
Preparing a solution by dilution—Example 4-11 illustrated.

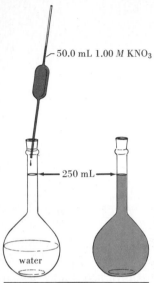

250 mL 0.200 M KNO$_3$

A pipet is used to dispense 50.0 mL of 1.00 M KNO$_3$ into a small quantity of water in a 250.0-mL volumetric flask. Following this, more water is added to bring the solution volume to the mark on the flask.

The basic principle of this method, illustrated through Figure 4-4, is that

all the solute in the initial, more concentrated solution appears in the final, diluted solution.   (4.19)

Statement (4.19) is all that is needed (together with the definition of molar concentration) in working dilution problems. However, some prefer a method based on a rearrangement of equations (4.17) and (4.18), that is,

no. mol solute = molarity (M) × volume (V, in liters);

and

no. mmol solute = molarity (M) × volume (V, in milliliters)

When a solution is diluted, the amount of solute *remains constant* between the initial (i) and final (f) solutions. Thus,

$$M_i \times V_i = \text{amount of solute (mol or mmol)} = M_f \times V_f$$

and

$$M_i \times V_i = M_f \times V_f \qquad (4.20)$$

Example 4-11   What volume of 1.000 M KNO$_3$ must be diluted with water to prepare 250.0 mL of 0.200 M KNO$_3$? (See Figure 4-4).

Solution.   Consider the two solutions separately. First, we calculate the amount of solute that must be present in the *final solution*.

$$\text{no. mol KNO}_3 = 0.250 \text{ L soln} \times \frac{0.200 \text{ mol KNO}_3}{1 \text{ L soln}}$$
$$= 0.0500 \text{ mol KNO}_3 = 50.0 \text{ mmol KNO}_3$$

Since all the solute in the final, diluted solution must come from the initial, more concentrated solution, let us now ask this question: What volume of 1.000 M KNO$_3$ must be taken to contain 0.0500 mol KNO$_3$?

$$\text{no. L soln} = 0.0500 \text{ mol KNO}_3 \times \frac{1 \text{ L soln}}{1.000 \text{ mol KNO}_3} = 0.0500 \text{ L soln}$$

or

$$\text{no. mL soln} = 50.0 \text{ mmol KNO}_3 \times \frac{1 \text{ mL soln}}{1.000 \text{ mmol KNO}_3} = 50.0 \text{ mL soln}$$

SIMILAR EXAMPLES: Review Problem 10; Exercises 13, 15.

Solution Stoichiometry.   Several of the ideas considered in this chapter can now be combined to answer the question posed in Example 4-12.

Example 4-12   What volume of 0.1060 M AgNO$_3$(aq) must react with 10.00 mL of 0.09720 M K$_2$CrO$_4$(aq) to precipitate all the chromate as Ag$_2$CrO$_4$?

$$2 \text{ AgNO}_3(\text{aq}) + \text{K}_2\text{CrO}_4(\text{aq}) \longrightarrow \text{Ag}_2\text{CrO}_4(\text{s}) + 2 \text{ KNO}_3(\text{aq})$$

Solution.   Because the amounts of reactants are small, the unit mmol is useful here (although, of course, the unit mol could be used as well). Also, a factor from the chemical equation can be on a millimole as well as on a mole basis.

This three-step approach is perhaps simplest: (1) Determine the amount of K$_2$CrO$_4$ that reacts. (2) Use the chemical equation to determine the amount of AgNO$_3$ required to react with the K$_2$CrO$_4$. (3) Calculate the volume of 0.1060 M AgNO$_3$ containing the required amount of AgNO$_3$.

FIGURE 4-5
An acid–base titration—Example 4-13 illustrated.

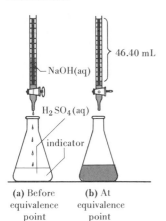

46.40 mL

NaOH(aq)

H_2SO_4(aq)

indicator

(a) Before equivalence point

(b) At equivalence point

The 5.00 mL H_2SO_4(aq) that is to participate in reaction (4.21) is added to a flask and diluted with water. A few drops of an appropriate acid–base indicator (phenolphthalein) are added. A solution of 0.875 *M* NaOH is contained in a long graduated tube from which the flow of liquid can be controlled with a stopcock. This device is called a buret. The buret is filled to the 0.00 mL mark. **(a)** Solution from the buret is added, first rapidly and then dropwise. **(b)** At the precise point where the reaction of the H_2SO_4 has been completed (the equivalence point) the acid–base indicator changes color. The buret reading at this point (46.40 mL) gives the volume of 0.875 *M* NaOH required for the titration.

no. mmol K_2CrO_4 = 10.00 mL soln $\times \dfrac{0.09720 \text{ mmol } K_2CrO_4}{1 \text{ mL soln}}$

$= 0.9720$ mmol K_2CrO_4

no. mmol $AgNO_3$ = 0.9720 mmol $K_2CrO_4 \times \dfrac{2 \text{ mmol } AgNO_3}{1 \text{ mmol } K_2CrO_4}$

$= 1.944$ mmol $AgNO_3$

no. mL $AgNO_3$(aq) = 1.944 mmol $AgNO_3 \times \dfrac{1 \text{ mL soln}}{0.1060 \text{ mmol } AgNO_3}$

$= 18.34$ mL $AgNO_3$(aq)

SIMILAR EXAMPLES: Exercises 16, 17(a).

A more common laboratory situation is that described in Example 4-13. A chemical reaction is carried out between two solutions, one of known concentration and the other unknown. Experimental data are used to determine the unknown concentration.

Example 4-13 **The electrolyte in a lead storage battery is H_2SO_4(aq). A 5.00-mL sample of a battery acid requires 46.40 mL of 0.875 *M* NaOH for its complete reaction (neutralization). What is the molar concentration of H_2SO_4 in the acid? (See Figure 4-5).**

$H_2SO_4(aq) + 2\,NaOH(aq) \longrightarrow Na_2SO_4(aq) + 2\,H_2O(l)$ (4.21)

Solution. **Let us again consider a three-step approach: (1) Determine the number of mmol NaOH in 46.40 mL of 0.875 *M* NaOH. (2) Determine the number of mmol H_2SO_4 that react with this NaOH. (3) Calculate the molarity of the H_2SO_4(aq).**

no. mmol NaOH = 46.40 mL $\times \dfrac{0.875 \text{ mmol NaOH}}{\text{mL}}$ = 40.6 mmol NaOH

no. mmol H_2SO_4 = 40.6 mmol NaOH $\times \dfrac{1 \text{ mmol } H_2SO_4}{2 \text{ mmol NaOH}}$ = 20.3 mmol H_2SO_4

Since the 20.3 mmol H_2SO_4 is derived from a 5.00-mL sample,

molar concentration = $\dfrac{20.3 \text{ mmol } H_2SO_4}{5.00 \text{ mL soln}}$ = 4.06 *M* H_2SO_4

SIMILAR EXAMPLES: Review Problem 11; Exercises 17, 18.

The calculation involved in Example 4-13 is not difficult, but the experimental procedure necessary to obtain the data for the calculation—a procedure called **titration**—is rather exacting. In Example 4-13, how can we be assured that as NaOH(aq), a colorless solution, is slowly added to the 5.00 mL of H_2SO_4(aq), also a colorless solution, the reaction is completed when exactly 46.40 mL has been added—and not 46.35 or 46.45 or 46.50 . . . ? This can be done by having present in the H_2SO_4(aq) a trace of a substance called an **indicator,** which changes color at the precise point (called the equivalence point) where all the H_2SO_4(aq) has reacted. Thus, the key to titration reactions is knowing how an indicator works and being able to select an appropriate indicator. We will return to this subject in Sections 18-4 and 18-5.

4-5 Some Additional Matters

Many situations dealing with the stoichiometry of chemical reactions can be handled by the methods introduced up to this point. There are, however, some complicating factors that need to be considered.

FIGURE 4-6

An analogy to determining the limiting reagent in a chemical reaction—assembling a hand-out experiment.

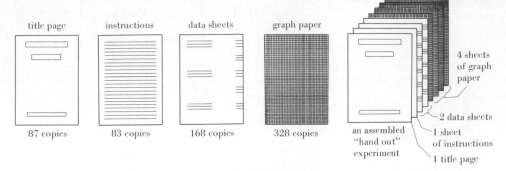

From the number of copies of each type of sheet (analogous to moles of reactants) calculate how many complete handouts (analogous to moles of products) can be assembled. Do you get 82? Which is the "limiting reagent?"

Based on the title page we would say that no more than 87 complete copies of the handout are possible, but based on the instruction page we would say no more than 83. Two data sheets are required per handout. There are enough data sheets for $168/2 = 84$ handouts, but no more than 83 handouts are possible because of the limited number of instructions sheets. Finally, because four sheets of graph paper are required per handout, we conclude that only $328/4 = 82$ handouts are possible. The excess pages are the title page (5 copies), the instructions sheet (one copy), and the data page (4 copies).

Determining the Limiting Reagent.   In examples presented previously we have stated which reactant (also called a reagent) was in excess. Some of the reactant in excess remains after a reaction is completed. The reactant that determines the outcome—the limiting reagent—is consumed completely. Situations may arise in which the limiting reagent is not indicated explicitly. In such cases the limiting reagent must be determined by calculation. The principle involved is illustrated in Figure 4-6.

Example 4-14   How many moles of $Fe(OH)_3(s)$ can be produced by allowing 1.0 mol Fe_2S_3, 2.0 mol H_2O, and 3.0 mol O_2 to react?

$$2\ Fe_2S_3(s) + 6\ H_2O(l) + 3\ O_2(g) \longrightarrow 4\ Fe(OH)_3(s) + 6\ S(s)$$

Solution.   The amount of Fe_2S_3 (1.0 mol) is less than of H_2O (2.0 mol) and O_2 (3.0 mol), but this *does not* automatically mean that Fe_2S_3 is the limiting reagent. *The amounts of reactants available must be compared to the proportions in which they combine.*

$$2\ \text{mol}\ Fe_2S_3 \approxeq 6\ \text{mol}\ H_2O \approxeq 3\ \text{mol}\ O_2$$

For example, the reaction of 3.0 mol O_2 requires this much H_2O.

$$\text{no. mol}\ H_2O = 3\ \text{mol}\ O_2 \times \frac{6\ \text{mol}\ H_2O}{3\ \text{mol}\ O_2} = 6\ \text{mol}\ H_2O$$

Only 2.0 mol H_2O is available. Some of the O_2 must remain unreacted. *$O_2(g)$ is in excess.* Is there a sufficient amount of Fe_2S_3 available to react with 2.0 mol H_2O?

$$\text{no. mol}\ Fe_2S_3 = 2.0\ \text{mol}\ H_2O \times \frac{2\ \text{mol}\ Fe_2S_3}{6\ \text{mol}\ H_2O} = 0.67\ \text{mol}\ Fe_2S_3$$

There is 1.0 mol Fe_2S_3 available—more than is required to react with the available H_2O. *Fe_2S_3 is also in excess, and H_2O is the limiting reagent.* The amount of $Fe(OH)_3$ that will be obtained then is determined by the amount of H_2O available.

$$\text{no. mol Fe(OH)}_3 = 2.0 \text{ mol H}_2\text{O} \times \frac{4 \text{ mol Fe(OH)}_3}{6 \text{ mol H}_2\text{O}} = 1.3 \text{ mol Fe(OH)}_3$$

An alternative approach takes the exact form outlined in Figure 4-6. Calculate the amount of $Fe(OH)_3$ produced from 1.0 mol Fe_2S_3, *assuming* an excess of H_2O and O_2. Next, calculate the amount of $Fe(OH)_3$ produced from 2.0 mol H_2O, *assuming* an excess of Fe_2S_3 and O_2. Finally, in a third calculation, *assume* that O_2 is the limiting reagent and that there is an excess of Fe_2S_3 and H_2O. Three results will be obtained, and the answer is the *smallest* of the three—1.3 mol $Fe(OH)_3$.

SIMILAR EXAMPLES: Review Problem 12; Exercise 22(a).

A more common situation is that of Example 4-15. Here the quantities of the available reactants must be converted to a mole basis before a comparison can be made to identify the limiting reagent.

Example 4-15 What mass of PbI_2 will precipitate if 2.85 g $Pb(NO_3)_2$ is added to 225 mL of 0.0550 *M* KI(aq)?

$$Pb(NO_3)_2(aq) + 2 \text{ KI(aq)} \longrightarrow PbI_2(s) + 2 \text{ KNO}_3(aq)$$

Solution

$$\begin{aligned} \text{no. mol Pb(NO}_3)_2 \text{ available} &= 2.85 \text{ g Pb(NO}_3)_2 \times \frac{1 \text{ mol Pb(NO}_3)_2}{331 \text{ g Pb(NO}_3)_2} \\ &= 8.61 \times 10^{-3} \text{ mol Pb(NO}_3)_2 \end{aligned}$$

$$\text{no. mol KI available} = 0.225 \text{ L} \times \frac{0.0550 \text{ mol KI}}{\text{L}} = 1.24 \times 10^{-2} \text{ mol KI}$$

Next, determine the amount of KI required to react with 8.61×10^{-3} mol $Pb(NO_3)_2$.

$$\begin{aligned} \text{no. mol KI required} &= 8.61 \times 10^{-3} \text{ mol Pb(NO}_3)_2 \times \frac{2 \text{ mol KI}}{1 \text{ mol Pb(NO}_3)_2} \\ &= 1.72 \times 10^{-2} \text{ mol KI} \end{aligned}$$

There is only 1.24×10^{-2} mol KI available. *KI is the limiting reagent.*

$$\text{no. g PbI}_2 = 1.24 \times 10^{-2} \text{ mol KI} \times \frac{1 \text{ mol PbI}_2}{2 \text{ mol KI}} \times \frac{461 \text{ g PbI}_2}{1 \text{ mol PbI}_2} = 2.86 \text{ g PbI}_2$$

SIMILAR EXAMPLES: Exercises 23, 24.

Theoretical Yield, Actual Yield, and Percent Yield. The quantity of product *calculated* to result from given quantities of initial reactants is called the **theoretical yield** of a reaction. The quantity of product that is *actually* produced in a chemical reaction is called the **actual yield.** The **percent yield** is defined as

$$\text{percent yield} = \frac{\text{actual yield}}{\text{theoretical yield}} \times 100 \tag{4.22}$$

There are many reactions for which the actual yield is almost exactly equal to the theoretical yield. Such reactions are said to be *quantitative,* i.e., they can be used in chemical analysis. On the other hand, for some reactions, particularly those involving organic compounds, the actual yield of a reaction is *less than* the theoretical yield, and the percent yield, *less than* 100%. This is because the reaction may not go to completion, competing reactions may reduce the yield of product, or material may be lost in handling.

Example 4-16   In the reaction of 1.00 mol CH_4 with an excess of Cl_2, 83.5 g CCl_4 is obtained. What is the **(a)** theoretical yield; **(b)** actual yield; and **(c)** percent yield of this reaction?

$$CH_4 + 4\ Cl_2 \longrightarrow CCl_4 + 4\ HCl$$

Solution

(a) From 1.00 mol CH_4 we would expect to obtain 1.00 mol CCl_4; or

$$\text{no. g } CCl_4 = 1.00 \text{ mol } CH_4 \times \frac{1 \text{ mol } CCl_4}{1 \text{ mol } CH_4} \times \frac{154 \text{ g } CCl_4}{1 \text{ mol } CCl_4} = 154 \text{ g } CCl_4$$

(b) The actual yield is 83.5 g CCl_4.

(c) The percent yield is obtained with equation (4.22).

$$\text{percent yield} = \frac{83.5 \text{ g } CCl_4}{154 \text{ g } CCl_4} \times 100 = 54.2\%$$

SIMILAR EXAMPLES: Review Problem 13; Exercises 28, 29.

Example 4-17   When heated in the presence of sulfuric or phosphoric acid, cyclohexanol, $C_6H_{12}O$, is converted to cyclohexene, C_6H_{10}.

$$C_6H_{12}O(l) \longrightarrow C_6H_{10}(l) + H_2O(l)$$

Additional procedures are required to obtain pure cyclohexene. The percent yield is 83%. What mass of cyclohexanol that is 91% pure must be used to obtain 25 g pure cyclohexene?

Solution.   The key to this problem is in recognizing that the actual yield is only 83% of the theoretical yield. The theoretical yield must be larger than the 25 g C_6H_{10} we wish to obtain in the reaction. In fact, the theoretical yield must be 25 g $\times$ (100/83) = 30 g. The remaining steps are to calculate the quantity of pure $C_6H_{12}O$ required to produce a theoretical 30 g C_6H_{10}; and, finally, the quantity of impure cyclohexanol required. This stepwise procedure is outlined below.

Step 1.   Calculate the theoretical yield needed.

$$\frac{\text{actual yield}}{\text{theoretical yield}} \times 100 = \text{percent yield}$$

$$\text{theoretical yield} = \frac{\text{actual yield} \times 100}{\text{percent yield}} = 25 \text{ g} \times \frac{100}{83} = 30 \text{ g}$$

Step 2.   Calculate the quantity of $C_6H_{12}O$ required to produce 30 g C_6H_{10}.

$$\text{no. g } C_6H_{12}O = 30 \text{ g } C_6H_{10} \times \frac{1 \text{ mol } C_6H_{10}}{82.1 \text{ g } C_6H_{10}} \times \frac{1 \text{ mol } C_6H_{12}O}{1 \text{ mol } C_6H_{10}} \times \frac{100.2 \text{ g } C_6H_{12}O}{1 \text{ mol } C_6H_{12}H}$$

$$= 37 \text{ g } C_6H_{12}O$$

Step 3.   Calculate the quantity of impure cyclohexanol required.

$$\text{no. g cyclohexanol (impure)} = 37 \text{ g } C_6H_{12}O \times \frac{100 \text{ g cyclohexanol (impure)}}{91 \text{ g } C_6H_{12}O}$$

$$= 41 \text{ g cyclohexanol (impure)}$$

SIMILAR EXAMPLE: Exercise 30.

Simultaneous and Consecutive Reactions.   Some stoichiometric calculations require that two or more chemical equations be used, each equation furnishing a conversion factor. In some cases the reactions occur at the same time (simultaneously), and in others, they occur in succession (consecutively). Example 4-18 deals with simultaneous reactions, and Example 4-19, with consecutive reactions.

Example 4-18   A 0.710-g sample of a magnalium alloy (70.0% Al–30.0% Mg) reacts with an excess of HCl(aq). What mass of H_2 is produced?

$$2\ Al(s) + 6\ HCl(aq) \longrightarrow 2\ AlCl_3(aq) + 3\ H_2(g)$$

$$Mg(s) + 2\ HCl(aq) \longrightarrow MgCl_2(aq) + H_2(g)$$

Solution
Step 1.   Use percent composition to determine the mass of each metal in the alloy. *Result:* 0.497 g Al; 0.213 g Mg.
Step 2.   Use molar masses to convert from mass to number of moles of each metal. *Result:* 0.0184 mol Al; 0.00877 mol Mg.
Step 3.   Determine the no. mol H_2 produced by each metal.

$$\text{no. mol } H_2 = 0.0184\ \text{mol Al} \times \frac{3\ \text{mol } H_2}{2\ \text{mol Al}} = 0.0276\ \text{mol } H_2$$

$$\text{no. mol } H_2 = 0.00877\ \text{mol Mg} \times \frac{1\ \text{mol } H_2}{1\ \text{mol Mg}} = 0.00877\ \text{mol } H_2$$

Step 4.   The total amount of H_2 produced is $0.0276 + 0.00877 = 0.0364$ mol H_2. Its mass is

$$\text{no. g } H_2 = 0.0364\ \text{mol } H_2 \times \frac{2.02\ \text{g } H_2}{1\ \text{mol } H_2} = 0.0735\ \text{g } H_2$$

SIMILAR EXAMPLES: Exercises 31, 32.

Example 4-19   Sodium chlorate, $NaClO_3$, can be produced as follows.

$$2\ KMnO_4 + 16\ HCl \longrightarrow 2\ KCl + 2\ MnCl_2 + 8\ H_2O + 5\ Cl_2$$

$$6\ Cl_2 + 6\ Ca(OH)_2 \longrightarrow Ca(ClO_3)_2 + 5\ CaCl_2 + 6\ H_2O$$

$$Ca(ClO_3)_2 + Na_2SO_4 \longrightarrow CaSO_4 + 2\ NaClO_3$$

How many moles of $NaClO_3$ are produced for every mole of HCl consumed? Assume an excess of all other reactants.

Solution
Step 1.   Determine the no. mol Cl_2 produced per mol HCl in the first reaction.

$$\text{no. mol } Cl_2 = 1.00\ \text{mol HCl} \times \frac{5\ \text{mol } Cl_2}{16\ \text{mol HCl}} = 0.312\ \text{mol } Cl_2$$

Step 2.   Determine the no. mol $Ca(ClO_3)_2$ formed from 0.312 mol Cl_2 in the second reaction.

$$\text{no. mol } Ca(ClO_3)_2 = 0.312\ \text{mol } Cl_2 \times \frac{1\ \text{mol } Ca(ClO_3)_2}{6\ \text{mol } Cl_2} = 0.0520\ \text{mol } Ca(ClO_3)_2$$

Step 3.   Determine the no. mol $NaClO_3$ formed from 0.0520 mol $Ca(ClO_3)_2$ in the third reaction.

$$\text{no. mol } NaClO_3 = 0.0520\ \text{mol } Ca(ClO_3)_2 \times \frac{2\ \text{mol } NaClO_3}{1\ \text{mol } Ca(ClO_3)_2} = 0.104\ \text{mol } NaClO_3$$

SIMILAR EXAMPLES: Exercises 34, 36.

Oxidation–Reduction Reactions. Before a stoichiometric calculation is possible, a balanced chemical equation is required. At times, however, the balanced condition may be very difficult to obtain by inspection. Often this is the case for reactions of a type called oxidation–reduction.

The key to identifying an oxidation–reduction reaction lies in assigning oxidation states to the atoms in the reactants and products. In the expression below oxidation states are denoted by small colored numerals.

$$\overset{0}{I_2} + \overset{0}{Br_2} + \overset{+1}{H_2}\overset{-2}{O} \longrightarrow \overset{+1}{H}\overset{+5}{I}\overset{-2}{O_3} + \overset{+1}{H}\overset{-1}{Br} \tag{4.23}$$

In an oxidation–reduction reaction certain atoms undergo changes in oxidation state (O.S.). Oxidation corresponds to an *increase* in O.S. and reduction to a *decrease* in O.S. In reaction (4.23) I atoms are *oxidized* (their O.S. increases from 0 in I_2 to +5 in HIO_3). Br atoms are *reduced* (their O.S. decreases from 0 in Br_2 to −1 in HBr). These changes are noted below.

$$\overset{0}{I_2} + \overset{0}{Br_2} + H_2O \longrightarrow \overset{+5}{HIO_3} + \overset{-1}{HBr}$$

increase of 5 in O.S. per I atom

decrease of 1 in O.S. per Br atom

In an oxidation–reduction reaction the total increase in O.S. for all atoms involved in oxidation must equal the total decrease in O.S. for all atoms involved in reduction. This fact requires that the coefficients of I_2, Br_2, HIO_3, and HBr be adjusted as follows.

$$I_2 + 5\,Br_2 + H_2O \longrightarrow 2\,HIO_3 + 10\,HBr$$

total increase of 10 in O.S.

total decrease of 10 in O.S.

The remaining coefficient (that of H_2O) is determined by inspection.

$$I_2 + 5\,Br_2 + 6\,H_2O \longrightarrow 2\,HIO_3 + 10\,HBr \tag{4.24}$$

The expression in Example 4-20 is written in ionic form, and in addition to achieving a balance in number of atoms of each type there must also be a balance of electric charge (recall Example 4-3). This charge balance results automatically when balancing is done by the method just outlined—the **oxidation state change method.**

> Some oxidation–reduction equations can be easily balanced by inspection, such as $2\,H_2 + O_2 \rightarrow 2\,H_2O$. Use the oxidation state change method when it is clear that balancing by inspection will be difficult or time consuming.

Example 4-20 Balance the following oxidation–reduction expression.

$$SO_3{}^{2-} + MnO_4{}^- + H^+ \longrightarrow SO_4{}^{2-} + Mn^{2+} + H_2O$$

Solution
Step 1. *Identify the atoms that undergo changes in oxidation state (O.S.) and indicate the increase or decrease in O.S. per atom.* (If there are no changes in O.S., it is not an oxidation–reduction reaction.)

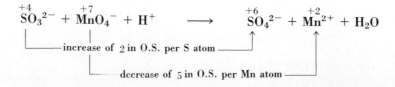

$$\overset{+4}{S}O_3{}^{2-} + \overset{+7}{Mn}O_4{}^- + H^+ \longrightarrow \overset{+6}{S}O_4{}^{2-} + \overset{+2}{Mn}^{2+} + H_2O$$

increase of 2 in O.S. per S atom

decrease of 5 in O.S. per Mn atom

Step 2. *Adjust the coefficients of species containing atoms that change in oxidation state so that the total increase in O.S. = total decrease in O.S.*

$$5\,SO_3^{2-} + 2\,MnO_4^- + H^+ \longrightarrow 5\,SO_4^{2-} + 2\,Mn^{2+} + H_2O$$

— total increase of 10 in O.S. —

— total decrease of 10 in O.S. —

Step 3. *Without changing coefficients previously established, complete the balancing by inspection.* As a result of Step 2 there are 23 O atoms on the left and only 21 on the right. The coefficient "3" for H_2O produces a balance of O atoms; the coefficient "6" for H^+ balances the H atoms.

$$5\,SO_3^{2-} + 2\,MnO_4^- + 6\,H^+ \longrightarrow 5\,SO_4^{2-} + 2\,Mn^{2+} + 3\,H_2O$$

Step 4. *Verify the electric charge balance.* The charges represented are

left: $(5 \times -2) + (2 \times -1) + (6 \times +1) = -6$
right: $(5 \times -2) + (2 \times +2) = -6$

SIMILAR EXAMPLES: Review Problem 14; Exercise 38(a)–(d).

There are several variations to the basic steps of balancing oxidation–reduction equations previously considered. Equations written with H^+ are conducted in acidic solution. The presence of OH^-, as in Example 4-21, signifies that the reaction is carried out in basic or alkaline solution. When balancing equations for oxidation–reduction reactions in basic solution it is often helpful to reverse steps 3 and 4 of Example 4-20, that is, to balance for electrical charge before completing the balance by inspection. This procedure is illustrated in Example 4-21. Also illustrated in Example 4-21 is that sometimes the same substance undergoes both oxidation and reduction.

Example 4-21 Balance the following oxidation–reduction expression.

$$Cl_2 + OH^- \longrightarrow Cl^- + ClO_3^- + H_2O$$

Solution
Step 1. Identify the atoms that undergo changes in oxidation state (O.S.) and indicate the increase or decrease in O.S. per atom.

$$\overset{0}{Cl_2} + OH^- \longrightarrow \overset{-1}{Cl^-} + \overset{+5}{ClO_3^-} + H_2O$$

— decrease of 1 in O.S. per Cl —

— increase of 5 in O.S. per Cl —

Step 2. Adjust the coefficients of the species containing atoms that change in O.S. so that the total increase in O.S. = total decrease in O.S.

$$3\,Cl_2 + OH^- \longrightarrow 5\,Cl^- + ClO_3^- + H_2O$$

— decrease of 5 in O.S. —

— increase of 5 in O.S. —

Step 3. Balance for electrical charge.

$$3\,Cl_2 + 6\,OH^- \longrightarrow 5\,Cl^- + ClO_3^- + H_2O$$

Step 4. Complete the balancing by inspection.

$$3\,Cl_2 + 6\,OH^- \longrightarrow 5\,Cl^- + ClO_3^- + 3\,H_2O \qquad (4.25)$$

SIMILAR EXAMPLES: Exercise 38(e), (f).

FIGURE 4-7
Identifying oxidizing
and reducing agents.

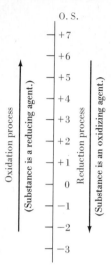

In an oxidation–reduction reaction the substance that is oxidized makes it possible for another substance to be *reduced*. The substance that is oxidized is called the **reducing agent.** Similarly, the substance that is reduced makes it possible for another substance to be *oxidized*. The substance that is reduced is the **oxidizing agent.** These statements are summarized diagramatically in Figure 4-7. Applied to reaction (4.24), I_2 is the reducing agent (it is oxidized) and Br_2 is the oxidizing agent (it is reduced). In reaction (4.25) Cl_2 is both the oxidizing and the reducing agent.

Predicting whether a particular combination of oxidizing and reducing agents actually results in an oxidation–reduction reaction requires the use of principles to be established in Chapter 20. At that time we will also consider another, more fundamental way of describing oxidation–reduction reactions. Until that time, our need is simply to be able to identify oxidizing and reducing agents and to have a way of balancing oxidation–reduction equations.

Chemical Equilibrium. With the methods presented in this chapter, we cannot answer this seemingly simple question: "How many moles of HI(g) are formed by the reaction of 0.10 mol I_2(g) with an excess of H_2(g)?"

$$H_2(g) + I_2(g) \rightleftharpoons 2\ HI(g) \tag{4.26}$$

The "expected" answer (0.20 mol HI) is *incorrect*. What is different here from previous situations is that while molecules of H_2 and I_2 are combining to form HI, molecules of HI are decomposing to re-form H_2 and I_2. There is both a forward (left-to-right) and a reverse (right-to-left) reaction, signified by the double arrow in equation (4.26).

In reaction (4.26) a point is reached where each reacting species is consumed and reformed at the same rate. This is a condition of **dynamic equilibrium**—*dynamic* because reaction among the molecules continues to occur, yet *equilibrium* because beyond this point there is no net change with time in the amounts of reacting species present. Since some of each species must be present at equilibrium, we cannot calculate the yield of an equilibrium reaction with factors from the chemical equation alone. The yield of HI in the hypothetical question posed above must be *less than* 0.20 mol—the I_2 is not completely consumed. In Chapter 15 we will learn some new principles which, when used in conjunction with factors from the chemical equation, will make possible an exact numerical answer.

In many chemical reactions the extent of the reverse reaction is negligible. This may be because a gaseous product escapes from the reaction mixture; because ions combine to form an insoluble precipitate; or because ions combine to form a substance which has very little tendency to ionize (as in the formation of H_2O by the combination of H^+ and OH^-). Also, many oxidation–reduction reactions are of such a nature that the forward reaction greatly predominates over the reverse reaction. In all these cases it can be assumed that the reaction "goes to completion." And it is to such reactions that the methods of calculation introduced in this chapter apply.

FOCUS ON
Industrial Chemistry

The typical apparatus of a chemical research laboratory (left) and of an industrial chemical plant (right) are rather dissimilar in appearance. The task of the chemical engineer is to convert a bench-scale laboratory process into a large-scale, economically viable operation. [Photos courtesy of BASF.]

Large-scale manufacturing processes for the production of chemicals are based on the same principles as those considered in this chapter and elsewhere in the text. Yet many factors beyond the usual textbook cases must be considered. We attempt here to capture the essence of industrial chemistry.

Hydrazine and Its Uses. Hydrazine, N_2H_4, is an oily, colorless liquid that freezes at 2.0°C and boils at 113.5°C. Its density at 25°C is 1.0045 g/cm^3. The nature of chemical bonding in hydrazine is discussed in Section 9-4, and its physical and chemical properties are explored further in Section 21-3.

The first important use of hydrazine was as a fuel in the rocket-powered ME-163 fighter airplane used by Germany in World War II, and it continues to find some use in rocket engines. Currently its largest use (40%) is in the synthesis of some 30 to 40 different pesticides. Next in importance (33%) is the production of hydrazine-based chemicals used as blowing agents in the polymer industry. (A blowing agent evolves gases on decomposing. The gases produce the holes or pores in products like sponge rubber and foamed plastics.) Hydrazine also finds use (15%) in water treatment. Hydrazine is easily oxidized by $O_2(g)$. When hydrazine is added to boilers or hot-water

heating systems, it scavenges (removes) dissolved oxygen from the water, thereby reducing the rate of corrosion of metal parts exposed to the hot water.

A method of manufacturing hydrazine (the Raschig process) is illustrated in Figure 4-8 in a form commonly used to represent an industrial process—a flow diagram. This flow diagram brings out several important aspects of industrial chemistry.

Stepwise Reactions, Intermediates, the Net Chemical Equation. An industrial process is usually carried out in stages or steps. The Raschig process involves three steps. Sodium hypochlorite (NaOCl), produced in the first reaction, and chloramine (NH_2Cl), produced in the second, are **intermediates** in the process. Their presence is crucial to the overall process, but they are consumed immediately after their formation. Hydrazine, the desired **end product,** is produced in the third reaction. The **net chemical equation,** representing the overall reaction, is obtained by adding together equations for the individual, consecutive reactions. In this addition intermediate species "cancel out."

By-products, Side Reactions. The net equation for the production of N_2H_4 shows that two other products are also

FIGURE 4-8
The commercial production of hydrazine (N_2H_4).

Production Scheme:

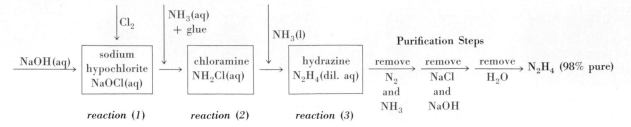

reaction (1)   reaction (2)   reaction (3)

The Net Chemical Reaction:

reaction (1) $2\,NaOH(aq) + Cl_2 \longrightarrow NaOCl(aq) + NaCl(aq) + H_2O$
reaction (2) $NaOCl(aq) + NH_3(aq) \longrightarrow NH_2Cl(aq) + NaOH(aq)$

reaction (3) $\underline{NH_2Cl(aq) + NH_3(l) + NaOH(aq) \xrightarrow{\Delta} N_2H_4(aq) + NaCl(aq) + H_2O}$

(1) + (2) + (3) $2\,NaOH(aq) + Cl_2 + NaOCl(aq) + NH_3(aq) + NH_2Cl(aq) + NH_3(l) + NaOH(aq) \longrightarrow$
$\qquad NaOCl(aq) + NaCl(aq) + H_2O + NH_2Cl(aq) + NaOH(aq) + N_2H_4(aq) + NaCl(aq) + H_2O$

cancellations: $2\,NaOH(aq) + Cl_2 + \cancel{NaOCl(aq)} + NH_3(aq) + \cancel{NH_2Cl(aq)} + NH_3(l) + \cancel{NaOH(aq)} \longrightarrow$
$\qquad \cancel{NaOCl(aq)} + NaCl(aq) + H_2O + \cancel{NH_2Cl(aq)} + \cancel{NaOH(aq)} + N_2H_4(aq) + NaCl(aq) + H_2O$

Net equation: $2\,NaOH(aq) + Cl_2 + 2\,NH_3 \longrightarrow N_2H_4(aq) + 2\,NaCl(aq) + 2\,H_2O$

Side Reaction: $2\,NH_2Cl(aq) + N_2H_4(aq) \xrightarrow{Cu^{2+}} 2\,NH_4Cl(aq) + N_2(g)$

---

formed—NaCl and H_2O. Substances formed along with the desired end product are called **by-products** of a process. Although they are of no value in the Raschig process, often the economic value of by-products may determine the feasibility of a process. (For example, about 90% of all hydrochloric acid is produced as a by-product of other manufacturing processes.)

In addition to their formation in the main chemical reaction, by-products may form through one or more competing **side reactions.** Some of the hydrazine formed in the third reaction combines with the intermediate chloramine to produce $NH_4Cl(aq)$ and $N_2(g)$. This side reaction, which reduces the yield of N_2H_4, is catalyzed (speeded up) by the presence of Cu^{2+} or other heavy metal ions. Removal of these ions helps to minimize the side reaction.

Purification of Product.   Rarely is the desired end product of an industrial chemical process in a state of sufficient purity for its intended uses—the product must be purified. Hydrazine produced in the Raschig process is in the form of a dilute water solution also containing NH_3, NaCl, and traces of NaOH. The substances that are obtainable as solids—NaCl and NaOH—are crystallized from solution. NH_3 and H_2O are removed by distillation. Ultimately a product containing 98% N_2H_4 or higher is obtainable. NH_3 recovered in the purification is reused in the main

reactions, illustrating still another principle of industrial chemistry: *Materials are recycled whenever possible.*

Reaction Conditions.   The precise conditions for a chemical process are not always indicated through chemical equations, nor are the reasons for these conditions. In the Raschig process the formation of chloramine (the second reaction) proceeds very rapidly, but the conversion of chloramine and ammonia to hydrazine (the third reaction) occurs more slowly. To speed up this conversion, elevated temperatures (about 130°C) are used for the third reaction.

At the same time that the main reaction is speeded up, the side reaction must be minimized. This is accomplished in two ways: (1) A large excess of NH_3 is used—perhaps 20 or 30 mol NH_3 per mole N_2H_4. This means that NH_2Cl is more likely to react with NH_3 than with N_2H_4. (2) A protein-based material–gelatin, albumin, glue—is added to the reaction mixture. Protein molecules bind with metal ions in solution (e.g., Cu^{2+}), reducing the ability of these ions to catalyze the side reaction.

In the third reaction NH_3 is introduced as the pure anhydrous (water-free) liquid, rather than in aqueous solution as in the second reaction. There are two reasons for this: (1) The heat given off when the NH_3 dissolves in the $NH_2Cl(aq)$ is sufficient to raise the reaction temperature to the desired 130°C (thus avoiding the need to consume fuel to heat the mixture). (2) The less water added to the reaction mixture, the less that will have to be removed in the purification steps.

Alternative Processes.   Modifications of the Raschig process are now coming into prominence. In one of these the hydrazine-forming reactions are carried out in the presence of acetone (an organic solvent). Acetone and hydrazine form a compound that is resistant to attack by NH_2Cl. The hydrazine-acetone compound is later decomposed into hydrazine and acetone. In this way, by eliminating the side reaction, the yield of hydrazine is close to the theoretical (i.e., about 100%) compared to a 60 to 80% yield in the original Raschig process. Currently about 25% of the hydrazine-producing capacity in the United States is by the original Raschig process and 75% by the modified process. Again two points are illustrated here: (1) Industrial methods of producing chemicals undergo constant change, but (2) the changeover from one process to another occurs over a period of time, as new plants are built or old ones converted.

Summary

Once the reactants and products have been identified, it is possible to represent a chemical reaction through a chemical equation. The physical states or forms in which reactants and products appear, and reaction conditions, can also be represented in the equation. A chemical equation must be balanced atomically, and, where necessary, for electrical charge as well. A wide variety of calculations (stoichiometric calculations) can be based on the chemical equation. The principal feature of such calculations is the use of conversion factors derived from the chemical equation. Molar masses, densities, and percent composition are often required as well.

Many chemical reactions are carried out in solution, and for this reason it proves convenient to describe the composition or concentration of solutions on a molar basis. The molarity or molar concentration indicates the number of moles of solute per liter of solution. From this definition it is possible to perform calculations relating molar concentration, solution volume, and amount of solute. This may be done for individual solutions or for situations in which solutions are mixed, diluted by adding more solvent, or concentrated by removing solvent. Molar concentrations can also serve as conversion factors in stoichiometric calculations. Of practical importance is the procedure for carrying out a reaction between two solutions—titration.

Additional features sometimes arise in stoichiometric calculations: It may be necessary to identify the single reactant, called the limiting reagent, that determines the amount of product formed (other reactants being in excess). The yield of product in a reaction may be less than the calculated value. Two or more reactions may occur at the same time, or a series of reactions may occur in succession. It may be difficult to balance certain chemical equations by inspection, in which case special methods have to be used. This is often the case with oxidation-reduction reactions. Finally, the methods of this chapter apply only to reactions that go to completion. (Reactions that reach a condition of dynamic equilibrium are considered in later chapters.)

Learning Objectives

As a result of studying Chapter 4, you should be able to

1. Write word and formula expressions for chemical reactions where the reactants and products are stated.

2. Balance a chemical equation atomically, and, where necessary, for electrical charge.

3. Write balanced equations for reactions involving the complete combustion of carbon-hydrogen compounds (hydrocarbons) and carbon-hydrogen-oxygen compounds.

4. Apply the terms *combustion, combination, decomposition, displacement,* and *metathesis* to describe types of reactions.

5. Solve a variety of problems based on the chemical equation, with quantities given and/or sought in units of moles, mass, volume, density, percent by mass, and so on.

6. Define the molar concentration (*M*) scale and perform calculations involving the amount of solute, solution volume, and the molarity of a solution.

7. Do calculations relating to the mixing of solutions or to the addition or removal of solvent from a solution.

8. Use solution volumes and molar concentrations in stoichiometric calculations for reactions involving solutions.

9. Use titration data to determine the composition of a sample of matter, such as the concentration of a solution.

10. Determine the limiting reagent in a chemical reaction.

11. Define the terms *actual yield*, *theoretical yield*, and *percent yield* of a reaction, and calculate these quantities.

12. Calculate the amount of product when two or more reactions occur simultaneously or consecutively.

13. Identify the oxidizing and reducing agents in oxidation–reduction reactions.

14. Balance oxidation–reduction equations by the oxidation state change method.

Some New Terms

Balancing an equation refers to placing numbers, **stoichiometric coefficients,** in front of the symbols and formulas in a chemical equation. In this way the numbers of atoms of each kind are made equal on the two sides of the equation. In **net ionic equations,** electrical charges must also be balanced.

By-products are those substances produced along with the principal reaction product in a chemical process.

A **chemical equation** is a symbolic representation of a chemical reaction; that is, symbols and formulas are substituted for the names of reactants and products.

A **chemical reaction** is a process in which one set of substances (reactants) is transformed into another set of substances (products).

Dilution is the process of reducing the concentration of a solution by adding more solvent.

An **intermediate** is the product of one reaction that is consumed in a following reaction in a process that proceeds through several steps.

The **limiting reagent** in a reaction is the reactant that is consumed completely. The amount of product(s) formed is dependent on the amount of the limiting reagent.

Molar concentration or molarity (M) refers to the composition or concentration of a solution expressed as number of moles of solute per liter of solution.

The **net chemical reaction** is the overall change that occurs in a process involving two or more steps.

An **oxidation–reduction reaction** is one in which certain atoms undergo changes in oxidation state. The substance containing atoms whose oxidation state *increases* is **oxidized** and is said to be the **reducing agent.** The substance containing atoms whose oxidation state *decreases* is **reduced** and is said to be an **oxidizing agent.**

A **side reaction** is a reaction that occurs at the same time as the principal reaction. Usually, the existence of side reactions reduces the yield of the desired product.

Titration is a procedure for carrying out a chemical reaction between two solutions by the controlled addition of one solution (from a buret) to the other.

The **yield** refers to the quantity of a desired product in a chemical process. The **theoretical yield** is the quantity *calculated* from the chemical equation. The **actual yield** is the *measured* quantity of product. **Percent yield** is the percent of the theoretical yield that is actually obtained.

Suggestions for Further Study

CARDULLA, F., ''Chemical of the Month: Hydrazine,'' *J. Chem. Educ.,* **60,** 505 (1983).

DAVIES, D. S., ''The Changing Nature of Industrial Chemistry,'' *Chem. & Eng. News,* **56**[8], 22 (1978).

KOLB, D., ''The Chemical Equation, Part I: Simple Reactions,'' *J. Chem. Educ.,* **55,** 184 (1978).

STRONG, L. E., ''Balancing Chemical Equations,'' *Chemistry,* **47**[1], 13 (1974).

Review Problems

1. Write balanced equations for each of the following.
  (a) magnesium + oxygen $\rightarrow$ magnesium oxide
  (b) sulfur + oxygen $\rightarrow$ sulfur dioxide
  (c) methane (CH_4) + oxygen $\rightarrow$ carbon dioxide + water
  (d) aqueous silver sulfate + aqueous barium iodide $\rightarrow$ solid barium sulfate + solid silver iodide

2. Balance the following equations by inspection.
  (a) $Na_2SO_4(s) + C(s) \rightarrow Na_2S(s) + CO_2(g)$
  (b) $HCl(g) + O_2(g) \rightarrow H_2O(l) + Cl_2(g)$
  (c) $PCl_3(l) + H_2O(l) \rightarrow H_3PO_3(aq) + HCl(aq)$
  (d) $PbO(s) + NH_3(g) \rightarrow Pb(s) + N_2(g) + H_2O(l)$
  (e) $Mg_3N_2(s) + H_2O(l) \rightarrow Mg(OH)_2(s) + NII_3(g)$

3. Balance the following equations written in *ionic* form.
  (a) $Zn(s) + Ag^+(aq) \rightarrow Zn^{2+}(aq) + Ag(s)$
  (b) $Mn^{2+}(aq) + H_2S(aq) \rightarrow MnS(s) + H^+(aq)$
  (c) $Al(s) + H^+(aq) \rightarrow Al^{3+}(aq) + H_2(g)$

4. Write a balanced chemical equation to represent
  (a) the complete combustion of C_5H_{12};
  (b) the complete combustion of $C_2H_6O_2$;
  (c) the neutralization of hydroiodic acid by aqueous sodium hydroxide, producing sodium iodide and water;
  (d) the precipitation of lead(II) iodide by the addition of potassium iodide to an aqueous solution of lead(II) nitrate.

5. Iron metal reacts with chlorine gas as follows.

$$2\ Fe(s) + 3\ Cl_2(g) \rightarrow 2\ FeCl_3(s)$$

How much $FeCl_3(s)$, in moles, is obtained when 3.15 mol Cl_2 reacts with excess Fe?

6. 0.382 mol PCl_3 is produced by the reaction

$$6\ Cl_2 + P_4 \rightarrow 4\ PCl_3$$

How many grams each of Cl_2 and P_4 are consumed?

7. The reaction of calcium hydride with water can be used to prepare small quantities of hydrogen gas.

$$CaH_2(s) + 2\ H_2O(l) \rightarrow Ca(OH)_2(s) + 2\ H_2(g)$$

(a) How much $H_2(g)$, in moles, results from the reaction of 312 g CaH_2 with an excess of water?
(b) What mass of H_2O is consumed in the reaction of 88.5 g CaH_2?
(c) What mass of CaH_2 must react with an excess of H_2O to produce 3.12×10^{25} molecules of $H_2(g)$?

8. What are the molar concentrations of the solutes listed below when dissolved in water?
(a) 2.17 mol C_2H_5OH in 5.12 L solution
(b) 12.35 mmol CH_3OH in 50.00 mL of solution
(c) 14.3 g $(CH_3)_2CO$ in 125 mL of solution
(d) 12.2 mL of pure glycerol, $C_3H_8O_3$ (d = 1.26 g/cm^3) in 375.0 mL solution

9. How much
(a) KCl, in moles, is required to prepare 3.55×10^3 L of 0.250 M KCl(aq)?
(b) Na_2SO_4, in grams, is required to produce 625 mL of 0.415 M Na_2SO_4(aq)?

(c) KOH, in mg, is present per mL of 0.105 M KOH(aq)?

10. What volume of 0.800 M KOH must be diluted with water to prepare 500.0 mL of 0.312 M KOH?

11. What volume of 0.1035 M NaOH is required for the titration (i.e., the neutralization) of 25.00 mL of 0.1132 M HNO_3?

$$HNO_3(aq) + NaOH(aq) \rightarrow NaNO_3(aq) + H_2O$$

12. A 0.500-mol sample of Cu is added to 125 mL of 6.0 M HNO_3(aq). Will the Cu dissolve completely? (*Hint:* Which is the limiting reagent?)

$$3\ Cu(s) + 8\ HNO_3(aq) \rightarrow$$
$$3\ Cu(NO_3)_2(aq) + 4\ H_2O + 2\ NO(g)$$

13. In the following reaction, 100.0 g $C_6H_{12}O$ yielded 69.0 g C_6H_{10}.

$$C_6H_{12}O \rightarrow C_6H_{10} + H_2O$$

(a) What is the theoretical yield of the reaction?
(b) What is the percent yield?
(c) What mass of $C_6H_{12}O$ should have been used to produce 100.0 g C_6H_{10}?

14. Balance the following oxidation-reduction equations by the oxidation state change method.
(a) $NO(g) + H_2(g) \rightarrow NH_3(g) + H_2O(g)$
(b) $Cu(s) + H^+ + NO_3^- \rightarrow Cu^{2+} + H_2O + NO(g)$
(c) $Zn(s) + H^+ + NO_3^- \rightarrow Zn^{2+} + H_2O + N_2O(g)$
(d) $Fe_2S_3(s) + H_2O + O_2(g) \rightarrow Fe(OH)_3(s) + S(s)$
(e) $H_2O_2 + MnO_4^- + H^+ \rightarrow Mn^{2+} + H_2O + O_2(g)$

15. For each of the reactions in Review Problem 14, identify the oxidizing agent and the reducing agent.

Exercises

Writing and balancing chemical equations

1. From the following descriptions of chemical reactions write appropriate balanced equations.
(a) Gaseous water reacts with carbon at high temperatures to produce carbon monoxide and hydrogen.
(b) Aluminum displaces copper(II) ion from aqueous solution, producing aluminum ion and copper.
(c) Zinc sulfide dissolves in hydrochloric acid, producing hydrogen sulfide gas and a water solution containing zinc ions.
(d) At high temperatures the gases chlorine and water react to form hydrogen chloride and oxygen gases.

2. Balance the following equations by inspection.
(a) $Cl_2(g) + H_2O(l) \rightarrow HCl(aq) + HOCl(aq)$
(b) $P_2H_4(l) \rightarrow PH_3(g) + P_4(s)$
(c) $NO_2(g) + H_2O(l) \rightarrow HNO_3(aq) + NO(g)$
(d) $S_2Cl_2 + NH_3 \rightarrow N_4S_4 + NH_4Cl + S_8$
(e) $SO_2Cl_2 + HI \rightarrow H_2S + H_2O + HCl + I_2$

3. Balance the following equations written in *ionic* form.
(a) $Fe^{3+}(aq) + H_2S(aq) \rightarrow Fe_2S_3(s) + H^+(aq)$
(b) $Al^{3+} + NH_3 + H_2O \rightarrow Al(OH)_3(s) + NH_4^+(aq)$
(c) $S_2O_3^{2-} + H^+ \rightarrow H_2O + S(s) + SO_2(g)$
(d) $MnO_2(s) + H^+ + Cl^- \rightarrow Mn^{2+} + H_2O + Cl_2(g)$

4. Write balanced equations to represent the complete combustion of (a) benzene, C_6H_6; (b) isopropyl alcohol, C_3H_7OH; (c) benzoic acid, C_6H_5COOH; (d) thiobenzoic acid, C_6H_5COSH (*Hint:* The sulfur is converted to sulfur dioxide).

Stoichiometry of chemical reactions

5. A laboratory method of preparing $O_2(g)$ involves the decomposition of $KClO_3(s)$.

$$2\ KClO_3(s) \xrightarrow{\Delta} 2\ KCl(s) + 3\ O_2(g)$$

A 25.5-g sample of $KClO_3(s)$ is decomposed. How many (a) mol $O_2(g)$; (b) molecules of $O_2(g)$; (c) g KCl(s) are produced?

6. A 128-g sample of a Na_2CO_3–$CaSO_4$ mixture containing 61.3% Na_2CO_3, by mass, is treated with excess HCl(aq). Only the Na_2CO_3 reacts. What mass of CO_2 is produced?

$$Na_2CO_3(s) + 2 HCl(aq) \rightarrow 2 NaCl(aq) + H_2O(l) + CO_2(g)$$

7. Iron ore is impure Fe_2O_3. When Fe_2O_3 is heated with an excess of carbon, iron metal is produced. From a sample of ore weighing 878 kg, 515 kg of pure iron is obtained. What is the % Fe_2O_3, by mass, in the ore sample? (*Hint:* How much Fe_2O_3 is required to produce 515 kg of iron?)

$$Fe_2O_3(s) + 3 C(s) \xrightarrow{\Delta} 2 Fe(l) + 3 CO(g)$$

8. Silver oxide decomposes at temperatures in excess of 300°C, yielding metallic silver and oxygen gas. A 2.95-g sample of *impure* silver oxide yields 0.183 g O_2. Assuming that Ag_2O is the only source of O_2, what is the percent, by mass, of Ag_2O in the sample?

9. A piece of Al foil measuring 5.11 in. × 3.23 in. × 0.0381 in. is dissolved in excess HCl(aq). What mass of H_2, in grams, is produced? (Density of Al = 2.70 g/cm^3.) (*Hint:* Use equation 4.16.)

10. How many mL $KMnO_4$(aq) containing 12.6 g $KMnO_4$/L must be used to convert 9.13 g KI to I_2?

$$2 KMnO_4 + 10 KI + 8 H_2SO_4 \rightarrow$$
$$6 K_2SO_4 + 2 MnSO_4 + 5 I_2 + 8 H_2O$$

Molar concentration

11. What are the molar concentrations of the solutes below?
  (a) urea, $CO(NH_2)_2$, if 132 g of the 98.3% pure solid is dissolved in 500.0 mL of aqueous solution.
  (b) diethyl ether, $(C_2H_5)_2O$, if 13.0 mg is dissolved in 3.00 gal of water (1 gal = 3.78 L).
  ★(c) NaCl, if a solution is found to have 1.52 ppm of Na. [Assume that NaCl is the only source of Na and that the solution density is 1.00 g/cm^3. Parts per million (ppm) can be taken to mean g Na per million g solution.]

12. How much
  (a) methanol, CH_3OH ($d = 0.792$ g/cm^3), in cm^3, must be dissolved in water to produce 4.80 L of 0.318 M CH_3OH?
  (b) ethanol, C_2H_5OH ($d = 0.789$ g/cm^3), in gal, must be dissolved in water to produce 55.0 gal of 1.50 M C_2H_5OH?
  ★(c) $Ca(NO_3)_2$, in mg, must be present in 50.0 L of a solution with 1.21 ppm Ca? [*Hint:* See Exercise 11(c).]

13. Two sucrose solutions—158 mL of 1.50 M $C_{12}H_{22}O_{11}$ and 273 mL of 1.25 M $C_{12}H_{22}O_{11}$—are mixed. What is the molarity of $C_{12}H_{22}O_{11}$ in the final solution? (Assume the solution volumes to be additive.)

14. What volume of a concentrated hydrochloric acid solution (36.0% HCl, by mass; $d = 1.18$ g/mL) is required to produce 20.0 L of 0.125 M HCl?

15. Water is evaporated from 85.0 mL of 0.512 M $MgSO_4$ solution until the solution volume becomes 61.5 mL. What is the molar concentration of $MgSO_4$ in the resulting solution?

Chemical reactions in solutions

16. What mass of $NaHCO_3$(s) must be added to 155 mL 0.245 M $Cu(NO_3)_2$(aq) to precipitate all the Cu^{2+} as $CuCO_3$(s)?

$$Cu(NO_3)_2(aq) + 2 NaHCO_3(s) \rightarrow$$
$$CuCO_3(s) + 2 NaNO_3(aq) + H_2O + CO_2(g)$$

17. For the neutralization reaction

$$Ca(OH)_2(s) + 2 HCl(aq) \rightarrow CaCl_2(aq) + 2 H_2O$$

  (a) What mass of $Ca(OH)_2$ is required to neutralize 325 mL of 0.410 M HCl?
  (b) What mass of $Ca(OH)_2$, in kilograms, is required to neutralize 215 L of an HCl solution that is 30.12% HCl, by mass, and has a density of 1.15 g/cm^3.

18. Household ammonia is NH_3(aq). 31.20 mL of 0.9918 M HCl is required to neutralize the NH_3 present in a 5.00-mL sample of household ammonia.

$$NH_3(aq) + HCl(aq) \rightarrow NH_4Cl(aq)$$

  (a) What is the molarity of NH_3 in the sample?
  (b) Assuming a density of 0.96 g/mL for the ammonia solution, what is its percent NH_3, by mass?

19. For use in titration reactions it is desired to prepare 20 L of HCl(aq) of a concentration known to four significant figures. A two-step approach is required. First, a solution having a concentration of about 0.25 M is prepared by dilution of concentrated HCl(aq). Then a sample of the diluted HCl(aq) is titrated with an NaOH solution of known concentration. The molar concentration of the HCl is calculated from the titration data.
  (a) What volume of concentrated HCl(aq) ($d = 1.19$ g/cm^3; 38% HCl, by mass) must be diluted to 20 L with water to prepare an 0.25 M HCl solution?
  (b) A 25.00-mL sample of the approximately 0.25 M HCl prepared in part (a) requires exactly 30.10 mL of 0.2000 M NaOH for its titration. What is the exact molar concentration of the diluted HCl(aq)?

$$HCl(aq) + NaOH(aq) \rightarrow NaCl(aq) + H_2O(l)$$

  (c) Why is a titration necessary? That is, why could not the final solution be prepared simply by an appropriate dilution of the concentrated hydrochloric acid?

20. An iron ore sample weighing 0.8313 g is dissolved in HCl(aq) and the iron is converted to $FeCl_2$(aq). This solution is then titrated with exactly 29.72 mL of 0.0410 M $K_2Cr_2O_7$. What must be the % Fe, by mass, in the ore sample?

$$6 FeCl_2(aq) + K_2Cr_2O_7(aq) + 14 HCl(aq) \rightarrow$$
$$6 FeCl_3(aq) + 2 CrCl_3(aq) + 2 KCl(aq) + 7 H_2O(l)$$

21. A method of adjusting the concentration of HCl(aq) is to allow the solution to react with a small quantity of Mg.

$$Mg(s) + 2\ HCl(aq) \rightarrow MgCl_2(aq) + H_2(g)$$

What mass of Mg must be added to 250.0 mL of 1.017 M HCl to reduce the concentration to exactly 1.000 M HCl?

Determining the limiting reagent

22. The following is encountered as a side reaction in the manufacture of rayon from wood pulp.

$$3\ CS_2 + 6\ NaOH \rightarrow 2\ Na_2CS_3 + Na_2CO_3 + 3\ H_2O$$

   (a) How many moles each of Na_2CS_3, Na_2CO_3, and H_2O are produced by allowing 1.00 mol each of CS_2 and NaOH to react?

   (b) What mass of Na_2CS_3 is produced in the reaction of 100.0 cm³ of liquid CS_2 (d = 1.26 g/cm³) and 3.50 mol NaOH?

23. What mass of H_2 is produced by the reaction of 1.75 g Al with 75.0 mL of 2.50 M HCl? (*Hint:* Use equation 4.16.)

24. Ammonia can be generated by heating together the solids NH_4Cl and $Ca(OH)_2$; $CaCl_2$ and H_2O are also formed. If a mixture containing 15.0 g each of NH_4Cl and $Ca(OH)_2$ is heated, what mass of NH_3 is formed? (*Hint:* Write a balanced equation for the reaction.)

25. What *total* mass of the mixed ZnS/BaSO₄ precipitate (a white pigment called lithopone) is obtained in the reaction of 85.1 mL of 0.150 M BaS with 43.2 mL of 0.355 M ZnSO₄?

$$BaS(aq) + ZnSO_4(aq) \rightarrow ZnS(s) + BaSO_4(s)$$

26. A mixture of 4.800 g H_2 and 36.40 g O_2 is allowed to react.
   (a) Write a balanced equation for this reaction.
   (b) What substances are present after the reaction, and in what quantities?
   (c) Show that the total mass of substances present before and after the reaction is the same.

27. A 60.7-g sample that is 96.2% $K_2Cr_2O_7$ is allowed to react with 318 mL of HCl(aq) with density 1.15 g/cm³ and 30.1% HCl, by mass. What mass of Cl_2 is produced?

$$K_2Cr_2O_7 + 14\ HCl \rightarrow 2\ KCl + 2\ CrCl_3 + 7\ H_2O + 3\ Cl_2(g)$$

Theoretical, actual, and percent yields

28. A laboratory manual calls for 13.0 g C_4H_9OH, 21.6 g NaBr, and 33.8 g H_2SO_4 as reactants in this reaction.

$$C_4H_9OH + NaBr + H_2SO_4 \rightarrow C_4H_9Br + NaHSO_4 + H_2O$$

A student following these directions obtains 16.8 g C_4H_9Br. What are **(a)** the theoretical yield, **(b)** the actual yield, and **(c)** the percent yield of this reaction?

29. Azobenzene, an intermediate in the manufacture of dyes, can be prepared from nitrobenzene by reaction with triethylene glycol in the presence of Zn and KOH. In one reaction 0.10 L of nitrobenzene (d = 1.20 g/cm³) and 0.30 L of triethylene glycol (d = 1.12 g/cm³) yielded 55 g of azobenzene. What was the percent yield of this reaction?

$$2\ C_6H_5NO_2 + 4\ C_6H_{14}O_4 \xrightarrow[\text{KOH}]{\text{Zn}} (C_6H_5N)_2 + 4\ C_6H_{12}O_4 + 4\ H_2O$$

nitro-   triethylene   azobenzene
benzene   glycol

30. How many grams of commercial acetic acid, which contains 97% $C_2H_4O_2$ by mass, must be allowed to react with an excess of PCl_3 to produce 50.0 g of acetyl chloride (C_2H_3OCl), if the reaction has a 70% yield?

$$C_2H_4O_2 + PCl_3 \rightarrow C_2H_3OCl + H_3PO_3 \quad \text{(not balanced)}$$

Simultaneous reactions

31. A mixture is known to contain 38.2% $MgCO_3$ and 61.8% $Mg(OH)_2$, by mass. How much HCl, in grams, is required to dissolve a 413-g sample of this mixture?

$$MgCO_3 + 2\ HCl \rightarrow MgCl_2 + H_2O + CO_2(g)$$
$$Mg(OH)_2(s) + 2\ HCl \rightarrow MgCl_2 + 2\ H_2O$$

32. A natural gas sample consists of 68.2% propane (C_3H_8) and 31.8% butane (C_4H_{10}), by mass. How many moles of $CO_2(g)$ would be produced by the complete combustion of 613 g of this gaseous mixture?

33. An organic liquid is either methyl alcohol (CH_3OH), ethyl alcohol (C_2H_5OH), or a mixture of the two. A 0.220-g sample of the liquid is burned in an excess of O_2 and yields 0.352 g CO_2. Is the liquid a pure alcohol or a mixture?

Consecutive reactions

34. Dichlorodifluoromethane, a widely used refrigerant, can be prepared by the following reactions.

$$CH_4 + 4\ Cl_2 \rightarrow CCl_4 + 4\ HCl$$
$$CCl_4 + 2\ HF \rightarrow CCl_2F_2 + 2\ HCl$$

How many moles of Cl_2 must be consumed to produce 18.2 mol CCl_2F_2?

35. A 25.00-mL sample of liquid benzene, C_6H_6 (d = 0.879 g/cm³), is burned completely, yielding CO_2 and H_2O as the only products. The CO_2 formed is passed into a barium hydroxide solution. What mass of $BaCO_3$ is obtained?

$$CO_2(g) + Ba(OH)_2(aq) \rightarrow BaCO_3(s) + H_2O$$

36. The following process has been used to obtain iodine from oil-field brines in California.

$$NaI + AgNO_3 \rightarrow AgI + NaNO_3$$
$$2\ AgI + Fe \rightarrow FeI_2 + 2\ Ag$$
$$2\ FeI_2 + 3\ Cl_2 \rightarrow 2\ FeCl_3 + 2\ I_2$$

What mass of $AgNO_3$ is required in the first step for every kg I_2 produced in the third step?

37. NaBr can be prepared as follows. How much Fe, in kg, is consumed to produce 5.00×10^3 kg NaBr?

$$Fe + Br_2 \rightarrow FeBr_2$$
$$FeBr_2 + Br_2 \rightarrow Fe_3Br_8 \qquad \text{(not balanced)}$$
$$Fe_3Br_8 + Na_2CO_3 \rightarrow NaBr + CO_2 + Fe_3O_4 \qquad \text{(not balanced)}$$

Oxidation–reduction

38. Balance the following oxidation–reduction equations by the oxidation state change method.

(a) $S_2O_3^{2-} + H_2O + Cl_2(g) \rightarrow SO_4^{2-} + Cl^- + H^+$

(b) $(NH_4)_2Cr_2O_7(s) \rightarrow Cr_2O_3(s) + N_2(g) + H_2O(g)$

(c) $P_4(s) + H^+ + NO_3^- + H_2O \rightarrow H_2PO_4^- + NO(g)$

(d) $MnO_4^- + NO_2^- + H^+ \rightarrow Mn^{2+} + NO_3^- + H_2O$

(e) $S_8(s) + OH^- \rightarrow S^{2-} + S_2O_3^{2-} + H_2O$

(f) $P_4(s) + OH^- + H_2O \rightarrow H_2PO_2^- + PH_3(g)$

★(g) $As_2S_3(s) + H^+ + NO_3^- + H_2O \rightarrow$
$$H_3AsO_4 + S(s) + NO(g)$$

★(h) $C_2H_5OH(aq) + MnO_4^-(aq) \rightarrow$
$$C_2H_3O_2^-(aq) + MnO_2(s) + H_2O + OH^-(aq)$$

39. Write a balanced oxidation–reduction equation for each of the following.

(a) The oxidation of $NH_3(g)$ to $NO(g)$ by $O_2(g)$. The $O_2(g)$ is reduced to $H_2O(g)$.

(b) The oxidation of Fe^{2+} to Fe^{3+} by MnO_4^- in acidic solution; the MnO_4^- is reduced to Mn^{2+}. An acidic solution means that H^+ and/or H_2O may appear as reactants or products.

(c) The reaction of H_2S—a reducing agent—with SO_2—an oxidizing agent. Both reactants are converted to elemental sulfur, and the other product is water.

Industrial chemistry

40. Sodium sulfate, a substance used extensively in the textile industry, can be prepared as follows.

$$2\,NaCl + H_2SO_4 \xrightarrow{\Delta} Na_2SO_4 + 2\,HCl(g)$$

The sulfuric acid is a concentrated aqueous solution having a density of 1.73 g/cm^3 and containing 80% H_2SO_4, by mass. What volume of the sulfuric acid solution must be used to react with 2.50×10^3 kg of NaCl?

41. In a particular plant using the Raschig process for N_2H_4 (refer to Figure 4-8), the mole ratio of NH_3 to NH_2Cl used in the third reaction is 30:1. What is the maximum quantity of NH_3, in kilograms, that is recoverable in the purification steps for every 1.00 kg Cl_2 that reacts?

42. Acrylonitrile, CH_2CHCN, is used in the production of synthetic fibers, plastics, and rubber goods. The Sohio process produces acrylonitrile from propylene, air, and ammonia. The unbalanced equation is

$$CH_2CHCH_3 + NH_3 + O_2 \rightarrow CH_2CHCN + H_2O$$

The process yields 0.73 lb of acrylonitrile (CH_2CHCN) per lb of propylene (CH_2CHCH_3).

(a) Balance the equation for the reaction.

(b) What is the percent yield of the process?

(c) At the percent yield calculated in part (b), what is the minimum mass of NH_3 required to produce 1.00 ton (2000 lb) of acrylonitrile by this process?

Additional Exercises

1. Hydrogen gas is passed over $Fe_2O_3(s)$ at 400°C. Water vapor is formed, together with a black residue—a compound consisting of 72.3% Fe and 27.7% O. Write a balanced equation for this reaction.

2. Balance the following equations by inspection.

(a) $SiCl_4(l) + H_2O(l) \rightarrow SiO_2(s) + HCl(g)$

(b) $CaC_2(s) + H_2O(l) \rightarrow Ca(OH)_2(s) + C_2H_2(g)$

(c) $Na_2HPO_4(s) \rightarrow Na_4P_2O_7(s) + H_2O(l)$

(d) $NCl_3(g) + H_2O(l) \rightarrow NH_3(g) + HOCl(aq)$

(e) $Al_2O_3(s) + H^+(aq) \rightarrow Al^{3+}(aq) + H_2O(l)$

3. Balance the following equations by the oxidation state change method.

(a) $Pb(NO_3)_2(s) \rightarrow PbO(s) + NO_2(g) + O_2(g)$

(b) $S_2O_3^{2-} + MnO_4^- + H^+ \rightarrow SO_4^{2-} + Mn^{2+} + H_2O$

(c) $HS^-(aq) + HSO_3^-(aq) \rightarrow S_2O_3^{2-}(aq) + H_2O$

(d) $Fe^{3+} + (NH_3OH)^+ \rightarrow Fe^{2+} + H^+ + H_2O + N_2O(g)$

(e) $O_2^-(aq) + H_2O \rightarrow OH^-(aq) + O_2(g)$

4. Synthesis gas, a mixture of CO(g) and H_2(g), is a starting material for the synthesis of a variety of organic compounds. It can be produced by allowing a hydrocarbon to react with steam— $H_2O(g)$—or with oxygen—O_2(g). Write a balanced equation to represent the formation of synthesis gas from (a) C_9H_{20}, using steam; (b) C_5H_{12}, using oxygen.

5. Use the terms from page 77, i.e., combustion, combination, decomposition, displacement, and metathesis, to describe the reactions encountered in Exercises 5, 6, 7, 17, 20, and 21. Comment on any difficulties that arise in applying these terms.

6. How many grams of Ag_2CO_3 must have been decomposed if 21.3 g Ag is obtained in this reaction?

$$Ag_2CO_3(s) \xrightarrow{\Delta} Ag(s) + CO_2(g) + O_2(g) \quad \text{(not balanced)}$$

7. Given the reaction

$$3\,Fe(s) + 4\,H_2O(g) \xrightarrow{\Delta} Fe_3O_4(s) + 4\,H_2(g)$$

(a) How many moles of H_2(g) can be produced from 35.8 g Fe and an excess of H_2O(g) [steam]?

(b) How many grams of H_2O would be consumed in the conversion of 315 g Fe to Fe_3O_4?

(c) If 2.18 mol H_2(g) is produced, what mass of Fe_3O_4 must also be produced?

8. Which of the following metals yields the maximum amount of H_2 *per gram* of metal reacting with HCl(aq): Na, Mg, Al, or Zn? [*Hint:* Write equations similar to (4.16).]

9. What volume of 0.750 *M* $CO(NH_2)_2$ solution must be diluted with water to produce 1.00 L of a solution with a concentration of 3.15 mg N per milliliter?

10. A seawater sample has a density of 1.03 g/cm^3 and 2.8% NaCl, by mass. A saturated solution of NaCl in water is 5.45 *M* NaCl. How much water would have to be evaporated from 5.00×10^6 L of the seawater before NaCl would precipitate? (A saturated solution contains the maximum amount of dissolved solute possible.)

11. 99.8 mL of 12.0% KI solution having a density of 1.093 g/mL is added to 96.7 mL of 14.0% $Pb(NO_3)_2$ solution having a density of 1.134 g/mL. What mass of PbI_2 is formed? (Both solution compositions are in percent, *by mass*.)

$$Pb(NO_3)_2(aq) + 2\ KI(aq) \rightarrow PbI_2(s) + 2\ KNO_3(aq)$$

12. Assuming that acetic acid, $HC_2H_3O_2$, is the only substance present that reacts with NaOH, calculate the molar concentration of acetic acid in ordinary vinegar if a 10.00-mL sample of the vinegar requires 19.32 mL of 0.500 *M* NaOH for its neutralization.

$$HC_2H_3O_2(aq) + NaOH(aq) \rightarrow NaC_2H_3O_2(aq) + H_2O(l)$$

13. A chemical plant using the Raschig process obtains 0.299 kg of 98.0% N_2H_4 for every 1.00 kg Cl_2 that reacts with excess NaOH and NH_3. What are the **(a)** theoretical, **(b)** actual, and **(c)** percent yields of *pure* N_2H_4? Refer to the net reaction in Figure 4-8.

14. Which of the following reactions would you expect to go to completion and which to reach a condition of equilibrium? Explain.
   (a) $FeS(s) + 2\ HCl(aq) \rightarrow FeCl_2(aq) + H_2S(g)$
   (b) $2\ NO(g) + O_2(g) \rightarrow 2\ NO_2(g)$
   (c) $Na_2SO_4(aq) + BaCl_2(aq) \rightarrow 2\ NaCl(aq) + BaSO_4(s)$
   (d) $Cl_2(aq) + H_2O(l) \rightarrow HCl(aq) + HOCl(aq)$

*__**15.**__ It is desired to determine the acetylsalicylic acid content of a series of aspirin tablets by titration with NaOH.

$$HC_9H_7O_4(aq) + NaOH(aq) \rightarrow NaC_9H_7O_4(aq) + H_2O(l)$$

Each of the tablets is expected to contain about 0.32 g of acetylsalicylic acid, $HC_9H_7O_4$. What molar concentration of NaOH must be used if titration volumes of about 22 mL are desired. (This procedure ensures good precision in measurements and allows the titration of two samples with the contents of a 50-mL buret.)

*__**16.**__ An 0.155-g sample of an aluminum–magnesium alloy is dissolved in an excess of hydrochloric acid, producing 0.0163 g H_2. What is the % Mg in the alloy? [*Hint:* Write equations similar to (4.16).]

*__**17.**__ The manufacture of ethyl alcohol, C_2H_5OH, yields diethyl ether, $(C_2H_5)_2O$, as a by-product. The *complete combustion* of a 1.005-g sample of the product of this process yields 1.963 g CO_2. What must be the percent, by mass, of C_2H_5OH and of $(C_2H_5)_2O$ in this sample?

*__**18.**__ Hydrogen produced by the decomposition of water has considerable potential as a fuel. A critical problem is in developing a reaction cycle—a series of chemical reactions—that has as its net reaction $2\ H_2O \rightarrow 2\ H_2 + O_2$. Demonstrate that this condition is met by the Fe/Cl cycle.

$$FeCl_2 + H_2O \xrightarrow{650°C} Fe_3O_4 + HCl + H_2$$

$$Fe_3O_4 + HCl + Cl_2 \xrightarrow{200°C} FeCl_3 + H_2O + O_2$$

$$FeCl_3 \xrightarrow{420°C} FeCl_2 + Cl_2$$

*__**19.**__ $CaCO_3(s)$ reacts with HCl(aq) to form H_2O, $CaCl_2(aq)$ and $CO_2(g)$. If a 45.0-g sample of $CaCO_3(s)$ is added to 1.25 L HCl(aq) that has a density of 1.13 g/cm^3 and contains 25.7% HCl, by mass, what will be the molarity of HCl in the solution remaining after the reaction is completed? (Assume that the solution volume remains constant.)

*__**20.**__ Under appropriate conditions, copper sulfate, potassium chromate, and water react to form a product containing Cu^{2+}, CrO_4^{2-}, and OH^- ions. Analysis of the compound yields 48.7% Cu^{2+}, 35.6% CrO_4^{2-}, and 15.7% OH^-.
   (a) Derive the empirical formula of the compound.
   (b) Write a plausible equation for the reaction.

Self-Test Questions

For questions 1 through 8 select the single item that best completes each statement.

1. For the reaction $2\ H_2S + SO_2 \rightarrow 3\ S + 2\ H_2O$
(a) 3 mol S is produced per mole of H_2S.
(b) 1 mol SO_2 is consumed per mole of H_2S.
(c) 1 mol H_2O is produced per mole of H_2S.
(d) the number of moles of products is independent of how many moles of reactants are used.

2. 1.0 mol of calcium cyanamide ($CaCN_2$) and 1.0 mol of water are allowed to react.

$$CaCN_2 + 3\ H_2O \rightarrow CaCO_3 + 2\ NH_3$$

The number of moles of NH_3 produced is (a) 3.0; (b) 2.0; (c) 1.0; (d) less than 1.0.

3. If the reaction of 1.00 mol $NH_3(g)$ and 1.00 mol $O_2(g)$

$$4\ NH_3(g) + 5\ O_2(g) \rightarrow 4\ NO(g) + 6\ H_2O(l)$$

is carried to completion, (a) all the $O_2(g)$ is consumed; (b) 4.0 mol NO(g) is produced; (c) 1.5 mol $H_2O(l)$ is produced; (d) none of these.

4. To prepare a solution that is 0.50 M KCl starting with 100 mL of 0.40 M KCl, (a) add 0.75 g KCl; (b) add 20 mL of water; (c) add 0.10 mol KCl; (d) evaporate 10 mL of water.

5. To complete the titration of 10.00 mL 0.0500 M NaOH

$$2\ NaOH(aq) + H_2SO_4(aq) \rightarrow Na_2SO_4(aq) + 2\ H_2O$$

requires (a) 50.0 mL of 0.0100 M H_2SO_4; (b) 25.0 mL of 0.0100 M H_2SO_4; (c) 100.0 mL of 0.0100 M H_2SO_4; (d) 10.0 mL of 0.100 M H_2SO_4.

6. The underlined atom with an oxidation state of -2 is (a) $\underline{S}O_4^{2-}$; (b) $\underline{O}_2$, (c) $\underline{N}_2H_4$; (d) $\underline{S}_2O_3^{2-}$.

7. In the reaction

$$4\ Fe^{2+} + O_2 + 4\ H^+ \rightarrow 4\ Fe^{3+} + 2\ H_2O,$$

the *reducing* agent is (a) Fe^{2+}; (b) O_2; (c) H^+; (d) Fe^{3+}.

8. In the reaction of 2.0 mol CCl_4 with an excess of HF, 1.70 mol CCl_2F_2 is obtained.

$$CCl_4 + 2\ HF \rightarrow CCl_2F_2 + 2\ HCl$$

(a) The theoretical yield is 1.70 mol CCl_2F_2.
(b) The theoretical yield is 1.0 mol CCl_2F_2.
(c) The percent yield of the reaction is 85%.
(d) The theoretical yield of the reaction depends on how large an excess of HF is used.

9. Write a balanced chemical equation to represent each of the following reactions.

  (a) The decomposition, by heating, of solid mercury(II) nitrate to produce pure liquid mercury, nitrogen dioxide gas, and oxygen gas.

  (b) The reaction of aqueous sodium carbonate solution with aqueous hydrochloric acid (hydrogen chloride) to produce water, carbon dioxide gas, and aqueous sodium chloride.

  (c) The complete combustion of malonic acid, a compound with 34.62% C, 3.88% H, and 61.50% O, by mass.

10. What volume of 0.0102 M $Ba(OH)_2(aq)$, in milliliters, is required to titrate 10.00 mL of 0.0526 M $HNO_3(aq)$?

$$2\ HNO_3(aq) + Ba(OH)_2(aq) \rightarrow Ba(NO_3)_2(aq) + 2\ H_2O$$

11. How many grams of Na must react with 125 mL H_2O to produce an NaOH solution that is 0.250 M? (Assume that the final solution volume is 125 mL.)

$$2\ Na(s) + 2\ H_2O(l) \rightarrow 2\ NaOH(aq) + H_2(g)$$

12. A nearly 100% yield is essential for a chemical reaction that is to be used to *analyze* a chemical compound, but is almost never expected for a reaction that is to be used to synthesize a compound. Explain why this is so.

5

Gases

Familiar samples of bulk matter exist as solids, liquids, or gases. This observation hardly needs mention; we understand it intuitively. However, we do need to describe the properties of these physical forms, or states, of matter in some detail.

The simplest of the three states of matter to understand is the gaseous state. Most of the ideas presented in this chapter predate the twentieth century, and the development of modern chemistry paralleled closely the growth of knowledge of the gaseous state. The behavior of gases figured prominently in the discovery of the laws of chemical combination and in the testing of Dalton's atomic theory. The stoichiometry of chemical reactions can be expanded in some interesting ways with a few key ideas about gases. Finally, the study of gases provides a basis for one of the great theories of science, the kinetic molecular theory. This theory will provide us with new insights, particularly of the concept of temperature.

5-1 Properties of a Gas

Gases may be characterized in many ways. All gases expand to fill and assume the shapes of their containers. All gaseous substances diffuse into one another and mix in all proportions; that is, all gaseous mixtures are homogeneous solutions. Gases are invisible in the sense that there are no visible particles of a gas. Some gases are colored, such as gaseous chlorine (greenish yellow), bromine (brownish red), and iodine (violet); some are combustible, such as hydrogen; and some are chemically inert, such as helium and neon.

Four basic properties determine the physical behavior of a gas: amount of gas, gas volume, temperature, and pressure. From numerical values of three of these it is possible to calculate a value of the fourth. This is done through a mathematical equation called an **equation of state.** In principle at least, many other properties of a gas can be calculated from an equation of state. We have already discussed to some extent the properties amount, volume, and temperature. A brief discussion of gas pressure is presented next.

5-2 Gas Pressure

That a rubber balloon expands when inflated with a gas is a familiar observation, but what keeps the balloon distended? A plausible hypothesis is that molecules of a gas are in constant motion, colliding with one another and with the walls of their container. In

FIGURE 5-1
The concept of liquid
pressure.

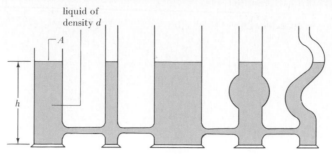

The pressure exerted by the liquid in the cylinder on the left is calculated in the text and shown, through equation (5.2), to depend only on the height of the liquid column (*h*) and the density of the liquid (*d*). All of the interconnected vessels pictured here fill to the same height; and the liquid pressures are the same, despite the different shapes and volumes of the containers.

collisions with the walls, a force is imparted. It is this force that keeps a balloon distended. To measure the total force imparted by a gas is not a simple matter, but gas *pressure* can be measured rather easily. Pressure is a force per unit area, that is, a force divided by the total area over which the force is exerted.

$$P = \frac{F}{A} \tag{5.1}$$

Liquid Pressure.   The pressure of a gas is most often measured *indirectly* by comparison with a liquid pressure. The concept of liquid pressure is illustrated in Figure 5-1 for a liquid with density *d,* contained in a cylinder with cross-sectional area *A* filled to height *h*. Equation (5.2) shows that *the pressure exerted by a liquid depends only on the height of the liquid column and the density of the liquid*. To establish this fact, recall the following: Weight (*W*) is a force. Weight (*W*) and mass (*m*) are proportional, with the proportionality constant being the acceleration due to gravity (*g*). The mass of a liquid is equal to the product of its density and volume (*m* = *d* · *V*). The volume (*V*) of a cylinder is equal to the product of its height (*h*) and cross-sectional area (*A*).

$$P = \frac{F}{A} = \frac{W}{A} = \frac{mg}{A} = \frac{gdV}{A} = \frac{dgAh}{A} = ghd \tag{5.2}$$

Measurement of Gas Pressure.   The most familiar gas is air. Actually, air is a mixture of several gases—principally nitrogen (78.08%), oxygen (20.95%), argon (0.93%), and carbon dioxide (0.03%). Life on the surface of earth exists at the bottom of a "sea" of air called the atmosphere, and all objects on earth are subjected to a pressure produced by this blanket of air. In 1643, Torricelli constructed the device pictured in Figure 5-2 to measure the pressure of the atmosphere. This device is called a **mercury barometer.**

As pictured in Figure 5-2a, if a tube having *both ends open* is placed upright in a container of mercury, the mercury levels inside and outside the tube are the same. To create the situation in Figure 5-2b, first a long glass tube (say, about 1 m long) is *sealed at one end* and filled with Hg(l). Next, the open end is covered while the tube is inverted into a container of mercury. Then this end is reopened. The mercury level in the tube does not drop to that in the outside container. Instead, it falls to a certain height and remains there. Something must maintain the mercury at a greater height inside the tube than outside. Originally, explanations of this phenomenon involved forces within the tube. We now understand that these forces exist *outside* the tube.

In the open-end tube (Figure 5-2a) the atmosphere exerts the same pressure on the

One early hypothesis
was that the top of the
mercury column was
attached to the top of
the tube by an invisible
thread.

FIGURE 5-2
Measurement of
atmospheric pressure
with a mercury
barometer.

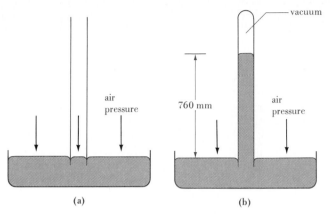

Black arrows represent the pressure exerted by the atmosphere.
(a) The mercury levels are equal inside and outside the open-end tube.
(b) A column of mercury 760 mm high is maintained in the closed-end tube.

surface of the mercury, both inside and outside the tube; the liquid levels are equal. With the closed-end tube (Figure 5-2b) there is no air inside the tube above the mercury. (Only a trace of mercury vapor is present.) The atmosphere exerts a force on the surface of the mercury that is transmitted through the liquid, holding mercury up in the tube. The column of liquid in the tube exerts a downward pressure that depends on its height (and the density of liquid mercury). For a particular height the pressure at the bottom of the mercury column and that of the atmosphere are equal and the column is maintained.

The height of mercury in a barometer is not constant, but varies with the location and atmospheric conditions. The **standard atmosphere** is defined as the pressure exerted by a mercury column of 760 mm height under conditions where the density of the mercury is 13.5951 g/cm^3 and the acceleration due to gravity $g = 9.80665$ m/s^2. This statement relates two useful units of pressure, the standard atmosphere (atm) and the millimeter of mercury (mmHg).

Can you see from equation (5.2) why it is necessary to specify the density of Hg and a value of g in the precise definition of a standard atmosphere?

1 atm = 760 mmHg

To honor Torricelli the pressure unit **torr** is also used. The torr is defined as exactly $\frac{1}{760}$ of a standard atmosphere.

760 torr = 1 atm

Thus, the pressure units torr and mmHg can be used interchangeably.

Mercury is a relatively rare, expensive, and rather poisonous liquid. Why use it rather than water as the liquid in a barometer? The answer lies in the extreme height necessary for a water barometer, as calculated in Example 5-1.

Example 5-1   What is the height of a water column, in meters, that could be maintained by standard atmospheric pressure?

Solution.   We can rephrase the question: What is the height of a column of water that exerts the same pressure as a column of mercury 76.0 cm (760 mm) high?

$$\text{pressure of Hg column} = gh_{Hg}d_{Hg} = g \times 76.0 \text{ cm} \times 13.6 \text{ g/cm}^3$$
$$\text{pressure of H}_2\text{O column} = gh_{H_2O}d_{H_2O} = g \times h_{H_2O} \times 1.00 \text{ g/cm}^3$$
$$\not{g} \times h_{H_2O} \times 1.00 \text{ g/cm}^3 = \not{g} \times 76.0 \text{ cm} \times 13.6 \text{ g/cm}^3$$
$$h_{H_2O} = 76.0 \text{ cm} \times \frac{13.6}{1.00} = 1.03 \times 10^3 \text{ cm} = 10.3 \text{ m}$$

SIMILAR EXAMPLE: Exercise 2.

FIGURE 5-3
Pumping water by
suction.

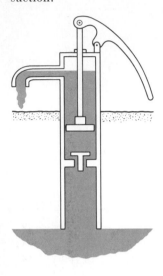

Example 5-1 helps us to understand the operation of an old-fashioned suction pump for drawing water from a well. As illustrated through Figure 5-3, the pump action is used to evacuate air from a cylindrical pipe in the well. Atmospheric pressure, acting on the surface of the water in the well, *pushes* a column of water up the evacuated pipe. Even if all the air in the pipe could be evacuated (which it cannot), the column of water could not be raised higher than 10.3 m. The use of a straw for drinking liquids is based on the same principle as the suction pump.

One cannot usually introduce a mercury barometer directly into a container of a gas. A device that is commonly used to measure gas pressures in the laboratory is a **manometer.** The principle of an open-end manometer is illustrated in Figure 5-4. As long as the gas pressure being measured and the prevailing atmospheric (barometric) pressure are equal, the heights of the mercury columns in the two arms of the manometer are equal. A difference in height of the two arms means a difference between the gas pressure and barometric pressure. The device pictured in Figure 5-5, a closed-end manometer, is useful for measuring low gas pressures.

Example 5-2 What is the gas pressure, P_{gas}, if the conditions in Figure 5-4b are that barometric pressure is 748.2 mmHg and $\Delta P = 25.0$ mm?

Solution

$$P_{gas} = P_{bar.} + \Delta P = 748.2 \text{ mmHg} + 25.0 \text{ mmHg} = 773.2 \text{ mmHg}$$

SIMILAR EXAMPLE: Exercise 3.

Example 5-3 What is the gas pressure, P_{gas}, if the manometer in Figure 5-4c is filled with glycerol ($d = 1.26$ g/cm^3), P_{bar} is 762.4 mmHg, and ΔP is 8.2 mm?

Solution.   First, we must convert ΔP, expressed as *8.2 mm of glycerol,* to an equivalent height of mercury. This is done in exactly the same way as in Example 5-1.

$$h_{Hg} = 8.2 \text{ mm} \times \frac{1.26}{13.6} = 0.76 \text{ mmHg}$$

In the remaining calculation, $\Delta P = -0.76$ mmHg. (P_{gas} is less than $P_{bar.}$.)

$$P_{gas} = P_{bar.} + \Delta P = 762.4 \text{ mmHg} - 0.76 \text{ mmHg} = 761.6 \text{ mmHg}$$

SIMILAR EXAMPLES: Exercises 3, 4.

FIGURE 5-4
Measurement of gas
pressure with an
open-end manometer.

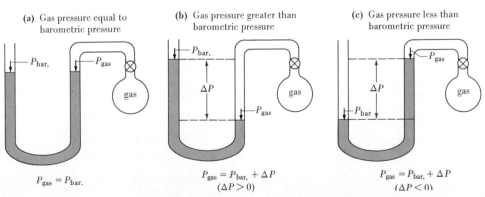

(a) Gas pressure equal to barometric pressure

(b) Gas pressure greater than barometric pressure

(c) Gas pressure less than barometric pressure

$P_{gas} = P_{bar.}$

$P_{gas} = P_{bar.} + \Delta P$
$(\Delta P > 0)$

$P_{gas} = P_{bar.} + \Delta P$
$(\Delta P < 0)$

The possible relationships between a measured gas pressure and barometric pressure are pictured here.

FIGURE 5-5
Measurement of gas
pressure with a
closed-end manometer.

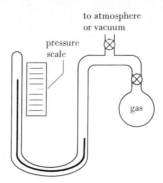

When the manometer is open to the atmosphere, air pressure pushes the mercury level to the top of the closed end. When the manometer is connected to a source of high vacuum, the mercury level in the closed end falls until the levels in the two arms of the U-tube become equal. This is the condition of "zero pressure." When the manometer is connected to a container of gas, the mercury level in the closed end rests at a level that can be read on a scale. This is the measured pressure. Manometers of this type are generally used in the pressure range from about 5 to 300 mmHg. Of course, the longer the closed arm of the manometer, the higher the pressure that can be measured.

Units of Pressure.   Many different units are used to express pressure. This is one quantity for which the adoption of a single unit, as proposed by the SI system, might eliminate some confusion. Some units of pressure are based on the height of a liquid column, some on a mass per unit area, and some on an actual force (weight) per unit area. In the set of equalities that follows, several different ways of expressing standard atmospheric pressure are given. The first three, shown in color, are the units we have introduced to this point and will continue to use throughout this text. The fourth and fifth units are based on a mass (rather than a force) per unit area. The unit pounds per square inch (psi) is employed most commonly in engineering work. The next three units, shown in boldface type, are the ones preferred in the SI system. The pascal is named after Blaise Pascal, a seventeenth-century scientist who made significant contributions to our understanding of pressure. In time, SI units may become the common ones for expressing pressure, but presently their use is limited. The unit millibar is commonly used by meteorologists.

$$\overset{(1)}{1 \text{ standard atmosphere (1 atm)}} = \overset{(2)}{760 \text{ mmHg}} = \overset{(3)}{760 \text{ torr}} = \overset{(4)}{14.7 \text{ lb/in.}^2 \text{ (psi)}}$$

$$= \overset{(5)}{1.0333 \text{ kg/cm}^2} = \overset{(6)}{\mathbf{101{,}325 \text{ newtons/m}^2 \text{ (N m}^{-2})}}$$

$$= \overset{(7)}{\mathbf{101{,}325 \text{ pascals (Pa)}}} = \overset{(8)}{\mathbf{101.325 \text{ kilopascals (kPa)}}}$$

$$= \overset{(9)}{1.01325 \text{ bars}} = \overset{(10)}{1013.25 \text{ millibars (mb)}} \qquad (5.3)$$

5-3 The Simple Gas Laws

Boyle's Law.   Of the several relationships among gas variables, the first to be discovered was the one between gas pressure and volume. This was accomplished in 1662 by Robert Boyle. Boyle found that

The volume of a fixed amount of gas maintained at a constant temperature is inversely proportional to the gas pressure.

The meaning of this statement is suggested through Figure 5-6.

FIGURE 5-6
Relationship between
gas volume and
pressure—Boyle's law.

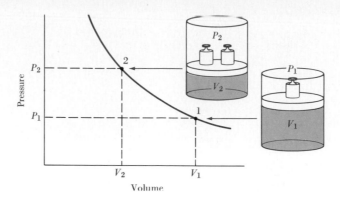

When temperature and the amount of gas are held constant, a doubling of the pressure causes the volume to decrease to one half its original value. The situation here is like operating a football pump with the needle plugged. The handle can be depressed slightly, and the air in the pump compressed to some extent. But it becomes increasingly difficult to reduce the gas volume further as more and more pressure (force per unit area) is required.

The gas pictured in Figure 5-6 is contained in a cylinder closed off by a freely moving "weightless" piston. The pressure of the gas is fixed by the total weight placed on top of the piston. [This weight (a force), divided by the area of the piston, yields the gas pressure.] If the weight on the piston is doubled, the pressure doubles and the gas volume decreases to one half of its original value; and so on.

Expressed mathematically, the inverse relationship between pressure and volume is

$$P \propto \frac{1}{V} \qquad \text{or} \qquad P = \frac{a}{V} \qquad \text{or} \qquad PV = a \quad \text{(a constant)} \tag{5.4}$$

Equation (5.4) shows that the product of the pressure and volume of a fixed amount of gas at a constant temperature is a constant (a). The graph of the relationship $PV = a$ shown in Figure 5-6 is of a form called an equilateral (or rectangular) hyperbola.

Example 5-4   The volume of a large, irregularly shaped, closed tank is determined as follows. The tank is first evacuated, and then it is connected to a 50.0-L cylinder of compressed nitrogen gas. The gas pressure in the cylinder, originally at 21.5 atm, falls to 1.55 atm after it is connected to the evacuated tank. What is the volume of the tank? (See Figure 5-7.)

FIGURE 5-7
An application of
Boyle's law—Example
5-4 visualized.

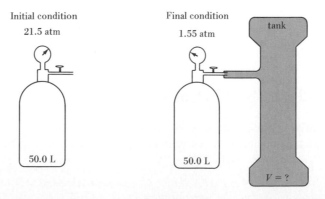

The final volume is the cylinder volume (50.0 L) plus that of the tank. The amount of gas remains constant when the cylinder is connected to the evacuated tank but the pressure drops from 21.5 to 1.55 atm.

Solution
Method 1.   Use equation (5.4).

Equation (5.4) is written for the initial (i) and final (f) conditions.

$$P_i V_i = a = P_f V_f$$

Now solve for the final volume, V_f.

$$V_f = V_i \times \frac{P_i}{P_f} = 50.0 \text{ L} \times \frac{21.5 \text{ atm}}{1.55 \text{ atm}} = 694 \text{ L}$$

Of this volume, 50.0 L is that of the cylinder. The volume of the tank is 694 L − 50.0 L = 644 L.

Method 2.   A "common sense" approach.

In the application of Boyle's law, a final gas volume will always be equal to an initial volume multiplied by a ratio of pressures (or a final pressure will be equal to an initial pressure multiplied by a ratio of volumes). We need to form the ratio of pressures that will cause the volume to change in the observed manner—to increase. Only the ratio 21.5 atm/1.55 atm will produce a V_f larger than V_i.

$$V_f = 50.0 \text{ L} \times \frac{21.5 \text{ atm}}{1.55 \text{ atm}} = 694 \text{ L}$$

Volume of tank = 694 L − 50.0 L = 644 L

SIMILAR EXAMPLES: Review Problems 3, 4; Exercises 5, 6.

Charles' Law.   The relationship between gas volume and temperature was discovered by the French physicist Charles in 1787 and, independently, by Gay-Lussac, who published it in 1802.

Figure 5-8 pictures a fixed amount of gas confined in a cylinder. The pressure is held

FIGURE 5-8
Gas volume as a
function of Celsius
temperature.

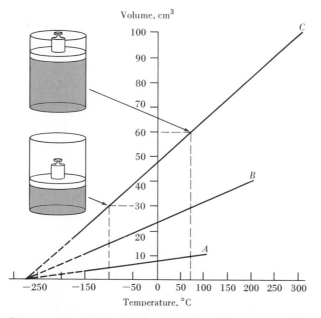

Of many possible starting conditions, three are represented here: A, 10 cm³ of gas at 1 atm and 100°C; B, 40 cm³ of gas at 1 atm and 200°C; C, 100 cm³ of gas at 1 atm and 300°C. When cooled from point C (300°C) to about 70°C, the gas volume decreases from 100 cm³ to 60 cm³. And, as pictured here, in the temperature interval from about 70°C to −100°C the gas volume decreases by half, i.e., from 60 cm³ to 30 cm³. The volume appears to become zero at about −270°C.

FIGURE 5-9
Gas volume as a function of kelvin temperature.

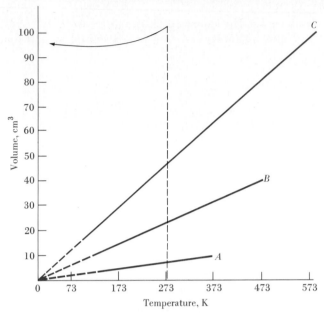

The vertical axis of Figure 5-8 (broken line) has been shifted 273.15° to the left. Note how the points *A*, *B*, and *C*, which were at 100, 200, and 300°C in Figure 5-8, now appear at 373, 473, and 573 K, respectively.

constant while the temperature is varied. The volume of gas increases as the temperature is raised or decreases as the temperature is lowered; the relationship is linear (straight line). Three possibilities are indicated in the figure.

A common feature of the lines in Figure 5-8 is the point of intersection with the temperature axis. Although different at every other temperature, the gas volumes for the three cases shown all appear to reach a value of zero at some temperature below −270°C (actually at −273.15°C). The temperature −273.15°C corresponds to that at which the volume of a hypothetical gas would become zero.* This is the **absolute zero of temperature.**

According to the SI convention, a kelvin temperature is referred to simply as kelvin and K, not degrees kelvin and °K. Thus, the kelvin temperature corresponding to 0°C is 273.15 K.

If the volume axis of Figure 5-8 is shifted 273.15°C to the left, as shown in Figure 5-9, the straight lines then pass through the origin of the new axes. The origin corresponds to the hypothetical zero volume at the absolute zero of temperature. The further effect of shifting the volume axis in this way is that we must add 273.15 degrees to each temperature value. This leads to the following relationship between Celsius and **kelvin** or **absolute** temperature.

$$T \text{ (K)} = t \text{ (°C)} + 273.15 \tag{5.5}$$

Thus, Charles' law may be stated in this way.

The volume of a fixed amount of gas at constant pressure is directly proportional to the kelvin (absolute) temperature.

Mathematically, this may be written as

$$V \propto T \quad \text{or} \quad V = bT \quad \text{(where } b \text{ is a constant)} \tag{5.6}$$

*All gases condense to liquids and solids before the temperature reaches absolute zero; and when we speak of the volume of a gas we mean the free volume among the gas molecules, not the volume of the molecules themselves. Thus, the hypothetical gas referred to here is one whose molecules are point masses and which does not condense to a liquid or solid.

From equation (5.6) we see that doubling the kelvin (absolute) temperature of a gas causes its volume to double. (Increasing the temperature of a gas from 1°C to 2°C or from 1°F to 2°F, of course, would not cause its volume to double.)

Example 5-5   A 177-cm^3 sample of a gas at 10.0°C is heated to 100.0°C while the pressure is held constant at 1 atm. What volume does the gas occupy at 100.0°C?

Solution
Method 1.   Use equation (5.6).
    Equation (5.6) is written for the initial (i) and final (f) conditions.

$$\frac{V_i}{T_i} = b = \frac{V_f}{T_f}$$

Now solve for the final volume, V_f.

$$V_f = V_i \times \frac{T_f}{T_i} = 177 \text{ cm}^3 \times \frac{(273 + 100) \text{ K}}{(273 + 10) \text{ K}}$$

$$= 177 \text{ cm}^3 \times \frac{373 \text{ K}}{283 \text{ K}} = 233 \text{ cm}^3$$

Method 2.   A "common sense" approach.
    The final gas volume is an initial volume multiplied by a ratio of kelvin temperatures. Only the ratio 373 K/283 K will produce a final volume larger than the initial volume.

$$V_f = 177 \text{ cm}^3 \times \frac{373 \text{ K}}{283 \text{ K}} = 233 \text{ cm}^3$$

SIMILAR EXAMPLES: Review Problems 5, 6.

Standard Conditions of Temperature and Pressure.   Because gas properties depend on temperature and pressure, it is convenient to specify a particular temperature and pressure at which comparisons can be made. The standard temperature for gases is defined as 0°C = 273.15 K and the standard pressure as 1 atm = 760 mmHg. Standard conditions are sometimes abbreviated as STP (or SC).

5-4 The Gas Laws and Development of the Atomic Theory

Avogadro's Hypothesis.   Verification of Dalton's atomic theory seemed to come in 1808 when Gay-Lussac reported on the combining volumes of gases. Gay-Lussac found that when gases react with one another they do so by volumes that are in the ratios of *small whole numbers*. For example, nitrogen and oxygen form three different compounds, with the combining ratios of N_2 to O_2 *by volume*, being 2:1, 1:1, and 1:2. These simple ratios do not exist for reactions involving solids and liquids, nor do such ratios exist, even for gases, if masses are compared rather than volumes.

One explanation of the **law of combining volumes** was that equal volumes of different gases, under identical conditions of temperature and pressure, contain equal numbers of particles (atoms). If chemical combination involves the union of atoms in simple numerical ratios, the combining volumes should also be in simple numerical ratios. There were some valid objections to this line of reasoning, however. Dalton argued that in the reaction of hydrogen and oxygen to form water the number of particles of water (OH) formed should be the same as the number of atoms of hydrogen (H) and of oxygen (O) reacting. If the "equal volumes–equal numbers" hypothesis were correct, the ratio of volumes of reactants and products should have been 1:1:1. By experiment the ratio proved to be *two* volumes of hydrogen to *one* of oxygen and *two* of steam—2:1:2.

FIGURE 5-10
Formation of water—
actual observation and
Avogadro's hypothesis.

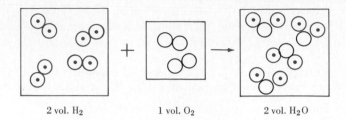

2 vol. H_2 1 vol. O_2 2 vol. H_2O

Avogadro published a paper in 1811 demonstrating that Gay-Lussac's law and Dalton's theory could be reconciled if one made *two* assumptions.

1. Equal volumes of different gases, under identical conditions of temperature and pressure, contain equal numbers of particles.
2. In many gases the ultimate particles are *molecules* consisting of a number of atoms joined together.

Avogadro proposed that hydrogen and oxygen both exist as *diatomic molecules,* that is, as H_2 and O_2, and that water has *two* H atoms for every O—H_2O! In the reaction of hydrogen and oxygen, the O_2 molecules split into half-molecules (atoms). The H_2 molecules and O half-molecules produce the same number of water molecules (H_2O) as H_2 molecules reacted. From two volumes of H_2 and one of O_2, two volumes of steam should form. This reasoning is outlined in Figure 5-10.

Cannizzaro's Work.   The scientific community was not ready for such bold assumptions as Avogadro's. His hypothesis was little used until it was promoted by Cannizzaro a half-century later. Cannizzaro reasoned as follows.

Take the atomic weight of hydrogen to be exactly 1. Assume that hydrogen exists as diatomic molecules, H_2. The molecular weight of hydrogen becomes exactly 2. Next, determine the volume of hydrogen gas that, under certain conditions of temperature and pressure, weighs exactly 2 g. The conditions chosen were 0°C and 1 atm (STP), and the volume proved to be 22.4 L. Now, 22.4 L of some other gas at STP should contain the same number of molecules as does 22.4 L of hydrogen. The ratio of the mass of 22.4 L of this gas to the mass of 22.4 L of hydrogen is the ratio of the molecular weight of the gas to that of H_2. This procedure is illustrated in Figure 5-11 for hydrogen and oxygen. By experiment, 22.4 L of oxygen at STP is found to weigh 32.00 g. The molecular weight of O_2 is 16 times as great as the molecular weight of H_2. Assuming the formula O_2, since the atomic weight of H is taken to be 1, the atomic weight of O is 16.

FIGURE 5-11
Cannizzaro's method illustrated.

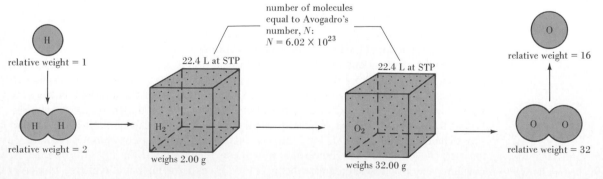

TABLE 5-1
Cannizzaro's method—the atomic weight of nitrogen

Substance	Mol. wt. (relative to H = 1)	Nitrogen, % by mass	Relative mass of N per molecule
hydrogen	2	—	—
ammonia	17	82.5	14
nitrous oxide	44	63.7	28
nitric oxide	30	46.7	14
nitrogen dioxide	46	30.4	14
nitrogen gas	28	100.0	28

This same procedure can be used to determine the atomic weights of other elements, as suggested through Table 5-1 for a series of gaseous nitrogen compounds. We may reason as follows.

1. 22.4 L of ammonia gas at STP weighs 17 g.
2. Ammonia consists of 82.5% N, by mass.
3. The relative mass of nitrogen in ammonia is $17 \times 0.825 = 14$.
4. 22.4 L of nitrous oxide at STP weighs 44 g.
5. Nitrous oxide consists of 63.7% N, by mass.
6. The relative mass of nitrogen in nitrous oxide is $44 \times 0.637 = 28$.
7. And so on.

The relative masses of nitrogen in all the molecules in Table 5-1 are either 14 or a multiple of 14. A plausible conclusion is that the atomic weight of nitrogen is 14, and that there is one N atom per molecule of ammonia, two per molecule of nitrous oxide, and so on.

Avogadro's hypothesis can be stated in two ways.

1. Equal volumes of different gases compared at the *same temperature and pressure* contain equal numbers of molecules.
2. Equal numbers of molecules of different gases compared at the *same temperature and pressure* occupy equal volumes.

Also, the volume of a gas at a fixed temperature and pressure is directly proportional to the amount of gas (i.e., to the number of molecules). If the amount of gas is doubled, the volume doubles; and so on. A mathematical statement of this fact is

$$V \propto n \quad \text{and} \quad V = cn \tag{5.7}$$

where c is a constant and n is the amount of gas.

At STP the number of molecules contained in 22.4 L of a gas is *1 mol* (see Figure 5-11). This quantity, 22.4 L of a gas at STP, is often referred to as the **molar volume of a gas.** Avogadro's law and statements derived from it, such as the molar volume of 22.4 L at STP, apply *only to gaseous substances*. There is no similar relationship dealing with liquids or solids.

Molar volume of a gas at STP visualized. The volume of the basketball is 7.5 L; the soccer ball, 6.0 L; and the football, 4.4 L. [See also, F. H. Jardine, "3 Basketballs = 1 Mole of Ideal Gas at STP," *J. Chem. Educ.*, **54**, 112 (1977).]

5-5 The Ideal Gas Equation

The simple gas laws are stated again below.

Boyle's law: $V \propto \dfrac{1}{P}$ (*n* and *T* constant)

Charles' law: $V \propto T$ (*n* and *P* constant)

Avogadro's law: $V \propto n$ (*P* and *T* constant)

Intuitively, it would seem that the volume of a gas should be directly proportional to the amount of gas and to temperature and inversely proportional to pressure. That is,

$$V \propto \frac{nT}{P} \qquad and \qquad V = \frac{RnT}{P} \qquad or \qquad PV = nRT \tag{5.8}$$

Mathematical proof of equation (5.8) is beyond the scope of this discussion,* but it has been amply demonstrated by experiment that any gas that obeys the three simple gas laws obeys equation (5.8) as well. Such a gas is said to be an **ideal gas,** and equation (5.8) is known as the **ideal gas equation.** Real gases can only approach the behavior implied by the ideal gas equation, as we shall learn shortly. Under suitable conditions, however, enough real gases do approach this behavior to make the equation very useful.

Before equation (5.8) can be applied to specific situations, a numerical value is needed for R, called the **gas constant.** One of the simplest means of establishing this value is to substitute into equation (5.8) the molar volume at STP, 22.414 L.

We will generally express R, which has the same value for all gases, to three significant figures.

0.0821 L atm mol^{-1} K^{-1}

Alternative units for R are required in some of its applications.

$$R = \frac{PV}{nT} = \frac{1 \text{ atm} \times 22.414 \text{ L}}{1 \text{ mol} \times 273.15 \text{ K}} = 0.082057 \frac{\text{L atm}}{\text{mol K}} \tag{5.9}$$

In using the ideal gas equation, you should note the following.

*Mathematically, the three proportionalities cannot simply be combined because each is stated for different conditions. For example, Boyle's law requires amount of gas and temperature to be held constant, but these are not the variables to be held constant in Charles' law. Several derivations of the ideal gas equation from the simple gas laws are possible.

1. There are five terms in the equation—*P, V, n, R,* and *T.* Four of them must be known; the equation is solved for the fifth. [In the examples that follow, all five terms are stated first (enclosed in brackets). This assists in identifying the unknown.]

2. Each term must be expressed in the proper units before substitution into the equation. The value of *R* serves as a guide in this. If, as is usually the case, its units are liter atmospheres per mole per kelvin (L atm mol^{-1} K^{-1}), the unit for pressure must be atmospheres; for volume, liters; for amount of gas, moles; and for temperature, kelvins.

Example 5-6   What is the volume occupied by 13.7 g Cl_2(g) at 45°C and 745 mmHg?

Solution

$$P = 745 \text{ mmHg} \times \frac{1 \text{ atm}}{760 \text{ mmHg}} = \frac{745}{760}\text{atm} = 0.980 \text{ atm}$$

$V = ?$ (this is the unknown)

$$n = 13.7 \text{ g Cl}_2 \times \frac{1 \text{ mol Cl}_2}{70.9 \text{ g Cl}_2} = 0.193 \text{ mol Cl}_2$$

$R = 0.0821$ L atm mol^{-1} K^{-1}

$T = 45°C + 273 = 318$ K

PV = nRT

Divide both sides by *P.*

$$P\frac{V}{P} = \frac{nRT}{P} \quad and \quad V = \frac{nRT}{P}$$

$$V = \frac{0.193 \text{ mol} \times 0.0821 \text{ L atm mol}^{-1} \text{ K}^{-1} \times 318 \text{ K}}{0.980 \text{ atm}}$$

$$= 5.14 \text{ L}$$

To check the cancellation of units we have generally looked for the same unit in the numerator and the denominator of a setup, for example, atm/atm = 1. Here we need to note that a unit such as mol^{-1} is the same as 1/mol. Thus, mol × mol^{-1} = 1. Also, K^{-1} × K = 1.

SIMILAR EXAMPLES: Review Problems 8, 9; Exercises 11, 12.

Example 5-7   A gas cylinder has a volume of 34.9 L. The cylinder is filled with nitrogen gas to a pressure of 5.1 atm at 20°C. What mass of N_2 is contained in the cylinder?

Solution

$P = 5.1$ atm

$V = 34.9$ L

$n = ?$ (this is the unknown)

$R = 0.0821$ L atm mol^{-1} K^{-1}

$T = 20°C + 273 = 293$ K

$$n = \frac{PV}{RT}$$

$$n = \frac{5.1 \text{ atm} \times 34.9 \text{ L}}{0.0821 \text{ L atm mol}^{-1} \text{ K}^{-1} \times 293 \text{ K}} = 7.4 \text{ mol}$$

$$\text{no. g N}_2 = 7.4 \text{ mol N}_2 \times \frac{28.0 \text{ g N}_2}{1 \text{ mol N}_2}$$

$$= 2.1 \times 10^2 \text{ g N}_2$$

SIMILAR EXAMPLES: Review Problems 8, 9; Exercises 11, 12.

FIGURE 5-12
Pressure of a fixed
amount of gas in a fixed
volume as a function of
temperature—Example
5-8 visualized.

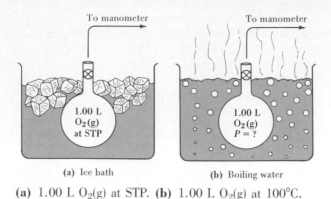

(a) 1.00 L O_2(g) at STP. (b) 1.00 L O_2(g) at 100°C.

In Examples 5-6 and 5-7 a single set of conditions—a single state—was involved. A single application of the ideal gas equation was required. At times a gas is described under *two* different sets of conditions—the initial and final states. Here two applications of the ideal gas equation are required. However, if one or more of the gas variables is held constant, the solution generally takes a simplified form. In fact, all the simple gas laws can be derived from the ideal gas equation. The simple law that pertains to Example 5-8 is a variation of Gay-Lussac's law that describes the relationship between gas pressure and temperature when the amount of gas and its volume are fixed.

Example 5-8   Pictured in Figure 5-12 is a 1.00-L flask of O_2(g), first at STP and then at 100°C. What is the pressure of the gas at 100°C?

Solution
Method 1.   Use of the ideal gas equation.
Since the volume (V), the amount of gas (n), and R remain constant, we can write the ideal gas equation in the form

$$\frac{nR}{V} = \frac{P}{T} = \frac{P_i}{T_i} = \frac{P_f}{T_f}$$

Then we can solve for P_f. (Note that P_i is standard pressure = 1.00 atm.)

$$P_f = P_i \times \frac{T_f}{T_i} = 1.00 \text{ atm} \times \frac{(100 + 273) \text{ K}}{273 \text{ K}} = 1.00 \text{ atm} \times \frac{373 \text{ K}}{273 \text{ K}} = 1.37 \text{ atm}$$

Method 2.   A "common sense" approach.
Gas pressure is directly proportional to kelvin temperature for a fixed amount of gas in a fixed volume. This means that

$$P_f = P_i \times \text{ ratio of kelvin temperatures}$$

$$P_f = 1.00 \text{ atm} \times \frac{373 \text{ K}}{273 \text{ K}} = 1.37 \text{ atm}$$

SIMILAR EXAMPLES: Review Problem 7; Exercises 10, 14.

5-6 Molecular Weight Determination

A more direct approach to establishing molecular weights than Cannizzaro's (see Figure 5-11) uses the ideal gas equation. For this purpose it is helpful to alter the equation slightly. The number of moles of gas, usually expressed as *n,* is also equal to the mass of gas, *m*, divided by the molar mass, *M* (whose units are g/mol). That is, $n = m/M$. The molecular weight (a dimensionless number) is numerically equal to the molar mass.

$$PV = \frac{mRT}{M} \qquad (5.10)$$

To determine the molecular weight of a gas with equation (5.10) requires measuring the volume (*V*) occupied by a known mass of gas (*m*) at a certain temperature (*T*) and pressure (*P*). The form of the ideal gas equation shown in equation (5.10) is not limited to the determination of molecular weights. It can be used in any application in which the quantity of gas is given or sought in grams rather than moles.

To establish that equation (5.10) is of the proper form, simply substitute the appropriate units for each term and check the cancellation of units.

$$PV = \frac{m}{M} RT$$

$$(\text{atm})(\text{L}) = \frac{(g)(\text{L atm mol}^{-1} \text{K}^{-1})(K)}{(g \text{ mol}^{-1})}$$

$$(\text{atm})(\text{L}) = (\text{L})(\text{atm})$$

This technique is known as *unit analysis*. It is especially useful in deriving and remembering equations.

Example 5-9 What is the molecular weight of a gas if 1.81 g of the gas occupies a volume of 1.52 L at 25°C and 737 mmHg?

Solution

$$\begin{cases} P = 737 \text{ mmHg} \times \dfrac{1 \text{ atm}}{760 \text{ mmHg}} = 0.970 \text{ atm} \\[2mm] V = 1.52 \text{ L} \\[2mm] m = 1.81 \text{ g} \\[2mm] M = ? \\[2mm] R = 0.0821 \text{ L atm mol}^{-1} \text{ K}^{-1} \\[2mm] T = 25 + 273 = 298 \text{ K} \end{cases}$$

Rearrange equation (5.10) as

$$M = \frac{mRT}{PV}$$

and substitute the known values.

$$M = \frac{1.81 \text{ g} \times 0.0821 \text{ L atm mol}^{-1} \text{ K}^{-1} \times 298 \text{ K}}{0.970 \text{ atm} \times 1.52 \text{ L}} = 30.0 \text{ g/mol}$$

The molecular weight of the gas is 30.0.

SIMILAR EXAMPLES: Review Problem 10; Exercise 17.

Example 5-10 A glass vessel weighs 40.1305 g when clean, dry, and evacuated; 138.2410 g when filled with water at 25.0°C (density of water = 0.9970 g/cm^3); and 40.2959 g when filled with propylene gas at 740.4 mmHg and 24.1°C. What is the molecular weight of propylene?

Solution. We must first determine the volume of the glass vessel (and hence the volume of the gas) and the mass of the gas. The gas constant, *R*, is expressed to four significant figures in this calculation to correspond to the other measured quantities.

mass of water to fill vessel = 138.2410 g − 40.1305 g = 98.1105 g

$$\text{volume of water (volume of vessel)} = 98.1105 \text{ g H}_2\text{O} \times \frac{1 \text{ cm}^3 \text{ H}_2\text{O}}{0.9970 \text{ g H}_2\text{O}}$$

$$= 98.41 \text{ cm}^3 = 0.09841 \text{ L}$$

mass of gas = 40.2959 g − 40.1305 g = 0.1654 g

temperature = 24.1°C + 273.15 = 297.2 K

$$\text{pressure} = 740.4 \text{ mmHg} \times \frac{1 \text{ atm}}{760.0 \text{ mmHg}} = 0.9742 \text{ atm}$$

$$\mathcal{M} = \frac{mRT}{PV} = \frac{0.1654 \text{ g} \times 0.08206 \text{ L atm mol}^{-1} \text{ K}^{-1} \times 297.2 \text{ K}}{0.9742 \text{ atm} \times 0.09841 \text{ L}} = 42.08 \text{ g/mol}$$

The molecular weight of propylene is 42.08.

SIMILAR EXAMPLE: Exercise 19.

The method of determining molecular weights outlined in Example 5-10 (the Dumas method) can be combined with an elemental analysis to yield the molecular formula of a gas. That is, if propylene is found to be 85.63% C and 14.37% H, by mass, what is its molecular formula?

5-7 Gas Densities

In Examples 5-9 and 5-10 we rearranged equation (5.10) to solve for molar mass, $\mathcal{M}$. A different rearrangement of the equation yields

$$\frac{m}{V} = \frac{\mathcal{M}P}{RT} \tag{5.11}$$

The term m/V is the mass of a gas divided by its volume—the **gas density.** Gas densities differ from those of solids and liquids in some important ways.

1. Gas densities are generally stated in g/L instead of g/cm^3.
2. Gas densities are strongly dependent on pressure and temperature, increasing as the gas pressure increases and decreasing as the temperature increases (see equation 5.11). Densities of liquids and solids do depend somewhat on temperature, but they are far less dependent on pressure.
3. The density of a gas is directly proportional to its molar mass. No simple relationship exists between density and molar mass for liquids and solids.

The density of a gas at STP can be easily calculated by dividing the molar mass of a gas by the molar volume (22.414 L/mol). For $O_2(g)$ at STP, for example, the density is 32.0 g mol^{-1}/22.4 L mol^{-1} = 1.43 g/L. Under other conditions of temperature and pressure, equation (5.11) is required.

Example 5-11   What is the density of oxygen gas (O_2) at 298 K and 0.987 atm?

Solution.   The terms for the right-hand side of equation (5.11) are readily obtainable for this example. The density is simply the left-hand side of the equation, m/V.

$$\frac{m}{V} = \frac{\mathcal{M}P}{RT} = \frac{32.0 \text{ g mol}^{-1} \times 0.987 \text{ atm}}{0.0821 \text{ L atm mol}^{-1} \text{ K}^{-1} \times 298 \text{ K}} = 1.29 \text{ g/L}$$

SIMILAR EXAMPLES: Review Problem 11; Exercise 20.

5-8 Gases in Chemical Reactions

We now have a new tool to apply to calculations dealing with gaseous reactants and/or products of a chemical reaction—the ideal gas equation. Specifically, information about gaseous species can be presented not only in grams and moles, but also in terms of gas volumes, temperatures, and pressures.

Example 5-12   What volume of $O_2(g)$, measured at 735 mmHg and 26°C, is produced when a 22.2-g sample of $KClO_3$ is decomposed?

$$2\ KClO_3(s) \xrightarrow{\Delta} 2\ KCl(s) + 3\ O_2(g)$$

Solution.   This problem is easiest to solve in two separate calculations. First, the number of moles of $O_2(g)$ is determined. Then the ideal gas equation is used.

$$\text{no. mol } O_2 = 22.2 \text{ g } KClO_3 \times \frac{1 \text{ mol } KClO_3}{123 \text{ g } KClO_3} \times \frac{3 \text{ mol } O_2(g)}{2 \text{ mol } KClO_3} = 0.271 \text{ mol } O_2$$

$$\left\{ \begin{array}{l} P = 735 \text{ mmHg} \times \dfrac{1 \text{ atm}}{760 \text{ mmHg}} = 0.967 \text{ atm} \\[2mm] V = ? \\[2mm] n = 0.271 \text{ mol} \\[2mm] R = 0.0821 \text{ L atm mol}^{-1} \text{ K}^{-1} \\[2mm] T = 26°C + 273 = 299 \text{ K} \end{array} \right.$$

$$V = \frac{nRT}{P} = \frac{0.271 \text{ mol} \times 0.0821 \text{ L atm mol}^{-1} \text{ K}^{-1} \times 299 \text{ K}}{0.967 \text{ atm}} = 6.88 \text{ L}$$

SIMILAR EXAMPLES: Review Problem 12; Exercises 23, 24.

Law of Combining Volumes.   The ideal gas equation, or more specifically Avogadro's hypothesis, is particularly useful in dealing with a second kind of calculation based on the balanced chemical equation. This relates to a situation in which either all the reactants and products are gases, or at least those involved in the particular calculation. Consider this reaction.

$$2\ NO(g) + O_2(g) \longrightarrow 2\ NO_2(g)$$

$$2 \text{ mol } NO(g) + 1 \text{ mol } O_2(g) \longrightarrow 2 \text{ mol } NO_2(g)$$

Suppose the gases are compared at the same T and P. Under these conditions 1 mol of gas occupies a volume of V liters; 2 mol of gas, $2V$ liters; and so on.

$$2V \text{ L } NO(g) + V \text{ L } O_2(g) \longrightarrow 2V \text{ L } NO_2(g)$$

Now divide each coefficient by V:

$$2 \text{ L } NO(g) + 1 \text{ L } O_2(g) \longrightarrow 2 \text{ L } NO_2(g)$$

From this description of the chemical equation, these statements may be written.

$$\frac{2 \text{ L } NO_2(g)}{2 \text{ L } NO(g)} \approx 1; \qquad \frac{2 \text{ L } NO_2(g)}{1 \text{ L } O_2(g)} \approx 1; \qquad \frac{1 \text{ L } O_2(g)}{2 \text{ L } NO(g)} \approx 1; \qquad \text{and so on.}$$

What we have just done is to develop, in modern terms, Gay-Lussac's law of combining volumes (recall page 111). This law is applied in Examples 5-13 and 5-14.

Example 5-13   Roasting of ZnS ore is the first step in the production of Zn.

$$2\ ZnS(s) + 3\ O_2(g) \longrightarrow 2\ ZnO(s) + 2\ SO_2(g)$$

What volume of $SO_2(g)$ forms *per* liter of $O_2(g)$ consumed? Both gases are measured at 740 mmHg and 25°C.

Solution.   Since the reactant and product being compared are *both gases*, and are *both measured at the same temperature and pressure*, a ratio of combining volumes can be derived from the balanced equation and used as follows.

$$\text{no. L } SO_2(g) = 1.00 \text{ L } O_2(g) \times \frac{2 \text{ L } SO_2(g)}{3 \text{ L } O_2(g)} = 0.667 \text{ L } SO_2(g)$$

SIMILAR EXAMPLES: Review Problem 13; Exercise 25(a).

You should note carefully the following points concerning a calculation of the kind performed in Example 5-13.

1. It was not necessary to use the specific temperature (25°C) and pressure (740 mmHg) at all. As long as a comparison is made at the *same T* and *P*, the relationship between volume and number of molecules (or moles) of a gas is the same for all gases.
2. If temperature and pressure are *not* identical for the gases being compared, the method of Example 5-13 will not work. Then it is best to convert information about the gases to a mole basis and to use mole rather than volume ratios. (See Example 5-14.)
3. If the relationship is between a *solid* (or *liquid*) and a gas, it is again necessary to use a mole ratio as a conversion factor (as in Example 5-12).

Example 5-14 A 20.1-L sample of $H_2(g)$, measured at 0°C and 750 mmHg, is mixed with 11.2 L of $O_2(g)$ measured at 27°C and 720 mmHg. The mixture is ignited and reacts to produce water. What amount of water is formed?

$$2 H_2(g) + O_2(g) \longrightarrow 2 H_2O(l)$$

Solution. Although there appears to be more than enough $O_2(g)$ to combine with all the $H_2(g)$, the gas volumes are *not* at the same temperature and pressure. The law of combining volume does *not* apply. To determine the limiting reagent, convert the quantities of the two reactants to moles and compare them.

$$n_{H_2} = \frac{PV}{RT} = \frac{750 \text{ mmHg} \times \dfrac{1 \text{ atm}}{760 \text{ mmHg}} \times 20.1 \text{ L}}{0.0821 \text{ L atm mol}^{-1} \text{ K}^{-1} \times 273 \text{ K}} = 0.885 \text{ mol}$$

$$n_{O_2} = \frac{PV}{RT} = \frac{720 \text{ mmHg} \times \dfrac{1 \text{ atm}}{760 \text{ mmHg}} \times 11.2 \text{ L}}{0.0821 \text{ L atm mol}^{-1} \text{ K}^{-1} \times 300 \text{ K}} = 0.431 \text{ mol}$$

The amount of O_2 required to react with the available H_2 is

$$\text{no. mol } O_2 \text{ required} = 0.885 \text{ mol } H_2 \times \frac{1 \text{ mol } O_2}{2 \text{ mol } H_2} = 0.442 \text{ mol } O_2$$

However, there is only 0.431 mol O_2 available. O_2 is the limiting reagent.

$$\text{no. mol } H_2O = 0.431 \text{ mol } O_2 \times \frac{2 \text{ mol } H_2O}{1 \text{ mol } O_2} = 0.862 \text{ mol } H_2O$$

SIMILAR EXAMPLES: Exercises 25(b), 26.

5-9 Mixture of Gases

Except to establish the number of moles of gas, at no point in our previous use of the simple gas laws or the ideal gas equation has it been necessary to identify the gas. This is because, as a first approximation at least, all gases behave pretty much alike. The ideal gas equation is applicable to all gases under the appropriate conditions. As a result the ideal gas equation may be applied to a *mixture of gases* just as it is to a single gas. To do this it is only necessary to use for the value of *n* the *total* number of moles of molecules in the gaseous mixture.

Example 5-15   What is the pressure exerted by a mixture of 1.0 g H_2 and 5.0 g He when confined to a volume of 5.0 L at 20°C?

Solution

$$n_{tot.} = \left(1.0 \text{ g } H_2 \times \frac{1 \text{ mol } H_2}{2.0 \text{ g } H_2}\right) + \left(5.0 \text{ g He} \times \frac{1 \text{ mol He}}{4.0 \text{ g He}}\right)$$

$$= 0.50 \text{ mol } H_2 + 1.25 \text{ mol He} = 1.75 \text{ mol gas}$$

$$P = \frac{n_{tot.}RT}{V} \tag{5.12}$$

$$= \frac{1.75 \text{ mol} \times 0.0821 \text{ L atm mol}^{-1} \text{ K}^{-1} \times 293 \text{ K}}{5.0 \text{ L}} = 8.4 \text{ atm}$$

SIMILAR EXAMPLES: Review Problem 14; Exercise 27.

In addition to his formulation of the atomic theory, John Dalton made an important contribution to the study of gaseous mixtures. Dalton considered that in a mixture of gases each gas expands to fill the container and exerts a **partial pressure** that is independent of the presence of other gases. *The sum of these partial pressures is equal to the total pressure of the mixture* (see Figure 5-13). For a mixture of gases, A, B, . . .

$$P_{tot.} = P_A + P_B + \cdots \tag{5.13}$$

That Dalton's law of partial pressures is equivalent to equation (5.12) can be easily demonstrated.

$$P_{tot.} = P_A + P_B + \cdots$$

$$= \frac{n_A RT}{V} + \frac{n_B RT}{V} + \cdots = \frac{RT}{V}(n_A + n_B + \cdots)$$

$$= \frac{n_{tot.}RT}{V} \quad \text{(where } n_{tot.} = n_A + n_B + \cdots)$$

An alternative expression known as Amagat's law is useful in dealing with gaseous mixtures whose compositions are expressed in percent by volume. Here we begin with the expression

$$V_{tot.} = \frac{n_{tot.}RT}{P_{tot.}} \tag{5.14}$$

and again note that $n_{tot.} = n_A + n_B + \cdots$. This allows us to write

FIGURE 5-13
Dalton's law of partial pressures illustrated.

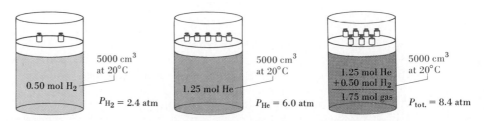

This figure indicates that the pressure of each gas is proportional to the number of moles of that gas. The total pressure of the mixture is the sum of the partial pressures of the individual gases.

$$V_{\text{tot.}} = \frac{n_A RT}{P_{\text{tot.}}} + \frac{n_B RT}{P_{\text{tot.}}} + \cdots$$

$$= V_A + V_B + \cdots \tag{5.15}$$

The terms $V_A, V_B, \ldots$ are called partial volumes. The **partial volume** of a component in a gaseous mixture is the volume that would be occupied by that component if it existed alone at the total pressure of the mixture. *The total volume of a gaseous mixture is the sum of the partial volumes of its components* (equation 5.15).

Still another useful expression for gaseous mixtures is obtained by taking the ratio of a partial pressure to a total pressure or a partial volume to a total volume.

$$\frac{P_A}{P_{\text{tot.}}} = \frac{n_A \dfrac{RT}{V_{\text{tot.}}}}{n_{\text{tot.}} \dfrac{RT}{V_{\text{tot.}}}} = \frac{n_A}{n_{\text{tot.}}} \quad \text{and} \quad \frac{V_A}{V_{\text{tot.}}} = \frac{n_A \dfrac{RT}{P_{\text{tot.}}}}{n_{\text{tot.}} \dfrac{RT}{P_{\text{tot.}}}} = \frac{n_A}{n_{\text{tot.}}}$$

which means that

$$\frac{n_A}{n_{\text{tot.}}} = \frac{P_A}{P_{\text{tot.}}} = \frac{V_A}{V_{\text{tot.}}} \tag{5.16}$$

The term $n_A/n_{\text{tot.}}$ is given a special name. It is the mole fraction of A in the gaseous mixture. **Mole fraction,** often denoted by the symbol χ, represents the fraction of all the molecules in a mixture that are of a given type. In a mixture of the gases A, B, . . . , the mole fractions are $\chi_A, \chi_B, \ldots$. The sum of the mole fractions of all the components in a mixture is 1, i.e., $\chi_A + \chi_B + \cdots = 1$.

As illustrated in Examples 5-16 and 5-17, expression (5.16) is an especially useful one for dealing with gaseous mixtures.

Example 5-16 What are the partial pressures of H_2 and He in the gaseous mixture described in Example 5-15?

Solution. From the number of moles of each gas and the conditions stated in Example 5-15, we may calculate the partial pressures directly.

$$P_{H_2} = \frac{n_{H_2} \cdot RT}{V} = \frac{0.50 \text{ mol} \times 0.0821 \text{ L atm mol}^{-1} \text{ K}^{-1} \times 293 \text{ K}}{5.0 \text{ L}} = 2.4 \text{ atm}$$

$$P_{He} = \frac{n_{He} \cdot RT}{V} = \frac{1.25 \text{ mol} \times 0.0821 \text{ L atm mol}^{-1} \text{ K}^{-1} \times 293 \text{ K}}{5.0 \text{ L}} = 6.0 \text{ atm}$$

As expected, these partial pressures, when added together, yield the total pressure calculated in Example 5-15—8.4 atm.

A second method makes use of expression (5.16), with mole fractions and total pressure obtained from Example 5-15.

$$P_{H_2} = \frac{n_{H_2}}{n_{\text{tot.}}} \times P_{\text{tot.}} = \frac{0.50}{1.75} \times 8.4 \text{ atm} = 2.4 \text{ atm}$$

$$P_{He} = \frac{n_{He}}{n_{\text{tot.}}} \times P_{\text{tot.}} = \frac{1.25}{1.75} \times 8.4 \text{ atm} = 6.0 \text{ atm}$$

SIMILAR EXAMPLE: Exercise 28.

Example 5-17 The major components of air are nitrogen, 78.08%; oxygen, 20.95%; argon, 0.93%; and carbon dioxide, 0.03%, by volume. What are the partial pressures of these four gases in a sample of air at standard atmospheric pressure (1.000 atm)?

Solution. Volume percentages are directly related to ratios of partial to total volumes. In a total volume of 100.0 L of air, the partial volume of $N_2(g)$ is 78.08L; $O_2(g)$, 20.95 L; and so on. We then substitute these values into equation (5.16).

$$P_{N_2} = \frac{V_{N_2}}{V_{tot.}} \times P_{tot.} = \frac{78.08 \text{ L}}{100.0 \text{ L}} \times 1.000 \text{ atm} = 0.7808 \text{ atm}$$

$$P_{O_2} = \frac{V_{O_2}}{V_{tot.}} \times P_{tot.} = \frac{20.95 \text{ L}}{100.0 \text{ L}} \times 1.000 \text{ atm} = 0.2095 \text{ atm}$$

$$P_{Ar} = \frac{V_{Ar}}{V_{tot.}} \times P_{tot.} = \frac{0.93 \text{ L}}{100.0 \text{ L}} \times 1.000 \text{ atm} = 0.0093 \text{ atm}$$

$$P_{CO_2} = \frac{V_{CO_2}}{V_{tot.}} \times P_{tot.} = \frac{0.03 \text{ L}}{100.0 \text{ L}} \times 1.000 \text{ atm} = 0.0003 \text{ atm}$$

SIMILAR EXAMPLE: Exercise 30.

A gaseous mixture is sometimes described by its **apparent molar mass**—the mass of one mole of molecules of the gaseous mixture. The apparent molar mass can be determined by adding together the contributions of each component to the mass of one mole of the mixture.

Example 5-18 From data in Example 5-17 calculate the apparent molar mass of air.

Solution. The key to this calculation again lies in equation (5.16). The ratio of the number of moles of a gaseous component to the total number of moles of gas (i.e., $n_A/n_{tot.}$) is the same as the volume ratio (i.e., $V_A/V_{tot.}$). In "one mole of air," $n_{tot.} = 1.000$ and the numbers of moles of the individual gases are

0.7808 mol N_2;   0.2095 mol O_2;   0.0093 mol Ar;   0.0003 mol CO_2

The apparent molar mass of air is

$$\left(0.7808 \text{ mol } N_2 \times \frac{28.01 \text{ g } N_2}{1 \text{ mol } N_2}\right) + \left(0.2095 \text{ mol } O_2 \times \frac{32.00 \text{ g } O_2}{1 \text{ mol } O_2}\right)$$
$$+ \left(0.0093 \text{ mol Ar} \times \frac{39.95 \text{ g Ar}}{1 \text{ mol Ar}}\right) + \left(0.0003 \text{ mol } CO_2 \times \frac{44.01 \text{ g } CO_2}{1 \text{ mol } CO_2}\right)$$
$$= 28.96 \text{ g/mol air}$$

FIGURE 5-14
Collection of a gas over water.

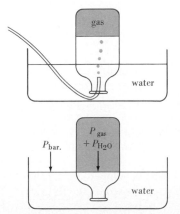

SIMILAR EXAMPLE: Exercise 30.

Collection of Gases over Water. The device pictured in Figure 5-14, called a pneumatic trough, represented a great breakthrough in the study of gases in the seventeenth century. It afforded a means of isolating gaseous products of chemical reactions. Since it is based on displacing water from a container, the method works only for gases that are insoluble in water, such as N_2, H_2, and O_2.

The gas that is collected is "wet." It is a mixture of the desired gas and water vapor. The gas being collected expands to fill the container and exerts its partial pressure: P_{gas}. Water vapor, produced by the evaporation of liquid water, also fills the container and exerts a partial pressure: P_{H_2O}. The pressure of the water vapor depends only on the temperature of the water. Water vapor pressure data are readily available in tabulated form (see Table 11-1). The concept of vapor pressure is explored more fully in Chapter 11.

According to Dalton's law the *total* pressure is the sum of the two partial pressures. Typically, the total pressure is measured simply by relating it to the prevailing pressure of the atmosphere (barometric pressure). If the container of gas is moved up or down until the water levels are equalized on the inside and outside, then the total gas pressure is made equal to the barometric pressure.

$$P_{bar.} = P_{tot.} = P_{gas} + P_{H_2O}$$

and

$$P_{gas} = P_{bar.} - P_{H_2O} \qquad (5.17)$$

Example 5-19   In the following reaction 81.2 cm³ of $O_2(g)$ is collected *over water* at 23°C and barometric pressure 751 mmHg. What must have been the mass of $Ag_2O(s)$ decomposed? (Vapor pressure of water at 23°C = 21.1 mmHg.)

$$2\ Ag_2O(s) \longrightarrow 4\ Ag(s) + O_2(g)$$

Solution.   First we need to calculate the number of moles of $O_2(g)$; this we can do with the ideal gas equation. The key to this calculation is the fact that the gas collected is "wet," i.e., a *mixture* of $O_2(g)$ and water vapor.

$$
\begin{cases}
P_{O_2} = P_{bar.} - P_{H_2O} = 751\ \text{mmHg} - 21.1\ \text{mmHg} \\[2mm]
\qquad = 730\ \text{mmHg} \times \dfrac{1\ \text{atm}}{760\ \text{mmHg}} = 0.961\ \text{atm} \\[3mm]
V = 81.2\ \text{cm}^3 = 0.0812\ \text{L} \\[2mm]
n = ? \\[2mm]
R = 0.0821\ \text{L atm mol}^{-1}\ \text{K}^{-1} \\[2mm]
T = 23°\text{C} + 273 = 296\ \text{K}
\end{cases}
$$

$$n = \frac{PV}{RT} = \frac{0.961\ \text{atm} \times 0.0812\ \text{L}}{0.0821\ \text{L atm mol}^{-1}\ \text{K}^{-1} \times 296\ \text{K}} = 0.00321\ \text{mol}$$

From the chemical equation we obtain a factor to convert from mol O_2 to mol Ag_2O. The molar mass of Ag_2O provides the final factor.

$$\text{no. g } Ag_2O = 0.00321\ \text{mol } O_2 \times \frac{2\ \text{mol } Ag_2O}{1\ \text{mol } O_2} \times \frac{232\ \text{g } Ag_2O}{1\ \text{mol } Ag_2O} = 1.49\ \text{g } Ag_2O$$

SIMILAR EXAMPLES: Review Problem 15; Exercise 31.

5-10 Kinetic Molecular Theory of Gases

The simple gas laws are empirical statements of the observed behavior of gases. These laws are reasonably accurate for most gases under normal conditions of temperature and pressure. As we have stated before, scientific laws express behavior through a correlation of observations or experiments. A scientific theory is an explanation of a law or a group of laws. A law is a statement of what will happen; a theory attempts to explain why this happens. A scientific theory is based on a model or concept from which various phenomena can be deduced logically.

The currently accepted theory for explaining gas behavior was developed during the middle nineteenth century. It is the kinetic molecular theory of gases, based on the following model.

1. A gas is comprised of extremely small particles called molecules (or atoms in some cases).

2. The molecules of a gas are usually separated by great distances. As a result they occupy only a very small fraction of the total gas volume. They are, in fact, assumed to be point masses.

3. There are assumed to be no intermolecular forces.

4. The molecules move constantly and randomly throughout the gas volume. As a result of their motion they undergo frequent collisions with one another and with the walls of their container.

5. Collisions between molecules are elastic. Individual molecules may gain or lose energy as a result of collisions; however, in a large collection of molecules at constant temperature, the total energy remains constant.

The basic equation of the kinetic molecular theory is obtained by totaling the forces exerted by the molecules of a gas as they collide with the walls of a container of volume V. The total force, divided by the area over which it is exerted, yields the gas pressure P. In terms of molecular properties, the force of molecular collisions depends on two factors. First is the frequency of molecular collisions with the container walls. In turn, this frequency depends on the speeds of the molecules; the faster they are moving, the more frequently they collide with the container walls. The second factor determining the forces of these collisions is the amount of translational kinetic energy (e_k) possessed by the gas molecules. Translational kinetic energy refers to the energy associated with the motion of a molecule through space: $e_k = \frac{1}{2}(mu^2)$, where m is the mass of the molecule and u is its speed (see also Appendix B). The faster they are moving, the more translational kinetic energy the molecules of a gas possess. Further analysis of the situation is complicated, however, by the fact that molecules move in all directions and at different speeds. The result obtained, offered here without proof, is

$$PV = \frac{n'm\overline{u^2}}{3} \tag{5.18}$$

where n' represents the number of molecules in the volume V. The term $\overline{u^2}$ represents the *average* of the *squares* of the molecular speeds. Based on this meaning of $\overline{u^2}$, we can represent the *average* translational kinetic energy of a collection of gas molecules as $\overline{e_k} = \frac{1}{2}(m\overline{u^2})$. By a slight rearrangement of equation (5.18), we obtain

$$PV = \tfrac{2}{3}n'(\tfrac{1}{2}m\overline{u^2}) \qquad \text{and} \qquad PV = \tfrac{2}{3}n'\overline{e_k} \tag{5.19}$$

Two more relationships can be derived from equation (5.19) by considering (a) 1 mol of gas, $n' = N_A$ (Avogadro's number), and (b) the ideal gas equation for 1 mol of gas, $PV = RT$.

$$PV = \frac{2}{3}N_A\overline{e_k} = RT \tag{5.20}$$

$$\overline{e_k} = \frac{3}{2}\frac{R}{N_A}T = \frac{3}{2}kT \tag{5.21}$$

The constant k is the gas constant *per molecule,* called the Boltzmann constant. Equation (5.21) provides a new definition of temperature. The kelvin temperature of a gas (T) is directly proportional to the average translational kinetic energy of its molecules ($\overline{e_k}$). Now we have a new conception of what changes in temperature mean—changes in the intensity of molecular motion. When heat flows from one body to another, this takes the form of molecules in the hotter body (higher temperature) giving up some of their kinetic energy through collisions with molecules in the colder body (lower temperature). The flow continues until the average molecular kinetic energies become equal, i.e., the temperatures become equalized. Equation (5.21) also provides a new way of looking at the absolute zero of temperature. It is the temperature at which molecular motion ceases.

Avogadro's Law. From equation (5.19), for two different gases, A and B,

$$P_A = \frac{2}{3}\frac{n'_A}{V_A}(\overline{e_k})_A \quad \text{and} \quad P_B = \frac{2}{3}\frac{n'_B}{V_B}(\overline{e_k})_B$$

If the two gases are compared at identical temperatures, $(\overline{e_k})_A = (\overline{e_k})_B$. At identical pressures, $P_A = P_B$. Under these conditions the number of molecules per unit volume must be the same for the two gases.

$$\frac{n'_A}{V_A} = \frac{n'_B}{V_B}$$

Thus, if equal volumes of gases are compared ($V_A = V_B$), the number of molecules of the two gases must be equal, $n'_A = n'_B$. If equal numbers of molecules are compared ($n'_A = n'_B$), the volumes must be equal, $V_A = V_B$.

In similar fashion, the other simple gas laws can also be derived from the kinetic molecular theory.

Distribution of Molecular Speeds. The statements we have made about an average kinetic energy imply that in a collection of molecules there is a *distribution* of energies from very high to very low. There is a distribution of speeds as well. Figure 5-15 shows a typical distribution.

Three different speeds are noted on the curve of Figure 5-15. These are the **most probable** or **modal speed, u_m**, the **average speed, $u_{av.} = \overline{u}$**, and the **root-mean-square speed, $u_{rms} = \sqrt{\overline{u^2}}$**. The root-mean-square speed is the square root of the average of the squares of the speeds of all the molecules in a sample. A simpler view is that u_{rms} is the speed of a molecule possessing the average kinetic energy. That is,

$$\overline{e_k} = \tfrac{1}{2}m\overline{u^2} = \tfrac{1}{2}m(\sqrt{\overline{u^2}})^2 = \tfrac{1}{2}m(u_{rms})^2$$

The root-mean-square speed can be calculated by substituting for the value of $\overline{e_k}$ in equation (5.21).

$$\overline{e_k} = \frac{1}{2}m\overline{u^2} = \frac{3}{2}\frac{R}{N_A}T \quad \text{and} \quad \overline{u^2} = \frac{3RT}{mN_A}$$

Since the product mN_A represents the mass of 1 mol of molecules, it can be replaced by the molar mass, $\mathcal{M}$.

$$u_{rms} = \sqrt{\overline{u^2}} = \sqrt{\frac{3RT}{\mathcal{M}}} \tag{5.22}$$

FIGURE 5-15
Distribution of
molecular speeds—
hydrogen gas at 0°C.

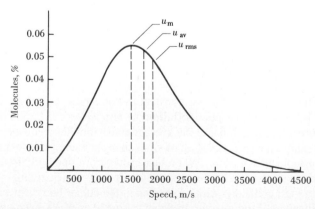

The ordinate values represent the percent of the molecules having a certain speed. The abscissas represent these speeds, based on an interval of 1 m/s. (For example, all molecules with speeds between 1499.5 and 1500.5 m/s are taken to have a speed of 1500 m/s.)

TABLE 5-2
Analogy to the distribution of molecular speeds

Consider ten automobiles on a highway traveling at these speeds:

	Speed, mi/h	**(Speed)^2**
	40	1,600
	42	1,764
	45	2,025
	48	2,304
most probable (modal)	50	2,500
speed = 50	50	2,500
	55	3,025
	57	3,249
	58	3,364
	60	3,600

sum of speeds = $\sum$ speed = 505 $\sum$ (speed)^2 = 25,931

average speed = $\overline{\text{speed}}$ = $\dfrac{\sum \text{speed}}{10}$ = 50.5 $\overline{(\text{speed})^2}$ = $\dfrac{\sum (\text{speed})^2}{10}$ = 2593.1

root-mean-square speed

$= \sqrt{\overline{(\text{speed})^2}} = \sqrt{2593.1} = 50.9$

modal speed: 50   average speed: 50.5   rms speed: 50.9

Derivations of u_m and $\bar{u}$ are more difficult than for u_{rms}, but these velocities are proportional to u_{rms}.

$$u_m = 0.816\, u_{rms} \qquad u_{av.} = \bar{u} = 0.921\, u_{rms} \tag{5.23}$$

An analogy is suggested in Table 5-2 that may help you to appreciate the distinction among most probable, average, and root-mean-square speed.

To calculate molecular speeds with expressions (5.22) and (5.23) requires that the gas constant be expressed as

$$R = 8.314 \text{ J mol}^{-1}\text{ K}^{-1}$$

Furthermore, to produce the proper cancellation of units, the joule (J) must be in terms of mass, length, and time. Since kinetic energy is expressed as K.E. $= \frac{1}{2}mv^2$, the joule must have the units of (mass) $\times$ (velocity)^2 = (kg)(m/s)^2. This leads to the value

$$R = 8.314 \text{ kg m}^2\text{ s}^{-2}\text{ mol}^{-1}\text{ K}^{-1}$$

Example 5-20   What is the root-mean-square speed (u_{rms}) of H_2 molecules at 50°C?

Solution.   Substitute into equation (5.22), noting that R must have the units described above and that the molar mass must be expressed in *kilograms* per mole.

$$u_{rms} = \sqrt{\frac{3 \times 8.314 \text{ kg m}^2\text{ s}^{-2}\text{ mol}^{-1}\text{ K}^{-1} \times 323 \text{ K}}{2.016 \times 10^{-3} \text{ kg/mol}}}$$

$$= \sqrt{4.00 \times 10^6 \text{ m}^2/\text{s}^2} = 2.00 \times 10^3 \text{ m/s}$$

This answer should seem reasonable when compared to Figure 5-15. At 0°C, u_{rms} for $H_2(g)$ is somewhat below 2000 m/s. Raising the temperature will increase u_{rms}.

SIMILAR EXAMPLES: Exercises 35, 36.

FIGURE 5-16
Effusion through an orifice.

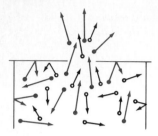

Average speeds of the two different types of molecules are suggested by the lengths of the arrows. The faster molecules (shown in color) effuse more rapidly.

Graham's Law. Molecular speeds are quite high. A speed of 1500 m/s corresponds to about 1 mi/s or 3000 mi/h. However, when gases are allowed to mix or diffuse into one another, they do not do so at nearly the rate implied by the speeds of the gas molecules. This is because the molecules collide with one another with great frequency and constantly change direction as a result of collisions. They do not follow straight-line paths over long distances. Nevertheless, gases do diffuse or mix, and the rate at which this occurs is dependent on the speeds of the gas molecules.

A concept related to diffusion rates is pictured in Figure 5-16. Here molecules are allowed to escape from their container through a tiny orifice or pin hole. This escape through an orifice is called **effusion.** The rates of effusion of molecules are directly proportional to their speeds. Thus, in comparing two different gases at the same temperature, we may write

$$\frac{\text{rate of effusion of A}}{\text{rate of effusion of B}} = \frac{\overline{u_A}}{\overline{u_B}} = \frac{0.921(u_{rms})_A}{0.921(u_{rms})_B} = \frac{(u_{rms})_A}{(u_{rms})_B}$$

A further substitution is possible using equation (5.22):

$$\frac{\text{rate of effusion of A}}{\text{rate of effusion of B}} = \sqrt{\frac{3RT/\mathcal{M}_A}{3RT/\mathcal{M}_B}} = \sqrt{\frac{\mathcal{M}_B}{\mathcal{M}_A}} \qquad (5.24)$$

The result shown in equation (5.24) is a kinetic-theory statement of a nineteenth-century law called Graham's law.

The rates of effusion (or diffusion) of two different gases are inversely proportional to the square roots of their molecular weights.

Equation (5.24) compares the rates of effusion of two different gases. *Lighter gases effuse faster than heavy ones.* The gas that effuses *fastest* takes the *shortest* time to do so. Also, the gas that effuses *fastest* travels *farthest* in a given period of time. An effective method of dealing with the several possible variations of equation (5.24) is to note that in every case a ratio of two terms (effusion rates, times, distances, . . .) is equal to the *square root* of a ratio of molar masses. That is,

$$\text{ratio of} \begin{cases} \text{molecular speeds} \\ \text{rates of effusion} \\ \text{effusion times} \\ \text{distance traveled by molecules} \\ \text{amount of gas effused} \end{cases} = \sqrt{\frac{\text{ratio of two}}{\text{molar masses}}} \qquad (5.25)$$

In using equation (5.25) to answer questions like those posed in Examples 5-21, 5-22, and 5-23, you need to establish whether the ratio of molar masses should be $\mathcal{M}_A/\mathcal{M}_B$ or $\mathcal{M}_B/\mathcal{M}_A$. This is most easily done by reasoning whether the ratio of properties sought (effusion rate, effusion time, etc.) should be greater or less than one.

Example 5-21 How does the average speed of a He atom compare to that of an H_2 molecule at the same temperature?

Solution. If we reason that lighter gases have higher speeds than heavier ones, we conclude that the average speed of an He atom should be *smaller* than for an H_2 molecule. This suggests the setup

$$\frac{\overline{u}_{He}}{\overline{u}_{H_2}} = \sqrt{\frac{\mathcal{M}_{H_2}}{\mathcal{M}_{He}}} = \sqrt{\frac{2.02}{4.00}} = 0.711$$

$$\overline{u}_{He} = 0.711\ \overline{u}_{H_2}$$

SIMILAR EXAMPLE: Exercise 35(b).

Example 5-22 A sample of Kr(g) escapes through a tiny hole in 87.3 s. How long would it take an identical sample of Ne(g) (i.e., one containing the same number of molecules per unit volume) to effuse under the same conditions of T and P?

Solution. Since Ne has a lower molar mass than Kr, it should effuse *faster* and take a *shorter* time to escape. The required ratio of molar masses should be less than one.

$$\frac{\text{effusion time for Ne}}{\text{effusion time for Kr}} = \sqrt{\frac{\mathcal{M}_{Ne}}{\mathcal{M}_{Kr}}} = \sqrt{\frac{20.2}{83.8}} = 0.491$$

effusion time for Ne = 0.491 × effusion time for Kr
= 0.491 × 87.3 s = 42.9 s

SIMILAR EXAMPLES: Review Problem 16; Exercises 38, 39.

Example 5-23 If molecules effusing from a sample of N_2 travel a distance of 60.2 cm in a certain period of time, how far will molecules of O_2 travel under the same conditions?

Solution. O_2 molecules are slightly heavier than N_2 molecules and should travel a *shorter* distance.

$$\frac{\text{distance for } O_2}{\text{distance for } N_2} = \sqrt{\frac{\mathcal{M}_{N_2}}{\mathcal{M}_{O_2}}} = \sqrt{\frac{28.0}{32.0}} = 0.935$$

distance for O_2 = 0.935 × 60.2 cm = 56.3 cm

SIMILAR EXAMPLES: Review Problem 16; Exercises 38, 39.

Applications of Diffusion. That gases effuse through openings and that they diffuse into one another are commonly experienced phenomena. A rubber balloon filled with H_2(g) or He(g) gradually deflates, no matter how tightly it is tied off. This is because gas molecules effuse through tiny, invisible holes in the rubber. As the H_2 or He effuses out of the balloon, air effuses in; but because the H_2 or He effuses more rapidly, the total number of molecules in the balloon decreases.

Natural gas is odorless, and for commercial use small quantities of a gaseous organic sulfur compound are added to it. The sulfur compound has an odor that can be detected in parts per billion (ppb) or less. When a leak occurs we rely on the diffusion of this odorous compound for detection of the leak.

In the Manhattan Project during World War II one of the methods developed for separating the desired isotope $^{235}_{92}U$ from the predominant species $^{238}_{92}U$ involved gaseous diffusion. In this process uranium is obtained as the gaseous hexafluoride, UF_6(g). Molecules of $^{235}UF_6$ diffuse a little faster than those of $^{238}UF_6$, producing a slight enrichment of the ^{235}U isotope. By repeating the process many times over, a separation of the isotopes can be achieved.

5-11 Nonideal Gases

We have stated on several occasions that real gases can be described by the ideal gas equation only under certain conditions. How serious are the departures from ideality displayed by real gases? An indication is given in Figure 5-17, where *PV/RT* is plotted as a function of *P*, for 1 mol of gas at a fixed temperature (0°C). If a gas is ideal, *PV/RT* = 1 for 1 mol of gas. The extent to which the measured value of *PV/RT*, called the *compressibility factor*, deviates from unity is a measure of the nonideality of a gas. Figure 5-17 suggests that all gases behave ideally at sufficiently low pressures, say below 1 atm, but

FIGURE 5-17
The behavior of real gases—compressibility factor for one mole of gas as a function of pressure at 0°C.

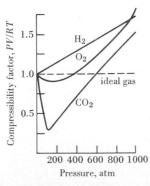

that deviations become significant at increased pressures. At very high pressures the compressibility factor is always greater than 1.

That real gases display behavior that deviates from the ideal is not difficult to rationalize. For example, Boyle's law predicts that at very high pressures a gas volume becomes extremely small, approaching zero. This cannot be, however, because the molecules themselves occupy space and are practically incompressible.

Another source of failure stems from the assumption of no intermolecular forces. If there is an attractive force among the molecules of a gas, the frequency and the force of the collisions of gas molecules with the container walls are reduced. The pressure exerted by a real gas on the container walls is less than would be predicted for an ideal gas. Intermolecular attractive forces are responsible for compressibility factors less than unity, and they become increasingly important at low temperatures where molecular motion is diminished in intensity.

To summarize, high pressures and low temperatures are the conditions that produce nonideal gas behavior. Conversely, gases tend to behave more ideally as the pressure is lowered and/or the temperature is raised.

The van der Waals Equation. A number of equations of state have been proposed for real gases, equations that apply over a wider range of temperatures and pressures than does the ideal gas equation. Such equations must correct for the effect of the two primary factors that cause deviations from ideal gas behavior: the volume associated with the molecules themselves and intermolecular forces of attraction. One equation that has found considerable application is the van der Waals equation.

$$\left(P + \frac{n^2 a}{V^2}\right)(V - nb) = nRT \tag{5.26}$$

In equation (5.26) V represents the volume of n moles of gas. The term $n^2 a / V^2$ is related to the intermolecular forces of attraction. It is added to the pressure because the measured pressure is lower than anticipated. The term b is related to the volume of the gas molecules and must be subtracted from the measured volume. Thus, $V - nb$ represents the *free* volume within the gas. The terms a and b have particular values for particular gases and vary somewhat with temperature and pressure (see Table 5-3). An equation of state for a nonideal gas is not as general as for an ideal gas; the identity of the gas enters into the equation. In Example 5-24 the pressure of a real gas is calculated with the van der Waals equation. Solving equation (5.26) for either n or V is a more difficult matter and, for those interested, is considered in Exercise 41. In Exercise 42 a still more general equation of state known as the *virial equation* is considered.

Example 5-24 Use the van der Waals equation to calculate the pressure exerted by 1.00 mol $Cl_2(g)$ when it is confined to a volume of 2.00 L at 273 K. Values of a and b are given in Table 5-3.

TABLE 5-3
Van der Waals constants for several gases

Gas	a, L^2 atm mol^{-2}	b, L mol^{-1}
Ar	1.35	0.0322
Cl_2	6.49	0.0562
CO	1.49	0.0399
CO_2	3.59	0.0427
H_2	0.244	0.0266
He	0.034	0.0237
N_2	1.39	0.0391
O_2	1.36	0.0318
SO_2	6.71	0.0564

Solution. Substitute the following values into equation (5.26).

$n = 1.00$ mol; $V = 2.00$ L; $T = 273$ K; $R = 0.0821$ L atm mol^{-1} K^{-1};

$n^2 a = (1.00)^2$ $mol^2 \times 6.49 \dfrac{L^2 \text{ atm}}{mol^2} = 6.49$ L^2 atm;

$nb = 1.00$ mol $\times 0.0562$ L/mol $= 0.0562$ L

$$P = \left(\frac{nRT}{V - nb}\right) - \frac{n^2 a}{V^2}$$

$$= \frac{1.00 \text{ mol} \times 0.0821 \text{ L atm } mol^{-1} K^{-1} \times 273 \text{ K}}{(2.00 - 0.0562) \text{ L}} - \frac{6.49 \text{ } L^2 \text{ atm}}{(2.00)^2 \text{ } L^2}$$

$$= 11.5 \text{ atm} - 1.62 \text{ atm} = 9.9 \text{ atm}$$

The pressure of the $Cl_2(g)$ calculated with the ideal gas equation is 11.2 atm. If only the b term is used in the van der Waals equation, the calculated pressure is 11.5 atm.

Including the *a* term reduces the calculated pressure by 1.62 atm. Under the conditions stated here, intermolecular forces are the main cause of the failure of $Cl_2(g)$ to behave ideally.

SIMILAR EXAMPLE: Exercise 40.

5-12 Postscript: The Atmosphere

If the outer limit of the atmosphere is taken to be the distance at which its composition becomes the same as that of interplanetary space (very, very low density atomic hydrogen gas, H), our atmosphere is about 10,000 km "thick." From another viewpoint the atmospheric blanket is very "thin." The total mass of the atmosphere is only about one-millionth that of the earth itself. Whether we think of the atmosphere as thick or thin, there is no question about the crucial role that it plays in the existence of life on earth.

Human beings and other animals depend on the oxygen content of the lower atmosphere to maintain their metabolic processes (see Section 27-3). Plants, through the process of photosynthesis, use carbon dioxide, a minor atmospheric component, and return oxygen to the atmosphere (see Section 27-2). Nitrogen, a vital element of life, circulates among organisms through a complex cycle, called the nitrogen cycle, that originates with atmospheric nitrogen (see Section 13-10). Carbon dioxide plays a key role in maintaining the heat balance of the earth (see Section 13-8), as does gaseous ozone (O_3) found in the stratosphere. Ozone also serves to screen the earth from harmful ultraviolet radiation (see Section 13-12).

The first 80 km or so of the atmosphere is a region known as the homosphere, so called because the composition of the gaseous mixture is essentially uniform (homogeneous) throughout this region. The portion of the atmosphere beyond the 80-km limit is called the heterosphere. It consists of four layers of gases—molecular nitrogen (N_2), atomic oxygen (O), helium (He), and atomic hydrogen (H) (Figure 5-18).

FIGURE 5-18
The atmosphere: structure, temperatures, composition, and other phenomena.

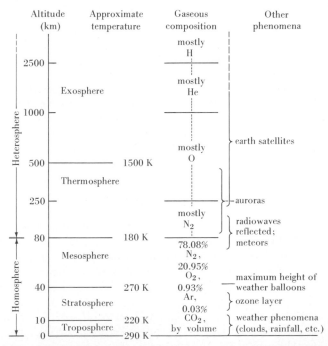

The values shown here are approximate. For example, the height of the troposphere varies from about 8 km at the poles to 16 km at the equator, and temperatures in the thermosphere vary greatly between day and night.

As suggested by Figure 5-18, the temperature of the atmosphere falls continuously for about the first 10 km above the earth's surface. This 10-km layer of air is that most familiar to us—the **troposphere.** Temperatures in the layer of air from about 10 to 40 km increase slowly from about 220 to 270 K. This is the region known as the **stratosphere.** (Supersonic transport aircraft fly in the lower regions of the stratosphere.) In the third atmospheric layer—the **mesosphere**—the temperature continues to rise to about 300 K and then falls to a minimum of about 180 K at 80 km. In the next layer of the atmosphere, the **thermosphere** (or **ionosphere**), temperatures rise continuously to 1500 K. In this region absorption of ultraviolet radiation from the sun causes gas molecules to ionize and/or dissociate. Thus, the atmosphere at these altitudes consists of positive and negative ions, free electrons, and neutral atoms and molecules. The term *ionosphere* is suggestive of this ionization process, and the term *thermosphere* suggests high temperatures associated with this ionized gas.

Example 5-25   At an altitude of 300 km the principal gaseous species is *atomic oxygen*, O. Oxygen atoms at this altitude have average speeds, $u_{av.} = 1.36 \times 10^3$ m/s. What is the approximate temperature corresponding to this molecular speed?

Solution.   We need to use two equations from the kinetic molecular theory described earlier. Equation (5.23) provides a relationship between $u_{av.}$ and u_{rms}, from which we determine

$$u_{rms} = \frac{u_{av.}}{0.921} = \frac{1.36 \times 10^3 \text{ m/s}}{0.921} = 1.48 \times 10^3 \text{ m/s}$$

Equation (5.22) provides a relationship between u_{rms} and kelvin temperature. As in Example 5-20, the gas constant must be expressed as 8.314 J mol^{-1} K^{-1}, and the molar mass of the gas in *kilograms* per mole. (Recall also that 1 joule = 1 kg m^2 s^{-2}.)

$$u_{rms} = \sqrt{\frac{3RT}{\mathcal{M}}} \qquad \text{and} \qquad (u_{rms})^2 = \frac{3RT}{\mathcal{M}} \qquad \text{and} \qquad T = \frac{\mathcal{M}(u_{rms})^2}{3R}$$

$$T = \frac{0.016 \text{ kg mol}^{-1} \times (1.48 \times 10^3)^2 \text{ m}^2 \text{ s}^{-2}}{3 \times 8.314 \text{ kg m}^2 \text{ s}^{-2} \text{ mol}^{-1} \text{ K}^{-1}} = 1.41 \times 10^3 \simeq 1400 \text{ K}$$

SIMILAR EXAMPLES: Exercises 35, 36.

An Additional Insight into the Meaning of Temperature.   The temperature of 1400 K calculated in Example 5-25, if achieved in bulk matter, would be considered very high (high enough to cause a bright red glow in iron). At high altitudes in the atmosphere there is a different significance to high temperature. The temperature is high in the sense that molecular speeds are high, but to experience the effects of high temperature requires that heat be transferred through *frequent* molecular collisions. Because the gas density at high altitudes is so low, molecular collisions occur only *infrequently*. A thermometer would give very low readings, even though in the midst of highly energetic, *but widely separated* molecules.

Meteors are extraterrestial chunks of matter that are trapped in the earth's gravitational field, disintegrate, and give off light as they fall through the atmosphere (hence the name "shooting stars"). Light emission from meteors is believed to be preceded by evaporation and ionization of atoms from the surface of the meteor. This evaporation of surface atoms in turn results from collisions with molecules of air. Meteors do not start to give off light until they fall to within 110 km of the earth's surface. The majority of them are completely vaporized in the range from about 80 to 110 km. Thus, meteors pass through the higher temperatures of the thermosphere without vaporizing. Instead, they vaporize in a lower-temperature region (about 220 K), but a region where gas densities are higher.

Summary

A gas is described through four variables—pressure, temperature, volume, and amount of gas. Gas pressure is most readily measured by comparing it to the pressure exerted by a liquid column, usually mercury. Atmospheric pressure is measured with a mercury barometer, other gas pressures, with various manometers.

Relationships between gas variables taken two at a time (with the remaining two held constant) are known as the simple gas laws. Most frequently encountered are Boyle's law, relating gas pressure and volume, Charles' law, relating gas volume and temperature, and Avogadro's law, relating volume and amount of gas. A number of important ideas originate with the simple gas laws. Among these are the concept of an absolute zero of temperature, a temperature scale (kelvin) based on this absolute zero, a standard condition of temperature and pressure (STP), and the molar volume of a gas at STP—22.4 L/mol.

By combining Boyle's, Charles', and Avogadro's laws, a more general statement of gas behavior is obtained, the ideal gas equation: $PV = nRT$. This equation can be solved for any one of the variables when values are known for the others. The ideal gas equation can also be applied to molecular weight and gas density determinations. The ideal gas equation is used in describing the gaseous reactants and/or products of a chemical reaction, and to describe mixtures of gases. Partial pressures and partial volumes are useful concepts for dealing with gaseous mixtures. A particularly important application of Dalton's law of partial pressures involves the collection of gases over water.

The kinetic molecular theory provides a theoretical basis for the ideal gas equation. With this theory it is possible to establish a relationship between average molecular kinetic energy and kelvin temperature. Another relationship involves the root-mean-square speed of molecules, the temperature, and the molar mass of a gas. Also, the effusion and diffusion of gases can be related to their molar masses through the kinetic molecular theory.

Real gases generally behave ideally only at high temperatures and low pressures. Nonideal behavior results chiefly from intermolecular attractions and the finite volumes occupied by gas molecules. Alternative equations of state have been developed for real gases. The best known of these, perhaps, is the van der Waals equation.

Learning Objectives

As a result of studying Chapter 5, you should be able to

1. Explain the operation of a mercury barometer, an open-end manometer, and a closed-end manometer; and be able to use data obtained with these instruments.

2. Calculate, for a fixed amount of gas at a fixed temperature, how the volume changes with pressure, and vice versa.

3. Calculate, for a fixed amount of gas at a fixed pressure, how the volume changes with temperature, and vice versa.

4. Discuss the significance of the absolute zero of temperature; and calculate kelvin temperatures from Celsius temperatures, and vice versa.

5. State the standard conditions of temperature and pressure (STP) for a gas.

6. State the relationship between the volume and amount of gas at a fixed temperature and pressure; and use the molar volume of a gas at STP in calculations.

7. Use the ideal gas equation to calculate one gas variable—pressure, volume, temperature, amount of gas—when the other three are known.

8. Rearrange and use the ideal gas equation to calculate molecular weights and gas densities from experimental data.

9. Use the ideal gas equation, together with other data, in stoichiometric calculations for reactions involving gases.

10. Calculate, for mixtures of gases, such quantities as partial pressures, total pressures, partial volumes, total volumes, compositions, and apparent molar masses.

11. Do calculations involving the collection of gases over water.

12. State the basic postulates of the kinetic molecular theory of gases.

13. Show how the results of the kinetic molecular theory can be used to deduce the simple gas laws.

14. Calculate the root-mean-square speed of the molecules of a gas of known molar mass at a given temperature.

15. Relate properties associated with the effusion or diffusion of gases to their molar masses.

16. Describe the conditions under which a gas is most likely to behave as an ideal gas, and use the van der Waals equation for nonideal behavior.

Some New Terms

One **atmosphere** (standard atmosphere) is the pressure exerted, under carefully specified conditions, by a column of mercury 760 mm high.

Avogadro's hypothesis states that equal volumes of different gases, compared under identical conditions of temperature and pressure, contain equal numbers of molecules.

A **barometer** is a device used to measure the pressure of the atmosphere.

Boyle's law states that the volume of a fixed amount of gas at a constant temperature is inversely proportional to the gas pressure.

Charles' law states that the volume of a fixed amount of gas at a constant pressure is directly proportional to the absolute (kelvin) temperature.

Dalton's law of partial pressures states that in a mixture of gases the total pressure is the sum of the partial pressures of the gases present.

The **gas constant, R,** is the numerical constant appearing in the ideal gas equation ($PV = nRT$).

Graham's law states that the rates of effusion (or diffusion) of two different gases are inversely proportional to the square roots of their molar masses. **Effusion** refers to the escape of a gas through a tiny hole. **Diffusion** refers to the spreading of a gas throughout a larger volume.

An **ideal gas** is one whose behavior can be predicted by the ideal gas equation. The behavior of a **nonideal gas** departs from that predicted by the ideal gas equation.

The **ideal gas equation** relates the pressure, volume, temperature, and number of moles of a gas (n) through the expression $PV = nRT$.

The **kinetic molecular theory of gases** is a model for describing gas behavior. It is based on a set of assumptions, and yields mathematical equations from which various properties of gases can be deduced.

A **manometer** is a device used to measure the pressure of a gas, either by comparing the gas pressure with barometric pressure (open-end manometer) or by measuring it directly (closed-end manometer).

Mole fraction describes a mixture (of gases, for example) in terms of the fraction of all the molecules that are of a particular type.

A **partial pressure** is the pressure exerted by an individual gas in a mixture, independently of the other gases.

Pressure is a force per unit area. Applied to gases, pressure is most easily understood in terms of the height of a liquid column that can be maintained by the gas.

Temperature (on the kelvin scale) is a measure of the average kinetic energy of the molecules of a substance.

The **van der Waals equation** is an equation of state for nonideal gases. It includes correction terms to account for intermolecular forces of attraction and for the volume occupied by the gas molecules themselves.

Suggestions for Further Study

Cohen, E. G. D., "Toward Absolute Zero," *American Scientist*, **65**, 752 (1977).

Davenport, D. A., M. Howe-Grant, and V. Srinivasan, "Musical Molecular Weights and Other Non-linear Properties of Gases," *J. Chem. Educ.*, **56**, 523 (1979).

Harris, A. D., "The Density and Apparent Molecular Weight of Air," *J. Chem. Educ.*, **60**, 74 (1984).

Herron, J. D., "Derivation of the Ideal Gas Law," *J. Chem. Educ.*, **57**, 201 (1980).

Vaitkunas, J. J., "Derivation of the Ideal Gas Law," *J. Chem. Educ.*, **56**, 530 (1979).

Whitaker, R. D., "The Early Development of Kinetic Theory," *J. Chem. Educ.*, **56**, 315 (1979).

Review Problems

1. Convert the following pressures to an equivalent pressure in standard atmospheres: **(a)** 737 mmHg; **(b)** 68.3 cm Hg; **(c)** 1215 torr; **(d)** 28 psi.

2. Calculate the height of a mercury column required to produce a pressure **(a)** of 1.35 atm; **(b)** of 618 torr; **(c)** equal to that of a column of water 138 ft high.

3. A sample of $O_2(g)$ occupies a volume of 31.7 L at 753 mmHg. What is the new gas volume if, while the temperature and amount of gas are held constant, the pressure is **(a)** lowered to 487 mmHg; **(b)** increased to 3.15 atm?

4. A sample of $N_2(g)$ at STP is compressed from a volume of 10.5 L to 832 cm^3 while the temperature and amount of gas are fixed. What is the final gas pressure?

5. A 138-cm^3 sample of Ne(g) is initially at 737 mmHg and 30°C. What will be the new volume if, while the pressure and amount of gas are held constant, the temperature is **(a)** increased to 85°C; **(b)** lowered to 0°C?

6. It is desired to increase the volume of a fixed amount of gas from 87.5 to 118 cm^3 while holding the pressure constant. What must be the final temperature if the initial temperature is 23°C?

7. Indicate how the final volume, V_f, is related to the initial volume, V_i, for a fixed amount of gas in each case.

(a) The pressure is decreased from 3 to 1 atm while the temperature is held at 25°C.
(b) The temperature is lowered from 400 to 100 K while the pressure is held constant at 1 atm.
(c) The temperature is raised from 200 to 300 K while the pressure is increased from 2 to 3 atm.

8. What is the volume occupied by 62.3 g CO(g) at 33°C and 728 mmHg?

9. A 23.5-L cylinder contains 45.7 g $SO_2(g)$ at 22°C. What is the pressure exerted by this gas?

10. A 0.341-g sample of gas has a volume of 355 cm^3 at 98.7°C and 743 mmHg. What is the molecular weight of this gas?

11. What is the density (in g/L) of $CO_2(g)$ at 28.7°C and 787 mmHg?

12. A method of removing $CO_2(g)$ from a spacecraft is to allow the CO_2 to react with NaOH. What volume of CO_2 at 25.8°C and 749 mmHg can be removed per kg NaOH?

$$2 \; NaOH(s) + CO_2(g) \rightarrow Na_2CO_3(s) + H_2O(l)$$

13. What volume of $O_2(g)$ is consumed in the combustion of 30.2 L $C_3H_8(g)$? Both gases are measured at STP.

$$C_3H_8(g) + O_2(g) \rightarrow CO_2(g) + H_2O(l) \quad \text{(unbalanced)}$$

14. What is the volume occupied by a mixture of 16.5 g Ne(g) and 34.5 g Ar(g) at 11.2 atm pressure and 37.8°C?

15. A 76.7-cm^3 sample of "wet" $O_2(g)$ is collected over water at 22°C and 752 mmHg barometric pressure. How much O_2, in moles, is present in the gas? Vapor pressure of H_2O at 22°C = 19.8 mmHg.

16. A sample of $Cl_2(g)$ effuses through a tiny hole in 32.5 s. How long would it take for an equivalent sample of $SO_2(g)$ to effuse under the same conditions?

Exercises

Pressure and its measurement

1. Convert each of the following pressures to the equivalent pressure in standard atmospheres: **(a)** 1387 mmHg; **(b)** 7.14 kg/cm^2; **(c)** 314 kPa; **(d)** 992 mb; **(e)** 2.53×10^5 N/m^2.

2. Calculate the following quantities.
(a) the height of a column of liquid glycerol ($d = 1.26$ g/cm^3) required to exert the same pressure as 2.85 m of $CCl_4(l)$ ($d = 1.59$ g/cm^3)
(b) the height of liquid benzene ($d = 0.879$ g/cm^3) required to exert a pressure of 3.14×10^4 N/m^2
(c) the density of a liquid if a 15.0-ft column is to exert a pressure of 12.5 lb/in.^{2}

3. The mercury level in the open arm of an open-end manometer is 283 mm above a reference point. In the arm connected to a container of gas, the level is 38 mm above the same reference point. Barometric pressure is 745 mmHg. What is the pressure of the gas, in atm?

4. A gas is collected over water as illustrated in Figure 5-14. The water level inside the gas container is 3.8 cm above the level outside. If barometric pressure is 753.5 mmHg, what is the total pressure of the gas in the container?

The simple gas laws

5. A sample of $N_2(g)$ that occupies a volume of 467 cm^3 at 746 mmHg is expanded, at constant temperature, to a volume of 569 cm^3. What is the new gas pressure?

6. A 15.5-L cylinder of Ar(g) is connected to an evacuated 2425-L tank. If the final gas pressure is 712 mmHg, what must have been the original gas pressure in the cylinder?

7. A fixed amount of gas, maintained in a constant volume of 247 cm^3, exerts a pressure of 814 mmHg at 23.2°C. At what temperature will the pressure of this gas become exactly 1.00 atm?

8. A 10.0-g sample of a gas has a volume of 4.62 L at 35°C and 762 mmHg. If to this *constant* 4.62-L volume is added 2.3 g of the same gas and the temperature raised to 51°C, what is the new gas pressure?

9. A 385-cm^3 sample of $O_2(g)$ is obtained at 22°C and 748 mmHg. **(a)** What would be the volume of this gas at STP? **(b)** How many moles of gas must be present?

10. A 12.5-g sample of gas is added to an evacuated, constant-volume vessel at 22°C. The pressure of the gas is to be held constant as the temperature is changed. How much of the gas must be allowed to escape when the temperature is raised to 195°C?

Ideal gas equation

11. A 47.3-L constant-volume cylinder containing 1.62 mol He(g) is heated until the pressure reaches 1.85 atm. What is the temperature of the gas?

12. Kr(g) in an 18.5-L cylinder exerts a pressure of 8.61 atm at 24.8°C. What is the mass of gas present?

13. A sample of gas has a volume of 4.18 L at 29.7°C and 732 mmHg. What is the volume of this gas at 24.8°C and 756 mmHg?

14. A 25.0-L cylinder contains 128 g $N_2(g)$ at 12°C. What mass of N_2 must be released to reduce the pressure in the cylinder to 1.65 atm?

$\star$**15.** A balloon is inflated with 1.00 ft^3 of He(g) at STP and released. What is the gas pressure in the balloon when it has expanded to a volume of 75.0 L? Assume a temperature of −20°C at this altitude.

*16. Use SI units and the ideal gas equation to
  (a) Express the gas constant R in the units $kPa\ dm^3\ mol^{-1}\ K^{-1}$.
  (b) Use the value found in (a), together with information from Appendix B, to obtain R in the units $J\ mol^{-1}\ K^{-1}$.
  (c) Calculate the pressure, in kPa, exerted by 1205 g CO(g) confined in a tank of $1.56\ m^3$ volume at 291 K.

Molecular weight determination

17. A gaseous hydrocarbon weighing 0.185 g occupies a volume of $110\ cm^3$ at 26°C and 743 mmHg. What is the molecular weight of this compound? What conclusion can you draw about its molecular formula?

18. A 2.650-g sample of a gas occupies a volume of $428\ cm^3$ at 742.3 mmHg and 24.3°C. Analysis of this compound shows it to be 15.5% C, 23.0% Cl, and 61.5% F, by mass. What is the molecular formula of this compound?

19. A glass vessel weighs 56.1035 g when evacuated; 264.2931 g when filled with Freon-113, a liquid with density $1.576\ g/cm^3$; and 56.2445 g when filled with acetylene gas at 749.3 mmHg and 20.02°C. What is the molecular weight of acetylene?

Gas densities

20. A particular application calls for $N_2(g)$ with a density of 1.50 g/L at 25.0°C. What must be the pressure of the $N_2(g)$?

21. The density of phosphorus vapor at 310°C and 775 mmHg is 2.64 g/L. What is the molecular formula of the phosphorus?

Cannizzaro's method

22. The following gases all contain the element X. Construct a table similar to Table 5-1 and determine the atomic weight of X. What element do you think X is?

Compound	Molecular weight	X, %
nitryl fluoride	65.01	49.4
nitrosyl fluoride	49.01	32.7
thionyl fluoride	86.07	18.6
sulfuryl fluoride	102.07	31.4

Gases in chemical reactions

23. A particular coal sample contains 2.32% S, by mass. When the coal is burned, the sulfur is converted to $SO_2(g)$. What volume of $SO_2(g)$, measured at 25.0°C and 749 mmHg, is produced by burning 2.0×10^6 lb of this coal?

24. A 2.71-g sample of a KCl–$KClO_3$ mixture is decomposed by heating and produces $90.2\ cm^3\ O_2(g)$, measured at 23.2°C and 741 mmHg. What is the percent $KClO_3$, by mass, in the mixture? (*Hint:* KCl in the mixture is unchanged.)

$$2\ KClO_3(s) \rightarrow 2\ KCl(s) + 3\ O_2(g)$$

25. The Haber process is the principal method for fixing nitrogen (converting N_2 to nitrogen compounds).

$$N_2(g) + 3\ H_2(g) \rightarrow 2\ NH_3(g)$$

  (a) How many liters of $NH_3(g)$ can be produced from 371 L $H_2(g)$ if the gases are measured at 525°C and 515 atm pressure?
  (b) How many liters of $NH_3(g)$, *measured at STP*, can be produced from 371 L $H_2(g)$ measured at 525°C and 515 atm pressure?

26. 1.50 L $H_2S(g)$, measured at 23.0°C and 735 mmHg, is mixed with 4.45 L $O_2(g)$, measured at 26.1°C and 750 mmHg, and burned.

$$2\ H_2S(g) + 3\ O_2(g) \rightarrow 2\ SO_2(g) + 2\ H_2O(g)$$

  (a) How much $SO_2(g)$, in moles, is produced?
  *(b) If the excess reactant and the products of the reaction are collected at 748 mmHg and 120.0°C, what volume will they occupy?

Mixtures of gases

27. A gas cylinder of 55.0 L volume contains $N_2(g)$ at a pressure of 32.5 atm and 23°C. What mass of Ne(g) must be introduced into this same cylinder to raise the total pressure to 65.0 atm?

28. A 1.85-L container of $H_2(g)$ at 777 mmHg and 25.0°C is connected to a 2.52-L container of He(g) at 742 mmHg and 25.0°C. What is the *total* gas pressure after the gases have mixed, with the temperature remaining at 25.0°C?

29. A mixture of 4.0 g $H_2(g)$ and an unknown quantity of He(g) is maintained at STP. If 10.0 g $H_2(g)$ is added to the mixture, while conditions are maintained at STP, the gas volume doubles. What mass of He is present?

30. Air that is exhaled (expired) by a human being differs from normal air. A typical analysis of expired air at 37°C and 760 mmHg, expressed as percent *by volume*, is 74.2% N_2, 15.2% O_2, 3.8% CO_2, 5.9% H_2O, and 0.9% Ar.
  (a) What is the apparent molar mass of this expired air? (Recall Example 5-18.)
  (b) Would you expect the density of expired air to be greater or less than that of ordinary air at the same temperature and pressure? Explain.
  (c) What is the ratio of the partial pressure of $CO_2(g)$ in expired air to that in ordinary air?

Collection of gases over water

31. A 1.93-g sample of Al reacts with excess HCl(aq) and the liberated H_2 is collected over water at 26°C at a barometric pressure of 738 mmHg. What volume of gas is collected? Vapor pressure of H_2O at 26°C = 25.2 mmHg.

$$2\ Al(s) + 6\ HCl(aq) \rightarrow 2\ AlCl_3(aq) + 3\ H_2(g)$$

32. A 243-cm^3 sample of Ar(g) at 26°C and at a barometric pressure of 755 mmHg is passed through water at 26°C. What is the volume of the gas when saturated with water vapor and again measured at 26°C and 755 mmHg barometric pressure? Water vapor pressure at 26°C = 25.2 mmHg.

33. A sample of O_2(g) is collected over water at 25°C. The volume of the gas is 1.28 L. In a subsequent experiment it is determined that the mass of O_2 present is 1.58 g. What must have been the barometric pressure at the time the gas was collected? Vapor pressure of water at 25°C = 23.8 mmHg.

Kinetic molecular theory

34. A kinetic theory verification of Avogadro's law was provided in the text. Use equations from Section 5-10 to verify Boyle's and Charles' laws.

35. The root-mean-square velocity, u_{rms}, of H_2 molecules at 273 K is 1.84×10^3 m/s.
   (a) At what temperature is u_{rms} for H_2 equal to 3.68×10^3 m/s?
   (b) What is u_{rms} for N_2 at 273 K?

36. Calculate u_{rms}, in m/s, for Cl_2(g) molecules at 25°C.

Effusion of gases

37. What are the ratios of the diffusion rates for the following pairs of gases? (a) H_2 and O_2; (b) H_2 and D_2 (D = deuterium, i.e., $^2_1$H); (c) $^{235}UF_6$ and $^{238}UF_6$

38. If 0.00251 mol NH_3(g) effuses through an orifice in a certain period of time, how much HCl(g) would effuse in the same time with the same initial conditions?

39. A sample of N_2(g) effuses through a tiny hole in 38 s. What must be the molecular weight of a gas that requires 55 s to effuse under identical conditions?

Nonideal gases

40. Calculate the pressure exerted by 1.00 mol CO_2(g) confined to a volume of 855 cm^3 at 30°C. Use (a) the ideal gas equation and (b) the van der Waals equation. (c) Compare the results and explain.

***41.** If the van der Waals equation is solved for volume, a cubic equation is obtained.
   (a) Derive the following equation by rearranging equation (5.26).

$$V^3 - n\left(\frac{RT + bP}{P}\right)V^2 + \left(\frac{n^2a}{P}\right)V - \frac{n^3ab}{P} = 0$$

   (b) What is the volume occupied by 132 g CO_2(g) at a pressure of 10.0 atm and 280 K? Use data from Table 5-3 as necessary.

***42.** The virial equation of state for O_2(g) has the form

$$P\bar{V} = RT\left\{1 + \frac{B}{\bar{V}} + \frac{C}{\bar{V}^2}\right\}$$

where $\bar{V}$ is the molar volume, $B = -21.89$ cm^3/mol, and $C = 1230$ cm^6/mol^2.
   (a) Use the equation to calculate the pressure exerted by 1 mol O_2(g) confined to a volume of 500 cm^3 (i.e., having $\bar{V} = 500$ cm^3/mol) at 273 K.
   (b) Is the result calculated in (a) consistent with that suggested for O_2(g) by Figure 5-17? Explain.

Additional Exercises

1. A gas occupies a volume of 323 cm^3 at 738 mmHg and 25°C. What *additional* pressure is required to reduce the gas volume to 275 cm^3?

2. A sample of N_2(g) occupies a volume of 58.0 cm^3 under the existing barometric pressure. Increasing the pressure by 125 mmHg reduces the volume to 49.6 cm^3. What is the prevailing barometric pressure?

3. Start with the conditions at points A, B, and C in Figure 5-8. Use Charles' law to calculate the volume of each gas at 0, −100, −200, −250, and −270°C; and show that indeed the volume of each gas becomes zero at −273.15°C.

4. A 12.1-L cylinder contains 38.7 g O_2(g) at 25°C. What is the pressure of this gas?

5. A 0.312-g sample of a gaseous compound occupies 185 cm^3 at 25.0°C and 745 mmHg. The compound consists of 85.6% C and 14.4% H, by mass. What is its molecular formula?

6. What mass of He(g) should be added to 2.15 L O_2(g) at 23°C

and 715 mmHg to increase the pressure to 1.25 atm? (The volume and temperature are held constant.)

7. A 1.85-g sample of NH_4NO_3(s) is introduced into an evacuated 2.12-L flask and then heated to 250°C. What is the total gas pressure in the flask at 250°C when the NH_4NO_3(s) has completely decomposed?

$$NH_4NO_3(s) \rightarrow N_2O(g) + 2\ H_2O(g)$$

8. What is the partial pressure of Cl_2(g) in a gaseous mixture at STP that consists of 50.0% N_2, 22.3% Ne, and 27.7% Cl_2, *by mass*?

9. Producer gas is a type of fuel gas made by passing air or steam through a bed of hot coal or coke. A typical producer gas has the following composition, in percent by volume: 8.0% CO_2, 23.2% CO, 17.7% H_2, 1.1% CH_4, and 50.0% N_2.
   (a) What is the apparent molar mass of this gas?
   (b) What is the density of this gas at 25°C and 752 mmHg?
   (c) What is the partial pressure of CO in this gaseous mixture at STP?

10. A *mixture* of $H_2(g)$ and $O_2(g)$ is prepared by electrolyzing 1.32 g of water, and the mixture of gases is collected over water at 30°C when the barometric pressure is 748 mmHg. The volume of "wet" gas obtained is 2.90 L. What must be the vapor pressure of water at 30°C?

$$2 H_2O(l) \xrightarrow{\text{electrolysis}} 2 H_2(g) + O_2(g)$$

11. At what temperature will u_{rms} for Ne(g) be the same as u_{rms} for He at 300 K?

12. Following the method in Table 5-2, determine $\bar{u}$ and u_{rms} for a group of six particles with the speeds: 9.8×10^3, 9.0×10^3, 8.3×10^3, 6.5×10^3, 3.7×10^3, and 1.8×10^3 m/s.

*13. Recall the composition of air (Example 5-17). What volume of air, measured at STP, is required to complete the combustion of 1.00×10^3 L of a natural gas (measured at 23°C and 741 mmHg) having the composition, 77.3% CH_4, 11.2% C_2H_6, 5.8% C_3H_8, 2.3% C_4H_{10} (and 3.4% noncombustible gases), by volume?

*14. Mixtures of the anesthetic gas cyclopropane, $(CH_2)_3$, and air with between 2.4 and 10.3% $(CH_2)_3$, by volume, are explosive. A sealed 1500-ml cylinder of $(CH_2)_3$ (g) at 2.50 atm and 25°C is placed in a fume hood of volume 72 ft³ containing air at 755 mmHg and 25°C. If the seal on the cylinder were to break and the $(CH_2)_3$ to mix with the air in the fume hood, would an explosive mixture result?

*15. A gaseous mixture of He and O_2 has a density of 0.518 g/L at 25°C and 720 mmHg. What is the % He, by mass, in the mixture?

*16. The equation

$$\frac{m/V}{P} = \frac{d}{P} = \frac{\mathcal{M}}{RT}$$

suggests that the ratio of gas density (d) to gas pressure (P), at constant temperature, should be a constant. The following gas density data were obtained for *oxygen* gas at various pressures at 273.15 K: **1.000 atm,** 1.428962 g/L; **0.750 atm,** 1.071485 g/L; **0.500 atm,** 0.714154 g/L; **0.250 atm,** 0.356985 g/L.

(a) Calculate values of d/P, and with a graph or by other means determine the best value of the term d/P for oxy-

gen at 273.15 K. (This is the value corresponding to oxygen as an ideal gas.)

(b) Use the value of d/P from part (a) to calculate a precise value of the atomic weight of oxygen and compare it with that listed in an atomic weight table.

*17. A particular limestone contains only $CaCO_3$ and $MgCO_3$. When a 0.4515-g sample of this limestone is decomposed by heating, 0.2398 g of mixed oxide (CaO and MgO) is obtained. What volume of $CO_2(g)$ at 752.0 mmHg and 285.3 K will be produced by decomposing 50.0 lb of this limestone?

$$MCO_3(s) \rightarrow MO(s) + CO_2(g) \quad \text{(where M = Ca or Mg)}$$

*18. A sounding balloon is a rubber bag, filled with $H_2(g)$ and carrying a set of instruments (the "payload"). Because this combination of bag, gas, and payload has a smaller mass than a corresponding volume of air, the balloon rises. As the balloon rises, it expands. From the following data estimate the maximum height to which the balloon can rise: mass of balloon, 1200 g; payload, 1700 g; quantity of $H_2(g)$ in balloon, 120 ft³ at STP; diameter of balloon at maximum height, 25 ft. Air pressure and temperature as a function of altitude are **0 km,** 1.0×10^3 mb, 288 K; **5 km,** 5.4×10^2 mb, 256 K; **10 km,** 2.7×10^2 mb, 223 K; **20 km,** 5.5×10^1 mb, 217 K; **30 km,** 1.2×10^1 mb, 230 K; **40 km,** 2.9×10^0 mb, 250 K; **50 km,** 8.1×10^{-1} mb, 250 K; **60 km,** 2.3×10^{-1} mb, 256 K.

*19. Atmospheric pressure as a function of altitude can be calculated with an equation known as the barometric formula.

$$P = P_0 \times 10^{-\mathcal{M}gh/2.303\ RT}$$

where P is the pressure, in atm, at an altitude of h meters. P_0 is the pressure at sea level (usually taken to be 1 atm), g is the acceleration due to gravity ($9.80\ m/s^2$); and $\mathcal{M}$ is the molar mass of air, expressed in kg per mole. R is expressed as $8.314\ J\ mol^{-1}\ K^{-1}$, and T is in kelvins.

(a) Estimate barometric pressure at the top of Mt. Whitney in California. (Altitude: 14,494 ft; assume a temperature of 10°C.)

(b) Use the barometric formula to show that barometric pressure decreases by one-thirtieth in value for every 900-ft increase in altitude.

Self-Test Questions

For questions 1 through 8 select the single item that best completes each statement.

1. The greatest pressure of the following is that exerted by (a) a column of Hg(l) 75.0 cm high ($d = 13.6\ g/cm^3$); (b) 10.0 g $H_2(g)$ at STP; (c) a column of air 10 mi high; (d) a column of $CCl_4(l)$ 60.0 cm high ($d = 1.59\ g/cm^3$).

2. For a fixed amount of gas at a fixed pressure, changing the temperature from *100°C* to *200 K* causes the gas volume (a) to

decrease; (b) to double; (c) to increase, but not to twice its original value; (d) not to change.

3. A sample of $O_2(g)$ is collected over water at 23°C at a barometric pressure of 751 mmHg (vapor pressure of water at 23°C = 21 mmHg). The *partial* pressure of $O_2(g)$ in the sample collected is (a) 21 mmHg; (b) 751 mmHg; (c) 0.96 atm; (d) 1.02 atm.

4. A comparison is made at standard temperature and pressure (STP) of 0.50 mol $H_2(g)$ and 1.0 mol He(g). The two gases will

(a) have equal average molecular kinetic energies; (b) have equal average molecular speeds; (c) occupy equal volumes; (d) have equal effusion rates.

5. At 0°C and 0.500 atm, 4.48 L $NH_3(g)$ (a) contains 0.20 mol NH_3; (b) weighs 1.70 g; (c) contains 6.02×10^{23} NH_3 molecules; (d) contains 0.40 mol NH_3.

6. In the reaction $2\,Al(s) + 6\,HCl(aq) \rightarrow 2\,AlCl_3(aq) + 3\,H_2(g)$
(a) 67.2 L $H_2(g)$ at STP is produced for every mol Al that reacts;
(b) 6 L $HCl(aq)$ is consumed for every 3 L $H_2(g)$ produced;
(c) 11.2 L $H_2(g)$ at STP is produced for every mol HCl consumed;
(d) 33.6 L $H_2(g)$ is produced, *regardless of temperature and pressure,* for every mol Al that reacts.

7. A mixture of 0.50 mol $H_2(g)$ and 0.50 mol $SO_2(g)$ is introduced into a 10.0-L container at 25°C. The container has a "pinhole" leak. After a period of time, the partial pressure of $H_2(g)$ in the *remaining mixture* (a) exceeds that of $SO_2(g)$; (b) is equal to that of $SO_2(g)$; (c) is less than that of $SO_2(g)$; (d) is the same as in the original mixture.

8. To establish a pressure of 2.00 atm in a 2.24-L cylinder containing 1.60 g $O_2(g)$ at 0°C, (a) add 1.60 g O_2; (b) release 0.80 g O_2; (c) add 2.00 g He; (d) add 0.60 g He.

9. 0.10 mol He(g) is added to 2.24 L $H_2(g)$ at standard temperature and pressure (STP). This is followed by an increase in temperature to 100°C while the pressure and amount of gas are held constant. What is the final gas volume?

10. Explain briefly why the height of the mercury column in a barometer is *independent* of the diameter of the barometer tube (i.e., whether the diameter is 1 mm, 1 cm, 10 cm, . . .).

11. Calculate the number of L $H_2(g)$ (measured at 22°C and 745 mmHg) required to react with 30.0 L CO(g) (measured at 0°C and 760 mmHg) in the reaction

$3\,CO(g) + 7\,H_2(g) \rightarrow C_3H_8(g) + 3\,H_2O(l)$

12. A particular gaseous hydrocarbon that is 82.7% C and 17.3% H, by mass, has a density of 2.35 g/L at 25°C and 752 mmHg. What is the molecular formula of this hydrocarbon?

6 Thermochemistry

In Chapter 4 we learned that chemical reactions can be used to transform matter from one chemical form (the reactants) to another (the products). We focused on calculating quantities of reactants consumed and products formed in a chemical reaction—reaction stoichiometry. Stoichiometric calculations were expanded in scope in Chapter 5 with consideration of gases in chemical reactions. Now we shift our attention to quantities of *energy* exchanged between a reaction mixture and its surroundings. Specifically, we will see how quantities of heat energy can be represented in chemical equations and incorporated into stoichiometric calculations.

6-1 Thermodynamics: A Preview

Thermodynamics, of which thermochemistry is one important aspect, deals with relationships between heat energy and other energy forms known as work. Our formal introduction to this branch of science will come in Chapter 16, but some of the terminology of thermodynamics is used so extensively in chemistry that we need to consider the meanings of a few terms at this point.

The portion of the universe selected for a thermodynamic study is called a **system,** and the portions of the universe with which the system interacts are called the **surroundings** of the system. A thermodynamic system may be as simple as a beaker of water or as complex as the contents of a blast furnace or a polluted lake. **Interactions** refer to the transfer of energy or matter between a system and its surroundings; these interactions are generally the focus of a thermodynamic study. Energy transfers can occur as **heat (q)** or in several other forms, known collectively as **work (w).** Energy transfers occurring as heat or work affect the total amount of energy contained within a system, its **internal energy (E).**

Thermodynamics is independent of any particular theory of the structure of matter. It was, in fact, fully developed as a science *before* modern atomic theory. Thus, the concept of the internal energy within a system can be handled by the methods of thermodynamics without ever describing where this energy comes from. We now recognize, however, that internal energy represents the total energy associated with the ultimate particles of matter in the system. This includes energy associated with chemical bonds between atoms, energies of intermolecular attractions, the kinetic energy of translational motion of molecules, and so on.

The concepts of heat and work are crucial to an understanding of thermodynamics, and we will need to say a good deal more about each. Our approach to thermodynamics

FIGURE 6-1
Pressure–volume work.

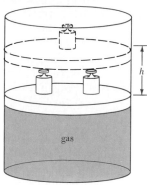

gas

When one of the weights is removed from the piston confining the gas, the remaining weight is lifted through the distance h. Work is performed.

will be in two parts. In this chapter we will focus on heat, the type of energy transfer that characterizes thermochemistry. In Chapter 16 we will present the three laws of thermodynamics. The first law establishes an important relationship among heat, work, and internal energy. The second law provides a criterion for predicting situations under which chemical reactions will occur and important ideas about chemical equilibrium (recall our introduction to this topic in Section 4-5). The third law gives us a useful basis for certain thermodynamic calculations. We will postpone a close look at the concept of work to Chapter 16. Nevertheless, a few words about this concept may be helpful here, mainly to establish a distinction between heat and work.

Work. Refer to Appendix B for the description of a succession of physical quantities leading to the notions of work and energy. Work is performed when a force acts through a distance, and the quantity of work, w, is the product of the force and the distance. The SI unit of work is the joule (J), which can also be expressed as

$$1 \text{ joule} = 1 \text{ kg m}^2 \text{ s}^{-2}$$

One way to think of work is as any form of energy transfer that can be expressed through the lifting or lowering of weights. Figure 6-1 suggests how work is performed when a gas is allowed to expand. Work associated with the expansion or compression of gases is called pressure–volume work.

6-2 Heat

Heat can be thought of as energy that is transferred as a result of a temperature difference. Energy, as heat, flows from a warmer body (higher temperature) to a colder body (lower temperature). At the molecular level this means that molecules of the warmer body lose kinetic energy to those in the colder body when the two bodies are brought into contact. As a result, the average translational kinetic energy of the molecules in the warmer body is lowered—its temperature is lowered. In the cooler body the temperature is raised. Energy is transferred—heat flows—between two bodies until they reach the same temperature.* It is important to emphasize, however, that a body contains energy, but it does not *contain* heat. Heat describes energy in transit across the boundary between a system and its surroundings.

The quantity of heat energy, q, required to change the temperature of a substance depends on how much the temperature is to be changed, the quantity of substance, and its identity (type of molecules). **Heat capacity** is the quantity of heat energy required to raise the temperature of a substance by 1°C. Heat capacity, of course, depends on the quantity of substance. **Specific heat capacity,** or more simply **specific heat,** is the quantity of heat energy required to raise the temperature of one *gram* of substance by 1°C. **Molar heat capacity** is the quantity of heat energy required to raise the temperature of one *mole* of substance by 1°C.

Historically, heat energy has been defined by the unit the calorie. One **calorie (cal)** is the quantity of heat energy required to raise the temperature of 1 g of water from 14.5 to 15.5°C. Thus, we can say that, at 15°C, the specific heat of water is 1.000 cal g^{-1} °C^{-1} (or 1.000 cal per g per °C). The molar heat capacity of water at the same temperature is 18.02 cal mol^{-1} °C^{-1} since there is 18.02 g H_2O in 1 mol. Specific heat is itself a function of temperature, and this is why a particular temperature interval is chosen in the definition of the calorie. For H_2O, the effect of temperature on specific heat is small, and

*Although the ultimate result of heat flow is to equalize the temperature of two bodies, the temperature of a body may remain constant for a time as heat enters or leaves it. This is what happens, for example, when a block of ice (0°C) absorbs heat from the surrounding air. Its temperature does not change until all the ice has melted. Such processes, called phase transitions, are described in Chapter 11.

It is unfortunate that in nutritional studies the unit called a Calorie is actually a kilocalorie. Thus, a "1 Calorie" soft drink actually has a food value of 1000 cal. Throughout this text, reference will always be to the defined calorie.

throughout the temperature interval from 0 to 100°C its specific heat generally can be taken as 1.00 cal g^{-1} °C^{-1}.

The calorie is a rather small quantity of energy, and in many applications the larger unit, the **kilocalorie (kcal),** is used. 1 kcal = 1000 cal. Also, it should be noted that the calorie is *not* an SI unit. We will introduce the appropriate SI unit shortly, and we will then redefine the calorie in terms of the SI unit.

Example 6-1   How much heat is required to raise the temperature of 735 g of water from 21.0 to 98.0°C? (Assume that the specific heat of water remains at 1.00 cal g^{-1} °C^{-1} throughout this temperature range.)

Solution.   The quantity of heat required to change the temperature of 735 g of water is 735 times as great as to change the temperature of 1 g of water. And the heat required to raise the temperature by 77.0°C (from 21.0 to 98.0°C) is 77 times as great as to raise the temperature 1°C. Thus, the specific heat of water must be multiplied by 735 and by 77.

$$\text{no. cal} = \frac{1.00 \text{ cal}}{\text{g water °C}} \times 735 \text{ g water} \times (98.0 - 21.0)°C = 5.66 \times 10^4 \text{ cal}$$

SIMILAR EXAMPLES: Review Problems 1, 2; Exercise 1.

Calculations involving specific heat can usually be performed by the line of reasoning in Example 6-1, but it is perhaps simpler to think in these terms.

$$\text{quantity of heat} = q$$
$$= \underbrace{\text{mass of substance} \times \text{specific heat}}_{\text{heat capacity}} \times \text{temperature change} \qquad (6.1)$$

Since the degree of temperature interval is the same on the two scales, either Celsius or Kelvin temperatures can be used in equation (6.2).

The temperature change in equation (6.1) can be expressed as

$$\Delta T = T_f - T_i \qquad (6.2)$$

where T_f is the final temperature, T_i is the initial temperature, and ΔT is the temperature change. The Greek letter Δ ("delta") is commonly used to represent the change in a property.

If the temperature of a substance is increased, the final temperature is greater than the initial temperature (noted symbolically as $T_f > T_i$) and ΔT is positive (i.e., $\Delta T > 0$). The quantity of heat energy calculated has a *positive* sign, signifying that heat is *absorbed* or *gained* when the temperature of a substance is increased. If the temperature of a substance is lowered, the final temperature is less than the initial temperature (noted symbolically as $T_f < T_i$). In this case ΔT is negative (i.e., $\Delta T < 0$). The quantity of heat carries a *negative* sign, signifying that heat is *evolved* or *lost* when a substance is cooled.

Development of the law of conservation of energy occurred over a long period of time. Count Rumford gave the essence of the law in explaining the heat evolved in the boring of cannons (1798). The law was proposed by Julius Mayer in 1842, and independently stated in more precise terms by Helmholtz in 1847. Credit for the law is now divided about equally among Mayer, Helmholtz, and James Joule.

An additional idea that enters into heat energy calculations is the **law of conservation of energy.** In interactions among objects or substances, the total energy remains constant. Thus, in an interaction between two objects, energy lost by one object must be gained by the other. The simple laboratory method of determining the specific heat of a metal illustrated in Figure 6-2 is based on the law of conservation of energy. In the exchange of heat energy between the lead (q_{lead}) and the water (q_{water}), the total quantity of heat energy must be zero.

$$q_{\text{lead}} + q_{\text{water}} = 0 \qquad (6.3)$$

or

$$q_{\text{lead}} = -q_{\text{water}} \qquad (6.4)$$

That is, the two terms must be equal in magnitude and opposite in sign. The heat lost by one object must be gained by the other.

FIGURE 6-2
Determination of
specific heat of lead—
Example 6-2 illustrated

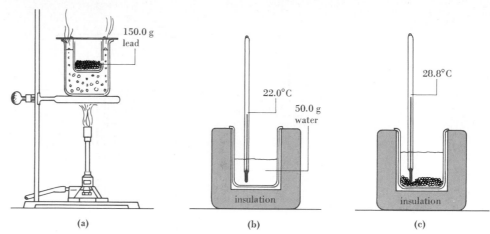

(a)   (b)   (c)

(a) 150.0 g of lead at the temperature of boiling water (100.0°C).
(b) 50.0 g of water in a thermally insulated beaker at 22.0°C.
(c) Final lead–water mixture at a temperature of 28.8°C.

Example 6-2   Use data presented in Figure 6-2 to calculate the specific heat of lead.

Solution.   First, let us use equation (6.1) to calculate q_{water}.

$$q_{water} = 50.0 \text{ g water} \times \frac{1.00 \text{ cal}}{\text{g water °C}} \times (28.8 - 22.0)°C = 340 \text{ cal}$$

From equation (6.4) we may write

$$q_{lead} = -q_{water} = -340 \text{ cal}$$

Now, from equation (6.1) again, we obtain

$$q_{lead} = 150.0 \text{ g lead} \times \text{sp. ht. lead} \times (28.8 - 100.0)°C$$
$$= -340 \text{ cal}$$

$$\text{sp. ht. lead} = \frac{-340 \text{ cal}}{150.0 \text{ g lead} \times (28.8 - 100.0)°C}$$
$$= \frac{-340 \text{ cal}}{150 \text{ g lead} \times (-71.2)°C} = 0.032 \frac{\text{cal}}{\text{g lead °C}}$$

SIMILAR EXAMPLES: Review Problem 3; Exercises 2, 4.

Mechanical Equivalent of Heat.   Figure 6-3 illustrates a significant experiment performed by James Joule in 1847. The work associated with the falling weights appeared as an increase in internal energy of the water or other liquid used, and resulted in a temperature rise in the liquid. In separate experiments the same temperature increases were produced by energy transfers through heat. The number of joules of work required to produce a given temperature increase proved to be about 4.15 times greater than the number of calories of heat required to produce the same temperature increase. Joule's studies established an equivalence between work and heat. Since the joule is the basic energy unit of the SI system, we will henceforth express both heat and work in joules. Based on modern measurements, the *defined* calorie and kilocalorie are

$$1 \text{ cal} = 4.184 \text{ J} \quad \text{and} \quad 1 \text{ kcal} = 4.184 \text{ kJ} \tag{6.5}$$

Here is a useful distinction between heat and work: Work implies organized molecular motion, and heat implies random or chaotic motion. In the work performed by the ex-

FIGURE 6-3
Establishing the
mechanical equivalent of
heat.

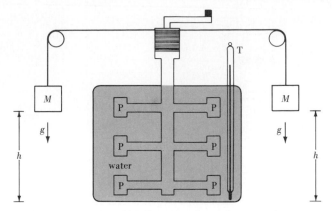

As the two weights M fall under the influence of gravity, g, through the distance h, a quantity of work is done equal to $2Mgh$. This work is first expressed in the rotational motion of the paddles, P. Then this rotational kinetic energy is converted to internal energy of the water as the paddles are slowed down and stopped by the water. The temperature of the water rises as a result of its increased internal energy. The process can be repeated many times by raising the weights by winding the crank. Separate experiments are performed to measure the quantity of heat required to produce the same temperature increase as the falling weights.

panding gas in Figure 6-1, all the atoms of the piston and weights moved in the same direction and by the same distance. In the work performed by the paddles in Figure 6-3, organized motion (falling weights, rotating paddles) was involved. When energy is transferred as heat, the average translational kinetic energy of molecules changes, but the molecules continue their random disorganized motion.

6-3 Heat of Reaction

When sucrose (ordinary cane sugar) is metabolized in the body, a complicated series of chemical reactions and energy conversions occurs (described in Section 27-3). The net result of these reactions, though, is the same as that obtained by the complete combustion of sucrose—production of $CO_2(g)$ and $H_2O(1)$.

$$C_{12}H_{22}O_{11}(s) + 12\ O_2(g) \longrightarrow 12\ CO_2(g) + 11\ H_2O(l) \tag{6.6}$$
(sucrose)

Lavoisier was the first to recognize that metabolism and combustion are closely related, that is, that the heat of reaction (6.6) is the same for both processes.

The caloric value of sucrose metabolized as a food is the same as the difference in internal energy between the products [12 mol $CO_2(g)$ and 11 mol $H_2O(l)$] and the reactants [1 mol $C_{12}H_{22}O_{11}(s)$ and 12 mol $O_2(g)$] in reaction (6.6). This combustion reaction can be carried out in such a way that the internal energy difference appears as a quantity of heat energy transferred between the reaction mixture (the system) and its surroundings. This quantity of heat energy can be called the **heat of reaction** and designated as q_{rxn}.

Calorimetry. The laboratory determination of a heat of reaction is carried out in a device called a **calorimeter.** The type shown in Figure 6-4, which is ideally suited for reaction (6.6), is called a **bomb calorimeter.** The thermodynamic system is the *contents* of the bomb, that is, just the reactants and the products. The steel bomb itself, the water in which the bomb is immersed, the thermometer, the stirrer, and so on, constitute the surroundings.

Heat evolved in the reaction goes largely toward raising the temperature of the water surrounding the bomb. Small quantities, however, are also required to raise the tempera-

FIGURE 6-4
Experimental
measurement of a heat
of reaction.

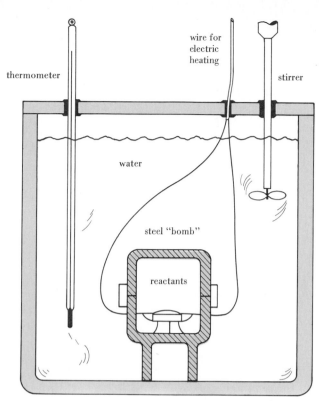

The reaction is initiated by momentary electric heating of a length of iron wire covered with the reactants. The heat of reaction is determined by measuring the total quantity of heat absorbed by the surroundings. The heat absorbed by the water is the product of the mass, specific heat, and temperature increase of the water. The heat absorbed by the rest of the surroundings—bomb, stirrer, thermometer, etc.—is the product of their combined heat capacity and the temperature increase. A separate experiment is required to establish this combined heat capacity, i.e., the heat capacity of the calorimeter assembly. Every calorimeter assembly has its own distinctive heat capacity.

ture of the bomb itself, of the stirrer, and of other parts of the calorimeter. Thus, we need to deal with three quantities of heat—q_{rxn}, q_{water}, and $q_{calorim.}$. The heat of reaction is q_{rxn}. The heat effect in the surroundings is separated into that affecting the water, q_{water}, and that affecting the rest of calorimeter assembly, $q_{calorim.}$. If we follow the reasoning used in developing equations (6.3) and (6.4), we conclude that

$$q_{rxn} + q_{water} + q_{calorim.} = 0 \tag{6.7}$$

and

$$q_{rxn} = -(q_{water} + q_{calorim.}) \tag{6.8}$$

To evaluate q_{water} and $q_{calorim.}$ we use equation (6.1). When equations (6.1) and (6.8) are combined, we obtain the final expression

$$q_{rxn} = -[(\text{mass of water} \times \text{sp. ht. water} \times \text{temp. change}) \tag{6.9}$$
$$+ (\text{heat capacity of calorimeter} \times \text{temp. change})]$$

Example 6-3   The burning of 1.010 g sucrose, $C_{12}H_{22}O_{11}$, in a bomb calorimeter causes the water temperature to rise from 24.92 to 28.33°C. The calorimeter contains 980.0 g water, and the heat capacity of the calorimeter is 785 J/°C. **(a)** What is the heat of combustion of sucrose, expressed in kJ/mol $C_{12}H_{22}O_{11}$? **(b)** Verify the claim of sugar manufacturers that one teaspoon of sugar (about 4.8 g) "contains only 18 Cal."

Solution

(a) Let us calculate q_{water} and $q_{calorim.}$ separately. Note that since 1 cal = 4.184 J, the specific heat of water may be written as 4.184 J (g $H_2O)^{-1}$ $°C^{-1}$.

$$q_{water} = 980.0 \text{ g } H_2O \times \frac{4.184 \text{ J}}{\text{g } H_2O \text{ °C}} \times (28.33 - 24.92)°C = 1.40 \times 10^4 \text{ J}$$

$$q_{calorim.} = \frac{785 \text{ J}}{°C} \times (28.33 - 24.92)°C = 2.68 \times 10^3 \text{ J} = 0.27 \times 10^4 \text{ J}$$

Now we use equation (6.8) to calculate q_{rxn} for the 1.010-g sample.

$$q_{rxn} = -(q_{water} + q_{calorim.}) = -(1.40 \times 10^4 \text{ J} + 0.27 \times 10^4 \text{ J}) = -1.67 \times 10^4 \text{ J}$$

per gram $C_{12}H_{22}O_{11}$:

$$q_{rxn} = \frac{-1.67 \times 10^4 \text{ J}}{1.010 \text{ g } C_{12}H_{22}O_{11}} = -1.65 \times 10^4 \text{ J/g } C_{12}H_{22}O_{11} = -16.5 \text{ kJ/g } C_{12}H_{22}O_{11}$$

per mol $C_{12}H_{22}O_{11}$:

$$q_{rxn} = \frac{-16.5 \text{ kJ}}{\text{g } C_{12}H_{22}O_{11}} \times \frac{342 \text{ g } C_{12}H_{22}O_{11}}{1 \text{ mol } C_{12}H_{22}O_{11}} = -5.64 \times 10^3 \text{ kJ/mol } C_{12}H_{22}O_{11}$$

(b) One of the quantities calculated in part (a) was the heat of combustion per gram of sucrose. We need to determine the heat of combustion of 4.8 g (one teaspoonful) of sucrose, and we need to convert this value to kcal.

$$\text{no. kcal} = \frac{-16.5 \text{ kJ}}{\text{g } C_{12}H_{22}O_{11}} \times 4.8 \text{ g } C_{12}H_{22}O_{11} \times \frac{1 \text{ kcal}}{4.184 \text{ kJ}} = -19 \text{ kcal}$$

Recall that a nutritional calorie, 1 Cal, is actually 1 kcal. The manufacturers' claim seems justified.

SIMILAR EXAMPLES: Review Problems 4, 5; Exercises 14, 15.

Example 6-4   The combustion of benzoic acid is often used in the laboratory to determine the heat capacity of a bomb calorimeter assembly. If the combustion of a 1.000-g sample of benzoic acid ($C_7H_6O_2$) causes a temperature increase of 4.96°C when a bomb calorimeter contains 1085 g of water, what is the heat capacity of the calorimeter? (A handbook lists the heat of combustion of benzoic acid as −26.42 kJ/g.)

In precise studies the heat capacity of a calorimeter is established with an energy input from an electrical source.

Solution.   All the data required in equation (6.9) are known except the heat capacity of the calorimeter. We substitute these data and solve for this heat capacity.

$$q_{rxn} = -26.42 \text{ kJ} = -2.642 \times 10^4 \text{ J}$$

$$= -\left[\left(1085 \text{ g water} \times \frac{4.184 \text{ J}}{\text{g water °C}} \times 4.96°C\right) + (\text{ht. cap.} \times 4.96°C)\right]$$

$$= -2.642 \times 10^4 \text{ J} = -2.25 \times 10^4 \text{ J} - (\text{ht. cap.} \times 4.96°C)$$

$$\text{heat capacity of calorimeter} = \frac{-2.25 \times 10^4 \text{ J} + 2.642 \times 10^4 \text{ J}}{4.96°C} = \frac{0.39 \times 10^4 \text{ J}}{4.96°C}$$

$$= 7.9 \times 10^2 \text{ J/°C}$$

SIMILAR EXAMPLES: Exercises 13, 17.

6-4 Enthalpy and Enthalpy Changes

In our discussion of bomb calorimetry we mentioned that energy evolved as heat in the combustion of sucrose (reaction 6.6) corresponds to the difference in internal energy between the products and reactants. We could have called this the change in internal energy for the reaction, ΔE. For any other combustion reaction in a bomb calorimeter (that is, for a reaction at constant volume) the heat of reaction is also ΔE. Most chemical reactions, however, are not carried out in bomb calorimeters. The metabolism of sucrose occurs under the conditions present in the human body. The combustion of methane (natural gas) in a water heater occurs in an open flame. This question then arises: "How does the heat of a reaction measured in a bomb calorimeter compare with the heat of reaction if the reaction is carried out in some other way?" The usual "other" way of conducting chemical reactions is in beakers, flasks, and other containers open to the atmosphere and under the constant pressure of the atmosphere.

In the combustion of sucrose the heat of reaction turns out to be the same, whether the reaction is carried out in a bomb calorimeter or in the open atmosphere. This is because, in either case, the only form of energy transfer between the reaction mixture and the surroundings is as heat. However, in many reactions carried out in the open atmosphere a small amount of pressure–volume work is done as the system expands or contracts. In these cases the measured heat of reaction is slightly different from ΔE for the reaction. Because of this fact, it proves useful to define a new thermodynamic property that is closely related to internal energy but has this important advantage: Its change corresponds to the measured heat of reaction for a reaction carried out in the open atmosphere (more precisely, a reaction carried out under constant pressure and with work limited to the pressure–volume type). The function that serves this purpose is called **enthalpy** and is denoted by the symbol H.*

In Chapter 16 we will restate enthalpy in mathematical terms and show how a heat of reaction determined under one set of conditions can be used to calculate the heat of reaction under different conditions. But these matters need not concern us now. We will limit our current discussion to enthalpy changes (ΔH), which you can consider to differ only slightly from changes in internal energy (ΔE). For the remainder of this section we explore some additional features of enthalpy that are important for its proper use.

Functions of State.   Internal energy, which we have described as the total energy associated with the ultimate particles of matter in a system, is something that *cannot be measured*. However, internal energy depends only on the conditions that characterize a system and not on how those conditions were achieved. The condition of a system is referred to as its **state,** and any property that depends only on the state of a system is called a **function of state** (or state function). This means that as long as its state is specified, a system has a *unique* value of E, even though we cannot measure the absolute value of the internal energy. Enthalpy, which is closely related to internal energy, also cannot be measured, but it is defined in such a way as to be a state function. For any particular state of a system there is a *unique* value of H. A further characteristic of a state function is that the difference in value of the function in two different states is a *unique* quantity.

An analogy to a state function is provided in Figure 6-5, which relates to climbing a mountain. Path (a) is short but steep; path (b) is longer and more gradual. The length of time to climb the mountain depends on the path chosen, but the total elevation gain is *fixed*. The elevation gain corresponds to ΔH. Furthermore, the loss of elevation in climbing back down the mountain is analogous to $-\Delta H$. Thus, following a round trip to the top of the mountain (ΔH) and back down ($-\Delta H$), the total elevation gain is *zero*. To state this

FIGURE 6-5
An analogy to a thermodynamic function of state.

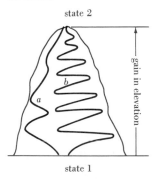

The gain in elevation in climbing from the base to the summit of the mountain is independent of the path chosen. This elevation gain is analogous to ΔH in a thermodynamic system.

---

*The term *heat content* has also been used for enthalpy, but there is an obvious objection to its use. A substance does not contain heat.

conclusion in thermodynamic terms, following a transition from state 1 to state 2 and back to the initial condition (state 1) all functions of state, including enthalpy, must regain their original values.

$$\text{state 1} \xrightarrow{\Delta H} \text{state 2} \xrightarrow{-\Delta H} \text{state 1} \tag{6.10}$$

We can think of the combustion of sucrose in this way.

$$\underbrace{C_{12}H_{22}O_{11}(s) + 12\ O_2(g)}_{\substack{\text{state 1,} \\ \text{having an enthalpy, } H_1 \\ H_1 \text{ cannot be} \\ \text{measured but has} \\ \text{a unique value.}}} \longrightarrow \underbrace{12\ CO_2(g) + 11\ H_2O(l)}_{\substack{\text{state 2,} \\ \text{having an enthalpy, } H_2 \\ H_2 \text{ cannot be} \\ \text{measured but has} \\ \text{a unique value.}}}$$

For the reaction: $\Delta H = H_2 - H_1$

ΔH has a unique value, which is the measured heat of reaction when the reaction is carried out at constant pressure.

Representing ΔH in a Chemical Reaction. From the result of Example 6-3 we can add this important bit of thermochemical information to the chemical equation (6.6).

The complete combustion of one mole of solid sucrose, producing gaseous carbon dioxide and liquid water as the sole products, is accompanied by a decrease of enthalpy equal to 5.64×10^3 kJ.

$$C_{12}H_{22}O_{11}(s) + 12\ O_2(g) \longrightarrow 12\ CO_2(g) + 11\ H_2O(l) \qquad \Delta\overline{H} = -5.64 \times 10^3 \text{ kJ/mol} \tag{6.11}$$

A line may be drawn above the symbol H (called an overbar) to signify that molar

FIGURE 6-6
Exothermic and endothermic reactions represented through an enthalpy diagram.

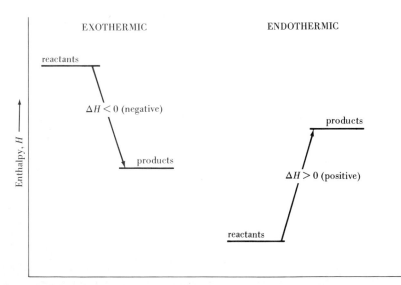

Because an absolute value of H cannot be measured, we cannot assign numerical values on the enthalpy axis. What we can say is that enthalpy increases in the direction shown by the arrow (upward). For an exothermic reaction, such as the combustion of sucrose (reaction 6.11), the reactants have a higher enthalpy than the products and the reaction is accompanied by a decrease in enthalpy. The enthalpy change, ΔH, is negative, i.e., has a value less than zero. For an endothermic reaction, such as the synthesis of NO(g) from its elements (reaction 6.12), the opposite situation prevails; i.e., ΔH is positive or greater than zero.

FIGURE 6-7
A calorimeter
constructed from
Styrofoam coffee cups.

The reaction mixture is in the inner cup. The outer cup provides additional thermal insulation from the surrounding air. The cup is closed off with a cork stopper through which a thermometer and a stirrer are immersed into the reaction mixture.

amounts of reactants and products are involved.* Thermochemical equations are sometimes written with fractional coefficients, as in the formation of one mole of NO(g) from its elements.

$$\tfrac{1}{2}N_2(g) + \tfrac{1}{2}O_2(g) \longrightarrow NO(g) \qquad \Delta\overline{H} = +90.37 \text{ kJ/mol} \tag{6.12}$$

Exothermic and Endothermic Reactions.   The negative sign of ΔH in equation (6.11) means that the enthalpy of the products is lower than that of the reactants. This *decrease* in enthalpy appears as energy evolved to the surroundings in the form of heat. A reaction which *gives off* heat is an **exothermic** reaction. The situation in reaction (6.12) is the opposite. The products have a higher enthalpy than the reactants; ΔH is positive. To produce this *increase* in enthalpy, heat is absorbed from the surroundings. A reaction which *absorbs* heat is an **endothermic** reaction. Figure 6-6 illustrates a useful method of indicating exothermic and endothermic reactions diagrammatically.

Experimental Determination of ΔH.   The simple calorimeter pictured in Figure 6-7 is much more commonly encountered in the general chemistry laboratory than is a bomb calorimeter. A chemical reaction is carried out in solution in a Styrofoam cup (generally in aqueous solution), and the temperature change is measured. Styrofoam is a good heat

---

*The unit kJ/mol appears in equation (6.11). We might ask, "per mole of what?" We can think of this as signifying -5.64×10^3 kJ of enthalpy change for every mole of $C_{12}H_{22}O_{11}$ consumed. But if we describe the reaction in terms of $O_2(g)$, the heat of reaction is $1/12 \times (-5.64 \times 10^3)$ kJ per mol O_2 consumed, since only $1/12$ mol $C_{12}H_{22}O_{11}$ is consumed per mol O_2. This difficulty would be compounded if we based the heat of reaction on one or another of the products formed. The way out of this difficulty is to base the heat of reaction, not on any particular reactant or product, but on the reaction as a whole. "One mole of reaction" is defined as involving the molar amounts of reactants and products represented in the balanced equation. Reaction (6.11) involves 1 mol $C_{12}H_{22}O_{11}$, 12 mol O_2, 12 mol CO_2, 11 mol H_2O, and -5.64×10^3 kJ of enthalpy change *per mole of reaction*.

This definition of a mole of reaction requires that a value of $\Delta\overline{H}$ always be linked with a particular equation; and because it is so linked, the overbar symbol and the /mol unit are frequently omitted. In Chapters 16 and 20 we will find the "per mole" terminology to be helpful.

insulator, so that there is very little heat transfer between the cup and the surrounding air. If a chemical reaction is exothermic, the heat energy released is retained within the solution and raises the temperature. If the reaction is endothermic, heat energy must be absorbed from the solution and its temperature falls. Because the reaction mixture is maintained under atmospheric pressure, the quantity of heat energy measured is at constant pressure: $q_P = \Delta H$.

Example 6-5   A 1.50-g sample of NH_4NO_3 is added to 35.0 g H_2O in a Styrofoam coffee cup and stirred until it dissolves. The temperature of the solution drops from 22.7 to 19.4°C. **(a)** Is the process endothermic or exothermic? **(b)** What is the heat of solution of NH_4NO_3 in water, expressed as kJ/mol NH_4NO_3?

Solution
(a) Since the water temperature decreases, the water must lose heat energy. This heat energy is absorbed by the NH_4NO_3 in order to dissolve. The process is endothermic.
(b) The two heat effects here can be designated, $q_{NH_4NO_3}$ and q_{water}. In the usual fashion we can write, $q_{NH_4NO_3} + q_{water} = 0$ and $q_{NH_4NO_3} = -q_{water}$.

$$q_{NH_4NO_3} = -\left[35.0 \text{ g } H_2O \times \frac{4.184 \text{ J}}{\text{g } H_2O \text{ °C}} \times (19.4 - 22.7)\text{°C} \right] = 4.8 \times 10^2 \text{ J}$$

The quantity of heat just calculated is for a 1.50-g sample. For 1.00 mol NH_4NO_3, that is, for

$$NH_4NO_3(s) \longrightarrow NH_4NO_3(aq) \qquad \Delta\overline{H} = ?$$

$$\Delta\overline{H} = \frac{4.8 \times 10^2 \text{ J}}{1.50 \text{ g } NH_4NO_3} \times \frac{80 \text{ g } NH_4NO_3}{1 \text{ mol } NH_4NO_3} \times \frac{1 \text{ kJ}}{1000 \text{ J}} = 26 \text{ kJ/mol } NH_4NO_3{}^*$$

SIMILAR EXAMPLES: Review Problem 6; Exercises 10, 11, 12.

To repeat what was mentioned earlier, a reaction carried out in a bomb calorimeter is often not at constant pressure. In these cases the heat of the reaction differs slightly from ΔH. The difference is usually so small, however, that it can be neglected. Until we encounter this point again in Chapter 16, we will consider all heat effects to be enthalpy changes, ΔH.

6-5 Relationships Involving ΔH

One of the most useful purposes served by the enthalpy concept is in providing the possibility of calculating large numbers of heats of reaction from a relatively small number of measurements. The following statements about enthalpy changes (ΔH) are crucial in this regard.

1. ΔH Is an Extensive Property. Enthalpy change is directly proportional to the amounts of substances involved in a process. If we double the amounts of reactants and products in reaction (6.12)—that is, if we double the equation—the enthalpy change is also doubled.

$$N_2(g) + O_2(g) \longrightarrow 2 \text{ NO}(g) \qquad \Delta\overline{H} = 2 \times (+90.37) = +180.74 \text{ kJ/mol}$$

2. ΔH Changes Sign When a Process is Reversed. Enthalpy (H) is a function of state. As described in the mountain-climbing analogy in Figure 6-5, if the direction of a process

---

*The heat of solution depends on the concentration of the solution produced. This value is for the production of about 0.5 M $NH_4NO_3(aq)$. The heat of solution for the production of 1 M $NH_4NO_3(aq)$, for example, would be slightly different.

is reversed, the change in property (ΔH) also reverses sign ($-\Delta H$). Thus, if for the formation of nitrogen oxide from its elements

$$\tfrac{1}{2} N_2(g) + \tfrac{1}{2} O_2(g) \longrightarrow NO(g) \qquad \Delta\overline{H} = +90.37 \text{ kJ/mol}$$

then for the decomposition of nitrogen oxide into its elements

$$NO(g) \longrightarrow \tfrac{1}{2} N_2(g) + \tfrac{1}{2} O_2(g) \qquad \Delta\overline{H} = -90.37 \text{ kJ/mol}$$

3. Hess's Law of Constant Heat Summation.

If a process can be considered to occur in stages or steps (either actually or hypothetically) the enthalpy change for the overall process can be obtained by summing the enthalpy changes for the individual steps.

This statement is again a consequence of the fact that enthalpy is a function of state. It can also be illustrated by the mountain-climbing analogy of Figure 6-5. Imagine that a trip from the base to the summit of the mountain is made in stages. The elevation gain (or loss) can be determined for each stage, and the total elevation gain is the sum of the changes for each stage (e.g., $+1000$ m, -200 m, $+400$ m, etc.).

Suppose that in the combination of $N_2(g)$ and $O_2(g)$, instead of stopping at $NO(g)$, the reaction is allowed to proceed to $NO_2(g)$.

$$\tfrac{1}{2} N_2(g) + O_2(g) \longrightarrow NO_2(g) \qquad \Delta\overline{H} = ? \qquad\qquad (6.13)$$

This reaction can be carried out in the two steps indicated below. When the two equations are summed, the net equation is (6.13). Hess's law states that the two enthalpy changes can also be summed, leading to $\Delta\overline{H}$ for reaction (6.13).

$$
\begin{array}{ll}
\tfrac{1}{2} N_2(g) + \tfrac{1}{2} O_2(g) \longrightarrow \cancel{NO(g)} & \Delta\overline{H} = +90.37 \text{ kJ/mol} \\
\cancel{NO(g)} + \tfrac{1}{2} O_2(g) \longrightarrow NO_2(g) & \Delta\overline{H} = -56.52 \text{ kJ/mol} \\
\hline
\tfrac{1}{2} N_2(g) + O_2(g) \longrightarrow NO_2(g) & \Delta\overline{H} = +90.37 - 56.52 \\
& \qquad\quad = +33.85 \text{ kJ/mol}
\end{array}
$$

Note that in this summation a species that would appear on both sides of the net equation (NO) cancels out (a feature previously illustrated in Figure 4-8). Figure 6-8 summarizes what we have just done through an enthalpy diagram.

FIGURE 6-8
Hess's law illustrated through an enthalpy diagram.

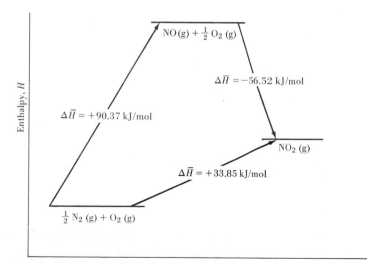

The conversion of $N_2(g)$ and $O_2(g)$ to $NO_2(g)$ is accompanied by the enthalpy change $\Delta\overline{H} = +33.85$ kJ/mol, whether the reaction is considered in a single step or in two steps [by way of $NO(g)$].

As another example, suppose that we wish to determine the enthalpy change ($\Delta \overline{H}$) for the reaction

$$3 \text{ C(graphite)} + 4 \text{ H}_2(g) \longrightarrow \text{C}_3\text{H}_8(g) \qquad \Delta \overline{H} = ? \tag{6.14}$$

How should we proceed? If graphite and $H_2(g)$ are introduced into a reaction vessel, some reaction may occur, but it will not go to completion. Neither will the products be limited to $C_3H_8(g)$; other hydrocarbons are formed as well. *$\Delta \overline{H}$ for reaction (6.14) cannot be measured directly.* Instead we must resort to an *indirect calculation* of the desired $\Delta \overline{H}$ from values of $\Delta \overline{H}$ that can be established by experiment. Heats of combustion are easily measured.

(a) $\text{C}_3\text{H}_8(g) + 5 \text{ O}_2(g) \longrightarrow 3 \text{ CO}_2(g) + 4 \text{ H}_2\text{O}(l)$ $\Delta \overline{H} = -2220.1$ kJ/mol

(b) $\text{C(graphite)} + \text{O}_2(g) \longrightarrow \text{CO}_2(g)$ $\Delta \overline{H} = -393.5$ kJ/mol

(c) $\text{H}_2(g) + \frac{1}{2} \text{O}_2(g) \longrightarrow \text{H}_2\text{O}(l)$ $\Delta \overline{H} = -285.8$ kJ/mol

The calculation of $\Delta \overline{H}$ for reaction (6.14) is completed in Example 6-6.

Example 6-6 Use equations (a), (b), and (c) to calculate $\Delta \overline{H}$ for reaction (6.14).

Solution. What we are trying to accomplish is the synthesis of one mole of $C_3H_8(g)$. Suppose we begin with a reaction in which C_3H_8 is produced from $CO_2(g)$ and $H_2O(l)$. This reaction can be represented by the reverse of equation (a).

$-$(a): $3 \text{ CO}_2(g) + 4 \text{ H}_2\text{O}(l) \longrightarrow \text{C}_3\text{H}_8(g) + 5 \text{ O}_2(g)$ $\begin{aligned}\Delta \overline{H} &= -(-2220.1 \text{ kJ/mol}) \\ &= +2220.1 \text{ kJ/mol}\end{aligned}$

Now suppose that the $CO_2(g)$ required in $-$(a) is produced by the combustion of graphite and the $H_2O(l)$, by the combustion of $H_2(g)$. To get the proper number of moles of each—3 of $CO_2(g)$ and 4 of $H_2O(l)$—we must multiply equation (b) by 3 and equation (c) by 4.

$3 \times$ (b): $3 \text{ C(graphite)} + 3 \text{ O}_2(g) \longrightarrow 3 \text{ CO}_2(g)$ $\begin{aligned}\Delta \overline{H} &= 3 \times (-393.5 \text{ kJ/mol}) \\ &= -1180.5 \text{ kJ/mol}\end{aligned}$

$4 \times$ (c): $4 \text{ H}_2(g) + 2 \text{ O}_2(g) \longrightarrow 4 \text{ H}_2\text{O}(l)$ $\begin{aligned}\Delta \overline{H} &= 4 \times (-285.8 \text{ kJ/mol}) \\ &= -1143.2 \text{ kJ/mol}\end{aligned}$

Let us think about the net change that we have described to this point: 3 mol C(graphite) and 4 mol $H_2(g)$ have been consumed and 1 mol $C_3H_8(g)$ has been produced. This is what is required in equation (6.14), and we are now able to combine the three modified equations, as follows:

$-$(a): $3 \text{ CO}_2(g) + 4 \text{ H}_2\text{O}(l) \longrightarrow \text{C}_3\text{H}_8(g) + 5 \text{ O}_2(g)$ $\Delta \overline{H} = +2220.1$ kJ/mol

$3 \times$ (b): $3 \text{ C(graphite)} + 3 \text{ O}_2(g) \longrightarrow 3 \text{ CO}_2(g)$ $\Delta \overline{H} = -1180.5$ kJ/mol

$4 \times$ (c): $4 \text{ H}_2(g) + 2 \text{ O}_2(g) \longrightarrow 4 \text{ H}_2\text{O}(l)$ $\Delta \overline{H} = -1143.2$ kJ/mol

$\overline{3 \text{ C(graphite)} + 4 \text{ H}_2(g) \longrightarrow \text{C}_3\text{H}_8(g)}$ $\Delta \overline{H} = -103.6$ kJ/mol

SIMILAR EXAMPLES: Review Problems 7, 8; Exercises 20, 21.

6-6 Standard Enthalpies of Formation

We have repeatedly noted that *absolute values of E and H do not exist*. Nevertheless, we have been successful in dealing with *changes* in these properties alone, e.g., ΔH.

We can return to our mountain analogy (Figure 6-5) for still another comparable situation. The difference in elevation between the summit and some fixed point at the base

TABLE 6-1

Some standard molar enthalpies (heats) of formation at 298 K

Substance	$\Delta \bar{H}^\circ_{f,\,298}$, kJ/mol[a]
$CH_4(g)$	−74.85
$C_2H_2(g)$	226.73
$C_2H_4(g)$	52.30
$C_2H_6(g)$	−84.68
$C_3H_8(g)$	−103.85
$CO(g)$	−110.54
$CO_2(g)$	−393.51
$HCl(g)$	−92.30
$H_2O(l)$	−285.85
$NH_3(g)$	−46.19
$NO(g)$	90.37
$SO_2(g)$	−296.90

[a]Values are for reactions in which one mole of substance is formed.

of the mountain can be determined very precisely, but what is the *absolute* elevation of the mountain? Do we mean by this the vertical distance between the mountaintop and the center of the earth? Do we mean the vertical distance between the mountaintop and the deepest trench in the ocean? No, by common agreement we mean the vertical distance between the mountaintop and mean sea level. If we arbitrarily assign to mean sea level an elevation of 0, all other points on earth can be assigned an elevation relative to this zero. The elevation of Mt. Everest is +8848 m; that of Badwater, Death Valley, California, is −86 m.

We can do the same thing with enthalpies. Here, *by convention,* we assign a value of **zero** to the enthalpies of the elements in their most stable forms at 1 atm pressure* at the specified temperature, a condition referred to as the **standard state.** The enthalpies of compounds can then be related to this arbitrary zero. The difference in enthalpy between *one mole* of a compound in its standard state and its elements in their standard states is called the **standard molar enthalpy of formation** (or simply the **molar heat of formation**) and denoted as $\Delta \bar{H}^\circ_f$. The superscript $^\circ$ signifies that the substances in a process are in their standard states.

Extensive tables of standard enthalpies of formation permit a variety of thermodynamic calculations. A few typical data are presented in Table 6-1. A more extensive tabulation is provided in Appendix D. Tabulated data are given most frequently for 25°C (298 K).

Let us apply Hess's law and other ideas from Section 6-5 to *calculate* the enthalpy change $\Delta \bar{H}^\circ_{rxn}$ (that is, the standard molar heat of reaction) for the combustion of 1 mol of ethane, $C_2H_6(g)$, with all reactants and products in their standard states.

$$C_2H_6(g) + \tfrac{7}{2} O_2(g) \longrightarrow 2\,CO_2(g) + 3\,H_2O(l) \qquad \Delta \bar{H}^\circ_{rxn} = ? \qquad (6.15)$$

Three equations that can be added together to yield equation (6.15) are

(a) $\qquad\qquad C_2H_6(g) \longrightarrow 2\,\cancel{C(graphite)} + \cancel{3\,H_2(g)} \qquad \Delta \bar{H} = -\Delta \bar{H}^\circ_f[C_2H_6(g)]$

(b) $\cancel{2\,C(graphite)} + 2\,O_2(g) \longrightarrow 2\,CO_2(g) \qquad\qquad \Delta \bar{H} = 2 \times \Delta \bar{H}^\circ_f[CO_2(g)]$

(c) $\qquad \cancel{3\,H_2(g)} + \tfrac{3}{2} O_2(g) \longrightarrow 3\,H_2O(l) \qquad\qquad \Delta \bar{H} = 3 \times \Delta \bar{H}^\circ_f[H_2O(l)]$

$$\overline{C_2H_6(g) + \tfrac{7}{2} O_2(g) \longrightarrow 2\,CO_2(g) + 3\,H_2O(l) \qquad \Delta \bar{H}^\circ_{rxn} = ? \qquad (6.15)}$$

Having just introduced the concept of enthalpies (heats) of formation, we should recognize that equation (a) is the *reverse* of the equation representing the formation of one mole of $C_2H_6(g)$ from its elements. $\Delta \bar{H}$ for equation (a) is the *negative* of the enthalpy of formation of $C_2H_6(g)$. For equations (b) and (c) the $\Delta \bar{H}$ values are two and three times the enthalpies of formation of $CO_2(g)$ and $H_2O(l)$, respectively. For reaction (6.15), then

$$\Delta \bar{H}^\circ_{rxn} = \{2 \times \Delta \bar{H}^\circ_f[CO_2(g)] + 3 \times \Delta \bar{H}^\circ_f[H_2O(l)]\} - \{\Delta \bar{H}^\circ_f[C_2H_6(g)]\} \qquad (6.16)$$

Equation (6.16) is simply a specific application of a more general relationship that is expressed as

$$\Delta \bar{H}^\circ_{rxn} = \left[\sum v_p \Delta \bar{H}^\circ_f(\text{products}) \right] - \left[\sum v_r \Delta \bar{H}^\circ_f(\text{reactants}) \right] \qquad (6.17)$$

The symbol Σ (Greek, sigma) means "the sum of." The terms that are added together are the products of the standard molar enthalpies of formation ($\Delta \bar{H}^\circ_f$) and their stoichiometric coefficients, v. One summation is required for the reaction products and another for the initial reactants. The enthalpy (heat) of reaction is the summation of terms for the products minus the summation of terms for the reactants.

---

*Recently, the International Union of Pure and Applied Chemistry (IUPAC) recommended that the standard-state pressure be changed from 1 atm (101,325 Pa) to 1 bar (1×10^5 Pa). Although several significant ramifications of this recommendation may develop over a period of time, the immediate effects of this change are minor (see Appendix D). We will continue to use 1 atm as the standard-state pressure in this text.

Example 6-7   Complete the calculation of $\Delta\overline{H}^{\circ}_{\text{rxn}}$ for reaction (6.15).

Solution.   The relationship of $\Delta\overline{H}^{\circ}_{\text{rxn}}$ to enthalpies of formation is expressed through equation (6.16). All that is required is the substitution of tabulated enthalpy of formation data (see Table 6-1) into this equation.

$$\Delta\overline{H}^{\circ}_{\text{rxn}} = 2 \times \Delta\overline{H}^{\circ}_{f}[\text{CO}_2(\text{g})] + 3 \times \Delta\overline{H}^{\circ}_{f}[\text{H}_2\text{O}(\text{l})] - \Delta\overline{H}^{\circ}_{f}[\text{C}_2\text{H}_6(\text{g})]$$

$$= 2 \times (-393.5 \text{ kJ/mol}) + 3 \times (-285.8 \text{ kJ/mol}) - (-84.7 \text{ kJ/mol})$$

$$\Delta\overline{H}^{\circ}_{\text{rxn}} = -787.0 \text{ kJ/mol} - 857.4 \text{ kJ/mol} + 84.7 \text{ kJ/mol} = -1559.7 \text{ kJ/mol}$$

SIMILAR EXAMPLES: Review Problem 9; Exercises 26, 27.

Example 6-8   The combustion of cyclopropane (an anesthetic) is represented as

$$(\text{CH}_2)_3(\text{g}) + \tfrac{9}{2}\,\text{O}_2(\text{g}) \longrightarrow 3\,\text{CO}_2(\text{g}) + 3\,\text{H}_2\text{O}(\text{l}) \qquad \Delta\overline{H}^{\circ}_{\text{rxn}} = -2091.4 \text{ kJ/mol}$$

Use this value of $\Delta\overline{H}^{\circ}_{\text{rxn}}$ and other data from Table 6-1 to calculate the standard enthalpy (heat) of formation of cyclopropane.

Solution.   This question requires an application of equation (6.17), but we solve for an unknown enthalpy (heat) of formation rather than for a heat of reaction. [Note that no $\Delta\overline{H}^{\circ}_{f}$ term appears for the $\tfrac{9}{2}\,\text{O}_2(\text{g})$, since the enthalpy (heat) of formation of a free element in its standard state is zero.]

$$\Delta\overline{H}^{\circ}_{\text{rxn}} = 3\,\Delta\overline{H}^{\circ}_{f}[\text{CO}_2(\text{g})] + 3\,\Delta\overline{H}^{\circ}_{f}[\text{H}_2\text{O}(\text{l})] - \Delta\overline{H}^{\circ}_{f}[(\text{CH}_2)_3(\text{g})]$$

$$-2091.4 \text{ kJ/mol} = 3 \times (-393.5 \text{ kJ/mol}) + 3 \times (-285.8 \text{ kJ/mol}) - \Delta\overline{H}^{\circ}_{f}[(\text{CH}_2)_3(\text{g})]$$

$$\Delta\overline{H}^{\circ}_{f}[(\text{CH}_2)_3(\text{g})] = -1180.5 \text{ kJ/mol} - 857.4 \text{ kJ/mol} + 2091.4 \text{ kJ/mol}$$

$$= +53.5 \text{ kJ/mol}$$

SIMILAR EXAMPLES: Review Problem 10; Exercise 28.

Since the enthalpy of formation data used in Examples 6-7 and 6-8 were for 298 K, the result of each calculation also applies only at 298 K. Generally, ΔH values are not significantly temperature dependent. This means that results obtained for 298 K are often applicable over a range of temperatures. (We discuss the temperature dependence of ΔH again in Chapter 16.)

FOCUS ON
Sources and Uses of Energy

Nineteen million cubic feet of gas was produced from underground coal in this experimental coal gasification site. [U.S. Department of Energy photo.]

In the United States energy consumption far exceeds the 10,000 to 12,000 kJ per capita daily requirement to sustain human life. With 6% of the world's population, the United States consumes about 30% of the world's energy production, for agriculture, industry, transportation, and material comforts. The current annual rate of energy consumption in the United States is about

$$8.4 \times 10^{16} \text{ kJ} = 2.0 \times 10^{16} \text{ kcal}$$
$$= 8.0 \times 10^{16} \text{ Btu}$$
$$= 80 \text{ quad}$$

FIGURE 6-9
Sources and uses of energy in the United States.

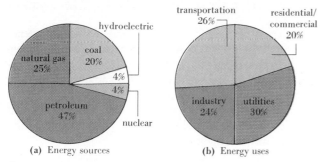

(a) Energy sources (b) Energy uses

Current energy consumption in the United States is about 8.4×10^{10} kJ/yr. The largest single consumer in the industrial sector is the chemical industry, followed closely by the primary metals industry.

[One British thermal unit (Btu) is the quantity of heat required to raise the temperature of one pound of water by one degree F. One quadrillion (10^{15}) Btu = 1 quad.]

The sources of this energy and the manner in which it is consumed are depicted in Figure 6-9.

Fossil Fuels.
As seen from Figure 6-9, the primary energy sources in use today are the so-called fossil fuels—coal, petroleum, and natural gas. These are carbon-containing materials that were formed from organisms that lived millions of years ago.

A direct consequence of the use of coal was the invention of the steam engine and the advent of the industrial revolution. The introduction of petroleum products as fuels in the past century ushered in the internal combustion engine and the jet age. Fossil fuels have served our needs well over the past few centuries, but what are the prospects for the future?

It is impossible to make accurate predictions regarding the ultimate fossil fuel resources of the world and the rate at which they will be consumed. The prospects of new discoveries, both major and minor, are quite uncertain. Also uncertain are the rate at which energy requirements will increase and the extent to which fossil fuels will be supplanted by other energy sources. Improvements in technology might allow for the extraction of fuels from lower-grade materials (e.g., oil from tar sands and oil shales). Nevertheless, it is possible to produce some very rough estimates of how much longer fossil fuels will be generally available, Figure 6-10 shows the production of fossil fuels as a function of time, from the approximate dates of their first general use to the anticipated dates of their exhaustion. As can be seen from this figure and its caption, reserves of coal could conceivably last for some

time into the future, but those of petroleum (including natural gas) seem likely to become unavailable much sooner.

Synthetic Fuels from Coal.
In the United States coal reserves have been variously estimated at between 5000 and 21,000 quads. Corresponding reserves of petroleum and natural gas do not exceed 1000 quads each. Despite its relative abundance, there has not been a significant increase in the production or use of coal in recent years. The difficulties with coal are twofold. First there are hazards and expense in the deep mining of coal. Strip mining, which is less hazardous and expensive, is also more damaging to the environment (although deep mines produce environmental damage as well). The second difficulty is in the burning of coal. The sulfur content of most coal is too high to allow for its direct combustion without exceeding environmental limits for atmospheric SO_2. It is necessary either to remove the $SO_2(g)$ from the flue gases or to remove the sulfur before the coal is burned.

One promising possibility for greater utilization of coal reserves is to convert the coal to gaseous or liquid fuels, either in surface installations or while the coal is still underground. The necessary conversions are from carbon (in coal) to carbon monoxide, methane, and other hydrocarbons; chemical reactions are involved.

Gasification of Coal.
Before the advent of cheap natural gas in the 1940s, gas produced from coal (variously called producer gas, town gas, or city gas) was widely used in the United States. This gas is manufactured by passing

FIGURE 6-10
Estimated world production of fossil fuels.

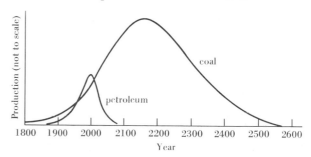

The exact shape and time of exhaustion for each curve depend on the estimate used for the total quantity of recoverable fuel and on the rate of production. These curves are based on the assumption of no increases from current rates of consumption, probably an unrealistic assumption. Even a growth rate of just 1% per year would cause the rate of consumption to double in 70 years and would greatly shorten the times to exhaustion.

steam and air through heated coal, leading to reactions such as

$$C(s) + H_2O(g) \longrightarrow CO(g) + H_2(g) \qquad (6.18)$$

$$CO(g) + H_2O(g) \longrightarrow CO_2(g) + H_2(g) \qquad (6.19)$$

$$2\ C(s) + O_2(g) \longrightarrow 2\ CO(g) \qquad (6.20)$$

$$C(s) + 2\ H_2(g) \longrightarrow CH_4(g) \qquad (6.21)$$

A typical producer gas consists of about 23% CO, 18% H_2, 8% CO_2, and 1% CH_4, by volume; but it also contains about 50% N_2, since air is used in the process. Because so much of the gas is noncombustible (i.e., the N_2 and CO_2), producer gas has only about 10 to 15% of the heat value of natural gas.

Modern gasification processes

1. Use $O_2(g)$ instead of air (thereby reducing $N_2(g)$ in the product to mere traces).
2. Provide for the removal of noncombustible CO_2 and of sulfur impurities.
3. Include a step to convert CO and H_2, in the presence of a catalyst, to CH_4.

$$CO(g) + 3\ H_2(g) \longrightarrow CH_4(g) + H_2O(g) \qquad (6.22)$$

With these modifications it is possible to obtain a substitute or **synthetic natural gas (SNG)**, a gaseous mixture with composition and heat value similar to natural gas.

Liquefaction of Coal.   Processes are currently being developed for producing liquid fuels from coal. These generally involve first the gasification of coal to water gas—a mixture of CO and H_2—by reaction (6.18). This is followed by catalytic reactions (Fischer–Tropsch process) in which liquid hydrocarbons are formed.

$$n\ CO + (2n + 1)H_2 \longrightarrow C_nH_{2n+2} + n\ H_2O \qquad (6.23)$$

In still another process water gas is converted to liquid methanol.

$$CO + 2\ H_2 \longrightarrow CH_3OH \qquad (6.24)$$

In 1942, some 32 million gallons of aviation fuel were made from coal in Germany; and currently, in South Africa, the Sasol process for coal liquefaction is used to produce gasoline and a variety of other petroleum products and chemicals.

Methanol, Ethanol, and Hydrogen.   Methanol, CH_3OH, can be obtained from coal by reaction (6.24). It can also be produced by thermal decomposition (pyrolysis) of wood, manure, sewage, or municipal waste. The heat of combustion of methanol is only about one half that of a typical gasoline on a mass basis, but methanol has a high octane number—106. It has been tested and used as a

fuel in internal combustion engines and found to burn cleaner than gasoline. Methanol can be used for space heating, electric power generation, fuel cells, and organic synthesis.

Ethanol, C_2H_5OH, is produced mostly from ethylene, C_2H_4, which in turn is derived from petroleum. Current interest centers on the production of ethanol by the fermentation of organic matter, a process known throughout recorded history. Ethanol production by fermentation is probably in its most advanced state in Brazil, where sugar cane and cassava (manioc) are the plant matter ("biomass") principally employed.

Another potentially useful fuel currently derived mostly from petroleum is hydrogen. Its heat of combustion, per gram, is twice that of methane and about three times that of gasoline. There are, of course, problems attendant to using hydrogen as a fuel. It can form explosive mixtures with air. It is bulky to transport as a gas because of its very low density. It is difficult to transport as a liquid because of its very low boiling point (20 K). It dissolves in metals, causing them to become brittle. The greatest problem, however, is the large quantity of energy required to produce hydrogen. If a cheap energy source can be developed (solar energy, nuclear fusion), abundant hydrogen can be obtained by the decomposition of water, either by electrolysis or thermally. Hydrogen might then become

FIGURE 6-11
Relative importance of different energy sources— present and future (speculative).

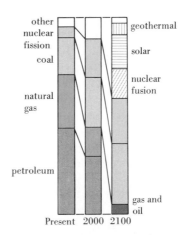

The projections for A.D. 2000 and 2100 are highly speculative. The actual situation might be much different. Solar energy might become an important energy source by 2000, or nuclear fusion may not be developed as an energy source even by 2100. Or perhaps biological sources may become increasingly important. One fact that seems quite certain for the time period shown is that natural gas and petroleum (oil) will decrease in importance.

a major resource of the future (see Section 13-3). A particular advantage of hydrogen over fossil fuels is that the product of its combustion is H_2O, not CO_2. Increasingly, the buildup of carbon dioxide in the atmosphere that accompanies the burning of fossil fuels is being recognized as an environmental problem (see Section 13-8).

Conclusion.   Although there is considerable uncertainty as to what the profile of future energy sources will be,

there is no doubt that it will differ greatly from that to which we are accustomed today. One possibility is suggested by Figure 6-11. (Nuclear fission and fusion are discussed in Chapter 25.)

Summary

Thermochemistry is concerned primarily with heat effects accompanying chemical reactions. In dealing with this subject, it is necessary to develop some ideas about the nature of heat and its measurement and the distinction between heat and work—the two forms of energy transfer in chemical reactions.

Combustion reactions can be carried out in a bomb calorimeter, a device in which a heat of reaction is used to change the temperature of a quantity of water and other objects in the surroundings. However, chemical reactions are generally not carried out under the conditions of bomb calorimetry. Most often they are conducted in containers open to the atmosphere and under the constant pressure of the atmosphere. For this purpose it is advantageous to define a thermodynamic function of state, enthalpy (H), having the property that its change (ΔH) corresponds to a heat effect at constant pressure.

For certain processes, ΔH values can be determined in a calorimeter of an especially simple design (see Figure 6-7). For many others it is possible to determine ΔH indirectly. Hess's law permits a process to be broken down into a series of steps and ΔH for the process to be obtained by summing ΔH values for the individual steps. The determination of enthalpy changes is also facilitated by establishing an *arbitrary* zero of enthalpy for the elements in their most stable forms at 1 atm. From this basis it is then possible to derive standard molar enthalpies (heats) of formation of compounds ($\Delta \overline{H}_f^\circ$). These data can be compiled into extensive listings (see Appendix D) and used in computations involving Hess's law.

Learning Objectives

As a result of studying Chapter 6, you should be able to

1. Distinguish between the energy forms heat and work.

2. Relate quantities of heat, temperature changes, and specific heats of substances (i.e., calculate one of these quantities from known values of the other two).

3. Explain the mechanical equivalent of heat, and convert calories to joules, and vice versa.

4. Explain the special feature of a function of state.

5. Calculate the heat of a reaction from bomb calorimetry data.

6. Explain the purposes served by the thermodynamic property of enthalpy (H).

7. Use data from a "Styrofoam coffee cup" calorimeter to determine heat effects at constant pressure.

8. Combine chemical equations with known ΔH values to obtain the net equation for a desired reaction and the heat of that reaction (ΔH).

9. Explain how standard molar enthalpies of formation are established.

10. Calculate heats of reaction from tabulated standard molar enthalpies (heats) of formation, or enthalpies of formation from experimentally determined heats of reaction.

Some New Terms

A **bomb calorimeter** is a device used to measure the heat of a combustion reaction.

The **calorie (cal)** is the quantity of heat required to raise the temperature of one gram of water by one degree Celsius (specifically, from 14.5 to 15.5°C); 1 cal = 4.184 J.

A **calorimeter** is a device (of which there are numerous types) used to measure the quantity of heat exchanged between a system and its surroundings.

Endothermic processes absorb heat from the surroundings as they occur. The quantity of heat carries a *positive* sign.

Enthalpy (*H*) is a thermodynamic function, related to internal energy, and useful in describing constant-pressure processes (see also page 498).

Enthalpy change (Δ*H*) is the difference in enthalpy between two states of a system. If the process by which the change occurs is a chemical reaction, the enthalpy change is called the **heat of reaction.**

Enthalpy (heat) of formation is the enthalpy change that accompanies the formation of a compound from the most stable forms of its elements.

Exothermic processes give off heat to the surroundings. The quantity of heat carries a *negative* sign.

A **function of state (state function)** is a property that depends only on the state or present condition of a system and not on how this state is attained.

Heat is a form of energy transfer resulting from a temperature difference.

Heat capacity is the quantity of heat energy required to change the temperature of an object or substance. This is often expressed as a **molar heat capacity,** e.g., cal mol^{-1} °C^{-1} or J mol^{-1} K^{-1}.

Hess's law states that the enthalpy change for an overall or net process is the sum of the enthalpy changes for individual steps in the process.

The **internal energy (*E*)** of a system is the total energy attributed to the particles of matter and their interactions within the system.

The **joule (J)** is the basic SI unit of energy. It is the quantity of work done when a force of one newton acts through a distance of one meter.

The **law of conservation of energy** states that energy can neither be created nor destroyed in processes.

Specific heat is the quantity of heat required to raise the temperature of one gram of substance by one degree Celsius.

The **standard state** of a substance refers to the most stable form of that substance at 1 atm pressure.

The **surroundings** represents that portion of the universe with which a system interacts.

A **system** is the portion of the universe selected for a thermodynamic study.

Work is a form of energy transfer that can be expressed as a force acting through a distance.

Suggestions for Further Study

BARTLETT, A. A., "Forgotten Fundamentals of the Energy Crisis," *Am. J. Physics,* **46,** 876 (1978).

BENT, H. A., "Energy and Exercise" (a six-part series), *J. Chem. Educ.,* **55,** 456, 526, 586, 659, 726, 796 (1978).

KERR, R. A., "Another Oil Resource Warning," *Science,* **223,** 382 (1984).

LLOYD, W. G., and D. A. DAVENPORT, "Applying Thermodynamics to Fossil Fuels: Heats of Combustion from Elemental Composition," *J. Chem. Educ.,* **57,** 56 (1980).

ROGERS, D. W., "An Informal History of the First Law of Thermodynamics," *Chemistry,* **49**[10], 11 (1976).

SCHRIESHEIM, A., and I. KIRSHENBAUM, "The Chemistry and Technology of Synthetic Fuels," *American Scientist,* **69,** 536 (1981).

STRANGES, A. N., "Synthetic Petroleum from Coal Hydrogenation," *J. Chem. Educ.,* **60,** 617 (1983).

TAYLOR, H. V., "Conservation Energy," *J. Chem. Educ.,* **58,** 185 (1981).

Review Problems

1. Express the quantity of heat in the unit indicated, and designate whether this heat is gained or lost by the liquid.

(a) The quantity of heat, in cal, when 325 g H_2O undergoes a temperature decrease of 2.92°C.

(b) The quantity of heat, in kcal, when 43.5 kg chloroform, $CHCl_3$(l), has its temperature changed from 16.8 to 22.3°C [sp. ht. of $CHCl_3$(l) = 0.232 cal g^{-1} °C^{-1}].

(c) The quantity of heat, in kJ, when 12.6 L ethylene dichloride, $C_2H_4Cl_2$(l), undergoes a temperature change from 48.3 to 24.7°C. [For $C_2H_4Cl_2$(l), d = 1.253 g/cm^3; sp. ht. = 0.310 cal g^{-1} °C^{-1}.]

2. Calculate the final temperature that results from the following processes.

(a) A 7.15-g sample of water at 18.2°C absorbs 107 cal of heat.

(b) A 3.56-kg sample of solid sulfur at 43.2°C gives off 167 kcal of heat (sp. ht. of S = 0.173 cal g^{-1} °C^{-1}).

(c) A 205-L sample of toluene, C_7H_8(l), at 32.2°C gives off 4.2 × 10^3 kJ of heat. [For C_7H_8(l), d = 0.866 g/cm^3; sp. ht. = 0.40 cal g^{-1} °C^{-1}.]

3. The experiment described in Example 6-2 is repeated with different metals substituting for lead. That is, the masses of metal and water and the initial temperatures of the metal and water are the same as in Figure 6-2. The final temperatures are given below. What is the specific heat of each metal? **(a)** Zn, final temperature 39.0°C; **(b)** Mg, 55.3°C; **(c)** Fe, 41.8°C.

4. Upon complete combustion in a bomb calorimeter, the following quantities of heat are evolved by the substances indicated. Express each heat of combustion as kJ/mol substance.

(a) 0.107 g of acetylene, C_2H_2(g), yields 5.36 kJ;

(b) 1.030 g of urea, $CO(NH_2)_2$(s), yields 2.601 kcal;

(c) 1.05 cm^3 of acetone, $(CH_3)_2CO$(l) (d = 0.791 g/cm^3), yields 6.22 kcal.

5. The following substances undergo complete combustion in a bomb calorimeter. The calorimeter assembly has a heat capacity of 827 J/°C and contains exactly 1000.0 g water. What is the final temperature if in each case the initial water temperature is 24.98°C? (Recall that the negative sign means that heat is given off by the substance that undergoes combustion.)

(a) 0.5060 g cyclohexanol, $C_6H_{12}O(l)$; heat of combustion = -890.7 kcal/mol

(b) 0.853 g thymol, $C_{10}H_{14}O(s)$; heat of combustion = -5.65×10^3 kJ/mol

(c) 1.25 mL of ethyl acetate, $C_4H_8O_2(l)$ ($d = 0.901$ g/cm^3); heat of combustion = -2246 kJ/mol

6. The following heats of solution are listed in a handbook. Indicate the final temperature you would expect to measure in a "coffee cup" calorimeter (see Figure 6-7), if in each of the following cases 1.00 g of the indicated solid is dissolved in 100.0 g H_2O that is initially at 24.8°C. (Assume a specific heat of 1.00 cal g^{-1} °C^{-1} for each solution.) (a) LiCl, $\Delta\overline{H}_{soln.} = -35.0$ kJ/mol; (b) MgSO$_4$, -84.9 kJ/mol; (c) Pb(NO$_3$)$_2$, $+31.8$ kJ/mol.

7. Use Hess's law to determine $\Delta\overline{H}$ for the reaction $CO(g) + \frac{1}{2} O_2(g) \rightarrow CO_2(g)$, given that

(a) C(graphite) $+ \frac{1}{2} O_2(g) \rightarrow CO(g)$ $\Delta\overline{H} = -110.54$ kJ/mol

(b) C(graphite) $+ O_2(g) \rightarrow CO_2(g)$ $\Delta\overline{H} = -393.51$ kJ/mol

8. Use Hess's law to determine $\Delta\overline{H}$ for the reaction: $C_3H_4(g) + 2 H_2(g) \rightarrow C_3H_8(g)$, given that

(a) $H_2(g) + \frac{1}{2} O_2(g) \rightarrow H_2O(l)$ $\Delta\overline{H} = -285.85$ kJ/mol

(b) $C_3H_4(g) + 4 O_2(g) \rightarrow 3 CO_2(g) + 2 H_2O(l)$ $\Delta\overline{H} = -1941$ kJ/mol

(c) $C_3H_8(g) + 5 O_2(g) \rightarrow 3 CO_2(g) + 4 H_2O(l)$ $\Delta\overline{H} = -2220$ kJ/mol

9. Use enthalpies of formation from Appendix D in equation (6.17) to determine the heats of the following reactions.

(a) $C_3H_8(g) + H_2(g) \rightarrow C_2H_6(g) + CH_4(g)$

(b) $2 H_2S(g) + 3 O_2(g) \rightarrow 2 SO_2(g) + 2 H_2O(l)$

(c) $3 Fe_2O_3(s) + H_2(g) \rightarrow 2 Fe_3O_4(s) + H_2O(l)$

10. The heats of combustion of the substances listed below were obtained from a handbook. Determine the enthalpy of formation ($\Delta\overline{H}_f^\circ$) of each substance. [*Hint:* Write an equation for each combustion reaction; assume that the heat of combustion is $\Delta\overline{H}^\circ$ for the combustion reaction; and use equation (6.17) and data from Table 6-1.]

(a) dimethyl ether, $(CH_3)_2O(g)$, -1.454×10^3 kJ/mol $(CH_3)_2O$

(b) benzaldehyde, $C_7H_6O(l)$, -3.520×10^3 kJ/mol C_7H_6O

(c) glucose, $C_6H_{12}O_6(s)$, -2.816×10^3 kJ/mol $C_6H_{12}O_6$

Exercises

Specific heat

1. An electric range burner weighing 583 g is turned off after reaching a temperature of 535°C. Assume an average specific heat of 0.4 J g^{-1} °C^{-1} for the burner.

(a) Approximately how much heat will the burner exchange with the surroundings in cooling back to room temperature (20°C)?

(b) Approximately how much water, in grams, could be heated from room temperature to the boiling point if all this heat could be transferred to the water?

2. A chunk of stainless steel (sp. ht. = 0.12 cal g^{-1} °C^{-1}) is transferred from an oven (152°C) to 125 mL of water at 24.8°C, into which it is immersed. The water temperature rises to 40.3°C. What was the mass of the steel? How accurate is this method of mass determination? Explain.

3. Magnesium metal has a specific heat of 1.04 J g^{-1} °C^{-1}. A 70.0-g sample of this metal, at a temperature of 99.8°C, is added to a beaker containing 50.0 g water at 30.0°C. The final water temperature is found to be 47.2°C. Is this result consistent with the law of conservation of energy? Explain.

4. A 74.8-g sample of copper metal at 143.2°C is added to an insulated vessel containing 165 cm^3 of glycerol, $C_3H_8O_3(l)$ ($d = 1.26$ g/cm^3), at 24.8°C. The final temperature is 31.1°C. The specific heat of copper is given as 0.393 J g^{-1} °C^{-1}. What is the *molar* heat capacity of glycerol?

⋆5. Brass has a density of 8.40 g/cm^3 and a specific heat of 0.385 J g^{-1} °C^{-1}. A cube of brass 6.80 mm on an edge, initially at a temperature of 92.1°C, is immersed into 17.5 g of water at 25.2°C in an insulated container. What is the final temperature of the mixture?

Heats of reaction

6. The reaction of quicklime (CaO) with water produces slaked lime [Ca(OH)$_2$], a substance widely used in the construction industry to make mortar and plaster. The reaction of quicklime and water is highly exothermic.

$$CaO(s) + H_2O(l) \rightarrow Ca(OH)_2(s) \qquad \Delta\overline{H} = -350 \text{ kJ/mol}$$

(a) What is the heat of reaction per gram of calcium oxide (CaO) reacted?

(b) How much heat, in kilojoules, is associated with the production of 1 ton (2000 lb) of slaked lime?

7. The combustion of methane gas in air is represented by the equation

$$CH_4(g) + 2 O_2(g) \rightarrow CO_2(g) + 2 H_2O(l) \quad \Delta\overline{H} = -890 \text{ kJ/mol}$$

(a) If the heat of this combustion reaction could be used with 100% efficiency, what volume of water could be heated from 21.8 to 60.3°C by burning 2.55×10^3 g $CH_4(g)$?

(b) How many liters of $CH_4(g)$ at 21.8°C and 748 mmHg must be burned to liberate 1.00×10^6 kJ of heat?

8. A particular natural gas consists, on a molar basis, of 83.0% CH_4, 11.2% C_2H_6, and 5.8% C_3H_8. The heats of combustion (ΔH) of these gases are -890 kJ/mol CH_4, -1559 kJ/mol C_2H_6, and -2219 kJ/mol C_3H_8. A 312-L sample of this gas, measured at 26.1°C and 753 mmHg, is burned at constant pressure in an excess of oxygen gas. How much heat is given off to the surroundings?

9. Thermite mixtures are used for certain types of welding. The thermite reaction is highly exothermic.

$$Fe_2O_3(s) + 2\ Al(s) \rightarrow Al_2O_3(s) + 2\ Fe(l)\quad \Delta\overline{H} = -850\ \text{kJ/mol}$$

1.00 mol Fe_2O_3 and 2.00 mol Al are mixed at room temperature (25°C) and a reaction is initiated. The liberated heat is retained within the products, whose combined specific heats over a broad temperature range are about 0.2 cal $g^{-1}\ °C^{-1}$. The melting point of iron is 1530°C. Show that the quantity of heat liberated is sufficient to raise the temperature of the products to the melting point of iron.

10. A pellet of potassium hydroxide, KOH, weighing 0.150 g is added to 45.0 g of water in a Styrofoam coffee cup. The water temperature rises from 24.1 to 24.9°C.
   (a) What is the approximate heat of solution of KOH, expressed in kJ/mol KOH?
   (b) How could the precision of the result be improved without modifying the apparatus in any way?

11. A lecture demonstration is being planned to illustrate an endothermic process. It is desired to lower the temperature of 1400 mL water in an insulated container from 25 to 10°C. What mass of $NH_4NO_3(s)$ should be dissolved in the water to achieve this result? Use the information in Example 6-5.

12. Care must be taken in preparing solutions of solutes that liberate heat on dissolving. The heat of solution of NaOH is -10 kcal/mol NaOH. To what approximate temperature will a sample of water, originally at 21°C, be raised in the preparation of 500 cm^3 of 7.0 *M* NaOH? Assume that no effort is made to remove heat from the solution.

Bomb calorimetry

13. A sample that is burned in a bomb calorimeter gives off 20.9 kJ of heat. The temperature of a 1155-g quantity of water increases by 3.68°C. Calculate the calorimeter's heat capacity.

14. What increase in temperature would you expect to occur in 1025 g of water in a bomb calorimeter if a 0.242-g sample of naphthalene is burned in an excess of $O_2(g)$ in the bomb? Assume a heat capacity of 802 J/°C for the calorimeter. The heat of combustion of naphthalene is -5.15×10^3 kJ/mol $C_{10}H_8$.

15. The burning of 2.051 g of glucose, $C_6H_{12}O_6$, in a bomb calorimeter causes the temperature of the water to increase from 24.92 to 31.41°C. The calorimeter contains 980.0 g of water. Assume a heat capacity of 812 J/°C for the calorimeter.
   (a) What is the heat of combustion of the glucose expressed in kJ/mol $C_6H_{12}O_6$?

   (b) Write a balanced equation for the combustion reaction, assuming that $CO_2(g)$ and $H_2O(l)$ are the sole products and that the heat of combustion can be represented as $\Delta\overline{H}$.

16. A 1.148-g sample of benzoic acid is burned in an excess of $O_2(g)$ in a bomb immersed in 1181 g of water. The temperature of the water rises from 24.96 to 30.25°C. In a second experiment, a 0.895-g powdered coal sample is burned in the same calorimeter assembly. The temperature of 1162 g of water rises from 24.98 to 29.81°C. How many metric tons (1 metric ton = 1000 kg) of this coal would have to be burned in a power plant to release 2.15×10^9 kJ of heat? The heat of combustion of benzoic acid is -26.42 kJ/g $C_7H_6O_2$.

17. Suppose that in the experiment described in Example 6-4 the combustion of the benzoic acid were carried out, in error, without immersing the bomb in water. What approximate final temperature would the bomb reach? (Neglect any loss of heat to the surrounding air.)

Functions of state

18. How can a thermodynamic function such as enthalpy (H) be used if its absolute value cannot be determined?

19. Both graphite and diamond are pure forms of carbon. The heat of combustion of graphite, yielding $CO_2(g)$ as the only product and with all reactants and the product being at 25°C and 1 atm pressure, is -393.51 kJ/mol. If diamond is substituted for graphite, would you expect the heat of combustion to be exactly the same or different? Explain.

Hess's law

20. Determine $\Delta\overline{H}$ for the reaction

$$N_2H_4(l) + 2\ H_2O_2(l) \rightarrow N_2(g) + 4\ H_2O(l)$$

from these data.

$$N_2H_4(l) + O_2(g) \rightarrow N_2(g) + 2\ H_2O(l)\quad \Delta\overline{H} = -622.33\ \text{kJ/mol}$$

$$H_2(g) + \tfrac{1}{2} O_2(g) \rightarrow H_2O(l)\qquad\qquad \Delta\overline{H} = -285.85\ \text{kJ/mol}$$

$$H_2(g) + O_2(g) \rightarrow H_2O_2(l)\qquad\qquad \Delta\overline{H} = -187.78\ \text{kJ/mol}$$

21. CCl_4, an important commercial solvent, is prepared by the reaction of $Cl_2(g)$ with a carbon compound. Determine $\Delta\overline{H}$ for the reaction

$$CS_2(l) + 3\ Cl_2(g) \rightarrow CCl_4(l) + S_2Cl_2(l)$$

given these data.

		kJ/mol
$CS_2(l) + 3\ O_2(g) \rightarrow CO_2(g) + 2\ SO_2(g)$	$\Delta\overline{H} =$	-1077
$2\ S(s) + Cl_2(g) \rightarrow S_2Cl_2(l)$	$\Delta\overline{H} =$	-60.2
$C(s) + 2\ Cl_2(g) \rightarrow CCl_4(l)$	$\Delta\overline{H} =$	-135.4
$S(s) + O_2(g) \rightarrow SO_2(g)$	$\Delta\overline{H} =$	-296.9
$C(s) + O_2(g) \rightarrow CO_2(g)$	$\Delta\overline{H} =$	-393.5

22. The heats of combustion ($\Delta\overline{H}$) per mole of 1,3-butadiene [$C_4H_6(g)$], normal butane [$C_4H_{10}(g)$], and $H_2(g)$ are -2543.5, -2878.6, and -285.85 kJ/mol, respectively. Use these data to calculate the heat of hydrogenation of 1,3-butadiene to normal butane. (*Hint:* Write equations for the combustion reactions.)

$$C_4H_6(g) + 2\ H_2(g) \rightarrow C_4H_{10}(g) \qquad \Delta\overline{H} = ?$$

23. Synthetic natural gas (SNG) is a gaseous mixture containing $CH_4(g)$ that is used in the synthesis of organic compounds. One reaction for the production of SNG is

$$4\ CO(g) + 8\ H_2(g) \rightarrow 3\ CH_4(g) + CO_2(g) + 2\ H_2O(l)$$
$$\Delta\overline{H} = ?$$

Use the following data, as necessary, to determine $\Delta\overline{H}$ for this SNG reaction.

$C(graphite) + \frac{1}{2} O_2(g) \rightarrow CO(g) \qquad \Delta\overline{H} = -110.54$ kJ/mol
$CO(g) + \frac{1}{2} O_2(g) \rightarrow CO_2(g) \qquad \Delta\overline{H} = -282.97$ kJ/mol
$H_2(g) + \frac{1}{2} O_2(g) \rightarrow H_2O(l) \qquad \Delta\overline{H} = -285.85$ kJ/mol
$C(graphite) + 2\ H_2(g) \rightarrow CH_4(g) \qquad \Delta\overline{H} = -74.85$ kJ/mol

24. Methanol, a potential fuel source, can be prepared by heating CO and H_2 under pressure in the presence of a catalyst.

$$CO(g) + 2\ H_2(g) \rightarrow CH_3OH(l) \qquad \Delta\overline{H} = ?$$

Use the fact that the heat of combustion of $CH_3OH(l)$ is $\Delta\overline{H} - -726.6$ kJ/mol $CH_3OH(l)$ and the following data to determine the heat of this reaction.

$C(graphite) + \frac{1}{2} O_2(g) \rightarrow CO(g) \qquad \Delta\overline{H} = -110.54$ kJ/mol
$C(graphite) + O_2(g) \rightarrow CO_2(g) \qquad \Delta\overline{H} = -393.51$ kJ/mol
$H_2(g) + \frac{1}{2} O_2(g) \rightarrow H_2O(l) \qquad \Delta\overline{H} = -285.85$ kJ/mol

25. A net reaction for a coal gasification process is given as

$$2\ C(s) + 2\ H_2O(g) \rightarrow CH_4(g) + CO_2(g)$$

Show that this net equation can be established by an appropriate combination of equations (6.18), (6.19), and (6.22).

Enthalpies (heats) of formation

26. Use standard enthalpies of formation from Table 6-1 to determine the heat of the reaction

$$2\ Cl_2(g) + 2\ H_2O(l) \rightarrow 4\ HCl(g) + O_2(g) \qquad \Delta\overline{H}° = ?$$

27. What is the heat of combustion of $C_2H_5OH(l)$ if the reactants and products are maintained at 25°C and 1 atm pressure? (*Hint:* Use data from Appendix D.)

28. Given the heat of the following reaction, determine the enthalpy of formation of $CCl_4(g)$ at 25°C and 1 atm. (*Hint:* Use data from Table 6-1.)

$$CH_4(g) + 4\ Cl_2(g) \rightarrow CCl_4(g) + 4\ HCl(g)$$
$$\Delta\overline{H}° = -402\ kJ/mol$$

29. What is the enthalpy change ($\Delta\overline{H}°_{rxn}$) for the net reaction if all reactants and products of the coal gasification process in Exercise 25 are measured at 25°C and 1 atm?

30. For the reaction

$$C_2H_4(g) + 3\ O_2(g) \rightarrow 2\ CO_2(g) + 2\ H_2O(l)$$
$$\Delta\overline{H}° = -1410.8\ kJ/mol$$

If the H_2O were obtained as a gas rather than a liquid, would the heat of reaction be greater (more negative) or smaller (less negative) than that indicated in the equation? Calculate the value of $\Delta\overline{H}$ in this case. (*Hint:* Refer to Appendix D.)

31. Use data from Appendix D, together with the fact that $\Delta\overline{H}°$ for the complete combustion of $C_5H_{12}(l)$ is -3534 kJ/mol, to calculate $\Delta\overline{H}°$ for the synthesis of 1 mol $C_5H_{12}(l)$ from CO(g) and $H_2(g)$.

$$5\ CO(g) + 11\ H_2(g) \rightarrow C_5H_{12}(l) + 5\ H_2O(l) \qquad \Delta\overline{H}° = ?$$

32. The decomposition of limestone, $CaCO_3(s)$, into quicklime, CaO(s), and $CO_2(g)$ is accomplished at 900°C in a gas-fired kiln. (Assume that heats of reaction under these conditions are the same as at 25°C and 1 atm pressure.)

  (a) Use data from Appendix D to determine how much heat is required to decompose 755 kg $CaCO_3(s)$.

  (b) If the heat energy calculated in part (a) is supplied by the combustion of methane, $CH_4(g)$, what volume of the gas, measured at 18.2°C and 737 mmHg, is required? [*Hint:* Again use data from Appendix D to determine the heat of combustion of $CH_4(g)$.]

33. Under the entry "H_2SO_4" a handbook lists several different values for the enthalpy of formation, $\Delta\overline{H}_f$. For example, for pure $H_2SO_4(l)$, $\Delta\overline{H}_f = -814$ kJ/mol H_2SO_4; for an aqueous solution that is 1.0 M H_2SO_4, $\Delta\overline{H}_f = -888$ kJ/mol H_2SO_4; for 0.25 M H_2SO_4, $\Delta\overline{H}_f = -890$ kJ/mol H_2SO_4; for 0.02 M H_2SO_4, $\Delta\overline{H}_f = -897$ kJ/mol H_2SO_4.

  (a) Explain why these values are not all the same.

  (b) When a concentrated aqueous solution of H_2SO_4 is diluted, does the solution temperature increase or decrease? Explain.

  ★**(c)** If 250 mL of 0.02 M H_2SO_4 is prepared by diluting pure $H_2SO_4(l)$ with water, estimate the change in temperature that occurs. [Assume that the $H_2SO_4(l)$ and the water used for its dilution are at the same temperature initially.]

★**34.** A 1.00-L sample (at STP) of a natural gas evolves, upon complete combustion at constant pressure, 43.6 kJ of heat. If the gas is a mixture of $CH_4(g)$ and $C_2H_6(g)$, what is its percent composition, by volume? [*Hint:* What are the heats of combustion of $CH_4(g)$ and $C_2H_6(g)$?]

★**35.** An alkane hydrocarbon has the formula C_nH_{2n+2}. Whatever the subscript for carbon, n, the subscript for hydrogen is "twice n plus 2." The enthalpies of formation of the alkanes decrease (become more negative) as the numbers of C atoms increase. Starting with propane, C_3H_8, for each additional CH_2 group in the formula the enthalpy of formation, $\Delta\overline{H}°_f$, changes by about -21 kJ/mol. Use this fact, and data from Appendix D, to estimate the heat of combustion of normal heptane, $C_7H_{16}(l)$.

Additional Exercises

In 1818, Dulong and Petit observed that the molar heat capacities of elements in their solid states are approximately constant. They derived the expression: atomic weight × specific heat (in cal g^{-1} $°C^{-1}$) = 6.4 (approximately). The specific heats of silicon, phosphorus, and lead are 0.17, 0.19, and 0.031 cal g^{-1} $°C^{-1}$, respectively. Use this information in the following three exercises.

1. Comment on the validity of the law of Dulong and Petit when applied to silicon, phosphorus, and lead.

2. To raise the temperature of 75.0 g of a particular metal by 15°C requires 107 cal of heat. What is the approximate atomic weight of the metal? What might the metal be?

*3. A sample of tin weighing 315 g and at a temperature of 65.0°C is added to 100.0 cm^3 of water at 25.0°C. Estimate the final water temperature.

*4. An old "trick" for cooling down a hot beverage quickly is to immerse a cold spoon into it. A silver spoon weighing 4.12 oz and at a temperature of 70°F is placed in a cup of hot coffee (8 oz) at 160°F. What will be the resulting liquid temperature? Assume that the specific heat of coffee is 1 cal g^{-1} $°C^{-1}$ and that of silver, 0.056 cal g^{-1} $°C^{-1}$. Comment on the effectiveness of this method of cooling a hot drink.

5. A British thermal unit (Btu) is defined as the quantity of heat required to change the temperature of 1 lb of water by 1°F. Assuming the specific heat of water to be independent of temperature, how much heat is required to raise the temperature of the water in a 40-gal water heater from 70 to 150°F? **(a)** in Btu; **(b)** in kcal; **(c)** in kJ. (1 L = 1.06 qt.)

6. A mixture of 187 g of copper and 415 g of water is heated in an open beaker from 24.0°C to 78.3°C. What is ΔH (in kJ) for the copper–water mixture? (Specific heats: water, 4.18 J g^{-1} $°C^{-1}$; copper, 0.393 J g^{-1} $°C^{-1}$.)

7. The combustion of normal octane in an excess of oxygen is represented by the equation

$$C_8H_{18}(l) + \tfrac{25}{2} O_2(g) \rightarrow 8 CO_2(g) + 9 H_2O(l)$$
$$\Delta\overline{H} = -5.48 \times 10^3 \text{ kJ/mol}$$

How much heat is liberated per gal $C_8H_{18}(l)$ burned at constant pressure? (density of octane = 0.703 g/cm^3)

8. The heat of solution of KI(s) in water is listed as $\Delta\overline{H}$ = +21.3 kJ/mol KI. If a quantity of KI is added to sufficient water at 23.5°C in a Styrofoam cup to produce 150.0 cm^3 of 2.50 *M* KI, what will the final temperature be? (Assume a density of 1.30 g/cm^3 and a specific heat of 4.18 J g^{-1} $°C^{-1}$ for 2.50 *M* KI.)

9. The heat of neutralization of HCl(aq) by NaOH(aq) is −55.90 kJ/mol H_2O produced. If 50.00 mL of 1.05 *M* NaOH is added to 25.00 mL of 1.86 *M* HCl, with both solutions originally at 24.72°C, what will be the final solution temperature?

Assume that no heat is lost to the surrounding air and that the aqueous solution produced in the neutralization reaction has a specific heat of 1.00 cal g^{-1} $°C^{-1}$.

$$HCl(aq) + NaOH(aq) \rightarrow NaCl(aq) + H_2O$$
$$\Delta\overline{H} = -55.90 \text{ kJ/mol}$$

10. The heats of combustion ($\Delta\overline{H}$) per mole of sucrose, fructose, and glucose are −1349.6, −675.6, and −673.0 kJ, respectively. Use these data to calculate $\Delta\overline{H}$ for the following reaction.

$$C_{12}H_{22}O_{11}(s) + H_2O(l) \rightarrow \underset{\text{fructose}}{C_6H_{12}O_6(s)} + \underset{\text{glucose}}{C_6H_{12}O_6(s)} \quad \Delta\overline{H} = ?$$

11. For the reaction $C_2H_4(g) + Cl_2(g) \rightarrow C_2H_4Cl_2(l)$, determine $\Delta\overline{H}$, given that

$$2 Cl_2(g) + 2 H_2O(l) \rightarrow 4 HCl(g) + O_2(g)$$
$$\Delta\overline{H} = +202.5 \text{ kJ/mol}$$

$$2 HCl(g) + C_2H_4(g) + \tfrac{1}{2} O_2(g) \rightarrow C_2H_4Cl_2(l) + H_2O(l)$$
$$\Delta\overline{H} = -319.6 \text{ kJ/mol}$$

$$\tfrac{1}{2} H_2(g) + \tfrac{1}{2} Cl_2(g) \rightarrow HCl(g)$$
$$\Delta\overline{H} = -92.30 \text{ kJ/mol}$$

$$H_2(g) + \tfrac{1}{2} O_2(g) \rightarrow H_2O(l)$$
$$\Delta\overline{H} = -285.85 \text{ kJ/mol}$$

12. A handbook lists two different values for the heat of combustion of hydrogen: 34.18 kcal/g H_2 if $H_2O(l)$ is formed, and 29.15 kcal/g H_2 if steam [$H_2O(g)$] is formed. Explain why these two values are different, and indicate what property this difference represents. Devise a means of verifying your conclusions with data from Appendix D.

13. A calorimeter that measures an exothermic heat of reaction by the quantity of ice that can be melted is called an **ice calorimeter.** Now consider that 0.100 L of methane gas, $CH_4(g)$, at 25.0°C and 744 mmHg is burned completely at constant pressure in an excess of air. The heat liberated is captured and used to melt 10.7 g of ice at 0°C. (The heat required to melt ice, called the heat of fusion, is 333.5 J/g.)
   (a) Write an equation for the combustion reaction.
   (b) What is ΔH for this reaction?

14. Which of the following gases has the greater fuel value, on a per liter (STP) basis. That is, which has the greater heat of combustion? (*Hint:* The only combustible gases are CH_4, C_3H_8, CO, and H_2.)
   (a) coal gas: 49.7% H_2, 29.9% CH_4, 8.2% N_2, 6.9% CO, 3.1% C_3H_8, 1.7% CO_2, and 0.5% O_2, by volume
   (b) sewage gas: 66.0% CH_4, 30.0% CO_2, and 4.0% N_2, by volume

*15. Joule published his definitive results on the mechanical equivalent of heat in 1850. One of these results was that "The quantity of heat capable of increasing the temperature of one pound of water by 1° Fahrenheit requires for its evolution the expenditure of a mechanical force represented by the fall of 772 lb through the space of one foot.'' Show that this result is equivalent to that presented in equation (6.5).

*16. Some of the butane, $C_4H_{10}(g)$, in a 200.0-L cylinder at 26.0°C is withdrawn and burned at constant pressure in an excess of air. As a result the pressure of the gas in the cylinder falls from 2.35 atm to 1.10 atm. The liberated heat is used to raise the temperature of 35.0 gal of water from 26.0°C to 62.2°C. Assuming that the combustion products are $CO_2(g)$ and $H_2O(l)$ exclusively, what is the efficiency of the water heater? (That is, what percent of the heat of combustion was absorbed by the water?)

*17. One of the advantages of modern coal gasification processes over earlier ones is said to be that some of the heat required to bring about gasification (equation 6.18) is supplied by the methanation reaction (equation 6.22). Use data from Appendix D to show that this is indeed possible.

*18. The metabolism of glucose, $C_6H_{12}O_6$, yields $CO_2(g)$ and $H_2O(l)$ as products. Heat released in the process is converted to useful work with about 70% efficiency. Calculate the mass of glucose metabolized by a 58.0-kg person in climbing a mountain with an elevation gain of 1450 m. Assume that the work performed in the climb is about four times that required simply to lift 58.0 kg by 1450 m. $\Delta \overline{H}_f^{\circ}$ of $C_6H_{12}O_6(s)$ is -1274 kJ/mol.

Self-Test Questions

For questions 1 through 8 select the single item that best completes each statement.

1. 1.00 *kcal* of heat is
(a) absorbed when 1.00 cm³ water is heated from 14.5 to 15.5°C.
(b) absorbed when 1.00 L water is heated from 20.0 to 30.0°C.
(c) given off when 100.0 cm³ water is cooled from 20.0 to 10.0°C.
(d) equal to 1.0×10^6 cal.

2. Each of four Styrofoam cups contains 75.0 g water at 25.0°C. To each cup is added 50.0 g of a different metal at 100°C. The final temperature is *highest* in the cup to which the added metal is
(a) Al (sp. ht. = 0.217 cal g^{-1} °C^{-1}); (b) Ag (sp. ht. = 0.056 cal g^{-1} °C^{-1}); (c) Fe (sp. ht. = 0.113 cal g^{-1} °C^{-1}); (d) Cu (sp. ht. = 0.093 cal g^{-1} °C^{-1}).

3. A 100-mL sample of water at 50°C is added to 50.0 mL of water at 30°C. The final temperature will be (a) 35.0°C; (b) 37.0°C; (c) 40.0°C; (d) 43.4°C.

4. The heat of solution of NaOH(s) is listed as -41.6 kJ/mol. When NaOH is dissolved in water, the solution temperature (a) increases; (b) decreases; (c) remains constant; (d) either increases or decreases, depending on how much NaOH is dissolved.

5. A handbook lists the heat of combustion of $CS_2(l)$ as -3.24 kcal/g CS_2. For the reaction, $CS_2(l) + 3 O_2(g) \rightarrow CO_2(g) + 2 SO_2(g)$, $\Delta \overline{H}$ is (a) -3.24 kcal/mol; (b) -1.03×10^3 kJ/mol; (c) -58.95 kJ/mol; (d) -13.6 kJ/mol.

6. The molar enthalpy of formation of $CO_2(g)$ is equal to (a) 0; (b) the molar heat of combustion of C(graphite); (c) the sum of the molar enthalpies of formation of CO(g) and $O_2(g)$; (d) the molar heat of combustion of CO(g).

7. A handbook lists the enthalpy of formation of $NH_3(g)$ as -46 kJ/mol. For the reaction

$$2 NH_3(g) \rightarrow N_2(g) + 3 H_2(g) \qquad \Delta \overline{H}^{\circ} = ?$$

(a) $\Delta \overline{H}^{\circ} = -46$ kJ/mol; (b) $\Delta \overline{H}^{\circ} = +46$ kJ/mol; (c) $\Delta \overline{H}^{\circ} = +92$ kJ/mol; (d) $\Delta \overline{H}^{\circ} = +138$ kJ/mol.

8. The enthalpy of formation of $CO_2(g)$ is -394 kJ/mol, and that of $H_2O(l)$ is -286 kJ/mol. The heat of combustion of $C_5H_{12}(l)$ is $\Delta \overline{H}^{\circ} = -3534$ kJ/mol.

$$C_5H_{12}(l) + 8 O_2(g) \rightarrow 5 CO_2(g) + 6 H_2O(l)$$
$$\Delta \overline{H}^{\circ} = -3534 \text{ kJ/mol}$$

The enthalpy of formation of $C_5H_{12}(l)$, in kJ/mol, is
(a) $+3534$
(b) $[-3534 - 5 \times (-394) - 6 \times (-286)]$
(c) $[5 \times (-394) + 6 \times (-286) - 3534]$
(d) $[5 \times (-394) + 6 \times (-286) + 3534]$

9. Explain briefly the difference in meaning between
  (a) specific heat and molar heat capacity of a substance;
  (b) endothermic and exothermic reaction;
  (c) enthalpy of formation and heat of combustion of the hydrocarbon $C_4H_{10}(g)$.

10. A 1.50-kg piece of iron (sp. ht. = 0.59 J g^{-1} °C^{-1}) is dropped into 0.755 kg of water and the water temperature is observed to rise from 21.3 to 38.6°C. What must have been the initial temperature of the iron?

11. The heat of combustion of phenol, $C_6H_5OH(s)$, is determined in a bomb calorimeter and found to be -32.55 kJ/g. Write a balanced equation for the reaction, including a value of $\Delta \overline{H}$ (which you can assume to be the molar heat of combustion).

12. The heats of combustion ($\Delta \overline{H}^{\circ}$) per mole of C(graphite) and CO(g) are -393.51 and -282.97 kJ/mol, respectively. In both cases $CO_2(g)$ is the sole product. For the formation of the poisonous gas, phosgene,

$$CO(g) + Cl_2(g) \rightarrow COCl_2(g) \qquad \Delta \overline{H}^{\circ} = -108 \text{ kJ/mol}$$

Use Hess's law to calculate the enthalpy of formation of $COCl_2(g)$.

7 Electrons in Atoms

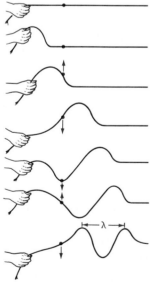

FIGURE 7-1
The simplest wave motion—traveling wave in a string.

The hypothetical string pictured here is infinitely long. Waves pass along the string only in the left-to-right direction. The wave is called a traveling wave.

For several reasons we need to acquire a more detailed knowledge of atomic and molecular structure. An understanding of atomic structure helps us to explain the forces between atoms that lead to the formation of molecules. With a knowledge of molecular structure we will be able to predict, among other things, whether a collection of molecules should exist as a solid, liquid, or gas and whether substances should form a solution on mixing. In this chapter we study the electronic structures of atoms. By the electronic structure of an atom we mean, basically, a description of where the electrons in an atom are most likely to be found.

Rutherford's nuclear atomic model (Section 2-6) became possible only after the principles had been established in two fundamental areas of physics—electricity and magnetism. To understand the behavior of electrons in atoms requires some appreciation of the nature of electromagnetic radiation. Our study will begin with this topic. Again, we will see how experiments and discoveries followed one another, for the most part in a systematic fashion. But there were a few dramatic breakthroughs as well.

7-1 Electromagnetic Radiation

Electric charges and magnetic poles exert forces over a distance, through electric and magnetic fields. Furthermore, these fields are complementary. A changing electric field induces a magnetic field, and vice versa. If electrically charged particles move with respect to one another, alternating electric and magnetic fields are produced and propagated through the space or medium surrounding the particles. The mode of propagation is called a **wave.** Energy is associated with the electric and magnetic fields, and the wave becomes a means of transmitting energy over distances. This energy transfer is called **electromagnetic radiation.**

Wave motion is complex, but certain characteristics of waves can be thought of in terms of vibrations in a string, pictured in Figure 7-1. The motion that is propagated through the string is the up-and-down motion of the hand holding the end of the string, starting at time $t = 0$ (top) and continuing through several intervals of time (bottom). The wave travels from left to right, but the vibrating medium (the string) moves up and down, that is, perpendicular to the direction of the wave itself. The position of a typical point on the string as a function of time is denoted by the colored dot. The arrows indicate the direction in which the dot is moving. At any instant of time the wave in the string consists of a number of regions in which the string is at a maximum or high point. These are called wave crests. There are a corresponding number of regions at a minimum or low point, called wave troughs. The distance between two successive crests (or troughs) is called the **wavelength,** usually designated by the Greek letter lambda, $\boldsymbol{\lambda}$.

Another characteristic property of a wave is its **frequency,** designated by the Greek letter nu, $\boldsymbol{\nu}$. This is the number of wave crests or troughs that pass through a given point in a fixed time. Frequency can be expressed by the unit s^{-1} (i.e., per second), meaning the number of occurrences, events, or cycles per second. For the vibrating string in Figure 7-1, this is just the number of times per second that the hand driving the wave goes through its up-and-down motion.

The product of the length of a wave (λ) and the number of cycles produced per second (i.e., the frequency, ν) indicates how far the wave front has traveled down the string in one second. This is *the velocity of the wave, c.*

$$c = \nu\lambda \qquad (7.1)$$

Now let us return to a discussion of electromagnetic waves. Propagation of these waves results from the oscillations of charged particles. This is analogous to the up-and-down hand motion in Figure 7-1. The electromagnetic waves are complicated by the fact that they travel in all directions simultaneously, that is, they are three-dimensional waves. An additional complication is that the waves actually represent two kinds of vibrations occurring simultaneously. These are an oscillating electric field and, perpendicular to this (at a 90° angle), an oscillating magnetic field. Still another interesting difference between electromagnetic radiation and other kinds of waves is that electromagnetic radiation requires no medium for its transmission. It can travel through vacuum or empty space. Despite these complications, an electromagnetic wave is often represented in the simplified manner shown in Figure 7-2. This is adequate for our purposes.

Frequency, Wavelength, and Velocity.   A variety of units are used to describe electromagnetic radiation. Frequency is usually expressed through the unit s^{-1}, referred to as cycles per second. The SI unit for cycles per second is the **hertz (Hz).** Wavelength must have a unit of length, and logically this should be the meter (m). However, because so many kinds of electromagnetic radiation are of very short wavelength, some smaller units are necessary. The units listed below are among those commonly encountered. [The angstrom is named for the Swedish physicist, Ångstrom (1814–1874), who used 1×10^{-10} m as a basic unit of wavelength in his studies of the solar spectrum; the angstrom is not an SI unit.]

1 centimeter (cm) = 1×10^{-2} m

1 nanometer (nm) = 1×10^{-9} m = 1×10^{-7} cm = 10 Å

1 angstrom (Å) = 1×10^{-10} m = 1×10^{-8} cm

A distinctive feature of electromagnetic radiation is that its velocity has a constant value of 2.997925×10^8 m s^{-1} in vacuum (usually rounded off to 3.00×10^8 m s^{-1}). Since ordinary light is a form of electromagnetic radiation, this characteristic velocity is

FIGURE 7-2
An electromagnetic wave.

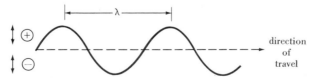

Although oversimplified, this representation does suggest that electromagnetic radiation results from the relative motion of electrically charged objects and denotes one characteristic of the radiation—the wavelength λ.

FIGURE 7-3
The electromagnetic spectrum.

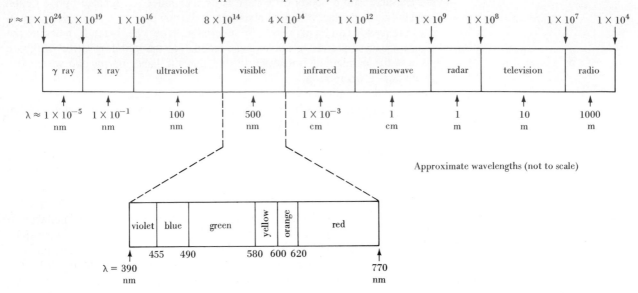

often called the **speed of light.** The frequencies and wavelengths of a number of different kinds of electromagnetic radiation are compared in Figure 7-3.

Example 7-1   A microwave oven produces electromagnetic radiation of wavelength 12.2 cm. What is this wavelength expressed in meters and in nanometers?

Solution

$$\text{no. m} = 12.2 \text{ cm} \times \frac{1 \text{ m}}{100 \text{ cm}} = 0.122 \text{ m}$$

$$\text{no. nm} = 12.2 \text{ cm} \times \frac{1 \text{ nm}}{1 \times 10^{-7} \text{ cm}} = 1.22 \times 10^{8} \text{ nm}$$

SIMILAR EXAMPLE: Review Problem 1.

Example 7-2   An FM radio station broadcasts on a frequency of 91.5 megahertz (MHz). What is the wavelength of these radio waves, in meters?

Solution.   The prefix *mega* denotes one million or 1×10^{6}. A frequency of 91.5 MHz = $91.5 \times 10^{6} \text{ s}^{-1}$.

Since radio waves are a form of electromagnetic radiation, $c = 3.00 \times 10^{8} \text{ m/s}$. Solve equation (7.1) for λ.

$$\lambda = \frac{c}{\nu} = \frac{3.00 \times 10^{8} \text{ m s}^{-1}}{91.5 \times 10^{6} \text{ s}^{-1}} = 3.28 \text{ m}$$

SIMILAR EXAMPLES: Review Problem 2; Exercise 2.

Example 7-3   Most of the light emitted by a sodium vapor lamp has a wavelength of 589 nm. What is the frequency of this radiation?

Solution

$$c = 3.00 \times 10^8 \text{ m/s}$$

$$\lambda = 589 \text{ nm} \times \frac{1 \times 10^{-9} \text{ m}}{1 \text{ nm}} = 5.89 \times 10^{-7} \text{ m}$$

$$\nu = ?$$

Equation (7.1) must be rearranged to the form $\nu = c/\lambda$.

$$\nu = \frac{c}{\lambda} = \frac{3.00 \times 10^8 \text{ m s}^{-1}}{5.89 \times 10^{-7} \text{ m}} = 5.09 \times 10^{14} \text{ s}^{-1}$$

SIMILAR EXAMPLE: Review Problem 3.

The Visible Spectrum.   The speed of light depends on the medium through which it travels. As a result, a light ray is refracted or bent as it passes from one medium to another. Because light waves of differing wavelengths are refracted differently, a light ray consisting of a large number of wavelength components is dispersed into a band or spectrum of colors as the ray passes through a medium. The shortest wavelength light that the human eye can detect corresponds to the color violet; the longest, red.

These points are illustrated in Figure 7-4. Here a beam of "white" light is passed through a narrow opening or slit and then dispersed into its colored components by a glass prism. A device in which light is dispersed and the intensity of each component is measured is called a **spectrometer.** If the spectrum is recorded photographically, the device is called a **spectrograph.**

The dispersion of white light is also illustrated in Color Section A.

White light is polychromatic (many-colored). If desired, after polychromatic light has been dispersed into a spectrum in a spectrometer, all the wavelength components except a narrow band can be blocked off. In this way monochromatic (one-color) light can be obtained.

7-2 Atomic Spectra

In the spectrum illustrated in Figure 7-4 the light source is "white" light. This could be sunlight or certain artificial light sources, for example, the heated filament of an ordinary electric light bulb. Each wavelength component of the white light, after passing through

FIGURE 7-4
The spectrum of white light.

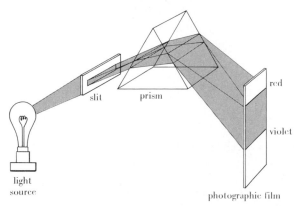

The wavelength components of white light have different speeds in a material medium. Because red light has the lowest frequency of the visible colors, it does not interact strongly with the medium. Its speed is not greatly reduced, and it is refracted least. Violet light, on the other hand, has the highest frequency of the visible colors. It has more opportunity to interact with the medium, is slowed down the most, and hence is refracted the most.

FIGURE 7-5
The production of an
atomic or line spectrum.

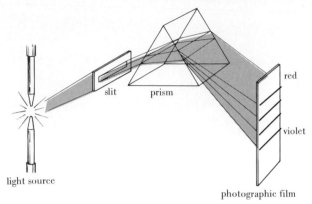

The light source depicted here is an electric arc between a pair of graphite electrodes. The substance whose spectrum is to be investigated is vaporized at the high temperature of the arc.

the slit and prism, produces an image of the slit on the photographic film in the form of a line. White light consists of so many wavelength components that its spectrum is a continuum of these lines. The spectrum of white light is continuous, displaying a gradual blending of colors from red, through orange, yellow, green, and blue, to violet.

On the other hand, with light emitted by most heated substances, only a discrete number of images of the entrance slit—a series of colored lines—is observed. These spectra are discontinuous, or line, spectra. The production of a line spectrum is illustrated in Figure 7-5.

The line spectrum obtained for each element differs from that of every other element—the line spectrum is like a fingerprint of the element. One of the first investigators to make extensive use of atomic spectra to identify chemical elements was the German chemist Robert Bunsen (1811–1899). During the latter half of the nineteenth century the spectra of most of the elements had been investigated, including even that of the element helium, which was discovered to exist on the sun before it was discovered on earth.

The visible spectrum of
hydrogen is also shown
in Color Section A.

Among the most extensively studied atomic spectra during the nineteenth century was that of the element hydrogen. The visible spectrum of hydrogen is rather simple, consisting of a red line, a green line, and a number of blue and violet lines that appear to converge to a limit in the ultraviolet region. The first four lines in the spectrum (starting with the red line at 656.3 nm) were investigated very carefully by Ångstrom. They correspond to the wavelengths listed in Figure 7-6. In 1885, the Swiss physicist-schoolteacher Johann Balmer, from just these four wavelength values, deduced, apparently by trial and error, the following formula for them.

FIGURE 7-6
The Balmer series for
hydrogen—a line
spectrum.

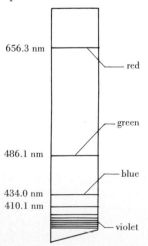

$$\lambda \text{ (in Å)} = 3645.6\left(\frac{n^2}{n^2 - 4}\right) \quad \text{where } n = 3, 4, 5, \ldots \tag{7.2}$$

If $n = 3$ is substituted into Balmer's formula, the wavelength obtained is 656.2 nm, in excellent agreement with Ångstrom's measured value for the red line. If $n = 4$ the wavelength of the green line is obtained, and so on. A more commonly encountered form of the Balmer equation is shown below, written in terms of the frequencies of the spectral lines.

$$\nu = Rc\left(\frac{1}{2^2} - \frac{1}{n^2}\right) = 3.2881 \times 10^{15} \text{ s}^{-1}\left(\frac{1}{2^2} - \frac{1}{n^2}\right) \tag{7.3}$$

R is a numerical constant, called the Rydberg constant, having a value of 10,967,800 m^{-1}; c is the velocity of light, 2.997925×10^8 m s^{-1}. To simplify calculations, the product $R \times c$ is also given in equation (7.3). To five significant figures, it is 3.2881×10^{15} s^{-1}.

Example 7-4 Use equation (7.3) to calculate the wavelength of the fourth line ($n = 6$) in the Balmer series of hydrogen. Compare this result with the value given in Figure 7-6.

Solution

$$\nu = 3.2881 \times 10^{15} \text{ s}^{-1} \times \left(\frac{1}{2^2} - \frac{1}{6^2}\right) = 3.2881 \times 10^{15} \text{ s}^{-1} \times \left(\frac{1}{4} - \frac{1}{36}\right)$$

$$= 3.2881 \times 10^{15} \text{ s}^{-1} \times (0.25000 - 0.02778)$$

$$= 3.2881 \times 10^{15} \text{ s}^{-1} \times 0.22222 = 7.3068 \times 10^{14} \text{ s}^{-1}$$

$$\lambda = \frac{c}{\nu} = \frac{2.9979 \times 10^8 \text{ m s}^{-1}}{7.3068 \times 10^{14} \text{ s}^{-1}} = 4.103 \times 10^{-7} \text{ m} = 410.3 \text{ nm}$$

The calculated value (410.3 nm) and the measured value (410.1 nm) agree within 0.2 nm. This is a good agreement.

SIMILAR EXAMPLES: Review Problem 4a, b; Exercise 4.

Example 7-5 A line is found in the Balmer series of hydrogen at 389.0 nm. To what value of n in equation (7.3) does this line correspond?

Solution. The wavelength 389.0 nm is 389.0×10^{-9} m $= 3.890 \times 10^{-7}$ m. Light of this wavelength has a frequency of

$$\nu = \frac{c}{\lambda} = \frac{2.998 \times 10^8 \text{ m s}^{-1}}{3.890 \times 10^{-7} \text{ m}} = 7.707 \times 10^{14} \text{ s}^{-1}$$

Now, rearrange equation (7.3) to the form

$$\frac{\nu}{3.2881 \times 10^{15} \text{ s}^{-1}} = \left(\frac{1}{2^2} - \frac{1}{n^2}\right)$$

and substitute the value of ν for the spectral line in question.

$$\frac{7.707 \times 10^{14} \text{ s}^{-1}}{3.2881 \times 10^{15} \text{ s}^{-1}} = 0.2344 = \frac{1}{4} - \frac{1}{n^2}$$

Solve for $1/n^2$,

$1/n^2 = 0.2500 - 0.2344 = 0.0156$

and then for its reciprocal,

$n^2 = 1/0.0156 = 64.1$

and then for the square root of the reciprocal.

$n = \sqrt{n^2} = \sqrt{64.1} = 8.01 \simeq 8.00$

The integral value of n in equation (7.3) is 8.

SIMILAR EXAMPLES: Review Problem 4c; Exercise 7c.

By the early nineteenth century a wave theory of light had been firmly established and was successful in explaining continuous spectra (like the rainbow). But the existence of atomic or line spectra could not be explained by this wave theory. Not even the electromagnetic theory of radiation, introduced by James Maxwell in the 1860s, was successful in explaining atomic spectra. That such a simple equation as Balmer's could be used to correlate atomic spectral data suggested that there was some basic principle underlying all atomic spectra. Yet this principle was never discovered by the methods of nineteenth-century physics known as classical physics.

FIGURE 7-7
An analogy between quantum mechanics and classical mechanics.

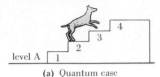

(a) Quantum case

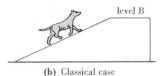

(b) Classical case

To climb from level A to level B the dog in (a) proceeds through 4 steps. It may change its position (and hence its potential energy) only in these 4 discrete steps of fixed magnitude. The dog in (b) can change its position and potential energy from level A to level B in any number of steps of any size whatsoever. Case (a) corresponds to quantum changes in energy, case (b) to classical changes in energy.

7-3 Quantum Theory

The key to the growing number of unsolved problems of nineteenth-century physics lay in a great breakthrough of modern science—the quantum theory. This theory was proposed in 1900 by Max Planck (1858–1947) to explain a phenomenon known as blackbody radiation—the emission of electromagnetic radiation by heated objects (such as the light from a "red-hot" iron poker). Planck's revolutionary hypothesis was that energy, like matter, is *discontinuous,* and consists of large numbers of tiny *discrete,* units called **quanta.** Whether a system gains or loses energy, it must do so in terms of these quanta. The energy associated with a quantum of electromagnetic radiation is proportional to the frequency of the radiation and is expressed by

$$E = h\nu \tag{7.4}$$

The proportionality constant, **h,** is called **Planck's constant** and has a value of 6.626×10^{-34} J s.

According to the laws of classical physics, the energy of a system should be able to assume any value and to change by any amount. By the principles of quantum theory, however, we find that the energy of a system can have only a unique set of values. This means that energy can change only by certain discrete amounts, called quantum jumps. A system may gain or lose one quantum of energy, or two quanta, or three, and so on; but it cannot gain or lose $\frac{1}{3}$, $\frac{1}{2}$, $1\frac{1}{3}$, or $2\frac{1}{2}$ quanta. Figure 7-7 suggests an analogy between classical and quantum theory that may prove helpful in understanding the essential difference between the two.

There have been instances in the history of science when a new hypothesis proved useful in explaining one phenomenon but was not generally applicable to any others. It was only in the discovery of other applications of the quantum hypothesis that it acquired status as a significant new theory of science. The first notable new success came in 1905 with Albert Einstein's (1879–1955) quantum explanation of the photoelectric effect.

The Photoelectric Effect. An interesting phenomenon, first observed by H. Hertz in 1887, is the photoelectric effect, depicted in Figure 7-8. A beam of light is shown striking a particular metal surface. This causes the emission of electrons and leaves the surface positively charged. According to classical physics, both the *number* of electrons ejected from the surface and their *energies* should depend on the *intensity* or *brightness* of the incident light. The number of ejected electrons does depend on the intensity of the inci-

FIGURE 7-8
The photoelectric effect.

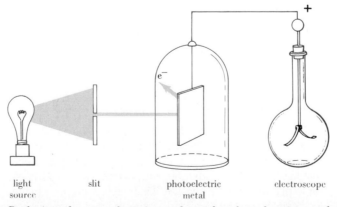

light source   slit   photoelectric metal   electroscope

By losing electrons from its surface, the photoelectric metal acquires a positive charge, as do the metal-foil leaves of the electroscope. Having like charges, the leaves repel one another. (See Appendix B for a further description of an electroscope.)

FIGURE 7-9
Dependence of
photoelectric effect on
the frequency of light.

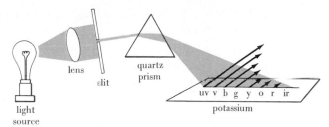

light
source

lens
slit
quartz
prism

uv v b g y o r ir
potassium

The lengths of the arrows represent the kinetic energies of the photoelectrons ejected by
different wavelength components of light. Color code: UV, ultraviolet; v, violet; b,
blue; g, green; y, yellow; o, orange; r, red; IR, infrared. Light with wavelength greater
than 710 nm (red) causes no photoelectric effect on potassium.

dent light, but the electron energies do not! These energies depend only on the frequency
(or wavelength) of the light. Thus, the kinetic energies of electrons ejected by a feeble
blue light are greater than the kinetic energies of electrons ejected by a bright red light (see
Figure 7-9). As with atomic spectra, here was another phenomenon that defied explanation by classical physics.

The photoelectric effect
has many practical ap-
plications, ranging from
automatic door openers
to light meters to light-
sensitive elements in television cameras.

Einstein proposed that electromagnetic radiation has particlelike characteristics and
that "particles" of light, called **photons,** possess a characteristic energy, given by
Planck's equation, $E = h\nu$. We can think of the energy of a light wave as being concentrated into photons. In the photoelectric effect these photons transfer energy during colli
sions with electrons. In each collision a photon gives up its entire energy—a quantum of
energy—to an electron. The more energetic the photon, the more energy it transfers to an
electron and the greater the kinetic energy of the ejected electron. So, the kinetic energy of
the electron should depend on the light frequency. The particlelike nature of light is
suggested by Figure 7-10.

The product of h and ν yields the energy of a single photon of electromagnetic radia-
tion in the unit joules. The energy of a typical photon is only a tiny fraction of a joule.
Often we deal with the much larger energy associated with a mole of photons, that is,
Avogadro's number (6.02205×10^{23}) of photons. Several alternative expressions for the
energy associated with ultraviolet radiation of wavelength 225 nm are presented in
Table 7-1.

FIGURE 7-10
Photons of light
visualized.

A light beam having this appearance

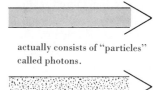

actually consists of "particles"
called photons.

Example 7-6 The lowest frequency light that can produce a photoelectric effect on
potassium metal is $4.2 \times 10^{14} \text{ s}^{-1}$. What is the energy of one photon of this light?

Solution

$E = h\nu = (6.626 \times 10^{-34} \text{ J s})(4.2 \times 10^{14} \text{ s}^{-1}) = 2.8 \times 10^{-19} \text{ J}$

SIMILAR EXAMPLES: Review Problem 5a; Exercise 8.

TABLE 7-1
Alternative expressions of energy associated with
ultraviolet radiation with $\lambda = 225$ nm

	Energy of	
	Single photon	Mole of photons
joule	8.81×10^{-19}	5.30×10^{5}
kilojoule	8.81×10^{-22}	5.30×10^{2}
calorie	2.11×10^{-19}	1.27×10^{5}
kilocalorie	2.11×10^{-22}	1.27×10^{2}

Example 7-7   What is the energy, in kJ/mol, associated with monochromatic radiation of wavelength 225 nm?

Solution.   First we must use equation (7.1) to determine the frequency of this radiation. Then we apply the Planck equation (7.4) to determine the energy per photon, in kJ/photon. Finally, we use Avogadro's number to convert from kJ/photon to kJ/mol photons.

$$\nu = \frac{c}{\lambda} = \frac{3.00 \times 10^8 \text{ m s}^{-1}}{225 \times 10^{-9} \text{ m}} = 1.33 \times 10^{15} \text{ s}^{-1}$$

$$\text{no. kJ/photon} = 6.626 \times 10^{-34} \frac{\text{J s}}{\text{photon}} \times 1.33 \times 10^{15} \text{ s}^{-1} \times \frac{1 \text{ kJ}}{1000 \text{ J}}$$

$$= 8.81 \times 10^{-22} \text{ kJ/photon}$$

$$\text{no. kJ/mol} = 8.81 \times 10^{-22} \frac{\text{kJ}}{\text{photon}} \times \frac{6.02 \times 10^{23} \text{ photons}}{1 \text{ mol}} = 5.30 \times 10^2 \text{ kJ/mol}$$

SIMILAR EXAMPLES: Review Problem 5; Exercises 8, 9.

7-4 The Bohr Atom

Even by the methods of classical physics, certain conclusions were possible concerning the behavior of electrons in atoms. It was obvious from the laws of electrostatics, for example, that negatively charged electrons could not remain at rest; otherwise, they would be attracted into the positively charged nucleus. Furthermore, the movement of electrons around the nucleus was a necessary condition to explain the emission of light. (Maxwell's theory stated that electromagnetic radiation resulted from the relative motion of electrically charged objects.) But a dilemma was posed for classical physics. According to classical physics, an electron must accelerate constantly in revolving about the nucleus of an atom. In doing so it would give off energy as light. Having lost energy, the electron should then be drawn closer to the nucleus. With each circuit the electron should lose more energy and be drawn still closer to the nucleus. This type of behavior suggests a spiraling motion in which the electron "falls" into the nucleus. But if electrons all were to suffer this fate, there could be no accounting for the fact that stable atoms consisting of electrically charged particles exist at all. The situation described here is pictured in Figure 7-11.

In 1913, the Danish physicist Niels Bohr (1885–1962) proposed that the dilemma just described could be resolved by introducing Planck's quantum of action (specifically, Planck's constant, *h*) into the description of atomic structure. His treatment of the hydrogen atom involved an interesting combination of classical and quantum theory, expressed through three ideas. The first two are the basic assumptions with which Bohr formulated his theory, and the third deals with specifying the allowable orbits for an electron in a hydrogen atom.

1. There is *only a certain set of allowable orbits* for an electron in a hydrogen atom. These orbits, referred to as stationary states of motion, are circular paths about the nucleus. The motion of an electron within a stationary state can be described by ordinary mechanics. However, even though classical theory would predict otherwise, *as long as an electron remains in a stationary state its energy remains constant and no light is emitted.*
2. An electron can pass only from one stationary state to another. In such transitions fixed, discrete quantities of energy (quanta) are involved, in accordance with Planck's equation, $E = h\nu$.

FIGURE 7-11
An unsatisfactory atomic model.

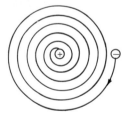

This model, based on classical physics, explains how an atom can emit light, though not why the light should show distinctive wavelength components. Furthermore, as energy is lost through light emission, the electron is drawn ever closer to the nucleus, eventually spiraling into it. This collapse of the atom would occur in a time interval much shorter than one second.

FIGURE 7-12
Bohr model of the
hydrogen atom.

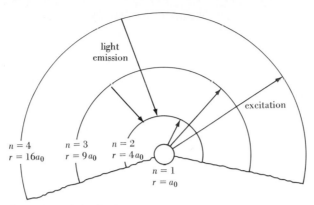

A portion of the hydrogen atom is pictured. The nucleus occupies a position at the center of the atom, and the electron is found in one of the discrete orbits $n = 1, 2, \ldots$. Electron transitions corresponding to excitation of the atom are shown in color, those to light emission in black.

3. The allowable stationary states are those in which certain properties of the electron have unique values. In particular, a property called the angular momentum must be an integral multiple of $h/2\pi$; that is, the angular momentum must be $nh/2\pi$ (where n is an integer and h is Planck's constant).

The atomic model for hydrogen based on these ideas is pictured in Figure 7-12. The allowable states for the electron are numbered, $n = 1$, $n = 2$, $n = 3$, and so on. These *integral* numbers are called **quantum numbers.** The letters, $K, L, M, N, \ldots$, are also used to designate the first few states. These designations by letter are based on terminology derived from spectroscopy.

Several properties of the electron in a hydrogen atom can be calculated with the Bohr theory. For one, the radii of the allowable orbits can be expressed as the multiples 1^2, 2^2, 3^2, 4^2, $\ldots$, n^2, of a certain orbit of minimum radius, designated as $a_0 = 0.53\text{Å}$. It is also possible to calculate the velocities associated with the electron in each of these orbits, and, most important of all, the energy. When the electron is free of the nucleus, its energy is taken to be zero. When the electron is attracted to the nucleus and bound in the orbit n, energy is emitted and the electron energy is lowered to the value

$$E_n = \frac{-B}{n^2} \tag{7.5}$$

B is a numerical constant with a value of 2.179×10^{-18} J.

One of the assumptions in Bohr's theory is that discrete quantities of energy are involved when an electron in a hydrogen atom passes from one allowable state to another. Let us use equation (7.5) to calculate the *difference in energy* between two states. As a case of special interest, we choose the states: $n = 3$ and $n = 2$.

$$\Delta E = E_3 - E_2 = \left(\frac{-B}{3^2}\right) - \left(\frac{-B}{2^2}\right) = \left(\frac{B}{2^2}\right) - \left(\frac{B}{3^2}\right) = B\left(\frac{1}{2^2} - \frac{1}{3^2}\right) \tag{7.6}$$

Equation (7.6) bears a striking resemblance to the Balmer equation (7.3)! Let us pursue this matter further. Normally, the electron in a hydrogen atom is believed to exist in the orbit closest to the nucleus ($n = 1$ or K state). Upon excitation a quantum of energy is absorbed and the electron jumps to a higher quantum state. Light emission occurs as the electron drops from a higher to a lower quantum state, and a unique quantity of energy is involved. Planck's equation (7.4) allows us to calculate the unique frequency (or wavelength) corresponding to this unique quantity of energy. Now let us return to our discussion of the transition of an electron from $n = 3$ to $n = 2$. The difference in energy between these two states, ΔE, is the energy of the emitted photon of light. We can write two equations for ΔE

$$\Delta E = h\nu \tag{7.4}$$

$$\Delta E = B\left(\frac{1}{2^2} - \frac{1}{3^2}\right) \tag{7.6}$$

and then set these equal to one another.

$$h\nu = B\left(\frac{1}{2^2} - \frac{1}{3^2}\right)$$

For the frequency of the photon of light, ν, we obtain

$$\nu = \frac{B}{h}\left(\frac{1}{2^2} - \frac{1}{3^2}\right) \tag{7.7}$$

The numerical value of the constant B/h can be determined readily.

$$\frac{B}{h} = \frac{2.179 \times 10^{-18}\ \text{J}}{6.626 \times 10^{-34}\ \text{J s}} = 3.289 \times 10^{15}\ \text{s}^{-1}$$

The constant B/h is practically identical to the product $R \times c$ in the Balmer equation (7.3)! We have succeeded in using results of Bohr's theory to derive the Balmer equation. If equation (7.7) is solved for a numerical result, the frequency obtained is that of the red line in the Balmer series. Every transition in which an electron moves from a higher-energy state to the state $n = 2$ produces a line in the Balmer series.

In Figure 7-13 the energies associated with the different allowable states for the electron in a hydrogen atom are shown as a group of lines. This representation is known as an **energy-level diagram.** When the electron is completely separated (ionized) from the atom ($n = \infty$), there is no attraction between the electron and the nucleus (proton). This,

FIGURE 7-13
Energy-level diagram
for the hydrogen atom.

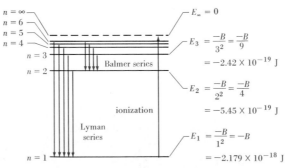

Among the features brought out by this diagram are

- The zero of energy corresponds to the completely ionized atom ($n = \infty$).
- Ionization of the normal hydrogen atom requires moving the electron from the level $n = 1$ to the level $n = \infty$, a process requiring 2.179×10^{-18} J.
- Electron transitions from higher quantum levels to the level $n = 2$ produce lines in the Balmer series; to the level $n = 1$, lines in the Lyman series.
- Energy differences between successive levels are smaller, the higher the values of n.

as we have already noted, is the zero of energy. Because the electron must absorb energy to become ionized and reach an energy state of zero, we designate the energy of the bound electron with a negative sign.

Example 7-8   What is the wavelength of the spectral line associated with the transition in which an electron in a hydrogen atom drops from the level $n = 2$ to the level $n = 1$?

Solution.   From Figure 7-13 we see that this line is in the Lyman series, which is in the ultraviolet region of the spectrum. We should expect our answer to be smaller than any of the wavelengths observed for the Balmer (visible) series. First, we need to use equation (7.5) to calculate the difference in energy between the levels $n = 2$ and $n = 1$.

$$E_1 = \frac{-B}{1^2} = -B \qquad \text{and} \qquad E_2 = \frac{-B}{2^2} = \frac{-B}{4}$$

$$\Delta E = E_2 - E_1 = \frac{-B}{4} - (-B) = \frac{3B}{4} = 0.7500\ B$$
$$= 0.7500 \times 2.179 \times 10^{-18}\ \text{J} = 1.634 \times 10^{-18}\ \text{J}$$

Now we can substitute this energy into Planck's equation (7.4) to obtain the frequency of the radiation.

$$\nu = \frac{E}{h} = \frac{1.634 \times 10^{-18}\ \text{J}}{6.626 \times 10^{-34}\ \text{J s}} = 2.466 \times 10^{15}\ \text{s}^{-1}$$

Finally, we use equation (7.1) to solve for the wavelength.

$$\lambda = \frac{c}{\nu} = \frac{2.998 \times 10^8\ \text{m s}^{-1}}{2.466 \times 10^{15}\ \text{s}^{-1}} \times \frac{1\ \text{nm}}{1 \times 10^{-9}\ \text{m}} = 121.6\ \text{nm}$$

SIMILAR EXAMPLES: Exercises 15, 16.

Shortcomings of the Bohr Theory.   The great success of the Bohr theory was in its ability to predict lines in the hydrogen atom spectrum. However, one of the discoveries of the time was that spectral lines have *fine structure,* especially in cases where excited atoms are placed in a magnetic field. That is, some principal lines were found actually to consist of a small number of very closed spaced lines. Fine structure in hydrogen spectra was explained through modification of the Bohr theory. However, the theory was never very successful in describing atomic spectra other than those of hydrogen, nor could it account for the ability of atoms to form molecules through chemical bonds.

7-5 Wave–Particle Duality

In the year 1905, Einstein set to rest a centuries-old dispute concerning the nature of light. Newton had advanced the proposition that light has a corpuscular or *particle* nature, that is, that it consists of a stream of energetic particles. Another theory was that of Huygens, who proposed that light consists of waves of energy.

To choose between the two theories required that accurate measurements be made of the speed of light in vacuum and in various media. Newton's view required that light travel *faster* in denser media, while Huygen's required that light travel *more slowly* in denser media. Accurate measurements of the speed of light showed that light does indeed travel *more slowly* in a denser medium. Thus, the wave model became firmly established. And along with this was also established the view that matter and energy are two distinctly different natural qualities governed by different laws. However, to explain the photoelectric effect Einstein was required to think of photons of light as if they were particles. Thus, there emerged the idea that light has a *dual* nature—in some instances its behavior is better understood in terms of waves and in other cases, particles.

In 1924, the French physicist Louis de Broglie, considering the nature of light and matter, offered a startling proposition: *Not only does light display particlelike characteristics, but small particles may at times display wavelike properties.* De Broglie's proposal received experimental verification in 1927—through experiments that led directly to the development of the electron microscope. De Broglie's description of matter waves was in mathematical terms. The de Broglie wavelength associated with a particle is related to the particle momentum, *p,* and Planck's constant, *h.* (Momentum is the product of mass, *m,* and velocity, *v.*)

$$\lambda = \frac{h}{p} = \frac{h}{mv} \tag{7.8}$$

In equation (7.8) wavelength is in meters, mass, in kilograms, and velocity, in meters per second. Planck's constant must be expressed in units of mass, length, and time, that is, by replacing the unit, joule, by the units $kg \ m^2 \ s^{-2}$ (see Appendix B, equation B.5).

Example 7-9   What is the wavelength associated with electrons traveling at one-hundredth the speed of light?

Solution.   The electron mass, expressed in kilograms, is 9.11×10^{-31} kg. The electron velocity v is $0.01 \times 3.00 \times 10^8$ m s^{-1}. Planck's constant $h = 6.626 \times 10^{-34}$ J s, and 1 J $= 1$ kg m^2 s^{-2}. Substituting these values into equation (7.8), we obtain

$$\lambda = \frac{6.626 \times 10^{-34} \ kg \ m^2 \ s^{-2} \ s}{(9.11 \times 10^{-31} \ kg)(3.00 \times 10^6 \ m \ s^{-1})} = 2.42 \times 10^{-10} \ m = 0.242 \ nm$$

SIMILAR EXAMPLES: Exercises 24, 25.

7-6 The Uncertainty Principle

The laws of classical physics are often thought of as universal truths. They tell us what physical behavior is permitted and what future events will follow from the present state of a system. When a rocket is fired, the exact point of impact can be calculated. Although errors in this calculation may arise from inaccuracies in measuring certain of the variables that affect the rocket trajectory, in principle these variables can be determined with the highest precision, leading to a result of any desired degree of accuracy. In classical physics nothing is left to chance—physical behavior can be predicted with certainty.

During the 1920s Niels Bohr and Werner Heisenberg thought about hypothetical experiments that would establish how precisely the behavior of subatomic particles could be determined. The two variables that determine this behavior are the position of a particle x and its momentum p. [Recall that momentum is the product of mass (m) and velocity (v); that is, $p = mv$.] The conclusion of these "thought" experiments was that there must always be uncertainties in measurement such that the product of the uncertainty in position, Δx, and in momentum, Δp, is

The symbol $\geqslant$ stands for equal to or greater than.

$$\Delta x \ \Delta p \geqslant \frac{h}{2\pi} \tag{7.9}$$

The significance of this expression, referred to as Heisenberg's uncertainty principle, is that position and momentum cannot both be measured with great precision simultaneously. If an experiment is designed to locate the position of a particle with great precision, it is not possible to measure its momentum precisely. Its future actions (trajectory) cannot

FIGURE 7-14
The uncertainty
principle.

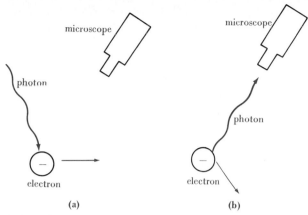

A free electron moves into focus of a hypothetical microscope (a). A photon of light strikes the electron and is reflected. In the collision the photon transfers momentum to the electron. The reflected photon is seen through the microscope, but the electron has moved out of focus (b). The exact position of the electron cannot be determined.

be predicted with certainty. Similarly, if the momentum is measured precisely, the position of the particle is not known with certainty. The uncertainty principle places a restriction on the precision with which we can make measurements and the certainty with which we can predict atomic events. But why should this be so?

Suppose that we wish to learn something of the behavior of an electron in a hydrogen atom by using a microscope with which to see the electron. What kind of microscope should this be? The resolving power of a microscope is limited to objects that are about the size of the wavelength of the light used. In an ordinary optical microscope using visible light, the resolving power is about 1000 nm. With an electron microscope the resolving power is about 1 nm.

The radius of the first Bohr orbit in a hydrogen atom is calculated to be 0.053 nm; thus, the diameter of a hydrogen atom is about 0.1 nm (10^{-10} m). Rutherford's experiments on the scattering of α particles by metal foils suggest that electrons are smaller still than the atoms containing them.

Suppose that the diameter of an electron is about 10^{-14} m. Light of this wavelength would have a frequency of 3×10^{22} s^{-1} ($\nu = c/\lambda$) and an energy per photon of 2×10^{-11} J ($E = h\nu$). But from Figure 7-13 we can see that this energy is far, far in excess of that required to ionize the electron in a hydrogen atom, that is, to strip the electron completely away from the atomic nucleus. Thus, in our attempt to see an electron in an atom, the measuring system (the light used) would interfere greatly with the measurement. We could not hope to determine the electron position and momentum with any precision. This point is illustrated further in Figure 7-14.

7-7 Wave Mechanics

The uncertainty principle is not easy for most people to accept philosophically. Einstein spent a good deal of time from the middle 1920s until his death in 1955 attempting, unsuccessfully, to disprove it.

The concepts introduced in the two preceding sections carry a number of implications regarding atomic structure. With the Bohr theory it is possible to calculate both the radius of an orbit and the velocity of an electron in this orbit. The former represents a precise definition of electron position and the latter of electron momentum. But, according to the uncertainty principle, it is impossible to measure both these quantities precisely. We should not adopt a theory that allows for the precise prediction of that which cannot be measured.

FIGURE 7-15
The electron as a
matter wave.

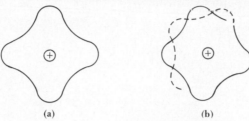

The wave pattern in (a), called a standing wave, is an acceptable representation of a matter wave. It has an integral number of wavelengths (four) about the nucleus, and successive waves reinforce one another. The pattern in (b) is unacceptable because the number of wavelengths is nonintegral (about 4.5), and successive waves overlap. Each maximum or crest in one part of the wave is cancelled by a minimum or trough in another part of the wave.

A second implication, explored with great success by Schrödinger in 1927, is that electrons in atoms can be treated as matter waves. Their motion can be likened to wave motion. But how are we to think of an electron in an atom according to this wave picture? Does the electron cease to exist as a particle? Is it literally smeared out into a wave? One answer to all these questions, of course, is that we do not know and cannot know. The Heisenberg uncertainty principle makes us ever aware of our inability to describe subatomic particles precisely.

The term *orbit* used in the Bohr theory suggests a definite two-dimensional pathway that electrons follow. The term *orbital* is intended only to outline the general three-dimensional region in which electrons are likely to be found.

The waves associated with electrons in atoms must correspond to certain allowable patterns (see Figure 7-15). These patterns can be described by mathematical equations, but this is a task well beyond the scope of our study. Let us just say that the acceptable solutions of these wave equations are known as **wave functions** (denoted by the Greek letter psi, ψ). Wave functions contain a set of three quantum numbers, and when specific values of these numbers are assigned, the result is called an **orbital.** The physical interpretation of an orbital is that it can be used to represent a region in space where an electron is likely to be found. The orbital description of an electron in an atom permits us to establish its energy state and also to think about the electron in either of two ways. First, we can think of the electron as being a cloud of negative electric charge, with the density of the charge varying from point to point. Or we can continue to think of an electron as a particle, with the probability of finding the electron varying from point to point. Although the wave function, ψ, has no physical significance, the *square* of the wave function, ψ^2, represents the electron charge density or the probability of finding an electron at any given point in the atom.

Perhaps the simplest orbital to describe is of the type known as $1s$. Figure 7-16a shows ψ^2 as a function of distance from the nucleus (r) along a line through the nucleus; it establishes that the highest electron charge density or electron probability in a $1s$ orbital

FIGURE 7-16
Three representations of
the 1s orbital.

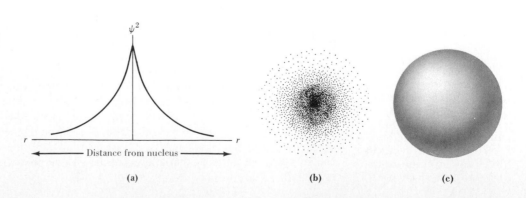

ψ^2

r —— Distance from nucleus —— r

(a)   (b)   (c)

FIGURE 7-17
Dart board analogy to a
1s orbital.

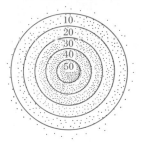

Summary of scoring

200 darts score	"50"
300	"40"
400	"30"
250	"20"
200	"10"
150	off the board

1500 darts	total

Imagine that a single dart (analogous to the single electron in a hydrogen atom) is thrown at a dart board 1500 times. The board itself contains 90% of all the dart holes (1350 out of 1500). It is analogous to the 1s orbital, though of course the region described by a 1s orbital is a sphere, not a circular disc. Where is the dart most likely to be found?

The density of dart holes (number of holes per unit area) is greatest in the "50" region. But if we ask what is the most likely score for a dart throw, it is "30" (400 throws out of 1500), not "50." Even though the density of dart holes is less in the "30" ring, the total area of the "30" ring is much greater than that of the "50" ring.

Similarly, although the probability of finding an electron (the value of ψ^2) is greatest in a small unit of volume around the nucleus, if we add up the probabilities for the volume units *equidistant* from the nucleus, the greatest *total* probability is in a spherical shell of radius 0.53 Å (0.053 nm). This proves to be the same as the radius of the first Bohr orbit!

exists at points in the vicinity of the nucleus. The pattern of dots in Figure 7-16b represents the distribution of electronic charge and electron probabilities in a plane with the nucleus at its center. Where the dots are closely spaced there is a higher charge density or a higher probability of finding an electron than where the spacing is greater. The pattern of dots in Figure 7-16b extends symmetrically in all directions and to all distances from the nucleus, with the spacing between dots constantly increasing. Because of this fact it is not possible to draw a picture that encompasses all such dots for all planes through the nucleus. Instead, we can only represent a region in which some portion (say 90%) of all the dots are found, that is, a region in which there is a 90% probability of finding an electron and 90% of the electron charge density is contained. The spherical envelope in Figure 7-16c portrays the 1s orbital in this way. The probability of finding an electron is the same at all points on the surface of this sphere. Figure 7-17 offers an analogy to the distribution of electron probabilities for the 1s orbital that may prove helpful. Figure 7-18 represents the electron charge distribution in an orbital of the type 2s.

FIGURE 7-18
The 2s orbital.

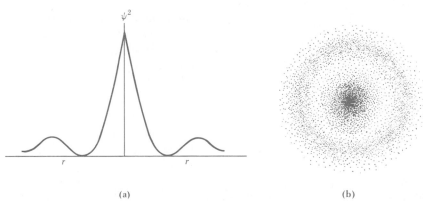

(a)   (b)

(a) As with the 1s orbital, the electron charge density is highest in the vicinity of the nucleus. It drops to zero at a particular distance from the nucleus (a condition called a node), rises to a second high value at a somewhat greater distance, and again gradually decreases, approaching zero at large distances from the nucleus.
(b) This diagram contains two regions with a high density of dots, corresponding to the two regions of high value of ψ^2 in (a). To encompass 90% of all the dots requires a spherical envelope that is larger than for the 1s orbital. Also, as discussed in the dartboard analogy of Figure 7-17, the most probable location of the electron is in a spherical shell corresponding to the second ring of dots.

7-8 Electron Orbitals and Quantum Numbers

To produce acceptable solutions to the Schrödinger wave equation, it is necessary to assign *integral* values to three different parameters, yielding *three* quantum numbers. Moreover, the allowable values for these quantum numbers are interrelated.

The first of these three is the **principal quantum number, *n*.** This quantum number may have only a *positive, nonzero integral* value.

$$n = 1, 2, 3, 4, \ldots \tag{7.10}$$

The second quantum number is the **orbital (azimuthal) quantum number, *l*,** which may be zero or a positive integer. It cannot be negative and it cannot be any larger than $n - 1$ (where *n* is the principal quantum number).

$$l = 0, 1, 2, 3, \ldots, n - 1 \tag{7.11}$$

The third quantum number is called the **magnetic quantum number, m_l.** Its value may be positive or negative, may include zero, and may range from $-l$ to $+l$ (where *l* is the orbital quantum number.)

$$m_l = -l, -l + 1, -l + 2, \ldots, 0, 1, 2, \ldots, +l \tag{7.12}$$

Example 7-10   What are the possible values of *l* and m_l for an electron with the principal quantum number, $n = 3$?

Solution.   From expression (7.11) we see that the allowable values of *l* are 0, 1, and 2. The allowable values of m_l depend on the value of *l* (expression 7.12).

If $l = 0$, there can be but a single value of m_l: 0.

If $l = 1$, there are three allowable values of m_l: −1, 0, +1.

If $l = 2$, there are five allowable values of m_l: −2, −1, 0, +1, +2.

SIMILAR EXAMPLES: Review Problem 6; Exercise 28.

Example 7-11   Can an electron have the quantum numbers $n = 2$, $l = 2$, and $m_l = 2$?

Solution.   No; the *l* quantum number cannot be greater than $n - 1$. Thus, if $n = 2$, *l* can be only 0 or 1. And if *l* can be only 0 or 1, m_l cannot be 2, because m_l can never be greater than *l* (expression 7.12).

SIMILAR EXAMPLES: Review Problem 8; Exercise 30.

Every combination of the three quantum numbers, *n*, *l*, and m_l corresponds to a different electron orbital. All orbitals having the same value of the quantum number *n* are said to be in the same **principal electronic shell** or **principal level,** and all orbitals having the same *l* value are in the same **subshell** or **sublevel.**

FIGURE 7-19
Three representations of a 2*p* orbital.

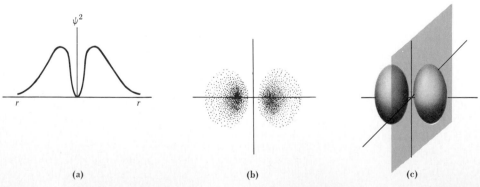

(a)   (b)   (c)

FIGURE 7-20
The three *p* orbitals.

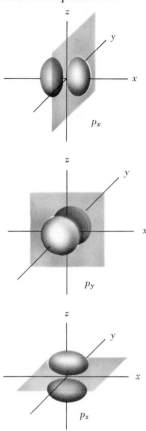

The principal shells are numbered in accordance with the value of *n*, but they may also be denoted by letter. The *first* principal electronic shell or the *K* shell consists of orbitals with *n* = 1; the *second* principal shell or *L* shell, of orbitals with *n* = 2; and so on. The value of the quantum number *n* relates to the energies of electrons and to their probable distances from the nucleus.

The value of the *l* quantum number determines the geometrical shape of the electron cloud or electron probability distribution. All orbitals with the value *l* = 0 are *s* orbitals. If the *s* orbital is in the first principal shell (*n* = 1), it is a 1*s* orbital, if in the second principal shell, 2*s*, and so on. The electron cloud or electron probability distribution for an *s* orbital is spherically symmetric, that is, has the shape of a sphere with the atomic nucleus at its center. Because, when *l* = 0, m_l must also be 0, there can be only one orbital of the *s* type for each principal shell.

The orbital type corresponding to *l* = 1 is the *p* orbital. Because when *l* = 1, m_l can have one of three values (−1, 0, +1), *p* orbitals occur in sets of three. That is, there are three *p* orbitals in the *p* subshell. As shown in Figure 7-19a, for a 2*p* orbital the electron charge density (ψ^2) is zero at the nucleus, rises to a maximum on either side of the nucleus, and then falls off with distance (*r*) along a line passing through the nucleus (i.e., the *x*, *y*, or *z* axis). The pattern of dots in Figure 7-19b represents the electron probability distribution in a plane passing through the nucleus (i.e., the *xy*, *xz*, or *yz* plane). The electron clouds or electron probability distributions for *p* orbitals are *not* spherically symmetric. The greatest probability of finding an electron in a 2*p* orbital is within the dumbbell-shaped region of Figure 7-19c, and this is the representation commonly used for a *p* orbital. The two lobes of high electron charge density or electron probability are separated by a nodal plane, a region in which the electron charge density is zero. Usually the three *p* orbitals are shown to be directed along the perpendicular axes through the nucleus, and are represented by the symbols p_x, p_y, and p_z. The set of three *p* orbitals is shown in Figure 7-20.

There is a set of five orbitals that have *l* = 2; these are the *d* orbitals, comprising the *d* subshell. The geometrical shapes corresponding to *d* orbitals, which are more complex than for *s* and *p* orbitals, are shown in Figure 7-21.

FIGURE 7-21
The five *d* orbitals.

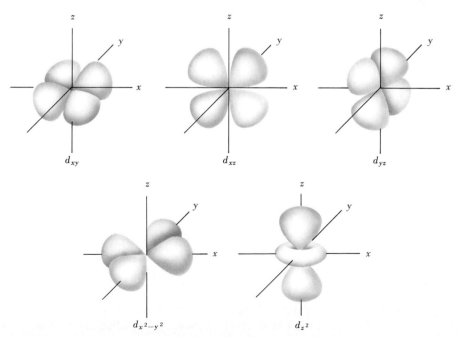

The designations *xy*, *xz*, *yz*, and so on are related to the values of the quantum number m_l (*l* = 2 for all *d* orbitals), but this is a detail not pursued further in the text.

TABLE 7-2
Electronic shells, orbitals, and quantum numbers[a]

Principal shell	K	L				M								
$n =$	1	2	2	2	2	3	3	3	3	3	3	3	3	3
$l =$	0	0	1	1	1	0	1	1	1	2	2	2	2	2
$m_l =$	0	0	-1	0	$+1$	0	-1	0	$+1$	-2	-1	0	$+1$	$+2$
orbital designation	1s	2s	2p	2p	2p	3s	3p	3p	3p	3d	3d	3d	3d	3d
number of orbitals in subshell	1	1		3		1		3				5		
total number of orbitals $= n^2$	1		4						9					

[a]When assigning the quantum number m_l, it is immaterial which value we assign first. We will follow the convention of beginning with the most negative of the permitted values and proceeding in the order: negative → zero → positive.

Some of the points discussed in the preceding paragraphs are illustrated through Table 7-2 and Example 7-12.

Example 7-12 Write an orbital designation for an electron with the quantum numbers $n = 4$, $l = 2$, and $m_l = 0$.

Solution.   The type of orbital is determined by the l quantum number. If $l = 2$, the orbital is of the d type. Because $n = 4$, the designation is $4d$.

SIMILAR EXAMPLES: Review Problem 7; Exercises 31, 32.

7-9 Electron Spin—A Fourth Quantum Number

In 1925, Uhlenbeck and Goudsmit proposed that some unexplained features of fine structure in the hydrogen spectra could be understood if it were assumed that electrons possess a fourth quantum number. The property of electrons associated with this fourth quantum number has become known as electron spin. The electron is pictured as spinning on its axis as it moves about the nucleus, much as the earth spins on its axis as it revolves about the sun. There appear to be two possibilities for electron spin. The quantum number describing electron spin, m_s, may have a value of $+\frac{1}{2}$ (also denoted by an arrow ↑) or $-\frac{1}{2}$ (denoted as ↓). Unlike the quantum numbers n, l, m_l, which have interrelated values, the value of m_s does not depend on the other three quantum numbers.

Although the concept of electron spin is useful, what proof is there that such a property exists, especially in view of the uncertainty principle? An experiment reported by Stern and Gerlach in 1922, although designed for a different purpose, seems to yield this proof. In the Stern-Gerlach experiment silver metal was vaporized in a furnace and a beam of silver atoms passed through an inhomogeneous (nonuniform) magnetic field. The beam was found to split in two. A simplified explanation is based on these points.

1. An electron, by virtue of its spin, has associated with it a magnetic field.

2. A pair of electrons with opposing spins has no associated magnetic field.

3. In a silver atom, 23 electrons have a spin of one type and 24 of the opposite type. The direction of deflection of a silver atom in a magnetic field depends on the type of spin on the "odd" electron.

4. In a collection of silver atoms there is an equal chance that the odd electron will have a spin of $+\frac{1}{2}$ or $-\frac{1}{2}$. Thus, the beam of atoms is split into two beams.

7-10 Many-Electron Atoms

To this point our description of electron orbitals has applied only to a hydrogen atom or hydrogenlike atoms—species containing a single electron: H, He$^+$, Li^{2+}, and so on. Exact solutions of the necessary mathematical equations for many-electron atoms are almost impossibly difficult. Nevertheless, the results obtained for the hydrogen atom are approximately correct for more complex atoms. By this we mean that the same types of quantum numbers and orbitals are assumed to exist for many-electron atoms as for the hydrogen atom. As we discover in subsequent chapters how this particular concept of electron orbitals allows us to explain so many chemical phenomena, our confidence in this assumption should grow.

Through equation (7.5) and Figure 7-13 we demonstrated that the energy levels in the Bohr hydrogen atom are a function only of the principal quantum number n. The same result is obtained with wave mechanics. That is, all the subshells in a given principal shell of the hydrogen atom (s, p, d, f, . . .) are at the same energy. But there are some important differences between the hydrogen atom and many-electron atoms with respect to orbital energies.

Nuclear Charge Effect. The attractive force of the nucleus for a given electron increases as the nuclear charge increases. As a result, we should expect the energy of interaction between the electron and the nucleus—the orbital energy—to become more negative with increasing atomic number.

Shielding Effect. In describing the attraction between a given electron and the nucleus, the presence of all the other electrons must also be considered. If the electron in question is the one farthest from the nucleus, the inner electrons reduce the effectiveness of the nucleus in attracting this outermost electron. They screen or shield the outermost electron from the full effects of the nucleus by partially neutralizing the nuclear charge. This shielding effect is illustrated in Figure 7-22.

The effectiveness of the shielding by inner electrons depends on the l value of the particular orbital for the outermost electron. For example, the p orbitals of an electronic shell extend to a greater distance from the nucleus than does the corresponding s orbital. Compared to being in an s orbital, an electron in a p orbital spends more time at distances from the nucleus where the shielding effect is strong (outside the dashed circle in Figure 7-22). The attractive force of the nucleus for an electron in a p orbital is less than for an electron in an s orbital, and the energy of the p orbital is higher than of the s orbital; in a d orbital the energy is higher still. The result of shielding is to cause a splitting of the energy level of each principal shell into separate levels for each subshell.

The ideas presented in this section are illustrated through Figure 7-23, which represents orbital energies for the first three shells of the hydrogen atom and some typical many-electron atoms.

FIGURE 7-22
The shielding effect.

The atom pictured is sodium. The charge on the nucleus is +11. The charge on the first electronic shell is −2, and on the second −8. If the nucleus and the two inner electronic shells are considered as a single unit (contained within the broken circle), the net charge on this unit is +1. For the time that the outer-shell electron spends outside the sphere of the inner electrons, the force on it is much the same as if the nucleus contained only a single unit of positive charge and there were only one electron in the atom.

FIGURE 7-23
Orbital energy diagram
for first three electronic
shells.

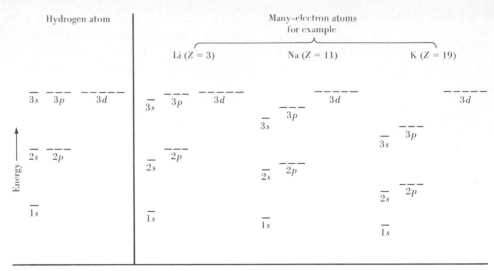

Energy levels for the subshells of the first three principal quantum levels are shown for a hydrogen atom and three typical many-electron atoms. Each many-electron atom has its own energy-level diagram. One of the distinctive features of the energy-level diagrams for many-electron atoms is the splitting of each quantum level into different energies for each subshell. Another feature is a steady decrease in the energies of these levels with increasing atomic number.

7-11 Electron Configurations

A designation of the most likely distribution of electrons among the orbitals of its electronic shells is called the **electron configuration** of an atom. We consider first three rules or principles that help in the assignment of a probable electron configuration to an atom, and then apply these principles to the various elements.

1. Electrons Occupy Orbitals in Such A Way as to Minimize the Energy of the Atom. Figure 7-23 implies an order in which electrons occupy orbitals, first the $1s$, then $2s$, $2p$, and so on. Actually, the energy of an atom is not minimized in most cases simply by filling the principal electronic shells in succession. One of the consequences of the splitting of energy levels for many-electron atoms is that at higher quantum numbers and for certain elements an overlapping of levels occurs, for example, with $4s$ filling before $3d$ in K and Ca. As a result, the order of filling of orbitals must be established by experiment. The order obtained is roughly the one listed below and also depicted in Figure 7-24.

$$1s, \ 2s, \ 2p, \ 3s, \ 3p, \ 4s, \ 3d, \ 4p, \ 5s, \ 4d, \ 5p, \ 6s, \ 4f, \ 5d, \ 6p, \ 7s, \ 5f, \ 6d, \ 7p \quad (7.13)$$

2. No Two Electrons in an Atom May Have All Four Quantum Numbers Alike—the Pauli Exclusion Principle. In 1926, Wolfgang Pauli studied the *absence* in emission spectra of lines which should have been present according to existing theory. His results led him to propose that any state or condition in which two electrons would have all four quantum numbers alike is *not* allowable. The first three quantum numbers, n, l, and m_l, determine a specific orbital. Two electrons may have these three quantum numbers alike, that is, they may have the same orbital designation; but if they do, they must have different values of m_s, the spin quantum number. *Only two electrons may exist in the same orbital and these electrons must have opposite spins.*

Because of this limit of two electrons per orbital, the maximum occupancy of a subshell can be obtained by *doubling* the number of orbitals in the subshell. That is, the s subshell consists of *one* orbital and has a capacity of 2 electrons; the p subshell consists of *three* orbitals and has a capacity of 6 electrons; and so on. The maximum occupancy of a

FIGURE 7-24
The order of filling of
electronic subshells.

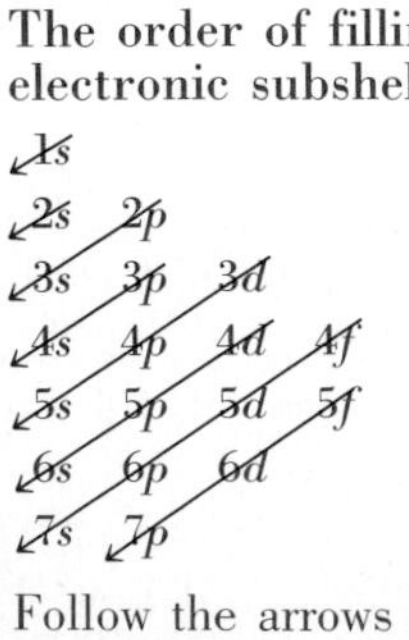

1s
2s 2p
3s 3p 3d
4s 4p 4d 4f
5s 5p 5d 5f
6s 6p 6d
7s 7p

Follow the arrows from top to bottom, and the order obtained is the same as that in expression (7.13).

principal shell is also *twice* the number of orbitals it contains (recall Table 7-2), leading to the expression

$$\begin{array}{l}\text{maximum number of electrons in the electronic}\\ \text{shell with principal quantum number } n\end{array} = 2n^2$$

3. The Principle of Maximum Multiplicity—Hund's Rule.   When orbitals of identical energy are available, electrons occupy these singly rather than in pairs. As a result, an atom tends to have as many unpaired electrons as possible. This behavior can be rationalized by saying that electrons, because they all carry the same electrical charge, seek out empty orbitals of similar energy in preference to pairing up with electrons in half-filled orbitals.

7-12 Electron Configurations of the Elements

To apply the principles of the preceding section, we first need a shorthand notation to designate electron configurations. That of an atom of carbon is represented below in two different ways.

spdf notation: C $1s^2 2s^2 2p^2$

orbital diagram: C   [↑↓] [↑↓] [↑ | ↑ |]
                         1s   2s   2p

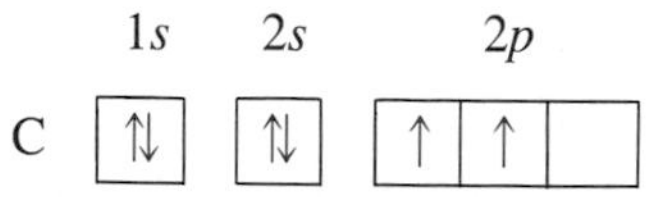

The total number of electrons to be assigned is six, the atomic number of carbon. Two of these electrons are in the 1s subshell, two in the 2s, and two in the 2p. The *spdf* notation only denotes the total number of electrons in each subshell. The orbital diagram breaks down each subshell into individual orbitals and indicates the number of electrons for each orbital. This is done through the use of arrows. An arrow pointing upward corresponds to an electron with one type of spin $(+\frac{1}{2})$ and an arrow pointing down to the other $(-\frac{1}{2})$.* Electrons with opposite (opposing) spins are said to be paired; the electrons in the 1s and the 2s orbitals of carbon are paired. Electrons with the same type of spin are said to have parallel spins, as in the 2p orbitals of carbon.

> The number of unpaired electrons in an atom can also be denoted with *spdf* notation if the subshells are broken down into individual orbtials, such as C $1s^2 2s^2 2p_x^1 2p_y^1$.

Two *incorrect* orbital diagrams for carbon are shown below. The first is incorrect because the two 2p electrons have been paired into a single 2p orbital in violation of Hund's rule. The second is incorrect because a configuration with singly occupied orbitals having electrons with opposing spins is not as energetically favorable as with parallel spins.

(incorrect) 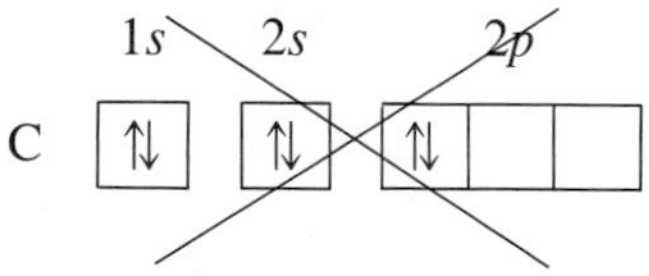   C   [↑↓] [↑↓] [↑↓ | |]   (incorrect)   C   [↑↓] [↑↓] [↑ | ↓ |]
                    1s   2s   2p   1s   2s   2p

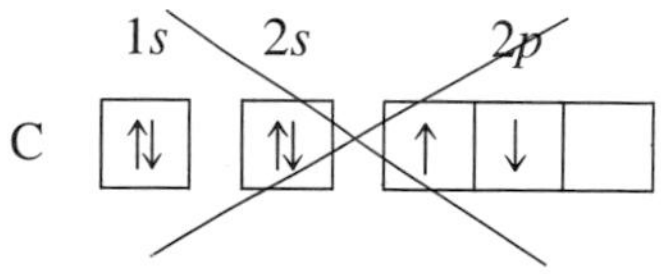

The Aufbau Process.   Let us consider the following *hypothetical* process—the building up of a more complex atom starting with the simplest atom, hydrogen. This hypothetical process is called the **Aufbau** process (meaning "building up" in German). In this process we proceed from an atom of one element to the next by adding a proton and the requisite number of neutrons to the nucleus and one electron to the appropriate orbital. It is to this differentiating electron that we pay particular attention.

Hydrogen, Z = 1.   The lowest energy state available to the electron in a hydrogen atom is the 1s orbital. The electron configuration is $1s^1$.

---

*It is customary to show half-filled orbitals with upward pointing arrows, but downward pointing arrows could be used as well.

Helium, $Z = 2$. In a helium atom a second electron goes into the $1s$ orbital. The two electrons must have opposing spins, $1s^2$.

Lithium, $Z = 3$. The differentiating electron cannot be accommodated in the $1s$ orbital (Pauli exclusion principle). It must be placed in the next available orbital, the $2s$. For lithium we may write, $1s^2 2s^1$.

Beryllium, $Z = 4$. The configuration is $1s^2 2s^2$.

Boron, $Z = 5$. The differentiating electron goes into the next available orbital, $2p$. Boron has the configuration $1s^2 2s^2 2p^1$.

Carbon, $Z = 6$. Here the rule of maximum multiplicity (Hund's rule) applies. The differentiating electron goes into a new $2p$ orbital and has a spin parallel to the first $2p$ electron.

	$1s$	$2s$	$2p$		
C	↑↓	↑↓	↑	↑	

Nitrogen, $Z = 7$. The atom has three unpaired electrons in its $2p$ orbitals.

	$1s$	$2s$	$2p$		
N	↑↓	↑↓	↑	↑	↑

Oxygen, $Z = 8$. The differentiating electron must go into an orbital already singly occupied. This reduces the number of unpaired electrons to two.

	$1s$	$2s$	$2p$		
O	↑↓	↑↓	↑↓	↑	↑

Fluorine, $Z = 9$, and Neon, $Z = 10$. In the fluorine atom there is one unpaired electron; in the neon atom all electrons are paired and the second principal shell is filled.

	$1s$	$2s$	$2p$		
Ne·	↑↓	↑↓	↑↓	↑↓	↑↓

Sodium, $Z = 11$, and Magnesium, $Z = 12$. In the sodium atom the differentiating electron must go into a new orbital in the third principal shell, the $3s$. In the magnesium atom there are two electrons in this orbital.

Na $1s^2 2s^2 2p^6 3s^1$ Mg $1s^2 2s^2 2p^6 3s^2$

Aluminum, $Z = 13$, to Argon, $Z = 18$. In this series of six elements the $3p$ orbitals fill, leading to

Al $1s^2 2s^2 2p^6 3s^2 3p^1$ through   Ar $1s^2 2s^2 2p^6 3s^2 3p^6$

Potassium, $Z = 19$, and Calcium, $Z = 20$. We now encounter the first "irregularity" in the order of filling of orbitals. According to expression (7.13) and Figure 7-24, the $4s$ orbital fills *before* the $3d$. The differentiating electrons for potassium and calcium go into the s orbital of the fourth electronic shell.

K $1s^2 2s^2 2p^6 3s^2 3p^6 4s^1$ Ca $1s^2 2s^2 2p^6 3s^2 3p^6 4s^2$

Since the first 18 electrons correspond to the configuration for argon, we may also write

K $[Ar]4s^1$ and   Ca $[Ar]4s^2$

Scandium, $Z = 21$, to Zinc, $Z = 30$.   This next series of elements is characterized by electrons filling the d orbitals of the third shell. The d subshell has a total capacity of 10 electrons, hence 10 elements are involved. A choice is necessary here as to whether to write the electron configuration of scandium as

(a) Sc $[Ar]3d^14s^2$ or   (b) Sc $[Ar]4s^23d^1$

Both methods are commonly used. Method (a) groups together all the subshells of a principal shell and places last subshells of the highest principal quantum level. Method (b) lists orbitals in the *apparent* order in which they fill. In this text we will use method (a). The electron configuration for zinc is

Zn $1s^22s^22p^63s^23p^63d^{10}4s^2$ or $[Ar]3d^{10}4s^2$

Gallium, $Z = 31$, to Krypton, $Z = 36$.   In this series of six elements the $4p$ subshell is filled. The configuration for krypton is

Kr $1s^22s^22p^63s^23p^63d^{10}4s^24p^6$

Rubidium, $Z = 37$, to Xenon, $Z = 54$.   In this series of 18 elements the subshells filled in succession are $5s$, $4d$, and $5p$.

Xe $1s^22s^22p^63s^23p^63d^{10}4s^24p^64d^{10}5s^25p^6$

Cesium, $Z = 55$, to Radon, $Z = 86$.   In this series of 32 elements, with a few exceptions, the subshells filled in succession are $6s$, $4f$, $5d$, and $6p$.

Rn $1s^22s^22p^63s^23p^63d^{10}4s^24p^64d^{10}4f^{14}5s^25p^65d^{10}6s^26p^6$

Francium, $Z = 87$, to ?.   Francium starts a series of elements in which the subshells filled are $7s$, $5f$, $6d$, and presumably $7p$, though elements in which the $7p$ subshell is occupied are not yet known.

Some Qualifying Remarks about the Aufbau Process.   There are several examples where the electron configuration predicted by the Aufbau process is not the one found by experiment. Among these, which can be seen in the listing in Appendix E, are Cr ($Z = 24$), Cu ($Z = 29$), and La ($Z = 57$). In some cases, even though the electron configuration obtained by the process is correct, the order of filling of orbitals implied by the process is not always in strict accordance with orbital energies. For example, the orbital energy of the $3d$ electron in Sc is lower than $4s$, even though in the Aufbau process we first fill the $4s$ subshell before adding electrons to the $3d$. Another complicating factor is that when considering the loss of electrons by an atom it is not always the last electrons added in the Aufbau process that are lost. The electron configuration of Sc^+ appears to be $[Ar]3d^14s^1$. The electron lost in forming Sc^+ from Sc is a $4s$ electron, not the $3d$ that was placed last in the Aufbau process. And we will find in later chapters that the electron configurations of ions are as useful as the electron configurations of the isolated atoms considered here.

Despite the qualifications just cited, the Aufbau process does provide a good overview of the assignment of electron configurations, especially when it is related to the periodic table of the elements, as we shall see in Section 8-4.

Example 7-13 Complete each of the following.
(a) ? $1s^22s^22p^63s^23p^5$
(b) Fe ($Z = 26$) $1s^22s^22p^63s^23p^63d^{(?)}4s^2$
(c) As ($Z = 33$)   ?

Solution

(a) All electrons must be accounted for in an electron configuration. Add up the superscript numerals $(2 + 2 + 6 + 2 + 5)$ to obtain the atomic number (17). Look up the correct element in a listing such as on the inside back cover or the periodic table on the inside front cover.

Cl $(Z = 17)$ $1s^2 2s^2 2p^6 3s^2 3p^5$

(b) The total number of electrons is 26. The number accounted for in the portion of the electron configuration given is $2 + 2 + 6 + 2 + 6 + 2 = 20$. There must be six $3d$ electrons.

Fe $(Z = 26)$ $1s^2 2s^2 2p^6 3s^2 3p^6 3d^6 4s^2$

(c) The first 18 electrons correspond to the configuration of Ar. The next two go into the $4s$ subshell; this accounts for 20 electrons. Following the $4s$ the next subshell to fill is the $3d$ (see Figure 7-24). The 10 $3d$ electrons bring the total to 30. The remaining three electrons go into the $4p$ subshell, resulting in the configuration

As $(Z = 33)$ $1s^2 2s^2 2p^6 3s^2 3p^6 3d^{10} 4s^2 4p^3$

SIMILAR EXAMPLES: Review Problems 10, 11, 12; Exercises 36, 37.

Summary

Ideas concerning the nature of electromagnetic radiation are a useful starting point in the discussion of atomic structure. The dispersion of white light produces a continuous spectrum—a rainbow. However, most light originating from excited atoms produces a discontinuous or line spectrum—a series of colored lines. The simplest line spectrum is that of hydrogen, which can be described through an empirical equation (the Balmer equation).

Theoretical explanations of atomic spectra required a breakthrough in our thinking about energy. Planck proposed that energy exists as tiny discrete units called quanta. Einstein used Planck's quantum theory to explain the photoelectric effect, and Bohr applied classical and quantum mechanics to develop a model of the hydrogen atom. Bohr's model permits a calculation of the permissible energy levels for the electron in a hydrogen atom. Energies of photons emitted by excited hydrogen atoms correspond to differences in these electron energy levels. Frequencies of lines in the hydrogen spectrum can be predicted by combining Bohr's theory and the Planck equation.

To provide explanations of such phenomena as the formation of chemical bonds between atoms, a model of atomic structure must be based on a new form of quantum theory—wave mechanics. The essential ideas contributing to this newer quantum mechanics are de Broglie's concept of wave-particle duality (i.e., the existence of matter waves) and Heisenberg's uncertainty principle. These ideas were used by Schrödinger to provide a new picture of the hydrogen atom.

The essential feature of the Schrödinger atom is to think of an electron as a cloud of negative electrical charge having a certain geometrical shape. Also, the electron can be viewed as a particle whose probability of being found extends throughout space, though the probability is highest in certain three-dimensional regions. The variation of charge density or electron probability from point to point is described through expressions known as orbitals. The key parameters that distinguish among orbitals are a set of three quantum numbers, n, l, and m_l. The specific orbital types described in this chapter are known as s, p, and d.

By considering the existence of a fourth quantum number (the spin quantum number) and a set of rules, the probable assignment of electrons to orbitals can be made. These assignments, called electron configurations, are made for the various elements as a conclusion to Chapter 7.

Learning Objectives

As a result of studying Chapter 7, you should be able to

1. Apply the basic expression relating the frequency, wavelength, and velocity of electromagnetic radiation, with appropriate regard for units.

2. List the various kinds of radiation and their approximate location in the electromagnetic spectrum.

3. Explain the essential difference between a continuous and a line spectrum.

4. Apply the Balmer equation to calculate the wavelengths of lines in the spectrum of hydrogen.

5. Use Planck's equation to relate the frequency and energy content of electromagnetic radiation.

6. State Bohr's assumptions and describe the picture of the hydrogen atom that results from them.

7. Calculate the energy of an electron in a hydrogen atom as a function of its principal quantum number, n.

8. Calculate the de Broglie wavelength of a particle.

9. Explain the basic differences between the Bohr and the Schrödinger descriptions of the hydrogen atom.

10. Apply the relationships among the three quantum numbers n, l, and m_l that result from wave mechanics.

11. Describe the shapes of the charge clouds or electron probability distributions for s, p, and d orbitals.

12. Explain how orbital energies for many-electron atoms differ from those of hydrogen.

13. Apply the three basic principles governing the assignment of electron configurations: the order of filling of electronic subshells, the Pauli exclusion principle, and Hund's rule of maximum multiplicity.

14. Use *spdf* notation or orbital diagrams to represent the electron configurations of different atoms.

Some New Terms

An **atomic (line) spectrum** is produced by dispersing light emitted by excited atoms. Only a discrete set of wavelength components (seen as colored lines) is present.

A **continuous spectrum** is one in which all wavelength components of the visible portion of the electromagnetic spectrum are present.

Electromagnetic radiation is a form of energy propagated through mutually perpendicular electric and magnetic fields.

An **electron configuration** is a representation showing the orbital designations of all the electrons in an atom.

Hund's rule (of maximum multiplicity) states that whenever orbitals of equal energies are available, electrons are assigned to these orbitals singly before any pairing of electrons occurs.

An **orbital** describes the electron charge density or the probability of finding an electron in an atom. The several kinds of orbitals ($s, p, d, f, \ldots$) differ from one another in the shapes of the electron clouds they describe.

An **orbital diagram** is a representation of an electron configuration in which the most probable orbital designation and the spin of each electron in the atom are indicated.

The **Pauli exclusion principle** states that no two electrons may have all four quantum numbers alike. This limits occupancy of an orbital to two electrons with opposing spins.

A **photon** is a "particle" of light. The energy of a beam of light is concentrated into these photons.

A **principal shell** refers to the collection of all orbitals having the same value of the principal quantum number, n. For example, the $3s$, $3p$, and $3d$ orbitals comprise the third principal shell ($n = 3$).

Quantum numbers are integral numbers whose values must be specified in order to solve the equations of wave mechanics. Three different quantum numbers are required: the *principal quantum number, n;* the *orbital quantum number, l;* and the *magnetic quantum number m_l.* The permitted values of these numbers are interrelated.

The **quantum theory** is based on the proposition that energy exists in the form of tiny, discrete units called quanta. Whenever an energy transfer occurs, it must involve an entire quantum.

spdf notation is a method of describing electron configurations in which the numbers of electrons assigned to each orbital are denoted as superscripts. For example, the electron configuration of Cl is $1s^2 2s^2 2p^6 3s^2 3p^5$.

A **subshell** refers to a collection of orbitals of the same type; e.g., the three $2p$ orbitals constitute the $2p$ subshell.

The **uncertainty principle** states that, when measuring the position and momentum of fundamental particles of matter, uncertainties in measurement are inevitable.

Wave mechanics is a form of quantum theory based on the concepts of wave–particle duality, the uncertainty principle, and the treatment of electrons as matter waves. Mathematical solutions of the equations of wave mechanics are known as **wave functions (ψ).**

Suggestions for Further Study

DAVIS, J. C., Jr., "Introduction to Spectroscopy Part III: Light and the Electromagnetic Spectrum," *Chemistry,* **48**[5], 18 (1975).

HAENDLER, B. L., "Presenting the Bohr Atom," *J. Chem. Educ.,* **59,** 372 (1982).

HERRON, J. D., "How Does the Electron Cross the Node?", *J. Chem. Educ.,* **57,** 651 (1980).

MASON, F. P., and R. W. RICHARDSON, "Why Doesn't the Electron Fall into the Nucleus?", *J. Chem. Educ.,* **60,** 40 (1983).

MAYBURY, R. H., "The Language of Quantum Mechanics," *J. Chem. Educ.,* **39,** 289 (1962).

MORWICK, J. J., "What is the Electron, Really?", *J. Chem. Educ.,* **55,** 662 (1978).

Review Problems

1. Restate each of the following wavelengths in the unit indicated.
  (a) 3015 Å = _____ nm   (b) 1.56 μm = _____ cm
  (c) 3.92 cm = _____ nm   (d) 376 nm = _____ m
  (e) 1812 nm = _____ μm   (f) 2.18 μm = _____ Å

2. What are the wavelengths, in meters, associated with radiation of the following frequencies? In what portion of the electromagnetic spectrum is each radiation found? (a) 6.2×10^{13} s^{-1}; (b) 4.5×10^{16} s^{-1}; (c) 5.50×10^{5} s^{-1}.

3. What is the frequency associated with radiation of each of the following wavelengths? (a) 1.3×10^{-4} cm; (b) 8.87 μm; (c) 371 Å; (d) 335 cm.

4. Use the Balmer equation (7.3) to determine
  (a) the frequency of the radiation corresponding to $n = 5$;
  (b) the wavelength of the line in the Balmer series corresponding to $n = 7$;
  (c) the value of n corresponding to the line in the Balmer series at 380 nm.

5. Use Planck's equation (7.4) to determine
  (a) the energy, in J/photon, of radiation of frequency, $\nu = 3.10 \times 10^{15}$ s^{-1};
  (b) the energy, in kJ/mol, of radiation of frequency, $\nu = 4.26 \times 10^{14}$ s^{-1};
  (c) the frequency of radiation having an energy of 3.54×10^{-20} J/photon;
  (d) the wavelength of radiation having an energy of 185 kJ/mol.

6. Give a possible value for the missing quantum number(s) in each of the following sets: (a) $n = 3$, $l = 0$, $m_l = $?; (b) $n = 3$, $l = $?, $m_l = -1$; (c) $n = $?, $l = 1$, $m_l = +1$; (d) $n = $?, $l = 2$, $m_l = $?

7. Write appropriate values of the n, l, and m_l quantum numbers for each of the following orbital designations: (a) 4s; (b) 3p; (c) 5f; (d) 3d.

8. Indicate which of the following sets of quantum numbers is *not* allowed.
  (a) $n = 3$, $l = 2$, $m_l = -1$ (b) $n = 2$, $l = 3$, $m_l = -1$
  (c) $n = 4$, $l = 0$, $m_l = -1$ (d) $n = 5$, $l = 2$, $m_l = -1$
  (e) $n = 3$, $l = 3$, $m_l = -3$ (f) $n = 5$, $l = 3$, $m_l = +2$

9. Arrange the following groups of orbitals in the order in which they fill with electrons in assigning electron configurations.
  5s, 3p, 3d, 4p, 5f, 6p, 6s

10. Complete the following. Use part (a) as an example.
  (a) Na $(Z = 11)$ $1s^2 2s^2 2p^6 3s^1$
  (b) _____ $1s^2 2s^2 2p^6 3s^2 3p^3$
  (c) Zr $(Z = 40)$ $[Kr]4d^{(?)}5s^2$
  (d) _____ $[Kr]4d^{(?)}5s^2 5p^4$
  (e) _____ $[Ar]3d^{(?)}4s^{(?)}4p^3$
  (f) Bi $(Z = 83)$ $[Xe]4f^{(?)}5d^{(?)}6s^{(?)}6p^{(?)}$

11. Without consulting Appendix E, identify the atoms with the following electron configurations.
  (a) $1s^2 2s^2 2p^1$
  (b) $[Ar]3d^3 4s^2$

	1s	2s	2p			3s	3p	
(c)	↑↓	↑↓	↑↓	↑↓	↑↓	↑↓	↑	↑

12. Without consulting Appendix E, use *spdf* notation to write the most probable electron configurations of the following atoms: (a) Al; (b) Rb; (c) Cd; (d) Sb; (e) Pb; (f) Xe

Exercises

Electromagnetic radiation

1. A certain radiation emitted by magnesium has a wavelength of 285.2 nm. Which of the following statements is (are) correct concerning this radiation?
  (a) It has a higher frequency than radiation with wavelength 315 nm.
  (b) It is visible to the eye.
  (c) It has a greater speed in vacuum than does red light of wavelength, 7100 Å.
  (d) Its wavelength is longer than that of x rays.

2. The current international standard of time, the second, is defined as 9,192,631,770 cycles of a particular radiation emitted by $^{133}_{55}$Cs atoms. What is the wavelength of this radiation?

3. How long does it take light from the sun, 93 million miles away, to reach the earth?

Atomic spectra

4. Use equation (7.3) to calculate the first four lines of the Balmer series of the hydrogen spectrum, starting with the *longest*-wavelength component.

5. A line is detected in the hydrogen spectrum at 1880 nm. Is this line in the Balmer series? Explain.

6. What is the wavelength limit to which the Balmer series for hydrogen converges; that is, what is the *shortest* possible wavelength in the series?

7. The Lyman series of the hydrogen spectrum can be represented by the equation

$$v = 3.2881 \times 10^{15} \ s^{-1} \left(\frac{1}{1^2} - \frac{1}{n^2} \right) \quad \text{where } n = 2, 3, \ \ldots$$

(a) Calculate the maximum and minimum wavelengths of lines in this series.
(b) In what portion of the electromagnetic spectrum will this series be found?
(c) What value of n corresponds to a spectral line at 95.0 nm?
★(d) Is there a line in the Lyman series at 108.5 nm? Explain.

Quantum theory

8. A certain radiation has a wavelength of 285 nm. What is the energy of (a) a single photon; (b) a mole of photons of this radiation?

9. What is the wavelength, in nm, of light that has an energy content of exactly 125 kcal/mol? In what portion of the electromagnetic spectrum is this light?

10. Figure 7-3 establishes regions of the electromagnetic spectrum in terms of frequency limits. What range of energies, in kJ/mol, corresponds to visible light?

The photoelectric effect

11. The lowest frequency light that will produce the photoelectric effect on a material is called the *threshold frequency*.
(a) The threshold frequency for platinum is $1.3 \times 10^{15} \ s^{-1}$. What is the energy of a quantum of this radiation?
(b) Will platinum display the photoelectric effect when exposed to ultraviolet light? infrared light? Explain.

12. In describing Einstein's quantum explanation of the photoelectric effect, Sir James Jeans made the following remark: "It not only prohibits killing two birds with one stone, but also the killing of one bird with two stones." Comment on the appropriateness of this analogy.

The Bohr atom

13. Use the description of the Bohr atom given in the text to determine (a) the radius, in nm, of the sixth Bohr orbit; (b) the energy of the electron when it is in this orbit (with the zero of energy being that of the free electron).

14. Calculate the increase in (a) distance from the nucleus and (b) energy when an electron in a hydrogen atom is excited from the first to the fourth Bohr orbit.

15. What are (a) the frequency and (b) the wavelength of the light emitted when the electron in a hydrogen atom drops from the energy level $n = 6$ to $n = 4$? (c) In what portion of the electromagnetic spectrum is this light?

16. Which of the following electron transitions requires that the greatest quantity of energy be *absorbed* by a hydrogen atom? From (a) $n = 1$ to $n = 2$; (b) $n = 2$ to $n = 4$; (c) $n = 3$ to $n = 6$; (d) $n = \infty$ to $n = 1$. Explain.

17. What electron transition in a hydrogen atom, starting from the orbit $n = 7$, will produce infrared light of wavelength 2170 nm? (*Hint:* In what orbit must the electron end up?)

★18. The Bohr theory can be extended to one-electron species other than the hydrogen atom, e.g., He^+, Li^{2+}, and Be^{3+}. In these cases the energies are related to the quantum number, n, through the expression

$$E_n = \frac{-Z^2 B}{n^2}$$

where Z is the atomic number of the species and $B = 2.179 \times 10^{-18} \ J$.
(a) What is the energy of the lowest level ($n = 1$) of a He^+ ion?
(b) What is the energy of the level $n = 3$ of a Li^{2+} ion?

★19. Both the Balmer series of the hydrogen spectrum and the energy-level diagram of Figure 7-13 feature lines that become very closely spaced in a certain region. Explain the relationship between these two phenomena.

The uncertainty principle

20. Describe the ways in which the Bohr model of the hydrogen atom violates the Heisenberg uncertainty principle.

21. Although Einstein himself made some early contributions to quantum theory, he was never able to accept the Heisenberg uncertainty principle. He stated, "God does not play dice with the world." What do you suppose that Einstein meant by this remark?

★22. Show that the uncertainty principle has little significance when applied to a large object like an automobile. (*Hint:* Assume that m is precisely known; assign a reasonable value to either Δx or Δv and estimate a value of the other.)

23. A proton is accelerated to one-tenth the velocity of light. Suppose that its velocity can be measured with a precision of $\pm 1\%$. What must be the uncertainty in its position?

Wave–particle quality

24. Which must possess a greater velocity to produce matter waves of the same wavelength (e.g., 1 nm), protons or electrons? Explain your reasoning.

25. What must be the velocity of a beam of electrons if they are to display a de Broglie wavelength of 1 nm?

Wave mechanics

26. Describe briefly the several differences between the orbits of the Bohr atom and the orbitals of the wave mechanical atom. Are there any similarities?

27. The greatest probability of finding the electron in a small volume element of the $1s$ orbital of the hydrogen atom is at the nucleus. Yet the most probable distance of the electron from the nucleus is 0.53 Å. How can you reconcile these two statements?

Quantum numbers and electron orbitals

28. Select the correct answer: An electron that has the quantum numbers $n = 3$ and $m_l = 2$ **(a)** must have the quantum number $m_s = +\frac{1}{2}$; **(b)** must have the quantum number $l = 1$; **(c)** may have the quantum number $l = 0, 1,$ or 2; **(d)** must have the quantum number $l = 2$.

29. With reference to Table 7-2, complete the entry for $n = 4$. The new subshell that arises is the f subshell. How many f orbitals are present in this subshell?

30. Which of the following sets of quantum numbers is not allowable? Why not?
   (a) $n = 2, l = 1, m_l = 0$
   (b) $n = 2, l = 2, m_l = -1$
   (c) $n = 3, l = 0, m_l = 0$
   (d) $n = 3, l = 1, m_l = -1$
   (e) $n = 2, l = 0, m_l = -1$
   (f) $n = 2, l = 3, m_l = -2$

31. What type of electron orbital (i.e., $s, p, d,$ or f) is designated: **(a)** $n = 2, l = 1, m_l = -1$; **(b)** $n = 4, l = 2, m_l = 0$; **(c)** $n = 5, l = 0, m_l = 0$.

32. What are the n and l quantum number designations for the subshells $3s, 4p, 5f,$ and $6d$?

33. How many orbitals can there be of each of the following types? Explain. **(a)** $2s$; **(b)** $3f$; **(c)** $4p$; **(d)** $5d$.

34. Which of the following statements is (are) correct for an electron that has the quantum numbers $n = 4$ and $m_l = -2$? Explain your reasoning.
   (a) The electron is in the fourth principal shell.
   (b) The electron may be in a d orbital.
   (c) The electron may be in a p orbital.
   (d) The electron must have a spin quantum number, $m_s = +\frac{1}{2}$.

Additional Exercises

1. In what region of the electromagnetic spectrum would you expect to find radiation emitted by a hydrogen atom when the electron in the atom falls from the orbit $n = 6$ to the orbit $n = 3$?

Electron configurations

35. Five electrons in an atom have the quantum numbers given below. Arrange these electrons in order of increasing energy. If any two have the same energy, so indicate.
   (a) $n = 5, l = 0, m_l = 0, m_s = +\frac{1}{2}$
   (b) $n = 3, l = 1, m_l = -1, m_s = -\frac{1}{2}$
   (c) $n = 3, l = 2, m_l = 0, m_s = +\frac{1}{2}$
   (d) $n = 3, l = 2, m_l = -2, m_s = -\frac{1}{2}$
   (e) $n = 3, l = 0, m_l = 0, m_s = -\frac{1}{2}$

36. State the basic idea(s) discussed in this chapter that is (are) violated by each of the following electron configurations, and replace each by the correct configuration. **(a)** B $1s^2 2s^3$ **(b)** Na $1s^2 2s^2 2p^6 2d^1$ **(c)** K $[Ar]3d^1$ **(d)** Ti $[Ar]4s^2 4p^2$ **(e)** Xe $[Kr]5s^2 5p^6 5d^{10}$ **(f)** Hg $[Xe]4f^{10} 5d^{10} 6s^2 6p^4$

37. Which of the following electron configurations is correct for phosphorus $(Z = 15)$? What is wrong with each of the others?

$$\qquad\qquad 3s \qquad\qquad 3p$$
(a) [Ne] ⇅ | ↑ | ↑ | ↑ |

$$\qquad\qquad 3s \qquad\qquad 3p$$
(b) [Ne] ⇅ | ↑ | ↑ | ↓ |

$$\qquad\qquad 3s \qquad\qquad 3p$$
(c) [Ne] ⇅ | ↑ | ↑ | ↑ |

$$\qquad\qquad 3s \qquad\qquad 3p$$
(d) [Ne] ⇅ | ⇅ | ↑ | |

38. On the basis of rules for electron configurations, indicate the number of **(a)** unpaired electrons in an atom of Si; **(b)** $3d$ electrons in an atom of S; **(c)** $4p$ electrons in an atom of As; **(d)** $3s$ electrons in an atom of Sr; **(e)** $4f$ electrons in an atom of Au.

39. The electron configurations described in the text are all for normal atoms in their ground states. An atom may absorb a quantum of energy and promote one or more electrons to a higher energy level; it becomes an "excited" atom. The following configurations represent excited states. Indicate why this is so. **(a)** $1s^2 2s^1 2p^1$; **(b)** $[Ne]3s^2 3p^2 3d^2$; **(c)** $[Ar]3d^{10} 4s^1 4p^3$.

*40. What would be the electron configuration of the element Cs in each case?
   (a) If there were *three* possibilities for electron spin.
   (b) If the quantum number, l, could have the value, n, and if all the rules governing electron configurations were otherwise valid.

*2. Between which two orbits of the Bohr hydrogen atom must an electron fall to produce light of wavelength 1876 nm?

*3. Balmer seems to have deduced his formula for the visible spectrum of hydrogen just by manipulating numbers. A more common scientific procedure is to graph experimental data and then find a mathematical equation to describe the graph. Show that equation (7.3) describes a straight line. Indicate which variables must be plotted and determine numerical values of the slope and intercept of this line.

4. What is the energy, in kJ/mol, of light having a wavelength of 825 nm?

5. Determine (a) the energy of an H atom when its electron is in the orbit $n = 5$; (b) the total energy required to ionize 1 mol of *normal* H atoms.

6. The subshell that arises after f is called the g subshell (i.e., s, p, d, f, g).
   (a) How many g orbitals are present in the g subshell?
   (b) In what principal electronic shell would the g subshell first occur, and what is the total number of orbitals in this principal shell?

7. Complete the following assignments by writing an acceptable value for the missing quantum number. What type of orbital is described by each set?
   (a) $n = ?$, $l = 2$, $m_l = 0$, $m_s = +\frac{1}{2}$
   (b) $n = 2$, $l = ?$, $m_l = -1$, $m_s = -\frac{1}{2}$
   (c) $n = 4$, $l = 2$, $m_l = 0$, $m_s = ?$
   (d) $n = ?$, $l = 0$, $m_l = ?$, $m_s = ?$

8. Which of the following electron configurations is correct for molybdenum ($Z = 42$)? Comment on the errors in each of the others.
   (a) $[Ar]3d^{10}3f^{14}$
   (b) $[Kr]4d^55s^1$
   (c) $[Kr]4d^55s^2$
   (d) $[Ar]3d^{14}4s^24p^8$
   (e) $[Ar]3d^{10}4s^24p^64d^6$

9. A light year is defined as the distance that electromagnetic radiation can travel in space in 1 year.
   (a) What is this distance, in km?
   (b) What is the distance, in km, to Alpha Centauri, the star closest to our solar system, if this distance is listed as 4.3 light years?

10. The *work function* of a photoelectric material is the energy that a photon of light must possess to just secure the release of an electron from the surface of a material. The corresponding frequency of the light is the threshold frequency. The higher the energy of the incident radiation, the more kinetic energy the ejected electrons have in moving away from the surface. The work function for the element mercury is equivalent to 435 kJ/mol of photons.
   (a) What is the threshold frequency?
   (b) What is the wavelength of light of this frequency?
   (c) Can the photoelectric effect be obtained with mercury using visible light?
   * (d) What is the maximum velocity of the electrons ejected when light of wavelength, 215 nm, strikes the surface of mercury?

*11. Use the equation given in Exercise 18 to calculate the wavelength of the spectral line for the transition of an electron from the orbit $n = 3$ to $n = 2$ in a Li^{2+} ion.

*12. The angular momentum of an electron in the Bohr hydrogen atom is mvr, where m is the mass of the electron, v, its velocity, and r, the radius of the Bohr orbit. Bohr established that the angular momentum only can have values equal to $nh/2\pi$ (where n is an integer). Combine these ideas and other data given in the text to obtain for an electron in the third orbit ($n = 3$) of a hydrogen atom: (a) its velocity; (b) the number of revolutions about the nucleus it makes per second.

*13. Radio signals from Voyager 1 spacecraft on its trip to Jupiter in the late 1970s were broadcast at a frequency of 8.4 gigahertz. On earth this radiation was received by a 64-m antenna capable of detecting signals as weak as 4×10^{-21} watt (1 watt = 1 J/s). Approximately how many photons per second did the antenna intercept from this signal?

*14. Certain metallic compounds, when heated in flames, impart characteristic colors to the flames: for example, sodium compounds, yellow; lithium, red; barium, green. "Flame tests" can be used to detect these elements.
   (a) If the flame temperature is 800°C, can collisions with other gaseous atoms or molecules possessing an average amount of kinetic energy supply the required energy to excite an atom to the point that visible light is emitted?
   (b) If not, how do you account for the excitation energy?

Self-Test Questions

For questions 1 through 8 select the single item that best completes each statement.

1. The *shortest* wavelength radiation of the following is
(a) 735 nm; (b) 6.3×10^{-5} cm; (c) 1.05 μm; (d) 3.5×10^{-6} m.

2. A particular electromagnetic radiation with wavelength 200 nm.
(a) has a higher frequency than radiation with wavelength 400 nm;
(b) is in the visible region of the electromagnetic spectrum;
(c) has a higher velocity in vacuum than does radiation of wavelength 400 nm;
(d) has a greater energy per photon than does radiation with wavelength 100 nm.

3. The set of quantum numbers, $n = 2$, $l = 2$, $m_l = 0$
(a) describes an electron in a $2d$ orbital; (b) describes an electron in a $2p$ orbital; (c) describes one of five orbitals of a similar type; (d) is not allowed.

4. If the quantum number, $n = 3$,
(a) the quantum, m_l, *must be* 0;
(b) the quantum number, l, *cannot be* larger than $+2$;
(c) the quantum number, m_s, must be $+\frac{1}{2}$;
(d) there are *three* possible values of m_l.

5. The m_l quantum number for an electron in a 5d orbital (a) can have any value less than 5; (b) may be 0; (c) may be $+\frac{1}{2}$ or $-\frac{1}{2}$; (d) is 3.

6. The number of 2p electrons in an atom of Cl is (a) 0; (b) 2; (c) 5; (d) 6.

7. The number of unpaired electrons in an atom of scandium ($Z = 21$) is (a) 3; (b) 2; (c) 1; (d) 0.

8. Of the following electron transitions in the Bohr hydrogen atom, the one for which light of the *longest* wavelength is *emitted* is that from (a) $n = 4$ to $n = 3$; (b) $n = 1$ to $n = 2$; (c) $n = 2$ to $n = 3$; (d) $n = 2$ to $n = 1$.

9. What is the energy content, in kJ/mol of photons, of a red light with frequency $4.00 \times 10^{14} \ s^{-1}$?

10. The atomic spectrum of sodium contains two bright yellow lines, one at 589.0 nm and the other at 589.6 nm. Which of the two lines represents the greater energy per photon? What is the *difference* in energy per photon between the two?

11. The line at 434 nm in the Balmer series of the hydrogen spectrum corresponds to a transition of an electron from the nth to the second Bohr orbit. What is the value of n?

$$\nu = 3.2881 \times 10^{15} \ s^{-1} \left(\frac{1}{2^2} - \frac{1}{n^2} \right)$$

12. Write out the complete electron configuration of (a) selenium ($Z = 34$), using *spdf* notation; (b) iodine ($Z = 53$), using an orbital diagram.

8 Atomic Properties and the Periodic Table

The properties of matter in bulk are determined by the properties of individual atoms. This is why chemists study atomic structure. One atomic property introduced in Chapter 7 was the electron configuration. We begin this chapter by describing the periodic table of the elements, which, in many ways, was the crowning achievement of nineteenth-century chemistry. Next, we turn to a discussion of the relationship between electron configurations and the periodic table. Finally, we will consider a number of other properties of individual atoms that provide a basis for understanding chemical bonding.

8-1 On the Idea of Order

One of the most important scientific activities is the search for order. If a large number of observations or objects can be arranged into categories according to some common features, it becomes easier to describe them. Moreover, it is often possible to discover an underlying cause for a particular order, and this discovery may lead, in turn, to a significant theory. Unfortunately, there is no way of knowing for certain what features to look for or the number of observations necessary to arrive at a classification scheme. Botanical observations, for example, were sufficiently numerous so that the task of ordering was completed in the eighteenth century. The field of chemistry was not ready for the establishment of order at the same time. The laws of chemical combination were not understood; the assignment of atomic weights was quite uncertain; too many elements remained undiscovered. Chemistry's turn came in the nineteenth century.

8-2 Periodic Law and the Periodic Table

A classification scheme of the elements similar to that used today was discovered independently and almost simultaneously by Dimitri Mendeleev and Lothar Meyer, in 1869. Their classifications were based on an early version of the periodic law.

If the elements are arranged in order of increasing atomic weight, certain sets of properties are found to recur periodically.

Atomic Volume. An early example of the periodic law involved a property known as **atomic volume**—the atomic weight of an element divided by its density. Meyer's plot of atomic volumes, published in 1870, displayed a periodicity based on atomic weights.

FIGURE 8-1
An illustration of the periodic law—atomic volume as a function of atomic number.

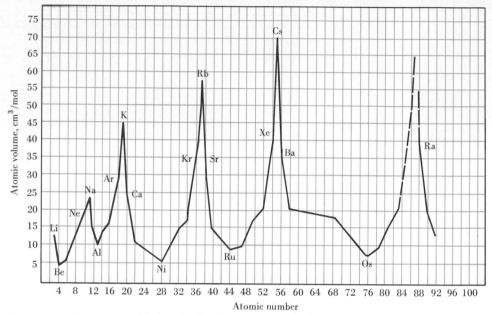

Meyer used the term *atomic volume* to refer to the property graphed here; molar volume—the volume of one mole of atoms—is a more accurate term.

Figure 8-1 is a later version based on atomic numbers. Very striking is the fact that atomic volumes rise to a maximum periodically, with the alkali metals Li, Na, K, Rb, and Cs. Several other properties, such as electrical conductivity, thermal conductivity, and hardness, when plotted as a function of atomic number (or atomic weight), yield similar curves.

By recognizing that atomic weight and molar mass are numerically equal, we can see the physical significance of the atomic volume. It is actually a molar volume—the volume occupied by one mole of atoms of an element.

$$\text{atomic (molar) volume (cm}^3/\text{mol)} = \text{molar mass (g/mol)} \times \frac{1}{d} \text{ (cm}^3/\text{g)} \qquad (8.1)$$

Example 8-1 Use data from Figure 8-1 to estimate the density of silver. Silver has an atomic number of 47 and an atomic weight of 108. From Figure 8-1, the atomic volume of the element with atomic number 47 is 10 cm^3/mol.

Solution

$$\text{no. g/cm}^3 = \frac{108 \text{ g}}{1 \text{ mol}} \times \frac{1 \text{ mol}}{10 \text{ cm}^3} = 11 \text{ g/cm}^3$$

Note that the result can be expressed only to two significant figures, since the graph of Figure 8-1 cannot be read any more precisely than this.

SIMILAR EXAMPLES: Exercises 1, 2.

Mendeleev's Periodic Table. A tabular arrangement of the elements based on the periodic law is called a **periodic table.** In Mendeleev's 1871 periodic table the elements were arranged in 12 horizontal rows and eight vertical columns or groups. The eight groups were further divided into subgroups. To achieve the objective of bringing similar elements into appropriate groups and subgroups, it was necessary to leave blank spaces for elements undiscovered at the time and to make assumptions about atomic weights that were not known with certainty. Mendeleev's periodic table is reproduced in Table 8-1.

The elements within a particular subgroup of Mendeleev's table have similar physical and chemical properties, and these properties change gradually from top to bottom in the group. For instance, the alkali metals in group I have low melting points that decrease in the order

Li (174°C) > Na (97.8°C) > K (63.7°C) > Rb (38.9°C) > Cs (28.5°C)

These elements also have high atomic volumes, as illustrated in Figure 8-1. In moving across one of the rows in Mendeleev's table, the properties change rather significantly from group to group.

At the top of Mendeleev's table are listed the formulas of the chlorides, hydrides, and oxides of the elements (R) in each group. Mendeleev was able to correlate these formulas with the group numerals, i.e., sodium chloride would have the formula $NaCl$; the arsenic-hydrogen compound arsine, AsH_3; and an oxide of molybdenum, MoO_3.

Correction of Atomic Weights.   To place them properly in his periodic table, Mendeleev made adjustments in the previously accepted atomic weights of a number of elements. One of these was indium. Indium was known to occur naturally in zinc ores and was assumed to form an oxide, InO, similar to that of zinc, ZnO. Based on the percent

TABLE 8-1
Mendeleev's periodic table of 1871

R	Group I	Group II	Group III	Group IV	Group V	Group VI	Group VII	Group VIII
O	R_2O	RO	R_2O_3	RO_2	R_2O_5	RO_3	R_2O_7	RO_4
W	RCl	RCl_2	RCl_3	RCl_4 RH_4	RH_3	RH_2	RH	
1	H = 1							
2	Li = 7	Be = 9.4	B = 11	C = 12	N = 14	O = 16	F = 19	
3	Na = 23	Mg = 24	Al = 27.3	Si = 28	P = 31	S = 32	Cl = 35.5	
4	K = 39	Ca = 40	____ = 44	Ti = 48	V = 51	Cr = 52	Mn = 55	Fe = 56, Co = 59, Ni = 59, Cu = 63
5	(Cu = 63)	Zn = 65	____ = 68	____ = 72	As = 75	Se = 78	Br = 80	
6	Rb = 85	Sr = 87	?Yt = 88	Zr = 90	Nb = 94	Mo = 96	____ = 100	Ru = 104, Rh = 104, Pd = 106, Ag = 108
7	(Ag = 108)	Cd = 112	In = 113	Sn = 118	Sb = 122	Te = 125	I = 127	
8	Cs = 133	Ba = 137	?Di = 138	?Ce = 140	. . .	. . .	. . .	
9	. . .	. . .	. . .	. . .	. . .	. . .	. . .	
10	. . .	. . .	?Er = 178	?La = 180	Ta = 182	W = 184	. . .	Os = 195, Ir = 197, Pt = 198, Au = 199
11	(Au = 199)	Hg = 200	Tl = 204	Pb = 207	Bi = 208	. . .	. . .	
12	. . .	. . .	. . .	Th = 231	. . .	U = 240	. . .	

composition of this oxide (82.5% In) indium had been assigned an atomic weight of approximately 76. This atomic weight would have placed indium, a metal, between arsenic and selenium, both nonmetals. Mendeleev proposed that indium formed the oxide, In_2O_3. From this formula the atomic weight obtained for indium was 113. As a result, Mendeleev placed indium in the space between cadmium and tin, both metals.

Other atomic weights corrected by Mendeleev were those of beryllium (from 13.5 to 9) and uranium (from 120 to 240).

Prediction of New Elements.   Mendeleev deliberately left blank spaces in his periodic table for elements yet to be discovered. Not only did he predict the existence of these elements, but he predicted what their properties would be. The blank space at atomic weight 72 was for an element in the same group as silicon. Mendeleev called this element, eka-silicon. The remarkable agreement between some measured properties of germanium and Mendeleev's predicted values of these properties is brought out in Table 8-2.

> The term "eka" is derived from Sanskrit and means "first." That is, eka-silicon means literally, first comes silicon (and then comes the unknown element).

Discovery of the Noble (Inert) Gases.   Cavendish, in 1785, reported that in reactions involving atmospheric gases a small portion, "not more than $\frac{1}{120}$ of the whole," remained completely unreacted. This gas was isolated by Rayleigh and Ramsay in 1894 and called argon ("the lazy one"). Another element, first observed in the sun's spectrum in 1868 and called helium, was isolated by Ramsay in 1895. Since these gases were unlike any other known elements, Ramsay assigned them to a new group of the periodic table (referred to as group 0 in this text). From the periodic law it was reasonable to expect that there were other members of this group, and the remaining members were discovered soon thereafter—neon, krypton, and xenon in 1898, and radon in 1900. As we shall learn in Chapter 9, knowledge of the noble gases played a key role in early theories of chemical bonding.

Atomic Number as the Basis for the Periodic Law.   In the early periodic table it was necessary to place certain pairs of elements out of order. For example, argon (at. wt. 39.9) was placed ahead of potassium (at. wt. 39.1). If this were not done, potassium, an active metal, would appear among the inert gases and argon, an inert gas, among a group of active metals. Once the concept of atomic number was introduced by Moseley in 1913, a reordering of the periodic table according to atomic number became possible. If the elements are arranged according to increasing atomic number, argon ($Z = 18$) naturally precedes potassium ($Z = 19$).

8-3 A Modern Periodic Table—The Long Form

Mendeleev's periodic table was a "short" form. Each main vertical group consisted of two subgroups. Most modern periodic tables are of the "long" form. The subgroups are separated from one another. Following are some of the features of the long form of the periodic table found on the inside front cover.

The horizontal rows of the table, which are arranged in order of increasing atomic number, are called **periods.** The vertical columns, which bring together similar elements, are called **groups** or **families.** The first period of the table consists of only two elements, hydrogen and helium. This is followed by two periods of eight elements each, lithium to neon, and sodium to argon. The fourth and fifth periods comprise 18 elements each, ranging from potassium to krypton and from rubidium to xenon. The sixth period is a long one of 32 members. To fit this period to a table which is held to a maximum width of 18 members requires that 14 members be extracted and placed at the bottom of the table. This series of 14 elements, which fits between lanthanum ($Z = 57$) and hafnium ($Z = 72$), is called the **lanthanoid** or rare earth series. The seventh and final period is incomplete but

> The "oid" ending means like or similar to (as in humanoid). The lanthanoids are like lanthanum, and the actinoids like actinium. The names lanthanide and actinide are also commonly used in place of lanthanoid and actinoid.

TABLE 8-2
Properties of germanium: predicted and observed

Property	Predicted: eka-silicon (1871)	Observed: germanium (1886)
atomic weight	72	72.6
density, g/cm^3	5.5	5.47
color	dirty gray	grayish white
density of oxide, g/cm^3	EsO_2: 4.7	GeO_2: 4.703
boiling point of chloride	$EsCl_4$: below 100°C	$GeCl_4$: 86°C
density of chloride, g/cm^3	$EsCl_4$: 1.9	$GeCl_4$: 1.887

is believed to be a long one. A 14-member series, extracted from the seventh period and placed at the bottom of the table, is called the **actinoid** series.

The groups in the periodic table are designated by Roman numerals and letters. The A group elements are known as **representative elements.** The B group elements, together with those in group VIII and the lanthanoid and actinoid series, comprise the **transition elements.** The group labeled 0 contains the noble (inert) gases.

Example 8-2   Refer to the periodic table on the inside front cover and indicate
(a) An element that is in group IVA and the fourth period.
(b) Two elements with properties similar to molybdenum.
(c) The known element that the unknown and undiscovered element, $Z = 114$, is most likely to resemble.

Solution
(a) The elements in the fourth period range from K ($Z = 19$) to Kr ($Z = 36$); those in group IVA, from C to Pb. The only element that is common to both of these groupings of elements is Ge ($Z = 32$).
(b) Molybdenum is in group VIB. Two other members of this group that it should resemble are chromium (Cr) and tungsten (W).
(c) Assuming that the seventh period has 32 members, the element $Z = 114$ should resemble lead, Pb ($Z = 82$), the element that it follows in group IVA.

SIMILAR EXAMPLES: Review Problem 1; Exercise 5.

8-4 Electron Configurations and the Periodic Table

In Table 8-3 three typical groups of elements have been extracted from the periodic table and their electron configurations noted. The similarity in configurations among the elements in a group is quite apparent. The noble gas atoms, with the exception of helium which has only two electrons, have outermost shells with eight electrons, in the configuration ns^2np^6 (where n is in the shell of highest principal quantum number). All the atoms of groups IA possess a single outer-shell electron in an s orbital, that is, ns^1. Atoms of the group VIIA elements have seven outer-shell electrons, in the configuration ns^2np^5.

Elements in the same group have similar physical and chemical properties and also similar electron configurations. We should begin to suspect that it is electron configurations that are principally responsible for the characteristic properties of the elements. Especially important is the electron configuration of the electronic shell of highest principal quantum number, that is, the *outermost* electronic shell. However, we should not lose sight of the fact that elements were first grouped according to similar properties; later the similarity of electron configurations—the theoretical basis—was established.

TABLE 8-3
Electron configurations of some groups of elements

Group	Element	Configuration
0	He	$1s^2$
	Ne	$1s^22s^22p^6$
	Ar	$1s^22s^22p^63s^23p^6$
	Kr	$1s^22s^22p^63s^23p^63d^{10}4s^24p^6$
	Xe	$1s^22s^22p^63s^23p^63d^{10}4s^24p^64d^{10}5s^25p^6$
	Rn	$1s^22s^22p^63s^23p^63d^{10}4s^24p^64d^{10}4f^{14}5s^25p^65d^{10}6s^26p^6$
IA	H	$1s^1$
	Li	$[He]2s^1$
	Na	$[Ne]3s^1$
	K	$[Ar]4s^1$
	Rb	$[Kr]5s^1$
	Cs	$[Xe]6s^1$
	Fr	$[Rn]7s^1$
VIIA	F	$[He]2s^22p^5$
	Cl	$[Ne]3s^23p^5$
	Br	$[Ar]3d^{10}4s^24p^5$
	I	$[Kr]4d^{10}5s^25p^5$
	At	$[Xe]4f^{14}5d^{10}6s^26p^5$

Although it is not correct in all details, Figure 8-2 shows a periodic table in which the order of filling of orbitals (the Aufbau process of Chapter 7) is summarized. Here it can be seen that atoms of *representative* elements are characterized by the filling of *s* or *p* subshells of the electronic shell of highest principal quantum number. For the *transition* elements it is the *d* or *f* subshells of an *inner* electronic shell (not the shell of highest principal quantum number) that are partially filled. A *d* subshell fills for transition elements in the main body of the table, and an *f* subshell for the lanthanoid and actinoid elements.

FIGURE 8-2
Electron configurations and the periodic table.

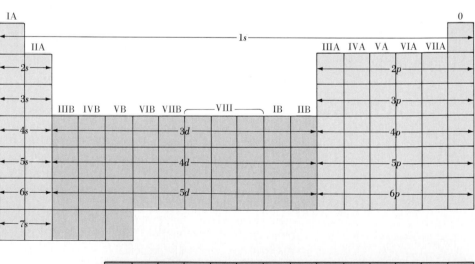

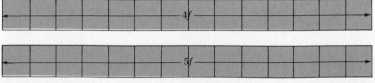

For the representative elements the group numeral (IA, IIA, . . .) is identical to the number of electrons in *s* and *p* orbitals of the outermost electronic shell. Atoms of the transition elements mostly have two outer-shell electrons in the configuration, ns^2; some, for example, Cr, Mo, Cu, Ag, and Au, have only one—ns^1. For the transition elements in *groups IB and IIB,* the group numerals do represent the number of outer-shell electrons— *s* electrons. For all other transition elements the group numeral is *not* the same as the number of outer-shell electrons, but it does represent the maximum number of electrons available for the formation of compounds—the maximum oxidation state of the element in its compounds.

The properties of an element are determined largely by the electron configuration of the *outermost* electronic shell. Adjacent members of a series of *representative* elements in the same period (such as P, S, and Cl) have rather different properties because of differences in their outer-shell electron configurations. Within a transition series differences in electron configurations are found mostly in *inner* shells. As a result, within a transition series there are similarities among adjacent members of the same period (e.g., Fe, Co, and Ni) *as well as* within the same vertical group.

Example 8-3 Based on the relationship between electron configurations and the periodic table, indicate how many **(a)** outer-shell electrons in an atom of bromine; **(b)** shells of electrons in an atom of strontium; **(c)** 5*p* electrons in atom of tellurium; **(d)** 3*d* electrons in an atom of zirconium; **(e)** unpaired electrons in an atom of indium.

Solution. Determine the atomic number of each element and the location of the element in the periodic table. Then establish the significance of each location.
(a) Bromine ($Z = 35$) is a representative element in group VIIA. There are *seven* outershell electrons in all the atoms in this group.
(b) The highest principal quantum number for atoms in the fifth period is $n = 5$. There are *five* electronic shells in the Sr atom.
(c) Tellurium ($Z = 52$) follows the *second* transition series of elements, in which the 4*d* subshell is filled. The next subshell to receive electrons after the 4*d* is the 5*p*. In the Aufbau process, Te is the fourth atom following Cd. The Te atom has *four* 5*p* electrons. (Alternatively, Te is in group VIA. All group VIA atoms have six outer-shell electrons, two *s* and four *p*. The outer-shell configuration of Te is $5s^2 5p^4$.)
(d) Zirconium ($Z = 40$) is in the *second* transition series, in which 4*d* orbitals fill. The filling of the 3*d* subshell occurs in the *first* transition series (from Sc to Zn). Thus, the 3*d* subshell of the Zr atom is filled. Zr has *ten* 3*d* electrons.
(e) Indium ($Z = 49$) is in group IIIA. The In atom has three outer-shell electrons with principal quantum number, $n = 5$, that is, $5s^2 5p^1$. The two 5*s* electrons are paired and the 5*p* electron is unpaired. The In atom has *one* unpaired electron.

SIMILAR EXAMPLES: Review Problem 5; Exercise 11.

8-5 Metals and Nonmetals

In Section 3-6, to assist in writing chemical names and formulas, we divided the elements into the categories, **metal** and **nonmetal.** At that time we also located these categories in the periodic table. Figure 8-3 extends this categorization of the elements to include the noble gases and a group of elements known as metalloids. **Metalloids** have the appearance of metals but display nonmetallic properties as well. Having just established a relationship between the electron configuration of the atoms of an element and the location of that element in the periodic table, we might wonder if there is not also a connection between electron configuration and the metallic/nonmetallic character of an element. Generally speaking there is.

FIGURE 8-3
Metals, nonmetals,
metalloids, and noble
gases.

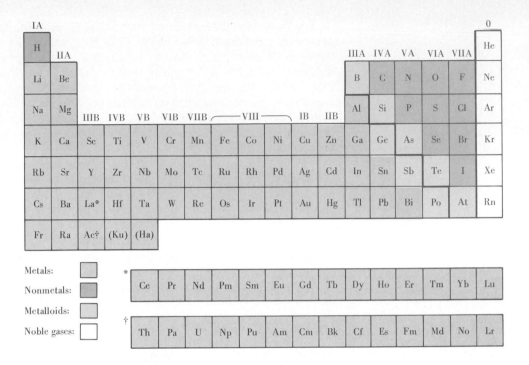

The noble gases are found in group 0 of the periodic table. Helium has the outer-shell electron configuration $1s^2$, and the other noble gases, ns^2np^6. These prove to be very stable electron configurations that can be altered only with great difficulty.

The electron configurations of the atoms of the group IA and IIA elements—the most active metals—differ from those of a noble gas atom by only one or two electrons in the *s* orbital of a new electronic shell. This fact is brought out clearly when we write electron configurations in the form

K $[Ar]4s^1$ and   Ca $[Ar]4s^2$

If a K atom is stripped of its outer-shell electron, it becomes the *positive ion* K^+, with the electron configuration [Ar]. A Ca atom acquires the [Ar] configuration following the removal of two electrons.

$$K([Ar]4s^1) \longrightarrow K^+([Ar]) + e^-$$

$$Ca([Ar]4s^2) \longrightarrow Ca^{2+}([Ar]) + 2\ e^-$$

A characteristic of the active metals of groups IA and IIA then is the tendency of their atoms to lose outer-shell electrons to produce positive ions with the electron configuration of a noble gas.

Atoms of the elements in groups VIIA and VIA, the most active nonmetals, have electron configurations with one and two electrons fewer than those of the corresponding noble gas. Atoms of these elements can acquire the electron configuration of a noble gas atom by *gaining* the appropriate number of electrons. The Cl atom becomes the *negative ion* Cl^- by gaining one electron, and the S atom becomes S^{2-} by gaining two electrons.

$$Cl([Ne]3s^23p^5) + e^- \longrightarrow Cl^-([Ar])$$

$$S([Ne]3s^23p^4) + 2\ e^- \longrightarrow S^{2-}([Ar])$$

Nonmetals are those elements whose atoms can acquire the electron configuration of a noble gas by gaining a small number of electrons.

Figure 8-3 shows that the transition elements are classified as metals. Atoms of these elements generally have two electrons in the s orbital of the electronic shell of highest principal quantum number (a few have a single electron in this orbital) and partially filled d and/or f orbitals of an inner electronic shell. Although an atom such as Sc acquires a noble gas electron configuration when it loses three electrons to become Sc^{3+}, most of the transition metal atoms do not acquire noble gas electron configurations following the loss of electrons. Furthermore, the electrons most readily lost are generally those of the s orbital of the outermost shell (recall the discussion of page 187), and several possibilities may exist for ion formation by a given metal atom. Thus, an atom of iron may lose two $4s$ electrons to form the ion Fe^{2+},

$$Fe([Ar]3d^64s^2) \longrightarrow Fe^{2+}([Ar]3d^6) + 2\ e^-$$

or it may lose the two $4s$ and one $3d$ electron to form the ion Fe^{3+}.

$$Fe([Ar]3d^64s^2) \longrightarrow Fe^{3+}([Ar]3d^5) + 3\ e^-$$

It is customary to separate the metals and nonmetals in a periodic table with a stair-step diagonal line. Some of the elements in groups IVA and VA are clear-cut nonmetals, some are predominantly metallic, and those adjacent to the line are metalloids, i.e., they exhibit both metallic and nonmetallic properties. Metalloids are also found at the top of group IIIA and the bottom of groups VIA and VIIA.

In later sections of this chapter we will consider some of the properties that can be studied to assess the relative metallic and nonmetallic characters of the elements.

8-6 Some Unresolved Issues Concerning the Periodic Table

The Placement of Hydrogen.   Although every other element has a definite place in the periodic table, the placement of hydrogen presents some difficulties. Its electron configuration, $1s^1$, suggests that it be placed in group IA, as has been done in this text. However, under normal conditions of temperature and pressure, hydrogen does not closely resemble the alkali metals. Later in the chapter we will consider two properties (ionization energy and electronegativity) whose values for H are markedly different from those of the alkali metals. Sometimes hydrogen is placed with the other elements whose outermost electronic shells contain just one electron less than that of a noble gas. This placement in group VIIA is not entirely satisfactory either, because hydrogen does not particularly resemble the halogen elements. Still another alternative is to place hydrogen by itself at the top of the periodic table above carbon, an element that it does resemble in some respects. The uniqueness of hydrogen regarding its location in the periodic table stems from the fact that the hydrogen atom has only one electron.

Group Designations in the Periodic Table.   The labeling of representative elements as A group and transition elements as B group has been widely used in the United States for many years. This labeling is followed consistently throughout this text. Another designation, found in some wall charts of the periodic table and in some of the chemical literature, labels all groups to the left of group VIII as A groups and all groups to the right of group VIII as B groups. Since no single acceptable method for designating groups has yet been devised, students of chemistry must be prepared to deal with discrepancies in the labeling of groups in the periodic table if the situation arises.

Predicting Properties of the Heavy Elements.   For certain elements in the sixth period (e.g., Au and Hg) the observed properties differ from those of the corresponding fifth-period element (i.e., Ag and Cd) in ways that do not seem to conform to the periodic law. A partial explanation of this observation, discussed in Section 23-1 and referred to as the lanthanoid contraction, is based on the filling of the 4f subshell, which occurs between elements Z = 57 and Z = 71. Other studies, however, suggest an even more fundamental basis to this behavior—Einstein's theory of relativity.

In Schrodinger's treatment of the hydrogen and hydrogenlike atoms, the values of certain properties, such as the mass and charge of an electron, enter into establishing the energies associated with electrons in various orbitals. According to Einstein's theory of relativity, the mass of a particle *increases* if the particle travels at speeds approaching the speed of light. The value listed for the mass of an electron in Table 2-1 is called the **rest mass.** Neglecting relativistic effects (i.e., effects based on the theory of relativity) and using only the rest mass of an electron in wave mechanical calculations is justified as long as the electron travels at moderate speeds. This procedure works well for atoms of low atomic number. In atoms of high atomic number, because of the large nuclear charge, electrons are accelerated to high speeds. This is especially so for electrons that have rather high probabilities of being found near the nucleus—*s* electrons. As the electron mass increases, the electron is drawn closer to the nucleus, the orbital energy decreases (becomes more negative), and properties related to this orbital energy are affected. Thus, since relativistic effects are encountered only with heavy elements, these effects may cause certain heavy elements to have properties that seem anomalous compared to lighter elements in the same group. Some examples will be considered in Section 22-6. The most significant departures from the periodic law are expected for the superheavy elements as larger numbers of these are synthesized and studied (see Section 25-4).

8-7 Atomic Radius

A number of physical and chemical properties are related to the sizes of atoms, but atomic size is somewhat difficult to define. We have seen that the probability of finding an electron falls off with increasing distance from the atomic nucleus, but nowhere does this probability reach zero. There is no precise outer boundary to an atom. Furthermore, atoms are generally observed not in isolation but in contact with other atoms, either of the same or of a different kind; and the size of an atom depends to a considerable extent on its environment. Given these facts, it is not difficult to understand why reproducible data on the sizes of atoms can be obtained only through carefully defined experimental measurements. Nevertheless, we can acquire some useful insights with the rough approximation that the size of an atom is measured by the most probable distance from the nucleus to the outer-shell electron(s). Let us call this distance the atomic radius and see what factors affect it, that is, the factors that affect atomic size.

1. Variation of Atomic Sizes Within a Group of the Periodic Table. We consider the distance of an electron from the nucleus of an atom to depend primarily on the principal quantum number, n, of the electron. As a result we should expect that the higher the quantum number of the outermost electronic shell, the larger the atom. This generalization holds well for the group members of lower atomic numbers, where the percent increase in size from one period of elements to the next is large (e.g., from Li to Na to K in group IA). The percent increase in size from one period to the next is much smaller for the elements of higher atomic number (e.g., from K to Rb to Cs). In these elements the outer-shell electrons are held more tightly by the nucleus than would otherwise be expected. This is because inner-shell electrons in *d* and *f* subshells are less effective than *s* and *p* electrons in screening outer-shell electrons from the nucleus (recall Figure 7-22). Nevertheless, as a useful generalization we can say that *the more electronic shells in an atom (the farther down a group of the periodic table), the larger the atom.*

2. Variation of Atomic Sizes Within a Period in the Periodic Table. Let us consider the hypothetical process of starting with sodium and adding one unit of positive charge to the nucleus and one outer-shell electron to build up, in succession, each atom in the third period. In this process the number of inner-shell electrons remains constant at 10 (in the configuration $1s^2 2s^2 2p^6$). To begin, let us *assume* that the inner-shell electron core is totally effective in shielding or screening the outer-shell electrons from the nucleus, and that the outer-shell electrons do not screen one another at all. In this case, in sodium, with an atomic number of 11, the effective charge on the combination of the nucleus and the inner-shell electron core would be +1. In magnesium, with atomic number 12, the effective charge would be +2; in aluminum, with atomic number 13, the effective charge experienced by the three outer-shell electrons would be +3, and so on. Actually neither assumption is correct. That is, the inner-shell core of electrons is not totally effective in screening outer-shell electrons from the nucleus. In sodium, the **effective nuclear charge (Z_{eff})**—the combined charge of the nucleus and the inner-shell electron core which acts on the outer-shell (3s) electron—is about +2 (not +1). And outer-shell electrons do screen one another somewhat from the nucleus; they are about one-third effective in doing so. As a result, the effective nuclear charge that each of the two outer-shell electrons in magnesium experiences is about $+2\frac{2}{3}$; in aluminum the effective nuclear charge is about $+3\frac{1}{3}$, and so on. In whatever way we look at the matter, outer-shell electrons experience a progressively stronger force of attraction to the nucleus as we move through a period of representative elements. The result is that *the atomic radius decreases from left to right through a period of elements.*

3. Variation of Atomic Sizes Within a Transition Series. For the fourth and higher periods the situation is somewhat different from that described in generalization 2 for the portions of the periods that include transition elements. In a series of transition elements, additional electrons go into an *inner* electron shell, where they participate quite effectively in screening outer-shell electrons. The number of electrons in the *outer* shell, on the other hand, tends to remain constant. Thus, the outer-shell electrons experience similar forces of attraction in these atoms. There is an initial sharp decrease in size for the first two or three members, but following that *atomic sizes change little in a transition series.* Consider Fe, Co, and Ni, for instance. Fe has 26 protons in the nucleus and 24 inner-shell electrons. In Co ($Z = 27$) there are 25 inner-shell electrons, and in Ni ($Z = 28$) there are 26. In each case the two outer shell electrons are under the influence of about the same effective nuclear charge.

Now let us be more specific about what we mean by an atomic radius, since, as we have already stated, the radius of an atom depends on the environment in which it is found. For *bonded* atoms we customarily speak of a covalent radius, ionic radius, and, in the case of metals, a metallic radius. For atoms that are *not* bonded together, the radius is known as the van der Waals radius. Three different radii for a sodium atom are compared in Figure 8-4.

The unit that has long been used to describe atomic dimensions, is the angstrom unit, Å. However, the angstrom is not a recognized SI unit. Appropriate SI units are either the nanometer (nm) or the picometer (pm).

$$1 \text{ Å} = 1 \times 10^{-10} \text{ m} = 1 \times 10^{-8} \text{ cm} = 0.10 \text{ nm} = 100 \text{ pm}$$

Even though the angstrom is not a recognized SI unit, it is still the unit preferred by most scientists who study atomic and molecular dimensions. We will use both the angstrom and the picometer in our discussion.

Covalent Radius In Chapter 10 we will picture covalent bonds as arising from the overlap of electron orbitals in the region between the centers of two atoms. The result is that the nuclei of bonded atoms approach each other more closely than do the nuclei of nonbonded ones. *The covalent radius is one half the distance between the nuclei of two identical atoms bonded together covalently.*

FIGURE 8-4
Covalent, ionic, and metallic radii compared.

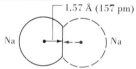

The **covalent radius** is one half the distance between the centers of two Na atoms in the gaseous molecule $Na_2(g)$. The **ionic radius** is based on the distance between centers of ions in an ionic compound, such as NaCl. Here, of course, the cation and anions are of different sizes. The **metallic radius** is taken as one half the distance between the centers of adjacent atoms in solid metallic sodium.

FIGURE 8-5
Covalent radii of atoms.

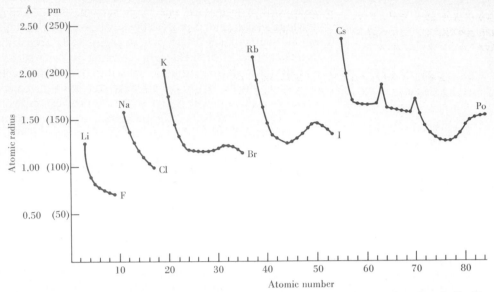

The values plotted here are for the bond type known as the single covalent bond. Radii are given both in angstroms and in picometers. Data for the noble gases are not shown in this graph because of the difficulty of measuring covalent radii for these atoms. (Only Kr, Xe, and Rn compounds are known.) Indirect methods yield estimated values for the noble gas atoms that seem to fit the trends discussed in this section. Also, explanations have been given for the small peaks at $Z = 63$ and $Z = 70$, but these are beyond the scope of this text.

Actually covalent radii themselves are not constant but vary with the character of the covalent bond between atoms, as we shall learn in Chapter 9. Figure 8-5 compares covalent radii based on the bond type called a single covalent bond. It illustrates the three generalizations about atomic sizes presented earlier in this section.

In the absence of statements to the contrary, we will take the term atomic radius to mean *covalent radius*.

Ionic Radius.   When electrons are removed from a metal atom to from a *positive* ion (cation), a significant reduction in size occurs. Usually the electrons lost are those of the shell of highest principal quantum number, and the resulting ion has one shell less than the metal atom.

Figure 8-6 compares five species: a Na atom, a Mg atom, a Na^+ ion, a Mg^{2+} ion, and a Ne atom. The Na atom is larger than Mg for the reason stated in generalization 2 on page 204. The ions are considerably smaller than the corresponding metal atoms, for the reason stated above. Na^+, Mg^{2+}, and Ne are said to be **isoelectronic**—they have the same number of electrons (10) in identical configurations ($1s^2 2s^2 2p^6$). Ne has a nuclear charge of $+10$. Na^+ is smaller than Ne because Na^+ has a nuclear charge of $+11$. Mg^{2+} is still smaller because it has a nuclear charge of $+12$.

When a nonmetal atom gains one or more electrons to form a *negative* ion (anion), there is an *increase* in size. The addition of electrons to an atom causes an increase in repulsions among the electrons. The electrons spread out more, and the size of the atom increases.

Figure 8-7 lists the radii of a number of ions, both negative and positive.

Example 8-4   **The following species are isoelectronic with the noble gas argon. Without reference to figures or tables in the text, arrange them in order of increasing size: Ar, K^+, Cl^-, S^{2-}, Ca^{2+}.**

FIGURE 8-6
A comparison of atomic and ionic sizes.

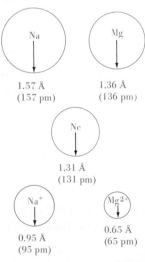

1.57 Å
(157 pm)

1.36 Å
(136 pm)

1.31 Å
(131 pm)

0.95 Å
(95 pm)

0.65 Å
(65 pm)

The radii shown for Na and Mg are covalent radii; for Na^+ and Mg^{2+}, ionic radii; and for Ne, the van der Waals radius. The variations in size can be explained by the factors discussed in the text.

FIGURE 8-7
Some representative ionic radii in picometers (pm).

Li^+ 60	Be^{2+} 31														B^{3+} 20	C —	N^{3-} 171	O^{2-} 140	F^- 136
Na^+ 95	Mg^{2+} 65														Al^{3+} 50	Si —	P^{3-} 212	S^{2-} 184	Cl^- 181
K^+ 133	Ca^{2+} 99	Sc^{3+} 81	Ti^{2+} 80	V^{2+} 88	Cr^{2+} 83	Mn^{2+} 80	Fe^{2+} 74	Co^{2+} 72	Ni^{2+} 69	Cu^{2+} 72	Zn^{2+} 74							Se^{2-} 198	Br^- 195
Rb^+ 148	Sr^{2+} 113																	Te^{2-} 221	I^- 216
Cs^+ 169	Ba^{2+} 135																		

Many of the elements form more than a single ion, and these different ions have different sizes. The data listed here are meant only to be representative.

Solution. The electron configuration that all five species have in common is $1s^2 2s^2 2p^6 3s^2 3p^6$. The greater the nuclear charge, the more tightly these electrons are held and the smaller the species. On this basis, Ca^{2+} is smallest, followed by K^+. If a species has an excess of electrons over protons, it becomes larger than the corresponding noble gas atom, the greater the negative charge the larger the size. Thus, S^{2-} is the largest species. The order is

$$Ca^{2+} < K^+ < Ar < Cl^- < S^{2-}$$

SIMILAR EXAMPLES: Review Problem 7; Exercises 17, 18.

8-8 Ionization Energy (Ionization Potential)

We have described several ways in which atoms can be caused to lose electrons. This may occur through shining light of an appropriate frequency on a suitable material (photoelectric effect), through heating certain materials (thermionic effect), and through collisions between an electron beam and gaseous atoms. In any event atoms do not lose electrons spontaneously under normal conditions. They must absorb energy in order for ionization to occur. The **ionization energy (ionization potential), *I*,** of an atom is the energy that the gaseous atom must absorb in order that its most loosely held electron may become completely separated from it.

Ionization energies can be measured in cathode ray tubes in which the atoms of interest are present as a gas under low pressure. Some typical values are

$$Mg(g) \longrightarrow Mg^+(g) + e^- \qquad I_1 = 7.65 \text{ eV/atom (738 kJ/mol)} \qquad (8.2)$$

$$Mg^+(g) \longrightarrow Mg^{2+}(g) + e^- \qquad I_2 = 15.04 \text{ eV/atom (1451 kJ/mol)} \qquad (8.3)$$

The symbol I_1 stands for first ionization energy, I_2, for second ionization energy, and so on. The loss of a second electron (as measured by I_2) occurs with greater difficulty than the first (as measured by I_1). This is because following ionization the ionized electron would have to move away from an ion with a charge of $+2$ (Mg^{2+}) rather than from an ion with a charge of $+1$ (Mg^+).

The energy unit used in equations (8.2) and (8.3) is the **electron volt (eV).** One electron volt is the energy acquired by an electron as it falls through an electrical potential difference of 1 volt (V). The electron volt is a very small unit of energy and quite suitable

TABLE 8-4
Ionization energies of
the alkali metal (group
IA) elements

	eV/atom	kJ/mol
Li	5.392	520.3
Na	5.139	495.9
K	4.341	418.9
Rb	4.177	403.0
Cs	3.894	375.7

to describe processes involving single atoms. When we consider atoms in large numbers, particularly in molar quantities, it is convenient to express ionization energies in terms of ionizing 1 mol of gaseous atoms. The conversion factor between electron volts per atom and kilojoules per mole (and kcal/mol) is given in equation (8.4).

$$1 \text{ eV/atom} = 96.49 \text{ kJ/mol} \ (=23.06 \text{ kcal/mol}) \tag{8.4}$$

The ease with which electrons can be removed from an atom requires consideration of several factors. It is reasonable to expect, however, that the farther an electron is from the nucleus, the smaller the force by which it is attracted to the nucleus and the more easily it should be extracted. The regions of greatest probability of finding outer-shell electrons are more distant from the nucleus in large atoms than in small atoms, and so *ionization energies decrease as the sizes of atoms increase.* This relationship is further illustrated through Table 8-4 for the alkali metal atoms and through Figure 8-8, in which ionization energies are plotted as a function of atomic number. The minima in Figure 8-8 come at the same atomic numbers as do the maxima in atomic volumes in Figure 8-1—the atomic numbers of the alkali metals. Another feature brought out by Figure 8-8 is the maxima in ionization energies that occur with the noble gases. The noble gas electron configuration is an exceptionally stable one that can be disrupted only with the expenditure of considerable energy.

If we consider the degree of metallic character of an element to be measured by the ease with which electrons can be removed from its atoms, then *the lower the ionization energy, the more metallic the element.* By this measure the atoms at the bottom of a group (larger atoms) are more metallic than those at the top (smaller atoms).

Table 8-5 lists ionization energies for the third period elements. With minor exceptions the trend in moving across a period (follow the colored stripe) is that ionization energies increase from group IA to group 0. (Recall that the effective nuclear charge

FIGURE 8-8
First ionization energies
as a function of atomic
number.

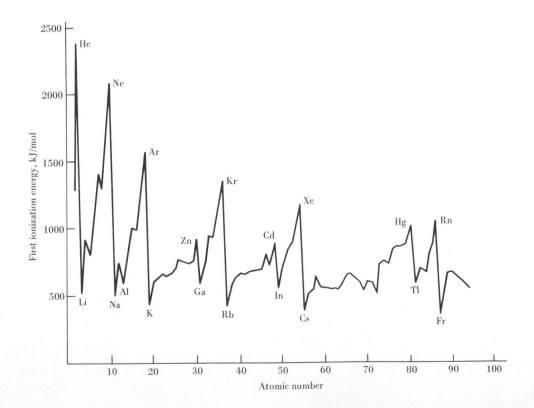

TABLE 8-5
Ionization energies of the third-row elements (in kJ/mol)

	Na	Mg	Al	Si	P	S	Cl	Ar
I_1	495.8	737.7	577.6	786.5	1012	999.6	1251.1	1520.5
I_2	4562	1451	1817	1577	1903	2251	2297	2666
I_3		7733	2745	3232	2912	3361	3822	3931
I_4			11580	4356	4957	4564	5158	5771
I_5				16090	6274	7013	6542	7238
I_6					21270	8496	9362	8781
I_7						27110	11020	12000

increases, and atoms get smaller and less metallic through the period.) Table 8-5 also lists stepwise ionization energies ($I_1, I_2, \ldots$). Note particularly the large breaks that occur as indicated by the zigzag diagonal line. Consider magnesium as an example. Although the second electron is removed with greater difficulty than the first, when two electrons are removed from a Mg atom it acquires the electron configuration of the noble gas Ne. Removal of a *third* electron requires breaking into the especially stable octet of electrons characteristic of the outer shell of a noble gas atom (ns^2np^6). This is an event that is not likely to occur in ordinary chemical processes. Similar reasoning would indicate that Na is not likely to occur as an ion with a charge greater than $+1$, nor Al with a charge greater than $+3$.

There are some exceptions to the regular increase of ionization energies in moving from left to right across the periodic table. For example, despite the fact that the Al atom is smaller than Mg, the first ionization energy is *lower* for Al (577.6 kJ/mol) than for Mg (737.7 kJ/mol). This is because the electron to be ionized in Al is in a higher energy orbital ($3p$) than is the electron ($3s$) to be ionized in Mg.

Example 8-5   How many joules of energy must be absorbed to convert to Li^+ all the atoms present in 1.00 mg Li(g)?

Solution.   We must convert from mg Li to mol Li, and then use the ionization energy of Li from Table 8-4.

$$\text{no. J} = 1.00 \times 10^{-3} \text{ g Li} \times \frac{1 \text{ mol Li}}{6.94 \text{ g Li}} \times \frac{520.3 \text{ kJ}}{1 \text{ mol Li}} \times \frac{1000 \text{ J}}{1 \text{ kJ}} = 75.0 \text{ J}$$

SIMILAR EXAMPLES: Review Problem 9; Exercise 24.

8-9 Electron Affinity

Electron affinity (EA) is the enthalpy change, ΔH, for the process in which an electron is brought from an infinite distance away up to a neutral gaseous atom and absorbed by it. For example,

$$Cl(g) + e^- \longrightarrow Cl^-(g) \qquad EA = -3.615 \text{ eV/atom } (-348.8 \text{ kJ/mol}) \qquad (8.5)$$

A few other values for the gain of one electron are -328.0, -324.6, -295.4, -141.1, and -200.43 kJ/mol for F, Br, I, O, and S, respectively.

TABLE 8-6
Electronegativities of selected elements

H 2.20										
Li 0.98	Be 1.57						B 2.04	C 2.55	N 3.04	O 3.44	F 3.98
Na 0.93	Mg 1.31						Al 1.61	Si 1.90	P 2.19	S 2.58	Cl 3.16
K 0.82	Ca 1.00	Sc 1.36	Ti 1.54	⋯	Cu 1.90	Zn 1.65	Ga 1.81	Ge 2.01	As 2.18	Se 2.55	Br 2.96
Rb 0.82	Sr 0.95	Y 1.22	Zr 1.33	⋯	Ag 1.93	Cd 1.69	In 1.78	Sn 1.96	Sb 2.05	Te 2.1	I 2.66
Cs 0.79	Ba 0.89									

The attraction of the nucleus of an atom for an additional electron results in a release of energy when a gaseous atom gains a single electron (EA < 0). The gain of a second electron requires the absorption of energy to overcome electron-electron repulsions (EA > 0). The affinities of O and S for two electrons (to form the ions O^{2-} and S^{2-}) are +704 and +386 kJ/mol, respectively.

Electron affinity is a property that until recently could not be measured easily by experiment; most electron affinities were derived indirectly from other measurements. Currently, methods do exist for their direct measurement.

Just as low ionization energy is a measure of metallic behavior, a low (very negative) value of electron affinity is a characteristic of active nonmetals*.

8-10 Electronegativity

Two criteria have been presented for expressing metallic and nonmetallic tendencies: ionization energy and electron affinity. These criteria are really quite adequate; yet it is useful to have a single criterion, especially when describing the bond type that results when atoms combine. The quantity we are seeking is called electronegativity. **Electronegativity** describes the ability of an atom to compete for electrons with another atom to which it is bonded. The electronegativity is related to ionization energy (I) and electron affinity (EA), since these quantities reflect the ability of an atom to lose or gain an electron. The most widely used electronegativity scale is based on an evaluation of bond energies and was devised by Linus Pauling. The basis of this scale is explored further in Chapter 9. Pauling's electronegativities are dimensionless (no units) numbers ranging from about 1 for very active metals to 3.98 for fluorine, the most active nonmetal. Numerical values of a few electronegativities are presented in Table 8-6.

As a rough rule, metals have electronegativities less than about 2; metalloids, about equal to 2; and nonmetals, greater than 2.

---

*In some sources the opposite sign convention is used. That is, the electron affinity of active nonmetals are highly positive quantities, such as EA = +3.615 eV/atom for Cl.

8-11 Magnetic Properties

On several occasions we have noted how atoms may display magnetic properties when placed in a magnetic field (recall, for example, the Stern–Gerlach experiment, Section 7-9). One basic interaction between *paired* electrons in atoms and a magnetic field results in atoms being repelled by the field. This phenomenon is called **diamagnetism.** Although all substances have diamagnetic properties, the forces associated with diamagnetism are quite weak and easily overcome by another type of magnetic property that exists in some substances. This is the property of **paramagnetism.** A paramagnetic substance is attracted into a magnetic field.

An electron in motion, whether it be orbital motion or by virtue of its spin, can be likened to a tiny electric current. By the laws of electromagnetism, an electric current is expected to induce a magnetic field around it. In a species having only filled electronic orbitals all electrons are paired, and these individual magnetic effects cancel out. Such a species is repelled by a magnetic field (diamagnetic). If a species has *unpaired* electrons, however, the individual magnetic effects do not cancel out and the species displays paramagnetism. The more unpaired electrons in a species the stronger the attractive force it experiences when it is placed in a magnetic field. A straightforward method for measuring magnetic properties of a substance, illustrated in Figure 8-9, involves weighing the substance "in" and "out" of a magnetic field. If the substance is diamagnetic, it weighs less in the magnetic field; if paramagnetic, it weighs more.

The measurement of magnetic properties is an experimental method that can assist in establishing electron configurations of atoms and ions. For example, an atom of iron is assigned the electron configuration

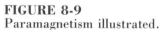

Fe [Ar] $\boxed{\uparrow\downarrow}\ \boxed{\uparrow}\ \boxed{\uparrow}\ \boxed{\uparrow}\ \boxed{\uparrow}$ $\boxed{\uparrow\downarrow}$

              3*d*   4*s*

If an iron atom loses its two 4*s* electrons to form the ion Fe^{2+}, we should expect a species with four unpaired electrons.

              3*d*

Fe^{2+} [Ar] $\boxed{\uparrow\downarrow}\ \boxed{\uparrow}\ \boxed{\uparrow}\ \boxed{\uparrow}\ \boxed{\uparrow}$

FIGURE 8-9
Paramagnetism illustrated.

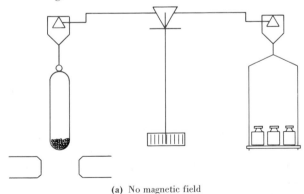
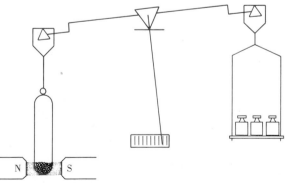

(a) No magnetic field   (b) Magnetic field turned on

(a) A sample is weighed in the absence of a magnetic field.
(b) When the field is turned on, the balanced condition is upset. The sample gains weight because it is attracted into the magnetic field.

The experimentally determined paramagnetism of solids containing Fe^{2+} does indeed correspond to four unpaired electrons. For the species Fe^{3+} the expectation, confirmed by experiment, is *five* unpaired electrons.

3d

Fe^{3+} [Ar] | ↑ | ↑ | ↑ | ↑ | ↑ |

The attraction of iron objects into a magnetic field is far stronger than can be accounted for by unpaired electrons alone. The special magnetic property possessed by iron and a few other metals and alloys is referred to as ferromagnetism (see Section 23-7).

Example 8-6 Which of the following species would you expect to be diamagnetic and which paramagnetic? **(a)** a Na atom; **(b)** a Mg atom; **(c)** a Cl^- ion; **(d)** a Ca^{2+} ion; **(e)** an Ag atom.

Solution

(a) Paramagnetic. The Na atom has a single 3s electron outside the Ne core. This electron is unpaired.

(b) Diamagnetic. The Mg atom has *two* 3s electrons outside the Ne core. They must be paired, as are all the other electrons.

(c) Diamagnetic. The Ar atom has all electrons paired ($1s^2 2s^2 2p^6 3s^2 3p^6$) and Cl^- is isoelectronic with Ar.

(d) Diamagnetic. Ca^{2+} is also isoelectronic with Ar.

(e) Paramagnetic. Even without considering the exact electron configuration of Ag, because the atom has 47 electrons—an odd number—at least one of the electrons must be unpaired.

SIMILAR EXAMPLES: Review Problem 12; Exercises 31, 32.

8-12 Using the Periodic Table to Compare Atomic Properties

Figure 8-10 summarizes the variation of atomic properties in groups and periods of the periodic table, and is used to answer some of the questions in Example 8-7.

Example 8-7 Of the following pairs of elements, which would you expect to have **(a)** the *lower* (first) ionization energy, Sr or Br? **(b)** a *more* nonmetallic character, Ga or P? **(c)** the *larger* atoms, Mg or I? **(d)** the *higher* electronegativity, Mg or I?

FIGURE 8-10
Atomic properties and the periodic table—a summary.

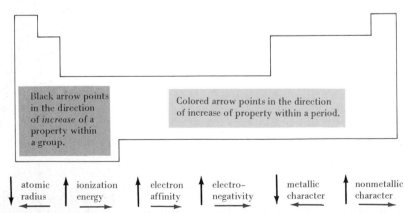

Black arrow points in the direction of *increase* of a property within a group.

Colored arrow points in the direction of increase of property within a period.

| ↓ atomic radius | ↑ ionization energy | ↑ electron affinity | ↑ electro-negativity | ↓ metallic character | ↑ nonmetallic character |

The atomic radius refers to covalent radius; ionization energy is the first ionization energy; electron affinity is based on *negative* values for active nonmetals, and the direction of increase is that in which electron affinities become more negative; metallic character refers generally to the ease of loss of electrons; and nonmetallic character to the ease of gain of electrons.

FIGURE 8-11
Comparisons of atomic properties—Example 8-7 illustrated.

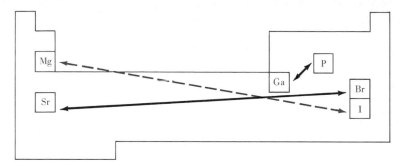

In general, qualitative comparisons of atomic properties can be made between pairs of atoms whose relative positions in the periodic table are as shown by the black arrows (from lower left to upper right or vice versa). Qualitative comparisons are sometimes more difficult (and at times impossible) in the direction of the colored broken arrow (from upper left to lower right or vice versa).

Solution

(a) Reference to Figure 8-10 reminds us that ionization energy *increases* from bottom to top in a group and from left to right in a period. Ionization energy *decreases* in the opposite direction, i.e., toward the *bottom* of a group and the *left* end of a period, a position where Sr is found. Sr has a lower ionization energy than Br.

(b) The direction of increasing nonmetallic character is toward the *top* of a group and the *right* end of a period, a condition met more closely by P than by Ga. P has more nonmetallic character than Ga.

(c) Here we have a problem! I has a greater number of electronic shells than does Mg (is farther down its group), suggesting that I is the larger atom. On the other hand, Mg is toward the extreme left of its period and I, the extreme right; this suggests that an Mg atom might be larger than an I atom. The question cannot be answered with Figure 8-10 alone. (Figure 8-5 indicates that the Mg atom is very slightly larger than I.) As suggested by Figure 8-11, certain comparisons cannot be made with the qualitative generalizations of this chapter.

(d) Although a comparison between Mg and I was not possible in part (c), it is here. Mg is a metal and I, a nonmetal (recall Figure 8-3). In general, electronegativities (EN) are greater for nonmetals (EN > 2) than for metals (EN < 2). I has a higher electronegativity than Mg.

SIMILAR EXAMPLES: Review Problems 6, 10; Exercises 17, 18, 19.

Summary

The periodic law is the historical basis of the tabular arrangement of the elements known as the periodic table. Mendeleev produced the first successful periodic table and used it to correct atomic weights and to predict the properties of undiscovered elements. Several modifications have been made to Mendeleev's original table, resulting in the currently used long form of the periodic table. Although the groupings of similar elements in Mendeleev's original table were based on *experimental* observations, a *theoretical* basis has also been established: Characteristic properties of the elements stem from their electron configurations. In the periodic table elements with similar electron configurations are grouped together.

To understand the physical and chemical properties of the elements, it is helpful to establish four categories: metals, nonmetals, metalloids, and noble gases. To compare the relative tendencies of atoms to behave as metals or nonmetals, it is necessary to study certain properties of the individual atoms—atomic properties. The atomic properties introduced in this chapter are atomic radius, ionization energy, electron affinity, and electronegativity. Magnetic properties of atoms and ions are helpful in establishing electron configurations.

Learning Objectives

As a result of studying Chapter 8, you should be able to

1. Illustrate the periodic law with graphs of selected properties of the elements as a function of atomic number.

2. Use the terms *periods, groups, families, representative elements,* and *transition elements* to describe individual elements and groupings of elements in the periodic table.

3. Use the periodic table as a guide to describe the filling of electron orbitals (Aufbau process).

4. Cite the basic features of the electron configurations of representative, transition, and inner transition elements.

5. Distinguish among metals, nonmetals, metalloids, and noble gases, and locate them in the periodic table.

6. State the factors that affect atomic size; distinguish among these terms: covalent, ionic, metallic, and van der Waals radius; and describe the relationship between atomic size and location of an element in the periodic table.

7. Use the properties of ionization energy, electron affinity, and electronegativity to evaluate the metallic/nonmetallic character of elements.

8. Relate the magnetic properties (diamagnetic or paramagnetic) of an atom or ion to its electron configuration.

Some New Terms

The **actinoids (actinides)** are a series of elements ($Z = 90$ to 103) characterized by partially filled $5f$ orbitals in their atoms. All the actinoid elements are radioactive.

Covalent radius is one half the distance between the centers of two atoms that are bonded together covalently. It is the atomic radius associated with an element in its covalent compounds.

Diamagnetism refers to the repulsion by a magnetic field of a species in which all electrons are paired.

Effective nuclear charge is the positive charge acting on a particular electron in an atom. Its value is the charge on the nucleus, reduced to the extent that other electrons screen the electron in question from the nucleus.

Electron affinity is the energy associated with the gain of an electron by a *gaseous* atom.

Electronegativity is a measure of the electron attracting power of an atom; metals have low electronegativities, nonmetals, high.

A **family** of elements is a numbered group from the periodic table, usually carrying a distinctive name: for example, group VIIA, the halogen family.

A **group** is a vertical column of elements in the periodic table; members of the group have similar properties.

Ionic radius is the radius of a spherical ion. It is the atomic radius associated with an element in its ionic compounds.

Ionization energy (ionization potential), I, is the energy required to remove the most loosely held electron from a *gaseous* atom.

Isoelectronic species have the same number of electrons (usually in the same configuration). Na^+ and Ne are isoelectronic.

The **lanthanoids (lanthanides)** are the series of elements ($Z = 58$ to 71) characterized by partially filled $4f$ orbitals in their atoms.

Metallic radius is one-half the distance between the centers of adjacent atoms in a metallic solid.

A **metalloid** is an element that may display both metallic and nonmetallic properties under the appropriate conditions (e.g., Si, Ge, As).

Metals are elements whose atoms have small numbers of electrons in the electronic shell of highest principal quantum number. Removal of an electron(s) from a metal atom occurs without great difficulty, producing a cation.

Noble gases are elements whose atoms have the electron configuration ns^2np^6 in the electronic shell of highest principal quantum number. (The noble gas He has the configuration, $1s^2$.)

Nonmetals are elements whose atoms tend to gain small numbers of electrons to form anions with the electron configuration of a noble gas.

Paramagnetism refers to the attraction of a magnetic field for a species containing unpaired electrons.

A **period** is a horizontal row of the periodic table. All members of a period have atoms with the same highest principal quantum number.

The **periodic law** refers to the periodic recurrence of certain physical and chemical properties when the elements are considered in terms of increasing atomic number.

The **periodic table** is an arrangement of the elements in which elements with similar physical and chemical properties are grouped together.

Representative elements are those whose atoms feature the filling of s or p orbitals of the electronic shell of highest principal quantum number.

Rest mass is the mass of a particle (e.g., an electron) when it is at rest. As its velocity approaches the speed of light, the mass of a particle increases.

Transition elements are those whose atoms feature the filling of d or f orbitals of an inner electronic shell.

Suggestions for Further Study

EDWARDS, P. P., and M. J. SIENKO, "On the Occurrence of Metallic Character in the Periodic Table of the Elements," *J. Chem. Educ.,* **60,** 691 (1983).

FERNELIUS, W. C., and W. H. POWELL, "Confusion in the Periodic Table of the Elements," *J. Chem. Educ.,* **59,** 504 (1982).

FIRSCHUNG, F. H., "Anomalies in the Periodic Table," *J. Chem. Educ.,* **58,** 478 (1981).

GRAHAM, L. R., "Textbook Writing and Scientific Creativity: The Case of Mendeleev," *National Forum* (Journal of Phi Kappa Phi), **63,** 22 (Winter 1983).

KELLER, O. L., Jr., "Predicted Properties of Elements 113 and 114," *Chemistry,* **43** [10], 8 (1970).

LOENING, K. L., Chairman, American Chemical Society Committee on Nomenclature, "Recommended Format for the Periodic Table of the Elements," *J. Chem. Educ.,* **61,** 136 (1984).

SANDERSON, R. T., "Ionization Energy and Atomic Structure," *Chemistry,* **46** [5], 12 (1973).

Review Problems

1. With reference to the periodic table, identify
(a) an element that is both in group IIIA and in the fifth period;
(b) an element similar to, and one unlike, sulfur;
(c) a highly reactive metal in the sixth period;
(d) the halogen element in the fifth period;
(e) an element with atomic number greater than 50 that is similar to the element with atomic number 18.

2. In what group of the periodic table are elements with the following electron configurations found?
(a) $1s^2 2s^2 2p^6 3s^2 3p^5$ (b) $[Ar]3d^{10}4s^2 4p^2$
(c) $1s^2 2s^2 2p^6 3s^2$ (d) $[Ar]3d^{10}4s^1$
(e) $[Xe]4f^{14}5d^4 6s^2$

3. Use the periodic table as a guide to write electron configurations for (a) In; (b) Y; (c) Sb; (d) Au.

4. Write electron configurations of the following ions: (a) Rb^+; (b) Br^-; (c) O^{2-}; (d) Ba^{2+}; (e) Zn^{2+}; (f) Ag^+; (g) Bi^{3+}.

5. Use Figure 8-2 as a guide to indicate the number of (a) $4s$ electrons in a K atom; (b) $5p$ electrons in an I atom; (c) $3d$ electrons in an atom of Zn; (d) $2p$ electrons in an atom of S; (e) $4f$ electrons in an atom of Pb; (f) $3d$ electrons in an atom of Ni.

6. For each of the following pairs, indicate the atom that has the *larger* size: (a) Br or As; (b) Sr or Mg; (c) Ca or Cs; (d) Ne or Xe; (e) C or O; (f) Hg or Cl.

7. Indicate the *smallest* and the *largest* species (atom or ion) in the following group: an Al atom, an Ar atom, an As atom, a Cs^+ ion, an F atom, an I^- ion, an N atom.

8. Use principles established in this chapter to arrange the following atoms in order of *increasing* value of the first ionization energy: Sr, Cs, S, F, As.

9. How such energy, in joules, must be absorbed to produce 1.86×10^{-6} mol of Mg^{2+} ions from gaseous Mg atoms?

10. Without referring to tables or figures in the text, indicate which of the atoms Bi, S, Ba, As, and Mg (a) is most metallic; (b) is most nonmetallic; (c) has the intermediate value when the five are arranged in order of increasing electronegativity.

11. Arrange the following elements in order of *decreasing* metallic character: Sc; Fe; Rb; Br; O; Ca; F; Te.

12. Which of the following species would you expect to be diamagnetic and which paramagnetic: K^+; Cr^{3+}; Zn^{2+}; Cd; Co^{3+}; Sn^{2+}; Br?

Exercises

The periodic law

1. The element francium occurs naturally in uranium ores, but it is estimated that there is less than 30 g present in the earth's crust at any one time. Because it is extremely rare, very little is known of its properties. Use Figure 8-1 and equation (8.1) to estimate its density.

2. Use data from Figure 8-1 and equation (8.1) to estimate the density that will be exhibited by element 114, if and when it is discovered. Assume a mass number of 298. What reason(s) can you think of why this estimate might be inaccurate?

3. The following melting points are in °C. Show that melting point is a periodic property of these elements: aluminum, 660; argon, −189; beryllium, 1278; boron, 2300; carbon, 3350; chlorine, −101; fluorine, −220; lithium, 179; magnesium, 651; neon, −249; nitrogen, −210; oxygen, −218; phosphorus, 590; silicon, 1410; sodium, 98; sulfur, 119.

The periodic table

4. There is every indication that the periodic table can be extended to elements of higher atomic numbers. What are the prospects of finding new elements within the existing periodic table at lower atomic numbers?

5. Assuming that the seventh period of elements is 32 members long, what would be the atomic number of the noble gas following radon (Rn)? of the alkali metal following francium (Fr)? What would you expect their approximate atomic weights to be?

6. Find the several pairs of elements that are "out of order" in the periodic table in terms of increasing atomic weights and explain why it is necessary to arrange them in inverse order by atomic weight.

7. Use Mendeleev's periodic table (Table 8-1) to predict formulas of the compounds listed below. Compare your results with the formulas you would have expected based on ideas introduced in Chapter 3 and explain any discrepancies.
   (a) the oxide of aluminum   **(b)** the oxide of sulfur
   (c) the chloride of silicon
   (d) the chloride of phosphorus   **(e)** the oxide of iron

Periodic table and electron configurations

8. Explain why the several periods in the periodic table do not all have the same number of members.

9. Sketch a periodic table that would permit inclusion of all members of each period into the main body of the table. How many "members" wide would the table be? Explain.

10. In Figure 8-2 no designation is made about three of the elements in the seventh period. **(a)** What are these three elements? **(b)** What distinctive features would you expect for their electron configurations?

11. Based on the relationship between electron configurations and the periodic table, give the number of **(a)** outershell electrons in an atom of Sb; **(b)** shells of electrons in an atom of Pt; **(c)** elements whose atoms have six outer-shell electrons; **(d)** unpaired electrons in an atom of Te; **(e)** transition elements in the sixth period of elements.

12. In Example 8-2 we concluded that the known element that the unknown and undiscovered element 114 would most closely resemble is Pb.
   (a) Write the electron configuration of Pb.
   (b) Propose a plausible electron configuration for element 114.

13. Write probable electron configurations for the following ions: **(a)** Sr^{2+}; **(b)** Y^{3+}; **(c)** Se^{2-}; **(d)** Cu^{2+}; **(e)** Ni^{2+}; **(f)** Ga^{3+}; **(g)** Ti^{2+}.

Atomic sizes

14. Explain why the sizes of atoms do not simply increase uniformly with increasing atomic number.

15. **(a)** Which is the smallest atom in group IIIA?
   (b) Which is the smallest of the following atoms: Te, In, Sr, Po, Sb?

16. How would you expect the sizes of the hydrogen ion, H^+, and the hydride ion, H^-, to compare with that of the He atom? Explain.

17. The following species are isoelectronic with the noble gas krypton. Arrange them in order of increasing size and comment on the principles involved in doing so: Rb^+; Y^{3+}; Br^-; Kr; Sr^{2+}; Se^{2-}.

18. Arrange the following species in expected order of increasing size: Y; Li^+; Se; Br^-.

***19.** Without reference to figures and tables in the text, arrange the following atomic and ionic species in the expected order of increasing size: B, Br, Br^-, Cl, Li^+, P, Rb, Y. Explain the basis of any uncertainties in your arrangement. Now use data from the text to establish the actual order of increasing size.

***20.** Refer to the caption of Figure 8-1 and explain why the volume (and hence the radius) of an individual atom cannot be determined from the experimentally measured molar volume.

Ionization energies, electron affinities

21. Are there any atoms for which the second ionization energy (I_2) is smaller than the first (I_1)? Explain.

22. Although the first ionization energy (I_1) of Na is smaller than for Mg, the second ionization energy (I_2) of Na is much greater than for Mg. Why is this so?

23. The ion Na^+ and the atom Ne are isoelectronic. The ease of loss of an electron by a gaseous Ne atom is measured by I_1 and has a value of 2081 kJ/mol. The ease of loss of an electron from a gaseous Na^+ ion is measured by I_2 for sodium and has a value of 4562 kJ/mol. Why are these values not the same?

24. How much energy, in eV, must be absorbed to ionize completely all the third-shell electrons in a gaseous phosphorus atom?

25. What is the maximum number of Cs^+ ions that can be produced per joule of energy absorbed by a sample of gaseous Cs atoms?

26. How much energy is involved when 1.00 mg of chlorine, in the form of Cl atoms, is converted completely to Cl^- ions in the gaseous state?

27. Use data from the energy level diagram in Figure 7-13 to determine the ionization energy of hydrogen, in eV/atom.

***28.** Without reference to figures or tables in the text, arrange the following ionization energies in order of *increasing* value. Explain the basis of any uncertainties in your arrangement. I_1 for F; I_2 for Ba; I_3 for Sc; I_2 for Na; I_3 for Mg

Electronegativities, metals and nonmetals

29. Plot the data of Table 8-6 as a function of atomic number and determine whether the property of electronegativity conforms to the periodic law. Do you think that it should?

30. Table 8-6 lists approximate electronegativity values for metals, nonmetals, and metalloids. Use relevant data from tables and figures in this chapter to establish rough ranges of first ionization energies for metals, nonmetals, and metalloids. Is the first ionization energy a satisfactory criterion for metallic/nonmetallic behavior? Explain.

Magnetic properties

31. Unpaired electrons are found in only one of the following species. Indicate which is that one and explain why: F^-; Ca^{2+}; Fe^{2+}; S^{2-}.

32. Write electron configurations consistent with the following data on number of unpaired electrons: V^{3+}, two; Cu^{2+}, one; Cr^{3+}, three.

33. Must all atoms with an odd atomic number be paramagnetic? Must all atoms with an even atomic number be diamagnetic? Explain.

* **34.** The number of unpaired electrons in an atom does not show the same periodicity with atomic number as do such properties as atomic radius and ionization energy. Give a reason(s) for this

difference. (*Hint:* You may find it useful to refer to the data in Appendix E.)

Predictions based on periodic relationships

35. In 1829, Dobereiner found that when certain similar elements are arranged in groups of three the atomic weight of the middle member of the group is roughly the average of the other two. Explain why Dobereiner's method works in the first two and fails in the other three cases for determining the atomic weight of **(a)** Na from those of Li and K; **(b)** Br from those of Cl and I; **(c)** Si from those of C and Ge; **(d)** Sb from those of As and Bi; **(e)** Ga from those of B and Tl.

36. Estimate the missing boiling point in the following series of compounds.
    (a) CH_4, $-164°C$; SiH_4, $-112°C$; GeH_4, $-90°C$; SnH_4, ?.
    (b) H_2O, ?; H_2S, $-61°C$; H_2Se, $-41°C$; H_2Te, $-2°C$.
Does your estimate of the boiling point of water, based on the data given here, agree with the known value? (An explanation is presented in Chapter 11.)

37. The element gallium (eka-aluminum) was unknown in Mendeleev's time, and he predicted properties of this element much as he did for eka-silicon. Predict the following for gallium: **(a)** its density; **(b)** the formula and percent composition of its oxide. [*Hint:* Use Figure 8-1, equation (8.1), and Table 8-1.]

Additional Exercises

1. The density of tellurium is 6.24 g/cm^3. Estimate its atomic weight using Figure 8-1 and equation (8.1).

2. Verify the statements on page 198 concerning Mendeleev's correction of the atomic weight of indium. That is, show that if indium oxide is assumed to have the formula, InO, its atomic weight must be 76; and if In_2O_3, 113. (Recall that the oxide is 82.5% In, by mass.)

* **3.** Studies conducted in 1880 showed that a chloride of uranium was 37.34% Cl, by mass, and had an approximate formula weight of 382. Uranium has a specific heat of 0.0276 cal g^{-1} $°C^{-1}$. Use these data, together with the law of Dulong and Petit given in the Additional Exercises of Chapter 6, to calculate the atomic weight of uranium and compare it with the value assigned by Mendeleev.

4. On the basis of the periodic table and rules for electron configurations, indicate the number of **(a)** $2p$ electrons in an atom of N; **(b)** $4s$ electrons in an atom of Rb; **(c)** $4d$ electrons in an atom

of As; **(d)** $4f$ electrons in an atom of Au; **(e)** unpaired electrons in an atom of Pb; **(f)** elements in group IVA of the periodic table; **(g)** elements in the sixth period of the periodic table.

5. With reference to the periodic table, indicate **(a)** the most active nonmetal; **(b)** the transition metal with lowest atomic number; **(c)** a metalloid whose atomic number places it exactly midway between two noble gas elements.

6. With reference to the periodic table, explain why
    (a) All nonmetals are representative elements, whereas among metals, some are representative and some are transition elements.
    (b) Metalloids are found only among the representative elements and not among the transition elements.

7. Use principles established in the text, but without reference to tables or figures, to arrange the following atoms in terms of
    (a) increasing first ionization energies: O; Rb; Br; Ca; Sc; Se; F; Cs; He.
    (b) decreasing metallic character: I; O; Cs; K; Te; F; Mg; Al.

8. Listed below are several atomic properties of the element germanium. With reference only to the periodic table, indicate probable values for each of the following elements either as greater than, about equal to, or less than the value for Ge.

	Covalent radius	First ionization energy	Electronegativity
Ge	122 pm	7.90 eV/atom	2.01
Al	?	?	?
In	?	?	?
Se	?	?	?

9. Which of the following species has the greatest number of unpaired electrons: **(a)** Ge; **(b)** Cl; **(c)** Cr^{3+}; **(d)** Br^-? Explain.

10. For the following groups of elements select the one that has the property noted.
   (a) The largest atom: H; Ar; Ag; Ba; Te; Au.
   (b) The lowest first ionization energy: B; Sr; Al; Br; Mg; Pb.
   (c) The greatest electron affinity: Na; I; Ba; Se; Cl; P.
   (d) The highest electronegativity: As; Ca; I; P; Ga; Se; Sn.
   (e) The largest number of unpaired electrons: F; N; S^{2-}; Mg^{2+}; Sc^{3+}; Ti^{3+}.

Self-Test Questions

For questions 1 through 8 select the single item that best completes each statement.

1. The element whose atoms have the electron configuration $[Kr]4d^{10}5s^25p^3$ **(a)** is in group IIIA of the periodic table; **(b)** bears a similarity to the element Bi; **(c)** is similar to the element Te; **(d)** is a transition element.

2. An atom of As has **(a)** 5 electrons in the $4p$ subshell; **(b)** 10 electrons in the $4d$ subshell; **(c)** 6 electrons in the $3p$ subshell; **(d)** 3 electrons in the $4s$ subshell.

3. The largest of the following species is **(a)** an Ar atom; **(b)** a K^+ ion; **(c)** a Ca^{2+} ion; **(d)** a Cl^- ion.

4. The *highest* (first) ionization energy of the following elements is that of **(a)** Cs; **(b)** Cl; **(c)** I; **(d)** Li.

5. The *greatest* electronegativity of the following elements is that of **(a)** Br; **(b)** Sn; **(c)** Ba; **(d)** Li.

6. The greatest quantity of energy is *released* when one electron is gained by an atom of **(a)** Sr; **(b)** Cs; **(c)** Si; **(d)** S.

7. The *most* metallic of the following elements is **(a)** Mg; **(b)** Li; **(c)** K; **(d)** Ca.

11. Each of the following formulas represents an actual compound, but one formula is inconsistent with the others given (i.e., not predictable from the others). Which is the one and what is the inconsistency? SiO_2; CO; H_2S; CCl_4; HCl; H_2O.

12. Use ideas presented in this chapter to indicate
   (a) three metals that you would expect to exhibit the photoelectric effect with visible light, and three that you would expect not to;
   (b) the noble gas element that should have the highest density when in the liquid state;
   (c) the approximate first ionization energy of fermium (Z = 100);
   (d) the approximate melting point of francium (Z = 87);
   (e) the approximate density of solid radium (Z = 88);
   (f) the approximate electronegativity of polonium (Z = 84).

*13. Refer to Exercise 18 in Chapter 7 and calculate the *second* ionization energy (I_2) for the helium atom. Compare your result with the tabulated value of 54.416 eV/atom.

*14. Various values are given for the radius of a hydrogen atom, including 37, 53, 120, and 210 pm. Why do you suppose there is such variability in the values given?

*15. Use values of basic physical constants and other data from the appendices to establish the relationship between eV/atom and kJ/mol in equation (8.4).

8. The number of *unpaired* electrons in an atom of Br is **(a)** 0; **(b)** 1; **(c)** 2; **(d)** 5.

9. For the atom $^{79}_{34}$Se indicate the number of **(a)** protons in the nucleus; **(b)** neutrons in the nucleus; **(c)** electrons in the third principal electronic shell; **(d)** electrons in the $2s$ orbital; **(e)** $4p$ electrons; **(f)** electrons in the shell of highest principal quantum number.

10. Which of the following would you expect to have the highest electronegativity: Pb; Sn; Br; As; Al? Explain.

11. Give the symbol of the element **(a)** in group IVA that has the smallest atoms; **(b)** in the fifth period that has the largest atoms; **(c)** in group VIIA that has the lowest electronegativity.

12. With reference to the periodic table, briefly explain why
   (a) There are 2 elements in the first period, 8 in the third, 18 in the fifth, and 32 in the seventh.
   (b) Argon (Ar), with an atomic weight of 39.948, is placed ahead of potassium (K), with an atomic weight of 39.102.

9 Chemical Bonding I: Basic Concepts

While the atomic theory was being developed various ideas were also entertained about the *combinations* of atoms that lead to chemical compounds. In compounds, atoms are held together by forces known as chemical bonds. Electrons play a key role in chemical bonding.

In Chapter 7 we proposed that within individual atoms regions exist where the probability of finding electrons is high. A logical extension of this view is to consider that within *combinations* of atoms there also exist regions in which electron probabilities are high. That is, electrons in atoms are described by atomic orbitals and in molecules by molecular orbitals. In this chapter, however, we will discuss chemical bonding in terms of simpler concepts. In Chapter 10 we will take a second look at some of these concepts from a more theoretical point of view, and we will also consider the molecular orbital approach.

9-1 Importance of Electrons in Chemical Bonding

It is interesting that one of the first real clues about chemical bonding came from a group of elements that show little tendency to form chemical compounds at all. These are the noble (inert) gases, whose discovery was described in Chapter 8.*

In 1916 several proposals about chemical bonding were made by two American chemists, Lewis and Langmuir, and a German, Kossel. Their basic thought was that if the inert gases do not combine with other elements, something unique about their electron configurations may prevent their doing so. And perhaps atoms that do unite experience changes in their electron configurations to make them more like inert gases. The theory developed around this model is referred to as the Lewis theory. It is based on these essential propositions.

1. Electrons, especially those of the outermost (valence) electronic shell, play a fundamental role in chemical bonding.
2. In some cases chemical bonding results from the *transfer* of one or more electrons from one atom to another. This leads to the formation of positive and negative ions and a bond type known as **ionic.**

---

*Belief in the uniqueness of noble gas electron configurations tended to discourage attempts to synthesize noble gas compounds. Since 1962, a number of compounds of Kr, Xe, and Rn have been prepared, but this fact does not alter the applicability of the ideas presented in this chapter.

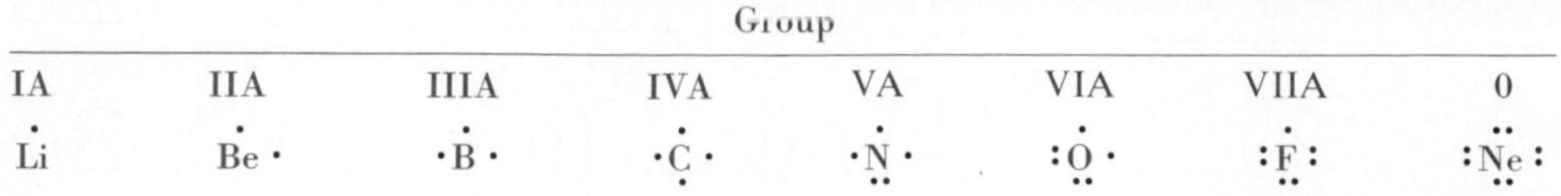

FIGURE 9-1
Lewis symbols of the
second row elements.

			Group				
IA	IIA	IIIA	IVA	VA	VIA	VIIA	0
Li	Be ·	·B ·	·C ·	·N ·	:O ·	:F :	:Ne :

3. In other cases chemical bonding results from a mutual *sharing* of pairs of electrons between atoms. This leads to molecules having a bond type called **covalent.**

4. The transfer or sharing of electrons occurs to the extent that each atom involved acquires an especially stable electron configuration. Often this configuration is that of a noble (inert) gas, that is, involving eight outer-shell electrons, an **octet.**

Lewis Symbols.   The Lewis symbol of an element consists of the common chemical symbol surrounded by a number of dots. The chemical symbol represents the kernel of the atom, consisting of the nucleus and *inner-shell* electrons. The dots represent the *outer-shell* or *valence* electrons. Thus, for silicon we can write the orbital diagram

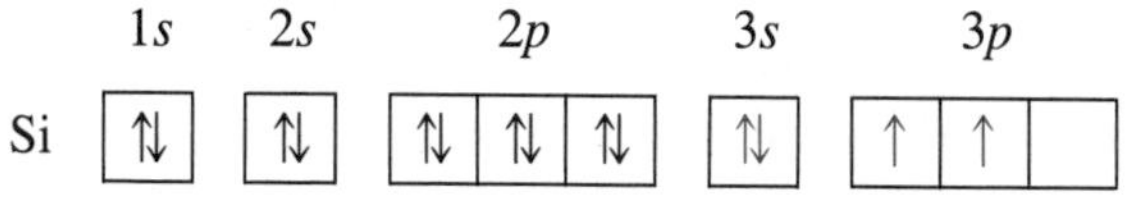

	1s	2s	2p			3s	3p		
Si	↑↓	↑↓	↑↓	↑↓	↑↓	↑↓	↑	↑	

A literal translation of this orbital diagram into a Lewis symbol produces

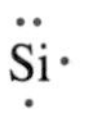

$$\overset{\cdot\cdot}{\underset{\cdot}{Si}}\cdot$$

where the pair of dots represents the pair of 3s electrons and the two lone dots represent the two unpaired 3p electrons. However, a representation that is more faithful to Lewis's original theory is

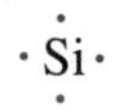

$$\cdot\overset{\cdot}{\underset{\cdot}{Si}}\cdot$$

The concept of electron spin had not been discovered at the time of Lewis's theory, and so there was no basis for thinking of electrons in the valence shell as existing in pairs. (In fact, in his original theory Lewis pictured the valence electrons as occupying the corners of a cubical atom.) The representation of Lewis symbols that we will find most useful is to place single dots on each side of the chemical symbol, up to a maximum of four, and then to pair up the dots until an octet is reached. This is the manner of writing Lewis symbols that is illustrated in Figure 9-1. This figure also illustrates that the number of outer-shell (valence) electrons equals the A group number.

Example 9-1   Refer to Figure 9-1 and write Lewis symbols for the following groups of elements: **(a)** N, P, As, Sb, Bi; **(b)** Ca, Ge, I.

Solution

(a) These are the elements of group VA of the periodic table. Their atoms all have five valence electrons (ns^2np^3). The Lewis symbols feature five dots.

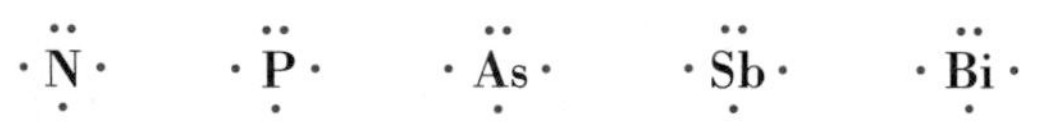

$$\cdot\overset{\cdot\cdot}{\underset{\cdot}{N}}\cdot \qquad \cdot\overset{\cdot\cdot}{\underset{\cdot}{P}}\cdot \qquad \cdot\overset{\cdot\cdot}{\underset{\cdot}{As}}\cdot \qquad \cdot\overset{\cdot\cdot}{\underset{\cdot}{Sb}}\cdot \qquad \cdot\overset{\cdot\cdot}{\underset{\cdot}{Bi}}\cdot$$

(b) Ca is in group IIA, Ge is in group IVA, and I is in group VIIA.

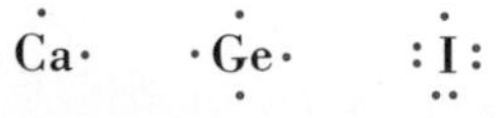

$$Ca\cdot \qquad \cdot\overset{\cdot}{Ge}\cdot \qquad :\overset{\cdot\cdot}{\underset{\cdot\cdot}{I}}:$$

SIMILAR EXAMPLE: Review Problem 1.

FIGURE 9-2
Formation of an ionic bond.

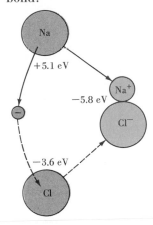

The processes depicted here have already been described in part in Chapter 8. The loss of an outer-shell electron by a sodium atom is accompanied by a decrease in size. There is essentially no change in size when an *isolated, nonbonded* chlorine atom gains an electron. The energy data refer to the loss of an electron by Na ($+5.1$ eV), the gain of an electron by Cl (-3.6 eV), and the energy of interaction of the two ions (-5.8 eV). The net energy change is -4.3 eV. The energy required to produce gaseous Na from solid Na and Cl atoms from Cl_2 molecules is not described here, but it is considerably less than 4.3 eV. A more complete analysis of the energetics of ionic crystal formation is provided in Section 11-12.

Lewis Structures.   Although Lewis's work dealt primarily with covalent bonding, we will use his ideas to depict both ionic and covalent bonds. A **Lewis structure** is a combination of Lewis symbols representing the transfer or sharing of electrons in a chemical bond.

$$\text{ionic bonding:} \qquad \text{Na}_\times \; + \; \cdot \overset{\cdot\cdot}{\underset{\cdot\cdot}{\text{Cl}}}: \;\longrightarrow\; [\text{Na}]^+ [\,\overset{\cdot\cdot}{\underset{\cdot\cdot}{\times}\text{Cl}}:\,]^- \qquad\qquad (9.1)$$

$$\underset{\text{Lewis symbols}}{} \qquad\qquad \underset{\text{Lewis structure}}{}$$

$$\text{covalent bonding:} \quad \text{H}_\times \; + \; \cdot \overset{\cdot\cdot}{\underset{\cdot\cdot}{\text{Cl}}}: \;\longrightarrow\; \text{H}\overset{\cdot\cdot}{\underset{\cdot\cdot}{\times}\text{Cl}}: \qquad\qquad (9.2)$$

$$\underset{\text{Lewis symbols}}{} \qquad\qquad \underset{\text{Lewis structure}}{}$$

In these two examples the electrons from one atom are denoted as $\times$ and from the other atom, as $\cdot$. However, it is impossible to distinguish among the electrons in bonded atoms. In all subsequent Lewis structures we will use only the dot symbol $\cdot$. Lewis structures are discussed more fully in the sections that follow.

FIGURE 9-3
Formation of an ionic crystal.

Each Na^+ ion (small sphere) is surrounded by six Cl^- ions (large spheres); in turn, each Cl^- is surrounded by six Na^+. An enormous number of ions cluster together into a crystalline solid.

9-2 Ionic Bonding

Figure 9-2 suggests the formation of a pair of ions, Na^+ and Cl^-, from the free, isolated atoms. Energy is required to remove the outer-shell ($3s^1$) electron from an Na atom—the first ionization energy, $I_1 = +5.1$ eV. Energy is released when a Cl atom absorbs an electron into its outer shell—the electron affinity, EA $= -3.6$ eV. Once formed, the ions attract each other and move into close proximity. This process is exothermic, with an energy change of -5.8 eV. The overall process is feasible energetically; the net energy change is $5.1 - 3.6 - 5.8 = -4.3$ eV/atom $= -415$ kJ/mol.

The ion pair $(\text{Na}^+)(\text{Cl}^-)$ pictured in Figure 9-2 exerts attractive forces on additional ion pairs. The result of such attractive forces is the clustering of *large* numbers of Na^+ and Cl^- ions into a solid crystal. (The additional energy change associated with the clustering of ion pairs into a crystal is about -2.4 eV per ion pair or -232 kJ/mol.) The formation of an ionic crystal is an integral part of the total process of ionic bonding. A portion of an ionic crystal of NaCl is pictured in Figure 9-3. The energetics of ionic compound formation and the structures of ionic crystals are discussed further in Chapter 11. For the present, let us simply note the following facts about ionic bonds.

1. An ionic bond results after the transfer of electrons between a *metal* and a *nonmetal* atom. In this transfer the metal atom becomes a positively charged ion (cation) and the nonmetal, a negatively charged ion (anion).

TABLE 9-1
Electron configurations for some metal ions

"Octet"		"18"		"18 + 2"	"Various"	
Na^+	Mg^{2+}	Cu^+	Zn^{2+}	In^+	Cr^{2+}:	$[Ar]3d^4$
K^+	Ca^{2+}	Ag^+	Cd^{2+}	Tl^+	Cr^{3+}:	$[Ar]3d^3$
Rb^+	Sr^{2+}	Au^+	Hg^{2+}	Sn^{2+}	Mn^{2+}:	$[Ar]3d^5$
Cs^+	Ba^{2+}		Ga^{3+}	Pb^{2+}	Mn^{3+}:	$[Ar]3d^4$
Fr^+	Ra^{2+}		In^{3+}	Sb^{3+}	Fe^{2+}:	$[Ar]3d^6$
	Al^{3+}		Tl^{3+}	Bi^{3+}	Fe^{3+}:	$[Ar]3d^5$
	Sc^{3+}				Co^{2+}:	$[Ar]3d^7$
	Y^{3+}				Co^{3+}:	$[Ar]3d^6$
	La^{3+}				Ni^{2+}:	$[Ar]3d^8$
					Ni^{3+}:	$[Ar]3d^7$

The octet configuration is that of a noble gas—ns^2np^6. Li^+ and Be^{2+} also have a noble gas electron configuration but not an octet of outer-shell electrons. Theirs is the configuration of helium: $1s^2$.

In the configuration labeled "18" all outer-shell electrons of the atom are ionized. The ion produced has a new outer shell with 18 electrons in the configuration $ns^2np^6nd^{10}$. For a few post-transition elements all outer-shell electrons but the two s electrons (an "inert pair") are ionized, producing ions with the "18 + 2" configuration, $(n-1)s^2(n-1)p^6(n-1)d^{10}ns^2$.

In ion formation among the transition metals, the outer-shell s electron(s) and some number of d electrons are ionized, producing the configurations listed as "various."

2. The nonmetal atom gains a sufficient number of electrons to produce an anion with a noble gas electron configuration. Several different types of electron configurations are found for metal ions; some are listed in Table 9-1.

3. Except in the gaseous state, ionic compounds are not composed of simple ion pairs or small clusters of ions. In the solid state each ion surrounds itself with ions of the opposite charge, producing an orderly array called a crystal.

4. A formula unit of an ionic compound is the smallest collection of ions that is electrically neutral. The formula unit is obtained automatically when the Lewis structure is written.

Example 9-2   Write Lewis structures for the following ionic compounds: **(a)** BaO; **(b)** $MgCl_2$; **(c)** potassium sulfide.

Solution

(a) Write the Lewis symbol and determine the number of electrons to be lost or gained by each atom in acquiring a noble gas electron configuration. The barium atom must lose two electrons and the oxygen atom must gain two.

$$Ba \quad + \quad \overset{\cdot}{\underset{\cdot\cdot}{:O:}} \quad \longrightarrow \quad [Ba]^{2+}[:\overset{\cdot\cdot}{\underset{\cdot\cdot}{O}}:]^{2-}$$

Lewis structure

(b) A Cl atom can accept but a single electron, whereas a Mg atom must lose two electrons. Two Cl atoms are required for each Mg atom.

$$Mg \quad + \quad \begin{matrix} \overset{\cdot\cdot}{\underset{\cdot\cdot}{:Cl:}} \\ \\ \overset{\cdot\cdot}{\underset{\cdot\cdot}{:Cl:}} \end{matrix} \quad \longrightarrow \quad [:\overset{\cdot\cdot}{\underset{\cdot\cdot}{Cl}}:]^-[Mg]^{2+}[:\overset{\cdot\cdot}{\underset{\cdot\cdot}{Cl}}:]^-$$

Lewis structure

(c) We are not given the formula of potassium sulfide, but as noted earlier, the formula follows directly from the principles of writing Lewis structures.

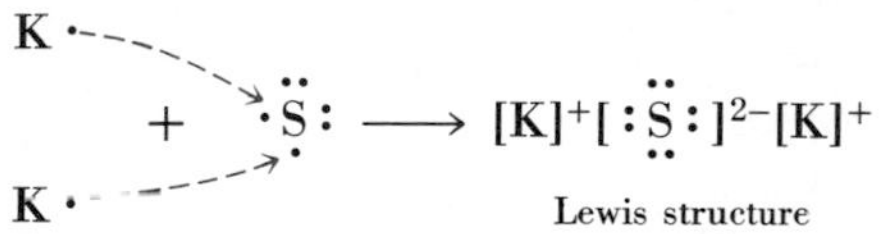

$$K \cdot \underset{}{\overset{}{\cdots}}$$
$$+ \quad \cdot \overset{\cdot\cdot}{\underset{\cdot\cdot}{S}} : \quad \longrightarrow \quad [K]^+ [: \overset{\cdot\cdot}{\underset{\cdot\cdot}{S}} :]^{2-} [K]^+$$
$$K \cdot \underset{}{\overset{}{\cdots}}$$

Lewis structure

To distinguish clearly between ionic and covalent compounds, Lewis structures for ionic compounds should show that positive and negative ions are present, as illustrated here.

SIMILAR EXAMPLES: Review Problem 2; Exercise 3.

9-3 Covalent Bonding

In the Lewis structures for NaCl and HCl, (9.1) and (9.2), the Cl atom acquires the electron configuration of a noble gas atom. The tendency for the Cl atom to gain an electron is equally strong in either case, but from what atom, Na or H, can an electron be extracted most readily? To begin, we note that neither an Na atom nor an H atom will give up an electron freely. However, the energy necessary to extract the valence electron from Na (I_1) is much smaller than from H—5.14 eV/atom compared to 13.6 eV/atom. *Sodium is much more metallic than hydrogen.* In fact, hydrogen is a *nonmetal* under normal conditions; it does not give up an electron to another nonmetal. Bonding between an H and a Cl atom involves the *sharing* of electrons, leading to a **covalent bond.**

We need to explore more fully the notion that a covalent bond involves the sharing of electrons between atoms. An in-depth treatment of the covalent bond is presented in Chapter 10. For the present let us see the kind of understanding of covalent bonding that is possible using the Lewis theory. First, we rewrite the Lewis structure for HCl to emphasize the sharing of electrons and the attainment of noble gas electron configurations.

$$\left(H \overset{\cdot}{\underset{\cdot}{:}} Cl \overset{\cdot\cdot}{\underset{\cdot\cdot}{:}} \right) \tag{9.3}$$

The dashed circles represent the outermost electronic shells of the bonded atoms. The effective configuration of the outermost shell is established by counting the number of electrons lying on each circle. For the H atom this is two, corresponding to the configuration of He. For the Cl atom it is eight, corresponding to the configuration of the Ar atom. Note that the two electrons between H and Cl (:) have been counted twice, a consequence of the fact that these are the shared electrons. Some additional simple structures are shown in Figure 9-4.

FIGURE 9-4
Some examples of covalent bonds.

Elements	H ·	· $\overset{\cdot\cdot}{O}$:	: $\overset{\cdot\cdot}{Cl}$ ·	· $\overset{\cdot\cdot}{N}$ ·
Compounds	H : $\overset{\cdot\cdot}{\underset{\cdot\cdot}{Cl}}$:	H : $\overset{\cdot\cdot}{\underset{\cdot\cdot}{O}}$: H	: $\overset{\cdot\cdot}{\underset{\cdot\cdot}{Cl}}$: $\overset{\cdot\cdot}{\underset{\cdot\cdot}{O}}$: : $\overset{\cdot\cdot}{\underset{\cdot\cdot}{Cl}}$:	H : $\overset{\cdot\cdot}{N}$: H H
Names	hydrogen chloride	hydrogen oxide (water)	chlorine oxide	hydrogen nitride (ammonia)
Molecules	HCl	H_2O	Cl_2O	NH_3
Molecular weights	36.46	18.02	86.91	17.03

We have already established and used the fact that several of the gaseous elements exist not as collections of isolated atoms but in *molecular* form. For example, we have noted these formulas: H_2, Cl_2, N_2, and O_2. Let us try now to establish these formulas through Lewis structures. The situation with H_2 and Cl_2 is relatively simple. Each atom acquires a noble gas electron configuration by sharing *one* pair of electrons with the atom to which it is bonded. The sharing of a single pair of electrons results in a **single** covalent bond, often represented by a dash sign (—). Electron pairs not involved in bond formation, **nonbonding pairs** or **lone pairs**, are sometimes also represented by dashes.

$$\text{H} \cdot + \cdot \text{H} \longrightarrow \text{H}:\text{H}   \text{or}   \text{H—H} \tag{9.4}$$

$$:\ddot{\text{Cl}}\cdot + \cdot\ddot{\text{Cl}}: \longrightarrow :\ddot{\text{Cl}}:\ddot{\text{Cl}}: \text{or} :\ddot{\text{Cl}}\text{—}\ddot{\text{Cl}}: \text{or} |\overline{\text{Cl}}\text{—}\overline{\text{Cl}}| \tag{9.5}$$

Multiple Covalent Bonds.   If we attempt to extend what we have just written for H_2 and Cl_2 to the molecule N_2, this is what we obtain.

$$:\dot{\text{N}}\cdot + \cdot\dot{\text{N}}: \longrightarrow :\dot{\text{N}}:\dot{\text{N}}:   \text{(incorrect)}$$

This structure violates the octet rule. The N atoms appear to have only six outer-shell electrons. The situation can be improved if we consider that more than a single pair of electrons may be shared between atoms. The Lewis structure we write for N_2 is

$$\text{N}:::\text{N}   \text{or}   :\text{N}\equiv\text{N}: \text{or} |\text{N}\equiv\text{N}| \tag{9.6}$$

The sharing of *three* pairs of electrons between two atoms, as in the N_2 molecule, is referred to as a **triple** covalent bond (≡).

In following this approach for O_2, we find a need to represent the sharing of *two* pairs of electrons between the O atoms. This sharing leads to a double covalent bond (=).

$$\text{O}::\text{O}   \text{or}   :\ddot{\text{O}}\text{=}\ddot{\text{O}}: \text{or} |\overline{\text{O}}\text{=}\overline{\text{O}}| \text{(incorrect)} \tag{9.7}$$

But why have we labeled the structures in (9.7) incorrect? We have done so because these structures fail to account for the fact that the O_2 molecule is paramagnetic (has unpaired electrons). We will return to this and other apparent failures of the octet rule shortly. For now we simply make this point: *The fact that a plausible Lewis structure can be written for a species is not proof that this structure is the true electronic structure.* Experimental verification is always required.

9-4 Covalent Lewis Structures—Some Examples

Despite the warning with which we closed the preceding section, the subject of Lewis structures is worth pursuing further. In considering some specific examples in this section, the following ideas will prove useful.

1. *All* the valence (outer-shell) electrons of the atoms in a Lewis structure must be accounted for.
2. *Usually,* each atom in a Lewis structure acquires an electron configuration with an outer-shell octet. [Hydrogen is limited to an outer-shell duet (two electrons).]
3. *Usually,* all the electrons in a Lewis structure are paired.
4. *Often* both atoms in a bonded pair contribute equal numbers of electrons to the covalent bond, but *sometimes* both electrons in a bonded pair are derived from a single atom. (Such a bond is referred to as a **coordinate covalent bond**.)

5. *Sometimes* it is necessary to represent double or triple covalent bonds in a Lewis structure.

6. *Sometimes* it is impossible to draw a single Lewis structure that is consistent with all the available data. In these instances the true structure can only be represented as a *composite* or *hybrid* of two or more plausible structures. This situation, called resonance, is discussed in Section 9-6.

Other useful ideas in writing a Lewis structure are to

(a) Start with a plausible skeleton structure. This is a representation of the order in which atoms are bonded together. The skeleton structure consists of one or more central atoms with other atoms (terminal atoms) bonded to the central atom(s). Employ the concept of formal charge (described on page 226) to assess the plausibility of a skeleton structure.
(b) Add up the total number of valence electrons for all the atoms in the structure. This a the number of electrons that must appear in the Lewis structure. (This rule is modified slightly when considering polyatomic anions or cations, as indicated in Section 9-5.)
(c) Place electron pairs around the terminal atoms to complete the valence shells (octets) of all the atoms. If this is not possible with the number of electrons available, then shift lone-pair electrons from the terminal atoms to form multiple bonds to the central atom.

Example 9-3 Write a plausible skeleton structure, and then the Lewis structure for the molecule hydrazine, N_2H_4.

Solution. Several skeleton structures for N_2H_4 are shown. Which is correct?

$$N—N—H—H—H—H \qquad H—H—N—N—H—H$$
$$\text{(a)} \qquad\qquad\qquad\qquad \text{(b)}$$

$$
\begin{array}{cc}
H & H \\
| & | \\
H—N—N—\!\!\!&\!\!\!II \\
\text{(c)} &
\end{array}
$$

There are times when the choice of a skeleton structure is difficult, requiring perhaps that actual experimental evidence be provided. Here we can reject structures (a) and (b) for this simple reason: If an H atom were to be covalently bonded to two other atoms simultaneously, this would place four electrons in the outer shell of the H atom. This is more electrons than the first shell can accommodate and certainly does not lead to the configuration of a noble gas. Or, to express this thought in another way, *a hydrogen atom can form only one single covalent bond.*

The total number of electrons to appear in the structure is 14 (*five* each from the *two* N atoms and *one* each from the *four* H atoms). The skeleton structure (c) already accounts for 10 of the electrons. The other four appear as two nonbonding or lone pairs, one pair on each N atom. The Lewis structure is

$$
\begin{array}{cc}
H\ H & H\ H \\
\ddot{\ }\ \ddot{\ } & |\ \ | \\
H\!:\!\ddot{N}\!:\!\ddot{N}\!:\!H \quad \text{or} \quad & H—N—N—H \\
\ddot{\ }\ \ddot{\ } & | \\
\end{array}
\qquad\qquad (9.8)
$$

SIMILAR EXAMPLES: Review Problem 3; Exercise 7.

Example 9-4 Write a plausible Lewis structure for the molecule hydrogen cyanide, which has the skeleton structure HCN.

Solution. The number of valence electrons in our structure must be 10 (*one* from H, *four* from C, and *five* from N). These electrons are all accounted for in the structure obtained by completing the valence shells for N and H.

$$H—C—\ddot{\overset{\displaystyle\cdot\cdot}{N}}\!: \quad \text{(incorrect)}$$

The structure just written is incorrect, however. The central C atom does not have an octet. The situation is corrected by moving two of the lone pairs of electrons of the N atom to form a triple covalent bond between the C and the N atoms.

$$\text{H} : \text{C} ::: \text{N} : \qquad \text{or} \qquad \text{H} - \text{C} \equiv \text{N} : \tag{9.9}$$

SIMILAR EXAMPLES: Review Problem 4; Exercise 8.

Example 9-5   Write a plausible Lewis structure for thionyl fluoride, SOF_2. (Experimental evidence indicates that the O and F atoms are bonded directly to the S atom.)

Solution.   To establish the skeleton structure of a molecule, we generally need to identify the central atom—the atom to which the others are bonded. We can eliminate the F atoms as possible central atoms in the skeleton structure because, having seven outer-shell electrons, they normally form only one single covalent bond apiece. The choice between S and O is not so easy. The valence-shell electron configurations of S and O are similar (both elements are in group VIA of the periodic table). It is for this reason that we were supplied with the statement that *sulfur is the central atom*.

The number of valence electrons in the structure is **26** (*six* from S, *six* from O, and *seven* each from the *two* F atoms). This number is sufficient to provide each atom with an octet. In the structure below, if we indicate the electrons originating from the S atom as ×, and from the other atoms as ·, we conclude that in the bond between the S and the O atom *both electrons are donated by the S atom*. This is a **coordinate covalent bond.**

$$: \overset{..}{\underset{..}{\text{F}}} \overset{\times}{\times} \overset{\overset{\times\times}{\text{S}}}{\underset{\times}{}} \overset{..}{\underset{..}{\text{O}}} :$$
$$\quad : \overset{}{\underset{..}{\text{F}}} :$$

Once formed, a coordinate covalent bond is no different from an ordinary covalent bond, and usually no attempt is made to differentiate among electrons in a Lewis structure. However, if desired, the presence of a coordinate covalent bond can be represented by an arrow pointing from the atom donating the electron pair to the one receiving it.

$$: \overset{..}{\underset{..}{\text{F}}} - \overset{..}{\underset{}{\text{S}}} - \overset{..}{\underset{..}{\text{O}}} : \qquad \text{or} \qquad : \overset{..}{\underset{..}{\text{F}}} - \overset{..}{\underset{}{\text{S}}} \to \overset{..}{\underset{..}{\text{O}}} : \tag{9.10}$$
$$\quad : \overset{}{\underset{..}{\text{F}}} : \qquad\qquad\qquad\quad : \overset{}{\underset{..}{\text{F}}} :$$

SIMILAR EXAMPLE: Exercise 8.

The Concept of Formal Charge.   Let us next consider a method of "electron bookkeeping" that can help us to choose among different possible skeleton structures. In Example 9-4 all the ordinary rules for Lewis structures would have been met equally well by either (a) or (b). Why did we choose structure (a)?

$$\text{H} - \text{C} \equiv \text{N} : \qquad \text{H} - \text{N} \equiv \text{C} : \tag{9.11}$$
$$\quad\; \text{(a)} \qquad\qquad\quad\;\; \text{(b)}$$

Suppose that in addition to the counting scheme that lets us establish that atoms acquire outer-shell octets, we count electron dots in Lewis structures in the following way: Count all *nonbonding* electrons as belonging entirely to the atom in which they are found. Count *bonding* electrons by dividing them equally between the bonded atoms. Figure 9-5 illustrates how this counting procedure is applied to structures (a) and (b) in (9.11). After having counted electrons in the manner indicated, determine whether any atoms in the Lewis structure have a **formal charge.**

FIGURE 9-5
Selecting a plausible
Lewis structure—the
concept of formal
charge illustrated.

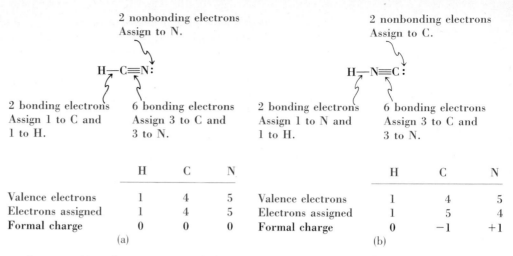

	H	C	N
Valence electrons	1	4	5
Electrons assigned	1	4	5
Formal charge	0	0	0

(a)

	H	C	N
Valence electrons	1	4	5
Electrons assigned	1	5	4
Formal charge	0	−1	+1

(b)

Structure (a) is the more plausible because there are no formal charges on any of the atoms represented in the structure.

*The formal charge is the number of outer-shell (valence) electrons in an isolated atom minus the number of electrons assigned to that atom in a Lewis structure.**

The rule that is used to establish structure (a) in (9.11) as the more plausible is this.

A Lewis structure in which there are no formal charges (i.e., where all formal charges are zero) is more plausible than one where formal charges are required. Where formal charges are necessary, these are generally of as small a magnitude as possible.

An important note about formal charges is in order. Even if we choose to denote formal charges, as in structure (b),

$$\overset{\oplus}{H}-\overset{\ominus}{N}\equiv C\!:$$ (9.12)

individual atoms within a covalent molecule do not carry actual net charges. To distinguish between formal charges and actual ionic charges, we will use small encircled charges for formal charges.

Example 9-6   Show that, even in the absence of the experimental data given in Example 9-5, structure (1) is more plausible than structure (2).

$$:\!\overset{\;\;\oplus}{\underset{\underset{(1)}{\overset{\displaystyle |}{\underset{\cdot\cdot}{F}\!:}}}{F}}\!-\!\overset{\cdot\cdot}{\underset{}{S}}\!-\!\overset{\;\;\ominus}{\underset{\cdot\cdot}{O}}\!: \qquad :\!\overset{\;\;\oplus}{\underset{\underset{(2)}{\overset{\displaystyle |}{\underset{\cdot\cdot}{F}\!:}}}{F}}\!-\!\overset{\cdot\cdot}{\underset{}{S}}\!-\!\overset{\;\;\ominus}{\underset{\cdot\cdot}{O}}\!:$$ (9.13)

Solution.   In both structures (1) and (2), the F atoms would be assigned seven electrons and would have formal charges of $7 - 7 = 0$. In structure (1) the S atom is assigned five electrons and has a formal charge of $6 - 5 = +1$. The formal charge of the O atom in structure (1) is $6 - 7 = -1$.

---

*An alternate definition that yields the same result and which some prefer is that
Formal charge = no. of valence electrons $-\frac{1}{2}$ (no. of bonding electrons)
                $-$ no. of nonbonding (lone pair) electrons

In structure (2) the S atom is assigned seven electrons and would have a formal charge of $6 - 7 = -1$. In turn, the O atom would have a formal charge of $6 - 5 = +1$.

Formal charges are found in both structures (1) and (2). However, recall the discussion of atomic properties in Chapter 8. Which atom, S or O, has the greater electronegativity; the more nonmetallic character? The answer to these questions is oxygen. We should choose the Lewis structure in which O, not S, *appears* to gain an electron and acquire a formal charge of -1. Structure (1) is the more plausible of the two.

SIMILAR EXAMPLES: Review Problem 6; Exercises 11, 12.

Example 9-6 suggests a second rule on formal charges.

In choosing among alternatives having similar distributions of formal charges, the most plausible Lewis structure is that in which negative formal charges are placed on the more electronegative atoms.

As a further result of applying this rule, it is found that *the central atom in a structure generally is the atom with lowest electronegativity*.

9-5 Covalent Bonding in Polyatomic Ions

As we learned in Section 3-8, a polyatomic ion consists of two or more atoms. The forces that operate *within* such ions to hold atoms together are covalent bonds. Consider, for example, the simple hydroxide ion, OH^-. Its Lewis structure must involve *eight* electrons: six from O, one from H, and one additional electron to account for the net charge of -1. (The "extra" electron is shown in color in structure 9.14, but of course there is no way of distinguishing among electrons.)

$$[\, :\ddot{\text{O}}: \text{H} \,]^- \tag{9.14}$$

Where does the OH^- ion gain its electron? Perhaps from a sodium atom that loses an electron to become a sodium ion. Thus, the *ionic* compound, sodium hydroxide, consists of simple Na^+ ions and *polyatomic* anions, OH^-. Within an OH^- ion the O and H atoms are joined by a single covalent bond.

$$\text{Na} \cdot + \cdot \ddot{\text{O}}: \text{H} \longrightarrow [\text{Na}]^+[\, :\ddot{\text{O}}: \text{H}]^- \tag{9.15}$$

Example 9-7   Write a plausible Lewis structure for the chlorite ion, ClO_2^-.

Solution
1. It is generally true for oxoanions (polyatomic anions containing oxygen) that the O atoms are all bonded directly to a central nonmetal atom. The expected skeleton structure is

O
|
Cl—O

2. The rule assessing the number of valence electrons in the Lewis structure (rule **b** of page 225) must be modified. That is, the number of valence electrons is **20** (*seven* from Cl, *six* each from the *two* O atoms, and *one extra electron to account for the net charge of* -1.)

3. The Lewis structure (including formal charges on the atoms) now follows directly.

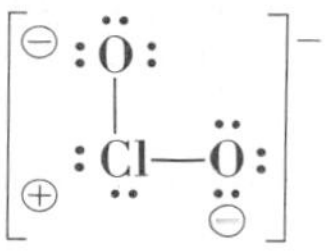

$$\left[\begin{array}{c} \ominus\ :\ddot{\text{O}}: \\ | \\ :\ddot{\text{Cl}}\!-\!\ddot{\text{O}}: \\ \oplus \quad \ominus \end{array} \right]^{-}$$
(9.16)

SIMILAR EXAMPLES: Review Problem 7; Exercises 15, 16.

We discover still another feature of the concept of formal charge when it is applied to the chlorite ion of Example 9-7. Each O atom has a formal charge of -1 and the Cl atom, $+1$. The sum of these formal charges is $-1 - 1 + 1 = -1$.

Where formal charges are assigned in a Lewis structure, these formal charges must total to zero for a neutral molecule, or to the net charge for a polyatomic ion.

FIGURE 9-6
Formation of the ammonium ion, NH_4^+.

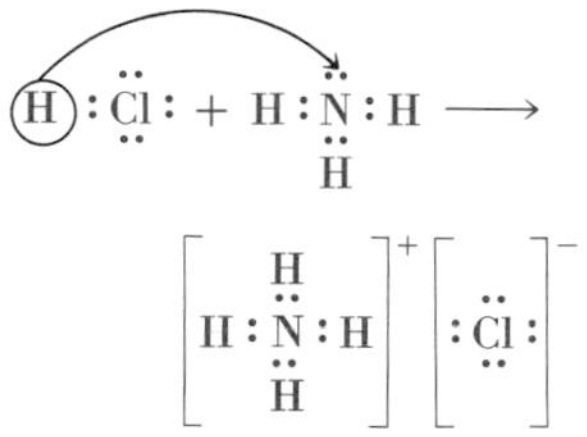

Figure 9-6 illustrates bonding in the polyatomic *cation* NH_4^+ and the formation of the ionic compound ammonium chloride by the reaction of hydrogen chloride and ammonia. In determining the number of valence shell electrons for the Lewis structure of the ammonium ion, we must *subtract* one electron to account for the net *positive* charge. That is, the number of valence electrons is **8** (*five* from N, *one* each from *four* H atoms, *minus one* electron to account for the net charge of $+1$). Note also that the formal charge on N is $+1$ and that the sum of the formal charges for all the atoms is equal to the ionic charge, $+1$.

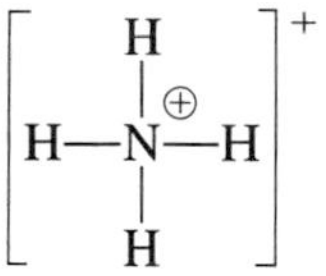

$$\left[\begin{array}{c} \text{H} \\ | \\ \text{H}\!-\!\overset{\oplus}{\text{N}}\!-\!\text{H} \\ | \\ \text{H} \end{array} \right]^{+}$$
(9.17)

9-6 Resonance

When the rules for Lewis structures are applied to the molecule SO_2, we obtain two equivalent structures.

$$:\ddot{\text{O}}\!=\!\ddot{\text{S}}\!-\!\ddot{\text{O}}: \qquad :\ddot{\text{O}}\!-\!\ddot{\text{S}}\!=\!\ddot{\text{O}}:$$

(a)   (b)

Which one is correct? Actually, neither one is. Both structures suggest that one sulfur-to-oxygen bond is single and the other double. The properties of the SO_2 molecule determined by experiment indicate, however, that the two bonds are the same. How shall we represent this fact? One method is suggested through structure (9.18).

A mundane analogy that has often been used is to consider a mule to be a resonance hybrid of a horse and a donkey. The mule has some of the qualities of both, but it is a very distinctive animal. It is neither a horse part of the time and a donkey the rest, nor is it half-horse and half-donkey.

$$:\ddot{\text{O}}\!=\!\ddot{\text{S}}\!-\!\ddot{\text{O}}: \longleftrightarrow :\ddot{\text{O}}\!-\!\ddot{\text{S}}\!=\!\ddot{\text{O}}:$$
(9.18)

The phenomenon we have been describing is called **resonance**. This is a situation in which more than one plausible structure can be written for a species and in which the true structure cannot be written at all. The true structure is considered to be a *hybrid* of the different plausible structures.

Example 9-8   Write the Lewis structure for the nitrate ion, NO_3^-. The O atoms are all bonded *in the same way* to the central N atom.

Solution.   The structure we are seeking must contain **24** valence electrons (*five* from N, *six* each from *three* O atoms, and *one* extra electron to produce the net charge of −1). Trial 1 provides each terminal atom with an octet, but it leaves the central atom with an incomplete octet. This situation is remedied by moving a lone pair of electrons from one of the O atoms to form a double bond between it and the N atom, as in trial 2.

trial 1: 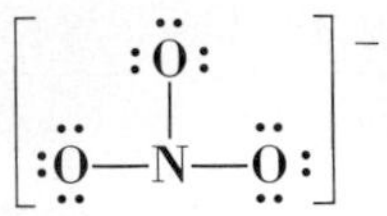
$$
\left[\begin{array}{c} :\!\overset{\cdot\cdot}{O}\!: \\[2pt] | \\[2pt] :\!\overset{\cdot\cdot}{\underset{\cdot\cdot}{O}}\!-\!N\!-\!\overset{\cdot\cdot}{\underset{\cdot\cdot}{O}}\!: \end{array}\right]^{-}
$$
   (incorrect)   trial 2:
$$
\left[\begin{array}{c} :\!\overset{\cdot\cdot}{O}\!: \\[2pt] | \\[2pt] :\!\overset{\cdot\cdot}{\underset{\cdot\cdot}{O}}\!-\!N\!=\!\overset{\cdot\cdot}{O}\!: \end{array}\right]^{-}
$$

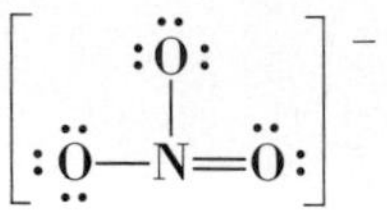

Trial 2 certainly produces a plausible structure, but it does not conform to the statement that all the nitrogen-to-oxygen bonds are *equivalent*. No single structure can be drawn to do this. Three equivalent structures can be written, and the true structure is a resonance hybrid of these.

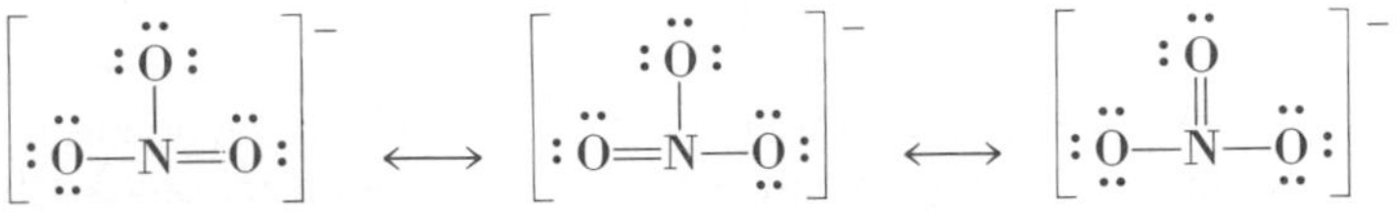

$$
\left[\begin{array}{c} :\!\overset{\cdot\cdot}{O}\!: \\ | \\ :\!\overset{\cdot\cdot}{\underset{\cdot\cdot}{O}}\!-\!N\!=\!\overset{\cdot\cdot}{O}\!: \end{array}\right]^{-} \longleftrightarrow \left[\begin{array}{c} :\!\overset{\cdot\cdot}{O}\!: \\ | \\ :\!\overset{\cdot\cdot}{O}\!=\!N\!-\!\overset{\cdot\cdot}{\underset{\cdot\cdot}{O}}\!: \end{array}\right]^{-} \longleftrightarrow \left[\begin{array}{c} :\!\overset{\cdot\cdot}{O} \\ \| \\ :\!\overset{\cdot\cdot}{\underset{\cdot\cdot}{O}}\!-\!N\!-\!\overset{\cdot\cdot}{\underset{\cdot\cdot}{O}}\!: \end{array}\right]^{-} \qquad (9.19)
$$

SIMILAR EXAMPLES: Exercises 17, 18, 19.

The need for the concept of resonance in dealing with the Lewis structure of SO_2 (9.18) was rather obvious. At times, even though a single Lewis structure can be written, better agreement with experiment is obtained if resonance is invoked. Such a case is illustrated through Figure 9-7, which also provides a few additional rules to help in identifying plausible contributing structures to a resonance hybrid.

9-7 Exceptions to the Octet Rule

Odd-Electron Species.   If the total number of valence electrons in a Lewis structure is an *odd* number, two immediate conclusions follow about the structure. There must be

1. At least one unpaired electron somewhere.
2. At least one atom lacking a completed octet of electrons.

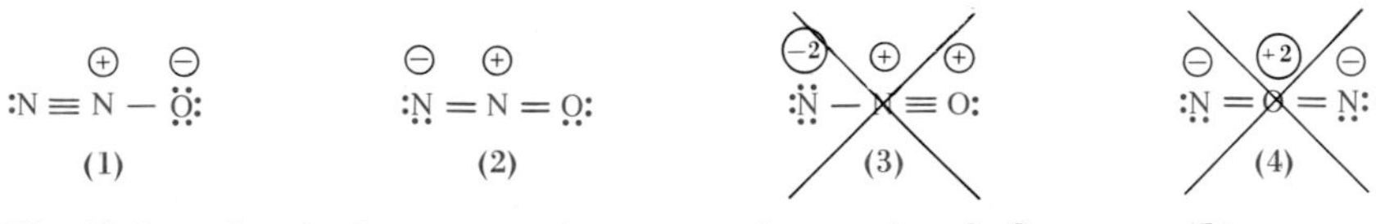

FIGURE 9-7
Writing a Lewis structure for N_2O.

$$
\underset{(1)}{:\!N\!\equiv\!\overset{\oplus}{N}\!-\!\overset{\ominus}{\underset{\cdot\cdot}{O}}\!:} \qquad \underset{(2)}{:\!\overset{\ominus}{N}\!=\!\overset{\oplus}{N}\!=\!\overset{\cdot\cdot}{O}\!:} \qquad \underset{(3)}{\overset{-2}{:\!N}\!-\!\overset{\oplus}{N}\!\equiv\!\overset{\oplus}{O}\!:} \qquad \underset{(4)}{:\!\overset{\ominus}{N}\!=\!\overset{+2}{O}\!=\!\overset{\ominus}{N}\!:}
$$

The N_2O molecule features a nitrogen-to-nitrogen bond. Structure (**1**) seems quite satisfactory. Experimental evidence, however, indicates that the N-to-N bond is intermediate between a double and a triple bond and the N-to-O bond, intermediate between a single and a double bond. This additional information suggests that structure (**2**) contributes to the resonance hybrid.

Structure (**3**) is not plausible because of the high formal charge on one of the atoms (N), because the most electronegative atom, O, has a positive rather than a negative formal charge, and because the structure violates this additional rule concerning formal charges: *A structure in which like formal charges (both positive or both negative) reside on adjacent atoms is very unlikely*.

Structure (**4**) and others like it are not possible contributing structures. *All contributing structures to a resonance hybrid must have the same arrangement of atoms*. In N_2O, an N atom, not O, is the central atom.

Take the case of the NO_2 molecule. The total number of valence electrons is 17. The Lewis structure must display the features cited above. Actually, there are two plausible structures and the true structure is a resonance hybrid of the two.

$$:\overset{..}{\underset{.}{O}}-\overset{..}{N}=\overset{..}{\underset{..}{O}}: \longleftrightarrow :\overset{..}{\underset{..}{O}}=\overset{..}{N}-\overset{..}{\underset{.}{O}}: \qquad (9.20)$$

Because of the presence of an unpaired electron, an odd-electron species must be paramagnetic. NO_2 is paramagnetic, and so is NO. Molecules with an *even* number of electrons are expected to have all electrons paired and to be diamagnetic. Yet the molecule O_2, with 12 valence electrons, is *paramagnetic. The O_2 molecule must have unpaired electrons.* This is why structure (9.7), which seemed to meet all the criteria of an acceptable Lewis structure for O_2, was said to be incorrect. It contained no unpaired electrons. Experimental evidence indicates that the bond in O_2 has some multiple-bond character. No single Lewis structure can be written to represent O_2.

$$:\overset{..}{\underset{.}{O}}-\overset{..}{\underset{.}{O}}: \longleftrightarrow :\overset{..}{\underset{..}{O}}=\overset{..}{\underset{..}{O}}: \qquad (9.21)$$

Incomplete Octets.   We have encountered Lewis structures in which one or more atoms has not had an outer-shell octet of electrons and have used this as a basis for rejecting a structure. But occasionally, situations arise where such structures appear, in fact, to be correct. This is the case with the molecule BF_3. (For emphasis the valence electrons of the B atom are shown in color.)

$$:\overset{..}{F}:B:\overset{..}{F}: \atop \qquad :\overset{..}{\underset{..}{F}}: \qquad (9.22)$$

That structure (9.22) is plausible for BF_3 is confirmed by the ease with which the compound $H_3N \cdot BF_3$ is formed. The N atom donates both electrons to the boron–nitrogen bond and the B atom completes its octet.

$$ \qquad (9.23)$$

The Expanded Octet.   Phosphorus forms two chlorides, PCl_3 and PCl_5. Covalent bonding in PCl_3 fulfills the basic criterion of Lewis structures—all atoms acquire an outer-shell octet. In PCl_5, because the five Cl atoms are bonded directly to the central P atom, ten electrons are found in the outer shell of the P atom. The octet has been "expanded" to ten electrons. In the molecule SF_6 it is expanded to 12. The situation may be depicted as follows.

$$ \qquad (9.24)$$

In Section 10-1 we shall see that when an atom acquires an octet of electrons the *s* and *p* subshells of the outermost electronic shell are filled. Expansion to 10 or 12 electrons requires additional orbitals. The energy difference between the 2*p* and 3*s* sublevels is too great to provide this additional bonding possibility to the nonmetals in the second period. Once the *d* subshell becomes available, expanded octets become possible. Thus, the phenomena described here are encountered with nonmetals in the third and higher periods.

Our usual method of resolving the difficulty of an incomplete octet on a central atom has been to shift a lone pair of electrons from a terminal atom to a bond with the central atom. Here this would produce a B-to-F double bond and formal charges of −1 on B and +1 on F—a less acceptable structure than (9.22).

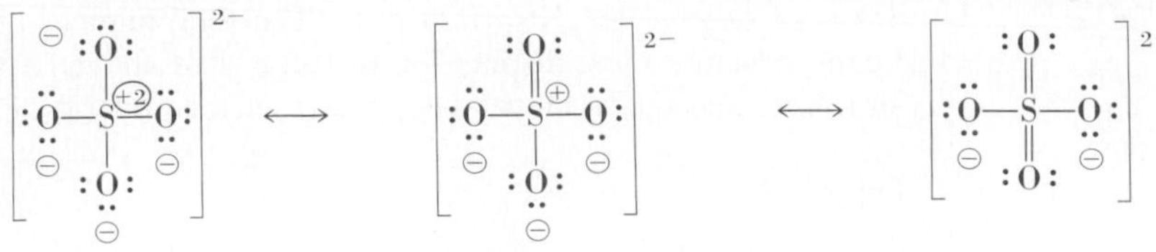

FIGURE 9-8
Representation of bonding in the sulfate ion.

$$\left[\begin{array}{c} \ominus\,:\!\ddot{O}: \\ :\!\ddot{O}-\!\overset{+2}{S}\!-\!\ddot{O}: \\ | \\ :O: \\ \ominus \end{array}\right]^{2-} \quad\longleftrightarrow\quad \left[\begin{array}{c} :O: \\ :\!\ddot{O}-\!\overset{+}{S}\!-\!\ddot{O}: \\ \| \\ :O: \\ \ominus \end{array}\right]^{2-} \quad\longleftrightarrow\quad \left[\begin{array}{c} :O: \\ :\!\ddot{O}-\!S\!-\!\ddot{O}: \\ \| \\ :O: \\ \ominus \end{array}\right]^{2-}$$

(four equivalent structures)   (six equivalent structures)

Application of the concepts of resonance and the expanded octet provides the best representation of the sulfate ion. The resonance hybrid of SO_4^{2-} involves contributions from all the structures shown here.

At times, by using expanded octets, we can write Lewis structures that correspond more closely to experimental observations. For example, if we attempt to write the Lewis structure of sulfate ion, SO_4^{2-}, *without* an expanded octet we obtain a structure in which all bonds are single bonds and all atoms carry a formal charge.

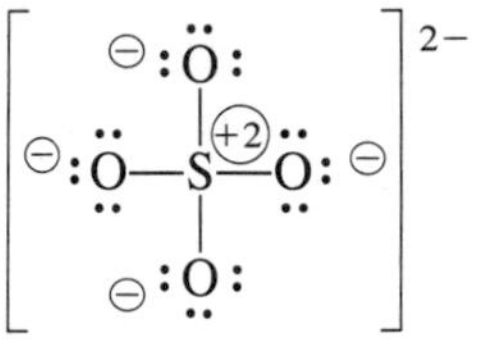

$$\left[\begin{array}{c} \ominus\,:\!\ddot{O}: \\ | \\ \ominus\,:\!\ddot{O}-\!\overset{+2}{S}\!-\!\ddot{O}:\,\ominus \\ | \\ \ominus\,:\!O: \end{array}\right]^{2-} \tag{9.25}$$

We can write a structure with fewer formal charges by employing some sulfur-to-oxygen double bonds, and experimental results indicate that there is some multiple bond character to these bonds. However, since the sulfur-to-oxygen bonds are all equivalent, it is necessary to invoke resonance in writing the Lewis structure of SO_4^{2-}, as shown in Figure 9-8.

Example 9-9   Which Lewis structures presented to this point could be improved by including a contributing structure(s) in which the central atom employs an expanded octet? Write these structures.

Solution.   Octet expansion can be considered for structures in which the central atom is a nonmetal of the third period or higher. We considered the contribution of expanded-octet structures to the Lewis structure of SO_4^{2-} in Figure 9-8. Other examples are

Contribution to (9.10)   Contribution to (9.16)   Contribution to (9.18)

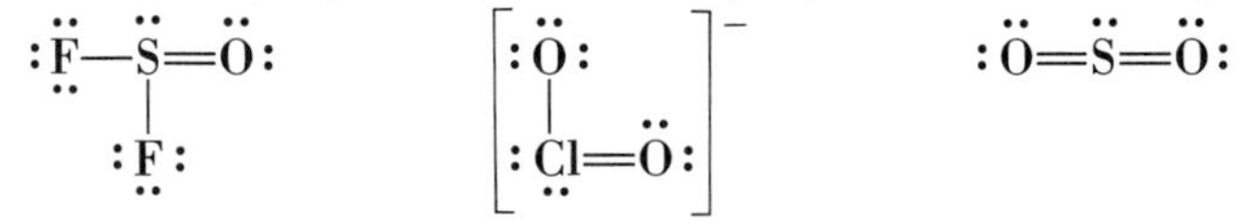

$$:\!\ddot{F}-\!\ddot{S}\!=\!\ddot{O}: \qquad \left[\begin{array}{c} :\!\ddot{O}: \\ | \\ :\!\ddot{Cl}\!=\!\ddot{O}: \end{array}\right]^{-} \qquad :\!\ddot{O}\!=\!\ddot{S}\!=\!\ddot{O}:$$
$$\qquad\quad | \\ \qquad :\!\ddot{F}:$$

SIMILAR EXAMPLES: Review Problem 9; Exercises 25, 26.

FIGURE 9-9
Geometrical shape of a molecule.

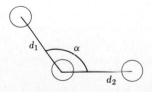

A hypothetical triatomic molecule is represented here. Although the atoms are actually in contact, for clarity only the centers of the atoms are shown (producing a so-called ball-and-stick model). To establish the shape of this molecule we must determine the distances between the centers of the bonded atoms (d_1 and d_2) and the angle between the adjacent bonds (α). Our primary concern is with the bond angle. For molecules with more than three atoms, additional bond distances and angles must be established, usually for a three-dimensional figure.

9-8 Molecular Shapes

By molecular shape we mean the geometrical figure that results if the nuclei of bonded atoms are joined by straight lines (see Figure 9-9). Because two points determine a straight line, all diatomic molecules are linear. Three points determine a plane; all triatomic molecules are planar. For molecules with more than three atoms (polyatomic molecules), planar and even linear shapes are sometimes encountered. Usually, however, the atoms define a three-dimensional figure. Molecular shapes cannot be predicted from empirical formulas; they must be determined experimentally.

Valence-Shell Electron-Pair Repulsion (VSEPR) Theory.   This theory of bonding states that *electron pairs, whether they are in chemical bonds or unshared (i.e., lone pairs), repel one another. Electron pairs tend to remain as far apart from each other as possible.* Or, in terms of the Pauli exclusion principle, if a pair of electrons occupies an orbital, another electron, regardless of its spin, cannot come into close proximity of the pair. VSEPR (pronounced "vesper") theory pictures *pairs* of electrons assuming a certain orientation with respect to the nucleus of an atom.

Consider a noble gas atom such as Ne, for example. What orientation will the four pairs of valence electrons ($2s^2 2p^6$) assume? As suggested by the "balloon" analogy in Figure 9-10, the electron pairs are farthest apart when they occupy the corners of a regular tetrahedron with the atomic nucleus at its center. Now consider the methane molecule, CH_4, in which the central C atom has acquired the Ne electron configuration by forming covalent bonds with four H atoms.

$$\begin{array}{c} H \\ \ddot{} \\ H:\underset{\displaystyle\ddot{}}{\overset{\displaystyle}{C}}:H \\ H \end{array}$$

The method predicts that CH_4 should be a *tetrahedral* molecule, with a C atom at the center of the tetrahedron and H atoms at the corners. This structure agrees with that established by experiment.

In NH_3 and H_2O the central atom is also surrounded by four pairs of electrons,

$$\begin{array}{ccc} H & & H \\ \ddot{} & & \ddot{} \\ H:\underset{\displaystyle\ddot{}}{\overset{\displaystyle}{N}}:H & \text{and} & :\underset{\displaystyle\ddot{}}{\overset{\displaystyle}{O}}:H \end{array}$$

but these molecules do not have a tetrahedral shape. The situation is this: VSEPR theory predicts the distribution of electron pairs. The geometrical shape of a molecule, however, is described in terms of the geometrical figure that results by *joining the appropriate atomic nuclei by straight lines.*

This and several other balloon analogies are pictured in Color Section A.

FIGURE 9-10
Balloon analogy to valence-shell electron-pair repulsion.

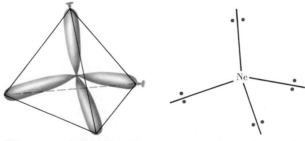

When two elongated balloons are twisted together at their centers, separation into four lobes occurs. To minimize mutual interferences, the lobes spread out into a tetrahedral shape. (A tetrahedron has four faces, each an equilateral triangle.) The lobes are analogous to valence-shell electron pairs. The distribution of the four pairs of valence-shell electrons of a neon atom is also shown.

FIGURE 9-11
Geometric shapes based on the tetrahedral distribution of four electron pairs—CH_4, NH_3, and H_2O.

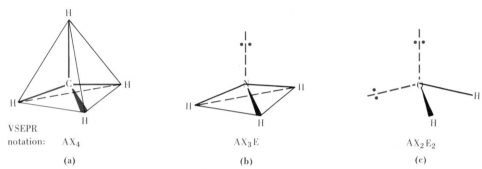

VSEPR notation: AX_4 AX_3E AX_2E_2
                   (a)   (b)   (c)

The shape of the molecule is established by the lines in black. Lone-pair electrons are shown as bold dots along broken lines originating at the central atom.
(a) All electron pairs about the central carbon atom are bond pairs. (The valence of the carbon atom is said to be *saturated*.) The lines that establish the shape of the CH_4 molecule (in black) are different from those representing carbon-to-hydrogen bonds (in color).
(b) The lone pair of electrons is directed toward the "missing" corner of the tetrahedron. The figure remaining is a trigonal pyramid. The nitrogen-to-hydrogen bonds form three of the edges of this pyramid.
(c) The H_2O molecule is an angular molecule outlined by the two oxygen-to-hydrogen bonds.

In the NH_3 molecule only *three* of the electron pairs are *bonding* pairs and the fourth pair is a *nonbonding* or **lone pair.** The geometrical figure obtained by joining the nuclei of the H atoms with the nucleus of the N atom is not a tetrahedron but a pyramid (called a trigonal pyramid). The pyramid has H atoms at the base and an N atom at the apex. In the H_2O molecule two of the four pairs about the O atom are *bonding* pairs and two are *lone* pairs. The figure obtained by joining the nuclei of the two H atoms with that of the O atom is a *planar* figure. We can say that the molecule is **angular** or **bent.** The geometrical shapes of CH_4, NH_3, and H_2O are shown in Figure 9-11. Also introduced in Figure 9-11 is a special notation in which the number of lone-pair electrons in a structure is specified. For example, the notation AX_2E_2 signifies that two atoms of X and two lone pairs of electrons (E) are distributed about the central atom, A.

The bond angles expected for the tetrahedral distribution of electron pairs are 109.5°—called the tetrahedral bond angle. In CH_4 the H—C—H bond angles are, in fact, 109.5°. The bond angles in NH_3 and H_2O are slightly smaller: 107° for the H—N—H bond and 104.5° for H—O—H. These less-than-tetrahedral bond angles can be explained by assuming that the charge cloud of the lone-pair electrons spreads out and forces the bonding electrons closer together, reducing the bond angles. In H_2O, with two lone pairs, the effect is greatest and the bond angle smallest.

We continue our discussion of the VSEPR theory by turning to Table 9-2. Close inspection of some of the examples there suggests the need for an additional rule in cases where different alternatives exist. For example, consider the case of five valence-shell electron pairs, with four bond pairs and one lone pair (i.e., AX₄E). Why is the lone pair shown in the triangular plane at the center of the bipyramid rather than at the top or bottom. The closer two pair of electrons are forced together, the stronger the repulsion between them. Repulsive forces become especially strong when the angle between two pair of electrons is of the order of 90° (a right angle). Moreover, repulsions involving lone-pair electrons are stronger than those involving bond pairs, leading to the following order for repulsive forces.

$$\text{lone pair–lone pair} > \text{lone pair–bond pair} > \text{bond pair–bond pair} \qquad (9.26)$$

Returning to the AX₄E case, placing the lone pair of electrons as indicated in Table 9-2 results in *two* interactions between a lone pair and a bond pair at 90°. If the lone pair were at the top or bottom of the bipyramid, there would be *three* such interactions. A similar situation is described in Example 9-12.

Example 9-10 Predict the geometrical shape of the molecule OCl_2.

Solution. Although VSEPR theory does not require it, let us begin with a Lewis structure of this molecule. Since the O atom has two unpaired electrons, we should expect it, and not one of the Cl atoms, to be the central atom.

$$:\overset{..}{\underset{..}{Cl}}:\overset{..}{O}:\overset{..}{\underset{..}{Cl}}:$$

This Lewis structure helps us to count valence-shell electrons around the central oxygen atom as follows:

from the O atom	= 6
from the Cl atoms, 2 × 1	= 2
total valence electrons	= 8
valence-shell electron pairs	= 4
number of bond pairs	= 2
number of lone pairs	= 2

For the geometrical shape corresponding to the distribution of two bond pairs and two lone pairs, that is, for AX_2E_2,

Conclusion: The molecule is angular or bent.

SIMILAR EXAMPLES: Review Problem 10; Exercise 32.

Example 9-11 The bonding scheme in hydrogen peroxide, H_2O_2, is H—O—O—H. Is this molecule linear in shape?

Solution. This molecule has *two* central atoms, that is, both O atoms.

$$H:\overset{..}{\underset{..}{O}}:\overset{..}{\underset{..}{O}}:H$$

Whichever O atom we look at, we see four pairs of valence-shell electrons—two bond pairs and two lone pairs. This means that each H—O—O bond angle must correspond to the following electron-pair distributions.

TABLE 9-2
Molecular geometry as a function of geometrical distribution of valence-shell electron pairs and number of lone pairs

Number of electron pairs	Geometrical distribution of electron pairs	Number of lone pairs	VSEPR notation	Molecular geometry	Ideal bond angles	Example
2	linear	0	AX_2	X—A—X (linear)	180°	$BeCl_2$
3	trigonal planar	0	AX_3	(trigonal planar)	120°	BF_3
	trigonal planar	1	AX_2E	(angular)	120°	SO_2[a]
4	tetrahedral	0	AX_4	(tetrahedral)	109.5°	CH_4
	tetrahedral	1	AX_3E	(trigonal pyramidal)	109.5°	NH_3
	tetrahedral	2	AX_2E_2	(angular)	109.5°	OH_2
5	trigonal bipyramidal	0	AX_5	(trigonal bipyramidal)	90°, 120°	PCl_5

TABLE 9-2 (Continued)

Number of electron pairs	Geometrical distribution of electron pairs	Number of lone pairs	VSEPR notation	Molecular geometry	Ideal bond angles	Example
	trigonal bipyramidal	1	AX_4E^b	(irregular tetrahedral)	90°, 120°	SF_4
	trigonal bipyramidal	2	AX_3E_2	(T-shaped)	90°	ClF_3
	trigonal bipyramidal	3	AX_2E_3	(linear)	180°	XeF_2
6	octahedral	0	AX_6	(octahedral)	90°	SF_6
	octahedral	1	AX_5E	(square pyramidal)	90°	BrF_5
	octahedral	2	AX_4E_2	(square planar)	90°	XeF_4

[a] For a discussion of the structure of SO_2, see page 238.
[a] For a discussion of the placement of the lone-pair electrons in this structure, see page 235.

For the molecule to be linear, both H—O—O bond angles would have to be 180°, but we see that they are more nearly tetrahedral.

Conclusion: H_2O_2 is not linear.

SIMILAR EXAMPLES: Review Problem 10; Exercise 32.

Example 9-12   Predict the shape of the polyatomic anion ICl_4^-.

Solution.   Instead of proceeding as in Example 9-10, let us consider a method that does not require drawing a Lewis structure.* This method, which applies to structures with *a single central atom*, is stated below in general terms (*italic*) and then applied to the case of ICl_4^- (**boldface**).

$$total\ no.\ electron\ pairs = \frac{no.\ of\ valence\ electrons \pm ionic\ charge}{2}$$

$$= \frac{(7 \times 5) + 1}{2} = 18$$

(Use + magnitude of ionic charge for polyatomic anions and − magnitude of ionic charge for polyatomic cations.)

$$no.\ bond\ pairs = no.\ of\ atoms - 1$$
$$= 5 - 1 = 4$$

$$no.\ central\ pairs = total\ no.\ pairs - [3 \times no.\ terminal\ atoms\ (excluding\ H)]$$
$$= 18 - (3 \times 4) = 6$$

(In ICl_4^-, one atom is the central atom and the other four are terminal atoms.)

$$no.\ lone\ pairs = no.\ central\ pairs - no.\ bond\ pairs$$
$$= 6 - 4 = 2$$

Six central electron pairs, with four as bond pairs and two as lone pairs, correspond to the structure AX_4E_2—square planar.

Figure 9-12 suggests two possible structures for ICl_4^-, but by applying relationship (9.26) we see that the square planar structure is indeed the correct one. Repulsions between the two lone pairs would be much greater if they were located at adjacent corners of the central square than when they are located above and below the square.

SIMILAR EXAMPLE: Exercise 32.

Structures with Multiple Covalent Bonds.   VSEPR theory considers all chemical bonds as negative charge centers, whether they are single or multiple bonds. One way to handle structures with multiple bonds is to treat multiple bonds *as if they were single bonds*, containing just a *single* electron pair in the bond. Let us apply this idea to the Lewis structures shown for SO_2 in (9.18). We count *three* electron pairs about the central S atom. Two of these pairs are bond pairs and one is a lone pair. The molecular geometry for this situation (i.e., corresponding to AX_2E) is an angular molecule. (See Table 9-2.)

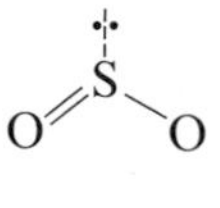

*The procedure outlined here is based on analyzing a Lewis structure without actually writing it. For example, each terminal atom is joined to the central atom by a bond pair and has three lone pairs of electrons (giving the terminal atom an outer-shell octet); hydrogen is an exception in having a bond pair but no lone-pair electrons. Thus, we have the statement that *no. central pairs = total no. pairs − [3 × no. terminal atoms (excluding H)]*.

FIGURE 9-12
Two predictions of the structure of ICl_4^- — Example 9-12 illustrated.

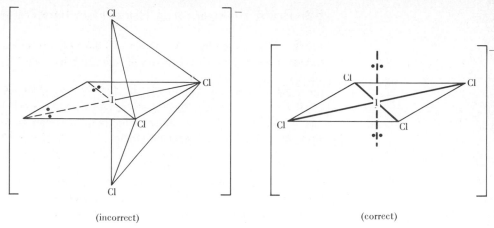

(incorrect)   (correct)

The observed structure is the square planar structure shown on the right.

A second way of handling structures with multiple bonds (illustrated in Example 9-13) is to use the method introduced in Example 9-12.

Example 9-13   What is the geometrical shape of the phosgene molecule $COCl_2$?

Solution

Method of Example 9-10.   A Lewis structure with C as the central atom is most plausible. This structure has no formal charges.

$$
:\overset{..}{\underset{..}{Cl}}: \qquad\qquad :\overset{..}{\underset{.}{Cl}}:
$$

$$
\overset{..}{\underset{..}{C}}::\overset{..}{\underset{..}{O}}: \qquad or \qquad \underset{}{C}=\overset{..}{\underset{..}{O}}:
$$

$$
:\overset{..}{\underset{..}{Cl}}: \qquad\qquad :\overset{..}{\underset{..}{Cl}}:
$$

Counting the carbon-to-oxygen double bond as if it were a single bond, the central atom has three pairs of electrons distributed around it. All are bond pairs. The geometrical distribution of these electron pairs and the geometrical shape of the molecule are both trigonal *planar*. The structure can be represented as

$$
\begin{array}{l} Cl \\ \diagdown \\ C=O \\ \diagup \\ Cl \end{array}
$$

Method of Example 9-12.   This method, by not requiring a Lewis structure, also does not require that either the central atom or multiple bonds be specifically identified.

$$
total\ no.\ electron\ pairs = \frac{4 + 6 + (2 \times 7)}{2} = 12
$$

$$
no.\ bond\ pairs = 4 - 1 = 3
$$

$$
no.\ central\ pairs = 12 - (3 \times 3) = 3
$$

$$
no.\ lone\ pairs = 3 - 3 = 0
$$

The structure corresponds to AX_3 of Table 9-2: trigonal planar.

SIMILAR EXAMPLE: Exercise 28.

9-9 Bond Energies and Bond Distances

Energy is *released* when atoms join together through a chemical bond and must be *absorbed* if bonded atoms are to be separated. Let us define **bond energy** as the quantity of energy required to break one mole of chemical bonds in a *gaseous* species. The SI units for bond energies are kilojoules per mole of bonds (kJ/mol).

It is reasonable to expect the strength of a chemical bond, as measured by bond energy, to depend on the nature of the bond between atoms. That is, a double bond is stronger than a single bond, and a triple bond, in turn, is stronger than a double bond. Furthermore, some correlation should be expected between bond energy and **bond distance (bond length)**—the distance between the nuclei of bonded atoms. The *stronger* a chemical bond, the *shorter* the bond distance. Table 9-3 lists values of bond energies and bond distances for a number of common chemical bonds. The term that was described in Chapter 8 as the single covalent radius is simply one half the distance between the nuclei of identical atoms joined by a single covalent bond. Thus, the single covalent radii of H and F are 37 pm (one half of 74 pm) and 64 pm (one half of 128 pm), respectively.

Bond energy is not difficult to conceptualize when considering diatomic molecules; there is but one bond per molecule. In the manner indicated in Chapter 6, we can think of bond energy (or **bond enthalpy**) as an enthalpy change or heat of reaction. For example,

bond breakage: $H_2(g) \longrightarrow 2\,H(g)$ $\Delta\overline{H} = +435$ kJ/mol

bond formation: $2\,H(g) \longrightarrow H_2(g)$ $\Delta\overline{H} = -435$ kJ/mol

With a polyatomic molecule, such as H_2O, the situation is somewhat different. The energy required to dissociate one mole of H atoms by breaking one O—H bond per molecule

$H—OH(g) \longrightarrow H(g) + OH(g)$ $\Delta\overline{H} = +492$ kJ/mol

is different from the energy required to dissociate a second mole of H atoms by breaking the remaining bonds in OH(g)

$O—H(g) \longrightarrow H(g) + O(g)$ $\Delta\overline{H} = +428$ kJ/mol

Because bond energies depend on the environment of the given bond—the portion of the

TABLE 9-3
Some representative bond energies and bond distances (lengths)

Bond	Bond energy, kJ/mol	Bond distance Å	pm	Bond	Bond energy, kJ/mol	Bond distance Å	pm
H—H	435	0.74	74	C—O	360	1.43	143
H—C	414	1.10	110	C=O	736	1.23	123
H—N	389	1.00	100	C—Cl	326	1.77	177
H—O	464	0.97	97	N—N	163	1.45	145
H—F	565	1.01	101	N=N	418	1.23	123
H—Cl	431	1.36	136	N≡N	946	1.09	109
H—Br	364	1.51	151	F—F	155	1.28	128
H—I	297	1.70	170	Cl—Cl	243	1.99	199
C—C	347	1.54	154	Br—Br	192	2.28	228
C=C	611	1.34	134	I—I	151	2.66	266
C≡C	837	1.20	120				
C—N	305	1.47	147				
C=N	615	1.28	128				
C≡N	891	1.16	116				

FIGURE 9-13
Bond breakage and formation in a chemical reaction—Example 9-14 illustrated.

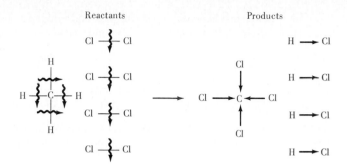

Reactants   Products

molecule close to the bond—it is customary to use *average* values. The bond energies in Table 9-3 are average values. For most molecules bond energies are approximately *additive,* i.e., the total bond energy is approximately the sum of the energies of the individual bonds.

Experimentally, spectroscopic methods can be used to determine bond distances and some bond energies. Other bond energies can be established indirectly, through Hess's law and data such as heats of formation and combustion. Example 9-16 provides an indication of how this is done.

Calculations Involving Bond Energies. For a reaction involving only *gaseous* species, we can visualize a hypothetical process in which, first, all the bonds in the reacting species are broken and then, the bonds in the products are formed. The sum of the enthalpy changes for bond breakage and formation yields the enthalpy change for the reaction, ΔH_{rxn}. Figure 9-13 depicts bond breakage and formation for the reaction

$$CH_4(g) + 4\ Cl_2(g) \longrightarrow CCl_4(g) + 4\ HCl(g) \tag{9.27}$$

Example 9-14   Calculate ΔH for the reaction

$$CH_4(g) + 4\ Cl_2(g) \longrightarrow CCl_4(g) + 4\ HCl(g) \qquad \Delta H_{rxn} = ?$$

Solution
ΔH *for bond breakage:*

 4 mol C—H bonds = 4 mol $\times$ (+414 kJ/mol) = +1656 kJ

4 mol Cl—Cl bonds = 4 mol $\times$ (+243 kJ/mol) = +972 kJ

ΔH *for bond formation:*

4 mol C—Cl bonds = 4 mol $\times$ (−326 kJ/mol) = −1304 kJ

4 mol H—Cl bonds = 4 mol $\times$ (−431 kJ/mol) = −1724 kJ

Heat of reaction:

$\Delta H_{rxn} = \Delta H_{\text{bond breakage}} + \Delta H_{\text{bond formation}}$
$\qquad = +1656\ kJ + 972\ kJ - 1304\ kJ - 1724\ kJ = -400\ kJ$

SIMILAR EXAMPLES: Exercises 37, 38.

Calculation of the heat of a reaction with tabulated bond energies generally offers no advantage over calculating this quantity from heat of formation data and expression (6.17). This is because heats of formation are usually known rather precisely, whereas bond energies are only average values. But there are times when heat-of-formation data are either not known or not available. Here bond energies can prove particularly useful.

The relationship between bond energies and enthalpy changes also permits a simple prediction of whether a reaction will be endothermic or exothermic: In general,

if weak bonds $\longrightarrow$ strong bonds $\Delta H < 0$ (*exothermic*)

and   (9.28)

if strong bonds $\longrightarrow$ weak bonds $\Delta H > 0$ (*endothermic*)

Example 9-15 applies this idea to a reaction involving some highly reactive, unstable species for which heats of formation are not normally tabulated.

Example 9-15   One of the reactions leading to photochemical smog involves the attack of a hydrocarbon molecule, RH, by a hydroxyl radical, OH. The products are a water molecule, H_2O, and a hydrocarbon radical, R· (which undergoes further reaction). Is this reaction endothermic or exothermic?

$$RH(g) + OH(g) \longrightarrow H_2O(g) + R\cdot(g)$$

Solution   In this reaction for every molecule of RH that reacts, one C—H bond is broken, requiring the absorption of 414 kJ/mol. An additional O—H bond is formed, converting OH(g) to $H_2O(g)$ and releasing 464 kJ/mol. Since more energy is released in forming new bonds than is absorbed in breaking old bonds, the reaction is exothermic.

SIMILAR EXAMPLE: Review Problem 12.

Example 9-16   The energy required to convert one mole of C(s) to C(g), called the heat of atomization, is 717 kJ/mol; the bond energy in H_2 is 435 kJ/mol; and the heat of formation of $CH_4(g)$ is −75 kJ/mol. Use these data to obtain an estimate of the C—H bond energy in $CH_4(g)$. Compare the result with the value listed in Table 9-3.

Solution.   The sum of reactions (1) and (2) represents the formation of $CH_4(g)$ from its elements in their standard states, for which $\Delta \overline{H}_f$ is given. The enthalpy change for reaction (1) is the heat of atomization of carbon. The heat of reaction (2) can be expressed in terms of the energy required to break two moles of H—H bonds and to form four moles of C—H bonds, but the value of the C—H bond energy is unknown. It is represented by x below.

(1)   C(s) $\longrightarrow$ C(g) $\Delta\overline{H} = +717$ kJ/mol
(2)   C(g) + 2 $H_2(g) \longrightarrow CH_4(g)$ $\Delta\overline{H} = (2 \times 435)$ kJ/mol $- 4x$
                                                                      $= 870$ kJ/mol $- 4x$

$\overline{}$

        C(s) + 2 $H_2(g) \longrightarrow CH_4(g)$ $\Delta\overline{H}_f = 717$ kJ/mol + 870 kJ/mol $- 4x$
                                                                          $= -75$ kJ/mol

$4x = (717 + 870 + 75)$ kJ/mol $= 1662$ kJ/mol
$x = 416$ kJ/mol

The value of the C—H bond energy listed in Table 9-3 is 414 kJ/mol. The agreement is quite good.

SIMILAR EXAMPLE: Exercise 39.

9-10 Partial Ionic Character of Covalent Bonds

We have established two fundamental bond types, ionic and covalent, and we have treated all bonds as if they were one or the other. But classification schemes generally involve borderline cases, and this is very much so with bond type.

FIGURE 9-14
Behavior of polar
molecules in the electric
field of a condenser.

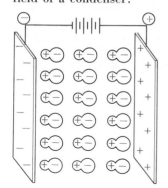

In the absence of an electric field the molecules are oriented randomly. However, when one plate is made negative and the other positive, polar molecules tend to orient themselves in the manner shown here. If the medium between the electrodes conducts electricity, a continuous flow of electrical charge—an electric current—results. In this case the device does not function as a condenser.

Polar Molecules and Dipole Moments.   The device depicted in Figure 9-14, consisting of a pair of electrodes separated by a *nonconducting* medium, is called a condenser or capacitor. Small quantities of electric charge can be stored on the electrodes, positive on one and negative on the other. The quantity of charge that can be stored depends on the medium between the electrodes. There is one class of substances that enhances this ability significantly. These are substances in whose molecules a separation of electric charge exists. Figure 9-14 suggests how such molecules line up in an electric field.

A separation into charge centers in a covalent molecule involves a shift of electrons toward one atom in a bond—the more electronegative (nonmetallic) atom. A molecule in which this charge separation exists is said to be **polar,** or it is termed a **dipole** because of the existence of two charge centers. For example, H_2 is a *nonpolar* molecule and HCl is *polar*.

$$H : H \qquad {}^{\delta+}H : \overset{..}{\underset{..}{Cl}} : {}^{\delta-}$$

The magnitude of the effect described here is denoted through the **dipole moment μ.** The dipole moment is the product of the magnitude of the charges (δ) and the distance separating them (d). (The δ symbols suggest a small magnitude of charge, less than the charge of an electron.)

$$\mu = \delta d \tag{9.29}$$

If the product of the charge and the distance of separation has a magnitude of 3.34×10^{-30} coulomb · meter (C · m), the dipole moment, μ, has a value of 1 **debye (D).** How significant is charge separation in a molecule like HCl? The measured dipole moment of HCl is 1.03 D, and the bond distance listed in Table 9-3 is 136 pm. That is,

$$\text{magnitude of charge} \times 1.36 \times 10^{-10} \text{ m} = 1.03 \text{ D} = 1.03 \times 3.34 \times 10^{-30} \text{ C} \cdot \text{m}$$

$$\text{magnitude of charge} = 3.44 \times 10^{-30} \text{ C} \cdot \text{m}/1.36 \times 10^{-10} \text{ m} = 2.53 \times 10^{-20} \text{ C}$$

This charge is about 16% of the charge on an electron (1.60×10^{-19} C), suggesting that HCl is about 16% ionic in character.

How does this new information change our outlook on the HCl molecule? Perhaps it is best to refer to HCl as a polar covalent molecule which is reasonably well represented by the Lewis structure

$$H : \overset{..}{\underset{..}{Cl}} : \tag{9.30}$$

but for which there is an ionic contribution

$$[H^+][: \overset{..}{\underset{..}{Cl}} : {}^-] \tag{9.31}$$

leading to the representation

$${}^{\delta+}H : \overset{..}{\underset{..}{Cl}} : {}^{\delta-} \tag{9.32}$$

Molecular Shapes and Dipole Moments.   There is an electronegativity difference between C and O atoms, and, as expected, the molecule CO is slightly polar. In the representation below the existence of a dipole moment in the carbon-to-oxygen bond is represented by a cross-base arrow ($\leftrightarrow$). The arrow points to the atom that attracts electrons more strongly—the more electronegative atom. (No attempt is made here to depict the multiple-bond character of the carbon-to-oxygen bond.)

$$C \overset{+}{\longrightarrow} O \qquad \mu = 0.11 \text{ D}$$

Even though there is an electronegativity difference and hence a **bond moment** along each carbon-to-oxygen bond, *the CO_2 molecule is nonpolar.* This can only mean that the effects of the two O atoms in attracting electrons cancel each other out. The O atoms lie

The result described here for CO_2 is analogous to a tug-of-war contest between equally matched teams. Although there is a strong pull in each direction, the knot at the center of the rope does not move.

along the same straight line through the central C atom.

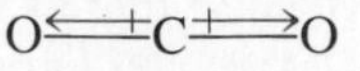

$$O \rightleftharpoons C \rightleftharpoons O \qquad \mu = 0$$

There is also an electronegativity difference between H and O atoms, leading to an O—H bond dipole moment of 1.51 D. If H_2O were a linear molecule, the bond moments would be in opposite directions and there would be no resultant dipole moment. But the measured dipole moment of water is 1.84 D. The molecule cannot be linear. The two bond moments combine to yield this resultant dipole moment for a particular bond angle—104°.

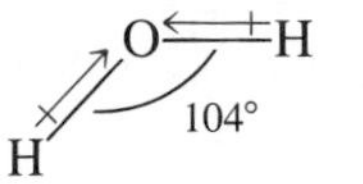

$$\underset{H}{\overset{O \rightleftharpoons H}{\diagdown}} \qquad \mu = 1.84 \text{ D} \qquad (9.33)$$

Verification of the tetrahedral shape of the molecule CCl_4 is possible through dipole moment measurements. Although the C—Cl bond moment is large (2.05 D), there is no resultant dipole moment in the molecule. The chlorine atoms must be arranged symmetrically about the carbon. Substituting a less electronegative atom for one of the Cl atoms (e.g., H) leads to an imbalance of attractive forces for electrons and a resulting dipole moment. Figure 9-15 illustrates this point.

Example 9-17 Which of the molecules listed would you expect to be polar and which nonpolar? Cl_2, ICl, BeF_2, NO, SO_2, XeF_4.

Solution

Polar: ICl, NO, SO_2. ICl and NO are diatomic molecules with an electronegativity difference between the bonded atoms. SO_2 is a bent molecule (nonlinear) with an electronegativity difference between S and O.

Nonpolar: Cl_2, BeF_2, XeF_4. Cl_2 is a homonuclear diatomic molecule. BeF_2 is linear. XeF_4 is a square planar molecule with the F atoms arranged symmetrically about the central Xe atom (recall Table 9-2).

SIMILAR EXAMPLES: Review Problem 13; Exercise 42.

Ionic Resonance Energy. If the polarity of the bond A—B is about the same as the bonds A—A and B—B, in particular if all the bonds are nonpolar, the bond energy of A—B may be taken as the average of A—A and B—B.

Example 9-18 Use data from Table 9-3 to calculate the energy of the Br—Cl bond.

Solution

$$E_{Br—Cl} = \tfrac{1}{2}(E_{Br—Br} + E_{Cl—Cl})$$
$$= \tfrac{1}{2}(192 + 243) = 218 \text{ kJ/mol}$$

(The experimental value = 218 kJ/mol.)

SIMILAR EXAMPLE: Exercise 44.

FIGURE 9-15
Dipole moments and molecular shapes.

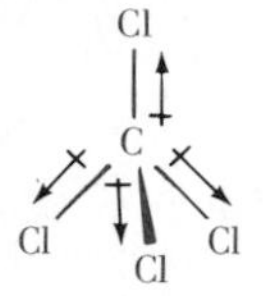

CCl_4: a nonpolar molecule
$\mu = 0$

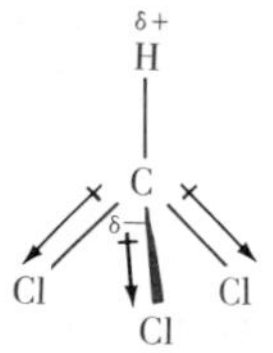

$CHCl_3$: a polar molecule
$\mu = 1.92$ debye

(a) The symmetrical distribution of the four carbon-to-chlorine bonds in CCl_4 causes a cancellation of all bond dipole moments; there is no resultant dipole moment in the molecule. CCl_4 is nonpolar.
(b) The carbon-to-hydrogen bond has a dipole moment of essentially zero because the electronegativities of C and H are quite similar. The three carbon-to-chlorine bond dipole moments cause the chlorine end of the molecule to develop a slight negative charge. The hydrogen end carries a slight positive charge. $CHCl_3$ is polar.

When the method of Example 9-18 is applied to the HCl molecule, it fails. The average of the bond energies for H—H and Cl—Cl is about 339 kJ/mol, but the experimentally determined value is 431 kJ/mol. The H—Cl bond is stronger by 92 kJ/mol than would be the case if the three types of bonds (H—H, Cl—Cl, H—Cl) all were of equal polarity. We have already seen one method of assessing the relative ionic character of the HCl molecule based on dipole moment; here is another. The energy difference of 92 kJ/mol is the additional stabilization contributed by the ionic structure (9.31) to the true structure (9.32). It is called the **ionic resonance energy.**

FIGURE 9-16
Percent ionic character
of a chemical bond
as a function of
electronegativity
difference.

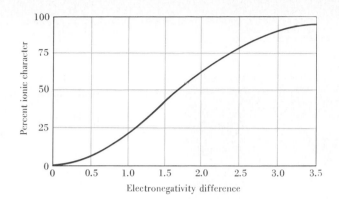

Electronegativity. Pauling's scale of electronegativity values presented in Section 8-10 was obtained through an extensive correlation of bond energies. Particular attention was given to ionic resonance energy. The equation developed for this purpose was

$$(\Delta EN)^2 = \frac{IRE}{96} \qquad (9.34)$$

where ΔEN is the electronegativity difference between the bonded atoms and IRE is the ionic resonance energy in kilojoules per mole.

From electronegativity values Pauling was able to assign a ''percent ionic character'' to a bond, as illustrated in Figure 9-16. If the electronegativity difference is very small, a bond is essentially covalent; and if large, essentially ionic. There is no particular electronegativity difference at which the bond type changes from covalent to ionic. As a very rough rule, if the electronegativity difference exceeds about 1.7, a bond is greater than 50% ionic.

Example 9-19 Use equation (9.34) to calculate the electronegativity difference between H and Cl and compare the result with data from Table 8-6. Use Figure 9-16 to estimate the percent ionic character in the H—Cl bond and compare the result with the estimate based on dipole moment measurements.

Solution. The ionic resonance energy for HCl stated above is 92 kJ/mol.

$$(\Delta EN)^2 = \frac{92}{96} = 0.96 \qquad \Delta EN = (0.96)^{1/2} = 0.98$$

From Table 8-6,

$$\Delta EN = EN_{Cl} - EN_H = 3.16 - 2.20 = 0.96$$

From Figure 9-16,

$$\Delta EN \approx 0.9 - 1.0 \qquad \text{% ionic character} \approx 20\%$$

From dipole moment measurement (page 243), the H—Cl bond is about 16% ionic.

SIMILAR EXAMPLES: Exercises 45, 46.

FOCUS ON
Polymers

In 1934, after four years of development work, Wallace Carrothers and his associates at E. I. du Pont de Nemours & Company succeeded in producing the first synthetic fiber—*nylon*. It is now possible for chemistry students to carry out a variation of this polymerization reaction in the general chemistry laboratory.

Based on the Lewis symbols of H and C (Figure 9-1), a first trial at writing a Lewis structure for ethylene, C_2H_4, might yield

$$\cdot \overset{\displaystyle H}{\underset{\displaystyle H}{C}} - \overset{\displaystyle H}{\underset{\displaystyle H}{C}} \cdot \qquad (9.35)$$

This structure is unacceptable because it leaves each C atom with an incomplete octet and an unpaired electron. If we apply the usual remedy—combining the unpaired electrons into an additional electron-pair bond between C and H—we obtain the accepted Lewis structure for ethylene.

$$\overset{\displaystyle H}{\underset{\displaystyle H}{C}} = \overset{\displaystyle H}{\underset{\displaystyle H}{C}} \qquad (9.36)$$

Another possibility for structure (9.35) is to join it with others of the same kind, as in this grouping of three.

$$\cdot C - C - C - C - C - C \cdot \qquad (9.37)$$

Here, the central C_2H_4 unit (1) has a satisfactory Lewis structure, but we still have a problem with the end units (2 and 3). Suppose we add two more of the units (9.35). Now, the Lewis structures of 1, 2, and 3 are taken care of and the problem of the unpaired electrons is shifted to the new end units (4 and 5).

$$\cdot C - C - C - C - C - C - C - C - C - C \cdot \qquad (9.38)$$

What we are describing is the plausibility of joining together a large number of simple molecules called **monomers** (Gr. *monos*, single, and *meros*, parts) into a complex long-chain molecule called a **polymer** (Gr. *polys*, many, and *meros*, parts). The chemical reaction that achieves this result is called a **polymerization reaction.** In the case in question the monomer is *ethylene* and the polymer is *polyethylene*, which is represented as

$$\left[\begin{array}{c} H \quad H \\ | \quad | \\ C - C \\ | \quad | \\ H \quad H \end{array} \right]_n \qquad (9.39)$$

The subscript n signifies that the monomer unit within the parentheses is repeated many times; n can have a range of values, say from several hundred to several thousand. A hypothetical polyethylene molecule is pictured in Figure 9-17.

Further aspects of the polymerization of ethylene that we will consider in Chapter 26 deal with how the double bond in C_2H_4 (structure 9.36) is "opened up" to initiate the polymerization reaction, how the polymer chains are propagated and terminated (to eliminate unpaired electrons), and the meaning of the term molecular weight when applied to a polymer.

Ubiquitous Polymers. The bulk of the polymer industry centers on *synthetic organic* polymers, like polyethylene. Ethylene is one of the principal starting materials for polymer production, and this accounts for its high ranking (sixth) among the key chemicals (recall Table 1-3). Another measure of the importance of polymers is that between 30 and 50% of all chemists and chemical engineers work with polymers.

In a general sense, any substance that can be thought of in terms of small units joined together into long chains or two- or three-dimensional networks is a polymer. This broad definition will allow us to consider certain *inorganic* materials as polymers, e.g., diamond and graphite in Chapter 11 and liquid and plastic sulfur, red phosphorus, and SiO_2 in Chapter 13. This definition also helps us to recognize *natural* as well as synthetic polymers. Among the natural polymers that we will encounter (Chapter 27) are carbohydrates, proteins, and nucleic acids. One of the most familiar of the natural polymers is probably natural rubber.

Rubber. The milky white fluid called **latex** is collected as an exudate of the tree *Hevea brasiliensis*. When this fluid is coagulated with salt and acetic acid, the product is crude rubber. This rubber is refined and processed according to its final use. The name "rubber" is attributed to the chemist Joseph Priestley (1770), who found that it could be used to rub out pencil marks. Its chief use for several decades was in pencil erasers. The repeating unit of rubber is shown in brackets below.

FIGURE 9-17
A hypothetical polyethylene molecule.

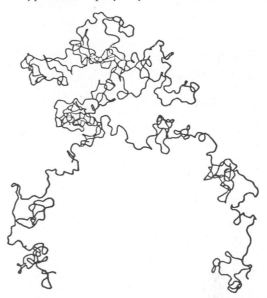

This is a randomly generated chain of 500 C_2H_4 monomer units. From L. R. G. Treloar, *Introduction to Polymer Science*, Wykeham Publications (London) Ltd., 1970.

$$
\begin{array}{c}
\mathrm{CH_3} \quad\quad \mathrm{H} \\
\mathrm{C}\!=\!\mathrm{C} \\
\mathrm{CH_2} \quad\quad \mathrm{CH_2}
\end{array}
\left[
\begin{array}{c}
\mathrm{CH_3} \quad\quad \mathrm{H} \\
\mathrm{C}\!=\!\mathrm{C} \\
\mathrm{CH_2} \quad\quad \mathrm{CH_2}
\end{array}
\right]
\begin{array}{c}
\mathrm{CH_3} \quad\quad \mathrm{H} \\
\mathrm{C}\!=\!\mathrm{C} \\
\mathrm{CH_2}
\end{array}
\qquad (9.40)
$$

rubber

Early rubber products had many undesirable properties: The rubber tended to creep or flow, and the products were sticky in hot weather and stiff in cold weather. In 1839, Charles Goodyear, a Connecticut inventor, accidentally spilled a sulfur–rubber mixture on a hot stove; the resulting rubber was stronger, more elastic, and more resistant to heat and cold. This treatment is now called **vulcanization** (after Vulcan, the Roman god of fire). The purpose of vulcanization is to form **crosslinks** between long polymer chains. In the illustration at the right the link is through two sulfur atoms.

$$
\begin{array}{c}
\mathrm{CH_3} \\
| \\
\text{\textasciitilde}\mathrm{CH_2}\!-\!\mathrm{C}\!=\!\mathrm{CH}\!-\!\mathrm{CH}\text{\textasciitilde} \\
| \\
\mathrm{S} \\
| \\
\mathrm{S} \\
| \\
\text{\textasciitilde}\mathrm{CH_2}\!-\!\mathrm{C}\!=\!\mathrm{CH}\!-\!\mathrm{CH}\text{\textasciitilde} \\
| \\
\mathrm{CH_3}
\end{array}
\qquad (9.41)
$$

TABLE 9-4
Some synthetic carbon-chain polymers

Name of polymer	Repeating unit	Some uses
Elastomers neoprene [polychloroprene]	$\left(\begin{array}{c} \text{H} \ \text{H} \ \text{Cl} \ \text{H} \\ \vert \quad \vert \quad \vert \quad \vert \\ -\text{C}-\text{C}=\text{C}-\text{C}- \\ \vert \qquad\qquad \vert \\ \text{H} \qquad\quad \text{H} \end{array}\right)_n$	wire and cable insulators, industrial hoses and belts, shoe soles and heels, gloves
silicone rubber	$\left(\begin{array}{c} \text{CH}_3 \\ \vert \\ -\text{Si}-\text{O}- \\ \vert \\ \text{CH}_3 \end{array}\right)_n$	gaskets, electrical insulation, surgical membranes, medical devices for use in the body
Fibers nylon 66	$\left(\begin{array}{c} \text{O} \qquad\qquad \text{O} \ \text{H} \qquad\qquad \text{H} \\ \Vert \qquad\qquad \Vert \ \vert \qquad\qquad \vert \\ -\text{C}-(\text{CH}_2)_4-\text{C}-\text{N}-(\text{CH}_2)_6-\text{N}- \end{array}\right)_n$	hosiery, rope, tire cord, fish line, parachutes, artificial blood vessels
Acrilan, Orlon	$\left(\begin{array}{c} \text{H} \ \ \text{C}\equiv\text{N} \\ \vert \qquad \vert \\ -\text{C}-\text{C}- \\ \vert \qquad \vert \\ \text{H} \quad\ \text{H} \end{array}\right)_n$	fabrics, carpets, drapes, upholstery, electrical insulation
Plastics polyethylene	$\left(\begin{array}{c} \text{H} \ \ \text{H} \\ \vert \quad \vert \\ -\text{C}-\text{C}- \\ \vert \quad \vert \\ \text{H} \ \ \text{H} \end{array}\right)_n$	bags, bottles, tubing, packaging film, paper coating
polypropylene	$\left(\begin{array}{c} \text{H} \ \ \text{CH}_3 \\ \vert \qquad \vert \\ -\text{C}-\text{C}- \\ \vert \qquad \vert \\ \text{H} \ \ \text{H} \end{array}\right)_n$	laboratory and household ware, storage battery cases, artificial turf, surgical casts, toys
PVC, "vinyl" [poly(vinyl chloride)]	$\left(\begin{array}{c} \text{H} \ \ \text{Cl} \\ \vert \quad \vert \\ -\text{C}-\text{C}- \\ \vert \quad \vert \\ \text{H} \ \ \text{H} \end{array}\right)_n$	bottles, records, floor tile, food wrap, piping, hoses, linings for ponds and reservoirs
Teflon [poly(tetrafluoroethylene)]	$\left(\begin{array}{c} \text{F} \ \ \text{F} \\ \vert \quad \vert \\ -\text{C}-\text{C}- \\ \vert \quad \vert \\ \text{F} \ \ \text{F} \end{array}\right)_n$	bearings, insulation, gaskets, nonstick surfaces (ovenware, frying pans), heat-resistant industrial plastics

Classification of Polymers. Some polymers have sites on their chains at which crosslinks between chains are formed when the polymer is heated for the first time. Such polymers can be heated to their softening points and molded into particular shapes. Because of crosslinking they retain these shapes on cooling. Reheating does not reverse the process; that is, instead of again becoming soft, the polymers degrade or decompose upon reheating. These are called **thermosetting** polymers. Other polymers, those unable to form crosslinks, also soften when heated, and they too can be molded into desired shapes that they retain on cooling. These polymers, however, are able to go through repeated cycles of heating, softening, reshaping, and cooling. These are called **thermoplastic** polymers. Another classification scheme is based on the properties of polymers described below. Examples are given in Table 9-4.

Elastomers. The chief characteristic of elastomers is their ability to be elongated under stress (stretched) and to regain their former shapes when the stress is relieved. In short, they are elastic. Rubber, whether natural or synthetic, is the best known example. Silicones are also elastomers (see Section 21-4). Current annual production of elastomers in the United States is about 3.5 million metric tons (1 metric ton = 1000 kg).

Fibers. Fibers are polymers oriented to provide optimum properties, such as high tensile strength, along one axis. These are threadlike polymers that can be woven into fabrics. Cotton, wool, and silk are natural fibers. Some synthetic fibers, such as nylon, Orlon, and Dacron, have additional desirable properties: increased tensile strength; lightness of weight; low moisture absorption; resistance to moths, mildew, rot, and fungus; and wrinkle resistance. Current annual United States production of fibers is about 10 million metric tons.

Plastics. Plastics, which are intermediate between elastomers and fibers, can be imparted with a variety of room-temperature properties. Thus polystyrene is stiff and brittle, whereas polypropylene is extremely tough, impact resistant, tear resistant, and flexible in thin sheets. Annually, about 15 million metric tons of plastics are produced in the United States.

Summary

The Lewis theory of chemical bonding is based on the behavior of electrons in the atoms being joined, especially the outer-shell or valence electrons. Although the Lewis theory can be used to predict formula units of ionic compounds, its greatest value is in describing covalently bonded species. The valence electrons of the bonded atoms are distributed throughout the framework or skeleton structure in accordance with a set of rules.

Sometimes experimental data must be used to establish the correct skeleton structure of a molecule. Lacking such data, the concept of formal charge may be used. At times, even though all the usual rules are followed, it is impossible to establish a single Lewis structure for a species. In these cases two or more plausible structures are obtained, and the true structure is said to be a resonance hybrid of these plausible structures. One of the basic assumptions of the Lewis theory is that all atoms in a Lewis structure acquire eight electrons in their valence shells (an octet). This assumption fails for odd-electron species, for electron-deficient compounds, and for compounds in which the central atom can accommodate an expansion of the octet, that is, 10 or 12 electrons.

A powerful method for predicting the shapes of molecules is the valence-shell electron-pair repulsion (VSEPR) theory. This method requires determining the number of pairs of valence electrons surrounding the central atom. The shape of a molecule depends both on the geometrical distribution of all valence-shell electron pairs and on whether these pairs are bonding pairs or lone (nonbonding) pairs.

Molecular properties such as bond energy and bond distance are not required in writing Lewis structures, but they are sometimes helpful in judging whether a structure is plausible. For example, they can be used to establish whether a covalent bond has multiple bond character. In addition, bond energies can be used to calculate enthalpy changes (heats) of reactions involving gaseous species.

In most of this chapter ionic and covalent bonds are treated as if they are distinctly different. In fact, most bonds have both partial ionic and partial covalent character. One indication of the partial ionic character of a covalent bond is the degree of separation of electrical charge that exists in the bond. This is measured through the dipole moment. Another measure of the partial ionic character of a bond is achieved through the use of electronegativities.

Learning Objectives

As a result of studying Chapter 9, you should be able to

1. State the basic assumptions of the Lewis theory of chemical bonding.

2. Relate the Lewis symbol of an element to its position in the periodic table.

3. Write Lewis structures for simple ionic compounds.

4. Use the basic rules of Lewis theory to propose a plausible skeleton structure for a molecule, and assign valence electrons to this structure.

5. Assign formal charges to the atoms in a Lewis structure.

6. Select the most plausible among alternative Lewis structures by applying rules pertaining to formal charges.

7. Recognize situations in which resonance occurs and write plausible structures contributing to a resonance hybrid.

8. Identify odd-electron species and write Lewis structures for them.

9. Recognize molecules in which expansion of the octet is required and write appropriate Lewis structures for them.

10. Relate the shape of a molecule to the distribution of valence-shell electron pairs around the central atom.

11. Use bond energies to calculate ΔH of a reaction involving gases.

12. Relate electronegativity data, molecular geometry, and the dipole moments of molecules.

13. Describe the relationship between electronegativity difference and the percent ionic character of a bond.

Some New Terms

Bond distance (bond length) is the distance between the nuclei of atoms joined by a chemical bond.

Bond energy (bond enthalpy) is the quantity of energy (usually expressed in kJ/mol) required to break one mole of chemical bonds in a gaseous species.

A **bonding pair** is a pair of electrons involved in bond formation.

In a **coordinate covalent bond** the electrons shared between atoms are contributed by just one of the atoms.

A **covalent bond** results from the sharing of electrons between atoms.

Dipole moment (μ) is a measure of the extent to which a separation of charges exists within a molecule. It is the product of the magnitude of the charge and the distance separating the charge centers. The unit used to measure dipole moment is the **debye**, 3.34×10^{-30} C m.

In a **double covalent bond** *two pairs* of electrons are shared between bonded atoms. The bond is represented by a double-dash sign (=).

Expanded octet is a term used to describe situations in which certain atoms in the third period or beyond are able to use 10 or 12 electrons in forming bonds.

Formal charge is the number of outer-shell (valence) electrons in an isolated atom minus the number of electrons assigned to that atom in a Lewis structure.

Incomplete octet is a term used to describe situations in which an atom fails to acquire eight outer-shell electrons when it forms bonds.

An **ionic bond** results from the transfer of electrons between metal and nonmetal atoms. Positive and negative ions are formed and held together by electrostatic attraction.

A **Lewis structure** is a combination of Lewis symbols that depicts the transfer or sharing of electrons in a chemical bond.

In the **Lewis symbol** of an element valence electrons are represented by dots placed around its chemical symbol.

A **lone pair** is a pair of electrons found in the valence shell of an atom and *not* involved in bond formation.

A **monomer** is a simple molecule that is capable of joining with others to form a complex long-chain molecule called a **polymer.**

A **multiple covalent bond** is a bond in which more than two electrons are shared between the bonded atoms.

In a **nonpolar molecule** the centers of positive and negative charge coincide. That is, there is no separation of charge within the molecule.

An **octet** refers to the presence of *eight* electrons in the outermost (valence) electronic shell of an atom.

An **odd-electron species** is one in which the total number of valence electrons is an *odd* number. At least one unpaired electron is present in the species.

In a **polar molecule** there exists a separation of electrical charge into positive and negative centers.

Resonance occurs when two or more plausible Lewis structures can be written for a species. The true structure is a composite or hybrid of these.

A **single covalent bond** results from the sharing of *one pair* of electrons between bonded atoms. It is represented by a single dash sign (—).

In a **triple covalent bond** *three pairs* of electrons are shared between the bonded atoms. The bond is represented by a triple-dash sign (≡).

The **valence-shell electron-pair repulsion (VSEPR) theory** relates the shape of a species to the geometrical distribution of electron pairs in the valence shell of the central atom.

Suggestions for Further Study

CARRAHER, C. E., Jr., "What Are Polymers?", *Chemistry,* **51** [5], 6 (1978).

GILLESPIE, R. J., "The Valence-Shell Electron-Pair Repulsion (VSEPR) Theory of Directed Valency," *J. Chem. Educ.,* **40,** 295 (1963).

GILLESPIE, R. J., "The Electron-Pair Repulsion Model for Molecular Geometry," *J. Chem. Educ.,* **47,** 18 (1970).

JENSEN, W. B., "Abegg, Lewis, Langmuir and the Octet Rule," *J. Chem. Educ.,* **61,** 191 (1984).

MARK, H. F., "The Development of Plastics," *American Scientist,* **72,** 156 (1984).

MICKEY, C. D., "Molecular Geometry," *J. Chem. Educ.,* **57,** 210 (1980).

SEARS, J. A., "Polymer Pioneers," *Chemistry,* **50** [7] 6 (1977).

WATKINS, K. W., "Heating in Microwave Ovens: An Example of Dipole Moments in Action," *J. Chem. Educ.,* **60,** 1043 (1983).

Review Problems

1. Write Lewis symbols for the following atoms and ions: **(a)** H; **(b)** Kr; **(c)** Ge; **(d)** Mg^{2+}; **(e)** Br^-; **(f)** Ga; **(g)** Sc^{3+}; **(h)** Cs; **(i)** S^{2-}.

2. Write Lewis structures for the following ionic compounds: **(a)** NaF; **(b)** MgO; **(c)** SrI_2.

3. Write plausible Lewis structures for the following molecules, which contain only single covalent bonds: **(a)** Br_2; **(b)** ICl; **(c)** OF_2 (FOF); **(d)** NI_3; **(e)** H_2Te.

4. The following molecules contain multiple covalent bonds. Give a plausible Lewis structure for each. **(a)** CS_2; **(b)** O_3; **(c)** H_2CO.

5. Indicate what is wrong with each of the following structures.

(a) H—H—N̈—Ö—H   **(b)** :Ö—C̈l—Ö:

(c) [·C̈=N̈:]⁻   **(d)** Ca—Ö:

6. Assign formal charges to the atoms in the species represented below. If there are no formal charges present for certain of these species, so indicate.

(a) :Ï—Ï: **(b)** (S with :O: :O:)

(c) [O-C-O]²⁻   **(d)** (C with O double bond, :Cl: :Cl:)

(e) [H—Ö—Ö:]⁻   **(f)** (N with :O: :O:)

7. Each of the following ionic compounds consists of a combination of monatomic and polyatomic ions. Represent these compounds with Lewis structures: **(a)** $Mg(OH)_2$; **(b)** NH_4I; **(c)** $Ca(ClO_2)_2$. (*Hint:* Each of the polyatomic ions is described in Section 9-5.)

8. Which of the following species would you expect to be diamagnetic and which paramagnetic? (*Note:* Some of these species are not especially stable.) **(a)** OH^-; **(b)** OH; **(c)** NO_3; **(d)** SO_3; **(e)** SO_3^{2-}; **(f)** HO_2.

9. Draw plausible Lewis structures for the following species, using the notion of expanded octets where necessary: **(a)** BrF_5; **(b)** PF_3; **(c)** ICl_3; **(d)** SF_4.

10. Use the valence-shell electron-pair repulsion theory to predict the geometrical shapes of the following species: **(a)** CO; **(b)** $SiCl_4$; **(c)** $SbCl_5$; **(d)** H_2Se; **(e)** ICl_3; **(f)** AlF_6^{3-}; **(g)** SO_3.

11. For each of the bonds shown in the structure below, indicate **(a)** the bond length and **(b)** the bond energy. How much energy, in joules, would be required to break all the bonds in *one molecule* of $H_2ClCCHO$?

(structure of $H_2ClCCHO$)

12. Without performing detailed calculations, indicate whether each of the following reactions is endothermic or exothermic.
  (a) $CH_4(g) + I(g) \rightarrow CH_3(g) + HI(g)$
  (b) $H_2(g) + I_2(g) \rightarrow 2\ HI(g)$
  (c) $C_2H_6(g) + Cl_2(g) \rightarrow C_2H_5Cl(g) + HCl(g)$

13. Which of the following molecules would you expect to have a dipole moment? **(a)** N_2; **(b)** NO; **(c)** BF_3; **(d)** HBr; **(e)** $HCBr_3$; **(f)** $SiCl_4$; **(g)** OCS.

14. Arrange the following in their expected order of *increasing* dipole moment: AsH_3; AsF_3; $AsCl_3$; AsI_3; $AsBr_3$.

15. Use electronegativity data and Figure 9-16 to arrange the following bonds in terms of *increasing* ionic character: C—H; F—H; Na—Cl; Br—H; K—F.

Exercises

Lewis theory

1. What are some of the essential differences in the ways in which Lewis structures are written for ionic and covalent bonds?

2. Give several examples for which the following statement proves to be incorrect.
"All atoms in a Lewis structure have an octet of electrons in their valence shells."

Ionic bonding

3. Derive the correct formulas for the following ionic compounds by writing Lewis structures: (a) lithium oxide; (b) sodium bromide; (c) strontium fluoride; (d) scandium chloride.

4. In all simple binary ionic compounds, the nonmetal atoms acquire the electron configurations of noble gas atoms. This is not always the case with the metal atoms. Explain why this is so.

5. Why is it inappropriate to use the term "molecules of NaCl" when describing *solid* sodium chloride? What would the term signify if one were describing *gaseous* sodium chloride?

Lewis structures

6. With reference to Lewis structures, what is meant by each of the following terms: (a) valence electrons; (b) octet; (c) unshared electron pairs; (d) multiple bonds; (e) coordinate covalent bonds; (f) resonance; (g) odd-electron species; (h) expanded octet?

7. By means of Lewis structures represent bonding between the following pairs of elements. Your structures should show clearly whether the bonding is essentially ionic or covalent. Give the name, formula, and formula weight of each. (a) Rb and Cl; (b) H and Se; (c) B and Cl; (d) Cs and S; (e) Sr and O; (f) F and O.

8. Write plausible Lewis structures for the following species: (a) H_2NOH; (b) N_2F_2; (c) HONO; (d) H_2NNO_2.

9. Indicate what is wrong with each of the following Lewis structures. Replace each by a more acceptable structure.

(a) $[: \overset{..}{S} — C \equiv \overset{..}{N} :]^-$

(b) $[: \overset{..}{\underset{..}{Cl}}]^+ [: \overset{..}{\underset{..}{O}} :]^{2-} [\overset{..}{\underset{..}{Cl}} :]^+$

(c) $: \overset{..}{O} = N = \overset{..}{O} :$

(d) $: \overset{..}{Cl} — \overset{..}{N} = \overset{..}{Cl} :$ with $: \overset{..}{\underset{..}{Cl}} :$ below

10. Suggest reasons why the following do not exist as stable molecules: (a) H_3; (b) HHe; (c) He_2; (d) H_3O.

Formal charge

11. Assign formal charges to the species represented below. If there are no formal charges present for certain of these species, so indicate.

(a) $\left[: \overset{..}{O} — \overset{\overset{..}{O}:}{\underset{..}{Cl}} — \overset{..}{O} : \right]^-$

(b) $: \overset{S}{\underset{\diagup \diagdown}{O}} \quad \overset{..}{O} :$

(c) $\left[: \overset{..}{F} — \overset{\overset{..}{F}:}{\underset{\underset{..}{F}:}{B}} — \overset{..}{F} : \right]^-$

(d) $H — \overset{..}{O} — \overset{\overset{:O:}{\parallel}}{\underset{\underset{H}{\overset{:O:}{|}}}{P}} — \overset{..}{O} — H$

(e) $[: \overset{..}{N} = N = \overset{..}{N} :]^-$

(f) $[: \overset{..}{N} — N \equiv N :]^-$

(g) $[: \overset{..}{O} = N = \overset{..}{O} :]^+$

12. Use the concept of formal charge to select the more likely skeleton structure for each of the following molecules: (a) H_2NOH or H_2ONH; (b) SCS or CSS; (c) NOCl or ONCl; (d) SCN^- or CNS^- or CSN^-.

*13. Although not specifically mentioned in the text, there is a relationship between the formation of a coordinate covalent bond and the presence of formal charges in a molecule. Describe this relationship. (*Hint:* You may find it helpful to use the alternate definition of formal charge in the footnote on page 227.)

Polyatomic ions

14. Propose Lewis structures for the following ionic species containing sulfur-to-sulfur bonds: (a) S_2^{2-}; (b) S_3^{2-}; (c) S_4^{2-}; (d) S_5^{2-}.

15. The polyatomic anions below involve covalent bonds between O atoms and the central nonmetal atom. Propose a plausible Lewis structure for each.
(a) OCl^-; (b) NO_2^-; (c) BrO_3^-.

16. Represent each of the following ionic compounds by an appropriate Lewis structure: (a) KIO_3; (b) $Ca(OCl)_2$; (c) NH_4ClO_4.

Resonance

17. In the manner used to establish the structures for SO_2 shown in (9.18), demonstrate that there are *three* equivalent structures that can be written for SO_3.

18. In Example 9-8 the phenomenon of resonance was illustrated for the nitrate ion. Resonance is also involved in the nitrite ion, NO_2^-. Represent this fact through appropriate Lewis structures.

19. With reference to the ozone molecule, O_3, show that no single structure can be written for this molecule if it is assumed that the two O—O bonds are equivalent.

20. Nitric acid, HNO_3, can be represented as a resonance hybrid of the structures shown below. Which structure(s) seems most plausible? Explain.

(a) ![structure a]

(b) ![structure b]

(c) ![structure c]

Odd-electron species

21. As with the case of NO_2 described in the text, the molecule NO is paramagnetic. Represent this molecule through a Lewis structure(s).

22. NO_2 may dimerize (two molecules join) to the molecule N_2O_4. Write a plausible Lewis structure for N_2O_4. Do you think that N_2O_4 is diamagnetic or paramagnetic?

23. Write plausible Lewis structures for the following odd-elec-tron species: **(a)** HO_2; **(b)** CH_3; **(c)** ClO_2; **(d)** NO_3.

Expanded octets

24. Phosphorus and sulfur atoms make use of expanded octets in many of their compounds. Nitrogen and oxygen never do. Would you expect As and Se to resemble more nearly P and S or N and O? Explain.

25. Draw plausible Lewis structures for the following species, using the notion of expanded octets where necessary: **(a)** $SO_3{}^{2-}$; **(b)** $HOClO_2$; **(c)** HONO; **(d)** $OP(OH)_3$; **(e)** O_2SCl_2; **(f)** XeO_3.

26. Indicate the nature of the sulfur-to-nitrogen bond in F_3SN (i.e., single, double, triple). (*Hint:* Use the notion of an ex-panded octet and ideas about formal charges.)

27. Exercise 17 refers to three *equivalent* structures for SO_3. Add to these *four* plausible structures based on an expanded octet for the sulfur atom. Of all seven structures (three from Exercise 17 and four here), which is the most plausible from the stand-point of formal charges? What experimental evidence would be helpful in assessing the relative importance of the various possi-ble structures?

Molecular shapes

28. Each of the following molecules contains one or more multi-ple covalent bonds. Draw plausible Lewis structures to represent this fact, and predict the shape of each molecule. **(a)** CO_2; **(b)** N_2O; **(c)** NSF; **(d)** $ClNO_2$.

29. Can you think of an example of a molecule in which the central atom has *one bonding pair* and *three lone pairs* of elec-trons? What must be the shape of this molecule?

30. The structure of BF_3 is shown in Table 9-2 to be planar. If a fluoride ion is attached to the B atom of BF_3 through a coordinate covalent bond, the ion $BF_4{}^-$ results. What is the geometrical shape of this ion?

31. Three possible Lewis structures for SO_3 were described in Exercise 17, and another four (based on an expanded octet for S) in Exercise 27. Why is the geometrical shape predicted for SO_3 *independent* of whichever of these Lewis structures is used?

32. Use the VSEPR theory to predict the geometrical shape of **(a)** the molecule OSF_2; **(b)** the molecule O_2SF_2; **(c)** the molecule XeF_4; **(d)** the ion $ClO_4{}^-$; **(e)** the ion $I_3{}^-$.

Bond distances

33. In the gaseous state, HNO_3 molecules have two nitrogen-to-oxygen bond distances of 121 pm and one of 140 pm. Draw a plausible Lewis structure(s) to represent this fact.

34. A relationship between bond distances and single covalent radii of atoms is suggested in Section 9-9. Use this relationship and appropriate data from Table 9-3 to calculate the bond dis-tances: **(a)** H—Cl; **(b)** C—N; **(c)** C—Cl; **(d)** C—F; **(e)** N—I.

35. Draw a sketch of the hydroxylamine molecule, H_2NOH, representing the shape of the molecule and, where possible, bond angles and distances.

Bond energies

36. Assuming that bond energies are additive, show that the total energy associated with the bonds in 1 mol of ethane, C_2H_6, is 2831 kJ. Would you expect the total bond energy in 1 mol of ethylene, C_2H_4, to be greater or less than for 1 mol of ethane? Explain.

37. Use bond energies from Table 9-3 to estimate the enthalpy change (ΔH) for the following reactions.

(a) $C_2H_6(g) + Cl_2(g) \rightarrow C_2H_5Cl(g) + HCl(g)$ $\Delta H = ?$
(b) $C_2H_4(g) + H_2(g) \rightarrow C_2H_6(g)$ $\Delta H = ?$

38. Use bond energies to calculate the heat of formation of $NH_3(g)$ and compare your result with the value listed in Ap-pendix D.

$$N_2(g) + 3\,H_2(g) \rightarrow 2\,NH_3(g)$$

39. The enthalpy (heat) of formation of ethane, $C_2H_6(g)$, is $\Delta \overline{H}_f^{\circ} = -84.61$ kJ/mol. Use this value, together with appropriate data from Table 9-3, to obtain a value of the enthalpy of atomization of carbon [i.e., $\Delta \overline{H}^{\circ}$ for the reaction, $C(s) \rightarrow C(g)$]. Compare your result with the value of the enthalpy of atomization of carbon used in Example 9-16.

*40. Use a value of 497 kJ/mol for the bond energy in $O_2(g)$ and other necessary data from the text to estimate the bond energy in $NO(g)$.

Polar molecules

41. Estimate the percent ionic character of the HBr molecule, given that the dipole moment is 0.79 D.

42. Predict the shapes of the following molecules, and then predict which you would expect to have resultant dipole moments: (a) SO_2; (b) NH_3; (c) H_2S; (d) C_2H_4; (e) SF_6; (f) CH_2Cl_2.

43. The molecule H_2O_2 has a dipole moment of 2.13 D. The bonding is H—O—O—H. Which of these bonds have bond dipole moments? Can the molecule be linear? Explain.

Partial ionic character of covalent bonds

44. Calculate the ionic resonance energies of HF and HBr. Do these values compare with that for HCl in the way you would expect? Explain.

45. Use the results of Exercise 44 and equation (9.34) to calculate the electronegativity differences between (a) H and F; (b) H and Br. Compare your results with values obtained from Table 8-6.

*46. Use the methods of Examples 9-18 and 9-19, together with appropriate data from the text, to estimate the O—O single bond energy. The measured N—O single bond energy is 201 kJ/mol.

Polymers

47. Describe the difference in meaning of the following pairs of terms: (a) monomer and polymer; (b) elastomer and fiber; (c) natural and synthetic rubber; (d) inorganic and organic polymer; (e) thermosetting and thermoplastic polymer.

48. For the polymer Teflon (see Table 9-4)
   (a) Draw the structure of a portion of this polymer chain consisting of four repeating units.
   (b) What is the % F in Teflon? Would you expect this to depend on the length of the polymer chains? Explain.

49. Formaldehyde, $\underset{H}{\overset{H}{>}}C{=}O$, is the monomer of polyformaldehyde, a polymer with carbon-to-oxygen bonds.
   (a) Draw the structure of a portion of this polymer chain consisting of ten repeating units.
   (b) What volume of $CO_2(g)$ measured at 25.0°C and 751 mmHg would be produced by the complete combustion of 1.05 g of this polymer?

50. A 315-cm^3 sample of propylene gas (C_3H_6) at 20°C and 748 mmHg is polymerized. If it were possible to produce polymer molecules that all had the formula

$$-\left(\begin{array}{cc} H & CH_3 \\ | & | \\ C & C \\ | & | \\ H & H \end{array}\right)_n-$$

where $n = 875$, how many polymer molecules would be formed?

Additional Exercises

1. Refer to Table 9-1 and write the complete electron configurations for the following ions: (a) Y^{3+}; (b) Cd^{2+}; (c) Sb^{3+}.

2. What are the principal assumptions made in writing Lewis structures for covalent molecules? Cite some exceptions to these assumptions.

3. A compound is found to consist of 47.5% S and 52.5% Cl, by mass. Write a Lewis structure for this compound and comment on its deficiencies. Write a different structure with the same ratio of S to Cl that is more plausible.

4. A 1.65-g sample of a hydrocarbon, when completely burned in an excess of $O_2(g)$, yields 5.37 g CO_2 and 1.65 g H_2O. Draw a plausible Lewis structure for the hydrocarbon molecule. (*Hint:* There is more than one possible arrangement of the C and H atoms.)

5. What is the formal charge on the indicated atom in each of the following?
   (a) oxygen in OH^- (structure 9.14)
   (b) sulfur in SO_2 (structure 9.18)
   (c) boron in $H_3N \cdot BF_3$ (structure 9.23)
   (d) phosphorus in PCl_5 (structure 9.24)
   (e) sulfur in SO_2 (Example 9-9)
   (f) iodine in ICl_4^- (Figure 9-12)

6. What is the relationship, if any, between the concepts of oxidation state and formal charge? Explain.

7. What is the relationship between the shapes of (a) the ammonia molecule, NH_3, and the ammonium ion, NH_4^+; (b) sulfur trioxide, SO_3, and the sulfate ion, SO_4^{2-}?

8. One each of the following species is linear, angular, planar, tetrahedral, and octahedral. Indicate the correct structure for each: (a) H_2Te; (b) C_2Cl_4; (c) CO_2; (d) $SbCl_6^-$; (e) SO_4^{2-}.

9. Some of these statements regarding molecular shape are always true and some are not. Identify those which are not always true and explain why they are not.
   (a) Diatomic molecules have a linear shape.
   (b) Molecules in which four atoms are bonded to the same central atom have a tetrahedral shape.
   (c) Molecules with a planar shape consist of three atoms (triatomic).
   (d) Molecules with a nonmetal of the second period as the central atom can not have an octahedral shape.

10. The bond energy in the carbon monoxide molecule, CO, is about 1070 kJ/mol. Use this fact, together with data from Table 9-3, to propose a plausible Lewis structure(s) for CO.

11. Indicate which of the following molecules you would expect to have a resultant dipole moment, and give reasons for your conclusions: (a) HCN; (b) SO_3; (c) CS_2; (d) OCS; (e) $SOCl_2$; (f) SiF_4; (g) POF_3; (h) XeF_2.

12. The oxide Cl_2O_7 has the structure shown below. The bonds labeled (a) have a length of 1.46 Å, and those labeled (b), 1.72 Å. Use this information to write a plausible Lewis structure(s) for the Cl_2O_7 molecule.

$$O \underset{(a)}{\overset{(a)(b)}{\underset{Cl}{\diagdown}}} \overset{(b)(a)}{\underset{O}{\overset{O}{\diagup}}} \overset{(a)}{\underset{Cl}{\diagdown}} \overset{(a)}{\underset{O}{\diagup}} O$$

13. A 0.212-g sample of a gaseous hydrocarbon occupies a volume of 127 cm^3 at 738 mmHg pressure and 24.7°C. Show that there is only one possible structure for this hydrocarbon and draw its Lewis structure.

*14. Carbon suboxide has the formula C_3O_2. The carbon-to-carbon distances are found to be 130 pm, and the carbon-to-oxygen distances are 120 pm. Propose plausible Lewis structures to account for these bond distances, and predict the geometrical shape of the molecule.

15. The aldehyde propynal has the formula HCCCHO. Draw a sketch that represents bonding in the molecule and its shape. Include bond distances and bond angles.

16. The total bond energy associated with all the bonds in
$$\overset{O}{\overset{\|}{}}$$
thiourethane, $H_2NCSCH_2CH_3$, is 4780 kJ/mol. Use this value, together with data from Table 9-3, to estimate the energy of the C—S bond. How would you expect this energy to compare in value with the bond energies of the carbon-to-sulfur bonds in CS_2?

17. Estimate the enthalpies (heats) of formation of the following species at 25°C and 1 atm: (a) $N_2H_4(g)$; (b) OH(g); (c) $CH_3(g)$. Use data from the text as necessary.

18. Use a value of 497 kJ/mol for the O-to-O bond energy in the O_2 molecule, together with other data in Table 9-3, to estimate the heat of combustion of dimethyl ether, H_3COCH_3. (*Hint:* All bonds in the dimethyl ether molecule are single covalent.)

*19. The bond energy in the $O_2(g)$ molecule is 497 kJ/mol; and the heat of formation of $H_2O_2(g)$ is -136 kJ/mol. Use these values, together with other appropriate data from the text, to estimate the oxygen-to-oxygen single bond energy.

*20. The text states that the *bond* dipole moment of the O—H bond is 1.51 D; the H—O—H bond angle is 104°; and the resultant dipole moment of the H_2O molecule is 1.84 D. (See expression 9.33.)
   (a) Show by an appropriate geometric calculation that the three statements made above for the H_2O molecule are mutually consistent.
   (b) Use the same method as developed in part (a) to estimate the bond angle in H_2S, given that the H—S bond moment is 0.67 D and the resultant dipole moment of the H_2S molecule is 0.93 D.

Self-Test Questions

For questions 1 through 6 select the single item that best completes each statement.

1. Of the following species, the one containing a triple covalent bond is (a) NO_3^-; (b) CN^-; (c) CO_2; (d) $AlCl_3$.

2. In the ammonium ion, NH_4^+; (a) the four H atoms are situated at the corners of a square; (b) all bonds are ionic; (c) all bonds are coordinate covalent; (d) the shape is that of a tetrahedron.

3. The formal charge of the O atoms in the ion $[:\overset{..}{O}{=}N{=}\overset{..}{O}:]^+$ is (a) -2; (b) -1; (c) 0; (d) $+1$.

4. All of the following molecules are linear except one. That one is (a) SO_2; (b) CO_2; (c) HCN; (d) C_2H_2.

5. Of the following molecules, all are polar except one. That one is (a) BCl_3; (b) CH_2Cl_2; (c) NO; (d) PCl_3.

6. All of the species indicated exist. Of the Lewis structures shown, one is plausible but the other three are much less so. The plausible one is
   (a) cyanate ion, $[:\overset{..}{\underset{..}{O}}{-}C{=}\overset{..}{N}:]^-$
   (b) carbide ion, $[C{\equiv}C:]^{2-}$
   (c) hypochlorite ion, $[:\overset{..}{\underset{..}{Cl}}{-}\overset{..}{\underset{..}{O}}:]^-$
   (d) nitrogen(II) oxide, $:\overset{..}{N}{=}\overset{..}{O}\cdot$

7. A chemical compound is found to have the following percent composition, by mass: 24.3% C, 71.6% Cl, and 4.1% H.
    (a) What is the *empirical* formula of this compound?
    (b) Draw a Lewis structure based on this empirical formula and comment on its inadequacies.
    (c) Propose a *molecular* formula for the compound that results in a more plausible Lewis structure.

8. Draw Lewis structures for two different molecules having the formula C_3H_4. Is either of these molecules linear? Explain.

9. In which of the following molecules is the nitrogen-to-nitrogen bond distance expected to be the shortest: (a) N_2H_4; (b) N_2; (c) N_2O_4; (d) N_2O? Explain.

10. Predict the shapes of the following sulfur-containing species: (a) SO_2; (b) SO_3; (c) SO_4^{2-}.

11. Given the bond energies: N-to-O bond in NO, 628 kJ/mol; H—H, 435 kJ/mol; N—H, 389 kJ/mol; O—H, 464 kJ/mol, calculate ΔH for the reaction

$$2\ NO(g) + 5\ H_2(g) \rightarrow 2\ NH_3(g) + 2\ H_2O(g)$$

12. The following statements are not made as carefully as they might be. Criticize each one.
    (a) Triatomic molecules have a planar shape.
    (b) Molecules in which there is an electronegativity difference between the bonded atoms are polar.
    (c) Lewis structures in which atoms carry formal charges are incorrect.

10 Chemical Bonding II: Additional Aspects

Ionic bonding is understandable in terms of the force of attraction between positive and negative ions. Our description of covalent bonding to this point has been less satisfactory. We have been successful in deriving formulas for a large number of covalent molecules by writing Lewis structures. Moreover, when combined with VSEPR theory, Lewis structures allow for a prediction of the shapes of molecules. However, they do not permit a prediction of bond energies or, in some cases, magnetic properties.

More adequate descriptions of the covalent bond require the methods of wave mechanics. As in Chapter 7 on the electronic structures of atoms, we limit ourselves to a few basic ideas applied in a qualitative way. The essential requirement of a covalent chemical bond is that it correspond to a region between bonded atoms where the probability of finding electrons or the electron charge density is high. In this chapter to describe these regions we consider two new approaches, known as the valence bond method and molecular orbital theory. Also considered are bonding in metals and in semiconductors.

10-1 The Valence Bond Approach to Chemical Bonding

This method of describing the covalent bond originated within a year or two of Schrödinger's work on the hydrogen atom. It views atoms involved in covalent bond formation as being largely unchanged from their isolated conditions. All that is considered to happen in bond formation is that electron orbitals centered on the individual atoms (atomic orbitals) overlap. A covalent bond, then, arises from the high electron charge density (high electron probability) in the region of atomic orbital overlap between bonded atoms. The overlap of two $1s$ orbitals in a hydrogen molecule is suggested in Figure 10-1.

FIGURE 10-1
Bonding in H_2 represented by atomic orbital overlap.

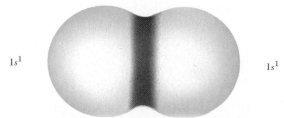

$1s^1$ $1s^1$

Each atomic orbital contains one unpaired electron. As a result of the overlap of the two orbitals, the electrons become paired and a region of high electron probability results—a covalent bond. Note how the characteristic features of the $1s$ atomic orbital are retained, except in the region of overlap (shown in black). (That is, compare this figure with Figure 7-16.)

FIGURE 10-2
Covalent bonding in
H₂S represented by
atomic orbital overlap.

Isolated atoms

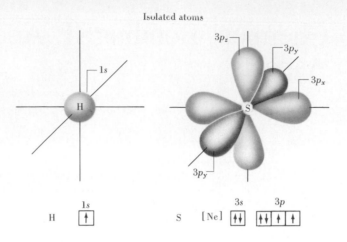

Covalent bonds

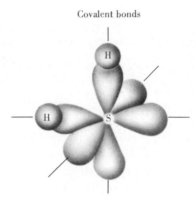

Orbitals containing a single electron are in color; those with an electron pair are grey. For S, only 3p orbitals are shown. The 1s orbitals of two hydrogen atoms overlap with the $3p_x$ and $3p_z$ orbitals of the sulfur atom. Here and elsewhere in this chapter the shapes of p orbitals have been elongated along the axes through the nucleus. This permits a better visualization of geometric structures.

The overlap of atomic orbitals involved in the formation of hydrogen-to-sulfur bonds in hydrogen sulfide is depicted in Figure 10-2. From this figure we should note the following.

1. The number of covalent bonds between atoms is such that, normally, all electrons become paired.
2. If valence electrons are counted in the same way as in Lewis structures, each atom normally acquires a noble gas electron configuration.
3. The shape of the molecule is determined by the geometrical orientation of the overlapping atomic orbitals of the bonded atoms.

Example 10-1 Describe the structure of the NH₃ molecule by the valence bond method.

Solution
Step 1. Draw orbital diagrams for the separate atoms that are to be bonded.

FIGURE 10-3
Bonding and structure
of NH_3 molecule—
Example 10-1
illustrated.

bonding orbitals of N atom

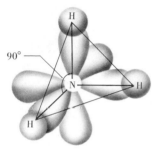

90°

covalent bonds formed

Orbitals with single electrons are shown in color;
those with an electron
pair are grey. Only bond-
ing orbitals are shown.
The 1s orbitals of three
hydrogen atoms overlap
with the three 2p orbitals
of the nitrogen atom.

Step 2. Sketch the orbitals of the central atom (N) that will be involved in the orbital overlap (see Figure 10-3).
Step 3. Complete the structure by bringing together the bonded atoms and representing the orbital overlap.
Step 4. Describe the structure. NH_3 is a trigonal pyramidal molecule. The three H atoms lie in the same plane. The N atom is situated at the apex of the pyramid above the plane of the H atoms. The three H—N—H bond angles are predicted to be 90°.

SIMILAR EXAMPLE: Review Problem 1.

If bonding involves two or all three p orbitals of the valence shell of the central atom, we should expect 90° bond angles; the p orbitals are mutually perpendicular. The measured bond angle in H_2S, for example, is 92°. But the measured bond angles are 104.5° in H_2O and 107° in NH_3, not the predicted 90°. The greater-than-90° bond angles in H_2O can be rationalized as follows. Because O is more electronegative than H, electrons are displaced toward the O atom in the O—H bonds in H_2O. This leaves the H atoms with a slight positive charge. The H atoms repel one another and cause an increase in the bond angle. The effect in NH_3 is similar, but it is less significant in H_2S because S is not as electronegative as N and O.

In addition to its inadequacies in dealing with H_2O and NH_3, the unmodified valence bond method fails for certain other molecules, such as CH_4 and CO_2. We need to develop a modification of the simple valence bond method.

10-2 Hybridization of Atomic Orbitals

In the examples in Section 10-1 we represented covalent bonding by starting with the normal or ground state electron configurations of the separated atoms. With this approach we would predict CH_2 rather than CH_4 as the simplest hydrocarbon molecule. To write an acceptable Lewis structure for CH_4 requires *four* unpaired electrons in the Lewis symbol of C. In terms of atomic orbitals this requirement can be met by postulating that a C atom acquires an "excited" electron configuration. One of the $2s$ electrons is promoted to the $2p$ subshell.

ground state: C $\begin{array}{ccc} 1s & 2s & 2p \\ \boxed{\uparrow\downarrow} & \boxed{\uparrow\downarrow} & \boxed{\uparrow\ \ \uparrow\ \ \ } \end{array}$

(10.1)

promotion: C $\begin{array}{ccc} 1s & 2s & 2p \\ \boxed{\uparrow\downarrow} & \boxed{\uparrow} & \boxed{\uparrow\ \ \uparrow\ \ \uparrow} \end{array}$

The molecular geometry predicted from the orbital diagram (10.1) would be a molecule with three mutually perpendicular C—H bonds (i.e., with bond angles of 90°). The fourth C—H bond would have no particular orientation with respect to the other three. However, we have already described the structure of CH_4 (see Table 9-2): The C—H bonds are directed in a tetrahedral fashion from the C atom; that is, the H—C—H bond angles are 109.5°. The orbital diagram (10.1) accounts for the correct number of bonds but not the correct orientation of these bonds.

The essential difficulty here is that the description of pure atomic orbitals (s, p, d, and f) is based on *isolated* atoms. There is no reason to assume, though, that the electron configurations of normal *isolated* atoms are applicable to *bonded* atoms. The fact that they are adequate in explaining some covalent bonds is fortunate. That they fail to apply in a large number of other cases is not an unreasonable finding either.

FIGURE 10-4
The sp^3 hybridization scheme.

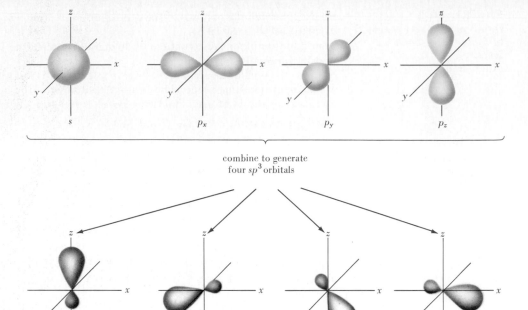

combine to generate
four sp^3 orbitals

which are represented
as the set

FIGURE 10-5
Bonding and structure
of CH_4.

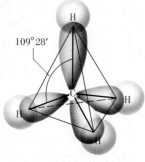

geometric structure

The four carbon orbitals involved in the bonding scheme (in grey) are $2sp^3$ hybrid orbitals. Those of the hydrogen atoms (in color) are $1s$. The structure is tetrahedral, with bond angles of about 109.5° (more exactly, 109°28′).

If the $2s$ and $2p$ orbitals are combined in an appropriate way, they generate a new set of orbitals. The set consists of four *identical* orbitals at exactly the tetrahedral bond angles, 109.5°. Orbitals obtained by this type of combination of simple atomic orbitals are called **hybrid orbitals.** The particular hybridization scheme described here and pictured in Figure 10-4 is called sp^3 hybridization. In a hybridization scheme the number of hybrid orbitals is equal to the total number of atomic orbitals involved; the symbol identifies these atomic orbitals. Thus, sp^3 signifies that one s and three p orbitals have been combined to produce a set of four new hybrid orbitals. A suitable orbital diagram for this hybridization scheme is

sp^3 hybridization: C $1s$ [↑↓] $2sp^3$ [↑ | ↑ | ↑ | ↑]   (10.2)

The use of sp^3 hybrid orbitals in bond formation in CH_4 is pictured in Figure 10-5.

The term "scheme" (a systematic plan for attaining some objective) is appropriate for describing hybridization. The objective is an after-the-fact attempt to account for the geometrical shape that is experimentally observed for a molecule. Hybridization is not an

FIGURE 10-6
sp^3 hybrid orbitals and bonding in H_2O and NH_3.

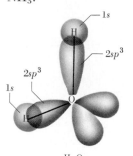

H_2O

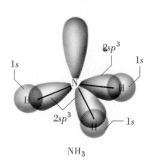

NH_3

actual physical phenomenon. There is no way during bond formation to see electron charge distributions rearranging from those described by simple atomic orbitals to those described by hybrid orbitals. Moreover, we must accept the fact that for some covalent bonds no single hybridization scheme works well.

Water (H_2O) and Ammonia (NH_3).   An sp^3 hybridization scheme produces good agreement with experiment and with VSEPR theory for the bond angles in H_2O and NH_3. In the orbital diagrams (10.3) and (10.4) the $2s$ and three $2p$ orbitals have been hybridized and the correct number of electrons placed in each orbital.

$$N \quad \boxed{\uparrow\downarrow}_{1s} \quad \boxed{\uparrow\downarrow \mid \uparrow \mid \uparrow \mid \uparrow}_{2sp^3} \tag{10.3}$$

$$O \quad \boxed{\uparrow\downarrow}_{1s} \quad \boxed{\uparrow\downarrow \mid \uparrow\downarrow \mid \uparrow \mid \uparrow}_{2sp^3} \tag{10.4}$$

As shown in Figure 10-6, in H_2O there are two bonds and two unshared (lone) pairs of electrons; the molecule is angular. In NH_3 there are three bonds and one unshared (lone) pair of electrons; the molecule has a trigonal pyramidal shape.

sp and sp^2 Hybrid Orbitals.   Two additional schemes for hybridizing s and p orbitals are shown in diagrams (10.5) and (10.6), one for B and one for Be.

$$B \quad \boxed{\uparrow\downarrow}_{1s} \ \boxed{\uparrow\downarrow}_{2s} \ \boxed{\uparrow \mid \ \mid \ }_{2p} \longrightarrow B \ \boxed{\uparrow\downarrow}_{1s} \ \boxed{\uparrow \mid \uparrow \mid \uparrow}_{2sp^2} \ \boxed{\ }_{2p} \tag{10.5}$$

ground state   hybridized

$$Be \quad \boxed{\uparrow\downarrow}_{1s} \ \boxed{\uparrow\downarrow}_{2s} \ \boxed{\ \mid \ \mid \ }_{2p} \longrightarrow Be \ \boxed{\uparrow\downarrow}_{1s} \ \boxed{\uparrow \mid \uparrow}_{2sp} \ \boxed{\ \mid \ }_{2p} \tag{10.6}$$

ground state   hybridized

As pictured in Figure 10-7, the *three sp^2* hybrid orbitals are directed *in a plane* at angles of 120°. The *two sp* hybrid orbitals are directed along a straight line, at a 180° angle. Thus, a molecule like BF_3 is *trigonal planar*, whereas $BeCl_2$ is *linear*.

d Hybrid Orbitals.   The concept of hybridization, if extended to include d orbitals, helps to explain the occurrence of expanded octets. Consider the molecule PCl_5, for example. The ground state electron configuration of phosphorus is

ground state: P   [Ne] $\boxed{\uparrow\downarrow}_{3s}$ $\boxed{\uparrow \mid \uparrow \mid \uparrow}_{3p}$ $\boxed{\ \mid \ \mid \ \mid \ \mid \ }_{3d}$

But to describe the formation of PCl_5 the following *hypothetical* process is proposed: (1) All the valence electrons are unpaired. (2) Electrons are placed successively and individually into $3s$, $3p$, and $3d$ orbitals. (3) All the orbitals that are occupied as a result are then hybridized.

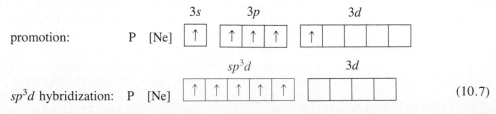

promotion: P   [Ne] $\boxed{\uparrow}_{3s}$ $\boxed{\uparrow \mid \uparrow \mid \uparrow}_{3p}$ $\boxed{\uparrow \mid \ \mid \ \mid \ }_{3d}$

sp^3d hybridization: P   [Ne] $\boxed{\uparrow \mid \uparrow \mid \uparrow \mid \uparrow \mid \uparrow}_{sp^3d}$ $\boxed{\ \mid \ \mid \ }_{3d}$ $\tag{10.7}$

FIGURE 10-7
The sp^2 and sp
hybridization schemes.

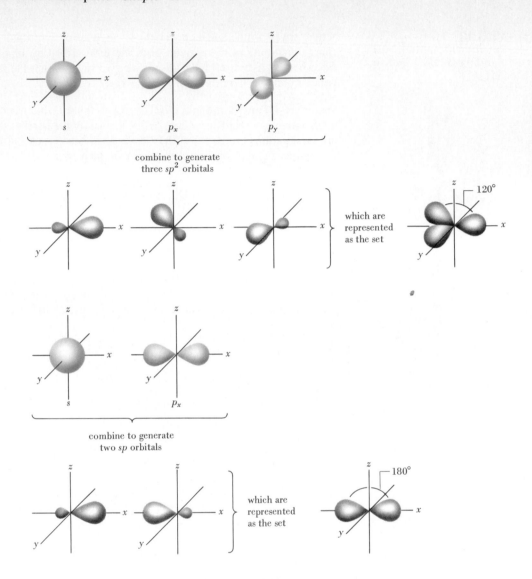

A set of sp^3d hybrid orbitals is pictured in Figure 10-8. A trigonal bipyramidal molecule results when these five orbitals are used in bond formation. In the molecule SF_6 there is a set of *six* bonding orbitals. These are the sp^3d^2 hybrid orbitals.

sp^3d^2 hybridization: S   [Ne] $\boxed{\uparrow}\boxed{\uparrow}\boxed{\uparrow}\boxed{\uparrow}\boxed{\uparrow}\boxed{\uparrow}$ sp^3d^2 $\boxed{}\boxed{}\boxed{}$ $3d$ (10.8)

The geometrical distribution of these orbitals, as shown in Figure 10-8, is octahedral.

Hybrid Orbitals and the Valence-Shell Electron-Pair Repulsion Theory.   In most cases the geometric structures of molecules predicted by the use of hybrid orbitals agree with predictions made by the valence-shell electron-pair repulsion (VSEPR) method. There is, of course, a connection between the two.

One hybrid orbital is produced for every atomic orbital involved in a hybridization scheme. In a molecule each of the hybrid orbitals of a central atom normally acquires an electron pair, either a bond pair or a lone pair. Thus, the number of hybrid orbitals equals the number of electron pairs. In most cases the orientation of hybrid orbitals is the same as

FIGURE 10-8
d hybrid orbitals—sp^3d
and sp^3d^2.

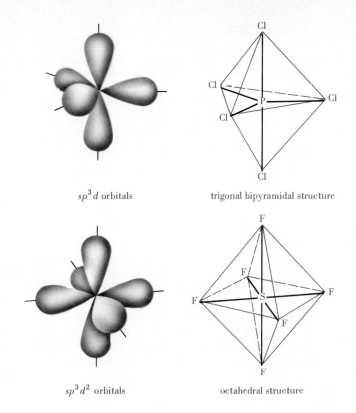

sp^3d orbitals

trigonal bipyramidal structure

sp^3d^2 orbitals

octahedral structure

that of electron pairs predicted by the VSEPR method, as shown in Table 10-1. There is one case, however, where the valence bond method is superior to the VSEPR method. This occurs when four pairs of valence electrons appear in the hybridization scheme dsp^2 rather than sp^3. The resulting structure has a square planar geometry not predicted by the VSEPR method. Some complex ions with this geometrical structure are described in Chapter 24.

TABLE 10-1
Hybrid orbitals and their geometric orientation

Atomic orbitals	Hybrid orbitals	Orientation	Example	Predicted bond angle
$s + p$	sp	linear	$BeCl_2$	180°
$s + p + p$	sp^2	trigonal planar	BF_3	120°
$s + p + p + p$	sp^3	tetrahedral	CH_4	109.5°
$d + s + p + p$	[a]dsp^2	square planar	$[Pt(NH_3)_4]^{2+}$	90°
$s + p + p + p + d$	[b]sp^3d	trigonal bipyramidal	PCl_5	120, 90°
$s + p + p + p + d + d$	[b]sp^3d^2	octahedral	SF_6	90°
$d + d + s + p + p + p$	[a]d^2sp^3	octahedral	$[Co(NH_3)_6]^{2+}$	90°

[a] These hybrid orbitals involve *d* orbital(s) from the next-to-outermost shell, together with *s* and *p* orbitals of the outermost shell. They are encountered commonly in the structures of complex ions (see Chapter 24).
[b] These hybrid orbitals involve *s*, *p*, and *d* orbitals, all from the outermost electronic shell. They are encountered in structures with an expanded octet that have nonmetals such as P, As, S, Cl, Br, and I as the central atom (see Chapter 21).

10-3 Multiple Covalent Bonds

Bonding in ethylene, C_2H_4 is represented by the following hybridization scheme. It results in both a set of hybrid orbitals (sp^2) and an *unhybridized p* orbital.

ground state: C

$1s$ $2s$ $2p$

promotion: C   (10.9)

$1s$ $2s$ $2p$

hybridization: C

$1s$ $2sp^2$ $2p$

The three sp^2 hybrid orbitals are directed in a plane and separated by 120° angles. Bonding through these orbitals leads to a molecule with all six atoms in the same plane. The molecule C_2H_4 is *planar*. The hybridization scheme, orbital overlap, and a space-filling model of the molecule are illustrated in Figure 10-9.

In the double covalent bond in Figure 10-9 we picture one of the bonds as resulting from the overlap of sp^2 hybrid orbitals along the line joining the nuclei of the two carbon atoms. Orbitals that overlap in this "end-to-end" fashion produce a **sigma bond,** designated **σ bond.** Another bond arises from the overlap of the unhybridized *p* orbitals. It has regions of high electron density above and below the plane of the carbon and hydrogen atoms. This "side-by-side" overlap of *p* orbitals produces a bond known as a **pi bond,** designated **π bond.**

FIGURE 10-9
sp^2 hybridization and bonding in C_2H_4.

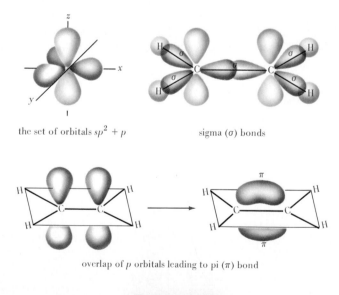

the set of orbitals $sp^2 + p$ sigma (σ) bonds

overlap of *p* orbitals leading to pi (π) bond

space–filling model

FIGURE 10-10
sp hybridization and
bonding in C_2H_2.

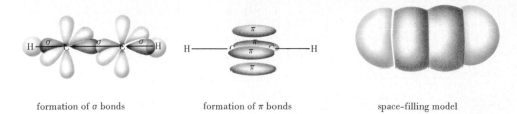

formation of σ bonds   formation of π bonds   space–filling model

Several aspects of bonding in C_2H_4 need further mention. First, the σ bond involves more extensive orbital overlap than does the π bond. As a result we should expect a carbon-to-carbon double bond (σ + π) to be stronger than a single bond (σ), but not twice as strong (from Table 9-3, C—C, 347 kJ/mol; C≡C, 611 kJ/mol). Second, the shape of a molecule is determined only by the orbitals leading to σ bonds (the σ-bond framework). Finally, rotation about a double bond is restricted; that is, the double bond is quite rigid. Consider the space-filling model of C_2H_4 in Figure 10-9. To twist one —CH_2 group out of the plane of the other would reduce the amount of overlap of the *p* orbitals. Rotation about a σ bond occurs more freely since the end-to-end overlap of orbitals is not disturbed by this type of motion.

In the molecule C_2H_2 (acetylene) the two carbon atoms are bonded through a triple covalent bond. The σ bonding scheme involves *sp* hybrid orbitals.

$$\text{C} \quad \boxed{\uparrow\downarrow} \quad \boxed{\uparrow \mid \uparrow} \quad \boxed{\uparrow \mid \uparrow} \qquad\qquad (10.10)$$

$$\begin{array}{ccc} 1s & 2sp & 2p \end{array}$$

In the triple bond in C_2H_2 one of the C≡C bonds is a σ bond and the other two are π bonds. These points are illustrated through Figure 10-10.

Illustrative Examples.   Starting with a Lewis structure it is possible to produce a three-dimensional sketch to represent the overlap of simple and hybrid orbitals in a molecule, as in Example 10-2.

Example 10-2   Describe the orbital overlap leading to bonding in the formaldehyde molecule, H_2CO, whose Lewis structure is shown below.

$$\begin{array}{c} \text{H} \\ | \\ \text{H}—\text{C}=\overset{\cdot\cdot}{\text{O}}: \end{array}$$

Solution
Step 1.   The shape of the molecule is established by the σ-bond framework. Since three electron pairs are found in σ bonds around the central C atom, the molecule is *trigonal planar* (120° bond angles).
Step 2.   Determine the simple and/or hybrid orbitals that will produce the predicted geometry. A trigonal planar structure is based on *sp²* *hybrid orbitals* of the central atom (recall Table 10-1).
Step 3.   Identify multiple bonds in the structure. A carbon-to-carbon double bond is a combination of one σ and one π bond. (A triple bond would signify one σ and two π bonds.)
Step 4.   Sketch the orbitals of the central atom that are involved in orbital overlap. The C atom (expression 10.9) uses two of its *sp²* hybrid orbitals to form σ bonds with two H atoms. The remaining *sp²* hybrid orbital is used to form a σ bond with oxygen. The simple *p* orbital of the C atom is used to form a π bond with O.

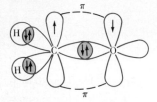

For simplicity, only bonding orbitals of the valence shells are shown. Specifically, the $2s$ and $2p_y$ orbitals of the oxygen atom, each of which contains an electron pair, have been omitted.

Step 5. Sketch the orbitals of the terminal atoms that are involved in orbital overlap. The H atoms employ their $1s$ orbitals. The situation with the O atom is less clear. Either of the orbital diagrams shown below for O will work. Each provides the simple p orbital needed for π-bond formation.

The sp^2 hybridization scheme places the lone-pair electrons of the valence shell of oxygen as far as possible from each other and from the σ bond pair. On the other hand, the chief purpose of employing hybridized atomic orbitals is to account for the observed geometrical structure of a species; and whether the lone-pair electrons of the O atom are in $2s$, $2p$, or $2sp^2$ orbitals does not alter our conclusions about the molecular geometry of H₂CO. We will adopt the practice of using hybridized atomic orbitals (where necessary) for central atoms and simple atomic orbitals for terminal atoms in a structure. A sketch of orbital overlap in H₂CO is shown in Figure 10-11.

SIMILAR EXAMPLES: Review Problem 2; Exercises 6, 7.

Orbital overlap is sometimes difficult to sketch in three dimensions. The more customary representation of a bonding scheme is that illustrated in Example 10-3 and Figure 10-12. In Figure 10-12 bonds between atoms are drawn as straight lines; bonds are labeled σ or π; and the orbital overlap involved in each bond is indicated, with the first orbital coming from the atom on the left and the second orbital, the atom on the right.

Example 10-3 Bond angles in the formic acid molecule, HCOOH, are indicated in the structure below. Propose a bonding scheme consistent with this structure.

Solution. The bond angles at the central C atom are essentially those for trigonal planar geometry (118° and 124° compared to an expected 120°). This suggests that the central C atom employs the orbital set, $sp^2 + p$. (The sp^2 orbitals are used in σ-bond formation and the p orbital for a π bond.) The observed C—O—H bond angle (108°) is very close to the tetrahedral angle (109.5°). This suggests that the O atom employs sp^3 hybrid orbitals in forming bonds to C and H (recall the hybridization scheme presented in equation 10.4). The situation with the terminal O atom in the C=O bond is the same as that described in Step 5 of Example 10-2.

SIMILAR EXAMPLES: Review Problem 4; Exercise 8.

A likely situation to be encountered is one in which only the formula of a molecule is given. In this case a plausible Lewis structure should be written; the σ-bond framework established; an appropriate hybridization scheme chosen for each central atom; and a bonding scheme represented. These are the steps required in Example 10-4.

Example 10-4 Propose a bonding scheme for hydrogen cyanate, HOCN.

Solution
Step 1. Write a plausible Lewis structure. The structure requires 16 valence electrons, leading to this first attempt.

$$\text{H—}\ddot{\text{O}}\text{—C—}\ddot{\text{N}}\text{:} \quad \text{(incorrect)}$$

FIGURE 10-13
Bonding and structure
of the HOCN
molecule—Example 10-4
illustrated.

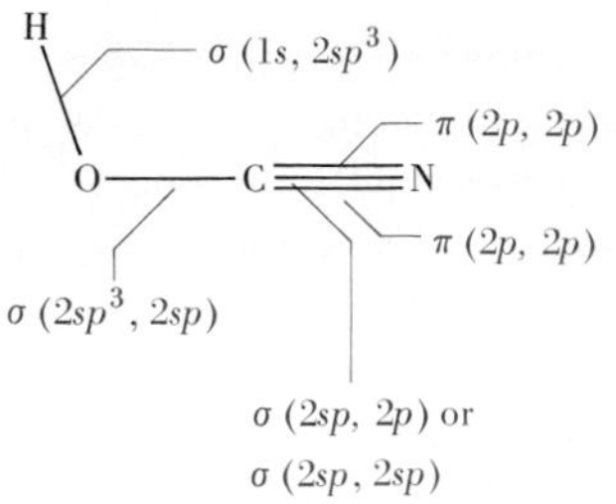

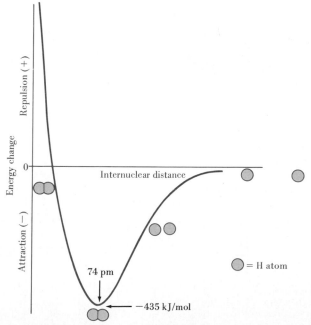

The lack of an octet on the C atom is remedied by moving two pairs of electrons from the terminal N atom, producing a triple bond between it and the C atom. (Other structures can be written, but the one here seems best because it has no formal charges.)

$$H—\overset{..}{\underset{..}{O}}—C{\equiv}N:$$

Step 2.   Determine the distribution of bond-pair and lone-pair electrons about the central atoms. *Four* electron pairs are distributed about the O atom. About the C atom there are two electron pairs in σ bonds and two pairs in π bonds.

Step 3.   Indicate the hybridization scheme for each central atom. According to Table 10-1, the scheme for O is sp^3 and for C, sp.

Step 4.   Similar to the situation described in Step 5 of Example 10-2, for the terminal nitrogen atom the choice is to use either the unmodified orbital diagram for N or an sp hybridization scheme.

Step 5.   Draw a diagram showing the orbital overlap scheme. This is done in Figure 10-13.

SIMILAR EXAMPLES: Review Problems 5, 6; Exercises 5, 9, 10.

10-4 Molecular Orbital Theory

Imagine that initially two hydrogen atoms are "infinitely" far apart. This is taken as the zero of energy. Now imagine that the two atoms are allowed to approach each other and that the energy change is plotted as a function of the distance between the nuclei of the two atoms—the **internuclear distance.** As the internuclear distance decreases (the atoms attract each other), more and more energy is released. The maximum energy release (435 kJ/mol), corresponding to the minimum in the curve of Figure 10-14, comes at an internuclear distance of 74 pm, the bond length in H_2. To force the atoms closer together

FIGURE 10-14
Energy of interaction of
two hydrogen atoms as
a function of
internuclear distance.

Two H atoms form the molecule H_2 at a particular internuclear distance (74 pm).

requires that energy again be absorbed, and the curve rises steeply from its minimum point.

The repulsion between H atoms in close proximity is readily understood. It results from the mutual repulsion of two positively charged nuclei. But what is the source of the attraction at intermediate distances?

Figure 10-15 shows two different arrangements of the protons and electrons in two H atoms brought into close proximity. In the arrangement where the electrons are located away from the internuclear region, there is a strong repulsion between protons. Energy is high, and the arrangement is unstable. The arrangement where the two electrons are located *between* the atomic nuclei is of *lower* energy than in the separated atoms. Thus, even classical theory (Coulomb's law) predicts that two H atoms should combine to form an H_2 molecule. However, the energy of interaction predicted by classical theory is much less than the measured value; the internuclear distance is also in error. Success in these predictions requires wave mechanics.

Let us think of electrons in terms of charge densities or probabilities extending over an entire molecule, that is, in terms of **molecular orbitals.** One method of deriving molecular orbitals is by an appropriate combination of atomic orbitals of the atoms being united into a molecule. Wave mechanics allows two possibilities. One combination of two $1s$ orbitals represented in Figure 10-15 produces a **bonding molecular orbital,** designated σ_{1s}^b. Another combination produces an **antibonding orbital,** σ_{1s}^*. Also depicted in Figure 10-15 is an energy level diagram for these two molecular orbitals: *The bonding molecular orbital is at a lower energy than the separate atomic orbitals, and the antibonding molecular orbital at a higher energy.*

In terms of electron probability or charge density, Figure 10-15 shows the bonding molecular orbital to correspond to a high electron probability or charge density *between*

FIGURE 10-15
The interaction of two hydrogen atoms.

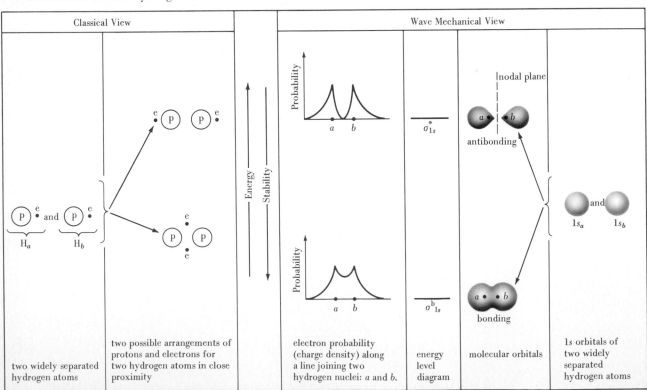

the atomic nuclei. Electron charge density concentrated in the internuclear region reduces repulsion between the positively charged nuclei. This permits bonding between the atoms—hence the term *bonding molecular orbital*. In the antibonding orbital the electron probability or charge density in the internuclear region is much lower. In fact, it falls to zero midway between the nuclei in a region called the *nodal plane*. Electron charge density in the antibonding orbital is concentrated away from the internuclear region, where it is ineffective in reducing internuclear repulsion—hence the term *antibonding molecular orbital*. The probability distributions for the bonding and antibonding molecular orbitals correspond roughly to the electron positions shown for the corresponding classical situations in Figure 10-15.

Basic Ideas Concerning Molecular Orbitals. To use molecular orbital theory to describe chemical bonding requires that we first establish some rules. These rules pertain to the particular molecular orbitals that arise when atomic orbitals are combined and the manner in which electrons are assigned to these molecular orbitals.

1. The number of molecular orbitals produced is equal to the number of atomic orbitals combined.
2. Of the two molecular orbitals produced when two atomic orbitals are combined, one is a *bonding* molecular orbital at a *lower* energy than the original atomic orbitals. The other is an *antibonding* orbital at a *higher* energy.
3. Electrons normally seek the lowest energy molecular orbitals available to them in a molecule.
4. The maximum number of electrons that can be assigned to a given molecular orbital is *two* (Pauli exclusion principle).
5. Electrons enter molecular orbitals of identical energies *singly* before they pair up (Hund's rule).
6. Formation of a bond between atoms requires that the number of electrons in bonding molecular orbitals exceed the number of electrons in antibonding orbitals.

First Period Elements. Figure 10-16 suggests four possibilities for assigning electrons to the molecular orbitals depicted in Figure 10-15. Let us consider them.

H_2^+: This species has a single electron. It enters the σ_{1s}^b orbital and produces a bond between the two H atoms. We might call this a one-electron or "half" bond.

FIGURE 10-16
Molecular orbital diagrams for the diatomic molecules (or ions) formed from first period elements.

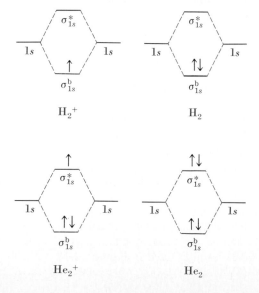

H_2: This molecule has two electrons, both in the σ_{1s}^b orbital. A regular single covalent bond is formed.

He_2^+: This ion has 3 electrons. Two are in the σ_{1s}^b orbital, and one in the σ_{1s}^*. The net number of bonding electrons is $2 - 1 = 1$. This species exists as a stable ion, but with only a one-electron or "half" bond.

He_2: Two electrons are in the σ_{1s}^b orbital and two in the σ_{1s}^*. The net number of bonding electrons is $2 - 2 = 0$. No bond is produced. We should not expect to encounter He_2 as a stable species.

An important idea used in the preceding examples is the following.

$$\text{bond order} = \tfrac{1}{2}(\text{no. } e^- \text{ in bonding M.O.} - \text{no. } e^- \text{ in antibonding M.O.}) \qquad (10.11)$$

Bond order indicates whether a bond is single, double, triple (or one half, three halves, five halves). Whereas Lewis theory and the valence bond method require electron-*pair* bonds, molecular orbital theory accounts quite nicely for *one*-electron bonds, which are known to exist.

Example 10-5   The bond energy of H_2 is 435 kJ/mol. Estimate the bond energies of H_2^+ and He_2^+.

Solution.   The bond order in H_2 is one, that is, a single bond. In H_2^+ and He_2^+ the bond order is $\tfrac{1}{2}$. We should expect the bonds in these two species to be only about one-half as strong as in H_2, that is, about 220 kJ/mol. (Actual values: H_2^+, 255 kJ/mol; He_2^+, 251 kJ/mol.)

SIMILAR EXAMPLE: Exercise 16.

Example 10-6   Which of the four species described in Figure 10-16 are paramagnetic?

Solution.   Paramagnetism requires the presence of unpaired electrons. One unpaired electron is found in H_2^+ and in He_2^+ ; these ions are paramagnetic.

SIMILAR EXAMPLE: Review Problem 8.

Second Period Elements.   To apply the molecular orbital method to elements of the second period requires that molecular orbitals be formed from atomic orbitals of the second principal electronic shell. We will limit our discussion to *diatomic* molecules. Also, we note again that two molecular orbitals—one bonding and one antibonding—are produced for every pair of atomic orbitals in the separated atoms. Because there are *four* orbitals in each atom ($2s$, $2p_x$, $2p_y$, $2p_z$) we need to deal with *eight* new molecular orbitals—four bonding and four antibonding. We must also have an energy level diagram for these orbitals.

The molecular orbitals formed from $2s$ orbitals have the same characteristics as those derived from $1s$ orbitals, but they are at a higher energy. As illustrated in Figure 10-17, there are *two* possibilities for combinations of p orbitals. Those that overlap along the same straight line (i.e., end to end) combine to produce σ orbitals: σ_{2p}^b and σ_{2p}^*. Those that overlap in a parallel or sidewise fashion produce π orbitals: π_{2p}^b and π_{2p}^*. There are *two* molecular orbitals of each π type because there are *two* pairs of p orbitals that are arranged in a parallel fashion. A distinction between bonding and antibonding molecular orbitals, first illustrated in Figure 10-15, is emphasized in Figure 10-17. In bonding molecular orbitals there is a high electron charge density between the atomic nuclei. In an antibonding orbital the electron charge density falls to zero in a plane perpendicular to the line joining the atomic nuclei, at a point midway between them. Energy level diagrams for these new orbitals are illustrated in Figure 10-18.

FIGURE 10-17
Representation of the molecular orbitals formed by a combination of $2p$ atomic orbitals.

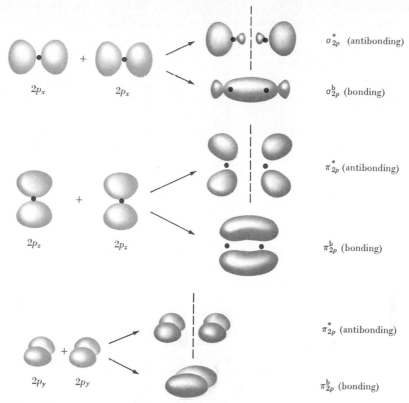

σ_{2p}^{*} (antibonding)

σ_{2p}^{b} (bonding)

π_{2p}^{*} (antibonding)

π_{2p}^{b} (bonding)

π_{2p}^{*} (antibonding)

π_{2p}^{b} (bonding)

$2p_x$ $2p_x$

$2p_z$ $2p_z$

$2p_y$ $2p_y$

These diagrams are meant simply to suggest the nature of the electron charge distribution for the several molecular orbitals. They are not exact in all details. Nodal planes for the antibonding orbitals are represented by the broken lines.

We can describe bonding in diatomic molecules of the second period elements (i.e., in Li_2, Be_2, B_2, . . .) in much the same way as for the first period elements. In this process we start with the σ_{1s}^{b} and the σ_{1s}^{*} orbitals filled. Then we add electrons, in order of increasing energy, to the available molecular orbitals formed from the second principal shells. Alternatively, we can think of the first shell (K shell) electrons as not involved in the bonding (i.e., as *nonbonding* electrons). This allows us simply to consider the assignment of the valence shell electrons of the two atoms. This assignment is depicted in Figure 10-18 (where the filled first electronic shells are denoted by the symbol KK).

Some of the previously unexplained features of the O_2 molecule can now be understood in terms of molecular orbitals. Each O atom brings *six* valence electrons to the diatomic molecule, O_2. There are *12* electrons to be assigned to molecular orbitals. The filling of these orbitals is depicted in Figure 10-18. The following features are brought out by the molecular orbital diagram for O_2.

1. The molecule has *two unpaired electrons*, even though the total number of electrons is even. Thus, the paramagnetism of O_2 is explained.
2. The total number of valence electrons in bonding orbitals is *eight*, and the number of electrons in antibonding orbitals is *four*. The excess of bonding over antibonding electrons is $8 - 4 = 4$. Counting two electrons per bond (i.e., using expression 10.11), this corresponds to a *double* covalent bond between oxygen atoms.

FIGURE 10-18
Molecular orbital diagrams for actual and hypothetical diatomic molecules of the second period elements.

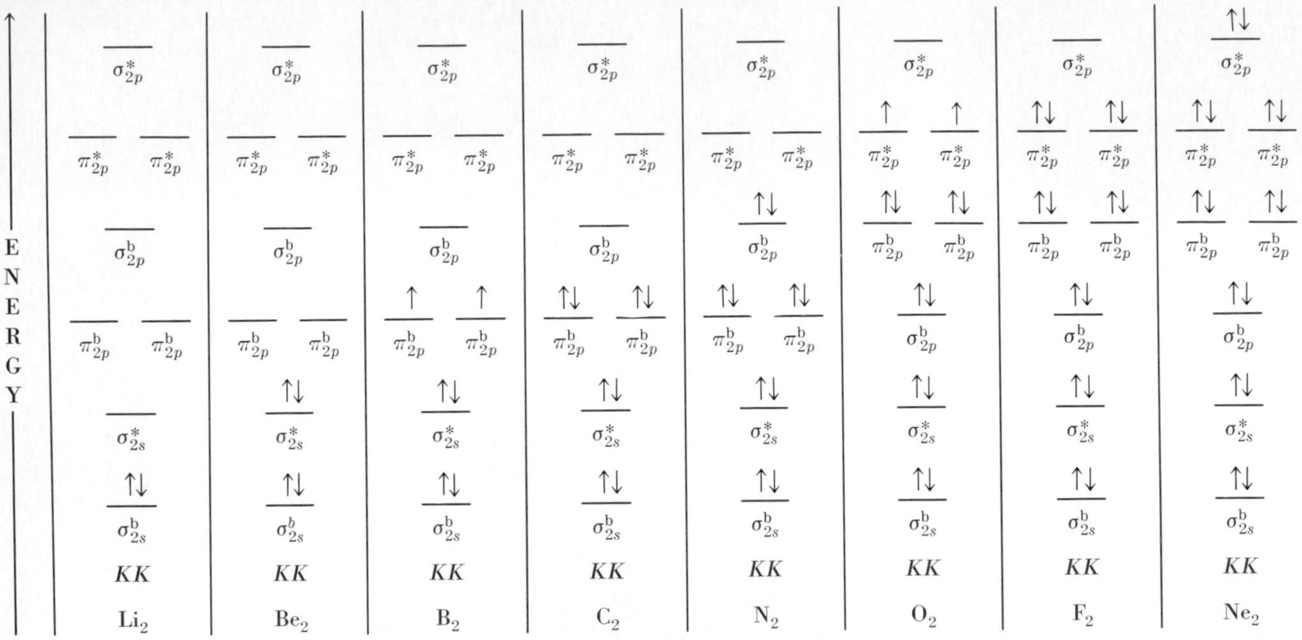

Because the $2p$ level is at a higher energy than $2s$ in the separated atoms, we expect the molecular orbitals formed by combinations of $2p$ atomic orbitals (σ_{2p}^b, π_{2p}^b, σ_{2p}^*, π_{2p}^b) to be at a higher energy than those formed from $2s$ orbitals (σ_{2s}^b, σ_{2s}^*). Concerning the relative placement of the σ_{2p}^b and π_{2p}^b molecular orbitals, more extensive overlap occurs when p atomic orbitals are combined end-to-end than side-by-side, suggesting that σ_{2p}^b lies at a lower energy than π_{2p}^b. This is the situation, confirmed by experiment, when the energy difference between $2s$ and $2p$ atomic orbitals is large (as in O, F, and Ne). When this difference is smaller (as in Li through N), there is a mixing of $2s$ and $2p$ atomic orbitals in the formation of molecular orbitals that results in a reversal of the σ_{2p}^b and π_{2p}^b levels. This variation in the energy-level diagrams of the diatomic molecules of the second period elements has little effect on the matters discussed in this chapter.

Example 10-7   Which of the species indicated in Figure 10-18 would you expect **(a)** *not* to exist as a stable molecule; **(b)** to have the highest bond energy?

Solution

(a) A stable diatomic molecule must have an excess of bonding over antibonding electrons. All the species in Figure 10-18 meet this criterion except two. These two—Be_2 and Ne_2—have equal numbers of bonding and antibonding electrons and a bond order of *zero* (recall expression 10.11). They do *not* exist as stable molecules.

(b) We would expect the greatest bond energy to exist in the species with the highest bond order. Inspection of Figure 10-18 shows that the species with the largest excess of bonding over antibonding electrons is N_2.

SIMILAR EXAMPLES: Review Problem 8; Exercises 14, 15.

Example 10-8   Represent bonding in O_2^+ with a molecular orbital diagram.

Solution.   The O_2 molecule has 12 valence electrons. In the *ion* O_2^+ there are 11 valence electrons. These are assigned to the available molecular orbitals in accordance with the principles established on page 269. The representation below resembles the orbital diagrams first developed for the electron configurations of atoms. The symbol *KK* means that electrons in the first electronic shells (*K* shells) are not involved in the bonding.

$$\sigma_{2s}^{b} \quad \sigma_{2s}^{*} \quad \sigma_{2p}^{b} \quad \pi_{2p}^{b} \quad \pi_{2p}^{*} \quad \sigma_{2p}^{*}$$

O_2^+ *KK*   ↑↓   ↑↓   ↑↓   ↑↓ ↑↓   ↑

SIMILAR EXAMPLES: Review Problem 9; Exercise 20.

10-5 Bonding in the Benzene Molecule

Earlier in this chapter we described some simple hydrocarbons—CH_4, C_2H_4, and C_2H_2—in terms of valence bond theory. These molecules can also be described through Lewis structures and VSEPR theory. But there are some organic compounds, notably aromatic hydrocarbons, that cannot be described adequately by any one of these approaches. They require instead use of the molecular orbital theory. Aromatic hydrocarbons have structures based on benzene, C_6H_6. The term "aromatic" originally referred to the fragrant aromas associated with many, but not all, compounds of this type.

Michael Faraday discovered benzene in 1825 in the gas lines of London. Benzene presented the field of organic chemistry with a problem lasting for 40 years: What is the structure of benzene? In 1834, it was shown to have the molecular formula C_6H_6. In 1865, Friedrich Kekule of the University of Bonn offered a structure.

Kekule's hypothesis was that the benzene molecule consists of a flat, cyclic, hexagonal structure of alternate carbon-to-carbon single and double bonds. Each carbon atom is bonded to two other carbon atoms and to only one hydrogen atom. Kekule accounted for the equivalence of the six carbon-to-carbon bonds by suggesting that the double bonds are not static, but instead oscillate from one position to another. To some chemists this suggested two discrete Kekule forms of benzene that were in equilibrium with one another. A more correct view is not of discrete Kekule structures but of a resonance hybrid structure toward which the Kekule forms are the two principal contributing structures. This view is suggested by Figure 10-19.

As depicted in Figure 10-20, the valence bond method of describing bonding in the benzene molecule requires the use of sp^2 and p orbitals by the carbon atoms. Overlap involving the sp^2 orbitals produces the σ bond framework and is consistent with the formula C_6H_6 and the hexagonal planar geometry (bond angles of 120°). The valence

FIGURE 10-19
Resonance in the benzene molecule and the Kekule structures.

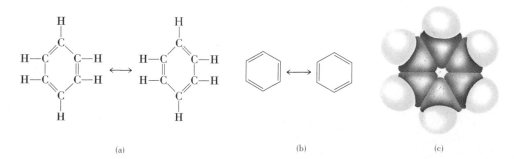

(a)   (b)   (c)

(a) Lewis structures for C_6H_6, showing alternate carbon-to-carbon single and double covalent bonds.
(b) Two equivalent Kekule structures for benzene. A carbon atom is at each corner of the hexagonal structure and a hydrogen atom is bonded to each carbon. (The symbols for carbon and hydrogen are customarily *not* written in these structures.)
(c) A space-filling model.

FIGURE 10-20
Bonding in benzene,
C_6H_6, by the valence
bond method.

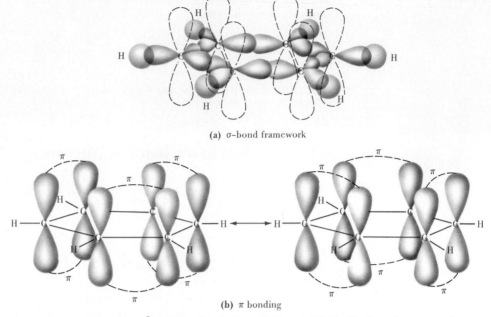

(a) σ–bond framework

(b) π bonding

(a) Carbon atoms use sp^2 and p orbitals (recall Figure 10-9). Each carbon atom forms three σ bonds, two with neighboring C atoms in the hexagonal ring and a third with an H atom.
(b) The overlap in sidewise fashion of $2p$ orbitals produces three π bonds. Thus, there are three double bonds (σ + π) between carbon atoms in the hexagonal ring. Two equivalent structures are possible.

bond method also accounts for double-bond formation, through the sidewise overlap of $2p$ orbitals yielding π bonds. However, the valence bond method fails to account for the *equivalence* of the carbon-to-carbon bonds (all are intermediate to single and double bonds) without invoking the phenomenon of resonance.

Delocalized Molecular Orbitals.   The key to a better explanation of bonding in benzene requires introduction of the concept of delocalized molecular orbitals. In a **delocalized molecular orbital** high electron probability or electron charge density extends over three

FIGURE 10-21
Molecular orbital
representation of
bonding in benzene,
C_6H_6.

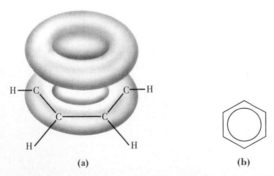

(a)   (b)

(a) A representation incorporating the σ-bond framework of Figure 10-20a and delocalized molecular orbitals (the doughnut-shaped regions) for the π bonds.
(b) A symbolic representation suggesting the delocalized nature of the π bonds (circle inscribed in the hexagon) that is often used in place of the Kekule structures of Figure 10-19b.

or more atoms instead of being limited (localized) to the internuclear region between two atoms.

The σ bond framework of Figure 10-20 describes adequately the bonds formed within the plane of the hexagonal ring of carbon and hydrogen atoms in C_6H_6. However, let us consider combining the six $2p$ atomic orbitals into molecular orbitals, a process that yields three bonding and three antibonding molecular orbitals of the π type. The three bonding orbitals fill with six electrons (one $2p$ electron from each C atom), and the three antibonding orbitals remain empty. The combination of the three bonding molecular orbitals describes the distribution of π electron charge in the molecule. This can be represented as two doughnut-shaped regions, one above and one below the plane of the C and H atoms (see Figure 10-21). Since they are spread out among six C atoms, these molecular orbitals are *delocalized*. The concept of delocalized electrons is carried over into the symbolic representation of the benzene molecule shown in Figure 10-21b. The circle inscribed within the hexagon represents the multiple bond character displayed by all six carbon atoms.

10-6 Other Structures with Delocalized Orbitals

The concept of delocalized orbitals that arises in molecular orbital theory is not limited to aromatic hydrocarbons. We should always look for this possibility in instances where a resonance hybrid is based on contributing structures in which multiple bonds appear. Consider, for example, the anion, NO_3^-, described in Chapter 9. In place of the resonance hybrid based on these structures,

we can write a single structure with a σ bond framework and delocalized electrons in π molecular orbitals. The situation is pictured in Figure 10-22.

In the σ bond framework we assume sp^2 hybridization for each atom. Of the 24 valence shell electrons, 18 are assigned to the sp^2 hybrid orbitals. Six of these 18 are shared in the regions of orbital overlap. The other 12 are found as six lone pairs, two pairs each on the O atoms. The sp^2 hybridization scheme leaves each atom with a simple p orbital. The combination of these four p orbitals produces four molecular orbitals of the π type. Two of the π orbitals are bonding molecular orbitals, and two are antibonding. The remaining six of the 24 valence shell electrons are now assigned to these π orbitals. Four electrons go into the bonding molecular orbitals and two into the antibonding. With four bonding and two antibonding electrons, the total number of bonds arising from the π orbitals is $\frac{2}{2} = 1$. This π bond is apportioned among the three nitrogen-to-oxygen bonds. Thus, each N—O bond is a $1\frac{1}{3}$ bond, exactly the conclusion we reach by averaging the three Lewis structures! With molecular orbital theory it is unnecessary to invoke the phenomenon of resonance.

One additional idea is needed to explain some situations where resonance occurs, e.g., SO_4^{2-}. Several contributing structures to the resonance hybrid were pictured in Figure 9-8. Those involving one or two sulfur-to-oxygen double bonds require octet expansion for the central S atom. However, VSEPR theory still predicts a tetrahedral distribution of bonds from the S atom. Valence bond theory provides for a tetrahedral distribution of bonds through sp^3 hybrid orbitals. But these orbitals participate only in σ-bond formation. What orbitals are involved in forming the π portions of the double bonds? These appear to involve a combination of $2p$ orbitals of the O atoms with $3d$ orbitals of the S atom, producing a type of bonding known as p_π–d_π.

FIGURE 10-22

Structure of the nitrate anion, NO_3^-.

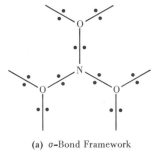

(a) σ-Bond Framework

(b) Delocalized π molecular orbital

10-7 Bonding in Metals

In nonmetal atoms the valence electronic shells generally contain more electrons than they do vacant or partially filled orbitals. To illustrate, in the valence shell ($n = 2$) of an atom of F, there are four orbitals ($2s$, $2p_x$, $2p_y$, $2p_z$) and seven electrons. By contrast, in lithium metal each atom has only one valence-shell electron ($2s^1$) and four valence-shell orbitals ($2s$, $2p_x$, $2p_y$, $2p_z$). In solid metallic lithium, each Li atom is bonded, somehow, to eight nearest neighbors. There appear to be too few electrons to hold these atoms together. This is a feature shared by all metals—more valence shell orbitals than electrons. A bonding scheme for metals must also account for these distinctive properties that all metals share, more or less.*

1. Ability to conduct electricity.
2. Ability to conduct heat.
3. Ease of deformation [i.e., ability to be flattened into sheets (malleability) and to be drawn into wires (ductility)].
4. Lustrous appearance.

One old and oversimplified model that can account for some of these properties is the **electron-sea model.** The metal is pictured as a network of positive ions immersed in a "sea of electrons." In lithium, for example, the ions would be Li^+ and one electron per atom would be contributed to the sea. These free electrons account for the characteristic metallic properties. If the ends of a bar of metal are connected to a source of electric current, electrons from the external source enter the bar at one end. Free electrons pass through the metal and leave the other end at the same rate. In thermal conductivity no electrons enter or leave the metal, but those in the region being heated gain kinetic energy and transfer this to other electrons. According to the electron-sea model the ease of deformation of metals can be thought of in this way: If one layer of metal ions is forced across another, perhaps by hammering, the internal structure remains essentially unchanged as the sea of electrons rapidly adjusts to the new situation.

Band Theory.   The term *electron sea* is not very specific about the region occupied by free electrons in a metal. A more exact description is possible with molecular orbital theory. For example, recall the formation of molecular orbitals by two Li atoms (Figure 10-18). Each Li atom contributes one $2s$ orbital to the production of two molecular orbitals—σ_{2s}^b and σ_{2s}^*. The electrons originally described as the $2s^1$ electrons of the Li atoms enter and half-fill these molecular orbitals; that is, they fill the σ_{2s}^b orbital and leave the σ_{2s}^* empty. If this combination of Li atoms is extended to a third Li atom, three molecular orbitals are formed, which contain three electrons; again the set of molecular orbitals is half-filled. This process can be extended to an enormously large number of atoms N, the total number of atoms in a crystal of lithium. Here is the result that is obtained: A set of N molecular orbitals is produced, with the difference in energy between the lowest and highest energy level in the set being not much greater than between the σ_{2s}^b and the σ_{2s}^* orbitals of Figure 10-18. Because the number of individual molecular orbitals in this set is so large (N), the energy separation between each pair of successive levels is extremely small. This collection of very closely spaced molecular orbital energy levels is called a **band.**

In the band just described, there are N electrons (a $2s^1$ electron from each Li atom)

---

*This classification of metallic properties is not perfect. The ability of metals to conduct electricity varies by a factor of about 100, from the best (Ag) to the poorest (Pu) conductor. Some metals (e.g., Mn and Bi) are quite brittle. And there are a few instances of nonmetals possessing some of the properties listed here. Carbon (as diamond) has good thermal conductivity; carbon (as graphite) conducts electricity (but only in two directions, not all three); and silicon has a high luster.

occupying, in pairs, the $N/2$ molecular orbitals of lowest energy. These are the electrons responsible for bonding the Li atoms together. They are valence electrons, and the band in which they are found is called a **valence band.** However, because the energy difference between the occupied and unoccupied levels in the valence band is so small, electrons can be easily excited from the highest of the filled levels to the unfilled levels that lie just slightly above them in energy. This excitation, which has the effect of producing mobile electrons, can be accomplished by heating the crystal or by applying a small electrical potential difference across the crystal. This is how the band theory explains thermal and electrical conductivity. The essential feature for electrical conductivity, then, is *an energy band that is only partly filled with electrons.* Such an energy band is called a **conduction band.** In lithium the $2s$ band is both a valence band and a conduction band.

If we extend this discussion to N atoms of beryllium, which has the electron configuration $1s^2 2s^2$, we would conclude that the $2s$ valence band is filled—N molecular orbitals and $2N$ electrons. But how can this be reconciled with the fact that beryllium is an electrical conductor? At the same time that $2s$ atomic orbitals are being combined into a $2s$ band, $2p$ orbitals combine to produce an *empty* $2p$ band. The lowest levels of the $2p$ band are at a lower energy than the highest levels of the $2s$ band—the bands overlap. As a consequence, empty molecular orbitals are available to the valence electrons in beryllium.

In an electrical insulator like diamond or silica (SiO_2), not only is the valence band filled, but there is a large **energy gap** between the valence band and a conduction band. Very few electrons are able to make the transition between the two, and no electrons are permitted in the **forbidden zone** that separates the two bands. Still another possibility, found in silicon and germanium, for example, is that a filled valence band and an empty conduction band are separated by only a small energy gap. Electrons in the valence band may acquire enough energy, e.g., thermal energy, to jump to a level in the conduction band. The greater the thermal energy, the more electrons that can make the transition. The electrical conductivity of a material of this type, a **semiconductor,** in contrast to that of metals, increases with temperature. In Figure 10-23 a simple comparison based on band theory is made of insulators, conductors, and semiconductors.

FIGURE 10-23
A comparison of electron energy levels in insulators, metallic conductors, and semiconductors.

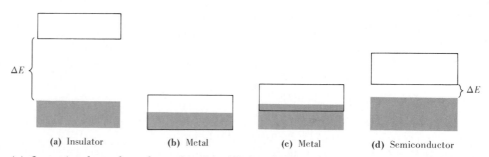

(a) Insulator   (b) Metal   (c) Metal   (d) Semiconductor

(a) In an insulator the valence band is filled with electrons, and a large energy gap, ΔE, separates the valence band and conduction band (outlined in black). Few electrons can make the transition when an electric field is applied, and the insulator does not conduct electric current.
(b) In some metals the valence band is only partially filled with electrons, and the valence band also serves as a conduction band (e.g., the half-filled $3s$ band in Na).
(c) In other metals the valence band is filled, but a conduction band overlaps it. In an electric field, electrons from the valence band can move through the conduction band (e.g., the empty $3p$ band of Mg overlaps with the filled $3s$ valence band.)
(d) In a semiconductor the valence band is filled and the conduction band is empty. The energy gap between the two, ΔE, is small enough, however, that some electrons make the transition simply by acquiring extra thermal energy.

FOCUS ON
Semiconductors

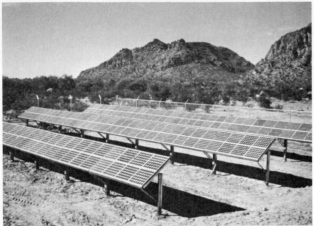

An array of thousands of photovoltaic cells. Such an installation can be used to convert solar to electric energy to operate pumps in irrigation systems. Similar arrays are used to produce electricity in spacecraft.

An alternative description of semiconductor behavior to that given in the preceding section is suggested by Figure 10-24. Here the normal condition is for all electrons to participate in electron-pair covalent bond formation. The electrons are in the valence band; they are localized. Thermal energy is sufficient, however, to cause some electrons to be displaced from electron-pair bonds into the crystal as a whole, that is, to become delocalized and enter the conduction band. For every electron that enters the conduction band (a conduction electron) a vacancy or "hole" is left at the site of the broken bond (which also means that a vacancy or hole is created in the valence band). An electron may leave a bond elsewhere in the crystal to fill this vacancy, in turn creating a vacancy at the bond site from which it originated, and so on. Because a deficiency of an electron at a bond site is equivalent to the creation of a positive charge, these vacancies are called **positive holes.** When an electric field is applied, both conduction electrons and positive holes migrate.

Semiconductors in which electrical conductivity involves the thermal promotion of valence electrons to the conduction band are called **intrinsic semiconductors.** Silicon and germanium are the two best known examples. More important, however, are **extrinsic semiconductors.** These are semiconductor materials to which have been added small and very carefully controlled quantities of impurity atoms—a process called **doping.** Consider, for example, the result of introducing a trace of phosphorus

(group VA) into a crystal of silicon (IVA). Because P atoms have five valence electrons compared to four for Si, one electron is promoted to the conduction band for every P atom introduced. This is equivalent to creating one immobile positive charge (P^+) and one mobile conduction electron for every P atom introduced as a dopant. Electrical conductivity in this type of semiconductor results primarily from the movement of conduction electrons; the semiconductor is said to be of the **n-type** ("n" referring to negative—the type of charge carried by electrons). Doping with a group IIIA element, such as boron, produces a semiconductor in which there is a deficiency of one electron for every B atom. Movement of electrons from bonds elsewhere in the crystal into vacancies at the sites of B atoms creates one immobile center of negative charge (B^-) and one positive hole for every B atom. Electrical conduction results primarily from the movement of positive holes, and the semiconductor is said to be of the **p-type.** The chemical constitutions of *n*- and *p*-type semiconductors are illustrated in Figure 10-25.

Appropriate combinations of *n*- and *p*-type semiconductors, called **transistors,** can be used to control electric currents. Transistors are the basic working components of modern solid-state electronic devices, whether they be

FIGURE 10-24
Formation of conduction electrons and positive holes in a semiconductor.

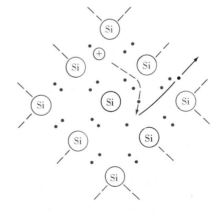

When an electron escapes from the bond between the two Si atoms shown in black, it moves into the crystal as a whole. It becomes a conduction electron and leaves a vacancy at the original bond site. An electron from elsewhere in the crystal (broken line) enters this vacancy, creating a vacancy or positive hole at the bond site from which it came.

FIGURE 10-25
N-type and *p*-type semiconductors.

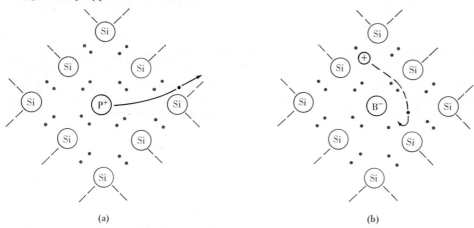

(a) (b)

(a) The presence of the group VA atom phosphorus in the crystal introduces an extra electron that is lost to the silicon crystal as a whole, that is, goes into the conduction band. This is an *n*-type semiconductor.
(b) The boron atom, having only three valence electrons, requires an additional electron from somewhere in the crystal in order to form bonds with each of its four neighboring Si atoms. As a result the boron atom becomes an immobile center of negative charge (B^-), and the bond vacancy created in the crystal becomes a positive hole. This is a *p*-type semiconductor.

found in television sets, electronic calculators, or high-speed computers. Another interesting use of semiconductors that is coming into prominence is in solar energy cells, such as the one diagrammed in Figure 10-26.

Figure 10-26 shows a *p*-type semiconductor, which might be Si doped with B, in contact with an *n*-type semiconductor, such as Si doped with P. Conduction electrons, the majority carriers of electricity in the *n*-type semiconductor, can migrate across the boundary or junction between the two semiconductors. Likewise, positive holes, the majority carriers in the *p*-type semiconductor, can migrate into the *n*-type semiconductor. These migrations can occur only to a very limited extent, however, because they tend to make the *p*-side of the junction, where the immobile B^- ions are found, acquire a net negative charge. Similarly, the *n*-side, where immobile P^+ ions are present, would acquire a net positive charge. Now imagine that the *p*-type semiconductor is struck by a beam of light. Electrons in the valence band can absorb some of this light energy and be promoted to the conduction band. Conduction electrons, unlike positive holes, can easily cross the junction into the *n*-type semiconductor. A flow of electrons, an electric current, is established. This device, known as a **photovoltaic cell,** converts solar to electrical energy. Because the conduction electrons produced in the *p*-type semiconductor can so readily be neutralized by positive holes, the layer of *p*-type semiconductor must

FIGURE 10-26
A photovoltaic (solar) cell using silicon-based semiconductors.

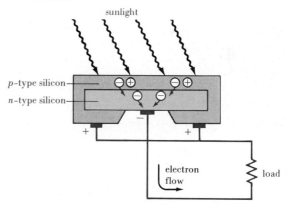

be kept very thin (about 10^{-4} cm). This way the electrons have a chance to cross the junction without being neutralized.

Summary

The valence bond method views a covalent bond as resulting from the overlap of atomic orbitals of the bonded atoms. This produces a high electron probability (electron charge density) in the region of overlap. Some simple covalent molecules (e.g., H_2, HCl, H_2S) can be described adequately in terms of the overlap of simple *s* and/or *p* orbitals. In the majority of cases, however, these simple atomic orbitals must be hybridized. That is, they must be replaced by a new set of hybrid orbitals whose properties depend on the number and types of simple atomic orbitals used to form them. The geometrical shape of a molecule is determined by the spatial distribution of the orbitals involved in bond formation, which for the most part, correspond to the distributions of electron pairs in VSEPR theory.

Two types of orbital overlap are encountered in the valence bond method. One type (σ) involves end-to-end overlap along the line joining the nuclei of the bonded atoms. The other (π) requires a "sidewise" overlap of two *p* orbitals. Single covalent bonds are σ bonds: a double bond consists of one σ and one π bond; and a triple bond, one σ and two π bonds.

In molecular orbital theory, when atoms join to form a molecule new regions of high electron probability—molecular orbitals—are established for the molecule as a whole. *Bonding* molecular orbitals correspond to high electron probability or electron charge density in the internuclear region between atoms. *Antibonding* molecular orbitals concentrate electron probability or charge density in regions away from the internuclear region. The numbers and kinds of molecular orbitals in a molecule are related to the corresponding atomic orbitals from which they arise. A procedure similar to the Aufbau process for the electron configurations of atoms can be employed to derive the electronic structure of a molecule. *Bond order* follows directly from the assignment of electrons to molecular orbitals: It is one-half the difference between the numbers of electrons in bonding molecular orbitals and in antibonding molecular orbitals.

A description of bonding in the benzene molecule (C_6H_6) requires the concept of delocalized molecular orbitals. These are regions of high electron probability that extend over *three or more* atoms in a molecule. Delocalized molecular orbitals also provide an alternative to the concept of resonance in dealing with species such as SO_2, SO_3, and NO_3^-. Finally, molecular orbital theory in the form called band theory can be applied to metals and semiconductors.

Learning Objectives

As a result of studying Chapter 10, you should be able to

1. Explain the basis of the valence bond method.

2. Write hybridization schemes for the formation of sp, sp^2, sp^3, sp^3d, and sp^3d^2 hybrid orbitals.

3. Predict geometrical shapes of molecules in terms of the simple and hybrid orbitals employed in bonding.

4. Discuss the relationship between VSEPR theory and the valence bond method of predicting molecular geometry.

5. Describe the conditions leading to σ and to π bond formation and the differences between these two types of bonds.

6. Propose plausible bonding schemes from Lewis structures or from experimental information about molecules (i.e., bond lengths, bond angles, and so on).

7. Explain the basis of molecular orbital theory.

8. Describe the differences between bonding and antibonding molecular orbitals.

9. Assign probable electron configurations, determine bond orders, and predict magnetic properties of the diatomic molecules and ions of the first and second period elements.

10. Describe bonding in the benzene molecule (C_6H_6) through Lewis structures, valence bond theory, and molecular orbital theory.

11. Use the concept of delocalized orbitals to write a *single* structure to represent a resonance hybrid (i.e., in cases such as NO_3^-, SO_2, SO_3, etc.).

12. Discuss some of the distinctive properties of metals and the way in which these properties can be explained with the electron-sea model and the band theory of metals.

13. Use the idea of electron bands to contrast insulators, metals, and semiconductors.

Some New Terms

An **antibonding molecular orbital** describes regions of high electron charge density located away from the internuclear region between two atoms.

Band theory is a form of molecular orbital theory applied to metals.

Bond order is one-half the difference between the numbers of electrons in bonding and in antibonding molecular orbitals in a molecule.

A **bonding molecular orbital** describes regions of high electron probability or charge density in the internuclear region between two atoms.

A **delocalized molecular orbital** describes regions of high electron probability or charge density that extend over three or more atoms.

A **hybrid orbital** is one of a set of equivalent orbitals used to replace simple atomic orbitals in describing certain covalent bonds.

Hybridization refers to the combining of simple atomic orbitals to generate new (hybrid) orbitals.

A **pi (π) bond** results from the "sidewise" overlap of p orbitals, producing a high electron charge density above and below the line joining the bonded atoms.

An **sp hybrid orbital** is one of two identical orbitals that result from the hybridization of one s and one p orbital. The angle between the two orbitals is 180°.

An **sp^2 hybrid orbital** is one of three identical orbitals that result from the hybridization of one s and two p orbitals. The angle between any two of the orbitals is 120°.

An **sp^3 hybrid orbital** is one of four identical orbitals that result from the hybridization of one s and three p orbitals. The angle between any two of the four orbitals is the tetrahedral angle—109.5°.

An **sp^3d hybrid orbital** is one of five orbitals that result from the hybridization of one s, three p, and one d orbital. The five orbitals are directed to the corners of a trigonal bipyramid.

An **sp^3d^2 hybrid orbital** is one of six orbitals that result from the hybridization of one s, three p, and two d orbitals. The six orbitals are directed to the corners of a regular octahedron.

A **semiconductor** is characterized by a small energy gap between a filled valence band and an empty conduction band. In an **intrinsic** semiconductor electrons acquire extra thermal energy and are promoted to the conduction band, leaving vacancies or **positive holes** in the valence band. Electrical conduction involves movement of both conduction electrons and positive holes. In an **extrinsic** semiconductor impurity atoms are added to create an excess of either conduction electrons (**n-type**) or positive holes (**p-type**).

A **sigma (σ) bond** results from the end-to-end overlap of simple or hybridized atomic orbitals along the straight line joining the nuclei of the bonded atoms.

The **valence bond method** treats a covalent bond in terms of the overlap of atomic orbitals. Electron probability (or charge density) is concentrated in the region of overlap.

Suggestions for Further Study

EBERLIN, D., and M. MONROE, "A Different Approach to Hybridization and Geometric Structure of Simple Molecules and Ions," *J. Chem. Educ.*, **59**, 285 (1982).

FERREIRA, R., "Molecular Orbital Theory: An Introduction," *Chemistry*, **41**[6], 8 (1968).

FINKLEA, H. O., "Photoelectrochemistry: Introductory Concepts," *J. Chem. Educ.*, **60**, 325 (1983).

MICKEY, C. D., "Solar Photovoltaic Cells," *J. Chem. Educ.*, **58**, 418 (1981).

MYERS, R. T., "Physical and Chemical Properties and Bonding of Metallic Elements," *J. Chem. Educ.*, **56**, 712 (1979).

SALEM, L., "A Faithful Couple: The Electron Pair," *J. Chem. Educ.*, **55**, 344 (1978).

WELLER, P. F., "An Introduction to Principles of the Solid State," *J. Chem. Educ.*, **47**, 501 (1970).

WELLER, P. F., "An Introduction to Principles of the Solid State: Extrinsic Semiconductors," *J. Chem. Educ.*, **48**, 831 (1971).

Review Problems

1. In the manner employed in Example 10-1, describe the structure and bonding in **(a)** HCl; **(b)** ICl; **(c)** H_2Se; **(d)** NI_3.

2. In the manner depicted in Figure 10-11, indicate the structures of the following simple molecules in terms of the overlap of simple atomic orbitals and hybrid orbitals: **(a)** H_2CCl_2; **(b)** $BeCl_2$; **(c)** BF_3.

3. Match each of the following species with one of these hybridization schemes: sp, sp^2, sp^3, sp^3d, sp^3d^2. **(a)** SF_6; **(b)** CS_2; **(c)** $SiCl_4$; **(d)** NO_3^-; **(e)** AsF_5. (*Hint:* You may find it useful to write Lewis structures and to refer to the VSEPR theory.)

4. Describe a bonding scheme, based on simple and hybridized atomic orbitals, to account for the structure of the hydroxylamine molecule.

5. For the following species label each σ and π bond. (*Hint:* You may find it helpful to write Lewis structures first.) **(a)** HCN; **(b)** C_2N_2; **(b)** $H_3CCHCHCCl_3$; **(d)** HONO.

6. Use the method of Figure 10-12 to represent bonding in the molecule, dimethyl ether, H_3COCH_3.

7. Which of the following molecules are linear? Which are planar? **(a)** $HC\equiv N$; **(b)** $N\equiv C—C\equiv N$; **(c)** $F_3C—C\equiv N$; **(d)** $H_2C=C=O$. (*Hint:* What type of hybrid orbitals are involved in the bonding?)

8. For each of the following indicate whether the species is diamagnetic or paramagnetic; if paramagnetic, indicate the number of unpaired electrons. **(a)** F_2; **(b)** N_2^+; **(c)** O_2^-.

Exercises

Valence bond method

1. Which of the following statements best describes the bond angle in H_2Se? Explain. **(a)** Greater than in H_2S; **(b)** less than in H_2S; **(c)** less than in H_2S, but not less than $90°$; **(d)** less than $90°$.

2. The Lewis structure of N_2 suggests a triple covalent bond. Describe this bonding in terms of atomic orbital overlap.

3. Indicate ways in which the valence bond method (atomic orbital overlap) is superior to Lewis structures in describing covalent bonds.

Hybridized atomic orbitals

4. Predict the shape of the ammonium ion, NH_4^+, and describe a bonding scheme that is consistent with this.

5. For each of the following species identify the central atom and indicate the hybridization scheme for that atom: **(a)** CO_2; **(b)** $ClNO_2$; **(c)** N_2O.

6. Acetic acid is a very common organic acid (5% by mass, in vinegar). Sketch a three-dimensional structure of the molecule that is consistent with this bonding scheme.

$$H_3C—\overset{\overset{\displaystyle O}{\|}}{C}—OH$$

7. Glycine is a simple amino acid with the formula H_2NCH_2COOH. Sketch a three-dimensional structure of the molecule, indicating the orbitals involved in the bonding scheme.

8. Propose a bonding scheme that is consistent with this structure for propynal. (*Hint:* You may find it useful to consult Table 9-3 to assess multiple bond character in some of the bonds.)

9. Use the method of Figure 10-12 to represent bonding in each of the following molecules: **(a)** CCl_4; **(b)** $ONCl$; **(c)** $HONO$; **(d)** H_3CCCH.

9. In the manner of Example 10-8, indicate the molecular orbital diagrams of **(a)** H_2^-; **(b)** N_2^+; **(c)** F_2^-; **(d)** Ne_2^+.

10. How many energy levels are present in the $3s$ conduction band of a single crystal of sodium weighing 25.3 mg? How many electrons are present in this band?

*10. Use the method of Figure 10-12 to represent bonding in the following species. **(a)** XeF_4; **(b)** C_3O_2; **(c)** XeF_2; **(d)** I_3^-; **(e)** $C_2O_4^{2-}$.

11. Describe a hybridization scheme for the central Cl atom in the molecule ClF_3 that is consistent with the geometrical structure pictured in Table 9-2. Which orbitals of the Cl atom are involved in overlap, and which are occupied by lone-pair electrons?

12. Although they have similar formulas, the hybridization schemes for PF_5 and BrF_5 are different. Explain why this is so.

13. Based on the distinction between σ and π bonds and the data given, estimate the bond strength of a carbon-to-carbon triple bond. Compare your result with the value listed in Table 9-3. Bond energies: C—C, 347 kJ/mol; C=C, 611 kJ/mol.

Molecular orbital method

14. With reference to the molecular orbital diagrams shown in Figure 10-18, which of the *stable* molecules are diamagnetic and which are paramagnetic?

15. Would you expect N_2^- and/or N_2^{2-} to be stable ionic species in the gaseous state? Explain.

16. The molecular orbital diagram of O_2 is shown in Figure 10-18, and for O_2^+ in Example 10-8. Which species, O_2 or O_2^+, has the stronger bond? Explain.

17. The paramagnetism of gaseous B_2 has been established. Explain how this observation confirms that the π_{2p}^b orbitals are at a lower energy than the σ_{2p}^b orbital for B_2.

18. Describe the bond order of diatomic carbon, C_2, in terms of Lewis theory and molecular orbital theory, and explain why the results are different.

19. In all of our discussion of bonding we have not encountered a bond order higher than triple. Use the energy level diagram of Figure 10-18 to show why this is to be expected.

20. The molecular orbital method can be applied to heteronuclear (different nuclei) diatomic molecules as well as to homonuclear (same nuclei) diatomic molecules. Based on the energy level diagram of Figure 10-18, suggest suitable molecular orbital diagrams for **(a)** NO; **(b)** NO^+; **(c)** CO; **(d)** CN; **(e)** CN^-; **(f)** CN^+; **(g)** BN.

21. We have used the term isoelectronic to refer to atoms with identical electron configurations. In the molecular orbital theory this term can be applied to molecules as well. Which of the species of Exercise 20 are isoelectronic?

22. One of the characteristics of an antibonding molecular orbital is the presence of a nodal plane. Which of the *bonding* molecular orbitals considered in this chapter have nodal planes? Explain how a molecular orbital can have a nodal plane and still be a bonding molecular orbital.

Delocalized molecular orbitals

23. Explain how it is possible to get around the concept of resonance by using molecular orbital theory.

24. Represent chemical bonding in the molecule SO_3 **(a)** by writing a Lewis structure(s); **(b)** by using a combination of localized and delocalized orbitals.

25. In which of the following species would you expect to find delocalized molecular orbitals? Explain. **(a)** C_2H_4; **(b)** CO_3^{2-}; **(c)** NO_2; **(d)** H_2CO.

★**26.** In a manner similar to that outlined in Section 10-6, propose a bonding scheme for SO_2 that is consistent with structures (9.18). To do so requires introducing the concept of a *nonbonding* molecular orbital, one in which the electron charge density neither adds nor detracts from bond formation. Explain why this is necessary.

Metallic bonding

27. Which of the following factors are especially important in determining whether a substance has metallic properties? Explain. **(a)** Atomic number; **(b)** atomic weight; **(c)** number of valence electrons; **(d)** number of empty orbitals; **(e)** total number of electronic shells in the atom.

28. Based just on the ground-state electron configurations of the atoms, how would you expect the melting points and hardnesses of sodium, iron, and zinc to compare?

Semiconductors

29. Which of the following would you expect to be a *p*-type, which an *n*-type, and which an intrinsic semiconductor? **(a)** germanium; **(b)** germanium doped with aluminum; **(c)** silicon doped with arsenic; **(d)** silicon.

30. Based on the discussion of semiconductors in the text, explain why
    (a) Even in an *n*-type semiconductor some of the electrical conductivity is due to positive holes.
    (b) The rate at which the conductivity of a semiconductor increases with temperature is greater for an intrinsic than for an extrinsic semiconductor.

31. The solar cell in Figure 10-26 is effective throughout the energy range of visible light.
    (a) Estimate the maximum value of the energy gap, in eV, separating the valence and conduction bands in silicon.
    (b) Why does the device operate over a broad range of wavelength components rather than at a single unique wavelength (as is so often the case when quantum effects are involved)?

★**32.** A solar cell that is 15% efficient in converting solar to electrical energy is exposed to full sunlight, which produces an energy flow of 1.00 kW/m^2. If the cell has an area of 40 cm^2 exposed to the sunlight
    (a) What is the power output of the cell, in watts?
    (b) If the power calculated in (a) is produced at 0.45 V, how much current does the cell deliver?
(*Hint:* Refer to Appendix B.)

Additional Exercises

1. Figure 10-14 represents the energy of interaction of two H atoms when the two electrons enter a bonding molecular orbital. Sketch a graph of energy vs. internuclear distance to represent the situation you would expect if the two electrons were to enter an antibonding orbital.

2. Show that both the valence bond method and molecular orbital theory provide an explanation for the existence of the covalent molecule Na_2 in the gaseous state.

3. The poisonous gas phosgene has the formula $COCl_2$. Propose an appropriate scheme of atomic orbital overlap for bonding in this molecule. What is its shape?

4. Lewis theory is satisfactory for explaining bonding in the ionic compound K_2O. However, it does not readily explain formation of the ionic compounds potassium superoxide, KO_2, and potassium peroxide, K_2O_2. **(a)** Show that molecular orbital theory can provide this explanation. **(b)** Write Lewis structures consistent with this explanation.

5. The structure of the molecule allene, $H_2C=C=CH_2$, is indicated below. Propose hybridization schemes for the three C atoms that are consistent with this structure.

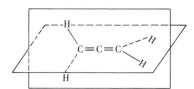

★**6.** Suppose that the σ_{2p}^b orbitals were to fill before the π_{2p}^b orbitals in **(a)** B_2 and **(b)** C_2. How would the number of unpaired electrons and the bond order in these molecules compare to what is shown in 10-18?

★**7.** In terms of concepts introduced in this chapter, explain why one would hesitate to stir a boiling aqueous solution with a metal rod but not with a rod made of glass or wood.

*8. Based on the plausible Lewis structure for HNO_3 in Exercise 33 of Chapter 9, propose a bonding scheme in the manner shown in Figure 10-12.

*9. He_2 does not exist as a stable molecule, but there is evidence that such a molecule can be formed between electronically excited He atoms. Suggest a bonding scheme based on molecular orbitals to account for this.

10. The molecule formamide, $HCONH_2$, has the following approximate bond angles: H—C—O, 123°; H—C—N, 113°; N—C—O, 124°; C—N—H, 119°; H—N—H, 119°. The C—N bond length is 138 pm. Two Lewis structures can be written for this molecule, with the true structure being a resonance hybrid of the two. Use the data given here (and refer to Table 9-3) to establish a bonding scheme for each structure.

Self-Test Questions

For questions 1 through 6 select the single item that best completes each statement.

1. A molecule in which sp^2 hybrid orbitals are employed for bond formation by the central atom is (a) NH_3; (b) CO; (c) SCl_2; (d) H_2CO.

2. A molecule containing π bonds is (a) PCl_5; (b) N_2; (c) OF_2; (d) He_2.

3. In carbon-hydrogen-oxygen compounds (a) all oxygen-to-hydrogen bonds are π bonds; (b) all carbon-to-hydrogen bonds are σ bonds; (c) all carbon-to-carbon bonds are π bonds; (d) all carbon-to-carbon bonds consist of a σ and one or more π bonds.

4. Of the following, the species with a bond order of 1 is (a) H_2^+; (b) H_2^-; (c) He_2; (d) Li_2.

5. Delocalized molecular orbitals are encountered in one of the following species. That one is (a) H_2; (b) HS^-; (c) CO_3^{2-}; (d) CH_4.

6. The best electrical conductor of the following materials is (a) Ge(s); (b) Br_2(l); (c) Li(s); (d) Si(s).

7. Propose a hybridization scheme to account for bonds formed by the carbon atom in each of the following molecules: (a) hydrogen cyanide, HCN; (b) chloroform, $CHCl_3$; (c) methyl alcohol, H_3COH; (d) carbamic acid,

8. What is the total number of (a) σ bonds and (b) π bonds in the molecule H_3CNCO? (*Hint:* Draw a plausible Lewis structure.)

9. Explain how well each of the following methods describes the shape of the water molecule. (a) Lewis theory; (b) valence bond method using simple atomic orbitals; (c) valence bond method using hybridized atomic orbitals; (d) VSEPR theory.

10. Is it correct to say that when a diatomic molecule loses an electron the bond energy always decreases, that is, that the bond is weakened? Explain.

11. Explain why the concept of delocalized molecular orbitals is essential to an understanding of bonding in the benzene molecule (C_6H_6).

12. Why does the hybridization scheme sp^3d *not* account for bonding in the molecule BrF_5? What hybridization scheme does work? Explain.

11

Liquids, Solids, and Intermolecular Forces

The covalent bond is an *intra*molecular force—a force between atoms *within* a molecule. Such forces influence molecular shapes, bond energies, and many aspects of chemical behavior. Physical properties of the condensed states of matter—liquids and solids—stem from *inter*molecular forces, that is, forces *between* molecules. Intermolecular forces are themselves closely related to intramolecular forces (i.e., to bond type). Thus, the present chapter follows naturally the material presented in the two preceding chapters.

We begin with an overview of some common properties of liquids and solids, and then show how these properties are related to intermolecular forces.

11-1 Comparison of the States of Matter

The atoms, ions, or molecules of a solid are in close contact. In many solids these structural particles exist in a highly ordered network called a **crystal.** Crystals have geometric shapes characterized by plane surfaces intersecting at definite angles. Not all solids are crystalline. If order among the structural particles does not prevail over long distances, the solid is said to be **amorphous.** Whether a solid is crystalline or amorphous, it occupies a definite volume and maintains a definite shape. Solids are practically incompressible. This observation supports the idea that the atoms, ions, or molecules of a solid are in close contact.

The structural particles of a liquid exist in close proximity, although they are usually thought of as not being in contact. Liquids are more compressible than solids because of the free volume that exists among the structural particles of a liquid. The intermolecular forces in a liquid are strong enough to maintain a liquid sample in a fixed volume, but not strong enough to maintain a fixed shape. A liquid tends to flow; it covers the bottom and assumes the shape of its container. Liquids are fluid.

Gases are also fluid. Gases expand to fill their containers and thus possess neither a definite shape nor volume. Also, because their atoms or molecules are ordinarily far apart, gases are highly compressible. (In a typical gas at STP the molecules themselves occupy less than 1% of the total gas volume.)

An additional comparison of the three states of matter is provided in Figure 11-1.

FIGURE 11-1
Comparison of the states of matter.

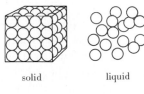

solid   liquid

gas

The structural particles in a solid are constrained to fixed points. They may vibrate about these points but, ordinarily, may not move from them. In a liquid there is some free volume among the structural particles; motions are more vigorous; and the structure is more random. In a gas there is a great deal of free volume; motion is chaotic; disorder is at a maximum.

11-2 Two Properties of Liquids

Surface Tension.   One property related to intermolecular forces in liquids is surface tension. Figure 11-2 suggests that the environment of molecules in the surface of a liquid is different from that of molecules in the interior. Special properties are associated with the surface of a liquid, such as an unbalanced force along the surface. This net force creates a measurable tension in the liquid surface—almost as if the liquid were covered with a tight skin. Another view of this matter is that, because they have greater numbers of neighboring molecules to which they are attracted, molecules in the interior of a liquid are in a lower energy state than those at the surface. As a consequence, as many molecules as possible take up positions in the interior of a liquid. The geometric shape having the smallest proportion of molecules at the surface is a sphere. So, liquids form spherical drops when they are allowed to fall freely, as in a rain shower.

From what has just been said, we see that to *increase* the surface of a liquid requires that energy be *expended*. Surface tension is a measure of the quantity of energy required; it has the units of energy per unit area, i.e., J/m^2. For water, the surface tension (λ) at 25°C is 7.20×10^{-2} J/m^2; for Hg(l), $\lambda = 47.2 \times 10^{-2}$ J/m^2. The effectiveness of intermolecular forces in creating a surface tension is lessened as the intensity of molecular motion increases. This means that surface tension decreases with increased temperature.

In order for a drop of liquid to spread into a thin film across the surface of another material, that is, to **wet** the surface, energy (albeit a very small amount) must be expended. And whether a drop of liquid will wet a surface or retain its spherical shape and stand on the surface depends on the comparative strengths of two types of intermolecular forces. **Cohesive forces** are the intermolecular forces between like molecules. **Adhesive forces** are intermolecular forces between *unlike* molecules, such as molecules of a liquid and molecules of an underlying surface. If cohesive forces are strong compared to adhesive forces, a liquid drop is maintained. If adhesive forces are sufficiently strong, the energy requirements for spreading the drop into a film are met; the liquid wets the surface.

Water wets many surfaces, such as glass and certain fabrics. This ability is essential if water is to be used as a cleaning agent. If a glass surface is coated with a film of oil or grease, water is no longer able to wet the glass and water droplets stand on the glass rather than form a film. Adding a detergent to water has two effects: The detergent solution dissolves grease, exposing the clean glass surface; and the detergent lowers the surface tension of the water. Lowering the surface tension means also lowering the energy required to spread the drop into a film. Other substances that reduce the surface tension of water, allowing it to spread more easily, are known as **wetting agents.** They are used in a variety of applications ranging from dishwashing to industrial processes. In some applications the *inability* of water to wet and penetrate a material is used to advantage. For

FIGURE 11-2
Intermolecular forces in a liquid.

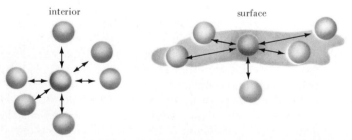

Molecules in the interior experience attractive forces from neighboring molecules in all directions. Molecules at the surface are attracted only by other surface molecules and by molecules below the surface.

FIGURE 11-3
Some phenomena related to surface tension.

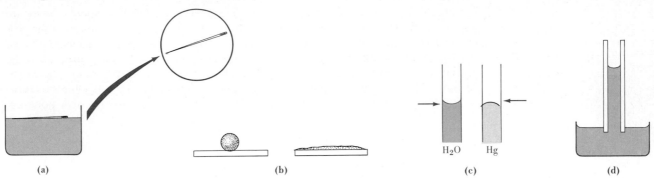

(a)   (b)   (c)   (d)

(a) Surface tension illustrated. That the surface of a liquid is under tension (as if covered by a skin) can be illustrated by carefully floating a clean steel needle on water. Steel is much more dense than water and would be expected to sink, but the surface tension of the water supports the needle.
(b) Wetting of a surface. Because of strong cohesive forces, a drop of Hg(l) does not wet glass and stands on the surface. By contrast, because of strong adhesive forces, a drop of water spreads into a thin film, i.e., wets glass.
(c) Meniscus formation. If a liquid (such as water) wets glass, the meniscus of a column of the liquid in a glass tube is concave. If the liquid does not wet glass (as with mercury), the meniscus is convex. (The effect is exaggerated here for emphasis.)
(d) Capillary action. A liquid wets the inside walls of a glass capillary and a thin film of liquid spreads up the tube. This spreading causes the pressure below the meniscus inside the tube to fall slightly below the pressure at the liquid–air interface outside the tube (atmospheric pressure). Atmospheric pressure then pushes a column of liquid up the tube to a sufficient height to eliminate the pressure difference. The *greater* the surface tension of the liquid and the *smaller* the diameter of the capillary tube, the higher the capillary rise. The action of a sponge in soaking up water depends on the rise of water into capillary pores of a fibrous material such as cellulose. The penetration of water into soils also depends in part on capillary action (e.g., a plant in a pot can be watered simply by placing the pot in a shallow tray of water).

example, liquid water will not penetrate through the microscopic holes of "Gore-Tex" film. Sandwiched between layers of nylon fabric, "Gore-Tex" is used in rainwear and boots. Its advantage over rubber is that the fabric "breathes," i.e., water vapor (perspiration) is able to escape through the microscopic holes.

When a glass tube is partially filled with water, the interface between the water and the air above it, called a meniscus, is curved upward (concave). The water is drawn slightly up the glass walls by adhesive forces between the water and the glass. With liquid mercury, the meniscus is curved downward (convex). The cohesive forces in mercury are strong, and the mercury does not wet glass.

Several phenomena related to surface tension are pictured in Figure 11-3.

Viscosity. Another property that is related, at least in part, to intermolecular forces of attraction is viscosity—a liquid's resistance to flow. One easy way to measure viscosity is to determine the time required for a liquid to drain through a fixed length of capillary tubing. Another method measures the time required for a steel ball to fall through a certain depth of liquid. As we know through personal experience, some liquids (water, ethanol, diethyl ether) pour or flow easily. Others, such as heavy motor oil, flow much more sluggishly (in fact, we say that they are *viscous*). Because the strengths of intermolecular forces of attraction decrease with temperature, viscosity also generally decreases with temperature.

Other factors that can significantly affect the viscosity of a liquid or liquid mixture are the sizes and shapes of the molecules present. At its melting point (119°C), sulfur produces a thin, mobile, straw-colored liquid; but at a *higher* temperature of about 160°C the liquid darkens and becomes very viscous. This results from the joining together of S_8 molecules into polymer chains with up to several thousand S atoms per chain. The entanglement of these chains produces great resistance to flow. Viscosity can sometimes give an indication of the degree of polymerization or the molecular weight of a polymer.

11-3 Vaporization

Molecules with kinetic energies sufficiently above average may overcome the attractive forces of neighboring molecules and escape from a liquid surface into the gaseous or vapor state. This phenomenon is called vaporization or evaporation. The tendency for a liquid to vaporize *increases* with an increase in liquid temperature. On the other hand, increased intermolecular forces in a liquid lead to a *decreased* tendency for vaporization.

Enthalpy (Heat) of Vaporization. The loss of more energetic molecules by vaporization reduces the average kinetic energy of the remaining molecules in a liquid. The liquid temperature falls. This is why a cooling effect is observed when a volatile liquid such as ether or acetone is allowed to evaporate from one's skin. Alternatively, the temperature of a liquid can be kept constant during vaporization. But this requires the absorption of heat from the surroundings to replace the energy carried away by the vaporized molecules. For the vaporization of one mole of liquid at constant temperature we can write

$$\Delta \overline{H}_{\text{vaporization}} = \overline{H}_{\text{vapor}} - \overline{H}_{\text{liquid}}$$

(The overbar on the symbol H, introduced in Chapter 6, signifies that a molar quantity is involved.)

As is the case with other absolute enthalpies, $\overline{H}_{\text{liquid}}$ and $\overline{H}_{\text{vapor}}$ cannot be measured; but since enthalpy is a function of state, these enthalpies have unique values. Their difference, $\Delta \overline{H}_{\text{vaporization}}$, or simply $\Delta \overline{H}_{\text{vap}}$, is a unique and *measurable* quantity. Moreover, the fact that $\Delta \overline{H}_{\text{vap}}$ is always positive does permit us to state that $\overline{H}_{\text{vapor}}$ is greater than $\overline{H}_{\text{liquid}}$. The appropriate SI units for an enthalpy (heat) of vaporization are J/mol and kJ/mol, but the units cal/g and kcal/mol are also commonly encountered.

Example 11-1 How much heat is required to convert 135 g of diethyl ether, $C_4H_{10}O$, from the liquid state at 20.0°C to the gaseous state at 30.0°C? The specific heat of liquid diethyl ether in the range 20 to 30°C is 2.30 J $g^{-1}°C^{-1}$. At 30.0°C $\Delta \overline{H}_{\text{vap}}$ is 27.8 kJ/mol.

Solution. The heat required is the sum of two quantities: that needed to raise the temperature of the liquid from 20.0°C to 30.0°C (calculated in the manner of Example 6-1) and that needed to vaporize the liquid at 30.0°C.

Energy to raise the temperature of the liquid ether:

$$\text{no. kJ} = 135 \text{ g ether} \times \frac{2.30 \text{ J}}{\text{g ether °C}} \times \frac{1 \text{ kJ}}{1000 \text{ J}} \times (30.0 - 20.0)°\text{C} = 3.10 \text{ kJ}$$

Energy to vaporize the ether:

$$\text{no. kJ} = 135 \text{ g } C_4H_{10}O \times \frac{1 \text{ mol } C_4H_{10}O}{74.1 \text{ g } C_4H_{10}O} \times \frac{27.8 \text{ kJ}}{1 \text{ mol } C_4H_{10}O} = 50.6 \text{ kJ}$$

Total energy required:

$$\text{no. kJ} = 3.10 \text{ kJ} + 50.6 \text{ kJ} = 53.7 \text{ kJ}$$

SIMILAR EXAMPLES: Review Problem 1; Exercises 4, 5.

Vapor Pressure. If the vapor produced by an evaporating liquid is not confined, evaporation will proceed until all the liquid has vaporized. On the other hand, if vaporization occurs into a *closed* vapor volume, as pictured in Figure 11-4, a different condition may result.

When a vapor is maintained in contact with a liquid, some molecules return from the vapor to the liquid. This process, which is the reverse of vaporization, is called **condensation.** The extent of condensation depends on the concentration of vapor molecules (number of molecules per unit volume) and on the area of contact between the liquid and its vapor. Also, as expected, the enthalpy of condensation is the negative of the enthalpy of vaporization. In a container with both liquid and vapor, vaporization and condensation occur simultaneously. Although molecules continue to pass back and forth between liquid and vapor, if sufficient liquid is present, eventually a condition is reached in which no *additional* vapor is formed. This condition is one of **dynamic equilibrium.** The term *dynamic equilibrium* always implies that two opposing processes are occurring simultaneously in a closed system, and in such a way as to offset one another. As a result there is no net change with time once equilibrium has been established.

The vapor in equilibrium with a liquid, like any gas, exerts a pressure. A special name is given to the characteristic pressure exerted by this vapor; it is called the **vapor pressure.** Magnitudes of vapor pressures, like so many other properties, vary widely. Liquids with high vapor pressures are said to be **volatile.** Those with very low vapor pressures are nonvolatile. Whether a liquid is volatile or nonvolatile at a given temperature is deter-

FIGURE 11-4
Establishing liquid–
vapor equilibrium at
constant temperature.

○→ molecules undergoing vaporization
○→ molecules undergoing condensation

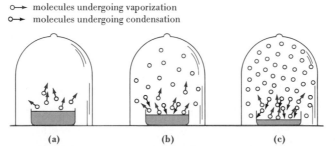

(a) (b) (c)

(a) A liquid is allowed to evaporate into a closed vapor volume. Initially only vaporization occurs.
(b) Condensation begins. However, because molecules are evaporating at a faster rate than they are condensing, the number of molecules in the vapor state continues to increase.
(c) The rate of condensation has become equal to the rate of vaporization. Dynamic equilibrium is established. The number of molecules present in the vapor state remains constant over time, as does the pressure exerted by the vapor.

mined primarily by intermolecular forces. Diethyl ether and acetone are highly volatile liquids. Water at ordinary temperatures is a moderately volatile liquid; at 25°C its vapor pressure is 23.8 mmHg. Gasoline is a mixture of hydrocarbons, most of which are somewhat more volatile than water. (A minor but nevertheless significant source of hydrocarbons in the production of smog is vaporized gasoline from oil refineries, automobile gas tanks, filling-station operations, etc.)

As an excellent first approximation, the vapor pressure of a liquid depends only on its temperature. The measurement of vapor pressure and its dependence on temperature are suggested by Figure 11-5.

A plot of vapor pressure as a function of temperature is known as a **vapor pressure curve.** Because vapor pressure increases with temperature, vapor pressure curves are always of the form shown in Figure 11-6. Vapor pressure data are also encountered in tabular form, as in Table 11-1 for water.

Boiling and the Boiling Point.   The temperature at which the vapor pressure of a liquid is equal to standard atmospheric pressure (1 atm = 760 mmHg) has special significance. It is called the **normal boiling point.** The condition of boiling results when a liquid is heated in a container *open to the atmosphere* and vaporization occurs *throughout* the liquid rather than simply at the surface. Pockets of vapor form within the bulk of the liquid, rise to the surface, and escape. During boiling, energy absorbed as heat is used only to convert molecules of liquid to vapor. The temperature remains constant until all the liquid has boiled away.

Figure 11-6 should help us to see that the boiling point of a liquid varies with atmospheric pressure. (Shift the line shown at $P = 760$ mmHg to higher or lower pressures and determine the temperatures of the new points of intersection with the vapor pressure curves.) Atmospheric pressures less than 1 atm are encountered quite commonly, notably at high altitudes. At an altitude of 1609 m (that of Denver, Colorado) atmospheric pressure is about 630 mmHg. The boiling point of water at this pressure is 95°C (203°F). To cook foods under these conditions of lower boiling temperatures, it is necessary to use longer cooking times. A "three-minute" boiled egg takes longer than 3 min to cook. The effect of high altitudes can be counteracted through a pressure cooker. In a pressure cooker the cooking water is maintained under higher pressure and its boiling point increases.

FIGURE 11-5
Measurement of vapor pressure.

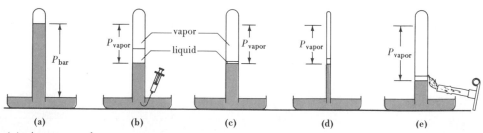

(a)   (b)   (c)   (d)   (e)

(a) A mercury barometer.
(b) A small volume of liquid is introduced to the top of the mercury column. The pressure of the vapor in equilibrium with the liquid depresses the level of the mercury in the barometer tube.
(c) By comparing this with situation (b) we see that vapor pressure is independent of the volume of liquid used to establish liquid–vapor equilibrium.
(d) The volume of vapor here is quite small, but vapor pressure is independent of the volume of vapor involved in the liquid–vapor equilibrium.
(e) An increase in temperature causes an increase in vapor pressure.

FIGURE 11-6
Vapor pressure curves
of several liquids.

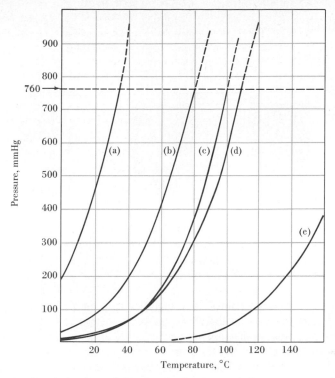

(a) Diethyl ether, $C_4H_{10}O$; (b) benzene, C_6H_6; (c) water, H_2O; (d) toluene, C_7H_8; (e) aniline, C_6H_7N.

TABLE 11-1
Vapor pressure of water at various temperatures

Temperature, °C	Pressure, mmHg	Temperature, °C	Pressure, mmHg
0	4.6	70	233.7
10	9.2	80	355.1
20	17.5	90	525.8
21	18.7	91	546.0
22	19.8	92	567.0
23	21.1	93	588.6
24	22.4	94	610.9
25	23.8	95	633.9
26	25.2	96	657.6
27	26.7	97	682.1
28	28.3	98	707.3
29	30.0	99	733.2
30	31.8	100	760.0
40	55.3	110	1074.6
50	92.5	120	1489.1
60	149.4		

FIGURE 11-7
Attainment of the
critical point.

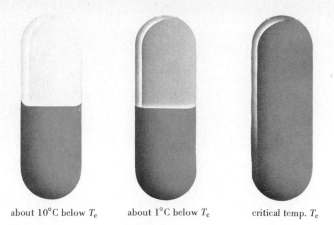

about 10°C below T_c about 1°C below T_c critical temp. T_c

The meniscus separating a liquid (bottom) from its vapor (top) disappears at the critical point. The liquid and vapor become indistinguishable.

The Critical Point. The qualification "in a container open to the atmosphere" is an important one in describing the phenomenon of boiling. If a liquid is heated in a *closed* container, boiling does not occur. Instead, the temperature and vapor pressure rise continuously. In some cases pressures many times greater than atmospheric may be attained.

If sufficient liquid is enclosed in a sealed tube and heated, as suggested by Figure 11-7, the following observations are made.

1. The density of the liquid decreases; that of the vapor increases; and eventually the two densities become equal.
2. The surface tension of the liquid approaches zero; the meniscus between the liquid and vapor becomes less distinct and eventually disappears.

The point at which these conditions are reached is called the **critical point** of the liquid. The temperature at this point is the **critical temperature** and the pressure, the **critical pressure.** The critical point is the highest temperature-pressure point on a vapor pressure curve.

Several critical temperatures and pressures are listed in Table 11-2. Substances having a critical temperature above room temperature can be liquefied at room temperature by

TABLE 11-2
Some critical temperatures and pressures

Substance	Critical temperature, K	Critical pressure, atm
H_2	33.2	12.8
N_2	126.0	33.5
O_2	154.3	49.7
CH_4	191.1	45.8
CO_2	304.1	72.9
HCl	324.5	81.6
NH_3	405.5	111.5
SO_2	430.3	77.7
H_2O	647.3	218.2

applying sufficient pressure to the gaseous state. Those with critical temperatures below room temperature require both the application of pressure *and* a lowering of temperature. A distinction that is sometimes made when describing the gaseous state of a substance is to call this a *vapor* if the temperature is below the critical temperature, and a *gas* if above the critical temperature. That is, a vapor is condensable by the application of pressure alone, and a gas requires that the temperature be lowered as well.

11-4 Some Calculations Involving Vapor Pressure and Related Concepts

Experimental Determination of Vapor Pressure.   The method suggested by Figure 11-5 makes an effective demonstration but it is not widely used. For one thing, the method would not work for vapor pressures that are either very low or very high. Also, the results obtained are not very accurate. A somewhat more useful method (called the transpiration method) is suggested in Example 11-2. An inert gas is saturated with the vapor under study, and the ideal gas equation is used to calculate the vapor pressure.

Example 11-2   113 L of He gas at 1360°C and prevailing barometric pressure is passed through molten silver at 1360°C. The gas becomes saturated with silver vapor. As a result, a loss of mass of 0.120 g is recorded in the liquid silver. What is the vapor pressure, in mmHg, of liquid silver at 1360°C? Neglect any change in gas volume due to the vaporization of the silver.

Solution.   Although the saturated gas is actually a mixture of He and Ag, we can deal with the Ag as if it were a single gas occupying a volume of 113 L. The necessary data follow.

P_{Ag} = ?   (this is the unknown)

V_{Ag} = 113 L

n_{Ag} = 0.120 g Ag × $\dfrac{1 \text{ mol Ag}}{108 \text{ g Ag}}$ = 0.00111 mol Ag

R = 0.0821 L atm mol^{-1} K^{-1}

T = 1360°C + 273 = 1633 K

$PV = nRT$ $P = \dfrac{nRT}{V}$

$P = \dfrac{(0.00111 \text{ mol})(0.0821 \text{ L atm mol}^{-1} \text{ K}^{-1})(1633 \text{ K})}{113 \text{ L}}$ = 1.32 × 10^{-3} atm

= 1.32 × 10^{-3} atm × $\dfrac{760 \text{ mmHg}}{1 \text{ atm}}$ = 1.00 mmHg

SIMILAR EXAMPLES: Review Problem 3; Exercise 9.

An Application of Vapor Pressure Data—Prediction of States of Matter.   To establish liquid–vapor equilibrium in a closed container requires that a certain quantity of substance be present as vapor, a quantity determined by the volume of vapor and its pressure (i.e., the vapor pressure of the liquid). The quantity of liquid present in the equilibrium is immaterial; it can be very small or quite large. What if we attempt to establish liquid–vapor equilibrium and the total quantity of substance available is less than sufficient to saturate the available volume with vapor? In this case the liquid will completely evaporate and vapor alone will exist, at a pressure less than the equilibrium vapor pressure. Various questions about the final condition that results from introducing a liquid and/or gaseous substance into a closed container can be answered with the ideal gas equation and vapor pressure data.

Example 11-3   As a result of a chemical reaction, 0.132 g H_2O is produced and maintained at a temperature of 50°C in a closed vessel of 525 cm³ volume. The vapor pressure of H_2O at 50°C is 92.5 mmHg. **(a)** Will the water in this vessel be present as liquid only, vapor only, or liquid and vapor in equilibrium? **(b)** If both liquid and vapor are present, what is the mass of each?

Solution

(a) Let us first eliminate the possibility that the water might be present as liquid only in this way: A 0.132-g sample of $H_2O(l)$ would occupy a volume of only about 0.132 cm³ [i.e., the density of $H_2O(l) \simeq 1.0$ g/cm³]. Something would have to fill the remaining volume (525 cm³) and this could only be vapor. The only possible answers are that the water exists as liquid and vapor in equilibrium, or as vapor only.

There are several ways that we can proceed at this point. One is to *assume* that liquid and vapor are present at equilibrium. In this case the pressure exerted by the vapor will be the vapor pressure of water at 50°C—92.5 mmHg. We can use the ideal gas equation to calculate the mass of water present in 525 cm³ at 92.5 mmHg at 50°C. If this calculated mass is *less than* the total mass of water available (0.132 g), then both liquid and water vapor must be present. On the other hand, if the calculated mass is greater than the total mass of water available, there is not enough water to saturate the vapor volume. In this case the water would be completely vaporized.

$$PV = \frac{m}{\mathcal{M}}RT \qquad \text{and} \qquad m = \frac{\mathcal{M}PV}{RT}$$

$$m = \frac{18.0 \text{ g mol}^{-1} \times (92.5/760) \text{ atm} \times 0.525 \text{ L}}{0.0821 \text{ L atm mol}^{-1} \text{ K}^{-1} \times 323 \text{ K}} = 0.0434 \text{ g}$$

Since the mass of water required to saturate the vapor is less than the total mass available, the condition must be one of liquid and vapor in equilibrium.

(b) Having established that the vessel contains both liquid and vapor, we can write

mass $H_2O(g) = 0.0434$ g

mass $H_2O(l) = 0.132 - 0.0434 = 0.089$ g

SIMILAR EXAMPLES: Review Problem 6; Exercise 18, 19.

Collection of Gases over Water.   We have already considered another situation where knowledge of vapor pressures is required—the collection of gases over water. This subject should be reviewed by rereading portions of Section 5-9, in particular, Example 5-19.

An Equation for Calculating Vapor Pressures.   Interpolating (estimating values) between points in a table or graph is difficult unless a linear relationship is involved. The vapor pressure curve is not linear—it becomes ever steeper as the temperature is increased. It is possible at times to convert a nonlinear relationship into a linear one by introducing a new function of the variables. Vapor pressure data are found to yield a straight line when *logarithm of the vapor pressure (log P)* is plotted against *reciprocal of the kelvin temperature (1/T)*. As described in Appendix A-4, a straight line can always be expressed through a simple mathematical equation. The equations for the lines in Figure 11-8 are of the form

$$\log P = -A \left(\frac{1}{T} \right) + B \tag{11.1}$$

equation of straight line: $\underbrace{y}_{} = \underbrace{m \cdot x}_{} + \underbrace{b}_{}$

Logarithmic functions that result from theoretical derivations are natural logarithmic functions, e.g.,

$$\ln P = \frac{-\Delta \overline{H}_{vap}}{RT} + B'$$

Natural and common (base-10) logarithms are related through the factor 2.303 (see Appendix A-2).

Whenever an equation such as (11.1) is found to describe a phenomenon, we can generally assume that there is a theoretical basis for the equation. In the present case the basis is that vapor pressure is a special example of a quantity called an equilibrium constant, and equilibrium constants vary with temperature in the manner indicated by equation (11.1). We will be able to derive equations like (11.1) following a further discussion of thermodynamics in Chapter 16. For now, we will simply use equation (11.1) and some of its variant forms.

The constant A in equation (11.1) is closely related to the enthalpy of vaporization of the liquid in question. As a reasonable approximation, $A = \Delta \overline{H}_{vap}(2.303R)$, where R is the gas constant. Substituting this value of A into equation (11.1) yields an expression known as the Clausius–Clapeyron equation.

$$\log P = \frac{-\Delta \overline{H}_{vap}}{2.303RT} + B \tag{11.2}$$

To calculate the vapor pressure (P) at some given temperature (T) with equation (11.2) requires that numerical values be available for $\Delta \overline{H}_{vap}$ and B. Such data are often tabulated

FIGURE 11-8
Vapor pressure data of Figure 11-6 replotted— $\log P$ versus $1/T$.

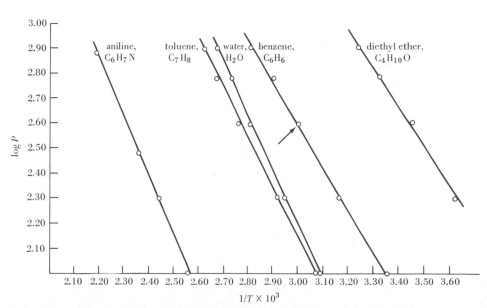

Data from Figure 11-6 have been recalculated and replotted as in the following example.

For benzene, at 60°C, the vapor pressure is 400 mmHg. $\log P = \log 400 = 2.60$. $T = 60°C = 333$ K; $1/T = 1/333 = 0.00300 = 3.00 \times 10^{-3}$; $1/T \times 10^3 = 3.00 \times 10^{-3} \times 10^3 = 3.00$.

The point corresponding to these data is marked by the arrow ($\rightarrow$).

in handbooks, but a more usual approach is to eliminate the constant B. Write equation (11.2) twice—once for vapor pressure P_1 at temperature T_1, and once for P_2 at T_2.

$$(1) \quad \log P_1 = \frac{-\Delta \overline{H}_{vap}}{2.303 \, RT_1} + B \qquad (2) \quad \log P_2 = \frac{-\Delta \overline{H}_{vap}}{2.303 \, RT_2} + B$$

Subtract equation (1) from equation (2).

$$\log P_2 - \log P_1 = \frac{-\Delta \overline{H}_{vap}}{2.303R} \left(\frac{1}{T_2} - \frac{1}{T_1} \right) + \not{B} - \not{B}$$

Use the equation in the above form or rearrange it to

$$\log \frac{P_2}{P_1} = \frac{\Delta \overline{H}_{vap}}{2.303R} \left(\frac{T_2 - T_1}{T_1 T_2} \right) \tag{11.3}$$

In equation (11.3) temperatures must be expressed in kelvins, the same units of pressure must be used for P_1 and P_2, and the same units of energy for $\Delta \overline{H}_{vap}$ and R. Generally this will require that R be expressed as 8.314 J mol^{-1} K^{-1}.

Example 11-4   Calculate the vapor pressure of water at 35°C with data from Table 11-1 and a value of 43.9 kJ/mol for $\Delta \overline{H}_{vap}$ of water.

Solution.   Let P_2 be the unknown vapor pressure at temperature $T_2 = 35°C = 308$ K. Choose for P_1 and T_1 known data (Table 11-1) at a temperature close to 35°C. At $T_1 = 30°C = 303$ K, $P_1 = 31.8$ mmHg. Substitute these values into equation (11.3) to obtain

> The temperature interval, $T_2 - T_1$, is kept small because $\Delta \overline{H}_{vap}$ varies somewhat with temperature.

$$\log \frac{P_2 \text{ (mmHg)}}{31.8 \text{ mmHg}} = \frac{43.9 \times 10^3 \text{ J mol}^{-1}}{2.303 \times 8.314 \text{ J mol}^{-1} \text{ K}^{-1}} \left[\frac{308 \text{ K} - 303 \text{ K}}{(303 \text{ K})(308 \text{ K})} \right]$$

$$= 2.29 \times 10^3 \text{ K} \left(\frac{5.4 \times 10^{-5}}{\text{K}} \right) = 1.2 \times 10^{-1} = 0.12$$

Next, determine the *antilogarithm* of 0.12. (What is the number whose logarithm is 0.12?) The antilogarithm is 1.3 (see Appendix A). Thus,

$$\frac{P_2}{31.8 \text{ mmHg}} = 1.3$$

$$P_2 = 1.3 \times 31.8 \text{ mmHg} = 41 \text{ mmHg}$$

(The measured vapor pressure of water at 35°C is 42.18 mmHg.)

SIMILAR EXAMPLES: Exercises 12, 13.

11-5 Transitions Involving Solids

Melting, Melting Point, and Heat of Fusion.   As the temperature of a crystalline solid is raised, its atoms, ions, or molecules vibrate increasingly vigorously. Eventually, a temperature is reached at which the crystalline structure is destroyed by these vibrations. The solid is converted to liquid. This process is called **melting.** The reverse process, conversion of liquid to solid, is called **freezing.** The temperature at which a pure liquid freezes and that at which the corresponding pure solid melts are identical. At this temperature, called either the **melting point** of the solid or the **freezing point** of the liquid, solid and liquid coexist in dynamic equilibrium.

FIGURE 11-9
Cooling curve for water.

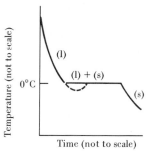

The broken line portion represents the condition of supercooling that occasionally occurs. (l) = liquid; (s) = solid.

If heat is added to a solid–liquid mixture at equilibrium, the solid is gradually converted to liquid *while the temperature remains constant*. When all the solid has melted, the temperature begins to rise. Conversely, the removal of heat from a solid–liquid mixture at equilibrium results in the constant-temperature conversion of liquid to solid. The quantity of energy required to melt a given amount of solid is called the **enthalpy (heat) of fusion.** When a liquid freezes it *gives off* the heat of fusion.

A well-known example of a melting point is that of ice, 0°C.* At this temperature ice and liquid water, in contact with air under standard atmospheric pressure, are in equilibrium. The heat of fusion of ice is 6.02 kJ/mol.

An easy method for determining the freezing point of a liquid is to measure the temperature of a liquid sample as it cools. The temperature decreases with time until the freezing point is reached. As solid begins to form, the temperature remains constant with time. When all the liquid has frozen, the temperature is again free to fall. A graphic representation of temperature versus time is called a **cooling curve.** An example is shown in Figure 11-9 for liquid water.

The behavior represented by the solid-line portions of Figure 11-9 is idealized. At times a liquid can be cooled *below* its normal freezing point before freezing occurs—a condition called **supercooling.** This happens when there are present few nuclei (such as suspended dust particles) on which solid crystals can form. When a supercooled liquid begins to freeze its temperature rises back to the normal freezing point, where freezing is completed. In these cases there is a dip in the cooling curve just before the straight-line portion.

With the ideas introduced here it is possible to include the solid state in calculations about the expected states of matter in a mixture, as in Example 11-5.

Example 11-5   A 25.0-g cube of ice at 0.0°C is added to 100.0 g of liquid water at 22.0°C in a thermally insulated container (e.g., a Styrofoam cup). The heat of fusion of ice is 6.02 kJ/mol and the specific heat of liquid water is 4.18 J g^{-1} °C^{-1}. **(a)** What is the final condition reached—liquid only, or ice and liquid? **(b)** What is the temperature in this final condition?

Solution

(a) Call the heat associated with melting the ice q_{ice}, and that for cooling the original liquid water, q_{water}.

$$q_{ice} = 25.0 \text{ g ice} \times \frac{1 \text{ mol ice}}{18.0 \text{ g ice}} \times \frac{6.02 \text{ kJ}}{1 \text{ mol ice}} = 8.36 \text{ kJ}$$

The maximum quantity of heat associated with cooling the liquid water would be for the temperature interval from 22.0 to 0.0°C.

$$q_{water} = 100.0 \text{ g water} \times \frac{4.18 \text{ J}}{\text{g water °C}} \times (0.0 - 22.0)°C = -9196 \text{ J} = -9.20 \text{ kJ}$$

The maximum quantity of heat that could be liberated in cooling the original liquid water is greater than that required to melt all the ice. The final condition is one of "all liquid" at a temperature somewhat above 0°C.

(b) Calculate the heat associated with (1) melting 25.0 g ice (as in part a), (2) warming the resulting 25.0 g liquid water from 0°C to a final, higher temperature T, and (3) cooling the original 100.0 g liquid water from 22.0°C to a final, lower temperature T. The sum of these three quantities of heat is zero, and this requires a specific value of T. Units have been eliminated from the expression that follows, but each of the three quantities of heat is expressed in joules.

*If air is excluded, and solid and liquid water are in equilibrium with their own vapor (at a pressure of 4.58 mmHg), the equilibrium temperature is slightly different, +0.01°C.

$$\underbrace{8360}_{(1)} + \underbrace{25.0 \times 4.18 \times (T - 0)}_{(2)} + \underbrace{100.0 \times 4.18 \times (T - 22.0)}_{(3)} = 0$$

$$8360 + 104T + 418T - 9200 = 0$$
$$522T = 840$$
$$T = 1.6°C$$

An alternative approach is to note from part (a) that for a temperature of 0°C the quantity of heat lost by the original liquid water (9200 J) exceeds that required to melt the ice (8360 J) by 840 J. Now determine the final temperature if 840 J were reabsorbed by 125.0 g liquid water at 0°C. The final temperature is 1.6°C.

$$125.0 \text{ g water} \times \frac{4.18 \text{ J}}{\text{g water °C}} \times (T - 0.0)°C = 840 \text{ J}$$

$$T = 1.6°C$$

SIMILAR EXAMPLES: Review Problem 5; Exercise 20.

Sublimation.   Vaporization of solids is also encountered, although in general the volatilities of solids are not as great as of liquids. The direct passage of molecules from a solid to a vapor is called **sublimation.** The reverse process, passage of molecules from the vapor to the solid state, is called **deposition.** When sublimation and deposition proceed at equal rates, a dynamic equilibrium is established between solid and vapor. The vapor exerts a characteristic pressure, called the sublimation pressure. A plot of sublimation pressure as a function of temperature is called a sublimation curve. The enthalpy (heat) of sublimation is the quantity of heat required to convert a certain quantity of solid to vapor. It is related to the enthalpies of fusion and vaporization in a simple way. At a given temperature,

The sublimation and deposition of solids are pictured in Color Section B.

$$\Delta \overline{H}_{\text{sub}} = \Delta \overline{H}_{\text{fus}} + \Delta \overline{H}_{\text{vap}} \tag{11.4}$$

Two very common solids with significant volatilities are ice and dry ice (solid carbon dioxide). People who live in cold climates are familiar with the fact that snow may disappear from the ground even though the temperature may fail to rise above 0°C. Under these conditions the snow does not melt; it sublimes.

11-6 Phase Diagrams

We have noted some general conditions under which a substance exists in three different states (phases) of matter. For instance, at high temperatures and low pressures we expect the gaseous state; at very low temperatures, the solid state. Moreover, we have described situations in which two states of matter coexist in equilibrium: liquid–vapor, solid–vapor, solid–liquid. All of this information can be summarized in a single graphic form called a phase diagram.

The phase diagram shown in Figure 11-10 for iodine is one of the simplest possible. Pressures and temperatures at which iodine exists as solid, liquid, and gas are denoted by areas in the diagram. Equilibrium conditions between two states of matter correspond to curves along which areas adjoin. The curve *OC* is the vapor pressure curve of liquid iodine (*C* is the critical point). *OB* is the sublimation curve of iodine. The effect of pressure on the melting point of iodine is represented by the curve *OD*; it is called the fusion curve. The point *O* has a special significance. It gives the unique temperature and

FIGURE 11-10
Phase diagram for
iodine.

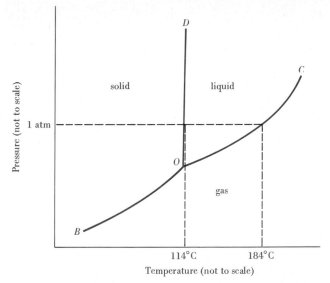

Note that the melting point and triple point temperatures for iodine are essentially the same. Generally, large pressure increases are required to produce even small changes in solid–liquid equilibrium temperatures.

pressure at which solid, liquid, and vapor coexist in equilibrium. It is called a **triple point;** for iodine this is at 114°C and 91 mmHg. The normal melting point (114°C) and boiling point (184°C) are the temperatures at which a line at $P = 1$ atm intersects the fusion and vapor pressure curves, respectively.

The behavior of carbon dioxide, shown in Figure 11-11, differs from that of iodine in one important respect—the pressure at the triple point is greater than 1 atm. A line at $P = 1$ atm intersects the sublimation curve, not the vapor pressure curve. If solid CO_2 (dry ice) is heated in a container open to the atmosphere, it sublimes away at a *constant* temperature of -78.5°C; it does not melt. The ability of dry ice to maintain a constant low temperature is the basis of its use in preserving frozen foods. Liquid CO_2 can exist only at pressures in excess of 5.1 atm.

FIGURE 11-11
Phase diagram for
carbon dioxide.

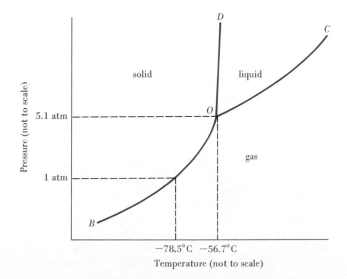

FIGURE 11-12
Phase diagram for water.

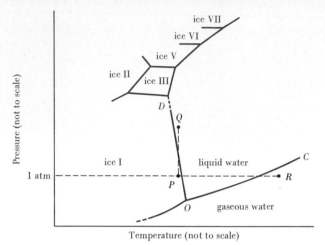

Point O, the triple point, is at $+0.0098°C$ and 4.58 mmHg. The critical point, C, is at 374.1°C and 218.2 atm. The negative slope of the fusion curve OD is exaggerated in this diagram. An increase in pressure of about 100 atm is required to produce a decrease of 1°C in the melting point of ice I. All of the high pressure forms of ice are more dense than liquid water. Ice I, ice III, and liquid water are at equilibrium (point D) at $-22.0°C$ and 2045 atm. Ice VII can be maintained at temperatures approximating the normal boiling point of water (100°C), but only at pressures in excess of 25,000 atm.

The phase diagram for water (Figure 11-12) presents several new features. The fusion curve OD has a *negative* slope, i.e., slopes toward the pressure axis. The melting point of ice *decreases* with increased pressure, unusual behavior for a solid. Another feature is that of **polymorphism,** the existence of a solid substance in more than one crystalline form. Ordinary ice (ice I) occurs under usual conditions of temperature and pressure. The other forms of ice exist only when maintained under very high pressures. Polymorphism is much more the rule than the exception among substances. Where polymorphism occurs a phase diagram displays triple points in addition to the familiar solid–liquid–vapor triple point. For example, ice I, ice III, and H$_2$O(l) are in equilibrium at $-22.0°C$ and 2045 atm.

> The polymorphism of mercury(II) iodide is illustrated in Color Section B.

Example 11-6 illustrates how a phase diagram may be used to explain the physical behavior of a substance.

Example 11-6   A sample of ice is maintained at 1 atm and at a temperature represented by point P in Figure 11-12. Describe what happens when **(a)** the temperature is raised, at constant pressure, to point R, and **(b)** the pressure is raised, at constant temperature, to point Q. The points P, Q, and R are shown in Figure 11-12, and the conditions that exist at these points are suggested by Figure 11-13.

Solution
(a) When the temperature reaches a point on the fusion curve OD (0°C), ice I begins to melt. The temperature remains constant as ice I is converted to liquid. When melting is complete, the temperature increases again. No vapor appears in the cylinder until the temperature reaches 100°C, the temperature at which the vapor pressure is 1 atm. When all the liquid has vaporized, the temperature is again free to rise, to a final value of R.
(b) Because of the very small compressibility of solids, essentially no change occurs until the pressure reaches a value corresponding to the point of intersection of

the constant-temperature line, *PQ*, with the fusion curve, *OD*. At this pressure melting occurs. Melting is accompanied by a significant decrease in volume (10% or so) as ice I is converted to liquid water. Following melting, additional increases in pressure produce very little effect since liquids are not very compressible.

SIMILAR EXAMPLES: Review Problem 7; Exercises 24, 25.

Le Châtelier's Principle: A Preview.   A useful aid in describing phase behavior is provided by Le Châtelier's principle, which we will consider in detail in Chapter 15 but which we preview here in a limited version. With respect to an equilibrium involving different phases of a substance,

An attempt to change the temperature or pressure of a system in phase equilibrium stimulates changes within the system that preserve the condition of equilibrium for as long as possible.

Consider again the phase equilibria encountered in Example 11-6. In Figure 11-12, at the point where the constant-pressure line *PR* intersects the fusion curve *OD*, ice I and $H_2O(l)$ are in equilibrium at constant temperature. At this point, an attempt to raise the temperature of the system by adding more heat stimulates the heat-absorbing (endothermic) process, the further melting of ice I. As more heat is added, more ice I melts and the equilibrium temperature remains constant. For a time the condition of equilibrium is preserved. Eventually, however, the ice I disappears; the two-phase equilibrium is destroyed; the system reverts to a single liquid phase; and the temperature is again free to rise. Another phase equilibrium is established when the line *PR* intersects the vapor pressure curve *OC*. An attempt to raise the temperature at this point again stimulates the heat-absorbing process—vaporization. The equilibrium condition and a constant temperature are maintained until all the liquid has vaporized.

Le Châtelier's principle can also be applied to the changes that occur as the pressure is slowly increased along the constant-temperature line *PQ* in Figure 11-12. When the system reaches a point on the fusion curve *OD* melting begins and equilibrium is established between ice I and $H_2O(l)$. An attempt to raise the pressure at this point stimulates the process in which the volume of the system decreases—the melting of ice I. Equilibrium is maintained until all the ice I has melted. Then the pressure is again free to rise.

FIGURE 11-13
Predicting phase changes—Example 11-6 illustrated.

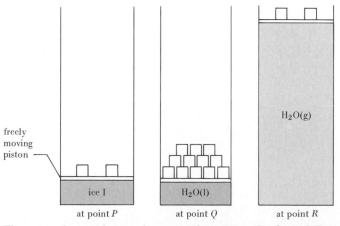

These are the conditions that prevail at points *P*, *Q*, and *R* in Figure 11-12.

The crossing of any two-phase curve in a phase diagram is called a transition, but there are special terms for the following transitions.

melting (S $\longrightarrow$ L)
freezing (L $\longrightarrow$ S)
vaporization (L $\longrightarrow$ V)
condensation (V $\longrightarrow$ L)
sublimation (S $\longrightarrow$ V)
deposition (V $\longrightarrow$ S)

The above discussion can be summarized and generalized by saying that in crossing a two-phase curve in a phase diagram

1 from lower to higher temperature along a constant-pressure line there is an *increase in enthalpy*, and

2 from lower to higher pressure along a constant-temperature line, there is a *decrease in volume*.

These statements help us to understand why a fusion curve generally has a positive slope (away from the pressure axis). Typical behavior is for a solid to be more dense than the corresponding liquid, and for increased pressure to favor the process, liquid $\rightarrow$ solid. These statements also suggest, for example, that for the transition, ice II $\rightarrow$ ice III, in Figure 11-12, $\Delta\bar{H}_{transition} > 0$; that the densities of all forms of ice except ice I are greater than the density of $H_2O(l)$; etc.

11-7 Evidence of Intermolecular Forces— Condensed States of the Noble Gases

The lightest of the group 0 elements, helium, forms no stable chemical bonds. Given this fact we might reason that helium would be a gas at all temperatures, right down to absolute zero. In fact, helium does condense to a liquid at 4 K and freeze to a solid (at 25 atm pressure) at 1 K. These data suggest the presence of weak intermolecular forces that overcome thermal agitation and cause the gas to condense if the temperature is lowered sufficiently. But what kind of force can this be?

11-8 Van der Waals Forces

Instantaneous and Induced Dipoles. In speaking of the electronic structure of an atom or molecule we refer to the *probability* of an electron being in a certain region at a given instant of time. One event that may occur is suggested by Figure 11-14—an instantaneous displacement of electrons toward one region of an atom or molecule. This displacement causes a normally nonpolar species to become polar; an instantaneous dipole is formed.

FIGURE 11-14
Instantaneous and induced dipoles.

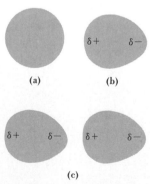

(a)   (b)

(c)

(a) *Normal condition*. A nonpolar species has a symmetrical charge distribution.
(b) *Instantaneous condition*. A displacement of the electronic charge produces an instantaneous dipole, with charges of $\delta+$ and $\delta-$.
(c) *Induced dipole*. The instantaneous dipole on the left induces a charge separation in the species on the right. The result is a dipole–dipole interaction; the two dipoles are attracted to each other.

FIGURE 11-15
Molecular shapes and polarizability.

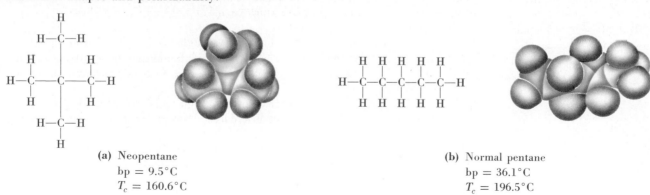

(a) Neopentane
bp = 9.5°C
$T_c = 160.6°C$

(b) Normal pentane
bp = 36.1°C
$T_c = 196.5°C$

The elongated normal pentane molecule is more easily polarized than the compact neopentane molecule. Intermolecular forces are stronger in normal pentane than in neopentane. As a result, normal pentane boils at a higher temperature than neopentane; its critical temperature is also higher than that of neopentane.

Following this, electrons in a neighboring atom or molecule may be displaced, also leading to a dipole. This is a process of induction, and the newly formed dipole is called an induced dipole.

> A commonplace example of induction is the attraction of a balloon to a wall. The balloon is charged by rubbing, and the charged balloon induces an opposite charge on the wall. (See also, Appendix B.)

Taken together these two events lead to an intermolecular force of attraction. Although it is proper to call this an instantaneous dipole–induced dipole interaction, the terms **dispersion force** and **London force** are generally used. (Fritz London offered a theoretical explanation of these forces in 1928.)

The ease with which an electron cloud is distorted by an external electric field (and hence the ease with which a dipole is induced) is called **polarizability.** In general, polarizability increases with the total number of electrons in a molecule. Since molecular weight is related in a general way to number of electrons, the polarizability of molecules and the strength of dispersion (London) forces increase with increased molecular weight. For example, radon (atomic weight, 222) has a much higher boiling point than does helium (atomic weight, 4): 211 K for Rn compared to 4 K for He. Another important factor in determining the strength of dispersion forces is molecular shape. Electrons in elongated molecules are more easily displaced than are those in small, compact, symmetrical molecules. Some physical properties of two substances with the identical numbers and kinds of atoms (isomers) but different molecular shapes are compared in Figure 11-15.

Dipole–Dipole Interactions.   In a *polar* substance molecules tend to become oriented with the positive end of one dipole directed toward the negative ends of neighboring dipoles. (An idealized situation is pictured in Figure 11-16.) This additional partial ordering of molecules can cause a substance to persist as a solid or liquid at temperatures higher than otherwise expected. For example, compare normal butane, C_4H_{10}, and acetone (dimethyl ketone), $(CH_3)_2CO$, two substances with the same molecular weight.

nonpolar: C_4H_{10} mol. wt. = 58   m.p. = −138.3°C   b.p. = −0.5°C

polar: $(CH_3)_2C\!\!=\!\!O$ mol. wt. = 58   m.p. = −94.8°C   b.p. = 56.2°C

The intermolecular forces described in this section are the forces that cause a gas to depart from ideal gas behavior. The van der Waals equation of state takes these forces into account, and, collectively, these forces are called **van der Waals forces.** In assessing relative contributions to van der Waals forces, the following factors need to be considered.

FIGURE 11-16
Dipole–dipole
interactions.

The arrangement shown
here is an idealized case.
Normally, thermal motion
upsets this orderly array,
cancelling out most of the
dipole–dipole attractions.
But this tendency for di-
poles to align themselves
can be of considerable
importance in affecting
the properties of sub-
stances, especially solids.

- Dispersion (London) forces are based on displacing all the electrons in a molecule; the strength of these forces increases with the number of electrons present (i.e., increased molecular weight) and depends to some extent on molecular shape; dispersion forces exist between all molecular species.
- Forces associated with permanent dipoles involve displacements of electron pairs in bonds rather than the molecule as a whole; they are found only for substances having resultant dipole moments; their existence represents an addition to the dispersion forces that are also present.
- In comparing substances of roughly comparable molecular weights, the presence of dipole forces can produce significant differences in properties [as in the comparison of C_4H_{10} and $(CH_3)_2CO$ above].
- In comparing substances of rather widely different molecular weights, dispersion forces are likely to be more important than dipole forces in determining relative values of physical properties.

Let us see how these statements are supported by the data in Table 11-3. HCl and F_2 have comparable molecular weights, but the presence of permanent dipoles in HCl causes it to have a significantly higher $\Delta\overline{H}_{vap}$ and boiling point than F_2. Within the series HCl, HBr, and HI, molecular weight increases sharply, and $\Delta\overline{H}_{vap}$ and boiling point increase in the order HCl < HBr < HI. The higher polarities of HBr and HCl relative to HI are not sufficient to reverse the trends produced by increasing strength of dispersion forces with increased molecular weight.

Example 11-7 Arrange the following substances in the order in which you would expect their boiling points to increase: CCl_4(M.W. 153.82); Cl_2(M.W. 70.91); ClNO (M.W. 65.46); N_2(M.W. 28.01).

Solution. Three of these are nonpolar substances for which we should expect only dispersion (London) forces. The strengths of these forces, and hence the boiling points, should increase with increasing molecular weight, i.e., $N_2 < Cl_2 < CCl_4$. ClNO has a molecular weight that is comparable to that of Cl_2, but the ClNO molecule is polar (bond angle ~120°). This should result in stronger intermolecular forces and a higher boiling point for ClNO than for Cl_2. However, we would not expect the boiling point of ClNO to be higher than that of CCl_4 because of the considerable difference in molecular weight between the two. The expected order is

$N_2 < Cl_2 < ClNO < CCl_4$

(The observed boiling points are 77.4, 239.1, 266.8, and 349.8 K, respectively.)

SIMILAR EXAMPLES: Review Problem 8; Exercises 27, 28.

TABLE 11-3
Intermolecular forces and properties of selected substances

| | Mol. wt. | Dipole moment, D | van der Waals forces | | $\Delta\overline{H}_{vap}$, kJ/mol | Boiling point, K |
			% dispersion	% dipole		
F_2	38.00	0	100	0	6.86	85.01
HCl	36.46	1.08	81.4	18.6	16.15	188.11
HBr	80.92	0.78	94.5	5.5	17.61	206.43
HI	127.91	0.38	99.5	0.5	19.77	237.80

FIGURE 11-17
Comparison of boiling points of some hydrides of the elements of groups IVA, VA, VIA, and VIIA.

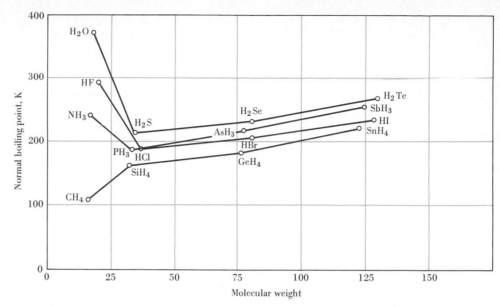

The values for NH_3, H_2O, and HF are unusually high compared to those of other members of their groups.

11-9 Hydrogen Bonds

Normal boiling points of a series of similar compounds are plotted as a function of molecular weight in Figure 11-17. Normal behavior is that displayed by the group IVA hydrides—boiling point increases regularly with increasing molecular weight. Three striking exceptions are noted in the figure—NH_3, H_2O, and HF.

Figure 11-18 illustrates for hydrogen fluoride three important ideas about the dipole–dipole interactions known as **hydrogen bonds.**

1. The H atom in HF is a center of positive charge; the F atom, of negative charge. Dipoles tend to align in the usual fashion, the positive end of one dipole directed toward the negative end of another. This alignment places an H atom between two F atoms. Because of the very small size of the H atom, the dipoles come into close proximity, producing a strong dipole–dipole interaction.

2. Together with being strongly bonded to one F atom, an H atom is weakly bonded to the F atom of a nearby HF molecule. This occurs through the lone pair of electrons on that F atom. In hydrogen bonding an H atom acts as a bridge between two nonmetal atoms.

3. An essential part of hydrogen bond formation is the tendency for hydrogen bonds to form throughout a cluster of molecules, producing so-called polymeric structures. The bond angle between two nonmetal atoms bridged by an H atom (i.e., the bond angle, X—H···X) is usually about 180°. However, there are preferred orientations for hydrogen-bonded molecules, such as in the pentagonal structure $(HF)_5$.

FIGURE 11-18
The hydrogen bond.

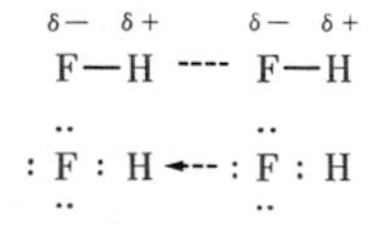

$$\overset{\delta-}{F}\!-\!\overset{\delta+}{H} \text{ ---- } \overset{\delta-}{F}\!-\!\overset{\delta+}{H}$$

$$: \overset{..}{F} : H \longleftarrow : \overset{..}{F} : H$$

two views of a hydrogen bond

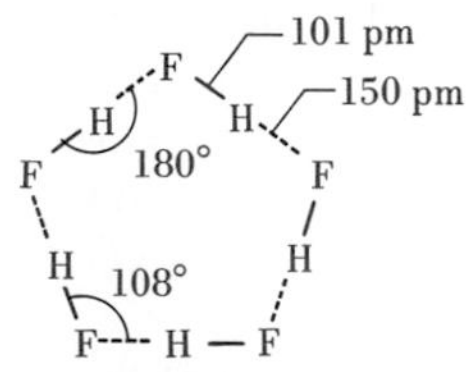

the species $(HF)_5$

In the gaseous state, several polymeric forms of the HF molecule exist in which the individual molecules (monomers) are held together through hydrogen bonds. A pentagonal arrangement of five HF molecules is shown here.

To summarize, the hydrogen bond is a rather strong intermolecular force, with energies of the order of 15 to 40 kJ/mol. (Van der Waals interactions correspond to energies of about 2 to 20 kJ/mol.) Strong hydrogen bonding is most likely to occur when an H atom in a molecule can be simultaneously attracted to a highly electronegative atom—F, O, or N—in a neighboring molecule. Weak hydrogen bonding may also occur between an H atom of one molecule and other nonmetal atoms of a neighboring molecule, e.g., Cl or S.

FIGURE 11-19
Hydrogen bonding in
water.

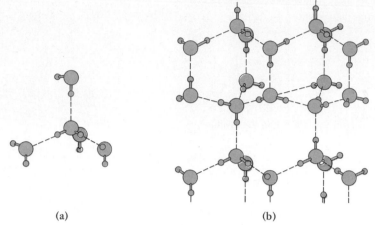

(a)   (b)

(a) Each water molecule is linked to four others through hydrogen bonds. The arrangement is tetrahedral. Each hydrogen atom is situated along a line joining two oxygen atoms, but somewhat closer to one oxygen atom (100 pm) than to the other (180 pm).
(b) The crystal structure of ice. Oxygen atoms are located in hexagonal rings arranged in layers. Positions available to the hydrogen atoms lie between pairs of oxygen atoms, again closer to one than to the other. This characteristic pattern is revealed at the macroscopic level in the hexagonal shapes of snowflakes.

The most common substance in which hydrogen bonding occurs is ordinary water. Figure 11-19 indicates how one water molecule is held to four neighbors in a tetrahedral arrangement by hydrogen bonds. This is the structural arrangement in crystalline water or ice. Hydrogen bonds hold the water molecules in a rigid but rather open structure. As ice melts, only a fraction of the hydrogen bonds are broken. Extensive hydrogen bonding persists in liquid water just above the melting point; liquid water retains an icelike structure. Evidence to this effect is provided by the low heat of fusion of water (6.02 kJ/mol), much less than would be expected if all the hydrogen bonds were to be broken.

The packing of molecules in liquid water at the melting point is closer than in ice; therefore, the liquid is more dense than ice. This is very unusual behavior for a substance. In all but a few cases, a liquid is less dense than the solid from which it is formed at the melting point. As liquid water is heated above the melting point, its density continues to *increase* as more hydrogen bonds are broken and further packing of the molecules occurs in the liquid state. Liquid water attains its maximum density at 3.98°C. Above this temperature the liquid behaves in a normal fashion; its density decreases with temperature. Through hydrogen bonding we can understand why a lake freezes from the top down and why the denser unfrozen water at the bottom of the lake is at a temperature somewhat above freezing (4°C = 39°F).

It is not an overstatement to say that hydrogen bonding makes life possible. Living organisms are maintained through a series of chemical reactions involving complex structures, such as DNA and proteins. Certain bonds in these structures must be capable of

FIGURE 11-20
Dimerization of gaseous
acetic acid.

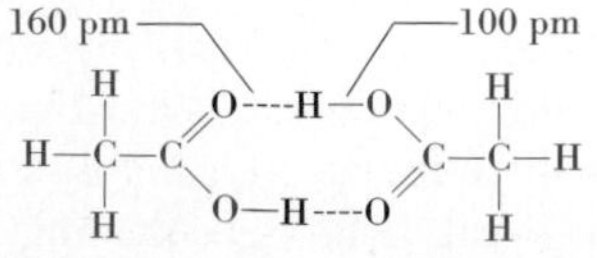

Hydrogen bonding permits molecules to exist in stable pairs (dimers).

FIGURE 11-21
The diamond structure.

(a)

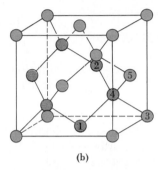

(b)

(a) A portion of the Lewis structure.
(b) Crystal structure. Each carbon atom is bonded to four others in a tetrahedral fashion. The segment of the entire crystal shown here is called a unit cell.

breaking and reforming with ease. Only hydrogen bonds have just the right energies to permit this. The importance of hydrogen bonding in molecules of biological significance is explored further in Chapter 27.

Liquids in which hydrogen bonding occurs exhibit stronger than usual intermolecular forces, and these liquids generally have high heats of vaporization. In acetic acid, H_3CCOOH, hydrogen bonding leads to the formation of dimers (double molecules) both in the liquid and *in the vapor state*. Not all the hydrogen bonds between molecules need to be broken to vaporize acetic acid and the heat of vaporization is abnormally *low*. Dimerization of acetic acid molecules is illustrated in Figure 11-20.

The type of hydrogen bonding described in this section is *inter*molecular. Another possibility is for a hydrogen atom to bridge two nonmetal atoms within the same molecule. This is called *intra*molecular hydrogen bonding (see Exercise 36).

11-10 Network Covalent Solids

In some substances covalent bonds are not limited to individual small molecules. They extend throughout a crystalline solid. In such cases inter- and intramolecular forces are indistinguishable since the entire crystal is held together by strong covalent bonds.

The Diamond Structure. A scheme to describe how carbon atoms can bond, one to another, is suggested in Figure 11-21. The two-dimensional Lewis structure (Figure 11-21a) is adequate only to establish the fact that this bonding involves ever-increasing numbers of carbon atoms and leads to a giant covalent molecule. It provides no information about the three-dimensional structure. The entire crystal structure can be inferred from the portion shown in Figure 11-21b. Each atom is bonded to four others. Atoms 1, 2, and 3 lie in a plane with atom 4 above the plane. Atoms 1, 2, 3, and 5 define a tetrahedron (sp^3 hybridization) with atom 4 inscribed in its center. When viewed from a particular direction, a nonplanar hexagonal arrangement of carbon atoms is seen (shown in grey).

The substance we have just described is diamond. If silicon atoms are substituted for one half the carbon atoms, the resulting structure is that of silicon carbide (carborundum). These two substances are both extremely hard. To break or scratch a crystal of diamond or silicon carbide requires that covalent bonds be broken. Both substances are widely used as abrasives, and diamond is the hardest substance known. They are nonconductors of electricity and do not melt or sublime until very high temperatures are reached. SiC sublimes at 2700°C, and diamond melts above 3500°C.

The Graphite Structure. Carbon atoms can be bonded together to produce a solid having properties very much different from diamond. This bonding involves sp^2 and p orbitals. The three orbitals of the sp^2 type are directed in a plane at angles of 120°; the p orbital is directed perpendicular to the plane, above and below. These are the same orbitals used by carbon atoms in C_6H_6 (described in Figure 10-20).

The crystal structure that results from this type of bonding is pictured in Figure 11-22; it is the graphite structure. Each carbon atom forms strong covalent bonds with three neighboring atoms in the same plane, giving rise to layers of carbon atoms in a hexagonal arrangement. The p electrons of the carbon atoms are *delocalized* (recall the discussion in Section 10-5). Bonding within layers is strong but between layers it is much weaker. Evidence of this is provided by bond distances. The C—C bond distance within a layer is 142 pm (1.42 Å); between layers it is 335 pm (3.35 Å).

The distinctive properties of graphite are derived from its unique crystal structure. Because bonding between layers is weak, the layers can be made to glide over one another readily. As a result graphite may be used as a lubricant, either in dry form or suspended in

FIGURE 11-22
The graphite structure.

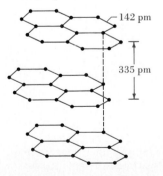

142 pm

335 pm

oil.* Layers flake off from a piece of graphite easily, making it valuable as the basic writing material in pencils. Because the *p* valence electrons in graphite are delocalized, they can be made to migrate through the planes of carbon atoms under the influence of an electric field. Graphite is a conductor of electric current and is widely used for electrodes in batteries and in industrial electrolysis processes. Diamond is not an electrical conductor because in its structure all of the valence electrons of the carbon atoms are localized or fixed permanently into single covalent bonds.

11-11 The Ionic Bond as an Intermolecular Force

Many of the physical properties of an ionic solid are determined by this one factor: How difficult is it to break up an ionic crystal and separate the ions it contains? In an ionic crystal each ion exerts attractive forces on several neighboring ions. In addition, ions with like charge exert a repulsive force on one another. Establishing the net strength of the forces within an ionic crystal is difficult, but this quantity can be calculated or determined by experiment. The quantity involved is called the lattice energy. **Lattice energy** is the quantity of energy liberated when separated ions, positive and negative, are allowed to come together to form an ionic crystal composed of 1 mol of formula units of a compound. Alternatively, lattice energy may be defined as the quantity of energy required to break up an ionic crystal, causing the complete separation of the ions in 1 mol of compound. The stronger the forces among ions, the more difficult it is to disrupt a crystal.

Any factor that contributes to strong attractive forces among ions will contribute to a high lattice energy. *The attractive force between a pair of oppositely charged ions increases with increased charge on the ions or with decreased ionic sizes,* as illustrated in Figure 11-23. The type of crystal structure is also a factor in establishing the magnitude of the lattice energy.

Lattice energies for most ionic compounds are sufficiently great that ions do not readily detach themselves from the crystal and pass into the gaseous state. Ionic solids do not sublime at ordinary temperatures. All ionic solids can be melted by supplying enough thermal energy to destroy the crystalline structure. In general the higher the lattice energy the higher the melting point.

---

*Apparently additional factors are involved in the lubricant properties of graphite, since these properties are greatly diminished when graphite is heated in vacuum to expel gases.

FIGURE 11-23
Interionic forces of attraction.

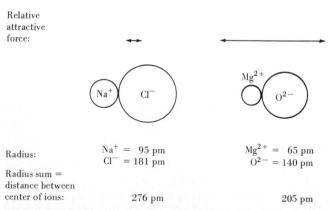

Relative attractive force:

Radius:
$$Na^+ = 95 \text{ pm}$$
$$Cl^- = 181 \text{ pm}$$
$$Mg^{2+} = 65 \text{ pm}$$
$$O^{2-} = 140 \text{ pm}$$

Radius sum = distance between center of ions:
276 pm   205 pm

Because of the higher charges on the ions and the closer proximity of their centers, the interionic attractive force between Mg^{2+} and O^{2-} is about seven times as great as between Na^+ and Cl^-.

Example 11-8   Would you expect KI or CaO to have the higher melting point?

Solution.   The higher lattice energy is expected for the combination of small, highly charged ions—Ca^{2+} and O^{2-}. The expected order of melting points is $KI < CaO$. The observed melting points are 677°C for KI and 2590°C for CaO.

SIMILAR EXAMPLES: Review Problem 9; Exercises 41, 42.

Two conditions are necessary for electrical conductivity: (1) Charged particles must be present and (2) the particles must be able to migrate in an electric field. In a solid ionic compound only the first requirement is met. However, whenever ions enter the liquid state, by melting an ionic crystal or by dissolving it in a suitable solvent, the liquid becomes a good electrical conductor. The energy required to break up an ionic crystal in the process of dissolving comes as a result of the interaction of ions with the solvent. Thus, the extent to which an ionic solid will dissolve is again determined, at least in part, by the lattice energy of the solid. That is, the lower the lattice energy the greater the quantity of an ionic solid that can be dissolved in a given quantity of solvent (see Section 12-1).

11-12 Calculation of Lattice Energy—The Born–Haber Cycle

The actual lattice energy of an ionic crystal need not be known to make qualitative predictions like those in the preceding section. However, calculation of lattice energy does offer interesting illustrations of Hess's law and two atomic properties, ionization energy and electron affinity. The familiar reaction for the formation of a mole of an ionic compound from its elements, for example NaCl(s), is

$$Na(s) + \tfrac{1}{2}Cl_2(g) \longrightarrow NaCl(s) \qquad \Delta\overline{H}_f^\circ = -411 \text{ kJ/mol} \tag{11.5}$$

But this is *not* the reaction that corresponds to the definition of lattice energy. Lattice energy, *U*, is based on the formation of a crystal from *gaseous ions*, step 5 in the following five-step process.

1. Sublimation of solid sodium:

$$Na(s) \longrightarrow Na(g) \qquad \Delta\overline{H}_1 = +108 \text{ kJ/mol}$$

2. Ionization of gaseous atomic sodium:

$$Na(g) \longrightarrow Na^+(g) + e^- \qquad \Delta\overline{H}_2 = +496 \text{ kJ/mol}$$

3. Dissociation of gaseous chlorine:

$$\tfrac{1}{2} Cl_2(g) \longrightarrow Cl(g) \qquad \Delta\overline{H}_3 = +121 \text{ kJ/mol}$$

4. Ion formation by gaseous atomic chlorine:

$$Cl(g) + e^- \longrightarrow Cl^-(g) \qquad \Delta\overline{H}_4 = -348 \text{ kJ/mol}$$

5. Combination of gaseous ions:

$$Na^+(g) + Cl^-(g) \longrightarrow NaCl(s) \qquad \Delta\overline{H}_5 = U = ?$$

The enthalpy changes listed above are indicated in Figure 11-24. From this figure it should also be clear that the sum of the five steps leads to the same result as the formation of one mole of NaCl(s) directly from its elements in their standard states. That is, $\Delta\overline{H}_f^\circ = \Delta\overline{H}_1 + \Delta\overline{H}_2 + \Delta\overline{H}_3 + \Delta\overline{H}_4 + \Delta\overline{H}_5$. Of these quantities, all are known except $\Delta\overline{H}_5 = U$. The value of $\Delta\overline{H}_2$ is the ionization energy of Na, and that of $\Delta\overline{H}_4$, the electron affinity of Cl. $\Delta\overline{H}_3$ is one half the Cl—Cl bond energy (recall Table 9-3).

FIGURE 11-24
Enthalpy diagram for
determining the lattice
energy, U, of NaCl(s).

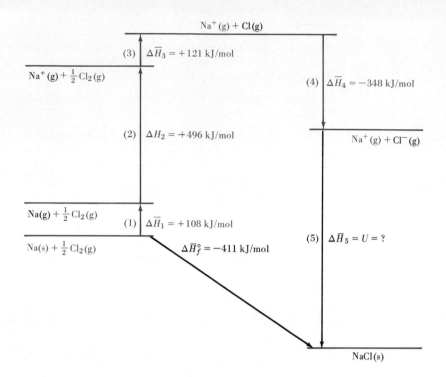

To proceed with the calculation of the lattice energy, $U = \Delta H_5$, we can write

(1)	$Na(s) \longrightarrow Na(g)$	$\Delta \overline{H}_1 = +108$ kJ/mol	
(2)	$Na(g) \longrightarrow Na^+(g) + e^-$	$\Delta \overline{H}_2 = +496$ kJ/mol	
(3)	$\frac{1}{2} Cl_2(g) \longrightarrow Cl(g)$	$\Delta \overline{H}_3 = +121$ kJ/mol	
(4)	$Cl(g) + e^- \longrightarrow Cl^-(g)$	$\Delta \overline{H}_4 = -348$ kJ/mol	
(5)	$Na^+(g) + Cl^-(g) \longrightarrow NaCl(s)$	$\Delta \overline{H}_5 = U = ?$	

$$Na(s) + \tfrac{1}{2}Cl_2(g) \longrightarrow NaCl(s) \qquad \Delta \overline{H}_f^\circ = -411 \text{ kJ/mol} \qquad (11.5)$$

$$\Delta \overline{H}_f^\circ = \Delta \overline{H}_1 + \Delta \overline{H}_2 + \Delta \overline{H}_3 + \Delta \overline{H}_4 + U$$

$$U = \Delta \overline{H}_f^\circ - \Delta \overline{H}_1 - \Delta \overline{H}_2 - \Delta \overline{H}_3 - \Delta \overline{H}_4$$

$$=(-411 - 108 - 496 - 121 + 348) \text{ kJ/mol} = -788 \text{ kJ/mol} \qquad (11.6)$$

The method illustrated here was developed by the German physicist Max Born and chemist Fritz Haber, and is known as the Born–Haber cycle.

11-13 Crystal Structures

Crystals—whether as ice, rock salt, quartz, or gemstones—have aroused human interest from earliest times. However, a fundamental understanding of the crystalline state has come only relatively recently, starting with the introduction of optical instruments and continuing into the present century with the development of x-ray diffraction. A key to this understanding has been the recognition that the observed regularity of crystals at the macroscopic level is due to an underlying microscopic regularity among groups of atoms, ions or molecules.

Crystal Lattices.   Dealing with regularity in one- and two-dimensional patterns is commonplace in human activities. The design shown in Figure 11-25a might be used as a

FIGURE 11-25
One- and two-dimensional patterns.

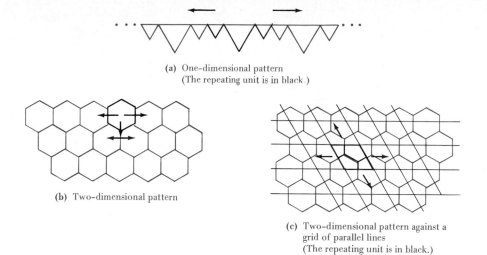

(a) One-dimensional pattern
(The repeating unit is in black.)

(b) Two-dimensional pattern

(c) Two-dimensional pattern against a
grid of parallel lines
(The repeating unit is in black.)

FIGURE 11-26

The cubic space lattice.

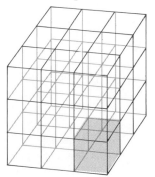

A typical parallelepiped formed by the intersection of mutually perpendicular parallel planes is shaded in grey. It is a cube.

Choosing a unit cell larger than the primitive unit cell is similar to choosing a group of several parallelograms in Figure 11-25c as the basis for describing the hexagonal array. For example, the design of the hexagon is clearly seen in a group of four parallelograms.

fringe or border on a piece of material. No matter how long this border, we can think of generating it simply by displacing the repeating unit in one dimension, that is, along a line to the left or to the right.

In the two-dimensional array of Figure 11-25b—a design sometimes used in ceramic tiling of floors and walls—we are accustomed to think of the hexagon as the repeating unit. However, it fails an important test for a repeating unit used by scientists who study crystal structures (crystallographers). *The repeating unit should generate neighboring units and the entire array by single straight displacements.* The hexagon outlined in black generates its neighbors in the same row by single displacements to the left and right, but to generate its neighbors in the next row requires *two* displacements, down and then to the left or to the right. The only two-dimensional figures that can meet the crystallographer's test for a repeating unit are those having parallel sides—squares, rectangles, or general parallelograms. The design of Figure 11-25b is redrawn in Figure 11-25c against a grid of parallel lines, and a repeating unit is identified. This repeating unit does possess the property of generating its neighbors and the entire array through single straight displacements. We might refer to the network of parallel lines in Figure 11-25c as a two-dimensional lattice and the points of intersection of the parallel lines as lattice points.

In the case of crystals we have to consider three dimensions. The basic lattice against which crystal structures can be described involves three sets of parallel planes. A special case is pictured in Figure 11-26; the planes are equidistant and mutually perpendicular (intersect at 90° angles). This is called the **cubic lattice.** It can be used to describe some, but by no means all, crystals. For some crystals the appropriate lattice may involve planes that are not equidistant or that do not intersect at 90° angles. In all there are seven possibilities, leading to seven crystallographic systems; but we will emphasize only the cubic system.

Intersections of the lattice planes produce three-dimensional figures having six faces arranged in three sets of parallel planes, figures called parallelepipeds. In Figure 11-26 these figures are cubes. We can identify a parallelepiped that can be used to generate the entire lattice by simple translation. This parallelepiped is called a **unit cell.** In describing a crystal it is convenient to arrange the three-dimensional space lattice so that structural particles of the crystal (i.e., atoms, ions or molecules) are situated at lattice points whenever possible. A unit cell having structural particles of the crystal only at its corners is called a primitive unit cell. It is the simplest unit cell that can be considered. Sometimes a unit cell involving more structural particles is chosen. In the **body-centered cubic (bcc)** structure, a structural particle of the crystal is found at the center of the cube as well as at

FIGURE 11-27
Unit cells in the cubic
crystal system.

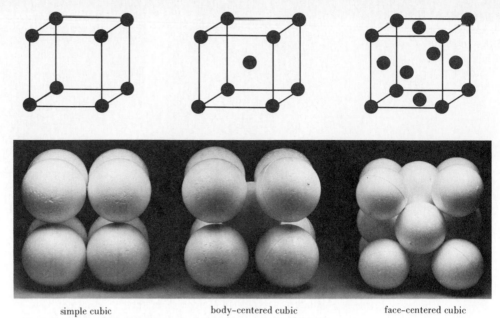

simple cubic   body–centered cubic   face–centered cubic

In the top row only the centers of spheres (atoms) are shown at their respective positions in the unit cells. In the space-filling models in the bottom row, contacts between spheres (atoms) are shown.

FIGURE 11-28
Closest packed
structures.

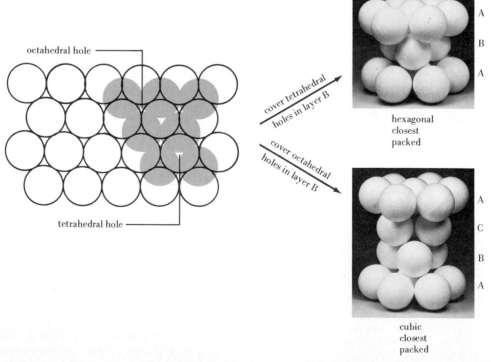

octahedral hole

tetrahedral hole

cover tetrahedral holes in layer B

cover octahedral holes in layer B

A
B
A

hexagonal
closest
packed

A
C
B
A

cubic
closest
packed

Spheres in layer A are outlined in color. Those added in layer B are shaded in grey.

FIGURE 11-29
A face-centered cubic unit cell for the cubic closest packing of spheres.

The 14 spheres on the left are extracted from a larger array of spheres in a cubic closest packed structure. The two middle layers each have six atoms; the top and bottom layers, one. Rotation of the 14 spheres reveals the *fcc* unit cell.

each corner. In a **face-centered cubic (fcc)** structure there is a structural particle at the center of each face as well as at each corner. These three unit cells are depicted in Figure 11-27.

FIGURE 11-30
The hexagonal closest packed (hcp) crystal structure.

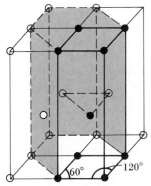

A unit cell is highlighted in heavy black and the atoms that are part of that cell are solid black. Note that the unit cell is not a cube. Three adjoining unit cells are also shown. The broken-line, color-shaded region, together with the highlighted unit cell, shows the layering (ABA) described in Figure 11-28.

Closest Packed Structures. Considerable insight into the structures of crystals can be gained by considering the way in which identical spheres—marbles, cannonballs, metal atoms—can be stacked. Unlike the case of cubes, there is no way of stacking spheres to fill all space, but there are certain arrangements in which the spheres come into as close contact as possible and the holes or voids are kept to a minimum. These are known as closest packed structures. Two such structures are presented in Figure 11-28.

Imagine a layer of identical spheres (layer A) in which each sphere is in contact with six others arranged in hexagonal fashion around it. Among the spheres there exist voids. Once the first sphere is placed into the next layer (layer B), the entire pattern for that layer is established. Again voids or holes are associated with layer B, but now the holes are of two different types. **Tetrahedral holes** fall directly over spheres in layer A and have this shape ▽. **Octahedral holes** fall directly over holes in layer A and have this shape ✿. Two possibilities now exist for the placement of the third layer. In one arrangement, the **hexagonal closest packed (hcp),** all the *tetrahedral* holes are covered. The third layer is identical to layer A and the structure begins to repeat itself. In the other arrangement, the **cubic closest packed,** all the *octahedral* holes are covered. The spheres in layer C are out of line with those of layer A. In this arrangement it is only when the fourth layer is added that the structure begins to repeat itself.

The cubic closest packed structure has a face-centered cubic unit cell which can be seen in the alternate view of closest packed spheres in Figure 11-29. The unit cell of the hexagonal closest packed structure is shown in Figure 11-30. In both the *hcp* and *fcc* structures voids account for only 25.96% of the total volume. Another arrangement in which the packing of spheres is not quite so close has a body-centered cubic (bcc) unit cell. In this structure voids account for 31.98% of the total volume. (A method of calculating the percent voids in crystal structures is outlined in Exercise 49.) The best examples of crystal structures based on the close packing of spheres are found among the metals. Some examples are listed in Table 11-4.

TABLE 11-4
Some features of close packed structures in metals

	Crystal coordination number	No. atoms per unit cell	Examples
hexagonal closest packed (hcp)	12	2	Cd, Mg, Ti, Zn
face-centered cubic (fcc)	12	4	Al, Cu, Pb, Ag
body-centered cubic (bcc)	8	2	Fe, K, Na, W

FIGURE 11-31
Apportioning atoms among bcc unit cells.

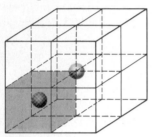

Eight unit cells are outlined. For clarity, only the centers of two atoms are pictured. Our attention is focused on the unit cell in color. The atom in the center of the cell belongs entirely to that cell. The corner atom is seen to be shared by all eight unit cells.

Crystal Coordination Number and Number of Atoms per Unit Cell.   In close packed structures of atoms, each atom is in contact with several others. To emphasize this fact we turn to the space-filling unit cells of Figure 11-27. For example, for the *bcc* unit cell we see that the center atom is in contact with each corner atom. The **crystal coordination number** is the number of neighboring atoms with which a given atom is in contact. For the *bcc* structure this is 8. For the *fcc* and *hcp* structures the crystal coordination number is 12. The easiest way to see this is from the layering of spheres described in Figure 11-28. Each sphere is in contact with six others in the same layer, three in the layer above, and three in the layer below.

Although we use 9 atoms to draw the *bcc* unit cell, it is incorrect to say that the unit cell contains or consists of 9 atoms. As shown in Figure 11-31, only the center atom belongs entirely to a *bcc* unit cell. The corner atoms are shared among eight adjoining unit cells. One eighth of each corner atom belongs to a given unit cell, and the eight corner atoms contribute the equivalent of one atom to the unit cell. Thus, the number of atoms in a *bcc* unit cell is *two*. For the *hcp* unit cell of Figure 11-30 an identical counting process indicates *two* atoms per unit cell. In the *fcc* unit cell the corner atoms account for $\frac{1}{8} \times 8 = 1$ atom and those in the center of the faces for $\frac{1}{2} \times 6 = 3$ atoms. The *fcc* unit cell contains *four* atoms.

X-Ray Diffraction.   Visible light, in combination with the human eye and brain, can provide knowledge of the structure of macroscopic objects. To reveal the manner in which atoms, ions, or molecules are arranged in a crystal, electromagnetic radiation of much shorter wavelengths is required. When a beam of x rays encounters atoms of a substance, x rays interact with electrons in the atoms and the original beam is reradiated, diffracted, or scattered in all directions. The scattering pattern is related to the distribution and density of electronic charge in the atoms. The scattered x rays must be made to produce a

FIGURE 11-32
Diffraction of x rays by a crystal.

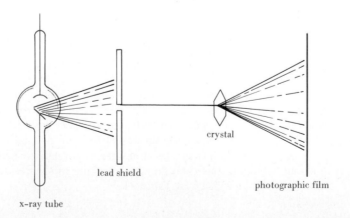

x-ray tube

lead shield

crystal

photographic film

FIGURE 11-33
A portion of a Laue
x-ray photograph of a
crystal.

The crystal producing this
pattern consists of Er^{3+}
and $HEDTA^{3-}$ ions. (Ethylenediaminetetraacetic
acid, EDTA, and ions
derived from it are de-
scribed in Chapter 24.)
[Courtesy of James E.
Benson, Ames Laboratory,
Iowa State University.]

visible pattern, as on a photographic film, and the microscopic structure of the object can
only be inferred from this pattern. The success of this inference depends on the amount of
scattered radiation that is recovered, that is, on how much "information" is gathered. The
power of the x-ray-diffraction method has been greatly increased by the use of high-speed
computers. Computers can process vast amounts of x-ray data (in much the same way the
human eye processes visible light).

Figure 11-32 suggests a method of scattering x rays from a crystal, and Figure 11-33
represents a photographic record of the scattered x rays called a Laue pattern (after Max
von Laue, who pioneered in the application of x rays to crystal-structure determination).
Some aspects of scattering patterns can be explained by a geometric analysis proposed by
W. H. Bragg and W. L. Bragg in 1912, illustrated in Figure 11-34.

Figure 11-34 pictures two rays in a monochromatic x-ray beam, labeled a and b. Wave
a is reflected from one plane of atoms or ions in a crystal and wave b from the next plane
below. Wave b travels a greater distance than wave a. The additional distance is $2d \sin \theta$.
In order to reinforce one another, the crests and troughs of the two waves must be in phase
(line up with one another) as they approach the detector. To satisfy this requirement, the
additional distance traveled by wave b must be an integral multiple of the wavelength of
the x rays.

$$n\lambda = 2d \sin \theta \qquad \text{(where } n = 1, 2, 3, \ldots \text{)} \tag{11.7}$$

The spacing between atomic planes can thus be determined by knowing λ and measuring θ. Different orientations of the crystal allow for atomic spacings and electron densities
to be determined along different directions through the crystal. With this type of information, the crystalline structure of a solid can be established.

Once a crystal structure has been established, certain properties can be determined by
calculation. In Example 11-9 a metallic radius is calculated, and in Example 11-10 Avogadro's number is estimated. In both of these calculations it is necessary to visualize a unit
cell of the crystal structure.

FIGURE 11-34
Determination of crystal
structure by x-ray
diffraction.

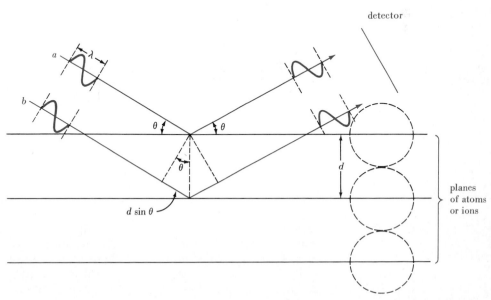

The two triangles outlined by the dotted lines are identical right triangles. The hypote-
nuse of each triangle is equal to the interatomic distance, d. The side opposite the
angle θ thus has a length of $d \sin \theta$. Wave b travels farther than wave a by the distance $2d \sin \theta$.

FIGURE 11-35
Determination of the
atomic radius of iron—
Example 11-9 illustrated.

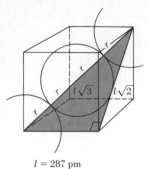

l = 287 pm

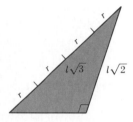

l = 287 pm

The right triangle must
conform to the Pythago-
rean formula: $a^2 + b^2 = c^2$. That is, $(l)^2 + (l\sqrt{2})^2 = (l\sqrt{3})^2$, or
$l^2 + 2(l^2) = 3(l^2)$.

Example 11-9 At room temperature iron crystallizes in a bcc structure. By x-ray diffraction, the edge of the cubic cell corresponding to Figure 11-35 is found to be 287 pm. What is the radius of an iron atom?

Solution. Nine atoms are associated with a *bcc* unit cell. One atom is located at each of the eight corners of the cube and one at the center of the cube. The three atoms along a cube diagonal are in contact. The length of the cube diagonal (the distance from the farthest upper-right corner to the nearest lower-left corner) is four times the atomic radius. But also shown in Figure 11-35 is that the diagonal of a cube is equal to $\sqrt{3}$ times the length. The length *l* is what is given.

$$4r = l\sqrt{3} \qquad r = \frac{\sqrt{3} \times 287 \text{ pm}}{4} = \frac{1.73 \times 287 \text{ pm}}{4} = 124 \text{ pm } (1.24 \text{ Å})$$

SIMILAR EXAMPLE: Review Problem 11.

Example 11-10 Use the fact that the density of iron is 7.86 g/cm³ and that its molar mass is 55.85 g Fe/mol Fe, together with data from Example 11-9, to estimate the value of Avogadro's number, N_A.

Solution. In Example 11-9 we are given the length of a unit cell of iron: $l = 287$ pm $= 287 \times 10^{-12}$ m $= 2.87 \times 10^{-8}$ cm. The volume of the unit cell is $V = l^3 = (2.87 \times 10^{-8})^3$ cm³ $= 2.36 \times 10^{-23}$ cm³. The mass of the unit cell is

$$m = V \times d = 2.36 \times 10^{-23} \text{ cm}^3 \times \frac{7.86 \text{ g}}{\text{cm}^3} = 1.85 \times 10^{-22} \text{ g}$$

If we divide the mass of a unit cell by the number of atoms that the cell contains, we obtain the mass of a single Fe atom. The molar mass of Fe, 55.85 g/mol, is the mass of Avogadro's number of atoms. The molar mass divided by the mass per atom is Avogadro's number. We have previously established that the number of atoms per unit cell for the *bcc* structure is *two* (see Table 11-4).

$$\text{mass of Fe atom} = \frac{1.85 \times 10^{-22} \text{ g}}{\text{unit cell}} \times \frac{1 \text{ unit cell}}{2 \text{ Fe atoms}} = 9.25 \times 10^{-23} \text{ g Fe/Fe atom}$$

$$N_A = \frac{55.85 \text{ g Fe}}{1 \text{ mol Fe}} \times \frac{1 \text{ Fe atom}}{9.25 \times 10^{-23} \text{ g Fe}} = 6.04 \times 10^{23} \text{ Fe atoms/mol Fe}$$

SIMILAR EXAMPLES: Review Problem 11; Exercises 48, 50.

11-14 Ionic Crystal Structures

When applied to an ionic crystal, the packing-of-spheres model is complicated by two factors: (1) Some of the ions are positively charged and some, negatively charged, and (2) the cations and anions are of different sizes. What we can expect, however, is that oppositely charged ions will come into very close proximity; generally we think of them as being in contact. Like-charged ions, because of mutual repulsions, generally are not in direct contact. Some ionic crystals can be pictured as a fairly close packed arrangement of ions of one type with voids filled by ions of the opposite charge. The relative sizes of cations and anions are important in establishing a particular packing arrangement. In defining a unit cell of an ionic crystal, the unit cell chosen must

1. by simple translation in three dimensions, generate the entire crystal;
2. indicate the crystal coordination numbers of the ions;
3. be consistent with the formula of the compound.

FIGURE 11-36
The unit cell of sodium chloride.

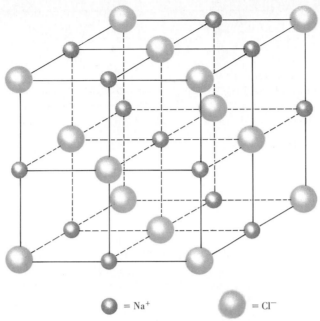

$\bigcirc$ = Na^+ $\bigcirc$ = Cl^-

For clarity, only the centers of the ions are shown. Oppositely charged ions are actually in contact.

Unit cells of crystalline NaCl and CsCl are pictured in Figures 11-36 and 11-37.

For an ionic crystal, the crystal coordination number is the number of nearest-neighboring ions of opposite charge to any given ion in the crystal. Note the Na^+ ion at the center of the unit cell in Figure 11-36. It is surrounded by *six* Cl^- ions. The crystal coordination numbers of both Na^+ and Cl^- are *six*. By contrast, the crystal coordination numbers of Cs^+ and Cl^- in Figure 11-37 are *eight*.

FIGURE 11-37
The unit cell of cesium chloride.

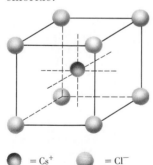

$\bigcirc$ = Cs^+ $\bigcirc$ = Cl^-

The Cs^+ ion is in the center of a cube with Cl^- ions at the corners. For clarity, only the centers of the ions are shown, though in reality each Cl^- is in contact with the Cs^+ ion. An alternative unit cell has Cl^- at the center and Cs^+ at the corners.

The 27 ions pictured in the unit cell of NaCl must be apportioned among the given and neighboring unit cells in the following way: Each Cl^- ion in a corner position is shared by eight unit cells, and each Cl^- in the center of a face, by two unit cells. This leads to a total number of Cl^- ions for a unit cell of $(8 \times \frac{1}{8}) + (6 \times \frac{1}{2}) = 1 + 3 = 4$. There are 12 Na^+ ions along edges of the unit cell, and each edge is shared by four unit cells. The Na^+ in the very center of the unit cell belongs entirely to that cell: $(12 \times \frac{1}{4}) + (1 \times 1) = 3 + 1 = 4$. The unit cell has the equivalent of 4 Na^+ and 4 Cl^- ions. The ratio of Na^+ to Cl^- is 1:1, corresponding to the formula NaCl.

Example 11-11 The ionic radii of Na^+ and Cl^- are 95 and 181 pm, respectively. What is the length of the unit cell of NaCl?

Solution. Again the key to solving this problem lies in understanding geometrical relationships in the unit cell. Along each edge of the unit cell (Figure 11-36) two Cl^- ions are in contact with one Na^+. The edge length is equal to the radius of one Cl^-, plus the diameter of Na^+, plus the radius of another Cl^-, that is,

$$\text{edge length} = (r_{Cl^-}) + (r_{Na^+}) + (r_{Na^+}) + (r_{Cl^-}) = 2(r_{Na^+}) + 2(r_{Cl^-})$$
$$= [(2 \times 95) + (2 \times 181)] \text{ pm} = 552 \text{ pm } (5.52 \text{ Å})$$

SIMILAR EXAMPLE: Exercise 52.

Ionic compounds of the type $M^{2+}X^{2-}$ (e.g., MgO, BaS, CaO) may form crystals of the NaCl type. For substances with formulas MX_2 or M_2X, the crystal structures are more complex. Because the cations and anions occur in unequal numbers, the crystals have two

FIGURE 11-38
Some unit cells of
greater complexity.

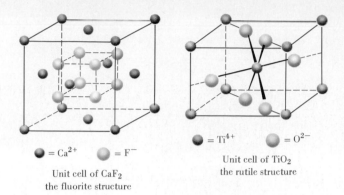

= Ca^{2+} = F^-

Unit cell of CaF_2
the fluorite structure

= Ti^{4+} = O^{2-}

Unit cell of TiO_2
the rutile structure

coordination numbers, one for the cation and another for the anion. Two typical structures of a more complex type are shown in Figure 11-38.

In CaF_2 (the fluorite structure) there are twice as many fluoride as calcium ions. The crystal coordination number of Ca^{2+} is eight, that of F^- is four. In TiO_2 (the rutile structure) Ti^{4+} has a crystal coordination number of six and O^{2-}, three. In this structure two of the O^{2-} ions are within the interior of the cell, two are in the top face, and two in the bottom face of the cell. Ti^{4+} ions are at the corners and the center of the cell.

11-15 Types of Solids: A Summary

The structural particles in a crystalline solid, we have learned, can be atoms, ions, or molecules. The intermolecular forces operating among these structural particles may be van der Waals forces, covalent bonds, interionic attractions, or the metallic bond. Table 11-5 lists the basic types of solid structures, the intermolecular forces within them, some of their characteristic properties, and typical examples of each.

TABLE 11-5
Types of solids

Type of solid	Structural particles	Intermolecular forces	Typical properties	Examples
molecular	molecules (atoms of noble gases)	London and/or dipole–dipole and/or hydrogen bonds	soft; low melting points; nonconductors of heat and electricity; sublime easily in many cases.	noble gas elements; CH_4; CO_2; P_4; S_8; I_2; H_2O.
network covalent	atoms	covalent bonds	very hard; very high melting points; nonconductors of electricity	C(diamond); SiC; SiO_2.
ionic	cations and anions	electrostatic attractions	hard; moderate to very high melting points; nonconductors of electricity (but good electrical conductors in the molten state).	NaCl; $NaNO_3$; MgO.
metallic	cations plus delocalized electrons	metallic bonds	hardness varies from soft to very hard; melting point varies from low to very high; lustrous; ductile; malleable; very good conductors of heat and electricity.	Na; Mg; Al; Fe; Zn; Cu; Ag; W.

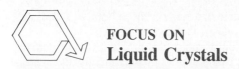

FOCUS ON
Liquid Crystals

A liquid crystal display screen on a portable microcomputer. [Courtesy Richard J. Botting.]

In 1888 the Austrian botanist Reinitzer discovered that the substance cholesteryl benzoate melted sharply at 145.5°C to produce a milky fluid, which in turn underwent a sharp transition to a clear liquid at 178.5°C. This substance, between 145.5 and 178.5°C, is in a phase that has the fluid properties of a liquid, the optical properties of a crystalline solid, and some unique properties of its own. The term *mesophase* has been used to denote this state of matter between liquid and solid. The more common term now used is **liquid crystal.** Figure 11-39 suggests how liquid crystals may be represented in a phase diagram.

Liquid crystals are observed most commonly in organic compounds comprised of cylindrically shaped (rod-like) molecules having molecular weights of 200 to 500 and having lengths that are 4 to 8 times their diameters. (Potentially, this is about 0.5% of all organic compounds.)

As illustrated in Figure 11-40, such molecules, when in the liquid state, have the expected random ordering. Liquid molecules are able to move (translate) in three directions and to rotate freely. In the liquid crystalline state molecules have both some mobility and some ordering among them. In the form known as **nematic** (meaning

FIGURE 11-40
The liquid crystalline state.

liquid

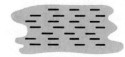
nematic liquid crystal

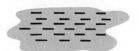

smectic liquid crystal

cholesteric liquid crystal

FIGURE 11-39
Representation of liquid crystals in a phase diagram.

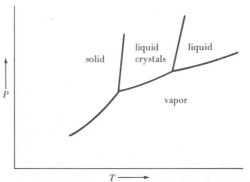

threadlike) the rodlike molecules are arranged in a parallel fashion; they are free to move in three directions but to rotate only about their long axes. (Imagine the possible ways in which you might move a particular pencil in a box of loosely packed pencils.) In the **smectic** form (meaning greaselike) the rodlike molecules are arranged in layers, with the long axes of the molecules perpendicular to the planes of the layers. The molecular motions possible here are translation in two directions (i.e., within a layer) and rotation about the long axis. The **cholesteric** form is related to the nematic form, but molecules are stratified into layers; and the direction in which the parallel molecules are oriented shifts from one plane to the next.

As suggested by Figure 11-40, each layer in a cholesteric structure has a different orientation of molecules from the layers immediately above and below it. Over a span of several layers, however, each particular orientation is repeated. A distinctive characteristic of a cholesteric liquid crystal is the distance between planes having molecules in the same orientation. When a film of cholesteric liquid crystals is struck by a beam of light, the properties of the reflected light depend on this distance. Since this distance is very temperature sensitive, the reflected light changes (e.g., in color) with changing temperature. This phenomenon serves as the basis of liquid-crystal temperature-sensing devices. Temperature changes as small as 0.01°C can be detected with such devices with ordinary white light. (See also, Color Section B.)

The orientation of molecules in a thin film of nematic liquid crystals is easily altered by pressure and by an electric field. The altered orientation affects the optical properties of the film, such as causing the film to become opaque. If, through electrodes arranged in certain patterns (say in the shape of numbers), an electric field is imposed on a thin film of liquid crystals, these patterns of electrodes are rendered visible. This is the principle that is used in liquid-crystal displays (LCD) in hand-held calculators and digital watches.

Liquid crystals have been found to occur widely in living matter. Cell membranes and certain tissues have structures that can be described as liquid crystalline; and hardening of the arteries is caused by a deposit of liquid crystalline compounds of cholesterol. Liquid crystalline properties have been identified in various polymers (e.g., Du Pont Kevlar fiber). As the range of materials displaying liquid crystalline properties is extended, so too will be the range of possible applications.

Summary

One result of the differing environments of surface and interior molecules in a liquid is surface tension and properties related to it. Another result is the tendency of energetic molecules at the surface to leave a liquid and pass into the vapor state—vaporization.

When vaporization and condensation occur simultaneously in a closed container, a condition of dynamic equilibrium is established. The vapor exerts a characteristic pressure called the vapor pressure of the liquid. A graph of vapor pressure versus temperature is known as a vapor pressure curve. A point of special interest on this curve is the temperature at which the vapor pressure equals barometric pressure—the boiling point. Another point of interest is the high-temperature terminus of the vapor pressure curve—the critical point. At this point the liquid and gaseous states of a substance become indistinguishable.

The phase diagram of a substance represents the phases or states of matter that exist at various temperatures and pressures. Characteristic points that can be located on a phase diagram are the normal melting and boiling points, the critical point, and the solid–liquid–vapor triple point. Phase diagrams can be used to describe the changes that occur when a sample of matter is heated, cooled, or subjected to a change in pressure.

Electrostatic attractions between instantaneous and induced dipoles are the most common intermolecular forces. The strength of these forces increases with increased molecular weight. For polar substances dipole–dipole attractions may also be significant. A hydrogen atom that is covalently bonded to one nonmetal atom may be attracted simultaneously to another nonmetal atom of high electronegativity in the same or a neighboring molecule. This hydrogen bonding is crucial to an understanding of the properties of solid and liquid water and of living matter.

In network covalent solids, bonds extend throughout a crystalline structure. Network covalent solids have high melting points compared to other covalent substances. If all the electrons in a network covalent solid are localized into bonds, the solid is an electrical insulator. However, if some of the electrons are *delocalized*, the solid is an electrical conductor.

The strength of intermolecular forces in an ionic solid is expressed through the lattice energy. Lattice energy, which is largely a function of ionic size and charge, can be calculated from thermochemical data and certain atomic and molecular properties.

To describe the structure of a solid, for example a

metal, it is useful to think in terms of a closely packed structure of spheres. The two closest-packed arrangements are the hexagonal and the cubic closest packed. A third arrangement, the body-centered cubic structure, is not so tightly packed. Experimental data required for a crystal-structure determination are acquired through the interaction of x rays with a crystal.

Ionic crystal structures can be described through a packing-of-spheres model, but the matter is complicated by the fact that the ions are not all of the same size or charge. A useful concept in describing crystals of all types is the unit cell. Knowledge of the unit cell of a crystal makes possible calculations involving densities, atomic radii, and other properties.

Learning Objectives

As a result of studying Chapter 11, you should be able to

1. Explain surface tension and related phenomena—drop formation, wetting of surfaces, and capillary action.

2. Describe dynamic equilibrium between a liquid and its vapor and state the factors that affect vapor pressure.

3. Derive useful data from plots of vapor pressure or log of vapor pressure as a function of temperature.

4. Calculate vapor pressures from experimental data, and use vapor pressure data to predict conditions for the existence of the vapor and/or liquid states.

5. Use the Clausius–Clapeyron equation to relate log of vapor pressure, temperature, and $\Delta \overline{H}_{vap}$ of a liquid.

6. Describe the significance of the critical point.

7. Interpret simple phase diagrams, and use phase diagrams to predict changes that will occur as a substance is heated, cooled, or subjected to a change in pressure.

8. Describe the common types of intermolecular forces and how these forces influence physical properties.

9. State conditions that lead to hydrogen bonds and some properties that result from hydrogen bond formation.

10. Give examples of network covalent solids and describe the bonding and characteristic properties of these solids.

11. Explain how lattice energy is related to ionic sizes and charges, and how physical properties of ionic solids are related to lattice energy.

12. Use the Born–Haber cycle to calculate lattice energies from thermochemical, atomic, and molecular data.

13. Describe how crystal structures are related to the close packing of spheres.

14. Explain the term, unit cell, and apply this concept to calculations of quantities such as atomic radii and densities.

15. Use the concept of the unit cell to establish ionic crystal coordination numbers and chemical formulas.

Some New Terms

Adhesive forces are intermolecular forces between unlike molecules, such as molecules of a liquid and of the surface with which it is in contact.

bcc is an abbreviation for body-centered cubic crystal structures.

Boiling is a process in which vaporization takes place throughout a liquid. It occurs when the vapor pressure of a liquid is equal to barometric pressure.

The **Born–Haber cycle** relates lattice energies of ionic solids to ionization energies, electron affinities, and heats of sublimation, dissociation, and formation.

Capillary action refers to the rise of a liquid in the pores of thin capillary tubes.

Cohesive forces are intermolecular forces between like molecules, such as in a drop of liquid.

Condensation is the passage of molecules from the gaseous state to the liquid state.

The **critical point** refers to the temperature and pressure where a liquid and its vapor become identical; it is the highest temperature point on the vapor pressure curve.

The **crystal coordination number** signifies the number of nearest neighboring atoms (or ions of opposite charge) to any given atom (or ion) in a crystal.

Deposition is the passage of molecules from the gaseous to the solid state.

fcc is an abbreviation for face-centered cubic crystal structures.

Freezing is the conversion of a liquid to a solid, and occurs at a fixed temperature known as the **freezing point.**

hcp is an abbreviation for hexagonal closest-packed crystal structures.

A **hydrogen bond** is an intermolecular attraction in which an H atom covalently bonded to one atom is attracted simultaneously to another strongly nonmetallic atom of the same or a nearby molecule.

An **induced dipole** is an atom or molecule in which a separation of charge is produced by a neighboring dipole.

An **instantaneous dipole** is an atom or molecule with a separation of charge produced by a momentary displacement of electrons from their normal distribution.

An **intermolecular force** is an attraction *between* molecules.

Lattice energy is the quantity of energy released in the formation of one mole of an ionic solid from its separated gaseous ions.

Liquid crystals are a form of matter with some of the properties of a liquid and some, of a crystalline solid.

London forces (dispersion forces) are intermolecular forces associated with instantaneous and induced dipoles.

Melting is the transition of a solid to a liquid.

A **network covalent solid** is a substance in which covalent bonds extend throughout a crystal.

Normal boiling point is the temperature at which the vapor pressure of a liquid is 1 atm.

Normal melting point is the temperature at which the melting of a solid occurs at 1 atm pressure. This is the same temperature as the **normal freezing point.**

A **phase diagram** is a graphical representation of the phases or states of matter of a substance that exist at various temperatures and pressures.

Sublimation is the passage of molecules from the solid to the gaseous state.

Surface tension is a property resulting from the differing environments of surface and interior molecules in a liquid.

Among the consequences of surface tension is the tendency of a free-falling liquid to minimize its surface area by forming spherical drops.

A **triple point** is the condition of temperature and pressure under which three phases of a substance (e.g., a solid, the liquid, and the vapor) coexist at equilibrium.

A **unit cell** is a small collection of atoms, ions, or molecules from which an entire crystal structure can be inferred.

van der Waals forces is a term used to describe, collectively, intermolecular forces of the London type and interactions between permanent dipoles.

Vaporization is the passage of molecules from the liquid to the gaseous state.

Vapor pressure is the pressure exerted by a vapor when it is in dynamic equilibrium with its liquid.

A **vapor pressure curve** is a graph of vapor pressure as a function of temperature.

Viscosity refers to a liquid's resistance to flow; its magnitude depends on intermolecular forces of attraction and, in some cases, on molecular sizes and shapes.

X-ray diffraction is a method of crystal structure determination based on the interaction of a crystal with x rays.

Suggestions for Further Study

BROWN, G. H., and P. P. CROOKER, "Liquid Crystals," *Chem. & Eng. News,* **61**[5], 24 (1983).

HOUSE, J. E., Jr., "Weak Intermolecular Interactions," *Chemistry,* **45**[4], 12 (1972).

HUGGINS, M. L., "The Hydrogen Bond and Other Reminiscences," *Chemtech,* **10**, 422 (1980).

LEVINSON, G. S., "A Simple Experiment for Determining Vapor Pressure and Enthalpy of Vaporization of Water," *J. Chem. Educ.,* **59**, 337 (1982).

LIVINGSTON, R. L., "The Teaching of Crystal Geometry in the Introductory Course," *J. Chem. Educ.,* **44**, 376 (1967).

PUPEZIN, J., G. JANSCO, and W. A. VAN HOOK, "The Vapor Pressure of Water: A Good Reference System?" *J. Chem. Educ.,* **48**, 114 (1971).

Special Issue: "Liquid Crystals," *Physics Today,* **35**[5] (1982).

WELLS, A. F., "Some Simple AX and AX_2 Structures," *J. Chem. Educ.,* **59**, 630 (1982).

Review Problems

1. At their normal boiling points, the enthalpies (heats) of vaporization of acetone, $(CH_3)_2CO$, chloroform, $CHCl_3$, and ethyl alcohol, C_2H_5OH, are 124.5, 59.0, and 204.3 cal/g, respectively.
  (a) How much heat is *evolved* when 1.25 kg $C_2H_5OH(g)$ condenses to $C_2H_5OH(l)$ at its normal boiling point?
  (b) What mass of acetone can be vaporized by the absorption of 12.5 kJ of heat?
  (c) What is $\Delta \overline{H}_{vap}$ of $CHCl_3$, in kJ/mol?

2. From Figure 11-6, estimate **(a)** the vapor pressure of C_6H_6 at 50°C; **(b)** the normal boiling point of $C_4H_{10}O$.

3. Equilibrium is established between $Br_2(l)$ and its vapor at 25.0°C. A 250.0-cm^3 sample of the vapor weighs 0.486 g. What is the vapor pressure of $Br_2(l)$ at 25.0°C, in mmHg?

4. Use data plotted in Figure 11-8 to estimate **(a)** the normal boiling point of aniline; **(b)** the vapor pressure of toluene at 75°C.

5. At their normal melting points, the enthalpies (heats) of fusion of Mg, Pb, and Cu are 2.140, 1.141, and 3.120 kcal/mol, respectively.
  (a) What mass of Mg will be melted if a sample of Mg(s), at its normal melting point, absorbs 985 cal of heat?
  (b) How much heat, in kcal, must be absorbed at the melting point to melt a bar of lead that is 8″ × 2″ × 1″? (Assume a density of 11 g/cm^3 for Pb.)
  (c) How much heat, in kJ, is *evolved* when a 2.12-kg sample of molten Cu freezes?

6. 0.180 g $H_2O(l)$ is sealed into an evacuated 2.50-L flask. What is the pressure of the vapor in the flask if the temperature is **(a)** 30°C; **(b)** 50°C; **(c)** 70°C? (Hint: Use data from Table 11-1. Does any of the water remain as liquid or does it vaporize completely?)

7. In the portion of the phase diagram for phosphorus shown below

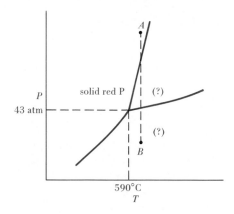

(a) Indicate the phases that exist in the regions labeled (?) and along each transition curve.
(b) A sample of solid red phosphorus cannot be melted by heating it in a container open to the atmosphere. Explain why this is so.
(c) Trace the phase changes that occur when the pressure on a sample of phosphorus is reduced from point A to B, at constant temperature.

8. In each of the following pairs, which would you expect to have the higher boiling point? Explain. **(a)** normal C_7H_{16} or normal $C_{10}H_{22}$; **(b)** C_3H_8 or H_3C—O—CH_3; **(c)** H_3CCH_2—S—H or H_3CCH_2—O—H. (The term "normal" means that all C atoms are in a straight chain; that is, there is no branching of the chain.

9. Arrange the following in the expected order of increasing lattice energy: CaO; $MgBr_2$; CsI.

10. The following data are given (all values are in kJ/mol):

$Li(s) \longrightarrow Li(g)$	$\Delta\bar{H} = +160.7$
$Na(s) \longrightarrow Na(g)$	$\Delta H = +107.8$
$Li(g) \longrightarrow Li^+(g) + e^-$	$\Delta\bar{H} = +520.5$
$Na(g) \longrightarrow Na^+(g) + e^-$	$\Delta\bar{H} = +495.4$
$F_2(g) \longrightarrow 2\ F(g)$	$\Delta\bar{H} = +157.8$
$F(g) + e^- \longrightarrow F^-(g)$	$\Delta\bar{H} = -328$
$Li(s) + \frac{1}{2}F_2(g) \longrightarrow LiF(s)$	$\Delta\bar{H} = -616.9$
$Na^+(g) + F^-(g) \longrightarrow NaF(s)$	$\Delta\bar{H} = -927.7$

(a) Calculate the lattice energy of LiF(s).
(b) Calculate the enthalpy (heat) of formation of NaF(s).

11. The fcc unit cell is a cube with atoms at each of the corners and in the center of each face, as shown below.

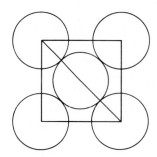

Assuming an atomic radius of 128 pm for a Cu atom,
(a) What is the length of the unit cell of Cu?
(b) What is the volume of the unit cell?
(c) How many atoms belong to the unit cell?
(d) What is the mass of a unit cell?
(e) Calculate the density of copper.

12. In the manner illustrated in the text for NaCl, show that the formula of CsCl is consistent with its unit cell pictured in Figure 11-37.

Exercises

Surface tension and related properties

1. Describe briefly the meaning of each term: **(a)** surface tension; **(b)** adhesive force; **(c)** capillary action; **(d)** wetting agent; **(e)** meniscus.

2. Silicone oils are used in water repellents for treating tents, hiking boots, and similar items. Explain how they function.

*3. When a wax candle is burned, the fuel consists of gaseous hydrocarbons that appear at the end of the candle wick. Describe the steps by which the solid wax is consumed.

Vaporization

4. When a sample of liquid benzene absorbs 1.00 kJ of heat at its normal boiling point of 80.1°C, the volume of $C_6H_6(g)$ produced (at 80.1°C and 1 atm) is 0.94 L. What is $\Delta\bar{H}_{vap}$ for C_6H_6?

5. For the combustion of $CH_4(g)$, $\Delta\bar{H} = -890$ kJ/mol CH_4. What volume of the gas, measured at 25.0°C and 748 mmHg, must be burned to provide for the evaporation of 4.55 L of liquid water at 100°C? The density of liquid water at 100°C = 0.958 g/cm³, and $\Delta\bar{H}_{vap} = +40.6$ kJ/mol.

6. A double boiler is used when a careful control of temperature is required in cooking. Water is boiled in an outer container to produce steam, and the steam condenses on the outside walls of an inner container in which cooking occurs. (A related laboratory device is called a steam bath.)

    (a) How is heat energy conveyed to the food to be cooked?

    (b) What is the maximum temperature that can be reached in the inside container?

7. If a volatile liquid vaporizes into the atmosphere from an ordinary container, the liquid temperature remains the same as that of the surroundings. If the same liquid vaporizes into the atmosphere from a thermally insulated container (a vacuum bottle or Dewar flask), its temperature falls below that of the surroundings. Explain this difference.

Vapor pressure and boiling point

8. Use data from Table 11-1 to estimate **(a)** the boiling point of water in Santa Fe, New Mexico, if the prevailing atmospheric pressure is 600 mmHg; **(b)** the prevailing atmospheric pressure in Leadville, Colorado, if the observed boiling point of water is 89°C.

9. A 25.0-L volume of He(g) at 30.0°C is passed through 6.220 g of liquid aniline (C_6H_7N) at 30.0°C. The liquid remaining after the experiment weighs 6.108 g. Assume that the He(g) becomes saturated with aniline vapor and that the total gas volume and temperature remain constant. What is the vapor pressure of aniline at 30.0°C?

10. 7.53 L of N_2(g) at 742 mmHg and 45.0°C is bubbled through CCl_4(l) at 45.0°C. Assuming the gas becomes saturated with CCl_4(g), what is the volume of the resulting gaseous mixture if the total pressure remains at 742 mmHg and the temperature, 45.0°C? The vapor pressure of CCl_4 at 45.0°C is 261 mmHg.

The Clausius–Clapeyron equation

11. By the method used to graph Figure 11-8, plot log P vs. $1/T$ for liquid yellow phosphorus, and estimate its normal boiling point. Vapor pressure data: 76.6°C, 1; 128.0°C, 10; 166.7°C, 40; 197.3°C, 100; 251.0°C, 400 mmHg.

12. Cyclohexanol, used in the manufacture of nylon, has a vapor pressure of 10.0 mmHg at 56.0°C and 100.0 mmHg at 103.7°C. Calculate its **(a)** $\Delta \overline{H}_{vap}$ and **(b)** normal boiling point.

13. The normal boiling point of normal octane, a constituent of gasoline, is 125.8°C and its $\Delta \overline{H}_{vap}$ is 33.9 kJ/mol. At what temperature does normal octane have a vapor pressure of 175 mmHg?

Critical point

14. Which substances listed in Table 11-2 can exist as liquids at room temperature (about 20°C)? Explain.

15. Can SO_2 be maintained as a liquid under a pressure of 100 atm at 0°C? Can liquid methane be obtained under the same conditions? (Refer to Table 11-2.)

Fusion

16. How much heat is required to melt a block of ice that measures 30.5 cm on an edge? The density of ice is 0.92 g/cm³ and the heat of fusion is 6.02 kJ/mol.

17. What is the total quantity of heat required to melt a 0.803-kg piece of Pb, starting with the sample at 25.0°C? Pb melts at 327.4°C; its heat of fusion is 4.774 kJ/mol; its average specific heat from 25 to 327.4°C is 0.134 J g⁻¹ °C⁻¹.

States of matter and phase diagrams

18. 2.50 g H_2O(l) is sealed in a 5.00-L flask at 120°C.

    (a) Show that the sample exists completely as vapor.

    (b) Estimate the temperature to which the flask must be cooled before liquid water condenses.

19. A 20.0-L vessel contains 0.100 mol H_2(g) and 0.050 mol O_2(g). The mixture is ignited with a spark and the reaction

$$2\ H_2(g) + O_2(g) \longrightarrow 2\ H_2O$$

goes to completion. The system is then cooled to 27°C. What is the final pressure in the vessel? (*Hint:* Is the H_2O formed present as a gas, a liquid, or a mixture of the two?)

20. A block of ice that is 8.0 cm × 2.5 cm × 2.7 cm is taken from a freezer at −25.0°C and added to 400.0 cm³ of H_2O(l) at 32.0°C. Assuming that the container is perfectly insulated from the surroundings, what will be the final temperature of the contents of the container? What state(s) of matter will be present? The specific heat of ice is 2.01 and that of H_2O(l) is 4.18 J g⁻¹ °C⁻¹. The heat of fusion of ice is 6.02 kJ/mol. Use 0.917 and 0.998 g/cm³ for the densities of ice and H_2O(l), respectively.

21. Explain why dry ice can be used to maintain frozen foods much more effectively than ordinary ice.

22. Why is the triple point of water (ice I–liquid–vapor) a better fixed point for establishing a thermometric scale than either the melting point of ice or the boiling point of water?

23. Is it likely that any of the following will occur naturally at or near the surface of the earth, anywhere on earth? Explain. **(a)** CO_2(s); **(b)** CH_4(l); **(c)** SO_2(g); **(d)** I_2(l); **(e)** O_2(l). (*Hint:* Use the appropriate phase diagrams and data from Table 11-2.)

24. Describe what happens to the following samples in a device like that pictured in Figure 11-13. Be as specific as you can about the temperatures and pressures at which changes occur.

    (a) A sample of H_2O is heated from −20°C to 200°C at a constant pressure of 600 mmHg.

    (b) The pressure on a sample of I_2 is increased from 90 mmHg to 100 atm at a constant temperature of 114.5°C.

    (c) A sample of CO_2 at 35°C is cooled to −100°C at a constant pressure of 50 atm. (*Hint:* Refer also to Table 11-2.)

25. Trace the phase changes that occur as a sample of H_2O(g), originally at 1.00 mmHg and −0.10°C, is compressed at constant temperature until the pressure reaches 100 atm.

26. Use the phase diagram in Figure 11-12 (and other information about water) to predict
  (a) the state in which water exists at 600 K and 225 atm;
  (b) whether the following transition at 3500 atm is endothermic or exothermic: ice V → ice II;
  (c) whether the density of ice III at −22.0°C and 2045 atm is greater or less than 1.00 g/cm³.

Van der Waals forces

27. One of the substances is out of order in the following list based on *increasing* boiling point. Identify it and put it in the proper place: N_2; O_3; F_2; Ar; Cl_2. Explain your reasoning.

28. Normal octane and isooctane are two components of gasoline having the formula C_8H_{18}. Normal octane has all eight C atoms in a straight chain. The chain length in isooctane is five, with three C atoms attached as side chains. Which of the two liquids has the higher boiling point? Explain.

29. A handbook lists the following normal boiling points for a series of normal (straight-chain) alkanes: propane, C_3H_8, −42.1°C; butane, C_4H_{10}, −0.5°C; pentane, C_5H_{12}, 36.1°C; hexane, C_6H_{14}, 68.7°C; heptane, C_7H_{16}, 98.4°C; octane, C_8H_{18}, 125.6°C. Estimate the normal boiling point of the normal alkane nonane, C_9H_{20}.

30. When another atom or group of atoms is substituted for one of the H atoms in benzene, C_6H_6, the boiling point changes. Explain the order of the following boiling points: C_6H_6, 80°C; C_6H_5Cl, 132°C; C_6H_5Br, 156°C; C_6H_5OH, 182°C.

Hydrogen bonding

31. Describe the conditions necessary for the formation of a hydrogen bond and how this bond differs from other intermolecular forces.

32. One of the following substances is a liquid at room temperature, whereas all the others are gaseous. Which do you think is the liquid? Explain. CH_3OH; C_3H_8; N_2; CO.

33. For each of the following substances describe the importance of London type forces, dipole–dipole interactions, and hydrogen bonds: (a) HCl; (b) Br_2; (c) ICl; (d) HF; (e) CH_4.

34. Figure 11-19 shows that one H_2O molecule can be bonded to four others through hydrogen bonds. What would you expect the situation to be with NH_3?

35. If water were a normal liquid, what would you expect to find for its (a) boiling point; (b) freezing point; (c) temperature of maximum density of the liquid; (d) relative densities of the solid and liquid states?

*__36.__ In some cases a hydrogen bond can be formed *within* a single molecule. This is called an *intra*molecular hydrogen bond. Do you think this type of hydrogen bonding is an important factor in (a) C_2H_6; (b) H_3CCH_2OH; (c) H_3CCOOH; (d) *o*-phthalic acid, $C_6H_4(COOH)_2$? (The structure of *o*-phthalic acid is pictured in Section 26-7.)

Network covalent solids

37. The text states that all electrons in diamond are localized, but that in graphite certain electrons are delocalized.
  (a) Explain the meaning of *localized* and *delocalized*.
  (b) Which electrons in graphite are delocalized?

38. Silicon carbide, SiC, crystallizes in a form similar to diamond, whereas the compound boron nitride, BN, crystallizes in a form similar to graphite.
  (a) Sketch the SiC structure as in Figure 11-21.
  (b) Propose a bonding scheme for BN.
  (c) BN can be obtained in a diamondlike form under high pressure, but SiC cannot be obtained in a graphitelike form. Why do you suppose that this is so?

39. Based on data presented in the text, would you expect diamond or graphite to have the greater density? Explain.

40. Diamond is often used as a cutting medium in glass cutters. What property of diamond makes this possible? Could graphite function as well?

Ionic properties and bonding

41. The melting points of NaF, NaCl, NaBr, and NaI are 988, 801, 755, and 651°C, respectively. Are these data consistent with ideas developed in Section 11-11? Explain.

42. Which compound in each of the following pairs would you expect to be the more water soluble? (a) MgF_2 or BaF_2; (b) MgF_2 or $MgCl_2$.

*__43.__ Use Coulomb's law to verify the conclusion concerning the relative strengths of the attractive forces in the ion pairs, Na^+Cl^- and $Mg^{2+}O^{2-}$, presented in Figure 11-23.

Born–Haber cycle

44. The enthalpy of formation of CsCl is −442.8 kJ/mol, and the enthalpy of sublimation of Cs(s) is 77.6 kJ/mol. Use these data, together with other data from the text, to calculate the lattice energy of CsCl.

*__45.__ The heat of formation of NaI(s) is −288 kJ/mol. Use this value, together with other data in the text, to calculate the lattice energy of NaI(s). (*Hint:* An extra step, in addition to those shown in Figure 11-24, and data from Appendix D are required.)

Crystal structures

46. Define what is meant by (a) closest packing of spheres; (b) tetrahedral holes; (c) octahedral holes.

47. Explain why there are two arrangements for the closest packing of spheres rather than a single one.

48. In a manner similar to Example 11-10, derive a value of Avogadro's number based on the fact that the crystal structure of Al is fcc; that the atomic radius of Al is 143.1 pm; and that the density of Al is 2.6984 g/cm³. (*Hint:* See also Review Problem 11.)

*49. Use the analyses of a *bcc* structure in Example 11-9 and the *fcc* structure in Review Problem 11 to determine the % voids in their respective packing-of-spheres arrangements, i.e., 31.98% for *bcc* and 25.96% for *fcc*. (*Hint:* What is the volume of the unit cell? How many atoms are present in the cell? What is their volume? Use the general case of the packing of spheres, without reference to a particular metal.)

*50. Magnesium crystallizes in the hcp arrangement shown in Figure 11-29. The dimensions of the unit cell are height, 520 pm

(5.20 Å); length of an edge, 320 pm (3.20 Å). Calculate the density of solid magnesium and compare with the measured value of 1.738 g/cm³.

Ionic crystal structures

51. Show that the unit cells for CaF_2 and TiO_2 shown in Figure 11-37 are consistent with their formulas.

52. Extend Example 11-11 by calculating the (a) volume of a unit cell; (b) mass of a unit cell; (c) density of NaCl.

Additional Exercises

1. The normal boiling point of acetone (a common solvent) is 56.5°C, and its heat of vaporization is 30.3 kJ/mol. What is the vapor pressure of acetone at 20.0°C?

*2. 100.0 L of $H_2O(g)$ at 100°C and 1 atm is passed into 1.00 kg of $H_2O(l)$. The $H_2O(l)$ is maintained in a thermally insulated container (such as a vacuum bottle or Dewar flask) and is initially at 18.0°C. What will be (a) the final mass of liquid; (b) the final temperature? $\Delta \overline{H}_{vap}$ of water at 100°C is 40.6 kJ/mol.

3. A 150.0 cm³ sample of $N_2(g)$ at 25.0°C and 750 mmHg, is passed through $C_6H_6(l)$ until the gas becomes saturated with $C_6H_6(g)$. The new volume of the gas is 172 cm³ at a total pressure of 750 mmHg. What is the vapor pressure of C_6H_6 at 25.0°C?

4. 525 cm³ of Hg(l) at 20°C is added to a large quantity of $N_2(l)$ kept at its boiling point in a thermally insulated container. What mass of $N_2(l)$ is vaporized as the Hg is brought to the temperature of the $N_2(l)$? For the specific heat of Hg(l) from 20°C to −39°C use 0.033 cal g⁻¹ °C⁻¹; and for Hg(s) from −39°C to −196°C, 0.030 cal g⁻¹ °C⁻¹. The density of Hg(l) is 13.6 g/cm³; the melting point of Hg is −39°C; and its heat of fusion is 2.30 kJ/mol. The boiling point of $N_2(l)$ is −196°C, and its $\Delta \overline{H}_{vap} = 5.58$ kJ/mol.

*5. The vapor pressure of $NH_3(l)$ can be expressed as

$$\log P \text{ (mmHg)} = 9.95028 - 0.003863T - \frac{1473.17}{T}$$

What is the normal boiling point of $NH_3(l)$?

6. The following data are given for CCl_4: normal melting point, −23°C; normal boiling point, 77°C; density of liquid, 1.59 g/cm³; heat of fusion, 30.5 kJ/mol; vapor pressure at 25°C, 110 mmHg.
   (a) How much heat must be absorbed to convert 10.0 g of solid CCl_4 to liquid at −23°C?
   *(b) How much heat is required to vaporize 20.0 L of $CCl_4(l)$ at its normal boiling point? (*Hint:* What is $\Delta \overline{H}_{vap}$?)
   (c) What is the volume occupied by 1.00 mol of the saturated vapor of CCl_4 at 77°C?
   (d) What phases—solid, liquid, and/or vapor—are present if 3.5 g CCl_4 is kept in an 8.21-L volume at 25°C?

*7. A cylinder containing 151 lb Cl_2 has a diameter of 10 in. and a height of 45 in. The gas pressure is 100 psi (1 atm = 14.7 psi) at 20°C. Cl_2 melts at −103°C, boils at −35°C, and has its critical point at 144°C and 76 atm. In what state(s) of matter does the Cl_2 exist in the cylinder?

8. A bottle of pop was placed in the freezer compartment of a refrigerator to cool quickly. When the bottle was taken out, the pop was still liquid; but when the cap was removed the pop froze instantaneously. Explain why this happened.

9. In their crystal structures, diamond, graphite, and ice all feature a hexagonal arrangement of atoms. How can this be reconciled with the facts that diamond is extremely hard and has a high melting point, graphite flakes easily and is a good electrical conductor, and ice has both a low density and low melting point.

10. Following are the values of enthalpies of vaporization of some typical liquids at their normal boiling points: H_2, 0.92 kJ/mol; CH_4, 8.16 kJ/mol; C_6H_6, 31.0 kJ/mol; H_2O, 40.7 kJ/mol. Explain the differences among these values.

11. In the manner of Figure 11-9, sketch a cooling curve to show how temperature varies with time when a sample of water is cooled from point *R* to point *P* in Figure 11-12. In a heating curve, temperature is plotted as a function of time while a sample is being heated. Sketch a heating curve for a sample of water from point *P* to *R*.

*12. In a similar fashion to Exercise 52, calculate the density of CsCl.

13. Place each of the following substances in the appropriate category in Table 11-5 and state your reason for each placement. (a) Si; (b) CCl_4; (c) $CaCl_2$; (d) Ag; (e) HCl.

*14. Because solid *p*-dichlorobenzene, $C_6H_4Cl_2$, sublimes rather easily, it has been used as a moth repellant. From the data given below, estimate the sublimation pressure of $C_6H_4Cl_2(s)$ at 25°C. For $C_6H_4Cl_2$: m.p. = 53.1°C; vapor pressure of $C_6H_4Cl_2(l)$ at 54.8°C is 10.0 mmHg; $\Delta \overline{H}_{fusion} = 17.88$ kJ/mol; $\Delta \overline{H}_{vap} = 72.22$ kJ/mol. [*Hint:* An equation like (11.3), with $\Delta \overline{H}_{subl}$ substituting for $\Delta \overline{H}_{vap}$, can be used for the sublimation curve.]

*15. In acetic acid vapor some molecules exist as monomers and some as dimers (see Figure 11-20). If the density of the vapor at 350 K and 1 atm is 3.23 g/L, what percent of the molecules must exist as dimers? Would you expect this percent to increase or decrease with increasing temperature?

*16. The second electron affinity of oxygen cannot be measured directly, that is,

$$O^-(g) + e^- \longrightarrow O^{2-}(g) \qquad EA_2 = ?$$

The O^{2-} ion can exist in the solid state, however, where the high energy requirement for its formation is offset by the large lattice energies of ionic oxides.

(a) Show that EA_2 can be calculated from the enthalpy of formation and lattice energy of MgO(s), enthalpy of sublimation of Mg(s), ionization energies of Mg, bond energy of O_2, and EA_1 for O(g).

(b) The enthalpy of sublimation of Mg(s) is 150 kJ/mol, the bond energy of $O_2(g)$ is 497 kJ/mol and the lattice energy of MgO is -3925 kJ/mol. Combine these data with other values in the text to calculate a value of EA_2 for oxygen.

Self-Test Questions

For questions 1 through 5 select the single item that best completes each statement.

1. Of the following, the one with the highest normal boiling point is (a) $O_2(l)$; (b) Ne(l); (c) $SO_3(l)$; (d) $Br_2(l)$.

2. The best electrical conductor of the following is (a) $CO_2(s)$; (b) Si(s); (c) NaCl(s); (d) $Br_2(l)$.

3. Of the compounds HF, CH_4, CH_3OH, and N_2H_4, hydrogen bonding as an important intermolecular force is expected in (a) none of these; (b) two of these; (c) all but one of these; (d) all of these.

4. Of the following properties, the magnitude of one must always increase with temperature. That property is (a) vapor pressure; (b) density; (c) $\Delta \overline{H}_{vap}$; (d) surface tension.

5. The form of carbon known as graphite (a) is harder than diamond; (b) contains a higher percentage of carbon than diamond; (c) is a better electrical conductor than diamond; (d) has equal carbon-to-carbon bond distances in all directions.

6. A television commercial claims that a product makes water "wetter." Is there any basis to this claim? Explain.

7. Which of the following factors would you expect to affect the vapor pressure of a liquid? Explain.
    (a) intermolecular forces in the liquid;
    (b) volume of liquid in the liquid–vapor equilibrium;
    (c) volume of vapor in the liquid–vapor equilibrium;
    (d) the size of the container in which the liquid–vapor equilibrium is established;
    (e) the temperature of the liquid.

8. 10.0 g of steam at 100°C and 100.0 g of ice at 0°C are added to 100.0 g of liquid water at 20°C in an insulated container. (For water: heat of fusion = 6.02 kJ/mol, heat of vaporization = 40.7 kJ/mol.) Which of these conditions will result? (a) The mixture will start to boil; (b) all of the liquid water will freeze; (c) all of the ice will melt; (d) a mixture of ice and liquid water will remain.

9. A dramatic lecture demonstration consists of continuously evacuating the water vapor produced from an open container of water in a bell jar. If the vacuum pump is sufficiently powerful, the water can be made to freeze. Indicate the principles involved in achieving this effect.

10. Explain what is wrong with the following approach used by a student to calculate the vapor pressure of a liquid at 50°C from a measured vapor pressure of 80 mmHg at 20°C.

$$\text{v.p. (at } 50°C) = 80 \text{ mmHg} \times \frac{323 \text{ K}}{293 \text{ K}} = 88 \text{ mmHg}$$

11. Argon, copper, sodium chloride, and carbon dioxide all crystallize in a fcc structure. How can this be and still have such a difference in their physical properties?

12. Summarize the characteristics of these four types of crystalline materials—ionic, molecular, network covalent, and metallic—in terms of intermolecular forces, structural particles of the crystal, and physical properties, such as melting point, boiling point, and electrical conductivity.

12 Solutions

The discussion of intermolecular forces in Chapter 11 was limited to *single* substances in the three states of matter. But intermolecular forces can exist among unlike molecules as well, and depending on the relative strengths of these forces a **heterogeneous** or **homogeneous** mixture results. The main body of this chapter is concerned with the properties of homogeneous mixtures or **solutions.**

12-1 Homogeneous and Heterogeneous Mixtures

A sample of matter having a fixed composition and uniform properties is called a **phase.** For example, water at 25°C and 1 atm pressure exists as a single liquid phase. All properties are uniform throughout this liquid phase. If a *small* quantity of salt (NaCl) is added to the water, the salt dissolves and the sample remains as a single liquid phase. The composition and properties of this new liquid phase, a salt solution, are different from those of pure water. This solution is a mixture, because it contains two substances. It is **homogeneous,** because its properties are uniform throughout the liquid. If some sand (SiO_2) is added to water, the sand settles to the bottom of the liquid as an undissolved solid. This water–sand mixture is a *two*-phase mixture (liquid + solid) and it is **heterogeneous.** Its composition and properties are not uniform. The composition and properties of the liquid phase are those of pure water; the composition and properties of the solid phase are those of sand.

Enthalpy of Solution. In a manner we have used before, this time for the formation of a solution, we can write for the enthalpy change

$\Delta H_{\text{soln}} = H_{\text{solution}} - H_{\text{pure components}}$

The enthalpy change is referred to as the enthalpy or heat of solution. The enthalpy change calculated in Example 6-5, for example, was the heat of solution of NH_4NO_3 in water.

Because a solution is a mixture of molecules (atoms or ions, in some cases), on average, solvent molecules are somewhat farther apart in a solution than in the pure solvent. The same is true of solute molecules. This suggests that solution formation can be thought of in terms of this hypothetical process: First, the distance between molecules of solvent is increased to the average distance that will prevail in the solution. This step requires that energy be absorbed to overcome cohesive intermolecular forces. This step is accompanied by an increase in enthalpy; it is endothermic. In a second endothermic step,

FIGURE 12-1
Enthalpy diagram for
solution formation.

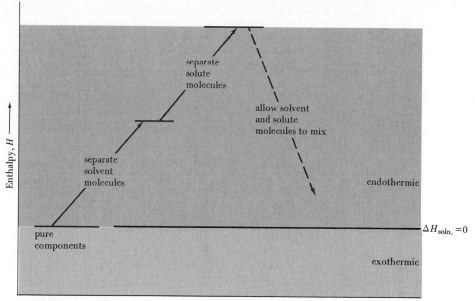

Depending on whether the broken arrow ends above, below, or on the line, the solution process is endothermic, exothermic, or has $\Delta H_{\text{soln}} = 0$, respectively.

a similar separation of solute molecules is achieved. The third, and final, step is to allow the solvent and solute molecules to mix. Intermolecular forces of attraction between the unlike molecules cause a release of energy; enthalpy decreases in this step. Depending on the comparative values of the enthalpy changes in the three steps just outlined, the overall enthalpy change may be positive (endothermic) or negative (exothermic) or in some cases, discussed below, $\Delta H_{\text{soln}} = 0$. As in earlier discussions of enthalpy, we can represent these enthalpy changes through an enthalpy diagram (see Figure 12-1).

Intermolecular Forces in Mixtures. To continue the discussion of the enthalpy of solution, we need to say more about intermolecular forces in mixtures. For the substances A and B let us represent the magnitude of the intermolecular forces between like molecules as $A \leftrightarrow A$ and $B \leftrightarrow B$ and between unlike molecules as $A \leftrightarrow B$ (see Figure 12-2). Four possibilities are described below. In this description comparative forces are denoted by: $\simeq$ approximately equal to, $>$ greater than, $<$ smaller than, $\ll$ much smaller than.

FIGURE 12-2
Representation of
intermolecular forces in
a solution.

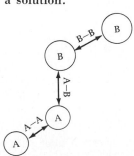

1. $A \leftrightarrow B \simeq A \leftrightarrow A \simeq B \leftrightarrow B$. If intermolecular forces between like and unlike molecules are of about the same strength, a random intermingling of molecules occurs and a homogeneous mixture or solution results. Properties of solutions of this type can generally be predicted from a knowledge of properties of the pure components. These solutions are said to be **ideal.** Specifically, the volume of an ideal solution is the sum of the volumes of the individual components ($\Delta V = 0$). The energies of interaction between like and unlike molecules are the same. There is no enthalpy change or heat effect on mixing the components ($\Delta H_{\text{soln}} = 0$). That is, the negative enthalpy change in the third step of Figure 12-1 is equal to the sum of the positive enthalpy changes for the first two steps. Most mixtures of liquid hydrocarbons (e.g., benzene–toluene) belong to this category.

2. $A \leftrightarrow B > A \leftrightarrow A, B \leftrightarrow B$. If intermolecular forces between unlike molecules exceed those between like molecules, a solution also forms. However, the properties of solutions of this type cannot be predicted from those of the pure components. These solutions are **nonideal.** The energy released in interactions between unlike molecules

FIGURE 12-3
Intermolecular force
between unlike
molecules leading to
nonideal solution.

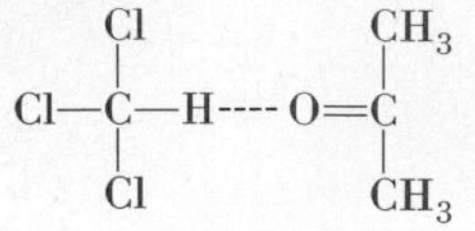

$$Cl-\underset{\underset{Cl}{|}}{\overset{\overset{Cl}{|}}{C}}-H\text{----}O=\underset{\underset{CH_3}{|}}{\overset{\overset{CH_3}{|}}{C}}$$

Weak hydrogen bonding
occurs between the H
atom of a $CHCl_3$ (chloroform) molecule and the O
atom of a $(CH_3)_2CO$ (acetone) molecule. This in-
termolecular force be-
tween unlike molecules
causes acetone–
chloroform solutions to be
nonideal.

exceeds that required to separate like molecules. Energy is given off to the surroundings
and the solution process is *exothermic* ($\Delta H_{soln} < 0$). Solutions of acetone and chloroform
are of this type (see Figure 12-3).

3. A $\leftrightarrow$ B < A $\leftrightarrow$ A, B $\leftrightarrow$ B. If forces of attraction between unlike molecules are
smaller than between like molecules, complete mixing may occur, but these solutions are
nonideal. The solution process is endothermic ($\Delta H_{soln} > 0$). Solutions of ethanol
(C_2H_5OH) and benzene (C_6H_6) are of this type.

4. A $\leftrightarrow$ B $\ll$ A $\leftrightarrow$ A, B $\leftrightarrow$ B. If intermolecular forces between unlike molecules are
much smaller than those between like molecules, the components remain segregated into a
heterogeneous mixture. For example, in a mixture of water and octane (a constituent of
gasoline) strong hydrogen bonds hold water molecules together in clusters. The nonpolar
octane molecules cannot exert strong attractive forces on polar water molecules, and the
two types of molecules do not mix.

As a further example of the process of solution formation, consider the dissolving of
an ionic solid in water. In Figure 12-4 water dipoles are pictured as clustering around ions
at the surface of a crystal. The negative ends of these dipoles are oriented toward the
positive ions, and the positive ends of dipoles, toward negative ions. If these ion–dipole
forces are strong enough to overcome the interionic attractions within the crystal, dissolving will occur. Moreover, these ion–dipole forces persist in the solution. An ion surrounded by a cluster of water molecules is said to be *hydrated*. Energy is released when
ions become hydrated. The greater the hydration energy in relation to the energy invest-
ment to separate ions from a crystal, the more energetically favorable is the solution
process.

Types of Solutions.   To repeat and extend some of the solution terminology introduced in
Chapter 4, the component present in greatest quantity, or that component which determines the state of matter in which the solution exists, is called the **solvent.** The compo-
nent(s) present in lesser quantity is (are) called the **solute(s).** A solution in which water is
the solvent is called an **aqueous** solution. A solution containing a relatively large quantity
of solute is said to be **concentrated.** If the quantity of solute is small, the solution is

FIGURE 12-4
Dissolving of an ionic
crystal in water.

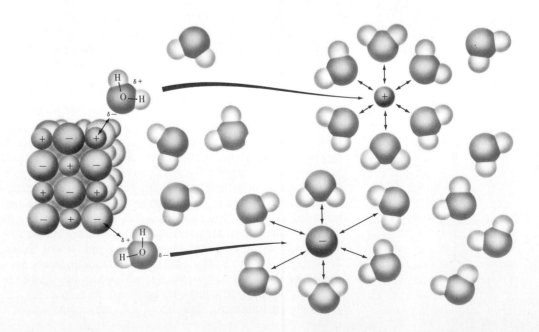

dilute. The term solution generally calls to mind a liquid solvent with either another liquid, a solid, or a gas as a solute. Three examples of solutions existing in the liquid state are

Gasoline: a mixture of a number of liquid hydrocarbons.
Seawater: an aqueous solution of sodium chloride and several other ionic solids.
Carbonated water: an aqueous solution of $CO_2(g)$.

Solutions may also exist in the gaseous and solid states. Since molecules of a gas are separated by great distances, molecules in a mixture of gases diffuse among one another randomly. *All gaseous mixtures are solutions*. The best known example of a gaseous solution is air, which consists of N_2, O_2, Ar and traces of other gases.

In a solid solution the solvent is a *solid* substance. The ability to form solid solutions is particularly common among metals, and such solid solutions are called **alloys.** In certain solid solutions, atoms of solute substitute for some of the solvent atoms in the crystalline lattice. These are called substitutional solid solutions and require that the solute and solvent atoms be of approximately the same size. Thus, copper (128 pm) and nickel (125 pm) form solid solutions in all proportions. In other solid solutions the solute atoms may take up positions in the interstices or holes in the solvent lattice. The formation of interstitial solid solutions requires that solute atoms be small enough to fit into the holes among the solvent atoms. Among the elements that often meet these requirements are carbon and hydrogen. Ordinary steels are alloys of iron and carbon.

> Not all mixtures of two or more metals (alloys) are solid solutions; some are heterogeneous, as is the lead-tin alloy known as solder. Also, in some mixtures a new intermetallic compound is formed.

12-2 Solution Concentration

Merely to state the components present in a solution does not describe the solution fully. The additional information needed is the *concentration* of the solution. There are many ways to describe solution concentration, but all of them must express the quantity of solute present in a given quantity of solvent (or solution). That is, in every system of concentration we must establish the following points:

1. The units used to measure the solute.
2. Whether the second quantity measured is the solvent or the total solution.
3. The units used to measure this second quantity.

Percent by Mass, Percent by Volume, and Related Quantities.   The statement, "5.00 g NaCl per 100.0 g of aqueous solution," has the following meaning: A solution is prepared by weighing out 5.00 g NaCl and dissolving it in 95.0 g H_2O, that is, a mass of water sufficient to produce 100.0 g of solution. The solution can be termed a 5.00% NaCl solution, *by mass*. This concentration unit, in which quantities of solute and solution are both measured by mass, is also referred to as **mass/mass percent** or as **% (mass/mass).**

When a liquid solute is used it is often convenient to prepare a solution on a volume basis, such as dissolving 5.00 mL of ethanol in a sufficient volume of water to produce 100.0 mL of solution. This ethanol–water solution is 5.00% ethanol, *by volume;* or, because both quantities are measured in volume units, the term **volume/volume percent** or **% (vol/vol)** can be used.

Still another possibility is that mass and volume units are mixed. For example, if the solute is measured by mass and the quantity of solution by volume, the term **mass/volume percent** or **% (mass/vol)** can be used. If a solution concentration is given on a percent basis, but with no further specification as to whether it is mass/mass, vol/vol, or mass/vol, one should assume that percent by mass is intended.

Solution concentrations expressed as percents have no theoretical significance, but they are commonly encountered and it is necessary to be familiar with them. Mass/volume percent is widely used in biological and medical laboratories and mass/mass percent is the solution concentration most often used in industrial chemistry. Still another concentration unit, used in clinical chemistry, for example, is mass per unit volume, e.g., g solute/cm^3 solution.

Molar Concentration (Molarity).   In Chapter 4 we noted that

1. The stoichiometry of chemical reactions is based on relative *numbers* of reacting atoms, ions, or molecules.
2. Many chemical reactions are conducted in solution.

For these reasons we introduced at that time a solution concentration unit based on *numbers of solute particles*—the molar concentration—which we presented in two ways.

$$\text{molar concentration } (M) = \frac{\text{no. mol solute}}{\text{no. L soln}} = \frac{\text{no. mmol solute}}{\text{no. mL soln}} \tag{12.1}$$

Molal Concentration (Molality).   Molarity is a function of temperature. This is because the quantity of solution is based on volume, and volume is a function of temperature. Suppose that a solution is prepared at 20°C using a volumetric flask calibrated at this same temperature but then the solution is used at 25°C. As the temperature increases from 20°C to 25°C, the amount of solute remains constant but the solution volume increases slightly. The number of moles of solute per liter (i.e., the molarity) *decreases* slightly.

Another situation in which molality concentration is useful is one in which the solvent, because it is a solid at room temperature, can be measured only by its mass not its volume.

For a variety of applications it is necessary to have a solution unit that is independent of temperature. The obvious unit is one in which *both* quantities, solute and solvent, are stated by mass. The mass of a substance is independent of temperature. A particularly useful unit is molality, in which the amount of solute is given in moles and the quantity of solvent (not solution) in kilograms. The units of molality then are *moles solute per kilogram solvent*. A solution in which 1 mol of NaCl is dissolved in 1000 g of water is described as a 1 molal solution and designated by the symbol 1 *m* NaCl. Molal concentration is defined by equation (12.2).

$$\text{molal concentration } (m) = \frac{\text{number of moles solute}}{\text{number of kilograms solvent}} \tag{12.2}$$

Mole Fraction.   The concentration units molality and molarity have the amount of solute expressed on a number basis (in moles), but the quantity of solvent or solution is on a mass or volume basis. To relate physical properties of solutions to solution concentration, it is sometimes necessary to use a concentration unit in which all solution components are described on a mole basis. This can be done through the mole fraction. The mole fraction of component i, designated χ_i, is the fraction of all the molecules in a solution that are of type i. The mole fraction of component j is χ_j, and so on. The sum of the mole fractions of all the solution components is 1.

The mole fraction of a solution component is defined by equation (12.3).

$$\chi_i = \frac{\text{moles of component, } i}{\text{total moles of all soln components}} \tag{12.3}$$

Another concentration unit related to the mole fraction is the **mole percent.** The mole percent of a solution component is the percent of all the molecules in a solution that are of a given type. Mole percents are mole fractions multiplied by 100.

12-3 Some Illustrative Examples Based on Solution Concentrations

Example 12-1   An ethanol–water solution is prepared by dissolving 10.00 cm³ of ethanol, C_2H_5OH ($d = 0.789$ g/cm³), in a sufficient volume of water to produce 100.0 cm³ of a solution with a density of 0.982 g/cm³ (see Figure 12-5). What is the concentration of this solution, expressed as **(a)** percent by volume; **(b)** percent by mass; **(c)** percent (mass/vol); **(d)** molarity; **(e)** molality; **(f)** mole fraction; and **(g)** mole percent of ethanol?

Solution
(a) Percent ethanol, by volume:

$$\% \text{ ethanol, by volume} = \frac{10.00 \text{ cm}^3 \text{ ethanol}}{100.0 \text{ cm}^3 \text{ soln}} \times 100 = 10.00\%$$

(b) Percent ethanol, by mass:

$$\text{no. g ethanol} = 10.00 \text{ cm}^3 \text{ ethanol} \times \frac{0.789 \text{ g ethanol}}{1.00 \text{ cm}^3 \text{ ethanol}} = 7.89 \text{ g ethanol}$$

$$\text{no. g soln} = 100.0 \text{ cm}^3 \text{ soln} \times \frac{0.982 \text{ g soln}}{1.00 \text{ cm}^3 \text{ soln}} = 98.2 \text{ g soln}$$

$$\% \text{ ethanol, by mass} = \frac{7.89 \text{ g ethanol}}{98.2 \text{ g soln}} \times 100 = 8.03\%$$

(c) Mass/volume percent ethanol:

$$\% \text{ ethanol (mass/vol)} = \frac{7.89 \text{ g ethanol}}{100.0 \text{ cm}^3 \text{ soln}} \times 100 = 7.89\%$$

(d) Molarity of ethanol:
To establish the various forms of percentage composition in parts (a), (b), and (c) did not require knowing the formula of ethanol. To express concentration on a molar basis does require that a mole of ethanol be identified.

$$\text{no. mol } C_2H_5OH = 10.00 \text{ cm}^3 \text{ ethanol} \times \frac{0.789 \text{ g ethanol}}{1.00 \text{ cm}^3 \text{ ethanol}} \times \frac{1 \text{ mol } C_2H_5OH}{46.1 \text{ g } C_2H_5OH}$$
$$= 0.171 \text{ mol } C_2H_5OH$$

$$\text{no. L soln} = 100.0 \text{ cm}^3 \text{ soln} \times \frac{1 \text{ L soln}}{1000 \text{ cm}^3 \text{ soln}} = 0.1000 \text{ L soln}$$

$$\text{molarity} = \frac{0.171 \text{ mol } C_2H_5OH}{0.1000 \text{ L soln}} = 1.71 \text{ } M \text{ } C_2H_5OH$$

(e) Molality of ethanol:
The key to determining molal concentration often is in establishing the mass of solvent.

mass soln = 98.2 g soln [see part (b)]

mass ethanol = 7.89 g ethanol [see part (b)]

mass H_2O = mass soln − mass ethanol = 98.2 g − 7.89 g = 90.3 g H_2O

$$\text{no. kg } H_2O = 90.3 \text{ g } H_2O \times \frac{1 \text{ kg } H_2O}{1000 \text{ g } H_2O} = 0.0903 \text{ kg } H_2O$$

$$\text{molality} = \frac{0.171 \text{ mol } C_2H_5OH}{0.0903 \text{ kg } H_2O} = 1.89 \text{ } m \text{ } C_2H_5OH$$

FIGURE 12-5
Preparation of an ethanol–water solution—Example 12-1 illustrated.

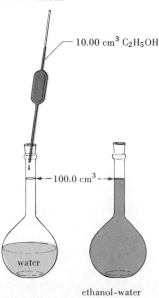

— 10.00 cm³ C_2H_5OH

—100.0 cm³—

water

ethanol–water solution: $d = 0.982$ g/cm³

FIGURE 12-6
Formation of a
saturated solution.

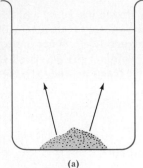

(a)

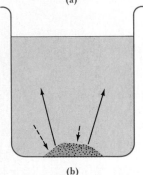

(b)

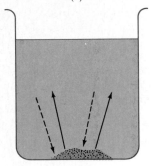

(c)

The lengths of the arrows
represent the rate of dissolving (⟶) and the
rate of precipitation
(⟵---).
(a) When solute and solvent are first brought together only the process of
dissolving occurs.
(b) After a time, though
dissolving continues at the
same rate, the rate of precipitation becomes significant.
(c) When dissolving and
precipitation occur at the
same rate, the solution is
saturated. There is no further change in solution
concentration with time.

(f) Mole fraction of ethanol:

The total moles of ethanol in the solution has already been calculated [see part (d)]. Now we must calculate the total moles of water present, based on the mass of water determined in part (e).

$$\text{no. mol } H_2O = 90.3 \text{ g } H_2O \times \frac{1 \text{ mol } H_2O}{18.0 \text{ g } H_2O} = 5.02 \text{ mol } H_2O$$

$$\chi_{C_2H_5OH} = \frac{0.171 \text{ mol } C_2H_5OH}{0.171 \text{ mol } C_2H_5OH + 5.02 \text{ mol } H_2O} = \frac{0.171}{5.19} = 0.0329$$

(g) Mole percent ethanol:

$$\text{mole percent } C_2H_5OH = 100\chi_{C_2H_5OH} = 100 \times 0.0329 = 3.29\%$$

SIMILAR EXAMPLES: Review Problems 2, 4, 5, 6; Exercises 8, 10, 11, 12.

More usual than the stepwise series of calculations illustrated in Example 12-1 is the need just to convert from one concentration unit to another. The general approach in these conversions is to identify (a) the quantities needed for the final concentration unit, (b) the quantities available through the original concentration unit, (c) any additional factors that may be needed, and (d) a useful quantity of the solution on which to base the calculation.

Example 12-2 Laboratory ammonia is 14.8 *M* NH_3(aq) and has a density of 0.8980 g/cm^3. What is the mole fraction of NH_3 in this solution?

Solution. To express solution concentration in mole fractions we need to determine the numbers of moles of solute (NH_3) and of solvent (H_2O). If we base our calculation on 1.000 L (1000 cm^3) of solution, the number of moles of solute is simply 14.8 mol NH_3. To determine the number of moles of H_2O in the solution we can proceed as follows.

$$\text{mass of solution} = 1000 \text{ cm}^3 \text{ soln} \times \frac{0.8980 \text{ g soln}}{\text{cm}^3 \text{ soln}} = 898.0 \text{ g soln}$$

$$\text{mass of } NH_3 = 14.8 \text{ mol } NH_3 \times \frac{17.0 \text{ g } NH_3}{1 \text{ mol } NH_3} = 252 \text{ g } NH_3$$

$$\text{mass of } H_2O = 898.0 \text{ g soln} - 252 \text{ g } NH_3 = 646 \text{ g } H_2O$$

$$\text{no. mol } H_2O = 646 \text{ g } H_2O \times \frac{1 \text{ mol } H_2O}{18.0 \text{ g } H_2O} = 35.9 \text{ mol } H_2O$$

$$\chi_{NH_3} = \frac{14.8 \text{ mol } NH_3}{14.8 \text{ mol } NH_3 + 35.9 \text{ mol } H_2O} = 0.292$$

SIMILAR EXAMPLE: Exercise 14.

12-4 Solubility Equilibrium

When a sufficiently large quantity of solute is maintained in contact with a limited quantity of solvent, dissolving occurs continuously. After a time, however, the reverse process
becomes increasingly important. This is the return of dissolved species (atoms, ions or
molecules) to the undissolved state, a process called **precipitation.** When dissolving and
precipitation occur at the same rate, the quantity of dissolved solute present in a given
quantity of solvent remains constant with time. The process is one of dynamic equilibrium
and the solution is said to be **saturated.** The formation of a saturated solution is suggested
by Figure 12-6. The concentration of the saturated solution is referred to as the **solubility**

FIGURE 12-7
Water solubility of
several salts as a
function of temperature.

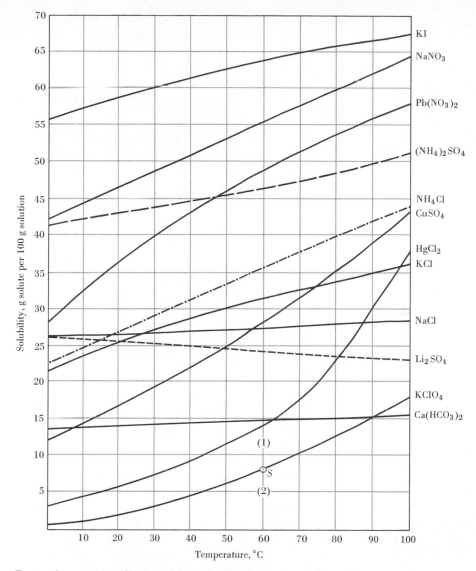

For each curve, as illustrated here for $KClO_4$, regions above the curves (1) represent
supersaturated solutions; points on the curves. *S*, saturated solutions; and regions below
the curves (2), unsaturated solutions.

In some solutions the
solute and solvent are
miscible in all propor-
tions. In these cases the
solution never becomes
saturated. Such is the
case with ethanol–water
solutions, for example.

of the solute in the given solvent. Solubility is generally a function of temperature, as
suggested by the **solubility curves** in Figure 12-7.

Suppose that a saturated solution is prepared at one temperature and then the temperature is changed to a value at which the solubility is lower (generally this is a lower
temperature). The usual result is that the excess solute precipitates from the solution. In
some cases, however, all the solute may remain in solution. Because the quantity of solute
in these cases is greater than in a normal saturated solution at the given temperature, such
a solution is said to be **supersaturated.** If a few crystals of solute are added to a supersaturated solution, the excess solute will usually precipitate. A solution containing less solute
than required for saturation is **unsaturated.** These terms are related to the solubility curve
for $KClO_4$ in Figure 12-7.

With but a few exceptions the water solubilities of ionic compounds increase with
temperature. Predicting those few exceptions is not so easily done. A useful generaliza-

tion for our purposes is that about 95% of all ionic compounds exhibit increased water solubility with temperature. Exceptions are found primarily among compounds containing the anions SO_3^{2-}, SO_4^{2-}, SeO_4^{2-}, AsO_4^{3-}, and PO_4^{3-}.

In our introduction of Le Châtelier's principle in Section 11-5, we learned that adding heat to a system at equilibrium (as in raising or attempting to raise the temperature) stimulates the heat-absorbing or endothermic process. If a solute has an endothermic enthalpy (heat) of solution, its solubility increases with temperature. Conversely, a solute with an exothermic enthalpy (heat) of solution exhibits a decreased solubility with temperature (i.e., the precipitation process is endothermic). There is a potential problem in applying this generalization, however, of which we need to be aware. The enthalpy of solution must be based on dissolving a small quantity of solute in a solution that is already saturated or nearly so, and this may be quite different from the heat effect observed on adding a solute to pure solvent. For example, when NaOH is dissolved in water the process is highly exothermic; but when additional NaOH is dissolved in NaOH(aq) that is already very nearly saturated, *heat is absorbed.** As a result we should expect the solubility of NaOH to *increase* with temperature, not decrease.

The increased solubility with temperature that is characteristic of most compounds can serve as a basis for purifying them. A saturated solution of the impure compound is prepared at an elevated temperature. Then the solution is cooled to a lower temperature where the excess solute crystallizes from solution. For this method to be most effective, it is necessary that the impurities not form a solid solution with the substance being crystallized. Usually they do not. Sometimes it is necessary to recrystallize the desired solute several times, especially if the original solution is saturated in one or more of the impurities.

Solubilities of Gases. When we think of the dissolving of gases in liquids in terms of the hypothetical three-step process of Figure 12-1, we discover an important difference from the cases considered earlier. The step which involves separating solute molecules is reversed. That is, in a gas the molecules are already widely separated, and when they enter a liquid solvent they must be brought much closer together, almost as if the gas were first condensed to a liquid and then dissolved. This step is exothermic, with a heat effect that is generally much greater than the energy requirement to increase the separation among solvent molecules. The overall solution process is generally exothermic, and the solubilities of gases generally *decrease* with increased temperature. The bubbles of gas that appear in water as it is heated, even at temperatures far below the boiling point, are expelled air. The decrease in solubility of air in water with increased temperature also accounts for the fact that most fish cannot live in warm water; there is an insufficient quantity of dissolved air present.

The effect of pressure on the solubility of a gas in a liquid is generally more significant than is the effect of temperature. The effect of pressure, as noted in Figure 12-8, is always the same: The solubility increases as the gas pressure is increased. Henry's law states that the concentration of a dissolved gas is proportional to the gas pressure above the solution.

$$C = k \cdot P_{gas} \tag{12.4}$$

The value of the proportionality constant, k, depends on the units of C and P.

Equilibrium between the gas above and the dissolved gas within a liquid is reached when the rates of evaporation and condensation of gas molecules become equal. The rate of condensation depends on the number of molecules per unit volume in the gaseous state. The rate of evaporation depends on the number of molecules of the dissolved gas per unit volume of solution. Thus, as the number of molecules per unit volume increases in the

FIGURE 12-8
Effect of pressure on the solubility of a gas.

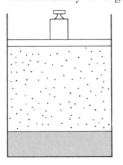

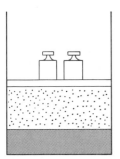

The concentration of dissolved gas is proportional to the pressure on the gas above the solution.

*The solid in equilibrium with saturated NaOH(aq) over a range of temperatures including 25°C is actually NaOH · H₂O. It is this hydrate of NaOH whose solubility dependence on temperature is predicted by Le Châtelier's principle.

gaseous state (through an increase in pressure) the number per unit volume must also increase in the liquid state (through an increase in the solution concentration). An assumption in Henry's law is that solute and solvent molecules do not interact appreciably, that the gas is *nonreactive*.

A practical application of Henry's law is encountered in soft drinks. The dissolved gas is CO_2 and the higher the gas pressure above the soda pop, the more CO_2 that can be kept dissolved. When a bottle is opened, the excess gas pressure is released and dissolved CO_2 escapes, usually rapidly enough to cause fizzing.

Unless otherwise specified, tabulated solubilities of gases are based on a gas pressure of 1 atm above a liquid.

Example 12-3 100.0 g H_2O dissolves 437 cm^3 $H_2S(g)$ measured at STP. What is the molal concentration of a saturated solution at 10.0 atm pressure?

Solution
Molality of saturated $H_2S(aq)$ at STP:

$$\text{no. mol } H_2S = 437 \text{ cm}^3 \text{ } H_2S \times \frac{1 \text{ L } H_2S}{1000 \text{ cm}^3 \text{ } H_2S} \times \frac{1 \text{ mol } H_2S}{22.4 \text{ L } H_2S}$$

$$= 0.0195 \text{ mol } H_2S$$

$$\text{no. kg } H_2O = 100.0 \text{ g } H_2O \times \frac{1 \text{ kg } H_2O}{1000 \text{ g } H_2O} = 0.1000 \text{ kg } H_2O$$

$$\text{molality of } H_2S(aq) = \frac{0.0195 \text{ mol } H_2S}{0.1000 \text{ kg } H_2O} = 0.195 \text{ } m$$

Evaluating k in equation (12.4):

$$k = \frac{\text{conc.}}{P_{gas}} = \frac{0.195 \text{ } m}{1 \text{ atm}}$$

At 10 atm: To determine the solubility at 10 atm pressure, we use equation (12.4) again, substituting the value of k just established.

$$\text{conc.} = k \times P_{gas} = \frac{0.195 \text{ } m}{1 \text{ atm}} \times 10 \text{ atm} = 1.95 \text{ } m$$

SIMILAR EXAMPLES: Review Problem 8; Exercises 20, 21.

12-5 Colligative Properties

There exist four related properties whose values in dilute solution, or to a first approximation in more concentrated solution, depend only on the number of solute particles present. That is, these properties do not depend on the identity of the solute. These four properties— vapor pressure lowering, boiling point elevation, freezing point depression, and osmotic pressure—are called colligative properties. Practical applications of colligative properties are numerous and varied. Also, the study of colligative properties has provided important methods of molecular weight determination and has contributed significantly to the development of solution theory.

Vapor Pressure Lowering.   In the discussion that follows we consider a solution of two components (binary solution), and call the solvent, A, and the solute, B. In the 1880s the French chemist F. M. Raoult found that a dissolved solute has the effect of *reducing* the vapor pressure of the solvent. The extent of the *lowering* of the vapor pressure (ΔP)

proves to be equal to the product of the mole fraction of *solute* (χ_B) and the vapor pressure of the pure solvent (P_A°). That is,

$$\Delta P = \chi_B P_A^\circ \tag{12.5}$$

In a solution of two components, $\chi_A + \chi_B = 1$, and $\chi_B = 1 - \chi_A$. Also, if the vapor pressure of the solvent above the solution is denoted as P_A, then $\Delta P = P_A^\circ - P_A$. Equation (12.5) can be rewritten as

$$P_A^\circ - P_A = (1 - \chi_A)P_A^\circ$$

and rearranged to the form in which Raoult's law usually is given.

$$\cancel{P_A^\circ} - P_A = \cancel{P_A^\circ} - \chi_A P_A^\circ$$

$$P_A = \chi_A P_A^\circ \tag{12.6}$$

Raoult's law states that

The vapor pressure of the solvent above a solution (P_A) is equal to the product of the vapor pressure of the pure solvent (P_A°) and its mole fraction in solution (χ_A).

If the solute(s) in a solution is volatile, we can also write

$$P_B = \chi_B P_B^\circ$$

In an ideal solution all components—solvent and solute(s) alike—adhere to Raoult's law over the entire concentration range. Solutions of benzene and toluene are essentially ideal. In all *dilute* solutions in which there is no chemical interaction among the components, Raoult's law will apply to the *solvent*. This is true whether the solution is ideal or nonideal. However, Raoult's law does not apply to the solute(s) in a dilute nonideal solution. This difference in behavior stems basically from this fact: Solvent molecules predominate in dilute solutions, and the solvent does not behave much differently than it would in the pure state. On the other hand, in dilute solutions solute molecules are surrounded by an overwhelming number of solvent molecules. This produces an environment for these solute molecules which is very much different from that in the pure solute. Although Raoult's law does not apply to the solute(s) in a dilute nonideal solution, Henry's law does.

To explain Raoult's law from fundamental principles presents some difficulties. Our first thought might be that the presence of solute molecules introduces intermolecular forces of attraction for the solvent molecules, thereby reducing the tendency for the solvent to vaporize and lowering its vapor pressure. This cannot be the case, however, because we defined an ideal solution as one in which intermolecular forces between like and unlike molecules are equal, and it is precisely to this type of solution that Raoult's law applies. Another attractive explanation is that the presence of solute molecules at the surface of a solution reduces the availability of solvent molecules for vaporization. If the rate of vaporization is reduced, not as many vapor molecules need to be present before the rate of condensation becomes equal to the rate of vaporization. This means that liquid–vapor equilibrium can be established at a lower vapor pressure. This explanation should help you to remember and to work with Raoult's law, although it too seems inconsistent with certain other principles and observations that are beyond the scope of this text to describe.* Fundamentally, all colligative properties are thermodynamic phenomena and they can best be explained in thermodynamic terms. We will return to this question briefly in Section 16-6.

---

*A discussion of these inconsistencies can be found in K. J. Mysels, *J. Chem. Educ.*, **32,** 179 (1955). For example, solutes of a certain type concentrate in the surface of a solution (surface-active solutes); but they are no more effective in lowering the vapor pressure of the solvent than are normal solutes.

Example 12-4 What are the partial and total vapor pressures at 25°C above a solution with equal numbers of molecules of benzene (C_6H_6) and toluene (C_7H_8) molecules? The vapor pressures of benzene and toluene at 25°C are 95.1 and 28.4 mmHg, respectively.

Solution. If the solution contains equal numbers of each component, the mole fractions must both be 0.500.

Partial pressures:

$$P_{\text{benz.}} = \chi_{\text{benz.}} P^\circ_{\text{benz.}} = 0.500 \times 95.1 \text{ mmHg} = 47.6 \text{ mmHg}$$

$$P_{\text{tol.}} = \chi_{\text{tol.}} P^\circ_{\text{tol.}} = 0.500 \times 28.4 \text{ mmHg} = 14.2 \text{ mmHg}$$

Total vapor pressure:

$$P_{\text{tot.}} = P_{\text{benz.}} + P_{\text{tol.}} = 47.6 \text{ mmHg} + 14.2 \text{ mmHg} = 61.8 \text{ mmHg}$$

SIMILAR EXAMPLES: Review Problem 9; Exercise 24.

Example 12-5 What is the composition of the vapor in equilibrium with the benzene–toluene solution of Example 12-4?

Solution. The ratio of each partial pressure to the total pressure yields the mole fraction of that component in the vapor. This conclusion follows from equation (5.16). (That is, $n_A/n_{\text{tot.}} = \chi_A = P_A/P_{\text{tot.}}$.) The composition of the vapor is

$$\chi_{\text{benz.}} = \frac{P_{\text{benz.}}}{P_{\text{tot.}}} = \frac{47.6 \text{ mmHg}}{61.8 \text{ mmHg}} = 0.770$$

$$\chi_{\text{tol.}} = \frac{P_{\text{tol.}}}{P_{\text{tot.}}} = \frac{14.2 \text{ mmHg}}{61.8 \text{ mmHg}} = 0.230$$

SIMILAR EXAMPLES: Review Problem 10; Exercise 26.

Liquid–Vapor Equilibrium—Ideal Solutions. Table 12-1 summarizes the data calculated in Examples 12-4 and 12-5 (shown in black) together with similar data for other benzene–toluene solutions. These data are presented graphically in Figure 12-9. This figure consists of four lines—three straight and one curved—spanning the entire concentration range. One straight line originates at $P = 0$ and increases to $P = 95.1$ mmHg at

TABLE 12-1
Vapor pressures and liquid-vapor compositions in benzene–toluene mixtures at 25°C

Liquid composition, expressed as mole fraction benzene	Vapor pressures, mmHg			Vapor composition, expressed as mole fraction benzene
	P_{benzene}	P_{toluene}	P_{total}	
0.000	0.0	28.4	28.4	0.000
0.100	9.5	25.6	35.1	0.271
0.200	19.0	22.7	41.7	0.456
0.300	28.5	19.9	48.4	0.589
0.400	38.0	17.0	55.0	0.691
0.500	47.6	14.2	61.8	0.770
0.600	57.1	11.4	68.5	0.834
0.700	66.6	8.5	75.1	0.887
0.800	76.1	5.7	81.8	0.930
0.900	85.6	2.8	88.4	0.968
1.000	95.1	0.0	95.1	1.000

FIGURE 12-9
Liquid–vapor
equilibrium for
benzene–toluene
mixtures at 25°C.

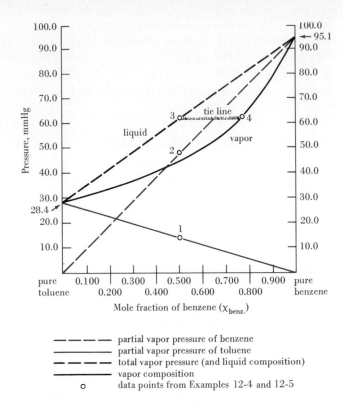

partial vapor pressure of benzene
partial vapor pressure of toluene
total vapor pressure (and liquid composition)
vapor composition
○   data points from Examples 12-4 and 12-5

$\chi_{benz.} = 1$. This straight line represents the partial vapor pressure of benzene as a function of solution composition. It has the equation $P_{benz.} = \chi \cdot P^{\circ}_{benz.}$, signifying that benzene follows Raoult's law. Another straight line originates at $P = 28.4$ mmHg and falls to $P = 0$ when $\chi_{benz.} = 1$. This line represents the partial vapor pressure of toluene, which also follows Raoult's law. The third straight line ranges from $P = 28.4$ mmHg when $\chi_{benz.} = 0$ to $P = 95.1$ mmHg when $\chi_{benz.} = 1$. This line shows how the *total* vapor pressure of benzene–toluene solutions varies with solution composition. It is obtained by adding together points from the two straight lines that lie below it. For example, the pressure at point 3 is the sum of the pressures at points 1 and 2. Point 3 then represents the total vapor pressure of a benzene-toluene solution in which $\chi_{benz.} = 0.500$. As shown in Example 12-5, the vapor in equilibrium with this solution is richer still in benzene; the vapor has $\chi_{benz.} = 0.770$ (point 4). We can think of a line segment joining points 3 and 4, called a **tie line,** in this way: The tie line is plotted at a constant pressure equal to the total vapor pressure of a solution. One end of the tie line represents the composition of the liquid solution and the other end, the composition of the vapor. Imagine establishing a series of tie lines throughout the composition range. The vapor ends of these tie lines can be joined by a smooth curve, the fourth curve in Figure 12-9 (shown in black). From the relative placement of the liquid and vapor curves we see that for ideal solutions of two components, *the vapor phase is richer in the more volatile component.*

Fractional Distillation.   Figure 12-10 shows a different way of looking at liquid–vapor equilibrium in the benzene–toluene system. Here we plot normal boiling point as a function of solution composition, and by normal boiling point we mean the temperature at which the *total* vapor pressure above a benzene–toluene solution is equal to one atmosphere. Since benzene is more volatile than toluene, it is reasonable to expect that adding benzene to toluene produces solutions with lower boiling points than that of toluene. Conversely, adding toluene to benzene produces solutions with higher boiling points than

FIGURE 12-10
Liquid–vapor
equilibrium for
benzene–toluene
mixtures: normal boiling
temperature vs.
composition.

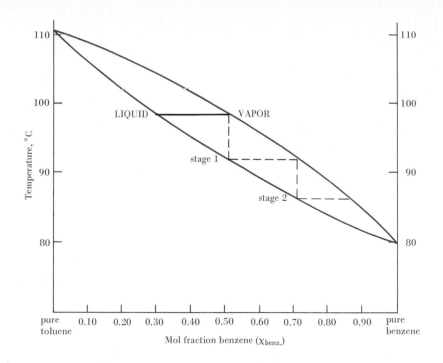

that of benzene. Boiling points in Figure 12-10, then, are expected to range from that of pure benzene to that of pure toluene. Another expectation is that above all benzene–toluene solutions, the vapor is richer in benzene (the more volatile component) than is the boiling solution. In Figure 12-10 the lower curve represents liquid solutions of varying composition and their boiling points. The upper curve represents the composition of vapor in equilibrium with boiling solutions. For example, a benzene–toluene solution with $\chi_{benz.}$ = 0.30 boils at a temperature of 98.6°C and is in equilibrium with a vapor in which $\chi_{benz.}$ = 0.51. These two points are joined by the solid tie line in Figure 12-10.

Imagine that the vapor with $\chi_{benz.}$ = 0.51 is removed from contact with the liquid solution and cooled to the point where the vapor completely condenses. The new liquid solution obtained from this vapor has $\chi_{benz.}$ = 0.51 (labeled stage 1 in Figure 12-10). When this new liquid solution is boiled, the vapor in equilibrium with it has $\chi_{benz.}$ = 0.71. If this vapor is removed and completely condensed, the resulting liquid will have $\chi_{benz.}$ = 0.71 (stage 2). This boiling/condensation process can be repeated several times more.

Now let us return to the original solution with $\chi_{benz.}$ = 0.30. Because the vapor is richer in benzene than is the boiling solution that produces it, as vapor is removed the remaining liquid solution becomes richer in the *less volatile* component, toluene. In the hypothetical boiling/condensation process outlined here, the ultimate vapor condensate is pure benzene and the liquid residue is pure toluene. This separation of substances based on their differing volatilities is called **fractional distillation.** Hydrocarbons in the gasoline range are separated from other petroleum components by fractional distillation (see Section 26-11). So is the separation of the constituents of liquid air achieved through fractional distillation (page 384).

Liquid–Vapor Equilibrium in Nonideal Solutions.   For a *nonideal* binary solution, the curves representing the partial vapor pressures of the two components and the total vapor pressure of the solution as a function of concentration are *not straight lines*. Neither is it always the case that the vapor phase above nonideal solutions is richer in the more volatile component.

FIGURE 12-11
Liquid–vapor
equilibrium with
positive deviations from
Raoult's law.

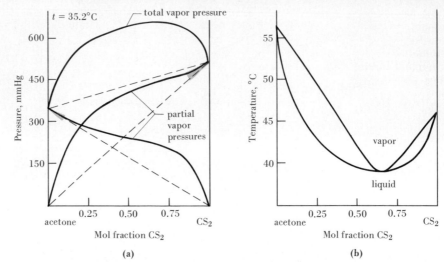

(a) Partial and total vapor pressure vs. composition (Raoult's law applies in the gray-shaded areas).
(b) Boiling point temperature vs. composition—a minimum boiling point azeotrope.

Departures from ideal solution behavior are sometimes classified according to whether partial vapor pressures of solution components and total vapor pressures are higher or lower than those expected for an ideal solution. For solutions of acetone and carbon disulfide (Figure 12-11a) these pressures are higher than predicted for an ideal solution; these solutions exhibit *positive* deviations from Raoult's law. In solutions of acetone and chloroform (Figure 12-12a) vapor pressures are lower than predicted and *negative* deviations from Raoult's law are exhibited. The nature of intermolecular forces in acetone–chloroform mixtures was illustrated in Figure 12-3: Intermolecular forces between unlike molecules reduce the tendency for each solution component to vaporize, accounting for the lower-than-ideal partial and total vapor pressures. Even though the departures from

FIGURE 12-12
Liquid–vapor
equilibrium with
negative deviations from
Raoult's law.

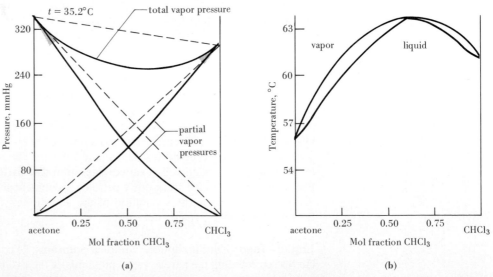

(a) Partial and total vapor pressures vs. composition (Raoult's law applies in the gray-shaded areas).
(b) Boiling point temperatures vs. composition—a maximum boiling azeotrope.

ideality for most solutions in Figures 12-11a and 12-12a are quite large, the partial vapor pressure curve for the *solvent* in *dilute* solutions does fall very nearly on the Raoult-law straight line. Thus, these graphs illustrate the statement made on page 338 that the solvent does conform to Raoult's law even in *dilute* nonideal solutions.

In considering graphs of boiling temperature versus composition for nonideal solutions, in the extreme cases where the total vapor pressure curve passes through a maximum or minimum, the corresponding boiling temperature graph passes through a minimum or maximum, respectively (see Figures 12-11b and 12-12b). That is, the maximum vapor pressure of an acetone–carbon disulfide solution at about $\chi_{CS_2} = 0.65$ corresponds to a minimum boiling point at about 38.5°C. In the acetone–chloroform system a minimum vapor pressure at about $\chi_{CHCl_3} = 0.61$ corresponds to a maximum boiling point at about 64°C. When it is boiled, a liquid solution with a composition corresponding to one of these maxima or minima produces a vapor of identical composition. Such solutions are called **constant-boiling solutions** or **azeotropes.** Fractional distillation of nonideal solutions displaying azeotropism yields one pure component and the azeotrope as the ultimate distillation products, rather than two pure liquids. One of the best known azeotropes is the minimum boiling azeotrope of ethanol (ethyl alcohol) and water with 95.6% ethanol, by mass, and a boiling point of 78.2°C. Since the usual material obtained in the industrial preparation of ethanol is an ethanol–water mixture with less than 95.6% ethanol, the products obtained by fractional distillation are pure water and the azeotrope. This is the reason why ethanol used in the laboratory is often only 95.6% pure. Special measures are required to remove the remaining 4.4% water if absolute ethanol is required.

Freezing Point Depression and Boiling Point Elevation. Up to this point we have assumed that both the solvent and the solute(s) are volatile. However, an important class of solutions is that in which the solutes are *nonvolatile*. For such solutions the nonvolatile solute still lowers the vapor pressure of the solvent; the higher its concentration the greater the vapor pressure lowering. This effect is pictured in Figure 12-13. Here the vapor

FIGURE 12-13
Vapor pressure lowering by a nonvolatile solute.

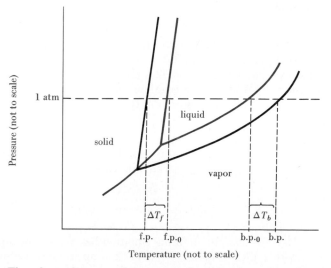

The phase diagram of the pure solvent is shown in color, and for the solvent containing a nonvolatile solute, in black. The freezing point and boiling point of the pure solvent are f.p.$_0$ and b.p.$_0$. The corresponding points for the solution are f.p. and b.p. The freezing point depression, ΔT_f, and boiling point elevation, ΔT_b, are indicated. Because the solute is nonvolatile and assumed to be insoluble in the solid solvent, the sublimation curve of the solvent is unaffected by the presence of solute in the solution phase.

TABLE 12-2
Cryoscopic and
ebullioscopic constants

Solvent	$K_f{}^a$	$K_b{}^a$
acetic acid	3.90	3.07
benzene	4.90	2.53
nitroben-zene	7.00	5.24
phenol	7.40	3.56
water	1.86	0.512

[a] Values correspond to freezing point depressions and boiling point elevations, in degrees Celsius, due to 1 mol of solute particles dissolved in 1 kg of solvent. Units: °C kg solvent (mol solute)$^{-1}$.

pressure curve and fusion curve for the solvent in a solution are superimposed on the phase diagram of the pure solvent. An important additional requirement in Figure 12-13 is that the solute be *insoluble* in the *solid* solvent, but there are many mixtures for which this requirement is met.

The intersection of the vapor pressure and sublimation curves for a solvent containing a nonvolatile solute comes at a lower temperature than for the pure solvent. Also displaced to lower temperatures is the fusion curve. Now recall how freezing points and boiling points are established. They are the temperatures at which a constant-pressure line at $P = 1$ atm intersects the fusion and vapor pressure curves, respectively. Four points of intersection are indicated in Figure 12-13—the freezing points and the boiling points of the pure solvent and of the solvent in a solution. The freezing point of the solvent is *depressed* and its boiling point is *elevated*.

Freezing point depression and boiling point elevation, like vapor pressure lowering, are proportional to mole fraction concentration. For *dilute* solutions this proportionality can be extended to molality.

$$\Delta T_f = K_f m \tag{12.7}$$
$$\Delta T_b = K_b m \tag{12.8}$$

ΔT_f and ΔT_b are the freezing point depression and boiling point elevation, respectively; m is the molality; K_f and K_b are proportionality constants. K_f is the **cryoscopic** or freezing point depression constant, and K_b the **ebullioscopic** or boiling point elevation constant. These constants are characteristic of the solvent and are based on a function of its melting point, enthalpy of fusion, and molecular weight (K_f) or of its boiling point, enthalpy of vaporization, and molecular weight (K_b). K_f and K_b represent the freezing point depression and the boiling point elevation for a 1 m solution. As a matter of fact, however, equations (12.7) and (12.8) often do not hold for solutions as concentrated as 1 m. Some typical values of K_f and K_b are listed in Table 12-2. Cooling curves for a pure solvent and solution are compared in Figure 12-14.

Historically, freezing point measurements have been used to establish molecular formulas. The required calculations can be thought of in terms of three questions, as illustrated through Example 12-6.

Example 12-6   (a) What is the molality of solute in an aqueous solution with a freezing point of $-0.450°C$? (b) If this solution is obtained by dissolving 2.12 g of an unknown compound in 48.92 g H_2O, what must be the molecular weight of the compound? (c) What is the true molecular formula of the compound if its analysis is 40.0% C, 53.3% O, and 6.7% H, by mass?

Solution
(a) The molality of solute is readily established by using equation (12.7) with the value of K_f listed for water in Table 12-2.

$$m = \frac{\Delta T_f}{K_f} = \frac{0.450°C}{1.86°C \text{ kg water (mol solute)}^{-1}} = 0.242 \frac{\text{mol solute}}{\text{kg water}}$$

(b) Here we use the defining equation for molal concentration (equation 12.2), but with a known molality (0.242) and an unknown molar mass ($\mathcal{M}$) of solute. The number of moles of solute is simply 2.12 g/$\mathcal{M}$.

$$m = \frac{2.12 \text{ g}/\mathcal{M}}{0.04892 \text{ kg water}} = 0.242 \frac{\text{mol}}{\text{kg water}}$$

$$\mathcal{M} = \frac{2.12 \text{ g}}{(0.04892 \times 0.242) \text{ mol}} = 179 \text{ g/mol}$$

The molecular weight is 179.

(c) **The empirical formula of the compound is determined from its percent composition by the method of Example 3-10. (This calculation is left as an exercise for the student.) The result obtained is CH_2O. This empirical formula leads to a formula weight of 30. The experimentally determined molecular weight—179—is almost exactly six times as large. The molecular formula is $C_6H_{12}O_6$.**

SIMILAR EXAMPLES: Review Problem 11; Exercises 27, 28.

Example 12-6 shows how measurement of a colligative property leads to a determination of molecular weight. There are limitations to this method, however, which must be understood. Remember that the boiling point of a liquid depends on atmospheric pressure. If boiling point elevation is used for molecular weight determination, it is necessary to maintain a constant barometric pressure. This is not particularly easy to do and, as a result, boiling point elevation is not commonly used. Because equation (12.7) is applicable only in dilute solutions (usually much less than 1 *m*), freezing points must be determined with considerable precision if water is the solvent ($K_f = 1.86$). In Example 12-6 the temperature measurement was made to $\pm 0.001°C$. Temperature measurements of this precision are not possible with ordinary laboratory thermometers. Greater precision is possible when a solvent with a larger value of K_f is used, such as cyclohexane ($K_f = 20$) or, better still, camphor ($K_f = 40$). If a solute has a high molecular weight, the number of moles in a sample may be too small to affect the freezing point appreciably. For these solutes the measurement of osmotic pressure is a better method.

Whatever solvent is used, it must be of high purity and have a freezing point that is conveniently measured. The melting point of camphor is rather high for usual laboratory operations, but camphor is still desirable as a solvent because of its large K_f.

The phenomenon of freezing point depression also has practical applications. Perhaps best known are methods used to lower the freezing point of water. An antifreeze (usually ethylene glycol), when added to the cooling system of an automobile, protects the coolant from freezing in cold weather. The use of $CaCl_2$ or $NaCl$ to lower the melting point of ice is also widely encountered, whether to de-ice roads or to prepare a freezing mixture for use in a home ice cream freezer.

Osmotic Pressure. Certain membranes, though they appear to be continuous sheets or films, actually contain a network of submicroscopic holes or pores. Small solvent molecules may pass through these pores, but the passage of dissolved solute molecules is severely restricted. Membranes having this property are said to be **semipermeable.** They may be of animal or vegetable origin and occur naturally, such as pig's bladder and parchment, or they may be synthetic materials, such as cellophane.

FIGURE 12-14
Cooling curves of a pure solvent and a solution compared.

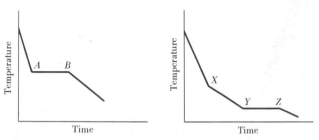

The cooling curve for the pure solvent has one horizontal break from *A* to *B* where complete freezing occurs. The cooling curve for the solution has a break at *X*, where the solvent begins to freeze from the solution (the freezing point). A second, horizontal break from *Y* to *Z* represents the freezing of both components from solution as a mixture of solids (the eutectic temperature). The freezing points of solutions referred to in this section correspond to the point *X*. (The behavior described here assumes that the solute is insoluble in the solid solvent. Not represented here is the fact that some supercooling is likely to occur at points *A*, *X*, and *Y*.)

FIGURE 12-15
An illustration of
osmosis.

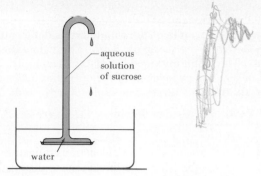

— aqueous
solution
of sucrose

water

Water molecules pass through the membrane, creating a pressure within the funnel.
This pressure causes the liquid level to rise and the solution to overflow. As this process continues, the solution inside the funnel becomes more dilute and the pure water
outside the funnel becomes a dilute sucrose solution. Liquid flow stops when the compositions of the solutions separated by the membrane become equal.

Figure 12-15 pictures an aqueous sucrose (sugar) solution in a long glass tube separated from pure water by a semipermeable membrane (permeable to water only). Water
molecules can pass through the membrane from either direction, and they do. But because
the concentration of water molecules is *greater* in the pure water than in the solution, there
is a net flow of water *from* the pure solvent *into* the solution. This net flow, called
osmosis, causes the solution level in the tube to rise. The more concentrated the sucrose
solution the higher the solution level rises. A 20% solution would be raised to more than
100 m!

The net flow of water into the sucrose solution can be reduced by applying a pressure
to the solution. This increases the flow of water in the reverse direction. The pressure
required to just stop the flow of water into the sucrose solution is known as the **osmotic
pressure** of the solution. For the 20% sucrose solution this pressure is about 15 atm.

Osmotic pressure is included among the colligative properties because its magnitude
depends only on the *number* of solute particles per unit volume of solution. It does not
depend on the identity of the solute. The expression written below (known as the van't
Hoff equation) works quite well for calculating osmotic pressures of *dilute* solutions. The
osmotic pressure is represented by the symbol π; R is the gas constant
(0.0821 L atm mol^{-1} K^{-1}); and T is the kelvin temperature. The term n represents the
moles of *solute* and V is the volume (in liters) of *solution;* the ratio, n/V, then, is the
molarity of the solution, M.

> There is a striking re-
> semblance between
> equation (12.9) and the
> ideal gas equation,
> $PV = nRT$. Think of π
> as being equivalent to a
> gas pressure exerted by
> n mol of gas confined to
> a volume of V L.

$$\pi = \left(\frac{n}{V}\right) RT = M \cdot RT \tag{12.9}$$

Example 12-7 What is the osmotic pressure at 25°C of an aqueous solution that is
0.0010 M $C_{12}H_{22}O_{11}$ (sucrose)?

Solution: Direct substitution into equation (12.9) leads to the result

$$\pi = \frac{0.0010 \text{ mol} \times 0.0821 \text{ L atm mol}^{-1} \text{ K}^{-1} \times 298 \text{ K}}{\text{L}} = 0.024 \text{ atm} \ (=18 \text{ mmHg})$$

SIMILAR EXAMPLES: Exercises 34, 35.

The 0.0010 M sucrose solution in Example 12-7 would have a molality of about
0.001 m. (In *dilute aqueous* solutions molarity and molality are essentially equal.) According to equation (12.7), we should expect a freezing point depression of about

0.00186°C for this solution. Such a small temperature difference is extremely difficult to measure with any precision. On the other hand, a pressure difference of 18 mmHg is rather easily measured. It corresponds to a solution height of about 0.25 m! This comparison suggests that measurement of osmotic pressure can be an important method of molecular weight determination when dealing with (a) very dilute solutions or (b) solutes of very high molecular weight.

Example 12-8 A solution is prepared by dissolving 1.08 g of the protein, human serum albumin, obtained from blood plasma, in 50.0 cm³ of aqueous solution. The solution has an osmotic pressure of 5.85 mmHg at 298 K. What is the molecular weight of the albumin?

Solution. First we need to express the osmotic pressure in atm.

$$\text{no. atm} = 5.85 \text{ mmHg} \times \frac{1 \text{ atm}}{760 \text{ mmHg}} = 7.70 \times 10^{-3} \text{ atm}$$

Now we apply equation (12.9) in a slightly modified form [i.e., with the number of moles of solute represented by mass of solute (m) divided by the molar mass ($\mathcal{M}$).]

$$\pi = \frac{(m/\mathcal{M}) \, RT}{V} \qquad \mathcal{M} = \frac{mRT}{\pi V}$$

$$\mathcal{M} = \frac{1.08 \text{ g} \times 0.0821 \text{ L atm mol}^{-1} \text{ K}^{-1} \times 298 \text{ K}}{7.70 \times 10^{-3} \text{ atm} \times 0.0500 \text{ L}} = 6.86 \times 10^4 \text{ g/mol}$$

The molecular weight of the albumin is 6.86×10^4.

SIMILAR EXAMPLES: Review Problem 13; Exercise 36.

FIGURE 12-16
Desalinization of seawater by reverse osmosis.

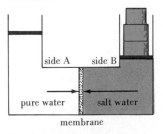

membrane

The membrane pictured here is permeable to water but not to sodium or chloride ions. The normal flow of water through the membrane, in the absence of external pressures, is from side A (pure water) to side B (salt water). If a pressure is exerted on side B that exceeds the osmotic pressure of the salt water, a net flow of water can be created in the *reverse* direction, that is, from the salt water (side B) to the pure water (side A). The magnitudes of the rates of flow of water molecules from each side are suggested by the lengths of the arrows.

Perhaps the most important examples of osmosis are those found in living organisms. Consider red blood cells, for instance. If red blood cells are placed in pure water, the cells expand and eventually rupture as a result of water entering the cells through osmosis. The osmotic pressure associated with the fluid inside the cell is equivalent to that of an 0.9% sodium chloride solution. Thus, if the cells are placed in a sodium chloride solution (saline solution) of this concentration, there is no net flow of water through the cell walls and the cells remain stable. This solution is said to be *iso*tonic. If the salt solution has a higher concentration than about 0.9%, water flows out of the cells and the cells shrink. The solution is *hyper*tonic. If the salt concentration is less than 0.9%, water flows into the cells and the solution is said to be *hypo*tonic. The walls (membranes) of red blood cells are approximately 10 nm thick and have pores (holes) about 0.8 nm in diameter. Water molecules have less than half this diameter and pass through easily. K^+ ions, which are found inside the cells, are also smaller than the pore diameters. But because the pore walls carry a positive electrical charge, K^+ ions are repelled. Thus, factors other than simple size may be involved in determining what species can pass through the pores of a semipermeable membrane.

An interesting practical application of the idea that an external pressure can be used to stop the osmotic flow of water is found in the process of **reverse osmosis.** It is currently being used to a limited extent as a method of desalinizing seawater or brackish water. As suggested by Figure 12-16, if a sufficiently high pressure is applied to a solution, the solvent can actually be forced to flow in the reverse direction, *from a solution into a pure solvent.* One of the problems associated with developing this scheme commercially is finding durable membrane materials with the required pore sizes and permeability properties.

12-6 Theory of Electrolytic Dissociation

Early investigators of the electrical properties of matter recognized that the ability to conduct electric current is not limited to metals. Some liquids and liquid solutions also conduct electric current. Pure liquid water is a very poor conductor of electric current, being essentially a nonconductor. The addition of certain solutes to water results in aqueous solutions that are excellent electrical conductors. Yet there are some solutes that do not enhance the electrical conductivity of water and still others that render it only weakly conducting. These three groups of solutes are termed **strong electrolytes, nonelectrolytes,** and **weak electrolytes,** respectively. Some representative examples are cited in Table 12-3 and illustrated in Figure 12-17.

For his doctoral dissertation (in 1884), a young Swedish chemist, Svante Arrhenius, undertook a careful investigation of the electrolytic conductivities of a variety of aqueous solutions. Prevailing opinion concerning ions in solution was that they form only as a result of the passage of electric current. Arrhenius, however, reached the conclusion that ions may exist in a solute and be dissociated from one another simply by dissolving the solute in water. The degree to which solute molecules are dissociated into ions he denoted by α, the **degree of dissociation.**

For a nonelectrolyte the electrolytic conductivity is extremely low; practically no ions exist in solution: $\alpha = 0$. For a weak electrolyte α is a small fractional number because in aqueous solution these solutes exist partly in ionic form and partly as undissociated molecules. In a strong electrolyte solution, especially at low concentrations, $\alpha = 1$. This value signifies that essentially complete dissociation of a solute into ions occurs in a strong electrolyte solution. Furthermore, from the measured electrolytic conductance it is possible to calculate the number of ions produced per mole of solute. For example, in NaCl,

FIGURE 12-17
Electrical conductivity
of aqueous solutions.

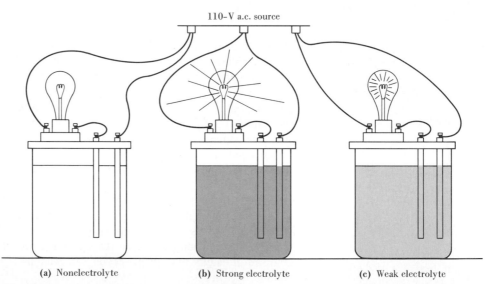

110–V a.c. source

(a) Nonelectrolyte   (b) Strong electrolyte   (c) Weak electrolyte

For electric current to flow requires that electrical contact be made between the two metal rods immersed in solution. This contact through the solution is possible only if the solution contains ions. The magnitude of the current is estimated by the brilliance of the incandescent lamp.
(a) The ionic concentration is essentially zero; no current flows.
(b) A solution having a high conductivity, even though the solute concentration may be low, is a strong electrolyte.
(c) If even at fairly high concentrations a solute imparts only low electrical conductivity to a solution, it is termed a weak electrolyte.

TABLE 12-3
Electrolytic properties of some aqueous solutions

Nonelectrolytes	Strong electrolytes		Weak electrolytes
	Ionic compounds	Covalent compounds	
H_2O (water)	NaCl	HCl	$HCHO_2$ (formic acid)
C_2H_5OH (ethanol)	$MgCl_2$	HBr	$HC_2H_3O_2$ (acetic acid)
$C_6H_{12}O_6$ (glucose)	KBr	HI	HOCl (hypochlorous acid)
$C_{12}H_{22}O_{11}$ (sucrose)	$KClO_4$	HNO_3	HNO_2 (nitrous acid)
$CO(NH_2)_2$ (urea)	KOH	H_2SO_4[a]	H_2SO_3 (sulfurous acid)
$C_2H_6O_2$ (ethylene glycol)	$Al_2(SO_4)_3$	$HClO_4$	NH_3 (ammonia)
$C_3H_8O_3$ (glycerol)	$CuSO_4$		$C_6H_5NH_2$ (aniline)
	$LiNO_3$		
plus		**plus**	**plus**
many	**plus**	**a few**	**many**
others	**many**	**others**	**others**
	others		

[a]H_2SO_4(aq) is completely ionized into H^+ and HSO_4^- ions, but HSO_4^- undergoes only partial ionization to H^+ and SO_4^{2-}. This solution behavior of H_2SO_4(aq) is discussed in Section 17-6.

$MgCl_2$, and K_2SO_4 the numbers of ions per mole of substance are found to be *two, three,* and *three moles,* respectively.

Concentrations are often expressed through bracket symbols, [ ]. The solutions being mixed in this example can be represented as $[Na_2SO_4] = 0.250\ M$ and $[NaCl] = 0.400\ M$. The resulting molarity of Na^+ is $[Na^+] = 0.467\ M$. If a symbol to denote the concentration unit is omitted in the bracket notation, molarity is understood.

Example 12-9   200.0 mL of 0.250 M Na_2SO_4 and 100.0 mL of 0.400 M NaCl are mixed. Assuming the solution volumes are additive, what is the resulting molarity of Na^+?

Solution   Both Na_2SO_4 and NaCl are strong electrolytes, completely dissociated in aqueous solutions. In Na_2SO_4(aq) there are twice as many moles of Na^+ ions as moles of Na_2SO_4. In NaCl(aq) the number of moles of Na^+ ions is the same as the number of moles of NaCl. We must determine the total number of moles of Na^+ ions present in a final solution of volume 300 mL = 0.300 L.

$$\text{no. mol } Na^+ \text{ in } Na_2SO_4(aq) = 200.0\text{ mL} \times \frac{1\text{ L}}{1000\text{ mL}} \times \frac{0.250\text{ mol } Na_2SO_4}{L} \times \frac{2\text{ mol } Na^+}{1\text{ mol } Na_2SO_4}$$
$$= 0.100\text{ mol } Na^+$$

$$\text{no. mol } Na^+ \text{ in } NaCl(aq) = 100.0\text{ mL} \times \frac{1\text{ L}}{1000\text{ mL}} \times \frac{0.400\text{ mol } NaCl}{L} \times \frac{1\text{ mol } Na^+}{1\text{ mol } NaCl}$$
$$= 0.0400\text{ mol } Na^+$$

$$\text{total mol } Na^+ = 0.100 + 0.0400 = 0.140\text{ mol } Na^+$$

$$\text{molarity of } Na^+ = \frac{0.140\text{ mol } Na^+}{0.300\text{ L}} = 0.467\ M$$

SIMILAR EXAMPLES: Review Problem 14; Exercise 37.

Arrhenius's theory of electrolytic dissociation was at first not accepted by his professors. Later, however, it was championed by such eminent chemists as van't Hoff and Ostwald. Arrhenius's theory marked the beginnings of the discipline of physical chemistry.

The value of a scientific theory lies in its ability to provide explanations of a variety of seemingly unrelated phenomena. Arrhenius's theory, though developed to explain electrolytic conductivity, also provided a basis for understanding chemical reactions and chemical equilibria in solutions. The relationship of his theory to these other phenomena is explored in later chapters. One of the immediate successes of the theory was in explaining certain anomalous values of colligative properties first investigated by van't Hoff.

Anomalous Behavior.   Certain solutes produce a greater effect on colligative properties than expected for a nonelectrolyte. The van't Hoff factor *i* is defined as

$$i = \frac{\text{measured value for solute}}{\text{expected value for a nonelectrolyte}} \qquad (12.10)$$

For a large group of solutes, nonelectrolytes such as urea, glycerol, and sucrose, *i* has a value of 1. For another equally large group of solutes, weak and strong electrolytes, *i* has values greater than 1.

Example 12-10   The following freezing points are observed for 0.010 *m* aqueous solutions. Calculate the van't Hoff factor for each solute and account for its value: urea, −0.0186°C; acetic acid, −0.0193°C; magnesium chloride, −0.054°C.

Solution.   To determine the van't Hoff factor, *i*, we must start with the freezing point depression expected for a nonelectrolyte. According to equation (12.7), this would be $\Delta T_f = 0.0186°C$.

urea $(CO(NH_2)_2)$: $i = \dfrac{0.0186°C}{0.0186°C} = 1.00$

Urea must be a nonelectrolyte that remains undissociated in aqueous solution.

acetic acid $(HC_2H_3O_2)$: $i = \dfrac{0.0193°C}{0.0186°C} = 1.04$

Acetic acid is a weak electrolyte. About 4% of the molecules are dissociated, producing two ions (H^+ and $C_2H_3O_2^-$) per molecule. ($\alpha = 0.04$.)

magnesium chloride $(MgCl_2)$: $i = \dfrac{0.054°C}{0.0186°C} = 2.9 \simeq 3$

A van't Hoff factor of $i \simeq 3$ suggests that $MgCl_2$ is dissociated in aqueous solution, producing three moles of ions per mole of compound. The reason why *i* is somewhat less than 3 is explained in Section 12-7.

SIMILAR EXAMPLES: Review Problem 15; Exercises 38, 40.

FIGURE 12-18
Interionic attractions in aqueous solution.

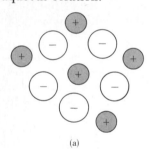

(a)

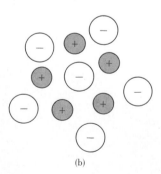

(b)

(a) A positive ion in aqueous solution is surrounded by a shell of negative ions.
(b) A negative ion attracts positive ions to its immediate surroundings.

12-7 Interionic Attractions

Once it had gained acceptance Arrhenius's theory stimulated great progress in physical chemistry. However, some observations made over the 40-year period from about 1880 to 1920 pointed to a need for refinements in this theory. Electrolytic conductances of concentrated solutions of strong electrolytes were found to be less than expected for complete dissociation into ions. These results suggest incomplete dissociation of a strong electrolyte in solution. Yet, x-ray diffraction studies indicate that salts exist in 100% ionic form in the solid state. Should they not also be completely ionized in solution?

The modern view of electrolyte solutions is based on a theory proposed by Debye and Hückel in 1923. The theory states that in aqueous solutions salts do exist in completely ionized form. The ions, however, do not behave independently of one another. Instead, each positive ion is surrounded by a cluster of predominantly negative ions, and each negative ion by a cluster in which positive ions predominate. That is, each ion is enveloped by an ionic atmosphere with a net charge opposite in sign to the central ion (see Figure 12-18).

In an electric field the mobility of each ion is reduced because of the attraction or drag of its neighbors in the ionic atmosphere. Similarly, the magnitudes of colligative properties are reduced (which explains why the value of *i* for $MgCl_2$ determined in Example 12-10 was 2.9 rather than 3). Thus, each type of ion in an aqueous solution has a total concentration based on the amount of solute dissolved, called the stoichiometric concentration. But the ion also has an "effective" concentration, called the **activity,** which takes into account interionic attractions. If the activity is used in place of stoichiometric concentration, solution properties can be predicted quite well, especially for dilute solutions. To relate activity to concentration requires the use of an **activity coefficient.** The Debye-Hückel theory provides a theoretical basis for calculating activity coefficients and activities.

12-8 Colloidal Mixtures

At the beginning of this chapter we chose sand in water as an example of a heterogeneous mixture. From common experience we expect sand to settle to the bottom of such a mixture, even if the quantity of sand (silica, SiO_2) is very small. SiO_2 is very insoluble in water. Yet it is possible to prepare mixtures in which large quantities of silica, up to 40% by mass, are dispersed in water and remain so for years! Such mixtures are clear, although with an opalescent or milky cast. Obviously, these dispersions do not involve ordinary grains of sand. Neither do they consist of dissolved ions or molecules. They are called colloidal mixtures.

The freezing points of colloidal mixtures of silica in water are only slightly below 0°C. We conclude that these mixtures contain *small* numbers of particles in comparison to true solutions with comparable solute concentrations. But if the numbers of particles are small, their masses and physical dimensions must be huge compared to typical solute particles. The molecular weights, or more correctly, particle weights, of colloids range into the hundreds of thousands or millions. Figure 12-19 compares colloidal particles of different sizes and shapes with the more familiar particles of chemistry—atoms, ions, and molecules. The production of colloidal particles may occur, as in the case of colloidal silica, by

FIGURE 12-19
A comparison of colloidal, molecular, and atomic dimensions.

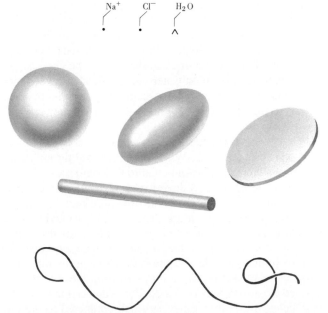

Some approximate sizes and shapes of colloidal particles are represented here. For comparison, particles with typical atomic and molecular dimensions are also shown.

FIGURE 12-20
Colloidal silica.

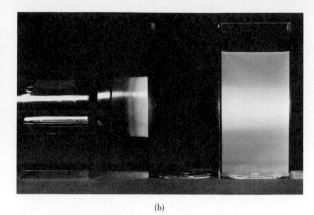

(a)

(b)

(a) The two containers have the same overall contents—30% SiO_2 in water, by mass. The one on the left is the expected heterogeneous mixture of sand in water. That on the right has SiO_2 in the colloidal state.
(b) The Tyndall effect. The flashlight beam is not visible (left) as it passes through water (or a true molecular solution), but it is readily visible as it passes through colloidal silica (right).
["Ludox" HS-30 Colloidal Silica: Courtesy of E. I. du Pont de Nemours & Company.]

the aggregation of large numbers of molecules through a process called *condensation* (see Section 13-9). A contrasting method that may sometimes be employed is that of *dispersion*. This involves breaking down larger particles mechanically, for example by grinding, until the particles are sufficiently small to remain suspended.

Figure 12-19 suggests that in order for a material to be classified as colloidal, one or more of its dimensions (length, width, or thickness) must fall in the approximate range 1 to 100 nm. If all the dimensions are less than 1 nm, the particles are in the molecular size range. If all the dimensions exceed 100 nm, the particles become of ordinary or macroscopic size (even if they are only visible under a microscope).

The colloidal particles in silica-water mixtures have a spherical shape; so do particles of bushy stunt virus. Some colloidal particles are rod shaped, for example, tobacco mosaic virus. Some have a disclike shape, like the gamma globulin in human blood plasma. Thin films, such as oil on water, are colloidal. And some colloids have the appearance of filaments or random coils, for example, cellulose fibers. As so aptly put years ago by Wilder Bancroft, "Colloid chemistry is the chemistry of bubbles, drops, grains, filaments, and films."

Determining whether a mixture is a true solution or colloidal is often possible by the method of Figure 12-20. When light is passed through a true solution, an observer viewing from a direction perpendicular to the beam sees no light. But in a colloidal suspension light is scattered in many directions and can be seen easily. This behavior, first studied by Tyndall in 1869, is known as the Tyndall effect. A common example is the scattering of light by dust particles in the light beam of a movie projector in a darkened room.

An important characteristic of colloidal particles is their high ratio of surface area to volume. It is well known that the atoms, ions, or molecules at the surface of a substance behave somewhat differently than do those in the interior. This is because species at the surface are subject to forces different from those in the interior (recall the discussion of surface tension in Section 11-2). For ordinary materials the proportion of atoms, ions or molecules at the surface is very small compared to the interior, and the distinctive phenomena associated with surfaces are masked. In colloidal materials these surface phenomena are often quite pronounced.

FIGURE 12-21
Surface of an SiO_2 particle in colloidal silica.

$$x\, SiO_2 \cdot y\, H_2O \begin{cases} OH^- \\ OH^- \\ OH^- \\ OH^- \\ OH^- \end{cases} \quad \begin{matrix} Na^+ \\ \\ Na^+ \\ \\ Na^+ \end{matrix}$$

The points made in this simplified drawing are

- The SiO_2 particles are hydrated.
- OH^- ions are preferentially adsorbed on the surface.
- In the immediate vicinity of the particle, negative ions outnumber positive ions, and the particle has a net negative charge.

Not illustrated here are the facts that

- Some of the negative charge on the surface comes from the formation of silicate anions, e.g., SiO_3^{2-}.
- As a whole, the solution in which these particles are found is electrically neutral.

Stability of Colloids. One of the properties of surfaces is that of being able to attach species to themselves, a phenomenon called **adsorption.** In their formation, some colloidal particles adsorb large numbers of ions from solution and become electrically charged. Silica particles in the colloidal silica mixtures mentioned previously adsorb hydroxide ions (OH^-) in preference to other ions, as suggested by Figure 12-21. As a result the silica particles all acquire a negative charge. Having like charges, the particles repel one another. It is these mutual repulsions that overcome the force of gravity and keep the particles suspended.

The coagulation of colloidal iron(III) oxide is illustrated in Color Section I.

Although the electrical charge factor can be an important one in stabilizing a colloid, a high concentration of ions can bring about the coagulation or precipitation of the colloid. The ions responsible for this are those carrying the opposite charge to the particles themselves. Thus, the colloidal silica particles pictured in Figure 12-21 are precipitated by cations, and the higher the charge on the cation the more effective it is (which explains why Al^{3+} ions are especially effective in precipitating negatively charged colloids). A common procedure for stabilizing a colloidal dispersion is to remove excess ions by **dialysis,** a process depicted in Figure 12-22. The process is similar to osmosis but is based

FIGURE 12-22
The principle of dialysis.

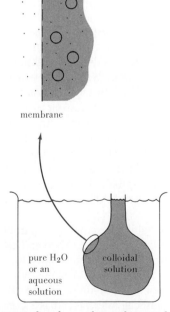

membrane

pure H_2O or an aqueous solution

colloidal solution

○ colloidal particles
• other solute particles

Water molecules, other solute molecules, and dissolved ions are free to pass through the pores of the membrane (e.g., cellophane) in either direction. The direction of net flow of these species depends on their relative concentrations on either side of the membrane. Colloidal particles cannot pass through the pores of the membrane.

FIGURE 12-23
The phenomenon of
electrophoresis.

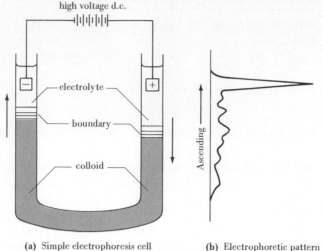

high voltage d.c.

electrolyte

boundary

colloid

Ascending

(a) Simple electrophoresis cell

(b) Electrophoretic pattern
for human plasma

(a) A colloidal suspension in water is covered by an electrolyte solution. Electrodes are immersed in the electrolyte and connected to a high-voltage direct current source. Colloidal particles, because of their adsorbed ions, carry electrical charges. As a result they are attracted to one electrode and repelled by the other; this causes movement of the boundaries. In the illustration here, the particles are positively charged. If colloidal particles of differing types are present, they migrate at different rates in an electric field; the initial sharp boundary separates into several boundaries.
(b) This is a representation of an electrophoresis boundary in which the separation of several proteins in human blood plasma occurs. The large peak corresponds to albumin, the smaller peaks to several globulins. The effectiveness of electrophoresis is limited by the effect of gravity, which causes colloidal particles to settle in a pattern determined by the particle masses. This limitation has now been essentially removed, on an experimental basis, in electrophoresis experiments conducted in the near zero-gravity environment of a space shuttle.

on the ability of small solute particles, especially ions, to pass through a semipermeable membrane together with solvent molecules. The membrane is impermeable to large colloidal particles. The process can be expedited by placing the container of colloidal solution in an electric field. In this electrodialysis, ions are attracted to the electrode carrying the opposite charge from themselves. In the functioning of a human kidney, blood, a colloidal solution, is dialyzed to remove excess electrolytes produced in metabolic processes and to restore a proper electrolyte balance. In certain kidney diseases, the kidney loses the ability to purify blood. In these instances the function of the kidney can be conducted through an external device known as an artificial kidney or a dialysis machine.

The rates at which colloidal particles migrate in an electric field depend on the magnitudes of the charges they carry, on their shapes, and on other factors. In the process of **electrophoresis,** illustrated in Figure 12-23, colloidal particles are separated according to these differences in mobility.

The electrostatic factor in stabilizing colloids is particularly important in a class referred to as lyophobic ("solvent fearing"). Another class of colloids called lyophilic ("solvent loving") owe their stabilities to an ability to swell in a solvent and remain suspended. If the suspending medium is water, the prefix "hydro" replaces "lyo" in these two terms (i.e., hydrophobic and hydrophilic). A large number of hydrophilic colloidal materials are of biochemical interest and are considered again in Chapter 27.

Types of Colloids. Colloidal mixtures can be categorized in part according to the phases of matter involved. A brief listing is provided in Table 12-4.

TABLE 12-4
Some common types of colloids

Dispersed phase	Dispersion medium	Type	Examples
solid	liquid	sol	clay sols,[a] colloidal gold
liquid	liquid	emulsion	oil in water, milk, mayonnaise
gas	liquid	foam	soap and detergent suds, whipped cream, meringues
solid	gas	aerosol[b]	smoke, dust-laden air[c]
liquid	gas	aerosol[b]	fog, mist (as in aerosol products)
solid	solid	solid sol	ruby glass, certain natural and synthetic gems, blue rock salt, black diamond
liquid	solid	solid emulsion	opal, pearl
gas	solid	solid foam	pumice, lava, volcanic ash

[a] In water purification it is sometimes necessary to precipitate clay particles or other suspended colloidal materials. This is often done by treating the water with an aluminum compound, such as $Al_2(SO_4)_3$. The negatively charged clay particles are neutralized by Al^{3+} ions and coagulate or settle from solution. Clay sols are also suspected of adsorbing organic substances, such as pesticides, and distributing them in the environment.

[b] Smogs are complex materials that are at least partly colloidal. The suspended particles are both solid (smoke) and liquid (fog): smoke + fog = smog. Other constituents of smog are molecular, such as sulfur dioxide, carbon monoxide, nitric oxide, and ozone.

[c] The bluish haze of tobacco smoke and the brilliant sunsets in desert regions are both attributable to the scattering of light by colloidal particles suspended in air.

FOCUS ON
Obtaining Pure Materials

A single crystal of ultrapure silicon. Purities of 99.999999 atom % or better are required in the starting materials used to manufacture semiconductor electronic devices. [Courtesy of Joel E. Arem.]

Fortunately, the high levels of purity required in semiconductors are not needed in every chemical substance and process, for these levels are difficult to achieve. Consider the most widely used chemical—water. Depending on whether water is to be used for agricultural, domestic, industrial, or laboratory purposes, its purity can be quite variable. Although we consider domestic drinking water to be pure, its impurity content is usually appreciable. Drinking water might contain up to 500 parts per million (ppm) or 0.05% dissolved solids, for example. This high a content of dissolved solids is unacceptable for most laboratory applications and for some industrial operations, such as in feedwater for boilers, which typically should not contain more than a few tenths ppm of dissolved solids. In the chemical laboratory, water is generally purified by distillation, a procedure discussed earlier in this chapter, or by deionization. Deionization through ion exchange resins can be used to remove essentially all the dissolved ionic solids in water (see Section 22-2). Although a fairly high mineral content (hardness) can be tolerated in drinking water, environmental regulations require that the content of certain organic materials be much lower— 0.10 ppm for trihalomethanes (e.g., chloroform, $CHCl_3$)

and only 0.2 part per billion (ppb) of the pesticide endrin. To test water samples for such trace impurities obviously requires that the reagents used in the testing themselves be ultrapure.

The usual product of an industrial chemical process is a chemical that may contain several percent impurities. For example, standard propellant-grade hydrazine produced by the Raschig process (recall Figure 4-8) is 98.0% N_2H_4. Such "technical grade" chemicals can often be used directly in other industrial processes. For use in laboratories and for the production of pharmaceuticals or as additives to foods, chemicals must generally be further purified, perhaps to a "reagent grade." And, for a few specialized uses, as previously mentioned, ultrapure chemicals are required. Very often the complexity of the purification process is reflected in the relative costs of the different grades of a chemical, as suggested by the data in Table 12-5.

The characteristics of a semiconductor are critically dependent on the level of dopant (B, P, As, . . .) used; generally this is of the order of a few tenths to one part per million. The purities of the basic semiconductor materials to which these dopant "impurities" are added must be 100 to 1000 times greater. That is, the dopants are added to a material that initially has no more than about 10 ppb of impurities. A useful method of preparing these ultrapure materials is zone refining.

Zone Refining.   Our discussion of the phenomenon of freezing point depression (Section 12-5) started with the assumption that a solute is soluble in a liquid solvent but not in the solid solvent that freezes from solution. If this

FIGURE 12-24
The principle of zone refining illustrated.

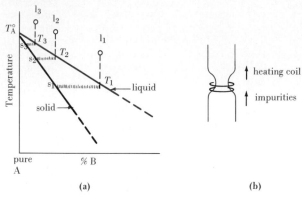

(a)   (b)

(a) The freezing point of pure A is T_A°. Addition of the impurity B lowers the freezing point of A; the greater the quantity of B, the greater the freezing point depression, as indicated by the line in color. The compositions of the solids that separate from the various freezing solutions lie along the line in black. When a liquid solution at point l_1 is cooled to the temperature T_1, it begins to freeze, initially producing a solid with composition s_1. Suppose that a small quantity of this solid could be removed from contact with the liquid and then reheated to point l_2. As the liquid l_2 is allowed to cool to point T_2, it begins to freeze, initially producing a solid with composition s_2. From this solid the liquid l_3 would be produced by remelting; liquid l_3 initially produces solid s_3 on freezing, and so on. With each melting/freezing cycle the melting (and freezing) point is increased and the point representing the composition of the solid moves closer to pure A. Different orientations of the liquid and solid composition lines are possible, but the technique of zone refining is generally applicable.
(b) As the heating coil moves up the rod of material, melting occurs. Impurities concentrate in the molten zone. The portion of the rod below the molten zone is purer than that in or above the zone. With each successive passage of the heating coil the rod becomes more pure. (See also Color Section D.)

TABLE 12-5
Specifications and prices of different grades of calcium carbonate

Grade	Specifications	Representative cost, \$/100 g
marble chips	for generating $CO_2(g)$[a]	0.70
light powder	98% $CaCO_3$, minimum	2.46
reagent	99% $CaCO_3$, approximate	6.25
reagent, analyzed	99.0% $CaCO_3$	11.28
ultrapure	99.999% $CaCO_3$	224.00

[a]For example, $CaCO_3(s) + 2\ HCl(aq) \rightarrow CaCl_2(aq) + H_2O + CO_2(g)$.

assumption were strictly valid, we would have, in principle, a method of preparing an ultrapure solid: Collect the solid that freezes from a solution, and the impurities will remain behind in the unfrozen solution. In actual practice the method is not so simple. Some impurities are entrapped in the frozen solvent, and this solid is wet with solution from which freezing has occurred, further contaminating it. Also, very likely some dissolving of the solute occurs in the solid solvent, even if this is slight. What we can say is that solute distributes itself between the liquid and solid solvent, as depicted in Figure 12-24a. Figure 12-24a bears a certain resemblance to the liquid–

vapor equilibrium curves presented in Figure 12-9 through 12-12. Those liquid–vapor equilibrium curves were used to establish how a substance can be purified by repeated cycles of vaporization and condensation. The caption to Figure 12-24 describes how repeated melting/freezing cycles can be used to obtain a pure solid. Figure 12-24b suggests that successive passes of a molten zone through a rod of material sweeps impurities to the end of the rod.

The end of the rod is cut off and the impurities are discarded. Wafers of semiconductor materials are obtained by cutting into thin slices a rod purified by zone refining.

Summary

Whether the mixing of two substances produces a solution depends largely on the strengths of intermolecular forces between like and unlike molecules. To predict whether an ionic solid will dissolve in water, forces of attraction between ions and water dipoles must be compared to interionic forces within the ionic crystal.

In describing a solution it is necessary to indicate the relative proportions of solvent and solute. This can be done through a variety of expressions for solution concentration. In most instances the ability of a solvent to dissolve a solute is limited. A solution containing this limiting quantity of solute is saturated. Solubility of a solute is a function of temperature, often represented through a solubility curve. Differences in solubility of a solute at different temperatures provide a basis for purifying substances known as recrystallization. The solubilities of gases depend on pressure as well as temperature. The concentration of a gas in solution is related to the pressure of the gas above the solution through Henry's law.

Colligative properties depend primarily on the number of solute particles in a solvent and not on the identity of the solute. One of these properties is the tendency of each solution component to lower the vapor pressure of other components. In ideal solutions this vapor pressure lowering conforms to Raoult's law. Graphs of boiling points of solutions as a function of composition help to describe how fractional distillation is used to separate solution components.

An important class of solutions is that in which solutes are nonvolatile. For these solutions freezing point depression and boiling point elevation can be used to determine the molecular weight of a solute. When dealing with solutes of high molecular weight (e.g., polymers) the preferred colligative property for molecular weight determination is osmotic pressure. Osmosis and osmotic pressure are important phenomena encountered in a number of biological systems.

Solutes in aqueous solution belong to one of three classes. Either they exist almost exclusively in molecular form (nonelectrolyte), partly in molecular and partly in ionic form (weak electrolyte), or in a completely ionized form (strong electrolyte). Depending on which of the three cases prevails, an aqueous solution is either a nonconductor of electricity, a weak conductor, or a good conductor. The expressions relating colligative properties to solution concentration must be modified if dissociation of a solute into ions occurs.

There are numerous instances, some of great practical importance, where a mixture does not fall neatly into the category of homogeneous (true solution) or heterogeneous. These intermediate cases are called colloidal mixtures. In general, if a material contains particles having one or more dimensions in the range 1 to 100 nm, it is called colloidal. Many distinctive features of colloidal materials stem from their high surface-to-volume ratios and their abilities to adsorb foreign molecules or ions on their surfaces.

Learning Objectives

As a result of studying Chapter 12, you should be able to

1. Explain the relationship between intermolecular forces of attraction and the formation of solutions, distinguishing between ideal and nonideal solutions.

2. Express solution concentration on a percent basis, or as molarity, molality, and mole fraction.

3. Distinguish among unsaturated, saturated, and supersaturated solutions, and describe how a solute may be purified by recrystallization.

4. Calculate the solubility of a gas in a liquid as a function of gas pressure.

5. Calculate partial vapor pressures of solution components, the total pressure, and the composition of vapor in equilibrium with an ideal solution.

6. Describe how solution components can be separated by fractional distillation.

7. Calculate, for dilute solutions of nonvolatile solutes, vapor pressures, freezing points, and boiling points.

8. Use measured freezing points or boiling points of solutions to calculate molecular weights of solutes.

9. Describe the process of osmosis, and use osmotic pressure data in molecular weight determinations.

10. Distinguish among strong, weak, and nonelectrolytes, and calculate ionic concentrations in aqueous solutions of strong electrolytes.

11. Describe the modifications of Arrhenius's theory of electrolytic dissociation made necessary because of interionic attractions in aqueous solutions.

12. Describe some distinctive properties of colloidal mixtures, and state how colloids differ from true solutions and from heterogeneous mixtures.

Some New Terms

Adsorption refers to the attachment of ions or molecules to the surface of a material.

An **alloy** is a mixture of two or more metals. Some alloys are solid solutions, some are heterogeneous mixtures, and some are intermetallic compounds.

An **azeotrope** is a solution which boils at a constant temperature, producing vapor of the same composition as the liquid. The boiling point of the azeotrope is, in some cases, lower than that of either solution component (minimum b. pt.), and in some cases higher (maximum b. pt.).

Colligative properties—vapor pressure lowering, freezing point depression, boiling point elevation and osmotic pressure—have values that depend only on the number of solute particles in the given solvent.

A **colloidal mixture** contains particles that are intermediate in size to those of a true solution and an ordinary heterogeneous mixture.

Dialysis is a process, similar to osmosis, in which ions or molecules in solution pass through a semipermeable membrane but colloidal particles do not.

Henry's law relates the solubility of a gas to the gas pressure maintained above a solution—$C = k \cdot P_{gas}$.

An **ideal solution** has $\Delta H_{soln} = 0$ and certain properties (notably vapor pressure) that are predictable from the properties of the solution components.

Molality (m) is a solution concentration expressed as number of moles of solute per kilogram of solvent.

The **mole fraction** of solution component i (χ_i) is the fraction of all the molecules in a solution which are of that type (i.e., type i).

A **nonelectrolyte** is a substance that is essentially un-ionized, both in the pure state and in solution.

Osmosis is the net flow of solvent molecules through a semipermeable membrane, from a more dilute solution (or from a pure solvent) into a more concentrated solution.

Osmotic pressure is the pressure that would have to be applied to a solution to stop the passage of molecules from the pure solvent through a semipermeable membrane into the solution.

Precipitation refers to the separation of a solid from a liquid solution.

Raoult's law states that the vapor pressure of a solution component is equal to the product of the vapor pressure of the pure liquid and its mole fraction in solution. Raoult's law applies to all volatile components in an ideal solution and to the solvent in a dilute nonideal solution.

Reverse osmosis is the passage through a semipermeable membrane of solvent molecules *from a solution into a pure solvent*. It can be achieved by applying to the solution a pressure in excess of its osmotic pressure.

A **saturated solution** is one that contains the maximum quantity of solute that is normally possible.

A **semipermeable membrane** permits the passage of some solution species but restricts the flow of others. It is a film of material containing submicroscopic pores.

A **solute(s)** is (are) the solution component(s) present in lesser amount(s) than the solvent.

The **solvent** is the solution component present in greatest quantity or the component that determines the state of matter in which a solution exists.

A **strong electrolyte** is a substance that exists completely in the form of ions in solution.

A **supersaturated solution** contains, because of its manner of preparation, more solute than normally expected.

An **unsaturated solution** contains less solute than the solution is capable of dissolving under the given conditions.

A **weak electrolyte** is a substance that is present in solution partly in the molecular form and partly as ions.

Zone refining is a purification process in which a rod of material is subjected to successive melting and freezing cycles. Impurities are swept by a moving molten zone to the end of the rod, which is cut off and discarded.

Suggestions for Further Study

HAWKINS, M. D., "Zone Melting," *Educ. in Chem.*, **9**, 94 (1972).

KANE, P. F., "Semiconductors: A Challenge in Trace Analysis," *Chemtech*, **1**, 532 (1971).

MATIJEVIC, E., "Colloids: The World of Neglected Dimensions," *Chemtech*, **3**, 656 (1973).

PFANN, W. G., "Zone Refining," *Scientific American*, **217**[12], 62 (1967).

SARQUIS, J., "Colloidal Systems," *J. Chem. Educ.*, **57**, 602 (1980).

TREPTOW, R. S., "Le Châtelier's Principle Applied to the Temperature Dependence of Solubility," *J. Chem. Educ.*, **61,**

499 (1984).

ZIEF, M. and A. J. BARNARD, Jr., "Re: High Purity Reagents and Their Uses," *Chemtech*, **3,** 440 (1973).

Review Problems

1. A saturated aqueous solution of KI at 20°C contains 144 g KI/100 g H_2O. Express this composition in the more conventional % (mass/mass), i.e., as g KI/100 g soln.

2. An aqueous solution with density 0.980 g/cm^3 at 20°C is prepared by dissolving 11.3 mL CH_3OH ($d = 0.793$ g/cm^3) in enough water to produce 75.0 mL of solution. What is the % CH_3OH, expressed as **(a)** % (vol/vol); **(b)** % (mass/vol); **(c)** % (mass/mass)?

3. It is desired to prepare 250.0 mL of 0.0250 M $AgNO_3$. What mass of a sample known to be 99.35% $AgNO_3$, by mass, is required for this purpose?

4. The concentration of an NaCl(aq) solution isotonic with blood is listed in the text as 0.90% NaCl, by mass. This solution has a density of 1.005 g/cm^3. What is the molarity of NaCl in the solution?

5. What is the molal concentration of *p*-dichlorobenzene in a solution prepared by dissolving 2.15 g $C_6H_4Cl_2$ in 25.0 cm^3 of benzene ($d = 0.879$ g/cm^3).

6. A solution is prepared by mixing the following hydrocarbons: 2.13 mol C_7H_{16}, 1.79 mol C_8H_{18}, and 3.11 mol C_9H_{20}. What are **(a)** the mole fraction and **(b)** the mole percent of each component of the solution?

7. A solution prepared by dissolving 0.80 mol NH_4Cl in 150.0 g H_2O is brought to a temperature of 25°C. Use Figure 12-7 to determine whether this solution is unsaturated or whether excess solute will precipitate.

8. Under an $O_2(g)$ pressure of 1.00 atm, 30.9 cm^3 of $O_2(g)$ at 25°C dissolves in 1.00 L H_2O at 25°C. Assuming that the solution volume remains at 1.00 L,

(a) What is the molarity of a saturated $O_2(aq)$ solution at 25°C when $P_{O_2} = 1$ atm?

(b) What is the aqueous solubility of O_2 at 25°C under the pressure of the atmosphere, which is 20.95% O_2, by volume? [*Hint:* Recall equations (5.16) and (12.4).]

9. What are the partial and total vapor pressures of a solution obtained by mixing 60.0 g benzene, C_6H_6, and 75.0 g toluene, C_7H_8, at 25°C? At 25°C the vapor pressure of pure C_6H_6 is 95.1 mmHg and that of C_7H_8 is 28.4 mmHg.

10. With reference to Review Problem 9, determine the vapor composition.

11. The addition of 1.10 g of an unknown compound reduces the freezing point of 75.22 g benzene from 5.51 to 4.90°C. What is the molecular weight of the compound?

12. The freezing point of an 0.01 m aqueous solution of a nonvolatile solute is -0.072°C. What would you expect the normal boiling point of this same solution to be?

13. Polyvinyl chloride (PVC) is a plastic widely used in the manufacture of food wrap and phonograph records. An 0.61-g sample of PVC is dissolved in 250.0 cm^3 of a suitable solvent at 25°C. The resulting solution has an osmotic pressure of 0.79 mmHg. What is the molecular weight of the PVC?

14. A solution is 0.110 M in KCl and 0.125 M in $MgCl_2$. What are the molarities of K^+, Mg^{2+}, and Cl^- in this solution?

15. Use your knowledge of strong, weak, and nonelectrolytes to arrange the following 0.0010 m aqueous solutions in order of *decreasing* freezing point: C_2H_5OH, NaCl, $MgBr_2$, $HC_2H_3O_2$, and $Al_2(SO_4)_3$.

Exercises

Homogeneous and heterogeneous mixtures

1. For each of the following solutions, indicate which component is the solvent and which is the solute. Comment on any difficulties in applying these terms. **(a)** 10 g C_2H_5OH(l) dissolved in 100 g H_2O(l); **(b)** 50 g CH_3OH(l) dissolved in 50 g H_2O(l); **(c)** 10 g CCl_4(l) dissolved in 50 g C_6H_6(l) + 50 g C_7H_8(l); **(d)** 1.0 M Na_2SO_4(aq).

2. Based on intermolecular forces, comment on the common phrases "like dissolves like" and "oil and water don't mix."

3. When 50.0 mL of ethanol and 50.0 mL of water are mixed, heat is evolved and the resulting solution has a volume of

96.0 mL. Which of the four situations described on page 329 do you think applies?

4. Explain the observation that all metal nitrates are water soluble, whereas many metal sulfides are not. Among metal sulfides, which would you expect to be most soluble?

Percent concentration

5. A sample of vinegar is 6.10% acetic acid ($HC_2H_3O_2$), by mass. What mass of $HC_2H_3O_2$ is contained in a 0.750-L bottle of the vinegar? Assume a density of 1.01 g/cm^3.

6. The calculations in Example 12-1 show that the percent ethanol, by mass, in a certain aqueous solution is less than the percent by volume in the same solution. Explain why you would expect this to be also true for all aqueous solutions of ethanol. Would it be true of all ethanol solutions, regardless of the other component? Explain.

7. Is either concentration term, percent by mass or percent by volume, independent of temperature? Explain.

Molar concentration

8. Commercial concentrated sulfuric acid is 94.0% H_2SO_4 and has a density of 1.831 g/cm^3. What is the molar concentration of H_2SO_4 in this solution?

9. How many mL of the ethanol-water solution described in Example 12-1 would have to be diluted with water to produce 1125 mL of 0.175 M C_2H_5OH?

10. A 10.00% by mass solution of ethanol, C_2H_5OH, in water has a density of 0.9831 g/cm^3 at 15°C and 0.9804 g/cm^3 at 25°C. What is the molarity of C_2H_5OH in this solution at each temperature?

Molal concentration

11. How many grams of iodine, I_2, must be dissolved in 125.0 mL of carbon tetrachloride, CCl_4 (density = 1.595 g/cm^3) to produce a 0.158 m I_2 solution?

12. An aqueous solution of hydrofluoric acid is 30.0% HF, by mass, and has a density of 1.101 g/cm^3. What are the molality and molarity of HF in this solution?

★13. A solution has a concentration described as 109.2 g KOH/L soln. The solution density is 1.09 g/cm^3. From 100.0 cm^3 of this solution it is desired to prepare 0.250 m KOH. What mass of which component, KOH or H_2O, must be added to the 100.0 cm^3 of solution?

Mole fraction, mole percent

14. Calculate the exact or approximate mole fraction of solute in the following aqueous solutions:
  (a) 12.2% C_2H_5OH, by mass;
  (b) 0.255 m $CO(NH_2)_2$ (urea);
  (c) 0.050 M $C_6H_{12}O_6$.
Which calculation(s) is (are) exact and which, approximate? Explain.

15. What mass of C_2H_5OH must be added to 100.0 mL of the solution described in Example 12-1(f) to increase the mole fraction of C_2H_5OH to 0.0500?

16. What volume of glycerol, $C_3H_8O_3$ ($d = 1.26$ g/cm^3), must be added per kilogram of water to produce a solution with 8.15 mole percent $C_3H_8O_3$?

Solubility equilibrium

17. Refer to Figure 12-7 and estimate the temperature at which a saturated aqueous solution of $KClO_4$ is 1.00 m.

18. A solution prepared by dissolving 26.0 g $KClO_4$ in 500.0 g of water is brought to a temperature of 20°C.
  (a) Refer to Figure 12-7 and determine whether the solution is unsaturated or supersaturated at 20°C.
  (b) Approximately what mass of $KClO_4$ must be added to make the solution saturated (if it is originally unsaturated) or what mass of $KClO_4$ can be crystallized from the solution (if it is originally supersaturated)?

19. A solid mixture consists of 95.0% NH_4Cl and 5.0% $(NH_4)_2SO_4$, by mass. A 50.0-g sample of this solid is added to 100.0 g of water at 90°C. With reference to Figure 12-7
  (a) Will all of the solid dissolve at 90°C?
  ★ (b) If the resulting solution is cooled to 0°C, approximately what mass of NH_4Cl can be crystallized?
  (c) Will $(NH_4)_2SO_4$ also crystallize at 0°C?

Solubility of gases

20. Certain natural waters contain dissolved $H_2S(g)$ [rotten-egg smell]. If such a water sample containing 0.5% by mass of H_2S is kept under a pressure of $H_2S(g)$ of 740 mmHg, will the sample dissolve more H_2S or lose some that is already dissolved? (*Hint:* Use data from Example 12-3.)

21. Most natural gases consist of about 90% methane, CH_4. Assume that the solubility of natural gas at 20°C and 1 atm gas pressure is about the same as that of CH_4, 0.02 g/kg water. If a sample of natural gas under a pressure of 20 atm is kept in contact with 1.00×10^3 kg of water, what mass of natural gas would you expect to dissolve?

22. Henry's law is often stated in this way: The mass of a gas dissolved by a given quantity of solvent at a fixed temperature is directly proportional to the pressure of the gas. Show how this statement is related to equation (12.4).

★23. Still another statement of Henry's law is this: A given quantity of liquid at a fixed temperature dissolves the same volume of gas at all pressures. What is the connection between this statement and the one given in Exercise 22? Under what conditions is this second statement not valid?

Raoult's law and liquid–vapor equilibrium

24. A benzene–toluene solution of $\chi_{benz.} = 0.300$ has a normal boiling point of 98.6°C. The vapor pressure of pure toluene at 98.6°C is 533 mmHg. What must be the vapor pressure of pure benzene at 98.6°C? (Assume ideal solution behavior.)

25. Calculate the vapor pressure at 25°C of a solution containing 21.8 g of the *nonvolatile* solute, urea, $CO(NH_2)_2$, in 525 g H_2O. The vapor pressure of water at 25°C is 23.8 mmHg.

*26. Calculate $\chi_{C_6H_6}$ in a benzene–toluene liquid solution that is in equilibrium at 25°C with a vapor phase that contains 62.0 mol % C_6H_6. (Use data from Review Problem 9.)

Freezing point depression and boiling point elevation

27. A compound is 42.4% C, 2.4% H, 16.6% N, and 37.8% O. The addition of 6.45 g of this compound to 50.0 mL of benzene ($d = 0.879$ g/cm^3) lowers the freezing point from 5.51 to 1.35°C. What is the molecular formula of this compound?

28. The addition of 1.00 g of benzene, C_6H_6, to 80.00 g of cyclohexane, C_6H_{12}, reduces the freezing point of the cyclohexane from 6.5 to 3.3°C.
   (a) What is the value of K_f for cyclohexane.
   (b) Which is the better solvent for molecular weight determinations by freezing point depression, benzene or cyclohexane? Explain.

29. What approximate proportions by volume of water ($d = 1.00$ g/cm^3) and ethylene glycol, $C_2H_6O_2$ ($d = 1.12$ g/cm^3), must be mixed to protect an automobile cooling system to $-10°C$? (Assume equation 12.7 applies.)

30. Citrus growers know that it is not necessary to fire their smudge (smoke) pots even if the temperature is expected to drop several degrees (Fahrenheit) below the normal freezing point of water for several hours.
   (a) Why doesn't the citrus fruit freeze at the normal freezing point (32°F)?
   (b) Why do you suppose that lemons freeze at a higher temperature than do oranges?

31. The boiling point of water at 756 mmHg is 99.85°C. What % $C_{12}H_{22}O_{11}$ should be present to raise the boiling point to 100°C at this pressure?

Osmotic pressure

32. When the stems of cut flowers are held in a concentrated salt solution, the flowers wilt. When a fresh cucumber is placed in a similar solution, it shrivels up (becomes pickled). Explain the basis of these phenomena.

33. Verify that a 20% sucrose solution would rise to a height of more than 100 m as a result of osmotic pressure.

34. At 25°C the average osmotic pressure of blood is 7.7 atm. What is the molar concentration of a glucose ($C_6H_{12}O_6$) solution that is isotonic with blood?

35. The mol. wt. of hemoglobin is 6.84×10^4. What mass of hemoglobin must be present per 100.0 cm^3 of a solution to exert an osmotic pressure of 6.15 mmHg at 25°C?

36. An 0.50-g sample of polyisobutylene in 100.0 cm^3 of benzene solution has an osmotic pressure at 25°C that supports a 5.1-mm column of the solution ($d = 0.88$ g/cm^3). What is the molecular weight of the polyisobutylene?

Strong electrolytes, weak electrolytes, and nonelectrolytes

37. Assuming no change in solution volume, what is the molarity of Cl^- in the solution obtained by adding 1.85 g $MgCl_2$ to 250.0 mL of 0.217 M $MgCl_2$?

38. Arrange the following aqueous solutions in order of increasing ability to conduct electric current. Comment on the reasons for this arrangement. 0.01 M NaCl; 1.0 M C_2H_5OH; 0.01 M $MgCl_2$; 0.01 M $HC_2H_3O_2$.

39. NH_3(aq) conducts electric current only weakly. The same is true for $HC_2H_3O_2$(aq). When these solutions are mixed, however, the resulting solution conducts electric current very well. Propose an explanation.

40. Predict the approximate freezing points of 0.10 m solutions of the following solutes dissolved in water: (a) urea; (b) NH_4NO_3; (c) $CaCl_2$; (d) $MgSO_4$; (e) ethanol; (f) HCl; (g) $HC_2H_3O_2$ (acetic acid).

41. Calculate the van't Hoff factors of the following weak electrolyte solutions:
   (a) 0.050 m $HCHO_2$, which freezes at $-0.0986°C$;
   (b) 0.100 M HNO_2, which has a hydrogen ion (and nitrite ion) concentration of 6.91×10^{-3} M.

Colloidal mixtures

42. Discuss some of the principal differences between colloidal mixtures and true solutions.

43. Describe what is meant by the terms (a) aerosol; (b) emulsion; (c) foam; (d) hydrophobic colloid; (e) electrophoresis.

44. The particles of a particular colloidal solution of arsenic trisulfide (As_2S_3) are negatively charged.
   (a) Sketch the results that would be obtained by the electrophoresis of this solution (see Figure 12-23).
   (b) Which 0.0005 M solution would be most effective in coagulating this colloidal solution: KCl, $MgCl_2$, $AlCl_3$, or Na_3PO_4? Explain.

*45. Suppose that 1.00 mg of gold is obtained in a colloidal dispersion in which the gold particles are assumed to be spherical, with a radius of 100 nm. (The density of gold is 19.3 g/cm^3.)
   (a) What is the total surface area of the particles?
   (b) What is the surface area of a single cube of gold weighing 1.00 mg?

Additional Exercises

1. A solution has a density of 1.235 g/cm^3 and is 90.0% glycerol, $C_3H_8O_3$, and 10.0% H_2O, by mass. Determine **(a)** the molarity of $C_3H_8O_3$ (with H_2O as the solvent); **(a)** the molarity of H_2O (with $C_3H_8O_3$ as the solvent); **(c)** the molality of H_2O in $C_3H_8O_3$; **(d)** the mole fraction of $C_3H_8O_3$; **(e)** the mole percent H_2O.

2. A brine solution contains 2.52% NaCl, by mass. If a 50.0 mL sample is found to weigh 51.1 g, how many liters of this brine would be required to extract 1000 kg NaCl?

3. Calculate the molality of the ethanol–water solution described in Exercise 10. Does the molality differ at the two temperatures (i.e., 15 and 25°C)? Explain.

★4. Water and phenol are only partially miscible at temperatures up to 66.8°C. In a mixture prepared at 29.6°C from 50.0 g of water and 50.0 g of phenol, 32.8 g of a phase consisting of 92.5% water and 7.50% phenol is obtained. This can be considered a saturated solution of phenol in water. What is the percent by mass of water in the second phase—a saturated solution of water in phenol?

5. Assuming the volumes are additive, what is the molarity of NO_3^- in a solution obtained by mixing 325 mL of 0.231 M KNO_3, 625 mL of 0.510 M $Mg(NO_3)_2$, and 825 mL of H_2O?

6. At 25°C and under 1 atm $N_2(g)$, the aqueous solubility of N_2 is 15.65 cm^3 $N_2(g)$ (measured at 25°C and 1 atm) per liter of solution. What would be the molarity of N_2 in a water solution that is in equilibrium with N_2 in the atmosphere? (Air contains 78.08% N_2, by volume.)

7. In a molecular weight determination it is desired to achieve a freezing point depression of between 2 and 3°C. If a 50.0-g sample of benzene is used as the solvent, what mass of unknown must be taken if the estimated molecular weight of the unknown is **(a)** 85; **(b)** 125?

8. Use the concentration of an isotonic saline solution given in the text to determine the osmotic pressure of blood at body temperature, 37.0°C. (*Hint:* Recall that NaCl is completely dissociated in aqueous solutions.)

9. What pressure is required in the reverse osmosis depicted in Figure 12-16 if the salt water contains 3.0% NaCl, by mass? (*Hint:* Recall that NaCl is completely dissociated in aqueous solutions. Also, assume a temperature of 25°C.)

10. HCl (b.p. = −84°C) and H_2O (b.p. = 100°C) form a maximum boiling azeotrope (b.p. = 110°C). A 5.00-mL sample of this azeotrope requires 30.32 mL of 1.006 M NaOH for its neutralization. The density of the azeotrope at 25°C is 1.099 g/cm^3.
   (a) Calculate the % HCl, by mass, in the azeotrope.
   (b) Sketch a boiling-point graph for the HCl–H_2O system in the manner of Figures 12-11 and 12-12.

$$HCl(aq) + NaOH(aq) \longrightarrow NaCl(aq) + H_2O$$

11. Solution A contains 0.515 g of urea, $CO(NH_2)_2$, dissolved in 85.0 g of water. Solution B contains 2.50 g of sucrose, $C_{12}H_{22}O_{11}$, dissolved in 92.5 g of water. Above which solution is the water vapor pressure greater?

★12. The two solutions described in Additional Exercise 11 are placed in separate containers but in an enclosure where their vapors may mix freely. Water evaporates from the solution of higher vapor pressure and condenses into the solution of lower vapor pressure until both solutions have the same water vapor pressure. What are the compositions of solutions A and B when this equilibrium is reached?

★13. At 20°C liquid benzene has a density of 0.879 g/cm^3; liquid toluene, 0.867 g/cm^3. Assuming ideal solutions
   (a) Calculate the densities of solutions containing 20, 40, 60, and 80 vol. % benzene.
   (b) Plot a graph of density vs. volume % composition.
   (c) Where V = vol. % benzene, establish that

$$d = \frac{1}{100}[0.879\, V + 0.867(100 - V)]$$

★14. The following data are given for the densities of ethanol-water solutions at 15°C as a function of *volume* percent ethanol: 0%, 0.999 g/cm^3; 20.0%, 0.977 g/cm^3; 40.0%, 0.952 g/cm^3; 60.0%, 0.914 g/cm^3; 80.0%, 0.864 g/cm^3; 100%, 0.794 g/cm^3. Are ethanol-water mixtures ideal?

★15. Demonstrate that for a *dilute aqueous* solution the molality is essentially equal to the molar concentration.

★16. Show that in a dilute solution the solute mole fraction is proportional to molality, and that in a dilute *aqueous* solution solute mole fraction is proportional to molarity.

★17. In Figure 12-10 (unlike the situation in Figure 12-9) the graph of boiling point versus liquid composition is not a straight line. Use the following data to demonstrate that this is indeed the situation expected for an ideal solution of two volatile components. For benzene, b.p. = 80.1°C, $\Delta\overline{H}_{vap}$ = 34.1 kJ/mol; for toluene, b.p. = 110.6°C, $\Delta\overline{H}_{vap}$ = 35.9 kJ/mol. (*Hint:* Pick a solution of any desired composition and determine the boiling point as if the graph were linear. For this temperature determine the partial and total vapor pressures above the solution. How does the total vapor pressure compare to 760 mmHg?)

★18. A saturated solution is prepared at 70°C containing 32.0 g $CuSO_4$ per 100.0 g soln. A 335-g sample of this solution is then cooled to 0°C and $CuSO_4 \cdot 5H_2O$ crystallizes out. If the concentration of a saturated solution at 0°C is 12.5 g $CuSO_4/100.0$ g soln, how many g $CuSO_4 \cdot 5H_2O$ would be obtained? (*Hint:* Note that the solution composition is stated in terms of $CuSO_4$ but that the solid that crystallizes is the hydrate, $CuSO_4 \cdot 5H_2O$.)

Self-Test Questions

For questions 1 through 8 select the single item that best completes each statement.

1. An *aqueous* solution is 0.01 M CH_3OH. The concentration of this solution is also very nearly (a) 0.01% CH_3OH (mass/vol); (b) 0.01 m CH_3OH; (c) $\chi_{CH_3OH} = 0.01$ (i.e., mole fraction $CH_3OH = 0.01$); (d) 0.99 M H_2O.

2. The most water soluble of the following compounds is (a) $C_6H_6(l)$; (b) $C(s)$; (c) $CH_3OH(l)$; (d) $C_{10}H_8(s)$.

3. The most likely of the following mixtures to be an ideal solution is (a) $NaCl-H_2O$; (b) $C_2H_5OH(l)-C_6H_6(l)$; (c) $C_7H_{16}(l)-H_2O$; (d) $C_7H_{16}(l)-C_8H_{18}(l)$.

4. The solubility of a nonreactive gas in water increases with (a) an increase in temperature; (b) an increase in gas pressure; (c) increases both in temperature and gas pressure; (d) an increase in the volume of water available.

5. The *best* electrical conductor of the following aqueous solutions is (a) 0.10 M NaCl; (b) 0.10 M C_2H_5OH (ethanol); (c) 0.10 M $HC_2H_3O_2$ (acetic acid); (d) 0.10 M $C_{12}H_{22}O_{11}$ (sucrose).

6. The aqueous solution with the *lowest* freezing point of the following group is (a) 0.01 m $MgSO_4$; (b) 0.01 m NaCl; (c) 0.01 m C_2H_5OH (ethanol); (d) 0.008 m MgI_2.

7. An ideal solution has equal mole fractions of two *volatile* components, A and B. In the *vapor* above the solution the mole fractions of A and B (a) are both 0.50; (b) are equal but not necessarily 0.50; (c) are not very likely to be equal; (d) are 1.00 and 0.00, respectively.

8. The best method for determining the molecular weight of a substance with high molecular weight generally involves measurement of (a) vapor density; (b) osmotic pressure; (c) freezing point depression; (d) boiling point elevation.

9. A solution contains 1.00 g of naphthalene, $C_{10}H_8$, in 44.0 g of benzene, C_6H_6.
    (a) What is the % $C_{10}H_8$, by mass, in this solution?
    (b) What is the molality of $C_{10}H_8$ in this solution?
    (c) What is the freezing point of the solution? [The freezing point of pure C_6H_6 is 5.51°C; K_f for C_6H_6 = 4.90°C kg solvent (mol solute)$^{-1}$.]

10. The boiling point of water at 735 mmHg is 99.07°C. What percent by mass of NaCl should be present in a water solution to raise the boiling point to 100.0°C? [K_b for water is 0.512°C kg solvent (mol solute)$^{-1}$.]

11. Pure liquid HCl is a poor electrical conductor. So is pure liquid water. When these two liquids are mixed, however, the resulting solution conducts electric current very well. How do you explain this?

12. How many mL of 0.250 M $MgCl_2$ must be added to 350.0 mL 0.250 M NaCl to produce a solution in which the concentration of chloride ion is 0.300 M Cl^-?

13

An Introduction to Descriptive Chemistry: The First 20 Elements

Chemical principles help us understand a variety of natural phenomena. These same principles have also provided a basis for many of the recent successes of applied chemistry, such as the creation of heretofore unknown materials.

An important bridge between theoretical and applied chemistry is descriptive chemistry. By this we mean a study of the elements and their compounds, based, where possible, on fundamental ideas about the structure and properties of matter. According to this definition certain aspects of descriptive chemistry have already been considered. What we will do in this chapter is approach descriptive chemistry in a more systematic way. We will focus on the first 20 elements, emphasizing topics that are related to principles that we have already studied, e.g., principles of atomic and molecular structure, periodic relationships, thermochemistry, phase equilibria, and intermolecular forces. Also, we will look at some of the typical reactions and uses of these elements and their compounds. Starting with Chapter 21, and for most of the text beyond that point, we will continue our discussion of the descriptive chemistry of these 20 and many of the other elements, employing at that time all of the principles developed in the text.

13-1 Occurrence and Preparation of the Elements

Of the 106 known elements, some 90 are obtained from natural sources. The remainder can only be synthesized by nuclear reactions (see Section 25-4). Moreover, most of the elements do not occur *free* in nature. Only about a fifth of the elements can be found in the elemental form; the rest occur exclusively in chemical combination.

Table 13-1 lists some of the more common elements and their abundances in the earth's crust. The earth's crust includes the atmosphere and terrestial waters together with the solid crust to a depth of approximately 50 km. Of the 12 elements listed in the body of Table 13-1, 10 are from among the first 20 elements, the elements surveyed in this chapter. Of course, the abundances of the elements in the earth's crust are considerably different from those of the entire earth, and even more different from those of the universe as a whole. For example, the core of the earth, with an approximate diameter of 3400 km, is thought to consist chiefly of iron and nickel. Hydrogen, although not a major component of the earth's crust, is estimated to account for 90% of all the atoms and three fourths of the mass of the universe, with helium accounting for most of the rest.

The data in Table 13-1 can be misleading. Aluminum, which is the most abundant of the metals, cannot be produced as cheaply as iron. Economic feasibility requires that an

TABLE 13-1
Abundance of the elements in the earth's crust[a]

Element	Abundance, mass %	Some commonly occurring forms
oxygen	49.3	water; silica; silicates; metallic oxides; molecular oxygen (in earth's atmosphere)
silicon	25.8	silica (sand, quartz, agate, flint); silicates (feldspar, clay, mica)
aluminum	7.6	silicates (clay, feldspar, mica); oxide (bauxite)
iron	4.7	oxide (hematite, magnetite)
calcium	3.4	carbonate (limestone, marble, chalk); sulfate (gypsum); fluoride (fluorite); silicates (feldspar, zeolites)
sodium	2.7	chloride (rock salt, ocean waters); silicates (feldspar, zeolite)
potassium	2.4	chloride; silicates (feldspar, mica)
magnesium	1.9	carbonate; chloride; sulfate (Epsom salts)
hydrogen	0.7	oxide (water); organic matter
titanium	0.4	oxide
chlorine	0.2	common salt (NaCl); sylvite (KCl); carnallite ($KCl \cdot MgCl_2 \cdot 6H_2O$)
phosphorus	0.1	phosphate rock [$Ca_3(PO_4)_2$]; organic matter
all others[b]	0.8	

[a] The earth's crust is taken to consist of the solid crust, terrestrial waters, and the atmosphere.

[b] This figure includes C, N, and S—all essential to life; less abundant, although common metals, such as Cu, Sn, Pb, Zn, and Cr; and some nonmetals with a variety of important uses, e.g., F, Br, I and B.

element be present in a concentrated form from which extraction is possible by physical or chemical means. These concentrated deposits are called **ores.** Aluminum is more widely distributed than iron, but Fe ores are more abundant than Al ores. Some elements found only in minor quantities in the earth's crust are widely known and used because of the availability of their ores (e.g., Cu, 0.005% abundance). Other elements are relatively abundant but difficult to isolate because they have no characteristic ores of their own. This is the case with rubidium, the sixteenth most abundant element.

For elements found in seawater, availability is based on their concentration in the water and on a chemically feasible and economically viable means of extracting them. The most abundant constituents of seawater are listed in Table 13-2. Seawater is already the principal source of two elements—Mg and Br. About one half the elements conceivably could be extracted from seawater.

Pure samples of certain elements can be obtained by physical separations; for example, the fractional distillation of liquid air is a source of the elements nitrogen, oxygen, and argon. The majority of the elements, however, require chemical reactions to release them from their combined forms. Since the objective of all such reactions is to obtain the element in the *zero* oxidation state, the required reactions are of the oxidation–reduction type.

13-2 Trends Among the First 20 Elements

From previous discussions we have come to expect similarities among the members of a vertical group of the periodic table. In this chapter we will also discover that the first member of a group differs somewhat from the heavier members; and among certain lighter elements "diagonal" similarities exist, specifically between Li and Mg, Be and Al, and B and Si.

TABLE 13-2
Principal constituents in seawater

Constituent	Concentration present, g/ton
Cl^-	18,980
Na^+	10,561
SO_4^{2-}	2,649
Mg^{2+}	1,272
Ca^{2+}	400
K^+	380
HCO_3^-	140
Br^-	65
H_3BO_3	26
Sr^{2+}	8
F^-	1

FIGURE 13-1
Two properties of the
second and third period
elements.

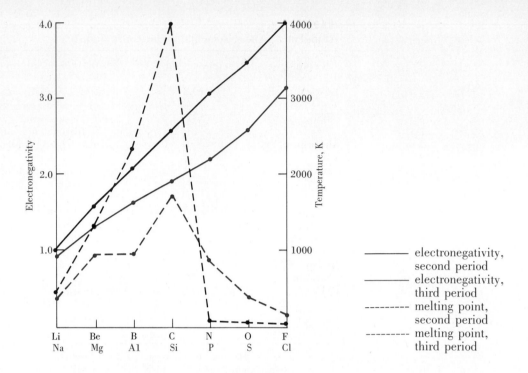

Horizontal trends are of various types, as suggested by Figure 13-1 for the second and third period elements. Some properties increase (or decrease) regularly from left to right in a period, e.g., electronegativity. Some properties may pass through a maximum (or minimum). Thus, melting point, which is related to the nature of intermolecular forces in a crystalline solid, rises to a maximum with C in the second period and with Si in the third. This is because C and Si form network covalent solids (recall Figure 11-21).

Example 13-1   Describe the trends that you would expect among the third period elements, E, **(a)** in electrical conductivity of the solid element; **(b)** in percent ionic character of the bond E—Cl.

Solution
(a) Electrical conductivity is closely related to metallic bonding. We should expect the three metals of the third period—Na, Mg, and Al—to be good electrical conductors. However, we have not studied the metallic state in sufficient detail to be able to make distinctions in the electrical conducting abilities among the three. Silicon is a metalloid and also a semiconductor. We should expect it to be not nearly so good an electrical conductor as the three metals preceding it. The remaining third-period elements—P, S, Cl (and Ar)—are all nonmetals and, therefore, nonconductors.

(b) We have learned that the greater the electronegativity difference between bonded atoms the greater the % ionic character of the bond (recall Section 9-10). EN differences can be obtained directly from Figure 13-1. They are the differences between the EN value for Cl and those of the other third-period elements. The greatest difference is between Cl and Na; the Na—Cl bond has the greatest % ionic character. This is followed by the Mg—Cl bond, Al—Cl bond, and so on through the third period. If we recall that an EN difference of about 2 or greater corresponds to a bond that is more than 50% ionic, we can see that only the Na—Cl and Mg—Cl bonds are essentially ionic and that the other bonds become progressively more covalent.

SIMILAR EXAMPLES: Exercises 1, 2.

13-3 Hydrogen

Since its identification by Cavendish in 1766, hydrogen, the simplest of the elements, has been at the center of the development of theories about the structure of matter. Dalton attempted to assign atomic weights to the various elements based on a relative mass of 1 for hydrogen; Prout (1815) suggested that hydrogen was the most fundamental atom and that all others could be formed from it; Davy (1810) suggested that hydrogen is the key element in common acids; Bohr (1913) chose the H atom for the first application of quantum mechanics to atomic structure; Schrödinger (1927) based his wave mechanics on the H atom; and theories of molecular structure use the H_2 molecule as a point of departure. But hydrogen is of practical importance too.

Occurrence and Preparation.   Hydrogen occurs in more compounds than any other element, but not all these compounds are equally suitable as starting materials for the preparation of $H_2(g)$. The most obvious compound to use, perhaps, is the most abundant one—water (H_2O). The required lowering of the oxidation state of H from +1 to 0 can be achieved with carbon, carbon monoxide, or methane as a reducing agent. [The term reforming, as used in reaction (13.3), means to reconstitute the elements of a hydrocarbon into a new compound(s).]

water gas reactions:
$$C(s) + H_2O(g) \xrightarrow{\Delta} CO(g) + H_2(g) \tag{13.1}$$

$$CO(g) + H_2O(g) \xrightarrow{\Delta} CO_2(g) + H_2(g) \tag{13.2}$$

reforming of methane: $CH_4(g) + H_2O(g) \xrightarrow{\Delta} CO(g) + 3\ H_2(g)$ (13.3)

Recall how Lavoisier decomposed HgO to obtain Hg(l) and O_2(g) (see Figure 2-1).

There are a few compounds that can be decomposed into their elements simply by heating to moderate temperatures, but such is not the case with H_2O. Even at temperatures as high as 2000°C, water is less than 1% dissociated into H_2 and O_2. A technique long used by chemists when decomposition is not possible thermally or by other chemical means is **electrolysis**—decomposition by electric current. The electrolysis of water is pictured in Figure 13-2, where a brief description of the process is also given. We will study the principles of electrochemistry in Chapter 20 and consider details of electrolysis reactions in that and subsequent chapters. For now, let us just note that when electric current is required to carry out a reaction this fact is indicated by writing "electrolysis" above the arrow.

$$2\ H_2O(l) \xrightarrow{\text{electrolysis}} 2\ H_2(g) + O_2(g) \tag{13.4}$$

Reaction (13.4) is simple and yields high-purity gases, but it does consume a great deal of electrical energy. Currently under investigation are ways of achieving the thermal decomposition of H_2O, but at temperatures far below the 2000°C cited above. The following cycle of reactions, for example, has H_2O as its only net reactant and $H_2(g)$ and $O_2(g)$ as its only net products. All other substances are recycled.

The concept of a net reaction for a multistep process was introduced in Figure 4-8.

$$6\ FeCl_2 + 8\ H_2O \xrightarrow{650°C} 2\ Fe_3O_4 + 12\ HCl + 2\ H_2$$
$$2\ Fe_3O_4 + 12\ HCl + 3\ Cl_2 \xrightarrow{200°C} 6\ FeCl_3 + 6\ H_2O + O_2$$
$$\underline{6\ FeCl_3 \xrightarrow{450°C} 6\ FeCl_2 + 3\ Cl_2}$$

net: $2\ H_2O \longrightarrow 2\ H_2 + O_2$ (13.5)

No temperature greater than 650°C is required in the process.

In the laboratory, $H_2(g)$ is most easily produced by the reaction of certain metals with an aqueous solution of an acid (e.g., HCl).

$$Zn(s) + 2\ H^+(aq) \longrightarrow Zn^{2+}(aq) + H_2(g) \tag{13.6}$$

FIGURE 13-2
The electrolysis of
water.

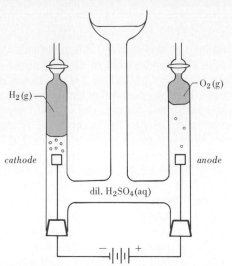

Electricity is passed through a dilute electrolyte solution, such as dilute $H_2SO_4(aq)$. H^+ ions are attracted to the negative electrode (cathode), where they gain electrons to form H atoms, which join to form H_2 molecules that escape from solution. Although SO_4^{2-} ions are attracted to the positive electrode (anode), they are not involved in the transfer of charge at the anode. Instead, a reaction occurs in which H_2O molecules decompose, replacing the H^+ ions lost at the cathode and releasing $O_2(g)$. The net reaction, then, is the production of the gases H_2 and O_2 at the expense of water molecules: $2 H_2O \rightarrow 2 H_2(g) + O_2(g)$.   Electrode processes are discussed in Chapter 20. In this chapter our interest is just on the net outcome of electrolysis reactions, not on the details of what happens at the electrodes.

If there is present in the acid solution NO_3^-, SO_4^{2-}, or any other oxidizing agent more powerful than H^+, $H_2(g)$ may not be produced. Instead a reduction product such as $NO(g)$, $N_2O(g)$, $NH_4^+(aq)$, or $SO_2(g)$ is obtained.

All but a few of the metals are capable of displacing $H_2(g)$ from a solution containing H^+; a means of identifying these exceptions is established in Chapter 20.

The most active of the metals, those of group IA and the heavier members of group IIA, are able to displace $H_2(g)$ from cold water.

$$2 M(s) + 2 H_2O(l) \longrightarrow 2 M^+(aq) + 2 OH^-(aq) + H_2(g) \qquad (13.7)$$
(M = group IA element)

$$M(s) + 2 H_2O(l) \longrightarrow M^{2+}(aq) + 2 OH^-(aq) + H_2(g) \qquad (13.8)$$
(M = Ca, Sr, Ba, Ra)

Reactions of Hydrogen. Hydrogen combines with the elements of the carbon (IVA), nitrogen (VA), oxygen (VIA), and halogen (VIIA) families, as well as with Be, Mg, B, Al, and Ga, to form **covalent hydrides.** For example,

$$H_2(g) + Cl_2(g) \longrightarrow 2 HCl(g) \qquad (13.9)$$

$$3 H_2(g) + N_2(g) \rightleftharpoons 2 NH_3(g) \qquad (13.10)$$

Means of influencing the equilibrium conditions in reaction (13.10) to maximize the production of NH_3 are discussed in Chapter 15.

Ionic hydrides are produced by the reaction of $H_2(g)$ with active metals, i.e., those of groups IA and IIA. In these compounds the hydrogen exists as hydride ion, $[H:]^-$.

$$2 M(s) + H_2(g) \longrightarrow 2 MH(s) \qquad M(s) + H_2(g) \longrightarrow MH_2(s) \qquad (13.11)$$
(M = any group IA element)   (M = Ca, Sr, Ba)

Ionic hydrides react with water to liberate $H_2(g)$. H^- is oxidized and H_2O is reduced.

CaH_2, a gray solid, is used as a portable source of $H_2(g)$, as for filling weather observation balloons.

$$H^- + H_2O \longrightarrow OH^-(aq) + H_2(g) \tag{13.12}$$

Uses. Large quantities of $H_2(g)$ are used in the synthesis of NH_3, HCl, and CH_3OH, and in the hydrogenation of fats and oils. It is used in cutting and welding torches, for filling lighter-than-air balloons, and, in liquid form, as a rocket fuel and for achieving low temperatures. $H_2(g)$ is a good reducing agent for producing metals from their oxides; for example,

$$Fe_2O_3(s) + 3\ H_2(g) \longrightarrow 2\ Fe(s) + 3\ H_2O(g) \tag{13.13}$$

One day hydrogen might largely replace gasoline as a fuel for transportation, natural gas as a fuel for space heating, and carbon (coke) as a metallurgical reducing agent. These uses of H_2 would constitute a fundamental change in our way of life, leading to **the hydrogen economy.** To achieve this hydrogen economy will require the production of vast quantities of $H_2(g)$ from water—either by electrolysis or a thermochemical cycle. In turn, this will require an almost unlimited source of energy, such as fusion energy (see Section 25-9).

Another problem that will have to be solved in the hydrogen economy is safe, economical storage of hydrogen. $H_2(g)$, because of its low density, requires large, bulky storage containers. Also, $H_2(g)$ forms explosive mixtures with $O_2(g)$ and air. $H_2(l)$ has a much higher density than $H_2(g)$ and represents a more efficient form of storage, although the liquid, of course, must be maintained at low temperatures. One of the more promising storage systems may be to dissolve $H_2(g)$ in metals (e.g., a Mg-Ni alloy). The gas could then be released by mild heating of the metal hydride.

13-4 Helium, Neon, and Argon

These are the three lightest members of group 0 of the periodic table. The other three members are Kr, Xe, and Rn. This group is variously known as the rare gases, inert gases, or noble gases, although He and Ar are not rare and Kr, Xe, and Rn are not inert. Helium was discovered to exist on the sun through spectroscopic observations of the solar eclipse of 1868. Argon was the first of the noble gases to be isolated from air, by John Rayleigh and William Ramsay in 1894. The isolation of He from a uranium mineral by Ramsay came in 1895. Based on Mendeleev's periodic table, of which the noble gases comprised a new group, Ramsay predicted that there should be an additional member with an atomic weight intermediate to that of He and Ar. This predicted gas, neon, was isolated from air by Ramsay and Morris Travers in 1898.

Until 1971, the U.S. government maintained a helium conservation program that called for He to be extracted from natural gas before the gas was sold. About 1×10^9 m^3 is currently being stored. When He is no longer available from natural gas, extraction from the atmosphere will make it a much more expensive commodity than it is today.

Occurrence. Air contains 0.000524% He, 0.001821% Ne, and 0.9340% Ar, by volume. Of the noble gases, Ar is the only one present in sufficient quantity to make its large-scale production from air practicable. In the case of helium, however, an alternate source is available. Certain natural gases in Texas, Oklahoma, and Kansas contain up to 2% He, by volume; and the extraction of He from these natural gases is feasible down to about 0.3% He. Alpha particles produced by radioactive decay give rise to He atoms in the earth's crust (recall Section 2-5).

Properties. The lighter noble gases are commercially important because of their lack of any significant chemical properties (i.e., their inertness) and the distinctiveness of some of their *physical* properties. Helium, in particular, possesses certain properties not shared by any other element. Properties of the common isotope ^{4}He include

If an empty beaker is partially submerged, open end up, in a liquid He-II bath, the liquid flows up the *outside* of the beaker, over the rim, and into the beaker, filling it to the same level as in the liquid He-II bath.

- A range of only one degree K between the boiling point (4.2 K) and the critical point (5.3 K).
- The existence of *two* forms of liquid He: a normal liquid, He-I, and a superfluid liquid, He-II. The superfluid has essentially no resistance to flow (viscosity) and a very high thermal conductivity.
- The existence of a liquid phase (He-II) down to 0 K.
- The lack of a triple point involving solid, liquid, and gaseous He. Solid He can only be obtained by applying pressure (about 25 atm) to He(l).

This unusual behavior can be represented through a phase diagram, as in Figure 13-3, but can be explained only in terms of quantum mechanics.

FIGURE 13-3
Phase diagram for helium-4.

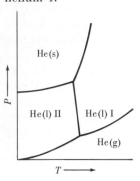

Uses.   Because of their inertness, both He and Ar can be used to blanket materials that need to be protected from N_2 and O_2 in the atmosphere, such as in certain types of welding, in metallurgical processes, and in the preparation of ultrapure Si and Ge in the manufacture of semiconductors. He is mixed with O_2 as a breathing mixture for deep-sea diving and in certain medical applications. Ar, mixed with N_2, is used to fill electric light bulbs, giving the filament longer life and increasing its electrical efficiency. Electric discharge through tubes containing Ne produces a distinctive red light (''neon lights'').

A large quantity of He is used to maintain materials at low temperatures (cryogenics). Metals essentially lose their electrical resistivity at liquid He temperatures and become superconductors. Powerful electromagnets can be made by immersing the electrical coils of the magnets in liquid He. Such magnets are used in particle accelerators and in fusion research (see Chapter 25). To improve the efficiency of electric power transmission, a possible future application might involve suspending metallic conductors in He(l) in underground transmission lines. Helium is also used to fill airships (blimps).

13-5 Lithium, Sodium, and Potassium

Two types of behavior are associated with the metallic state. Chemical behavior, including the ability to form ionic compounds with nonmetals, is associated with such atomic properties as low ionization energies and low electronegativities. Physical behavior, including hardness, high melting point, and the ability to conduct heat and electricity, is a reflection of metallic bonding. The group IA atoms, having but a single *s* valence electron that is rather easily lost, display the chemical behavior of metals to the highest degree. At the same time, because of their large sizes and limited numbers of valence electrons, bonding between alkali metal atoms is not as strong as in most metals, accounting for the low densities, melting points, and hardnesses listed in Table 13-3.

TABLE 13-3
Some properties of the alkali metals

	Li	Na	K	Rb	Cs
Density, g/cm³ at 20°C	0.534	0.971	0.862	1.532	1.873
Melting point, °C	179	97.8	63.7	38.9	28.5
Hardness[a]	0.6	0.4	0.5	0.3	0.3
Electrical conductivity[b]	17.4	35.2	23.1	13.0	8.1
Flame color	carmine	yellow	violet	bluish red	blue

[a]Hardness measures the ability of substances to scratch, abrade, or indent one another. On the Mohs scale, hardnesses of 10 minerals range from that of talc (0) to diamond (10). Other values: wax (0°C), 0.2; asphalt, 1–2; fingernail, 2.5; copper, 2.5–3; iron, 4–5; chromium, 9. Each substance is capable of scratching only others of hardness values less than its own.
[b]On a scale relative to silver, 100; copper, 95.9; gold, 67.5.

The low ionization energies of the group IA atoms (recall Table 8-4) suggest that the outer-shell electrons of these atoms can be promoted to higher energy levels relatively easily. This excitation might occur, for example, as a result of collisions between atoms in a gas flame. When these excited atoms revert to their normal (ground) states, specific amounts of energy are emitted in the form of light, imparting a characteristic color to the flame. The yellow color of the sodium flame is produced by the transition $[Ne]3p^1 \rightarrow [Ne]3s^1$.

This characteristic yellow color is shown in Color Section C.

Preparation and Uses of the Metals. Because an alkali metal atom is easily oxidized to the metal ion, the reverse process—reduction of the metal ion to the free metal—is difficult to accomplish by chemical means. The principal method of preparing Li and Na involves electrolysis of a molten salt, usually the chloride. For example,

$$2\ NaCl(l) \xrightarrow{\text{electrolysis}} 2\ Na(l)\ +\ Cl_2(g) \tag{13.14}$$

Because the alkali metals are very good reducing agents (i.e., easily oxidized), one alkali metal can be used to reduce another, as in the preparation of K.

$$KCl(l)\ +\ Na(g) \rightleftharpoons NaCl(l)\ +\ K(g) \tag{13.15}$$

Normally, reaction (13.15) would reach a condition of equilibrium in which most of the KCl(l) remains unreacted. However, the reaction can be conducted in such a way that the K(g) is carried away from the reaction mixture. Removal of one of the reacting species in a chemical equilibrium constitutes a stress which, according to Le Châtelier's principle, can be relieved by the production of more of that species. Sometimes a reaction can be carried to completion in this way.

The addition of small quantities of Li imparts high-temperature strength to Al and ductility to Mg. Li is also used in the production of specialized electrochemical cells and batteries. The principal use of Na has been in the production of tetramethyllead and tetraethyllead as antiknock additives for gasoline. Na is also used as a reducing agent in the production of certain metals, e.g., titanium, and as a light-emitting substance in lamps for street and highway lighting. Liquid Na is used as the heat exchange medium in fast-breeder nuclear reactors (see Section 25-8). Uses of potassium metal are limited to a few special applications where sodium, the cheaper metal, will not suffice.

A sodium vapor lamp is shown in Color Section A.

Reactions of Li, Na, and K. The chemistry of the IA metals is largely a reflection of the ease of oxidation of the metal atoms to the metal ions. The direct action of a halogen (X_2) on a group IA metal (M) results in a binary halide.

$$2\ M(s)\ +\ X_2(g) \longrightarrow 2\ MX(s) \tag{13.16}$$

Oxides result from the direct action of $O_2(g)$ on the metals; the varied products that are possible are described in Section 13-12. Ionic hydrides of the alkali metals were described in Section 13-3. Of the alkali metals, only Li forms a nitride by direct union with $N_2(g)$.

$$6\ Li(s)\ +\ N_2(g) \longrightarrow 2\ Li_3N(s) \tag{13.17}$$

The nitride reacts with water to form $NH_3(g)$ and LiOH. Li, Na, and K are such active metals that they will displace $H_2(g)$ from acid solutions (reaction 13.6) and from water (reaction 13.7).

Important Compounds of Li, Na, and K. Among the most important compounds of lithium are the chloride, sulfate, and carbonate. Li_2CO_3 is used in the manufacture of some types of glass and ceramic ware. In high purity it is used in the treatment of certain mental disorders. It also serves as a starting material in the preparation of other Li compounds, e.g.,

$$Li_2CO_3(s)\ +\ Ca(OH)_2(aq) \longrightarrow CaCO_3(s)\ +\ 2\ LiOH(aq) \tag{13.18}$$

An important use of LiOH(s) is in atmospheric regeneration (i.e., removal of CO_2) in submarines and spacecraft.

$$2 \text{ LiOH(s)} + CO_2(g) \longrightarrow Li_2CO_3(s) + H_2O \tag{13.19}$$

Sodium chloride is one of the most important mineral substances. Present annual consumption in the United States is about 50 million tons. Salt is used in the dairy industry, in the treatment of hides, the preservation of meat and fish, the control of ice on roads, and the regeneration of water softeners. In the chemical industry, sodium chloride is a source of sodium metal, chlorine gas, sodium hydroxide, hydrochloric acid, sodium carbonate, sodium sulfate, and other sodium and chlorine compounds.

Sodium hydroxide, which is produced by the electrolysis of NaCl(aq), is used in petroleum refining and in the manufacture of soaps, textiles, plastics, and other chemicals. Annual U.S. consumption is about 12 million tons.

Sodium sulfate is obtained both from natural sources and from the following process devised by J. R. Glauber (1604–1670).

$$H_2SO_4(\text{conc. aq}) + 2 \text{ NaCl(s)} \xrightarrow{\Delta} Na_2SO_4(s) + 2 \text{ HCl(g)} \tag{13.20}$$

The method works well because HCl is volatile and H_2SO_4 is not. The major use of Na_2SO_4 is in the paper industry (about 70% of the annual U.S. consumption of 1 million tons). In the kraft process for papermaking, undesirable lignin is removed from wood by digesting the wood in an alkaline solution of Na_2S. Na_2S is produced by reducing Na_2SO_4 with carbon.

The manufacture of glass accounts for slightly over one half of the Na_2CO_3 consumed in the United States. Other uses include the manufacture of chemicals, about 20%; pulp and paper, 6%; soap and detergents, 5%; and water treatment, 3%.

$$Na_2SO_4 + 4 \text{ C} \xrightarrow{\Delta} Na_2S + 4 \text{ CO} \tag{13.21}$$

About 100 lb of Na_2SO_4 is required for every ton of paper produced.

Sodium carbonate (soda ash) is widely used, especially in the manufacture of glass. Of the several million tons produced annually in the United States, somewhat more than half comes from natural sources—dry lakes in California and immense deposits in western Wyoming. The remainder is produced by a process developed by the Belgian chemist Solvay in 1863, described in Section 22-1.

Potassium compounds have some uses similar to those of sodium, e.g., K_2CO_3 in

FIGURE 13-4
Generalized Born–Haber cycle for alkali metal halides.

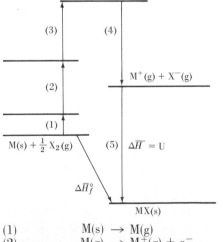

(1) $M(s) \rightarrow M(g)$
(2) $M(g) \rightarrow M^+(g) + e^-$
(3) $\frac{1}{2} X_2(g) \rightarrow X(g)$
(4) $X(g) + e^- \rightarrow X^-(g)$
(5) $M^+(g) + X^-(g) \rightarrow MX(s)$ $\Delta\overline{H} = U$
See also Figure 11-24.

glass and ceramics; but the most important use by far is in fertilizers. Potassium is one of the three main nutrients required by plants (N and P are the other two). KCl is commonly used for this purpose because this is the form in which most potassium is obtained from natural sources.

Properties of Alkali Metal Compounds.   The lattice energies of alkali metal compounds are generally lower than those of many other ionic compounds. This means also that the alkali metal compounds have melting points that are not exceptionally high (e.g., LiCl, 605°C; NaCl, 801°C; KCl, 770°C). Figure 13-4 helps us to explain some of the properties of alkali metal compounds. If the metal M is lithium, the enthalpy of sublimation (1) and the ionization energy (2) are considerably larger than for the other alkali metals because of the small size of the Li atom. But the lattice energy (step 5) also tends to be larger because of the small size of the Li^+ ion. The magnitude of the enthalpy of formation of a lithium compound relative to other alkali metal compounds depends, then, on the enthalpy changes of steps 1, 2, and 5, and most particularly on step 5. In the formation of fluorides, for example, the lattice energy of LiF(s) is so large (because of the small sizes of both Li^+ and F^- ions) that $\Delta \overline{H}_f^\circ$ is more negative for LiF than for any other alkali metal fluoride. The high lattice energy of LiF makes it much less water soluble than the other alkali metal fluorides.

	LiF	NaF	KF	RbF	CsF
lattice energy, kJ/mol	−1043	−928	−826	−789	−758
solubility, g/100 g H_2O	0.27	4.22	92.3	130.6	366.5

The situation just described for LiF does not hold for all lithium compounds, however. Another factor affecting the solubilities of lithium compounds is the high energy of hydration of the Li^+ ion in aqueous solution (again because of the small size of the Li^+ ion). This effect, which causes the solubilities of some lithium compounds to be unusually high, is discussed further in Section 22-1.

Diagonal Relationships.   As we have seen throughout this section, Li has several properties that set it apart from the rest of the alkali metals. In its ability to form a nitride, and in the very low solubility of its fluoride, carbonate, and phosphate, Li bears a strong resemblance to Mg. This similarity is referred to as a diagonal relationship. Such relationships also exist between Be and Al and between B and Si (see Figure 13-5). The Li–Mg similarity is thought to result from the approximately equal sizes of the Li and Mg atoms and the Li^+ and Mg^{2+} ions.

Example 13-2   Write equations to represent the reaction of **(a)** Na with H_2O; **(b)** K with HCl(aq); **(c)** Li_3N with H_2O.

FIGURE 13-5
Diagonal relationships.

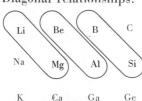

The encircled pairs of elements exhibit many similar properties.

Solution

(a) Equation (13.7) represents the general reaction of a IA metal with H_2O. Since Na is a IA metal, all we need to do is substitute Na for M in equation (13.7).

$$2\ Na(s) + 2\ H_2O(l) \longrightarrow 2\ Na^+(aq) + 2\ OH^-(aq) + H_2(g)$$

(b) Equation (13.6) represents the dissolving of a metal (Zn) in an acid (HCl). This equation is followed by the statement that all but a few metals are capable of displacing $H_2(g)$ from an aqueous solution of H^+. We certainly would not expect K to be one of the exceptions; it is among the most active (most easily oxidized) metals.

$$2\ K(s) + 2\ H^+(aq) \longrightarrow 2\ K^+(aq) + H_2(g)$$

(c) Often reactions are described verbally and not by an equation. Here we are required to supply an equation for a reaction that has already been described. On page 371 we find the statement, "The nitride (meaning Li_3N) reacts with water to form NH_3 and LiOH."

$$Li_3N(s) + 3\ H_2O \longrightarrow 3\ LiOH(aq) + NH_3(g)$$

SIMILAR EXAMPLES: Review Problems 4, 5; Exercise 8.

13-6 Beryllium, Magnesium, and Calcium

From a chemical standpoint, in their abilities to react with water and acids and to form ionic compounds, the heavier IIA elements—Ca, Sr, Ba, and Ra—are nearly as active as the alkali metals (IA). In terms of certain physical properties (e.g., density, hardness, and melting point), the IIA elements are more typically metallic than the IA elements, as can be seen by comparing Table 13-4 and Table 13-3.

The Special Case of Beryllium.   Be (and to some extent Mg) is rather different from the heavier numbers of group IIA. It can be seen from Table 13-4 that Be has a considerably higher melting point and is much harder than the other group IIA elements. Also, its chemical properties differ significantly. Some of its distinctive properties are

- Be is quite unreactive toward air and water.
- BeO does not react with water. [For the other oxides, $MO + H_2O \rightarrow M(OH)_2$.]
- Be and BeO dissolve in strongly basic solution to form the ion BeO_2^{2-}.
- $BeCl_2$ and BeF_2 in the molten state are poor conductors of electricity.

The chemical behavior of Be is best understood in terms of the small size and high ionization energy of the Be atom. The tendency to form the Be^{2+} ion is limited, and the ability of the Be atom to form covalent bonds is more pronounced. The reactivity of BeO with strongly basic solutions indicates that the oxide has acidic properties. This in turn is associated with small ionic size and high charge (discussed in Section 17-8). Covalent bond formation by beryllium appears to involve hybridized atomic orbitals. Figure 13-6 depicts bonding through sp hybrid orbitals in $BeCl_2(g)$ and through sp^3 hybrid orbitals in $BeCl_2(s)$.

Preparation and Uses of Be, Mg, and Ca.   Reduction of compounds to the free metals is not quite so difficult to achieve with Be, Mg, and Ca as with the alkali metals. Still, however, the preferred methods of preparation of Mg and Ca involve electrolysis. The molten salt chosen is generally the chloride, as in the Dow process for Mg (described further in Section 22-2).

TABLE 13-4
Some properties of the alkaline earth metals

	Be	Mg	Ca	Sr	Ba
Density, g/cm³ at 20°C	1.85	1.74	1.55	2.54	3.5
Melting point, °C	1278	651	845	769	725
Hardness[a]	ca. 5	2.0	1.5	1.8	ca. 2
Electrical conductivity[a]	8.8	36.3	35.2	7.0	—
Flame color	none	none	orange-red	scarlet	green

[a] See footnotes to Table 13-3.

FIGURE 13-6
Covalent bonds in $BeCl_2$.

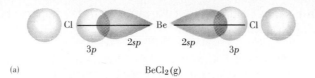

(a) $BeCl_2(g)$

(b) $BeCl_2(s)$

(a) In gaseous $BeCl_2$ discrete molecules exist with the bonding scheme shown in the figure.
(b) In solid $BeCl_2$ two Cl atoms are bonded to a Be atom through normal covalent bonds. Two others are bonded by coordinate covalent bonds, using lone-pair electrons of the Cl atoms (bonds shown as arrows). The arrangement is essentially tetrahedral. Of course, once formed, these two types of bonds cannot be distinguished from one another. $BeCl_2$ units are linked into long chain-like polymeric molecules—$(BeCl_2)_n$.

$$MgCl_2(l) \xrightarrow{\text{electrolysis}} Mg(l) + Cl_2(g) \tag{13.22}$$

Beryllium is produced from its fluoride, with Mg as a reducing agent.

Because of its ability to withstand metal fatigue, an alloy of Cu containing about 2% Be is used in making springs, clips, and electrical contacts. Other Be alloys are used for structural purposes where light weight is a primary requirement. Because the Be atom has so little stopping power for x rays or neutrons, beryllium is used in fabricating "windows" for x ray tubes and for various components in nuclear reactors.

Magnesium has a lower density than any other structural metal and is widely used in fabricating lightweight objects, such as aircraft parts, generally as an alloy with Al and other metals. Magnesium is also used as a reducing agent in certain metallurgical processes.

The primary use of calcium is as a reducing agent in the preparation, from their oxides or fluorides, of other, less common, metals such as Sc, W, Th, U, Pu, and most of the lanthanoids. Calcium is also used in the manufacture of batteries, the production of alloys, and in deoxidizing and degassing metals.

Reactions of Be, Mg, and Ca. Some typical reactions, illustrated with Mg, are

$$Mg + X_2 \longrightarrow MgX_2 \tag{13.23}$$
$$(X = F, Cl, Br, I)$$

$$2\,Mg + O_2 \longrightarrow 2\,MgO \tag{13.24}$$

$$3\,Mg + N_2 \longrightarrow Mg_3N_2 \tag{13.25}$$

Calcium reacts with cold water as shown in equation (13.8); but in the case of Mg, an impervious film of $Mg(OH)_2$ covers the surface and immediately stops the reaction. Mg does react with steam, however.

$$Mg(s) + H_2O(g) \xrightarrow{\Delta} MgO(s) + H_2(g) \tag{13.26}$$

Beryllium fails to react with either cold water or steam.

Important Compounds of Mg and Ca.   A few important compounds of magnesium and some of their uses are the

- carbonate (paint, printing inks, fireproofing and polishing compounds)
- chloride (production of Mg metal, fireproofing agents, cements)
- oxide (refractory bricks for furnaces and cement kilns)
- sulfate (fertilizers, animal-feed supplement, various applications in the textile industry)

Limestone is a naturally occurring form of $CaCO_3$ containing some clay and other impurities. It is the most widely used type of rock. Limestone's primary use is as a building stone (about 70%). Other uses include the manufacture of cement (15%), as a flux in metallurgical processes (5%), as a source of quicklime and slaked lime (5%), and as an ingredient in glass.

To obtain pure $CaCO_3$ from limestone requires (1) the thermal decomposition of limestone (calcination), (2) the reaction of CaO with water, and (3) the conversion of $Ca(OH)_2$ to precipitated $CaCO_3$.

calcination: $$CaCO_3(s) \xrightarrow{900°C} CaO(s) + CO_2(g) \tag{13.27}$$

slaking: $$CaO(s) + H_2O(l) \longrightarrow Ca(OH)_2(s) \tag{13.28}$$

carbonation: $$Ca(OH)_2(aq) + CO_2(g) \longrightarrow CaCO_3(s) + H_2O(l) \tag{13.29}$$

The principal use of precipitated $CaCO_3$ is as a coating to impart brightness, opacity, smoothness, and good ink-absorbing qualities to paper.

The term "lime" is used for two different calcium compounds. CaO is called **quicklime** and is formed in reaction (13.27); $Ca(OH)_2$ is called **slaked lime** and is formed in reaction (13.28). Reaction (13.27) is reversible, and at room temperature the reverse reaction occurs almost exclusively. As the temperature is increased, the forward reaction becomes more favorable, especially if $CO_2(g)$ is continuously exhausted from the furnace in which the reaction is conducted.

Slaked lime is the cheapest commercial alkaline (basic) substance, and it is used in all applications where high water solubility is not essential. $Ca(OH)_2$ is used in the manufacture of other alkalis and bleaching powder, in the purification of sugar, in tanning hides, and in water softening. A mixture of slaked lime, sand, and water is the familiar mortar used in brick laying. The initial setting of the mortar involves absorption of excess water by the bricks and its loss by evaporation. Final hardening of the mortar involves absorption of $CO_2(g)$ and conversion of $Ca(OH)_2$ back to $CaCO_3$.

Reaction (13.28), by which slaked lime is produced, is highly exothermic ($\Delta \overline{\overline{H}}° = -65$ kJ/mol). The heat of the reaction is sufficient to convert some excess water to steam.

$$Ca(OH)_2(s) + CO_2(g) \longrightarrow CaCO_3(s) + H_2O(g) \tag{13.30}$$

Another important calcium-containing mineral is **gypsum,** $CaSO_4 \cdot 2H_2O$, of which about 50 million tons is consumed annually in the United States. About one half of this quantity is converted to the hemihydrate ($\frac{1}{2}$-hydrate) known as **plaster of Paris.**

$$\underset{\text{gypsum}}{CaSO_4 \cdot 2H_2O(s)} \xrightarrow{\Delta} \underset{\text{plaster of Paris}}{CaSO_4 \cdot \tfrac{1}{2}H_2O(s)} + \tfrac{3}{2} H_2O(g) \tag{13.31}$$

When mixed with water, plaster of Paris reverts back to gypsum. Because it expands as it sets, a plaster of Paris–water mixture is useful in making castings where sharp details of an object must be retained. It is widely used in jewelry making and in dental work. The most important application of plaster of Paris, however, is in producing gypsum wall board, which has all but supplanted other interior wall covering materials in the construction industry.

13-7 Boron and Aluminum

Because they are both members of group IIIA we expect similarities between boron and aluminum. Similarities do exist, but there are perhaps more differences than similarities. Furthermore, from the diagonal relationship (Figure 13-5) we expect B to resemble Si and Al to resemble Be.

The tendency to lose electrons (three) to acquire a noble gas electron configuration is much less pronounced in B and Al than in the group IA and IIA elements. In fact, although we postulate the ion B^{3+} in descriptions of certain phenomena, the energy required to produce it is too high for B^{3+} to exist. Even the ion Al^{3+} is not commonly encountered, except perhaps in $AlF_3(s)$. In aqueous solution Al^{3+} is found as the hydrated ion $[Al(H_2O)_6]^{3+}$ or in other complex forms.

Boric acid, $B(OH)_3$, is a weak *acid* (produces H^+ rather than OH^- in aqueous solution). Aluminum hydroxide, $Al(OH)_3$, is primarily basic, although it does exhibit acidic properties under some circumstances. Whereas aluminum forms a single, ill-defined hydride, boron forms a series of hydrides. In this, B resembles Si, although boron hydrides are more complex than those of silicon. And because the bonded pair B—N is isoelectronic with the bonded pair C—C, there exist a number of covalent boron-nitrogen compounds with properties similar to organic compounds (Chapter 26).

Covalent Bonding.   Compounds of B and Al are characterized by (1) bonding through sp^2 hybrid orbitals and (2) a deficiency of electrons in the valence shell of the bonded B or Al atom. Let us consider some of the consequences of these conditions as they relate to the halides.

In the formula unit AlX_3 (X = Cl, Br, I), the $3s$ and all $3p$ orbitals of the Al atom are hybridized. Three of these sp^3 hybrid orbitals are occupied with electron pairs (one electron each from the three halogen atoms and three from Al); the fourth orbital is vacant. In the *dimer* Al_2X_6 the Al atom in each AlX_3 unit acquires an octet by sharing a pair of electrons donated by an X atom in the other AlX_3 unit. Dimerization through this process of "back bonding" is illustrated in Figure 13-7. (Dimerization does not occur in the halides BX_3, possibly because the small size of the B atom prevents it from being simultaneously bonded to four X atoms.)

In the presence of other atoms with strong electron-donating properties, the halides of B and Al may form species called adducts. In structure (9.23) we saw that the incomplete octet associated with B in BF_3 could be completed through formation of a coordinate covalent bond with N in NH_3, giving rise to the adduct $F_3B:NH_3$. In the presence of diethyl ether, for example, Al_2Cl_6 dimers split and each $AlCl_3$ unit forms an adduct with an ether molecule—$(C_2H_5)_2O:AlCl_3$. The arrangement of three Cl atoms and the O atom around the Al atom is tetrahedral. The halides of B and Al are used in organic chemistry where, through adduct formation, they can catalyze certain reactions.

Physical and Chemical Properties and Uses of Aluminum.   Pure aluminum is a malleable, ductile, silvery-colored metal of low density. Its density is only about one third that of steel. The metal is not very strong; but its strength increases considerably when it is alloyed with Cu, Mg, Mn, or Si. Al–Mg alloys, because of their low densities, are extensively used in the aircraft industry. They also find application in the construction industry.

Another important use of Al is as an electrical conductor. For a given diameter wire, Al has only about 60% of the conductivity of Cu, but because of its low density it is a better conductor than Cu on a mass basis. For this reason Al has found considerable use in recent years in electrical transmission lines. Aluminum in home electric wiring is now regarded as a fire hazard, however. Among the more familiar uses of Al is the fabrication of pots, pans, and other kitchen utensils. Both the metal and its ion (Al^{3+}) are non-poisonous.

AlF_3 has considerable ionic character. Its melting point is 1040°C, compared to 194°C for $AlCl_3$ (at 5.2 atm), 97.5°C for $AlBr_3$, and 191°C for AlI_3.

FIGURE 13-7
Bonding in Al_2Cl_6.

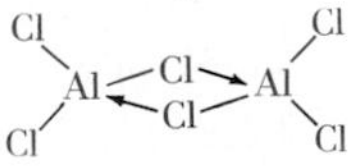

Two Cl atoms bridge the $AlCl_3$ units, producing the dimer Al_2Cl_6. Electrons donated by these Cl atoms to Al atoms are denoted by arrows.

Aluminum is a good reducing agent, meaning that it is fairly easily oxidized. For example, it dissolves in acids to produce $H_2(g)$.

Al does not dissolve in $HNO_3(aq)$. In this property it resembles Cr, for reasons given in Section 23-5.

$$2 \text{ Al(s)} + 6 \text{ H}^+(aq) \longrightarrow 2 \text{ Al}^{3+}(aq) + 3 \text{ H}_2(g) \tag{13.32}$$

If there is a good oxidizing agent present in the acidic solution, some species other than H^+ is reduced, such as in $H_2SO_4(aq)$ where SO_4^{2-} is reduced to $SO_2(g)$.

$$2 \text{ Al(s)} + 12 \text{ H}^+(aq) + 3 \text{ SO}_4^{2-}(aq) \longrightarrow 2 \text{ Al}^{3+}(aq) + 6 \text{ H}_2O + 3 \text{ SO}_2(g) \tag{13.33}$$

Aluminum is one of a small group of metals that dissolve in alkaline (basic) as well as acidic solution. This behavior, which results from acidic properties of $Al(OH)_3$, is explained in Section 22-3.

$$2 \text{ Al(s)} + 2 \text{ OH}^-(aq) + 6 \text{ H}_2O \longrightarrow 2 \text{ Al(OH)}_4^-(aq) + 3 \text{ H}_2(g) \tag{13.34}$$

The powdered metal is easily oxidized in air in a highly exothermic reaction.

$$2 \text{ Al(s)} + 3 \text{ O}_2(g) \longrightarrow \text{Al}_2O_3(s) \qquad \Delta \overline{H}^\circ = -1670 \text{ kJ/mol} \tag{13.35}$$

The source of oxygen for the conversion of Al to Al_2O_3 can be a substance other than air. In the thermite reaction (also known as the Goldschmidt reaction), Al combines with oxygen from another metal oxide, reducing that metal to its free state.

$$\text{Fe}_2O_3(s) + 2 \text{ Al(s)} \longrightarrow \text{Al}_2O_3(s) + 2 \text{ Fe(l)} \tag{13.36}$$

The thermite reaction can be used in on-site welding of large metal objects. Also it can be used for the preparation of small quantities of metals (e.g., chromium, if Cr_2O_3 is substituted for Fe_2O_3).

A widely used reducing agent, especially in organic chemistry, is lithium aluminum hydride, $LiAlH_4$. The AlH_4^- ion can be thought of as an adduct of AlH_3 and H^-. The reduction of water by AlH_4^- produces $H_2(g)$.

$$\text{AlH}_4^- + 4 \text{ H}_2O \longrightarrow \text{Al(OH)}_4^-(aq) + 4 \text{ H}_2(g) \tag{13.37}$$

Acid–base reactions of $Al(OH)_3$, complex ion formation by Al^{3+}, the metallurgy of Al, and other aspects of the chemistry of aluminum are described later in the text.

13-8 Carbon

Carbon occupies a special place among the elements in the variety and complexity of the compounds that it can form. This can be attributed to the great strength of carbon-to-carbon bonds, making it possible for C atoms to bond together into long chains, into rings, or combinations of rings and chains. Moreover, the ground-state electron configuration of carbon ($1s^2 2s^2 2p^2$) is easily hybridized to produce the orbital sets sp^3, or $sp^2 + p$, or $sp + p^2$. This means that carbon chain and ring structures may have multiple as well as single bonds between C atoms. The study of carbon atoms in combination with H, O, N, S, and a few other elements is such a vast undertaking that it constitutes a specialized branch of chemistry—organic chemistry. Organic chemistry and closely related topics in biochemistry are surveyed in Chapters 26 and 27. For the present we will just look at a few aspects of what might be called the inorganic chemistry of carbon.

Physical Forms of Carbon.   In Section 11-10 we used the two principal forms of carbon—diamond and graphite—as examples of network covalent solids. There we discussed the nature of the bonding and some of the physical properties of each form. Of the two forms, graphite is the more stable at ordinary temperatures and pressures (i.e., at 298 K and 1 atm). Thus, it is assigned an enthalpy of formation of zero, and there is an enthalpy difference between graphite and diamond.

$$\text{C(graphite)} \longrightarrow \text{C(diamond)} \qquad \Delta \overline{H}^\circ = +1.88 \text{ kJ/mol} \tag{13.38}$$

That graphite is the stable form of carbon means that in a phase diagram of carbon the point (1 atm, 298 K) falls in the area representing graphite. The area representing diamond is at much higher pressures. All of this suggests that, contrary to jewelers' advertising claims that "diamonds last forever," in time a diamond should convert to graphite. However, phase changes that require rearrangement of a crystalline structure often occur extremely slowly at room temperature (essentially not at all), and, fortunately, this is the case with the diamond → graphite transition. Of course, it also follows that diamonds cannot be synthesized from graphite under ordinary conditions. Conditions that existed on earth for the creation of natural diamonds were probably similar to those that are now used to synthesize artificial diamonds in the laboratory—pressures from 5000 to 100,000 atm, temperatures from 1200 to 2400 K, and the presence of catalysts.

In addition to the graphite and diamond modifications, carbon can be obtained in a variety of forms known as **amorphous carbon** (although most of these seem to consist of microcrystalline graphite). When coal is heated in the absence of air, various volatile substances are driven off (coal gas), leaving a high-carbon residue known as **coke.** This same type of destructive distillation of wood and other organic matter yields **charcoal.** Incomplete combustion of natural gas produces a smoky flame, and this smoke can be deposited as a finely divided soot called **carbon black.** Finally, a very pure form of carbon can be obtained by heating sucrose in the absence of air.

$$C_{12}H_{22}O_{11}(s) \xrightarrow{\Delta} 12\ C(s) + 11\ H_2O(g) \tag{13.39}$$

Coke is the principal metallurgical reducing agent in use today, and it is also used in the manufacture of fuel gases (recall reaction 6.18). The principal uses of carbon black are as a filler in rubber tires and as a pigment in printing inks.

Oxides of Carbon.   When carbon or any of a variety of organic compounds is burned in a limited quantity of air, the principal carbon-containing product is **carbon monoxide.**

$$2\ C(s) + O_2(g) \longrightarrow 2\ CO(g) \tag{13.40}$$

Because of a very high ratio of surface area to volume, certain forms of charcoal (called activated charcoal) exhibit strong surface properties, specifically an ability to adsorb substances from liquid solutions or from the gaseous state. Activated charcoal is used in gas masks, for example.

A battery of modern coke ovens at a steelmaking plant. The 80 slot-type ovens, commissioned in 1982, are designed to produce 850,000 tons of coke per year. [Courtesy of Bethlehem Steel Corporation.]

Lavoisier was first to demonstrate (1772) that diamonds consist of carbon; that they burn in air to produce $CO_2(g)$; and that they do not burn in the absence of air.

In an excess of air, **carbon dioxide** is formed.

$$C(s) + O_2(g) \longrightarrow CO_2(g) \tag{13.41}$$

Also, an excess of air completes the conversion of CO to CO_2.

$$2\ CO(g) + O_2(g) \longrightarrow 2\ CO_2(g) \tag{13.42}$$

Incomplete combustion of fossil fuels is a principal source of CO as an air pollutant. About 80% of this CO is believed to originate in automotive internal combustion engines. CO is an inhalation poison because CO molecules bond to Fe atoms in hemoglobin in blood, displacing the O_2 molecules normally carried by the hemoglobin. This action of CO is similar to that involved in the formation of metal carbonyls, discussed in Section 23-7. On the beneficial side, CO mixed with H_2 in various proportions, known as **synthesis gas,** is used to synthesize a host of organic compounds, such as

$$CO(g) + 2\ H_2(g) \longrightarrow CH_3OH(l) \tag{13.43}$$
$$\text{methanol}$$

Also, CO is a good reducing agent. In a blast furnace, for example, coke is converted to CO by reaction (13.40) and the CO reduces iron oxide to iron.

$$Fe_2O_3(s) + 3\ CO(g) \xrightarrow{\Delta} 2\ Fe(l) + 3\ CO_2(g) \tag{13.44}$$

Some scientists are concerned about the "synfuel" (synthetic fuel) approach to solving future energy problems. Such an approach implies large increases in the combustion of fossil fuels and a corresponding buildup of atmospheric $CO_2(g)$.

$CO_2(g)$ is not normally considered to be an air pollutant because it is essentially nontoxic. However, its potential effects on the environment are disturbing. The problem is that $CO_2(g)$ absorbs infrared radiation from the earth and reradiates it back to earth. This "greenhouse effect" (see Figure 13-8) could have the effect of raising the average temperature of the earth. With an increase in temperature could come changes in climate and agricultural productivity and, possibly, even melting of the polar ice caps.

$CO_2(g)$ in the atmosphere increased from 296 to 315 ppm (parts per million) in the first half of this century; and from 1958 to 1979, by another 20 ppm, bringing the current total to a figure in excess of 335 ppm.* These increases are attributed to the burning of fossil fuels and slash-and-burn agricultural methods. Increases in low cloud coverage and particulate concentrations in the atmosphere could offset some of the effects of a $CO_2(g)$ buildup. Best estimates are that the earth's average temperature might increase by as much as 6°C over the next 200 years. Scientific opinion on the significance of a CO_2 buildup varies, primarily because the oceans serve as a vast reservoir for dissolved CO_2, and in ways that are not completely understood.

$CO_2(g)$ dissolves in water and comes into equilibrium with carbonic acid, a weak acid that ionizes in two stages. Neutralization of the acid through the first stage (e.g., with NaOH) produces a salt variously called a hydrogen carbonate, an acid carbonate, or a bicarbonate. Neutralization through the second stage (i.e., neutralization of the acid carbonate) yields a carbonate.

$$CO_2(g) + H_2O \rightleftharpoons H_2CO_3(aq) \tag{13.45}$$

$$H_2CO_3(aq) + Na^+ + OH^- \longrightarrow \underset{\text{sodium hydrogen carbonate}}{Na^+ + HCO_3^-} + H_2O \tag{13.46}$$

$$Na^+ + HCO_3^- + Na^+ + OH^- \longrightarrow \underset{\text{sodium carbonate}}{2\ Na^+ + CO_3^{2-}} + H_2O \tag{13.47}$$

Reversal of the above reactions through the action of an acid yields $CO_2(g)$; that is,

$$Na_2CO_3(aq) + 2\ H^+(aq) \longrightarrow 2\ Na^+(aq) + H_2O + CO_2(g) \tag{13.48}$$

In later chapters we will discuss phenomena associated with carbon dioxide, carbonic acid, and carbonates, ranging from their action in blood to the formation of limestone caves to hardness in water and water softening.

*R. Revelle, "Carbon Dioxide and World Climate," *Scientific American,* **247**[2], 35 (1982).

FIGURE 13-8
The "greenhouse effect."

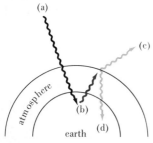

(a) Sunlight received by earth.
(b) Infrared radiation (heat) from surface of earth absorbed by $CO_2(g)$ in atmosphere.
(c) Infrared radiation emitted to space by $CO_2(g)$.
(d) Infrared radiation emitted back to earth by $CO_2(g)$.

FIGURE 13-9
Bonding in silica—SiO_2.

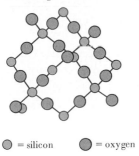

○ = silicon   ● = oxygen

13-9 Silicon

A silicon atom, like a carbon atom, can form four bonds simultaneously, arranged in a tetrahedral fashion. The element Si crystallizes in the same type of fcc structure as diamond (Figure 11-21). Whereas diamond is an electrical insulator and graphite, a moderately good conductor, silicon is a *semiconductor*. In silica (SiO_2) each Si atom is bonded to *four* O atoms and each O atom to *two* Si atoms. This structural arrangement extends throughout a very large network, as suggested by Figure 13-9. Certain properties of silica resemble those of diamond. Quartz, a form of silica, has a Mohs hardness of 7 compared to 10 for diamond (see again, footnote to Table 13-3). Quartz has a high melting point (about 1700°C) and is a nonconductor of electric current. Quartz is a common form of silica, but there are more than a dozen forms of silica altogether.

Silicate Minerals.   If C is the key element of the living world, Si is no less the key element of the inanimate or mineral world. Only the simple alkali metal silicates can be obtained as water-soluble compounds. The vast majority of silicates are highly insoluble. A common feature of silicate minerals is the complexity of the silicate anions. Nevertheless, within these complex anions the basic structural unit is a simple tetrahedral arrangement of four O atoms about a central Si atoms. These tetrahedra may exist (a) as separate units; (b) joined into chains or rings in groups of 2, 3, 4, or 6; (c) joined together into long single chains or (d) double chains; (e) arranged in sheets; (f) linked into a three-dimensional network. Two of these possibilities are pictured in Figure 13-10, and several examples are listed in Table 13-5. To establish the charge on a silicate anion in this table, it is convenient to think of each Si atom as carrying a charge of +4, and each O, −2. Thus, the charge on the anion SiO_4 is −4, and on Si_2O_7, −6.

Colloidal Silica.   Just as CO_2 dissolves in water and reacts with bases to produce carbonates, SiO_2 dissolves slowly in strong bases. It forms a variety of silicates, such as Na_2SiO_3 (sodium metasilicate) and Na_2SiO_4 (sodium orthosilicate). Aqueous sodium silicate is often encountered under the name "water glass."

Acidification of an aqueous carbonate solution produces H_2CO_3, which decomposes to H_2O and CO_2 (reaction 13.48). The silicic acids resulting from the acidification of silicate solutions are also unstable; they decompose to silica. Depending on the acidity of the solution, the silica may be obtained as a colloidal dispersion, a gelatinous precipitate, or a solidlike gel in which all of the liquid is entrapped. Water molecules are eliminated

This is the process referred to in Section 12-8 for the preparation of colloidal silica.

FIGURE 13-10
Silicate anions.

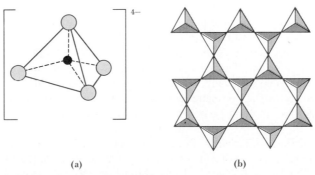

(a)   (b)

(a) The simple anion $SiO_4{}^{4-}$ consists of a central Si atom surrounded by four O atoms in a tetrahedral arrangement.
(b) A portion of the anion structure of mica. The cations required to balance the electric charge of the sheet anions are located above and below each sheet. Usually these are Al^{3+} and K^+. Talc and clay have similar structures.

TABLE 13-5
Some representative silicate minerals

Mineral type	Anion type	Mineral	Composition
(a) orthosilicate	SiO_4^{4-}	zircon	$ZrSiO_4$
		olivine	Mg_2SiO_4
(b) polysilicate	$Si_2O_7^{6-}$	thortveitite	$Sc_2Si_2O_7$
	$Si_6O_{18}^{12-}$	beryl	$Be_3Al_2Si_6O_{18}$
(c) pyroxene	Si—O chains	diopside	$CaMg(SiO_3)_2$
		spodumene	$LiAl(SiO_3)_2$
(d) amphibole	Si—O double chains	tremolite	$Ca_2Mg_5(Si_4O_{11})_2(OH)_2$
		hornblende	$(Ca,Na,K)_{2-3}(Mg,Fe,Al)_5$ $(Si,Al)_2Si_6O_{22}(OH)_2$
(e) mica	Si—O sheets	muscovite	$KAl_2(AlSi_3O_{10})(OH)_2$
(f) zeolite	Si—O three-dimensional network	natrolite	$Na_2Al_2Si_3O_{10} \cdot 2\,H_2O$

between neighboring silicic acid molecules, producing a polymer of ever increasing size, as suggested below.

$$SiO_4^{4-}(aq) + 4\,H^+(aq) \longrightarrow Si(OH)_4 \qquad (13.49)$$

which is followed by

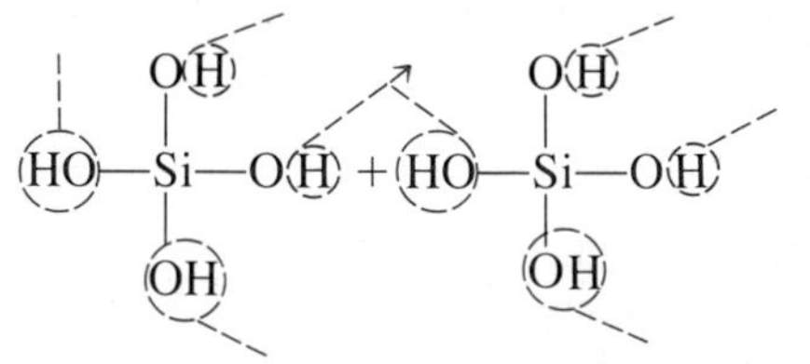

$$+ \cdots \longrightarrow \longrightarrow (x\ SiO_2 \cdot yH_2O) \qquad (13.50)$$

colloidal silica

Glass. If a mixture of sodium and calcium carbonates is fused with sand at about 1500°C, the product is a mixture of sodium and calcium silicates. This mixture is a liquid which, on cooling, becomes so viscous that it ceases to flow. The product has the appearance of a solid, and even in sheets of considerable thickness, it is transparent to visible light. What has just been described is ordinary soda-lime glass. Variations in the proportions of the three basic ingredients, together with the admixture of other substances, can be used to alter the properties of glass. For example, borosilicate glass (Pyrex) contains about 13% B_2O_3.

Any substance in a glasslike state is referred to as a "glass," whether it is of silicate origin or not. This terminology is based on the uniqueness of the glass structure. Because there is no long-range order in the arrangements of their atoms, glasses are noncrystalline. X ray diffraction patterns of glasses resemble those of liquids; and glasses do not have definite melting points.

Portland Cement. Portland cement is a complex mixture of calcium silicates and aluminates. It is formed by heating limestone at about 1500°C with materials rich in silica (SiO_2) and alumina (Al_2O_3). Minor components may also be present. The product is pulverized before use. The reactions of cement with water and its subsequent hardening are complicated. Certain of the reactions occur almost immediately upon mixing cement and water. Others progress over a period of months or years. Because only water is

Cement manufacture. The production of cement occurs in long, firebrick-lined rotary kilns. A carefully proportioned mixture of powdered materials, including limestone, clay, and sand, is heated to progressively higher temperatures as it moves down the inclined kiln. First moisture, and then chemically bound water, are driven off. This is followed by calcination of $MgCO_3$ and $CaCO_3$ (reaction 13.27), and finally by the reaction of oxides to form silicates, aluminates, and other cement components. The pile of black material in the background is powdered coal, the fuel used to heat the rotary kilns. [Courtesy of Jerry Farr, Riverside Cement Co.]

required to set portland cement, it is referred to as a hydraulic cement. It will set even when completely submerged in water, as in the construction of bridge piers. Pure cement does not have a great deal of strength, but when mixed with sand and gravel it sets into a hard mass. This mixture is common concrete.

13-10 Nitrogen

Nitrogen has the electron configuration $1s^2 2s^2 2p^3$. In forming compounds with other atoms, the N atom is able to gain, or more likely share, three electrons to acquire a valence shell octet, $1s^2 2s^2 2p^6$. The oxidation state of N in its compounds can range from -3 to $+5$. The maximum oxidation state corresponds to its periodic group number, VA.

Although the variability of oxidation states results in an unusually rich chemistry of nitrogen compounds, the precursor of all nitrogen compounds, elemental nitrogen, N_2, is rather inert. This lack of reactivity can be attributed to the great strength of the bond between N atoms in N_2; 946.4 kJ of energy is required to rupture 1 mol of these bonds.

$$N \equiv N(g) \longrightarrow 2 \, N(g) \qquad \Delta \overline{H}^\circ = +946.4 \text{ kJ/mol} \qquad (13.51)$$

Because of this high bond energy, many nitrogen-containing compounds have positive enthalpies of formation.

Nitrogen in the Atmosphere.   As we have noted on several occasions, $N_2(g)$ is the primary component of air (78.08%, by volume). Moreover, with the exception of deposits

FIGURE 13-11
The separation of
atmospheric gases.

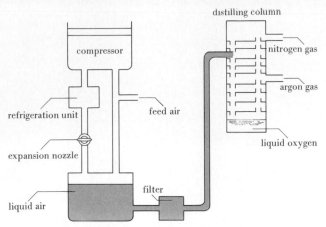

Clean air is fed into a compressor and then cooled by refrigeration. The cold air expands through a nozzle, and as a result is cooled still further—sufficiently to cause it to liquefy. The liquid air is filtered to remove CO_2(s) and hydrocarbons and then distilled. Liquid air enters the top of the column where nitrogen, the most volatile component, passes off as a gas. In the middle of the column gaseous argon is removed, and liquid oxygen, the least volatile component, collects at the bottom. The normal boiling points of nitrogen, argon, and oxygen are -195.8, -185.7, and $-183.0°C$, respectively.

of $NaNO_3$ in Chile and Peru, nitrogen compounds do not occur in significant quantities in the earth's crust. This means that the principal source of N_2(g) and manufactured nitrogen compounds is the atmosphere. The separation of liquid air into its components by fractional distillation is described in Figure 13-11.

Current consumption of N_2 in the United States is about 15 million tons annually. One important use of N_2(g) is to provide inert (blanketing) atmospheres in the metals, electronics, and chemical process industries. Liquid N_2 is used as a freezing agent in the food processing industry. Another important use, of course, is in the production of nitrogen compounds, principally through the manufacture of NH_3.

Chemically combined nitrogen is called "fixed" nitrogen, and any process that converts N_2 into its compounds is referred to as the fixation of nitrogen. Nitrogen is one of the elements essential to living matter. Since animals and most plants can only use fixed nitrogen, natural nitrogen fixation processes are of extreme importance. Normally, nitrogen consumed by plants and animals is returned to the environment. A natural cycle exists by which nitrogen passes from one form to another, the **nitrogen cycle,** pictured in Figure 13-12.

The delicate balance of the nitrogen cycle can be easily upset by human activities. When land is extensively cultivated, fixed nitrogen is removed at a greater rate than it can be replenished naturally. This requires that nitrogen compounds be added to the soil as fertilizers. Also, the need for explosives, plastics, fibers, and industrial chemicals requires the artificial fixation of nitrogen. It has been estimated that the introduction of fixed nitrogen into the nitrogen cycle through human activities now equals or exceeds that from natural sources. Some estimates place the quantity of nitrogen being fixed annually at 92 million metric tons (1 metric ton = 1000 kg). The rate of return of nitrogen to the atmosphere (denitrification) is estimated to be 83 million metric tons. The difference, 9 million metric tons annually, is accumulating as nitrogen compounds in soil, groundwater, surface water, and the oceans.*

*C. C. Delwiche "The nitrogen cycle," In *The Biosphere*, W. H. Freeman, San Francisco, 1970.

FIGURE 13-12
The nitrogen cycle in nature.

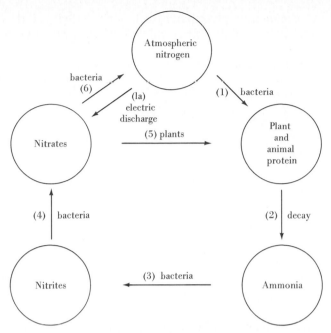

(1) Certain bacteria that reside as parasites in the root nodules of leguminous plants (beans, peas, clover, and alfalfa) can fix atmospheric nitrogen directly for conversion into plant proteins. Animal requirements are met by feeding on these and other plants. **(2)** The decay of plant and animal protein leads to the formation of ammonia. **(3)** and **(4)** Successive bacterial actions convert ammonia into nitrates. **(5)** The natural forms of fixed nitrogen that most plants require are the nitrates. This consumption again leads to plant and animal protein. **(6)** Certain denitrifying bacteria are capable of decomposing nitrates into elemental nitrogen, which returns to the atmosphere. **(1a)** As a result of electric discharges in rainstorms (lightning), a series of chemical reactions results in the direct production of nitrates as nitric acid, HNO_3.

Ammonia and Ammonium Compounds.   The main process for the manufacture of ammonia is the Haber process, described in detail in Chapter 15.

$$N_2(g) + 3\ H_2(g) \rightleftharpoons 2\ NH_3(g) \tag{13.52}$$

Under the name "anhydrous ammonia" $NH_3(l)$ can be applied directly to fields as a fertilizer. Alternatively, aqueous solutions of NH_3, NH_4NO_3, and urea may be used. $NH_3(aq)$ is also found in a variety of household cleaning agents. The most important use of NH_3, however, is in synthesizing a host of other nitrogen compounds. The reaction of ammonia (a base) with acids produces ammonium salts, such as

$$NH_3(aq) + HCl(aq) \longrightarrow \underset{\text{ammonium chloride}}{NH_4Cl(aq)} \tag{13.53}$$

Ammonium chloride is used in the manufacture of dry-cell batteries, in cleaning metal surfaces, and as a flux in soldering metals.

   Ammonium sulfate is the most widely used solid fertilizer. **Ammonium nitrate** has a higher percentage of N than does $(NH_4)_2SO_4$ and is also used as a fertilizer. NH_4NO_3 is unstable toward heat, however, making it hazardous for certain applications and accounting for its use in explosives.

$$NH_4NO_3(s) \xrightarrow{\Delta} N_2O(g) + 2\ H_2O(g) \tag{13.54}$$

Ammonium phosphates [such as $NH_4H_2PO_4$ and $(NH_4)_2HPO_4$] are good fertilizers because they supply both N and P for plant growth. They are also used as fire retardants.

Nitrogen Oxides.   Nitrogen forms a series of oxides in which the oxidation state can have every value from $+1$ (N_2O) to $+5$ (N_2O_5). **Dinitrogen oxide** [nitrogen(I) oxide, nitrous oxide] has the structure

$$:N\equiv N-\overset{\cdot\cdot}{\underset{\cdot\cdot}{O}}: \longleftrightarrow :\overset{\cdot\cdot}{\underset{\cdot\cdot}{N}}=N=\overset{\cdot\cdot}{\underset{\cdot\cdot}{O}}: \qquad (13.55)$$

and can be prepared by reaction (13.54). Its main use is as an anesthetic ("laughing gas").

Nitrogen oxide [nitrogen(II) oxide, nitric oxide] is an odd-electron molecule; it is paramagnetic. A suitable Lewis structure would appear to be $:\overset{\cdot}{N}=\overset{\cdot}{O}:$. With molecular orbital theory we can think of NO as being similar to N_2, but with an O atom substituting for one of the N atoms. The extra electron brought by the O atom to the molecular orbital diagram of N_2 in Figure 10-18 would enter an antibonding π_{2p}^* orbital, suggesting a bond order of 2.5 for NO.

The molecular orbital diagram for NO is the same as that shown for O_2^+ in Example 10-8.

Laboratory methods for preparing NO include the reaction of Cu with cold dilute $HNO_3(aq)$.

$$3\ Cu(s) + 8\ H^+ + 2\ NO_3^- \longrightarrow 3\ Cu^{2+} + 4\ H_2O + 2\ NO(g) \qquad (13.56)$$

Commercially, NO is produced by the catalytic oxidation of NH_3 (the Ostwald process). This oxidation constitutes the first step in the conversion of NH_3 to other nitrogen compounds.

$$4\ NH_3(g) + 5\ O_2(g) \xrightarrow{Pt} 4\ NO(g) + 6\ H_2O(g) \qquad (13.57)$$

Nitrogen dioxide [nitrogen(IV) oxide] is another paramagnetic odd-electron molecule. NO_2 molecules tend to pair up to produce diamagnetic N_2O_4, **dinitrogen tetroxide** [dimer of nitrogen(IV) oxide].

As we might expect, the O atoms in these structures are equivalent. A more complete description would show contributing structures to a resonance hybrid.

$$(13.58)$$

There are several unusual features about the N_2O_4 structure. One of these, an N—N bond distance that is about 20% longer than the usual N—N single bond, helps us to rationalize the existence of positive formal charges on the adjacent N atoms. $NO_2(g)$ can be prepared by the reaction of Cu with warm concentrated $HNO_3(aq)$.

This reaction is pictured in Color Section C.

$$Cu(s) + 4\ H^+ + 2\ NO_3^- \longrightarrow Cu^{2+} + 2\ H_2O + 2\ NO_2(g) \qquad (13.59)$$

Often, however, when brown $NO_2(g)$ is observed in a reaction involving $HNO_3(aq)$, the actual product may be colorless NO(g). NO is readily oxidized to NO_2 in air (as happens to NO produced in an automobile engine).

$$2\ NO(g) + O_2(g) \longrightarrow 2\ NO_2(g) \qquad (13.60)$$

Nitric Acid and Nitrates.   Reaction (13.57), followed by (13.60), followed by the reaction of $NO_2(g)$ with water describes the commercial preparation of nitric acid.

$$3\ NO_2(g) + H_2O(l) \longrightarrow 2\ HNO_3(aq) + NO(g) \qquad (13.61)$$

Pure nitric acid is a colorless liquid ($d = 1.50\ g/cm^3$). It dissociates at temperatures slightly above its melting point ($-42°C$) to N_2O_4, O_2, and water. Ordinary concentrated nitric acid is an aqueous solution with a density of $1.41\ g/cm^3$ and a concentration of about 15 M HNO_3. It generally has a yellow color due to the presence of dissolved oxides

FIGURE 13-13
Air temperature as a
function of altitude.

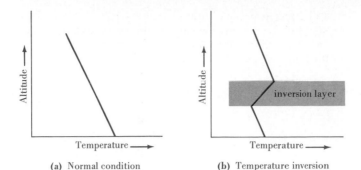

(a) Normal condition   (b) Temperature inversion

(a) Under normal conditions, air temperature decreases with altitude.
(b) Under conditions of temperature inversion, a layer of warmer air is suspended aloft.
At times the bottom of the inversion layer may be at ground level.

of nitrogen. Nitrate salts are produced by neutralizing nitric acid. In addition to its acidic properties, HNO_3 is a good oxidizing agent. With dilute $HNO_3(aq)$ and an active metal, the reduction product is N_2O (or even NH_4^+ in some cases).

$$4 \text{ Zn(s)} + 10 \text{ H}^+ + 2 \text{ NO}_3^- \longrightarrow 4 \text{ Zn}^{2+} + 5 \text{ H}_2\text{O} + \text{N}_2\text{O(g)} \qquad (13.62)$$

With a less active metal, the reduction product is $NO(g)$, as shown in equation (13.56). With concentrated $HNO_3(aq)$, the product is $NO_2(g)$, as in reaction (13.59). Nitric acid also reacts with nonmetallic elements, generally with the formation of an oxoacid and $NO(g)$.

In cold concentrated $HNO_3(aq)$:

$$\text{S(s)} + 2 \text{ HNO}_3(aq) \longrightarrow \text{H}_2\text{SO}_4(aq) + 2 \text{ NO(g)} \qquad (13.63)$$

In hot concentrated $HNO_3(aq)$:

$$3 \text{ I}_2(s) + 10 \text{ HNO}_3(aq) \longrightarrow 6 \text{ HIO}_3(aq) + 2 \text{ H}_2\text{O} + 10 \text{ NO(g)} \qquad (13.64)$$

Air Pollution and Oxides of Nitrogen. Air pollution that results in reduced visibility and in minor or major irritations to most people is commonly called "smog." Although revealed most dramatically in southern California, air pollution is a worldwide phenomenon.

Local smog conditions depend on the rate at which pollutants are produced and the rate of their dispersal. Dispersal by winds carries pollutants in a horizontal direction along the earth's surface. In the absence of winds mixing must occur vertically. Air at the surface of the earth is heated, expands, and rises. Cooler air aloft descends, itself becomes heated, and rises again. Thus, a constant circulation occurs through a large air mass and pollutants at ground level are dispersed.

Dispersal of pollutants can be seriously impaired by phenomena known as **temperature inversions.** Air temperature normally decreases regularly with increased altitude; the normal temperature-altitude profile is shown in Figure 13-13a. At times a different situation may prevail. For example, a mass of cool ocean air may move onshore and displace a layer of warmer air, holding it aloft; the temperature–altitude profile takes the form of Figure 13-13b. Temperature inversions are particularly common in latitudes of about 30 to 35° in summer months when a belt of high pressure air encircles the earth.

Smog formation is illustrated in Color Section E.

Under conditions of temperature inversion the volume of air into which pollutants can be dispersed is reduced. The inversion layer acts as a lid on the smoggy air below. In the Los Angeles basin, which is surrounded by mountains on three sides, there is little horizontal displacement of pollutants and temperature inversions assume a special importance.

TABLE 13-6
Simplified reaction scheme for the production of photochemical smog

nitrogen–oxygen reactions:	$NO_2 + h\nu \longrightarrow NO + O$	rxn. 1
	$O + O_2 \longrightarrow O_3$	2
	$O_3 + NO \longrightarrow NO_2 + O_2$	3
hydrocarbon[a] reactions:	$RH + O \longrightarrow RO\cdot$	4
	$RO\cdot + O_2 \longrightarrow RO_3\cdot$	5
	$RO_3\cdot + RH \longrightarrow$ aldehydes + ketones	6
	(RCHO) (RCOR)	
	$RO_3\cdot + NO \longrightarrow RO_2\cdot + NO_2$	7
	$RO_3\cdot + O_2 \longrightarrow RO_2\cdot + O_3$	8
	$RO_3\cdot + NO_2 \longrightarrow$ peroxyacyl nitrates	9
	(PAN)	

[a] R = a hydrocarbon chain, e.g., C_8H_{17}, or ring, e.g., C_6H_5. Thus, RH = C_8H_{18}, C_6H_6, . . . ; $RO\cdot$, $RO_2\cdot$, $RO_3\cdot$ refer to highly reactive, odd-electron species called radicals.

Although the direct combination of $N_2(g)$ and $O_2(g)$ to $NO(g)$ is very limited at low temperatures, it becomes more significant at the high temperatures encountered in internal combustion engines and, to some extent, in electric power plants. As we have already seen, once $NO(g)$ is formed in air it is readily oxidized to $NO_2(g)$ (reaction 13.60). $NO_2(g)$ may then trigger the sequence of reactions outlined in Table 13-6. The process is thought to begin when an NO_2 molecule absorbs a quantum ($h\nu$) of near-ultraviolet radiation and dissociates to NO and O (rxn. 1). This is followed by the reaction of O and O_2 to produce ozone, O_3 (rxn. 2). The O_3 in turn reacts with NO to reform NO_2 and O_2 (rxn. 3). If these were the only reactions occurring, the O_3 level would not be nearly so high as it becomes under smog conditions. Reactions 4, 5, and 8 in Table 13-6 account for most of the O_3 production. These reactions involve unburned hydrocarbons (RH), another product of the internal combustion engine.

Because of the important role played by sunlight in its formation, this type of smog is referred to as **photochemical smog.** The characteristic eye irritation associated with photochemical smog is probably due to formaldehyde (HCOOH) and acrolein ($H_2C\!=\!CHCHO$) produced in rxn. 6 and peroxyacyl nitrates (PAN) produced in rxn. 9. The incidence of all respiratory diseases (e.g., bronchitis and emphysema) is much higher than normal under smog conditions. Also photochemical smog causes heavy crop losses and the deterioration of such materials as rubber goods.

Catalytic systems have been developed to reduce the emission of hydrocarbons and CO by promoting their complete oxidation in exhaust fumes. For these catalysts to function properly they must be used with lead-free gasoline. This, in turn, has required changes in petroleum-refining processes. To remove oxides of nitrogen from automotive exhaust requires their reduction, preferably to $N_2(g)$. However, the optimum catalyst composition to promote the reduction of oxides of nitrogen is not the same as for the oxidation of hydrocarbons and CO. Dual catalyst systems are now being developed to deal with both types of pollutants.

Control of hydrocarbons has reduced smog levels in central Los Angeles. However, in the absence of effective control of the oxides of nitrogen, the smog-forming reactions simply occur more slowly and lead to increased smog in cities to the east of Los Angeles.

13-11 Phosphorus

Phosphorus, the 12th most abundant element in the earth's crust, occurs principally as deposits of phosphate rock (phosphorite) found in several states in the U.S. (principally Florida), in Morocco, and in the U.S.S.R. Phosphate rock has a variable composition but generally its constituents occur in the same proportions found in the mineral fluorapatite, $[3Ca_3(PO_4)_2\cdot CaF_2]$. For the past 100 years, elemental phosphorus has been prepared by a method that involves heating phosphate rock, silica, and coke in an electric furnace.

$$2 \text{ Ca}_3(\text{PO}_4)_2(s) + 10 \text{ C}(s) + 6 \text{ SiO}_2(s) \xrightarrow{1500°\text{C}} 6 \text{ CaSiO}_3 (l) + 10 \text{ CO}(g) + \text{P}_4(g)$$

(13.65)

$\text{P}_4(g)$ is condensed and collected under water as a solid.

Allotropy of Phosphorus.   The existence of an element in two or more forms is called **allotropy.** The form of solid phosphorus obtained from reaction (13.65) is a waxy, white, phosphorescent solid. (A phosphorescent material glows in the dark.) **White P** has a melting point of 44.1°C, producing a liquid that boils at 287°C. White P is a nonconductor of electricity, ignites spontaneously on contact with air (hence the reason for storing it under water), is essentially insoluble in water, very soluble in $\text{CS}_2(l)$, and slightly soluble in many organic solvents.

The basic structural units of white phosphorus are P_4 molecules. These molecules, pictured in Figure 13-14, are tetrahedral; a P atom is found at each corner of the tetrahedron. The P-to-P bonds in P_4 appear to involve the overlap of $3p$ orbitals almost exclusively. Such overlap normally produces 90° bond angles (recall Figure 10-2), but in P_4 the P—P—P bond angles are 60°. The bonds are said to be strained, and species with strained bonds are generally quite reactive, as is white P.

When heated to 400°C or upon long standing under exposure to light, white P transforms into amorphous **red P.** What is believed to occur is the opening up of bonds in P_4 tetrahedra and the joining together of these fragments into long chains. In the process the P—P—P bond angle increases. Red P, a more stable solid modification than white P, is correspondingly less reactive. [For example, it combines with atmospheric $\text{O}_2(g)$ only very slowly.] The triple point of red P is 590°C and 43 atm. Thus, red P sublimes without melting (at about 420°C).

The most stable form of phosphorus appears to be black P, which can be formed from white P under high pressures, or by heating white P in the presence of a catalyst (Hg) and "seed" crystals of black P. Black P has a layered crystalline structure, similar to graphite, but the layers are buckled. Black P is a semiconductor.

Oxides and Oxoacids of Phosphorus.   Many oxides of P have been reported; the two most important are oxides having P in the oxidation states of $+3$ and $+5$. The simplest formulas that can be written for these are P_2O_3 and P_2O_5, with the corresponding names, phosphorus trioxide and phosphorus pentoxide. However, since each of these has a molecular structure based on the P_4 tetrahedron, more appropriate molecular formulas are P_4O_6 and P_4O_{10}. As shown in Figure 13-15, in P_4O_6 one O atom bridges each pair of P atoms, hence six O atoms in the molecule. In P_4O_{10} there is an additional O atom bonded to each P atom. Both oxides are prepared by the direct action of $\text{O}_2(g)$ on white P. As should seem reasonable (recall the case of the oxides of carbon), the oxide with P in the lowest oxidation state—P_4O_6—is produced when the available quantity of O_2 is limited.

FIGURE 13-14
The P_4 molecule.

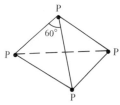

We have also used the term polymorphism to describe the existence of different solid modifications of a substance (Section 11-6). Allotropy is a more general term.

FIGURE 13-15
Molecular structures of P_4O_6 and P_4O_{10}.

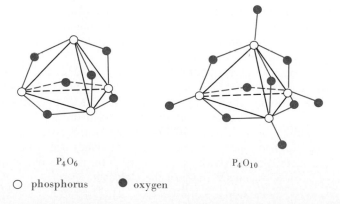

P_4O_6 P_4O_{10}

○ phosphorus   ● oxygen

In an excess of $O_2(g)$, P_4O_{10} is formed. P_4O_6 is a low-melting solid (mp 23.9°C; bp 175.4°C) and P_4O_{10}, a solid that melts at about 580°C. Of the several oxides, only P_4O_{10} is of commercial importance.

P_4O_6 and P_4O_{10} each reacts with water to produce an oxoacid.

$$P_4O_6 + 6\ H_2O \longrightarrow \underset{\text{phosphorous acid}}{4\ H_3PO_3}$$

$$P_4O_{10} + 6\ H_2O \longrightarrow \underset{\text{phosphoric acid}}{4\ H_3PO_4}$$

(13.66)

Phosphoric acid is used in treating metals to provide them with a corrosion-resistant coating or an undercoating to which paint can be applied. In the food industry it is used to impart tartness to soft drinks. Calcium phosphates are used as dietary supplements, in baking powders, and as polishing agents in dentifrices. Sodium and potassium phosphates are used in dairy products, hams, sausage, canned seafood, cereals, and processed potatoes. Other uses of phosphates as fertilizers and in detergents are discussed in Section 21-3.

Other Compounds of Phosphorus.   Phosphorus forms a hydride somewhat analogous to NH_3. The hydride, PH_3, called **phosphine,** is a gas at room temperature and is somewhat soluble in water. Like NH_3, PH_3 forms phosphonium ion (PH_4^+) and phosphonium salts. Unlike NH_3, PH_3 has a positive enthalpy of formation, is thermally unstable, burns in air, and is extremely poisonous. The PH_3 molecule has a pyramidal shape with H—P—H bond angles of 93° (close to the 90° predicted for bonding through $3p$ orbitals). PCl_3, a colorless liquid (bp 76.1°C), is the most important of the phosphorus halides, and is prepared by the action of $Cl_2(g)$ on white P. Like PH_3, the PCl_3 molecule has a pyramidal shape, although the bonds involve some sp^3 hybridization of the P atom. The chief use of PCl_3 is as a reaction intermediate in the synthesis of organic phosphorus compounds, which have uses ranging from motor oil additives to defoliants. PCl_5 is produced by the further reaction of PCl_3 with $Cl_2(g)$. It is a greenish white solid that sublimes at 160°C (but can be melted at 167°C under 1.2 atm pressure). The structure of the PCl_5 molecule is a trigonal bipyramid (see Table 9-2).

13-12 Oxygen

Oxygen, in group VIA, is both one of the most nonmetallic of the elements (electronegativity = 3.44) and one of the most important of the nonmetals. $O_2(g)$ is obtained by the fractional distillation of liquid air (Figure 13-11), and is used chiefly in the manufacture of steel. Mixtures of $O_2(g)$ and acetylene or $H_2(g)$ are used to generate high temperatures for metal welding. Liquid oxygen (''lox'') is used as an oxidizer in rocket systems, and $O_2(g)$ is used in various breathing mixtures. Anticipated future uses include direct injection into streams and sewage systems as a means of oxidizing wastes and as a reactant in coal gasification processes.

The electron configuration of oxygen is $1s^2 2s^2 2p^4$. Two of the $2p$ electrons are unpaired. As noted in Chapters 9 and 10, bonding in the O_2 molecule is not easily explained. In particular, with a Lewis structure it is difficult to describe both the bond order and the fact that the molecule is paramagnetic. A satisfactory description is possible with molecular orbital theory (Figure 10-18). Oxygen forms compounds with all the elements except the light noble gases. In general, its compounds with metals are ionic and with nonmetals, covalent. Reaction with oxygen is one of the aspects of chemical behavior usually considered in a systematic study of the elements.

Ozone. An important allotrope of ordinary oxygen (dioxygen) is **ozone, O_3.** Ozone can be described more successfully with a Lewis structure than can O_2, although even here a resonance hybrid must be proposed. The O—O bonds are equivalent and have a bond order halfway between a single and double bond.

The natural abundance of $O_3(g)$ at sea level is only 0.04 ppm (parts per million). In the stratosphere the ozone content is appreciably greater, although still limited to trace amounts. It reaches about 10 ppm at altitudes of 25 to 30 km. Atmospheric ozone plays a vital role in maintaining life on earth because of the ability of O_3 molecules to absorb ultraviolet radiation in the range 210 to 290 nm. The effect is twofold. First, ultraviolet radiation of this wavelength is damaging to tissues. Many forms of life would not be possible if this radiation were not screened out. Also, in absorbing ultraviolet radiation O_3 molecules undergo dissociation with the evolution of heat, thereby helping to maintain a heat balance in the atmosphere.

The principal chemical reactions that produce $O_3(g)$ in the upper atmosphere are

$$O_2 + h\nu \longrightarrow O + O \tag{13.67}$$

$$O_2 + O + M \longrightarrow O_3 + M \tag{13.68}$$

Equation (13.67) describes the dissociation of O_2 following the absorption of ultraviolet radiation. The reaction of atomic and molecular oxygen to produce ozone is represented by equation (13.68). The "third body" M [e.g., $N_2(g)$] is needed to carry off excess collision energy, else the O_3 produced would be so energetic as to decompose spontaneously.

The absorption of ultraviolet radiation by O_3 molecules is represented by equation (13.69). The energy released in reaction (13.70) accounts for the heating effect referred to above.

$$O_3 + h\nu \longrightarrow O_2 + O \tag{13.69}$$

$$O_3 + O \longrightarrow 2 O_2 \tag{13.70}$$

The production of $O_3(g)$ directly from $O_2(g)$ is highly endothermic and does not occur under normal conditions in the lower atmosphere.

$$3 O_2(g) \longrightarrow 2 O_3(g) \qquad \Delta\overline{H}^\circ = +285 \text{ kJ/mol}$$

This reaction will proceed to some extent under high-energy conditions as in electrical discharges. Thus, some ozone is produced during electric storms, and ozone formation is the cause of the sharp pungent odor sometimes encountered around electric machinery. The principal method of preparing laboratory ozone is by passing a silent electric discharge through oxygen gas.

$O_3(g)$ is an excellent oxidizing agent. Its oxidizing ability is surpassed by very few substances [e.g., $F_2(g)$ and $OF_2(g)$]. An important oxidation–reduction reaction used to determine trace amounts of O_3 and other oxidants in polluted air involves oxidation of iodide ion.

$$2 I^-(aq) + O_3(g) + H_2O \longrightarrow 2 OH^-(aq) + I_2(aq) + O_2(g)$$

After acidification of the solution, I_2 is titrated with sodium thiosulfate solution, using a starch indicator (see Section 21-1).

A measure of the severity of photochemical smog is the **oxidant level.** Oxidants in smog are those components capable of oxidizing I^- to I_2. In addition to O_3 the principal oxidants are NO_2 and organic peroxides.

Water and Hydrogen Peroxide. Certainly the most important compound of oxygen is water, H_2O. Probably more properties of water have been measured than of any other chemical substance. Among the properties commented upon elsewhere in the text are its molecular structure; its crystal structure; its vapor pressure, heat of vaporization, specific heat, heat of fusion, melting point, triple point, boiling point, and critical point; its ionization constant; and its solvent properties.

Although not as important as water, **hydrogen peroxide, H_2O_2,** also has interesting properties. The molecule features an O—O bond as indicated by the Lewis structure

$$H:\ddot{O}:\ddot{O}:H$$

The geometric structure of the molecule is shown in Figure 13-16.

Hydrogen peroxide enters into a wide variety of oxidation–reduction reactions. In reaction (13.71) it acts as an oxidizing agent (reduced to H_2O). In reaction (13.72) it assumes the role of a reducing agent (oxidized to O_2).

> If a substance is to act as either an oxidizing or a reducing agent, it must contain an element in an oxidation state between the highest and lowest values possible. The observed oxidation states for oxygen are -2, -1, and 0. In H_2O_2 the oxidation state is -1.

$$H_2O_2 + 2\ I^- + 2\ H^+ \longrightarrow 2\ H_2O + I_2 \tag{13.71}$$

$$5\ H_2O_2 + 2\ MnO_4^- + 6\ H^+ \longrightarrow 2\ Mn^{2+} + 8\ H_2O + 5\ O_2(g) \tag{13.72}$$

Hydrogen peroxide can be prepared by a variety of methods. For the laboratory preparation of small quantities, the addition of barium peroxide to cold, dilute, aqueous sulfuric acid is convenient.

$$BaO_2(s) + H_2SO_4(aq) \longrightarrow BaSO_4(s) + H_2O_2(aq)$$

Pure H_2O_2 is a pale-blue liquid with a freezing point of $-0.46°C$. The liquid is considerably more dense than water (1.47 g/cm^3 at $0°C$). The pure compound is unstable. The decomposition of H_2O_2 is an exothermic reaction which is catalyzed by light and a variety of materials (such as iron and copper).

$$2\ H_2O_2(l) \longrightarrow 2\ H_2O(l) + O_2(g) \qquad \Delta\overline{H}° = -197 \text{ kJ/mol} \tag{13.73}$$

Aside from its industrial uses, such as in bleaching wood pulp, dilute $H_2O_2(aq)$ has been used in the household as a mild antiseptic and as a bleaching agent. A desirable feature of H_2O_2 is that its reaction products are harmless—$H_2O(l)$ and $O_2(g)$.

Oxides of the Elements. There are several ways in which oxides may be classified. For example, among the ionic oxides we may speak of normal oxides, peroxides, and super-oxides. **Normal oxides,** such as Li_2O and CaO, involve the simple oxide anion. The **peroxide** ion, such as in Na_2O_2, features an O—O bond and oxidation state of -1 for oxygen. The **superoxide** ion, such as in KO_2, also features an O—O bond, an unpaired electron, and an oxidation state of $-\frac{1}{2}$ for oxygen. Lewis structures of these ions are

FIGURE 13-16
Geometric structure of H_2O_2.

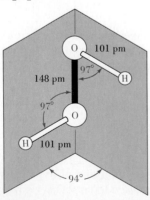

oxide ion   peroxide ion   superoxide ion

$$\left[:\ddot{O}:\right]^{2-} \qquad \left[:\ddot{O}:\ddot{O}:\right]^{2-} \qquad \left[:\ddot{O}:\ddot{O}:\right]^{-}$$

All three ions are unstable toward water and can only be found in the solid state. The decompositions of O_2^{2-} and O_2^- involve oxidation and reduction. Both of these ions, especially O_2^-, are good oxidizing agents.

oxide: $O^{2-} + H_2O \longrightarrow 2\ OH^-(aq)$
peroxide: $2\ O_2^{2-} + 2\ H_2O \longrightarrow O_2(g) + 4\ OH^-(aq)$
superoxide: $4\ O_2^- + 2\ H_2O \longrightarrow 3\ O_2(g) + 4\ OH^-(aq)$

Another classification scheme, in which oxides are viewed as being acidic, basic, or amphoteric, is discussed in Section 17-8.

13-13 Sulfur

Based on periodic relationships and electron configurations, a similarity is to be expected between S and O. Both elements form ionic compounds with active metals and both form similar covalent compounds, such as H_2S and H_2O, CS_2 and CO_2, SCl_2 and Cl_2O. However, there are factors that produce some differences between oxygen and sulfur compounds. The O atom has a single covalent bond radius of 74 pm, whereas that of an S atom is 104 pm. The comparative electronegativities are 3.44 for O and 2.58 for S. Hydrogen bonding is not a significant feature in sulfur compounds as it is in some oxygen compounds. Although H_2O has a very high boiling point ($+100°C$) for a compound of such low molecular weight (18), the boiling point of H_2S (molecular weight, 34) is more nearly normal ($-61°C$). Compared to the O atom, the S atom has a greater capacity to be bonded simultaneously to other atoms (as in SF_6) because of the availability of $3d$ orbitals (i.e., an expanded octet).

Allotropy of Sulfur. Figure 13-17 represents a few of the better known molecular species of sulfur. Among the different physical forms of sulfur that can be observed are

- **rhombic sulfur** (S_α), which has sixteen S_8 rings in a unit cell and converts at 95.5°C to
- **monoclinic sulfur** (S_β). Monoclinic sulfur is thought to have six S_8 rings in its unit cell. It melts at 119°C, yielding
- **liquid sulfur** (S_λ) comprised of S_8 molecules. This is a yellow, transparent, mobile liquid. At about 160°C, however, the S_8 rings open up and join together into long spiral-chain molecules, resulting in a
- **liquid sulfur** (S_μ) that is dark in color and very thick and viscous. This liquid boils at 445°C, producing
- **sulfur vapor,** S_8, which dissociates into progressively smaller molecular species as the temperature is raised.
- **Plastic sulfur** forms if liquid S_μ is poured into cold water. It consists of chainlike molecules and has rubberlike qualities when first formed. On standing, however, it becomes brittle and eventually converts to rhombic sulfur.

The allotropy of sulfur as a function of temperature can be summarized as

$$S_\alpha \xrightarrow{95.5°C} S_\beta \xrightarrow{119} S_\lambda \xrightarrow{160} S_\mu \xrightarrow{445} S_8(g) \longrightarrow S_6 \longrightarrow S_4 \xrightarrow{1000} S_2 \xrightarrow{2000} S$$

Because of the sluggishness of some of the transitions, additional phenomena may be observed. If rhombic sulfur is heated rapidly, for example, it fails to convert to monoclinic sulfur and melts at 113°C.

Production and Uses of Sulfur. Sulfur occurs abundantly in the earth's crust—as elemental sulfur, as mineral sulfides and sulfates, as $H_2S(g)$ in natural gas, and as organic

FIGURE 13-17
Different molecular forms of sulfur.

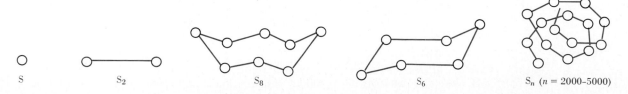

S S_2 S_8 S_6 S_n (n = 2000–5000)

FIGURE 13-18
The Frasch process for
mining sulfur.

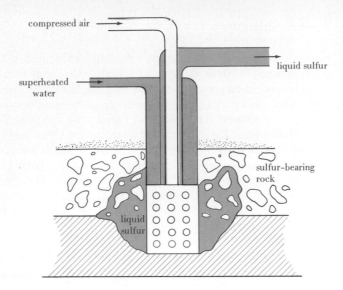

sulfur compounds in coal and oil. Extensive deposits of elemental sulfur are found in the U.S. in Texas and Louisiana, some of them at offshore sites. In the Frasch process (Figure 13-18) a mixture of superheated water and steam (at 160°C and 16 atm) is forced down the outermost of three concentric pipes into an underground bed of sulfur-containing rock. The sulfur melts and forms a liquid pool. Compressed air (20–25 atm) is passed through the innermost pipe, forcing liquid sulfur up the remaining pipe. At the surface the liquid sulfur is collected in large bins and allowed to freeze, yielding solid sulfur about 99.5% pure.

For a time in the U.S. the Frasch process was essentially the exclusive source of sulfur; but now with a need to control sulfur emissions from industrial operations, about one half of the U.S. production is as by-product sulfur. For example, high-sulfur natural gases can be desulfurized by the reaction

$$2\ H_2S(g) + SO_2(g) \xrightarrow{\Delta} 3\ S(g) + 2\ H_2O(g) \tag{13.74}$$

Although significant quantities of sulfur are used in vulcanizing rubber (page 247), by far the most important use is in the manufacture of sulfuric acid. Sulfuric acid, in turn, ranks first among all manufactured chemicals, with a typical annual U.S. production exceeding 40 million tons. Its production and uses are described at the end of the next chapter.

FIGURE 13-19
Structures of some
sulfur oxides.

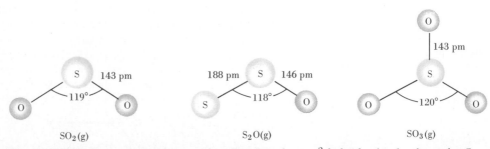

The SO_2 molecule owes its shape to bonding based on sp^2 hybrid orbitals about the S atom. The molecule S_2O has a similar structure but with an S atom substituted for one O atom. In the gaseous state SO_3 molecules are planar with 120° bond angles.

Oxides and Oxoacids of Sulfur.   A number of oxides of sulfur have been reported, but only sulfur dioxide, SO_2, and sulfur trioxide, SO_3, are commonly encountered. The structures of some oxides of sulfur are outlined in Figure 13-19. $SO_2(g)$ can be commercially prepared by the direct oxidation (burning) of sulfur or by "roasting" metal sulfides.

$$S(s) + O_2(g) \longrightarrow SO_2(g)$$

$$2\ ZnS(s) + 3\ O_2(g) \longrightarrow 2\ ZnO(s) + 2\ SO_2(g)$$

The principal source of SO_3 is the oxidation of $SO_2(g)$. A catalyst (Pt or V_2O_5) helps to increase the rate at which equilibrium is attained.

$$2\ SO_2(g) + O_2(g) \rightleftharpoons 2\ SO_3(g) \tag{13.75}$$

The reaction of SO_2 with water produces $H_2SO_3(aq)$, but this acid, sulfurous acid, has never been isolated in pure form. By contrast, H_2SO_4, sulfuric acid, resulting from the reaction of SO_3 and water, is a pure liquid ($d = 1.86$ g/cm^3; mp = 10°C).

Aqueous sulfurous acid and its salts (sulfites) are good reducing agents,

$$Cl_2(g) + SO_3^{2-}(aq) + H_2O \longrightarrow 2\ Cl^-(aq) + SO_4^{2-}(aq) + 2\ H^+(aq) \tag{13.76}$$

but they can also act as oxidizing agents, as in this reaction with H_2S.

$$2\ H_2S(g) + 2\ H^+(aq) + SO_3^{2-}(aq) \longrightarrow 3\ H_2O + 3\ S(s) \tag{13.77}$$

Dilute $H_2SO_4(aq)$ enters into all the common reactions of a strong mineral acid, such as neutralizing bases, dissolving active metals to liberate $H_2(g)$, and dissolving carbonates to liberate $CO_2(g)$.

Concentrated H_2SO_4 has some distinctive properties. It has a very strong affinity for water, strong enough even to remove chemically bonded water from certain compounds. When concentrated H_2SO_4 is dropped onto ordinary cane sugar, the sugar becomes charred or carbonized.

$$C_{12}H_{22}O_{11}(s) \xrightarrow{H_2SO_4\ (conc.)} 12\ C(s) + 11\ H_2O \tag{13.78}$$

Although all acids produce a stinging sensation on the skin, concentrated H_2SO_4 produces severe burns as a result of reactions like that shown in equation (13.78).

The concentrated acid is a moderately good oxidizing agent and can dissolve some metals that are insoluble in acids whose anions are not oxidizing agents (e.g., HCl).

$$Cu(s) + 2\ H_2SO_4 \longrightarrow Cu^{2+} + SO_4^{2-} + 2\ H_2O + SO_2(g) \tag{13.79}$$

Air Pollution and Oxides of Sulfur.   **Industrial smog** consists primarily of particles (ash and smoke), SO_2, and H_2SO_4 mist. Among the sources of SO_2 are industrial operations such as occur in oil refineries, smelters, coke plants, and sulfuric acid plants. The main contributors, however, are power plants burning coal or high-sulfur fuel oils. The SO_2 can undergo air oxidation to SO_3, especially when catalyzed on the surface of airborne particles or through reaction with NO_2. Then SO_3 reacts with water vapor to produce H_2SO_4 mist. A further reaction may occur with NH_3 to produce particles of $(NH_4)_2SO_4$. The exact physiological effects of SO_2 and H_2SO_4 mist at low concentrations are not well understood, but these substances are respiratory irritants and levels above 0.10 ppm are generally considered unhealthful.

H_2SO_4 mist produced in this way is one of the most important constituents of acid rain.

The control of industrial smog hinges on the removal of sulfur from fuels and the control of $SO_2(g)$ emissions where they do occur. Dozens of processes have been proposed for the removal of SO_2 from stack gases. None has yet proved superior to all others, and none, in fact, has proved totally effective in removing SO_2 under all circumstances. One simple method involves the reactions

$$SO_2(g) + MgO(s) \underset{750°C}{\overset{150°C}{\rightleftharpoons}} MgSO_3(s)$$

The $SO_2(g)$ released when $MgSO_3(s)$ is heated can be used in the manufacture of H_2SO_4.

Pollution control unit at an electric power plant. The pollution control unit (scrubber) at the far right was added to an existing power plant. In the scrubber, flue gases from the combustion of coal in the power plant are mixed with a lime–water mixture. This treatment removes about 83–85% of the $SO_2(g)$ and more than 99% of the particulate matter before the gases are vented through the stack. The magnitude of the pollution control operation is seen through both the physical size of the unit and the fact that 3–4% of the power plant's output of electricity is used to operate motors, pumps, and filters in the pollution control unit. [Photograph courtesy of Dusquesne Light Company.]

Removal of SO_2 from exhaust gases adds considerable expense to the burning of fossil fuels. It may be that processes to remove S from fuels before burning will prove to be more feasible.

The role of sulfur pollutants in the natural sulfur cycle is suggested by Figure 13-20.

FIGURE 13-20
The sulfur cycle in the environment.

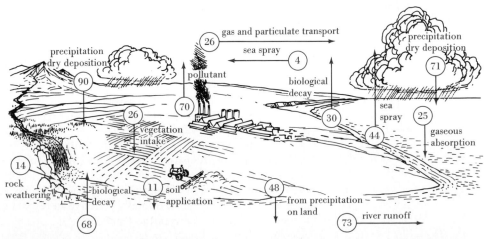

The figures are in millions of tons per year and were calculated on a worldwide basis. Sulfur circulates among land, sea, and air with a net transfer of about 95 million tons of sulfur to the oceans each year. [Source: E. Robinson and R. C. Robbins, Where does it all go?, *Stanford Res. Inst. J.*, No. 23, 8 (1968).]

13-14 Fluorine and Chlorine

These two elements belong to periodic group VIIA, a group of nonmetals called the halogens, meaning "salt formers." Formation of ionic compounds with metals (salts) is a typical reaction of members of this group. No matter which of several criteria we use, the halogens stand out as a group of very active nonmetals; and fluorine is the most nonmetallic of all the elements. The halogen atoms have valence-shell electron configurations that are one electron short of the stable octet of the noble gases—ns^2np^5. Their ionization energies are high—electrons are lost with difficulty. Their electron affinities are large and negative—electrons are gained readily. And they have high electronegativities.

Some Physical Properties of Fluorine, Chlorine, and Other Halogens.   Table 13-7 lists some properties of the halogens. Two halogen atoms, X, form a strong bond between them in the molecule, X_2. However, attractive forces between molecules are rather weak. They are of the instantaneous dipole–induced dipole type (London forces), and increase in strength with increasing molecular weight. We can combine ideas from Chapters 8 and 11 to answer questions of the following sort concerning properties of the halogens.

Example 13-3   Estimate values of the missing entry for Br in Table 13-7.

Solution.   The basic principle here is that, in general, properties vary continuously within a group in the periodic table. All of the missing values for Br should fall between those of Cl and I. Because the atomic number and the atomic weight of Br are midway between those of Cl and I, a rough approximation would be that the missing values for Br are the averages of the values listed for Cl and I. The actual values are given in parentheses.

$$\text{predicted electronegativity} = \frac{3.16 + 2.66}{2} = 2.91 \quad \text{(actual value, 2.96)}$$

$$\text{predicted covalent radius} = \frac{99 + 133}{2} = 116 \text{ pm} \quad \text{(actual value, 114 pm)}$$

$$\text{predicted ionic radius} = \frac{181 + 216}{2} = 198 \text{ pm} \quad \text{(actual value, 196 pm)}$$

SIMILAR EXAMPLES: Exercises 37, 38.

Example 13-4   Estimate values of the missing entries for F_2 in Table 13-7.

Solution.   Because the strengths of intermolecular forces of the London type decrease with decreasing molecular weight, our expectation should surely be that F_2 is a gas. (It is a yellow gas.) Based on an extrapolation of the data for Cl, Br, and I, very rough estimates of the melting and boiling points of F_2 might be of the order of $-180°C$ and $-120°C$, respectively. (The actual melting point is $-220°C$, and the boiling point, $-188°C$.)

SIMILAR EXAMPLES: Exercises 37, 38.

The estimated values of the melting and boiling points of F_2 in Example 13-4 deviate rather sharply from the actual values. This fact underscores an observation made on several occasions in this chapter—the first member of a group differs in some respects from other group members. These differences can often be attributed to the small size and compact electron clouds of the first period atoms and, in matters of compound formation, to the nonavailability of d orbitals for bonding. Several other differences between F and Cl are noted in this section.

TABLE 13-7
Some properties of the halogen elements

	Atomic Properties				
	At. no.	At. wt.	Electro-negativity	Covalent radius, X, pm	Ionic radius, X^-, pm
F	9	19.00	3.98	72	136
Cl	17	35.45	3.16	99	181
Br	35	79.90	a	a	a
I	53	126.90	2.66	133	216

	Physical Properties		
	Melting point, °C	Boiling point, °C	Physical form at room temperature
F_2 [b]			
Cl_2	-101	-34.6	greenish yellow gas
Br_2	-7.2	$+58.8$	red-brown liquid
I_2	$+114$	$+184.4$	grayish black solid

[a] See Example 13-3.
[b] See Example 13-4.

Preparation and Uses of Fluorine and Chlorine.   Because F and Cl show such a strong tendency to gain electrons, naturally occurring forms of these elements involve the ions F^- (e.g., fluorspar, CaF_2) and Cl^- (e.g., NaCl). To obtain the free halogen element (X_2) requires oxidation of the halide ion (X^-). There is no common oxidizing agent strong enough to oxidize F^-; electrolytic methods must be employed. Generally this involves HF in molten KHF_2 as the electrolyte.

The hydrogen difluoride ion in KHF_2 features a strong hydrogen bond, with an H^+ midway between two F^- ions, $[F-H-F]^-$. Chlorine does not form such an ion.

$$2\ H^+ + 2\ F^- \xrightarrow{\text{electrolysis}} H_2(g) + F_2(g) \tag{13.80}$$

$Cl_2(g)$ can be prepared by oxidation of $Cl^-(aq)$ with a strong oxidizing agent like $KMnO_4$.

$$2\ MnO_4^- + 16\ H^+ + 10\ Cl^- \longrightarrow 2\ Mn^{2+} + 8\ H_2O + 5\ Cl_2(g) \tag{13.81}$$

Most commercially prepared Cl_2, however, is obtained by electrolysis, either of a molten chloride such as NaCl or $MgCl_2$ (recall reaction 13.14) or an aqueous solution.

$$2\ Cl^- + 2\ H_2O \xrightarrow{\text{electrolysis}} 2\ OH^- + H_2(g) + Cl_2(g) \tag{13.82}$$

This reaction, which is the basis of the chlor-alkali industry, is one of the most important of all electrolysis reactions. It is a source of three valuable substances, NaOH, H_2, and Cl_2. The electrolysis of NaCl(aq) is explored more fully in Chapter 20.

Fluorine is used in the preparation of uranium hexafluorides, the compounds that are separated in gaseous diffusion plants to isolate [235]U as a nuclear fuel (recall page 129). Among other important fluorine-containing materials are

- chlorofluorocarbons [e.g., $CCl_2F_2(g)$, dichlorodifluoromethane, Freon-12], used as refrigerants and propellants for spray cans;
- sodium fluorosilicate, Na_2SiF_6, used to fluoridate water and as an additive to tooth-pastes to prevent tooth decay;

- synthetic cryolite, Na_3AlF_6, used as the electrolyte in the electrolytic extraction of Al from Al_2O_3 (see Section 22-3).
- Teflon, an inert, temperature-resistant carbon-fluorine polymer (see Table 9-4).

Chlorine and its compounds have myriad uses in chemical and related industries. Chlorine is used to purify drinking water and in the treatment of swimming pools, sewage effluents, and industrial wastes. These applications are based on the oxidizing power of Cl_2 and OCl^-, which forms when Cl_2 is dissolved in basic solutions. Hypochlorite solutions are used in household bleaches and in the production of paper and rayon. Chlorine is used in the production of CCl_4, ethylene glycol (an automotive antifreeze), and a host of other organic compounds. Typical annual production of Cl_2 in the United States exceeds 12 million tons.

Environmental Concerns over Fluorine and Chlorine Compounds.   In recent years it has been discovered that a number of widely used compounds of F and Cl pose environmental hazards. Perhaps the first such compound identified was the popular insecticide DDT (dichlorodiphenyltrichloroethane, $C_{14}H_9Cl_5$). Concern now centers on chlorofluorocarbons and polychlorinated biphenyls (PCBs).

There are no natural processes by which chlorofluorocarbons are destroyed at the earth's surface. In the stratosphere, however, they decompose by absorbing ultraviolet radiation. One of the products of this decomposition, atomic Cl, enters into reactions with stratospheric O_3 that have the effect of converting O_3 to O_2 (reaction 13.70). Partial destruction of the ozone layer would permit more ultraviolet radiation to reach the surface of the earth, increasing the risk of skin cancers. In the United States, chlorofluorocarbons have been phased out as propellants for spray cans; CO_2 and other gases are used instead.

PCBs, used at one time in inks, plastics, and paper coatings and now used only in electrical transformers and capacitors, are among the most persistent synthetic chemicals released to the environment. They can withstand very high temperatures and are not readily degraded by natural agents. Controls on the production and use of these materials in the United States are now quite stringent.

An interesting consequence of the chlorination of water has recently come to light. When a municipal water source is river water, the water may contain traces of industrial wastes—hydrocarbons, for instance. Chlorination of such water not only kills microorganisms but chlorinates these pollutants. A total of 66 chemical pollutants has been detected in drinking water in the lower Mississippi Valley (New Orleans). These include known toxic materials, such as $CHCl_3$, CCl_4, and PCBs. Water purification with ozone, widely used in Europe, is now attracting attention in the United States. Oxygen is also being tested as an agent in water purification. Still another method involves adsorption of impurities on activated charcoal.

Reactions of Fluorine and Chlorine. F_2 reacts with most of the elements, many inorganic compounds, and most organic compounds. Cl_2 also reacts with most of the elements and many compounds. An indication of the large number of reactions of these elements is provided in Table 13-8. An important and interesting compound of fluorine is HF. It is the reagent most often used in preparing other fluorine compounds. Its own preparation involves the reaction of a nonvolatile acid (H_2SO_4) with the salt (CaF_2) of a volatile acid (HF).

$$CaF_2(s) + H_2SO_4(\text{conc. aq}) \xrightarrow{\Delta} CaSO_4(s) + 2\ HF(g) \tag{13.83}$$

One well-known property of HF is its ability to etch (and ultimately to dissolve) glass.

$$SiO_2(s) + 4\ HF(aq) \xrightarrow{\Delta} 2\ H_2O + SiF_4(g) \tag{13.84}$$

This requires that HF be stored in special containers, such as Teflon.

TABLE 13-8
Selected reactions of fluorine and chlorine (X = F or Cl)

Reaction with	Reaction equation
alkali metals	$2\ M + X_2 \longrightarrow 2\ MX$
alkaline earth metals	$M + X_2 \longrightarrow MX_2$
other metals (e.g., Fe)	$2\ Fe + 3\ X_2 \longrightarrow 2\ FeX_3$
hydrogen	$H_2 + X_2 \longrightarrow 2\ HX$
sulfur	$S_8 + 4\ Cl_2 \longrightarrow 4\ S_2Cl_2$
	$S_8 + 24\ F_2 \longrightarrow 8\ SF_6$
phosphorus[a]	$P_4 + 6\ X_2 \longrightarrow 4\ PX_3$
other halogens[b]	$Cl_2 + F_2 \longrightarrow 2\ ClF$
	$Cl_2 + Br_2 \longrightarrow 2\ BrCl$
water	$2\ X_2 + 2\ H_2O \longrightarrow 4\ HX + O_2$
	also: $Cl_2 + H_2O \longrightarrow HCl + HOCl$

[a]PF_5 and PCl_5 are also formed.

[b]Other diatomic interhalogen compounds that are known are BrF, IF, ICl, and IBr. Depending on reaction conditions, more complex interhalogen compounds can be obtained; most of these contain fluorine, e.g., ClF_3, BrF_3, IF_3, ICl_3, ClF_5, BrF_5, IF_5, and IF_7.

HCl has a melting point of $-115°C$ and a boiling point of $-85°C$. Our first thought might be that HF, having a lower molecular weight, should have lower melting and boiling points than HCl; but it does not. HF melts at $-83°C$ and boils at $20°C$. This pronounced difference in properties results from strong hydrogen bonding in HF, which does not occur in HCl. This difference is similar to that found between H_2O and H_2S.

Summary

The first 20 elements are among the most abundant in the earth's crust. They and their compounds have many important uses. $H_2(g)$ is easily obtainable from a variety of sources and has important applications in the synthesis of other materials. Its possible future uses as a fuel and a metallurgical reducing agent hinge on the development of cheaper means of producing it. The lighter noble gases are most noted for their chemical inertness. In addition liquid He has a number of very special uses based on its unique properties.

Li, Na, K, and the other elements of group IA are very active metals. Their atoms lose electrons easily to form the ions M^+. This oxidation occurs when the metals react with the halogens, oxygen, hydrogen, acids, or water. Most alkali metal compounds are highly water soluble, but there are some exceptions among Li compounds. In this and certain other respects Li bears a resemblance to Mg, behavior referred to as a diagonal relationship.

Be, Mg, Ca, and the other members of group IIA are also metals, but here again, as in group IA, the lightest member of the group, Be, differs in properties from the heavier members and resembles Al. Mg and Ca engage in most of the same reactions that the IA metals do. Many of their compounds, however, are insoluble in water. A number of important calcium compounds are derived from the mineral limestone ($CaCO_3$).

The chemistry of boron is essentially that of a nonmetal. Although Al exhibits some properties similar to boron (e.g., in forming a covalent chloride), its most important uses are based on its metallic properties.

The inorganic chemistry of carbon deals primarily with the physical forms of carbon, the oxides of carbon, carbonic acid, and various carbonates. The inorganic chemistry of silicon focuses on the properties of SiO_2 and the variety of silicates that are the backbone of the mineral world.

A practical aspect of the chemistry of nitrogen concerns the means of converting the stable N_2 molecule into nitrogen-containing compounds needed as fertilizers and for other purposes. The oxides of nitrogen are important because of their role in these conversions, for example, in the production of nitric acid. Also, however, oxides of nitrogen are implicated in the formation of photochemical smog. Much of the chemistry of phosphorus centers on its oxides and materials that can be produced from phosphoric acid, H_3PO_4.

The chemistry of oxygen is based primarily on dioxygen, O_2, and water, but ozone, O_3, and hydrogen peroxide, H_2O_2, also have interesting and important properties, particularly in regard to their oxidation-reduction chemistry. Sulfur is of interest because of its variety of molecular and physical forms and because of its use in the production of sulfuric acid, H_2SO_4, the most important manufactured chemical. Oxides of sulfur are common air pollutants produced by power plants and other industrial operations.

Fluorine and chlorine are the first two members of the halogen group (VIIA), the most active nonmetals. Their chemistry includes forming ionic halides with metals, a variety of covalent halides with nonmetals, and compounds among themselves (interhalogens). Chlorine and fluorine compounds have important uses, but some of them pose special environmental problems.

Learning Objectives

As a result of studying Chapter 13, you should be able to

1. Describe and explain physical properties of a number of representative elements in terms of their atomic properties and positions in the periodic table.

2. Use periodic relationships to predict certain properties of the elements and their compounds (e.g., melting point and boiling point).

3. Explain the origin of flame colors that are characteristic of some of the elements.

4. Cite examples of ways in which the first member of a group differs from other members of the same family, including a description of the "diagonal relationship."

5. Describe methods that are used to obtain the first 20 elements from naturally occurring sources, some typical reactions of these elements, and some of their important uses.

6. Name several important compounds of the first 20 elements, write equations for their preparation, and describe some of their uses.

7. Write equations for the reactions of various metals with water, acids, and acids whose anions are oxidizing agents.

8. Describe the different molecular or physical forms of helium, carbon, oxygen, sulfur, and phosphorus and the physical behavior associated with them.

9. Outline the natural processes involved in the nitrogen cycle.

10. Explain the functions of O_3 and CO_2 in relation to heating of the earth.

11. Discuss some of the factors involved in the formation of smog.

12. List some common air pollutants and describe their sources and methods used to control them.

Some New Terms

An **adduct** is a compound formed by the joining together of two simpler molecules through a coordinate covalent bond.

Allotropy refers to the existence of a substance in two or more molecular or physical forms, such as O_2 and O_3, and white, red, and black P.

Diagonal relationships refer to similarities that exist between certain pairs of elements in different groups and periods of the periodic table, i.e., Li/Mg, Be/Al, and B/Si.

Electrolysis refers to the decomposition of a substance, either in the molten state or in an electrolyte solution, by means of electric current.

Industrial smog is air pollution in which the chief pollutants are $SO_2(g)$, $SO_3(g)$, H_2SO_4 mist, and smoke.

An **interhalogen** compound is a binary covalent compound between two halogen elements, such as ClF and BrF_3.

The **nitrogen cycle** is a series of processes by which atmospheric N_2 is fixed, enters into the food chain of animals, and eventually is returned to the atmosphere by bacteria.

The **peroxide** ion has the structure $[:\overset{..}{\underset{..}{O}}\!-\!\overset{..}{\underset{..}{O}}:]^{2-}$.

Photochemical smog is air pollution resulting from reactions involving sunlight, oxides of nitrogen, ozone, and hydrocarbons.

The **superoxide** ion has the structure $[:\overset{..}{\underset{..}{O}}\!-\!\overset{..}{\underset{.}{O}}:]^{-}$.

Suggestions for Further Study

ALLEN, W. M., "The Diagonal Periodic Relationship," *Chemistry,* **43**[4], 22 (1970).

BAMBERGER, C. E., and J. BRAUNSTEIN, "Hydrogen: A Versatile Element," *American Scientist,* **63**, 438 (1975).

COMPANION, A. and K. SCHUG, "Ceramics and Glass," *Chemistry,* **46**[9], 27 (1973).

HANST, P. L., "Noxious Gases in the Air, Part I. Photochemical Smog," *Chemistry,* **51**[1], 8 (1978); "Part II. Halogenated Products," **51**[2], 6 (1978).

HOUSE, J. E., Jr., "Beryllium," *Chemistry,* **44**[11], 10 (1971).

LIKENS, G. E., et al., "Acid Rain," *Scientific American,* **241**[4], 43 (1979).

NAVRATIL, J. D., "Fluorine, a Hostile Element," *Chemistry,* **42**[2], 11 (1969).

REVELLE, R., "Carbon Dioxide and World Climate," *Scientific American,* **247**[2], 35 (1982).

WATKINS, K. W., "Chemical of the Month: Lime," *J. Chem. Educ.,* **60,** 60 (1983).

ZIMMERMAN, J., "The Strange World of Helium," *Chemistry,* **43**[2], 14 (1970).

Review Problems

1. Describe briefly what is meant by each of the following terms, citing specific examples wherever possible: **(a)** allotropy; **(b)** an ore; **(c)** diagonal relationship; **(d)** nitrogen fixation; **(e)** thermite reaction; **(f)** interhalogen compound; **(g)** noble gas.

2. Supply a name or formula for each of the following: **(a)** sodium peroxide; **(b)** magnesium nitride; **(c)** $Ca(OH)_2$; **(d)** sodium perchlorate; **(e)** $LiAlH_4$; **(f)** O_2^-; **(g)** hydrogen difluoride ion; **(h)** ammonium dihydrogen phosphate; **(i)** H_2O_2; **(j)** $NaHSO_3$; **(k)** ozone; **(l)** N_2O_4; **(m)** calcium hydrogen carbonate; **(n)** bromine trifluoride.

3. What chemical substances constitute these common industrial materials? **(a)** coke; **(b)** limestone; **(c)** quicklime; **(d)** slaked lime; **(e)** silica; **(f)** gypsum; **(g)** water glass; **(h)** plaster of Paris; **(i)** synthesis gas.

4. Complete the following equations for the reactions of substances with water.

(a) $LiH(s) + H_2O \rightarrow$ **(b)** $Na_2O_2(s) + H_2O \rightarrow$

(c) $Ca(s) + H_2O \rightarrow$ **(d)** $CaO(s) + H_2O \rightarrow$

(e) $P_4O_{10}(s) + H_2O \rightarrow$ **(f)** $C(s) + H_2O \xrightarrow{\Delta}$

5. Complete the following equations for reactions of substances with acids.

(a) $Mg(s) + HCl(aq) \rightarrow$

(b) $NH_3(g) + HNO_3(aq) \rightarrow$

(c) $CaCO_3(s) + HCl(aq) \rightarrow$

(d) $NaF(s) + H_2SO_4(\text{conc. aq}) \xrightarrow{\Delta}$

(e) $Ag(s) + HNO_3(\text{dil. aq}) \rightarrow$

6. Write equations to show how H_2O_2 **(a)** oxidizes NO_2^- to NO_3^- in acidic solution; **(b)** oxidizes $SO_2(g)$ to $SO_4^{2-}(aq)$ in basic solution; **(c)** reduces MnO_4^- to Mn^{2+} in acidic solution; **(d)** reduces $Cl_2(g)$ to $Cl^-(aq)$ in basic solution. (In an acidic solution you may write H^+ and/or H_2O on either side of the equation, as necessary; for a basic solution, OH^- and/or H_2O.)

7. Write the formula of a chemical compound mentioned in this chapter that has **(a)** N in the oxidation state (O.S.) +4; **(b)** O in the O.S. −1; **(c)** Cl in the O.S. +1; **(d)** S in the O.S. −2; **(e)** P in the O.S. −3; **(f)** H in the O.S. −1.

8. Identify the oxidizing and reducing agents in the following reactions from the chapter: (13.3); (13.26); (13.56); (13.61); (13.64); (13.65); (13.72).

9. Use data from Appendix D to establish whether each of the following reactions is endothermic or exothermic at 298 K and 1 atm: (13.2); (13.3); (13.13); (13.27); (13.30); (13.43). [*Hint:* Recall equation (6.17).]

10. State which of the first 20 elements

(a) is the most abundant nonmetal, and which, the most abundant metal;

(b) are gases at STP;

(c) are semiconductors in the solid state;

(d) is the best oxidizing agent;

(e) exhibit allotropy;

(f) has the highest melting point;

(g) has the lowest critical point temperature;

(h) form compounds with oxygen.

Exercises

Group and period trends

1. For the following groups of substances, select

(a) the most metallic of the group K, Be, Ca;

(b) the most soluble of the compounds Li_2CO_3, Na_2CO_3, $CaCO_3$;

(c) the best oxidizing agent of the group H_2O_2, O_2, O_3;

(d) the most volatile of the liquids $H_2O(l)$, $H_2S(l)$, $HF(l)$;

(e) the hardest of the substances $Ca(s)$, $C(\text{graphite})$, $SiO_2(s)$;

(f) the best electrical conductor among $LiF(l)$, $BeF_2(l)$, $CF_4(l)$, $F_2O(l)$;

(g) the highest melting point among $LiCl(s)$, $NaCl(s)$, $LiF(s)$, $NaF(s)$.

2. Use data from the text to plot, in a manner similar to Figure 13-1, the following properties of the third period elements: **(a)** single covalent radius; **(b)** first ionization energy; **(c)** number of unpaired electrons in the neutral atoms. Explain the significance of each graph.

Hydrogen

3. Write equations to show how each of the following substances can be used in the preparation of $H_2(g)$: **(a)** H_2O; **(b)** $HI(aq)$; **(c)** $Mg(s)$; **(d)** $CO(g)$; **(e)** $NaOH(aq)$. Use other reactants as necessary, e.g., H_2O, acids, metals.

4. What mass of $CaH_2(s)$ would be required to generate the $H_2(g)$ to fill a 225-L observation balloon at 748 mmHg and 18°C?

Helium, neon, argon

5. A 55-L cylinder contains Ar at 125 atm and 25°C. What minimum volume of air at STP must have been liquefied and distilled to produce this Ar? Air contains 0.934% Ar, by volume.

6. A typical natural gas from which He can be extracted contains 0.3% He, by volume. Air contains 5 ppm He, by volume. How much more abundant is He in the natural gas than in air?

7. Suppose that a breathing mixture is prepared in which He is substituted for N_2, i.e., a gas with 79% He and 21% O_2, by volume. What is
   (a) the density of this mixture, in g/L, at STP?
   (b) the apparent molar mass of the mixture (recall page 123)?

Lithium, sodium, potassium

8. Write equations to represent the reactions of the following with H_2O: (a) K(s); (b) KH(s); (c) KO_2(s).

9. A pure white solid is either LiCl or KCl. Describe a simple test(s) to determine which it is.

*10. Use the lattice energy of LiF given in Section 13-5, together with an enthalpy of sublimation of Li(s) = 160.7 kJ/mol and other data from the text, to calculate $\Delta \overline{H}_f^\circ$ for LiF(s).

Beryllium, magnesium, calcium

11. Write equations for the reactions that you would expect to occur when
    (a) $MgCO_3$(s) is heated to a high temperature;
    (b) molten $CaCl_2$ is electrolyzed;
    (c) Be(s) is dissolved in concentrated NaOH(aq);
    (d) Ca(s) is added to cold dilute HCl(aq).

12. Magnesium oxide is produced by heating Mg in air. From 0.200 g Mg, 0.305 g of product is obtained.
    (a) Can the compound be pure MgO? Explain.
    (b) What other substance is probably present?
    (c) How would you test for the presence of the substance referred to in part (b)?

Boron, aluminum

13. Write equations for the reactions that you would expect to occur when Al(s) is allowed to react with (a) HI(aq); (b) H_2SO_4(aq); (c) KOH(aq); (d) Cr_2O_3 at high temperatures; (e) O_2(g) at high temperatures; (f) Cl_2(g) at high temperatures.

14. Explain what is meant by the terms (a) electron-deficient molecule; (b) dimer; (c) adduct.

15. The enthalpies of formation of Al_2O_3(s), Fe_2O_3(s), MnO_2(s), and MgO(s) are −1670, −824, −519, and −602 kJ/mol, respectively.
    (a) Determine the heat of the thermite reaction (13.36) per mole of Fe produced.
    (b) Determine the heat of reaction per mole of Mn, if MnO_2 is substituted for Fe_2O_3.
    (c) Show that if MgO were substituted for Fe_2O_3 the reaction would be endothermic. (Al does not reduce MgO to Mg.)

Carbon

16. Write equations for the reactions that you would expect to occur when
    (a) C_8H_{18} is burned in a limited quantity of air;
    (b) CO(g) is heated with PbO(s);
    (c) CO_2(g) is bubbled into KOH(aq);
    (d) $MgCO_3$(s) is added to HCl(aq).

17. The reaction of methane and sulfur vapor produces carbon disulfide and hydrogen sulfide. Carbon disulfide reacts with Cl_2(g) to form carbon tetrachloride and S_2Cl_2. Further reaction of carbon disulfide and S_2Cl_2 produces additional carbon tetrachloride and sulfur. Write a series of equations for the reactions described here.

18. The combustion of 5 billion metric tons of fossil fuels (the current annual rate, worldwide) raises the atmospheric content of CO_2 by 0.7 ppm. Starting with the current level of 335 ppm, and assuming a fossil fuel reserve of 10,000 billion metric tons, what is the limit to which the CO_2 content of air could be raised by burning all this fuel?

Silicon

19. With reference to Table 13-5, show that the formula given for mica is consistent with the usual oxidation states and ionic charges of its constituents.

20. Write chemical equations for the following reactions.
    (a) The reduction of silica to elemental silicon by aluminum metal.
    (b) The preparation of potassium metasilicate by the high-temperature fusion of silica and potassium carbonate.
    (c) The formation of monosilane (SiH_4) by the reaction of silicon tetrachloride with lithium aluminum hydride.
    (d) The combustion of trisilane (Si_3H_8), yielding silica.
    (*Hint*: The essential substances have been named. You may have to supply an additional reactant and/or product to complete each equation.)

21. Describe and explain the similarities and differences between the reaction of a silicate with an acid and that of a carbonate with an acid.

Nitrogen

22. Write balanced equations to represent
    (a) the complete neutralization of H_2SO_4(aq) with NH_3(aq);
    (b) the dissolving of Ag in conc. HNO_3(aq);
    (c) the reaction of hot conc. HNO_3(aq) with carbon;
    (d) the role of NO_2(g) in the formation of O_3(g) in photochemical smog.

23. Step 1a in the nitrogen cycle in Figure 13-12 involves the production of nitric acid during a lightning storm. Suggest a series of reactions by which HNO_3 could form under these conditions.

24. In 1968, before any pollution controls existed on automotive emissions, over 75 billion gal of gasoline were consumed in the U.S. as a motor fuel. Assuming an emission of oxides of nitrogen of 5 g per vehicle mile and an average mileage of 15 mi/gal of gasoline, what mass of nitrogen oxides, in tons, was released to the atmosphere?

*25. Zn can reduce NO_3^- to $NH_3(g)$ in basic solution. (The following equation is *not* balanced.)

$$NO_3^- + Zn(s) + OH^- + H_2O \rightarrow Zn(OH)_4^{2-} + NH_3(g)$$
$$(13.85)$$

The NH_3 can be neutralized with an excess of HCl(aq), and the unreacted HCl titrated with NaOH. In this way a quantitative determination of NO_3^- is achieved. A 25.00-mL sample of solution was treated according to (13.85). The $NH_3(g)$ was passed into 50.00 mL of 0.1500 *M* HCl. The excess HCl required 32.10 mL of 0.1000 *M* NaOH for its titration. What was the molarity of NO_3^- in the original sample?

Phosphorus

26. A certain phosphate rock is 58.0% $Ca_3(PO_4)_2$. What mass of this rock is required to produce 135 kg of phosphorus, assuming no loss of material in reaction (13.65)?

27. The following process could be used to obtain solid red P. Liquid white P is boiled at 287°C and phosphorus vapor condensed to solid red P at 350°C. Explain how it is possible for the vapor to deposit as a solid at a temperature *higher* than the boiling point of the liquid.

Oxygen

28. Use Lewis structures or other information from this chapter to explain the facts that
   (a) H_2S is gaseous at room temperature while H_2O is liquid;
   (b) O_3 is diamagnetic;
   (c) The O—O bond lengths in O_2, O_3, and H_2O_2 are 121, 128, and 148 pm, respectively.

29. In a disproportionation reaction the same substance is both oxidized and reduced. Write an equation for a plausible disproportionation of H_2O_2.

30. $O_3(g)$ is a powerful oxidizing agent. Write equations to represent oxidation of (a) I^- to I_2 in acidic solution; (b) sulfur in the presence of moisture to sulfuric acid; (c) $[Fe(CN)_6]^{4-}$ to $[Fe(CN)_6]^{3-}$ in basic solution. In each case $O_3(g)$ is reduced to $O_2(g)$, and H^+ and/or H_2O and OH^- and/or H_2O can be included as necessary.

Additional Exercises

1. Complete each of the following equations. If no reaction occurs, so state.
   (a) $CaH_2(s) + H_2O \rightarrow$
   (b) $Mg(s) + HC_2H_3O_2(aq) \rightarrow$
   (c) $BeO(s) + H_2O \rightarrow$
   (d) $Ca(s) + H_2O \rightarrow$
   (e) $Cl_2(g) + H_2O \rightarrow$
   (f) $NaCl(aq) \xrightarrow{\text{electrolysis}}$
   (g) $CO_2(g) + Na^+(aq) + OH^-(aq) \rightarrow$

31. Show that the structure of the O_3 molecule given in the text is consistent with the prediction made with VSEPR theory.

*32. It has been estimated that if all the O_3 in the atmosphere were brought to sea level at STP, the gas would form a layer 0.3 cm thick. Estimate the number of O_3 molecules in the earth's atmosphere. (Assume that the radius of the earth is 4000 mi.)

Sulfur

33. Use Lewis structures or other information from this chapter to explain the fact that
   (a) S_2Cl_2 has a structure similar to H_2O_2.
   (b) SO_2 possesses a dipole moment but SO_3 does not.
   (c) There exists a metastable, paramagnetic, purple solid, S_2.

34. Sulfur can occur naturally as sulfates, but not as sulfites. Explain why this is so.

35. 25.0 L of a natural gas, measured at 25°C and 740 mmHg, is bubbled through Pb^{2+}(aq); a precipitate weighing 0.535 g is obtained. What is the % H_2S, by volume, in the natural gas? $[Pb^{2+} + H_2S \rightarrow PbS + 2 H^+.]$

36. The Frasch process for mining sulfur uses superheated water at 160°C.
   (a) What is superheated water?
   *(b) Under what minimum pressure must this water be maintained? (*Hint:* Use relevant information from Chapter 11.)

Fluorine, chlorine

37. Use data from Table 13-7 to predict the melting points and boiling points of BrCl and ICl. What principle is involved?

38. In a similar manner to Exercise 37, use data from Example 13-4 and Table 13-7 to estimate the melting and boiling points of ClF.

39. Write balanced equations for the following reactions.
   (a) $Ca(s) + Cl_2(g) \rightarrow$ (b) $P_4(s) + Cl_2(g) \rightarrow$
   (c) $F_2(g) + H_2O \rightarrow$ (d) $Al(s) + F_2(g) \rightarrow$

2. Explain why the compound SF_6 exists but OF_6 does not; PCl_5 but not NCl_5.

3. Suppose that 90% of $SO_2(g)$ emissions of an electric power plant are converted to H_2SO_4. What volume of sulfuric acid (98% H_2SO_4, $d = 1.84$ g/cm³) could be produced from the 2.2×10^6 tons of coal (3.5% S) used by the plant annually?

4. Suppose that, unlike in Figure 13-2, no attempt is made to separate the $H_2(g)$ and $O_2(g)$ produced by the electrolysis of water (13.4).

(a) What volume of an H_2/O_2 mixture at 23°C and 748 mmHg could be obtained as a result of electrolyzing 10.0 g H_2O?

(b) Recalculate the volume in part (a) by taking into account the fact that the H_2/O_2 mixture is saturated with $H_2O(g)$. Assume a total pressure of 748 mmHg and a vapor pressure of 20.5 mmHg for the dilute electrolyte solution.

(c) Express the composition of the gas in part (b) in mole fractions of $H_2(g)$, $O_2(g)$, and $H_2O(g)$.

5. With reference to Figure 13-1, give a reasonable explanation of the following observations.

(a) The melting point of C is higher than that of Si.

(b) The group IA, IIA, and IIIA metals of the third period have *lower* melting points than do those of the second period. On the other hand, the third period elements of groups VA, VIA, and VIIA have *higher* melting points than do the second period elements.

6. Oxidation of IO_3^- by Cl_2 in basic solution yields paraperiodate ion.

$$IO_3^- + Cl_2 + OH^- \rightarrow H_3IO_6^{2-} + Cl^-$$

A precipitate is formed with Ag^+ ion.

$$H_3IO_6^{2-} + Ag^+ \rightarrow Ag_3IO_5(s) + H_2O + H^+$$

A solution of paraperiodic acid is formed when an aqueous suspension of Ag_3IO_5 is treated with $Cl_2(g)$.

$$Ag_3IO_5(s) + Cl_2(g) + H_2O \rightarrow H_5IO_6(aq) + AgCl(s) + O_2(g)$$

The paraperiodic acid can be crystallized from this solution.

(a) Balance all the equations given.

(b) What mass of H_5IO_6 can be prepared from 113 g $NaIO_3$?

*7. The composition of a phosphate mineral can be expressed as % P, % P_2O_5, or % BPL [bone phosphate of lime, $Ca_3(PO_4)_2$].

(a) Show that % P = $0.436 \times$ (% P_2O_5) and % BPL = $2.185 \times$ (% P_2O_5).

(b) What is the significance of a % BPL greater than 100?

(c) What is the % BPL of a typical phosphate rock?

*8. Propose a plausible bonding scheme for O_3 involving hybrid orbitals.

*9. Assuming a similar packing of spherical atoms in the crystalline alkali metals, account for the fact that Na has a higher density than both Li and K. (*Hint:* Use data from Table 13-3.)

*10. The normal boiling point of liquid sulfur is 445°C, and the following data are given for triple points in the system sulfur: $S_\alpha S_\beta V$, 95.5°C and 0.004 mmHg; $S_\beta LV$, 119°C and 0.05 mmHg; $S_\alpha S_\beta L$, 151°C and 1288 atm. At pressures above 1288 atm the only solid phase is S_α. Use these data to sketch a phase diagram for sulfur (not to scale).

Self-Test Questions

For questions 1 through 8 select the single item that best completes each statement.

1. All of the following metals react with cold water except (a) K; (b) Al; (c) Ca; (d) Li.

2. All of the following molten compounds are good electrical conductors except (a) $BeCl_2$; (b) KF; (c) CsI; (d) NaCl.

3. Under the appropriate conditions, each of the following can act as an oxidizing agent except (a) H_2O_2; (b) Cl^-; (c) F_2; (d) SO_2.

4. Of the following group of liquids, the highest boiling point is expected for (a) HCl; (b) CS_2; (c) H_2S; (d) CCl_4.

5. All of the following substances react with water. The pair that yield the same gaseous product are (a) CaO and CaH_2; (b) Na and Na_2O_2; (c) F_2 and Na_2O_2; (d) Li_3N and LiH.

6. The % H, by mass, is greatest in (a) the earth's crust; (b) the earth's core; (c) the universe as a whole; (d) the compound H_2O.

7. The *cheapest* raw material from which an alkaline (basic) medium can be prepared is (a) coke; (b) limestone; (c) phosphate rock; (d) gypsum.

8. Of the following, the one that is *unimportant* in the production of fertilizers is (a) phosphate rock; (b) H_2SO_4; (c) NH_3; (d) Na_2CO_3.

9. Supply a name or formula for each of the following: (a) lithium hydride; (b) $KHSO_4$; (c) ClO_2^-; (d) SiO_2; (e) potassium perchlorate; (f) sodium nitrite; (g) N_2O_5; (h) BaO_2.

10. Write balanced equations to represent (a) the decomposition of $MgCO_3$ by heating; (b) the action of HCl(aq) on $MgCO_3(s)$; (c) the formation of H_3PO_4 from P; (d) the reduction in acidic solution of Cl_2 to Cl^- by H_2O_2; (e) the oxidation of NH_3 to NO by O_2; (f) the reaction of Cu with dilute $HNO_3(aq)$.

11. A chemistry magazine gives the 1982 U.S. production of $O_2(g)$ as 3.58×10^{11} ft^3 at STP. What is this quantity of $O_2(g)$ expressed in kg?

12. Explain why the air pollution control measures required for automobiles are not the same as those required for fossil-fuel electric power plants.

14

Chemical Kinetics

The principles of stoichiometry permit us to calculate the amounts of substances that can be produced by a chemical reaction. They tell us nothing, however, about *how long* it takes for a reaction to occur. For an industrial process one might actually choose a reaction that gives a lower yield but proceeds *faster* than an alternative reaction yielding the same product. On the other hand, certain reactions that proceed extremely rapidly might not be desirable either—they might lead to explosions! And then there are circumstances where chemical reactions are unwanted. Here we prefer that whatever reaction does occur do so as slowly as possible. This is the objective of adding a rust inhibitor to the coolant in an automobile radiator or storing milk in a refrigerator.

The cases just cited suggest a need to be able to measure, control, and where possible *predict* the rates of chemical reactions. These topics are all part of the study of **chemical kinetics.** Also, chemical kinetics sometimes helps us to deduce something about a **mechanism** for a reaction. This is a detailed, step-by-step suggestion of how initial reactants are converted to final products. Predictions of the rates of chemical reactions are based on mathematical equations called rate laws. Methods of deriving and using rate laws are the central topics of this chapter.

14-1 The Rate of a Chemical Reaction

Since our principal concern will be with the rates of chemical reactions, we should begin with an explanation of what we mean by this term. Rate (or speed or velocity) suggests something that happens in a unit of time, e.g., per second, per minute, ... What happens in chemical reactions is that amounts of reactants and products change. These changes are most commonly expressed through changes in *molar concentrations*. Thus, by the rate of the hypothetical reaction

$$A + 3\,B \longrightarrow 2\,C + 2\,D \tag{14.1}$$

we might mean the rate of decrease of molar concentration of A. This gives the rate of reaction the units mol L^{-1} s^{-1} (mol per liter per second), for example.

We can also describe the rate of reaction (14.1) in terms of the disappearance of B or the formation of C or D, but now we have a problem: These rates are not the same as the rate of disappearance of A. From the coefficients in the equation we see that *three* moles of B are consumed for every mole of A. That is, B disappears three times as fast as A, or

rate of disappearance of B = 3 × rate of disappearance of A.

Similarly, we can write that

rate of formation of C = 2 × rate of disappearance of A
rate of formation of D = 2 × rate of disappearance of A

406

Another problem with these varying expressions for the rate of a reaction is that rates of disappearance are *negative* (concentrations decrease with time) and rates of formation are *positive* (concentrations increase with time). All of these matters are taken care of, resulting in a single, positive expression for the rate of reaction (14.1), for example, if we express it as

$$\text{rate of reaction} = - \text{ rate of disappearance of A}$$
$$= -\tfrac{1}{3} \text{ rate of disappearance of B}$$
$$= \tfrac{1}{2} \text{ rate of formation of C} \qquad (14.2)$$
$$= \tfrac{1}{2} \text{ rate of formation of D}$$

If the above system of dividing by the coefficients in the chemical equation is not used, then we need to make clear the species whose concentration is being followed when we speak of the rate of a chemical reaction.

We need to become even more specific about the meaning of the rate of a chemical reaction and will do so in the next section. First let us illustrate some of these introductory ideas through Example 14-1, where we also reintroduce the bracket notation first mentioned on page 349. That is, the symbol [A] stands for the molar concentration of A—the number of moles of A per liter of solution.

Example 14-1 Suppose that at some particular point in the hypothetical reaction (14.1) it is found that the concentration of A is 1.0000 *M*, i.e., [A] = 1.0000 *M*. Suppose that exactly one minute later [A] = 0.9982 *M*. **(a)** What is the rate of the reaction at this point, expressed in mol L^{-1} min^{-1}? **(b)** What is the rate of formation of C at this point? **(c)** What is the rate of reaction expressed in mol L^{-1} s^{-1}?

Solution

(a) The rate of disappearance of A is given by the *change* in molar concentration, $\Delta[A]$, divided by the time interval, Δt, over which this change occurs. The change in molar concentration is $\Delta[A] = 0.9982\ M - 1.0000\ M = -0.0018\ M$. The time interval is $\Delta t = 1.00$ min.

$$\text{rate of disappearance of A} = \frac{\Delta[A]}{\Delta t} = \frac{-0.0018\ M}{1.00\ \text{min}}$$
$$= -1.8 \times 10^{-3}\ M\ min^{-1}$$
$$= -1.8 \times 10^{-3}\ \text{mol } L^{-1}\ min^{-1}$$

$$\text{rate of reaction} = - \text{ rate of disappearance of A}$$
$$= -(-1.8 \times 10^{-3}\ \text{mol } L^{-1}\ min^{-1})$$
$$= 1.8 \times 10^{-3}\ \text{mol } L^{-1}\ min^{-1}$$

(b) From expression (14.2) we see that the rate of reaction = $\tfrac{1}{2}$ rate of formation of C, or

$$\text{rate of formation of C} = 2 \times \text{ rate of reaction}$$
$$= 2 \times 1.8 \times 10^{-3}\ \text{mol } L^{-1}\ min^{-1}$$
$$= 3.6 \times 10^{-3}\ \text{mol } L^{-1}\ min^{-1}$$

(c) The decrease in concentration used in the rate calculation in (a) occurs over a 1-min time interval. A much smaller change in concentration will occur over a 1-s interval. Specifically, we must convert min^{-1} to s^{-1}. (Note that $min^{-1} \times min = 1$.)

$$\text{rate of reaction} = 1.8 \times 10^{-3}\ \text{mol } L^{-1}\ min^{-1} \times \frac{1\ \text{min}}{60\ \text{s}}$$
$$= 3.0 \times 10^{-5}\ \text{mol } L^{-1}\ s^{-1}$$

SIMILAR EXAMPLES: Review Problems 1, 2; Exercise 1.

14-2 Experimental Determination of Reaction Rates

In our continuing exploration of rates of reactions, we will consider as a specific illustration the decomposition of hydrogen peroxide in aqueous solution.

$$H_2O_2(aq) \longrightarrow H_2O + \tfrac{1}{2}O_2(g) \tag{14.3}$$

In the course of this reaction, $O_2(g)$ escapes from the reaction mixture and the reaction eventually goes to completion.* The progress of the reaction can easily be followed by removing small samples of the solution from time to time and titrating them with $KMnO_4$ in aqueous acidic solution.

$$2\,MnO_4^- + 5\,H_2O_2 + 6\,H^+ \longrightarrow 2\,Mn^{2+} + 8\,H_2O + 5\,O_2(g) \tag{14.4}$$

Typical data for the concentration of H_2O_2 as a function of time are listed in Table 14-1 and plotted in Figure 14-1. How data from the titrations with $KMnO_4$ are converted to $[H_2O_2]$ is suggested through Example 14-2.

Example 14-2   When exactly 300 s had elapsed after the start of reaction (14.3), a 5.00-mL portion of the reaction mixture was removed and immediately titrated with 0.1000 M MnO_4^- in acidic solution (reaction 14.4); 37.1 mL of the 0.1000 M $MnO_4^-(aq)$ was required. What must have been $[H_2O_2]$ at this 300-s point in the reaction?

Solution. $[H_2O_2]$ in the 5.00-mL portion was the same as $[H_2O_2]$ in the larger sample that was undergoing decomposition. For this 5.00-mL portion

$$\text{no. mmol } H_2O_2 = 37.1 \text{ mL soln.} \times \frac{0.1000 \text{ mmol } MnO_4^-}{\text{mL soln.}} \times \frac{5 \text{ mmol } H_2O_2}{2 \text{ mmol } MnO_4^-}$$

$$= 9.28 \text{ mmol } H_2O_2$$

$$[H_2O_2] = \frac{9.28 \text{ mmol } H_2O_2}{5.00 \text{ mL}} = 1.86 \ M$$

SIMILAR EXAMPLE: Review Problem 3.

Since chemical reaction continues in the portion of a reaction mixture that is removed, either the analysis of this portion must be carried out very quickly or the reaction must be effectively stopped (quenched), e.g., by dilution or by cooling.

TABLE 14-1
Decomposition of H_2O_2

Time, s	$[H_2O_2]$, M
0	2.32
300	1.86
600	1.49
1200	0.98
1800	0.62
3000	0.25

Rate of Reaction: A Variable Quantity.   From Figure 14-1 we see that the concentration of H_2O_2 decreases with time at a rate that is initially rapid but then slows down; the reaction *decelerates*. For example, in the first 600 s $[H_2O_2]$ decreases from about 2.30 M to 1.50 M, a decrease of 0.80 mol H_2O_2/L. In the next 600 s the decrease is only about 0.50 mol H_2O_2/L. Ordinarily, the reaction rate varies during the course of a reaction. We need to specify further exactly how the rate of a reaction is to be expressed.

Rate of Reaction Expressed as $-\Delta[H_2O_2]/\Delta t$.   One approach to describing the rate of decomposition of H_2O_2 is to extract data from Figure 14-1 and tabulate these data (shown in color in Table 14-2). Column III of Table 14-2 lists the molar concentrations of H_2O_2 at the times shown in column I. Column II states the arbitrary time interval that we have chosen between data points—400 s. Column IV reports the changes in concentration that occur for each 400-s interval. The figures in column V represent the reaction rates; their values decrease continuously with time.

Starting with $[H_2O_2] = 2.32$ M at $t = 0$, with each succeeding time interval the concentration becomes smaller; $\Delta[H_2O_2]$ is a *negative* quantity. But according to the conven-

---

*Although reaction (14.3) goes to completion, it does so very slowly. Generally a catalyst is used to speed up the reaction. The function of a catalyst in this reaction is described in Section 14-10.

FIGURE 14-1
Graphical determination
of the rate of the
reaction

$$H_2O_2 \longrightarrow H_2O + \tfrac{1}{2} O_2(g)$$

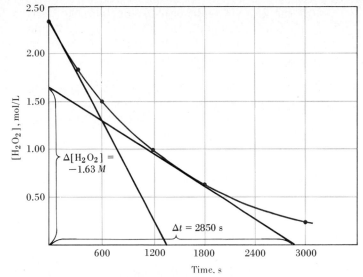

Reaction rates are determined from the slopes of the tangent lines
(see Examples 14-3 and 14-4).

tion established through expression (14.2), the rate of reaction is the *negative* of the rate of
disappearance of H_2O_2. We should write

$$\text{rate of reaction} = \frac{-\Delta[H_2O_2]}{\Delta t} \tag{14.5}$$

Rate of Reaction Expressed as the Slope of a Tangent Line. When expressed as
$-\Delta[H_2O_2]/\Delta t$, the rate of reaction is simply an average value over the time interval
chosen. For example, the rate of decomposition of H_2O_2 averages $15.0 \times$

TABLE 14-2
Decomposition of H_2O_2—derived rate data

I	II	III	IV	V
Time, s	Δt, s	$[H_2O_2]$, mol/L	$\Delta[H_2O_2]$, mol/L	Reaction rate = $-\Delta[H_2O_2]/\Delta t$, mol L^{-1} s^{-1}
0		2.32		
	400		−0.60	15.0×10^{-4}
400		1.72		
	400		−0.42	10.5×10^{-4}
800		1.30		
	400		−0.32	8.0×10^{-4}
1200		0.98		
	400		−0.25	6.2×10^{-4}
1600		0.73		
	400		−0.19	4.8×10^{-4}
2000		0.54		
	400		−0.15	3.8×10^{-4}
2400		0.39		
	400		−0.11	2.8×10^{-4}
2800		0.28		

In calculus notation, the ratio $-\Delta[H_2O_2]/\Delta t$ in the limit where $\Delta t \to 0$ can be replaced by the derivative $-d[H_2O_2]/dt$. That is,

$$\lim_{\Delta t \to 0} \frac{-\Delta[H_2O_2]}{\Delta t}$$

$$= \frac{-d[H_2O_2]}{dt}$$

10^{-4} mol L^{-1} s^{-1} in the interval from 0 to 400 s (see the first entry of Table 14-2). We might think of this as the rate of reaction at the middle of the interval—200 s. We could also describe the rate of reaction at 200 s with concentration data at 100 s and 300 s and $\Delta t = 200$ s, but the result would differ slightly from 15.0×10^{-4} mol L^{-1} s^{-1}. A unique value of the rate of reaction, called an **instantaneous reaction rate,** is obtained only in the limit where the time interval is allowed to approach zero, that is, $\Delta t \to 0$. Under these circumstances the rate of reaction becomes equal to the *negative of the slope of the tangent line* to the graph of $[H_2O_2]$ as a function of time. Two tangent lines are drawn in Figure 14-1, one at $t = 0$ and one at $t = 1500$ s.

Example 14-3   From Figure 14-1 determine the rate of decomposition of H_2O_2 at 1500 s.

Solution.   From a graph like Figure 14-1, a rate of reaction is most readily established through the slope of a tangent line, in this case at $t = 1500$ s.

Even though we can attempt to read points from the graph to three significant figures, because of uncertainty in constructing the tangent line we probably are not justified in carrying more than two figures in the slope.

$$\text{rate of reaction} = -\text{ slope of tangent line} = \frac{-\Delta[H_2O_2]}{\Delta t} = \frac{1.63 \text{ mol/L}}{2850 \text{ s}}$$

$$= 5.7 \times 10^{-4} \text{ mol } L^{-1} \text{ s}^{-1}$$

SIMILAR EXAMPLES: Review Problems 4, 15; Exercise 23d, e.

Initial Rate of Reaction.   To determine rates of reaction by the graphical method of Example 14-3 requires that data be collected over an extended period of time. Occasionally all that is required is the rate of reaction immediately after the reactants are brought together—the **initial rate of reaction.** This rate can be obtained by dividing the change in a reactant concentration that occurs in a brief time interval following the start of the reaction ($\Delta[\text{reactant}]$) by the time interval (Δt). The initial rate is also equal to the slope of the tangent line at $t = 0$ in a graph such as Figure 14-1. Calculations based on the initial rate work only if the time interval chosen is short enough that the rate remains essentially constant during the interval. This interval corresponds to the time span over which the tangent line and the "concentration vs. time" curves practically coincide. In Figure 14-1 this is about 200 s. Put in another way, initial rate calculations work over the time span in which only a few percent of the available reactant(s) are consumed.

Example 14-4   For the decomposition of H_2O_2 represented by the data in Table 14-1, (a) determine the initial rate of reaction, and (b) $[H_2O_2]$ at $t = 100$ s.

Solution
(a) Two methods for determining the initial rate are suggested in the above discussion. The first, which involves dividing a change in reactant concentration by a time interval, is not likely to work well here. The shortest time interval in Table 14-1 is 300 s, and this exceeds the time interval over which we expect the initial rate to remain fairly constant. The second method involves determining the slope of the tangent line at $t = 0$. This we can do from Figure 14-1, based on the intersections of the tangent line with the axes: $t = 0$, $[H_2O_2] = 2.32$ M; $t = 1330$ s, $[H_2O_2] = 0$.

$$\text{initial rate} = -\text{ slope of tangent} = \frac{-(0 - 2.32) \text{ mol/L}}{1330 \text{ s}}$$

$$= 1.7 \times 10^{-3} \text{ mol } L^{-1} \text{ s}^{-1}$$

(b) Let us assume that the rate determined in (a) remains essentially constant for at least 100 s. Since

$$\text{rate of reaction} = \frac{-\Delta[H_2O_2]}{\Delta t} \qquad \text{then} \qquad \Delta[H_2O_2] = - \text{ rate of reaction} \times \Delta t$$

$$\Delta[H_2O_2] = -1.7 \times 10^{-3} \text{ mol L}^{-1} \text{ s}^{-1} \times 100 \text{ s} = -1.7 \times 10^{-1} \text{ mol L}^{-1}$$

$$[H_2O_2]_{t=100\ s} = [H_2O_2]_{t=0} + \Delta[H_2O_2]$$
$$= 2.32\ M - 0.17\ M = 2.15\ M$$

SIMILAR EXAMPLES: Review Problem 14; Exercises 2, 8, 9, 23d.

14-3 The Rate Law for Chemical Reactions

For many reactions the reaction rate can be expressed through a mathematical equation known as the **rate law** or **rate equation.** Consider the reaction

$$a\,A + b\,B + \cdots \longrightarrow g\,G + h\,H + \cdots \tag{14.6}$$

where a, b, . . . stand for coefficients in the balanced equation. The rate of reaction can be represented as

$$\text{rate of reaction} = k[A]^m[B]^n \cdots \tag{14.7}$$

In this expression the symbols [A], [B], . . . represent molar concentrations. The exponents m, n, . . . are generally small integral numbers, although in some cases they may be fractional or negative. It is important to note that there is *no* relationship between the exponents m, n, . . . , and the corresponding coefficients in the balanced equation, a, b, ... If in some cases they happen to be identical (i.e., $m = a$, $n = b$), this is just a matter of chance; it is *not* to be expected.

The exponents in the rate equation determine the **order of the reaction.** If $m = 1$, the reaction is said to be *first order in A*. If $n = 2$, the reaction is *second order in B,* and so on. The total of the exponents $m + n + \cdots$ is the overall order of the reaction. The term k in equation (14.7) is called the **rate constant.** It is a proportionality constant that is characteristic of the particular reaction and is significantly dependent only on temperature or the presence of catalysts. The larger the value of k the faster a reaction proceeds. The rate of reaction, as we have seen, is usually expressed in the units *moles per liter per unit time*, e.g., *mol L^{-1} s^{-1}*. The units of k, on the other hand, depend on the order of the reaction.

The significance of the rate law expression seems to have first been recognized by the Norwegian mathematician Cato Guldberg and his brother-in-law chemist, Peter Waage. In 1865 they proposed that the force (rate) of a chemical reaction is equal to the product of the active masses (concentrations) of the reactants and an affinity coefficient (rate constant), with each active mass raised to some power. Moreover, they seemed to be fully aware that these definite powers were not necessarily integral numbers and not deducible from the balanced chemical equation. The Guldberg and Waage formulation is generally referred to as the **law of mass action.**

The manner in which the initial rate of reaction (14.8) is experimentally determined is described in Additional Exercise 13 and pictured in Color Section C.

Method of Initial Rates.   In this method of establishing the exponents in a rate equation the initial rate of reaction for different sets of initial concentrations is measured. Consider the following reaction between peroxodisulfate and iodide ions in aqueous solution. (To simplify algebraic expressions in Example 14-5, we let A = $S_2O_8^{2-}$, B = I^-, and so on.)

$$S_2O_8^{2-} + 3\,I^- \longrightarrow 2\,SO_4^{2-} + I_3^- \tag{14.8}$$
$$(A \quad + 3\,B \longrightarrow 2\,C \quad\quad + D)$$

The rate equation for reaction (14.8) has the form

$$\text{rate of reaction} = k[S_2O_8^{2-}]^m[I^-]^n = k[A]^m[B]^n \tag{14.9}$$

The method of initial rates is applied in Example 14-5.

Example 14-5   Use data from Table 14-3 to establish **(a)** the order of reaction (14.8) with respect to $S_2O_8^{2-}$ and I^- and **(b)** the overall order.

Solution

(a)   Our task is to determine values of m and n in equation (14.9). In comparing Expt. 2 to Expt. 1, $[B]$ is held constant while $[A]$ is doubled. The rate of reaction also increases by a factor of 2. That is, $[B]_2 = [B]_1$, $[A]_2 = 2 \times [A]_1$, and $R_2 = 2 \times R_1$. These observations require that the exponent $m = 1$, as established in the ratios below.

$$\frac{R_2}{R_1} = 2 = \frac{k[A]_2^m[B]_2^n}{k[A]_1^m[B]_1^n} = \frac{(2 \times [A]_1)^m}{[A]_1^m} = \frac{2^m [A]_1^m}{[A]_1^m} = 2^m$$

In order that $2^m = 2$, $m = 1$. The reaction is first order in $S_2O_8^{2-}$.

In comparing Expt. 2 to Expt. 3, $[A]$ is held constant, $[B]_2 = 2 \times [B]_3$, and $R_2 = 2 \times R_3$. These observations require that the exponent $n = 1$.

$$\frac{R_2}{R_3} = 2 = \frac{k[A]_2^m[B]_2^n}{k[A]_3^m[B]_3^n} = \frac{(2 \times [B]_3)^n}{[B]_3^n} = \frac{2^n [B]_3^n}{[B]_3^n} = 2^n$$

If $2^n = 2$, then $n = 1$; and the reaction is first order in I^-.

(b)   The overall order of the reaction is $m + n = 1 + 1 = 2$—second order.

SIMILAR EXAMPLES: Review Problem 6; Exercises 13, 14, 15.

In Example 14-5 we established that if the initial reaction rate *doubles* with a doubling of the initial concentration of a reactant, the reaction is first order in that reactant. A similar analysis would show that if the initial reaction rate *quadruples* with the doubling of a reactant concentration, the reaction is *second order* in that reactant. For *third order*, the initial rate would increase *eightfold* with a doubling of the reactant concentration.

Now that we know the exponents in the peroxodisulfate–iodide rate equation (14.9), we can determine the value of the rate constant, k. What is required is that for a given reaction mixture we simultaneously establish $[S_2O_8^{2-}]$, $[I^-]$, and the rate of reaction.

Example 14-6   Use the results of Example 14-5 and data from Table 14-3 to determine the value of k in the rate equation (14.9).

TABLE 14-3
Experimental data for the reaction

$$S_2O_8^{2-} + 3\,I^- \longrightarrow 2\,SO_4^{2-} + I_3^-$$
$$(A \quad + 3\,B \longrightarrow 2\,C \quad + D)$$

Exper-iment	Initial concentrations, M		Initial rate of reaction[a] mol L^{-1} s^{-1}
	$[A] = [S_2O_8^{2-}]$	$[B] = [I^-]$	
1	$[A]_1 = 0.038$	$[B]_1 = 0.060$	$R_1 = 1.4 \times 10^{-5}$
2	$[A]_2 = 0.076$	$[B]_2 = 0.060$	$R_2 = 2.8 \times 10^{-5}$
3	$[A]_3 = 0.076$	$[B]_3 = 0.030$	$R_3 = 1.4 \times 10^{-5}$

[a]For example, this might be − rate of disappearance of $S_2O_8^{2-}$ or rate of formation of I_3^-.

FIGURE 14-2
The purpose of a rate law.

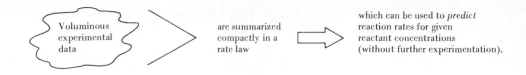

Voluminous experimental data → are summarized compactly in a rate law → which can be used to *predict* reaction rates for given reactant concentrations (without further experimentation).

Solution. We can use the data for any one of the three experiments of Table 14-3, together with the values $m = n = 1$. Equation (14.9) is solved for k.

$$k = \frac{\text{rate of reaction}}{[S_2O_8^{2-}]^m[I^-]^n} = \frac{R_1}{[S_2O_8^{2-}][I^-]} = \frac{1.4 \times 10^{-5} \text{ mol L}^{-1} \text{ s}^{-1}}{0.038 \text{ mol L}^{-1} \times 0.060 \text{ mol L}^{-1}}$$
$$= 6.1 \times 10^{-3} \text{ L mol}^{-1} \text{ s}^{-1}$$

SIMILAR EXAMPLE: Exercise 14.

It is now also possible for us to calculate the rate of a reaction at some given point in the reaction. For this we use the rate equation and known values of k and reactant concentrations. The result will be an *instantaneous* rate of reaction (similar to the tangents to a curve like Figure 14-1). Once the reaction proceeds beyond this given point, reactant concentrations change and so does the rate of reaction.

Example 14-7 What is the rate of reaction (14.8) at a point where $[S_2O_8^{2-}] = 0.050$ *M* and $[I^-] = 0.025$ *M*?

Solution. From the results of Examples 14-5 and 14-6 we can write

$$\text{rate} = k[S_2O_8^{2-}][I^-] = 6.1 \times 10^{-3} \text{ L mol}^{-1} \text{ s}^{-1} \times 0.050 \text{ mol/L} \times 0.025 \text{ mol/L}$$
$$= 7.6 \times 10^{-6} \text{ mol L}^{-1} \text{ s}^{-1}$$

SIMILAR EXAMPLE: Exercise 14.

Example 14-7 illustrates the main purpose served by rate laws. They permit us to predict rates of chemical reactions. Figure 14-2 is offered as a brief summary of how this comes about.

FIGURE 14-3
A straight-line plot for the zero-order reaction

$A \rightarrow$ products

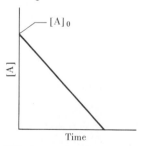

[A] decreases from a maximum value of $[A]_0$ at time $t = 0$ to $[A] = 0$ at a time $t = [A]_0/k$. The rate constant $k = -(\text{slope})$.

14-4 Zero-Order Reactions

Occasionally, the rate of a reaction appears not to depend on the concentration of the reactant(s) at all. This situation is encountered when some other variable controls the rate of reaction, e.g., light intensity in a photochemical reaction or availability of enzyme in an enzyme-catalyzed reaction. In such cases the reaction proceeds at a *constant rate.*

rate of reaction = k = constant   (14.10)

Equation (14.10) is a special case of (14.7) in which each of the exponents $m, n, \ldots$, equals zero. The reaction is **zero order,** and the units of k are the same as those of the rate itself. We have seen that the rate of a reaction is given by the slope of the tangent line to a graph of concentration of a reactant as a function of time. Because the rate of a zero-order reaction is constant there is but one tangent line: The graph of concentration of a reactant as a function of time must itself be a *straight line* (see Figure 14-3).

14-5 First-Order Reactions

The decomposition of H_2O_2 (reaction 14.3) is a first-order reaction in H_2O_2. This means that $[H_2O_2]$ appears in the rate equation to the first power.

rate of reaction $= k[H_2O_2]$

From the measured rate of reaction at a particular $[H_2O_2]$, it is possible to calculate the rate constant, k.

Example 14-8   When $[H_2O_2] = 0.78 \, M$, the rate of reaction (14.3) is 5.7×10^{-4} mol L^{-1} s^{-1}. What is the value of k for this first-order reaction?

Solution.   We rearrange the rate equation and solve for k.

$$k = \frac{\text{rate of reaction}}{[H_2O_2]} = \frac{5.7 \times 10^{-4} \text{ mol } L^{-1} \text{ s}^{-1}}{0.78 \text{ mol } L^{-1}} = 7.3 \times 10^{-4} \text{ s}^{-1}$$

SIMILAR EXAMPLE: Exercise 23f.

The method used to determine k in Example 14-8 is deceptively simple. The problem is twofold: How do we know that the reaction is first order? How do we determine the rate of the reaction at any given point without making additional measurements at other points? You may recognize the data in Example 14-8 as being based on a tangent line in Figure 14-1. This line can be drawn only after a significant portion of the concentration vs. time graph is plotted. This requires a number of experimental data points.

Consider this hypothetical first-order reaction: $A \rightarrow B + C$. The rate law for the reaction is

$$\text{rate of reaction} = - \text{ rate of disappearance of A} = k[A] \qquad (14.11)$$

This rate law is the starting point of a derivation that is based on the integration concept from calculus and yields this important result.*

$$\ln \frac{[A]_t}{[A]_0} = -kt \qquad or \qquad \ln [A]_t - \ln [A]_0 = -kt \qquad (14.12)$$

Expressed in common logarithms, equation (14.12) becomes

$$\log \frac{[A]_t}{[A]_0} = \frac{-kt}{2.303} \qquad or \qquad \log [A]_t - \log [A]_0 = \frac{-kt}{2.303} \qquad (14.13)$$

In equations (14.12) and (14.13) the term $[A]_t$ represents the concentration of A at some time t. $[A]_0$ represents the initial concentration of A, that is, at $t = 0$. k is the rate constant for the reaction. Only the time t and the corresponding concentration $[A]_t$ are

---

*The steps in this derivation are
(1) Replace the rate of disappearance of A in equation (14.11) by $d[A]/dt$.
(2) Rearrange (14.11) to the form, $d[A]/[A] = -kdt$.
(3) Integrate between the limits $[A]_0$ at time $t = 0$ and $[A]_t$ at time t.
$$\int_{[A]_0}^{[A]_t} \frac{d[A]}{[A]} = -k \int_0^t dt, \quad \text{yielding } \ln \frac{[A]_t}{[A]_0} = -kt$$

FIGURE 14-4
Test for a first-order
reaction: decomposition
of H_2O_2.

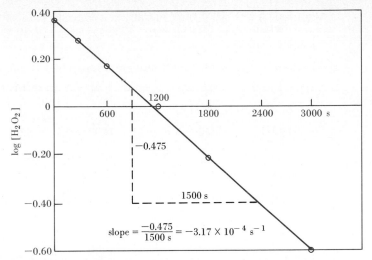

The data from Table 14-1 are

t, s	$[H_2O_2]$, M	$\log[H_2O_2]$
0	2.32	0.365
300	1.86	0.270
600	1.49	0.173
1200	0.98	−0.009
1800	0.62	−0.21
3000	0.25	−0.60

variables in these equations. A slight rearrangement of these equations shows them to be equations of a straight line. For example, if $\log[A]_t$ is plotted as a function of t,

$$\log[A]_t = \left(\frac{-k}{2.303}\right)t + \log[A]_0 \tag{14.14}$$

equation of
straight line: $y \quad = \quad m \cdot x \; + \quad b$

A test for a first-order reaction, then, is to plot the logarithm of a reactant concentration vs. time and determine if the graph is linear. Data from Table 14-1 for the first-order decomposition of H_2O_2 are plotted in Figure 14-4. The value of the rate constant k can be derived from the slope of the line.

$$\frac{-k}{2.303} = m = \text{slope} = -3.17 \times 10^{-4} \text{ s}^{-1}$$

$$k = 2.303 \times 3.17 \times 10^{-4} \text{ s}^{-1} = 7.30 \times 10^{-4} \text{ s}^{-1}$$

Knowing that a reaction is first order and having a numerical value of the rate constant k, allow us to use equation (14.12) or (14.13) to do calculations of the type outlined in Examples 14-9 and 14-10. In Example 14-9 we will use both equations (14.12) and (14.13) to show that they always yield identical results. From this point on you may use whichever equation you prefer, which may depend on how particular functions are handled on your electronic calculator. Beyond this point the text will continue to show the two logarithmic alternatives from time to time. Calculations, however, will be based on the common logarithmic form, since scientific notation is in terms of the base 10 (i.e., in powers of ten).

Example 14-9 $H_2O_2(aq)$, initially at a concentration of 2.32 M, is allowed to decompose. What will be $[H_2O_2]$ at $t = 1200$ s? Use $k = 7.30 \times 10^{-4}$ s^{-1} for this first-order decomposition.

Solution.   For substitution into equations (14.12) and (14.13) we have

$[H_2O_2]_t = ?$ $[H_2O_2]_0 = 2.32$ M $k = 7.30 \times 10^{-4}$ s^{-1} $t = 1200$ s

Use of equation (14.12)

$\ln [H_2O_2]_t = -7.30 \times 10^{-4}$ s$^{-1} \times 1200$ s $+ \ln 2.32$

$\qquad\qquad = \qquad\quad -0.876 \qquad\qquad + 0.842 \quad = -0.034$

$[H_2O_2]_t = \text{antiln}(-0.034) = e^{-0.034} = 0.97$ M

(To find the number whose natural logarithm is -0.034, raise e to the -0.034 power.)

Use of equation (14.13)

$$\log [H_2O_2]_t = \frac{-7.30 \times 10^{-4} \text{ s}^{-1} \times 1200 \text{ s}}{2.303} + \log 2.32$$

$\qquad\qquad = \qquad\quad -0.380 \qquad\qquad + \quad 0.365 \quad = -0.015$

$[H_2O_2]_t = \text{antilog}(-0.015) = 10^{-0.015} = 0.97$ M

(To find the number whose common logarithm is -0.015, raise 10 to the -0.015 power.)

SIMILAR EXAMPLES: Review Problem 8; Exercise 18.

The value listed in Table 14-1 for $[H_2O_2]$ at 1200 s is 0.98 M. Note how well the calculated value in Example 14-9 agrees with the experimental value, and also note this fact: From the initial rate of a reaction we can calculate changes in concentration only over a brief time interval following the start of the reaction (as in Example 14-4). With equations (14.12) and (14.13), called *integrated* rate equations, we are able to calculate, from initial conditions, quantities for times much further into the reaction.

The data used in the equations of first-order kinetics need not always be in terms of molar concentrations. Sometimes masses of reactants, or just the fraction of reactant consumed, are sufficient, as in Example 14-10.

Example 14-10   Use a value of $k = 7.30 \times 10^{-4}$ s^{-1} for the first-order decomposition of $H_2O_2(aq)$ to determine **(a)** the % H_2O_2 that has decomposed in the first 500 s after the reaction begins and **(b)** the time required for one-half of the sample to decompose.

Solution

(a) Regardless of the value of $[H_2O_2]_0$, the ratio, $[H_2O_2]_t/[H_2O_2]_0$, represents the fractional part of the initial quantity of A that remains *unreacted* at time t. Our problem essentially is to determine this ratio at $t = 500$ s. The following form of equation (14.13) is most useful for this purpose.

$$\log \frac{[H_2O_2]_t}{[H_2O_2]_0} = \frac{-kt}{2.303} = \frac{-7.30 \times 10^{-4} \text{ s}^{-1} \times 500 \text{ s}}{2.303} = -0.158$$

$$\frac{[H_2O_2]_t}{[H_2O_2]_0} = \text{antilogarithm} \ (-0.158) = 0.695 \quad \text{and} \quad [H_2O_2]_t = 0.695[H_2O_2]_0$$

The fractional part of the original H_2O_2 that remains *undecomposed* is 0.695 (69.5%). The fractional part that must have *decomposed* is $1.000 - 0.695 = 0.305$. The % H_2O_2 decomposed is 30.5%.

(b) What we are seeking here is the time at which $[H_2O_2]_t = \frac{1}{2}[H_2O_2]_0$. Note that in the expression below $\log \frac{1}{2} = \log 1 - \log 2 = 0 - \log 2$.

$$\log \frac{[H_2O_2]_t}{[H_2O_2]_0} = \log \frac{\frac{1}{2}[H_2O_2]_0}{[H_2O_2]_0} = \log \frac{1}{2} = -\log 2 = -0.3010 = \frac{-kt}{2.303}$$

$$t = \frac{0.3010 \times 2.303}{k} = \frac{0.3010 \times 2.303}{7.30 \times 10^{-4} \text{ s}^{-1}}$$

$$= 9.50 \times 10^2 \text{ s} = 950 \text{ s}$$

SIMILAR EXAMPLES: Review Problem 8; Exercise 19.

Half-life of a Reaction. The time calculated in Example 14-10(b) is known as the **half-life** for the reaction. This is the time required for the concentration (or quantity) of a reactant to decrease to one-half of a previous value. The half-life, $t_{1/2}$, is obtained by the substitutions $[A]_t = \frac{1}{2}[A]_0$ and $t = t_\frac{1}{2}$ into equation (14.13). These are the substitutions that we made in Example 14-10(b). Regardless of the identity or initial concentration of reactant A, *if the reaction is first order, the half-life is constant and depends only on* k. From Example 14-10(b) we see that the relationship between $t_{1/2}$ and k, expressed to three significant figures, is

$$t_{1/2} = \frac{0.3010 \times 2.303}{k} = \frac{0.693}{k} \tag{14.15}$$

In Example 14-10(b), the time required for the quantity of H_2O_2 to be reduced to $\frac{1}{4}$ of its original value would be $(950 + 950)$ s; to $\frac{1}{8}$ of its original value, $(950 + 950 + 950)$ s, and so on. This test of constancy of half-life (which holds only for first-order reactions) can be applied to a simple plot of concentration against time. Try this with Figure 14-1. That is, starting with $[H_2O_2] = 2.32 \, M$ at $t = 0$, at what time is $[H_2O_2] \simeq 1.16 \, M$? $[H_2O_2] \simeq 0.58 \, M$? $[H_2O_2] \simeq 0.29 \, M$?

Reactions Involving Gases. In reactions involving gases the reaction rate is often measured in terms of gas pressure. It is not difficult to derive equations similar to (14.12) and (14.13) for first-order gas-phase reactions. Consider the reaction $A(g) \rightarrow$ products to be first order and replace $[A]$ by the partial pressure, P_A, as follows.

$$P_A V = n_A RT$$

$$\frac{n_A}{V} = [A] = \frac{P_A}{RT}$$

Now substitute $[A] = P_A/RT$ in equation (14.13) to obtain

$$\log \frac{[A]_t}{[A]_0} = \log \frac{(P_A)_t/RT}{(P_A)_0/RT} = \log \frac{(P_A)_t}{(P_A)_0} = \frac{-kt}{2.303} \tag{14.16}$$

The first-order decomposition of di-*t*-butyl peroxide (DTBP) to acetone and ethane is represented by the equation

$$C_8H_{18}O_2(g) \longrightarrow 2 \, C_3H_6O(g) + C_2H_6(g) \tag{14.17}$$

and the partial pressure of DTBP as a function of time is plotted in Figure 14-5.

FIGURE 14-5
Decomposition of di-*t*-butyl peroxide (DTBP).

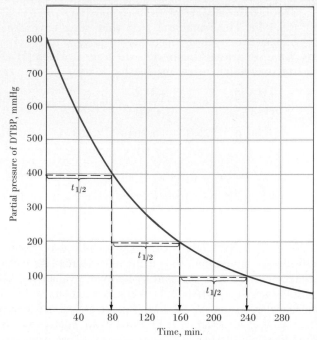

The rate of a gas-phase reaction can be followed by measuring the partial pressure of a gaseous species. The decomposition of di-*t*-butyl peroxide (DTBP) is described through equation (14.17). Three successive half-life intervals of 80 min each are indicated.

Example 14-11   Reaction (14.17) has a half-life of 80 min at 147°C. A reaction is started with pure DTBP at 147°C and 800 mmHg in a flask of constant volume. **(a)** At what time will the partial pressure of DTBP be 100 mmHg? **(b)** What will be the *total* gas pressure when the partial pressure of DTBP is 700 mmHg?

Solution

(a) In one half-life period both the number of moles and the partial pressure of DTBP decrease to one-half of their initial values: $P_{DTBP} = 400$ mmHg. In a second half-life period, P_{DTBP} falls to 200 mmHg, and in a third, to 100 mmHg. The time required is three half-life periods, $3 \times t_{1/2} = 3 \times 80$ min = 240 min. (See also Figure 14-5.)

Other calculations involving the decomposition of DTBP might require use of equation (14.16). The value of k for such uses would be
$k = 0.693/t_{1/2}$
$= 0.693/80$ min
$= 8.7 \times 10^{-3}$ min^{-1}.

(b) Three moles of product gases appear for every mole of DTBP consumed. A sufficient amount of DTBP is consumed to cause a decrease in its partial pressure of 100 mmHg (i.e., from 800 mmHg to 700 mmHg). Products are formed in sufficient amount to produce a partial pressure of 300 mmHg.

total pressure $= P_{DTBP} + P_{prod.} = 700 + 300 = 1000$ mmHg

SIMILAR EXAMPLES: Exercises 21, 22.

Radioactive decay processes are nuclear, not chemical, processes; they are considered in detail in Chapter 25.

Examples of First-Order Reactions and the Significance of k.   One of the best known examples of first-order processes is the decay of radioactive atoms. For example, the isotope $^{32}_{15}P$, which is used in biochemical studies, has a half-life of 14.3 days. Whatever number of $^{32}_{15}P$ atoms are at hand at the moment, there will be half this number present in 14.3 days; one quarter this number in 14.3 + 14.3 = 28.6 days; etc. The rate constant for the decay is $k = 0.693/t_{1/2}$ and equations (14.12) and (14.13) apply, with numbers of

TABLE 14-4
Some typical first-order processes

Process	Half-life, $t_{1/2}$	Rate constant k, s^{-1}
radioactive decay of $^{238}_{92}U$	4.51×10^9 years	4.87×10^{-18}
radioactive decay of $^{14}_{6}C$	5.73×10^3 years	3.83×10^{-12}
radioactive decay of $^{32}_{15}P$	14.3 days	5.61×10^{-7}
radioactive decay of $^{26}_{11}Na$	1.0 s	6.9×10^{-1}
$(CH_2)_2O(g) \xrightarrow{415°C} CH_4(g) + CO(g)$ ethylene oxide	56.3 min	2.05×10^{-4}
$2 N_2O_5 \xrightarrow[45°C]{in\ CCl_4} 2 N_2O_4 + O_2(g)$	18.6 min	6.21×10^{-4}
$C_{12}H_{22}O_{11}(aq) + H_2O \xrightarrow{15°C} C_6H_{12}O_6(aq) + C_6H_{12}O_6(aq)$ sucrose glucose fructose	8.4 h	2.3×10^{-5}
$HC_2H_3O_2(aq) \longrightarrow H^+(aq) + C_2H_3O_2^-(aq)$	8.9×10^{-7} s	7.8×10^5

atoms substituting for concentrations, i.e., N_t for $[A]_t$ and N_0 for $[A]_0$. A few radioactive decay processes, together with several other examples of first-order kinetics, are summarized in Table 14-4.

Although we have considered a number of quantitative questions concerning first-order kinetics, we should not lose sight of the fact that some questions can be answered qualitatively. When comparing two first-order processes, the one with the *larger* value of k or the *shorter* half-life proceeds more rapidly. One interpretation of the rate constant k is that it represents the fractional part of a reactant that is consumed per unit time. If k for the reaction A $\rightarrow$ products has a value of 1.0×10^{-3} s^{-1}, then at all times one-thousandth of the reactant present is being consumed every second.

rate of reaction = $k[A]$ = $(0.001 \times [A])$ s^{-1}

A first-order reaction proceeding ten times as fast would have $k = 1.0 \times 10^{-2}$ s^{-1}; one-hundredth (or 1%) of the reactant present would be consumed every second. If a first-order reaction has less than a 1-s half life, more than half of the remaining reactant is consumed every second. From the data in Table 14-4 we see that the rates of first-order processes can vary greatly.

14-6 Second-Order Reactions

If the hypothetical reaction A $\rightarrow$ B + C is second order in A, the rate law is

rate of reaction = $k[A]^2$ (14.18)

If the hypothetical reaction A + B $\rightarrow$ C + D is first order in A and first order in B, the overall order is second, and the rate law is

The peroxodisulfate–iodide ion reaction (14.8) fits the rate law expression (14.19).

rate of reaction = $k[A][B]$ (14.19)

However, these facts about the reaction order *cannot* be deduced from the balanced equation. Reaction order can be determined *only from experimental rate data*. How must these data be treated to indicate if a reaction is second order? The situation is more complex than with first-order reactions, and we limit ourselves to a single case: a reaction involving a *single* reactant and following equation (14.18). As in the case of first-order

FIGURE 14-6
A straight-line plot for the second-order reaction
A → products

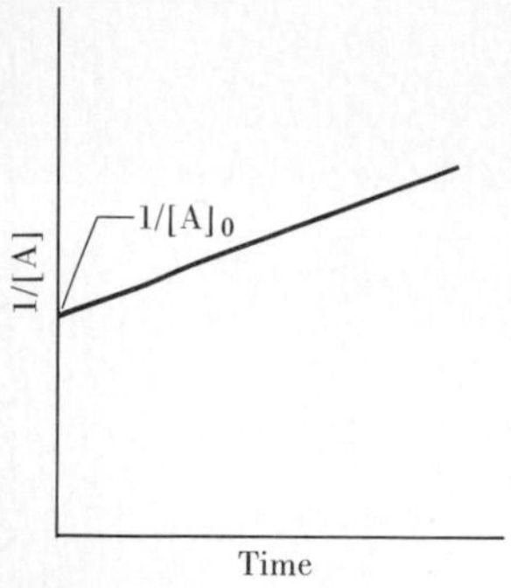

The reciprocal of the concentration, 1/[A], has its lowest value at the start of the reaction. As the reaction proceeds, [A] decreases and 1/[A] increases, in a straight-line fashion. The slope of the line is the rate constant k.

reactions, converting equation (14.18) to a more useful form requires the use of integration, a calculus procedure that we will not pursue. The result obtained, in the recognizable form of a straight-line graph, is

$$\underbrace{\frac{1}{[A]_t}}_{} = \underbrace{k \cdot t}_{} + \underbrace{\frac{1}{[A]_0}}_{} \tag{14.20}$$

equation of straight line: $\quad y \quad = \quad mx \quad + \quad b$

A plot of $1/[A]_t$ as a function of time yields a straight line with a slope of k (see Figure 14-6). Each of the three terms in equation (14.20) must have the same units—L/mol. Since the product kt has the units L/mol, the units of k must be $L \ mol^{-1} \ (time)^{-1}$, that is, $L \ mol^{-1} \ s^{-1}$, $L \ mol^{-1} \ min^{-1}$, and so on. To establish the half-life for a second-order reaction of this type, substitute $t = t_{1/2}$ and $[A]_t = \frac{1}{2}[A]_0$ into equation (14.20) to obtain

$$t_{1/2} = \frac{1}{k[A]_0} \tag{14.21}$$

The half-life for the reaction is *not* a constant. Its value depends on the concentration of reactant at the start of each half-life interval (i.e., depends on $[A]_0$).

Example: 14-12 The data listed in Table 14-5 were obtained for the decomposition reaction: A → 2 B + C. **(a)** Establish the order of the reaction. **(b)** What is the rate constant k? **(c)** What is the half-life, $t_{1/2}$, if the initial $[A] = 1.00 \ M$?

Solution
(a) Plot the following three graphs.

 1. [A] vs. time. (If a straight line, reaction is zero order.)
 2. log [A] vs. time. (If a straight line, reaction is first order.)
 3. 1/[A] vs. time. (If a straight line, reaction is second order.)

These graphs are shown in Figure 14-7; the reaction is second order.
(b) The slope of graph 3 in Figure 14-7 is

$$k = \frac{(4.00 - 1.00) \ \text{L/mol}}{25 \ \text{min}} = 0.12 \ \text{L mol}^{-1} \ \text{min}^{-1}$$

(c) According to equation (14.21)

$$t_{1/2} = \frac{1}{k[A]_0} = \frac{1}{0.12 \ \text{L mol}^{-1} \ \text{min}^{-1} \times 1.00 \ \text{mol/L}} = 8.3 \ \text{min}$$

SIMILAR EXAMPLES: Review Problem 12; Exercises 24, 25, 26.

TABLE 14-5
Kinetic data for Example 14-12

Time, min	[A], *M*	log [A]	1/[A]
0	1.00	0.00	1.00
5	0.63	−0.20	1.6
10	0.46	−0.34	2.2
15	0.36	−0.44	2.8
25	0.25	−0.60	4.0

FIGURE 14-7
Testing for the order of
a reaction—Example
14-12 illustrated.

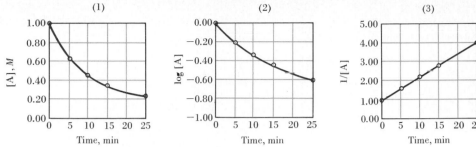

The straight-line plot is that shown in (3). The reaction is second order.

14-7 Theoretical Treatment of Rates of Chemical Reactions

The statement that *chemical reactions occur as a result of collisions between molecules* may seem so obvious as not to require mention. Yet a theory to explain chemical reactions in terms of molecular collisions did not become firmly established until the early decades of the present century. The kinetic-molecular theory had to be developed first. Recall for a moment a number of ideas in our discussion of this theory in Section 5-10. We noted that there is a distribution of kinetic energies and speeds of the molecules of a gas. Also, we were able to write an expression for the average molecular speed. Although beyond the scope of this text to do so, the number of collisions between molecules per unit time can be derived from kinetic-molecular theory. This number is called the **collision frequency.**

For example, consider 0.01 mol of a gas in a 1-L vessel at 25°C, that is, a gas at an initial concentration of 0.01 M. Assuming a molecular weight of about 100 for the gas, it can be shown that the frequency of molecular collisions is of the order of 10^{30} collisions per second. If each of these collisions led to chemical reaction, the rate of reaction would be of the order of 10^6 mol L^{-1} s^{-1}.

$$\frac{10^{30} \text{ collisions } L^{-1} \text{ } s^{-1}}{6.02 \times 10^{23} \text{ collisions/mol}} \simeq 10^6 \text{ mol } L^{-1} \text{ } s^{-1} \tag{14.22}$$

The rate of reaction suggested by equation (14.22) is extremely fast. There are, in fact, reactions that proceed at these fast rates, e.g., some reactions involving ions in aqueous solution. If the hypothetical gas underwent reaction at the rate of (14.22), the gas would be completely consumed in 10^{-8} s. A more typical rate of a gas-phase reaction is of the order of 10^{-4} mol L^{-1} s^{-1}. Thus, typically, only a fraction of the collisions among gaseous molecules lead to chemical reaction, an observation that can be accounted for by two factors: (1) Only the more energetic molecules in a mixture undergo reaction as a result of collisions. (2) The probability of a particular collision resulting in chemical reaction depends on the orientation of the colliding molecules.

The extra energy that molecules must possess in order to react is called the **activation energy.** With the kinetic-molecular theory it is possible to establish what fraction of all the molecules in a collection possess energies in excess of any particular value. A hypothetical activation energy and the fraction of molecules possessing energies in excess of this value are indicated on a distribution curve of molecular energies in Figure 14-8. Think of the rate of a reaction as depending on the product of the collision frequency *and* the fraction of activated molecules. Because the fraction of activated molecules is generally so small, the rate of a reaction is usually much smaller than the collision frequency. Moreover, the *larger* the activation energy, the *smaller* the fraction of activated molecules and the more slowly a reaction proceeds.

FIGURE 14-8
Distribution of
molecular energies.

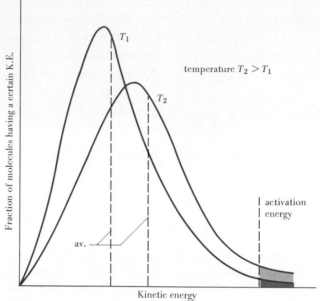

At the higher temperature, T_2, the distribution of energies is broadened, the average molecular kinetic energy increases, and many more molecules possess energies greater than the activation energy.

To visualize the reaction

$$A_2(g) + B_2(g) \longrightarrow 2\ AB(g) \tag{14.23}$$

in terms of collision theory, assume that during a collision between a molecule of A_2 and B_2 the bonds A—A and B—B break and the bonds A—B form. The result is the conversion of reactants A_2 and B_2 to the product AB. But as pictured in Figure 14-9, this assumption does not hold for every collision. A particular orientation of the molecules may be required if the collision is to be effective in producing chemical reaction. Figure 14-9 suggests that the number of unfavorable orientations often exceeds the number of

FIGURE 14-9
Molecular collisions and
chemical reactions.

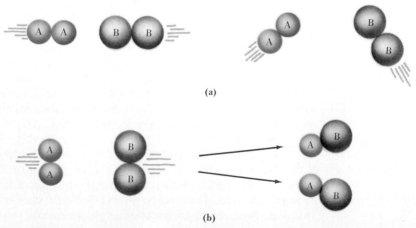

(a) Unfavorable collisions for chemical reaction.
(b) A favorable collision for chemical reaction.

favorable ones. This means that the probability of a particular collision being favorable to reaction is generally small (i.e., the probability is less than 1).

If we denote collision frequency as Z, the fraction of activated molecules as f, and the probability factor as p, the rate of a chemical reaction has the form

$$\text{rate of reaction} = p \cdot f \cdot Z \tag{14.24}$$

Collision frequency is proportional to the concentrations of the molecular species involved in collisions (say that these species are A and B). That is, Z can be replaced by $[A] \times [B]$, and this more familiar rate expression can be written

$$\text{rate of reaction} \propto pf[A][B] = k[A][B] \tag{14.25}$$

The collision theory seems to lead to a general rate equation for chemical reactions, but there are some serious shortcomings to the result we have just stated. Equation (14.25) describes a reaction that is second order overall, yet we know that other reaction orders are possible. At times the probability factor is much smaller than can be explained just in terms of collision orientations; in a few cases it is very large, that is, much greater than 1. This discussion has assumed that chemical reaction occurs as a result of the collision of two molecules, a situation that we will refer to in Section 14-10 as a bimolecular process. In a few reactions a simultaneous collision of three bodies is required. And, more important, in some reactions, although molecules acquire activation energy through collisions, the actual reaction (e.g., the rearrangement of chemical bonds) does not occur at the time of a collision. The theoretical basis of chemical kinetics must involve more than just the concepts associated with simple collision theory.

An important extension of collision theory has been developed by the American chemist Henry Eyring (1901–81), and others. It focuses on the details of a collision, in particular on a hypothesized intermediate species, called an **activated complex,** that forms during an energetic collision. This species exists very briefly, and then dissociates— either back to the original reactants (in which case there is no reaction) or to product molecules. An activated complex might be represented as follows.

$$
\begin{array}{ccccc}
\text{A} & \text{B} & \text{A}\cdots\text{B} & & \text{A——B} \\
| & + \ | & \vdots \quad \vdots & \longrightarrow & \\
\text{A} & \text{B} & \text{A}\cdots\text{B} & & \text{A——B} \\
\text{\small reactants} & & \text{\small activated} & & \text{\small products} \\
& & \text{\small complex} & &
\end{array}
\tag{14.26}
$$

An activated complex has old bonds stretched to the breaking point and new bonds only partially formed. Only if colliding molecules possess a large quantity of kinetic energy to invest in producing this strained species can an activated complex form. The energy required is the activation energy.

Another way of looking at activation energy is presented in Figure 14-10, which is a **reaction profile** for the reaction of $H_2(g)$ and $I_2(g)$ to form $HI(g)$. In Figure 14-10 energies of the species involved are plotted on the vertical axis and a quantity called the reaction coordinate, on the horizontal axis. The reaction coordinate can be thought of as representing the extent of the reaction. That is, the reaction starts with reactants on the left, passes through an activated complex, and ends with products on the right. The difference in energies between the reactants and products is $\Delta \overline{H}$ for the reaction. The formation of $HI(g)$ is a slightly exothermic reaction. The difference in energy between that of the activated complex and that of the reactants—171 kJ/mol—is the activation energy. Thus, a large energy barrier separates the reactants from the products. Only especially energetic reactant molecules can cross this barrier (by forming an activated complex which then dissociates into product molecules).

Figure 14-10 can also be used to describe the reverse process, the decomposition of $HI(g)$ into $H_2(g)$ and $I_2(g)$. The activation energy for the reverse reaction is 184 kJ/mol.

FIGURE 14-10
Energy profile for the reaction

$H_2(g) + I_2(g) \rightarrow 2\ HI(g)$

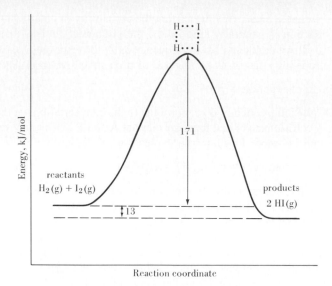

Figure 14-10 suggests the following important relationship between activation energy, E_a, and enthalpy (heat) of reaction, $\Delta\overline{H}$.

$$\Delta\overline{H} = E_a(\text{forward}) - E_a(\text{reverse}) \tag{14.27}$$

for HI formation: $\Delta\overline{H} = 171$ kJ/mol $- 184$ kJ/mol $= -13$ kJ/mol

Also, it should be apparent that for an *endothermic* reaction the activation energy must be equal to or exceed the endothermic heat of reaction (and usually it is greater).

An Analogy to Activation Energy. It may be helpful to think of activation energy in terms of this analogy: Imagine that a hike is being planned between two mountain valleys separated by a ridge through which there is a pass. The elevations of the valleys correspond to the energies of the reactants and products, and the elevation of the pass above the valleys to activation energy. The ease or difficulty of the hike is not so much a matter of the difference in elevation of the two valleys as it is of the elevation of the pass above the starting point. If the climb from the starting point to the pass is very long and steep, not too many individuals may care to take the hike, no matter how much "downhill" there is on the other side.

Attempts at purely theoretical predictions of reaction rate constants have not been very successful. The principal value of reaction rate theories is in providing concepts to facilitate the discussion of experimentally observed reaction rate data. In the next section we see how the concept of activation energy enters into a discussion of the effect of temperature on reaction rates.

14-8 The Effect of Temperature on Reaction Rates— The Arrhenius Equation

As a practical matter we know that chemical reactions tend to go faster at higher temperatures. We speed up certain biochemical reactions by raising the temperature, for example in cooking foods. On the other hand, we slow down some reactions by lowering the temperature, such as in refrigerating or freezing cooked foods to prevent spoilage. And now we have an explanation for the profound effect of temperature on reaction rates:

TABLE 14-6
Specific rate constant, k, at several temperatures for the
reaction $N_2O_5(\text{in CCl}_4) \longrightarrow N_2O_4(\text{in CCl}_4) + \frac{1}{2}O_2(g)$

t, °C	T, K	$1/T$, K^{-1}	k, s^{-1}	$\log k$
0	273	3.66×10^{-3}	7.87×10^{-7}	-6.10
25	298	3.36×10^{-3}	3.46×10^{-5}	-4.46
35	308	3.25×10^{-3}	1.35×10^{-4}	-3.87
45	318	3.14×10^{-3}	4.98×10^{-4}	-3.30
55	328	3.05×10^{-3}	1.50×10^{-3}	-2.82
65	338	2.96×10^{-3}	4.87×10^{-3}	-2.31

Collision frequency increases with temperature and we might expect this also to be a factor in speeding up a chemical reaction. However, collision frequency (which is proportional to $\sqrt{T}$) accounts for less than 1% of the increase described here.

Increasing the temperature increases the fraction of the molecules that have energies in excess of the activation energy (recall Figure 14-8). This factor is so important that for many chemical reactions it can lead to a doubling or tripling of the reaction rate for a temperature increase of only 10°C. Data to illustrate the effect of temperature on the decomposition of N_2O_5 are presented in Table 14-6.

Even without plotting the data in Table 14-6, we see that a graph of rate constant k against temperature would increase sharply with temperature. The graph would not be linear. This situation is reminiscent of that encountered with vapor pressure. In that case a steeply rising curve (Figure 11-6) was converted to a straight line (Figure 11-8) by plotting $\log P$ vs. $1/T$. Let us try a similar plot in this instance, that is, $\log k$ vs. $1/T$. The necessary data are given in Table 14-6 and plotted in Figure 14-11. The graph is indeed linear! Its equation has the form

$$\log k = \left(\frac{-E_a}{2.303R}\right)\frac{1}{T} + B \tag{14.28}$$

equation of
straight line: $y \;=\; m \;\cdot\; x + b$

FIGURE 14-11
Temperature dependence of the rate constant k for the reaction

N_2O_5 (in CCl$_4$) $\rightarrow$ N_2O_4 (in CCl$_4$) $+ \frac{1}{2} O_2(g)$

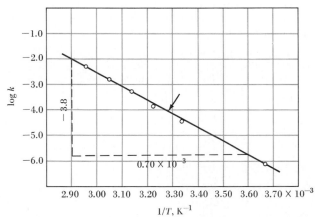

This graph can be used to establish the activation energy E_a for the reaction (see equation 14-28).

$$\text{slope of line} = \frac{-E_a}{2.303\,R} = \frac{-3.8}{0.70 \times 10^{-3}\,\text{K}^{-1}} = -5.4 \times 10^3\,\text{K}$$

$E_a = 2.303 \times 8.314\ \text{J mol}^{-1}\,\text{K}^{-1} \times 5.4 \times 10^3\ \text{K} = 1.0 \times 10^5\ \text{J/mol} = 1.0 \times 10^2\ \text{kJ/mol}$

(A more precise plot of the data of Table 14-6 yields a value of $E_a = 106$ kJ/mol.) The colored arrow is referred to in Example 14-13.

Equation (14.28) can be used for a variety of purposes. One is to establish the activation energy of a reaction graphically, as shown in Figure 14-11. Variations of this equation are also frequently encountered. For example, in the manner outlined on page 296, the constant B can be eliminated from equation (14.28) to obtain the following useful equation.

$$\log \frac{k_2}{k_1} = \frac{E_a}{2.303R}\left(\frac{T_2 - T_1}{T_2 T_1}\right) \tag{14.29}$$

In equation (14.29) T_2 and T_1 are two kelvin temperatures. k_2 and k_1 are the rate constants at these temperatures. E_a is the activation energy in J/mol. R is the gas constant, expressed as 8.314 J mol^{-1} K^{-1}.

Example 14-13   Use data from Table 14-6 and Figure 14-11 to determine the temperature at which the half-life for the decomposition of N_2O_5 is 2 h.

Solution.   Since both Table 14-6 and Figure 14-11 are based on rate constants, k, our first step must be to convert the half-life of 2 h to a corresponding value of k. For a first-order reaction

$$k = \frac{0.693}{t_{1/2}} = \frac{0.693}{2\text{ h}} = \frac{0.693}{7200\text{ s}} = 9.62 \times 10^{-5}\text{ s}^{-1}$$

We may now proceed in either of two ways.
Graphical method
We are seeking the temperature at which $k = 9.62 \times 10^{-5}$ and $\log k = \log 9.62 \times 10^{-5} = -4.02$. This point can be located directly on the straight-line graph of Figure 14-11 (marked by the colored arrow). Corresponding to $\log k = -4.02$, $1/T = 3.28 \times 10^{-3}$ K^{-1}; $T = (1/3.28 \times 10^{-3})$ K $= 305$ K $= 32°$C.
With equation (14.29)
Suppose that we denote as T_2 the temperature at which $k = 9.62 \times 10^{-5}$ s^{-1}. For T_1, we must use some other temperature at which k is known. Suppose that we take $T_1 = 298$ K and $k_1 = 3.46 \times 10^{-5}$ s^{-1}. The activation energy is 106 kJ/mol $= 1.06 \times 10^5$ J/mol (the more precise value given in Figure 14-11). Solve equation (14-29) for T_2. (For simplicity units have been omitted below. The temperature is obtained in kelvins.)

$$\log \frac{k_2}{k_1} = \log \frac{9.62 \times 10^{-5}}{3.46 \times 10^{-5}} = \log 2.78$$

$$0.444 = \frac{1.06 \times 10^5}{2.303 \times 8.314}\left(\frac{T_2 - 298}{298 \times T_2}\right)$$

$$= 5.54 \times 10^3\left(\frac{T_2 - 298}{298 \times T_2}\right)$$

$$132\, T_2 = 5.54 \times 10^3\, T_2 - 1.65 \times 10^6$$

$$5.41 \times 10^3\, T_2 = 1.65 \times 10^6$$

$$T_2 = 305\text{ K}$$

SIMILAR EXAMPLES: Review Problem 9; Exercises 32, 33.

14-9 Catalysis

A reaction can be speeded up by increasing the fraction of the molecules that have energies in excess of the activation energy. Raising the temperature is one way to increase this fraction. Another way not requiring an increase in temperature is to find an alternate reaction pathway of *lower* activation energy. The function of a catalyst in a chemical

To continue the analogy introduced on page 424, a catalyst is like a guide who can ease the climb for a party of hikers by showing them an easier route (less steep) to their objective.

reaction is to provide this alternate pathway. A catalyst enters into a chemical reaction in such a way that it undergoes no permanent change. As a result its formula does not appear in the net chemical equation (although its presence is generally indicated by an appropriate notation above the arrow sign). The success or failure of a commercial process for producing a substance often hinges on finding appropriate catalysts for the reactions involved, The ranges of temperatures and pressures that can be employed in industrial processes is not possible with biochemical reactions. The availability of appropriate catalysts for these reactions is absolutely crucial to the existence of living matter.

Mechanism of Catalysis.   Figure 14-12 represents the decomposition of formic acid (HCOOH). In the uncatalyzed reaction a hydrogen atom must be transferred from one part of the formic acid molecule to another before the breaking of a C—O bond can occur. The energy requirement for this atom transfer is large, resulting in a high activation energy and a slow reaction.

The acid-catalyzed decomposition of formic acid can be represented by

$$\underset{\text{H—C—O—H}}{\overset{\overset{\text{O}}{\|}}{}} \xrightarrow{\text{H}^+} \text{H}_2\text{O} + \text{CO} \tag{14.30}$$

In this reaction a hydrogen ion from solution attaches itself to the oxygen atom that is singly bonded to the carbon atom. The activated complex $(\text{HCOOH}_2)^+$ is formed. The C—O bond ruptures, and a hydrogen atom attached to a *carbon* atom in the intermediate species $(\text{HCO})^+$ is released to the solution as a hydrogen ion. This reaction pathway does not require an atom transfer within the activated complex. Thus, it has a substantially lower activation energy than does the uncatalyzed reaction; it proceeds at a faster rate.

In the acid-catalyzed decomposition of formic acid the reactant and catalyst are present in a single phase. This type of catalysis is known as **homogeneous catalysis.** If the reaction of ethylene and hydrogen to form ethane gas is attempted in the gaseous state, the reaction rate is extremely low. The probable activated complex for this reaction is a four-membered cyclic ring, a structure of very high energy.

$$\begin{array}{c}\text{H} \\ | \\ \text{H}\end{array} + \begin{array}{c}\text{CH}_2 \\ \| \\ \text{CH}_2\end{array} \longrightarrow \begin{array}{c}\text{H}\cdots\text{CH}_2 \\ \vdots \quad \vdots \\ \text{H}\cdots\text{CH}_2\end{array} \longrightarrow \begin{array}{c}\text{H—CH}_2 \\ | \\ \text{H—CH}_2\end{array} \tag{14.31}$$

FIGURE 14-12
An example of homogeneous catalysis.

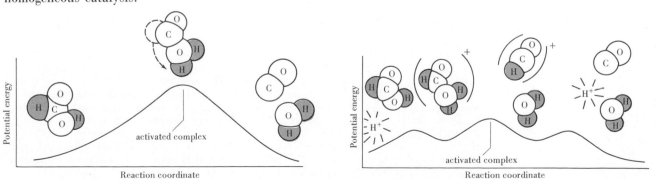

(a) Uncatalyzed Reaction   (b) Catalyzed Reaction

The potential energy of the activated complex—the activation energy—is lowered in the presence of a catalyst. [From George C. Pimentel, ed., *Chemistry: An Experimental Science.* Freeman, San Francisco, 1963. Reproduced by permission of the Chemical Education Material Study.]

There is no practical way to catalyze this reaction homogeneously in the gaseous state. A different method of catalysis is required.

Heterogeneous Catalysis. Reaction (14.31) can be changed from a homogeneous gaseous reaction to a reaction occurring on a surface—a heterogeneous reaction. If this surface material is properly chosen, the reaction rate can be increased significantly; the catalytic action is referred to as **heterogeneous catalysis.** The precise mechanism of heterogeneous catalysis is imperfectly understood. It appears, though, that the availability of d electrons and d orbitals in surface atoms of the catalyst plays an important role. Catalytic activity is associated with a large number of transition elements and their compounds.

The key requirement in heterogeneous catalysis is that reactants be *adsorbed* from a gaseous or solution phase onto the surface of the catalyst. Not all surface atoms are equally effective as catalysts; those that are constitute the **active sites** of a catalyst. Basically, then, heterogeneous catalysis involves (1) adsorption of reactants, (2) diffusion of reactants along the surface, (3) reaction at an active site to form adsorbed product, and (4) *desorption* of the product. These steps are described for the catalytic hydrogenation of ethylene to ethane in Figure 14-13. The net reaction is simply that described in equation (14.31), and the catalyst is unchanged.

Heterogeneous catalysts must be carefully prepared and maintained, for they are easily "poisoned." Trace amounts of certain impurities may become bound to active sites and destroy their catalytic activity. This situation occurs if arsenic is present in platinum, for example, forming platinum arsenide at the active sites. It may also occur if gasoline containing lead compounds as additives is used in automobiles equipped with catalytic mufflers. [These catalysts are designed to promote the combustion of carbon monoxide and hydrocarbons (see Section 13-8).]

FIGURE 14-13
Heterogeneous catalysis and the reaction $C_2H_4(g) + H_2(g) \rightarrow C_2H_6(g)$.

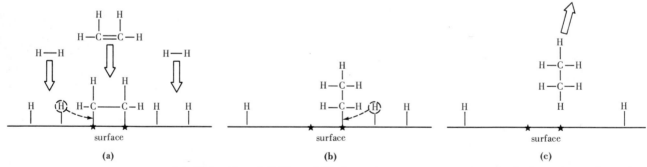

(a) Molecules of C_2H_4 and H_2 from the gaseous state are adsorbed on the surface of a catalyst (e.g., a finely divided form of nickel). The adsorption of H_2 involves dissociation of the molecules and formation of metal–hydrogen (M—H) bonds. The adsorbed H atoms are able to skip about the surface from one site to another. One of the H atoms is shown about to move from its site to a neighboring one where a molecule of C_2H_4 is adsorbed at an active site ($\star$).
(b) A C—H bond is formed between a C_2H_4 molecule and an H atom. The ethyl group (—C_2H_5) that is formed remains adsorbed. A second H atom is shown about to migrate from a nearby site to the active site of the —C_2H_5 group.
(c) Attachment of a second H atom results in the formation of a molecule of C_2H_6. This molecule, because its bonding capacity is saturated through C—H bonds, is easily desorbed and escapes into the gaseous state.

TABLE 14-7
Some applications of catalysis

Process	Product(s)	Typical catalyst(s)	Text reference
polymerization of propylene	polypropylene	titanium(IV) halides	Table 9-4
methane reforming with steam	$H_2(g)$ and $CO(g)$	Ni on K_2O/Al_2O_3	reaction (13.3)
methanol synthesis	$CH_3OH(g)$	ZnO/Cr_2O_3	reaction (4.15)
oxidation of $SO_2(g)$	$SO_3(g)$	V_2O_5 plus K_2SO_4 on SiO_2	reaction (14.42)
ammonia synthesis	$NH_3(g)$	Fe with Al_2O_3, MgO, CaO, and K_2O	reaction (15.35)
ammonia oxidation	NO(g)	90% Pt–10% Rh wire gauze	reaction (13.57)
alkylation of benzene	$C_6H_5C_2H_5$	H_3PO_4 on SiO_2	Figure 26-6

Table 14-7 lists a few important industrial catalysts, some for processes that we have already encountered and others for processes to be studied later.

Enzymes as Catalysts. Some catalysts, for example platinum metal, can catalyze a wide variety of reactions. Unlike platinum, the catalysts associated with chemical reactions in a living organism must be very specific. These catalysts, called **enzymes,** are high-molecular-weight proteins. Many enzymes catalyze one particular reaction and no others. For example, in alcoholic fermentation the six-carbon compound glucose is broken down into two molecules of ethanol and two of carbon dioxide.

$$C_6H_{12}O_6 \longrightarrow 2\ C_2H_5OH + 2\ CO_2$$

This process requires 12 enzymatic steps. In the last of these acetaldehyde is reduced to ethanol through the action of the enzyme alcohol dehydrogenase. (The necessary hydrogen atoms are furnished by other species in the reaction.)

$$
\begin{array}{ccc}
\text{H} & \text{O} & \\
| & \| & \\
\text{H—C—C—H} & \xrightarrow[\text{dehydrogenase}]{\text{alcohol}} & \text{H—C—C—O—H} \\
| & & \\
\text{H} & &
\end{array}
$$

The simplest mechanism of enzyme action, known as the Michaelis–Menten mechanism, involves a reactant species, called the **substrate** (S), attaching itself to an active site on the enzyme (E). The result is an enzyme–substrate complex (ES). This complex dissociates to produce a product species (P) and the original enzyme (E). Thus, a two-step mechanism can be written, with each step being reversible (denoted by the double arrow $\rightleftharpoons$).

FIGURE 14-14
The Michaelis–Menton mechanism of enzyme action.

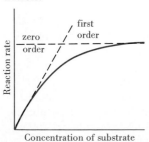

Concentration of substrate

$$S + E \rightleftharpoons ES$$

$$ES \rightleftharpoons E + P$$

Figure 14-14, which shows how the reaction rate varies with substrate concentration, is consistent with this mechanism. Along the ascending portion of the curve the reaction is first order in S, because the rate at which the complex ES is formed is proportional to [S].

rate of reaction = $k[S]$

At high concentrations of the substrate the reaction is zero order. The enzyme is saturated, and adding more substrate cannot accelerate the reaction.

rate of reaction = $k'[S]^0 = k'$

14-10 Reaction Mechanisms

One important reason for studying chemical kinetics is the fact that knowledge of the rate of a reaction can provide insight into the succession of steps by which the reaction proceeds—**the reaction mechanism.**

Analogies to a Reaction Mechanism.   Consider this analogy to a reaction mechanism: A person in Los Angeles meets an old acquaintance from New York City and asks, "When did you leave New York?" The answer: "This morning." The friend must have made the principal part of the journey by airplane. If the friend had answered, "Two weeks ago," the conclusion on the mode of travel would not have been obvious. Further questioning (experimentation) would have been required.

At times, the fact that a reaction occurs very rapidly suggests a plausible reaction mechanism, as in the preceding analogy. More often, however, we gain insight into a reaction mechanism by identifying the *slow* or rate-determining step, as in this analogy: A business traveler lands at an airport, rents a car, and then takes one of two possible routes to visit a potential customer. The trip normally requires 20 minutes by either route, but it takes the traveler 50 minutes. When he learns of this, the customer remarks, "You must have come by Route 330 and been delayed by the one-way construction traffic." The slow or rate-determining step in a reaction mechanism is analogous to a "bottleneck" in a flow of traffic.

If the business traveler had offered a ride into the city to a seat companion who was a native of the area, he might have been guided by the companion (catalyst) to the faster route.

Elementary Processes.   Each molecular event that significantly alters a molecule's energy or geometry is called an elementary process. The combined result of all the elementary processes is the net reaction. Unlike for the net reaction, however, concentration-term exponents in the rate law for an *elementary* process *are* the same as the corresponding coefficients in the balanced equation for the elementary process. Moreover, by writing rate equations for the elementary steps and combining them in the appropriate fashion, a rate equation can be determined for the net reaction. A test of the *plausibility* of a mechanism is that it yield the same rate equation as that determined experimentally. The word "plausible" is emphasized because there is no way to *prove* a reaction mechanism. It is often possible to propose several mechanisms that are consistent with the observed rate law.

Other ideas essential to developing a plausible reaction mechanism are

1. Elementary processes in which a single molecule dissociates—**unimolecular**—or two molecules collide—**bimolecular**—are much more probable than a process requiring the simultaneous collision of three bodies—**termolecular.**

2. All elementary processes are reversible and may reach a **steady-state condition.** In the steady state the rates of the forward and reverse processes become equal. The concentration of some intermediate becomes constant with time.

3. One elementary process may occur much more slowly than all the others, and in some cases may determine the rate at which the overall reaction proceeds. Such a process is called the **rate-determining step.**

The decomposition of H_2O_2 can also be catalyzed heterogeneously, e.g., by colloidal platinum.

The Catalyzed Decomposition of Hydrogen Peroxide.   The decomposition of H_2O_2 (reaction 14.3), which we studied earlier in this chapter, is catalyzed by a number of substances. In the presence of HBr, for example, this two-step process seems to occur.

(slow) $H_2O_2 + H^+ + Br^- \longrightarrow HOBr + H_2O$

(fast) $\underline{\hspace{0.8cm} H_2O_2 + HOBr \longrightarrow H_2O + H^+ + Br^- + O_2(g) \hspace{0.8cm}}$

net: $2\,H_2O_2 \longrightarrow 2\,H_2O + O_2(g)$

As required for a catalyzed reaction, the formula of the catalyst does not appear in the net equation. The intermediate species HOBr is consumed in the second step just as rapidly as it can be formed in the first step. The rate of disappearance of H_2O_2, then, is determined by the rate of the slow first step in the mechanism.

rate of reaction $= -$ rate of disappearance of $H_2O_2 = k[H_2O_2][H^+][Br^-]$ (14.32)

Because HBr, a catalyst in the reaction, is constantly regenerated, the concentration terms $[H^+]$ and $[Br^-]$ are unchanged (constant) throughout a given reaction. The observed reaction rate is

rate of reaction $= k'[H_2O_2]$ (14.33)

Equation (14.33) is the rate law that we used in Section 14-5. However, equation (14.32) does suggest that the rate of decomposition of H_2O_2 will be affected by the concentration of HBr introduced initially. That is, for each different initial concentration of HBr the rate constant in (14.33) has a different value.

The mechanism described here helps to explain why the reaction is first order in H_2O_2 and not second order as the net equation might suggest. Only half of the H_2O_2 consumed participates in the slow, rate-determining step. The other half disappears in the fast reaction that comes *after* the rate-determining step (after the reaction bottleneck has been passed).

The Hydrogen–Iodine Reaction.   Is there a reaction for which the mechanism consists of a single elementary step? That is, is there a reaction for which the mechanism is that implied by the net equation? For the better part of the twentieth century the classic example of such a reaction had been that of gaseous hydrogen and iodine.

$$H_2(g) + I_2(g) \longrightarrow 2 HI(g)$$ (14.34)

rate of reaction $= k[H_2][I_2]$

The rate equation for a simple one-step bimolecular process based on the collision of an H_2 and an I_2 molecule is the same as that for the net reaction.

However, as a result of a detailed experimental investigation of this reaction, it was concluded that the hydrogen–iodine reaction is probably more complex than the one-step mechanism of (14.34).* In particular the proposal was made of a two-step mechanism. In the first step, which occurs rapidly, iodine molecules are believed to dissociate into iodine atoms. The second step is believed to involve the simultaneous collision of *two* iodine atoms and a hydrogen molecule. We should expect this *termolecular* step to occur much more slowly and to be the *rate-determining step*. This step is illustrated in Figure 14-15.

(fast) $\qquad\qquad\qquad I_2(g) \underset{k_2}{\overset{k_1}{\rightleftharpoons}} 2\ I(g)$ (14.35)

(slow) $\quad \underline{2\ I(g) + H_2(g) \overset{k_3}{\longrightarrow} 2\ HI(g)}$ (14.36)

net: $\qquad\qquad I_2(g) + H_2(g) \longrightarrow 2\ HI(g)$ (14.34)

The net equation obtained by adding together the two elementary processes is indeed that expected.

For the *rate-determining step* (14.36) we can write

rate of reaction $= k_3[I]^2[H_2]$ (14.37)

However, since atomic I is an intermediate in the reaction, we must eliminate its concentration term from the final rate equation. This we can do by assuming that the fast reversible step (14.35) reaches a *steady-state condition*, in which I_2 is consumed and

FIGURE 14-15
Rate-determining step in the reaction

$H_2(g) + I_2(g) \rightarrow 2\ HI(g)$

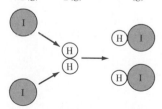

The rate-determining step for this reaction appears to involve the simultaneous collision of two I atoms with an H_2 molecule.

*J. Sullivan, *J. Chem. Phys.* **46,** 73 (1967).

produced at equal rates and from which I is withdrawn only very slowly through the rate-determining step.

$-$ rate of disappearance of I_2 = rate of formation of I_2

$$k_1[I_2] = k_2[I]^2$$

and $\qquad [I]^2 = \dfrac{k_1}{k_2}[I_2] \qquad\qquad\qquad\qquad\qquad\qquad\qquad (14.38)$

Now, if we substitute into the rate equation (14.37) the expression for $[I]^2$ from the steady-state condition (14.38), we obtain for the rate of the net reaction

$$\text{rate of reaction} = \frac{k_1 k_3}{k_2}[H_2][I_2] = k[H_2][I_2] \qquad (\text{where } k = k_1 k_3/k_2) \qquad (14.39)$$

Equation (14.39) agrees with the observed rate law. Whether the mechanism outlined here is the actual mechanism of the reaction, we cannot say. All that we can say is that it is *plausible*.

The Hydrogen–Bromine Reaction—A Chain Reaction. Next we might ask if the mechanism of the hydrogen–bromine reaction is similar to that of the hydrogen–iodine reaction. The answer is no, because the observed rate law is much more complicated than expected (that is, it is *not* rate = $k[H_2][Br_2]$). One mechanism that leads to the observed rate of reaction is

initiation: $\qquad\qquad Br_2 \longrightarrow 2\ Br \qquad\qquad\qquad\qquad\qquad\qquad (14.40a)$

propagation: $\quad Br + H_2 \longrightarrow HBr + H \qquad\qquad\qquad\qquad\quad (14.40b)$

$\qquad\qquad\qquad H + Br_2 \longrightarrow HBr + Br$

$$\text{and so on}$$

inhibition: $\quad H + HBr \longrightarrow H_2 + Br \qquad\qquad\qquad\qquad\quad (14.40c)$

termination: $\quad Br + Br \longrightarrow Br_2 \qquad\qquad\qquad\qquad\qquad\qquad (14.40d)$

Chain reactions are involved in the formation of photochemical smog (Section 13-10), in certain reactions between halogens and hydrocarbons (Section 26-2) and in polymerization reactions (Section 26-3).

A Br atom formed in the initiation reaction (14.40a) attacks an H_2 molecule to produce a molecule of HBr and an H atom. This H atom, in turn, attacks a Br_2 molecule to produce a second HBr molecule and a Br atom, and so on through thousands of steps in the propagation reactions (14.40b). As the product HBr accumulates its participation in reaction (14.40c) slows the rate of build up of product. This is an inhibition reaction. Finally, the combination of two Br atoms terminates the chain. The collection of steps described here is called a **chain reaction.** The net change represented by all the elementary steps in reaction (14.40b) is

$$H_2(g) + Br_2(g) \longrightarrow 2\ HBr(g) \qquad\qquad\qquad\qquad\qquad (14.41)$$

Example 14-14 The reaction $2\ NO + Cl_2 \rightarrow 2\ NOCl$ has the observed rate law: rate of reaction = $k[NO]^2[Cl_2]$. Show that the following mechanism is consistent with this rate law.

(fast) $\quad NO + Cl_2 \underset{k_2}{\overset{k_1}{\rightleftharpoons}} NOCl_2$

(slow) $\quad NO + NOCl_2 \overset{k_3}{\longrightarrow} 2\ NOCl$

Solution. Assume that the second step of the mechanism, the slow step, is rate determining.

rate of reaction = $k_3[NO][NOCl_2]$

Also assume that a fast reversible reaction (the first step above) reaches a steady-state condition.

rate of formation of $NOCl_2 = k_1[NO][Cl_2] = k_2[NOCl_2]$
$$= -\text{ rate of disappearance of } NOCl_2$$

Determine the steady-state concentration of $NOCl_2$

$$[NOCl_2] = \frac{k_1[NO][Cl_2]}{k_2}$$

and substitute this concentration into the rate-determining step.

$$\text{rate of reaction} = k_3[NO]\frac{k_1[NO][Cl_2]}{k_2} = \frac{k_1k_3[NO]^2[Cl_2]}{k_2} = k[NO]^2[Cl_2]$$

SIMILAR EXAMPLES: Review Problems 10, 11; Exercises 43, 45, 46.

FOCUS ON
Sulfuric Acid

Large complexes for the manufacture of phosphate fertilizers often include a sulfuric acid plant (outlined in the photograph). The principal use of sulfuric acid is in the manufacture of fertilizers. [Courtesy of W. R. Grace & Co.]

Sulfuric acid has a long history; it has been produced for at least 500 and possibly 1000 years. One method of production, dating back to the fifteenth century, involved heating the hydrate $FeSO_4 \cdot 7H_2O$. The sulfuric acid industry can be dated from about the middle eighteenth century, fol-

lowing the invention of a process in which a mixture of S and KNO_3 was burned and the gaseous products dissolved in water in lead-lined chambers, yielding aqueous sulfuric acid. Later it was discovered that oxides of nitrogen formed by heating KNO_3 acted as intermediates in the process, and that they could be recovered and reused. We now consider intermediates of this type to be catalysts. The manufacture of H_2SO_4 by the "lead chamber" process is one of the earliest examples of *homogeneous catalysis*.

The key step in the manufacture of H_2SO_4 is the conversion of $SO_2(g)$ to $SO_3(g)$.

$$SO_2(g) + \tfrac{1}{2}O_2(g) \longrightarrow SO_3(g) \tag{14.42}$$

Reaction (14.42) occurs very slowly unless it is catalyzed. A simplified description of catalysis by the oxides of nitrogen is

$$NO(g) + \tfrac{1}{2}O_2(g) \longrightarrow NO_2(g) \tag{14.43}$$

$$NO_2(g) + SO_2(g) \longrightarrow SO_3(g) + NO(g) \tag{14.44}$$

H_2SO_4 produced by the lead chamber process is impure and diluted with water to the point that it is never more than 80% H_2SO_4, by mass. The method is now obsolete. It is interesting to speculate, though, whether reactions such as (14.43) and (14.44) might catalyze the conversion of atmospheric $SO_2(g)$ to $SO_3(g)$ in the production of acid rain.

A method of converting $SO_2(g)$ to $SO_3(g)$ in the presence of platinum metal was patented in England in 1831. Here *heterogeneous catalysis* is involved. Adsorption of $SO_2(g)$ and $O_2(g)$ on the Pt is followed by reaction at active sites and desorption of the $SO_3(g)$ (recall Figure

14-13). The net reaction is (14.42). The advantages of this alternate route to sulfuric acid, the **contact process,** were not fully appreciated until later in the century when the need arose in the synthetic organic chemical industry for a material variously called *oleum* and *fuming sulfuric acid.* Oleum is a solution of excess SO_3 in pure sulfuric acid (in a sense it is "greater than 100% H_2SO_4"). Oleum cannot be produced by the lead chamber process. In the contact process pure $SO_3(g)$ is passed into 98 or 99% sulfuric acid to form oleum, which can then be diluted with water to the exact strength of H_2SO_4 desired. For example, if we use the formula $H_2S_2O_7$ (pyrosulfuric acid) for a particular oleum, the reactions are

$$SO_3(g) + H_2SO_4(l) \longrightarrow H_2S_2O_7(l) \qquad (14.45)$$
$$\text{(oleum)}$$

$$H_2S_2O_7(l) + H_2O(l) \longrightarrow 2 H_2SO_4(l) \qquad (14.46)$$

$$H_2SO_4(l) \xrightarrow{H_2O} H_2SO_4(aq) \qquad (14.47)$$

Platinum metal catalyst, in a finely divided form called platinum black, is easily "poisoned," e.g., by as little as 1×10^{-8} g/L of arsenic compounds in the gases brought into contact with the catalyst. Because of this difficulty other catalysts have been developed for the contact process. The principal one in use today is vanadium pentoxide (V_2O_5) mixed with alkali metal sulfates.

Other modifications in the process have been stimulated by the need to greatly reduce the quantities of noxious gases (principally SO_2) emitted in sulfuric acid manufacture. By a double absorption of $SO_3(g)$ in sulfuric acid and by recycling the reactants through the catalyst several times, it is now possible to obtain about 99.7% conversion of $SO_2(g)$ to $SO_3(g)$. The chief source of $SO_2(g)$ for the contact process comes from burning pure sulfur. In the past this sulfur has been mainly from the Frasch mining process (Section 13-13). Now, with the need to remove sulfur compounds from natural gas and from stack gases in power plants and smelters, much of the sulfur used in the manufacture of sulfuric acid is recovered sulfur (recall re-

TABLE 14-8
Relative importance of some uses of sulfuric acid in the United States

Use	1960	1970	1980
fertilizers	38.4%	51.0%	68.1%
manufacture of titanium dioxide (a white pigment)	9.1	5.5	1.8
petroleum refining	8.8	7.0	4.5
iron and steel pickling	5.6	1.6	0.3
manufacture of cellulose products	3.6	2.2	0.9
ore processing	1.8	3.6	4.8
batteries	0.5	0.4	0.4
Total consumption (in millions of kg)	17,500	28,500	41,200

action 13.74). It is likely that by the end of the century most of the sulfur used will be recovered sulfur.

Because the potential uses of sulfuric acid are so varied, it was once said that the quantity of sulfuric acid produced is a good indicator of the degree of industrialization of a nation and of general economic conditions. This statement is no longer applicable in the United States, however, since the bulk of H_2SO_4 production is now used for a single purpose—the manufacture of fertilizers. Table 14-8 illustrates how dramatic this shift has been in the past few decades. The use of sulfuric acid in the production of phosphate fertilizers is discussed in Section 21-3.

Summary

The rate of a chemical reaction is related to how fast the concentration of a reactant decreases or that of a product increases with time. An *initial* rate of reaction is determined by dividing the change in concentration of a reactant at the start of a reaction by the short time interval over which this change occurs. Beyond this initial stage, the *instantaneous* rate of reaction is given by the slope of a tangent line to a concentration vs. time graph. One goal of a kinetic study is to be able to express the reaction rate through a rate law having the form

reaction rate $= k[A]^m[B]^n \cdots$

The order of a reaction is related to the exponents in the rate law. Reactions that proceed at a constant rate, *independent* of the concentrations of the reactants, are zero-order reactions. A first-order reaction most often has a single concentration term appearing in the rate law and that term is raised to the first power. The most common forms of a second-order rate law have either a single concentration term raised to the second power or two concentration terms each raised to the first power. One method of establishing the order of a reaction requires measuring the *initial* reaction rate in a series of experiments. A second method requires plotting appropriate functions of reactant concentration against time so as to obtain a straight-line graph. A third method involves substituting experimentally determined concentrations and times into appropriate mathematical equations.

An important characteristic associated with reaction rates is the half-life of a reaction. For a first-order reaction the half-life is *independent* of the concentration of the reactant. For other reaction orders the half-life is concentration dependent.

The theoretical basis of chemical kinetics includes these essential ideas: Chemical reactions occur as a result of collisions between molecules. Only collisions in which molecules possess sufficient energy and a proper geometrical orientation are effective in yielding a product. The course of a chemical reaction can be depicted through an energy diagram, called a reaction profile, in which the energies of reactants, products, and activated complex(es) are represented. Such a profile permits a visualization of the enthalpy change and activation energy of a reaction.

Raising the temperature of a reaction mixture increases the fraction of the molecules possessing energies in excess of the activation energy—the reaction speeds up. The increase in rate of reaction with temperature can be expressed through a mathematical equation. Another method of speeding up a reaction is to employ a catalyst. Some catalyst molecules participate as intermediates that are regenerated in a homogeneous reaction. In another type of catalysis—heterogeneous catalysis—the catalyst provides a surface on which the desired reaction proceeds at a faster rate. Biochemical reactions are promoted by catalysts called enzymes.

It is sometimes possible to propose a mechanism for a reaction. This is done by postulating a series of elementary processes, writing rate equations for them, and combining these elementary rate equations into a rate law for the net reaction.

Learning Objectives

As a result of studying Chapter 14, you should be able to

1. Describe how the rate of a reaction is related to the rate of disappearance of a reactant or formation of a product.

2. Obtain the data needed for a kinetic study from the results of simple chemical analyses.

3. Establish the exact rate of a chemical reaction from the slope of a tangent line to a concentration vs. time graph.

4. Determine the *initial* rate and an *instantaneous* rate of a reaction, either from a graph or by calculation.

5. Apply the method of initial rates to derive the rate law for a reaction.

6. Use a rate law and rate data to calculate a rate constant, k, or use the rate law and rate constant to calculate rate data.

7. Establish, through rate data, equations, and graphs, whether a reaction is zero order, first order, or second order.

8. Use the concept of the half-life of a reaction for zero-order, first-order, and second-order reactions.

9. Describe the collision theory of reactions, stating the factors that affect collision frequency and those that lead to favorable collisions.

10. Describe a reaction in terms of a reaction profile, activated complex, and activation energy.

11. Use the Arrhenius equation in calculations involving rate constants, temperatures, and activation energies.

12. Describe the role of a catalyst and explain the difference between homogeneous and heterogeneous catalysis.

13. Describe a reaction mechanism, and distinguish between elementary processes and a net chemical reaction.

14. Use the concepts of steady-state condition and rate-determining step in testing a plausible mechanism for a reaction.

Some New Terms

An **activated complex** is an intermediate species in the mechanism of a chemical reaction.

Activation energy is the energy that molecules must possess so that collisions between them will lead to chemical reaction. It is the energy required to form an activated complex.

Active sites are the locations at which catalysis occurs, whether on the surface of a heterogeneous catalyst or an enzyme.

The **Arrhenius equation** relates the rate of a reaction to temperature and activation energy.

A **bimolecular process** is an elementary process involving the collision of two molecules.

Catalysis is the speeding up of a reaction in the presence of an agent **(catalyst)** that changes the reaction mechanism to one of lower activation energy.

A **chain reaction** is a reaction mechanism in which certain elementary processes occur many times in the overall reaction mechanism.

Collision frequency is the number of collisions occurring between molecules in a unit of time.

Collision theory describes reactions in terms of molecular collisions—the frequency of collisions and the probability that they will lead to chemical reaction.

An **elementary process** is a molecular event representing a single step in a reaction mechanism.

A **first-order** reaction is one for which the sum of the concentration-term exponents in the rate law is 1.

The **half-life** is the time required for one half of a reactant to be consumed in a chemical reaction.

The **initial rate of a reaction** is the rate of a reaction immediately after the reactants are brought together.

The **method of initial rates** establishes the order of a reaction from the initial rates of reaction.

The **order of a reaction** relates to the exponents of the concentration terms in the rate law for the reaction.

The **rate constant, k,** is the proportionality constant in the rate law of a chemical reaction.

A **rate-determining step** in a reaction mechanism is an elementary process that is much slower than other steps and determines the rate of the overall reaction.

The **rate law (rate equation)** for a reaction relates the reaction rate to the concentrations of reactants. It has the form: reaction rate = $k[A]^m[B]^n \cdots$.

The **rate of a chemical reaction** is related to how fast reactants are consumed or products are formed.

A **reaction mechanism** is a set of elementary steps or processes by which a reaction is proposed to occur.

A **reaction profile** is a graphical representation of a chemical reaction in terms of the energies of the reactants, activated complex(es), and products.

A **second-order** reaction is one for which the sum of the concentration-term exponents in the rate law is 2.

A **steady-state condition** is reached in an elementary process when a species is formed and consumed at equal rates and its concentration remains constant with time.

A **termolecular process** is an elementary process in which three molecules collide simultaneously.

A **unimolecular process** is an elementary process in which a single molecule dissociates.

A **zero-order** reaction proceeds at a rate that is *independent* of reactant concentrations.

Suggestions for Further Study

BLAND, W. J., "Sulphuric Acid—Modern Manufacture and Uses," *Educ. in Chem.*, **21,** 7 (1984).

KOLB, D., "Catalysis," *J. Chem. Educ.*, **56,** 743 (1979).

LABUZA, T. P., "Application of Chemical Kinetics to Deterioration of Foods," *J. Chem. Educ.*, **61,** 348 (1984).

LOGAN, S. R., "The Origin and Status of the Arrhenius Equation," *J. Chem. Educ.*, **59,** 279 (1982).

MICKEY, C. D., "Chemical Kinetics: Reaction Rates," *J. Chem. Educ.*, **57,** 659 (1980).

SATTERFIELD, C. N., "Industrial Heterogeneous Catalysis: An Overview," *Chemtech*, **11,** 618 (1981).

SCHOFIELD, M., "Early Days of Sulfuric Acid," *Chemistry*, **45**[9], 11 (1972).

Review Problems

1. In the reaction A → products, the initial concentration of A is 0.1205 *M*, and 53 s later, 0.1168 *M*. What is the initial rate of this reaction, expressed in **(a)** mol L^{-1} s^{-1}; **(b)** mol L^{-1} min^{-1}?

2. In the reaction 2 A + B → C + 3 D, reactant A is found to be disappearing at the rate of 2.6×10^{-4} mol L^{-1} s^{-1}. **(a)** What is the rate of the reaction? **(b)** What is the rate of formation of D?

3. Example 14-2 illustrates how $[H_2O_2]$ in reaction (14.3) can be followed by titration with MnO_4^-(aq). **(a)** As the reaction proceeds, does the volume of MnO_4^-(aq) required for each successive titration increase or decrease? **(b)** Determine the volume of 0.1000 *M* MnO_4^-(aq) that must have been required for each of the remaining titrations in Table 14-1, i.e., at 600 s, 1200 s, etc.

4. From Figure 14-1 estimate the rate of decomposition of H_2O_2 at **(a)** $t = 1000$ s; **(b)** the point in the reaction where $[H_2O_2] = 0.50$ *M*.

5. The decomposition of acetaldehyde is found to be second order. Write the rate equation for this decomposition: $CH_3CHO \rightarrow CH_4 + CO$.

6. The initial rate of the reaction 2 A + 2 B → C + D is determined for several different initial conditions. The results obtained are tabulated below. **(a)** What is the order of the reaction with respect to A and to B? **(b)** What is the overall reaction order?

Expt.	[A], *M*	[B], *M*	Initial rate, mol L^{-1} s^{-1}
1	0.210	0.115	6.30×10^{-4}
2	0.210	0.230	1.25×10^{-3}
3	0.420	0.115	2.51×10^{-3}
4	0.420	0.230	5.13×10^{-3}

7. Substance A decomposes by a first-order reaction. Starting initially with [A] = 2.00 M, after 201 min [A] = 0.250 M. For this reaction what is **(a)** $t_{1/2}$; **(b)** k?

8. The first-order reaction   A → products   has $k = 1.0 \times 10^{-3}$ s^{-1}. Starting with [A] = 0.85 M,
  (a) What is [A] 10.0 min later?
  (b) At what time will [A] = 0.25 M?
  (c) How long will it take for 90% of the A originally present to be consumed?

9. The rate constant for the reaction $H_2(g) + I_2(g) \rightarrow 2\ HI(g)$ has been determined at the following temperatures: 556 K, $k = 1.2 \times 10^{-4}$ L mol^{-1} s^{-1}; 666 K, $k = 3.8 \times 10^{-2}$ L mol^{-1} s^{-1}.
  (a) Calculate the activation energy for the reaction.
  (b) At what temperature will the rate constant have the value $k = 1.0 \times 10^{-3}$ L mol^{-1} s^{-1}?

10. For the reaction   A + B → C + D   the following mechanism is proposed.

(slow)   2 A → I + D
(fast)   I + B → A + C

  (a) Show that this mechanism is consistent with the net equation.
  (b) Write the rate equation expected for this mechanism.

11. The reaction, $H_2 + 2\ ICl \rightarrow I_2 + 2\ HCl$, is found to be first order in H_2 and in ICl. A proposed mechanism for the reaction involves the following first step.

(slow) $H_2 + ICl \rightarrow HI + HCl$

  (a) Write a plausible second step in a two-step mechanism.
  (b) Is the second step slow or fast? Explain.

Three different sets of data of [A] vs. time are given in the table below for the reaction   A → products.   (*Hint:* There are several ways of arriving at answers for each of the following five questions.)

I		II		III	
Time, s	[A], M	Time, s	[A], M	Time, s	[A], M
0	1.00	0	1.00	0	1.00
25	0.78	25	0.75	25	0.80
50	0.61	50	0.50	50	0.67
75	0.47	75	0.25	75	0.57
100	0.37	100	0.00	100	0.50
150	0.22			150	0.40
200	0.14			200	0.33
250	0.08			250	0.29

12. Which of these sets of data corresponds to a **(a)** zero-order, **(b)** first-order, **(c)** second-order reaction?

13. What is the approximate half-life of the first-order reaction?

14. What is the approximate initial rate of the second-order reaction?

15. What is the approximate rate of the reaction at $t = 75$ s for the **(a)** zero-order, **(b)** first-order, **(c)** second-order reaction?

16. What is the approximate concentration of A remaining after 110 s in the **(a)** zero-order, **(b)** first-order, **(c)** second-order reaction?

Exercises

Rates of reactions

1. In the reaction   A → products,   the initial concentration of A is 0.1503 M. After 1.00 min, [A] = 0.1455 M, and after 2.00 min, [A] = 0.1409 M.
  (a) Calculate the rate of the reaction during the first minute and during the second minute.
  (b) Why are these two rates not equal?

2. If the reaction   A + 2 B → 2 C   has a rate of 2.50×10^{-5} mol L^{-1} s^{-1}, at a time when [A] = 0.5000 M,
  (a) What is the rate of formation of C?
  (b) What will be [A] 1.00 min later?
  (c) Assuming that the rate remains at 2.50×10^{-5} mol L^{-1} s^{-1}, how long would it take for [A] to decrease from 0.5000 M to 0.4900 M?

3. In Example 14-3 we determined the rate of decomposition of $H_2O_2(aq)$ to be 5.7×10^{-4} mol L^{-1} s^{-1}. What is the rate of production of $O_2(g)$ from 1.00 L of the $H_2O_2(aq)$ at this point, expressed as **(a)** mol O_2 s^{-1}; **(b)** mol O_2 min^{-1}; **(c)** cm^3 O_2 at STP per minute.

4. A reaction involves the decomposition   A → B + C.   What is the meaning of each of the following terms with respect to this reaction? **(a)** [A]$_0$; **(b)** [A]$_t$; **(c)** Δ[A]; **(d)** Δt; **(e)** $-$Δ[A]/Δt; **(f)** Δ[B]/Δt; **(g)** $t_{1/2}$.

5. Describe briefly the meaning of each of the following terms. **(a)** initial rate of reaction; **(b)** instantaneous rate of reaction; **(c)** zero-order reaction; **(d)** half-life of a reaction.

6. Explain the difference in meanings of the terms *rate of a reaction* and *rate constant of a reaction*. Is there any type of reaction for which the two are the same? Explain.

7. For the reaction   A → products,   what are the units of the rate constant, k, if the reaction is **(a)** zero order; **(b)** first order; **(c)** second order in A.

8. The initial rate of decomposition of H_2O_2 (reaction 14.3) in a particular experiment is 1.7×10^{-3} mol L^{-1} s^{-1}. Assume that this rate holds for about 2 min. Starting with 175 mL of 1.55 M H_2O_2(aq) at $t = 0$,

   (a) What is $[H_2O_2]$ after 10 s?
   (b) At what time after the start of the experiment would you expect $[H_2O_2] = 1.50$ M?
   ★(c) What volume of O_2(g), measured at STP, is released from solution in the first minute of the reaction? (*Hint:* Determine the decrease in $[H_2O_2]$. Then determine the actual decrease in amount of H_2O_2 (in moles), to which the amount of O_2 can be related.)

9. Refer to Experiment 2 of Table 14-3. Exactly 1 min after the reaction is started, what are (a) $[S_2O_8^{2-}]$ and (b) $[I^-]$ in the mixture?

10. In the hypothetical reaction $A(g) \rightarrow 2 B(g) + C(g)$,   the *total* pressure increases while the *partial* pressure of $A(g)$ decreases. If the initial pressure of $A(g)$ in a vessel of constant volume is 1000 mmHg

   (a) What is the total pressure when the reaction has gone to completion?
   (b) What is the total gas pressure when the partial pressure of $A(g)$ has fallen to 800 mmHg?

★**11.** Use the volumes of MnO_4^-(aq) calculated in Review Problem 3 and show that the number of mL MnO_4^-(aq) required for each titration plotted as a function of time yields the same type of curve as Figure 14-1; and that from the tangent to the curve at $t = 1500$ s one can obtain the same value for the rate of reaction as that determined in Example 14-3.

★**12.** A statement is made in the text (page 410) that the value obtained for the reaction rate through the expression $-\Delta[H_2O_2]/\Delta t$ depends on the time interval chosen for Δt. Use data from Figure 14-1 to show that this is indeed the case. (*Hint:* Estimate the reaction rate at 1500 s using values of Δt that range from quite large, e.g., 3000 s, to very small.)

Method of initial rates

13. What would you expect for the initial rate of reaction for a fourth experiment in Table 14-3 in which $[A] = [S_2O_8^{2-}] = 0.019$ M and $[B] = [I^-] = 0.015$ M?

14. The rate of the reaction $2 HgCl_2 + C_2O_4^{2-} \rightarrow 2 Cl^- + 2 CO_2(g) + Hg_2Cl_2(s)$ is followed by measuring the number of moles of Hg_2Cl_2 that precipitate per liter per minute.

Expt.	$[HgCl_2]$, M	$[C_2O_4^{2-}]$, M	Initial rate, mol L^{-1} min^{-1}
1	0.105	0.15	1.8×10^{-5}
2	0.105	0.30	7.1×10^{-5}
3	0.052	0.30	3.5×10^{-5}
4	0.052	0.15	8.9×10^{-6}

   (a) Determine the order of the reaction with respect to $HgCl_2$, with respect to $C_2O_4^{2-}$, and overall.
   (b) What is the value of the rate constant, k?
   (c) What would be the initial rate of reaction if $[HgCl_2] = 0.020$ M and $[C_2O_4^{2-}] = 0.22$ M?

15. Listed below are *initial rates*, expressed as the rate of decrease of partial pressure of a reactant, at 826°C.

$$2 \ NO(g) + 2 \ H_2(g) \rightarrow N_2(g) + 2 \ H_2O(g)$$

With initial $P_{H_2} =$ 400 mmHg		With initial $P_{NO} =$ 400 mmHg	
Initial P_{NO}, mmHg	Rate, mmHg/s	Initial P_{H_2}, mmHg	Rate, mmHg/s
359	0.750	289	0.800
300	0.515	205	0.550
152	0.125	147	0.395

   (a) What is the order of the reaction with respect to NO, with respect to H_2, and overall?
   (b) Write the rate equation for this reaction.

★**16.** Hydroxide ion is involved in the mechanism of the following reaction but is not consumed in the net reaction.

$$OCl^- + I^- \xrightarrow{\ OH^-\ } OI^- + Cl^-$$

   (a) From the data given, determine the order of the reaction with respect to OCl^-, I^-, and OH^-.
   (b) What is the overall reaction order?
   (c) Write the rate equation and determine the value of the rate constant, k.

$[OCl^-]$, M	$[I^-]$, M	$[OH^-]$, M	Rate formation OI^-, mol L^{-1} s^{-1}
0.0040	0.0020	1.00	4.8×10^{-4}
0.0020	0.0040	1.00	5.0×10^{-4}
0.0020	0.0020	1.00	2.4×10^{-4}
0.0020	0.0020	0.50	4.6×10^{-4}
0.0020	0.0020	0.25	9.4×10^{-4}

First-order reactions

17. Some of the following statements are true and some are not, regarding the *first-order* reaction $2 A \rightarrow B + C$.   Indicate which are true and which are false. Explain your reasoning.

   (a) The rate of the reaction decreases as more and more of B and C are formed.
   (b) The time required for one half of substance A to react is directly proportional to the quantity of A.
   (c) A plot of $[A]$ vs. time yields a straight line.
   (d) The rate of formation of C is one-half the rate of disappearance of A.

18. A certain first-order decomposition has a value of $k = 6.5 \times 10^{-4}$ s^{-1}. Starting with a concentration of reactant of 0.50 M, how long will it take for the reactant to be 80% decomposed?

19. In the first-order reaction   A $\rightarrow$ products,   it is found that 99% of the original amount of reactant A decomposes in 185 min. What is the half-life, $t_{1/2}$, of this decomposition reaction?

20. The following first-order reaction is conducted in CCl$_4$(l) at 45°C.

$$N_2O_5 \rightarrow N_2O_4 + \tfrac{1}{2}O_2(g)$$

The rate constant $k = 6.2 \times 10^{-4}$ s^{-1}. An 80.0-g sample of N$_2$O$_5$ is dissolved in CCl$_4$(l) and allowed to decompose at 45°C.
   (a) How long will it take for the quantity of N$_2$O$_5$ to be reduced to 2.5 g?
   (b) What total volume of O$_2$ (at STP) is produced up to this point?

21. Concerning the decomposition of di-*t*-butyl peroxide (DTBP) described in Example 14-11,
   (a) Determine the time at which the *partial* pressure of DTBP = 700 mmHg.
   *(b) Determine the times at which the *total* gas pressure is 2000 mmHg and 2100 mmHg.

22. The decomposition of dimethyl ether at 504°C is

$$(CH_3)_2O(g) \rightarrow CH_4(g) + H_2(g) + CO(g)$$

The following data are partial pressures of dimethyl ether (DME) as a function of time: $t = 0$ s, $P_{DME} = 312$ mmHg; 390 s, 264 mmHg; 777 s, 224 mmHg; 1195 s, 187 mmHg; 3155 s, 78.5 mmHg.
   (a) Show that the reaction is first order.
   (b) What is the value of the rate constant, k?
   (c) What is the total gas pressure at 390 s?
   (d) What is the total gas pressure when the reaction has gone to completion?
   (e) What is the total gas pressure at $t = 1000$ s?

23. Benzenediazonium chloride decomposes by a first-order reaction in water, yielding N$_2$(g) as a product.

$$C_6H_5N_2Cl \rightarrow C_6H_5Cl + N_2(g)$$

The reaction can be followed by measuring the volume of N$_2$(g) evolved as a function of time. The following data were obtained for the decomposition of an 0.071 M solution at 50°C. ($t = \infty$ corresponds to the completed reaction.) To convert from volume of N$_2$(g) to [C$_6$H$_5$N$_2$Cl], note that after 3 min 10.8 cm^3 of a total 58.3 cm^3 N$_2$(g) is evolved, corresponding to this fraction of the total reaction: 10.8/58.3 = 0.185. The fraction of C$_6$H$_5$N$_2$Cl remaining is $1.00 - 0.185 = 0.815$. [C$_6$H$_5$N$_2$Cl] at 3 min = 0.815×0.071 $M = 0.058$ M.

Time, min	N$_2$(g), cm^3	Time, min	N$_2$(g), cm^3
0	0	18	41.3
3	10.8	21	44.3
6	19.3	24	46.5
9	26.3	27	48.4
12	32.4	30	50.4
15	37.3	∞	58.3

   (a) Determine [C$_6$H$_5$N$_2$Cl] remaining after 21 min.
   (b) Construct a table similar to Table 14-2, with an interval of time, $\Delta t = 3$ min. That is, determine [C$_6$H$_5$N$_2$Cl] at 3, 6, 9, . . . min; Δ[C$_6$H$_5$N$_2$Cl] over every 3-min interval; and Δ[C$_6$H$_5$N$_2$Cl]/Δt for each 3-min interval.
   (c) Plot a graph similar to Figure 14-1 showing both the formation of N$_2$(g) and the disappearance of C$_6$H$_5$N$_2$Cl as a function of time.
   (d) What is the initial rate of the reaction?
   (e) From the graph of part (c), estimate the rate of the reaction through the slope of the tangent to the curve at $t = 21$ min. Compare with the reported value of 1.1×10^{-3} mol C$_6$H$_5$N$_2$Cl L^{-1} min^{-1}.
   (f) Write a rate-law expression for the first-order decomposition of C$_6$H$_5$N$_2$Cl and use the method of Example 14-8 to estimate a value of k based on the rate determined in parts (d) and (e).
   (g) Determine $t_{1/2}$ for this reaction by calculation (equation 14.15) and by estimation from the graph of the rate data.
   (h) At what time should the decomposition of the sample be three fourths completed?
   (i) Plot log [C$_6$H$_5$N$_2$Cl] vs. time (as in Figure 14-4) and show that the reaction is indeed first order.
   (j) Determine k from the slope of the log plot of part (i).

Establishing the order of a reaction

24. For the reaction   A $\rightarrow$ 2 B + C,   the following data are obtained for [A] as a function of time: $t = 0$ min, [A] = 0.80 M; 8 min, 0.60 M; 24 min, 0.35 M; 40 min, 0.20 M.
   (a) By suitable means establish the order of the reaction.
   (b) What is the value of the rate constant, k?
   (c) Calculate the rate of formation of B at $t = 30$ min.

25. In three different experiments the following results were obtained for the reaction   A $\rightarrow$ products: [A]$_0$ = 1.00 M, $t_{1/2}$ = 50 min; [A]$_0$ = 2.00 M, $t_{1/2}$ = 25 min; [A]$_0$ = 0.50 M, $t_{1/2}$ = 100 min. Write the rate equation for this reaction and indicate the value of k.

26. For the reaction   A $\rightarrow$ products,   the following data were obtained. $t = 0$ s,   [A] = 0.715 M;   22 s,   0.605 M;   74 s, 0.345 M; 132 s, 0.055 M. (a) What is the order of this reaction? (b) What is the half-life of the reaction?

Collision theory; activation energy

27. Explain why
(a) A reaction rate cannot be calculated from collision frequency alone.
(b) The rate of a chemical reaction may increase so dramatically with temperature while the collision frequency increases much more slowly.
(c) The addition of a catalyst to a reaction mixture can have such a pronounced effect on the rate of a reaction, even if the temperature is held constant.

28. For the reversible reaction $A + B \rightleftharpoons C + D$, the enthalpy change of the forward reaction is $+21$ kJ/mol. The activation energy of the forward reaction is 84 kJ/mol.
(a) What is the activation energy of the reverse reaction?
(b) In the manner of Figure 14-10 sketch a graph of the energy profile of this reaction.

29. By an appropriate sketch indicate why there is some relationship between the enthalpy change and the activation energy for an *endothermic* reaction but not for an exothermic reaction.

30. If even a tiny spark is introduced into a mixture of $H_2(g)$ and $O_2(g)$, a highly exothermic explosive reaction occurs. Without the spark the mixture remains unreacted indefinitely.
(a) Explain this difference in behavior.
(b) Why is the nature of the reaction independent of the size of the spark?

31. Under certain conditions the rates of gas-phase reactions are found to be affected by the presence of inert gases. How do you account for this?

Effect of temperature on reaction rate

32. Experiment 3 of Table 14-3 is repeated at several temperatures and the following rate constants are established: 3°C, $k = 1.4 \times 10^{-3}$ L mol^{-1} s^{-1}; 13°C, 2.9×10^{-3}; 24°C, 6.2×10^{-3}; 33°C, 1.20×10^{-2}.
(a) Construct a graph of log k vs. $1/T$.
(b) What is the activation energy, E_a, of the reaction?
(c) Calculate a value of the rate constant, k, at 40°C.
(d) What would be the *initial rate* of the reaction for Experiment 3 at 50°C?

33. The reaction $2 NO_2(g) \rightarrow 2 NO(g) + O_2(g)$ has a rate constant $k = 1.0 \times 10^{-10}$ at 300 K and an activation energy of 111 kJ/mol. At what temperature will this reaction have a rate constant $k = 1.0 \times 10^{-5}$? (Assume no change in the reaction mechanism.)

34. The text states that as a rule of thumb reaction rates double for a temperature increase of about 10°C.
(a) What must be the approximate activation energy for this statement to be true for reactions at about room temperature?
(b) Would you expect this rule of thumb to apply to the formation of HI from its elements at room temperature? (Recall Figure 14-10.) Explain.

35. Concerning the "rule of thumb" stated in Exercise 34, estimate how much faster cooking will occur in a pressure cooker in which the vapor pressure of water is 2.00 atm than in water under normal boiling conditions. (*Hint:* Refer to Table 11-1.)

36. With reference to the data in Table 14-6 and Figure 14-11, to what temperature must a solution in which $[N_2O_5] = 0.15\ M$ be heated to have an *initial rate* of decomposition equal to that of a $1.25\ M$ solution at 0°C?

Catalysis

37. The following statements are sometimes encountered with reference to catalysis, but they are not stated as carefully as they might be. What slight modifications should be made in each of them?
(a) A catalyst is a substance that speeds up a chemical reaction but does not take part in the reaction.
(b) The function of a catalyst is to lower the activation energy for a chemical reaction.

38. What is the principal difference between the catalytic activity of platinum metal and an enzyme?

39. In a particular enzyme reaction the following data are obtained on the rate of disappearance of substrate S: $t = 0$ min, $[S] = 1.00\ M$; 20 min, $0.90\ M$; 60 min, $0.70\ M$; 100 min, $0.50\ M$; 160 min, $0.20\ M$. What is the order of this reaction with respect to S in the concentration range studied?

Reaction mechanisms

40. We have used the terms order of a reaction and molecularity of an elementary process (i.e., unimolecular, bimolecular, termolecular). What is the relationship, if any, between these two terms?

41. The rate-determining step in a reaction mechanism is sometimes referred to as a bottleneck. Comment on the appropriateness of this analogy.

42. The collision theory states that chemical reactions occur as a result of collisions between molecules. A unimolecular elementary process in a reaction mechanism involves dissociation of a *single* molecule. How can these two ideas be compatible? Explain.

43. For this reversible elementary process, determine the steady-state concentration of N_2O_2, that is, $[N_2O_2]$.

$$2\ NO \underset{k_2}{\overset{k_1}{\rightleftharpoons}} N_2O_2$$

44. For the reaction $2 NO(g) + O_2(g) \rightarrow 2 NO_2(g)$, the rate law, expressed as the rate of formation of NO_2, is found to be rate $= k[NO]^2[O_2]$.
(a) Is the rate law consistent with the one-step mechanism $2 NO + O_2 \rightarrow 2 NO_2$?
(b) Why is this one-step mechanism not very plausible?

45. Show that the rate law in Exercise 44 is consistent with the following mechanism.

(fast) $2\ NO \underset{k_2}{\overset{k_1}{\rightleftharpoons}} N_2O_2$

(slow) $N_2O_2 + O_2 \xrightarrow{k_3} 2\ NO_2$

*46. Propose a two-step mechanism for the reaction $2\ O_3(g) \rightarrow 3\ O_2(g)$ that is consistent with the observed rate law.

$$\text{rate} = k\frac{[O_3]^2}{[O_2]}$$

Additional Exercises

1. The first-order reaction $A \rightarrow$ products has $t_{1/2} = 150$ s.
(a) What percent of a sample of A remains *unreacted* 600 s after a reaction has been started?
(b) What is the reaction rate when $[A] = 0.50\ M$?

2. What is the initial rate of disappearance of I^- in reaction (14.8) if the initial concentrations are $[S_2O_8{}^{2-}] = 0.15\ M$ and $[I^-] = 0.010\ M$? Use data from Examples 14-6 and 14-7.

3. No specific equation is given in the text for the half-life of a zero-order reaction. Use Figure 14-3 to
(a) Show that the half-life depends on the initial concentration $[A]_0$.
(b) Derive an equation for the half-life in terms of $[A]_0$ and k.

4. The following data were obtained for the decomposition of N_2O_5 in $CCl_4(l)$ at 45°C: $t = 0$ s, $[N_2O_5] = 1.46\ M$; 423 s, $1.09\ M$; 753 s, $0.89\ M$; 1116 s, $0.72\ M$; 1582 s, $0.54\ M$; 1986 s, 0.43 M; 2343 s, 0.35 M. Determine a value of k for the reaction $N_2O_5 \rightarrow N_2O_4 + \frac{1}{2} O_2(g)$.

5. The half-life of the radioactive isotope $^{32}_{15}P$ is 14.3 days. How long would it take for a sample of $^{32}_{15}P$ to lose 99% of its radioactivity?

6. The half-life for the first-order decomposition of nitramide, $NH_2NO_2(aq) \rightarrow N_2O(g) + H_2O(l)$, is 123 min at 15°C. If 165 mL of a 0.105 M NH_2NO_2 solution is allowed to decompose, how long must the reaction proceed to produce 50.0 cm^3 of "wet" $N_2O(g)$ measured at 15°C and a barometric pressure of 756 mmHg? (The vapor pressure of water at 15°C is 12.8 mmHg.)

7. The rate-determining step in the hydrogen–iodine reaction is thought to be $2\ I(g) + H_2(g) \rightarrow 2\ HI(g)$. The rate constant for this step has been determined at 520 and 710 K, yielding values of 3.96×10^{-5} and 1.61×10^{-4} $L^2\ mol^{-2}\ s^{-1}$, respectively. What is the activation energy, E_a, for this elementary process?

8. A statement is made in the text that in enzyme-catalyzed reactions the reaction is first order at low substrate concentrations and becomes zero order at high concentrations. Certain gas-phase reactions on a heterogeneous catalyst are found to be first order at low gas pressures and zero order at high pressures. What is the connection between these two situations?

9. From the result of Example 14-4a, determine the initial rate of evolution of $O_2(g)$, in cm^3(STP)/s, from 0.100 L of 2.32 M $H_2O_2(aq)$.

*10. The decomposition of ethylene oxide at 690 K is followed by measuring the *total* gas pressure as a function of time. The data obtained are $t = 10$ min, $P_{tot.} = 139.14$ mmHg; 20 min, 151.67 mmHg; 40 min, 172.65 mmHg; 60 min, 189.15 mmHg; 100 min, 212.34 mmHg; 200 min, 238.66 mmHg; ∞, 249.88 mmHg. For the reaction $(CH_2)_2O(g) \rightarrow CH_4(g) + CO(g)$,
(a) What must be the initial total pressure (i.e., the pressure of pure ethylene oxide at $t = 0$)?
(b) What is the order of the reaction?

*11. The following data were obtained for the reaction $2\ A + B \rightarrow$ products. Use appropriate methods introduced in this chapter to establish the order of this reaction with respect to A and to B.

Experiment 1, [B] = 1.00 M		Experiment 2, [B] = 0.50 M	
Time, min	[A], M	Time, min	[A], M
0	1.000×10^{-3}	0	1.000×10^{-3}
1	0.951×10^{-3}	1	0.975×10^{-3}
5	0.779×10^{-3}	5	0.883×10^{-3}
10	0.607×10^{-3}	10	0.779×10^{-3}
20	0.368×10^{-3}	20	0.607×10^{-3}

*12. These data were obtained for the decomposition of acetaldehyde at 518°C, $CH_3CHO(g) \rightarrow CH_4(g) + CO(g)$: $t = 0$ s, $P_{tot.} = 363$ mmHg; 42 s, 397 mmHg; 73 s, 417 mmHg; 105 s, 437 mmHg; 190 s, 477 mmHg; 310 s, 517 mmHg; 480 s, 557 mmHg; 665 s, 587 mmHg.
(a) By an appropriate graphical method, establish the order of the reaction.
(b) Obtain a value of the rate constant k.

*13. The peroxodisulfate-iodide reaction (14.8) can be followed by the series of reactions

$$S_2O_8^{2-}(aq) + 3\,I^-(aq) \rightarrow 2\,SO_4^{2-}(aq) + I_3^-(aq)$$

$$2\,S_2O_3^{2-}(aq) + I_3^-(aq) \rightarrow S_4O_6^{2-}(aq) + 3\,I^-(aq)$$

$$I_3^-(aq) + \text{starch}(aq) \rightarrow \text{blue complex}$$

Solutions of $S_2O_8^{2-}$ and I^- are mixed in the presence of a fixed amount of $S_2O_3^{2-}$ and starch indicator. The I_3^- produced in the main reaction reacts very rapidly with $S_2O_3^{2-}$ until all the $S_2O_3^{2-}$ is consumed. Then the I_3^- combines with the starch to produce a deep blue color. The rate of reaction can be related to the appearance of this blue color. In an experiment, 25.0 mL of 0.20 M $(NH_4)_2S_2O_8$, 25.0 mL of 0.20 M KI, 10.0 mL of 0.010 M $Na_2S_2O_3$, and 5.0 mL of starch solution are mixed. How long after mixing will the blue color appear? (*Hint:* Use the rate constant from Example 14-6.)

*14. Show that the following mechanism is consistent with the rate law established for the iodide–hypochlorite reaction in Exercise 16.

(fast) $OCl^- + H_2O \underset{k_2}{\overset{k_1}{\rightleftharpoons}} HOCl + OH^-$

(slow) $I^- + HOCl \xrightarrow{k_3} HOI + Cl^-$

(fast) $HOI + OH^- \underset{k_5}{\overset{k_4}{\rightleftharpoons}} H_2O + OI^-$

Self-Test Questions

For questions 1 through 8 select the single item that best completes each statement.

1. For the reaction $A \rightarrow$ products, a plot of [A] vs. time is found to be a straight line. The order of this reaction is (a) zero; (b) first; (c) second; (d) impossible to determine from this graph.

2. A *first-order* reaction $A \rightarrow$ products has a half-life of 100 s. Whatever the quantity of substance A involved in a particular reaction, (a) the reaction goes to completion in 200 s; (b) the quantity of A remaining after 200 s is half of what remains after 100 s; (c) the same quantity of A is consumed for every 100 s of the reaction; (d) 100 s elapses before the reaction begins.

3. The rate equation for the reaction $2\,A + B \rightarrow C$ is found to be: rate $= k[A][B]$. (a) The unit of k must be s^{-1}; (b) $t_{1/2}$ is a constant; (c) the value of k depends on the initial concentrations of A and B; (d) the rate of formation of C is one half the rate of disappearance of A.

4. The decomposition of substance A is *second order:* $A \rightarrow$ products. The initial rate of decomposition when $[A]_0 = 0.50\ M$ is (a) the same as the initial rate for any other value of $[A]_0$; (b) half as great as when $[A]_0 = 1.00\ M$; (c) five times as great as when $[A]_0 = 0.10\ M$; (d) four times as great as when $[A]_0 = 0.25\ M$.

5. The rate of a chemical reaction generally increases rapidly, even for small temperature increases, because of a rapid increase with temperature in (a) the collision frequency; (b) the fraction of molecules with energies in excess of the activation energy; (c) the activation energy; (d) the average kinetic energy of gas molecules.

6. The reaction $A + B \rightarrow C + D$ has $\Delta\overline{H} = +25$ kJ/mol, and an activation energy, $E_a = $ (a) -25 kJ/mol; (b) less than $+25$ kJ/mol; (c) more than $+25$ kJ/mol; (d) either less than $+25$ kJ/mol or more than $+25$ kJ/mol, which can only be determined by experiment.

7. A catalyst speeds up a chemical reaction by increasing (a) the average kinetic energy of molecules; (b) the frequency of molecular collisions; (c) the activation energy of the reaction; (d) the proportion of molecules with energies in excess of the activation energy.

8. For the net reaction $A + B \rightarrow 2\,C$, which proceeds by a *single-step bimolecular mechanism,* the following equation is applicable: (a) $t_{1/2} = 0.693/k$; (b) rate of reaction $= k[A][B]$; (c) rate of appearance of $C = $ rate disappearance A; (d) $\log [A] = (-k/2.303)t + \log [A]_0$.

9. A kinetic study of the reaction $A \rightarrow$ products yields the data $t = 0$ s, $[A] = 2.00\ M$; 500 s, 1.00 M; 1500 s, 0.50 M; 3500 s, 0.25 M. In the simplest way possible determine the order of this reaction.

10. In the *first-order* decomposition of substance A the following concentrations are found to exist at the indicated times following the start of the reaction: $t = 0$ s, $[A] = 0.88\ M$; 50 s, 0.62 M; 100 s, 0.44 M; 150 s, 0.31 M. Calculate the *instantaneous* rate of reaction at 100 s.

11. The reaction $A \rightarrow$ products is first order in A.
  (a) If 1.60 g A is allowed to decompose for 20 min, the mass of A remaining undecomposed is found to be 0.40 g. What is the half-life, $t_{1/2}$, of this reaction?
  (b) Starting with 1.60 g A, what is the mass of A remaining undecomposed after 33.2 min?

12. The decomposition of acetaldehyde, CH_3CHO, can be catalyzed by $I_2(g)$.
uncatalyzed reaction: $E_a = 190$ kJ/mol

$$CH_3CHO(g) \rightarrow CH_4(g) + CO(g)$$

catalyzed reaction: $E_a = 136$ kJ/mol

$$CH_3CHO(g) + I_2(g) \rightarrow CH_3I(g) + HI(g) + CO(g)$$
$$CH_3I(g) + HI(g) \rightarrow CH_4(g) + I_2(g)$$

The following enthalpies of formation are also given: $\Delta\overline{H}_f^\circ[CH_3CHO(g)] = -166$ kJ/mol; $\Delta\overline{H}_f^\circ[CH_4(g)] = -74.9$ kJ/mol; $\Delta\overline{H}_f^\circ[CO(g)] = -110.5$ kJ/mol. Sketch the reaction profiles for the catalyzed and uncatalyzed reactions, representing $\Delta\overline{H}$ and E_a.

15 Principles of Chemical Equilibrium

Dynamic equilibrium is a condition in which two opposing processes occur at equal rates. As a result, no further *net* change occurs in a system at equilibrium. We have already encountered several equilibrium situations. Let us review two of them briefly.

1. When a liquid vaporizes into a closed container, there comes a time when molecules return to the liquid state at the same rate at which they leave it. That is, vapor *condenses* at the same rate at which liquid *vaporizes*. Even though molecules continue to pass back and forth between the liquid and vapor, at equilibrium the pressure exerted by the vapor remains constant with time.
2. When a solute dissolves in a solvent, a point is reached where the rate at which additional solute particles dissolve is just matched by the rate at which dissolved solute precipitates. The solution becomes saturated and its concentration remains constant with time.

One of the characteristics of a system at equilibrium, then, is that certain properties acquire values that remain unchanged with time. In this chapter we turn our attention to dynamic equilibrium in chemical reactions and focus on the property known as the equilibrium constant. We will learn how to use this constant in qualitative and quantitative discussions of equilibrium. In the next chapter we will study the theoretical basis of the equilibrium constant.

15-1 The Condition of Chemical Equilibrium

The situation pictured by the three graphs in Figure 15-1 is unlike anything encountered in our discussion of stoichiometry in Chapter 4. In experiment 1, 0.00150 mol each of H_2 and I_2 are allowed to react. In the usual fashion we can write

forward reaction: $H_2(g) + I_2(g) \longrightarrow 2 HI(g)$ (15.1)

However, as soon as some HI has formed it begins to dissociate back to H_2 and I_2.

reverse reaction: $2 HI(g) \longrightarrow H_2(g) + I_2(g)$ (15.2)

Thus, two reactions occur simultaneously—a forward and a reverse reaction. When it reaches 0.00234 mol, the amount of HI ceases to increase. The amounts of HI, H_2, and I_2 present all remain constant with time. The two opposing reactions continue to occur but

FIGURE 15-1
Three approaches to
equilibrium in the
reaction

$$H_2(g) + I_2(g) \rightleftharpoons 2\ HI(g)$$

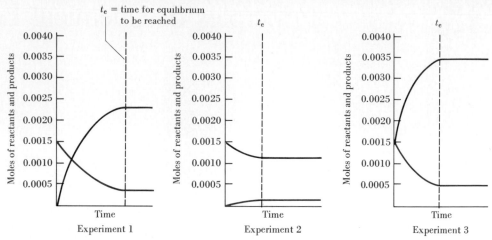

The data plotted here are from Table 15-1.

——— mol H_2 = mol I_2
——— mol HI

now at equal rates. A condition of dynamic equilibrium is achieved. The forward and reverse reactions can be written together by using a double arrow ($\rightleftharpoons$).

$$H_2(g) + I_2(g) \rightleftharpoons 2\ HI(g) \tag{15.3}$$

Experiment 2 represents a different approach to equilibrium in the same reaction—starting with pure HI and forming H_2 and I_2. Once more a time is reached following which there is no further net change, because the rate of re-formation of HI from H_2 and I_2 has become equal to the rate of dissociation of HI. Experiment 3 represents a situation in which all three reactants are present initially.

Two points should be noted in Figure 15-1: (1) In no case is any reacting species completely consumed, and (2) based only on the amounts of reactants and products at equilibrium, there is no apparent common feature in the three situations.

The Hydrogen–Iodine–Hydrogen Iodide Equilibrium.   Let us use data from Table 15-1 to explore the equilibrium condition in reaction (15.3) a little more fully. In particular, let us seek the constant property of a chemical reaction at equilibrium referred to in the introduction to this chapter.

In Table 15-2 three different attempts are summarized for the three experiments of Table 15-1. One of these expressions does, in fact, give almost identical numerical values in all three cases.

$$K_c = \frac{[HI]^2}{[H_2][I_2]} = 50.2 \quad \text{(at } 445°C) \tag{15.4}$$

The symbol K_c denotes an expression based on molar concentrations at equilibrium.

Three experiments are not enough to establish the constant value of expression (15.4), but repeated experimentation at 445°C would yield the same value. The significance of expression (15.4) is that whenever equilibrium is established among $H_2(g)$, $I_2(g)$, and HI(g) at 445°C, the particular ratio of molar concentrations must have a value of 50.2. We need to say more about the quantity K_c, called an **equilibrium constant.** Before we do, let us illustrate its meaning in the following example.

TABLE 15-1

Three approaches to equilibrium in the reaction $H_2(g) + I_2(g) \rightleftharpoons 2\ HI(g)^{a,b}$

Exper-iment	Initial amount,[c] mol $\times 10^3$			Equilibrium amount,[c] mol $\times 10^3$			Equilibrium concentration,[c] $M \times 10^3$		
	H_2	I_2	HI	H_2	I_2	HI	$[H_2]$	$[I_2]$	$[HI]$
1	1.50	1.50	—	0.330	0.330	2.34	0.412	0.412	2.92
2	—	—	1.50	0.165	0.165	1.17	0.206	0.206	1.46
3	1.50	1.50	1.50	0.495	0.495	3.51	0.619	0.619	4.39

[a]Temperature = 445°C; volume of reaction mixture = 0.8000 L.

[b]Known initial amounts of H_2, I_2, and/or HI are sealed into glass vessels and maintained at a constant temperature until equilibrium is established. Then the vessels are quickly chilled and the contents transferred to aqueous solutions at room temperature. These solutions are titrated with $Na_2S_2O_3(aq)$ using starch indicator. The titration data are used to establish the equilibrium amount of I_2. Other equilibrium amounts can be related to this quantity (see Exercise 10).

[c]Figures in these columns have been multiplied by 10^3. In experiment 1, for example, initial amount of $H_2 = 1.50 \times 10^{-3}$ mol; equilibrium amount of $H_2 = 0.330 \times 10^{-3}$ mol; equilibrium concentration of $H_2 = 0.330 \times 10^{-3}$ mol $H_2/0.8000$ L $= 0.412 \times 10^{-3}\ M$.

TABLE 15-2

In search of a constant ratio of concentrations to describe equilibrium in the reaction $H_2(g) + I_2(g) \rightleftharpoons 2\ HI(g)^{a,b}$

Exper-iment	*Try:* $\dfrac{[HI]}{[H_2][I_2]}$	*Try:* $\dfrac{2 \times [HI]}{[H_2][I_2]}$	*Try:* $\dfrac{[HI]^2}{[H_2][I_2]}$
1	$\dfrac{2.92 \times 10^{-3}}{(0.412 \times 10^{-3})^2} = 1.72 \times 10^4$	$\dfrac{2 \times 2.92 \times 10^{-3}}{(0.412 \times 10^{-3})^2} = 3.44 \times 10^4$	$\dfrac{(2.92 \times 10^{-3})^2}{(0.412 \times 10^{-3})^2} - 50.2$
2	$\dfrac{1.46 \times 10^{-3}}{(0.206 \times 10^{-3})^2} = 3.44 \times 10^4$	$\dfrac{2 \times 1.46 \times 10^{-3}}{(0.206 \times 10^{-3})^2} = 6.88 \times 10^4$	$\dfrac{(1.46 \times 10^{-3})^2}{(0.206 \times 10^{-3})^2} = 50.2$
3	$\dfrac{4.39 \times 10^{-3}}{(0.619 \times 10^{-3})^2} = 1.15 \times 10^4$	$\dfrac{2 \times 4.39 \times 10^{-3}}{(0.619 \times 10^{-3})^2} = 2.29 \times 10^4$	$\dfrac{(4.39 \times 10^{-3})^2}{(0.619 \times 10^{-3})^2} = 50.3$

[a]Equilibrium concentrations are from Table 15-1 at 445°C.

[b]Since in each experiment $[H_2] = [I_2]$ at equilibrium, the denominator in each expression has been written as a squared term, e.g., $(0.412 \times 10^{-3})(0.412 \times 10^{-3}) = (0.412 \times 10^{-3})^2$.

Example 15-1   If the equilibrium concentrations of H_2 and I_2 in reaction (15.3) at 445°C are found to be $[H_2] = 4.84 \times 10^{-5}\ M$ and $[I_2] = 1.68 \times 10^{-3}\ M$, what must be the equilibrium concentration of HI?

Solution.   The three equilibrium concentrations are related through expression (15.4). We need to substitute the known molar concentrations into (15.4) and solve for [HI].

$$[HI]^2 = K_c[H_2][I_2] = 50.2 \times 4.84 \times 10^{-5} \times 1.68 \times 10^{-3} = 4.08 \times 10^{-6}$$

$$[HI] = \sqrt{[HI]^2} = \sqrt{4.08 \times 10^{-6}} = 2.02 \times 10^{-3}\ M$$

SIMILAR EXAMPLES: Review Problems 5, 8; Exercises 12, 13.

The Equilibrium Constant, K_c.   From the specific example of the hydrogen–iodine–hydrogen iodide reaction, we now turn to the general case of a reversible reaction at a condition of equilibrium.

For the generalized reaction

$$a \text{ A} + b \text{ B} + \cdots \rightleftharpoons g \text{ G} + h \text{ H} + \cdots \tag{15.5}$$

The equilibrium constant expression has the form

$$\frac{[G]^g [H]^h \cdots}{[A]^a [B]^b \cdots} = K_c \tag{15.6}$$

In Section 16-7 we will introduce a quantity known as the thermodynamic equilibrium constant, a dimensionless number. In anticipation of this later development, we will not attach units to the equilibrium constant values in this chapter.

The numerator is the product of the concentrations of the species written on the right side of the equation ([G], [H], . . .), each concentration being raised to a power given by the coefficient in the balanced equation (g, h, . . .). The denominator is the product of the concentrations of the species written on the left side of the equation ([A], [B], . . .), again with each concentration raised to a power given by the coefficient in the balanced equation (a, b, . . .).

The numerical value of the **equilibrium constant, K_c,** depends uniquely on the particular reaction and on the temperature.

15-2 Additional Relationships Involving Equilibrium Constants

Relationship of K_c to the Balanced Chemical Equation. The reversible reaction involving $SO_2(g)$, $O_2(g)$, and $SO_3(g)$ is described in three different ways below.

$$2 \text{ SO}_2(g) + \text{O}_2(g) \rightleftharpoons 2 \text{ SO}_3(g) \qquad K_c(a) = 2.8 \times 10^2 \text{ at } 1000 \text{ K} \tag{15.7a}$$

$$2 \text{ SO}_3(g) \rightleftharpoons 2 \text{ SO}_2(g) + \text{O}_2(g) \qquad K_c(b) = ? \tag{15.7b}$$

$$\text{SO}_2(g) + \tfrac{1}{2} \text{O}_2(g) \rightleftharpoons \text{SO}_3(g) \qquad K_c(c) = ? \tag{15.7c}$$

The expressions obtained for the equilibrium constants are

$$K_c(a) = \frac{[SO_3]^2}{[SO_2]^2[O_2]} = 2.8 \times 10^2 \text{ at } 1000 \text{ K} \tag{15.8a}$$

$$K_c(b) = \frac{[SO_2]^2[O_2]}{[SO_3]^2} = ? \tag{15.8b}$$

$$K_c(c) = \frac{[SO_3]}{[SO_2][O_2]^{1/2}} = ? \tag{15.8c}$$

For a given set of initial conditions, the equilibrium concentrations of SO_2, O_2, and SO_3 acquire a unique set of values. This is true regardless of which of the three expressions in (15.8) we choose to represent equilibrium. The K_c values for the expressions in equations (15.8a–c) must be related in some way. Since

$$\frac{[SO_2]^2[O_2]}{[SO_3]^2} = \frac{1}{[SO_3]^2/([SO_2]^2[O_2])}$$

then

$$K_c(b) = \frac{1}{K_c(a)} = \frac{1}{2.8 \times 10^2} = 3.6 \times 10^{-3}$$

and since

$$\frac{[SO_3]}{[SO_2][O_2]^{1/2}} = \left\{ \frac{[SO_3]^2}{[SO_2]^2[O_2]} \right\}^{1/2}$$

then

$$K_c(c) = \{K_c(a)\}^{1/2} = \sqrt{2.8 \times 10^2} = 1.7 \times 10^1$$

To summarize, we note that

1. Whatever expression is used for K_c, it must be matched to the corresponding balanced chemical equation.
2. If an equation is reversed, the value of K_c is inverted; that is, the new equilibrium constant is the reciprocal of the old one.
3. If the coefficients in a balanced equation are multiplied by a common factor (2, 3, . . .), the new equilibrium constant will be the old one raised to the corresponding power (2, 3, . . .).
4. If the coefficients in a balanced equation are divided by a common factor (2, 3, . . .), the new equilibrium constant will be the corresponding root of the old one (square root, cube root, . . .).

Example 15-2   For the reaction $NH_3 \rightleftharpoons \frac{1}{2} N_2 + \frac{3}{2} H_2$, $K_c = 5.2 \times 10^{-5}$ at 298 K. What is the value of K_c at 298 K for the reaction $N_2 + 3 H_2 \rightleftharpoons 2 NH_3$?

Solution.   To obtain the desired equation, the original equation must be (1) reversed and (2) doubled. Thus,

(1) $\frac{1}{2} N_2 + \frac{3}{2} H_2 \rightleftharpoons NH_3$ $\qquad K_c(1) = \dfrac{[NH_3]}{[N_2]^{1/2}[H_2]^{3/2}} = \dfrac{1}{5.2 \times 10^{-5}} = 1.9 \times 10^4$

(2) $N_2 + 3 H_2 \rightleftharpoons 2 NH_3$ $\qquad K_c(2) = ?$

$$K_c(2) = \frac{[NH_3]^2}{[N_2][H_2]^3} = \left\{ \frac{[NH_3]}{[N_2]^{1/2}[H_2]^{3/2}} \right\}^2 = \{K_c(1)\}^2 = (1.9 \times 10^4)^2 = 3.6 \times 10^8$$

SIMILAR EXAMPLES: Review Problem 2; Exercise 5.

Combining Equilibrium Constant Expressions.   Suppose that we are given the following equilibrium constant data at 25°C.

$$N_2(g) + O_2(g) \rightleftharpoons 2 NO(g) \qquad K_c = 4.1 \times 10^{-31} \tag{15.9}$$

$$N_2(g) + \tfrac{1}{2} O_2(g) \rightleftharpoons N_2O(g) \qquad K_c = 2.4 \times 10^{-18} \tag{15.10}$$

and that we wish to establish K_c for the reaction

$$N_2O(g) + \tfrac{1}{2} O_2(g) \rightleftharpoons 2 NO(g) \qquad K_c = ? \tag{15.11}$$

We may obtain (15.11) by this combination of equations (15.9) and (15.10).

(1) $\qquad N_2(g) + O_2(g) \rightleftharpoons 2 NO(g) \qquad\qquad K_c(1) = 4.1 \times 10^{-31}$
(2) $\qquad\qquad N_2O(g) \rightleftharpoons N_2(g) + \tfrac{1}{2} O_2(g) \qquad K_c(2) = 1/(2.4 \times 10^{-18})$
$\qquad\qquad\qquad\qquad\qquad\qquad\qquad\qquad\qquad\qquad = 4.2 \times 10^{17}$

net: $N_2O(g) + \tfrac{1}{2} O_2(g) \rightleftharpoons 2 NO(g) \qquad\qquad K_c = ?$

The product, $K_c(1) \times K_2(2)$, when simplified, yields K_c for the net reaction.

$$\underbrace{\frac{[NO]^2}{[N_2][O_2]}}_{K_c(1)} \times \underbrace{\frac{[N_2][O_2]^{1/2}}{[N_2O]}}_{K_c(2)} = \frac{[NO]^2}{[N_2O][O_2]^{1/2}} = K_c(net)$$

$$K_c(net) = K_c(1) \times K_c(2) = 4.1 \times 10^{-31} \times 4.2 \times 10^{17} = 1.7 \times 10^{-13} \tag{15.12}$$

The important generalization established by (15.12) is that

The equilibrium constant for a net reaction is the product of the equilibrium constants for the individual reactions being added.

The Equilibrium Constant Expressed as K_p. Equilibrium constants for gaseous systems can be based on the partial pressures of gases rather than on molar concentrations. An equilibrium constant written in this way is called a **partial pressure equilibrium constant** and is denoted by the symbol K_p. To illustrate the relationship between K_p and K_c for a reaction, let us consider again reaction (15.7a).

$$2\ SO_2(g) + O_2(g) \rightleftharpoons 2\ SO_3(g) \qquad K_c = 2.8 \times 10^2 \text{ at } 1000\ K \tag{15.7a}$$

$$K_c = \frac{[SO_3]^2}{[SO_2]^2[O_2]}$$

Also, according to the ideal gas law, $PV = nRT$ and

$$[SO_3] = \frac{n_{SO_3}}{V} = \left(\frac{P_{SO_3}}{RT}\right) \qquad [SO_2] = \frac{n_{SO_2}}{V} = \left(\frac{P_{SO_2}}{RT}\right) \qquad [O_2] = \frac{n_{O_2}}{V} = \left(\frac{P_{O_2}}{RT}\right)$$

Substituting the circled terms for concentrations in K_c, we obtain the expression

$$K_c = \frac{(P_{SO_3}/RT)^2}{(P_{SO_2}/RT)^2(P_{O_2}/RT)} = \frac{(P_{SO_3})^2}{(P_{SO_2})^2(P_{O_2})} \times RT \tag{15.13}$$

The ratio of partial pressures shown in color in (15.13) is the equilibrium constant, K_p. The relationship between K_p and K_c for reaction (15.7a) is

$$K_c = K_p \times RT \qquad \text{and} \qquad K_p = \frac{K_c}{RT} = K_c(RT)^{-1} \tag{15.14}$$

If a similar derivation were carried out for the general reaction

$$a\ A(g) + b\ B(g) + \cdots \rightleftharpoons g\ G(g) + h\ H(g) + \cdots \tag{15.5}$$

the result would be

$$K_p = K_c(RT)^{\Delta n} \tag{15.15}$$

where Δn is the difference in the stoichiometric coefficients of *gaseous* products and reactants; that is, $\Delta n = (g + h + \cdots) - (a + b + \cdots)$. In reaction (15.7a), $\Delta n = 2 - (2 + 1) = -1$, just as noted in equation (15.14).

Although we have not attached units to equilibrium constants (see again, marginal note on page 446), to use equation (15.15) requires that specific units be chosen for the terms in K_c and K_p expressions. As we have already seen, K_c is based on molar concentrations. In this text we will choose the unit *atm* for partial pressures. In equation (15.15), then, a value of $R = 0.0821$ L atm mol^{-1} K^{-1} is consistent with the units chosen for the terms in K_c and K_p expressions. Occasionally, in handbooks or elsewhere, K_p values are listed for units other than atm (e.g., mmHg).

Example 15-3 Complete the calculation of K_p for reaction (15.7a) from the data given.

Solution. If concentrations in K_c are expressed on a molar basis and partial pressures in K_p as atm, the solution to equation (15.14) becomes

$$K_p = K_c(RT)^{-1} = 2.8 \times 10^2(0.0821 \times 1000)^{-1} = \frac{2.8 \times 10^2}{0.0821 \times 1000} = 3.4$$

SIMILAR EXAMPLES: Review Problem 4; Exercise 3.

Example 15-4   What is the value of K_p for the hydrogen–iodine–hydrogen iodide reaction at 445°C?

Solution.   For the reaction $H_2(g) + I_2(g) \rightleftharpoons 2\ HI(g)$, $\Delta n = 0$. This means that in the expression $K_p = K_c(RT)^{\Delta n}$, $K_p = K_c$ (since any number raised to the "0" power has a value of 1). Referring to expression (15.4), we conclude that

$$K_p = K_c = 50.2$$

SIMILAR EXAMPLES: Review Problem 4; Exercise 3.

Equilibria Involving Pure Liquids and Solids (Heterogeneous Reactions).   An equilibrium constant expression contains terms only for substances whose concentrations or partial pressures can take on different values during the course of a chemical reaction. Since their compositions are not variable, even while participating in a chemical reaction, *pure* solids and *pure* liquids are not represented in equilibrium constant expressions.* Reactions (15.3) and (15.7a) are *homogeneous* reactions (occurring within a single phase) and their equilibrium constant expressions (15.4) and (15.8a) contain a term for each reactant. For the following *heterogeneous* reaction

$$C(s) + H_2O(g) \rightleftharpoons CO(g) + H_2(g)$$

the equilibrium constant expression contains terms only for the species present in the homogeneous gaseous phase—H_2O, CO, and H_2.

$$K_c = \frac{[CO][H_2]}{[H_2O]}$$

Another example of a heterogeneous reaction is the decomposition of calcium carbonate.

$$CaCO_3(s) \rightleftharpoons CaO(s) + CO_2(g) \tag{15.16}$$

for which the equilibrium constant expression contains but a single term.

$$K_c = [CO_2(g)] \tag{15.17}$$

K_p can be written in a similar fashion, and the relationship between K_p and K_c is that derived from equation (15.15), with $\Delta n = 1$.

$$K_p = P_{CO_2} \qquad K_p = K_c(RT) \tag{15.18}$$

According to (15.18), the equilibrium pressure of $CO_2(g)$ in contact with $CaO(s)$ and $CaCO_3(s)$ is in itself a value of the equilibrium constant, K_p. And, as expected, the pressure of the CO_2 does not depend on the quantities of $CaO(s)$ and $CaCO_3(s)$ present (but *both* solids must be present).

A liquid–vapor equilibrium is a *physical* equilibrium (no chemical reaction is involved), but the principles just presented also apply. For the vaporization equilibrium of water we can write

$$H_2O(l) \rightleftharpoons H_2O(g)$$

$$K_c = [H_2O(g)] \qquad K_p = P_{H_2O} \qquad K_p = K_c(RT) \tag{15.19}$$

Thus, equilibrium vapor pressures can be viewed as equilibrium constants, K_p, at different temperatures, and again, their values do not depend on the quantity of liquid present.

---

*An alternate statement is that in a *thermodynamic* equilibrium constant every species involved in the reaction is represented through its *activity*. However, since the activities of pure solids and liquids are exactly 1.000, these terms have no effect on equilibrium constant expressions. (This matter is discussed further in Section 16-7.)

Example 15-5 Equilibrium is established in the following reaction at 60°C, and the gas partial pressures are found to be $P_{HI} = 3.65 \times 10^{-3}$ atm and $P_{H_2S} = 9.96 \times 10^{-1}$ atm. What is the value of K_p for the reaction?

$$H_2S(g) + I_2(s) \rightleftharpoons 2 HI(g) + S(s) \qquad K_p = ?$$

Solution. Recall that terms for pure solids do not appear in an equilibrium constant expression. The value of K_p is

$$K_p = \frac{(P_{HI})^2}{(P_{H_2S})} = \frac{(3.65 \times 10^{-3})^2}{9.96 \times 10^{-1}} = 1.34 \times 10^{-5}$$

SIMILAR EXAMPLES: Review Problem 5; Exercise 4.

15-3 Significance of the Magnitude of an Equilibrium Constant

In principle we can write an equilibrium constant expression and establish the numerical value of an equilibrium constant for every chemical reaction, but it is only in certain situations that these constants are significant. Table 15-3 lists equilibrium constant values for several reactions that we have previously encountered in the text.

The first of these reactions represents the synthesis of water from its elements; it is the reaction with which we introduced fundamental ideas about stoichiometry in Section 4-3. At that time we assumed that reaction (4.8) proceeds only in the forward direction; that the reaction continues until one of the reactants is consumed; that the reaction *goes to completion*. For these assumptions to hold, an equilibrium constant expression must have at least one term in the *denominator* that is very small (approaching zero).

A very large numerical value of K_c or K_p signifies that the forward reaction, as written, goes to completion or very nearly so.

At 298 K, the value of K_p for reaction (4.8) is 1.4×10^{83}, and we are entirely justified in assuming that the reaction goes to completion.

The second reaction in Table 15-3 deals with the synthesis of NO(g) from N_2(g) and O_2(g). Here we find that, at 298 K, the value of K_p (or K_c) is very small (5.3×10^{-31}). To account for a very small numerical value for an equilibrium constant expression, the *numerator* must be very small (approaching zero).

A very small numerical value of K_c or K_p signifies that the forward reaction, as written, does not occur to any significant extent.

At 1800 K the value of K_p for the synthesis of NO(g) is much larger than at 298 K, suggesting that at high temperatures the forward reaction does occur to some extent before equilibrium is established. It is for this reason that high-temperature combustion processes carried out in the presence of air always produce some NO(g) as an air pollutant. When

TABLE 15-3
Some equilibrium reactions

Reaction	Equilibrium constant, K_p	Text reference
$2 H_2(g) + O_2(g) \rightleftharpoons 2 H_2O(l)$	1.4×10^{83} at 298 K	reaction (4.8)
$N_2(g) + O_2(g) \rightleftharpoons 2 NO(g)$	5.3×10^{-31} at 298 K	Section 6-5
	1.3×10^{-4} at 1800 K	
$2 NO(g) + O_2(g) \rightleftharpoons 2 NO_2(g)$	1.6×10^{12} at 298 K	reaction (13.60)
$H_2(g) + I_2(g) \rightleftharpoons 2 HI(g)$	50.2 at 718 K	reaction (15.3)
$2 SO_2(g) + O_2(g) \rightleftharpoons 2 SO_3(g)$	3.4 at 1000 K	reaction (15.7a)
$C(s) + H_2O(g) \rightleftharpoons CO(g) + H_2(g)$	1.6×10^{-21} at 298 K	reaction (13.1)
	10 at 1100 K	

$NO(g)$ comes into contact with $O_2(g)$ at 298 K, it is converted to $NO_2(g)$ in a reaction (reaction 13.60) that goes nearly to completion ($K_p = 1.6 \times 10^{12}$).

For the synthesis of $HI(g)$ from its elements at 718 K (445°C), we see that K_p (or K_c) is neither very small nor very large. Both the forward and the reverse reactions occur to a significant extent, and appreciable amounts of $H_2(g)$, $I_2(g)$, and $HI(g)$ are found at equilibrium, as we have already seen through Figure 15-1 and Table 15-1. We reach a similar conclusion about the conversion of $SO_2(g)$ to $SO_3(g)$ at 1000 K and the water gas reaction (13.1) at 1100 K. At 298 K, however, we would not expect the water gas reaction to occur to any significant extent because of the very small value of K_p.

We comment further on the direction and extent of a reaction in the next section, and in Section 15-6 we will calculate actual quantities of reactants and products present when a reaction reaches equilibrium. At times, however, the simple ideas presented here can help you to make a general assessment of an equilibrium situation before proceeding with detailed calculations.

15-4 Predicting the Direction and Extent of a Reaction

At each point in the progress of a reaction it is possible to formulate a ratio of concentrations having the same form as the equilibrium constant expression. This generalized ratio is called the **reaction quotient,** often designated by the symbol Q. For the generalized reversible reaction (15.5) the reaction quotient is

$$Q = \frac{[G]^g[H]^h \cdots}{[A]^a[B]^b \cdots} \tag{15.20}$$

If the values substituted into the reaction quotient, Q, are a valid set of equilibrium concentrations, then Q will be found to equal K_c.

What is the likelihood that any chosen set of *initial* concentrations for the reactants and products in a reversible reaction will in fact be equilibrium concentrations? The likelihood is exceedingly small! Further reaction must occur in which *all* of the reactant and product concentrations change until the reaction quotient, Q, becomes equal to K_c. Depending on the relationship of Q to K_c, a net reaction proceeds either in the forward direction (to the right) or in the reverse direction (to the left).

The three experiments pertaining to reaction (15.21) were described through Figure 15-1 and Tables 15-1 and 15-2. They are presented again in Table 15-4.

TABLE 15-4
Predicting the direction of change in a reversible chemical reaction

$H_2(g) + I_2(g) \rightleftharpoons 2\,HI(g)$ $K_c = 50.2^a$

Exper- iment	Initial concentration,[b] $M \times 10^3$			Initial reaction quotient $Q = \dfrac{[HI]^2}{[H_2][I_2]}$	Comparison of Q and K_c	Direction of net chemical reaction
	$[H_2]$	$[I_2]$	$[HI]$			
1	1.88	1.88	0	$Q = \dfrac{0}{(1.88 \times 10^{-3})^2} = 0$	$Q < K_c$	to the right
2	0	0	1.88	$Q = \dfrac{(1.88 \times 10^{-3})^2}{0} = \infty$	$Q > K_c$	to the left
3	1.88	1.88	1.88	$Q = \dfrac{(1.88 \times 10^{-3})^2}{(1.88 \times 10^{-3})^2} = 1$	$Q < K_c$	to the right

[a]Temperature = 445°C; reaction volume = 0.8000 L.
[b]Initial concentrations are obtained by dividing initial amounts of reacting species (from Table 15-1) by the reaction volume of 0.8000 L. Concentrations have been multiplied by 10^3 in this column. That is, in experiment 1, $[H_2] = 1.88 \times 10^{-3}$ M, etc.

$$H_2(g) + I_2(g) \rightleftharpoons 2\,HI(g) \qquad K_c = 50.2 \text{ at } 445^\circ C \tag{15.21}$$

Let us focus on the *initial* concentrations of the reactants. In experiment 1 only $H_2(g)$ and $I_2(g)$ are present initially. This means that $[HI(g)] = 0$ and the reaction quotient, $Q = 0$; whereas K_c for the reaction is 50.2. We know that in order for equilibrium to be established in experiment 1 some $HI(g)$ must be produced. A net reaction proceeds in the forward direction or to the right. As $[HI(g)]$ increases, $[H_2(g)]$ and $[I_2(g)]$ decrease. The reaction quotient, Q, increases in value until it becomes equal to K_c.

*A net reaction proceeds from left
to right* (*the forward reaction*) *if* $\quad Q < K_c \tag{15.22}$

In experiment 2 of Table 15-4, only $HI(g)$ is present initially, no $H_2(g)$ and $I_2(g)$. If $[H_2(g)] = [I_2(g)] = 0$, the reaction quotient is infinitely large, $Q = \infty$. Again, the value of $K_c = 50.2$. In this situation we know that in order for equilibrium to be established, a net reaction must proceed in the *reverse* direction, that is, to the left. In this way the concentrations of $H_2(g)$ and $I_2(g)$ increase while $[HI(g)]$ decreases. Ultimately, the value of Q becomes equal to K_c and equilibrium is established.

*A net reaction proceeds from
right to left* (*the reverse direction*) *if* $\quad Q > K_c \tag{15.23}$

In experiment 3 of Table 15-4, all three reactants are present initially and the direction of the net reaction is not immediately obvious. However, consideration of the criteria just established shows that (15.22) applies; that is, $Q = 1$ is less than $K_c = 50.2$. The net reaction proceeds to the right.

The criteria for predicting the *direction* of chemical change in a reversible reaction are illustrated in Figure 15-2 and applied in Example 15-6. Predicting the *extent* of reaction (i.e., actual equilibrium concentrations from initial concentrations) requires additional algebraic calculations and is the subject of Example 15-14.

Example 15-6 For the reaction $CO(g) + H_2O(g) \rightleftharpoons CO_2(g) + H_2(g)$, $K_c = 1.00$ at about 1100 K. The following amounts of substances are brought together at this temperature and allowed to react: 1.00 mol CO 1.00 mol H_2O, 2.00 mol CO_2, and 2.00 mol H_2. Relative to their initial amounts, which of the reactants will be present in greater amount and which, in lesser amount, when equilibrium is established?

Solution. Basically, all that is required here is to determine the direction in which the net reaction proceeds; and for this we use the criteria (15.22) and (15.23). To substitute concentrations into the reaction quotient, we assume an arbitrary reaction volume, V. Its value is immaterial, since in this case volume cancels out.

Volume terms will cancel from a reaction quotient or an equilibrium constant expression only if the total of the exponents of the concentration terms in the numerator equals that in the denominator.	$Q = \dfrac{[CO_2][H_2]}{[CO][H_2O]} = \dfrac{(2.00/V)(2.00/V)}{(1.00/V)(1.00/V)} = 4.00$

$$4.00 > K_c = 1.00$$

Because $Q > K_c$ reaction proceeds to the left. When equilibrium is established, the amounts of CO_2 and H_2 will have *decreased* from their initial values and the amounts of CO and H_2O will have *increased*.

SIMILAR EXAMPLES: Review Problem 9; Exercises 14, 18.

15-5 Altering Equilibrium Conditions—Le Châtelier's Principle

Our ultimate goal in this chapter is to do calculations that give us detailed information about the condition of equilibrium in a reversible chemical reaction. Yet, there are times when *qualitative* statements about equilibrium are sufficient. Moreover, in cases where necessary data are not available, a qualitative statement is all that is possible. The French

FIGURE 15-2
Predicting the direction
of change in a
reversible reaction.

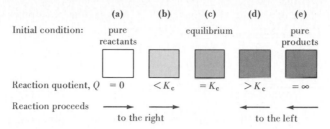

From Table 15-4 experiment 1 corresponds to initial condition **(a)**, experiment 2 to condition **(e)**, and experiment 3 to **(b)**. The situation in Example 15-6 corresponds to condition **(d)**.

chemist Le Châtelier (1884) formulated a statement that is useful in describing the condition of equilibrium. Le Châtelier's principle is difficult to state unambiguously, but its essence is that

Recall also our previous introduction to Le Châtelier's principle in Section 11-6.

Actions that tend to change the temperature, pressure, or concentrations of reactants in a system at equilibrium stimulate a net reaction that restores equilibrium in the system.

The way in which the system responds to changes in these variables involves, in some cases, a shift of the equilibrium condition "to the right" (meaning favoring the forward reaction) and in others, a shift "to the left" (favoring the reverse reaction). Generally, it is not difficult to predict the outcome of changing a system variable. At times, however, it is easy to overlook some secondary effects of changing a system variable that may cause a prediction to be at variance with actual experience (e.g., see Exercise 44).

Effect of Changing the Amounts of Reacting Species.   Let us return to the equilibrium

$$2\ SO_2(g) + O_2(g) \rightleftharpoons 2\ SO_3(g) \qquad K_c = 2.8 \times 10^2\ \text{at 1000 K} \qquad (15.7a)$$

Figure 15-3a depicts a particular equilibrium mixture, and Figure 15-3b, a disturbance in the form of adding 1.00 mol SO_3 while the volume of the system is held constant. How will the amounts of the reactants change to reestablish equilibrium? One approach is to calculate the reaction quotient, Q, immediately after the addition of the 1.00 mol SO_3. The addition of any quantity of SO_3 to a constant-volume equilibrium mixture makes the

FIGURE 15-3
Changing equilibrium
conditions by increasing
the amount of one of
the reactants in the
reaction

$2\ SO_2(g) + O_2(g) \rightleftharpoons$
$\qquad\qquad 2\ SO_3(g)$
$K_c = 2.8 \times 10^2$ at 1000 K

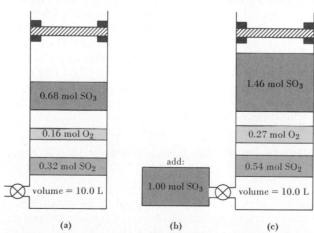

(a) The original equilibrium condition.   **(b)** Disturbance caused by the addition of 1.00 mol SO_3.   **(c)** The new equilibrium condition. Calculation of the new equilibrium amounts is the subject of Exercise 22.

value of Q larger than that of K_c. A net reaction must proceed in the direction that reduces $[SO_3]$, that is, to the left, the reverse direction.

Original equilibrium

$$Q = \frac{[SO_3]^2}{[SO_2]^2[O_2]} = K_c$$

Following disturbance

$$Q = \frac{[SO_3]^2}{[SO_2]^2[O_2]} > K_c$$

The amount of each species in the new equilibrium condition can be calculated by methods to be introduced in the next section. These amounts are represented in Figure 15-3c.

An alternative, *qualitative* approach, based on Le Châtelier's principle, is to note that if the system is to resist an action that increases the equilibrium concentration of one of its reactants, it must do so by favoring the reaction in which that reactant is consumed. Here, this means the reverse reaction—conversion of some of the added SO_3 to SO_2 and O_2. In the new equilibrium there will be greater amounts of all the reactants than in the initial equilibrium, although the additional amount of SO_3, of course, will be less than the added 1.00 mol.

Example 15-7   Predict the effect of introducing additional $H_2(g)$ into a constant-volume equilibrium mixture of $N_2(g)$, $H_2(g)$, and $NH_3(g)$.

$$N_2(g) + 3\ H_2(g) \rightleftharpoons 2\ NH_3(g)$$

Solution.   The action of increasing the concentration of $H_2(g)$ stimulates a shift of the equilibrium condition to the right. However, only a portion of the added $H_2(g)$ is consumed in this reaction. When equilibrium is reestablished, there will still be more $H_2(g)$ than present originally. The amount of $NH_3(g)$ will also be greater, but the amount of $N_2(g)$ will be *smaller*. [$N_2(g)$ must be consumed along with $H_2(g)$ if a net reaction is to occur from left to right. Note that additional $H_2(g)$ was added to the original mixture but no additional $N_2(g)$.]

SIMILAR EXAMPLES: Review Problems 14, 15; Exercises 35, 36.

Example 15-8   What is the effect on equilibrium in the reaction $CaCO_3(s) \rightleftharpoons CaO(s) + CO_2(g)$ produced by **(a)** adding more $CaCO_3(s)$ and **(b)** removing some $CO_2(g)$.

Solution
(a) The addition or removal of substances will affect an equilibrium condition only if they produce changes in the concentration terms that appear in an equilibrium constant expression. $CaCO_3(s)$, as we noted in Section 15-2, does not appear in expression (15.18). Adding more $CaCO_3(s)$ will not affect the equilibrium condition. That is, the partial pressure of $CO_2(g)$ in the mixture is unchanged.
(b) Removal of $CO_2(g)$ from an equilibrium mixture has the effect of reducing its concentration (or partial pressure). To resist this change, more $CaCO_3(s)$ will decompose to replace the $CO_2(g)$ that has been removed and to restore equilibrium. If $CO_2(g)$ is removed *continuously* from the reaction mixture, equilibrium will never be established and the reaction will go to completion (as described in the discussion of reaction 13.27).

SIMILAR EXAMPLES: Review Problems 14, 15; Exercises 35, 36.

Effect of Change of Pressure.   The equilibrium mixture of Figure 15-4a has its volume reduced to one tenth of its original value by increasing the external pressure on the mixture. Once more an adjustment of equilibrium amounts of the reactants must occur in accord with the expression for K_c.

$$K_c = \frac{[SO_3]^2}{[SO_2]^2[O_2]} = \frac{(n_{SO_3}/V)^2}{(n_{SO_2}/V)^2(n_{O_2}/V)} = \frac{(n_{SO_3})^2}{(n_{SO_2})^2(n_{O_2})} \cdot V = 2.8 \times 10^2 \qquad (15.24)$$

FIGURE 15-4
Effect of pressure change on equilibrium condition in the reaction

$$2\,SO_2(g) + O_2(g) \rightleftharpoons 2\,SO_3(g)$$
$$K_c = 2.8 \times 10^2 \text{ at } 1000 \text{ K}$$

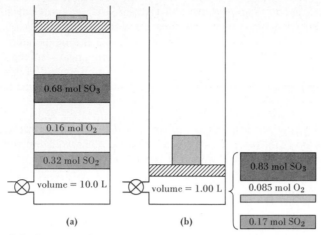

Calculation of the equilibrium amounts in **(b)** is the subject of Exercise 22.

From equation (15.24) it follows that, if V is reduced by a factor of 10, the ratio

$$\frac{(n_{SO_3})^2}{(n_{SO_2})^2(n_{O_2})}$$

must increase by a factor of 10. The equilibrium amount of SO_3 must increase and the amounts of SO_2 and O_2 must decrease. A quantitative description of the new equilibrium amounts is given in Figure 15-4b.

An equilibrium system responds to an increase in external pressure by shrinking into the smallest volume available to it. In the reaction $2\,SO_2(g) + O_2(g) \rightleftharpoons 2\,SO_3(g)$, 3 mol of *gases* on the left produces 2 mol of *gases* on the right. The product of the reaction, SO_3, occupies a smaller volume than the reactants from which it is formed. Thus, an increase in pressure results in the production of additional SO_3.

*When the pressure on an equilibrium mixture involving **gases** is increased, a net reaction proceeds in the direction in which the number of moles of **gases** becomes smaller. If the pressure is decreased, a net reaction proceeds in the direction producing a larger number of moles of **gases**.*

The term, gases, is stressed in this statement because the effect of pressure on reaction equilibria involving condensed phases is generally negligible. Solids and liquids are practically incompressible.

Effect of an Inert Gas. Since an inert gas does not participate in the equilibrium reaction, we should expect its presence not to affect the equilibrium condition. Actually, whether an inert gas does affect an equilibrium condition depends on precisely how the inert gas is introduced. If a quantity of helium is added to the equilibrium mixture pictured in Figure 15-4a, *while the volume is held constant,* the *total* gas pressure will increase. The partial pressures of $SO_2(g)$, $O_2(g)$, and $SO_3(g)$ will remain *constant,* however, as will their equilibrium amounts. Under these conditions the presence of the inert gas does not affect the equilibrium condition. On the other hand, if helium is added to the mixture of Figure 15-4a *at constant pressure,* the reaction volume will *increase.* The effect on the equilibrium will be the same as if the volume increase had been brought about by reducing the external pressure. The equilibrium condition will shift to the side involving the greater number of moles of gas—to the left. Another way to look at this matter is that the presence of an inert gas affects an equilibrium condition only if this produces changes in concentrations (or partial pressures) of the reactants.

Effect of Temperature. In Section 15-7 we will consider a quantitative relationship between the equilibrium constant and temperature. A qualitative statement is possible from Le Châtelier's principle. Changing the temperature of an equilibrium mixture is accomplished by adding heat to or removing heat from the system. Addition of heat favors the heat-absorbing (endothermic) reaction. Removal of heat favors the heat-evolving (exothermic) reaction. (The system attempts to replace the heat that is being removed.) In summary,

Raising the temperature of an equilibrium mixture causes the equilibrium condition to shift in the direction of the endothermic reaction. Lowering the temperature causes a shift in the direction of the exothermic reaction.

Example 15-9 Is the conversion of $SO_2(g)$ to $SO_3(g)$ favored at high or low temperatures?

$$2\ SO_2(g) + O_2(g) \rightleftharpoons 2\ SO_3(g) \qquad \Delta\overline{H}^\circ = -180 \text{ kJ/mol}$$

Solution. Raising the temperature favors the endothermic reaction, the reverse reaction above. To favor the forward (exothermic) reaction requires that the temperature be lowered. Therefore, conversion of $SO_2(g)$ to $SO_3(g)$ is favored at *low* temperatures.

SIMILAR EXAMPLES: Review Problem 16; Exercises 37, 38.

15-6 Equilibrium Calculations—Some Illustrative Examples

Calculations relating to the condition of chemical equilibrium are among the most important encountered in chemistry. Such calculations are the subject of the next several chapters. In this section we consider some examples that employ the general equilibrium principles established earlier in the chapter. Each example includes a brief section labeled "comments," which describes the special features illustrated. The collection of these "comments" constitutes the basic methodology of equilibrium calculations.

Example 15-10 Equilibrium is established in the reaction $N_2O_4(g) \rightleftharpoons 2\ NO_2(g)$ at 25°C. The quantities of reactant and product present in a 3.00-L vessel are 7.64 g N_2O_4 and 1.56 g NO_2. What is the value of K_c for this reaction?

Solution

Equilibrium amounts, mol	Equilibrium concentrations, mol/L
$7.64 \text{ g } N_2O_4 \times \dfrac{1 \text{ mol } N_2O_4}{92.0 \text{ g } N_2O_4} = 0.0830 \text{ mol } N_2O_4$	$[N_2O_4] = \dfrac{0.0830 \text{ mol } N_2O_4}{3.00 \text{ L}} = 0.0277\ M$
$1.56 \text{ g } NO_2 \times \dfrac{1 \text{ mol } NO_2}{46.0 \text{ g } NO_2} = 0.0339 \text{ mol } NO_2$	$[NO_2] = \dfrac{0.0339 \text{ mol } NO_2}{3.00 \text{ L}} = 0.0113\ M$

$$K_c = \frac{[NO_2]^2}{[N_2O_4]} = \frac{(1.13 \times 10^{-2})^2}{2.77 \times 10^{-2}} = 4.61 \times 10^{-3}$$

SIMILAR EXAMPLES: Review Problems 6, 10; Exercise 11.

Comments. Correct substitutions must be made into an equilibrium constant expression, K_c. To ensure that this is done, it is helpful to tabulate the data and carefully label each item. Equilibrium concentrations in mol/L must be used, *not* equilibrium amounts in moles.

Example 15-11 A 0.0200-mol sample of SO_3 is introduced into an evacuated 1.52-L vessel and heated to 900 K, where equilibrium is established. The amount of SO_3 present at equilibrium is found to be 0.0142 mol. What are the values of (a) K_c and (b) K_p at 900 K for the reaction

$$2 \, SO_3(g) \rightleftharpoons 2 \, SO_2(g) + O_2(g)?$$

Solution. In the table of data below, the key item is the change in amount of SO_3: $(0.0142 - 0.0200)$ mol $SO_3 = -0.0058$ mol SO_3. (The negative sign signifies that this amount of reactant is consumed to establish equilibrium.) In the row labeled "change" we must relate the changes in amounts of SO_2 and O_2 to this change in amount of SO_3. For this we use the balanced equation, in particular the stoichiometric coefficients 2, 2, and 1. That is, 1 mol SO_2 and $\frac{1}{2}$ mol O_2 are produced for every mole of SO_3 consumed.

the reaction:	$2 \, SO_3(g)$	$\rightleftharpoons$	$2 \, SO_2(g)$	$+$	$O_2(g)$
initial amounts:	0.0200 mol		0.00 mol		0.00 mol
change:	-0.0058 mol		$+0.0058$ mol		$+0.0029$ mol
equilibrium amounts:	0.0142 mol		0.0058 mol		0.0029 mol
equilibrium concentrations:	$[SO_3] = 0.0142 \text{ mol}/1.52 \text{ L}$ $= 9.34 \times 10^{-3} \, M$		$[SO_2] = 0.0058 \text{ mol}/1.52 \text{ L}$ $= 3.8 \times 10^{-3} \, M$		$[O_2] = 0.0029 \text{ mol}/1.52 \text{ L}$ $= 1.9 \times 10^{-3} \, M$

(a) $K_c = \dfrac{[SO_2]^2[O_2]}{[SO_3]^2} = \dfrac{(3.8 \times 10^{-3})^2(1.9 \times 10^{-3})}{(9.34 \times 10^{-3})^2} = 3.1 \times 10^{-4}$

(b) $K_p = K_c(RT)^{\Delta n} = 3.1 \times 10^{-4}(0.0821 \times 900)^{(2+1)-2}$
$= 3.1 \times 10^{-4}(0.0821 \times 900)^1 = 2.3 \times 10^{-2}$

SIMILAR EXAMPLES: Review Problems 6, 10; Exercises 7, 8, 9.

Comments

1. The chemical equation for a reversible reaction can serve *both* to establish the form of the equilibrium constant expression *and* to provide the conversion factors to relate the equilibrium amounts and concentrations to the specified initial conditions.

2. Whether working with K_c or K_p or the relationship between them, these expressions must be based on the chemical equation that is given, not on what may have been used in other situations: K_p and K_c were related for the sulfur dioxide–oxygen–sulfur trioxide reaction in Example 15-3, but the result here is different because the form of the chemical equation is different.

Example 15-12 Ammonium hydrogen sulfide, NH_4HS, dissociates appreciably, even at room temperature.

$$NH_4HS(s) \rightleftharpoons NH_3(g) + H_2S(g) \qquad K_p(\text{atm}) = 1.08 \times 10^{-1} \text{ at } 25°C$$

A sample of $NH_4HS(s)$ is introduced into an evacuated flask and allowed to establish equilibrium at 25°C. What is the total gas pressure at equilibrium?

Solution. K_p for this reaction is simply the product of the partial pressures of $NH_3(g)$ and $H_2S(g)$, each stated in atm. Moreover, since these gases are produced in equimolar amounts by the dissociation of NH_4HS, $P_{NH_3} = P_{H_2S}$; and $P_{tot.} = P_{NH_3} + P_{H_2S} = 2 \times P_{NH_3}$.

$K_p = (P_{NH_3})(P_{H_2S}) = (P_{NH_3})(P_{NH_3}) = (P_{NH_3})^2 = 1.08 \times 10^{-1} = 10.8 \times 10^{-2}$

$P_{NH_3} = \sqrt{10.8 \times 10^{-2}} = 3.29 \times 10^{-1} \text{ atm}$

$P_{tot.} = 2 \times P_{NH_3} = 2 \times 3.29 \times 10^{-1} \text{ atm} = 0.658 \text{ atm}$

SIMILAR EXAMPLES: Exercises 26, 27, 28.

Comments. When writing a K_p equilibrium constant expression look for relationships among partial pressures of the reactants. If the total gas pressure needs to be related to reactant partial pressures, this can be done through equations presented in Chapter 5 (e.g., equations 5.12, 5.13, and 5.16).

Example 15-13   An 0.0240-mol sample of $N_2O_4(g)$ is allowed to dissociate and come to equilibrium with $NO_2(g)$ in an 0.372-L flask at 25°C. What is the percent dissociation of the N_2O_4?

$$N_2O_4(g) \rightleftharpoons 2\,NO_2(g) \qquad K_c = 4.61 \times 10^{-3} \text{ at } 25°C$$

Solution.   By percent dissociation we mean the percent of the N_2O_4 molecules present initially that are converted to NO_2 (see Figure 15-5). This requires a determination of the number of moles of reactant and product at equilibrium. But now for the first time we must introduce an algebraic unknown, x. Suppose that we let $x =$ the number of moles of N_2O_4 that dissociate. We enter this value into the row labeled "change" in the table below. The amount of NO_2 produced is $2x$.

the reaction:	$N_2O_4(g)$	$\rightleftharpoons$	$2\,NO_2(g)$
initial amounts:	0.0240 mol		0.00 mol
change:	$-x$ mol		$+2x$ mol
equilibrium amounts:	$(0.0240 - x)$ mol		$2x$ mol
equilibrium concentrations:	$[N_2O_4]$ $= (0.0240 - x\,\text{mol})/0.372\text{ L}$		$[NO_2]$ $= 2x\,\text{mol}/0.372\text{ L}$

$$K_c = \frac{[NO_2]^2}{[N_2O_4]} = \frac{\left(\dfrac{2x}{0.372}\right)^2}{\dfrac{0.0240 - x}{0.372}} = \frac{4x^2}{0.372(0.0240 - x)} = 4.61 \times 10^{-3}$$

$$4x^2 = 4.12 \times 10^{-5} - (1.71 \times 10^{-3})x$$
$$x^2 + (4.28 \times 10^{-4})x - 1.03 \times 10^{-5} = 0$$

For a discussion of quadratic equations see Appendix A.

$$x = \frac{-4.28 \times 10^{-4} \pm \sqrt{(4.28 \times 10^{-4})^2 + 4 \times 1.03 \times 10^{-5}}}{2}$$

$$x = \frac{-4.28 \times 10^{-4} \pm \sqrt{(1.83 \times 10^{-7}) + (4.12 \times 10^{-5})}}{2}$$

$$x = \frac{-4.28 \times 10^{-4} \pm \sqrt{4.14 \times 10^{-5}}}{2}$$

$$x = \frac{-4.28 \times 10^{-4} \pm 6.43 \times 10^{-3}}{2}$$

The symbol $\pm$ in this equation signifies that there are two possible roots. One can decide between the two by considering their physical meaning. In this problem x must be a *positive* quantity, smaller than 0.0240.

$$= \frac{-4.28 \times 10^{-4} + 6.43 \times 10^{-3}}{2} = \frac{6.00 \times 10^{-3}}{2}$$

$$x = 3.00 \times 10^{-3} \text{ mol } N_2O_4 \quad \text{(dissociated)}$$

The percent dissociation of the N_2O_4 is given by the expression

$$\% \text{ dissoc. of } N_2O_4 = \frac{3.00 \times 10^{-3} \text{ mol } N_2O_4 \text{ dissoc.}}{0.0240 \text{ mol } N_2O_4 \text{ initially}} \times 100$$

$$= 12.5\%$$

SIMILAR EXAMPLES: Exercises 29, 30, 31.

FIGURE 15-5
Equilibrium in the reaction

$N_2O_4(g) \rightleftharpoons 2\ NO_2(g)$

at 25°C—Example 15-13 illustrated.

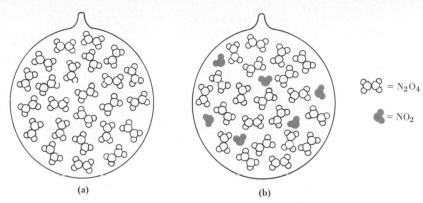

(a)   (b)

$\bowtie$ = N_2O_4

$\blacktriangleleft$ = NO_2

Each "molecule" illustrated represents 0.001 mol.
(a) Initially, pure N_2O_4 is introduced into an evacuated glass bulb and the bulb is sealed. The illustration shows 24 "molecules" corresponding to 0.024 mol N_2O_4.
(b) At equilibrium, some molecules of N_2O_4 have dissociated to NO_2 (shown in gray). The illustration contains 21 N_2O_4 and 6 NO_2 "molecules," corresponding to 0.021 mol N_2O_4 and 0.006 mol NO_2.

Comments.   When one or more of the quantities in an equilibrium constant expression must be stated in terms of the algebraic unknown, x, there are usually several ways in which x can be defined. No hard-and-fast rules need be given, except to note that it is necessary to establish through this definition whether x is

1. An *amount* of substance (mol) or a *concentration* (mol/L).
2. Stated in terms of a reactant *consumed* or a *product* formed.
3. A positive or a negative quantity. (A definition that makes x a positive quantity is generally preferred.)

Example 15-14   A solution is prepared having the initial concentrations: $[Fe^{3+}] = [Hg_2^{2+}] = 0.5000\ M$; $[Fe^{2+}] = [Hg^{2+}] = 0.03000\ M$. The following reaction occurs among these ions.

$$2\ Fe^{3+}(aq) + Hg_2^{2+}(aq) \rightleftharpoons 2\ Fe^{2+}(aq) + 2\ Hg^{2+}(aq)\quad K_c = 9.14 \times 10^{-6}\ \text{at 25°C}$$

What will be the ionic concentrations when equilibrium is established?

Solution.   Although it is not necessary to do so, comparing Q and K_c may help us to visualize the solution to this problem.

$$Q = \frac{[Fe^{2+}]^2[Hg^{2+}]^2}{[Fe^{3+}]^2[Hg_2^{2+}]} = \frac{(0.03000)^2(0.03000)^2}{(0.5000)^2(0.5000)} = \frac{8.10 \times 10^{-7}}{1.25 \times 10^{-1}}$$

$$= 6.48 \times 10^{-6} < K_c = 9.14 \times 10^{-6}$$

Since Q is smaller than K_c, a reaction must proceed to the right (recall criterion 15.22). Let us define x as the number of moles per liter of Fe^{3+} that are converted to Fe^{2+}. The several equilibrium concentrations can then be expressed in terms of x, as shown below.

the reaction:	$2\ Fe^{3+}(aq)$	$+$	$Hg_2^{2+}(aq)$	$\rightleftharpoons$	$2\ Fe^{2+}(aq)$	$+$	$2\ Hg^{2+}(aq)$
initial concentrations:	0.5000 *M*		0.5000 *M*		0.03000 *M*		0.0300 *M*
change:	$-x\ M$		$-x/2\ M$		$+x\ M$		$+x\ M$
equilibrium concentrations:	$(0.5000 - x)\ M$		$(0.5000 - x/2)\ M$		$(0.03000 + x)\ M$		$(0.03000 + x)\ M$

$$K_c = \frac{(0.03000 + x)^2(0.03000 + x)^2}{(0.5000 - x)^2(0.5000 - x/2)} = 9.14 \times 10^{-6}$$

Solving this equation can be simplified greatly if the following assumption is made, and *if the assumption proves valid.* If x is much smaller than 0.5000, then $(0.5000 - x) \simeq 0.5000$, and $(0.5000 - x/2) \simeq 0.5000$. This assumption leads to the expression

$$K_c = \frac{(0.03000 + x)^2(0.03000 + x)^2}{(0.5000)^2(0.5000)} = 9.14 \times 10^{-6}$$

$$(0.03000 + x)^4 = 1.14 \times 10^{-6} = 114 \times 10^{-8}$$

Take the *fourth* root of each side of this equation (i.e., take the square root twice).

$$(0.03000 + x)^2 = 10.7 \times 10^{-4}$$

$$(0.03000 + x) = 3.27 \times 10^{-2}$$

$$x = 3.27 \times 10^{-2} - 0.03000 = 2.7 \times 10^{-3}$$

The simplifying assumption appears to be valid: 2.7×10^{-3} is considerably smaller than 0.5000.

Equilibrium concentrations

$$[Fe^{2+}] = 0.03000 + x = 0.03000 + 2.7 \times 10^{-3} = 3.27 \times 10^{-2} \ M$$

$$[Hg^{2+}] = 0.03000 + x = 0.03000 + 2.7 \times 10^{-3} = 3.27 \times 10^{-2} \ M$$

$$[Fe^{3+}] = 0.5000 - x = 0.5000 - 2.7 \times 10^{-3} = 4.973 \times 10^{-1} \ M$$

$$[Hg_2^{2+}] = 0.5000 - x/2 = 0.5000 - 1.4 \times 10^{-3} = 4.986 \times 10^{-1} \ M$$

SIMILAR EXAMPLES: Review Problem 12; Exercises 16, 17, 23, 24.

Comments

1. It is sometimes useful to compare the reaction quotient, Q, to the equilibrium constant, K_c, to determine the direction in which a net reaction will proceed.

2. In some equilibrium calculations—often those involving species in aqueous solution—one can work with molar concentrations exclusively. No specific reference is made to moles of reactants and solution volumes.

3. Algebraic solutions can often be greatly simplified if one recognizes certain relationships among the terms in the equilibrium constant expression. Usually, these simplifications take the form of x being much smaller than some other numerical value to which it is added or from which it is subtracted.

15-7 The Effect of Temperature on Equilibrium

In general, the equilibrium constant for a reaction is temperature-dependent. Values of K_p for the sulfur dioxide–oxygen–sulfur trioxide reaction at several temperatures are tabulated in Table 15-5, together with familiar functions of these data—$\log K$ and $1/T$. Figure 15-6 shows that a plot of $\log K$ vs. $1/T$ yields a straight line. The equation of this straight line is

$$\underbrace{\log K}_{y} = \underbrace{\frac{-\Delta \overline{H}^\circ}{2.303R}}_{m} \cdot \underbrace{\frac{1}{T}}_{x} + \underbrace{\text{constant}}_{b} \tag{15.25}$$

equation of straight line: y $=$ m $\cdot x +$ b

TABLE 15-5
Equilibrium constants, K_p, for the reaction
$2 SO_2(g) + O_2(g) \rightleftharpoons 2 SO_3(g)$
at several temperatures

T, K	$1/T$, K^{-1}	K_p	$\log K_p$
800	12.5×10^{-4}	9.1×10^2	2.96
850	11.8×10^{-4}	1.7×10^2	2.23
900	11.1×10^{-4}	4.2×10^1	1.62
950	10.5×10^{-4}	1.0×10^1	1.00
1000	10.0×10^{-4}	3.2×10^0	0.51
1050	9.52×10^{-4}	1.0×10^0	0.00
1100	9.09×10^{-4}	3.9×10^{-1}	-0.41
1170	8.55×10^{-4}	1.2×10^{-1}	-0.92

Moreover, as illustrated twice previously (in establishing equations 11.3 and 14.29), the constant term can be eliminated from equation (15.25) to yield a result with a familiar form (called the van't Hoff equation).

$$\log \frac{K_2}{K_1} = \frac{\Delta \overline{H}^\circ}{2.303\ R} \left(\frac{T_2 - T_1}{T_2 T_1} \right) \tag{15.26}$$

K_2 and K_1 are the equilibrium constants at the kelvin temperatures T_2 and T_1. $\Delta \overline{H}^\circ$ is the standard molar enthalpy (heat) of reaction. Both positive and negative values of $\Delta \overline{H}^\circ$ are possible; and the assumption is made that $\Delta \overline{H}^\circ$ is independent of temperature, which is valid in most cases.

According to Le Châtelier's principle, if $\Delta \overline{H}^\circ > 0$ (endothermic), the forward reaction is favored with increased temperature, suggesting that the value of K increases with temperature. If $\Delta \overline{H}^\circ < 0$ (exothermic), the reverse reaction is favored by an increase in

FIGURE 15-6
Temperature dependence of the equilibrium constant K_p for the reaction

$$2 SO_2(g) + O_2(g) \rightleftharpoons 2 SO_3(g)$$

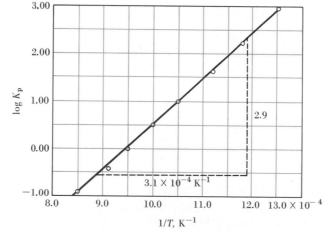

This graph can be used to establish the heat of reaction, $\Delta \overline{H}^\circ$ (see equation 15.25).

$$\text{slope} = \frac{-\Delta \overline{H}^\circ}{2.303\ R} = \frac{2.9}{3.1 \times 10^{-4}\ \text{K}^{-1}} = 9.4 \times 10^3\ \text{K}$$

$$\Delta \overline{H}^\circ = -2.303 \times 8.314\ \text{J mol}^{-1}\ \text{K}^{-1} \times 9.4 \times 10^3\ \text{K}$$
$$= -1.8 \times 10^5\ \text{J/mol} = -1.8 \times 10^2\ \text{kJ/mol}$$

temperature and the value of K decreases with temperature. Equation (15.26) yields *quantitative* results in agreement with *qualitative* observations from Le Châtelier's principle, as confirmed by Example 15-15.

Example 15-15 Use data from Table 15-5 and Figure 15-6 to estimate the temperature at which $K_p = 1.0 \times 10^6$ for the reaction

$$2\ SO_2(g) + O_2(g) \rightleftharpoons 2\ SO_3(g)$$

Solution. For substitution into equation (15.26), use a known value from Table 15-5 and the heat of reaction established through Figure 15-6. That is,

$$\begin{cases} T_2 = 800\ K \\ T_1 = ? \end{cases} \quad \begin{matrix} K_2 = 9.1 \times 10^2 \\ K_1 = 1.0 \times 10^6 \end{matrix} \quad \Delta \overline{H}^\circ = -1.8 \times 10^5\ \text{J/mol}$$

$$\log \frac{9.1 \times 10^2}{1.0 \times 10^6} = \log 9.1 \times 10^{-4} = -3.04 = \frac{-1.8 \times 10^5}{2.303 \times 8.314} \left(\frac{800 - T_1}{800 T_1} \right)$$

$$\frac{2.303 \times 8.314 \times 3.04 \times 800}{1.8 \times 10^5} T_1 = 800 - T_1$$

$$0.26\, T_1 = 800 - T_1$$

$$1.26\, T_1 = 800 \qquad T_1 = \frac{800}{1.26} = 635\ K$$

SIMILAR EXAMPLES: Review Problem 13; Exercises 40, 41, 43.

Here is a final noteworthy point: The Clausius–Clapeyron equation (11.3) for the variation of vapor pressure with temperature is just a special form of equation (15.26). Vapor pressures are values of K_p (recall equation 15.19), and $\Delta \overline{H}^\circ{}_{\text{vap}}$ is equivalent to $\Delta \overline{H}^\circ{}_{\text{rxn}}$.

15-8 Kinetic Basis of the Equilibrium Constant

Several times we have referred to the fact that when equilibrium is established, the rates of a forward and a reverse reaction become equal. Can this statement serve as the basis of a theoretical derivation of the equilibrium constant expression? The first attempts to relate reaction rates and the equilibrium constant are generally attributed to Guldberg and Waage (see again page 411). They proposed that the equilibrium condition could be described simply by equating the rate laws for the forward and reverse reactions. However, this formulation did not always lead to the equilibrium constant expression. Guldberg and Waage did not always use coefficients from the balanced equation as exponents in rate law expressions. Other investigators, including van't Hoff, seem to have arrived at the correct form of the equilibrium constant expression before Guldberg and Waage.

Let us pursue Guldberg and Waage's lead a bit further. For this we use as an illustrative example the hydrogen–iodine–hydrogen iodide reaction

$$H_2(g) + I_2(g) \underset{k_2}{\overset{k_1}{\rightleftharpoons}} 2\ HI(g) \tag{15.27}$$

Suppose that the *mechanism* of this reaction involves simple one-step bimolecular processes of the type first described in Section 14-10. We would write the following simple rate laws and equate them for the equilibrium condition. Rearrangement of equation

(15.28) then leads to an expression that has the same form as the equilibrium constant established in (15.4).

$$\text{forward rate} = k_1[H_2][I_2] = k_2[HI]^2 = \text{reverse rate} \tag{15.28}$$

$$K_c = \frac{k_1}{k_2} = \frac{[HI]^2}{[H_2][I_2]} \tag{15.29}$$

But there is a serious objection to this derivation! An alternate mechanism was presented in Section 14-10 for reaction (15.27). This was the two-step mechanism outlined through equations (15.30) and (15.31). What is the equilibrium condition when described by this alternate mechanism?

$$(\text{fast}) \qquad\qquad I_2(g) \underset{k_2}{\overset{k_1}{\rightleftharpoons}} 2\ I(g) \tag{15.30}$$

$$(\text{slow}) \quad 2\ I(g) + H_2(g) \underset{k_4}{\overset{k_3}{\rightleftharpoons}} 2\ HI(g) \tag{15.31}$$

Let us first establish the *steady-state condition* for each of these equilibria.

$$(\text{fast}) \qquad k_1[I_2] = k_2[I]^2 \qquad and \qquad \frac{k_1}{k_2} = \frac{[I]^2}{[I_2]} \tag{15.32}$$

$$(\text{slow}) \quad k_3[I]^2[H_2] = k_4[HI]^2 \qquad and \qquad \frac{k_3}{k_4} = \frac{[HI]^2}{[H_2][I]^2} \tag{15.33}$$

Now, solve equation (15.32) for $[I]^2$ and substitute into equation (15.33) to obtain

$$\frac{k_3}{k_4} = \frac{[HI]^2}{\dfrac{k_1}{k_2}[I_2][H_2]} \qquad and \qquad \frac{k_1 k_3}{k_2 k_4} = \frac{[HI]^2}{[I_2][H_2]} = K_c \tag{15.34}$$

Again we obtain an expression having the same form as the equilibrium constant expression first written in equation (15.4)! Actually, any plausible mechanism of a reversible reaction can be used to derive an equilibrium constant expression of the expected form (i.e., as written in equation 15.6).

Effect of a Catalyst on Equilibrium.   The presence of a catalyst in a reversible reaction has the effect of speeding up *both* the forward and reverse reactions. The condition of equilibrium is reached more quickly, but the equilibrium amounts of the reacting species are *unchanged* by the catalyst. Thus, for a given set of reaction conditions the equilibrium amounts of $SO_2(g)$, $O_2(g)$, and $SO_3(g)$ have fixed values. This is true whether the reaction is carried out as a slow homogeneous gas-phase reaction or a faster heterogeneous reaction on the surface of a catalyst. Or, stated in another way, the presence of a catalyst does not change the numerical value of the equilibrium constant.

$$2\ SO_2(g) + O_2(g) \rightleftharpoons 2\ SO_3(g) \qquad K_c = 2.8 \times 10^2 \text{ at } 1000 \text{ K} \tag{15.7a}$$

The role of a catalyst is to change the mechanism of a chemical reaction to one involving a lower activation energy (recall Figure 14-12). Also, a catalyst has no effect on the condition of equilibrium in a reversible reaction. Taken together these facts mean that an equilibrium condition must be *independent* of the reaction mechanism. Thus, as stated above, the kinetic derivation of the equilibrium constant should not depend on the particular mechanism chosen. Moreover, we could always conceive of a *hypothetical* catalyst that changes the mechanism of a reversible reaction to the simple one-step process suggested by the balanced chemical equation. The equilibrium constant expression derived from such a mechanism would then always be of the form written in equation (15.6).

FOCUS ON
The Synthesis of Ammonia and Other Nitrogen Compounds

Nobel Prize: 1909

Wilhelm Ostwald (1853–1932)

"For his work on catalysis and on the conditions of chemical equilibrium and velocities of chemical reactions."

Nobel Prize: 1918

Fritz Haber (1868–1934)

"For the synthesis of ammonia from its elements, nitrogen and hydrogen."

Nobel Prize: 1931

Carl Bosch (1870–1940)

(with Friedrich Bergius)

"In recognition of their contributions to the invention and development of chemical high-pressure methods."

The development of methods of obtaining nitrogen compounds from atmospheric nitrogen was directly or indirectly responsible for three Nobel Prizes. [Courtesy of German Information Center.]

Elemental nitrogen occurs in the atmosphere to the extent of 78% by volume, but because of nitrogen's relative inertness, nitrogen-containing compounds do not occur extensively in nature. The only significant natural source is Chilean saltpeter ($NaNO_3$). Methods of synthesizing nitrogen compounds, referred to as artificial nitrogen fixation, are industrial processes of utmost importance. The chief method is the reaction of N_2 and H_2 to form ammonia. The ammonia can be oxidized to oxides of nitrogen, which in turn can be converted to nitric acid and nitrate salts. Urea,

another important nitrogen-containing chemical, is formed by the reaction of NH_3 and CO_2. The production scheme for these several chemicals is outlined in Figure 15-7.

Ammonia.   The theoretical basis of the ammonia synthesis reaction and its laboratory testing were the work of Fritz Haber (1908). Haber's laboratory apparatus was capable of producing about 1 kg NH_3 per day. The development effort required to convert Haber's process into a commercial operation was one of the most challenging

FIGURE 15-7
Production scheme for some important nitrogen-containing chemicals.

Those shown in black have important uses as fertilizers.

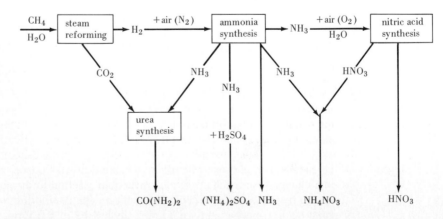

chemical engineering problems of its time, requiring that methods be devised to conduct gaseous chemical reactions at high temperatures and pressures in the presence of an appropriate catalyst. This work was done under the leadership of Carl Bosch at Badische Anilin und Soda Fabrik (BASF). By 1913 a plant was in operation that could produce 30,000 kg NH_3 per day. Modern ammonia plants have about 50 times this capacity.

Some relevant data on the ammonia synthesis reaction are

$$N_2(g) + 3 H_2(g) \rightleftharpoons 2 NH_3(g) \qquad (15.35)$$

at 298 K: $K_p = 6.2 \times 10^5$ and $\Delta \overline{H}^\circ = -92.38$ kJ/mol

For every 4 mol of gases that react [1 mol $N_2(g)$ and 3 mol $H_2(g)$], 2 mol NH_3 is produced. Increasing the pressure forces the reaction mixture into a smaller volume and favors the reaction producing the smaller number of moles of gas—the production of $NH_3(g)$. The forward reaction is *exothermic*. The exothermic reaction is favored if the temperature is *lowered*. Thus, the optimum conditions for the production of NH_3 appear to be *high pressures* and *low temperatures*.

These "optimum" conditions, however, do not take into account the rate of reaction. Although the equilibrium production of NH_3 is favored at low temperatures, the rate of its formation is so slow as to make the method unfeasible. One way of speeding up the reaction is to raise the temperature (even though the equilibrium concentration of NH_3 decreases in doing so). Another way is to use a catalyst. The usual operating conditions for the Haber–Bosch process are about 550°C, pressures ranging from 150 to 350 atm, and a catalyst—usually iron in the presence of Al_2O_3, MgO, CaO, and K_2O. The dramatic difference between the theoretical optimum conditions and the actual operating conditions is suggested through Figure 15-8.

Another way to increase the rate of production of NH_3 is to remove NH_3 continuously as it is formed. This is done by liquefying the NH_3 and recycling the N_2 and H_2, which are not easily liquefied. The disturbance caused by the continuous removal of NH_3 displaces the equilibrium toward the production of more NH_3. In fact, the mixture need not be allowed to come to equilibrium at all. In this way practically 100% conversion of N_2 and H_2 to NH_3 is possible. A schematic representation of a portion of the ammonia synthesis reaction is provided in Figure 15-9.

A critical step in the ammonia synthesis reaction is the production of $H_2(g)$. This may be done in several ways. In the original Haber–Bosch process, $H_2(g)$ was produced from coke and steam by the water gas reactions (13.1 and 13.2). The common method now used is steam reforming of natural gas (methane) or other hydrocarbons.

FIGURE 15-8
Equilibrium conversion of $N_2(g)$ and $H_2(g)$ to $NH_3(g)$ as a function of temperature and pressure.

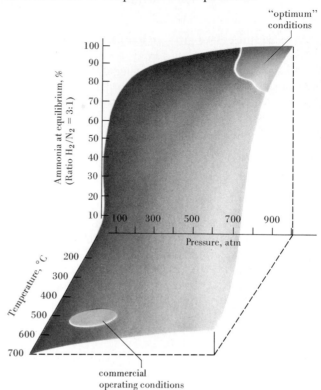

FIGURE 15-9
Ammonia synthesis reaction.

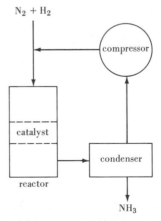

The gaseous N_2–H_2 mixture is introduced into a reactor at high temperature and pressure in the presence of a catalyst. The gaseous N_2–H_2–NH_3 mixture leaves the reactor and is cooled as it passes through a condenser. Liquefied NH_3 is removed, and the remaining N_2–H_2 mixture is compressed and returned to the reactor.

$$CH_4(g) + H_2O(g) \longrightarrow CO(g) + 3\ H_2(g) \qquad (13.3)$$

followed by

$$CO(g) + H_2O(g) \longrightarrow CO_2(g) + H_2(g) \qquad (13.2)$$

As a result, the cost of nitrogen fertilizers is closely linked to the cost of hydrocarbon fuels.

In addition to its use in producing nitric acid and urea, ammonia is used in the preparation of various nitrogen-containing monomers employed in the manufacture of nylon, acrylic polymers, and polyurethane foams. Ammonia is also used in the manufacture of pharmaceuticals, various organic and inorganic chemicals, detergents, and cleansers. Another important use is its direct application to fields as a fertilizer (under the name "anhydrous ammonia").

Nitric Acid. The key step in the production of nitric acid is the conversion of $NH_3(g)$ to $NO(g)$. The conditions developed by Wilhelm Ostwald early in the present century—brief contact (0.01 s) of $NH_3(g)$ and an excess of $O_2(g)$ with a Pt catalyst at about 1000°C—are essentially those in use today in the Ostwald process.

$$4\ NH_3(g) + 5\ O_2(g) \xrightarrow[900°C]{Pt/Rh} 4\ NO(g) + 6\ H_2O(g)$$

$$(15.36)$$

The $NO(g)$ is cooled to near room temperature, where its conversion to $NO_2(g)$ is favorable (recall Table 15-3).

$$2\ NO(g) + O_2(g) \rightleftharpoons 2\ NO_2(g) \qquad (15.37)$$

The $NO_2(g)$ is dissolved in water, and the $NO(g)$ is recycled.

$$3\ NO_2(g) + H_2O \longrightarrow 2\ HNO_3(aq) + NO(g) \qquad (15.38)$$

Nitric acid is used to prepare nitrate salts (chiefly NH_4NO_3), the explosives nitroglycerine and nitrocellulose, and various nitrogen-containing organic compounds.

Urea. The reaction of $NH_3(g)$ and $CO_2(g)$ at high temperatures (190°C) yields an intermediate, ammonium carbamate ($NH_4CO_2NH_2$), which dehydrates (loses H_2O) to form urea. The urea and H_2O form a concentrated liquid solution. The net equilibrium reaction, for which no catalyst is required, is

$$2\ NH_3(g) + CO_2(g) \rightleftharpoons \underset{\text{urea}}{CO(NH_2)_2} + H_2O \qquad (15.39)$$

As expected, the forward reaction is favored by high pressures; the usual operating condition is about 200 atm. A small excess of NH_3 beyond the 2:1 molar proportions of equation (15.39) is also used.

Urea contains 46% nitrogen, by mass. It is an excellent fertilizer, used as the pure solid, as the solid in mixtures with ammonium salts, or as a very concentrated aqueous solution mixed with NH_4NO_3 and/or NH_3. Urea is also used as a feed supplement for cattle and in the production of polymers and pesticides.

Summary

When two competing reactions—a forward and a reverse reaction—occur simultaneously, reaction does not go to completion. The *reversible* reaction proceeds to the point where the rates of the forward and reverse reactions become equal. At this point of dynamic equilibrium no further net change occurs and the amounts of the reacting species remain constant with time. The chemical equation describes the proportions in which reactants participate in a reversible reaction. A quantitative description of the equilibrium condition is provided through the equilibrium constant expression.

The *form* of the equilibrium constant expression, K_c, is established through the balanced chemical equation. The *numerical value* of the constant is determined by experiment. If a reaction involves gases, an equilibrium constant expression, K_p, can be based on partial pressures of gases. K_p and K_c are related through a simple mathematical equation. The equilibrium constant expression is inverted when a chemical equation is written in the reverse direction. If two or more equations are added together, the equilibrium constant for the resulting net reaction is the product of the constants for the individual reactions.

A ratio of initial reactant concentrations can be formed in the same manner as the equilibrium constant expression. By comparing the numerical value of this ratio, called the reaction quotient, Q, to K_c, one can determine the *direction* in which a net reaction proceeds. An algebraic equation can be written and solved to determine the *extent* of reaction, that is, to relate the final equilibrium to the initial conditions. A variety of possibilities exists for equilibrium calculations, but the basic principles and algebraic techniques involved are few in number.

One variable that generally has a substantial effect on the value of an equilibrium constant is temperature. The van't Hoff equation relates log K to the standard molar enthalpy change for a reaction and to kelvin temperature. The qualitative effect of temperature on a reversible reaction is that raising the temperature favors the endothermic reaction and lowering the temperature, the exothermic reaction. The addition of a catalyst to a reversible reaction speeds up the forward and reverse reactions equally. Equilibrium is attained more rapidly but without any change in the equilibrium concentrations. Le Châtelier's principle states that an equilibrium condition undergoes modification or "shifts" whenever an equilibrium

mixture is disturbed. These disturbances may take the form of the addition or removal of reacting species or of changes in temperature or pressure.

Chemical equilibrium has theoretical links to other areas of chemistry. In this chapter the relationship of the equilibrium constant expression to chemical kinetics is established.

Learning Objectives

As a result of studying Chapter 15, you should be able to

1. Describe the condition of equilibrium in a reversible reaction.

2. Write the equilibrium constant expression, K_c, from the chemical equation for a reversible reaction.

3. Derive K values for situations where chemical equations are reversed, multiplied through by constant coefficients or added together.

4. Assess the relative importance of the forward and reverse reactions from the magnitude of an equilibrium constant.

5. Write an equilibrium constant expression in terms of partial pressures of gases, K_p; and relate a value of K_p to the corresponding value of K_c.

6. Calculate a numerical value of an equilibrium constant if equilibrium conditions are given.

7. Predict the direction in which a reaction proceeds toward equilibrium by comparing the reaction quotient, Q, to K_c.

8. Make qualitative predictions of how equilibrium conditions change when an equilibrium mixture is disturbed.

9. Calculate the final equilibrium condition in a reversible reaction from a given set of initial conditions.

10. Relate the equilibrium constant to the standard molar enthalpy of reaction, $\Delta \overline{H}^\circ$, and to kelvin temperature, both graphically and algebraically.

11. Demonstrate that the form of an equilibrium constant expression is consistent with the rate laws and mechanisms of chemical reactions.

Some New Terms

Equilibrium refers to a condition where a forward and reverse process proceed at equal rates and no further net change occurs (e.g., amounts of reactants and products remain constant with time).

An **equilibrium constant** describes the relationship among the concentrations (or partial pressures in some cases) of the substances within an equilibrium system. The numerical value of the constant does not depend on how the equilibrium condition is attained.

K_c is a relationship that exists among the equilibrium concentrations of reactants and products in a reversible reaction at a given temperature.

K_p, the partial pressure equilibrium constant, is a relationship that exists among the partial pressures of gaseous reactants and products in a reversible reaction at equilibrium at a given temperature.

Le Châtelier's principle states that actions that tend to change the temperature, pressure, or concentrations of reactants in a system at equilibrium stimulate a net reaction that restores equilibrium in the system.

The **reaction quotient, Q,** is a ratio of concentration terms having the same form as the equilibrium constant expression, but usually applied to *nonequilibrium* conditions. It is used to determine the direction in which a net reaction occurs to establish equilibrium.

Suggestions for Further Study

ALLSOP, R. T., and N. H. GEORGE, "Le Châtelier—A Redundant Principle?" *Educ. in Chem.*, **21**, 54 (1984).

FELDMAN, M. R., and M. L. TARVER, "Profiles in Chemistry: Fritz Haber," *J. Chem. Educ.*, **60**, 463 (1983).

GALLAGHER, J. T., and F. M. TAYLER, "The Physical Chemistry of Ammonia Manufacture and Nitrogenous Fertilizer Production," *Educ. in Chem.*, **4**, 30 (1967).

GARRETT, G. W., and D. SALLADAY, "Fluid Fertilizers," *Chemtech*, **14**, 250 (1984).

MICKEY, C. D., "Chemical Equilibrium," *J. Chem. Educ.*, **57**, 801 (1980).

MILLER, G. C., "Chemical of the Month: Ammonia," *J. Chem. Educ.*, **58**, 424 (1981).

TREPTOW, R. S., "Le Châtelier's Principle: A Reexamination and Method of Graphic Illustration," *J. Chem. Educ.*, **57**, 417 (1980).

Review Problems

1. Write an equilibrium constant expression, K_c, for each of the following reactions.
   (a) $CO(g) + Cl_2(g) \rightleftharpoons COCl_2(g)$
   (b) $2 NO(g) + O_2(g) \rightleftharpoons 2 NO_2(g)$
   (c) $CO_2(s) \rightleftharpoons CO_2(g)$
   (d) $CS_2(g) + 4 H_2(g) \rightleftharpoons CH_4(g) + 2 H_2S(g)$
   (e) $2 NaHCO_3(s) \rightleftharpoons Na_2CO_3(s) + CO_2(g) + H_2O(g)$

2. From the values of K_c given

$CO(g) + H_2O(g) \rightleftharpoons CO_2(g) + H_2(g)$ $K_c = 23.2$ at 600 K

$SO_2(g) + \frac{1}{2} O_2(g) \rightleftharpoons SO_3(g)$ $K_c = 56$ at 900 K

$2 H_2S(g) \rightleftharpoons 2 H_2(g) + S_2(g)$ $K_c = 2.3 \times 10^{-4}$ at 1405 K

$2 NO_2(g) \rightleftharpoons 2 NO(g) + O_2(g)$ $K_c = 1.8 \times 10^{-6}$ at 457 K

determine values of K_c for the following reactions.
   (a) $CO_2(g) + H_2(g) \rightleftharpoons CO(g) + H_2O(g)$
   (b) $2 SO_2(g) + O_2(g) \rightleftharpoons 2 SO_3(g)$
   (c) $H_2S(g) \rightleftharpoons H_2(g) + \frac{1}{2} S_2(g)$
   (d) $NO(g) + \frac{1}{2} O_2(g) \rightleftharpoons NO_2(g)$

3. Determine K_c for the reaction $\frac{1}{2} N_2(g) + \frac{1}{2} O_2(g) + \frac{1}{2} Br_2(g) \rightleftharpoons NOBr(g)$ from the following information (at 298 K).

$2 NO(g) \rightleftharpoons N_2(g) + O_2(g)$ $K_c = 2.4 \times 10^{30}$

$NO(g) + \frac{1}{2} Br_2(g) \rightleftharpoons NOBr(g)$ $K_c = 1.4$

4. Determine numerical values of K_p corresponding to the values of K_c listed for the four reactions in Review Problem 2.

5. Equilibrium is established in the reversible reaction $A + B \rightleftharpoons 2 C$. Following are the equilibrium concentrations: $[A] = 0.47 M$, $[B] = 0.55 M$, $[C] = 0.36 M$. What is the value of K_c for this reaction?

6. The reaction $2 A(g) + B(g) \rightleftharpoons C(g)$ was allowed to come to equilibrium. The *initial* amounts of the reactants present in a 1.80-L vessel were 1.18 mol A and 0.78 mol B. The amount of A, at equilibrium, was found to be 0.92 mol. What is the value of K_c for this reaction?

7. For the reaction $CO(g) + H_2O(g) \rightleftharpoons CO_2(g) + H_2(g)$, $K_c = 1.00$ at about 1100 K. Which of the following statements must always be true concerning this reaction at equilibrium at 1100 K?
   (a) $[CO] = [H_2O] = [CO_2] = [H_2]$
   (b) $[CO] \times [H_2O] = [CO_2] \times [H_2]$
   (c) $[CO] = [H_2O]$ and $[CO_2] = [H_2]$
   (d) $[CO] \times [H_2O] = [CO_2] \times [H_2] = 1.00$

8. An equilibrium mixture of SO_2, SO_3, and O_2 gases is maintained in a 11.5-L flask at a temperature at which $K_c = 55.2$ for the reaction $2 SO_2(g) + O_2(g) \rightleftharpoons 2 SO_3(g)$.
   (a) If the numbers of moles of SO_2 and SO_3 in the flask are equal, how much O_2 is present?
   (b) If the number of moles of SO_3 in the flask is twice the number of moles of SO_2, how much O_2 is present?

9. For the reaction $3 A(g) + B(g) \rightleftharpoons 2 C(g)$,
   (a) Write the equilibrium constant expression for K_c.
   (b) If, at a given temperature, $K_c = 8.8$, can a mixture of 1.55 mol each of A, B, and C exist at equilibrium in a 1.15-L flask?
   (c) If the mixture in (b) is not at equilibrium, in what direction will a net chemical reaction occur, i.e., to the left or to the right?

10. When 1.00 mol $I_2(g)$ is introduced into an evacuated 1.00-L flask at 1200°C, it is 5% dissociated into I atoms. For the reaction $I_2(g) \rightleftharpoons 2 I(g)$, what is the value of (a) K_c; (b) K_p?

11. 0.100 mol H_2 and 0.100 mol I_2 are sealed in a 1.50-L vessel and the mixture is allowed to come to equilibrium at 445°C. What are the equilibrium concentrations of H_2, I_2, and HI?

$H_2(g) + I_2(g) \rightleftharpoons 2 HI(g)$ $K_c = 50.2$ at 445°C

12. 0.100 mol H_2 and 0.100 mol HI are sealed in a 1.50-L vessel and the mixture is allowed to come to equilibrium at 445°C. How many moles of I_2 will be present when equilibrium is established?

$H_2(g) + I_2(g) \rightleftharpoons 2 HI(g)$; $K_c = 50.2$ at 445°C

13. Estimate the value of K_p for the reaction $2 SO_2(g) + O_2(g) \rightleftharpoons 2 SO_3(g)$ at 25°C. Use data from Table 15-5 and Figure 15-6.

14. What effect would increasing the external pressure on the reaction mixture have on the equilibrium condition in each of the following reactions?
   (a) $C(s) + H_2O(g) \rightleftharpoons CO(g) + H_2(g)$
   (b) $CO(g) + H_2O(g) \rightleftharpoons CO_2(g) + H_2(g)$
   (c) $4 HCl(g) + O_2(g) \rightleftharpoons 2 H_2O(g) + 2 Cl_2(g)$

15. A mixture of $HCl(g)$, $O_2(g)$, $H_2O(g)$, and $Cl_2(g)$ is brought to equilibrium at 200°C.

$4 HCl(g) + O_2(g) \rightleftharpoons 2 H_2O(g) + 2 Cl_2(g)$

What would be the effect on the equilibrium amount of $HCl(g)$ if
   (a) additional $O_2(g)$ were added to the mixture at constant volume?
   (b) $Cl_2(g)$ were removed from the reaction mixture at constant volume?
   (c) the volume of the reaction mixture were increased to twice its original value?
   (d) a catalyst were added to the reaction mixture?

16. For which of the following reactions would you expect the percent dissociation to increase with increasing temperature? Explain.
   (a) $NO(g) \rightleftharpoons \frac{1}{2} N_2(g) + \frac{1}{2} O_2(g)$ $\Delta \overline{H}° = -90.2$ kJ/mol
   (b) $SO_3(g) \rightleftharpoons SO_2(g) + \frac{1}{2} O_2(g)$ $\Delta \overline{H}° = +98.9$ kJ/mol
   (c) $N_2H_4(g) \rightleftharpoons N_2(g) + 2 H_2(g)$ $\Delta \overline{H}° = -95.4$ kJ/mol
   (d) $COCl_2(g) \rightleftharpoons CO(g) + Cl_2(g)$ $\Delta \overline{H}° = +108.3$ kJ/mol

Exercises

Writing equilibrium constant expressions

1. Based on the following descriptions of reversible reactions, write a balanced equation and the K_c expression for each.
   (a) Oxygen gas oxidizes gaseous ammonia to gaseous nitrogen and water vapor.
   (b) Hydrogen gas reduces gaseous nitrogen dioxide to gaseous ammonia and water vapor.
   (c) Liquid acetone [$(CH_3)_2CO$] is in equilibrium with its vapor.
   (d) Chlorine gas reacts with liquid carbon disulfide to produce the liquids CCl_4 and S_2Cl_2.
   (e) Nitrogen gas reacts with the solids, sodium carbonate and carbon, to produce solid sodium cyanide and carbon monoxide gas.

2. Write an equilibrium constant expression, K_c, for the formation of *1 mol* of each of the following *gaseous* compounds from its *gaseous* elements. (a) NO; (b) HCl; (c) NH_3; (d) ClF_3; (e) NOCl.

3. Determine values of K_c from the K_p values given.
   (a) $SO_2Cl_2(g) \rightleftharpoons SO_2(g) + Cl_2(g)$
   $$K_p = 2.9 \times 10^{-2} \text{ at } 303 \text{ K}$$
   (b) $2 NO(g) + O_2(g) \rightleftharpoons 2 NO_2(g)$
   $$K_p = 1.48 \times 10^4 \text{ at } 184°C$$
   (c) $Sb_2S_3(s) + 3 H_2(g) \rightleftharpoons 2 Sb(s) + 3 H_2S(g)$
   $$K_p = 0.429 \text{ at } 713 \text{ K}$$

4. The vapor pressure of water at 25°C is 23.8 mmHg. Write K_p for the vaporization of water, in the unit, atm. What is the value of K_c for the vaporization process?

5. Given the equilibrium constant values

$N_2(g) + \frac{1}{2} O_2(g) \rightleftharpoons N_2O(g) \qquad K_c = 3.4 \times 10^{-18}$

$N_2O_4(g) \rightleftharpoons 2 NO_2(g) \qquad K_c = 4.6 \times 10^{-3}$

$\frac{1}{2} N_2(g) + O_2(g) \rightleftharpoons NO_2(g) \qquad K_c = 4.1 \times 10^{-9}$

Determine a value of K_c for the reaction

$2 N_2O(g) + 3 O_2(g) \rightleftharpoons 2 N_2O_4(g)$

6. Use the following equilibrium data at 1200 K to estimate a value of K_p at 1200 K for the reaction $2 H_2(g) + O_2(g) \rightleftharpoons 2 H_2O(g)$. (*Hint:* Note that some of the values are K_p and some, K_c.)
   (a) $C(graphite) + CO_2(g) \rightleftharpoons 2 CO(g) \qquad K_c = 0.64$
   (b) $CO_2(g) + H_2(g) \rightleftharpoons CO(g) + H_2O(g) \qquad K_p = 1.4$
   (c) $C(graphite) + \frac{1}{2} O_2(g) \rightleftharpoons CO(g) \qquad K_c = 1 \times 10^8$

Experimental determination of equilibrium constants

7. 1.00 g PCl_5 is introduced into a 250-mL flask and the flask heated to 250°C, where the dissociation of PCl_5 is allowed to reach equilibrium: $PCl_5(g) \rightleftharpoons PCl_3(g) + Cl_2(g)$. The quantity of $Cl_2(g)$ present at equilibrium is found to be 0.25 g. What is the value of K_c for the dissociation reaction at 250°C?

8. A mixture of 1.00 g $H_2(g)$ and 1.06 g $H_2S(g)$ is introduced into a 0.500-L flask and the mixture is allowed to come to equilibrium at 1670 K: $2 H_2(g) + S_2(g) \rightleftharpoons 2 H_2S(g)$.   The equilibrium amount of $S_2(g)$ is found to be 8.00×10^{-6} mol. Determine the value of K_p.

9. An experiment used to establish principles of chemical equilibrium was the homogeneous reaction of ethanol (C_2H_5OH) and acetic acid (CH_3COOH) to produce ethyl acetate and water.

$$C_2H_5OH + CH_3CO_2H \rightleftharpoons CH_3CO_2C_2H_5 + H_2O \qquad (15.40)$$

The reaction can be followed by analyzing the equilibrium mixture for its acetic acid content.

$$2 CH_3CO_2H(aq) + Ba(OH)_2(aq) \rightarrow$$
$$Ba(CH_3CO_2)_2(aq) + 2 H_2O \quad (15.41)$$

An experiment is performed in which 1.000 mol of acetic acid and 0.500 mol of ethanol are mixed and allowed to come to equilibrium. A sample representing exactly one-hundredth of the total equilibrium mixture requires 28.85 mL of 0.1000 M $Ba(OH)_2$ for its titration. Show that the equilibrium constant, K_c, for reaction (15.40) is 4.0. (*Hint:* It is not necessary to know the volume of the reaction mixture.)

*10. The decomposition of HI(g) is represented by the equation $2 HI(g) \rightleftharpoons H_2(g) + I_2(g)$. HI(g) is introduced into five identical 400-cm^3 glass bulbs, and the five bulbs are maintained at 623 K. Each bulb is opened after a period of time and analyzed for I_2 by titration with 0.0150 M $Na_2S_2O_3(aq)$.

$$I_2(aq) + 2 Na_2S_2O_3(aq) \rightarrow Na_2S_4O_6(aq) + 2 NaI(aq)$$

What is the value of K_c at 623 K?

Bulb number	Original amount of HI(g), g	Bulb opened after, h	Volume of 0.0150 M $Na_2S_2O_3$ required for titration, mL
1	0.300	2	20.96
2	0.320	4	27.90
3	0.315	12	32.31
4	0.406	20	41.50
5	0.280	40	28.68

Equilibrium relationships

11. An equilibrium mixture at 1000 K contains 0.276 mol H_2, 0.276 mol CO_2, 0.224 mol CO, and 0.224 mol H_2O.

$$CO_2(g) + H_2(g) \rightleftharpoons CO(g) + H_2O(g)$$

   (a) Show that for this reaction K_c is independent of the reaction volume, V.
   (b) Determine the value of K_c.

12. Equilibrium is established at a temperature at which $K_c = 375$ for the reaction $2 NO(g) + O_2(g) \rightleftharpoons 2 NO_2(g)$. The equilibrium amount of $O_2(g)$ found in a 0.755-L vessel is 0.0148 mol. What is the ratio of [NO] to [NO$_2$] in this equilibrium mixture?

13. For the dissociation of $I_2(g)$ at about 1200°C, $I_2(g) \rightleftharpoons 2 I(g)$, $K_c = 1.1 \times 10^{-2}$. What volume vessel is required if it is desired that 1.00 mol I_2 and 0.50 mol I be present at equilibrium?

Direction and extent of chemical change

14. Can a mixture of 3 mol O_2, 2 mol SO_2, and 6 mol SO_3 be maintained indefinitely in a 8.50-L flask at a temperature at which $K_c = 100$ in the reaction $2 SO_2(g) + O_2(g) \rightleftharpoons 2 SO_3(g)$? If not, in what direction will a net reaction occur?

15. Starting with 1.00 mol each of $CO(g)$ and $COCl_2(g)$ in a 1.75-L reaction vessel at 668 K, what is the number of moles of $Cl_2(g)$ produced at equilibrium?

$$CO(g) + Cl_2(g) \rightleftharpoons COCl_2(g) \qquad K_c = 1.2 \times 10^3 \text{ at } 668 \text{ K}.$$

16. With reference to Example 15-6, what will be the amounts of $CO(g)$, $H_2O(g)$, $CO_2(g)$, and $H_2(g)$ when equilibrium is established?

17. 3.00 mol $SbCl_3$ and 1.00 mol Cl_2 are introduced into an evacuated 5.00-L vessel and equilibrium is established at 248°C. How many moles of $SbCl_5$, $SbCl_3$, and Cl_2 are present at equilibrium?

$$SbCl_5(g) \rightleftharpoons SbCl_3(g) + Cl_2(g) \quad K_c = 2.5 \times 10^{-2} \text{ at } 248°C$$

18. If 0.390 mol SO_2, 0.156 mol O_2, and 0.657 mol SO_3 are introduced simultaneously into a 1.90-L reaction vessel at 1000 K,
   (a) Is this mixture at equilibrium?
   (b) If not, in what direction must a net reaction proceed to establish equilibrium?

$$2 SO_2(g) + O_2(g) \rightleftharpoons 2 SO_3(g) \quad K_c = 2.8 \times 10^2 \text{ at } 1000 \text{ K}$$

★19. Calculate the actual equilibrium amounts of SO_2, O_2, and SO_3 in Exercise 18.

20. 1.00 g *each* of $CO(g)$, $H_2O(g)$, and $H_2(g)$ are sealed in a 1.41-L vessel and brought to equilibrium at 600 K. What mass of $CO_2(g)$ is present in the equilibrium mixture?

$$CO(g) + H_2O(g) \rightleftharpoons CO_2(g) + H_2(g) \qquad K_c = 23.2$$

21. 0.250 mol *each* of $H_2(g)$ and $I_2(g)$ are allowed to establish equilibrium with HI(g) in a 4.10-L reaction flask at 445°C. What will be the mole percent HI in the equilibrium mixture?

$$H_2(g) + I_2(g) \rightleftharpoons 2 HI(g) \qquad K_c = 50.2$$

★22. Derive, by calculation, the equilibrium amounts of SO_2, O_2, and SO_3 listed in **(a)** Figure 15-3(c); **(b)** Figure 15-4(b).

23. An aqueous solution is made 1.00 *M* in $AgNO_3$ and 1.00 *M* in $Fe(NO_3)_2$ and allowed to come to equilibrium. What are the values of [Ag$^+$], [Fe^{2+}], and [Fe^{3+}] when equilibrium is established?

$$Ag^+(aq) + Fe^{2+}(aq) \rightleftharpoons Fe^{3+}(aq) + Ag(s) \qquad K_c = 2.98$$

★24. Solid iron metal is added to a solution having the concentrations $[Cr^{3+}] = 0.250 \, M$, $[Cr^{2+}] = 0.0500 \, M$, and $[Fe^{2+}] = 0.00100 \, M$. What are the concentrations of these ions when equilibrium is established?

$$2 Cr^{3+}(aq) + Fe(s) \rightleftharpoons 2 Cr^{2+}(aq) + Fe^{2+}(aq) \quad K_c = 10.34$$

Partial pressure equilibrium constant, K_p

25. 1.00 mol *each* of $SO_2(g)$ and $Cl_2(g)$ are introduced into an evacuated 2.50-L flask and the following equilibrium is established at 303 K.

$$SO_2Cl_2(g) \rightleftharpoons SO_2(g) + Cl_2(g) \qquad K_p = 2.9 \times 10^{-2}$$

For this equilibrium, calculate **(a)** the partial pressure of SO_2Cl_2; **(b)** the total gas pressure.

26. A sample of air with an original mole ratio of nitrogen to oxygen of 79:21 is heated to 2500 K. When equilibrium is established, the mole percent NO present is found to be 1.8%. Calculate K_p for the reaction

$$N_2(g) + O_2(g) \rightleftharpoons 2 NO(g) \qquad K_p \text{ at } 2500 \text{ K} = ?$$

(*Hint:* The result is independent of both volume and total pressure. The presence of other gases in air can be neglected.)

27. A sample of $NH_4HS(s)$ is introduced into a 1.60-L flask containing 0.170 g NH_3. What is the total gas pressure when equilibrium is established in the flask at 25°C?

$$NH_4HS(s) \rightleftharpoons NH_3(g) + H_2S(g) \qquad K_p = 0.108 \text{ at } 25°C$$

28. In the manufacture of sodium carbonate by the Solvay process, $NaHCO_3(s)$ is decomposed by heating.

$$2 NaHCO_3(s) \rightleftharpoons Na_2CO_3(s) + CO_2(g) + H_2O(g)$$
$$K_p = 0.23 \text{ at } 100°C$$

   (a) If a sample of $NaHCO_3(s)$ is brought to a temperature of 100°C in a closed container, what will be the total gas pressure (in atm) at equilibrium?
   (b) A mixture of 1.00 mol each of $NaHCO_3(s)$ and $Na_2CO_3(s)$ is introduced into a 2.50-L flask in which $P_{CO_2} = 2.10$ atm and $P_{H_2O} = 715$ mmHg. When equilibrium is established (at 100°C), will the partial pressures of $CO_2(g)$ and $H_2O(g)$ be greater or less than their initial partial pressures? Explain.
   ★(c) Starting with the initial conditions of part (b), what will be the partial pressures of $CO_2(g)$ and $H_2O(g)$ when equilibrium is established at 100°C?

Dissociation reactions

29. If the reaction mixture described in Example 15-13 were transferred to a 10.0-L vessel, would the % dissociation increase, decrease, or remain the same? Explain.

30. Calculate the % dissociation referred to in Exercise 29.

31. What is the % dissociation of HI(g) into its gaseous elements at 340°C?

$$H_2(g) + I_2(g) \rightleftharpoons 2\ HI(g) \qquad K_p = 6.9 \times 10^1$$

★32. A sample of pure $PCl_5(g)$ is introduced into an evacuated flask and allowed to dissociate. If the fraction of the PCl_5 molecules that dissociate is denoted by α, and if the total gas pressure at equilibrium is P, show that for the reaction

$$PCl_5(g) \rightleftharpoons PCl_3(g) + Cl_2(g) \qquad K_p = \frac{\alpha^2 P}{1 - \alpha^2}$$

33. With reference to the equation established in Exercise 32, if $K_p = 1.78$ at 250°C,
  (a) What is the % dissociation of PCl_5 at 250°C and 1 atm total pressure?
  (b) Under what total pressure must the gaseous mixture be maintained to limit dissociation of PCl_5 to 10.0%?

Le Châtelier's principle

34. Continuous removal of one of the products of a chemical reaction has the effect of causing the reaction to go to completion. Explain.

35. The *endothermic* reaction $A(g) + B(g) \rightleftharpoons 2\ C(g)$ proceeds to an equilibrium condition at 200°C. Which of the following statements is true? Explain. (*Hint:* There may be more than one correct statement.)
  (a) If the mixture is transferred to a reaction vessel of twice the volume, the amounts of reactants and products will remain unchanged.
  (b) Addition of an appropriate catalyst will result in the formation of a greater amount of C(g).
  (c) Lowering the reaction temperature to 100°C will result in the formation of a greater amount of C(g).
  (d) Addition of an inert gas, such as helium, will have little or no effect on the equilibrium.

36. Show that the % dissociation in reaction (1) depends on the volume of the reaction vessel and in reaction (2) it does not. Explain this difference from the standpoint of Le Châtelier's principle.

(1) $SO_2Cl_2(g) \rightleftharpoons SO_2(g) + Cl_2(g)$

(2) $CS_2(g) \rightleftharpoons C(s) + S_2(g)$

37. Explain why all dissociation reactions of the type $A_2(g) \rightleftharpoons 2\ A(g)$ proceed to a greater extent at higher temperatures [e.g., $I_2(g) \rightleftharpoons 2\ I(g)$].

38. The reaction $N_2(g) + O_2(g) \rightleftharpoons 2\ NO(g)$, $\Delta \overline{H}° = +181$ kJ/mol, occurs whenever a substance is burned in air. This reaction occurs in internal combustion engines, leading to the formation of oxides of nitrogen that are involved in the production of photochemical smog. High-compression engines, characteristic of large automobiles, operate at high temperatures.

  (a) What effect do these high temperatures have on the equilibrium production of NO(g)?
  (b) What effect does high temperature have on the rate of this reaction?

39. The freezing of $H_2O(l)$ at 0°C can be represented as

$$H_2O(l,\ d = 1.00\ \text{g/cm}^3) \rightleftharpoons H_2O(s,\ d = 0.92\ \text{g/cm}^3)$$

Explain why the application of pressure to ice at 0°C causes the ice to melt. Is this behavior to be expected of solids in general?

Effect of temperature on equilibrium constants

40. For the reaction $N_2O_4(g) \rightleftharpoons 2\ NO_2(g)$, $\Delta \overline{H}° = +61.5$ kJ/mol and $K_p = 0.113$ at 298 K.
  (a) What is the value of K_p at 0°C?
  (b) At what temperature will $K_p = 1.00$?

41. The following equilibrium constants have been determined for the reaction $H_2(g) + I_2(g) \rightleftharpoons 2\ HI(g)$: $K_c = 50.0$ at 448°C and 66.9 at 350°C. Use these data to estimate $\Delta \overline{H}°$ for the reaction and compare your result with that given in the reaction profile of Figure 14-10.

42. Use data from Table 15-3 and Appendix D to estimate the value of K_p at 100°C for the reaction $2\ NO(g) + O_2(g) \rightleftharpoons 2\ NO_2(g)$.

43. The following data are given for the temperature variation of K_p (partial pressures expressed in atm) for the reaction

$$2\ NaHCO_3(s) \rightleftharpoons Na_2CO_3(s) + CO_2(g) + H_2O(g)$$

t, °C	K_p
30	1.66×10^{-5}
50	3.90×10^{-4}
70	6.27×10^{-3}
100	2.31×10^{-1}

  (a) Plot a graph similar to Figure 15-6 and determine $\Delta \overline{H}°$ for the reaction.
  (b) Calculate the temperature at which the total gas pressure above a mixture of $NaHCO_3(s)$ and $Na_2CO_3(s)$ is 2.00 atm.

★44. The reaction $A(s) \rightleftharpoons B(s) + 2\ C(g) + \frac{1}{2} D(g)$ is found to have $\Delta \overline{H}° = 0$.
  (a) Will K_p increase, decrease, or remain constant with temperature? Explain.
  (b) If a constant-volume mixture originally at equilibrium at 298 K is heated to 400 K, will the amount of D(g) present increase, decrease, or remain constant? Explain.

Additional Exercises

1. Assume that when equilibrium is established in a certain mixture of sulfur dioxide, oxygen, and sulfur trioxide, $[SO_2] = [SO_3]$. Show that $[O_2]$ has the same value regardless of which of the three expressions (15.8a), (15.8b), or (15.8c) is used to describe the equilibrium.

2. Use data from Review Problem 3 to determine K_p at 298 K for the reaction

$$\tfrac{1}{2} N_2(g) + \tfrac{1}{2} O_2(g) + \tfrac{1}{2} Br_2(g) \rightleftharpoons NOBr(g) \qquad K_p = ?$$

3. The high-temperature dissociation of salicylic acid is represented by the equation

$$C_7H_6O_3(g) \rightleftharpoons C_6H_6O(g) + CO_2(g)$$

As a result of an experiment carried out at 200°C, an initial sample of 0.300 g $C_7H_6O_3$ in a 50.0-cm³ vessel yielded an equilibrium mixture in which the partial pressure of $CO_2(g)$ was found to be 1.50 atm. What are **(a)** K_c and **(b)** K_p for this reaction at 200°C?

4. Formamide is used as an intermediate and solvent in the manufacture of pharmaceuticals, dyes, and agricultural chemicals. At elevated temperatures it decomposes to $NH_3(g)$ and $CO(g)$.

$$HCONH_2(g) \rightleftharpoons NH_3(g) + CO(g) \qquad K_c = 4.84 \text{ at } 400 \text{ K}$$

If 0.100 mol $HCONH_2(g)$ is allowed to dissociate in a 1.50-L flask at 400 K, what will be the *total* pressure at equilibrium?

5. With reference to the reaction described in Exercise 9,

$$C_2H_5OH + CH_3CO_2H \rightleftharpoons CH_3CO_2C_2H_5 + H_2O \qquad K_c = 4.0$$

15.5 g C_2H_5OH, 25.0 g CH_3CO_2H, 45.5 g $CH_3CO_2C_2H_5$, and 52.0 g H_2O are mixed and allowed to react.
(a) In what direction will a net reaction occur?
(b) What will be the equilibrium quantities of each of the reacting species?

6. A solution is prepared having $[Fe^{3+}] = 0.4000\ M$ and $[Hg_2^{2+}] = 0.2500\ M$. What are the values of $[Fe^{3+}]$, $[Fe^{2+}]$, $[Hg_2^{2+}]$, and $[Hg^{2+}]$ when equilibrium is established?

$$2\ Fe^{3+}(aq) + Hg_2^{2+}(aq) \rightleftharpoons 2\ Fe^{2+}(aq) + 2\ Hg^{2+}(aq)$$
$$K_c = 9.14 \times 10^{-6}$$

7. Use data from Appendix D to establish if the forward reaction is favored by high or low temperatures.

$$2\ NO(g) + O_2(g) \rightleftharpoons 2\ NO_2(g)$$

8. An equilibrium condition is attained in the reaction

$$Fe_3O_4(s) + 4\ H_2(g) \rightleftharpoons 3\ Fe(s) + 4\ H_2O(g)$$

at 150°C. What would be the effect on $[H_2O(g)]$ in a constant-volume reaction mixture if
(a) additional $H_2(g)$ were introduced?
(b) more $Fe(s)$ were added?
(c) a catalyst were employed?

9. Use Le Châtelier's principle to make qualitative predictions about
(a) the effect on the amount of $Cl_2(g)$ at equilibrium if the volume of the reaction vessel in Exercise 15 is increased from 1.75 L to 2.50 L;
(b) the effect on the equilibrium amounts of $SbCl_5$, $SbCl_3$, and Cl_2 if a catalyst is used in the reaction in Exercise 17;
(c) the effect on the % dissociation of $HI(g)$ in Exercise 31 if an inert gas is added until the total pressure exerted by the gaseous mixture is increased from 1.0 atm to 10.0 atm;
(d) the effect on the total pressure exerted by $CO_2(g)$ and $H_2O(g)$ in equilibrium with $NaHCO_3(s)$ and $Na_2CO_3(s)$ if the temperature is raised from 25°C to 200°C (see Exercise 43).

10. What is the % dissociation of $H_2S(g)$ if 1.00 mol H_2S is introduced into an evacuated 1.10-L vessel at 1000 K?

$$2\ H_2S(g) \rightleftharpoons 2\ H_2(g) + S_2(g) \qquad K_c = 1.0 \times 10^{-6}$$

11. With reference to the data in Exercise 40, at what temperature is a sample of N_2O_4 50% dissociated into NO_2, if the total pressure is maintained at 1 atm?

12. At 2000 K the reaction $2\ CH_4(g) \rightleftharpoons C_2H_2(g) + 3\ H_2(g)$ has $K_c = 0.154$. If a reaction mixture at equilibrium at 2000 K contains 0.10 mol each of $CH_4(g)$ and $H_2(g)$ in a 1.00-L vessel, **(a)** what is the mole fraction of $C_2H_2(g)$ present? **(b)** Is the conversion of $CH_4(g)$ to $C_2H_2(g)$ favored by high or low pressures? Explain.

13. Show that in terms of mole fractions of gases and *total* gas pressure, the equilibrium constant expression for $N_2(g) + 3\ H_2(g) \rightleftharpoons 2\ NH_3(g)$ is

$$K_p = \frac{(\chi_{NH_3})^2}{(\chi_{N_2})(\chi_{H_2})^3} \times \frac{1}{(P_{tot.})^2}$$

⋆14. For the synthesis of ammonia at 500 K, $N_2(g) + 3\ H_2(g) \rightleftharpoons 2\ NH_3(g)$, $K_p = 9.06 \times 10^{-2}$. If N_2 and H_2 are allowed to react in the mole ratio 1:3 and the total pressure is maintained at 1.00 atm, what is the mol % NH_3 at equilibrium? (*Hint:* Use the result of the preceding exercise and a method of successive approximations.)

⋆15. A mixture of $H_2S(g)$ and $CH_4(g)$ in the mole ratio 2:1 was allowed to come to equilibrium at 700°C and a total gas pressure of 1 atm. The *equilibrium* mixture was analyzed for the amount of H_2S present; 9.54×10^{-3} mol H_2S was found. The CS_2 present at equilibrium was converted successively to H_2SO_4 and then to $BaSO_4$; 1.42×10^{-3} mol $BaSO_4$ was obtained. Use these data to determine K_p at 700°C for the reaction

$$2\ H_2S(g) + CH_4(g) \rightleftharpoons CS_2(g) + 4\ H_2(g) \qquad K_p \text{ at } 700°C = ?$$

⋆16. For the loss of water by the trihydrate $CuSO_4 \cdot 3H_2O$, the following data are given at 298 K.

$CuSO_4 \cdot 3H_2O(s) \rightleftharpoons CuSO_4 \cdot H_2O(s) + 2\ H_2O(g)$
$K_p = 5.43 \times 10^{-5}$ (pressures in atm); $\Delta \overline{H}° = +113$ kJ/mol

(a) Write an equation similar to equation (15.26) to show how the partial pressure of water vapor varies with temperature above a mixture of the two hydrates.

(b) At what temperature is the partial pressure of the water vapor above the hydrates equal to 0.100 atm?

*17. The decomposition of the poisonous gas phosgene is represented by the equation $COCl_2(g) \rightleftharpoons CO(g) + Cl_2(g)$. Values of K_p for this reaction are listed as $K_p = 6.7 \times 10^{-9}$ at 100°C and $K_p = 4.44 \times 10^{-2}$ at 395°C. At what temperature is $COCl_2$ 15% dissociated when the total gas pressure is maintained at 1.00 atm?

*18. What is the apparent molar mass of the gaseous mixture that results when $COCl_2(g)$ is allowed to dissociate at 395°C and a total pressure of 3.00 atm?

$COCl_2(g) \rightleftharpoons CO(g) + Cl_2(g)$ $K_p = 4.44 \times 10^{-2}$ at 395°C

Self-Test Questions

For questions 1 through 8 select the single item that best completes each statement.

1. In the reaction $H_2(g) + I_2(g) \rightleftharpoons 2\ HI(g)$, a mixture that initially contains 2 mol H_2 and 1 mol I_2 produces, at equilibrium, (a) 1 mol HI; (b) 2 mol HI; (c) more than 2 but less than 4 mol HI; (d) less than 2 mol HI.

2. Equilibrium is established in the reaction $2\ SO_2(g) + O_2(g) \rightleftharpoons 2\ SO_3(g)$ at a temperature at which $K_c = 100$. If the number of moles of SO_3 in the equilibrium mixture is equal to the number of moles of SO_2,
(a) the number of moles of O_2 is also equal to the number of moles of SO_2;
(b) the number of moles of O_2 is half the number of moles of SO_2;
(c) $[O_2] = 0.01\ M$;
(d) $[O_2]$ may have any of several different values.

3. The volume of the reaction vessel containing an equilibrium mixture in the reaction $SO_2Cl_2(g) \rightleftharpoons SO_2(g) + Cl_2(g)$ is increased. When equilibrium is reestablished,
(a) the amount of $Cl_2(g)$ will have increased;
(b) the amount of $SO_2(g)$ will have decreased;
(c) the amount of $Cl_2(g)$ will have remained unchanged;
(d) the amount of $SO_2Cl_2(g)$ will have increased.

4. Which of the following statements is true when equilibrium is established in the reaction $A + B \rightleftharpoons C + D$; $K_c = 10.0$?
(a) $[C][D] = [A][B]$;   (b) $[C] = [A]$ and $[B] = [D]$;
(c) $[A][B] = 0.10 \times [C][D]$;   (d) $[A] = [B] = [C] = [D] = 10.0\ M$.

5. Equilibrium in a mixture of $CO(g)$, $H_2O(g)$, $CO_2(g)$, and $H_2(g)$ is established in a 1.00-L container at 1000 K. The following data are given.

$CO(g) + H_2O(g) \rightleftharpoons CO_2(g) + H_2(g)$
$$\Delta \overline{H}° = -42\ \text{kJ/mol} \qquad K_c = 0.66$$

The equilibrium amount of $H_2(g)$ can be increased by (a) adding a catalyst; (b) increasing the temperature; (c) transferring the mixture to a 10.0-L container; (d) none of the methods described in (a), (b), or (c).

6. 1.00 mol *each* of $CO(g)$, $H_2O(g)$, and $CO_2(g)$ are introduced into a 10.0-L flask at a temperature at which $K_c = 10.0$ for the reaction

$CO(g) + H_2O(g) \rightleftharpoons CO_2(g) + H_2(g)$ $K_c = 10.0$

When the reaction reaches a state of equilibrium,
(a) the amount of H_2 will be 1.00 mol.
(b) the amount of $CO_2(g)$ will be greater than 1.00 mol, and the amounts of $CO(g)$, $H_2O(g)$, and $H_2(g)$ will each be less than 1.00 mol.
(c) the amounts of all reactants and products will be greater than 1.00 mol.
(d) the amounts of $CO_2(g)$ and $H_2(g)$ will each be greater than 1.00 mol, and the amounts of $CO(g)$ and $H_2O(g)$ will each be less than 1.00 mol.

7. For the reaction $2\ NO_2(g) \rightleftharpoons 2\ NO(g) + O_2(g)$, $K_c = 1.8 \times 10^{-6}$ at 184°C. At 184°C, the value of K_c for the reaction $NO(g) + \frac{1}{2}\ O_2(g) \rightleftharpoons NO_2(g)$ is (a) 0.9×10^6; (b) 7.5×10^2; (c) 5.6×10^5; (d) 2.8×10^5.

8. For the dissociation reaction $2\ H_2S(g) \rightleftharpoons 2\ H_2(g) + S_2(g)$, $K_p = 1.2 \times 10^{-2}$ at 1065°C. For this same reaction at 298 K (a) K_c is less than K_p; (b) K_c is greater than K_p; (c) $K_c = K_p$; (d) whether K_c is less than, equal to, or greater than K_p depends on the total gas pressure.

9. Describe how the balanced chemical equation for a reversible reaction is used in equilibrium calculations. Explain why the balanced equation *alone* cannot be used for determining the composition of an equilibrium mixture.

10. Explain briefly the relationship between
(a) the rates of chemical reactions and the condition of equilibrium.
(b) the reaction quotient, Q, and the equilibrium constant expression, K_c.
(c) the equilibrium constants K_c and K_p.

11. A 0.0010-mol sample of $S_2(g)$ is allowed to dissociate in an 0.500-L flask at 1000 K. When equilibrium is reached, 1.0×10^{-11} mol $S(g)$ is present. What is K_c for the reaction $S_2(g) \rightleftharpoons 2\ S(g)$ at 1000 K?

12. Into a 1.00-L vessel at 1000 K are introduced 0.100 mol each of $NO(g)$ and $Br_2(g)$ and 0.0100 mol $NOBr(g)$.

$2\ NO(g) + Br_2(g) \rightleftharpoons 2\ NOBr(g)$ $K_c = 1.32 \times 10^{-2}$ at 1000 K

(a) In what direction must a net reaction occur?
*(b) What is the partial pressure of $NOBr(g)$ in the vessel when equilibrium is established?

16 Thermodynamics and Chemistry

Thermodynamics deals with the relationship between heat and other energy forms. Its development, one of the most significant scientific achievements of the nineteenth century, was largely through the efforts of physicists and engineers seeking to attain higher efficiencies in heat engines. An interest in improving heat engines has again come into prominence because of the need to utilize fossil fuels more effectively. But for the past 75 years the most important applications of thermodynamics have been in chemistry. The laws of thermodynamics provide powerful tools for studying chemical reactions. In Chapter 6 we discussed one important aspect of thermodynamics—thermochemistry, i.e., the heat effects accompanying chemical reactions. In this chapter we will formally introduce the first law of thermodynamics, which is the basis of thermochemistry. Also, we will consider the second law of thermodynamics, which allows us to deal with the *direction* and *extent* of a chemical reaction in a more satisfactory manner than was possible in Chapter 15. Specifically, the second law provides a basis for deriving values of equilibrium constants from certain thermodynamic properties introduced in this chapter. Finally, in the third law of thermodynamics we will discover a starting point for the experimental evaluation of these new thermodynamic properties.

16-1 The First Law of Thermodynamics

In Chapter 6 we made repeated use of the law of conservation of energy. It is the underlying principle of calorimetry, as expressed, for example, in the fundamental equation of bomb calorimetry.

$$q_{rxn} + q_{water} + q_{calorim.} = 0 \qquad (6.7)$$

Equation (6.7) states that the sum of all the heat effects in a process is zero. Or, all the heat energy lost by the system (the reaction mixture) is gained by the surroundings (the water in which the bomb is immersed, together with the rest of the calorimeter assembly). Or, energy can neither be created nor destroyed in a process.

The first law of thermodynamics is a restatement of the law of conservation of energy in a form that considers the internal energy of a system and distinguishes between the two fundamental forms of energy transfer—heat and work. In an isolated system the total energy remains constant. Or, if a system exchanges heat and/or work with its surroundings, this must occur in such a way that the total energy of the system and its surroundings remains constant. In terms of internal energy (E), heat (q), and work (w),

$$\Delta E = q - w \tag{16.1}$$

In applying equation (16.1) we will use the following conventions.

ΔE is the change in internal energy of the system for some process.

q is a *positive* quantity if heat is *absorbed* by the system ($q > 0$).

q is a *negative* quantity if heat is *lost* by the system ($q < 0$).

w is a *positive* quantity if work is done *by* the system ($w > 0$).

w is a *negative* quantity if work is done *on* the system ($w < 0$).

Suppose that a system absorbs some heat energy ($q > 0$) and does work on the surroundings at the same time ($w > 0$). The change in internal energy, ΔE, is equal to the heat absorbed less the work done ($q - w$). If the quantity of heat absorbed is greater than the work done, ΔE is positive. If more work is done than heat absorbed, ΔE is negative.

Example 16-1   A gas does 243 J of work while expanding (as previously illustrated in Figure 6-1). During the expansion the gas absorbs 225 J of heat from the surroundings. What is ΔE for the gas?

Solution.   Heat absorbed is a *positive* quantity: $q = +225$ J. The work *done* by the system (gas) is also a positive quantity: $w = +243$ J.

$$\Delta E = q - w = +225 \text{ J} - (+243 \text{ J}) = -18 \text{ J}$$

SIMILAR EXAMPLE: Review Problem 1.

In Section 6-4 we pointed out that the absolute value of the internal energy of a system cannot be measured. For a given state or condition of the system, however, the internal energy has a *unique* value: *internal energy is a function of state*. When a system undergoes a change between two states, the *difference* in internal energy, ΔE, also has a *unique* value. That is, the value of ΔE is *independent* of the means or path by which the change between the two states is made. The values of q and w, on the other hand, do depend on the path taken—there are many combinations of q and w that could lead to the value of $\Delta E = -18$ J in Example 16-1.

Change in Internal Energy for a Chemical Reaction.   In a chemical reaction we can think of the reactants as representing one state of a thermodynamic system, state 1, with an internal energy E_1. The products represent a different state—state 2, with an internal energy, E_2.

reactants $\longrightarrow$ products
  (state 1)   (state 2)
   E_1 E_2

Accompanying the reaction there occurs a change in internal energy,

$$\Delta E = E_2 - E_1$$

which according to the first law of thermodynamics can also be represented as

$$\Delta E = q - w$$

To evaluate ΔE we need to measure q and w, and these quantities, of course, depend on how the reaction is carried out.

Authors vary in their thermodynamic notation. Sometimes the symbol U is used instead of E; a different sign convention may be adopted for w; and a different form used for equation (16.1).

If a reaction is carried out in a bomb calorimeter (recall Figure 6-4), the thermodynamic system is the *contents* of the bomb. The rest of the assembly represents the surroundings. The bomb confines the system to a *constant volume: There is no opportunity for the system to do work or to have work done on it.* As a result, $w = 0$, and $\Delta E = q_V$ (where the subscript V stands for constant volume). A bomb calorimetry experiment, then, measures a quantity known as the **heat of reaction at constant volume (q_V),** which is equal to ΔE for the reaction.

The Thermodynamic Function Enthalpy, *H*.   If a reaction is carried out in a container *open to the atmosphere,* it will, in most instances, have a slightly different heat of reaction than if carried out in a bomb calorimeter. In such cases, rather than the volume of the system remaining constant, the external pressure is constant (essentially 1 atm). The **heat of reaction at constant pressure** is designated as q_p. If the only type of work considered is pressure–volume work, and if the work is performed at constant pressure, the quantity of work is $w = P\Delta V$. Now we can write

$$\Delta E = q_p - P\Delta V \quad \text{(at constant pressure)}$$

and

$$q_p = \Delta E + P\Delta V \quad \text{(at constant pressure)} \tag{16.2}$$

At this point let us reintroduce the thermodynamic property, enthalpy, H, and let us define it as the sum of the internal energy and the pressure–volume product of a system.

$$H = E + PV \tag{16.3}$$

For a change at *constant pressure,*

$$\Delta H = \Delta E + P\Delta V = q_p \tag{16.4}$$

Since E and the PV product are both functions of state, their sum, H, is also a state function.

Expressions (16.2) and (16.4) are equivalent. In Chapter 6 we described enthalpy change as being the heat effect accompanying a process carried out at constant pressure. Here we see why this is so, based on the first law of thermodynamics and the definition given in (16.3). A comparison of internal energy and enthalpy is made in Table 16-1.

Relationship Between ΔH and ΔE.   If only liquids and solids are involved in a reaction, very little volume change occurs. According to equation (16.4), in such reactions ΔH and ΔE should have essentially the same value (since $P\Delta V \simeq 0$). A greater change in volume occurs when gases are involved, as in the combustion of naphthalene.

$$C_{10}H_8(s) + 12\ O_2(g) \longrightarrow 10\ CO_2(g) + 4\ H_2O(l) \quad \Delta\overline{E} = -5151.9 \text{ kJ/mol} \tag{16.5}$$

Here 12 mol *gaseous* O_2 is replaced by 10 mol *gaseous* CO_2.

TABLE 16-1
Internal energy (E) and enthalpy (H) compared

	E	$H = E + PV$
nature	fundamental	invented for convenience
most useful at:	constant volume (e.g., reaction in a bomb calorimeter)	constant pressure (e.g., reaction in an open beaker)
first-law statement under these conditions	$q_V = \Delta E$	$q_P = \Delta H$

If we treat $O_2(g)$ and $CO_2(g)$ as ideal gases and compare them at the same T and P, we can write

$$V_{CO_2} = \frac{n_{CO_2}RT}{P} \qquad \text{and} \qquad V_{O_2} = \frac{n_{O_2}RT}{P}$$

The change in volume in the reaction is thus

$$\Delta V = V_{CO_2} - V_{O_2} = (n_{CO_2} - n_{O_2})\frac{RT}{P}$$

or

$$\Delta V = \Delta n_g \left(\frac{RT}{P}\right) \qquad \text{where } \Delta n_g = \text{number of moles of } \textit{gaseous} \qquad (16.6)$$
$$\textit{products} \text{ minus the number of}$$
$$\text{moles of } \textit{gaseous reactants}$$

By substituting equation (16.6) into equation (16.4), we obtain

$$\Delta H = \Delta E + \cancel{P}\left(\frac{\Delta n_g RT}{\cancel{P}}\right)$$

and

$$\Delta H = \Delta E + \Delta n_g RT \tag{16.7}$$

Equation (16.7) is generally applied at 25°C (298 K). If the values of ΔH and ΔE are in kJ/mol, this requires that R be expressed as 8.314×10^{-3} kJ mol^{-1} K^{-1}. The RT product becomes 8.314×10^{-3} kJ mol^{-1} K$^{-1} \times 298$ K $= 2.48$ kJ/mol. Δn_g represents the change in number of moles of gas per mole of reaction.* Combining these features into equation (16.7) we obtain the useful expression

$$\textit{at 298 K:} \quad \Delta \overline{H} = \Delta \overline{E} + 2.48\Delta \overline{n}_g \tag{16.8}$$

Example 16-2   Use expressions (16.5) and (16.8) to calculate $\Delta \overline{H}$ for the combustion of $C_{10}H_8(s)$.

Solution.   A value of $\Delta \overline{E}$ is given in equation (16.5): -5151.9 kJ/mol. Also, from equation (16.5) we see that $\Delta \overline{n}_g = n_{CO_2} - n_{O_2} = -2$ mol gas per mol reaction.

$$\Delta \overline{H} = \Delta \overline{E} + 2.48\Delta \overline{n}_g$$

$$\Delta \overline{H} = -5151.9 \text{ kJ/mol} + 2.48 \times (-2) \text{ kJ/mol}$$

$$\Delta \overline{H} = -5151.9 \text{ kJ/mol} - 4.96 \text{ kJ/mol} = -5156.9 \text{ kJ/mol}$$

The magnitude of the heat of combustion of $C_{10}H_8(s)$ at constant pressure ($\Delta \overline{H} = q_p$) is just slightly greater than that at constant volume ($\Delta \overline{E} = q_V$). It is greater by about 5 parts in 5000, or by about 0.1%.

SIMILAR EXAMPLES: Review Problem 2; Exercises 5, 6.

Two points are noteworthy about Example 16-2: (1) The difference between ΔH and ΔE is very small, just as we asserted would usually be the case when we introduced enthalpy in Section 6-4. (2) The essential difference between ΔH and ΔE for a reaction is the pressure–volume work associated with the expansion or compression of gases. In

*As pointed out on page 149, per mole of reaction means that the numbers of moles of reactants and products correspond to the stoichiometric coefficients in the balanced equation. For reaction (16.5) this means that $\Delta \overline{n}_g = 10$ mol $CO_2 - 12$ mol $O_2 = -2$ mol gas per mol reaction.

FIGURE 16-1
Heat evolved in the
constant-pressure
combustion of
naphthalene—Example
16-2 visualized.

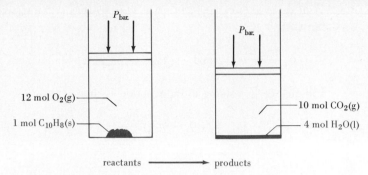

In the combustion reaction $C_{10}H_8(s) + 12\ O_2(g) \rightarrow 10\ CO_2(g) + 4\ H_2O(l)$ the volume of the reactants is essentially that of 12 mol $O_2(g)$, and the volume of the products, 10 mol $CO_2(g)$. The total quantity of energy evolved as heat at constant pressure is that which would have been evolved at constant volume ($\Delta \overline{E}$) *plus* an additional small quantity corresponding to the work of compression done by the surroundings. (The volumes occupied by $C_{10}H_8(s)$ and $H_2O(l)$ are negligible, as is the presence of a small amount of water vapor in the products.)

reaction (16.5) the volume is compressed from that occupied by 12 mol of gas to that occupied by 10. Work is done *by* the surroundings *on* the system when the reaction is conducted at constant pressure, as suggested by Figure 16-1.

16-2 In Search of a Criterion for Spontaneous Change

On occasion in Chapter 15 we used the phrase "the direction of chemical change." By this we meant whether a net reaction would proceed in the forward or the reverse direction, based on a comparison of the reaction quotient, Q, and the equilibrium constant, K. To be more precise we should add one word to this phrase and speak of "the direction of *spontaneous* chemical change."*

A **spontaneous** or **natural process** is one that occurs in a system left to itself; no external action is required to make it happen. A spontaneous change continues until a system reaches a state of equilibrium; then no further net change occurs. For example, the running down of a tightly wound clock occurs spontaneously, but the clock cannot rewind itself. The winding of a clock is a *nonspontaneous* process: External action (winding by human hands) is required to make this process happen. A more interesting example from a chemical standpoint is the corrosion ("rusting") of an iron pipe exposed to the atmosphere. Although the process may occur quite slowly, it does so *continuously* and in the *same* direction. The amount of iron decreases and the amount of rust increases until a final state of equilibrium is reached where practically all of the iron has been consumed. To reverse this process, that is, to convert the rust back into pure iron (and to fabricate the iron into a pipe essentially identical to the original one) is *not* impossible. However, the process is certainly not spontaneous. In fact, this nonspontaneous reverse process is essentially what occurs in the manufacture of iron from iron ore. From these examples we might draw the following conclusions.

1. If a process is found to be spontaneous, the reverse process is nonspontaneous.
2. Both spontaneous and nonspontaneous processes are *possible,* but only spontaneous

*In a way "spontaneous" is not an ideal term for what we are attempting to describe here because the word has a practical implication of something that occurs rapidly. Spontaneous processes may in fact occur quite slowly (such as the rusting of iron). What is intended by the term spontaneous is more along the lines of a *precise* dictionary definition: "acting in accordance with or resulting from natural feeling, temperament, or disposition, or from a native internal proneness, readiness, or tendency, without compulsion, constraint, or premeditation."

FIGURE 16-2
Direction of spontaneous change in a mechanical system.

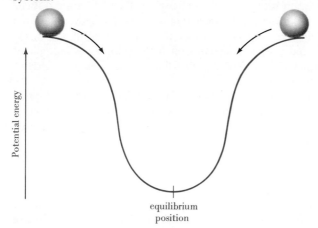

Whether we consider the ball on the left or the one on the right, the direction of spontaneous change is downhill. The ball reaches a position of equilibrium when it comes to rest at the bottom, the point of lowest potential energy.

FIGURE 16-3
Search for a criterion for spontaneous chemical change.

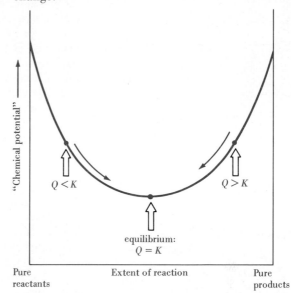

We are looking for some thermodynamic property, here called the "chemical potential," that has a minimum value at a point in the reaction somewhere between the pure reactants and the pure products. At this point the reaction is at equilibrium, with the reaction quotient Q equal to the equilibrium constant K. For a condition on either side of the equilibrium point, spontaneous reaction will occur in the direction of the equilibrium point.

processes will occur *naturally*. Nonspontaneous processes require the system to be acted upon in some way.

Useful as these two conclusions are, there is a third one that is more useful still. This is a conclusion that permits us to predict whether the forward or the reverse direction is the direction of spontaneous change. We will refer to this as a criterion for spontaneous change. To begin, we might look to mechanical systems for a clue: A ball rolls downhill and water flows to a lower level. Figure 16-2 illustrates a common feature of these processes—a decrease in potential energy.

In a scientific paper written in 1875, Berthelot stated this conclusion in the form of a *principle of maximum work:* "All chemical changes occurring without the intervention of outside energy tend toward the production of bodies, or a system of bodies, which liberate more heat."

For a chemical system the property analogous to the potential energy of a mechanical system is the internal energy or the closely related property of enthalpy. In the 1870s, the French chemist P. Berthelot and the Danish chemist J. Thomsen proposed that the direction of spontaneous change was that in which the enthalpy of a system decreases. An enthalpy decrease means that heat is given off by the system to the surroundings. Their conclusion then was that *exothermic* reactions should be spontaneous. In fact, exothermic processes generally are spontaneous, but so are some endothermic ones! The melting of ice at room temperature, the evaporation of liquid ether from an open beaker, and the dissolving of ammonium nitrate in water are all examples of *spontaneous, endothermic* processes. In abandoning enthalpy as our criterion for spontaneous change, let us agree that it would be useful to find some other thermodynamic function having the properties implied by Figure 16-3: We seek a function whose value falls to a minimum at the point where a chemical reaction has reached equilibrium.

FIGURE 16-4
The mixing of ideal gases.

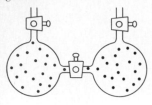

(a) Before mixing

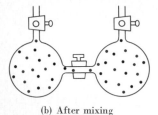

(b) After mixing

• gas A   gas B •

The total volume of the system and the total gas pressure are the same in each case pictured above. The net change that occurs is this.
(a) Before mixing, each gas is confined to one half of the total volume (a single bulb) at a pressure of 1.00 atm.
(b) After mixing, each gas has expanded into the total volume (both bulbs) yielding a partial pressure of 0.500 atm.

16-3 Entropy and Disorder

In continuing our search for a criterion for spontaneous change, perhaps we should focus on endothermic processes. After all, these are the ones that occur in contradiction to the Berthelot–Thomsen principle of maximizing the quantity of heat evolved in a process. The three spontaneous, endothermic processes mentioned in the closing paragraph of Section 16-2 share a common characteristic. To see what this is, let us first consider a simpler case—the mixing of ideal gases.

Figure 16-4 depicts a situation in which one ideal gas, labeled A, is introduced into a glass bulb at a pressure of 1.00 atm. A second ideal gas, B, is introduced into a second bulb, which is identical to the first. Again the pressure is brought to 1.00 atm. The two bulbs are joined by a stopcock valve. This initial situation is pictured in Figure 16-4a. Assume that no chemical reaction is possible between the two gases, and now imagine that the valve between the two bulbs is opened. Intuitively, what would we expect to happen?

We know that the molecules of a gas are in constant motion and that they will move into whatever space is available to them—gases expand to fill their containers. In this case each gas expands into the bulb containing the other—the gases mix. The mixing will continue until the partial pressure of each gas becomes 0.500 atm in each bulb, as illustrated in Figure 16-4b. Since the mixing of ideal gases is a spontaneous process, we might next inquire as to what property of the system has changed.

The internal energy and enthalpy of an ideal gas depend only on temperature, not on the gas pressure or volume. For the mixing of ideal gases at constant temperature, because there are no intermolecular forces, $\Delta E = \Delta H = 0$. One characteristic of the system, however, is greatly altered by the mixing process: the degree of *order* that prevails. In the initial condition of Figure 16-4a there is some degree of order, at least to the extent that all the molecules of A are found on one side of the valve and all those of B on the other side. After mixing, half of the molecules of A and half of those of B are found in each bulb. The molecules have reached the maximum state of mixing or *disorder* possible. A thermodynamic property that relates to the degree of disorder in a system is called **entropy** and denoted by the symbol S.

The higher the degree of randomness or disorder in a system, the greater its entropy.

With reference to the mixing of gases in Figure 16-4,

A(g) + B(g) $\longrightarrow$ mixture of A(g) and B(g)

$$\Delta S = S_{\text{final state}} - S_{\text{initial state}}$$

$$\Delta S = S_{\text{mixt. of gases}} - [S_{A(g)} + S_{B(g)}]$$

$$\Delta S > 0$$

In the mixing of gases, since disorder *increases,* we expect entropy to *increase,* and ΔS is a *positive* quantity, that is, $\Delta S > 0$. Now let us see what we can conclude about the entropy changes for the three endothermic processes described earlier.

In the melting of ice, a crystalline solid (recall Figure 11-19) is replaced by a less structured liquid. Disorder and entropy increase in the process of melting. Molecules in the gaseous state, because of the large free volume in which they can move, have a much higher entropy than in the corresponding liquid state: The process of evaporation is accompanied by an increase in entropy. In the dissolving of an ionic solid such as ammonium nitrate in water, a crystalline solid and a pure liquid are replaced by a mixture of ions and water molecules in the liquid (solution) state (recall Figure 12-4). There is a tendency for some ordering of water molecules to occur around the ions in solution (referred to as hydration of the ions). However, this ordering tendency is not as great as the disorder produced by destroying the original crystalline solid. Again, disorder and entropy increase

FIGURE 16-5
Entropy and disorder—
three processes in which
entropy increases.

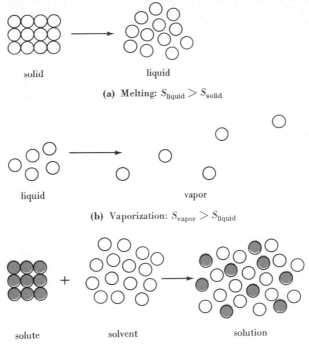

(a) Melting: $S_{liquid} > S_{solid}$

(b) Vaporization: $S_{vapor} > S_{liquid}$

(c) Dissolving: $S_{soln} > (S_{solvent} + S_{solute})$

Each of the processes pictured here results in greater disorder and an increased entropy.

in the dissolving process. The increased disorder accompanying the types of processes described here is pictured in Figure 16-5. In general, entropy *increases* are expected to accompany processes in which

- pure liquids or liquid solutions are formed from solids;
- gases are formed, either from solids or liquids;
- the number of molecules of gases increases in the course of a chemical reaction; (16.9)
- the temperature of a substance is increased. (Increased temperature means increased molecular motion, whether it be vibrational motion of atoms or ions in a solid or translational motion of molecules in a liquid or gas.)

Example 16-3 **Predict whether each of the following processes involves an increase or decrease in entropy.**
(a) The explosive decomposition of NH_4NO_3:

$$2\ NH_4NO_3(s) \longrightarrow 2\ N_2(g) + 4\ H_2O(g) + O_2(g)$$ *increase*

(b) The conversion of SO_2 to SO_3:

$$2\ SO_2(g) + O_2(g) \longrightarrow 2\ SO_3(g)$$ *decrease*

(c) The extraction of NaCl(s) from seawater:

$$NaCl(aq) \longrightarrow NaCl(s)$$ *decrease*

(d) The "water gas" reaction:

$$CO(g) + H_2O(g) \longrightarrow CO_2(g) + H_2(g)$$

Solution

(a) Here a solid yields a large quantity of gas. An *increase* in entropy is expected.

(b) Three moles of gaseous reactants produce 2 mol of gaseous product. The *loss* of 1 mol of gas in the course of the reaction suggests a greater degree of order in the $SO_3(g)$ than in the gases from which it is formed. A *decrease* in entropy is expected.

(c) The ions Na^+ and Cl^- achieve a high degree of order when they leave the solution state and arrange themselves into the crystalline state. A *decrease* in entropy is expected.

(d) The entropies of the four gases are all likely to be different because of differences in their molecular structures. There should be an entropy change in the reaction. However, because no change occurs in the number of gaseous molecules in the course of the reaction, the entropy change is likely to be small. Moreover, there is no way for us to conclude from the relationships stated in (16.9) whether the entropy increases or decreases.

SIMILAR EXAMPLES: Review Problem 3; Exercises 8, 9.

Our use of the entropy concept will be mostly qualitative, as in Example 16-3, but we should comment briefly on how entropy changes can be experimentally measured. Entropy, like enthalpy, must be defined in such a way as to be a function of state. It must have a unique value for each state or condition of a system so that the difference in entropy between two states, ΔS, will also have a unique value. A further aspect of the definition of entropy, of course, is that it be based on measurable quantities. Two measurable quantities that affect the amount of disorder in a system are a quantity of heat and temperature. For example, addition of a quantity of heat to a solid at its melting point produces some of the more disordered liquid state; and the more heat absorbed, the greater the amount of disorder produced. The ability of a given quantity of heat to produce disorder is greater if the heat is absorbed by a well-ordered system (low temperature) than by a system which is already highly disordered (high temperature). The following defining equation for entropy change is consistent with these statements; that is, ΔS is directly proportional to q and inversely proportional to T.

$$\Delta S = \frac{q_{rev}}{T} \tag{16.10}$$

Equation (16.10) is deceptively simple in appearance. The difficulty in using it is that ΔS for a process must have a unique value (again, because S is a function of state), but the value of q depends on the path chosen. Thus, equation (16.10) holds only when the path or means by which a process is carried out is carefully defined. The path must be of a type called reversible, for which $q = q_{rev}$. Fortunately, we do not need to deal with the subtleties of equation (16.10) in this text. Equation (16.10) does help us, however, to establish the units of entropy and entropy change; they are J/K or $J \, K^{-1}$.

The Meaning of $\Delta S_{universe}$.   A moment's reflection reveals that we still have not found a suitable *single* criterion for spontaneous change. For example, based on the entropy change of the water alone, how do we explain the *spontaneous* freezing of water at $-10°C$? If the entropy of the system increases when ice melts, it must *decrease* when water freezes. The answer to this puzzle lies in the fact that there are always *three* entropy changes that must be assessed. These are the entropy change of the system, of the surroundings, and the total of the two—the so-called entropy change of the "universe." The relationship among the three is

$$\Delta S_{total} = \Delta S_{universe} = \Delta S_{system} + \Delta S_{surroundings} \tag{16.11}$$

It is beyond the scope of this text to pursue this matter, but for all *spontaneous* processes it is true that

$$\Delta S_{\text{univ}} = \Delta S_{\text{syst}} + \Delta S_{\text{surr}} > 0 \tag{16.12}$$

Equation (16.12) is in fact a statement of the second law of thermodynamics.

All spontaneous or natural processes produce an increase in the entropy of the universe.

The freezing of water is accompanied by a decrease in entropy of the *system*. But as long as the temperature is below 0°C, the entropy of the *surroundings* increases to a greater extent. The *total* entropy change is *positive* and the process is indeed spontaneous.

16-4 Free Energy and Spontaneous Change

Total entropy change is a valid criterion for spontaneity (spontaneous change), but is it practical? Assessment of the entropy change of the surroundings can be a very tedious process, even impossible at times if the interactions between a system and its surroundings are not completely known. What we need is a criterion that eliminates reference to the surroundings and that can be applied *just to the system itself*.

Consider a spontaneous process at constant temperature and pressure occurring within a system that is limited to doing only pressure–volume work. This process is accompanied by a heat effect, q_p, which, as we have previously seen, is equal to ΔH for the system (ΔH_{syst}). The heat effect experienced by the surroundings is the *negative* of that for the system. If q in equation (16.10) is taken to be equal to $-\Delta H_{\text{syst}}$, then for the entropy change of the surroundings we have $\Delta S_{\text{surr}} = -\Delta H_{\text{syst}}/T$.* Now substitute this value of ΔS_{surr} into equation (16.11).

$$\Delta S_{\text{univ}} = \Delta S_{\text{syst}} - \frac{\Delta H_{\text{syst}}}{T}$$

Multiply by T to obtain

$$T\Delta S_{\text{univ}} = T\Delta S_{\text{syst}} - \Delta H_{\text{syst}} = -(\Delta H_{\text{syst}} - T\Delta S_{\text{syst}})$$

and then change signs.

$$-T\Delta S_{\text{univ}} = \Delta H_{\text{syst}} - T\Delta S_{\text{syst}} \tag{16.13}$$

Equation (16.13) has, on its right side, terms involving *only the system*. On the left side appears the term ΔS_{univ}, which for a spontaneous process must be a positive quantity (recall equation 16.12).

Equation (16.13) can serve as our basic criterion for spontaneous change, but it is customary to cast it in a slightly different form, through the introduction of a new thermodynamic property, the **Gibbs free energy, *G*.** The Gibbs free energy for a system is defined as

$$G = H - TS$$

and for a change at constant T and P,

$$\Delta G = \Delta H - T\Delta S \tag{16.14}$$

By comparing equations (16.13) and (16.14) we see that

$$\Delta G = -T\Delta S_{\text{univ}}$$

This free energy function was formulated by the American mathematical physicist J. Willard Gibbs in 1876.

*In case you are wondering why $\Delta H_{\text{syst}}/T$ cannot also be substituted for ΔS_{syst}, this is one of those subtleties in the use of equation (16.10) referred to on page 482. Because the spontaneous process occurring in the system is a natural process, it cannot be reversible, and q_p for an irreversible process cannot be used in equation (16.10).

Because the criterion for a spontaneous process is that $\Delta S_{uniuv} > 0$, *for a process occurring at constant T and P in a closed system, spontaneous change is accompanied by a decrease in free energy, that is,*

$$\Delta G < 0 \tag{16.15}$$

That free energy carries an energy unit is not difficult to establish from equation (16.14). The unit of ΔH is joules (J), and so is that of $T\Delta S$ (K × J/K = J). ΔG is the difference between these two quantities of energy.

Application of the New Criterion for Spontaneous Change. In considering H and S separately we concluded that decreases in H (ΔH negative) and increases in S (ΔS positive) both tend to favor *spontaneous* change. For a process in which ΔH is negative and ΔS is positive, equation (16.14) indicates a *negative* value of ΔG. It should come as no surprise, then, that our new criterion (16.15) also states that for a spontaneous process at constant temperature and pressure, $\Delta G < 0$.

Altogether there are four possibilities for ΔG, based on the signs of ΔH and ΔS. These possibilities are presented in Table 16-2. An important conclusion can be drawn from the data in this table: At low temperatures the enthalpy change ΔH is more significant in determining the sign of ΔG, whereas at high temperatures ΔS assumes this role.

Example 16-4 Is a dissociation reaction of the type $AB(g) \rightarrow A(g) + B(g)$, favored at high or low temperatures?

Solution. This reaction involves the breaking of a bond between A and B. Energy must be absorbed; $\Delta H > 0$. Because 2 mol of gas is produced from 1 mol, increased disorder results from the reaction; $\Delta S > 0$.

$$\Delta G = \underset{+}{\underline{\Delta H}} \underset{-}{\underline{- T\Delta S}} = \underset{+ \text{ or } -}{\underline{?}}$$

Whether the dissociation process is spontaneous at some given temperature depends on the relative values of the ΔH term and the $T\Delta S$ term, and this we do not know. But what we can say is that the higher the temperature, the more negative the term $-T\Delta S$ becomes. At some temperature the magnitude of $T\Delta S$ will just exceed that of ΔH. At this temperature, $\Delta G < 0$. Dissociation reactions of this type are favored at *high* temperatures. An example is $I_2(g) \rightarrow I(g) + I(g)$.

SIMILAR EXAMPLES: Review Problem 4; Exercise 12.

An interesting conclusion can be drawn from the result of Example 16-4: There should exist an upper temperature limit for the stabilities of chemical compounds. No matter how positive the value of ΔH for dissociation of a compound into its atoms, the term $T\Delta S$ will

TABLE 16-2
Criterion for spontaneous change: $\Delta G = \Delta H - T\Delta S$

Case	ΔH	ΔS	ΔG	Result	Example
1	−	+	−	spontaneous at all temp.[a]	$2\,N_2O(g) \longrightarrow 2\,N_2(g) + O_2(g)$
2	−	−	$\begin{cases} - \\ + \end{cases}$	spontaneous at low temp. nonspontaneous at high temp.	$H_2O(l) \longrightarrow H_2O(s)$
3	+	+	$\begin{cases} + \\ - \end{cases}$	nonspontaneous at low temp. spontaneous at high temp.	$2\,NH_3(g) \longrightarrow N_2(g) + 3\,H_2(g)$
4	+	−	+	nonspontaneous at all temp.[a]	$3\,O_2(g) \longrightarrow 2\,O_3(g)$

[a]If either ΔH or ΔS changes sign in the temperature range under consideration, then one of the other cases applies.

eventually exceed it as temperature is increased. If we consider the complete range of temperatures from absolute zero to the interior temperatures of the stars (about 3×10^7 K) we find that molecules exist over only a very small portion of this range (up to about 1×10^4 K or about 0.03% of this range).

16-5 Standard Free Energy Change, $\Delta G°$

As with other thermodynamic functions, the free energies of chemical substances depend on their state or condition. To facilitate the use of this function in calculations, we need to establish standard state conditions, as was done with enthalpy in Chapter 6. The standard state conventions we shall use are

for a solid: the pure substance at 1 atm pressure.

for a liquid: the pure substance at 1 atm pressure.

for a gas: an ideal gas at 1 atm partial pressure.   (16.16)

for a solute: an ideal solution at 1 *M* concentration.

The free energy change when reactants and products are all in their standard states is referred to as the **standard free energy change, $\Delta G°$.** For a reaction in which a compound is formed from its elements, $\Delta G°_{rxn}$ is referred to as the **standard free energy of formation** of the compound, represented by $\Delta G°_f$. If the amount of compound formed is *one mole,* as is usually the case in tabulated data, it is customary to use an overbar on the symbol, that is, $\Delta \overline{G}°_f$. And by convention, standard free energies of formation of the elements in their most stable forms at 1 atm pressure are defined as *zero* at the specified temperature. Selected values of $\Delta \overline{G}°_f$ at 298 K are tabulated in Appendix D.

Additional relationships needed in applying the free energy function are similar to the three presented in Section 6-5 for enthalpy.

1. ΔG is an extensive property.
2. ΔG changes sign when a process is reversed.
3. ΔG for a net or overall process can be obtained by summing the ΔG values for the individual steps.

Example 16-5   The standard molar enthalpy of formation of silver oxide at 298 K is -30.59 kJ/mol. The standard molar free energy change, $\Delta \overline{G}°$, for the dissociation of silver oxide at 298 K is given below. What is $\Delta \overline{S}°$ for this reaction?

$$2 \text{ Ag}_2\text{O(s)} \longrightarrow 4 \text{ Ag(s)} + \text{O}_2\text{(g)}   \qquad \Delta \overline{G}° = +22.43 \text{ kJ/mol}$$

Solution.   In applying equation (16.14) we note that the reactants and products are in their standard states. Therefore, $\Delta \overline{G}° = \Delta \overline{H}° - T \Delta \overline{S}°$, and

$$\Delta \overline{S}° = \frac{\Delta \overline{H}° - \Delta \overline{G}°}{T}$$

Before we can solve for $\Delta \overline{S}°$, it is necessary to have a value of $\Delta \overline{H}°$. This can be obtained in the familiar manner employed in Section 6-6.

$$\Delta \overline{H}° = 4\{\Delta \overline{H}°_f[\text{Ag(s)}]\} + \Delta \overline{H}°_f[\text{O}_2\text{(g)}] - 2\{\Delta \overline{H}°_f[\text{Ag}_2\text{O(s)}]\}$$

$$= \quad\quad 0 \quad\quad + \quad\quad 0 \quad\quad - 2(-30.59 \text{ kJ/mol})$$

$$= +61.18 \text{ kJ/mol}$$

$$\Delta \overline{S}° = \frac{61.18 \text{ kJ/mol} - 22.43 \text{ kJ/mol}}{298 \text{ K}} = 0.130 \text{ kJ mol}^{-1} \text{ K}^{-1}$$

That $\Delta \overline{S}°$ is a positive quantity is a reasonable finding—from a solid reactant both a solid and a *gaseous* product are formed (recall Example 16-3a).

SIMILAR EXAMPLES: Review Problem 5; Exercises 15, 16.

Example 16-6 What is $\Delta \overline{G}^{\circ}$ at 298 K for the reaction $C(s) + CO_2(g) \rightleftharpoons 2\ CO(g)$?

Solution. Here we can use tabulated data from Appendix D to determine $\Delta \overline{G}^{\circ}$ directly.

$$\Delta \overline{G}^{\circ} = 2\{\Delta \overline{G}_{f}^{\circ}[CO(g)]\} - \Delta \overline{G}_{f}^{\circ}[C(s)] - \Delta \overline{G}_{f}^{\circ}[CO_2(g)]$$
$$= 2(-137.28 \text{ kJ/mol}) - 0 - (-394.38 \text{ kJ/mol})$$
$$= -274.56 + 394.38 = +119.82 \text{ kJ/mol}$$

SIMILAR EXAMPLE: Review Problem 6.

16-6 Free Energy and Equilibrium

We have established that for a spontaneous process $\Delta G < 0$, and for a nonspontaneous process $\Delta G > 0$. Some mention should now be made of the situation in which $\Delta G = 0$. This is the condition of equilibrium. When a system is at equilibrium, there is an equal tendency for a process to proceed in either the forward or the reverse direction. Even an infinitesimal change in one of the experimental variables (e.g., temperature or pressure) will cause a net change to occur. As long as a system at equilibrium is left undisturbed, however, there will be no net change with time.

A hypothetical condition of equilibrium is presented in Figure 16-6 as the intersection of two graph lines. One line represents the temperature variation of the enthalpy change and the other, the product $T\Delta S$. The ΔH line may have a positive slope in some cases and negative in others, but the slope is generally quite gradual—ΔH is not especially temperature sensitive. The slope of the $T\Delta S$ line is expected to be rather steep, because of the presence of the temperature factor, T, in the product, $T\Delta S$. The vertical distance from the $T\Delta S$ line to the ΔH line represents $\Delta G = \Delta H - T\Delta S$. Starting at the left side of the figure, ΔG is positive and large. The magnitude of ΔG decreases with increasing temperature. At the right side of the figure, $T\Delta S$ exceeds ΔH in value and ΔG is negative. At the point of intersection of the two lines, $\Delta G = 0$ and the system is at equilibrium.

FIGURE 16-6
Free energy change as a function of temperature.

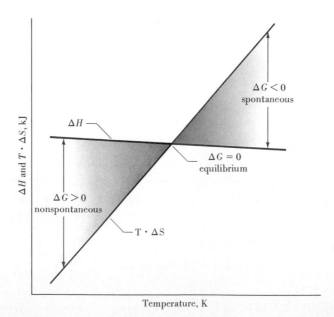

Phase Transitions. Consider for a moment the application of Figure 16-6 to a description of the melting of ice. If we deal with the solid and liquid in their standard states (i.e., under 1 atm pressure), the intersection of the two lines comes at 273.15 K. This is the normal melting point of ice. At temperatures above 273.15 K the melting of ice is a spontaneous process. Below 273.15 K it is nonspontaneous. At 273.15 K ice and liquid water are at equilibrium. This equilibrium condition provides us with a method for determining the entropy change, $\Delta \overline{S}^\circ$, for the melting of ice. (The molar heat of fusion of ice at 273.15 K is 6.02 kJ/mol.)

For the process $H_2O(s, 1 \text{ atm}) \rightleftharpoons H_2O(l, 1 \text{ atm})$ *at* 273.15 K:

$$\Delta \overline{G}^\circ = \Delta \overline{H}^\circ - T\Delta \overline{S}^\circ = 0 \qquad \Delta \overline{H}^\circ = T\Delta \overline{S}^\circ \qquad \Delta \overline{S}^\circ = \frac{\Delta \overline{H}^\circ}{T}$$

$$\Delta \overline{S}^\circ = \frac{6.02 \text{ kJ/mol}}{273.15 \text{ K}} = 2.20 \times 10^{-2} \text{ kJ mol}^{-1} \text{ K}^{-1} = 22.0 \text{ J mol}^{-1} \text{ K}^{-1}$$

The calculation just performed is a specific application of the general equation that can be written to describe any process at equilibrium.

$$\left. \begin{aligned} \Delta G &= \Delta H - T\Delta S = 0 \\ \Delta S &= \frac{\Delta H}{T} \end{aligned} \right\} \text{ at equilibrium} \qquad (16.17)$$

When the process involves a transition between phases at constant temperature—melting, freezing, vaporization, condensation, and so on—descriptive subscripts may be used. That is, the subscripts "tr," "fus," "vap," and so on, may be attached to ΔG, ΔH, ΔS, and T. If we wish to describe a transition involving molar quantities of substances in their standard states, equation (16.17) is modified in the usual fashion: overbars and superscript degree signs are placed on the thermodynamic symbols.

$$\Delta \overline{S}^\circ_{tr} = \frac{\Delta \overline{H}^\circ_{tr}}{T_{tr}} \qquad (16.18)$$

Example 16-7 What is the standard molar entropy of vaporization of water at 100°C? The standard molar enthalpy of vaporization at 100°C is 40.7 kJ/mol.

Solution. Let us first translate this verbal description of the process into a chemical equation.

$$H_2O(l, 1 \text{ atm}) \rightleftharpoons H_2O(g, 1 \text{ atm}) \qquad \Delta \overline{H}^\circ_{vap} = 40.7 \text{ kJ/mol} \qquad \Delta \overline{S}^\circ_{vap} = ?$$

and then apply equation (16.18).

$$\Delta \overline{S}^\circ_{vap} = \frac{\Delta \overline{H}^\circ_{vap}}{T_{bp}} = \frac{40.7 \text{ kJ/mol}}{373.15 \text{ K}} = 0.109 \text{ kJ mol}^{-1} \text{ K}^{-1} = 109 \text{ J mol}^{-1} \text{ K}^{-1}$$

SIMILAR EXAMPLES: Review Problem 8; Exercises 24, 25.

A useful generalization, known as Trouton's rule, is that for many liquids at their normal boiling points the standard molar *entropy of vaporization* has a value of about 88 J mol^{-1} K^{-1}.

$$\Delta \overline{S}^\circ_{vap} = \frac{\Delta \overline{H}^\circ_{vap}}{T_{bp}} \simeq 88 \text{ J mol}^{-1} \text{ K}^{-1} \qquad (16.19)$$

If the degree of disorder produced in transferring 1 mol of molecules from liquid to vapor at 1 atm pressure is roughly comparable for different liquids, then we should expect

TABLE 16-3
Some standard molar entropy changes for phase transitions

Substance	T_{tr}, K	$\Delta \overline{H}^{\circ}_{tr}$, kJ/mol	$\Delta \overline{S}^{\circ}_{tr}$, J mol^{-1} K^{-1}
Vaporization			
acetone	329	30.25	91.9
ethanol	351	39.33	112
n-hexane	342	28.58	83.6
lead	2024	179.41	88.64
Fusion			
benzene	279	9.83	35.2
camphor	452	6.86	15.2
lead	601	4.77	7.94
mercury	234	2.30	9.83

similar values of $\Delta \overline{S}^{\circ}_{vap}$. Instances where Trouton's rule fails are also understandable. In water and ethanol, hydrogen bonding among molecules produces a greater degree of order than otherwise expected in the liquid state. The degree of disorder produced in the vaporization process is greater than normal, and $\Delta \overline{S}^{\circ}_{vap} > 88$ J mol^{-1} K^{-1}.* There is no regularity in the values of entropies of fusion, other than that they are generally smaller for metals than for most other substances. Selected values for a few phase transitions are presented in Table 16-3.

16.7 Relationship of $\Delta \overline{G}^{\circ}$ to K

For the vaporization of water, $H_2O(l) \rightleftharpoons H_2O(g)$, the standard molar free energy change at 25° C is $\Delta \overline{G}^{\circ} = +8.58$ kJ/mol. The positive sign of $\Delta \overline{G}^{\circ}$ cannot mean that water is incapable of vaporizing at 25°C, for practical experience tells us that it can. The difficulty here is that $\Delta \overline{G}^{\circ}$ refers to liquid water and water vapor in their *standard states*. Its positive value simply tells us that $H_2O(l)$ at 1 atm pressure will not spontaneously produce $H_2O(g)$ at *1 atm pressure* at 25°C, that is,

$$H_2O(l, 1 \text{ atm}) \rightleftharpoons H_2O(g, 1 \text{ atm}) \qquad \Delta \overline{G}^{\circ} = +8.58 \text{ kJ/mol} \qquad (16.20)$$

The equilibrium vapor pressure of water at 25°C is found experimentally to be 23.8 mmHg = 0.0313 atm. The condition of equilibrium is

$$H_2O(l, 0.0313 \text{ atm}) \rightleftharpoons H_2O(g\ 0.0313 \text{ atm}) \qquad \Delta \overline{G} = 0 \qquad (16.21)$$

Figure 16-7 suggests the difference between the $\Delta \overline{G}^{\circ}$ of (16.20) and $\Delta \overline{G}$ of (16.21).

The Expression $\Delta \overline{G}^{\circ} = -2.303RT \log K$. $\Delta \overline{G}^{\circ}$ alone can be used to determine the direction of spontaneous change only if the reactants and products are in their standard states. Because reaction conditions of interest to us will often be nonstandard ones, we should pursue the criterion for spontaneous change further. To do so we need to relate the molar free energy change for nonstandard conditions $\Delta \overline{G}$, to the corresponding molar free

---

*The description of vaporization in terms of producing disorder helps to explain Raoult's law. The molar entropy of vaporization of a solvent should be the same whether vaporization occurs from the pure solvent or from an ideal solution (say, one containing a nonvolatile solute). However, since the solution state is more disordered than the pure solvent, the equilibrium vapor produced from the solution would also have to be more disordered than that produced by the pure solvent. The vapor molecules above a solution must be more widely separated or at a lower pressure than those above the pure solvent.

FIGURE 16-7
Liquid–vapor
equilibrium and the
direction of spontaneous
change.

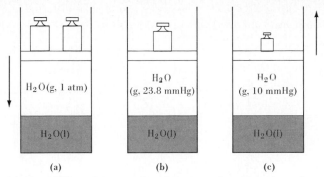

(a) (b) (c)

(a) At 25°C and 1 atm, the direction of spontaneous change is the condensation of H$_2$O(g). For the vaporization process H$_2$O(l, 1 atm) $\rightarrow$ H$_2$O(g, 1 atm), $\Delta\overline{G} = \Delta\overline{G}^\circ = +8.58$ kJ/mol.
(b) At 25°C and 23.8 mmHg, the equilibrium H$_2$O(l) $\rightleftharpoons$ H$_2$O(g) is established; $\Delta\overline{G} = 0$.
(c) At 25°C and 10 mmHg, the vaporization of water is spontaneous; for H$_2$O (l, 10 mmHg) $\rightarrow$ H$_2$O (g, 10 mmHg), $\Delta\overline{G} < 0$.
In each case the black arrow suggests the direction of spontaneous change.

energy change for a process in which all reactants and products are in their standard states, $\Delta\overline{G}^\circ$. The key term for doing this is the reaction quotient, Q, written for the specified states of reactants and products. Unfortunately, a derivation of this relationship is beyond the scope of the present discussion, and the equation below is given without proof.

$$\Delta\overline{G} = \Delta\overline{G}^\circ + RT \ln Q \qquad (16.22)$$

or, in terms of common logarithms,

$$\Delta\overline{G} = \Delta\overline{G}^\circ + 2.303\, RT \log Q \qquad (16.23)$$

The value of equations (16.22) and (16.23) lies in applying them to situations in which the specified states of the reactants and products are *nonstandard* states, including *equilibrium* states. Thus, if we apply them to a system at equilibrium, the value of ΔG is *zero* and the reaction quotient, Q, is the *equilibrium constant, K*. This leads to the following expressions.

At equilibrium:

$$\Delta\overline{G} = \Delta\overline{G}^\circ + RT \ln K = 0$$

and

$$\Delta\overline{G}^\circ = -RT \ln K \qquad (16.24)$$

or, in terms of common logarithms,

$$\Delta\overline{G}^\circ = -2.303\, RT \log K \qquad (16.25)$$

The most important implication of equations (16.24) and (16.25) is that from tabulated thermodynamic data (as in Appendix D) we can derive $\Delta\overline{G}^\circ$ values, and from these, in turn, equilibrium constants, K. As a result, many of the calculations considered in Chapter 15 now become possible without the need for direct experimental equilibrium measurements. We will attempt such calculations shortly, but first we need to consider two other matters briefly.

FIGURE 16-8
Free energy change, equilibrium, and the direction of spontaneous change.

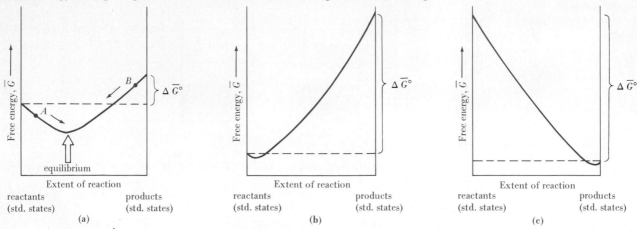

(a) Free energy is plotted as a function of the extent of reaction. The difference between the standard molar free energies of reactants and products is the standard molar free energy change, $\Delta \overline{G}°$. The equilibrium point lies somewhere between pure reactants and pure products. The free energy of the equilibrium mixture is lower than those of the pure reactants, of the pure products, and of any other mixture of the two. Mixtures A and B will each undergo spontaneous change in the direction of the equilibrium mixture.

Only if $\Delta \overline{G}°$ has a small magnitude (e.g., 10 to 20 kJ/mol), either positive or negative, will the equilibrium mixture have appreciable amounts of both reactants and products.

(b) If $\Delta \overline{G}°$ is large and positive, the minimum in free energy lies very close to the pure reactant side, and very little reaction occurs before equilibrium is reached.

(c) If $\Delta \overline{G}°$ is large and negative, the minimum in free energy lies very close to the product side, and the reaction goes essentially to completion.

The first of these matters is that we have completed our search for a criterion for spontaneous change. We can replace the hypothetical graphical representation of Figure 16-3 with our final result, as is done in Figure 16-8. The second matter is discussed in the following paragraphs.

The Thermodynamic Equilibrium Constant. To be used in equations (16.24) and (16.25), equilibrium constants must be modified to a form called the thermodynamic equilibrium constant. This is a constant that is written in terms of activities or "effective concentrations" (see Section 12-7). For the general reaction

$$a \, A + b \, B + \cdots \rightleftharpoons g \, G + h \, H + \cdots \tag{16.26}$$

the thermodynamic equilibrium constant is

$$K = \frac{(a_G)^g (a_H)^h \cdots}{(a_A)^a (a_B)^b \cdots} \tag{16.27}$$

In applying equation (16.27), only numerical values of the activities are substituted for the a symbols: Each a symbol is actually a ratio of the equilibrium activity of a reactant to its activity in its standard state. However, standard states have been defined to have unit activity—the denominator in each ratio is 1. Thus, as long as the units used to express equilibrium activities are those used to define standard states, activity units in the a ratios cancel. This leaves a unitless but numerically equal to the equilibrium activity. The following are useful approximations in establishing activities.

For pure solids and liquids: The activity is taken as unity, $a = 1$.

For gases: Assume ideal gas behavior. Then the activity of a gas is equal to its pressure in atm.

For components in solution: Assume ideal behavior (e.g., no interionic attractions). Then if the standard state is taken to be 1 M, the activity is equal to the molar concentration.

Whether K_c or K_p can be used in equation (16.25) depends on whether either of these expressions results from writing a thermodynamic equilibrium constant expression. One of the cases cited in Example 16-8 results in a value of K_p because partial pressures are substituted for activities. Another yields a value of K_c, with dilute concentrations substituting for activities. In the remaining case, because both concentrations and partial pressures appear in the equilibrium constant expression, the symbol K is used.

Example 16-8 Write thermodynamic equilibrium constant expressions for the following reversible processes, making appropriate substitutions for activities.

(a) $C(s) + H_2O(g) \rightleftharpoons CO(g) + H_2(g)$

(b) $PbI_2(s) \rightleftharpoons Pb^{2+}(aq) + 2\ I^-(aq)$

(c) $O_2(g) + 2\ S^{2-}(aq) + 2\ H_2O(l) \rightleftharpoons 4\ OH^-(aq) + 2\ S(s)$

Solution

(a) $K = \dfrac{(a_{CO(g)})(a_{H_2(g)})}{(a_{C(s)})(a_{H_2O(g)})} = \dfrac{(P_{CO})(P_{H_2})}{(1)(P_{H_2O})} = \dfrac{(P_{CO})(P_{H_2})}{(P_{H_2O})} = K_p$

(b) $K = \dfrac{(a_{Pb^{2+}})(a_{I^-})^2}{a_{PbI_2(s)}} = \dfrac{[Pb^{2+}][I^-]^2}{1} = [Pb^{2+}][I^-]^2 = K_c$

(c) $K = \dfrac{(a_{S(s)})^2(a_{OH^-})^4}{(a_{O_2(g)})(a_{S^{2-}})^2(a_{H_2O})^2} = \dfrac{(1)^2[OH^-]^4}{(P_{O_2})[S^{2-}]^2(1)^2} = \dfrac{[OH^-]^4}{(P_{O_2})[S^{2-}]^2}$

SIMILAR EXAMPLE: Review Problem 10.

Illustrative Examples. The first two examples make use of equation (16.25) to provide information about equilibrium in the reaction described in Example 16-5.

$$2\ Ag_2O(s) \rightleftharpoons 4\ Ag(s) + O_2(g) \tag{16.28}$$

Example 16-9 Use data from Example 16-5, as needed, to determine the equilibrium partial pressure of $O_2(g)$ above a mixture of Ag(s) and $Ag_2O(s)$ at 298 K. Treat the $O_2(g)$ as an ideal gas.

Solution. The equilibrium in question is described by the expression

$$K = \frac{(a_{Ag(s)})^4(a_{O_2(g)})}{(a_{Ag_2O(s)})^2} = \frac{(1)^4(P_{O_2})}{(1)^2} = P_{O_2} = K_p$$

To determine the value of $K = K_p$ at 298 K we use $\Delta \overline{G}^\circ$ from Example 16-5 in equation (16.25). In equation (16.25), the units of R must be J mol^{-1} K^{-1} and temperature must be in kelvins.

$\Delta \overline{G}^\circ = -2.303RT \log K = +22.43$ kJ/mol

$\quad = -2.303 \times 8.314\ \text{J mol}^{-1}\ \text{K}^{-1} \times 298\ \text{K} \times \log K = 22.43 \times 10^3\ \text{J/mol}$

$\log K = \dfrac{-22.43 \times 10^3}{2.303 \times 8.314 \times 298} = -3.931 = 0.069 - 4.00$

$\quad K = 1.17 \times 10^{-4}$

But we have already seen that $K = K_p = P_{O_2}$. Therefore, $P_{O_2} = 1.17 \times 10^{-4}$ atm

SIMILAR EXAMPLES: Review Problems 9, 11; Exercise 30.

Here we discover a reason for specifying "a mole of reaction." The "mol^{-1}" term in the value of $\Delta \overline{G}^\circ$ provides for the cancellation of "mol^{-1}" in the units of R, making log K a dimensionless quantity. A balanced chemical equation must always be given for a reaction. From this, the "mole of reaction," the $\Delta \overline{G}^\circ$ value, and the equilibrium constant expression, K, can all be established in a consistent way.

Example 16-10   At what approximate temperature will the dissociation pressure of $Ag_2O(s)$ become equal to 0.10 atm? Use data from Example 16-5, and indicate any assumptions that are made in this estimation.

Solution

Step 1.   If the dissociation pressure is to be 0.10 atm, this means that $P_{O_2} = 0.10$ atm, and $K = K_P = P_{O_2} = 0.10$ for reaction (16.28).

Step 2.   If we know a value of K, we can use equation (16.25) to determine a value of $\Delta\overline{G}°$ in terms of the unknown temperature, T.

$$\begin{aligned} \Delta\overline{G}° &= -2.303RT \log K \\ &= -2.303 \times 8.314 \times 10^{-3}\ T\ \text{kJ mol}^{-1}\ \text{K}^{-1} \times \log (0.10) \\ &= -0.0191T \times (-1.00) = 0.0191T\ \text{kJ mol}^{-1}\ \text{K}^{-1} \end{aligned}$$

Step 3.   A temperature at which we do have some data for this reaction is 298 K. From Example 16-5 we obtain the values $\Delta\overline{H}° = +61.18$ kJ/mol and $\Delta\overline{S}° = 0.130$ kJ mol^{-1} K^{-1}.

Step 4.   Let us *assume* that at the temperature T the values of $\Delta\overline{H}°$ and $\Delta\overline{S}°$ are about the same as at 298 K. In substituting data into the Gibbs equation below, the value used for $\Delta\overline{G}°$ is that established in Step 2.

$$\Delta\overline{G}° = \Delta\overline{H}° - T\Delta\overline{S}°$$

$$0.0191T\ \text{kJ mol}^{-1}\ \text{K}^{-1} = 61.18\ \text{kJ mol}^{-1} - T \times 0.130\ \text{kJ mol}^{-1}\ \text{K}^{-1}$$

$$(0.130 + 0.0191)T\ \text{kJ mol}^{-1}\ \text{K}^{-1} = 61.18\ \text{kJ mol}^{-1}$$

$$T = \frac{61.18\ \text{kJ mol}^{-1}}{0.149\ \text{kJ mol}^{-1}\ \text{K}^{-1}} = 411\ \text{K}$$

SIMILAR EXAMPLES: Exercises 36 through 39.

In Example 16-9 we were asked to calculate an equilibrium partial pressure and in Example 16-10, an equilibrium temperature. Sometimes the question being asked does not require calculating an equilibrium concentration, pressure, or temperature, or any other numerical quantity. In these cases we should look for some qualitative means to answer the question, as in Example 16-11.

Example 16-11   The reaction $C(s) + CO_2(g) \rightleftharpoons 2\ CO(g)$ was the subject of Example 16-6. If $C(s)$ and $CO_2(g)$ at 1 atm pressure are maintained in contact at 298 K, will there be any significant reaction to produce $CO(g)$?

Solution.   In Figure 16-8 we saw that for a reaction with a large positive $\Delta\overline{G}°$ very little product(s) is formed before the reaction reaches equilibrium. The $\Delta\overline{G}°$ calculated in Example 16-6 certainly qualifies as a large positive $\Delta\overline{G}°$—nearly +120 kJ/mol. We can safely conclude that very, very little $CO(g)$ will be produced in this reaction at 298 K. (Note also that $\Delta\overline{G}° = +120$ kJ/mol corresponds to log $K = -120 \times 10^3/2.303RT \simeq -21$; or $K \simeq 1 \times 10^{-21}$. This is a very small equilibrium constant and also suggests that very little product would be formed.)

SIMILAR EXAMPLES: Review Problem 14; Exercise 31.

$\Delta\overline{G}°$ **as a Function of Temperature.**   Quite often, values of $\Delta\overline{G}°$ and K are desired at temperatures other than those at which values of $\Delta\overline{H}°$ and $\Delta\overline{S}°$ are available. Exact expressions that incorporate the temperature dependence of $\Delta\overline{H}°$ and $\Delta\overline{S}°$ are available to relate $\Delta\overline{G}°$ and T. However, it is beyond the scope of this text to consider these. Fortunately, the *approximation* introduced in Example 16-10 often works: $\Delta\overline{H}^\circ$ and $\Delta\overline{S}^\circ$ are taken to be temperature-independent. The Gibbs equation can then be written as follows.

Regardless of the temperature in question, assume that these quantities have the same values as at some other temperature at which they are known or can be determined.

$$\Delta\overline{G}^\circ = \Delta\overline{H}^\circ - T \times \Delta\overline{S}^\circ$$

An ability to describe $\Delta\overline{G}^\circ$ as a function of temperature should provide a means of describing the equilibrium constant, K, as a function of temperature as well. This is because of the relationship $\Delta\overline{G}^\circ = -2.303RT \log K$. In fact, equation (15.26), which relates K to temperature, can be derived from the Gibbs equation, as follows.

$$\Delta\overline{G}^\circ = \Delta\overline{H}^\circ - T\Delta\overline{S}^\circ$$

$$\frac{\Delta\overline{G}^\circ}{T} = \frac{\Delta\overline{H}^\circ}{T} - \Delta\overline{S}^\circ$$

$$\frac{-2.303R\cancel{T} \log K}{\cancel{T}} = \frac{\Delta\overline{H}^\circ}{T} - \Delta\overline{S}^\circ$$

$$\log K = \frac{-\Delta\overline{H}^\circ}{2.303RT} + \frac{\Delta\overline{S}^\circ}{2.303R}$$

By assuming that $\Delta\overline{H}^\circ$ and $\Delta\overline{S}^\circ$ are independent of temperature, we can write the above equation for two different temperatures, T_2 and T_1,

$$\log K_2 = \frac{-\Delta\overline{H}^\circ}{2.303RT_2} + \frac{\Delta\overline{S}^\circ}{2.303R} \quad \text{and} \quad \log K_1 = \frac{-\Delta\overline{H}^\circ}{2.303RT_1} + \frac{\Delta\overline{S}^\circ}{2.303R}$$

take the difference between the two (which eliminates the term $\Delta S^\circ/2.303R$),

$$\log K_2 - \log K_1 = \frac{-\Delta\overline{H}^\circ}{2.303R}\left(\frac{1}{T_2} - \frac{1}{T_1}\right)$$

and rearrange to the final result.

$$\log \frac{K_2}{K_1} = \frac{\Delta\overline{H}^\circ}{2.303R}\left(\frac{T_2 - T_1}{T_2 T_1}\right) \tag{15.26}$$

16-8 The Third Law of Thermodynamics

Although absolute values of most thermodynamic properties cannot be determined, it is possible to obtain *absolute* values of the entropy for many substances. This is made possible by the third law of thermodynamics.

The entropy of a pure perfect crystal at the absolute zero of temperature is zero.

The pure perfect crystal at the absolute zero of temperature represents the greatest order possible for a thermodynamic system. As the temperature is raised from values near 0 K, the entropy increases. Absolute entropies are always positive quantities.

If more than one form of a solid substance exists, there will be a small entropy of transition at the temperature at which the low-temperature form is converted to a new solid phase. Another sharp increase in entropy comes at the melting point of the solid, and a still larger increase at the boiling point of the liquid. Entropies of transition can be calculated with equation (16.18). The determination of entropy changes in temperature ranges where there are no phase transitions requires the calorimetric determination of specific

FIGURE 16-9
Absolute entropy as a
function of temperature.

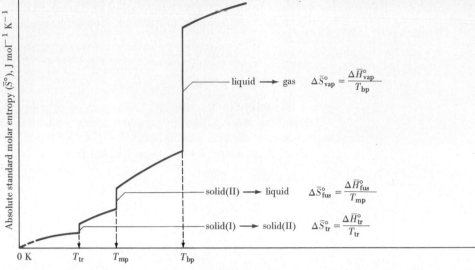

The absolute standard molar entropy of a hypothetical substance is represented here.
By the third law of thermodynamics, an absolute entropy of zero is expected at 0 K.
However, experimental measurements cannot be carried to this temperature—
extrapolation techniques are required (broken-line portion of graph). A transition at T_{tr}
is noted from the solid(I) to solid (II) form of the substance. Melting occurs at the normal melting point, T_{mp}, and vaporization at the normal boiling point, T_{bp}.

heats. The variation of entropy with temperature is illustrated through Figure 16-9. Typical entropy data are included in Appendix D; and such data are used in Example 16-12.

Example 16-12   Is the dissociation of NO(g) favored at high or low temperatures?

$$2\ NO(g) \rightleftharpoons N_2(g) + O_2(g)$$

Solution.   We can choose among the four possibilities listed in Table 16-2 if we know
the signs of ΔH and ΔS. But what are the signs of ΔH and ΔS for the dissociation of
NO(g)? We cannot answer this question in the simple manner that worked for the
dissociation reaction $AB(g) \rightarrow A(g) + B(g)$ in Example 16-4. There, ΔH had to be
positive because chemical bonds were broken and none were formed; and ΔS was
positive because of the increase in number of moles of gases in the forward reaction.
Here, since new chemical bonds are formed (in N_2 and O_2) at the same time that old
bonds are broken (in NO), we cannot say that ΔH should be a positive quantity. Also,
none of the ideas about entropy changes offered through (16.9) applies here; we
cannot assess the sign of ΔS either.

Let us take the following approach: Determine the standard molar entropy and
enthalpy changes at 298 K from data in Appendix D, and then refer to Table 16-2.

At 298 K:

$$\Delta \overline{S}^\circ = \overline{S}^\circ[N_2(g)] + \overline{S}^\circ[O_2(g)] - 2\{\overline{S}^\circ[NO(g)]\}$$

$$\Delta \overline{S}^\circ = 191.50 + 205.02 - 2(210.62) = -24.72\ \text{J mol}^{-1}\ \text{K}^{-1}$$

$$\Delta \overline{H}^\circ = \Delta \overline{H}_f^\circ[N_2(g)] + \Delta \overline{H}_f^\circ[O_2(g)] - 2\{\Delta \overline{H}_f^\circ[NO(g)]\}$$

$$\Delta \overline{H}^\circ = \quad 0 \quad + \quad 0 \quad - 2(+90.37) \quad = -180.74\ \text{kJ/mol}$$

A process with $\Delta H < 0$ and $\Delta S < 0$ corresponds to case 2 in Table 16-2. The forward
process is spontaneous at low temperatures ($\Delta \overline{G} < 0$). Thus, the dissociation of
NO(g) is favored at *low* temperatures.

SIMILAR EXAMPLE: Review Problem 13.

Since the dissociation of
NO(g) is favored at low
temperatures, its forma-
tion is favored at high
temperatures. As a re-
sult, the production of
NO(g) from $N_2(g)$ and
$O_2(g)$ occurs in any
combustion process in
air. The higher the
combustion tempera-
ture, the more impor-
tant the production of
NO(g) becomes, and the
more serious the attend-
ant air pollution prob-
lems that are created.

AN EARLY CATHODE RAY EXPERIMENT. One of the early observations on cathode rays (William Crookes, 1879) was their deflection in a magnetic field. The rays are themselves invisible but are observed through the green fluorescence they produce when they strike a zinc sulfide-coated screen. The magnetic field is produced by the large magnet to the right and slightly behind the screen. [Photograph by Carey B. Van Loon.]

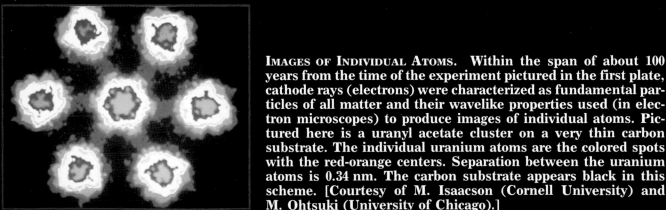

IMAGES OF INDIVIDUAL ATOMS. Within the span of about 100 years from the time of the experiment pictured in the first plate, cathode rays (electrons) were characterized as fundamental particles of all matter and their wavelike properties used (in electron microscopes) to produce images of individual atoms. Pictured here is a uranyl acetate cluster on a very thin carbon substrate. The individual uranium atoms are the colored spots with the red-orange centers. Separation between the uranium atoms is 0.34 nm. The carbon substrate appears black in this scheme. [Courtesy of M. Isaacson (Cornell University) and M. Ohtsuki (University of Chicago).]

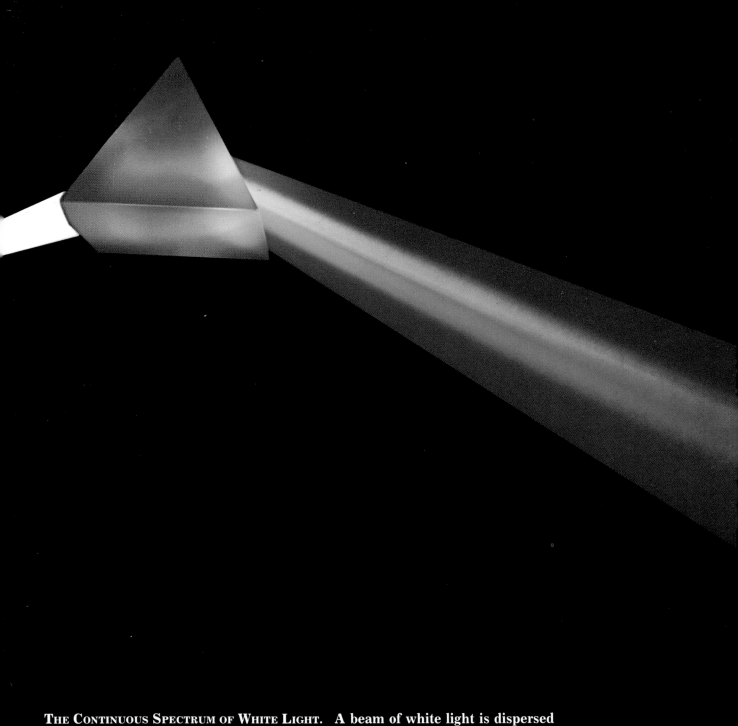

THE CONTINUOUS SPECTRUM OF WHITE LIGHT. A beam of white light is dispersed into its individual components, which extend over all wavelengths from red to violet (the rainbow). [Courtesy of Bausch and Lomb Company.]

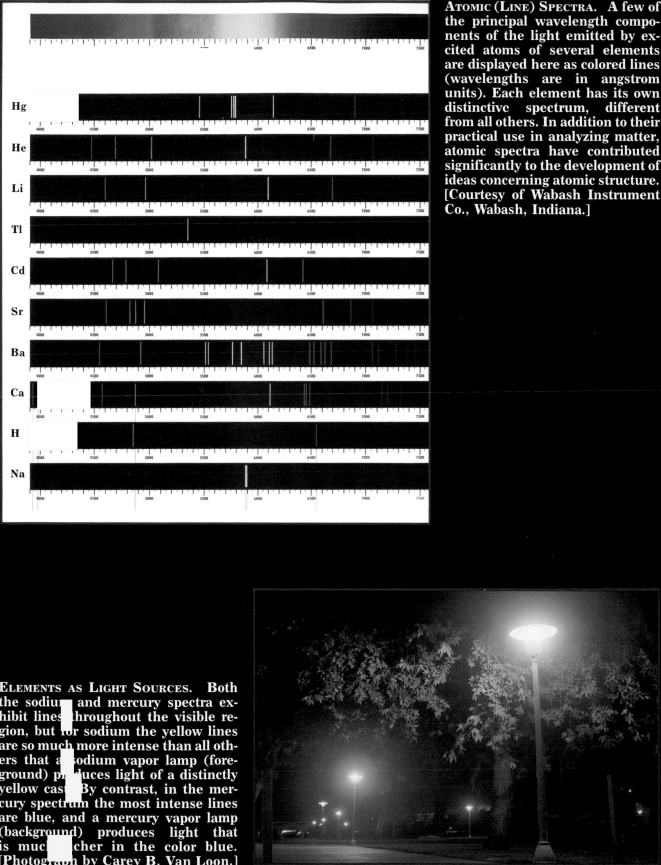

ATOMIC (LINE) SPECTRA. A few of the principal wavelength components of the light emitted by excited atoms of several elements are displayed here as colored lines (wavelengths are in angstrom units). Each element has its own distinctive spectrum, different from all others. In addition to their practical use in analyzing matter, atomic spectra have contributed significantly to the development of ideas concerning atomic structure. [Courtesy of Wabash Instrument Co., Wabash, Indiana.]

ELEMENTS AS LIGHT SOURCES. Both the sodium and mercury spectra exhibit lines throughout the visible region, but for sodium the yellow lines are so much more intense than all others that a sodium vapor lamp (foreground) produces light of a distinctly yellow cast. By contrast, in the mercury spectrum the most intense lines are blue, and a mercury vapor lamp (background) produces light that is much richer in the color blue. [Photograph by Carey B. Van Loon.]

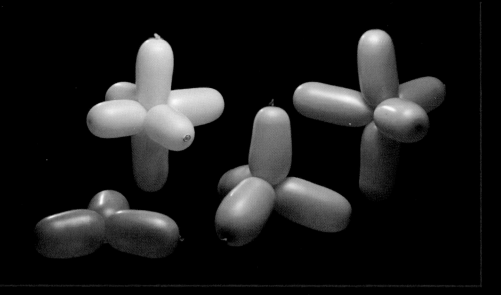

THE VALENCE-SHELL ELECTRON-PAIR REPULSION THEORY ILLUSTRATED.
Electron pairs distribute themselves in the valence shell of the central atom in
a molecule or polyatomic ion in much the same way as the lobes formed by
twisting together elongated balloons. The relationship between electron pair
distributions and molecular shapes is summarized in Table 9-2. The electron
pair distributions pictured here are octahedral (yellow), trigonal bipyramidal
(pink), tetrahedral (green), and trigonal planar (orange). [Courtesy of Arlo
Harris; photograph by Carey B. Van Loon.]

B. Physical Changes

SUBLIMATION OF IODINE. Even at temperatures well below its melting point of 114°C,
solid iodine exhibits an appreciable vapor
pressure. Here, purple iodine vapor is produced at about 70°C and redeposits as solid
iodine on the colder walls of the flask.
[Photograph by Carey B. Van Loon.]

PHASE TRANSITION IN SOLID MERCURY(II) IODIDE. The stable form of mercury(II) iodide at room temperature is a red solid (left). At 126°C the red solid undergoes a phase change to a yellow solid. Mercury(II) iodide also exhibits an appreciable sublimation pressure, and the vapor condenses to a mixture of the yellow and red solids on the colder walls of the flask (right). [Photographs by Carey B. Van Loon.]

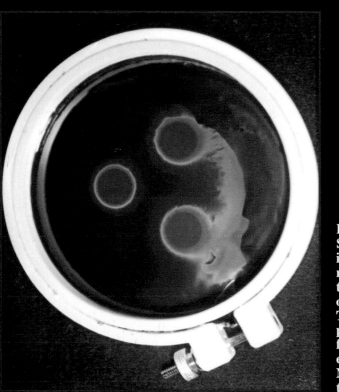

EFFECT OF TEMPERATURE ON LIQUID CRYSTALS. Some liquid crystalline materials undergo a change in color with temperature. Here a liquid crystalline material has been coated on the surface of a polyester film supported on an embroidery hoop. A heated object is placed under the film and allowed to cool. The colors of the liquid crystals establish the temperature distribution in the object. Liquid crystals find many applications in nondestructive testing, e.g., in the electronics industry. [Courtesy of Edward N. Sharpless, President, Liquid Crystal Applications, Inc.]

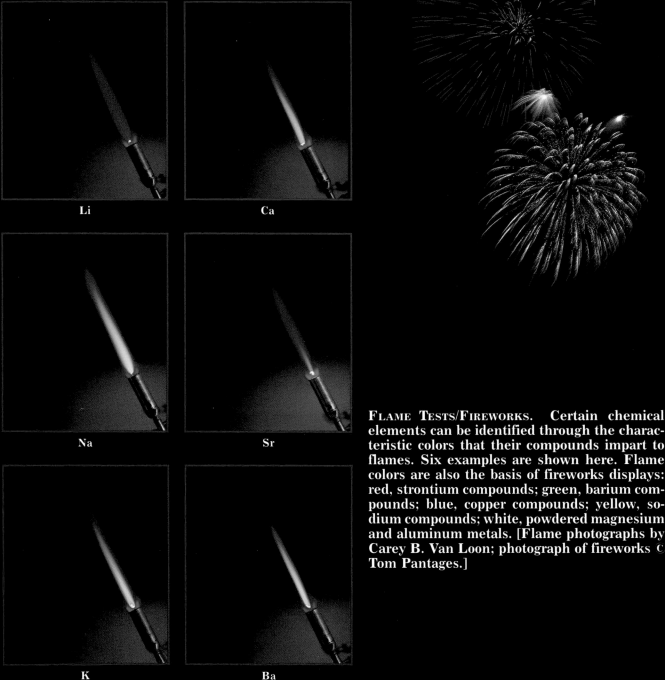

Li

Ca

Na

Sr

K

Ba

FLAME TESTS/FIREWORKS. Certain chemical elements can be identified through the characteristic colors that their compounds impart to flames. Six examples are shown here. Flame colors are also the basis of fireworks displays: red, strontium compounds; green, barium compounds; blue, copper compounds; yellow, sodium compounds; white, powdered magnesium and aluminum metals. [Flame photographs by Carey B. Van Loon; photograph of fireworks © Tom Pantages.]

QUALITATIVE TESTS FOR Co^{2+} AND Fe^{3+}.
In the presence of thiocyanate ion, SCN^-,
Co^{2+}(aq) forms a blue complex ion,
$[Co(SCN)_4]^{2+}$ (left). In a mixture of
Co^{2+}(aq) and Fe^{3+}(aq), even if the quantity of Fe^{3+} is slight, the blood-red complex ion $[Fe(H_2O)_5(SCN)]^{2+}$ masks the
color of the cobalt(II) complex ion (center). However, if the solution is treated
with fluoride ion, Fe^{3+} is converted to the
extremely stable, pale yellow complex
$[FeF_6]^{3-}$; and the Co^{2+} can be detected
through its thiocyanato complex ion,
yielding a blue-green solution (right).
[Photographs by Carey B. Van Loon.]

TEST FOR IODIDE ION. When a colorless aqueous solution containing I^- is treated
with Cl_2(aq), some of the I^- is oxidized to I_2, which in the presence of excess I^-
produces red-brown triiodide ion, I_3^- (left). I_2 is considerably more soluble in nonpolar organic solvents (e.g., CS_2) than in aqueous KI and is extracted into the
organic layer, producing a violet color (right). [Photographs by Carey B. Van
Loon.]

An "Iodine Clock" Reaction. One of the products of the reaction illustrated here is iodine, in the form of triiodide ion, I_3^- (see reaction 14.8). Also present in the reaction mixture are small quantities of thiosulfate ion, $S_2O_3^{2-}$, and starch indicator. Iodine reacts with $S_2O_3^{2-}$ as long as there is any $S_2O_3^{2-}$ present, and then it forms a complex with starch; the solution changes from colorless to a deep blue color. Here, the appearance of the blue color is being used to indicate the rate of reaction (14.8)—the sooner the color appears, the faster the reaction. At room temperature, for a given set of initial concentrations, the blue color appears after 49.8 s (left). For the same initial concentrations the reaction proceeds much more slowly in an ice bath, requiring about 4.5 min (center). When the room-temperature reaction is repeated in the presence of a drop of Cu^{2+}(aq) as a catalyst, the reaction rate is much faster; the reaction time is 17.7 s (right). [Photographs by Carey B. Van Loon.]

Dissolving Action of Nitric Acid on Brass. Except for tin, the several metals in a typical brass sample—Cu, Zn, Sn, Pb, Fe—are readily soluble in concentrated HNO_3(aq). The red-brown gaseous product is nitrogen dioxide, NO_2 (left). Tin is oxidized to a hydrated SnO_2, which remains as a white precipitate in the solution made blue by Cu^{2+} (right). The precipitate can be filtered off, purified, dried and weighed, thereby providing a basis for determining the percent Sn in the brass sample (see page 57). [Photographs by Carey B. Van Loon.]

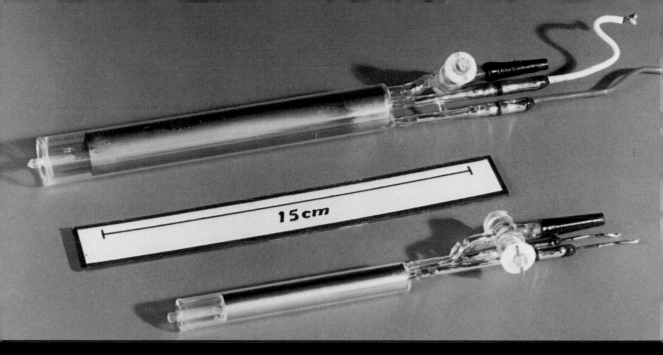

RADIOCARBON DATING. Pictured here are special radiation counters used in radiocarbon dating (see Section 25-5). A small sample (as little as 10 mg) of an organic material is burned and the evolved $CO_2(g)$ passed through the counter, where beta particles emitted by carbon-14 are counted. Recently such a counter was used to authenticate a gold and silver studded belt from about A.D. 700. The sample taken for analysis was flax fibers from the belt. [Courtesy of Brookhaven National Laboratory.]

D. Producing Modern Materials

ZONE REFINING. For certain uses, such as in modern semiconductor devices, ultrapure materials are required. Here a single crystal of silicon is being purified by zone refining (see page 356). [Photograph by Sol Mednick

Photovoltaic Cells. One of the applications of ultrapure materials produced by zone refining or similar techniques is in fabricating photovoltaic cells. A small array of such cells is illustrated here. [Courtesy of Exxon Corporation.]

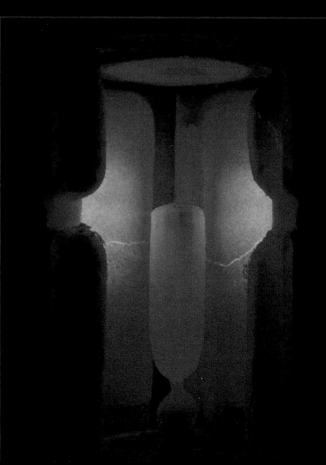

Manufacture of Artificial Gemstones. Many natural gemstones have aluminum oxide, Al_2O_3, as their principal constituent, together with small quantities of transition metal oxide "impurities" (see Table 22-5). Similar gemstones can be artificially produced. In the furnace pictured here, a mixed powder is sprayed from above, melts in the hottest regions of the furnace, and deposits as a liquid layer that then solidifies. The solidified material is gradually withdrawn from the furnace as layer after layer is added to it. Gemstones produced in this way have many industrial applications, such as the use of rubies in lasers. [Courtesy of Hrand Djevahirdjian, S.A.]

NEW LIGHTWEIGHT MATERIALS. The development of aircraft to fly at high altitudes and high speeds with significantly improved fuel economies has dictated the replacement of traditional construction materials (e.g., aluminum) by strong, lightweight materials, such as a composite of graphite and epoxy resin in the prototype Beechcraft Starship 1 airplane pictured here. [Courtesy of Beech Aircraft Corporation.]

REPLACING ENVIRONMENTALLY HAZARDOUS MATERIALS. The typical painter's palette features a number of toxic materials. For example, a variety of popular yellow pigments are based on CdS, and red pigments, on CdSe and HgS. Both Cd and Hg are toxic metals (see page 700). For ordinary use, as in house paints, these pigments have been largely replaced by iron oxides, which are environmentally safe. [Photograph by Barbara Cushing.]

**FIRST PRODUCT OF A SPACE-BASED MANUFAC-
TURING PROCESS.** The objects pictured here are uniform spheres of a polystyrene plastic, 10.0 μm in diameter, manufactured in the low-gravity environment of space shuttle flights. When produced on earth, polystyrene spheres of this size are nonuniform in diameter and often egg-shaped. Uniform spheres are used in calibrating certain medical research instruments, in monitoring environmental particulate pollutants, and in conjunction with the manufacture of finely powdered materials such as pigments and inks. [Courtesy of NASA.]

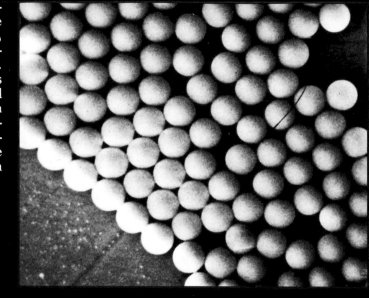

LIMESTONE CAVERN AND FORMATIONS. Carbon dioxide from the atmosphere, when dissolved in rainwater, participates in a series of reactions that can ultimately lead to the formation of limestone caverns with their familiar stalactites and stalagmites (see also, page 681). [Photograph of Temple of the Sun, Carlsbad Cavern, New Mexico, by Fred E. Mang, Jr.; courtesy of the National Park Service.]

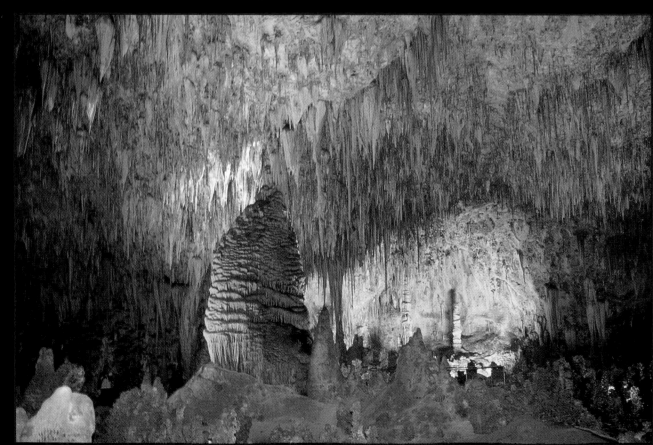

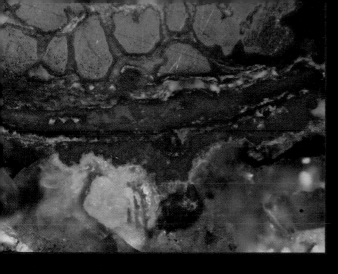

CORROSION OF BRONZE. Just as limestone cavern formation is a natural chemical effect, so is the atmospheric corrosion of copper and copper alloys such as bronze. The familiar green patina that forms on these metals is a basic copper carbonate (reaction 13.50). In this photomicrograph (approximately $100\times$) of a fragment of ancient bronze found in Thailand, the interface between the metal and its corrosion layer is clearly visible. [Courtesy of William Marin, Jr., Brookhaven National Laboratory.]

EUTROPHICATION OF A LAKE. Still another natural chemical effect in the environment is the eutrophication of a lake, but this effect can be greatly accelerated by human activities, e.g., through phosphates in waste water (see page 658). [Courtesy of The Soil Conservation Service, Tom McCabe.]

INDUSTRIAL SMOG. This form of air pollution is derived from $SO_2(g)$ in coal-fired power plants and other industrial operations (see page 395). Until the advent of strict pollution control measures, sights like this one were not uncommon. (Contrast this photograph with the fully controlled power plant pictured on page 396.) [Courtesy of Environmental Protection Agency.]

PHOTOCHEMICAL SMOG. This form of air pollution results from the action of sunlight on oxides of nitrogen, which are emitted by automobiles and other high-temperature combustion sources (see page 387). Its hallmark is a colored haze, as seen here. [Photograph by Sepp Seitz © 1980; Woodfin Camp & Associates.]

SMOG EFFECTS. The documented and suspected effects of air pollution are numerous, e.g., respiratory ailments, crop losses, and acidification of lakes. A particularly striking effect is the action of acidic air pollutants on marble, as this statue shows. [Courtesy of CNRI/French Information Services.]

pH INDICATORS. Color changes in acid–base indicators can be used to determine the pH of a solution (see page 545). The indicators shown here and the pH ranges over which they change color are methyl violet (0–2); thymol blue, acid range (1–3); methyl orange (3–5); phenol red (6–8); thymol blue, base range (8–10); alizarin yellow R (10–12). [Photographs by Carey B. Van Loon.]

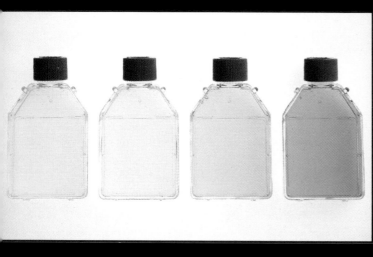

HYDROLYSIS OF SALT SOLUTIONS. Bromothymol blue indicator undergoes a color change at about the neutral point of pH 7. Its colors in buffer solutions at pH 6, 7, and 8 are shown in the top row. In the bottom row bromothymol blue is present in four different salt solutions. The first solution contains $Al_2(SO_4)_3$(aq), and its yellow color demonstrates the acidic properties of $[Al(H_2O)_6]^{3+}$. The second solution contains NH_4Cl(aq), and its yellow color indicates that NH_4^+ is also acidic. In the third solution, $NaCl$(aq), no hydrolysis occurs and the pH is 7. In the fourth solution, $NaC_2H_3O_2$(aq), the blue color indicates that $C_2H_3O_2^-$ acts as a base. [Photographs by Carey B. Van Loon.]

PREPARATION AND ACTION OF A BUFFER SOLUTION. Thymol blue indicator assumes a red color in 0.20 M HCl (top row, left). An excess of $NaC_2H_3O_2$ converts the HCl to $HC_2H_3O_2$, resulting in a $HC_2H_3O_2$–$NaC_2H_3O_2$ buffer solution with pH ≈ 5; the thymol blue changes to a yellow color (top row, center). Addition of a pellet of NaOH to this buffer solution has practically no effect on the pH (top row, right). Thymol blue in distilled water is also yellow (bottom row, left); but the addition of a pellet of NaOH causes a large increase in pH, indicated by the blue color (bottom row, right). Pure water has no buffer capacity. [Photographs by Carey B. Van Loon.]

AMPHOTERISM. When an excess of $OH^-(aq)$ is added to a solution containing $Al^{3+}(aq)$ and $Fe^{3+}(aq)$, the Fe^{3+} precipitates as $Fe(OH)_3(s)$ and the $Al(OH)_3(s)$ first formed redissolves to produce $[Al(OH)_4]^-(aq)$ (left). $Al(OH)_3(s)$ is amphoteric. The precipitate of $Fe(OH)_3(s)$ is removed by filtration and the $[Al(OH)_4]^-(aq)$ is made slightly acidic through the action of CO_2, here added as dry ice (center). The precipitated $Al(OH)_3(s)$ collects at the bottom of a clear, colorless solution (right). A method similar to that outlined here is used to purify bauxite ore (hydrated Al_2O_3) as the first step in the manufacture of aluminum (see page 688). [Photographs by Carey B. Van Loon.]

G. Electrochemistry

DISPLACEMENT OF $Ag^+(aq)$ BY $Cu(s)$. A coil of copper wire is placed in an aqueous solution of silver nitrate (left) and immediately the displacement reaction begins: $2 Ag^+(aq) + Cu(s) \rightarrow Cu^{2+}(aq) + 2 Ag(s)$. After about 30 min a crystalline deposit of silver and a blue coloration of the $Cu^{2+}(aq)$ can be clearly seen. The photograph at the right was taken after about 2 h; in time the reaction goes essentially to completion. [Photographs by Carey B. Van Loon.]

DISPLACEMENT OF Cu²⁺(aq) BY Zn(s).
$Cu^{2+}(aq)$ produced in the displacement reaction of $Cu(s)$ and $Ag^+(aq)$ can itself be displaced from solution by a more active metal such as zinc. $Cu^{2+}(aq)$ and $Zn(s)$ are shown in the photograph at the left. The products of the displacement reaction (right) are a red-brown precipitate of $Cu(s)$ and a colorless solution of $Zn^{2+}(aq)$:
$$Cu^{2+}(aq) + Zn(s) \rightarrow Zn^{2+}(aq) + Cu(s).$$
[Photographs by Carey B. Van Loon.]

THE ELECTROCHEMICAL MECHANISM OF CORROSION. In the corrosion of an iron nail (top left) oxidation of the iron to $Fe^{2+}(aq)$ occurs at the strained regions of the nail—the head and the tip. The presence of $Fe^{2+}(aq)$ is marked by the formation of a blue precipitate in the presence of $K_3[Fe(CN)_6]$. Reduction of dissolved O_2 to $OH^-(aq)$ occurs on other portions of the nail. The presence of $OH^-(aq)$ is signaled by the pink coloration of phenolphthalein indicator. With a bent nail oxidation occurs at the additional point of strain, the bend (top right). An iron nail can be protected from corrosion by the more active metal zinc (bottom left). Here, the zinc corrodes and no blue precipitate is seen, indicating the absence of $Fe^{2+}(aq)$. Coating an iron nail with the less active metal copper (bottom right) provides protection only to the regions fully covered by the copper. The exposed portions of the nail corrode even more rapidly than otherwise. Corrosion reactions are discussed in Section 20-7. [Photograph by Carey B. Van Loon.]

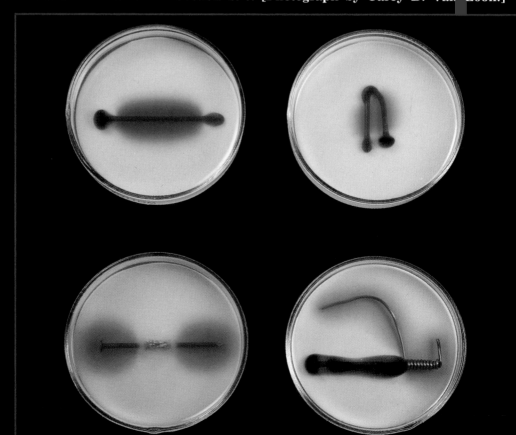

SOME MINERALS OF THE METALS OF THE FIRST TRANSITION SERIES. With the principal exception of some native copper, the metals of the first transition series must be extracted from their mineral ores. The production of the metals ranges from that of scandium, produced only in kilogram quantities, to that of iron, with annual world-wide production approaching one billion tons. Eight of the minerals are arranged in clockwise fashion, starting with thortveitite (the ice-cream-cone shaped specimen at "nine o'clock"). The ninth mineral, chalcopyrite, is at the center.

1.	Thortveitite	$ScSi_2O_7$	from Ireland
2.	Ilmenite	$FeTiO_3$	from Norway
3.	Vanadinite	$3Pb_3(VO_4)_2 \cdot PbCl_2$	from Arizona
4.	Chromite	$Fe(CrO_2)_2$	from New Zealand
5.	Pyrolusite	MnO_2	from New Mexico
6.	Hematite	Fe_2O_3	from Great Britain
7.	Cobaltite	$CoAsS$	from USSR
8.	Pentlandite	$(Ni,Fe)_9S_8$	from Czechoslovakia
9.	Chalcopyrite	$CuFeS_2$	from Pennsylvania

OXIDATION–REDUCTION REACTIONS.   A general feature of the transition elements is variability of oxidation states, and this means that many of the reactions of these elements and their compounds are oxidation–reduction reactions. Pictured on the left is ammonium dichromate, which contains both an oxidizing agent ($Cr_2O_7^{2-}$) and a reducing agent (NH_4^+). The products of the reaction between these ions (reaction 23.12) are $Cr_2O_3(s)$, $N_2(g)$, and $H_2O(g)$. Considerable heat and light are also evolved (center). The product on the right is pure $Cr_2O_3(s)$. [The further reduction of $Cr_2O_3(s)$ with $Al(s)$ as a reducing agent can be carried out in the thermite reaction (23.13).] [Photographs by Carey B. Van Loon.]

STABILITY OF OXIDATION STATES.   Of the various oxidation states that a transition metal may display in its compounds, some may not be readily attainable. For example, a freshly prepared blue solution of $Cr^{2+}(aq)$, obtained by dissolving chromium metal in dilute $HCl(aq)$ (left), is oxidized within a few minutes to green $Cr^{3+}(aq)$ (right). (See also, equation 23.16.) [Photographs by Carey B. Van Loon.]

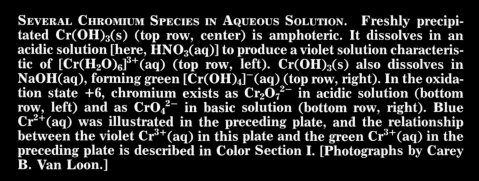

SEVERAL CHROMIUM SPECIES IN AQUEOUS SOLUTION. Freshly precipitated $Cr(OH)_3(s)$ (top row, center) is amphoteric. It dissolves in an
acidic solution [here, $HNO_3(aq)$] to produce a violet solution characteristic of $[Cr(H_2O)_6]^{3+}(aq)$ (top row, left). $Cr(OH)_3(s)$ also dissolves in
$NaOH(aq)$, forming green $[Cr(OH)_4]^-(aq)$ (top row, right). In the oxidation state +6, chromium exists as $Cr_2O_7^{2-}$ in acidic solution (bottom
row, left) and as CrO_4^{2-} in basic solution (bottom row, right). Blue
$Cr^{2+}(aq)$ was illustrated in the preceding plate, and the relationship
between the violet $Cr^{3+}(aq)$ in this plate and the green $Cr^{3+}(aq)$ in the
preceding plate is described in Color Section I. [Photographs by Carey
B. Van Loon.]

SOME VANADIUM SPECIES IN SOLUTION.
The element vanadium is shown in
four different oxidation states. The
yellow solution has vanadium in the
+5 oxidation state, as VO_2^+. In the
blue solution the oxidation state is +4,
in VO^{2+}. The green solution contains
the ion V^{3+}, and the violet solution,
V^{2+}. [Photograph courtesy of Teledyne
Wah Chang, Albany, Oregon.]

1. Complex Ions and Coordination Compounds

TWO COORDINATION COMPOUNDS OF COBALT(III).   Yellow $[Co(NH_3)_6]Cl_3$
and purple $[Co(NH_3)_5Cl]Cl_2$ figured prominently in the development of
Werner's coordination theory (see Section 24-1). [Courtesy of Arlo Harris;
photograph by Carey B. Van Loon.]

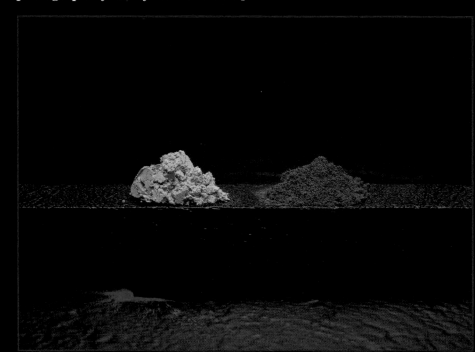

EFFECT OF LIGANDS ON THE COLORS OF COORDINATION COMPOUNDS. These compounds all consist of a six-coordinate cobalt complex ion in combination with nitrate ions. In each case the complex ion has five NH_3 molecules and one other group as ligands. From left to right the compounds are $[Co(NH_3)_5Cl](NO_3)_2$, $[Co(NH_3)_5Br](NO_3)_2$, $[Co(NH_3)_5I](NO_3)_2$, $[Co(NH_3)_5NO_2](NO_3)_2$, $[Co(NH_3)_5SO_4]NO_3$, and $[Co(NH_3)_5CO_3]NO_3$. [Courtesy of Arlo Harris; photograph by Carey B. Van Loon.]

"INERT" COMPLEX IONS. The solution on the left is obtained by dissolving $CrCl_3 \cdot 6H_2O$ in water. The green color is due to $[Cr(H_2O)_4Cl_2]^+$. A slow exchange of H_2O for Cl^- ligands leads to a violet solution of $[Cr(H_2O)_6]^{3+}$ in one or two days (right). (See also, Section 24-12.) [Photograph by Carey B. Van Loon.]

"LABILE" COMPLEX IONS. The exchange of ligands in the coordination sphere of Cu^{2+} occurs as rapidly as solutions can be mixed. The solution at the extreme left is produced by dissolving anhydrous $CuSO_4$ in concentrated HCl(aq). Its yellow color is caused by $[CuCl_4]^{2-}$. Dilution of this solution with a small volume of water produces a yellow-green solution, resulting from the replacement of some Cl^- ligands by H_2O molecules. When dissolved in water, $CuSO_4$ produces a light blue solution of $[Cu(H_2O)_4]^{2+}(aq)$. If this solution is treated with $NH_3(aq)$, the H_2O molecules are displaced as ligands by NH_3 molecules. The solution acquires the very deep blue color characteristic of $[Cu(NH_3)_4]^{2+}(aq)$ (extreme right). [Photograph by Carey B. Van Loon.]

COAGULATION OF A COLLOIDAL DISPERSION. The red-brown mixture on the left is a colloidal dispersion of hydrous Fe_2O_3, obtained by adding a few drops of concentrated $FeCl_3(aq)$ to boiling water. The colloid is formed by a condensation reaction involving complex ions such as the one pictured in Figure 24-15. That the mixture is colloidal is easily demonstrated by adding a few drops of $Al_2(SO_4)_3(aq)$. Rapid coagulation of the colloidal particles leads to a precipitate of Fe_2O_3 (right). [Photographs by Carey B. Van Loon.]

Although very dissimilar in appearance, the two devices pictured here work on the same principle. At the left is a model of an early Watt steam engine featuring an all-important external condenser to convert steam back to liquid water after the steam has performed some work. The ship at the right contains equipment designed to produce electricity from heat extracted from ocean water, a process known as ocean thermal energy conversion (OTEC). [Courtesy of Smithsonian Institution (left); U.S. Department of Energy photo (right).]

The underlying principle of the second law of thermodynamics was deduced by a French military engineer, Sadi Carnot, in 1824 in a study of the efficiencies of heat engines. William Thomson (Lord Kelvin) recognized the significance of Carnot's work and saw in it the basis of the second law of thermodynamics and an absolute temperature scale.

The basic principle of a heat engine is that heat (q_h) is absorbed by the working substance of the engine at a high temperature (T_h). This heat is partly converted to work (w), and the remainder (q_l) is released to the surroundings at a lower temperature (T_l). The process is pictured in Figure 16-10.

The efficiency of the engine is governed by the ratio w/q_h. If all the heat absorbed could be converted to work, the engine would be 100% efficient. The second law of thermodynamics places an absolute limit on the efficiency of a heat engine, *and it is never 100%*. The expression obtained is

$$\text{efficiency} = \frac{w}{q_h} = \frac{T_h - T_l}{T_h} \qquad (16.29)$$

where temperature is in kelvins.

The following rearrangement of equation (16.29) may provide added insight into the matter of converting heat to work.

$$w = q_h\left(\frac{T_h - T_l}{T_h}\right) = \left(q_h \times \frac{T_h}{T_h}\right) - \left(q_h \times \frac{T_l}{T_h}\right)$$
$$= q_h - \left(q_h \times \frac{T_l}{T_h}\right) \qquad (16.30)$$

The work obtainable from a given quantity of heat (q_h) is equal to the quantity of heat *less* a fraction of that quantity. The second term of equation (16.30)—$q_h \times (T_l/T_h)$—is the fraction of the heat energy absorbed at the high tem-

FIGURE 16-10
Schematic representation of a heat engine.

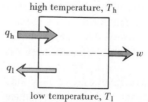

The efficiency of the engine is determined by the quantity of heat q_h that is converted to work w. The smaller the quantity of heat released to the surroundings at the lower temperature q_l, the more efficient the engine. The second law of thermodynamics places a theoretical maximum efficiency on every heat engine; it is always less than 100% (see equation 16.29).

perature that is *unavailable* for work. To reduce this fraction to its lowest possible value—*zero*—would require that the low temperature (T_l) be 0 K. This observation provides a new way of looking at the absolute zero of temperature: It is the temperature at which heat would have to be discharged by a heat engine if the engine were to have a theoretical efficiency of 100%. For a given low temperature (T_l), another way to reduce the value of the fraction of heat unavailable for work is to *raise* the temperature at which a heat engine absorbs heat (T_h). That is, the *higher* the temperature at which heat is absorbed, the greater the proportion of that heat that can be converted to work—the higher the "quality" of the heat. Heat rejected to a low temperature sink by a heat engine has been premanently degraded in quality.

According to equation (16.29), if a steam engine operates between 100°C (the boiler temperature) and 25°C (the condenser temperature), it can have an efficiency of only 0.20, that is, 20%; 80% of the heat supplied to the boiler is given off to the surroundings.

$$\text{efficiency} = \frac{373 - 298}{373} = \frac{75}{373} = 0.20$$

The ocean thermal energy conversion (OTEC) technology makes use of the temperature difference between the warm surface layer of the ocean and a much colder, deeper layer of water. In the Hawaiian Islands this difference amounts to about 36°C. The warm water is used to vaporize a working substance, such as ammonia. This vapor then drives a turbine that turns an electric generator. After performing work the working substance is condensed back to the liquid state by the colder, bottom water. The cycle is repeated continuously. The efficiencies of current OTEC devices are only about 2%, but the attractiveness of the method is that there is no fuel cost. The temperature gradient of the ocean water is produced by solar energy.

Conventional electric power plants are based on the burning of a fossil fuel, which provides the heat energy to convert water to steam. The steam powers a turbine that drives an electric generator. The process is not highly efficient. As we have just noted, the efficiency can be improved by using a higher working temperature for the turbine (T_h). This can be accomplished by operating the system under high pressure, at temperatures well in excess of 100°C, using superheated steam. Still, however, most electric power plants operate at efficiencies of less than 40%. Thus, more than half the heat required to produce electric power is waste heat. The implications of this fact are twofold: wasted heat means wasted fuel, and as the waste heat enters the environment it leads to thermal pollution.

FIGURE 16-11
Efficiency of a fossil-fuel electric power plant and the origin of thermal pollution.

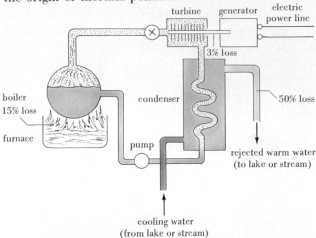

Water is heated in a boiler and converted to superheated steam. The steam is allowed to expand in the turbine where it does work by turning the rotor blades. The turbine shaft drives the electric generator, which produces electric power. Conversion of mechanical to electrical energy in the turbine-generator combination is nearly 100% efficient.

If the steam were to be rejected to the surroundings, all the makeup water in the boiler would have to be cold water. This would require more fuel for heating than if the steam is returned to the boiler. However, because of its greatly expanded volume after performing work in the turbine, the steam must be condensed back to liquid before being returned to the boiler. Condensation of the steam occurs in the condenser. The water that is used for cooling carries away up to 50% of the heat energy that was released in burning the fuel. It is this waste heat that produces thermal pollution.

Thermal Pollution. The origin of thermal pollution in a power plant is suggested by Figure 16-11. The best known effects of thermal pollution are those produced on fish and other aquatic life. The differential growth rate of algae with changing temperatures can cause one type of algae, which is ideal food for fish, to be displaced by another, which is a poor food or even toxic to fish. Also, because their metabolic rate goes up with temperature, fish need more food and oxygen as the temperature of their water rises; but the solubility of air in water *decreases* with increasing temperature. Temperature changes affect other physiological processes too. Some fish are killed by the thermal shock of even relatively small temperature changes.

Even if all other pollution sources could be controlled, release of heat energy to the surroundings by heat engines would still be unavoidable. Means must be found to minimize the amount of waste heat or to put the waste heat to useful purposes. The first objective can be met through more efficient heat engines and through the direct conversion of other energy forms to electricity (such as with fuel cells and solar energy devices). The second objective can be met by using waste heat from power plants to supply hot water for industrial, commercial, or residential purposes, or using this water for agricultural purposes. None of these methods to deal with thermal pollution is well established currently, and the problem is likely to continue to grow before solutions are found.

Summary

The first law of thermodynamics relates the internal energy change (ΔE) in a system to exchanges of heat (q) and work (w) between a system and its surroundings. A chemical reaction can be treated as a thermodynamic system, with ΔE for the system equal to the heat of reaction at constant volume, q_V. This particular heat of reaction is what is measured in a bomb calorimeter.

Chemical reactions are generally carried out at constant pressure, not at constant volume; the enthalpy function (H) has been devised to describe constant-pressure processes. In particular, $\Delta H = q_p$, the heat of a reaction carried out at constant pressure and in which only pressure-volume work is performed. For reactions involving only condensed phases, $\Delta H \simeq \Delta E$. Differences between ΔH and ΔE are larger if gases are involved, though still negligible in most cases. ΔH and ΔE can be related through a simple mathematical equation.

Although most exothermic reactions are spontaneous, many *endothermic* reactions are spontaneous too. A common feature shared by systems in which spontaneous endothermic processes occur is that they undergo an increase in disorder or randomness. The thermodynamic function related to the degree of disorder or randomness in a system is *entropy, S*. If entropy change alone is used as a criterion for spontaneous change, it is necessary to assess the *total* entropy change accompanying a process. This is the sum of the entropy changes of the system *and* the surroundings. Such an assessment is inconvenient and sometimes very difficult to make.

The thermodynamic property devised to provide a criterion for spontaneous change based just on the system itself is the *Gibbs free energy, G*. The criterion for spontaneous change in a closed system at constant temperature and pressure is that there be a *decrease in free energy:* $\Delta G < 0$. The free energy criterion for *equilibrium* is that $\Delta G = 0$. At equilibrium, $\Delta H = T \Delta S$, an expression that is particularly useful in dealing with phase transitions. For liquid–vapor equilibria at the normal boiling point, Trouton's rule is often applicable: $\Delta \overline{S}^\circ_{vap} = \Delta \overline{H}^\circ_{vap}/T_{bp} \simeq 88 \text{ J mol}^{-1} \text{ K}^{-1}$.

The standard free energy change, $\Delta \overline{G}^\circ$, accompanies the conversion of reactants in their standard states to products in their standard states. Tabulated free energy data are usually standard molar free energies of formation, $\Delta \overline{G}^\circ_f$. A useful relationship exists between standard molar free energy change and the equilibrium constant: $\Delta \overline{G}^\circ = -2.303RT \log K$. If a value of $\Delta \overline{G}^\circ$ can be obtained, say by the appropriate combination of tabulated thermodynamic data, a value of K can be calculated. The K values obtained from $\Delta \overline{G}^\circ$ data are thermodynamic equilibrium constants. These equilibrium constant expressions are based on the *activities* of reactants and products. Activities can be related to concentrations and partial pressures through a few simple conventions.

A key to successful thermodynamic calculations is in being able to determine entropy changes from tabulated *absolute* molar entropies. The third law of thermodynamics provides a basis upon which absolute entropies may be experimentally established.

Learning Objectives

As a result of studying Chapter 16, you should be able to

1. State the first law of thermodynamics and the sign conventions used for heat, q, and work, w.

2. Calculate the value of one of the following from known values of the other two: ΔE, q, and w.

3. Explain the purpose served by the thermodynamic property of enthalpy (H), describe how ΔH is related to ΔE, and calculate one from the other for reactions involving gases.

4. Explain the meaning of the term "spontaneous change" as it applies to chemical reactions.

5. Discuss the significance of entropy and its relationship to the degree of disorder within a system.

6. Predict for certain processes whether entropy increases or decreases.

7. Write the principal equations that define free energy and free energy change and state the basic criterion for spontaneous change, $\Delta G < 0$.

8. Relate $\Delta \overline{G}^\circ$, $\Delta \overline{H}^\circ$, and $\Delta \overline{S}^\circ$ through the equation: $\Delta \overline{G}^\circ = \Delta \overline{H}^\circ - T \Delta \overline{S}^\circ$.

9. Calculate entropy and enthalpy changes for phase transitions, using the condition of equilibrium ($\Delta G = 0$).

10. Write thermodynamic equilibrium constant expressions for chemical reactions and relate these to K_c and K_p.

11. Calculate equilibrium constants using tabulated data and the relationship between $\Delta \overline{G}^\circ$ and K.

12. Explain how the third law of thermodynamics makes possible an evaluation of *absolute* entropies of substances.

Some New Terms

Enthalpy, H, is a thermodynamic function used to describe constant-pressure processes. $H = E + PV$, and at constant pressure, $\Delta H = \Delta E + P \Delta V$.

Entropy, S, is a measure of the degree of *disorder* in a system; the greater the disorder, the greater the entropy.

Entropy change, ΔS, expresses the extent to which the degree of order changes as the result of some process. A positive ΔS means an increase in *disorder*.

Entropy change of the universe, ΔS_{univ}, is the total entropy change (system and surroundings) for a process. For every spontaneous change, $\Delta S_{univ} > 0$.

The **first law of thermodynamics** states that in interactions between a system and its surroundings energy is neither created nor destroyed. The difference in the quantities of heat and/or work exchanged must be reflected as a change in the internal energy of the system: $\Delta E = q - w$.

Free energy, G, is a thermodynamic function designed to provide a criterion for spontaneous changes. It is defined through the equation: $G = H - TS$.

Free energy change, ΔG, indicates the direction of spontaneous change. For a spontaneous process at constant temperature and pressure in a closed system, $\Delta G < 0$.

A **heat engine** is a device for converting heat into work. The engine absorbs heat at a high temperature, converts part of it to work, and discharges the remaining heat to the surroundings at a lower temperature.

The **second law of thermodynamics** relates to the direction of spontaneous change. All spontaneous processes produce an increase in the entropy of the universe.

A **spontaneous or natural process** is one that is able to take place in a system left to itself. No external action is required to make the process go, although in some cases the process may take a very long time to occur.

Standard free energy change, ΔG°, is the free energy change of a process when reactants and products are all in their standard states. The equation relating the standard molar free energy change and the equilibrium constant is $\Delta \overline{G}^\circ = -2.303RT \log K$.

The **standard free energy of formation, $\Delta \overline{G}_f^\circ$,** is the standard free energy change associated with the formation of 1 mol of compound from its elements.

The **thermodynamic equilibrium constant, K,** is an equilibrium constant expression based on activities. In dilute solutions activities can be replaced by molar concentrations; in ideal gases, by partial pressures in atm.

The **third law of thermodynamics** postulates that the entropy of a pure perfect crystal at 0 K is zero.

Trouton's rule states that at their normal boiling points the entropies of vaporization of many liquids are about 88 J mol^{-1} K^{-1}.

Suggestions for Further Study

ANON., "How Hot Is Solar Energy?" *Env. Sci. & Tech.*, **11,** 651 (1977).

BICKFORD, F. R., "Entropy and Its Role in Introductory Chemistry," *J. Chem. Educ.*, **59,** 317 (1982).

COUTANT, C. C., "How to Put Waste Heat to Work," *Env. Sci. & Tech.*, **10,** 868 (1976).

HARRIS, W. F., "Clarifying the Concept of Equilibrium in Chemically Reacting Systems," *J. Chem. Educ.*, **59,** 1034 (1982).

KLEIN, M. J., "Carnot's Contribution to Thermodynamics," *Physics Today*, **27**[8], 23 (1974).

LEVIN, A. A., et. al., "Thermal Discharge: Ecological Effects," *Env. Sci. & Tech.*, **6,** 224 (1972).

LOWE, J. P., "Heat-Fall and Entropy," *J. Chem. Educ.*, **59,** 353 (1982).

RAMAN, V. V., "Evolution of the Second Law of Thermodynamics," *J. Chem. Educ.*, **47,** 331 (1970).

Review Problems

1. What are the changes in internal energy of a system, ΔE, if the system

(a) absorbs 58 J of heat and does 58 J of work?

(b) absorbs 125 J of heat and does 687 J of work?

(c) loses 22 J of heat and has 111 J of work done on it?

(d) absorbs no heat and does 117 J of work?

2. For each of the reactions listed below, indicate whether $\Delta \overline{H}$ is equal to, greater than, or less than $\Delta \overline{E}$. (*Hint:* Recall that "greater than" means more positive or less negative.)

(a) $C(graphite) + H_2O(g) \rightarrow CO(g) + H_2(g)$
$$\Delta \overline{H} = +130 \text{ kJ/mol}$$

(b) $3\ CO(g) + 7\ H_2(g) \rightarrow C_3H_8(g) + 3\ H_2O(l)$
$$\Delta \overline{H} = -628 \text{ kJ/mol}$$

3. Indicate whether you would expect the entropy to increase or decrease in each of the following reactions. If a determination cannot be made simply by inspecting the equation, state why this is so.

(a) $(CH_3CH_2)_2O(l) \rightarrow (CH_3CH_2)_2O(g)$

(b) $CuSO_4 \cdot 3\ H_2O(s) + 2\ H_2O(g) \rightarrow CuSO_4 \cdot 5\ H_2O(s)$

(c) $SO_3(g) + H_2(g) \rightarrow SO_2(g) + H_2O(g)$

(d) $H_2S(g) + O_2(g) \rightarrow H_2O(g) + SO_2(g)$ [not balanced]

4. From the data given, indicate which of the four cases in Table 16-2 applies for each reaction.

(a) $H_2(g) \rightarrow 2\ H(g)$

(b) $2\ SO_2(g) + O_2(g) \rightarrow 2\ SO_3(g)$ at 298 K,
$$\Delta \overline{H}^\circ = -197.8 \text{ kJ/mol}$$

(c) $N_2H_4(g) \rightarrow N_2(g) + 2\ H_2(g)$ at 298 K,
$$\Delta \overline{H}^\circ = -95.4 \text{ kJ/mol}$$

(d) $N_2(g) + 3\ Cl_2(g) \rightarrow 2\ NCl_3(l)$ $\Delta \overline{H}^\circ = +230 \text{ kJ/mol}$

5. A particular handbook lists values of $\Delta \overline{G}_f^\circ$ and $\Delta \overline{H}_f^\circ$ but not molar entropies. From the data given below, determine $\Delta \overline{S}^\circ$ for the reaction $NH_3(g) + HCl(g) \rightarrow NH_4Cl(s)$. All data are at 298 K.

$\Delta \overline{H}_f^\circ$ values (kJ/mol): $NH_3(g)$, -46.2; $HCl(g)$, -92.3; $NH_4Cl(s)$, -315.4.

$\Delta \overline{G}_f^\circ$ values (kJ/mol): $NH_3(g)$, -16.6; $HCl(g)$, -95.3; $NH_4Cl(s)$, -203.9.

6. Use data from Appendix D to determine values at 298 K of $\Delta \overline{G}^\circ$ for the following reactions.

(a) $N_2(g) + 3\ H_2(g) \rightarrow 2\ NH_3(g)$

(b) $C_2H_2(g) + 2\ H_2(g) \rightarrow C_2H_6(g)$

(c) $Fe_3O_4(s) + 4\ H_2(g) \rightarrow 3\ Fe(s) + 4\ H_2O(g)$

(d) $MgO(s) + 2\ HCl(g) \rightarrow MgCl_2(s) + H_2O(g)$

7. If a graph similar to Figure 16-6 were drawn for the process

$H_2O(l, 1\ atm) \rightarrow H_2O(g, 1\ atm)$

(a) At what temperature would the two lines intersect?

(b) What would be the value of ΔG at this point?

8. From the following data, determine $\Delta \overline{S}_{tr}^\circ$ (in J mol^{-1} K^{-1}) for each transition.

(a) The boiling of HCl(l) at $-85.05°C$, with $\Delta \overline{H}^\circ{}_{vap} = 3.86$ kcal/mol.

(b) The melting of Na(s) at $97.82°C$, with $\Delta H^\circ{}_{fusion} = 27.05$ cal/g.

(c) The transition of rhombic to monoclinic sulfur at $95.5°C$, with $\Delta \overline{H}^\circ{}_{tr} = 96$ cal/mol.

9. For the reaction $Cl_2(g) \rightleftharpoons 2\ Cl(g)$, $K_p = 2.45 \times 10^{-7}$ at 1000 K. What is $\Delta \overline{G}^\circ$ for this reaction at 1000 K?

10. Write thermodynamic equilibrium constant expressions for the following reactions. Do any of these expressions correspond to K_c or K_p?

(a) $2\ NO(g) + O_2(g) \rightleftharpoons 2\ NO_2(g)$

(b) $MgSO_3(s) \rightleftharpoons MgO(s) + SO_2(g)$

(c) $HC_2H_3O_2(aq) \rightleftharpoons H^+(aq) + C_2H_3O_2^-(aq)$

(d) $2\ NaHCO_3(s) \rightleftharpoons Na_2CO_3(s) + H_2O(g) + CO_2(g)$

(e) $MnO_2(s) + 4\ H^+(aq) + 2\ Cl^-(aq) \rightleftharpoons$
$$Mn^{2+}(aq) + 2\ H_2O(l) + Cl_2(g)$$

11. For the reaction $2\ SO_2(g) + O_2(g) \rightleftharpoons 2\ SO_3(g)$, $K_c = 2.8 \times 10^2$ at 1000 K.

(a) What is K_p for this reaction?

(b) What is $\Delta \overline{G}^\circ$ at 1000 K?

(c) Use the concept of the reaction quotient from Section 15-4 to determine in what direction a reaction will occur if one mixes in a 2.50-L vessel at 1000 K 0.40 mol SO_2, 0.18 mol O_2, and 0.72 mol SO_3.

(d) Show that the same result as in part (c) is obtained with equation (16.22) or (16.23).
(*Hint:* What units must be used for the terms in Q?)

12. Use data from Appendix D to determine K_p at 298 K for the reaction $2\ NO(g) + Cl_2(g) \rightleftharpoons 2\ NOCl(g)$.

13. In Example 16-3 we were unable to conclude, by inspection, whether $\Delta \overline{S}^\circ$ for the reaction, $CO(g) + H_2(g) \rightarrow CO_2(g) + H_2O(g)$ should be positive or negative. Use absolute molar entropies from Appendix D to obtain a value of $\Delta \overline{S}^\circ$ at 298 K.

14. Without performing detailed calculations, indicate whether any of the following reactions is expected to occur to a significant extent at 298 K.

(a) Conversion of dioxygen (O_2) to ozone (O_3): $3\ O_2(g) \rightarrow 2\ O_3(g)$

(b) Dissociation of N_2O_4 to NO_2: $N_2O_4(g) \rightarrow 2\ NO_2(g)$

(c) Formation of BrCl: $Br_2(l) + Cl_2(g) \rightarrow 2\ BrCl(g)$

Exercises

First law of thermodynamics

1. *The internal energy of a fixed quantity of an ideal gas depends only on its temperature.* A sample of an ideal gas is allowed to expand at a *constant temperature* (isothermal expansion). **(a)** Does the gas do work? **(b)** Does the gas exchange heat with its surroundings? **(c)** What happens to the temperature of the gas? **(d)** What is ΔE for the gas?

2. In an *adiabatic* process a system is thermally insulated from its surroundings such that there is no exchange of heat ($q = 0$). For the adiabatic expansion of an ideal gas **(a)** does the gas do work? **(b)** does the internal energy of the gas increase, decrease, or remain constant? **(c)** What happens to the temperature of the gas? (*Hint:* Also refer to Exercise 1.)

3. State whether you think each of the following observations is in any way possible and indicate your reasons. (*Hint:* Also refer to the statements made in Exercises 1 and 2.)
   (a) An ideal gas is expanded at *constant temperature* and is observed to do twice as much work as the heat it absorbs from its surroundings.
   (b) A gas absorbs a quantity of heat from its surroundings while being compressed.

★4. An experiment, reported to have been performed by Joule while on his honeymoon in Switzerland, involved measuring the temperature of the water at the top of a waterfall and again at the bottom. The temperature of the water at the bottom was found to be slightly higher than at the top. Explain this observation.

Enthalpy and internal energy

5. For each of the reactions described below, indicate whether $\Delta \overline{H}$ is equal to, greater than, or less than $\Delta \overline{E}$. (*Hint:* Write equations for the reactions; assume that all reactants and products are at 298 K; and recall that "greater than" means more positive, or less negative.)
   (a) The complete combustion of butanol, $C_4H_9OH(l)$.
   (b) The complete combustion of glucose, $C_6H_{12}O_6(s)$.
   (c) The decomposition of $NH_4NO_3(s)$ into liquid water and gaseous dinitrogen oxide.

6. An experimental determination of the heat of combustion of isopropyl alcohol in a bomb calorimeter yields a value of -33.41 kJ/g C_3H_7OH.

$$C_3H_7OH(l) + \tfrac{9}{2} O_2(g) \rightarrow 3\ CO_2(g) + 4\ H_2O(l)$$

   (a) What is $\Delta \overline{E}$ for this reaction (in kJ/mol)?
   (b) What is $\Delta \overline{H}$ for the reaction (in kJ/mol)?

7. Only one of the following expressions can be used to describe the heat of a chemical reaction *regardless of how the reaction is carried out*. Which is the correct expression and why? **(a)** q_V; **(b)** q_P; **(c)** $\Delta E + w$; **(d)** ΔE; **(e)** ΔH.

Entropy and disorder

8. Based on the relationship of entropy to the degree of order in a system, indicate whether each of the following changes represents an increase or decrease in entropy in a system: **(a)** the freezing of ethanol; **(b)** the sublimation of dry ice; **(c)** the burning of a rocket fuel.

9. Which of each pair of substances would you expect to have the greater entropy? Explain your reasoning.
   (a) 1 mol $H_2O(l,\ 1$ atm, $50°C)$ *or* 1 mol $H_2O(g,\ 1$ atm, $50°C)$
   (b) 50.0 g $Fe(s,\ 1$ atm, $20°C)$ *or* 0.80 mol $Fe(s,\ 1$ atm, $20°C)$
   (c) 1 mol $Br_2(l,\ 1$ atm, $58°C)$ *or* 1 mol $Br_2(s,\ 1$ atm, $-10°C)$
   (d) 0.10 mol $O_2(g,\ 0.10$ atm, $25°C)$ *or* 0.10 mol $O_2(g,\ 10.0$ atm, $25°C)$

10. Explain why
   (a) Some exothermic reactions do not occur spontaneously.
   (b) Some reactions in which the entropy of the system increases also do not occur spontaneously.

★11. Use ideas from this chapter to comment on the meaning of a famous remark attributed to Rudolf Clausius (1865): "Die Energie der Welt ist konstant; die Entropie der Welt strebt einem Maximum zu. [The energy of the world is constant; the entropy of the world increases toward a maximum.]"

Free energy and spontaneous change

12. The decomposition of nitrosyl bromide is represented by the equation

$$NOBr(g) \rightarrow NO(g) + \tfrac{1}{2} Br_2(g)\ \ \Delta \overline{H}° = +8.54\ \text{kJ/mol at 298 K}$$

Do you expect this decomposition to occur to a greater extent at low or high temperatures? Explain.

13. For the process pictured in Figure 16-4, what are the values (positive, negative, or zero) for ΔH, ΔS, and ΔG? Explain your reasoning.

14. What values of ΔH, ΔS, and ΔG would you expect for the formation of an ideal solution of liquid components (i.e., are these values positive, negative, or zero)? Explain.

Standard free energy change

15. For the reaction $2\ PCl_3(l) + O_2(g) \rightarrow 2\ POCl_3(l)$, $\Delta \overline{H}° = -555$ kJ/mol at 298 K. The absolute molar entropies at 298 K (in J mol^{-1} K^{-1}) are PCl_3, 217; O_2, 205; and $POCl_3$, 222. What is $\Delta \overline{G}°$ for this reaction at 298 K?

16. At what temperature will the following reaction have the values of $\Delta \overline{G}°$, $\Delta \overline{H}°$, and $\Delta \overline{S}°$ given?

$$2\ PbS(s) + 3\ O_2(g) \rightarrow 2\ PbO(s) + 2\ SO_2(g)$$

$$\Delta \overline{G}° = -777.8\ \text{kJ/mol} \qquad \Delta \overline{H}° = -843.7\ \text{kJ/mol}$$

$$\Delta \overline{S}° = -0.165\ \text{kJ mol}^{-1}\ \text{K}^{-1}$$

17. The absolute entropies of $F_2(g)$ and $F(g)$ at 298 K are given as 203.3 and 158.7 J mol^{-1} K^{-1}, respectively. Use these data, together with that listed below, to estimate the bond energy of the F_2 molecule. Compare your result with that listed in Table 9-3.

$$F_2(g) \rightarrow 2\ F(g) \qquad \Delta\overline{G}° = 123.85 \text{ kJ/mol at 298 K}\cdot$$

*18. The standard free energy of combustion of *n*-octane is indicated below. What would be the free energy change if the water were produced as a gas instead of as a liquid? (*Hint:* Use data from Appendix D.)

$$C_8H_{18}(l) + \tfrac{25}{2}\ O_2(g) \rightarrow 8\ CO_2(g) + 9\ H_2O(l)$$
$$\Delta\overline{G}° = -5.28 \times 10^3 \text{ kJ/mol at 298 K}$$

Free energy and equilibrium

19. At how many different temperatures and in what temperature range can this equilibrium be established?

$$H_2O(l,\ 0.50 \text{ atm}) \rightleftharpoons H_2O(g,\ 0.50 \text{ atm})$$

20. Refer to Figures 11-10 and 16-6. Does solid or liquid iodine have the lower free energy at 110°C and 1 atm pressure?

21. Refer to Figures 11-11 and 16-6. Which has the lowest free energy at 1 atm and −60°C, solid, liquid, or gaseous carbon dioxide? Explain.

Phase transitions

22. Example 16-7 dealt with the standard molar enthalpy and entropy of vaporization of water at 100°C.
   (a) Use data from Appendix D to determine corresponding values at 25°C.
   *(b) Explain the differences in values of $\Delta\overline{H}°_{vap}$ and $\Delta\overline{S}°_{vap}$ between these two temperatures.

23. Which of the following substances would you expect to obey Trouton's rule most closely? Explain your reasoning.
   (a) HF;
   (b) $C_6H_5CH_3$;
   (c) CH_3OH.

24. Estimate the normal boiling point of bromine, Br_2, in the following two-step procedure, and compare your result with the measured value of 58.8°C.
   (a) Determine $\Delta\overline{H}°_{vap}$ for Br_2 with data from Appendix D.
   (b) Estimate the normal boiling point, assuming that Trouton's rule is obeyed. (You will also be assuming that $\Delta\overline{H}°_{vap}$ is independent of temperature.)

25. Use the method outlined in Exercise 24 to estimate the normal boiling point of mercury.

Relationship of $\Delta\overline{G}$, $\Delta\overline{G}°$, Q, and K

26. Refer to situation (c) in Figure 16-7. Calculate $\Delta\overline{G}$ at 298 K for the vaporization process described, that is,

$$H_2O(l,\ 10 \text{ mmHg}) \rightarrow H_2O(g,\ 10 \text{ mmHg}) \qquad \Delta\overline{G} = ?$$

(*Hint:* Recall equations 16.22 and 16.23.)

27. The following data are given at 298 K.

$$CCl_4(l,\ 1 \text{ atm}) \rightarrow CCl_4(g,\ 1 \text{ atm})$$

$$\Delta\overline{S}° = 94.98 \text{ J mol}^{-1}\text{ K}^{-1} \qquad \Delta\overline{H}°_f[CCl_4(l)] = -139.3 \text{ kJ/mol}$$
$$\Delta\overline{H}°_f[CCl_4(g)] = -106.7 \text{ kJ/mol}$$

   (a) Show that $CCl_4(l)$ does not spontaneously vaporize to produce $CCl_4(g)$ at *1 atm pressure*.
   (b) Calculate the equilibrium vapor pressure of CCl_4 at 298 K.

The thermodynamic equilibrium constant

28. Why must the thermodynamic equilibrium constant, K, be used in the expression $\Delta\overline{G}° = -2.303RT \log K$, rather than simply K_c? What are the circumstances under which K_c and/or K_p may be used in place of K?

29. A method of preparing $H_2(g)$ involves passing steam over hot iron: $3\ Fe(s) + 4\ H_2O(g) \rightleftharpoons Fe_3O_4(s) + 4\ H_2(g)$
   (a) Write an expression for the thermodynamic equilibrium constant for this reaction.
   (b) Explain why the partial pressure of $H_2(g)$ is independent of the amounts of $Fe(s)$ and $Fe_3O_4(s)$ present.
   (c) Can we conclude that the production of $H_2(g)$ from $H_2O(g)$ could be accomplished regardless of what proportions of $Fe(s)$ and $Fe_3O_4(s)$ are used? Explain.

$\Delta\overline{G}°$ and K

30. For the following equilibrium reactions discussed in Chapter 15, calculate the value of $\Delta\overline{G}°$ at the indicated temperature. (*Hint:* How is each equilibrium constant related to a thermodynamic equilibrium constant?)
   (a) $H_2(g) + I_2(g) \rightleftharpoons 2\ HI(g) \qquad K_c = 50.2$ at 445°C
   (b) $N_2O(g) + \tfrac{1}{2}\ O_2(g) \rightleftharpoons 2\ NO(g)$
   $$K_c = 1.7 \times 10^{-13} \text{ at 25°C}$$
   (c) $N_2O_4(g) \rightleftharpoons 2\ NO_2(g) \qquad K_c = 4.61 \times 10^{-3}$ at 25°C
   (d) $2\ SO_2(g) + O_2(g) \rightleftharpoons 2\ SO_3(g)$
   $$K_p = 9.1 \times 10^2 \text{ at 800°C}$$
   (e) $2\ Fe^{3+}(aq) + Hg_2^{2+}(aq) \rightleftharpoons 2\ Fe^{2+}(aq) + 2\ Hg^{2+}(aq)$
   $$K_c = 9.14 \times 10^{-6} \text{ at 25°C}$$

31. For the decomposition of $CaCO_3(s)$ to $CaO(s)$ and $CO_2(g)$ (a reaction in the manufacture of portland cement) the standard molar free energy and enthalpy changes at 298 K are

$$CaCO_3(s) \rightleftharpoons CaO(s) + CO_2(g)$$

$$\Delta\overline{H}° = +178.45 \text{ kJ/mol} \qquad \Delta\overline{G}° = +130.30 \text{ kJ/mol}$$

   (a) Does the decomposition of $CaCO_3(s)$ occur to any appreciable extent at room temperature?
   (b) Does $CaCO_3(s)$ decompose more readily by raising or lowering the temperature from 298 K?

32. The equilibrium constant at 298 K for the reaction $CO(g) + Cl_2(g) \rightleftharpoons COCl_2(g)$ is $K_p = 5.64 \times 10^{35}$. What is the standard free energy of formation of $COCl_2(g)$ at 298 K, that is, $\Delta\overline{G}°_f[COCl_2(g)]$? (*Hint:* You will need to use some data from Appendix D.)

33. In Example 16-11 the statement was made that only a trace of $CO(g)$ would be produced in the reaction $C(s) + CO_2(g) \rightleftharpoons 2\,CO(g)$ at 298 K. Show by calculation that this is in fact the case.

34. Use data from Appendix D to calculate the vapor pressure of mercury (in mmHg) at 298 K.

★35. A statement is made in the caption to Figure 16-8 that only if $\Delta\overline{G}°$ has a small magnitude, either positive or negative, will an equilibrium mixture contain appreciable amounts of all reactants and products. Illustrate this statement using hypothetical data.

The variation of $\Delta\overline{G}°$ with temperature

36. Use data from Appendix D to determine for the reaction

$$CO(g) + H_2O(g) \rightleftharpoons CO_2(g) + H_2(g)$$

(a) $\Delta\overline{H}°$, $\Delta\overline{S}°$, and $\Delta\overline{G}°$ at 298 K;
(b) K_p at 1100 K. (*Hint:* Recall the assumptions made in Examples 16-10 and 16-12.)
(c) Compare the result calculated in part (b) with that found in Example 15-6.

37. In Example 15-15 the approximate temperature at which $K_p = 1.0 \times 10^6$ for the reaction $2\,SO_2(g) + O_2(g) \rightleftharpoons 2\,SO_3(g)$ was determined by calculation. Obtain another estimate of this temperature with data from Appendix D and equation (16-14). Compare your result to that of Example 15-15.

38. For the reaction $C(s) + CO_2(g) \rightleftharpoons 2\,CO(g)$ described in Example 16-6 and Exercise 33.

(a) Is conversion of $CO_2(g)$ to $CO(g)$ favored at high or low temperatures?
(b) If the equilibrium partial pressure of $CO_2(g)$ is maintained at 1.00 atm, at approximately what temperature does the equilibrium partial pressure of $CO(g)$ become equal to that of $CO_2(g)$?

39. For the reaction $N_2(g) + O_2(g) \rightleftharpoons 2\,NO(g)$, at what approximate temperature will an equimolar mixture of $N_2(g)$ and $O_2(g)$ be 1% converted to $NO(g)$? (*Hint:* Use data from Appendix D or Example 16-12.)

Heat engines

40. Why is it not possible to develop a heat engine that is 100% efficient in converting heat to work?

41. What is the maximum efficiency of a heat engine that absorbs heat at 300°C and discharges waste heat at 20°C?

42. If a steam electric power plant discharges condensate at 40°C and is found to operate at 36% efficiency,

(a) What is the minimum temperature of the steam used in the plant?
(b) Why is the actual steam temperature probably higher than that calculated in part (a)?

Additional Exercises

1. A statement is made in Exercise 1 that for a fixed quantity of an ideal gas the internal energy depends only on temperature (i.e., not on pressure or volume). Can the same kind of statement be made concerning (a) the enthalpy and (b) the entropy of an ideal gas? Explain.

2. Although we would not expect a large difference in value between ΔH and ΔE for the melting of ice, there is a slight difference.
(a) Are the values of ΔH and ΔE for this process positive or negative?
(b) Which is the larger of the two? Explain.

3. Would you expect the reaction $Cl_2(g) \rightleftharpoons 2\,Cl(g)$ to occur to a greater extent in the forward direction by raising or lowering the temperature? Explain your reasoning.

4. Use data from Appendix D to determine values at 298 K of $\Delta\overline{G}°$ and K for the following reactions.
(a) $NO(g) + O_2(g) \rightleftharpoons NO_2(g)$ (not balanced)
(b) $HCl(g) + O_2(g) \rightleftharpoons H_2O(g) + Cl_2(g)$ (not balanced)
(c) $Fe_2O_3(s) + H_2(g) \rightleftharpoons Fe_3O_4(s) + H_2O(g)$ (not balanced)

5. An *equilibrium* mixture at 1000 K contains 0.276 mol H_2, 0.276 mol CO_2, 0.224 mol CO, and 0.224 mol H_2O.

$$CO_2(g) + H_2(g) \rightleftharpoons CO(g) + H_2O(g)$$

(a) What is the equilibrium constant, K_p, for this reaction at 1000 K?
(b) Calculate $\Delta\overline{G}°$ at 1000 K.
(c) In what direction will a spontaneous reaction occur if one brings together at 1000 K: 0.0500 mol CO_2, 0.070 mol H_2, 0.0400 mol CO, and 0.0850 mol H_2O?

6. Use data from Appendix D and other information from this chapter to determine the temperature at which the dissociation of $I_2(g)$ becomes appreciable [e.g., with the $I_2(g)$ being 50% dissociated into $I(g)$ at 1 atm total pressure].

7. Indicate for each of the following whether a net reaction will occur to the left or to the right at 298 K? [*Hint:* Use data from Appendix D and equation (16.22) or (16.23).]
(a) $SO_2(g, 0.010\ atm) + \frac{1}{2}\,O_2(g, 0.0010\ atm) \rightleftharpoons SO_3(g, 2.0\ atm)$
(b) $C_2H_2(g, 1.2\ atm) + 2\,H_2(g, 0.020\ atm) \rightleftharpoons C_2H_6(g, 3.0\ atm)$
(c) $Br_2(1, 1\ atm) \rightleftharpoons Br_2(g, 0.10\ atm)$

★8. Many statements have been made about the first and second laws of thermodynamics that draw upon everyday life. One such statement, in gambler's parlance, is "You can't win and you can't even break even." Explain the basis of this statement.

*9. At its normal boiling point of 78.4°C, $\Delta \overline{S}^{\circ}_{vap}$ of ethanol is $+112.1$ J mol^{-1} K^{-1}. Use these data, together with values from Appendix D, to estimate the following at 298 K for $C_2H_5OH(g)$: (a) $\Delta \overline{H}^{\circ}_f$; (b) $\Delta \overline{G}^{\circ}_f$; (c) $\overline{S}^{\circ}$.

*10. The normal boiling point of cyclohexane, C_6H_{12}, is 80.7°C. Estimate the temperature at which the vapor pressure of cyclohexane is 100 mmHg.

*11. From the data given in Exercise 43 of Chapter 15, estimate a value of $\Delta \overline{S}^{\circ}$ at 298 K for the reaction

$$2 \; NaHCO_3(s) \rightarrow Na_2CO_3(s) + H_2O(g) + CO_2(g)$$

*12. One of the steps that appears to be involved in smog formation is the reaction of atomic oxygen, $O(g)$, with molecular oxygen, $O_2(g)$. [The atomic oxygen is produced by the action of sunlight on $NO_2(g)$.]

$$O(g) + O_2(g) \rightleftharpoons O_3(g)$$

A sample of air is found to contain 10 parts per million of $O_3(g)$

by volume. If the $O_3(g)$ is assumed to be in equilibrium with $O_2(g)$ and $O(g)$, what would have to be the approximate partial pressure of atomic oxygen present in the air, that is, $P_{O(g)}$?

*13. Use appropriate data from Appendix D to estimate the % dissociation of $H_2O(g)$ at 2000 K if the total pressure is maintained at 1.00 atm

$$H_2O(g) \rightleftharpoons H_2(g) + \tfrac{1}{2} O_2(g)$$

*14. Assume that the steam [i.e., $H_2O(g)$] in the power plant referred to in Exercise 42 is in equilibrium with liquid water. What is the steam pressure corresponding to the minimum steam temperature calculated in Exercise 42(a)?

*15. Two identical steel springs are provided. Each is dissolved in 500 mL of 6 MHCl(aq). One is dissolved in its relaxed condition. The other is compressed to one half its normal length and dissolved. How would you expect the magnitudes of the heat of reaction to compare in the two cases? Explain. (The reaction is exothermic.)

Self-Test Questions

For questions 1 through 8 select the single item that best completes each statement.

1. $\Delta E = +100$ J for a system that *gives off* 100 J of heat and (a) does 200 J of work; (b) has 200 J of work done on it; (c) does no work; (d) has 100 J of work done on it.

2. For the reaction $N_2(g) + O_2(g) \rightleftharpoons 2 \; NO(g)$, (a) ΔE is less than ΔH; (b) ΔE is greater than ΔH; (c) $\Delta E = 0$; (d) $\Delta E = \Delta H$.

3. For a process to occur spontaneously,
(a) the entropy of the system must increase.
(b) the entropy of the surroundings must increase.
(c) both the entropy of the system and of the surroundings must increase.
(d) the entropy of the universe must increase.

4. The free energy change of a reaction is a measure of
(a) the quantity of heat given off to the surroundings.
(b) the direction in which a net reaction occurs.
(c) the increased molecular disorder that occurs in the system.
(d) how rapidly the reaction occurs.

5. If a reaction has $\Delta H < 0$ and $\Delta S < 0$, the reaction proceeds furthest in the forward direction at (a) low temperatures; (b) high temperatures; (c) all temperatures; (d) no temperature.

6. If it is necessary to employ electric current (electrolysis) to carry out a chemical reaction, then for that reaction (a) $\Delta H > 0$; (b) $\Delta G = \Delta H$; (c) $\Delta G > 0$; (d) $\Delta S > 0$.

7. For the reaction $Br_2(g) \rightarrow 2 \; Br(g)$, we should expect that at all temperatures (a) $\Delta H < 0$; (b) $\Delta S > 0$; (c) $\Delta G < 0$; (d) $\Delta S < 0$.

8. If $\Delta \overline{G}^{\circ} = 0$ for a reaction, then (a) $\Delta \overline{H}^{\circ} = 0$; (b) $\Delta \overline{S}^{\circ} = 0$; (c) $K = 0$; (d) $K = 1$.

9. Explain briefly why
(a) The change in entropy in a system is not always a suitable criterion for spontaneous change.
(b) $\Delta \overline{G}^{\circ}$ is so important in dealing with the question of spontaneous change, even though the conditions employed in a reaction are often nonstandard.

10. Explain the relationships among $\Delta \overline{G}$, $\Delta \overline{G}^{\circ}$, and the equilibrium constant, K, for a reversible chemical reaction.

11. A handbook lists the following standard enthalpies of formation at 298 K for cyclopentane, C_5H_{10}: $\Delta \overline{H}^{\circ}_f[C_5H_{10}(l)] = -105.9$ kJ/mol and $\Delta \overline{H}^{\circ}_f[C_5H_{10}(g)] = -77.2$ kJ/mol.
(a) Estimate the normal boiling point of cyclopentane.
(b) Estimate $\Delta \overline{G}^{\circ}$ for the vaporization of cyclopentane at 298 K.
(c) Comment on the significance of the sign of $\Delta \overline{G}^{\circ}$ at 298 K.

12. The following data are given at 298 K.

	$\Delta \overline{H}^{\circ}_f$, kJ/mol	$\overline{S}^{\circ}$, J mol^{-1} K^{-1}
$NH_4NO_3(s)$	-365.56	151.08
$N_2O(g)$	$+81.55$	219.99
$H_2O(l)$	-285.85	69.96

For the reaction $NH_4NO_3(s) \rightarrow N_2O(g) + 2 \; H_2O(l)$
(a) Is the forward reaction endothermic or exothermic?
(b) What is the value of $\Delta \overline{G}^{\circ}$ at 298 K?
(c) What is the value of the equilibrium constant, K?
(d) Is the forward reaction favored at temperatures above 25°C, below 25°C, both, or neither? Explain.

17 Acids and Bases

In this chapter we extend our discussion of equilibrium to two types of substances that figure prominently in many equilibria—acids and bases. We begin with an overview of acid–base theories; and then proceed to describe a number of topics of special interest, some qualitatively and some quantitatively.

17-1 Acid–Base Theories

That acids and bases (alkalis) have been known since ancient times is reflected in their very names. The term *acid* is derived from the Latin *acetum,* which means vinegar. The acid constituent in vinegar is acetic acid, H_3CCOOH. The term *alkali* is derived from the Arabic word for ashes. Until relatively recent times, the principal source of bases or alkalis was wood ashes. That the product of the reaction of an acid and a base (neutralization) is a *salt* has also been understood for at least three centuries.

Theoretical attempts to explain acid–base behavior form an important chapter in the history of chemistry. Lavoisier proposed (1777) that all acids contain a common element—oxygen. (The name oxygen, proposed by Lavoisier, is derived from Greek and means "acid former.") By showing that muriatic acid (hydrochloric acid) contains only hydrogen and chlorine and no oxygen, Davy (1810) established that hydrogen, not oxygen, was the common element in acids.

Arrhenius's Theory. In his theory of electrolytic dissociation Svante Arrhenius (1884) proposed that an electrolyte dissociates into ions upon dissolving in water; a strong electrolyte dissociates completely; a weak electrolyte, only partially. A substance that dissociates to produce hydrogen ions (H^+) is an acid, e.g., HCl.

$$HCl(aq) \longrightarrow H^+(aq) + Cl^-(aq) \tag{17.1}$$

A base dissociates to produce hydroxide ions (OH^-).

$$NaOH(aq) \longrightarrow Na^+(aq) + OH^-(aq) \tag{17.2}$$

The reaction of an acid and a base—a neutralization reaction—can be represented in one of three ways:

complete equation:
$$\underset{\text{an acid}}{HCl} + \underset{\text{a base}}{NaOH} \longrightarrow \underset{\text{a salt}}{NaCl} + \underset{\text{water}}{H_2O} \tag{17.3}$$

ionic equation:
$$\underset{\text{an acid}}{H^+ + Cl^-} + \underset{\text{a base}}{Na^+ + OH^-} \longrightarrow \underset{\text{a salt}}{Na^+ + Cl^-} + \underset{\text{water}}{H_2O} \tag{17.4}$$

net ionic equation:
$$H^+ + OH^- \longrightarrow H_2O \tag{17.5}$$

The net ionic equation is an especially appropriate representation of a neutralization reaction according to the Arrhenius theory. It brings out this essential point: *A neutralization reaction involves the combination of hydrogen and hydroxide ions to form water.*

There is another way in which the Arrhenius theory explains the process of neutralization better than any prior theory. The enthalpies of neutralization of strong acids and bases are found to be essentially constant: -55.90 kJ/mol water formed. That $\Delta\overline{H}_{neu}$ is independent of the identity of the strong acid and base is readily understood: The essential reaction is always

$$H^+ + OH^- \longrightarrow H_2O \tag{17.5}$$

The Arrhenius theory also met with success in explaining the catalytic activity of acids in certain reactions. The acids that prove to be the most effective catalysts are also those that have the best electrical conductivity, that is, strong acids. The stronger the acid, the higher the H^+ concentration in solution. It is the H^+ ion that is the actual catalyst in these reactions. An example of an acid-catalyzed reaction, the decomposition of formic acid, was considered in Section 14-9.

Brønsted–Lowry Theory. Despite its successes and continued usefulness, the Arrhenius theory does have limitations. For one, it does not recognize any constituent other than OH^- as imparting basic properties to a substance. This requirement led to representing the ionization of aqueous solutions of ammonia as

$$NH_4OH(aq) \rightleftharpoons NH_4{}^+(aq) + OH^-(aq) \tag{17.6}$$

Yet the substance NH_4OH (ammonium hydroxide) appears not to exist. That is, it cannot be isolated in pure form as can sodium hydroxide (NaOH).

Moreover, even in Arrhenius's time, reactions were being conducted in *nonaqueous* solvents, such as liquid ammonia. Some of these reactions seemed to have the characteristics of acid–base reactions. Clearly, however, OH^- was not present, since there were no oxygen atoms in the system at all. For example, ammonium chloride and sodium amide react in liquid ammonia as follows.

complete: $\qquad\qquad NH_4Cl + NaNH_2 \longrightarrow NaCl + 2\ NH_3 \tag{17.7}$

ionic: $\qquad NH_4{}^+ + Cl^- + Na^+ + NH_2{}^- \longrightarrow Na^+ + Cl^- + 2\ NH_3 \tag{17.8}$

net ionic: $\qquad\qquad\qquad NH_4{}^+ + NH_2{}^- \longrightarrow 2\ NH_3 \tag{17.9}$

Reaction (17.9) can be considered an acid–base reaction with $NH_4{}^+$ analogous to H^+ and $NH_2{}^-$ to OH^-. The reaction can be explained through a theory of acids and bases proposed independently by J. N. Brønsted in Denmark and T. M. Lowry in Great Britain in 1923. According to the Brønsted–Lowry theory, an acid is a **proton* donor** and a base, a **proton acceptor,** as suggested in reaction (17.10).

$$\underset{\text{acid(1)}}{NH_4{}^+} + \underset{\text{base(2)}}{NH_2{}^-} \rightleftharpoons \underset{\text{acid(2)}}{NH_3} + \underset{\text{base(1)}}{NH_3} \tag{17.10}$$

A number of ideas are implicit in equation (17.10). An acid, call it acid(1), loses a proton and becomes base(1). Similarly, base(2) gains a proton and becomes acid(2). In general proton transfer reactions are reversible. When base(1) regains a proton, it forms acid(1). Base(1) is said to be the **conjugate base** of acid(1). Similarly, acid(2) is the **conjugate acid** of base(2). Figure 17-1 may help you to visualize the proton transfer involved in reaction (17.10).

The direction in which an acid–base reaction tends to occur, that is, the direction of proton transfer, depends on the relative strengths of the species involved. If the acid is strong, its conjugate base is weak, and vice versa. The net reaction goes from strong acid

NH_3 (the solvent) is both the conjugate base of the acid $NH_4{}^+$ and the conjugate acid of the base $NH_2{}^-$ in reaction (17.10).

---

*In acid–base theory the term proton refers to a particular H atom that has lost its electron, that is, the nucleus of a hydrogen atom, H^+.

FIGURE 17-1
Brønsted–Lowry acid–
base reaction.

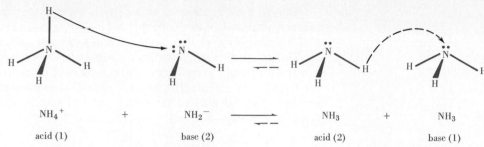

NH_4^+	+	NH_2^-	
acid (1)		base (2)	
NH_3	+	NH_3	
acid (2)		base (1)	

This figure depicts the proton transfer in reaction (17.10). The solid arrows represent the forward reaction and the broken arrows, the reverse reaction. NH_4^+/NH_2^- is a strong acid/strong base combination, whereas NH_3/NH_3 is a weak acid/weak base combination. The forward direction is favored in this acid–base reaction.

and base to weak acid and base. We will explore some of these relationships between conjugate acids and bases more fully in Section 17-8.

Example 17-1   For each of the following identify the acids and bases involved in both the forward and reverse reactions.
(a) $HClO_2 + H_2O \rightleftharpoons H_3O^+ + ClO_2^-$
(b) $OCl^- + H_2O \rightleftharpoons HOCl + OH^-$
(c) $NH_3 + H_2PO_4^- \rightleftharpoons NH_4^+ + HPO_4^{2-}$
(d) $HCl + H_2PO_4^- \rightleftharpoons H_3PO_4 + Cl^-$

Solution.   Consider $HClO_2$ in reaction (a). It is converted to ClO_2^- by losing a proton (H^+). $HClO_2$ must be an acid, and ClO_2^- is its conjugate base. H_2O accepts the proton lost by $HClO_2$. Then H_2O must be a base and H_3O^+, its conjugate acid. In reaction (b), H_2O acts an an acid and OH^- is its conjugate base. Reactions (c) and (d) illustrate another species that can either donate or accept a proton—an **amphiprotic** species. In this case the amphiprotic species is $H_2PO_4^-$.

(a) $HClO_2 + H_2O \rightleftharpoons H_3O^+ + ClO_2^-$
    acid(1)   base(2)   acid(2)   base(1)

(b) $H_2O + OCl^- \rightleftharpoons HOCl + OH^-$
    acid(1)   base(2)   acid(2)   base(1)

(c) $H_2PO_4^- + NH_3 \rightleftharpoons NH_4^+ + HPO_4^{2-}$
    acid(1)   base(2)   acid(2)   base(1)

(d) $HCl + H_2PO_4^- \rightleftharpoons H_3PO_4 + Cl^-$
    acid(1)   base(2)   acid(2)   base(1)

SIMILAR EXAMPLES: Review Problem 1; Exercises 1, 4.

Acid–base reactions in solvents containing neither H^+ nor OH^- can also be described by the "solvent system" concept: Associated with every solvent there is a characteristic cation and anion, e.g.,

$$2\ SO_2 \rightarrow SO^{2+} + SO_3^{2-}$$

Any substance that produces the characteristic cation (SO^{2+}) in the solvent is an acid, and any substance that produces the characteristic anion (SO_3^{2-}) is a base.

Two additional features of the Brønsted–Lowry theory are illustrated in Example 17-1.

• Any species that is an acid by the Arrhenius theory remains an acid in the Brønsted–Lowry theory. The same is true of bases.
• Certain species that would not be classified as bases by the Arrhenius theory are so classified by the Brønsted–Lowry theory, e.g., OCl^- and $H_2PO_4^-$.

Lewis Theory.   G. N. Lewis developed an alternative to the Arrhenius theory of acids and bases at about the same time as Brønsted and Lowry (1923). Lewis's theory has some additional advantages over the Brønsted–Lowry theory in that it permits the acid-base classification to be used for some reactions in which *neither* H^+ *nor* OH^- is present.

In Lewis theory an acid is an **electron pair acceptor** and a base is an **electron pair donor.** From what we know about chemical bonding, then, we should expect acids to be species with *available orbitals and a lack of electrons.* Bases are species with *lone-pair electrons available for sharing.* In addition, an acid–base reaction leads to the formation of a covalent bond between the acid and the base.

With the definitions just given, we would classify H^+ as an acid because of the presence of an empty orbital ($1s$) that can accept a pair of electrons. OH^- and NH_3 are bases because of the presence in these species of available lone-pair electrons.

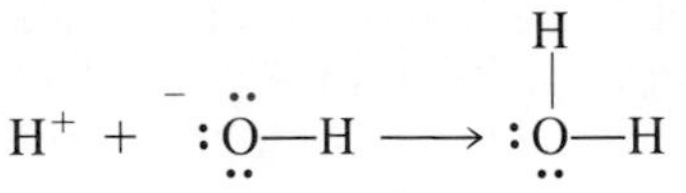

$$H^+ + \ ^-:\!\ddot{O}\!-\!H \longrightarrow \ :\!\ddot{O}\!-\!H \quad \text{(with } H \text{ above)} \tag{17.11}$$

$$H^+ + H\!-\!\ddot{N}\!-\!H \longrightarrow \left[H\!-\!N\!-\!H \right]^+ \tag{17.12}$$

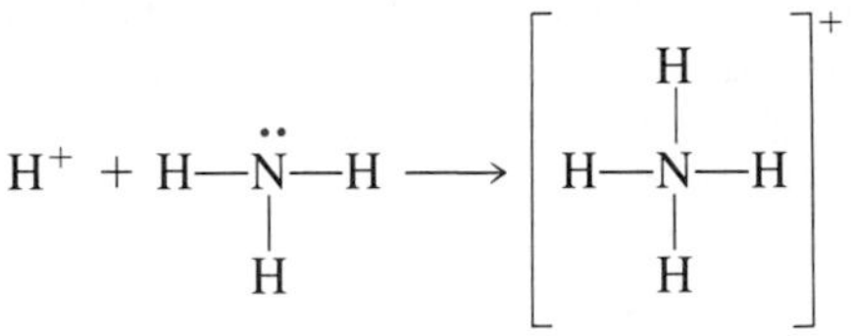

In the following reaction, which we first encountered in Section 9-7, BF_3 acts as a Lewis acid and NH_3 as a Lewis base. An electron-pair bond (a coordinate covalent bond) is formed between the B and the N atom.

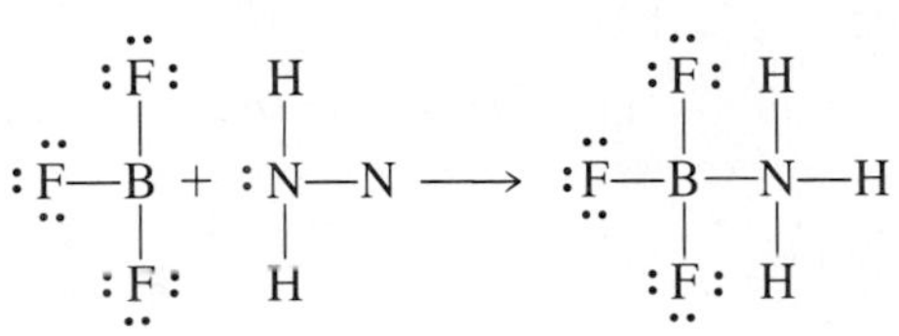

$$\text{(17.13)}$$

Example 17-2 Each of the following is an acid–base reaction in the Lewis sense. Which species is the acid and which is the base?
(a) $BF_3 + F^- \longrightarrow BF_4^-$
(b) $Zn^{2+}(aq) + 4\,NH_3(aq) \longrightarrow [Zn(NH_3)_4]^{2+}$

(c)

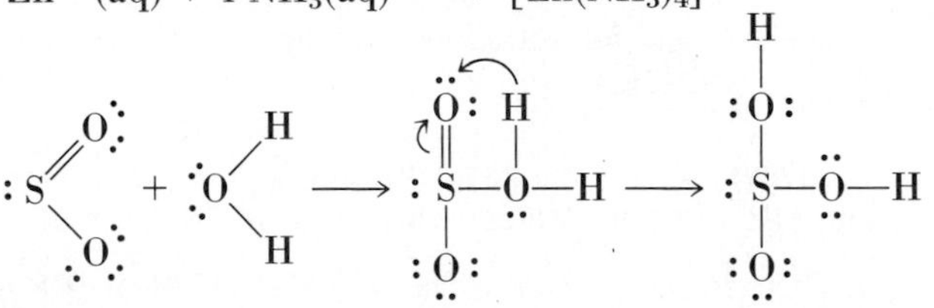

Solution
(a) Note in equation (17.13) that BF_3 is an electron-deficient molecule, with a vacant orbital on the B atom. The fluoride ion has an outer-shell octet of electrons. BF_3 is the electron pair acceptor—the acid. F^- is the electron pair donor—the base.
(b) Note in equation (17.13) that NH_3 is an electron pair donor—a base. Although from the equation written it is not clear just how the Zn^{2+} ion accepts electrons from the NH_3, it must indeed do so. Altogether four pairs of electrons are accepted by the Zn^{2+} to form the *complex ion* $[Zn(NH_3)_4]^{2+}$. Complex ion formation is discussed in Chapter 24.
(c) Here the key to identifying the acid and the base is through the structure of the product molecule (sulfurous acid, H_2SO_3). This molecule has three O atoms bonded to sulfur, whereas SO_2 has two. The third O atom is the O atom from H_2O, which supplies an electron pair. H_2O is the Lewis base and SO_2 is the Lewis acid. Note also that a rearrangement of a pair of electrons and one H atom occurs in this reaction.

SIMILAR EXAMPLES: Review Problem 2; Exercise 6.

17-2 Self-ionization (Autoionization) of Water

Water is a very poor electrical conductor. Yet that it does conduct electric current feebly indicates that some ions are present. According to the Arrhenius theory, these ions, which arise from the ionization of water molecules themselves, are H^+ and OH^-.

$$H_2O \rightleftharpoons H^+ + OH^- \tag{17.14}$$

The self-ionization of water is also an acid–base reaction in the Brønsted–Lowry sense (equation 17.15). One water molecule acts as an acid; it loses a proton. Another water molecule acts as a base. It accepts the proton, with which it forms a coordinate covalent bond through an unshared pair of electrons on the oxygen atom. The resulting ions are **H_3O^+**, called **hydronium ion,** and **OH^-, hydroxide ion.** The ionization reaction is reversible, and in the reverse reaction a hydronium ion loses a proton to a hydroxide ion. In fact, since acid(2) and base(1) are *much* stronger than acid(1) and base(2), the reverse reaction is much more significant than the forward reaction. *Equilibrium is displaced far to the left.*

$$\text{:O:H} + \text{:O:H} \rightleftharpoons \left(\text{:O:H}\right)^+ + \left(\text{:O:H}\right)^- \tag{17.15}$$

acid (1)   base (2)   acid (2)   base (1)

We can describe equilibrium in the self-ionization of water through a thermodynamic equilibrium constant expression.

$$K = \frac{(a_{H_3O^+})(a_{OH^-})}{(a_{H_2O})^2}$$

Since the activity of the water—a pure liquid—is 1,

$$K = (a_{H_3O^+})(a_{OH^-})$$

Furthermore, because the ion concentrations are very small, we can substitute molar concentrations for activities, leading to the final expression

$$K_w = [H_3O^+][OH^-]$$

There are several experimental methods of determining the concentrations of H_3O^+ and OH^- in pure water. All lead to the result that

at 25°C in pure water: $[H_3O^+] = [OH^-] = 1.0 \times 10^{-7}\ M$

That $[H_3O^+]$ and $[OH^-]$ in pure water must be equal can be seen from the balanced equation (17.15). The equilibrium constant for the self-ionization of water is called the **ion product of water,** and represented as **K_w.** At 25°C,

$$K_w = [H_3O^+][OH^-] = 1.0 \times 10^{-14} \tag{17.16}$$

Like all other equilibrium constants, the ion product of water is temperature-dependent. At 60°C, $K_w = 9.6 \times 10^{-14}$; at 100°C, 5.5×10^{-13}.

Calculations of $[H^+]$ in water yield results as low as $10^{-130}\ M$. As described by N. V. Sidgwick (1950), this corresponds to *one* free proton in 10^{70} universes filled with a 1 *M* acid solution!

The Nature of the Proton in Aqueous Solution.   One reason for favoring equation (17.15) over (17.14) for the self-ionization of water is that it conforms to the Brønsted–Lowry theory, which is more general than the Arrhenius theory. There is another equally important reason. The Arrhenius theory (i.e., through equation 17.14) postulates the existence of H^+ ions in aqueous solution. Recall that the H^+ ion is just a lone proton (the nucleus of a hydrogen atom). Because of their very small size and high positive charge density, we should expect H^+ ions to seek out centers of negative charge with which to form bonds. Thus, hydrogen ions, H^+, are not expected to exist in water solution.

What is the situation with respect to the hydronium ion, H_3O^+, whose existence in water solution is postulated by the Brønsted–Lowry theory? Does it exist? For many years this seemed an unanswerable question, and hydronium ion was thought just to be the simplest of a series of species called *hydrated* protons: $[H(H_2O)_n]^+$. That is, if $n = 1$, the species obtained is $[H(H_2O)]^+ = H_3O^+$—the hydronium ion. However, there is experimental evidence, established through x-ray diffraction studies, for the existence of the hydronium ion in the solid state. What was once thought to be a monohydrate of perchloric acid, $HClO_4 \cdot H_2O$, is now known to be the ionic compound $H_3O^+ClO_4^-$. This salt, which we might call hydronium perchlorate, is structurally quite similar to ammonium perchlorate, $NH_4^+ClO_4^-$. In addition, recent experimental evidence has established the presence of H_3O^+ in aqueous solution. A current view is of the existence of structures such as the one depicted in Figure 17-2—a central hydronium ion hydrogen-bonded to three H_2O molecules.

The formula of the structure shown in Figure 17-2 could be written as $H_3O^+ \cdot 3H_2O$ or $H_9O_4^+$, but we will not do so. Generally, all ions are hydrated in aqueous solution, and we have found it satisfactory just to use the symbol (aq) to represent this fact. Throughout this chapter we refer to hydronium ion in solution as H_3O^+ or H_3O^+(aq).

Hydroxide ion appears to form a similar species: $OH^- \cdot 3H_2O$ or $H_7O_4^-$. We will continue simply to speak of OH^- or OH^-(aq).

17-3 Strong Acids and Strong Bases

When an acid is added to water, as in aqueous solutions of hydrochloric acid, in addition to the self-ionization of water

$$H_2O + H_2O \rightleftharpoons H_3O^+ + OH^- \qquad (17.17)$$
$$\text{acid} \quad \text{base} \qquad \text{acid} \quad \text{base}$$

the acid also ionizes.

$$HCl + H_2O \longrightarrow H_3O^+ + Cl^- \qquad (17.18)$$
$$\text{acid} \quad \text{base} \qquad \text{acid} \quad \text{base}$$

In very concentrated HCl(aq) it is not proper to assume that ionization of HCl goes to completion—we can smell HCl(g) in the vapor above such solutions. In all the situations described in this text, however, the assumption of complete ionization will be valid.

The self-ionization of water, the forward reaction in equation (17.17), occurs only to a slight extent. By contrast, the ionization of HCl, a strong acid, goes essentially to completion (equation 17.18). In calculating the concentration of H_3O^+ in an aqueous HCl solution, unless the HCl(aq) is extremely dilute, it is generally acceptable to consider the ionization of HCl to be the sole source of H_3O^+.

Example 17-3 Calculate $[H_3O^+]$, $[Cl^-]$, and $[OH^-]$ in a 100.0-mL sample of 0.015 M HCl(aq).

Solution. We should first note that molar concentrations are independent of solution volume. That is, $[H_3O^+]$ in 0.015 M HCl(aq) is the same whether we are describing 1.00 L, 10.0 L, or 100.0 mL of solution. The volume of solution does not enter into this calculation (but see Example 17-5 for a calculation in which it does).

If we assume that HCl is the sole source of H_3O^+ and that it is completely ionized in aqueous solution,

$$[H_3O^+] = 0.015 \ M$$

Furthermore, as indicated by equation (17.18), one Cl^- is produced for every H_3O^+. Therefore,

$$[Cl^-] = 0.015 \ M$$

To calculate $[OH^-]$ we must recognize (a) that all the OH^- is derived from the self-ionization of water (equation 17.17) and (b) that in any aqueous solution $[H_3O^+]$

FIGURE 17-2
The hydronium ion in aqueous solution.

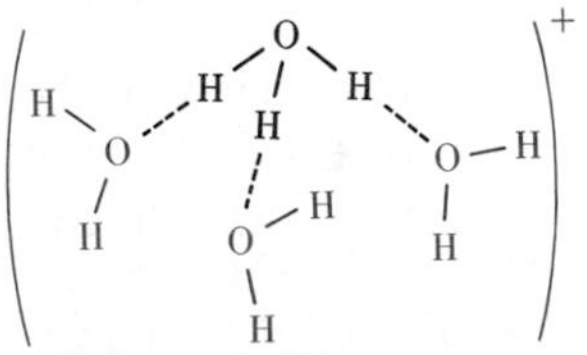

Represented here is a probable species that exists in aqueous solution. The central H_3O^+ is hydrogen-bonded to three H_2O molecules.

and $[OH^-]$ must have values consistent with the ion product of water.

$$K_w = [H_3O^+][OH^-] = 1.0 \times 10^{-14}$$
$$(0.015)[OH^-] = 1.0 \times 10^{-14}$$

$$[OH^-] = \frac{1.0 \times 10^{-14}}{1.5 \times 10^{-2}} = 0.67 \times 10^{-12} = 6.7 \times 10^{-13} \ M$$

SIMILAR EXAMPLES: Review Problem 3; Exercise 8.

The concentration of OH^- in 0.015 *M* HCl (Example 17-3) is much smaller than in pure water ($6.7 \times 10^{-13} \ M$ compared to $1.0 \times 10^{-7} \ M$). The addition of an acid *represses* the ionization of water. That is, it favors the *reverse* of reaction (17.17). This result is in accord with Le Châtelier's principle: The self-ionization equilibrium of water is disturbed by increasing the concentration of H_3O^+ present, and the equilibrium shifts in the direction that removes some (though not very much) of the added H_3O^+. In the new equilibrium $[H_3O^+]$ is greater than it was in pure water and $[OH^-]$ is smaller. Another way to view the situation is that in an aqueous solution the expression, $K_w = [H_3O^+][OH^-] = 1.0 \times 10^{-14}$, must be obeyed. If $[H_3O^+]$ is increased above that found in pure water, then $[OH^-]$ must decrease to the point where the product of the two ion concentrations equals K_w.

Example 17-4   Calculate $[H_3O^+]$, $[OH^-]$, and $[Ba^{2+}]$ in a 50.0-mL sample of 0.010 *M* $Ba(OH)_2$(aq).

Solution.   Again, the volume of the solution does not enter into these calculations. In this solution we assume that $Ba(OH)_2$ is completely dissociated and that it is the sole source of OH^-, leading to

$$[OH^-] = \frac{0.010 \ mol \ Ba(OH)_2}{L} \times \frac{2 \ mol \ OH^-}{1 \ mol \ Ba(OH)_2} = 0.020 \ M$$

The molar concentration of Ba^{2+} is

$$[Ba^{2+}] = \frac{0.010 \ mol \ Ba(OH)_2}{L} \times \frac{1 \ mol \ Ba^{2+}}{1 \ mol \ Ba(OH)_2} = 0.010 \ M$$

And we calculate $[H_3O^+]$ by using K_w for water.

$$K_w = [H_3O^+][OH^-] = [H_3O^+](0.020) = 1.0 \times 10^{-14}$$
$$[H_3O^+] = 5.0 \times 10^{-13} \ M$$

SIMILAR EXAMPLE: Review Problem 3.

TABLE 17-1
The common strong acids and bases

Acids	Bases
HCl	NaOH
HBr	KOH
HI	RbOH
$HClO_4$	CsOH
HNO_3	$Ca(OH)_2$
H_2SO_4[a]	$Sr(OH)_2$
	$Ba(OH)_2$

[a] H_2SO_4 ionizes in two distinct steps. It is a strong acid only in its first ionization step (see page 520).

To use the method of Examples 17-3 and 17-4 requires knowledge of what compounds are strong acids and what compounds are strong bases. The common strong acids and bases are listed in Table 17-1.

Neutralization Reactions.   Suppose we were to mix the two solutions that were the subjects of Examples 17-3 and 17-4. First, we would expect a final solution volume of 100.0 mL + 50.0 mL = 150.0 mL. If no reaction were to occur in the mixed solution, $[H_3O^+]$ and $[OH^-]$ would be

$$[H_3O^+] = \frac{no. \ mol \ H_3O^+}{no. \ L \ soln.} = \frac{0.1000 \ L \times 0.015 \ mol \ H_3O^+/L}{0.1500 \ L} = 0.010 \ M$$

and

$$[OH^-] = \frac{no. \ mol \ OH^-}{no. \ L \ soln.} = \frac{0.0500 \ L \times 0.020 \ mol \ OH^-/L}{0.1500 \ L} = 0.0067 \ M$$

The product of $[H_3O^+]$ and $[OH^-]$ would be $(0.010)(0.0067) = 6.7 \times 10^{-5}$. However, in any aqueous solution at 25°C, the product of these ion concentrations must be equal to $K_w = 1.0 \times 10^{-14}$. *A solution cannot be simultaneously 0.010 M in H_3O^+ and 0.0067 M in OH^-.* A chemical reaction must occur in which H_3O^+ and OH^- combine to form water.

$$H_3O^+ + OH^- \longrightarrow 2\,H_2O \qquad\qquad (17.19)$$

Here, then, is a new way to look at a neutralization reaction. It occurs in order to maintain the required value of an equilibrium constant (K_w). We will encounter other situations where a reaction occurs for similar reasons in the next two chapters. In Section 18-4 we will examine some practical matters concerning neutralization reactions, such as selecting an appropriate indicator for an acid–base titration.

Example 17-5   When the solutions of Examples 17-3 and 17-4 are mixed **(a)** is the final solution acidic or basic? **(b)** what is $[H_3O^+]$?

Solution
(a) From equation (17.19) we see that H_3O^+ and OH^- combine in a 1:1 mol ratio. We need to compare the no. mol H_3O^+ present in 100.0 mL 0.015 *M* HCl with the no. mol OH^- present in 50.0 mL 0.010 *M* Ba(OH)$_2$. One of these reactants is the limiting reagent and the other is in excess. The excess reagent determines whether the final solution is acidic or basic.

$$\text{no. mol } H_3O^+ = 0.1000\text{ L} \times \frac{0.015\text{ mol } H_3O^+}{L} = 0.0015\text{ mol } H_3O^+$$

$$\text{no. mol } OH^- = 0.0500\text{ L} \times \frac{0.020\text{ mol } OH^-}{L} = 0.0010\text{ mol } OH^-$$

OH^- is the limiting reagent; H_3O^+ is in excess; and the final solution is acidic.
(b) We need to calculate the amount of H_3O^+ that remains in excess.

$$\text{no. mol } H_3O^+ \text{ consumed} = 0.0010\text{ mol } OH^- \times \frac{1\text{ mol } H_3O^+}{1\text{ mol } OH^-} = 0.0010\text{ mol } H_3O^+$$

$$\text{no. mol } H_3O^+ \text{ in excess} = 0.0015 - 0.0010 = 0.0005\text{ mol } H_3O^+$$

The excess H_3O^+ is found in 150 mL of solution, leading to

$$[H_3O^+] = \frac{0.0005\text{ mol } H_3O^+}{0.150\text{ L soln.}} = 3 \times 10^{-3}\ M$$

SIMILAR EXAMPLES: Review Problem 8; Exercises 13, 14.

17-4 pH and pOH

The concentrations of H_3O^+ and OH^- found in aqueous solutions are variable over an extreme range. In 0.015 *M* HCl, $[H_3O^+] = 0.015\ M$. In 0.010 *M* Ba(OH)$_2$, $[H_3O^+] = 5.0 \times 10^{-13}\ M$. The advantage of writing the exponential form $[H_3O^+] = 5.0 \times 10^{-13}$ over the decimal form $[H_3O^+] = 0.00000000000050\ M$ is obvious; but a further simplification is also possible. This is accomplished through the pH notation introduced in 1909 by Søren Sørensen, a Danish biochemist. (Sørensen meant the symbol pH to stand for "potential of hydrogen.") If a solution has

A more precise definition of pH requires that it be related to the activity of H_3O^+ ($a_{H_3O^+}$) rather than simply its molarity.

$$[H_3O^+] = 10^{-x} \qquad\qquad (17.20)$$

$$pH = x \qquad\qquad (17.21)$$

pH = 2.30 is expressed to two significant figures. The number "2" locates the decimal point in the pH, just as does the exponent "−3" in $[H_3O^+] = 5.0 \times 10^{-3}$ (see also, Appendix A).

The mathematical operation in transforming equation (17.20) to (17.21) involves the use of logarithms. The pH is the *negative* of the logarithm of $[H_3O^+]$.

$$pH = -\log [H_3O^+] \tag{17.22}$$

In 0.010 M HCl, $[H_3O^+] = 1.0 \times 10^{-2}$ M, and pH $= -\log (1.0 \times 10^{-2}) = -(-2.00) = 2.00$. For 0.0050 M HCl, $[H_3O^+] = 5.0 \times 10^{-3}$ M, and

$$pH = -\log (5.0 \times 10^{-3}) = -(\log 5.0 + \log 10^{-3}) = -(0.70 - 3.00) = 2.30$$

To determine $[H_3O^+]$ from a known pH value requires the taking of an antilogarithm. For example, if pH = 5.30,

$$[H_3O^+] = 10^{-5.30} = 10^{0.70} \times 10^{-6.00}$$
$$= 5.0 \times 10^{-6} \ M$$

But you have probably already discovered that a modern electronic calculator performs these logarithm and antilogarithm calculations with your hardly having to give a second thought to the matter.

For dealing with $[OH^-]$, pOH is defined in an analogous fashion to pH.

$$pOH = -\log [OH^-] \tag{17.23}$$

Since in any dilute aqueous solution at 25°C $[H_3O^+][OH^-] = 1.0 \times 10^{-14}$, we obtain the useful expression (17.24) by taking negative logarithms.

$$-(\log [H_3O^+][OH^-]) = -\log (1.0 \times 10^{-14})$$
$$-\log [H_3O^+] - \log [OH^-] = -(-14.00)$$
$$pH + pOH = 14.00 \tag{17.24}$$

The pH values associated with a number of common materials are listed in Figure 17-3. As indicated in the figure, pH = 7 represents a neutral solution; pH < 7, acidic; and pH > 7, basic.

Conversions between $[H_3O^+]$ and pH are often encountered as part of a larger problem, as illustrated through Example 17-6.

Example 17-6 **What is the pH of the aqueous solution that results from the reaction of a 3.50-mg sample of Na(s) with 275 mL H_2O?**

$$2 \ Na(s) + 2 \ H_2O(l) \longrightarrow 2 \ Na^+(aq) + 2 \ OH^-(aq) + H_2(g)$$

Solution. The calculation can best be done in several steps:
Step 1. Calculate the amount of Na(s) that reacts.

$$\text{no. mol Na} = 3.50 \text{ mg} \times \frac{1 \text{ g Na}}{1000 \text{ mg Na}} \times \frac{1 \text{ mol Na}}{23.0 \text{ g Na}} = 1.52 \times 10^{-4} \text{ mol Na}$$

(Without question, Na is the limiting reagent since there is far more than 1.52×10^{-4} mol H_2O in 275 mL H_2O.)
Step 2. Calculate the amount of OH^- that is produced.

$$\text{no. mol OH}^- = 1.52 \times 10^{-4} \text{ mol Na} \times \frac{2 \text{ mol OH}^-}{2 \text{ mol Na}} = 1.52 \times 10^{-4} \text{ mol OH}^-$$

Step 3. Calculate the molar concentration of OH^-.

$$[OH^-] = \frac{1.52 \times 10^{-4} \text{ mol OH}^-}{0.275 \text{ L solution}} = 5.53 \times 10^{-4} \ M$$

Step 4. Use expression (17.23) to calculate pOH.

$$pOH = -\log [OH^-] = -\log (5.53 \times 10^{-4}) - (0.743 - 4.000)$$
$$= -(0.743 - 4.000) = 3.257$$

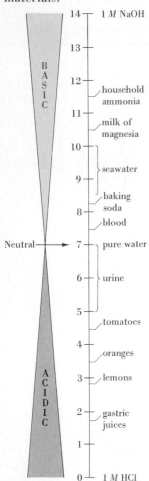

FIGURE 17-3
The pH scale and pH values of some common materials.

14 — 1 M NaOH

13 —

B
A   12 —
S   household
I   11 — ammonia
C
            milk of
            magnesia
10 —

            seawater
9 —
            baking
8 — soda
            blood

Neutral → 7 — pure water

6 — urine

5 —
            tomatoes
4 —
            oranges
A 3 — lemons
C
I 2 — gastric
D   juices
I
C 1 —

0 — 1 M HCl

It is important to note that a change of pH of one unit represents a ten-fold change in $[H_3O^+]$. For example, orange juice is about 10 times more acidic than tomato juice.

Step 5. Use expression (17.24) to calculate pH.

$pH = 14.00 - pOH = 14.00 - 3.26 = 10.74$

SIMILAR EXAMPLES: Review Problem 7; Exercise 9.

17-5 Weak Acids and Weak Bases

Most acids and bases are weak, and the situation that results when they are dissolved in water is more complex than that encountered in Example 17-3. Again two ionization reactions must be considered, and again the ionization of the weak acid (or base) generally occurs to a greater extent than that of water.

$$H_2O + H_2O \rightleftharpoons H_3O^+ + OH^- \qquad (17.17)$$
$$\text{acid} \quad \text{base} \qquad \text{acid} \quad \text{base}$$

$$HOCl + H_2O \rightleftharpoons H_3O^+ + OCl^- \qquad (17.25)$$
$$\text{acid} \quad \text{base} \qquad \text{acid} \quad \text{base}$$

The ionization of HOCl represented by equation (17.25) differs from that of HCl represented by (17.18) in a highly significant way: The ionization of HOCl is a reversible process and must be represented by an equilibrium constant expression.

$$K_a = \frac{[H_3O^+][OCl^-]}{[HOCl]} = 2.95 \times 10^{-8}$$

The term K_a is called the **ionization constant** of hypochlorous acid. Its value, 2.95×10^{-8}, must be determined by experiment. Ionization constants for a few typical weak acids and bases are listed in Table 17-2. The symbol K_a is commonly used to represent the ionization constant of a weak acid and K_b for a weak base. pK is a commonly

TABLE 17-2
Ionization constants for some weak acids and weak bases in water at 25°C

	Ionization equilibrium	Ionization constant, K	pK[a]
Acid		$K_a =$	$pK_a =$
acetic	$HC_2H_3O_2 + H_2O \rightleftharpoons H_3O^+ + C_2H_3O_2^-$	1.74×10^{-5}	4.76
benzoic	$HC_7H_5O_2 + H_2O \rightleftharpoons H_3O^+ + C_7H_5O_2^-$	6.3×10^{-5}	4.20
chlorous	$HClO_2 + H_2O \rightleftharpoons H_3O^+ + ClO_2^-$	1.2×10^{-2}	1.92
formic	$HCHO_2 + H_2O \rightleftharpoons H_3O^+ + CHO_2^-$	1.8×10^{-4}	3.74
hydrocyanic	$HCN + H_2O \rightleftharpoons H_3O^+ + CN^-$	4.0×10^{-10}	9.40
hydrofluoric	$HF + H_2O \rightleftharpoons H_3O^+ + F^-$	6.7×10^{-4}	3.17
hypochlorous	$HOCl + H_2O \rightleftharpoons H_3O^+ + OCl^-$	2.95×10^{-8}	7.53
monochloroacetic	$HC_2H_2ClO_2 + H_2O \rightleftharpoons H_3O^+ + C_2H_2ClO_2^-$	1.35×10^{-3}	2.87
nitrous	$HNO_2 + H_2O \rightleftharpoons H_3O^+ + NO_2^-$	5.13×10^{-4}	3.29
phenol	$HOC_6H_5 + H_2O \rightleftharpoons H_3O^+ + C_6H_5O^-$	1.6×10^{-10}	9.80
Base		$K_b =$	$pK_b =$
ammonia	$NH_3 + H_2O \rightleftharpoons NH_4^+ + OH^-$	1.74×10^{-5}	4.76
aniline	$C_6H_5NH_2 + H_2O \rightleftharpoons C_6H_5NH_3^+ + OH^-$	4.30×10^{-10}	9.37
ethylamine	$C_2H_5NH_2 + H_2O \rightleftharpoons C_2H_5NH_3^+ + OH^-$	4.4×10^{-4}	3.36
hydroxylamine	$HONH_2 + H_2O \rightleftharpoons HONH_3^+ + OH^-$	9.1×10^{-9}	8.04
methylamine	$CH_3NH_2 + H_2O \rightleftharpoons CH_3NH_3^+ + OH^-$	4.2×10^{-4}	3.38
pyridine	$C_5H_5N + H_2O \rightleftharpoons C_5H_5NH^+ + OH^-$	2.0×10^{-9}	8.70

[a] Although some of these pK values could be expressed with an additional significant figure, the circumstances of a calculation often do not warrant this.

used shorthand designation for an equilibrium constant: $pK = -\log K$. Thus, for hypochlorous acid, $pK_a = -\log (2.95 \times 10^{-8}) = -(-7.530) = 7.53$.

Ionization vs. Dissociation.   The terms dissociation and ionization are often used synonymously, but there is a slight difference in their meanings. Dissociation means "coming apart," as in the dissociation of $N_2O_4(g)$.

$$N_2O_4(g) \rightleftharpoons 2\ NO_2(g)$$

It is also appropriate to speak of the dissociation of NaCl upon dissolving.

$$NaCl(s) \longrightarrow Na^+(aq) + Cl^-(aq)$$

Ionization of Na and Cl atoms occurs in the formation of crystalline NaCl, not when it dissolves. With HOCl, however, dissociation and ionization (ion formation) occur simultaneously. In our discussion of weak acids and bases we will emphasize proton transfer (and hence ion formation). For this reason we will use the term ionization.

Example 17-7   The pH of 1.10 *M* HCN solution is found to be 4.7. What is the value of K_a for HCN?

Solution.   Suppose we let x be the number of moles per liter of HCN that ionize. The relevant data are summarized below the equation. Note that $[H_3O^+] = [CN^-]$ since these ions are formed in equal numbers in the following reaction (and the ionization of water can be considered negligible).

	HCN	+ H₂O ⇌	H₃O⁺	+ CN⁻
placed in solution:	1.10 *M*		—	—
changes:	$-x$ *M*		$+x$ *M*	$+x$ *M*
at equilibrium:	$(1.10 - x)$ *M*		x *M*	x *M*

But $x = [H_3O^+]$ and is directly related to the measured pH.

$$\log [H_3O^+] = -pH = -4.7 = 0.3 - 5.0$$

$$x = [H_3O^+] = (\text{antilog } 0.3) \times (\text{antilog} -5) = 2 \times 10^{-5}$$

The ionization constant, K_a, is

$$K_a = \frac{[H_3O^+][CN^-]}{[HCN]} = \frac{x \cdot x}{1.10 - x} = \frac{(2 \times 10^{-5})(2 \times 10^{-5})}{1.10 - (2 \times 10^{-5})}$$

$$= \frac{(2 \times 10^{-5})^2}{1.10} = 3.6 \times 10^{-10} = 4 \times 10^{-10}$$

SIMILAR EXAMPLES: Review Problem 9; Exercises 16,17.

Example 17-8   A saturated aqueous solution of the weak base aniline contains 36.0 g $C_6H_5NH_2$/L. What is the pH of this solution?

$$C_6H_5NH_2 + H_2O \rightleftharpoons C_6H_5NH_3^+ + OH^- \qquad K_b = 4.30 \times 10^{-10}$$

Solution.   Even though it is a weak base, let us assume that the aniline produces practically all of the OH^- in solution. This is equivalent to saying that aniline is a stronger base than is H_2O. As in Example 17-7, the relevant data are summarized below the ionization equilibrium equation. A preliminary calculation is required in this case, though: the *molar* concentration of aniline in a saturated aqueous solution.

$$[C_6H_5NH_2] = \frac{36.0 \text{ g } C_6H_5NH_2}{\text{L soln.}} \times \frac{1 \text{ mol } C_6H_5NH_2}{93.1 \text{ g } C_6H_5NH_2} = 0.387 \text{ } M$$

$$C_6H_5NH_2 + H_2O \rightleftharpoons C_6H_5NH_3^+ + OH^-$$

placed in solution:	0.387 M	—	—
changes:	$-x$ M	$+x$ M	$+x$ M
at equilibrium:	$(0.387 - x)$ M	x M	x M

$$K_b = \frac{[C_6H_5NH_3^+][OH^-]}{[C_6H_5NH_2]} = \frac{x \cdot x}{0.387 - x} = 4.30 \times 10^{-10}$$

To solve this equation, let us assume that x is very small, that is, $x \ll 0.387$.

$$\frac{x^2}{0.387} = 4.30 \times 10^{-10} \qquad x^2 = 1.66 \times 10^{-10} \qquad [OH^-] = x = 1.29 \times 10^{-5} \ M$$

Our assumption is valid since $x = 1.29 \times 10^{-5} = 0.0000129 \ll 0.387$.

$$pOH = -\log[OH^-] = -(\log 1.29 \times 10^{-5}) = 4.889$$

$$pH = 14.00 - pOH = 14.00 - 4.889 = 9.11$$

SIMILAR EXAMPLES: Review Problems 10, 11; Exercises 18, 19.

Example 17-9 What is $[NO_2^-]$ in 0.00250 M HNO$_2$(aq)?

Solution. The initial steps to describe the ionization equilibrium are the same here as they were for HCN in Example 17-7 and $C_6H_5NH_2$ in Example 17-8.

$$HNO_2 + H_2O \rightleftharpoons H_3O^+ + NO_2^- \qquad K_a = 5.13 \times 10^{-4}$$

placed in solution:	0.00250 M	—	—
changes:	x M	$+x$ M	$+x$ M
at equilibrium:	$(0.00250 - x)$ M	x M	x M

$$K_a = \frac{[H_3O^+][NO_2^-]}{[HNO_2]} = \frac{x \cdot x}{0.00250 - x} = 5.13 \times 10^{-4} \qquad (17.26)$$

Let us make the usual assumption, that is, $x \ll 0.00250$ and $(0.00250 - x) \simeq 0.00250$.

$$\frac{x^2}{0.00250} = 5.13 \times 10^{-4} \qquad x^2 = 1.28 \times 10^{-6} \qquad [NO_2^-] = x = 1.13 \times 10^{-3} \ M$$

The value of x is about 40% as large as 0.00250. This is much too large to neglect. That is, if $x = 1.13 \times 10^{-3}$, then $0.00250 - x = 0.00250 - 0.00113 = 0.00137 \neq 0.00250$.

Since our assumption failed, we must return to equation (17.26) and seek an exact solution. This means solving a quadratic equation.

$$\frac{x^2}{0.00250 - x} = 5.13 \times 10^{-4}$$

$$x^2 + 5.13 \times 10^{-4} \ x - 1.28 \times 10^{-6} = 0$$

$$x = \frac{-5.13 \times 10^{-4} \pm \sqrt{(5.13 \times 10^{-4})^2 + 4 \times 1.28 \times 10^{-6}}}{2}$$

$$= \frac{-5.13 \times 10^{-4} \pm 2.32 \times 10^{-3}}{2}$$

$$x = [NO_2^-] = 9.04 \times 10^{-4} \ M$$

SIMILAR EXAMPLES: Exercises 22, 23.

The simplifying assumption in Examples 17-7 and 17-8, which failed in Example 17-9, was equivalent to saying this: treat the weak acid or weak base as if all of it remains essentially nonionized. When can such an assumption be made? The validity of this assumption depends on the magnitude of K_a (or K_b) and the total concentration of weak acid (or weak base). The following is a very rough generalization.

Value to consider	The assumption that most of a weak acid or base remains essentially nonionized usually	
	Works if value is	Does not work if value is
K_a or K_b	less than about 1×10^{-5}	greater than about 1×10^{-5}
total concentration of weak acid or weak base	greater than about 0.01 M	less than about 0.01 M

In Example 17-9, the value of K_a for HNO_2 is seen to be greater than 1×10^{-5}, and the concentration of HNO_2 placed in solution, less than 0.01 M. The usual simplifying assumption failed. In the final analysis, the best approach is always to test the validity of any assumptions made in an equilibrium calculation.

Simplifying assumptions are also more likely to be valid if only two significant figures are carried in a calculation rather than three or four.

Degree of Ionization. In a similar manner to the percent dissociation of gases discussed in Chapter 15 (recall Example 15-13), we can establish a percent ionization of a weak acid or base. In Example 17-10 we calculate the percent ionization of acetic acid at three different concentrations and reach this conclusion: *The degree of ionization (percent ionization) of a weak electrolyte increases as the solution becomes more dilute*. This conclusion is pictured and contrasted to the behavior of a strong acid in Figure 17-4.

FIGURE 17-4
Percent ionization of an acid as a function of concentration.

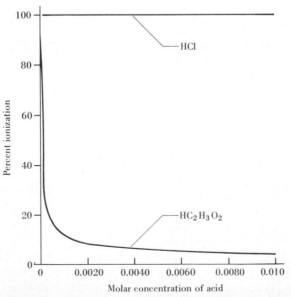

Over the concentration range shown, HCl(aq) is considered to be completely ionized— 100% ionized. The percent ionization of $HC_2H_3O_2$(aq) increases from about 4% in 0.010 M $HC_2H_3O_2$ to essentially 100% when the solution becomes extremely dilute.

Example 17-10   Calculate the % ionization of acetic acid in **(a)** 1.0 M; **(b)** 0.10 M; **(c)** 0.010 M HC$_2$H$_3$O$_2$(aq).

$$HC_2H_3O_2 + H_2O \rightleftharpoons H_3O^+ + C_2H_3O_2^- \qquad K_a = 1.74 \times 10^{-5}$$

Solution.   In each case let us assume that the weak acid is essentially nonionized, so that we can write [HC$_2$H$_3$O$_2$] = 1.0 $- x \approx$ 1.0 M in (a), 0.10 M in (b), and 0.010 M in (c).

(a) $K_a = \dfrac{[H_3O^+][C_2H_3O_2^-]}{[HC_2H_3O_2]} = \dfrac{x \cdot x}{1.0} = 1.74 \times 10^{-5}$

$x = [H_3O^+] = 4.2 \times 10^{-3} \ M$

% ionization of 1.0 M HC$_2$H$_3$O$_2$ $= \dfrac{4.2 \times 10^{-3} \text{ mol } H_3O^+/L}{1.0 \text{ mol } HC_2H_3O_2/L} \times 100 = 0.42\%$

To complete the calculations of parts (b) and (c), substitute the appropriate value of [HC$_2$H$_3$O$_2$]. Solve for [H$_3$O$^+$] and then determine the percent ionization.
(b) 0.10 M HC$_2$H$_3$O$_2$(aq) is 1.3% ionized.
(c) 0.010 M HC$_2$H$_3$O$_2$(aq) is 4.2% ionized.

SIMILAR EXAMPLES: Review Problem 12; Exercise 24.

The inverse of the calculation illustrated in Example 17-10 is also of importance because from the degree of ionization (α) of a weak acid or base one can calculate a value of K_a or K_b. (Percent ionization is $100 \times \alpha$.) The degree of ionization can be established by experimental measurements, e.g., of the electrical conductivity or of the freezing point depression of the weak electrolyte solution (see Exercise 27).

17-6 Polyprotic Acids

All of the acids listed in Table 17-2 are weak *monoprotic* acids. They produce only one proton per acid molecule, even if at times there is more than one hydrogen atom in the molecule. For example, in acetic acid only the hydrogen atom shown in color is ionizable.

$$(17.27)$$

There are some acids, however, that contain *more than one* ionizable hydrogen atom per molecule. These are called **polyprotic acids.** Ionization constants for several polyprotic acids are listed in Table 17-3. An important polyprotic acid is hydrogen sulfide (hydrosulfuric acid), H$_2$S. Its ionization occurs in two steps.

first ionization step: $H_2S + H_2O \rightleftharpoons H_3O^+ + HS^-$ (17.28)
second ionization step: $HS^- + H_2O \rightleftharpoons H_3O^+ + S^{2-}$ (17.29)

An ionization constant expression can be written for each step in the usual manner.

$$K_{a_1} = \frac{[H_3O^+][HS^-]}{[H_2S]} = 1.1 \times 10^{-7} \qquad K_{a_2} = \frac{[H_3O^+][S^{2-}]}{[HS^-]} = 1.0 \times 10^{-14} \quad (17.30)$$

One generalization about polyprotic acids is that the second ionization constant, K_{a_2}, is always *smaller* than the first, K_{a_1}. This is a reasonable finding: We would expect greater difficulty in separating a proton from a double negatively charged ion, S^{2-} (the second step) than from a single negatively charged ion, HS$^-$ (the first step). These additional observations about the H$_2$S equilibria are generally applicable to weak polyprotic acids (with the qualification noted in item 3).

TABLE 17-3
Ionization constants of some common polyprotic acids

Acid	Ionization equilibria	Ionization constants, K	pK
carbonic[a]	$H_2CO_3 + H_2O \rightleftharpoons H_3O^+ + HCO_3^-$	$K_{a_1} = 4.2 \times 10^{-7}$	$pK_{a_1} = 6.38$
	$HCO_3^- + H_2O \rightleftharpoons H_3O^+ + CO_3^{2-}$	$K_{a_2} = 5.6 \times 10^{-11}$	$pK_{a_2} = 10.25$
hydrosulfuric	$H_2S + H_2O \rightleftharpoons H_3O^+ + HS^-$	$K_{a_1} = 1.1 \times 10^{-7}$	$pK_{a_1} = 6.96$
	$HS^- + H_2O \rightleftharpoons H_3O^+ + S^{2-}$	$K_{a_2} = 1.0 \times 10^{-14}$	$pK_{a_2} = 14.00$
oxalic	$H_2C_2O_4 + H_2O \rightleftharpoons H_3O^+ + HC_2O_4^-$	$K_{a_1} = 5.4 \times 10^{-2}$	$pK_{a_1} = 1.27$
	$HC_2O_4^- + H_2O \rightleftharpoons H_3O^+ + C_2O_4^{2-}$	$K_{a_2} = 5.4 \times 10^{-5}$	$pK_{a_2} = 4.27$
phosphoric	$H_3PO_4 + H_2O \rightleftharpoons H_3O^+ + H_2PO_4^-$	$K_{a_1} = 5.9 \times 10^{-3}$	$pK_{a_1} = 2.23$
	$H_2PO_4^- + H_2O \rightleftharpoons H_3O^+ + HPO_4^{2-}$	$K_{a_2} = 6.2 \times 10^{-8}$	$pK_{a_2} = 7.21$
	$HPO_4^{2-} + H_2O \rightleftharpoons H_3O^+ + PO_4^{3-}$	$K_{a_3} = 4.8 \times 10^{-13}$	$pK_{a_3} = 12.32$
phosphorous	$H_3PO_3 + H_2O \rightleftharpoons H_3O^+ + H_2PO_3^-$	$K_{a_1} = 5.0 \times 10^{-2}$	$pK_{a_1} = 1.30$
	$H_2PO_3^- + H_2O \rightleftharpoons H_3O^+ + HPO_3^{2-}$	$K_{a_2} = 2.5 \times 10^{-7}$	$pK_{a_2} = 6.60$
sulfurous[b]	$H_2SO_3 + H_2O \rightleftharpoons H_3O^+ + HSO_3^-$	$K_{a_1} = 1.3 \times 10^{-2}$	$pK_{a_1} = 1.89$
	$HSO_3^- + H_2O \rightleftharpoons H_3O^+ + SO_3^{2-}$	$K_{a_2} = 6.3 \times 10^{-8}$	$pK_{a_2} = 7.20$
sulfuric[c]	$H_2SO_4 + H_2O \longrightarrow H_3O^+ + HSO_4^-$	$K_{a_1} =$ very large	$pK_{a_1} < 0$
	$HSO_4^- + H_2O \rightleftharpoons H_3O^+ + SO_4^{2-}$	$K_{a_2} = 1.29 \times 10^{-2}$	$pK_{a_2} = 1.89$

[a] H_2CO_3 is also in equilibrium with dissolved $CO_2(g)$ and H_2O.

$$CO_2(aq) + H_2O \rightleftharpoons H_2CO_3(aq)$$

H_2CO_3 cannot be isolated in pure form.

[b] H_2SO_3 is a hypothetical, nonisolable species produced in the reaction

$$SO_2(aq) + H_2O \rightleftharpoons H_2SO_3(aq)$$

[c] H_2SO_4 is completely ionized in the first step.

1. All species involved in the ionization equilibria exist together in a single solution phase. For any given equilibrium condition, each concentration—$[H_2S]$, $[H_3O^+]$, $[HS^-]$, $[S^{2-}]$—has a fixed value.

2. The concentrations of all species present in the solution must be consistent with both ionization constant expressions, K_{a_1} and K_{a_2}.

3. Because HS^- is much weaker an acid than H_2S, H_3O^+ is produced almost exclusively in the *first* ionization step. [This assumption is valid only if $K_{a_1} \gg K_{a_2}$ (say by a factor of 10^3 or 10^4). Where the difference between K_{a_1} and K_{a_2} is not so great, the second ionization may also be significant as a source of H_3O^+.]

4. Even though further ionization in the second step occurs to a very limited extent, this ionization is the *only* source of S^{2-}.

Example 17-11   A saturated aqueous solution of $H_2S(g)$ at 25°C and 1 atm is 0.10 M H_2S. Calculate $[H_2S]$, $[H_3O^+]$, $[HS^-]$, and $[S^{2-}]$ in this solution.

Solution.   To describe initial and equilibrium concentrations in the usual manner, we need to write *two* expressions, one for each ionization step. The number of moles per liter of H_2S ionized in the first step is represented by x. The number of moles per liter of HS^- that undergo further ionization in the second step is represented by y. The final equilibrium concentrations are shown in color.

$$H_2S \quad + \; H_2O \; \Longleftrightarrow \; H_3O^+ \; + \; HS^-$$

placed in solution:	$0.10 \, M$	—	—
changes:	$-x \, M$	$+x \, M$	$+x \, M$
after first ionization:	$(0.10 - x) \, M$	$x \, M$	$x \, M$

$$HS^- \quad + \; H_2O \; \Longleftrightarrow \; H_3O^+ \quad + \quad S^{2-}$$

from first ionization:	$x \, M$	$x \, M$	—
changes:	$-y \, M$	$+y \, M$	$+y \, M$
after second ionization:	$(x - y) \, M$	$(x + y) \, M$	$y \, M$

Equilibrium concentrations

$$[H_2S] = 0.10 - x \qquad [H_3O^+] = x + y \qquad [HS^-] = x - y \qquad [S^{2-}] = y$$

Equilibrium constant expressions

$$K_{a_1} = \frac{[H_3O^+][HS^-]}{[H_2S]} = \frac{(x + y)(x - y)}{0.10 - x} = 1.1 \times 10^{-7} \tag{17.31}$$

$$K_{a_2} = \frac{[H_3O^+][S^{2-}]}{[HS^-]} = \frac{(x + y)y}{x - y} = 1.0 \times 10^{-14} \tag{17.32}$$

To calculate the equilibrium concentration of each species is not nearly so difficult as first appearances might suggest. The key to a simple solution is in Statement 3 above. The concentration of H_3O^+ produced in the second ionization, y, is very much smaller than that produced in the first ionization, x. That is, $y \ll x$. This means that $(x + y) \simeq (x - y) \simeq x$. Furthermore, because H_2S is quite a weak acid, most of it remains nonionized. This means that $x \ll 0.10$ and $(0.10 - x) \simeq 0.10$. With these assumptions we obtain the much simpler algebraic expressions:

$$K_{a_1} = \frac{x \cdot x}{0.10} = 1.1 \times 10^{-7} \qquad and \qquad K_{a_2} = \frac{\cancel{x} \cdot y}{\cancel{x}} = 1.0 \times 10^{-14}$$

$$x^2 = 1.1 \times 10^{-8}$$

$$x = 1.0 \times 10^{-4} \qquad\qquad\qquad y = 1.0 \times 10^{-14}$$

Equilibrium concentrations

$$[H_2S] = 0.10 \, M \qquad [H_3O^+] = 1.0 \times 10^{-4} \, M$$

$$[HS^-] = 1.0 \times 10^{-4} \, M \qquad [S^{2-}] = 1.0 \times 10^{-14} \, M$$

SIMILAR EXAMPLES: Review Problem 13; Exercises 28, 31.

As long as Statement 3 of page 518 is valid, the general ideas illustrated by Example 17-11 are that

1. $[H_3O^+]$ in a weak polyprotic acid solution can be calculated through the K_{a_1} expression alone.
2. The molar concentration of the anion produced in the second ionization step of a weak diprotic acid (e.g., $[S^{2-}]$) is numerically equal to K_{a_2}.

Where Statement 3 is not valid, calculation of the concentrations of species in solution is more difficult, because it requires the simultaneous solution of two or more algebraic equations. Additional aspects of equilibria involving polyprotic acids are considered in Chapters 18 and 19.

A Somewhat Different Case—H_2SO_4.

H_2SO_4 differs from the other polyprotic acids in Table 17-3 in this important respect: It is a strong acid in its first ionization step and weak only in the second. Ionization is complete in the first step, and this means that in most $H_2SO_4(aq)$ solutions we can assume that $[H_2SO_4] \simeq 0$. Thus, if a solution is 0.50 M H_2SO_4, we can treat it as if it were 0.50 M H_3O^+ and 0.50 M HSO_4^- initially and then determine how much further ionization of HSO_4^- occurs to produce additional H_3O^+ and SO_4^{2-}.

Example 17-12   What are $[H_3O^+]$, $[HSO_4^-]$, and $[SO_4^{2-}]$ in 0.50 M H_2SO_4?

Solution.   Using the same formulation as for H_2S in Example 17-11, we can write

$$H_2SO_4 \; + \; H_2O \longrightarrow \; H_3O^+ \; + \; HSO_4^-$$

placed in solution:	0.50 M	—	—
changes:	$-0.50\ M$	$+0.50\ M$	$+0.50\ M$
after first ionization:	$\simeq 0$	0.50 M	0.50 M

$$HSO_4^- \; + \; H_2O \rightleftharpoons \; H_3O^+ \; + \; SO_4^{2-}$$

from first ionization:	0.50 M	0.50 M	—
changes:	$-x$	$+x$	$+x$
after second ionization:	$(0.50-x)\ M$	$(0.50+x)\ M$	$x\ M$

We need to deal with just one ionization equilibrium expression, K_{a_2}. If we assume that $x \ll 0.50$, then $(0.50+x) \simeq (0.50-x) \simeq 0.50$ and

$$K_{a_2} = \frac{[H_3O^+][SO_4^{2-}]}{[HSO_4^-]} = \frac{(0.50+x)x}{(0.50-x)} = 1.29 \times 10^{-2}$$

$[H_3O^+] = 0.50 + x = 0.51\ M$ $[HSO_4^-] = 0.50 - x = 0.49$
$[SO_4^{2-}] = x = 0.0129\ M$

The assumption made in this calculation is similar to that made in Example 17-11 for H_2S and leads to $[SO_4^{2-}] = K_{a_2}$. Notice, however, that this assumption would fail if the original $H_2SO_4(aq)$ were much more dilute than 0.50 M H_2SO_4 (see Exercise 33).

SIMILAR EXAMPLES: Exercises 32, 33.

17-7 Cations and Anions as Acids and Bases

The acids and bases emphasized in our discussion to this point have been neutral molecular species. However, proton donors and acceptors are not limited to neutral molecules. Many ions can act in these capacities as well. For example, the second and subsequent ionization steps of a polyprotic acid involve an anion acting as an acid, as in the ionization

$$H_2PO_4^- + H_2O \rightleftharpoons H_3O^+ + HPO_4^{2-} \qquad K = K_{a_2} = 6.2 \times 10^{-8}$$

Each of the following is also an acid–base reaction. Let us think about how they might be described.

$$NH_4^+ + H_2O \rightleftharpoons NH_3 + H_3O^+ \tag{17.33}$$

$$C_2H_3O_2^- + H_2O \rightleftharpoons HC_2H_3O_2 + OH^- \tag{17.34}$$

Reaction (17.33) suggests that NH_4^+ is an *acid*, able to donate a proton to water, a *base*. The equilibrium constant for this reaction can be referred to as the acid ionization constant, K_a, of the ammonium ion, NH_4^+. Reaction (17.34) shows $C_2H_3O_2^-$ acting as a *base* by accepting a proton from water, an *acid*. Here the applicable equilibrium constant is a base ionization constant, K_b, for the acetate ion, $C_2H_3O_2^-$. Table 17-4 presents ionization constant data for several ions.

Hydrolysis.   In pure water at 25°C, $[H_3O^+] = [OH^-] = 1.0 \times 10^{-7}\ M$. When a salt such as NaCl is added to water, complete dissociation into Na^+ and Cl^- ions occurs, but

TABLE 17-4
Acid and base ionization constants for several ions at 25°C

	Ionization equilibrium	Ionization constant, K	pK
Acid		$K_a =$	p$K_a =$
ammonium ion	$NH_4^+ + H_2O \rightleftharpoons H_3O^+ + NH_3$	5.7×10^{-10}	9.24
anilinium ion	$C_6H_5NH_3^+ + H_2O \rightleftharpoons H_3O^+ + C_6H_5NH_2$	2.3×10^{-5}	4.64
ethylammonium ion	$C_2H_5NH_3^+ + H_2O \rightleftharpoons H_3O^+ + C_2H_5NH_2$	2.3×10^{-11}	10.64
methylammonium ion	$CH_3NH_3^+ + H_2O \rightleftharpoons H_3O^+ + CH_3NH_2$	2.4×10^{-11}	10.62
Base		$K_b =$	p$K_b =$
acetate ion	$C_2H_3O_2^- + H_2O \rightleftharpoons OH^- + HC_2H_3O_2$	5.7×10^{-10}	9.24
chlorite ion	$ClO_2^- + H_2O \rightleftharpoons OH^- + HClO_2$	8.3×10^{-13}	12.08
cyanide ion	$CN^- + H_2O \rightleftharpoons OH^- + HCN$	2.5×10^{-5}	4.60
fluoride ion	$F^- + H_2O \rightleftharpoons OH^- + HF$	1.5×10^{-11}	10.82
hypochlorite ion	$OCl^- + H_2O \rightleftharpoons OH^- + HOCl$	3.4×10^{-7}	6.47
nitrite ion	$NO_2^- + H_2O \rightleftharpoons OH^- + HNO_2$	1.9×10^{-11}	10.72

these ions do not influence the ionization of water. The pH of the solution remains at 7.

$$Na^+ + Cl^- + H_2O \longrightarrow \text{no reaction}$$

We can say that neither Na^+ nor Cl^- has acidic or basic properties. They are neutral ions.

When NH_4Cl is added to water, the pH falls below 7. This means that $[H_3O^+]$ in the solution increases and $[OH^-]$ decreases. A reaction producing H_3O^+ must occur between the added ions and water molecules. Cl^- cannot act as an acid; it has neither a proton to donate nor an ability to accept an electron pair. And Cl^- is too weak a base to accept a proton from H_2O. However, as we have already seen, a reaction does occur between NH_4^+ and H_2O.

$$Cl^- + H_2O \longrightarrow \text{no reaction}$$

$$NH_4^+ + H_2O \rightleftharpoons NH_3 + H_3O^+ \tag{17.33}$$

Although reaction (17.33) is fundamentally no different from other acid–base reactions, a reaction between an ion and water is often called a **hydrolysis** reaction. The ammonium ion hydrolyzes and the chloride ion does not.

When sodium acetate is added to water, the pH rises above 7. This means that $[OH^-]$ in the solution increases and $[H_3O^+]$ decreases. Sodium ion has neither acidic nor basic properties, but acetate ion hydrolyzes.

$$Na^+ + H_2O \longrightarrow \text{no reaction}$$

$$C_2H_3O_2^- + H_2O \rightleftharpoons HC_2H_3O_2 + OH^- \tag{17.34}$$

From the preceding discussion let us first examine some qualitative statements about hydrolysis.

Hydrolyses of several ions are illustrated in Color Section F.

- In general, salts are completely dissociated into ions in aqueous solutions.
- Salts of strong acids and strong bases (such as NaCl) do not hydrolyze.
- Salts of *weak* acids and strong bases (such as $NaC_2H_3O_2$) undergo hydrolysis, producing a basic solution: pH > 7. It is the *anion* in such a salt which acts as a base.
- Salts of strong acids and *weak* bases (such as NH_4Cl) undergo hydrolysis to produce an acidic solution: pH < 7. In such a salt it is the *cation* that acts as an acid.
- Salts of *weak* acids and *weak* bases (such as $NH_4C_2H_3O_2$) hydrolyze, but whether the resulting solution is neutral, acidic, or basic depends on the relative values of K_a and K_b for the ions that hydrolyze.

Example 17-13 Predict whether you would expect each of the following solutions to be acidic, basic, or neutral: **(a)** NaCN(aq); **(b)** KCl(aq); **(c)** NH₄CN(aq).

Solution

(a) The ions in solution are Na^+, which does not hydrolyze, and CN^-, which does: $CN^- + H_2O \rightleftharpoons HCN + OH^-$. Because the hydrolysis reaction produces OH^-, the solution is basic.

(b) Neither K^+ nor Cl^- undergoes hydrolysis. KCl(aq) is neutral—pH = 7.

(c) *Both* NH_4^+ and CN^- hydrolyze in aqueous solution, one to produce H_3O^+ and the other, OH^-. With values from Table 17-4 we can write

$$NH_4^+ + H_2O \rightleftharpoons NH_3 + H_3O^+ \qquad K_a = 5.7 \times 10^{-10}$$

$$CN^- + H_2O \rightleftharpoons HCN + OH^- \qquad K_b = 2.5 \times 10^{-5}$$

Because K_b is larger than K_a, we should expect CN^- to hydrolyze to a greater extent than NH_4^+. This means that $[OH^-] > [H_3O^+]$ and the solution will be basic.

SIMILAR EXAMPLES: Review Problem 14; Exercise 37.

In Example 17-13 we predicted that a solution of NaCN should be basic (pH > 7). In Example 17-14 we use the ionization constant, K_b, for CN^- to *calculate* the pH of a NaCN solution.

Example 17-14 What is the pH of an 0.50 *M* NaCN solution?

Solution. The concentrations of the several species involved in the hydrolysis reaction are summarized below, where $x = [OH^-]$.

$$CN^- + H_2O \rightleftharpoons HCN + OH^-$$

placed in solution:	0.50 *M*	—	—
changes:	− *x M*	+ *x M*	+ *x M*
equilibrium concentrations:	(0.50 − *x*) *M*	*x M*	*x M*

$$K_b = \frac{[OH^-][HCN]}{[CN^-]} = 2.5 \times 10^{-5}$$

$$\frac{x \cdot x}{0.50 - x} = \frac{x^2}{0.50 - x} = 2.5 \times 10^{-5}$$

Next a familiar assumption can be made. If $x \ll 0.50$, then $0.50 - x \simeq 0.50$.

$$x^2 = (0.50)(2.5 \times 10^{-5}) = 1.2 \times 10^{-5} = 12 \times 10^{-6}$$

$$x = [OH^-] = (12 \times 10^{-6})^{1/2} = 3.5 \times 10^{-3}$$

$$[H_3O^+] = \frac{K_w}{[OH^-]} = \frac{1.0 \times 10^{-14}}{3.5 \times 10^{-3}} = 2.9 \times 10^{-12}$$

$$pH = -\log [H_3O^+] = -\log (2.9 \times 10^{-12}) = 11.54$$

SIMILAR EXAMPLES: Review Problem 15; Exercises 34, 35.

To predict that NH₄CN(aq) is basic (pH > 7) is easily done, as in Example 17-13(c). To calculate the *exact* pH of NH₄CN(aq) is much more difficult because several algebraic equations must be solved simultaneously. This calculation is left as an exercise for the interested student (see Additional Exercise 15).

The Relationship Between Ionization Constants of Weak Acids (or Bases) and Their Conjugates. Ionization constants for cations and anions are not often encountered in tabulations such as Table 17-4. Usually they must be derived from other data. This derivation is a straightforward application of the method of combining equilibrium constant expressions introduced in Section 15-2.

Written below are equations for the ionization of (a) acetic acid as a weak acid and (b) acetate ion as a base. The sum of these two equations represents the self-ionization of water. When equations are added, we learned in Section 15-2, the equilibrium constant of the net reaction is the *product* of the equilibrium constants of the reactions that are combined.

(a) $HC_2H_3O_2 + H_2O \rightleftharpoons H_3O^+ + C_2H_3O_2^-$ K_a(acetic acid)
(b) $C_2H_3O_2^- + H_2O \rightleftharpoons OH^- + HC_2H_3O_2$ K_b(acetate ion)

net: $2 H_2O \rightleftharpoons H_3O^+ + OH^-$ $K_w = K_a$(acetic acid) (17.35) $\times K_b$(acetate ion)

The result in equation (17.35) can be rearranged to show that *the ionization constant of the conjugate base (the anion) of a weak acid is*

$$K_b = \frac{K_w}{K_a} \tag{17.36}$$

We could similarly demonstrate that *the ionization constant of the conjugate acid (the cation) of a weak base is*

$$K_a = \frac{K_w}{K_b} \tag{17.37}$$

If we wished to calculate the pH of an aqueous solution of sodium benzoate ($NaC_7H_5O_2$), we would first have to use data from Table 17-2 and equation (17.36) to determine K_b for benzoate ion. This is done in Example 17-15.

The constants K_a (for cations) and K_b (for anions) are sometimes referred to as hydrolysis constants, K_h, since they are associated with hydrolysis reactions. As noted previously, however, there is no real distinction between hydrolysis reactions and other acid–base reactions.

Example 17-15 What is the value of K_b for benzoate ion, $C_7H_5O_2^-$?

$C_7H_5O_2^- + H_2O \rightleftharpoons HC_7H_5O_2 + OH^-$; $K_b = ?$

Solution. From equation (17.36) we can write

$$K_b(C_7H_5O_2^-) = \frac{K_w}{K_a(HC_7H_5O_2)} = \frac{1.0 \times 10^{-14}}{6.3 \times 10^{-5}} = 1.6 \times 10^{-10}$$

SIMILAR EXAMPLE: Review Problem 16.

Hydrated Metal Ions as Acids. With the exception of NH_4^+, the cations in Table 17-4 are not too familiar (they are derived from organic molecules). Among simple metal ions, we have learned that Na^+ does not hydrolyze. Neither do other familiar cations of the group IA and IIA metals. However, with many other metal ions, especially of the transition metals, hydrolysis is important. These cations ionize as acids—proton donors. They do this through the ionization of water molecules attached to the metal ion. Thus, it is the *hydrated* metal ions that hydrolyze, as in the case of Fe^{3+}(aq).

$$[Fe(H_2O)_6]^{3+} + H_2O \rightleftharpoons [Fe(H_2O)_5OH]^{2+} + H_3O^+ \tag{17.38}$$

One of the six H_2O molecules attached to Fe^{3+} in the complex ion $[Fe(H_2O)_6]^{3+}$ loses a proton to a free H_2O molecule, converting it to H_3O^+. The H_2O molecule that has ionized is converted to OH^-, which remains attached to the Fe^{3+} ion and reduces the charge of the complex ion from 3+ to 2+.

We will explore the ionization of hydrated cations more fully in Chapters 22–24, where this phenomenon will be frequently encountered.

17-8 Molecular Structure and Acid–Base Behavior

For the quantitative descriptions of aqueous solutions that we have been emphasizing in this chapter, we have had to consider two questions primarily: Do the ionization processes in solution go to completion or do they reach a state of equilibrium? If an equilibrium condition is involved, what are the relevant equilibrium constant expressions? If we had answers to these questions, we generally were able to calculate any concentrations of interest to us.

There are times, however, when we seek only some qualitative answers: Why is HOCl a weak acid whereas $HClO_4$ is a strong acid? Why is ethanol, C_2H_5OH, so much weaker an acid than phenol, C_6H_5OH? Although some situations concerning acid (or base) strength are quite complex, in most cases conclusions based on a consideration of molecular structure prove to be correct.

Strengths of Brønsted–Lowry Acids and Bases. Before attempting to answer questions like the two just posed, we need to say a bit more about acid and base strength in relation to the Brønsted–Lowry theory. If we study the following two reactions, we find that the first goes essentially to completion and the other proceeds hardly at all in the forward direction.

$$HClO_4 \; + \; OH^- \; \rightleftharpoons \; H_2O \; + \; ClO_4^-$$

acid(1)	base(2)	acid(2)	base(1)
very strong	strong	weak	very weak

$$HCO_3^- \; + \; Br^- \; \rightleftharpoons \; HBr \; + \; CO_3^{2-}$$

acid(1)	base(2)	acid(2)	base(1)
weak	very weak	very strong	strong

Each reaction illustrates a statement first made on page 505: *A Brønsted–Lowry acid–base reaction is always favored in the direction of the stronger to the weaker acid/base combination.* This statement in turn suggests the need for a listing of Brønsted–Lowry acids and bases according to their relative strengths. Such a list is provided in Table 17-5. A point to note in Table 17-5 is that the stronger an acid, the weaker its conjugate base.

The relative placement of some of the entries in Table 17-5 follows directly from what we have learned elsewhere in this chapter. For example, acetic acid ($K_a = 1.74 \times 10^{-5}$) is stronger than carbonic acid ($K_{a_1} = 4.2 \times 10^{-7}$); and carbonic acid is a stronger acid than its anion, HCO_3^- ($K_{a_2} = 5.6 \times 10^{-11}$). Why do we rank $HClO_4$ ahead of HCl, however? In water solution both of these acids are so strong that they are completely ionized. Water is said to have a **leveling effect** on these two acids. Water is a strong enough base to accept protons from either acid, blurring whatever differences may exist between them.

To distinguish between the strengths of $HClO_4$ and HCl, we need to use a **differentiating solvent.** This is a solvent that is itself such a weak base that it will accept protons from the stronger of the two acids more readily than from the weaker one. Diethyl ether, $C_2H_5OC_2H_5$, has less affinity for protons than does water. In diethyl ether, $HClO_4$ is still essentially completely ionized, but HCl is only partially ionized. In considering the relative strengths of acids and bases, then, the solvent chosen for these comparisons plays an important role. Our attention, nevertheless, will continue to be focused on aqueous solutions.

Bond Strength and Acid Strength. The strength of an acid or base must relate ultimately to the ease with which a proton is lost or gained.

It is an oversimplification to relate the acidities of the binary acids HX to the mere breaking of the bond H—X. For one thing, bond energies are given for the dissociation of gaseous species, and here we are dealing with species in solution. Nevertheless, it would seem that the stronger the H—X bond, the weaker the acid. This generalization does hold

TABLE 17-5

Relative strengths of some common Brønsted–Lowry acids and bases

	Acid		Conjugate base	
perchloric acid	$HClO_4$	perchlorate ion	ClO_4^-
hydroiodic acid	HI	iodide ion	I^-
hydrobromic acid	HBr	bromide ion	Br^-
hydrochloric acid	HCl	chloride ion	Cl^-
nitric acid	HNO_3	nitrate ion	NO_3^-
sulfuric acid	H_2SO_4	hydrogen sulfate ion	HSO_4^-
hydronium ion[a]	H_3O^+	water[a]	H_2O
hydrogen sulfate ion	HSO_4^-	sulfate ion	SO_4^{2-}
nitrous acid	HNO_2	nitrite ion	NO_2^-
acetic acid	$HC_2H_3O_2$	acetate ion	$C_2H_3O_2^-$
carbonic acid	H_2CO_3	bicarbonate ion	HCO_3^-
ammonium ion	NH_4^+	ammonia	NH_3
bicarbonate ion	HCO_3^-	carbonate ion	CO_3^{2-}
water	H_2O	hydroxide ion	OH^-
methanol	CH_3OH	methoxide ion	CH_3O^-
ammonia	NH_3	amide ion	NH_2^-

increasing acid strength (left margin arrow, pointing up)

increasing base strength (right margin arrow, pointing down)

[a] The hydronium ion/water combination refers to the ease with which a proton is passed from one water molecule to another; that is, $H_3O^+ + H_2O \rightleftharpoons H_3O^+ + H_2O$.

true for the series of acids, HF, HCl, HBr, HI, for which the bond energies increase in the order

$$HI < HBr < HCl < HF$$

whereas the acid strengths decrease in the order

$$\underbrace{HI > HBr > HCl}_{\text{strong}} > \underbrace{HF}_{\text{weak}}$$

An alternative view leading to the same ordering for these acids is that, as the anion decreases in size, its affinity for a proton increases. The anion is the conjugate base of the acid, and as its base strength increases, the acid strength decreases.

Acidic, Basic, and Amphoteric Oxides. Most of the elements form compounds with oxygen, and one classification scheme for oxides uses the terms, acidic, basic, and amphoteric. To see how the scheme works, consider the hypothetical oxide, E_xO_y, formed by the element E. Consider also that this oxide reacts with water to form a hydroxo compound containing one or more bonds, E—O—H.

$$E_xO_y + H_2O \longrightarrow \;E{-}O{-}H$$

Any factor that draws electrons toward the atom E strengthens the E—O bond, weakens the O—H bond, and causes the hydroxo compound to ionize as an acid. The ability of the compound to donate protons is increased in the presence of a strong base such as OH^-.

$$E{-}O{-}(H + OH^-) \longrightarrow \left(E{-}O\right)^- + H_2O$$

This type of behavior is favored by small size and high charge (as indicated by high

(margin note:) The fact that HF is a weak acid whereas the other hydrogen halides are very strong has always seemed an anomaly. One recent proposal suggests that in HF(aq) ion pairs are held together by strong hydrogen bonds. This keeps the concentration of free H_3O^+ from being as large as otherwise expected.

$$HF + H_2O \longrightarrow$$
$$(^-F\text{---}H_3O^+) \rightleftharpoons$$
$$H_3O^+ + F^-$$

oxidation state or high formal charge) on the atom E, and is expected if the element E is nonmetallic. In these situations the oxide E_xO_y is called an **acid anhydride** (from the Greek, "without water"), and the hydroxo compound is an **oxoacid.**

A large size and small charge on the atom E favors rupture of the E—O bond. The hydroxo compound ionizes to produce OH^- and the atom E becomes a cationic species. This tendency to ionize as a base is increased when the hydroxo compound is placed in an acidic solution. Behavior of this type is associated with metallic character in the element E. In these instances the oxide E_xO_y is referred to as a **base anhydride.**

$$\overset{\diagdown}{\underset{\diagup}{E}}\text{--}O\text{--}H + H_3O^+ \longrightarrow \left(\overset{\diagdown}{\underset{\diagup}{E}}\right)^+ + 2\,H_2O$$

In some cases the hydroxo compound may act either as an acid or a base. This behavior is known as **amphoterism** and is associated with elements having both metallic and nonmetallic properties.

A more quantitative description of the effect of size and charge on the acid–base character of a hydroxo compound is based on charge density—the ratio of the charge of the cation to its radius. The higher the charge density, the more acidic the compound. Conversely, the lower the charge density the more basic the compound. For the following hydroxo compounds the charge densities and order of increasing acidity are

$$Mg(OH)_2 < Be(OH)_2 \simeq Al(OH)_3 < B(OH)_3$$

	basic	amphoteric		acidic
ionic radius, pm:	Mg^{2+}, 65	Be^{2+}, 31	Al^{3+}, 50	B^{3+}, 20
charge density:	2/65 =	2/31 =	3/50 =	3/20 =
	0.031 $<$	0.065 $\simeq$	0.060 $<$	0.15

This classification scheme for element oxides is summarized in Table 17-6.

Strengths of Oxoacids. An oxoacid has at least one E—O—H bond and other bonds that are either E—O—H or E—O (or occasionally, E—H). The formation and acidic behavior of an oxoacid are pictured in Figure 17-5.

Factors that promote the withdrawal of electrons from the O—H bond toward the nonmetal atom, E, favor breakage of the O—H bond and increased acid strength. These factors include a high electronegativity of the central atom, E, and an increased number of

TABLE 17-6
Classification of some oxides

Acidic		Basic	Amphoteric	
Representative elements				
Cl_2O	P_4O_{10}	Na_2O	BeO	GeO
SO_2	CO_2	K_2O	Al_2O_3	Sb_2O_3
SO_3	SiO_2	MgO	SnO	
N_2O_5	B_2O_3	CaO	PbO	
Transition elements				
CrO_3		Sc_2O_3	ZnO	
MoO_3		TiO_2	Cr_2O_3	
WO_3		ZrO_2		
Mn_2O_7				

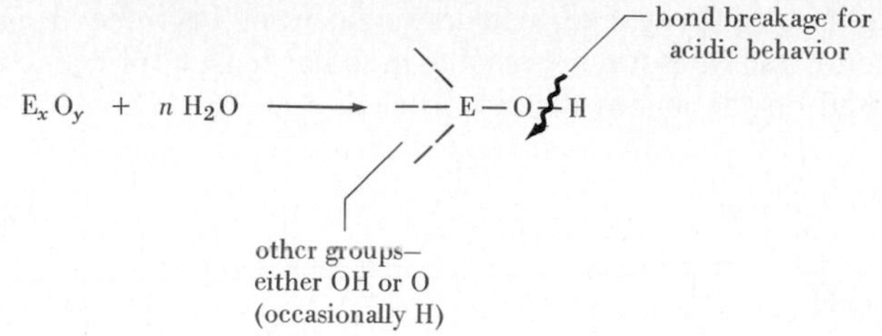

FIGURE 17-5
Formation and acidic behavior of an oxoacid.

$$E_x O_y \ + \ n\,H_2O \ \longrightarrow \ \text{—}\text{—}E\text{—}O\text{—}H$$

bond breakage for acidic behavior

other groups— either OH or O (occasionally H)

oxygen atoms (not OH groups) bonded directly to the E atom. The fact that HNO_3 is a strong acid whereas H_3PO_4 is weak is consistent with these statements.

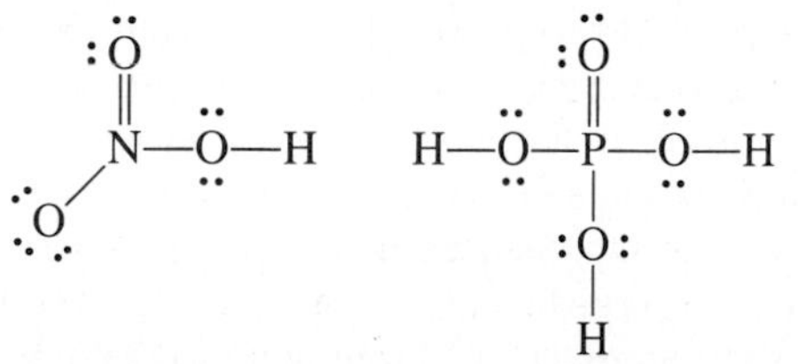

strong   weak

In the series of oxoacids of chlorine, as the oxidation state and formal charge on the Cl atom increase, this atom becomes more of a "positive charge" center. Electron density is drawn away from the O—H bond, leading to increased acid strength. The variation in acid strengths is from HOCl, quite a weak acid, to $HOClO_3$ (perchloric acid), one of the strongest known.

O.S. = +1   O.S. = +3   O.S. = +5   O.S. = +7
F.C. = 0   F.C. = +1   F.C. = +2   F.C. = +3

These Lewis structures suggest that the formulas HOCl, HOClO, . . . , are more accurate representations of these oxoacids than HClO, $HClO_2$, ...

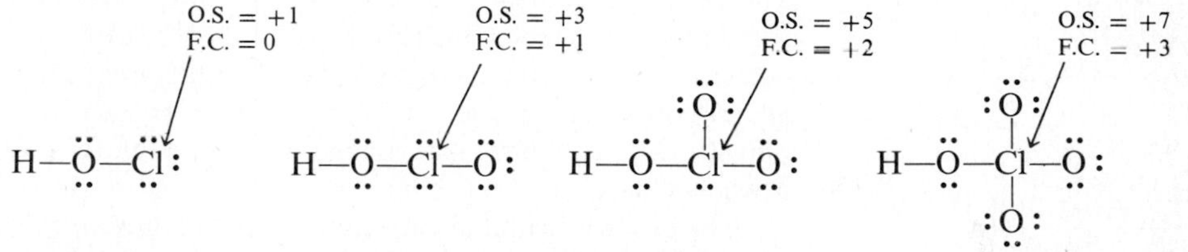

$K_a = 3 \times 10^{-8}$ $K_a = 1.2 \times 10^{-2}$ $K_a = 1 \times 10^3$ $K_a = 1 \times 10^8$
hypochlorous acid   chlorous acid   chloric acid   perchloric acid
HOCl (HClO)   HOClO (HClO_2)   HOClO_2 (HClO_3)   HOClO_3 (HClO_4)

The following generalization often can be applied to oxoacids with the formula, $EO_m(OH)_n$ (where E is the central atom). If

$m = 0$, $K_{a_1} \simeq 10^{-7}$; $m = 1$, $K_{a_1} \simeq 10^{-2}$; $m = 2$, K_{a_1} is large; $m = 3$, K_{a_1} is very large.

Note how well these approximate values of K_a agree with the data listed for the oxoacids of chlorine.

Strengths of Organic Acids. Ethanol and acetic acid both have an O—H group bonded to a carbon atom, but acetic acid is a much stronger acid than is ethanol.

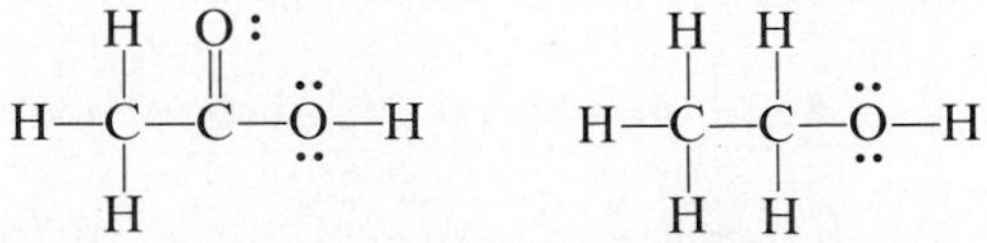

acetic acid, $pK_a = 4.76$ ethanol, $pK_a = 15.9$

(17.39)

We might rationalize that the carbonyl oxygen atom in acetic acid ($\!>\!C\!\!=\!\!O$), being highly electronegative, withdraws electrons from the O—H bond. The result is that the proton can be lost more readily. Another reason for acetic acid being the stronger acid is based on the anions that are formed.

$$(17.40)$$

acetate ion   ethoxide ion

In acetate anion resonance occurs. Two plausible structures can be written in which the double bond is shifted from one O atom to the other. The net effect is that each carbon-to-oxygen bond is a "$\frac{3}{2}$" bond and each O atom carries "$\frac{1}{2}$" negative charge. In short, the excess unit of negative charge is spread out. This reduces the ability of either O atom to attract a proton and makes acetate ion a moderately weak Brønsted–Lowry base. Ethoxide ion, on the other hand, has no resonance possibilities. The unit of negative charge is centered on a single O atom. The ion is a much stronger base than acetate ion. If a conjugate base is strong, then the corresponding acid is weak (recall Table 17-5).

The substitution of groups can also have an effect on the strength of an organic acid. Replacement of one H atom by a Cl atom in acetic acid produces monochloroacetic acid.

$$(17.41)$$

monochloroacetic acid, $pK_a = 2.87$

The highly electronegative Cl atom helps to draw electrons away from the O—H bond. The bond is weakened; the proton is lost more readily; and the acid is a stronger acid than acetic acid. Viewed from the standpoint of the anion, the electronegative Cl atom causes an additional spreading out of the unit of negative charge beyond that described in (17.40). Monochloroacetate ion has less attraction for a proton than does acetate ion; it is a weaker base.

The electron-withdrawing power of certain groups in molecules is called the **inductive effect.** The inductive effect is strongest when the substituent group is adjacent to the carboxyl group ($-C\overset{O}{\underset{O-H}{\diagup}}$). The effect falls off rapidly as the substituent is moved farther away on a hydrocarbon chain. The benzene ring system (phenyl group) exhibits an inductive effect and accounts for the greater acid strength of phenol (17.42) relative to ethanol.

$$(17.42)$$

phenol, $pK_a = 10.0$

Electronegative substituents on the phenyl group may also exhibit an inductive effect, causing the acid to become somewhat stronger.

Example 17-16   Which member of each pair of acids is the stronger?

(a) $CH_2FC\overset{O}{\underset{OH}{\diagup}}$ or $CH_2ClC\overset{O}{\underset{OH}{\diagup}}$

(I)   (II)

(b) $ClCH_2CH_2C$ $\overset{O}{\underset{OH}{}}$ or CH_3CHClC $\overset{O}{\underset{OH}{}}$

(I)   (II)

(c) (benzene ring)—OH   or   (benzene ring)—OH with Br

(I)   (II)

Solution
(a) Because F is more electronegative than Cl, it should exert a stronger pull on the electrons in the O—H bond than does Cl. Monofluoroacetic acid (I) is a stronger acid than monochloroacetic acid (II).
(b) The location from which the Cl atom exerts the strongest inductive effect is directly adjacent to the carboxyl group. Compound II (2-chloropropanoic acid) is more acidic than compound I (3-chloropropanoic acid).
(c) The electronegative Br atom draws electrons away from the O—H group. o-Bromophenol (II) is a stronger acid than phenol (I).

SIMILAR EXAMPLES: Review Problem 17; Exercises 40, 41.

Summary

Much of solution chemistry can be explained by acid–base theories. The Brønsted–Lowry theory describes acid–base reactions in terms of proton transfer. The tendencies for substances to lose or gain protons, that is, their relative acid and base strengths, determine the direction in which an acid–base reaction will occur—from the stronger to the weaker acid/base combination. The Lewis acid–base theory views an acid–base reaction in terms of covalent bond formation between an electron pair acceptor (acid) and an electron pair donor (base). Its greatest use is in situations where reactions occur in the absence of a solvent or in which an acid contains no hydrogen atoms.

In the self-ionization of water a proton is transferred from one water molecule to another, producing the ions H_3O^+ and OH^-. The equilibrium constant for this self-ionization is the ion product of water, K_w. Constancy of the ion product permits a calculation of $[OH^-]$ from a known value of $[H_3O^+]$, and vice versa. Useful shorthand designations for $[H_3O^+]$ and $[OH^-]$ are the symbols pH and pOH. In any aqueous solution at 25°C, pH + pOH = 14.00.

Strong acids ionize completely in aqueous solutions, producing H_3O^+. In aqueous solution, a strong base ionizes completely, producing OH^-. With weak acids and weak bases ionization of the molecular acid or base does not go to completion. Equilibrium exists between the nonionized acid or base and its ions, and is described through the ionization constant, K_a or K_b. Some weak acids produce more than one proton per molecule, and they do so in a stepwise fashion. That is, polyprotic acid molecules first lose one proton; acid anions then lose a second proton; and so on. Distinct ionization constants K_{a_1}, K_{a_2}, . . . apply to each step. Calculations involving ionization equilibria are similar to those introduced for gas-phase equilibria in Chapter 15, although some additional considerations are necessary for polyprotic acids.

In reactions between ions and water—sometimes called hydrolysis reactions—the ions can be treated as weak acids or bases. Values of K_a and K_b for these reactions can be established from the fact that the product of the ionization constant of a weak acid (or base) and that of its conjugate base (or acid) is equal to K_w for water. Hydrolysis accounts for the observation that many salt solutions are not pH neutral (pH $\neq$ 7).

Molecular structure establishes whether a substance will have acidic, basic, or amphoteric properties. In addition, molecular structure affects whether an acid or base is strong or weak. Differences in molecular structure among similar types of compounds often can be used to predict relative values of K_a or K_b.

Learning Objectives

As a result of studying Chapter 17, you should be able to

1. Describe the similarities and differences among the Arrhenius, Brønsted–Lowry, and Lewis theories of acids and bases.

2. Identify Brønsted–Lowry conjugate acids and bases and write equations to represent acid–base reactions.

3. Identify Lewis acids and bases and write equations for acid-base reactions involving them.

4. Describe the nature of the proton in aqueous solution, with special attention to the hydronium ion, H_3O^+.

5. Name the common strong acids and bases.

6. Calculate ionic concentrations in solutions of strong electrolytes, and relate $[H_3O^+]$ and $[OH^-]$ through K_w.

7. Use the basic relationships among $[H_3O^+]$, $[OH^-]$, pH, and pOH in numerical calculations.

8. Identify a weak acid or weak base, write a chemical equation to represent its ionization, and set up its ionization constant expression.

9. Calculate, from appropriate data, ionization constants, concentrations of nonionized weak acids or bases, or concentrations of their ions in solution.

10. Describe the ionization of a polyprotic acid in aqueous solution and calculate the concentrations of the different species present in such a solution.

11. Predict which ions hydrolyze and whether aqueous salt solutions are acidic, basic, or neutral.

12. Calculate values of K_a for cations and K_b for anions from ionization constants of their conjugates and K_w of water.

13. Calculate the pH values of aqueous salt solutions in which hydrolysis occurs.

14. Predict the direction of acid–base reactions from the relative strengths of Brønsted–Lowry acids and bases.

15. Predict whether certain oxides and hydroxo compounds are acidic, basic, or amphoteric.

16. Predict the relative strengths of acids and bases from a knowledge of their molecular structures.

Some New Terms

An **acid anhydride** is an oxide that reacts with water to form an acid.

Amphoterism refers to the ability of certain oxides and hydroxo compounds to act either as acids or bases.

In the **Arrhenius acid–base theory** an acid produces H^+ and a base produces OH^- in aqueous solution.

A **base anhydride** is an oxide that reacts with water to form a base.

Brønsted–Lowry theory describes acids as proton donors and bases as proton acceptors.

A **conjugate base** remains after a Brønsted–Lowry acid has lost a proton. A **conjugate acid** is formed when a Brønsted–Lowry base gains a proton. Every acid has its conjugate base and every base, its conjugate acid.

Degree of ionization (or percent ionization) refers to the extent to which molecules of a weak acid or base have ionized. The degree of ionization increases as the weak electrolyte solution is diluted.

Hydrolysis is a special name given to acid–base reactions in which the acids and bases are ions and water molecules. As a result of hydrolysis, for many salt solutions pH $\neq$ 7.

Hydronium ion, H_3O^+, is the principal form in which protons are found in aqueous solution. The terms ''hydrogen ion'' and ''hydronium ion'' are often used synonymously.

An **ionization constant** is an equilibrium constant describing the ionization of a weak acid or weak base. The symbol K_a is used for weak-acid ionizations and K_b for weak-base ionizations.

The **ion product of water**, K_w, is the product of $[H_3O^+]$ and $[OH^-]$ in pure water or an aqueous solution. This product has a unique value which depends only on temperature. At 25°C, $K_w = 1.0 \times 10^{-14}$.

Lewis acid–base theory considers an acid to be an electron pair acceptor and a base, an electron pair donor. An acid–base reaction consists of the formation of a covalent bond between the acid and the base.

An **oxoacid** is an acid in which the ionizable hydrogen atom(s) is bonded through an oxygen atom to a central nonmetal atom, that is, E—O—H. Other groups bonded to the central atom are either additional —OH groups or O atoms (or occasionally, H atoms).

pH is a shorthand designation for $[H_3O^+]$ in a solution. It is defined as pH $= -\log [H_3O^+]$.

pK is a shorthand designation for an ionization constant; p$K = -\log K$. pK values are useful when comparing the relative strengths of acids or bases.

pOH relates to the $[OH^-]$ in an aqueous solution; pOH $= -\log [OH^-]$.

A **polyprotic acid** is capable of losing more than a single proton per molecule in acid–base reactions. Protons are lost in a stepwise fashion, with the first proton being the most readily lost.

Self-ionization is an acid–base reaction in which one solvent molecule acts as an acid and donates a proton to another solvent molecule acting as a base.

Suggestions for Further Study

BURKE, J. D., "On Calculating [H$^+$]," *J. Chem. Educ.*, **53**, 79 (1976).

GIGUÈRE, P. A., "The Great Fallacy of the H$^+$ Ion and the True Nature of H$_3$O$^+$," *J. Chem. Educ.*, **56**, 571 (1979).

HOUSE, J. E., and R. C. REITER, "Errors in Calculating Hydrogen Ion Concentrations," *J. Chem. Educ.*, **45**, 679 (1968).

KOLB, D., "Acids and Bases," *J. Chem. Educ.*, **55**, 459 (1978).

KOLB, D., "The pH Concept," *J. Chem. Educ.*, **56**, 49 (1979).

LESSLEY, S. D., and R. O. RAGSDALE, "Trends in the Acidities of Some Binary Hydrides in Aqueous Solution," *J. Chem. Educ.*, **53**, 19 (1976).

MYERS, R. T., "Strength of the Hydrohalic Acids," *J. Chem. Educ.*, **53**, 17 (1976).

Review Problems

1. For each of the following identify the acids and bases involved in both the forward and reverse directions.
- (a) $HOBr + H_2O \rightleftharpoons H_3O^+ + OBr^-$
- (b) $HSO_4^- + H_2O \rightleftharpoons H_3O^+ + SO_4^{2-}$
- (c) $HS^- + H_2O \rightleftharpoons H_2S + OH^-$
- (d) $C_6H_5NH_3^+ + OH^- \rightleftharpoons C_6H_5NH_2 + H_2O$

2. Indicate whether each of the following is a Lewis acid or base: (a) OH^-; (b) $B(OH)_3$; (c) $AlCl_3$; (d) CH_3NH_2. (*Hint:* You may find it helpful to draw Lewis structures.)

3. Calculate $[H_3O^+]$ and $[OH^-]$ for each solution.
- (a) 0.0030 M HCl
- (b) 0.045 M NaOH
- (c) 0.0015 M Sr(OH)$_2$
- (d) 7.2×10^{-3} M HNO$_3$

4. What is the pH of each of the following solutions? (a) 1.0×10^{-3} M HCl; (b) 0.000180 M HBr; (c) 1.15×10^{-3} M NaOH; (d) 4.1×10^{-4} M NaOH.

5. What is the pOH of each of the following solutions?
- (a) 1.0×10^{-2} M NaOH
- (b) 0.0068 M LiOH
- (c) 0.00520 M Ba(OH)$_2$
- (d) 3.51×10^{-4} M HCl

6. What is $[H_3O^+]$ in a solution with (a) pH = 6.0; (b) pH = 3.15; (c) pH = 0.65; (d) pOH = 4.10; (e) pOH = 11.15?

7. What is the pH of a water solution containing 1.06 g Ba(OH)$_2 \cdot$8H$_2$O in 575 mL of solution?

8. What is $[H_3O^+]$ in a solution obtained by mixing 25.00 mL 0.150 M HNO$_3$ and 10.00 mL 0.412 M KOH? (*Hint:* Is the solution acidic, basic, or neutral?)

9. In an aqueous solution prepared by dissolving 0.355 mol butyric acid (HC$_4$H$_7$O$_2$) in 715 mL of solution, it is found that $[H_3O^+] = [C_4H_7O_2^-] = 2.73 \times 10^{-3}$ M. What is the value of K_a for the ionization of butyric acid?

$$HC_4H_7O_2 + H_2O \rightleftharpoons H_3O^+ + C_4H_7O_2^- \qquad K_a = ?$$

10. Ionization constants of three acids are listed below.

$$HC_2H_3O_2 + H_2O \rightleftharpoons H_3O^+ + C_2H_3O_2^- \qquad K_a = 1.74 \times 10^{-5}$$
acetic acid

$$HC_8H_7O_2 + H_2O \rightleftharpoons H_3O^+ + C_8H_7O_2^- \qquad K_a = 4.9 \times 10^{-5}$$
phenylacetic acid

$$HC_6H_4ClO + H_2O \rightleftharpoons H_3O^+ + C_6H_4ClO^- \qquad K_a = 3.2 \times 10^{-9}$$
o-chlorophenol

- (a) What is $[C_2H_3O_2^-]$ in 0.815 M HC$_2$H$_3$O$_2$?
- (b) What is the pH of 0.105 M HC$_8$H$_7$O$_2$?
- (c) What molar concentration of *o*-chlorophenol is necessary to produce a solution with pH = 4.86? (*Hint:* What are the values of $[H_3O^+]$ and $[C_6H_4ClO^-]$ in this solution?)

11. Calculate $[(CH_3)_3NH^+]$ in an aqueous solution that is 1.52 M (CH$_3$)$_3$N (trimethylamine).

$$(CH_3)_3N + H_2O \rightleftharpoons (CH_3)_3NH^+ + OH^- \qquad K_b = 6.2 \times 10^{-5}$$

12. What is the (a) degree of ionization and (b) percent ionization of propionic acid in a solution that is 0.25 M HC$_3$H$_5$O$_2$?

$$HC_3H_5O_2 + H_2O \rightleftharpoons H_3O^+ + C_3H_5O_2^- \qquad K_a = 1.34 \times 10^{-5}$$

13. For an 0.015 M solution of the weak diprotic acid H$_2$CO$_3$, calculate (a) $[H_3O^+]$, (b) $[HCO_3^-]$, and (c) $[CO_3^{2-}]$. (Use data from Table 17-3.)

14. Predict whether each of the following aqueous solutions is acidic, basic, or neutral: (a) KCl; (b) NH$_4$NO$_3$; (c) NaNO$_2$; (d) NaI; (e) Ca(OCl)$_2$.

15. Calculate the pH of an aqueous solution that is 0.25 M NH$_4$Cl.

$$NH_4^+ + H_2O \rightleftharpoons H_3O^+ + NH_3 \qquad K_a = 5.75 \times 10^{-10}$$

16. From data in Table 17-2 determine a value of (a) K_a for C$_5$H$_5$NH$^+$; (b) K_b for CHO$_2^-$; (c) K_b for C$_6$H$_5$O$^-$.

17. Which is the more acidic of each of the following pairs of acids? (a) HBr *or* HI; (b) HOClO *or* HOI; (c) H$_3$CCH$_2$COOH *or* Cl$_3$CCH$_2$COOH; (d) I$_3$CCH$_2$CH$_2$COOH *or* H$_3$CCH$_2$CF$_2$COOH.

Exercises

Brønsted–Lowry theory of acids and bases

1. According to the Brønsted–Lowry theory, which of the following would you expect to be acidic and which basic? **(a)** HNO_2; **(b)** OCl^-; **(c)** NH_2^-; **(d)** NH_4^+; **(e)** $CH_3NH_3^+$.

2. What are the conjugate acids of the following bases? **(a)** OH^-; **(b)** Cl^-; **(c)** OCl^-; **(d)** CN^-.

3. The following acids are all *monoprotic* (yielding a single proton). Write the formula of the conjugate base of each acid; **(a)** HIO_4; **(b)** $HC_3H_5O_2$; **(c)** C_6H_5COOH; **(d)** $C_6H_5NH_3^+$.

4. Substances that can either lose or gain protons are said to be *amphiprotic*. Which of the following are amphiprotic? **(a)** OH^-; **(b)** NH_3; **(c)** H_2O; **(d)** HS^-; **(e)** NO_3^-; **(f)** HCO_3^-; **(g)** HSO_4^-; **(h)** HNO_3.

5. The Brønsted–Lowry theory can be applied to acid–base reactions in nonaqueous solvents. Indicate whether each of the following would be an acid, a base, or either in pure liquid acetic acid, $HC_2H_3O_2$, as a solvent: **(a)** $C_2H_3O_2^-$; **(b)** H_2O; **(c)** $HC_2H_3O_2$; **(d)** $HClO_4$. (*Hint:* Think of analogous situations in water or ammonia.)

Lewis theory of acids and bases

6. Each of the following is a Lewis acid–base reaction. Which is the acid and which is the base? Explain.
  (a) $SO_3 + H_2O \rightarrow H_2SO_4$
  (b) $Al(OH)_3(s) + OH^-(aq) \rightarrow [Al(OH)_4]^-$

7. Show that in each of the following practical cases a Lewis acid–base reaction is involved. Identify the acid and the base. (*Hint:* Draw Lewis electronic structures as necessary.)
  (a) $SO_2(g)$ can be removed from the exhaust gases of a power plant by allowing it to combine with lime, $CaO(s)$. The result is the ionic compound $CaSO_3(s)$.
  (b) $CO_2(g)$ can be removed from confined quarters (such as spacecraft) by allowing it to combine with an alkali metal hydroxide, e.g.,
    $CO_2(g) + LiOH(s) \rightarrow LiHCO_3(s)$.

Strong acids, strong bases, and pH

8. What is $[H_3O^+]$ in a solution obtained by dissolving 312 cm^3 $HCl(g)$, measured at 30°C and 740 mmHg, in 3.25 L of water solution?

9. What is the pH of a solution obtained by mixing 215 mL of 3.00×10^{-3} M HI and 475 mL of 6.40×10^{-2} M HCl?

10. What volume of concentrated HCl(aq) that is 36.0% HCl, by mass, and has a density of 1.18 g/cm^3 is required to produce 6.55 L of solution with pH = 1.85?

★11. A saturated aqueous solution of $Mg(OH)_2$ has a pH of 10.53. What is the solubility of $Mg(OH)_2$, expressed in mg/L?

★12. Can a solution of pH 8 be prepared by dissolving HCl in water? If it is possible to do so, indicate how. If it is not possible, indicate why not.

Neutralization

13. What volume of 0.151 M HCl is required to exactly neutralize 25.1 mL 0.218 M NaOH?

14. A 21.22-mL sample of a KOH(aq) solution is required to exactly neutralize 25.00 mL 0.1051 M HBr. What is the molar concentration of the KOH(aq)?

15. 25.00 mL of a HNO_3(aq) solution with a pH of 2.52 is mixed with 25.00 mL of a KOH(aq) solution with a pH of 12.05. What is the pH of the final solution? (*Hint:* Is the solution acidic, basic, or neutral?)

Weak acids, weak bases, and pH (Use data from Table 17-2 as necessary.)

16. Normal caproic acid, $HC_6H_{11}O_2$, found in small amounts in coconut and palm oils, is used in making artificial flavors. A saturated aqueous solution of the acid contains 11 g/L and has pH = 2.94. Calculate K_a for the acid.

17. The pH of an 0.250 M aqueous solution of β-picoline, an organic solvent used in the chemical industry, is found to be 10.86. What is K_b for this base?

$$CH_3C_5H_4N + H_2O \rightleftharpoons CH_3C_5H_4NH^+ + OH^- \qquad K_b = ?$$

18. The compound *o*-nitrophenol, $HC_6H_4NO_3$, is slightly soluble in water and ionizes as a weak acid. A saturated solution of *o*-nitrophenol has pH = 4.53. What is the solubility of *o*-nitrophenol in water, expressed in g/L?

$$HC_6H_4NO_3 + H_2O \rightleftharpoons H_3O^+ + C_6H_4NO_3^- \qquad K_a = 5.9 \times 10^{-8}$$

19. The active ingredient in aspirin is acetylsalicylic acid.

$$HC_9H_7O_4 + H_2O \rightleftharpoons H_3O^+ + C_9H_7O_4^- \qquad K_a = 2.75 \times 10^{-5}$$

What is the pH of the solution obtained by dissolving two aspirin tablets in 250 mL water? Assume that each tablet contains 0.32 g of acetylsalicylic acid.

20. An aqueous solution of dichloroacetic acid, prepared by dissolving 0.10 mol $HC_2HCl_2O_2$ per liter of solution, has pH = 1.30. What is K_a for this acid?

21. A particular vinegar is found to contain 6.0% acetic acid by mass. What mass of this vinegar should be added to 1.00 L of water to produce a solution with pH = 4.62?

$$HC_2H_3O_2 + H_2O \rightleftharpoons H_3O^+ + C_2H_3O_2^- \qquad K_a = 1.74 \times 10^{-5}$$

22. What is $[C_2H_2ClO_2^-]$ in a solution containing 4.5×10^{-3} mol of monochloroacetic acid per liter?

$$HC_2H_2ClO_2 + H_2O \rightleftharpoons H_3O^+ + C_2H_2ClO_2^-$$
$$K_a = 1.35 \times 10^{-3}$$

23. The organic base piperidine is found in small amounts in black pepper. What is the pH of 287 mL of a water solution containing 118 mg of piperidine?

$$C_5H_{11}N + H_2O \rightleftharpoons C_5H_{11}NH^+ + OH^- \qquad K_b = 1.6 \times 10^{-3}$$

Degree of ionization (Use data from Table 17-2 as necessary.)

24. What must be the total molarity of $HC_2H_3O_2$ in an aqueous solution if the acid is to be 1.00% ionized?

25. The % ionization found for acetic acid in Example 17-10 was 0.42% in 1.0 M $HC_2H_3O_2$, 1.3% in 0.10 M $HC_2H_3O_2$, and 4.2% in 0.010 M $HC_2H_3O_2$. Should we expect to find that the percent ionization is 13% in 0.0010 M $HC_2H_3O_2$ and 42% in 0.00010 M $HC_2H_3O_2$? Explain.

26. What is the % ionization of trichloroacetic acid in an 0.050 M $HC_2Cl_3O_2$ solution?

$$HC_2Cl_3O_2 + H_2O \rightleftharpoons H_3O^+ + C_2Cl_3O_2^- \qquad K_a = 2.0 \times 10^{-1}$$

★27. From the observation that 0.0500 M vinylacetic acid has a freezing point of $-0.096°C$, determine K_a for this acid.

$$HC_4H_5O_2 + H_2O \rightleftharpoons H_3O^+ + C_4H_5O_2^- \qquad K_a = ?$$

Polyprotic acids (Use data from Table 17-3 as necessary.)

28. For the following solutions of $H_2S(aq)$, determine $[H_3O^+]$, $[HS^-]$, and $[S^{2-}]$: **(a)** 0.075 M H_2S; **(b)** 0.0050 M H_2S; **(c)** 1.0×10^{-5} M H_2S.

29. The assumptions on page 518 refer to a *di*protic acid. Consider the *tri*protic acid H_3A that ionizes as follows.

$$H_3A + H_2O \rightleftharpoons H_3O^+ + H_2A^- \qquad K_{a_1}$$
$$H_2A^- + H_2O \rightleftharpoons H_3O^+ + HA^{2-} \qquad K_{a_2}$$
$$HA^{2-} + H_2O \rightleftharpoons H_3O^+ + A^{3-} \qquad K_{a_3}$$

   (a) What relationship would you expect among K_{a_1}, K_{a_2}, and K_{a_3}; that is, which is largest, smallest?
   (b) Under what conditions will $[HA^{2-}] = K_{a_2}$?
   (c) Will $[A^{3-}] = K_{a_3}$? Explain.

30. Phosphoric acid, H_3PO_4, ionizes in three steps. Use assumptions similar to those on page 518 to calculate the following in 0.100 M $H_3PO_4(aq)$: $[H_3O^+]$, $[H_2PO_4^-]$, $[HPO_4^{2-}]$, and $[PO_4^{3-}]$. (*Hint:* Refer to Exercise 29.)

31. The antimalarial drug quinine, $C_{20}H_{24}O_2N_2$, a *diprotic base*, has a water solubility of 1.00 g per 1900 mL of solution. $K_{b_1} = 1.08 \times 10^{-6}$ and $K_{b_2} = 1.5 \times 10^{-10}$.
   (a) Write equations to represent the ionization equilibria corresponding to K_{b_1} and K_{b_2}.
   (b) What is the pH of saturated aqueous quinine?

32. What is **(a)** $[H_3O^+]$ in 0.750 M $H_2SO_4(aq)$; **(b)** $[SO_4^{2-}]$ in 1.10 M $H_2SO_4(aq)$?

33. Calculate $[H_3O^+]$, $[HSO_4^-]$, and $[SO_4^{2-}]$ in 0.0100 M $H_2SO_4(aq)$. (*Hint:* Do the usual simplifying assumptions hold?)

Hydrolysis (Use data from Tables 17-2, 17-3, and 17-4, as necessary.)

34. Sodium benzoate, $NaC_7H_5O_2$, is a commonly used food preservative, at a concentration of about 0.10%, by mass. Determine the pH of a 0.10%, by mass, solution of sodium benzoate in water.

35. It is desired to produce an aqueous solution of pH = 8.75 by dissolving one of the following salts in water. Which salt would you use and at what molar concentration? **(a)** NH_4Cl; **(b)** $KHSO_4$; **(c)** KNO_2; **(d)** $NaNO_3$.

36. Pyridine, C_5H_5N, forms a salt, pyridinium hydrochloride, as a result of a reaction with HCl. Write an ionic equation to represent the hydrolysis of the pyridinium ion, and calculate the pH of 0.0482 M $C_5H_5NH^+Cl^-$.

37. For each of the following ions, write equations to represent ionization and hydrolysis. Then use data from Table 17-3 to predict whether each ion makes the solution acidic or basic. **(a)** HSO_3^-; **(b)** HS^-; **(c)** $HC_2O_4^-$.

Molecular structure and acid strength (Use data from tables in this chapter, as necessary.)

38. Based on the Brønsted–Lowry theory, explain why
   (a) Acetic acid is a strong acid in $NH_3(l)$ but a weak acid in $H_2O(l)$.
   (b) Ammonia is a strong base in $HC_2H_3O_2(l)$ but a weak base in $H_2O(l)$.

39. Predict the direction favored in each of the following acid–base reactions. That is, does the reaction tend to go more in the forward or the reverse direction?
   (a) $HBr + OH^- \rightleftharpoons H_2O + Br^-$
   (b) $HSO_4^- + NO_3^- \rightleftharpoons HNO_3 + SO_4^{2-}$
   (c) $CH_3OH + C_2H_3O_2^- \rightleftharpoons HC_2H_3O_2 + CH_3O^-$
   (d) $HC_2H_3O_2 + CO_3^{2-} \rightleftharpoons HCO_3^- + C_2H_3O_2^-$
   (e) $HNO_2 + ClO_4^- \rightleftharpoons HClO_4 + NO_2^-$
   (f) $H_2CO_3 + CO_3^{2-} \rightleftharpoons HCO_3^- + HCO_3^-$

40. Arrange the following in the order of increasing acid strength.

(a) HI

(b) HCl

(c) $H{-}\overset{\displaystyle H}{\underset{\displaystyle H}{C}}{-}\overset{\displaystyle O}{C}{-}O{-}H$

(d) $Cl{-}\overset{\displaystyle H}{\underset{\displaystyle H}{C}}{-}\overset{\displaystyle O}{C}{-}O{-}H$

(e) $F{-}\overset{\displaystyle F}{\underset{\displaystyle F}{C}}{-}\overset{\displaystyle O}{C}{-}O{-}H$

(f) $I{-}\overset{\displaystyle H}{\underset{\displaystyle H}{C}}{-}\overset{\displaystyle O}{C}{-}O{-}H$

41. Which is the stronger base, propylamine, $CH_3CH_2CH_2NH_2$, or aniline, ⬡$-NH_2$? (*Hint:* Compare the conjugate acids and recall the situation involving ethanol and phenol.)

42. In comparing the acid strengths of acetic acid and ethanol, we reasoned from the standpoint of the anions. That is, because of the increased spread of the negative charge, acetate ion is a weaker base than ethoxide ion. Show that this same line of reasoning can be applied in describing the strengths of oxoacids.

43. Phosphorous acid has the formula H_3PO_3, but it is listed in Table 17-3 as a *diprotic* acid (two ionizable hydrogen atoms). Also, notice that K_{a_1} for phosphorous acid is about the same as K_{a_1} for phosphoric acid. Propose a Lewis structure for H_3PO_3 that is consistent with these facts.

44. Use the general rule on page 527 **(a)** to estimate the value of K_{a_1} for H_3AsO_4; **(b)** to write a Lewis structure for hypophosphorous acid, H_3PO_2, for which $pK_{a_1} = 1.1$.

Additional Exercises

1. Calculate $[H_3O^+]$ and pH in **(a)** $0.0087\ M$ HI(aq); **(b)** $0.0018\ M$ $Ca(OH)_2$(aq); **(c)** saturated $Ba(OH)_2$(aq) [containing 39 g $Ba(OH)_2 \cdot 8\ H_2O$ per liter].

2. What volume of $0.467\ M$ NaOH must be diluted to 1.00 L to produce a solution with pH $= 11.81$?

3. In the manner used in equation (17.15), represent the self-ionization of the following liquid solvents: **(a)** NH_3; **(b)** HF; **(c)** CH_3OH; **(d)** $HC_2H_3O_2$; **(e)** H_2SO_4.

4. Draw Lewis structures corresponding to the cations and anions formed in each of the self-ionization reactions of the preceding exercise.

5. What are the pH values of the following solutions of benzoic acid?
   (a) $0.40\ M$ $HC_7H_5O_2$
   (b) $1.0 \times 10^{-3}\ M$ $HC_7H_5O_2$
   (c) $1.4 \times 10^{-5}\ M$ $HC_7H_5O_2$

6. The pH of a saturated solution of phenol at room temperature is 4.90. What is the solubility of phenol in grams per liter of saturated solution?

$$HC_6H_5O + H_2O \rightleftharpoons H_3O^+ + C_6H_5O^- \qquad K_a = 1.6 \times 10^{-10}$$

7. With information from this chapter, but without performing calculations, arrange the following $0.01\ M$ solutions in order of *increasing* pH: NH_3; KOH; H_2SO_4; $HC_2H_3O_2$; $Ba(OH)_2$; HCl.

8. For $0.165\ M$ $H_2C_2O_4$(aq), calculate $[H_3O^+]$, $[HC_2O_4^-]$, and $[C_2O_4^{2-}]$. For oxalic acid, $K_{a_1} = 5.4 \times 10^{-2}$; $K_{a_2} = 5.4 \times 10^{-5}$.

9. Which compound is the more *acidic* in each of the following pairs? Explain your reasoning.
   (a) $ClCH_2CH_2OH$ *or* CH_3CH_2OH

(b) CH_3NH_2 *or* $ClCH_2NH_2$
(c) $ClCH_2CH_2COOH$ *or* CH_2FCOOH.

10. The following generalization is sometimes used in assessing the acidic, basic, or amphoteric character of a representative element oxide: If the quantity $\sqrt{\text{cation charge}/\text{cation radius}}$ (pm) is less than 0.22 the oxide is basic, greater than 0.32, acidic, and of an intermediate value, amphoteric. Show that this generalization leads to the same conclusion as stated about MgO, BeO, B_2O_3, and Al_2O_3 in Section 17-8.

★11. Show that when $[H_3O^+]$ of a solution is reduced to one-half of its original value, the pH value increases by 0.30 unit, *regardless of the initial pH*. Can it also be said that when any solution is diluted to one-half of its original concentration its pH value increases by 0.30 unit? Explain.

★12. What total concentration of $HC_2H_3O_2$ must be placed in aqueous solution if the solution is to have the same freezing point as $0.150\ M$ $HC_2H_2ClO_2$ (monochloroacetic acid)?

★13. What is the pH of a solution that is $0.315\ M$ $HC_2H_3O_2$ and $0.250\ M$ $HCHO_2$?

★14. Data are given in Section 17-2 on K_w as a function of temperature. Use these data to verify that the heat of neutralization of strong acids by strong bases is about -56 kJ/mol H_2O produced.

★15. Calculate the pH of $1.0\ M$ NH_4CN(aq). (*Hint:* Identify the six species [excluding H_2O] whose concentrations in this solution are "unknown," and find six equations relating these unknowns. Three equations are equilibrium constant expressions—K_w, K_a for HCN, and K_b for NH_3. Another two conditions are that $[NH_3] + [NH_4^+] = 1.0\ M = [HCN] + [CN^-]$. A simplifying assumption is that $[NH_4^+] = [CN^-]$.)

Self-Test Questions

For questions 1 through 8 select the single item that best completes each statement.

1. The number of moles of hydroxide ion, OH^-, in 0.300 L of $0.0050\ M$ $Ba(OH)_2$ is (a) 0.0075; (b) 0.0015; (c) 0.0030; (d) 0.0050.

2. A solution has a pH of 5.00. In this solution $[OH^-]$ must be (a) $1.0 \times 10^{-9}\ M$; (b) $1.0 \times 10^{-7}\ M$; (c) greater than $1.0 \times 10^{-5}\ M$; (d) $1.0 \times 10^{-5}\ M$.

3. $[H_3O^+]$ in $0.10\ M$ $HC_3H_5O_2$ (propionic acid) is
(a) equal to $[H_3O^+]$ in $0.10\ M$ HNO_2 (nitrous acid)
(b) less than $[H_3O^+]$ in $0.10\ M$ HI (hydroiodic acid)
(c) greater than $[H_3O^+]$ in $0.10\ M$ HBr (hydrobromic acid)
(d) equal to 0.10 mol H_3O^+/L

4. An aqueous solution is $0.10\ M$ in the weak base, CH_3NH_2. In this solution (a) $[H_3O^+] = 0.10\ M$; (b) $[OH^-] = 0.10\ M$; (c) pH < 7; (d) pH < 13.

5. Of the following solutions the one that is most alkaline (basic) is (a) $NaCl(aq)$; (b) $NH_4Cl(aq)$; (c) $KNO_3(aq)$; (d) $KC_2H_3O_2(aq)$.

6. Of the following solutions the one that is most acidic is (a) $NaHSO_4(aq)$; (b) $NaCl(aq)$; (c) $NaC_2H_3O_2(aq)$; (d) $Na_2S(aq)$.

7. One of the following ions is amphiprotic, that is, has the ability to lose or gain a proton in aqueous solution. That one is (a) HCO_3^-; (b) CO_3^{2-}; (c) Cl^-; (d) NH_4^+.

8. The reaction of acetic acid, $HC_2H_3O_2$, with a base will proceed furthest toward completion (to the right) when that base is (a) H_2O; (b) NH_3; (c) Cl^-; (d) $HClO_4$.

9. Arrange the following 0.01 M aqueous solutions in order of *decreasing* $[H_3O^+]$: $NH_3(aq)$; $HNO_3(aq)$; $NaNO_2(aq)$; $H_2SO_4(aq)$; $NaOH(aq)$; $NaCl(aq)$; $Ba(OH)_2(aq)$; $NH_4ClO_4(aq)$; $HC_2H_3O_2(aq)$.

10. What mass of benzoic acid, $HC_7H_5O_2$, must be dissolved in 250.0 mL of water solution to produce a pH = 2.60?

$$HC_7H_5O_2 + H_2O \rightleftharpoons H_3O^+ + C_7H_5O_2^- \qquad K_a = 6.3 \times 10^{-5}$$

11. Explain why you would expect
  (a) $HClO_4$ to be a stronger acid than HNO_3.
  (b) trifluoroacetic acid, $HC_2F_3O_2$, to be a stronger acid than acetic acid, $HC_2H_3O_2$.
  (c) *o*-chloroaniline to be a weaker base than aniline.

o-chloroaniline   aniline

12. Explain the following statements.
  (a) $[H_3O^+]$ in a strong acid solution doubles as the total concentration of the acid is doubled, whereas for a weak acid $[H_3O^+]$ increases only by $\sqrt{2}$.
  (b) Even though two H_3O^+ ions are produced for every S^{2-} ion produced by the ionization of H_2S in aqueous solution, $[S^{2-}]$ is not simply $\frac{1}{2}[H_3O^+]$; it is *much* smaller than this.

18

Additional Aspects of Acid–Base Equilibria

In Chapter 17 we limited our discussion to equilibrium approached from the standpoint of a pure acid or base dissolved in water (that is, equilibrium approached "from the left"). Many practical applications of acid–base equilibria involve the ionization of acids and bases in the presence of one of the products of the ionization, with a resultant effect known as the common ion effect. The common ion effect—the focus of this chapter—will help us understand how certain solutions, buffer solutions, are able to resist changes in their pH. The common ion effect is also important in the action of acid–base indicators. An understanding of these indicators, together with other equilibrium principles, makes possible a detailed description of acid–base titrations.

18-1 The Common Ion Effect in Acid–Base Equilibria

The ionization of HCl(aq) produces H_3O^+, an ion also produced by the self-ionization of water. H_3O^+ is an ion common to both ionization processes; it is a **common ion.** All H_3O^+ ions in HCl(aq), regardless of their source, participate in the self-ionization equilibrium of H_2O. That is, in the expression, $K_w = [H_3O^+][OH^-] = 1.0 \times 10^{-14}$, $[H_3O^+]$ is the total concentration of hydronium ion. In HCl(aq), because of the high concentration of H_3O^+ produced by the strong acid, $[OH^-]$ is *very much smaller* than in pure water. The self-ionization equilibrium is shifted far to the left. A strong base also represses the self-ionization of water by increasing the concentration of the common ion, OH^-. Actually, the self-ionization of water occurs to such a slight extent that even weak acids and bases exhibit a common ion effect on this equilibrium.

$$H_2O + H_2O \rightleftharpoons H_3O^+ + OH^-$$

◀ In the presence of an acid or base, equilibrium condition shifts.

Solutions of Weak Acids and Strong Acids.   Just as it represses the self-ionization of water, a strong acid represses the ionization of a weak acid, through the common ion H_3O^+. For example,

$$HC_2H_3O_2 + H_2O \rightleftharpoons H_3O^+ + C_2H_3O_2^- \qquad K_a = 1.74 \times 10^{-5} \qquad (18.1)$$

◀ In the presence of a strong acid, equilibrium shifts

Example 18-1 (a) Determine $[H_3O^+]$ and $[C_2H_3O_2^-]$ in 0.100 M $HC_2H_3O_2$. (b) Then determine these same quantities in a solution that is 0.100 M in both $HC_2H_3O_2$ and HCl.

Solution

(a) This calculation is performed in the same way as those of Chapter 17. That is,

$$HC_2H_3O_2 \; + \; H_2O \; \rightleftharpoons \; H_3O^+ \; + \; C_2H_3O_2^-$$

placed in solution:	0.100 M	—	—
changes:	$- x$ M	$+ x$ M	$+ x$ M
at equilibrium:	$(0.100 - x)$ M	x M	x M

If we assume that x is quite small, then $0.100 - x \approx 0.100$ and

$$K_a = \frac{[H_3O^+][C_2H_3O_2^-]}{[HC_2H_3O_2]} = \frac{x^2}{0.100} = 1.74 \times 10^{-5}$$

$$x^2 = 1.74 \times 10^{-6} \qquad x = [H_3O^+] = [C_2H_3O_2^-] = 1.32 \times 10^{-3} \; M$$

(b) Here we arrange the equilibrium data from part (a) and the information about the common ion, H_3O^+, below equation (18.1) in the following fashion.

$$HC_2H_3O_2 \; + \; H_2O \; \rightleftharpoons \; H_3O^+ \; + \; C_2H_3O_2^-$$

from weak acid:	$(0.100 - x)$ M	x M	x M
from 0.100 M HCl:	—	0.100 M	—
at equilibrium:	$(0.100 - x)$ M	$(0.100 + x)$ M	x M

Because of the repression of the ionization of $HC_2H_3O_2$ by the strong acid, HCl, we should expect x to be very small. This means that $(0.100 - x) \approx (0.100 + x) \approx 0.100$, and leads to the expression

$$K_a = \frac{[H_3O^+][C_2H_3O_2^-]}{[HC_2H_3O_2]} = \frac{(0.100 + x)(x)}{0.100 - x} = \frac{0.100(x)}{0.100} = 1.74 \times 10^{-5}$$

$$x = [C_2H_3O_2^-] = 1.74 \times 10^{-5} \; M \qquad \begin{aligned} 0.100 + x &= [H_3O^+] = 0.100 \; M \\ 0.100 - x &= [HC_2H_3O_2] = 0.100 \; M \end{aligned}$$

SIMILAR EXAMPLES: Review Problem 1; Exercise 2a, d.

To summarize the effect of HCl on the ionization of $HC_2H_3O_2$ calculated in Example 18-1,

1. All the H_3O^+ in a mixture of a strong acid and a weak acid is assumed to come from the strong acid. (This assumption will not be valid, however, if the strong acid is very dilute and/or if K_a of the weak acid is large.)

2. In the *absence* of any additional solute, $[C_2H_3O_2^-] = [H_3O^+]$. In the *presence* of a strong acid, $[C_2H_3O_2^-] \ll [H_3O^+]$. In Example 18-1 the common ion from the strong acid, H_3O^+, caused nearly a 100-fold decrease in $[C_2H_3O_2^-]$—from 1.32×10^{-3} M to 1.74×10^{-5} M.

The ideas presented here are also applicable to mixtures of weak and strong bases. In these solutions $[OH^-]$ is established by the strong base, and ionization of the weak base is repressed.

Solutions of Weak Acids and Salts of Weak Acids. The salt of a weak acid is a strong electrolyte—it is completely dissociated into its ions in aqueous solution. One of the ions, the *anion,* is a common ion in the ionization equilibrium of the weak acid. The presence of this common ion represses the ionization of the weak acid. For example,

$$NaC_2H_3O_2(aq) \longrightarrow Na^+ + C_2H_3O_2^-$$

$$HC_2H_3O_2 + H_2O \rightleftharpoons H_3O^+ + C_2H_3O_2^- \qquad K_a = 1.74 \times 10^{-5} \qquad (18.2)$$

⬅ In the presence of acetate salts, equilibrium shifts.

In mixtures of acetic acid and sodium acetate, $[C_2H_3O_2^-]$ has a "high" value and $[H_3O^+]$ is diminished by the presence of this common ion, as illustrated in Example 18-2.

Example 18-2 Calculate $[H_3O^+]$ and $[C_2H_3O_2^-]$ in a solution that is 0.100 M both in $HC_2H_3O_2$ and in $NaC_2H_3O_2$.

Solution. The setup below is very similar to that used in Example 18-1(b), except that $NaC_2H_3O_2$ provides the common ion.

$$HC_2H_3O_2 + H_2O \rightleftharpoons H_3O^+ + C_2H_3O_2^-$$

from weak acid:	$(0.100 - x)\,M$	$x\,M$	$x\,M$
from 0.100 M NaC$_2$H$_3$O$_2$:	—	—	0.100 M
at equilibrium:	$(0.100 - x)\,M$	$x\,M$	$(0.100 + x)\,M$

Again we should expect x to be very small, because of the repression of the ionization of $HC_2H_3O_2$ caused by the anion, $C_2H_3O_2^-$. $(0.100 - x) \simeq (0.100 + x) \simeq 0.100$, and

$$K_a = \frac{[H_3O^+][C_2H_3O_2^-]}{[HC_2H_3O_2]} = \frac{(x)(0.100 + x)}{0.100 - x} = \frac{(x)\cancel{0.100}}{\cancel{0.100}} = 1.74 \times 10^{-5}$$

$$x = [H_3O^+] = 1.74 \times 10^{-5}\,M \qquad 0.100 + x = [C_2H_3O_2^-] = 0.100\,M$$

SIMILAR EXAMPLES: Review Problem 2; Exercises 2b, e, 3.

As in Example 18-1, the assumption made about the magnitude of x in Example 18-2 is equivalent to saying that all the anion in a mixture of a weak acid and its salt comes from the salt. Again, this assumption is valid *unless* the salt concentration is very low and/or K_a of the weak acid is large.

Once more we should add that the same ideas illustrated in Example 18-2 apply to mixtures of weak bases and their salts. (In this case the salt provides a common cation.) A summary of the common ion effect on the ionization of weak acids and bases is provided in Table 18-1.

TABLE 18-1
Summary of the common ion effect on the ionization of weak acids and bases

Ionization equilibrium	In the *absence* of common ion	Common ion	Its effect	Common ion	Its effect
weak acid (e.g., HA = HC$_2$H$_3$O$_2$) $HA + H_2O \rightleftharpoons H_3O^+ + A^-$	$[H_3O^+] = [A^-]$	H_3O^+	reduces $[A^-]$; $[H_3O^+] > [A^-]$	A^-	reduces $[H_3O^+]$; $[A^-] > [H_3O^+]$
weak base (e.g., B = NH$_3$) $B + H_2O \rightleftharpoons BH^+ + OH^-$	$[BH^+] = [OH^-]$	OH^-	reduces $[BH^+]$; $[OH^-] > [BH^+]$	BH^+	reduces $[OH^-]$; $[BH^+] > [OH^-]$

18-2 Buffer Solutions

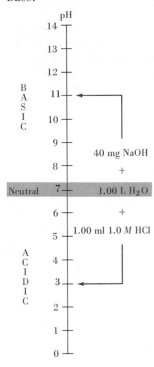

If even a very small quantity of either an acid or a base is added to pure water, the pH changes dramatically. Pure water, as outlined below and illustrated in Figure 18-1, has no resistance to a change in pH—it has no buffer capacity.

Pure water has a $pH = 7$. Addition of 0.001 mol HCl (e.g., 1.00 mL of 1.0 M HCl) to 1.00 L of pure water produces $[H_3O^+] = 10^{-3}\ M$ and $pH = 3$. Addition of 0.001 mol NaOH (e.g., 40 mg NaOH) to 1.00 L of pure water produces $[OH^-] = 10^{-3}\ M$, and $pH = 11$. (18.3)

The acetic acid–sodium acetate solution of Example 18-2 *does* have a capacity to resist a change in pH—it is a **buffer** solution. Let us see how this fact can be established, first qualitatively and then quantitatively. In any solution of acetic acid and sodium acetate, the K_a expression for acetic acid must apply.

$$K_a = \frac{[H_3O^+][C_2H_3O_2^-]}{[HC_2H_3O_2]} = 1.74 \times 10^{-5} \tag{18.4}$$

Suppose we take a solution that has equal concentrations of $HC_2H_3O_2$ and $NaC_2H_3O_2$ (as in Example 18-2). In such a solution $[H_3O^+] = 1.74 \times 10^{-5}\ M$. Now consider adding a small quantity of either an acid (H_3O^+) or a base (OH^-) to this solution. The following reactions will occur in the buffer solution.

$$\underset{\text{(small added amt.)}}{C_2H_3O_2^- +\qquad H_3O^+} \longrightarrow HC_2H_3O_2 + H_2O \tag{18.5}$$

$$\underset{\text{(small added amt.)}}{HC_2H_3O_2 +\qquad OH^-} \longrightarrow C_2H_3O_2^- + H_2O \tag{18.6}$$

In the first instance (18.5) a small amount of the salt ($C_2H_3O_2^-$) is converted to the acid ($HC_2H_3O_2$). In the second instance (18.6) a small amount of the acid is converted to the salt. In both cases, essentially all that has occurred is a very slight change in the ratio $[C_2H_3O_2^-]/[HC_2H_3O_2]$ in equation (18.4). This change is so slight that $[H_3O^+]$ remains essentially constant. The acetic acid–sodium acetate buffer maintains a constant pH at a value of about 5.

To obtain a buffer in basic solution, a weak base and its salt can be used, as illustrated in Example 18-3.

Example 18-3   Show that an aqueous solution of NH_3 and NH_4Cl is a basic buffer solution.

Solution.   The key to buffer action, as we have just seen, is the presence in solution of a component that reacts with small amounts of added acid and a component that reacts with small amounts of added base. In the present case these components are NH_3 and NH_4^+, respectively.

$$\underset{\text{(small added amt.)}}{NH_3 +\qquad H_3O^+} \longrightarrow NH_4^+ + H_2O$$

$$\underset{\text{(small added amt.)}}{NH_4^+ +\qquad OH^-} \longrightarrow NH_3 + H_2O$$

In any aqueous solution containing NH_3 and NH_4^+ the ionization equilibrium of NH_3 is established $(NH_3 + H_2O \rightleftharpoons NH_4^+ + OH^-)$ and the K_b expression must be satisfied.

$$K_b = \frac{[NH_4^+][OH^-]}{[NH_3]} = 1.74 \times 10^{-5}$$

If approximately equal concentrations of the base (NH_3) and its salt (NH_4^+) are maintained in solution, then $[OH^-] \simeq 1 \times 10^{-5}$, $pOH \simeq 5$, and $pH \simeq 9$. Thus, ammonia–ammonium chloride solutions are basic buffer solutions.

SIMILAR EXAMPLES: Review Problem 3; Exercises 6, 7a.

Calculating the pH of Buffer Solutions.   To calculate the pH of a buffer solution requires that equilibrium concentrations of the weak acid (or weak base) and its salt be substituted into an ionization constant expression, as illustrated in Example 18-4.

Example 18-4.   Calculate the pH of a buffer solution that is $0.15\ M\ HC_2H_3O_2$–$0.50\ M\ NaC_2H_3O_2$.

Solution.   In the usual fashion we can write

$$HC_2H_3O_2 + H_2O \rightleftharpoons H_3O^+ + C_2H_3O_2^- \qquad K_a = 1.74 \times 10^{-5}$$

placed in solution:	$0.15\ M$	—	$0.50\ M$
changes:	$-x\ M$	$+x\ M$	$+x\ M$
at equilibrium:	$(0.15 - x)\ M$	$x\ M$	$(0.50 + x)\ M$

Assume that $x \ll 0.15$ (which means also that $x \ll 0.50$).

$$\frac{[H_3O^+][C_2H_3O_2^-]}{[HC_2H_3O_2]} = \frac{x(0.50 + x)}{(0.15 - x)} = \frac{x(0.50)}{(0.15)} = 1.74 \times 10^{-5}$$

$$x = [H_3O^+] = 5.2 \times 10^{-6} \qquad pH = -\log(5.2 \times 10^{-6}) = 5.28$$

SIMILAR EXAMPLES: Review Problem 4; Exercises 7c, 14a.

A Useful Equation for Buffer Solutions.   A slightly different approach to Example 18-4 involves making some modifications to the ionization constant expression for $HC_2H_3O_2$ before substituting equilibrium concentrations. That is, start with the familiar equation

$$K_a = \frac{[H_3O^+][C_2H_3O_2^-]}{[HC_2H_3O_2]} \tag{18.4}$$

and express the right side as a product of $[H_3O^+]$ and the *ratio* of concentrations, $[C_2H_3O_2^-]/[HC_2H_3O_2]$.

$$K_a = [H_3O^+] \times \frac{[C_2H_3O_2^-]}{[HC_2H_3O_2]} \tag{18.7}$$

Take the *negative logarithm* of each side of equation (18.7).

$$-\log K_a = -\log \left([H_3O^+] \times \frac{[C_2H_3O_2^-]}{[HC_2H_3O_2]}\right) = -\log [H_3O^+] - \log \frac{[C_2H_3O_2^-]}{[HC_2H_3O_2]}$$

Recall that $pH = -\log [H_3O^+]$ and $pK_a = -\log K_a$.

$$pK_a = pH - \log \frac{[C_2H_3O_2^-]}{[HC_2H_3O_2]} \tag{18.8}$$

Finally, solve equation (18.8) for pH.

$$pH = pK_a + \log \frac{[C_2H_3O_2^-]}{[HC_2H_3O_2]} \tag{18.9}$$

For acetic acid, $pK_a = 4.76$.

$$pH = 4.76 + \log \frac{[C_2H_3O_2^-]}{[HC_2H_3O_2]} \tag{18.10}$$

If we make the same substitutions into equation (18.10) that we made in Example 18-4, that is, $[C_2H_3O_2^-] = 0.50 + x \simeq 0.50\ M$, and $[HC_2H_3O_2] = 0.15 - x \simeq 0.15\ M$, we obtain the same pH value as we did there.

$$pH = 4.76 + \log \frac{0.50}{0.15} = 4.76 + \log 3.3 = 4.76 + 0.52 = 5.28$$

An expression similar to equation (18.9) can be derived for a solution of a weak base (B) and its cation (BH^+). That is,

$$pOH = pK_b + \log \frac{[BH^+]}{[B]} \tag{18.11}$$

Applied to the NH_3/NH_4^+ buffer, and with $pK_b = 4.76$, equation (18.11) becomes

$$pOH = 4.76 + \log \frac{[NH_4^+]}{[NH_3]} \tag{18.12}$$

In applying equations such as (18.9) and (18.11), it is customary to substitute stoichiometric concentrations (the concentrations of substances placed in solution) for equilibrium concentrations. That is, the weak acid or base is assumed to be essentially nonionized in the presence of its salt. This assumption is valid only if (a) neither buffer component is too dilute and (b) K_a (or K_b) is not too large. To avoid misapplication of these equations, it is necessary to test this assumption in some cases. A situation in which these equations greatly facilitate matters is when a series of related calculations is necessary (see Section 18-4).

Preparing Buffer Solutions. According to equation (18.10), to prepare a buffer solution with pH = 4.76 should be an easy matter. Dissolve acetic acid and sodium acetate in water in equimolar concentrations (such as $[HC_2H_3O_2] = [C_2H_3O_2^-] = 1.00\ M$ or $[HC_2H_3O_2] = [C_2H_3O_2^-] = 0.50\ M$). Then, since $[C_2H_3O_2^-]/[HC_2H_3O_2] = 1.00$ and $\log 1.00 = 0$, pH = 4.76.

What if we wish to prepare a buffer solution with pH = 5.10? One method would be to find a weak acid having $pK_a = 5.10$. A solution with equimolar concentrations of this acid and its salt would have pH = 5.10. Finding such an acid is not always possible. An alternative method, illustrated in Example 18-5, is to use an acid with $pK_a \simeq 5$ and an appropriate ratio of concentrations of anion and acid (but not an equimolar ratio).

Example 18-5 What mass of $NaC_2H_3O_2$ must be dissolved in 0.300 L of 0.250 M $HC_2H_3O_2$ to produce a solution with pH = 5.10? (Assume that the solution volume remains constant at 0.300 L.)

Solution. In this problem we are dealing with 0.250 M $HC_2H_3O_2$, a fairly concentrated solution of a weak acid for which K_a is quite small (i.e., $K_a = 1.74 \times 10^{-5}$). If $[C_2H_3O_2^-]$ in this solution is also fairly high, then all the conditions for substituting stoichiometric concentrations for equilibrium concentrations in equation (18.10) have been met. By "fairly high" concentrations of $HC_2H_3O_2$ and $C_2H_3O_2^-$ we mean about 1000 times larger than $[H_3O^+]$, which for pH = 5.10 is about $1 \times 10^{-5}\ M$. The value of $[C_2H_3O_2^-]$ calculated with equation (18.10) below is seen to meet this requirement.

$$pH = 4.76 + \log \frac{[C_2H_3O_2^-]}{[HC_2H_3O_2]}$$

$$5.10 = 4.76 + \log \frac{[C_2H_3O_2^-]}{[HC_2H_3O_2]} \quad \text{and} \quad \log \frac{[C_2H_3O_2^-]}{0.250} = 5.10 - 4.76 = 0.34$$

$$\frac{[C_2H_3O_2^-]}{0.250} = \text{antilogarithm } 0.34 = 2.2$$

$$[C_2H_3O_2{}^-] = 0.250 \times 2.2 = 0.55 \; M$$

$$\text{no. g NaC}_2\text{H}_3\text{O}_2 = 0.300 \text{ L} \times \frac{0.55 \text{ mol NaC}_2\text{H}_3\text{O}_2}{\text{L}} \times \frac{82.0 \text{ g NaC}_2\text{H}_3\text{O}_2}{1 \text{ mol NaC}_2\text{H}_3\text{O}_2}$$

$$= 14 \text{ g NaC}_2\text{H}_3\text{O}_2$$

SIMILAR EXAMPLES: Review Problems 5, 6; Exercises 8, 15a.

Reaction (18.13) is between a strong acid (HCl) and the salt of a weak acid ($NaC_2H_3O_2$). Acetate ion, a base, accepts a proton from H_3O^+ to form molecules of the weak acid, $HC_2H_3O_2$. Equilibrium is displaced far to the right.

$$H_3O^+ + Cl^- + Na^+ + C_2H_3O_2{}^- \rightleftharpoons HC_2H_3O_2 + Na^+ + Cl^- + H_2O \qquad (18.13)$$

This method of preparing a buffer is illustrated in Color Section F and Exercise 15.

If HCl is in excess, the result of reaction (18.13) is a mixture of a weak acid and strong acid. If $NaC_2H_3O_2$ is in excess, the result is a mixture of a weak acid and its salt—*a buffer solution*. An alternative method of preparing a buffer solution, then, is to mix a strong acid with an excess of a salt of a weak acid (or a strong base with an excess of a salt of a weak base).

Calculating pH Changes in Buffer Solutions.   We should make explicit here an idea that we have used implicitly before: A calculation based on an equilibrium constant expression often must be preceded by a stoichiometric calculation. In Example 17-5, in which we calculated $[H_3O^+]$ in a solution following the reaction of a strong acid and a strong base, the stoichiometric calculation was of a limiting and an excess reagent; the equilibrium calculation was based on K_w. To calculate the change in pH produced by adding a small amount of acid or base to a buffer solution, buffer action is represented through chemical equations (such as 18.5 or 18.6). Stoichiometric principles are applied to establish how much of one buffer component is consumed and how much of the other is produced. From these results a new set of equilibrium concentrations is established and the ionization constant expression or an equation derived from it (such as 18.9 or 18.11) is used to calculate $[H_3O^+]$ or pH. These two aspects of the pH calculation are highlighted in Example 18-6.

Example 18-6   What is the effect on the pH of 1.00 L of 0.100 M $HC_2H_3O_2$–0.100 M $NaC_2H_3O_2$ buffer solution by adding (a) 0.001 mol H_3O^+? (b) 0.001 mol OH^-?

Solution
(a) First we must calculate the concentrations of the buffer components before and after addition of the 0.001 mol H_3O^+.
  Step 1.   Stoichiometric calculation.

	$C_2H_3O_2{}^-$	+	H_3O^+	$\longrightarrow$	$HC_2H_3O_2$	+	H_2O
in original buffer:	0.100 M		—		0.100 M		
add:			+ 0.001 M				
changes:	− 0.001 M		− 0.001 M		+ 0.001 M		
in final buffer:	0.099 M		?		0.101 M		

The pH of the buffer solution after addition of the acid can now be calculated with equation (18.10).
  Step 2.   Equilibrium calculation.

$$\text{pH} = 4.76 + \log \frac{0.099}{0.101} = 4.76 + \log 0.98 = 4.76 - 0.01 = 4.75$$

The effect of adding 0.001 mol H_3O^+ to 1.00 L of the original buffer solution is to lower its pH, but only from 4.76 to 4.75.
(b) The same two-step approach is used as in part (a).

Step 1.   Stoichiometric calculation.

$$HC_2H_3O_2 + OH^- \longrightarrow C_2H_3O_2^- + H_2O$$

in original buffer:	0.100 M	—	0.100 M
add:		+ 0.001 M	
changes:	− 0.001 M	− 0.001 M	+ 0.001 M
in final buffer:	0.099 M	≃ 0	0.101 M

Step 2.   Equilibrium calculation.

$$\text{pH} = 4.76 + \log \frac{0.101}{0.099} = 4.76 + \log 1.02 = 4.76 + 0.01 = 4.77$$

The effect of adding 0.001 mol OH^- to 1.00 L of the original buffer solution is to raise its pH, but only from 4.76 to 4.77.

SIMILAR EXAMPLES: Exercises 13, 14b, c, 15b.

The differing effects of acid or base on pure water and on a buffered solution are illustrated in Color Section F.

When added to 1.00 L of pure water the amounts of acid and base used in Example 18-6 (i.e., 0.001 mol) produce pH changes of *4 units* (recall Figure 18-1). In the acetic acid–sodium acetate buffer, the corresponding change in pH is only *0.01 unit!*

If a buffer solution is diluted, both $[A^-]$ and $[HA]$ are reduced by the same factor. The *ratio* $[A^-]/[HA]$ remains constant, and so does the pH (pH = pK_a + log $[A^-]/[HA]$). *Resistance to pH changes upon dilution* is another important property of buffer solutions.

Buffer Capacity and Buffer Range.   In Example 18-6 addition of more than 0.100 mol H_3O^+/L would cause essentially complete conversion of $C_2H_3O_2^-$ to $HC_2H_3O_2$. The result would be a mixture of a weak acid ($HC_2H_3O_2$) and a strong acid (the unreacted H_3O^+). The solution would become much more acidic than the original buffer, and the buffering action would be destroyed. In a similar manner, addition of more than 0.100 mol OH^-/L would convert all the $HC_2H_3O_2$ to $C_2H_3O_2^-$, resulting in a mixture of unreacted base (OH^-) and the salt of a weak acid ($NaC_2H_3O_2$). Again the buffering action would be destroyed.

Buffer capacity refers to the amount of acid or base that may be added to a buffer solution before its pH changes appreciably. In general, the maximum capacity to resist pH changes exists when the concentrations of weak acid (or base) and its salt are kept large and approximately equal to one another. The buffer has its maximum capacity at pH = pK_a (or pOH = pK_b). Whenever the ratio of salt to weak electrolyte concentration is either less than about 0.10 or greater than about 10, the buffer loses its effectiveness. Since log 0.10 = −1 and log 10 = +1, this means that the effective buffer range is about one pH unit on either side of the value of pK. For acetic acid–sodium acetate buffers the effective range is from about pH 3.76 to 5.76; for ammonia–ammonium chloride, about pH 8.24 to 10.24.

The Importance of pH Maintenance in Blood.   Maintenance of the proper pH in blood and in intracellular fluids is absolutely crucial to the processes that occur in living organisms. This is primarily because the functioning of enzymes—the catalysts for these processes—is sharply pH-dependent. The normal pH value of blood is 7.4. Severe illness or death can result from sustained variations of a few tenths of a pH unit.

Among the factors that can lead to a condition of acidosis, in which there is a decrease in the pH of blood, are heart failure, kidney failure, diabetes mellitus, persistent diarrhea, or a long-term high-protein diet. A temporary condition of acidosis may result from prolonged, intensive exercise.

Alkalosis, characterized by an increase in the pH of blood, may occur as a result of severe vomiting, hyperventilation (overbreathing, sometimes caused by anxiety or hysteria), or exposure to high altitudes (altitude sickness). In recent studies performed on

Mount Everest on climbers who reached the summit (8848 m = 29,028 ft) without supplemental oxygen, the pH of arterial blood was found to be between 7.7 and 7.8.* Extreme hyperventilation is required to compensate for the very low partial pressures of O_2 (about 43 mmHg) at this altitude.

Blood as a Buffered Solution. A measure of the buffer capacity of human blood is the fact that addition of 0.01 mol HCl to one liter of blood lowers the pH only from 7.4 to 7.2. The same amount of HCl lowers the pH of a saline (NaCl) solution isotonic with blood from 7.0 to 2.0.

Several factors are involved in the control of the pH of blood. A particularly important one is the ratio of dissolved HCO_3^- (bicarbonate ion) to H_2CO_3 (carbonic acid). $CO_2(g)$ is moderately soluble in water, and in aqueous solution reacts only to a limited extent to produce H_2CO_3. Nevertheless, in using K_{a_1} we treat the dissolved CO_2 as if it were completely converted to H_2CO_3. Moreover, although H_2CO_3 is a weak diprotic acid, in the carbonic acid–bicarbonate ion buffer system we deal only with the first ionization step: H_2CO_3 is the weak acid and HCO_3^- is the conjugate base (salt).

$$CO_2(aq) + H_2O \rightleftharpoons H_2CO_3(aq) \tag{18.14}$$

$$H_2CO_3 + H_2O \rightleftharpoons H_3O^+ + HCO_3^- \qquad K_{a_1} = 4.2 \times 10^{-7} \tag{18.15}$$

$$HCO_3^- + H_2O \rightleftharpoons H_3O^+ + CO_3^{2-} \qquad K_{a_2} = 5.6 \times 10^{-11} \tag{18.16}$$

Carbon dioxide enters the blood from tissues as the by-product of metabolic reactions. In the lungs $CO_2(g)$ is exchanged for $O_2(g)$, which is transported throughout the body by the blood.

In blood, the concentration of HCO_3^- is about 20 times greater than that of H_2CO_3. With an equation similar to (18.9) and a value of $pK_{a_1} = -\log (4.2 \times 10^{-7}) = 6.38$, we can write

$$pH = pK_{a_1} + \log \frac{[HCO_3^-]}{[H_2CO_3]} = 6.38 + \log \left(\frac{20}{1}\right) = 6.38 + \log 20$$
$$= 6.38 + 1.30 = 7.68$$

This H_2CO_3/HCO_3^- buffer does not seem to establish the proper pH for blood (pH = 7.4). Furthermore, the large ratio of $[HCO_3^-]$ to $[H_2CO_3]$ seems to place this buffer well outside the range of its maximum buffer capacity. The situation is rather complex, but some of the factors involved are

1. The need to neutralize excess acid (lactic acid produced by exercise) is generally greater than the need to neutralize excess base. The high proportion of HCO_3^- helps in this regard.

2. If additional H_2CO_3 is needed to neutralize excess alkalinity, $CO_2(g)$ in the lungs can be reabsorbed to build up the H_2CO_3 content of the blood.

3. Important contributions to maintaining the pH of blood are made by other components in blood, such as the phosphate buffer, $H_2PO_4^-/HPO_4^{2-}$, with a pH of 7.2.

Additional Applications of Buffer Solutions. Buffer solutions have other important applications, albeit less dramatic than maintaining the pH of blood. Certain chemical reactions may consume, produce, or be catalyzed by H_3O^+ (recall Figure 14-12). To study the kinetics of these reactions, or simply to control their reaction rates, requires that the pH be controlled. This control can be achieved by conducting reactions in buffered solutions. Enzyme reactions are particularly sensitive to pH changes. Protein studies must often be performed in buffered media because the magnitude and kind of electrical charge carried by protein molecules depend on the pH (see Section 27-2). An example of this pH

*J. B. West, "Human Physiology at Extreme Altitudes on Mount Everest," *Science*, **223**, 784 (1984).

FIGURE 18-2
pH ranges for some common indicators.

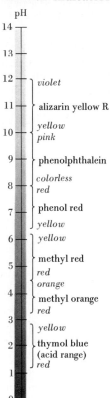

Color changes of thymol blue and several other indicators can be seen in Color Section F.

dependence is the separation of proteins by electrophoresis (recall Figure 12-23). The solubilities of many substances, e.g., hydroxides, carbonates, and sulfides, depend on $[H_3O^+]$; the importance of buffer solutions in controlling the pH of solubility/precipitation processes should become evident in the next chapter.

18-3 Acid–Base Indicators

We have discussed at some length the meaning of pH, how it can be calculated, and how it can be controlled; but we have not yet commented on how it may be measured experimentally. One simple method involves the use of indicators. An **acid–base indicator** is a *weak acid* for which the nonionized acid (HIn) has one color [color(1)] and the anion another [color(2)]. When a small amount of indicator is placed in a solution, depending on whether the ionization equilibrium of the indicator is displaced toward the acid or anion form, the solution acquires either color(1) or color(2). The direction of displacement of equilibrium in reaction (18.17) depends on $[H_3O^+]$, and hence, on pH.

$$\underset{\text{color(1)}}{HIn} + H_2O \rightleftharpoons H_3O^+ + \underset{\text{color(2)}}{In^-} \tag{18.17}$$

To assess the pH range in which an indicator will work, we can again write an expression similar to equation (18.9), that is,

$$pH = pK_a + \log \frac{[In^-]}{[HIn]} \tag{18.18}$$

In general, if 90% or more of an indicator is in the form HIn, the solution in which it is found will assume color(1). If 90% or more is in the form In^-, the solution acquires color(2). These two conditions correspond roughly to the ratios $[In^-]/[HIn] \simeq 0.10$ and $[In^-][HIn] \simeq 10$. The logarithms of these ratios are about -1 and $+1$, respectively. Thus, an indicator changes from color(1) to color(2) over a pH range of about *2 units*. At the middle of this range, that is, with $[HIn] = [In^-]$, $pH = pK_a$. At this midpoint the solution color is a "mixture" of color(1) and color(2). The indicator is said to be undergoing a color change. The colors and pH ranges for several common indicators are illustrated in Figure 18-2.

Thymol blue is a weak *diprotic* acid and undergoes color changes in *two* pH ranges. One range is from pH 1.2 to 2.8, where the color changes from red to yellow. The other range is from pH 8.0 to 9.6, where the color changes from yellow to blue. (Only the acid range is included in Figure 18-2.) Ionization equilibria for thymol blue can be represented as

$$\underset{\text{(color 1)}}{H_2In} + H_2O \rightleftharpoons H_3O^+ + \underset{\text{(color 2)}}{HIn^-}$$

$$\underset{\text{(color 2)}}{HIn^-} + H_2O \rightleftharpoons H_3O^+ + \underset{\text{(color 3)}}{In^{2-}}$$

An acid–base indicator is usually prepared as a solution (in water, ethanol, or some other solvent). In acid–base titrations a small volume (a few drops) of the indicator solution is added to the solution being titrated. In another form, porous paper is impregnated with an indicator solution and dried. When this paper is moistened with the solution being tested, it acquires a color determined by the pH of the solution. This paper is usually called pH paper.

Acid–base indicators find their greatest use in applications where a precise determination of pH is not necessary. For example, they are used in soil-testing kits to establish the pH of soils. Soils are usually acidic in regions of high rainfall and heavy vegetation and alkaline in more arid regions, but the pH can vary considerably with local conditions. If

The color change of phenol red indicator can be seen in Color Section F.

a soil is found to be too acidic for a certain crop, its pH can be raised by adding lime (CaO). To reduce the pH of a soil, the addition of gypsum ($CaSO_4 \cdot 2\ H_2O$) or organic matter is generally effective. In swimming pools, a particular pH is required for the most effective action of chlorinating agents, to prevent growth of algae, to avoid corrosion of the pool plumbing, etc. The preferred pH is generally about 7.4 and phenol red is a common indicator used in testing pool water. If chlorination is accomplished with $Cl_2(g)$, the pool water becomes acidic because of the reaction of Cl_2 with H_2O (see Table 13-8). In this case alkaline substances, such as lime or sodium carbonate, are used to raise the pH. If NaOCl(aq) is used as the chlorinating agent, the pool water becomes alkaline and the pH is adjusted by adding an acid (e.g., HCl or H_2SO_4).

More precise than the use of indicators is the measurement of pH with an electrical measuring instrument called a pH meter, described in Section 20-5.

18-4 Neutralization Reactions and Titration Curves

We have been using the term neutralization to mean the reaction of an acid and a base to from a salt and water; and we have seen that the fundamental reaction that occurs during neutralization is

$$H_3O^+ + OH^- \longrightarrow 2\ H_2O \qquad (18.19)$$

In Example 17-5 we applied stoichiometric principles to determine the excess reagent in the reaction of a strong acid and strong base, so that we could then calculate $[H_3O^+]$ in the resulting solution. An equally important situation is one in which both the acid and base are consumed in the neutralization and *neither* is present in excess. This condition is called the **equivalence point** of the neutralization. To locate the equivalence point in a neutralization requires that careful control be exercised over the addition of base to acid (or acid to base). This, as we first learned in Section 4-4, can be accomplished through the procedure known as titration. But how is the precise equivalence point in a titration to be found? We are now in a position to answer this question.

In a titration one of the solutions to be neutralized, say the acid, is placed in a flask or beaker. The other solution, the base, is contained in a buret and is added to the acid, first rapidly and then dropwise, until the equivalence point of the titration is reached. One attempts to locate the equivalence point through the color change of an acid–base indicator. The point in a titration where the indicator changes color is called the **end point** of the indicator. What is needed is to match the indicator end point with the equivalence point of the neutralization. This can be done if we can find an indicator whose color change occurs over a pH range that includes the pH corresponding to the equivalence point.

Titration of a Strong Acid by a Strong Base.   For the titration of 25.00 mL 0.1000 *M* HCl (a strong acid) by 0.1000 *M* NaOH (a strong base) we can calculate the pH of the solution at various points during the titration. These data can be plotted in the form of solution pH versus volume of added base, a form called a **titration curve.** From this curve we can establish the pH at the equivalence point and thus select a suitable indicator for the titration. The necessary calculations are of the type involved in Example 18-7. Note that the calculations of parts (b) and (d) are similar to the calculation of Example 17-5.

Example 18-7   What is the pH at each of the following points in the titration of 25.00 mL 0.1000 *M* HCl by 0.1000 *M* NaOH? **(a)** Before the addition of any NaOH; **(b)** after the addition of 24.00 mL 0.1000 *M* NaOH; **(c)** at the equivalence point; **(d)** after the addition of 26.00 mL 0.1000 *M* NaOH.

Solution

(a) Before the addition of NaOH we are dealing only with 0.1000 M HCl. This solution has $[H_3O^+] = 0.1000$ M and a pH = 1.00.

(b) The total number of moles of H_3O^+ to be titrated is

$$\text{no. mol } H_3O^+ = 0.02500 \text{ L} \times \frac{0.1000 \text{ mol } H_3O^+}{L} = 2.500 \times 10^{-3} \text{ mol } H_3O^+$$

The number of moles of OH^- present in 24.00 mL 0.1000 M NaOH is

$$\text{no. mol } OH^- = 0.02400 \text{ L} \times \frac{0.1000 \text{ mol } OH^-}{L} = 2.400 \times 10^{-3} \text{ mol } OH^-$$

This point in the neutralization reaction can be represented by

$$H_3O^+ \quad + \quad OH^- \quad \longrightarrow \quad 2 H_2O$$

initial:	2.500×10^{-3} mol	—	
add:		2.400×10^{-3} mol	
after reaction:	0.100×10^{-3} mol	—	

The 0.100×10^{-3} mol H_3O^+ is present in 49.00 mL of solution (the original 25.00 mL of acid plus the added 24.00 mL of base).

$$[H_3O^+] = \frac{0.100 \times 10^{-3} \text{ mol } H_3O^+}{0.04900 \text{ L}} = 2.04 \times 10^{-3} M$$

$$\text{pH} = -\log [H_3O^+] = -\log (2.04 \times 10^{-3}) = 2.69$$

(c) At the equivalence point 2.500×10^{-3} mol NaCl has been produced and is present in 50.00 mL of solution. The solution is 0.05000 M NaCl. Since neither Na^+ nor Cl^- hydrolyzes, this solution has a pH = 7.00.

(d) Beyond the equivalence point excess OH^- is present. For example, following the addition of 26.00 mL 0.1000 M NaOH.

$$H_3O^+ \quad + \quad OH^- \quad \longrightarrow \quad 2 H_2O$$

initial:	2.500×10^{-3} mol	—	
add:		2.600×10^{-3} mol	
after reaction:	—	0.100×10^{-3} mol	

$$[OH^-] = \frac{0.100 \times 10^{-3} \text{ mol } OH^-}{0.05100 \text{ L}} = 1.96 \times 10^{-3} M$$

$$\text{pOH} = -\log(1.96 \times 10^{-3}) = 2.71 \qquad \text{pH} = 14.00 - 2.71 = 11.29$$

SIMILAR EXAMPLES: Review Problem 8: Exercises 26a, 29a.

The titration curve for Example 18-7 is shown in Figure 18-3. The most interesting feature of this NaOH–HCl titration curve is that the pH changes quite slowly until just before the equivalence point is reached. At the equivalence point the pH rises very sharply, by approximately 6 units for an addition of only 0.10 mL of base (corresponding to 2 drops). Beyond the equivalence point the pH again changes quite slowly as excess NaOH is added. Any indicator whose color changes in the pH range from about 4 to 10 is suitable for this titration.

Titration of a Weak Acid by a Strong Base. The neutralization of a weak acid by a strong base can be thought of in a slightly different way from that of a strong acid by a strong base. Initially most of the weak acid is present in solution as nonionized molecules, HA, rather than as H_3O^+ and A^-. In the presence of a strong base, protons are transferred directly from the nonionized HA molecules to OH^-. For the neutralization of $HC_2H_3O_2$ by NaOH, the net equation is

$$HC_2H_3O_2 + OH^- \longrightarrow H_2O + C_2H_3O_2^- \tag{18.20}$$

FIGURE 18-3
Titration curve for
strong acid by a strong
base—25.00 mL of
0.1000 *M* HCl by
0.1000 *M* NaOH.

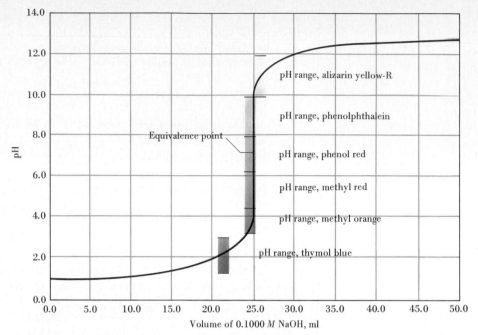

The indicators whose color-change ranges fall along the steep portion of the titration curve are all suitable for this titration. Thymol blue changes color too soon (that is, before 25.00 mL of base has been added); alizarin yellow-R, too late.

Example 18-8 considers the titration of 25.00 mL 0.1000 *M* $HC_2H_3O_2$ by 0.1000 *M* NaOH. Since it takes just as much 0.1000 *M* NaOH to react with 2.500×10^{-3} mol $HC_2H_3O_2$ in reaction (18.20) as it does to react with 2.500×10^{-3} mol H_3O^+ in reaction (18.19), the equivalence point in Example 18-8 comes at the same titrant volume as in Example 18-7. However, the pH at the equivalence point is different in the two cases, as are other details of the titration curves.

Example 18-8 What is the pH at each of the following points in the titration of 25.00 mL 0.1000 *M* $HC_2H_3O_2$ by 0.1000 *M* NaOH? **(a)** Before the addition of any NaOH; **(b)** after the addition of 10.00 mL 0.1000 *M* NaOH; **(c)** after the addition of 12.50 mL 0.1000 *M* NaOH; **(d)** at the equivalence point; **(e)** after the addition of 26.00 mL 0.1000 *M* NaOH.

Solution

(a) The initial pH—that of 0.1000 *M* $HC_2H_3O_2$—can be obtained from the calculation in Example 18-1(a): $pH = -\log (1.3 \times 10^{-3}) = 2.88$.

(b) The total number of moles of acetic acid to be neutralized is

$$\text{no. mol } HC_2H_3O_2 = 0.02500 \text{ L} \times \frac{0.1000 \text{ mol } HC_2H_3O_2}{L}$$

$$= 2.500 \times 10^{-3} \text{ mol } HC_2H_3O_2$$

At this point in the titration the number of moles of OH^- added $= 0.01000$ L $\times$ 0.1000 mol OH^-/L $= 1.000 \times 10^{-3}$ mol OH^-. The amount of acid neutralized is 1.000×10^{-3} mol, and the amount of acetate ion produced is also $1.000 \times$

10^{-3} mol. The amount of unreacted acid is 1.500×10^{-3} mol. The solution components are present in $25.00 + 10.00 = 35.00$ mL $= 0.03500$ L.

$$HC_2H_3O_2 \quad + \quad OH^- \quad \longrightarrow \quad CH_2H_3O_2^- \quad + \quad H_2O$$

initial:	2.500×10^{-3} mol	—		
add:		1.000×10^{-3} mol		
after reaction:	1.500×10^{-3} mol	—	1.000×10^{-3} mol	
concentrations:	$\dfrac{1.500 \times 10^{-3} \text{ mol}}{0.03500 \text{ L}}$	—	$\dfrac{1.000 \times 10^{-3} \text{ mol}}{0.03500 \text{ L}}$	
	$= 4.286 \times 10^{-2} M$		$= 2.857 \times 10^{-2} M$	

The concentrations of the buffer components can be substituted directly into equation (18.10) for a calculation of the pH of the solution.

$$pH = 4.76 + \log \frac{[C_2H_3O_2^-]}{[HC_2H_3O_2]} = 4.76 + \log \frac{2.857 \times 10^{-2}}{4.286 \times 10^{-2}}$$
$$= 4.76 + \log 0.6666 = 4.76 - 0.18 = 4.58$$

(c) The addition of 12.50 mL 0.1000 M NaOH converts one half of the initial $HC_2H_3O_2$ to $C_2H_3O_2^-$. An analysis of the solution composition as in part (b) would indicate that $[HC_2H_3O_2] = [C_2H_3O_2^-]$. $\log([C_2H_3O_2^-]/[HC_2H_3O_2]) = \log 1.00 = 0$. At this half-neutralization point, $pH = pK_a = 4.76$.

(d) At the equivalence point neutralization has just been completed, and the solution is the same as one would get by dissolving 2.500×10^{-3} mol $NaC_2H_3O_2$ in 50.00 mL. In this solution $[C_2H_3O_2^-] = 2.500 \times 10^{-3}/0.05000 = 0.05000$ M. The question is: What is the pH of 0.05000 M $NaC_2H_3O_2$? The answer lies in recognizing that $C_2H_3O_2^-$ hydrolyzes, and then using the method of Example 17-14.

$$C_2H_3O_2^- \quad + \quad H_2O \rightleftharpoons HC_2H_3O_2 + OH^-$$

placed in solution:	0.05000 M		
changes:	$-x$ M	$+x$ M	$+x$ M
equilibrium:	$(0.05000 - x)$ M	x M	x M

Assume that x is very small and that $0.05000 - x \simeq 0.05000$.

$$K_b = \frac{[HC_2H_3O_2][OH^-]}{[C_2H_3O_2^-]} = \frac{x \cdot x}{0.05000} = 5.75 \times 10^{-10}$$

$$x^2 = 2.88 \times 10^{-11} \qquad x = [OH^-] = 5.37 \times 10^{-6} \qquad pOH = -\log(5.37 \times 10^{-6})$$

$$pOH = 5.270 \qquad pH = 14.00 - 5.27 = 8.73$$

(e) After the addition of 26.00 mL 0.1000 M NaOH, the total number of moles of OH^- added $= 0.02600$ L $\times$ 0.1000 mol $OH^-/$L $= 2.600 \times 10^{-3}$ mol OH^-. Of this total amount, 2.500×10^{-3} mol OH^- is consumed to neutralize the $HC_2H_3O_2$. This leaves 0.100×10^{-3} mol OH^- in 51.00 mL of solution.

$$[OH^-] = \frac{0.100 \times 10^{-3} \text{ mol } OH^-}{0.05100 \text{ L}} = 1.96 \times 10^{-3} M$$

$$pOH = 2.708 \qquad pH = 14.00 - 2.71 = 11.29$$

SIMILAR EXAMPLES: Review Problem 9; Exercises 26b, 27, 29b.

The principal features to note in the titration curve of a weak acid by a strong base, illustrated in Figure 18-4, are these.

FIGURE 18-4
Titration curve for
weak acid by a strong
base—25.00 mL of
0.1000 M HC$_2$H$_3$O$_2$ by
0.1000 M NaOH.

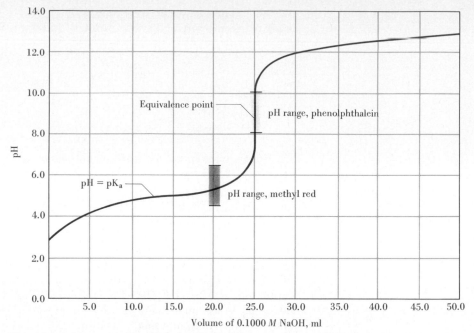

Phenolphthalein is a suitable indicator for this titration, but methyl red is not. When exactly one half of the acid is neutralized, [HC$_2$H$_3$O$_2$] = [C$_2$H$_3$O$_2^-$] and pH = pK_a = 4.76.

1. The initial pH is higher than in the titration curve of a strong acid by a strong base (because the weak acid is only partially ionized).

2. There is a fairly sharp increase in pH at the start of the titration. [The acetate ion resulting from the neutralization reaction (18.20) acts as a common ion and represses the ionization of acetic acid.]

3. Over a long section of the curve preceding the equivalence point, the pH changes quite gradually. (Solutions represented by this portion of the curve contain appreciable concentrations of both HC$_2$H$_3$O$_2$ and C$_2$H$_3$O$_2^-$. These are *buffer* solutions.)

4. The pH at the point where a weak acid is half-neutralized is pH = pK_a. (At half-neutralization, [HC$_2$H$_3$O$_2$] = [C$_2$H$_3$O$_2^-$].)

5. The pH at the equivalence point is greater than 7. (This results from the hydrolysis of C$_2$H$_3$O$_2^-$.)

6. Beyond the equivalence point the titration curve for a weak acid by a strong base is identical to that of a strong acid by a strong base. (In this portion of the titration, the pH is determined only by the concentration of free OH$^-$.)

7. The steep portion of the titration curve at the equivalence point occurs over a shorter pH range (from about pH 7 to 10) than for the titration of a strong acid by a strong base (about pH 4 to 10).

8. The selection of indicators suitable for the titration of a weak acid by a strong base is more limited than for a strong acid by a strong base.

Titration of a Weak Polyprotic Acid.   The most striking visual evidence that a polyprotic acid ionizes in distinctive steps is provided in the neutralization of the acid. For example, in the neutralization of phosphoric acid, essentially all H$_3$PO$_4$ molecules are first converted to the salt, NaH$_2$PO$_4$; then NaH$_2$PO$_4$ is converted to Na$_2$HPO$_4$; and finally, Na$_2$HPO$_4$ is converted to Na$_3$PO$_4$. That is,

$$H_3PO_4 + OH^- \longrightarrow H_2PO_4^- + H_2O$$

is followed by

$$H_2PO_4^- + OH^- \longrightarrow HPO_4^{2-} + H_2O$$

is followed by

$$HPO_4^{2-} + OH^- \longrightarrow PO_4^{3-} + H_2O$$

Corresponding to these three distinctive stages there are three equivalence points. For every mole of H_3PO_4, 1 mol NaOH is required to reach the first equivalence point. At this first equivalence point, the solution is essentially one of NaH_2PO_4. This is an acidic solution because the further ionization of $H_2PO_4^-$

$$H_2PO_4^- + H_2O \rightleftharpoons H_3O^+ + HPO_4^{2-} \qquad K_{a_2} = 6.2 \times 10^{-8} \tag{18.21}$$

occurs to a greater extent than its hydrolysis.

$$H_2PO_4^- + H_2O \rightleftharpoons H_3PO_4 + OH^- \qquad K_b = 1.7 \times 10^{-12} \tag{18.22}$$

An additional mole of NaOH is required to titrate the acid to its second equivalence point. At this second equivalence point, the solution is *basic* because the hydrolysis of HPO_4^{2-}

$$HPO_4^{2-} + H_2O \rightleftharpoons H_2PO_4^- + OH^- \qquad K_b = 1.6 \times 10^{-7} \tag{18.23}$$

occurs to a greater extent than the further ionization of HPO_4^{2-},

$$HPO_4^{2-} + H_2O \rightleftharpoons H_3O^+ + PO_4^{3-} \qquad K_{a_3} = 4.8 \times 10^{-13} \tag{18.24}$$

The complete neutralization of H_3PO_4 to Na_3PO_4 cannot normally be accomplished in a titration, even though a third equivalence point might be expected. This is because the pH of the strongly hydrolyzed Na_3PO_4 solution at the third equivalence point is higher than can normally be attained in an acid–base titration, approaching pH = 13. An Na_3PO_4 solution is nearly as basic as the NaOH solution that would be used to form it in a titration. Thus, the titration curve (see Figure 18-5) has two breaks instead of three, with the breaks equally spaced in terms of volume of titrant (NaOH) used.

FIGURE 18-5
Titration of a weak polyprotic acid by a strong base—10.00 mL of 0.1000 *M* H_3PO_4 by 0.1000 *M* NaOH.

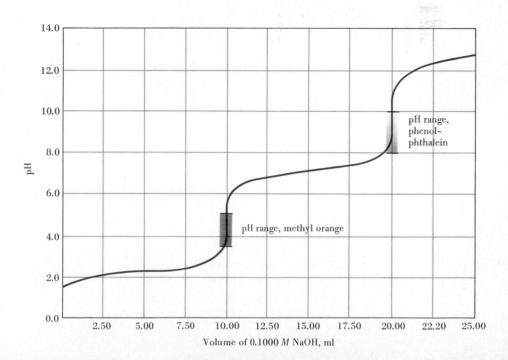

18-5 The pH of Solutions of Salts of Polyprotic Acids

In our discussion of the titration of H_3PO_4 by a strong base we established through qualitative reasoning the approximate pH at the three equivalence points expected in the titration curve. The precise pH values at these three points can all be calculated, but the calculations are more difficult than for the equivalence points of monoprotic acids. Of the three calculations, the most straightforward is that of the pH of $Na_3PO_4(aq)$.

Na_3PO_4 cannot ionize further and the pH of its aqueous solution is determined by the hydrolysis of PO_4^{3-}.

$$PO_4^{3-} + H_2O \rightleftharpoons HPO_4^{2-} + OH^- \qquad \begin{aligned} K_b &= K_w/K_a(HPO_4^{2-}) \\ &= 1.00 \times 10^{-14}/4.8 \times 10^{-13} \\ &= 2.1 \times 10^{-2} \end{aligned} \qquad (18.25)$$

HPO_4^{2-} can also hydrolyze to $H_2PO_4^-$; but K_b for HPO_4^{2-}, which is 1.6×10^{-7}, is much smaller than K_b for PO_4^{3-}. We can assume that hydrolysis of PO_4^{3-} occurs only through the single step represented by equation (18.25).

Example 18-9 What is the pH of $1.0\ M\ Na_3PO_4$?

Solution. In the usual fashion we can write,

$$PO_4^{3-} + H_2O \rightleftharpoons HPO_4^{2-} + OH^- \qquad K_b = 2.1 \times 10^{-2}$$

placed in solution:	$1.0\ M$	—	—
changes:	$-x\ M$	$+x\ M$	$+x\ M$
at equilibrium:	$(1.0 - x)\ M$	$x\ M$	$x\ M$

$$K_b = \frac{[HPO_4^{2-}][OH^-]}{[PO_4^{3-}]} = \frac{x \cdot x}{1.0 - x} = 2.1 \times 10^{-2}$$

Because K_b is relatively large, the usual assumption that $x \ll 1.0$ is not likely to work well here. Solution by the quadratic formula leads to $x = [OH^-] = 0.13\ M$.

$$pOH = -\log[OH^-] = -\log 0.13 = +0.89$$

$$pH = 14.00 - 0.89 = 13.11$$

SIMILAR EXAMPLE: Exercise 33.

From the result of Example 18-9 we can understand the significance of the statement made in the preceding section that $Na_3PO_4(aq)$ is such an alkaline solution that generally it cannot be produced in a titration of $H_3PO_4(aq)$. For this reason a third equivalence point did not appear in the titration curve of Figure 18-5. The result of Example 18-9 also helps to explain why Na_3PO_4 (known commercially as trisodium phosphate or TSP) is a commonly used cleaning agent. One of the properties generally required of a cleaning agent is that it produce an alkaline medium, which helps to solubilize grease and oils.

To calculate the *exact* pH of $NaH_2PO_4(aq)$ or $Na_2HPO_4(aq)$ is more complicated a task than the calculation of Example 18-9. This is because in each case two equilibria must be considered simultaneously—one an ionization and the other a hydrolysis—that is, equations (18.21) and (18.22) for $NaH_2PO_4(aq)$ and (18.23) and (18.24) for $Na_2HPO_4(aq)$. These calculations are left as an exercise for the interested student (Additional Exercise 15), but the results are sufficiently important for us to state them without proof. For solutions that are reasonably concentrated (say $0.10\ M$ or greater), the pH values of these solutions are *independent* of concentration and are given by the following expressions. (pK_a values are from Table 17-3.)

for NaH₂PO₄: $pH = \frac{1}{2}(pK_{a_1} + pK_{a_2}) = \frac{1}{2}(2.23 + 7.21) = 4.72 \qquad (18.26)$

for Na₂HPO₄: $pH = \frac{1}{2}(pK_{a_2} + pK_{a_3}) = \frac{1}{2}(7.21 + 12.32) = 9.76 \qquad (18.27)$

Expressions like (18.26) and (18.27) are generally applicable to salts of polyprotic acids and permit us, for example, to calculate the pH of $NaHCO_3(aq)$.

$$\text{for } NaHCO_3: \quad pH = \tfrac{1}{2}(pK_{a_1} + pK_{a_2}) = \tfrac{1}{2}(6.38 + 10.25) = 8.32 \tag{18.28}$$

$NaHCO_3$ is only a mildly basic substance, which accounts for its use in antacids. Na_2CO_3, on the other hand, produces rather strongly alkaline solutions and is often used commercially as an inexpensive base. Calculation of the pH of $Na_2CO_3(aq)$ is done in the same way as illustrated for $Na_3PO_4(aq)$ in Example 18-9.

18-6 Acid–Base Equilibrium Calculations—A Summary

Between this and the preceding chapter we have considered a wide variety of types of equilibrium calculations, which inevitably seems to lead to the question of what method to employ when facing a given problem-solving situation. Here are some points to consider.

1. What are the equilibrium expressions that must be obeyed? Which are the most significant?
    One equilibrium expression that must be obeyed in every aqueous solution is $K_w = [H_3O^+][OH^-] = 1.00 \times 10^{-14}$, but we have not found it necessary to apply it in every aqueous equilibrium calculation. One situation where it will be significant is if we are asked to calculate $[OH^-]$ in an *acidic* solution or $[H_3O^+]$ in a *basic* solution. [After all, an acid does not produce $OH^-(aq)$, nor does a base produce $H_3O^+(aq)$.] Another situation is one in which the solution is likely to be near pH = 7.
    The ionization of H_3PO_4, a triprotic acid, must be represented through three ionization steps; but if a question concerns $[H_3O^+]$ in $H_3PO_4(aq)$, only the first ionization step is important because this is the step in which essentially all the H_3O^+ is produced. On the other hand, if we are asked to calculate the pH of a solution containing $H_2PO_4^-$ and HPO_4^{2-} ions, K_{a_2} of H_3PO_4 would seem to be the important expression, since this is the equilibrium constant that describes formation of one ion from the other.
2. Are there reactions possible among any of the solution components, and what is their stoichiometry?
    If asked to calculate $[OH^-]$ in a solution that is 0.10 M NaOH and 0.20 M NH_4Cl, before answering that $[OH^-] = 0.10\ M$, we should consider if a solution can be simultaneously 0.10 M in OH^- and 0.20 M in NH_4^+. It cannot, of course, since any solution containing both OH^- and NH_4^+ must also contain NH_3, and in concentrations that conform to K_b for NH_3. What happens is that the NaOH and NH_4Cl react in a 1 : 1 mol ratio to produce NH_3. The question then becomes one of describing a *buffer* solution that is 0.10 M NH_3–0.10 M NH_4^+.
3. What are the species potentially present in the solution, and how significant are their concentrations likely to be?
    In principle, any solution containing phosphoric acid or a phosphate salt contains the following species (in addition to H_2O): H_3PO_4, $H_2PO_4^-$, HPO_4^{2-}, PO_4^{3-}, OH^-, H_3O^+, and possibly other cations. But as we have already seen, in $H_3PO_4(aq)$ the only species present in significant concentrations are H_3PO_4, H_3O^+, and $H_2PO_4^-$. And, if the solution is described as $Na_3PO_4(aq)$, the significant species are Na^+, PO_4^{3-}, HPO_4^{2-}, and OH^-. In a solution of HCl and $HC_2H_3O_2$, the significant ionic species are H_3O^+ and Cl^-. $HC_2H_3O_2$ would be present mostly in nonionized form because of the repression of its ionization by the strong acid. $[C_2H_3O_2^-]$ would be quite low. On the other hand, in the mixture of the two weak acids, $HC_2H_3O_2$ and $HC_3H_5O_2$, $[HC_2H_3O_2]$, $[HC_3H_5O_2]$, $[C_2H_3O_2^-]$, $[C_3H_5O_2^-]$, and $[H_3O^+]$ would probably all enter significantly into calculations.
4. Can the situation be *readily* described as of a certain type?
    If the given situation can be recognized as corresponding to one of the important types

discussed in these chapters, then it can usually be treated by the method introduced for that type (sometimes with and sometimes without simplifying assumptions being justified). But if the calculation is not *readily* seen to be of a certain type, the other points summarized here should be carefully investigated. Some of the basic categories considered in these two chapters are listed below. The examples refer to cases in which the essential question was to determine the concentrations of various species present in solution.

- ionization of a strong acid (Example 17-3).
- ionization of a strong base (Example 17-4).
- ionization of a weak monoprotic acid (Example 17-9).
- ionization of a weak monoprotic base (Example 17-8).
- ionization of a weak polyprotic acid (or base) (Example 17-11).
- hydrolysis of a salt of a monoprotic acid or base (Example 17-14).
- ionization of a weak acid (or base) in the presence of a strong acid (or base)—common ion effect (Example 18-1).
- ionization of a weak acid (or base) in the presence of its salt—common ion effect, buffer solution (Examples 18-2, 18-4).
- reaction of strong acid and base, followed by calculation of ionic concentrations (Examples 17-5, 18-7).
- reaction of weak acid (or base) with strong base (or acid), followed by calculation of ionic concentrations (Example 18-8).
- hydrolysis of the salt of a polyprotic acid (Example 18-9).

18-7 Postscript: Equivalent Weight and Normality

Because of the difficulties they had in establishing molecular weights, nineteenth century chemists developed the concept of equivalent weights instead. It is a concept that is still used to some extent in analytical chemistry. In this section we give brief consideration to equivalent weight and to a concentration scale based on equivalent weight—normality.

Let us define equivalent weight (or equivalent), for example, of substances A and B, such that

One equivalent of A reacts completely with one equivalent of B,

that is,

Recall the significance of the symbol $\rightleftharpoons$, described on page 15.

1 equiv. A $\rightleftharpoons$ 1 equiv. B   (18.29)

This definition of an equivalent obviously requires that we know the kind of reaction in which substances participate. For the present we will limit our discussion to acid–base reactions.

Equivalent Weight in Acid–Base Reactions. *An equivalent is a quantity of substance that will liberate or react with one mole of hydrogen ions.* The equivalent weight of HCl is equal to its molar mass, that is, 36.46 g HCl/equiv HCl. The reaction of HCl with NaOH can be represented by the ionic equation

$$\underbrace{H^+(aq) + Cl^-(aq)}_{1\ mol\ =\ 1\ equiv.} + \underbrace{Na^+(aq) + OH^-(aq)}_{1\ mol\ =\ 1\ equiv.} \longrightarrow HOH + Na^+(aq) + Cl^-(aq) \quad (18.30)$$

An equivalent of NaOH is identical to a mole; the equivalent weight of NaOH is 40.00 g NaOH/equiv. NaOH.

When completely neutralized with NaOH, sulfuric acid (H_2SO_4) yields *two* H^+ ions per molecule. That is,

$$\underbrace{2\ H^+(aq) + SO_4^{2-}(aq)}_{1\ mol\ =\ 2\ equiv.} + \underbrace{2\ Na^+(aq) + 2\ OH^-(aq)}_{2\ mol\ =\ 2\ equiv.} \longrightarrow$$

$$2\ H_2O + 2\ Na^+(aq) + SO_4^{2-}(aq) \qquad (18.31)$$

There are *two* equivalents in 1 mol of H_2SO_4. The equivalent weight is *one half* the molar mass—49.04 g H_2SO_4/equiv. H_2SO_4.

The situation with phosphoric acid, H_3PO_4, is more complicated. If H_3PO_4 participates in reaction (18.32) its equivalent weight is equal to its molar mass. In reaction (18.33) the equivalent weight is *one half* the molar mass; and in reaction (18.34), *one third* the molar mass.

$$H_3PO_4(aq) + NaOH(aq) \longrightarrow NaH_2PO_4(aq) + H_2O \qquad (18.32)$$

$$H_3PO_4(aq) + 2\ NaOH(aq) \longrightarrow Na_2HPO_4(aq) + 2\ H_2O \qquad (18.33)$$

$$H_3PO_4(aq) + 3\ NaOH(aq) \longrightarrow Na_3PO_4(aq) + 3\ H_2O \qquad (18.34)$$

Example 18-10 Determine the equivalent weight of the following substances: **(a)** HNO_3; **(b)** $Ba(OH)_2$.

Solution
(a) Since only 1 mol H^+ can be produced per mole of HNO_3,

1 equiv. HNO_3 = 1 mol HNO_3

equiv. wt. HNO_3 = molar mass HNO_3 = 63.0 g HNO_3

(b) In water solution $Ba(OH)_2$ produces *2* mol OH^- for every mole of compound, and this 2 mol OH^- is capable of reacting with 2 mol H^+. One mole of $Ba(OH)_2$ is *2* equivalents.

2 equiv. $Ba(OH)_2$ = 1 mol $Ba(OH)_2$

1 equiv. $Ba(OH)_2$ = $\frac{1}{2}$ mol $Ba(OH)_2$

equiv. wt. $Ba(OH)_2$ = $\frac{1}{2}$ molar mass $Ba(OH)_2$ = $\frac{1}{2} \times 171.4$ = 85.7 g $Ba(OH)_2$

SIMILAR EXAMPLE: Review Problem 10.

Normality. Normality concentration is defined in a similar fashion to molarity.

$$\text{normality } (N) = \frac{\text{number equiv. solute}}{\text{number liters soln.}} \qquad (18.35)$$

A 1 *normal* solution of HCl can be prepared by dissolving 36.5 g HCl (1 equiv.) in 1 L of a water solution. Since this mass of HCl is also 1 mol, the solution is also 1 *molar*.

$$1.00\ N\ HCl = \frac{36.5\ g\ HCl}{1.00\ L\ soln.} = 1.00\ M\ HCl$$

Normality and molarity are not equal for an aqueous solution of $Ba(OH)_2$, however.

$$0.01\ M\ Ba(OH)_2 = \frac{0.01\ mol\ Ba(OH)_2}{1.00\ L\ soln.} = \frac{0.02\ equiv.\ Ba(OH)_2}{1.00\ L\ soln.}$$

$$= 0.02\ N\ Ba(OH)_2$$

The relationship between normality and molarity concentration is

$$\text{normality} = n \times \text{molarity} \qquad (18.36)$$

where n represents the number of moles of hydrogen ions per mole of compound that a solute is capable of releasing (acid) or reacting with (base).

Example 18-11   What are the normalities of the following solutions?
(a) 0.50 M NaOH; **(b)** 0.02 M H_2SO_4.

Solution
(a) One mole of OH^- is produced for every mole of NaOH. The value of n in equation (18.36) is 1. Normality and molarity are identical for NaOH solutions. 0.50 M NaOH = 0.50 N NaOH.
(b) Two moles of H^+ are produced in the complete neutralization of one mole of H_2SO_4; $n = 2$. Normality = 2 × molarity = 2 × 0.02 M = 0.04 N H_2SO_4.

SIMILAR EXAMPLES: Review Problem 11; Exercises 35, 36, 37.

Solution Stoichiometry and Normality Concentration.   Equation (18.35) can be recast in a slightly different form.

$$\text{no. equiv. solute} = \text{normality} \times \text{no. L soln.} \tag{18.37}$$

Another common expression is obtained by dividing both sides of this equation by 1000. One-thousandth of a liter is a milliliter (mL) and one-thousandth of an equivalent is a **milliequivalent (meq).** Thus, normality concentration can be expressed either as equiv./L or meq/mL.

$$\text{no. meq solute} = \text{normality} \times \text{no. mL soln.} \tag{18.38}$$

According to the definition of the equivalent (equation 18.29), we can write

no. equiv. A $\eqsim$ no. equiv. B

$$V_A \cdot N_A \eqsim V_B \cdot N_B \tag{18.39}$$

In equation (18.39) the symbols N refer to normality concentration, and the subscripts indicate whether the solute is reactant A or B. The symbols V refer to the corresponding solution volumes. These volumes can be expressed either in liters or in milliliters, as long as the same unit is used for each.

Example 18-12   The concentration of an HCl(aq) solution is established by titration with a standard solution of $Ba(OH)_2$(aq). The HCl(aq) is then used to determine the normality of an NaOH(aq) solution. A 25.00-mL sample of the HCl(aq) is titrated with 30.08 mL 0.1000 N $Ba(OH)_2$. A 10.00-mL sample of the NaOH(aq) requires 25.10 mL of the HCl(aq) for its titration. What is the normality of the NaOH(aq)?

Solution
Standardization of HCl(aq):

$$V_{HCl} \times N_{HCl} = V_{Ba(OH)_2} \times N_{Ba(OH)_2}$$

$$25.00 \text{ mL} \times N_{HCl} = 30.08 \text{ mL} \times \frac{0.1000 \text{ meq}}{\text{mL}}$$

$$N_{HCl} = 0.1203 \frac{\text{meq}}{\text{mL}}$$

Determination of NaOH(aq):

$$V_{NaOH} \times N_{NaOH} = V_{HCl} \times N_{HCl}$$

$$10.00 \text{ mL} \times N_{NaOH} = 25.10 \text{ mL} \times \frac{0.1203 \text{ meq}}{\text{mL}}$$

$$N_{NaOH} = 0.3020 \frac{\text{meq}}{\text{mL}}$$

SIMILAR EXAMPLES: Review Problem 12; Exercise 38.

The concepts of equivalent weight and normality do not offer any important new possibilities for problem solving. Example 18-12 could have been done just as well with moles and molar concentrations.

You will notice that in Example 18-12 we wrote no chemical equations. This represents one of the principal advantages of normality concentration. When expressed in equivalents, reactants always combine in a 1:1 ratio and equation (18.39) is applicable. Even if the chemical formula of a substance is not known, its equivalent weight and normality concentrations with respect to certain types of reactions can still be established. These advantages, however, are offset by this principal disadvantage: *The equivalent weight of a substance and the normality concentration of its solutions depend on the particular reaction in which the substance participates* (recall equations 18.32 through 18.34). Equivalent weight and normality often *do not* have unique values, as do formula (molecular) weight and molarity. Although equivalent weight and normality will continue to be used for some time, it is likely that their use will decrease in importance. This is especially true now that the mole has been established as the base unit for amount of substance in the SI system.

Summary

The degree of ionization of a weak acid in aqueous solution can be reduced by adding to the solution either a strong acid or a salt of the weak acid. In each case this occurs through the common ion effect. The weak acid/salt combination is especially important: a solution with these two components resists changes in its pH—it is a buffer solution. A weak base/salt combination is also a buffer.

The typical buffer solution has one component to react with small added amounts of acid and another to react with bases. The pH at which a buffer solution functions is determined by the pK value of the weak acid or base and the *ratio* of molar concentration of salt to that of weak acid or base. The molar concentrations of salt and weak acid or weak base also establish the buffer capacity—the amount of added acid or base that the buffer is capable of neutralizing. Buffer action plays a critical role in the functioning of blood and other fluids in living organisms.

In a titration, solutions of an acid and a base are allowed to react to the point where exact neutralization occurs. At this point (the equivalence point) there is neither excess acid nor base in solution but simply the salt produced by their neutralization. Whether this salt solution is acidic, basic, or neutral is determined by whether the salt can hydrolyze or undergo further ionization. A titration curve represents the pH of the solution being titrated as a function of the volume of titrating agent (titrant) added. Generally, a titration curve shows a sharp change in solution pH at the equivalence point. The object in choosing an indicator for a titration is to find one that changes color as close to the equivalence point as possible. An acid–base indicator is a weak acid, and the selection of an appropriate indicator for a titration requires matching its pK_a value to the pH at the equivalence point.

Learning Objectives

As a result of studying Chapter 18, you should be able to

1. Describe the effect of common ions on the ionization of weak acids and bases, and calculate the concentrations of species present in solutions of weak acids or bases and their common ions.

2. Explain why pure water cannot resist changes in pH and how buffer solutions work to maintain a constant pH.

3. Derive and use the basic equations that relate to the pH of buffer solutions.

4. Calculate the pH of a buffer solution from concentrations of the buffer components and a value of K_a or K_b, and describe how to prepare a buffer having a specific pH.

5. Determine the changes in pH of a buffer solution that result from the addition of acids or bases.

6. Define the terms "buffer range" and "buffer capacity" and establish numerical values for them.

7. Describe how the blood buffer system works.

8. Explain how an acid–base indicator works.

9. Calculate pH values for various combinations of weak and strong acids and bases.

10. Plot titration curves and extract significant information from such curves—initial pH, buffer region, equivalence point, indicator selection.

11. Calculate the pH of certain solutions of salts of polyprotic acids.

12. Apply the concepts of equivalent weight and normality to solution stoichiometry problems, especially those involving titrations.

Some New Terms

An **acid–base indicator** is a substance that can be used to measure the pH of a solution. The indicator takes on one color when in its nonionized acid form and a different color in its anion form. Its color in a particular solution depends on which form predominates.

A **buffer** solution resists a change in pH. It contains components capable of reacting with (neutralizing) small added amounts of acid or base.

Buffer capacity refers to the amount of acid and/or base that a buffer solution can neutralize and still maintain an essentially constant pH.

Buffer range is the range of pH values over which a particular buffer system will function.

The **common ion effect** describes the effect on an equilibrium by a substance that furnishes ions that can participate in the equilibrium. For example, sodium acetate, $NaC_2H_3O_2$, furnishes the common ion, $C_2H_3O_2^-$, to the ionization equilibrium of acetic acid, $HC_2H_3O_2$.

The **end point of a titration** is the point in the titration where the indicator used changes color. A properly chosen indicator for a titration must have its end point correspond as closely as possible to the equivalence point of the titration reaction.

The **equivalence point of a neutralization reaction** is the condition in which an acid and a base neutralize one another.

Equivalent weight is a definition of a quantity of substance based on the extent to which it enters into certain kinds of reactions. For example, an equivalent weight of acid liberates 1 mol H^+, and an equivalent weight of base reacts with 1 mol H^+.

Normality (*N*) is a concentration unit used in conjunction with the concept of equivalent weight. Normality is the number of equivalents of solute per liter of solution.

A **titration curve** is a graph of solution pH versus volume of titrant. It outlines how pH changes during an acid–base titration, and it can be used to establish such features as the equivalence point of the titration.

Suggestions for Further Study

LOTT, J. A., "Hydrogen Ions in Blood," *Chemistry*, **51**[4], 6 (1978).

NAKAYAMA, F. S., "Hydrolysis of Sodium Carbonate," *J. Chem. Educ.*, **47**, 67 (1970).

STAIRS, R. A., "Unified Calculation of Titration Curves," *J. Chem. Educ.*, **55**, 99 (1978).

WIGER, G. R., and DE LA CAMP, U., "Conjugate Acid–Base Mixtures in the General Chemistry Laboratory," *J. Chem. Educ.*, **55**, 401 (1978).

WILLIS, C. J., "Another Approach to Titration Curves," *J. Chem. Educ.*, **58**, 659 (1981).

YINGST, A., "Evaluation of Titration Analyses with Logarithmic Concentration Diagrams," *J. Chem. Educ.*, **44**, 601 (1967).

Review Problems

1. For a solution that is 0.355 *M* $HC_3H_5O_2$ and 0.106 *M* HI, calculate **(a)** $[H_3O^+]$; **(b)** $[OH^-]$; **(c)** $[C_3H_5O_2^-]$; **(d)** $[I^-]$.

$$HC_3H_5O_2 + H_2O \rightleftharpoons H_3O^+ + C_3H_5O_2^- \qquad K_a = 1.34 \times 10^{-5}$$

2. For a solution that is 0.143 *M* NH_3 and 0.0875 *M* NH_4Cl, calculate **(a)** $[OH^-]$; **(b)** $[NH_4^+]$; **(c)** $[Cl^-]$; **(d)** $[H_3O^+]$

$$NH_3 + H_2O \rightleftharpoons NH_4^+ + OH^- \qquad K_b = 1.74 \times 10^{-5}$$

3. Write equations to show how each of the following buffer solutions reacts with a small added amount of acid or base.
   (a) $HCHO_2/NaCHO_2$ **(b)** $C_6H_5NH_2/C_6H_5NH_3^+Cl^-$
   (c) NaH_2PO_4/Na_2HPO_4

4. Calculate the pH of a buffer solution that is
   (a) 0.0552 *M* $HC_7H_5O_2$ and 0.132 *M* $NaC_7H_5O_2$

$$HC_7H_5O_2 + H_2O \rightleftharpoons H_3O^+ + C_7H_5O_2^-$$
$$K_a = 6.3 \times 10^{-5}$$

   (b) 0.085 *M* NH_3 and 0.17 *M* NH_4Cl

$$NH_3 + H_2O \rightleftharpoons NH_4^+ + OH^- \qquad K_b = 1.74 \times 10^{-5}$$

5. What concentration of formate ion, $[CHO_2^-]$, should be present in 0.472 *M* $HCHO_2$ to produce a buffer solution with pH = 4.12?

$$HCHO_2 + H_2O \rightleftharpoons H_3O^+ + CHO_2^- \qquad K_a = 1.8 \times 10^{-4}$$

6. What concentration of ammonia, $[NH_3]$, should be present in a solution that has $[NH_4^+] = 0.812$ *M* to produce a buffer solution with pH = 9.15?

$$NH_3 + H_2O \rightleftharpoons NH_4^+ + OH^- \qquad K_b = 1.74 \times 10^{-5}$$

7. Phenol red indicator changes from yellow to red in the pH range from 6.6 to 8.0. *Without making detailed calculations,* state what color the indicator will assume in each of the following solutions. **(a)** 0.10 *M* KOH; **(b)** 0.10 *M* $HC_2H_3O_2$; **(c)** 0.10 *M* NH_4NO_3; **(d)** 0.10 *M* HBr; **(e)** 0.10 *M* NaCN.

8. Calculate the pH at the points in the titration of 20.00 mL 0.350 *M* KOH at which **(a)** 15.00 mL and **(b)** 20.00 mL 0.425 *M* HCl have been added.

$$H_3O^+ + OH^- \rightleftharpoons 2 H_2O$$

9. Calculate the pH at the points in the titration of 25.00 mL 0.108 M HNO$_2$ at which **(a)** 10.00 mL and **(b)** 20.00 mL 0.162 M NaOH have been added. For HNO$_2$, $K_a = 5.1 \times 10^{-4}$.

$$HNO_2 + OH^- \rightarrow H_2O + NO_2^-$$

10. Calculate the equivalent weights of the following substances for use in acid–base reactions. **(a)** HClO$_4$; **(b)** Mg(OH)$_2$; **(c)** HC$_3$H$_5$O$_2$ (propionic acid).

11. What are the normality concentrations corresponding to the following molarities? Indicate any cases in which more than one normality seems possible. **(a)** 0.24 M KOH; **(b)** 0.15 M H$_2$C$_2$O$_4$; **(c)** 2×10^{-3} M Ca(OH)$_2$.

12. What volume of 0.1090 N NaOH is required for the complete neutralization of **(a)** 25.00 mL 0.1471 N H$_2$SO$_4$? **(b)** 15.00 mL 0.08511 M H$_2$SO$_4$?

Exercises

The common ion effect in acid–base equilibria (Use data from Table 17-2, as necessary.)

1. Describe the effect on the pH of the solution produced by adding **(a)** NaNO$_2$ to HNO$_2$(aq); **(b)** NaNO$_3$ to HNO$_3$(aq). Why are the effects not the same? Explain.

2. Calculate the concentration of the ionic species indicated.
- **(a)** [H$_3$O$^+$] in a solution that is 0.100 M HCl and 0.100 M HOCl
- **(b)** [NO$_2^-$] in a solution that is 0.100 M NaNO$_2$ and 0.0100 M HNO$_2$.
- **(c)** [Cl$^-$] in a solution that is 0.200 M HCl and 0.0500 M HC$_2$H$_3$O$_2$
- **(d)** [C$_2$H$_3$O$_2^-$] in a solution that is 0.100 M HCl and 0.300 M HC$_2$H$_3$O$_2$
- **(e)** [OH$^-$] in a solution that is 0.200 M (NH$_4$)$_2$SO$_4$ and 0.500 M NH$_3$

3. Lactic acid, HC$_3$H$_5$O$_3$, is found in sour milk. A solution containing 10.0 g NaC$_3$H$_5$O$_3$ in 100.0 mL of 0.0500 M HC$_3$H$_5$O$_3$ has pH = 4.11. What is K_a of lactic acid?

4. What is the pOH of a solution obtained by adding 1.37 mg of aniline hydrochloride (C$_6$H$_5$NH$_3^+$Cl$^-$) to 3.25 L of 0.105 M aniline (C$_6$H$_5$NH$_2$)?

$$C_6H_5NH_2 + H_2O \rightleftharpoons C_6H_5NH_3^+ + OH^- \qquad K_b = 4.30 \times 10^{-10}$$

*5. You are given 250.0 mL of 0.100 M HC$_3$H$_5$O$_2$(aq) (propionic acid, $K_a = 1.35 \times 10^{-5}$). You wish to adjust the pH of this aqueous propionic acid by adding an appropriate solution to it. What volume would you add of **(a)** 1.00 M HCl to lower the pH to 1.00; **(b)** 1.00 M NaC$_3$H$_5$O$_2$ to raise the pH to 4.00; **(c)** water to raise the pH by 0.15 unit?

Buffer solutions (Use data from Tables 17-2 and 17-3 as necessary.)

6. Indicate which of the following aqueous solutions are buffer solutions. Explain your reasoning. (*Hint:* You must also consider whether any reactions occur between the solution components listed.)
- **(a)** 0.100 M NaCl
- **(b)** 0.100 M NaCl–0.100 M NH$_4$Cl
- **(c)** 0.100 M CH$_3$NH$_2$–0.150 M CH$_3$NH$_3^+$Cl$^-$
- **(d)** 0.100 M HCl-0.050 M NaNO$_2$
- **(e)** 0.100 M HCl–0.200 M NaC$_2$H$_3$O$_2$
- **(f)** 0.100 M HC$_2$H$_3$O$_2$–0.125 M NaC$_3$H$_5$O$_2$

7. In the text the H$_2$PO$_4^-$/HPO$_4^{2-}$ combination is mentioned as playing a role in maintaining the pH of blood.
- **(a)** Write equations to show how a solution containing these ions functions as a buffer.
- **(b)** Verify that this buffer is most effective at pH 7.2.
- **(c)** Calculate the pH of a buffer solution in which [H$_2$PO$_4^-$] = 0.050 M and [HPO$_4^{2-}$] = 0.150 M. (*Hint:* Recall that phosphoric acid ionizes in three distinct steps and focus on the second step.)

8. What mass of (NH$_4$)$_2$SO$_4$ must be added to 320.0 mL of 0.105 M NH$_3$ to yield a solution with pH 9.35?

9. You are given the task of preparing a buffer solution with pH 3.50 and have available the following solutions, all 0.100 M: HCHO$_2$, HC$_2$H$_3$O$_2$, H$_3$PO$_4$, NaCHO$_2$, NaC$_2$H$_3$O$_2$, and NaH$_2$PO$_4$. Describe how you would prepare this buffer solution. (*Hint:* What volumes of which solutions would you use?)

*10. Suppose that the task in the preceding exercise were modified to require exactly 1.00 L of the buffer solution with pH 3.50 and that the solutions available were 0.100 M NaCHO$_2$, 0.100 M NaC$_2$H$_3$O$_2$, 0.100 M NaH$_2$PO$_4$, and 1.00 M HCl. Describe how you would prepare this buffer solution.

11. In blood the ratio [HCO$_3^-$]/[H$_2$CO$_3$] $\simeq$ 20. What is the pH of a buffer solution with [C$_2$H$_3$O$_2^-$]/[HC$_2$H$_3$O$_2$] = 20? How effective would this buffer be? Explain.

12. Compare the following buffers with respect to their **(a)** pH and **(b)** buffer capacities: 0.010 M HC$_2$H$_3$O$_2$–0.010 M NaC$_2$H$_3$O$_2$ and 1.0 M HC$_2$H$_3$O$_2$–0.50 M NaC$_2$H$_3$O$_2$.

13. With respect to the buffer solution described in Example 18-4, that is, 0.15 M HC$_2$H$_3$O$_2$–0.50 M NaC$_2$H$_3$O$_2$,
- **(a)** What is the pH if 1.00 mL of 6.00 M HCl is added to 100.0 mL of the buffer?
- **(b)** What is the pH if 10.00 mL of 6.00 M HCl is added instead?

14. A buffer solution is prepared by dissolving 1.51 g NH$_3$ and 3.85 g (NH$_4$)$_2$SO$_4$ in 0.500 L of water solution.
- **(a)** What is the pH of this solution?
- **(b)** If 1.00 g NaOH is added to this solution, what is the pH?
- **(c)** How many mL of 12 M HCl must be added to 0.500 L of the original buffer solution to change its pH to 9.00?

15. An acetic acid–sodium acetate buffer can be prepared by allowing an *excess* of $NaC_2H_3O_2$ to react with HCl. The strong acid HCl is converted to the weak acid $HC_2H_3O_2$.

$$C_2H_3O_2^- \quad + \quad H_3O^+ \quad \rightarrow HC_2H_3O_2 + H_2O$$
(from $NaC_2H_3O_2$)   (from HCl)

    (a) If 10.0 g $NaC_2H_3O_2$ is added to 0.300 L of 0.200 *M* HCl, what is the pH of the resulting solution?
    (b) If 1.00 g $Ba(OH)_2$ is added to the solution in part (a), what happens to the pH?
    (c) What is the capacity of the buffer solution in part (a) toward $Ba(OH)_2$?
    (d) What is the pH of the solution in part (a) following the addition of 5.2 g $Ba(OH)_2$?

Acid–base indicators (Use data from Tables 17-2 and 17-3 as necessary.)

16. In the use of acid–base indicators,
    (a) Why is it generally sufficient to use a *single* indicator in an acid–base titration but often necessary to use *several* indicators to establish the approximate pH of a solution?
    (b) Why must the amount of indicator used in an acid–base titration be kept as small as possible?

17. A handbook lists the following data for some acid–base indicators.

		Color change
bromphenol blue	$K_a = 1.41 \times 10^{-4}$	yellow → blue
		(acid) (anion)
bromcresol green	$K_a = 2.09 \times 10^{-5}$	yellow → blue
bromthymol blue	$K_a = 7.9 \times 10^{-8}$	yellow → blue
2,4-dinitrophenol	$K_a = 1.26 \times 10^{-4}$	colorless → yellow
chlorophenol red	$K_a = 1.0 \times 10^{-6}$	yellow → red
thymolphthalein	$K_a = 1.0 \times 10^{-10}$	colorless → blue

    (a) Which of these indicators change color in acidic solution which in basic solution, and which near the neutral point?
    (b) What is the approximate pH of a solution if bromcresol green indicator assumes a green color; if chlorophenol red assumes an orange color?

18. With reference to the indicators listed in Exercise 17, what would be the color of each combination?
    (a) 2,4-dinitrophenol when placed in 0.100 *M* HCl(aq)
    (b) chlorophenol red in 1.00 *M* NaCl
    (c) thymophthalein in 1.00 *M* NH_3
    (d) bromthymol blue in 1.00 *M* NH_4NO_3
    (e) bromcresol green in seawater (recall Figure 17-3)
    (f) bromphenol blue in saturated CO_2(aq) [Consider CO_2(aq) to be 0.034 *M* H_2CO_3.]

19. The indicator methyl red has a $pK_a = 4.95$. It changes color from red to yellow over the pH range 4.4 to 6.2.
    (a) If the indicator is placed in a buffer solution that is 0.10 *M* $HC_2H_3O_2$–0.10 *M* $NaC_2H_3O_2$, what percent of the indicator will be in the acid form and what percent in the anion form?
    (b) Which form of the indicator do you think has the "stronger" (more visible) color, the acid form (red) or the anion form (yellow)? Explain your reasoning.

Neutralization reactions

20. A 18.67-mL sample of 0.07152 *M* HI is required to titrate a 25.00-mL sample of an unknown strong base. What is $[OH^-]$ in the base?

21. Excess $Ca(OH)_2$(s) is shaken with water to produce a saturated solution. A 50.00-mL sample of the clear, saturated solution is withdrawn and titrated. 10.7 mL 0.1032 *M* HCl is required for this titration. What is the solubility of $Ca(OH)_2$, expressed as g $Ca(OH)_2$ per L soln.?

22. Two solutions are mixed: (A) 100.0 mL of an HCl(aq) with pH 2.50 and (B) 100.0 mL of an NaOH(aq) with pH 11.00. What is the pH of the solution that results from mixing A and B?

★23. With reference to Exercise 22, if solution A were 100.0 mL of a weak acid solution with pH 2.50 and solution B, 100.0 mL of a weak base solution with pH 11.00,
    (a) Would the final pH on mixing the two solutions be greater or less than that calculated in Exercise 22? Explain.
    (b) What additional information would be required to calculate the actual pH of the mixed solution?

★24. The neutralization of NaOH by HCl is represented in equation (1) below, and the neutralization of NH_3 by HCl, in equation (2).

    (1) $OH^- + H_3O^+ \rightleftharpoons 2\ H_2O \qquad K = ?$
    (2) $NH_3 + H_3O^+ \rightleftharpoons NH_4^+ + H_2O \qquad K = ?$
    (a) Determine the equilibrium constant K for each reaction.
    (b) Explain why each neutralization reaction can be considered to go to completion.

Titration curves

25. In the text several differences were pointed out between the titration curves for a strong acid by a strong base and a weak acid by a strong base (i.e., between Figures 18-3 and 18-4). One point that is identical in the two curves, however, is the volume of 0.1000 *M* NaOH required to reach the equivalence point. Explain why this should be so.

26. Sketch the following titration curves. Indicate the initial pH and the pH corresponding to the equivalence point. Indicate the volume of titrant required to reach the equivalence point, and select a suitable indicator from Figure 18-2.
    (a) 25.0 mL 0.100 *M* KOH by 0.200 *M* HI
    (b) 10.0 mL 1.00 *M* NH_3 by 0.250 *M* HCl

27. Sketch a series of titration curves for the following three hypothetical weak acids when titrated with 0.1000 M NaOH. Select suitable indicators for the titrations. (*Hint:* Select a few key points at which to estimate the pH of the solution, for example, the initial pH, the condition pH = pK_a, and the pH at the equivalence point.)
  (a) 10.00 mL of 0.1000 M HX; $K_a = 1.0 \times 10^{-3}$
  (b) 10.00 mL of 0.1000 M HY; $K_a = 1.0 \times 10^{-5}$
  (c) 10.00 mL of 0.1000 M HZ; $K_a = 1.0 \times 10^{-7}$

28. A 10.00-mL solution that is 0.0400 M H_3PO_4 and 0.0150 M NaH_2PO_4 is titrated with 0.0200 M NaOH. Sketch the titration curve obtained. (*Hint:* How many equivalence points are there? Are they equally spaced along the volume axis as in Figure 18-5?)

29. Determine, by calculation, the pH at the point where the original acid is 90% neutralized in the titration of
  (a) 25.00 mL of 0.1000 M HCl by 0.1000 M NaOH (see Figure 18-3);
  (b) 25.00 mL of 0.1000 M $HC_2H_3O_2$ by 0.1000 M NaOH (see Figure 18-4).

★**30.** Thymol blue in its acid range is not a suitable indicator for the titration of HCl by NaOH. Suppose that a student uses thymol blue by mistake in the titration of Figure 18-3, and suppose that the indicator end point is taken to be at pH = 2.
  (a) Would there be a sharp color change [i.e., produced by a single drop of NaOH(aq)]?
  (b) What percent of the original HCl remains unneutralized at this point?

★**31.** Calculate the volume of 0.1000 M NaOH that must be added to just reach the following.
  (a) a pH of 3.00 in the titration of 25.00 mL 0.1000 M HCl (Figure 18-3).
  (b) a pH of 5.25 in the titration of 25.00 mL 0.1000 M $HC_2H_3O_2$ (Figure 18-4).

  (c) a pH of 7.50 in the titration of 10.00 mL 0.1000 M H_3PO_4 (Figure 18-5).

Hydrolysis of salts of polyprotic acids

32. Is a solution that is 0.10 M Na_2S(aq) likely to be acidic, basic, or neutral? Explain.

33. Calculate the pH of **(a)** 1.0 M Na_2CO_3(aq) and **(b)** 0.010 M Na_2CO_3(aq).

34. Arrange the following 0.10 M solutions in order of decreasing acidity: $NaC_2H_3O_2$, $NaHC_2O_4$, $NaHSO_4$, NaH_2PO_4. Explain your reasoning.

Equivalent weight and normality concentration

35. In aqueous solutions carbonate ion, CO_3^{2-}, can react with *two* H^+ ions yielding one molecule of H_2O and one of CO_2. What mass of $Na_2CO_3 \cdot 10H_2O$, in grams, is required to produce 2.00 L of 0.175 N Na_2CO_3(aq)?

36. It is desired to prepare a standard solution of $Ba(OH)_2$ for use in acid–base titrations. What is the approximate maximum *normality* solution that can be prepared if the solubility of barium hydroxide is 3.89 g $Ba(OH)_2$ per 100 g of solution at 20°C?

37. A 25.00-mL sample of H_3PO_4(aq) requires 31.15 mL of 0.242 N KOH for its titration in reaction (18.33). What is the normality of this H_3PO_4(aq) if it is always to be used **(a)** in reaction (18.32); **(b)** in reaction (18.33); **(c)** in reaction(18.34)?

38. A sample of battery acid is to be analyzed for its sulfuric acid content. A 1.00-mL sample weighs 1.239 g. This sample is diluted to 250.0 mL and 10.00 mL of the diluted acid requires 32.40 mL 0.0100 N $Ba(OH)_2$ for its titration. What is the % H_2SO_4, by mass, in the battery acid?

Additional Exercises

1. In Example 17-10, the percent ionization of $HC_2H_3O_2$ was calculated for **(a)** 1.0 M, **(b)** 0.10 M, and **(c)** 0.010 M $HC_2H_3O_2$. Recalculate the percent ionization if each of the three solutions is also made 0.10 M in $NaC_2H_3O_2$. Explain why these results differ from those of Example 17-10.

2. Although we say that a buffered solution has a constant pH and [H_3O^+], small changes in pH do translate into significant changes in [H_3O^+]. What is the *percent* increase in [H_3O^+] when the pH of blood drops from 7.4 to 7.3?

3. A buffer solution is prepared by dissolving 1.50 g each of benzoic acid, $HC_7H_5O_2$, and sodium benzoate, $NaC_7H_5O_2$, in 150.0 mL of water solution.

$$HC_7H_5O_2 + H_2O \rightleftharpoons H_3O^+ + C_7H_5O_2^- \quad K_a = 6.3 \times 10^{-5}$$

  (a) What is the pH of this buffer solution?
  (b) Which buffer component, and in what quantity, must be added to the solution to change its pH to 4.00?

4. A solution is 0.318 M $HCHO_2$ (formic acid). What mass of sodium formate ($NaCHO_2$) must be added to 250.0 mL of this solution to produce a buffer solution of pH = 3.62? [$K_a(HCHO_2) = 1.8 \times 10^{-4}$.]

5. If to 100.0 mL of the buffer solution prepared in the preceding exercise is added 0.35 mL 15 M NH_3, what will be the pH of the resulting solution?

6. In what approximate pH range would you expect each of the following buffer solutions to be most effective? **(a)** HNO_2–$NaNO_2$; **(b)** CH_3NH_2–$CH_3NH_3^+Cl^-$; **(c)** NH_3–$(NH_4)_2SO_4$.

7. It is desired to bring the pH of 0.500 L 0.500 M $NH_4Cl(aq)$ to a value of 7.00. How many drops (1 drop = 0.05 mL) of which of the following solutions would you use: 10.0 M HCl, 10.0 M NH_3?

8. What aqueous concentrations of the following substances are required to obtain solutions with the pH values indicated?
   (a) $Ba(OH)_2$ for pH 12.44
   (b) Aniline, $C_6H_5NH_2$, for pH 8.91
   (c) $HC_2H_3O_2$ in 0.313 M $NaC_2H_3O_2$ for pH 4.56
   (d) NH_4Cl for pH 5.05

9. Use appropriate values of equilibrium constants to determine whether a solution can be simultaneously
   (a) 0.10 M NH_3 *and* 0.10 M NH_4Cl, with pH 6.07;
   (b) 0.10 M $NaC_2H_3O_2$ *and* 0.058 M HI;
   (c) 0.10 M KNO_2 *and* 0.25 M KNO_3.

10. The single equilibrium equation written below can be applied to different phenomena described in this or the preceding chapter.

$$HC_2H_3O_2 + H_2O \rightleftharpoons H_3O^+ + C_2H_3O_2^- \qquad K_a = 1.74 \times 10^{-5}$$

Indicate the phenomenon—ionization of pure acid, common ion effect, buffer action, hydrolysis—for each of the following combinations of concentrations.
   (a) $[H_3O^+]$ and $[HC_2H_3O_2]$ are high; $[C_2H_3O_2^-]$ is very low.
   (b) $[C_2H_3O_2^-]$ is high; $[HC_2H_3O_2]$ and $[H_3O^+]$ are very low.
   (c) $[HC_2H_3O_2]$ is high; $[H_3O^+]$ and $[C_2H_3O_2^-]$ are low.
   (d) $[HC_2H_3O_2]$ and $[C_2H_3O_2^-]$ are high; $[H_3O^+]$ is low.

★**11.** The titration of a weak acid by a weak base is not a particularly satisfactory procedure because the pH does not increase sharply at the equivalence point. Demonstrate this fact by sketching a titration curve for the neutralization of 10.00 mL of 0.100 M $HC_2H_3O_2$ by 0.100 M NH_3.

★**12.** At times a salt of a weak base can be titrated by a strong base. Use appropriate data from the text to sketch a titration curve for the titration of 20.00 mL of 0.0500 M $C_6H_5NH_3^+Cl^-$ by 0.1000 M NaOH.

★**13.** Carbonic acid is a weak diprotic acid (H_2CO_3), having $K_{a_1} = 4.2 \times 10^{-7}$ and $K_{a_2} = 5.6 \times 10^{-11}$. The equivalence points for the titration of this acid come at approximately pH 4

and pH 9. Suitable indicators for use in titrating carbonic acid or carbonate solutions are methyl orange and phenolphthalein.
   (a) Sketch the titration curve that would be obtained in titrating a sample of $NaHCO_3(aq)$ with 1.00 M HCl.
   (b) Sketch the titration curve for $Na_2CO_3(aq)$ by 1.00 M HCl.
   (c) What volume of 0.100 M HCl is required for the neutralization of 1.00 g $NaHCO_3(s)$?
   (d) What volume of 0.100 M HCl is required for the complete neutralization of 1.00 g $Na_2CO_3(s)$?
   (e) A sample of NaOH contains a small amount of Na_2CO_3. For titration to the phenolphthalein end point, 0.1000 g of this sample requires 23.98 mL 0.1000 M HCl. An additional 0.78 mL is required to the methyl orange end point. What is the % Na_2CO_3, by mass, in the sample?

★**14.** Thymol blue indicator has *two* pH ranges. It changes color from red to yellow in the pH range 1.2 to 2.8, and from yellow to blue in the pH range 8.0 to 9.6. What is the color of the indicator in the following situations?
   (a) The indicator is placed in 350 mL 0.205 M HCl.
   (b) To the solution in part (a) is added 250 mL 0.500 M $NaNO_2$.
   (c) To the solution in part (b) is added 150 mL 0.100 M NaOH.
   (d) To the solution in part (c) is added 5.00 g $Ba(OH)_2$.

★**15.** Consider a solution of $NaH_2PO_4(aq)$ of molar concentration, M, and derive equation (18.26) by showing that the pH is independent of M.

★**16.** A solution is prepared that is 0.150 M $HC_2H_3O_2$ and 0.250 M $NaCHO_2$.
   (a) Show that this is a buffer solution.
   (b) Calculate the pH of this buffer solution.
   (c) What is the final pH if to 1.00 L of this buffer solution is added 1.00 L 0.100 M HCl?
(*Hint:* Write equilibrium constant expressions for the two acids, $HC_2H_3O_2$ and $HCHO_2$. In parts (b) and (c), the total acetate concentration = $[HC_2H_3O_2] + [C_2H_3O_2^-]$ and total formate concentration = $[HCHO_2] + [CHO_2^-]$. Because each solution must be electrically neutral, in (b), $[Na^+] + [H_3O^+] = [C_2H_3O_2^-] + [CHO_2^-] + [OH^-]$; and in (c); $[Na^+] + [H_3O^+] = [C_2H_3O_2^-] + [CHO_2^-] + [OH^-] + [Cl^-]$. Because the buffer solution is acidic, a useful simplifying assumption is that $[OH^-] \approx 0$.)

Self-Test Questions

For questions 1 through 8 select the single item that best completes each statement.

1. To *repress* the ionization of formic acid, $HCHO_2(aq)$, add to the solution (a) NaCl; (b) NaOH; (c) $NaCHO_2$; (d) $NaNO_3$.

2. To *increase* the ionization of formic acid, $HCHO_2(aq)$, add to the solution (a) NaCl; (b) $NaCHO_2$; (c) H_2SO_4; (d) $NaHCO_3$.

3. To raise the pH of 1.00 L of 0.50 M HCl(aq) *significantly*, add (a) 0.50 mol $HC_2H_3O_2$; (b) 1.00 mol NaCl; (c) 0.60 mol $NaC_2H_3O_2$; (d) 0.40 mol NaOH.

4. To convert NH_4^+(aq) to NH_3(aq), (a) add H_3O^+; (b) raise the pH; (c) add KNO_3(aq); (d) add NaCl.

5. The effect of adding 0.001 mol KOH to 1.00 L of a solution that is 0.10 M NH_3–0.10 M NH_4Cl is to (a) raise the pH very slightly; (b) lower the pH very slightly; (c) raise the pH by several units; (d) lower the pH by several units.

6. The most acidic of the following 0.10 M salt solutions is that of (a) Na_2S; (b) $NaHSO_4$; (c) $NaHCO_3$; (d) Na_2HPO_4.

7. If an indicator is to be used in an acid–base titration having an equivalence point in the pH range 8 to 10, the indicator must (a) be a weak base; (b) have $K_a \simeq 1 \times 10^{-9}$; (c) ionize in two steps; (d) be added to the solution only after the solution has become alkaline.

8. When a solution of a weak monoprotic acid has been *half-neutralized* by a strong base, (a) the pH = $\frac{1}{2}pK_a$; (b) pH = $\frac{1}{2}$ of the pH value at the equivalence point; (c) pH = twice the initial pH value; (d) pH = pK_a.

9. Explain the difference in meaning between an indicator *end point* and the *equivalence point* of a titration.

10. Indicate whether you would expect the equivalence point of each of the following titrations to be below, above, or at pH 7. Explain your reasoning. **(a)** $NaHCO_3$(aq) is titrated with NaOH(aq); **(b)** HCl(aq) is titrated with NH_3(aq); **(c)** KOH(aq) is titrated with III(aq).

11. A $HCHO_2$–$NaCHO_2$ buffer solution is to be prepared; $K_a(HCHO_2$, formic acid) = 1.8×10^{-4}.
   (a) What mass of $NaCHO_2$ must be dissolved in 0.500 L 0.650 M $HCHO_2$ to produce a pH of 3.90?
   (b) If to the 0.500 L of buffer solution produced in part (a) is added one small pellet of NaOH (about 0.20 g), what will the new pH value be?

12. 25.0 mL 0.0100 M $HC_7H_5O_2$ (the monoprotic acid, benzoic acid, $K_a = 6.3 \times 10^{-5}$) is titrated by 0.0100 M $Ba(OH)_2$. Calculate the pH **(a)** of the initial acid solution; **(b)** after the addition of 6.25 mL of 0.0100 M $Ba(OH)_2$; **(c)** at the equivalence point; **(d)** after the addition of a total of 15.0 mL 0.0100 M $Ba(OH)_2$.

19

Solubility and Complex Ion Equilibria

We open this chapter by considering the nature of equilibrium between a slightly soluble ionic solid and its ions in aqueous solution—solubility equilibrium. The solubility of a solute is given by the concentration of the saturated solution, usually expressed as number of moles of solute per liter of saturated solution. As with acid–base equilibria, we shall find solubility equilibria, and hence solute solubilities, to be strongly affected by the presence of common ions. The solubility equilibria of certain solutes are also affected by the simultaneous occurrence of acid–base reactions. This explains why, for example, some solutes that are insoluble in water are readily soluble in acidic solutions. And still another factor that can greatly increase solute solubilities is complex ion formation, a subject introduced here and discussed more fully in Chapter 24.

An interesting context in which all the equilibrium situations of this and the preceding two chapters are encountered is the qualitative analysis of common cations. The qualitative analysis scheme is presented later in the chapter, with particular emphasis on the role of H_2S ionization equilibria. Most of the concepts of this chapter are amenable to quantitative treatment, and to be able to calculate the concentrations of solution species is important in applied chemistry. Yet, considerable insight into solution equilibria is also possible through qualitative expressions, chiefly net ionic equations and solubility rules. Both are featured in this chapter.

19-1 The Solubility Product Constant, K_{sp}

Silver chromate is slightly soluble (sparingly soluble) in water. The equilibrium existing in a saturated solution is

$$Ag_2CrO_4(s) \rightleftharpoons 2\,Ag^+(aq) + CrO_4^{2-}(aq) \tag{19.1}$$

for which the thermodynamic equilibrium constant expression is

$$K = \frac{(a_{Ag^+})^2(a_{CrO_4^{2-}})}{(a_{Ag_2CrO_4(s)})} \tag{19.2}$$

This expression can be simplified considerably by applying the conventions introduced in Section 16-7. The activity of a pure solid = 1; and in *dilute* solutions molar concentrations may be substituted for activities of solutes. The result obtained is

$$K = K_c = \frac{[Ag^+]^2[CrO_4^{2-}]}{(1)} = [Ag^+]^2[CrO_4^{2-}] \tag{19.3}$$

Because only molar concentration terms appear in expression (19.3), it is appropriate to refer to this equilibrium constant by the symbol K_c. However, a special term and symbolism generally are used instead. An equilibrium constant expression representing equilibrium between a slightly soluble ionic compound and its ions in aqueous solution is called a **solubility product constant,** designated K_{sp}. For a saturated aqueous solution of Ag_2CrO_4 at 25°C,

$$K_{sp} = [Ag^+]^2[CrO_4^{2-}] = 2.4 \times 10^{-12} \tag{19.4}$$

Some typical solubility product constants are listed in Table 19-1.

TABLE 19-1
Solubility product constants at 25°C

Solute	Solubility equilibrium	K_{sp}
aluminum hydroxide	$Al(OH)_3(s) \rightleftharpoons Al^{3+}(aq) + 3\,OH^-(aq)$	1.3×10^{-33}
barium carbonate	$BaCO_3(s) \rightleftharpoons Ba^{2+}(aq) + CO_3{}^{2-}(aq)$	5.1×10^{-9}
barium hydroxide	$Ba(OH)_2(s) \rightleftharpoons Ba^{2+}(aq) + 2\,OH^-(aq)$	5×10^{-3}
barium sulfate	$BaSO_4(s) \rightleftharpoons Ba^{2+}(aq) + SO_4{}^{2-}(aq)$	1.1×10^{-10}
bismuth(III) sulfide	$Bi_2S_3(s) \rightleftharpoons 2\,Bi^{3+}(aq) + 3\,S^{2-}(aq)$	1×10^{-97}
cadmium sulfide	$CdS(s) \rightleftharpoons Cd^{2+}(aq) + S^{2-}(aq)$	8.0×10^{-27}
calcium carbonate	$CaCO_3(s) \rightleftharpoons Ca^{2+}(aq) + CO_3{}^{2-}(aq)$	2.8×10^{-9}
calcium fluoride	$CaF_2(s) \rightleftharpoons Ca^{2+}(aq) + 2\,F^-(aq)$	2.7×10^{-11}
calcium hydroxide	$Ca(OH)_2(s) \rightleftharpoons Ca^{2+}(aq) + 2\,OH^-(aq)$	5.5×10^{-6}
calcium sulfate	$CaSO_4(s) \rightleftharpoons Ca^{2+}(aq) + SO_4{}^{2-}(aq)$	9.1×10^{-6}
chromium(III) hydroxide	$Cr(OH)_3(s) \rightleftharpoons Cr^{3+}(aq) + 3\,OH^-(aq)$	6.3×10^{-31}
cobalt(II) sulfide	$CoS(s) \rightleftharpoons Co^{2+}(aq) + S^{2-}(aq)$	4.0×10^{-21}
copper(II) sulfide	$CuS(s) \rightleftharpoons Cu^{2+}(aq) + S^{2-}(aq)$	6.3×10^{-36}
iron(II) sulfide	$FeS(s) \rightleftharpoons Fe^{2+}(aq) + S^{2-}(aq)$	6.3×10^{-18}
iron(III) hydroxide	$Fe(OH)_3(s) \rightleftharpoons Fe^{3+}(aq) + 3\,OH^-(aq)$	4×10^{-38}
lead(II) chloride	$PbCl_2(s) \rightleftharpoons Pb^{2+}(aq) + 2\,Cl^-(aq)$	1.6×10^{-5}
lead(II) chromate	$PbCrO_4(s) \rightleftharpoons Pb^{2+}(aq) + CrO_4{}^{2-}(aq)$	2.8×10^{-13}
lead(II) iodide	$PbI_2(s) \rightleftharpoons Pb^{2+}(aq) + 2\,I^-(aq)$	7.1×10^{-9}
lead(II) sulfate	$PbSO_4(s) \rightleftharpoons Pb^{2+}(aq) + SO_4{}^{2-}(aq)$	1.6×10^{-8}
lead(II) sulfide	$PbS(s) \rightleftharpoons Pb^{2+}(aq) + S^{2-}(aq)$	8.0×10^{-28}
lithium phosphate	$Li_3PO_4(s) \rightleftharpoons 3\,Li^+(aq) + PO_4{}^{3-}(aq)$	3.2×10^{-9}
magnesium carbonate	$MgCO_3(s) \rightleftharpoons Mg^{2+}(aq) + CO_3{}^{2-}(aq)$	3.5×10^{-8}
magnesium fluoride	$MgF_2(s) \rightleftharpoons Mg^{2+}(aq) + 2\,F^-(aq)$	3.7×10^{-8}
magnesium hydroxide	$Mg(OH)_2(s) \rightleftharpoons Mg^{2+}(aq) + 2\,OH^-(aq)$	1.8×10^{-11}
magnesium phosphate	$Mg_3(PO_4)_2(s) \rightleftharpoons 3\,Mg^{2+}(aq) + 2\,PO_4{}^{3-}(aq)$	1×10^{-25}
manganese(II) sulfide	$MnS(s) \rightleftharpoons Mn^{2+}(aq) + S^{2-}(aq)$	2.5×10^{-13}
mercury(I) chloride	$Hg_2Cl_2(s) \rightleftharpoons Hg_2{}^{2+}(aq) + 2\,Cl^-(aq)$	1.3×10^{-18}
mercury(II) sulfide	$HgS(s) \rightleftharpoons Hg^{2+}(aq) + S^{2-}(aq)$	1.6×10^{-52}
nickel(II) sulfide	$NiS(s) \rightleftharpoons Ni^{2+}(aq) + S^{2-}(aq)$	3.2×10^{-19}
silver bromide	$AgBr(s) \rightleftharpoons Ag^+(aq) + Br^-(aq)$	5.0×10^{-13}
silver carbonate	$Ag_2CO_3(s) \rightleftharpoons 2\,Ag^+(aq) + CO_3{}^{2-}(aq)$	8.1×10^{-12}
silver chloride	$AgCl(s) \rightleftharpoons Ag^+(aq) + Cl^-(aq)$	1.6×10^{-10}
silver chromate	$Ag_2CrO_4(s) \rightleftharpoons 2\,Ag^+(aq) + CrO_4{}^{2-}(aq)$	2.4×10^{-12}
silver iodide	$AgI(s) \rightleftharpoons Ag^+(aq) + I^-(aq)$	8.5×10^{-17}
silver sulfate	$Ag_2SO_4(s) \rightleftharpoons 2\,Ag^+(aq) + SO_4{}^{2-}(aq)$	1.4×10^{-5}
silver sulfide	$Ag_2S(s) \rightleftharpoons 2\,Ag^+(aq) + S^{2-}(aq)$	6.3×10^{-50}
strontium carbonate	$SrCO_3(s) \rightleftharpoons Sr^{2+}(aq) + CO_3{}^{2-}(aq)$	1.1×10^{-10}
strontium sulfate	$SrSO_4(s) \rightleftharpoons Sr^{2+}(aq) + SO_4{}^{2-}(aq)$	3.2×10^{-7}
tin(II) sulfide	$SnS(s) \rightleftharpoons Sn^{2+}(aq) + S^{2-}(aq)$	1.0×10^{-25}
zinc sulfide	$ZnS(s) \rightleftharpoons Zn^{2+}(aq) + S^{2-}(aq)$	1.0×10^{-21}

Example 19-1 Write a solubility product constant expression (K_{sp}) for the formation of a saturated aqueous solution of Bi_2S_3.

Solution. Unless otherwise indicated, a K_{sp} expression is based on the chemical equation written per mole of the solid solute. The ionic concentration terms appearing in the expression are raised to powers equal to the coefficients on the right side of the chemical equation. No term appears in the denominator of a K_{sp} expression. (Why?)

$$Bi_2S_3(s) \rightleftharpoons 2\ Bi^{3+}(aq) + 3\ S^{2-}(aq)$$

$$K_{sp} = [Bi^{3+}]^2[S^{2-}]^3$$

SIMILAR EXAMPLES: Review Problems 1, 2.

19-2 Relationship Between Solubility and K_{sp}

It is possible to relate K_{sp} to an experimentally determined solubility, as in Example 19-2, or to calculate the solubility from a tabulated value of K_{sp}, as in Example 19-3. However, there is an implicit assumption in these calculations—that the dissolved solute exists only as simple, free cations and anions and that these ions do not associate into more complex species. Reasons why this assumption is not always valid are discussed on page 569.

Example 19-2 A 100.0-mL sample is removed from a water solution saturated in MgF_2 at 18°C. The water is completely evaporated from the sample and a 7.6-mg deposit of $MgF_2(s)$ is obtained. What is K_{sp} for MgF_2 at 18°C?

$$MgF_2(s) \rightleftharpoons Mg^{2+}(aq) + 2\ F^-(aq) \qquad K_{sp} = ?$$

Solution. We must first determine the molar solubility of MgF_2 and then relate the ionic concentrations, $[Mg^{2+}]$ and $[F^-]$, to it.

$$\text{no. mol } MgF_2 \text{ per L satd. soln.} = \frac{7.6\ \text{mg } MgF_2}{0.100\ \text{L soln.}} \times \frac{1\ \text{g } MgF_2}{1000\ \text{mg } MgF_2} \times \frac{1\ \text{mol } MgF_2}{62.3\ \text{g } MgF_2}$$

$$= 1.2 \times 10^{-3}\ \text{mol } MgF_2/\text{L}$$

Key factors in the expression below are those (shown in color) which relate the number of moles of ions in solution to the number of moles of solute dissolved.

$$[Mg^{2+}] = 1.2 \times 10^{-3}\ \frac{\text{mol } MgF_2}{\text{L}} \times \frac{1\ \text{mol } Mg^{2+}}{1\ \text{mol } MgF_2} = 1.2 \times 10^{-3}\ M$$

$$[F^-] = 1.2 \times 10^{-3}\ \frac{\text{mol } MgF_2}{\text{L}} \times \frac{2\ \text{mol } F^-}{1\ \text{mol } MgF_2} = 2 \times 1.2 \times 10^{-3} = 2.4 \times 10^{-3}\ M$$

$$K_{sp} = [Mg^{2+}][F^-]^2 = (1.2 \times 10^{-3})(2.4 \times 10^{-3})^2 = 6.9 \times 10^{-9}$$

SIMILAR EXAMPLES: Review Problem 4; Exercise 3.

Example 19-3 Calculate the molar solubility of Ag_2CrO_4 in water at 25°C.

Solution. The formation of a saturated solution is represented by

$$Ag_2CrO_4(s) \rightleftharpoons 2\ Ag^+(aq) + CrO_4^{2-}(aq) \qquad K_{sp} = 2.4 \times 10^{-12}$$

Two moles of Ag^+ and *one* mole of CrO_4^{2-} ion appear in the saturated solution for every mole of Ag_2CrO_4 dissolved. If S represents the number of moles of Ag_2CrO_4 that have dissolved per liter of saturated solution, then at equilibrium

$$[Ag^+] = 2S \qquad [CrO_4^{2-}] = S$$

The solubility product relationship must be satisfied by these concentrations.

$$K_{sp} = [Ag^+]^2[CrO_4^{2-}] = (2S)^2(S) = 2.4 \times 10^{-12}$$

$$4S^3 = 2.4 \times 10^{-12}$$

$$S^3 = 0.60 \times 10^{-12}$$

$$S = (0.60)^{1/3} \times 10^{-4} = 0.84 \times 10^{-4}$$

$$S = \text{molar solubility} = 8.4 \times 10^{-5} \text{ mol Ag}_2\text{CrO}_4/\text{L}$$

SIMILAR EXAMPLES: Review Problem 3; Exercise 2.

Examples 19-2 and 19-3 demonstrate that molar solubilities and solubility product constants are related, but they are by no means identical. Each can be used as a basis for *calculating* the other. Their numerical values will never be equal.

Limitation of K_{sp} to Slightly Soluble Solutes.   We have used the terms "slightly soluble solute" and "sparingly soluble solute" in discussing the solubility product constant. Cannot similar expressions be written for saturated solutions of moderately or highly soluble ionic compounds, for example, NaCl, KNO_3, or NaOH? For these solutes we can indeed write solubility equilibrium equations similar to (19.1) and *thermodynamic* equilibrium constant expressions similar to (19.2). What we cannot do is substitute ionic concentrations for ionic activities, as we did in deriving equations (19.3) and (19.4) from (19.2). Saturated solutions of moderately or highly soluble ionic compounds are simply much too concentrated to permit the assumption that activities and molar concentrations are equal. Without this assumption much of the value of the solubility product concept is lost. Even if the qualifier, "slightly soluble," is omitted in describing a solubility equilibrium, whenever a K_{sp} value is given it will be for a slightly soluble ionic compound.

The Common Ion Effect.   In the situations discussed to this point, the saturated solution has contained ions derived from a *single* source, the pure solid solute. What will be the effect on equilibrium in the saturated solution if some of these ions are introduced from a *second* source as well? As an example, suppose that to the saturated solution of Ag_2CrO_4 described in Example 19-3 is added some CrO_4^{2-} ion—*a common ion*—from a source such as $K_2CrO_4(aq)$.

According to Le Châtelier's principle, a system at equilibrium responds to an increase in concentration of one of its reactants by shifting in the direction in which that reactant is consumed. In this case, to an original equilibrium mixture

$$Ag_2CrO_4(s) \rightleftharpoons 2\ Ag^+(aq) \quad + CrO_4^{2-}(aq)$$

is added CrO_4^{2-}.

The reverse reaction is favored,

leading to a new equilibrium in which

additional $Ag_2CrO_4(s)$ precipitates,	$[Ag^+]$ is less than in the original equilibrium,	$[CrO_4^{2-}]$ is greater than in the original equilibrium.

The solubility of a slightly soluble ionic compound is lowered by the presence of a second solute that furnishes a common ion. The common ion effect in solubility equilibrium is pictured in Figure 19-1.

Most ionic compounds for which K_{sp} values are given are termed "insoluble." By this we simply mean that their solubilities are very limited.

FIGURE 19-1
The common ion effect in solubility equilibrium.

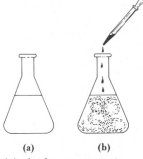

(a)   **(b)**

(a) A clear saturated solution.
(b) Addition of a small volume of a solution containing a common ion. The common ion reduces the solubility of the solute, and the excess solid precipitates.

This is our second encounter with the common ion effect. Our first was in Section 18-1. A review of some of the conclusions of that section will establish certain similarities to solubility equilibrium. In Section 18-1 we found that the presence of the common ions, H_3O^+ and $C_2H_3O_2^-$, *repressed* the ionization of $HC_2H_3O_2$. Here we find that the effect of adding the common ion CrO_4^{2-} (or Ag^+) to a saturated solution is to *reduce* the solubility of Ag_2CrO_4. We will discover other similarities to situations in Chapter 18. For example, there we dealt with ionization constants (K_w, K_a, and K_b) and studied the effect of changing the concentrations of various solution components on a property like pH. Here we will deal with solubility product constants (K_{sp}) and see how changing the concentrations of various solution components affects the property of solute solubility. In Example 19-4 we recalculate the solubility of Ag_2CrO_4 in the presence of the common ion, CrO_4^{2-}.

Example 19-4   What is the molar solubility of Ag_2CrO_4 in 0.10 M K_2CrO_4(aq)?

Solution.   The ion concentrations at equilibrium, *derived from the dissolved* Ag_2CrO_4, can be written as they were in Example 19-3. That is, if we let $S =$ the molar solubility of Ag_2CrO_4, then $[Ag^+] = 2S$ and $[CrO_4^{2-}] = S$. However, an *additional* 0.10 mol CrO_4^{2-}/L is derived from the K_2CrO_4(aq). The *total* ion concentrations at equilibrium are $[Ag^+] = 2S$ and $[CrO_4^{2-}] = 0.10 + S$. This kind of information can generally be summarized effectively in the following manner.

$$Ag_2CrO_4(s) \rightleftharpoons 2\,Ag^+(aq) + CrO_4^{2-}(aq)$$

from $Ag_2CrO_4(s)$:	$2S$	S mol/L
from 0.10 M K_2CrO_4(aq):	—	0.10 mol/L
equilibrium concentrations:	$2S$	$(0.10 + S)$ mol/L

The usual K_{sp} relationship must be satisfied, that is,

$$K_{sp} = [Ag^+]^2[CrO_4^{2-}] = (2S)^2(0.10 + S) = 2.4 \times 10^{-12}$$

Because we know that S in this case will be even smaller than it was in Example 19-3, we can safely assume that $S \ll 0.10$ and $(0.10 + S) \simeq 0.10$.

$$(2S)^2(0.10) = 4S^2(0.10) = 2.4 \times 10^{-12}$$

$$S^2 = 6.0 \times 10^{-12}$$

$$S = 2.4 \times 10^{-6} = 2.4 \times 10^{-6}\ M$$

Since we defined S to be the number of moles of Ag_2CrO_4 dissolved per liter of 0.10 M K_2CrO_4(aq), the value of S is, in fact, the molar solubility we are seeking: 2.4×10^{-6} mol Ag_2CrO_4/L.

SIMILAR EXAMPLES: Review Problem 5; Exercises 11, 12.

The molar solubility of Ag_2CrO_4 in the presence of 0.10 M CrO_4^{2-} calculated in Example 19-4 (2.4×10^{-6} mol Ag_2CrO_4/L) is 35 times less than its value in pure water calculated in Example 19-3 (8.4×10^{-5} mol Ag_2CrO_4/L). Calculation of the solubility of Ag_2CrO_4 in the presence of 0.10 M Ag^+ would show an even more striking effect produced by Ag^+ as the common ion.

The Diverse ("Uncommon") Ion Effect—The Salt Effect.   Having just established the effect of common ions, we might next ask: Do ions different from those involved in the solubility equilibrium ("uncommon" ions) have an effect on the solubilities of sparingly soluble ionic compounds? They do, but in a different way from common ions. The effect is significant but not as striking as the common ion effect. Moreover, the presence of "uncommon" ions tends to *increase* rather than decrease solubility. As the total ionic concentration of a solution increases, interionic attractions become more important (recall Section 12-7). Activities (effective concentrations) become smaller than the stoichiomet-

FIGURE 19-2
Comparison of the
common ion effect and
the salt effect.

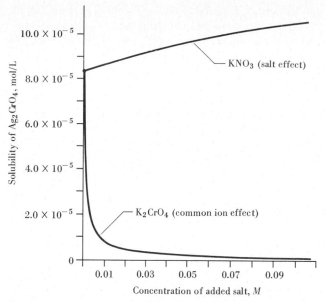

The presence of the common ion CrO_4^{2-}, derived from $K_2CrO_4(aq)$, reduces the solubility of Ag_2CrO_4 by a factor of about 35 over the concentration range shown (from 0 to 0.10 M added salt). Over this same range, the solubility of Ag_2CrO_4 is increased by the presence of the "uncommon" or diverse ions from KNO_3, but only by about 25% or so.

ric or measured concentrations. For the ions involved in the solution process this means that higher concentrations must appear in solution before equilibrium is established—*the solubility increases*. Figure 19-2 suggests the different effects of common and "uncommon" ions.

The "uncommon" or diverse ion effect is more commonly referred to as the **salt effect.** In all further discussion of solubility equilibrium, we will neglect the salt effect. This means that we must limit our discussion to very dilute solutions with all ions derived from a sparingly soluble solute, or to more concentrated solutions with a common ion present.

Other Factors Affecting the Solubilities of Sparingly Soluble Ionic Compounds. In each calculation performed to this point we have *assumed* that all of the dissolved solute appears in solution as separated cations and anions. In many cases this assumption is *not* valid. For example, in a saturated solution of magnesium fluoride **ion pairs** consisting of a Mg^{2+} and an F^- ion, that is, MgF^+, might be found. To the extent that ion-pair formation occurs in a solution, the free ion concentrations tend to be reduced. This means that the amount of solute that must dissolve to maintain the required *free* ion concentrations to satisfy the K_{sp} expression increases: *Solubility increases when ion-pair formation occurs in solution*. Although ion-pair formation can be significant in some cases (especially for a moderately soluble solute yielding ions with high charge), we will not specifically consider its effect on solubility equilibrium. We must accept the fact, however, that certain of our calculations may be somewhat in error as a consequence.

A more significant factor than ion-pair formation is that an ionic species participating in a solubility equilibrium may be *simultaneously* involved in an acid-base or complex ion equilibrium. We will consider these possibilities later in the chapter.

Constancy of K_{sp}. When considering such phenomena as the salt effect, the value of K_{sp} based on molar concentrations varies depending on the ionic atmosphere. But for all the applications considered in this text we will assume that the form of the solubility product constant expression and the value of K_{sp} remain unchanged. As with other equilibrium constants, however, values of K_{sp} are temperature dependent.

19-3 Precipitation Reactions

We have used the term *solubility equilibrium* to describe the phenomena encountered to this point. But, as we have seen on previous occasions, an equilibrium condition can be approached from *either* direction. If the equilibria of the preceding sections are approached by starting with ions in solution and producing pure, undissolved solute, then the process involved is a *precipitation reaction*. And there is a great deal that we can say about precipitation reactions from the standpoint of solubility product constants.

Criterion for Precipitation from Solution. The most fundamental question we can ask about a precipitation reaction is whether it will in fact occur for a given set of conditions. Suppose that a solution is made simultaneously 0.10 M in Ag^+ and 0.10 M in Cl^-. Should a precipitate of AgCl(s) form? To answer this question we begin with a chemical equation to represent equilibrium between the slightly soluble solute and its ions, together with a value of K_{sp} for this equilibrium.

$$AgCl(s) \rightleftharpoons Ag^+(aq) + Cl^-(aq) \tag{19.5}$$

$$K_{sp} = [Ag^+][Cl^-] = 1.6 \times 10^{-10} \tag{19.6}$$

Now recall how we dealt with the question of the direction of net change in Section 15-4. We formulated a quantity called the reaction quotient, Q, and compared its value with that of the equilibrium constant, K. In this case the reaction quotient is just the product, $[Ag^+][Cl^-]$, based on the initial concentrations of these ions. For precipitation reactions, Q is sometimes called the ion product.

$$Q = (0.10)(0.10) = 1 \times 10^{-2} > K_{sp} = 1.6 \times 10^{-10}$$

We conclude that reaction should occur to the left or in the reverse direction in equation (19.5)—*precipitation should occur*.

More general conclusions about precipitation from solution are

Precipitation should occur if $Q > K_{sp}$.

Precipitation does not occur if $Q < K_{sp}$. $\tag{19.7}$

A solution is just saturated if $Q = K_{sp}$.

Illustrative Examples. In each of the following three examples a different idea important to a successful quantitative description of precipitation reactions is presented.

The point illustrated by Example 19-5 is that the criterion for precipitation must be applied to ion concentrations *after* mixing of solutions. That is, any possible dilution effects must be taken into account.

Example 19-5 Should a precipitate form if 10.0 mL 0.0010 M AgNO₃(aq) is added to 500.0 mL 0.0020 M K₂CrO₄(aq)?

$$Ag_2CrO_4(s) \rightleftharpoons 2\,Ag^+(aq) + CrO_4^{2-}(aq) \qquad K_{sp} = 2.4 \times 10^{-12}$$

Solution
Initial amounts:

$$\text{no. mol Ag}^+ = 0.0100 \text{ L} \times \frac{0.0010 \text{ mol AgNO}_3}{1 \text{ L}} \times \frac{1 \text{ mol Ag}^+}{1 \text{ mol AgNO}_3}$$

$$= 1.0 \times 10^{-5} \text{ mol Ag}^+$$

$$\text{no. mol CrO}_4{}^{2-} = 0.500 \text{ L} \times \frac{0.0020 \text{ mol K}_2\text{CrO}_4}{1 \text{ L}} \times \frac{1 \text{ mol CrO}_4{}^{2-}}{1 \text{ mol K}_2\text{CrO}_4}$$

$$= 1.0 \times 10^{-3} \text{ mol CrO}_4{}^{2-}$$

Total solution volume $= 0.0100 \text{ L} + 0.500 \text{ L} = 0.510 \text{ L}$.

After mixing:

$$[\text{Ag}^+] = \frac{1.0 \times 10^{-5} \text{ mol Ag}^+}{0.510 \text{ L}} = 2.0 \times 10^{-5} \text{ M}$$

$$[\text{CrO}_4{}^{2-}] = \frac{1.0 \times 10^{-3} \text{ mol CrO}_4{}^{2-}}{0.510 \text{ L}} = 2.0 \times 10^{-3} \text{ M}$$

Application of criterion:

$$Q = [\text{Ag}^+]^2[\text{CrO}_4{}^{2-}] = (2.0 \times 10^{-5})^2(2.0 \times 10^{-3}) = 8.0 \times 10^{-13}$$
$$= 8.0 \times 10^{-13} < K_{sp}(= 2.4 \times 10^{-12})$$

Precipitation of $\text{Ag}_2\text{CrO}_4(s)$ will *not* occur.

SIMILAR EXAMPLES: Review Problem 6; Exercises 17, 18.

Note that if Q had been based on the ion concentrations as given, rather than after mixing, we would have concluded, *erroneously*, that precipitation should occur. That is,

$$(1 \times 10^{-3})^2$$
$$\times (2 \times 10^{-3}) > K_{sp}$$

In Example 19-6 our interest centers on whether precipitation goes to completion, that is, on the concentration of an ion (Mg^{2+}) *remaining* in solution after precipitation has occurred.

Example 19-6 The first step in the extraction of magnesium metal from seawater involves precipitating Mg^{2+} as $\text{Mg(OH)}_2(s)$. The magnesium ion concentration in seawater is about 0.059 M. If a seawater sample is treated so that its $[\text{OH}^-]$ is maintained at $2.0 \times 10^{-3} \text{ M}$, **(a)** what will be the concentration of Mg^{2+} remaining in solution after precipitation has occurred? **(b)** Can we say that precipitation is complete?

$$\text{Mg(OH)}_2(s) \rightleftharpoons \text{Mg}^{2+}(aq) + 2 \text{ OH}^-(aq) \qquad K_{sp} = 1.8 \times 10^{-11}$$

Solution

(a) There is no question that precipitation will occur since the product, $[\text{Mg}^{2+}][\text{OH}^-]^2 = (0.059)(2.0 \times 10^{-3})^2 = 2.4 \times 10^{-7}$, exceeds K_{sp}. We need to determine the $[\text{Mg}^{2+}]$ remaining in a solution from which some solid Mg(OH)_2 has precipitated. For the saturated solution at equilibrium, we substitute $[\text{OH}^-]$ (maintained constant at $2.0 \times 10^{-3} \text{ M}$) into the solubility product constant expression and solve for $[\text{Mg}^{2+}]$.

$$K_{sp} = [\text{Mg}^{2+}][\text{OH}^-]^2 = [\text{Mg}^{2+}] \times (2.0 \times 10^{-3})^2 = 1.8 \times 10^{-11}$$

$$[\text{Mg}^{2+}] = \frac{1.8 \times 10^{-11}}{4.0 \times 10^{-6}} = 4.5 \times 10^{-6} \text{ M}$$

A useful rule of thumb is that precipitation is complete if less than one part per thousand (0.1%) of the original solute is left unprecipitated.

(b) $[\text{Mg}^{2+}]$ in the seawater is reduced from 0.059 M to $4.5 \times 10^{-6} \text{ M}$ as a result of the precipitation reaction. We can conclude that precipitation is complete.

SIMILAR EXAMPLES: Exercises 22, 23a.

1 line short

In the example just considered the concentration of one of the precipitating ions (OH^-) was kept constant while the other ion (Mg^{2+}) was removed from solution. In some precipitation reactions *both* ion concentrations change during the precipitation, as illustrated in Example 19-7.

Example 19-7 A 50.0-mL sample of 0.0152 *M* $Na_2SO_4(aq)$ is added to 50.0 mL of 0.0125 *M* $Ca(NO_3)_2(aq)$. **(a)** Should precipitation of $CaSO_4(s)$ occur? **(b)** Will precipitation of Ca^{2+} be complete?

$$CaSO_4(s) \rightleftharpoons Ca^{2+}(aq) + SO_4^{2-}(aq) \qquad K_{sp} = 9.1 \times 10^{-6}$$

Solution

(a) The concentrations of ions present *after* mixing (i.e., in the final 100.0 mL solution) are

$$[Ca^{2+}] = 0.0500 \text{ L} \times \frac{0.0125 \text{ mol } Ca^{2+}}{L} \times \frac{1}{0.1000 \text{ L}} = 6.25 \times 10^{-3} \ M$$

$$[SO_4^{2-}] = 0.0500 \text{ L} \times \frac{0.0152 \text{ mol } SO_4^{2-}}{L} \times \frac{1}{0.1000 \text{ L}} = 7.60 \times 10^{-3} \ M$$

$$\begin{aligned} Q = [Ca^{2+}][SO_4^{2-}] &= (6.25 \times 10^{-3})(7.60 \times 10^{-3}) \\ &= 4.75 \times 10^{-5} > K_{sp} = 9.1 \times 10^{-6} \end{aligned}$$

Precipitation of $CaSO_4(s)$ should occur.

(b) The course of the precipitation reaction is outlined below. In this outline the molar concentration of Ca^{2+} in equilibrium with $CaSO_4(s)$ is set equal to x, for which we must solve.

	$CaSO_4(s) \rightleftharpoons$	$Ca^{2+}(aq) +$	$SO_4^{2-}(aq)$
initial concentrations from (a):		0.00625 *M*	0.00760 *M*
consumed in the precipitation:		0.00625 − x	0.00625 − x
at equilibrium:		x	0.00760 − (0.00625 − x)
			0.00760 − 0.00625 + x
			0.00135 + x

The essential difference between a problem of this type and the one considered in Example 19-6 is in the last column of figures. In Example 19-6, $[OH^-]$ was the same, initially and after precipitation was completed. Here, $[SO_4^{2-}]$ decreases from its initial to its equilibrium value.

What we have indicated above is that if $[Ca^{2+}]$ falls from 0.00625 *M* to x, then the precipitation of $CaSO_4(s)$ must have consumed a concentration of Ca^{2+} equal to 0.00625 − x. One SO_4^{2-} ion is removed from solution for every Ca^{2+} ion. The decrease in $[SO_4^{2-}]$ must also be 0.00625 − x. The $[SO_4^{2-}]$ *remaining* in solution at equilibrium is the initial concentration minus that which is consumed: 0.00760 − (0.00625 − x) = 0.00135 + x.

We can now substitute the equilibrium concentrations into the K_{sp} expression and solve for x.

The usual simplifying assumption that x is very small compared to a number to which it is added or from which it is subtracted does not work here. The number 0.00135 is itself quite small. The calculated value of x turns out to be larger, not smaller, than 0.00135.

$$K_{sp} = [Ca^{2+}][SO_4^{2-}] = (x)(0.00135 + x) = 9.1 \times 10^{-6}$$

The result is a quadratic equation; $x^2 + 0.00135x - 9.1 \times 10^{-6} = 0$. Solution of this equation by the quadratic formula leads to the result

$$x = \frac{-0.00135 \pm \sqrt{(0.00135)^2 + 4 \times 9.1 \times 10^{-6}}}{2} = \frac{-0.00135 \pm 6.2 \times 10^{-3}}{2}$$

$$[Ca^{2+}] = x = \frac{4.8 \times 10^{-3}}{2} = 2.4 \times 10^{-3} \ M$$

The percentage of calcium ion left in solution can be expressed as

$$\% \ Ca^{2+} \ \text{remaining} = \frac{2.4 \times 10^{-3} \ M \ Ca^{2+}}{6.25 \times 10^{-3} \ M \ Ca^{2+}} \times 100 = 38\%$$

Precipitation of Ca^{2+} is incomplete.

SIMILAR EXAMPLES: Exercises 23b, 24, 25.

Completeness of Precipitation. Example 19-7 suggests the following generalization concerning the completeness of a precipitation reaction.

Value to consider	Complete precipitation is likely if value is	Precipitation may not be complete if value is
K_{sp} of precipitate	small	large
concentration of common ion in saturated solution	large	small

Unfortunately, there is no simple definition of ''small'' and ''large.'' We can say, though, that in Example 19-7, K_{sp} was large compared to most of the K_{sp} values in Table 19-1, and the concentration of common ion (SO_4^{2-}) in the saturated solution was small. These are both conditions which lead to *incompleteness* of precipitation.

19-4 Precipitation Reactions in Quantitative Analysis

In Chapter 3 we described how the precipitation of a solid can be used to determine the exact composition of a sample of matter (see, e.g., Example 3-13). What does it take to conduct a successful precipitation analysis? One factor that we assessed through Examples 19-6 and 19-7 is that precipitation must be as complete as possible (so that very little of the desired substance is left in solution). In purifying a precipitate by washing, it is sometimes necessary to do so with a solution containing a common ion rather than with pure water. This is done to reduce the solubility of the precipitate. Figure 19-3 outlines these ideas as applied to the gravimetric determination (determination by weighing) of calcium.

FIGURE 19-3
Outline of a gravimetric analysis for calcium.

Ca^{2+}(aq) [from sample being analyzed]

add excess
$(NH_4)_2C_2O_4$(aq)

$CaC_2O_4 \cdot H_2O$(s) [impure]

wash with dilute $(NH_4)_2C_2O_4$(aq);
filter and dry

$CaC_2O_4 \cdot H_2O$(s) [pure]

heat to 500°C

$CaCO_3$(s) [pure]

A weighed sample is dissolved [usually in HCl(aq)] to obtain Ca^{2+}(aq). The solution is treated with excess $(NH_4)_2C_2O_4$(aq) and the hydrate, $CaC_2O_4 \cdot H_2O$, precipitates. The precipitate is purified by washing with $(NH_4)_2C_2O_4$(aq). When heated strongly, $CaC_2O_4 \cdot H_2O$(s) is converted to $CaCO_3$(s), in which form the calcium is finally weighed.

In the washing process it is important to use a solution that provides a common ion to reduce the solubility of $CaC_2O_4 \cdot H_2O$(s) but one that does not leave a residue that would contaminate the final $CaCO_3$(s). Upon heating, $(NH_4)_2C_2O_4$ undergoes decomposition to gaseous products—NH_3, H_2O, CO, and CO_2.

Fractional Precipitation.   Another technique that can be better understood through principles of solubility equilibrium is that of fractional precipitation. This term refers to a situation in which two or more ions in solution, each capable of being precipitated by the same reagent, are *separated* by the use of that reagent: *One ion is precipitated and the other(s) remains in solution.* The primary condition for a successful fractional precipitation is that there be a significant difference in the solubilities of the substances to be separated. (Usually this means a significant difference in their K_{sp} values, say of the order of 10^5 or greater.)

Example 19-8 considers the separation of $Cl^-(aq)$ and $I^-(aq)$ through the use of $Ag^+(aq)$. The data needed to describe this separation are the solubility equilibrium equations and solubility product constants for AgCl and AgI.

$$AgCl(s) \rightleftharpoons Ag^+(aq) + Cl^-(aq) \qquad K_{sp} = 1.6 \times 10^{-10}$$

$$AgI(s) \rightleftharpoons Ag^+(aq) + I^-(aq) \qquad K_{sp} = 8.5 \times 10^{-17}$$

Example 19-8   To a solution that has $[Cl^-] = 0.010 \ M$ and $[I^-] = 0.010 \ M$, is slowly added $0.10 \ M \ AgNO_3(aq)$ (see Figure 19-4).
(a) Show that AgI(s) precipitates before AgCl(s).
(b) At the point AgCl(s) begins to precipitate, what is $[I^-]$ remaining in solution?
(c) Is separation of $I^-(aq)$ and $Cl^-(aq)$ by fractional precipitation feasible?

Solution

(a) As a drop of the $AgNO_3(aq)$ enters the solution, $[Ag^+]$ builds up from a value of zero to a point where one of the ion products, Q, exceeds the corresponding K_{sp}. Then precipitation occurs. The required values of $[Ag^+]$ for precipitation are

for AgI(s) to ppt: $Q = [Ag^+][I^-] = [Ag^+](0.010) = 8.5 \times 10^{-17} = K_{sp}$
$$[Ag^+] = 8.5 \times 10^{-15} \ M$$

for AgCl(s) to ppt: $Q = [Ag^+][Cl^-] = [Ag^+](0.010) = 1.6 \times 10^{-10} = K_{sp}$
$$[Ag^+] = 1.6 \times 10^{-8} \ M$$

Since the required $[Ag^+]$ to start the precipitation of AgI(s) is less than that for AgCl(s), AgI(s) will be first to precipitate. As long as a significant quantity of AgI(s) is forming, the free silver ion concentration in solution is not able to attain the value required for the precipitation of AgCl(s).

(b) As more and more AgI(s) precipitates, $[I^-]$ gradually decreases; and this permits $[Ag^+]$ to increase. When $[Ag^+]$ reaches $1.6 \times 10^{-8} \ M$, precipitation of AgCl(s) begins. Next, we need to answer the question: What is $[I^-]$ at the point where $[Ag^+] = 1.6 \times 10^{-8} \ M$? For this we use K_{sp} for AgI, and solve for $[I^-]$.

$$K_{sp} = [Ag^+][I^-] = (1.6 \times 10^{-8})[I^-] = 8.5 \times 10^{-17}$$

$$[I^-] = 5.3 \times 10^{-9} \ M$$

(c) Before AgCl(s) begins to precipitate, $[I^-]$ will have been reduced from 1.0×10^{-2} to $5.3 \times 10^{-9} \ M$. Essentially all of the I^- will have precipitated from solution as AgI(s) while the Cl^- remains in solution. Fractional precipitation is feasible for separating mixtures of Cl^- and I^-.

SIMILAR EXAMPLES: Review Problem 7; Exercises 26, 27.

How might we conduct the titration illustrated in Figure 19-4 and Example 19-8? That is, how can we stop the titration before AgCl(s) starts to precipitate—before $[Ag^+]$ reaches $1.6 \times 10^{-8} \ M$? A particularly effective method is to follow $[Ag^+]$ during the titration. $[Ag^+]$ increases very rapidly between the point where AgI has finished precipitating and AgCl is about to begin (see Additional Exercise 12). Methods of determining very low ion concentrations in solution are discussed in Section 20-5.

FIGURE 19-4
Fractional
precipitation—Example
19-8 illustrated.

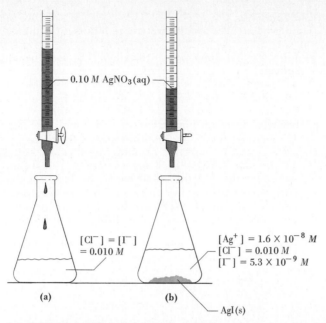

0.10 M AgNO$_3$(aq)

$[Cl^-] = [I^-]$
$= 0.010\ M$

$[Ag^+] = 1.6 \times 10^{-8}\ M$
$[Cl^-] = 0.010\ M$
$[I^-] = 5.3 \times 10^{-9}\ M$

(a)   (b)

AgI(s)

(a) 0.10 M AgNO$_3$(aq) is slowly added to a solution that is 0.010 M in Cl$^-$ and 0.010 M in I$^-$.
(b) This is the condition at the point where AgCl(s) would just begin to precipitate. Essentially all of the I$^-$ has precipitated as AgI(s), leaving $[I^-] = 5.3 \times 10^{-9}\ M$ in the solution. The Cl$^-$ and I$^-$ have been separated.

19-5 Writing Net Ionic Equations

Suppose that we are asked to predict whether a chemical reaction will occur when the following solutions are mixed.

$$\text{Pb(NO}_3)_2(aq) + \text{KI(aq)} \longrightarrow\ ? \tag{19.8}$$

An appropriate start would be to write (19.8) in the *ionic* form, that is, to show the actual ionic species that are brought together.

$$\text{Pb}^{2+}(aq) + 2\ \text{NO}_3{}^-(aq) + \text{K}^+(aq) + \text{I}^-(aq) \longrightarrow \tag{19.9}$$

Now we can use our chemical knowledge. Is there any combination of cations and anions in (19.9) that might produce a precipitate?

The potential products are KNO$_3$, KI, Pb(NO$_3$)$_2$, and PbI$_2$. Of these possible ion combinations only PbI$_2$ is listed in Table 19-1. We might conclude that it is "insoluble" and the other potential products are all soluble. As long as [Pb^{2+}] and [I$^-$] in the mixed solution are large enough that $[\text{Pb}^{2+}][\text{I}^-]^2 > K_{sp}(\text{PbI}_2)$, a precipitate should form.

$$\text{Pb}^{2+}(aq) + 2\ \text{I}^-(aq) \longrightarrow \text{PbI}_2(s) \tag{19.10}$$

The method used here is not entirely satisfactory, however. The available listing of solubility product constants may not be extensive enough to be certain that the absence of a substance from the list means that it is soluble. A better approach is to employ a set of **solubility rules,** as in Table 19-2. From Table 19-2 we note that all nitrates are soluble— neither Pb(NO$_3$)$_2$ nor KNO$_3$ is expected to precipitate. Among the common iodides, all are soluble *except* those of Ag$^+$, Hg$_2{}^{2+}$, and Pb^{2+}. A precipitate of PbI$_2$ should form.

TABLE 19-2
Some general rules for the water solubilities of common ionic compounds[a]

1. All common compounds of the alkali (IA) metals and the ammonium ion (NH_4^+) are **soluble.**
2. All nitrates (NO_3^-), chlorates (ClO_3^-), perchlorates (ClO_4^-), and acetates ($C_2H_3O_2^-$) are **soluble.** Silver acetate is only **moderately soluble.**
3. The chlorides (Cl^-), bromides (Br^-), and iodides (I^-) of most metals are **soluble.** The principal *exceptions* are those of Pb^{2+}, Ag^+, and Hg_2^{2+}.
4. All sulfates (SO_4^{2-}) are **soluble** *except* for those of Sr^{2+}, Ba^{2+}, Pb^{2+}, and Hg_2^{2+}. Sulfates of Ca^{2+} and Ag^+ are only **moderately soluble.**
5. All carbonates (CO_3^{2-}), chromates (CrO_4^{2-}), and phosphates (PO_4^{3-}) are **insoluble** *except* for those of the alkali metals (including ammonium).
6. The group IA metal hydroxides (OH^-) are **soluble.** The hydroxides of Ca^{2+}, Sr^{2+}, and Ba^{2+} are **moderately soluble.** The rest of the hydroxides are **insoluble.**
7. The sulfides of all metals are **insoluble** *except* for those of NH_4^+ and the IA and IIA metals.

[a] Generally speaking, if a saturated solution of an ionic solute is greater than about 0.10 *M*, the solute is said to be soluble; if less than about 0.01 *M*, insoluble; and if of an intermediate concentration (i.e., between about 0.01 *M* and 0.10 *M*), moderately soluble.

Example 19-9   Predict whether a chemical reaction should occur in each of the following cases. If so, write a net ionic equation to represent the reaction.
(a) $NaOH(aq) + MgCl_2(aq) \longrightarrow$?
(b) $(NH_4)_2SO_4(aq) + CuCl_2(aq) \longrightarrow$?

Solution
(a) The ions present in the mixed solution are Na^+, Mg^{2+}, OH^-, and Cl^-. Since all common sodium compounds are soluble in water (see Table 19-2), the Na^+ remains in solution. Magnesium hydroxide is *insoluble*. The reaction is

$$Mg^{2+}(aq) + 2\ OH^-(aq) \longrightarrow Mg(OH)_2(s) \qquad (19.11)$$

Note that it was possible to go directly to the net ionic equation without first having to write an equation with the "spectator" ions—Na^+ and Cl^-.
(b) An analysis of the solubility relationships for all the possible ionic combinations in this case indicates that there are no insoluble combinations. All ammonium compounds are soluble, and $CuCl_2$ and $CuSO_4$ are not among the few insoluble chlorides and sulfates cited in Table 19-2. No chemical reaction occurs, a fact that can be indicated as follows

$$(NH_4)_2SO_4(aq) + CuCl_2(aq) \longrightarrow \text{no reaction} \qquad (19.12)$$

SIMILAR EXAMPLE: Review Problem 8.

We will write net ionic equations on several occasions in the remaining sections of this chapter.

19-6 Solubility and pH

If the ions derived from a slightly soluble solute are able to enter into acid–base reactions with H_3O^+ or OH^-, the solubility of the solute will be affected by pH. Consider $Mg(OH)_2$, for example. OH^- ions derived from the solubility equilibrium can react with H_3O^+ to form H_2O.

$$Mg(OH)_2(s) \rightleftharpoons Mg^{2+}(aq) + 2\ OH^-(aq) \qquad K_{sp} = 1.8 \times 10^{-11} \qquad (19.13)$$

$$OH^-(aq) + H_3O^+(aq) \longrightarrow 2\ H_2O \qquad (19.14)$$

According to Le Châtelier's principle, reaction (19.14) disturbs the equilibrium represented by (19.13) through removal of OH^-. Equilibrium in (19.13) is displaced to the right as $Mg(OH)_2(s)$ dissolves in an attempt to replace OH^- drawn off by reaction (19.14). In fairly acidic solutions reactions (19.13) and (19.14) go to completion and $Mg(OH)_2$ is highly soluble. The net reaction is

$$Mg(OH)_2(s) + 2\ H_3O^+(aq) \longrightarrow Mg^{2+}(aq) + 4\ H_2O \tag{19.15}$$

More on Net Ionic Equations.   Although introduced in the preceding section as a qualitative means of describing precipitation reactions, net ionic equations can be written for a variety of reaction types. Equation (19.15), for example, is a net ionic equation showing that $Mg(OH)_2(s)$ is soluble in acidic solutions. Example 19-10 illustrates how solubility equilibrium, acid–base equilibrium, and gas formation may all be involved in a net ionic equation.

Example 19-10   Write a net ionic equation to represent the dissolving of $CaCO_3(s)$ in $HC_2H_3O_2(aq)$.

Solution.   Carbonate ions are produced by the reaction

$$CaCO_3(s) \rightleftharpoons Ca^{2+}(aq) + CO_3^{2-}(aq) \tag{19.16}$$

Hydronium ion is furnished by the ionization of acetic acid.

$$HC_2H_3O_2(aq) + H_2O \rightleftharpoons H_3O^+(aq) + C_2H_3O_2^-(aq) \tag{19.17}$$

Carbonate ion, CO_3^{2-}, is a stronger base than acetate ion, $C_2H_3O_2^-$ (recall Table 17-5) and accepts protons from H_3O^+.

$$CO_3^{2-}(aq) + H_3O^+(aq) \longrightarrow HCO_3^-(aq) + H_2O \tag{19.18}$$

Reaction (19.18) promotes both the ionization of $HC_2H_3O_2$ (19.17) and the dissolving of $CaCO_3$ (19.16). Further reaction occurs between HCO_3^- and H_3O^+.

$$HCO_3^-(aq) + H_3O^+(aq) \longrightarrow H_2CO_3(aq) + H_2O \tag{19.19}$$

Finally, H_2CO_3 decomposes.

$$H_2CO_3(aq) \longrightarrow H_2O + CO_2(g) \tag{19.20}$$

   The net reaction that occurs when calcium carbonate dissolves in acetic acid, obtained by combining equations (19.16) through (19.20), is

$$CaCO_3(s) + 2\ HC_2H_3O_2(aq) \longrightarrow Ca^{2+}(aq) + 2\ C_2H_3O_2^-(aq) + H_2O + CO_2(g) \tag{19.21}$$

SIMILAR EXAMPLE: Review Problem 8.

   These further ideas about writing net ionic equations are suggested by the way in which equation (19.21) is written. Ionic formulas are used only for strong electrolytes (salts and strong acids and bases). In equation (19.21) the salt, calcium acetate, is written in ionic form. Molecular formulas are used for *nonelectrolytes* (H_2O), *weak electrolytes* ($HC_2H_3O_2$), *gases* (CO_2), and *solids* ($CaCO_3$). If the equilibrium condition is displaced far to the right, a single arrow is used to signify that the reaction goes to completion. If a significant amount of all reactants and products coexist at equilibrium, a double arrow is used.

Illustrative Examples.   There are essentially three types of calculations that are encountered in situations where both a solubility equilibrium and an acid–base equilibrium are involved. The points illustrated by Examples 19-11 through 19-13 are

Example 19-11: Determining whether a precipitate will form.
Example 19-12: Controlling the concentration of a solute species either to cause precipitation or to prevent it.
Example 19-13: Determining the solubility of a solute.

In the first two types of calculations we can proceed by considering first one equilibrium (solubility or acid–base) and then the other. The third type of calculation can be more difficult because at times two (or more) equilibria must be considered *simultaneously*.

Example 19-11   Should $Mg(OH)_2$ precipitate from a solution that is $0.010\ M\ MgCl_2$ if the solution is also made $0.10\ M\ NH_3$?

Solution
Step 1.   Consider the ionization equilibrium in $NH_3(aq)$.

$$NH_3 + H_2O \rightleftharpoons NH_4^+ + OH^- \qquad K_b = 1.74 \times 10^{-5}$$

If we let $x = [NH_4^+] = [OH^-]$ and $[NH_3] = (0.10 - x) \simeq 0.10$, we obtain

$$K_b = \frac{[NH_4^+][OH^-]}{[NH_3]} = \frac{x \cdot x}{0.10} = 1.74 \times 10^{-5}$$

$$x^2 = 1.7 \times 10^{-6} \qquad x = [OH^-] = 1.3 \times 10^{-3}\ M$$

Step 2.   Now we can rephrase the question as: Should $Mg(OH)_2(s)$ precipitate from a solution in which $[Mg^{2+}] = 1.0 \times 10^{-2}\ M$ and $[OH^-] = 1.3 \times 10^{-3}\ M$? We must compare the ion product, Q, with K_{sp}.

$$Q = [Mg^{2+}][OH^-]^2 = (1.0 \times 10^{-2})(1.3 \times 10^{-3})^2$$

$$= 1.7 \times 10^{-8} > K_{sp} = 1.8 \times 10^{-11}$$

Precipitation should occur.

SIMILAR EXAMPLES: Review Problem 9; Exercise 30.

Example 19-12   What $[NH_4^+]$ must be maintained to prevent the precipitation of $Mg(OH)_2$ from a solution that is $0.010\ M\ MgCl_2$ and $0.10\ M\ NH_3$?

Solution.   The maximum value of the ion product, Q, before precipitation occurs is 1.8×10^{-11}. This allows us to determine the maximum $[OH^-]$ that can be tolerated.

$$[Mg^{2+}][OH^-]^2 = (1.0 \times 10^{-2})[OH^-]^2 = 1.8 \times 10^{-11}$$

$$[OH^-]^2 = 1.8 \times 10^{-9} \qquad [OH^-] = 4.2 \times 10^{-5}\ M$$

Next we determine what $[NH_4^+]$ must be present in $0.10\ M\ NH_3$ to maintain $[OH^-] = 4.2 \times 10^{-5}\ M$.

$$K_b = \frac{[NH_4^+][OH^-]}{[NH_3]} = \frac{[NH_4^+](4.2 \times 10^{-5})}{0.10} = 1.74 \times 10^{-5}$$

$$[NH_4^+] = 0.041\ M$$

SIMILAR EXAMPLES: Review Problem 10; Exercise 31.

The solution described in Example 19-12, since it contains both NH_3 and its cation, NH_4^+, is a *buffer solution*. What we have illustrated is how a buffer solution can be used to control a precipitation reaction.

At one point in the qualitative analysis scheme (described in Section 19-9), it is necessary to precipitate Ca^{2+}, Sr^{2+}, and Ba^{2+} as carbonates while Mg^{2+} remains in solution. $MgCO_3$ is more soluble than $CaCO_3$, $SrCO_3$, and $BaCO_3$ (note K_{sp} values in Table 19-1). The pH of the solution in which the carbonate precipitation occurs is critical. If the pH is too low, the carbonates will not precipitate completely. (Recall Example 19-10 describing the solubility of $CaCO_3$ in acidic solution.) If the pH is too high, $Mg(OH)_2$ will precipitate along with the carbonates of Ca, Sr, and Ba. The NH_3–NH_4^+ buffer maintains just the proper pH to accomplish the desired separation.

Example 19-13 Calculate the molar solubility of $CaF_2(s)$ in a buffer solution with pH = 3.00.

Solution. CaF_2 dissolves to the extent that its K_{sp} expression is obeyed. At the same time the ionization equilibrium of HF also must be satisfied. The most direct approach in situations where two or more equilibria must be considered simultaneously is to look for a way of combining these into a single equilibrium constant expression that can be solved. In combining the expressions below we make use of ideas first introduced in Section 15-2.

$$CaF_2(s) \rightleftharpoons Ca^{2+}(aq) + \underline{2\ F^-(aq)} \qquad\qquad K_{sp} = 2.7 \times 10^{-11}$$
$$\underline{2\ H_3O^+(aq) + \underline{2\ F^-(aq)}} \rightleftharpoons 2\ HF(aq) + 2\ H_2O \qquad K = 1/(K_a)^2 = 1/(6.7 \times 10^{-4})^2$$
$$CaF_2(s) + 2\ H_3O^+(aq) \rightleftharpoons Ca^{2+}(aq) + 2\ HF(aq) + 2\ H_2O$$
$$K = K_{sp}/(K_a)^2 = 6.0 \times 10^{-5}$$
$$(19.22)$$

We can now deal with expression (19.22) in a familiar fashion. That is, if we let the solubility of $CaF_2 = x$ mol/L and note that $[H_3O^+]$ remains constant at 1.0×10^{-3} (corresponding to pH = 3.00), we obtain

$$CaF_2(s) + 2\ H_3O^+(aq) \rightleftharpoons Ca^{2+}(aq) + 2\ HF(aq) + 2\ H_2O$$

initial concentrations:	$1.0 \times 10^{-3}\ M$	—	—
changes:	—	$+ x\ M$	$+ 2x\ M$
at equilibrium:	$1.0 \times 10^{-3}\ M$	$x\ M$	$2x\ M$

$$K = \frac{[Ca^{2+}][HF]^2}{[H_3O^+]^2} = \frac{x \cdot (2x)^2}{(1.0 \times 10^{-3})^2} = 6.0 \times 10^{-5}$$

$$4x^3 = 6.0 \times 10^{-11} \qquad x^3 = 1.5 \times 10^{-11} \qquad x = 2.5 \times 10^{-4}$$

The molar solubility of CaF_2 in a buffer solution of pH 3.00 is $2.5 \times 10^{-4}\ M$.

SIMILAR EXAMPLES: Exercises 32, 33.

19-7 Complex Ions and Coordination Compounds—An Introduction

Cobalt forms a simple ionic chloride, $CoCl_3$, in which three electrons are transferred from a Co atom to Cl atoms. But in the presence of $NH_3(aq)$ cobalt(III) chloride can form a series of coordination compounds with such formulas as

$$CoCl_3 \cdot 6NH_3 \qquad CoCl_3 \cdot 5NH_3 \qquad CoCl_3 \cdot 4NH_3 \qquad\qquad (19.23)$$
$$\text{(a)} \qquad\qquad\qquad \text{(b)} \qquad\qquad\qquad \text{(c)}$$

That these are three different compounds is seen in the fact that when treated with excess $AgNO_3(aq)$ compound **(a)** yields 3 mol $AgCl(s)$ per mole of compound; compound **(b)** yields only 2 mol $AgCl(s)$; and compound **(c)** only 1 mol.

Our present conception of coordination compounds is based on ideas proposed by the Swiss chemist Alfred Werner in 1893. Werner's main proposal was that certain metal atoms (primarily those of the transition metals) have *two types of valence*. One, the *primary* valence, is based on the number of electrons lost in forming the metal ion. A *secondary* or auxiliary valence is responsible for bonding coordinated groups, called **ligands,** to the central metal ion. In Werner's theory better representations of the compounds listed in (19.23) are

$$[Co(NH_3)_6]Cl_3 \qquad [Co(NH_3)_5Cl]Cl_2 \qquad [Co(NH_3)_4Cl_2]Cl \qquad\qquad (19.24)$$
$$\text{(a)} \qquad\qquad\qquad \text{(b)} \qquad\qquad\qquad \text{(c)}$$

These representations show that the ligands are bonded directly to the central Co^{3+} ion. The combination of a central metal ion and its ligands is called a **complex ion,** and a neutral compound containing complex ions is called a **coordination compound.** The region surrounding the central metal ion where ligands are found is called the **coordination sphere.** The number of positions in the coordination sphere at which ligand attachment can occur is the **coordination number** of the central metal ion. Some of these terms are further illustrated below.

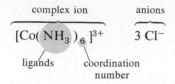

a coordination compound

The coordination number of Co^{3+} in its complex ions is *six,* that is, there are always six ligands attached to the Co^{3+} ion. In compound **(a)** of (19.24) all six ligands are NH_3 molecules and the three Cl^- ions are free anions. In compound **(b)** five NH_3 molecules and one Cl^- ion are the ligands and *two* Cl^- ions are free anions. In compound **(c),** four NH_3 molecules and two Cl^- ions comprise the ligands, leaving only *one* free Cl^- anion. The structures in (19.24) easily account for the observed difference in behavior of these compounds toward $AgNO_3(aq)$.

Bonding, structure, properties, and uses of complex ions and coordination compounds are explored in some detail in Chapter 24. For the present our interest is in how complex ion formation affects other equilibrium processes in aqueous solutions.

19-8 Equilibria Involving Complex Ions

If a moderately concentrated solution of $NH_3(aq)$ is added to solid silver chloride, the solid dissolves.

$$AgCl(s) + 2\ NH_3(aq) \longrightarrow [Ag(NH_3)_2]^+(aq) + Cl^-(aq) \tag{19.25}$$

Ag^+ from AgCl combines with NH_3 to form the complex ion $[Ag(NH_3)_2]^+$. The coordination compound $[Ag(NH_3)_2]Cl$ is soluble in $NH_3(aq)$. $AgBr(s)$ is only slightly soluble in $NH_3(aq)$ and $AgI(s)$ is essentially insoluble.

To understand these observations, we need to think of reaction (19.25) as involving two equilibria simultaneously.

$$AgCl(s) \rightleftharpoons Ag^+(aq) + Cl^-(aq) \tag{19.26}$$

$$Ag^+(aq) + 2\ NH_3(aq) \rightleftharpoons [Ag(NH_3)_2]^+(aq) \tag{19.27}$$

Because $[Ag(NH_3)_2]^+$ is a stable complex ion, equilibrium in reaction (19.27) is displaced to the right and the equilibrium concentration of Ag^+ is very low. K_{sp} for AgCl is not exceeded even in the presence of moderately high Cl^- concentrations. Thus, AgCl(s) is soluble in $NH_3(aq)$. Additional predictions concerning the aqueous chemistry of $[Ag(NH_3)_2]^+$ can be made with Le Châtelier's principle, as illustrated in Example 19-14.

Example 19-14 Predict the effect of adding $HNO_3(aq)$ to a saturated solution of $[Ag(NH_3)_2]Cl$ in $NH_3(aq)$.

Solution. HNO_3 neutralizes the free NH_3 in solution.

$$H^+(aq) + NO_3^-(aq) + NH_3(aq) \longrightarrow NH_4^+(aq) + NO_3^-(aq)$$

Loss of NH_3 upsets the equilibrium in reaction (19.27). To replace the neutralized NH_3, $[Ag(NH_3)_2]^+$ dissociates—the reverse of reaction (19.27). This, in turn, produces an increase in $[Ag^+]$—a disturbance that is relieved by the reverse of reaction (19.26). $AgCl(s)$ precipitates.

SIMILAR EXAMPLES: Exercises 35, 36.

Formation Constants of Complex Ions.   To provide a *quantitative* description of equilibria involving complex ions requires knowledge of the **formation constant, K_f.** For reaction (19.27)

$$Ag^+(aq) + 2\,NH_3(aq) \rightleftharpoons [Ag(NH_3)_2]^+(aq) \tag{19.27}$$

the equilibrium constant describing equilibrium among the complex ion and the central metal ion and ligands from which it is formed is

$$K_f = \frac{[[Ag(NH_3)_2]^+]}{[Ag^+][NH_3]^2} = 1.6 \times 10^7 \tag{19.28}$$

A more complete description of the formation of a complex ion views this to occur in a *stepwise* fashion, as described in Section 24-10; but for our present purposes we will use only overall formation constants. Selected data are listed in Table 19-3.

Sometimes complex ion equilibria are written in the reverse of the method used in equation (19.27), i.e., $[Ag(NH_3)_2]^+ \rightleftharpoons Ag^+ + 2\,NH_3$. In this case the equilibrium constant is the reciprocal of the type shown in equation (19.28). Written in this way the constant is called a dissociation constant, K_D, or an instability constant, K_i.

$$K_D = \frac{[Ag^+][NH_3]^2}{[[Ag(NH_3)_2]^+]} = \frac{1}{K_f} = \frac{1}{1.6 \times 10^7} = 6.2 \times 10^{-8}$$

TABLE 19-3
Formation constants for some complex ions

Complex ion	Equilibrium reaction	K_f
$[AlF_6]^{3-}$	$Al^{3+} + 6\,F^- \rightleftharpoons [AlF_6]^{3-}$	6.7×10^{19}
$[Cd(CN)_4]^{2-}$	$Cd^{2+} + 4\,CN^- \rightleftharpoons [Cd(CN)_4]^{2-}$	7.1×10^{18}
$[Co(NH_3)_6]^{3+}$	$Co^{3+} + 6\,NH_3 \rightleftharpoons [Co(NH_3)_6]^{3+}$	4.5×10^{33}
$[Cu(CN)_3]^{2-}$	$Cu^+ + 3\,CN^- \rightleftharpoons [Cu(CN)_3]^{2-}$	2×10^{27}
$[Cu(NH_3)_4]^{2+}$	$Cu^{2+} + 4\,NH_3 \rightleftharpoons [Cu(NH_3)_4]^{2+}$	1.1×10^{13}
$[Fe(CN)_6]^{4-}$	$Fe^{2+} + 6\,CN^- \rightleftharpoons [Fe(CN)_6]^{4-}$	1×10^{37}
$[Fe(CN)_6]^{3-}$	$Fe^{3+} + 6\,CN^- \rightleftharpoons [Fe(CN)_6]^{3-}$	1×10^{42}
$[PbCl_3]^-$	$Pb^{2+} + 3\,Cl^- \rightleftharpoons [PbCl_3]^-$	2.4×10^1
$[HgCl_4]^{2-}$	$Hg^{2+} + 4\,Cl^- \rightleftharpoons [HgCl_4]^{2-}$	1.2×10^{15}
$[HgI_4]^{2-}$	$Hg^{2+} + 4\,I^- \rightleftharpoons [HgI_4]^{2-}$	1.9×10^{30}
$[Ni(CN)_4]^{2-}$	$Ni^{2+} + 4\,CN^- \rightleftharpoons [Ni(CN)_4]^{2-}$	1×10^{22}
$[Ag(NH_3)_2]^+$	$Ag^+ + 2\,NH_3 \rightleftharpoons [Ag(NH_3)_2]^+$	1.6×10^7
$[Ag(CN)_2]^-$	$Ag^+ + 2\,CN^- \rightleftharpoons [Ag(CN)_2]^-$	5.6×10^{18}
$[Ag(S_2O_3)_2]^{3-}$	$Ag^+ + 2\,S_2O_3^{2-} \rightleftharpoons [Ag(S_2O_3)_2]^{3-}$	1.7×10^{13}
$[Zn(NH_3)_4]^{2+}$	$Zn^{2+} + 4\,NH_3 \rightleftharpoons [Zn(NH_3)_4]^{2+}$	4.1×10^8
$[Zn(CN)_4]^{2-}$	$Zn^{2+} + 4\,CN^- \rightleftharpoons [Zn(CN)_4]^{2-}$	1×10^{18}
$[Zn(OH)_4]^{2-}$	$Zn^{2+} + 4\,OH^- \rightleftharpoons [Zn(OH)_4]^{2-}$	4.6×10^{17}

Illustrative Examples. As with the illustrative examples in Section 19-6, basically there are three types of calculations that combine solubility and complex ion formation equilibria. These are

Example 19-15: Determining whether a precipitate will form.
Example 19-16: Controlling the concentration of a solute species (e.g., that of a ligand) to cause precipitation or to prevent it.
Example 19-17: Determining the solubility of a solute.

And as with the examples in Section 19-6, the first two types are easiest because they permit equilibrium expressions to be used successively. The third type again may require that two (or more) equilibria be considered simultaneously.

Example 19-15 A 0.10-mol sample of $AgNO_3$ is dissolved in 1.00 L of 1.00 M NH_3. If 0.010 mol NaCl is added to this solution, will a precipitate form?

Solution. The total silver concentration in the solution is 0.10 mol/L, found partly as Ag^+ and partly as $[Ag(NH_3)_2]^+$. In the setup shown we assume that practically all the silver is complexed. If 0.10 mol Ag^+ is complexed, then twice this amount of NH_3 must also be complexed.

$$Ag^+ \quad + 2\, NH_3 \rightleftharpoons [Ag(NH_3)_2]^+$$

dissolve: $\qquad$ 0.10 M $\qquad$ 1.00 M
at equilibrium
(assume $x \ll 0.10$): $\quad x = [Ag^+]$ $\quad$ 0.80 M $\qquad\qquad$ 0.10 M

$$\frac{[[Ag(NH_3)_2]^+]}{[Ag^+][NH_3]^2} = \frac{0.10}{[Ag^+](0.80)^2} = 1.6 \times 10^7$$

$$[Ag^+] = 9.8 \times 10^{-9}$$

We must compare $[Ag^+][Cl^-]$ with $K_{sp}(AgCl) = 1.6 \times 10^{-10}$.

$$[Ag^+] = 9.8 \times 10^{-9} \qquad [Cl^-] = 1.0 \times 10^{-2}$$

$$(9.8 \times 10^{-9})(1.0 \times 10^{-2}) < 1.6 \times 10^{-10}$$

AgCl will not precipitate.

SIMILAR EXAMPLES: Review Problem 11; Exercise 37.

Example 19-16 What is the minimum concentration of aqueous NH_3 required to prevent AgCl(s) from precipitating from 1.00 L of a solution containing 0.10 mol $AgNO_3$ and 0.010 mol NaCl?

Solution. The $[Cl^-]$ that must be maintained in solution is 1.0×10^{-2}. If no precipitation is to occur, $[Ag^+][Cl^-] \leq K_{sp}$.

$$[Ag^+](1.0 \times 10^{-2}) \leq K_{sp} = 1.6 \times 10^{-10} \qquad [Ag^+] \leq 1.6 \times 10^{-8}$$

The maximum concentration of *free, uncomplexed* Ag^+ permitted in solution is 1.6×10^{-8} M. This means that essentially all the Ag^+ (0.10 mol/L) must be in the form of the complex ion, $[Ag(NH_3)_2]^+$. We need to solve the following expression for $[NH_3]$.

$$K_f = \frac{[[Ag(NH_3)_2]^+]}{[Ag^+][NH_3]^2} = \frac{1.0 \times 10^{-1}}{1.6 \times 10^{-8}[NH_3]^2} = 1.6 \times 10^7$$

$$[NH_3]^2 = 0.39 \qquad [NH_3] = 0.62\ M$$

The concentration calculated above is that of *free, uncomplexed* NH_3. Considering as well the 0.20 mol NH_3/L consumed to produce $[Ag(NH_3)_2]^+$, the total concentration of $NH_3(aq)$ required is

$$[NH_3]_{tot} = 0.62 + 0.20 = 0.82 \; M$$

SIMILAR EXAMPLES: Review Problem 12; Exercise 38

Example 19-17   What is the molar solubility of AgCl(s) in 0.100 M $NH_3(aq)$?

Solution.   As in Example 19-13, the most direct approach is to try to write a single equilibrium constant expression by combining the separate equilibrium reactions involved. In this case (19.25) is obtained by adding together (19.26) and (19.27). The equilibrium constant for reaction (19.25) is the product of K_{sp} for AgCl and K_f for $[Ag(NH_3)_2]^+$.

$$AgCl(s) \rightleftharpoons Ag^+(aq) + Cl^-(aq) \qquad\qquad K_{sp} = 1.6 \times 10^{-10}$$
$$Ag^+(aq) + 2\;NH_3(aq) \rightleftharpoons [Ag(NH_3)_2]^+(aq) \qquad\qquad K_f = 1.6 \times 10^7$$

$$AgCl(s) + 2\;NH_3(aq) \rightleftharpoons [Ag(NH_3)_2]^+(aq) + Cl^-(aq) \qquad \begin{aligned} K &= K_{sp} \times K_f \\ &= 2.6 \times 10^{-3} \end{aligned}$$

According to equation (19.25), if x mol AgCl(s) dissolves per liter of solution (the molar solubility), the expected concentrations of $[Ag(NH_3)_2]^+$ and Cl^- are also equal to x.

$$AgCl(s) + \quad 2\;NH_3(aq) \quad \rightleftharpoons \quad [Ag(NH_3)_2]^+(aq) + Cl^-(aq)$$

	NH_3	$[Ag(NH_3)_2]^+$	Cl^-
initial concentrations:	0.100 M		
changes:	$-2x$ M	$+x$ M	$+x$ M
at equilibrium:	$(0.100 - 2x)$ M	x M	x M

$$K = \frac{[[Ag(NH_3)_2]^+][Cl^-]}{[NH_3]^2} = \frac{x \cdot x}{(0.100 - 2x)^2} = \left\{ \frac{x}{(0.100 - 2x)} \right\}^2 = 2.6 \times 10^{-3}$$

At this point, to avoid using the quadratic formula we have usually made an assumption such as, if x is very small $0.100 - x \approx 0.100$. Here a simple solution to the equation is possible by taking the square root of both sides.

$$\frac{x}{(0.100 - 2x)} = \sqrt{2.6 \times 10^{-3}} = 5.1 \times 10^{-2} \qquad x = 5.1 \times 10^{-3} - 0.10x$$

$$1.10x = 5.1 \times 10^{-3} \qquad and \qquad x = 4.6 \times 10^{-3}$$

The solubility of AgCl(s) in 0.100 M $NH_3(aq)$ is 4.6×10^{-3} mol AgCl/L.

SIMILAR EXAMPLE: Exercise 39.

One disadvantage of combining equilibrium constant expressions instead of dealing with them separately is that occasionally we may lose sight of the physical situation involved. In Example 19-17 the molar solubility of AgCl is x and so is the chloride ion concentration. However, it is the *total* of $[Ag^+]$ and $[[Ag(NH_3)_2]^+]$ that is equal to x. By setting $[[Ag(NH_3)_2]^+] = x$ in Example 19-17 we were in effect assuming that $[Ag^+]$ was so small as to be negligible. We can expect this assumption to be valid for stable complex ions, that is, for complex ions with large values of K_f. (See Additional Exercise 15 for a situation where this assumption is not valid.)

19-9 Qualitative Analysis

Qualitative analysis refers to a set of laboratory procedures that can be used to *separate* and *test* for the presence of ions in solution. The analysis is said to be *qualitative* because it reveals only which ions are present in a mixture. The analysis does not necessarily indicate the compounds from which these ions are derived, or the actual quantities present

FIGURE 19-5
A qualitative analysis
scheme for cations.

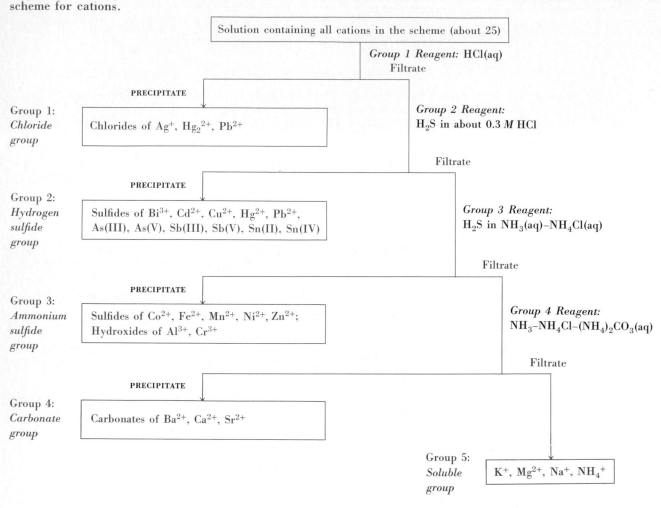

Solution containing all cations in the scheme (about 25)

Group 1 Reagent: HCl(aq)
Filtrate

PRECIPITATE

Group 1:
*Chloride
group*

Chlorides of Ag^+, Hg_2^{2+}, Pb^{2+}

Group 2 Reagent:
H_2S in about 0.3 *M* HCl

Filtrate

PRECIPITATE

Group 2:
*Hydrogen
sulfide
group*

Sulfides of Bi^{3+}, Cd^{2+}, Cu^{2+}, Hg^{2+}, Pb^{2+},
As(III), As(V), Sb(III), Sb(V), Sn(II), Sn(IV)

Group 3 Reagent:
H_2S in NH_3(aq)–NH_4Cl(aq)

Filtrate

PRECIPITATE

Group 3:
*Ammonium
sulfide
group*

Sulfides of Co^{2+}, Fe^{2+}, Mn^{2+}, Ni^{2+}, Zn^{2+};
Hydroxides of Al^{3+}, Cr^{3+}

Group 4 Reagent:
NH_3–NH_4Cl–$(NH_4)_2CO_3$(aq)

Filtrate

PRECIPITATE

Group 4:
*Carbonate
group*

Carbonates of Ba^{2+}, Ca^{2+}, Sr^{2+}

Group 5:
*Soluble
group*

K^+, Mg^{2+}, Na^+, NH_4^+

(i.e., not the masses or concentrations of substances). Perhaps more so than any other set of common laboratory procedures, qualitative analysis illustrates the complete range of solution equilibrium concepts considered in this and preceding chapters.

Qualitative analysis can be applied both to anions and cations. A particular set of procedures used to perform an analysis is called a **qualitative analysis scheme.** Our interest will center on the analysis of cations by the scheme outlined in Figure 19-5 (in a form called a flowchart or flow diagram).

As indicated in Figure 19-5, the basic approach is to separate the common cations into groups (usually five) by *precipitation*. For example, only the members of group 1, that is, only Ag^+, Hg_2^{2+}, and Pb^{2+}, form insoluble chlorides. All other common metal chlorides are water soluble. Group 2 cations are precipitated as sulfides from acidic solution. Cations of group 3 form insoluble sulfides in basic solution. The group 4 cations are precipitated as carbonates. The cations of group 5 remain in solution throughout the separation of the other four groups.

Following the separation of cations into the five major groups, further separation and testing must be done within each group. The end result is to establish for an unknown mixture the presence or absence of each ion in the scheme. Detailed flow diagrams can be written for this further separation and testing.

Other approaches to the qualitative analysis scheme are often encountered. For example, analysis can begin with the soluble group cations (group 5), followed by analysis of groups 1 through 4 on a separate sample.

Example 19-18 You are given an unknown *colorless* aqueous solution and told that it contains *none, one, two,* or *all three* of the following ions: Ag$^+$, Ba^{2+}, Cu^{2+}. You treat the unknown with (NH$_4$)$_2$CO$_3$(aq), and obtain a *white* precipitate. What conclusions can you draw from this observation? (Use data from Figure 19-5 and Table 19-2, together with general laboratory knowledge.)

Solution. All three of the possible ions produce an insoluble carbonate (Table 19-2, item 5). The formation of a precipitate only allows you to conclude that at least one of the three ions is present. The general laboratory knowledge expected of you is some familiarity with the colors of ionic compounds and of ions in solution. Of the three ions, only Cu^{2+} displays a color in solution (blue). Compounds containing Cu^{2+} are also colored. The observation that neither the solution nor the precipitate is colored *suggests* that Cu^{2+} is *absent*. (It is still possible—although probably not likely—that Cu^{2+} is present in the solution. At very low concentrations its color cannot be detected.) Whether the white precipitate is Ag$_2$CO$_3$(s), BaCO$_3$(s), or a mixture of the two cannot be determined from this single observation.

SIMILAR EXAMPLE: Exercise 44.

Can you see how an unambiguous answer to Example 19-18 could be obtained by following the qualitative analysis scheme of Figure 19-5? If the sample were first treated with HCl(aq), formation of a precipitate would prove the presence of Ag$^+$ (the only one of the three ions that forms an insoluble chloride). Lack of a precipitate would establish the *absence* of Ag$^+$. The filtrate—the solution remaining after removal of any AgCl(s)—could now be treated with H$_2$S. Here, formation of a precipitate would prove the presence of Cu^{2+}, and lack of a precipitate, the absence of Cu^{2+}. *After* the removal of any Ag$^+$ or Cu^{2+} present in the unknown, a test could be performed for Ba^{2+} using the group 4 reagent.

19-10 H$_2$S Equilibria Revisited—Precipitation and Solubilities of Metal Sulfides

As we learned in Section 17-6, hydrogen sulfide in aqueous solutions (hydrosulfuric acid) is a weak *di*protic acid. One of the conclusions we reached in our previous discussion of H$_2$S is that in an aqueous solution of H$_2$S, [S^{2-}] = K_{a_2} = 1.0×10^{-14} M. With this value of [S^{2-}] we can determine which metal ions can be precipitated as sulfides—an important question in the qualitative analysis scheme.

Example 19-19 Which of the following ions will precipitate from a solution that is kept saturated in H$_2$S and 0.010 M in each: Pb^{2+}, Zn^{2+}, and Mn^{2+}?

K_{sp}(PbS) = 8×10^{-28} K_{sp}(ZnS) = 1.0×10^{-21} K_{sp}(MnS) = 2.5×10^{-13}

Solution. In a saturated H$_2$S solution, [S^{2-}] = K_{a_2} = 1.0×10^{-14} M. The following comparisons must be made.

[Pb^{2+}][S^{2-}] = $(1.0 \times 10^{-2})(1.0 \times 10^{-14})$ = 1.0×10^{-16} > K_{sp} = 8×10^{-28}
PbS precipitates.

[Zn^{2+}][S^{2-}] = $(1.0 \times 10^{-2})(1.0 \times 10^{-14})$ = 1.0×10^{-16} > K_{sp} = 1.0×10^{-21}
ZnS precipitates.

[Mn^{2+}][S^{2-}] = $(1.0 \times 10^{-2})(1.0 \times 10^{-14})$ = 1.0×10^{-16} < K_{sp} = 2.5×10^{-13}
MnS does not precipitate.

SIMILAR EXAMPLES: Review Problem 13; Exercise 40a.

The Common Ion Effect in H₂S Solutions.   Selective precipitation of sulfides cannot be accomplished very effectively in a solution containing only H_2S. In the qualitative analysis scheme the H_2S equilibria are altered by the presence of other substances. For example, adding a strong acid (HCl) to an aqueous solution of H_2S reduces $[S^{2-}]$ in solution.

When $[H_3O^+]$ in solution is determined by some species other than the H_2S itself, calculations pertaining to H_2S equilibria are more readily made by using a *combination* of ionization constants. In a manner we have employed before, we can write

$$H_2S + H_2O \rightleftharpoons H_3O^+ + HS^- \qquad K_{a_1} = 1.1 \times 10^{-7}$$
$$HS^- + H_2O \rightleftharpoons H_3O^+ + S^{2-} \qquad K_{a_2} = 1.0 \times 10^{-14}$$

$$\overline{H_2S + 2\,H_2O \rightleftharpoons 2\,H_3O^+ + S^{2-} \qquad K = K_{a_1} \times K_{a_2}} \qquad (19.29)$$

and

$$K_{a_1} \times K_{a_2} = \frac{[H_3O^+]^2[S^{2-}]}{[H_2S]} = (1.1 \times 10^{-7})(1.0 \times 10^{-14}) = 1.1 \times 10^{-21} \qquad (19.30)$$

Equation (19.29) states that for every molecule of H_2S that undergoes *complete* ionization, two H_3O^+ ions and one S^{2-} ion are formed. Equation (19.30) indicates that if any *two* of the three concentration terms it relates are known, the remaining one can be calculated. Remember, however, that in aqueous solutions containing only H_2S, $[H_3O^+]$ is *very much larger* than $[S^{2-}]$. It is not just twice as large, as equation (19.29) might seem to suggest. In such solutions practically all the H_3O^+ is produced in the first ionization step, together with HS^-. Very little of the HS^- ionizes further to produce S^{2-}. This is another example of the tendency for physical insight into an equilibrium condition to become obscured when equilibrium constant expressions are combined. (The other example given was on page 583.)

Selective Precipitation of Metal Sulfides.   In the qualitative analysis scheme (recall Figure 19-5) the *least* soluble of the metal sulfides of cation group 3 is ZnS and the *most* soluble of those of cation group 2 is CdS (see Table 19-1). To achieve an effective separation of these two groups of cations, precipitation must be carried out under conditions where CdS precipitates and ZnS does not. The necessary conditions are established in Example 19-20.

Example 19-20   What $[H_3O^+]$ must be maintained in a saturated H_2S solution (0.10 *M* H_2S) to precipitate CdS, but not ZnS, if Cd^{2+} and Zn^{2+} are each present initially at a concentration of 0.10 *M*?

$$CdS(s) \rightleftharpoons Cd^{2+}(aq) + S^{2-}(aq) \qquad K_{sp} = 8.0 \times 10^{-27}$$
$$ZnS(s) \rightleftharpoons Zn^{2+}(aq) + S^{2-}(aq) \qquad K_{sp} = 1.0 \times 10^{-21}$$

Solution.   If ZnS is *not* to precipitate,

$$[Zn^{2+}][S^{2-}] < K_{sp} = 1.0 \times 10^{21}$$
$$(0.10)[S^{2-}] < 1.0 \times 10^{-21}$$
$$[S^{2-}] < 1.0 \times 10^{-20}$$

The maximum value of $[S^{2-}]$ before ZnS will precipitate is 1.0×10^{-20} *M*. The $[H_3O^+]$ required to maintain this $[S^{2-}]$ is

$$K_{a_1} \times K_{a_2} = \frac{[H_3O^+]^2[S^{2-}]}{[H_2S]} = \frac{[H_3O^+]^2(1.0 \times 10^{-20})}{0.10} = 1.1 \times 10^{-21}$$

$$[H_3O^+]^2 = 1.1 \times 10^{-2} \qquad [H_3O^+] = 1.0 \times 10^{-1} = 0.10\ M$$

That CdS should precipitate under these conditions is easily demonstrated.

$$[Cd^{2+}][S^{2-}] = (0.10)(1.0 \times 10^{-20}) = 1.0 \times 10^{-21} > K_{sp} = 8.0 \times 10^{-27}$$

Any concentration of $[H_3O^+]$ greater than 0.10 M ensures that no ZnS precipitates. In actual practice the cation group 2 precipitating reagent is kept at about pH = 0.5 ($[H_3O^+]$ = 0.3 M).

SIMILAR EXAMPLES: Review Problem 14; Exercise 40b.

Dissolving Metal Sulfides.   The qualitative analysis scheme requires that sulfides be both precipitated and redissolved. To discuss the dissolving of metal sulfides it is again helpful to combine two equilibrium reactions. Consider, for example, the dissolving of PbS(s) in HCl(aq).

$$PbS(s) \rightleftharpoons Pb^{2+}(aq) + S^{2-}(aq) \qquad\qquad K_{sp} = 8.0 \times 10^{-28} \quad (19.31)$$

$$\underset{\text{(from HCl)}}{2\,H_3O^+} + \underset{\text{(from PbS)}}{S^{2-}} \rightleftharpoons H_2S(aq) + 2\,H_2O \qquad\qquad K = \frac{1}{K_{a_1} \times K_{a_2}}$$
$$= 9.1 \times 10^{20} \quad (19.32)$$

$$PbS(s) + \underset{\text{(from HCl)}}{2\,H_3O^+} \rightleftharpoons Pb^{2+}(aq) + H_2S(aq) + 2\,H_2O \quad K = \frac{K_{sp}}{K_{a_1} \times K_{a_2}}$$
$$= 7.3 \times 10^{-7} \quad (19.33)$$

Equation (19.31) describes the solubility equilibrium for PbS; K_{sp} is the applicable equilibrium constant. Sulfide ion from PbS combines with H_3O^+ from HCl to produce H_2S. This equilibrium reaction is described by the *reciprocal* of $K_{a_1} \times K_{a_2}$. The net equation (19.33) indicates that as S^{2-} from PbS is converted to $H_2S(aq)$, Pb^{2+} appears in solution—the PbS dissolves. Because of the small value of K for reaction (19.33), however, we should not expect very much PbS to dissolve. In fact, we know that the reverse of reaction (19.33) is strongly favored in 0.3 M HCl: Pb^{2+} precipitates with the group 2 cations in the qualitative analysis scheme. Nevertheless, high concentrations of H_3O^+ will promote some dissolving of PbS, as illustrated in Example 19.21.

Example 19-21   What is the molar solubility of PbS in 1.0 M HCl(aq)?

Solution.   The relevant data are listed below equation (19.33), where the molar solubility = $[Pb^{2+}]$ = x.

$$PbS(s) + 2\,H_3O^+ \rightleftharpoons Pb^{2+} + H_2S(aq) + 2\,H_2O \quad (19.33)$$

initial concentrations:	1.0 M		
changes:	$-2x\ M$	$+x\ M$	$+x\ M$
at equilibrium:	$(1.0 - 2x)\ M$	$x\ M$	$x\ M$

$$K = \frac{[Pb^{2+}][H_2S]}{[H_3O^+]^2} = \frac{x \cdot x}{(1.0 - 2x)^2} = 7.3 \times 10^{-7}$$

If, as we suspect, x is very small, then $(1.0 - 2x) \simeq 1.0$.

$$\frac{x^2}{1.0} = 7.3 \times 10^{-7} \qquad x = [Pb^{2+}] = 8.5 \times 10^{-4}\ M$$

SIMILAR EXAMPLES: Exercises 42, 43.

It is perhaps worth noting that situations in which both solubility equilibrium and the H_2S equilibria are involved fall into the same general categories mentioned in Sections 19-6 and 19-8. That is,

- Determining if a sulfide precipitate will form, given a cation concentration and information for use in equation (19.30).
- Controlling the precipitation of sulfides by controlling $[H_3O^+]$ in the H_2S equilibria (as in Example 19-20).
- Determining the solubility of a sulfide (as in Example 19-21).

Summary

Equilibrium between a slightly soluble ionic compound and its ions in aqueous solution is expressed through the *solubility product constant*, K_{sp}. In using the K_{sp} expression two situations are commonly encountered: (1) The ions in a saturated solution may be derived *solely* from the slightly soluble solute, or (2) an additional salt(s) may be present that contributes ions either *common to* or *different from* those of the slightly soluble solute. The solubility of the solute is *greatly reduced* in the presence of common ions derived from a dissolved salt. A salt whose ions are different from those of the slightly soluble solute causes a *small increase* in the solubility of the solute. The formation of ion pairs in solution may also cause the solubility of a solute to be somewhat higher than predicted from the K_{sp} value.

A comparison of the reaction quotient (ion product) Q with K_{sp} provides a criterion for precipitation: If $Q > K_{sp}$, precipitation should occur; if $Q < K_{sp}$, the solution remains unsaturated. Another matter of interest concerns the completeness of a precipitation reaction. If a precipitation is carried out in the presence of a high concentration of a common ion, it generally goes to completion. Combinations of factors, such as lack of sufficient common ion and a moderately high value of K_{sp}, can result in incomplete precipitation. At times, ions in solution can be separated by *fractional precipitation*. One type of ion is removed by precipitation while the others remain in solution.

By using the common ion effect to alter equilibria of weak acids and bases, other solution equilibria can be affected as well. For example, by controlling pH through the use of buffers, solution concentrations of such ions as OH^-, CO_3^{2-}, and S^{2-} can be maintained over an extreme range. Such control allows for the selective precipitation or dissolving of ionic compounds.

The formation of a complex ion from a central metal ion and ligands can be viewed as an equilibrium process with an equilibrium constant called the *formation constant*, K_f. In general, if the formation constant of a complex ion is large, the concentration of *free* metal ion in equilibrium with the complex ion is very small. Complex ion formation can render certain insoluble materials quite soluble in appropriate aqueous solutions, such as AgCl(s) in NH_3(aq).

Precipitation reactions find extensive application in *qualitative analysis*. In the qualitative analysis scheme for cations, the cations are first separated into groups based on differing solubilities of their compounds. Further separation and testing of ions is then done within each group. Aqueous solutions of H_2S are particularly useful in this scheme.

Learning Objectives

As a result of studying Chapter 19, you should be able to

1. Write the solubility product constant expression, K_{sp}, for a slightly soluble ionic compound.

2. Calculate K_{sp} from the experimentally determined solubility of a slightly soluble ionic compound, or the solubility from a known value of K_{sp}.

3. Calculate the effect of common ions on the aqueous solubilities of slightly soluble ionic compounds.

4. Describe how the presence of ''uncommon'' ions or the formation of ion pairs affects solute solubilities.

5. Apply the criterion for precipitation from solution.

6. Determine, by calculation, whether the precipitation of a slightly soluble solute will be complete for a given set of conditions.

7. Explain how fractional precipitation works and the conditions under which it can be used.

8. State the general rules that apply to the water solubilities of ionic compounds, and write net ionic equations based on these solubility rules.

9. Describe, through net ionic equations and through calculations, the effect of pH on the precipitation and dissolving of certain substances.

10. Write equations showing the effect of complex ion formation on other equilibrium processes, such as solubility equilibrium.

11. Calculate the concentrations of ligands and free metal ion in equilibrium with a complex ion.

12. Use K_f values in conjunction with K_{sp} values to make predictions about the precipitation of solutes in the presence of complexing ligands.

13. Calculate the solubilities of certain solutes in the presence of complexing ligands.

14. Relate $[H_3O^+]$ and $[S^{2-}]$ in H_2S(aq) solutions through the combined equilibrium constant expression, $K_{a_1} \times K_{a_2}$.

15. Predict conditions under which metal sulfides will precipitate from solution and conditions under which they will dissolve.

16. Describe how precipitation reactions are employed in a qualitative analysis scheme, and draw conclusions about the presence or absence of ions in an unknown from experimental observations.

Some New Terms

A **complex ion** is a combination of a central metal ion and attached groups called ligands.

A **coordination compound** is a substance containing complex ions.

The **formation constant** K_f is the equilibrium constant describing equilibrium among a complex ion, the free metal ion, and ligands.

Fractional precipitation is a technique in which ions in solution are separated by the addition of a precipitating reagent.

Ion-pair formation refers to the association of cations and anions in solution. Such combinations, when they occur to a significant extent, can have an effect on solution equilibria.

Ligands are the groups that are coordinated (bonded) to the central metal ion in a complex ion.

Molar solubility is the molar concentration of solute (mol/L) in a saturated solution.

A **net ionic equation** is a chemical equation in which only the species actually participating in a chemical reaction are shown. The form in which each reactant occurs is also represented, that is, as ions in solution, insoluble precipitate, gas, and so on.

Qualitative analysis is a laboratory method, based on a variety of solution equilibrium concepts, for determining the presence or absence of certain cations or anions in a sample.

Quantitative analysis refers to the analysis of substances or mixtures to determine the *quantities* of the various components rather than their mere presence or absence.

The **salt effect** is that of ions *different* from those directly involved in a solution equilibrium. The salt effect is also known as the diverse or "uncommon" ion effect.

The **solubility product constant** K_{sp} describes equilibrium in a saturated solution of a slightly soluble ionic compound. It is the product of ionic concentration terms, with each term raised to an appropriate power.

Suggestions for Further Study

COLE, G. M., and W. H. WAGGONER, "Qualitative Analysis: The Current Status," *J. Chem. Educ.,* **60,** 135 (1983).

COOPER, J. N., "Solubility of Lead Bromide in Nitrate Media," *J. Chem. Educ.,* **49,** 282 (1972).

HAIGHT, G. P., "The Relation Between Solubility and Solubility Product Is a Limiting Case," *J. Chem. Educ.,* **55,** 452 (1978).

MICKEY, C. D., "Using the Equilibrium Concept," *J. Chem. Educ.,* **58,** 56 (1981).

PAPE, W., "More on the Common Ion Effect," *J. Chem. Educ.,* **58,** 1019 (1981).

SHAKASHIRI, B. Z., G. E. DIRREEN, and F. JUERGENS, "Solubility and Complex Ion Equilibria of Silver(I) Species in Aqueous Solution," *J. Chem. Educ.,* **57,** 813 (1980).

Review Problems

(Use data from tables in Chapters 17, 18 and 19, as necessary.)

1. Write K_{sp} expressions for the following equilibria. For example, $AgCl(s) \rightleftharpoons Ag^+(aq) + Cl^-(aq)$ $K_{sp} = [Ag^+][Cl^-]$

  (a) $Ag_2SO_4(s) \rightleftharpoons 2\,Ag^+(aq) + SO_4^{2-}(aq)$
  (b) $Ra(IO_3)_2(s) \rightleftharpoons Ra^{2+}(aq) + 2\,IO_3^-(aq)$
  (c) $Ni_3(PO_4)_2(s) \rightleftharpoons 3\,Ni^{2+}(aq) + 2\,PO_4^{3-}(aq)$
  (d) $Hg_2C_2O_4(s) \rightleftharpoons Hg_2^{2+}(aq) + C_2O_4^{2-}(aq)$
  (e) $PuO_2CO_3(s) \rightleftharpoons PuO_2^{2+}(aq) + CO_3^{2-}(aq)$

2. Write solubility equilibrium equations that are described by the following K_{sp} expressions. For example, $K_{sp} = [Ag^+][Cl^-]$ represents $AgCl(s) \rightleftharpoons Ag^+(aq) + Cl^-(aq)$.

  (a) $K_{sp} = [Fe^{3+}][OH^-]^3$ **(b)** $K_{sp} = [BiO^+][OH^-]$
  (c) $K_{sp} = [Hg_2^{2+}][I^-]^2$ **(d)** $K_{sp} = [Pb^{2+}]^3[AsO_4^{3-}]^2$
  (e) $K_{sp} = [Cu^{2+}]^2[Fe(CN)_6^{4-}]$
  (f) $K_{sp} = [Mg^{2+}][NH_4^+][PO_4^{3-}]$

3. Calculate the water solubility, in mol/L, of each of the following: **(a)** $BaCrO_4(s)$, $K_{sp} = 1.2 \times 10^{-10}$; **(b)** $PbBr_2$, $K_{sp} = 4.0 \times 10^{-5}$; **(c)** CeF_3, $K_{sp} = 8 \times 10^{-16}$; **(d)** $Mg_3(AsO_4)_2$, $K_{sp} = 2.1 \times 10^{-20}$.

4. The following aqueous solubility data, expressed in mol solute/L, are derived from a handbook. What are the values of K_{sp} for these solutes? **(a)** $CsMnO_4$, $3.8 \times 10^{-3}\ M$; **(b)** $Pb(ClO_2)_2$, $2.8 \times 10^{-3}\ M$; **(c)** Li_3PO_4, $2.9 \times 10^{-3}\ M$.

5. Calculate the molar solubility of $Mg(OH)_2(s)$ in **(a)** pure water; **(b)** $0.015\ M$ $MgCl_2(aq)$; **(c)** $0.217\ M$ $KOH(aq)$.

6. Predict whether a precipitate is expected to form in a solution with the ion concentrations listed.
  (a) $[Mg^{2+}] = 0.015\ M$, $[CO_3^{2-}] = 0.0072\ M$
  (b) $[Ag^+] = 0.0038\ M$, $[SO_4^{2-}] = 0.0105\ M$
  (c) $[Cr^{3+}] = 0.041\ M$, $[H_3O^+] = 0.0016\ M$
  (*Hint:* What is $[OH^-]$?)

7. $KI(aq)$ is slowly added to a solution that is $0.10\ M$ in both Pb^{2+} and Ag^+.
  (a) Which precipitate should form first, PbI_2 or AgI?
  (b) What $[I^-]$ is required for the *second* cation to begin to precipitate?
  (c) What concentration of the first ion to precipitate remains in solution at the point where the second ion begins to precipitate?
  (d) Can these two cations be effectively separated by fractional precipitation?

8. Predict whether a reaction is likely to occur in each of the following cases. If so, write a net ionic equation.
   (a) $NaI(aq) + ZnSO_4(aq) \rightarrow$
   (b) $CuSO_4(aq) + Na_2CO_3(aq) \rightarrow$
   (c) $AgNO_3(aq) + CuCl_2(aq) \rightarrow$
   (d) $BaS(aq) + CuSO_4(aq) \rightarrow$
   (e) $Al(OH)_3(s) + HCl(aq) \rightarrow$
   (f) $CaC_2O_4(s) + HCl(aq) \rightarrow$
   (g) $CdS(s) + HC_2H_3O_2(aq) \rightarrow$
       (*Hint:* Recall Example 19-20.)

9. A buffer solution is prepared which is $0.50\ M$ $HC_2H_3O_2$ and $0.25\ M$ $NaC_2H_3O_2$. Which of the following ions can be maintained at a concentration of $0.10\ M$ or greater without precipitating as the hydroxide from this solution? **(a)** Ca^{2+}; **(b)** Al^{3+}; **(c)** Cr^{3+}.

10. To $0.350\ L$ $0.100\ M$ NH_3 is added $0.150\ L$ $0.100\ M$ $MgCl_2$.
   (a) Show that $Mg(OH)_2$ should precipitate.
   (b) What mass of $(NH_4)_2SO_4$ should be added to cause the $Mg(OH)_2$ to redissolve. (*Hint:* What mass of $(NH_4)_2SO_4$ would have prevented precipitation in the first place?)

Exercises

(Use data from tables in Chapters 17, 18, and 19, as necessary.)

K_{sp} and solubility

1. A statement is made in the text that although K_{sp} and molar solubility of a slightly soluble solute are related, they are not equal. Demonstrate why this is so.

2. Which of the following saturated aqueous solutions has the highest $[Mg^{2+}]$? **(a)** $MgCO_3$; **(b)** MgF_2; **(c)** $Mg_3(PO_4)_2$.

3. A handbook lists the solubility of barium oxalate in water as 9 mg per 100 mL. What is K_{sp} for BaC_2O_4?

4. How many parts per million (ppm) of fluoride ion (i.e., g F^- per 10^6 g of solution) are present in a water solution that is saturated in CaF_2?

★5. One of the substances sometimes responsible for the "hardness" of water is $CaSO_4$. A particular water sample has 131 ppm of $CaSO_4$ (131 g $CaSO_4$ per 10^6 g of water). If this water is boiled in a tea kettle, approximately what fraction of the water must be evaporated before $CaSO_4(s)$ begins to deposit? Assume that the solubility of $CaSO_4$ does not change with temperature in the range 0 to 100°C.

6. A 25.00-mL sample of a clear *saturated* solution of PbI_2 requires 13.3 mL of a certain $AgNO_3(aq)$ for its titration. What is the molarity of this $AgNO_3(aq)$?

$$I^-(aq) \quad + \quad Ag^+(aq) \quad \rightarrow AgI(s)$$
[from satd. $PbI_2(aq)$]   [from $AgNO_3(aq)$]

7. Saturated $CaC_2O_4(aq)$ is prepared and 250.0 mL of the solution is withdrawn and titrated with 4.9 mL 0.00131 M

11. Which of the following ion concentrations can be maintained in the same solution *without* a precipitate forming? (*Hint:* What is the concentration of the *free* cation in each case?)
   (a) $[[Ag(CN)_2]^-] = 0.012\ M$, $[CN^-] = 1.05\ M$,   and $[I^-] = 2.0\ M$
   (b) $[[Ag(S_2O_3)_2]^{3-}] = 0.012\ M$, $[S_2O_3^{2-}] = 1.05\ M$, and $[I^-] = 2.0\ M$
   (c) $[[Cu(NH_3)_4]^{2+}] = 0.055\ M$, $[NH_3] = 1.8\ M$,   and $[S^{2-}] = 3.2 \times 10^{-5}\ M$

12. What concentration of *free* CN^- must be maintained in a solution that is $1.8\ M$ $AgNO_3$ and $0.327\ M$ $NaCl$ to prevent $AgCl(s)$ from precipitating?

13. A solution is $0.10\ M$ in Cd^{2+}, Cu^{2+}, and Fe^{2+}. Which ions will precipitate as sulfides if the solution is also made to be **(a)** $0.010\ M$ $H_2S(aq)$; **(b)** $0.01\ M$ $H_2S(aq)$–$0.010\ M$ $HCl(aq)$?

14. Should FeS precipitate from a solution that is saturated in H_2S ($0.10\ M$) and $0.0022\ M$ in Fe^{3+} at pH = 3.55?

15. A solution is saturated in H_2S ($0.10\ M$ H_2S) and $0.015\ M$ $FeSO_4(aq)$. What is the minimum pH that can be maintained and still have FeS(s) precipitate from this solution?

$KMnO_4(aq)$. What is the value of K_{sp} for CaC_2O_4 obtained from these data? The titration reaction is

$$5\ C_2O_4^{2-}(aq) + 2\ MnO_4^-(aq) + 16\ H^+(aq) \rightarrow$$
$$2\ Mn^{2+}(aq) + 8\ H_2O + 10\ CO_2(g)$$

★8. To precipitate as $Ag_2S(s)$ all the Ag^+ present in 338 mL of a saturated solution of $AgBrO_3$ requires $30.4\ cm^3$ of $H_2S(g)$ measured at 23°C and 748 mmHg. What is K_{sp} for $AgBrO_3$?

$$2\ Ag^+(aq) + H_2S(g) \rightarrow Ag_2S(s) + 2\ H^+(aq)$$

★9. S is the molar solubility of a slightly soluble ionic compound in pure water. Derive an algebraic relationship between S and K_{sp} in each of the following cases. Then calculate the molar solubility. For example, K_{sp} of $AgCl = 1.6 \times 10^{-10}$. $S = \sqrt{K_{sp}} = 1.3 \times 10^{-5}$ mol AgCl/L.
   (a) $Mg(OH)_2$, $K_{sp} = 1.8 \times 10^{-11}$
   (b) Ag_2CO_3, $K_{sp} = 8.1 \times 10^{-12}$
   (c) $Al(OH)_3$, $K_{sp} = 1.3 \times 10^{-33}$
   (d) Li_3PO_4, $K_{sp} = 3.2 \times 10^{-9}$
   (e) Bi_2S_3, $K_{sp} = 1 \times 10^{-97}$

The common ion effect

10. The salts KI and KNO_3 are found to have different effects on the solubility of AgI in water. Describe the effect of each and explain why the effects are different.

11. A $0.200\ M$ Na_2SO_4 solution is saturated with Ag_2SO_4 and is found to have $[Ag^+] = 9.2 \times 10^{-3}\ M$. What is the value of K_{sp} for Ag_2SO_4 obtained from these data?

12. Demonstrate that the effect of Ag^+ ion on reducing the water solubility of Ag_2CrO_4 is even greater than the effect of CrO_4^{2-} described in Example 19-4. [*Hint:* What is the solubility of Ag_2CrO_4 in 0.10 M $AgNO_3(aq)$?]

13. Plot a graph similar to Figure 19-2 to show how the solubility of lead iodate varies with concentration of $KIO_3(aq)$ ranging from pure water to 0.10 M $KIO_3(aq)$. For $Pb(IO_3)_2$, $K_{sp} = 3.2 \times 10^{-13}$.

14. A large excess of $Mg(OH)_2(s)$ is maintained in contact with 500.0 mL of its saturated solution. What will be $[Mg^{2+}]$ at equilibrium in the final solution in each case?

   (a) 500.0 mL of pure water is added to the mixture and equilibrium is reestablished.

   (b) 100.0 mL of the clear saturated solution is removed from the original mixture and added to 500.0 mL of pure water.

   (c) 25.00 mL of clear saturated solution is removed from the original mixture and added to 250.0 mL 0.065 M $MgCl_2(aq)$.

   (d) 50.00 mL of clear saturated solution is removed from the original mixture and added to 150.0 mL 0.150 M $KOH(aq)$.

15. A particular water sample is saturated in CaF_2 at the same time that it has a Ca^{2+} content of 115 ppm (i.e., 115 g Ca^{2+} per 10^6 g of water sample). What is the fluoride ion content of the water in ppm?

*16. Calculate the solubility of MgF_2 in 5.50×10^{-4} M $MgCl_2(aq)$. How effective is the common ion (Mg^{2+}) in reducing the solubility of MgF_2 in this case? Explain.

Criterion for precipitation from solution

17. Should precipitation of MgF_2 occur if 17.5 mg $MgCl_2 \cdot 6H_2O$ is added to 325 mL 0.045 M KF?

18. Should precipitation of $PbCl_2(s)$ occur when 155 mL 0.016 M $KCl(aq)$ is added to 245 mL 0.175 M $Pb(NO_3)_2(aq)$?

19. What must be the pH in a solution that is 0.17 M in Fe^{3+} to just cause the precipitation of $Fe(OH)_3(s)$?

20. Should precipitation occur in the following cases?

   (a) 1.0 mg $NaCl$ is added to 1.00 L 0.10 M $AgNO_3(aq)$.

   (b) One drop (0.05 mL) of 0.20 M KBr is added to 200 mL of a saturated solution of $AgCl$.

   (c) One drop (0.05 mL) of 0.0150 M $NaOH(aq)$ is added to 5.0 L of a solution with 2.0 mg Mg^{2+} per liter.

21. If 0.025 g KCl is added to 0.75 L of a solution saturated in Ag_2CO_3, will $AgCl$ precipitate?

Completeness of precipitation

22. When 200 mL of 0.350 M $K_2CrO_4(aq)$ is added to 200 mL of 0.100 M $AgNO_3(aq)$,

   (a) Should a precipitate form?

   (b) What $[Ag^+]$ is left unprecipitated?

23. A certain sample of water is found to have $[Mg^{2+}] = 1.0 \times 10^{-2}$ M.

   (a) What $[OH^-]$ must be maintained in this water to remove essentially all the magnesium by precipitation as $Mg(OH)_2(s)$? (*Hint:* Assume that all the Mg^{2+} has been removed when $[Mg^{2+}] = 1 \times 10^{-6}$ M.)

   (b) What $[OH^-]$ should be maintained to effect the removal of 90% of the dissolved magnesium?

24. What percent of the Ba^{2+} in solution is precipitated as $BaCO_3(s)$ if *equal volumes* of 0.0020 M $Na_2CO_3(aq)$ and 0.0010 M $BaCl_2(aq)$ are mixed?

*25. What is $[Pb^{2+}]$ remaining in solution if 225 mL 0.15 M $KCl(aq)$ is added to 135 mL 0.12 M $Pb(NO_3)_2(aq)$? (*Hint:* Look for a simplifying assumption but not the usual one.)

Fractional precipitation

26. Assume that the seawater sample described in Example 19-6 contains approximately 440 g Ca^{2+} per metric ton of seawater (1 metric ton = 1000 kg; density of seawater $\simeq$ 1.03 g/cm^3).

   (a) Should $Ca(OH)_2(s)$ precipitate from seawater under the conditions stated, that is, with $[OH^-]$ maintained at 2.0×10^{-3} M?

   (b) Is the separation of Ca^{2+} and Mg^{2+} from seawater by fractional precipitation feasible?

27. To a solution that is 0.250 M in $NaCl(aq)$ and 0.0022 M in $KBr(aq)$ is slowly added $AgNO_3(aq)$.

   (a) Which should precipitate first, $AgCl(s)$ or $AgBr(s)$?

   (b) Can the Cl^- and Br^- be separated effectively by this fractional precipitation?

28. Which of the following reagents would work best to separate Ba^{2+} and Ca^{2+} from a solution in which both are present at a concentration of 0.05 M? **(a)** 0.10 M $NaCl(aq)$; **(b)** 0.50 M $Na_2SO_4(aq)$; **(c)** 0.001 M $NaOH(aq)$; **(d)** 0.50 M $Na_2CO_3(aq)$.

29. Refer to Example 19-8 and the discussion following it. $AgI(s)$ is yellow and $AgCl(s)$ is white. Could this fact be used as a basis for determining the point at which to stop the titration of Figure 19-4? Explain.

Solubility and pH

30. Should the following precipitates form under the given conditions?

   (a) $PbI_2(s)$, from a solution that is 1.05×10^{-3} M HI, 1.05×10^{-3} M NaI, and 1.1×10^{-3} M $Pb(NO_3)_2$.

   (b) $Mg(OH)_2(s)$, from 2.50 L of 0.0150 M $Mg(NO_3)_2$ to which is added 1 drop (0.05 mL) of 1.00 M NH_3.

   (c) $Al(OH)_3(s)$, from a solution that is 1.0×10^{-2} M in Al^{3+}, 0.01 M $HC_2H_3O_2$, and 0.01 M $NaC_2H_3O_2$.

31. The solubility of $Mg(OH)_2$ in a particular buffer solution is found to be 0.95 g/L. What must be the pH of the buffer solution?

32. Calculate the molar solubility of $Mg(OH)_2$ in 1.00 M $NH_4Cl(aq)$. (*Hint:* Consider the reaction $Mg(OH)_2(s) + 2 NH_4^+(aq) \rightleftharpoons Mg^{2+}(aq) + 2 NH_3(aq) + 2 H_2O$.)

33. The solubility of MgF_2 in a particular buffer solution is found to be 0.049 g/L. What is the pH of the buffer solution? (*Hint:* What are the two equilibrium expressions that should be combined?)

34. A handbook lists the solubility of $CaHPO_4$ as 0.32 g $CaHPO_4 \cdot 2H_2O/L$ and lists a K_{sp} value of

$$CaHPO_4(s) \rightleftharpoons Ca^{2+}(aq) + HPO_4^{2-}(aq) \qquad K_{sp} = 1 \times 10^{-7}$$

  (a) Are these data consistent? (That is, are the molar solubilities the same when derived in two different ways?)
  (b) How do you account for the "discrepancy"? (*Hint:* Recall the nature of phosphate species in solution.)

Complex ion equilibria

35. $PbCl_2(s)$ is found to be somewhat soluble in HCl(aq) but not in HNO_3(aq). Explain this difference in behavior.

36. Write equations to represent the following observations: $Zn(OH)_2(s)$ is insoluble in pure water but is readily soluble in dilute HCl(aq), $HC_2H_3O_2$(aq), NH_3(aq), and NaOH(aq).

37. A solution is prepared that is 0.10 M in *free* NH_3, 0.10 M NH_4Cl, and 0.15 M in $[Cu(NH_3)_4]^{2+}$. Should $Cu(OH)_2(s)$ precipitate from this solution? K_{sp} for $Cu(OH)_2$ is 1.6×10^{-19}.

38. Refer to Example 19-15. What mass of KI could be dissolved in 1.00 L of the $AgNO_3$–NH_3 solution before AgI(s) would precipitate?

39. What mass of AgBr can be dissolved in 1.00 L of each of the following solutions? **(a)** 1.50 M NH_3; **(b)** 0.10 M NaCN; **(c)** 0.50 M $Na_2S_2O_3$.

Precipitation and solubilities of metal sulfides

40. A buffer solution is 0.25 M $HC_2H_3O_2$–0.15 M $NaC_2H_3O_2$, saturated in H_2S (0.10 M) and has $[Mn^{2+}] = 0.015$ M.
  (a) Show that MnS will not precipitate from this solution.
  (b) Which buffer component would you increase in concentration, and to what minimum value, to ensure that precipitation of MnS would begin? (*Note:* The concentration of the other buffer component is held constant.)

41. In the qualitative analysis of a mixture of cations, Pb^{2+} is first precipitated as $PbCl_2$. Later, PbS is precipitated when the saturated $PbCl_2$(aq) is saturated with H_2S (0.10 M) and its pH adjusted to about 0.5. Show that in fact the precipitation of PbS should occur under these conditions.

42. What is the solubility of FeS, in g/L, in a buffer solution that is 0.500 M $HC_2H_3O_2$–0.250 M $NaC_2H_3O_2$? (*Hint:* Assume that $[H_3O^+]$ remains constant in the dissolving reaction.)

★43. Calculate the molar solubility of CoS(s) in 0.100 M $HC_2H_3O_2$. (*Hint:* Combine the appropriate equilibrium constant expressions.)

Qualitative analysis

44. Addition of HCl(aq) to a solution containing several different cations produces a white precipitate. The filtrate is removed and treated with H_2S(aq) in 0.3 M HCl. No precipitate forms. Which of the following conclusions is(are) valid? Explain.
  (a) Ag^+ and/or Hg_2^{2+} probably present
  (b) Mg^{2+} probably not present
  (c) Pb^{2+} probably not present
  (d) Fe^{2+} probably not present

45. Why is Pb^{2+} found in qualitative analysis groups 1 *and* 2? Why do you suppose that none of the other common cations is found in more than one group?

46. What reagent solution might you use to separate the cations in the following pairs, that is, with one ion appearing in solution and the other in a precipitate? (*Hint:* Refer to Figure 19-5; consider water also to be a reagent.)
  (a) $BaCl_2(s)$ and NaCl(s)
  (b) $MgCO_3(s)$ and $Na_2CO_3(s)$
  (c) $AgNO_3(s)$ and $KNO_3(s)$
  (d) $PbSO_4(s)$ and $Pb(NO_3)_2(s)$

47. In the qualitative analysis scheme for the silver group, $PbCl_2(s)$ is separated from AgCl(s) by dissolving it in hot water. Given the K_{sp} values for $PbCl_2$ listed below, assess the water solubility of $PbCl_2(s)$ at low and high temperatures. Use the definitions of soluble, insoluble, and moderately soluble given in the footnote to Table 19-2. K_{sp} for $PbCl_2 = 1.6 \times 10^{-5}$ at 25°C; 3.3×10^{-3} at 80°C.

★48. Explain the following restrictions in the qualitative analysis scheme for cations.
  (a) An unknown solution cannot simply be divided into about two dozen samples and a test performed for a single different ion on each sample.
  (b) The separation of ions into groups cannot be done in just any order desired, such as groups 4, 3, 1, 2, and 5.

Additional Exercises

1. A handbook lists the solubilities 1.0×10^{-5} mol $BaSO_4$/L and 7.1×10^{-5} mol $BaCO_3$/L. When 0.50 M Na_2CO_3(aq) is added to a saturated solution of $BaSO_4$, a precipitate of $BaCO_3$(s) forms. How do you account for this fact, given that $BaCO_3$ has a larger K_{sp} than does $BaSO_4$?

2. A solution is saturated with $PbSO_4$ at 50°C ($K_{sp} = 2.3 \times 10^{-8}$). How many mg $PbSO_4$ will precipitate from 815 mL of this solution if it is cooled to 25°C? ($K_{sp} = 1.6 \times 10^{-8}$.)

3. A 25.00-mL sample of a saturated solution of $SrCrO_4$ requires 25.5 mL of 0.0138 M $Fe(NO_3)_2$(aq) for its titration according to the reaction

$$CrO_4{}^{2-}(aq) + 3\ Fe^{2+}(aq) + 8\ H^+(aq) \rightarrow$$
$$Cr^{3+}(aq) + 3\ Fe^{3+}(aq) + 4\ H_2O$$

Calculate K_{sp} for $SrCrO_4$.

4. If 10.0 mg $CaCl_2$ is added to 250.0 mL 0.03500 M Na_2SO_4,
  (a) Should precipitation occur?
  (b) Will precipitation of Ca^{2+} be complete?

5. The electrolysis of $MgCl_2$(aq) can be represented as

$$Mg^{2+}(aq) + 2\ Cl^-(aq) + 2\ H_2O \rightarrow$$
$$Mg^{2+}(aq) + 2\ OH^-(aq) + H_2(g) + Cl_2(g)$$

The electrolysis of a 315-mL sample of 0.220 M $MgCl_2$ is continued until 1.04 L H_2(g) at 23°C and 748 mmHg has been collected. Will $Mg(OH)_2$(s) precipitate when electrolysis is carried to this point?

6. You are provided with the following: NaOH(aq), K_2SO_4(aq), $Mg(NO_3)_2$(aq), $BaCl_2$(aq), NaCl(aq), $Ca(NO_3)_2$(aq), Ag_2SO_4(s), and $BaSO_4$(s). Write net ionic equations to show how you would use these reagents to obtain the following: **(a)** $CaSO_4$(s); **(b)** $Mg(OH)_2$(s); **(c)** KCl(aq); **(d)** AgCl(s).

7. In Example 19-18 a statement was made that the distinctive blue color of Cu^{2+}(aq) might not be detectable if a solution were too dilute. Suppose that you had an unknown solution in which $[Cu^{2+}] = 0.02\ M$.
  (a) Show that you would be able to precipitate CuS from 10 mL of this solution by adding a few drops of 0.50 M Na_2S (1 drop = 0.05 mL).
  (b) How much CuS would be obtained, and do you think that it would be visible to the unaided eye?

8. The following pertain to the precipitation or dissolving of metal sulfides. Predict whether each reaction proceeds to a significant extent in the forward direction.
  (a) $Cu^{2+}(aq) + H_2S(\text{satd. aq}) \rightarrow$
  (b) $Mg^{2+}(aq) + H_2S(\text{satd. aq}) \xrightarrow{0.3\ M\ HCl}$
  (c) PbS(s) + HCl(0.3 M) $\rightarrow$
  (d) $Ag^+(aq) + H_2S(\text{satd. aq}) \xrightarrow{0.3\ M\ HCl}$

9. Determine if 1.50 g $H_2C_2O_4$ (oxalic acid) can be dissolved in 0.200 L of 0.150 M $CaCl_2$ without the formation of CaC_2O_4(s) (for which $K_{sp} = 1.3 \times 10^{-9}$). (*Hint:* You will also need data from Table 17-3.)

10. It is desired to carry out the separation of Fe^{2+} and Mn^{2+} by precipitating FeS and leaving Mn^{2+} in solution.
  (a) Show that this can be accomplished in a solution that is 0.50 M $HC_2H_3O_2$ and 0.40 M $NaC_2H_3O_2$ and saturated in H_2S.
  (b) What is the pH of the solution *after* precipitation of the FeS? (*Hint:* Why doesn't the pH remain constant?)
The following conditions apply: $[Fe^{2+}] = 0.10\ M$, $[Mn^{2+}] = 0.10\ M$, and $[H_2S] = 0.10\ M$.

*11. Write net ionic equations to represent the following observations.
  (a) When concentrated $CaCl_2$(aq) is added to Na_2HPO_4(aq), a white precipitate is formed that is 38.7% Ca, by mass.
  (b) When a piece of dry ice [CO_2(s)] is placed into a clear dilute solution of "lime water" [$Ca(OH)_2$(aq)], bubbles of gas are evolved. At first a white precipitate forms, but then it redissolves.

*12. For the titration that is the subject of Example 19-8 and illustrated in Figure 19-4, verify the assertion on page 574 that $[Ag^+]$ increases very rapidly between the point where AgI has finished precipitating and AgCl is about to begin.

*13. The solubility of AgCN(s) in 0.200 M NH_3(aq) is 8.8×10^{-6} mol/L. What is the value of K_{sp} for AgCN?

*14. The solubility of $CdCO_3$(s) in 1.00 M KI(aq) is 1.2×10^{-3} mol/L. Given that K_{sp} for $CdCO_3$ is 5.2×10^{-12}, what is K_f for $CdI_4{}^{2-}$?

*15. Use K_{sp} for $PbCl_2$ and K_f for $[PbCl_3]^-$ to determine the molar solubility of $PbCl_2$ in 0.10 M HCl(aq). (*Hint:* Recall the discussion following Example 19-17.)

*16. In the Mohr titration, Cl^- is determined by titration with $AgNO_3$(aq). The solution also contains a small volume of K_2CrO_4(aq) as an indicator. After sufficient $AgNO_3$(aq) has been added to precipitate all of the Cl^-, a red precipitate of Ag_2CrO_4 forms. Explain the basis of this titration. That is, why does the red color not appear while AgCl(s) is still precipitating? Why does the red precipitate appear *immediately* after the AgCl(s) has finished precipitating?

*17. A mixture of $PbSO_4$(s) and PbS_2O_3(s) is shaken with pure water until a saturated solution is formed. Both solids remain in excess. What is $[Pb^{2+}]$ in the saturated solution? (*Hint:* Both of the following equilibrium expressions are required, and a third equation as well.)

$$PbSO_4(s) \rightleftharpoons Pb^{2+}(aq) + SO_4{}^{2-}(aq) \qquad K_{sp} = 1.6 \times 10^{-8}$$

$$PbS_2O_3(s) \rightleftharpoons Pb^{2+}(aq) + S_2O_3{}^{2-}(aq) \qquad K_{sp} = 4.0 \times 10^{-7}$$

*18. Use the method of the preceding exercise to determine $[Pb^{2+}]$ in a saturated solution in contact with a mixture of $PbCl_2$(s) and $PbBr_2$(s).

$$PbCl_2(s) \rightleftharpoons Pb^{2+}(aq) + 2\ Cl^-(aq) \qquad K_{sp} = 1.6 \times 10^{-5}$$

$$PbBr_2(s) \rightleftharpoons Pb^{2+}(aq) + 2\ Br^-(aq) \qquad K_{sp} = 4.0 \times 10^{-5}$$

*19. 2.50 g Ag_2SO_4(s) is added to a beaker containing 0.150 L 0.025 *M* $BaCl_2$.
   (a) Write an equation for any reaction that occurs.
   (b) Describe the final contents of the beaker, that is, the masses of any precipitates present and the concentrations of the ions in solution.

*20. Calculate the molar solubility of MnS in water (a) based only on the solubility product expression, K_{sp}, for MnS and (b) taking into account the hydrolysis of S^{2-} to HS^-. (c) Explain the difference in the results obtained.

Self-Test Questions

For questions 1 through 8 select the single item that best completes each statement.

1. Pure water is saturated with the slightly soluble solute, PbI_2. In this saturated solution, (a) $[Pb^{2+}] = [I^-]$; (b) $[Pb^{2+}] = K_{sp}$ of PbI_2; (c) $[Pb^{2+}] = \sqrt{K_{sp}}$ of PbI_2; (d) $[Pb^{2+}] = 0.5 [I^-]$.

2. The addition of 1.85 g Na_2SO_4 to 500.0 mL of saturated aqueous $BaSO_4$ has the following effect on the saturated solution: (a) reduces $[Ba^{2+}]$; (b) reduces $[SO_4^{2-}]$; (c) increases the solubility of $BaSO_4$; (d) has no effect.

3. The slightly soluble solute Ag_2CrO_4 is expected to be *most* soluble in (a) pure water; (b) 0.10 *M* K_2CrO_4; (c) 0.25 *M* KNO_3; (d) 0.40 *M* $AgNO_3$.

4. Cu^{2+} and Pb^{2+} are both present in an aqueous solution. To precipitate one of the ions and leave the other in solution, add (a) H_2S(aq); (b) H_2SO_4(aq); (c) HNO_3(aq); (d) NH_4NO_3(aq).

5. The addition of K_2CO_3(aq) to the following solutions is expected to yield a precipitate in every case but one. That one is (a) $BaCl_2$(aq); (b) $CaBr_2$(aq); (c) $(NH_4)_2SO_4$(aq); (d) $Pb(NO_3)_2$(aq).

6. A *large* excess of MgF_2(s) is maintained in contact with 1.00 L of pure water to produce a saturated solution of MgF_2. When an additional 1.00 L of pure water is added to the mixture and equilibrium reestablished, compared to its value in the original saturated solution, $[Mg^{2+}]$ will be (a) the same; (b) twice as large; (c) half as large; (d) some unknown fraction of the original $[Mg^{2+}]$.

7. To increase the molar solubility of $CaCO_3$(s) in a saturated aqueous solution, add (a) $NaHSO_4$; (b) Na_2CO_3; (c) NH_3; (d) more water.

8. The best way to ensure complete precipitation from *saturated* H_2S(aq) of a metal ion, M^{2+}, as its sulfide, MS(s), is to (a) add an acid; (b) increase $[H_2S]$ in the solution; (c) raise the pH; (d) heat the solution.

9. The solubility product constant of PbI_2 is

$$PbI_2(s) \rightleftharpoons Pb^{2+}(aq) + 2\,I^-(aq) \qquad K_{sp} = 7.1 \times 10^{-9}$$

Determine the mg Pb^{2+} per milliliter in an 0.065 *M* KI solution that is also saturated with PbI_2.

10. Which of the following solids are likely to be more soluble in acidic solution, and which in basic solution? Which are likely to have a solubility that is independent of pH? Explain. (a) $H_2C_2O_4$; (b) $MgCO_3$; (c) CdS; (d) KCl; (e) $NaNO_3$; (f) $Ca(OH)_2$.

11. A solution is 0.010 *M* in K_2CrO_4 and 0.010 *M* in K_2SO_4. To this solution is slowly added 0.10 *M* $Pb(NO_3)_2$(aq).
   (a) What should be the first substance to precipitate from the solution?
   (b) What is $[Pb^{2+}]$ at the point where the second substance precipitates?
   (c) Are the two substances effectively separated by this fractional precipitation? Explain. (K_{sp} for $PbCrO_4 = 2.8 \times 10^{-13}$; K_{sp} for $PbSO_4 = 1.6 \times 10^{-8}$.)

12. Without performing detailed calculations, indicate whether either of the following compounds is expected to be appreciably soluble in NH_3(aq): (a) CuS, $K_{sp} = 6.3 \times 10^{-36}$; (b) $CuCO_3$, $K_{sp} = 1.4 \times 10^{-10}$. Also use the fact that K_f for $[Cu(NH_3)_4]^{2+}$ is 1.1×10^{13}.

20 Oxidation–Reduction and Electrochemistry

We introduced the concept of oxidation state in Chapter 3, and in Chapter 4 we first described a type of reaction characterized by changes in oxidation state—oxidation–reduction reactions. Classifying certain reactions as oxidation–reduction provided a useful basis for discussing some aspects of descriptive chemistry in Chapter 13. Before returning to a discussion of descriptive chemistry, we consider in this chapter the theoretical basis of oxidation–reduction.

In the preceding three chapters we considered another important reaction type, acid–base reactions, and found that their fundamental nature involved *proton* transfer. The fundamental nature of oxidation–reduction reactions, as we will soon see, involves *electron* transfer. We will also reexamine the concepts of spontaneous changes, equilibrium, and free energy changes in relation to a new property—the electrode potential. Still other topics explored here are the electrolytic mechanism of corrosion and some practical uses of electrochemistry.

20-1 Oxidation–Reduction: Some Definitions

When an iron object is exposed to the atmosphere, it rusts. A simplified equation is

$$4 \text{ Fe(s)} + 3 \text{ O}_2(g) \longrightarrow 2 \text{ Fe}_2\text{O}_3(s) \tag{20.1}$$

In this reaction iron combines with oxygen. Originally, the term "oxidation" was applied to reactions in which a substance combines with oxygen.

Iron rust is an oxide of iron and so are most iron ores. In simplified fashion, the production of iron from iron ore is described by

$$\text{Fe}_2\text{O}_3(s) + 3 \text{ CO}(g) \longrightarrow 2 \text{ Fe(l)} + 3 \text{ CO}_2(g) \tag{20.2}$$

Reaction (20.2) involves the oxidation of $CO(g)$ to $CO_2(g)$. The oxygen atoms required for this oxidation come from the $Fe_2O_3(s)$, which is reduced. Originally, the term "reduction" was used to denote reactions in which oxygen is removed from a substance. Both oxidation and reduction occur in reactions (20.1) and (20.2), and they can be called oxidation–reduction reactions. A definition of oxidation–reduction based solely on a transfer of oxygen atoms is too restrictive; it would limit us to reactions in which oxygen atoms are involved. We need broader definitions.

One type of reaction that we should include in the category "oxidation–reduction" is the displacement of silver ion from aqueous solution by copper metal (see Figure 20-1). In addition, it is useful to separate the net reaction into an **oxidation half-reaction** and a

This reaction is also illustrated in Color Section G.

FIGURE 20-1
Displacement of Ag^+
from aqueous solution
by copper metal.

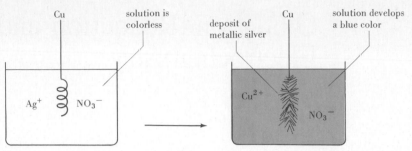

Copper metal displaces Ag^+ ion from solution, producing a feathery deposit of metallic silver (a silver tree).

$$Cu(s) + 2 Ag^+(aq) \longrightarrow Cu^{2+}(aq) + 2 Ag(s)$$

reduction half-reaction. Finally, our definitions should assign the term oxidation to describe what happens to the copper, and reduction to the silver ion. The result of these requirements is

oxidation:	$Cu(s) \longrightarrow Cu^{2+}(aq) + 2 e^-$	(20.3)
reduction:	$Ag^+(aq) + e^- \longrightarrow Ag(s)$	(20.4)
net reaction:	$Cu(s) + 2 Ag^+(aq) \longrightarrow Cu^{2+}(aq) + 2 Ag(s)$	(20.5)

The oxidation state of copper *increases* from 0 to $+2$ (corresponding to the loss of two electrons by each copper atom). The oxidation state of silver *decreases* from $+1$ to 0 (corresponding to the gain of one electron by each silver ion). What equations (20.3), (20.4), and (20.5) seem to require by way of definitions is:

- Oxidation is a process in which the oxidation state of some element *increases* and in which electrons appear on the right-hand side of the oxidation half-equation.
- Reduction is a process in which the oxidation state of some element *decreases* and in which electrons appear on the left-hand side of the reduction half-equation.
- Oxidation and reduction half-reactions must always occur together. Moreover, the total number of electrons associated with the oxidation process must be equal to the total number associated with the reduction process.

20-2 More on Balancing Oxidation–Reduction Equations

Before adding half-equations (20.3) and (20.4) to obtain the net equation (20.5), it was necessary to *multiply (20.4) by the factor "2."* This was done because equal numbers of electrons must appear in the oxidation and in the reduction half-equations. One method of balancing oxidation–reduction equations is based on this requirement. It is called the half-reaction or ion electron method and is considered below. Another approach is based on the definitions of oxidation and reduction in terms of oxidation states. This is the oxidation state change method that we learned in Section 4-5.

Ion Electron or Half-reaction Method. In this method oxidation and reduction half-equations are written separately and then combined into a balanced net equation. Consider the reaction of sulfite and permanganate ions in acidic solution.*

$$SO_3^{2-} + H^+ + MnO_4^- \longrightarrow SO_4^{2-} + Mn^{2+} + H_2O$$

---

*As noted in Section 17-2, protons in aqueous solutions are hydrated. In Chapters 17, 18, and 19 we chose to represent them as hydronium ions, H_3O^+, because of the importance of this species in a discussion of acids and bases. Here we will represent them as H^+, to simplify the appearance of oxidation–reduction equations.

Step 1. *Identify the species involved in oxidation state changes and write "skeleton" half-equations based on them.* In comparing the two oxoanions of sulfur, the oxidation state of S in SO_3^{2-} is +4, and in SO_4^{2-} it is +6. The oxidation half-reaction involves the conversion of sulfite to sulfate ion. The oxidation state of Mn decreases from +7 to +2 in the net reaction. The conversion of MnO_4^- to Mn^{2+} occurs in the reduction half-reaction.

oxidation: $SO_3^{2-} \longrightarrow SO_4^{2-}$
reduction: $MnO_4^- \longrightarrow Mn^{2+}$

Step 2. *Balance each half-equation "atomically."* To show the same number of atoms of each type on both sides of the half-equation, it is often necessary to add H_2O and either H^+ (for acidic solutions) or OH^- (for basic solutions). For an acidic solution add *one* H_2O molecule for every O atom needed on the side that is deficient in O atoms. To the *opposite* side of the half-equation, add *two* H^+ for every H_2O molecule that was used.

oxidation: $SO_3^{2-} + H_2O \longrightarrow SO_4^{2-} + 2\,H^+$
reduction: $MnO_4^- + 8\,H^+ \longrightarrow Mn^{2+} + 4\,H_2O$

Step 3. *Balance each half-equation "electrically."* To the *right-hand* side of the *oxidation* half-equation, add the number of electrons necessary to achieve the same net charge on both sides of the half-equation. Do the same to the *reduction* half-equation by adding electrons to the *left-hand* side.

oxidation: $SO_3^{2-} + H_2O \longrightarrow SO_4^{2-} + 2\,H^+ + 2\,e^-$
(net charge on each side, −2)
reduction: $MnO_4^- + 8\,H^+ + 5\,e^- \longrightarrow Mn^{2+} + 4\,H_2O$
(net charge on each side, +2)

Step 4. *Obtain the net oxidation–reduction equation by combining the half-equations.* Multiply through the oxidation half-equation by *5* and through the reduction half-equation by *2*. This will result in 10 e^- on *each* side of the net equation. These terms will then cancel out. Electrons must not appear in the final net equation.

$5\,SO_3^{2-} + 5\,H_2O \longrightarrow 5\,SO_4^{2-} + 10\,H^+ + \cancel{10\,e^-}$
$2\,MnO_4^- + 16\,H^+ + \cancel{10\,e^-} \longrightarrow 2\,Mn^{2+} + 8\,H_2O$

$5\,SO_3^{2-} + 5\,H_2O + 2\,MnO_4^- + 16\,H^+ \longrightarrow 5\,SO_4^{2-} + 10\,H^+ + 2\,Mn^{2+} + 8\,H_2O$

Step 5. *Simplify.* If the net equation contains the same species on both sides of the equation, cancel it from the side where it appears in lesser amount. Subtract *five* H_2O from each side of the net equation of step 4. This leaves *three* H_2O on the right. Also subtract *ten* H^+ from each side, leaving *six* on the left.

$$5\,SO_3^{2-} + 2\,MnO_4^- + 6\,H^+ \longrightarrow 5\,SO_4^{2-} + 2\,Mn^{2+} + 3\,H_2O \qquad (20.6)$$

Step 6. *Verify.* Check the final net equation to ensure that it is balanced both "atomically" and "electrically." For example, show that in equation (20.6) the net charge on each side of the equation is − 6.

Illustrative Examples. Some of the additional aspects of balancing oxidation–reduction equations illustrated through the examples are

Example 20-1
- Half-equations can be written without a prior assessment of oxidation state changes.
- More than a single oxidation and/or reduction process may occur in the same oxidation–reduction reaction.
- In the balanced net equation, H^+ and H_2O do not necessarily appear on the same side(s) of the equation as in the original unbalanced equation.

Example 20-2
- Oxidation–reduction reactions may occur in basic solutions.

Example 20-3
- The same substance may undergo both oxidation and reduction, called a **disproportionation reaction.**
- A combination of the half-reaction and the oxidation state change methods often provides the simplest means of balancing an equation.

Example 20-1   Balance the oxidation–reduction equation

$$As_2S_3(s) + NO_3^- + H^+ \longrightarrow H_3AsO_4 + S(s) + NO(g) + H_2O$$

Solution.   The two half-equations that trace the fate of each atomic species (other than H and O) are

(1) $As_2S_3 \longrightarrow H_3AsO_4 + S$

(2) $NO_3^- \longrightarrow NO$

We can balance half-equation (1) through the following steps, ultimately identifying it as an oxidation half-equation because electrons appear on the *right*. Even though both arsenic and sulfur atoms are oxidized, no initial determination of oxidation states is required.

$$As_2S_3 \longrightarrow H_3AsO_4 + S$$
$$As_2S_3 \longrightarrow 2\ H_3AsO_4 + 3\ S$$
$$As_2S_3 + 8\ H_2O \longrightarrow 2\ H_3AsO_4 + 3\ S$$
$$As_2S_3 + 8\ H_2O \longrightarrow 2\ H_3AsO_4 + 3\ S + 10\ H^+$$

oxidation: $As_2S_3 + 8\ H_2O \longrightarrow 2\ H_3AsO_4 + 3\ S + 10\ H^+ + 10\ e^-$

Next, half-equation (2) is balanced through a series of steps and is identified as a reduction half-equation.

$$NO_3^- \longrightarrow NO$$
$$NO_3^- \longrightarrow NO + 2\ H_2O$$
$$NO_3^- + 4\ H^+ \longrightarrow NO + 2\ H_2O$$

reduction: $NO_3^- + 4\ H^+ + 3\ e^- \longrightarrow NO + 2\ H_2O$

The oxidation half-equation is multiplied by *three* and the reduction half-equation by *ten.* The half-equations are combined and the appropriate numbers of H^+ and OH^- are canceled to obtain the balanced net equation.

oxidation: $3\ As_2S_3 + 24\ H_2O \longrightarrow 6\ H_3AsO_4 + 9\ S + 30\ H^+ + \cancel{30\ e^-}$
reduction: $10\ NO_3^- + 40\ H^+ + \cancel{30\ e^-} \longrightarrow 10\ NO + 20\ H_2O$

net: $3\ As_2S_3 + 10\ NO_3^- + 10\ H^+ + 4\ H_2O \longrightarrow 6\ H_3AsO_4 + 9\ S + 10\ NO$

SIMILAR EXAMPLE: Review Problem 2.

Example 20-2   Balance the oxidation–reduction equation

$$Cr(OH)_3(s) + OCl^- + OH^- \longrightarrow CrO_4^{2-} + Cl^- + H_2O$$

Solution.   The oxidation state of Cr in $Cr(OH)_3$ is +3, and in CrO_4^{2-} it is +6. In OCl^- the oxidation state of Cl is +1, and in Cl^- it is −1.

oxidation: $Cr(OH)_3(s) \longrightarrow CrO_4^{2-}$
reduction: $OCl^- \longrightarrow Cl^-$

TABLE 20-1
Achieving a balance of H_2O and OH^- in oxidation and reduction half-equations in basic solutions

	To the side deficient in oxygen	To the other side
to balance O atoms:	for every O atom required, add two OH^-	add one H_2O
to balance H atoms:	To the side deficient in hydrogen	To the other side
	for every H atom required, add one H_2O	add one OH^-

In basic solutions, whether we add OH^- or H_2O to a half-equation, we find ourselves adding *both* H and O atoms; however, in OH^- there is *one* H atom per O atom and in H_2O, *two* H atoms per O atom. In general, we should add OH^- to the side deficient in O atoms and H_2O to the side deficient in H atoms. The method of Table 20-1 can be quite useful.*

In the skeleton oxidation half-equation there are *three* O atoms on the left and *four* on the right. According to Table 20-1, to gain an O atom on the left we should add *two* OH^- to the left and *one* H_2O to the right.

oxidation: $Cr(OH)_3(s) + 2\ OH^- \longrightarrow CrO_4{}^{2-} + H_2O$ (not balanced)

The H atoms are still out of balance—*five* on the left and *two* on the right. Again according to Table 20-1, we should add *three* H_2O to the right and *three* OH^- to the left.

oxidation: $Cr(OH)_3(s) + 5\ OH^- \longrightarrow CrO_4{}^{2-} + 4\ H_2O$ (not balanced)

Next we add *three* electrons to the right to balance electric charge.

oxidation: $Cr(OH)_3(s) + 5\ OH^- \longrightarrow CrO_4{}^{2-} + 4\ H_2O + 3\ e^-$

Completing the reduction half-equation is a bit simpler because only O atoms are out of balance in the skeleton half-equation.

reduction: $OCl^- + H_2O \longrightarrow Cl^- + 2\ OH^-$ (not balanced)

Charge balance is achieved by adding *two* electrons to the left.

reduction: $OCl^- + H_2O + 2\ e^- \longrightarrow Cl^- + 2\ OH^-$

Finally, we multiply through the oxidation half-equation by 2 and through the reduction half-equation by 3. The two half-equations are combined and the net equation simplified by canceling the appropriate number of H_2O and OH^-.

oxidation: $2\ Cr(OH)_3(s) + 10\ OH^- \longrightarrow 2\ CrO_4{}^{2-} + 8\ H_2O + \cancel{6\ e^-}$
reduction: $3\ OCl^- + 3\ H_2O + \cancel{6\ e^-} \longrightarrow 3\ Cl^- + 6\ OH^-$
_____
net: $2\ Cr(OH)_3(s) + 3\ OCl^- + 4\ OH^- \longrightarrow 2\ CrO_4{}^{2-} + 3\ Cl^- + 5\ H_2O$

SIMILAR EXAMPLE: Review Problem 3.

Example 20-3   Balance the oxidation–reduction equation

$$P_4(s) + OH^- + H_2O \longrightarrow H_2PO_2{}^- + PH_3(g)$$

*Another commonly used method is to treat the reaction *as if* it occurred in acidic solution (i.e., by using H^+ and H_2O to balance the half-equations). A number of OH^- ions equal to the number of H^+ ions in the net equation is added to *both* sides of the net equation. H^+ and OH^- are combined into H_2O and the net equation is simplified.

Solution. Although the method of Example 20-2 works here too, the final result can be reached a little more easily in the following way: P_4 *is both oxidized and reduced.* In the oxidation process the oxidation state increases from 0 (in P_4) to $+1$ (in $H_2PO_2^-$). This corresponds to *one* electron on the right side of the half-equation, *per P atom,* or, for *four* P atoms,

oxidation: $\quad P_4(s) \longrightarrow 4\ H_2PO_2^- + 4\ e^-$ (not balanced)

The *eight* units of negative charge on the right side can be offset by *eight* OH^- ions on the left. This leads to a half-equation that also is balanced atomically.

oxidation: $\quad P_4(s) + 8\ OH^- \longrightarrow 4\ H_2PO_2^- + 4\ e^-$

In the reduction half-equation the oxidation state decreases from 0 (in P_4) to -3 (in PH_3). This corresponds to *three* electrons on the left side of the half-equation, *per P atom,* or, for *four* P atoms,

reduction: $\quad P_4(s) + 12\ e^- \longrightarrow 4\ PH_3(g)$ (not balanced)

The *twelve* units of negative charge on the left side can be offset by *twelve* OH^- ions on the right.

reduction: $\quad P_4(s) + 12\ e^- \longrightarrow 4\ PH_3(g) + 12\ OH^-$ (not balanced)

A final balance is achieved through *twelve* H_2O molecules on the left.

reduction: $\quad P_4(s) + 12\ H_2O + 12\ e^- \longrightarrow 4\ PH_3(g) + 12\ OH^-$

Finally, the oxidation half-equation is multiplied by 3; the half-equations are combined; and the net equation is simplified.

oxidation: $\quad 3\ P_4(s) + 24\ OH^- \longrightarrow 12\ H_2PO_2^- + 12\ e^-$
reduction: $\quad P_4(s) + 12\ H_2O + 12\ e^- \longrightarrow 4\ PH_3(g) + 12\ OH^-$
***
net: $\quad 4\ P_4(s) + 12\ H_2O + 12\ OH^- \longrightarrow 12\ H_2PO_2^- + 4\ PH_3(g)$

divide all coefficients by 4:

net: $\quad P_4(s) + 3\ H_2O + 3\ OH^- \longrightarrow 3\ H_2PO_2^- + PH_3(g)$

SIMILAR EXAMPLE: Review Problem 4.

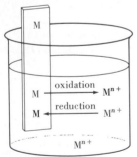

FIGURE 20-2
An electrochemical half-cell.

The half-cell consists of a metal electrode, M, immersed in an aqueous solution of its ions, M^{n+}. (The anions required to maintain electrical neutrality in the solution are not shown.) The situation illustrated here is limited to metals that do not react with water.

20-3 Measurement of Oxidation and Reduction Tendencies

If a solution of $Zn(NO_3)_2$ is substituted for the $AgNO_3$ solution in Figure 20-1, no reaction occurs. Why does copper behave differently toward Ag^+ and Zn^{2+}? To answer this and other fundamental questions, it is helpful to think in terms of a device in which an oxidation–reduction reaction can actually be separated into two distinct half-reactions.

Measurement of Electromotive Force (EMF). Figure 20-2 pictures a metal strip, M, called an **electrode,** immersed in a solution containing the metal ions, M^{n+}. The entire assembly is called a **half-cell.** Three kinds of interactions can take place between the metal atoms on the electrode and the metal ions in solution.

- A metal ion M^{n+} may collide with the electrode and undergo no change.
- A metal ion may collide with the electrode, gain n electrons and be converted to a metal atom M. *The ion is reduced.*
- A metal atom M on the electrode may lose n electrons and enter the solution as the ion M^{n+}. *The metal atom is oxidized.*

Equilibrium between the metal and its ions, which is quickly established, can be represented as

$$M(s) \underset{\text{reduction}}{\overset{\text{oxidation}}{\rightleftharpoons}} M^{n+}(aq) + n\,e^{-} \tag{20.7}$$

The net numbers of electrons on the electrode before and after equilibrium is established will be slightly different. As a result, the electrode acquires a very slight electric charge; the solution acquires the opposite charge.

The magnitude of the charge on an electrode, when it is in equilibrium with its ions in solution, would seem to be directly related to the tendency of the metal atoms to undergo oxidation and of the metal ions to undergo reduction. That is, the stronger the oxidation tendency the more negative would be the charge on the electrode (because of the electrons left behind by atoms that are oxidized). Or, the stronger the tendency for reduction, the more positive would be the charge on the electrode (because of electrons being extracted from the metal surface by ions as they are reduced). But there is a difficulty in trying to use the magnitude of the charge on an electrode as a criterion for oxidation and reduction tendencies. Surely the magnitude of this charge would depend on the size of the electrode, or more precisely, its surface area. The greater the surface area of the electrode in contact with the solution, the greater the magnitude of the charge that should have accumulated when equilibrium is reached. This problem could be overcome by assessing the charge density, the *charge per unit area* at the electrode surface. This quantity would be independent of the total surface area. Charge density, in turn, establishes the electric potential on the electrode surface. The electric potential at some point is defined as the quantity of work associated with bringing a fundamental unit of electric charge (say an electron) from an infinite distance away up to the point in question. Thus, the quantity of work required to push an electron (a negatively charged particle) onto the surface of a negatively charged electrode would be greater, the greater the negative charge density on the electrode.

The hypothetical process just described could be used to evaluate a quantity known as a *single electrode potential*. Unfortunately, no one has ever been able to design an experiment to measure a single electrode potential. In some previous cases where we encountered a process that cannot be carried out directly, we devised an indirect method that produced the same result (such as in the use of Hess's law in Chapter 6). In Section 22-1 a thermochemical cycle that leads to an indirect *estimation* of single electrode potentials is discussed, but that method is not capable of yielding precise and accurate values.

Fortunately, there is a straightforward experimental method that does yield very precise results, but it is based on measuring a *difference* in potential between *two* electrodes. If an electrical connection is made between two regions of differing electric charge density, electric charge will flow from the region of higher charge density or higher electric potential to the region of lower charge density or lower electric potential. This flow of electric charge is called an electric current, and the greater the difference in potential between two points, the greater the current that flows between them. Even the slightest difference in electric potential is enough to set up an electric current. This is analogous to the observation that water will always flow from a higher to a lower level, no matter how slight the difference in levels, if a connection is provided between the two levels.

In conclusion, then, we must shift our attention from individual half-cells, as pictured in Figure 20-2, to a combination of two half-cells, as pictured in Figure 20-3. A combination of two half-cells is called an **electrochemical cell.**

The electrical connection of two half-cells must be done in a special way. *Both* the metal electrodes *and* the solutions have to be connected, so that a continuous circuit is formed through which charged particles flow. The electrodes can simply be connected by a metal wire which permits the flow of electrons.

The flow of electric current between the solutions must be in the form of a migration of ions. This cannot occur through a wire but only through another solution which "bridges" the two half-cells; this connection is called a **salt bridge.**

FIGURE 20-3
Measurement of the
electromotive force of a
cell.

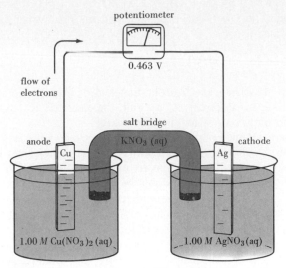

The cell consists of two half-cells with electrodes joined by a wire and solutions by a salt bridge. (The ends of the salt bridge are plugged with a porous material that allows ions to migrate but prevents the bulk flow of liquid.) The potentiometer measures the difference in electric potential between the two electrodes; this difference is 0.463 volt (V).

When these connections have been made in Figure 20-3, the following changes occur. Copper atoms lose electrons at the copper electrode and enter the solution as Cu^{2+} ions. The electrons lost by the copper atoms pass through the wire and the electrical measuring circuit to the silver electrode. Here Ag^+ ions from solution gain electrons and deposit as silver metal. Without a salt bridge, the solution in the copper half-cell would acquire excess Cu^{2+} and a net positive charge. In the silver half-cell there would be a deficiency of Ag^+, an excess of anions, and a negative charge buildup in the solution. Electric current could not continue to flow. The salt bridge allows for passage of electric current between the solutions. Consider the copper half-cell. Excess Cu^{2+} ions in this half-cell enter the salt bridge and migrate toward the silver half-cell. Also, anions from the salt bridge ($NO_3{}^-$) migrate into the copper half-cell. In the silver half-cell, $NO_3{}^-$ ions migrate out of the half-cell and K^+ ions from the salt bridge migrate in. The net reaction that occurs is

oxidation: $Cu(s) \longrightarrow Cu^{2+}(aq) + 2\ e^-$
reduction: $\underline{2\{Ag^+(aq) + e^- \longrightarrow Ag(s)\}}$
net: $Cu(s) + 2\ Ag^+(aq) \longrightarrow Cu^{2+}(aq) + 2\ Ag(s)$ (20.5)

The reading on the meter in the electrical circuit (0.463 V) is also of significance. It represents the **potential difference** between the two half-cells. Since this potential difference is the "driving force" for electrons, it is often called the **electromotive force (emf)** of the cell or the **cell potential.** The unit used to measure electric potential is the **volt,** so the cell potential is also referred to as the **cell voltage.** One definition of the unit, volt, helps to relate it to other units: The passage of one coulomb of electric charge through a potential difference of one volt produces a quantity of work equal to one joule.

1 joule (J) = 1 volt (V) × 1 coulomb (C)   (20.8)

Returning to the opening question of this section of why copper does not displace zinc ion from solution, the answer can be found in constructing the electrochemical cell pictured in Figure 20-4. Here we see that zinc shows a greater tendency to be oxidized than

FIGURE 20-4
The reaction
$Zn(s) + Cu^{2+}(aq) \longrightarrow$
$\qquad Zn^{2+}(aq) + Cu(s)$
occurring in an
electrochemical cell.

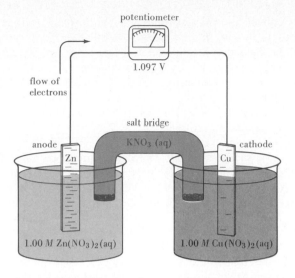

Reaction (20.9) is illus-
trated in Color Sec-
tion G.

does copper. That is, electrons flow from the zinc to the copper electrode. The reaction
that occurs spontaneously in the electrochemical cell is

oxidation: $Zn(s) \longrightarrow Zn^{2+}(aq) + 2\ e^-$
reduction: $Cu^{2+}(aq) + 2\ e^- \longrightarrow Cu(s)$

net: $\overline{Zn(s) + Cu^{2+}(aq) \longrightarrow Zn^{2+}(aq) + Cu(s)}$ (20.9)

Displacement of $Zn^{2+}(aq)$ by $Cu(s)$—the reverse of reaction (20.9)—does not occur
spontaneously.

Precision Measurement of Cell Potentials.   The emf of an electrochemical cell can be
measured with great precision, but only if the measurement is made in a certain way. The
measured potential difference between the anode and cathode depends on the quantity of
electric charge that passes between them, that is, on the amount of electric current drawn
from the cell. Some of the emf of the cell is expended in overcoming the internal electrical
resistance of the cell; and, as current is drawn from the cell, the concentrations of the
species in the half-cell compartments change. This change in concentrations causes the
electrode potentials to change, in a manner that reduces the potential difference. This is
analogous to the difference in water levels becoming smaller as water flows between two
levels or the difference in temperature of two bodies becoming smaller as heat flows
between them.

The simplest device for measuring a difference in electric potential is a common
voltmeter, but a voltmeter must draw a significant electric current to register a meter
reading. A voltmeter will not give a highly precise value of a cell emf. In a device known
as a **potentiometer** the flow of electric current from the electrochemical cell under investigation is countered by current of the same magnitude flowing in the *opposite* direction
and drawn from another electrochemical cell of known emf. When this condition of
essentially zero current is attained, the two cells have equal and opposite emf values.

Cell Diagrams and Terminology.   Sketching an electrochemical cell as in Figures 20-3
and 20-4 is somewhat tedious and cumbersome. A symbolic representation is often used
to describe a cell. This representation is called a **cell diagram.** For the electrochemical
cell of Figure 20-4 it takes the form

anode ⌐   salt
         bridge ⌐   ⌐cathode
$Zn(s)|Zn^{2+}(aq)||Cu^{2+}(aq)|Cu(s)$ (20.10)
   half-cell   half-cell

Alessandro Volta (1745–1827) made the first cells of the type being described here. He generated electric currents (in a device called a voltaic pile) by stacking up, alternately, pieces of copper and zinc (i.e., Cu, Zn, Cu, . . .) separated by pieces of paper soaked in NaCl(aq).

By convention, the electrode shown *at the left* is the one at which *oxidation* occurs; it is called the **anode.** At the electrode shown *at the right, reduction* occurs; this electrode is called the **cathode.** A single vertical line represents the boundary between an electrode and another phase (e.g., an aqueous solution). A double vertical line signifies that solutions are joined by a salt bridge. The parenthetical expressions in equation (20.10) are just the familiar (s) and (aq), but we can be more specific about electrode and solution conditions. For example, the symbol $Zn^{2+}(0.10\ M)$ would indicate that a solution is 0.10 molar in Zn^{2+}. Combinations expressed as Zn/Zn^{2+} and Cu^{2+}/Cu are often called **couples.** Zn/Zn^{2+} is the oxidation couple, and Cu^{2+}/Cu is the reduction couple.

The electrochemical cells considered to this point are all of a type that *produce* electricity as a result of *spontaneous* chemical change. They are all called **galvanic** or **voltaic** cells. Another possibility that we will consider later (Section 20-8) is the production of a *nonspontaneous* chemical change through the *consumption* of electricity.

Standard Electrode Potentials. We have been describing the *measurement* of potential *differences*. However, if we had a way to assign numerical values to metal–metal ion combinations or couples, we might then be able to *calculate* cell potentials. The way to do this is to choose a particular couple and assign it a value of *zero*. Other couples can then be compared to this reference electrode.

The reference electrode for potential measurements is the **standard hydrogen electrode (S.H.E.),** pictured in Figure 20-5. The standard hydrogen electrode involves H^+ ions in solution at unit activity ($a = 1$); for simplicity we will take this to be essentially $1\ M\ H^+$. H_2 molecules in the gaseous state are at a pressure of 1 atm. The oxidized (H^+) and reduced (H_2) forms of hydrogen come into contact on an inert platinum metal surface and impart a characteristic potential to the surface. The temperature is taken to be exactly 25°C. The conditions specified here can be written in the form of an equation. Also, they can be represented through a half-cell couple.

$$2\ H^+(a = 1) + 2\ e^- \xrightleftharpoons{\text{on Pt}} H_2(g,\ 1\ atm) \qquad E° = 0.0000\ \text{volt (V)} \qquad (20.11)$$

$$H^+(a = 1)\,|\,H_2(g,\ 1\ atm),\ Pt \qquad\qquad\qquad\qquad (20.12)$$

Another symbol commonly used for standard electrode potentials is $\mathscr{E}°$.

By international agreement, a **standard electrode potential, $E°$,** is based on the tendency for a **reduction** process to occur at the electrode.* To represent other standard electrodes, we can write expressions of this sort.

$$Cu^{2+}(1\ M) + 2\ e^- \rightleftharpoons Cu(s) \qquad E° = ? \qquad (20.13)$$

$$Cl_2(g,\ 1\ atm) + 2\ e^- \rightleftharpoons 2\ Cl^-(1\ M) \qquad E° = ? \qquad (20.14)$$

In all cases the ionic species are present in aqueous solution at unit activity (approximately $1\ M$); gases are at 1 atm pressure. Where no solid substance is indicated, the potential is established on an inert platinum electrode.

To determine values of $E°$ for electrodes such as those of (20.13) and (20.14), we need to measure the potential difference between *two* electrodes. This can be done through an electrochemical cell with a S.H.E. as one electrode and the standard electrode in question as the other. In the following voltaic cell the measured potential difference is 0.337 V, with electrons flowing from the H_2 to the Cu electrode. Since this is the emf of the cell formed from two standard electrodes, it is referred to as the **standard cell potential, $E°_{cell}$.**

$$Pt,\ H_2(g,\ 1\ atm)\,|\,H^+(1\ M)\,\|\,Cu^{2+}(1\ M)\,|\,Cu(s) \qquad E°_{cell} = 0.337\ V \qquad (20.15)$$

---

*By an earlier convention, standard electrode potentials were based on oxidation processes. One must be careful in using tabulated data to determine which convention applies.

The reaction that occurs in the voltaic cell of (20.15) is

oxidation: $H_2(g, 1 \text{ atm}) \longrightarrow 2 H^+(1 M) + 2 e^-$
reduction: $Cu^{2+}(1 M) + 2 e^- \longrightarrow Cu(s)$

net: $H_2(g, 1 \text{ atm}) + Cu^{2+}(1 M) \longrightarrow 2 H^+(1 M) + Cu(s)$ $E^\circ_{cell} = 0.337 \text{ V}$

(20.16)

According to reaction (20.16) $Cu^{2+}(1 M)$ must be reduced more easily than is $H^+(1 M)$. The standard electrode potential representing the reduction of $Cu^{2+}(aq)$ to $Cu(s)$ is $+0.337$ V.

$$Cu^{2+}(1 M) + 2 e^- \rightleftharpoons Cu(s) \qquad E^\circ = +0.337 \text{ V} \qquad (20.17)$$

When a standard hydrogen electrode is combined with a standard zinc electrode, electrons are found to flow in the opposite direction from the cell in (20.15), that is, *from the zinc to the hydrogen electrode*. The S.H.E. acts as the *cathode* and the standard zinc electrode as the *anode*. The measured value of E°_{cell} is 0.760 V.

$$Zn(s)|Zn^{2+}(1 M)\|H^+(1 M)|H_2(g, 1 \text{ atm}), Pt \qquad E^\circ_{cell} = 0.760 \text{ V} \qquad (20.18)$$

The reaction that occurs in the voltaic cell (20.18) is

oxidation: $Zn(s) \longrightarrow Zn^{2+}(1 M) + 2 e^-$
reduction: $2 H^+(1 M) + 2 e^- \longrightarrow H_2(g, 1 \text{ atm})$

net: $Zn(s) + 2 H^+(1 M) \longrightarrow Zn^{2+}(1 M) + H_2(g, 1 \text{ atm})$ $E^\circ_{cell} = 0.760 \text{ V}$

(20.19)

Here reduction of $Zn^{2+}(1 M)$ must occur with *greater difficulty* than that of $H^+(1 M)$, since oxidation, not reduction, occurs at the zinc electrode. E°_{cell} of equation (20.19) describes the tendency for zinc to become *oxidized. If we consider the reduction tendency to be the opposite of the oxidation tendency,* then

$$Zn^{2+}(1 M) + 2 e^- \rightleftharpoons Zn(s) \qquad E^\circ = -0.760 \text{ V} \qquad (20.20)$$

In summary:

- The potential of the standard hydrogen electrode is set at zero.
- Any electrode at which a *reduction* half-reaction shows a *greater* tendency to occur than does $2 H^+(1 M) + 2 e^- \rightarrow H_2(g, 1 \text{ atm})$ has a *positive* electrode potential, E°.
- Any electrode at which a *reduction* half-reaction shows a *lesser* tendency to occur than does $2 H^+(1 M) + 2 e^- \rightarrow H_2(g, 1 \text{ atm})$ has a *negative* electrode potential, E°.
- If the tendency for a reduction process is given by E°, the oxidation tendency is simply the negative of this value, that is, $-E^\circ$.
- With these ideas it is possible to develop extensive tabulations of standard electrode potentials, as suggested by Table 20-2.

In Figure 20-6 some of the reduction half-reactions that we have been describing are located on a vertical scale where the zero corresponds to the standard hydrogen electrode. Reduction half-reactions in the colored region occur *more readily* than the reduction of H^+ to $H_2(g)$ ($E^\circ > 0$), and reduction half-reactions in the gray region, *less readily* ($E^\circ < 0$).

In the voltaic cell (20.18), H^+ is reduced at the S.H.E. and the measured cell potential is 0.760 V. Since Cu^{2+} is reduced even more readily than H^+, we should expect the cell potential difference to be greater if the S.H.E. in (20.18) were replaced by the Cu^{2+}/Cu couple. In fact, we should predict the cell emf to be $0.337 - (-0.760) = 1.097$ V, just as measured in Figure 20-4. This prediction of E°_{cell} corresponds to the following three-step process, which is also illustrated through Examples 20-4 and 20-5.

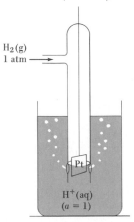

FIGURE 20-5
The standard hydrogen electrode (S.H.E.).

$H_2(g)$
1 atm

Pt

$H^+(aq)$
($a = 1$)

TABLE 20-2
Some selected standard electrode potentials

Reduction half-reaction	E^o, V
Acidic solution	
$F_2(g) + 2\,e^- \longrightarrow 2\,F^-(aq)$	$+2.87$
$O_3(g) + 2\,H^+(aq) + 2\,e^- \longrightarrow O_2(g) + H_2O$	$+2.07$
$S_2O_8^{2-}(aq) + 2\,e^- \longrightarrow 2\,SO_4^{2-}(aq)$	$+2.01$
$H_2O_2(aq) + 2\,H^+(aq) + 2\,e^- \longrightarrow 2\,H_2O$	$+1.77$
$MnO_4^-(aq) + 8\,H^+(aq) + 5\,e^- \longrightarrow Mn^{2+}(aq) + 4\,H_2O$	$+1.51$
$PbO_2(s) + 4\,H^+(aq) + 2\,e^- \longrightarrow Pb^{2+}(aq) + 2\,H_2O$	$+1.455$
$Cl_2(g) + 2\,e^- \longrightarrow 2\,Cl^-(aq)$	$+1.360$
$Cr_2O_7^{2-}(aq) + 14\,H^+(aq) + 6\,e^- \longrightarrow 2\,Cr^{3+}(aq) + 7\,H_2O$	$+1.33$
$MnO_2(s) + 4\,H^+(aq) + 2\,e^- \longrightarrow Mn^{2+}(aq) + 2\,H_2O$	$+1.23$
$O_2(g) + 4\,H^+(aq) + 4\,e^- \longrightarrow 2\,H_2O$	$+1.229$
$2\,IO_3^-(aq) + 12\,H^+(aq) + 10\,e^- \longrightarrow I_2(s) + 6\,H_2O$	$+1.195$
$Br_2(l) + 2\,e^- \longrightarrow 2\,Br^-(aq)$	$+1.065$
$NO_3^-(aq) + 4\,H^+(aq) + 3\,e^- \longrightarrow NO(g) + 2\,H_2O$	$+0.96$
$Ag^+(aq) + e^- \longrightarrow Ag(s)$	$+0.800$
$Fe^{3+}(aq) + e^- \longrightarrow Fe^{2+}(aq)$	$+0.771$
$O_2(g) + 2\,H^+(aq) + 2\,e^- \longrightarrow H_2O_2(aq)$	$+0.682$
$I_2(s) + 2\,e^- \longrightarrow 2\,I^-(aq)$	$+0.535$
$Cu^+(aq) + e^- \longrightarrow Cu(s)$	$+0.52$
$H_2SO_3(aq) + 4\,H^+(aq) + 4\,e^- \longrightarrow S(s) + 3\,H_2O$	$+0.45$
$Cu^{2+}(aq) + 2\,e^- \longrightarrow Cu(s)$	$+0.337$
$SO_4^{2-}(aq) + 4\,H^+(aq) + 2\,e^- \longrightarrow 2\,H_2O + SO_2(g)$	$+0.17$
$Sn^{4+}(aq) + 2\,e^- \longrightarrow Sn^{2+}(aq)$	$+0.154$
$S(s) + 2\,H^+(aq) + 2\,e^- \longrightarrow H_2S(g)$	$+0.141$
$2\,H^+(aq) + 2\,e^- \longrightarrow \mathbf{H_2(g)}$	0.0000
$Pb^{2+}(aq) + 2\,e^- \longrightarrow Pb(s)$	-0.126
$Sn^{2+}(aq) + 2\,e^- \longrightarrow Sn(s)$	-0.136
$Fe^{2+}(aq) + 2\,e^- \longrightarrow Fe(s)$	-0.440
$Zn^{2+}(aq) + 2\,e^- \longrightarrow Zn(s)$	-0.763
$Al^{3+}(aq) + 3\,e^- \longrightarrow Al(s)$	-1.66
$Mg^{2+}(aq) + 2\,e^- \longrightarrow Mg(s)$	-2.375
$Na^+(aq) + e^- \longrightarrow Na(s)$	-2.714
$Ca^{2+}(aq) + 2\,e^- \longrightarrow Ca(s)$	-2.76
$K^+(aq) + e^- \longrightarrow K(s)$	-2.925
$Li^+(aq) + e^- \longrightarrow Li(s)$	-3.045
Basic solution	
$O_3(g) + H_2O + 2\,e^- \longrightarrow O_2(g) + 2\,OH^-$	$+1.24$
$OCl^-(aq) + H_2O + 2\,e^- \longrightarrow Cl^- + 2\,OH^-$	$+0.89$
$O_2(g) + 2\,H_2O + 4\,e^- \longrightarrow 4\,OH^-(aq)$	$+0.401$
$CrO_4^{2-}(aq) + 4\,H_2O + 3\,e^- \longrightarrow Cr(OH)_3(s) + 5\,OH^-$	-0.13
$S(s) + 2\,e^- \longrightarrow S^{2-}(aq)$	-0.48
$2\,H_2O + 2\,e^- \longrightarrow H_2(g) + 2\,OH^-(aq)$	-0.828
$SO_4^{2-}(aq) + H_2O + 2\,e^- \longrightarrow SO_3^{2-}(aq) + 2\,OH^-(aq)$	-0.93

FIGURE 20-6
Representation of
standard electrode
potentials.

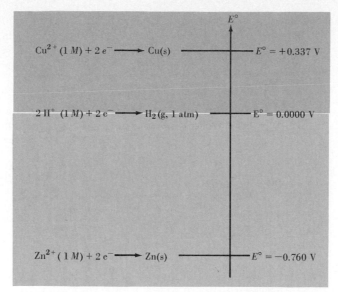

Reduction half-reactions that occur more readily than the reduction of $H^+(1\ M)$ to $H_2(g, 1\ atm)$ have positive values of $E°$ and appear in the colored region. Those with a lesser tendency for reduction have negative values of $E°$ and appear in the gray region.

1. Write the proposed reduction half-equation and a standard reduction potential, $E°_{red}$, to describe it. This will be a value of $E°$ from Table 20-2.
2. Write the proposed oxidation half-equation and a standard oxidation potential, $E°_{ox}$, to describe it. This will be the *negative* of the $E°$ value listed in Table 20-2.
3. Combine the half-equations into a net oxidation–reduction equation. *Add* together the oxidation and reduction potentials to obtain $E°_{cell}$.

In completing Step 3 this important point must be noted: Since electrode potential is an *intensive* property of an electrode system, its value does not depend on the amounts of substances involved. $E°$ values are *unaffected* by multiplying half-equations by constant coefficients.

Example 20-4   Calculate $E°_{cell}$ for the reaction that occurs in the voltaic cell

$Ag(s)|Ag^+(1\ M)\|H^+(1\ M)|O_2(g, 1\ atm), Pt(s)$

Solution.   According to the conventions for writing cell diagrams, the electrode on the left is the anode, where oxidation occurs.

oxidation: $Ag(s) \longrightarrow Ag^+(1\ M) + e^-$ $E°_{ox} = -(+0.800) = -0.800\ V$

The reduction half-reaction occurring at the cathode is

reduction: $O_2(g, 1\ atm) + 4\ H^+(1\ M) + 4\ e^- \longrightarrow 2\ H_2O$ $E°_{red} = +1.229\ V$

$E°_{cell} = -0.800 + 1.229 = +0.429\ V$

To write the net oxidation–reduction equation, multiply the oxidation half-equation by *four* and combine the oxidation and reduction half-equations.

net: $4\ Ag(s) + O_2(g, 1\ atm) + 4\ H^+(1\ M) \longrightarrow 4\ Ag^+(1\ M) + 2\ H_2O$

SIMILAR EXAMPLES: Review Problem 5; Exercises 13, 14.

Although the standard electrode potentials in Table 20-2 are based on the arbitrarily assigned value of 0.0000 V for the standard hydrogen electrode, it is not necessary to use a S.H.E. in an experimental measurement designed to determine an unknown electrode potential. Use of some other electrode system with a precisely known value of $E°$ will do, as illustrated in Example 20-5.

Example 20-5   The cell potential of the following voltaic cell is measured. Use this measured value, together with a value from Table 20-2, to obtain the standard reduction potential for the Cd^{2+}/Cd electrode.

$$Cd(s)|Cd^{2+}(1\ M)\|Cu^{2+}(1\ M)|Cu(s) \qquad E°_{cell} = 0.740\ V$$

Solution.   The half-equations are written and combined below.

oxidation:	$Cd(s) \longrightarrow Cd^{2+}(1\ M) + 2\ e^-$	$E°_{ox} = ?$
reduction:	$Cu^{2+}(1\ M) + 2\ e^- \longrightarrow Cu(s)$	$E°_{red} = +0.337\ V$
net:	$Cd(s) + Cu^{2+}(1\ M) \longrightarrow Cd^{2+}(1\ M) + Cu(s)$	$E°_{cell} = E°_{ox} + 0.337\ V$ $= 0.740\ V$

$$E°_{ox} = 0.740\ V - 0.337\ V = 0.403\ V.$$

For the oxidation of $Cd(s)$ to $Cd^{2+}(1\ M)$, $E° = 0.403\ V$. For the reduction

$$Cd^{2+}(1\ M) + 2\ e^- \longrightarrow Cd(s) \qquad E° = -0.403\ V$$

SIMILAR EXAMPLES: Review Problem 6; Exercise 7.

20-4 Electrical Work and Free Energy Change

In Chapter 16 we learned two things about *nonspontaneous* processes. One is that it takes work to make a nonspontaneous process occur, and the other is that for such a process, $\Delta G > 0$. Conversely, for a *spontaneous* process, $\Delta G < 0$. We did not state this point at that time, but a *spontaneous process is capable of doing work;* and $-\Delta G$ is the maximum amount of work that can be done.

$$-\Delta G = w_{max} \tag{20.21}$$

When a reaction is carried out in a voltaic cell, it also does work. Moreover, this is the maximum amount of work obtainable for the process. The quantity of electrical work done in a voltaic cell is

$$w_{elec.} = n\mathscr{F}E_{cell} \tag{20.22}$$

where n is the number of moles of electrons per mole of reaction. $\mathscr{F}$, the **Faraday constant,** is the electric charge per mole of electrons, 96,500 C/mol e^-. E_{cell} is the emf of the voltaic cell, in volts.

If we equate (20.21) and (20.22), we obtain the expression

A more precise value of $\mathscr{F}$ is 9.6487×10^4 C/mol e^-.

$$\Delta\overline{G} = -n\mathscr{F}E_{cell} \tag{20.23}$$

and if the reactants and products in the half-cells are in their standard states,

$$\Delta\overline{G}° = -n\mathscr{F}E°_{cell} \tag{20.24}$$

To establish the value of n to be used in equations (20.23) and (20.24) requires expressing a net oxidation–reduction equation as the sum of the two half-equations, in which electrons appear. Thus, for the net reaction in Example 20-4, $n = 4$ moles of electrons per mole of reaction, i.e., 4 mol e^-/mol rxn, or more simply, 4 mol e^-/mol. $\Delta\overline{G}°$ for that reaction is

$$\Delta\overline{G}° = \frac{-4\ \text{mol}\ e^-}{mol} \times \frac{96,500\ C}{mol\ e^-} \times 0.429\ V = -1.66 \times 10^5\ \frac{V \times C}{mol}$$

$$= -1.66 \times 10^5\ \text{J/mol} = -166\ \text{kJ/mol}$$

Spontaneous Change in Oxidation–Reduction Reactions. The condition for spontaneous change in oxidation–reduction reactions remains that established in Chapter 16: $\Delta G < 0$. However, according to equation (20.23), this can also be stated as $E_{cell} > 0$. (That is, E_{cell} must be positive if ΔG is to be negative.) If we deal with reactants and products in their standard states, then, according to equation (20.24), for spontaneous change $E°_{cell} > 0$. In order to predict the direction of spontaneous change in an oxidation–reduction reaction, we need simply to add a couple of steps to the three listed on page 607.

4. If $E°_{cell}$ is *positive*, a reaction will occur spontaneously in the forward direction. If $E°_{cell}$ is *negative*, the reaction will proceed spontaneously in the reverse direction.
5. If a cell reaction is reversed, $E°_{cell}$ changes sign.

Example 20-6 **Will aluminum metal displace Cu^{2+} ion from aqueous solution? That is, does spontaneous change occur in the forward direction?**

$$2\ Al(s) + 3\ Cu^{2+}(1\ M) \longrightarrow 3\ Cu(s) + 2\ Al^{3+}(1\ M)$$

Solution. The net equation is obtained by adding together (20.25) and (20.26). As usual coefficients must be adjusted to eliminate electrons, e^-, from the net equation.

oxidation: $\quad 2\{Al(s) \longrightarrow Al^{3+}(1\ M) + 3\ e^-\} \qquad E°_{ox} = -(-1.66)$
$$= +1.66\ V \qquad (20.25)$$

reduction: $\quad \underline{3\{Cu^{2+}(1\ M) + 2\ e^- \longrightarrow Cu(s)\} \qquad E°_{red} = +0.337\ V} \qquad (20.26)$

net: $\quad 2\ Al(s) + 3\ Cu^{2+}(1\ M) \longrightarrow 3\ Cu(s) + 2\ Al^{3+}(1\ M)$
$$E°_{cell} = +2.00\ V \qquad (20.27)$$

Since $E°_{cell}$ is *positive*, the direction of spontaneous change is the forward direction.

SIMILAR EXAMPLES: Review Problem 7; Exercise 9.

Example 20-7 **Will oxygen gas oxidize sulfate ion to peroxodisulfate ion in acidic solution? That is, will the following reaction occur to any significant extent in the forward direction?**

$$4\ SO_4^{2-}(aq) + O_2(g) + 4\ H^+(aq) \longrightarrow 2\ S_2O_8^{2-}(aq) + 2\ H_2O \qquad (20.28)$$

Solution. We begin by splitting equation (20.28) into two half-equations. Electrode potential data are obtained from Table 20-2.

oxidation: $\quad 2\ SO_4^{2-}(aq) \longrightarrow S_2O_8^{2-}(aq) + 2\ e^- \qquad E°_{ox} = -(+2.01\ V) = -2.01\ V$
reduction: $\quad O_2(g) + 4\ H^+(aq) + 4\ e^- \longrightarrow 2\ H_2O \qquad E°_{red} = +1.229\ V$

and

$$E°_{cell} = E°_{ox} + E°_{red} = -2.01\ V + 1.23\ V = -0.78\ V$$

The large negative value of $E°_{cell}$ suggests that $O_2(g)$ will not oxidize SO_4^{2-} to $S_2O_8^{2-}$ to any significant extent.

SIMILAR EXAMPLES: Review Problem 8; Exercise 12.

Let us examine the result of Example 20-7 more closely to establish its full significance: What we determined in Example 20-7 was $E°_{cell}$ for the electrochemical cell written as

$$Pt|SO_4^{2-}(1\ M),\ S_2O_8^{2-}(1\ M)\|H^+(1\ M)|O_2(g,\ 1\ atm),\ Pt$$

The fact that $E°_{cell} < 0$ means that electrons actually flow *from* the oxygen electrode (the anode) *to* the sulfate-peroxodisulfate electrode (the cathode). The direction of spontaneous change is the reverse of reaction (20.28)—peroxodisulfate ion oxidizes water to produce $O_2(g)$, itself being reduced to SO_4^{2-}.

We can use equation (20.24) to convert $E_{cell}^{\circ} = -0.78$ V to a value of $\Delta\overline{G}^{\circ}$ for reaction (20.28). In this calculation $n = 4$ mol e$^-$/mol. [To obtain equation (20.28) we multiply the oxidation half-equation by two and add to the reduction half-equation.]

$$\Delta\overline{G}^{\circ} = -n\mathscr{F}E_{cell}^{\circ} = \frac{-4 \text{ mol e}^-}{\text{mol}} \times \frac{96,500 \text{ C}}{\text{mol e}^-} \times (-0.78) \text{ V} = 3.0 \times 10^5 \text{ J/mol}$$

$$= 3.0 \times 10^2 \text{ kJ/mol}$$

$\Delta\overline{G}^{\circ}$ for reaction (20.28) is very large and positive. This corresponds to the situation depicted in Figure 16-8b, where for a large positive $\Delta\overline{G}^{\circ}$ the equilibrium point in a reaction is seen to lie very close to the pure reactants. Again we are led to the conclusion that the forward reaction in (20.28) will occur hardly at all.

Although we used electrochemical cells as the bases for identifying the direction of spontaneous change in Examples 20-6 and 20-7, it is not necessary actually to conduct these reactions in electrochemical cells. A strip of aluminum metal will displace Cu^{2+} just by being added directly to a solution of Cu^{2+}(aq); a strip of copper metal will displace Ag^+ by being added to a solution of Ag^+(aq) (as in Figure 20-1); and so on.

Relationship Between E_{cell}° and K. One of the most significant applications of equation (20.24) comes in combining it with equation (16.25).

$$\Delta\overline{G}^{\circ} = -2.303RT \log K = -n\mathscr{F}E_{cell}^{\circ}$$

and

$$E_{cell}^{\circ} = \frac{2.303RT}{n\mathscr{F}} \log K \tag{20.29}$$

If we use a value of 8.314 J mol^{-1} K^{-1} for R; the unit mol e$^-$/mol for n; and 298.15 K for the temperature (the temperature at which E° values are usually tabulated), we find that

$$\frac{2.303RT}{n\mathscr{F}} = \frac{2.303 \times 8.314 \text{ J mol}^{-1} \text{ K}^{-1} \times 298.15 \text{ K}}{n \text{ (mol e}^-/\text{mol)} \times 96,487 \text{ C/mol e}^-} = \left(\frac{0.0592}{n}\right)\frac{\text{J}}{\text{C}} = \left(\frac{0.0592}{n}\right)\text{V}$$

With the foregoing stipulations we may write for E_{cell}° (in volts, V)

$$E_{cell}^{\circ} = \frac{0.0592}{n} \log K \tag{20.30}$$

Example 20-8 What is K for the following reaction at 25°C?

$$Cu^{2+}(aq) + Sn^{2+}(aq) \longrightarrow Sn^{4+}(aq) + Cu(s)$$

Solution. First we must determine E_{cell}° for this reaction.

oxidation: $\quad$ $Sn^{2+}(aq) \longrightarrow Sn^{4+}(aq) + 2 \text{ e}^-$		$E_{ox}^{\circ} = -(+0.154)$
		$= -0.154$ V
reduction: $\quad$ $Cu^{2+}(aq) + 2 \text{ e}^- \longrightarrow Cu(s)$		$E_{red}^{\circ} = +0.337$ V
net: $\quad$ $Cu^{2+}(aq) + Sn^{2+}(aq) \longrightarrow Sn^{4+}(aq) + Cu(s)$		$E_{cell}^{\circ} = 0.183$ V

The number of moles of electrons per mole of cell reaction is 2.

$$E_{cell}^{\circ} = \frac{0.0592}{2} \log K = 0.183 \qquad \log K = \frac{2 \times 0.183}{0.0592} = 6.18$$

$$K = \text{antilog } 6.18 = 1.5 \times 10^6$$

SIMILAR EXAMPLES: Review Problem 9; Exercise 16.

20-5 E_{cell} as a Function of Concentrations

If the galvanic cell pictured in Figure 20-4 is operated at different concentrations of Zn^{2+} and Cu^{2+}, E_{cell} is found to vary in the manner suggested by Table 20-3 and Figure 20-7. The equation of the straight line in Figure 20-7 is

$$E_{cell} = 1.10 - 0.03 \log \frac{[Zn^{2+}]}{[Cu^{2+}]} \tag{20.31}$$

Relationships of this type were first studied by Walter Nernst (1864–1941). Equation (20.31) is a specific example of the general equation now known as the **Nernst equation.** We have shown how this equation may be established by experiment, but it can also be derived from thermodynamics. For the reaction

$$a\,A + b\,B + \cdots \rightleftharpoons g\,G + h\,H + \cdots \tag{20.32}$$

we can write

$$\Delta\overline{G} = \Delta\overline{G}^\circ + 2.303RT \log Q \tag{16.23}$$

Substituting equations (20.23) and (20.24) into (16.23), we obtain

The terms $\Delta\overline{G}^\circ$ and E_{cell}° both refer to a reaction in which reactants and products are in their standard states. By contrast, $\Delta\overline{G}$ and E_{cell} refer to nonstandard conditions.

$$-n\mathscr{F}E_{cell} = -n\mathscr{F}E_{cell}^\circ + 2.303RT \log Q \qquad and \qquad E_{cell} = E_{cell}^\circ - \frac{2.303RT}{n\mathscr{F}} \log Q$$

Q is the reaction quotient and has the form established in Chapters 15 and 16. At 25°C, $2.303RT/n\mathscr{F}$ has the value $(0.0592/n)$V. Again, n is the number of moles of electrons transferred per mole of cell reaction. At 25°C the Nernst equation becomes

$$E_{cell} = E_{cell}^\circ - \frac{0.0592}{n} \log \frac{(a_G)^g(a_H)^h \cdots}{(a_A)^a(a_B)^b \cdots} \tag{20.33}$$

In equation (20.33) we can make the usual substitutions for activities: $a = 1$ for pure solids and liquids; $a =$ partial pressures (atm) for gases; $a =$ molar concentrations for solution components.

FIGURE 20-7
Variation of E_{cell} with ion concentrations for the cell reaction

$$Zn(s) + Cu^{2+}(aq) \longrightarrow Zn^{2+}(aq) + Cu(s)$$

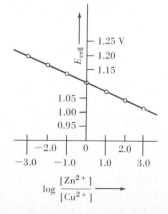

Example 20-9   Calculate E_{cell} for the voltaic cell pictured in Figure 20-8.

$$Pt|Fe^{2+}(0.10\ M),\ Fe^{3+}(0.20\ M)\|Ag^+(1.0\ M)|Ag(s)$$

Solution.   We begin by using data from Table 20-2 to determine E_{cell}°. Then we apply the Nernst equation.

TABLE 20-3
Variation of E_{cell} with ion concentrations for the cell reaction
$Zn(s) + Cu^{2+}(aq) \longrightarrow Zn^{2+}(aq) + Cu(s)$

$[Zn^{2+}]$, M	$[Cu^{2+}]$, M	$\dfrac{[Zn^{2+}]}{[Cu^{2+}]}$	$\log\dfrac{[Zn^{2+}]}{[Cu^{2+}]}$	E_{cell}°, V
1.0	1.0×10^{-3}	1.0×10^3	3.0	1.01
1.0	1.0×10^{-2}	1.0×10^2	2.0	1.04
1.0	1.0×10^{-1}	1.0×10^1	1.0	1.07
1.0	1.0	1.0	0	1.10
1.0×10^{-1}	1.0	1.0×10^{-1}	-1.0	1.13
1.0×10^{-2}	1.0	1.0×10^{-2}	-2.0	1.16
1.0×10^{-3}	1.0	1.0×10^{-3}	-3.0	1.19

FIGURE 20-8
A voltaic cell with nonstandard conditions—Example 20-9 illustrated.

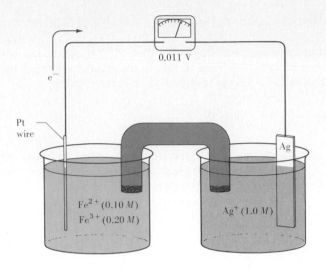

Determining E°_{cell}:

oxidation:	$Fe^{2+}(aq) \longrightarrow Fe^{3+}(aq) + e^-$	$E^{\circ}_{\text{ox}} = -(+0.771)$
		$= -0.771$ V
reduction:	$Ag^+(aq) + e^- \longrightarrow Ag(s)$	$E^{\circ}_{\text{red}} = +0.800$ V
net:	$Fe^{2+}(aq) + Ag^+(aq) \longrightarrow Fe^{3+}(aq) + Ag(s)$	$E^{\circ}_{\text{cell}} = +0.029$ V

$$(20.34)$$

Nernst equation:

$$E_{\text{cell}} = E^{\circ}_{\text{cell}} - \frac{0.0592}{n} \log \frac{[Fe^{3+}]}{[Fe^{2+}][Ag^+]}$$

Substitute: $E^{\circ}_{\text{cell}} = +0.029$ V; $n = 1$; $[Fe^{2+}] = 0.10$ M; $[Fe^{3+}] = 0.20$ M; $[Ag^+] = 1.0$ M.

$$E_{\text{cell}} = 0.029 - \frac{0.0592}{1} \log \frac{0.20}{0.10 \times 1.0}$$

$$= 0.029 - (0.0592 \log 2.0)$$

$$E_{\text{cell}} = 0.029 - (0.0592 \times 0.30)$$

$$= 0.029 - 0.018 = 0.011 \text{ V}$$

SIMILAR EXAMPLES: Review Problem 10; Exercise 18.

The Nernst Equation and Le Châtelier's Principle. The fact that $E^{\circ}_{\text{cell}} > 0$ signifies that the reaction

$$Fe^{2+}(aq) + Ag^+(aq) \longrightarrow Fe^{3+}(aq) + Ag(s) \qquad E^{\circ}_{\text{cell}} = 0.029 \text{ V} \qquad (20.34)$$

occurs spontaneously if reactants and products are in their standard states. According to Le Châtelier's principle the *forward* reaction is favored even more strongly as the concentration of Fe^{2+} and/or Ag^+ is *increased* and that of Fe^{3+}, *decreased*. To favor the *reverse* reaction, the concentration of Fe^{2+} and/or Ag^+ should be *decreased* and that of Fe^{3+}, *increased*. The Nernst equation permits us to *calculate* the effect of these varying conditions on E_{cell}.

Example 20-10 What is the *minimum* $[Ag^+]$ at which the cell of Figure 20-8 is able to function as a voltaic cell with Fe^{2+}/Fe^{3+} as the anode?

$$Pt|Fe^{2+}(0.10\ M),\ Fe^{3+}(0.20\ M)\|Ag^+(?\ M)|Ag(s)$$

Solution. According to the above discussion, if $[Ag^+]$ is lowered sufficiently, $E_{cell} = 0$. If $E_{cell} = 0$, the cell reaction is at equilibrium (with neither the forward nor the reverse reaction favored). The $[Ag^+]$ we are seeking is for $E_{cell} = 0$.

$$E_{cell} = 0 = 0.029 - \frac{0.0592}{1} \log \frac{0.20}{0.10 \times [Ag^+]}$$

$$= 0.029 - 0.0592(\log 2.0 - \log [Ag^+])$$

$$0.029 = 0.0592(0.30 - \log [Ag^+]) \qquad 0.0592 \log [Ag^+] = 0.018 - 0.029$$

$$\log [Ag^+] = \frac{-0.011}{0.0592} = -0.19 \qquad [Ag^+] = 0.65 \ M$$

Note that if $[Ag^+]$ is reduced below $0.65 \ M$, $E_{cell} < 0$ and the Ag/Ag^+ electrode becomes the anode.

SIMILAR EXAMPLES: Exercises 26, 27.

Measurement of pH. Suppose that we set up the electrochemical cell illustrated in Figure 20-9. This consists of *two* hydrogen electrodes. One is a *standard hydrogen electrode* (S.H.E.) and the other is a hydrogen electrode immersed in a solution of unknown $[H^+]$. The two half-cells are joined by a salt bridge, resulting in the voltaic cell,

$$\text{Pt, } H_2(g, 1 \text{ atm})|H^+(x \ M)\|H^+(1 \ M)|H_2(g, 1 \text{ atm}), \text{ Pt} \qquad (20.35)$$

The reaction occurring in this cell is

oxidation: $\quad \cancel{H_2(g, 1 \text{ atm})} \longrightarrow 2 H^+(x \ M) + \cancel{2 e^-}$

reduction: $\quad 2 H^+(1 \ M) + \cancel{2 e^-} \longrightarrow \cancel{H_2(g, 1 \text{ atm})}$

net: $\qquad\quad 2 H^+(1 \ M) \longrightarrow 2 H^+(x \ M) \qquad (20.36)$

Any electrochemical cell in which the net cell reaction is simply the change in concentration of some species (here $[H^+]$) is called a concentration cell. A **concentration cell** consists of two half-cells with *identical electrodes* but differing ion concentrations. Because the electrodes are identical, $E°$ for the oxidation will be numerically equal and opposite in sign to $E°$ for the reduction. As a result, $E°_{cell} = 0$. However, because the ion concentrations do differ, there is a potential difference between the two half-cells. The Nernst equation for reaction (20.36) takes the form

$$E_{cell} = E°_{cell} - \frac{0.0592}{2} \log \frac{(x)^2}{(1)^2}$$

which simplifies to

$$E_{cell} = 0 - \frac{0.0592}{2} \times 2 \log \frac{x}{1} = -0.0592 \log x$$

FIGURE 20-9
A concentration cell, consisting of two hydrogen electrodes, for measuring pH.

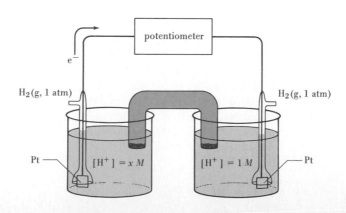

FIGURE 20-10
A glass electrode for pH measurements.

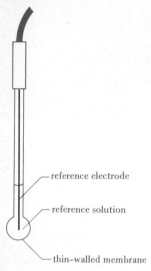

— reference electrode

— reference solution

— thin-walled membrane

Since x is $[H^+]$ in the unknown solution, and $-\log x = -\log [H^+] = $ pH, our final result is

$$E_{cell} = 0.0592 \text{ pH} \tag{20.37}$$

If an unknown solution has a pH of 3.50, for example, the measured cell voltage in Figure 20-9 will be $E_{cell} = 0.0592 \times 3.50 = 0.207$ V.

The Glass Electrode.   Constructing and using a hydrogen electrode is difficult. The Pt metal surface must be specially prepared and maintained, gas pressures must be controlled, and the electrode cannot be used in the presence of strong oxidizing or reducing agents. A better approach to pH measurement than that of Figure 20-9 replaces the S.H.E. with some other reference electrode of precisely known $E°$ value. The second hydrogen electrode is replaced by a **glass electrode.** The basic feature of this electrode is a thin glass membrane of a carefully regulated chemical composition. When the electrode is dipped into a solution, depending on the type and concentration of ions present, a potential is established on the outer surface of the membrane. This potential is registered through a reference electrode immersed in a solution inside the membrane. The glass electrode pictured in Figure 20-10, immersed in an unknown solution, serves as a half-cell. When combined with another reference half-cell, the assembly functions as a voltaic cell.

The most commonly used glass electrodes are those whose potentials are determined by $[H^+]$ in solution. These are the glass electrodes used in common laboratory pH meters. Other glass electrodes have been developed that can function with Na^+, K^+, or other cations. These so-called **ion-selective electrodes** are especially valuable because ordinary half-cells cannot use very active metals as electrodes. These metals react with water to liberate $H_2(g)$.

Potentiometric Titrations.   In Chapter 18 we demonstrated, by *calculation*, how pH varies with titrant volume in an acid–base titration, and we displayed the results in a graph known as a titration curve. We found that the pH changes very sharply with titrant volume at the equivalence point. An acid–base indicator whose color changes in the pH range where a titration curve is ascending (or descending) very steeply is a suitable indicator for the titration. But what if we want to titrate an acid or base in a solution that itself is strongly colored or very turbid? An acid–base indicator would not be suitable because its color change would be obscured.

Equation (20.37) and the device to which it applies (Figure 20-9) gives us a totally different approach to titration. According to equation (20.37), the measured value of E_{cell} in Figure 20-9 is directly proportional to pH. Suppose that we place an unknown acid in the anode half cell (the colored compartment) in Figure 20-9, titrate it with a strong base, and measure E_{cell} after each addition of titrant. A plot of E_{cell} (or pH) against volume of titrant will have the same shape as one of the titration curves of Chapter 18. From the graph we can determine the volume of titrant corresponding to the equivalence point and thus complete the analysis.

Since the electrical device required to measure E_{cell} is called a potentiometer (see again, Figure 20-9), this type of titration is called a **potentiometric titration.** The great utility of potentiometric titrations is that they can be used for any titration where the concentration of one of the reactants changes extremely rapidly at the equivalence point, and where an electrode exists whose potential depends on the concentration of that reactant. For example, the titration reaction pictured in Figure 19-4 can be followed in an electrochemical cell consisting of a reference electrode and a second electrode (an indicator electrode) whose potential depends on $[Ag^+]$ (see Additional Exercise 15). A further aspect of potentiometric titrations is that such titrations can be performed automatically, that is, by machine.

Measurement of K_{sp}. One of the best ways to measure K_{sp} of a slightly soluble salt is electrochemically. Consider this concentration cell.

$$Pb(s)|Pb^{2+}(sat'd.\ PbCl_2(aq))\|Pb^{2+}(0.100\ M)|Pb(s) \qquad E_{cell} = 0.0237\ V \qquad (20.38)$$

In the anode half-cell lead is immersed in a saturated aqueous solution of $PbCl_2$. In the cathode half-cell a second lead strip is dipped into a solution with $[Pb^{2+}] = 0.100\ M$. The two half-cells are connected by a salt bridge and the difference in potential between the two electrodes is measured. It is 0.0237 V. The cell reaction occurring in this concentration cell is

oxidation: $\cancel{Pb(s)} \longrightarrow Pb^{2+}(sat'd.\ PbCl_2(aq)) + \cancel{2e^-}$
reduction: $\underline{Pb^{2+}(0.100\ M) + \cancel{2e^-} \longrightarrow \cancel{Pb(s)}}$
net: $Pb^{2+}(0.100\ M) \longrightarrow Pb^{2+}(sat'd.\ PbCl_2(aq)) \qquad E_{cell} = 0.0237\ V \quad (20.39)$

Example 20-11 With the data listed in (20.39), calculate K_{sp} for $PbCl_2$.

$$PbCl_2(s) \rightleftharpoons Pb^{2+}(aq) + 2\ Cl^-(aq) \qquad K_{sp} = ?$$

Solution. Let us represent $[Pb^{2+}]$ in saturated $PbCl_2(aq)$ by x. Then we apply the Nernst equation to the cell reaction (20.39). (Recall that $E^\circ_{cell} = 0$ for a concentration cell.)

$$E_{cell} = E^\circ_{cell} - \frac{0.0592}{2} \log \frac{x}{0.10}$$

$$0.0237 = 0 - 0.0296(\log x - \log 0.100)$$

$$0.0237 = -0.0296 \log x + 0.0296 \log 0.100$$

$$0.0296 \log x = 0.0296 \times (-1.000) - 0.0237$$

$$\log x = \frac{-0.0296 - 0.0237}{0.0296} = -1.80$$

$$x = [Pb^{2+}] = 1.6 \times 10^{-2}\ M \qquad [Cl^-] = 2\ [Pb^{2+}] = 3.2 \times 10^{-2}\ M$$

$$K_{sp} = [Pb^{2+}][Cl^-]^2 = (1.6 \times 10^{-2})(3.2 \times 10^{-2})^2 = 1.6 \times 10^{-5}$$

SIMILAR EXAMPLES: Exercises 25, 27.

FIGURE 20-11
The Leclanché (dry) cell.

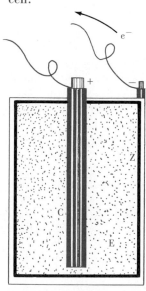

C: carbon rod serving as cathode—reduction occurs.
Z: zinc container serving as anode—oxidation occurs.
E: electrolyte—a moist paste of MnO_2, $ZnCl_2$, NH_4Cl, and carbon black.

20-6 Production of Electric Energy by Chemical Change

An important use of voltaic cells is the production of electric energy by chemical change. Four rather different devices are considered in this section.

Leclanché (Dry) Cell. A familiar flashlight cell is pictured in Figure 20-11. Oxidation occurs at a zinc anode and reduction at an inert carbon cathode. The electrolyte is a moist paste of MnO_2, $ZnCl_2$, NH_4Cl, and carbon black. The difference in potential of the two electrodes is about 1.5 V. Because there is no free liquid in the cell, it is called a "dry" cell.

The anode reaction is simple—the oxidation of zinc atoms to Zn^{2+}.

oxidation: $Zn(s) \longrightarrow Zn^{2+}(aq) + 2\ e^-$

The reduction is more complex. Essentially, it involves the reduction of MnO_2 to a series of compounds having Mn in a $+3$ oxidation state, for example, Mn_2O_3.

reduction: $2\ MnO_2(s) + H_2O + 2\ e^- \longrightarrow Mn_2O_3(s) + 2\ OH^-(aq)$

An acid–base reaction occurs between the OH^- and NH_4^+.

acid–base reaction: $\quad NH_4^+(aq) + OH^-(aq) \longrightarrow NH_3(g) + H_2O$

A buildup of a layer of $NH_3(g)$ around the cathode cannot be allowed; it would disrupt the electric current. This is prevented by a reaction between Zn^{2+} and $NH_3(g)$ leading to the formation of complex ions, such as $[Zn(NH_3)_4]^{2+}$.

complex ion formation: $\quad Zn^{2+}(aq) + 4\, NH_3(aq) \longrightarrow [Zn(NH_3)_4]^{2+}(aq)$

The Leclanché cell is called a **primary** cell. The electrode reactions cannot be reversed. The cell is not rechargeable. Rechargeable cells, called **secondary** cells, use oxidation–reduction reactions that can be reversed by an external electric energy source. The most familiar one is probably the lead storage cell.

Lead Storage Cell. The electrodes in this cell are plates of a lead-antimony alloy. The anodes are impregnated with spongy lead metal, the cathodes with red-brown lead dioxide. The electrolyte is a dilute sulfuric acid solution. When the cell pictured in Figure 20-12 is allowed to discharge, the following reactions occur. [Think of these as involving the oxidation of Pb^0 and the reduction of Pb^{4+}, both to Pb^{2+}, followed by the precipitation of $PbSO_4(s)$.]

oxidation: $\quad Pb(s) + SO_4^{2-}(aq) \longrightarrow PbSO_4(s) + 2\, e^-$
reduction: $\quad PbO_2(s) + 4\, H^+(aq) + SO_4^{2-} + 2\, e^- \longrightarrow PbSO_4(s) + 2\, H_2O$
net: $\quad Pb(s) + PbO_2(s) + 4\, H^+ + 2\, SO_4^{2-} \longrightarrow 2\, PbSO_4(s) + 2\, H_2O$
$$E_{cell} \simeq 2.0 \text{ V} \qquad (20.40)$$

When the plates become partially coated with $PbSO_4(s)$ and the electrolyte has been diluted by the water produced, the cell is in a discharged condition. To recharge it, electrons are made to flow in the opposite direction using an external electric source. The net reaction that occurs is the reverse of (20.40).

To prevent short circuiting, alternating anode and cathode plates are separated by sheets of insulating material. A group of anodes is connected together electrically, and a group of cathodes is similarly connected. This "parallel" connection increases the electrode area in contact with the electrolyte solution and increases the current-delivering capacity of the cell. Cells are then joined together in "series" fashion, + to −, to produce a battery. In a 6-V battery there are three cells; in a 12-V battery, six cells.

FIGURE 20-12
A lead storage cell.

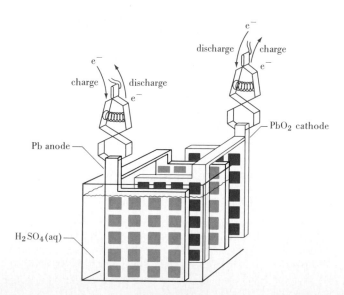

$PbSO_4(s)$ is less dense than $Pb(s)$ and $PbO_2(s)$, and its formation can cause the grid structure of the lead plates to expand and rupture. Also, $PbSO_4$ obstructs the flow of $H_2SO_4(aq)$ to the electrodes. As a consequence a lead-storage battery should not be allowed to discharge to the point where more than about 25 to 30% conversion of Pb and PbO_2 to $PbSO_4$ has occurred. Beyond this point it may be difficult or impossible to reverse the cell reaction (a condition known as sulfation). In recharging, the process should be stopped at the point where all the $PbSO_4$ has been converted back to Pb and PbO_2. Beyond this point continued passage of electric current causes the electrolysis of water, and the evolution of $H_2(g)$ and $O_2(g)$ can interfere with the freshly formed deposits of Pb and PbO_2. In addition, the mixture of $H_2(g)$ and $O_2(g)$ is an explosive hazard.

The Silver–Zinc Cell.   A checklist of the desirable features of a storage cell system is quite lengthy. It includes such factors as a cell reaction that is completely reversible, that involves highly insoluble reactants of the lowest possible molar masses, and that has the highest E_{cell} and number of moles of electrons per mole of reaction possible. Construction of the cell should be strong, lightweight, and should minimize energy-wasting internal electrical resistance by having electrodes as close together as possible and separated by the best electrolytic conductor possible. No currently available cell has all these features, though some are much better than others. The silver–zinc cell, which first became available in the 1950's, satisfies a number of these criteria rather well. Its construction is as follows.

Zn, ZnO(s)|KOH(sat'd)|AgO(s), Ag

The half-reactions on discharging are

<table>
<tr><td>anode:</td><td>$Zn(s) + 2\ OH^-(aq) \longrightarrow ZnO(s) + H_2O + 2\ e^-$</td></tr>
<tr><td>cathode:</td><td>$AgO(s) + H_2O + 2\ e^- \longrightarrow Ag(s) + 2\ OH^-(aq)$</td></tr>
<tr><td>net:</td><td>$Zn(s) + AgO(s) \longrightarrow ZnO(s) + Ag(s)$</td></tr>
</table>

The half-reactions and net reaction are reversed on charging.

Because no solution species is involved in the net reaction, the quantity of electrolyte can be kept very small, so small that the electrodes can be maintained in very close proximity and the cell kept nearly dry. The electric storage capacity of the silver–zinc cell is the greatest of all commercially available storage cells, about six times as great as a lead storage cell of the same size. Unfortunately, the cell can only be recharged a few hundred times before the electrodes deteriorate; and the high cost of silver is another disadvantage to its use.

Fuel Cells.   A voltaic cell produces electric energy with a high efficiency, as high as 90%. This is in contrast to the 30 to 40% efficiencies encountered in the combustion–steam turbine–electric generator method. Cannot fuels be consumed more efficiently by a direct conversion of chemical energy to electricity we might ask? The objective of a fuel cell is to achieve this conversion. Presently, fuel cells have achieved their most publicized successes as energy devices for space vehicles. Their uses will undoubtedly multiply in the future as conventional fuel sources dwindle and alternative means of energy production are explored more fully. The essential process involved in a fuel cell is

fuel + oxygen $\longrightarrow$ oxidation products

The essential requirement is an electrode system with which this reaction can be carried out. The free energy change of the reaction is released as electric energy.

A fuel cell consists of a compartment containing electrolyte (a concentrated aqueous solution or a molten salt) and porous electrodes into which the gases and electrolyte may diffuse and enter into reaction. One of the simplest and most successful fuel cells, shown schematically in Figure 20-13, involves the reaction of $H_2(g)$ and $O_2(g)$ to form water. In an alkaline solution (e.g., 25% KOH) these reactions occur.

Note that in this cell the ionic form of silver is Ag^{2+} rather than the more common Ag^+.

FIGURE 20-13
Schematic representation of a hydrogen-oxygen fuel cell.

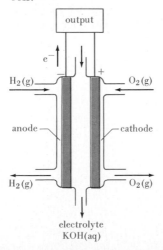

oxidation: $2 H_2(g) + 4 OH^-(aq) \longrightarrow 4 H_2O + 4 e^-$
reduction: $O_2(g) + 2 H_2O + 4 e^- \longrightarrow 4 OH^-(aq)$

net: $2 H_2(g) + O_2(g) \longrightarrow 2 H_2O(l)$ (20.41)

A fuel cell is more properly described as an energy conversion device rather than as an electrical battery. As long as fuel and $O_2(g)$ are available, it will produce electricity. It does not have a limited capacity, as does a primary cell. However, neither can it store electric energy, as does a secondary cell.

The first fuel cell based on reaction (20.41) was devised in 1842, but its efficiency was very poor. Development of the internal combustion engine seems to have been favored over that of electrochemical energy converters because of this poor efficiency. On the other hand, although the efficiency of converting heat to work is subject to thermodynamic limitations (recall page 495), no such limitations exist for electrochemical energy conversion. Early problems in carrying out chemical reactions on electrode surfaces (due to a phenomenon called polarization) are now being resolved, and the full potential of electrochemical energy converters is coming to be realized.

20-7 Electrochemical Mechanism of Corrosion

An important group of oxidation–reduction processes are those involved in corrosion. The fact that the combined cost of corrosion protection and corrosion losses amount to billions of dollars annually lends practical as well as theoretical importance to this subject.

The processes involved in the corrosion of iron can be demonstrated in a particularly graphic manner as pictured in Figure 20-14. The object undergoing corrosion is an iron nail. The nail is embedded in a gel of agar in water. Incorporated in the gel are the acid–base indicator phenolphthalein and the substance $K_3[Fe(CN)_6]$ (potassium ferricyanide).

Following are the observations that can be made within hours of starting the experiment. At the head of the nail and at the tip a deep blue precipitate forms. Along the body of the nail the agar gel acquires a pink color. The blue precipitate, known as Turnbull's blue, establishes the presence of iron(II). The pink color, of course, is characteristic of phenolphthalein in a basic solution. From these observations we can write two half-equations.

oxidation: $2 Fe(s) \longrightarrow 2 Fe^{2+}(aq) + 4 e^-$
reduction: $O_2 + 2 H_2O + 4 e^- \longrightarrow 4 OH^-(aq)$

The results of this and other corrosion processes are pictured in Color Section G.

Thus, in the corrosion of the nail oxidation occurs at the two ends. Electrons given up in the oxidation pass along the body of the nail where they are used to reduce dissolved O_2. The reduction product, OH^-, is detected by the phenophthalein. The overall corro-

FIGURE 20-14
Corrosion of an iron nail.

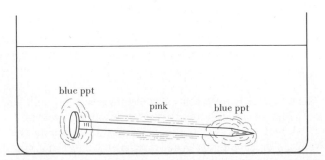

The nail is imbedded in an agar gel that is impregnated with phenolphthalein and $K_3[Fe(CN)_6]$.

FIGURE 20-15
Protection of iron
against electrolytic
corrosion.

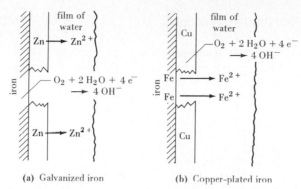

(a) Galvanized iron   (b) Copper-plated iron

In the anodic reaction (oxidation), the metal that is more easily oxidized loses electrons to produce metal ions. In case (a) this is zinc; in case (b), iron. In the cathodic reaction (reduction), oxygen gas, which is dissolved in a thin film of adsorbed water, is reduced to hydroxide ion. Rusting of iron does not occur in (a), but it does in (b).

$$Fe^{2+} + 2\ OH^- \longrightarrow Fe(OH)_2(s)$$

$$4\ Fe(OH)_2(s) + O_2 + 2\ H_2O \longrightarrow 4\ Fe(OH)_3(s)$$

$$2\ Fe(OH)_3(s) \longrightarrow \underset{\text{rust}}{Fe_2O_3 \cdot H_2O} + 2\ H_2O$$

sion reaction is an electrochemical one. With a bent nail, oxidation occurs at three points: the head, the tip, and the bend. The nail is preferentially corroded at these points because the strained metal is more active (more anodic) than the unstrained metal.

With some metals, such as aluminum, the corrosion products (Al_2O_3) form a tough adherent coating that protects the underlying metal from further corrosion. But iron oxide (rust) flakes off an object, constantly exposing fresh surface which corrodes. It is this very difference in behavior of the corrosion products that explains why cans made of iron deteriorate rapidly under environmental conditions, whereas aluminum cans have an almost unlimited lifetime. A number of methods, of varying degrees of effectiveness, have been devised to protect a metal from corrosion. The simplest involves coating the surface with paint or some other protective coating. An iron surface is protected in this way only as long as the coating does not chip or peel off.

Another method of protecting an iron surface is to plate it with a thin layer of a second metal. Iron can be coated with copper by electroplating or with tin by dipping into the molten metal. In both cases protection of the underlying iron is achieved only as long as the coating remains intact. If the coating is cracked, as when a "tin" can is dented for example, the underlying iron is exposed and corrodes. Iron, being more active than copper and tin, undergoes oxidation; the reduction half-reaction occurs on the plating. When iron is coated with zinc (galvanized iron) the situation is different. Zinc is more active than iron. If a break occurs in the zinc plating, the iron is still protected. Zinc is oxidized in place of iron, and corrosion products protect zinc from further corrosion. The difference in these two types of protective action is brought out in Figure 20-15. Still another method can be used to protect iron and steel objects—ships, storage tanks, pipelines, plumbing systems. This involves connecting to the object, either directly or through a wire, a chunk of magnesium or other active metal. Oxidation occurs at the active metal and it gradually dissolves. The iron surface acquires electrons from the oxidation of the active metal; the iron acts as a cathode and supports a *reduction* half-reaction. As long as some of the active metal remains the iron is protected. This type of corrosion protection is called cathodic protection and the magnesium or other active metal is called, appropriately, a "sacrificial anode." The method is illustrated in Figure 20-16.

FIGURE 20-16
Protection of an iron
tank from corrosion
with a sacrificial
magnesium anode.

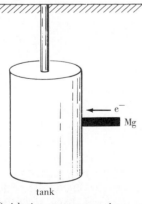

Oxidation occurs at the
Mg anode, and reduction
occurs on the iron tank.

20-8 Electrolysis and Nonspontaneous Chemical Change

Let us return to a consideration of Figure 20-4, which already has provided us with insights into several electrochemical phenomena. If the cell is allowed to function spontaneously, electrons flow from zinc to copper, and the net chemical change is that of equation (20.9).

voltaic cell: $Zn(s) + Cu^{2+}(aq) \longrightarrow Cu(s) + Zn^{2+}(aq)$ $E^\circ_{cell} = +1.097$ V

By connecting the electrodes to an external energy source—either a generator or a voltaic cell of sufficient emf, such as a lead storage battery—electrons can be made to flow in the opposite direction. The chemical reaction in this case is the reverse of (20.9). In an electrolysis reaction electric energy is used to produce a chemical change that will not occur spontaneously: E_{cell} is negative.

electrolysis: $Cu(s) + Zn^{2+}(aq) \longrightarrow Zn(s) + Cu^{2+}(aq)$ $E^\circ_{cell} = -1.097$ V (20.42)

Predicting Electrode Reactions.   If a potential difference exceeding 1.097 V is applied to the cell of Figure 20-4, with zinc as the cathode and copper the anode, the electrolysis reaction (20.42) will occur. Similar calculations can be made regarding other electrolysis reactions. However, what actually happens may not always correspond to these calculations.

In many cases the voltage necessary to bring about a particular electrode reaction may exceed that which is calculated theoretically. Interactions called polarization may occur between the electrode surface and the species involved in an electrode reaction. This may require that an overpotential be applied in order for the electrode reaction to occur. An overpotential is the potential difference in excess of that calculated theoretically required to produce electrolysis. Overpotentials are particularly common when gases are involved. For example, the overpotential for the discharge of $H_2(g)$ at a mercury cathode is approximately 1.5 V, whereas on a platinum cathode it is practically zero.

A second complicating factor is that if the material being electrolyzed contains several species capable of undergoing oxidation and reduction, competing electrode reactions may occur. In the electrolysis of *molten* sodium chloride only one oxidation and one reduction are possible.

oxidation: $2 Cl^- \longrightarrow Cl_2(g) + 2 e^-$
reduction: $2 Na^+ + 2 e^- \longrightarrow 2 Na(l)$

In the electrolysis of *aqueous* sodium chloride *two* oxidation and *two* reduction half-reactions must be considered.

oxidation: $2 Cl^- \longrightarrow Cl_2(g) + 2 e^-$ $E^\circ_{ox} = -1.36$ V   (20.43)
            $2 H_2O \longrightarrow O_2(g) + 4 H^+ + 4 e^-$ $E^\circ_{ox} = -1.23$ V   (20.44)

reduction: $2 Na^+ + 2 e^- \longrightarrow 2 Na(s)$ $E^\circ_{red} = -2.71$ V   (20.45)
            $2 H_2O + 2 e^- \longrightarrow H_2(g) + 2 OH^-$ $E^\circ_{red} = -0.83$ V   (20.46)

The electrode potentials for half-reactions (20.43) and (20.44) are similar in magnitude. Exact values depend on $[Cl^-]$ in the one case and $[H^+]$ in the other. If the NaCl solution is concentrated, the oxidation half-reaction (20.43) is favored; if it is quite dilute, (20.44).

As far as the reduction half-reaction is concerned, the reduction of water occurs much more readily than that of Na^+. Generally, only the half-reaction (20.46) occurs. The principal exception is if liquid mercury is used as a cathode. In this case, because of the high overpotential of hydrogen on mercury and the solubility of sodium metal in liquid mercury, the half-reaction (20.45) actually is observed.

FIGURE 20-17
Predicting electrode reactions in electrolysis—Example 20-12 illustrated.

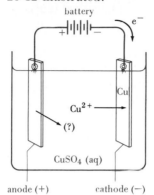

Electrons are forced onto the copper cathode by the external source (battery). Cu^{2+} ions are attracted to the cathode and are reduced to Cu(s). The oxidation half-reaction at the anode depends on the metal used for the anode.

Example 20-12   With reference to Figure 20-17, predict the electrode reactions and the net electrolysis reaction that will occur when the anode is made of (a) copper and (b) platinum.

Solution.   In both cases the cathode reaction is the same, the deposition of copper from the $CuSO_4(aq)$.

reduction: $Cu^{2+}(aq) + 2 e^- \longrightarrow Cu(s)$ $E^\circ_{red} = +0.34$ V

(a)   The half-reaction occurring at the anode is the oxidation of copper.

oxidation: $Cu(s) \longrightarrow Cu^{2+}(aq) + 2 e^-$ $E^\circ_{ox} = -0.34$ V

The net electrolysis reaction is

$$Cu(s)[\text{anode}] \longrightarrow Cu(s)[\text{cathode}] \qquad (20.47)$$

Copper from the anode dissolves at the same rate that Cu^{2+} ions are deposited at the cathode. The solution concentration remains unchanged.

(b)   Platinum metal is much too difficult to oxidize in an electrolysis process. Also, the tendency for SO_4^{2-} to be oxidized to $S_2O_8^{2-}$ is very low ($E^\circ_{ox} = -2.01$ V). The oxidation that occurs most readily is that of water, as shown by equation (20.44).

oxidation: $2 H_2O \longrightarrow O_2(g) + 4 H^+(aq) + 4 e^-$ $E^\circ_{ox} = -1.23$ V

The overall electrolysis reaction is

$$2 Cu^{2+}(aq) + 2 H_2O \longrightarrow 2 Cu(s) + 4 H^+(aq) + O_2(g) \qquad E^\circ_{cell} = -0.89 \text{ V}$$
$$(20.48)$$

SIMILAR EXAMPLES: Review Problem 12; Exercise 34.

In Example 20-12(b) the oxidation half-reaction involved the oxidation of H_2O to $O_2(g)$. In some electrolysis reactions the reduction half-reaction is the reduction of H_2O to H_2. A point worth noting is that in any case where other oxidation and reduction half-reactions are not feasible, the electrolysis of an aqueous solution will result in the decomposition of H_2O to $H_2(g)$ and $O_2(g)$. This is the combination of half-reactions (20.44) and (20.46).

Faraday's Laws of Electrolysis.   The relationship between quantities of electric energy consumed and chemical change produced in electrolysis is one of the many important questions to which Michael Faraday (1791–1867) sought answers. Faraday's first law of electrolysis notes that

the amount of chemical change produced is proportional to the quantity of electric charge that passes through an electrolytic cell.

Faraday's second law of electrolysis states that

a given quantity of electricity produces the same number of equivalents of any substance in an electrolysis.

An equivalent of substance is associated with *1 mol of electrons* in a half-reaction (see also, Section 20-9). Let us rewrite the electrolysis reaction (20.48) in terms of half-equations based on the transfer of *1 mol of electrons* between the anode and the cathode.

oxidation: $\frac{1}{2} H_2O \longrightarrow \frac{1}{4} O_2(g) + H^+(aq) + e^-$
reduction: $\frac{1}{2} Cu^{2+}(aq) + e^- \longrightarrow \frac{1}{2} Cu(s)$

From these half-equations we would define one (electrochemical) equivalent as equal to $\frac{1}{2}$ mol H_2O, $\frac{1}{4}$ mol O_2, 1 mol H^+, $\frac{1}{2}$ mol Cu^{2+}, and $\frac{1}{2}$ mol Cu(s). Thus, the passage of

1 mol of electrons through the electrolysis cell of Figure 20-17 is signaled by the deposition of $\frac{1}{2}$ mol Cu (31.78 g) at the cathode. One ampere (A) of electric current represents the passage of 1 coulomb of charge per second (C/s). Thus, the product, current × time (s), yields the total quantity of charge transferred, in coulombs (C). The Faraday constant, 96,500 C/mol e, allows for a conversion between coulombs of charge and moles of electrons. Two types of calculation are possible with this kind of information. One may make electrical measurements and calculate the extent of chemical change (as in Example 20-13). Alternatively, by determining the extent of chemical change one can establish the quantity of electricity involved in an electrolysis. Methods of determining chemical change include weighing a deposit on an electrode or titrating a product of the electrolysis. An electrolytic cell designed for the purpose of determining quantities of electric charge is called a **coulometer** (see Exercise 38).

Example 20-13 What mass of copper, in grams, is deposited by a current of 1.50 A in 1.00 h in the electrolysis of a $CuSO_4$ solution?

Solution. The electrode reaction is $Cu^{2+}(aq) + 2\ e^- \rightarrow Cu(s)$, which yields the conversion factor 1 mol Cu $\approx$ 2 mol e^-. The calculation is best done in three steps.

$$\text{no. C} = 1.00\ \text{h} \times \frac{60\ \text{min}}{1\ \text{h}} \times \frac{60\ \text{s}}{1\ \text{min}} \times \frac{1.50\ \text{C}}{1\ \text{s}} = 5.40 \times 10^3\ \text{C}$$

$$\text{no. mol } e^- = 5.40 \times 10^3\ \text{C} \times \frac{1\ \text{mol } e^-}{9.65 \times 10^4\ \text{C}} = 5.60 \times 10^{-2}\ \text{mol } e^-$$

$$\text{no. g Cu} = 5.60 \times 10^{-2}\ \text{mol } e^- \times \frac{1\ \text{mol Cu}}{2\ \text{mol } e^-} \times \frac{63.55\ \text{g Cu}}{1\ \text{mol Cu}} = 1.78\ \text{g Cu}$$

SIMILAR EXAMPLES: Review Problem 13; Exercise 36.

Industrial Electrolysis Processes. An important use of electrolysis is in refining metals. The usual metallurgical smelting process produces copper metal that is too impure for most of its intended uses. For example, the presence of arsenic lowers the electrical conductivity of copper, making it unfit for the manufacture of wire and other electrical conductors. The electrolysis described by equation (20.47) in Example 20-12(a) is used to refine copper to purities of 99.95% or higher. A large chunk of impure copper is taken as

Cathodes of 99.98% copper being lifted from an electrolytic refining tank. [Courtesy of Anaconda Copper Company.]

the anode and a strip of pure copper, as the cathode. During electrolysis, copper is transported continuously through the solution (as Cu^{2+}) from anode to cathode. Gold and silver are commonly found as impurities in copper. These metals are less active than copper, that is, less easily oxidized. They do not enter into the anode reaction, but simply deposit at the bottom of the electrolysis tank in a sludge called anode mud. The economic value of anode mud is often enough to cover the cost of the electrolytic refining of copper.

Among the common substances that are produced almost exclusively by electrolytic processes are the alkali metals, magnesium, aluminum, chlorine, fluorine, hydrogen peroxide, and sodium hydroxide. It is not an overstatement that modern industry and modern society in general could not function without the availability of electrolysis reactions. One of the most important of industrial electrolysis reactions—the chlor-alkali process—is described at the end of this chapter.

20-9 Postscript: Equivalent Weight and Normality in Oxidation–Reduction Reactions

A balanced oxidation–reduction equation can be used in stoichiometric calculations just like any other balanced equation. Sometimes, however, oxidation–reduction reactions are treated from the standpoint of equivalent weight and normality rather than the mole and molar concentration. Let us explore briefly how this is done.

For oxidation–reduction reactions *an equivalent is the amount of substance associated with 1 mol of electrons in a half-reaction*. Consider the equation

$$5\ Fe^{2+} + MnO_4^- + 8\ H^+ \longrightarrow 5\ Fe^{3+} + Mn^{2+} + 4\ H_2O \qquad (20.49)$$

Expressed as balanced half-equations, it becomes

oxidation: $\quad 5\ Fe^{2+} \longrightarrow 5\ Fe^{3+} + 5\ e^-$
reduction: $\quad MnO_4^- + 8\ H^+ + 5\ e^- \longrightarrow Mn^{2+} + 4\ H_2O$

We conclude that 5 mol Fe^{2+} is 5 equiv Fe^{2+}, or that 1 mol Fe^{2+} is 1 equiv Fe^{2+}. The situation with MnO_4^- is that 1 mol MnO_4^- is 5 equiv MnO_4^-, or $\frac{1}{5}$ mol MnO_4^- is 1 equiv MnO_4^-. Stated in another way:

$$1\ \text{mol}\ Fe^{2+} \backsimeq \tfrac{1}{5}\ \text{mol}\ MnO_4^- \qquad \text{and} \qquad 1\ \text{equiv}\ Fe^{2+} \backsimeq 1\ \text{equiv}\ MnO_4^-$$

The end point of a permanganate titration is signaled by the first lasting pink color in solution.

Example 20-14 A piece of pure iron weighing 0.1568 g is dissolved in acidic solution and titrated with 26.24 mL of a $KMnO_4(aq)$ solution. What is the normality of the $KMnO_4(aq)$?

Solution

$$\text{no. equiv}\ Fe^{2+} = \text{no. equiv}\ Fe = \text{no. mol}\ Fe = 0.1568\ \text{g Fe} \times \frac{1\ \text{mol Fe}}{55.85\ \text{g Fe}}$$

$$= 2.808 \times 10^{-3}\ \text{equiv}\ Fe^{2+}$$

$$\text{no. equiv}\ KMnO_4 = \text{no. equiv}\ MnO_4^- = \text{no. equiv}\ Fe^{2+} = 2.808 \times 10^{-3}$$

$$\text{normality}\ KMnO_4 = \frac{2.808 \times 10^{-3}\ \text{equiv}\ KMnO_4}{0.02624\ \text{L}} = 0.1070\ N\ KMnO_4$$

SIMILAR EXAMPLES: Review Problem 14; Exercise 40.

Example 20-15 25.8 mL of the 0.1070 N $KMnO_4$ solution described in Example 20-14 is used to titrate 50.0 mL of a saturated solution of sodium oxalate, $Na_2C_2O_4$. What is the solubility of $Na_2C_2O_4$ in g/L?

$$5\ C_2O_4^{2-} + 2\ MnO_4^- + 16\ H^+ \longrightarrow 2\ Mn^{2+} + 8\ H_2O + 10\ CO_2(g) \qquad (20.50)$$

Solution. The number of equivalents of MnO_4^- used in the titration is

$$\text{no. equiv } MnO_4^- = 0.0258 \text{ L} \times \frac{0.1070 \text{ equiv } MnO_4^-}{L} = 2.76 \times 10^{-3} \text{ equiv } MnO_4^-$$

Using the basic idea that 1 equiv $Na_2C_2O_4 \backsimeq 1$ equiv $C_2O_4^{2-} \backsimeq 1$ equiv MnO_4^-, we can express the solubility of $Na_2C_2O_4$ in equiv/L, that is, in *normality* concentration.

$$\frac{2.76 \times 10^{-3} \text{ equiv } C_2O_4^{2-}}{0.0500 \text{ L}} = 5.52 \times 10^{-2} \text{ } N \text{ } Na_2C_2O_4$$

The final step is to convert from equiv $Na_2C_2O_4$ to g $Na_2C_2O_4$. For this we turn to equation (20.50). The reduction of 2 mol of MnO_4^- to Mn^{2+} involves 10 mol of electrons (recall the reduction half-equation in 20.49). Ten moles of electrons must also be associated with the oxidation of 5 mol $Na_2C_2O_4$ to $CO_2(g)$. The amount of $Na_2C_2O_4$ associated with *1 mol of electrons* is 0.500 mol $Na_2C_2O_4$: The equivalent weight of $Na_2C_2O_4$ is one half its molar mass, or $0.500 \times 134 = 67.0$ g $Na_2C_2O_4$/equiv $Na_2C_2O_4$.

$$\text{solubility} = \frac{5.52 \times 10^{-2} \text{ equiv } Na_2C_2O_4}{L} \times \frac{67.0 \text{ g } Na_2C_2O_4}{1 \text{ equiv } Na_2C_2O_4} = 3.70 \text{ g } Na_2C_2O_4/\text{L}$$

SIMILAR EXAMPLES: Review Problem 15; Exercise 42.

If the $KMnO_4$ solution of Examples 20-14 and 20-15 were used in a reaction in which MnO_4^- is reduced to $MnO_2(s)$ instead of Mn^{2+}, its normality concentration would *not* be 0.1070 *N*. This is because reduction of MnO_4^- to MnO_2 involves *3* mol of electrons per mole of MnO_4^-, whereas reduction to Mn^{2+} involves *5* mol of electrons per mole of MnO_4^-. *Molarity* is *independent* of the reaction in which a solution participates. At times, *normality* may not be. One of the drawbacks of equivalent weight and normality, then, is that one must have prior knowledge of the type of reaction in which a solution is to be used.

FOCUS ON
The Chlor-Alkali Process

Current U.S. annual production of Cl_2 and NaOH is approximately 20 billion pounds of each, and most of that production is carried out in electrolysis cells such as those shown here. [Photo courtesy of PPG Industries, Inc.]

In Section 20-8 we noted that the combination of half-equations (20.43) and (20.46) describes the electrolysis of concentrated aqueous solutions of chloride ion.

oxid: $2 \text{ Cl}^-(aq) \longrightarrow Cl_2(g) + 2 \text{ e}^-$
$$E^\circ_{ox} = -1.36 \text{ V} \qquad (20.43)$$

red: $2 \text{ H}_2O + 2 \text{ e}^- \longrightarrow 2 \text{ OH}^-(aq) + H_2(g)$
$$E^\circ_{red} = -0.83 \text{ V} \qquad (20.46)$$

net: $2 \text{ Cl}^-(aq) + 2 \text{ H}_2O \longrightarrow$
$$2 \text{ OH}^-(aq) + H_2(g) + Cl_2(g)$$
$$E^\circ_{cell} = -2.19 \text{ V} \qquad (20.51)$$

If the source of chloride ion is NaCl, the products of reaction (20.51), in addition to $H_2(g)$, are chlorine gas and the alkali NaOH. A process yielding these products is called a **chlor-alkali process.**

Diaphragm Cell. One type of chlor-alkali electrolysis cell, called a diaphragm cell, is pictured in Figure 20-18. The cell consists of an anode compartment, where $Cl_2(g)$ is produced, and a cathode compartment, where the production of $H_2(g)$ and NaOH(aq) occurs. The Cl_2 and H_2 must be kept separated because they form explosive mixtures. Contact between $Cl_2(aq)$ and NaOH(aq) must also be prevented, and this is the purpose served by the diaphragm. (The reaction of Cl_2 with NaOH to form NaOCl and $NaClO_3$ is described in Section 21-1.) A difference in solution levels is maintained so that NaCl(aq) flows slowly from the anode to the cathode compartment. This minimizes the back flow of NaOH(aq) into the anode compartment; but it also means that the product in the cathode compartment is a mixture, e.g., 10 to 12% NaOH(aq) and 14 to 16% NaCl(aq). The solution from the cathode compartment must then be concentrated by evaporating water and purified by crystallizing NaCl(s). Typically, the final product in this chlor-alkali process is 50% NaOH(aq) with about 1% NaCl as impurity. The $Cl_2(g)$ may contain up to 1.5% $O_2(g)$ because of the alternate oxidation process described by equation (20.44).

From the $E°_{cell}$ value in (20.51) we conclude that the electrolysis of NaCl(aq) requires a source of direct electric current of voltage exceeding 2.19 V. Actually, because of the internal resistance of the electrolysis cell and overpotentials at the electrodes (recall page 620), a somewhat higher voltage is required, about 3.5 V. If a current of 1.00 A were to pass through a diaphragm cell continuously for 24 hr, the quantity of Cl_2 produced would be about

$$\text{no. g } Cl_2 = 24 \text{ h} \times \frac{60 \text{ min.}}{1 \text{ h}} \times \frac{60 \text{ s}}{1 \text{ min}} \times \frac{1.00 \text{ C}}{\text{s}}$$

$$\times \frac{1 \text{ mole e}^-}{96,500 \text{ C}} \times \frac{1 \text{ mol } Cl_2}{2 \text{ mole e}^-} \times \frac{70.9 \text{ g } Cl_2}{1 \text{ mol } Cl_2}$$

$$= 32 \text{ g } Cl_2$$

This is a minuscule rate of production for a commercial process. In order for the cell to produce about 1 metric ton (1000 kg) of Cl_2 per day, a current of approximately 31,000 A is required. It should come as no surprise, then, to learn that the chlor-alkali industry consumes about 0.5% of all the electric power produced annually in the United States.

Mercury Cell. The products of the diaphragm cell process are suitable for most purposes, but there are some applications (e.g., rayon manufacture) where high purity

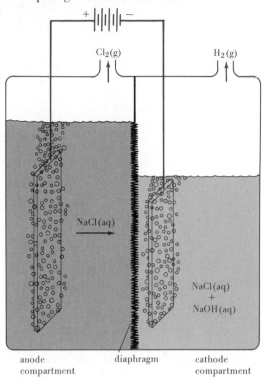

FIGURE 20-18
A diaphragm chlor-alkali cell.

The anode may be made of graphite or, in more modern technology, of specially treated titanium metal. The diaphragm and cathode are generally fabricated as a composite unit consisting of asbestos or an asbestos–polymer mixture deposited on a steel wire mesh or perforated steel cathode.

NaOH(aq) is required. An electrolytic process that produces this higher purity NaOH(aq) is pictured in Figure 20-19. It is based on the fact that a mercury cathode has a high overpotential for the reduction of H_2O to OH^- and $H_2(g)$. The reduction that occurs instead is that of $Na^+(aq)$ to Na, which dissolves in Hg(l) to form an amalgam with about 0.5% Na, by mass.

oxid: $2 Cl^-(aq) \longrightarrow Cl_2(g) + 2 e^-$
$$E°_{ox} = -1.36 \text{ V}$$

red: $2 Na^+(aq) + 2 e^- \longrightarrow 2 Na(\text{in Hg})$
$$E°_{red} = -1.77 \text{ V}$$

net: $2 Na^+(aq) + 2 Cl^-(aq) \longrightarrow$
$$2 Na(\text{in Hg}) + Cl_2(g)$$
$$E°_{cell} = -3.13 \text{ V} \qquad (20.52)$$

When the Na amalgam is removed from the cell and treated with water, NaOH(aq) is formed,

$2 Na(\text{in Hg}) + 2 H_2O \longrightarrow$
$$2 Na^+(aq) + 2 OH^-(aq) + H_2(g) + Hg(l) \qquad (20.53)$$

FIGURE 20-19
The mercury-cell chlor-alkali process.

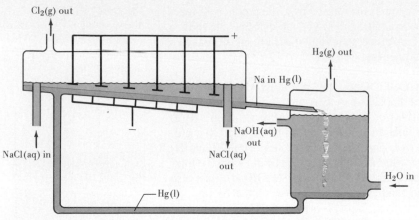

The cathode is a layer of Hg(l) that flows along the bottom of the tank. Anodes, at which $Cl_2(g)$ forms, are immersed in NaCl(aq) just above the Hg(l). Sodium formed at the cathode dissolves in the Hg(l), and the sodium amalgam is decomposed with water, producing NaOH(aq) and $H_2(g)$. The liberated Hg(l) is recycled.

and the liquid mercury is recycled back to the electrolysis cell. Notice that in the mercury-cell chlor-alkali process NaOH(aq) and Cl_2 never come in contact.

The mercury cell would seem to have many advantages over the diaphragm cell, particularly in being able to produce concentrated high-purity NaOH without extensive follow-up procedures. One important disadvantage, though, is that the mercury cell requires a higher voltage (about 4.5 V) than the diaphragm cell and consumes more electric energy, e.g., 3100 kWh/ton Cl_2 in a mercury cell compared to 2700 in a diaphragm cell. Another serious drawback of the mercury cell is the need to control mercury effluents into the environment. Prior to the establishment of environmental regulations, mercury losses were about 200 g Hg per metric ton of Cl_2 produced. Mercury losses are now limited to 0.28 g Hg per metric ton of Cl_2 in existing plants and half this amount in new plants. About 25% of chlor-alkali production in the United States is by the mercury-cell process, but this percentage is not likely to increase because of the difficulty in controlling mercury losses.

Membrane Cells.   The ideal chlor-alkali process is one that is energy efficient and that does not use mercury. A type of cell that offers these advantages is one in which the porous diaphragm of Figure 20-18 is replaced with a cation exchange membrane that permits hydrated cations (Na^+ and H_3O^+) to pass between the anode and cathode compartments but severely restricts the flow of Cl^-. The preparation of polymeric materials to function as cation exchange membranes is one of the most significant modern developments in chlor-alkali technology, and in time the membrane cell may predominate in the industry.

Summary

The essential changes that occur in an oxidation–reduction reaction are most readily seen by separating the net reaction into two half-reactions. In the oxidation half-reaction certain atoms undergo an increase in oxidation state, and electrons appear on the right side of the half-equation. In the reduction half-reaction the oxidation states of certain atoms decrease, and electrons appear on the left side of the half-equation. In a net oxidation–reduction equation, the same number of electrons must appear in each half-equa-

tion. This requirement is the basis of balancing oxidation–reduction equations.

To establish in what direction an oxidation–reduction reaction occurs spontaneously, it is necessary to have a measure of the relative tendencies for oxidation and reduction. These tendencies are described through standard electrode potentials, $E°$. Standard electrode potentials are expressed relative to the reduction of H^+(aq) to H_2(g) at the standard hydrogen electrode (S.H.E.), which is assigned a value of $E° = 0.0000$ V. Tabulated standard electrode potentials can be used to design voltaic cells, which produce electricity from chemical change. If an electrochemical cell has $E_{cell} > 0$, the cell reaction proceeds spontaneously in the forward direction.

Cell voltages predicted from tabulations of *standard* electrode potentials, $E°_{cell}$ values, are based on standard state conditions for reactants and products. For nonstandard state conditions, the Nernst equation permits calculation of E_{cell} from $E°_{cell}$ and the activities of reactants and products. Among the many applications of cell potential measurements are potentiometric titrations and the determination of pH and K_{sp} values. $E°_{cell}$ values can also be used to determine free energy changes ($\Delta \overline{G}°$) and equilibrium constants, K. Voltaic cells are encountered in such practical devices as flashlight and lead storage batteries. Less desirable are the voltaic cell reactions that occur in electrochemical corrosion.

In electrolysis, an external source of electricity is used to force electrons to flow in the direction opposite to that in which they would flow spontaneously. The amount of chemical change produced in an electrolysis cell is directly proportional to the quantity of electric charge passing through the cell, as stated through Faraday's laws of electrolysis. Numerous important industrial processes make use of electrolysis.

Learning Objectives

As a result of studying Chapter 20, you should be able to

1. Separate an oxidation–reduction equation into half-equations, complete and balance the half-equations, and combine them into a balanced net oxidation–reduction equation.

2. Describe a voltaic (galvanic) cell in terms of the electrodes, salt bridge, half-cell reactions, net cell reaction, and cell diagram.

3. Describe the standard hydrogen electrode (S.H.E.) and explain how other standard electrode potentials are related to it.

4. Use tabulated $E°$ values to determine $E°_{cell}$ for an oxidation–reduction reaction, and predict the direction of spontaneous change.

5. Describe the effect of varying conditions (concentration, gas pressures) on E_{cell} values, both qualitatively and quantitatively.

6. Use the relationships that exist among $\Delta \overline{G}°$, $E°_{cell}$, and K.

7. Describe some common voltaic cells—the dry cell, lead storage cell, silver-zinc cell, and fuel cell.

8. Explain the corrosion of metals in electrochemical terms and describe methods of corrosion protection.

9. Describe the essential features of an electrolytic cell and the way in which it differs from a voltaic (galvanic) cell.

10. Identify oxidation and reduction processes that might be involved in an electrolysis reaction and choose among them to predict the most probable reaction.

11. Calculate relationships between amount of chemical change and quantity of electric charge involved in an electrolysis.

12. Apply the concepts of equivalent weight and normality to stoichiometric calculations involving oxidation–reduction reactions.

Some New Terms

The **anode** is an electrode at which an oxidation half-reaction occurs.

The **cathode** is an electrode at which a reduction half-reaction occurs.

Cell potential, E_{cell}, is the term used to describe the difference in potential between two electrodes in an electrochemical cell. If the reactants and products at each electrode are in their standard states, $E_{cell} = E°_{cell}$.

A **chlor-alkali process** is an industrial process for the electrolysis of NaCl(aq) to produce Cl_2(g) ("chlor") and NaOH(aq) ("alkali").

In a **disproportionation reaction** the same substance is both oxidized and reduced.

An **electrochemical cell** is a device in which an oxidation–reduction reaction is carried out in the form of separate half-reactions for oxidation and reduction.

An **electrode potential** is the electric potential that exists on an electrode in contact with the oxidized and reduced forms of some substance.

An **electrolytic cell** is an electrochemical cell in which an external source of electric energy is used to carry out a *nonspontaneous* reaction.

The **Faraday, $\mathscr{F}$**, is the quantity of electric charge associated with 1 mol of electrons, 96,487 C/mol e⁻.

Faraday's laws of electrolysis describe the relationship between the quantity of electric charge that passes through an electrolytic cell and the amount of chemical change that occurs.

A **half-reaction** describes one portion of a net oxidation–reduction reaction, either the oxidation or the reduction.

The **Nernst equation** is used to relate E_{cell}, $E°_{cell}$, and the activities of the reactants and products in a cell reaction.

In an **oxidation half-reaction** certain atoms undergo an *increase* in oxidation state and electrons appear on the right side of a half-equation.

An **oxidation–reduction** reaction is the net reaction that results from oxidation and reduction processes occurring simultaneously.

In a **reduction half-reaction** certain atoms undergo a *decrease* in oxidation state and electrons appear on the left side of a half-equation.

A **voltaic cell** is an electrochemical cell that produces electricity from a spontaneous chemical reaction.

Suggestions for Further Study

ALKIRE, R. C., "Electrochemical Engineering," *J. Chem. Educ.*, **60**, 274 (1983).

DOUGLAS, D. L., and J. R. BIRK, "Batteries for Energy Storage" (a two-part series), *Chemtech*, **13**, 58, 120 (1983).

FISCHER, R. B., "Ion-Selective Electrodes," *J. Chem. Educ.*, **51**, 387 (1974).

LEDDY, J. J., "The Chlor-Alkali Industry," *J. Chem. Educ.*, **57**, 640 (1980).

SAMMELLS, A. F., "Fuel Cells and Electrochemical Energy Storage," *J. Chem. Educ.*, **60**, 320 (1983).

SAPIO, J. P., and R. D. BRAUN, "Ion Selective Electrodes," *Chemistry*, **46**[6], 14 (1973).

SLABAUGH, W. H., "Corrosion," *J. Chem. Educ.*, **51**, 218 (1974).

SMITH, W. L., "Corrosion" (a three-part series), *Chemistry*, **49**[1], 14; [5], 7; [8], 10 (1976).

VENKATESH, S., and B. V. TILAK, "Chlor-Alkali Technology," *J. Chem. Educ.*, **60**, 276 (1983).

Review Problems

1. Complete and balance the following half-equations, and indicate whether oxidation or reduction is involved.

(a) $S_2O_8^{2-} \rightarrow SO_4^{2-}$

(b) $HNO_3 \rightarrow N_2O$ (acidic sol.)

(c) $CH_4 \rightarrow CO_2$ (acidic sol.)

(d) $Br^- \rightarrow BrO_3^-$ (basic sol.)

(e) $NO_3^- \rightarrow NH_3$ (basic sol.)

2. Balance the following equations by the half-reaction (ion electron) method.

(a) $Cu(s) + H^+ + NO_3^- \rightarrow Cu^{2+} + NO(g) + H_2O$

(b) $Zn(s) + H^+ + NO_3^- \rightarrow Zn^{2+} + NH_4^+ + H_2O$

(c) $H_2O_2 + MnO_4^- + H^+ \rightarrow Mn^{2+} + H_2O + O_2(g)$

(d) $S_2O_3^{2-} + Cl_2(g) + H^+ \rightarrow HSO_4^- + Cl^- + H_2O$

(e) $P(s) + H^+ + NO_3^- \rightarrow H_2PO_4^- + NO(g) + H_2O$

3. Balance, by the half-reaction method, the following oxidation–reduction equations for reactions occurring in basic solution.

(a) $CN^- + MnO_4^- + OH^- \rightarrow MnO_2(s) + CNO^- + H_2O$

(b) $[Fe(CN)_6]^{3-} + N_2H_4(g) + OH^- \rightarrow$
$$[Fe(CN)_6]^{4-} + N_2(g) + H_2O$$

(c) $Fe(OH)_2(s) + O_2(g) + OH^- \rightarrow Fe(OH)_3(s) + H_2O$

(d) $C_2H_5OH(aq) + MnO_4^- + OH^- \rightarrow$
$$C_2H_3O_2^- + MnO_2(s) + H_2O$$

4. Use the half-reaction method to balance the following equations for disproportionation reactions.

(a) $Cl_2(g) + H_2O \rightarrow H^+ + Cl^- + HOCl$

(b) $Br_2(l) + OH^- \rightarrow Br^- + BrO_3^- + H_2O$

(c) $S_2O_4^{2-} + H_2O \rightarrow S_2O_3^{2-} + HSO_3^-$

(d) $MnO_4^{2-} + H_2O \rightarrow MnO_2(s) + MnO_4^- + OH^-$

5. Write the cell reactions for the electrochemical cells diagrammed below, and use data from Table 20-2 to calculate $E°_{cell}$ for each reaction.

(a) $Zn(s)|Zn^{2+}(aq)\|Sn^{2+}(aq)|Sn(s)$

(b) $Pt(s)|Fe^{2+}(aq), Fe^{3+}(aq)\|Sn^{4+}(aq), Sn^{2+}(aq)|Pt(s)$

(c) $Cu(s)|Cu^{2+}(aq)\|Cl^-(aq)|Cl_2(g), Pt(s)$

6. From the data listed below, together with data from Table 20-2, determine the quantity indicated.

(a) $E°_{cell}$ for the cell

$$Pt(s), Cl_2(g)|Cl^-(aq)\|Pb^{2+}(aq), H^+(aq)|PbO_2(s)$$

(b) $E°_{red}$ for the couple, Sc^{3+}/Sc, given that

$$Mg(s)|Mg^{2+}(aq)\|Sc^{3+}(aq)|Sc(s) \qquad E°_{cell} = +0.35 \text{ V}$$

(c) $E°_{red}$ for the couple, Cu^{2+}/Cu^+, given that

$$Pt(s)|Cu^+(aq), Cu^{2+}(aq)\|Ag^+(aq)|Ag(s) \quad E°_{cell} = +0.647 \text{ V}$$

7. Assume that all reactants and products are in their standard states and use data from Table 20-2 to predict whether a spontaneous reaction will occur in the forward direction in each case.

(a) $Sn(s) + Zn^{2+} \rightarrow Sn^{2+} + Zn(s)$

(b) $2 Fe^{3+} + 2 I^- \rightarrow 2 Fe^{2+} + I_2(s)$

(c) $4 NO_3^- + 4 H^+ \rightarrow 3 O_2(g) + 4 NO(g) + 2 H_2O$

(d) $O_2(g) + 2 Cl^- \rightarrow 2 OCl^-$ (basic soln.)

(e) $2 H_2O_2(aq) \rightarrow 2 H_2O + O_2(g)$

8. Use data from Table 20-2 to predict whether, to any significant extent,

(a) $Mg(s)$ will displace Sn^{2+} from aqueous solution;

(b) lead metal will dissolve in 1 M HCl;

(c) SO_4^{2-} will oxidize Fe^{2+} to Fe^{3+} in acidic solution;

(d) SO_4^{2-} will oxidize Fe^{2+} to Fe^{3+} in basic solution;

(e) $Cl_2(g)$ will displace I^- from aqueous solution to produce I_2.

9. Write the equilibrium constant expression for each of the following reactions and determine the numerical value of K at 25°C. Use data from Table 20-2 as necessary.

(a) $Fe^{3+}(aq) + Ag(s) \rightarrow Fe^{2+}(aq) + Ag^+(aq)$

(b) $MnO_2(s) + 4\ H^+(aq) + 2\ Cl^-(aq) \rightarrow$
$$Mn^{2+}(aq) + 2\ H_2O + Cl_2(g)$$

(c) $2\ OCl^-(aq) \rightarrow 2\ Cl^-(aq) + O_2(g)$ (basic solution)

10. Write a Nernst-equation expression (20.33) for the following oxidation–reduction reactions in the text: (20.5); (20.27); (20.49).

11. For the cell pictured in Figure 20-9, what is E_{cell} if the unknown solution in the half-cell on the left (a) has a pH = 5.12; (b) is 0.00185 M HCl (aq); (c) is 0.357 M $HC_2H_3O_2$?

12. Use data from Table 20-2 to predict the probable products when Pt electrodes are used in the electrolysis of (a) $CuCl_2(aq)$; (b) HCl(aq); (c) $Na_2SO_4(aq)$; (d) $BaCl_2(l)$; (e) KI(aq); (f) KOH(aq).

13. What mass of metal would be deposited by the passage of 1.56 A of current for 2.25 h in the electrolysis of a solution containing (a) Zn^{2+}; (b) Al^{3+}; (c) Ag^+; (d) Ni^{2+}.

14. Calculate the equivalent weights of the following substances when they are used in the reaction indicated.

(a) Na_2SO_3, reaction (20.6)

(b) Al, reaction (20.27)

(c) PbO_2, reaction (20.40)

(d) O_2, reaction (20.41)

15. What is the % Fe, by mass, in an iron ore if the Fe^{2+} derived from a 0.8312-g ore sample requires 25.13 mL of 0.2821 N $KMnO_4$(aq) for its titration by reaction (20.49)?

Exercises

(Use data from Table 20-2, as necessary.)

Definitions and terminology

1. Indicate the essential *differences* in meanings of the following pairs of terms: (a) oxidation and reduction; (b) oxidizing agent and reducing agent; (c) half-reaction and net reaction; (d) voltaic (galvanic) cell and electrolytic cell; (e) anode and cathode; (f) E°_{cell} and E_{cell}.

Oxidation–reduction reactions

2. Balance the following oxidation–reduction equations by an appropriate method. [*Hint:* Add H^+ (or OH^-) and/or H_2O as necessary.]

(a) $Fe_2S_3(s) + H_2O + O_2(g) \rightarrow Fe(OH)_3(s) + S(s)$

(b) $IBr + BrO_3^- + H^+ \rightarrow IO_3^- + Br^- + H_2O$

(c) $As_2S_3(s) + OH^- + H_2O_2 \rightarrow AsO_4^{3-} + H_2O + SO_4^{2-}$

(d) $CrI_3(s) + H_2O_2 + OH^- \rightarrow CrO_4^{2-} + IO_4^- + H_2O$

(e) $F_5SeOF + OH^- \rightarrow SeO_4^{2-} + F^- + O_2(g) + H_2O$

(f) $Ag(s) + CN^- + O_2(g) + OH^- \rightarrow [Ag(CN)_2]^- + H_2O$

(g) $B_2Cl_4 + OH^- \rightarrow BO_2^- + Cl^- + H_2O + H_2(g)$

(h) $C_2H_5NO_3 + Sn + H^+ \rightarrow$
$$NH_2OH + C_2H_5OH + Sn^{2+} + H_2O$$

3. With reference to Table 20-2, arrange the following oxidizing agents in order of increasing power in acidic solution: $I_2(s)$, $IO_3^-(aq)$, $F_2(g)$, $Na^+(aq)$, $Zn^{2+}(aq)$, $PbO_2(s)$.

4. Certain substances can act only as oxidizing agents, others only as reducing agents. But there are some substances that can act in either capacity. Refer to Table 20-2 and indicate the situation for each of the following: (a) $Zn(s)$; (b) $S^{2-}(aq)$; (c) $Fe^{2+}(aq)$; (d) $Al^{3+}(aq)$; (e) $I_2(s)$; (f) $S(s)$; (g) $Cr_2O_7^{2-}(aq)$.

Standard electrode potentials

5. From the observations listed, estimate the approximate value of the standard electrode potential for the half-reaction $M^{2+}(aq) + 2\ e^- \rightarrow M(s)$.

(a) The metal M dissolves in HNO_3(aq) but not in HCl(aq); it displaces Ag^+(aq) but not Cu^{2+}(aq).

(b) The metal M dissolves in HCl(aq) producing $H_2(g)$, but displaces neither Zn^{2+}(aq) nor Fe^{2+}(aq).

6. You are given the task of estimating the standard electrode potential of indium.

$$In^{3+}(aq) + 3\ e^- \rightarrow In(s) \qquad E^\circ = ?$$

You do not have any electrical equipment but you do have all of the metals listed in Table 20-2 and water solutions of their ions. Describe the experiments you would perform and indicate the accuracy you would expect in your results.

★7. Two electrochemical cells are assembled in which these reactions occur.

$$V^{2+} + VO^{2+} + 2\ H^+ \rightarrow 2\ V^{3+} + H_2O \qquad E^\circ_{cell} = 0.616\ V$$

$$V^{3+} + Ag^+ + H_2O \rightarrow VO^{2+} + 2\ H^+ + Ag(s)$$
$$E^\circ_{cell} = 0.439\ V$$

Use these data and other values from Table 20-2 to calculate the standard electrode potential for the half-reaction

$$V^{3+} + e^- \rightarrow V^{2+}$$

Predicting oxidation–reduction reactions

8. According to standard electrode potentials, Na metal seemingly should displace Mg^{2+} from aqueous solution. Yet if this reaction is attempted, it is found not to occur. Explain. (*Hint:* What reaction does occur?)

9. Copper does not dissolve in HCl(aq), but it does dissolve in HNO_3(aq), producing Cu^{2+}(aq).
  (a) Explain this difference in behavior.
  (b) Write an equation to represent the reaction of Cu and HNO_3(aq).

10. Zinc will react with 1 M HCl to displace H_2(g), but copper will not. If a piece of copper metal is joined to one of zinc and the pair of metals immersed in an HCl solution, bubbles of H_2(g) appear at the copper metal.
  (a) Does this mean that copper reacts with HCl(aq)?
  (b) What is the reaction that occurs?
  (c) What is the function of the copper metal?

11. What observations would you expect to make if copper and silver metal are joined together in the manner described in Exercise 10 and the metals immersed in (a) HCl(aq); (b) HNO_3(aq)?

12. Write a balanced equation to represent the spontaneous disproportionation of H_2O_2(aq).

Voltaic (galvanic) cells

13. In each of the following examples, diagram a voltaic cell that utilizes the given reaction. Label the anode and cathode; indicate the direction of electron flow; write a balanced equation for each cell reaction; and calculate E°_{cell}.
  (a) $Cu(s) + Fe^{3+} \rightarrow Cu^{2+} + Fe^{2+}$
  (b) $Sn^{2+} + Cr_2O_7^{2-} + H^+ \rightarrow Sn^{4+} + Cr^{3+} + H_2O$
  (c) $Cl_2(g) + H_2O \rightarrow Cl^- + O_2(g) + H^+$

14. Diagram the cell and, if possible, calculate the value of E°_{cell} for a voltaic cell in which
  (a) Cl_2(g) is reduced to Cl^- and Fe(s) is oxidized to Fe^{2+};
  (b) Zn^{2+} is displaced from solution as Zn(s);
  (c) The net cell reaction is $2 Cu^+ \rightarrow Cu^{2+} + Cu(s)$.

$\Delta\bar{G}^\circ$, E°_{cell}, and K

15. Use data from Table 20-2 and appropriate equations from the text to determine $\Delta\bar{G}^\circ$ for the following reactions. (*Hint:* The equations are not balanced.)
  (a) $Al(s) + Zn^{2+} \rightarrow Al^{3+} + Zn(s)$
  (b) $Pb^{2+} + MnO_4^- + H^+ \rightarrow PbO_2(s) + Mn^{2+} + H_2O$
  (c) $H^+ + Cl^- + MnO_2(s) \rightarrow Mn^{2+} + H_2O + Cl_2(g)$

16. Determine the value of K at 298 K for reaction (a) in Exercise 15. Does the displacement of Zn^{2+} by Al(s) go to completion? (*Hint:* What is $[Zn^{2+}]$ remaining at equilibrium if Al metal is added to a solution with $[Zn^{2+}] = 1.0$ M?)

⋆17. The standard electrode potential corresponding to the reduction $Cr^{3+} + e^- \rightarrow Cr^{2+}$ is $E^\circ = -0.407$ V. If excess Fe(s) is added to a solution in which $[Cr^{3+}] = 1.00$ M, what will be $[Fe^{2+}]$ when equilibrium is established at 298 K?

$$Fe(s) + 2 Cr^{3+} \rightleftharpoons Fe^{2+} + 2 Cr^{2+}$$

Concentration dependence of E_{cell}—the Nernst equation

18. Use the Nernst equation and data from Table 20-2 to calculate E_{cell} for each of the following electrochemical cells. (*Hint:* Some of these have positive and some, negative E_{cell} values.)
  (a) $Fe(s)|Fe^{2+}(0.20\ M)\|Sn^{2+}(0.050\ M)|Sn(s)$
  (b) $Pt(s), Cl_2(g, 0.50\ atm)|Cl^-(1.2\ M)\|Ag^+(0.35\ M)|Ag(s)$
  (c) $Mg(s)|Mg^{2+}(0.012\ M\|OH^-(0.75\ M)|H_2(g,\ 0.50\ atm)$, Pt(s)

19. Write an equation to represent the oxidation of Mn^{2+} to MnO_2(s) by O_2(g) in acidic solution. Will this reaction occur spontaneously as written if all other reactants and products are in their standard states and (a) $[H^+] = 5.5$ M; (b) $[H^+] = 1.0$ M; (c) $[H^+] = 0.050$ M; (d) pH = 9.13?

20. Show that the oxidation of Cl^- to Cl_2(g) by $Cr_2O_7^{2-}$ in acidic solution will not occur spontaneously with reactants and products in their standard states. Nevertheless, this method can be used to produce Cl_2(g) in the laboratory. Explain why this is so. (*Hint:* What experimental conditions would you use?)

21. The following oxidizing agents are all listed in Table 20-2: Cl_2(g), O_2(g), MnO_4^-(aq), H_2O_2(aq), F_2(g).
  (a) For which of them is the oxidizing power dependent on pH?
  (b) For which is the oxidizing power independent of pH?
  (c) Where the oxidizing power is pH dependent, which are better oxidizing agents in acidic and which in basic solution?

22. If $[Zn^{2+}]$ is maintained at 1.0 M
  (a) What is the minimum $[Cu^{2+}]$ for which reaction (20.9) is still spontaneous in the forward direction?
  (b) Does the displacement of Cu^{2+}(aq) by Zn(s) go to completion?

23. A galvanic cell is constructed of two hydrogen electrodes, one immersed in a solution with H^+ at 1.0 M and the other in 0.85 M KOH.
  (a) Determine E_{cell} for the reaction that occurs.
  (b) Compare your result with E° for the reduction of H_2O to H_2(g) in basic solution and explain.

24. If the 0.85 M KOH solution of Exercise 23 is replaced by 0.85 M NH_3
  (a) Will E_{cell} be higher or lower than in 0.85 M KOH?
  (b) What will be the value of E_{cell}?

25. A voltaic cell is constructed as follows.

Ag(s)|Ag$^+$(sat'd. Ag$_2$SO$_4$(aq))‖Ag$^+$(0.125 M)|Ag(s)

What is the value of E_{cell}?

26. The following voltaic cell is constructed.

Sn(s)|Sn^{2+}(0.150 M)‖Pb^{2+}(0.550 M)|Pb(s)

(a) What is E_{cell} initially?
(b) If the cell is allowed to operate spontaneously, will E_{cell} increase, decrease, or remain constant with time? Explain.
(c) What will be E_{cell} when [Pb^{2+}] has fallen to 0.500 M?
(d) What will be [Sn^{2+}] at the point where E_{cell} = 0.020 V?
(e) What are the concentrations of each of the reacting species when E_{cell} = 0?

27. It is desired to construct the following voltaic cell to have E_{cell} = 0.110 V. What [Cl$^-$] must be present in the cathode half-cell to achieve this result?

Ag(s)|Ag$^+$(sat'd. AgI(aq))‖Ag$^+$(sat'd. AgCl, x M Cl$^-$)|Ag(s)

Batteries and fuel cells

28. The nickel-cadmium cell features one electrode of Cd(OH)$_2$(s) on cadmium serving as the anode on discharge. The second electrode is coated with Ni(OH)$_2$ and Ni(OH)$_3$. The electrolyte is KOH(aq). Write electrode reactions and the net cell reaction for the discharge of this cell.

29. One type of fuel cell (see Figure 20-13) uses a mixture of molten carbonates as an electrolyte. In this mixture some decomposition of carbonate ion occurs.

CO$_3^{2-}$ → CO$_2$ + O^{2-}

Write the electrode reactions corresponding to the net cell reaction CH$_4$(g) + 2 O$_2$(g) → CO$_2$(g) + 2 H$_2$O(g). (*Hint:* Use O^{2-} in writing electrode reactions.)

Electrochemical mechanism of corrosion

30. Comment on the appropriateness of the statement "A corroding metal is like a galvanic cell."

31. Write half-equations and a net equation to correspond to the observations one would make in Figure 20-14 if
(a) a copper wire were wrapped around the iron nail;
(b) the nail were driven through a piece of zinc.

32. Natural gas transmission pipes are sometimes protected against corrosion by maintaining a small potential difference between the pipe and an inert electrode buried in the ground. Describe how the method works. (*Hint:* Is the inert electrode maintained positively or negatively charged with respect to the pipe?)

Electrolysis reactions

33. The commercial production of magnesium involves the electrolysis of molten MgCl$_2$. Why cannot the simpler electrolysis of MgCl$_2$(aq) be used instead?

34. If a lead storage battery is charged at too high a voltage, gases are produced at each electrode. (It is possible to recharge a lead storage battery only because of the high overpotential for gas formation on the electrodes.)
(a) What are these gases?
(b) Write a cell reaction to describe their formation.

35. It is sometimes possible to separate two metal ions from each other through electrolysis. One ion is reduced to the free metal at the cathode and the other remains in solution. Comment on the effectiveness of this method in the following cases. (That is, how complete would be the separation of the two ions?)
(a) Cu^{2+} and K$^+$; (b) Cu^{2+} and Ag$^+$; (c) Pb^{2+} and Sn^{2+}.

Faraday's laws of electrolysis

36. Calculate the quantity indicated for each of the following electrolyses:
(a) The mass of Zn deposited at the cathode in 756 s when 1.05 A of electric current is passed through a solution of Zn^{2+}(aq).
(b) The time required to produce 2.18 g I$_2$ at the anode if a current of 4.28 A is passed through KI(aq).
(c) [Cu^{2+}] remaining in 335 mL of a solution that was originally 0.215 M CuSO$_4$, after the passage of 2.17 A for 235 s and the deposition of Cu at the cathode.
(d) The time required to reduce [Ag$^+$] in 415 mL of AgNO$_3$(aq) from 0.185 to 0.175 M by electrolyzing the solution with Pt electrodes and a current of 3.12 A.

37. Concentrated Na$_2$SO$_4$(aq) is electrolyzed with a current of 1.75 A for 1.32 h. Assuming that the anode reaction is exclusively (20.44), what volume of O$_2$(g) is produced at 25°C and 747 mmHg barometric pressure if (a) the gas is dry; (b) the gas is saturated with water vapor?

38. In a silver coulometer, Ag$^+$ is reduced to Ag(s) at a Pt cathode. If 1.96 g Ag is deposited in 787 s by a certain quantity of electricity, (a) how much electric charge (in C) must have passed, and (b) what was the magnitude (in A) of the electric current?

★**39.** A test for completeness of electrodeposition of Cu from a solution of Cu^{2+}(aq) is to add NH$_3$(aq). A blue color signifies the formation of the complex ion [Cu(NH$_3$)$_4$]$^{2+}$ (K_f = 1.1 × 10^{13}). 250.0 mL of 0.1000 M CuSO$_4$(aq) is electrolyzed with a 3.512 A current for 1368 s. A sufficient quantity of NH$_3$(aq) is added to complex any remaining Cu^{2+} and to maintain a free [NH$_3$] = 0.10 M. If [Cu(NH$_3$)$_4$]$^{2+}$ is detectable at concentrations as low as 1 × 10^{-5} M, would you expect the blue color to form in this case? (*Hint:* Use the more precise value of $\mathscr{F}$ given in the marginal note on page 608.)

Stoichiometry of oxidation–reduction reactions

40. 40.10 mL K$_2$Cr$_2$O$_7$(aq) is required for the titration of an 0.2050-g sample of FeSO$_4$ that is 99.72% pure. What are (a) the molarity and (b) the normality of the K$_2$Cr$_2$O$_7$(aq)?

6 Fe^{2+} + Cr$_2$O$_7^{2-}$ + 14 H$^+$ → 6 Fe^{3+} + 2 Cr^{3+} + 7 H$_2$O

*41. 25.00 mL of the $K_2Cr_2O_7$(aq) described in Exercise 40 is added to 25.00 mL of an acidic solution that is 0.0101 M H_2O_2(aq). What is $[Cr_2O_7^{2-}]$ following chemical reaction. (*Hint:* What is the reaction?)

42. The $KMnO_4$(aq) solution described in Example 20-14 is used in the following titration. What volume of the solution is required to react with 0.0417 g $Na_2S_2O_3$?

$$S_2O_3^{?-} \; | \; MnO_4^- + H^+ \rightarrow MnO_2(s) + SO_4^{2-} + H_2O$$
$$\text{(not balanced)}$$

Additional Exercises

1. Predict whether, to any significant extent,
   (a) Zn(s) will displace Al^{3+} from solution;
   (b) MnO_4^- will oxidize Cl^- to Cl_2(g) in acidic solution;
   (c) O_2(g) will oxidize S^{2-} to S in basic solution;
   (d) Cu(s) will dissolve in 1 M HCl;
   (e) the reaction $Mn^{2+} + O_2(g) + H_2O \rightarrow MnO_2(s) + H_2O_2 + H^+$ will occur as written.

2. E_{cell}° is measured for the reaction

$$3\ U^{4+} + 2\ NO_3^- + 2\ H_2O \rightarrow 3\ UO_2^{2+} + 2\ NO(g) + 4\ H^+$$
$$E_{cell}^{\circ} = 0.63 \text{ V}$$

What is E° for the reduction of UO^{2+} to U^{4+} in acidic solution?

3. A $KMnO_4$(aq) solution is to be standardized by titration against As_2O_3(s). A 0.1097-g sample of As_2O_3 requires 26.10 mL of the $KMnO_4$(aq) for its titration. What are the molarity and normality of the $KMnO_4$(aq)?

$$As_2O_3 + MnO_4^- + H^+ \rightarrow H_3AsO_4 + Mn^{2+} + H_2O$$
$$\text{(not balanced)}$$

4. Mn^{2+}(aq) can be determined by titration with MnO_4^-(aq).

$$Mn^{2+} + MnO_4^- + OH^- \rightarrow MnO_2(s) + H_2O \text{ (not balanced)}$$

A 15.00-mL sample of Mn^{2+}(aq) requires 20.06 mL of the $KMnO_4$(aq) of Additional Exercise 3 for its titration. What are the molarity and normality of the Mn^{2+}(aq)?

5. In a similar fashion to Exercise 22, would you say that the displacement of Pb^{2+} from a 1.0 M $Pb(NO_3)_2$ solution can be carried to completion by tin metal?

6. Why do you suppose that copper plumbing is preferred to cast iron in modern construction?

7. Estimate the voltage of the fuel cell in Figure 20-13.

8. SCN^-(aq) is oxidized by I_2 to SO_4^{2-}(aq). Write a balanced equation for this oxidation–reduction reaction at (a) low pH, where the other products are I^- and HCN; (b) pH $\simeq$ 7, where the other products are I^- and ICN.

*9. A solution is prepared by saturating 100.0 mL 1.00 M NH_3(aq) with AgBr. A silver electrode is immersed in this solution, which is joined by a salt bridge to a S.H.E. What will be the measured E_{cell}? Is the S.H.E. the anode or cathode? (*Hint:* Review Example 19-17.)

*10. The electrolysis of Na_2SO_4(aq) is conducted in two separate half cells joined by a salt bridge, as suggested below.

$$Pt|Na_2SO_4(aq)\|Na_2SO_4(aq)|Pt$$

   (a) In one experiment the solution in the anode half-cell is found to become more acidic and that in the cathode compartment, more basic during the electrolysis. When the electrolysis is discontinued and the two solutions mixed, the resulting solution has pH = 7. Write half-equations and the net electrolysis equation.
   (b) In a second experiment, a 10.00-mL sample of H_2SO_4(aq) of unknown concentration and phenolphthalein indicator are added to the Na_2SO_4(aq) in the cathode compartment. Electrolysis is carried out with a current of 21.5 mA (milliamperes) for 683 s, at which point the solution in the cathode compartment acquires a lasting pink color. What is the molarity of the unknown H_2SO_4(aq)?

*11. An Ni anode and a Fe cathode are placed in a solution with $[Ni^{2+}] = 1.0\ M$ and then connected to a battery.
   (a) Write the equation for the electrolysis reaction.
   (b) The iron has a surface area of 165 cm^2. How long must electrolysis be continued with a current of 1.50 A to build a 0.050-mm-thick deposit of nickel on the iron? (Density of nickel = 8.90 g/cm^3.)

*12. 100.0-mL solutions with ion concentrations of 1.000 M were placed in each of the half-cell compartments of the cell pictured in Figure 20-4. The cell was operated as an *electrolysis* cell, with copper as the anode, and zinc as the cathode. A current of 0.500 A was used. Assume that the only electrode reactions occurring were those involving Cu/Cu^{2+} and Zn/Zn^{2+}. Electrolysis was stopped after 10.00 h and the cell was allowed to function as a *voltaic* cell. What was E_{cell}?

*13. A common reference electrode consists of a silver wire coated with AgCl(s) and immersed in 1 M KCl.

$$AgCl(s) + e^- \rightarrow Ag(s) + Cl^-(1\ M) \qquad E^\circ = +0.2223\ V$$

   (a) What is E°_{cell} when this electrode is a *cathode* in combination with a zinc electrode as an *anode?*

   (b) Cite several reasons why this electrode is easier to use than a standard hydrogen electrode.

   (c) By comparing the potential of the silver-silver chloride electrode with that of the standard silver electrode, determine K_{sp} for AgCl.

*14. Recovery as a by-product in the metallurgy of lead is an important source of Ag. The percentage of Ag in lead was determined as follows. A 1.050-g sample was dissolved in nitric acid to produce Pb^{2+}(aq) and Ag^+(aq). The solution was diluted to 355 mL with water; an Ag electrode was immersed in the solution; and the potential difference between this electrode and a S.H.E. was found to be 0.503 V. What was the % Ag, by mass, in the lead metal?

*15. The course of a precipitation titration can be followed by measuring the emf of a cell. The resulting curve of E_{cell} vs. volume of titrant shows a sharp change in E_{cell} at the equivalence point. The following cell is set up.

$$Hg(l)|Hg_2Cl_2(s)|Cl^-(0.10\ M)\|Ag^+(x\ M)|Ag(s)$$

The potential of the anode half-reaction remains constant at $E^\circ_{ox} = -0.2802$ V.

   (a) Write an equation for the cell reaction.

   (b) Calculate E_{cell} at various points in the titration of 50.0 mL of 0.0100 M AgNO$_3$ carried out in the cathode half-cell compartment by adding 0.01000 M KI.

   (c) Sketch a titration curve for the titration.

Self-Test Questions

For questions 1 through 8 select the single item that best completes each statement.

1. In the reaction $Cu(s) + 2\ H_2SO_4(aq) \rightarrow CuSO_4(aq) + 2\ H_2O + SO_2(g)$ (a) H_2SO_4(aq) is oxidized; (b) Cu(s) is reduced; (c) SO_2(g) is the reducing agent; (d) H_2SO_4(aq) is reduced.

2. The process in which NpO_2^+ is converted to Np^{4+} (a) is an oxidation half-reaction; (b) can only occur in an electrochemical cell; (c) is a reduction half-reaction; (d) is an oxidation–reduction reaction.

3. For the half-reaction Hg^{2+}(aq) $+ 2\ e^- \rightarrow Hg(l)$, $E^\circ = 0.851$ V, which means that (a) Hg is more readily oxidized than H_2; (b) Hg^{2+} is more readily reduced than H^+; (c) Hg(l) will dissolve in 1 M HCl; (d) Hg(l) will displace Zn^{2+} from aqueous solution.

4. The reaction Cu^{2+}(aq) $+ 2\ Cl^-$(aq) $\rightarrow Cu(s) + Cl_2(g)$ has $E^\circ_{cell} = -1.02$ V. This reaction (a) can be made to produce electricity in a voltaic cell; (b) can be made to occur in an electrolysis cell; (c) occurs whenever Cu^{2+} and Cl^- are brought together in aqueous solution; (d) can occur in acidic solution but not in basic solution.

5. The value of E°_{cell} for the oxidation–reduction reaction $Zn(s) + Pb^{2+}(1.0\ M) \rightarrow Zn^{2+}(1.0\ M) + Pb(s)$ is +0.66 V. For the reaction $Zn(s) + Pb^{2+}(0.10\ M) \rightarrow Zn^{2+}(0.10\ M) + Pb(s)$, $E_{cell} =$ (a) +0.63 V; (b) +0.66 V; (c) +0.69 V; (d) +0.72 V.

6. For the reaction $Co(s) + Ni^{2+} \rightarrow Co^{2+} + Ni(s)$, $E^\circ_{cell} = +0.03$ V. If cobalt metal is added to a water solution in which $[Ni^{2+}] = 1.0\ M$, (a) the reaction will not proceed in the forward direction at all; (b) the displacement of Ni^{2+} from solution by Co will go to completion; (c) the displacement of Ni^{2+} from solution by Co will proceed to a considerable extent, but the reaction will stop before Ni^{2+} is completely displaced; (d) only the reverse reaction will occur.

7. The gas evolved at the *anode* when K$_2$SO$_4$(aq) is electrolyzed between Pt electrodes is most likely (a) O_2; (b) H_2; (c) SO_2; (d) SO_3.

8. A quantity of electrical charge that brings about the deposition of 4.5 g Al from Al^{3+} at a cathode will also produce the following volume (STP) of H_2(g) from H^+ at a cathode: (a) 5.6 L; (b) 11.2 L; (c) 22.4 L; (d) 44.8 L.

9. Write half-equations to represent the oxidation of PbO(s) to PbO_2(s) and the reduction of MnO_4^-(aq) to MnO_2(s), both in basic solution. Combine the two half-equations into a balanced net equation.

10. Diagram a voltaic (galvanic) cell in which the following reaction occurs. Label the anode and cathode; use a table of standard electrode potentials to determine E°_{cell}; and balance the equation for the cell reaction.

$$Zn(s) + H^+ + NO_3^- \rightarrow Zn^{2+} + H_2O + NO(g)$$

11. The voltaic (galvanic) cell indicated below registers an $E_{cell} = +0.108$ V. What is the pH of the unknown solution?

$$Pt, H_2(g,\ 1\ atm)|H^+(x\ M)\|H^+(0.10\ M)|H_2(g,\ 1\ atm),\ Pt$$

12. For the reaction $2\ Cu^+ + Sn^{4+} \rightarrow 2\ Cu^{2+} + Sn^{2+}$, $E^\circ_{cell} = -0.0050$ V. A solution is prepared that is 1.00 M in each of the four ions. Estimate the concentration of each of the four ions when equilibrium is established at 298 K.

21

Chemistry of the Representative Elements I: Nonmetals

We encountered most of the important nonmetals in our study of the first 20 elements in Chapter 13. In this chapter we will extend our discussion of those nonmetals to include aspects of their chemical behavior that can best be understood with the full range of chemical principles introduced in the preceding 20 chapters. We will also briefly consider some of the heavier nonmetals and the noble gases.

21-1 Group VIIA—The Halogen Elements

In Section 13-14 we commented on the pronounced nonmetallic character of fluorine and chlorine (and to a lesser extent, bromine and iodine), as evidenced by their atomic and physical properties. We spoke of the strong tendency of F and Cl atoms to gain electrons, accounting for their natural occurrence in the ionic forms F^- and Cl^- and the difficulty of preparing the pure elements from these ionic forms. Perhaps the most practical measure of this tendency to gain electrons is the one introduced in Chapter 20—the standard reduction potential. For the four halogens these potentials are

$$X_2 + 2\,e^- \longrightarrow 2\,X^-(aq) \qquad E° \text{ for} \begin{cases} F_2(g) = +2.87 \text{ V} \\ Cl_2(g) = +1.360 \text{ V} \\ Br_2(l) = +1.065 \text{ V} \\ I_2(s) = +0.535 \text{ V} \end{cases} \qquad (21.1)$$

Preparation of the Halogens. By comparing expression (21.1) with a table of standard reduction potentials (Table 20-2), we see that the reduction of $F_2(g)$ to $F^-(aq)$ has the greatest tendency to occur of all reduction half-reactions. Conversely, the oxidation of $F^-(aq)$ to $F_2(g)$ has the least tendency to occur among all possible oxidation half-reactions ($E°_{ox} = -2.87$ V). There is no reduction half-reaction with which this oxidation half-reaction can be combined to yield a spontaneous oxidation–reduction reaction. This is the reason why $F^-(aq)$ cannot be oxidized to $F_2(g)$ by chemical means and why an electrolytic method is required to prepare the free element from its naturally occurring sources (recall equation 13.80).

Chemical oxidation of $Cl^-(aq)$ to $Cl_2(g)$, on the other hand, is possible. What is required is a reduction half-reaction with an $E° > 1.360$ V. A laboratory method described in Chapter 13 was the oxidation of $Cl^-(aq)$ by $MnO_4^-(aq)$ in acidic solution (reaction 13.81). For this reaction, $E°_{ox} = -1.360$ V, $E°_{red} = +1.51$ V, and $E°_{cell} =$

+0.15 V. Another common laboratory method of preparing $Cl_2(g)$ involves heating a mixture of $MnO_2(s)$ and concentrated $HCl(aq)$.

oxid:	$2\ Cl^-(aq) \longrightarrow Cl_2(g) + 2\ e^-$	$E^\circ_{ox} = -1.36\ V$
red:	$MnO_2(s) + 4\ H^+ + 2\ e^- \longrightarrow Mn^{2+} + 2\ H_2O$	$E^\circ_{red} = +1.23\ V$
net:	$MnO_2(s) + 4\ H^+ + 2\ Cl^- \longrightarrow Mn^{2+} + 2\ H_2O + Cl_2(g)$	$E^\circ_{cell} = -0.13\ V$

Why does this method work even though $E^\circ_{cell} < 0$? It works because heating the solution drives off $Cl_2(g)$ and because the $HCl(aq)$ is concentrated. According to Le Châtelier's principle, both of these conditions should favor the forward reaction.

Oxidation of $Cl^-(aq)$ to $Cl_2(g)$ by chemical means is practicable in the laboratory but not in large-scale commercial processes. The required oxidizing agents ($KMnO_4$, MnO_2) are too expensive. The method of choice in chemical industry is again electrolysis, of a molten chloride (e.g., reaction 13.14) or, more likely, $NaCl(aq)$. The electrolysis of $NaCl(aq)$ is the basis of the chlor-alkali industry discussed in Chapter 20.

Bromine is commercially produced by extraction from seawater, where it occurs to an extent of about 70 parts per million (ppm) as Br^-. First the seawater is adjusted to pH 3.5 and treated with $Cl_2(g)$, which oxidizes Br^- to Br_2.

$$Cl_2(g) + 2\ Br^-(aq) \longrightarrow Br_2(l) + 2\ Cl^-(aq) \tag{21.2}$$

Do you recall why solutions containing CO_3^{2-} are strongly basic? (See page 553.)

The liberated Br_2 is swept from the seawater with a current of air. Various possibilities exist for concentrating the Br_2 obtained. One method involves passing the bromine-laden air into an alkaline solution [usually $Na_2CO_3(aq)$] where a disproportionation reaction occurs.

$$3\ Br_2 + 6\ OH^- \rightleftharpoons 5\ Br^- + BrO_3^- + 3\ H_2O \tag{21.3}$$

Reaction (21.3) is reversed by acidifying the solution.

$$5\ Br^- + BrO_3^- + 6\ H^+ \longrightarrow 3\ Br_2(l) + 3\ H_2O \tag{21.4}$$

The most important use of Br_2 has been in the manufacture of ethylene dibromide ($C_2H_4Br_2$), a gasoline additive. The function of the ethylene dibromide is to react with lead from the antiknock additive, tetraethyllead, to produce volatile lead(II) bromide, which is carried off with the automotive exhaust. Other uses of bromine include the manufacture of organic compounds used as dyes, pharmaceuticals, fumigants, and pesticides. An important inorganic compound of bromine is $AgBr$, the primary light-sensitive agent used in photographic film.

Iodine is obtainable in small quantities from dried seaweed, since certain marine plants absorb and concentrate I^- selectively in the presence of Cl^- and Br^-. Low concentrations of $I^-(aq)$ are also found in some natural brines (salt solutions) associated with oil fields. From such sources oxidation of I^- by a variety of oxidizing agents is possible. A more abundant natural source of iodine is $NaIO_3$, found in large deposits in Chile. To convert IO_3^- to I_2 requires the use of a *reducing* agent, e.g., sodium hydrogen sulfite (bisulfite). A two-step procedure is used in which $IO_3^-(aq)$ is first reduced to $I^-(aq)$.

Environmental restrictions, which have caused a decline in the use of leaded gasoline, have also reduced the demand for ethylene dibromide. Another use that has been banned in the United States is as a fumigant for grain products and fruit. Ethylene dibromide (EDB) is a known carcinogen in rats.

$$IO_3^- + 3\ HSO_3^- \longrightarrow I^- + 3\ SO_4^{2-} + 3\ H^+ \tag{21.5}$$

The reaction of $I^-(aq)$ with additional $IO_3^-(aq)$ in acidic solution produces I_2 (similar to the $Br^--BrO_3^-$ reaction of 21.4).

$$5\ I^- + IO_3^- + 6\ H^+ \longrightarrow 3\ I_2 + 3\ H_2O \tag{21.6}$$

Iodine is of less commercial importance than chlorine and bromine, although it and its compounds do have applications as catalysts, in medicine, and in the preparation of photographic emulsions (as AgI). Iodine and its compounds also have important uses in the analytical chemistry laboratory (see page 642).

FIGURE 21-1
Standard electrode
potential diagram for
chlorine.

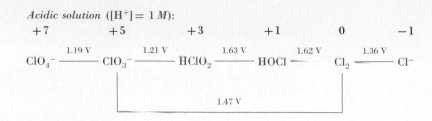

Acidic solution ([H$^+$] = 1 *M*):

The numbers in black are the oxidation states of the chlorine. The potentials listed are for reduction processes with reactants and products at unit activity. The data for acidic solution are at [H$^+$] = 1 *M*; in basic solution [OH$^-$] = 1 *M*. Because of the basic properties of ClO$_2^-$ and OCl$^-$, the weak acids HClO$_2$ and HOCl are formed in acidic solution.

Electrode (Reduction) Potential Diagrams.   As we have been noting, much of the chemical behavior of the halogens and their compounds involves oxidation–reduction reactions; and a discussion of this behavior can be facilitated by the diagrammatic method introduced in Figure 21-1. In this diagram several chlorine-containing species are represented. Various pairs of these species are joined by line segments. Written above each line is the standard electrode potential for the reduction of one species to the other. These two species, a couple, form the basis of a reduction half-equation. For example,

the symbolism   ClO$_4^-$ $\xrightarrow{1.19\ V}$ ClO$_3^-$

signifies

$$ClO_4^- + 2\ H^+ + 2\ e^- \longrightarrow ClO_3^- + H_2O \qquad E^\circ_{red} = 1.19\ V$$

Example 21-1   Use data from Figure 21-1 to predict whether the following reaction will occur to any significant extent in the forward direction.

$$Cl_2(g) + H_2O \longrightarrow Cl^- + ClO_3^- + H^+$$

Solution. E° values are taken from the lines joining Cl$_2$ with Cl$^-$ and Cl$_2$ with ClO$_3^-$. Since all values in Figure 21-1 are reduction potentials, E°_{ox} is the negative of the corresponding E°_{red} listed in Figure 21-1.

oxid:	Cl$_2$ + 6 H$_2$O $\longrightarrow$ 2 ClO$_3^-$ + 12 H$^+$ + 10 e$^-$	$E^\circ_{ox} = -1.47$ V
red:	5{Cl$_2$ + 2 e$^-$ $\longrightarrow$ 2 Cl$^-$}	$E^\circ_{red} = +1.36$ V
net:	3 Cl$_2$(g) + 3 H$_2$O $\longrightarrow$ 5 Cl$^-$ + ClO$_3^-$ + 6 H$^+$	$E^\circ_{cell} = -0.11$ V

The negative value of E°_{cell} suggests that the reaction does not occur to a significant extent in the forward direction.

SIMILAR EXAMPLES: Review Problems 5, 6.

We have had to combine oxidation and reduction half-equations into a net oxidation–reduction equation on numerous occasions, and in these instances we have found that E°_{ox}

and E°_{red} can be added together to yield E°_{cell} for the net reaction. The combination of half-equations called for in Example 21-2 is of a different nature. Here the objective is to establish the electrode potential for an unknown *half-reaction* from known values for related half-reactions. When half-reactions of the same kind are combined (i.e., either both oxidation or both reduction), electrons *do not* cancel out and the half-cell potentials *are not* additive; but free energy changes *are* additive. The key to the method of Example 21-2, then, is multiple use of expression (20.24), $\Delta\overline{G}^\circ = -n\mathscr{F}E^\circ$.

Example 21-2   Refer to Figure 21-1 and establish the value of the standard electrode potential for the reduction of ClO_3^- to OCl^- in basic solution.

Solution.   The desired half-reaction is the sum of the two half-reactions that follow. Electrode potentials are converted to free energy changes through the expression $\Delta\overline{G}^\circ = -n\mathscr{F}E^\circ$ or $-\Delta\overline{G}^\circ = n\mathscr{F}E^\circ$.

$$
\begin{array}{ll}
ClO_3^- + H_2O + 2\,e^- \longrightarrow ClO_2^- + 2\,OH^- & -\Delta\overline{G}^\circ = 2\,\mathscr{F}(0.35) \\
ClO_2^- + H_2O + 2\,e^- \longrightarrow OCl^- + 2\,OH^- & -\Delta\overline{G}^\circ = 2\,\mathscr{F}(0.65) \\
\hline
ClO_3^- + 2\,H_2O + 4\,e^- \longrightarrow OCl^- + 4\,OH^- & -\Delta\overline{G}^\circ = 2\,\mathscr{F}(0.35 + 0.65)
\end{array}
$$

But for the desired half-reaction we may also write that $-\Delta\overline{G}^\circ = n\mathscr{F}E^\circ$, and in this half-reaction $n = 4$, whereas it was 2 in each of the others.

$$ E^\circ = \frac{-\Delta\overline{G}^\circ}{n\mathscr{F}} = \frac{2\mathscr{F}(0.35 + 0.65)}{4\mathscr{F}} = 0.50 \text{ V} $$

SIMILAR EXAMPLES: Review Problem 7; Exercise 4.

Hydrogen Halides.   We have encountered the hydrogen halides from time to time throughout the text. In aqueous solution they are typical mineral acids whose acid strengths decrease in the direction of increasing bond energy, that is

$$ \underbrace{HI > HBr > HCl}_{\substack{\text{strong} \\ \text{acids}}} > \underset{\substack{\text{weak} \\ \text{acid}}}{HF} $$

Strong hydrogen bonding associated with HF, both in the gaseous state and in aqueous solution, seems to account for some of its rather unusual properties relative to the other hydrogen halides.

The hydrogen halides can be prepared by direct combination of the elements

$$ H_2(g) + X_2(g) \longrightarrow 2\,HX(g) \tag{21.7} $$

but in reactions that differ in their speeds and degree of completion. The reaction of $H_2(g)$ and $F_2(g)$ is very fast, occurring with explosive violence under some conditions. The reaction of $H_2(g)$ and $Cl_2(g)$ also proceeds rapidly (explosively) in the presence of light. With Br_2 and I_2 reaction occurs much more slowly. The reaction of $H_2(g)$ with the lighter halogens occurs through a chain mechanism (reaction 14.40), but with $I_2(g)$ the mechanism involves a termolecular step (reaction 14.36) and a high activation energy (Figure 14-10). From the data in Table 21-1 we see that the free energies of formation of HF, HCl, and HBr are large and negative, suggesting that for them reaction (21.7) goes to completion. For HI(g) the value of $\Delta\overline{G}_f^\circ$ is small and positive. The value of K_p for the decomposition of HI (calculated in Example 21-3) is large enough to suggest that the dissociation into its elements should occur to a significant extent even at room temperature (Additional Exercise 14). However, because of its high activation energy (184 kJ/mol) the dissociation reaction occurs very slowly in the absence of a catalyst. As a result, HI(g) is quite stable at room temperature.

TABLE 21-1
Free energy of formation of hydrogen halides at 298 K

	$\Delta\overline{G}_f^\circ$, kJ/mol
HF(g)	-270.70
HCl(g)	-95.27
HBr(g)	-53.22
HI(g)	$+1.30$

Example 21-3   What is the value of K_p for the dissociation of HI(g) into its elements at 298 K?

Solution.   The dissociation reaction at 298 K is

$$HI(g) \rightleftharpoons \tfrac{1}{2} H_2(g) + \tfrac{1}{2} I_2(s) \qquad \Delta \overline{G}^\circ = -1.30 \text{ kJ/mol}$$

Since this is the reverse of the formation reaction, $\Delta \overline{G}^\circ$ is the negative of the value listed in Table 21-1.

$$\Delta \overline{G}^\circ = -2.303RT \log K_p$$

$$\log K_p = \frac{-\Delta \overline{G}^\circ}{2.303RT} = \frac{-(-1.30 \times 10^3 \text{ J mol}^{-1})}{2.303 \times 8.314 \text{ J mol}^{-1} \text{ K}^{-1} \times 298 \text{ K}} = 0.228$$

$$K_p = \text{antilog} (0.228) = 1.69$$

SIMILAR EXAMPLE: Exercise 5.

Another method of preparing hydrogen halides is to heat halide salts with a concentrated nonvolatile acid. The volatile hydrogen halide escapes from the reaction mixture. As we learned in Chapter 13, this is the principal commercial method of preparing HF.

$$CaF_2(s) + H_2SO_4(\text{conc. aq}) \xrightarrow{\Delta} CaSO_4(s) + 2 \text{ HF}(g) \tag{21.8}$$

Applied to the preparation of HCl, the halide salt is generally NaCl(s).

$$NaCl(s) + H_2SO_4(\text{conc. aq}) \xrightarrow{\Delta} NaHSO_4(s) + HCl(g) \tag{21.9}$$

Further heating of NaCl(s) in the presence of $NaHSO_4$ produces additional HCl(g).

$$NaCl(s) + NaHSO_4(s) \xrightarrow{\Delta} Na_2SO_4(s) + HCl(g) \tag{21.10}$$

The method of reactions (21.9) and (21.10) is unsuitable for the preparation of HBr and HI, because these compounds are rather easily oxidized to the free elements and H_2SO_4(conc. aq) is a mild oxidizing agent. That is,

$$2 \text{ HBr} + H_2SO_4(\text{conc. aq}) \xrightarrow{\Delta} 2 \text{ H}_2O + Br_2(g) + SO_2(g) \tag{21.11}$$

This difficulty can be overcome by using a nonvolatile acid that has very little oxidizing power, such as H_3PO_4.

$$NaBr(s) + H_3PO_4(\text{conc. aq}) \xrightarrow{\Delta} NaH_2PO_4 + HBr(g) \tag{21.12}$$

Other methods are also possible, such as reaction of Br_2(1) with red P

$$P_4(s) + 6 \text{ Br}_2(l) \longrightarrow 4 \text{ PBr}_3 \tag{21.13}$$

followed by the action of water on PBr_3.

$$PBr_3 + 3 \text{ H}_2O \longrightarrow H_3PO_3 + 3 \text{ HBr}(g) \tag{21.14}$$

Oxoacids and Oxoanions of the Halogens.   An important class of compounds of the halogens are the oxygen containing acids and their salts. A summary of the known oxoacids is presented in Table 21-2. Chlorine forms a complete set of acids; but for bromine and iodine the acids $HBrO_2$ and HIO_2 do not exist or, at least, have not been observed. A system of nomenclature of oxoacids and oxoanions was discussed in Section 3-8. Lewis structures were presented on page 527 and used to relate the strengths of the oxoacids to their structures. These Lewis structures conformed to the idea encountered in

TABLE 21-2
Oxoacids of the halogens[a]

Oxidation state of the halogen	Chlorine	Bromine	Iodine
+1	HOCl	HOBr	HOI
+3	$HClO_2$	—	—
+5	$HClO_3$	$HBrO_3$	HIO_3
+7	$HClO_4$	$HBrO_4$	HIO_4; H_5IO_6

[a] In all these acids H atoms are bonded to O atoms, not to the central halogen atom. More accurate representations would be HOClO (instead of $HClO_2$), HOClO$_2$ (instead of $HClO_3$), and so on.

Chapter 9 that in a molecular structure the central atom is generally *less* electronegative than the atoms bonded to it. If we carry this principle over to the hypothetical oxoacids of fluorine, such acids would have F, the *most electronegative* of all atoms, as the central atom. Also they would have the F atom in positive oxidation states and carrying positive formal charges. These facts suggest that F should not form oxoacids, that in aqueous solution F should exist in the oxidation state -1. The compound HOF has been reported and its properties studied in the solid and liquid states. In water, however, it rapidly decomposes to HF and $O_2(g)$. Our statement about the nonexistence of oxoacids of fluorine is essentially correct. This is still another example of how fluorine differs from the other halogens, a theme explored in several ways in Chapter 13.

Hypochlorous acid is formed in the reaction of Cl_2 with water.

$$Cl_2(g) + H_2O \rightleftharpoons HOCl(aq) + H^+(aq) + Cl^-(aq) \tag{21.15}$$

A solution of "chlorine water," commonly used as an oxidizing agent in the laboratory, is thus a mixture of $Cl_2(aq)$, HOCl(aq), and HCl(aq). The concentration of HOCl(aq) can be increased by adding a reactant to remove Cl^-, displacing equilibrium to the right in reaction (21.15). This can be accomplished with mercury(II) oxide.

$$2 HgO(s) + 2 H^+(aq) + 2 Cl^-(aq) \longrightarrow HgO \cdot HgCl_2(s) + H_2O \tag{21.16}$$

Although obtainable in aqueous solutions, HOCl cannot be isolated in the pure state.

If $Cl_2(g)$ is dissolved in an alkaline solution, equilibrium is also displaced far to the right in reaction (21.15). The HCl(aq) and HOCl(aq) are neutralized and an aqueous solution of a **hypochlorite** salt is formed.

$$Cl_2(g) + 2 OH^-(aq) \longrightarrow OCl^-(aq) + Cl^-(aq) + H_2O \tag{21.17}$$

Reaction (21.17) is carried out commercially in a chlor-alkali cell in which, rather than keeping the electrolysis products separated, $Cl_2(g)$ from the anode is deliberately mixed with NaOH(aq) from the cathode (recall Figure 20-18). Common household bleaches and swimming pool "chlorine" are aqueous alkaline solutions of NaOCl obtained by reaction (21.17). The product of the reaction of $Cl_2(g)$ with $Ca(OH)_2(s)$ is a complex mixture of salts containing calcium hypochlorite, $Ca(OCl)_2$, and known as **bleaching powder.** Unlike HOCl, hypochlorite salts can be obtained in the pure state. Hypochlorous acid and its salts are strong oxidizing agents, capable of oxidizing I^- to I_2, Fe^{2+} to Fe^{3+}, Mn^{2+} to MnO_4^-, and Pb^{2+} to PbO_2.

Chlorous acid can be prepared by treating an aqueous suspension of barium chlorite with $H_2SO_4(aq)$. $BaSO_4(s)$ precipitates and $HClO_2(aq)$ remains.

$$Ba(ClO_2)_2 + H_2SO_4(aq) \longrightarrow BaSO_4(s) + 2 HClO_2(aq) \tag{21.18}$$

FIGURE 21-2
Structures of the chlorine oxoanions.

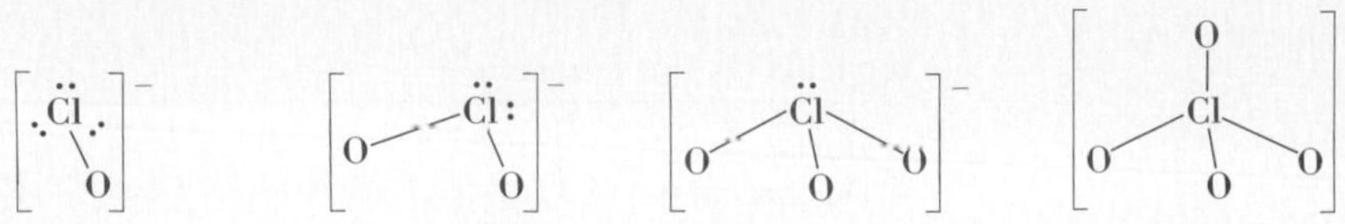

hypochlorite, OCl^-
(linear)

chlorite, ClO_2^-
(angular)

chlorate, ClO_3^-
(trigonal pyramidal)

perchlorate, ClO_4^-
(tetrahedral)

Lone-pair electrons are shown only for the central Cl atom.

FIGURE 21-3
Structure of the H_5IO_6 molecule.

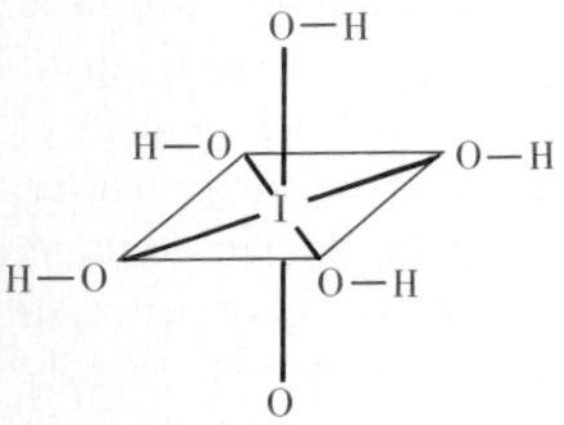

If we represent H_5IO_6 in the form $IO_m(OH)_n$, $m = 1$ and $n = 5$, that is, $IO(OH)_5$. According to the generalization on page 527, we should expect $K_{a_1} \approx 10^{-2}$; the measured value is $K_{a_1} = 5.1 \times 10^{-4}$.

Chlorite salts are generally produced by reducing chlorine dioxide with peroxide ion in aqueous solution.

$$2 ClO_2(g) + O_2^{2-}(aq) \longrightarrow 2 ClO_2^-(aq) + O_2(g) \tag{21.19}$$

Sodium chlorite is used as a bleaching agent for textiles. $ClO_2(g)$ for reaction (21.19) can be obtained by the reduction of $NaClO_3$ with $SO_2(g)$ in sulfuric acid.

Chloric acid is prepared by treating barium chlorate with $H_2SO_4(aq)$. As in reaction (21.18), $BaSO_4$ precipitates.

$$Ba(ClO_3)_2(aq) + H_2SO_4(aq) \longrightarrow BaSO_4(s) + 2 HClO_3(aq) \tag{21.20}$$

Chloric acid cannot be isolated in the pure state. **Chlorate** salts are formed when hot alkaline solutions are treated with $Cl_2(g)$.

$$3 Cl_2(g) + 6 OH^- \longrightarrow 5 Cl^- + ClO_3^- + 3 H_2O \tag{21.21}$$

The chloride and chlorate salts are separated by fractional crystallization. Chlorates are good oxidizing agents. In addition, solid chlorates yield oxygen directly, accounting for their use in matches and fireworks. A simple laboratory method of producing $O_2(g)$ involves heating $KClO_3(s)$ in the presence of a catalyst (MnO_2).

$$2 KClO_3(s) \longrightarrow 2 KCl(s) + 3 O_2(g) \tag{21.22}$$

Perchloric acid can be prepared by distilling a mixture of $KClO_4$ and H_2SO_4 under reduced pressure.

$$KClO_4 + H_2SO_4(conc. aq) \longrightarrow KHSO_4 + HClO_4 \tag{21.23}$$

Perchlorate salts are mainly prepared by electrolyzing chlorate solutions. Oxidation of ClO_3^- occurs at a Pt anode through the half-reaction

$$ClO_3^-(aq) + H_2O \longrightarrow ClO_4^-(aq) + 2 H^+(aq) + 2 e^- \tag{21.24}$$

$O_2(g)$, **which is often an oxidation product in electrolysis, does not form in half-reaction (21.24) because of the high overvoltage of $O_2(g)$ on Pt (see again, page 620).**

Perchlorates are only mild oxidizing agents, but the solid salts do lose $O_2(g)$ on strong heating. An interesting laboratory use of perchlorate salts is in solution studies where complex ion formation is to be avoided. ClO_4^- has the least tendency of any anion to act as a ligand in complex ion formation.

Geometric structures of the oxoanions of chlorine are easily predictable from their Lewis structures and the VSEPR theory, as indicated in Figure 21-2.

Because of its large size, the I atom is able to surround itself with and bond to six other atoms. This accounts for the existence of an oxoacid such as H_5IO_6, **paraperiodic acid,** whose structure is shown in Figure 21-3. Although the I atom is in the oxidation state $+7$, H_5IO_6 is not a strong acid (as is HIO_4). Five of the O atoms in H_5IO_6 have H atoms bonded to them. The relationship between acid strength and number of O atoms bonded to the central atom (recall Section 17-8) applies only to O atoms *with no attached H atoms*. In this case there is only one such O atom. H_5IO_6 is a weak polyprotic acid, and only the first H atom ionizes with comparative ease: $pK_{a_1} = 3.29$; $pK_{a_2} = 8.3$; $pK_{a_3} = 11.6$.

FIGURE 21-4
Structures of some
interhalogen molecules.

Type:	XY	XY$_3$	XY$_5$	XY$_7$
Shape:	linear	T-shaped	square pyramidal	pentagonal bipyramidal

FIGURE 21-5
Structure of the I$_3^-$ ion.

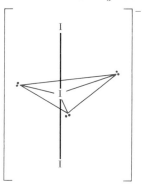

The bond-pair electrons
are shown as black lines.
The lone-pair electrons of
the central I atom are
represented by bold dots.

**Triiodide ion is pictured
in Color Section C.**

Interhalogen Compounds.   Two different halogen elements can unite to produce interhalogen compounds. Many compounds of this type are known. These compounds and their physical states at 25°C and 1 atm are listed in Table 21-3. The molecular structures of the interhalogen compounds feature the larger, less electronegative halogen as the central atom and the smaller halogen atoms as the terminal atoms. Molecular shapes of the interhalogens agree quite well with predictions based on the VSEPR theory. Some representative structures are shown in Figure 21-4, including one type (IF$_7$) that we have not considered previously. This is a structure with *seven* electron pairs distributed about the central atom in the form of a pentagonal bipyramid (sp^3d^3 hybridization).

All the interhalogen compounds, with the exception of IF$_5$, are very reactive. ClF$_3$ and BrF$_3$, for example, react with explosive violence with water, organic materials, and some inorganic materials. These two trifluorides are used to fluorinate compounds, as in the preparation of UF$_6$ for use in the separation of uranium isotopes by gaseous diffusion (recall page 129). ICl is used as an iodination reagent in organic chemistry.

Polyhalide Ions.   Iodine is only slightly soluble in pure water (0.3 g/L) but rather highly soluble in KI(aq). This enhanced solubility results from the formation of the triiodide ion.

$$I_2(s) + I^-(aq) \rightleftharpoons I_3^-(aq) \qquad K_c = 710 \text{ (at 298 K)} \tag{21.25}$$

Triiodide ion is one of a fairly large group of species, called polyhalide ions, that are produced by the reaction of a halide ion with a halogen or interhalogen molecule. In this reaction the halide ion acts as a Lewis base (electron pair donor) and the molecule as a Lewis acid (electron pair acceptor). The structure of the I$_3^-$ ion is shown in Figure 21-5. Iodine solutions in aqueous KI (and hence triiodide ion) are widely used in analytical chemistry.

Analytical Chemistry of the Halogens.   Detection of Cl$^-$, Br$^-$, and I$^-$ in aqueous solutions can be accomplished with AgNO$_3$(aq). The silver halides are all insoluble.

TABLE 21-3
Some interhalogen compounds^a

	Type		
XY	XY$_3$	XY$_5$	XY$_7$
ClF(g)	ClF$_3$(g)	ClF$_5$	
BrF(g)	BrF$_3$(l)	BrF$_5$(l)	
BrCl(g)			
ICl(s)	ICl$_3$(s)	IF$_5$(l)	IF$_7$(g)
IBr(s)			

^{a}Some interhalogen compounds, particularly of the type XY, are unstable (e.g., at 298 K $\Delta\overline{G}_f^\circ$ for BrCl(g) is only -0.88 kJ/mol).

Moreover, because of differences in their K_{sp} values, first AgI, then AgBr, and finally AgCl precipitate from an aqueous solution containing Cl^-, Br^-, and I^- as $AgNO_3(aq)$ is slowly added (recall Section 19-4).

Br^- and I^- can also be detected by reaction with $Cl_2(aq)$. The Cl_2 is reduced to Cl^-, and Br^- or I^- is oxidized to Br_2 or I_2. The liberated Br_2 or I_2 is extracted from aqueous solution with a solvent, such as CS_2 (see Figure 21-6). In CS_2 iodine has a violet color; bromine, red. If Br^- and I^- occur together in aqueous solution, I^- can be oxidized to IO_3^- with an excess of $Cl_2(aq)$. Simultaneously, Br^- is oxidized only to Br_2, which is extracted by the $CS_2(l)$. To detect the oxoanions of Cl, Br, and I, oxidation–reduction reactions can be used to reduce the oxoanion to the free halogen or the halide ion. Then the methods just outlined can be used.

> **Br₂, of course, is not a strong enough oxidizing agent to oxidize $Cl^-(aq)$ to $Cl_2(g)$, nor can I_2 oxidize either $Cl^-(aq)$ or $Br^-(aq)$.**

Iodine and iodide ion are among the most versatile reagents used in oxidation–reduction reactions in analytical chemistry. For example, a standard solution of I_2 can be used to titrate certain reducing agents, such as sulfite ion.

$$SO_3^{2-} + I_2 + H_2O \longrightarrow SO_4^{2-} + 2\,I^- + 2\,H^+ \tag{21.26}$$

Iodide ion can be used in the titration of strong oxidizing agents.

$$2\,MnO_4^- + 10\,I^- + 16\,H^+ \longrightarrow 2\,Mn^{2+} + 5\,I_2 + 8\,H_2O \tag{21.27}$$

Sometimes, a reaction such as (21.27) is conducted by adding an *excess* of $I^-(aq)$ to the oxidizing agent and then titrating the liberated I_2 with sodium thiosulfate. Even when iodine is present only in trace amounts, it forms a deep blue complex with starch. The end point of reaction (21.28) is marked by the disappearance of the blue color associated with the starch–iodine complex.

> **This starch–iodine complex is featured in Color Section C.**

$$I_2 + 2\,S_2O_3^{2-} \longrightarrow 2\,I^- + S_4O_6^{2-} \tag{21.28}$$

$$I_2 + starch \longrightarrow blue\ complex \tag{21.29}$$

In reactions in which iodine is used as an analytical reagent, it is usually in the presence of I^- and therefore exists mostly as triiodide ion, I_3^-, as noted in equation (21.25).

FIGURE 21-6
A test for iodide ion.

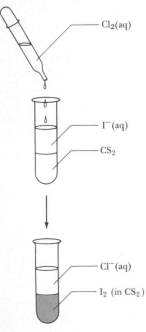

Cl₂(aq)

I⁻(aq)

CS₂

Cl⁻(aq)

I₂ (in CS₂)

21-2 The Group VIA Elements

One of the important points made in Chapter 13 about oxygen, the first member of this group, was that while it does bear some resemblance to other members of the group, it also differs from them in several respects. Similarities are attributable to the similar valence-shell electron configurations of the atoms of the group VIA elements. Differences result from the (a) unavailability of *d* orbitals for bonding, (b) small size, and (c) high electronegativity associated with O atoms. The data in Table 21-4, which compare a few of the properties of oxygen and sulfur, summarize some similarities and differences.

Oxygen.   Oxygen is so central to a study of chemistry that inevitably we find ourselves considering physical and chemical properties of this element and its compounds as we attempt to gain an understanding of chemical principles. Some of the first examples in our discussion of stoichiometry dealt with combustion reactions, in which substances react with $O_2(g)$ to yield products such as $CO_2(g)$, $H_2O(l)$, and $SO_2(g)$. Combustion reactions again figured prominently in the study of thermochemistry. Many of the molecules and ions used as examples in the chapters on chemical bonding were oxygen-containing species. Water was a primary subject in the discussion of liquids, solids, and intermolecular forces; as it was again in the study of acid–base and other solution equilibria. Ozone and hydrogen peroxide were considered in Chapter 13, and the acidic, basic, and amphoteric properties of element oxides in Chapter 17. To illustrate the frequency with which aspects

TABLE 21-4
Some comparisons of oxygen and sulfur

Oxygen	Sulfur
$O_2(g)$ at 298 K and 1 atm	$S_8(s)$ at 298 K and 1 atm
Two allotropes, $O_2(g)$ and $O_3(g)$	Two solid allotropes and many different species in liquid and gaseous states
Possible oxidation states: -2, -1, 0	Possible oxidation states: all values from -2 to $+6$[a]
$O_2(g)$ and $O_3(g)$ are very good oxidizing agents	$S(s)$ is a poor oxidizing agent
Forms, with metals, oxides that are mostly ionic in character	Forms ionic sulfides with the most active metals but many metal sulfides have partial covalent character
O^{2-} completely decomposes in water, producing OH^-	S^{2-} strongly hydrolyzes in water to HS^- (and OH^-) but some free S^{2-} remains
O not often the central atom in a structure, and can never have more than 4 atoms bonded to it. More commonly has two (as in H_2O) or three (as in H_3O^+)	S the central atom in many structures. Can easily accommodate up to 6 atoms around itself (e.g., SO_3, SO_4^{2-}, SF_6)
Can form only two- and three-atom chains, as in H_2O_2 and O_3. Compounds with O—O bonds decompose readily	Can form molecules with up to six S atoms per chain in compounds such as H_2S_n, Na_2S_n, $H_2S_nO_6$
Forms the oxide CO_2, which reacts with $NaOH(aq)$ to produce $Na_2CO_3(aq)$	Forms the sulfide CS_2, which reacts with $NaOH(aq)$ to produce $Na_2CS_3(aq)$

Forms the hydride H_2O, which
  is a liquid at 298 K and 1 atm
  is extensively hydrogen-bonded
  has a large dipole moment
  is an excellent ionizing solvent
  forms hydrates and aqua complexes
  is oxidized with difficulty
	Froms the hydride H_2S, which
	  is a (poisonous) gas at 298 K and 1 atm
	  is not hydrogen-bonded
	  has a small dipole moment
	  is a poor solvent
	  forms no complexes
	  is easily oxidized

[a] For an assessment of the oxidation state of S in the ion $S_2O_8^{2-}$, see page 648.

of oxygen chemistry are encountered in other contexts, consider the following methods of preparing $O_2(g)$. Each appeared in the context cited.

Heating oxides of metals of low reactivity:

$$2\ HgO(s) \xrightarrow{\Delta} 2\ Hg(l) + O_2(g) \qquad \text{(Figure 2-1)}$$

$$2\ Ag_2O(s) \xrightarrow{\Delta} 4\ Ag(s) + O_2(g) \qquad \text{(reaction 16.28)}$$

Heating certain oxygen-containing salts:

$$2\ KClO_3 \xrightarrow{\Delta} 2\ KCl(s) + 3\ O_2(g) \qquad \text{(Example 5-12; reaction 21.22)}$$

Decomposition of water by a thermochemical cycle:

$$2\ H_2O(g) \xrightarrow{\Delta} 2\ H_2(g) + O_2(g) \qquad \text{(net reaction 13.5)}$$

Decomposition of water by electrolysis:

$$2\ H_2O(l) \xrightarrow{\text{electrolysis}} 2\ H_2(g) + O_2(g) \qquad \text{(reaction 13.4)}$$

$$2\ H_2O \longrightarrow 4\ H^+(aq) + O_2(g) + 4\ e^- \qquad \text{(half-reaction 20.44)}$$

Reaction of an ionic peroxide with water:

$$2\ O_2^{2-} + 2\ H_2O \longrightarrow 4\ OH^-(aq) + O_2(g) \qquad \text{(page 392)}$$

Reaction of a superoxide with water:

$$4\ O_2^- + 2\ H_2O \longrightarrow 4\ OH^-(aq) + 3\ O_2(g) \qquad \text{(page 392)}$$

FIGURE 21-7
Electrode potential
diagram for oxygen.

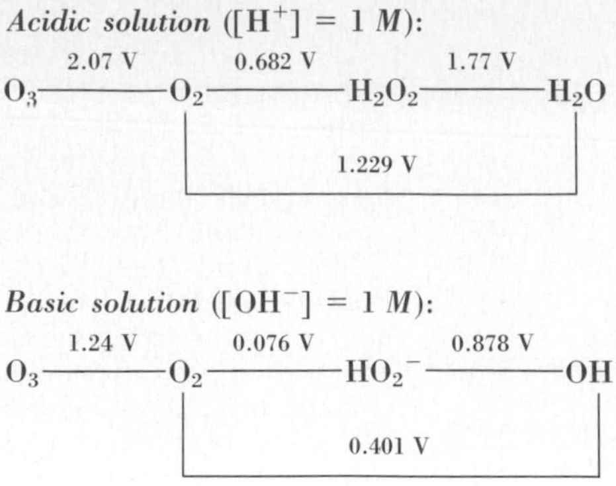

Acidic solution ([H^+] = 1 *M*):

$$O_3 \xrightarrow{\ 2.07\ V\ } O_2 \xrightarrow{\ 0.682\ V\ } H_2O_2 \xrightarrow{\ 1.77\ V\ } H_2O$$

1.229 V

Basic solution ([OH^-] = 1 *M*):

$$O_3 \xrightarrow{\ 1.24\ V\ } O_2 \xrightarrow{\ 0.076\ V\ } HO_2^- \xrightarrow{\ 0.878\ V\ } OH^-$$

0.401 V

The ion HO_2^- results from the acid ionization of H_2O_2 when it is placed in a strongly alkaline solution.

Catalylic decomposition of hydrogen peroxide:

$$2\ H_2O_2(aq) \longrightarrow 2\ H_2O + O_2(g) \qquad \text{(reactions 13.73, 14.3)}$$

Oxidation of hydrogen peroxide:

$$5\ H_2O_2 + 2\ MnO_4^- + 6\ H^+ \longrightarrow 2\ Mn^{2+} + 8\ H_2O + 5\ O_2(g) \qquad \text{(reaction 13.72)}$$

Oxidation–Reduction Reactions of Oxygen, Ozone, and Hydrogen Peroxide. Several reactions of this type were presented in Chapter 13, but now we can also look at these reactions from the standpoint of the electrode potential diagrams shown in Figure 21-7. These diagrams make clear the facts that $O_3(g)$ can only act as an oxidizing agent and that H_2O_2 may act either as an oxidizing or reducing agent.

Electrode Potential Diagram for Sulfur. As with the halogens and oxygen, the oxidation–reduction behavior of sulfur-containing species can be summarized through an electrode potential diagram, as in Figure 21-8.

Example 21-4 Zinc dissolves in HCl(conc. aq) to liberate $H_2(g)$; but from H_2SO_4(conc. aq), $SO_2(g)$ is evolved. Explain this difference in behavior.

Solution. The dissolving process is the half-reaction $Zn(s) \rightarrow Zn^{2+}(aq) + 2\ e^-$, for which $E^\circ_{ox} = +0.76$ V (see Table 20-2). This oxidation half-reaction can be combined with the reduction half-reaction $2\ H^+(aq) + 2\ e^- \rightarrow H_2(g)$, for which $E^\circ_{red} = 0.00$ V. The result is a spontaneous net oxidation–reduction reaction with

FIGURE 21-8
Electrode potential
diagram for sulfur.

Acidic solution ([H^+] = 1 *M*):

+6		+5		+4		+2.5		+2		0		−2
SO_4^{2-}	−0.22 V	$S_2O_6^{2-}$	0.57 V	SO_2	0.51 V	$S_4O_6^{2-}$	0.08 V	$S_2O_3^{2-}$	0.50 V	S	0.14 V	H_2S

0.17 V   0.45 V

Basic solution ([OH^-] = 1 *M*):

+6		+4		+2.5		+2		0		−2
SO_4^{2-}	−0.93 V	SO_3^{2-}	−0.79 V	$S_4O_6^{2-}$	0.08 V	$S_2O_3^{2-}$	−0.74 V	S	−0.48 V	S^{2-}

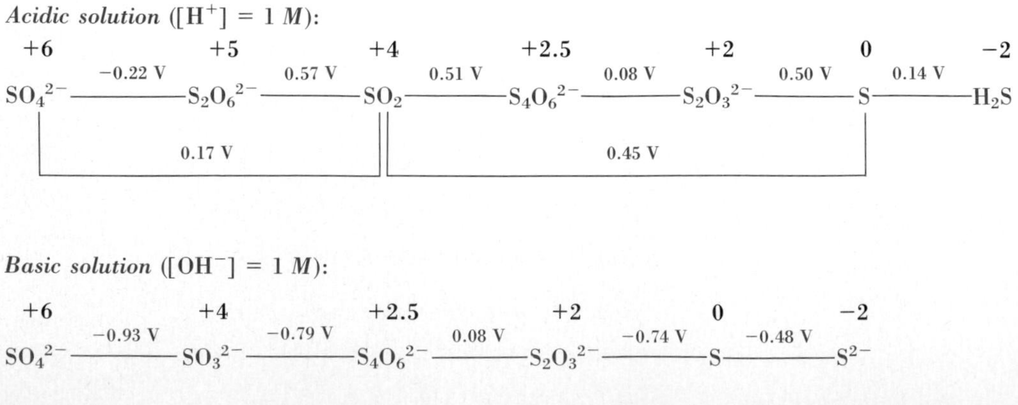

FIGURE 21-9
Structures of the sulfite (SO_3^{2-}) and sulfate (SO_4^{2-}) ions.

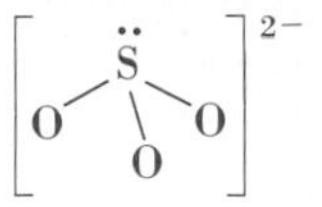

trigonal pyramidal

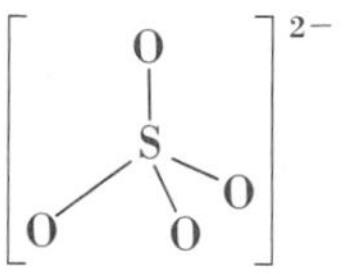

tetrahedral

$E_{cell}^{\circ} = +0.76$ V. This is what occurs when Zn dissolves in HCl(aq) and accounts for the liberated $H_2(g)$. Although this same combination of half-reactions could be written for the dissolving of Zn in H_2SO_4(conc. aq), a different reduction is more likely to occur.

$$SO_4^{2-}(aq) + 4\,H^+(aq) + 2\,e^- \rightarrow 2\,H_2O + SO_2(g) \qquad E_{red}^{\circ} = +0.17\ V$$

For the net spontaneous reaction

$$Zn(s) + 4\,H^+(aq) + SO_4^{2-}(aq) \rightarrow Zn^{2+}(aq) + 2\,H_2O + SO_2(g)$$

$E_{cell}^{\circ} = +0.93$ V, and this is the reaction that occurs.

SIMILAR EXAMPLE: Review Problem 9.

Sulfites and Sulfates. Some of the properties of sulfurous acid, H_2SO_3(aq), and sulfuric acid, H_2SO_4, were discussed in Section 13-13. The salts of these acids are sulfites and sulfates, respectively. Since each acid is a diprotic acid, two types of salts are possible for each.

$$H_2SO_3(aq) + H_2O \rightleftharpoons H_3O^+ + HSO_3^- \qquad K_{a_1} = 1.3 \times 10^{-2}$$
hydrogen sulfite (bisulfite) ion

$$HSO_3^- + H_2O \rightleftharpoons H_3O^+ + SO_3^{2-} \qquad K_{a_2} = 6.3 \times 10^{-8}$$
sulfite ion

$$H_2SO_4 + H_2O \rightleftharpoons H_3O^+ + HSO_4^- \qquad K_{a_1} = large$$
hydrogen sulfate (bisulfate) ion

$$HSO_4^- + H_2O \rightleftharpoons H_3O^+ + SO_4^{2-} \qquad K_{a_2} = 1.29 \times 10^{-2}$$
sulfate ion

FIGURE 21-10
Structures of some thio ions.

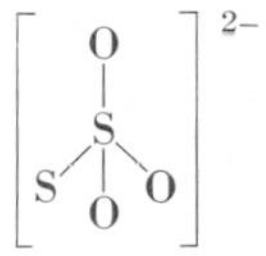

$S_2O_3^{2-}$ thiosulfate ion

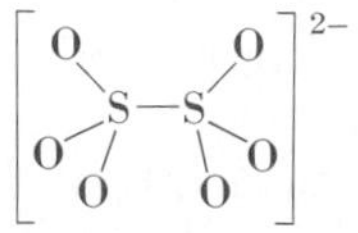

$S_2O_6^{2-}$ dithionate ion

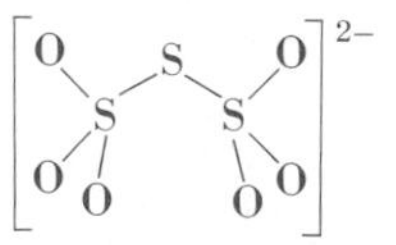

$S_3O_6^{2-}$ trithionate ion

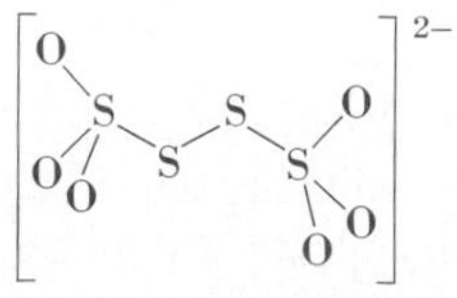

$S_4O_6^{2-}$ tetrathionate ion

We generally consider H_2SO_4 in dilute aqueous solution to be completely ionized in the first ionization step.

Sulfate ion in acidic solution, i.e., H_2SO_4(aq), is a mild oxidizing agent (recall Example 21-4), whereas sulfite ion is a good reducing agent (reaction 13.76) and, under certain conditions, an oxidizing agent (reaction 13.77). The structures of these two ions, which have been noted on previous occasions, are shown again in Figure 21-9. In each structure the central S atom uses an expanded octet and the sulfur-to-oxygen bonds have some multiple bond character.

Except for ammonium and alklai metal sulfites, sulfites are insoluble. Most sulfates, on the other hand, are water soluble. The principal exceptions are $SrSO_4$, $BaSO_4$, $PbSO_4$, and Hg_2SO_4. $CaSO_4$ and Ag_2SO_4 are only moderately soluble. The presence of SO_3^{2-}(aq) or SO_4^{2-}(aq) is generally detected by precipitation of the barium salt. Following precipitation it is easy to establish whether the (white) precipitate is $BaSO_3$(s) or $BaSO_4$(s). $BaSO_3$, as do all sulfites, decomposes in acidic solution with the evolution of $SO_2(g)$.

$$BaSO_3(s) + 2\,H_3O^+(aq) \longrightarrow Ba^{2+}(aq) + 3\,H_2O + SO_2(g) \qquad (21.30)$$

$BaSO_4$(s) is unaffected by acids.

Thio Compounds. In thio compounds an S atom replaces an O atom. Replacement of one of the O atoms in sulfate ion, SO_4^{2-}, leads to thiosulfate ion, $S_2O_3^{2-}$. The formal oxidation state of S in $S_2O_3^{2-}$ is +2 although, as can be seen from Figure 21-10, the two S atoms are not equivalent. (The central S atom is in the oxidation state +6 and the terminal S atom, −2.) Also shown in Figure 21-10 are anions derived from the series of acids called thionic acids, $H_2S_nO_6$ (where $n = 2, 3, 4, 5$, or 6). Thiosulfate ion is formed when an alkaline solution of sodium sulfite is boiled with elemental sulfur. The sulfur is oxidized and the sulfite ion is reduced.

$E°$ in alkaline solution for the couple $SO_3^{2-}/S_2O_3^{2-}$ is not listed in the electrode potential diagram of Figure 21-8. Neither is $E°$ in acidic solution for the couple $SO_2(g)/S_2O_3^{2-}$. These values can be obtained by the method of Example 21-2.

$oxid$: $2\,S + 6\,OH^- \longrightarrow S_2O_3^{2-} + 3\,H_2O + 4\,e^-$ $E°_{ox} = +0.74\ V$
red: $2\,SO_3^{2-} + 3\,H_2O + 4\,e^- \longrightarrow S_2O_3^{2-} + 6\,OH^-$ $E°_{red} = -0.57\ V$

net: $SO_3^{2-} + S \longrightarrow S_2O_3^{2-}$ $E°_{cell} = +0.17\ V$ (21.31)

Equation (21.31) establishes that the formation of $S_2O_3^{2-}$ occurs spontaneously in *basic* solutions and that $S_2O_3^{2-}$ should be stable in such solutions. If we repeat the calculation for an *acidic* solution, we obtain the result

$oxid$: $2\,S + 3\,H_2O \longrightarrow S_2O_3^{2-} + 6\,H^+ + 4\,e^-$ $E°_{ox} = -0.50\ V$
red: $2\,SO_2(g) + 2\,H^+ + 4\,e^- \longrightarrow S_2O_3^{2-} + H_2O$ $E°_{red} = +0.40\ V$

net: $S(s) + SO_2(g) + H_2O \longrightarrow 2\,H^+ + S_2O_3^{2-}$ $E°_{cell} = -0.10\ V$ (21.32)

The negative value of $E°_{cell}$ signifies that in acidic solution the spontaneous reaction is the decomposition of $S_2O_3^{2-}$, the reverse of (21.32).

$$S_2O_3^{2-} + 2\,H^+ \longrightarrow H_2O + SO_2(g) + S(s) \tag{21.33}$$

In the acidic decomposition of $S_2O_3^{2-}$, the sulfur is obtained, first in a colloidal state, and then as an allotropic modification, rho sulfur, composed of the S_6 rings pictured in Figure 13-17. Since $S_2O_3^{2-}$ is both oxidized and reduced in reaction (21.33), the reaction is a disproportionation reaction.

Thiosulfate ion is a reducing agent widely used in analytical chemistry in conjunction with iodine. For example, in the analysis of Cu^{2+} an excess of iodide ion is added, and the liberated iodine is titrated with a standard solution of $Na_2S_2O_3$. The iodine–thiosulfate titration was described through equations (21.28) and (21.29).

Another important use of $Na_2S_2O_3$ is as a "fixing agent" in the photographic process. It functions through the ability of $S_2O_3^{2-}$ to form complex ions with Ag^+ (see Section 24-13).

$$2\,Cu^{2+} + 4\,I^- \longrightarrow 2\,CuI(s) + I_2 \tag{21.34}$$

The thionate ions depicted in Figure 21-10 can be formed from SO_3^{2-} or $S_2O_3^{2-}$ by oxidation–reduction reactions that use oxidizing or reducing agents of the appropriate strengths. For example, tetrathionate ion is formed in the reduction of I_2 by thiosulfate ion (reaction 21.28).

Sulfides.   All metals form sulfides, and most sulfides are insoluble in water. Most are soluble in acidic solution, however, and a few are soluble in basic solution. The differing solubilities of the sulfides allow metal ions to be separated into different groups. This, as we have already learned, is the basis of a qualitative analysis scheme for cations (see Figure 19-5).

Whether a given sulfide will precipitate from solution depends on the pH of the solution. Predictions require the use of the ionization constants for H_2S and solubility product constants for metal sulfides, as previously described in Section 19-10. For the simple metal ion M^{2+}, which precipitates from an aqueous solution saturated in H_2S, we may write

$$H_2S(aq) + M^{2+}(aq) \longrightarrow MS(s) + 2\,H^+(aq) \tag{21.35}$$

Dissolution of a metal sulfide in an acid can be represented by the reverse of equation (21.35), that is,

$$MS(s) + 2\,H^+(aq) \longrightarrow M^{2+}(aq) + H_2S(g) \tag{21.36}$$

Some metal sulfides have such low values of K_{sp} that they will not dissolve in ordinary mineral acids. However, they can be partially dissolved in a strong oxidizing acid. The sulfide ion is oxidized to insoluble elemental sulfur, and the metal ion appears in solution.

$$3\,CuS(s) + 8\,H^+ + 2\,NO_3^- \longrightarrow 3\,Cu^{2+} + 4\,H_2O + 2\,NO(g) + 3\,S(s) \tag{21.37}$$

Mercuric sulfide, HgS, which has the smallest solubility of all metal sulfides (K_{sp} = 1.6×10^{-52}), can be dissolved in a mixture of nitric and hydrochloric acids called **aqua regia.** Here dissolution involves oxidation of the sulfide ion and formation of the complex ion $[HgCl_4]^{2-}$.

$$HgS(s) + 4\ H^+ + 4\ Cl^- + 2\ NO_3{}^- \longrightarrow [HgCl_4]^{2-}(aq) + 2\ H_2O + 2\ NO_2(g) + S(s) \tag{21.38}$$

Polysulfides. A few metal sulfides have an acidic nature. They dissolve in basic solutions containing high concentrations of S^{2-} or $S_n{}^{2-}$, polysulfide ion. **Polysulfide ions,** which can be prepared by boiling a solution of a soluble sulfide with free sulfur, have S atoms bonded together into chains.

$$(:\ddot{S}\!-\!\ddot{S}:)^{2-} = \text{disulfide ion} \qquad (:\ddot{S}\!-\!\ddot{S}\!-\!\ddot{S}:)^{2-} = \text{trisulfide ion}$$

$S_n{}^{2-}$ = polysulfide ion (where n is an integer from 2 to about 9)

Polysulfide solutions contain all these ions in dynamic equilibrium. Tin(II) sulfide is one of the sulfides that is soluble in a polysulfide solution.

$$SnS(s) + S_2{}^{2-}(aq) \longrightarrow SnS_3{}^{2-}(aq) \tag{21.39}$$

In reaction (21.39) oxidation of tin(II) to tin(IV) occurs along with dissolution of the sulfide; $SnS_3{}^{2-}$, called thiostannate ion, is the sulfur analog of $SnO_3{}^{2-}$, stannate ion. Upon acidification of a solution containing $SnS_3{}^{2-}$, insoluble SnS_2 precipitates.

$$SnS_3{}^{2-}(aq) + 2\ H^+(aq) \longrightarrow SnS_2(s) + H_2S(g) \tag{21.40}$$

Some of the points discussed in the preceding few paragraphs are further illustrated through Table 21-5.

$H_2S(g)$ for the precipitation of metal sulfides can be obtained in several ways. A traditional method involves the action of HCl(aq) on FeS(s).

$$FeS(s) + 2\ HCl(aq) \longrightarrow FeCl_2(aq) + H_2S(g) \tag{21.41}$$

The generally preferred laboratory method is the hydrolysis of thioacetamide. This hydrolysis is quite slow at room temperature, but it proceeds at an appreciable rate at higher temperatures.

$$\underset{\text{thioacetamide}}{CH_3\overset{\overset{\textstyle S}{\|}}{C}NH_2} + H_2O \longrightarrow \underset{\text{acetamide}}{CH_3\overset{\overset{\textstyle O}{\|}}{C}NH_2} + H_2S(g) \tag{21.42}$$

Other Sulfur Compounds. The variety and complexity of sulfur compounds is impressive, resulting largely from the possibility of octet expansion for a central S atom mentioned in Table 21-4 and elsewhere in the text. Consider, for example, combinations of

TABLE 21-5
Solubilities of some metal sulfides

H_2O	Soluble in 0.3 M HCl (K_{sp})	Soluble in 3 M HNO_3 (K_{sp})	Soluble in aqua regia (K_{sp})	Soluble in KOH(aq) or $(NH_4)_2S_n$(aq) (K_{sp})
K_2S	MnS (2.5×10^{-13})	CdS (8.0×10^{-27})	HgS (1.6×10^{-52})	SnS (1.0×10^{-25})
Na_2S	FeS (6.3×10^{-18})	PbS (8×10^{-28})		As_2S_3
CaS	CoS (4.0×10^{-21})	CuS (6.3×10^{-36})		Sb_2S_3
	ZnS (1.0×10^{-21})			

sulfur and fluorine. There are two different compounds with the formula S_2F_2, as well as the compounds SF_4, SF_6, and S_2F_{10}. Substitution of other atoms for some of the F atoms accounts for compounds such as SF_5Cl, SF_5Br, and $(SF_5)_2O$.

Compounds with the characteristic group $OS{\displaystyle \overset{\diagup}{\diagdown}}$ can be thought of as derived from sulfurous acid $[OS(OH)_2]$ by substituting other atoms for the —OH groups. These are known as **thionyl** compounds, e.g., $OSCl_2$ = thionyl chloride. Compounds with the group $O_2S{\displaystyle \overset{\diagup}{\diagdown}}$, in turn, are derived from sulfuric acid $[O_2S(OH)_2]$. These are known as **sulfuryl** compounds, e.g., O_2SCl_2 = sulfuryl chloride.

We would not expect to be able to oxidize $SO_4{}^{2-}$ in an electrolysis reaction because other oxidation half-reactions (e.g., $2\,H_2O \rightarrow O_2(g) + 4\,H^+ + 4\,e^-$) occur more readily. However, electrolysis can be carried out in such a way as to create a high overvoltage for $O_2(g)$ formation. In the electrolysis of H_2SO_4 with Pt electrodes, oxidation of $HSO_4{}^-$ at the anode occurs by the half-reaction

$$2\,HSO_4{}^- \longrightarrow H_2S_2O_8 + 2\,e^- \tag{21.43}$$

(H_2 is produced in the accompanying reduction half-reaction.)

The product of reaction (21.43) is called **peroxodisulfuric acid** and has the structure

$$\text{HO}\!-\!\overset{\displaystyle O}{\underset{\displaystyle O}{\overset{\|}{\underset{\|}{S}}}}\!-\!O\!-\!O\!-\!\overset{\displaystyle O}{\underset{\displaystyle O}{\overset{\|}{\underset{\|}{S}}}}\!-\!\text{OH} \tag{21.44}$$

The presence of the —O—O— bond is characteristic of **peroxo** compounds. Salts of $H_2S_2O_8$ contain the ion $S_2O_8{}^{2-}$ and are called **peroxodisulfates** (often referred to more simply as persulfates).

The ion $S_2O_8{}^{2-}$ readily hydrolyzes in aqueous solution, yielding H_2O_2 and $HSO_4{}^-$.

$$S_2O_8{}^{2-} + 2\,H_2O \longrightarrow H_2O_2 + 2\,HSO_4{}^- \tag{21.45}$$

In fact, the hydrolysis of peroxodisulfates is the chief method used in the commercial production of H_2O_2.

Peroxodisulfate ion is one of the most powerful oxidizing agents.

$$S_2O_8{}^{2-} + 2\,e^- \longrightarrow 2\,SO_4{}^{2-} \qquad E° = +2.01\ \text{V} \tag{21.46}$$

The formal oxidation state of S in $S_2O_8{}^{2-}$ is $+7$, although a more accurate representation of the situation is to consider that the six terminal O atoms are in the oxidation state -2 and the two atoms in the —O—O— linkage, -1. This leaves the oxidation state of S as $+6$ and indicates that the reduction involves breaking the —O—O— bond and converting these two O atoms from the oxidation state -1 to -2.

Selenium, Tellurium, and Polonium.   These elements are similar to sulfur in many respects, yet they also have properties that illustrate the progressive changes that occur in moving down a group of the periodic table. Polonium has the physical appearance of a metal, and it has other metallic characteristics as well, such as an ability to conduct electric current and to form cations. These observations are in keeping with the position of Po adjacent to the stair-step diagonal line in the periodic table.

Selenium and tellurium form unpleasant smelling gaseous hydrides, H_2Se and H_2Te, that are similar to but less stable than, H_2S. For example, the heats of formation of H_2Se and H_2Te are positive, whereas that of H_2S is negative. H_2Se and H_2Te ionize in aqueous solution somewhat more readily that does H_2S. They produce HSe^-, Se^{2-}, HTe^-, and Te^{2-} ions, and they precipitate metal ions as selenides and tellurides. Se and Te also form acids in which the nonmetal is in a higher oxidation state—H_2SeO_3, H_2TeO_3, H_2SeO_4, and H_6TeO_6. Selenic acid, H_2SeO_4, resembles H_2SO_4 very closely, e.g., the ionization constants of the two acids are almost identical.

Whereas sulfur is a nonconductor, the allotropic form of selenium known as grey selenium is a semiconductor; and because its electrical conductivity increases in the presence of light, it finds use in photoelectric devices and in photocopying machines. Grey selenium is also used in the manufacture of rectifiers (used to convert alternating to direct electric current). Se and Te find some use in the preparation of alloys, and their compounds are used as additives to control the color of glass.

Se and Te are found as selenides and tellurides in sulfide ores, and the principal source of these elements is the anode mud obtained in the electrolytic refining of copper (see again, page 623). Polonium is a rare, radioactive element and interest in this element is centered on properties associated with its radioactivity.

21-3 The Group VA Elements

Perhaps the most interesting feature of the group VA elements is that, even more so than with the group VIA elements just considered, properties of both nonmetals and metals are displayed within the group. (The stepwise diagonal line in the periodic table passes through group VA.) The electron configurations of the elements provide only a limited clue to their metallic-nonmetallic behavior. Their outer shell electron configurations are ns^2np^3. There are a number of ways in which this electron configuration may be altered when a group VA atom enters into compound formation. Several possibilities are worthy of special mention.

One possibility is the gain or, more likely, sharing of three electrons in the outer shell to produce the noble gas configuration ns^2np^6 and an oxidation state of -3. This is especially so for the smaller atoms N and P. For the larger atoms—As, Sb, and Bi—the p^3 set of electrons may be lost. This leads to an electron configuration involving a next-to-outermost shell of 18 and an outer shell of 2. This so-called "18 + 2" configuration is adopted by a number of ionic species derived from metals. In some cases, all five outer shell electrons may be involved in compound formation, leading to the oxidation state of $+5$.

Assessment of Metallic–Nonmetallic Character in Group VA. To aid in this discussion a number of properties have been listed in Table 21-6. For group VA the usual decrease of ionization energy with increasing atomic number is noted. This establishes the order of metallic character within the group. Nitrogen is least metallic and bismuth is most metallic. Of course, all these ionization energies are high compared to the IA and IIA elements. The electronegativities indicate a high degree of nonmetallic character for nitrogen and less so for the remaining members of the group. None of the elements in group VA is highly metallic, however.

TABLE 21-6
Selected properties of group VA elements

Element	Atomic radius, pm	Electro-negativity	First ionization energy, kJ/mol	Common physical form(s)	Density of solid, g/cm^3	Comparative electrical conductivity[a]
N	150	3.04	1402	gas	1.03 ($-252°$C)	—
P	190	2.19	1012	waxlike white solid	1.82	
				red-colored solid	2.36	10^{-15}
As	200	2.18	947	yellow solid	2.03	
				gray solid with metallic luster	5.73	4.6
Sb	220	2.05	834	yellow solid	5.3	
				silvery white metallic solid	6.69	4.2
Bi	—	1.9	703	pinkish white metallic solid	9.75	1.4

[a] These values are relative to an assigned value of 100 for silver.

TABLE 21-7
Some thermodynamic properties of group VA hydrides

Hydride	Free energy of formation $\Delta \bar{G}_f^\circ$ (298 K), kJ/mol	Standard electrode potential, V
ammonia, NH_3	-16.6	$N_2 \xrightarrow{+0.27} NH_3$
phosphine, PH_3	$+8.8$	$P_4 \xrightarrow{-0.03} PH_3$
arsine, AsH_3	$+157.7$	$As \xrightarrow{-0.54} AsH_3$
stibine, SbH_3	$+147.7$	$Sb \xrightarrow{-0.51} SbH_3$
bismuthine, BiH_3	$+230(?)$	$Bi \xrightarrow{-0.80(?)} BiH_3$

Three of the elements—phosphorus, arsenic, and antimony—exhibit allotropy. For phosphorus the stable form at room temperature is red phosphorus. Its physical properties are those of a nonmetal. For example, its triple point is at 590°C and 43 atm pressure; red phosphorus sublimes without melting. For arsenic and antimony the stable allotropic forms are the "metallic" ones. They have high densities, fair thermal conductivities, and limited abilities to conduct electricity. Bismuth has no nonmetallic allotropic forms.

The increase in metallic behavior from top to bottom in group VA is reflected by the data listed in Table 21-7. Covalent bond formation with hydrogen is expected for nonmetallic elements. The negative value of the free energy of formation of $NH_3(g)$ suggests it is a stable molecule that forms spontaneously from its elements in their standard states. The values for the other hydrides are positive and increase in magnitude with increasing atomic number, suggesting decreasing stabilities. In fact, BiH_3 is so unstable that its properties have not been measured with any accuracy.

Another indication of the order of stability of the covalent hydrides is provided by electrode potentials. A decrease in electrode potential signifies a decreasing tendency toward reduction of the free element to the hydride.

The group VA elements form a number of different oxides. Table 21-8 describes the solubilities of the oxides X_2O_3 or X_4O_6 in acidic and basic solutions. The oxides of nitrogen, phosphorus, and arsenic are acidic; antimony oxide is amphoteric; and the oxide of bismuth has only basic properties. The acid–base behavior of the oxides establishes the gradation of properties within group VA, from nonmetallic to metallic, as well as any other criterion.

Electrode Potential Diagram for Nitrogen.   Figure 21-11 lists some important nitrogen-containing species through an electrode potential diagram. The variability possible in the oxidation state of N suggests the existence of a large number of nitrogen compounds.

TABLE 21-8
Acid–base behavior of some group VA oxides

Oxide	Nature of oxide	Principal product(s) obtained when dissolved in	
		Acidic soln.	Basic soln.
N_2O_3	acidic	HNO_2	NO_2^-
P_4O_6	acidic	H_3PO_3	$H_2PO_3^-$, HPO_3^{2-}
As_4O_6	acidic	$H_3AsO_3(HAsO_2)$	AsO_3^{3-} (AsO_2^-)
Sb_2O_3	amphoteric	SbO^+	SbO_2^-
Bi_2O_3	basic	BiO^+ and Bi^{3+}	insoluble

FIGURE 21-11
Electrode potential diagram for nitrogen.

Acidic solution ($[H^+] = 1\,M$):

$$NO_3^- \xrightarrow{+0.81\,V} NO_2 \xrightarrow{+1.07\,V} HNO_2 \xrightarrow{+0.99\,V} NO \xrightarrow{+1.59\,V} N_2O \xrightarrow{+1.77\,V} N_2 \xrightarrow{-1.87\,V} NH_3OH^+ \xrightarrow{+1.46\,V} N_2H_5^+ \xrightarrow{+1.24\,V} NH_4^+$$

Basic solution ($[OH^-] = 1\,M$):

$$NO_3^- \xrightarrow{-0.85\,V} NO_2 \xrightarrow{+0.88\,V} NO_2^- \xrightarrow{-0.46\,V} NO \xrightarrow{+0.76\,V} N_2O \xrightarrow{+0.94\,V} N_2 \xrightarrow{-3.04\,V} NH_2OH \xrightarrow{+0.74\,V} N_2H_4 \xrightarrow{+0.10\,V} NH_3$$

Hydrides of Nitrogen. There are three nitrogen hydrides, NH_3, N_2H_4, and HN_3. A fourth compound, NH_2OH, though containing oxygen, is closely related to NH_3.

Ammonia, NH_3, is produced by the Haber–Bosch process (described in Chapter 15) and is the ultimate source from which other nitrogen compounds are synthesized. NH_3 is the commonest weak base.

$$NH_3(aq) + H_2O \rightleftharpoons NH_4^+(aq) + OH^-(aq) \qquad K_b = 1.74 \times 10^{-5} \qquad (21.47)$$

Neutralization of NH_3 by acids is the source of several ammonium compounds. The properties of the NH_4^+ ion are similar to those of the alkali metal ions, which means that all common ammonium salts are water soluble. Because N exists in its lowest possible oxidation state (-3) in NH_3, in oxidation–reduction reactions NH_3 is always a reducing agent (it is oxidized).

When in the presence of a very strong base, such as H^- or O^{2-}, NH_3 may act as an *acid,* i.e., a proton donor.

$$NaH + NH_3(l) \longrightarrow NaNH_2 + H_2(g) \qquad (21.48)$$

The NH_2^- ion is the **amide ion.** Sodium amide (sodamide) is used in a variety of reactions designed to synthesize organic molecules.

If an H atom in NH_3 is replaced by the group $-NH_2$, the resulting molecule is H_2N-NH_2 or N_2H_4, **hydrazine.** Replacement of an H atom in NH_3 by $-OH$ produces NH_2OH, **hydroxylamine.** Both these compounds are weak bases. Because of its two N atoms, N_2H_4 can accept two protons in a stepwise fashion. However, the ion $N_2H_6^{2+}$ can be obtained in appreciable concentrations only in strongly acidic solutions.

$$NH_2OH(aq) + H_2O \rightleftharpoons NH_3OH^+ + OH^- \qquad K_b = 9.1 \times 10^{-9}$$

$$N_2H_4(aq) + H_2O \rightleftharpoons N_2H_5^+ + OH^- \qquad K_{b_1} = 8.5 \times 10^{-7}$$

$$N_2H_5^+(aq) + H_2O \rightleftharpoons N_2H_6^{2+} + OH^- \qquad K_{b_2} = 8.9 \times 10^{-16}$$

Hydrazine and hydroxylamine form salts analogous to ammonium salts, that is, salts such as $[NH_3OH]Cl$, $N_2H_5NO_3$, and $N_2H_6SO_4$. These salts all hydrolyze in water to yield acidic solutions.

Hydrazine and some of its derivatives burn in air with the evolution of much heat; they are used as rocket fuels.

Reaction (21.49) has also been used as the basis of a fuel cell (recall Figure 20-13).

$$N_2H_4(l) + O_2(g) \longrightarrow N_2(g) + 2\,H_2O(l) \qquad (21.49)$$

Both N_2H_4 and NH_2OH can act either as oxidizing or reducing agents (usually the latter) depending on the pH and the substances with which they react (see Figure 21-11). For example, NH_2OH reduces Fe(III) to Fe(II) in acidic solutions (reaction 21.50), whereas it oxidizes Fe(II) to Fe(III) in basic solutions (reaction 21.51).

$$4\ Fe^{3+} + 2\ NH_3OH^+ \longrightarrow 4\ Fe^{2+} + 6\ H^+ + H_2O + N_2O(g)\quad E°_{cell} = +0.72\ V \quad (21.50)$$

$$2\ Fe(OH)_2(s) + NH_2OH + H_2O \longrightarrow 2\ Fe(OH)_3(s) + NH_3 \quad E°_{cell} = +0.98\ V \quad (21.51)$$

The oxidation of hydrazine in acidic solution by nitrite ion produces **hydrogen azide, HN_3**.

$$N_2H_5^+ + NO_2^- \longrightarrow HN_3 + 2\ H_2O \tag{21.52}$$

The structure of the HN_3 molecule is

Pure HN_3 is a colorless liquid that boils at 37°C. It is very unstable and will detonate when subjected to shock. In aqueous solution HN_3 is a weak acid, called **hydrazoic acid;** its salts are called **azides.** Azides resemble chlorides in some properties (e.g., AgN_3 is insoluble in water), but they are extremely unstable. Some azides, e.g., lead azide, are used to make detonators.

Oxides, Oxoacids, and Oxoanions of Nitrogen.   Listed in Table 21-9 are six oxides of nitrogen. The enthalpies (heats) of formation of these oxides are mostly positive. This is because energy released in the formation of nitrogen-to-oxygen bonds is not enough to compensate for the large quantity of energy required to break the very strong $N\equiv N$ bond in $N_2(g)$.

Example 21-5   The bond energies of N_2, O_2, and NO are 946, 499, and 632 kJ/mol, respectively. Use these data to estimate the enthalpy (heat) of formation of NO(g).

Solution.   The formation of one mole of NO(g) can be represented as

$$\tfrac{1}{2}\ N_2(g) + \tfrac{1}{2}\ O_2(g) \longrightarrow NO(g) \qquad \Delta\overline{H}°_f = ?$$

In this reaction $\tfrac{1}{2}$ mol of bonds in N_2 and $\tfrac{1}{2}$ mol of bonds in O_2 must be broken. One mole of bonds in NO is formed.

$$\Delta\overline{H}°_f = \Delta\overline{H}°_{\text{bond breakage}} + \Delta\overline{H}°_{\text{bond formation}}$$
$$= [\tfrac{1}{2}\ (946) + \tfrac{1}{2}\ (499)]\ kJ/mol - 632\ kJ/mol = +91\ kJ/mol$$

SIMILAR EXAMPLES: Review Problem 10; Exercise 11.

TABLE 21-9
Oxides of nitrogen

Oxidation state	Oxide	At 298 K and 1 atm		
		Physical state	$\Delta\overline{H}°_f$, kJ/mol	$\Delta\overline{G}°_f$, kJ/mol
+1	dinitrogen oxide, N_2O (nitrous oxide)	colorless gas	+81.55	+103.6
+2	nitrogen oxide, NO (nitric oxide)	colorless gas	+90.37	+86.69
+3	dinitrogen trioxide, N_2O_3	—	—	—
+4	nitrogen dioxide, NO_2	brown gas	+33.85	+51.84
+4	dinitrogen tetroxide, N_2O_4	colorless gas	+9.67	+98.28
+5	dinitrogen pentoxide, N_2O_5	colorless solid	−41.8	+113.8

Free energies of formation of the oxides are also positive, suggesting that some of these oxides should decompose rather easily, such as

$$2 N_2O(g) \longrightarrow 2 N_2(g) + O_2(g) \qquad \Delta \overline{G}° \text{ (at 298 K)} = -207 \text{ kJ/mol} \qquad (21.53)$$

Actually, N_2O is quite stable at room temperature. This is because the activation energy for the decomposition reaction is very high—about 250 kJ/mol. At elevated temperatures (about 600°C) the rate of decomposition becomes appreciable. Reaction (21.53) accounts for the ability of $N_2O(g)$ to support combustion. Once a high enough temperature has been reached, the material undergoing combustion combines with the $O_2(g)$ produced in reaction (21.53).

$$H_2(g) + N_2O(g) \xrightarrow{\Delta} H_2O(l) + N_2(g) \qquad (21.54)$$

$$Cu(s) + N_2O(g) \xrightarrow{\Delta} CuO(s) + N_2(g) \qquad (21.55)$$

Some common methods of preparing the oxides of nitrogen are outlined in Table 21-10. All nitrates decompose on heating, but only NH_4NO_3 produces $N_2O(g)$. Nitrates of active metals (e.g., $NaNO_3$) yield the corresponding nitrite and $O_2(g)$. Thermal decomposition of the nitrates of less active metals [e.g., $Pb(NO_3)_2$] yields the metal oxide, $NO_2(g)$, and $O_2(g)$. $NO_2(g)$ is also produced at low temperatures by the reaction of NO and O_2. N_2O_3 can be produced by cooling and condensing an equimolar mixture of NO and NO_2. N_2O_3 is stable only in the solid state and as a liquid below about 250 K. Above this temperature it decomposes to $NO(g)$ and $NO_2(g)$. N_2O_4 is a dimer of NO_2 and the two species are in equilibrium in the gaseous state up to about 150°C. Above this temperature dissociation of N_2O_4 to NO_2 is complete. Pure N_2O_4 can be obtained as a pale yellow liquid (bp 21°C) or colorless solid (m.p. −11°C). N_2O_5 can be prepared by dehydrating (removing water from) HNO_3, e.g., with P_4O_{10}. Gaseous N_2O_5 is unstable, but the solid is stable below 0°C. Decomposition of N_2O_5 produces N_2O_4 and O_2.

K_p for the formation of NO_2 from NO and O_2, which is 1.6×10^{12} at 25°C, falls to about 0.1 at 600°C.

The N_2O_4–NO_2 equilibrium is discussed in Examples 15-10 and 15-13.

N_2O_3 is the acid anhydride of nitrous acid

$$N_2O_3 + H_2O \longrightarrow 2 HNO_2 \qquad (21.56)$$

and N_2O_5, of nitric acid.

$$N_2O_5 + H_2O \longrightarrow 2 HNO_3 \qquad (21.57)$$

NO_2 produces both HNO_2 and HNO_3 when it reacts with water.

$$2 NO_2(g) + H_2O \longrightarrow H^+ + NO_3^- + HNO_2 \qquad (21.58)$$

Reaction (21.58) is an oxidation–reduction reaction for which, from Figure 21-11, we can calculate that $E°_{cell} = +0.26$ V.

TABLE 21-10
Preparation of oxides of nitrogen

Oxide	A method of preparation
N_2O	$NH_4NO_3(s) \xrightarrow{\Delta} N_2O(g) + 2 H_2O(l)$
NO	$3 Cu(s) + 8 H^+ + 2 NO_3^- \longrightarrow 3 Cu^{2+} + 2 NO(g) + 4 H_2O$
N_2O_3	$NO(g) + NO_2(g) \rightleftharpoons N_2O_3(g) \qquad$ at 298 K: $\qquad K_p = 0.48$
NO_2	$2 Pb(NO_3)_2(s) \xrightarrow{\Delta} 2 PbO(s) + 4 NO_2(g) + O_2(g)$
	$2 NO(g) + O_2(g) \rightleftharpoons 2 NO_2(g) \qquad$ at 298 K: $\qquad \Delta \overline{H}° = -113$ kJ/mol; $\qquad K_p = 1.6 \times 10^{12}$
N_2O_4	$2 NO_2(g) \rightleftharpoons N_2O_4(g) \qquad$ at 298 K: $\qquad \Delta \overline{H}° = -58$ kJ/mol; $\qquad K_p = 8.84$
N_2O_5	$4 HNO_3(l) + P_4O_{10}(s) \longrightarrow 4 HPO_3 + 2 N_2O_5(s)$

TABLE 21-11
Nitric acid as an oxidizing agent

Nitric acid concentration	Reducing agent	Principal reduction product[a]
15 M	Cu; Ag	$NO_2(g)$
8 M	Cu; Ag	$NO(g)$
dilute[b]	Zn; Fe	$N_2O(g)$

[a] The oxidation products are Cu^{2+}, Ag^+, Zn^{2+}, and Fe^{3+}, respectively.
[b] If $[H_3O^+]$ is increased by adding $H_2SO_4(aq)$, NH_3OH^+ or NH_4^+ might be obtained. In *basic* solution, reduction of NO_3^- by Zn produces $NH_3(g)$.

$HNO_3(aq)$ is both a strong acid and a strong oxidizing agent capable of yielding a variety of reduction products depending on the reducing agent chosen and the concentration of the acid. Some of the many possibilities are listed in Table 21-11. $HNO_2(aq)$ is a weak acid that can function either as an oxidizing or reducing agent. Thus, it is reduced to $NO(g)$ by I^- and oxidized to NO_3^- by MnO_4^-.

Structures of some of the oxides of nitrogen were considered in Section 13-10. Figure 21-12 presents structures of nitrous and nitric acids and their anions. From these structures it is clear that although we write the formulas HNO_2 and HNO_3, more appropriate representations would be HONO and $HONO_2$.

Oxoacids of Phosphorus.   In Section 13-11 we discussed the two principal oxides of phosphorus, P_4O_6 and P_4O_{10}, and noted that they are the acid anhydrides of H_3PO_3 and

FIGURE 21-12
Structures of some oxoacids and oxoanions of nitrogen.

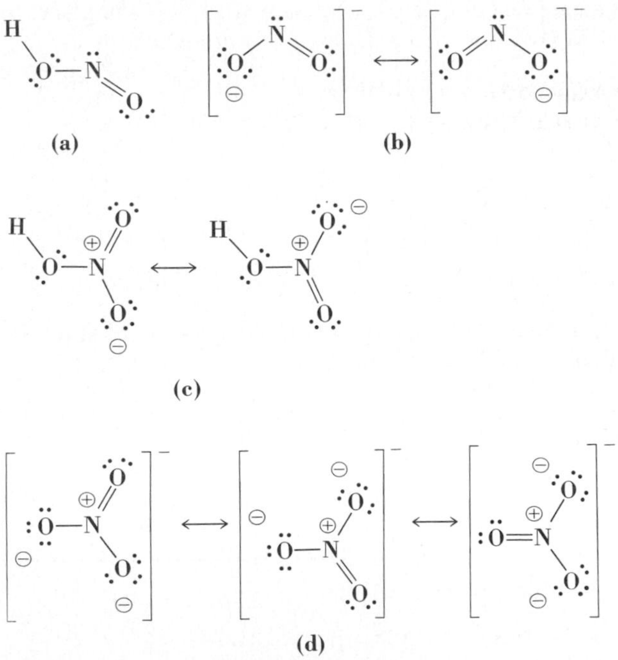

(a) Nitrous acid, HONO: The N and O atoms are in the same plane and the H atom is out of the plane.
(b) Nitrite ion, NO_2^-: The ion has an angular shape.
(c) Nitric acid, $HONO_2$: The N and O atoms are in the same plane and the H atom is out of the plane.
(d) Nitrate ion, NO_3^-: The ion has a trigonal planar shape.

TABLE 21-12
Oxoacids of phosphorus

Oxidation state	Name	Formula	Structure
+1	hypophosphorous acid	H_3PO_2	$\begin{array}{c} O \\ \parallel \\ HO{-}P{-}H \\ \mid \\ H \end{array}$
+3	orthophosphorous acid	H_3PO_3	$\begin{array}{c} O \\ \parallel \\ HO{-}P{-}OH \\ \mid \\ H \end{array}$
+3	pyrophosphorous acid	$H_4P_2O_5$	$\begin{array}{c} O\quad\quad O \\ \parallel\quad\quad\parallel \\ HO{-}P{-}O{-}P{-}OH \\ \mid\quad\quad\mid \\ H\quad\quad H \end{array}$
+4	hypophosphoric acid	$H_4P_2O_6$	$\begin{array}{c} O\quad\quad O \\ \parallel\quad\quad\parallel \\ HO{-}P{-}\!\!-\!\!P{-}OH \\ \mid\quad\quad\mid \\ OH\quad OH \end{array}$
+5	orthophosphoric acid	H_3PO_4	$\begin{array}{c} O \\ \parallel \\ HO{-}P{-}OH \\ \mid \\ OH \end{array}$
+5	pyrophosphoric acid	$H_4P_2O_7$	$\begin{array}{c} O\quad\quad O \\ \parallel\quad\quad\parallel \\ HO{-}P{-}O{-}P{-}OH \\ \mid\quad\quad\mid \\ OH\quad OH \end{array}$
+5	metaphosphoric acid	$(HPO_3)_n$	(see expression 21.62)

H_3PO_4, respectively. These and other oxoacids are listed in Table 21-12. A complication suggested in this table is explored more fully in Figure 21-13—the existence of ortho, meta, and pyro acids.

Orthophosphoric acid forms when P_4O_{10} reacts with an excess of H_2O.

$$P_4O_{10}(s) + 6\ H_2O \longrightarrow 4\ H_3PO_4(aq) \tag{21.59}$$

When orthophosphoric acid is heated to temperatures in excess of 215°C, **pyrophosphoric acid** results.

$$2\ H_3PO_4 \xrightarrow{\Delta} H_4P_2O_7 + H_2O \tag{21.60}$$

When either the ortho or pyro acid is heated to temperatures in excess of about 300°C, a glassy product is formed. This is probably a polymerized form of **metaphosphoric acid,** HPO_3, i.e., $(HPO_3)_n$, where $n = 2, 3, 4, 6$. To apply the scheme of Figure 21-13 to other acids, proceed in the following way.

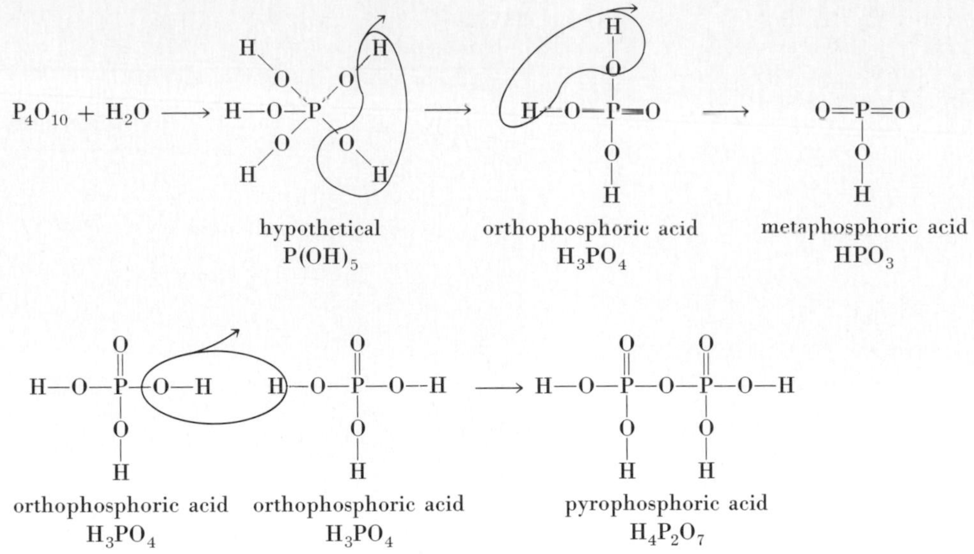

FIGURE 21-13
Some oxoacids of
phosphorus(V).

$$P_4O_{10} + H_2O \longrightarrow$$

hypothetical
$P(OH)_5$

orthophosphoric acid
H_3PO_4

metaphosphoric acid
HPO_3

orthophosphoric acid
H_3PO_4

orthophosphoric acid
H_3PO_4

pyrophosphoric acid
$H_4P_2O_7$

1. Start with an oxide in which the element E is in the oxidation state $+n$.
2. Consider that when the oxide reacts with water, the hydroxo compound $E(OH)_n$ is formed. (n is the oxidation number of E.)
3. If the hydroxo compound $E(OH)_n$ actually exists, it is the *ortho* acid (i.e., H_nEO_n). When this hydroxo compound does not exist, the loss of one water molecule yields the ortho acid.
4. The *meta* acid is formed from the ortho acid by the loss of one water molecule.

Example 21-6 Supply names and/or formulas for the following: (a) H_3AsO_4; (b) metaarsenous acid; (c) magnesium pyroarsenate.

Solution
(a) Arsenic is in the oxidation state $+5$. The hypothetical hydroxo compound would be $As(OH)_5$. Loss of one molecule of water leads to H_3AsO_4. The acid is called **orthoarsenic acid**; it is analogous to H_3PO_4.
(b) This is an "ous" acid and so must have arsenic in the oxidation state $+3$. The hydroxo compound corresponding to this oxidation state is $As(OH)_3$ or H_3AsO_3 and does actually exist. It is the ortho acid. The loss of one molecule of water leads to **$HAsO_2$**; this is the meta acid.
(c) The pyroacid is formed by the loss of one H_2O from two H_3AsO_4 molecules and has the formula $H_4As_2O_7$. The magnesium salt can be thought of as the combination of Mg^{2+} and $As_2O_7{}^{4-}$, i.e., **$Mg_2As_2O_7$**.

SIMILAR EXAMPLE: Review Problem 11.

Ionization and Neutralization of Oxoacids of Phosphorous. Orthophosphoric acid is a weak polyprotic acid; the molecule H_3PO_4 has three ionizable H atoms. Ionization equilibria for H_3PO_4 were discussed in Section 17-6, and its titration curve was presented in Section 18-4. Because only H atoms attached to an O atom are ionizable, not H atoms linked directly to the central atom, we see from Table 21-12 that hypophosphorous acid is monoprotic and phosphorous acid, diprotic.

Polyphosphoric Acids and Polyphosphates. Pyrophosphoric acid, $H_4P_2O_7$, as indicated in Figure 21-13, results from the elimination of one molecule of H_2O from two

Pure H_3PO_4 and $H_4P_2O_7$ are colorless crystalline solids. H_3PO_4 melts at 29°C; $H_4P_2O_7$, at 61°C.

molecules of H_3PO_4; it has 34.8% P, by mass. If a H_2O molecule is split out from between a molecule of H_3PO_4 and one of $H_4P_2O_7$, the resulting molecule is $H_5P_3O_{10}$, with 36.0% P. This process of eliminating H_2O can be continued until the % P reaches about 39%, and the species obtained are called **polyphosphoric acids.** Of the phosphoric acids only H_3PO_4 and $H_4P_2O_7$ can be obtained as pure substances. Commercial syrupy phosphoric acid is a mixture of H_3PO_4, $H_4P_2O_7$, and the polyphosphoric acids illustrated below.

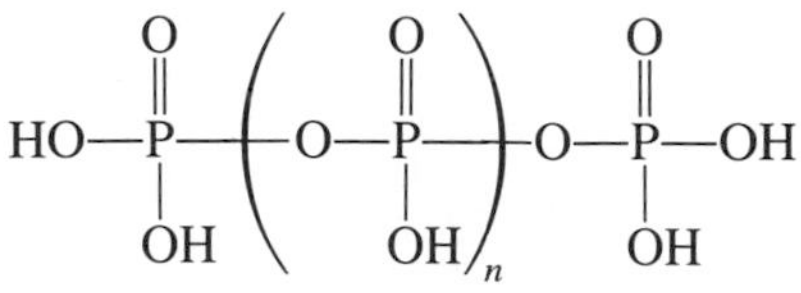

$$\text{structure} \qquad or \qquad H_2PO_3(HPO_3)_nPO_4H_2 \qquad (21.61)$$

repeating unit = HPO_3; if
$n = 1$, $H_5P_3O_{10}$ (tripolyphosphoric acid);
$n = 2$, $H_6P_4O_{13}$ (tetrapolyphosphoric acid); etc.

Salts of these acids are called **polyphosphates;** e.g., $Na_5P_3O_{10}$ is sodium tripolyphosphate.

Removal of an additional H_2O from a molecule of a polyphosphoric acid results in a chain or ringlike structure with the formula, $(HPO_3)_n$. These acids are called **metaphosphoric acids** and are named according to the number of P atoms they contain. A ring structure is illustrated below.

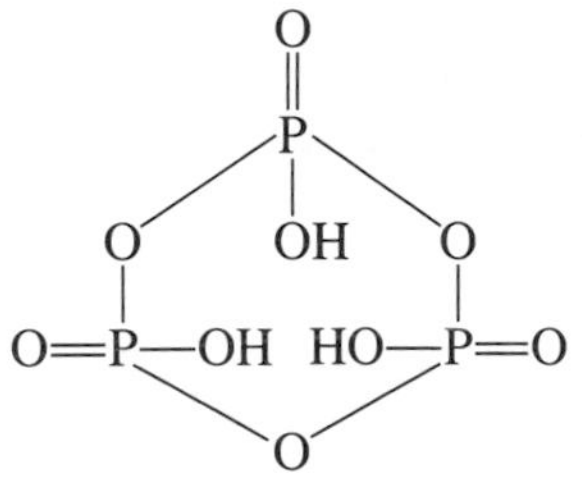

$$\text{structure}$$

metaphosphoric acid = $(HPO_3)_n$
$n = 3$, trimetaphosphoric acid;
$n = 4$, tetrametaphosphoric acid; etc.

$$(21.62)$$

FIGURE 21-14
Sequestering action of polyphosphates on metal ions.

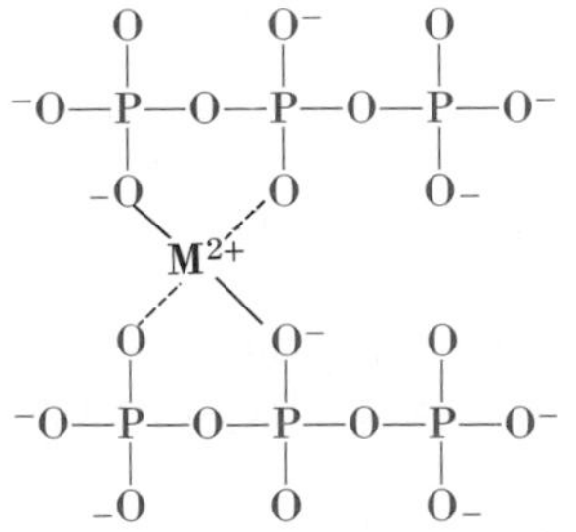

Salts of these acids are called **metaphosphates;** e.g., $Na_3P_3O_9$ is sodium trimetaphosphate.

Although it is easiest to think about polyphosphoric acids in terms of eliminating H_2O from H_3PO_4 and to think of polyphosphates as salts formed in the neutralization of these poly acids, commercial methods of preparing polyphosphates involve heating simple phosphates. For example, $Na_5P_3O_{10}$ is prepared by heating a powdered mixture containing Na_2HPO_4 and NaH_2PO_4 in a 2:1 mole ratio.

$$2\ Na_2HPO_4 + NaH_2PO_4 \xrightarrow{\Delta} Na_5P_3O_{10} + 2\ H_2O \qquad (21.63)$$

Sodium metaphosphate is obtained as a glassy solid when NaH_2PO_4 is heated to temperatures above 620°C and the melt rapidly cooled.

$$n\ NaH_2PO_4 \xrightarrow{\Delta} (NaPO_3)_n + n\ H_2O \qquad (21.64)$$

The commercial product "Calgon" (meaning calcium gone) is a sodium metaphosphate with a formula weight of 1500 to 2000, corresponding to the formula $(NaPO_3)_n$ with $n = 15{-}20$.

Polyphosphoric acids containing 36 to 37% P are used to produce high-strength fertilizers and as catalysts in petroleum refining. Sodium tripolyphosphate is used in cement manufacturing and in oil well drilling, but its most extensive use has been as a builder in detergents. Detergents lose their effectiveness when used in water containing dipositive metal ions such as Ca^{2+}. The function of a builder is to complex or **sequester** these ions, reducing the concentration of the free ions to the point where they do not interfere with the detergent action. The sequestering action of $P_3O_{10}^{5-}$ ions is illustrated in Figure 21-14. Metaphosphates exhibit a similar action on metal ions and they are used in water softening (see Section 22-2).

Phosphate Rock.   The principal source of phosphorus compounds is phosphate rock, a complex material containing fluorapatite [$3Ca_3(PO_4)_2 \cdot CaF_2$]. $Ca_3(PO_4)_2$ can be extracted from fluorapatite and used in the preparation of P_4 by the method outlined in Section 13-11. If P_4 (usually as a liquid) is burned in air, the product is P_4O_{10}, and P_4O_{10} is the acid anhydride of H_3PO_4. An impure form of phosphoric acid can be prepared by the direct action of H_2SO_4 on phosphate rock.

[$3Ca_3(PO_4)_2 \cdot CaF_2$] + 10 H_2SO_4 + 20 H_2O $\longrightarrow$

   fluorapatite

$$6 \ H_3PO_4 + 10 \ [CaSO_4 \cdot 2H_2O] + 2 \ HF \qquad (21.65)$$

   gypsum

The principal use of phosphorus compounds is in fertilizers. A mixture of $CaSO_4$ and the more soluble calcium dihydrogen phosphate, called **normal superphosphate,** has a phosphorus content equivalent to 7 to 9% P, by mass. It is produced by treating phosphate rock with H_2SO_4 in a reaction similar to (21.65) but employing different proportions of the rock and acid.

[$3Ca_3(PO_4)_2 \cdot CaF_2$] + 7 H_2SO_4 + 3 H_2O $\longrightarrow$

$$\underbrace{3 \ [Ca(H_2PO_4)_2 \cdot H_2O] + 7 \ CaSO_4} + 2 \ HF \qquad (21.66)$$

   normal superphosphate

If phosphate rock is treated with H_3PO_4 (derived from reaction 21.65) instead of H_2SO_4, the product is known as **triple superphosphate.** This process eliminates $CaSO_4$, and the product has a much higher phosphorus content than normal superphosphate, about 20 to 21% P.

[$3Ca_3(PO_4)_2 \cdot CaF_2$] + 14 H_3PO_4 + 10 H_2O $\longrightarrow$ $\underbrace{10 \ [Ca(H_2PO_4)_2 \cdot H_2O]} + 2 \ HF$

$$(21.67)$$

   triple superphosphate

Environmental Problems of Phosphorus.   HF is a byproduct of superphosphate production (reactions 21.66 and 21.67). In some instances it is recovered, but in the past it was mostly released into waterways. Now, because of strict environmental controls it has become necessary to neutralize the HF, usually with lime. Large settling ponds are required for this reaction. Because two thirds of phosphate rock is waste, deposits of waste rock are accumulated in fertilizer manufacture. The handling of this waste adds to the cost and complexity of the operation.

Changes that occur in freshwater bodies as a result of enrichment by nutrients are referred to by the term **eutrophication.** This is a natural process that occurs over geologic time periods, but it is greatly accelerated by human activities. A body of water that receives large quantities of nutrients such as nitrates and phosphates experiences excessive growth of algae. This is followed by oxygen depletion, fish kills, growth of anerobic bacteria, and other undesirable effects. Natural sources of these nutrients include animal wastes, decomposition of dead organic matter, and natural nitrogen fixation. Human sources include industrial wastes, municipal sewage plant effluents, and fertilizer runoff; but one of the main contributors has been the builders from detergents (e.g., $Na_5P_3O_{10}$). Currently many municipalities have adopted ordinances limiting the phosphate content of detergents. This action appears to be improving environmental conditions, although the case against phosphorus is not conclusive. It may be that in some cases, even in the presence of high concentrations of phosphates, the eutrophication process is limited by some other nutrient or the availability of CO_2. (Photosynthetic activity by algae is dependent on the quantity of CO_2 produced by microorganisms that metabolize organic compounds.)

The method of Section 13-11 is called the thermal process. The method outlined here is the wet process. Over 90% of H_3PO_4 is produced by the wet process, both in the United States and worldwide.

Eutrophication of a lake is pictured in Color Section E.

Another means of reducing phosphate discharges into the environment is through their removal in sewage treatment plants. In the processing of sewage, polyphosphates are decomposed to orthophosphates, and organic phosphorus compounds are degraded to orthophosphates by bacterial action. The orthophosphates may then be precipitated, either as iron(III) phosphates, aluminum(III) phosphates, or as $Ca_3(PO_4)_2$ or $Ca_5OH(PO_4)_3$. The precipitating agents are generally aluminum(III) sulfate, iron(III) chloride, or calcium hydroxide. In a fully equipped modern sewage treatment plant, up to 98% of the phosphates in sewage can be removed.

21-4 The Group IVA Elements

The differences between carbon and silicon, as outlined in Table 21-13, are perhaps the most striking to be found between the second and third period elements of a group in the periodic table. As suggested by the approximate bond energies, strong C—C and C—H bonds account for the central role of carbon-atom chains and rings in establishing the chemical behavior of carbon. A study of these chains and rings and their attached atoms (e.g., H, O, N, and S) is the focus of organic chemistry (Chapter 26) and biochemistry (Chapter 27). The relatively weaker Si—Si and Si—H bonds imply a less important "organic chemistry" of silicon, and the strength of the Si—O bond accounts for the predominance of the silicates among silicon compounds.

In this section we consider a few carbon compounds which, in addition to the oxides and carbonates discussed in Section 13-8, constitute the inorganic chemistry of carbon. To complement the inorganic chemistry of silicon, which was presented in Section 13-9, we will briefly consider some organosilicon compounds.

Carbon is strictly nonmetallic in its behavior. Silicon is also essentially nonmetallic though its semiconductor properties have been noted. The semiconductor behavior of the metalloid germanium has also been explored (Chapter 10). Tin and lead, although having certain properties resembling those of the lighter elements in group IVA, are essentially metallic elements, and they are studied in Chapter 22.

TABLE 21-13
Some comparisons of carbon and silicon

Carbon	Silicon
Two principal allotropes, graphite and diamond	One stable, diamond-type crystalline modification
Forms two stable *gaseous* oxides, CO and CO_2, and several less stable ones, such as C_3O_2	Forms only one *solid* oxide (SiO_2) that is stable at room temperature; a second oxide (SiO) is stable only in the temperature range 1180–2480°C
Insoluble in alkaline medium	Dissolves in alkaline medium and forms $H_2(g)$ and $SiO_4^{4-}(aq)$
Principal oxoanion is CO_3^{2-}, which has a planar shape	Principal oxoanion is SiO_4^{4-}, which has a tetrahedral shape
Strong tendency for catenation, with straight and branched chains and rings containing up to hundreds of C atoms[a]	Less tendency for catenation, with silicon atom chains limited to about six Si atoms[a]
Readily forms multiple bonds through use of the orbital sets $sp^2 + p$ and $sp + p^2$	Multiple bond formation much less common than with carbon and limited to $p_\pi - d_\pi$ type (recall Section 10-6)
Approximate single bond energies: C—C, 347 kJ/mol C—H, 414 C—O, 360 C—Cl, 326	Approximate single bond energies: Si—Si, 226 kJ/mol Si—H, 318 Si—O, 368 Si—Cl, 250–335

[a]Catenation refers to the joining together of like atoms into chains.

Some Additional Inorganic Carbon Compounds.   Carbon combines with metals to form carbides. With the transition metals these are interstitial carbides—carbon atoms take up positions in the holes or voids in the crystalline structure of the metal. With active metals the carbides are ionic, containing, for example, the ion $(:C\equiv C:)^{2-}$. **Calcium carbide** is formed by the high-temperature reaction of lime and coke.

$$CaO(s) + 3\ C(s) \xrightarrow{2000°C} CaC_2(s) + CO(g) \tag{21.68}$$

Calcium carbide is an important product because its reaction with water produces acetylene, a gas widely used in the synthesis of organic compounds.

$$\underset{\text{acetylene}}{CaC_2(s) + 2\ H_2O \longrightarrow Ca(OH)_2(s) + C_2H_2(g)} \tag{21.69}$$

Two other important inorganic compounds of carbon are **carbon disulfide,** CS_2, and **carbon tetrachloride,** CCl_4. The modern method of preparing CS_2 involves the reaction of methane and sulfur vapor in the presence of a catalyst.

$$CH_4(g) + 4\ S(g) \xrightarrow{\Delta} CS_2(l) + 2\ H_2S(g) \tag{21.70}$$

CS_2 is a highly flammable, volatile liquid useful as a solvent for sulfur, phosphorus, bromine, iodine, fats, and oils. Its uses as a solvent are decreasing, however, because of its poisonous nature. Other important uses are in the manufacture of rayon and cellophane. CCl_4 can be prepared by the direct chlorination of methane.

$$CH_4(g) + 4\ Cl_2(g) \longrightarrow CCl_4(l) + 4\ HCl(g) \tag{21.71}$$

Although CCl_4 has found extensive use as a solvent, drycleaning agent, and fire extinguisher, these uses are now declining. CCl_4 causes liver and kidney damage and is a known carcinogen. Another use of CCl_4 is in the production of chlorofluorocarbons (Section 13-14).

Pseudohalogens.   Certain groupings of atoms, several containing C atoms, have some of the characteristics of a halogen atom. These are called pseudohalogens and include

—CN (cyanide)   —OCN (cyanate)   —SCN (thiocyanate)   —N₃ (azide)

The **cyanide** ion, CN^-, is similar to the halide ions, X^-, in that it forms an insoluble silver salt, AgCN, and in that it forms the acid, HCN, though the acid is very weak. CN^- differs from the halide ions in being extremely poisonous. Compounds such as CNCl and CNBr, which result from the action of Cl_2 and Br_2 on HCN, are examples of pseudohalogen-halogen compounds.

Organosilicon Compounds.   Single bonds of the type Si—Si are not strong enough for very long chains to exist. Nevertheless, a series of **silanes** can be prepared, up to a limit of six silicon atoms per chain.

monosilane   disilane   trisilane   hexasilane

Monosilane results from the reaction of lithium aluminum hydride and silicon tetrachloride.

$$LiAlH_4 + SiCl_4 \longrightarrow LiCl + AlCl_3 + SiH_4(g) \tag{21.72}$$

The silanes are thermally unstable. Moderate heating of the higher silanes causes decomposition to the lower silanes, and above 500°C, to the elements. Like the hydrocarbons, the silanes are combustible. In fact, they ignite spontaneously in air.

$$SiH_4 + 2 O_2 \longrightarrow SiO_2 + 2 H_2O \tag{21.73}$$

Other atoms can be substituted for hydrogen atoms in the silanes rather easily. For example, the series of compounds SiH_3Cl, SiH_2Cl_2, $SiHCl_3$, and $SiCl_4$ results from the successive substitution of Cl for H atoms in SiH_4.

The reaction of $(CH_3)_2SiCl_2$ with water produces a compound, $(CH_3)_2Si(OH)_2$, dimethyl silanol, which undergoes a reaction (polymerization) by the successive elimination of water from among large numbers of the silanol molecules. The result is a material consisting of long-chain molecules, **silicones.**

$$(CH_3)_2SiCl_2 + 2 H_2O \longrightarrow (CH_3)_2Si(OH)_2 + 2 HCl$$

a silicone

Depending on the length of the chains and the degree of crosslinking between chains, silicones are obtained either as oils or as rubberlike materials. Silicone oils are not volatile; they may be heated without decomposition; they can be cooled to low temperatures without becoming viscous or solidifying. (Hydrocarbon oils become very viscous at low temperatures.) Silicone rubbers retain their elasticity at low temperatures; they are chemically resistant and thermally stable.

21-5 Boron

Of the group IIIA elements the only one that is almost exclusively nonmetallic in its physical and chemical properties is boron. A factor that governs much of its chemical behavior is the electron deficiency associated with many of its compounds, making them strong Lewis acids, e.g., BF_3. A class of boron compounds in which this electron deficiency leads to bonding of a type not heretofore considered in this text is the boron hydrides.

Boron Hydrides. The molecule BH_3 (borane or borine) is believed to exist as a reaction intermediate, but it cannot be isolated as a stable species. The simplest boron hydride that can be isolated is **diborane,** B_2H_6. Resolution of the structure of B_2H_6 has resulted in fundamental contributions to bonding theory, particularly to molecular orbital theory. The problem is this: In the molecule B_2H_6 there are 14 valence shell atomic orbitals but only 12 valence electrons. The molecule appears to be *electron deficient.*

The currently accepted structure is shown in Figure 21-15. The two B atoms and four of the H atoms lie in the same plane. The orbitals used by the B atoms to bond these particular H atoms are sp^2 (the H—B—H bond angles in the plane are 121.8°). Eight electrons are involved in these bonds. This leaves four electrons and six atomic orbitals to be accounted for: an sp^2 and a p orbital on each B atom and a $1s$ on each of the two remaining H atoms. The two B atoms and two remaining H atoms are joined together through two bonds variously called, "three-center," "bridge," or "banana" bonds. The six atomic orbitals just mentioned can be combined into six molecular orbitals, two of which are bonding orbitals. Placement of the four remaining electrons into these two orbitals completes the bonding picture. Thus, each bridge bond unites *three* atoms, but with only *two* electrons. If the concept of bridge bonds is extended to bonds of the type, B—B—B, structures of the higher boranes, such as that of B_5H_9 shown in Figure 21-16, can also be described.

An artificial heart made of silicone–urethane plastics. [Courtesy of R. Ward, Mercor, Incorporated.]

FIGURE 21-15
Structure of diborane, B_2H_6.

FIGURE 21-16
Structure of pentaborane, B_5H_9.

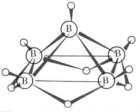

The boron atoms are joined through multicenter B—B—B bonds. Five of the H atoms are bonded directly to B atoms. The other four H atoms bridge pairs of B atoms.

A borate mining and processing installation in the Mojave Desert of California.
[© Steve Strickland, 1980.]

Other Boron Compounds. Boron compounds are widely distributed in the earth's crust, but concentrated ores are found only in a few locations—in Italy, U.S.S.R., Tibet, Turkey, and the desert regions of California. Typical of these ores is the hydrated borate **borax,** $Na_2B_4O_7 \cdot 10H_2O$. Figure 21-17 suggests how borax can be converted to a wide variety of boron compounds.

Although elemental boron can be prepared by the reduction of B_2O_3 with an active metal, the product is rather impure. Higher purity boron can be obtained by the reduction of boron halides (usually BBr_3) with $H_2(g)$. This higher purity boron can be brought to the ultrapure levels required for semiconductor applications by zone refining (recall Figure 12-24).

One of the key compounds used in the synthesis of other boron compounds is **boric acid,** $B(OH)_3$. As discussed in Section 17-8, the high ratio of ionic charge to ionic radius for B^{3+} causes $B(OH)_3$ to ionize as a weak acid. The ionization reaction involves formation of the ion $B(OH)_4^-$.

$$B(OH)_3(aq) + 2\,H_2O \rightleftharpoons H_3O^+ + B(OH)_4^- \qquad K_a = 5.6 \times 10^{-10} \qquad (21.74)$$

In solutions more concentrated than about 0.1 M, $B(OH)_3$ and $B(OH)_4^-$ combine to form a series of polyborate ions (similar to the formation of polyphosphate ions described in Section 21-3).

As expected of the salts of a weak acid, borate salts produce basic solutions by

FIGURE 21-17
Preparation of some
boron compounds.

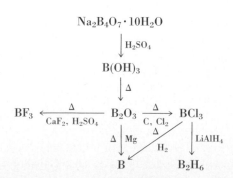

hydrolysis, and this accounts for their use in cleaning agents. However, of the several million tons of borates produced annually, worldwide, only a small proportion is used in this way. Boron compounds are also used in products as varied as adhesives, cement, disinfectants, fertilizers, fire retardants, glass, herbicides, metallurgical fluxes, and textile bleaches and dyes.

Example 21-7   Based on Figure 21-17 write chemical equations for the successive conversions of borax to **(a)** boric acid, **(b)** B_2O_3, and **(c)** impure boron metal.

Solution.   Figure 21-17 lists the key substances involved in each reaction. Our task is to identify other plausible reactants and/or products.

(a) Conversion of a salt to the corresponding acid is an acid–base reaction and does not involve changes in oxidation states. Balancing should be possible by inspection. The additional substances required in the equation are Na_2SO_4 and H_2O.

$$Na_2B_4O_7 \cdot 10\ H_2O + H_2SO_4 \longrightarrow 4\ B(OH)_3 + Na_2SO_4 + 5\ H_2O$$

(b) Conversion of a hydroxo compound to an oxide requires that H_2O be driven off.

$$2\ B(OH)_3 \xrightarrow{\Delta} B_2O_3 + 3\ H_2O$$

(c) Since the boron need not be of high purity, direct reduction of B_2O_3 with Mg is possible.

$$3\ Mg + B_2O_3 \xrightarrow{\Delta} 2\ B + 3\ MgO$$

SIMILAR EXAMPLES: Exercises 32, 36.

21-6 Group 0—The Noble Gases

Sources of the lighter noble gases and their physical properties and uses were discussed in Section 13-4. For a long time these were the only aspects of the noble gases that could be considered, since no chemical compounds of these elements were known. In fact, the inertness of the noble gases helped to provide a theoretical framework for the Lewis theory of bonding: Other atoms tend to form bonds so that they may acquire noble gas electron configurations (ns^2np^6). Moreover, the unquestioning attitude of most chemists toward the stability of noble gas electron configurations tended to retard rather than to stimulate the search for noble gas compounds.

In 1962, Bartlett and Lohmann discovered that O_2 and PtF_6 would join in a $1:1$ mole ratio to form the compound O_2PtF_6. Properties of this compound (e.g., its paramagnetism) suggested the bonding to be ionic: $(O_2)^+(PtF_6)^-$. The quantity of energy required to extract an electron from O_2 is 12.2 eV/molecule (or 1177 kJ/mol). Bartlett noted that the first ionization energy of Xe (1170 kJ/mol) is almost identical to that of O_2 and reasoned that the compound $XePtF_6$ might also exist. He was able to prepare a yellow crystalline solid with a composition corresponding to this formula.* Soon thereafter a considerable number of noble gas compounds were synthesized by chemists around the world. For example, XeF_4 was prepared by the reaction of Xe and F_2 in the mole ratio $1:5$ in a nickel vessel at 400°C and at about 6 atm. By varying the Xe/F_2 ratio and other reaction conditions, XeF_2 and XeF_6 can also be obtained. All are colorless, crystalline solids that sublime easily.

A fluoride of krypton, KrF_2, has been prepared, and compounds of radon also exist, though experiments with Rn are difficult to perform because of the intense radioactivity of all its isotopes. No compounds of He, Ne, or Ar are known. In general the conditions necessary for the formation of noble gas compounds are

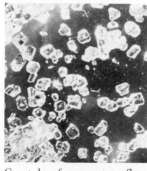

Crystals of xenon tetrafluoride, XeF_4. [Courtesy of Argonne National Laboratory.]

*It has since been established that this solid has the formula $Xe(PtF_6)_n$, where n is between 1 and 2.

- a readily ionizable (and therefore, heavy) noble gas atom and
- highly electronegative groups (e.g., F or O) to bond to the noble gas atom.

Compounds have been synthesized with Xe in the oxidation states of

+2	+4	+6	+8
e.g., XeF_2	XeF_4, $XeOF_2$	XeF_6, XeO_3	XeO_4, H_4XeO_6

Because it is difficult to oxidize Xe to these positive oxidation states, we should expect Xe compounds to be very strong oxidizing agents (i.e., to be easily reduced). For example, in aqueous acidic solution

$$XeF_2(aq) + 2 H^+ + 2 e^- \longrightarrow Xe + 2 HF \qquad E° = +2.64 \text{ V} \qquad (21.75)$$

The hydrolysis of XeF_2 in basic solution is an oxidation–reduction reaction in which OH^- is oxidized to $O_2(g)$.

$$2 XeF_2 + 4 OH^- \longrightarrow 2 Xe + 4 F^- + 2 H_2O + O_2(g) \qquad (21.76)$$

The fluorides XeF_2, XeF_4, and XeF_6 are stable if kept from contact with water. XeO_3 is an explosive white solid and XeO_4 is an explosive colorless gas.

Bonding in Noble Gas Compounds. The valence bond theory permits bonding schemes for Xe compounds that are consistent with their observed molecular shapes. These require sp^3d hybridization for XeF_2 and sp^3d^2, for XeF_4.

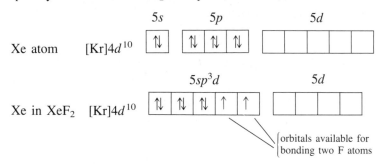

distribution of sp^3d orbitals: trigonal bipyramidal
geometric structure of XeF_2: linear

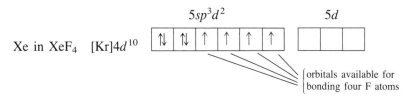

distribution of sp^3d^2 orbitals: octahedral
geometric structure of XeF_4: square planar

FIGURE 21-18
Molecular shapes of
(a) XeF_2 and (b) XeF_4.

(a)

(b)

Only the lone-pair electrons of Xe are shown.

Based on the high promotion energy for an electron from a $5p$ to a $5d$ orbital and on the observed bond strengths and bond distances in these fluorides, there is some doubt, however, whether d orbitals are involved in bond formation. A molecular orbital description of XeF_2 seems to provide better agreement with observed properties by proposing that a single electron pair bonds both F atoms to Xe (similar to the B—H—B bonds in B_2H_6 described in the preceding section).

What is clear is that VSEPR theory does do a satisfactory job of predicting molecular shapes for most Xe compounds and this is the approach that we will follow. The molecular shapes of XeF_2 and XeF_4 predicted by VSEPR theory are shown in Figure 21-18.

Example 21-8 Predict the molecular shapes of (a) XeF_2 and (b) XeF_4.

Solution. Here we can proceed by the method introduced in Example 9-12.

	(a) XeF_2	(b) XeF_4
total number of electron pairs:	$\dfrac{8 + (2 \times 7)}{2} = 11$	$\dfrac{8 + (4 \times 7)}{2} = 18$
no. bond pairs: [no. atoms − 1]	$3 - 1 = 2$	$5 - 1 = 4$
no. central pairs: [no. electron pairs − 3 × no. terminal atoms (excluding H)]	$11 - (3 \times 2) = 5$	$18 - (3 \times 4) = 6$
no. lone pairs: [no. central pairs − no. bond pairs]	$5 - 2 = 3$	$6 - 4 = 2$
distribution of electron pairs:	5: trigonal bipyramidal	6: octahedral
molecular shape:	linear	square planar

SIMILAR EXAMPLES: Review Problem 12; Exercise 37.

Summary

In this chapter principles developed throughout the first 20 chapters of the text are applied to various aspects of the descriptive inorganic chemistry of the nonmetals. For example, oxidation–reduction reactions are discussed with the aid of electrode potential diagrams, and electrode potential diagrams are given for Cl, O, S, and N.

Oxoacids and oxoanions are studied in terms of the methods required to prepare them, their acid–base properties, their strengths as oxidizing or reducing agents, and their structures. Some additional possibilities for oxoacids of sulfur are the substitution of S for O atoms (thiosulfates) and the presence of —O—O— bonds (peroxosulfates). With phosphorus an additional possibility is the elimination of H_2O from simpler acid molecules to produce polyphosphoric acids, polyphosphates, metaphosphoric acids, and metaphosphates.

The solubility characteristics of a number of compounds are noted in this chapter. Particular attention is given to the effect of acids, bases, and oxidizing agents on the solubilities of metal sulfides and the relationship of these solubilities to the qualitative analysis scheme for cations. Also noted at several points is how Le Châtelier's principle is used to control equilibrium processes.

The differences between the first and higher members of a group of the periodic table is encountered again in this chapter, as it was in Chapter 13. Examples cited are the failure of fluorine to form oxoacids and the numerous differences between O and S and between C and Si. Also, by considering several criteria, the progression of properties from nonmetallic to metallic within group VA is established. Bond energies and thermodynamic stabilities are applied to a study of nitrogen oxides, and additional insights into chemical bonding are provided through a study of the boron hydrides, interhalogen compounds, and noble gas compounds.

Throughout the chapter practical uses of the nonmetals and their compounds are described. Quite often it is found that a particular use of a substance is a consequence of some special property that it possesses.

Learning Objectives

As a result of studying Chapter 21, you should be able to

1. Use electrode (reduction) potential diagrams as a means of writing half equations and predicting oxidation–reduction reactions.

2. Outline the principal methods of preparing the halogen elements, the hydrogen halides, and the oxoacids and oxoanions of the halogens.

3. Describe the structures of the oxoanions of the halogens, sulfur, and nitrogen.

4. Predict the molecular geometry of interhalogen compounds, polyhalide ions, and noble gas compounds.

5. Cite the similarities and differences between the first and second members of a group of the periodic table (e.g., O and S in group VIA).

6. List several methods of preparing $O_2(g)$.

7. Describe a number of thio compounds by name, formula, and structure.

8. Predict the solubilities of metal sulfides in acidic, basic, and oxidizing solutions.

9. Describe the acid–base and oxidation–reduction properties of NH_3, N_2H_4, and NH_2OH.

10. Outline methods of preparing the oxides of nitrogen.

11. Describe the simple oxoacids of phosphorus and a number of polyphosphoric acids, metaphosphoric acids, polyphosphates, and metaphosphates, by name and formula.

12. Name ortho, meta, and pyro oxoacids according to a scheme based on the loss of H_2O molecules.

13. Indicate the unique features of chemical bonding in the boron hydrides.

14. Discuss the oxidizing and/or reducing power of the halogen elements and their oxoacids and oxoanions, oxygen, ozone, hydrogen peroxide, the oxoacids and oxoanions of sulfur and nitrogen, and the noble gas compounds.

15. Cite uses of some of the compounds discussed in the chapter, e.g., $NaOCl$, $NaIO_3$, $Na_2S_2O_3$, NaH_2PO_4, $Na_2B_4O_7 \cdot 10H_2O$.

Some New Terms

An **electrode (reduction) potential diagram** lists standard electrode (reduction) potentials for oxidation–reduction couples of an element and various of its ionic and compound forms.

Eutrophication is the deterioration of a freshwater body, caused by nutrients such as nitrates and phosphates, that results in growth of algae, oxygen depletion, and fish kills.

An **interhalogen compound** is a compound involving different halogen elements, e.g., ICl or BrF_3.

A **meta acid** is formed by the elimination of two H atoms and one O atom (i.e., H_2O) from the ortho acid.

Metaphosphoric acids have the formula $(HPO_3)_n$ and their salts are called **metaphosphates.**

An **ortho acid** is an oxoacid containing the maximum number of OH groups possible.

Peroxo compounds contain the characteristic group —O—O—, such as HO_3S—O—O—SO_3H ($H_2S_2O_8$, peroxodisulfuric acid).

In a **polyhalide ion** two or more halogen atoms are covalently bonded into a polyatomic anion, e.g., I_3^-.

Polyphosphoric acids have the formula $H_2PO_3(HPO_3)_nPO_4H$ and their salts are called **polyphosphates.**

Polysulfide ions, S_n^{2-}, have sulfur atoms bonded together into chains of length $n = 2$ to about $n = 9$.

Pseudohalogens are groupings of atoms (e.g., —CN, —OCN, —SCN, —N_3) that have some of the characteristics of a halogen atom.

A **pyro acid** is formed by the elimination of two H atoms and one O atom (i.e., H_2O) from between two molecules of the ortho acid.

A **silicone** is an organosilicon polymer containing —O—Si—O— bonds.

Sulfuryl compounds contain the group $O_2S{<}$.

Superphosphate is a mixture of $Ca(H_2PO_4)_2$ and $CaSO_4$ produced by the action of H_2SO_4 on phosphate rock.

A **thio compound** is one in which an S atom replaces an O atom. For example replacement of an O by S converts SO_4^{2-} to $S_2O_3^{2-}$ (thiosulfate ion).

Thionyl compounds contain the group $OS{<}$.

A **three-center (bridge, banana) bond** unites *three* atoms through *two* electrons in a delocalized molecular orbital, as with the B—H—B bonds in diborane, B_2H_6.

Suggestions for Further Study

ARKLES, B., "Look What You Can Make Out of Silicones," *Chemtech,* **13,** 542 (1983).

GREENWOOD, N. N., and J. SMITH, "Models of Boron Hydrides," *Educ. in Chem.,* **1,** 25 (1964).

HUNT, C. B., "Noble Gas Compounds—In the Beginning," *Educ. in Chem.,* **20,** 177 (1983).

MOODY, G. J., "A Decade of Xenon Chemistry," *J. Chem. Educ.,* **51,** 628 (1974).

PANNU, S. S., "Chemical of the Month: Nitric Acid," *J. Chem. Educ.,* **61,** 174 (1984).

SARQUIS, M., "Arsenic and Old Myths," *J. Chem. Educ.,* **56,** 815 (1979).

SHELDON, R. P., "Phosphate Rock," *Scientific American,* **246**[6], 45 (1982).

STEIN, L., "Noble Gas Compounds," *Chemistry,* **47**[9], 15 (1974).

Review Problems

1. Provide an acceptable name or formula for each of the following.

(a) $KBrO_3$ (b) ClF_3
(c) sodium hypoiodite   (d) trisilane
(e) $KOCN$ (f) sodium dithionate
(g) silver azide   (h) NaH_2PO_4

2. Complete and balance equations for the following reactions. If no reaction occurs, so state.

(a) $CaCl_2(s) + H_2SO_4(conc.\ aq) \xrightarrow{\Delta}$
(b) $I_2(s) + Cl^-(aq) \rightarrow$
(c) $PbS(s) + HCl(aq) \rightarrow$
(d) $NH_3(aq) + HClO_4(aq) \rightarrow$
(e) $NO(g) + O_2(g) \rightarrow$
(f) $Hg(NO_3)_2(s) \xrightarrow{\Delta}$

3. Give a practical method that could be used in the laboratory to prepare (a) $O_2(g)$; (b) $HCl(aq)$; (c) $N_2O(g)$; (d) $BaSO_3(s)$.

4. Write a chemical equation to represent the reaction of (a) $Cl_2(g)$ with cold $NaOH(aq)$; (b) $NaI(s)$ with hot H_2SO_4 (conc. aq); (c) $Cl_2(g)$ with $Br^-(aq)$; (d) $CdS(s)$ with $HNO_3(aq)$; (e) $Fe(s)$ with very hot $N_2O(g)$; (f) $Pb(s)$ with 8 M $HNO_3(aq)$.

5. Write half-equations for the following half-reactions.

acidic solution: *basic solution:*

$H_5IO_6 \xrightarrow{+1.60\ V} IO_3^-$ $OCl^- \xrightarrow{+0.88\ V} Cl^-$

$H_3PO_2 \xrightarrow{-0.51\ V} P_4$ $B_2H_6 \xrightarrow{+0.78\ V} BH_4^-$

$Sb_2O_5 \xrightarrow{+0.58\ V} SbO^+$ $HXeO_4^- \xrightarrow{+1.24\ V} Xe$

6. Use data from the electrode potential diagrams indicated to predict whether the stated reaction is likely to occur to a significant extent in the forward direction.

(a) $H_2O_2 + HClO_2 \rightarrow H^+ + ClO_3^- + H_2O$
                      (Figures 21-1, 21-7)

(b) $2\ Cl^- + 2\ NO_3^- + 4\ H^+ \rightarrow$
     $2\ H_2O + 2\ NO_2(g) + Cl_2(g)$ (Figures 21-1, 21-11)

(c) $Cl_2(g) + S^{2-} \rightarrow 2\ Cl^- + S(s)$ in basic solution
                        (Figures 21-1, 21-8)

(d) $3\ OCl^- + H_2O \rightarrow ClO_2^- + Cl_2(g) + 2\ OH^-$
                             (Figure 21-1)

7. Use data from Figure 21-8 to obtain the standard electrode potential for the reduction of (a) $SO_2(g)$ to $S_2O_3^{2-}$ in acidic solution and (b) SO_3^{2-} to $S_2O_3^{2-}$ in basic solution.

8. Use data from this chapter and Table 20-2 to predict which of the following are strong enough oxidizing agents in acidic solutions to oxidize H_2O_2 to $O_2(g)$. Explain the basis of your predictions. (a) $Cl_2(g)$; (b) $H_2SO_4(aq)$; (c) $Cr_2O_7^{2-}(aq)$; (d) $MnO_2(s)$; (e) $I_2(s)$

9. Use electrode potential data from this chapter or Table 20-2 to predict which of the following outcomes is the more likely. Explain your reasoning in each case.

(a) When $Cl_2(g)$ is added to an aqueous solution containing I^-, which is more likely to be produced, $O_2(g)$ or I_2?
(b) When added to an acidic solution containing NH_4^+, is H_2O_2 more likely to be oxidized to $O_2(g)$ or reduced to H_2O?
(c) When $Ag(s)$ is added to a solution that is 6 M in both H_2SO_4 and HNO_3, is the gaseous product most likely to be NO, SO_2, or H_2?

10. Given the bond energies at 298 K: O_2, 499; N_2, 946; F_2, 159; Cl_2, 243; ClF, 251; OF (in OF_2), 213; OCl (in OCl_2), 205; and NF (in NF_3), 280 kJ/mol, respectively, calculate $\Delta \overline{H}_f^\circ$ at 298 K for 1 mol of (a) $ClF(g)$; (b) $OF_2(g)$; (c) $OCl_2(g)$; (d) $NF_3(g)$.

11. Use the scheme of Figure 21-13 to supply plausible names or formulas for the following (a) calcium orthophosphate; (b) potassium pyrophosphate; (c) $NaSbO_2$; (d) $NaBiO_3$; (e) sodium orthobismuthate.

12. Use VSEPR theory to predict the probable geometric structures of the molecules (a) XeO_3; (b) XeO_4; (c) $OXeF_4$.

Exercises

The Halogens

1. Make a general statement about which of the elements, Cl_2, Br_2, and I_2, displaces other halogens from a solution of halide ions. That is, will the reaction $Br_2 + 2\ I^- \rightarrow 2\ Br^- + I_2$ occur? Will the reaction $Br_2 + 2\ Cl^- \rightarrow 2\ Br^- + Cl_2$ occur? And so on.

2. Fluorine was not mentioned in the halogen displacement series in Exercise 1.

(a) In principle would you expect F_2 to be able to displace Cl^-, Br^- and I^- from solution (producing Cl_2, Br_2, and I_2)?

(b) What difficulty would be encountered in attempting these displacements with F_2?
(c) Is there any reagent that can displace $F^-(aq)$, producing $F_2(g)$? Explain.

3. In Example 21-1 we concluded that this reaction will not occur spontaneously in acidic solution: $3\ Cl_2 + 3\ H_2O \rightarrow 5\ Cl^- + ClO_3^- + 6\ H^+$. Will this disproportionation occur in a strongly basic solution? Explain.

4. Use the methods of Examples 21-1 and 21-2 to determine (a) the standard electrode potential for the reduction of $HOCl$ to Cl^-; (b) whether the reaction $2\ HOCl \rightarrow HClO_2 + H^+ + Cl^-$ will go essentially to completion as written.

5. Use data from Table 21-1 to determine for the dissociation of $HCl(g)$ into its elements at 298 K (a) K_p and (b) the % dissociation.

6. You have available these materials: H_2O, $CaO(s)$, $NaCl(s)$, $NaBr(s)$, $KOH(aq)$, H_2SO_4(conc. aq), and H_3PO_4(conc. aq). How would you use these, together with common laboratory equipment, to prepare (a) $CaCl_2$; (b) KBr; (c) $KBrO_3$?

7. Use VSEPR theory to predict the geometric structures of (a) BrF_3; (b) IF_5; (c) ICl_2^-; (d) Cl_3IF^-.

*8. The following data are given.

$$IO_3^- + 3\ H_2SO_3(aq) \rightarrow I^- + 3\ SO_4^{2-} + 6\ H^+$$
$$E^\circ_{cell} = 0.92\ V$$

$$HIO(aq) + H^+ + I^- \rightarrow I_2(s) + H_2O \qquad E^\circ_{cell} = 0.91\ V$$

Use these data together with values from Table 20-2 to complete the standard electrode potential diagram shown.

$$IO_3^- \underset{\underset{\text{(?)}}{\underline{\hspace{5cm}}}}{\overset{\text{(?)}}{\rule{1.5cm}{0.4pt}}} HIO \overset{\text{(?)}}{\rule{1.5cm}{0.4pt}} I_2(s) \overset{0.54\ V}{\rule{1.5cm}{0.4pt}} I^-$$

Oxygen

9. Each of the following compounds produces $O_2(g)$ when strongly heated. Write a plausible equation for the reactions that occur. (a) $HgO(s)$; (b) $KClO_4(s)$; (c) $Hg(NO_3)_2$; (d) $H_2O_2(l)$.

10. Use Figure 21-7 to write equations for the disproportionation of H_2O_2 in (a) acidic and (b) basic solutions. (*Hint:* Review the meaning of disproportionation on page 598.)

11. For the conversion of $O_2(g)$ to $O_3(g)$, which can be accomplished in an electric discharge, $3\ O_2(g) \rightarrow 2\ O_3(g)$, $\Delta \overline{H}^\circ = +285$ kJ/mol. The bond energy in O_2, which is essentially the $O{=}O$ double bond energy, is 499 kJ/mol. The $O{-}O$ single bond energy is 142 kJ/mol.
   (a) Calculate the average $O{-}O$ bond energy in $O_3(g)$.
   (b) Estimate the average bond energy in $O_3(g)$ from the structure on page 391 and compare this result with that of part (a).

*12. Refer to Figure 10-18 and arrange the following species in the expected order of increasing (a) bond distance and (b) bond strength (energy). State the basis of your prediction. O_2, O_2^+, O_2^-, O_2^{2-}.

Sulfur

13. Give an appropriate name to each of the following: (a) ZnS; (b) $KHSO_3$; (c) S_4^{2-}; (d) $K_2S_4O_6$; (e) OSF_2.

14. Through a chemical equation give a specific example that illustrates
   (a) the dissolving of a metal sulfide in $HCl(aq)$;
   (b) the action of a *nonoxidizing* acid on a metal sulfite;
   (c) the oxidation of $SO_2(aq)$ to $SO_4^{2-}(aq)$ by $MnO_2(s)$ in acidic solution;
   (d) a plausible reaction of $O_2(g)$ with a basic solution containing S^{2-}.

15. Available in a chemical laboratory are elemental sulfur, chlorine gas, metallic sodium, and water. Show how you would use these substances (and air) to produce aqueous solutions containing (a) Na_2SO_3; (b) Na_2SO_4; (c) $Na_2S_2O_3$. (*Hint:* You will have to use information from other chapters as well as this one, e.g., Chapter 13.)

16. Based on the discussion of peroxo compounds propose a plausible formula and structure for *peroxomonosulfuric acid*. (*Hint:* Recall structure 21.44.)

17. A statement is made in the text that HgS is the least soluble of all metal sulfides; yet K_{sp} for Bi_2S_3 is 1×10^{-96} compared to 1.6×10^{-52} for HgS. Explain this apparent discrepancy.

18. What mass of Na_2SO_3 must have been present in a sample that required 26.50 mL of 0.0510 M $KMnO_4$ for its oxidation to Na_2SO_4 in an acidic solution? MnO_4^- is reduced to Mn^{2+}.

19. A 1.100-g sample of copper ore is dissolved and the $Cu^{2+}(aq)$ is treated with excess KI. The liberated iodine requires 12.12 mL of 0.1000 M $Na_2S_2O_3$ for its titration. What is the % copper, by mass, in the ore? (*Hint:* Recall equation 21-34.)

Nitrogen

20. Write chemical equations to represent the following.
   (a) equilibrium between nitrogen dioxide and dinitrogen tetroxide in the gaseous state
   (b) the decomposition of $NaNO_3(s)$ by heating
   (c) the neutralization of $NH_3(aq)$ by $H_2SO_4(aq)$
   (d) the dissolving of silver metal in 8 M $HNO_3(aq)$
   (e) the complete combustion of the rocket fuel, dimethyl hydrazine, $(CH_3)_2NNH_2$

21. What is the pH of an aqueous solution that is (a) 0.032 M in NH_2OH; (b) 0.018 M in $[NH_3OH]Cl$?

22. NH_3OH^+ can act as an oxidizing agent in acidic solutions.
   (a) Write a half-equation to represent the reduction of NH_3OH^+ to NH_4^+.
   (b) Use data from Figure 21-11 to determine E° for the half-reaction described in part (a).
   (c) Will the oxidation of Fe^{2+} to Fe^{3+} occur in acidic solution with NH_3OH^+ as the oxidizing agent?

23. The heat of formation of hydrazine $(\Delta \overline{H}_f^\circ)$ is +50.63 kJ/mol. What is the heat of combustion of hydrazine as represented in equation (21.49)?

24. Draw plausible Lewis structures for (a) N_2O_3 (which has an $N{-}N$ bond) and (b) N_2O_5 (which has an $N{-}O{-}N$ bond).

25. Trace the changes in molecular species (i.e., oxides of nitrogen) that you would observe by slowly heating a sample of $N_2O_4(s)$ from $-20°C$ to $700°C$. (*Hint:* Recall the discussion of Table 21-10.)

*26. Use data from Figure 21-11 and equation (21.51) to determine E° for the reduction couple $Fe(OH)_3/Fe(OH)_2$ in basic solution.

Phosphorus

27. Write chemical equations to show why
   (a) A solution of Na_3PO_4 is strongly basic.
   (b) The first equivalence point in the titration of H_3PO_4 is on the acid side of pH 7.

28. Supply an appropriate name for each of the following:
(a) HPO_4^{2-}; **(b)** $Ca_2P_2O_7$; **(c)** $H_6P_4O_{13}$; **(d)** $(NaPO_3)_4$.

29. You have available concentrated aqueous solutions of H_3PO_4 and NaOH. With these solutions as starting materials, indicate how you would prepare the solids **(a)** $Na_5P_3O_{10}$ and **(b)** $(NaPO_3)_n$.

30. Various glassy metaphosphates have been called sodium metaphosphate. Explain why this name is less specific than most chemical names in describing the composition of a substance.

31. Write a series of equations to show how triple superphosphate could be produced without the use of H_2SO_4 anywhere in the process.

Carbon, silicon

32. Write chemical equations for the reactions that you would expect to occur when
   (a) KCN(aq) is added to $AgNO_3$(aq);
   (b) Al_4C_3 reacts with water to produce CH_4(g);
   (c) S_3H_8 is burned in an excess of air.

33. Describe what is meant by the terms *silane* and *silanol*. What is their role in the preparation of silicones?

34. In a manner similar to that outlined on page 661,
   (a) write equations to represent the reaction of $(CH_3)_3SiCl$ with water, followed by the elimination of H_2O from the resulting silanol molecules.

   (b) Does a silicone polymer form?
   (c) What would be the corresponding product obtained from H_3CSiCl_3?

Boron

35. The molecule tetraborane has the formula B_4H_{10}.
   (a) Show that this is an electron deficient molecule.
   (b) How many bridge bonds must occur in the molecule?
   (c) Show that the carbon analog, butane, C_4H_{10}, is not electron deficient.

36. Write chemical equations to represent
   (a) the preparation of boron from BBr_3;
   (b) the formation of BF_3 from B_2O_3;
   (c) the reaction at high temperatures of boron with N_2O(g).

Noble gas compounds

37. Predict plausible molecular shapes for **(a)** O_2XeF_2; **(b)** O_3XeF_2; **(c)** O_2XeF_4; **(d)** XeF_5^+.

★38. Use VSEPR theory to predict a plausible structure for XeF_6, and comment on the difficulty in applying the valence bond method of page 664 to a description of this structure.

★39. The bond energies of Cl_2 and F_2 are 243 and 155 kJ/mol, respectively. Use these data to explain why XeF_2 is a much more stable compound than $XeCl_2$. (*Hint:* Recall that Xe exists as a monatomic gas.)

★40. Write plausible half-equations and a balanced oxidation–reduction equation for the disproportionation of XeF_4 to Xe and XeO_3 in aqueous acidic solution. Xe and XeO_3 are produced in a $2:1$ mol ratio and O_2(g) is also produced.

Additional Exercises

1. Supply a name for each of the following. **(a)** $H_2S_2O_8$; **(b)** K_2HPO_4; **(c)** $Sr(ClO_4)_2$; **(d)** HIO_3; **(e)** BaO_2; **(f)** $Mg_2P_2O_7$; **(g)** $Hg(SCN)_2$; **(h)** MgTe; **(i)** $Ba(N_3)_2$; **(j)** $K_2S_3O_6$; **(k)** As_2S_3; **(l)** CSe_2; **(m)** $Pb_3(AsO_4)_2$; **(n)** $Ag_2S_2O_3$.

2. Supply a formula for each of the following. **(a)** magnesium nitrite; **(b)** bromine pentafluoride; **(c)** calcium hydrogen sulfide; **(d)** potassium cyanate; **(e)** phosphorus dichloride trifluoride; **(f)** cesium trisulfide; **(g)** hydrazinium chloride; **(h)** lithium dithionate; **(i)** calcium telluride; **(j)** iron(II) orthophosphate; **(k)** lead(II) metaarsenite; **(l)** calcium telluride; **(m)** copper(I) azide.

3. Complete and balance each of the following equations. If no reaction occurs, so state.
   (a) $KI(s) + H_3PO_4(conc. aq) \rightarrow$
   (b) $NaClO_3(s) \xrightarrow{\Delta}$
   (c) $K_2O_2(s) + H_2O \rightarrow$
   (d) $CuS(s) + HCl(aq) \rightarrow$
   (e) $I_2(s) + KI(aq) \rightarrow$
   (f) $SO_3^{2-}(aq) + MnO_4^-(aq) + H^+(aq) \rightarrow$
   (g) $Br_2(l) + Cl^-(aq) \rightarrow$

4. Write equations to show how H_2O_2 **(a)** oxidizes NO_2^- to NO_3^- in acidic solution; **(b)** oxidizes SO_2(g) to SO_4^{2-} in basic solution; **(c)** reduces MnO_4^- to Mn^{2+} in acidic solution; **(d)** reduces Cl_2(g) to Cl^- in basic solution.

5. Show that the hypothetical process pictured in Figure 21-13 also leads to correct formulas for **(a)** carbonic acid (H_2CO_3); **(b)** nitrous acid (HNO_2); **(c)** nitric acid (HNO_3); **(d)** sulfuric acid (H_2SO_4); **(e)** periodic acid (HIO_4).

6. Nitramide and hyponitrous acid both have the formula $H_2N_2O_2$. Hyponitrous acid is a weak diprotic acid; nitramide contains the amide group (—NH_2). Based on this information write plausible Lewis structures for these two substances.

7. *Peroxonitrous acid* is an unstable intermediate formed in the oxidation of HNO_2 by H_2O_2. It has the same formula as *nitric acid*, HNO_3. Show how you would expect these two acids to differ in structure.

8. The structure of $N(SiH_3)_3$ involves a planar arrangement of N and Si atoms, whereas that of the related compound $N(CH_3)_3$ has a pyramidal arrangement of N and C atoms. Propose bonding schemes for these molecules that are consistent with this observation.

*9. The unstable species HO_2 has O in the oxidation state $-1/2$. For the half-reaction $O_2 + H^+ + e^- \rightarrow HO_2$, $E° = -0.131$ V. Reconstruct the electrode potential diagram for oxygen in acidic solution (Figure 21-7) to include the species HO_2 and fill in any missing electrode potential values.

*10. Use the data developed in Additional Exercise 9 to show that HO_2 undergoes spontaneous disproportionation in acidic solution.

*11. Use data from Table 21-10 and appropriate equation(s) from elsewhere in the text to verify the statements in the text that
   (a) $N_2O_4(g)$ is essentially completely dissociated into $NO_2(g)$ at 150°C.
   (b) The reverse of the reaction $2 NO(g) + O_2(g) \rightarrow NO_2(g)$ proceeds essentially to completion at 600°C.

*12. The total solubility of $Cl_2(g)$ in water is 6.4 g/L at 25°C. At this temperature the hydrolysis reaction

$$Cl_2(aq) + H_2O \rightleftharpoons HOCl(aq) + H^+(aq) + Cl^-(aq)$$

has a value of $K_c = 4.4 \times 10^{-4}$. For a saturated aqueous solution of Cl_2 in water, calculate $[Cl_2]$, $[HOCl]$, $[H^+]$, and $[Cl^-]$.

*13. The concentration of a saturated solution of I_2 in water is 1.33×10^{-3} M. Also

$$I_2(aq) \rightleftharpoons I_2(CCl_4) \qquad K = \frac{[I_2]_{CCl_4}}{[I_2]_{aq}} = 85.5$$

A 10.0-mL sample of saturated $I_2(aq)$ is shaken with 10.0 mL CCl_4. After equilibrium is established, the two liquid layers are separated. (*Hint:* Recall Figure 21-6.)
   (a) What mass of I_2, in mg, remains in the water layer?
   (b) If the 10.0-mL water layer in (a) is extracted with a second 10.0-mL portion of CCl_4, what will be the number of mg I_2 remaining in the water?
   (c) If the 10.0-mL sample of saturated $I_2(aq)$ had originally been extracted with 20.0 mL CCl_4, would the quantity of I_2 remaining in the aqueous solution have been less than, equal to, or greater than in part (b)? Explain.

*14. Use data from Example 21-3 to calculate the % dissociation of $HI(g)$ into its elements at 298 K.

*15. Estimate the % dissociation of $Cl_2(g)$ into $Cl(g)$ at 1 atm total pressure and 1000 K. Use data from Appendix D and equations established elsewhere in the text, as necessary.

Self-Test Questions

For questions 1 through 8 select the single item that best completes each statement.

1. To displace Br_2 from an aqueous solution containing Br^-, add (a) $I_2(aq)$; (b) $Cl_2(aq)$; (c) $Cl^-(aq)$; (d) $I_3^-(aq)$.

2. All of the following compounds yield $O_2(g)$ when heated strongly (e.g., to about 1000 K) except (a) $KClO_3$; (b) HgO; (c) N_2O; (d) $CaCO_3$.

3. The expected gaseous product when Cu is dissolved in concentrated $HNO_3(aq)$ is (a) NO_2; (b) H_2; (c) N_2; (d) NH_3.

4. All of the following are bases except (a) N_2H_4; (b) NH_2OH; (c) NH_3; (d) HN_3.

5. To dissolve mercuric sulfide, HgS ($K_{sp} = 1.6 \times 10^{-52}$) use (a) $HNO_3(aq)$; (b) $HCl(aq)$; (c) a mixture of $HNO_3(aq)$ and $HCl(aq)$; (d) $NaOH(aq)$.

6. The best reducing agent of the following is (a) H_2S; (b) $Cl^-(aq)$; (c) $SO_4^{2-}(aq)$; (d) O_3.

7. Al of the following have a tetrahedral shape except (a) SO_4^{2-}; (b) XeF_4; (c) ClO_4^-; (d) XeO_4.

8. The term "thio" is used in the names of all of the following compounds except (a) $Na_2S_2O_3$; (b) $NaCS_3$; (c) $NaSCN$; (d) Na_2S_3.

9. Write chemical equations to represent
   (a) the thermal decomposition of $Pb(NO_3)_2(s)$;
   (b) the reaction of $Cl_2(g)$ with cold $NaOH(aq)$;
   (c) the neutralization of $H_3PO_4(aq)$ to the second equivalence point with $KOH(aq)$;
   (d) the action of hot concentrated $H_2SO_4(aq)$ on $KBr(s)$;
   (e) the formation of pyrophosphoric acid from orthophosphoric acid.

10. Use principles from this chapter and elsewhere in the text to explain why
   (a) Ag will dissolve in HNO_3(conc. aq) but not in HCl(conc. aq).
   (b) I_2 is much more soluble in KI(aq) than it is in pure water.
   (c) H_2S, which has nearly twice the molecular weight of H_2O, is a gas at room temperature whereas H_2O is a liquid.

11. The abundance of F^- in seawater is 1 g F^- per ton of seawater. Suppose that a commercially feasible method could be found to extract fluorine from seawater.

   (a) What mass of F_2 could be obtained from 1 km^3 of seawater ($d = 1.03$ g/cm^3)?

   (b) Do you think the process would resemble that for extracting bromine from seawater? Explain.

12. A portion of the standard electrode (reduction) potential diagram of selenium is given below. What is the $E°$ value for the reduction of H_2SeO_3 to H_2Se?

$$SeO_4^{2-} \xrightarrow{\text{1.15 V}} H_2SeO_3 \xrightarrow{\text{0.74 V}} Se \xrightarrow{-0.35 \text{ V}} H_2Se$$
$$\underset{(?)}{\underline{\hspace{6cm}}}$$

22 Chemistry of the Representative Elements II: Metals

The representative metals range from the most metallic—the heavier members of group IA—to the rather noble mercury. Thus, in our study of the representative metals we will note wide variations in certain properties, such as reactivity toward water, acids, and alkalis. We will find that the salts of some metals have high water solubility, whereas other metal salts are insoluble. Some oxides are easily reduced by carbon; others are not, requiring that the metals be prepared electrolytically. To sort out this rich variety of observations, to the extent possible we will use fundamental principles (e. g., electrode potentials, free energy data, equilibrium constants). In developing an understanding of why metals behave the way they do in laboratory situations, we should also gain insights into why certain industrial methods have evolved for producing metals and their compounds, and why the metals and their compounds are used in some of the ways that they are.

22-1 Group IA—The Alkali Metals

Several properties of the alkali metals were listed in Table 13-3 and discussed at that time. Additional properties are given in Table 22-1. The principal conclusion that we can draw from these two tables is that the alkali metals are a group of active (the most active) metals. They have low ionization energies, large negative electrode potentials, etc. Another conclusion is that, in general, properties vary uniformly within the group as is expected from the periodic law. There are some irregularities, though, and these are found mostly with the first member, Li.

Lithium and its compounds differ from the other alkali metals in several ways, including

- low solubility of its carbonate, fluoride, hydroxide, and phosphate;
- ability to form a nitride (Li_3N);
- formation of a normal oxide (Li_2O) rather than a peroxide or superoxide;
- on heating, decomposition of its carbonate and hydroxide to the oxide.

This behavior, which resembles that of Mg and its compounds, was cited as an example of

TABLE 22-1
Properties of the alkali metals[a]

	Li	Na	K	Rb	Cs
metallic radius, pm	155	190	235	248	267
ionic radius, pm	60	95	133	148	169
ionic charge density [ionic charge/(ionic radius, Å)]	+1.67	+1.05	+0.75	+0.68	+0.59
sublimation energy, kJ/mol [$M(s) \longrightarrow M(g)$; $\Delta \overline{H}^{\circ}_{subl}$]	155	109	90	86	79
first ionization energy, kJ/mol [$M(g) \longrightarrow M^+(g) + e^-$; $\Delta \overline{H}^{\circ}_{ioniz}$]	520	496	419	403	376
hydration energy, kJ/mol [$M^+(g) \longrightarrow M^+(aq)$; $\Delta \overline{H}^{\circ}_{hydr}$]	−506	−397	318	−289	−259
the sum: $\{\Delta \overline{H}^{\circ}_{subl} + \Delta \overline{H}^{\circ}_{ioniz} + \Delta \overline{H}^{\circ}_{hydr}\}^{b}$	169	208	191	200	196
electrode potential E°, V [$M^+(aq) + e^- \longrightarrow M(s)$]	−3.045	−2.714	−2.925	−2.925	−2.923

[a]Several physical properties are listed in Table 13-3.
[b]The significance of this sum of terms is discussed on page 674.

FIGURE 22-1
Hydration of a Li^+ ion.

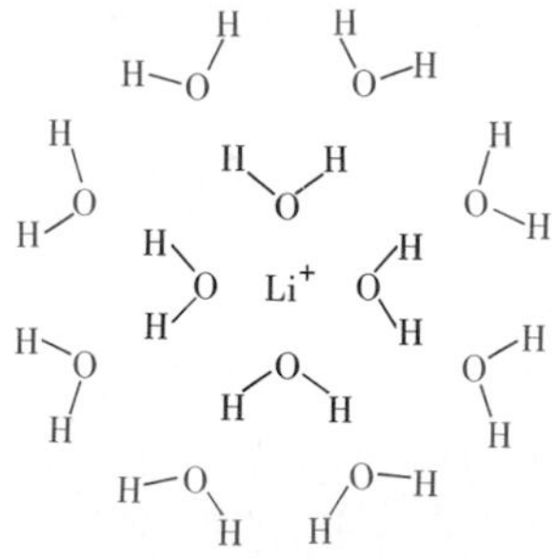

The Li^+ ion has a small number of H_2O molecules held to it by electrostatic forces in a primary hydration sphere (in black). These molecules, in turn, hold other H_2O molecules, but more weakly, in a secondary hydration sphere. In all, the Li^+ ion may have about 25 H_2O molecules surrounding it in a three-dimensional envelope (not the planar arrangement suggested here).

the diagonal relationship on page 373. It also illustrates a point made repeatedly in Chapters 13 and 21—the first member differs rather significantly from heavier members of a group in the periodic table. Among the alkali metals this difference can be attributed to the high charge density—the high ratio of cation charge to cation radius—for Li^+ compared to the other alkali metal ions (see Table 22-1).

One consequence of the high charge density of Li^+ is its large negative enthalpy of hydration, $\Delta \overline{H}^{\circ}_{hydr}$. In the hypothetical process in which gaseous Li^+ ions are added to water, the ions surround themselves with a sheath of H_2O molecules, as pictured in Figure 22-1. Energy is released in this interaction because of the attraction of the negative ends of H_2O dipoles to the positive ion. Other alkali metal ions also are hydrated in aqueous solutions, but Li^+ is more heavily hydrated than they because of its especially high positive charge density. That Li^+(aq) is heavily hydrated shows up in the fact that Li^+, despite its small size, has a lower mobility in an electric field than the larger alkali metal cations. Lithium salts produce abnormally high vapor pressure lowering and freezing point depression in aqueous solutions. Also, LiCl has an exothermic heat of solution, whereas those of NaCl and KCl are endothermic.

Electrode Potentials. Large negative values of electrode potentials mean that the tendencies for reduction processes to occur are very slight. In turn, these large negative values signify a strong tendency for the reverse process—oxidation. For the alkali metals the oxidation potentials are

$$M(s) \longrightarrow M^+(aq) + e^- \qquad E^{\circ}_{ox} = +3.045, +2.714, +2.925, +2.925, +2.923 \text{ V,}$$
$$\text{for Li, Na, K, Rb, and Cs, respectively.}$$

Ease of oxidation is the mark of an active metal, and Li shows the greatest tendency among the alkali metals to undergo oxidation to a unipositive ion in aqueous solution. Yet, by other criteria (e.g., first ionization energy), Li is the *least* metallic of the alkali metals. How can this apparent discrepancy be explained?

One way to analyze the situation is to consider, in a stepwise fashion, factors that determine the magnitude of E°_{ox}, and in doing so perhaps to discover what is unusual about

TABLE 22-2
Thermochemical cycle for evaluating E_{ox}° for a metal

Oxidation of alkali metal, M	Reduction of H^+
(1) Sublimation $\quad$ M(s) $\longrightarrow$ ~~M(g)~~ $\quad \Delta H_1$	(4) Reverse of hydration $\quad$ H^+(aq) $\longrightarrow$ ~~H^+(g)~~ $\quad \Delta H_4$
(2) Ionization $\quad$ ~~M(g)~~ $\longrightarrow$ ~~M^+(g)~~ $+$ e^- $\quad \Delta H_2$	(5) Reverse of ionization $\quad$ ~~H^+(g)~~ $+$ e^- $\longrightarrow$ ~~H(g)~~ $\quad \Delta H_5$
(3) Hydration $\quad$ ~~M^+(g)~~ $\longrightarrow$ M^+(aq) $\quad \Delta H_3$	(6) Recombination of H atoms $\quad$ ~~H(g)~~ $\longrightarrow$ $\frac{1}{2}$ H_2(g) $\quad \Delta H_6$

(1) + (2) + (3):
M(s) $\longrightarrow$ M^+(aq) + e^-
$\Delta \overline{H}_{ox}^\circ = \Delta H_1 + \Delta H_2 + \Delta H_3$
$\qquad$ = values listed in
$\qquad \qquad$ Table 22-1

(4) + (5) + (6):
H^+(aq) + e^- $\longrightarrow$ $\frac{1}{2}$ H_2(g)
$\Delta \overline{H}_{red}^\circ = \Delta H_4 + \Delta H_5 + \Delta H_6$
$\qquad$ = +1079 − 1312 − 216
$\qquad$ = −449 kJ/mol

$$M(s) + H^+(aq) \longrightarrow M^+(aq) + \tfrac{1}{2} H_2(g)$$
$$\Delta \overline{H}^\circ = \Delta \overline{H}_{ox}^\circ + \Delta \overline{H}_{red}^\circ$$
$$\Delta \overline{H}^\circ = (\Delta \overline{H}_{ox}^\circ - 449) \text{ kJ/mol}$$

Li. As a first step, represented through expression (22.1), we can equate E_{ox}° to the value of E_{cell}° for the displacement of H_2(g) from H^+(aq) by an alkali metal, M.

$$
\begin{array}{ll}
M(s) \longrightarrow M^+(aq) + e^- & E_{ox}^\circ = ? \\
H^+(aq) + e^- \longrightarrow \tfrac{1}{2}H_2(g) & E_{red}^\circ = 0.0000 \text{ V} \\
\hline
M(s) + H^+(aq) \longrightarrow M^+(aq) + \tfrac{1}{2} H_2(g) & E_{cell}^\circ = E_{ox}^\circ = ?
\end{array}
\qquad (22.1)
$$

Next we can relate E_{cell}° to $\Delta \overline{G}^\circ$ through the expression

$$\Delta \overline{G}^\circ = -n\mathscr{F}E_{cell}^\circ \qquad (20.24)$$

followed by this familiar equation,

$$\Delta \overline{G}^\circ = \Delta \overline{H}^\circ - T\Delta \overline{S}^\circ \qquad (16.14)$$

If we assume that the value of $\Delta \overline{G}^\circ$ is essentially that of $\Delta \overline{H}^\circ$, we can see that whatever makes $\Delta \overline{H}^\circ$ of reaction (22.1) more negative also makes $\Delta \overline{G}^\circ$ more negative. The more negative the value of $\Delta \overline{G}^\circ$, the more positive the value of E_{cell}° (expression 20.24); and the more positive the value of E_{cell}°, the more positive the value of E_{ox}° (expression 22.1).

With data from Table 22-1 we can assess, through the thermochemical cycle presented in Table 22-2, the factors that determine the magnitude of $\Delta \overline{H}^\circ$. Oxidation of the alkali metal M involves the steps

1. subliming the solid metal to the gaseous state;
2. ionizing the gaseous metal atoms to gaseous metal ions;
3. hydrating the ions by dissolving M^+(g) in water to produce M^+(aq).

These steps are pictured in Figure 22-2. The sum of the enthalpy changes for these three steps (given in the next-to-last line of Table 22-1) is $\Delta \overline{H}^\circ$ for the oxidation half-reaction, i.e., $\Delta \overline{H}_{ox}^\circ$.

$$\Delta \overline{H}_{ox}^\circ = \Delta H_1 + \Delta H_2 + \Delta H_3 = \Delta \overline{H}_{subl}^\circ + \Delta \overline{H}_{ioniz}^\circ + \Delta \overline{H}_{hydr}^\circ \qquad (22.2)$$

FIGURE 22-2
Thermochemical cycle for $\Delta \overline{H}_{ox}^\circ$ of an alkali metal,

$$M(s) \rightarrow M^+(aq) + e^-$$

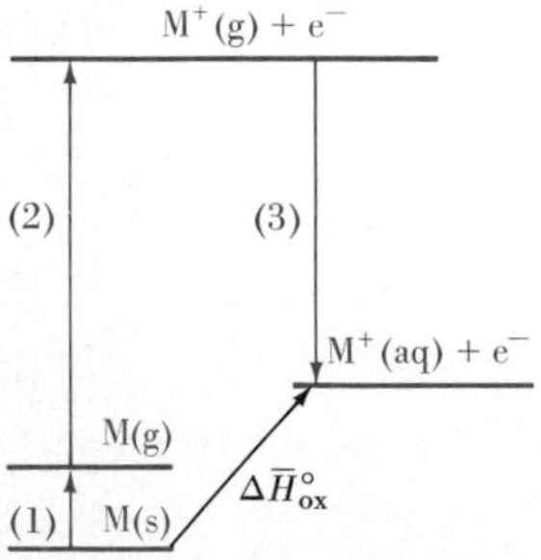

(1) Sublimation of metal atoms. (2) Ionization of gaseous metal atoms. (3) Hydration of gaseous metal ions.

Displacement of $H_2(g)$ from $H^+(aq)$ is a reduction half-reaction, which can be viewed through steps 4, 5, and 6 of Table 22-2 and for which $\Delta\overline{H}^\circ_{red} = -449$ kJ/mol. For the overall reaction,

$$\Delta\overline{H}^\circ = \Delta\overline{H}^\circ_{ox} + \Delta\overline{H}^\circ_{red} \quad \text{or}$$
$$\Delta\overline{H}^\circ = (\Delta\overline{H}^\circ_{ox} - 449) \text{ kJ/mol} \tag{22.3}$$

Our final conclusion, then, is that the *smaller* the value of $\Delta\overline{H}^\circ_{ox}$, the *more negative* the value of $\Delta\overline{H}^\circ$ in (22.3), the *more negative* the value of $\Delta\overline{G}^\circ$ in (16.18), the more positive the value of E°_{cell} in (20.24), and the *more positive* the value of E°_{ox} in (22.1).

For Li, the large values of $\Delta\overline{H}^\circ_{subl}$ and $\Delta\overline{H}^\circ_{ioniz}$ are offset by an unusually large (negative) value of $\Delta\overline{H}^\circ_{hydr}$. The sum of these three quantities, $\Delta\overline{H}^\circ_{ox}$, is *smaller* for Li (169 kJ/mol) than for any of the other alkali metals. As a consequence, the *largest* value of E°_{ox} is that of Li. Li is the most easily oxidized of the alkali metals in *aqueous solution*; but for reactions in the *gaseous* state, because of its high ionization energy, Li reverts back to its expected position of least metallic in the group.

Preparation of Alkali Metal Compounds.   Although applied to sodium compounds, the outline in Figure 22-3 is generally applicable to other alkali metal compounds as well. NaCl, as the most abundant naturally occurring sodium compound, occupies a central position in the figure. Since the end of the nineteenth century, when large-scale electrolytic processes were introduced, it has been possible to produce the full range of sodium compounds from NaCl as the starting material. Prior to the advent of electrolysis, however, chemical methods of producing NaOH from NaCl were crucial to the chemical industry. The most important chemical route to NaOH is from the carbonate. A slurry of

FIGURE 22-3
Preparation of sodium compounds.

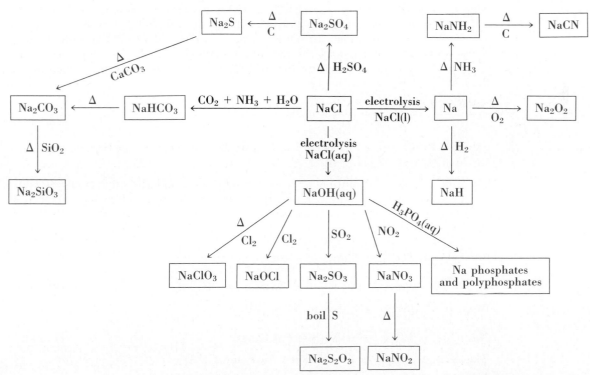

Most of these reactions are described in the text (e.g., Chapters 13, 21, and 22). Alternative methods of preparation are possible for a number of these compounds.

$Ca(OH)_2(s)$ is treated with $Na_2CO_3(aq)$; $Ca(OH)_2(s)$ is converted to the more insoluble $CaCO_3(s)$; and an aqueous solution of NaOH is obtained.

$$Ca(OH)_2(s) + Na_2CO_3(aq) \rightleftharpoons CaCO_3(s) + 2\ NaOH(aq) \tag{22.4}$$

Expressed in net ionic form this reaction is

$$Ca(OH)_2(s) + CO_3^{2-}(aq) \rightleftharpoons CaCO_3(s) + 2\ OH^-(aq) \tag{22.5}$$

Example 22-1   Show that, in principle, $Na_2CO_3(aq)$ can be converted almost completely to NaOH(aq) by reaction (22.5). Use solubility product data from Table 19-1.

Solution.   We begin by combining the following equilibria to obtain K for the net reaction (22.5).

$$Ca(OH)_2(s) \rightleftharpoons Ca^{2+}(aq) + 2\ OH^-(aq) \qquad\qquad K = K_{sp} = 5.5 \times 10^{-6}$$
$$CO_3^{2-}(aq) + Ca^{2+}(aq) \rightleftharpoons CaCO_3(s) \qquad\qquad K = 1/K_{sp} = 1/2.8 \times 10^{-9}$$

$$\overline{Ca(OH)_2(s) + CO_3^{2-}(aq) \rightleftharpoons CaCO_3(s) + 2\ OH^-(aq)\quad K = 5.5 \times 10^{-6}/2.8 \times 10^{-9}}$$

For the net equilibrium,

$$K = \frac{[OH^-]^2}{[CO_3^{2-}]} = \frac{5.5 \times 10^{-6}}{2.8 \times 10^{-9}} = 2.0 \times 10^3$$

Our first clue that the reaction goes essentially to completion is the large value of K. We could now proceed by assuming some reasonable initial concentration of Na_2CO_3 (e.g., 1.00 or 2.00 M) and calculating $[CO_3^{2-}]$ and $[OH^-]$ at equilibrium.

Alternatively, we can note that at equilibrium

$$[OH^-]^2 = 2.0 \times 10^3 \times [CO_3^{2-}] \quad\text{\textit{or}}\quad [OH^-] = \sqrt{2.0 \times 10^3} \times \sqrt{[CO_3^{2-}]}$$
$$= 45 \times \sqrt{[CO_3^{2-}]}$$

Suppose that as a result of reaction (22.5) $[CO_3^{2-}]$ is reduced from its initial value to an equilibrium value of about 0.01 M. At equilibrium, $[OH^-] = 45 \times \sqrt{0.01} = 4.5\ M$. A reaction in which an initial $[CO_3^{2-}]$ is reduced to 0.01 M while $[OH^-]$ increases to 4.5 M must surely be one that has gone essentially to completion. (Can you see that the *initial* concentration of CO_3^{2-} in this case must have been about 2.25 M?)

SIMILAR EXAMPLES: Review Problem 5; Exercise 8a.

Example 22-2   Use information from Figure 22-3 to propose a method of synthesizing Na_2CO_3 from NaCl.

Solution.   Figure 22-3 outlines a series of reactions through their principal reactants and products. In some cases we must supply formulas for other plausible products. A possible first step is the conversion of NaCl to Na_2SO_4 by reactions (21.9) and (21.10).

$$NaCl(s) + H_2SO_4\ (\text{conc. aq}) \xrightarrow{\Delta} NaHSO_4(s) + HCl(g) \tag{21.9}$$

$$NaCl(s) + NaHSO_4(s) \xrightarrow{\Delta} Na_2SO_4(s) + HCl(g) \tag{21.10}$$

Next, the Na_2SO_4 is reduced to Na_2S with carbon.

$$Na_2SO_4(s) + 4\ C(s) \xrightarrow{\Delta} Na_2S(s) + 4\ CO(g) \tag{13.21}$$

The final step is a reaction between Na_2S and $CaCO_3$.

$$Na_2S(s) + CaCO_3(s) \xrightarrow{\Delta} CaS(s) + Na_2CO_3(s) \tag{22.6}$$

SIMILAR EXAMPLE: Exercise 1.

Leblanc and Lavoisier, two contemporary chemists, were both victims of the French Revolution. Lavoisier was guillotined on May 8, 1794. Leblanc, because of conflicting claims to his process and the intervention of the revolution, never received his prize money. He was reduced to poverty and committed suicide in 1806.

The Leblanc Process. Until the end of the eighteenth century the principal source of Na_2CO_3 was from salt lakes (e.g., in Egypt). The capacity to produce NaOH by reaction (22.5) was thus limited by the availability of Na_2CO_3, and the demand for NaOH was rapidly increasing, particularly for the manufacture of soap. In 1775, in order to gain independence from the importation of Na_2CO_3, the French government, through its Academy of Sciences, offered a prize to anyone who could devise a process for the preparation of NaOH from NaCl. Nicolas Leblanc responded with the process outlined in Example 22-2.

The Solvay Process. In the early decades of the nineteenth century several scientists and engineers tried, unsuccessfully, to develop a commercial process based on the reaction of NH_4HCO_3 and NaCl to form $NaHCO_3$. The Belgian engineer, Ernest Solvay, was successful in doing so in 1865, and within three decades the Solvay process had completely supplanted the Leblanc process.

In the Solvay process cold concentrated NaCl(aq) is treated with NH_3(g) and CO_2(g). $NaHCO_3$(s), being the least soluble of all the possible products, precipitates from solution.

$$Na^+ + Cl^- + NH_3 + CO_2 + H_2O \longrightarrow NaHCO_3(s) + NH_4^+ + Cl^- \tag{22.7}$$

Sodium carbonate is produced from the bicarbonate by heating.

Ordinary baking soda is $NaHCO_3$. When dough containing $NaHCO_3$ is baked, reaction (22.8) occurs. The released CO_2(g) raises the dough, creating a "light and airy" baked product, i.e., one filled with holes.

$$2\ NaHCO_3(s) \xrightarrow{\Delta} Na_2CO_3(s) + H_2O(g) + CO_2(g) \tag{22.8}$$

The Solvay process is outlined in Figure 22-4.

The early success of the Solvay process stemmed from the efficient use of raw materials through recycling. Only one by-product results, $CaCl_2$. The demand for $CaCl_2$ is quite limited, however, and only a small percentage of the several million tons produced annually is consumed. The rest has commonly been dumped, generally into streams, but this is no longer permitted. Ways must be found to recover and reuse $CaCl_2$ if the Solvay process is to continue in use. Here is an example where economic feasibility is no longer the critical factor. Environmental concerns are overriding. Fortunately, natural deposits from which Na_2CO_3 can be extracted, such as those found in Wyoming, should be sufficient to meet demands for several thousand years.

FIGURE 22-4
The Solvay process for the manufacture of $NaHCO_3$.

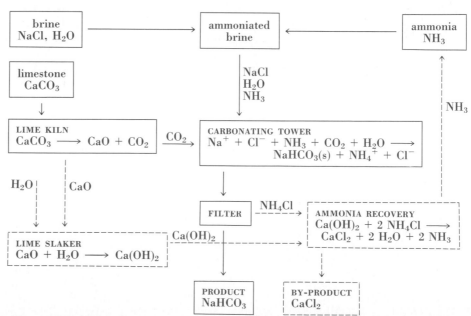

The main reaction sequence is traced by solid arrows. Recycling reactions are shown by broken arrows.

22-2 Group IIA—The Alkaline Earth Metals

Table 22-3 lists some of the same properties for the alkaline earth metal atoms and ions as did Table 22-1 for the alkali metals. These data allow us to make certain general statements about the IIA metals and comparisons to group IA.

Metallic radii increase with increasing atomic number in the manner we have come to expect. Ionic radii are much smaller than the metallic radii because the ions carry a net charge of +2. The IIA metal ions (M^{2+}) are also considerably smaller than the corresponding IA metal ions (M^+) because of this additional positive charge. Because of their high charges and small radii, the IIA metal ions carry a high positive charge density (e.g., +3.08 for Mg^{2+} compared to +1.05 for Na^+). The energy required to strip an alkaline earth metal atom of its two valence electrons—the sum of the first and second ionization energies—is very large, e.g., +2188 kJ/mol Mg. Removal of just a single electron can be accomplished with a much smaller expenditure of energy, e.g., +737 kJ/mol Mg. Why do the alkaline earth metals form compounds such as MCl_2 instead of MCl? The answer comes through an analysis of all the steps involved in ionic compound formation (recall the Born–Haber cycle in Section 11-12). The additional energy required to produce dipositive from unipositive alkaline earth metal ions is more than offset by the increased lattice energies of MCl_2(s) over MCl(s) and, for species in aqueous solution, by the much larger hydration energies associated with M^{2+}(aq) compared to M^+(aq). This comparison is further explored in Additional Exercise 12.

The group IIA ions, because of their large negative reduction potentials, are difficult to reduce to the free metal. So, as with the group IA metals, we find that electrolysis is an important commercial method for their preparation. The commercial production of magnesium is described next.

Production of Magnesium. Although some magnesium is prepared by thermal reduction (generally using silicon as a reducing agent), the principal methods for its industrial production are based on electrolysis of the molten chloride. In the **Dow process** magnesium is precipitated from seawater as the hydroxide; Mg^{2+} is the only common cation in seawater that forms an insoluble hydroxide. The source of OH^- is $Ca(OH)_2$, which is derived by heating limestone or oyster shells and treating the resulting CaO with water (recall equations 13.27 and 13.28). The precipitated $Mg(OH)_2$(s) is washed, filtered, and dissolved in HCl(aq). The concentrated $MgCl_2$(aq) that results is evaporated nearly to dryness. The $MgCl_2$ is then melted and electrolyzed, yielding pure Mg metal and Cl_2(g). The Cl_2(g) is recycled by conversion to HCl. The Dow process is outlined in Figure 22-5, and the electrolysis of $MgCl_2$(l) is pictured in Figure 22-6.

TABLE 22-3
Properties of the alkaline earth metals[a]

	Be	Mg	Ca	Sr	Ba
metallic radius, pm	111	160	197	215	217
ionic radius, pm	31	65	99	113	135
ionic charge density [ionic charge/(ionic radius, Å)]	+6.45	+3.08	+2.02	+1.77	+1.48
1st + 2nd ionization energies [$M(s) \longrightarrow M^{2+}(g) + 2\ e^-$], kJ/mol	+2657	+2188	+1735	+1614	+1468
hydration energy, kJ/mol [$M^{2+}(g) + aq \longrightarrow M^{2+}(aq)$]	−2385	−1940	−1600	−1460	−1320
electrode potential $E°$, V [$M^{2+}(aq) + 2\ e^- \longrightarrow M(s)$]	−1.70	−2.375	−2.76	−2.89	−2.90

[a]Several physical properties are listed in Table 13-4.

FIGURE 22-5
The Dow process for the production of Mg.

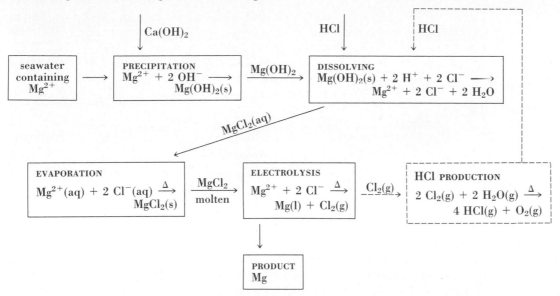

The main reaction sequence is traced by solid arrows. Recycling of $Cl_2(g)$ is shown by broken arrows.

Of the major structural metals, Fe and Al occur more abundantly in the earth's crust than does Mg, but their production will become more difficult as high-grade ores are depleted. Magnesium, on the other hand, occurs to the extent of 0.14% by mass in seawater and will remain at this concentration indcfinitely. All the Mg that has yet been produced could have been obtained from just a few km^3 of seawater. In addition to its many applications in lightweight alloys, magnesium is used in batteries, in fireworks and flash photography, in the formation of important reagents for organic chemical reactions, and as a reducing agent in the production of such metals as Ti and Zr.

FIGURE 22-6
The electrolysis of molten $MgCl_2$.

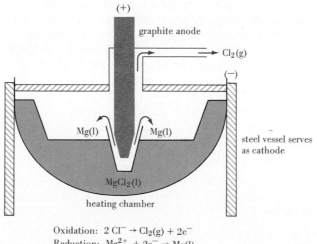

Oxidation: $2 Cl^- \rightarrow Cl_2(g) + 2e^-$
Reduction: $Mg^{2+} + 2e^- \rightarrow Mg(l)$

The actual electrolyte is a mixture of molten NaCl, $CaCl_2$, and $MgCl_2$. This mixture has a lower melting point and higher electrical conductivity than does $MgCl_2$ alone, but only Mg^{2+} is reduced under the conditions employed in the electrolysis.

The precipitation of $Mg(OH)_2$ from seawater is carried out in large circular vats. This is the first step in the Dow process for extracting magnesium from seawater. [Courtesy of The Dow Chemical Company.]

Industrial operations that involve electrolytic reactions and that require materials to be heated consume large quantities of energy. Their feasibility depends on abundant sources of energy and its efficient use. Presently, energy consumption in the manufacture of magnesium averages about 300 MJ/kg Mg (300 MJ = 300×10^6 joules). Process modifications under development may reduce this requirement to about 200 MJ/kg Mg. By contrast, the energy requirement to melt and recast recycled Mg is about 7 MJ/kg Mg. Recycling, where possible, is an important consideration in the production and use of materials.

Alkaline Earth Metal Compounds.   As we have already noted, alkaline earth cations have high positive charge densities. In combination with certain anions this leads to very high lattice energies and salts that are either insoluble or only moderately soluble in water (e.g., carbonates, fluorides, hydroxides). In other cases, where the lattice energy is not quite so high (e.g., in combination with large uninegative anions), the large hydration energies of the cations may result in high water solubility and salts that crystallize from solution as hydrates. Typical hydrates are $MX_2 \cdot 6H_2O$ (where M = Mg, Ca, or Sr, and X = Cl or Br).

Salts with high water solubilities produce a large lowering of the vapor pressure in their concentrated aqueous solutions. In some cases this vapor pressure lowering is so great that water vapor in the air may exist at a high enough pressure to condense on a salt and dissolve it. The salt is said to **deliquesce.** All the hydrates $MX_2 \cdot 6H_2O$ mentioned above are deliquescent. In fact, $CaCl_2 \cdot 6H_2O$ (the by-product of the Solvay process) has been used to control dust on dirt roads. The hydrate deliquesces and the $CaCl_2(aq)$ produced keeps the road wet.

An important feature of the alkaline earths not shared by the alkali metals (except Li) is the instability of their carbonates at high temperatures.

$$MCO_3(s) \xrightarrow{\Delta} MO(s) + CO_2(g) \qquad (22.9)$$

Decomposition of the carbonate affords a means of preparing the metal oxide other than direct combination of the metal and oxygen. And from the oxide the hydroxide can easily be formed.

$$MO(s) + H_2O \longrightarrow M(OH)_2(s) \qquad (22.10)$$

Important uses of $CaCO_3$, CaO, and $Ca(OH)_2$ in the construction industry were discussed in Section 13-6.

The ready availability of calcium carbonate and the simplicity of reactions (22.9) and (22.10) make $Ca(OH)_2$ the cheapest alkaline substance for use in the chemical industry.

Calcium Ion in Natural Waters. Carbonates, especially $CaCO_3$, are involved in chemical processes that are responsible for a number of natural phenomena. One process begins as rainwater dissolves atmospheric $CO_2(g)$. This makes the water slightly acidic, through the formation and ionization of carbonic acid.

$$CO_2(g) + H_2O \rightleftharpoons H_2CO_3 \tag{22.11}$$

$$H_2CO_3 + H_2O \rightleftharpoons H_3O^+ + HCO_3^- \qquad K_{a_1} = 4.2 \times 10^{-7} \tag{22.12}$$

$$HCO_3^- + H_2O \rightleftharpoons H_3O^+ + CO_3^{2-} \qquad K_{a_2} = 5.6 \times 10^{-11} \tag{22.13}$$

As CO_2-charged rainwater seeps through limestone beds, insoluble $CaCO_3$ is converted to soluble $Ca(HCO_3)_2$.

$$CaCO_3(s) + H_2O + CO_2 \rightleftharpoons Ca(HCO_3)_2(aq) \tag{22.14}$$

Over time this dissolving action can produce a large cavity in the limestone bed—a limestone cave. Reaction (22.14) is reversible, however, and the evaporation of a solution of $Ca(HCO_3)_2$ causes a loss of both water and CO_2 and conversion of $Ca(HCO_3)_2$ back to $CaCO_3(s)$. This process occurs very slowly; but over a period of many years, as $Ca(HCO_3)_2(aq)$ drips from the ceiling of a cave, $CaCO_3(s)$ remains as icicle-like deposits called **stalactites.** Some of the dripping $Ca(HCO_3)_2(aq)$ may hit the floor of the cave before decomposition occurs and limestone ($CaCO_3$) deposits build up from the floor in formation called **stalagmites.** Eventually, some stalactites and stalagmites grow together into limestone columns.

Example 22-3 *Without performing detailed calculations,* show how the data of equations (22.12) and (22.13), together with K_{sp} for $CaCO_3$ (2.8×10^{-9}), can be used to establish that reaction (22.14) describes the dissolving action of rainwater on limestone ($CaCO_3$).

Solution. We need to use two ideas considered earlier in the text.

1. In $H_2CO_3(aq)$, a diprotic acid, since $K_{a_1} \gg K_{a_2}$, $[H_3O^+] = [HCO_3^-]$ and $[CO_3^{2-}] = K_{a_2} = 5.6 \times 10^{-11} M$ (see again, page 519).
2. When pure water is saturated with $CaCO_3$, $[Ca^{2+}] = [CO_3^{2-}]$; $K_{sp} = [Ca^{2+}][CO_3^{2-}] = [CO_3^{2-}]^2 = 2.8 \times 10^{-9}$. In this solution $[CO_3^{2-}] = \sqrt{2.8 \times 10^{-9}} = 5.3 \times 10^{-5} M$.

$[CO_3^{2-}]$ in saturated $CaCO_3(aq)$ is much higher than $[CO_3^{2-}]$ normally produced by the ionization of H_2CO_3. Carbonate ion derived from $CaCO_3$ acts as a common ion and displaces equilibrium (22.13) *to the left,* converting CO_3^{2-} to HCO_3^- and consuming H_3O^+. Removal of H_3O^+ in this way stimulates equilibrium (22.12) to shift *to the right* to produce more H_3O^+. Equilibrium (22.11) also shifts *to the right* to replace the H_2CO_3 that has ionized. The net reaction is that CO_2, H_2O, and CO_3^{2-} (from $CaCO_3$) are consumed and HCO_3^- [as $Ca(HCO_3)_2$] is produced, just as indicated by equation (22.14).

SIMILAR EXAMPLES: Review Problem 6; Exercises 8a, 9.

As we have just noted, rainwater, because of its solvent action on atmospheric gases, soil, and rocks, is not chemically pure. It may acquire anywhere from a few parts to perhaps 1000 parts per million (ppm) of dissolved substances. Water containing ions capable of yielding significant quantities of a precipitate is said to be **hard.** Water that owes its hardness primarily to HCO_3^- and associated cations is said to be **temporary hard water.**

The process responsible for limestone formations in natural caves—the reverse of reaction (22.14)—is greatly accelerated when water is heated. Equilibria are displaced in the reverse direction from that discussed in Example 22-3: H_2CO_3 breaks down into H_2O and CO_2, which escape. Some bicarbonate ions combine with H_3O^+ to replace the

Section of a pipe with boiler scale derived from hard water. [Courtesy of Nalco Chemical Company.]

H_2CO_3; and to replenish the H_3O^+ consumed in this way, an equal number of HCO_3^- ions ionize further to H_3O^+ and CO_3^{2-}. The consequence of these shifts in equilibria is a net reaction in which HCO_3 decomposes to CO_3^{2-}, H_2O, and CO_2.

$$\cancel{H_2CO_3} \xrightarrow{\Delta} H_2O + CO_2(g)$$
$$HCO_3^- + \cancel{H_3O^+} \longrightarrow \cancel{H_2O} + \cancel{H_2CO_3}$$
$$\underline{HCO_3^- + \cancel{H_2O} \longrightarrow \cancel{H_3O^+} + CO_3^{2-}}$$
$$2\ HCO_3^- \xrightarrow{\Delta} CO_3^{2-} + H_2O + CO_2 \tag{22.15}$$

Reaction (22.15) represents the principal deleterious effect of temporary hard water. CO_3^{2-} regenerated when the water is heated reacts with multivalent cations in the water to form a mixed precipitate of $MgCO_3$, $CaCO_3$, and $FeCO_3$ called **boiler scale.** The formation of boiler scale is a very serious problem in many industries. For example, in steam power plants the formation of scale can cause a boiler to overheat, presenting an explosion hazard.

Water Softening. Hard water is acceptable for some applications, but not for others. The term "water softening" refers to the removal of natural mineral impurities, and there are several ways in which this can be accomplished. Simple boiling will soften temporary hard water, but with the formation of unwanted boiler scale. Water containing significant concentrations of anions other than HCO_3^-, e.g., SO_4^{2-}, together with associated cations, is called **permanent hard water.** This is because such water cannot be softened simply by heating.

Another suitable method for softening temporary hard water is to treat it with a base and filter off the precipitated metal carbonate.

$$HCO_3^- + OH^- \longrightarrow H_2O + CO_3^{2-} \tag{22.16}$$

$$CO_3^{2-} + M^{2+} \longrightarrow MCO_3(s) \tag{22.17}$$

The source of OH^- may be **slaked lime,** $Ca(OH)_2$, or **washing soda,** Na_2CO_3. (Can you write an equation to show how Na_2CO_3 produces OH^- by hydrolysis?) Permanent hard water can also be softened with Na_2CO_3. Cations such as Ca^{2+} and Mg^{2+} precipitate as carbonates, and Na_2SO_4 remains in solution.

(a)

(b)

(c)

(d)

Reactions of Ca^{2+} and aqueous solutions of CO_2. **(a)** CO_2 (dry ice) is added to dilute $Ca(OH)_2(aq)$. **(b)** $CaCO_3(s)$ precipitates. **(c)** The precipitate redissolves in the presence of excess $CO_2(aq)$ to form $Ca(HCO_3)_2(aq)$. **(d)** $Ca(HCO_3)_2(aq)$ decomposes on heating and reprecipitates as $CaCO_3(s)$. (See also, Exercise 9 and Additional Exercise 11.)

FIGURE 22-7
Ion exchange process.

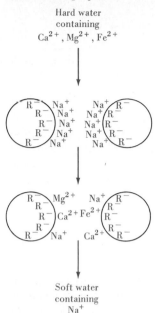

Hard water
containing
Ca^{2+}, Mg^{2+}, Fe^{2+}

Soft water
containing
Na^+

The resin pictured here is a cation exchange resin: multivalent cations (Ca^{2+}, Mg^{2+}, Fe^{2+}) are exchanged for univalent cations (Na^+). This resin can be represented as RNa. Other ion exchange resins, designated as ROH, exchange OH^- for other anions; these are anion exchange resins.

Ion Exchange. One of the best methods of softening water is through **ion exchange.** The ion exchange medium may be a natural sodium aluminosilicate, called a **zeolite,** or a synthetic resinous material. Ion exchange materials consist of macromolecular (polymer) particles capable of ionizing to produce fixed ions, which remain attached to the particle surfaces, and free or mobile counterions. It is the counterions that may exchange positions with other ions in a solution passed through a bed of these particles.

In the resin pictured in Figure 22-7, the fixed ions, R, are negatively charged and the counterions are positive. At the start the counterions in the resin bed are Na^+. When hard water is passed through the bed, Ca^{2+}, Mg^{2+}, and Fe^{2+}, because of their higher charge, displace Na^+ as counterions. To regenerate the resin concentrated NaCl(aq) is passed through. In high concentration Na^+ ions displace the multivalent ions from the resin particles, restoring the resin to its original condition. The ion exchange material has an indefinite lifetime. The only material consumed in softening water by this method is the sodium chloride used in the regenerating process. However, water softening by this method is not without some disadvantages. Replacement of multivalent cations by Na^+ would be detrimental to anyone on a low-sodium diet. Ion exchange processes can be represented through simple chemical equations. If a zeolite is involved, the symbol Z is used; if a synthetic resin, R.

$$Na_2Z + M^{2+} \longrightarrow MZ + 2\ Na^+ \tag{22.18}$$

$$Na_2R + M^{2+} \longrightarrow MR + 2\ Na^+ \tag{22.19}$$

An ion exchange resin can have H^+ instead of Na^+ as its counterions, for example, by flushing the resin with concentrated HCl(aq) instead of NaCl(aq). When water containing multivalent cations is passed through this H^+-charged resin, the exchange reaction is

$$H_2R + M^{2+} \longrightarrow MR + 2\ H^+ \tag{22.20}$$

This reaction presents two possibilities: (1) Titration with a standard base of the H^+ liberated when a sample of hard water is passed through the resin gives a measure of the hardness of the water, as illustrated in Example 22-4. (2) Water in which all cations have been replaced by H^+ can next be passed through a second ion exchange resin in which all *anions* are replaced by OH^-. The H^+ and OH^- combine to form H_2O. Because the water sample is now essentially free of all ions, it is called **deionized water.** Deionized water is commonly used in the chemical laboratory because ions present in ordinary tap water generally interfere with chemical reactions.

Example 22-4 After being passed through a column containing the ion exchange resin H_2R, a 100.0-mL sample of hard water requires 15.17 mL of 0.0265 *M* NaOH for its titration. What is the hardness of the water, expressed as ppm Ca^{2+}?

Solution
Step 1. Determine the no. mmol H^+ present in the 100.0-mL water sample. The titration reaction is simply $H^+ + OH^- \to H_2O$. The no. mmol H^+ is equal to the no. mmol OH^- used in the titration.

$$\text{no. mmol}\ H^+ = 15.17\ \text{mL NaOH soln.} \times \frac{0.0265\ \text{mmol}\ OH^-}{\text{mL NaOH soln.}} \times \frac{1\ \text{mmol}\ H^+}{1\ \text{mmol}\ OH^-}$$
$$= 0.402\ \text{mmol}\ H^+$$

Step 2. Determine the no. mmol Ca^{2+} originally in the 100.0-mL water sample. According to equation (22.20), each Ca^{2+} in the hard water releases two H^+ from the ion exchange resin.

$$\text{no. mmol}\ Ca^{2+} = 0.402\ \text{mmol}\ H^+ \times \frac{1\ \text{mmol}\ Ca^{2+}}{2\ \text{mmol}\ H^+} = 0.201\ \text{mmol}\ Ca^{2+}$$

Step 3. Determine the mass of Ca^{2+} in the 100.0-mL water sample.

$$\text{no. g } Ca^{2+} - 0.201 \text{ mmol } Ca^{2+} \times \frac{1 \text{ mol } Ca^{2+}}{1000 \text{ mmol } Ca^{2+}} \times \frac{40.08 \text{ g } Ca^{2+}}{1 \text{ mol } Ca^{2+}}$$
$$= 8.06 \times 10^{-3} \text{ g } Ca^{2+}$$

Many U.S. government agencies report water analyses in terms of individual ions (such as ppm Ca^{2+}). Another widely used method is to report hardness in terms of a $CaCO_3$ equivalent: 80.6 ppm Ca^{2+} is equivalent to $(80.6 \times 100.1/40.08) = 201$ ppm $CaCO_3$.

Step 4. Express the result of Step 3 as parts per million (ppm). The sample used in Steps 1, 2, and 3 was 100.0 mL = 100 g (density of water = 1.00 g/mL). That is,

$$\frac{8.06 \times 10^{-3} \text{ g } Ca^{2+}}{100 \text{ g water}} = \frac{8.06 \times 10^{-5} \text{ g } Ca^{2+}}{\text{g water}}$$

The quantity of Ca^{2+} present is 8.06×10^{-5} g Ca^{2+} *per g water*. In one million (1.00×10^6) g water the mass of Ca^{2+} is

$$\text{no. g } Ca^{2+} = 1.00 \times 10^6 \text{ g water} \times \frac{8.06 \times 10^{-5} \text{ g } Ca^{2+}}{\text{g water}} = 80.6 \text{ g } Ca^{2+}$$

But this is the answer we are seeking: 80.6 g Ca^{2+} in 1.00×10^6 g water is 80.6 ppm Ca^{2+}.

SIMILAR EXAMPLES: Review Problems 7, 8.

Soaps and Detergents. One of the most frequently encountered effects of hard water in the household is its action on soaps and detergents. Some fundamental ideas about the molecular structures of these common materials are presented in Figure 22-8. We can

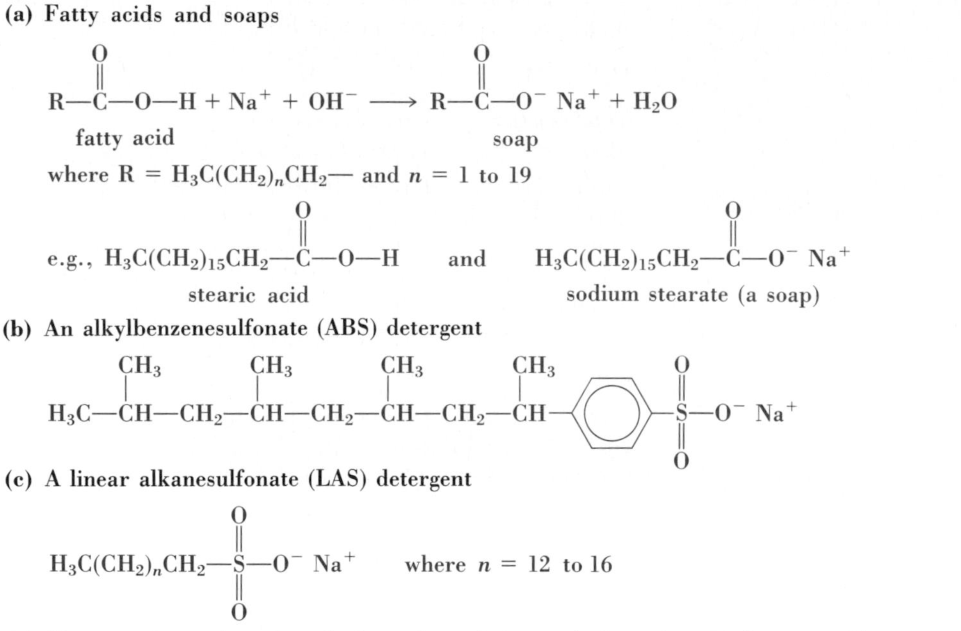

FIGURE 22-8
Structural features of fatty acids, soaps, and detergents.

(a) Fatty acids and soaps

$$R-\overset{\overset{\textstyle O}{\|}}{C}-O-H + Na^+ + OH^- \longrightarrow R-\overset{\overset{\textstyle O}{\|}}{C}-O^- \ Na^+ + H_2O$$

fatty acid   soap

where R = $H_3C(CH_2)_nCH_2-$ and n = 1 to 19

e.g., $H_3C(CH_2)_{15}CH_2-\overset{\overset{\textstyle O}{\|}}{C}-O-H$ and $H_3C(CH_2)_{15}CH_2-\overset{\overset{\textstyle O}{\|}}{C}-O^- \ Na^+$

stearic acid   sodium stearate (a soap)

(b) An alkylbenzenesulfonate (ABS) detergent

$$H_3C-\underset{\underset{\textstyle CH_3}{|}}{CH}-CH_2-\underset{\underset{\textstyle CH_3}{|}}{CH}-CH_2-\underset{\underset{\textstyle CH_3}{|}}{CH}-CH_2-\underset{\underset{\textstyle CH_3}{|}}{CH}-\bigcirc\!\!\!\!\bigcirc-\overset{\overset{\textstyle O}{\|}}{\underset{\underset{\textstyle O}{\|}}{S}}-O^- \ Na^+$$

(c) A linear alkanesulfonate (LAS) detergent

$$H_3C(CH_2)_nCH_2-\overset{\overset{\textstyle O}{\|}}{\underset{\underset{\textstyle O}{\|}}{S}}-O^- \ Na^+ \qquad \text{where } n = 12 \text{ to } 16$$

(a) Fatty acids are based on hydrocarbon chains with from 4 to 22 C atoms. The carbon atom at one end of the chain is in a carboxyl group, $-\overset{\overset{\textstyle O}{\|}}{C}-O-H$. Ordinary soaps are the sodium or potassium salts of fatty acids.

(b) An alkylbenzenesulfonate (ABS) detergent can be thought of as a derivative of sulfuric acid, $HO-SO_2-OH$. A benzene ring with its attached hydrocarbon chain substitutes for one of the $-OH$ groups, and Na replaces the remaining H atom. Because of branching in the hydrocarbon chain, these molecules are resistant to the action of microorganisms—they are not biodegradable.

(c) A linear alkanesulfonate (LAS) detergent can also be thought of as a derivative of sulfuric acid, but the hydrocarbon portion of the molecule has no chain branching. These molecules are biodegradable.

FIGURE 22-9
Representation of the cleaning action of soap.

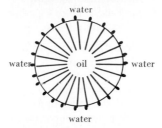

The cleaning action results from the structure of the soap molecules. Each molecule has both a long nonpolar portion (shown in color) and a polar carboxyl group. The interface between an oil droplet and the aqueous medium in which it is suspended is lined with soap molecules having their nonpolar ends in the oil and polar ends in the water. The oil droplet is solubilized.

represent a typical soap as the ionic compound $RCOO^- Na^+$, and note that alkali metal soaps are soluble in water. However, soaps of multivalent cations are not water soluble, as seen in this all-too-familiar reaction.

$$2\ Na^+ + 2\ RCOO^- + Ca^{2+} \longrightarrow Ca(RCOO)_2(s) + 2\ Na^+ \qquad (22.21)$$

soap   ''bathtub ring''

Soaps are effective (though expensive) water softeners, but unfortunately the object being cleansed becomes coated with the precipitated heavy metal soaps. One solution to this problem is to use softened water, that is, water that has been treated according to reactions (22.16) and (22.17) or (22.18) or (22.19). Another solution is to use a synthetic detergent. **Synthetic detergents** are salts of certain organic sulfonic acids, i.e, $RSO_3^- Na^+$. The great advantage of a detergent over a soap is that its calcium and other heavy metal salts are soluble. [Although detergents do not form precipitates in hard water, they do not function effectively unless a builder is present to sequester multivalent cations (see again, page 657).]

Soaps and detergents function by solubilizing or emulsifying oils with water. Figure 22-9 suggests how soap molecules bring this about.

22-3 The Group IIIA Metals

Boron, the first element in group IIIA, is a metalloid. Aluminum, in its physical appearance and physical properties and in much of its chemical behavior, is metallic. So are the remaining members of group IIIA—gallium, indium, and thallium. Some properties of the group IIIA metals are listed in Table 22-4.

Negative reduction potentials indicate that these elements are more metallic than hydrogen. The ionization energies of the group IIIA metals are comparable to one another, but the hydration energy of Al^{3+} is largest among the group IIIA cations. This accounts for the fact that Al^{3+} has the most negative reduction potential of the group IIIA cations and that Al is the most active of the group IIIA metals.

An interesting feature of the elements Ga, In, and Tl not found in Al is the ability to form unipositive ions. This ability is usually ascribed to the existence of an inert pair of electrons, **ns^2**, in atoms of post-transition elements. Thus, a Ga atom can lose its $4p$ electron and retain its $4s$ electrons to form the ion, Ga^+, with the electron configuration $[Ar]3d^{10}4s^2$. This possibility becomes more pronounced with heavier members of the group. In fact, with thallium the +1 oxidation state is more stable in aqueous solution than is the +3 oxidation state.

$$Tl^{3+}(aq) + 2\ e^- \longrightarrow Tl^+(aq) \qquad E^\circ = +1.25\ V \qquad (22.22)$$

TABLE 22-4
Some properties of the group IIIA metals

	Al	Ga	In	Tl
melting point, °C	660.4	29.8	156.6	303.5
boiling point, °C	2467	2403	2080	1457
electron configuration	$[Ne]3s^23p^1$	$[Ar]3d^{10}4s^24p^1$	$[Kr]4d^{10}5s^25p^1$	$[Xe]4f^{14}5d^{10}6s^26p^1$
metallic radius, pm	143	141	166	171
ionic radius, pm				
M^+	—	113	132	140
M^{3+}	50	62	81	95
electrode potential, E°, V				
$[M^{3+}(aq) + 3\ e^- \longrightarrow M(s)]$	−1.66	−0.56	−0.34	+0.72
$[M^+(aq) + e^- \longrightarrow M(s)]$	—	—	−0.25	−0.34

The chemistry of gallium is similar to that of aluminum. Thallium bears some similarity to lead, e.g., in its high density (11.85 g/cm^3), softness, and the poisonous nature of its compounds. Thallium(I) compounds resemble alkali metal compounds in some respects—TlOH is a strong base and the Tl^+ ion (radius 140 pm) can crystallize together with the K^+ ion (radius 133 pm) in chlorates, perchlorates, sulfates, and various phosphates. In other respects Tl^+ resembles Ag^+, as in forming a light-sensitive, insoluble chloride, TlCl.

Gallium remains a liquid over one of the longest temperature ranges of any substance (over two thousand degrees Celsius), and because of this property it finds some use in high-temperature thermometry. Ga, In, and Tl and their compounds have a number of commercial uses—alloys, transistors, photoconductors, specialty glasses—but none of these requires large quantities of the metals.

Aluminum Ion in Solution.   Two factors must be considered in assessing the water solubilities of aluminum compounds: the small size and high charge of the Al^{3+} ion and its very high hydration energy (-4613 kJ/mol). When Al^{3+} is combined with small, highly charged anions, the high lattice energies of the resulting solids render them insoluble in water. Such is the case with Al_2O_3. Even AlF_3, a combination of Al^{3+} with the univalent anion F^-, has only a low water solubility (about 0.07 M). $AlCl_3$, $AlBr_3$, and AlI_3 have considerable covalent character (see again, Section 13-7). These compounds are quite water soluble.

A number of aluminum salts, like those of the group IIA metals, crystallize from solution as hydrates. Some of these hydrates are highly soluble and therefore deliquescent, such as $AlX_3 \cdot 6H_2O$ (where X = Cl, Br, I, or ClO_3) and $Al(NO_3)_3 \cdot 9H_2O$. Moreover, certain aspects of the aqueous chemistry of aluminum compounds derive from the nature of the hydrated aluminum ion—$[Al(H_2O)_6]^{3+}$. One distinctive feature is that aqueous solutions of aluminum salts are *acidic*, which comes about in this way: Because of the attractive power of the small, highly charged Al^{3+} ion for electrons, an O—H bond in a ligand H_2O molecule breaks. A proton is released to an H_2O molecule outside the coordination sphere. The original H_2O ligand is converted to OH^-, and the complex ion to $[Al(H_2O)_5OH]^{2+}$.

The acidic nature of $[Al(H_2O)_6]^{3+}$ is illustrated in Color Section F.

$$[Al(H_2O)_6]^{3+} + H_2O \rightleftharpoons H_3O^+ + [Al(H_2O)_5OH]^{2+} \qquad (22.23)$$

Reaction (22.23) is reversible. In strongly acidic solutions the ionization of $[Al(H_2O)_6]^{3+}$ is repressed, but in alkaline solution its ionization is favored. If a solution is made sufficiently alkaline, reaction (22.23) can be carried to completion and be followed by further ionization of other H_2O ligand molecules.

$$[Al(H_2O)_5OH]^{2+} + H_2O \rightleftharpoons H_3O^+ + [Al(H_2O)_4(OH)_2]^+ \qquad (22.24)$$

$$[Al(H_2O)_4(OH)_2]^+ + H_2O \rightleftharpoons H_3O^+ + Al(H_2O)_3(OH)_3 \qquad (22.25)$$

The product of reaction (22.25) is **hydrated aluminum hydroxide,** $Al(OH)_3 \cdot 3H_2O$. Actually, what we have demonstrated through the preceding three equations is the precipitation of Al^{3+} as its hydroxide. If we return to these three equations, add an OH^- to both sides of each equation to combine with the H_3O^+ produced in the ionization, add the three equations, and cancel out some H_2O molecules, we obtain

$$[Al(H_2O)_6]^{3+} + 3\ OH^- \longrightarrow Al(OH)_3 \cdot 3H_2O(s) + 3\ H_2O \qquad (22.26)$$

Of course, simpler ways to represent this precipitation are

$$[Al(H_2O)_6]^{3+} + 3\ OH^- \longrightarrow Al(OH)_3(s) + 6\ H_2O \qquad (22.27)$$

and

$$Al^{3+}(aq) + 3\ OH^-(aq) \longrightarrow Al(OH)_3(s) \qquad (22.28)$$

TABLE 22-5
Some gemstones based on Al_2O_3

Gem	Impurity
white sapphire	none
blue sapphire	Fe, Ti
green sapphire	Co
yellow sapphire	Ni, Mg
star sapphire	Ti
ruby	Cr

In a strongly basic solution, such as in NaOH(aq), reaction (22.25) can be carried a step further.

$$Al(H_2O)_3(OH)_3(s) + H_2O \rightleftharpoons [Al(H_2O)_2(OH)_4]^-(aq) + H_3O^+ \qquad (22.29)$$

Now the Al^{3+} becomes the central ion of a complex *anion* called **aluminate ion.** For simplicity this ion is usually represented as $[Al(OH)_4]^-$ by eliminating the two H_2O ligand molecules (and sometimes even as AlO_2^- by eliminating still two more H_2O molecules from $[Al(OH)_4]^-$). The fact that $Al(OH)_3(s)$ is soluble in alkaline solution can be represented somewhat more simply by the equation

$$Al(OH)_3(s) + OH^- \longrightarrow [Al(OH)_4]^-(aq) \qquad (22.30)$$

To summarize this discussion, the principal species one expects to encounter in the aqueous chemistry of Al^{3+} are

$$[Al(H_2O)_6]^{3+} \text{ (or } Al^{3+}) \underset{H_3O^+}{\overset{OH^-}{\rightleftharpoons}} Al(OH)_3(s) \underset{H_3O^+}{\overset{OH^-}{\rightleftharpoons}} [Al(OH)_4]^- \qquad (22.31)$$

Equation (22.31) demonstrates the amphoteric nature of $Al(OH)_3$ (and also Al_2O_3). It acts either as an acid or a base and dissolves in either an acid or a base, a fact that we established in a somewhat different way in Section 17-8. Equation (22.31) can also help us to understand a statement made in Chapter 13 that aluminum is one of the few metals that is soluble in alkaline as well as acidic solutions. With its strong tendency to be oxidized to Al^{3+}, we should expect Al(s) to displace $H_2(g)$ from water, just as do the IA and the heavier IIA metals.

$$2\,Al(s) + 6\,H_2O \longrightarrow 2\,Al(OH)_3(s) + 3\,H_2(g) \qquad (22.32)$$

This reaction probably does occur on a clean aluminum surface, but the surface film of $Al(OH)_3$ (or hydrated Al_2O_3) protects the underlying metal. However, in acidic solution or in the presence of a strong base this surface film dissolves and the metal undergoes further attack.

$$2\,Al(OH)_3(s) + 2\,OH^-(aq) \longrightarrow 2[Al(OH)_4]^-(aq) \qquad (22.33)$$

The net reaction when Al is brought into contact with a strongly basic solution is the sum of reactions (22.32) and (22.33), and this proves to be identical to the reaction that we first described in Chapter 13.

$$2\,Al(s) + 2\,OH^-(aq) + 6\,H_2O \longrightarrow 2[Al(OH)_4]^-(aq) + 3\,H_2(g) \qquad (13.34)$$

Aluminum Compounds. The naturally occurring forms from which most aluminum compounds are derived are the **oxide,** Al_2O_3, and various **hydrated oxides,** e.g., $Al_2O_3 \cdot H_2O$ and $Al_2O_3 \cdot 3H_2O$. The mineral **corundum,** which is used as an abrasive, is Al_2O_3. The ore **bauxite,** which is used in the manufacture of aluminum metal, is a mixture of the mono- and trihydrates. Many familiar gemstones are composed of Al_2O_3 with small quantities of other metal oxides as impurities (see Table 22-5). Artificial gemstones can be prepared by fusing corundum with the appropriate metal oxides. Other uses of Al_2O_3 (alumina) are as a refractory lining for high-temperature furnaces and as a catalyst support in industrial chemical processes.

Although some forms of Al_2O_3 are resistant to attack by acids and alkalis, it is possible to dissolve Al_2O_3 in both acidic and basic media to obtain the forms of aluminum ion described in the preceding section. For example, aluminum sulfate is prepared by the action of hot, concentrated $H_2SO_4(aq)$ on Al_2O_3.

$$Al_2O_3(s) + 3\,H_2SO_4(\text{conc. aq}) \longrightarrow Al_2(SO_4)_3(aq) + 3\,H_2O \qquad (22.34)$$

It crystallizes from solution as $Al_2(SO_4)_3 \cdot 18H_2O$.

Aluminum cookware should never be used to contain strongly alkaline substances.

Production of artificial gemstones is illustrated in Color Section D.

Modern foam-type fire-extinguishing systems produce foams by mechanical means rather than by chemical reaction.

One interesting use of $Al_2(SO_4)_3$ is in foam-type fire extinguishers. When the extinguisher is activated, $Al_2(SO_4)_3$(aq) and $NaHCO_3$(aq) are mixed in the presence of an emulsifying agent. The reaction is an acid–base reaction. $Al(H_2O)_6^{3+}$ goes through the series of ionizations represented in equations (22.23), (22.24), and (22.25), with HCO_3^- accepting the protons released. Aluminum precipitates as the hydroxide, and HCO_3^- is converted to CO_2(g) and water. The CO_2(g)–$Al(OH)_3$(s) mixture is produced as a foam, stabilized by the emulsifying agent. The foam blankets the burning material and extinguishes the fire. The net reaction is

$$Al_2(SO_4)_3(aq) + 6\ HCO_3^-(aq) \longrightarrow 2\ Al(OH)_3(s) + 3\ SO_4^{2-}(aq) + 6\ CO_2(g) \quad (22.35)$$

The use of $Al_2(SO_4)_3$ in water purification is described in Table 12-4.

Commercial applications consume over 1×10^6 tons of $Al_2(SO_4)_3$ annually in the United States, with about one half of this in various aspects of papermaking. Other uses include the dyeing, waterproofing, and fireproofing of fabrics; as a food additive; and in the treatment of industrial wastes, municipal water supplies, and municipal sewage.

An aqueous solution containing equimolar amounts of $Al_2(SO_4)_3$ and K_2SO_4 crystallizes as potassium aluminum sulfate, $KAl(SO_4)_2 \cdot 12H_2O$. This salt, which is variously known as potash alum, common alum, or ordinary alum, belongs to a large general class of substances called alums. **Alums** have the formula $M(I)M(III)(SO_4)_2 \cdot 12H_2O$ [where M(I) is almost any unipositive cation, except Li^+, and M(III) is a tripositive cation—Al^{3+}, Ti^{3+}, V^{3+}, Cr^{3+}, Mn^{3+}, Fe^{3+}, Co^{3+}, Ga^{3+}, In^{3+}, Re^{3+}, and Ir^{3+}]. Alums contain the ions $[M(H_2O)_6]^+$, $[M(H_2O)_6]^{3+}$, and SO_4^{2-} in the ratio 1:1:2. The most common alums have M(I) = K^+ or NH_4^+ and M(III) = Al^{3+}. Li^+ does not form alums because the ion is too small to meet the structural requirements of the crystal.

Alums are used for some of the same purposes as the simpler salts from which they are derived. One important use of potash alum is as a mordant in dyeing. The fabric to be dyed is dipped in a solution of the alum and heated with steam. Hydrolysis of $[Al(H_2O)_6]^{3+}$ deposits $Al(OH)_3$(s) into the fibers of the material and the dye is adsorbed on the $Al(OH)_3$.

A number of aluminum compounds feature covalent bonds. These compounds and some of their uses were noted in Section 13-7.

Production of Aluminum. Friedrich Wöhler isolated pure aluminum in 1827 by heating aluminum chloride with potassium. Wöhler's method was improved in the 1850s when Na was substituted for K as the reducing agent. This development lowered the price of Al from about \$90 to \$5 per pound; but it still remained a semiprecious metal used chiefly in jewelry and artwork. Napoleon III's elegant set of flatware was made of aluminum. So were the crown worn by Christian X of Denmark and the metal cap placed atop the Washington Monument in 1884. A major breakthrough in aluminum manufacture came in 1886 with the Hall–Heroult process, developed simultaneously but independently by Charles Martin Hall in the United States and Paul Heroult in France. Hall and Heroult were of the same age, and Hall was a student at Oberlin College at the time of his invention.

These insoluble impurities, called red mud, present a disposal problem. More than 10 million tons of red mud is produced annually in the United States.

A number of interesting principles are involved in the manufacture of aluminum. First, the chief ore, bauxite, contains Fe_2O_3, SiO_2, and TiO_2 as impurities in Al_2O_3. Because it is amphoteric, Al_2O_3 dissolves in NaOH(aq) and the impurities do not.

$$Al_2O_3(s) + 2\ OH^-(aq) + 3\ H_2O \longrightarrow 2[Al(OH)_4]^-(aq) \quad (22.36)$$

When the solution containing $[Al(OH)_4]^-$ is diluted with water or slightly acidified, $Al(OH)_3$(s) precipitates.

Reaction (22.37) is illustrated in Color Section F.

$$[Al(OH)_4]^-(aq) + H_3O^+ \longrightarrow Al(OH)_3(s) + 2\ H_2O \quad (22.37)$$

Pure Al_2O_3 is obtained by heating $Al(OH)_3$.

$$2\ Al(OH)_3(s) \xrightarrow{\Delta} Al_2O_3(s) + 3\ H_2O(g) \quad (22.38)$$

Al_2O_3 has an extremely high melting point—too high to make its direct electrolysis practicable. Instead, a small percentage of Al_2O_3 is dissolved in molten **cryolite,** Na_3AlF_6, which has a much lower melting point. The electrolysis cell pictured in Figure 22-10 is operated at about 950°C. Aluminum of 99.6 to 99.8% purity is obtained. The electrode reactions are not known with certainty, but the net result is

oxid: $\quad 3\{C(s) + 2\ O^{2-} \longrightarrow CO_2(g) + 4\ e^-\}$
red: $\quad\ \ 4\{Al^{3+} + 3\ e^- \longrightarrow Al(l)\}$

net: $\quad\ \ 3\ C(s) + 4\ Al^{3+} + 6\ O^{2-} \longrightarrow 4\ Al(l) + 3\ CO_2(g)$ $\hspace{3em}$ (22.39)

We learned in Section 20-7 that aluminum metal is coated with an oxide that protects the metal from corrosion. Its maximum resistance to corrosion is exhibited between pH 4.5 and 8.5. Much of the aluminum used commercially is treated in such a way as to build up its oxide coating. In one method, called **anodizing,** an aluminum object is made the anode and a graphite rod, the cathode, in an electrolyte bath of $H_2SO_4(aq)$. The anode half-reaction is

$$2\ Al(s) + 3\ H_2O \longrightarrow Al_2O_3(s) + 6\ H^+ + 6\ e^- \hspace{3em} (22.40)$$

Al_2O_3 coatings of varying porosities and thicknesses can be obtained. Also, the oxide can be made to absorb coloring matter or other additives.

Despite the success of the Hall–Heroult process, problems remain. When this process was first introduced, cryolite was obtained from natural sources; but this material is quite rare and a *synthetic* cryolite has had to be developed, such as by the reaction

$$12\ HF + Al_2O_3 \cdot 3H_2O + 6\ NaOH \longrightarrow 2\ Na_3AlF_6 + 12\ H_2O \hspace{2em} (22.41)$$

With but one or two exceptions, all the Al produced in the world is derived from bauxite. Although Al is the most abundant metal in the earth's crust, not much of it is found as bauxite. For the past fifty years research has been in progress to develop methods of extracting alumina from clay and other mineral sources, but these methods are not yet economically feasible. And finally there is the matter of energy consumption. About five times as much energy is required to make a ton of aluminum as to make a ton of steel. Various efforts are underway to improve the energy efficiency of aluminum manufacturing, with the most effective being recycling of used aluminum. The energy required to recycle Al is only about 5% of that to produce the metal from bauxite. Currently about

FIGURE 22-10
Electrolysis cell for aluminum production.

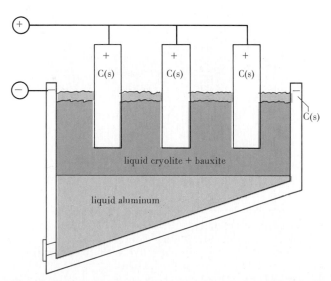

The cathode is a carbon lining in a steel tank. The anodes are also made of carbon. Liquid aluminum is more dense than the electrolyte medium and collects at the bottom of the tank. A crust of frozen electrolyte forms at the top of the cell.

20% of the Al consumed in the United States is recycled, and by the year 2000 this may reach 30%.

22-4 Extractive Metallurgy

One of the common features that we discovered in our study of the more active metals—groups IA, IIA, and Al—was that the favored commercial method of producing them is electrolysis. To produce them by chemical reduction requires expensive reducing agents (e.g., K or Na to reduce aluminum compounds). Less active metals usually are prepared by a series of steps that does include chemical reduction, often with carbon (coke or coal) as the reducing agent.

We use the general term metallurgy to refer to the study of metals and **extractive metallurgy** for the winning of pure metals from their ores. There is no single method of extractive metallurgy that works for all metals, but there are some basic operations that generally apply. We consider these next and apply them to the extractive metallurgy of zinc as a specific example.

Concentration.   In mining operations the desired mineral from which a metal is to be extracted often constitutes only a few percent (or even just a fraction of a percent) of the material mined. It is necessary to separate the desired ore from waste rock before proceeding with other metallurgical operations. One useful method that works in the few cases where an ore contains magnetic materials (e.g., Fe_3O_4) is to pass the crushed mixture of ore and rock through a magnetic field. The ore and rock are obtained in separate piles. This process is called **magnetic beneficiation.** Another method that is more generally applicable, **flotation,** is described in Figure 22-11.

Roasting.   The purpose of roasting an ore is to convert the metal to its oxide, which can then be reduced. The commercially important ores of zinc are the sulfide (sphalerite) and

FIGURE 22-11
Concentration of an ore by flotation.

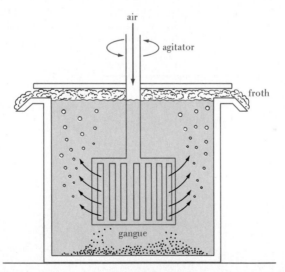

Powdered ore is suspended in water in a large vat, together with suitable additives, and the mixture is agitated with air. Particles of ore become attached to air bubbles, rise to the top of the vat, and are collected in the overflow froth. Particles of the undesired waste rock (gangue) fall to the bottom.

The success of this method depends on the use of proper additives—a material that will produce a stable foam (frother) and a substance (collector) that coats the particles of ore but does not "wet" the particles to be rejected. Pine oil is widely used as a frother and sodium ethyl xanthate as a collector.

the carbonate (smithsonite). When strongly heated, sulfides liberate $SO_2(g)$; carbonates liberate $CO_2(g)$.

$$2\ ZnS(s) + 3\ O_2(g) \xrightarrow{\Delta} 2\ ZnO(s) + 2\ SO_2(g) \tag{22.42}$$

$$ZnCO_3(s) \xrightarrow{\Delta} ZnO(s) + CO_2(g) \tag{22.43}$$

Reduction. Carbon, in the form of coke or powdered coal, is generally the preferred reducing agent where its use is possible. The reduction usually involves several reactions that occur simultaneously. Both $C(s)$ and $CO(g)$ function as reducing agents. The reduction of ZnO is carried out at about 1100°C, a temperature above the boiling point of zinc. The zinc is obtained as a vapor and condensed to the liquid.

$$ZnO(s) + C(s) \longrightarrow Zn(g) + CO(g) \tag{22.44}$$

$$ZnO(g) + CO(g) \longrightarrow Zn(g) + CO_2(g) \tag{22.45}$$

$$C(s) + CO_2(g) \rightleftharpoons 2\ CO(g) \tag{22.46}$$

Refining. The metal produced in the reduction step is usually not pure enough for its intended uses. Impurities must be removed, that is, the metal must be refined. The impurities found in zinc are mostly Cd and Pb. They are removed by fractional distillation of the liquid zinc.

About one-half of the zinc produced worldwide is refined electrolytically, mostly in a process that combines reduction and refining. ZnO from the roasting step is dissolved in $H_2SO_4(aq)$.

$$ZnO(s) + 2\ H^+(aq) + SO_4^{2-}(aq) \longrightarrow Zn^{2+}(aq) + SO_4^{2-}(aq) + H_2O \tag{22.47}$$

Powdered Zn is added to the solution to displace less active metals (e.g., Cu), and the solution is electrolyzed between lead anodes and aluminum cathodes. $H_2(g)$ does not form on the cathodes because of its high overpotential. The electrode reactions are

cathode: $\quad Zn^{2+}(aq) + 2\ e^- \longrightarrow Zn(s)$
anode: $\quad H_2O \longrightarrow \frac{1}{2} O_2(g) + 2\ H^+(aq) + 2\ e^-$
unchanged: $\quad SO_4^{2-}(aq) \longrightarrow SO_4^{2-}(aq)$
net: $\quad Zn^{2+} + SO_4^{2-} + H_2O \longrightarrow Zn(s) + 2\ H^+ + SO_4^{2-} + \frac{1}{2} O_2(g)$ (22.48)

Note that in the net electrolysis reaction Zn^{2+} is reduced to pure metallic zinc and sulfuric acid is regenerated. The acid is reused in step (22.47).

Example 22-5 Write chemical equations to represent the following metallurgical processes: **(a)** roasting of galena, PbS; **(b)** reduction of $Cu_2O(s)$, using charcoal as a reducing agent; **(c)** deposition of pure silver from an aqueous solution of Ag^+.

Solution
(a) We expect this process to be essentially the same as reaction (22.42).

$$2\ PbS(s) + 3\ O_2(g) \xrightarrow{\Delta} 2\ PbO(s) + 2\ SO_2(g)$$

(b) The simplest possible equation is

$$Cu_2O(s) + C(s) \xrightarrow{\Delta} 2\ Cu(l) + CO(g)$$

(c) This process involves a reduction half-reaction. The accompanying oxidation half-reaction is not specified. Neither is it specified whether this is an electrolysis process or whether silver is displaced by a more active metal. In either case the half-reaction is

$$Ag^+(aq) + e^- \longrightarrow Ag(s)$$

SIMILAR EXAMPLE: Review Problem 9.

Thermodynamics of Extractive Metallurgy.   To understand why carbon is able to reduce ZnO(s) at 1100°C, it is helpful to think in terms of a competition for O atoms between Zn and C. The equations below indicate that C has a greater tendency to become oxidized at 1100°C than does Zn(g). [$\Delta\overline{G}^\circ$ is more negative for reaction (a) than for (b).]

(a) $2\ C(s) + O_2(g) \longrightarrow 2\ CO(g)$ $\Delta\overline{G}^\circ \approx -460$ kJ/mol
(b) $2\ Zn(g) + O_2(g) \longrightarrow 2\ ZnO(s)$ $\Delta\overline{G}^\circ \approx -360$ kJ/mol

If we reverse (b) and add the two equations we obtain equation (22.44) and its $\Delta\overline{G}^\circ$ value.

(a) $2\ C(s) + \cancel{O_2(g)} \longrightarrow 2\ CO(g)$ $\Delta\overline{G}^\circ \approx -460$ kJ/mol
−(b) $2\ ZnO(s) \longrightarrow 2\ Zn(g) + \cancel{O_2(g)}$ $\Delta\overline{G}^\circ \approx +360$ kJ/mol

       $2\ ZnO(s) + 2\ C(s) \longrightarrow 2\ Zn(g) + 2\ CO(g)$ $\Delta\overline{G}^\circ \approx -100$ kJ/mol
or $ZnO(s) + C(s) \longrightarrow Zn(g) + CO(g)$ $\Delta\overline{G}^\circ \approx -50$ kJ/mol   (22.44)

With its large negative $\Delta\overline{G}^\circ$ value, we expect equilibrium in reaction (22.44) greatly to favor formation of the products.

   How does the competition for O atoms between C and Zn proceed at other temperatures? An especially useful tool to help answer questions like this is the graphical representation of $\Delta\overline{G}^\circ$ as a function of temperature shown in Figure 22-12. The lines   (2 C + O$_2$ → 2 CO)   and   (2 Zn + O$_2$ → 2 ZnO)   cross at about 950°C. At about this temperature Zn and C have an equal affinity for O atoms; $\Delta\overline{G}^\circ$ for reaction (22.44) is 0; and the equilibrium constant for the reaction is K = 1. At temperatures above about 950°C the magnitude of the negative $\Delta\overline{G}^\circ$ value of reaction (a) exceeds the positive value of $\Delta\overline{G}^\circ$ of the reverse of reaction (b). For reaction (22.44), $\Delta\overline{G}^\circ < 0$ (negative) and the reduction reaction proceeds spontaneously. As the temperature falls below about 950°C, the reverse situation holds. $\Delta\overline{G}^\circ$ becomes progressively more positive and the equilibrium yield of products becomes progressively smaller.

   In Section 16-8, when considering $\Delta\overline{G}^\circ$ as a function of temperature, we noted that $\Delta\overline{H}^\circ$ does not change appreciably with temperature and that the change in $\Delta\overline{G}^\circ$ is due primarily to the $T\Delta\overline{S}^\circ$ term in the expression

$$\Delta\overline{G}^\circ = \Delta\overline{H}^\circ - T\Delta\overline{S}^\circ \tag{16.14}$$

For the reaction $2\ Zn(s) + O_2(g) \to 2\ ZnO(s)$, $\Delta\overline{H}^\circ < 0$, and we expect a decrease in entropy because of the disappearance of one mole of gas: $\Delta\overline{S}^\circ < 0$. As the temperature increases the $T\Delta\overline{S}^\circ$ term becomes more negative; $-T\Delta\overline{S}^\circ$ becomes more positive; and

FIGURE 22-12
$\Delta\overline{G}^\circ$ as a function of temperature for some reactions of extractive metallurgy.

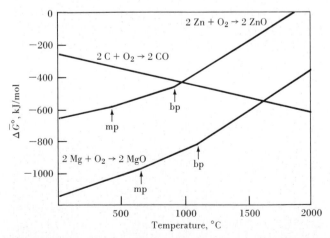

The points noted by arrows are the melting and boiling points of the metals zinc and magnesium.

$\Delta \overline{H}^\circ - T\Delta \overline{S}^\circ$ becomes less negative (or more positive). That is, $\Delta \overline{G}^\circ$ increases with temperature, just as shown in Figure 22-12. For the reaction $2\ C(s) + O_2(g) \rightarrow 2\ CO(g)$, $\Delta \overline{H}^\circ$ is also negative, but we expect an increase in entropy because *two* moles of gas are formed from *one* mole of gas and two of solid: $\Delta \overline{S}^\circ > 0$. As temperature increases, $T\Delta \overline{S}^\circ$ becomes more positive; $-T\Delta \overline{S}^\circ$, more negative; and $\Delta \overline{H}^\circ - T\Delta \overline{S}^\circ$, more negative. $\Delta \overline{G}^\circ$ decreases with temperature.

One of the reasons why carbon is such a good reducing agent is that its *descending* line in Figure 22-12 eventually intersects the *ascending* lines for most metals. Of course, for some metals the intersection comes at too high a temperature to make reduction of its oxide by carbon a practicable process. And in some cases where reduction of a metal oxide by carbon is possible, carbon may still not be the reducing agent chosen because of its tendency to form metal carbides, which may be undesirable in the final product.

Example 22-6   Would you expect the reaction $MgO(s) + C(s) \rightarrow Mg(g) + CO(g)$ to occur to any significant extent at 1500°C?

Solution.   We can extract these approximate data for 1500°C from Figure 22-12.

$$2\ C(s) + \cancel{O_2(g)} \longrightarrow 2\ CO(g) \qquad\qquad \Delta \overline{G}^\circ \simeq -530 \text{ kJ/mol}$$
$$\underline{2\ MgO(s) \longrightarrow 2\ Mg(g) + \cancel{O_2(g)} \qquad\qquad \Delta \overline{G}^\circ \simeq +610 \text{ kJ/mol}}$$
$$2\ MgO(s) + 2\ C(s) \longrightarrow 2\ Mg(g) + 2\ CO(g) \qquad \Delta \overline{G}^\circ \simeq +80 \text{ kJ/mol}$$

The positive value of $\Delta \overline{G}^\circ$ indicates that the reduction of MgO(s) by C(s) does not occur to a significant extent at 1500°C (i.e., the value of K is much smaller than 1). At temperatures above 1600°C, however, $\Delta \overline{G}^\circ$ becomes negative.

SIMILAR EXAMPLES: Review Problem 10; Exercises 19, 20.

22-5 Tin and Lead

Both tin and lead have been known from ancient times and were thought to be related. At one time the Romans called tin *plumbum album,* and lead *plumbum nigrum.*

Carbon, the first member of group IVA, is a nonmetal. The next two members, silicon and germanium, are metalloids. Among the more interesting properties of Si and Ge is semiconductor behavior, which we have discussed previously. Tin and lead are mostly metallic in their behavior.

The data in Table 22-6 suggest that tin and lead are rather similar to each other. Both are soft, malleable, and low melting. The ionization energies of the two metals are about the same, as are their standard electrode potentials. This means that their tendencies to be

TABLE 22-6
Some properties of tin and lead

	Tin	Lead
density, g/cm^3	5.77 (α)	11.34
	7.29 (β)	
melting point, °C	232	327
electron configuration	[Kr]$4d^{10}5s^25p^2$	[Xe]$4f^{14}5d^{10}6s^26p^2$
principal oxidation states	+2, +4	+2, +4
ionization energy, kJ/mol		
first	709	716
second	1412	1450
electrode potential E°, V		
[$M^{2+}(aq) + 2\ e^- \longrightarrow M(s)$]	−0.136	−0.126
[$M^{4+}(aq) + 2\ e^- \longrightarrow M^{2+}(aq)$]	+0.15	+1.5

oxidized to the +2 oxidation state are comparable; both are slightly more metallic than hydrogen.

The existence of the principal oxidation states of +2 and +4 is a consequence of the inert pair effect described in Section 22-3. In the +2 oxidation state the inert pair, ns^2, is not involved in bond formation and in the +4 oxidation state it is. The tendency to use the inert pair, that is, to exist in the +4 oxidation state, is much greater for tin than for lead, as seen in this way: The reduction of Sn(IV) to Sn(II) has only a slightly positive $E°$ value ($+0.15$ V), whereas for the reduction of Pb(IV) to Pb(II) it is large ($E° = +1.5$ V). Sn(II) can act as a reducing agent and Pb(IV) forms strong oxidizing agents. [Recall how the reduction of $PbO_2(s)$ to $PbSO_4(s)$ is used as the cathode reaction in the lead storage battery (page 616).]

Another difference between tin and lead is that tin exists in two allotropic forms (α and β) and lead has but a single form. The lack of allotropism in lead, situated at the bottom of group IVA, is similar to what we found for bismuth, at the bottom of group VA. Lack of allotropism is an important indicator that metallic properties increase in progressing from top to bottom in both groups.

The α (grey) form of tin is stable below 13°C and β (white) tin, above 13°C. Ordinarily, when a sample of β (white) tin is cooled, it must be maintained at temperatures below 13°C for a long time before the transition to α (grey) tin occurs. Once it does begin, however, the transformation takes place rather rapidly and with dramatic results. The tin expands and crumbles to a powder. This is because the α (grey) tin is less dense than the β (white) tin. This transformation of tin has resulted in the disintegration of organ pipes, buttons, medals, and various other tin objects, causing the transformation to be called, variously, tin disease, tin pest, and the tin plague.

Metallurgy and Uses of Tin and Lead. Tin is found mainly as **cassiterite, SnO₂.** The ore is concentrated by flotation and then roasted. Since the ore is already in the form of the oxide, the purpose of roasting is to oxidize metallic impurities and to remove sulfur and arsenic as their volatile oxides. Next, the oxide is reduced with carbon (coal).

$$SnO_2(s) + 2\ C(s) \xrightarrow{\Delta} Sn(1) + 2\ CO(g) \tag{22.49}$$

Tin from reaction (22.49) is purified by remelting. The easily melted tin is poured off from unmelted metals. Impurities that remain dissolved in the liquid tin are oxidized and removed by skimming the oxide film from the liquid surface.

Lead is found chiefly as **galena, PbS.** The ore is concentrated by flotation. Some quartz (SiO_2) is added and the ore is then roasted. The main reaction is

$$2\ PbS(s) + 3\ O_2(g) \xrightarrow{\Delta} 2\ PbO(s) + 2\ SO_2(g) \tag{22.50}$$

Reduction is carried out with coke and limestone, and involves the reactions

$$PbO(s) + C(s) \xrightarrow{\Delta} Pb(l) + CO(g) \tag{22.51}$$

$$PbO(s) + CO(g) \xrightarrow{\Delta} Pb(l) + CO_2(g) \tag{22.52}$$

[The purpose of the SiO_2 in reaction (22.50) is to convert any $PbSO_4$ that may form at the high roasting temperature to lead silicate, $PbSiO_3$. Limestone in the reduction reactions decomposes to CaO and CO_2. The CaO converts $PbSiO_3$ to PbO by forming a $CaSiO_3$ slag.]

The refining of lead requires several steps since a number of impurities are present. First, the lead is melted and maintained for a time at a temperature below the melting point of copper. The Cu crystallizes and is skimmed off the liquid lead. Air is now blown through the liquid and a surface scum of lead arsenate and lead antimonate is skimmed off. Finally, 1 to 2% Zn is added to the molten lead. Silver passes from the Pb(l) to the Zn(l), in which it is more soluble. The Zn–Ag alloy crystallizes and is skimmed off.

About one half of all tin produced is used in tin plate, especially in plating iron for use in cans for storing foods (Section 20-7). The next most important use is in the manufacture of **solders**—low-melting alloys used to join together wires or pieces of metal. Other important alloys of tin are **pewter** (85% Sn, 7% Cu, 6% Bi, 2% Sb) and **bronze** (90% Cu, 10% Sn).

The chief uses of lead are in the manufacture of lead storage batteries (Section 20-6), solders and other alloys, cable coverings, ammunition, tetraethyl lead (as an antiknock additive for gasoline), and radiation shields (to protect against x rays and λ rays).

Reactions of Tin and Lead.   Although tin and lead are quite similar in physical properties, some important differences are found in their chemical behaviors. These differences stem from the stability of the +4 oxidation state for tin in contrast to lead, and from the fact that so many lead compounds are insoluble.

Tin dissolves in HCl(aq), slowly in dilute acid and more rapidly in concentrated acid.

$$Sn(s) + 2 HCl(aq) \longrightarrow Sn^{2+}(aq) + 2 Cl^-(aq) + H_2(g) \tag{22.53}$$

In concentrated HNO_3(aq), tin is oxidized to SnO_2(s).

Reaction (22.54) is illustrated in Color Section C.

$$Sn(s) + 4 HNO_3(\text{conc. aq}) \longrightarrow SnO_2(s) + 2 H_2O + 4 NO_2(g) \tag{22.54}$$

Tin also dissolves in concentrated NaOH(aq) through a reaction similar to that described for aluminum (see page 687). In this reaction tin is oxidized to the +4 oxidation state in the complex ion $[Sn(OH)_6]^{2-}$; $H_2(g)$ is the other product.

The reactions of lead with dilute HCl(aq) and H_2SO_4(aq) stop after an initial brief reaction because the products, $PbCl_2$(s) and $PbSO_4$(s), protect the metal from further attack. However, $PbCl_2$(s) slowly dissolves in concentrated HCl(aq) through the formation of the complex ion $[PbCl_3]^-$. In turn, this makes it possible for lead to dissolve.

$$Pb(s) + 3 HCl(\text{conc. aq}) \longrightarrow [PbCl_3]^-(aq) + H^+(aq) + H_2(g) \tag{22.55}$$

Lead is unattacked by concentrated H_2SO_4(aq) at temperatures up to 200°C. The products of its reaction with HNO_3(aq) are $Pb(NO_3)_2$(aq) and various oxides of nitrogen, depending on the reaction conditions (recall page 654).

When tin is heated in air, the product is SnO_2(s). The air oxidation of lead, on the other hand, produces PbO(s). Metallic tin reacts with Cl_2(g) to form $SnCl_4$ (a reaction used to recover tin from scrap tin plate). Lead reacts with Cl_2(g) to form $PbCl_2$. When heated with S, both metals yield sulfides. With tin the SnS formed reacts further to yield SnS_2. Lead forms only PbS.

Compounds of Tin and Lead.   Although SnO exists, it is unstable; and on heating in air it converts to SnO_2. In addition to PbO (yellow lead oxide, litharge), two other oxides of lead are brown lead dioxide, PbO_2, and "red lead," Pb_3O_4. Pb_3O_4 is obtained by the air oxidation of PbO at 400°C, but the reverse reaction occurs at higher temperatures.

$$6 PbO(s) + O_2(g) \rightleftharpoons 2 Pb_3O_4(s) \tag{22.56}$$

Pb_3O_4 has lead in two oxidation states, Pb(II) and Pb(IV), perhaps as Pb_2PbO_4. When treated with HNO_3(aq), the Pb(II) forms $Pb(NO_3)_2$(aq) and the Pb(IV) precipitates as PbO_2(s).

$$Pb_3O_4(s) + 4 HNO_3(aq) \longrightarrow 2 Pb(NO_3)_2(aq) + PbO_2(s) + 2 H_2O \tag{22.57}$$

Simple hydroxides of tin and lead do not appear to exist. Acid-base properties must be described in terms of their oxides. SnO and PbO, by forming salts with acids, display basic properties; but they also have an acidic character. They dissolve in NaOH(aq).

$$SnO(s) + OH^-(aq) + H_2O \longrightarrow [Sn(OH)_3]^-(aq) \tag{22.58}$$

PbO forms $[Pb(OH)_3]^-$. The higher oxides, SnO_2 and PbO_2, are also amphoteric, dis-

FIGURE 22-13
Structures of the
(a) stannite and
(b) stannate ions.

(a)

(b)

playing acidic properties by dissolving in NaOH(aq).

$$SnO_2(s) + 2\ OH^-(aq) + 2\ H_2O \longrightarrow [Sn(OH)_6]^{2-} \tag{22.59}$$

Anions having Sn and Pb as central atoms are best named as complex ions (see Section 24-4), but they are often named like oxoanions. If the central atom has the oxidation state $+2$, the *ite* ending is used; if $+4$, *ate*. There appear to be a variety of *ite* and *ate* species; some exist in solution and some only in solids. The best characterized are

stannite ion $[Sn(OH)_3]^-$ (also written as $[Sn(OH)_4]^{2-}$, $HSnO_2^-$, and SnO_2^{2-})

stannate ion $[Sn(OH)_6]^{2-}$ (also written as SnO_4^{4-} and SnO_3^{2-})

plumbite ion $[Pb(OH)_3]^-$ (also written as $[Pb(OH)_4]^{2-}$)

plumbate ion $[Pb(OH)_6]^{2-}$ (also written as PbO_4^{4-} and PbO_3^{2-})

Structures of the stannite and stannate ions are shown in Figure 22-13.

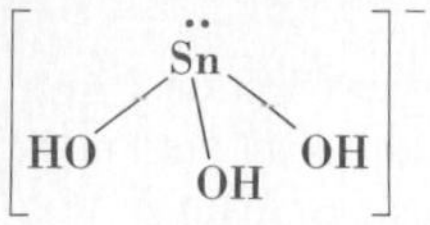

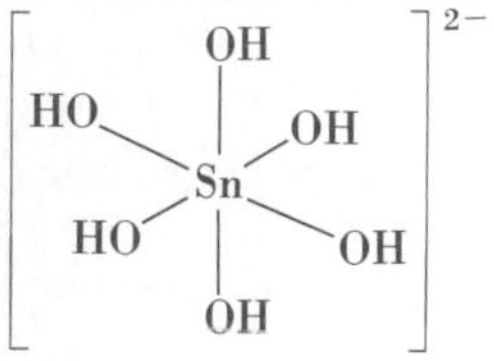

Stannite ion is a good reducing agent because of the ease with which it is oxidized to stannate ion.

$$[Sn(OH)_3]^-(aq) + 3\ OH^-(aq) \longrightarrow [Sn(OH)_6]^{2-}(aq) + 2\ e^-   \qquad E^\circ_{ox} = +0.96\ V \tag{22.60}$$

This property is not shared by plumbite ion, as we might expect from previous comments on the oxidation states of lead. The sulfur analogs of stannite and stannate ions (i.e., SnS_2^{2-} and SnS_3^{2-}) were discussed in connection with the solubilities of metal sulfides in Section 21-2.

Lead oxides are used in the manufacture of lead storage batteries, glass, ceramic glazes, cements (PbO), metal-protecting paints (Pb_3O_4), matches (PbO_2), and explosives (PbO_2). Lead oxides also serve as sources of other lead compounds. The reaction of PbO with acetic acid (CH_3COOH) yields lead(II) acetate, $Pb(CH_3COO)_2$. Lead(II) acetate is essentially a covalent compound. It is highly soluble in water and is used when high aqueous concentrations of lead are required. A reaction between Pb_3O_4 and acetic acid produces lead(IV) acetate or lead tetraacetate, $Pb(CH_3COO)_4$, which is an important oxidizing agent in organic chemistry. Reaction (22.57) produces $Pb(NO_3)_2(aq)$, one of the few soluble lead compounds. Addition of $CrO_4^{2-}(aq)$ to $Pb(NO_3)_2(aq)$ precipitates insoluble $PbCrO_4(s)$, a paint pigment known as **chrome yellow.** Another lead-based pigment, used in ceramic glazes and once used extensively in the manufacture of paint, is basic lead carbonate, $2\ PbCO_3 \cdot Pb(OH)_2$, also known as **white lead.**

The chlorides of tin—$SnCl_2$ and $SnCl_4$—both have a number of industrial uses. In addition, $SnCl_2$ is a good reducing agent and is used in the chemical laboratory to reduce Fe(III) to Fe(II), Hg(II) to Hg(I) and Hg, and Cu(II) to Cu(I). The fluoride, SnF_2, has an important use as an anti-cavity additive to toothpaste. SnO_2 is used as a jewelry abrasive, and SnS_2 (tin bronze) is used as a pigment and for imitation gilding.

22-6 The Group IIB Metals

To this point in the text we have discussed only elements that belong to A groups in the periodic table. These are the main group or representative elements. In the next chapter we will consider transition elements, all of which belong either to B subgroups or group VIII. There is one B subgroup, however, which does not fit the definition of transition elements that we will give at that time. This is subgroup IIB—Zn, Cd, and Hg.

We might expect these three elements to have some similarities to the transition elements which they immediately follow; and they do in such properties as the ability to form a large variety of complex ions. Also we might expect some similarity to the group IIA metals, since the IIA and IIB elements are found in the same group in Mendeleev's

TABLE 22-7
Some properties of the IIB metals

	Zn	Cd	Hg
density, g/cm^3	7.14	8.64	13.59 (ℓ)
melting point, °C	419.6	320.9	−38.87
boiling point, °C	907	765	357
electron configuration	[Ar]$3d^{10}4s^2$	[Kr]$4d^{10}5s^2$	[Xe]$4f^{14}5d^{10}6s^2$
atomic radius, pm	133	149	150
ionization energy, kJ/mol			
first	906	867	1006
second	1703	1631	1809
principal oxidation state(s)	+2	+2	+1, +2
electrode potential $E°$, V			
[M^{2+}(aq) + 2 e$^-$ ⟶ M]	−0.763	−0.403	+0.851
[M_2^{2+}(aq) + 2 e$^-$ ⟶ 2 M]	—	—	+0.796

periodic table (page 197). These similarities do exist, mostly for Zn and Cd, in that they display a single principal oxidation state (+2), their ions are colorless in solution, their solid compounds are mostly without strong color, and Zn^{2+} and Cd^{2+} can replace (are isomorphous with) Mg^{2+} in some of its compounds, e.g., $Zn(Mg)SO_4 \cdot 7H_2O$.

As seen from Table 22-7, the IIB metals have low melting and boiling points. This can probably be attributed to the relatively weak metallic bonding associated with the "18 + 2" electron configuration. Mercury is the only metal that exists as a liquid at room temperature (although gallium can easily be supercooled to room temperature). Mercury differs from Zn and Cd in a number of ways in addition to its physical appearance.

- Mercury has very little tendency to combine with oxygen; the oxide, HgO, is thermally unstable.
- Very few mercury compounds are water soluble and most are not hydrated.
- Many mercury compounds are covalent. The stability of the Hg—C bond makes possible a large number of organic mercury compounds. Mercury halides, except HgF_2, are only slightly ionized in aqueous solution.
- Mercury forms a diatomic ion with a metal–metal covalent bond, Hg_2^{2+}.
- Mercury will not displace $H_2(g)$ from H^+(aq).

Some of these differences found in mercury can probably be attributed to the fact that $4f$ electrons are not as effective in shielding outer-shell electrons as are electrons in other inner subshells. This leads to a higher effective nuclear charge and a smaller size than we would otherwise expect for the Hg atom. As a result, ionization energies for Hg are somewhat higher than for Zn and Cd. Hydration energies for Hg_2^{2+} and Hg^{2+} are also not as large as for Zn^{2+} and Cd^{2+}. The net result of these factors is that the reduction potentials for Hg_2^{2+} and Hg^{2+} are positive, whereas those of Zn^{2+} and Cd^{2+} are negative. Other explanations take into account the increased importance of the inert pair effect with the bottom member of a group. However, perhaps the most significant factor of all in explaining differences between the fifth and sixth period members of a group is relativistic effects on orbital energies in heavy atoms (see again, Section 8-6).

Metallurgy and Uses of Zn, Cd, and Hg. The production of zinc was used as a specific example of extractive metallurgy in Section 22-4. Cadmium generally occurs together with zinc in its ores, so that cadmium is obtained as a by-product in the production of zinc. Cadmium, because of its lower boiling point, can be separated from zinc by fractional distillation. If electrolytic reduction and refining of zinc is the chosen method, just prior to

FIGURE 22-14
$\Delta \overline{G}°$ as a function of temperature for the reaction

$2 \text{ Hg} + \text{O}_2 \rightarrow 2 \text{ HgO}$

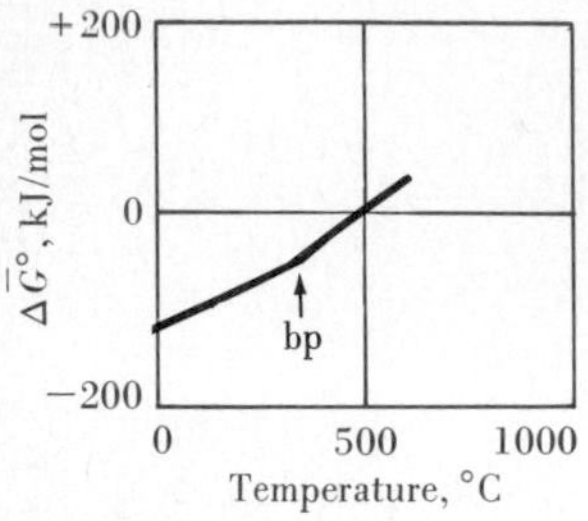

At about the temperature where $\Delta \overline{G}° = 0$, spontaneous decomposition of HgO occurs. [The arrow marks the boiling point of Hg(l).]

reaction (22.48) the solution containing Zn^{2+} is treated with powdered zinc, which goes into solution as Zn^{2+} and displaces Cd^{2+}.

$$\text{Zn}(s) + \text{Cd}^{2+}(aq) \longrightarrow \text{Zn}^{2+}(aq) + \text{Cd}(s) \tag{22.61}$$

The Cd(s) is filtered off, dissolved in acidic solution, and electrolyzed to yield pure cadmium.

Mercury is one on the simplest metals to extract from its ores. When cinnabar (HgS) is heated it decomposes directly to Hg(g), which condenses to Hg(l).

$$\text{HgS}(s) + \text{O}_2(g) \xrightarrow{\Delta} \text{Hg}(g) + \text{SO}_2(g) \tag{22.62}$$

Alternative processes that eliminate the emission of $SO_2(g)$ involve roasting HgS with Fe or with CaO.

$$\text{HgS}(s) + \text{Fe}(s) \xrightarrow{\Delta} \text{FeS}(s) + \text{Hg}(g) \tag{22.63}$$

$$4 \text{ HgS}(s) + 4 \text{ CaO}(s) \xrightarrow{\Delta} 3 \text{ CaS}(s) + \text{CaSO}_4(s) + 4 \text{ Hg}(g) \tag{22.64}$$

The roasting of HgS does not yield HgO. HgO is unstable at high temperatures, decomposing to Hg(g) and $O_2(g)$, as suggested by Figure 22-14.

Mercury is purified by treatment with dilute $HNO_3(aq)$, which oxidizes most of the impurities. Insoluble products float to the surface of the liquid and are removed. Final purification is by distillation. Mercury can be easily obtained with a purity exceeding that of most metals (e.g., 99.9998% Hg or better).

About one third of the metallic zinc produced is used in coating iron to give it corrosion protection (Section 20-7). An iron object may be coated with zinc by dipping or spraying it with Zn(l), by electroplating it from a solution of $Zn^{2+}(aq)$, or by mixing it with powdered Zn and heating. By whatever method this coating is achieved, the product is called **galvanized iron.** Large quantities of Zn are consumed in the manufacture of alloys. For example, about 20% of the production of Zn is used in **brass,** an alloy having Cu as its major component, 20 to 45% Zn, and small quantities of Sn, Pb, and Fe. Brass is a good electrical conductor and it is corrosion resistant. Zinc is also used in the manufacture of dry cells, in printing (lithography), in the construction industry (roofing materials), and as sacrificial anodes in corrosion protection (Section 20-7).

Cadmium is used instead of zinc to protect iron in special applications. It is used in bearing alloys, in low-melting solders, in aluminum solders, and as an additive to impart strength to copper. Another important use, based on its neutron absorbing capacity, is in control rods and shielding for nuclear reactors (see Section 25-8).

The principal uses of mercury take advantage of the combination of its metallic and liquid properties and its high density. Thus, it is used in thermometers, barometers, gas pressure regulators, and electrical relays and switches. Liquid mercury is used for electrodes in electrochemical cells, such as in the chlor-alkali process (Chapter 20). Mercury vapor is used in fluorescent tubes and street lamps. Mercury forms alloys, called **amalgams,** with almost all metals, and some of these amalgams are of commercial importance. Dental amalgam consists of 70% Hg and 30% Cu. Mercury compounds are used as pharmaceuticals, germicides, and fungicides.

Iron is one of the few metals that does not form an amalgam. Iron containers are used to store and ship mercury.

Reactions and Compounds of Zn, Cd, and Hg. We saw from Table 22-7 that Zn and Cd are reasonably active metals, but Hg is not. Zn and Cd dissolve in HCl(aq); Hg does not. However, all three metals react with concentrated $H_2SO_4(aq)$, as exemplified by Hg.

$$2 \text{ Hg}(l) + 2 \text{ H}_2\text{SO}_4(\text{conc. aq}) \longrightarrow \text{Hg}_2\text{SO}_4(s) + 2 \text{ H}_2\text{O} + \text{SO}_2(g) \tag{22.65}$$

The three metals also dissolve in $HNO_3(aq)$, and with a variety of products possible. Hg(l) is oxidized to Hg_2^{2+} in dilute $HNO_3(aq)$; the reduction product is NO(g). In excess HNO_3(conc. aq), Hg^{2+} and $NO_2(g)$ are produced. With zinc and $HNO_3(aq)$ the reduction products may have nitrogen in any oxidation state from $+4$ (NO_2) to -3 (NH_4^+).

Hydrated zinc ion in solution ionizes in the same manner as described for $[Al(H_2O)_6]^{3+}$ on page 686, and this makes solutions of $Zn^{2+}(aq)$ somewhat acidic. Addition of small quantities of OH^- to $Zn^{2+}(aq)$ produces $Zn(OH)_2(s)$, and an excess of OH^- yields zincate ion, $[Zn(OH)_4]^{2-}$. In the same summary fashion as used for aluminum in expression (22.31), we may write

$$[Zn(H_2O)_6]^{3+}(aq) \underset{H_3O^+}{\overset{OH^-}{\rightleftharpoons}} Zn(OH)_2(s) \underset{H_3O^+}{\overset{OH^-}{\rightleftharpoons}} [Zn(OH)_4]^{2-}(aq) \tag{22.66}$$

Not only is $Zn(OH)_2(s)$ amphoteric but so too is $ZnO(s)$. That is,

$$ZnO(s) + 2\ H_3O^+(aq) \longrightarrow Zn^{2+}(aq) + 3\ H_2O \tag{22.67}$$

$$ZnO(s) + 2\ OH^-(aq) + H_2O \longrightarrow [Zn(OH)_4]^{2-}(aq) \tag{22.68}$$

Zn dissolves in NaOH(aq), a fact that can also be described in the manner outlined for Al on page 687.

$$Zn(s) + 2\ OH^-(aq) + 2\ H_2O \longrightarrow [Zn(OH)_4]^{2-}(aq) + H_2(g) \tag{22.69}$$

$Cd(OH)_2$ displays basic properties by dissolving in acids. Its acidic character is much weaker than that of $Zn(OH)_2$, and it dissolves only in very concentrated NaOH(aq). $Zn(OH)_2$ and $Cd(OH)_2$ are both soluble in $NH_3(aq)$ because of complex ion formation. For example,

$$Zn(OH)_2(s) + 4\ NH_3(aq) \longrightarrow [Zn(NH_3)_4]^{2+}(aq) + 2\ OH^-(aq) \tag{22.70}$$

An interesting aspect of the chemistry of mercury centers on the disproportionation reaction

$$Hg_2^{2+}(aq) \longrightarrow Hg^{2+}(aq) + Hg(l) \qquad E^\circ_{cell} = -0.15\ V \tag{22.71}$$

Although this reaction does not occur spontaneously under standard-state conditions, many situations arise in which $[Hg^{2+}]$ is brought to a very low value. In these cases, in accordance with Le Châtelier's principle, we can expect a displacement of equilibrium to the right, perhaps enough so as to drive the reaction essentially to completion. Such is the result, for example, when $Hg_2^{2+}(aq)$ is treated with $S^{2-}(aq)$. HgS(s) is so insoluble that reaction (22.71) does in fact occur. That is,

$$Hg_2^{2+}(aq) + S^{2-}(aq) \longrightarrow HgS(s) + Hg(l) \tag{22.72}$$

Table 22-8 lists a few important compounds of the IIB metals and some of their uses.

TABLE 22-8
Some important compounds of the IIB metals

Compound	Uses
ZnO	reinforcing agent in rubber; pigment; cosmetics; dietary supplement; photoconductors in copying machines.
ZnS	phosphors in x ray and television screens; pigment; luminous paints.
$ZnSO_4$	rayon manufacture; animal feeds; wood preservative.
CdO	electroplating; batteries; catalyst; nematocide.
CdS	solar cells; photoconductor in xerography; phosphors; pigment.
$CdSO_4$	electroplating; standard voltaic cells (Weston cell).
HgO	polishing compounds; dry cells; antifouling paints; fungicide; pigment.
$HgCl_2$	manufacture of Hg compounds; disinfectant; fungicide; insecticide; wood preservative.
Hg_2Cl_2	electrodes; pharmaceuticals; fungicide.

Excerpts from a Letter to Benjamin Vaughan

Philada, July 31, 1786

Dear Friend,

I recollect that, when I had the great Pleasure of seeing you at Southampton, now a 12 month since, we had some Conversation on the bad Effects of Lead taken inwardly . . . .

The First Thing I remember of this kind was a general Discourse in Boston, when I was a Boy, of a Complaint from North Carolina against New England Rum, that it poison'd their People, giving them the Dry Bellyach, with Loss of the Use of their Limbs. The Distilleries being examin'd on the occasion, it was found that several of them used Leaden Still-heads and Worms, and the Physicians were of Opinion, that the Mischief was occasioned by the Use of Lead. The Legislature of the Massachusetts there-upon pass'd an Act, prohibiting under severe Penalties the Use of such Still-heads and Worms thereafter. Inclos'd I send you a Copy of the Acct, taken from my Printed Law-Book.

. . . . .

When I was in Paris with Sir John Pringle in 1767, he visited La Charité, a Hospital particularly famous for the Cure of that Malady, and brought from thence a Pamphlet containing a List of the Names of Persons, specifying their Professions or Trades, who had been cured there. I had the curiosity to examine that List, and found that all the Patients were of Trades, that, some way or other, use or work in Lead; such as Plumbers, Glaziers, Painters, &c., excepting only two kinds, Stonecutters and Soldiers. These I could not reconcile to my Notion, that Lead was the cause of that Disorder. But on my mentioning this Difficulty to a Physician of that Hospital, he inform'd me that the Stonecutters are continually using

Benjamin Franklin (1706–1790). [Courtesy of The Franklin Mint.]

melted Lead to fix the Ends of Iron Balustrades in Stones; and that the Soldiers had been employ'd by Painters, as Labourers, in Grinding of Colours.

This, my dear Friend, is all I can at present recollect on the Subject. You will see by it, that the Opinion of this mischievous Effect from Lead is at least above Sixty Years old; and you will observe with Concern how long a useful Truth may be known and exist, before it is generally receiv'd and practis'd on.

I am, ever, yours most affectionately,
B. Franklin

Certain metal ions (e.g., Na^+, K^+, and Ca^{2+}) are found in high concentrations in living organisms. Other metals (e.g., Fe, Co, Mo, Mn, Cu, and Zn) normally are present only in small quantities, but they are known to perform vital functions. Some metal ions, though not present in organisms to any appreciable extent, seem to be harmless. Such is the case with Al^{3+}. Finally, there is a group of metals that to varying degrees are toxic to living organisms. The best known examples are Pb, Sn, As, Sb, Hg, and Cd. Their effects seem to be cumulative since the human body has no good mechanisms for eliminating them.

Lead. The metal poisoning to which humans have been longest exposed is probably lead poisoning. Beginning with the ancient Romans and continuing to modern times, lead has been used in plumbing systems, including those designed to transport water. Exposure to lead has also come through cooking and eating utensils and pottery glazes. Of more recent origin is exposure to lead-based paints and lead compounds associated with the use of tetraethyllead in gasoline. The health effects of lead are not fully known, but mild forms of lead poisoning produce nervousness and mental depression. More severe cases can lead to permanent nerve, brain, and kidney damage.

Lead poisoning involves interference with the biochemical reactions that produce the iron-containing heme group in hemoglobin. In some cases of lead poisoning it is possible to reduce lead levels in the body through the use of chelating agents (see page 761). It has been estimated that millions of people in the United States have elevated levels of lead in their bodies, with perhaps several hundred thousand young children having hazardous lead levels.

Benjamin Franklin's letter to Benjamin Vaughn indicates that knowledge of the toxic effects of lead have been known for at least 200 to 300 years. The graph in Figure 22-15 shows that it was precisely during Franklin's lifetime—the early industrial revolution—that lead began to accumulate significantly in the environment. A very sharp increase in the environmental deposition of lead beginning about 1940 is generally believed to correspond to the rapid growth in the use of lead compounds as antiknock additives to gasoline. Also, before 1940 lead levels in the Antarctic ice were so low as to be undetectable; since 1940 they have risen to above 0.02 μg Pb/kg ice. More recent evidence of the correlation between the use of lead in gasoline and the body burden of lead carried by humans is provided by Figure 22-16. As the use of lead in gasoline has declined, so too has the level of lead in the blood of the general population. Figure 22-16 suggests that phasing out the use of leaded gasoline should be a very significant control measure.

Mercury. Accumulations of mercury in the body affect the nervous system and cause brain damage. "Hatter's disease" (which afflicted the mad Hatter in *Alice's Adventures in Wonderland*) was a form of chronic mercury poisoning. Mercury compounds were used to convert fur to felt for making hats. One mechanism of mercury poisoning seems to be interference with the functioning of sulfur-containing enzymes; Hg has a high affinity for sulfur. All mercury compounds are poisonous, but in varying degrees. Organic mercury compounds are generally much more poisonous than inorganic ones. An insidious aspect of mercury poisoning is that certain microorganisms possess the ability to convert mercury to methylmercury (CH_3Hg^+) compounds, which then concentrate in the food chains of fish and other aquatic life.

In the free metallic state, mercury is most poisonous as a vapor. Levels of mercury that exceed 10 μg Hg/m^3 air are considered unsafe. Even though we think of mercury as having a low vapor pressure, the concentration of Hg in its saturated vapor far exceeds this limit. Spilled mercury is a special health hazard, particularly if in cleanup operations some tiny droplets go undetected. Although mercury levels in remote locations generally do not exceed a few hundredths microgram per m^3 of air, in certain industrialized settings—chlor-alkali plants, thermometer factories, smelters—mercury levels may exceed the safety limit. Unsafe levels may even be found in dental laboratories.

FIGURE 22-15
Deposition of lead in Greenland ice cap.

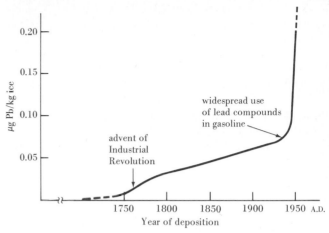

Fallen snow is converted to ice, in which the deposited lead is trapped. In 800 B.C. the level was less than 0.001 μg Pb/kg ice. Deposition of lead in the ice appears to be closely linked to human activities. [Data from M. Murozumi, T. J. Chow, and C. C. Patterson, *Geochim. Cosmochim. Acta*, **33**, 1247 (1969).]

FIGURE 22-16
Lead in gasoline (—) and lead in blood (—).

The level of lead in the blood of a representative population has declined with the decline in use of lead in gasoline. Medical experts generally consider a lead level of 30 μg Pb/dL blood to pose a significant health risk, but recent evidence suggests measurable effects at lower levels (e.g., changes in brain-wave patterns of children have been detected at levels of 10–15 μg Pb/dL blood). [Data from Environmental Protection Agency Office of Policy Analysis, 1984.]

Cadmium.   Although Zn is an essential element in trace amounts, in higher concentrations it is toxic. Much more toxic, however, even at low levels, is cadmium. One effect of cadmium poisoning is an extremely painful skeletal disorder known as ''itai-itai kyo'' (Japanese for ''ouch-ouch'' disease). This effect was discovered following World War II in an area of Japan where effluents from a zinc mine became mixed with irrigation water used in rice fields. Cadmium poisoning was discovered in residents of the area who ate the rice. Cadmium poisoning can also cause liver damage, kidney failure, and pulmonary disease; and Cd appears to be a contributing factor in high blood pressure. The mechanism of Cd poisoning may involve substitution in certain enzymes of Cd, a poison, for Zn, an essential element. Concern over cadmium poisoning has increased with an awareness that some Cd is almost always found in zinc and zinc compounds, materials that have many commercial uses (recall Table 22-8).

Summary

The alkali (IA) metals are the most active of the metals, as indicated by their low ionization energies and large negative reduction potentials. Li is rather different from the other IA metals, and these differences for the most part can be explained in terms of the high positive charge density on Li^+. The alkali metals are prepared by the electrolysis of their molten salts. Electrolysis of NaCl(aq) produces NaOH(aq), a substance from which other Na compounds can be prepared. NaOH can also be made from Na_2CO_3. In turn, Na_2CO_3 can be produced from NaCl, NH_3, and CO_2 by the Solvay process.

The alkaline earth (IIA) metals are also very active and are generally prepared by electrolysis of a molten salt. In the case of magnesium the salt is $MgCl_2$, with the Mg^{2+} derived from sea water. The comparative strengths of lattice energies and cation hydration energies are important in establishing the solubilities of IIA metal salts. A large number of these salts are insoluble in water, but others are quite soluble and crystallize from aqueous solution as deliquescent hydrates. Among the most important of the insoluble salts are the carbonates, especially $CaCO_3$. Reversible reactions involving CO_3^{2-}, HCO_3^-, $CO_2(g)$, and H_2O account for the formation of both limestone caves and temporary hard water. Water may also have noncarbonate or permanent hardness through the presence of such ions as SO_4^{2-}. One of the objections to hard water is its action on soaps. Water can be softened either through chemical reactions or by ion exchange. Alternatively, synthetic detergents can be used even in hard water.

The principal metal of group IIIA is aluminum, whose large-scale use has been made possible through an effective method of production. Advantage is taken of the amphoteric nature of Al_2O_3 to separate it from impurities, and advantage is taken of the solubility of Al_2O_3 in the much lower melting Na_3AlF_6 to produce a liquid medium in which electrolysis is possible. Much of the solution chemistry of aluminum centers on the hydrated aluminum ion, $[Al(H_2O)_6]^{3+}$, the amphoteric hydroxide, $Al(OH)_3$, and aluminate ion, $[Al(OH)_4]^-$, which forms in basic solution. For example, this solution behavior accounts for the fact that Al is soluble in NaOH(aq).

The remaining representative metals can be obtained by more traditional methods of extractive metallurgy. Generally this begins with concentration of the ore, often by flotation. Next the ore is roasted to convert it to the metal oxide. This is followed by reduction, with carbon being the reducing agent generally used. The final step is that of refining the impure metal. A useful tool in determining the required reduction temperature is a graph of free energy change for oxide formation as a function of temperature.

Tin and lead in group IVA have some similarities, but differences in their behaviors are also found. The chief difference is that Sn acquires the oxidation state +4 rather easily whereas with lead the +2 oxidation state is preferred. This means that lead(IV) compounds are easily reduced and are, therefore, powerful oxidizing agents.

In subgroup IIB the heaviest member, Hg, is rather different from the other two—Zn and Cd. These differences can be explained from several standpoints. The solution behavior of Zn^{2+} is similar to that of Al^{3+}, e.g., $Zn(OH)_2$ is amphoteric and Zn is soluble in NaOH(aq).

Learning Objectives

As a result of studying Chapter 22, you should be able to

1. Describe how standard electrode potentials are related to enthalpies of sublimation, ionization, and hydration; and how lattice energies and enthalpies of hydration affect the solubilities of ionic compounds.

2. Outline the methods that are used to produce Na, Mg, and Al.

3. Outline the Solvay process for the production of $NaHCO_3$, showing how substances are recycled in the process.

4. Outline the series of reactions whereby rainwater acquires temporary hardness and show how this hardness is removed by heating.

5. Write chemical equations for reactions that can be used to soften temporary and permanent hard water.

6. Write formulas for some typical alums.

7. Describe the four fundamental processes of traditional extractive metallurgy—concentration, roasting, reduction, and refining—with specific reference to metals such as Sn, Pb, Zn, Cd, and Hg.

8. Establish from free energy data (as in Figure 22-12) which reducing agents will bring about reduction of a metal oxide at a given temperature.

9. Write equations to show the effect of H_2O, of dilute and concentrated HCl(aq), H_2SO_4(aq), and HNO_3(aq), and of NaOH(aq) on metals, such as those of groups IA and IIA, Al, Sn, Pb, Zn, Cd, and Hg.

10. Write equations to represent the amphoteric nature of the oxides and hydroxides of Al, Sn, and Zn.

11. Explain ways in which Li differs from other members of group IA; Tl differs from other members of group IIIA; Pb differs from Sn; and Hg differs from Zn and Cd.

12. State some of the principal uses of the representative metals.

13. Describe some of the more important reactions and uses of compounds of the representative elements.

Some New Terms

Alums are double salts having the general formula, $M(I)M(III)(SO_4)_2 \cdot 12H_2O$. M(I) is a unipositive cation (except Li^+) and M(III) is one of a group of tripositive cations, e.g., Al^{3+}, Fe^{3+}, Cr^{3+}.

Amalgams are metal alloys containing mercury. Depending on their compositions, some are liquid and some are solid.

Anodizing refers to the electrodeposition of Al_2O_3 on Al to further protect it against corrosion.

Deionized water is water that has been freed of most of its impurity ions by passing it successively through a cation and an anion exchange resin.

Deliquescence is a process in which atmospheric water vapor condenses on a highly soluble solid and the solid dissolves in the water to produce an aqueous solution.

Detergents are salts of organic sulfonic acids, $RSO_3^-Na^+$, in which R is a hydrocarbon chain or a combination of a benzene ring and hydrocarbon chain.

Extractive metallurgy refers to the process of extracting a metal from its ores. Generally this occurs in four steps. **Concentration** separates the ore from waste rock (gangue). **Roasting** converts the ore to the metal oxide. **Reduction** (usually with carbon) converts the oxide to the metal. **Refining** removes impurities from the metal.

Galvanizing is the name given to any process in which an iron surface is coated with zinc to give it corrosion protection.

Hard water contains dissolved minerals in significant concentrations. If the hardness is primarily due to HCO_3^- and associated cations, the water is said to have **temporary hardness.** If hardness is due to anions other than HCO_3^- (e.g., SO_4^{2-}), it is referred to as **permanent hardness.**

The **inert pair effect** refers to the effects on the properties of certain post-transition elements that result from a pair of electrons in the s orbital of the valence shell, that is **ns^2.**

Ion exchange is a process in which ions in solution are exchanged for corresponding ions held to the surface of an ion exchange material. For example, Ca^{2+} and Mg^{2+} may be exchanged for Na^+; or SO_4^{2-} for OH^-.

Soaps are the salts of fatty acids, e.g., $RCOO^-Na^+$, where the R group is a hydrocarbon chain containing from 3 to 21 C atoms.

Solders are low melting alloys used for joining wires or pieces of metal. They usually contain metals such as Sn, Pb, Bi, and Cd.

Stalactites and **stalagmites** are limestone ($CaCO_3$) formations in limestone caves produced by the slow decomposition of $Ca(HCO_3)_2$(aq).

Suggestions for Further Study

CARTER, D. E. and Q. FERNANDO, "Chemical Toxicology II: Metal Toxicity," *J. Chem. Educ.*, **56**, 490 (1979).

HABASHI, F., "Chemistry and Metallurgy," *Chemistry*, **45**[9], 6 (1972).

HABASHI, F., "Hydrometallurgy," *Chem. Eng. News*, **60**[6], 46 (1982).

HAUPIN, W. E., "Electrochemistry of the Hall–Heroult Process for Aluminum Smelting," *J. Chem. Educ.*, **60**, 279 (1983).

LAYMAN, P. L., "Detergent Report," *Chem. Eng. News*, **62**[4], 17 (1984).

MORRIS, D. L., "Stress, Collisions, and Constants: Part III. The Solvay Process," *Chemistry*, **44**[6], 15 (1971).

MELLON, E. F., "Alfred E. Stock and the Insidious 'Quecksilbervergiftung'," *J. Chem. Educ.*, **54**, 211 (1977).

NAVRATIL, J. D., "Magnesium," *Chemistry*, **44**[5], 6 (1971).

PATTERSON, S. H., "Aluminum from Bauxite: Are There Alternatives?", *American Scientist*, **65**, 345 (1977).

TACKETT, S. L., "The Franklin Letter on Lead Poisoning," *J. Chem. Educ.*, **58**, 274 (1981).

Review Problems

1. Provide an acceptable name or formula for each of the following.

(a) $Pb(CH_3COO)_2$
(b) magnesium hydrogen carbonate
(c) Hg_2Br_2
(d) $CaCl_2 \cdot 6H_2O$
(e) aluminum ammonium alum
(f) zincate ion
(g) $NaAl(OH)_4$
(h) $K_2Sn(OH)_6$

2. Complete and balance the following equations. If no reaction occurs, so state.

(a) $MgCO_3(s) \overset{\Delta}{\rightarrow}$
(b) $CaO(s) + HCl(aq) \rightarrow$
(c) $Al(s) + KOH(aq) + H_2O \rightarrow$
(d) $Pb(s) + HNO_3(aq) \rightarrow$
(e) $Hg(l) + HCl(dil.\ aq) \rightarrow$
(f) $ZnO(s) + CO(g) \overset{\Delta}{\rightarrow}$

3. Assuming the availability of water, common reagents (acids, bases, salts), and simple laboratory equipment, give a practical method that could be used to prepare (a) $MgCl_2 \cdot 6H_2O$ from $MgCO_3(s)$; (b) $NaAl(OH)_4$ from $Na(s)$ and $Al(s)$; (c) $ZnS(s)$ from $ZnO(s)$; (d) $HgS(s)$ from $Hg(l)$.

4. Write a chemical equation to represent the reaction of (a) $K_2CO_3(aq)$ and $Ba(OH)_2(aq)$; (b) $Mg(HCO_3)_2(aq)$ upon heating; (c) tin(II) oxide when heated with carbon; (d) HNO_3(conc. aq) and metallic tin; (e) $H_2SO_4(aq)$ and $CdO(s)$; (f) $PbO_2(s)$ and $HI(aq)$.

5. For each of the following indicate if equilibrium is displaced either far to the left or far to the right. Detailed calculations are not required.

(a) $Ba(OH)_2(s) + SO_4^{2-}(aq) \rightleftharpoons BaSO_4(s) + 2\ OH^-(aq)$
(b) $Mg(OH)_2(s) + CO_3^{2-}(aq) \rightleftharpoons MgCO_3(s) + 2\ OH^-(aq)$
(c) $Ca(OH)_2(s) + 2\ F^-(aq) \rightleftharpoons CaF_2(s) + 2\ OH^-(aq)$

6. Without performing detailed calculations, indicate why you would expect each of the following reactions to occur to a significant extent as written. Use equilibrium constants from earlier chapters, as necessary.

(a) $SrCO_3(s) + 2\ HC_2H_3O_2(aq) \rightarrow$
$$Sr(C_2H_3O_2)_2(aq) + H_2O + CO_2(g)$$
(b) $Ba(OH)_2(s) + 2\ NH_4^+(aq) \rightarrow$
$$Ba^{2+}(aq) + 2\ NH_3(aq) + 2\ H_2O$$
(c) $ZnS(s) + 2\ H^+(conc.\ aq) \rightarrow Zn^{2+}(aq) + H_2S(g)$

7. A sample of water whose hardness is expressed as 185 ppm Ca^{2+} is passed through an ion exchange column and the Ca^{2+} is replaced by Na^+. What is $[Na^+]$ in the water that has been so treated? (*Hint:* Base your calculation on a 1000-L sample, which weighs 1.0×10^6 g.)

8. A sample of water has a hardness expressed as 77.5 ppm Ca^{2+}. This sample is passed through an ion exchange column and the Ca^{2+} is replaced by H^+. What is the pH of the water after it has been so treated? [*Hint:* Base your calculation on 1000 L $(1.0 \times 10^6$ g) of water.]

9. Write chemical equations to represent the most probable outcome in each of the following. If no reaction is likely to occur, so state.

(a) $CdCO_3(s) \overset{\Delta}{\rightarrow}$
(b) $MgO(s) \overset{\Delta}{\rightarrow}$
(c) $SnO_2(s) + CO(g) \overset{\Delta}{\rightarrow}$
(d) $Cd^{2+}(aq) + SO_4^{2-}(aq) \xrightarrow{\text{electrolysis}}$
(e) $HgO(s) \overset{\Delta}{\rightarrow}$
(f) $MgO(s) + Zn(s) \overset{\Delta}{\rightarrow}$

10. Use data from Figures 22-12 and 22-14 to answer the following.

(a) Will $C(s)$ reduce $HgO(s)$ to Hg at 100°C?
(b) Will $Hg(l)$ reduce $ZnO(s)$ to $Zn(s)$ at 200°C?
(c) Will $Mg(s)$ reduce $ZnO(s)$ to $Zn(s)$ at 200°C?

Exercises

Alkali (IA) metals

1. Refer to Figure 22-3 (as well as to appropriate information from Chapter 21) and write balanced equations for the preparation of all the compounds indicated there as being derived from $NaOH(aq)$.

2. Apply the method of Example 22-1 to the specific case where a slurry of $Ca(OH)_2(s)$ in water is also made 1.000 M $Na_2CO_3(aq)$ and calculate the concentrations, *at equilibrium*, of CO_3^{2-} and OH^-.

3. Suppose that in reaction (22.4) $Na_2SO_4(aq)$ is substituted for $Na_2CO_3(aq)$.

(a) Will the reaction go essentially to completion as was the case in Example 22-1? Explain.
(b) What will be $[SO_4^{2-}]$ and $[OH^-]$, *at equilibrium*, if a slurry of $Ca(OH)_2(s)$ is mixed with 1.00 M $Na_2SO_4(aq)$?

4. An analysis of a Solvay plant shows that for every 1.00 ton of NaCl consumed 1.03 tons of $NaHCO_3$ is obtained. Only 1.5 lb NH_3 is consumed in the overall process.

(a) What is the % efficiency of this process for converting NaCl to $NaHCO_3$?
(b) Why is so little NH_3 required?

5. What is the pH of the resulting solution if 0.445 L NaCl(aq) is electrolyzed for 137 s with a current of 1.08 A? Does the result depend on the concentration of NaCl present? Explain.

Magnesium

6. From these two statements estimate the quantity of Mg that has been produced since the metal was discovered: All the Mg produced in the world could have been made from as little as 4 km³ of sea water. The quantity of Mg^{2+} in sea water is 1272 g Mg^{2+}/ton sea water. (The density of sea water is 1.03 g/cm³.)

*7. The electrolysis of 0.250 L of 0.220 M MgCl$_2$ is conducted until 104 cm^3 of gas (a mixture of H$_2$ and water vapor) is collected at 23°C and 748 mmHg. Will Mg(OH)$_2$(s) precipitate if electrolysis is carried to this point? (Use 21 mmHg as the vapor pressure of the solution.)

8. The dissolving of MgCO$_3$(s) in NH$_4{}^+$(aq) can be represented as

$$MgCO_3(s) + NH_4{}^+(aq) \rightleftharpoons Mg^{2+}(aq) + HCO_3{}^-(aq) + NH_3(aq)$$

(a) Arrange the following three solubilities in the expected order, from most to least soluble: MgCO$_3$(s) in 1.00 M NH$_4$Cl(aq); MgCO$_3$(s) in the solution 1.00 M NH$_3$–1.00 M NH$_4$Cl; MgCO$_3$(s) in the solution 0.100 M NH$_3$–1.00 M NH$_4$Cl.

*(b) Calculate each of the solubilities referred to in part (a).

Hard Water

9. Write chemical equations for the reactions represented by the photographs on page 682.

10. A particular hard water contains 180.0 ppm HCO$_3{}^-$. What mass of CaO is required to soften 1.00×10^6 gal of this water. [*Hint:* Recall equations (22.16) and (22.17).]

11. Supose that all the cations associated with HCO$_3{}^-$ in Exercise 10 are Ca^{2+}.

(a) What is the total mass of CaCO$_3$ that would be precipitated in softening 1.00×10^6 gal of this water?

(b) Show that in the CaCO$_3$, half the Ca^{2+} is derived from the CaO used in the water softening and half from the water itself.

12. A particular water sample has a hardness of 131 ppm SO$_4{}^{2-}$ (as CaSO$_4$).

(a) Show how this water can be softened with Na$_2$CO$_3$.

(b) What mass of Na$_2$CO$_3$ is required to soften 385 L of this water?

13. Describe, using equations similar to (22.19), how a sample of hard water containing Fe^{3+}, Ca^{2+}, Mg^{2+}, HCO$_3{}^-$, and SO$_4{}^{2-}$ can be deionized by passing it first through a cation exchange resin and then an anion exchange resin.

Aluminum

14. The maximum resistance to corrosion of Al metal is between pH 4.5 and 8.5. Explain how this observation is consistent with other facts about the behavior of Al presented in the text.

15. Describe a series of *simple* chemical reactions with which you could determine whether a particular metal sample is Aluminum 2S (99.2% Al) or magnalium (70% Al, 30% Mg). You are permitted to destroy the metal sample in the testing.

16. In the purification of bauxite ore as a preliminary step in the production of aluminum, [Al(OH)$_4$]$^-$(aq) can be converted to Al(OH)$_3$(s) by passing CO$_2$(g) through it. Write an equation for the reaction that you would expect to occur.

17. Concerning the compound NaAl(SO$_4$)$_2 \cdot$12H$_2$O

(a) Why is ''sodium alum'' not a sufficient name for the compound?

(b) What is a more appropriate name for the compound?

(c) Is an aqueous solution of this compound acidic, basic, or neutral? Explain.

18. Use information from the chapter to explain why neither the compound Al(HCO$_3$)$_3$ nor the compound Al$_2$(CO$_3$)$_3$ exists.

Extractive metallurgy

19. The following approximate data are given for the reaction 2 H$_2$(g) + O$_2$(g) → 2 H$_2$O(g) $\Delta \overline{G}°$ = −465 kJ/mol at 100°C; −420 kJ/mol at 500°C; −360 kJ/mol at 1000°C; −240 kJ/mol at 2000°C.

(a) Sketch $\Delta \overline{G}°$ as a function of temperature on the same axes as used in Figure 22-12.

(b) At what approximate temperature does the water gas reaction   C(s) + H$_2$O(g) → CO(g) + H$_2$(g)   become spontaneous?

20. Calcium metal will reduce MgO(s) to Mg at all temperatures from 0 to 2000°C. Use this fact to sketch a plausible graph of $\Delta \overline{G}°$ versus temperature for the reaction   2 Ca + O$_2$ → 2 CaO.   For calcium, mp = 839°C; bp = 1484°C. (*Hint:* Relate your sketch to Figure 22-12.)

*21. According to Figure 22-12, $\Delta \overline{G}°$ decreases with temperature for the reaction   2 C(s) + O$_2$(g) → 2 CO(g).   For the following reactions would you expect $\Delta \overline{G}°$ to increase, decrease, or remain essentially constant with temperature?

(a) C(s) + O$_2$(g) → CO$_2$(g)

(b) 2 CO(g) + O$_2$(g) → 2 CO$_2$(g)

*22. Explain why (a) breaks in the straight lines of Figure 22-12 occur at the melting points and boiling points of the metals; (b) the slopes of the lines become more positive at these breaks; (c) the break at the boiling point is sharper than at the melting point.

Tin and lead

23. The names dilead(II)lead(IV) oxide and plumbous orthoplumbate have both been used to describe red lead, Pb$_3$O$_4$. Show that these names are consistent with remarks made in the text about this oxide of lead.

24. Write chemical equations to represent the following reactions.

(a) The reaction of Pb with hot concentrated HNO$_3$(aq).

(b) The oxidation of HCl(aq) to Cl$_2$(g) by PbO$_2$(s).

(c) The reduction of Fe^{3+} to Fe^{2+} by Sn^{2+} in aqueous solution.

(d) The production of basic lead carbonate (white lead) by the action of H$_2$O and CO$_2$ on PbO.

25. With the reagents (acids, bases, salts) and equipment commonly available in a chemical laboratory indicate how you would prepare (a) PbCrO$_4$(s) from Pb$_3$O$_4$(s); (b) PbCl$_2$(s) from PbS(s).

26. Scrap tin plate can be recycled by dissolving the tin in NaOH(aq), followed by electrolysis. Write plausible equations for the reactions involved.

★27. Use information from this chapter and elsewhere in the text to explain why neither the compound $PbBr_4$ nor the compound PbI_4 exists.

★28. To prevent the air oxidation of aqueous solutions of Sn^{2+} to Sn^{4+}, sometimes metallic tin is kept in contact with the Sn^{2+}(aq). Suggest how this helps to prevent the oxidation.

Group IIB metals

29. Write chemical equations to represent
  (a) the dissolving of Cd in H_2SO_4(conc. aq);
  (b) the dissolving of Hg in dilute HNO_3(aq);
  (c) the dissolving of ZnO(s) in $HC_2H_3O_2$(aq);
  (d) the action of excess NaOH(aq) on $ZnSO_4$(aq).

30. Use data from Table 22-7 to establish the standard potential for the reduction $2\,Hg^{2+}(aq) + 2\,e^- \rightarrow Hg_2^{2+}(aq)$.

31. The vapor pressure of Hg as a function of temperature is

$$\log p \text{ (mmHg)} = \frac{-0.05223a}{T} + b$$

where $a = 61{,}960$; $b = 8.118$; T = kelvin temperature

Show that the concentration of Hg(g) in equilibrium with Hg(l) greatly exceeds the maximum permissible level of 0.05 mg Hg/m^3 air. (Assume a temperature of 25°C.)

32. The text notes that in small quantities Zn is an essential metal (though it is toxic in higher concentrations). It is also noted that Sn is among the poisonous metals. Give reasons why tin plate is used rather than galvanized iron in fabricating cans for food storage.

Additional Exercises

1. Write chemical equations for the following reactions.
  (a) The decomposition of $Mg(HCO_3)_2$ on heating.
  (b) The recovery of NH_3 in the Solvay process.
  (c) Production of an alkaline solution for the precipitation of $Mg(OH)_2$(s) from seawater.
  (d) The softening of temporary hard water with NH_3(aq).
  (e) The separation of Fe_2O_3 inpurity from bauxite ore.
  (f) The formation of lead(II) sulfate during the high temperature roasting of lead(II) sulfide.
  (g) The successive conversion of KCl to K_2SO_4 to K_2S to K_2CO_3.

2. Write chemical equations to represent the dissolving of **(a)** Zn in HCl(aq); **(b)** Pb in HCl(conc. aq); **(c)** cadmium oxide in dilute HNO_3(aq); **(d)** cadmium hydroxide in NH_3(aq); **(e)** Al in KOH(aq); **(f)** sodium aluminum alum in NH_3(aq).

3. Oersted (1825) produced aluminum chloride by passing Cl_2(g) over a heated mixture of aluminum oxide and carbon. Wöhler (1827) prepared Al by heating aluminum chloride with potassium. Write plausible equations for these reactions.

4. Exercise 16 describes how $Al(OH)_3$(s) can be obtained by treating $[Al(OH)_4]^-$(aq) with CO_2(aq). Could HCl(aq) be used instead of CO_2(g)? Explain.

5. Predict whether each of the following is expected to produce an aqueous solution that is acidic, basic, or neutral: **(a)** Na_2SO_4; **(b)** $KAl(SO_4)_2 \cdot 12H_2O$; **(c)** $KAl(OH)_4$; **(d)** $ZnSO_4$; **(e)** NH_4Cl.

6. A description for preparing potassium aluminum alum calls for dissolving aluminum foil in KOH(aq). The solution obtained is treated with H_2SO_4(aq) and the alum is crystallized from the resulting solution. Write plausible equations for the reactions just described.

7. In this and the preceding chapter several examples were cited of the first member of a group differing from the second. Also in this chapter differences were cited between the fifth period and the sixth period members of a group. What are the fundamental reasons for these differences?

8. What mass of "bathtub ring" will form if 15.5 L of water having 78 ppm Ca^{2+} is treated with an excess of the soap potassium stearate, $CH_3(CH_2)_{16}COO^-K^+$?

9. A particular hard water contains 96 ppm SO_4^{2-} and 183 ppm HCO_3^-, with Ca^{2+} as the only cation. How many ppm Ca^{2+} does the water contain?

★10. Refer to the hard water sample described in Additional Exercise 9.
  (a) What mass of CaO is required to remove the HCO_3^- from a 150-L sample of this water?
  (b) How many ppm of SO_4^{2-} and Ca^{2+} remain after the treatment described in part (a)?
  (c) What mass of Na_2CO_3 is required to remove the remaining Ca^{2+} found in part (b)?

11. Concerning the reactions represented by the photographs on page 682,
  (a) If the $Ca(OH)_2$(aq) were replaced by $CaCl_2$(aq), would a precipitate form in the presence of CO_2(aq)? Explain.
  ★(b) The $CaCO_3$(s) precipitate redissolves if the $Ca(OH)_2$(aq) used is about 0.005 M, but fails to redissolve if the $Ca(OH)_2$(aq) is saturated. By calculation, demonstrate that this difference in behavior is to be expected.

*12. The enthalpy of sublimation of Mg is +150 kJ/mol and the lattice energy of $MgCl_2(s)$ is −2526 kJ/mol. Use these facts, together with data from this and earlier chapters, to demonstrate that the enthalpy of formation of $MgCl_2(s)$ is much larger (more negative) than that of $MgCl(s)$. [*Hint:* Assume that the lattice energy of $MgCl(s)$ is about the same as that of $NaCl(s)$.]

*13. Use Figure 22-12 to estimate for the reaction

$$ZnO(s) + C(s) \rightleftharpoons Zn(l) + CO(g) \qquad K = ?$$

(a) a value of K at the boiling point of zinc;
(b) the equilibrium pressure of $CO(g)$ at the temperature of part (a).

Self-Test Questions

For questions 1 through 8 select the single item that best completes each statement.

1. Production of the metal by reduction with carbon at moderate temperatures is not feasible with (a) CdO; (b) Al_2O_3; (c) PbO; (d) HgO.

2. All but one of the following can be used to soften temporary hard water. That one is (a) NH_3; (b) Na_2CO_3; (c) NH_4Cl; (d) NaOH.

3. Of the following sulfides the one that produces the free metal directly on roasting is (a) HgS; (b) PbS; (c) Na_2S; (d) SnS_2.

4. Of the following oxides, all are soluble in NaOH(aq) except (a) ZnO; (b) Al_2O_3; (c) Fe_2O_3; (d) SnO_2.

5. The most difficult of the following ions to reduce to the free metal in aqueous solution is (a) Zn^{2+}; (b) Cd^{2+}; (c) Pb^{2+}; (d) Hg^{2+}.

6. Of the following solutions the one that is acidic is (a) $ZnSO_4(aq)$; (b) $NaAl(OH)_4(aq)$; (c) $NaHCO_3(aq)$; (d) $KNO_3(aq)$.

7. The best oxidizing agent of the following oxides is (a) SnO_2; (b) PbO_2; (c) HgO; (d) MgO.

*14. An Al production cell of the type pictured in Figure 22-10 operates at a current of 1.00×10^5 A and a voltage of 4.5 V. The cell is 38% efficient in converting electric energy to chemical change. (The rest of the electric energy is dissipated as heat energy in the cell.)

(a) What mass of Al can be produced by this cell in 8.00 h?
(b) If the electric energy required to power this cell is produced by burning coal (85% C; heat of combustion of C = 32.8 kJ/g) in a power plant with 35% efficiency, what mass of coal must be burned to produce the mass of Al determined in part (a)?

8. All of the following find appreciable use in low melting alloys except (a) Sn; (b) Pb; (c) Cd; (d) Mg.

9. Write chemical equations to represent
(a) the dissolving of tin in concentrated $HNO_3(aq)$;
(b) the reaction of $MgCO_3(s)$ with $CO_2(aq)$;
(c) the dissolving of zinc in concentrated NaOH(aq);
(d) the roasting of cadmium sulfide followed by reduction with carbon.

10. Why is (a) the melting point of MgO (2800°C) so much higher than that of BaO (1920°C)? (b) the water solubility of $MgCl_2$ much greater than that of MgF_2?

11. Indicate how you would establish through simple laboratory tests whether a particular metal sample is a Pb–Sn or a Pb–Cd alloy.

12. A sample of water has its hardness due only to $CaSO_4$. When this water is passed through an *anion* exchange resin, SO_4^{2-} ions are replaced by OH^-. A 25.00-mL sample of the water so treated requires 21.58 mL of 1.00×10^{-3} M H_2SO_4 for its titration. What is the hardness of the water, expressed in ppm of $CaSO_4$?

23 Chemistry of the Transition Elements

Elements of the main transition series (also called the "*d* block" elements) have atoms or ions with partially filled *d* orbitals. Partially filled *f* orbitals are characteristic of atoms and ions of the inner transition ("*f* block") elements. All the elements in the middle section of the periodic table fit one or another of these descriptions. More than half the elements belong either to a transition or inner transition series. The chemistry of these elements has both theoretical and practical significance.

We begin with a survey of the general properties of transition elements. This is followed by discussions—some brief and some of greater length—of the individual members of the first transition series. The coinage metals—Cu, Ag, Au—are considered as a group, as is the first series of inner transition elements, the lanthanoids. The chapter closes with a discussion of the cation group of the traditional qualitative analysis scheme that contains several elements of the first transition series.

One important characteristic of transition elements is the ability to form complex ions. This tendency is mentioned in several instances in this chapter and is explored more fully in the next.

23-1 General Properties of the Transition Elements

A transition element must have partially filled *d* orbitals in either the neutral atom or the atom in one of its oxidation states. Copper does not have *d* orbital vacancies in the neutral atom nor in Cu(I); but it does in Cu(II). Copper is a transition element. Zinc does not have *d* orbital vacancies in either the neutral atom or in Zn(II). Zinc is a representative element (as are Cd and Hg); group IIB was considered in Chapter 22.

The properties listed in Table 23-1 for elements in the first transition series ($Z = 21$ to $Z = 29$) are clearly those of a group of metallic elements. High melting points, good electrical conductivity, and moderate to extreme hardness result from the ready availability of electrons and orbitals for metallic bonding.

Atomic Radii. In a transition series the essential difference in atomic structure between successive elements involves one unit of positive charge on the nucleus and one electron in an orbital of an *inner* electronic shell. This is a rather small difference and does not produce significant changes in atomic size, especially in the middle of the series. As a result, in the first transition series one finds strong similarities among *horizontal* groupings of elements. For example, the three elements, Fe, Co, and Ni—the iron triad—are usually discussed as a single group (see Section 23-7).

Some ores of these metals are featured in Color Section H.

708

TABLE 23-1
Selected properties of elements of the first transition series

	Sc	Ti	V	Cr	Mn	Fe	Co	Ni	Cu
atomic number	21	22	23	24	25	26	27	28	29
electron config.[a]	$3d^1 4s^2$	$3d^2 4s^2$	$3d^3 4s^2$	$3d^5 4s^1$	$3d^5 4s^2$	$3d^6 4s^2$	$3d^7 4s^2$	$3d^8 4s^2$	$3d^{10} 4s^1$
metallic radius, pm	161	145	132	127	124	124	125	125	128
ionization energy, kJ/mol									
first	631	658	650	653	718	759	758	737	746
second	1235	1310	1414	1592	1509	1561	1646	1753	1958
third	2389	2653	2828	2987	3249	2957	3232	3394	3554
electrode potential,[b] V	-2.08	-1.63	-1.18	-0.91	-1.19	-0.44	-0.28	-0.23	$+0.34$
common oxidation states	3	2, 3, 4	2, 3, 4, 5	2, 3, 6	2, 3, 4, 7	2, 3	2, 3	2	1, 2
melting point, °C	1397	1672	1710	1900	1244	1530	1495	1455	1083
density, g/cm³	2.99	4.49	5.96	7.20	7.20	7.86	8.90	8.91	8.92
hardness[c]	—	—	—	9.0	5.0	4.5	—	—	—
electrical conductivity[d]	—	2	3	10	2	17	24	24	97

[a] Each atom has an argon inner core configuration.
[b] For the reduction process $M^{2+}(aq) + 2\,e^- \rightarrow M(s)$ [except for scandium, where the ion is $Sc^{3+}(aq)$].
[c] Hardness values are on the Mohs scale (see Table 13-3).
[d] Compared to an arbitrarily assigned value of 100 for silver.

The terms "lanthanoid" and "lanthanide" are often used interchangeably.

Lanthanoid Contraction.   When an element in the first transition series is compared with those of the second and third series within the same group, some important differences are noted. Table 23-2 lists representative data for the members of group VIB—Cr, Mo, and W. The most notable feature of the table is that the atomic (metallic) radii of Mo and W are the same. Along with the usual filling of *s*, *p*, and *d* sublevels, in the interval of elements separating Mo and W the 4*f* sublevel is also filled. Electrons in an *f* subshell are not very effective in screening outershell electrons from the nucleus. As a result these outer-shell electrons are held more tightly than would otherwise be the case and atomic size does not increase as expected. The limited shielding ability of 4*f* electrons means that in the series of elements in which the 4*f* sublevel is being filled, an actual decrease in atomic size occurs. This filling occurs in the lanthanoid series ($Z = 58$ to 71), so the phenomenon is called the lanthanoid contraction. An example of ways in which Cr differs from Mo and W is that Cr can exist in aqueous solution as the simple hydrated ions $Cr^{2+}(aq)$ and $Cr^{3+}(aq)$. Ionic forms of Mo and W are polyatomic (e.g., oxoanions). The

TABLE 23-2
Some properties of group VIB—Cr, Mo, W

Transition series	Element	Atomic number	Electron configuration	Atomic radius, pm	Standard electrode potential,[a] V	Oxidation states[b]
first	Cr	24	$[Ar]3d^5 4s^1$	127	-0.744	2, **3**, 6
second	Mo	42	$[Kr]4d^5 5s^1$	139	-0.20	2, 3, 4, **5**, **6**
third	W	74	$[Xe]4f^{14} 5d^4 6s^2$	139	-0.11	2, 3, 4, 5, **6**

[a] For the reduction process $M^{3+} + 3\,e^- \rightarrow M(s)$.
[b] The most common oxidation state(s) is shown in boldface.

oxidation state $+3$ is very common for Cr, whereas higher oxidation states ($+5$, $+6$) are favored with Mo and W.

Electron Configurations and Oxidation States.   All the elements of the first transition series have an electron configuration with the following characteristics:

1. An inner core of electrons in the argon configuration.
2. Two electrons in the $4s$ orbital for seven members and one $4s$ electron for the remaining two (Cr and Cu).
3. A number of electrons in $3d$ orbitals ranging from one in Sc to ten in Cu.

Sc has the electron configuration $[Ar]3d^14s^2$. In combination with certain other elements, the loss of three electrons by Sc atoms is energetically favorable and Sc^{3+} ions are formed. Ti atoms, with the electron configuration $[Ar]3d^24s^2$, can use four electrons in compound formation and display the oxidation state $+4$. It is also possible for Ti atoms to use a smaller number of electrons, such as through the loss of the $4s^2$ electrons to form the ion Ti^{2+}. With Ti, then, we note two features: (a) a variety of possible oxidation states and (b) a maximum oxidation state corresponding to the group number IVB. These two features continue with V, Cr, and Mn, for which the maximum oxidation states are $+5$, $+6$, and $+7$, respectively. A shift in behavior is noted in group VIII, however. Although Fe, Co, and Ni can all exist in more than one oxidation state, they do not display the variety found in the earlier members of the first transition series. Neither do they exhibit a maximum oxidation state corresponding to their group number (VIII). As we traverse the first transition series, the nuclear charge, number of d electrons, and energy requirement for the successive ionization of d electrons all increase. The use of a large number of d electrons becomes energetically unfavorable, and only the lower oxidation states are commonly encountered. Because of its single $4s$ electron in the configuration $[Ar]3d^{10}4s^1$, copper can exhibit an oxidation state of $+1$ as well as $+2$.

Although the transition elements display a variety in their oxidation states, the possible oxidation states differ in the ease with which they can be attained and in their stabilities. These stabilities depend on a number of factors—other atoms to which the transition metal is bonded, whether the compound is in crystalline form or in solution, the pH of an aqueous solution, etc. For example, $TiCl_2$ is a well-characterized compound, but the ion Ti^{2+} is unstable in aqueous solution; $Co^{3+}(aq)$ is unstable, but can be stabilized in complex ions with the appropriate ligands. Generalizations are difficult, beyond noting that stable higher oxidation states are generally found for oxoanions, fluorides, and oxofluorides. (Stabilization of higher oxidation states by F is probably due to the combined effects of the weak F—F bond, the high electronegativity of fluorine, and the small size of F^-, which leads to high lattice energies.) In Table 23-1 and elsewhere, when the term "common" is used to describe an oxidation state, this can be taken to mean an oxidation state found in a number of compounds and/or in aqueous solution.

Ionization Energies and Electrode Potentials.   The ionization energies are fairly constant across the first transition series of elements. The values of the first ionization energies are about the same as for the group IIA metals. Standard electrode potentials increase in value gradually across the transition series. However, with the exception of the oxidation of Cu to Cu^{2+}, all these elements are more readily oxidized than hydrogen. This means that they displace $H^+(aq)$, with the production of $H_2(g)$.

Ionic and Covalent Compounds.   We tend to think of metals as forming ionic compounds with nonmetals. This is certainly the case with IA metal compounds and most IIA metal compounds. On the other hand, we have seen that some metal compounds have significant covalent character, e.g., $BeCl_2$ and $AlCl_3$ (Al_2Cl_6). The transition metal compounds display both ionic and covalent character. In general, compounds with the transition metal in lower oxidation states are essentially ionic, and those in higher oxidation states (such as $TiCl_4$) have covalent character.

Color and Magnetism.   As explained more fully in Section 24-9, electronic transitions that occur within partially filled d subshells impart color to solid transition metal compounds and their solutions. Absence of these transitions, in turn, accounts for the fact that so many representative metal compounds are without color. Because individual orbitals in d subshells are half-filled before electron pairing begins, many transition metal compounds are paramagnetic. A special form of magnetism is displayed by Fe, Co, Ni, and some of their alloys (see Section 23-7).

Catalytic Activity.   Another consequence of d orbital availability in transition metals and transition metal compounds is their catalytic activity. An unusual ability to adsorb gaseous species makes the transition metals (e.g., Ni and Pt) good heterogeneous catalysts. The possibility of multiple oxidation states accounts for the catalytic effect of some transition metal ions on certain oxidation–reduction reactions. Many homogeneous chemical reactions seem to involve the exchange of ligands in complex ions, and complex ion formation is one of the distinctive features of the transition metals.

In Table 14-7 we noted the catalytic activity of Ni, Fe, Pt, Rh, Cr_2O_3, V_2O_5, and the titanium halides. Further examples in this chapter include $TiCl_4$, V_2O_5, and MnO_2. Catalysis is a key phenomenon in about 90% of all chemical manufacturing processes, and the transition elements are often the key elements in the catalysts used.

Comparison of Transition and Representative Elements.   With the representative elements the s and p orbitals of the outermost electronic shell are the most important in establishing the nature of chemical bonding. Participation by d orbitals is essentially nonexistent for the second period elements and the group IA and IIA metals and of limited importance in the heavier nonmetals. With the transition elements d orbitals are of primary importance in chemical bond formation and s and p orbitals are of less consequence. Most of the observed behavioral differences between the transition and representative elements—multiple versus single oxidation states, complex ion formation, color, magnetic properties, and catalytic activity—can be traced to this issue of which orbitals are most involved in bond formation.

23-2 Scandium

Scandium is found in nature with the lanthanoid elements and is similar to them (Section 23-10). On the other hand, Sc^{3+} ion displays many similarities to Al^{3+}. Among these are: hydrolysis of $[Sc(H_2O)_6]^{3+}(aq)$ to yield acidic solutions; formation of an amphoteric gelatinous hydroxide, $Sc(OH)_3$; formation in alkaline solution of such ions as $[Sc(H_2O)_2(OH)_4]^-$; formation of a fluoro compound, Na_3ScF_6 (similar to cryolite, Na_3AlF_6), in which Sc_2O_3 can be dissolved and Sc obtained by electrolysis.

Because of its noble gas electron configuration, the ion Sc^{3+} lacks some of the characteristic properties of transition metal ions. In particular, the ion is colorless and diamagnetic, as are most of its salts.

Scandium is a rare metal. Its presence in the earth's crust has been estimated at from 5 to 30 ppm, and it is found only in a few mineral deposits. The commercial uses of Sc are very limited and its production is measured in gram or kilogram quantities, not tonnages. One recent use has been as a component in high-intensity lamps. The pure metal is usually prepared by the electrolysis of a fused mixture of $ScCl_3$ with other chlorides.

23-3 Titanium

Titanium is the ninth most abundant element, comprising 0.6% of the earth's crust. Three physical properties underlie most current uses of the metal: low density, high structural strength, and corrosion resistance. The first two properties account for its extensive use in the aircraft industry, and the third for its applications in the chemical industry, where it is used in pipes, component parts of pumps, and reaction vessels.

Titanium Tetrachloride. $TiCl_4$ is one of the most important compounds of titanium. It is the starting material for preparing other Ti compounds; it plays a central role in the metallurgy of titanium; and it is used in formulating catalysts for the production of polyethylene and other plastics. The usual method of preparing $TiCl_4$ involves reaction of the naturally occurring ore **rutile** (TiO_2) with carbon and $Cl_2(g)$.

$$TiO_2(s) + 2\ C(s) + 2\ Cl_2(g) \xrightarrow{\Delta} TiCl_4(g) + 2\ CO(g) \tag{23.1}$$

$TiCl_4$ is a colorless liquid (m.p. $-24°C$; b.p. $136°C$). Molecules of $TiCl_4$ have a tetrahedral shape with Cl—Ti—Cl bond angles of $109.5°$. The hydrolysis of $TiCl_4$,

$$TiCl_4(l) + 2\ H_2O \longrightarrow TiO_2(s) + 4\ HCl(aq) \tag{23.2}$$

when carried out in moist air, is the basis of certain types of smoke grenades.

In the $+4$ oxidation state all the valence shell electrons of Ti atoms are employed in bond formation. In this oxidation state Ti bears a strong resemblance to the group IVA elements. The physical properties and molecular shape of $TiCl_4$ are similar to those of CCl_4 and $SiCl_4$. $SiCl_4$ also fumes in moist air in a reaction like (23.2).

Titanium Dioxide. One of the important reactions of $TiCl_4$ is the production of pure TiO_2. A gaseous mixture of $TiCl_4$ and O_2 is passed through a silica tube at about $700°C$, and reaction (23.3) occurs.

$$TiCl_4(g) + O_2(g) \xrightarrow{\Delta} TiO_2(s) + 2\ Cl_2(g) \tag{23.3}$$

TiO_2 is amphoteric, though it is actually not very soluble in either acids or alkalis. It dissolves slowly in hot concentrated $H_2SO_4(aq)$, from which a sulfate can be crystallized. With molten alkalis it forms *titanates,* e.g., K_2TiO_3.

Because of its whiteness, opacity, inertness, nontoxicity, and relative cheapness, TiO_2 is now the most widely used white pigment for paints. In this use TiO_2 has largely displaced toxic basic lead carbonate (see page 696). TiO_2 is also used as a paper whitener and in glass, ceramics, floor coverings, and cosmetics.

Metallurgy of Titanium. Extensive production of Ti is a recent development, spurred at first by the needs of the military and then by the aircraft industry. Ti is a good alternative to Al and steel in aircraft because Al loses its strength at high temperature and steel is too dense.

The first step in the production of Ti is the conversion of rutile ore to $TiCl_4$ (reaction 23.1). The purified $TiCl_4$ is next reduced to Ti by using a good reducing agent. The **Kroll process** uses Mg.

$$TiCl_4(g) + 2\ Mg(l) \xrightarrow[\text{He}]{850°C} Ti(s) + 2\ MgCl_2(l) \tag{23.4}$$

The reaction is carried out in a steel vessel. The $MgCl_2(l)$ is removed and electrolyzed to produce Mg and Cl_2, both of which are recycled (reactions 23.1 and 23.4). The Ti is

Vacuum-distilled metallic titanium sponge produced by magnesium reduction. The temperature in the Kroll process (23.4) is kept below the melting point of titanium. The metal is obtained as chunks of solid called titanium sponge. [Courtesy of Teledyne Wah Chang, Albany, Oregon.]

obtained as a sintered mass called titanium sponge. The sponge must be subjected to further treatment and alloying with other metals before it can be used. One of the most difficult problems in the commercial development of Ti was that of devising new metallurgical techniques for fabricating the metal.

In 1947, the United States production of Ti was only 2 tons. Today it is measured in the thousands of tons. Now there is concern that continued widespread use of the metal will deplete the world's known reserves of rutile (estimated to be equivalent to 10–70 million tons of Ti). Fortunately, methods are available for extracting TiO_2 from the more abundant ore **ilmenite** ($FeTiO_3$), although they are currently not cost competitive with extraction from rutile. The story of titanium and its oxide illustrates how materials practically unknown and unused in one decade may become major production items in the next.

23-4 Vanadium

Vanadium is a fairly abundant element (0.02% of the earth's crust) and is found in several dozen ores. One commercially important ore is V_2S_5; the others are more complex. The metallurgy of vanadium is not simple, but vanadium of high purity (99.99%) is obtainable. For most of its applications, though, V is prepared as an iron–vanadium alloy, **ferrovanadium,** containing from 35 to 95% V. Ferrovanadium is produced by reducing V_2O_5 with silicon in the presence of iron. SiO_2 combines with CaO to form a liquid slag of calcium silicate.

$$2\ V_2O_5 + 5\ Si\ (+\ Fe)\ \xrightarrow{\Delta}\ \underset{\text{ferrovanadium}}{4\ V\ (+\ Fe)} + 5\ SiO_2 \qquad (23.5)$$

$$SiO_2(s) + CaO(s)\ \xrightarrow{\Delta}\ CaSiO_3(l) \qquad (23.6)$$

About 80% of the vanadium produced is used in the manufacture of steel. Vanadium-containing steels are used in applications requiring strength and toughness, such as in springs and high-speed machine tools.

TABLE 23-3
Characteristics of oxides and some ions of vanadium

O.S.	Oxide	Behavior	Ion[a]	Name of ion	Color of ion
+2	VO	basic	V^{2+}	vanadium(II) (vanadous)	violet
+3	V_2O_3	basic	V^{3+}	vanadium(III) (vanadic)	green
+4	VO_2	amphoteric	VO^{2+}	oxovanadium(IV) (vanadyl)	blue
			[b]	hypovanadate (vanadite)	brown
+5	V_2O_5	amphoteric	VO_2^+	dioxovanadium(V)[c]	yellow
			VO_4^{3-}	orthovanadate[d]	colorless

[a] Some of these ions are hydrated in aqueous solution, e.g., $[V(H_2O)_6]^{2+}$, $[V(H_2O)_6]^{3+}$, $[VO(H_2O)_4]^{2+}$, and $[VO_2(H_2O)_4]^+$.
[b] There is no simple anionic species of vanadium(IV). One formulation of this ion is $V_4O_9^{2-}$.
[c] This ion is obtained only in strongly acidic solutions (pH <1.5).
[d] Orthovanadates can be obtained only in highly alkaline solutions (pH >13). At lower pH values the anionic species are more complex, e.g., pyrovanadates ($V_2O_7^{4-}$) from pH 10 to 13 and metavanadates, $(VO_3^-)_n$, from pH 7 to 10.

Vanadium Oxides.   As outlined in Table 23-3, in each of its oxidation states vanadium forms an oxide. These oxides behave in accordance with the factors discussed in Section 17-8: If the metal is in a low oxidation state (having a low charge density) the oxide acts as a base. With a higher oxidation state (and charge density) on the central atom, acidic properties increase. In the case of vanadium, the oxides with V in the oxidation states +2 and +3 are basic; in the +4 and +5 oxidation states, amphoteric.

The most important oxide is V_2O_5, and its most important use is as a catalyst, such as in the conversion of $SO_2(g)$ to $SO_3(g)$ in the contact method for the manufacture of sulfuric acid (recall Chapter 14). The activity of V_2O_5 as an oxidation catalyst may be linked to the reversible loss of oxygen that occurs from 700 to 1100°C.

Oxidation States.   Figure 23-1 summarizes the oxidation–reduction relationships among some of the species listed in Table 23-3. In general, compounds with V in its highest oxidation state (+5) are good oxidizing agents; and in its lowest oxidation state, vanadium (as V^{2+}) is a good reducing agent.

Some of these species are also featured in Color Section H.

Example 23-1   It is desired to reduce $VO_2^+(aq)$ to $V^{2+}(aq)$ in a stepwise fashion. Can this be accomplished using as a reducing agent **(a)** Zn(s), **(b)** $Sn^{2+}(aq)$, **(c)** $I^-(aq)$? Use data from Figure 23-1 and Table 20-2.

Solution.   The reduction potentials are written in a stepwise fashion in Figure 23-1. We need to write the oxidation half-reaction for each of the proposed reducing agents and see if for each step of the reduction E°_{cell} is positive or negative.
(a) $Zn(s) \rightarrow Zn^{2+}(aq) + 2\ e^-$ $E^\circ_{ox} = +0.763$ V
In adding +0.763 V to the numerical values in Figure 23-1, a positive E°_{cell} is obtained for each step. That is, (+0.763 V + 1.00 V) = +1.763 V; (+0.763 + 0.361 V) = +1.124 V; etc. Zinc will reduce VO_2^+ to V^{2+}.

FIGURE 23-1
Electrode potential diagram for vanadium.

$$VO_2^+(aq) \xrightarrow{+1.00\ V} VO^{2+}(aq) \xrightarrow{+0.361\ V} V^{3+}(aq) \xrightarrow{-0.255\ V} V^{2+}(aq) \xrightarrow{-1.18\ V} V(s)$$

(yellow)   (blue)   (green)   (violet)

(b) $Sn^{2+}(aq) \rightarrow Sn^{4+}(aq) + 2 e^-$ $E°_{ox} = -0.154$ V

Sn^{2+} will reduce VO_2^+ to VO^{2+}; and this will be followed by the reduction of VO^{2+} to V^{3+}, for which $E°_{cell} = -0.154$ V $+ 0.361$ V $= +0.207$ V. The process stops here since for the next step $E°_{cell} < 0$. Sn^{2+} will not reduce VO_2^+ to V^{2+}.

(c) $2 I^-(aq) \rightarrow I_2(s) + 2 e^-$ $E°_{ox} = -0.535$ V

I^- will reduce VO_2^+ to VO^{2+}, but the reduction will not go beyond this point. I^- will not reduce VO_2^+ to V^{2+}.

SIMILAR EXAMPLES: Review Problem 7; Exercises 6, 9.

23-5 Chromium

Although it is found only to the extent of 122 ppm in the earth's crust, chromium is one of the most important industrial metals. The principal ore is **chromite,** $Fe(CrO_2)_2$, from which an alloy of Fe and Cr called **ferrochrome** is obtained by the reduction

$$Fe(CrO_2)_2 + 4 C \xrightarrow{\Delta} \underbrace{Fe + 2 Cr}_{\text{ferrochrome}} + 4 CO(g) \tag{23.7}$$

Ferrochrome may be added directly to iron, together with other metals, to produce steel. Chromium metal is very hard and maintains a bright surface through the protective action of an invisible oxide coating. It is extensively used in plating other metals.

Chrome Plating.   Chrome plating of steel is accomplished from an aqueous solution containing CrO_3 and H_2SO_4 in a mass ratio of about 100:1. The plating obtained is thin and porous and tends to develop cracks. Usual practice is first to plate the steel with copper or nickel, which is the true protective coating. Then chromium is plated over this for decorative purposes. The technical art of chrome plating is well understood, but the mechanism of the electrodeposition has not been established. The efficiency of chrome plating is limited by the fact that reduction of $Cr(VI)$ to $Cr(0)$ produces only $\frac{1}{6}$ mol Cr per faraday: Large quantities of electric energy are required for chrome plating relative to other types of metal plating.

Oxides of Chromium.   The oxides of chromium (like those of V) illustrate the general principles of acid–base properties of element oxides discussed in Section 17-8. Their behavior is summarized in Table 23-4. The amphoterism of Cr_2O_3 and $Cr(OH)_3$ can be represented by a sequence of reactions similar to that used for Al in expression (22.31), that is, starting with the ionization of hydrated $Cr^{3+}(aq)$.

Some of the species listed in Table 23-4 are also shown in Color Section H.

$$[Cr(H_2O)_6]^{3+}(aq) + H_2O \rightleftharpoons [Cr(H_2O)_5OH]^{2+}(aq) + H_3O^+(aq) \tag{23.8}$$

Other Compounds of Chromium.   Pure chromium dissolves in dilute $HCl(aq)$ and $H_2SO_4(aq)$ to produce Cr^{2+} ion. Nitric acid and other oxidizing agents alter the surface of the metal (perhaps by formation of an oxide coating) and render the metal resistant to further attack—the metal becomes *passive*. The best source of chromium compounds, then, is not the pure metal but alkali metal chromates, which can be obtained from chromite ore by reactions such as

$$4 Fe(CrO_2)_2 + 8 Na_2CO_3 + 7 O_2(g) \xrightarrow{\Delta} 2 Fe_2O_3 + 8 Na_2CrO_4 + 8 CO_2 \tag{23.9}$$

A pure dichromate can be crystallized from an acidified solution of the chromate.

$$2 Na_2CrO_4(aq) + H_2SO_4(aq) \longrightarrow Na_2Cr_2O_7(aq) + Na_2SO_4(aq) + H_2O \tag{23.10}$$

Reduction of a dichromate with carbon yields Cr_2O_3.

$$K_2Cr_2O_7 + 2 C \xrightarrow{\Delta} Cr_2O_3 + K_2CO_3 + CO(g) \tag{23.11}$$

TABLE 23-4
Characteristics of oxides and some ions of chromium

O.S.	Oxide[a]	Hydroxo compound	Behavior	Ion	Name of ion	Color of ion
+2	CrO	$Cr(OH)_2$	basic	Cr^{2+b}	chromium(II) (chromous)	light blue
+3	Cr_2O_3	$Cr(OH)_3{}^c$	amphoteric	Cr^{3+d}	chromium(III) (chromic)	violet
				$Cr(OH)_4{}^{-e}$	chromite[f]	green
+6	CrO_3	$H_2CrO_4{}^g$ $H_2Cr_2O_7{}^h$	acidic	$CrO_4{}^{2-}$	chromate	yellow
				$Cr_2O_7{}^{2-i}$	dichromate	orange

[a] The oxide CrO_2 is also well known. It is a ferromagnetic electrical conductor used in high-quality recording tape.
[b] The hydrated ion is $[Cr(H_2O)_6]^{2+}$.
[c] The hydroxide is probably a hydrated oxide, $Cr_2O_3 \cdot xH_2O$.
[d] The hydrated ion $[Cr(H_2O)_6]^{3+}$ is violet. Substitution of other ligands for H_2O molecules causes a change in color, e.g., to green (see Section 24-9).
[e] This ion is actually $[Cr(H_2O)_2(OH)_4]^-$ and is also sometimes represented in dehydrated form as $CrO_2{}^-$.
[f] A more systematic name, diaquatetrahydroxochromate(III), is based on complex ion nomenclature (see Section 24-4).
[g] The hydroxo compound is $CrO_2(OH)_2$.
[h] The hydroxo compound is $Cr_2O_5(OH)_2$.
[i] See page 717 for the relationship between $CrO_4{}^{2-}$ and $Cr_2O_7{}^{2-}$.

Reaction (23.12) is illustrated in Color Section H.

Ammonium dichromate yields Cr_2O_3 simply upon being heated.

$$(NH_4)_2Cr_2O_7(s) \xrightarrow{\Delta} Cr_2O_3(s) + N_2(g) + 4\,H_2O(g) \qquad (23.12)$$

In applications where pure Cr is required, $Cr_2O_3(s)$ can be reduced by aluminum in the thermite reaction.

$$Cr_2O_3(s) + 2\,Al(s) \longrightarrow Al_2O_3(s) + 2\,Cr(l) \qquad (23.13)$$

Other Cr(III) compounds can be obtained from $Cr_2O_3(s)$, by dissolving it in acids or bases, or by the reduction of $Cr_2O_7{}^{2-}$, as in the preparation of **chrome alum.**

$$K_2Cr_2O_7(aq) + H_2SO_4(aq) + 3\,SO_2(g) \xrightarrow{\Delta} \underbrace{K_2SO_4(aq) + Cr_2(SO_4)_3(aq)}_{\substack{\text{crystallizes as}\\ KCr(SO_4)_2 \cdot 12H_2O}} + H_2O \qquad (23.14)$$

Chromium(II) compounds can be prepared by the reduction of Cr(III) compounds, with zinc in acidic solution or electrolytically at a lead cathode. The most distinctive feature of chromium(II) compounds is their reducing power.

$$Cr^{3+}(aq) + e^- \longrightarrow Cr^{2+}(aq) \qquad E° = -0.41\,V \qquad (23.15)$$

The relationship between $Cr^{2+}(aq)$ and $Cr^{3+}(aq)$ is shown in Color Section H.

The reverse of (23.15)—oxidation of $Cr^{2+}(aq)$—has $E_{ox}° = +0.41\,V$. An important use of Cr(II) compounds is in purging gases of traces of $O_2(g)$, through the reaction

$$4\,Cr^{2+}(aq) + O_2(g) + 4\,H^+(aq) \longrightarrow 4\,Cr^{3+}(aq) + 2\,H_2O \quad E_{cell}° = +1.64\,V \quad (23.16)$$

The Chromate–Dichromate Equilibrium   The red oxide CrO_3 dissolves in water to produce a strongly acidic solution. Although chromic acid H_2CrO_4 might be postulated as a product of the reaction, such a compound has never been isolated in the pure state. The observed reaction is

$$2\,CrO_3(s) + H_2O \longrightarrow 2\,H^+ + Cr_2O_7{}^{2-}$$

It is possible to crystallize a dichromate salt, such as $Na_2Cr_2O_7$ or $K_2Cr_2O_7$, from a water solution of CrO_3. If the solution is made basic, the color turns from red-orange to yellow. From basic solutions only chromate salts can be crystallized, for example, Na_2CrO_4 or K_2CrO_4. CrO_3 can be obtained by the action of concentrated sulfuric acid on a chromate or dichromate. This red solid is a powerful oxidizing agent and, in conjunction with concentrated sulfuric acid, is commonly used as a cleaning solution for laboratory glassware. Its principal mode of action is to oxidize grease.

Whether a solution contains $Cr(VI)$ as dichromate or chromate ion is a function of pH, since the equilibrium between these ions depends on the concentration of H^+.

$$2\ CrO_4^{2-} + 2\ H^+ \rightleftharpoons Cr_2O_7^{2-} + H_2O \tag{23.17}$$

$$K_c = \frac{[Cr_2O_7^{2-}]}{[CrO_4^{2-}]^2[H^+]^2} = 3.2 \times 10^{14} \tag{23.18}$$

$$\frac{[Cr_2O_7^{2-}]}{[CrO_4^{2-}]^2} = 3.2 \times 10^{14}[H^+]^2 \tag{23.19}$$

Equation (23.17) is actually the sum of two equilibrium expressions. The first is an acid–base reaction.

$$H^+ + CrO_4^{2-} \rightleftharpoons HCrO_4^-$$

The second involves the elimination of a water molecule from between two $HCrO_4^-$ ions (a dehydration reaction).

$$2\ HCrO_4^- \rightleftharpoons Cr_2O_7^{2-} + H_2O$$

Quantitative calculations of the relative amounts of the two ions as a function of $[H^+]$ can be made with equation (23.19), but a qualitative prediction can be made as well by applying Le Châtelier's principle to equation (23.17). Clearly, $Cr_2O_7^{2-}$ is the predominant species in acidic solution and CrO_4^{2-} in basic solution. Control of the chromate–dichromate equilibrium through the control of pH is important in applications where dichromate solutions are used as oxidizing agents and chromate solutions to precipitate metal chromates.

$K_2Cr_2O_7$ is one of the most extensively used oxidizing agents in the analytical laboratory. For example, it can be used to determine the quantity of iron present in a sample. The sample is dissolved in acid; any iron that appears as Fe^{3+} is reduced back to Fe^{2+} [using $Zn(s)$ or $Sn^{2+}(aq)$ as a reducing agent]; and the following titration is performed.

$$6\ Fe^{2+} + Cr_2O_7^{2-} + 14\ H^+ \longrightarrow 6\ Fe^{3+} + 2\ Cr^{3+} + 7\ H_2O \tag{23.20}$$

Dichromates are also used as oxidizing agents in chemical industry. In the chrome tanning process, hides are immersed in $Na_2Cr_2O_7(aq)$, which is then reduced by $SO_2(g)$ to soluble basic chromic sulfate, $Cr(OH)SO_4$. Collagen, a protein in hides, reacts to form an insoluble complex chromium compound. The hides become tough, pliable, and resistant to biological decay. They are converted to **leather.**

Chromate ion in basic solution is not a good oxidizing agent.

$$CrO_4^{2-}(aq) + 4\ H_2O + 3\ e^- \longrightarrow Cr(OH)_3(s) + 5\ OH^-(aq) \quad E° = -0.13\ V \tag{23.21}$$

In fact, $Cr(OH)_3(s)$ is rather easily oxidized to $CrO_4^{2-}(aq)$; H_2O_2 is a suitable oxidizing agent. With CrO_4^{2-} precipitation reactions are more important than oxidation-reduction reactions. Insoluble $PbCrO_4$ and $ZnCrO_4$ both find use as yellow pigments.

The structures of the CrO_4^{2-} and $Cr_2O_7^{2-}$ ions are suggested by Figure 23-2. The Cr atom is at the center of a tetrahedron with O atoms at the corners. In $Cr_2O_7^{2-}$ two tetrahedra share an O atom. The Cr—O distance in the Cr—O—Cr link is somewhat greater than the other Cr—O distances.

FIGURE 23-2
Structures of CrO_4^{2-} and $Cr_2O_7^{2-}$.

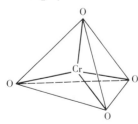

23-6 Manganese

Like V and Cr, the most important uses of Mn are in steel production, and for this the iron–manganese alloy, **ferromanganese,** is generally used. Ferromanganese is produced by reducing a mixture of iron and manganese oxides with carbon. The principal manganese ore is **pyrolusite,** MnO_2.

$$MnO_2 + Fe_2O_3 + 5\ C \xrightarrow{\Delta} \underline{Mn + 2\ Fe} + 5\ CO(g) \tag{23.22}$$

ferromanganese

In steel production, Mn participates in the purification of iron by reacting with sulfur and oxygen and removing them through slag formation. An added function is to increase the hardness of steel. Steel containing high proportions of Mn is extremely tough and wear resistant and is used in such applications as railroad rails and earth moving machinery.

Oxidation States.   The electron configuration of Mn is $[Ar]3d^54s^2$. By employing first the two $4s$ electrons and then, consecutively, up to all five of its unpaired $3d$ electrons, manganese exhibits all oxidation states from $+2$ to $+7$. The important reactions of manganese compounds then are oxidation–reduction reactions, which can be summarized through the electrode potential diagrams in Figure 23-3.

A number of important conclusions can be drawn from Figure 23-3. For example,

1. $Mn^{3+}(aq)$ is unstable. Its disproportionation is spontaneous.

$$2\ Mn^{3+} + 2\ H_2O \longrightarrow Mn^{2+} + MnO_2(s) + 4\ H^+ \qquad E^\circ_{cell} = +0.54\ V \qquad (23.23)$$

2. Manganate ion, MnO_4^{2-}, is unstable in acidic solution. Reaction (23.24) is spontaneous.

$$3\ MnO_4^{2-} + 4\ H^+ \longrightarrow MnO_2(s) + 2\ MnO_4^- + 2\ H_2O \qquad E^\circ_{cell} = +1.70\ V \quad (23.24)$$

3. But MnO_4^{2-} can be obtained in alkaline solution.

$$3\ MnO_4^{2-} + 2\ H_2O \longrightarrow MnO_2(s) + 2\ MnO_4^- + 4\ OH^- \qquad E^\circ_{cell} = +0.04\ V \quad (23.25)$$

That is, if $[OH^-]$ is kept sufficiently high, the reverse of reaction (23.25) becomes important, and significant concentrations of MnO_4^{2-} can be maintained in solution.

Manganese Compounds.   The principal source of manganese compounds is MnO_2. When MnO_2 is heated in the presence of an alkali and an oxidizing agent, a manganate salt is produced.

$$3\ MnO_2 + 6\ KOH + KClO_3 \longrightarrow 3\ K_2MnO_4 + KCl + 3\ H_2O(g) \qquad (23.26)$$

K_2MnO_4 is extracted from the fused mass with water, and can then be oxidized to $KMnO_4$ (e.g., with Cl_2 as an oxidizing agent). Alternatively, if $MnO_4^{2-}(aq)$ is acidified, MnO_4^- is produced by reaction (23.24).

Potassium permanganate, $KMnO_4$, is an important oxidizing agent. For chemical analyses it is generally used in acidic solutions, where it is reduced to $Mn^{2+}(aq)$. In the

FIGURE 23-3
Electrode potential diagram for manganese.

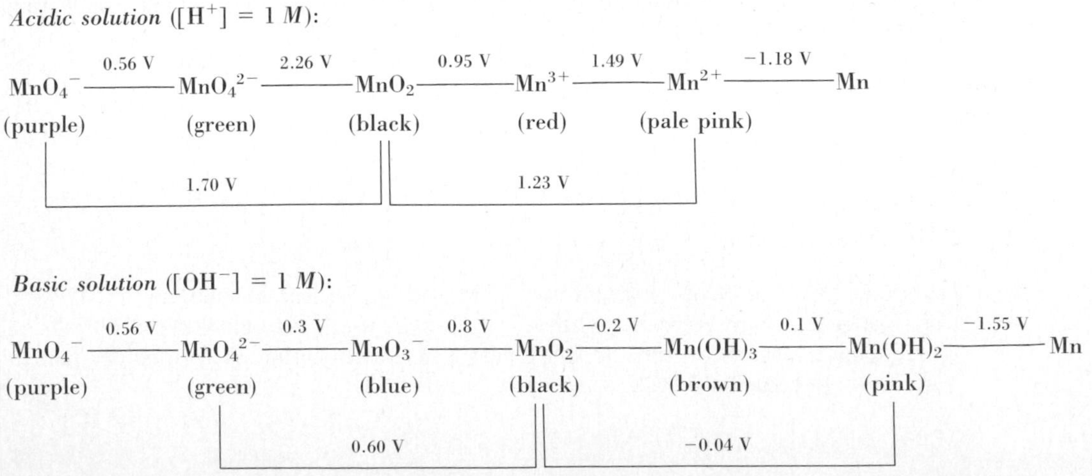

Acidic solution ($[H^+] = 1\ M$):

Basic solution ($[OH^-] = 1\ M$):

analysis of iron by MnO_4^-, a sample is prepared in the same manner as described for reaction (23.20) and titrated with $MnO_4^-(aq)$.

$$5\ Fe^{2+} + MnO_4^- + 8\ H^+ \longrightarrow 5\ Fe^{3+} + Mn^{2+} + 4\ H_2O \qquad (23.27)$$

$Mn^{2+}(aq)$ has a very pale pink color that is barely discernible. $MnO_4^-(aq)$ is intensely colored (purple). At the end point of titration (23.27) the solution being titrated acquires a lasting deep pink color with just a single drop of excess $MnO_4^-(aq)$. $MnO_4^-(aq)$ is less satisfactory for titrations in alkaline solutions because the insoluble reduction product $MnO_2(s)$ obscures the end point. Other titrations performed with MnO_4^- include the determination of nitrites, H_2O_2, and calcium (after precipitation as the oxalate). In organic chemistry $MnO_4^-(aq)$ is used to oxidize alcohols and unsaturated hydrocarbons (see Chapter 26). Manganese dioxide, MnO_2, is used in dry cells (recall Figure 20-11), in glass and ceramic glazes, and as a catalyst.

Manganese Nodules.   These are rocklike objects found on the ocean floor. They are composed of layers of oxides of Mn and Fe, with small quantities of other metals such as Co, Cu, and Ni. The nodules are roughly spherical in shape with diameters ranging from a few millimeters to about 15 cm. They are believed to grow at a rate of a few millimeters per million years. It has been proposed that marine organisms may play a role in their formation. Estimates of the total quantity of these nodules are very great, perhaps billions of tons. However, numerous challenges exist to developing manganese nodules as a significant raw material. Methods must be perfected to explore the seabed, dredge for the nodules, and transport them through several thousand meters of sea water. Also, new metallurgical processes must be devised for extracting the desired metals. Nontechnical, but problems nevertheless, are the political and legal issues involved in mining in international waters. (Who owns the nodules?) The largest deposits currently known are in an area southeast of the Hawaiian Islands.

Manganese nodules on the ocean floor. [Courtesy of Lawrence Sullivan, Lamont-Doherty Geological Observatory of Columbia University.]

FIGURE 23-4
The phenomenon of ferromagnetism.

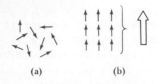

(a)   (b)

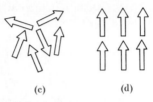

(c)   (d)

(a) In ordinary paramagnetism the magnetic moments of the atoms or ions are randomly distributed.
(b) In a ferromagnetic material the magnetic moments are aligned into domains.
(c) In an unmagnetized piece of the material the domains are randomly oriented.
(d) In a magnetic field the domains are oriented in the direction of the field and the material becomes magnetized.

23-7 The Iron Triad—Fe, Co, and Ni

Iron, with annual worldwide production approaching one billion tons, is, of course, the most important metal in modern civilization. It is found widely distributed in the earth's crust in an abundance of 4.7%. Cobalt is among the rarer metals. It constitutes only 20 ppm of the earth's crust, but it occurs in sufficiently concentrated deposits (ores) so that its annual production runs into the millions of pounds. Cobalt is used primarily in alloys with other metals. Nickel ranks twenty-fourth in abundance among the elements in the earth's crust. Its ores are mainly the sulfides, oxides, silicates, and arsenides. Particularly large deposits are found in Canada. Of the 300 million pounds or so of nickel consumed annually in the United States, about 80% is used in the production of alloys. Another 15% is used for electroplating, and the remainder for miscellaneous purposes.

As expected, there are some similarities between Fe, Co, and Ni and other elements in the first transition series. Various combinations of these metals form alloys with one another, and all of the simple metal ions have orbitals available for complex ion formation. But there are differences between the Fe–Co–Ni group and the elements preceding them in the first transition series. For example, Fe, Co, and Ni do not form stable oxoanions like VO_3^-, CrO_4^{2-}, and MnO_4^-. Also, they do not exhibit the same variability of oxidation state. The strongest similarities for elements of group VIII in the periodic table occur horizontally among the three elements in each period of the group. For this reason Fe, Co, and Ni are generally considered a group of three—a triad.

Ferromagnetism.   One unique property possessed only by Fe, Co, and Ni among the common chemical elements is that of ferromagnetism. Although Fe^{2+}, Co^{2+}, and Ni^{2+} all have unpaired electrons, the property of ferromagnetism cannot be accounted for by the paramagnetism of these ions alone. In the solid state the metal ions are thought to be grouped together into small regions containing rather large numbers of metal ions. These regions are called **domains.** Instead of the individual magnetic moments of the ions within a domain being randomly oriented, they are all directed in the same way. In an unmagnetized piece of iron the domains are oriented in several directions and their magnetic effects cancel. When the metal is placed in a magnetic field, however, the domains are lined up and a strong resultant magnetic effect is produced. This alignment of domains may actually involve the growth of domains with favorable orientations at the expense of those with unfavorable orientations. The ordering of domains persists when the object is removed from the magnetic field, and thus permanent magnetism results (see Figure 23-4).

The key to ferromagnetism involves two basic factors: that the species involved have unpaired electrons (a property possessed by many species) and that interionic distances be of just the right magnitude to make possible the ordering of ions into domains. If atoms are too large, interactions among them are too weak to produce this ordering. With small atoms the tendency is for atoms to pair and their magnetic moments to cancel. This critical factor of atomic size is just met in Fe, Co, and Ni. However, it is possible to prepare alloys of metals other than these three in which this condition is met (e.g., Al–Cu–Mn, Ag–Al–Mn, and Bi–Mn). Also, certain rare earth elements, for example, gadolinium (Gd) and dysprosium (Dy), are ferromagnetic at low temperatures.

Metal Carbonyls.   The transition metals, with few exceptions, form compounds with carbon monoxide. These compounds are called metal carbonyls. In the simple metal carbonyls listed in Table 23-5,

1. Each CO molecule contributes an electron pair to an empty orbital of the metal atom.
2. All electrons are paired (most metal carbonyls are diamagnetic).
3. The metal atom acquires the electron configuration of the noble gas Kr.

The structures of the simple carbonyls in Figure 23-5 are those that one might predict from

TABLE 23-5
Three metal carbonyls

	Number of e		
	From metal	From CO	Total
$Cr(CO)_6$	24	12	36
$Fe(CO)_5$	26	10	36
$Ni(CO)_4$	28	8	36

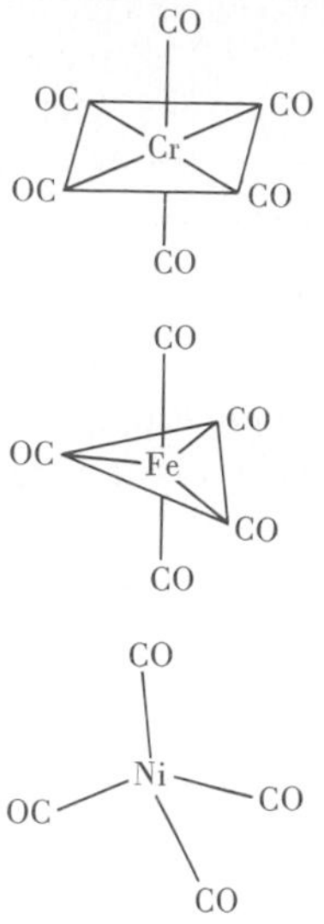

FIGURE 23-5
Structures of some
simple carbonyls.

VSEPR theory (i.e., based on a number of electron pairs equal to the number of CO molecules).

In the species $Mn(CO)_5$ the number of electrons associated with the Mn atom would be 35 (25 of its own and two each from the five CO molecules). This would make it an odd-electron (paramagnetic) species. When two $Mn(CO)_5$ units join by forming an Mn—Mn bond between them, each Mn atom acquires the equivalent of 36 electrons (a noble gas electron configuration) and all electrons are paired. The *binuclear* carbonyl $Mn_2(CO)_{10}$ is shown in Figure 23-6.

Metal carbonyls are produced in several ways. Nickel metal combines with CO(g) at ordinary temperatures and pressures in a reversible reaction.

$$Ni(s) + 4\ CO(g) \rightleftharpoons Ni(CO)_4(l) \tag{23.28}$$

With iron it is necessary to use higher temperatures (200°C) and CO pressures (100 atm).

$$Fe(s) + 5\ CO(g) \rightleftharpoons Fe(CO)_5(g) \tag{23.29}$$

In other cases the carbonyl is obtained by reducing a metal compound in the presence of CO(g).

Carbon monoxide poisoning results from an action similar to carbonyl formation. CO molecules coordinate with Fe atoms in hemoglobin, displacing the O_2 molecules normally carried by the hemoglobin. The metal carbonyls themselves are also very poisonous.

Oxidation States.   Variability of oxidation states is encountered within the iron triad elements, even if not to the same degree as with certain other transition elements like vanadium and manganese. The oxidation state +2 is commonly encountered with all three elements.

$$Fe^{2+}\quad [Ar]3d^6 \qquad Co^{2+}\quad [Ar]3d^7 \qquad Ni^{2+}\quad [Ar]3d^8$$

For cobalt and nickel the oxidation state +2 is the most stable, but for iron the most stable is the oxidation state +3.

$$3d$$
$$Fe^{2+}\quad [Ar]\ \boxed{\uparrow\downarrow\ \vert\ \uparrow\ \vert\ \uparrow\ \vert\ \uparrow\ \vert\ \uparrow}$$

$$3d$$
$$Fe^{3+}\quad [Ar]\ \boxed{\uparrow\ \vert\ \uparrow\ \vert\ \uparrow\ \vert\ \uparrow\ \vert\ \uparrow}$$

An electron configuration in which the *d* subshell is half-filled, with all electrons unpaired, has a special stability. This fact suggests that Fe(II) can be oxidized to Fe(III) without great difficulty; for example, in the presence of oxygen at 1 atm pressure and with $[H^+] = 1\ M$,

$$4\ Fe^{2+} + O_2(g) + 4\ H^+ \longrightarrow 4\ Fe^{3+} + 2\ H_2O \qquad E^\circ_{cell} = +0.44\ V \tag{23.30}$$

FIGURE 23-6
Structure of a binuclear
carbonyl: $Mn_2(CO)_{10}$.

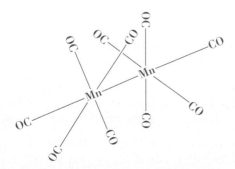

But even at lower partial pressures of oxygen, as in the atmosphere, and in less acidic media, reaction (23.30) may still be spontaneous.

For Co(II) and Ni(II) the loss of an additional electron does not lead to an electron configuration with half-filled d orbitals.

$$3d$$

Co^{3+} [Ar] [↿⇂ ↑ ↑ ↑ ↑]

$$3d$$

Ni^{3+} [Ar] [↿⇂ ↿⇂ ↑ ↑ ↑]

As a result the conversion of Co(II) to Co(III) and Ni(II) to Ni(III) is accomplished not nearly so easily as in the case of iron. Consider, for example, this standard electrode potential for cobalt.

$$Co^{3+}(aq) + e^- \longrightarrow Co^{2+}(aq) \qquad E° = 1.82 \text{ V} \tag{23.31}$$

The *reduction* of Co^{3+} to Co^{2+} occurs readily; Co(III) compounds tend to be very good oxidizing agents. For reasons explained in Section 24-13, the +3 oxidation state of cobalt can be stabilized, however, if the Co^{3+} is part of a complex ion. The oxidation state +3 is also difficult to achieve with nickel and is not commonly encountered. Nickel compounds in higher oxidation states do find use, however, as electrode materials in the nickel–cadmium cell (see Exercise 28, Chapter 20).

Because of the stability attributed to the electron configuration of Fe^{3+}, iron does not commonly occur in higher oxidation states. The oxidation state +6 is attainable, however, under very strong oxidizing conditions.

$$2 \text{ Fe(OH)}_3(s) + 3 \text{ OCl}^- + 4 \text{ OH}^- \rightleftharpoons 2 \text{ FeO}_4^{2-} + 3 \text{ Cl}^- + 5 \text{ H}_2O \tag{23.32}$$

The ion FeO_4^{2-} is called **ferrate ion,** and a few ferrate salts have been prepared. For example, barium ferrate, $BaFeO_4$, is a purple colored solid. As might be expected the ferrates are unstable and are powerful oxidizing agents.

Some Reactions of the Iron Triad Elements. The reactions of the iron triad elements are many and varied. The metals are all more active than hydrogen and liberate $H_2(g)$ from an acidic solution.

$$Ni(s) + 2 \text{ H}^+ + 2 \text{ Cl}^- \longrightarrow Ni^{2+} + 2 \text{ Cl}^- + H_2(g) \tag{23.33}$$

Hydrated, colored ions are characteristic of the iron triad elements. Co^{2+} and Ni^{2+} are red and green, respectively. In aqueous solution hydrated Fe^{2+} is pale green, and fully hydrated Fe^{3+} is pale purple. Generally, solutions of Fe^{3+} are yellow to brown in color, but this color is probably due to the presence of species formed in the hydrolysis of $Fe^{3+}(aq)$. As we have seen before with $Al^{3+}(aq)$, $Zn^{2+}(aq)$, and $Cr^{3+}(aq)$, this hydrolysis proceeds in a stepwise fashion, beginning with

$$[Fe(H_2O)_6]^{3+} + H_2O \rightleftharpoons [Fe(H_2O)_5OH]^{2+} + H_3O^+ \tag{23.34}$$

However, with Fe^{3+} the reaction does not proceed beyond $Fe(OH)_3(s)$, even in the presence of excess $OH^-(aq)$—$Fe(OH)_3$ is not amphoteric.

Salts of the iron triad elements usually crystallize from solution as hydrates. The hydrate of Co(II) chloride has an interesting application. When exposed to atmospheric moisture, depending on the partial pressure of $H_2O(g)$, the hydrate assumes different colors. In dry air, water of hydration is lost and the solid acquires a blue color; as the humidity increases, the solid undergoes a gradual color change to pink. This reaction has been used as an inexpensive, if somewhat crude, method of moisture determination.

$$CoCl_2 \cdot 6H_2O(s) \rightleftharpoons CoCl_2 \cdot 2H_2O(s) + 4 \text{ H}_2O(g) \tag{23.35}$$
$$\text{(pink)} \qquad\qquad \text{(blue)}$$

TABLE 23-6
Some qualitative tests for iron(II) and iron(III)

Reagent	Iron(II)	Iron(III)
NaOH(aq)	green precipitate	red-brown precipitate
$K_4[Fe(CN)_6]$	white precipitate, turning blue rapidly	Prussian blue precipitate
$K_3[Fe(CN)_6]$	Turnbull's blue precipitate	red-brown coloration (no precipitate)
KSCN	no coloration	deep red coloration

The basis for writing formulas and systematic names of complex ions such as these is taken up in Chapter 24.

The **hexacyanoferrates** are compounds containing the complex ions $[Fe(CN)_6]^{4-}$ and $[Fe(CN)_6]^{3-}$. Their systematic names are hexacyanoferrate(II) and hexacyanoferrate(III), respectively, but they are also commonly called **ferro**cyanide and **ferri**cyanide. Iron(III) in aqueous solution yields a dark blue precipitate, **Prussian blue,** when treated with potassium hexacyanoferrate(II). The exact structure of Prussian blue is not known, but it is probably quite complex. It has the empirical formula $Fe_7C_{18}N_{18} \cdot 10H_2O$. A simplified representation of this reaction is

$$4\,Fe^{3+} + 3[Fe(CN)_6]^{4-} \longrightarrow Fe_4[Fe(CN)_6]_3 \qquad (23.36)$$

The formation of Turnbull's blue is the basis of the test for iron illustrated in Color Section G.

Iron(II) compounds yield a blue precipitate, **Turnbull's blue,** when treated with potassium hexacyanoferrate(III). The reaction appears to proceed in two stages. The first is an oxidation–reduction reaction in which iron(II) is oxidized to iron(III) and hexacyanoferrate(III) is reduced to hexacyanoferrate(II).

$$Fe^{2+} + [Fe(CN)_6]^{3-} \longrightarrow Fe^{3+} + [Fe(CN)_6]^{4-} \qquad (23.37)$$

This is followed by reaction (23.36). Turnbull's blue is a lighter shade than Prussian blue and this may be caused by the admixture of some white $K_2\{Fe[Fe(CN)_6]\}$.

The qualitative tests for iron just described, together with two others, are summarized in Table 23-6. The dark red coloration resulting from the reaction of thiocyanate ion, SCN^-, with Fe^{3+} is the basis of an extremely sensitive test for iron. The composition of the product appears to be $[Fe(H_2O)_5SCN]^{2+}$.

This test for Fe^{3+} is illustrated in Color Section C.

A very distinctive reaction of Ni^{2+} that can be used both for its qualitative detection and its quantitative determination is the formation of a neutral complex with dimethylglyoxime, which precipitates from solution as a brilliant scarlet precipitate. In addition to the coordination bonds between N atoms and the Ni^{2+} ion, this complex features hydrogen bonds.

$$(23.38)$$

23-8 Metallurgy of Iron and Steel

A type of steel called wootz steel was first produced in India about 3000 years ago. This is the same steel that became famous in ancient times as Damascus steel, renowned for its suppleness, its ability to hold a cutting edge, and its use in making swords. Many technological advances have been made since ancient times. These include introduction of the

FIGURE 23-7
Typical blast furnace.

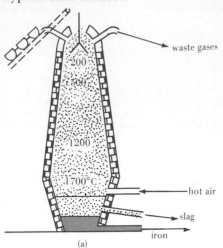

(a)

(b)

(a) Iron ore, coke, and limestone are added at the top of the furnace, and hot air is introduced through the bottom. Maximum temperatures are attained near the bottom of the furnace where molten iron and slag collect.
(b) The blast furnace shown here is capable of producing nearly 1×10^7 kg of pig iron per day. It stands about 90 m high, is computer-controlled, and is equipped with the latest environmental control devices. [Courtesy of Bethlehem Steel Corporation.]

blast furnace in about A.D. 1300, the Bessemer converter in 1856, the open hearth furnace in the 1860s, and most recently, the basic oxygen furnace. However, a true understanding of the iron and steel making process has developed only within the past few decades, based on concepts of thermodynamics, equilibrium, and kinetics.

Pig Iron. The reduction of iron ore, which is accomplished in a blast furnace, involves an impressive array of reactions. In the following simplified scheme, approximate temperatures are given so that these reactions may be keyed to regions of the blast furnace pictured in Figure 23-7.

formation of reducing agents, principally $CO(g)$ and $H_2(g)$:

$$2\ C + O_2 \longrightarrow 2\ CO\ (1700°C)$$
$$C + CO_2 \longrightarrow 2\ CO\ (>1000°C) \tag{23.39}$$
$$C + H_2O \longrightarrow CO + H_2\ (>600°C)$$

reduction of iron oxide:

$$3\ CO + Fe_2O_3 \longrightarrow 2\ Fe + 3\ CO_2\ (900°C)$$
$$3\ H_2 + Fe_2O_3 \longrightarrow 2\ Fe + 3\ H_2O\ (900°C) \tag{23.40}$$

slag formation to remove impurities from ore:

$$CaCO_3 \longrightarrow CaO + CO_2\ (800–900°C)$$
$$CaO + SiO_2 \longrightarrow CaSiO_3(l)\ (1200°C) \tag{23.41}$$
$$3\ CaO + P_2O_5 \longrightarrow Ca_3(PO_4)_2(l)\ (1200°C)$$

impurity formation in the iron:

$$MnO + C \longrightarrow Mn + CO\ (1400°C)$$
$$SiO_2 + 2\ C \longrightarrow Si + 2\ CO\ (1400°C) \tag{23.42}$$
$$P_2O_5 + 5\ C \longrightarrow 2\ P + 5\ CO\ (1400°C)$$

TABLE 23-7
Some typical iron ores

Type of ore	Formula
oxide	
magnetite	Fe_3O_4
hematite	Fe_2O_3
ilmenite	$FeTiO_3$
limonite	$HFeO_2$
carbonate	
siderite	$FeCO_3$
sulfide	
pyrite	FeS_2
pyrrhotite	FeS

TABLE 23-8
Some common types of iron and steel

Composition, mass %	Trade name
98.5 Fe	wrought iron
99 Fe; 1 C	steel
97 Fe; 3 C	white cast iron
94 Fe; 3.5 C; 2.5 Si	gray cast iron
95.1 Fe; 3 Ni; 1.5 Cr; 0.4 C	nickel-chrome steel
94.5 Fe; 5 W; 0.5 C	tungsten steel
84.3 Fe; 14.5 Si; 0.85 C; 0.35 Mn	duriron
74 Fe; 18 Cr; 8 Ni; 0.18 C	stainless N

The blast furnace charge consists of iron ore, coke, a slag-forming flux, and perhaps some scrap iron. The exact proportions used depend on the composition of the iron ore and its impurities. The formulas of some typical ores are listed in Table 23-7. The purpose of the flux is to maintain the proper ratio of acidic oxides (SiO_2, Al_2O_3, and P_2O_5) to basic oxides (CaO, MgO, and MnO) to obtain an easily liquefied silicate, aluminate, or phosphate slag. Since in most ores the acidic oxides predominate, the flux generally employed is limestone, $CaCO_3$, or dolomite, $CaCO_3 \cdot MgCO_3$.

The iron obtained from a blast furnace is called **pig iron.** It contains about 95% Fe, 3 to 4% C, and varying quantities of other impurities. **Cast iron** can be obtained by pouring pig iron directly into molds of the desired shape. Cast iron is very hard and brittle and can be used only in applications where it will not be subjected to mechanical or thermal shock. For example, it is used in engine blocks, brake drums, and transmission housings in automobiles.

Steel. Until the 1850s the method of producing steel had not changed significantly from ancient times. Iron, as cast iron and wrought iron, was still the principal metal of commerce. (Wrought iron is a purified iron low in carbon content and easily worked, that is, malleable.) In 1856, Henry Bessemer, in England, introduced a new, rapid, inexpensive method of converting iron to steel. The iron and steel industry was completely transformed by this and subsequent inventions.

The fundamental changes that must be accomplished in any steel-making process are these: (1) reduction of the carbon content from an original 3 to 4 in pig iron to 0 to 1.5% in steel; (2) removal through slag formation of Si, Mn, and P (each present in pig iron to the extent of 1% or so), together with other minor impurities; and (3) addition of alloying elements (such as Cr, Ni, Mn, V, Mo, and W) to give the steel its desired end properties. Table 23-8 lists a few of the many different types of iron and steel in use today.

The **Bessemer converter,** shown in Figure 23-8, is a steel vessel with a refractory lining. The refractory lining may be either silica-based (acid Bessemer) or made of dolomite (basic Bessemer), depending on the type of impurities to be removed from the iron. The vessel can be rotated on a pair of trunions, one of which is hollow to allow for the passage of air. A blast of air is injected into the molten iron through a series of holes (tuyeres) in the bottom lining. A vigorous exothermic process occurs, involving a variety of reactions. Some of these are presented in simplified form in Table 23-9.

In the 1860s the **open hearth furnace** was developed, principally through the efforts of William Siemens in England. This furnace, pictured in Figure 23-9, works on a regenerative principle. The high temperatures required to maintain iron in a molten condition are achieved by burning a gaseous fuel (such as natural gas) over the metal. The hot gaseous products are exhausted through a network of firebricks, called checkers. Periodically, the direction of gas and air flow are reversed, so that the entering gases pass

FIGURE 23-8
A Bessemer converter.

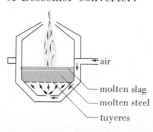

air

molten slag
molten steel
tuyeres

FIGURE 23-9
An open hearth furnace.

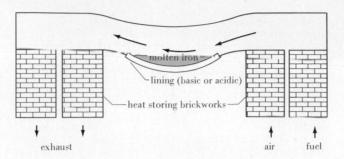

through the hot checkers and become preheated before combustion. The firebrick lining of the open hearth furnace can be either of the acid or basic type, although the basic type is the only one commonly used. The usual furnace charge consists of scrap iron, molten pig iron, iron oxide, and limestone. The reactions are similar to those listed in Table 23-9. Typically, a batch of steel can be produced in about 12 hours.

The most important method in use today is the **oxygen steel-making process,** This process is a logical extension of earlier practice. It uses pure oxygen gas rather than air to support the oxidation reactions required in refining iron. The process is carried out in a vessel much like a Bessemer converter. Oxygen gas, at about 10 atm, and a stream of powdered limestone are fed through a water-cooled lance and discharged above the molten metal (see Figure 23-10). A typical reaction time is 22 minutes.

Steel making is an activity that has been undergoing rapid technological change. In 1962 the quantity of steel made in the United States by the basic oxygen method was only about 4% of that produced in open hearth furnaces. Today, the basic oxygen method is the principal steel-making method; the Bessemer converter is obsolete; and more steel is being produced in electric arc furnaces (e.g., from scrap steel) than in open hearth furnaces. The ultimate goal may be to produce iron and steel directly from iron ore in a single-step (perhaps continuous) process. Methods currently exist for the direct reduction of iron ore to iron at temperatures below the melting point of any of the materials used in the process, that is, without the need of blast furnaces. The most successful of these direct reduction processes use as reducing agents $CO(g)$ and $H_2(g)$ obtained in the reaction of steam with natural gas (reaction 13.3). The economic viability of these methods depends on an abundant supply of natural gas. Currently only a few percent of the world's iron production is by direct reduction, but this is a fast-growing component of the steel industry, particularly in the Middle East and South America.

FIGURE 23-10
A basic oxygen furnace.

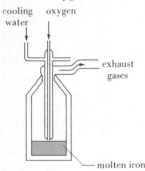

TABLE 23-9
Reactions occurring in steel-making processes

$$2\,C + O_2 \longrightarrow 2\,CO$$
$$2\,CO + Si \longrightarrow SiO_2 + 2\,C$$
$$CO + Mn \longrightarrow MnO + C$$
$$C + FeO \longrightarrow Fe + CO$$
$$Fe_3C + FeO \longrightarrow 4\,Fe + CO$$
$$2\,Fe + O_2 \longrightarrow 2\,FeO$$
$$2\,FeO + Si \longrightarrow 2\,Fe + SiO_2$$
$$FeO + Mn \longrightarrow Fe + MnO$$
$$\left. \begin{array}{l} FeO + SiO_2 \longrightarrow FeO \cdot SiO_2 \\ MnO + SiO_2 \longrightarrow MnO \cdot SiO_2 \end{array} \right\} slag$$
$$4\,P + 5\,O_2 \longrightarrow 2\,P_2O_5$$
$$3\,CaO + P_2O_5 \longrightarrow Ca_3(PO_4)_2 \quad (slag)$$

A basic oxygen furnace being charged with molten iron from a blast furnace. [Courtesy of Bethlehem Steel Corporation.]

Direct reduction of iron ore yields solid pellets rather than liquid pig iron. The pellets are handled with industrial magnets. [Courtesy of Midrex Corporation.]

23-9 The Coinage Metals—Cu, Ag, and Au

Some properties of these metals are listed in Table 23-10. The alkali metals (IA) and the coinage metals (IB) appear together in group I of Mendeleev's periodic table, but the only similarity between the two is in having a single *s* electron in the valence shells of their atoms. Regarding this electron, the combination of increased nuclear charge and reduced effectiveness of *d* (and *f*) electrons in shielding outer-shell *s* electrons causes the ns^1 electron to be held more tightly in IB than in IA atoms. Loss of this electron, as measured by ionization energies, is almost twice as difficult for a IB atom as for a IA atom. In terms of electrode potentials, the IB metals are less active than hydrogen and among the metals most difficult to oxidize (their reduction potentials are positive). It is this resistance toward oxidation that imparts "nobility" to the IB metals.

The similarity of the IB elements to the other transition elements that precede them in the periodic table is in being able to employ *d* electrons in chemical bonding. Thus, the IB elements can exist in different oxidation states, exhibit paramagnetism and color in some

TABLE 23-10
Some properties of Cu, Ag, and Au

	Cu	Ag	Au
electron configuration	$[Ar]3d^{10}4s^1$	$[Kr]4d^{10}5s^1$	$[Xe]4f^{14}5d^{10}6s^1$
metallic radius, pm	128	144	144
first ionization energy, kJ/mol	745	731	890
electrode potential, V			
$M^+(aq) + e^- \rightarrow M(s)$	+0.522	+0.800	+1.68
$M^{2+}(aq) + 2\ e^- \rightarrow M(s)$	+0.337	+1.39	—
$M^{3+}(aq) + 3\ e^- \rightarrow M(s)$	—	—	+1.42
oxidation states[a]	+1, **+2**	**+1**, +2	+1, **+3**

[a]The most common oxidation state is shown in boldface.

of their compounds, and form complex ions. They also possess to a high degree some distinctive physical properties of the metallic state—malleability, ductility, and excellent electrical and thermal conductivity.

In comparing the IB metals among themselves, we find the same trends noted for the IIB metals in Section 22-6, and for essentially the same reasons. The lightest metal (Cu) is the most active of the three; the heaviest (Au), the least active. Bonding in most Au(III) compounds is covalent.

Oxidation States.   Much can be learned about the stabilities of coinage metal compounds by considering electrode potential data. For the different oxidation states of copper we can write

The value of $E°$ for half-reaction (23.43) is obtained by the method introduced in Example 21-2.

$$Cu^+(aq) \longrightarrow Cu^{2+}(aq) + e^- \qquad E°_{ox} = -0.152 \text{ V} \qquad (23.43)$$
$$Cu^+(aq) + e^- \longrightarrow Cu(s) \qquad E°_{red} = +0.522 \text{ V} \qquad (23.44)$$
$$2 \; Cu^+(aq) \longrightarrow Cu^{2+}(aq) + Cu(s) \qquad E°_{cell} = +0.370 \text{ V} \qquad (23.45)$$

According to equation (23.45), $Cu^+(aq)$ spontaneously disproportionates under standard-state conditions. This does not mean, however, that aqueous solutions of Cu(I) compounds are not possible. To assess the circumstances under which they are likely to be encountered, let us establish the equilibrium constant for (23.45) through the relationship

$$E°_{cell} = \frac{0.0592}{n} \log K \qquad (20.30)$$

By substituting $n = 1$ and $E°_{cell} = +0.370$ into equation (20.30), we obtain

$$\log K = \frac{1 \times 0.370}{0.0592} = 6.25 \quad and \quad K = 1.8 \times 10^6$$

Thus, for the disproportionation reaction (23.45)

$$K = \frac{[Cu^{2+}]}{[Cu^+]^2} = 1.8 \times 10^6 \quad and \quad [Cu^{2+}] = 1.8 \times 10^6 \times [Cu^+]^2 \qquad (23.46)$$

This is how we can interpret equation (23.46): If we attempt to maintain any appreciable $[Cu^+]$ in aqueous solution, $[Cu^{2+}]$ would be a large number (because it would have to be about 2 million times the square of $[Cu^+]$). Disproportionation would go to completion. On the other hand, if $[Cu^+]$ is kept very low (as in a very slightly soluble solute or a stable complex ion), $[Cu^{2+}]$ is exceedingly small and the copper(I) species is stable. (Remember that the square of any number less than one is *smaller* than the number.)

Example 23-2   Can a solution be prepared that is **(a)** 0.020 M in Cu^+; **(b)** 1.0 $\times$ 10^{-10} M in Cu^+?

Solution
(a) In order to maintain $[Cu^+] = 0.020 \; M$ in a solution, we see from equation (23.46) that the solution would simultaneously need to have $[Cu^{2+}] = 1.8 \times 10^6 \times (0.020)^2 = 7.2 \times 10^2 \; M$. This is an impossibly high concentration of Cu^{2+}. In a solution prepared just by dissolving 0.020 mol/L of some copper(I) salt, the Cu^+ ion undergoes essentially complete disproportionation.
(b) In a solution with $[Cu^+] = 1.0 \times 10^{-10} \; M$, the $[Cu^{2+}]$ that simultaneously needs to be present is $[Cu^{2+}] = 1.8 \times 10^6 \times (1.0 \times 10^{-10})^2 = 1.8 \times 10^{-14} \; M$. In this instance $[Cu^{2+}]$ is about 6000 times *smaller* than $[Cu^+]$. In other words, disproportionation of Cu^+ occurs hardy at all. A solution with $[Cu^+] = 1.0 \times 10^{-10} \; M$ can be prepared (and this is essentially the situation one finds in a saturated solution of CuCN, which has $K_{sp} = 3.2 \times 10^{-20}$).

SIMILAR EXAMPLES: Review Problem 9; Exercises 26, 27.

An analysis of the oxidation states of Ag shows that

$$2\ Ag^+(aq) \longrightarrow Ag^{2+}(aq) + Ag(s) \qquad E^\circ_{cell} = -1.18\ V, \tag{23.47}$$

The disproportionation of $Ag^+(aq)$ is a very unfavorable reaction, and this accounts for the fact that silver is found mostly in Ag(I) compounds. Ag(II) compounds are formed only under very strong oxidizing conditions, and Ag(II) compounds themselves are good oxidizing agents because of their strong tendency to be reduced to Ag(I).

The situation with gold is that

$$3\ Au^+(aq) \longrightarrow Au^{3+}(aq) + 2\ Au(s) \qquad E^\circ_{cell} = +0.39\ V \tag{23.48}$$

$Au^+(aq)$ tends to disproportionate to $Au^{3+}(aq)$. Although the $+3$ oxidation state is most common for gold, this is actually not encountered in the form of Au^{3+} ion. Instead the Au atom is covalently bonded to other atoms, producing a complex species in which Au has the oxidation state $+3$, e.g., $[AuCl_4]^-$.

Compounds of Cu, Ag, and Au. As expected, none of the IB metals will dissolve in HCl(aq). Cu and Ag both dissolve in concentrated H_2SO_4(aq) and HNO_3(aq). In H_2SO_4(aq) the products are Cu^{2+}, Ag^+, and SO_2(g). In HNO_3(aq) they are Cu^{2+}, Ag^+, and either NO_2(g) or NO(g), depending on the concentration of the acid (recall Table 21-11). Au does not dissolve in either acid, but it does dissolve in aqua regia (1 part HNO_3 and 3 parts HCl). Oxidation is accomplished by the HNO_3 and the reaction is promoted by the formation of the stable complex ion $[AuCl_4]^-$.

The dissolution of Cu in concentrated HNO_3(aq) is shown in Color Section C.

$$Au(s) + 4\ H^+ + NO_3^- + 4\ Cl^- \longrightarrow [AuCl_4]^- + 2\ H_2O + NO(g) \tag{23.49}$$

The metals are generally resistant to air oxidation, although in moist air copper corrodes to produce green basic copper carbonate. This is the green color associated with copper roofing and gutters and bronze statues. Fortunately this corrosion product forms a tough adherent coating that protects the underlying metal. The corrosion reaction is complex but may be summarized as follows.

The tarnish that forms when silver metal is exposed to air is due to Ag_2S produced by the reaction of Ag with atmospheric sulfur compounds.

$$2\ Cu + H_2O + CO_2 + O_2 \longrightarrow \underset{\text{basic copper carbonate}}{Cu_2(OH)_2CO_3} \tag{23.50}$$

Addition of OH^-(aq) to an aqueous solution of Cu^{2+} precipitates $Cu(OH)_2$(s), which is soluble in NH_3(aq) owing to the formation of the deep blue complex ion $[Cu(NH_3)_4]^{2+}$.

$$Cu^{2+}(aq) + 2\ OH^-(aq) \longrightarrow Cu(OH)_2(s) \tag{23.51}$$

$$Cu(OH)_2(s) + 4\ NH_3(aq) \longrightarrow [Cu(NH_3)_4]^{2+}(aq) + 2\ OH^-(aq) \tag{23.52}$$

Addition of OH^-(aq) to Ag^+(aq) precipitates silver(I) oxide.

$$2\ Ag^+(aq) + 2\ OH^-(aq) \longrightarrow Ag_2O(s) + H_2O \tag{23.53}$$

Trace amounts of Cu are essential to life, but larger quantities are toxic, especially to bacteria, algae, and fungi. Among the many copper compounds used as pesticides are the basic acetate, carbonate, chloride, hydroxide, and sulfate. Commercially, the most important copper compound is $CuSO_4 \cdot 5H_2O$. In addition to its agricultural applications, $CuSO_4$ is used in batteries and electroplating, in preparing other copper salts, and in the petroleum, rubber, and steel industries.

Silver nitrate is the principal silver compound of commerce and is also an important laboratory reagent for the precipitation of anions (most anions form insoluble silver salts). These precipitation reactions can be used for the quantitative determination of anions, either gravimetrically (by weighing precipitates) or volumetrically (by titration). $AgNO_3$ is the source from which most other Ag compounds are derived. Ag compounds are used in electroplating, in the manufacture of batteries, in medicinal chemistry, as catalysts, and in cloud seeding (AgI). Their most important use by far (mostly as silver halides) is in photography (see Section 24-13).

Copper ore being concentrated by the flotation method. [Courtesy of Kennecott.]

Gold compounds are used in electroplating, photography, medicinal chemistry, and the manufacture of special glasses and ceramics.

Metallurgy and Uses of Copper.   The extraction of Cu from its ores (generally as sulfides) is considerably more complicated than the general scheme illustrated for zinc in Section 22-4. This complexity arises because of the presence of iron sulfides in copper ores. The usual procedures would result in iron being produced together with copper. To avoid this, iron must be removed before the final reduction to copper metal takes place. Five steps are required altogether: (1) concentration, (2) roasting, (3) smelting, (4) converting, and (5) refining.

> Copper was once rather widely available in the free state. Now free copper is being mined in the United States only in Michigan.

Currently most copper-bearing ores being mined have only about 0.5% Cu. Concentration of these ores is essential. This is usually done by the flotation process (recall Figure 22-11), yielding an ore concentrate with about 20 to 40% Cu.

The function served by roasting the concentrated ore (where necessary) is to convert some iron sulfide to iron oxide, with the copper remaining as the sulfide. This can be done if the temperature is kept below 800°C.

> Almost 200 kg of waste is produced for every kg Cu obtained.

$$2\ FeS(s) + 3\ O_2(g) \xrightarrow{\Delta} 2\ FeO(s) + 2\ SO_2(g) \tag{23.54}$$

The roasted ore is then transferred to a smelting furnace at 1400°C. In this furnace the charge melts and separates into two layers. The bottom layer is copper matte, consisting chiefly of the molten sulfides of copper and iron. The top layer is a silicate slag formed by the reaction of oxides of Fe, Ca, and Al with SiO_2 (which either is present in the ore or is added).

$$FeO(s) + SiO_2(s) \xrightarrow{\Delta} FeSiO_3(l) \tag{23.55}$$

After smelting, the copper matte is transferred to another furnace (the converter), where air is blown through the molten mass. First the remaining iron sulfide is converted to the oxide (reaction 23.54), followed by slag formation (reaction 23.55). The slag is poured off and air is again blown through the furnace. Now the following reactions occur, producing a product that is about 98 to 99% Cu.

$$2\ Cu_2S + 3\ O_2(g) \xrightarrow{\Delta} 2\ Cu_2O + 2\ SO_2(g) \tag{23.56}$$

$$2\ Cu_2O + Cu_2S \xrightarrow{\Delta} 6\ Cu(l) + SO_2(g) \tag{23.57}$$

The product of reaction (23.57) is called blister copper because of the presence of frozen bubbles of $SO_2(g)$. It can be used where high purity is not required (e.g., plumbing). High purity is essential in electrical applications, and purification is accomplished by electrolysis (Section 20-8).

About one half of the copper used in the United States is for electrical applications. Another 20% is used in plumbing and other aspects of the construction industry. The third important use of copper is in the manufacture of alloys, chiefly brass and bronze.

Metallurgy and Uses of Silver and Gold. Silver and gold both can be found free in nature, but easily accessible deposits have all but disappeared. A typical gold ore may contain only about 10 g Au per ton. In the **cyanidation** process $O_2(g)$ oxidizes the free metal to Au^+, which complexes with CN^-.

$$4\ Au(s) + 8\ CN^- + O_2(g) + 2\ H_2O \longrightarrow 4[Au(CN)_2]^- + 4\ OH^- \tag{23.58}$$

The pure metal is then displaced from solution by an active metal.

$$2\ [Au(CN)_2]^-(aq) + Zn(s) \longrightarrow 2\ Au(s) + [Zn(CN)_4]^{2-}(aq) \tag{23.59}$$

The extraction of silver generally requires dissolving an ore in $CN^-(aq)$, followed by displacement of the silver.

$$Ag_2S(s) + 4\ CN^-(aq) \longrightarrow 2[Ag(CN)_2]^-(aq) + S^{2-}(aq) \tag{23.60}$$

Ag_2S is highly insoluble, and to repress the reverse of reaction (23.60) air is blown through the mixture to oxidize S^{2-} to SO_4^{2-}.

Ag and Au are also obtained during the refining of other metals. They occur in the anode sludge produced in the electrolytic refining of Cu (Section 20-8), and Ag is obtained in the Parkes process for desilverizing lead (page 694).

Metallic silver and gold have uses in addition to being a source of their compounds. They both have served monetary purposes since ancient times. Millions of kg Au are currently held in the monetary reserves of nations throughout the world; and although Ag is much less used in the production of coins than in the past, a reserve of millions of kg Ag exists in old U.S. coins alone. In the past the most significant use of Ag was in silverware and jewelry. These uses are now declining relative to the use of Ag (the best electrical conductor among metals) in electrical contacts, printed circuits, and batteries. Silver is also widely used in alloys, such as silver solder and dental amalgams. About three fourths of the world's consumption of gold is for jewelry. Other important uses are in dentistry, coins and medals, and specialized industrial applications. Gold is important in the electronics industry because of its excellent electrical and thermal conductivity, its resistance to oxidation, and the ease with which it can be electroplated onto other metals.

23-10 The Lanthanoid Elements

The elements from cerium ($Z = 58$) through lutetium ($Z = 71$) are *inner transition* elements—their electron configurations feature the filling of $4f$ orbitals. These elements, together with lanthanum ($Z = 57$), which closely resembles them, are variously called the **lanthanoid, lanthanide,** or **rare earth** elements. The last-mentioned name is a misnomer, though, because these elements are not that rare. La, Ce, and Nd are more abundant than lead; and Tm is about as abundant as iodine. The lanthanoids occur primarily as oxides, and mineral deposits containing them are found in various locations. Large deposits near the California–Nevada border are being developed to provide rare earth oxides used in phosphors in color television sets.

Some Properties of the Lanthanoids. Because their differences in electron configuration are mainly in $4f$ orbitals, and because $4f$ electrons play a minor role in chemical bonding, strong similarities are found among the lanthanoids. For example, for the reduction process

$$M^{3+}(aq) + 3 e^- \longrightarrow M(s) \qquad (23.61)$$

$E°$ values do not show much variation; all fall between -2.52 V (La) and -2.25 V (Lu). The differences in properties that do exist among the lanthanoids mostly arise from the lanthanoid contraction discussed in Section 23-1. This contraction is best illustrated in the radii of the ions M^{3+}. These decrease regularly by about 1 to 2 pm for each unit increase in atomic number, from a radius of 106 pm for La^{3+} to 85 pm for Lu^{3+}.

Separation of the Lanthanoids. The lanthanoid elements are extremely difficult to extract from their natural sources and to separate from one another. All the methods for doing so are based on this principle: Species that are strongly *dissimilar* can often be completely separated in a one-step process, such as separating $Ag^+(aq)$ and $Cu^{2+}(aq)$ by adding $Cl^-(aq)$—AgCl is insoluble. Species that are very *similar* can at best be *fractionated* in a one-step process. That is, the ratio of the concentration of one species to that of another can be altered slightly. To achieve a complete separation may require repetition of the same basic step hundreds, or even thousands, of times.

Procedures that have been used to separate the lanthanoids include fractional crystallization, fractional precipitation, solvent extraction, and ion exchange. The details of these procedures are beyond the scope of this text, but it should be noted that a number of advances in chemical technology came about through development of these procedures.

Reactions of the Lanthanoids. The lanthanoids (Ln) are active metals that liberate $H_2(g)$ from hot water and from dilute acids by undergoing oxidation to $Ln^{3+}(aq)$. The lanthanoids combine with $O_2(g)$, sulfur, the halogens, $N_2(g)$, $H_2(g)$, and carbon, in much the same way as expected for metals about as active as the alkaline earths. Preparations of the pure metals can be achieved by electrolytic reduction of Ln^{3+} in a molten salt.

The most common oxidation state for the lanthanoids is +3. About half the lanthanoids can also be obtained in the oxidation state +2, and about half, +4. The special stability associated with an electron configuration involving half-filled f orbitals (f^7) may account for some of the observed oxidation states, but the reason for the predominance of the +3 oxidation state is less clear. Most of the lanthanoid ions are paramagnetic and colored in aqueous solution.

The chlorides, bromides, iodides, nitrates, and perchlorates of the lanthanoids are all water soluble; the sulfates are of variable solubility. The oxides and hydroxides are basic.

23-11 Qualitative Analysis of Some Transition Metal Ions

In this return to the subject of qualitative analysis, we will consider (a) how cations are separated and identified following precipitation of a cation group, and (b) how the procedures of qualitative analysis can be used to illustrate some descriptive chemistry of the metals.

Aspects of qualitative analysis that have already been considered are (a) the solubility relationships used in cation group separations (Section 19-9), (b) the use of H_2S as a precipitating agent (Section 19-10), and (c) the importance of buffering action in qualitative analysis procedures (Section 19-6). For the present discussion we choose the group of eight cations that constitute the ammonium sulfide group (cation group 3 of Figure 19-5). The scheme is described in the numbered items listed below and through the flow chart in Figure 23-11. The notes provide additional detail on a few interesting points.

1. A solution containing the cations of groups 3, 4, and 5 (see Figure 19-5), is treated with $(NH_4)_2S$ in an NH_3–NH_4Cl buffer solution. Under these conditions Fe^{3+}, Cr^{3+}, and

FIGURE 23-11
Qualitative analysis of the ammonium sulfide group.

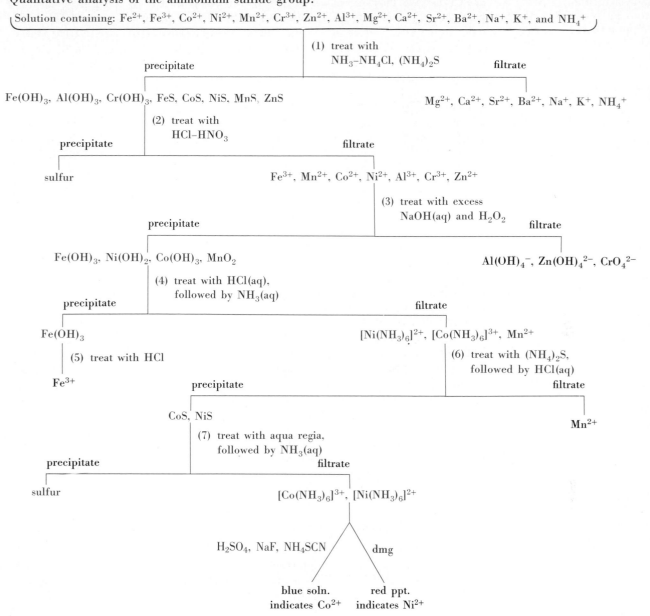

Al^{3+} precipitate as hydroxides and Fe^{2+}, Co^{2+}, Ni^{2+}, Mn^{2+}, and Zn^{2+} as sulfides. The solution contains cations of groups 4 and 5.

Note: Since the sulfides of the ammonium sulfide group are more soluble than most sulfides, a basic solution is needed to maintain a high $[S^{2-}]$ for their precipitation (recall equation 19.30). But the pH of the solution must not be too high or Mg^{2+} will precipitate as $Mg(OH)_2$. The use of a NH_3–NH_4Cl buffer to regulate the precipitation of $Mg(OH)_2$ was the subject of Example 19-12.

2. The precipitate from step 1 is treated with HCl–HNO_3(aqua regia), and any precipitated sulfur is filtered off.

Note: All components of the precipitate in step 1, except NiS and CoS, are readily soluble in HCl(aq). NiS and CoS require aqua regia for their dissolution (recall reaction 21.38).

The separation of Al^{3+} from Fe^{3+} based on the amphoterism of $Al(OH)_3(s)$ is illustrated in Color Section F.

3. The solution from step 2 is treated with an excess of NaOH. The hydroxides of Al, Zn, and Cr dissolve; they are *amphoteric*. The hydroxides of Fe, Co, Ni, and Mn are basic and precipitate. The mixture of solution and precipitate is next treated with H_2O_2, which oxidizes $[Cr(OH)_4]^-$ to CrO_4^{2-} in the solution and Co(II) to Co(III) and Mn(II) to Mn(IV) in the precipitate. The solution is subjected to further separations and tests for Al, Zn, and Cr. A yellow color in the solution indicates the presence of chromium (as CrO_4^{2-}). The precipitate consists of $Fe(OH)_3$, $Ni(OH)_2$, $Co(OH)_3$, and MnO_2.

Note: Because CrO_4^{2-} in alkaline solution is not a good oxidizing agent (reaction 23.21), oxidation of $[Cr(OH)_4]^-$ to CrO_4^{2-} is not difficult to achieve.

4. The precipitate from step 3 may be analyzed in a number of ways. One method involves dissolving the precipitate in HCl(aq), followed by treatment with NH_3(aq). This causes $Fe(OH)_3$ to reprecipitate and Co^{2+} and Ni^{2+} to be converted to $[Co(NH_3)_6]^{3+}$ and $[Ni(NH_3)_6]^{2+}$. Manganese also appears in solution as Mn^{2+}.

Note: MnO_2(s) is reduced to Mn^{2+}(aq) by HCl(aq). In an alkaline solution Co(II) is oxidized to Co(III) by O_2(g) and the presence of NH_3 stabilizes the Co(III) through the formation of the complex ion $[Co(NH_3)_6]^{3+}$. (The stabilization of oxidation states as a result of complex ion formation is discussed in Chapter 24.)

5. The $Fe(OH)_3$(s) is filtered off and dissolved in HCl(aq). The presence of Fe^{3+} is confirmed by the tests outlined in Table 23-6.

6. The filtrate containing $[Co(NH_3)_6]^{3+}$, $[Ni(NH_3)_6]^{2+}$, and Mn^{2+} is treated with $(NH_4)_2S$ to reprecipitate the sulfides. Because MnS is considerably more soluble than CoS and NiS, it is dissolved in HCl(aq), leaving a residue of CoS and NiS. Tests for Mn^{2+} are performed on the filtrate.

7. The sulfide precipitate is dissolved in aqua regia, followed by treatment with NH_3(aq). Thus, the species $[Co(NH_3)_6]^{3+}$ and $[Ni(NH_3)_6]^{2+}$ are obtained once again. A portion of the solution containing these ions is treated with dimethylglyoxime (dmg). Formation of a scarlet precipitate indicates the presence of nickel (see structure 23.38). A test for cobalt is performed on a second portion of the solution which is treated with H_2SO_4, NaF, and NH_4SCN. The presence of cobalt is disclosed by the formation of the blue complex ion $[Co(SCN)_4]^{2-}$.

This test for cobalt is shown in Color Section C.

Note: Acidification with H_2SO_4 causes Co(III) to decompose back to Co(II). The presence of NaF is required to complex Fe^{3+} as $[FeF_6]^{3-}$, so that any traces of Fe^{3+} present in the solution will not react with SCN^- and interfere with the test for cobalt.

Summary

More than half the elements are transition elements. The transition elements are all metals (most are more active than hydrogen). They tend to be found in several different oxidation states in their compounds, and they readily form complex ions (discussed further in Chapter 24).

Within a group of *d* block elements, the members of the *second* and *third* transition series resemble each other more than they do the group member from the first transition series. This is a consequence of the phenomenon known as the lanthanoid contraction. Another type of resemblance is that which occurs among certain adjacent members of the same transition series. One such group considered in this chapter is the iron triad—Fe, Co, Ni. A particular property that these elements share is ferromagnetism.

The possibility of a variety of oxidation states for the transition metals means that oxidation and reduction are involved in many of their reactions. Some of the more important oxidizing agents used in the laboratory are transition metal compounds, e.g., dichromates and permanganates. Dichromate ion participates in acid–base and hydration–dehydration equilibria with chromate ion. Chromate ion is a good precipitating agent. Most of the oxides and hydroxides of the transition metals are basic. This is typical of fairly active metals. However, certain oxides and hydroxides are amphoteric, and several examples are presented in the chapter.

The transition elements are among the most widely used metals. Their desirable properties range from the ready availability, low cost, and structural strength of iron to the excellent electrical conductivity of copper. Several of the transition metals—V, Cr, Mn, Co, Ni, Mo, W—are added to iron in steelmaking. The metallurgies of several transition metals are discussed in this chapter.

Several of the first transition series metals fall within the same group of cations in the qualitative analysis scheme. Many of the acid–base, oxidation–reduction, and precipitation reactions of this chapter are essential to the experimental procedures for separating and identifying these cations.

Learning Objectives

As a result of studying Chapter 23, you should be able to

1. Cite ways in which the transition elements differ from the representative elements.

2. Describe the lanthanoid contraction and explain how it affects the properties of transition elements.

3. Describe sources and uses of the transition elements and some of their compounds, e.g., $TiCl_4$, V_2O_5, K_2CrO_4, $K_2Cr_2O_7$, $KMnO_4$, MnO_2, Fe_2O_3, $CuSO_4$, and $AgNO_3$.

4. Outline the metallurgies of Ti, iron and steel, Cu, Ag, and Au.

5. State which are the most common oxidation states of the first transition series and group IB elements, which of these are stable, and which are unstable.

6. Use electrode potential data to predict disproportionation reactions.

7. Write equations for some important oxidation–reduction reactions, especially those involving permanganate and dichromate ions.

8. Discuss the chromate–dichromate equilibrium and the effect of pH on the concentrations of these ions in aqueous solution.

9. Write chemical equations to illustrate the amphoteric nature of certain of the transition metal oxides and hydroxides, especially those of Cr(III).

10. Describe the property of ferromagnetism and the features of atomic structure that lead to it.

11. Describe the formation of metal carbonyls.

12. Describe qualitative tests for identifying Fe^{2+} and Fe^{3+}.

13. Discuss separations and tests for ions in the ammonium sulfide group of the qualitative analysis scheme.

14. Draw conclusions about the presence or absence of ions in a qualitative analysis "unknown," based on the results of laboratory tests.

Some New Terms

Ferromagnetism is a property that permits certain materials (notably Fe, Co, and Ni) to be made into permanent magnets. The magnetic moments of individual atoms are aligned into domains. In the presence of a magnetic field, these domains orient themselves to produce a permanent magnetic moment.

Iron triad is a term used for the group, Fe, Co, Ni, to emphasize similarities in their physical and chemical properties.

Lanthanoid contraction describes the decrease in atomic size in a series of elements in which an f subshell fills (an inner transition series). It results from the ineffectiveness of f electrons in shielding outershell electrons from the nuclear charge of an atom.

The **lanthanoid elements** are those of the first inner transition series ($Z = 58$ to 71). Because lanthanum resembles them, La ($Z = 57$) is generally considered together with them.

Metal carbonyls, e.g., nickel carbonyl, $Ni(CO)_4$, are compounds formed between certain metal atoms and CO molecules.

Pig iron is an impure form of iron (about 95% Fe, 3–4% C) produced in a blast furnace.

Steel is a term used to describe iron alloys containing from 0 to 1.5%C together with other key elements, such as V, Cr, Mn, Ni, W, and Mo.

Suggestions for Further Study

DAVIS, K. A., "Chemical of the Month: Titanium Dioxide," *J. Chem. Educ.*, **59**, 158 (1982).

HOUSE, J. E., Jr., "Carbon Monoxide Complexes of Uncharged Metals," *Chemistry*, **48**[5], 12 (1975).

HUXLEY, J. V., "Recent Developments in Copper Production," *Educ. in Chem.*, **10**, 94 (1973).

LONG, G. J., and H. P. LEIGHLY, Jr., "The Iron–Iron Carbide Phase Diagram," *J. Chem. Educ.*, **59**, 948 (1982).

LUDI, A., "Prussian Blue, an Inorganic Evergreen," *J. Chem. Educ.*, **58**, 1013 (1981).

MENARD, H. W., "Time, Chance, and The Origin of Manganese Nodules," *American Scientist*, **64**, 519 (1976).

MILLER, J. R., "The Direct Reduction of Iron Ore," *Scientific American*, **235**[1], 68 (1976).

SELLERS, N., "Chemistry of Steelmaking," *J. Chem. Educ.*, **57**, 139 (1980).

Review Problems

1. Provide an acceptable name or formula for each of the following.
    (a) barium dichromate   **(b)** $Sc(OH)_3$
    (c) chromium trioxide   **(d)** MnO
    (e) iron(II) silicate   **(f)** $Fe(CO)_5$

2. Describe the chemical composition of the materials described as **(a)** ferromanganese; **(b)** cast iron; **(c)** chromite ore; **(d)** chrome alum; **(e)** aqua regia; **(f)** Prussian blue.

3. Complete and balance the following equations. If no reaction occurs, so state.

    (a) $TiCl_4(g) + Na(l) \overset{\Delta}{\rightarrow}$

    (b) $Cr_2O_3(s) + Al(s) \overset{\Delta}{\rightarrow}$
    (c) $Ag(s) + HCl(aq) \rightarrow$
    (d) $K_2Cr_2O_7(aq) + KOH(aq) \rightarrow$

    (e) $MnO_2(s) + C(s) \overset{\Delta}{\rightarrow}$
    (f) $Fe(OH)_3(s) + NaOH(aq) \rightarrow$

4. Balance the following oxidation–reduction equations.
    (a) $Fe_2S_3(s) + H_2O + O_2(g) \rightarrow Fe(OH)_3(s) + S(s)$
    (b) $Mn^{2+} + S_2O_8^{2-} + H_2O \rightarrow MnO_4^- + SO_4^{2-} + H^+$
    (c) $Ag(s) + CN^- + O_2(g) + H_2O \rightarrow [Ag(CN)_2]^- + OH^-$

5. Suggest chemical reactions that might be used to obtain the following compounds from Na_2CrO_4: **(a)** $Na_2Cr_2O_7$; **(b)** Cr_2O_3; **(c)** $CrCl_3$; **(d)** $NaCr(OH)_4$.

6. Through a chemical equation, give an example to represent the dissolving of
    (a) a transition metal in an oxidizing acid;
    (b) a transition metal oxide in $NaOH(aq)$;
    (c) a transition metal sulfide in $HCl(aq)$;
    (d) an iron triad metal hydroxide in $NH_3(aq)$;
    (e) an inner transition metal in $HCl(aq)$.

7. Using data from the chapter and from Table 20-2,
    (a) Construct an electrode potential diagram which includes $Cr_2O_7^{2-}$, Cr^{3+}, and Cr^{2+} in acidic solution.
    (b) Assess the abilities of the species in Example 23-1 (Zn, Sn^{2+}, and I^-) to reduce $Cr_2O_7^{2-}$ to Cr^{3+} and then to Cr^{2+}.

8. What are the relative concentrations of $Cr_2O_7^{2-}$ and CrO_4^{2-} in a solution with a pH of **(a)** 5.0; **(b)** 9.12?

9. Equilibrium is established in reaction (23.45) at 25°C with all three reactants present, i.e., Cu, Cu^+, and Cu^{2+}.
    (a) What must be the molar concentrations of Cu^+ and Cu^{2+} if they are found to be equal to each other?
    (b) What percent of the copper ion is present as Cu^+ if the *total* copper ion concentration at equilibrium is 1.00×10^{-4} M?

10. What single reagent solution (including water) could be used to effect the separation of the following pairs of solids? **(a)** NaOH and $Fe(OH)_3$; **(b)** $Ni(OH)_2$ and $Fe(OH)_3$; **(c)** Cr_2O_3 and $Fe(OH)_3$; **(d)** MnS and CoS.

Exercises

Properties of the transition elements

1. Describe how the transition elements compare with representative metals (e.g., group IIA) with respect to oxidation states, formation of complexes, colors of compounds, and magnetic properties.

2. Why do the atomic radii vary so much more for two representative elements that differ by one unit in atomic number than they do for two transitions elements that differ by one unit?

3. With but minor irregularities, the melting points of the first series of transition metals rise from that of Sc to Cr and then fall to that of Cu (also note that the melting point of Zn = 420°C). Give a plausible reason for this observation, based on atomic structure.

4. With but minor exceptions the standard reduction potentials of the first transition series metals increase regularly, from that of Sc (-2.08 V) to that of Cu ($+0.34$ V). In the lanthanoid series the potentials increase by less than 0.3 V, from that of La (-2.52 V) to that of Lu (-2.25 V).
    (a) Why is the variation of $E°$ values so much smaller for the lanthanoids than for the first transition series?
    (b) Why do you suppose that the lanthanoids are more active metals than those of the first transition series?

Oxidation–reduction

5. Write plausible half-equations to represent **(a)** $VO^{2+}(aq)$ as an oxidizing agent in acidic solution; **(b)** $Cr^{2+}(aq)$ as a reducing agent; **(c)** the oxidation of $Fe(OH)_3(s)$ to FeO_4^{2-} in basic solution; **(d)** the reduction of $[Ag(CN)_2]^-$ to silver metal.

6. Use electrode potential data from this chapter and from Table 20-2 to predict whether each of the following reactions will occur to any significant extent as written.
    (a) $2\ VO_2^+ + 6\ Br^- + 8\ H^+ \rightarrow 2\ V^{2+} + 3\ Br_2 + 4\ H_2O$
    (b) $VO_2^+ + Fe^{2+} + 2\ H^+ \rightarrow VO^{2+} + Fe^{3+} + H_2O$
    (c) $MnO_2(s) + H_2O_2 + 2\ H^+ \rightarrow Mn^{2+} + 2\ H_2O + O_2(g)$

7. When Zn is added to an acidified solution of $K_2Cr_2O_7$, the color of the solution changes from orange to green; then to blue; and, over a period of time, back to green. Write equations for this series of reactions.

8. The electrode potential diagram of Figure 23-3 does not include a value of $E°$ for the reduction of MnO_4^- to Mn^{2+} in acidic solution. Use other data in the figure to establish this value, and compare your result with the value listed in Table 20-2.

★9. Select four reducing agents to carry out the series of reductions given in Figure 23-1. The first reducing agent should reduce VO_2^+ to V^{3+}, but no further; and so on.

Chromium and chromium compounds

10. Use data from Table 23-1 and equation (23.15) to determine $E°$ for the reduction of $Cr^{3+}(aq)$ to $Cr(s)$.

11. Explain why it is reasonable to expect the principal chemistry of dichromate ion to involve oxidation–reduction reactions and that of chromate ion, precipitation reactions.

12. If $CO_2(g)$ under pressure is passed into $Na_2CrO_4(aq)$, $Na_2Cr_2O_7(aq)$ is formed. What is the function of the $CO_2(g)$? Write a plausible equation for the net reaction.

13. The ionization constant for the species $HCrO_4^-$ is listed as $K_a = 3.2 \times 10^{-7}$. Calculate a value of K for the hydration–dehydration equilibrium.

$$2\ HCrO_4^- \rightleftharpoons Cr_2O_7^{2-} + H_2O \qquad K = ?$$

14. If a solution is prepared by dissolving 0.100 mol Na_2CrO_4 in 1.00 L of a buffer solution at pH 7.00, what will be $[CrO_4^{2-}]$ and $[Cr_2O_7^{2-}]$ in the solution?

15. Show that in a solution that is 0.10 M in Ba^{2+}, 0.10 M in Sr^{2+}, 0.10 M in Ca^{2+}, 1.0 M in $HC_2H_3O_2$, 1.0 M in $NH_4C_2H_3O_2$, and 0.0010 M in $Cr_2O_7^{2-}$, $BaCrO_4$ will precipitate but not $SrCrO_4$ or $CaCrO_4$. Use data from this chapter and Table 19-1, as necessary.

16. How long would an electric current of 4.2 A have to pass through a chrome-plating bath to produce a deposit 0.0010 mm thick on an object with a surface area of 22.7 cm^2? (The density of Cr is 7.14 g/cm^3.)

The iron triad

17. Explain why the iron triad elements resemble each other so strongly.

18. How does the property of ferromagnetism differ from ordinary paramagnetism? Why is this property so limited in its occurrence among the elements?

19. Which of the iron triad ions, M^{2+}, would you expect to be oxidized to M^{3+} by $O_2(g)$ in acidic solution?

$$4\ M^{2+} + O_2(g) + 4\ H^+ \rightarrow 4\ M^{3+} + 2\ H_2O$$

Carbonyls

20. Use methods outlined in the chapter to predict formulas and structures of the carbonyls of **(a)** molybdenum; **(b)** osmium; **(c)** rhenium.

21. Iron and nickel carbonyls are liquids at room temperature, but that of cobalt is a solid. Why should this be so?

22. The compound $Na[V(CO)_6]$ has been reported. Discuss the probable nature of chemical bonding in this compound.

The coinage metals

23. In the metallurgical extraction of Ag and Au often an alloy of the two metals is obtained. They can be separated either with concentrated HNO_3 or boiling concentrated H_2SO_4, in a process called *parting*. Write chemical equations to show how these separations work. What minimum further treatment would be required to obtain each pure metal?

24. In acidic solution silver(II) oxide first dissolves to produce $Ag^{2+}(aq)$. This is followed by the oxidation of H_2O to $O_2(g)$ and the reduction of Ag^{2+} to Ag^+.
  (a) Write equations for the dissolution and oxidation–reduction reactions.
  * **(b)** Show that the second reaction is indeed spontaneous. (*Hint:* You will have to use the method of Example 21-2 for part of the solution.)

25. Determine the values of K for reactions (23.47) and (23.48).

* **26.** If one attempts to make CuI_2 by the reaction of $Cu^{2+}(aq)$ and $I^-(aq)$, $CuI(s)$ and I_2 are obtained instead. Without performing detailed calculations, show why this reaction should occur as written.

$$2\ Cu^{2+} + 4\ I^- \rightarrow 2\ CuI(s) + I_2$$

[*Hint:* Use electrode potentials and $K_{sp}(CuI) = 1.1 \times 10^{-12}$.]

* **27.** Without performing detailed calculations, show that significant decomposition of AuCl occurs if one attempts to make a saturated aqueous solution. Use equation (23.48) and $K_{sp}(AuCl) = 2.0 \times 10^{-13}$.

Qualitative analysis

28. With respect to the qualitative analysis scheme for the ammonium sulfide group outlined in the text, write equations to represent **(a)** dissolving FeS in HCl(aq); **(b)** dissolving CoS in aqua regia; **(c)** the action of excess NaOH(aq) and H_2O_2 on $Cr^{3+}(aq)$; **(d)** the action of $NH_3(aq)$ on a mixture of $Fe(OH)_3(s)$ and $Ni(OH)_2(s)$; **(e)** a qualitative test for $Fe^{2+}(aq)$; **(f)** the action of HCl(aq) on $MnO_2(s)$.

29. Describe what might happen in the qualitative analysis of the ammonium sulfide group if one did the following, in error.
  (a) failed to include NH_4Cl in the reagent used to treat the original solution in the flow sheet of Figure 23-11.
  (b) failed to add H_2O_2 to the reagent used in step 3.
  (c) used NaOH(aq) in step 4 rather than $NH_3(aq)$.
  (d) used aqua regia in step 6 instead of HCl(aq).
  (e) neglected to use NaF in the test for cobalt in step 7.

30. A particular stainless steel sample is known to contain some combination of Ni, Cr, and Mn alloyed with Fe. Outline a simplified qualitative analysis scheme that would enable you to determine which of these metals are present.

31. An unknown solution contains only group 3 ions as its possible cations. The solution is first treated with an $NH_3–NH_4Cl$ buffer solution and no precipitate is observed. A light-colored precipitate does form when $(NH_4)_2S$ is added. This precipitate is completely soluble in $HCl(aq)$. Indicate for each of the group 3 cations whether it is present or absent or whether its presence is uncertain.

32. A metal sample, when dissolved and subjected to qualitative analysis, yields a cation group 3 precipitate that is partially soluble in $NaOH(aq)$. The remainder dissolves in aqua regia. The solution from the aqua regia reaction again yields a precipitate when treated with $H_2O_2(aq)$ and $NaOH(aq)$. This precipitate is completely soluble in $NH_3(aq)$. From these observations indicate what conclusions are possible concerning cation group 3 metals that may have been present in the original sample.

Quantitative analysis

33. Nickel can be determined by the precipitation of nickel dimethylglyoxime.
  (a) What is the formula of this compound (see structure 23.38)?
  (b) A 1.502-g sample of steel yields 0.259 g of nickel dimethylglyoxime. What is the percent Ni in the steel?

34. A 0.589-g sample of pyrolusite ore (impure MnO_2) is treated with 1.651 g of oxalic acid ($H_2C_2O_4 \cdot 2H_2O$) in an acidic medium. Following this reaction the excess oxalic acid is titrated with 0.1000 M $KMnO_4$, 30.06 mL being required. What is the % MnO_2 in the ore?

$$H_2C_2O_4 + MnO_2 + 2\,H^+ \rightarrow Mn^{2+} + 2\,H_2O + 2\,CO_2$$

$$5\,H_2C_2O_4 + 2\,MnO_4^- + 6\,H^+ \rightarrow$$
$$2\,Mn^{2+} + 8\,H_2O + 10\,CO_2$$

Additional Exercises

1. Describe a simple chemical test to distinguish between $Fe^{2+}(aq)$ and $Fe^{3+}(aq)$.

2. Suggest a series of reactions, using common chemicals, by which each of the following syntheses can be performed: **(a)** $Fe(OH)_3(s)$ from $FeS(s)$; **(b)** $BaCrO_4(s)$ from $BaCO_3(s)$ and $K_2Cr_2O_7(aq)$; **(c)** $CrCl_3(aq)$ from $(NH_4)_2Cr_2O_7(s)$; **(d)** $MnCO_3(s)$ from $MnO_2(s)$.

3. When a soluble lead compound is added to a solution containing primarily *orange* dichromate ion, *yellow* lead chromate precipitates. Describe the equilibria involved.

4. When *yellow* $BaCrO_4$ is dissolved in $HCl(aq)$, a *green* solution is obtained. Write a chemical equation to account for the color change.

5. In the text three conclusions were reached about certain ions of manganese by using the electrode potential diagrams of Figure 23-3. Show that the following statements are true as well.
  (a) MnO_4^- in acidic solution oxidizes Mn^{2+} to $MnO_2(s)$; in basic solution MnO_4^- oxidizes $MnO_2(s)$ to MnO_4^{2-}.
  (b) MnO_4^- will slowly liberate $O_2(g)$ from water, either in acidic or basic solution.

6. Although Au is soluble in aqua regia (3 parts HCl + 1 part HNO_3), Ag is not. What is the likely reason(s) for this difference?

7. The only important compounds of Ag(II) are AgF_2 and AgO. Explain why you would expect these two compounds to be stable, but not others such as $AgCl_2$, $AgBr_2$, AgS, etc.

8. Write a plausible equation for the disproportionation reaction brought about by passing $CO_2(g)$ into $MnO_4^{2-}(aq)$.

★9. Show that the corrosion reaction in which Cu is converted to its basic carbonate (reaction 23.50) can be thought of in terms of a combination of oxidation–reduction, acid–base, and precipitation reactions.

★10. In an atmosphere with industrial smog Cu corrodes to a basic sulfate, $Cu_2(OH)_2SO_4$. Propose a series of chemical reactions to describe this corrosion.

★11. Calculate $[Au^+]$, $[Au^{3+}]$, and $[Cl^-]$ in a saturated solution obtained by dissolving AuCl in water. (*Hint:* Neglect complex ion formation and use data from Exercise 27.)

★12. Suppose a solution is 0.50 M in OH^- and 0.10 M in MnO_4^- in the presence of $MnO_2(s)$. Show that a significant concentration of MnO_4^{2-} exists in the solution.

★13. Use data from Figure 23-3, together with equations from elsewhere in the text, to estimate K_{sp} for $Mn(OH)_2$.

★14. A steel sample is to be analyzed for Cr and Mn, simultaneously. By suitable treatment the Cr is oxidized to $Cr_2O_7^{2-}$ and the Mn to MnO_4^-. A 10.000-g sample of steel is used to produce 250.0 mL of a solution containing $Cr_2O_7^{2-}$ and MnO_4^-. A 10.00-mL portion (aliquot) of this solution is added to a $BaCl_2$ solution, and by proper adjustment of the acidity, the chromium is completely precipitated as $BaCrO_4$; 0.0549 g is obtained. A second 10.00-mL aliquot of this solution requires exactly 15.95 mL of 0.0750 M standard Fe^{2+} solution for its titration (in acid solution). Calculate the % Mn and % Cr in the steel sample. (*Hint:* Both MnO_4^- and $Cr_2O_7^{2-}$ are reduced by Fe^{2+}.)

Self-Test Questions

For questions 1 through 8 select the single item that best completes each statement.

1. A property generally expected of the transition elements is (a) low melting points; (b) high ionization energies; (c) variable oxidation states; (d) positive standard electrode (reduction) potentials.

2. Only one of the following ions is diamagnetic. That ion is (a) Cr^{2+}; (b) Fe^{3+}; (c) Cu^{2+} (d) Sc^{3+}.

3. Of the following elements the one that is *not* expected to display an oxidation state of +6 in any of its compounds is (a) Ti; (b) Cr: (c) Mn; (d) Fe.

4. One might expect $Cl_2(g)$ to be produced if an HCl(aq) solution is heated strongly in the presence of (a) CuO; (b) MnO_2; (c) Cr^{3+}; (d) Fe^{2+}.

5. Of the following ions the one most likely to disproportionate in aqueous solution is (a) Fe^{3+}; (b) Ag^+; (c) Cr^{2+}; (d) Cu^+.

6. The best oxidizing agent among the following ions is (a) Au^+; (b) Ag^+; (c) Ag^{2+}; (d) Cu^{2+}.

7. To increase the water solubility of $BaCrO_4(s)$, add to the water (a) NH_4Cl; (b) $Ba(OH)_2$; (c) Na_2CrO_4; (d) $BaCl_2$.

8. Of the following materials, the highest % C is found in (a) wrought iron; (b) cast iron; (c) steel; (d) stainless steel.

9. Why is +3 the most stable oxidation state for Fe, whereas it is +2 for Co and Ni?

10. What are the products obtained when $Mg^{2+}(aq)$ and $Cr^{3+}(aq)$ are each treated with a limited amount of NaOH(aq)? With an excess of NaOH(aq)? Why are the results different in these two cases?

11. Which of these reagents—H_2O, NaOH(aq), HCl(aq)—can be used to dissolve the following compounds? **(a)** $CuSO_4$; **(b)** $Ni(OH)_2$; **(c)** $AgNO_3$; **(d)** FeS; **(e)** $Cr(OH)_3$?

12. A solution is believed to contain one or more of the following ions: Cr^{3+}, Zn^{2+}, Fe^{3+}, Ni^{2+}. When the solution is treated with NaOH(aq) and H_2O_2, a precipitate is formed. The solution separated from this precipitate is colorless. The precipitate is dissolved in HCl(aq) and the resulting solution is treated with NH_3(aq). No precipitation occurs. Based solely on these observations, what conclusions can you draw about the ions in the solution? That is, are any proved to be present? to be absent? Are further tests needed?

24 Complex Ions and Coordination Compounds

In Section 19-7, to describe substances such as $CoCl_3 \cdot 6NH_3$, $CoCl_3 \cdot 5NH_3$, and $CoCl_3 \cdot 4NH_3$, we introduced the term coordination compound. This term suggests that two simpler substances, $CoCl_3$ and NH_3, are joined together or coordinated into more complex substances. Better representations of these compounds, we learned, are

$$[Co(NH_3)_6]Cl_3 \qquad [Co(NH_3)_5Cl]Cl_2 \qquad [Co(NH_3)_4Cl_2]Cl \qquad (24.1)$$
$$\text{(a)} \qquad\qquad\qquad \text{(b)} \qquad\qquad\qquad \text{(c)}$$

where the groups attached to the central metal ion are called ligands, and the combination of central ion and attached ligands is a complex ion. These representations helped us to understand the experimental observation that, when treated with excess $AgNO_3(aq)$, the three compounds produce 3, 2, and 1 mol AgCl per mol of coordination compound, respectively. However, these formulas alone do not suggest why compound (a) is yellow in color, why compound (b) is purple, and why compound (c) exists in *two* forms (isomers), one violet and one green. To explain these observations we will have to acquire some understanding of the bonding and structure of complex ions, two of the topics discussed in this chapter.

Compounds (a) and (b) are shown in Color Section I.

We will also look more closely at the question of the stabilities of complex ions and how these stabilities provide bases for phenomena as diverse as the separation of cations in qualitative analysis, the fixing of photographic film, and the sequestering of ions. Also, we will mention some important complexes of the living state.

24-1 Early Theories of Coordination Compounds

Prior to work published by Alfred Werner in 1891, one of the commonly held beliefs about bonding in coordination compounds was that the valence or combining capacity of the central atom had the same fixed value as exhibited in its simple salts. For Co in compounds (a), (b), and (c) above, this was postulated to be three. The three groups attached directly to the metal atom, it was believed, would be either Cl atoms or NH_3 molecules. It was further postulated that NH_3 molecules could form chainlike structures (similar to the carbon-chain molecules that were being so actively studied at the time).

TABLE 24-1
Some Pt(IV) coordination compounds

Compound	No. mol ions per mol compound
$[Pt(NH_3)_6]Cl_4$	5
$[Pt(NH_3)_5Cl]Cl_3$	4
$[Pt(NH_3)_4Cl_2]Cl_2$	3
$[Pt(NH_3)_3Cl_3]Cl$	2
$[Pt(NH_3)_2Cl_4]$	0
$Na[Pt(NH_3)Cl_5]$	2
$Na_2[PtCl_6]$	3

Finally, the only ionizable Cl atoms would be those at the ends of NH_3 chains, i.e., in an environment similar to the ionic salt NH_4Cl. The structures proposed for (a), (b), and (c) according to this "chain" theory were

$$
\begin{array}{c}
\text{Co}\overset{\displaystyle NH_3\,Cl}{\underset{\displaystyle NH_3\,Cl}{-NH_3-NH_3-NH_3-NH_3\,Cl}} \\[1em]
(a)
\end{array}
$$

$$
\begin{array}{c}
\text{Co}\overset{\displaystyle Cl}{\underset{\displaystyle NH_3\,Cl}{-NH_3-NH_3-NH_3-NH_3\,Cl}} \\[1em]
(b)
\end{array}
\qquad (24.2)
$$

$$
\begin{array}{c}
\text{Co}\overset{\displaystyle Cl}{\underset{\displaystyle Cl}{-NH_3-NH_3-NH_3-NH_3\,Cl}} \\[1em]
(c)
\end{array}
$$

The postulated structures (24.2) are consistent with the previously mentioned abilities of these compounds to precipitate AgCl. And they are also consistent with another observation: that the electrical conductivity in aqueous solution is greatest for (a) and decreases progressively for (b) and (c). That is, these structures suggest the presence of four ions per formula unit of (a)—three Cl^- and $Co(NH_3)_6^{3+}$—three ions per formula unit of (b), and two of (c).

The alternative theory proposed by Werner, as we have already seen (Section 19-7), was that of a primary valence and a secondary valence for the central metal atom. The primary valence is exhibited in the initial formation of an ion (Co^{3+} in expression 24.1), and the secondary valence, by the attachment of a fixed number of ligands directly to this ion (six in the case of Co^{3+}). Werner's formulations (24.1) could be easily extended to predict the existence of a compound that should yield no precipitate with $AgNO_3(aq)$ and which should be a nonelectrolyte.

$$[Co(NH_3)_3Cl_3] \qquad (24.3)$$

Moreover, it should also be possible to obtain compounds such as

$$Na[Co(NH_3)_2Cl_4] \qquad (24.4)$$

FIGURE 24-1
Molar conductivities of some Pt(IV) coordination compounds.

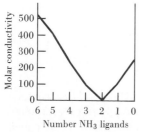

The compounds listed in Table 24-1 have the conductivities given by the points on this graph. The compounds are identified by the number of NH_3 molecules coordinated to the Pt(IV).

Molar conductivity is the electrical conductivity, under precisely defined conditions, of an aqueous solution containing one mole of compound.

in which the complex ion acquires a net *negative* charge. Compound (24.4) would again be expected to conduct electric current, but not to yield a precipitate with $AgNO_3(aq)$. There was no way that the chain theory could account for such possibilities as compounds (24.3) and (24.4) if the valency of Co was fixed at three. Werner was not able to synthesize these compounds, but his line of reasoning was supported through some related compounds (e.g., the nonelectrolyte $IrCl_3 \cdot 3NH_3$).

Table 24-1 lists several coordination compounds of Pt(IV), and Figure 24-1 plots the electrical conductivities of these compounds. $[Pt(NH_3)_2Cl_4]$, as we would expect from Werner's theory, is a nonelectrolyte; it does not ionize. We cannot appropriately call it a complex ion, but it clearly belongs to the series of species listed in Table 24-1. This difficulty is overcome by referring to all coordinated species, whether they exist as cations, anions, or neutral molecules, as **complexes.**

FIGURE 24-2
Structures of some
complex ions.

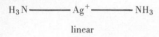

linear

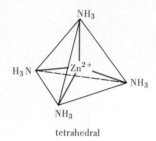

tetrahedral

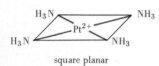

square planar

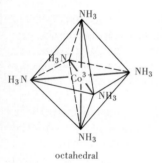

octahedral

These are among the most
commonly observed struc-
tures for complex ions.

24-2 Coordination Number and Structures of Complexes

An additional postulate of Werner's theory was that the secondary valencies of a central metal ion are directed to specific positions in the coordination sphere. Among the implications of this postulate is that complex ions should possess distinctive geometrical shapes. The four most commonly observed structures are depicted in Figure 24-2.

The number of secondary valencies displayed by a central ion is the coordination number of the ion. Coordination numbers ranging from 2 to 12 have been observed in complexes, although the number 6 is by far the most common, followed by 4. Coordination number 2 is mostly limited to complexes of Cu(I), Ag(I), and Au(I). Coordination numbers greater than 6 are not often found in members of the first transition series but are more common in those of the second and third series. Stable complexes with coordination numbers 3 and 5 are rarely encountered. The coordination number(s) exhibited by an ion depends on a number of factors, such as the ratio of the radius of the central metal ion to that of the attached ligands.

Coordination numbers of some common ions are listed in Table 24-2. One practical use of the coordination number is to assist in writing and interpreting formulas of complexes.

Example 24-1   What are the coordination number and oxidation state of Al in the complex ion $[Al(H_2O)_4(OH)_2]^+$?

Solution.   The complex ion has *four* H_2O molecules and *two* OH^- ions as ligands. The coordination number is *six*. Of these six ligand groups, *two* carry a charge of -1 each (the OH^- ions) and four are *neutral* (the H_2O molecules). The total contribution of the OH^- ions to the net charge on the complex ion is -2. Since the net charge on the complex ion is $+1$, the charge of the central aluminum ion, and hence its oxidation state, must be $+3$. Diagrammatically, we can write

$$\text{charge} = x \quad\quad\quad \overset{\text{charge of } -1 \text{ on OH}^-;}{\text{total negative charge} = -2}$$

$$[Al(H_2O)_4(OH)_2]^+$$

$$\underset{\text{coordination number} = 6}{\quad} \quad \overset{\text{net charge on complex ion:}}{\quad}$$
$$x - 2 = +1$$
$$x = +3$$

SIMILAR EXAMPLE: Review Problem 1.

In the structure of $[Co(NH_3)_6]^{3+}$ depicted in Figure 24-2, all NH_3 ligands are equidistant from the Co^{3+} ion and the structure is octahedral. Figure 24-3 represents the structure of the complex ion $[Cu(NH_3)_4(H_2O)_2]^{2+}$. Here the two H_2O molecules in the so-called axial positions are more distant from the Cu^{2+} ion than are the four NH_3 molecules in the square planar (equatorial) arrangement. The ion has a distorted octahedral structure (called a tetragonal structure). If the axial ligands in such a structure are sufficiently far from the central ion, the octahedral structure with coordination number 6 gives way to the square planar structure with coordination number 4. The ion in Figure 24-3 is generally written as $[Cu(NH_3)_4]^{2+}$ and described as having square planar geometry.

24-3 Ligands

A ligand is a species having an atom(s) capable of donating an electron pair(s) to a central metal ion at a particular site(s) in the coordination sphere. Thus, a ligand is a Lewis base and the metal ion is a Lewis acid. A ligand capable of donating a single electron pair (as

TABLE 24-2
Coordination numbers of
some common metal ions

Cu^+	2, 4	Zn^{2+}	4, 6
Ag^+	2		
Au^+	2, 4	Al^{3+}	4, 6
		Sc^{3+}	6
Ca^{2+}	6	Cr^{3+}	6
Fe^{2+}	6	Fe^{3+}	6
Co^{2+}	4, 6	Co^{3+}	6
Ni^{2+}	4, 6	Au^{3+}	4
Cu^{2+}	4, 6		

TABLE 24-3
Some common unidentate ligands

Formula	Name as ligand	Formula	Name as ligand	Formula	Name as ligand
Neutral molecules		*Anions*		*Anions*	
H_2O	aqua[a]	F^-	fluoro	SO_4^{2-}	sulfato
NH_3	ammine	Cl^-	chloro	$S_2O_3^{2-}$	thiosulfato
CO	carbonyl	Br^-	bromo	NO_2^{-b}	nitro
NO	nitrosyl	I^-	iodo	ONO^{-b}	nitrito
CH_3NH_2	methylamine	O^{2-}	oxo	SCN^{-c}	thiocyanato
C_5H_5N	pyridine	OH^-	hydroxo	NCS^{-c}	isothiocyanato
		CN^-	cyano		

[a]Until recently, the term aquo was used.
[b]If the nitrite ion is attached through the N atom (—NO_2), the name *nitro* is used; if attached through an O atom (—ONO), *nitrito*.
[c]If the thiocyanate ion is attached through the S atom (—SCN), the name *thiocyanato* is used; if attachment is through the N atom (—NCS), *isothiocyanato*.

FIGURE 24-3
Structure of
$[Cu(NH_3)_4(H_2O)_2]^{2+}$.

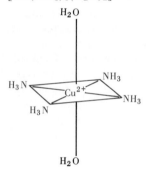

is NH_3 through its N atom) is called a **unidentate** ligand. These ligands may be monoatomic anions (but not neutral atoms) such as halide ions, polyatomic anions such as NO_2^-, simple molecules like NH_3, or more complex molecules like pyridine, C_5H_5N. A listing of some unidentate ligands is provided in Table 24-3.

Chelates.   Some ligands are capable of donating more than a single electron pair, from different atoms in the ligand and to different sites in the geometric structure of a complex. These are called **multidentate** ligands. The molecule **ethylenediamine (en)** can donate *two* electron pairs, one from each N atom. It is a **bidentate** ligand.

$$H{-}\overset{\uparrow}{\underset{H}{N}}{-}CH_2CH_2{-}\overset{\uparrow}{\underset{H}{N}}{-}H \tag{24.5}$$

Three common multidentate ligands are shown in Table 24-4.

TABLE 24-4
Some common multidentate ligands (chelating agents)

Abbreviation	Name	Formula
en	ethylenediamine	$H_2N{\cdot\cdot}\overset{\displaystyle CH_2{-}CH_2}{}{\cdot\cdot}NH_2$
ox	oxalato	oxalate structure with C—C bonded, each C double-bonded to O, each single-bonded to O^-
EDTA	ethylenediaminetetraacetato	EDTA structure

FIGURE 24-4
Two representations of
the chelate $[Pt(en)_2]^{2+}$.

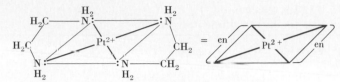

The ligands attach at adjacent corners along an edge of the square. They do *not* bridge
the square by attaching to opposite corners.

The attachment of two ethylenediamine ligands to a Pt^{2+} ion is depicted in Figure
24-4. That multiple attachment of ligands does occur can be established in two ways.

1. The complex ion $[Pt(en)_2]^{2+}$ has no capacity to coordinate additional ligands, that is,
not NH_3, H_2O, Cl^-, and so on. Each en ligand must be attached at two points to account
for the coordination number of four for Pt^{2+}.
2. The en ligands exhibit no further basic properties once attached to the Pt^{2+} ion. That
is, they do not accept protons from water to produce OH^- ions in solution, as would be
expected for a weak base. Both —NH_2 groups of each en molecule must be tied up in the
complex ion.

Note that in $[Pt(en)_2]^{2+}$ there exist two five-membered rings of Pt, N, and C atoms
(see Figure 24-4). When bonding between a metal ion and multidentate ligands results in
ring formation (usually five- or six-membered), the process is called **chelation.** The
species produced is called a **chelate,** and the multidentate ligand is a **chelating agent.**
The origins of words used to describe ligands are interesting. Ligand comes from the
Latin word *ligare,* meaning to bind. Dentate is derived from the Latin word *dens,* meaning tooth. Figuratively speaking, a unidentate ligand has one tooth, a bidentate ligand two
teeth, and so on. The ligand attaches itself to the central ion in accordance with the
number of teeth it possesses. Chelate is derived from the Greek word *chela,* meaning a
crab's claw. The mode of attachment of a chelating agent to a metal ion bears a certain
resemblance to a crab's claw.

24-4 Nomenclature

Prior to Werner's time no attempt was made to develop a systematic nomenclature for
complexes. The colors of some compounds were used as a basis for naming them. The
compound $[Co(NH_3)_5Cl]Cl_2$ is purple in color and was called purpureocobaltic chloride.
By analogy, any other complex ion with the formula $[M(NH_3)_5Cl]^{n+}$ was referred to as a
"purpureo" complex ion (even if its color was not purple). Another approach was simply
to name the compound after the investigator who first prepared it. For example,
$NH_4[Cr(NH_3)_2(NCS)_4]$ was called Reinecke's salt. Werner developed a system of nomenclature that has long been used; but in recent years, through international agreements,
Werner's system has been modified. Although usage still varies somewhat, all the complexes encountered in this text can be named by the eight rules listed below and illustrated
in Example 24-2.

Rules for Naming Complex Ions

1. Cations are named before anions. This is the same rule that applies to simple salts
like NaCl.
**2. For a complex ion or neutral complex, the names of the ligands are given first,
followed by the name of the central ion.** This order is the *opposite* of that in which
formulas are written (where the symbol of the metal appears first in the formula).

3. In naming ligands an "o" ending is used for anions. Specifically, "ide" is changed to "*o*," "ate" to "*ato*," and "ite" to "*ito*."

4. If a ligand is a neutral molecule, the unmodified name of the molecule is used. However, four important exceptions to this rule are noted in Table 24-3—the names for H_2O, NH_3, CO, and NO.

5. The Greek prefixes mono = 1, di = 2, tri = 3, tetra = 4, penta = 5, and hexa = 6 are used to designate the number of ligands of a given type. For example, *dichloro* signifies two Cl^- ions as ligands, and *pentaaqua*, five H_2O molecules. **Also used are the prefixes bis = 2, tris = 3, tetrakis = 4, ...** This is done to avoid confusion if the ligand name itself contains prefixes. The name **bis**(ethylene*di*amine) signifies two ethylenediamine molecules as ligands. The prefix *di* within the parentheses refers to an aspect of the structure of the ligand, not to the number of ligands in the complex. This series of prefixes is also used for ligands that are more complicated than the simple unidentate ligands of Table 24-3. For example, **tris**(oxalato) indicates that three oxalate ions, $C_2O_4^{2-}$, are ligands in a complex.

By an older rule anions are named before neutral ligands, i.e., dichlorotetraaqua.

6. Ligands are named in alphabetical order. For example, if four H_2O molecules and two Cl^- ions are ligands in a complex ion, they are named in the order *tetraaquadichloro*. Prefixes are not considered in establishing alphabetical order (aqua precedes chloro). In formulas for complexes, neutral ligands are generally written before anions, e.g., $[Co(en)_2Cl_2]^+$.

7. The oxidation state of the central metal ion in a complex cation is denoted by a Roman numeral placed in parentheses following the name of the ion. The metal name remains unchanged. The ion $[Co(NH_3)_6]^{3+}$, for example, is called *hexaamminecobalt(III) ion*.

8. In complex anions the oxidation state of the central ion is denoted by a Roman numeral, and the name of the central ion is modified to carry an "ate" ending. The ion $[Al(H_2O)_2(OH)_4]^-$ is called the *diaquatetrahydroxoaluminate(III) ion*. For the following metals the English name is replaced by the Latin name, to which the "ate" ending is added: iron → **ferrate**; tin → **stannate**; lead → **plumbate**; copper → **cuprate**; silver → **argentate**; gold → **aurate**. Thus, $[CuCl_4]^{2-}$ is *tetrachlorocuprate(II) ion*.

Example 24-2 (a) What is the formula of the compound pentaaquachlorochromium(III) chloride? (b) What is the name of the compound $K_3[Fe(CN)_6]$? (c) What is the name of the complex ion $[Pt(en)_2Cl_2]^{2+}$?

Solution
(a) The central metal ion is Cr^{3+}. There are one Cl^- ion and five H_2O molecules as ligands. The complex ion carries a net charge of +2. Two Cl^- ions are required outside the coordination sphere to neutralize this charge. The formula of the coordination compound is $[Cr(H_2O)_5Cl]Cl_2$.
(b) This compound consists of K^+ cations and complex anions having the formula $[Fe(CN)_6]^{3-}$. Each cyanide ion carries a charge of -1, so the oxidation state of the iron must be +3. The Latin name "ferrum" is changed to the "ate" ending because the complex ion is an anion. The name of the anion is hexacyanoferrate(III) ion. The coordination compound is potassium hexacyanoferrate(III).
(c) The ethylenediammine ligands are neutral and they are *bidentate*. The ligand Cl^- is uninegative, and two Cl^- ions are present in the complex ion. The coordination number is *six*; the net charge on the complex ion is +2; and the charge on the central metal ion is +4. The ligands are named in alphabetical order, and the prefix "bis" is required for (en).

$[Pt(en)_2Cl_2]^{2+}$ = dichlorobis(ethylenediamine)platinum(IV) ion.

SIMILAR EXAMPLES: Review Problems 2, 3, 4; Exercises 4, 5.

Although most complexes are named in the manner just outlined, some common or "trivial" names are still in use. Chief among these are the ferrocyanide, $[Fe(CN)_6]^{4-}$, and ferricyanide, $[Fe(CN)_6]^{3-}$, ions. These common names suggest the oxidation states of the central metal ions through the "o" and "i" designations, but they do not indicate coordination numbers. Clearly, systematic names—hexacyanoferrate(II) and hexacyanoferrate(III)—are superior.

24-5 Isomerism

Two or more species having the same composition (i.e., the same formula) but different structures and properties are said to be **isomers.** Isomerism arises among complexes for a variety of reasons which fall into two broad categories. **Structural isomers** differ in basic structure or bond type—what ligands are bonded to the central ion, through which atoms? **Stereoisomers** are alike at the bonding level but differ in geometrical or spatial arrangements among the ligands. Of the five kinds of isomerism that follow, the first three represent structural isomerism and the remaining two, stereoisomerism.

Ionization Isomerism.   The same numbers and types of groups are present in each of the two coordination compounds (24.6). Thus, they have the same empirical formulas. However, the compounds are not identical. In compound (24.6a) $SO_4{}^{2-}$ is part of the octahedral complex ion; the Cl^- is the free anion of the coordination compound. In compound (24.6b) the roles of the $SO_4{}^{2-}$ and Cl^- are reversed.

$$[Cr(NH_3)_5SO_4]Cl \qquad\qquad [Cr(NH_3)_5Cl]SO_4 \qquad\qquad (24.6)$$

pentaamminesulfatochromium(III)   pentaamminechlorochromium(III)
chloride   sulfate
(a)   (b)

Coordination Isomerism.   A similar situation to that just described arises when a coordination compound is made up of both complex cations and complex anions. The ligands may be distributed differently between the complex ions, yet the compositions and empirical formulas of the coordination compounds remain the same.

$$[Co(en)_3][Cr(ox)_3] \qquad\qquad [Cr(en)_3][Co(ox)_3] \qquad\qquad (24.7)$$

tris(ethylenediamine)cobalt(III)   tris(ethylenediamine)chromium(III)
tris(oxalato)chromate(III)   tris(oxalato)cobaltate(III)
(a)   (b)

Linkage Isomerism.   Some ligands may attach to the central metal ion of a complex ion in different ways. For example, the nitrite ion has electron pairs available for coordination both on the N and O atoms.

$$\left[\; \overset{\displaystyle \ddot{N}}{:\!\ddot{O}\diagup \;\;\diagdown \ddot{O}\!:} \;\right]^{-} \qquad\qquad (24.8)$$

Whether attachment of this ligand is through the N or the O atom, the formula of the complex ion is unaffected. However, the properties of the complex ion may be affected. When attachment occurs through the N atom, the ligand is referred to as "nitro." Coordination through the O atom produces a "nitrito" complex.

$$[Co(NH_3)_4(NO_2)Cl]^{+} \qquad\qquad [Co(NH_3)_4(ONO)Cl]^{+} \qquad\qquad (24.9)$$

tetraamminechloronitrocobalt(III) ion   tetraamminechloronitritocobalt(III) ion
(a)   (b)

FIGURE 24-5
Geometric isomerism
illustrated.

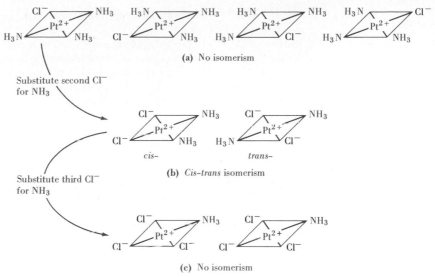

(a) No isomerism

Substitute second Cl⁻
for NH₃

cis-   *trans-*

(b) *Cis-trans* isomerism

Substitute third Cl⁻
for NH₃

(c) No isomerism

For the square planar complexes shown here, isomerism exists only when two Cl⁻ ions have replaced NH₃ molecules.

FIGURE 24-6
Cis-trans isomers of an
octahedral complex.

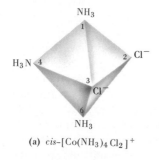

(a) *cis*-[Co(NH₃)₄Cl₂]⁺

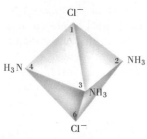

(b) *trans*-[Co(NH₃)₄Cl₂]⁺

The Co³⁺ ion is at the center of the octahedron and one NH₃ ligand is on the far corner (corner 5), out of view.

Geometric Isomerism. If a single Cl⁻ ion is substituted for an NH₃ molecule in the complex ion [Pt(NH₃)₄]²⁺, the point at which substitution occurs is immaterial. All four possibilities are alike, as shown in Figure 24-5a. Substitution of a second Cl⁻ produces two distinct possibilities, however (Figure 24-5b). The two Cl⁻ ions can either be along the same edge of the square planar structure (**cis**) or on opposite corners (**trans**). To distinguish between the two isomers, one must either draw a structure or refer to the appropriate name. The formula alone does not distinguish between them. (Note also that this complex is a neutral species, not an ion.)

$$[Pt(NH_3)_2Cl_2] \qquad\qquad (24.10)$$

cis-diamminedichloroplatinum(II)
or
trans-diamminedichloroplatinum(II)

With the substitution of a third Cl⁻, again there is but a single structure; isomerism disappears (Figure 24-5c).

Geometric isomerism is very common among octahedral complexes, but here the situation is a bit more complicated. Take the complex ion [Co(NH₃)₆]³⁺ as an example. Substitution of one Cl⁻ for an NH₃ produces no isomers. Substitution of two Cl⁻ ions leads to *cis-trans* isomerism. The *cis* isomer has Cl⁻ ions along the same edge of the octahedron. The *trans* isomer has Cl⁻ ions on opposite corners, that is, at opposite ends of a line drawn through the central metal ion. These two isomers are pictured in Figure 24-6. One of the differences in their properties is that the *cis* isomer is a blue-violet color and the *trans* is a bright green.

Substitution of a third Cl⁻ for one of the NH₃ molecules again leads to two possibilities. Refer to Figure 24-6a. If substitution occurs either at position 1 or 6, the result is the same—three Cl⁻ on the same face of the octahedron. This isomer may also be called *cis*.* If the substitution is made either at position 4 or 5, the result is three Cl⁻ around a perimeter of the octahedron. This isomer may also be called *trans*.* Substitution of a fourth Cl⁻ leads to two isomers, named this time based on the positions of the two remaining NH₃ molecules.

*In more precise nomenclature, this type of *cis* isomer is called a facial (*fac*) isomer, and this type of *trans* isomer, a meridional (*mer*) isomer.

Example 24-3   Sketch all the possible isomers of $[Co(ox)(NH_3)_3Cl]$.

Solution.   The Co^{3+} ion exhibits a coordination number of 6. The structure is octahedral. Recall that ox (oxalate ion) is a bidentate ligand carrying a double-negative charge (Table 24-4). Recall also that such a ligand must be attached in *cis* positions, not *trans* (Figure 24-4). Once the ox ligand is placed, we see that there are two possibilities. The three NH_3 molecules can be situated (1) on the same face of the octahedron (*cis* isomer) or (2) around a perimeter of the octahedron (*trans* isomer).

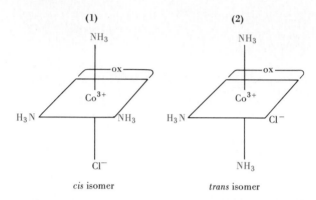

SIMILAR EXAMPLES: Review Problems 8, 9; Exercise 13.

Optical Isomerism.   The isomers of $[Co(en)_3]^{3+}$ depicted in Figure 24-7 are not as obviously different from each other as those that we have considered previously. Each of the six positions has the same ligand attachment, through the N atom of an ethylenediamine. The two isomers are related to each other as are an object and its image in a mirror. Features on the right side of an object appear on the left side of its mirror image, and vice versa. As a familiar example, a left hand produces as its mirror image a right hand.

If an object has certain symmetrical features to its structure, the object and its mirror image are superimposable and indistinguishable. That is, starting with a particular orientation of the object and its mirror image, the object can be rotated in such a way as to assume a new orientation that is identical to its original mirror image. If an object is lacking in these symmetry features (is asymmetric), it and its mirror image are nonsuperimposable and remain distinguishable. Consider, for example, an open-top cubical cardboard box. There are a number of ways (hypothetical, of course) in which the box can be superimposed on its mirror image. Now imagine that a distinctive label is placed at a corner of one side of the box. In this case there is no way that the box and its mirror image can be superimposed; they are nonsuperimposable and clearly different. There is no way that a form-fitting left glove can be worn on the right hand (turning it inside out is disallowed). Likewise, as suggested through Figure 24-7, the two complex ion structures are nonsuperimposable; no combination of rotation or inversion of one structure will make it identical to its mirror image.

Structures that are nonsuperimposable are called **enantiomers** and are said to be **chiral** (pronounced kye · rull). (Structures that *are* superimposable are *achiral*.) Whereas other types of isomers may differ rather significantly in their physical and chemical properties, enantiomers have identical properties, except in a few specialized situations. These exceptions involve phenomena that are closely linked to chirality or "handedness" at the molecular level. One such phenomenon is that of optical activity, pictured in Figure 24-8. Interactions between a beam of polarized light and the electrons in an enantiomer cause the plane of the polarized light to be rotated. One enantiomer rotates the plane of polarized

FIGURE 24-7
Optical isomers.

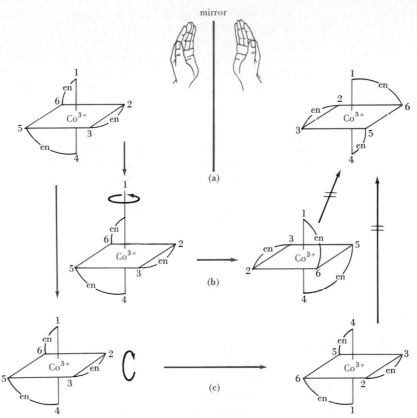

(a) The two structures shown, like a left and a right hand, are mirror images and are nonsuperimposable.
(b) The structure on the left is rotated about its vertical axis (180°). This places the en group in the central plane in the same orientation as in the structure on the right in (a), but the rest of the structure is nonsuperimposable.
(c) The structure on the left is inverted (top to bottom). Again, this leads to an arrangement that is nonsuperimposable with the structure on the right in (a).

light to the right (clockwise) and is said to be **dextrorotatory** (designated + or *d*). The other enantiomer rotates the plane of polarized light to the same extent but to the left (counterclockwise). It is said to be **levorotatory** (− or *l*). (The terms dextro and levo are derived from the Latin words *dexter* = right and *laevus* = left.)

FIGURE 24-8
Optical activity.

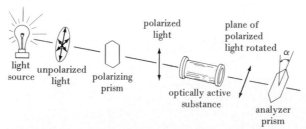

Light from an ordinary source consists of electromagnetic waves vibrating in all planes; it is unpolarized. Some substances (e.g., polaroid) possess the ability to screen out all light waves except those vibrating in a particular plane. Other substances—optically active materials—have the ability to rotate the plane of polarized light. In the illustration the light is rotated to the right by the angle α.

We might call this a chiral situation: choosing a matched pair of gloves from a bin of gloves of identical sizes, 99 made for the right hand and 1 for the left hand.

In the synthesis of an optically active complex ion, usually a mixture of equal amounts of each enantiomer is obtained. The optical rotation of one isomer just cancels that of the other, and the mixture, called a **racemic mixture,** produces no net rotation of the plane of polarized light at all. Separating the *d* and *l* isomers of a racemic mixture is called **resolution.** This can sometimes be done by chemical reactions based on chirality, that is, with the two enantiomers behaving differently. Many phenomena of the living state, such as the activity of an enzyme or the ability of a microorganism to promote a reaction, involve chirality (see Chapter 27).

Isomerism and Werner's Theory. The study of isomerism played a central role in establishing Werner's theory of coordination compounds. An important early question in the field involved the structure of complexes with coordination number 6. Werner proposed that they are octahedral, but other possibilities were also suggested. One of these, pictured in Figure 24-9, was a hexagonal structure (resembling the structure of benzene, which had recently been established). The choice of the octahedral over the hexagonal structure was strongly suggested by the observation that there are but *two* isomers of complex ions like $[Co(NH_3)_4Cl_2]^+$ (Figure 24-6). The hexagonal structure would require *three,* and a third one was never found.

More direct evidence came when Werner synthesized optical isomers like $[Co(en)_3]^{3+}$ (Figure 24-7). Neither the hexagonal structure nor any other structure considered as alternatives to the octahedron accounts for optical activity. However, this evidence was still not accepted as conclusive proof by some of Werner's critics. They associated optical activity with carbon atoms and argued that the optical activity that Werner observed was caused by C atoms in his ligands and not by the structure of the complex ion. Werner succeeded (in 1908) in preparing an optically active coordination compound that was totally inorganic (no C atoms were present). Werner was awarded the Nobel prize in 1913 for his prodigious efforts in developing the fundamentals of coordination chemistry.

FIGURE 24-9
Hypothetical structure for $[Co(NH_3)_4Cl_2]^+$.

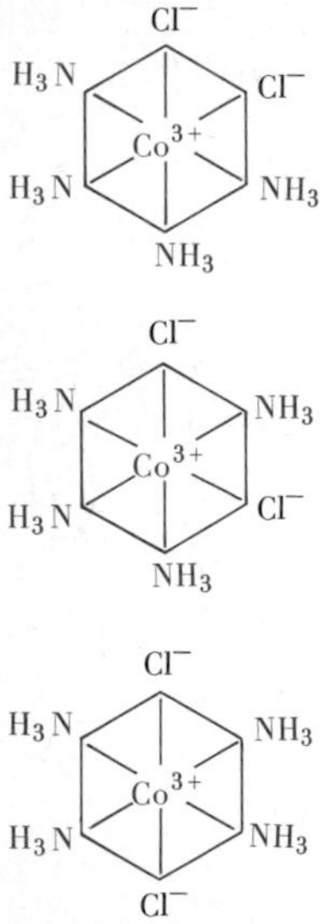

If the distribution of ligands were in the planar hexagonal structure shown here, there should be *three* distinct isomers, but only *two* isomers are found.

24-6 Bonding in Complex Ions—Valence Bond Theory

One view of the metal ion-ligand bond is based on coordinate covalency—the central ion furnishes the orbitals and the ligands furnish electron pairs. Although this view is now considered inadequate, it does account for the observed coordination number of the central ion and describe the geometric structure of a complex ion.

A scheme consistent with the formula and structure of the complex ion $[Co(NH_3)_6]^{3+}$ involves the hybridization of two 3*d* orbitals with a 4*s* and three 4*p* orbitals, yielding six d^2sp^3 orbitals with an octahedral distribution.

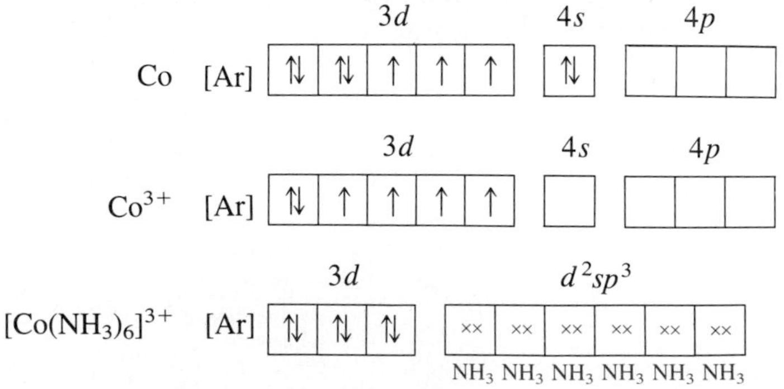

Three additional hybridization schemes are given on the next page for the tetrahedral $[Zn(NH_3)_4]^{2+}$, the square planar $[Pt(NH_3)_4]^{2+}$, and the linear $[Ag(NH_3)_2]^+$, respectively. Geometric structures of all four complex ions were presented in Figure 24-2.

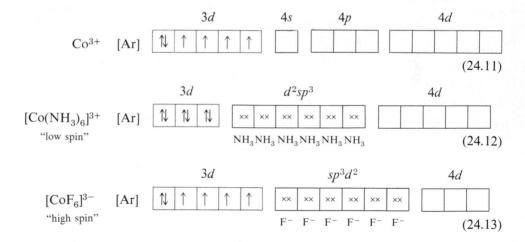

24-7 Inner and Outer Orbital Complexes

The species $[Co(NH_3)_6]^{3+}$ is diamagnetic; $[CoF_6]^{3-}$ has a similar octahedral structure, but it is paramagnetic and appears to have *four* unpaired electrons. One explanation of this observation assumes that there are two ways in which an equivalent set of hybrid orbitals can be formed. In the first case, exemplified by $[Co(NH_3)_6]^{3+}$, the d orbitals come from an inner electronic shell (the third), and the resulting hybrid orbitals are d^2sp^3. The complex ion is called an inner orbital complex. Furthermore, because all electrons are paired and the complex ion is diamagnetic, it is referred to as a "low spin" complex. In $[CoF_6]^{3-}$ it is assumed that the d orbitals come from the outer electronic shell (the fourth), and the resulting hybrid orbitals are sp^3d^2. The complex ion is called an outer orbital complex and is a "high spin" complex. These bonding schemes are consistent with the observed magnetic data.

24-8 Bonding in Complex Ions—Crystal Field Theory

The valence bond theory of bonding in complex ions has some shortcomings. For instance, it does not provide insight into the origin of the characteristic colors of complex ions. Neither does it explain why $[Co(NH_3)_6]^{3+}$ is an inner orbital complex and $[CoF_6]^{3-}$ is an outer orbital complex. An alternative that is more successful in doing so is the crystal field theory. In the crystal field model, bonding in a complex ion is considered to be an

FIGURE 24-10
Approach of six anions to a metal ion to form a complex ion with octahedral structure.

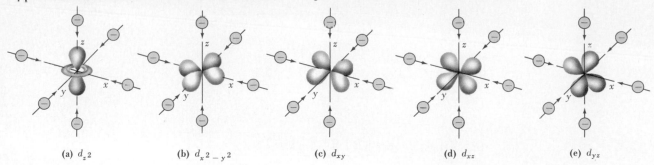

(a) d_{z^2} (b) $d_{x^2-y^2}$ (c) d_{xy} (d) d_{xz} (e) d_{yz}

The ligands (anions in this case) approach the central metal ion along the x, y, and z axes. Maximum interference occurs with the d_{z^2} and $d_{x^2-y^2}$ orbitals, and their energies are raised. Interference with the other d orbitals is not as great. A difference in energy results between the two sets of d orbitals.

Modifications of the simple crystal field theory that take into account such factors as the partial covalency of the metal–ligand bond are called ligand field theory. Often the single term "ligand field theory" is used to signify both the purely electrostatic crystal field theory and its modifications.

electrostatic attraction between the positively charged nucleus of the central metal ion and electrons in the ligands. Repulsion occurs between the ligand electrons and the electrons of the central ion. The theory focuses on these repulsive forces, particularly as they affect d electrons of the central metal ion.

The d orbitals first presented in Figure 7-21 are not all alike in their spatial orientations, but for an isolated atom or ion they do have equal energies. One of them, d_{z^2}, is directed along the z axis and another, $d_{x^2-y^2}$, has lobes along the x and y axes. The remaining three have lobes extending into regions between the perpendicular x, y, and z axes.

Figure 24-10 pictures six anions (ligands) approaching a central metal ion along the x, y, and z axes. This direction of approach leads to an octahedral complex. As a result of electrostatic repulsion between the negatively charged ligands and the d electrons of the metal ion, the d energy level is split into two groups. One group, consisting of $d_{x^2-y^2}$ and d_{z^2}, has its energy raised with respect to a hypothetical free ion in the field of the ligands. The other group, consisting of d_{xy}, d_{xz}, and d_{yz}, has its energy lowered. The difference in energy between the two groups is represented by the symbol Δ, as shown in Figure 24-11.

FIGURE 24-11
Splitting of d energy levels in the formation of an octahedral complex ion.

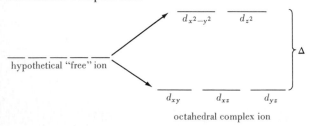

FIGURE 24-12
Crystal field splitting in a tetrahedral complex ion.

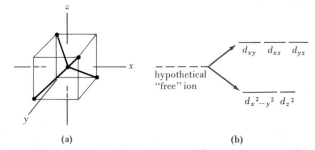

(a)   (b)

(a) The positions of attachment of ligands to a metal ion leading to the formation of a tetrahedral complex ion.
(b) Interference with the d orbitals directed along the x, y, and z axes is not as great as with those that lie between the axes (again, see Figure 24-10). As a result, the pattern of crystal field splitting is reversed from that of an octahedral complex.

The pattern of splitting of the *d* energy levels is different for complex ions with other geometric shapes. The pattern for tetrahedral complexes is described in Figure 24-12, and that of square planar complexes in Figure 24-13. If the same combinations of ligands, metal ions, and metal–ligand distances are compared for hypothetical complexes of different structures, the greatest energy separation of the *d* levels is found for the square planar complex and the least for the tetrahedral complex. The relative magnitudes of these separations, compared to that of the octahedral complex (Δ), are tetrahedral (0.44Δ) and square planar (1.74Δ).

Ligands differ in ability to produce a splitting of the *d* energy levels. Strong Lewis bases, such as CN^- and NH_3, produce a "strong" field. They repel *d* electrons most strongly and cause a greater separation of the *d* energy levels than do weak bases like H_2O and F^-. Concerning the distribution of *d* electrons of the metal ion, we should expect orbitals to be singly occupied, as in the complex ion $[CoF_6]^{3-}$ shown in (24.14). However, in some cases, the energy separation between the lower and higher energy states (Δ) is larger than the energy required to pair electrons. Then the energetically favored arrangement is for electrons to remain paired in the lower-energy *d* orbitals, as in $[Co(NH_3)_6]^{3+}$.

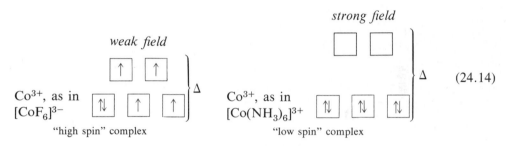

(24.14)

We have just seen how crystal field theory offers an explanation of the magnetic properties of the inner and outer complexes described by structures (24.12) and (24.13).

FIGURE 24-13
Comparison of crystal field splitting in a square planar and an octahedral complex.

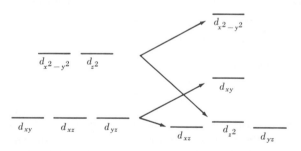

Consider an octahedral complex in which the two ligands along the *z* axis move progressively farther away from the central ion, with the final result that the remaining four ligands are in a square planar configuration about the central ion. (This process can be visualized in terms of Figure 24-3 and the accompanying discussion of the structure of $[Cu(NH_3)_4(H_2O)_2]^{2+}$.)

Splitting of the *d* energy level in a square planar complex can be related to that of the octahedral complex. Because there are no ligands along the *z* axis in a square planar complex, we should expect the repulsion between ligands and d_{z^2} electrons to be much less than in an octahedral complex. The d_{z^2} energy level is lowered considerably from that in an octahedral complex. The energy of the $d_{x^2-y^2}$ orbital is raised, since the *x* and *y* axes represent the direction of approach of four ligands to the central ion. The energy of the d_{xy} orbital is also raised because this orbital lies in the plane of the ligands in the square planar complex. The d_{xz} and d_{yz} orbitals are concentrated in planes perpendicular to that of the square planar complex and their energies are lowered slightly.

Predictions of this same sort can be made for other complex ions using the following listing, called the **spectrochemical series,** to describe the degree of splitting produced by ligands.

strong field
$$CN^- > -NO_2^- > en > py \simeq NH_3 > SCN^- -$$

weak field
$$>H_2O > OH^- > F^- > Cl^- > Br^- > I^- \qquad (24.15)$$

(When NO_2^- is attached through an O atom or SCN^- through the S atom, a different placement in the spectrochemical series is found than the one given here.)

Example 24-4   How many unpaired electrons are present in the octahedral complex $[Fe(CN)_6]^{3-}$?

Solution.   The Fe atom has the electron configuration $[Ar]3d^64s^2$. The Fe^{3+} ion has the configuration $[Ar]3d^5$. CN^- is a strong-field ligand. Because of the large energy separation produced by this ligand in the d levels of the metal ion, all the electrons are found in the lowest energy levels. There is *one* unpaired electron.

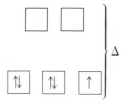

SIMILAR EXAMPLES: Exercises 17, 18, 19.

Example 24-5   The complex ion $[Ni(CN)_4]^{2-}$ is found to be diamagnetic. Use ideas from the crystal field theory to speculate on its probable structure.

Solution.   We can eliminate an octahedral structure because the coordination number is 4 (not 6). Our choice appears to be between a tetrahedral and a square planar structure.

The electron configuration of Ni is $[Ar]3d^84s^2$ and of Ni(II), $[Ar]3d^8$. Since the complex ion is found to be diamagnetic, all $3d$ electrons must be paired. Now compare the d energy level diagrams for a tetrahedral (Figure 24-12) and a square planar complex (Figure 24-13). For the tetrahedral complex ion the d orbitals are found in two groups. In $[Ni(CN)_4]^{2-}$ the lower group would be fully occupied with four electrons, and the remaining four electrons would be distributed among the three orbitals in the upper group. There would be two unpaired electrons and the complex ion would be paramagnetic. In the square planar complex the three lowest energy orbitals would be filled. Because of the large energy separation between the d_{xy} and the $d_{x^2-y^2}$ orbitals, we should expect the remaining two electrons to be paired in the d_{xy} orbital and the $d_{x^2-y^2}$ to be empty, leading to a diamagnetic species. The probable structure of $[Ni(CN)_4]^{2-}$ is square planar.

SIMILAR EXAMPLES: Exercises 21, 22.

24-9 Color of Complex Ions

The absorption of electromagnetic radiation by an ionic species in solution requires that electrons within the ion undergo a transition from one energy state to another. The light

TABLE 24-5
Some coordination compounds of Cr^{3+} and their colors

Isomer	Color	Isomer	Color
$[Cr(H_2O)_6]Cl_3$	violet	$[Cr(NH_3)_6]Cl_3$	yellow
$[Cr(H_2O)_5Cl]Cl_2$	blue-green	$[Cr(NH_3)_5Cl]Cl_2$	purple
$[Cr(H_2O)_4Cl_2]Cl$	green	$[Cr(NH_3)_4Cl_2]Cl$	violet

quanta absorbed must have energies just equal to the difference in the two energy states involved in the transition. Where this transition energy corresponds to a wavelength component in visible light, that component is absorbed and the transmitted light is colored. The color of the transmitted light—that is, the "color" of the solution—is the *complement* of the color that is absorbed. A solution containing $[Cu(H_2O)_4]^{2+}$ absorbs yellow light (about 580 nm), and the transmitted light is blue. A solution with $[CuCl_4]^{2-}$ absorbs blue light, and the transmitted light is yellow. The principal combinations of colors and their complements are blue $\leftrightarrow$ yellow; red $\leftrightarrow$ blue-green; green $\leftrightarrow$ purple.

Ions having a noble gas electron configuration, an outer shell of 18 electrons, or the "18 + 2" configuration do not possess electronic transitions in the energy range corresponding to visible light. "White" light passes through solutions of these ions without being absorbed; these ions are colorless in solution. Examples are the alkali and alkaline earth metal ions, the halide ions, Zn^{2+} and Bi^{3+}.

Crystal field splitting of the d energy levels produces the energy difference, Δ, that accounts for the colors of complex ions. Promotion of an electron from a lower to a higher d level results from the absorption of the appropriate components of white light; the transmitted light is colored.

The colors of some complex ions of chromium are given in Table 24-5. The complex ions in question are all octahedral, so the pattern of the crystal field splitting of the d energy levels is that of Figure 24-11. Chromium(III) complexes all have a d^3 configuration. In each case there are three unpaired electrons in the lower energy group.

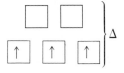

The effect of different ligands on the colors of coordination compounds is illustrated in Color Section I.

The magnitude of Δ, the crystal field splitting, is determined by the particular ligands attached to the Cr^{3+} ion. This magnitude determines what wavelength light must be absorbed to produce an electronic transition, and, in turn, what color light is transmitted by the solution. For example, in $[Cr(NH_3)_6]^{3+}$ the crystal field splitting (Δ) is greater than in $[Cr(H_2O)_6]^{3+}$. As a result, $[Cr(NH_3)_6]^{3+}$ absorbs light of a shorter wavelength (violet) than does $[Cr(H_2O)_6]^{3+}$ (yellow). The light *transmitted* by $[Cr(NH_3)_6]^{3+}$—and hence its color—is yellow; that of $[Cr(H_2O)_6]^{3+}$ is violet.

24-10 Additional Aspects of Complex Ion Equilibria

In Chapter 19 we considered how equilibria involved in complex ion formation affect other equilibrium processes, such as the dissolving of a slightly soluble solute. That discussion was pursued both qualitatively and quantitatively. For quantitative applications we introduced an equilibrium constant called the formation constant of the complex ion, K_f (Section 19-8). In this section we reconsider complex ion equilibria in somewhat more detail.

Displacement Reactions and Stepwise Formation (Stability) Constants. By the method of Chapter 19 we would represent the formation of $[Zn(NH_3)_4]^{2+}$ with the equations

$$Zn^{2+}(aq) + 4\ NH_3(aq) \rightleftharpoons [Zn(NH_3)_4]^{2+}(aq) \qquad (24.16)$$

$$K_f = \frac{[[Zn(NH_3)_4]^{2+}]}{[Zn^{2+}][NH_3]^4} = 4.1 \times 10^8 \qquad (24.17)$$

However, we have since learned that cations in aqueous solution exist mostly in hydrated form, which for $Zn^{2+}(aq)$ can be written as $[Zn(H_2O)_4]^{2+}$. As a consequence, NH_3 molecules do not enter an empty coordination sphere. They must displace H_2O molecules in a stepwise fashion. The reaction

$$[Zn(H_2O)_4]^{2+} + NH_3 \rightleftharpoons [Zn(H_2O)_3NH_3]^{2+} + H_2O \qquad (24.18)$$

for which

$$K_1 = \frac{[[Zn(H_2O)_3NH_3]^{2+}]}{[[Zn(H_2O)_4]^{2+}][NH_3]} = 3.9 \times 10^2 \qquad (24.19)$$

is followed by

$$[Zn(H_2O)_3NH_3]^{2+} + NH_3 \rightleftharpoons [Zn(H_2O)_2(NH_3)_2]^{2+} + H_2O \qquad (24.20)$$

for which

$$K_2 = \frac{[[Zn(H_2O)_2(NH_3)_2]^{2+}]}{[[Zn(H_2O)_3NH_3]^{2+}][NH_3]} = 2.1 \times 10^2 \qquad (24.21)$$

and so on.

The value of K_1 in equation (24.19) can also be designated as β_1 and called the formation constant for the complex ion $[Zn(H_2O)_3NH_3]^{2+}$. The formation of $[Zn(H_2O)_2(NH_3)_2]^{2+}$ is represented by the *sum* of equations (24.18) and (24.20),

$$[Zn(H_2O)_4]^{2+} + 2\ NH_3 \rightleftharpoons [Zn(H_2O)_2(NH_3)_2]^{2+} + 2\ H_2O \qquad (24.22)$$

and the formation constant β_2, in turn, is given by the *product* of equations (24.19) and (24.21).

$$\beta_2 = \frac{[[Zn(H_2O)_2(NH_3)_2]^{2+}]}{[[Zn(H_2O)_4]^{2+}][NH_3]^2} = K_1 \times K_2 = 8.2 \times 10^4 \qquad (24.23)$$

TABLE 24-6
Stepwise and overall formation (stability) constants for several complex ions[a]

Metal ion	Ligand	K_1	K_2	K_3	K_4	K_5	K_6	β_n (or K_f)[b]
Ag^+	NH_3	2.0×10^3	7.9×10^3					1.6×10^7
Zn^{2+}	NH_3	3.9×10^2	2.1×10^2	1.0×10^2	5.0×10^1			4.1×10^8
Cu^{2+}	NH_3	1.9×10^4	3.9×10^3	1.0×10^3	1.5×10^2			1.1×10^{13}
Ni^{2+}	NH_3	6.3×10^2	1.7×10^2	5.4×10^1	1.5×10^1	5.6	1.1	5.3×10^8
Cu^{2+}	en	5.2×10^{10}	2.0×10^9					1.0×10^{20}
Ni^{2+}	en	3.3×10^7	1.9×10^6	1.8×10^4				1.1×10^{18}
Ni^{2+}	EDTA	4.2×10^{18}						4.2×10^{18}

[a]In many tabulations in the chemical literature, formation constant data are presented as logarithms, i.e., $\log K_1$, $\log K_2, \dots$, and $\log \beta_n$.
[b]The β_n listed is for the number of steps shown, i.e., for $[Ag(NH_3)_2]^+$, $\beta_2 = K_f = K_1 \times K_2$; for $[Ni(en)_3]^{2+}$, $\beta_3 = K_f = K_1 \times K_2 \times K_3$; and for $[Ni(EDTA)]^{2-}$, $\beta_1 = K_f = K_1$.

For the next ion in the series, $[Zn(H_2O)(NH_3)_3]^{2+}$, $\beta_3 = K_1 \times K_2 \times K_3$. For the final member, $[Zn(NH_3)_4]^{2+}$, $\beta_4 = K_1 \times K_2 \times K_3 \times K_4$, and this is the value that was called the formation constant in Section 19-8 and listed as K_f in Table 19-3. Additional stepwise formation constant data are presented in Table 24-6.

The large numerical value of K_1 for reaction (24.18) indicates that Zn^{2+} has a greater affinity for NH_3 (a stronger Lewis base) than it does for H_2O. Displacement of ligand H_2O molecules by NH_3 occurs even if the number of NH_3 molecules present in aqueous solution is much smaller than the number of H_2O molecules, i.e., in dilute $NH_3(aq)$. The fact that the magnitudes of successive K values decrease regularly in the displacement process can be explained in statistical terms: An NH_3 molecule has a better chance of replacing an H_2O molecule in $[Zn(H_2O)_4]^{2+}$, where each coordination position is occupied by H_2O, than in $[Zn(H_2O)_3NH_3]^{2+}$, where one of the positions is already occupied by NH_3. For other complexes, if irregularities in the K values arise, it is often because of a change in structure of the complex ion at some point in the series of displacement reactions.

Chelation. If the ligand in a substitution process is multidentate, it simultaneously displaces as many H_2O molecules as there are points of attachment. Thus, ethylenediamine (en) displaces H_2O molecules in $[Ni(H_2O)_6]^{2+}$ two at a time, in three steps. The first step is

$$[Ni(H_2O)_6]^{2+} + en \rightleftharpoons [Ni(en)(H_2O)_4]^{2+} + 2 H_2O \tag{24.24}$$

There is an interesting difference between this first step and the first two steps in the displacement of H_2O by NH_3 in $[Ni(H_2O)_6]^{3+}$. Reaction (24.24), with two H_2O molecules being replaced by a single en, is accompanied by a larger increase in entropy than is the succession of two one-for-one displacements leading to $[Ni(H_2O)_4(NH_3)_2]^{2+}$. A more positive ΔS in the expression: $\Delta G = \Delta H - T\Delta S$ leads to a more negative ΔG, a larger value of K, and a more thermodynamically stable complex ion. This additional stability of the complex ion is referred to as the **chelation effect.** Note in Table 24-6 how much larger is β_3 for $[Ni(en)_3]^{2+}$ than β_6 for $[Ni(NH_3)_6]^{2+}$. The formation constant for $[Ni(EDTA)]^{2-}$, involving the hexadentate ligand EDTA, is also much larger.

Use of Formation Constants in Calculations. Our purpose in this section has been to increase our understanding of complex ions by focusing on the stepwise formation of complex ions (or more correctly the stepwise displacement of H_2O by other ligands). A point that we should specifically emphasize, though, is that in calculations of the type encountered in this text (such as Examples 19-15, 19-16, and 19-17) only overall formation reactions and overall formation constants, K_f, are required. That is, we assume that the concentrations of intermediate species are negligible, even though there are obviously situations where this is not the case.

24-11 Acid–Base Reactions of Complex Ions

We have described complex ion formation in terms of Lewis acids and bases (Section 24-3). Complex ions may also exhibit acid–base properties in the Brønsted–Lowry sense, i.e., act as proton donors and acceptors. Figure 24-14 illustrates the ionization of $[Fe(H_2O)_6]^{3+}$ as an acid. A proton from a *ligand* water molecule in hexaaquairon(III) ion is transferred to a *solvent* water molecule. The H_2O ligand is converted to OH^-.

$$[Fe(H_2O)_6]^{3+} + H_2O \rightleftharpoons [Fe(H_2O)_5OH]^{2+} + H_3O^+ \qquad K_{a_1} = 9 \times 10^{-4} \tag{24.25}$$

The second ionization step is

$$[Fe(H_2O)_5OH]^{2+} + H_2O \rightleftharpoons [Fe(H_2O)_4(OH)_2]^+ + H_3O^+ \qquad K_{a_2} = 5 \times 10^{-4} \tag{24.26}$$

FIGURE 24-14
Ionization of
$[Fe(H_2O)_6]^{3+}$.

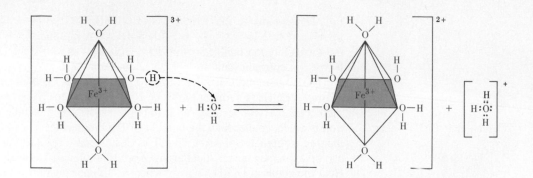

From these K_a values we see that Fe^{3+}(aq) is fairly acidic. To repress ionization (hydrolysis) of $[Fe(H_2O)_6]^{3+}$ it is necessary to maintain a low pH by the addition of acids such as HNO_3 or $HClO_4$. (Recall that increasing $[H_3O]^+$ displaces reactions 24.25 and 24.26 to the left.) The ion $[Fe(H_2O)_6]^{3+}$ is violet in color, but aqueous solutions of Fe^{3+}(aq) are generally yellow owing to the presence of hydroxo complex ions.

The ionization of $[Fe(H_2O)_4(OH)_2]^+$ can proceed a step further, particularly in the presence of an alkaline substance, leading to a precipitate of $[Fe(H_2O)_3(OH)_3]$. Another reaction that can occur simultaneously with (24.25) and (24.26) is the formation of a bridged structure by the joining of two simple complex ions into a binuclear complex ion, as illustrated in Figure 24-15. More units may add to the binuclear structure, resulting in aggregates having colloidal dimensions. Colloidal ferric hydroxide (hydrous ferric oxide) can be prepared by means of these reactions by adding an iron(III) salt to boiling water.

Formation of this colloid is pictured in Color Section I.

Cr^{3+} and Al^{3+} behave in a similar manner to Fe^{3+}, except that with them hydroxo complex ion formation continues until complex anions are produced. $Cr(OH)_3$ and $Al(OH)_3$, as we have previously noted, are soluble in alkaline as well as acidic solutions; they are amphoteric.

Regarding the acid strengths of aqua complex ions, a critical factor is the charge-to-radius ratio of the central metal ion, just as we found in an earlier discussion of acid strength (Section 17-8). Thus, the small, highly charged Fe^{3+} attracts electrons away from an O—H bond in a ligand water molecule, causing it to ionize as an acid ($K_{a_1} = 9 \times 10^{-4}$). The attraction of Fe^{2+} for electrons is weaker, as is its acid strength ($K_{a_1} = 1 \times 10^{-7}$).

24-12 Some Kinetic Considerations

When NH_3(aq) is added to a solution containing Cu^{2+}, there is an immediate change in color from pale blue to very deep blue. The reaction involves NH_3 molecules displacing H_2O molecules as ligands.

$$[Cu(H_2O)_4]^{2+} + 4\ NH_3 \longrightarrow [Cu(NH_3)_4]^{2+} + 4\ H_2O \qquad (24.27)$$

    (pale blue)   (very deep blue)

FIGURE 24-15
A binuclear complex of
Fe(III).

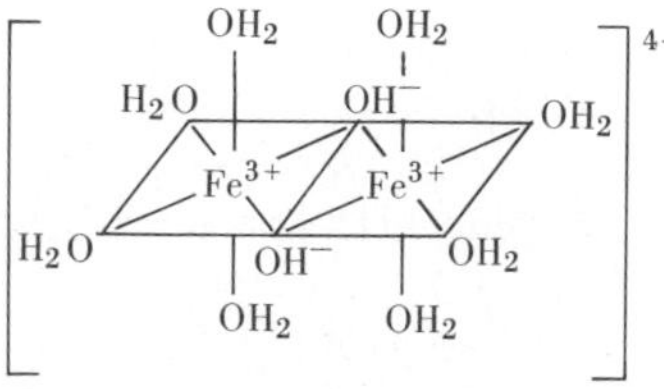

$$\begin{bmatrix} & OH_2 & OH_2 & \\ H_2O & | & OH^- & OH_2 \\ & \diagdown Fe^{3+} \diagdown Fe^{3+} \diagup & \\ H_2O & | & OH_2 & \\ & OH^- & OH_2 & \\ & OH_2 & OH_2 & \end{bmatrix}^{4+}$$

This binuclear structure is produced by the reaction

$$2\ [Fe(H_2O)_6]^{3+} \rightleftharpoons [Fe_2(H_2O)_8(OH)_2]^{4+} + 2\ H_3O^+ \qquad K = 1.2 \times 10^{-3}$$

This reaction occurs very rapidly—as rapidly as the two reactants can be brought to-gether. The addition of HCl(aq) to an aqueous solution of Cu^{2+} produces an immediate color change from pale blue to green, or even yellow if the HCl(aq) is sufficiently concentrated.

$$[Cu(H_2O)_4]^{2+} + 4\ Cl^- \longrightarrow [CuCl_4]^{2-} + 4\ H_2O \qquad (24.28)$$
$$\text{(pale blue)} \qquad\qquad\qquad \text{(yellow)}$$

Complex ions in which ligands can be interchanged rapidly are said to be **labile.** $[Cu(H_2O)_4]^{2+}$, $[Cu(NH_3)_4]^{2+}$, and $[CuCl_4]^{2-}$ are all labile.

In freshly prepared $CrCl_3$(aq) the ion $[Cr(H_2O)_4Cl_2]^+$ produces a green color, but the color gradually turns to violet. This color change results from the very slow exchange of H_2O for Cl^- as ligands. A complex ion that exchanges ligands slowly is said to be nonlabile or **inert.** In general, complex ions of the first transition series, except for those of Cr(III) and Co(III), are kinetically labile. Those of the second and third transition series are generally kinetically inert. Whether a complex ion is labile or inert affects the ease with which it can be studied. That the inert ones are easiest to obtain and characterize may explain why so many of the early studies of complex ions were based on Cr(III) and Co(III).

The terms "labile" and "inert" are not related to the thermodynamic stabilities of complex ions, nor to the equilibrium constants for ligand-substitution reactions. The terms are *kinetic* terms, referring to the *rates* at which ligands may be exchanged. (See also, Color Section I.)

24-13 Applications of Coordination Chemistry

The number and variety of applications of coordination chemistry are impressive, ranging from analytical chemistry to biochemistry. The examples given here are intended simply to convey an idea of this diversity.

Hydrates.   Often when a compound is crystallized from an aqueous solution of its ions the crystals obtained are hydrated. A hydrate is a substance that has associated with each formula unit a certain number of water molecules. In some cases the water molecules are ligands bonded directly to a metal ion. The coordination compound $[Co(H_2O)_6](ClO_4)_2$ may be represented as the hexahydrate, $Co(ClO_4)_2 \cdot 6H_2O$. In the hydrate $CuSO_4 \cdot 5H_2O$, four H_2O molecules are associated with copper in the complex ion $[Cu(H_2O)_4]^{2+}$, and the fifth with the SO_4^{2-} anion through hydrogen bonding. Another possibility for hydrate formation is that the water molecules may be incorporated into definite positions in the solid crystal but not associated with any particular cations or anions. This is referred to as lattice water, as in $BaCl_2 \cdot 2H_2O$. Finally, part of the water may be coordinated to an ion and part of it may be lattice water. Apparently, this is the case with the alums, such as $KAl(SO_4)_2 \cdot 12H_2O$.

Stabilization of Oxidation States.   The standard electrode potential for the reduction of Co(III) to Co(II) is

$$Co^{3+}(aq) + e^- \longrightarrow Co^{2+}(aq) \qquad E° = +1.82\ V \qquad (24.29)$$

This suggests that Co^{3+}(aq) is a strong oxidizing agent, strong enough in fact to oxidize water to O_2(g).

$$4\ Co^{3+}(aq) + 2\ H_2O \longrightarrow 4\ Co^{2+}(aq) + 4\ H^+ + O_2(g) \qquad E°_{cell} = +0.59\ V \quad (24.30)$$

Yet, one of the complex ions featured in this chapter has been $[Co(NH_3)_6]^{3+}$. This ion is stable in water solution, even though it contains cobalt in the oxidation state +3. Reaction (24.30) will not occur if the concentration of Co^{3+} is sufficiently low, and $[Co^{3+}]$ is kept very low because of the great stability of the complex ion.

$$Co^{3+}(aq) + 6\ NH_3(aq) \rightleftharpoons [Co(NH_3)_6]^{3+}(aq) \qquad K_f = 4.5 \times 10^{23} \qquad (24.31)$$

In fact the concentration of free Co^{3+} is so low that for the half-reaction

$$[Co(NH_3)_6]^{3+} + e^- \longrightarrow [Co(NH_3)_6]^{2+} \tag{24.32}$$

E_{red}° is only $+0.10$ V. As a consequence, not only is $[Co(NH_3)_6]^{3+}$ stable, but $[Co(NH_3)_6]^{2+}$ is rather easily oxidized to the Co(III) complex.

The formation of stable complexes affords a means of attaining certain oxidation states that might otherwise be difficult or impossible.

The Photographic Process. A photographic film is basically an emulsion of silver bromide in gelatin. When the film is exposed to light, silver bromide granules become activated according to the intensity of the light striking them. When the exposed film is placed in a developer solution [a mild reducing agent such as hydroquinone, $C_6H_4(OH)_2$], the activated granules of silver bromide are reduced to black metallic silver. The unactivated granules in the unexposed portions of the film are practically unaffected. This developing process produces the photographic image.

The photographic process cannot be terminated at this point, however. The unactivated granules of silver bromide would eventually be reduced to black metallic silver upon exposing the film to light again. The image on the film must be "fixed." This requires that the black metallic silver that results from the developing be left on the film and the remaining silver bromide be removed. The "fixer" commonly employed is sodium thiosulfate (also known as sodium hyposulfite or "hypo"). In the fixing process AgBr(s) is dissolved and the complexed silver ion is washed away.

$$AgBr(s) + 2 S_2O_3^{2-} \longrightarrow [Ag(S_2O_3)_2]^{3-} + Br^- \tag{24.33}$$

Qualitative Analysis. In the separation and detection of cations in the qualitative analysis scheme, Ag^+, Pb^{2+}, and Hg_2^{2+} are first precipitated as chlorides (recall Figure 19-5). All other common cations form soluble chlorides. $PbCl_2(s)$ is removed from AgCl(s) and $Hg_2Cl_2(s)$ by its greater solubility in hot water. AgCl(s) is separated from $Hg_2Cl_2(s)$ by its solubility in $NH_3(aq)$, described by equation (19.25).

At another point in the qualitative analysis scheme it is desired to precipitate CdS in the presence of Cu^{2+}. Normally, Cu^{2+} would precipitate along with the Cd^{2+} since K_{sp} for CuS is smaller than for CdS, 6.3×10^{-36} compared to 8.0×10^{-27}. However, by treating the solution of Cu^{2+} and Cd^{2+} with an excess of CN^- prior to saturation with H_2S, this separation can be achieved. The following reactions occur.

$$Cd^{2+} + 4 CN^- \rightleftharpoons [Cd(CN)_4]^{2-} \qquad K_f = 7.1 \times 10^{18} \tag{24.34}$$

$$2 Cu^{2+} + 10 CN^- \rightleftharpoons 2 [Cu(CN)_4]^{3-} + C_2N_2(g) \tag{24.35}$$

Reaction (24.35) is an oxidation–reduction reaction in which Cu^{2+} is reduced to Cu^+ and complexed with CN^- as well. The complex ion $[Cu(CN)_4]^{3-}$ is very stable; its K_f value is 1×10^{28}. The concentration of *free* Cu^+ in equilibrium with the complex ion is very low. When a solution containing this complex ion is later saturated with H_2S, K_{sp} for Cu_2S is not exceeded. By contrast, $[Cd^{2+}]$ in equilibrium with $[Cd(CN)_4]^{2-}$ is sufficiently great that K_{sp} of CdS is exceeded under these same conditions.

Electroplating. Electrolyte solutions used in commercial electroplating are quite complex. Each component plays a role in achieving the final objective of a bright smooth fine-grained metal deposit. A number of metals, e.g., Cu, Ag, and Au, are generally plated from solutions of their cyano complex ions. In the electrolysis reaction below the object to be plated is made the cathode and a piece of copper metal is the anode.

anode: $Cu + 4 CN^- \longrightarrow [Cu(CN)_4]^{3-} + e^-$ $\qquad$ (24.36)

cathode: $[Cu(CN)_4]^{3-} + e^- \longrightarrow Cu + 4 CN^-$ $\qquad$ (24.37)

The net change simply involves the transfer of Cu metal from the anode to the cathode through the formation, migration, and decomposition of the complex ion $[Cu(CN)_4]^{3-}$. An additional advantage of electroplating Cu from a solution of $[Cu(CN)_4]^{3-}$ is that 1 mol of Cu is obtained per Faraday, in contrast to $\frac{1}{2}$ mol per Faraday that would be obtained from a solution of Cu^{2+}.

The need to complex Cu^{2+} in the Raschig process for the production of hydrazine, N_2H_4, was mentioned in Chapter 4.

Sequestering Metal Ions.   Metal ions may act as catalysts in promoting undesirable chemical reactions in a manufacturing process, or they may alter in some way the properties of the material being manufactured. Thus, for many industrial purposes it is imperative to remove mineral impurities from water. Often these impurities are present only in trace amounts, e.g., Cu^{2+}. Precipitation of metal ions is feasible only if K_{sp} for the precipitate is very small.

One method of water treatment involves chelation. Among the chelating agents widely employed are the salts of **ethylenediaminetetraacetic acid (EDTA),** e.g., the sodium salt.

$$4\,Na^+ \begin{bmatrix} {}^-OOCCH_2 & & CH_2COO^- \\ & NCH_2CH_2N & \\ {}^-OOCCH_2 & & CH_2COO^- \end{bmatrix} \tag{24.38}$$

As an example, the formation constants of $[Ca(EDTA)]^{2-}$ and $[Mg(EDTA)]^{2-}$ are large enough ($K_f = 4 \times 10^{10}$ and 4×10^8, respectively) that the concentrations of $Ca^{2+}(aq)$ and $Mg^{2+}(aq)$ can be reduced to the point that these ions will not precipitate in the presence of any common reagents (including soaps).

Chelation with EDTA can also be used in the treatment of metal poisoning. If a person suffering from lead poisoning is fed $[Ca(EDTA)]^{2-}$, the following exchange occurs because $[Pb(EDTA)]^{2-}$ ($K_f = 1 \times 10^{18}$) is even more stable than $[Ca(EDTA)]^{2-}$ ($K_f = 4 \times 10^{10}$).

$$Pb^{2+} + [Ca(EDTA)]^{2-} \longrightarrow [Pb(EDTA)]^{2-} + Ca^{2+} \tag{24.39}$$

The lead complex is excreted by the body and the Ca^{2+} remains as a body nutrient. The structure of $[Pb(EDTA)]^{2-}$ is shown in Figure 24-16. The high degree of stability of this and other EDTA complexes can be attributed to the presence of five, five-membered chelate rings in the complex.

FIGURE 24-16
The structure of $[Pb(EDTA)]^{2-}$.

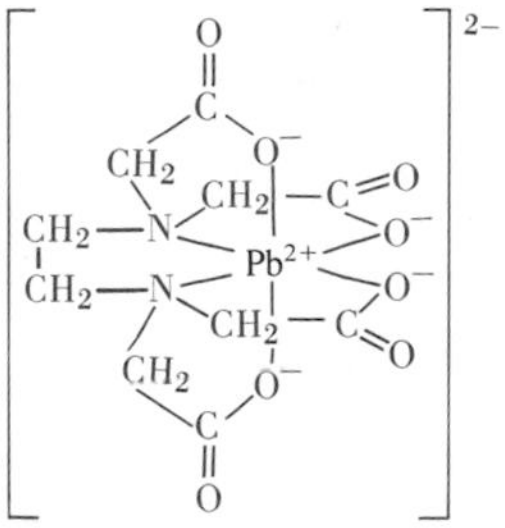

The structures of other metal–EDTA complexes have a metal ion M^{n+} in place of the Pb^{2+}.

Biological Applications: Porphyrins.   The structure illustrated in Figure 24-17 is commonly found in both plant and animal matter. If the groups substituted at the eight bonds shown in black are all H atoms, the molecule is called prophin. By substituting other groups at these eight positions, the structures obtained are called porphyrins. Metal ions can replace the two H atoms on the central N atoms and coordinate simultaneously with all four N atoms. The porphyrin is a tetradentate ligand or chelating agent for the central metal ion. In chlorophyll the central ion is Mg^{2+}, and in hemoglobin, Fe^{2+}. (Structures of these complexes are pictured in Figures 27-9 and 27-16, respectively.)

FIGURE 24-17
The porphyrin structure.

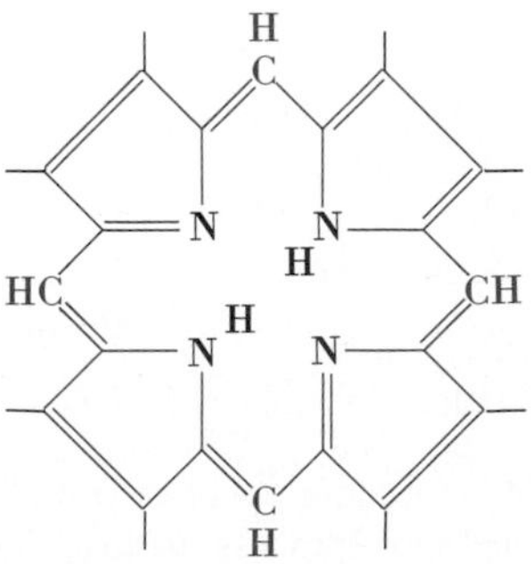

Summary

Many metal ions, particularly those of the transition elements, have the ability to form bonds with electron-donor groups (ligands). The number of ligands coordinated and the geometrical distribution of these ligands about the metal ion are distinctive features of a complex ion. Some ligands have more than one electron pair for donation. These multidentate ligands are able to attach simultaneously to two or more positions in the coordination sphere of the central metal ion. This multiple attachment produces complexes with five- or six-membered rings of atoms—chelates.

To name complexes requires application of a set of rules. These rules deal with such matters as denoting the number and kind of ligands, the oxidation state of the central metal ion, and whether the complex is neutral, a cation, or an anion. The positions in the coordination sphere of the central ion at which attachment of ligands may occur are not always equivalent. In geometric isomerism different structures with different properties result depending on where this attachment occurs. Optical isomers differ only in certain properties that depend on chirality, e.g., the rotation of plane polarized light. Other forms of isomerism depend on such factors as the atom of the ligand through which linkage occurs, and whether groups in a coordination compound are present within or outside the coordination sphere of the metal ion.

The structures of complex ions can be rationalized by appropriate orbital hybridization schemes. More successful in explaining the magnetic properties and characteristic colors of complex ions is the crystal field theory. This theory emphasizes the splitting of the d energy level as a result of repulsions between electrons of the central ion and of the ligands. Different splitting patterns are obtained for different geometric structures. A prediction of the magnitude of d-level splitting produced by a ligand can be made through the spectrochemical series.

The formation of a complex ion can be viewed as a stepwise equilibrium process in which other ligands displace H_2O molecules from aqua complex ions. There is an equilibrium constant for each step, and these stepwise constants can be combined into an overall formation constant for the complex ion, K_f. The ability of ligand water molecules to ionize results in acidic properties for some aqua complexes and helps to explain amphoterism. Also important is the rate at which a complex ion exchanges ligands between its coordination sphere and the solution. Exchange is rapid in a labile complex and slow in an inert complex.

The formation of complex ions can be used to stabilize certain oxidation states, such as Co(III). Other applications include dissolving precipitates, such as AgCl by $NH_3(aq)$ in the qualitative analysis scheme and AgBr by $Na_2S_2O_3(aq)$ in the photographic process, and sequestering ions by chelation, as with EDTA.

Learning Objectives

As a result of studying Chapter 24, you should be able to

1. Identify the central ion and ligands, determine the coordination number and oxidation state of the central ion, and establish the net charge on a complex ion.

2. Give the coordination numbers of some common metal ions and write the names and formulas of some common unidentate and multidentate ligands.

3. Write distinctive names based on formulas of complexes, and distinctive formulas based on names.

4. Draw plausible structures for complex ions from information conveyed by their names and formulas.

5. Describe the types of isomerism found among complex ions and identify the possible isomers in specific cases.

6. Use the valence bond method to describe bonding and structures of complex ions.

7. Explain the basis of the crystal field theory of bonding in complex ions.

8. Use the spectrochemical series to make predictions about d-level splitting and the number of unpaired electrons in complex ions.

9. Explain the origin of color in aqueous solutions of complex ions.

10. Write equations to show the formation of complex ions by the stepwise displacement of H_2O by other ligands, and relate the overall formation constant, K_f, to stepwise equilibrium constants.

11. Describe how an aqua complex ion ionizes as an acid, and explain amphoterism from this standpoint.

12. Explain how complex ion formation can be used to stabilize oxidation states.

13. Cite ways in which complex ion equilibria are used in the qualitative analysis scheme.

14. Describe applications of complex ion formation in the photographic process, in electroplating, and in water treatment.

Some New Terms

A **chelate** results from the attachment of multidentate ligands to a metal ion. Chelates are five- or six-membered rings that include the central metal ion and atoms of the ligands.

A **chelating agent** is a **multidentate ligand.** It simultaneously attaches to two or more positions in the coordination sphere of the central metal ion.

Chiral refers to the nonsuperimposability of enantiomers and situations affected by this property.

Coordination isomerism arises in certain coordination compounds having both a complex cation and a complex anion. An interchange of ligands between the two complex ions leaves the composition of the compound unchanged.

Coordination number is the number of positions available for the attachment of ligands to a central metal ion.

The **coordination sphere** is the region around a metal ion where linkage to ligands can occur to produce a complex ion.

Crystal field theory describes bonding in complex ions in terms of electrostatic attractions between ligands and the nucleus of the central metal ion. Particular attention is focused on the splitting of the d energy level of the central metal ion that results from electron repulsions.

Dextrorotatory means the ability to rotate the plane of polarized light to the right, designated d or $+$.

Enantiomers (optical isomers) are molecules whose structures are not superimposable. The structures are mirror images of one another and the molecules are optically active.

The **(overall) formation constant, K_f,** of a complex ion is obtained by combining equilibrium constants for the stepwise displacement by other ligands of H_2O molecules from the coordination sphere of a metal ion.

Geometric isomerism refers to the formation of nonequivalent structures based on the positions at which ligands are attached to a central metal ion in a complex ion.

Inert is the term used to describe a complex ion in which the exchange of ligands occurs very slowly.

Ionization isomerism arises when a ligand from the coordination sphere of a metal ion is exchanged for an ion outside the coordination sphere.

Labile is the term used to describe a complex ion in which rapid exchange of ligands occurs.

Levorotatory means the ability to rotate the plane of polarized light to the left, designated l or $-$.

Linkage isomerism applies to complex ions with the same compositions but with one or more ligands bonded differently.

Optical isomerism refers to the existence of enantiomers (which differ in their abilities to rotate the plane of polarized light).

A **racemic mixture** is a mixture containing equal amounts of the d and l isomers of an optically active substance.

Resolution refers to the separation of the optically active isomers from a racemic mixture.

The **spectrochemical series** is a ranking of ligand abilities to produce a splitting of the d energy level of a central metal ion in a complex ion.

A **unidentate ligand** is one that has but a single pair of electrons available for donation to a central metal ion. It becomes attached at only a single position in the coordination sphere.

Suggestions for Further Study

KAUFFMAN, G. B., and J. F. BAXTER, Jr., "Hydrated Cations in the General Chemistry Course," *J. Chem. Educ.*, **58,** 349 (1981).

KELLER, E. L., "Photography, Part I: Images on Silver," *Chemistry,* **43**[9], 6 (1970).

LOEFFLER, B. M., and R. G. BURNS, "Shedding Light on the Color of Gems and Minerals," *American Scientist,* **64,** 636 (1976).

MICKEY, C. D., "Optical Activity," *J. Chem. Educ.*, **57,** 442 (1980).

MICKEY, C. D., "Some Aspects of Coordination Chemistry," *J. Chem. Educ.*, **58,** 257 (1981).

SOLOV'EV, Y. I., "D. I. Mendeleev's Conceptions Concerning the Structure of Complex Compounds." *J. Chem. Educ.*, **55,** 494 (1978).

ZIPP, A. D., and S. G. ZIPP, "Pt(NH$_3$)$_2$Cl$_2$ and Cancer," *J. Chem. Educ.*, **54,** 739 (1977).

Review Problems

1. What is the coordination number and the oxidation state of the metal ion in each of the following complexes?
- **(a)** $[Ni(NH_3)_6]^{2+}$
- **(b)** $[AlF_6]^{3-}$
- **(c)** $[Cu(CN)_4]^{2-}$
- **(d)** $[Cr(NH_3)_3Br_3]$
- **(e)** $[Fe(ox)_3]^{3-}$
- **(f)** $[Cr(EDTA)]^-$

2. Name the following complex ions. (Do not attempt to distinguish among isomers in this problem.)
- **(a)** $[Ag(NH_3)_2]^+$
- **(b)** $[Fe(H_2O)_5OH]^{2+}$
- **(c)** $[ZnCl_4]^{2-}$
- **(d)** $[Pt(en)_2]^{2+}$
- **(e)** $[Co(NH_3)_4(NO_2)Cl]^+$

3. Name the following coordination compounds.
- **(a)** $[Co(NH_3)_5Br]SO_4$
- **(b)** $[Co(NH_3)_5SO_4]Br$
- **(c)** $[Cr(NH_3)_6][Co(CN)_6]$
- **(d)** $Na_3[Co(NO_2)_6]$
- **(e)** $[Co(en)_3]Cl_3$

4. Write appropriate formulas for the following species. (Do not attempt to distinguish among isomers in this problem.)
- **(a)** dicyanosilver(I) ion
- **(b)** diamminetetrachloronickelate(II) ion
- **(c)** tris(ethylenediamine)copper(II) sulfate
- **(d)** sodium diaquatetrahydroxoaluminate(III)

5. Write formulas for the following hydrates.
   (a) iron(III) chloride hexahydrate
   (b) cobalt(II) hexachloroplatinate(IV) hexahydrate

6. Draw a structure to represent the complex ion *trans*-$[Cr(NH_3)_4ClOH]^+$.

7. Draw structures to represent these three complex ions:
   (a) $[PtCl_4]^{2-}$ (b) $[Fe(en)Cl_4]^-$
   (c) *cis*-$[Fe(en)(ox)Cl_2]^-$

8. How many different structures are possible for each of the following complex ions?

(a) $[Co(NH_3)_5H_2O]^{3+}$ (b) $[Co(NH_3)_4(H_2O)_2]^{3+}$
(c) $[Co(NH_3)_3(H_2O)_3]^{3+}$ (d) $[Co(NH_3)_2(H_2O)_4]^{3+}$

9. Indicate what type of isomerism may be found in each of the following cases. If no isomerism is possible, so indicate.
   (a) $[Zn(NH_3)_4][CuCl_4]$ (b) $[Fe(CN)_5SCN)]^{4-}$
   (c) $[Ni(NH_3)_5Cl]^+$ (d) $[Pt(py)Cl_3]^-$
   (e) $[Cr(NH_3)_3(OH)_3]^-$

10. (a) Draw an orbital diagram to represent bonding in the complex ion $[AuBr_4]^-$.
    (b) Would you expect this complex ion to be diamagnetic or paramagnetic? Explain.

Exercises

Definitions and terminology

1. The following terms relate to the structure of a complex ion. What is the meaning of each? (a) octahedral geometry; (b) dsp^2 hybrid bonding; (c) aqua complex.

2. Describe the difference in meaning of the terms in each of the following pairs. (a) coordination number and oxidation number; (b) unidentate and multidentate ligand; (c) structural isomerism and stereoisomerism; (d) *cis* and *trans* isomer; (e) *d* and *l* isomer; (f) nitro and nitrito complex ion; (g) low and high spin complex.

3. What characteristics distinguish a chelate complex from an ordinary complex ion (such as an aqua complex ion)?

Nomenclature

4. Supply acceptable names for the following. (Do not attempt to distinguish among isomers in this exercise.)
   (a) $[Co(NH_3)_4(H_2O)(OH)]^{2+}$ (b) $[Co(NH_3)_3(-NO_2)_3]$
   (c) $[Pt(NH_3)_4][PtCl_6]$ (d) $[Fe(ox)_2(H_2O)_2]^-$
   (e) $[Fe(py)(CN)_5]^{3-}$ (f) $Ag_2[HgI_4]$

5. Write appropriate formulas for the following. (Do not attempt to distinguish among isomers in this exercise.)
   (a) potassium hexacyanoferrate(II)
   (b) bis(ethylenediamine)copper(II) ion
   (c) tetraaquadihydroxoaluminum(III) chloride
   (d) amminechlorobis(ethylenediamine)chromium(III) sulfate
   (e) tris(ethylenediamine)iron(II) hexacyanoferrate(III)

Bonding and structure in complex ions

6. What type of geometric structure would you predict for $[Au(CN)_2]^-$? Explain.

7. The complex ion $[FeCl_4]^-$ has a tetrahedral structure. Use the valence bond method to propose a bonding scheme for this complex ion. How many unpaired electrons are present in the structure?

8. In what way would you expect the bonding scheme for the high-spin complex $[Co(H_2O)_6]^{2+}$ to differ from that given in the text for $[Co(NH_3)_6]^{3+}$?

9. Draw a plausible structure to represent
   (a) $[PtCl_4]^{2-}$ (b) *cis*-$[Zn(NH_3)(OH)_3]^-$
   (c) $[Cr(H_2O)_5Cl]^{2+}$

10. Draw plausible structures of the following chelate complexes.
    (a) $[Pt(ox)_2]^{2-}$ (b) $[Cr(ox)_3]^{3-}$
    (c) $[Fe(EDTA)]^{2-}$

Isomerism

11. Would you expect *cis-trans* isomerism to occur in a complex ion with a (a) tetrahedral; (b) square planar; (c) linear structure? Explain.

12. Draw structures for each of the isomers indicated in expressions (24.6), (24.7), and (24.9).

13. Which of these octahedral complexes would you expect to exhibit *geometric* isomerism? Explain.
    (a) $[Cr(NH_3)_5OH]^{2+}$ (b) $[Cr(NH_3)_3(H_2O)(OH)_2]^+$
    (c) $[Cr(en)_2Cl_2]^+$ (d) $[Cr(en)Cl_4]^-$
    (e) $[Cr(en)_3]^{3+}$

14. Draw a structure for *cis*-dichlorobis(ethylenediamine)cobalt(III) ion. Is this ion optically active? Is the *trans* isomer optically active? Explain.

15. If A, B, C, and D are four different ligands,
    (a) how many geometric isomers will be found for $[PtABCD]^{2+}$?
    (b) will tetrahedral $[ZnABCD]^{2+}$ display optical isomerism?

Crystal field theory

16. Describe how the crystal field theory makes possible an explanation of the fact that so many transition metal compounds are colored.

17. One of the following ions is paramagnetic and one is diamagnetic; which is which? **(a)** $[MoCl_6]^{3-}$; **(b)** $[Co(en)_3]^{3+}$

18. If the ion Cr^{2+} is linked with strong-field ligands to produce an octahedral complex, the complex has *two* unpaired electrons. If Cr^{2+} is linked with weak-field ligands, the complex has *four* unpaired electrons. How do you account for this difference? What hybridization schemes should be used to describe bonding by the valence bond method for the two types of complexes?

19. In contrast to the case of Cr^{2+} considered in Exercise 18, no matter what ligand is linked to Cr^{3+} to form an octahedral complex, the complex always has *three* unpaired electrons. Explain this fact.

20. Cyano complexes of transition metal ions (e.g., Fe^{2+} and Cu^{2+}) are often yellow in color, whereas aqua complexes are often green or blue. Why is there this difference?

21. Predict
  (a) the number of unpaired electrons expected for the tetrahedral complex ion $[CoCl_4]^{2-}$;
  (b) whether the square planar complex ion $[Cu(py)_4]^{2+}$ is diamagnetic or paramagnetic;
  (c) whether octahedral $[Mn(CN)_6]^{3-}$ or tetrahedral $[FeCl_4]^-$ has a greater number of unpaired electrons.

22. In Example 24-5 we chose between a tetrahedral and a square planar structure for $[Ni(CN)_4]^{2-}$ based on magnetic properties. Could we similarly use magnetic properties to establish whether the ammine complex of Ni(II) is $[Ni(NH_3)_6]^{2+}$ or tetrahedral $[Ni(NH_3)_4]^{2+}$? Explain.

Complex ion equilibria

23. Write equations to represent the following observations.
  (a) A mixture of $Mg(OH)_2(s)$ and $Zn(OH)_2(s)$ is treated with $NH_3(aq)$. The $Zn(OH)_2$ dissolves but the $Mg(OH)_2(s)$ is left behind.
  (b) When $NaOH(aq)$ is added to $CuSO_4(aq)$, a pale blue precipitate forms. If $NH_3(aq)$ is added, the precipitate redissolves, producing a solution with an intense deep-blue color. If this solution is made acidic with $HNO_3(aq)$, the color is converted back to pale blue.
  (c) A quantity of $CuCl_2(s)$ is dissolved in concentrated $HCl(aq)$ and produces a yellow solution. The solution is diluted to twice its volume with water and assumes a green color. Upon dilution to 10 times its original volume, the solution becomes pale blue in color.

24. Write a series of equations to show the stepwise displacement of H_2O ligands in $[Fe(H_2O)_6]^{3+}$ by ethylenediamine, for which $\log K_1 = 4.34$; $\log K_2 = 3.31$; and $\log K_3 = 2.05$. What is the overall formation constant, $\beta_3 = K_f$, for $[Fe(en)_3]^{3+}$?

25. Without performing detailed calculations, show why you would expect the concentrations of the various ammine-aqua

complex ions to be negligible compared to that of $[Cu(NH_3)_4]^{2+}$ in a solution having a total Cu(II) concentration of $0.10\ M$ and a total concentration of NH_3 of $1.0\ M$. Under what conditions would the concentrations of these ammine-aqua complex ions (e.g., $[Cu(H_2O)_3NH_3]^{2+}$) become more significant relative to the concentration of $[Cu(NH_3)_4]^{2+}$? Explain.

26. Without performing detailed calculations, verify the statement on page 761 that neither Ca^{2+} nor Mg^{2+} found in natural waters is likely to precipitate from the water upon the addition of other reagents, if the ions are complexed with EDTA. (*Hint:* Use data from this chapter and Chapter 19. Assume some reasonable values for the total metal ion concentration and that of *free* EDTA, e.g., $0.10\ M$ each.)

Acid–base properties

27. Write simple chemical equations to show how the complex ion $[Cr(H_2O)_5OH]^{2+}$ acts as **(a)** an acid; **(b)** a base.

28. For a solution that is made up to be $0.100\ M$ in $[Fe(H_2O)_6]^{3+}$,
  (a) Calculate the pH of the solution assuming that ionization of the aqua complex ion proceeds only through the first step (24.25).
  (b) Calculate $[Fe(H_2O)_5OH]^{2+}$ if the solution is also $0.100\ M\ HClO_4$. (ClO_4^- does not complex with Fe^{3+}.)
  (c) Can the pH of the solution be mantained so that $[[Fe(H_2O)_5OH]^{2+}]$ does not exceed $1 \times 10^{-6}\ M$? Explain.

Applications

29. From data in Chapter 19
  (a) Derive an equilibrium constant for reaction (24.33) and explain why this reaction (the fixing of photographic film) is expected to go essentially to completion.
  (b) Explain why $NH_3(aq)$ cannot be used in the fixing of photographic film.

30. Show that the oxidation of $[Co(NH_3)_6]^{2+}$ to $[Co(NH_3)_6]^{3+}$ referred to on page 760 should occur spontaneously in alkaline solution with H_2O_2 as an oxidizing agent. (*Hint:* Refer also to Figure 21-7.)

31. A current of 2.13 A is passed for 0.347 h between a pair of Cu electrodes. What mass of Cu is deposited at the cathode if the electrolyte is **(a)** $[Cu(H_2O)_4]^{2+}$; **(b)** $[Cu(CN)_4]^{3-}$? Why are these quantities different?

32. Use data provided in Section 24-13 to demonstrate that if a solution is $0.10\ M$ in total copper, $0.10\ M$ in total cadmium, $0.10\ M$ in *free* CN^-, and saturated with H_2S, $CdS(s)$ will precipitate but not $Cu_2S(s)$. [Assume a pH of 7. $K_{sp}(CdS) = 8.0 \times 10^{-27}$; $K_{sp}(Cu_2S) = 1.2 \times 10^{-49}$.]

Additional Exercises

1. The following compounds were discovered before Werner's time. Rewrite each *empirical* formula to be consistent with Werner's formulation for coordination compounds. Name each compound.

   (a) Zeise's salt: $PtCl_2 \cdot KCl \cdot C_2H_4$ (This compound consists of a simple unipositive cation and a uninegative complex anion.)

   (b) Magnus' green salt: $PtCl_2 \cdot 2\ NH_3$ (This compound consists of a dipositive complex cation and a dinegative complex anion.)

2. Figure 24-9 presents a structure that was at one time thought to be an alternative to the octahedral structure. Another was that shown below.

   (a) Show that for the complex ion $[Co(NH_3)_4Cl_2]^+$ the above structure predicts three geometrical isomers.

   (b) Show that neither the above structure nor Figure 24-9 would account for optical isomerism in $[Co(en)_3]^{3+}$.

3. Sketch all the possible isomers of $[Co(ox)(NH_3)_2Cl_2]^-$.

4. The *cis* and *trans* isomers of $[Co(en)_2Cl_2]^+$ can be distinguished through a displacement reaction with oxalate ion. What difference in reactivity toward oxalate ion would you expect between the *cis* and *trans* isomers? (*Hint:* Which ligands would you expect oxalate ion to displace more readily?)

5. Propose a bonding scheme that is consistent with the square planar structure of $[Cu(NH_3)_4]^{2+}$. Is this complex ion diamagnetic or paramagnetic?

6. A tabulation of formation constant data lists the following log K values for the formation of $[CuCl_4]^{2-}$: log K_1 = 2.80; log K_2 = 1.60; log K_3 = 0.49; log K_4 = 0.73.

   (a) What is the overall formation constant $\beta_4 = K_f$ for $[CuCl_4]^{2-}$?

   (b) In connection with reaction (24.28), estimate the concentration of HCl(aq) required in a solution that is 0.10 M $CuSO_4$ to produce a visible yellow color? Assume that 99% conversion of $[Cu(H_2O)_4]^{2+}$ to $[CuCl_4]^{2-}$ is sufficient for this to happen, and neglect the presence of any mixed aqua-chloro complex ions.

★7. (a) Draw a diagram similar to (24.14) to represent the distribution of d electrons in the octahedral complex ion $[Ti(H_2O)_6]^{3+}$.

   (b) The ion $[Ti(H_2O)_6]^{3+}$ absorbs light of wavelength 490 nm. What is the energy separation, Δ, expressed in kJ/mol?

★8. With reference to the stability of $[Co(NH_3)_6]^{3+}$(aq),

   (a) Verify that E°_{cell} for reaction (24.30) is +0.59 V.

   (b) Calculate $[Co^{3+}]$ in a solution that has a total concentration of cobalt of 1.0 M and $[NH_3]$ = 0.10 M.

   (c) Show that for the value of $[Co^{3+}]$ calculated in part (b), reaction (24.30) will not occur. (*Hint:* What is $[H_3O^+]$ in 0.10 M NH_3? Assume a low, but reasonable, concentration of Co^{2+}, say 1×10^{-4} M, and a partial pressure of O_2(g) of 0.2 atm.)

★9. A Cu electrode is immersed in a solution that is 1.00 M NH_3 and 1.00 M in $[Cu(NH_3)_4]^{2+}$. If a standard hydrogen electrode is the anode, E_{cell} is found to be -0.08 V. What is the value obtained by this method for the formation constant, K_f, of $[Cu(NH_3)_4]^{2+}$?

★10. The following concentration cell is constructed.

$$Ag|Ag^+(0.10\ M\ [Ag(CN)_2]^-, 0.10\ M\ KCN)\|Ag^+(0.10\ M)|Ag$$

If K_f for $[Ag(CN)_2]^-$ is 5.6×10^{18}, what value would you expect for E_{cell}. (*Hint:* Recall that the electrode on the left is the anode.)

Self-Test Questions

For questions 1 through 8 select the single item that best completes each statement.

1. The oxidation state of Ni in the complex ion $[Ni(CN)_4I]^{3-}$ is (a) -3; (b) -2; (c) $+2$; (d) $+5$.

2. The coordination number of Pt in the complex ion $[Pt(en)_2Cl_2]^{2+}$ is (a) 3; (b) 4; (c) 5; (d) 6.

3. Of the following complex ions one exhibits isomerism. That one is (a) $[Ag(NH_3)_2]^+$; (b) $[Co(NH_3)_5NO_2]^{2+}$; (c) $[Pt(en)Cl_2]$; (d) $[Co(NH_3)_5Cl]^{2+}$.

4. Of the following, the one that exhibits optical isomerism is (a) *cis*-$[Co(en)_2Cl_2]^+$; (b) $[Co(NH_3)_4Cl_2]^+$; (c) $[Co(NH_3)_2Cl_4]^-$; (d) *trans*-$[Co(en)_2Cl_2]^+$.

5. The number of unpaired electrons expected for the complex ion $[Cr(NH_3)_6]^{2+}$ is (a) 2; (b) 3; (c) 4; (d) 5.

6. Of the following complex ions, one is a Brønsted–Lowry acid. That one is (a) $[Cu(NH_3)_4]^{2+}$; (b) $[FeCl_4]^-$; (c) $[Fe(H_2O)_6]^{3+}$; (d) $[Zn(OH)_4]^{2-}$.

7. Of the following complex ions, the one that probably has the largest overall formation constant, K_f, is (a) $[Co(NH_3)_6]^{3+}$; (b) $[Co(H_2O)_6]^{3+}$; (c) $[Co(H_2O)_4(NH_3)_2]^{3+}$; (d) $[Co(en)_3]^{3+}$.

8. The most soluble of the following compounds in $NH_3(aq)$ is (a) $BaSO_4$; (b) $Cu(OH)_2$; (c) SiO_2; (d) $MgCO_3$.

9. Name the following complex ions and coordination compounds.

   (a) $[Cr(NH_3)_4(OH)_2]Br$ **(b)** $K_3[Co(-NO_2)_6]$
   (c) $[Fe(en)_2(H_2O)_2]^{2+}$ **(d)** $[Pt(en)_2Cl_2]SO_4$

10. Sketch a plausible geometric structure for
   (a) $[Co(NH_3)_5Cl]^{2+}$
   (b) $[Cr(en)_2(ox)]^+$
   (c) *cis*-diammincdinitroplatinum(II)
   (d) *trans*-triamminetrichlorocobalt(III)

11. Explain the following observations in terms of complex ion formation.
   (a) $Al(OH)_3(s)$ is soluble in $NaOH(aq)$ but insoluble in $NH_3(aq)$.
   (b) $ZnCO_3(s)$ is soluble in $NH_3(aq)$ but $ZnS(s)$ is not.
   (c) $CoCl_3$ is unstable in water solution, being reduced to $CoCl_2$ and liberating $O_2(g)$. On the other hand, $[Co(NH_3)_6]Cl_3$ can be easily maintaincd in aqueous solution.

12. In both $[Fe(H_2O)_6]^{2+}$ and $[Fe(CN)_6]^{4-}$ the iron is present as Fe^{2+}; yet $[Fe(H_2O)_6]^{2+}$ is paramagnetic, whereas $[Fe(CN)_6]^{4-}$ is diamagnetic. Explain this difference.

25 Nuclear Chemistry

The ordinary chemistry of the elements is based on phenomena in which electrons play a major role. Both chemical properties and those physical properties related to intermolecular forces are derived from the electronic structures of atoms, ions, and molecules. The primary functions of atomic nuclei in the phenomena we have considered to this point are in establishing the masses of atoms and molecules and in furnishing a center of positive charge to keep electrons in position.

In this chapter we consider a variety of phenomena stemming *directly* from the nuclei of atoms. We will refer to these phenomena, collectively, as nuclear chemistry. Nuclear chemical phenomena are of primary importance in dealing with the actinoid (actinide) elements. However, these phenomena are also encountered with certain nuclides of the lighter elements, in particular, with artificial nuclides. Yet another part of nuclear chemistry is a study of the effects of ionizing radiation on matter—one of the central issues in the current "nuclear debate."

25-1 The Phenomenon of Radioactivity

The term "radioactivity" was proposed by Marie Curie to describe the most readily observable phenomenon that accompanies transformations of certain atomic nuclei—the emission of ionizing radiation. Ionizing radiation, as the name implies, interacts with matter to produce ions. This means that the radiation is sufficiently energetic to break chemical bonds. Some ionizing radiation is particulate (consists of particles) and some is nonparticulate. We will return to a discussion of the effects of ionizing radiation on matter in Section 25-10. For the present, let us comment briefly on the types of ionizing radiation associated with radioactivity, depicted in Figure 25-1. This will take the form of a review and extension of ideas first encountered in Chapter 2.

FIGURE 25-1
Three types of radiation from radioactive materials.

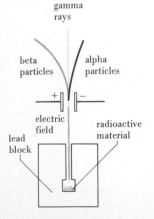

The radioactive material is enclosed in a lead block with a narrow opening. All the radiation except for that passing through the opening is absorbed by the lead. When this radiation is passed through an electric field, it splits into three beams. One beam is undeflected—these are gamma (γ) rays. A second beam is attracted toward the negatively charged plate; these are the positively charged alpha (α) particles. The third beam is deflected toward the positive plate. This is the beam of beta (β) particles—electrons. Because of their greater momentum (product of mass and velocity), α particles are deflected to a smaller extent than β particles. A similar situation exists if a magnetic field is substituted for an electric field.

FIGURE 25-2

Production of gamma rays.

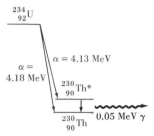

The transition of a $^{230}_{90}$Th nucleus between the two energy states shown results in the emission of 0.05 MeV of energy in the form of γ rays.

Alpha (α) Rays. This radiation consists of a stream of alpha (α) particles. These particles are identical to the nuclei of helium-4 atoms. Alpha rays produce large numbers of ions as they penetrate matter, even though their penetrating power is low. (They can generally be stopped by a sheet of paper.) Because of their positive charge, α particles are deflected by magnetic and electric fields (see Figure 25-1). The production of α particles by a radioactive nucleus can be represented through a nuclear equation. A **nuclear equation** is written in a fashion similar to a chemical equation. However, the condition of balance in a nuclear equation requires the sum of atomic numbers and the sum of mass numbers on the left side of the equation to equal the corresponding sums on the right. In equation (25.1), mass numbers are written as superscript numerals; atomic numbers are subscript numerals; and the α particle is represented as $^4_2$He. Mass numbers total 238, and atomic numbers total 92. The loss of an α particle by a nucleus causes a decrease of two in its atomic number and four in its mass number.

$$^{238}_{92}\text{U} \longrightarrow {}^{234}_{90}\text{Th} + {}^4_2\text{He} \tag{25.1}$$

Beta (β^-) Rays. These rays are also comprised of particles, and β^- particles are identical to electrons. Beta rays have a greater penetrating power but a lower ionizing power than α rays. (They can pass through aluminum foil up to about 2 to 3 mm thick.) Beta particles are also deflected by electric and magnetic fields but in the *opposite* direction from α particles (see Figure 25-1). Moreover, β^- particles suffer greater deflections in these fields than do α particles, because the β^- particles have so much smaller a mass than do α particles. The nucleus of $^{234}_{90}$Th, which was a product of reaction (25.1), is itself unstable. $^{234}_{90}$Th undergoes radioactive decay by β^- emission. When a nucleus loses a β^- particle, represented as $_{-1}^0$e, the atomic number of the nucleus *increases* by one unit and its mass number is unchanged. Thus, a β^- particle is treated as if it had an atomic number of -1 and a mass number of 0.

$$^{234}_{90}\text{Th} \longrightarrow {}^{234}_{91}\text{Pa} + {}^{\ 0}_{-1}\text{e} \tag{25.2}$$

Gamma (γ) Rays. Some radioactive decay processes yielding α or β particles leave a nucleus in an energetic state. The nucleus then loses energy in the form of electromagnetic radiation—a gamma (γ) ray. Gamma rays are a highly penetrating form of radiation. They are undeflected by electric and magnetic fields. In the radioactive decay of $^{234}_{92}$U, 77% of the nuclei emit α particles having an energy of 4.18 MeV. The remaining 23% of the $^{234}_{92}$U nuclei produce α particles with energies of 4.13 MeV. In the latter case the $^{230}_{90}$Th nuclei are left with an excess energy of 0.05 MeV. This energy is released as γ rays. If we denote the unstable, energetic Th nucleus as $^{230}_{90}$Th*, we can write

$$^{234}_{92}\text{U} \longrightarrow {}^{230}_{90}\text{Th}^* + {}^4_2\text{He} \tag{25.3}$$

$$^{230}_{90}\text{Th}^* \longrightarrow {}^{230}_{90}\text{Th} + \gamma \tag{25.4}$$

This γ emission process is represented diagrammatically in Figure 25-2.

The energy unit electron volt, eV, was introduced in Section 8-8. The unit MeV is a million electron volts. This unit and its relationship to other energy units are discussed further in Section 25-6.

Positrons (β^+). The emission of β^- rays, as we shall discover in Section 25-7, is characteristic of nuclei in which the ratio of number of neutrons to number of protons is too large for stability. If this ratio is too small for stability, radioactive decay may occur by positron emission. A positron is a *positively* charged particle that is otherwise identical to a β^- particle or electron. It is designated as β^+ or as $_{+1}^0$e. Positron emission is commonly encountered with artificially radioactive nuclei of the lighter elements (see Section 25-3). For example,

$$^{30}_{15}\text{P} \longrightarrow {}^{30}_{14}\text{Si} + {}^0_{+1}\text{e} \tag{25.5}$$

Electron Capture (E.C.). A second process that achieves the same effect as positron emission is electron capture (E.C.). In this process an electron from an inner electron shell (usually the K or L shell) is absorbed by the nucleus. (Inside the nucleus the electron is used to convert a proton into a neutron.) The dropping of an electron from a higher quantum level into that vacated by the captured electron results in the emission of x radiation. For example,

$$^{202}_{81}\text{Tl} \xrightarrow{\text{E.C.}} {}^{202}_{80}\text{Hg} \quad \text{(followed by x radiation)} \tag{25.6}$$

Example 25-1 Write nuclear equations to represent (a) α particle emission by ^{222}Rn; (b) radioactive decay of bismuth-215 to polonium; (c) decay of a radioactive nucleus to produce ^{58}Ni and a positron.

Solution

(a) Two of the species involved in this process can be written directly from the information given. The remaining species is identified by using the basic principle for balancing nuclear equations. (Note that as long as a name or chemical symbol is given, e.g., Rn, the atomic number of the element follows directly.)

$$^{222}_{86}\text{Rn} \longrightarrow (?) + {}^{4}_{2}\text{He} \quad \textit{and} \quad {}^{222}_{86}\text{Rn} \longrightarrow {}^{218}_{84}\text{Po} + {}^{4}_{2}\text{He}$$

(b) Bismuth has the atomic number 83, and polonium, 84. The type of emission that leads to an increase of one unit in atomic number is the β^- ray.

$$^{215}_{83}\text{Bi} \longrightarrow {}^{215}_{84}\text{Po} + {}^{0}_{-1}\text{e}$$

(c) We are given the products of the radioactive decay process. The only radioactive nucleus that can produce them is $^{58}_{29}\text{Cu}$.

$$^{58}_{29}\text{Cu} \longrightarrow {}^{58}_{28}\text{Ni} + {}^{0}_{+1}\text{e}$$

SIMILAR EXAMPLES: Review Problem 2; Exercise 5.

25-2 Naturally Occurring Radioactive Nuclides

$^{209}_{83}\text{Bi}$ is the nuclide of highest atomic and mass number that is stable. All known nuclides beyond it in atomic and mass numbers are radioactive. Naturally occurring $^{238}_{92}\text{U}$ is radioactive and disintegrates by the loss of α particles.

$$^{238}_{92}\text{U} \longrightarrow {}^{234}_{90}\text{Th} + {}^{4}_{2}\text{He} \tag{25.7}$$

$^{234}_{90}\text{Th}$ is also radioactive; its decay is by β^- emission.

$$^{234}_{90}\text{Th} \longrightarrow {}^{234}_{91}\text{Pa} + {}^{0}_{-1}\text{e} \tag{25.8}$$

$^{234}_{91}\text{Pa}$ also decays by β^- emission.

$$^{234}_{91}\text{Pa} \longrightarrow {}^{234}_{92}\text{U} + {}^{0}_{-1}\text{e} \tag{25.9}$$

$^{234}_{92}\text{U}$ is radioactive also.

Radioactive Decay Series. The chain of radioactive decay begun with $^{238}_{92}\text{U}$ continues through a number of steps of α and β emission until it eventually terminates with a stable isotope of lead—$^{206}_{82}\text{Pb}$. The entire scheme is outlined in Figure 25-3. All naturally occurring radioactive nuclides of high atomic number belong to one of three decay series—the **uranium series** just described, the **thorium series,** or the **actinium series.**

One application of these radioactive decay schemes is determining the ages of rocks and thereby the age of the earth (see Section 25-5). The appearance of certain radioactive substances in the environment can also be explained through radioactive decay series.

FIGURE 25-3
The natural radioactive decay series for $^{238}_{92}$U.

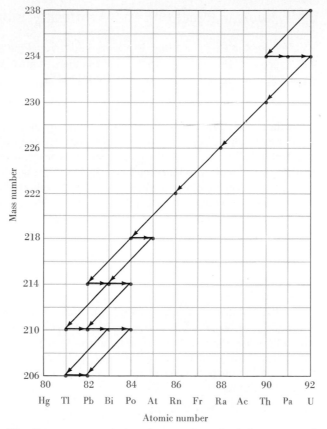

The long arrows pointing down and to the left correspond to α-particle emissions. The short horizontal arrows represent β^- emissions. Other natural decay series originate with isotopes of thorium, actinium, and neptunium (see Exercise 7 and Additional Exercise 5).

The first person to associate lung diseases in miners with the mining environment was the Swiss physician-chemist Paracelsus (1493–1541).

^{222}Rn produced by the decay of ^{238}U is now known to be the cause of a fatal lung disease afflicting miners in central Europe for centuries. This disease was lung cancer and the uranium responsible for it was present in the gold, silver, and platinum ores that were being mined. ^{210}Po and ^{210}Pb have been detected in cigarette smoke. These radioactive nuclides are derived from ^{238}U, found in trace amounts in the phosphate fertilizers used in tobacco fields. It has been established that these α-emitting nuclides are implicated in the link between cigarette smoking and cancer and heart disease.

Radioactivity, which is so common among the nuclides of high atomic number, is a relatively rare phenomenon among the *naturally occurring* lighter nuclides. ^{40}K is a radioactive nuclide, as are ^{50}V and ^{138}La. ^{40}K decays by β^- emission and by electron capture.

$$^{40}_{19}K \longrightarrow {}^{40}_{20}Ca + {}^{0}_{-1}e \qquad and \qquad {}^{40}_{19}K \xrightarrow{E.C.} {}^{40}_{18}Ar$$

At the time that the earth was formed ^{40}K was much more abundant than it is now. It is believed that the high argon content of the atmosphere (0.934%, by volume, and almost all of it as ^{40}Ar) has been derived from the radioactive decay of ^{40}K. Aside from ^{40}K, the most important radioactive nuclides of the lighter elements are those that are *artificially* produced.

25-3 Nuclear Reactions and Artificially Induced Radioactivity

We described in Chapter 2 how Rutherford discovered the existence of the proton outside the nucleus of an atom. He found that bombardment of nitrogen nuclei by α particles produced protons, the other product being O atoms.

$$^{14}_{7}N + ^{4}_{2}He \longrightarrow ^{17}_{8}O + ^{1}_{1}H \tag{25.10}$$

Although the principles involved in writing equation (25.10) are similar to those used to represent radioactive decay, there is this difference: Two "reactants" are written on the left. Instead of a nucleus disintegrating spontaneously, it must be struck by another small particle to induce a **nuclear reaction.** In the more condensed representation given by (25.11), the target and product nuclei are represented on the left and right of a parenthetical expression. Within the parentheses the bombarding particle is written first, followed by the ejected particle.

$$^{14}N(\alpha,p)^{17}O \tag{25.11}$$

$^{17}_{8}O$ is a naturally occurring *nonradioactive* nuclide of oxygen (0.037% natural abundance). The situation with $^{30}_{15}P$, which can also be produced by a nuclear reaction, is somewhat different.

In 1934, when bombarding aluminum with α particles, Irene Curie and her husband Frédéric Joliot observed the emission of two types of particles—neutrons and positrons. The Joliots observed that when bombardment by α particles was terminated, the emission of neutrons also stopped; that of positrons, however, continued. Their conclusion was that the nuclear bombardment produces $^{30}_{15}P$, which undergoes radioactive decay by the emission of positrons.

$$^{27}_{13}Al + ^{4}_{2}He \longrightarrow ^{30}_{15}P + ^{1}_{0}n \qquad \text{or} \qquad ^{27}Al(\alpha,n)^{30}P \tag{25.12}$$

$$^{30}_{15}P \longrightarrow ^{30}_{14}Si + ^{0}_{+1}e \tag{25.13}$$

$^{30}_{15}P$ was the first radioactive nuclide produced by artificial means. Since the time of its discovery over 1000 other radioactive nuclides have been produced. The number of known radioactive nuclides now exceeds considerably the number of nonradioactive nuclides (about 280). A few of the many applications of artificially induced radioactivity are considered in Section 25-11.

Example 25-2   Write **(a)** a condensed representation of the nuclear reaction, $^{14}_{7}N + ^{1}_{0}n \rightarrow ^{14}_{6}C + ^{1}_{1}H$; **(b)** an expanded representation of the nuclear reaction, $^{59}Co(n,?)^{56}Mn$.

Solution
(a) The bombarding particle is simply a neutron, n, and the emitted particle is a proton, p. The condensed equation is $^{14}N(n,p)^{14}C$.
(b) The missing particle in parentheses must have $A = 4$ and $Z = 2$; it is an α particle.

$$^{59}_{27}Co + ^{1}_{0}n \longrightarrow ^{56}_{25}Mn + ^{4}_{2}He$$

SIMILAR EXAMPLES: Review Problems 3, 4; Exercise 9.

25-4 Transuranium Elements

Until 1940 the only known elements were those that occur naturally. In 1940, the first synthetic element was produced by bombarding $^{238}_{92}U$ atoms with neutrons. First the unstable nucleus $^{239}_{92}U$ is formed. This nucleus then undergoes β^{-} decay, yielding the element neptunium, with $Z = 93$.

FIGURE 25-4
A charged-particle
accelerator—the
cyclotron.

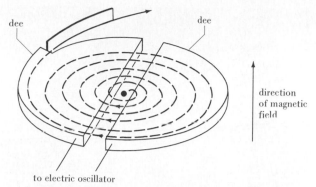

to electric oscillator

direction
of magnetic
field

The accelerator consists of two hollow, flat, semicircular boxes, called dees, that are kept electrically charged. The entire assembly is maintained within a magnetic field. The particles to be accelerated, in the form of positive ions, are produced at the center of the opening between the dees. They are then attracted into the negatively charged dee and forced into a circular path by the magnetic field. When the particles leave the dee and enter the gap, the electric charges on the dees are reversed, so that the particles are attracted into the opposite dee. The particles are accelerated as they pass the gap, and travel a wider circular path in the new dee. This process is repeated many times over until the particles are brought to the required energy to induce the desired nuclear reaction. They are then brought out of the accelerator and made to strike the target.

A simple analogy to the nuclear accelerator is found in the action of a playground swing. If the rider is given a push each time the swing reaches the end of its arc, the swing accelerates and the arc grows wider with each push.

$$^{238}_{92}\text{U} + ^{1}_{0}\text{n} \longrightarrow ^{239}_{92}\text{U} \tag{25.14}$$

$$^{239}_{92}\text{U} \longrightarrow ^{239}_{93}\text{Np} + ^{0}_{-1}\text{e} \tag{25.15}$$

Bombardment by neutrons is a particularly effective way to produce nuclear reactions because these heavy, uncharged particles are not repelled as they approach a nucleus.

Since 1940 all the elements from $Z = 93$ to 106 have been synthesized. For example, an isotope of the element, $Z = 105$, was produced in 1970 by bombarding atoms of $^{249}_{98}\text{Cf}$ with $^{15}_{7}\text{N}$ nuclei.

$$^{249}_{98}\text{Cf} + ^{15}_{7}\text{N} \longrightarrow ^{260}_{105}\text{Ha} + 4^{1}_{0}\text{n} \tag{25.16}$$

Elements 104 and 105 are the first in a series of transition elements following the actinoid series; they might be called transactinoid elements. From our knowledge of the periodic table, we should predict these elements to resemble hafnium and tantalum, respectively. Moreover, there is every reason to believe that more elements will be synthesized.

Charged-Particle Accelerators.   To bring about nuclear reactions such as (25.16) requires that energetic particles be used as projectiles for bombarding atomic nuclei. Such energetic particles can be obtained in an accelerator. A type of accelerator known as a cyclotron is described in Figure 25-4.

A charged-particle accelerator, as the name implies, can only produce beams of charged particles as projectiles (e.g., $^{1}_{1}\text{H}^{+}$). In many cases neutrons are more effective as projectiles for nuclear bombardment. The neutrons required can themselves be generated through a nuclear reaction produced by a charged-particle beam. In the following reaction $^{2}_{1}\text{H}$ represents a beam of deuterons (actually $^{2}_{1}\text{H}^{+}$) from an accelerator.

$$^{9}_{4}\text{Be} + ^{2}_{1}\text{H} \longrightarrow ^{10}_{5}\text{B} + ^{1}_{0}\text{n}$$

Another important source of neutrons for nuclear reactions is a nuclear reactor (see Section 25-8).

The 400 GeV particle accelerator at Fermilab, Batavia, Illinois. (1 GeV = 1×10^9 eV = 1000 MeV.) [Courtesy of the Fermi National Accelerator Laboratory.]

25-5 Rate of Radioactive Decay

In time, we can expect every atomic nucleus of a radioactive nuclide to disintegrate, but it is impossible to predict when a given nucleus will do so. Radioactivity is a random process. However, this all-important observation has been made: *The rate of disintegration of a radioactive material, that is, the decay rate, is directly proportional to the number of atoms present in a sample.* Consider a case where, *on the average,* 100 atoms undergo disintegration each second in a 1-million-atom sample, we would expect the average decay rate to be 200 atoms/s in a 2×10^6 atom sample, 50 atoms/s in a 5×10^5 atom sample, and so on. In mathematical terms,

$$\text{rate of decay} \propto N \qquad and \qquad \text{rate of decay} = \lambda \cdot N \tag{25.17}$$

The decay rate is expressed in atoms per unit time, such as atoms/s. N is the number of atoms in the sample being observed. λ is the **decay constant;** its unit is $(\text{time})^{-1}$.

Equation (25.17) represents a first-order process. By simply substituting N for [A] and λ for k in the several equations presented in Chapter 14, we establish the following useful equations.

$$\log N_t - \log N_0 = \log \frac{N_t}{N_0} = \frac{-\lambda t}{2.303} \tag{25.18}$$

$$t_{1/2} = \frac{0.693}{\lambda} \tag{25.19}$$

Here N_t represents the number of atoms remaining at time t and N_0, the number at some initial time ($t = 0$). Instead of characterizing radioactive decay through the decay constant λ, it is customary to do so through the **half-life, $t_{1/2}$.** The shorter the half-life, the larger the value of λ and the faster the decay process. Half-lives of radioactive nuclides range over periods of time from extremely short to very long, as suggested by the representative data in Table 25-1.

TABLE 25-1
Some representative half-lives

Nuclide	Half-life[a]	Nuclide	Half-life[a]
$^{3}_{1}\text{H}$	12.26 y	$^{90}_{38}\text{Sr}$	28.1 y
$^{14}_{6}\text{C}$	5730 y	$^{131}_{53}\text{I}$	8.070 d
$^{13}_{8}\text{O}$	8.7×10^{-3} s	$^{137}_{55}\text{Cs}$	30.23 y
$^{28}_{12}\text{Mg}$	21 h	$^{214}_{84}\text{Po}$	1.64×10^{-4} s
$^{32}_{15}\text{P}$	14.3 d	$^{222}_{86}\text{Rn}$	3.823 d
$^{35}_{16}\text{S}$	88 d	$^{226}_{88}\text{Ra}$	1600 y
$^{40}_{19}\text{K}$	1.28×10^{9} y	$^{234}_{90}\text{Th}$	24.1 d
$^{80}_{35}\text{Br}$	17.6 min	$^{238}_{92}\text{U}$	4.51×10^{9} y

[a] s, second; min, minute; h, hour; d, day; y, year.

Example 25-3 $^{238}_{92}\text{U}$ has a half-life of 4.51×10^{9} y. It decays by the process shown in equation (25.1). How many α particles are produced per second in a sample containing 1.00×10^{20} atoms of $^{238}_{92}\text{U}$?

Solution. From $t_{1/2}$ we obtain λ, using equation (25.19).

$$\lambda = \frac{0.693}{4.51 \times 10^{9} \text{ y}} = 1.54 \times 10^{-10} \text{ y}^{-1}$$

We use equation (25.17) to solve for the rate of decay.

$$\text{rate} = (1.54 \times 10^{-10} \text{ y}^{-1}) \times (1.00 \times 10^{20} \text{ atoms}) = 1.54 \times 10^{10} \text{ atoms/y} \qquad (25.20)$$

One α particle is produced for every atom that disintegrates. The rate of α-particle production is expressed by equation (25.20). On a *per second* basis, this becomes

$$\text{rate of decay} = 1.54 \times 10^{10} \ \alpha \ \text{particles/y} \times \frac{1 \text{ y}}{365 \text{ d}} \times \frac{1 \text{ d}}{24 \text{ h}} \times \frac{1 \text{ h}}{60 \text{ min}} \times \frac{1 \text{ min}}{60 \text{ s}}$$

$$= 488 \ \alpha \ \text{particles/s}$$

Radioactive decay is subject to statistical fluctuations, and the result obtained here is simply the *average* decay rate. It should not be taken to mean, literally, that for each and every second exactly 488 α particles are produced.

SIMILAR EXAMPLES: Review Problem 6; Exercises 11, 12.

Radiocarbon Dating. Carbon-containing compounds in living organisms maintain an equilibrium with ^{14}C in the atmosphere. The ^{14}C nuclide is radioactive and has a half-life of 5730 y. The activity associated with carbon in this equilibrium is about 15 disintegrations per minute (dis/min) per gram of carbon. When an organism ceases to live (e.g., a tree is felled), this equilibrium is destroyed and the disintegration rate falls off. From the measured disintegration rate at some later time, an estimate of the age (i.e., elapsed time since death of the organism) can be made.

$^{14}_{6}\text{C}$ is formed at a constant rate in the upper atmosphere by the bombardment of $^{14}_{7}\text{N}$ with neutrons.

$$^{14}_{7}\text{N} + ^{1}_{0}\text{n} \longrightarrow ^{14}_{6}\text{C} + ^{1}_{1}\text{H}$$

The neutrons are produced by cosmic rays. $^{14}_{6}\text{C}$ disintegrates by β^{-} emission.

Example 25-4 A wooden object is found in an Indian burial mound and subjected to radiocarbon dating. The decay rate associated with its ^{14}C content is 10 dis min^{-1} per g C. What is the age of the object (i.e., time elapsed since the tree was cut down)?

Solution. In this example three equations, (25.17), (25.18), and (25.19), are required. Equation (25.19) is again used to determine the decay constant.

$$\lambda = \frac{0.693}{5730 \text{ y}} = 1.21 \times 10^{-4} \text{ y}^{-1}$$

Next we use equation (25.17) to represent the actual number of atoms: N_0 at $t = 0$ (the time when the ^{14}C equilibrium was destroyed) and N_t at time t (the present time). The rate of decay just prior to the ^{14}C equilibrium being destroyed is 15 dis min^{-1} per g C, and at the time of the measurement, 10 dis min^{-1} per g C. The corresponding numbers of atoms are proportional to these decay rates divided by λ.

$$N_0 = \frac{\text{decay rate (at } t = 0)}{\lambda} = \frac{15}{\lambda}$$

$$N_t = \frac{\text{decay rate (at time } t)}{\lambda} = \frac{10}{\lambda}$$

Finally, we substitute into equation (25.18).

$$\log \frac{N_t}{N_0} = \log \frac{10/\lambda}{15/\lambda} = \log \frac{10}{15} = \frac{-(1.21 \times 10^{-4} \text{ y}^{-1})t}{2.303}$$

$$-0.18 = -(5.25 \times 10^{-5} \text{ y}^{-1})t$$

$$t = \frac{0.18}{5.25 \times 10^{-5} \text{ y}^{-1}} = 3.4 \times 10^3 \text{ y}$$

SIMILAR EXAMPLES: Review Problem 7; Exercises 18, 19.

Age of the Earth. The natural radioactive decay scheme of Figure 25-3 suggests the eventual fate awaiting all the $^{238}_{92}U$ found in nature—conversion to lead. Naturally occurring uranium minerals always have associated with them some nonradioactive lead formed by radioactive decay. From the mass ratio of $^{206}_{82}Pb$ to $^{238}_{92}U$ in such a mineral it is possible to estimate the age of the rock containing the mineral. By the age of the rock we mean the time elapsed since molten magma froze to become a rock. One assumption of this method is that the initial radioactive nuclide, the final stable nuclide, and all the products of a decay series remain in the rock. Another assumption is that any lead present in the rock initially consisted of the several isotopes of lead in their present naturally occurring abundances.

An exact treatment of this subject would require some discussion of the relationship between the rates of decay of a radionuclide called a "parent" and the product nuclide called a "daughter." A discussion of these relationships is beyond the scope of this text, but an indication of the method is still possible.

The half-life of $^{238}_{92}U$ is 4.5×10^9 y. According to the natural decay scheme of Figure 25-3, the basic changes that occur as atoms of $^{238}_{92}U$ and its daughters pass through the entire sequence of steps is

$$^{238}_{92}U \longrightarrow {}^{206}_{82}Pb + 8 \, {}^{4}_{2}He + 6 \, {}^{0}_{-1}e \qquad (25.21)$$

Discounting the mass associated with the β^- particles, we can see that for every 238 g of uranium that undergoes complete decay, 206 g of lead and 32 g of helium are produced.

Suppose that in a rock containing no lead initially, 1.000 g $^{238}_{92}U$ had disintegrated through one half-life period, 4.5×10^9 y. At the end of that time there would be present in the sample

0.500 g $^{238}_{92}$U undisintegrated

and

$$0.500 \times \frac{206}{238} = 0.433 \text{ g } ^{206}_{82}\text{Pb}$$

with the ratio

$$\frac{^{206}_{82}\text{Pb}}{^{238}_{92}\text{U}} = \frac{0.433}{0.500} = 0.866 \qquad (25.22)$$

A $^{206}_{82}$Pb/$^{238}_{92}$U ratio smaller that that shown in (25.22) would suggest the solid mineral had not been in existence for as long as one half-life period of $^{238}_{92}$U. A higher ratio would suggest a greater age for the rock. The best estimates of the age of the oldest rocks and presumably of the earth itself are in fact about 4.5×10^9 y. These estimates are based on the $^{206}_{82}$Pb to $^{238}_{92}$U ratio and on ratios for other pairs of nuclides from natural radioactive decay series.

Example 25-5   The thorium radioactive decay series produces one atom of ^{208}Pb as the final disintegration product of an atom of ^{232}Th. The half-life of ^{232}Th is 1.4×10^{10} y. A certain rock is found to have a ^{208}Pb/^{232}Th mass ratio of $0.14:1.00$. Use these data to estimate the age of the rock.

Solution.   The decay constant, λ, is obtained in the usual fashion.

$$\lambda = \frac{0.693}{1.4 \times 10^{10} \text{ y}} = 5.0 \times 10^{-11} \text{ y}^{-1}$$

Let us base our calculation on a quantity of mineral containing 1.00 g ^{232}Th at the present time, t. The total mass of ^{232}Th present in the sample of mineral when it was formed must have been the 1.00 g present currently plus the mass of ^{232}Th required to produce 0.14 g ^{208}Pb.

$$\text{no. g } ^{232}\text{Th} = 0.14 \text{ g } ^{208}\text{Pb} \times \frac{232 \text{ g } ^{232}\text{Th}}{208 \text{ g } ^{208}\text{Pb}} = 0.16 \text{ g } ^{232}\text{Th}$$

The total mass of ^{232}Th present initially was $1.00 + 0.16 = 1.16$ g ^{232}Th. Since the number of atoms in a sample of an element is directly proportional to the mass of the sample, we can substitute 1.00 g for N_t and 1.16 g for N_0. We now have the necessary data to substitute into equation (25.18) and solve for t.

$$\log \frac{N_t}{N_0} = \log \frac{1.00}{1.16} = \frac{-5.0 \times 10^{-11} \text{ y}^{-1} \, t}{2.303}$$

$$-0.064 = -2.2 \times 10^{-11} \text{ y}^{-1} \, t$$

$$t = \frac{-0.064}{-2.2 \times 10^{-11} \text{ y}^{-1}} = 2.9 \times 10^9 \text{ y}$$

SIMILAR EXAMPLES: Exercises 16, 17.

25-6 Energetics of Nuclear Reactions

A complete assessment of a nuclear reaction requires use of the mass-energy equivalence given by Albert Einstein.

$$E = mc^2 \qquad (25.23)$$

In chemical reactions energy changes are so small that the equivalent mass changes are undetectable (though real nevertheless). We say that mass is conserved in a chemical reaction. In nuclear reactions, energies involved are orders of magnitude greater. Perceptible changes in mass do occur.

If the exact masses of nuclides are known, it is possible to calculate the energy of a nuclear reaction with equation (25.23). The term m corresponds to the net change in mass, in kg, and c, the velocity of light, is expressed in m/s. The resulting energy is in joules. Another common unit for expressing nuclear energy is the MeV (million electron volt).

$$1 \text{ MeV} = 1.602 \times 10^{-13} \text{ J} \tag{25.24}$$

Example 25-6   What is the energy associated with the α decay of ^{238}U **(a)** in MeV; **(b)** in kJ/mol?

$$^{238}_{92}\text{U} \longrightarrow {}^{234}_{90}\text{Th} + {}^{4}_{2}\text{He}.$$

The nuclidic masses in atomic mass units (u) are

$$^{238}_{92}\text{U} = 238.0508 \text{ u} \qquad {}^{234}_{90}\text{Th} = 234.0437 \text{ u} \qquad {}^{4}_{2}\text{He} = 4.0026 \text{ u}$$

Solution
(a) The net change in mass that accompanies the decay of a single nucleus of ^{238}U is $234.0437 + 4.0026 - 238.0508 = -0.0045$ u. This loss of mass appears as kinetic energy carried away by the α particle. In the setup below, the relationship between the units u and g is established most readily by noting that 1 u is exactly $\frac{1}{12}$ of the mass of a carbon-12 atom.

$$1 \text{ u} = \frac{1}{12} \times \frac{12.00 \text{ g}}{6.02 \times 10^{23}} = 1.66 \times 10^{-24} \text{ g}$$

$$E = 0.0045 \text{ u} \times \frac{1.66 \times 10^{-24} \text{ g}}{\text{u}} \times \frac{1 \text{ kg}}{1000 \text{ g}} \times (3.00 \times 10^8)^2 \frac{\text{m}^2}{\text{s}^2}$$

$$= 6.7 \times 10^{-13} \text{ J}$$

$$E = 6.7 \times 10^{-13} \text{ J} \times \frac{1 \text{ MeV}}{1.602 \times 10^{-13} \text{ J}} = 4.2 \text{ MeV}$$

(b) The calculation in part (a) is for a single disintegration. The energy in kJ/mol is based on the disintegration of 1 mol of atoms.

$$E = \frac{6.7 \times 10^{-13} \text{ J}}{\text{atom}} \times \frac{6.02 \times 10^{23} \text{ atoms}}{1 \text{ mol}} \times \frac{1 \text{ kJ}}{1000 \text{ J}}$$

$$= 4.0 \times 10^8 \text{ kJ/mol}$$

SIMILAR EXAMPLES: Review Problem 8; Exercises 21, 22.

If in the calculation of Example 25-6a we had been dealing with a mass of exactly 1.000 u (instead of 0.0045 u), the calculated energy would have been 931.2 MeV. This provides a useful conversion factor between mass and energy.

$$1 \text{ atomic mass unit (u)} = 931.2 \text{ MeV} \tag{25.25}$$

Nuclear Binding Energy.   Figure 25-5 depicts a process in which the nucleus of a $^{4}_{2}$He atom is produced from two protons and two neutrons. In the formation of this nucleus there is a **mass defect** of 0.0305 u. That is, the experimentally determined mass of a $^{4}_{2}$He nucleus is 0.0305 u *less* than the combined masses of two protons and two neutrons. This "lost" mass is liberated as energy. With expression (25.25) it can be shown that 0.0305 u of mass is equivalent to an energy of 28.4 MeV. Since this is the energy released in forming an $^{4}_{2}$He nucleus, it is referred to as the **binding energy** of the nucleus. (Viewed in another way, an $^{4}_{2}$He nucleus would have to absorb 28.4 MeV to cause its protons and neutrons to become separated.) If we consider the binding energy to be apportioned

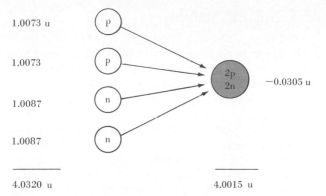

The mass of a helium nucleus ($^4_2$He) is 0.0305 u (atomic mass unit) less than the combined masses of two protons and two neutrons. The energy equivalent to this loss of mass (called the mass defect) is the nuclear energy that binds the nuclear particles together.

equally among the two protons and two neutrons in $^4_2$He, we obtain a binding energy per nuclear particle (nucleon) of 7.10 MeV. Similar calculations can be made for other nuclei, leading to the graph shown in Figure 25-6.

Figure 25-6 indicates that the maximum binding energy per nucleon is found in a nucleus with a mass number of approximately 60. This leads to two interesting conclusions: (a) If small nuclei are combined into a heavier one (up to about $A = 60$), the binding energy per nucleon increases and a certain quantity of mass must be converted to energy. The nuclear reaction is highly exothermic. This is a fusion process, and serves as the basis of the hydrogen bomb. (b) For nuclei having mass numbers above 60, the addition of extra nucleons to the nucleus would require the expenditure of energy (since the binding energy per nucleon decreases). On the other hand, the *disintegration* of heavier nuclei into lighter ones is accompanied by the release of energy. This is a nuclear fission process and serves as the basis of the atomic bomb and conventional nuclear power reactors. Nuclear fission and fusion are considered in greater detail in Sections 25-8 and 25-9. But first let us see what insights Figure 25-6 provides into the question of nuclear stability.

FIGURE 25-6
Average binding energy
per nucleon as a
function of atomic
number.

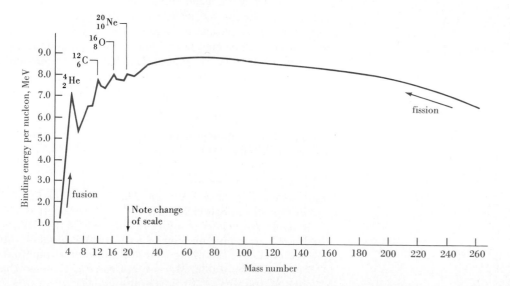

TABLE 25-2
Magic numbers for nuclear stability

Number of protons	Number of neutrons
2	2
8	8
20	20
28	28
50	50
82	82
114	126
	184
	196

25-7 Nuclear Stability

A number of basic questions have probably occurred to you as we have been describing nuclear decay processes: Why do some radioactive nuclei decay by α emission, some by β^- emission, and so on? Why do the lighter elements have so few naturally occurring radioactive nuclides, whereas those of the heavier elements all seem to be radioactive?

Our first clue to answers for such questions comes from Figure 25-6, where several nuclides have been specifically noted. These nuclides have higher binding energies per nucleon than those of their neighbors. Their nuclei are especially stable. This observation is consistent with a theory of nuclear structure known as the **shell theory.** In the formation of a nucleus, protons and neutrons are believed to occupy a series of nuclear shells. This process is analogous to building up of the electronic structure of an atom by the successive addition of electrons to electronic shells. Just as the Aufbau process produces, periodically, electron configurations of exceptional stability, so do certain nuclei acquire a special stability as nuclear shells are closed. This condition of special stability of an atomic nucleus comes for certain numbers of protons and/or neutrons known as **magic numbers.** These magic numbers are listed in Table 25-2.

Analogous to the situation with electrons, nucleons possess the property of spin, and a pairing of spins also occurs in the filling of nuclear shells. A nucleus that contains an odd number of nucleons will have a resultant nuclear spin. There are fewer stable nuclides with odd numbers of nucleons than even numbers. This situation is illustrated by the distribution of numbers of protons (Z) and neutrons (N) among the known stable nuclides.

Z even–N even: 163 nuclides
Z even–N odd: 55
Z odd–N even: 50
Z odd–N odd: 4

The Z odd–N odd combination is found only in the lighter elements: ^2_1H, ^6_3Li, $^{10}_5B$, $^{14}_7N$.

Another manifestation of the pairing of nucleons is that elements of *odd* atomic number generally have only one or two stable isotopes, whereas those of even atomic number have several. When all the protons are paired (even atomic number), the nucleus is able to accommodate a greater range in the number of neutrons. This leads to a greater variety of isotopes. Some representative data are shown in Table 25-3.

For the lighter elements (up to about $Z = 20$) the common stable nuclides have equal numbers of protons and neutrons, for example, ^4_2He, $^{12}_6C$, $^{16}_8O$, $^{28}_{14}Si$, $^{40}_{20}Ca$. For higher atomic numbers, because of increasing repulsive forces between protons, larger numbers of neutrons must be present to stabilize a nucleus and the n/p ratio increases. For bismuth the ratio is about 1.5:1. Figure 25-7 indicates an approximate range of n/p ratios for stable nuclides as a function of atomic number.

TABLE 25-3
Stable isotopes of a few elements

Element	Z	Number of stable isotopes
H	1	2
O	8	3
F	9	1
Ne	10	3
Cl	17	2
Ca	20	6
Cu	29	2
Sn	50	10
I	53	1
Hg	80	7

Example 25-7 Which of the following nuclides would you expect to be stable and which, radioactive? (a) ^{76}As; (b) ^{120}Sn; (c) ^{214}Po.

Solution
(a) ^{76}As has $Z = 33$ and $N = 43$. This is an odd–odd combination that is found only in four of the lighter elements. ^{76}As is radioactive. (Note also that this nuclide is outside the belt of stability in Figure 25-7.)
(b) Sn has an atomic number of 50—a magic number. The neutron number is 70 in the nuclide ^{120}Sn. This is an even–even combination and we should expect the nucleus to be stable. Moreover, Figure 25-7 shows that this nuclide is within the belt of stability. ^{120}Sn is a stable nuclide.

FIGURE 25-7
Neutron-to-proton ratio
and the stability of
nuclides.

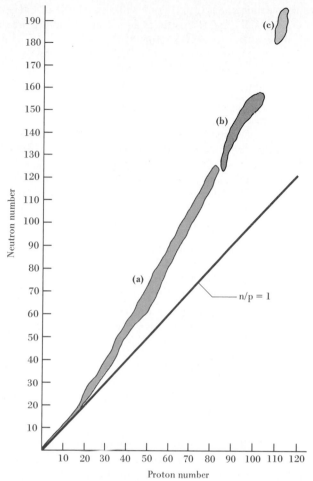

Neutron number

Proton number

n/p = 1

(a)

(b)

(c)

(a) The belt of naturally occurring stable nuclides, ranging from $_1^1$H to $_{83}^{209}$Bi.
(b) Naturally occurring and man-made radioactive nuclides of the heavier elements.
(c) Possible heavy nuclides of high stability (long radioactive half-lives).

(c) ^{214}**Po has an atomic number of 84. All known nuclides with** $Z > 83$ **are radioac-**
tive. ^{214}**Po is radioactive.**

SIMILAR EXAMPLES: Review Problem 10; Exercises 23, 25.

Using the ideas presented here (i.e., nuclear shell theory, magic numbers, etc.),
nuclear scientists have predicted the possible existence of nuclides of high atomic number.
These nuclides would have very long half-lives. Currently, a search is on to find such
nuclides, either naturally or by creating them in an accelerator. (In order for a nuclide still
to be present following the creation of the earth, its half-life would have to be greater than
about 10^8 y.) Figure 25-7 suggests the general range of proton and neutron numbers for
these nuclides.

Mechanism of Radioactive Decay. The emission of an α particle by a nucleus is not
difficult to visualize. A bundle of four nucleons (two protons, two neutrons) is ejected
from a nucleus and the nucleus becomes energetically more stable. Alpha particle emission is pretty much limited to nuclides with $Z > 82$, though there are a few α emitters
among the lanthanoid nuclides. Gamma ray emission can be visualized simply in terms of

TABLE 25-4
Radioactive properties
of isotopes of fluorine

Isotope	Mode of decay	Half-life
^{17}F	β^+	66 s
^{18}F	β^+, E.C.	109.7 min
^{19}F	stable	—
^{20}F	β^-	11.4 s
^{21}F	β^-	4.4 s
^{22}F	β^-	4.0 s

some rearrangement of nucleons with the release of energy as electromagnetic radiation. Emission of β^- and β^+ particles and electron capture are harder to picture. However, we can think in the following terms.*

$$\beta^- \text{ emission:} \qquad {}^1_0n \longrightarrow {}^1_1p + {}^0_{-1}e \qquad (25.26)$$

$$\beta^+ \text{ emission:} \qquad {}^1_1p \longrightarrow {}^1_0n + {}^0_{+1}e \qquad (25.27)$$

$$\text{electron capture:} \quad {}^1_1p + {}^0_{-1}e \longrightarrow {}^1_0n \qquad (25.28)$$

In general, if a nuclide lies above the belt of stability in Figure 25-7 the n/p ratio is too high. A neutron is converted to a proton and a β^- particle is emitted (25.26). If a nuclide lies below the belt of stability, the n/p ratio is too low. Either a proton is converted to a neutron, followed by β^+ emission (25.27) or electron capture occurs (25.28). The situation for a series of isotopes of fluorine is presented in Table 25-4.

Example 25-8   By what mode will the nuclide ^{82}Y decay?

Solution.   That this nuclide is radioactive can be seen in two ways. It lies below the belt of stability in Figure 25-7, and it has an odd–odd combination of protons and neutrons. The nuclide has too few neutrons to be stable. A proton must be converted to a neutron, either by positron (β^+) emission (25.27) or by electron capture (25.28).

SIMILAR EXAMPLE: Exercise 24.

25-8 Nuclear Fission

In 1934, the Italian physicist Enrico Fermi proposed that transuranium elements might be produced by the bombardment of uranium with neutrons. He reasoned that the successive loss of β^- particles would cause the atomic number to increase, perhaps as high as to 96. When such experiments were carried out, it was found that in fact the product did emit β^- particles. But in 1938, two chemists, Hahn and Strassman, found by chemical analysis that the products of the neutron bombardment of uranium did not correspond to elements with $Z > 92$. Neither were they the neighboring elements of uranium—Ra, Ac, Th, and Pa. Instead, the products consisted of radioisotopes of much lighter elements, such as strontium and barium. Neutron bombardment of uranium nuclei causes certain of them to undergo **fission** into smaller fragments. A fission process is depicted in Figure 25-8.

The energy equivalent of the mass destroyed in a fission process is somewhat variable because a variety of fission fragments is possible. However, the average energy for each fission event is approximately 3.20×10^{-11} J (200 MeV).

$$^{235}_{92}U + n \longrightarrow {}^{236}_{92}U \longrightarrow \text{fission fragments} + \text{neutrons} + 3.20 \times 10^{-11} \text{ J}$$

An energy of 3.20×10^{-11} J may seem small, but this energy is for the fission of a *single* $^{235}_{92}U$ nucleus. What if 1.00 g $^{235}_{92}U$ were to undergo fission?

$$\text{no. kJ} = 1.00 \text{ g } ^{235}U \times \frac{1 \text{ mol } ^{235}U}{235 \text{ g } ^{235}U} \times \frac{6.02 \times 10^{23} \text{ atoms } ^{235}U}{1 \text{ mol } ^{235}U} \times \frac{3.20 \times 10^{-11} \text{ J}}{1 \text{ atom } ^{235}U}$$

$$= 8.20 \times 10^{10} \text{ J} = 8.20 \times 10^7 \text{ kJ}$$

This is an enormous quantity of energy! By contrast, to release this same quantity of energy would require the complete combustion of nearly three tons of coal.

---

*Equations (25.26) and (25.27) are oversimplifications of the actual case. In order that certain properties be conserved, it is necessary to postulate the presence of other extremely tiny particles (about 0.0004 times the mass of an electron) in β decay processes. These are the neutrino in equation (25.26) and the antineutrino in (25.27). The existence of these particles has been confirmed.

FIGURE 25-8
Nuclear fission of $^{235}_{92}$U
with thermal neutrons.

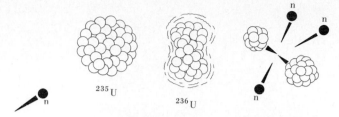

A $^{235}_{92}$U nucleus is struck by a neutron possessing ordinary thermal energy. First the unstable nucleus $^{236}_{92}$U is produced; this then breaks up into a light and a heavy fragment and several neutrons. A variety of nuclear fragments is possible, but the most probable mass number for the light fragment is 97, and for the heavy one 137.

Nuclear Reactors.   In the fission of $^{235}_{92}$U, on average, 2.5 neutrons are released per fission event. These neutrons, on average, produce two or more fission events. The neutrons produced by the second round of fission produce another four or five events, and so on. The result is a **chain reaction.** If the reaction is uncontrolled, the total released energy causes an explosion; this is the basis of the atomic bomb. Spontaneous fission resulting in an uncontrolled explosion occurs only if the quantity of ^{235}U exceeds the **critical mass.** With subcritical masses, neutrons escape from the ^{235}U at too great a rate to sustain a chain reaction.

In a nuclear reactor, fission energy is released in a controlled manner. One common design for a nuclear reactor, called the light water reactor (LWR), is pictured in Figure 25-9. In the core of the reactor, rods of uranium-rich fuel are suspended in liquid water maintained under a pressure of from 70 to 150 atm. The water serves a dual purpose. First, it slows down the neutrons given off in the fission process so that they possess only normal thermal energy. These so-called thermal neutrons are more able to induce fission than highly energetic ones. In this capacity the water is said to act as a **moderator.** The second function of the water is as a heat-transfer medium. The energy of the fission

FIGURE 25-9
Light water nuclear
reactor.

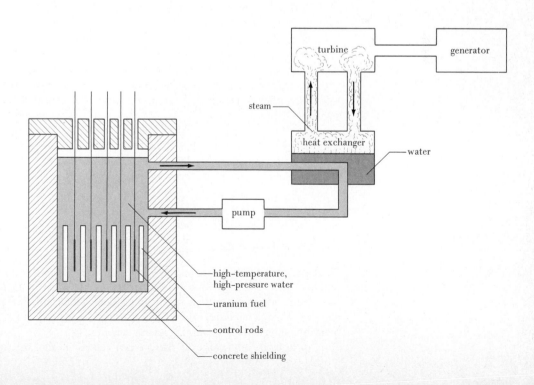

The core of a nuclear reactor in the process of being refueled. [Courtesy of Atomic Industrial Forum, Inc.]

reaction maintains the water at a high temperature (about 300°C). The high-temperature water is brought in contact with colder water in a heat exchanger. The colder water is converted to steam, which drives a turbine, which in turn drives an electric generator. A final component of the nuclear reactor is a set of **control rods,** usually cadmium metal, whose function is to absorb neutrons. When the rods are lowered into the reactor, the fission process is slowed down. When the rods are raised, the density of neutrons and the rate of fission increase.

A nuclear reactor based on the fission of $^{235}_{92}\text{U}$ is referred to as a nuclear burner. In the nuclear burner the fissionable nuclide is consumed (perhaps at the rate of 1 to 3 kg/d), and highly radioactive waste products accumulate. The separation, concentration, and disposal of these radioactive wastes is a difficult problem, requiring a considerable amount of chemical technology. A completely satisfactory method for the disposal of radioactive wastes is yet to be found.

Another important aspect of nuclear burning is the high rate of consumption of a relatively rare fissionable material. $^{235}_{92}\text{U}$ accounts for only approximately 0.71% of naturally occurring uranium. To extract pure $^{235}_{92}\text{U}$ from uranium ores requires that high-grade ores be employed, usually U_3O_8. Estimated world reserves of U_3O_8 are not extensive, and conventional nuclear reactors are of limited potential in the long-term production of energy.

Breeder Reactors. All that is required to initiate the fission of $^{235}_{92}\text{U}$ are neutrons of ordinary thermal energies. Nuclei of $^{238}_{92}\text{U}$, the abundant nuclide of uranium (99.28%), undergo the following reactions when struck by energetic neutrons.

$$^{238}_{92}\text{U} + ^{1}_{0}\text{n} \longrightarrow {}^{239}_{92}\text{U}$$

$$^{239}_{92}\text{U} \longrightarrow {}^{239}_{93}\text{Np} + {}^{0}_{-1}\text{e}$$

$$^{239}_{93}\text{Np} \longrightarrow {}^{239}_{94}\text{Pu} + {}^{0}_{-1}\text{e}$$

A fissionable nuclide such as $^{235}_{92}\text{U}$ is called a fissile nuclide; $^{239}_{94}\text{Pu}$ is also fissile. A nuclide such as $^{238}_{92}\text{U}$, which can be converted into a fissile nuclide, is said to be fertile. In a breeder nuclear reactor a small quantity of fissile nuclide provides the neutrons that convert a large quantity of fertile nuclide into a fissile one. (The fissile nuclide then participates in a self-sustaining chain reaction.)

An obvious advantage of the breeder reactor is that the amount of uranium ''fuel'' available would immediately jump by a factor of about 100. This is the ratio of naturally occurring $^{238}_{92}U$ to $^{235}_{92}U$. But the advantage is even greater than this. Breeder reactors may be able to use as nuclear fuels materials that have even very low uranium contents. For example, shale deposits exist in the western Appalachian Mountains that contain about 0.006% U by weight. These deposits extend for several hundred square miles, and all of this material is potential nuclear fuel.

There are, however, important disadvantages to the use of breeder reactors. This is especially true of the type that has been pursued most vigorously—the liquid-metal-cooled fast breeder reactor (LMFBR). Systems must be perfected for handling a liquid metal, such as sodium, which becomes highly radioactive in the reactor. Also, the rate of heat and neutron production are both greater in the LMFBR than in the LWR, resulting in a more rapid deterioration of materials. These factors will complicate greatly the design and operation of the reactor. Perhaps the greatest unsolved problems are those of handling radioactive wastes and reprocessing plutonium fuel. Plutonium is one of the most toxic substances known. It can cause lung cancer when inhaled even in microgram (10^{-6} g) amounts. Federal health standards limit exposures to this substance to a total body burden of only 0.6 μg. Furthermore, because of its long half-life (24,000 y), any accident involving plutonium could leave an affected area almost permanently contaminated.

25-9 Nuclear Fusion

The fusion of atomic nuclei is the process whereby energy is produced on the sun. An uncontrolled fusion reaction is the basis of the hydrogen bomb. If a fusion reaction can be controlled, this will provide an almost unlimited source of energy. The nuclear reaction that holds the most immediate promise is the deuterium–tritium reaction.

$$^2_1H + ^3_1H \longrightarrow ^4_2He + ^1_0n \tag{25.29}$$

The difficulties in developing a fusion reaction are probably without parallel in the history of technology. In fact, the feasibility of a controlled fusion reaction has yet to be fully demonstrated. The basic problems are these:

In order to permit their fusion, the nuclei of deuterium and tritium must come into close proximity. Because atomic nuclei repel one another, this close approach requires the

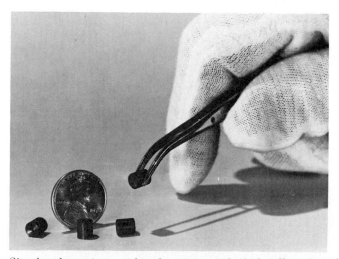

Simulated uranium oxide–plutonium oxide fuel pellets for a breeder nuclear reactor test facility. One such pellet would produce energy equivalent to three tons of coal. [Courtesy of Atomic Industrial Forum, Inc.]

In the hydrogen bomb these high temperatures are attained by exploding an atomic bomb, which triggers the fusion reaction.

nuclei to have very high thermal energies. At the temperatures necessary to initiate a fusion reaction, gases are completely ionized into a mixture of atomic nuclei and electrons known as a **plasma.** Still higher plasma temperatures—over 40,000,000 K—are required to initiate a *self-sustaining* reaction (one that releases more energy than is required to get it started). Obviously, there is no container that can withstand these fantastically high temperatures. A method must be devised to confine the plasma out of contact with other materials and at a sufficiently high density and for a sufficient period of time to permit the fusion reaction to occur. The two methods receiving greatest attention currently are confinement in a magnetic field and the heating of a frozen deuterium-tritium pellet with a laser beam. Another series of technical problems that must be solved involves the handling of liquid lithium, which is the anticipated heat transfer medium and tritium (^3_1H) source.

$$\underset{\text{(fast)}}{^7_3\text{Li}} + \,^1_0\text{n} \longrightarrow \,^4_2\text{He} + \underset{\text{(slow)}}{^3_1\text{H}} + \,^1_0\text{n} \tag{25.30}$$

Finally, for the magnetic containment method the magnetic field must be produced by superconducting magnets maintained at temperatures near absolute zero. Thus, the fusion reactor must have regions in which the temperature ranges from near 0 K to tens of millions of degrees kelvin, separated by distances of perhaps only 2 m.

The advantages of fusion over fission should be very great. Since deuterium comprises about one in every 6500 H atoms, the oceans of the world can supply an almost limitless amount of nuclear fuel. It is estimated that there is sufficient lithium on the earth to provide a source of tritium for about 1 million years.

25-10 Effect of Radiation on Matter

We now turn our attention briefly to the fate of the radiation or emanations of radioactive nuclei. Although there are substantial differences in the way in which α, β, and γ rays interact with atoms and molecules as they pass through samples of matter, they share an

FIGURE 25-10
Some interactions of radiation with matter.

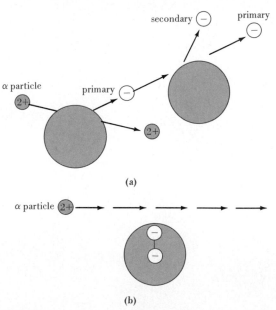

(a)

(b)

(a) The production of primary and secondary electrons by collisions.
(b) The excitation of an atom by the passage of an α particle. An electron is raised to a higher energy level within the atom. The excited atom reverts to its normal state by emitting radiation.

important feature: They tend to dislodge electrons from atoms and molecules to produce ions. The ionizing power of radiation may be described in terms of the number of ion pairs it forms. An ion pair consists of an ionized electron and the resulting positive ion. Alpha particles have the greatest ionizing power, followed by β particles and then γ rays. The ionized electrons produced directly by the collisions of particles of radiation with atoms are called primary electrons. These electrons may themselves possess sufficient energies to cause secondary ionizations. Thus, even though some radiation, such as γ rays, may produce few primary electrons, the total ionization associated with its passage through matter may be considerable.

Not all interactions between radiation and matter cause ion formation. In some cases electrons may simply be raised to higher atomic or molecular energy levels. The return of these electrons to their normal states is then accompanied by radiation—x rays, ultraviolet light, or visible light, depending on the energies involved. A practical example of this effect is found in luminescent watch dials. The dial numerals are painted with a mixture of a fluorescent material, such as zinc sulfide, and traces of a radioactive material, such as radium (an α emitter). Excitation of the zinc sulfide by α particles is accompanied by light emission; the dial numerals are visible in the dark.

Some of the possibilities described here are pictured in Figure 25-10.

Radiation Detectors.   The interactions of radiation with matter can serve as bases for the detection of radiation and the measurement of its intensity. One of the simplest methods is that used by Becquerel in his discovery of radioactivity—the exposure of a photographic plate. The effect of α, β, and γ rays on a photographic emulsion is similar that of x rays.

A device that has played an important role in the development of knowledge about radioactivity is the **cloud chamber** pictured in Figure 25-11. The principle involved is quite simple. If a vapor is brought to a saturated condition, say by expansion and cooling, and if there are no nuclei (such as dust particles) on which condensation can occur, the vapor may become supersaturated. If ionizing radiation passes through such a vapor, the

FIGURE 25-11
Wilson cloud chamber.

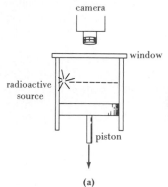

(a)

(b)

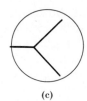

(c)

(d)

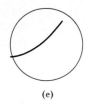

(e)

(f)

(a) Saturated vapor (say, ethanol) is cooled by sudden expansion. As a result it becomes supersaturated. Ion pairs produced in the supersaturated vapor serve as nuclei for condensation tracks.
(b) α particle tracks.
(c) An α particle in collision with an atomic nucleus. Two tracks emanate from the point where the original track disappears (one is the α particle track, and the other is that of the atom that is struck).
(d) α particle deflected in a magnetic field.
(e) β^- particle deflected in a magnetic field.
(f) Positrons (β^+) deflected in a magnetic field.

FIGURE 25-12
Geiger–Müller counter.

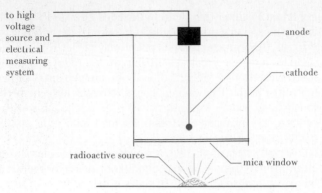

Radiation enters the G–M tube through the mica window. Ions produced by the radiation cause an electrical breakdown of the gas in the tube (usually argon). A pulse of electric current passes through the electric circuit and is counted.

ions produced become nuclei for the production of liquid. Droplets are produced in the form of a cloudlike track. The type of ionizing radiation and some of its characteristics can be inferred from such tracks. For instance, α particles produce short, straight thick tracks, whereas β^- particles produce long, thin meandering tracks. If the cloud chamber is placed in a magnetic field, the tracks are curved—one direction for α particles and the opposite for β^- particles.

A modification of the cloud chamber that is particularly useful in detecting high-energy radiation such as γ rays is the **bubble chamber.** In this device a substance, usually hydrogen, is kept just at its boiling point. As ion pairs are produced by the transit of an ionizing ray, bubbles of vapor form around the ions. Again tracks result that can be photographed and analyzed.

The most common device for detecting and measuring ionizing radiation is the **Geiger-Müller counter** pictured in Figure 25-12. The G-M counter consists of a cylindrical cathode with a wire anode running along its axis. The anode and cathode are sealed in a gas-filled glass tube. The tube is operated in such a way that ions produced by radiation passing through the tube trigger pulses of electric current. It is these pulses that are counted.

Effect of Ionizing Radiation on Living Matter.   All life exists against a background of naturally occurring ionizing radiation—cosmic rays, ultraviolet light, and emanations from radioactive elements such as uranium in rocks. The level of this radiation varies from point to point on earth, being greater, for instance, at higher elevations. Only in recent times have human beings been able to create situations where living organisms might be exposed to radiation at levels significantly higher than natural background.

The interactions of radiation with living matter are the same as with other forms of matter—ionization, excitation, and dissociation of molecules. There is no question of the effect of large dosages of ionizing radiation on living organisms—the organisms are killed. But even slight exposures to ionizing radiation can cause changes in cell chromosomes. Thus it is commonly held that even at low dosage rates ionizing radiation can result in birth defects, leukemia, bone cancer, and other forms of cancers. The nagging question that has eluded any definitive answers is how great an increase in the incidence of birth defects and cancers is caused by certain levels of radiation.

Radiation Dosage.   Several different units are used to describe radiation dosage, that is, the amount of radiation to which matter is exposed. A summary is provided in Table 25-5.

A special counter used in radiocarbon dating is pictured in Color Section C.

TABLE 25-5
Units of radiation dosage[a]

Unit	Definition
Curie	An amount of radioactive material decaying at the same rate as 1 g of radium (3.7×10^{10} dis/s).
Rad	A dosage of radiation able to deposit 1×10^{-2} J of energy per kilogram of matter.
Rem	A unit related to the rad, but taking into account the varying effects of different types of radiation of the same energy on biological matter. This relationship is through a "quality factor," which may be taken as equal to one for x rays, γ rays, and β particles. For protons and slow neutrons the factor has a value of 5; and for α particles, 10. Thus, an exposure to 1 rad of x rays is about equal to 1 rem, but 1 rad of α particles is about equal to 10 rem.

[a]Sources of α radiation are relatively harmless when external to the body and extremely hazardous when taken internally, as in the lungs or stomach. Other forms of radiation (x rays, γ rays), because they are highly penetrating, are hazardous even when external to the body.

It is thought that exposure in a single short time interval to 1000 rem of radiation would kill 100% of the population exposed. Exposure to 450 rem would probably result in death within 30 days in about 50% of the population. A single dosage of 1 rem delivered to 1 million people would probably produce about 100 cases of cancer within 20 to 30 years. The total body radiation received by most of the world's population from normal background sources is about 0.13 rem [130 millirem (mrem)] per year. The exposure in a chest x ray examination is about 20 mrem; and the estimated maximum exposure to the general population associated with the production of nuclear power is about 5 mrem/y.

Some of the foregoing statements about radiation exposures and their anticipated effects are based on (1) medical histories of the survivors of the Hiroshima and Nagasaki atomic blasts, (2) the incidence of leukemia and other cancers in children whose mothers received diagnostic radiation during pregnancy, and (3) the occurrence of lung cancers among uranium miners in the United States. What does all of this tell us about a ''safe'' level of radiation exposure? One approach has been to extrapolate from these high dosage rates to the lower dosages affecting the general population. This has led the United States National Council on Radiation Protection and Measurements to recommend that the dosage rate for the general population be limited to 0.17 rem (170 mrem) per year from all sources above background level. However, experts disagree on how the data observed for high-dosage exposures should be extrapolated to low dosages. According to some, the 0.17 rem/y figure is too high, by perhaps a factor of 10. If so, exposure to an additional 0.17 rem/y above normal background levels may cause statistically significant increases in the incidence of birth defects and cancers.

25-11 Applications of Radioisotopes

Both the potential of nuclear reactions to provide new sources of energy and the destructive capacity of these same nuclear reactions have been cited. Less heralded but also important are a variety of practical applications of radioisotopes. We close this chapter with a brief survey of a few of these applications.

Cancer Therapy.   We have noted how ionizing radiation in low dosages can induce cancers, but this same radiation, particularly γ rays, can also be used in the treatment of cancer. The basis of such treatment is that, although the radiation tends to destroy all cells, cancerous cells are more easily destroyed than normal ones. Thus, a carefully directed beam of γ rays or high-energy x rays of the appropriate dosage may be used to arrest the growth of cancerous cells. Also coming into use for some forms of cancer is radiation therapy using beams of neutrons.

Radioactive Tracers.   Radioactive isotopes are different from nonradioactive ones only in the instability of their nuclei, not in physical or chemical properties. Thus, in any physical or chemical process radioactive and nonradioactive isotopes are expected to behave in the same way. This fact serves as the fundamental principle in the use of radioactive tracers or "tagged" atoms. For example, if a small quantity of artificially produced radioactive ^{32}P (as a phosphate) is added to a nutrient solution that is fed to plants, the uptake of the phosphorus can be followed by charting the regions of the plant that become radioactive. Similarly, the fate of iodine in the human body can be determined by having an individual drink a solution of dissolved iodides containing a small quantity of radioactive iodine as a tracer. Abnormalities in the thyroid gland can be detected in this way.

Industrial applications of tracers are also numerous. The fate of a catalyst in a chemical plant can be followed by incorporating a radioactive tracer in the catalyst, for example, ^{192}Ir in a Pt–Ir catalyst. By monitoring the activity of the ^{192}Ir one can determine the rate at which the catalyst is being carried away and to which parts of the plant.

Structures and Mechanisms.   Often detailed knowledge of the mechanism of a chemical reaction or the structure of a species can be inferred from experiments using radioisotopes as tracers. Consider the following experimental proof of the statement made in Section 21-2 that the two S atoms in the thiosulfate ion, $S_2O_3^{2-}$, are not equivalent.

$S_2O_3^{2-}$ is prepared from radioactive sulfur (^{35}S) and sulfite ion containing the nonradioactive isotope ^{32}S.

$$^{35}S + {}^{32}SO_3^{2-} \longrightarrow {}^{35}S^{32}SO_3^{2-} \tag{25.31}$$

When the thiosulfate ion is decomposed by acidification, all the radioactivity appears in the precipitated sulfur and none in the $SO_2(g)$. The ^{35}S atoms must be bonded in a different way than the ^{32}S atoms (see Figure 21-10).

$$^{35}S^{32}SO_3^{2-} + 2\ H^+ \longrightarrow H_2O + {}^{32}SO_2(g) + {}^{35}S(s) \tag{25.32}$$

In reaction (25.33) nonradioactive KIO_4 is added to a solution containing iodide ion labeled with the radioisotope, ^{128}I. All the radioactivity appears in the I_2 and none in the IO_3^-. This proves that all the IO_3^- is produced by reduction of IO_4^- and none by oxidation of I^-.

$$IO_4^- + 2\ {}^{128}I^- + H_2O \longrightarrow {}^{128}I_2 + IO_3^- + 2\ OH^- \tag{25.33}$$

If a chromium(III) salt containing radioactive ^{51}Cr is added to a chrome-plating bath, no radioactivity shows up in the chrome plate. This proves that the reduction of Cr(VI) to Cr(0) at the cathode does not proceed through the oxidation state Cr(III). (Chrome plating was discussed in Section 23-5.)

Analytical Chemistry.   We have learned how a substance can be analyzed by precipitation from solution. The usual procedure involves filtering, washing, drying, and weighing a pure precipitate. An alternative is to allow the substance to be analyzed to react with a reagent containing a radioisotope. By measuring the activity of the precipitate and comparing it with that of the original solution, it is possible to calculate the amount of precipitate without having to purify, dry, and weigh it. (For example, Ag^+ in solution can be precipitated as radioactive AgCl by treatment with a solution containing radioactive Cl^-.)

Another method of importance in analytical chemistry is **neutron activation analysis.** In this procedure the sample to be analyzed, normally nonradioactive, is bombarded with neutrons; the element of interest is converted to a radioisotope. The radioactivity of this radioisotope is measured. This measurement is combined with a knowledge of such factors as the rate of neutron bombardment, the half-life of the radioisotope, and the efficiency of the radiation detector to calculate the quantity of the element in the sample. The method is especially attractive because (1) trace quantities of elements can be determined (sometimes in parts per billion or less); (2) a sample can be tested without destroying it; and (3) the sample can be in any state of matter, including biological materials. Among its many uses, neutron activation analysis has been used to study archeological artifacts and to determine the authenticity of old paintings. (Old masters formulated their own paints. Differences between formulations are easily detected through the trace elements they contain.)

Radiation Processing. This term describes industrial applications of ionizing radiation—γ rays from ^{60}Co or electron beams from electron accelerators. The ionizing radiation is used in the production of certain materials or to modify their properties. Its most extensive current use is in breaking, reforming, and cross-linking polymer chains to affect the physical and mechanical properties of plastics used in foamed products, electrical insulation, and packaging materials. Exposure to ionizing radiation is used to sterilize medical supplies such as sutures, syringes, and hospital garb. In sewage treatment plants radiation processing has been used to decrease the settling time of sewage sludge and to kill pathogens. An important future use may be in the preservation of foods, that is, as an alternative to canning, freeze-drying or refrigeration. A significant aspect of radiation processing is that, unlike in neutron activation analysis, for example, the irradiated material is *not* rendered radioactive. Also of interest is the fact that radiation processing can produce desirable effects at a very low energy cost.

Summary

Radioactivity refers to the ejection of particles (α, β^-, β^+), the capture of electrons from an inner shell, or the emission of electromagnetic radiation (γ) by unstable nuclei. Often these processes are part of a radioactive decay series. That is, they occur through several steps until a stable (nonradioactive) nucleus is finally produced. With the exception of γ ray emission, radioactive decay processes lead to the transformation (transmutation) of one element into another.

All nuclides having atomic number greater than 83 are radioactive. Although a few occur naturally, most radioactive nuclides of lower atomic number are produced artificially, by bombarding appropriate target nuclei with energetic particles. The high-energy particles induce a nuclear reaction that yields the desired nuclide. Equations can be written for nuclear reactions, whether they occur through particle bombardment or by spontaneous radioactive decay. In either case the sum of the atomic numbers and the sum of the mass numbers must be constant between the two sides of the equation.

The rate of radioactive decay is directly proportional to the number of atoms in a sample. Each radioactive nuclide has a characteristic decay constant and half-life, and calculations of decay rates can be made using equations similar to those for first-order chemical kinetics. Measurements of decay rates of radioactive nuclides have a number of practical applications, ranging from determining the ages of rocks to the dating of wooden objects (carbon-14 dating).

Large quantities of energy are associated with nuclear reactions, sufficiently large that mass changes occur. The quantity of energy released in the formation of a nucleus from protons and neutrons can be calculated; and when these nuclear binding energies per nucleon are plotted as a function of mass number, a distinctive graph is obtained (Figure 25-6). From this graph one can establish that fission of heavy nuclei and fusion of lighter nuclei yield large quantities of energy. The fission process is the basis of the atomic bomb and nuclear reactors. Fusion is the energy-producing process of the stars, the hydrogen bomb, and the, as yet unperfected, thermonuclear reactor.

The stability of a nucleus depends on several factors, among them being whether the numbers of protons and neutrons are odd or even, and whether either of these is a "magic number." The most important factor is the neutron-to-proton ratio in the nucleus and whether this ratio lies within the belt of stable nuclides (Figure 25-7). Nuclides outside the belt of stability are radioactive. Their placement with respect to this belt can generally serve to indicate whether radioactive decay will occur by α, β^-, or β^+ emission or by electrons capture.

One of the principal effects of the interaction of radiation with matter is the production of ions. This phenomenon can be used to detect radiation, and it is also the basis of radiation damage to living matter. Several methods have been developed to measure radiation dosages and to predict the biological effects of these dosages, but much uncertainty remains. Despite the hazards associated with radioactivity, many beneficial applications exist as well. Radioactive nuclides are used in cancer therapy, in basic studies of chemical structures and mechanisms, in analytical chemistry, and in chemical industry.

Learning Objectives

As a result of studying Chapter 25, you should be able to

1. Name the different types of radioactive decay processes and describe the characteristics of their radiation.

2. Write nuclear equations for radioactive decay processes.

3. Describe the three natural radioactive decay series, using the uranium series as an example.

4. Write equations for nuclear reactions produced artificially.

5. Name some of the transuranium elements and describe how they are made.

6. Describe the principles involved in the operation of a charged-particle accelerator.

7. Calculate the rate of radioactive decay, the half-life, or the number of atoms in a sample of a radioactive nuclide if two of the three quantities are known.

8. Determine the ages of rocks from a measured mass ratio of a stable nuclide to a radioactive one (such as $^{206}Pb/^{238}U$), and the ages of carbon-containing materials from the decay rate of carbon-14.

9. Calculate the energies associated with nuclear reactions.

10. Calculate the average nuclear binding energy per nucleon for a nuclide.

11. Describe the factors that determine nuclear stability, establish whether a particular nuclide is likely to be stable or radioactive, and predict the type of decay process expected for a radioactive nuclide.

12. Describe the processes of nuclear fission and nuclear fusion, including the problems with using them as energy sources.

13. Explain the effects of ionizing radiation on matter and describe several radiation-detection devices based on these effects.

14. Discuss methods of expressing radiation dosages, some biological hazards of ionizing radiation, and sources of radiation to which the general population is exposed.

15. Discuss some practical, beneficial uses of radioisotopes.

Some New Terms

An **alpha (α) particle** is a combination of two protons and two neutrons identical to the nucleus of an ordinary helium atom, that is, $^4_2He^{2+}$.

A **beta (β^-) particle** is an electron emitted as a result of the conversion of a neutron to a proton in a radioactive nucleus.

A **breeder reactor** is a nuclear reactor that creates more nuclear fuel than it consumes, for example by converting ^{238}U to ^{239}Pu.

A **charged-particle accelerator** is a device that imparts high energies to charged particles for use in nuclear reactions.

A **cloud chamber** is a device used to detect ionizing radiation through the formation of a trail of droplets along the path traveled by radiation through a supersaturated vapor.

A **curie** is a quantity of radioactive material decaying at the same rate as 1 g of radium (3.7×10^{10} dis/s).

Electron capture (E.C.) is a form of radioactive decay in which an electron from an inner electronic shell is absorbed by a nucleus. In the nucleus the electron is used to convert a proton to a neutron.

A **gamma (γ) ray** is a form of electromagnetic radiation emitted by certain radioactive nuclei.

A **Geiger-Müller counter** is a device used to detect ionizing radiation. Each ionizing event that occurs in the counter produces an electric discharge that can be recorded.

The **half-life** of a radioactive nuclide is the time required for one half of the atoms present in a sample to undergo radioactive decay.

Magic numbers is a term used to describe numbers of protons and neutrons that confer a special stability to an atomic nucleus.

Nuclear binding energy is the energy released when nucleons (protons and neutrons) are fused into an atomic nucleus. This energy replaces an equivalent quantity of matter.

Nuclear fission is a radioactive decay process in which a heavy nucleus breaks up into two lighter nuclei and several neutrons.

In **nuclear fusion** small atomic nuclei are fused into larger ones, with some of their mass being converted to energy.

A **nuclear reactor** is a device in which nuclear fission is carried out as a controlled chain reaction. That is, neutrons produced in one fission event trigger the fission of another nucleus, and so on.

A **positron (β^+)** is a *positive* electron emitted as a result of the conversion of a proton to a neutron in a radioactive nucleus.

A **rad** is a quantity of radiation able to deposit 1×10^{-2} J of energy per kilogram of matter.

A **radioactive decay series** is a succession of individual steps whereby an initial radioactive nuclide (e.g., ^{238}U) is ultimately converted to a stable nuclide (e.g., ^{206}Pb).

Radiocarbon dating is a method of determining the age of a carbon-containing material based on the rate of decay of radioactive carbon-14.

A **rem** is a unit of radiation related to the rad, but taking into account the varying effects on biological matter of different types of radiation of the same energy.

A **transuranium element** is one with an atomic number $Z > 92$.

Suggestions for Further Study

COTTER, M. J., "Neutron Activation Analysis of Paintings," *American Scientist,* **69,** 17 (1981).

KULCINSKI, G. L., et al., "Energy for the Long Run: Fission or Fusion," *American Scientist,* **67,** 78 (1979).

League of Women Voters Education Fund, *A Nuclear Power Primer: Issues for Citizens* (Washington, DC: League of Women Voters of the United States, 1982).

MAMMANO, N. J., "A Chemistry Lesson at Three Mile Island," *J. Chem. Educ.,* **57,** 286 (1980).

MYERS, H. G., "Radioisotopes in Plant Operations," *Chemtech,* **11,** 489 (1981).

SEABORG, G. T., "The New Elements," *American Scientist,* **68,** 279 (1980).

SILVERMAN, J., "Radiation Processing: The Industrial Applica-

tions of Radiation Chemistry," *J. Chem. Educ.,* **58,** 168 (1981).*

UPTON, A. C., "The Biological Effects of Low-Level Ionizing Radiation," *Scientific American,* **246**[2], 41 (1982).

WILSON, R. R., "U.S. Particle Accelerators at Age 50," *Physics Today,* **34**[11], 86 (1981).

YALOW, R. S., "Radioactivity in the Service of Man," *J. Chem. Educ.,* **59,** 735 (1982).

ZURER, P. S., "U.S. Charts Plans for Nuclear Waste Disposal," *Chem. Eng. News,* **61**[29], 20 (1983).

*Several other articles on radiation chemistry will be found in the same issue of this journal.

Review Problems

1. Which of the following—α, β, or γ—generally has the greatest **(a)** penetrating power through matter; **(b)** ionizing power in matter; **(c)** deflection in a magnetic field?

2. Supply the missing information in each of the following nuclear equations representing a radioactive decay process.
   (a) $^{32}_{16}S \rightarrow ^{?}_{17}Cl + _{-1}^{0}e$ **(b)** $^{14}_{8}O \rightarrow ^{14}_{7}N + ?$
   (c) $^{235}_{?}U \rightarrow ^{?}_{?}Th + ?$ **(d)** $^{214}? \rightarrow ^{?}_{?}Po + _{-1}^{0}e$

3. Complete the following nuclear equations.
   (a) $^{23}_{11}Na + ^{2}_{1}H \rightarrow ? + ^{1}_{1}H$
   (b) $^{59}_{27}Co + ? \rightarrow ^{56}_{25}Mn + ^{4}_{2}He$
   (c) $^{238}_{92}U + ^{2}_{1}H \rightarrow ? + _{-1}^{0}e$
   (d) $^{246}_{96}Cm + ^{13}_{6}C \rightarrow ^{254}_{102}No + ?$
   (e) $^{238}_{92}U + ^{14}_{7}N \rightarrow ^{246}_{?}Es + ? \, ^{1}_{0}n$

4. Write an equation to represent each of the following nuclear processes.
   (a) The reaction of two deuterium nuclei (deuterons) to produce a nucleus of ^{3}He.
   (b) The production of $^{243}_{97}$Bk by the α particle bombardment of $^{241}_{95}$Am.
   (c) The bombardment of $^{121}_{51}$Sb by α particles to produce $^{124}_{53}$I followed by its radioactive decay by positron emission.

5. For the radioactive nuclides in Table 25-1,
   (a) Which one has the largest value of the decay constant, γ?
   (b) Which one would display a 75% reduction in radioactivity from its current value in approximately two days?

   (c) Which ones would have lost more than 99% of their radioactivity in one month?

6. A sample of radioactive $^{35}_{16}$S is found to disintegrate at a rate of 1.00×10^3 atoms/min. The half-life of $^{35}_{16}$S is 87.9 d. How long will it take for the activity of this sample to decrease to the point of producing **(a)** 115; **(b)** 86; and **(c)** 43 dis/min?

7. A wooden art object is claimed to have been found in an Egyptian pyramid and is offered for sale to an art museum. Nondestructive radiocarbon dating of the object reveals a disintegration rate of 12 dis min^{-1} per g C. Do you think the object is authentic?

8. With appropriate equations in the text, determine
   (a) The energy in joules corresponding to the destruction of 1.05×10^{-23} g of matter.
   (b) The energy in MeV that would be released if one α particle was completely destroyed.
   (c) The number of neutrons that could conceivably be created from 1.50×10^6 MeV of energy.

9. The measured mass of the nuclide $^{20}_{10}$Ne is 19.99244 u. Determine the binding energy per nucleon (in MeV) in this atom. (*Hint:* The nuclidic mass includes the mass of electrons as well as that of the nucleus. Also, recall Figure 25-5.)

10. Two of the following isotopes do not occur naturally. Which are they? **(a)** ^{2}H; **(b)** ^{32}S; **(c)** ^{80}Br; **(d)** ^{132}Cs; **(e)** ^{184}W.

Exercises

Definitions and terminology

1. Describe briefly the meaning of each of the following concepts or terms introduced in this chapter: (a) neutron-to-proton ratio; (b) nucleon; (c) mass-energy relationship; (d) background radiation; (e) radioactive decay series; (f) nuclear accelerator.

2. Explain the difference in meaning between the following pairs of terms: (a) naturally occurring and artificially produced radioisotope; (b) electron and positron; (c) primary and secondary ionization; (d) transuranium and transactinoid element; (e) nuclear fission and nuclear fusion.

3. What is meaning of each of these symbols in describing nuclear phenomena? (a) α; (b) γ; (c) $t_{1/2}$; (d) γ; (e) β^+.

4. Supply a name or symbol for each of the following nuclear particles: (a) ^4_2He; (b) beta particle; (c) neutron; (d) ^1_1H; (e) $^0_{+1}\text{e}$; (f) tritium.

Radioactive processes

5. What is the nucleus obtained in each process?
   (a) $^{234}_{94}\text{Pu}$ decays by α emission.
   (b) $^{248}_{97}\text{Bk}$ decays by β^- emission.
   (c) $^{196}_{82}\text{Pb}$ goes through two successive E.C. processes.
   (d) $^{214}_{82}\text{Pb}$ decays through two successive β^- emissions.
   (e) $^{226}_{88}\text{Ra}$ decays through three successive α emissions.
   (f) $^{69}_{33}\text{As}$ decays by β^+ emission.

6. Both β^- (electron) and β^+ (positron) emission are observed for artificially produced radioisotopes of low atomic numbers, but only β^- (electron) emission is observed with naturally occurring radioisotopes of high atomic number. What is the reason for this observation?

Radioactive decay series

7. The natural decay series starting with the radionuclide $^{232}_{90}\text{Th}$ follows the sequence represented below. Construct a graph of this series, similar to Figure 25-3.

$$^{232}_{90}\text{Th}-\alpha-\beta-\beta-\alpha-\alpha-\alpha \begin{array}{c} \alpha-\beta \quad \alpha-\beta \\ \diagup \diagdown \diagup \diagdown \\ \diagdown \diagup \diagdown \diagup \\ \beta-\alpha \quad \beta-\alpha \end{array} {}^{208}_{82}\text{Pb}$$

8. The uranium series described in Figure 25-3 is also known as the "4n + 2" series because the mass number of each nuclide in the series can be expressed by the equation $A = 4n + 2$, where n is an integer. Show that this equation is indeed applicable to the uranium series.

Nuclear reactions

9. Write out the nuclear equations represented by the following symbolic notation: (a) $^7\text{Li}(p,\gamma)^8\text{Be}$; (b) $^{33}\text{S}(n,p)^{33}\text{P}$; (c) $^{239}\text{Pu}(\alpha,n)^{242}\text{Cm}$; (d) $^{238}\text{U}(\alpha,3n)^{239}\text{Pu}$.

10. Write an equation for each of the nuclear reactions represented by Figure 25-3.

Rate of radioactive decay

11. The disintegration rate for a sample containing $^{60}_{27}\text{Co}$ as the only radioactive nuclide is found to be 185 atoms/min. The half-life of $^{60}_{27}\text{Co}$ is 5.2 y. Estimate the number of atoms of $^{60}_{27}\text{Co}$ in the sample.

12. How long must the radioactive sample of Exercise 11 be maintained before the disintegration rate falls to 101 dis/min?

13. The radioisotope $^{32}_{15}\text{P}$ is used extensively in biochemical studies. Its half-life is 14.2 d. Suppose that a sample containing this isotope has an activity 1000 times the detectable limit. For how long a time could an experiment be run with this sample before the radioactivity could no longer be detected? (*Hint:* Use $N_0 = 1.00 \times 10^3$ and $N_t = 1.00$.)

14. A sample containing $^{234}_{88}\text{Ra}$, which decays by α particle emission, is observed to disintegrate at the following rate, expressed as disintegrations per minute or counts per minute (cpm). What is the half-life of this nuclide? $t = 0$, 1000 cpm; $t = 1$ h, 992 cpm; $t = 10$ h, 924 cpm; $t = 100$ h, 452 cpm; $t = 250$ h, 138 cpm.

15. The unit **curie** is defined as a disintegration rate of 3.7×10^{10} dis/s. What mass of ^{226}Ra, with a half-life of 1602 y, is required to produce 1.00 millicurie of radiation?

16. If a meteorite is approximately 4.5×10^9 y old, what should be the mass ratio $^{208}\text{Pb}/^{232}\text{Th}$ in the meteorite? The half-life of ^{232}Th is 1.39×10^{10} y.

17. One method of dating rocks is based on their $^{87}\text{Sr}/^{87}\text{Rb}$ ratio. The ^{87}Rb is a β^- emitter with a half-life of 5×10^{11} y. A certain rock is found to have a mass ratio $^{87}\text{Sr}/^{87}\text{Rb}$ of $0.004 : 1.00$. What is the age of the rock?

Radiocarbon dating

18. What should be the current rate of disintegration of ^{14}C, expressed in dis min^{-1} per g C, for a wooden object that is believed to have been made in 1100 B.C.?

19. The lowest level of ^{14}C activity that seems possible for experimental detection is 0.03 dis min^{-1}/g C. What is the maximum age of an object that can be determined by the carbon-14 method?

⋆20. The carbon-14 dating method is based on the assumption that the rate of production of ^{14}C by cosmic ray bombardment has remained constant for thousands of years and that the ratio of ^{14}C to ^{12}C has also remained constant. Can you think of any effects of human activities that could invalidate this assumption in the future?

Energetics of nuclear reactions

21. Calculate the energy (in MeV) released in the nuclear reaction

$$^{10}_{5}B + ^{4}_{2}He \rightarrow ^{13}_{6}C + ^{1}_{1}H$$

The nuclidic masses are $^{10}_{5}B$ = 10.01294 u; $^{4}_{2}He$ = 4.00260 u; $^{13}_{6}C$ = 13.00335 u; $^{1}_{1}H$ = 1.00783 u.

22. You are given the following exact atomic masses; $^{6}_{3}Li$ = 6.01513 u; $^{4}_{2}He$ = 4.00260 u; $^{3}_{1}H$ = 3.01604 u; $^{1}_{0}n$ = 1.008665 u. How much energy is released in the nuclear reaction $^{6}_{3}Li + ^{1}_{0}n \rightarrow ^{4}_{2}He + ^{3}_{1}H$, expressed in MeV?

Nuclear stability

23. Which member of the following pairs of nuclides would you expect to be most abundant in natural sources? Explain your reasoning. **(a)** $^{20}_{10}Ne$ or $^{22}_{10}Ne$; **(b)** $^{17}_{8}O$ or $^{18}_{8}O$; **(c)** $^{6}_{3}Li$ or $^{7}_{3}Li$.

24. One member each of the following pairs of radioisotopes decays by β^- emission and the other by positron (β^+) emission. Which is which? Explain your reasoning. **(a)** $^{29}_{15}P$ and $^{33}_{15}P$; **(b)** $^{120}_{53}I$ and $^{134}_{53}I$.

25. Sometimes the most abundant isotope of an element can be established by rounding off the atomic weight to the nearest whole number, for example, ^{39}K, ^{85}Rb, and ^{88}Sr. But at other times the isotope corresponding to the ''rounded-off'' atomic weight does not even occur naturally, for example, ^{64}Cu. Explain the basis of this observation.

26. Some nuclides are said to be ''doubly magic.'' What do you suppose this term means? Postulate some nuclides that might be doubly magic and locate them in Figure 25-7.

Fission and fusion

27. Describe briefly what is meant by the following types of nuclear reactors: **(a)** burner; **(b)** breeder; **(c)** thermonuclear.

28. Based on Figure 25-6, can you explain why more energy is released in a fusion than in a fission process?

29. Use data from the text to determine how many metric tons (1 metric ton = 1000 kg) of bituminous coal (85% C) would have to be burned to release as much energy as is produced by the fission of 1.00 kg $^{235}_{92}U$.

Effect of radiation on matter

30. Explain why the rem is more satisfactory than the rad as a unit for measuring radiation dosage.

31. The Geiger–Müller counter is a much more efficient device for detecting and measuring γ rays than is a cloud chamber. Why do you think this is so?

32. Discuss briefly the basic difficulties in establishing the physiological effects of low-level radiation.

33. ^{90}Sr is both a product of radioactive fallout and a radioactive waste in a nuclear reactor. This radioisotope is a β^- emitter with a half-life of 27.7 y. Suggest reasons why ^{90}Sr is such a potentially hazardous substance.

Application of radioisotopes

34. Describe how you might go about finding a leak in the $H_2(g)$ supply line in an ammonia synthesis plant by using radioactive materials.

35. Explain why neutron activation analysis is so useful in determining trace elements in a sample, in contrast to ordinary methods of quantitative analysis such as precipitation or titration.

36. A small quantity of NaCl containing radioactive $^{24}_{11}Na$ is added to an aqueous solution of $NaNO_3$. The solution is cooled and $NaNO_3$ is crystallized from the solution. Would you expect the $NaNO_3$ to be radioactive? Explain.

37. The following reactions are carried out using HCl(aq) containing some tritium ($^{3}_{1}H$) as a tracer. Would you expect any of the tritium radioactivity to appear in the $NH_3(g)$? In the H_2O? Explain.

$$NH_3(aq) + HCl(aq) \rightarrow NH_4Cl(aq)$$
$$NH_4Cl(aq) + NaOH(aq) \rightarrow NaCl(aq) + H_2O + NH_3(g)$$

Additional Exercises

1. Each of the following isotopes is radioactive. Which would you expect to decay by β^- emission and which by positron (β^+) emission? **(a)** $^{28}_{15}P$; **(b)** $^{45}_{19}K$; **(c)** $^{72}_{30}Zn$.

2. The half-life of tritium is 12.26 y. What would be the rate of decay of tritium atoms, per second, in 1.00 L of hydrogen gas at STP containing 0.15% tritium atoms?

3. Explain the similarities and differences between a cloud chamber and a bubble chamber.

4. Use data from Table 2-1, together with the measured mass of the nuclide $^{16}_{8}O$, 15.99491 u, to determine the binding energy per nucleon (in MeV) in this atom.

5. If you follow the same description as in Exercise 8, the thorium series may be called the ''4n'' series and the actinium series the ''4n + 3'' series. A ''4n + 1'' series has also been established with $^{237}_{93}Np$ as the parent nuclide. To which radioactive series does each of the following nuclides belong? **(a)** $^{214}_{83}Bi$; **(b)** $^{216}_{84}Po$; **(c)** $^{215}_{85}At$; **(d)** $^{235}_{92}U$.

★6. Use the definition of Exercise 15 to determine how many millicuries of radiation are produced by a sample containing 5.10 mg ^{229}Th, which has the half-life of 7340 y?

★7. Calculate the minimum kinetic energy (in MeV) that α particles must possess to produce the nuclear reaction

$$^4_2\text{He} + {}^{14}_7\text{N} \rightarrow {}^{17}_8\text{O} + {}^1_1\text{H}$$

The nuclidic masses are $^4_2\text{He} = 4.00260$ u; $^{14}_7\text{N} = 14.00307$ u; $^1_1\text{H} = 1.00783$ u; $^{17}_8\text{O} = 16.99913$ u. (*Hint:* What is the increase in mass in this process?)

★8. The packing fraction of a nuclide is related to the fraction of the total mass of a nuclide that is converted to nuclear binding energy. It is defined as the fraction $(M - A)/A$, where M is the actual nuclidic mass and A is the mass number. Use data from a handbook (such as *The Handbook of Chemistry and Physics*, published by the CRC Press) to determine the packing fractions of some representative nuclides. Plot a graph of packing fraction versus mass number and compare it to Figure 25-6. Explain the relationship between the two.

★9. ^{40}K undergoes radioactive decay both by electron capture to ^{40}Ar and β^- emission to ^{40}Ca. The fraction of the decay that occurs by electron capture is 0.110. The half-life of ^{40}K is 1.27×10^9 y. Assuming that a rock in which ^{40}K has undergone decay retains all of the ^{40}Ar produced, what would be the ^{40}Ar/^{40}K mass ratio in a rock that is 1.5×10^9 y old?

★10. Reference is made in the text to using a certain shale deposit in the Appalachian Mountains containing 0.006% U as a potential fuel in a breeder reactor. Assuming a density of 2.5 g/cm^3, how much energy could be released from 1.00×10^3 cm^3 of this material? Assume a fission energy of 3.20×10^{-11} J per fission event (i.e., per U atom).

Self-Test Questions

For questions 1 through 8 select the single item that best completes each statement.

1. Of the following types of radiation, the only one to be deflected in a magnetic field is (a) x ray; (b) γ ray; (c) β ray; (d) neutrons.

2. A process that produces a one-unit increase in atomic number is (a) electron capture; (b) β^- emission; (c) α emission; (d) γ ray emission.

3. Of the following nuclides, the one most likely to be radioactive is (a) ^{31}P; (b) ^{66}Zn; (c) ^{37}Cl; (d) ^{108}Ag.

4. One of the following elements has eight naturally occurring *stable* isotopes. We should expect that one to be (a) Ra; (b) Au; (c) Cd; (d) Br.

5. Of the following nuclides, the one most likely to decay by positron (β^+) emission is (a) ^{59}Cu; (b) ^{63}Cu; (c) ^{67}Cu; (d) ^{68}Cu.

6. Among the following nuclides, the highest nuclear binding energy per nucleon is found for (a) $^3_1$H; (b) $^{16}_8$O; (c) $^{56}_{26}$Fe; (d) $^{235}_{92}$U.

7. The most radioactive of the isotopes of an element is the one with the largest value of its (a) half-life, $t_{1/2}$; (b) neutron number, N; (c) mass number, Z; (d) radioactive decay constant, λ.

8. Given a radioactive nuclide with $t_{1/2} = 1.00$ h and a current disintegration rate of 1000 atoms s^{-1}, three hours from now the disintegration rate will be (a) 1000 atoms s^{-1}; (b) 333 atoms s^{-1}; (c) 250 atoms s^{-1}; (d) 125 atoms s^{-1}.

9. Write nuclear equations to represent (a) the decay of ^{230}Th by α particle emission; (b) the decay of ^{54}Co by positron emission; (c) the nuclear reaction ^{232}Th $(\alpha,4n)^{232}$U.

10. Iodine-129 is a product of nuclear fission, whether from an atomic bomb or a nuclear power plant. It is a β^- emitter with a 1.7×10^7 y half-life. How many disintegrations per second would occur in a sample containing 1.00 mg ^{129}I?

11. ^{223}Ra has a half-life of 11.4 d. How long would it take for the radioactivity associated with a sample of ^{223}Ra to decrease to 1% of its current value?

12. Explain why

(a) Radioactive nuclides with intermediate half-lives are generally more hazardous than those with extremely short or extremely long half-lives.

(b) Some radioactive substances are hazardous from a distance, whereas others must be taken internally to constitute a hazard.

(c) Argon is the most abundant of the noble gases in the atmosphere.

(d) Francium is such a rare element (less than about 30 g present in the earth's crust at any one time), and it cannot be extracted from minerals containing the alkali metals.

(e) Such extremely high temperatures will be required to develop a self-sustaining thermonuclear (fusion) process as an energy source.

26 Organic Chemistry

Several million compounds exist containing carbon atoms in combination with hydrogen, oxygen, nitrogen, or certain other elements. Organic chemistry is the chemistry of these compounds. The element carbon is singled out for special study because of the ability of carbon atoms to form strong covalent bonds with one another. Carbon atoms may join together into straight chains, branched chains, and rings. The nearly infinite number of possible bonding arrangements of carbon atoms into these chains and rings accounts for the vast number and variety of carbon compounds.

Originally, organic chemistry dealt only with compounds derived from living matter. Living matter was thought to possess a ''vital force'' necessary for the synthesis of these compounds. In 1828, the German chemist Friedrich Wöhler heated ammonium cyanate, derived from inorganic substances, and obtained the organic compound urea.

$$KOCN + NH_4Cl \longrightarrow KCl + NH_4OCN$$

$$\underset{\text{ammonium cyanate}}{NH_4OCN} \xrightarrow{\text{heat}} \underset{\text{urea}}{H_2NCONH_2}$$

The urea formed in this way proved to be identical to urea isolated from urine.

26-1 The Nature of Organic Compounds and Structures

The simplest organic compounds are those of carbon and hydrogen—hydrocarbons. The simplest of the hydrocarbons is methane, CH_4, the principal constituent of natural gas. Shown below for methane are three ways of representing an organic molecule. A Lewis structure shows the distribution of all valence electrons in a molecule. A structural formula focuses on the electrons involved in bond formation, using a dash to represent a single bond (double and triple dashes for double and triple bonds). A condensed formula conveys essentially the same information as a structural formula but in a single line.

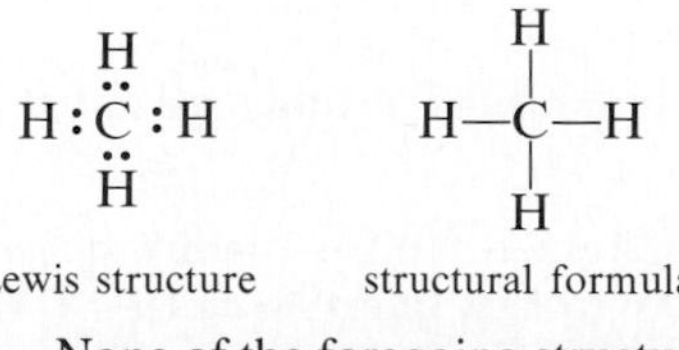

Lewis structure	structural formula	condensed formula

None of the foregoing structures describes the geometrical shape of the CH_4 molecule.

> "I must tell you that I can make urea without the use of kidneys, either man or dog. Ammonium cyanate is urea."
>
> F. Wöhler
> to J. J. Berzelius
> Feb. 22, 1828

797

FIGURE 26-1
Structural representation of the methane molecule.

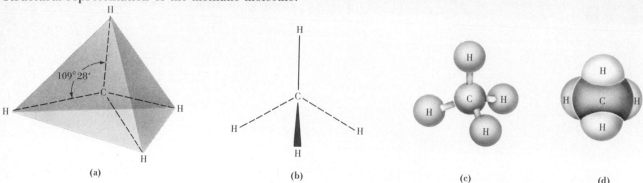

(a)   (b)   (c)   (d)

(a) Tetrahedral structure showing bond angle.
(b) Convention used to suggest a three-dimensional structure through a structural formula. The solid line represents a bond in the plane of the page. The dashed lines project *away* from the viewer and the heavy wedge projects *toward* the viewer.
(c) Ball-and-stick model.   **(d)** Space-filling model.

However, both from VSEPR theory (Section 9-8) and valence bond theory (Section 10-2) we expect the distribution of the four electrons pairs around the central carbon atom to be tetrahedral. In a CH_4 molecule the four H atoms are equivalent: They are equidistant from the C atom and attached to it by covalent bonds of equal strength. The angle between any two C—H bonds is $109°28'$. Molecular models are often used to represent organic molecules. Two widely used forms are illustrated in Figure 26-1. By increasing the number of carbon atoms in a molecule, other members of a hydrocarbon series can be represented, as in Figure 26-2.

Skeletal Isomerism.   From Figure 26-2 we see that there are *two* ways of assembling a hydrocarbon molecule with four carbon and ten hydrogen atoms. There are *two* different compounds with the formula C_4H_{10}. One is called butane and the other, isobutane. As we have already learned, compounds having the same molecular formula but different structural formulas are called **isomers.** Numerous possibilities for isomerism are found among organic compounds. In the case considered here the isomers differ in their carbon chains— one is a straight chain and the other, a branched chain. This type of isomerism is called **chain** or **skeletal isomerism.**

   The names given to the first four members of the hydrocarbon series in Figure 26-2 are common names. As the length of the carbon chain increases, a root name is used that reflects the number of carbon atoms in the chain. To this root is attached a characteristic "ane" ending. Certain branched chain compounds are often referred to by the prefix "iso." The hydrocarbon series initiated with methane continues beyond the four-carbon compound with these characteristic names: pentane (C_5H_{12}), hexane (C_6H_{14}), heptane (C_7H_{16}), octane (C_8H_{18}), nonane (C_9H_{20}), and decane ($C_{10}H_{22}$). All of the longer chain hydrocarbons display skeletal isomerism, and the more carbon atoms present the greater the number of possible isomers. There are 18 isomers of octane, 35 of nonane, 75 of decane, and so on.

Example 26-1   Write structural formulas for all the possible isomers of hexane, C_6H_{14}.

Solution.   The basic question is: In how many different ways can six C atoms be bonded together? The key to this question is the word *different*. For example, the following formulas are not different. (Think of all the C atoms as if they were "hinged." Each structure can be rearranged into a six-carbon straight chain.)

$$\begin{array}{ccc} & C-C & C-C-C \\ C-C-C-C-C-C & C-C\quad C-C & C-C-C \end{array} \quad \text{and so on}$$

We start with the *one* straight chain molecule, and for simplicity show only the carbon skeleton. (The complete structure requires adding the appropriate number of H atoms to produce four bonds at each C atom.)

$$\underset{(1)}{C-C-C-C-C-C}$$

Next we look for the possibilities involving a five-carbon chain with one C atom as a side chain. There are only *two* possibilities.

$$\underset{(2)}{\overset{\displaystyle C}{\underset{|}{C-C-C-C-C}}} \qquad \underset{(3)}{\overset{\displaystyle C}{\underset{|}{C-C-C-C-C}}}$$

FIGURE 26-2
Representation of some additional hydrocarbons.

Ethane:

$$\begin{array}{c} \;\;\overset{H}{|}\;\;\overset{H}{|} \\ H-C-C-H \\ \;\;\underset{H}{|}\;\;\underset{H}{|} \end{array} \qquad CH_3-CH_3$$

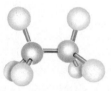 

Propane:

$$\begin{array}{c} \;\;\overset{H}{|}\;\;\overset{H}{|}\;\;\overset{H}{|} \\ H-C-C-C-H \\ \;\;\underset{H}{|}\;\;\underset{H}{|}\;\;\underset{H}{|} \end{array} \qquad CH_3-CH_2-CH_3$$

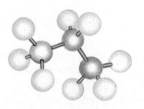 

Butane:

$$\begin{array}{c} \;\;\overset{H}{|}\;\;\overset{H}{|}\;\;\overset{H}{|}\;\;\overset{H}{|} \\ H-C-C-C-C-H \\ \;\;\underset{H}{|}\;\;\underset{H}{|}\;\;\underset{H}{|}\;\;\underset{H}{|} \end{array} \qquad CH_3-(CH_2)_2-CH_3$$

Isobutane:

$$\begin{array}{c} \;\;\overset{H}{|}\;\;\overset{H}{|}\;\;\overset{H}{|} \\ H-C-C-C-H \\ \;\;\underset{H}{|}\;\;\underset{|}{}\;\;\underset{H}{|} \\ \quad\;\; H-C-H \\ \quad\;\;\;\; \underset{H}{|} \end{array} \qquad HC(CH_3)_3$$

(a)   (b)

(a) Structural formulas.   (b) Condensed formulas.   (c) Ball-and-stick models.
(d) Space-filling models.

For example, if the following structure is "flopped" from left to right, it is seen to be identical to (2).

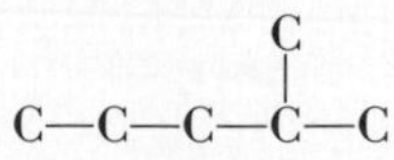

```
          C
          |
C—C—C—C—C
```

Now let us consider four-carbon chains with two one-carbon side chains. There is *one* possibility for side chains attached to *different* C atoms of the main chain (structure 4), and *one* possibility for side chains attached to the *same* C atom (structure 5).

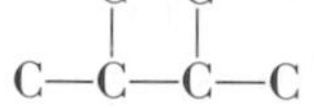

```
    C   C              C
    |   |              |
C—C—C—C          C—C—C—C
                       |
                       C
   (4)                (5)
```

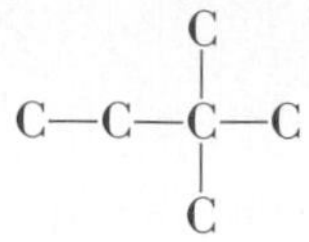

Any additional possibilities for a carbon skeleton prove to be identical to others already encountered. Thus, these two are identical to (5).

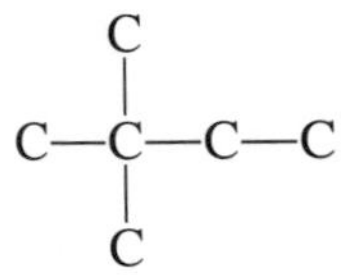

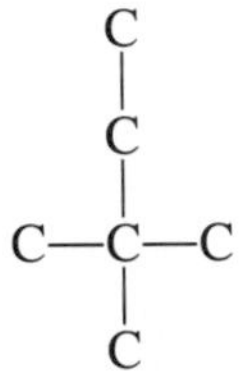

```
                    C
                    |
    C               C
    |               |
C—C—C—C        C—C—C
    |               |
    C               C
```

The number of isomers of hexane is 5.

SIMILAR EXAMPLES: Review Problems 2, 4; Exercise 7.

Nomenclature. Early in the history of organic chemistry, chemists assigned names of their own choosing to new compounds. Often these names were related to the origin or certain properties of the compounds, and some of these names are still in common use. Citric acid is found in citrus fruit; uric acid is present in urine; formic acid is found in ants (from the Latin word for ant, *formica*); and morphine induces sleep (from *Morpheus,* the ancient Greek god of sleep). As thousands upon thousands of new compounds were synthesized it became apparent that a system of nomenclature based on common names would be unworkable. Following several interim systems, one recommended by the International Union of Pure and Applied Chemistry (IUPAC or IUC) was adopted. In saturated hydrocarbons all carbon-to-carbon bonds are single bonds. A few of the more important rules for naming saturated hydrocarbons of the type, C_nH_{2n+2}, are these.

1. The generic (family) name of a saturated hydrocarbon is *alkane.*
2. Select the *longest* continuous carbon chain in the molecule and use the parent hydrocarbon name of this chain as the base name.
3. Every branch of the main chain is considered to be a substituent derived from another hydrocarbon. For these substituents the ending of the base name is changed from "ane" to "yl."
4. Number the carbon atoms of the continuous base chain so that the substituents appear *at the lowest numbers* possible.
5. Each substituent receives a name and number. For identical substituents use di, tri, tetra, and so on, and *repeat the numbers.*
6. Numbers are separated from other numbers by commas and from letters by dashes.
7. Arrange the substituents alphabetically by name, regardless of the numbers they carry or their complexity.
8. Whenever alternative base chains of equal length are possible, always name a compound so as to have the maximum number of side chains.

In applying rule 3, hydrocarbon substituents or alkyl groups should be named as follows.

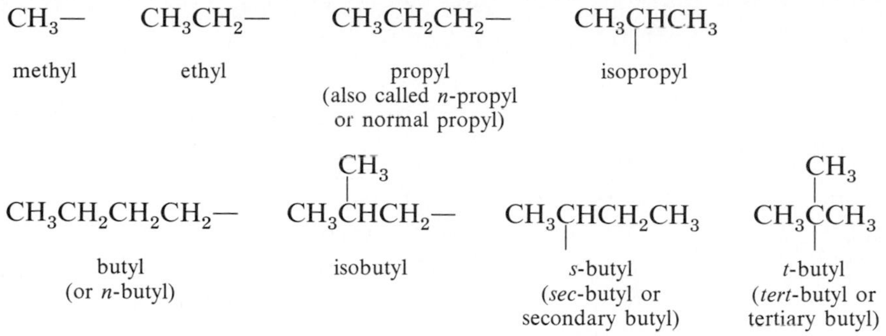

CH$_3$—   CH$_3$CH$_2$—   CH$_3$CH$_2$CH$_2$—   CH$_3$CHCH$_3$

methyl   ethyl   propyl   isopropyl
                              (also called *n*-propyl
                               or normal propyl)

CH$_3$CH$_2$CH$_2$CH$_2$—   CH$_3$CHCH$_2$—   CH$_3$CHCH$_2$CH$_3$ CH$_3$CCH$_3$

                              CH$_3$ CH$_3$

butyl   isobutyl   *s*-butyl   *t*-butyl
(or *n*-butyl)   (*sec*-butyl or   (*tert*-butyl or
                                  secondary butyl)   tertiary butyl)

Example 26-2   Give appropriate IUPAC names for the following compounds.

(a)
      CH$_3$ CH$_3$
       | |
CH$_3$—C—CH$_2$—CH—CH$_3$
 1 |2 3   4   5
      CH$_3$

(b)
                    CH$_3$
                     |
CH$_3$—CH$_2$—CH—CH$_2$
  1   2   3   4|
                    ^{5}CH$_2$
                     |
                    ^{6}CH$_3$

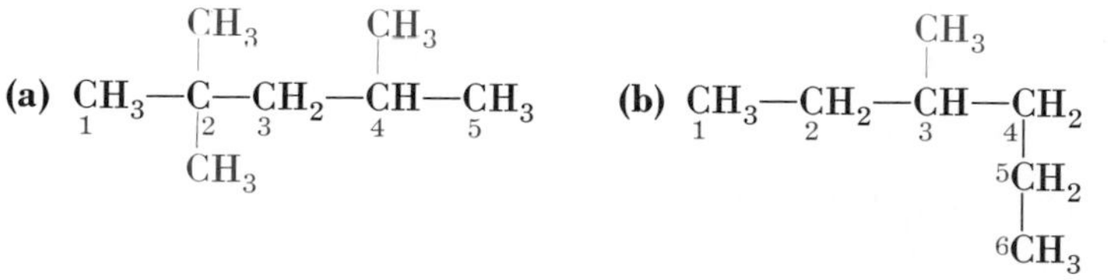

Solution

(a) The side chain substituents to be named are shown in color. Each is a methyl group, —CH$_3$. Two methyl groups are on the second carbon atom and one methyl, on the fourth. The main carbon chain has five atoms. The correct name is

2,2,4-trimethylpentane

If the carbon atoms in structure (a) were numbered from right to left, the name obtained would be 2,4,4-trimethylpentane. However, this is *not* an acceptable name. It does not use the smallest numbers possible.

(b) The chain length is six, not four. The methyl group is on the third carbon atom. The correct name is

3-methylhexane

SIMILAR EXAMPLES: Review Problem 6; Exercises 10, 12, 13.

Example 26-3   Write structural formulas and condensed formulas for the following compounds: (a) 4-*t*-butyl-2-methylheptane; (b) 2,6-dimethyl-3-ethylheptane.

Solution

(a) The substituent group, *t*-butyl, is attached to the fourth C atom in a seven-carbon chain. A methyl group is attached to the second C atom.

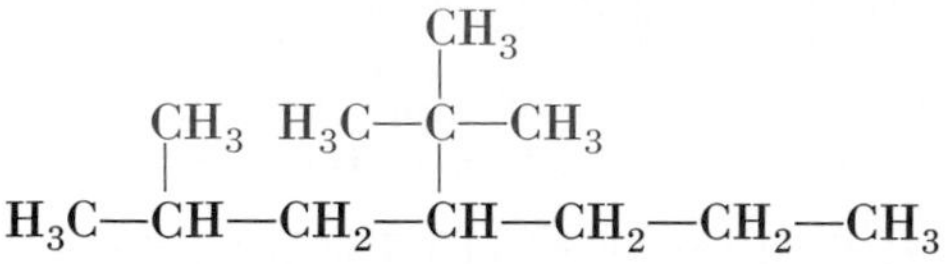

                  CH$_3$
                   |
      CH$_3$ H$_3$C—C—CH$_3$
       | |
H$_3$C—CH—CH$_2$—CH—CH$_2$—CH$_2$—CH$_3$

or

(CH$_3$)$_2$CHCH$_2$CH[C(CH$_3$)$_3$]CH$_2$CH$_2$CH$_3$
                    (condensed formula)

In arranging the substituent groups alphabetically as required in nomenclature rule 7, the symbols *n*, *s*, and *t* are not considered. Thus, butyl (even though *t*-butyl) precedes methyl in naming this structure.

(b) Methyl groups are substituted at the second and sixth C atoms and an ethyl group at the third.

$$
\begin{array}{ccc}
& CH_3 & \\
& | & \\
CH_3 & CH_2 & CH_3 \\
| & | & | \\
H_3C\text{---}CH\text{---}CH\text{---}CH_2\text{---}CH_2\text{---}CH\text{---}CH_3
\end{array}
$$

or

$(CH_3)_2CHCH(C_2H_5)CH_2CH_2CH(CH_3)_2$

(condensed formula)

It appears that this structure could also have been named 5-isopropyl-2-methyl-heptane. This is *not* done because only two side chain substituents would be involved instead of three (see rule 8).

SIMILAR EXAMPLES: Review Problems 7, 8; Exercises 11, 13, 14.

Positional Isomerism.   A variety of atoms or groups of atoms can be substituents on carbon chains, for example, Br. The three monobromopentanes are isomeric but they possess the same carbon skeleton. Because they differ in the position of the bromine atom on the carbon chain, they are called **positional isomers.**

$$
CH_3CH_2CH_2CH_2CH_2Br \qquad
\underset{\underset{\displaystyle Br}{|}}{CH_3CH_2CH_2CHCH_3} \qquad
\underset{\underset{\displaystyle Br}{|}}{CH_3CH_2CHCH_2CH_3}
$$

1-bromopentane   2-bromopentane   3-bromopentane

Example 26-4   Represent all the possible isomers of C_4H_9Cl.

Solution.   Perhaps the simplest approach is to consider the number of positional isomers for each of the skeletal isomers of butane. There are *two* possibilities based on *n*-butane (structures 1 and 2) and *two* based on isobutane (structures 3 and 4).

$$
CH_3CH_2CH_2CH_2Cl \qquad
\underset{\underset{\displaystyle Cl}{|}}{CH_3CH_2CHCH_3} \qquad
\underset{\overset{\displaystyle CH_3}{|}}{CH_3CHCH_2Cl} \qquad
\overset{\overset{\displaystyle CH_3}{|}}{\underset{\underset{\displaystyle Cl}{|}}{CH_3CCH_3}}
$$

(1)   (2)   (3)   (4)

SIMILAR EXAMPLES: Review Problem 4; Exercise 7.

Functional Groups.   Elements other than carbon and hydrogen are typically found in organic compounds as distinctive groupings of one or several atoms. In some cases these groupings are substituted for H atoms in hydrocarbon chains or rings, and in others, for C atoms themselves. These groupings of atoms are called **functional groups,** and the remainder of the molecule is referred to by the symbol **R.** Table 26-1 lists some of the functional groups most frequently encountered.

Example 26-5   Use information from Table 26-1 and elsewhere in this section to derive an acceptable name for the compound $(CH_3)_2CHOCH_2CH_2CH_3$.

Solution.   First we convert the condensed formula into a structural formula.

$$
\begin{array}{ccccccc}
& & H & & & & \\
& & | & & & & \\
H & H\text{---}C\text{---}H & H & H & H \\
| & | & | & | & | \\
H\text{---}C\text{------}C\text{---}O\text{---}C\text{---}C\text{---}C\text{---}H \\
| & | & | & | & | \\
H & H & H & H & H
\end{array}
$$

TABLE 26-1
Some classes of organic compounds[a]

Type of compound	General structural formula	Example	Name
alkanes	R—H	$CH_3CH_2CH_2CH_2CH_3$	pentane
alkenes	$\underset{R}{\overset{R}{C}}{=}\underset{R}{\overset{R}{C}}$	$CH_3CH_2CH_2CH{=}CH_2$	1-pentene
alkynes	R—C≡C—R	$CH_3CH_2C{\equiv}CCH_3$	2-pentyne
alcohols	R—OH	$CH_3CH_2CH_2CH_2CH_2OH$	1-pentanol
alkyl halides	R—X	$CH_3CH_2\underset{Br}{CH}CH_2CH_3$	3-bromopentane
ethers	R—O—R	$CH_3CH_2OCH_2CH_2CH_3$	ethyl propyl ether
aldehydes	$R{-}\overset{O}{\overset{\|}{C}}{-}H$	$CH_3CH_2CH_2CH_2\overset{O}{\overset{\|}{C}}H$	pentanal
ketones	$R{-}\overset{O}{\overset{\|}{C}}{-}R$	$CH_3CH_2\overset{O}{\overset{\|}{C}}CH_2CH_3$	3-pentanone
acids	$R{-}\overset{O}{\overset{\|}{C}}{-}OH$	$CH_3CH_2CH_2CH_2\overset{O}{\overset{\|}{C}}OH$	pentanoic acid
esters	$R{-}\overset{O}{\overset{\|}{C}}{-}O{-}R$	$CH_3CH_2CH_2CH_2\overset{O}{\overset{\|}{C}}OCH_3$	methyl pentanoate
amines	$R{-}NH_2$	$CH_3CH_2CH_2CH_2CH_2NH_2$	pentylamine

[a] Some of the functional groups appearing here and discussed later in the chapter have distinctive names: —OH, hydroxyl; $\overset{\diagdown}{\diagup}C{=}O$, carbonyl; $\overset{O}{\overset{\|}{-C}}$ —OH, carboxyl; —NH$_2$, amino.

The presence of the group C—O—C signifies that this is an ether. The group shown on the left is isopropyl; the one on the right, *n*-propyl or simply propyl. The compound can be called isopropyl propyl ether.

SIMILAR EXAMPLE: Review Problem 5.

26-2 Alkanes

In this section we explore further some properties of the alkanes. The essential characteristic of alkane hydrocarbon molecules is that only single covalent bonds are present. In these compounds the bonds are said to be saturated; the alkanes are known as **saturated hydrocarbons.**

The alkanes range in complexity from methane, CH_4 (accounting for over 90% of natural gas) to molecules containing 50 carbon atoms or more (found in petroleum). Each alkane differs from the preceding one by a —CH_2— or methylene group. This constant unit difference forms the basis of a **homologous series.** Members of such a series are usually closely related in chemical and physical properties. For example, a gradual increase in boiling point is noted with an increase in molecular weight. This trend of boiling points is understandable in terms of London-type forces introduced in Chapter 11. These forces increase in strength with increasing molecular weight. Also understandable from

TABLE 26-2
Boiling points of some isomeric alkanes

Family	Isomer	Boiling point, °C	Family	Isomer	Boiling point, °C
butane	*n*-butane	−0.5	hexane	*n*-hexane	68.7
	isobutane	−11.7		3-methylpentane	63.3
pentane	*n*-pentane	36.1		isohexane	60.3
	isopentane	27.9		2,3-dimethylbutane	58.0
	2,2-dimethylpropane	9.5		2,2-dimethylbutane	49.7
	(neopentane)			(neohexane)	

ideas presented in Chapter 11 are the data of Table 26-2, which show that the more branching on a carbon chain the lower the boiling point of the isomer (recall Figure 11-15).

Conformations. An important type of motion in alkane molecules is suggested by ball-and-stick models. This is a motion in which one group rotates with respect to the other. Two of the many possible orientations of the two —CH_3 groups in ethane are shown in Figure 26-3.

In one of these configurations, when the molecule is viewed head-on along the C—C bond, one set of C—H bonds is directly behind the other. This structure is referred to as the **eclipsed conformation.** In this conformation the distance between H atoms on adjacent C atoms is at a minimum, leading to a condition of maximum repulsion between H atoms. This conformation is slightly less stable than the staggered conformation. In the **staggered conformation** the hydrogen atoms are located a maximum distance apart. Although we expect the staggered conformation to be the more stable, at room temperature the thermal energy of an ethane molecule is sufficient that free rotation of the methyl groups about the C—C bond takes place. At lower temperatures, however, ethane does occur mostly in the staggered conformation. A similar situation is also encountered with higher alkanes.

Ring Structures. Alkanes in chain structures have the formula C_nH_{2n+2} and are called **aliphatic.** Alkanes can also exist in ring or cyclic structures called **alicyclic.** These rings can be thought of as resulting from the joining together of two ends of an aliphatic chain

FIGURE 26-3
Rotation about the
C—C bond in ethane.

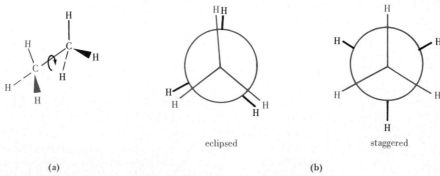

eclipsed   staggered

(a)   (b)

(a) Rotation about the C—C bond.
(b) Conformations of C_2H_6, in a representation known as a Newman projection. The view is along the C—C bond, end on.

by the elimination of a hydrogen atom from each end. Simple alicyclic compounds have the formula C_nH_{2n}.

$$\underset{\substack{\text{cyclopropane}\\ C_3H_6}}{\overset{\displaystyle CH_2}{\underset{\displaystyle H_2C\text{———}CH_2}{}}} \qquad \underset{\substack{\text{cyclobutane}\\ C_4H_8}}{\overset{\displaystyle H_2C—CH_2}{\underset{\displaystyle H_2C—CH_2}{|\quad\;\;|}}} \qquad \underset{\substack{\text{cyclopentane}\\ C_5H_{10}}}{\overset{\displaystyle CH_2}{\underset{\displaystyle H_2C—CH_2}{H_2C\quad\;CH_2}}} \qquad \underset{\substack{\text{cyclohexane}\\ C_6H_{12}}}{\overset{\displaystyle CH_2}{\underset{\displaystyle CH_2}{\underset{\displaystyle CH_2\quad CH_2}{CH_2\quad CH_2}}}}$$

The naming of alicyclic compounds follows the rules established in the preceding section. Thus the name of

is 1,3-dimethylcyclopentane. By convention, when a ring structure is drawn, usually neither the ring C atoms nor the H atoms bonded to them are written out.

The bond angles in cyclopropane are 60° compared to the normal 109.5°; the bonds are quite strained. As a result there are numerous reactions in which the ring will break open to yield a chain molecule—propane or a propane derivative. Cyclopropane is more reactive than most alkanes.

If the four carbon atoms in cyclobutane were to exist in the same plane, the C—C bond angles would be 90°. In reality the molecule buckles slightly to relieve the bond strain. If the cyclopentane molecule existed as a planar structure, the bond angles would be 108°, which is quite close to the normal 109.5°. However, in this planar structure all the H atoms would be eclipsed. A more stable arrangement is for one of the carbon atoms to be buckled out of the plane of the other four and for the H atoms to adopt a more staggered conformation.

With ball-and-stick models it can be seen that there are two possible conformations of cyclohexane. These are the **"boat"** and the **"chair"** conformations, pictured in Figure 26-4. The models demonstrate clearly that H atoms on the first and fourth C atoms in the boat form come close enough together to repel one another. In the chair form all the H atoms are staggered. The chair form is the more stable conformation of cyclohexane. Also evident from Figure 26-4 is the fact that the twelve H atoms in cyclohexane are not quite equivalent. Six of them extend outward from the ring and are called **equatorial** H atoms. Of the other six, three are directed above and three below the ring; these are the six **axial** H atoms. When another group (say —CH₃) is substituted for an H atom on the ring, the

FIGURE 26-4
Conformations of cyclohexane.

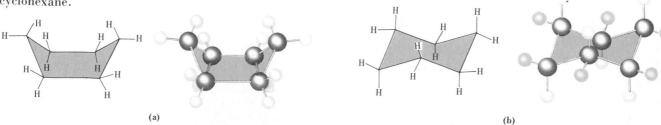

(a)

(b)

(a) Boat form. **(b)** Chair form; the equatorial H atoms are shown in color; the other H atoms are axial.

preferred position is an equatorial one. This causes a minimum interference with other groups on the ring.

Preparation of Alkanes. The primary source of alkanes is petroleum, but several laboratory methods have also been developed for their preparation. Some of these use organic substances of types that are described later in this chapter. Unsaturated hydrocarbons, whether containing double or triple bonds, may be converted to saturated hydrocarbons by the addition of hydrogen to the multiple bond system in the presence of a metal catalyst (equations 26.1 and 26.2). In the Würtz reaction halogenated hydrocarbons react with alkali metals to produce alkanes of double the carbon content (26.3). Alkali metal salts of carboxylic acids may be fused with alkali hydroxides. Here a hydrocarbon is obtained containing one carbon atom less than the original acid salt (26.4).

$$CH_2{=}CH_2 + H_2 \xrightarrow[\text{heat/pressure}]{\text{Pt or Pd}} CH_3{-}CH_3 \tag{26.1}$$
$$\text{ethane}$$

$$HC{\equiv}CH + 2\,H_2 \xrightarrow[\text{heat/pressure}]{\text{Pt or Pd}} CH_3{-}CH_3 \tag{26.2}$$
$$\text{ethane}$$

$$2\,CH_3CH_2Br + 2\,Na \longrightarrow 2\,NaBr + CH_3{-}CH_2{-}CH_2{-}CH_3 \tag{26.3}$$
$$\underset{\substack{\text{ethyl bromide}\\ \text{or bromoethane}}}{} \qquad\qquad \underset{\text{butane}}{}$$

$$\underset{\text{sodium acetate}}{CH_3\overset{\displaystyle O}{\overset{\displaystyle \|}{C}}ONa} + NaOH \xrightarrow{\text{heat}} Na_2CO_3 + \underset{\text{methane}}{CH_4} \tag{26.4}$$

The preceding equations illustrate that organic reactions are generally represented in a manner different from inorganic reactions. Frequently, only the organic reactant is shown on the left. Inorganic reagents and reaction conditions (and sometimes even organic reagents) are written above the arrow. Only the important products are shown on the right. The numerous by-products of organic reactions are usually omitted. Equations may or may not be written in balanced form.

Ordinary paraffin wax, used in candles, waxed paper, and home canning, is a mixture of long-chain paraffin hydrocarbons.

Substitution Reactions. Saturated hydrocarbons have little affinity for most chemical reagents. Because of this they have become known as paraffin hydrocarbons (Latin, *parum*, little; *affinis*, reactivity). Paraffin hydrocarbons are insoluble in water and do not react with aqueous solutions of acids, bases, or oxidizing agents. Halogens react only slowly with alkanes at room temperature; but at higher temperatures, particularly in the presence of light, halogenation occurs. In this reaction a halogen atom *substitutes* for a hydrogen atom. The mechanism of this substitution reaction appears to be by a chain reaction, written as follows for the chlorination of methane. (For emphasis, only the electrons involved in bond breakage or formation are shown in this reaction mechanism.)

initiation: $Cl{:}Cl \xrightarrow[\text{light}]{\text{heat or}} 2\,Cl\cdot$

propagation: $H_3C{:}H + Cl\cdot \longrightarrow H_3C\cdot\ + H{:}Cl$
$H_3C\cdot\ + Cl{:}Cl \longrightarrow H_3C{:}Cl + Cl\cdot$

termination: $Cl\cdot\ + Cl\cdot \longrightarrow Cl{:}Cl$
$H_3C\cdot\ + Cl\cdot \longrightarrow H_3C{:}Cl$
$H_3C\cdot\ + H_3C\cdot \longrightarrow H_3C{:}CH_3$

The reaction is initiated when some Cl_2 molecules absorb sufficient energy to dissociate into Cl atoms (represented above as Cl·). Cl atoms collide with CH_4 molecules to produce methyl radicals ($H_3C\cdot$), which combine with Cl_2 molecules to form the product, CH_3Cl. When any or all of the last three reactions proceed to the extent of consuming the radicals present, the reaction stops. This free-radical chain reaction usually yields a mixture of products. The net equation for the formation of chloromethane is

$$CH_4 + Cl_2 \xrightarrow[\text{light}]{\text{heat or}} CH_3Cl + HCl \tag{26.5}$$

Polyhalogenation can also occur to form CH_2Cl_2, dichloromethane (methylene dichloride, a solvent); $CHCl_3$, trichloromethane (chloroform, a solvent and anesthetic); and CCl_4, tetrachloromethane (carbon tetrachloride, a solvent).

Oxidation is the reaction of hydrocarbons underlying their important use as fuels. For example,

$$C_7H_{16}(l) + 11\,O_2(g) \longrightarrow 7\,CO_2(g) + 8\,H_2O(l) \quad \Delta \bar{H}^\circ = -4812\,\text{kJ/mol} \tag{26.6}$$

Hydrocarbon fuels are discussed further in Section 26-11.

26-3 Alkenes and Alkynes

Unsaturated hydrocarbons contain some multiple bonds between carbon atoms. The simple **alkenes** or **olefins** contain one double bond and have the general formula C_nH_{2n} in their straight chain or branched chain forms. Simple **alkynes** or **acetylenes** have one triple bond between carbon atoms and can be represented by the general formula C_nH_{2n-2}.

With slight modifications, the basic rules for naming alkanes apply also to alkenes and alkynes. The base chain is taken to be the longest chain containing the multiple bond. The ending "ene" is used for alkenes and "yne" for alkynes. Common names are established by considering alkenes to be derivatives of ethylene and alkynes, of acetylene. The carbon atoms of a chain are numbered in such a way as to place the multiple bond at the lowest possible number, and only that number is referred to in locating the multiple bond. (It is understood that the multiple bond is between this and the next-highest-numbered carbon atom.) Thus, 4-methyl-2-pentyne has a triple bond between the second and third C atoms of a five-carbon chain.

$$CH_2{=}CH_2 \qquad CH_3CH_2CH{=}CH_2$$

ethene   1-butene
(ethylene)   (ethylethylene)

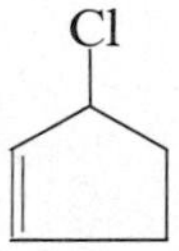

3-chlorocyclopentene

$$HC{\equiv}CH \qquad CH_3CH_2C{\equiv}CH \qquad CH_3CHC{\equiv}CCH_3$$
$$\hspace{7.5cm} | $$
$$\hspace{7.5cm} CH_3$$

ethyne   1-butyne   4-methyl-2-pentyne
(acetylene)   (ethylacetylene)   (isopropylmethylacetylene)

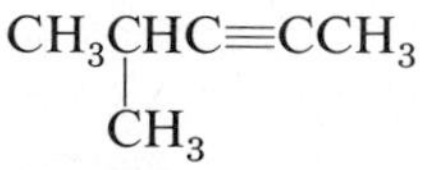

Example 26-6   What is the systematic name of the following structure?

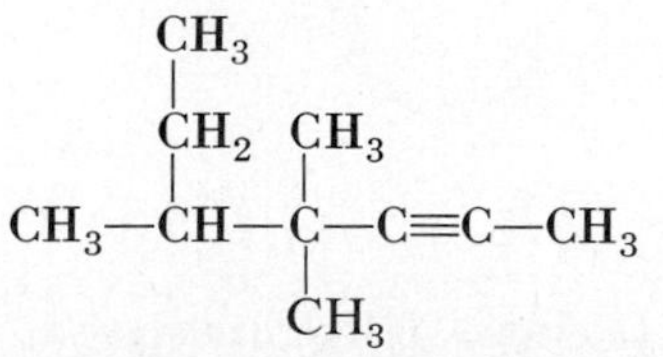

$$CH_3$$
$$|$$
$$CH_2 \quad CH_3$$
$$| \qquad |$$
$$CH_3{-}CH{-}C{-}C{\equiv}C{-}CH_3$$
$$\hspace{3cm} |$$
$$\hspace{3cm} CH_3$$

Solution. The longest chain is numbered to place the triple bond at the lowest possible number—2. Having made this decision, we now establish that there are three methyl groups at the positions 4, 4, and 5. The name of the compound is

4,4,5-trimethyl-2-heptyne

The name 4,4-dimethyl-5-ethyl-2-hexyne is unacceptable since it is not based on the longest possible hydrocarbon chain.

SIMILAR EXAMPLES: Review Problems 7, 8; Exercises 10, 13, 14.

The alkenes are similar to the alkanes in physical properties. At room temperature, those containing two to four carbon atoms are gases; those with 5 to 18 are liquids; and those with more than 18 are solids. In general, alkynes have higher boiling points than their alkane and alkene counterparts.

Geometric Isomerism. The compounds 2-butene, $CH_3CH=CHCH_3$, and 1-butene, $CH_2=CHCH_2CH_3$, are isomers. The difference between these two isomers is in the position of the double bond. This is another example of a positional isomer. Further consideration of the structure of 2-butene suggests another kind of isomerism.

H\
C=C
CH₃   CH₃
(a)

CH₃   H
C=C
H   CH₃
(b)

The terms *cis* and *trans* have the same meaning here as in the geometric isomerism of complex ions (Section 24-5).

In a double bond there occurs an overlap of $2p$ orbitals to form a π bond, *in addition to* the formation of a σ bond. Rotation about the double bond at room temperature is severely restricted. As a result the foregoing molecules (a) and (b) are distinct and different (see Figure 26-5). To distinguish them by name, (a) is designated *cis*-2-butene (*cis,* Latin, on the same side) and (b) is called *trans*-2-butene (*trans,* Latin, across). This type of isomerism is called **geometric isomerism.**

FIGURE 26-5
Geometric isomerism in 2-butene.

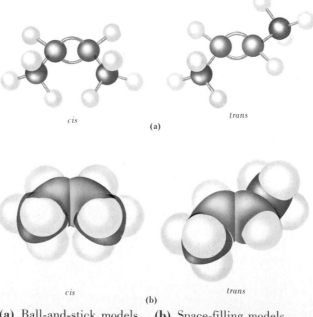

cis   *trans*
(a)

cis   *trans*
(b)

(a) Ball-and-stick models. **(b)** Space-filling models.

As we learned in Section 24-5, geometric isomerism is only one type of a general kind of isomerism called **stereoisomerism** (from the Greek word *stereos,* meaning solid or three-dimensional in nature). In stereoisomerism the number and types of atoms and bonds are the same, but certain atoms are oriented differently in space. Another type of stereoisomerism that was introduced in Chapter 24 is optical isomerism. Stereoisomerism, as we shall see in the next chapter, has profound implications in the specificity of the chemical reactions giving uniqueness to living organisms.

Preparation. Two general reactions for the preparation of olefins use alcohols and alkyl halides as starting materials (substrates). These are **elimination reactions,** processes in which atoms are removed from adjacent positions on a carbon chain. A small molecule is produced and an additional bond is formed between the C atoms. H_2O is eliminated in equation (26.7) and HBr in (26.8).

$$CH_3-\underset{\underset{HO}{|}}{\overset{\overset{H}{|}}{C}}-\underset{\underset{H}{|}}{\overset{\overset{H}{|}}{C}}-H \xrightarrow[\text{heat}]{H_2SO_4} CH_3CH{=}CH_2 + H_2O \tag{26.7}$$

$$CH_3\underset{\underset{H}{|}}{\overset{\overset{H}{|}}{C}}-\underset{\underset{Br}{|}}{\overset{\overset{H}{|}}{C}}-H \xrightarrow{\text{alcoholic KOH}} CH_3CH{=}CH_2 + KBr + HOH \tag{26.8}$$

Acetylene, the simplest alkyne, has been one of the most important organic raw materials used in industry. As we learned in Section 21-4, it is prepared from coal, water, and limestone.

$$CaCO_3 \xrightarrow{\text{heat}} CaO + CO_2 \tag{26.9}$$

$$CaO + 3\,C \xrightarrow[2000°C]{\substack{\text{electric} \\ \text{furnace,}}} CaC_2 + CO \tag{26.10}$$
calcium
acetylide
(calcium
carbide)

$$CaC_2 + 2\,H_2O \longrightarrow HC{\equiv}CH + Ca(OH)_2 \tag{26.11}$$

Most other alkynes are prepared from acetylene itself by taking advantage of the acidity of the C—H bond. In the presence of a very strong base, such as sodium amide, the acetylene donates protons to amide ions and forms a sodium salt, sodium acetylide (equation 26-12). This acetylide can then react with an alkyl halide (26.13).

$$H-C{\equiv}C-H + Na^+\,NH_2^- \longrightarrow NH_3 + H-C{\equiv}C^-\,Na^+ \tag{26.12}$$

$$H-C{\equiv}C^-\,Na^+ + CH_3Br \longrightarrow HC{\equiv}C-CH_3 + NaBr \tag{26.13}$$

By continuing this reaction, the triple bond can be positioned as desired in the chain, as in the synthesis of 2-pentyne.

$$H-C{\equiv}C-CH_3 \xrightarrow{\text{NaNH}_2} Na^+\,{}^-C{\equiv}C-CH_3 + NH_3 \tag{26.14}$$

$$Na^+\,{}^-C{\equiv}C-CH_3 + CH_3CH_2Br \longrightarrow CH_3CH_2C{\equiv}CCH_3 + NaBr \tag{26.15}$$

Addition Reactions.   The most significant distinction between alkanes and alkenes is that alkanes react by *substitution* and alkenes by *addition*.

alkane: $CH_3-CH_3 + Br_2 \longrightarrow CH_3-CH_2-Br + HBr$ (26.16)

alkene: $CH_2{=}CH_2 + Br_2 \longrightarrow \begin{array}{c} CH_2-CH_2 \\ | \quad\quad | \\ Br \quad\; Br \end{array}$ (26.17)

The splitting of a Br_2 molecule produces two identical Br atoms; $CH_2{=}CH_2$ yields two $-CH_2$ groups. Br_2 and $CH_2{=}CH_2$ are symmetrical reagents. The two parts that result when HBr or $CH_2CH{=}CH_2$ are split are not identical. These are unsymmetrical reagents.

When *unsymmetrical* HBr is added to *unsymmetrical* propene, a problem arises. Which of the following products will form?

$$CH_3CH{=}CH_2 + H-Br \longrightarrow \begin{array}{c} CH_3CH-CH_2 \\ | \quad\quad | \\ Br \quad\; H \end{array} \quad \text{or} \quad \begin{array}{c} CH_3CH-CH_2 \\ | \quad\quad | \\ H \quad\; Br \end{array}$$

2-Bromopropane is the sole product obtained. This fact can be explained in terms of the electron-donating abilities of H atoms and alkyl groups. However, rather than go into further detail on this matter, we simply present an empirical rule proposed by Markovnikov in 1871.

In the addition of an unsymmetrical reagent (HX, HOH, HCN, $HOSO_3H$) *to an unsymmetrical olefin, such as* $CH_3CH{=}CH_2$, *the more positive fragment of the reagent (usually hydrogen) adds to the carbon atom with the greater number of attached hydrogen atoms.*

Note how Markovnikov's rule applies to reactions (26.18) and (26.19).

$$\begin{array}{c} CH_3 \\ | \\ CH_3-C{=}CH_2 \end{array} + H_2O \xrightarrow{\text{10\% } H_2SO_4} \begin{array}{c} CH_3 \\ | \\ CH_3-C-CH_3 \\ | \\ OH \end{array}$$ (26.18)

t-butyl alcohol

$$CH_3-C{\equiv}CH + HCl \longrightarrow \begin{array}{c} H \\ | \\ CH_3-C{=}C-H \\ | \\ Cl \end{array} \xrightarrow{HCl} \begin{array}{c} Cl \;\; H \\ | \quad\; | \\ CH_3-C-C-H \\ | \quad\; | \\ Cl \;\; H \end{array}$$ (26.19)

methylacetylene   2-chloropropene   2,2-dichloropropane
(propyne)

$$HC{\equiv}CH + HCN \longrightarrow \begin{array}{c} CN \\ | \\ H-C{=}C-H \\ | \\ H \end{array}$$ (26.20)

cyanoethylene
(acrylonitrile)

$$CH_2{=}CHCH_3 \xrightarrow{MnO_4^-, \; H_2O} \begin{array}{c} CH_2-CH-CH_3 \\ | \quad\quad | \\ OH \quad OH \end{array}$$ (26.21)

propylene   1,2-propanediol
(propene)   (propylene glycol)

The addition of H_2O to a double bond (as in reaction 26.18) is the reverse of the reaction in which a double bond is formed by the elimination of H_2O (as in reaction 26.7). An equilibrium is involved that favors the addition reaction in *dilute* acid and the elimination reaction in *concentrated* H_2SO_4(aq). Purple-colored permanganate is decolorized by olefins (reaction 26.21). This is in contrast to the nonreactivity of alkanes and provides a qualitative test for distinguishing between alkanes and alkenes (Baeyer test). The addition reactions of HCl and HCN to alkynes (reactions 26.19 and 26.20) are used commercially to synthesize intermediates for polymer production.

We have already considered the fundamental nature of high polymers (Chapter 9), and, in Section 21-4, the formation of silicones by a process known as *condensation polymerization*. The mechanism described here is known as *free-radical addition polymerization*. Additional examples of both types are discussed in the special feature at the end of this chapter.

Polymerization.   Ethylene and other olefins can enter into reactions in which the net effect is the opening up of the C=C double bond and the formation of giant molecules. The key to the reaction is a free-radical initiator. In the first step of the process outlined in equations (26.22) through (26.25), an organic peroxide dissociates into two radicals (equation 26.22). The radicals add to the C=C double bonds of ethylene molecules, forming radical intermediates (26.23). These radical intermediates successively attack more ethylene molecules, forming new intermediates of longer and longer length (26.24). The reaction proceeds by a chain mechanism. The chains terminate as a result of reactions such as (26.25).

initiation:
$$R\!-\!O:O\!-\!R \longrightarrow 2\,R\!-\!O\cdot \tag{26.22}$$
$$\underset{\text{organic peroxide}}{}$$

followed by
$$CH_2\!=\!CH_2 + RO\cdot \longrightarrow R\!-\!O\!-\!CH_2\!-\!CH_2\cdot \tag{26.23}$$

propagation:
$$ROCH_2CH_2\cdot + CH_2\!=\!CH_2 \longrightarrow ROCH_2CH_2CH_2CH_2\cdot \tag{26.24}$$

$$RO(CH_2)_3CH_2\cdot + CH_2\!=\!CH_2 \longrightarrow RO(CH_2)_5CH_2\cdot \longrightarrow \longrightarrow$$

termination:
$$RO(CH_2)_xCH_2\cdot + RO\cdot \longrightarrow RO(CH_2)_xCH_2OR \tag{26.25}$$

or
$$2\,RO(CH_2)_xCH_2\cdot \longrightarrow RO(CH_2)_xCH_2CH_2(CH_2)_xOR$$

26-4 Aromatic Hydrocarbons

Aromatic hydrocarbon molecules have structures based on the molecule, benzene, C_6H_6. In Section 10-5 we discussed chemical bonding in benzene in some detail and arrived at various representations of the benzene molecule, including

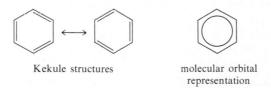

Kekule structures   molecular orbital
                           representation

Of these two possibilities we will choose the molecular orbital representation.
    Other aromatic hydrocarbons can be viewed as derivatives of benzene.

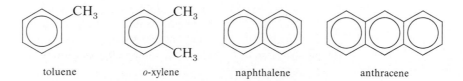

toluene   *o*-xylene   naphthalene   anthracene

Toluene and *o*-xylene are substituted benzenes and naphthalene and anthracene feature fused benzene rings. Whenever two rings are fused together, there is a loss of two carbon and four hydrogen atoms. Thus, naphthalene has the formula $C_{10}H_8$ and anthracene, $C_{14}H_{10}$.
    When one of the six equivalent H atoms of a benzene molecule is removed, the species that results is called a **phenyl** group. Two phenyl groups may bond together as in biphenyl, or phenyl groups may be substituents on aliphatic hydrocarbon chains.

phenyl group   biphenyl   triphenylmethane

Other groups may be substituted for H atoms, and this raises problems in nomenclature. These problems are handled by a numbering system for the C atoms in the ring. If the name of an aromatic compound is to be based on a common name other than benzene, e.g., toluene, the characteristic substituent group ($-CH_3$) is assigned position "1" on the benzene ring.

3-bromotoluene
(*m*-bromotoluene)

2-bromochlorobenzene
(*o*-bromochlorobenzene)

1,4-dichlorobenzene
(*p*-dichlorobenzene)

2-chlorotoluene
(*o*-chlorotoluene)

DDT was at one time a widely used pesticide. Its use in the United States has been discontinued, however, because of environmental hazards.

2,2-di-(*p*-chlorophenyl)-1,1,1-trichloroethane (DDT)

The terms "ortho," "meta," and "para" (*o*-, *m*-, *p*-) may be used when there are two substituents on the benzene ring. **Ortho** refers to substituents on adjacent carbon atoms, **meta** to substituents with one carbon atom between them, and **para** to substituents opposite to one another on the ring.

Benzene and its homologs are similar to other hydrocarbons in that they are insoluble in water but soluble in organic solvents. The boiling points of the aromatic hydrocarbons (arenes) are slightly higher than those of the alkanes of similar carbon content. For example, *n*-hexane, C_6H_{14}, boils at 69°C, whereas benzene boils at 80°C. The planar structure and highly delocalized electron density in aromatic hydrocarbons increases the attractive forces acting between molecules, resulting in higher boiling points. The symmetrical structure of benzene permits closer packing in the crystalline state, resulting in a higher melting point than for *n*-hexane. Benzene melts at 5.5°C and *n*-hexane melts at −95°C.

Aromatic hydrocarbons are highly flammable and should always be handled with care. Prolonged inhalation of benzene vapor results in a decreased production of both red and white blood corpuscles, and this can prove fatal. Also, benzene is a carcinogen. Benzene should be used only under well-ventilated conditions. Fused ring systems, such as 3,4-benzpyrene,

3,4-benzpyrene

are commonly encountered when organic materials are heated to high temperatures in

FIGURE 26-6
Some substitution reactions of benzene illustrated.

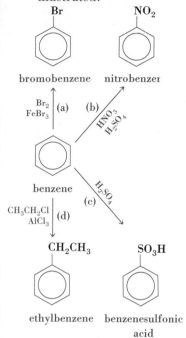

(a) Halogenation.
(b) Nitration.
(c) Sulfonation.
(d) Alkylation.

limited contact with air, a process known as **pyrolysis** (thermal decomposition). 3,4-Benzpyrene has been isolated in the tar formed by burning cigarettes and as a decomposition product of grease in the charcoal grilling of meat. 3,4-Benzpyrene is one of the most active hydrocarbon carcinogens.

Aromatic Substitution Reactions. Alkenes and alkynes have *localized* regions of high electron density (multiple bonds). The electron density associated with the unsaturation in an aromatic ring is *delocalized* into a π electron cloud (recall the doughnut-shaped regions in Figure 10-21). Simple reactions of the aromatic ring involve not the addition of a reagent but the substitution of other atoms or groups for H atoms. Among the reactions of this type are halogenation, nitration, sulfonation, and alkylation. Examples of these are presented in Figure 26-6.

The substitution of a single group, X, can occur at any of the six positions of the benzene ring; they are all equivalent. For the introduction of a second substituent, however, the question arises as to which of the remaining five sites the new group will occupy. If all of these sites were equally preferred, the distribution of products would be this purely statistical one.

$$\underbrace{\qquad\qquad}_{\text{40\% ortho } (^2/_5)} \quad \underbrace{\qquad\qquad}_{\text{40\% meta } (^2/_5)} \quad \underbrace{\qquad}_{\text{20\% para } (^1/_5)}$$

The following scheme describes the products resulting from nitration followed by halogenation (equation 26.26) and halogenation followed by nitration (equation 26.27). It shows that *the substitution is not random*. The —NO$_2$ group directs Cl to a meta position and the —Cl group is an ortho,para director.

$$(26.26)$$

$$(26.27)$$

Whether a group is an ortho,para or a meta director depends on how the presence of one substituent alters the electron distribution in the benzene ring. This makes attack by a second group more likely at one type of position than another. Examination of a large number of reactions leads to the following order.

ortho, para directors: —NH$_2$, —OR, —OH, —OCOR, —R, —X
  (from strongest to weakest)

meta directors: —NO$_2$, —CN, —SO$_3$H, —CHO, —COR, —COOH, —COOR
  (from strongest to weakest)

Example 26-7 Predict the products of the mononitration of

(a)  (b)

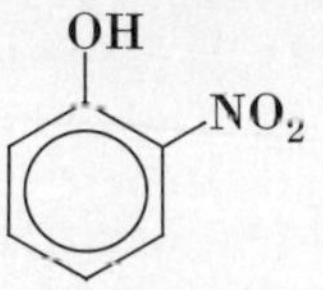

Solution

(a) Since —CHO is a meta director, we should expect the nitration of benzaldehyde (a) to produce 3-nitrobenzaldehyde almost exclusively.

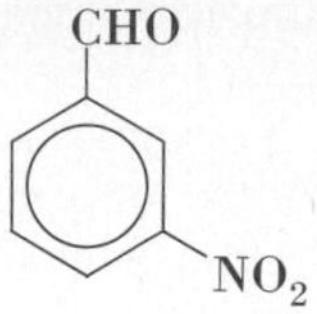

(b) The —OH group is an ortho,para director and we should expect the products

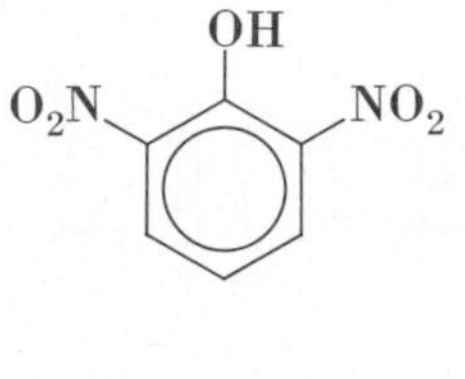

2,6-dinitrophenol 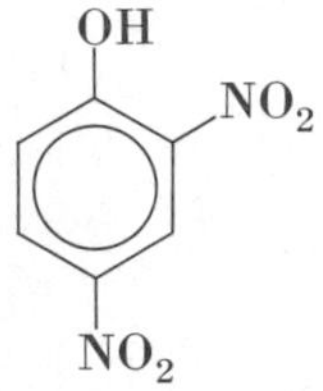   2,4-dinitrophenol

Considering the —NO_2 group as a meta director leads to the same conclusion. Exercise 27 at the end of the chapter illustrates a case where a conclusion is not quite as simple.

SIMILAR EXAMPLES: Exercises 26, 27.

26-5 Alcohols, Phenols, and Ethers

The presence of —OH, the **hydroxyl** group, characterizes the alcohols and phenols. Depending on the nature of the carbon atom to which the hydroxyl group is attached, three classes of alcohols are obtained.

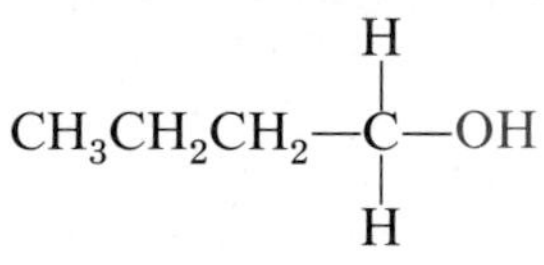

1-butanol (*n*-butyl alcohol) (a *primary* alcohol)   2-butanol (*s*-butyl alcohol) (a *secondary* alcohol)   2-methyl-2-propanol (*t*-butyl alcohol) (a *tertiary* alcohol)

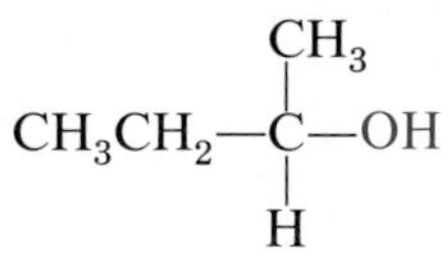

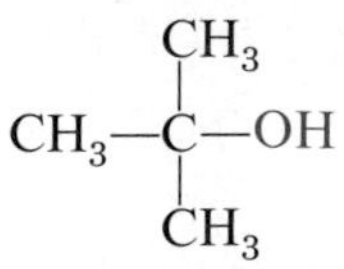

There may be more than one —OH group present in a molecule, resulting in a polyhydric alcohol.

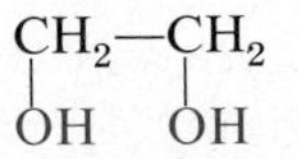

1,2-ethanediol (ethylene glycol)   1,2,3-propanetriol (glycerol)

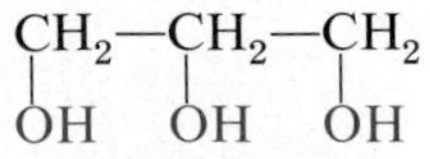

Hexachlorophene, a phenol, was once used as an antiseptic in toothpastes, soaps, and deodorants. However, in view of laboratory observations that small concentrations of hexachlorophene produce brain damage in rats, its use is currently restricted to certain medical applications.

In phenols the hydroxyl group is attached to an aromatic ring.

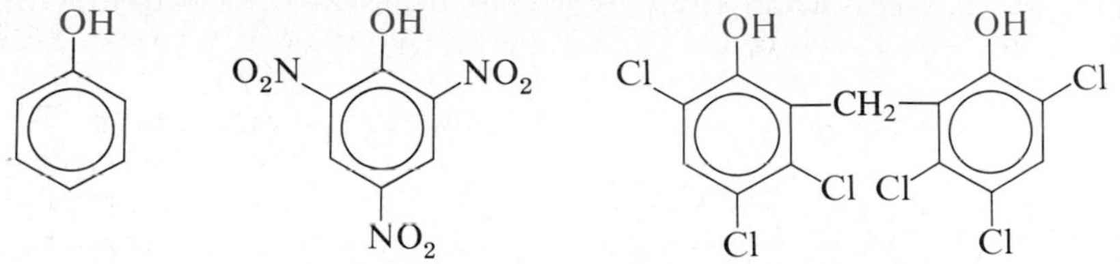

phenol (carbolic acid) — 2,4,6-trinitrophenol (picric acid) — hexachlorophene

Properties of the aliphatic alcohols are strongly influenced by hydrogen bonding. As the chain length increases, the influence of the polar hydroxyl group on the properties of the molecule diminishes. The molecule becomes less like water and more like a hydrocarbon. As a consequence, low-molecular-weight alcohols tend to be water soluble; high-molecular-weight alcohols are not. The boiling points and solubilities of the phenols vary widely, depending on the nature of the other substituents on the benzene ring.

Preparation and Uses. Alcohols can be obtained by the hydration of alkenes or the hydrolysis of alkyl halides.

$$\text{hydration:}\quad \underset{\text{propene (propylene)}}{CH_3CH{=}CH_2} + H_2O \xrightarrow{H_2SO_4} \underset{\text{2-propanol (isopropyl alcohol)}}{CH_3\overset{\overset{\displaystyle OH}{|}}{C}HCH_3} \qquad (26.28)$$

$$\text{hydrolysis:}\quad \underset{n\text{-propyl bromide}}{CH_3CH_2CH_2Br} + OH^- \longrightarrow \underset{n\text{-propyl alcohol}}{CH_3CH_2CH_2OH} + Br^- \qquad (26.29)$$

Methanol is known as wood alcohol because it can be produced by the destructive distillation of wood. This substance is highly toxic and can lead to blindness or death if ingested. Most methanol is manufactured synthetically from carbon monoxide and hydrogen.

$$CO(g) + 2\,H_2(g) \xrightarrow[\substack{200\text{ atm}\\ ZnO,\ Cr_2O_3}]{350\,^\circ C} CH_3OH(g) \qquad (26.30)$$

Ethanol, CH_3CH_2OH, is the common ''alcohol'' to the layperson. It is obtainable by the fermentation of blackstrap molasses, the residue from the purification of sugar cane, or from other materials containing natural sugars. The principal synthetic method is the hydration of ethylene with sulfuric acid (similar to reaction 26.18).

Ethylene glycol, CH_2OHCH_2OH, is water soluble and has a higher boiling point (197°C) than water. Because of these properties it makes an excellent permanent nonvolatile antifreeze for use in automobile radiators. It is also used in the manufacture of solvents, paint removers, and plasticizers (softeners). Propylene glycol is used in suntan lotions and, in conjunction with propellants, to produce nonaqueous foams in aerosol products.

Glycerol (glycerin), $CH_2OHCHOHCH_2OH$, is obtained commercially as a by-product in the manufacture of soap. It is a sweet, syrupy liquid that is miscible with water in all proportions. Because it has the ability to take up moisture from the air, it can be used to keep skin moist and soft, accounting for its use in lotions and cosmetics. It is also used to maintain the moisture content of tobacco and candy.

Reactions of the OH Group. The reactivity of the —OH group results from (1) the unshared electron pairs on the O atom, making the molecule basic in the Lewis sense, or

(2) the polarity of the O—H bond, causing the molecule to act as a proton donor, to be acidic in the Bronsted–Lowry sense. Reactions (26.31) and (26.32) illustrate the first point and (26.33), the second.

$$CH_3CH—\overset{\cdot\cdot}{\underset{\cdot\cdot}{O}}H \underset{CH_3}{|} + CH_3\overset{\overset{O}{\|}}{C}—Cl \longrightarrow CH_3\overset{\overset{O}{\|}}{C}OCH(CH_3)_2 + HCl \qquad (26.31)$$

isopropyl alcohol   acetyl chloride   isopropyl acetate
(an acid halide)   (an ester)

$$\begin{matrix} CH_2—OH \\ | \\ CH—OH \\ | \\ CH_2—OH \end{matrix} + 3\,HONO_2 \longrightarrow 3\,H_2O + \begin{matrix} CH_2ONO_2 \\ | \\ CHONO_2 \\ | \\ CH_2ONO_2 \end{matrix} \qquad (26.32)$$

glycerol   glyceryl trinitrate
(nitroglycerine)

$$CH_3—O—H \xrightarrow{\text{NaOH}} CH_3O^- Na^+ + H_2O \qquad (26.33)$$
$K_a = 1 \times 10^{-16}$

Nitroglycerine was first prepared in 1846 by Sobreno, an Italian, but it remained for Alfred Nobel, a Swedish chemist (1861), to mix it with diatomaceous earth and produce a material less sensitive to shock. This material is called dynamite.

Although commonly used, nitroglycerine is an incorrect name for the structure in (26.32). The compound is an ester and should be so named (i.e., glyceryl trinitrate). A nitro compound has a —NO_2 group bonded directly to a carbon atom, as in nitrobenzene.

Ethers. Ethers are compounds with the general formula R—O—R. Structurally, they can be pure aliphatic, pure aromatic, or mixed.

$$CH_3—O—CH_3$$

dimethyl ether   diphenyl ether   methyl phenyl ether
(anisole)

Ethers can be prepared by the elimination of water from between two alcohol molecules using a strong dehydrating agent, such as concentrated H_2SO_4.

$$CH_3CH_2OH + HOCH_2CH_3 \xrightarrow[\text{conc.}]{H_2SO_4} CH_3CH_2OCH_2CH_3 + H_2O \qquad (26.34)$$

diethyl ether

Chemically, the most notable property of ethers is their comparative lack of reactivity. The ether linkage is stable to most oxidizing and reducing agents and to action by dilute acids and alkalies.

Diethyl ether has been used extensively as a general anesthetic. It is easy to administer and produces excellent relaxation of the muscles. Also, the pulse rate, rate of respiration, and blood pressure are affected only slightly. However, it is somewhat irritating to the respiratory passages and produces a nauseous after effect. More recently methyl propyl ether (neothyl) has been used, and it is reported to be less irritating. Methyl ether, a gas at room temperatures, is used as a propellant for aerosol sprays. Higher ethers have found extensive use as solvents for varnishes and lacquers.

26-6 Aldehydes and Ketones

Aldehydes and ketones contain the carbonyl group.

$$>C=O$$

If both groups attached to a carbonyl group are carbon residues, the resulting compound is called a ketone. If one of the groups is a carbon residue and the other a hydrogen atom, the compound is called an aldehyde.

methanal
(formaldehyde)

3-chlorobutanal
(β-chlorobutyraldehyde)

benzaldehyde

propanone
(acetone)

3-pentanone
(diethyl ketone)

methyl phenyl ketone
(acetophenone)

Preparation and Uses. Partial oxidation of a primary alcohol produces an aldehyde and further oxidation yields a carboxylic acid. This sequence of oxidations is illustrated in the photograph below. Oxidation of a secondary alcohol produces a ketone.

$$CH_3CH_2OH \xrightarrow[H^+]{Cr_2O_7^{2-}} CH_3CHO \xrightarrow[H^+]{Cr_2O_7^{2-}} CH_3CO_2H \qquad (26.35)$$

ethanol
(a primary alcohol)

acetaldehyde
(an aldehyde)

acetic acid
(an acid)

A balanced equation for (26.36) can be obtained with the half-reaction method introduced in Chapter 20. 2-Propanol is oxidized to 2-propanone, and $Cr_2O_7^{2-}$ is reduced to Cr^{3+}.

$$CH_3CHOHCH_3 \xrightarrow[H^+]{Cr_2O_7^{2-}} CH_3\overset{O}{\overset{\|}{C}}CH_3 \qquad (26.36)$$

2-propanol
(a secondary alcohol)

propanone
(a ketone)

The successive oxidation of ethanol (left) to acetaldehyde (center) to acetic acid (right) illustrated through molecular models (see equation 26.35).

Formaldehyde, a colorless gas, dissolves readily in water. A 40% solution in water, called formalin, is used as an embalming fluid and a tissue preservative. Formaldehyde is also used in the manufacture of synthetic resins. A polymer of formaldehyde, called paraformaldehyde, is used as an antiseptic and an insecticide. Acetaldehyde is an important raw material for the production of acetic acid, acetic anhydride, and the ester, ethyl acetate.

Acetone is the most important of the ketones. It is a volatile liquid (boiling point, 56°C) and highly flammable. Acetone is a good solvent for a variety of organic compounds; and because of this it is widely used in solvents for varnishes, lacquers, and plastics. Unlike many common organic solvents, acetone is miscible with water in all proportions. This property, combined with its volatility, makes acetone a useful drying agent for laboratory glassware. Residual water is removed through several rinses with acetone, and the remaining liquid acetone film quickly evaporates. One method of producing acetone involves the *dehydrogenation* of isopropyl alcohol in the presence of a copper catalyst.

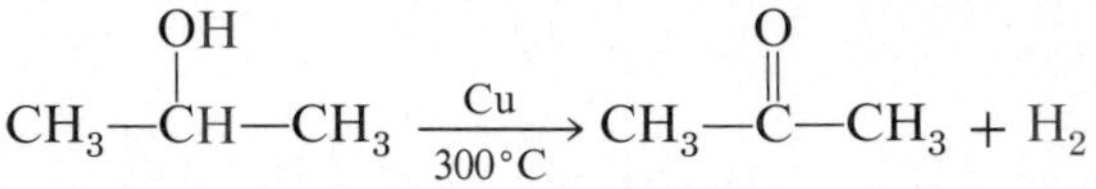

$$CH_3-\overset{\overset{\displaystyle OH}{|}}{CH}-CH_3 \xrightarrow[300°C]{Cu} CH_3-\overset{\overset{\displaystyle O}{||}}{C}-CH_3 + H_2 \qquad (26.37)$$

Aldehydes and ketones occur widely in nature. Typical natural sources are

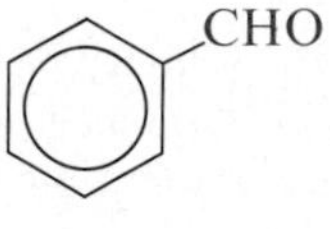

benzaldehyde
(almonds)

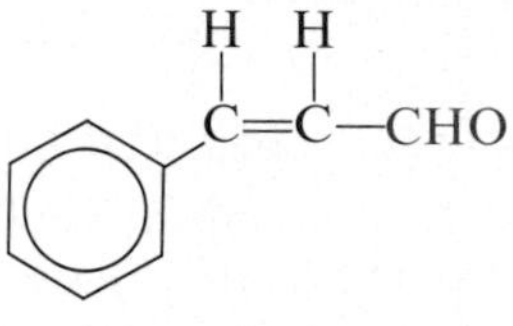

cinnamaldehyde
(cinnamon)

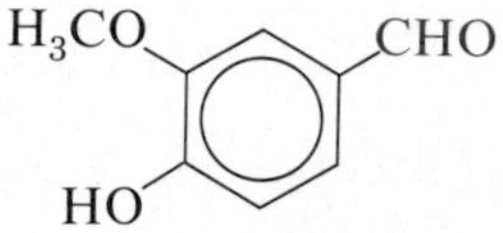

vanillin
(vanilla)

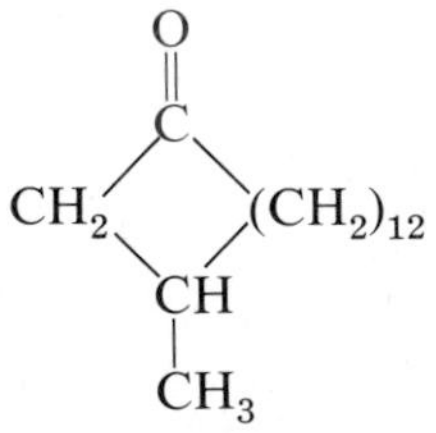

muscone
(obtained from musk deer;
used in perfumes)

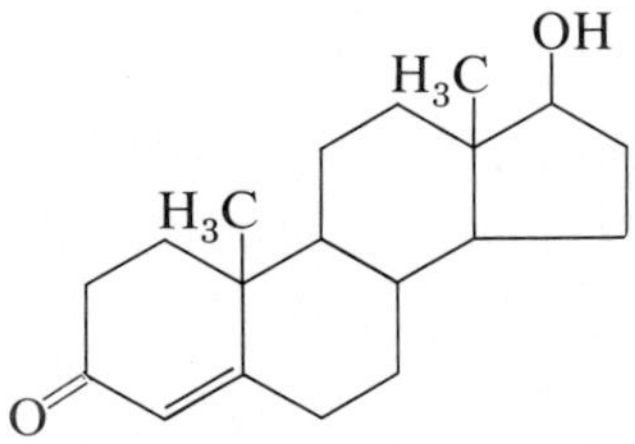

testosterone
(male sex hormone)

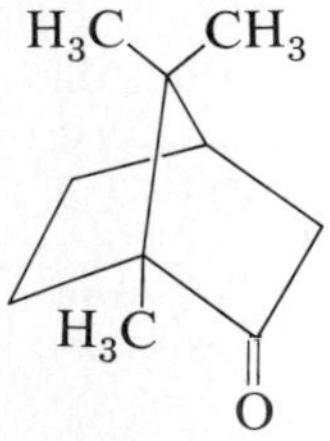

camphor
(obtained from
camphor tree)

26-7 Carboxylic Acids and Their Derivatives

Compounds that contain the carboxyl group (*carb*onyl and hydr*oxyl*)

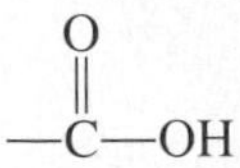

$$-\overset{\overset{\displaystyle O}{||}}{C}-OH$$

are called carboxylic acids; they have the general formula $R-CO_2H$. Many compounds are known where R is an aliphatic residue. These are called fatty acids since compounds of this type are readily available from naturally occurring fats and oils. The carboxyl

TABLE 26-3
Some common carboxylic acids

Structural formula	Common name	IUPAC name	K_a
HCO_2H	formic acid	methanoic	1.78×10^{-4}
CH_3CO_2H	acetic acid	ethanoic	1.74×10^{-5}
$CH_3CH_2CO_2H$	propionic acid	propanoic	1.35×10^{-5}
$CH_3(CH_2)_2CO_2H$	butyric acid	butanoic	1.48×10^{-5}
$CH_3(CH_2)_{16}CO_2H$	stearic acid	octadecanoic	
$CH_3(CH_2)_7CH{=}CH(CH_2)_7CO_2H$	oleic acid	9-octadecenoic	
$C_6H_5CO_2H$	benzoic	benzoic	6.4×10^{-5}
$O_2NC_6H_4CO_2H$	p-nitrobenzoic	4-nitrobenzoic	3.8×10^{-4}
HO_2CCO_2H	oxalic	ethanedioic	$\begin{cases}(K_{a_1})3.5 \times 10^{-2}\\(K_{a_2})6.1 \times 10^{-5}\end{cases}$

group can also be found attached to the benzene ring. If two carboxyl groups are found on the same molecule the acid is called a dicarboxylic acid.

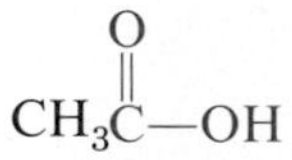

acetic acid
(an aliphatic acid)

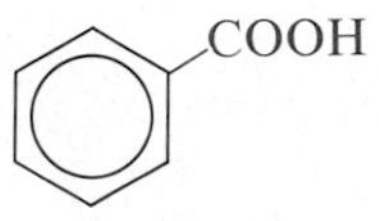

benzoic acid
(an aromatic acid)

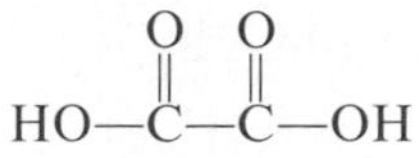

oxalic acid
(an aliphatic
dicarboxylic acid)

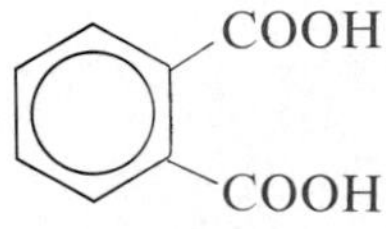

phthalic acid
(an aromatic
dicarboxylic acid)

The carboxylic acids have widely used common names as well as systematic names. Some examples are given in Table 26-3.

Substituted aliphatic acids can be named either by their IUPAC names or by using Greek letters in conjunction with common names. Aromatic acids are named as derivatives of benzoic acid.

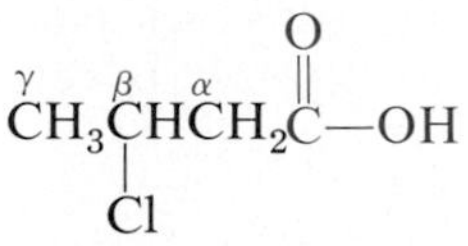

3-chlorobutanoic acid
β-chlorobutyric acid

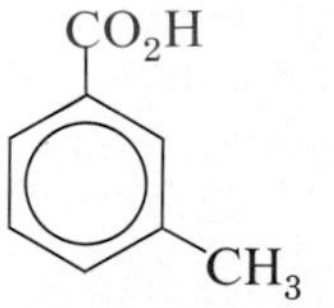

3-methylbenzoic acid
m-methylbenzoic acid
(also m-toluic acid)

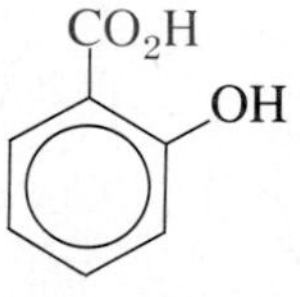

2-hydroxybenzoic acid
o-hydroxybenzoic acid
(also salicyclic acid)

o-Hydroxybenzoic acid or salicylic acid occurs in nature in the willow tree (genus *Salix*). The acetyl derivative of this acid is aspirin, an analgesic (pain killer) and antipyretic (fever reducer).

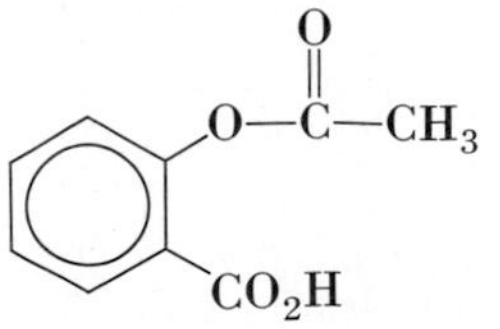

acetylsalicylic acid
aspirin

Because many derivatives of the carboxylic acids involve simple replacement of the hydroxyl groups, special names have been developed for the remaining portion of the molecule, $R{-}\overset{O}{\underset{}{C}}{-}$. The group —COR is given the general name **acyl.** Some specific examples of its use are

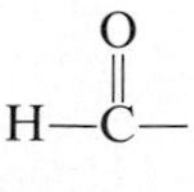

formyl

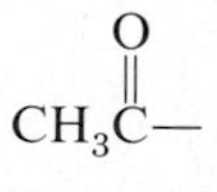

acetyl

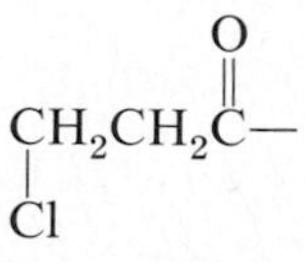

β-chloropropionyl

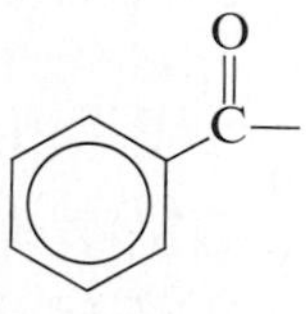

benzoyl

Among the important acid derivatives are

$$R-\overset{\displaystyle O}{\overset{\|}{C}}-X \qquad R-\overset{\displaystyle O}{\overset{\|}{C}}-O-R' \qquad R-\overset{\displaystyle O}{\overset{\|}{C}}-O-\overset{\displaystyle O}{\overset{\|}{C}}-R' \qquad R-\overset{\displaystyle O}{\overset{\|}{C}}-NH_2$$

acyl halide   an ester   an acid anhydride   an amide
(an acid halide)
(X = halogen)

Unlike the pungent odors of the carboxylic acids from which they are derived, esters have very pleasant aromas. The characteristic fragrances of many flowers and fruits can be traced to the esters they contain. Esters are used in perfumes and in the manufacture of flavoring agents for the confectionery and soft drink industries. Most esters are colorless liquids, insoluble in water. Their melting points and boiling points are generally lower than those of alcohols and acids of comparable carbon content. This is because of the absence of hydrogen bonding in esters.

Preparation. Two methods for the preparation of the carboxylic acids are illustrated through equations (26.38) and (26.39).

oxidation of an alcohol:

$$CH_3CH_2OH \xrightarrow[H^+]{K_2Cr_2O_7} CH_3CO_2H \tag{26.38}$$

oxidation of an aldehyde:

$$CH_3CH_2CHO \xrightarrow[OH^-]{KMnO_4} CH_3CH_2CO_2K \xrightarrow{H^+} CH_3CH_2CO_2H + K^+ \tag{26.39}$$

Reactions of the Carboxyl Group. The carboxyl group displays the chemistry of both the carbonyl and the hydroxyl group. Donation of a proton to a base leads to salt formation. The sodium and potassium salts of long-chain fatty acids are known as **soaps,** for example, sodium stearate (see Chapter 27).

Heating the ammonium salt of a carboxylic acid causes the elimination of water and the formation of an **amide.** Further loss of water occurs if the amide is heated with a strong dehydrating agent such as P_2O_5. The final product, which contains the group —C≡N, is called a **nitrile.** These reactions can be reversed by the stepwise treatment with water.

$$CH_3-\overset{\displaystyle O}{\overset{\|}{C}}-O^- NH_4^+ \xrightarrow{heat} CH_3-\overset{\displaystyle O}{\overset{\|}{C}}-NH_2 + H_2O \tag{26.40}$$

ammonium acetate   acetamide

$$heat \downarrow P_2O_5$$

$$CH_3C\equiv N \tag{26.41}$$

acetonitrile

The product of the reaction of an acid and an alcohol is called an **ester.** It, too, can be viewed as forming by the elimination of H_2O through the respective hydroxyl groups of an acid and an alcohol. The mechanism of this reaction is such that the —OH of the resulting water comes from the *acid* and the —H from the *alcohol.*

$$CH_3\underset{\underset{\displaystyle CH_3}{|}}{CH}-\overset{\displaystyle O}{\overset{\|}{C}}-OH + CH_3(CH_2)_3OH \longrightarrow H_2O + CH_3\underset{\underset{\displaystyle CH_3}{|}}{CH}-\overset{\displaystyle O}{\overset{\|}{C}}-O(CH_2)_3CH_3 \tag{26.42}$$

methylpropionic acid   1-butanol   butyl methylpropionate
(isobutyric acid)   (*n*-butyl alcohol)   (*n*-butyl isobutyrate)

FIGURE 26-7
A classification scheme for amines.

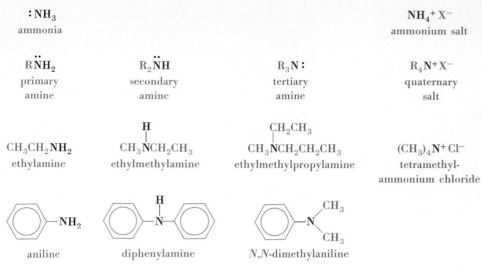

Amines are derivatives of ammonia. Quaternary salts are analogous to ammonium salts.

26-8 Amines

Amines are organic derivatives of ammonia in which one or more organic residues (R) are substituted for H atoms. Their classification is based on the number of R groups bonded to the nitrogen atom—one for primary amines, two for secondary, and three for tertiary. This classification scheme is illustrated in Figure 26-7.

Amines of low molecular weight are gases that are readily soluble in water, yielding basic solutions. The volatile members have odors similar to ammonia but more "fish-like." Amines form hydrogen bonds, but these bonds are weaker than are those in water because nitrogen is less electronegative than oxygen. As in ammonia (see Figure 10-6), the nitrogen atoms in the amines use sp^3 hybrid orbitals, resulting in pyramidal structures. The lone pair electrons occupy one of the sp^3 orbitals. Amines, like ammonia, owe their basicity to these lone pair electrons. In aromatic amines, because of unsaturation in the benzene ring, electrons are drawn into the ring and this reduces the electron density on the nitrogen atom. As a result, aromatic amines are weaker bases than ammonia. Aliphatic amines are somewhat stronger bases than ammonia. Some properties of amines are listed in Table 26-4.

Dimethylamine is used as an accelerator in the removal of hair from hides in the processing of leather. Butyl- and amylamines are used as antioxidants, corrosion inhibi-

TABLE 26-4
Some properties of selected amines

Name	Formula	Boiling point, °C	K_b
ammonia	NH_3	−33.4	1.8×10^{-5}
methylamine	CH_3NH_2	−6.5	44×10^{-5}
ethylamine	$CH_3CH_2NH_2$	16.6	47×10^{-5}
butylamine	$CH_3CH_2CH_2CH_2NH_2$	77.8	40×10^{-5}
aniline	$C_6H_5NH_2$	184	4.2×10^{-10}
N-methylaniline[a]	$C_6H_5NHCH_3$	196	7.1×10^{-10}

[a]The designation "*N*" in *N*-methylaniline signifies that the methyl group is attached to the N atom and not the benzene ring.

FIGURE 26-8
Pyridine.

(a)

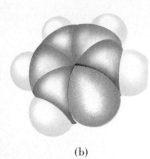

(b)

(a) Structural formula.
(b) Space-filling model.

tors, and in the manufacture of oil-soluble soaps. Dimethyl- and trimethylamines are used in the manufacture of ion exchange resins. Additional uses are found in the manufacture of disinfectants, insecticides, herbicides, drugs, dyes, fungicides, soaps, cosmetics, and photographic developers.

Several methods may be used to synthesize amines, but we limit out concern to an especially important one—the reduction of nitro compounds.

$$\langle\!\langle\bigcirc\rangle\!\rangle-NO_2 \xrightarrow[\text{HCl}]{\text{Fe}} \langle\!\langle\bigcirc\rangle\!\rangle-NH_3^+ \; Cl^- \xrightarrow{\text{NaOH}} \langle\!\langle\bigcirc\rangle\!\rangle-NH \qquad (26.43)$$

26-9 Heterocylic Compounds

In the ring structures considered to this point, all the ring atoms have been carbon; these structures are said to be carbocyclic. Many compounds are encountered, both naturally and synthetically, in which one or more of the atoms in a ring structure is not carbon. These ring structures are said to be heterocyclic. The heterocyclic systems most commonly encountered contain N, O, and S atoms, and the rings are of various sizes.

Pyridine is a nitrogen analog of benzene (see Figure 26-8), but unlike benzene it is water soluble and it has basic properties (the unshared pair of electrons on the N atom is not part of the π electron cloud of the ring system). Pyridine was once obtained exclusively from coal tar, but it is now used so extensively that several synthetic methods have been developed for its production. It is a liquid with a disagreeable odor used in the production of pharmaceuticals such as sulfa drugs and antihistamines, as a denaturant for ethyl alcohol, as a solvent for organic chemicals, and in the preparation of waterproofing agents for textiles. Other examples of heterocyclic compounds will be encountered in Chapter 27.

26-10 Synthesis of Organic Compounds

Originally, all organic compounds were isolated from natural sources. However, as the chemical behavior of these compounds came to be better understood, chemists began to devise *synthetic* methods of producing them from simple starting materials. Now, organic synthesis is one of the most important aspects of organic chemistry. Equipped with a knowledge of a wide variety of reaction types, together with an understanding of the mechanisms of organic reactions, the organic chemist can devise schemes for assembling simple molecular species into more complex structures. A simple example follows:

Example 26-8   Suggest a synthesis for ethyl acetate, $CH_3CO_2CH_2CH_3$, using only inorganic substances and carbon (coke) as starting materials.

Solution.   The compound in question is an ester. In equation (26.42) we noted that acids react with alcohols to produce esters. Our problem is to devise syntheses for *ethyl alcohol* and *acetic acid*. Recall that

$$CaCO_3 \xrightarrow{\text{heat}} CaO + CO_2(g) \qquad (26.9)$$

$$CaO + 3 \; C \xrightarrow[2000°]{\text{electric furnace}} CaC_2 + CO(g) \qquad (26.10)$$

$$CaC_2 + 2 \; H_2O \longrightarrow HC\equiv CH + Ca(OH)_2 \qquad (26.11)$$

Ethylene is produced by adding H_2 to C_2H_2,

$$HC\equiv CH + H_2 \xrightarrow[\text{heat/pressure}]{\text{Pt or Pd}} H_2C=CH_2$$

Ethanol is obtained by the addition of H_2O to C_2H_4.

$$H_2C{=}CH_2 + H_2O \xrightarrow{H_2SO_4} CH_3CH_2OH$$

Some of the ethanol is oxidized to acetic acid.

$$CH_3CH_2OH \xrightarrow[H^+]{K_2Cr_2O_7} CH_3CO_2H$$

Finally, ethanol and acetic acid are combined to form ethyl acetate.

$$CH_3COOH + HOCH_2CH_3 \longrightarrow H_2O + CH_3-\overset{\overset{\textstyle O}{\|}}{C}-O-CH_2CH_3$$

SIMILAR EXAMPLES: Exercises 35, 36.

26-11 Raw Materials for the Organic Chemical Industry

We have just seen through Example 26-8 how an organic compound may be synthesized largely from inorganic substances. However, we did need a source of carbon and we chose coke (derived from coal). The principal sources of carbon for the industrial synthesis of organic chemicals are petroleum, coal, and plant products. Currently, petroleum is the most important of these three sources, although there is now renewed interest in coal because reserves of coal are much more plentiful than those of petroleum. Plant products (biomass) may develop as an important future source.

Coal. Coal is an organic, rocklike material with a high ratio of carbon to hydrogen and other elements. (One proposed formula for a "molecule" of bituminous coal is $C_{153}H_{115}N_3O_{13}S_2$.) To synthesize hydrocarbons or other desired organic compounds from coal requires decreasing the C/H ratio.

In the method of **pyrolysis,** coal (usually bituminous coal) is heated to a high temperature (350 to 1000°C) in the absence of air. Volatile products are formed and an impure carbon residue called **coke** remains. Condensation of the volatile products of this destructive distillation yields black viscous **coal tar.**

$$\text{coal} \xrightarrow[\substack{(absence \\ of\ air)}]{heat} \text{coke + coal tar + coal gas}$$

One ton of bituminous coal yields about 1500 lb of coke, 8 gal of coal tar, and 10,000 ft^3 of coal gas. Coal gas is a mixture of H_2, CH_4, CO, C_2H_6, NH_3, CO_2, H_2S, and other components. At one time coal gas was used as a fuel. Coal tar can be distilled to yield the fractions listed in Table 26-5. From these fractions, in turn, other organic chemicals can be produced.

TABLE 26-5
Coal tar fractions

Boiling range	Name	Tar, mass %	Primary constituents
below 200°C	light oil	5	benzene, toluene, xylenes
200–250	middle oil (carbolic oil)	17	naphthalene, phenol, pyridine
250–300	heavy oil (creosote oil)	7	naphthalenes and methylnaphthalenes, cresols, quinoline
300–350	green oil	9	anthracene, carbazole
residue	—	62	pitch or tar

TABLE 26-6
Principal petroleum fractions

Boiling range, °C	Composition	Fractions	Uses
0–30	C_1–C_4	gas	gaseous fuel
30–60	C_5–C_7	petroleum ether	solvents
60–100	C_6–C_8	ligroin	solvents
70–150	C_6–C_9	gasoline	motor fuel
175–300	C_{10}–C_{16}	kerosene	jet fuel, diesel oil
over 300	C_{16}–C_{18}	gas-oil	diesel fuel, cracking stock
—	C_{18}–C_{20}	wax-oil	lubricating oil, mineral oil, cracking stock
—	C_{21}–C_{40}	paraffin wax	candles, wax paper
—	above C_{40} plus C	residuum	roofing tar, road materials, waterproofing

Pyrolysis can be thought of as a carbon-removal process. Coke is formed and the remaining products are correspondingly enriched in hydrogen and other elements. Coal gasification or liquefaction schemes involve the addition of hydrogen (and usually also oxygen). In general these schemes are based on chemical reactions that have been known for 75 years of more, but they have been updated by new technology, particularly new catalyst systems. One approach, for example, is to burn a coal–water slurry to obtain a mixture of $CO(g)$ and $H_2(g)$. This gaseous mixture is converted to methanol, and with the proper catalysts, the methanol is converted to acetic acid. Heat evolved in burning the coal is used to meet heat requirements in other parts of the process. Sulfur is removed from the coal and converted to $H_2SO_4(aq)$. The process is nonpulluting, energy efficient, and produces only the desired end products (together with some CO_2) from coal and water as starting materials.

Petroleum. The principal constituents of crude oil are aliphatic hydrocarbons. Certain low-molecular-weight hydrocarbons are found dissolved in crude oil or are produced in the manufacture of gasoline. These compounds are removed and compressed into liquid form in cylinders. Propane and butane sold in this form are known as **liquefied petroleum gas (LPG).**

Crude oil is indeed a complex mixture. It has been estimated that in petroleum boiling up to 200°C there are at least 500 compounds, some aliphatic, some alicyclic, and some aromatic. Petroleum is refined by distillation into various fractions. A typical fractionation yields the products listed in Table 26-6.

Fuel Production. Not all of the gasoline components listed in Table 26-6 are equally desirable as fuels. Some of them burn more smoothly than others. (Explosive burning results in engine "knocking.") The octane hydrocarbon, **2,2,4-trimethylpentane,** has excellent engine performance; it is given an octane rating of 100. **n-Heptane** has poor engine performance; its octane rating is set at 0. These two hydrocarbons serve as a basis for establishing the quality of automotive fuels, which are mixtures of a large number of hydrocarbons. In general, branched chain hydrocarbons have higher octane numbers than their straight chain counterparts.

2,2,4-trimethylpentane
(isooctane)
octane rating: 100

n-heptane
octane rating: 0

The fractional distillation of petroleum is carried out in columns such as these. The one at the left operates at atmospheric pressure and the one at the right, under vacuum.

Gasoline obtained by the fractional distillation of petroleum has an octane number of 50 to 55 and is not acceptable for use in automobiles. Extensive modifications of its composition are required. The principal methods employed are of three types—thermal and catalytic cracking, reforming, and alkylation. The chemical changes involved are represented by the equations in Figure 26-9.

In **thermal cracking** large hydrocarbon molecules are broken down into molecules in the gasoline range. The presence of special catalysts promotes the production of branched chain hydrocarbons. The process known as **reforming** or isomerization converts straight

FIGURE 26-9
Some reactions associated with the production of gasoline.

(a) Cracking

$$C_{15}H_{32} \xrightarrow[\text{catalyst}]{\text{heat}} C_8H_{18} + C_7H_{14}$$

$$C_8H_{18} \xrightarrow[\text{catalyst}]{\text{heat}} \underset{\text{butane}}{CH_3CH_2CH_2CH_3} + \underset{\text{2-butene}}{CH_3CH=CHCH_3}$$

(b) Reforming

$$n\text{-}C_4H_{10} \xrightarrow[\text{catalyst}]{\text{heat}} \text{iso-}C_4H_{10} \text{ or } C_4H_8 + H_2$$

$$\underset{\text{1-butene}}{CH_3CH_2CH=CH_2} \xrightarrow[\text{catalyst}]{\text{heat}} \underset{\text{isobutene}}{CH_3\overset{\overset{\displaystyle CH_3}{|}}{C}=CH_2}$$

(c) Alkylation

$$\underset{\text{isobutane}}{CH_3\overset{\overset{\displaystyle CH_3}{|}}{\underset{\underset{\displaystyle CH_3}{|}}{CH}}} + \underset{\text{isobutene}}{CH_2=\overset{\overset{\displaystyle CH_3}{|}}{C}CH_3} \xrightarrow[\text{catalyst}]{\text{heat}} \underset{\substack{\text{2,2,4-trimethylpentane}\\ \text{(isooctane)}}}{CH_3\overset{\overset{\displaystyle CH_3}{|}}{\underset{\underset{\displaystyle CH_3}{|}}{C}}-CH_2-\overset{\overset{\displaystyle CH_3}{|}}{\underset{\underset{\displaystyle H}{|}}{C}}CH_3}$$

A catalytic cracking unit at a petroleum refinery.

chain to branched chain hydrocarbons. Also, alicyclic hydrocarbons are converted to aromatic hydrocarbons, which possess higher octane numbers. In thermal and catalytic cracking, some of the products are low molecular weight, unsaturated hydrocarbons or olefins. In the **alkylation** process these unsaturated compounds are polymerized to higher molecular weight olefins. These can be used directly as fuel components or hydrogenated to produce saturated hydrocarbons.

The octane rating of gasoline has been further improved by the addition of certain "antiknock" compounds which prevent premature combustion. Most widely used in the past have been tetraethyllead, $(C_2H_5)_4Pb$, and tetramethyllead, $(CH_3)_4Pb$. For example, the addition of 6 mL of tetraethyllead to 1 gal of 2,2,4-trimethylpentane would raise its octane rating to 120.3. In order to combine with the lead that would otherwise be deposited in the automobile engine, ethylene dibromide, $BrCH_2CH_2Br$, or ethylene dichloride, $ClCH_2CH_2Cl$, is also added to leaded gasoline; but the lead halides thus formed are exhausted to the atmosphere, creating an environmental hazard. In recent years lead-free gasolines have been appearing on the market for use in automobiles with catalytic converters. Their production requires modifications in the gasoline refining process to increase octane ratings in another way, such as increasing the proportion of aromatic hydrocarbons. These modifications account for the fact that lead-free gasoline is more expensive to produce than a gasoline with a lead additive.

Other additives to prevent fuel-line freeze-up, carburetor icing, spark-plug fouling, engine corrosion, and engine deposits make gasoline as complex a mixture as the crude oil from which it comes.

Petrochemicals.   Although the principal products of the petroleum industry are fuels, chemicals produced from petroleum—petrochemicals—are essential to modern society. Current annual production of benzene in the United States exceeds 11 billion lb. Over 90% of this is produced from petroleum. The process involves cyclization and dehydrogenation of *n*-hexane to the aromatic hydrocarbon. Of the petroleum-produced benzene, about 40% is used to manufacture ethylbenzene for the production of styrene plastics, 18% to manufacture phenol, 6% to synthesize dodecylbenzene (for detergents), and 2% to make aniline. The production of aromatic compounds by dehydrogenation of alkanes yields large amounts of hydrogen gas. An important use of this hydrogen is in the Haber–Bosch synthesis of ammonia.

FOCUS ON
Polymerization Reactions

Plastics are replacing traditional materials (e.g., metals) in many applications. Plastic pipe, such as the polyethylene piping shown here, is employed in over 50% of new gas distribution systems. Key discoveries of the 1950s dealing with the mechanism of polymerization reactions have been largely responsible for the modern plastics era. [Courtesy Du Pont Company.]

The special feature at the end of Chapter 9 dealt with some fundamental ideas about polymers, and throughout the text we have encountered specific examples of polymeric materials. Here we consider in somewhat more detail the types of reactions that are employed to produce polymers.

Chain-Reaction Polymerization. This is the type of polymerization reaction typically encountered with monomers that have carbon-to-carbon double bonds. The net result is that the double bonds "open up" and monomer units add to growing chains. As with other chain reactions the mechanism involves three characteristic steps: initiation, propagation, and termination.

The formation of polystyrene is illustrated in Figure 26-10. In this polymerization a small amount of benzoyl peroxide is present as an initiator. A molecule of this substance decomposes at 70°C to form two benzoyloxy radicals, which in turn lose CO_2 molecules to become phenyl radicals. A phenyl radical adds to a molecule of styrene to produce the more stable diphenylethane radical, which in turn adds another styrene molecule, and so on. Termination of the chain occurs when two free radicals combine.

Because it is initiated by a free radical, the polymerization reaction in Figure 26-10 is sometimes called a free-radical addition polymerization. However, chain-reaction polymerization can also be initiated by cationic and anionic species or by a coordination complex. Chain-reaction polymerization tends to form high molecular weight polymers (mol. wt. up to 10^7) by rapid exothermic reactions. Some typical polymers produced by chain-reaction polymerization are listed in Table 26-7.

Copolymers. One of the polymers listed in Table 26-7, SBR rubber, consists of two different types of monomers—1,3-butadiene and styrene. It is called a copolymer. Many synthetic polymers are copolymers, and it is possible to control the formation of copolymers so that the different monomers (noted as X and Y below) are joined in different patterns.

random
X—Y—Y—X—Y—X—X—X—Y—X—Y—Y

block
X—X—X—X—Y—Y—Y—Y—X—X—X—X

alternating
X—Y—X—Y—X—Y—X—Y—X—Y—X—Y

graft

```
                        Y
                        |
                        Y
                        |
X—X—X—X—X—X—X—X—X—X—X—X
  |         |           |
  Y         Y           Y
  |         |           |
  Y         Y           Y
  |                     |
  Y                     Y
                        |
                        Y
```

FIGURE 26-10
Chain-reaction polymerization—formation of polystyrene.

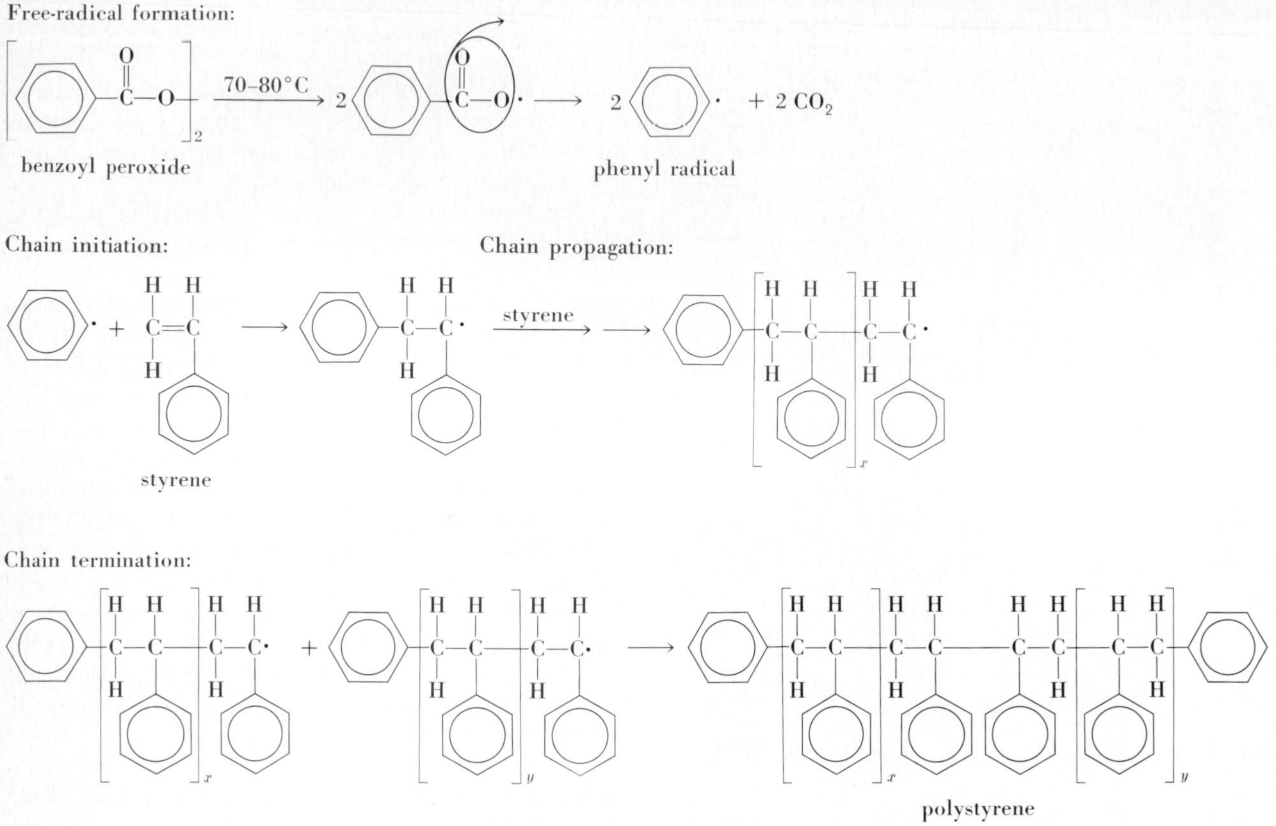

Free-radical formation:

benzoyl peroxide

phenyl radical

Chain initiation: Chain propagation:

styrene

Chain termination:

polystyrene

The method of chain termination illustrated here—the joining of two free-radical chains—
is just one of several possibilities. Termination will also occur as a result of the reaction of a chain with a radical initiator (phenyl radical).

Step-Reaction Polymerization. In this type of polymerization the monomers typically have two or more functional groups that undergo a reaction that joins the two
monomers together. Usually this involves the elimination
of a small molecule, such as H_2O; and this type of polymerization is also referred to as condensation polymerization. Unlike chain-reaction polymerization, where reaction of a monomer can occur only on a growing polymer
chain, in step-reaction polymerization any pair of monomers is free to join into a dimer; a dimer may join with a
monomer to form a trimer; two dimers may join into a
tetramer; etc. Step-reaction polymerization tends to occur
slowly and to produce polymers of only moderately high
molecular weights (mol. wt. less than about 10^5). The
formation of Dacron and nylon 66 is illustrated in Figure 26-11.

Stereospecific Polymers. The physical properties of a
polymer are determined by a number of factors, such as

the average length of the polymer chains and the strength
of intermolecular forces between chains. Another important factor is whether the polymer chains display any crystallinity. In general, amorphous polymers are glasslike or
rubbery. A high-strength fiber, on the other hand, must
possess some crystallinity. Crystals, as we learned in
Chapter 11, are characterized by long-range order among
their structural units.

The usual designation of a polymer, when applied to
polypropylene

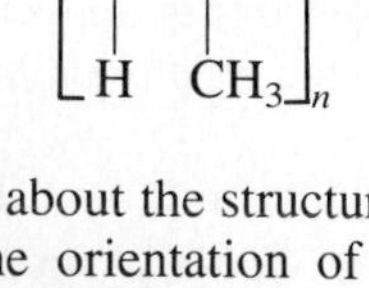

is not very revealing about the structure of the polymer. It
does not indicate the orientation of the $-CH_3$ groups
along the polymer chain.

TABLE 26-7
Some polymers produced by chain-reaction polymerization

Name	Monomer(s)	Polymer
polyethylene	$CH_2{=}CH_2$	$\left[\!CH_2{-}CH_2\!\right]_n$
polypropylene	$CH_2{=}CHCH_3$	$\left[CH_2{-}\underset{\underset{CH_3}{\vert}}{CH}\right]_n$
poly(vinyl chloride)	$CH_2{=}CHCl$	$\left[CH_2{-}\underset{\underset{Cl}{\vert}}{CH}\right]_n$
polyacrylonitrile	$CH_2{=}CHCN$	$\left[CH_2{-}\underset{\underset{CN}{\vert}}{CH}\right]_n$
polystyrene	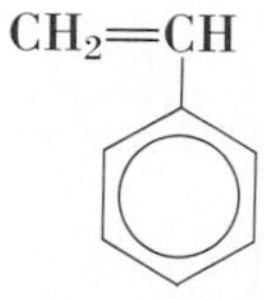 $CH_2{=}CH$ (phenyl)	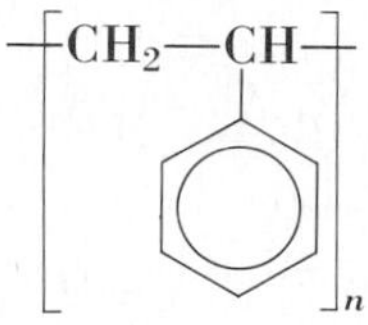 $\left[CH_2{-}CH\right]_n$ (phenyl)
poly(butadiene-co-styrene) SBR, Buna S	$CH_2{=}CHCH{=}CH_2$ 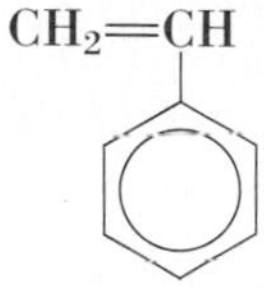 + $CH_2{=}CH$ (phenyl)	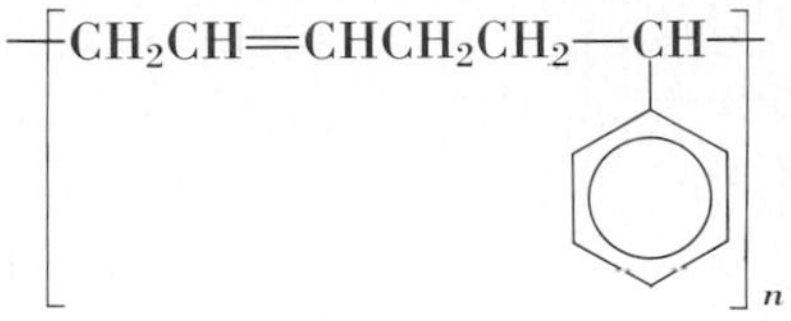 $\left[CH_2CH{=}CHCH_2CH_2{-}CH\right]_n$ (phenyl)

FIGURE 26-11
Step-reaction
polymerization

terephthalic acid   ethylene glycol

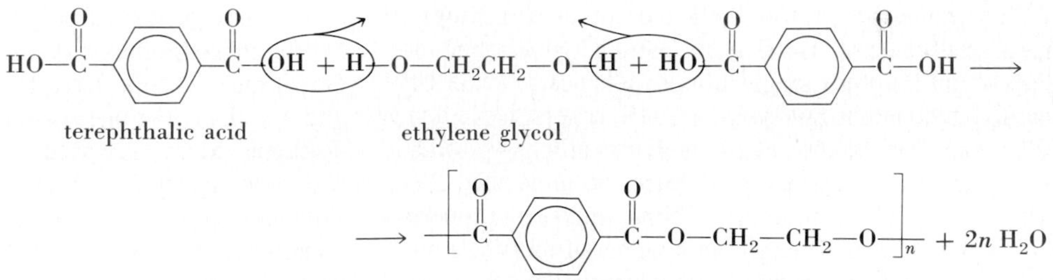

poly(ethylene glycol terephthalate)
(Dacron)

(a)

$HO_2C(CH_2)_4CO_2H + H_2N(CH_2)_6NH_2 \xrightarrow{-H_2O} \left[\underset{\underset{O}{\parallel}}{C}{-}(CH_2)_4{-}\underset{\underset{O}{\parallel}}{C}{-}NH{-}(CH_2)_6{-}NH\right]_n$

adipic acid   1,6-hexanediamine   poly(hexamethyleneadipamide)
(nylon 66)

(b)

(a) Formation of Dacron. The reaction is carried out at elevated temperatures, reduced pressures, and in the presence of sodium methoxide, CH_3ONa.
(b) Formation of nylon 66.

If propylene (CH_3—CH=CH_2) is polymerized by a method similar to that outlined in Figure 26-10, the orientation of the —CH_3 groups on the polymer chain is random. A polymer of this type is called **atactic.** Because the structure of the polymer chains is not regular, atactic polymers are amorphous.

atactic

In the **isotactic** polymer, however, all —CH_3 groups have the same orientation, and in the **syndiotactic** polymer the —CH_3 groups alternate in their positions along the chain.

isotactic

Because of their structural regularity, these two types of polymers do possess crystallinity, which makes them stronger and more resistant to chemical attack than the atactic polymer.

syndiotactic

In the 1950s Ziegler and Natta developed procedures for controlling the spatial orientation of substituent groups on a polymer chain through the use of special catalysts [e.g., $(CH_3CH_2)_3Al + TiCl_4$]. This discovery revolutionized polymer chemistry, for through stereospecific polymerization it is literally possible to "tailor make" large molecules.

A chemist setting out to build a giant molecule is in the same position as an architect designing a building. He has a number of building blocks of certain shapes and sizes, and his task is to put them together in a structure to serve a particular purpose.

GUILIO NATTA

Summary

Organic chemistry deals with compounds of carbon. Simplest among these are carbon-hydrogen compounds—hydrocarbons. In hydrocarbons the C atoms are bonded to one another in straight or branched chains or in rings; H atoms are bonded to the C atoms. Some hydrocarbon molecules contain only single bonds (alkanes); others have some double bonds (alkenes); and still others, triple bonds (alkynes). Yet another class of hydrocarbons—aromatic hydrocarbons—is based on the benzene molecule, C_6H_6. For a given class of hydrocarbons, physical properties (e.g., melting and boiling points) generally follow a regular pattern with increasing molecular weight.

Greater variety among organic compounds results with the incorporation of certain atoms or groupings of atoms (functional groups) into hydrocarbon structures. These substituent groups are introduced through chemical reactions. With alkanes and aromatic hydrocarbons these reactions are based on *substitution:* A functional group replaces an H atom in the hydrocarbon. With alkenes and alkynes chemical reaction occurs by *addition:* Functional group atoms are joined to the C atoms at points of unsaturation (double or triple bonds).

Isomerism is frequently encountered among organic compounds. One form of isomerism is based on molecules with the same total number of C atoms but with different branching of the carbon chain. Another stems from the different positions on a hydrocarbon chain or ring at which functional groups may be attached. Still other isomers (e.g., *cis-trans*) arise from different orientations of substituent groups in space.

Among the reactions that can be used for preparing alkanes are the addition of H_2 to an alkene or alkyne and the reaction of an alkyl halide with sodium. Alkenes can be prepared by *elimination* reactions. For example, the elimination of H_2O through the removal of an H atom and an —OH group from adjacent C atoms leaves a double bond between the C atoms. The principal alkyne, acetylene, is produced by the reaction of calcium carbide with water. Other alkynes can be prepared from acetylene based on the acidity of the C—H bond in HC≡CH.

Compounds of the general formula ROH are alcohols (phenols if R is C_6H_5). Alcohols can be prepared by the *hydration* of an alkene (addition of HOH) or the hydrolysis of an alkyl halide. Ethers (R'OR) result from the elimination of HOH from between two alcohol molecules. Aldehydes, RCHO, and ketones, RCOR′, feature the

carbonyl group, $\diagdown$C$=$O. They can be prepared by the

controlled oxidation of alcohols. Carboxylic acids are weak acids having the general formula RCOOH and featuring the carboxyl group, $-\text{C} \diagup^{O} \diagdown_{OH}$. In addition to their

typical acid–base reactions, carboxylic acids react with alcohols to form esters. Other reactions include the formation of amides and nitriles. Carboxylic acids can be prepared by the oxidation of an alcohol or an aldehyde.

Amines are organic derivatives of ammonia, and, like ammonia, they have basic properties. They can be prepared by the reduction of nitro compounds.

The substitution of other atoms (such as N, O, or S) for C atoms in ring structures yields heterocyclic compounds. These are widely encountered among molecules of the living state (see Chapter 27).

Organic chemicals can be derived from petroleum, coal, or biomass; petroleum is the chief source. In the manufacture of motor fuels, the composition of petroleum is greatly altered, with extensive use being made of the fundamental reactions of organic chemistry. Increasingly, chemicals obtained from petroleum—petrochemicals—are being used in the manufacture of polymers.

Learning Objectives

As a result of studying Chapter 26, you should be able to

1. Give examples of alkane, alkene, alkyne, and aromatic hydrocarbons.

2. Draw structural and condensed formulas for hydrocarbons for which systematic (IUPAC) names are given.

3. Name hydrocarbon molecules for which structural or condensed formulas are given.

4. Determine all the possible skeletal isomers of simple hydrocarbons of given formulas (e.g., all the isomers of C_5H_{12}).

5. Discuss the physical properties of aliphatic and aromatic hydrocarbons in relation to their bonding, structures, and molecular weights.

6. Write equations for several reactions used in the preparation of alkanes, alkenes, and alkynes.

7. Explain why alkanes and aromatic hydrocarbons react by substitution and alkenes and alkynes by addition.

8. Write equations for the reactions of alkanes with halogens and with oxygen, emphasizing the radical chain nature of the alkane-halogen reaction.

9. Predict the products of an addition reaction at a multiple bond.

10. Name the common functional groups and give examples of compounds containing them.

11. Write structural formulas, identify possible isomers, and name organic compounds containing functional groups.

12. Describe methods of preparing alcohols, ethers, aldehydes, ketones, acids, and amines; and write equations for some typical reactions of these functional groups.

13. Propose schemes for synthesizing some simple organic compounds.

14. Describe the processes that are used to increase the yield of gasoline from petroleum and to improve the performance of gasoline in internal combustion engines.

Some New Terms

The **acyl** group is $-\overset{\displaystyle O}{\overset{\|}{C}}-\mathbf{R}.$ If R = H, this is the **formyl** group; R = CH_3, **acetyl;** and R = C_6H_5, **benzoyl.**

In **addition reactions** functional group atoms are joined to the C atoms at points of unsaturation in alkene and alkyne hydrocarbon molecules. **Markonikov's rule** helps to establish the way in which addition of a reagent occurs.

Alcohols contain the functional group —OH and have the general formula ROH.

Aldehydes have the general formula $\mathbf{R}-\overset{\displaystyle O}{\overset{\|}{\mathbf{C}}}-\mathbf{H}.$

Alicyclic hydrocarbon molecules have their carbon atom skeletons arranged in rings and resemble aliphatic (rather than aromatic) hydrocarbons.

Aliphatic hydrocarbon molecules have their carbon atom skeletons arranged in straight or branched chains.

Alkane hydrocarbon molecules have only single covalent bonds between C atoms. In their chain structures alkanes have the general formula C_nH_{2n+2}.

Alkene hydrocarbons have one or more carbon-to-carbon double bonds in their molecules. The simple alkenes have the general formula C_nH_{2n}.

Alkyne hydrocarbons have one or more carbon-to-carbon triple bonds in their molecules. The simple alkynes have the general formula C_nH_{2n-2}.

An **amide** has the general formula $R-\overset{\overset{O}{\|}}{C}-NH_2$.

An **amine** is an organic base having the formula RNH_2 (primary), R_2NH (secondary), or R_3N (tertiary), depending on the number of H atoms of an NH_3 molecule that are replaced by R groups.

Aromatic hydrocarbon molecules have carbon atoms arranged in hexagonal rings, based on benzene, C_6H_6.

Chain or **skeletal isomers** have the same number of C and H atoms in their hydrocarbon chains but a different pattern of branching in the chain.

A **condensed formula** is a simplified representation of a structural formula.

Conformations refer to the different spatial arrangements possible in a hydrocarbon molecule. Examples are the eclipsed and staggered conformation of hydrocarbon chains and the "boat" and "chair" forms of cyclohexane.

An **elimination reaction** is one in which atoms are removed from adjacent positions on a hydrocarbon chain, producing a small molecule (e.g., H_2O) and an additional bond between C atoms.

An **ester** is the product of the elimination of H_2O from between an acid and an alcohol molecule. Esters have the general formula $R-\overset{\overset{O}{\|}}{C}-O-R'$.

Ethers have the general formula ROR'.

A **functional group** is an atom or grouping of atoms attached to a hydrocarbon residue, R. The functional group often confers specific properties to an organic molecule.

Heterocyclic compounds are based on hydrocarbon ring structures in which one or more of the C atoms is replaced by atoms such as N, O, or S.

A **homologous series** is a group of compounds that differ in composition by some constant unit ($-CH_2$ in the case of alkanes).

IUPAC (or **IUC**) refers to a system of relating the names and structural formulas of organic compounds.

Ketones have the general formula $R-\overset{\overset{O}{\|}}{C}-R'$.

A **meta** (*m-*) **isomer** has two substituents on a benzene ring separated by one C atom.

A **nitrile** has the general formula $R-C\equiv N$.

An **ortho** (*o-*) **isomer** has two substituents attached to adjacent C atoms in a benzene ring.

A **para** (*p-*) **isomer** has two substituents located opposite to one another on a benzene ring.

Phenols have the functional group $-OH$ as part of an aromatic hydrocarbon structure.

The **phenyl group** is a benzene ring from which one H atom has been removed: $-C_6H_5$.

A **polycyclic aromatic hydrocarbon** is obtained whenever two or more benzene rings are fused together (with an appropriate loss of C and H atoms).

Polymerization is the process of producing a giant molecule (polymer) from simpler molecular units (monomers).

Positional isomers differ in the position on a hydrocarbon chain or ring where a functional group(s) is attached.

Saturated hydrocarbon molecules contain only single bonds between carbon atoms.

In **stereoisomerism** the number and types of atoms and bonds in molecules are the same, but certain atoms are oriented differently in space. *Cis-trans* **isomerism** is a form of stereoisomerism.

A **structural formula** for a compound indicates which atoms in a molecule are bonded together, and whether by single, double, or triple bonds.

Substitution reactions are typical of those involving alkane and aromatic hydrocarbons. In such a reaction a functional group replaces an H atom on a chain or ring.

Unsaturated hydrocarbon molecules contain one or more carbon-to-carbon multiple bonds.

Suggestions for Further Study

EVANS, E. B., "Catalytic Processes in the Petroleum Industry," *Educ. in Chem.*, **8**, 55 (1971).

FERNELIUS, W. C., H. WITTCOFF, and R. E. VARNERIN, "Ethylene: The Organic Chemical Industry's Most Important Building Block," *J. Chem. Educ.*, **56**, 385 (1979).

HARRIS, F. W., "Introduction to Polymer Chemistry," *J. Chem. Educ.*, **58**, 837 (1981).*

*Several other articles on polymers will be found in the same issue of this journal.

KOLB, D., "The Aromatic Ring," *J. Chem. Educ.*, **56**, 334 (1979).

KOLB, D. and K. KOLB, "Petroleum Chemistry," *J. Chem. Educ.*, **56**, 465 (1979).

LIPELES, E. S., "Friedrich August Kekule," *J. Chem. Educ.*, **58**, 624 (1981).

WITTCOFF, H., "Propylene—A Basis for Creative Chemistry," *J. Chem. Educ.*, **57**, 707 (1980).

WITTCOFF, H., "Benzene—The Polymer Former," *J. Chem. Educ.*, **58**, 270 (1981).

Review Problems

1. Draw Lewis structures of the following simple organic molecules: **(a)** $CH_3CHClCH_3$; **(b)** $HOCH_2CH_2OH$; **(c)** CH_3CHO.

2. Draw structural formulas for all the isomers of **(a)** pentane; **(b)** heptane.

3. Which of the following pairs of molecules are isomers and which are not? Explain.

(a) $CH_3CH_2CH_2CH_3$ and $CH_3CH=CHCH_3$
(b) $CH_3(CH_2)_5CH(CH_3)_2$ and
$CH_3(CH_2)_4CH(CH_3)CH_2CH_3$
(c) $CH_3CHClCH_2CH_3$ and $CH_3CH_2CHClCH_3$
(d)

$$\underset{H}{\overset{H}{\diagdown}}C=C\underset{H}{\overset{Cl}{\diagup}} \quad \text{and} \quad \underset{Cl}{\overset{H}{\diagdown}}C=C\underset{H}{\overset{H}{\diagup}}$$

(e)

[benzene ring with NO_2 substituent]   and   [benzene ring with NO_2 substituent]

(f)

[benzene ring with OH and NO_2]   and   [benzene ring with OH and NO_2]

4. By drawing suitable structural formulas, establish that there are **(a)** 5 isomers of $C_3H_5Cl_3$ and **(b)** 9 isomers of $C_4H_8Cl_2$.

5. Identify the functional group in each compound (i.e., whether an alcohol, amine, etc.).

(a) $CH_3CHBrCH_2CH_3$ **(b)** CH_3CH_2COOH
(c) $C_6H_5CH_2CHO$ **(d)** $(CH_3)_2CHCH_2OCH_3$
(e) $CH_3COCH_2CH_3$ **(f)** $CH_3CH(NH_2)CH_2CH_3$

(g) [benzene ring]—CH_2CH_3 **(h)** $CH_3CO_2CH_3$

6. Give an acceptable name for each of the following structures.

(a)
$$CH_3CH_2CH_2CH_2CH_2\overset{\overset{\displaystyle CH_3}{|}}{\underset{\underset{\displaystyle CH_3}{|}}{C}}-CH_2CH_3$$

(b)
$$CH_3-\overset{\overset{\displaystyle CH_3}{|}}{\underset{\underset{\displaystyle CH_3}{|}}{C}}-CH_3$$

(c)
$$\overset{\overset{\displaystyle CH_3}{|}}{CH_2}CH-CH_2-\overset{\overset{}{\underset{\underset{\displaystyle Cl}{|}}{CH}}}{}-\overset{\underset{\underset{\displaystyle Cl}{|}}{CH}}{}-CH_3 \atop \underset{\displaystyle C_2H_5}{|}$$

(d)
$$CH_3-CH_2-CH-CH_2-CH_3 \atop \underset{\displaystyle H_3C-\overset{\overset{}{|}}{\underset{\underset{\displaystyle CH_3}{|}}{C}}-CH_3}{}$$

(e) $CH_3CH_2\overset{\overset{\displaystyle Cl}{|}}{CH}-CH=CHCH_2CH_3$

7. Draw a structural formula to correspond to each of the following names.
(a) 3-bromo-2-methylpentane
(b) 3-isopropyloctane
(c) 2-pentene
(d) ethyl *n*-propyl ether

8. Write a condensed formula for each of the following chemical substances.
(a) isopropyl alcohol (rubbing alcohol)
(b) tetraethyllead (an antiknock component of gasoline)
(c) 1,1,1-chlorodifluoroethane (a refrigerant)
(d) 2-methyl-1,3-butadiene (used in the manufacture of elastomers)
(e) 2-butenal (crotonaldehyde, used in organic syntheses)
(f) 1,3-cyclopentadiene (used in organic syntheses)

9. Supply a correct name or formula for each of the following aromatic compounds.

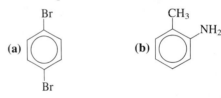

(c) 1,3,5-trimethylbenzene   **(d)** *p*-nitrophenol
(e) 3-amino-2,5-dichlorobenzoic acid (amiben, a plant growth regulator)

10. Indicate the principal product of each of these reactions.

(a) $CH_3CH_3 + Cl_2 \xrightarrow{\text{heat}}$

(b) $CH_3CH_2CH=CH_2 + H_2 \xrightarrow[\text{heat/pressure}]{\text{Pt}}$

(c) $CH_3CH_2CHOHCH_2CH_3 \xrightarrow[\text{H}^+]{Cr_2O_7^{2-}}$

(d) $CH_3CH_2CH=CH_2 + H_2O \xrightarrow[\text{H}_2SO_{4(aq)}]{10\%}$

(e) $CH_3CH_2OH + CH_3C\overset{\displaystyle O}{\underset{\displaystyle OH}{\diagup\!\!\!\diagdown}} \longrightarrow$

(f) [benzene ring]$-C\overset{\displaystyle O}{\underset{\displaystyle OH}{\diagup\!\!\!\diagdown}}$ + KOH(aq) $\longrightarrow$

Exercises

Definitions and terminology

1. What are the essential features that characterize (a) an organic compound; (b) an alkane; (c) an aromatic hydrocarbon?

2. What is the difference in meaning of the following terms? (a) aliphatic and alicyclic; (b) aliphatic and aromatic; (c) paraffin and olefin; (d) alkane and alkyl; (e) normal (*n*-) and iso-; (f) primary, secondary, tertiary; (g) axial and equatorial.

3. Give a definition and a well-chosen example of each of the following: (a) condensed formula; (b) homologous series; (c) olefin; (d) free radical; (e) ortho,para director.

Organic structures

4. Write structural formulas corresponding to these condensed formulas.
   (a) $CH_3CH_2CH_2CHBrCH_3$
   (b) $(CH_3)_2CHCH_2CH_2CH(CH_3)CH_2CH_3$
   (c) $(CH_3)_3CCH_2CH(CH_3)CH_2CH_2CH_3$
   (d) $CH_3CH_2CH(CH_3)C(C_2H_5){=}CH_2$

5. With appropriate sketches represent chemical bonding in terms of the overlap of pure or hybridized atomic orbitals in the following molecules.
   (a) C_2H_6
   (b) $H_2C{=}CHCl$
   (c) $CH_3C{\equiv}CH$
   (d) $CH_3\overset{\overset{\displaystyle O}{\|}}{C}CH_3$
   (e) $CH_3CH_2NH_2$

Isomers

6. Indicate the difference in these three types of isomers: skeletal, positional, and geometrical. Which term best describes each of the following pairs of isomers?
   (a) $CH_3CH_2CH_2Cl$ and $CH_3CHClCH_3$
   (b) $CH_3CH(CH_3)CH_2CH_3$ and $CH_3(CH_2)_3CH_3$
   (c) $CHCl{=}CHCl$ and $CH_2{=}CCl_2$
   (d) [structure: 2-methylaniline] and [structure: 3-methylaniline]
   (e) [structure: maleic acid] and [structure: fumaric acid]

7. By drawing suitable structural formulas, establish that there are 17 isomers of $C_6H_{13}Cl$. (*Hint:* Refer to Example 26-1).

Functional Groups

8. The functional groups in each of the following pairs have certain features in common, but what is the essential difference between them?
   (a) carbonyl and carboxyl
   (b) amine and amide
   (c) acid and acid anhydride
   (d) aldehyde and ketone

9. Give one example of each of the following types of compounds: (a) aliphatic nitro compound; (b) aromatic amine; (c) chlorophenol; (d) aliphatic diol; (e) unsaturated aliphatic alcohol; (f) alicyclic ketone; (g) halogenated alkane; (h) aromatic dicarboxylic acid.

Nomenclature and formulas

10. Give an acceptable name for each of the following structures.
    (a) $CH_3CH_2C(CH_3)_3$ (b) $C(CH_3)_2{=}CH_2$
    (c) $\underset{\underset{\displaystyle CH_2}{|}}{CH_2}{-}CH{-}CH_3$ (d) $CH_3C{\equiv}CCH(CH_3)_2$
    (e) $CH_3CH(C_2H_5)CH(CH_3)CH_2CH_3$
    (f) $CH_3CH(CH_3)CH(CH_3)C(C_3H_7){=}CH_2$

11. Draw a structure to correspond to each of the following names.
    (a) isopentane   (b) cyclohexene
    (c) 3-hexyne   (d) 2-butanol
    (e) isopropyl methyl ether (f) propionaldehyde
    (g) *t*-butyl chloride   (h) diethylmethylamine
    (i) isobutyric acid   (j) isobutyl propionate

12. Does each of the following names convey sufficient information to suggest a specific structure? Explain.
    (a) pentene   (b) butanone
    (c) butyl alcohol   (d) methylaniline
    (e) methylcyclopentane

13. Is each of the following names correct? If not, indicate why not and give a correct name.
    (a) 3-pentene   (b) pentadiene
    (c) 1-propanone   (d) bromopropane
    (e) 2,6-dichlorobenzene (f) 2-methyl-3-pentyne
    (g) 2-methyl-4-*n*-butyloctane
    (h) 4,4-dimethyl-5-ethyl-1-hexyne
    (i) 1,3-dimethylcyclohexane
    (j) 3,4-dimethyl-2-pentene

14. Supply condensed or structural formulas for the following substances.
    (a) 2,2,4-trimethylpentane (isooctane—a constituent of gasoline)
    (b) 2,4,6-trinitrotoluene (TNT—an explosive)
    (c) methyl salicylate (oil of wintergreen) (*Hint:* Recall that salicylic acid is *o*-hydroxybenzoic acid.)

(d) 2-hydroxyl-1,2,3-propanetricarboxylic acid (citric acid, $C_6H_8O_7$)

(e) 1,5-cyclooctadiene (an intermediate in the manufacture of resins)

(f) *o-t*-butylphenol (an antioxidant in aviation gasoline)

(g) 1-phenyl-2-aminopropane (benzedrine—an amphetamine, ingredient in "pep pills")

(h) 2-methylheptadecane (a scx pheromone of tiger moths—a chemical used for communication among members of the species) (*Hint:* "Heptadeca" means 17.)

(i) 3,7,11-trimethyl-2,6,10-dodecatriene-1-ol (farnesol—odor of lilly-of-the-valley) (*Hint:* "Dodeca" means 12.)

(j) 2,6-dimethyl-5-hepten-1-al (used in the manufacture of perfume)

Experimental determination of formulas

15. Combustion of 184 mg of a hydrocarbon gave 577 mg CO_2 and 236 mg H_2O. Calculate the empirical formula of the compound. What is the molecular formula if the molecular weight is subsequently found to be 56?

Alkanes

16. Draw structural formulas for all the isomers listed in Table 26-2 and show that, indeed, the more compact structures yield lower boiling points.

17. What is the most stable conformation of the molecule *t*-butylcyclohexane? (*Hint:* Is the ring in the boat or chair form? Is the *t*-butyl group in an axial or equatorial position?)

18. Write the structure of each alkane.

(a) Molecular weight = 44; forms two different monochlorination products.

(b) Molecular weight = 58; forms two different monobromination products.

19. Name the principal products obtained in the reaction of

(a) $CH_3CH_2CH=CH_2$ with H_2 in the presence of a catalyst

(b) *n*-propyl bromide with sodium

(c) sodium butyrate with sodium hydroxide

(d) propane with chlorine gas in the presence of ultraviolet light

20. In the chlorination of CH_4 some CH_3CH_2Cl is obtained as a product. Explain why this should be so.

Alkenes

21. Why is it not necessary to refer to ethene and propene as 1-ethene and 1-propene? Can the same be said for butene?

22. Alkenes (olefins) and cyclic alkanes (alicyclics) each have the generic formula C_nH_{2n}. In what important ways do these types of compounds differ structurally?

23. Draw the structures of the products of each of the following reactions.

(a) propylene + hydrogen (Pt, heat)

(b) 2-butanol + heat (in the presence of sulfuric acid)

(c) sodium acetylide with *t*-butyl bromide

24. Use Markovnikov's rule to predict the product of the reaction of

(a) HCl with $CH_3CCl=CH_2$

(b) HCN with $CH_3C\equiv CH$

(c) HCl with $CH_3CH=C(CH_3)_2$

(d) 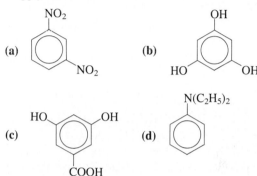 with HBr

Aromatic compounds

25. Supply a name or formula for each of the following.

(e) *p*-phenylphenol (a fungicide)

(f) phenylacetylene

(g) 2-hydroxy-4-isopropyltoluene (thymol—flavor constituent of the herb thyme)

26. Predict the products of the monobromination of (a) *m*-dinitrobenzene; (b) aniline; (c) *p*-bromoanisole.

27. Write the isomers to be expected from the mononitration of *m*-methoxybenzaldehyde.

In actual fact no 3-methoxy-5-nitrobenzaldehyde is obtained. What does this fact imply about the strength of meta and ortho,para directors?

28. What principal product would you expect to obtain when toluene is allowed to react with (a) $HNO_3 + H_2SO_4$; (b) $Cl_2 + FeCl_3$; (c) Cl_2 without $FeCl_3$ but in the presence of ultraviolet light?

★29. The symbol

which is often used to represent the benzene molecule, is also the structural formula of cyclohexatriene. Are benzene and cyclohexatriene the same substance? Explain.

*30. In the representation for benzene

the inscribed circle represents electrons in a π bonding system (recall Figure 10-21). How many electrons are represented by the circles in this representation of naphthalene?

Organic reactions

31. Describe what is meant by each of the following reaction types and illustrate with an example from the text: **(a)** Aliphatic substitution reaction; **(b)** aromatic substitution reaction; **(c)** addition reaction; **(d)** elimination reaction.

32. Draw a structure to represent the principal product of each of the following reactions.
   (a) 1-pentanol + excess dichromate (acid catalyst)
   (b) butyric acid + ethanol (acid catalyst)
   (c) *o*-nitrophenol + sodium hydroxide
   (d) $CH_3CH_2C(CH_3)=CH_2 + H_2O$ (in the presence of H_2SO_4)

33. Use the half-reaction method to balance the following oxidation–reduction equations.

   (a) ⬡—NO_2 + Fe + H^+ ⟶

   ⬡—NH_3^+ + Fe^{3+} + H_2O

   (b) $CH_3CH=CH_2 + MnO_4^- + H_2O$ ⟶
   $CH_3\overset{\text{OH}}{CH}-\overset{\text{OH}}{CH_2} + MnO_2 + OH^-$

   (c) ⬡—OH + $Pb(C_2H_3O_2)_4$ ⟶

   ⬡$\overset{\text{CHO}}{}$CHO + $Pb(C_2H_3O_2)_2$ + $HC_2H_3O_2$

   (d) H_3C—⬡—$CH_3 + H^+ + Cr_2O_7^{2-}$ ⟶

   HO_2C—⬡—$CO_2H + Cr^{3+} + H_2O$

34. A 10.6-g sample of benzaldehyde was allowed to react with 5.9 g $KMnO_4$ in an excess of KOH(aq). After filtration of the MnO_2(s) and acidification of the solution, 6.1 g of benzoic acid was isolated. What was the percent yield of this reaction?

Organic synthesis

35. Starting with benzene and any aliphatic or inorganic reagents required, how would you synthesize **(a)** *m*-bromonitrobenzene; **(b)** *p*-aminotoluene?

36. Starting with acetylene as the only source of carbon, together with any inorganic reagents desired, devise syntheses for **(a)** acetaldehyde; **(b)** 1,1,2,2-tetrabromoethane; **(c)** acetonitrile; **(d)** isopropyl acetate.

Polymerization reactions

37. Explain why Dacron is called a polyester. What is the % O, by mass, in Dacron?

38. Nylon 66 is produced by the reaction of 1,6-hexanediamine with adipic acid. A different nylon polymer is obtained if sebacyl chloride

$$\overset{O}{\overset{\|}{Cl C}}(CH_2)_8\overset{O}{\overset{\|}{C Cl}}$$

is substituted for the adipic acid. What is the basic repeating unit of this nylon structure?

39. Could a polymer be formed by the reaction of terephthalic acid with ethyl alcohol in place of ethylene glycol? Explain.

40. On page 247 the phenomenon of crosslinking was discussed in connecton with the vulcanization of rubber. Polymers can be formed through the reaction of *o*-phthalic acid

⬡$\overset{\text{COOH}}{\underset{\text{COOH}}{}}$

either with ethylene glycol or with glycerol. One of these polymers involves crosslinking; the other does not. One of these polymers is soft and tacky; the other is hard and brittle. Represent the structures of these two polymers and indicate the expected properties of each.

41. Draw structures of the following copolymers, using monomer formulas given in this chapter or in Table 9-4.
   (a) the copolymer (called Saran) of vinyl chloride and vinylidine chloride, $CH_2=CCl_2$;
   (b) poly(styrene-coacrylonitrile).

42. In referring to the molecular weight of a polymer, we can speak only of the "average molecular weight." Explain why the molecular weight of a polymet is not a unique quantity, as it is for a substance like benzene.

*43. Explain why a polymer formed by free-radical addition generally has a higher molecular weight than a corresponding polymer formed by condensation polymerization.

Additional Exercises

1. Write structural formulas for the following compounds.
   (a) ethylisobutylmethylamine
   (b) 3-chloro-2,3-dimethylbutanoic acid
   (c) 1-phenyl-2-butanone
   (d) *t*-butyl isobutyrate
   (e) 1-chloro-2,4-octadiene
   (f) *trans*-1,4-dibromobutadiene

2. Give an acceptable name for each of the following.
   (a) $(CH_3)_2CHOCH_2CH_2CH_3$
   (b) $(CH_3)_2CHCOCH_3$
   (c) $CH_2{=}CHCH{=}CH_2$
   (d) $(CH_3)_3CCH_2CH(CH_3)CH_2CH_2CH_3$

   (e)

   (f)

3. Draw and name all the isomers for (a) C_6H_{14}; (b) C_4H_8; (c) C_4H_6. (*Hint:* Do not forget rings, double bonds, and combinations of these.)

4. Write the structure of each alkane: (a) molecular weight = 72; forms four monochlorination products; (b) molecular weight = 72; forms a single monochlorination product.

5. Why is *cis-trans* isomerism encountered with olefins but not with paraffins?

6. Methanol is a weaker acid than water, whereas phenol is stronger. Methylamine is a stronger base than ammonia, whereas aniline is weaker. Explain these observations.

7. Outline a series of reactions that could be used to synthesize ethanol from coal, water, and other inorganic materials.

8. Draw a structure to represent the principal product of each of the following reactions.
   (a) dimethylamine + hydrochloric acid
   (b) isopropanol + sodium
   (c) *t*-butyl bromide + NaOH(aq)

★9. Combustion of a 0.1908-g sample of a compound gave 0.2895 g CO_2 and 0.1192 g H_2O. Combustion of a second sample, weighing 0.1825 g, yielded 40.2 mL of $N_2(g)$, collected over 50% KOH(aq) (vapor pressure = 9 mmHg) at 25°C and 735 mmHg barometric pressure. When 1.082 g of compound was dissolved in 26.00 g benzene (m.p. 5.50°C, K_f = 5.12), the solution had a freezing point of 3.66°C. What is the molecular formula of the compound?

★10. The three isomeric tribromobenzenes, I, II, and III, when nitrated, form three, two, and one mononitrotribromobenzenes, respectively. Assign correct structures to I, II, and III.

★11. Write the name and structure of each aromatic hydrocarbon.
   (a) Formula: C_8H_{10}; forms three ring monochlorination products.
   (b) Formula: C_9H_{12}; forms one ring mononitration product.
   (c) Formula: C_9H_{12}; forms four ring mononitration products.

★12. In the molecule 2-methylbutane, the organic chemist distinguishes the different types of hydrogen and carbon atoms as being primary (1°), secondary (2°), and tertiary (3°). For the monochlorination of hydrocarbons the following ratio of reactivities has been found; 3°/2°/1° = 4.3:3:1. How many different monochloro derivatives of 2-methylbutane are possible and what percent of each would you expect to find?

Self-Test Questions

For questions 1 through 8 select the single item that best completes each statement.

1. The compound isoheptane has the formula (a) C_7H_{14}; (b) $(CH_3)_2CH(CH_2)_3CH_3$; (c) $CH_3(CH_2)_5CH_3$; (d) $C_6H_5CH_3$.

2. *Three* isomers exist of the hydrocarbon (a) C_3H_8; (b) C_4H_{10}; (c) C_6H_6; (d) C_5H_{12}.

3. The hydrocarbon cyclobutane has the same carbon-to-hydrogen ratio as (a) C_4H_{10}; (b) $CH_3CH{=}CHCH_3$; (c) $CH_3C{\equiv}CCH_3$; (d) C_6H_6.

4. *Cis-trans* isomerism is expected in the compound (a) $ClCH{=}CHCl$; (b) $CH_2{=}CCl_2$; (c) $ClCH_2CH_2Cl$; (d) $Cl_2C{=}CCl_2$.

5. The compound 2-chloro-3-methyl-1-butanol has the formula
(a) $CH_2ClC(CH_3)_2CH_2OH$
(b) $CH_3CHOHCH(CH_3)CH_2Cl$
(c) $CH_3CH(CH_3)CHClCH_2OH$
(d) $CH_3CHClCH(CH_3)CH_2OH$

6. The compound

is named (a) *o*-aminotoluene; (b) *p*-methylaniline; (c) *m*-methyl-benzene; (d) 3-methylaniline.

7. The most acidic of the following substances is

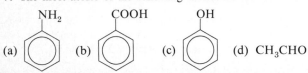

(a)   (b)   (c)   (d) CH_3CHO

8. To prepare methyl ethyl ketone one should oxidize (a) 2-propanol; (b) 1-butanol; (c) 2-butanol; (d) *t*-butyl alcohol.

9. Draw structural formulas for the following compounds:
(a) dichlorodifluoromethane (Freon 12—a refrigerant);
(b) *p*-bromophenol; **(c)** 3-hydroxy-2-butanone; **(d)** MTBE (methyl *t*-butyl ether, an antiknock gasoline additive).

10. Draw structural formulas for all the possible isomers of $C_5H_{11}Br$ and name them.

11. Indicate the principal organic product(s) that you would expect to obtain by
  (a) treating $CH_3CH_2CH{=}CH_2$ with *dilute* H_2SO_4(aq);
  (b) exposing a mixture of chlorine and propane gases to ultraviolet light;
  (c) heating a mixture of isopropyl alcohol and benzoic acid;
  (d) oxidizing *s*-butyl alcohol with $Cr_2O_7^{2-}$ in acidic solution.

12. Give a *simple* test that you might use to determine whether an organic substance is **(a)** C_2H_6 or C_8H_{18}; **(b)** C_2H_6 or C_2H_4; **(c)** C_2H_5OH or $C_6H_{13}OH$;

(d)   CHO   *or*   COOH

27 Chemistry of the Living State

Even though not stated in scientific terms, the biblical commentary "Dust you are, to dust you shall return" certainly suggests our relationship to nature and its laws. But there exists a fascinating interlude between dust and dust—that which we call life. A living organism maintains a low entropy within itself and thus resists, for a time, the universal tendency to approach equilibrium, where entropy and disorder reach a maximum. To maintain the high degree of order characteristic of the living state requires the information of heredity, the energy of biochemical reactivity, and raw materials to build cells. These are some of the topics considered in this chapter.

27-1 Structure and Composition of the Cell

From one-celled plants and animals to the form of greatest complexity, *Homo sapiens,* there is a bewildering variety of life forms. Yet these forms share many characteristics. Of the known elements, about 50 occur in living matter in measurable concentrations. Of these, about 22 have functions that are definitely known. The 11 most abundant elements in living organisms and their percentages in the human body are listed in Table 27-1. Four elements—oxygen, carbon, hydrogen, and nitrogen—together account for 96% of human body mass.

In addition to water, which is the most abundant compound in all living organisms, the important constituents of the cell are compounds of three types: **lipids, carbohydrates,** and **proteins.**

The cell is the fundamental unit of all life. Cells combine to form tissues; tissues may be grouped into organs; and organs combine into organisms. Two views of a typical animal cell are presented in Figures 27-1 and 27-2. Figure 27-1 describes the types and complexity of molecules found in cells and Figure 27-2 pictures the substructure of a cell. You may find it helpful to refer back to these figures from time to time as you proceed through this chapter.

27-2 Principal Constituents of the Cell

Lipids. A precise definition of lipids is not possible. They are simply those constituents of plant and animal tissue that are soluble in solvents of low polarity, such as chloroform, carbon tetrachloride, diethyl ether, or benzene. Many compounds fit this description. The following categories, though arbitrary, are widely accepted.

TABLE 27-1
Chemical elements found in the human body

Element	Percent, by mass
O	65
C	18
H	10
N	3
Ca	2
P	1.2
K	0.20
S	0.20
Cl	0.20
Na	0.11
Mg	0.04

trace elements: Mn, Fe, Co, Cu, Zn, B, Al, V, Mo, I, Si

FIGURE 27-1
Cellular organization.

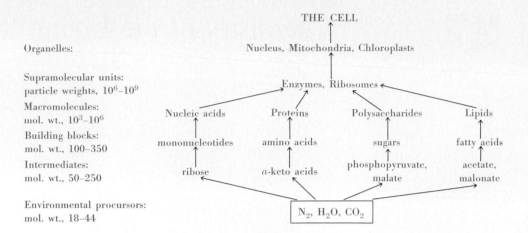

THE CELL

Organelles: Nucleus, Mitochondria, Chloroplasts

Supramolecular units:
particle weights, 10^6–10^9 Enzymes, Ribosomes

Macromolecules:
mol. wt., 10^3–10^6 Nucleic acids   Proteins   Polysaccharides   Lipids

Building blocks:
mol. wt., 100–350   mononucleotides   amino acids   sugars   fatty acids

Intermediates:
mol. wt., 50–250   ribose   α-keto acids   phosphopyruvate,   acetate,
                                                                  malate   malonate

Environmental procursors:
mol. wt., 18–44 N_2, H_2O, CO_2

FIGURE 27-2
A typical animal cell.

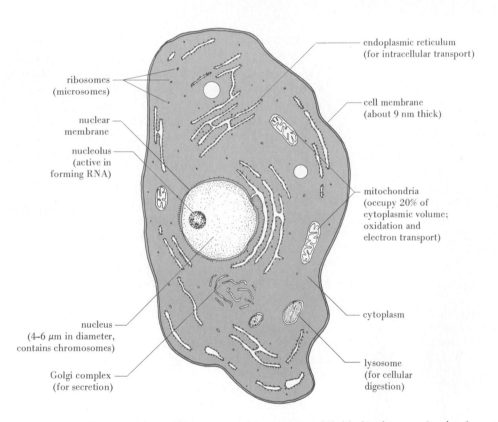

endoplasmic reticulum
(for intracellular transport)

ribosomes
(microsomes)

nuclear
membrane

cell membrane
(about 9 nm thick)

nucleolus
(active in
forming RNA)

mitochondria
(occupy 20% of
cytoplasmic volume;
oxidation and
electron transport)

cytoplasm

nucleus
(4–6 μm in diameter,
contains chromosomes)

lysosome
(for cellular
digestion)

Golgi complex
(for secretion)

1. The Triglycerides. The most common group of lipids in plants and animals are esters of glycerol (1,2,3-propanetriol) with long-chain monocarboxylic acids (fatty acids). Glycerol provides the three-carbon backbone

$$\underset{\underset{H}{|}}{\overset{\overset{H}{|}}{HO-C}}-\underset{\underset{H}{|}}{\overset{\overset{OH}{|}}{C}}-\underset{\underset{H}{|}}{\overset{\overset{H}{|}}{C}}-OH$$

and the acids provide acyl groups

$$R-\overset{\overset{O}{\|}}{C}-$$

TABLE 27-2
Some common fatty acids

Common name	IUPAC name	Formula
Saturated acids		
lauric acid	dodecanoic acid	$C_{11}H_{23}CO_2H$
myristic acid	tetradecanoic acid	$C_{13}H_{27}CO_2H$
palmitic acid	hexadecanoic acid	$C_{15}H_{31}CO_2H$
stearic acid	octadecanoic acid	$C_{17}H_{35}CO_2H$
Unsaturated acids		
oleic acid	9-octadecenoic acid	$C_{17}H_{33}CO_2H$
linoleic acid	9,12-octadecadienoic acid	$C_{17}H_{31}CO_2H$
linolenic acid	9,12,15-octadecatrienoic acid	$C_{17}H_{29}CO_2H$
elcosteoric acid	9,11,13-octadecatrienoic acid	$C_{17}H_{29}CO_2H$

The systematic names for these esters is triacylglycerol, but the common name that has long been used is triglyceride, the name we use here. If all acid groups are the same, the triglyceride is called a *simple* glyceride; otherwise the triglyceride is a *mixed* glyceride. Some long-chain or fatty acids commonly encountered in triglycerides are listed in Table 27-2.

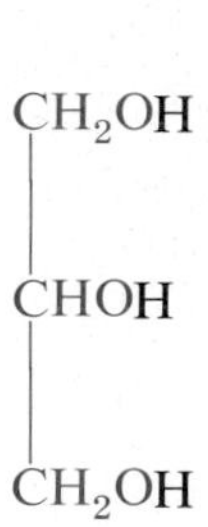

$$CH_2OH$$
$$CHOH$$
$$CH_2OH$$
glycerol

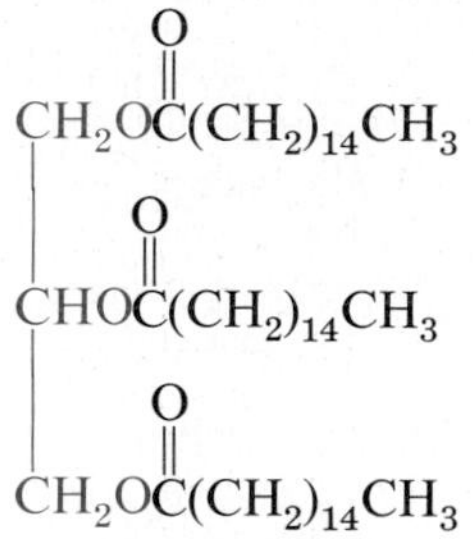

$$CH_2O\overset{O}{\overset{\|}{C}}(CH_2)_{14}CH_3$$
$$CHO\overset{O}{\overset{\|}{C}}(CH_2)_{14}CH_3$$
$$CH_2O\overset{O}{\overset{\|}{C}}(CH_2)_{14}CH_3$$
glyceryl tripalmitate,
tripalmitin
(a simple glyceride; a fat)

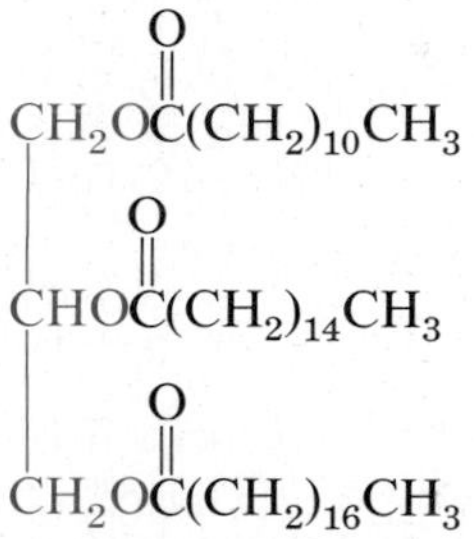

$$CH_2O\overset{O}{\overset{\|}{C}}(CH_2)_{10}CH_3$$
$$CHO\overset{O}{\overset{\|}{C}}(CH_2)_{14}CH_3$$
$$CH_2O\overset{O}{\overset{\|}{C}}(CH_2)_{16}CH_3$$
glyceryl lauropalmitostearate
(a mixed glyceride; a fat)

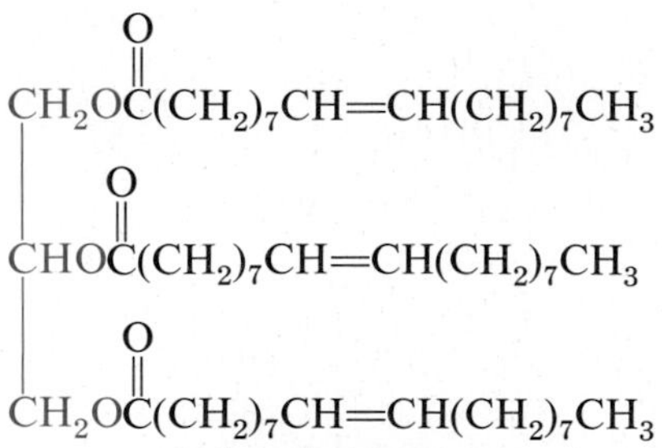

$$CH_2O\overset{O}{\overset{\|}{C}}(CH_2)_7CH{=}CH(CH_2)_7CH_3$$
$$CHO\overset{O}{\overset{\|}{C}}(CH_2)_7CH{=}CH(CH_2)_7CH_3$$
$$CH_2O\overset{O}{\overset{\|}{C}}(CH_2)_7CH{=}CH(CH_2)_7CH_3$$
glyceryl trioleate,
triolein
(a simple glyceride; an oil)

Example 27-1 **Write a structural formula for glyceryl butyropalmitooleate.**

Solution. This is a mixed glyceride in which the acid groups are

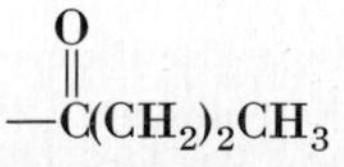

$$-\overset{O}{\overset{\|}{C}}(CH_2)_2CH_3$$
butyric

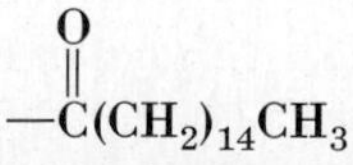

$$-\overset{O}{\overset{\|}{C}}(CH_2)_{14}CH_3$$
palmitic

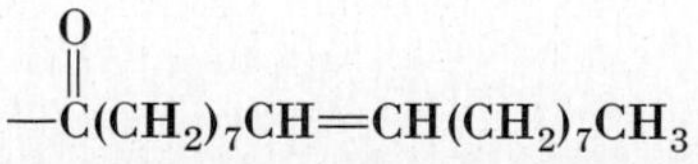

$$-\overset{O}{\overset{\|}{C}}(CH_2)_7CH{=}CH(CH_2)_7CH_3$$
oleic

The complete structure is thus

$$CH_2O\overset{O}{\overset{||}{C}}(CH_2)_2CH_3$$

$$CHO\overset{O}{\overset{||}{C}}(CH_2)_{14}CH_3$$

$$CH_2O\overset{O}{\overset{||}{C}}(CH_2)_7CH{=}CH(CH_2)_7CH_3$$

SIMILAR EXAMPLES: Review Problems 1, 2.

The **fats** are glyceryl esters in which *saturated* acid components predominate; they are solids at room temperature. **Oils** have a predominance of *unsaturated* fatty acids and are liquids at room temperature. The composition of fats and oils is variable and depends not only on the particular plant or animal species involved but also on dietary and climatic factors.

When pure, fats and oils are colorless, odorless, and tasteless. The characteristic colors, odors, and flavors commonly associated with them are imparted by other organic substances that are present in the impure materials. The yellow color of butter is caused by the presence of β-carotene (a yellow pigment also found in carrots and marigolds). The taste of butter is attributed to these two compounds,

$$CH_3{-}\overset{O}{\overset{||}{C}}{-}\underset{\underset{OH}{|}}{C}HCH_3 \qquad CH_3{-}\overset{O}{\overset{||}{C}}{-}\overset{O}{\overset{||}{C}}{-}CH_3$$

3-hydroxy-2-butanone   diacetyl

both produced in the aging of cream.

Glycerides can be hydrolyzed in alkaline solution to produce glycerol and the alkali metal salts of the fatty acids. These salts are commonly known as **soaps,** and the hydrolysis process is called **saponification.**

An early process for making soap required first that wood ashes be soaked in a ceramic pot to produce KOH. These "pot ashes" gave the element "potassium" its name.

$$CH_2O\overset{O}{\overset{||}{C}}(CH_2)_{16}CH_3$$
$$CHO\overset{O}{\overset{||}{C}}(CH_2)_{16}CH_3 \;+\; 3\,KOH \;\longrightarrow\; CHOH \;+\; 3\,CH_3(CH_2)_{16}\overset{O}{\overset{||}{C}}OK \qquad (27.1)$$
$$CH_2O\overset{O}{\overset{||}{C}}(CH_2)_{16}CH_3$$

tristearin   glycerol   potassium stearate
(MW = 890)   (a soap)

Saponification reactions can be used in the laboratory to yield information about the structures of glycerides. This is done through the **saponification value**—the number of milligrams of KOH required to saponify 1.00 g of the glyceride.

Example 27-2 What is the saponification value of tristearin?

Solution. The balanced equation for the saponification reaction is given in (27.1). We need simply to calculate the number of moles of tristearin in 1.00 g, and then, successively, the number of moles, grams, and milligrams of KOH required.

$$\text{no. mg KOH} = 1.00 \text{ g tristearin} \times \frac{1 \text{ mol tristearin}}{890 \text{ g tristearin}} \times \frac{3 \text{ mol KOH}}{1 \text{ mol tristearin}}$$

$$\times \frac{56.1 \text{ g KOH}}{1 \text{ mol KOH}} \times \frac{1000 \text{ mg KOH}}{1.00 \text{ g KOH}}$$

$$= 189 \text{ mg KOH}$$

SIMILAR EXAMPLES: Review Problems 3a, 4; Exercise 9.

Another useful quantity in characterizing a glyceride is the **iodine number**—the number of grams of I_2 reacting with 1.00×10^2 g of a glyceride as a result of the addition of iodine to any double bonds that are present.

Example 27-3   What is the iodine number of triolein?

Solution

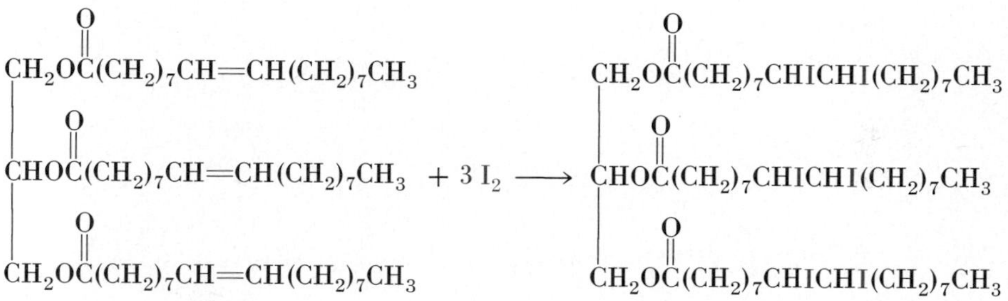

triolein
(MW = 884)

$$\text{no. g } I_2 = 1.00 \times 10^2 \text{ g triolein} \times \frac{1 \text{ mol triolein}}{884 \text{ g triolein}} \times \frac{3 \text{ mol } I_2}{1 \text{ mol triolein}} \times \frac{254 \text{ g } I_2}{1 \text{ mol } I_2}$$

$$= 86.2 \text{ g } I_2$$

SIMILAR EXAMPLES: Review Problem 3b; Exercise 10.

The compositions, saponification values, and iodine numbers of some common fats and oils are listed in Table 27-3.

Unsaturation in a fat or oil may be removed by the catalytic addition of hydrogen (hydrogenation). Thus, oils or low-melting fats can be changed to higher melting fats.

TABLE 27-3
Some common fats and oils

| Lipid | Component acids, %, by mass | | | | | | Saponification value | Iodine number |
| | Saturated | | | Unsaturated | | | | |
	Myristic	Palmitic	Stearic	Oleic	Linoleic	Linolenic		
Fats								
butter	7–10	24–26	10–13	28–31	1–3	0.2–0.5	210–230	26–28
lard	1–2	28–30	12–18	40–50	7–13	0–1	195–203	46–70
Edible oils								
corn	1–2	8–12	2–5	19–49	34–62	—	187–196	109–133
safflower	—	6–7	2–3	12–14	75–80	0.5–1.5	188–194	140–156

These fats, when mixed with skim milk, fortified with vitamin A, and artificially colored, are known as margarines. Edible fats and oils both hydrolyze and cleave at the double bonds by oxidation on exposure to heat, air, and light. The low-molecular-weight fatty acids produced give off offensive odors, the condition known as rancidity. Antioxidants, such as 3-*t*-butyl-4-hydroxyanisole (BHA), retard this oxidative rancidity. They are commonly added to oils in the high-temperature cooking of potato chips and other foods. Medical evidence suggests a relationship between a high intake of saturated fats and the incidence of coronary heart disease. For this reason many diets call for the substitution of unsaturated for saturated fatty acids in foods. In general, mammal fats are saturated, whereas those derived from vegetables, seafood, and poultry are unsaturated.

2. Phosphatides.   The phosphatides (phospholipids) occur in all vegetable and animal cells and are especially prevalent in nerve tissue. They are derived from glycerol, fatty acids, phosphoric acid, and a nitrogen-containing compound. (In the following structures, R and R′ are long-chain alkyl groups.)

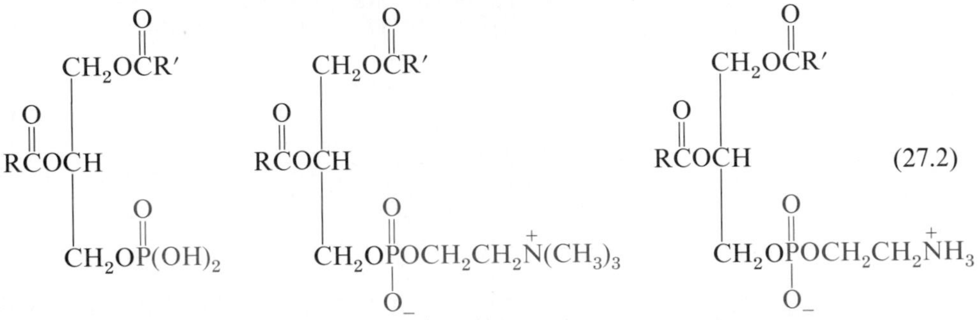

$$
\begin{array}{ccc}
\underset{\substack{\\ \\ \text{a phosphatidic acid}}}{
\begin{array}{l}
\text{CH}_2\text{O}\overset{\displaystyle O}{\overset{\|}{\text{C}}}\text{R}' \\[4pt]
\text{R}\overset{\displaystyle O}{\overset{\|}{\text{C}}}\text{OCH} \\[4pt]
\text{CH}_2\text{O}\overset{\displaystyle O}{\overset{\|}{\text{P}}}(\text{OH})_2
\end{array}}
&
\underset{\substack{\text{phosphatidylcholine}\\ \text{(a lecithin)}}}{
\begin{array}{l}
\text{CH}_2\text{O}\overset{\displaystyle O}{\overset{\|}{\text{C}}}\text{R}' \\[4pt]
\text{R}\overset{\displaystyle O}{\overset{\|}{\text{C}}}\text{OCH} \\[4pt]
\text{CH}_2\text{O}\overset{\displaystyle O}{\overset{\|}{\text{P}}}\text{OCH}_2\text{CH}_2\overset{+}{\text{N}}(\text{CH}_3)_3 \\[2pt]
\quad\overset{|}{\text{O}}_-
\end{array}}
&
\underset{\substack{\text{phosphatidylethanolamine}\\ \text{(a cephalin)}}}{
\begin{array}{l}
\text{CH}_2\text{O}\overset{\displaystyle O}{\overset{\|}{\text{C}}}\text{R}' \\[4pt]
\text{R}\overset{\displaystyle O}{\overset{\|}{\text{C}}}\text{OCH} \\[4pt]
\text{CH}_2\text{O}\overset{\displaystyle O}{\overset{\|}{\text{P}}}\text{OCH}_2\text{CH}_2\overset{+}{\text{N}}\text{H}_3 \\[2pt]
\quad\overset{|}{\text{O}}_-
\end{array}}
\end{array} \tag{27.2}
$$

Choline, which is a quaternary ammonium compound, has the formula $[\text{HOCH}_2\text{CH}_2\text{N}^+(\text{CH}_3)_3]\text{OH}^-$. Its basicity in aqueous solution is comparable to that of KOH. When the alcohol group of choline is esterified with a phosphatidic acid, the product is a phosphatidylcholine or a **lecithin.** The lecithins are found in brain and nerve tissue and in egg yolk. In contrast to simple fats and oils, lecithins form stable colloidal suspensions in water. They are obtained from soybeans and are used as emulsifiers in the dairy and confectionery industries.

Lecithins have both a highly polar and a nonpolar component. [The polar portion of the molecule is shown in color (27.2).] Lecithins are associated with membranes enclosing cell nuclei and mitochondria. Their physiological role is apparently to associate water-insoluble lipids and water-soluble components of an organism, such as in the transport of lipids in the bloodstream or in the movement of fats from one tissue to another.

Removal of the fatty acid residue on the central carbon atom of a lecithin produces a lysolecithin. If this compound comes in contact with red blood cells, disintegration of the cells (hemolysis) occurs. The venom of poisonous snakes contains an enzyme that converts lecithins to lysolecithins. This accounts for the sometimes fatal effects of snakebites. Some spiders and insects also produce toxic results by the same mechanism.

Phosphatidylethanolamines or phosphatidylserines are known as **cephalins.** They are found in brain tissue, and they are also intimately involved in the blood-clotting process. Phospholipids are involved in the transport of ions across cell membranes, in certain secretory processes, and in the electron transport processes of respiration.

3. Waxes.   When fatty acids form esters with long-chain *mono*hydric alcohols, the products are rather high-melting solids (35 to 100°C) called waxes. Beeswax is largely ceryl myristate, $C_{13}H_{27}CO_2C_{26}H_{53}$. It melts between 62 and 65°C and is used in shoe polish, candles, and wax coatings. Carnauba wax, a plant wax from a Brazilian palm tree, is largely myricyl cerotate, $C_{25}H_{51}CO_2C_{31}H_{63}$. It melts between 80 and 87°C and is used in polishes and to coat mimeograph stencils. Spermaceti wax (also called whale oil al-

Ethanolamine has the formula $\text{HOCH}_2\text{CH}_2\text{NH}_2$ and serine, $\text{HOCH}_2\text{CHNH}_2\text{CO}_2\text{H}$.

though not really an oil) consists mainly of cetyl palmitate, $C_{15}H_{31}CO_2C_{16}H_{33}$, and cetyl alcohol, $C_{16}H_{33}OH$. This material, obtained from the head cavity of a whale, has been used as a softening agent in ointments and in cosmetics.

Carbohydrates.   The literal meaning of the term "carbohydrate" is hydrate of carbon: $C_x(H_2O)_y$. Thus, sucrose or cane sugar, with the formula $C_{12}H_{22}O_{11}$, might be represented as $C_{12}(H_2O)_{11}$. A more useful definition, however, is that carbohydrates are polyhydroxy aldehydes, polyhydroxy ketones, their derivatives, and substances that yield them upon hydrolysis. Carbohydrates that are ketones are called ketoses; those that are aldehydes are called aldoses. If the compound contains five carbon atoms it is a pentose, six carbon atoms, a hexose, and so on.

The simplest carbohydrates are the monosaccharides. Oligosaccharides contain from two to ten monosaccharide units bonded together. Names can be assigned to reflect the actual number of such units present, such as *di*saccharide and *tri*saccharide. Mono- and oligosaccharides are also called **sugars.** Polysaccharides contain more than ten monosaccharide units. The general term for all carbohydrates is glycoses. In summary,

$$
\text{Glycoses}\begin{cases}
\text{Monosaccharides} \\
\quad \text{aldose (aldotriose, aldotetrose, . . .)} \\
\quad \text{ketoses (ketotriose, ketotetrose, . . .)} \\
\text{Oligosaccharides (from 2 to 10 monosaccharide units)} \\
\quad \text{disaccharides (e.g., sucrose)} \\
\quad \text{trisaccharides (e.g., raffinose)} \\
\quad \text{and so on.} \\
\text{Polysaccharides (more than 10 monosaccharide units)} \\
\quad \text{(e.g., starch and cellulose)}
\end{cases}
$$

The simplest glycose is 2,3-dihydroxypropanal (glyceraldehyde), an *aldotriose*. If a ball-and-stick model of this molecule is made, an interesting form of stereoisomerism becomes evident—optical isomerism. Figure 27-3 illustrates that there are *two* nonsuperimposable structures for glyceraldehyde. As we learned in Section 24-5, such structures are related to each other like a right and a left hand, or like an object and its mirror image; they are called enantiomers. Also, as we learned in Section 24-5, one enantiomer rotates the plane of polarized light to the right and is said to be *dextro*rotatory (designated + or

FIGURE 27-3
Optical isomerism in glyceraldehyde.

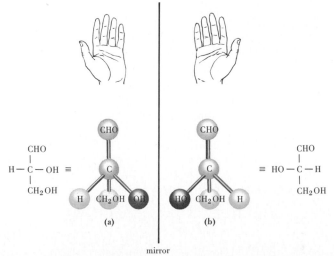

The structure in **(a)** is not superimposable on **(b),** just as a right and a left hand are not superimposable. (A right-handed glove cannot be worn on a left hand.)

d); the other rotates the plane of polarized light to the left and is *levo*rotatory (designated − or *l*). Almost all molecules exhibiting optical isomerism possess at least one carbon atom with four different groups attached to it. Such a carbon atom is said to be **asymmetric** or **chiral.**

Which arrangement of groups at the asymmetric carbon atom in glyceraldehyde, that is, which **absolute configuration,** is associated with dextrorotatory and which with levorotatory properties? In 1930, Rosanoff made a bold guess with a 50:50 chance of being correct. He proposed that the structure in Figure 27-3a was the dextrorotatory (+) one. To this species he assigned a small capital letter D, and to the structure in Figure 27-3b, the letter L. Nineteen years later, an x-ray structure determination of a closely related compound proved that the guess was correct. D-Glyceraldehyde is indeed dextrorotatory; that is, it is D-(+)-glyceraldehyde.*

It is difficult to represent a three-dimensional structure in a plane (two-dimensional) drawing. A useful convention (the Fischer convention) is to place the structural formula on the page so that the backbone of the molecule is arranged from top to bottom, with the most oxidized portion of the molecule (—CHO) at the top and the least oxidized (—CH$_2$OH), at the bottom. Attached groups (—H and —OH) are written to the sides. The end groups of the backbone extend *behind* the plane of the page, *away* from the viewer. The glyceraldehyde enantiomers are written

$$
\begin{array}{ccc}
\text{CHO} & & \text{CHO} \\
| & & | \\
\text{H—C—OH} & & \text{HO—C—H} \\
| & & | \\
\text{CH}_2\text{OH} & & \text{CH}_2\text{OH}
\end{array}
\tag{27.3}
$$

D-(+)-glyceraldehyde   L-(−)-glyceraldehyde

and establish the D-L convention to be used for other sugars: The —H and —OH groups on the next-to-last (penultimate) carbon atom extend in *front* of the page, *toward* the viewer. If the —OH group on this penultimate carbon atom is to the *right,* the configuration is D. If the —OH is to the left, the configuration is L. This convention is applied below to the four-carbon aldoses. The penultimate carbon atoms are shown in color. Figure 27-4 is offered as a further aid in picturing the relationship between a three-dimensional structure and its two-dimensional representation.

$$
\begin{array}{ccc}
\text{CHO} & & \text{CHO} \\
| & & | \\
\text{H—C—OH} & & \text{HO—C—H} \\
| & & | \\
\text{H—C—OH} & & \text{HO—C—H} \\
| & & | \\
\text{CH}_2\text{OH} & & \text{CH}_2\text{OH}
\end{array}
\tag{27.4}
$$

mirror
D-(−)-erythrose   L-(+)-erythrose

But there are two more possibilities for 2,3,4-trihydroxybutanal—D-threose and L-threose. They too are enantiomers.

$$
\begin{array}{ccc}
\text{CHO} & & \text{CHO} \\
| & & | \\
\text{HO—C—H} & & \text{H—C—OH} \\
| & & | \\
\text{H—C—OH} & & \text{HO—C—H} \\
| & & | \\
\text{CH}_2\text{OH} & & \text{CH}_2\text{OH}
\end{array}
\tag{27.5}
$$

mirror
D-(−)-threose   L-(+)-threose

*A more generalized system for absolute configurations uses *R* (Latin, *rectus,* right) in place of D and *S* (Latin, *sinister,* left) for L. However, D and L symbols are still widely used for carbohydrates.

FIGURE 27-4
The structure of D-(−)-erythrose.

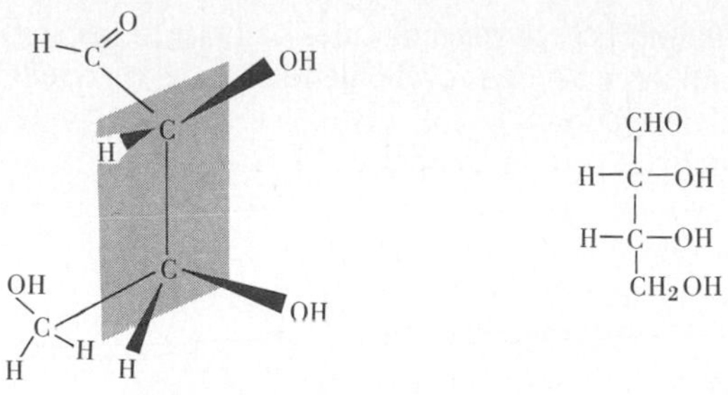

(a) (b)

The three-dimensional structure **(a)** is represented in two dimensions by **(b)**.

If we compare the configurations of D-erythrose and D-threose, we note that these two molecules are *not* mirror images. Molecules that are optical isomers but *not* mirror images of one another are called **diastereomers.**

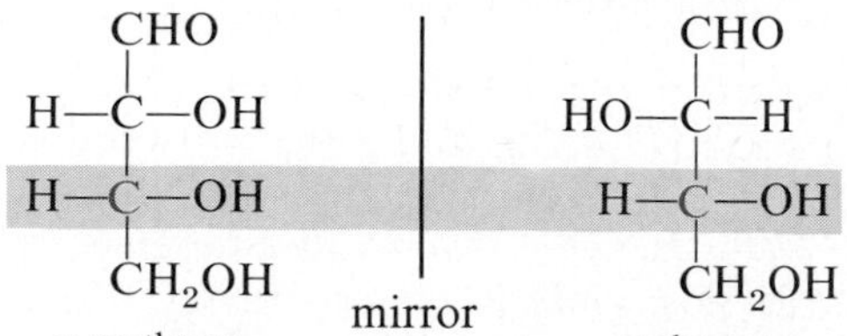

not mirror images

D-erythrose D-threose

A final note of importance is that the designations D and L refer to the configuration of the groups bonded to a chiral carbon atom. The terms (+) and (−) represent the actual direction of rotation of polarized light. Thus a molecule can have a D configuration and be levorotatory (−), or an L configuration and be dextrorotatory (+), as we see in structures (27.4) and 27.5).

Enantiomers (shortened form, antiomers) have the same physical and chemical properties. They differ only in the *direction,* not the extent to which they rotate plane-polarized light. Diastereomers (shortened form, diamers) do differ both in physical and chemical properties. Also they differ in the extent to which they rotate plane-polarized light.

A mixture of equal amounts of the D and L configurations of a substance, called a *dl* or racemic mixture, does not rotate the plane of polarized light either to the left or to the right. The designation DL-erythrose, for example, signifies a racemic mixture. Usually, when molecules with chiral centers are synthesized, the product is a racemic mixture. This is because the creation of these centers is a random process like flipping a coin (an equal probability for "heads" or "tails"). If optically pure isomers are desired, the racemic mixture must be separated into the component enantiomers by a process called **resolution.** Sometimes this is carried out through an enzyme reaction, with a particular enzyme that reacts with one enantiomer but not the other.

Monosaccharides. Of the 16 possible aldohexoses only three occur widely in nature: D-glucose, D-galactose, and D-mannose.

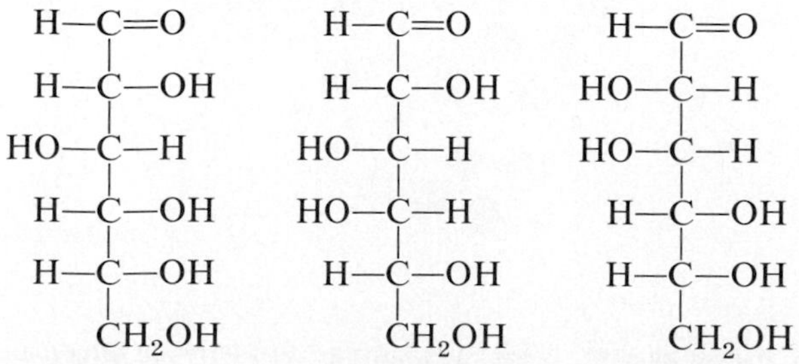

D-glucose D-galactose D-mannose

To some extent these sugar molecules do exist in the straight chain forms shown, but, in the main, all three occur as cyclic **hemiacetals.** We need to consider briefly the subject of hemiacetal formation.

An aldehyde and an alcohol can react to produce a hemiacetal.

$$R-\overset{\overset{\displaystyle O}{\|}}{C}-H + (R'O)(H) \Longrightarrow R-\overset{\overset{\displaystyle OH}{|}}{\underset{\underset{\displaystyle OR'}{|}}{C}}-H \tag{27.6}$$

hemiacetal

Straight chain hemiacetals are unstable and can undergo further reaction with a second alcohol molecule to form an **acetal.**

$$R-\overset{\overset{\displaystyle OH}{|}}{\underset{\underset{\displaystyle OR'}{|}}{C}}-H + R'OH \xrightarrow{H^+} R-\overset{\overset{\displaystyle OR'}{|}}{\underset{\underset{\displaystyle OR'}{|}}{C}}-H + H_2O \tag{27.7}$$

acetal

Reactions (27.6) and (27.7) involved two different molecules. With long-chain poly-hydroxy aldehydes like glucose, the opportunity exists for *intra*molecular hemiacetal formation. The —OH group of the fifth carbon atom (C-5) adds to the carbonyl of the C-1 atom and produces a ring structure composed of five C atoms and one O atom, as illustrated in Figure 27-5. The configuration of this six-membered ring is of the "chair" type. Moreover, unlike the straight chain molecule, the ring hemiacetal is stable.

Closure of the ring in hemiacetal formation creates a new chiral center (at the C-1 atom), with two possible orientations to consider. In the α form the OH at C-1 is axial (directed down); in the β form it is equatorial (extends out from the ring). The α and β forms of D-glucose are pictured in Figure 27-6.

The naming of monosaccharides is complicated by the fact that ring formation occurs. However, each term in a name conveys precise information about the molecule. Thus, D-(+)-glucose refers to the straight chain form of glucose in the D configuration; this form is dextrorotatory (+). The name α-D-(+)-glucose denotes the ring form derived from D-glucose in which the α configuration is found at the C-1 atom.

With some sugars a sufficient amount of the straight chain form is in equilibrium with the cyclic form so that the sugar engages in an oxidation–reduction reaction with $Cu^{2+}(aq)$. The $Cu^{2+}(aq)$ is reduced to red Cu_2O, and the aldehyde portion of the sugar is

FIGURE 27-5
Models of the glucose molecule.

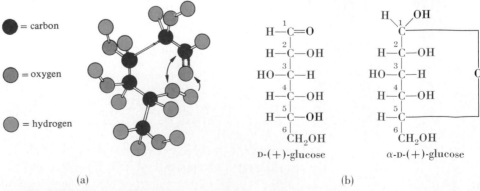

(a)   (b)

(a) Ball-and-stick model indicating the atoms involved in ring closure.
(b) Ring closure represented through Fischer projection formulas.

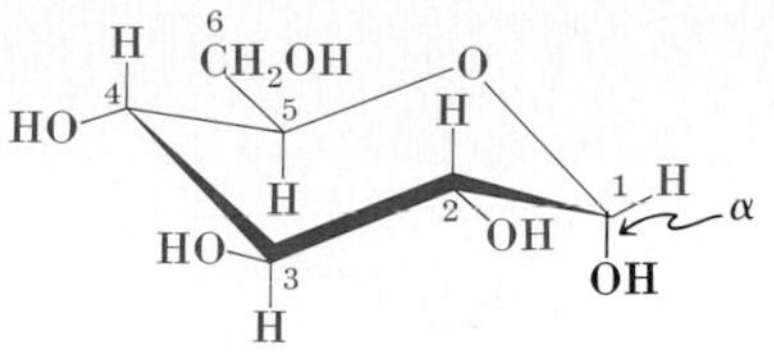

FIGURE 27-6
α and β forms of
D-glucose.

α-D-glucopyranose

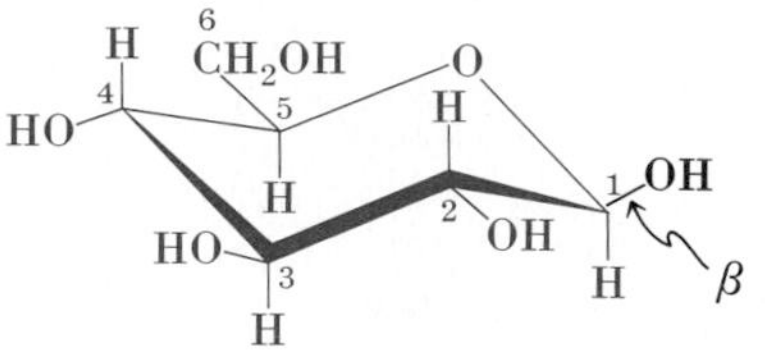

β-D-glucopyranose

A still more precise name for α-D-(+)-glucose is α-D-(+)-glucopyranose. The term *pyranose* signifies a six-membered, oxygen-containing heterocycle of the pyran type.

oxidized (to an acid). These sugars are known as **reducing sugars.** This test for a reducing sugar is conducted with alkaline copper ion complexed as the tartrate (Fehling's solution) or the citrate (Benedict's solution).

$$\text{cyclic sugar as hemiacetal} \rightleftharpoons \underset{\text{open chain form}}{\overset{\text{CHO}}{\underset{|}{\overset{|}{\text{CHOH}}}}} \xrightarrow{\text{Cu}^{2+}} \underset{\text{red ppt.}}{\overset{\text{COOH}}{\underset{|}{\overset{|}{\text{CHOH}}}}} + \text{Cu}_2\text{O(s)} \qquad (27.8)$$

Disaccharides. In equations (27.6) and (27.7) we saw that one molecule of an alcohol converts an aldehyde to a hemiacetal and a second molecule of alcohol converts the hemiacetal to an acetal. Acetals of the glycoses are called glycosides. D-Glucose, for example, can form a glycoside (more specifically for glucose, glucoside) with the —OH group supplied by a second monosaccharide unit. In this case, because two monosaccharide units are joined together, the product is a disaccharide. Hydrolysis of a disaccharide by acid or enzymatic catalysis yields two molecules of monosaccharides.

In considering disaccharides, attention is focused on three question.

1. What are the component monosaccharide units?
2. Is the configuration of the linkage between the monosaccharide units α or β?
3. What is the ring size in each monosaccharide unit?

The important naturally occurring disaccharides—maltose, cellobiose, lactose, and sucrose—are presented in Figure 27-7. In maltose the monosaccharide unit on the left, originally in the hemiacetal form, is converted to the acetal by reacting with the hydroxyl group on the C-4 of a second unit. Linkage of the two units is in the α manner. The glucose unit on the right is present as a hemiacetal in its α form. This hemiacetal exists in equilibrium with the open chain form and can be oxidized by Cu^{2+}(aq). Maltose is a reducing sugar. [The full-acetal glucose unit on the left, however, is stable to Cu^{2+}(aq).] Maltose is produced by the action of malt enzyme on starch. It undergoes fermentation, in the presence of yeast, to glucose, and then to ethanol and CO_2(g).

Cellobiose is obtained by careful hydrolysis of cellulose. It is a glucose-glucose disaccharide with β linkages; it is also a reducing sugar. Lactose is the reducing sugar present in milk (4 to 6% in cow's milk and 5 to 8% in human milk). It is a galactose-glucose disaccharide having β linkages. Sucrose is ordinary table sugar (cane or beet sugar). It is

FIGURE 27-7
Some common
disaccharides.

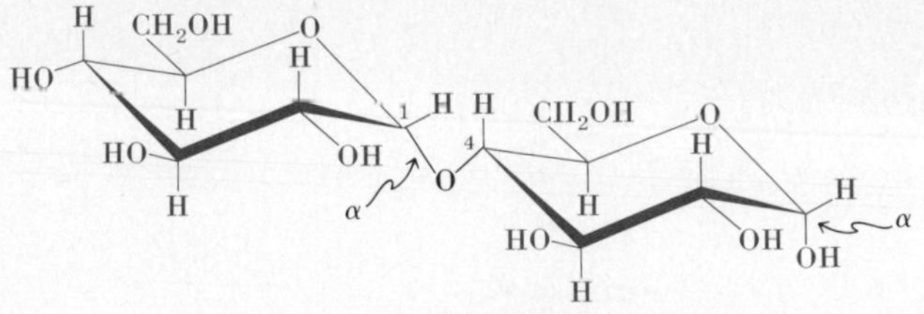

maltose (α-form)

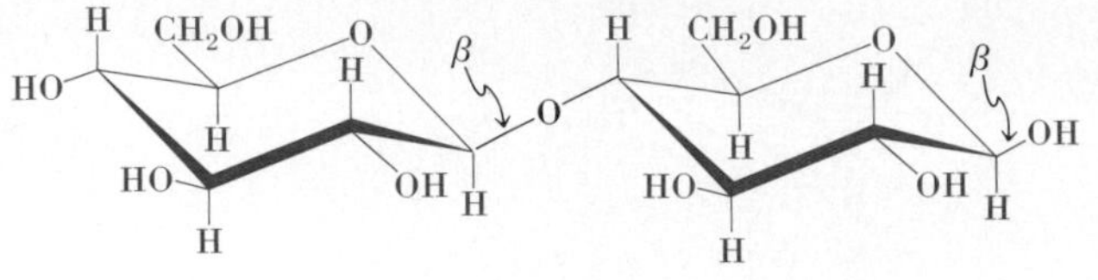

cellobiose

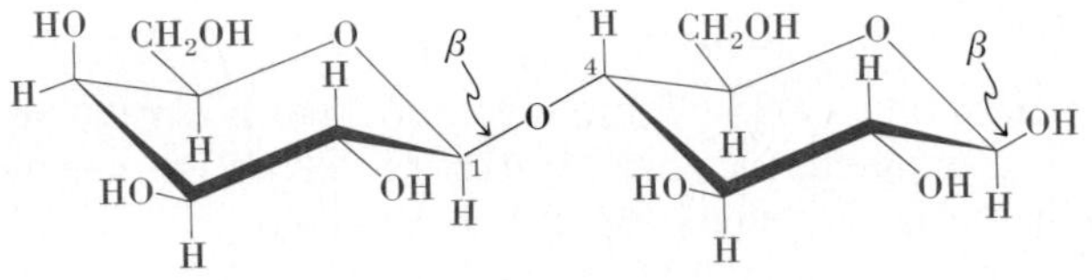

lactose (β-form)

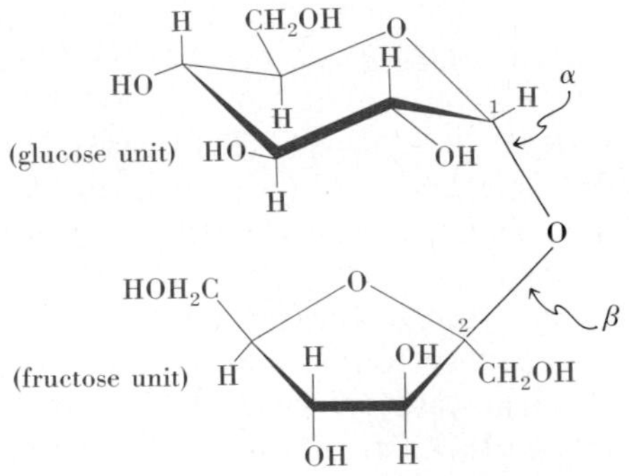

sucrose

a glucose-fructose disaccharide linked 1α, 2β. Since the glucose unit is tied up as an acetal and the fructose as a ketal, sucrose is a *nonreducing* sugar.

Polysaccharides.   Polysaccharides are composed of monosaccharide units joined into long chains by oxygen linkages. **Starch,** with a molecular weight between 20,000 and 1,000,000, is the reserve carbohydrate of many plants and is the bulk constituent of cereals, rice, corn, and potatoes. Its structural features are brought out in Figure 27-8. **Glycogen** is the reserve carbohydrate of animals, occurring in liver and muscle tissue. It has a higher molecular weight than starch, and the polysaccharide chains are more branched. **Cellulose** is the main structural material of plants. It is the chief component of wood pulp, cotton, and straw. Complete hydrolysis gives glucose. Partial hydrolysis yields cellobiose, indicating that cellulose is a β-linked polymer (see Figure 27-8). Cellulose has a molecular weight between 300,000 and 500,000, corresponding to 1800 to 3000 glucose units. Most animals, including human beings, do not possess the necessary enzymes to hydrolyze β linkages. As a result they cannot digest cellulose. Certain bacteria

FIGURE 27-8
Two common polysaccharides.

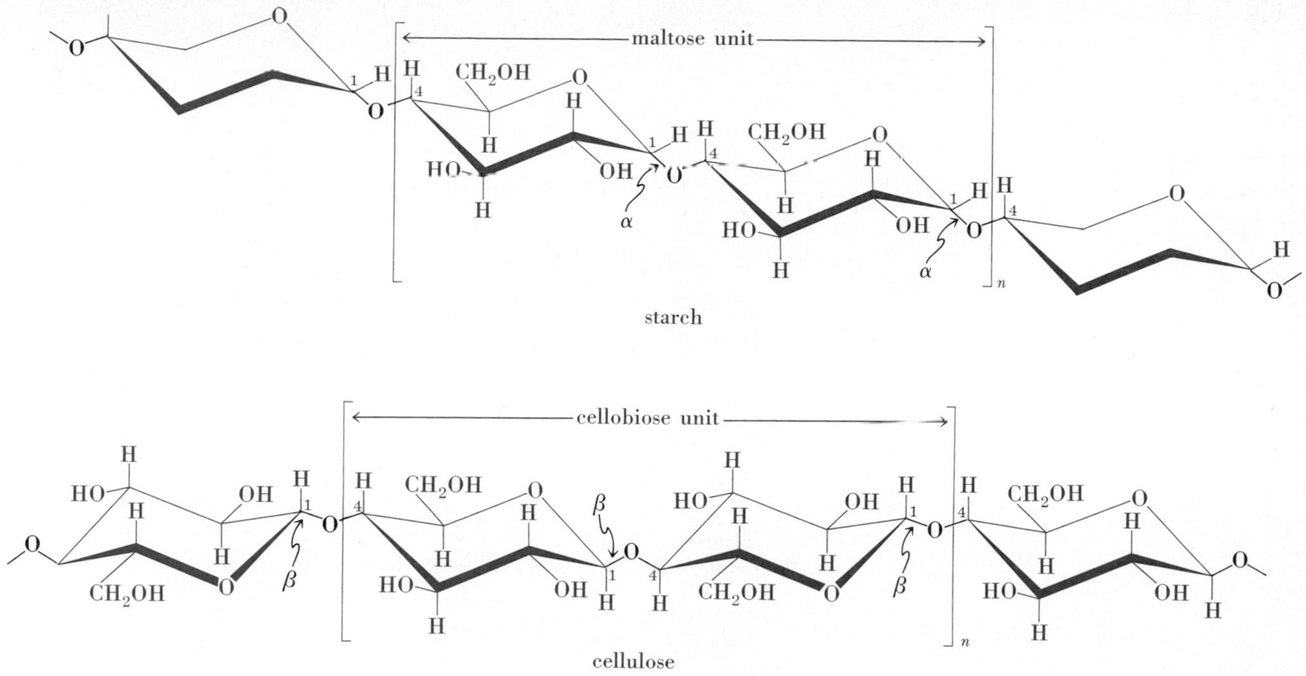

starch

cellulose

in ruminants (cows, sheep, horses) and termites can hydrolyze cellulose, allowing them to use it as food. Termites, as we know, subsist on a diet of wood.

Photosynthesis.   Green plants conduct the process of photosynthesis in the presence of light. They use carbon dioxide as the only source of carbon, together with water, inorganic salts, and a catalytic agent called chlorophyll.

$$n \, CO_2 + n \, H_2O \xrightarrow[\text{chlorophyll}]{\text{sunlight}} (CH_2O)_n + n \, O_2 \qquad (27.9)$$

Photosynthesis has been estimated to account for the annual conversion of 200 billion tons of carbon, as carbon dioxide, to carbohydrates. Concurrently, about 400 billion tons of $O_2(g)$ are released. Between 10 and 20% of this photosynthetic production occurs on land and the remainder in the oceans.

The photosynthesis reaction just given is greatly oversimplified. The currently accepted mechanism, proposed by Melvin Calvin (Nobel prize, 1963), involves as many as 100 sequential steps for the conversion of 6 mol of carbon dioxide to 1 mol of glucose. The elucidation of this mechanism was aided greatly by the use of the radioactive tracer carbon-14. For simplicity, the overall photosynthetic process is divided into two phases: (1) the conversion of solar energy to chemical energy, the light reaction; and (2) the synthesis, promoted by enzymes, of carbohydrate intermediates. This latter reaction can occur in the absence of light and is called the dark reaction. A key role in the light reaction is served by the chlorophyll. This substance, whose structure is shown in Figure 27-9, is a chelate complex. The four nitrogen atoms form a square about the central magnesium ion and the entire structure is planar. Structures of this type, first considered in Section 24-13, are called **porphyrins** and are commonly found in both plants and animals.

Biomass.   As an energy source "biomass" is any material produced by photosynthesis, that is, plants or their principal components (cellulose, starch, sugars). Some biomass

FIGURE 27-9
Structure of
chlorophyll a.

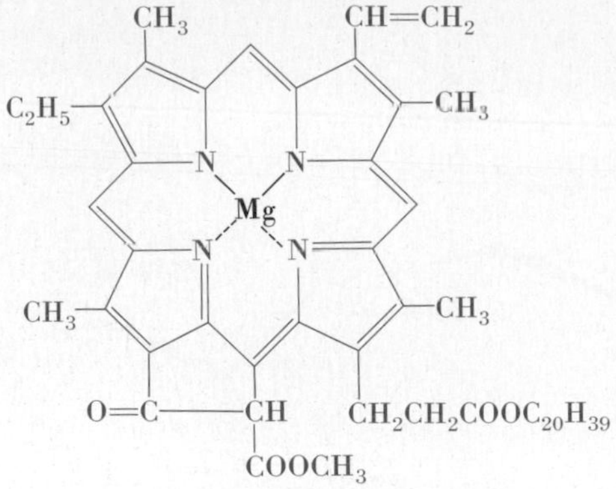

chlorophyll a

(e.g., wood) may be used directly as a fuel. Some may be converted to other gaseous, liquid, or solid materials for use as fuels or chemical raw materials.

Perhaps the best known and most widely used biomass conversion method involves the fermentation of sugars to produce ethanol. A fermentation process involves the decomposition of organic matter in the absence of air through the action of a microorganism.

$$\text{hexose sugar} \xrightarrow[\text{in yeast}]{\text{microorganisms}} 2\,C_2H_5OH + 2\,CO_2(g)$$

Disaccharides such as sucrose and polysaccharides (e.g., starch) can be hydrolyzed into monosaccharides by enzymes and then fermented to ethanol. The principal raw material for the industrial production of ethanol by fermentation is corn. Ethanol from this source is currently finding some use in the fuel "gasohol," a mixture of 10% ethanol and 90% gasoline. The biomass material that may find increasing future use as a chemical raw material is cellulose.

Plants are no longer being converted in significant quantities to fossil fuels (coal, petroleum, natural gas) by geologic processes. In principle some of the same compounds now being produced from petroleum could be made directly from cellulose. Methanol (wood alcohol) is formed in the destructive distillation (pyrolysis) of wood. Cellulose can be hydrolyzed to glucose and then converted to ethanol by fermentation. Also, fermentation processes might be used to produce a series of oxygenated compounds—alcohols and ketones. These could then be converted to hydrocarbons. Thus, the entire spectrum of organic chemicals could be produced from the simple molecules CO_2 and H_2O. The required energy would be mostly solar. Combustion of the organic chemicals or products made from them would simply return CO_2 and H_2O to the environment.

The name "protein" is derived from the Greek word *proteios*, meaning "of first importance" (similar to the derivation of "proton").

Proteins. Proteins are probably the most complex organic materials found in nature. They are the basis of protoplasm and are found in all living organisms. Proteins, as muscle, skin, hair, and other tissue, make up the bulk of the body's nonskeletal structure. As enzymes they catalyze biochemical reactions. As hormones they regulate metabolic processes; and as antibodies they counteract the effect of invading species and substances.

When a protein is hydrolyzed by dilute acids, alkalis, or hydrolytic enzymes, a mixture of α-amino acids results. [The term α means that the amino group ($—NH_2$) is on the carbon atom adjacent to the carboxyl group.] Proteins are high-molecular-weight long-chain polymers composed of these α-amino acids. Of the known α-amino acids, about 20 have been identified as building blocks of most plant and animal proteins. These are listed in Table 27-4.

TABLE 27-4
Some common amino acids

Name	Symbol	Formula	pI
		Neutral amino acids	
glycine	Gly	H CH(NH$_2$)CO$_2$H	5.97
alanine	Ala	CH$_3$ CH(NH$_2$)CO$_2$H	6.00
valine[a]	Val	(CH$_3$)$_2$CH CH(NH$_2$)CO$_2$H	5.96
leucine[a]	Leu	(CH$_3$)$_2$CHCH$_2$ CH(NH$_2$)CO$_2$H	6.02
isoleucine[a]	Ileu or Ile	CH$_3$CH$_2$CH(CH$_3$) CH(NH$_2$)CO$_2$H	5.98
serine	Ser	HOCH$_2$ CH(NH$_2$)CO$_2$H	5.68
threonine[a]	Thr	CH$_3$CHOH CH(NH$_2$)CO$_2$H	5.6
phenylalanine[a]	Phe	C$_6$H$_5$CH$_2$ CH(NH$_2$)CO$_2$H	5.48
methionine[a]	Met	CH$_3$SCH$_2$CH$_2$ CH(NH$_2$)CO$_2$H	5.74
cysteine	Cys	HSCH$_2$CH(NH$_2$)CO$_2$H	5.05
cystine	(Cys)$_2$	$+$SCH$_2$CH(NH$_2$)CO$_2$H]$_2$	4.8
tyrosine	Tyr	4-HOC$_6$H$_4$CH$_2$CH(NH$_2$)CO$_2$H	5.66
tryptophan[a]	Trp		5.89
proline	Pro		6.30
hydroxyproline	Hyp		
		Acidic amino acids	
aspartic acid	Asp	HO$_2$CCH$_2$CH(NH$_2$)CO$_2$H	2.77
glutamic acid	Glu	HO$_2$CCH$_2$CH$_2$CH(NH$_2$)CO$_2$H	3.22
		Basic amino acids	
lysine[a]	Lys	H$_2$N(CH$_2$)$_4$CH(NH$_2$)CO$_2$H	9.74
arginine	Arg	H$_2$NCNHNH(CH$_2$)$_3$CH(NH$_2$)CO$_2$H	10.76
histidine	His		

[a] Essential amino acids. In addition to these, arginine and glycine are required by the chick, arginine by the rat, and histidine by human infants.

Other than glycine ($H_2NCH_2CO_2H$), naturally occurring amino acids are optically active, mostly with an L configuration.

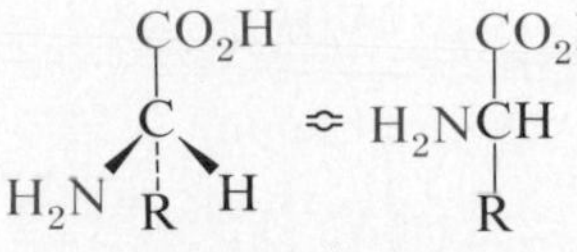

$$
\underset{H_2N \quad R \quad H}{\overset{CO_2H}{\underset{|}{C}}} \quad \approx \quad H_2N\underset{R}{\overset{CO_2H}{\underset{|}{CH}}}
$$

The reference structure for establishing the absolute configurations of amino acids is again glyceraldehyde, with the $-NH_2$ group substituting for $-OH$ and $-CO_2H$, for $-CHO$. The molecule shown above has an L configuration because the $-NH_2$ group appears on the *left* of the penultimate carbon atom.

Certain amino acids are required for proper health and growth in human beings, yet the body is unable to synthesize them. These must be ingested through foods, and are called **essential** amino acids. Eight are known to be essential; the case of three others is less certain (see Table 27-4).

The amino acids are colorless, crystalline, high-melting solids that are moderately soluble in water. In an acidic solution the amino acid exists as a cation, with a proton attaching itself to the unshared pair of electrons on the nitrogen atom in the group $-NH_2$. In a basic solution an anion is formed, through the loss of a proton by the $-CO_2H$ group. At the neutral point a proton is transferred from $-CO_2H$ to $-NH_2$. The product is a dipolar ion or a "zwitterion."

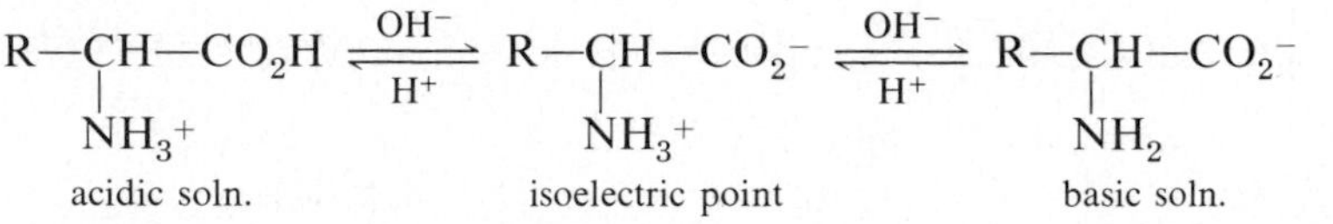

$$
\underset{\substack{NH_3^+ \\ \text{acidic soln.}}}{R-CH-CO_2H} \underset{H^+}{\overset{OH^-}{\rightleftarrows}} \underset{\substack{NH_3^+ \\ \text{isoelectric point}}}{R-CH-CO_2^-} \underset{H^+}{\overset{OH^-}{\rightleftarrows}} \underset{\substack{NH_2 \\ \text{basic soln.}}}{R-CH-CO_2^-} \qquad (27.10)
$$

Amino acids are amphoteric. The pH at which the dipolar structure predominates is called the **isoelectric point** or pI. At this pH the molecule does not migrate in an electric field in an electrophoresis apparatus (recall Figure 12-23). At a pH above the pI the molecule migrates to the anode (positive electrode), below the pI to the cathode (negative electrode). Basic amino acids have a pI above 7, acidic ones below 7, and neutral ones near 7. In most amino acids the basicity of the amino group is about equal to the acidity of the carboxyl group. The largest group of amino acids are essentially pH neutral.

Peptides. Amino acid molecules can be joined together by the elimination of water molecules between them. Two amino acids thus joined form a dipeptide. The bond between the two amino acid units (shown in color below) is called a peptide linkage or bond.

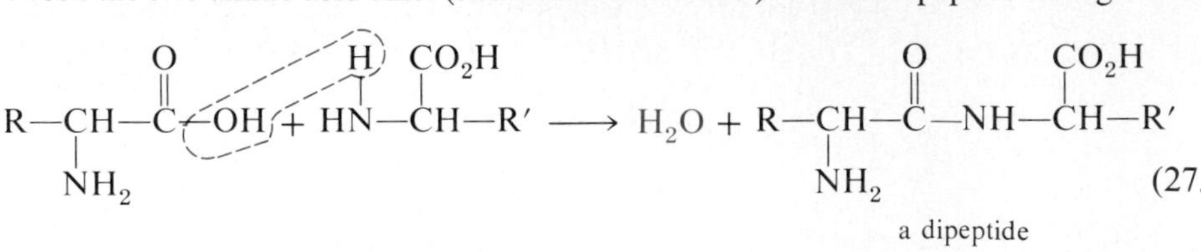

$$
\underset{NH_2}{R-CH-\overset{O}{\overset{\|}{C}}-OH} + HN-\underset{}{\overset{CO_2H}{\underset{|}{CH}}}-R' \longrightarrow H_2O + \underset{NH_2}{R-CH-\overset{O}{\overset{\|}{C}}-NH-\overset{CO_2H}{\underset{|}{CH}}-R'}
$$

$$(27.11)$$

a dipeptide

A tripeptide has three amino acid residues and two peptide linkages. A large number of amino acid units may join to form a *poly*peptide.

A simple convention is used to write the structures and names of polypeptides. The amino acid unit present at one end of the polypeptide chain has a free $-NH_2$ group; this is the "N-terminal" end. The other end of the chain has a free $-CO_2H$ group; it is the "C-terminal" end. The structure is written with the N-terminal end to the left and C-terminal to the right. The base name of the polypeptide is that of the C-terminal amino acid. All the other amino acid units in the chain are named as substituents of this acid; as such their names are changed from the "ine" to the "yl" ending. Abbreviations are also commonly used in polypeptide names, as illustrated in Example 27-4.

FIGURE 27-10
Amino acid sequence in beef insulin—primary structure of a protein.

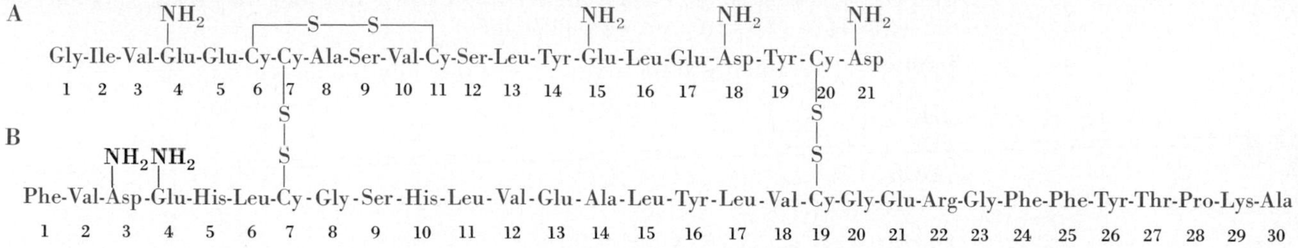

A

Gly-Ile-Val-Glu-Glu-Cy-Cy-Ala-Ser-Val-Cy-Ser-Leu-Tyr-Glu-Leu-Glu-Asp-Tyr-Cy-Asp

1   2   3   4   5   6   7   8   9   10 11 12 13 14 15 16 17 18 19 20 21

B

Phe-Val-Asp-Glu-His-Leu-Cy-Gly-Ser-His-Leu-Val-Glu-Ala-Leu-Tyr-Leu-Val-Cy-Gly-Glu-Arg-Gly-Phe-Phe-Tyr-Thr-Pro-Lys-Ala

1   2   3   4   5   6   7   8   9   10 11 12 13 14 15 16 17 18 19 20 21 22 23 24 25 26 27 28 29 30

There are two polypeptide chains joined by disulfide (—S—S—) linkages. One chain has 21 amino acids, and the other 30. In chain A the Gly at the left end is N-terminal and the Asp is C-terminal. In chain B Phe is N-terminal, and Ala is C-terminal.

Example 27-4 What is the name of the polypeptide whose structure is shown?

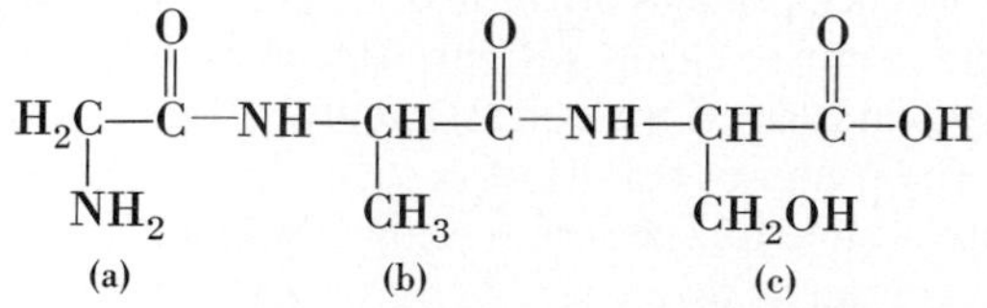

$$H_2C-C(=O)-NH-CH-C(=O)-NH-CH-C(=O)-OH$$

with substituents NH_2 (a), CH_3 (b), CH_2OH (c)

Solution. The three amino acids in this tripeptide are identified through Table 27-4. (a) = glycine; (b) = alanine; (c) = serine. The C-terminal amino acid is serine. The name is

glycylalanylserine (Gly-Ala-Ser)

SIMILAR EXAMPLES: Review Problems 7, 8; Exercise 21.

Suppose that a tripeptide is known to consist of the three amino acids: A, B, and C. What is the correct structure: ABC?, ACB?, . . . Can you see that there are six possibilities? For longer chains, of course, the number of possibilities is enormous. Determining the sequence of amino acids in a polypeptide chain is one of the most significant problems in all of biochemistry. The Nobel prize in 1958 was awarded to Frederick Sanger for elucidation of the structure of beef insulin (see Figure 27-10). The method employed is outlined in Figure 27-11.

FIGURE 27-11
Experimental determination of amino acid sequence.

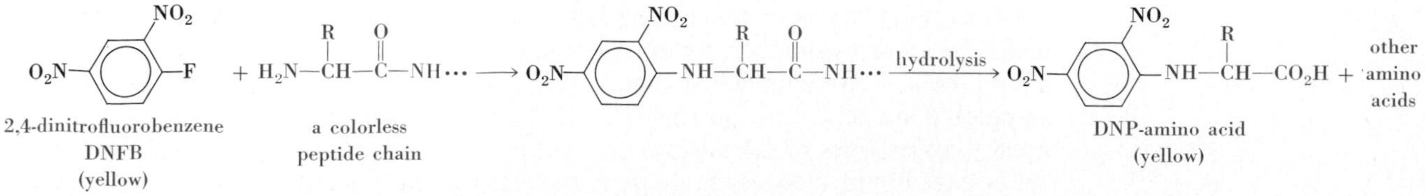

2,4-dinitrofluorobenzene
DNFB
(yellow)

a colorless
peptide chain

DNP-amino acid
(yellow)

In the reaction outline the N-terminal amino acid ends up with the yellow "marker" (a dinitrophenyl group, DNP) attached to it. By gentle hydrolysis and repeated use of the marker, a polypeptide chain can be broken down and the sequence of the individual units determined.

Example 27-5   A polypeptide, on complete hydrolysis, yielded the amino acids A, B, C, D, and E. Partial hydrolysis and sequence proof gave single amino acids together with the following larger fragments: AD, DC, DCB, BE, and CB. What must be the sequence of amino acids in the polypeptide?

Solution.   By arranging the fragments in the following manner,

```
AD
 DC
 DCB
   BE
 CB
```

we see that only the sequence A-D-C-B-E is consistent with the fragments observed.

SIMILAR EXAMPLE: Exercise 22.

The distinction between large polypeptides and proteins is arbitrary. It is generally accepted that if the molecular weight is over 10,000 (roughly 50 to 75 amino acid units) the substance is a protein. Proteins, like amino acids, are amphoteric. They possess characteristic isoelectric points, and their acidity or basicity depends on their amino acid composition. When proteins are heated, treated with salts, or exposed to ultraviolet light, profound and complex changes occur. This **denaturation** usually brings about lowering of solubility and loss of biological activity. Examples of denaturation of proteins are numerous. The frying or boiling of an egg involves the denaturation (coagulation) of the egg albumin, a protein. The beauty shop "permanent wave" takes advantage of a denaturation process that is reversible. The proteins found in hair (e.g., keratin) contain disulfide linkages (—S—S—). Treatment of hair with a reducing agent causes cleavage of these linkages—a denaturation process. Following this step the hair is set into the desired shape. Next, the hair is treated with a mild oxidizing agent. The disulfide linkages are reestablished and the hair remains in the style in which it was set.

Structure of proteins.   The **primary structure** of a protein, as we have already seen, refers to the exact sequence of amino acids in the polypeptide chains that make up the protein. Among the most complex proteins that have been analyzed is gamma globulin. It consists of four polypeptide chains, contains 1320 amino acid units (19,996 atoms), and has a molecular weight of 100,000.

What are the shapes of the long polymeric chains themselves? Are they simply limp and entangled like a plate of spaghetti or is there some order within chains and among chains? The structure or shape of an individual protein chain is referred to as **secondary structure.** The first work on this subject was published in 1951 by Pauling and Corey, describing x ray diffraction studies on polylysine, a synthetic polypeptide. They postulated that the orientation of this polypeptide and thus of protein chains is *helical.* A spiral, helical, or springlike shape can be either left- or right-handed; but because proteins are composed of L-amino acids their helical structure is right-handed (see Figure 27-12).

As we have learned previously, highly ordered materials produce distinctive x-ray-diffraction patterns, and from these patterns much can be learned about their structure. Conversely, materials in which there is little order—amorphous materials—do not have distinctive x-ray-diffraction patterns. From x ray studies it is clear that not all proteins have a helical configuration. Gamma globulin is one that does not. Hemoglobin and myoglobin are helical in some portions of their chains and randomly oriented in others. Finally, other types of orientations are possible. For example, β-keratin and silk fibroin are arranged in pleated sheets. In these proteins the side chains extend above and below the pleated sheets and hydrogen bonding is between *different* molecules (interpeptide bonding) lying next to each other and about 0.47 nm apart in the same sheet. These sheets are stacked on top of one another about 1.0 nm apart, rather like a pile of sheets of corrugated roofing (see Figure 27-13).

FIGURE 27-12
An alpha helix—
secondary structure of a
protein.

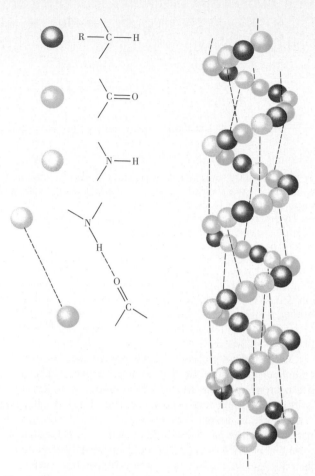

The helical structure is stabilized through formation of hydrogen bonds between carboxyl oxygen atoms in one turn and amide hydrogen atoms in the next turn above. The
bulky R groups are directed outward from the atoms in the spiral.

FIGURE 27-13
Pleated-sheet model of
β-keratin.

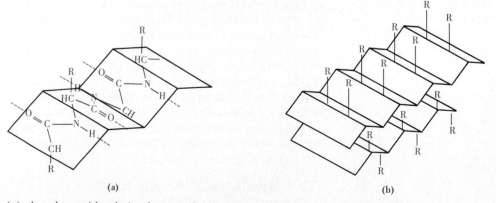

(a)

(b)

(a) A polypeptide chain showing the direction of interpolypeptide hydrogen bonds
(other polypeptide chains lie to the left and to the right of the chain shown). Bulky R
groups extend above and below the pleated sheet.
(b) The stacking of pleated sheets.

FIGURE 27-14
Linkages contributing to the tertiary structure of proteins.

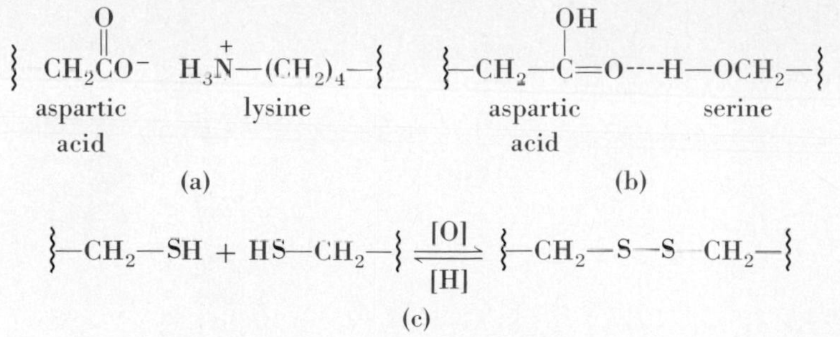

$$\}\,CH_2\overset{\displaystyle O}{\overset{\displaystyle \|}{C}}O^- \quad H_3\overset{+}{N}-(CH_2)_4-\{ \qquad \}-CH_2-\overset{\displaystyle OH}{\overset{\displaystyle |}{C}}{=}O\,{\cdots}\,H-OCH_2-\{$$

aspartic   lysine   aspartic   serine
acid   acid

(a)   (b)

$$\}-CH_2-SH + HS-CH_2-\{ \; \underset{[H]}{\overset{[O]}{\rightleftharpoons}} \; \}-CH_2-S-S\;\;CH_2-\{$$

(c)

(a) Salt linkages. Acid-base interactions between different coils. In the example shown, the carboxyl group of an aspartic acid unit on one coil donates a proton to the free amine group of a lysine unit on another.
(b) Hydrogen bonding. Interactions between side chains of certain amino acids, for example, aspartic acid and serine.
(c) Disulfide linkages. Oxidation of the highly reactive thioalcohol (—SH) group of cysteine to a disulfide (—S—S—) can occur (as in beef insulin).

The folding of polypeptide chains into a tertiary structure is influenced by an additional factor. Hydrophobic hydrocarbon portions of the chains (R groups) tend to be drawn into close proximity in the interior of the structure, leaving ionic groups at the exterior.

Does a helical protein molecule possess any additional structural features? That is, are the helices elongated, twisted, knotted, or what? The final statement regarding the shape of a protein molecule lies in a description of its tertiary structure. Because the internal hydrogen bonding that occurs between atoms in successive turns of a protein helix is weak, these hydrogen bonds ought to be easily broken. In particular, we should expect them to be replaced by hydrogen bonds to water molecules when the protein is placed in water. That is, the α helix should open up and become a randomized structure when placed in water (recall the analogy of the limp spaghetti). But experimental evidence indicates that this does not happen. We are led to the conclusion that other forces must be involved in compressing the long α-helical chains into definite geometric shapes. Each protein has its own three-dimensional shape or **tertiary structure.** Three types of linkages involved in tertiary structures are described in Figure 27-14.

By x-ray-diffraction studies Kendrew and Perutz (Nobel prize, 1963) were able to elucidate the primary, secondary, and tertiary structures of myoglobin. The primary structure is that of a peptide of 153 units in a single chain. Secondary structure involves 70% coiling of the chain into an α helix. The tertiary structure is depicted in Figure 27-15.

FIGURE 27-15
Representation of the tertiary structure of myoglobin.

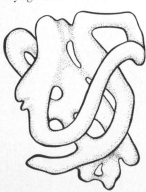

The hemoglobin molecule consists of four separate polypeptide chains or subunits. The arrangement of these four subunits constitutes a still higher order of structure referred to as the **quaternary structure** (see Figure 27-16). Because it is a single polypeptide, myoglobin has no quaternary structure.

Even minor changes in the structure of a protein can have profound effects. Hemoglobin contains four polypeptide chains, each with 146 amino acid units. The substitution of valine for glutamic acid at one site in two of these chains gives rise to the sometimes fatal blood disease known as sickle cell anemia. Apparently, the altered hemoglobin has a reduced ability to transport oxygen through the blood.

27-3 Biochemical Reactivity

Although the outward appearances of organisms vary remarkably, the similarities that exist in the chemical reactions that occur within organisms are equally striking. The totality of these reactions is referred to as **metabolism.** The part of this process in which

FIGURE 27-16
Quaternary protein structure—the structures of heme and hemoglobin.

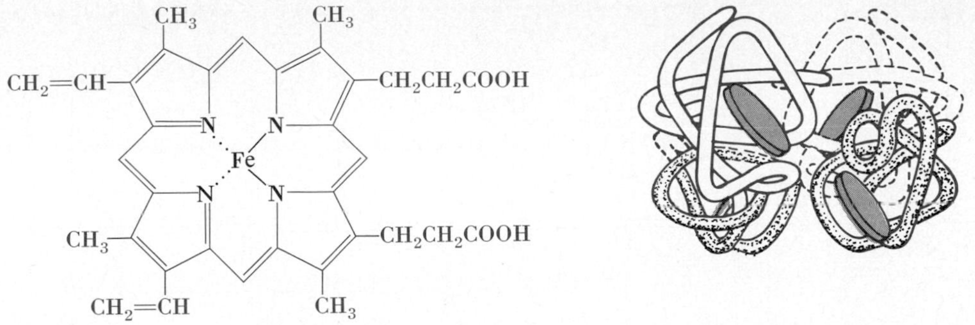

(a)

(b)

Hemoglobin is a conjugated protein. It consists of nonprotein groups (called prosthetic groups) bonded to a protein portion. **(a)** The structure of heme. **(b)** Four heme units and four polypeptide coils are bonded together in a molecule of hemoglobin.

molecules are broken down or degraded is called catabolism, and that part in which molecules are synthesized is called anabolism. Reactions for which the standard free energy change is positive are endergonic, and those for which it is negative are exergonic. The chemical substances involved in metabolism are called metabolites.

The raw materials for metabolic transformations are organic molecules, but the reactions involved are very dissimilar to those considered in the preceding chapter. In fragile biological systems, heat and pressure cannot be used to force chemical reactions to go, nor can strongly acidic or basic catalysts. The substances that do control the processes of metabolism are the catalysts known as enzymes. Herein lies the basis of biological uniqueness. An organism develops into what it is because of the particular set of enzymes it possesses.

Metabolism is a complex subject and anything more than a brief overview is much beyond the scope of this text. The discussion that follows centers on the summary of metabolic processes outlined in Figure 27-17.

Carbohydrate Metabolism. Foods containing starch are the principal sources of carbohydrates for humans and many animals. Digestion of starch begins in the mouth through the action of salivary enzymes, the amylases. Starch is converted to maltose and polysaccharides known as dextrins. This process continues in the acidic medium of the stomach, and the maltose and polysaccharides pass on to the small intestine. Here, amylase from the pancreas completes the conversion of polysaccharides to maltose, and the enzyme maltase converts maltose to glucose. Glucose is absorbed through the wall of the small intestine into the bloodstream, from which it is distributed to other organs.

Glucose is ultimately oxidized to carbon dioxide and water with the liberation of energy; the principal intermediate in this process is glucose 6-phosphate (glucose-6P). Its formation is controlled by the pancreatic hormone, insulin. Once formed glucose-6P may be converted to glycogen (a polysaccharide stored in the liver), back to glucose, or it may be metabolized. The major route for this metabolism involves the anaerobic (absence of air) Embden–Meyerhof pathway, followed by an aerobic cycle (Krebs cycle). These interrelationships are suggested diagrammatically in Figure 27-17.

In glucose 6-phosphate a phosphate group replaces the —OH group on the C-6 atom of a pyranose (see Figure 27-6).

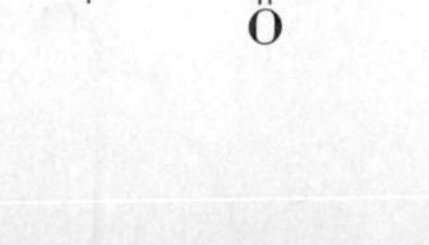

Lipid Metabolism. Fats stored in the body represent a rich source of energy. The digestion of fats and oils occurs primarily in the small intestine, through the action of a combination of lipase enzymes. The products of this enzyme hydrolysis are glycerol,

FIGURE 27-17
Metabolism outline.

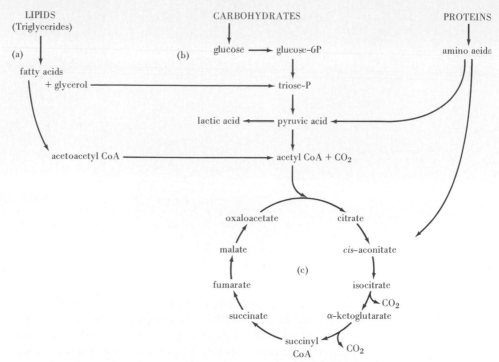

(a) Fatty acid section. Fatty acids are degraded two carbon atoms at a time. Acetyl units are fed into the citric acid cycle (c) as acetyl CoA.

(b) Glycolysis section (Embden–Meyerhof pathway). These reactions are anaerobic (no oxygen required). Carbohydrates are degraded to the six-carbon sugar glucose, and then to the three-carbon triose-P (glyceraldehyde-3-phosphate). Next the three-carbon acid, pyruvic acid, is formed from triose-P. Pyruvic acid loses a molecule of CO_2, yielding the two-carbon acetyl unit, which combines with coenzyme A (CoA) to form acetyl CoA.

(c) Citric acid cycle (Krebs cycle). A two-carbon acetyl unit from acetyl CoA joins with the four-carbon oxaloacetate unit to produce the six-carbon tricarboxylic acid, citric acid (designated here as citrate). A two-step conversion to isocitrate occurs, followed by the loss of a molecule of CO_2 and the formation of the five-carbon α-ketoglutarate. Another CO_2 molecule is lost in the formation of succinyl CoA. The remainder of the cycle involves a succession of four-carbon acids leading to oxaloacetate. The oxaloacetate regenerated at the end of the cycle now joins with another acetyl unit and the cycle is repeated. The net change occurring in the cycle, then, is that a two-carbon acetyl unit enters the cycle and two molecules of CO_2 leave.

Emphasis in this outline is on the degradation of large molecules to small ones (catabolism). Some of the species shown in the citric acid cycle are also involved in the biosynthesis of larger molecules (anabolism). For example, the amino acid asparagine is synthesized from oxaloacetate.

The structure of glyceraldehyde 3-phosphate is

mixtures of mono- and diglycerides, and fatty acids. These are absorbed into the bloodstream through the wall of the intestine. Glycerol is converted to glyceraldehyde-3-phosphate (triose phosphate) and joins into the glucose metabolism route previously described. Fatty acids are oxidized to carbon dioxide and water, with the release of energy, in a series of reactions known as β oxidation. In this process oxidation occurs at the β carbon atom of a fatty acid, followed by cleavage. This means that two-carbon pieces (acetic acid) are split off. The process requires the presence of coenzyme A (CoA). For example, with palmitic acid ($C_{15}H_{31}CO_2H$) the process must be repeated seven times, with the formation of eight molecules of acetyl coenzyme A, which enter the Krebs cycle (Figure 27-17).

Protein Metabolism. In the stomach, hydrochloric acid and the enzyme pepsin hydrolyze about 10% of the amide linkages in proteins and produce polypeptides in the molecular weight range of 500 to several thousand. In the small intestine peptidases such as trypsin and chymotrypsin (from the pancreas) cleave the polypeptides into very small fragments. These are then acted upon by aminopeptidase and carboxypeptidase. The resulting free amino acids pass through the wall of the intestine, into the bloodstream, and on to various organs. Each amino acid has its own characteristic metabolic reactions, but in general each is converted to an intermediate that enters the Krebs cycle. Proteins may also be synthesized from amino acids in body cells under directions supplied by nucleic acids (see Section 27-4). However, the eight essential amino acids mentioned previously cannot be synthesized in the body and must be obtained from digested proteins.

Energy Relationships in Metabolism. Reactions in which molecules are synthesized in an organism, anabolic reactions, must acquire energy from reactions in which molecules are degraded, catabolic reactions. The fundamental agents responsible for these energy exchanges are adenosine triphosphate (ATP) and adenosine diphosphate (ADP). The energy released in exergonic reactions is stored by the conversion of ADP to ATP.

$$\text{ADP} + \text{inorganic phosphate (P}_i\text{)} + 30\text{--}50 \text{ kJ} \underset{\substack{\text{energy} \\ \text{released}}}{\overset{\substack{\text{energy} \\ \text{absorbed}}}{\rightleftharpoons}} \text{ATP} + \text{H}_2\text{O} \qquad (27.12)$$

Energy is released from foods by oxidation processes. The energy released in the oxidation is picked up by ADP, which is converted to ATP. Enzymes catalyze each conversion every step along the way. ADP, ATP, and two important intermediates, nicotinamide adenine dinucleotide (NAD) and flavin adenine dinucleotide (FAD), are pictured in Figure 27-18.

The exact way in which ADP and ATP enter into metabolic processes was not detailed in Figure 27-17. However, it is known that the conversion of 1 mol of glucose to CO_2 and H_2O is accompanied by the conversion of 38 mol ADP to ATP.

$$C_6H_{12}O_6 + 6\,O_2 + 38\,\text{ADP} + 38\,P_i \longrightarrow 6\,CO_2 + 6\,H_2O + 38\,\text{ATP} \qquad (27.13)$$

Assuming that 33.5 kJ of energy is absorbed for each mole of ADP converted to ATP, the energy stored in ATP as a result of the metabolism of 1 mol glucose is $38 \times 33.5 = 1270$ kJ. The total energy released when 1 mol glucose is converted to CO_2 and H_2O is 2870 kJ. Thus, the efficiency of energy storage in the high-energy bonds of ATP is $(1270/2870) \times 100 = 44\%$. Compared to the efficiency of heat engines in converting heat to work (recall Chapter 16), the metabolic process is seen to be an efficient one. Nearly half the energy of glucose can be stored by the body for later use. Perhaps the complexity of the metabolic process can be better appreciated in terms of the data just given. If the metabolism of glucose occurred in a single step, with one ADP converted to ATP in that step, only $(33.5/2870) \times 100 = 1\%$ of the available energy would be conserved. Because metabolism occurs in many steps there is an opportunity for a much greater quantity of energy to be stored.

Enzymes. An enzyme is a biological catalyst that contains protein. Some enzymes are made up only of protein; some need cofactors or coenzymes to function (e.g., FAD and NAD). Enzymes are specific for each biological transformation and catalyze a reaction without requiring a change in temperature or pH. Originally, enzymes were assigned common or trivial names, such as pepsin and catalase. Present practice, however, is to name them after the processes they catalyze, usually employing an "ase" ending.

In 1913, Michaelis and Menten proposed a model of enzyme reactivity based on the formation of an enzyme-substrate complex. According to this model, an enzyme can exert its catalytic activity only after combining with the reacting substance, the **substrate,** to

FIGURE 27-18
Some important
chemical intermediates
in metabolism.

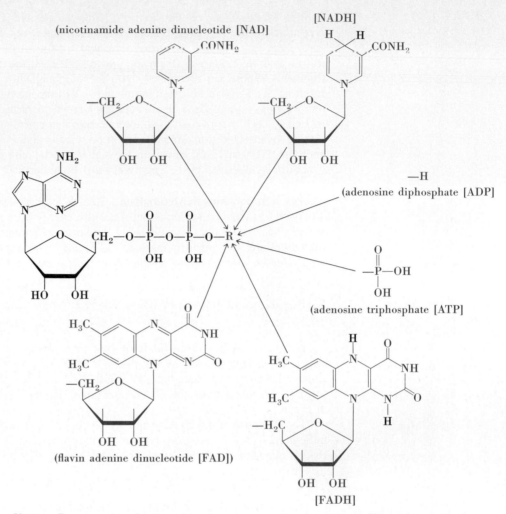

(nicotinamide adenine dinucleotide [NAD]

[NADH]

(adenosine diphosphate [ADP]

(adenosine triphosphate [ATP])

(flavin adenine dinucleotide [FAD])

[FADH]

Various R groups can be joined to the structure shown in black. The reduced form of NAD, called NADH, contains one additional H atom, shown in black. The reduced form of FAD, called FADH, contains two additional H atoms, also shown in black.

form a complex. There appears to be a definite site on the enzyme where the substrate combines; for some enzymes there is evidence that more than one active site exists. Reaction of the substrate (S) with the enzyme (E) to form a complex (ES) permits the reaction to proceed via a path of lower activation energy than the noncatalyzed path. When the complex decomposes, products (P) are formed and the enzyme is regenerated (see Figure 27-19). This general reaction scheme and a specific example are presented below.

$$E + S \rightleftharpoons ES \longrightarrow E + P$$

$$\text{sucrase} + \text{sucrose} \rightleftharpoons \frac{\text{sucrase-sucrose}}{\text{complex}} \xrightarrow{H_2O} \text{sucrase} + \text{glucose} + \text{fructose}$$

The implications of the Michaelis–Menton mechanism for the rates of enzyme-catalyzed reactions were considered in Section 14-9.

FIGURE 27-19
The lock-and-key model
of enzyme action.

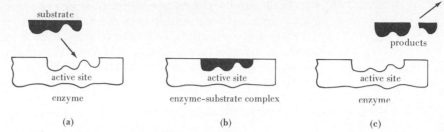

(a) The substrate attaches itself to an active site on the enzyme molecule. **(b)** Reaction occurs. **(c)** Product species detach themselves from the site, freeing the enzyme molecule to attach another substrate molecule. The substrate and enzyme must have complementary structures to produce a complex, hence the term *lock-and-key*.

Generally, a 10°C temperature rise produces an approximate doubling of a reaction rate, but with enzymes a certain temperature is reached beyond which a decrease in rate sets in. The optimum temperature for enzyme activity is about 37°C (98°F) for the enzymes present in warm-blooded animals. The decrease in rate beyond this temperature results from the fact that enzymes are proteins and proteins are denatured by heat. This denaturation disrupts the secondary and tertiary structure and distorts the active site on the enzyme.

Protein behavior is extremely sensitive to changes in pH. At high and low pH values complete denaturation of enzymes occurs, but even milder changes in pH cause drastic changes in the rate of enzyme action. Most enzymes in the body have their maximum activity between pH 6 and 8, with gastric enzymes being notable exceptions. In addition to the effects of temperature and pH, specific inhibition can occur when a molecule other than the substrate competes for an active enzyme site. Also heavy metal ions (Hg^{2+}, Pb^{2+}, and Ag^+) may combine in a nonreversible way with active site groups, such as —OH, —SH, —CO_2^-, and —NH_3^+, and deactivate the enzyme. Oxalic and citric acids inhibit blood clotting by competing for the calcium ions necessary for the activation of the enzyme thromboplastin. It has also been suggested that antibiotics function by inhibiting enzyme-coenzyme reactions in microorganisms.

Hormones. A hormone is a secretion of a ductless or endocrine gland, such as the thyroid, the pituitary, the pancreas (in part), the adrenals, and parts of the testes and ovaries. These glands secrete their products directly into the blood stream through which these products reach all parts of the body. The hormones appear to aid in the control of biological reactions, but their exact role is not clearly understood. Hormones are sometimes referred to as "chemical messengers."

A well-known protein hormone is insulin. Its function is to lower the blood sugar level by increasing the rate of conversion of glucose into muscle and liver glycogen. There is considerable evidence that insulin acts by controlling the phosphorylation of glucose. In the absence of a sufficient amount of insulin, the condition of diabetes mellitus results. Among its clinical symptoms are high levels of sugar in the blood and urine and the formation of excess ketones, giving a distinctive odor to the breath.

Vitamins. Vitamins are substances necessary to maintain normal health, growth, and nutrition; however, they are not used in building cells or as an energy source. Their apparent function is as catalysts for biological processes. The vitamins are sometimes classified as fat soluble (vitamins A, D, E, and K) and water soluble (vitamins B and C). Vitamin A, although not found in plants itself, can be formed from β-carotene, a yellow pigment found in plants. A deficiency of vitamin A causes night blindness and xerophthalmia, a disease of the eyes in which the tear glands cease to function. Vitamin D is associated with the proper deposition of calcium phosphate, which in turn is related to normal teeth and bone development and the prevention of rickets.

Vitamin E is sometimes called the fertility factor and is involved in the proper func tioning of the reproductive system. It is found in vegetable oils, such as corn germ oil, cottonseed oil, peanut oil, and wheat germ oil. There is more than one form of vitamin E; the structure for α-tocopherol, the most active, is shown below.

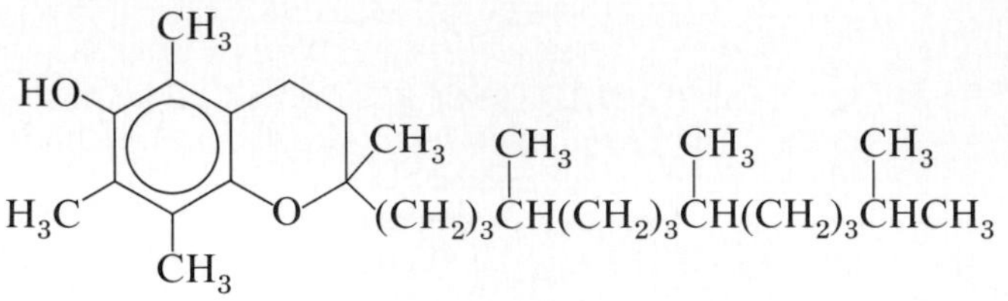

vitamin E
α-tocopherol

Vitamin K is the antihemorrhagic factor involved in bloodclotting. There are two K vitamins. Vitamin K_1 is obtained from alfalfa and vitamin K_2 by bacterial action in the intestine.

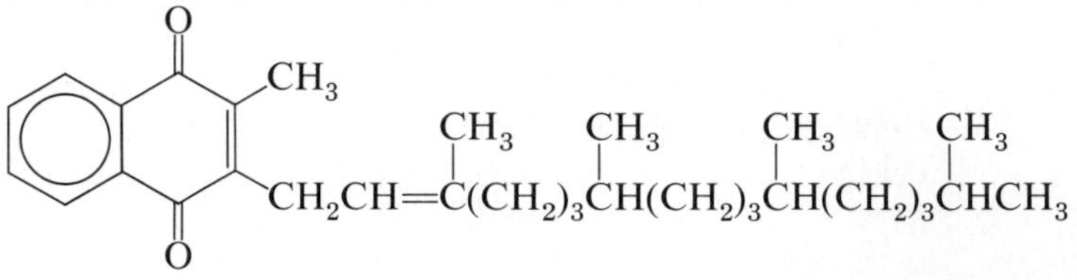

vitamin K_1

Vitamin C, ascorbic acid, is the vitamin that prevents scurvy. It is found in citrus fruits, green peppers, parsley, and tomatoes. The vitamin B complex has been shown to consist of many substances, most of which seem to be involved in energy transformations related to metabolism. A deficiency of vitamin B_1, thiamine, leads to the disease called beriberi. Lack of vitamin B_2, riboflavin, causes inflammation of the lips, dermatitis, a dryness and burning of the eyes, and sensitivity to light. Both of these B vitamins are distributed widely in nature, in lean meat, nuts, and leafy vegetables.

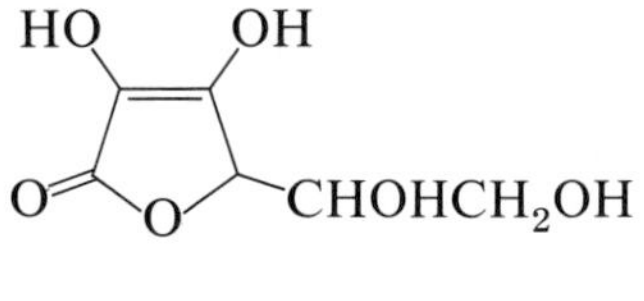

vitamin C, ascorbic acid

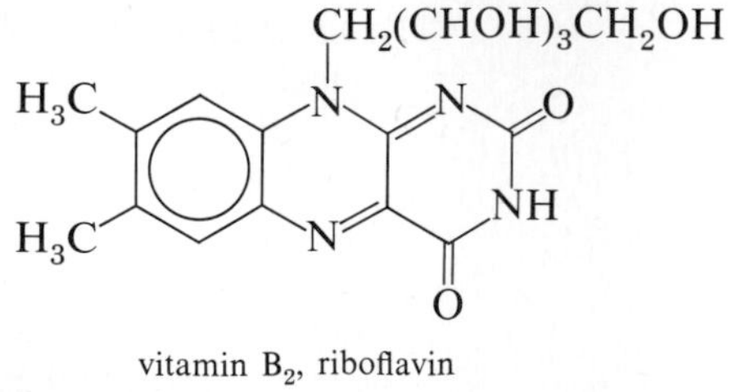

vitamin B_2, riboflavin

27-4 The Nucleic Acids

As we noted earlier, lipids, carbohydrates, and proteins, taken together with water, constitute about 99% of most living organisms. The remaining 1% includes some compounds of vital importance to the existence, development, and reproduction of all forms of life. Among these are the nucleic acids. Nucleic acids carry the information that directs the metabolic activity of cells.

FIGURE 27-20
Hydrolysis products of nucleic acids.

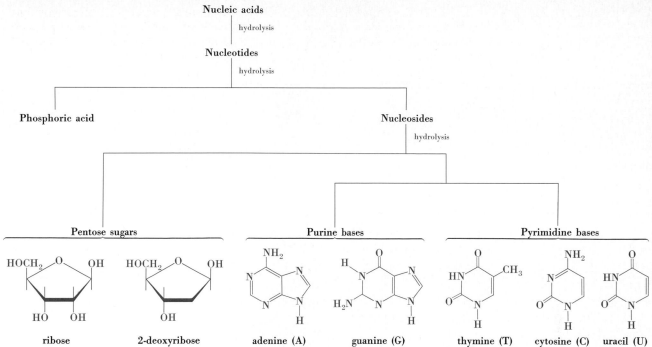

Tracing the hydrolysis reactions in the reverse direction, the combination of a pentose sugar and a purine or pyrimidine base yields a nucleoside. A nucleoside, in combination with phosphoric acid, yields a nucleotide. A nucleic acid is a polymer of nucleotides.

If the sugar is 2-deoxyribose and the bases A, G, T, and C, the nucleic acid is DNA. If the sugar is ribose and the bases A, G, U, and C, the nucleic acid is RNA. (The term 2-deoxy means without an oxygen atom on the second carbon atom.)

The nucleus of the cell contains **chromosomes** that cause replication from one generation to the next. The individual portions of the chromosomes that carry specific traits are known as genes. It has now been established that DNA or **deoxyribonucleic acid** is the actual substance constituting the genes. Figure 27-20 traces the steps that may be followed in degrading nucleic acids into their simpler constituents—heterocyclic amines known as purines and pyrimidines, a five-carbon sugar (ribose or 2-deoxyribose), and phosphoric acid. Figure 27-21 represents a portion of a nucleic acid chain.

The usual form for DNA is a **double helix.** The postulation of this structure of DNA by Francis Crick and James Watson in 1953 was one of the great scientific breakthroughs of modern times. Their work was critically dependent on two other contemporary scientific achievements. One was the precise x ray diffraction studies on DNA by Maurice Wilkins and Rosalind Franklin. The other was the discovery by Erwin Chargaff of a set of regularities regarding the occurrence of the purine and pyrimidine bases in nucleic acids. For DNA derived from a particular organism, these regularities, known as the **base-pairing rules,** require the following.

1. The amount of adenine is equal to the amount of thymine (A = T).
2. The amount of guanine is equal to the amount of cytosine (G = C).
3. The total amount of purine bases is equal to the total amount of pyrimidine bases (G + A = C + T).

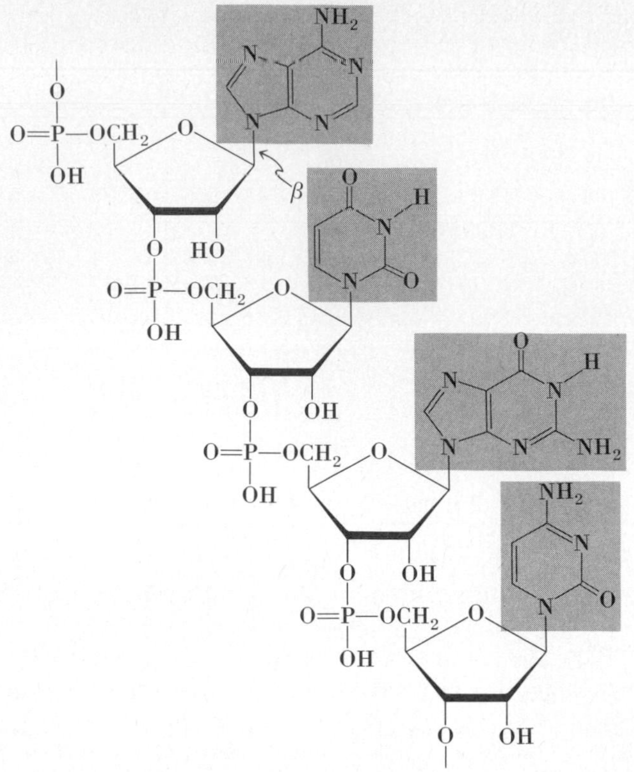

FIGURE 27-21
A portion of a nucleic
acid chain.

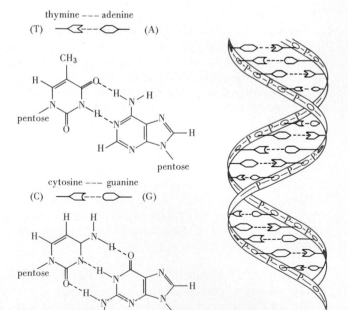

FIGURE 27-22
DNA model.

thymine --- adenine
(T) ◁--◇ (A)

cytosine --- guanine
(C) ◁--◇ (G)

◯ deoxyribose
−P− phosphate ester
--- hydrogen bond
◇ adenine (A)
▷ thymine (T)
◯ guanine (G)
▷ cytosine (C)

Since large measures of intuition and inspiration were involved in Crick and Watson's work, it is not easy to show how their model was established from the facts just described. Instead, let us simply demonstrate how their model, shown in Figure 27-22, is consistent with the base-pairing rules.

In order to maintain the structure of a double helix, it is necessary that a force exist between the two single strands. As in the α-helical structure of a protein, the postulated force is based on hydrogen bonds, involving hydrogen, nitrogen, and oxygen atoms on the purine and pyrimidine bases. The necessary conditions for hydrogen bonding will exist only if an A on one strand appears opposite a T on the other, or if a G is bonded to a C. No other combinations will work. For example, C cannot be paired with T. Both are relatively small molecules (single ring) and would not approach each other closely enough between the strands. The combination of G and A cannot occur because the molecules are too large (double rings). With the combinations C + A and G + T, the conditions for hydrogen bonding are not right. Thus, it follows rather directly that the total amount of A must equal the total amount of T, and so on. The Chargoff rules are explained.

Proposal of the DNA structure was the important first step in the development of the theory of DNA, but at least three other questions must be explained by this theory: (1) How does a DNA molecule reproduce itself during cell division? (2) How does DNA direct the synthesis of proteins in the cell? (3) How is the information required to obtain the exact sequence of amino acids in a protein coded into the DNA structure? We cannot go into detail on these questions, but let us elaborate a bit.

The critical step in the replication of a DNA molecule requires the molecule to unwind into single strands. As the unwinding occurs, nucleotides present in the cell nucleus, through the action of enzymes, become attached to the exposed portions of the two single strands, converting each to a new double helix of DNA. As suggested by Figure 27-23, when the original DNA (parent) molecule is completely unwound, two new molecules (daughters) appear in its place!

There are two pieces of evidence, each very convincing, that the process outlined here does indeed occur. First, electron micrographs of the DNA molecule have now been obtained, including some that capture DNA in the act of replication. Another elegant experiment involves growing bacteria in a medium containing ^{15}N atoms, so that all the N atoms of the bases of the DNA molecules are ^{15}N. The bacteria are then transferred to a nutrient with nucleotides containing normal ^{14}N. Here the bacteria are allowed to divide and reproduce themselves. The DNA of the offspring cells are then analyzed. Those of the first generation, for example, consist of DNA molecules with one strand having ^{15}N and the other ^{14}N atoms. This is exactly the result to be expected if replication occurs by the unzipping process described in Figure 27-23.

The puzzle of protein synthesis is that this occurs in the cytoplasm of the cell, not in the nucleus. Yet the molecule that directs the synthesis, DNA, is found only in the

FIGURE 27-23
Replication of a DNA molecule visualized.

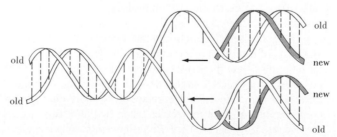

As the unzipping process occurs, hydrogen bonds between the old DNA strands are broken. The new strands grow in the direction of the arrows by attaching nucleotides which then form hydrogen bonds to the old DNA strands.

FIGURE 27-24
A representation of protein synthesis.

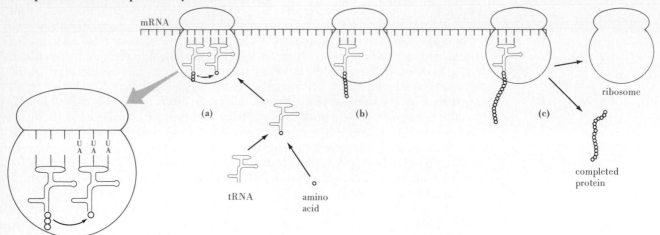

(a) Through the action of an enzyme, a tRNA molecule brings a single amino acid to a site on a ribosome. The anticodon of the tRNA (AAA for the example shown) must be complementary to the codon of the mRNA (UUU). (The amino acid carried by this tRNA is phenylalanine.) The amino acid is added to the chain in the manner shown; the chain moves from the tRNA on the left to the one on the right.
(b) As the ribosome moves along the mRNA strand, more and more amino acid units are added through the proper matching of tRNA molecules with the code on the mRNA.
(c) When the ribosome reaches the end of the mRNA strand, it and the completed protein are released. The ribosome is free to repeat the process.

nucleus. Here is where the different forms of RNA play a fundamental role. The several steps involved in this process are represented, collectively, in Figure 27-24.

First, a molecule of **messenger RNA, mRNA,** is synthesized on a portion of a DNA strand in the nucleus. This probably involves an unzipping of the DNA molecule similar to the process of DNA replication described above. The mRNA migrates out of the nucleus into the cytoplasm. mRNA has a high affinity for ribosomes and gathers them up along the chain. The combination of the mRNA and its ribosomes is called a **polysome. Transfer RNA, tRNA,** refers to a variety of rather short RNA chains, each of which is capable of attaching only a specific amino acid. The function of tRNA is to bring a specific amino acid to a site on the ribosome where the amino acid can form a polypeptide bond and become part of a growing polypeptide chain. The ribosome moves along the mRNA chain, attaching different tRNA molecules and incorporating their amino acids into the polypeptide chain. When the ribosomes reaches the end of the mRNA chain it falls off and releases the protein molecule that has been synthesized. The entire process is like the stringing of beads.

The code (set of directions) that determines the exact sequence of amino acids in the synthesis of a protein is incorporated in the chromosomal DNA. It is found in the particular pattern of base molecules on the double helix. Since there are only four different bases possible in an mRNA molecule—U, A, G, and C—but 20 different amino acids, it is clear that the code cannot correspond to individual base molecules. There are $4^2 = 16$ combinations of base molecules taken two at a time (i.e., UU, UA, UG, UC, etc.). But these could only account for 16 amino acids. When the base molecules are taken three at a time, there are $4^3 = 64$ possible combinations. A group of three base molecules in a DNA strand, called a **triplet,** causes a complementary set of base molecules to appear in the mRNA formed on it. This triplet or **codon** on the mRNA must be matched by a complementary triplet, called an **anticodon,** in a tRNA molecule. The particular tRNA with this anticodon carries a specific amino acid to the site of protein synthesis.

TABLE 27-5
The genetic code

Second base

First base		U	C	A	G
U		UUU⎫ Phe UUC⎭ UUA⎫ Leu UUG⎭	UCU⎫ UCC⎪ Ser UCA⎪ UCG⎭	UAU⎫ Tyr UAC⎭ UAA Nonsense UAG Nonsense	UGU⎫ Cys UGC⎭ UGA Nonsense UGG Trp
C		CUU⎫ CUC⎪ Leu CUA⎪ CUG⎭	CCU⎫ CCC⎪ Pro CCA⎪ CCG⎭	CAU⎫ His CAC⎭ CAA⎫ Gln CAG⎭	CGU⎫ CGC⎪ Arg CGA⎪ CGG⎭
A		AUU⎫ AUC⎪ Ile AUA⎭ AUG Met	ACU⎫ ACC⎪ Thr ACA⎪ ACG⎭	AAU⎫ Asn AAC⎭ AAA⎫ Lys AAG⎭	AGU⎫ Ser AGC⎭ AGA⎫ Arg AGG⎭
G		GUU⎫ GUC⎪ Val GUA⎪ GUG⎭	GCU⎫ GCC⎪ Ala GCA⎪ GCG⎭	GAU⎫ Asp GAC⎭ GAA⎫ Glu GAG⎭	GGU⎫ GGC⎪ Gly GGA⎪ GGG⎭

Initiation of a polypeptide chain appears to require the presence of the codon AUG, and termination of the chain, either UAA or UAG or UGA.

Through an ingenious set of experiments, which led to a Nobel prize in 1968 for Nirenberg, Holley, and Khorana, the genetic code (Table 27-5) has now been cracked. The significant features of the genetic code are the following.

- There is more than one triplet code for most amino acids.
- The first two letters of the codon are most significant. There is considerable variation in the third.
- There are three codons that do not correspond to any amino acids. Although they are referred to as nonsense codons, these seem to play a role in stopping protein synthesis (rather like the word "STOP" used to separate phrases in a telegram).
- The various codons direct the *same* protein synthesis, whether in bacteria, plants, lower animals, or humans.

Summary

Four categories of substances found in living organisms are considered in this chapter—lipids, carbohydrates, proteins, and nucleic acids.

One familiar group of lipids are the triglycerides. These are esters of glycerol with long-chain monocarboxylic (fatty) acids. If saturated fatty acids predominate, the triglyceride is a fat. If unsaturation (in the form of double bonds) occurs in some of the fatty acid components, the triglyceride is an oil. The catalytic addition of hydrogen to double bonds in a triglyceride converts an oil to a fat. Hydrolysis of a triglyceride with a strong base (saponification) yields glycerol and metal salts of the fatty acids—soaps. In phosphatides (phospholipids) phosphoric acid and a nitrogen-containing compound are substituted for one of the fatty acid components of a triglyceride.

The simplest carbohydrate molecules are five- and six-carbon-chain polyhydroxy aldehydes and ketones. However, these molecules convert readily into cyclic structures, that is, five- and six-membered rings. The subject is further complicated by the existence of chiral carbon atoms in these molecules, rendering the molecules optically active, that is, able to rotate the plane of polarized light. Moreover, optical isomers are encountered—molecules of identical composition that differ in their spatial orientations and in their optical activity. Monosaccharides or simple sugars can be readily joined into polysaccharides containing from a few to a few thousand monosaccharide units.

The basic building blocks of proteins are some 20 different α-amino acids. Two amino acid molecules may join by eliminating a H_2O molecule between them. The result is a peptide bond, —CONH—. Long polypeptide chains are formed as this process is repeated. Additional structural features include the twisting of a polypeptide chain into a helical coil and bonding between coils.

Lipids, carbohydrates, and proteins are complex molecules. Their metabolism by the body involves breaking them down into their simplest units. Carbohydrates are broken down into monosaccharides; proteins, into amino acids; and lipids, into glycerol and two-carbon-chain acids. Ultimately, these products are decomposed into still simpler molecules, such as CO_2, H_2O, NH_3, and urea. Energy released in these processes is stored in the substance adenosine triphosphate (ATP).

DNA molecules found in cell nuclei are able to synthesize RNA molecules, the framework on which protein synthesis occurs. Proteins acquire the correct sequence of amino acids because each RNA molecule carries a code imparted to it in its synthesis by DNA. DNA molecules also have the ability to replicate themselves during cell division, which ensures that the correct protein-building code is passed on from one generation of cells to the next.

Learning Objectives

As a result of studying Chapter 27, you should be able to

1. List the four principal types of substances found in cells—lipids, carbohydrates, proteins, and nucleic acids—and describe the basic chemical composition of each.

2. Write structural formulas and names of triglycerides and indicate whether their constituent fatty acids are saturated or unsaturated.

3. Relate the structures, saponification values, and iodine numbers of triglycerides.

4. Explain how phosphatides and waxes, which are also lipids, differ from triglycerides.

5. Classify carbohydrates as monosaccharides, oligosaccharides, and polysaccharides; and give examples of each.

6. Describe the phenomenon of optical activity and the structural features of a molecule that produce it.

7. Outline the conversion of a straight chain monosaccharide into its cyclic form, and the joining of monosaccharide units into polysaccharides.

8. Show how the form adopted by an amino acid varies with solution pH.

9. Describe the formation of peptide bonds between amino acid molecules, and assign systematic names and structures to polypeptides.

10. Explain how the sequence of amino acids in a polypeptide chain is determined from information acquired in the hydrolysis of the polypeptide.

11. Explain the meaning of secondary, tertiary, and, where applicable, quaternary structure of a protein.

12. Describe metabolism and how the energy released in metabolism is stored.

13. Explain enzyme action in terms of the "lock-and-key" model.

14. Name the principal constituents of the nucleic acids and indicate how these constituents are linked together into chains and, in the case of DNA, a double helix.

15. Explain how DNA replicates itself during cell division; and outline the process of protein synthesis, indicating the role of DNA, mRNA, and tRNA.

Some New Terms

Absolute configuration refers to the spatial arrangement of the groups attached to a chiral carbon atom. The two possibilities are D and L.

An **α-amino acid** has a amino group (—NH_2) attached to the C atom adjacent to a carboxyl group (—COOH).

A **carbohydrate** is a polyhydroxy aldehyde, a polyhydroxy ketone, a derivative of these, or a substance that yields them upon hydrolysis.

Denaturation refers to the loss of biological activity of a protein brought about by changes in its secondary and tertiary structure.

Deoxyribonucleic acid (DNA) is the substance that makes up the genes of the chromosomes in the nuclei of cells.

Diastereomers are optically active isomers of a compound, but their structures are *not* mirror images (as are enantiomers).

An **enzyme** is a substance containing protein that catalyzes biological reactions.

Fats are triglycerides in which saturated fatty acid components predominate.

The **genetic code** describes the sequences of bases in DNA molecules which determine complementary sequences in mRNA and, ultimately, sequences of amino acids in proteins.

The **iodine number** of a triglyceride is the number of grams of I_2 reacting with 1.00×10^2 g of the triglyceride. The iodine number indicates the degree of unsaturation in the fatty acid components.

The **isoelectric point** or **pI** of an amino acid or a protein is the pH at which the dipolar structure or "zwitterion" predominates.

Lipids include a variety of substances sharing the property of solubility in solvents of low polarity [$CHCl_3$, CCl_4, C_6H_6, $(C_2H_5)_2O$].

Metabolism refers to the totality of chemical reactions occurring within an organism, reactions in which large molecules are broken down (catabolism) or small ones are synthesized into larger ones (anabolism).

A **monosaccharide** is a single, simple molecule having the structural features of a carbohydrate. It is called a **sugar.**

Nucleic acids are cell components comprised of purine and pyrimidine bases, pentose sugars, and phosphoric acid.

Oils are triglycerides in which unsaturated fatty acid components predominate.

Oligosaccharides are carbohydrates consisting of from two to ten monosaccharide units. They are also called **sugars.**

A **peptide bond** is formed by the elimination of a water molecule from between two amino acids. The H atom comes from the —NH_2 group of one amino acid, and the —OH group from the —COOH group of the other acid.

A **polypeptide** is formed by the joining together of a large number of amino acid units through peptide bonds.

A **polysaccharide** is a carbohydrate consisting of more than ten monosaccharide units (e.g., starch and cellulose).

Primary structure refers to the sequence of amino acids in the polypeptide chains that make up a protein.

A **protein** is a large polypeptide, that is, having a molecular weight of 10,000 or more.

A **reducing sugar** is one that is able to reduce Cu^{2+}(aq) to red Cu_2O. The sugar must have available an aldehyde group (which is oxidized to an acid).

Ribonucleic acid (RNA), through its **messenger RNA (mRNA)** and **transfer RNA (tRNA)** forms, is involved in the synthesis of proteins in cells.

Saponification is the hydrolysis of a triglyceride by a strong base. The products of saponification are glycerol and a soap. The number of mg KOH required to saponify 1.00 g of the triglyceride is known as the **saponification value.**

Secondary structure of a protein describes the structure or shape of a polypeptide chain, for example, coiling into a helix.

A **soap** is a metal salt of a fatty acid (e.g., sodium stearate).

A **substrate** is a substance that undergoes a biological reaction through the action of an enzyme.

Tertiary structure of a protein describes the types of linkages between polypeptide chains that give a protein its three-dimensional structure.

Triglycerides are esters of glycerol (1,2,3-propanetriol) with long-chain monocarboxylic (fatty) acids.

Suggestions for Further Study

CALVIN, M., "Photosynthesis as a Source of Energy and Materials," *American Scientist,* **64,** 270 (1976).

DICKERSON, R. E., "The DNA Helix and How It Is Read," *Scientific American,* **249**[6], 94 (1983).

DUNNE, C. P., "Biosynthesis of Plant and Animal Foods," *J. Chem. Educ.,* **61,** 271 (1984).*

FRIEDEN, E., "The Chemicals of Life," *Scientific American,* **227**[1], 52 (1972).

GOHEEN, D. W., "Chemicals from Wood and Other Biomass,

Part II. The Chemistry of Conversion," *J. Chem. Educ.,* **58,** 544 (1981).

HINKLE, P. C., and R. E. MCCARTY, "How Cells Make ATP," *Scientific American,* **238**[3], 104 (1978).

KRISHNAMURTY, S., "The Intriguing Biological Role of Vitamin E," *J. Chem. Educ.,* **60,** 465 (1983).

RAW, I., and G. W. HOLLEMAN, "Water—Energy for Life," *Chemistry,* **46**[5], 6 (1973).

SENOZAN, N. M., and R. L. HUNT, "Hemoglobin: Its Occurrence, Structure, and Adaptation," *J. Chem. Educ.,* **59,** 173 (1982).

SHARON, N., "Carbohydrates," *Scientific American,* **243**[5], 90 (1980).

*Several other articles on the chemistry of the food cycle will be found in the same issue of this journal.

Review Problems

1. Name the following compounds.

(a)
$$H_2CO-\overset{\displaystyle O}{\overset{\displaystyle \|}{C}}-C_{17}H_{35}$$
$$HCO-\overset{\displaystyle O}{\overset{\displaystyle \|}{C}}-C_{17}H_{33}$$
$$H_2CO-\overset{\displaystyle O}{\overset{\displaystyle \|}{C}}-C_{11}H_{23}$$

(b)
$$H_2CO-\overset{\displaystyle O}{\overset{\displaystyle \|}{C}}-C_{17}H_{31}$$
$$HCO-\overset{\displaystyle O}{\overset{\displaystyle \|}{C}}-C_{17}H_{31}$$
$$H_2CO-\overset{\displaystyle O}{\overset{\displaystyle \|}{C}}-C_{17}H_{31}$$

(c) $C_{13}H_{27}CO_2^-Na^+$

2. Write structural formulas for the following.
   (a) glyceryl lauromyristolinoleate
   (b) trilaurin
   (c) potassium palmitate
   (d) cetyl linoleate [cetyl alcohol = $CH_3(CH_2)_{14}CH_2OH$]

3. For glyceryl butyropalmitooleate, whose structure is written in Example 27-1, determine its (a) saponification value and (b) iodine number.

4. A simple glyceride is found to have a saponification value of 209. What is this simple glyceride? [*Hint:* Will the simple glyceride have a higher or lower molecular weight than tristearin (Example 27-2)?]

5. The structure of L-(+)-arabinose is given below. From this structure derive the structure of (a) D-(−)-arabinose; (b) a diastereomer of L-(+)-arabinose.

H—C=O
H—C—OH
HO—C—H
HO—C—H
    CH₂OH

L-(+)-arabinose

Exercises

Structure and composition of the cell

Exercises 1 to 5 refer to a typical *E. coli* bacterium. This is a cylindrical cell about 2 μm long and 1 μm in diameter, weighing about 2×10^{-12} g, and containing about 80% water by volume. (See Figure 27-2.)

1. The intracellular pH is 6.4 and $[K^+] = 1.5 \times 10^{-4}$ M. Determine the number of (a) H^+ ions and (b) K^+ ions in a typical cell.

2. The *E. coli* cell contains about 1.5×10^4 ribosomes. Assuming a ribosome to be a sphere with a diameter of 18 nm, what percentage of the cell volume do the ribosomes occupy?

3. Calculate the number of lipid molecules present, assuming their average molecular weight to be 700 and the lipid content to be 2%.

4. The cell is about 15% protein, by mass, with 90% of this protein in the cytoplasm. Assuming an average molecular weight of 3×10^4, how many protein molecules are present in the cytoplasm?

5. A single chromosomal DNA molecule contains about 4.5 million mononucleotide units. If this molecule were extended so that the mononucleotide units were 450 pm apart, what would be the length of the molecule? How does this compare with the length of the cell itself? What does this result suggest about the shape of the DNA molecule?

Lipids

6. Describe briefly what is meant by each of the following terms, using specific examples where appropriate: (a) lipid; (b) triglyceride; (c) simple glyceride; (d) mixed glyceride; (e) fatty acid; (f) soap.

7. Explain the essential distinction between the following pairs of materials: (a) a fat and a lipid; (b) a fat and an oil; (c) a fat and a wax; (d) butter and margarine.

6. Write the formulas of the species expected if the amino acid phenylalanine is maintained in (a) 1.0 M HCl; (b) 1.0 M NaOH; (c) a buffer solution with pH 5.5.

7. Write the structures of (a) glycylmethionine; (b) isoleucylleucylserine.

8. For the polypeptide Gly-Ala-Ser-Thr, (a) write the structural formula; (b) name the polypeptide. (*Hint:* Which is the N-terminal and which is the C-terminal amino acid?)

9. With reference to Figure 27-21, identify the purine bases, the pyrimidine bases, the pentose sugars, and the phosphate groups. Is this a chain of DNA or RNA? Explain.

10. What polypeptide would be synthesized by the following coding on a mRNA strand? ACCCAUCCCUUGGCGAGUGGUAUGUAA

8. Explain why phospholipids are more water soluble than simple or mixed glycerides.

9. What is the saponification value of glyceryl lauropalmitostearate?

10. What simple triglyceride has a saponification value of 193 and an iodine number of 174 (see Table 27-2).

11. Oleic acid is a moderately unsaturated fatty acid. Linoleic acid belongs to a group called **polyunsaturated.** What structural feature characterizes polyunsaturated fatty acids? Is stearic acid polyunsaturated? Is eleostearic acid? Why do you suppose safflower oil is so highly recommended in dietary programs?

12. In light of present medical knowledge, which is a more desirable lipid for human consumption, one with a high saponification value or a high iodine number? Which is the "best" of those listed in Table 27-3 from this standpoint?

Carbohydrates

13. Describe what is meant by each of the following terms, using specific examples where appropriate: (a) monosaccharide; (b) disaccharide; (c) oligosaccharide; (d) polysaccharide; (e) sugar; (f) glycose; (g) aldose, (h) ketose; (i) pentose; (j) hexose.

14. The following terms are all related to stereoisomers and their optical activity. Explain the meaning of each. (a) dextrorotatory; (b) levorotatory; (c) racemic mixture; (d) diastereomers; (e) (+); (f) (−); (g) D configuration; (h) *l*; (i) *d*.

15. Write the structure for the straight chain form of L-glucose. Is this isomer dextrorotatory or levorotatory?

16. The pure α and β forms of D-glucose rotate the plane of polarized light to the right by 112° and 18.7°, respectively (denoted as +112° and +18.7°). Are these two forms of glucose enantiomers or diastereomers? (Consider also the structures shown in Figure 27-6.)

17. When a mixture of the pure α and β forms of glucose is allowed to reach equilibrium in solution, the rotation changes to $+52.7°$ (a phenomenon known as mutarotation). What are the percentages of the α and β forms in the equilibrium mixture? [*Hint:* Refer to Exercise 16. If the mixture were 50:50, the rotation would be $0.50(+112) + 0.50(+18.7) = +65.4°$.]

18. Why is fruit sugar (D-fructose) a reducing sugar while sucrose is not? (*Hint:* See Figure 27-7.)

Amino acids, polypeptides, and proteins

19. Describe what is meant by each of the following terms, using specific examples where appropriate: **(a)** α-amino acid; **(b)** zwitterion; **(c)** isoelectric point; **(d)** peptide bond; **(e)** polypeptide; **(f)** protein; **(g)** N-terminal amino acid; **(h)** α helix; **(i)** denaturation.

20. A mixture of the amino acids lysine, proline, and glutamic acid is subjected to electrophoresis at a pH of 6.3. In what direction will each amino acid migrate?

21. Write the structures of **(a)** the different tripeptides that can be obtained from a combination of alanine, serine, and lysine; **(b)** the tetrapeptides containing two serine and two alanine amino acid units.

22. Upon complete hydrolysis a polypeptide yields the following amino acids: Gly, Leu, Ala, Val, Ser, Thr. Partial hydrolysis yields the following fragments: Ser-Gly-Val, Thr-Val, Ala-Ser, Leu-Thr-Val, Gly-Val-Thr. An experiment using a marker establishes that Ala is the N-terminal amino acid.
    (a) Establish the amino acid sequence in this polypeptide.
    (b) What is the name of the polypeptide?

23. Describe what is meant by the primary, secondary, and tertiary structure of a protein. What is the quaternary structure? Do all proteins have a quaternary structure? Explain.

24. A 1.00-mL solution containing 1.0 mg of an enzyme was deactivated by the addition of 0.346 μmol AgNO$_3$ (1 μmol = 1×10^{-6} mol.) What is the *minimum* molecular weight of the enzyme? Why does this calculation yield only a minimum value? (*Hint:* How many active sites are present in each enzyme molecule?)

25. Sickle cell anemia is sometimes referred to as a "molecular" disease. Comment on the appropriateness of this term.

Biochemical reactivity

26. Describe briefly the meaning of each of the following terms as they apply to metabolism: **(a)** metabolite; **(b)** anabolism; **(c)** catabolism; **(d)** endergonic; **(e)** ADP; **(f)** ATP.

27. The metabolism of a particular metabolite has a theoretical free energy change of -837 kJ/mol. The metabolism of 1 mol of this material in a living organism results in the conversion of 15 mol ADP to ATP. What is the percent efficiency of this metabolism?

28. Calculate the equilibrium constant for the hydrolysis of glucose 6-phosphate to glucose and phosphoric acid if $\Delta \overline{G}° = -13.8$ kJ/mol.

29. Explain why the action of an enzyme is so dependent on pH.

Nuclei acids

30. What are the two major types of nucleic acids? List their principal components.

31. DNA has been called the "thread of life." Comment on the appropriateness of this expression.

32. What are the principal functions of each of the following in protein synthesis? **(a)** DNA; **(b)** mRNA; **(c)** tRNA.

33. A ribosome is sometimes said to *read* an mRNA strand. Suggest a meaning for this expression?

34. Propose a plausible polypeptide sequence on a DNA strand that would code for the synthesis of the polypeptide Ser-Gly-Val-Ala. Why is there more than one possible sequence for the DNA strand?

*35. If one strand of a DNA molecule has the sequence of bases AGC, what must be the sequence on the opposite strand? Draw a structure of this portion of the double helix, showing all hydrogen bonds.

Additional Exercises

1. Write structural formulas for the following.
    (a) hydrogenation product of triolein;
    (b) saponification products of trilaurin;
    (c) iodination product of glyceryl lauromyristolinoleate.

2. Castor oil is a mixture of triglycerides having about 90% of its fatty acid content as the unsaturated hydroxy aliphatic acid, ricinoleic acid.

$$CH_3(CH_2)_5CHOHCH_2CH{=}CH(CH_2)_7COOH.$$

Estimate the saponification value and iodine number of castor oil. (*Hint:* What triglyceride should you assume?)

3. There are eight aldopentoses. Draw their structures and indicate which are enantiomers.

4. The term **epimer** is used to describe diastereomers that differ in the configuration about a *single* carbon atom. Which pairs of the three naturally occurring aldohexoses shown on page 847 are epimers (i.e., are D-galactose and D-mannose epimers, etc.)?

5. The protein molecule hemoglobin contains four atoms of iron (recall Figure 27-16). The mass percent of iron in hemoglobin is 0.34%. What is the molecular weight of hemoglobin?

★6. Draw complete structures for each of the chemical intermediates shown in Figure 27-18, that is, ADP, ATP, NAD, NADH, FAD, and FADH.

★7. What is a nucleoside? Draw structures of the following nucleosides using information from Figure 27-20.
(a) cytidine   (b) uridine   (c) guanosine
(d) deoxycytidine   (e) deoxyadenosine

★8. In the experiment described on page 867, the first generation offspring of DNA molecules each contained one strand with ^{15}N atoms and one with ^{14}N. If the experiment were carried through a second, third, and fourth generation, what fractions of the DNA molecules would still have strands with ^{15}N atoms?

★9. Bradykinin is a nonapeptide that is obtained by the partial hydrolysis of blood serum protein. It causes a lowering of the blood pressure and an increase in capillary permeability. Complete hydrolysis of bradykinin yields three proline (Pro), two arginine (Arg), two phenylalanine (Phe), one glycine (Gly), and one serine (Ser) amino acid units. The N-terminal and C-terminal units are both arginine (Arg). In a hypothetical experiment partial hydrolysis and sequence proof reveals the following fragments: Gly-Phe-Ser-Pro, Pro-Phe-Arg; Ser-Pro-Phe; Pro Pro Gly; Pro-Gly-Phe; Arg-Pro-Pro; Phe-Arg. Deduce the sequence of amino acid units in bradykinin.

★10. A pentapeptide was isolated from a cell extract and purified. A portion of the compound was treated with 2,4-dinitrofluorobenzene (DNFB) and the resulting material hydrolyzed. Analysis of the hydrolysis products revealed 1 mol of DNP-methionine, 2 mol of methionine, and 1 mol each of serine and glycine. A second portion of the original compound was partially hydrolyzed and separated into four products. Separately, the four products were hydrolyzed further, giving the following four sets of compounds. (a) 1 mol of DNP-methionine, 1 mol of methionine, and 1 mol of glycine; (b) 1 mol of DNP-methionine and 1 mol of methionine; (c) 1 mol of DNP-serine and 1 mol of methionine; (d) 1 mol of DNP-methionine, 1 mol of methionine, and 1 mol of serine. What is the structural formula of the pentapeptide?

Self-Test Questions

For questions 1 through 8 select the single item that best completes each statement.

1. The substance *glyceryl trilinoleate* (linoleic acid: $C_{17}H_{31}COOH$) is best described as a (a) fat; (b) oil; (c) wax; (d) fatty acid.

2. One can most easily distinguish between *glyceryl tristearate* (stearic acid: $C_{17}H_{35}COOH$) and *glyceryl trioleate* (oleic acid: $C_{17}H_{33}COOH$) by measuring their (a) molecular weights; (b) saponification values; (c) iodine numbers; (d) hydrolysis constants.

3. The mixture of sugars referred to as DL (or *dl*)-erythrose rotates the plane of polarized light (a) to the left; (b) to the right; (c) first to the left and then to the right; (d) neither to the left nor to the right.

4. Of the following names, the one that refers to a simple sugar in its cyclic (ring) form is (a) β-galactose; (b) L-(−)-glyceraldehyde; (c) D-(+)-glucose; (d) DL-erythrose.

5. The coagulation of egg whites by boiling is an example of (a) saponification; (b) inversion of a sugar; (c) hydrolysis of a protein; (d) denaturation of a protein.

6. A molecule in which the energy of metabolism is stored is (a) ATP; (b) glucose; (c) CO_2; (d) glycerol.

7. Of the following, the one that is not a constituent of a nucleic acid chain is (a) purine base; (b) phosphate group; (c) glycerol; (d) pentose sugar.

8. The structure of the DNA molecule is best described as (a) a random coil; (b) a double helix; (c) a pleated sheet; (d) partly coiled.

9. Calculate the maximum mass of a sodium soap that could be prepared from 125 g of the triglyceride *glyceryl tripalmitate* (palmitic acid: $C_{15}H_{31}COOH$).

10. Upon complete hydrolysis of a pentapeptide the following amino acids are obtained: valine (Val), phenylalanine (Phe), glycine (Gly), cysteine (Cys), and tyrosine (Tyr). Partial hydrolysis yields the following fragments: Val-Phe, Cys-Gly, Cys-Val-Phe, Tyr-Phe. Glycine is found to be the N-terminal acid. What is the sequence of the amino acids in the polypeptide?

11. Explain why enzyme action is so dependent on factors such as temperature, pH, and the presence of metal ions. Why are enzymes so specific in the reactions they catalyze? (That is, why don't they catalyze a variety of reactions, as does platinum metal, for example?)

12. Explain what is meant by the primary, secondary, and tertiary structure of a protein.

APPENDIX A

Mathematical Operations

A-1 Exponential Arithmetic

The numerical quantities encountered in this text range from very small to very large in value. For example, the mass of an individual hydrogen atom is 0.00000000000000000000000167 g; the number of molecules in 18.016 g H_2O is 602,205,000,000,000,000,000,000. In the form expressed, both of these numbers would be very cumbersome to handle in computations. To express such numbers with greater ease, the methods of exponential arithmetic may be employed.

positive powers

$$10^0 = 1$$

$$10^1 = 10$$

$$10^2 = 10 \times 10 = 100$$

$$10^3 = 10 \times 10 \times 10 = 1000$$

negative powers

$$10^0 = 1$$

$$10^{-1} = \frac{1}{10} = 0.1$$

$$10^{-2} = \frac{1}{10 \times 10} = 0.01$$

$$10^{-3} = \frac{1}{10 \times 10 \times 10} = \frac{1}{10^3} = 0.001$$

A number is said to be written in exponential form when it is expressed as a coefficient multiplied by a power of ten; $a \cdot 10^y$. For example, $120 = 1.2 \times 100 = 1.2 \times 10^2$.

Illustrative Examples

$24,100 = 2.41 \times 10,000 = 2.41 \times 10^4$
$0.0038 = 3.8 \times 0.001 = 3.8 \times 10^{-3}$
$6.1 \times 10^6 = 6.1 \times 1,000,000 = 6,100,000$
$4.7 \times 10^{-3} = 4.7 \times 0.001 = 0.0047$

Addition and Subtraction. To add or subtract numbers written in exponential form, the numbers must first be expressed to the same power of 10. Thus,

$$0.0560 + 0.0038 - 0.0152 = 5.60 \times 10^{-2} + 0.38 \times 10^{-2} - 1.52 \times 10^{-2}$$
$$= (5.60 + 0.38 - 1.52) \times 10^{-2} = 4.46 \times 10^{-2}$$

Multiplication. Consider the numbers $a \cdot 10^y$ and $b \cdot 10^z$. Their product is $a \cdot b \cdot 10^{(y+z)}$. Coefficients are multiplied and exponents are added.

A1

$$0.0220 \times 0.0040 \times 750 = (2.20 \times 10^{-2})(4.0 \times 10^{-3})(7.5 \times 10^2)$$
$$= (2.20 \times 4.0 \times 7.5) \times 10^{(-2-3+2)} = 66 \times 10^{-3}$$
$$= 6.6 \times 10^1 \times 10^{-3} = 6.6 \times 10^{-2}$$

Division.   Consider the two numbers, $a \cdot 10^y$ and $b \cdot 10^z$. Their quotient is $a \cdot 10^y/b \cdot 10^z = (a/b) \times 10^{(y-z)}$. The coefficients are divided and the exponent of the denominator is subtracted from the exponent in the numerator.

$$\frac{20.0 \times 636 \times 0.150}{0.0400 \times 1.80} = \frac{(2.00 \times 10^1)(6.36 \times 10^2)(1.50 \times 10^{-1})}{4.00 \times 10^{-2} \times 1.80}$$

$$= \frac{2.00 \times 6.36 \times 1.50 \times 10^{(1+2-1)}}{4.00 \times 1.80 \times 10^{-2}} = \frac{19.1 \times 10^2}{7.20 \times 10^{-2}}$$

$$= 2.65 \times 10^{(2-(-2))} = 2.65 \times 10^4$$

Raising a Number to a Power.   To "square" the number $a \cdot 10^y$ means to determine the value $(a \cdot 10^y)^2$ or the product $(a \cdot 10^y)(a \cdot 10^y)$. According to the rule for multiplication, this product is $(a \times a) \times 10^{(y+y)} = a^2 \cdot 10^{2y}$. When an exponential number is raised to a power, the coefficient is raised to that power and the exponent is multiplied by the power.

$$(0.0034)^3 = (3.4 \times 10^{-3})^3 = (3.4)^3 \times 10^{3 \times (-3)} = 39 \times 10^{-9} = 3.9 \times 10^{-8}$$

Extracting the Root of an Exponential Number.   To extract the root of a number is the same as raising the number to a fractional power. This means that the square root of a number is the number to the one-half power; the cube root is the number to the one-third power; and so on. Thus,

$$\sqrt{a \cdot 10^y} = (a \cdot 10^y)^{1/2} = a^{1/2} \cdot 10^{y/2}$$
$$\sqrt{156} = \sqrt{1.56 \times 10^2} = (1.56)^{1/2} \times 10^{2/2} = 1.25 \times 10^1 = 12.5$$

In the following example, where the cube root is extracted, the exponent (-5) is not divisible by 3; the number is rewritten so that the new exponent will be divisible by 3.

$$(3.52 \times 10^{-5})^{1/3} = (35.2 \times 10^{-6})^{1/3} = (35.2)^{1/3} \times 10^{-6/3} = 3.28 \times 10^{-2}$$

A-2 Logarithms

The common logarithm (log) of a number (N) is that exponent (x) to which the base 10 must be raised to yield the number N.

$$\log N = x \quad \text{means that} \quad N = 10^x = 10^{\log N}$$

For simple powers of ten:

$$\log 1 = \log 10^0 = 0$$

$$\log 10 = \log 10^1 = 1 \qquad \log 0.10 = \log 10^{-1} = -1$$

$$\log 100 = \log 10^2 = 2 \qquad \log 0.01 = \log 10^{-2} = -2$$

Multiplication and Division.   From the definition of a logarithm we can write: $M = 10^{\log M}$ and $N = 10^{\log N}$. This means that

$$M \times N = 10^{\log M} \times 10^{\log N} = 10^{(\log M + \log N)}$$

But, again from the definition of a logarithm, we have

$$M \times N = 10^{\log (M \times N)}$$

which means that

$$\log (M \times N) = \log M + \log N$$

Similarly,

$$\log \frac{M}{N} = \log M - \log N$$

Illustrative Examples. Most numbers encountered in measurements and calculations are not simple powers of 10. To allow for the determination of the logarithms of such numbers, tables of logarithms have been developed (see Table A-1). In the example below log 7.34 is obtained from such a table.

$$\log 734 = \log (7.34 \times 10^2) = \log 7.34 + \log 10^2$$
$$= 0.866 + 2 = 2.866$$

The logarithms of numbers smaller than 1 can also be obtained easily.

$$\log 0.00130 = \log (1.30 \times 10^{-3}) = \log 1.30 + \log 10^{-3}$$
$$= 0.114 - 3 = -2.886$$

A number has a logarithm of 4.350. What is the number? The number we are seeking is said to be the antilogarithm of 4.350, that is, the number whose logarithm is 4.350. In exponential form, the unknown number is $N = a \cdot 10^x$.

$$\log N = 4.350 = 0.350 + 4$$
$$a = \text{antilog } 0.350 = 2.24$$
$$x = 4$$
$$N = 2.24 \times 10^4$$

Another commonly encountered example is this text is of the form: $\log N = -4.350$; what is N? The key operation here involves stating -4.350 as a difference of two numbers.

$$\log N = -4.350 = 0.650 - 5$$
$$N = (\text{antilog } 0.650) \times 10^{-5}$$
$$N = 4.47 \times 10^{-5}$$

With a modern electronic calculator, of course, logarithms and antilogarithms can be obtained directly, without the use of a table of logarithms or the multistep procedures outlined here.

Questions sometimes arise about the appropriate number of significant figures to use in expressing a logarithm or antilogarithm. The fundamental rule to follow is that the digit(s) to the left of the decimal point in a logarithm is used to establish a power of ten. All digits to the right of the decimal point in a logarithm are significant. Thus, the logarithm -2.08 is expressed to *two* significant figures. The antilogarithm of -2.08 should also be expressed to *two* significant figures: 8.3×10^{-3}. To help settle this point, take the antilogarithms of -2.07, -2.08, and -2.09. You will find these antilogarithms to have the same first digit and slight differences in the second digit, i.e., 8.5×10^{-3}, 8.3×10^{-3}, and 8.1×10^{-3}, respectively.

Natural Logarithms. The definition of a logarithm is not limited to the base 10. For example, corresponding to $2^3 = 8$ we can write, $\log_2 8 = 3$ (read as "the logarithm of 8 to the base 2 is equal to 3"); $\log_2 10 = 3.322$; and so on. Several equations encountered in this text are derived using the methods of calculus and involve logarithmic functions. These derivations require that the logarithm be a "natural" one. This is the case only if for the base we use the quantity $e = 2.71828. \ldots$ A logarithm to the base "e" is usually denoted by the symbol "ln"; that is, $\log_e x = \ln x$. The relationship between $\ln x$ and $\log_{10} x$ is a simple one. It involves the factor $\log_e 10 = 2.303$.

TABLE A-1
Four place logarithms

N	0	1	2	3	4	5	6	7	8	9
10	0000	0043	0086	0128	0170	0212	0253	0294	0334	0374
11	0414	0453	0492	0531	0569	0607	0645	0682	0719	0755
12	0792	0828	0864	0899	0934	0969	1004	1038	1072	1106
13	1139	1173	1206	1239	1271	1303	1335	1367	1399	1430
14	1461	1492	1523	1553	1584	1614	1644	1673	1703	1732
15	1761	1790	1818	1847	1875	1903	1931	1959	1987	2014
16	2041	2068	2095	2122	2148	2175	2201	2227	2253	2279
17	2304	2330	2355	2380	2405	2430	2455	2480	2504	2529
18	2553	2577	2601	2625	2648	2672	2695	2718	2742	2765
19	2788	2810	2833	2856	2878	2900	2923	2945	2967	2989
20	3010	3032	3054	3075	3096	3118	3139	3160	3181	3201
21	3222	3243	3263	3284	3304	3324	3345	3365	3385	3404
22	3424	3444	3464	3483	3502	3522	3541	3560	3579	3598
23	3617	3636	3655	3674	3692	3711	3729	3747	3766	3784
24	3802	3820	3838	3856	3874	3892	3909	3927	3945	3962
25	3979	3997	4014	4031	4048	4065	4082	4099	4116	4133
26	4150	4166	4183	4200	4216	4232	4249	4265	4281	4298
27	4314	4330	4346	4362	4378	4393	4409	4425	4440	4456
28	4472	4487	4502	4518	4533	4548	4564	4579	4594	4609
29	4624	4639	4654	4669	4683	4698	4713	4728	4742	4757
30	4771	4786	4800	4814	4829	4843	4857	4871	4886	4900
31	4914	4928	4942	4955	4969	4983	4997	5011	5024	5038
32	5051	5065	5079	5092	5105	5119	5132	5145	5159	5172
33	5185	5198	5211	5224	5237	5250	5263	5276	5289	5302
34	5315	5328	5340	5353	5366	5378	5391	5403	5416	5428
35	5441	5453	5465	5478	5490	5502	5514	5527	5539	5551
36	5563	5575	5587	5599	5611	5623	5635	5647	5658	5670
37	5682	5694	5705	5717	5729	5740	5752	5763	5775	5786
38	5798	5809	5821	5832	5843	5855	5866	5877	5888	5899
39	5911	5922	5933	5944	5955	5966	5977	5988	5999	6010
40	6021	6031	6042	6053	6064	6075	6085	6096	6107	6117
41	6128	6138	6149	6160	6170	6180	6191	6201	6212	6222
42	6232	6243	6253	6263	6274	6284	6294	6304	6314	6325
43	6335	6345	6355	6365	6375	6385	6395	6405	6415	6425
44	6435	6444	6454	6464	6474	6484	6493	6503	6513	6522
45	6532	6542	6551	6561	6571	6580	6590	6599	6609	6618
46	6628	6637	6646	6656	6665	6675	6684	6693	6702	6712
47	6721	6730	6739	6749	6758	6767	6776	6785	6794	6803
48	6812	6821	6830	6839	6848	6857	6866	6875	6884	6893
49	6902	6911	6920	6928	6937	6946	6955	6964	6972	6981
50	6990	6998	7007	7016	7024	7033	7042	7050	7059	7067
51	7076	7084	7093	7101	7110	7118	7126	7135	7143	7152
52	7160	7168	7177	7185	7193	7202	7210	7218	7226	7235
53	7243	7251	7259	7267	7275	7284	7292	7300	7308	7316
54	7324	7332	7340	7348	7356	7364	7372	7380	7388	7396
N	0	1	2	3	4	5	6	7	8	9

N	0	1	2	3	4	5	6	7	8	9
55	7404	7412	7419	7427	7435	7443	7451	7459	7466	7474
56	7482	7490	7497	7505	7513	7520	7528	7536	7543	7551
57	7559	7566	7574	7582	7589	7597	7604	7612	7619	7627
58	7634	7642	7649	7657	7664	7672	7679	7686	7694	7701
59	7709	7716	7723	7731	7738	7745	7752	7760	7767	7774
60	7782	7789	7796	7803	7810	7818	7825	7832	7839	7846
61	7853	7860	7868	7875	7882	7889	7896	7903	7910	7917
62	7924	7931	7938	7945	7952	7959	7966	7973	7980	7987
63	7993	8000	8007	8014	8021	8028	8035	8041	8048	8055
64	8062	8069	8075	8082	8089	8096	8102	8109	8116	8122
65	8129	8136	8142	8149	8156	8162	8169	8176	8182	8189
66	8195	8202	8209	8215	8222	8228	8235	8241	8248	8254
67	8261	8267	8274	8280	8287	8293	8299	8306	8312	8319
68	8325	8331	8338	8344	8351	8357	8363	8370	8376	8382
69	8388	8395	8401	8407	8414	8420	8426	8432	8439	8445
70	8451	8457	8463	8470	8476	8482	8488	8494	8500	8506
71	8513	8519	8525	8531	8537	8543	8549	8555	8561	8567
72	8573	8579	8585	8591	8597	8603	8609	8615	8621	8627
73	8633	8639	8645	8651	8657	8663	8669	8675	8681	8686
74	8692	8698	8704	8710	8716	8722	8727	8733	8739	8745
75	8751	8756	8762	8768	8774	8779	8785	8791	8797	8802
76	8808	8814	8820	8825	8831	8837	8842	8848	8854	8859
77	8865	8871	8876	8882	8887	8893	8899	8904	8910	8915
78	8921	8927	8932	8938	8943	8949	8954	8960	8965	8971
79	8976	8982	8987	8993	8998	9004	9009	9015	9020	9025
80	9031	9036	9042	9047	9053	9058	9063	9069	9074	9079
81	9085	9090	9096	9101	9106	9112	9117	9122	9128	9133
82	9138	9143	9149	9154	9159	9165	9170	9175	9180	9186
83	9191	9196	9201	9206	9212	9217	9222	9227	9232	9238
84	9243	9248	9253	9258	9263	9269	9274	9279	9284	9289
85	9294	9299	9304	9309	9315	9320	9325	9330	9335	9340
86	9345	9350	9355	9360	9365	9370	9375	9380	9385	9390
87	9395	9400	9405	9410	9415	9420	9425	9430	9435	9440
88	9445	9450	9455	9460	9465	9469	9474	9479	9484	9489
89	9494	9499	9504	9509	9513	9518	9523	9528	9533	9538
90	9542	9547	9552	9557	9562	9566	9571	9576	9581	9586
91	9590	9595	9600	9605	9609	9614	9619	9624	9628	9633
92	9638	9643	9647	9652	9657	9661	9666	9671	9675	9680
93	9685	9689	9694	9699	9703	9708	9713	9717	9722	9727
94	9731	9736	9741	9745	9750	9754	9759	9763	9768	9773
95	9777	9782	9786	9791	9795	9800	9805	9809	9814	9818
96	9823	9827	9832	9836	9841	9845	9850	9854	9859	9863
97	9868	9872	9877	9881	9886	9890	9894	9899	9903	9908
98	9912	9917	9921	9926	9930	9934	9939	9943	9948	9952
99	9956	9961	9965	9969	9974	9978	9983	9987	9991	9996
N	0	1	2	3	4	5	6	7	8	9

$$\log_e x = (\log_e 10) \times (\log_{10} x)$$

$$\ln x = 2.303 \log_{10} x - 2.303 \log x$$

In equations in this text where the factor "2.303" appears, you can assume that this was introduced to relate a natural logarithm to a logarithm to the base 10 (a common logarithm). Either use the equation as written, based on common logarithms (log), or omit the factor "2.303" and use natural logarithms (ln). Most electronic calculators can handle both of these functions. One application where common logarithms *must* be used, however, is in calculations involving pH, since pH is defined in terms of a logarithm to the base 10 (see Section 17-4).

A-3 Algebraic Operations

An algebraic equation is solved when one of the quantities, the unknown, is expressed in terms of all the other quantities in the equation. This effect is achieved when the unknown is present, alone, on one side of the equation, and the rest of the terms are on the other side. To solve an equation a rearrangement of terms may be necessary. The basic principle governing these rearrangements is quite simple. *Whatever is done to one side of the equation must be done as well to the other.*

$$(x^2 \times y) + 6 = z \qquad \textbf{\textit{Solve for x.}}$$

$$(x^2 \times y) + \cancel{6} - \cancel{6} = z - 6 \qquad \text{(1) Subtract 6 from each side.}$$

$$(x^2 \times y) = z - 6$$

$$\frac{x^2 \times \cancel{y}}{\cancel{y}} = \frac{z - 6}{y} \qquad \text{(2) Divide each side by } y.$$

$$x^2 = \frac{z - 6}{y}$$

$$\sqrt{x^2} = \sqrt{\frac{z - 6}{y}} \qquad \text{(3) Extract the square root of each side. (Find the quantity } \sqrt{N} \text{ that, when multiplied by itself, yields the number } N.\text{)}$$

$$x = \sqrt{\frac{z - 6}{y}} \qquad \text{(4) Simplify. The square root of } x^2 \text{ is simply } x.$$

Quadratic Equations. A quadratic equation has the form $ax^2 + bx + c = 0$, where a, b, and c are constants (a cannot be equal to 0). A number of calculations encountered in the text require that a quadratic equation be solved. The solution of such an equation is

$$x = \frac{-b \pm \sqrt{b^2 - 4ac}}{2a}$$

In Example 15-13 the following quadratic equation is obtained.

$$x^2 + 4.28 \times 10^{-4}\, x - 1.03 \times 10^{-5} = 0$$

Its solution is

$$x = \frac{-4.28 \times 10^{-4} \pm \sqrt{(4.28 \times 10^{-4})^2 + 4 \times 1.03 \times 10^{-5}}}{2}$$

$$= \frac{-4.28 \times 10^{-4} \pm \sqrt{(1.83 \times 10^{-7}) + (4.12 \times 10^{-5})}}{2}$$

$$= \frac{-4.28 \times 10^{-4} \pm \sqrt{4.14 \times 10^{-5}}}{2} = \frac{-4.28 \times 10^{-4} \pm 6.43 \times 10^{-3}}{2}$$

$$= \frac{-4.28 \times 10^{-4} + 6.43 \times 10^{-3}}{2} = \frac{6.00 \times 10^{-3}}{2} = 3.00 \times 10^{-3}$$

Note that only the (+) value of the (±) sign was used in solving for x. If the (−) value had been used, a negative value of x would have resulted. However, for the given situation a negative value of x is meaningless.

Higher Degree Equations. The highest power of x in a quadratic equation is 2; a quadratic is a second-degree equation. Some equations have the unknown appearing to a higher degree. Exact solutions of algebraic equations of higher degree than second are also possible, but a simple method that usually works quite well is the *method of successive approximations*. The following equation is obtained in the solution of Exercise 15-19; it is a cubic equation (third degree in x).

$$\frac{(0.657 + 2x)^2 \times 1.90}{(0.390 - 2x)^2(0.156 - x)} = 280$$

It can be reduced to the simpler cubic equation

$$x^3 - 0.537x^2 + 0.103x - 0.00520 = 0$$

Suppose we try the solution $x = 0.10$ and substitute into the above equation.

$$(1 \times 10^{-3}) - (5.37 \times 10^{-3}) + (1.03 \times 10^{-2}) - (5.20 \times 10^{-3})$$
$$= (11.3 \times 10^{-3}) - (10.6 \times 10^{-3}) > 0$$

The sum of terms on the left side > 0.
   Now try $x = 1 \times 10^{-2}$.

$$(1 \times 10^{-6}) - (5.37 \times 10^{-5}) + (1.03 \times 10^{-3}) - (5.20 \times 10^{-3}) < 0$$

The sum of terms on the left side <0.
   The value of x we are seeking has a value $0.01 < x < 0.10$. By successive approximations the value $x = 0.076$ can be established, and these approximations can be made easily if an electronic calculator or computer is used.

A-4 Graphs

Suppose the following sets of numbers are obtained for two quantities x and y by laboratory measurement.

$x = 0, 1, 2, 3, 4, \ldots$
$y = 2, 4, 6, 8, 10, \ldots$

The relationship between these sets of numbers is not difficult to establish.

$y = 2x + 2$

Ideally, the results of experimental measurements are best expressed through a mathematical equation. Sometimes, however, an exact equation cannot be written or its form is not clear from the experimental data. The graphing of data is very useful in such cases. In Figure A-1 the points listed above are located on a coordinate grid, in which x values are placed along the horizontal axis (abscissa) and y values along the vertical axis (ordinate). For each point the x and y values are indicated in parentheses.

The data points are seen to define a straight line. A mathematical equation for a straight line always has the form

$$y = mx + b \qquad\qquad (A.1)$$

FIGURE A-1
A straight line graph:
$y = mx + b$.

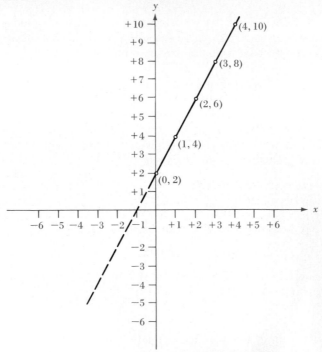

Values of m, the **slope** of the line, and b, the **intercept,** can be obtained from the straight line graph.

When $x = 0$, $y = b$. The intercept is the point where the straight line intersects the y-axis. The slope can be obtained from two points on the graph.

$$y_2 = mx_2 + b \qquad \text{and} \qquad y_1 = mx_1 + b$$

$$y_2 - y_1 = m(x_2 - x_1)$$

$$m = \frac{y_2 - y_1}{x_2 - x_1}$$

From the straight line in Figure A-1 can you establish that $m = b = 2$?

APPENDIX B

Some Basic Physical Concepts

B-1 Velocity and Acceleration

Time elapses as an object is displaced from one point to another. The **velocity** of the object is defined as the distance traveled per unit of time. An automobile that travels a distance of 60.0 km in exactly one hour has a velocity of 16.7 m/s.

Table B-1 contains experimental data on the velocity of a free-falling body. For this type of motion velocity is not constant; it increases with time—the falling body "speeds up" continuously. The rate of change of velocity with time is called **acceleration.** It has the units of distance per unit time per unit time. The constant acceleration in Table B-1, called the *acceleration due to gravity,* is 9.8 m/s^2.

TABLE B-1
Velocity and acceleration of a free-falling body

Time elapsed, s	Total distance, m	Velocity, m/s	Acceleration, m/s^2
0	0		
1	4.9	4.9	
2	19.6	14.7	9.8
3	44.1	24.5	9.8
4	78.4	34.3	9.8

B-2 Force and Work

Newton's first law of motion states that an object at rest remains at rest, and that an object in motion remains in uniform motion, unless acted upon by an external force. The tendency for an object to remain at rest or in uniform motion is called **inertia;** a **force** is that which is required to overcome inertia. Since the application of a force to an object either gives it motion or changes its motion, the actual effect of a force is to change the velocity of an object. Change in velocity is an acceleration, so force is that which provides an object with acceleration.

Newton's second law of motion describes the force, F, required to produce an acceleration, a, in an object of mass, m.

$$F = ma \qquad (B.1)$$

A9

The basic unit of force in the SI system is the **newton (N).** It is the force required to provide a one kilogram mass with an acceleration of one meter per second per second.

$$1 \text{ N} = 1 \text{ kg} \times 1 \text{ m s}^{-2} \tag{B.2}$$

The acceleration listed in Table B-1 is often denoted by the symbol g. The force of gravity on an object (its weight) is the product of the mass of the object and the acceleration of gravity.

$$F = mg \tag{B.3}$$

Work is performed when a force acts through a distance. The **joule (J)** is the amount of work associated with 1 newton (N) acting through a distance of 1 m.

$$1 \text{ J} = 1 \text{ N} \times 1 \text{ m} \tag{B.4}$$

From the definition of the newton in (B.2), we can also write

$$1 \text{ J} = 1 \text{ kg} \times 1 \text{ m s}^{-2} \times 1 \text{ m} = 1 \text{ kg m}^2 \text{ s}^{-2} \tag{B.5}$$

B-3 Energy

Energy is defined as the capacity to do work, but there are further ways of categorizing energy beyond this simple statement. For example, an object in motion has the immediate capacity to do work, and its energy is called **kinetic** energy. An object at rest may also have the capacity to do work by changing its position. The energy it possesses, which can be transformed to actual work, is called **potential** energy. As a ball rolls down a hill, some of its potential energy is converted to kinetic energy.

The **kinetic** energy of an object is given by one-half the product of its mass and the square of its velocity.

$$KE = \tfrac{1}{2}mv^2 \tag{B.6}$$

Mathematical expressions for potential energy are also possible, but their exact forms depend on the manner in which this energy is "stored."

B-4 Magnetism

Attractive and repulsive forces associated with a magnet are centered at regions called **poles.** A magnet has a north and a south pole. If two magnets are aligned such that the north pole of one is directed toward the south pole of the second, an attractive force develops. If the alignment brings like poles into proximity, either both north or both south, a repulsive force develops. *Unlike poles attract; like poles repel.*

A magnetic field exists in that region surrounding a magnet in which the influence of the magnet can be felt. Internal changes produced within an iron object by a magnetic field, not produced in a field-free region, are responsible for the attractive force that the object experiences.

B-5 Static Electricity

Another property with which certain objects may be endowed is electric charge. Analogous to the case with magnetism, *unlike charges attract and like charges repel* (see Figure B-1). In Coulomb's law, stated below, a *positive* force is *repulsive;* and a *negative* force is *attractive.*

$$F = \frac{Q_1 Q_2}{\epsilon r^2} \tag{B.7}$$

FIGURE B-1
Forces between
electrically charged
objects.

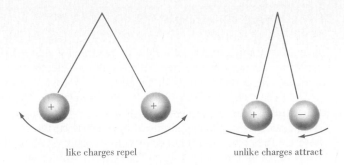

like charges repel   unlike charges attract

where Q_1 is the magnitude of the charge on object 1.
   Q_2 is the magnitude of the charge on object 2.
   r is the distance between the objects.
   ϵ is a proportionality constant called the **dielectric constant** of the medium.
   The numerical value of this constant reflects the effect that the medium separating the two charged objects has on the force existing between them. For
   vacuum, $\epsilon = 1$, and for other media ϵ is greater than 1 (e.g., for water
   $\epsilon = 78.5$).

   An **electric field** exists in that region surrounding an electrically charged object in
which the influence of the electric charge is felt. If an uncharged object is brought into the
field of a charged object, the uncharged object may undergo internal changes that it would
not experience in a field-free region. These changes may lead to the production of electric
charges in the formerly uncharged object, a phenomenon called **induction** (illustrated in
Figure B-2).

FIGURE B-2
Production of electric
charges by induction in
a gold leaf electroscope.

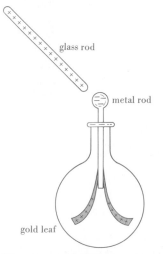

glass rod

metal rod

gold leaf

The glass rod acquires a positive electric charge by being rubbed with a silk cloth. As
the rod is brought near the electroscope a separation of charge occurs in the electroscope. The leaves become positively charged and repel one another. Negative charge is
attracted to the spherical terminal at the end of the metal rod. If the glass rod is removed, the charges on the electroscope redistribute themselves and the leaves collapse.
If before the glass rod is removed the spherical ball is touched by an electric conductor, negative charge is removed from the ball, the electroscope retains a net positive
charge, and the leaves remain outstretched.

B-6 Current Electricity

Current electricity consists of a flow of electrically charged particles. In electric currents in metallic conductors, the charged particles are electrons; in molten salts or in aqueous solutions, the particles are both negatively and positively charged ions.

The unit of electric charge is called a **coulomb (C).** The unit of electric current known as the **ampere (A)** is defined as a flow of 1 C/s through an electrical conductor. Two variables determine the magnitude of the electric current, I, flowing through a conductor. These are the potential difference or voltage drop, E, along the conductor and the electrical resistance of the conductor, R. The units of voltage and resistance are the **volt (V)** and **ohm,** respectively. The relationship of electric current, voltage, and resistance is given through Ohm's law.

$$I = \frac{E}{R} \tag{B.8}$$

One joule of energy is associated with the passage of one coulomb of electric charge through a potential difference (voltage) of one volt. That is, one joule = one volt-coulomb. Electric **power** refers to the rate of production (or consumption) of electric energy. It has the unit, **watt (W).**

$$1 \text{ W} = 1 \text{ J s}^{-1} = 1 \text{ V C s}^{-1}$$

Since one C s^{-1} is a current of one ampere,

$$1 \text{ W} = 1 \text{ V} \times 1 \text{ A} \tag{B.9}$$

Thus, a 100-watt light bulb operating at 110 V draws a current of 100 W/110 V = 0.91 A.

B-7 Electromagnetism

The relationship between electricity and magnetism is an intimate one. Interactions of electric and magnetic fields result in (1) magnetic fields associated with the flow of electric current (as in electromagnets), (2) forces experienced by current-carrying conductors when placed in a magnetic field (as in electric motors), and (3) electric current being induced when an electric conductor is moved through a magnetic field (as in electric generators). Numerous observations described in this text can be understood in terms of electromagnetic phenomena.

APPENDIX C

SI Units

The system of units that will in time be used universally for expressing all measured quantities is Le Système International d'Unités (The International System of Units), adopted in 1960 by the Conference Générale des Poids et Measures (General Conference of Weights and Measures). A summary of some of the provisions of the SI convention is provided here.

C-1 SI Base Units

A single unit has been established for each of the basic quantities involved in measurement. These are

Physical quantity	Unit	Symbol
length	meter	m
mass	kilogram	kg
time	second	s
electric current	ampere	A
temperature	kelvin	K
luminous intensity	candela	cd
amount of substance	mole	mol
plane angle	radian	rad
solid angle	steradian	sr

C-2 SI Prefixes

Distinctive prefixes are attached to the base unit to express quantities that are **multiples** (greater than) or **submultiples** (less than) of the base unit. The multiples and submultiples are obtained by multiplying the base unit by powers of ten.

Multiple	Prefix	Symbol	Submultiple	Prefix	Symbol
10^{12}	tera	T	10^{-1}	deci	d
10^9	giga	G	10^{-2}	centi	c
10^6	mega	M	10^{-3}	milli	m
10^3	kilo	k	10^{-6}	micro	μ
10^2	hecto	h	10^{-9}	nano	n
10^1	deka	da	10^{-12}	pico	p
			10^{-15}	femto	f
			10^{-18}	atto	a

C-3 Derived SI Units

A number of quantities must be derived from measured values of the SI base quantities [e.g., volume has the unit (length)^3]. Two sets of derived units are given, those whose names follow directly from the base units and those that are given special names.

Two other SI conventions are illustrated through this table: (a) Units are written in the singular form—meter or m (*not* meters or ms); (b) negative exponents are preferred to the shilling bar or solidus (/)—m s^{-1} and m s^{-2} (*not* m/s and m/s/s).

Physical quantity	Unit	Symbol
area	square meter	m^2
volume	cubic meter	m^3
velocity	meter per second	$m\,s^{-1}$
acceleration	meter per second squared	$m\,s^{-2}$
density	kilogram per cubic meter	$kg\,m^{-3}$
molar mass	kilogram per mole	$kg\,mol^{-1}$
molar volume	cubic meter per mole	$m^3\,mol^{-1}$
molar concentration	mole per cubic meter	$mol\,m^{-3}$

Physical quantity	Unit	Symbol	In terms of SI units
frequency	hertz	Hz	s^{-1}
force	newton	N	$kg\,m\,s^{-2}$
pressure	pascal	Pa	$N\,m^{-2}$
energy	joule	J	$kg\,m^2\,s^{-2}$
power	watt	W	$J\,s^{-1}$
electric charge	coulomb	C	$A\,s$
electric potential difference	volt	V	$J\,A^{-1}\,s^{-1}$
electric resistance	ohm	Ω	$V\,A^{-1}$

C-4 Units to Be Discouraged or Abandoned

There are several commonly used units whose use is to be discouraged and ultimately abandoned. Their gradual disappearance is to be expected, though each is used in this text. A few such units are listed.

Another SI convention is implied here. No commas are used in expressing large numbers but spaces are left between groupings of three digits (that is, 101 325 instead of 101,325). (Decimal points are written either as periods or commas.)

Physical quantity	Unit	Symbol	Definition in SI units
length	angstrom	Å	1×10^{-10} m
force	dyne	dyn	1×10^{-5} N
energy	erg	erg	1×10^{-7} J
energy	calorie	cal	4.184 J
pressure	atmosphere	atm	101 325 Pa
pressure	millimeter of mercury	mmHg	$(13.5951)(980.665) \times 10^{-2}$ Pa
pressure	torr	torr	133.322 Pa

C-5 Fundamental Constants

The fundamental constants introduced in this text, such as the speed of light, acceleration due to gravity, gas constant, Planck's constant, and the Faraday constant, continue to be used, but their units should be expressed as SI units. (See the inside back cover for a listing.)

APPENDIX D

Thermodynamic Properties of Substances[a]

	$\Delta \bar{H}_f^\circ$, kJ/mol	$\Delta \bar{G}_f^\circ$, kJ/mol	$\bar{S}^\circ$, J mol^{-1} K^{-1}
$Al_2O_3(s)$	−1669.79	−1576.41	51.00
$BaCO_3(s)$	−1218.8	−1138.9	112.1
$B_2H_6(g)$	31.4	82.8	232.88
$B_2O_3(s)$	−1263.6	−1184.1	54.02
$Br(g)$	111.75	82.38	174.93
$Br_2(g)$	30.71	3.14	245.35
$Br_2(l)$	0	0	152.30
$BrCl(g)$	14.7	− 0.88	239.9
$C(g)$	718.39	672.95	157.99
$C(diamond)$	1.88	2.85	2.43
$C(graphite)$	0	0	5.69
$CCl_4(g)$	− 106.7	− 64.0	309.4
$CO(g)$	− 110.54	− 137.28	197.90
$CO_2(g)$	− 393.51	− 394.38	213.64
$CH_4(g)$	− 74.85	− 50.79	186.19
$CH_2Cl_2(g)$	− 82.0	− 58.6	234.2
$C_2H_2(g)$	226.73	209.20	200.83
$C_2H_4(g)$	52.30	68.12	219.45
$C_2H_6(g)$	− 84.68	− 32.89	229.49
$C_3H_8(g)$	− 103.85	− 23.47	269.9
$C_6H_6(g)$	82.93	129.66	269.20
$C_6H_6(l)$	49.04	124.52	172.80
$CH_3OH(g)$	− 200.67	− 162.00	239.70
$CH_3OH(l)$	− 238.66	− 166.36	126.8
$C_2H_5OH(l)$	− 277.65	− 174.77	160.67
$CaCO_3(s)$	−1207.1	−1128.76	92.88
$CaO(s)$	− 635.5	− 604.17	39.75
$Ca(OH)_2(s)$	− 986.6	− 898.8	76.15
$CaSO_4(s)$	−1432.7	−1320.3	106.7
$Cl(g)$	121.38	105.39	165.10
$Cl_2(g)$	0	0	222.97
$CuO(s)$	− 155.2	− 127.2	43.5
$Cu_2O(s)$	− 166.69	− 146.36	100.8
$Fe_2O_3(s)$	− 822.16	− 741.0	90.0
$Fe_3O_4(s)$	−1117.13	−1014.20	146.4

[a] Data are for 298.15 K and 1 atm pressure. To illustrate the difference produced in changing standard pressure from 1 atm to 1 bar (10^5 Pa) (see page 153), for $CO_2(g)$: $\Delta \bar{H}_f^\circ$ is unchanged; $\Delta \bar{G}_f^\circ$ becomes −394.41 kJ/mol; and $\bar{S}^\circ$ becomes 213.75 J mol^{-1} K^{-1}.

	$\Delta \bar{H}_f^\circ$, kJ/mol	$\Delta \bar{G}_f^\circ$, kJ/mol	$\bar{S}^\circ$, J mol^{-1} K^{-1}
H(g)	217.94	203.26	114.60
H$_2$(g)	0	0	130.58
HBr(g)	− 36.23	− 53.22	198.49
HCl(g)	− 92.30	− 95.27	186.69
HF(g)	− 268.61	− 270.70	173.51
Hl(g)	25.94	1.30	206.31
H$_2$O(g)	− 241.84	− 228.61	188.74
H$_2$O(l)	− 285.85	−237.19	69.96
H$_2$S(g)	− 20.17	− 33.01	205.64
HCHO(g)	− 115.9	− 110.0	218.7
He(g)	0	0	126.06
Hg(g)	60.84	31.76	174.89
Hg(l)	0	0	77.4
I(g)	106.61	70.17	180.67
I$_2$(g)	62.26	19.37	260.58
I$_2$(s)	0	0	116.7
KCl(s)	− 435.89	− 408.32	82.68
MgCl$_2$(s)	− 641.83	− 592.33	89.5
MgO(s)	− 601.83	− 569.57	26.8
MnO$_2$(s)	− 519.7	− 464.8	53.1
N(g)	472.71	455.55	153.22
N$_2$(g)	0	0	191.50
NH$_3$(g)	− 46.19	− 16.65	192.51
NH$_4$Cl(s)	− 315.38	− 203.89	94.6
NO(g)	90.37	86.69	210.62
N$_2$O(g)	81.55	103.60	219.99
NO$_2$(g)	33.85	51.84	240.45
N$_2$O$_4$(g)	9.67	98.28	304.30
NOCl(g)	52.59	66.36	264
NaCl(s)	− 410.99	− 384.01	72.38
O(g)	247.53	230.12	160.96
O$_2$(g)	0	0	205.02
O$_3$(g)	142.3	163.43	237.7
PCl$_3$(g)	− 306.4	− 286.3	311.6
PCl$_5$(g)	− 398.9	− 324.6	353
S(rhombic)	0	0	31.88
S(monoclinic)	0.30	0.096	32.55
SO$_2$(g)	− 296.90	− 300.37	248.53
SO$_3$(g)	− 395.18	− 370.37	256.23
SO$_2$Cl$_2$(l)	− 389	− 314	207
UO$_2$(s)	−1130	−1075	77.8
ZnO(s)	− 347.98	− 318.19	43.9

APPENDIX E

Electron Configurations of the Elements

Atomic number	Element	Electron configuration
1	H	$1s^1$
2	**He**	$1s^2$
3	Li	[He] $2s^1$
4	Be	___ $2s^2$
5	B	___ $2s^2 2p^1$
6	C	___ $2s^2 2p^2$
7	N	___ $2s^2 2p^3$
8	O	___ $2s^2 2p^4$
9	F	___ $2s^2 2p^5$
10	**Ne**	$1s^2 2s^2 2p^6$
11	Na	[Ne] $3s^1$
12	Mg	___ $3s^2$
13	Al	___ $3s^2 3p^1$
14	Si	___ $3s^2 3p^2$
15	P	___ $3s^2 3p^3$
16	S	___ $3s^2 3p^4$
17	Cl	___ $3s^2 3p^5$
18	**Ar**	$1s^2 2s^2 2p^6 3s^2 3p^6$
19	K	[Ar] $4s^1$
20	Ca	___ $4s^2$
21	Sc	___ $3d^1 4s^2$
22	Ti	___ $3d^2 4s^2$
23	V	___ $3d^3 4s^2$
24	Cr	___ $3d^5 4s^1$
25	Mn	___ $3d^5 4s^2$
26	Fe	___ $3d^6 4s^2$
27	Co	___ $3d^7 4s^2$
28	Ni	___ $3d^8 4s^2$
29	Cu	___ $3d^{10} 4s^1$
30	Zn	___ $3d^{10} 4s^2$
31	Ga	___ $3d^{10} 4s^2 4p^1$

Atomic number	Element	Electron configuration
32	Ge	___ $3d^{10}4s^24p^2$
33	As	___ $3d^{10}4s^24p^3$
34	Se	___ $3d^{10}4s^24p^4$
35	Br	___ $3d^{10}4s^24p^5$
36	**Kr**	$1s^22s^22p^63s^23p^63d^{10}4s^24p^6$
37	Rb	[Kr] $5s^1$
38	Sr	___ $5s^2$
39	Y	___ $4d^15s^2$
40	Zr	___ $4d^25s^2$
41	Nb	___ $4d^45s^1$
42	Mo	___ $4d^55s^1$
43	Tc	___ $4d^55s^2$
44	Ru	___ $4d^75s^1$
45	Rh	___ $4d^85s^1$
46	Pd	___ $4d^{10}$
47	Ag	___ $4d^{10}5s^1$
48	Cd	___ $4d^{10}5s^2$
49	In	___ $4d^{10}5s^25p^1$
50	Sn	___ $4d^{10}5s^25p^2$
51	Sb	___ $4d^{10}5s^25p^3$
52	Te	___ $4d^{10}5s^25p^4$
53	I	___ $4d^{10}5s^25p^5$
54	**Xe**	$1s^22s^22p^63s^23p^63d^{10}4s^24p^64d^{10}5s^25p^6$
55	Cs	[Xe] $6s^1$
56	Ba	___ $6s^2$
57	La	___ $5d^16s^2$
58	Ce	___ $4f^26s^2$
59	Pr	___ $4f^36s^2$
60	Nd	___ $4f^46s^2$
61	Pm	___ $4f^56s^2$
62	Sm	___ $4f^66s^2$
63	Eu	___ $4f^76s^2$
64	Gd	___ $4f^75d^16s^2$
65	Tb	___ $4f^96s^2$
66	Dy	___ $4f^{10}6s^2$
67	Ho	___ $4f^{11}6s^2$
68	Er	___ $4f^{12}6s^2$

Atomic number	Element	Electron configuration
69	Tm	_____ $4f^{13}6s^2$
70	Yb	_____ $4f^{14}6s^2$
71	Lu	_____ $4f^{14}5d^16s^2$
72	Hf	_____ $4f^{14}5d^26s^2$
73	Ta	_____ $4f^{14}5d^36s^2$
74	W	_____ $4f^{14}5d^46s^2$
75	Re	_____ $4f^{14}5d^56s^2$
76	Os	_____ $4f^{14}5d^66s^2$
77	Ir	_____ $4f^{14}5d^76s^2$
78	Pt	_____ $4f^{14}5d^96s^1$
79	Au	_____ $4f^{14}5d^{10}6s^1$
80	Hg	_____ $4f^{14}5d^{10}6s^2$
81	Tl	_____ $4f^{14}5d^{10}6s^26p^1$
82	Pb	_____ $4f^{14}5d^{10}6s^26p^2$
83	Bi	_____ $4f^{14}5d^{10}6s^26p^3$
84	Po	_____ $4f^{14}5d^{10}6s^26p^4$
85	At	_____ $4f^{14}5d^{10}6s^26p^5$
86	**Rn**	$1s^22s^22p^63s^23p^63d^{10}4s^24p^64d^{10}4f^{14}5s^25p^65d^{10}6s^26p^6$
87	Fr	[Rn] $7s^1$
88	Ra	_____ $7s^2$
89	Ac	_____ $6d^17s^2$
90	Th	_____ $6d^27s^2$
91	Pa	_____ $5f^26d^17s^2$
92	U	_____ $5f^36d^17s^2$
93	Np	_____ $5f^46d^17s^2$
94	Pu	_____ $5f^67s^2$
95	Am	_____ $5f^77s^2$
96	Cm	_____ $5f^76d^17s^2$
97	Bk	_____ $5f^86d^17s^2$
98	Cf	_____ $5f^{10}7s^2$
99	Es	_____ $5f^{11}7s^2$
100	Fm	_____ $5f^{12}7s^2$
101	Md	_____ $5f^{13}7s^2$
102	No	_____ $5f^{14}7s^2$
103	Lr	_____ $5f^{14}6d^17s^2$
104	(Ku)	_____ $5f^{14}6d^27s^2$
105	(Ha)	_____ $5f^{14}6d^37s^2$

Answers to Selected Exercises

Numerical answers may differ slightly from those listed here, depending on the number of steps used to solve a problem and the rounding off of results at each step. Values used for atomic weights and other constants should permit final answers of three or four significant figures in most cases. Occasionally more precise values are required.

Chapter 1

Review Problems: **1.** (a) 1.17×10^3 g; (b) 7.115 m; (c) 0.000621 kg; (d) 0.673 L; (e) 173 mm; (f) 482 cm^3; (g) 2.07×10^3 mg; (h) 481 m. **2.** (a) 254.2 in.; (b) 26.0 lb; (c) 504 in.; (d) 9.0×10^3 s; (e) 907.0 yd; (f) 1.95×10^4 ft. **3.** (a) 53 cm; (b) 6.7 m; (c) 3.1×10^2 g; (d) 1.4×10^2 lb; (e) 103 ft; (f) 0.19 oz. **4.** (a) 1×10^6; (b) 2.79×10^7; (c) 2.59×10^6; (d) 1728; (e) 2.83×10^4; (f) 1×10^{21}. **5.** (a) 95°F; (b) 30°C; (c) 444°C; (d) −261°F. **6.** 1.26 g/cm^3. **7.** (a) 3.88×10^2 g; (b) 2.25 L; (c) 37.0 lb. **8.** 7.5×10^2 g. **9.** 136 g. **10.** (a) 3.8×10^3; (b) 4.82×10^5; (c) 1.00×10^{-1}; (d) 6.212×10^3; (e) 2.111×10^5; (f) 6.5×10^{-6}; (g) 8.7×10^{-3}; (h) 6.00×10^{-2}; (i) 2.2×10^1. **11.** (a) 6180; (b) 0.812; (c) 0.0004613; (d) 0.043; (e) 0.00000347; (f) 6.70; (g) 6.2; (h) 29,800,000,000; (i) 0.000000168. **12.** (a) 3; (b) 2 or 3; (c) 2; (d) 5; (e) 5; (f) 5; (g) 2 to 5; (h) 4; (i) 1. **13.** (a) 3162; (b) 3.200×10^3; (c) 218.5; (d) 0.06504; (e) 6.000×10^{-5}; (f) 327.3; (g) 1.860×10^5; (h) 2.200×10^5; (i) 14.70. **14.** (a) 8×10^5; (b) 5.9×10^6; (c) 3.5×10^{-4}; (d) 4.3; (e) 7.6×10^5; (f) 8.5×10^4. **15.** (a) 1.000 g/cm^3; (b) 0.9997 g/cm^3.
Exercises: **1.** (a) physical; (b) chemical; (c) physical; (d) chemical. **2.** (a) substance; (b) heterogeneous mixture; (c) homogeneous mixture; (d) heterogeneous mixture; (e) heterogeneous mixture; (f) homogeneous mixture; (g) substance; (h) heterogeneous mixture. **3.** (a) chemical; (b) physical; (c) physical. **5.** (a) extensive; (b) intensive; (c) extensive; (d) intensive. **9.** (a) 1.86×10^5 mi/s; (b) 5×10^{15} to 6×10^{15} t; (c) 1.73×10^{17} W; (d) 1×10^{-6} m; (e) 1×10^{-5} m. **10.** (a) 9.1×10^{-2}; (b) 2.1×10^3; (c) 3.6×10^{-5}; (d) 4.6×10^2. **12.** (a) 7.52×10^4; (b) 4.3×10^{10}; (c) 3.0×10^{-3}; (d) 72.6; (e) 2.16×10^4. **13.** 21.877 g. **15.** 5.03 m. **16.** $7.26/3 lb.

17. 10.2 s. **18.** 7.92 in. **19.** (a) 6.7×10^2 mg; (b) 10 mg/kg. **20.** (a) 0.25 m^3; (b) 2.5×10^4 cm^2. **21.** (a) 43,560 ft^2; (b) 0.405 hm^2. **22.** 1.71×10^3 km/h. **24.** 47.8°C; −8.3°C. **25.** no. **26.** −459.67°F. **27.** 0.792 g/cm^3. **28.** 0.849 g/cm^3. **29.** 102 cm^3. **30.** 2 qt, 1 pt, 0.3 cup. **31.** (a) < (c) < (b). **32.** 0.04 g. **33.** 13% A, 25% B, 44% C, 13% D, 5% F. **34.** 5.4×10^2 g. **35.** 12.2 L. **36.** 0.8688 g/cm^3. **37.** 35°C. **38.** 1.61.
Self-Test Questions: **1.** (c) **2.** (a) **3.** (d) **4.** (b) **5.** (c) **6.** (b) **7.** (d) **8.** (a) **10.** 1.11×10^4. **11.** 436 lb. **12.** $603.

Chapter 2

Review Problems: **1.** 4.01 g Zn. **2.** (a) yes; (b) 27.3% C, 72.7% O. **4.** (a) 0; (b) -2.40×10^{-4} C; (c) $+4.81 \times 10^{-4}$ C. **6.** (a) $^{40}_{18}Ar < ^{39}_{19}K < ^{58}_{27}Co < ^{59}_{29}Cu < ^{120}_{48}Cd < ^{112}_{50}Sn < ^{122}_{52}Te$; (b) $^{39}_{19}K < ^{40}_{18}Ar < ^{59}_{29}Cu < ^{58}_{27}Co < ^{112}_{50}Sn < ^{122}_{52}Te < ^{120}_{48}Cd$; (c) $^{39}_{19}K < ^{40}_{18}Ar < ^{58}_{27}Co < ^{59}_{29}Cu < ^{112}_{50}Sn < ^{120}_{48}Cd < ^{122}_{52}Te$. **7.** (a) 58.6%; (b) 41%. **8.** (a) 0.751015; (b) 2.581147; (c) 7.82551. **9.** 80.916. **10.** 238.0.
Exercises: **5.** HO_2. **6.** 24. **8.** N_2O; NO_2. **9.** HgO and Hg_2O. **15.** (a) 2.4×10^5; (b) 1.1×10^8. **21.** (a) 47 p, 60 n, 47 e; (b) 8.9087575; (c) 6.68374. **24.** 1.661×10^{-24} g. **25.** (a) 1.31×10^{-7} g; (b) 1.34×10^{-7} g; (c) 1.33×10^{-7} g. **28.** (a) 78.70%; (b) 24; (c) 23.98 u. **29.** (c) 7.5% ^6_3Li, 92.5% ^7_3Li. **30.** 0.36% $^{15}_7N$. **31.** about 4% each of ^{29}Si and ^{30}Si. **32.** (a) six; (b) 36, 37, 38, 38, 39, 40; (c) most abundant, $^1H^{35}Cl$; second, $^1H^{37}Cl$. **33.** 200.6.
Self-Test Questions: **1.** (d) **2.** (c) **3.** (b) **4.** (a) **5.** (c) **6.** (b) **7.** (d) **8.** (c) **11.** -2.609×10^3 C/g. **12.** 108.9 u.

Chapter 3

Review Problems: **1. (a)** 2.32×10^{24}; **(b)** 9.81×10^{21}; **(c)** 2.0×10^{15}.   **2. (a)** 13.6 mol Al; **(b)** 296 g Cl_2; **(c)** 668 kg Zn; **(d)** 2.99×10^{24} Fe atoms; **(e)** 3.96×10^{25} Li^+ ions.   **3. (a)** 149.2; **(b)** 34.9 mol H; **(c)** 4.61×10^{24} C atoms; **(d)** 2.29 g O/g N.   **4. (a)** 21; **(b)** 1.67:1; **(c)** 1.14 g O/g C; **(d)** 18.5% N.   **5.** 74.04% C, 7.46% H, 8.64% N, 9.86% O.   **6.** Co_2O_3.   **7.** C_3H_8O. **8.** $C_8H_6O_4$.   **9.** K.   **10. (a)** potassium iodide; **(b)** calcium chloride; **(c)** potassium cyanide; **(d)** magnesium nitrate; **(e)** iodine trichloride; **(f)** chlorine dioxide; **(g)** phosphorus pentachloride.   **11. (a)** stannous fluoride; **(b)** PbO_2; **(c)** Co_2S_3; **(d)** potassium periodate; **(e)** aurous chloride.   **12. (a)** 0; **(b)** -2; **(c)** $+4$; **(d)** $+5$; **(e)** $+5$; **(f)** $+6$; **(g)** $+3$; **(h)** $+6$.   **13. (a)** CaO; **(b)** SrF_2; **(c)** $Mg(OH)_2$; **(d)** Cs_2CO_3; **(e)** $Hg(NO_3)_2$; **(f)** Fe_2S_3; **(g)** $Mg(ClO_4)_2$; **(h)** $KHSO_4$.   **14. (a)** HBr; **(b)** chlorous acid; **(c)** HIO_3; **(d)** nitrous acid; **(e)** H_2SO_4; **(f)** hydrosulfuric acid.   **15.** 43.86% H_2O.

Exercises: **2. (a)** 2.32×10^{24}; **(b)** 7.85×10^{22}; **(c)** 2.0×10^{15}; **(d)** 1.95×10^{23}.   **3. (a)** 6.89; **(b)** 20.7; **(c)** 55.1; **(d)** 6.89.   **4. (a)** 0.333; **(b)** 0.698; **(c)** 4.27×10^{23}. **5.** 1.797×10^4 g.   **6.** 3.36×10^{23}.   **7.** 4.38×10^{22}. **8.** 4.36×10^{12} Pb atoms.   **9. (a)** 4.96×10^{-5} mol S_8; **(b)** 2.39×10^{20} S atoms.   **10. (a)** 1.26×10^{15} $CHCl_3$ molecules.   **13. (a)** 69; **(b)** 0.45:1.00; **(c)** 0.565 g Ge/g S; **(d)** 128 g S; **(e)** 6.68 g; **(f)** 6.203×10^{23} C atoms. **14.** 36.17%.   **17.** 4.7 lb.   **18.** $C_{10}H_8$.   **19.** SeO_2, SeO_3.   **20. (a)** $C_{19}H_{16}O_4$; **(b)** $C_6H_8O_7$.   **21.** CF_2Cl_2. **22. (a)** -4; **(b)** $+4$; **(c)** -1; **(d)** 0; **(e)** $+6$; **(f)** $+2.5$. **24.** N_2O, NO, N_2O_3, NO_2, N_2O_5.   **28.** $MgCl_2 \cdot 6\,H_2O$. **29.** $MgSO_4 \cdot 7\,H_2O$.   **30.** 1.27 g.   **31. (a)** 90.53% C, 9.50% H; **(b)** C_4H_5; **(c)** C_8H_{10}.   **32.** C_7H_8O. **33.** CH_4N.   **34.** C_2H_5.   **35.** 0.4976 g AgI. **36.** 78.0% Cu, 11.3% Sn, 6.13% Zn, 4.59% Pb. **37.** $ZnSO_4 \cdot 7H_2O$.   **38.** compounds are PCl_3 and PCl_5. **39.** 24.3.   **40.** 26.9.   **41.** 56.

Self-Test Questions: **1. (c)**   **2. (b)**   **3. (d)**   **4. (a)** **5. (c)**   **6. (c)**   **7. (c)**   **8.** 494 cm^3.   **9. (a)** calcium iodide; **(b)** $Fe_2(SO_4)_3$; **(c)** SO_3; **(d)** BrF_5; **(e)** ammonium cyanide; **(f)** calcium chlorite; **(g)** $LiHCO_3$. **10. (a)** 57.49% Cu; **(b)** 720 g CuO.   **11.** $C_{13}H_6Cl_6O_2$. **12.** $Na_2SO_3 \cdot 7\,H_2O$.

Chapter 4

Review Problems: **1. (a)** $2\,Mg + O_2 \rightarrow 2\,MgO$; **(b)** $S + O_2 \rightarrow SO_2$; **(c)** $CH_4 + 2\,O_2 \rightarrow CO_2 + 2\,H_2O$; **(d)** $Ag_2SO_4(aq) + BaI_2(aq) \rightarrow BaSO_4(s) + 2\,AgI(s)$. **2. (a)** 1, 2 → 1, 2; **(b)** 4, 1 → 2, 2; **(c)** 1, 3 → 1, 3; **(d)** 3, 2 → 3, 1, 3; **(e)** 1, 6 → 3, 2.   **3. (a)** 1, 2 → 1, 2; **(b)** 1, 1 → 1, 2; **(c)** 2, 6 → 2, 3.   **4. (a)** $C_5H_{12} + 8\,O_2 \rightarrow 5\,CO_2 + 6\,H_2O$; **(b)** $2\,C_2H_6O_2 + 5\,O_2 \rightarrow 4\,CO_2 + 6\,H_2O$; **(c)** $HI(aq) + NaOH(aq) \rightarrow NaI(aq) + H_2O$; **(d)** $Pb(NO_3)_2(aq) + 2\,KI(aq) \rightarrow PbI_2(s) + 2\,KNO_3(aq)$. **5.** 2.10 mol $FeCl_3$.   **6.** 11.8 g P_4, 40.6 g Cl_2. **7. (a)** 14.8 mol H_2; **(b)** 75.7 g H_2O; **(c)** 1.09×10^3 g CaH_2.

8. (a) 0.424 M C_2H_5OH; **(b)** 0.2470 M CH_3OH; **(c)** 1.97 M $(CH_3)_2CO$; **(d)** 0.445 M $C_3H_8O_3$. **9. (a)** 888 mol KCl; **(b)** 36.8 g Na_2SO_4; **(c)** 5.89 mg KOH. **10.** 0.195 L.   **11.** 27.34 mL.   **12.** no. **13. (a)** 81.9 g C_6H_{10}; **(b)** 84.2%; **(c)** 145 g $C_6H_{12}O$. **14. (a)** 2, 5 → 2, 2; **(b)** 3, 8, 2 → 3, 4, 2; **(c)** 4, 10, 2 → 4, 5, 1; **(d)** 2, 6, 3 → 4, 6; **(e)** 5, 2, 6 → 2, 8, 5. **15.** oxidizing agents: (a) NO, (b) NO_3^-, (c) NO_3^-, (d) O_2, (e) MnO_4^-; reducing agents: (a) H_2, (b) Cu, (c) Zn, (d) Fe_2S_3, (e) H_2O_2.

Exercises: **5. (a)** 0.311 mol O_2; **(b)** 1.87×10^{23} molecules O_2; **(c)** 15.5 g KCl.   **6.** 32.6 g CO_2. **7.** 84.1% Fe_2O_3.   **8.** 89.8% Ag_2O.   **9.** 3.12 g H_2. **10.** 138 mL.   **11. (a)** 4.32 M $CO(NH_2)_2$; **(b)** 1.55×10^{-5} M $(C_2H_5)_2O$; **(c)** 6.61×10^{-5} M NaCl. **12. (a)** 61.7 mL CH_3OH; **(b)** 4.82 gal C_2H_5OH; **(c)** 247 mg $Ca(NO_3)_2$.   **13.** 1.34 M $C_{12}H_{22}O_{11}$.   **14.** 215 mL. **15.** 0.707 M $MgSO_4$.   **16.** 6.38 g $NaHCO_3$. **17. (a)** 4.94 g $Ca(OH)_2$; **(b)** 75.6 kg $Ca(OH)_2$. **18. (a)** 6.18 M NH_3; **(b)** 10.9% NH_3.   **19. (a)** 4.0×10^2 mL; **(b)** 0.2408 M HCl.   **20.** 49.1% Fe. **21.** 51 mg Mg.   **22. (a)** 0.33 mol Na_2CS_3, 0.17 mol Na_2CO_3, 0.50 mol H_2O; **(b)** 1.70×10^2 g Na_2CS_3. **23.** 0.190 g H_2.   **24.** 4.76 g NH_3.   **25.** 4.23 g. **26. (b)** 0.214 g H_2, 0.0 g O_2, 41.00 g H_2O.   **27.** 42.3 g Cl_2.   **28. (a)** 24.0 g C_4H_9Br; **(b)** 16.8 g C_4H_9Br; **(c)** 70.0%.   **29.** 62%.   **30.** 56 g $C_2H_4O_2$.   **31.** 456 g HCl.   **32.** 41.8 mol CO_2.   **33.** a mixture. **34.** 72.8 mol Cl_2.   **35.** 333 g $BaCO_3$.   **36.** 1.34 kg $AgNO_3$.   **37.** 1.02×10^3 kg Fe.   **40.** 1.5×10^3 L. **41.** 6.95 kg NH_3.   **42. (b)** 58%; **(c)** 1.1×10^3 lb NH_3. *Self-Test Questions:*   **1. (c)**   **2. (d)**   **3. (a)**   **4. (a)** **5. (b)**   **6. (c)**   **7. (a)**   **8. (c)** **9. (a)** $Hg(NO_3)_2(s) \rightarrow Hg(l) + 2\,NO_2(g) + O_2(g)$; **(b)** $Na_2CO_3(aq) + 2\,HCl(aq) \rightarrow 2\,NaCl(aq) + H_2O + CO_2(g)$; **(c)** $C_3H_4O_4 + 2\,O_2(g) \rightarrow 3\,CO_2(g) + 2\,H_2O(l)$. **10.** 25.8 mL.   **11.** 0.719 g Na.

Chapter 5

Review Problems: **1. (a)** 0.970 atm; **(b)** 0.899 atm; **(c)** 1.599 atm; **(d)** 1.9 atm.   **2. (a)** 103 cm; **(b)** 618 mm; **(c)** 3.09 m.   **3. (a)** 49.0 L; **(b)** 9.97 L.   **4.** 12.6 atm. **5. (a)** 163 cm^3; **(b)** 124 cm^3.   **6.** 126°C.   **7. (a)** $V_f = 3V_i$; **(b)** $v_f = 0.250V_i$; **(c)** $V_f = V_i$.   **8.** 58.2 L. **9.** 0.735 atm.   **10.** 30.0.   **11.** 1.84 g/L.   **12.** 311 L. **13.** 151 L.   **14.** 3.83 L.   **15.** 0.00305 mol O_2. **16.** 30.9 s.

Exercises: **1. (a)** 1.825 atm; **(b)** 6.91 atm; **(c)** 3.10 atm; **(d)** 0.979 atm; **(e)** 2.50 atm.   **2. (a)** 3.60 m; **(b)** 3.65 m; **(c)** 1.92 g/cm^3.   **3.** 1.30 atm.   **4.** 750.7 mmHg. **5.** 612 mmHg.   **6.** 148 atm.   **7.** 4°C. **8.** 986 mmHg.   **9. (a)** 351 cm^3; **(b)** 0.0157 mol. **10.** 4.6 g.   **11.** 385°C.   **12.** 546 g.   **13.** 3.98 L. **14.** 79 g.   **15.** 0.349 atm.   **16. (c)** 66.7 kPa. **17.** 42.2.   **18.** C_2ClF_5.   **19.** 26.05.   **20.** 1.31 atm. **21.** P_4.   **23.** 1.6×10^7 L.   **24.** 11.0% $KClO_3$.

25. (a) 247 L $NH_3(g)$; **(b)** 4.37×10^4 L $NH_3(g)$.
26. (a) 0.0597 mol $SO_2(g)$; **(b)** 6.82 L. **27.** 1.5×10^3 g Ne. **28.** 758 mmHg. **29.** 12 g He. **30. (a)** 28.7; **(b)** less; **(c)** about 130:1. **31.** 2.80 L. **32.** 251 cm^3.
33. 741 mmHg. **35. (a)** 1.09×10^3 K; **(b)** 494 m/s.
36. 324 m/s. **37. (a)** 3.98:1.0; **(b)** 1.4:1.0;
(c) 1.004:1.000. **38.** 0.00171 mol HCl. **39.** 59.
40. (a) 29.1 atm; **(b)** 25.7 atm. **41. (b)** 6.54 L.
42. (a) 43.1 atm.
Self-Test Questions: **1.** (b) **2.** (a) **3.** (c) **4.** (a)
5. (b) **6.** (c) **7.** (c) **8.** (d) **9.** 6.1 L.
11. 77.3 L. **12.** C_4H_{10}.

Chapter 6

Review Problems: **1. (a)** -949 cal; **(b)** $+56$ kcal;
(c) -483 kJ. **2. (a)** 33.2°C; **(b)** -228°C; **(c)** 18°C.
3. (a) 0.0929 cal g^{-1} °C^{-1}; **(b)** 0.248 cal g^{-1} °C^{-1};
(c) 0.113 cal g^{-1} °C^{-1}. **4. (a)** -1.30×10^3 kJ/mol
$C_2H_2(g)$; **(b)** -634.6 kJ/mol $CO(NH_2)_2(s)$; **(c)** -1.82×10^3 kJ/mol $(CH_3)_2CO(l)$. **5. (a)** 28.74°C; **(b)** 31.39°C;
(c) 30.71°C. **6. (a)** 26.8°C; **(b)** 26.5°C; **(c)** 24.6°C.
7. -282.97 kJ/mol. **8.** -293 kJ/mol.
9. (a) -55.68 kJ/mol; **(b)** -1125.16 kJ/mol;
(c) -53.63 kJ/mol. **10. (a)** -191 kJ/mol $(CH_3)_2O(g)$;
(b) -93 kJ/mol $C_7H_6O(l)$; **(c)** -1260 kJ/mol $C_6H_{12}O_6(s)$.
Exercises: **1. (a)** -1×10^2 kJ; **(b)** 3×10^2 g H_2O.
2. 1.4×10^2 g. **4.** 2.3×10^2 J mol^{-1} °C^{-1}. **5.** 26.1°C.
6. (a) -6.24 kJ/g CaO; **(b)** -4.29×10^6 kJ.
7. (a) 882 L; **(b)** 2.76×10^4 L. **8.** $\Delta H = -1.313 \times 10^4$ kJ. **10.** $\Delta \overline{H} = -6 \times 10^1$ kJ/mol KOH.
11. 0.27 kg NH_4NO_3. **12.** approx. 90°C. **13.** 8.4×10^2 J/°C. **14.** 1.91°C. **15.(a)** -2.80×10^3 kJ/mol
$C_6H_{12}O_6$. **16.** 70.5 metric tons. **17.** about 33°C above
room temperature. **20.** $\Delta \overline{H} = -818.47$ kJ/mol.
21. $\Delta \overline{H} = -284$ kJ/mol. **22.** $\Delta \overline{H} = -236.6$ kJ/mol.
23. $\Delta \overline{H} = -747.60$ kJ/mol. **24.** $\Delta \overline{H} = -128.1$ kJ/mol. **26.** $\Delta \overline{H}° = +202.5$ kJ/mol.
27. $\Delta \overline{H}° = -1366.92$ kJ/mol $C_2H_5OH(l)$.
28. $\Delta \overline{H}°_f[CCl_4(g)] = -108$ kJ/mol. **29.** $\Delta \overline{H}° = +15.32$ kJ/mol. **30.** $\Delta \overline{H}° = -1322.8$ kJ/mol.
31. $\Delta \overline{H}° = -1026$ kJ/mol. **32. (a)** 1.34×10^6 kJ;
(b) 3.72×10^4 L $CH_4(g)$. **33. (b)** increase; **(c)** $\Delta t = +0.4$°C. **34.** 87.0% CH_4, by volume. **35.** $\Delta \overline{H}° = -4.853 \times 10^3$ kJ/mol C_7H_{16}.
Self-Test Questions: **1.** (c) **2.** (a) **3.** (d) **4.** (a)
5. (b) **6.** (b) **7.** (c) **8.** (d) **10.** 101°C.
11. $C_6H_5OH(s) + 7 O_2(g) \rightarrow 6 CO_2(g) + 3 H_2O(l)$; $\Delta \overline{H} = -3.063 \times 10^3$ kJ/mol. **12.** $\Delta \overline{H}°_f[COCl_2(g)] = -219$ kJ/mol.

Chapter 7

Review Problems: **1. (a)** 301.5 nm; **(b)** 1.56×10^{-4} cm;
(c) 3.92×10^7 nm; **(d)** 3.76×10^{-7} m; **(e)** 1.812 μm;
(f) 2.18×10^4 Å. **2. (a)** 4.8×10^{-6} m; **(b)** 6.7×10^{-9} m; **(c)** 5.45×10^2 m. **3. (a)** 2.3×10^{14} s^{-1};
(b) 3.38×10^{13} s^{-1}; **(c)** 8.09×10^{15} s^{-1}; **(d)** 8.96×10^7 s^{-1}. **4. (a)** $\nu = 6.9050 \times 10^{14}$ s^{-1}; **(b)** $\lambda = 397.12$ nm; **(c)** $n = 10$. **5. (a)** 2.05×10^{-18} J/photon;

(b) 1.70×10^2 kJ/mol; **(c)** $\nu = 5.34 \times 10^{13}$ s^{-1}; **(d)** $\lambda = 6.47 \times 10^{-7}$ m. **6. (a)** $m_l = 0$; **(b)** $l = 2$ or 1; **(c)** $n = 2$ or larger; **(d)** $n = 3$ or larger, $m_l = -2, -1, 0, +1$, or $+2$.
7. (a) $n = 4$, $l = 0$, $m_l = 0$; **(b)** $n = 3$, $l = 1$, $m_l = -1$, 0, or $+1$; **(c)** $n = 5$, $l = 3$, $m_l = -3, -2, -1, 0, +1, +2$, or $+3$;
(d) $n = 3$, $l = 2$, $m_l = -2, -1, 0, +1$, or $+2$. **8.** not
allowed: (b), (c), and (e). **9.** $3p$, $3d$, $4p$, $5s$, $6s$, $6p$, $5f$.
10. (b) P $(Z = 15)$; **(c)** $\ldots 4d^2 \ldots$; **(d)** Te $(Z = 52)$,
$\ldots 4d^{10} \ldots$; **(e)** As $(Z = 53)$, $\ldots 3d^{10}4s^2 \ldots$;
(f) $\ldots 4f^{14}5d^{10}6s^26p^3$. **11. (a)** B; **(b)** V; **(c)** Si.
12. (a) $[Ne]3s^23p^1$; **(b)** $[Kr]5s^1$; **(c)** $[Kr]4d^{10}5s^2$;
(d) $[Kr]4d^{10}5s^25p^3$; **(e)** $[Xe]4f^{14}5d^{10}6s^26p^2$;
(f) $[Kr]4d^{10}5s^25p^6$.
Exercises: **1. (a)** and (d). **2.** 3.261226 cm.
3. 8.3 min. **4.** 656.21 nm, 486.08 nm, 434.00 nm,
410.13 nm. **6.** 364.56 nm. **7. (a)** 121.56 nm and
91.174 nm; **(b)** ultraviolet; **(c)** $n = 5$. **8. (a)** 6.96×10^{-19} J/photon; **(b)** 4.19×10^2 kJ/mol. **9.** 229 nm.
10. from 2.58×10^{-19} to 5.10×10^{-19} J/photon.
11. (a) 8.6×10^{-19} J. **13. (a)** 19 Å; **(b)** -6.053×10^{-20} J. **14. (a)** 8.0 Å; **(b)** $\Delta E = 2.043 \times 10^{-18}$ J.
15. (a) $\nu = 1.142 \times 10^{14}$ s^{-1}; **(b)** $\lambda = 2.625 \times 10^{-6}$ m;
(c) infrared. **17.** $n = 4$. **18. (a)** -8.716×10^{-18} J;
(b) -2.179×10^{-18} J. **23.** $\Delta x \geq 2.10 \times 10^{-13}$ m.
25. 7×10^5 m/s. **28.** (d). **30.** not allowed: (b), (e),
and (f). **31. (a)** $2p$; **(b)** $4d$; **(c)** $5s$. **32.** $n = 3$, $l = 0$;
$n = 4$, $l = 1$; $n = 5$, $l = 3$; $n = 6$, $l = 2$. **33. (a)** 1; **(b)** 0;
(c) 3; **(d)** 5. **34. (a)** and **(b)**. **35.** (e) $<$ (b) $<$ (c) $=$
(d) $<$ (a). **38. (a)** 2; **(b)** 0; **(c)** 3; **(d)** 2; **(e)** 14.
40. (a) $1s^32s^32p^93s^33p^93d^{15}4s^34p^95s^1$;
(b) $1s^21p^62s^22p^62d^{10}3s^23p^63d^{10}3f^14s^24p^65s^2$.
Self-Test Questions: **1.** (b) **2.** (a) **3.** (d) **4.** (b)
5. (b) **6.** (d) **7.** (c) **8.** (a) **9.** 1.60×10^2 kJ/mol. **10.** greater energy: 589.0 nm; energy
difference: 4×10^{-22} J/photon. **11.** $n = 5$.
12. (a) $1s^22s^22p^63s^23p^63d^{10}4s^24p^4$; **(b)** orbital diagram
corresponds to $[Kr]4d^{10}5s^25p^5$.

Chapter 8

Review Problems: **1. (a)** In; **(b)** similar: Se, Te; unlike,
many others (e.g., metals); **(c)** Cs (or Ba); **(d)** I; **(e)** Xe or
Rn. **2. (a)** VIIA; **(b)** IVA; **(c)** IIA; **(d)** IB; **(e)** VIB.
3. (a) $[Kr]4d^{10}5s^25p^1$; **(b)** $[Kr]4d^15s^2$; **(c)** $[Kr]4d^{10}5s^25p^3$;
(d) $[Xe]4f^{14}5d^{10}6s^1$. **4. (a)** $[Ar]3d^{10}4s^24p^6$; **(b)** same
as (a); **(c)** $1s^22s^22p^6$; **(d)** $[Kr]4d^{10}5s^25p^6$; **(e)** $[Ar]3d^{10}$;
(f) $[Kr]4d^{10}$; **(g)** $[Xe]4f^{14}5d^{10}6s^2$. **5. (a)** 1; **(b)** 5; **(c)** 10;
(d) 6; **(e)** 14; **(f)** 8. **6. (a)** As; **(b)** Sr; **(c)** Cs; **(d)** Xe;
(e) C; **(f)** Hg. **7.** smallest: F atom; largest: I^- ion.
8. Cs $<$ Sr $<$ As $<$ S $<$ F. **9.** 4.07 J. **10. (a)** Ba;
(b) S; **(c)** Bi. **11.** Rb $>$ Ca $>$ Sc $>$ Fe $>$ Te $>$ Br $>$ O $>$
F. **12.** diagmagnetic: K^+, Zn^{2+}, Cd, Sn^{2+}; paramagnetic:
Cr^{3+}, Co^{3+}, Br.
Exercises: **1.** 2.6 g/cm^3. **2.** about 15 g/cm^3. **5.** $Z = 118$; $Z = 119$, at. wt. about 310. **11. (a)** 5; **(b)** 6; **(c)** 5;
(d) 2; **(e)** 24. **15. (a)** B; **(b)** Te. **18.** $Li^+ <$ Se $<$ Y $<$
Br^-. **24.** 176.8 eV. **25.** 1.603×10^{18} Cs^+ ions.

26. -9.84 J.   **27.** $I_1 = 13.60$ eV/atom.   **31.** Fe^{2+}.
32. V^{3+}, $[Ar]3d^2$; Cu^{2+}, $[Ar]3d^9$; Cr^{3+}, $[Ar]3d^3$.
37. (a) 5.7 g/cm³; **(b)** Ga_2O_3, 74% Ga.
Self-Test Questions: **1. (b)   2. (c)   3. (d)   4. (b)**
5. (a)   6. (d)   7. (c)   8. (b)   9. (a) 34; **(b)** 45;
(c) 18; **(d)** 2; **(e)** 4; **(f)** 6.   **10.** Br.   **11. (a)** C; **(b)** Rb;
(c) At.

Chapter 9
Review Problems: **1. (a)** H· **(b)** :K̈r: **(c)** ·Ge· **(d)** Mg^{2+}

(e) [:B̈r:]⁻ **(f)** ·Ga· **(g)** Sc^{3+} **(h)** Cs· **(i)** [:S̈:]²⁻

2. (a) Na⁺ [:F̈:]⁻ **(b)** Mg^{2+}[:Ö:]²⁻

(c) ⁻[:Ï:] Sr^{2+} [:Ï:]⁻   **3. (a)** :B̈r—B̈r: **(b)** :Ï—C̈l:

(c) :F̈—Ö: **(d)** :N̈—Ï: **(e)** :T̈e—H   **4. (a)** :S̈=C=S̈:
(with :Ï: above and :F̈: below in (c); with Ï: above and :Ï: below in (d))

(b) :Ö—Ö=Ö: **(c)** H—C̈=Ö: **6. (a)** none; **(b)** +1 on

S, −1 on singly bonded O; **(c)** −1 on each singly bonded O;
(d) none; **(e)** −1 on terminal O; **(f)** +1 on N, −1 on singly
bonded O.   **7. (a)** Mg^{2+} and two [:Ö—H]⁻

(b) [H—N̈—H]⁺ (with H above and H below)   [:Ï:]⁻ **(c)** Ca^{2+} and two [:Ö—Cl—Ö:]⁻ (with :Ö: above)

8. diamagnetic: (a), (d), (e); paramagnetic: (b), (c), (f).
10. (a) linear; **(b)** tetrahedral; **(c)** trigonal bipyramidal;
(d) angular; **(e)** T-shaped; **(f)** octahedral; **(g)** trigonal planar.
11. (b) 4.40×10^{-18} J to break all bonds in one molecule.
12. (a) endothermic; **(b)** exothermic; **(c)** exothermic.
13. dipole moments expected: (b), (d), (e), (g).
14. $AsH_3 < AsI_3 < AsBr_3 < AsCl_3 < AsF_3$.
15. C—H < Br—H < F—H < Na—Cl < K—F.
Exercises: **7. (a)** RbCl, ionic, f. wt. = 120.9; **(b)** H_2Se,
covalent, 80.98; **(c)** BCl_3, covalent, 117.2; **(d)** Cs_2S, ionic,
297.9; **(e)** SrO, ionic, 103.6; **(f)** OF_2, covalent, 54.00.
11. (a) +2 on Cl, −1 on each O; **(b)** none; **(c)** −1 on B;
(d) none; **(e)** +1 on central N, −1 on each terminal N;
(f) +1 on central N, −2 on singly bonded N; **(g)** +1 on N.
12. (a) H_2NOH; **(b)** SCS; **(c)** ONCl; **(d)** SCN⁻.   **20.** most
plausible: (a) and (b).   **26.** sulfur-to-nitrogen triple bond.
29. AXE_3, linear (e.g., HF).   **30.** tetrahedral.
32. (a) trigonal pyramid; **(b)** tetrahedral; **(c)** square planar;
(d) tetrahedral; **(e)** linear.   **34. (a)** 137 pm; **(b)** 149 pm;
(c) 177 pm; **(d)** 141 pm; **(e)** 205 pm.
37. (a) −100 kJ/mol; **(b)** −129 kJ/mol.
38. $\Delta H_f^\circ[NH_3(g)] = -4.0 \times 10^1$ kJ/mol.
40. 631 kJ/mol.   **41.** 11% ionic.   **42.** expected dipole
moments: (a), (b), (c), (f).   **44.** ionic resonance energies:

HF, 270 kJ/mol; HBr, 50 kJ/mol.   **45. (a)** 1.68;
(b) 0.72.   **46.** 209 kJ/mol.   **48. (b)** 76.0% F.
49. (b) 867 cm³ CO_2(g).   **50.** 8.88×10^{18}.
Self-Test Questions: **1. (b)   2. (d)   3. (c)   4. (a)**
5. (a)   6. (c)   7. (a) CH_2Cl; **(c)** $C_2H_4Cl_2$.

8.   H—C=C=C—H   and   H—C≡C—C—H   **9. (b)**
(with H above and H below each terminal C in first; with H above, H, H around right C in second)

10. (a) angular; **(b)** trigonal planar; **(c)** tetrahedral.
11. -7.5×10^2 kJ/mol.

Chapter 10
Review Problems: **1.** orbital overlap: **(a)** 1s of H with 3p
of Cl; **(b)** 3p of Cl with 5p of I; **(c)** 1s of each H with two
of the 4p of Se; **(d)** each 2p of N with a 5p of each I.
2. (a) two C—H bonds: 1s of H with $2sp^3$ of C, two C—Cl
bonds: $2sp^3$ of C with 3p of Cl; **(b)** two Be—Cl bonds: 2sp
of Be with 3p of Cl; **(c)** three B—F bonds: $2sp^2$ of B with
2p of F.   **3. (a)** sp^3d^2; **(b)** sp; **(c)** sp^3; **(d)** sp^2; **(e)** sp^3d.
4. sp^3 hybridization at N and O atoms; $2sp^3$–$2sp^3$ overlap for
N—O bond; $2sp^3$–1s overlap for N—H and O—H bonds.
5. (a) H—C, σ; C≡N, one σ, two π; **(b)** C—C, one σ;
C≡N, one σ, two π; **(c)** C=C, one σ, one π; all others, σ
bonds; **(d)** H—O, σ; O—N, σ; N=O, one σ, one π.
6. bonds in H_3C—all σ with 1s–$2sp^3$ overlap; bonds in
C—O—C, both σ with $2sp^3$–$2sp^3$ overlap.   **7.** linear: (a),
(b); planar: (d).   **8.** diamagnetic: (a); paramagnetic: (b), (c).

9. (a) σ_{1s}^b [↑↓] σ_{1s}^* [↑]

(b) KK σ_{2s}^b [↑↓] σ_{2s}^* [↑↓] π_{2p}^b [↑↓][↑↓] σ_{2p}^b [↑] π_{2p}^* [ ][ ] σ_{2p}^* [ ]

(c) KK σ_{2s}^b [↑↓] σ_{2s}^* [↑↓] σ_{2p}^b [↑↓] π_{2p}^b [↑↓][↑↓] π_{2p}^* [↑↓][↑↓] σ_{2p}^* [↑]

10. 6.62×10^{20} energy levels and 6.62×10^{20} electrons.
Exercises: **1. (c).   5. (a)** C atoms: 2sp; **(b)** N atom:
$2sp^2$; **(c)** N atom: 2sp.   **13.** 875 kJ/mol.   **14.** stable
diamagnetic: Li_2, C_2, N_2, F_2; stable paramagnetic: B_2, O_2.
15. both N_2^- and N_2^{2-}.   **16.** O_2^+ has a stronger bond.
21. isoelectronic: NO⁺, CO, CN⁻; isoelectronic: CN⁺, BN.
25. (b), (c).   **29. (a)** intrinsic; **(b)** p-type; **(c)** n-type;
(d) intrinsic.   **31. (a)** energy gap is less than 1.7 eV.
32. (a) 0.60 watt; **(b)** 1.3 A.
Self-Test Questions: **1. (d)   2. (b)   3. (b)   4. (d)**
5. (c)   6. (c)   7. (a) sp; **(b)** sp^3; **(c)** sp^3 **(d)** sp^2.
8. (a) 6 σ; **(b)** 2 π.

Chapter 11
Review Problems: **1. (a)** 2.55×10^5 cal; **(b)** 24.0 g;
(c) 29.4 kJ/mol.   **2. (a)** 300 mmHg; **(b)** 35°C.
3. 226 mmHg.   **4. (a)** 184°C; **(b)** 2.6×10^2 mmHg.
5. (a) 11.2 g Mg; **(b)** 16 kcal; **(c)** 435 kJ.

6. (a) 31.8 mmHg; **(b)** 80.6 mmHg; **(c)** 85.6 mmHg.
7. (a) liquid and gas; **(c)** solid → liquid → gas.
8. (a) $C_{10}H_{22}$; **(b)** H_3C—O—CH_3; **(c)** H_3CCH_2—O—H.
9. $CsI < MgBr_2 < CaO$.   **10. (a)** $U = -1049$ kJ/mol;
(b) $\Delta \overline{H}_f^{\circ} = -574$ kJ/mol.   **11. (a)** 362 pm; **(b)** 4.74×10^{-23} cm^3; **(c)** 4; **(d)** 4.22×10^{-22} g; **(e)** 8.90 g/cm^3.
12. unit cell contains 1 Cs^+ and $(8 \times 1/8)$ Cl^-, leading to the formula CsCl.
Exercises: **4.** 31 kJ/mol.   **5.** 273 L CH_4(g).
8. (a) 93.5°C; **(b)** 507 mmHg.   **9.** 0.904 mmHg.
10. 11.6 L.   **11.** 280°C.   **12. (a)** 49.8 kJ/mol;
(b) 160°C.   **13.** 75°C.   **14.** CO_2, HCl, NH_3, SO_2, H_2O.   **16.** 8.7×10^3 kJ.   **17.** 51.0 kJ.
18. (b) 97°C.   **19.** 26.7 mmHg.   **20.** 18°C.
26. (a) liquid; **(b)** exothermic; **(c)** greater than 1.00 g/cm^3.
27. $N_2 < F_2 < Ar < O_3 < Cl_2$.   **42. (a)** BaF_2;
(b) $MgCl_2$.   **44.** $U = -669$ kJ/mol.   **45.** $U = -703$ kJ/mol.   **50.** 1.75 g/cm^3.   **52. (a)** 1.68×10^{-22} cm^3; **(b)** 3.89×10^{-22} g; **(c)** 2.32 g/cm^3.
Self-Test Questions: **1.** (d)   **2.** (b)   **3.** (c)   **4.** (a)
5. (c)   **7.** (a) and (e).   **8.** (c).

Chapter 12

Review Problems: **1.** 59.0% KI.   **2. (a)** 15.1% vol/vol;
(b) 11.9% mass/vol; **(c)** 12.2% mass/mass.   **3.** 1.07 g.
4. 0.15 M NaCl.   **5.** 0.664 m.   **6. (a)** $\chi_{C_7H_{18}} = 0.303$, $\chi_{C_8H_{18}} = 0.255$, $\chi_{C_9H_{20}} = 0.442$; **(b)** 30.3, 25.5, and 44.2 mol %, respectively, of C_7H_{18}, C_8H_{18}, and C_9H_{20}.
7. unsaturated.   **8. (a)** 1.26×10^{-3} M; **(b)** 2.64×10^{-4} M.   **9.** $P_{C_6H_6} = 46.1$ mmHg; $P_{C_7H_8} = 14.6$ mmHg; $P_{tot.} = 60.7$ mmHg.   **10.** $\chi_{C_6H_6} = 0.759$; $\chi_{C_7H_8} = 0.241$.
11. mol. wt. $= 1.2 \times 10^2$.   **12.** 100.02°C.
13. mol. wt. $= 6.1 \times 10^4$.   **14.** 0.110 M K^+, 0.125 M Mg^{2+}, and 0.360 M Cl^-.   **15.** order of freezing points of 0.001 m aqueous solutions: $C_2H_5OH > HC_2H_3O_2 > NaCl > MgBr_2 > Al_2(SO_4)_3$.
Exercises: **5.** 46.2 g $HC_2H_3O_2$.   **8.** 17.5 M H_2SO_4.
9. 115 mL.   **10.** at 15°C: 2.134 M; at 25°C: 2.128 M.
11. 8.00 g I_2.   **12.** 21.4 m HF; 16.5 M HF.   **13.** 682 g H_2O.   **14. (a)** $\chi_{C_2H_5OH} = 0.0516$; **(b)** $\chi_{CO(NH_2)_2} = 4.57 \times 10^{-3}$; **(c)** $\chi_{C_6H_{12}O_6} \simeq 9 \times 10^{-4}$.   **15.** 4.3 g C_2H_5OH.
16. 3.60×10^2 mL $C_3H_8O_3$.   **17.** 80°C.
18. (a) supersaturated; **(b)** about 16 g $KClO_4$ crystallizes.
19. (a) yes; **(b)** 18 g NH_4Cl; **(c)** no.   **20.** sample will dissolve more H_2S.   **21.** 4×10^2 g.   **24.** 1.29×10^3 mmHg.   **25.** 23.5 mmHg.   **26.** $\chi_{C_6H_6} = 0.328$.
27. $C_6H_4N_2O_4$.   **28. (a)** 20°C kg solv. (mol solute)$^{-1}$.
29. 3 parts $C_2H_6O_2$ to 10 parts H_2O, by volume.
31. 9.0% $C_{12}H_{22}O_{11}$.   **34.** 0.31 M $C_6H_{12}O_6$.   **35.** 2.26 g.
36. mol. wt. $= 2.8 \times 10^5$.   **37.** 0.588 M Cl^-.
40. (a) -0.19°C; **(b)** -0.38°C; **(c)** -0.57°C; **(d)** -0.38°C;
(e) -0.19°C; **(f)** -0.38°C; **(g)** -0.20°C.   **41. (a)** $i = 1.06$;
(b) $i = 1.07$.   **44. (b)** $AlCl_3$.   **45. (a)** 16 cm^2; **(b)** 8.35×10^{-3} cm^2.
Self-Test Questions: **1.** (b)   **2.** (c)   **3.** (d)   **4.** (b)
5. (a)   **6.** (d)   **7.** (c)   **8.** (b)   **9. (a)** 2.22% $C_{10}H_8$;

(b) 0.178 m $C_{10}H_8$; **(c)** 4.64°C.   **10.** 5.0% NaCl.
12. 8×10^1 mL.

Chapter 13

Review Problems: **2. (a)** Na_2O_2; **(b)** Mg_3N_2; **(c)** calcium hydroxide; **(d)** $NaClO_4$; **(e)** lithium aluminum hydride;
(f) superoxide ion; **(g)** HF_2^-; **(h)** $NH_4H_2PO_4$; **(i)** hydrogen peroxide; **(j)** sodium hydrogen sulfite; **(k)** O_3; **(l)** dinitrogen tetroxide [nitrogen(IV) oxide dimer]; **(m)** $Ca(HCO_3)_2$;
(n) BrF_3.   **3. (a)** C; **(b)** $CaCO_3$; **(c)** CaO; **(d)** $Ca(OH)_2$;
(e) SiO_2; **(f)** $CaSO_4 \cdot 2$ H_2O; **(g)** Na_2SiO_3(aq);
(h) $CaSO_4 \cdot \frac{1}{2}$ H_2O; **(i)** CO(g) + H_2(g).   **4. (a)** LiH + H_2O → LiOH + H_2; **(b)** 2 Na_2O_2 + 2 H_2O → 4 NaOH + O_2;
(c) Ca + 2 H_2O → $Ca(OH)_2$ + H_2; **(d)** CaO + H_2O → $Ca(OH)_2$; **(e)** P_4O_{10} + 6 H_2O → 4 H_3PO_4; **(f)** C + H_2O → CO + H_2.   **5. (a)** Mg + 2 HCl → $MgCl_2$ + H_2;
(b) NH_3 + HNO_3 → NH_4NO_3; **(c)** $CaCO_3$ + 2 HCl → $CaCl_2$ + H_2O + CO_2; **(d)** 2 NaF + H_2SO_4 → Na_2SO_4 + 2 HF; **(e)** 3 Ag + 4 HNO_3 → 3 $AgNO_3$ + 2 H_2O + NO.   **6. (a)** H_2O_2 + NO_2^- → H_2O + NO_3^-; **(b)** H_2O_2 + SO_2 + 2 OH^- → 2 H_2O + SO_4^{2-}; **(c)** 5 H_2O_2 + 2 MnO_4^- + 6 H^+ → 2 Mn^{2+} + 8 H_2O + 5 O_2; **(d)** H_2O_2 + Cl_2 + 2 OH^- → 2 H_2O + 2 Cl^- + O_2.   **7. (a)** NO_2 and N_2O_4;
(b) peroxides, such as H_2O_2; **(c)** hypochlorites, such as HOCl and NaOCl; **(d)** sulfides, such as H_2S and Na_2S; **(e)** PH_3;
(f) ionic hydrides; such as NaH and CaH_2 **8.** (13.3): oxidizing agent, H_2O, reducing agent, CH_4; (13.26): H_2O, Mg; (13.56): NO_3^-, Cu; (13.61): NO_2, NO_2; (13.64): HNO_3, I_2; (13.65): $Ca_3(PO_4)_2$, C; (13.72): MnO_4^-, H_2O_2.
9. exothermic: (13.30), (13.43); endothermic: (13.2), (13.3), (13.13), (13.27).   **10. (a)** O, Al; **(b)** H_2, He, N_2, O_2, F_2, Ne, Cl_2, and Ar; **(c)** B and Si; **(d)** F_2; **(e)** C, O, P, and S;
(f) C; **(g)** He; **(h)** all except He, Ne, and Ar.
Exercises: **1. (a)** K; **(b)** Na_2CO_3; **(c)** O_3; **(d)** H_2S(l);
(e) SiO_2(s); **(f)** LiF(l); **(g)** LiF(s).   **4.** 195 g CaH_2.
5. 6.7×10^5 L.   **6.** 600 times.   **7. (a)** 0.44 g/L;
(b) mol. wt. $= 9.9$.   **10.** -612 kJ/mol.
15. (a) -423 kJ/mol Fe; **(b)** -593 kJ/mol Mn.   **18.** about 1700 ppm.   **24.** 6×10^6 t.   **25.** $[NO_3^-] = 0.1716$ M.
26. 1.17×10^3 kg.   **29.** 2 H_2O_2 → 2 H_2O + O_2.
30. (a) O_3 + 2 I^- + 2 H^+ → O_2 + I_2 + H_2O; **(b)** 3 O_3 + S + H_2O → H_2SO_4 + 3 O_2; **(c)** O_3 + 2 $[Fe(CN)_6]^{4-}$ + H_2O → 2 $[Fe(CN)_6]^{3-}$ + 2 OH^- + O_2.   **32.** 5×10^{37} O_3 molecules.   **35.** 0.225% H_2S, by volume.
Self-Test Questions: **1.** (b)   **2.** (a)   **3.** (b)   **4.** (d)
5. (c)   **6.** (c).   **7.** (b)   **8.** (d)   **9. (a)** LiH;
(b) potassium hydrogen sulfate; **(c)** chlorite ion; **(d)** SiO_2;
(e) $KClO_4$; **(f)** $NaNO_2$; **(g)** dinitrogen pentoxide [nitrogen(V) oxide]; **(h)** barium superoxide.
10. (a) $MgCO_3$(s) $\overset{\Delta}{\rightarrow}$ MgO(s) + CO_2(g); **(b)** $MgCO_3$ + 2 HCl → $MgCl_2$ + H_2O + CO_2; **(c)** 4 P + 5 O_2 → P_4O_{10}, followed by P_4O_{10} + 6 H_2O → 4 H_3PO_4; **(d)** H_2O_2 + Cl_2 → 2 H^+ + 2 Cl^- + O_2; **(e)** 4 NH_3 + 5 O_2 → 4 NO + 6 H_2O; **(f)** 3 Cu + 8 H^+ + 2 NO_3^- → 3 Cu^{2+} + 4 H_2O + 2 NO.   **11.** 1.44×10^{10} kg O_2.

Chapter 14

Review Problems: **1. (a)** 7.0×10^{-5} mol L^{-1} s^{-1}; **(b)** 4.2×10^{-3} mol L^{-1} min^{-1}. **2. (a)** 1.3×10^{-4} mol L^{-1} s^{-1}; **(b)** 3.9×10^{-4} mol L^{-1} s^{-1}. **3. (a)** decrease; **(b)** 600 s, 29.8 mL; 1200 s, 19.6 mL; 1800 s, 12.4 ml: 3000 s, 5.0 mL. **4. (a)** 7.7×10^{-4} mol L^{-1} s^{-1}; **(b)** 3.7×10^{-4} mol L^{-1} s^{-1}. **5.** rate of reaction = $k[CH_3CHO]^2$. **6. (a)** 2nd order in A, 1st order in B; **(b)** 3rd order overall. **7. (a)** 67.0 min.; **(b)** 1.03×10^{-2} min^{-1}. **8. (a)** 0.47 M; **(b)** 1.2×10^3 s; **(c)** 2.3×10^3 s. **9. (a)** $E_a = 1.61 \times 10^2$ kJ/mol; **(b)** 5.93×10^2 K. **10. (b)** rate = $k[A]^2$. **11. (a)** ICl + HI $\rightarrow$ I$_2$ + HCl; **(b)** fast. **12. (a)** II; **(b)** I; **(c)** III. **13.** 70 s. **14.** 1.0×10^{-2} mol L^{-1} s^{-1}. **15. (a)** 1.0×10^{-2} mol L^{-1} s^{-1}; **(b)** 4.6×10^{-3} mol L^{-1} s^{-1}; **(c)** 3.7×10^{-3} mol L^{-1} s^{-1}. **16. (a)** [A] = 0; **(b)** [A] $\simeq$ 0.33 M; **(c)** [A] $\simeq$ 0.48 M.

Exercises: **1. (a)** 4.8×10^{-3} mol L^{-1} min^{-1} and 4.6×10^{-3} mol L^{-1} min^{-1}. **2. (a)** 5.00×10^{-5} mol L^{-1} s^{-1}; **(b)** [A] = 0.4985 M; **(c)** $\Delta t = 4.00 \times 10^2$ s. **3. (a)** 2.8×10^{-4} mol O$_2$ s^{-1}; **(b)** 1.7×10^{-2} mol O$_2$ min^{-1}; **(c)** 3.8×10^2 cm^3 O$_2$ min^{-1}. **7. (a)** mol L^{-1} s^{-1}; **(b)** s^{-1}; **(c)** L mol^{-1} s^{-1}. **8. (a)** [H$_2$O$_2$] = 1.53 M; **(b)** 3×10^1 s; **(c)** 0.20 L O$_2$(g). **9. (a)** [S$_2$O$_8^{2-}$] = 0.074 M; **(b)** [I$^-$] = 0.055 M. **10. (a)** 3000 mmHg; **(b)** 1400 mmHg. **13.** initial rate = 1.8×10^{-6} mol L^{-1} s^{-1}. **14. (a)** 1st order in HgCl$_2$, 2nd order in C$_2$O$_4^{2-}$, 3rd order overall; **(b)** $k = 7.6 \times 10^{-3}$ L^2 mol^{-2} min^{-1}; **(c)** initial rate = 7.4×10^{-6} mol L^{-1} min^{-1}. **15. (a)** 1st order in H$_2$, 2nd order in NO, 3rd order overall; **(b)** rate = $k \times P_{H_2} \times (P_{NO})^2$. **16. (a)** 1st order in I$^-$ and in OCl$^-$, -1 order in OH$^-$; **(b)** 1st order overall; **(c)** $k = 60$ s^{-1}. **18.** 2.5×10^3 s. **19.** $t_{1/2}$ = 28 min. **20. (a)** 5.5×10^3 s; **(b)** 8.04 L O$_2$(g). **21. (a)** 15 min; **(b)** 160 min, 193 min. **22. (b)** $k = 4.42 \times 10^{-4}$ s^{-1}; **(c)** $P_{tot.}$ = 4.0×10^2 mmHg; **(d)** $P_{tot.}$ = 936 mmHg; **(e)** $P_{tot.}$ = 536 mmHg. **23. (a)** 0.017 M; **(d)** 4.3×10^{-3} mol L^{-1} min^{-1}; **(f)** $k = 6.5 \times 10^{-2}$ min^{-1} **(g)** 11 min; **(h)** 22 min. **24. (a)** 1st order; **(b)** k = 0.035 min^{-1}; **(c)** 2.0×10^{-2} mol L^{-1} s^{-1}. **25.** rate of reaction = $k[A]^2$, k = 0.020 L mol^{-1} min^{-1}. **26. (a)** zero order; **(b)** $t_{1/2} = 7.0 \times 10^1$ s. **32. (b)** E_a = 52 kJ/mol; **(c)** $k = 1.9 \times 10^{-2}$ L mol^{-1} s^{-1}; **(d)** initial rate = 8.0×10^{-5} mol L^{-1} s^{-1}. **33.** 405 K. **34. (a)** 53 kJ/mol. **36.** 2.9×10^2 K. **39.** zero order.

Self-Test Questions: **1. (a)** **2. (b)** **3. (d)** **4. (d)** **5. (b)** **6. (c)** **7. (d)** **8. (b)** **9.** 2nd order. **10.** 3.0×10^{-3} mol L^{-1} s^{-1}. **11. (a)** $t_{1/2}$ = 10 min; **(b)** 0.160 g A.

Chapter 15

Review Problems: **1. (a)** $K_c = [COCl_2]/[CO][Cl_2]$; **(b)** $K_c = [NO_2]^2/[NO]^2[O_2]$; **(c)** $K_c = [CO_2(g)]$; **(d)** $K_c = [H_2S]^2[CH_4]/[CS_2][H_2]^4$; **(e)** $K_c = [CO_2(g)][H_2O(g)]$. **2. (a)** 4.31×10^{-2}; **(b)** 3.1×10^3; **(c)** 1.5×10^{-2}; **(d)** 7.5×10^2. **3.** $K_c = 9.1 \times 10^{-16}$. **4.** K_p = 23.2, 6.5, 2.7 $\times 10^{-2}$, and 6.8×10^{-5}, respectively. **5.** K_c = 0.50.

6. K_c = 0.77. **7.** statement (b). **8. (a)** 0.208 mol O$_2$; **(b)** 0.834 mol O$_2$. **9. (a)** $K_c = [C]^2/[A]^3[B]$; **(b)** no; **(c)** to the right. **10. (a)** $K_c = 1.1 \times 10^{-2}$; **(b)** K_p = 1.3. **11.** [H$_2$] = [I$_2$] = 0.015 M; [HI] = 0.104 M. **12.** 2.0×10^{-3} mol I$_2$. **13.** $K_p \simeq 1 \times 10^{23}$. **14. (a)** displace equilibrium to the left; **(b)** none; **(c)** displace equilibrium to the right. **15. (a)** decrease; **(b)** decrease; **(c)** increase; **(d)** none. **16.** dissociation increases with temperature for (b) and (d).

Exercises: **3. (a)** $K_c = 1.2 \times 10^{-3}$; **(b)** $K_c = 5.55 \times 10^5$; **(c)** K_c = 0.429. **4.** K_p = 0.0313; $K_c = 1.28 \times 10^{-3}$. **5.** $K_c = 1.2 \times 10^6$. **6.** $K_p = 5 \times 10^{14}$. **7.** $K_c = 3.8 \times 10^{-2}$. **8.** K_p = 1.79. **10.** $K_c = 1.49 \times 10^2$. **11.** K_c = 0.659. **12.** [NO]/[NO$_2$] = 0.369. **13.** 23 L. **14.** no; a net reaction occurs to the right. **15.** 1.5×10^{-3} mol Cl$_2$. **16.** 1.50 mol each of CO, H$_2$O, CO$_2$, and H$_2$. **17.** 0.94 mol SbCl$_5$, 2.06 mol SbCl$_3$, 0.06 mol Cl$_2$. **18. (a)** no; **(b)** reaction proceeds to the right. **19.** 0.240 mol SO$_2$; 0.081 mol O$_2$; 0.807 mol SO$_3$. **20.** 0.92 g CO$_2$. **21.** 78.0 mol% HI. **23.** [Ag$^+$] = [Fe^{2+}] = 0.44 M; [Fe^{3+}] = 0.56 M. **24.** [Cr^{3+}] = 0.03 M; [Cr^{2+}] = 0.27 M; [Fe^{2+}] = 0.11 M. **25. (a)** $P_{SO_2Cl_2}$ = 9.33 atm; **(b)** $P_{tot.}$ = 10.5 atm. **26.** $K_p = 2.1 \times 10^{-3}$. **27.** 0.673 atm. **28. (a)** 0.96 atm; **(b)** less; **(c)** P_{CO_2} = 1.33 atm; P_{H_2O} = 0.17 atm. **29.** increase. **30.** 5.0×10^1%. **31.** 19%. **33. (a)** 80.0%; **(b)** 176 atm. **35. (a)** true; **(b)** false; **(c)** false; **(d)** true. **40. (a)** $K_p = 1.2 \times 10^{-2}$; **(b)** T = 326 K. **41.** $\Delta \overline{H}°$ = -11 kJ/mol. **42.** $K_p = 2 \times 10^8$. **43. (b)** 3.8×10^2 K.

Self-Test Questions: **1. (d)** **2. (c)** **3. (a)** **4. (c)** **5. (d)** **6. (b)** **7. (b)** **8. (a)** **11.** $K_c = 2.0 \times 10^{-19}$. **12. (a)** reverse direction; **(b)** 0.30 atm.

Chapter 16

Review Problems: **1. (a)** 0; **(b)** -562 J; **(c)** $+89$ J; **(d)** -117 J. **2. (a)** greater; **(b)** less. **3. (a)** increase; **(b)** decrease; **(c)** uncertain; **(d)** decrease. **4. (a)** case 3; **(b)** case 2; **(c)** case 1; **(d)** case 4. **5.** $\Delta \overline{S}°$ = -0.285 kJ mol^{-1} K^{-1}. **6. (a)** -33.30 kJ/mol; **(b)** -242.09 kJ/mol; **(c)** $+99.76$ kJ/mol; **(d)** -60.8 kJ/mol. **7. (a)** 100°C; **(b)** ΔG = 0. **8. (a)** 85.9 J mol^{-1} K^{-1}; **(b)** 7.014 J mol^{-1} K^{-1}; **(c)** 1.09 J mol^{-1} K^{-1}. **9.** $+127$ kJ/mol. **10. (a)** $K = K_p = (P_{NO_2})^2/(P_{NO})^2(P_{O_2})$; **(b)** $K = K_p = P_{SO_2}$; **(c)** $K = K_c = [H^+][C_2H_3O_2^-]/[HC_2H_3O_2]$; **(d)** $K = K_p = (P_{CO_2})(P_{H_2O})$; **(e)** $K = (P_{Cl_2})[Mn^{2+}]/[H^+]^4[Cl^-]^2$. **11. (a)** K_p = 3.4; **(b)** $\Delta \overline{G}° = -1.0 \times 10^4$ J/mol; **(c)** to the right. **12.** $K_p = 1.32 \times 10^7$. **13.** $\Delta \overline{S}°$ = $+73.90$ J mol^{-1} K^{-1}. **14. (a)** no; **(b)** yes; **(c)** yes. *Exercises:* **5. (a)** $\Delta \overline{H} < \Delta \overline{E}$; **(b)** $\Delta \overline{H} = \Delta \overline{E}$; **(c)** $\Delta \overline{H} > \Delta \overline{E}$. **6. (a)** $\Delta \overline{E}$ = -2008 kJ/mol C$_3$H$_7$OH; **(b)** $\Delta \overline{H}$ = -2012 kJ/mol C$_3$H$_7$OH. **7.** (c). **8. (a)** decrease; **(b)** increase; **(c)** increase. **9. (a)** 1 mol H$_2$O(g, 1 atm, 50°C); **(b)** 50.0 g Fe(s, 1 atm, 20°C); **(c)** 1 mol Br$_2$(1, 1 atm, 58°C); **(d)** 0.10 mol O$_2$(g, 0.10 atm, 25°C). **13.** ΔH = 0; $\Delta S > 0$; $\Delta G < 0$. **14.** ΔH = 0; $\Delta S > 0$; $\Delta G < 0$. **15.** -497 kJ/mol. **16.** 399 K. **17.** $+157.87$ kJ/mol.

18. -5.20×10^3 kJ/mol.   **20.** $I_2(s)$.   **21.** $CO_2(g)$
22. (a) $\Delta \overline{H}^\circ_{vap} = 44.01$ kJ/mol.   **23. (b).**
24. (a) $\Delta \overline{H}^\circ_{vap} = 30.71$ kJ/mol; **(b)** 3.5×10^2 K.   **25.** 6.9×10^2 K.   **26.** $\Delta \overline{G} = -2.2$ kJ/mol.   **27. (b)** 1.4×10^2 mmHg.   **30. (a)** -23.4 kJ/mol; **(b)** $+68.9$ kJ/mol;
(c) $+5.5$ kJ/mol; **(d)** -61 kJ/mol; **(e)** $+28.8$ kJ/mol.
31. (a) no; **(b)** raising the temperature.
32. -341 kJ/mol.   **34.** 0.00207 mmHg.
36.(a) $\Delta \overline{H}^\circ = -41.13$ kJ/mol, $\Delta \overline{S}^\circ = -42.42$ J mol^{-1} K^{-1}, $\Delta \overline{G}^\circ = -28.49$ kJ/mol; **(b)** $K_p = 0.55$.   **38. (a)** high;
(b) 977 K.   **39.** about 2000 K.   **41.** 48.9%.
42. (a) 4.9×10^2 K.
Self-Test Questions: **1.** (b)   **2.** (d)   **3.** (d)   **4.** (b)
5. (a)   **6.** (c)   **7.** (b)   **8.** (d)   **11. (a)** 3.3×10^2 K;
(b) $\Delta \overline{G}^\circ_{vap} = 3$ kJ/mol.   **12. (a)** exothermic; **(b)** $\Delta \overline{G}^\circ = -186.81$ kJ/mol; **(c)** $K_p = 5 \times 10^{32}$;   **(d)** all temperatures.

Chapter 17
Review Problems: **1. (a)** acids: HOBr, H_3O^+; bases: H_2O, OBr^-; **(b)** HSO_4^-, H_3O^+; H_2O, SO_4^{2-}; **(c)** H_2O, H_2S; HS^-, OH^-; **(d)** $C_6H_5NH_3^+$, H_2O; OH^-, $C_6H_5NH_2$.   **2. (a)** base;
(b) acid; **(c)** acid; **(d)** base.   **3. (a)** $[H_3O^+] = 3.0 \times 10^{-3}$ M, $[OH^-] = 3.3 \times 10^{-12}$ M; **(b)** 2.2×10^{-13}, 4.5×10^{-2}; **(c)** 3.3×10^{-12}, 3.0×10^{-3}; **(d)** 7.2×10^{-3}, 1.4×10^{-12}.   **4. (a)** pH = 3.00; **(b)** 3.745; **(c)** 11.06;
(d) 10.61.   **5. (a)** pOH = 2.00; **(b)** 2.17; **(c)** 1.983;
(d) 10.54.   **6. (a)** $[H_3O^+] = 1 \times 10^{-6}$ M; **(b)** 7.1×10^{-4};
(c) 0.22; **(d)** 1.3×10^{-10}; **(e)** 1.4×10^{-3}.   **7.** pH = 12.07.   **8.** $[H_3O^+] = 9.1 \times 10^{-13}$ M.   **9.** $K_a = 1.51 \times 10^{-5}$.   **10. (a)** $[C_2H_3O_2^-] = 3.77 \times 10^{-3}$ M; **(b)** pH = 2.64; **(c)** $[HC_6H_4ClO] = 0.061$ M.   **11.** $[(CH_3)_3NH^+] = 9.7 \times 10^{-3}$ M.   **12. (a)** $\alpha = 0.0072$; **(b)** 0.72%.
13. (a) $[H_3O^+] = 7.9 \times 10^{-5}$ M; **(b)** $[HCO_3^-] = 7.9 \times 10^{-5}$ M; **(c)** $[CO_3^{2-}] = 5.6 \times 10^{-11}$ M.   **14. (a)** neutral;
(b) acidic; **(c)** basic; **(d)** neutral; **(e)** basic.   **15.** pH = 4.92.   **16. (a)** $K_a = 5.0 \times 10^{-6}$; **(b)** $K_b = 5.6 \times 10^{-11}$;
(c) $K_b = 6.2 \times 10^{-5}$.   **17. (a)** HI; **(b)** HOClO;
(c) Cl_3CCH_2COOH; **(d)** $H_3CCH_2CF_2COOH$.
Exercises: **1.** acidic: (a), (d), (e); basic: (b), (c).
2. (a) H_2O; **(b)** HCl; **(c)** HOCl; **(d)** HCN.   **3. (a)** IO_4^-;
(b) $C_3H_5O_2^-$; **(c)** $C_6H_5COO^-$; **(d)** $C_6H_5NH_2$.   **4.** (b), (c), (d), (f), (g).   **5. (a)** base; **(b)** either; **(c)** either; **(d)** acid.
8. $[H_3O^+] = 3.75 \times 10^{-3}$ M.   **9.** pH = 1.348.
10. 7.9 mL.   **11.** 9.9 mg $Mg(OH)_2/L$.   **13.** 36.2 mL.
14. 0.1238 M KOH.   **15.** pH = 11.60.   **16.** $K_a = 1.3 \times 10^{-5}$.   **17.** $K_b = 2.1 \times 10^{-6}$.   **18.** 2.1 g/L.   **19.** pH = 3.21.   **20.** $K_a = 5 \times 10^{-2}$.   **21.** 3.4 mg.
22. $[C_2H_2ClO_2^-] = 1.9 \times 10^{-3}$ M.   **23.** pH = 11.32.
24. 0.172 M.   **26.** 8×10^1%.   **27.** $K_a = 8 \times 10^{-5}$.
28. (a) $[H_3O^+] = [HS^-] = 9.1 \times 10^{-5}$ M, $[S^{2-}] = 1.0 \times 10^{-14}$ M; **(b)** $[H_3O^+] = [HS^-] = 2.3 \times 10^{-5}$ M, $[S^{2-}] = 1.0 \times 10^{-14}$ M; **(c)** $[H_3O^+] = [HS^-] = 1.0 \times 10^{-6}$ M, $[S^{2-}] = 1.0 \times 10^{-14}$ M.   **30.** $[H_3O^+] = [H_2PO_4^{2-}] = 0.022$ M; $[HPO_4^{2-}] = 6.2 \times 10^{-8}$ M; $[PO_4^{3-}] = 1.4 \times 10^{-18}$ M.   **31. (b)** pH = 9.62.   **32. (a)** $[H_3O^+] = 0.763$ M; **(b)** $[SO_4^{2-}] = 0.0129$ M.   **33. (a)** $[H_3O^+] =$

0.0147 M; $[HSO_4^-] = 0.0053$ M; $[SO_4^{2-}] = 4.68 \times 10^{-3}$ M.
34. pH = 8.0.   **35.** 1.6 M KNO_2.   **36.** pH = 3.31.
39. (a) forward; **(b)** reverse; **(c)** reverse; **(d)** forward;
(e) reverse; **(f)** forward.   **40.** (c) < (f) < (d) < (e) < (b) < (a).   **41.** propylamine.   **44. (a)** $K_a \approx 1 \times 10^{-2}$.
Self-Test Questions: **1.** (c)   **2.** (a)   **3.** (b)   **4.** (d)
5. (d)   **6.** (a)   **7.** (a)   **8.** (b)   **9.** $H_2SO_4 > HNO_3 > HC_2H_3O_2 > NH_4ClO_4 > NaCl > NaNO_2 > NH_3 > NaOH > Ba(OH)_2$.   **10.** 3.0 g $HC_7H_5O_2$.

Chapter 18
Review Problems: **1. (a)** $[H_3O^+] = 0.106$ M; **(b)** $[OH^-] = 9.4 \times 10^{-14}$ M; **(c)** $[C_3H_5O_2^-] = 4.49 \times 10^{-5}$ M; **(d)** $[I^-] = 0.106$ M.   **2. (a)** $[OH^-] = 2.84 \times 10^{-5}$ M; **(b)** $[NH_4^+] = 0.0875$ M; **(c)** $[Cl^-] = 0.0875$ M; **(d)** $[H_3O^+] = 3.5 \times 10^{-10}$ M.   **3. (a)** $HCHO_2 + OH^- \rightarrow H_2O + CHO_2^-$ and $CHO_2^- + H_3O^+ \rightarrow HCHO_2 + H_2O$;
(b) $C_6H_5NH_2 + H_3O^+ \rightarrow C_6H_5NH_3^+ + H_2O$ and $C_6H_5NH_3^+ + OH^- \rightarrow C_6H_5NH_2 + H_2O$;
(c) $H_2PO_4^- + OH^- \rightarrow HPO_4^{2-} + H_2O$ and $HPO_4^{2-} + H_3O^+ \rightarrow H_2PO_4^- + H_2O$.   **4. (a)** pH = 4.59; **(b)** pH = 8.94;   **5.** $[CHO_2^-] = 1.12$ M.   **6.** $[NH_3] = 0.66$ M.
7. (a) red; **(b)** yellow; **(c)** yellow; **(d)** yellow; **(e)** red.
8. (a) pH = 12.26; **(b)** pH = 1.426.   **9. (a)** 3.47;
(b) pH = 12.08.   **10. (a)** 100.5; **(b)** 29.2; **(c)** 74.1.
11. (a) 0.24 N KOH; **(b)** 0.15 N and 0.30 N $H_2C_2O_4$; **(c)** 4×10^{-3} N $Ca(OH)_2$.   **12. (a)** 33.74 mL; **(b)** 23.42 mL.
Exercises: **2. (a)** 0.100 M; **(b)** 0.100 M; **(c)** 0.200 M;
(d) 5.22×10^{-5} M; **(e)** 2.18×10^{-5} M.   **3.** $K_a = 1.4 \times 10^{-3}$.   **4.** pOH = 5.277.   **5. (a)** 28 mL; **(b)** 3.5 mL;
(c) 250.0 mL.   **6.** (c), (e), (f).   **7. (c)** pH = 7.69.
8. 1.8 g $(NH_4)_2SO_4$.   **9.** 1.00 L 0.100 M $HCHO_2$ + 0.58 L 0.100 M $NaCHO_2$.   **10.** 59.5 mL 1.00 M HCl in 0.100 M $NaCHO_2$.   **13. (a)** pH = 5.08; **(b)** pH = 1.04.
14. (a) pH = 9.42; **(b)** pH = 9.78; **(c)** 2.9 mL.
15. (a) pH = 4.77; **(b)** pH = 4.95; **(c)** 5.1 g $Ba(OH)_2$;
(d) pH = 11.6.   **18. (a)** colorless; **(b)** red; **(c)** blue;
(d) yellow; **(e)** blue; **(f)** green.   **19. (a)** 39% in anion form; **(b)** acid form (red).   **20.** $[OH^-] = 0.05340$ M.
21. 0.815 g/L.   **22.** pH = 2.96.   **24. (a)** rxn 1: $K = 1 \times 10^{14}$, rxn 2: $K = 1.7 \times 10^9$.   **29. (a)** pH = 2.28; **(b)** pH = 5.71.   **30. (a)** no; **(b)** 18.16%.   **31. (a)** 24.50 mL;
(b) 19 mL; **(c)** 16.6 mL.   **33. (a)** pH = 12.11; **(b)** pH = 11.11.   **34.** $NaHSO_4 > NaHC_2O_4 > NaH_2PO_4 > NaC_2H_3O_2$.   **35.** 50.1 g.   **36.** 0.46 N.
37. (a) 0.151 N; **(b)** 0.302 N; **(c)** 0.453 N.   **38.** 32.1% H_2SO_4.
Self-Test Questions: **1.** (c)   **2.** (d)   **3.** (c)   **4.** (b)
5. (a)   **6.** (b)   **7.** (b)   **8.** (d)   **10. (a)** above 7;
(b) below 7; **(c)** pH = 7.   **11. (a)** 31 g; **(b)** pH = 3.89.
12. (a) pH = 3.10; **(b)** pH = 4.20; **(c)** pH = 8.00; **(d)** pH = 11.08.

Chapter 19
Review Problems: **1. (a)** $K_{sp} = [Ag^+]^2[SO_4^{2-}]$; **(b)** $K_{sp} = [Ra^{2+}][IO_3^-]^2$; **(c)** $K_{sp} = [Ni^{2+}]^3[PO_4^{3-}]^2$; **(d)** $K_{sp} = [Hg_2^{2+}][C_2O_4^{2-}]$; **(e)** $K_{sp} = [PuO_2^{2+}][CO_3^{2-}]$.

2. (a) $Fe(OH)_3(s) \rightleftharpoons Fe^{3+}(aq) + 3\ OH^-(aq)$;
(b) $BiOOH(s) \rightleftharpoons BiO^+(aq) + OH^-(aq)$;
(c) $Hg_2I_2(s) \rightleftharpoons Hg_2^{2+}(aq) + 2\ I^-(aq)$;
(d) $Pb_3(AsO_4)_2(s) \rightleftharpoons 3\ Pb^{2+}(aq) + 2\ AsO_4^{3-}(aq)$.
(e) $Cu_2[Fe(CN)_6](s) \rightleftharpoons 2\ Cu^{2+}(aq) + [Fe(CN)_6]^{4-}(aq)$;
(f) $MgNH_4PO_4(s) \rightleftharpoons Mg^{2+}(aq) + NH_4^+(aq) + PO_4^{3-}(aq)$.
3. (a) 1.1×10^{-5} mol/L; **(b)** 2.2×10^{-2} mol/L; **(c)** 7×10^{-5} mol/L; **(d)** 4.5×10^{-5} mol/L. **4. (a)** 1.4×10^{-5};
(b) 8.8×10^{-8}; **(c)** 1.9×10^{-9}. **5. (a)** 1.7×10^{-4} M;
(b) 1.7×10^{-5} M; **(c)** 3.8×10^{-10} M. **6. (a)** yes; **(b)** no;
(c) no. **7. (a)** AgI; **(b)** $[I^-] = 2.7 \times 10^{-4}$ M; **(c)** $[Ag^+] =$
3.1×10^{-13} M; **(d)** yes. **8. (a)** no reaction; **(b)** $Cu^{2+}(aq) +$
$CO_3^{2-}(aq) \rightarrow CuCO_3(s)$; **(c)** $Ag^+(aq) + Cl^-(aq) \rightarrow AgCl(s)$;
(d) $Ba^{2+}(aq) + S^{2-}(aq) + Cu^{2+}(aq) +$
$SO_4^{2-}(aq) \rightarrow BaSO_4(s) + CuS(s)$; **(e)** $Al(OH)_3(s) +$
$3\ H^+(aq) \rightarrow Al^{3+}(aq) + 3\ H_2O(l)$; **(f)** $CaC_2O_4(s) +$
$2\ H^+(aq) \rightarrow Ca^{2+}(aq) + H_2C_2O_4(aq)$; **(g)** no reaction.
9. only Ca^{2+}. **10. (b)** 1.7 g $(NH_4)_2SO_4$; **11. (a)** no
precipitate; **(b)** precipitate forms; **(c)** precipitate forms.
12. $[CN^-] = 2.6 \times 10^{-5}$ M. **13. (a)** all precipitate;
(b) Cd^{2+} and Cu^{2+} but not Fe^{2+} **14.** no. **15.** pH =
3.29.
Exercises: **2.** (b). **3.** $K_{sp} = 2 \times 10^{-7}$. **4.** 7.2 ppm
F^-. **5.** 0.68 or 68%. **6.** 4.5×10^{-3} M $AgNO_3$.
7. $K_{sp} = 4.1 \times 10^{-9}$. **8.** $K_{sp} = 5.30 \times 10^{-5}$. **11.** $K_{sp} =$
1.7×10^{-5}. **14. (a)** 1.7×10^{-4} M; **(b)** 2.8×10^{-5} M;
(c) 0.058 M; **(d)** 1.4×10^{-9} M. **15.** 1.8 ppm F^-.
16. 1.9×10^{-3} mol MgF_2/L. **17.** yes. **18.** no.
19. pH = 1.8. **20. (a)** yes; **(b)** yes; **(c)** no. **21.** yes.
22. (a) yes; **(b)** $[Ag^+] = 4.0 \times 10^{-6}$ M. **23. (a)** $[OH^-] =$
4.2×10^{-3} M; **(b)** $[OH^-] = 1.3 \times 10^{-4}$ M. **24.** 98%.
25. $[Pb^{2+}] = 0.016$ M. **26. (a)** no; **(b)** yes.
27. (a) AgBr; **(b)** no. **28.** (b). **30. (a)** no; **(b)** no;
(c) yes. **31.** pH = 9.53. **32.** 0.18 M. **33.** pH =
3.80. **37.** no. **38.** 1.4×10^{-6} g KI.
39. (a) 0.66 g AgBr/L; **(b)** 9.4 g AgBr/L; **(c)** 39 g AgBr/L.
40. (b) increase $[C_2H_3O_2^-]$ to 1.7 M. **42.** 0.23 g FeS/L.
43. 1×10^{-3} mol CoS/L.
Self-Test Questions: **1.** (d) **2.** (a) **3.** (c) **4.** (b)
5. (c) **6.** (a) **7.** (a) **8.** (c) **9.** $3.5 \times$
10^{-4} mg Pb^{2+}/mL. **10.** more soluble in acidic solution;
(b), (c), (f); in basic solution: (a), independent of pH: (d),
(e). **11. (a)** $PbCrO_4$; **(b)** $[Pb^{2+}] = 1.6 \times 10^{-6}$ M;
(c) yes. **12.** $CuCO_3$, but not CuS.

Chapter 20
Review Problems: **1. (a)** *red.* $S_2O_8^{2-} + 2\ e^- \rightarrow 2\ SO_4^{2-}$;
(b) *red.* $2\ HNO_3 + 8\ H^+ + 8\ e^- \rightarrow N_2O + 5\ H_2O$; **(c)** *oxid.*
$CH_4 + 2\ H_2O \rightarrow CO_2 + 8\ H^+ + 8\ e^-$; **(d)** *oxid.* $Br^- +$
$6\ OH^- \rightarrow BrO_3^- + 3\ H_2O + 6\ e^-$; **(e)** *red.* $NO_3^- +$
$6\ H_2O + 8\ e^- \rightarrow NH_3 + 9\ OH^-$. **2.** coefficients are
(a) 3, 8, 2 → 3, 2, 4; **(b)** 4, 10, 1 → 4, 1, 3; **(c)** 5, 2,
6 → 2, 8, 5; **(d)** $S_2O_3^{2-} + 4\ Cl_2(g) + 5\ H_2O \rightarrow 2\ HSO_4^- +$
$8\ Cl^- + 8\ H^+$; **(e)** $3\ P + 2\ H_2O + 2\ H^+ +$
$5\ NO_3^- \rightarrow 3\ H_2PO_4^- + 5\ NO$. **3. (a)** $3\ CN^- +$
$2\ MnO_4^- + H_2O \rightarrow 2\ MnO_2 + 3\ CNO^- + 2\ OH^-$; **(b)** 4

$[Fe(CN)_6]^{3-} + N_2H_4 + 4\ OH^- \rightarrow 4\ [Fe(CN)_6]^{4-} + N_2 +$
$4\ H_2O$; **(c)** $4\ Fe(OH)_2 + O_2 + 2\ H_2O \rightarrow 4\ Fe(OH)_3$;
(d) $3\ C_2H_5OH + 4\ MnO_4^- \rightarrow 3\ C_2H_3O_2^- + 4\ MnO_2 +$
$4\ H_2O + OH^-$. **4.** coefficients are **(a)** 1, 1 → 1, 1, 1;
(b) 3, 6 → 5, 1, 3; **(c)** 2, 1 → 1, 2; **(d)** 3, 2 → 1, 2, 4.
5. (a) $Zn(s) + Sn^{2+}(aq) \rightarrow Zn^{2+}(aq) + Sn(s)$, $E°_{cell} =$
$+0.627$ V; **(b)** $2\ Fe^{2+}(aq) + Sn^{4+}(aq) \rightarrow 2\ Fe^{3+}(aq) +$
$Sn^{2+}(aq)$, $E°_{cell} = -0.617$ V; **(c)** $Cu(s) +$
$Cl_2(g) \rightarrow Cu^{2+}(aq) + 2\ Cl^-(aq)$, $E°_{cell} = +1.023$ V.
6. (a) $E°_{cell} = +0.095$ V; **(b)** $E°_{red} = -2.03$ V; **(c)** $E°_{red} =$
$+0.153$ V. **7. (a)** no; **(b)** yes; **(c)** no; **(d)** no; **(e)** yes.
8. (a) yes; **(b)** yes; **(c)** no; **(d)** no; **(e)** yes. **9. (a)** $K =$
0.32; **(b)** $K = 4 \times 10^{-5}$; **(c)** $K = 1 \times 10^{33}$.

10. (a) $E_{cell} = 0.463 - \dfrac{0.0592}{2} \log \dfrac{[Cu^{2+}]}{[Ag^+]^2}$; **(b)** $E_{cell} =$
$2.00 - \dfrac{0.0592}{6} \log \dfrac{[Al^{3+}]^2}{[Cu^{2+}]^3}$; **(c)** $E_{cell} = 0.74 -$
$\dfrac{0.0592}{5} \log \dfrac{[Fe^{3+}]^5[Mn^{2+}]}{[Fe^{2+}]^5[MnO_4^-][H^+]^8}$. **11. (a)** 0.303 V;
(b) 0.162 V; **(c)** 0.154 V. **12. (a)** $Cu(s), Cl_2(g)$;
(b) $H_2(g), Cl_2(g)$; **(c)** $H_2(g), O_2(g)$; **(d)** $Ba(l), Cl_2(g)$;
(e) $H_2(g), OH^-(aq), I_2$; **(f)** $H_2(g), O_2(g)$.
13. (a) 4.28 g Zn; **(b)** 1.18 g Al; **(c)** 14.1 g Ag; **(d)** 3.85 g
Ni. **14. (a)** 63.0; **(b)** 8.99; **(c)** 119.6; **(d)** 8.00.
15. 47.63% Fe.
Exercises: **3.** $Na^+ < Zn^{2+} < I_2 < IO_3^- < PbO_2 < F_2$.
4. oxidizing agents only: (d), (g); reducing agents only: (a),
(b); either: (c), (e), (f). **5. (a)** $0.337 < E° < 0.800$ V;
(b) $-0.440 < E° < 0.000$ V. **7.** $E° = -0.255$ V.
12. $2\ H_2O_2 \rightarrow 2\ H_2O + O_2$.
13. (a) $Cu(s)|Cu^{2+}(aq)||Fe^{3+}(aq), Fe^{2+}(aq)|Pt(s)$;
(b) $Pt|Sn^{2+}(aq), Sn^{4+}(aq)||Cr_2O_7^{2-}(aq), Cr^{3+}(aq)|Pt$; **(c)** Pt,
$O_2(g)|H_2O, H^+(aq)||Cl^-(aq)|Cl_2(g)$, Pt. **15. (a)** $\Delta\overline{G}° =$
-5.2×10^2 kJ/mol; **(b)** $\Delta\overline{G}° = -5 \times 10^1$ kJ/mol; **(c)** $\Delta\overline{G}° =$
$+25$ kJ/mol. **16.** $K = 1 \times 10^{91}$. **17.** $[Fe^{2+}] =$
0.42 M. **18. (a)** $+0.286$ V; **(b)** -0.573 V;
(c) $+1.620$ V. **22. (a)** $[Cu^{2+}] = 6 \times 10^{-38}$ M; **(b)** yes.
23. (a) $+0.824$ V. **24. (a)** lower; **(b)** $+0.686$ V.
25. $+0.037$ V. **26. (a)** $+0.027$ V; **(b)** decrease;
(c) $+0.022$ V; **(d)** $[Sn^{2+}] = 0.218$ M; **(e)** $[Sn^{2+}] = 0.48$ M,
$[Pb^{2+}] = 0.22$ M. **27.** $[Cl^-] = 2.4 \times 10^{-4}$ M.
36. (a) 0.269 g Zn; **(b)** 388 s; **(c)** $[Cu^{2+}] = 0.207$ M;
(d) 1.3×10^2 s. **37. (a)** 0.538 L; **(b)** 0.556 L.
38. (a) 1.75×10^3 C; **(b)** 2.22 A. **39.** yes.
40. (a) 5.594×10^{-3} M; **(b)** 3.356×10^{-2} N.
41. $[Cr_2O_7^{2-}] = 1.12 \times 10^{-3}$ M. **42.** 32.9 mL.
Self-Test Questions: **1.** (d) **2.** (c) **3.** (b) **4.** (b)
5. (b) **6.** (c) **7.** (a) **8.** (a) **9.** $3\ PbO(s) +$
$2\ MnO_4^- + H_2O \rightarrow 3\ PbO_2(s) + 2\ MnO_2(s) + 2\ OH^-$.
10. $Zn(s)|Zn^{2+}(aq)||H^+(aq), NO_3^-(aq)|NO(g)$, Pt(s).
11. pH = 2.82. **12.** $[Cu^+] = 1.08$ M; $[Sn^{4+}] = 1.04$ M;
$[Cu^{2+}] = 0.92$ M; $[Sn^{2+}] = 0.96$ M.

Chapter 21
Review Problems: **1. (a)** potassium bromate; **(b)** chlorine
trifluoride; **(c)** NaOI; **(d)** Si_3H_8; **(e)** potassium cyanate:

(f) $Na_2S_2O_6$; (g) AgN_3; (h) sodium dihydrogen phosphate.
2. (a) $CaCl_2(s) + H_2SO_4(conc. aq) \rightarrow CaSO_4(s) + 2\ HCl(g)$;
(b) no reaction; (c) no reaction; (d) $NH_3(aq) +$
$HClO_4(aq) \rightarrow NH_4ClO_4(aq)$; (e) $2\ NO(g) +$
$O_2(g) \rightarrow 2\ NO_2(g)$; (f) $2\ Hg(NO_3)_2(s) \rightarrow 2\ HgO(s) +$
$4\ NO_2(g) + O_2(g)$,   followed by $2\ HgO(s) \rightarrow 2Hg(l) +$
$O_2(g)$.   4. (a) $Cl_2(g) + 2\ OH^-(aq) \rightarrow OCl^-(aq) +$
$Cl^-(aq) + H_2O$; (b) $2\ NaI(s) + 3\ H_2SO_4(conc.$
$aq) \xrightarrow{\Delta} 2\ NaHSO_4(s) + 2\ H_2O + SO_2(g) + I_2(g)$; (c) $Cl_2(g) +$
$2\ Br^-(aq) \rightarrow 2\ Cl^-(aq) + Br_2(aq)$; (d) $3\ CdS(s) + 8\ H^+ +$
$2\ NO_3^- \rightarrow 3\ Cd^{2+} + 4\ H_2O + 3\ S(s) + 2\ NO(g)$;
(e) $Fe(s) + N_2O(g) \xrightarrow{\Delta} FeO(s) + N_2(g)$ or $2\ Fe(s) +$
$3\ N_2O(g) \xrightarrow{\Delta} Fe_2O_3(s) + 3\ N_2(g)$; (f) $3\ Pb(s) + 8\ H^+ +$
$2\ NO_3^- \rightarrow 3\ Pb^{2+} + 4\ H_2O + 2\ NO(g)$.   5. $H_5IO_6 +$
$H^+ + 2\ e^- \rightarrow IO_3^- + 3\ H_2O$; $4\ H_3PO_2 + 4\ H^+ +$
$4\ e^- \rightarrow P_4 + 8\ H_2O$; $Sb_2O_5 + 6\ H^+ + 4\ e^- \rightarrow 2\ SbO^+ +$
$3\ H_2O$; $OCl^- + H_2O + 2\ e^- \rightarrow Cl^- + 2\ OH^-$; $B_2H_6 +$
$2\ H_2O + 4\ e^- \rightarrow 2\ BH_4^- + 2\ OH^-$; $HXeO_4^- + 3\ H_2O +$
$6\ e^- \rightarrow Xe + 7\ OH^-$.   6. (a) yes; (b) no; (c) yes;
(d) no.   7. (a) $E° = +0.40$ V; (b) $E° = -0.57$ V.
8. strong enough: (a), (c), (d).   9. (a) I_2; (b) reduced to
H_2O; (c) NO.   10. (a) -49 kJ/mol ClF; (b) -17 kJ/mol
OF_2; (c) $+83$ kJ/mol OCl_2; (d) -129 kJ/mol NF_3.
11. (a) $Ca_3(PO_4)_2$; (b) $K_4P_2O_7$; (c) sodium metaantimonite;
(d) sodium metabismuthate; (e) Na_3BiO_4.   12. (a) trigonal
pyramidal; (b) tetrahedral; (c) square-based pyramidal.
Exercises: 3. yes.   4. (a) $E° = +1.49$ V; (b) no.
5. (a) $HCl(g) \rightleftharpoons \frac{1}{2} H_2(g) + \frac{1}{2} Cl_2(g)$, $K_p = 2 \times 10^{-17}$; (b) $4 \times$
$10^{-15}\%$ dissociated.   7. (a) T-shaped; (b) square-based
pyramidal; (c) linear; (d) square planar.
11. (a) 303 kJ/mol.   13. (a) zinc sulfide; (b) potassium
hydrogen sulfite; (c) tetrasulfide ion; (d) potassium
tetrathionate; (e) thionyl fluoride.   16. H_2SO_5.
18. 0.426 g Na_2SO_3.   19. 7.002% Cu.   21. (a) pH =
9.23; (b) pH = 3.85.   22. (a) $NH_3OH^+ + 2\ H^+ +$
$2\ e^- \rightarrow NH_4^+ + H_2O$; (b) $E° = +1.35$ V; (c) yes.
23. -622.33 kJ/mol.   26. $E°_{red} = -0.56$ V.
28. (a) (mono)hydrogen phosphate ion; (b) calcium
pyrophosphate; (c) tetrapolyphosphoric acid; (d) sodium
tetrametaphosphate.   37. (a) irregular tetrahedral;
(b) trigonal bipyramidal; (c) octahedral; (d) square-based
pyramidal.   38. pentagonal-based pyramidal.
Self-Test Questions: 1. (b)   2. (d)   3. (a)   4. (d)
5. (c)   6. (a)   7. (b)   8. (d)
9. (a) $2\ Pb(NO_3)_2 \rightarrow 2\ PbO + 4\ NO_2 + O_2$; (b) $Cl_2 +$
$2\ OH^- \rightarrow ClO^- + Cl^- + H_2O$; (c) $H_3PO_4 +$
$2\ KOH \rightarrow K_2HPO_4 + 2\ H_2O$; (d) $2\ KBr +$
$3\ H_2SO_4 \xrightarrow{\Delta} 2\ KHSO_4 + 2\ H_2O + SO_2 + Br_2$;
(e) $2\ H_3PO_4 \rightarrow H_4P_2O_7 + H_2O$.   11. 1×10^6 kg F_2.
12. $E°_{red} = +0.38$ V.

Chapter 22

Review Problems: 1. (a) lead acetate; (b) $Mg(HCO_3)_2$;
(c) mercury(I) bromide; (d) calcium chloride hexahydrate;

(e) $NH_4Al(SO_4)_2 \cdot 12H_2O$; (f) $[Zn(OH)_4]^{2-}$; (g) sodium
aluminate; (h) potassium stannate.
2. (a) $MgCO_3(s) \xrightarrow{\Delta} MgO(s) + CO_2(g)$; (b) $CaO(s) +$
$2\ HCl(aq) \rightarrow CaCl_2(aq) + H_2O$; (c) $2\ Al(s) + 2\ KOH(aq) +$
$6\ H_2O \rightarrow 2\ KAl(OH)_4(aq) + 3\ H_2(g)$; (d) $3\ Pb(s) +$
$8\ HNO_3(aq) \rightarrow 3\ Pb(NO_3)_2(aq) + 4\ H_2O + 2\ NO(g)$; (e) no
reaction; (f) $ZnO(s) + CO(g) \xrightarrow{\Delta} Zn(g) + CO_2(g)$.
4. (a) $Ba^{2+}(aq) + CO_3^{2-}(aq) \rightarrow BaCO_3(s)$;
(b) $Mg(HCO_3)_2 \xrightarrow{\Delta} MgCO_3(s) + H_2O + CO_2(g)$;
(c) $SnO(s) + C(s) \xrightarrow{\Delta} Sn(l) + CO(g)$; (d) $Sn(s) +$
$4\ HNO_3 \rightarrow SnO_2(s) + 2\ H_2O + 4\ NO_2(g)$; (e) $CdO(s) +$
$H_2SO_4(aq) \rightarrow CdSO_4(aq) + H_2O$; (f) $PbO_2(s) + 4\ H^+ +$
$4\ I^- \rightarrow PbI_2(s) + 2\ H_2O + I_2$.   5. (a) to the right; (b) to
the left; (c) to the right.   7. $[Na^+] = 9.2 \times 10^{-3}$ M.
8. pH = 2.42.   9. (a) $CdCO_3(s) \xrightarrow{\Delta} CdO(s) + CO_2(g)$;
(b) no reaction; (c) $SnO_2(s) + CO(g) \xrightarrow{\Delta} SnO(s) + CO_2(g)$,
followed by $SnO(s) + CO(g) \xrightarrow{\Delta} Sn(l) + CO_2(g)$;
(d) $2\ Cd^{2+} + 2\ H_2O \xrightarrow{electrolysis} 2\ Cd(s) + O_2(g) + 4\ H^+(aq)$;
(e) $2\ HgO(s) \rightarrow 2\ Hg(l) + O_2(g)$; (f) no reaction.
10. (a) yes; (b) no; (c) yes.
Exercises: 2. $[CO_3^{2-}] = 0.002$ M and $[OH^-] \simeq 2$ M.
3. (a) no; (b) $[SO_4^{2-}] = 0.68$ M and $[OH^-] = 0.64$ M.
4. (a) 71.5%.   5. pH = 11.54.   6. 6×10^6 t Mg.
7. $Mg(OH)_2(s)$ will precipitate.
8. (a) solubility in 1.00 M $NH_4Cl >$ in 0.100 M NH_3–1.00 M
$NH_4Cl >$ in 1.00 M NH_3–1.00 M NH_4Cl; (b) $[Mg^{2+}] =$
7.1×10^{-3} M in 1.00 M NH_4Cl; $[Mg^{2+}] = 1.9 \times 10^{-3}$ M in
0.100 M NH_3–1.00 M NH_4Cl; $[Mg^{2+}] = 6.0 \times 10^{-4}$ M in
1.00 M NH_3–1.00 M NH_4Cl.   10. 313 kg CaO.
11. (a) 1.12×10^3 kg $CaCO_3$.   12. (b) 55.6 g Na_2CO_3.
21. (a) essentially constant; (b) increase.   30. $E°_{red} =$
$+0.90$ V.
Self-Test Questions: 1. (b)   2. (c)   3. (a)   4. (c)
5. (a)   6. (a)   7. (b)   8. (d)   9. (a) $Sn(s) +$
$4\ HNO_3(conc. aq) \rightarrow SnO_2(s) + 2\ H_2O + 4\ NO_2(g)$;
(b) $MgCO_3(s) + H_2O + CO_2 \rightarrow Mg^{2+} + 2\ HCO_3^-$; (c) $Zn +$
$2\ OH^- + 2\ H_2O \rightarrow [Zn(OH)_4]^{2-} + H_2(g)$; (d) $CdS +$
$O_2 \rightarrow CdO + SO_2$,   followed by $CdO + C \rightarrow Cd + CO$.
12. 34.6 ppm Ca^{2+}

Chapter 23

Review Problems: 1. (a) $BaCr_2O_7$; (b) scandium
hydroxide; (c) CrO_3; (d) manganese(II) oxide; (e) $FeSiO_3$;
(f) iron (penta)carbonyl.   2. (a) Fe + Mn; (b) Fe with
more than 1% C; (c) $Fe(CrO_2)_2$; (d) $KCr(SO_4)_2 \cdot 12\ H_2O$;
(e) one part HNO_3 and three parts HCl; (f) $Fe_4[Fe(CN)_6]_3$.
3. (a) $TiCl_4 + 4\ Na \rightarrow 4\ NaCl + Ti$; (b) $Cr_2O_3 +$
$2\ Al \rightarrow Al_2O_3 + 2\ Cr$; (c) no reaction; (d) $K_2Cr_2O_7 +$
$2\ KOH \rightarrow 2\ K_2CrO_4 + H_2O$; (e) $MnO_2 + 2\ C \rightarrow Mn +$
$2\ CO$; (f) no reaction.   4. coefficients are (a) 2, 6, 3 → 4,
6; (b) 2, 5, 8 → 2, 10, 16; (c) 4,8,1,2 → 4,4.
6. (a) $3\ Cu + 8\ H^+ + 2\ NO_3^- \rightarrow 3\ Cu^{2+} + 4\ H_2O + 2\ NO$;
(b) $Cr_2O_3 + 2\ OH^-(aq) + 3\ H_2O \rightarrow 2\ [Cr(OH)_4]^-(aq)$;

(c) $FeS + 2 H_3O^+(aq) \rightarrow Fe^{2+}(aq) + 2 H_2O + H_2S(g)$;
(d) $Ni(OH)_2 + 6 NH_3(aq) \rightarrow [Ni(NH_3)_6]^{2+}(aq) + 2 OH^-(aq)$;
(e) $2 Ln + 6 H_3O^+(aq) \rightarrow 2 Ln^{3+}(aq) + 6 H_2O + 3 H_2(g)$.
8. (a) $[Cr_2O_7^{2-}] = 3.2 \times 10^4 \times [CrO_4^{2-}]^2$; (b) $[Cr_2O_7^{2-}] = 1.8 \times 10^{-4} \times [CrO_4^{2-}]^2$.   **9.** (a) $[Cu^+] = [Cu^{2+}] = 5.6 \times 10^{-7} M$; (b) about 7% Cu^+.   **10.** (a) water; (b) $NH_3(aq)$;
(c) $NaOH(aq)$; (d) $HCl(aq)$.
Exercises: **5.** (a) $VO^{2+} + 2 H^+ + e^- \rightarrow V^{3+} + H_2O$;
(b) $Cr^{2+} \rightarrow Cr^{3+} + e^-$; (c) $Fe(OH)_3 + 5 OH^- \rightarrow FeO_4^{2-} + 4 H_2O + 3 e^-$; (d) $[Ag(CN)_2]^- + e^- \rightarrow Ag(s) + 2 CN^-$.
6. (a) no; (b) yes; (c) yes.   **10.** $E°_{red} = -0.73$ V.
13. $K = 33$.   **14.** $[CrO_4^{2-}] = 0.070 M$; $[Cr_2O_7^{2-}] = 0.015 M$.   **16.** 43 s.   **24.** dissolution reaction: $AgO(s) + 2 H_3O^+ \rightarrow Ag^{2+}(aq) + 3 H_2O$; oxidation–reduction reaction: $4 Ag^{2+} + 2 H_2O \rightarrow 4 Ag^+ + 4 H^+ + O_2(g)$.   **25.** for reaction (23.47), $K = 1 \times 10^{-20}$; for reaction (23.48), $K = 1 \times 10^{13}$.   **33.** (a) $NiC_8H_{14}N_4O_4$; (b) 3.50% Ni.
34. 82.3% MnO_2.
Self-Test Questions: **1.** (c)   **2.** (d)   **3.** (a)   **4.** (b)
5. (d)   **6.** (c)   **7.** (a)   **8.** (b)   **10.** limited $OH^-(aq)$: $Mg(OH)_2(s)$ and $Cr(OH)_3(s)$; excess $OH^-(aq)$: $Mg(OH)_2(s)$ and $[Cr(OH)_4]^-(aq)$.   **11.** (a) H_2O, HCl; (b) HCl; (c) H_2O;
(d) HCl; (e) NaOH or HCl.   **12.** ion present: Ni^{2+}; ions absent: Cr^{3+}, Fe^{3+}; further tests needed: Zn^{2+}.

Chapter 24

Review Problems: **1.** (a) coordination number: 6, oxidation state: +2; (b) 6, +3; (c) 4, +2; (d) 6, +3; (e) 6, +3; (f) 6, +3;   **2.** (a) diamminesilver(I);
(b) pentaaquahydroxoiron(III); (c) tetrachlorozincate(II);
(d) bis(ethylenediamine)platinum(II);
(e) tetraamminechloronitrocobalt(III).
3. (a) pentaamminebromocobalt(III) sulfate;
(b) pentaamminesulfatocobalt(III) bromide;
(c) hexaamminechromium(III) hexacyanocobaltate(III);
(d) sodium hexanitrocobaltate(III);
(e) tris(ethylenediamine)cobalt(III) chloride.
4. (a) $[Ag(CN)_2]^-$; (b) $[Ni(NH_3)_2Cl_4]^{2-}$; (c) $[Cu(en)_2]SO_4$;
(d) $Na[Al(H_2O)_2(OH)_4]$.   **5.** (a) $FeCl_3 \cdot 6 H_2O$;
(b) $Co[PtCl_6] \cdot 6 H_2O$.   **8.** (a) one; (b) two; (c) two;
(d) two.   **9.** (a) coordination isomerism; (b) linkage isomerism; (c) none; (d) none; (e) *cis-trans (fac-mer)*.

	$5d$				dsp^2			
10. (a) [Xe]$4f^{14}$	↑↓	↑↓	↑↓	↑↓	··	··	··	··

·· represents electron pairs from Br^-; (b) diamagnetic.
Exercises: **4.** (a) tetraammineaquahydroxocobalt(III) ion;
(b) triamminetrinitrocobalt(III); (c) tetraammineplatinum(II) hexachloroplatinate(IV); (d) diaquabis(oxalato)ferrate(III) ion;
(e) pentacyanopyridineferrate(II) ion; (f) silver tetraiodomercurate(II).   **5.** (a) $K_4[Fe(CN)_6]$;
(b) $[Cu(en)_2]^{2+}$; (c) $[Al(H_2O)_4(OH)_2]Cl$;
(d) $[Cr(en)_2(NH_3)Cl]SO_4$; (e) $[Fe(en)_3]_3[Fe(CN)_6]_2$.   **13.** (b),
(c).   **17.** (a) paramagnetic; (b) diamagnetic.   **21.** (a) 3;
(b) paramagnetic; (c) $[FeCl_4]^-$.   **24.** $K_f = 5.0 \times 10^9$.
28. (a) pH = 2.0; (b) $[Fe(H_2O)_5OH]^{2+} = 9 \times 10^{-4} M$;

(c) no.   **29.** (a) $K = 8.5$.   **31.** (a) 0.876 g Cu;
(b) 1.75 g Cu.
Self-Test Questions: **1.** (c)   **2.** (d)   **3.** (b)   **4.** (a)
5. (a)   **6.** (c)   **7.** (d)   **8.** (b)   **9.** (a) *cis-* or *trans-*tetraamminedihydroxochromium(III) bromide;
(b) potassium hexanitrocobaltate(III); (c) *cis-* or *trans-*diaquabis(ethylenediamine)iron(II) ion; (d) *cis-* or *trans-*dichlorobis(ethylenediamine)platinum(IV) sulfate.

Chapter 25

Review Problems: **1.** (a) γ; (b) α; (c) β.   **2.** (a) $^{32}_{17}Cl$;
(b) $^0_{+1}e$; (c) $^{235}_{92}U$, $^{231}_{90}Th$, ^4_2He; (d) $^{214}_{83}Bi$, $^{214}_{84}Po$.
3. (a) $^{24}_{11}Na$; (b) ^1_0n; (c) $^{240}_{94}Pu$; (d) 5 ^1_0n; (e) $^{246}_{99}Es$, 6 ^1_0n.
4. (a) $^2_1H + ^2_1H \rightarrow ^3_2He + ^1_0n$; (b) $^{241}_{95}Am + ^4_2He \rightarrow ^{243}_{97}Bk + 2 ^1_0n$; (c) $^{121}_{51}Sb + ^4_2He \rightarrow ^{124}_{53}I + ^1_0n$.   **5.** (a) $^{214}_{84}Po$; (b) $^{28}_{12}Mg$;
(c) $^{13}_8O$, $^{28}_{12}Mg$, $^{80}_{35}Br$, $^{214}_{84}Po$, $^{222}_{86}Rn$.   **6.** (a) 274 d; (b) 3.2 × 10^2 d; (c) 4.1×10^2 d.   **7.** no.   **8.** (a) 9.45×10^{-10} J;
(b) 3.726×10^3 MeV; (c) 1.60×10^3 neutrons.
9. 8.10 MeV.   **10.** (c) and (d).
Exercises: **5.** (a) $^{230}_{92}U$; (b) $^{248}_{98}Cf$; (c) $^{196}_{80}Hg$; (d) $^{214}_{84}Po$;
(e) $^{214}_{82}Pb$; (f) $^{69}_{32}Ge$.   **11.** 7.4×10^8 atoms.   **12.** 4.5 y.
13. 142 d.   **14.** 3.63 d.   **15.** 1.0×10^{-3} g.
16. 0.22 g ^{208}Pb/1.00 g ^{232}Th.   **17.** 4×10^9 y.
18. 1.0×10^1 dis min^{-1} per g C.   **19.** 5.1×10^4 y.
21. 4.06 MeV.   **22.** 4.80 MeV.   **23.** (a) $^{20}_{10}Ne$; (b) $^{18}_8O$;
(c) ^7_3Li.   **24.** β^-: $^{33}_{15}P$ and $^{134}_{53}I$; positron: $^{29}_{15}P$ and $^{120}_{53}I$.
29. 2.94×10^3 metric tons.
Self-Test Questions: **1.** (c)   **2.** (b)   **3.** (d)   **4.** (c)
5. (a)   **6.** (c) **7.** (d)   **8.** (d)   **9.** (a) $^{230}_{90}Th \rightarrow ^{226}_{88}Ra + ^4_2He$; (b) $^{54}_{27}Co \rightarrow ^{54}_{26}Fe + ^0_{+1}e$; (c) $^{232}_{90}Th + ^4_2He \rightarrow ^{232}_{92}U + 4 ^1_0n$.   **10.** 6.1×10^3 dis s^{-1}.   **11.** 76 d.

Chapter 26

Review Problems:

2. (a)

(b)

3. isomers: (b) and (f).

4. (a) positions for the three Cl atoms: 1,1,1; 1,1,2; 1,1,3; 1,2,2; 1,2,3; **(b)** positions for the two Cl atoms based on *n*-butane: 1,1; 1,2; 1,3; 1,4; 2,2; 2,3; positions based on isobutane (2-methylpropane): 1,1; 1,2; 1,3.   **5. (a)** halide (bromide); **(b)** carboxylic acid; **(c)** aldehyde; **(d)** ether; **(e)** ketone; **(f)** amine; **(g)** phenyl; **(h)** ester.

6. (a) 3,3-dimethyloctane; **(b)** 2,2-dimethylpropane (neopentane); **(c)** 2,3-dichloro-5-ethylheptane; **(d)** 2,2-dimethyl-3-ethylpentane; **(e)** 5-chloro-3-heptene.

7. (a) $CH_3CH(CH_3)CHBrCH_2CH_3$;
(b) $CH_3CH_2CH[CH(CH_3)_2]CH_2CH_2CH_2CH_2CH_3$;
(c) $CH_3CH{=}CHCH_2CH_3$; **(d)** $CH_3CH_2OCH_2CH_2CH_3$.

8. (a) $CH_3CH(OH)CH_3$; **(b)** $Pb(C_2H_5)_4$; **(c)** $ClCF_2CH_3$; **(d)** $CH_2{=}C(CH_3)CH{=}CH_2$; **(e)** $CH_3CH{=}CHCHO$.

9. (a) *p*-dibromobenzene (1,4-dibromobenzene); **(b)** *o*-methylaniline or *o*-aminotoluene;

(c) **(d)** **(e)**

10. (a) CH_3CH_2Cl; **(b)** $CH_3CH_2CH_2CH_3$;

(c) **(d)** $CH_3CH_2CH(OH)CH_3$;

(e) **(f)** $K^+ +$

Exercises: **6. (a)** positional; **(b)** skeletal; **(c)** positional; **(d)** positional; **(e)** geometric.   **10. (a)** 2,2-dimethylbutane; **(b)** 2-methyl-1-propene or, simply, methylpropene; **(c)** methylcyclopropane; **(d)** 4-methyl-2-pentyne; **(e)** 3,4-dimethylhexane; **(f)** 3,4-dimethyl-2-*n*-propyl-1-pentene.   **12. (a)** no; **(b)** yes; **(c)** no; **(d)** no; **(e)** yes.   **13.** (i) and (j) are correct.

14. (a) $(CH_3)_2CCH_2CH(CH_3)_2$; **(b)**

(c) **(d)** $CH_2(CO_2H)C(CO_2H)CH_2(CO_2H)$;

(e) **(f)**

(g)

(h) $(CH_3)_2CH(C_{15}H_{31})$;
(i) $(CH_3)_2C{=}CH(CH_2)_2C(CH_3){=}CH(CH_2)_2C(CH_3){=}CHCH_2OH$;
(j) $(CH_3)_2C{=}CHCH_2CH_2CH(CH_3)CHO$.   **15.** C_4H_8.

18. (a) $CH_3CH_2CH_3$; **(b)** $CH_3CH_2CH_2CH_3$ or $CH(CH_3)_3$.
19. (a) *n*-butane; **(b)** *n*-hexane and NaBr; **(c)** *n*-propane and Na_2CO_3; **(d)** 1-chloropropane and 2-chloropropane.
24. (a) $CH_3CCl_2CH_3$; **(b)** $CH_3C(CN){=}CH_2$

(c) $CH_3CH_2CCl(CH_3)_2$; **(d)** —$CHBrCH_3$.

25. (a) *m*-dinitrobenzene; **(b)** 1,3,5-trihydroxybenzene; **(c)** 3,5-dihydroxybenzoic acid; **(d)** N,N-diethylaniline;

(e) $HO-$ **(f)** $-C{\equiv}CH$

(g)

26. (a) 1-bromo-3,5-dinitrobenzene; **(b)** 2-bromoaniline, 4-bromoaniline; **(c)** 2,4-dibromoanisole.   **30.** 10.
33. (a) 1, 2, 7 → 1, 2, 2; **(b)** 3, 2, 4 → 3, 2, 2; **(c)** 1, 1 → 1, 1, 2; **(d)** 1, 16, 2 → 1, 4, 10.   **34.** $9.0 \times 10^1\%$.
37. 33.3% O.   **38.** $[CO(CH_2)_8CO(NH)(CH_2)_6(NH)]_n$
Self-Test Questions: **1.** (b)   **2.** (d)   **3.** (b)   **4.** (a)   **5.** (c)   **6.** (d)   **7.** (b)   **8.** (c)

9. (a) **(b)** $HO-$$-Br$

(c) **(d)**

10. the isomers are 1-bromopentane, 2-bromopentane, 3-bromopentane, 1-bromo-3-methylbutane, 2-bromo-3-methylbutane, 1-bromo-2-methylbutane, 2-bromo-2-methylbutane, 1-bromo-2,2-dimethylpropane.

11. (a) **(b)** $CH_3CH_2CH_2Cl$ and

(c) **(d)** $CH_3CH_2CCH_3$.

12. (a) C_2H_6 is a gas at room temperature; C_8H_{18} is a liquid; **(b)** C_2H_4, an olefin, will decolorize MnO_4^-(aq); C_2H_6, an alkane, will not. **(c)** C_2H_5OH is miscible with water in all proportions; $C_6H_{13}OH$ has very limited water solubility; **(d)** determine the pH of a dilute aqueous solution; C_6H_5COOH is acidic, and C_6H_5CHO is not.

Chapter 27

Review Problems·   **1. (a)** glyceryl laurooleosterate;
(b) glyceryl trilinoleate; **(c)** sodium myristate.
3. (a) saponification value = 253; **(b)** iodine number = 38.2.
4. tripalmitin.   **6. (a)** $C_6H_5CH_2CHCO_2H$
$$NH_3^+Cl^-$$
(b) $C_6H_5CH_2CHCO_2^-Na^+$ **(c)** $C_6H_5CH_2CHCO_2^-$
$$NH_2 \qquad\qquad NH_3^+$$
8. (b) glycylalanylserylthreonine.
10. Thr-His-Pro-Leu-Ala-Ser-Gly-Met.
Exercises: **1. (a)** 5×10^2 H$^+$ ions; **(b)** 2×10^5 K$^+$ ions.
2. 2%.   **3.** 3×10^7.   **4.** 5×10^6.   **5.** length of

extended DNA molecule: 2.0×10^3 μm.   **9.** 216.
10. trilinolein.   **16.** diastereomers.   **17.** 37% α,
63% β.   **20.** lysine to cathode; glutamic acid to anode;
proline does not migrate.
22. (a) Ala-Ser-Gly-Val-Thr-Leu;
(b) alanylserylglycylvalylthreonylleucine.   **24.** minimum
mol. wt. = 2.9×10^3.   **27.** about 60.0% efficiency.
28. $K = 2.6 \times 10^2$.   **34.** possible DNA base sequence;
AGA-CCA-CAA-CGA.
Self-Test Questions: **1.** (b)   **2.** (c)   **3.** (d)   **4.** (a)
5. (d)   **6.** (a)   **7.** (c)   **8.** (b)   **9.** 129 g soap.
10. Gly-Cys-Val-Phe-Tyr.

Index

Boldface page numbers indicate end-of-chapter definitions. Abbreviations in italics after page numbers indicate exercise (*ex.*), figure (*f.*), marginal note or footnote (*n.*), and table (*t.*).

A

Å, 165, 205, A15 *t.*
Absolute configuration, 846, **870**
Absolute entropy, 494 *f.*
  table of, A16–17
Absolute temperature, 110–11
Absolute zero, 110
Acceleration, A9
  due to gravity, A9
Acetal, 848
Acetaldehyde, 817
Acetamide, 820
Acetate ion, 66 *t.*, 521, 528
Acetic acid, 513 *t.*, 527, 817, 819, 860
  in buffer solutions, 539–42
  hydrogen bonding in, 306 *f.*, 307
  titration of, by sodium hydroxide, 547–50
Acetone, 818
Acetone–chloroform solutions, 330 *f.*, 342–43
Acetophenone, 817
Acetyl chloride, 816
Acetylene, 807
  bonding in, 265
  in organic synthesis, 836 *ex.*
Acetylsalicylic acid, 532 *ex.*, 819 *n.*
Achiral, 748
Acid(s), **69**, 504–35
  amino, 852–56, 873 *ex.*
  anions as, 520–23
  Arrhenius, 504–505, **530**
  binary, 65, **69**
  Brønsted–Lowry, 505–506, 524, 525 *t.*, 530 *ex.*
  carboxylic, 818–20
  cations as, 520–23
  conjugate, 505, 522–23, **530**
  dicarboxylic, 819
  electrolytic properties of, 349 *t.*
  fatty, 684, 842
  hydrated metal ions as, 523, 686, 699, 722, 757–58
  ionization constants of, 513, 514, 518 *t.*, 521 *t.*
  Lewis, 506–507, **530**, 532 *ex.*
  naming of, 65–67
  neutralization of, 504–505, 510–11, 532 *ex.*, 546–51
  organic, 527–29
  polyprotic, 517–20, **530**, 550–53
  saturated, 841 *t.*
  strong, 509–11, 532 *ex.*, 536–37, 546–47
  structure of, and strength, 524–29, 533–34 *ex.*
  unsaturated, 841 *t.*
  weak, 513–17, 532 *ex.*, 536–38, 547–51
Acid anhydride, 526, **530**
Acid–base indicator, 545–46, **558**, 560 *ex.*
Acidosis, 543
Acid rain, 395 *n.*, 681 *n.*
Acrilan, 248 *t.*
Acrolein, in smog, 388
Acrylonitrile, 810
Actinide elements. *See* Actinoid elements
Actinium series, 770
Actinoid elements, 199, **214**
  See also Transuranium elements
Activated charcoal, 379 *n.*
Activated complex, 423, 427 *f.*, **435**
Activation energy, 421, 422 *f.*, 423–24, **435**
Active site, 428, **435**
Activity, 351, 490–91
Activity coefficient, 351
Acyl group, 819, **831**
Acyl halide, 820
Addition reactions, 810, **831**
Adduct, 377, **401**
Adenine, 865 *f.*, 866 *f.*
Adenosine diphosphate, 861, 862 *f.*

Selected Physical Constants

acceleration
due to gravity: g 9.8067 m s^{-2}

speed of light
(in vacuum): c 2.99792458 × 10^8 m s^{-1}

gas constant: R 0.082057 L atm mol^{-1} K^{-1}
            8.3143 J mol^{-1} K^{-1}

electronic charge: e 1.6021 × 10^{-19} C

electronic rest mass: m 9.1091 × 10^{-28} g

Planck's constant: h 6.6256 × 10^{-34} J s

Faraday constant: $\mathfrak{F}$ 9.6487 × 10^4 C/mol e$^-$

Avogadro's number: N 6.02205 × 10^{23} mol^{-1}

Multiples and Submultiples in the SI System

(See also Appendix C)

Multiple	Prefix	Symbol	Submultiple	Prefix	Symbol
10^{12}	tera	T	10^{-1}	deci	d
10^9	giga	G	10^{-2}	centi	c
10^6	mega	M	10^{-3}	milli	m
10^3	kilo	k	10^{-6}	micro	μ
10^2	hecto	h	10^{-9}	nano	n
10^1	deka	da	10^{-12}	pico	p
			10^{-15}	femto	f
			10^{-18}	atto	a

Some Common Conversion Factors

Length
1 meter (m) = 39.37 inches (in.)
1 in. = 2.54 centimeters (cm)

Mass
1 kilogram (kg) = 2.205 pounds (lb)
1 lb = 453.5 grams (g)

Force
1 newton (N) = 1 kg m s^{-2}

Volume
1 liter (L) = 1000 mL = 1000 cm^3
1 L = 1.057 quart (qt)

Energy
1 joule (J) = 1 N m = 1 kg m^2 s^{-2}
1 calorie (cal) = 4.1840 J
1 electron volt (eV) = 1.6021 × 10^{-19} J
1 eV/atom = 96.49 kJ/mol
1 kilowatt hour (kW hr) = 3600.0 kJ

mass–energy equivalence: 1 atomic mass unit (amu) = 1.66 × 10^{-24} g = 931 MeV

Some Useful Geometric Formulas

perimeter of a rectangle = $2l + 2w$; area of a rectangle = $l \times w$;
volume of a parallelepiped = $l \times w \times h$; area of a triangle = $\frac{1}{2}$(base × height);
circumference of a circle = $2\pi r$; area of a circle = πr^2; area of a sphere = $4\pi r^2$;
volume of a sphere = $\frac{4}{3}\pi r^3$; volume of a cylinder or prism = (area of base) × (height)